Bacteria and Fungi from Fish and Other Aquatic Animals, 2nd Edition

A Practical Identification Manual

Bacteria and Fungi from Fish and Other Aquatic Animals, 2nd Edition

A Practical Identification Manual

Nicky B. Buller

Department of Agriculture and Food Western Australia
c/o Animal Health Laboratory, Locked Bag 4, Bentley Delivery Centre,
Bentley, Western Australia 6983

www.cabi.org

CABI is a trading name of CAB International

CABI
Nosworthy Way
Wallingford
Oxfordshire OX10 8DE
UK

Tel: +44 (0)1491 832111
Fax: +44 (0)1491 833508
E-mail: info@cabi.org
Website: www.cabi.org

CABI
38 Chauncy Street
Suite 1002
Boston, MA 02111
USA

Tel: +1 800 552 3083 (toll free)
E-mail: cabi-nao@cabi.org

A catalogue record for this book is available from the British Library, London, UK.

Library of Congress Cataloging-in-Publication Data

Buller, Nicky B., author.
Bacteria and fungi from fish and other aquatic animals : a practical identification manual / Nicky B. Buller, Department of Agriculture and Food Western Australia. – 2nd edition.
pages cm
Includes bibliographical references and index.
ISBN 978-1-84593-805-5 (alk. paper)
1. Aquatic animals–Microbiology. I. Title.

QR106.B85 2014
578.76–dc23

2014011552

ISBN-13: 978 1 84593 805 5

Commissioning editor: Rachel Cutts
Editorial assistant: Emma McCann
Production editor: James Bishop

Typeset by SPi, Pondicherry, India
Printed and bound in the UK by Halstan Printing Group, Amersham

Contents

List of Tables and Figures

Tables

Figures

Acknowledgements

For the second edition

Dr Jeremy Carson, Department of Primary Industries, Water and Environment, Tasmania kindly provided about 30 type strains of *Vibrio* species, enabling me to test their reactions in the conventional biochemical media and the API 20E kit, which are reported in this second edition.

Dr Bruno Gomez-Gil (CIAD/Mazatián Unit for Aquaculture, Mexico) provided test results for *V. rotiferianus* not listed in the literature, and supplied the full list of reactions for the API 20E for inclusion in this manual.

Dr Fran Stephens, Department of Fisheries Western Australia, Fish Health Unit and Professor Louis Evans, Aquatic Science, Curtin University provided many of the early references on *Aphanomyces astaci*, some of which were translations from the original papers.

Thank you to Dr Brian Jones, Dr Fran Stephens, Paul Hillier and all staff at the Fish Health Unit, Department of Fisheries Western Australia, which is co-located at Animal Health Laboratories, Department of Agriculture and Food Western Australia, for their encouragement for this manual and their infectious enthusiasm when investigating diseases from aquatic animals. To Sam Hair and all staff in the bacteriology laboratory, thank you for your support.

For the first edition

I would like to acknowledge Dr Jeremy Carson and Nick Gudkovs (Australian Animal Health Laboratories, Geelong, Victoria) for allowing me to visit their respective laboratories, particularly in the early part of my career, for discussions on fish bacterial diseases and isolation techniques. In addition, Dr Jeremy Carson and his laboratory who, over the years, have confirmed the identification of some of our *Vibrio* and *Flavobacterium* strains isolated from diagnostic cases. In particular, for confirmation of the identity of Animal Health Laboratory Department of Agriculture (AHLDA) diagnostic strains mentioned in this manual (*Flavobacterium columnare*, *Listonella anguillarum*, *Vibrio agarivorans*, *V. halioticoli* and *V. mediterranei*). To Dr Annette Thomas (Department of Primary Industries, Queensland), thank you for the gift of cultures of *V. alginolyticus*, *V. (carchariae) harveyi* and *Streptococcus iniae* and for the many discussions over the years on veterinary and fish pathogens. Thank you to Dr Bruno Gomez-Gil for test results on *V. rotiferianus* not listed in the literature, and for supplying the paper on *V. pacinii* before publication, for inclusion in this manual. To Dr Fabiano Thompson (Laboratory for Microbiology, Ghent University, Belgium) for providing further API 20E reactions for *V. brasiliensis*, *V. neptunius* and *V. xuii* not listed in the journal article. To the fish pathologists who I have worked with, in particular Dr Brian Jones and in memorandum Dr Jeremy Langdon, who have helped me to a better understanding of fish diseases. I am also grateful to the Fisheries Research and Development Council for funding to attend workshops on fish diseases held at the University of Tasmania in 1996 and the fish bacteriologist's workshops held at the Australian Animal Health Laboratories, Geelong in 2000 and 2001.

The photograph of *Renibacterium salmoninarum* was from the Animal Health Laboratory (AHL) archives. There was no documentation as to the source of this photograph and thus I am unable to acknowledge the origin. The rest of the photographs were taken by the author and were of cultures obtained as diagnostic submissions at AHL, type strains or from Dr A. Thomas. The photographs of *Flavobacterium columnare* adhering to gill tissue are courtesy of Dr Brian Jones.

A great many microbiologists and other scientists have been responsible throughout the years for developing and refining the media and techniques used for isolation and identification of bacteria. The media quoted from the literature in this book are referenced to include the originator and refiners of those media. My apologies if I am remiss in leaving anybody out.

Due to the expense of producing the photographic plates, assistance towards their cost was achieved from sponsorship from the following organizations. Their assistance is gratefully acknowledged.

Principle sponsor:

Agriculture, Fisheries and Forestry-Australia (AFFA), Aquatic Animal Health Subprogram, funded through Fisheries Research and Development Corporation (FRDC).

Other sponsors:

Department of Fisheries, Western Australia

Oxoid Australia

Animal Health Laboratories, Department of Agriculture and Food, Western Australia.

I would also like to acknowledge Mr Tim Hardwick, editor at CABI, for his assistance in bringing this manuscript to publication.

Introduction

Many of the organisms and aquatic animal species covered in this manual reflect those we see in the Bacteriology Laboratory at Animal Health Laboratories, Department of Agriculture and Food Western Australia. Over the years we have tested many different fin fish, aquarium species, abalone, cultured and captive seahorses, prawns, crabs, lobsters, sea hares, sea urchins, coral, sea stars or starfish, penguins, whales, dolphins, seals, seabirds and many others. All these samples have a myriad of bacteria associated with them and it is a challenge to culture, isolate and identify the pathogenic bacteria. The diversity of bacterial species covered in this manual reflects the diverse range of samples we receive into our diagnostic laboratory, which includes not only samples from farmed and wild aquaculture, marine mammals and wildlife, but also samples from terrestrial animals.

Many countries with aquaculture industries send students, particularly postgraduate students, to universities all over the world. Quite often the postgraduate study they undertake, especially masters and doctorate studies, involves microbiology. In many cases the students have a basic understanding of microbiology, but require further knowledge for their studies. This manual attempts to provide such students with a good resource for undertaking microbiological techniques so that they can not only apply these to their studies, but apply their knowledge in their own countries when investigating disease outbreaks. The intention is that well-equipped laboratories may find this manual an additional resource. The world of farmed and wild aquatic animals presents many interesting microbiological challenges and the knowledge of that particular microbiological world is increasing daily and is being unravelled with an array of technical methods. Many of these technical methods are included in this manual.

Due to the increase in aquaculture research, intensive fish farming systems and the international trade of live aquatic animals and products, and the emergence of new diseases, laboratories must keep abreast of techniques for the isolation and identification of bacteria and fungi in either a diagnostic or research capacity. In this manual there is an emphasis on bacteria of interest to the aquaculture industry, be they pathogens, normal flora or strains that may be used as probiotics. Some bacteria that have been isolated from diverse habitats are also included. This manual attempts to provide laboratories with an up-to-date and standardized database of methods and biochemical identification tables that can be used to isolate and identify bacteria from aquatic sources. Molecular diagnostics has become routine in many laboratories and some of the methods available are included in this manual.

A vast number of samples from aquatic animal species are received into our laboratory from both cool and warm climates. This manual includes as many of those bacteria as possible that may be found during examination of these samples from this diverse range of hosts and habitats. Not only pathogens but also environmental and saprophytic organisms are included to aid in the understanding of the microflora that may be found in such samples. Many bacteria from the more extreme environments also have been included, because with the increase in aquaculture throughout the world, and the increasing knowledge about the microflora of such habitats, these organisms may find their way into some laboratories via the samples submitted. Those isolates that are capable of growing on the isolation media recommended for aquatic organisms, such as ZoBell's or Marine agar 2216 (Difco), are included in this manual. In addition, I have included some of the Antarctic organisms, as being rich in omega-3 polyunsaturated fatty acids which are significant in the diet of marine finfish, they have been suggested as a low cost food source (Nichols *et al.*, 1996), and therefore there is a likelihood they may be present in gastrointestinal samples that come into a laboratory.

Medical laboratories are also required to identify an increasing number of bacteria from aquatic habitats that may be involved in clinical infections. This book may also assist in the identification of such bacteria that are not normally listed in the commercial databases such as API (bioMérieux).

Results of phenotypic test results reported in the literature can be confusing. It is important to perform tests using the methods that have been used by the reporting literature. In this book, the majority of biochemical tests have been performed according to West and Colwell (1984), Cowan and Steel (1970), and these methods are listed. Results are also included from the commercial identification kits available, namely API 20E, API 50CH and API-ZYM from bioMérieux. Results from these tests are listed in the appropriate tables. Phenotypic tests that may produce different results between biochemical tube media and commercial identification kits include: citrate reaction, decarboxylases indole and some carbohydrates. Where different strains have been used in the literature and different phenotypic results reported, the results of these organisms have been listed separately. This is an attempt to enable microbiologists to provide the best possible identification of an organism isolated from a diagnostic or research sample.

A defined set of biochemical tests is used as much as possible in this manual and from this the majority of pathogens and non-pathogens encountered from aquatic sources can be identified to genus level and, for the more commonly known bacteria, to species level. The aim of having a defined biochemical set is so that laboratories who make in-house media can prepare media that will enable as many bacteria as possible to be cultured and identified in the routine laboratory without having to prepare an excessive number of tests and growth media. Some of the problems with variations in biochemical reactions reported in the literature are due to different methods being used. This manual provides a standardized set of biochemical identification methods for aquatic organisms and the reactions reported here are based on this defined set.

Bacteria that require specialized media and identification tests are included here to assist laboratories in identification of these organisms. In the case of *Brucella*, *Mycoplasma* and *Mycobacteria*, these methods are intended as a guide, because these organisms should be sent to a laboratory that specializes in their identification. Other media, such as alternative methods for the detection of carbohydrate fermentation by *Flavobacterium* species, for example, are also included.

The terms fermentation and utilization are often confused in the literature, and in some cases it has been difficult to assess by which method the test was done. It is important to distinguish between fermentation and utilization because they are separate methodologies and a bacterium may show a positive test for fermentation of a carbohydrate, yet negative for utilization of that same carbohydrate when tested as a sole carbon growth source. Basically, fermentation refers to the fermentation or breakdown of a carbohydrate, commonly called a 'sugar'. The breakdown products are detected by a pH change in the medium according to a colour change in the pH indicator, usually phenol red. Utilization refers to a test where a carbon source is assessed as a sole growth source for that bacterium. No other nutrients are present in the medium and growth is observed macroscopically, seen as an increase in the turbidity of the test medium. pH indicators are not included in utilization tests. The exception to this is citrate utilization using Simmons' method.

Layout of the Manual

This manual details the steps taken to isolate and identify an unknown bacterium. Experienced microbiologists may find some sections basic, but these are included for students and newly graduated microbiologists or for researchers who are unfamiliar with the techniques of bacteriology, particularly those used in a diagnostic laboratory.

In the phenotypic identification section a flow chart (Fig. 3.1) directs the microbiologist to the appropriate biochemical identification table required for the identification of the unknown organism. The biochemical identification tables are named according to genus such as *Aeromonas* and *Vibrio*, or to Gram reaction and cell shape, or oxidase reaction.

In the conventional identification tables, the organisms are listed in alphabetical order (with the exception of the *Vibrio* tables, Tables 3.22 and 3.23) under headings of Pathogenic or Environmental. The Pathogenic and Environmental differences refer to the pathogenicity for fish and aquatic animals, and generally not to humans or terrestrial animals. The inclusion of biochemical reactions for saprophytic and other species is intended to assist with the identification and ensure that the correct identification is made where species have similar results. In the *Vibrio* tables, the organisms are listed according to their groupings based on ODC, LDC and ADH reactions. The intention is to use these groupings as the starting point for identification, similar to a flow chart. Tables for the API databases have the organisms listed alphabetically.

Significance

Fish and other aquatic animals (farmed and wild) are prone to bacterial infections in the same way as terrestrial animals, especially when stressed. Disease may occur systemically or be confined to external surfaces such as the skin or gills. In many instances, the pathogenic bacteria are ubiquitous in the environment, or may form part of the normal internal bacterial flora of an aquatic animal. One study suggested that up to 28 different *Vibrio* species may be found in the hepatopancreas (10^4 CFU/g), intestine and stomach (10^6 CFU/g) of healthy shrimp. *Vibrio* species identified included *V. alginolyticus*, *V. parahaemolyticus*, *V. cholerae* as well as *Photobacterium damselae*. In diseased states only one or two *Vibrio* species are found (Gomez-Gil *et al.*, 1998). Therefore, many factors need to be considered in making a disease diagnosis, such as clinical signs and symptoms, pathology, amount of growth of the cultured bacteria, the numbers of different species cultured, the tissue site from which it was isolated, and the sterility of the specimen collection (Lightner and Redman, 1998).

Bacterial microflora on the surface of fish are heterogeneous in their salt requirement for growth. This requirement for salt (halophilic) is usually retained after serial subculture. Likewise, the gut of fish can be composed almost exclusively of halophilic vibrios (Simidu and Hasuo, 1968; Liston, 1957). Therefore, when attempting culture for pathogens, their salt requirement needs to be taken into account. This also applies to the biochemical identification tests.

Bacterial diseases affecting aquatic animals are detailed in texts such as Austin and Austin (2012), and Woo and Bruno (2011) (see Further Reading and Other Information Sources section).

Photographs of Culture and Microscopic Appearance of Organisms

The microscopic and cultural appearance of 31 species of bacteria and some of the biochemical test results are detailed here. Obviously not all bacteria can be presented, however, genera and species have been photographed on commonly used media to show their cultural appearance on that medium. Some species such as *Vibrio mimicus* and *Vibrio cholerae* are photographed to indicate how similar they can be in cultural appearance and how similar they may appear to motile *Aeromonas* species. Likewise the motile *Aeromonas* species all appear similar on blood agar media, whereas the non-motile *Aeromonas salmonicida* has a distinctive colony appearance, slow-growing, with pigment production after a few days of incubation.

Reactions of some biochemical tests are included for those not familiar with these reactions.

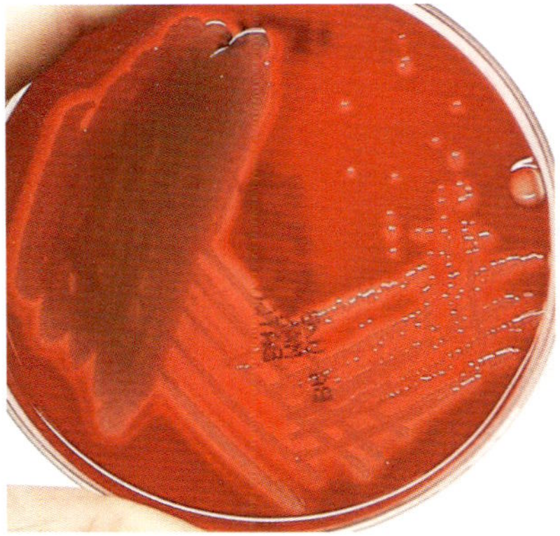

Fig. 1. *Aeromonas hydrophila* on BA, 24 h.

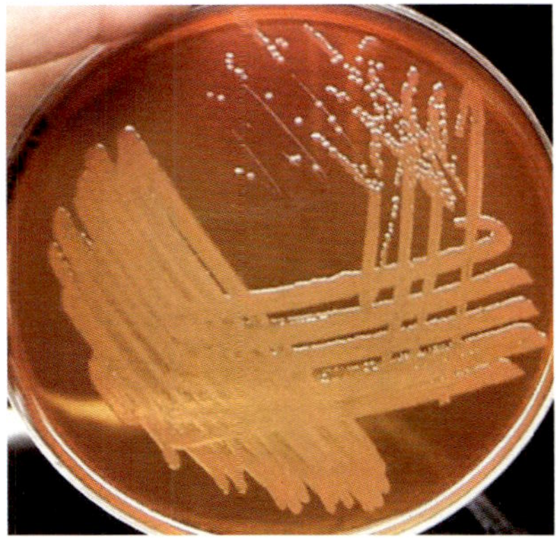

Fig. 2. *Aeromonas hydrophila* on MCA, 24 h.

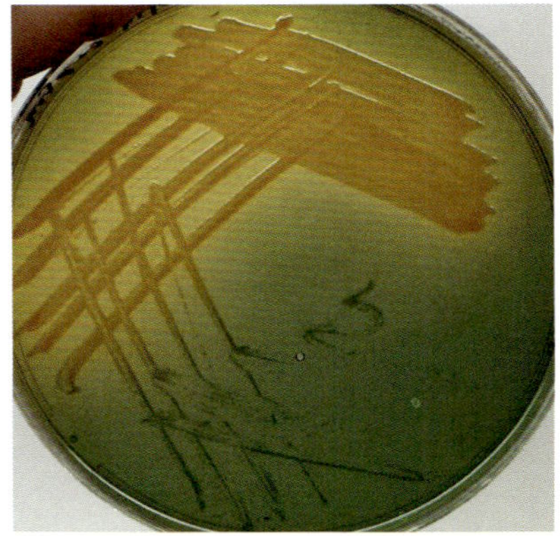

Fig. 3. *Aeromonas hydrophila* on TCBS, 24 h.

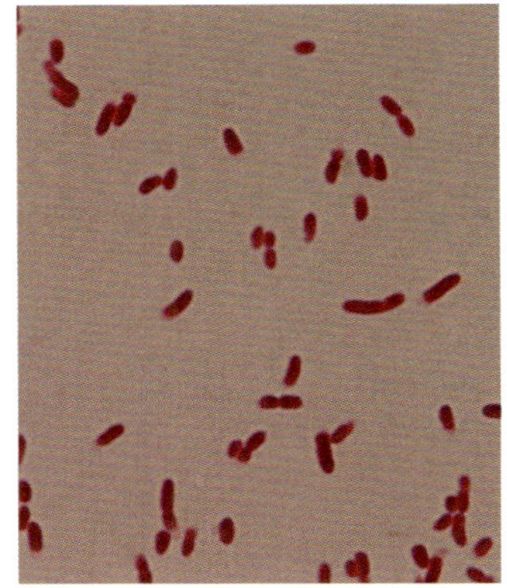

Fig. 4. *Aeromonas hydrophila*, Gram stain.

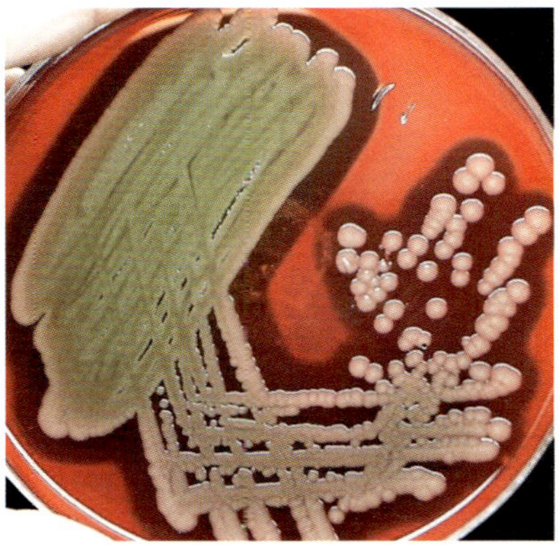

Fig. 5. *Aeromonas jandaei* on BA, 48 h.

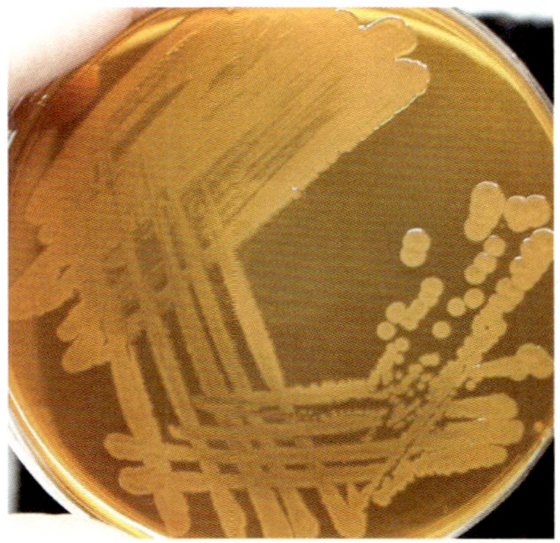

Fig. 6. *Aeromonas jandaei* on MCA, 3 days.

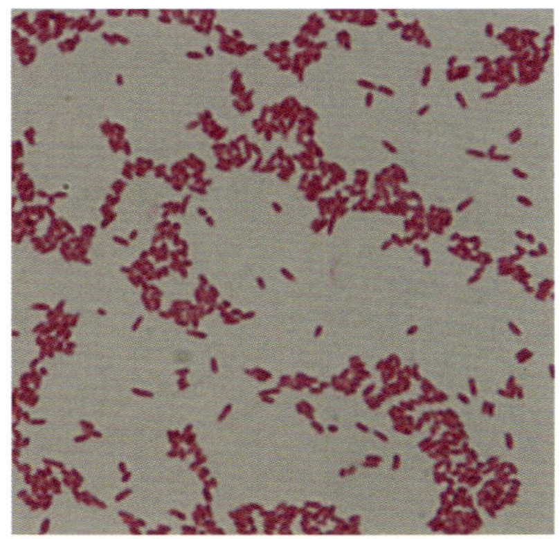

Fig. 7. *Aeromonas jandaei*, Gram stain.

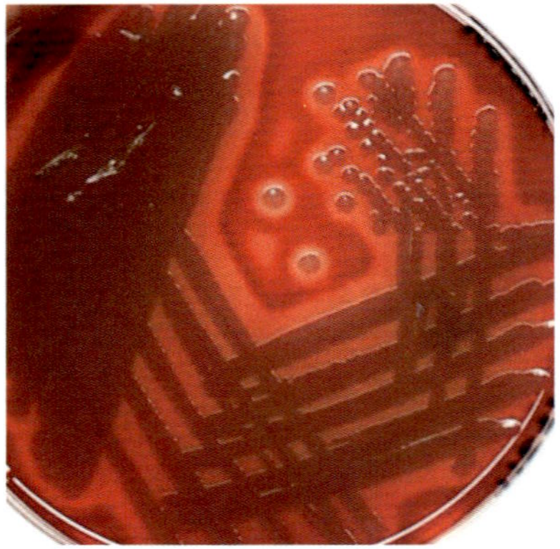

Fig. 8. *Aeromonas veronii* biovar Sobria on BA, 48 h, 25°C.

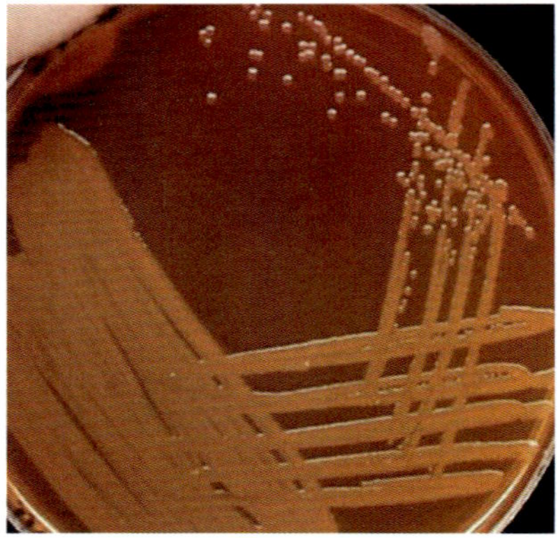

Fig. 9. *Aeromonas veronii* biovar Sobria on MCA, 24 h, 25°C.

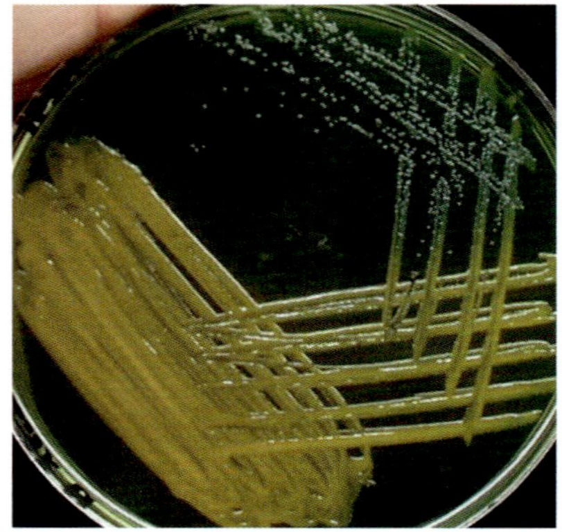

Fig. 10. *Aeromonas veronii* biovar Sobria on TCBS, 24 h, 25°C.

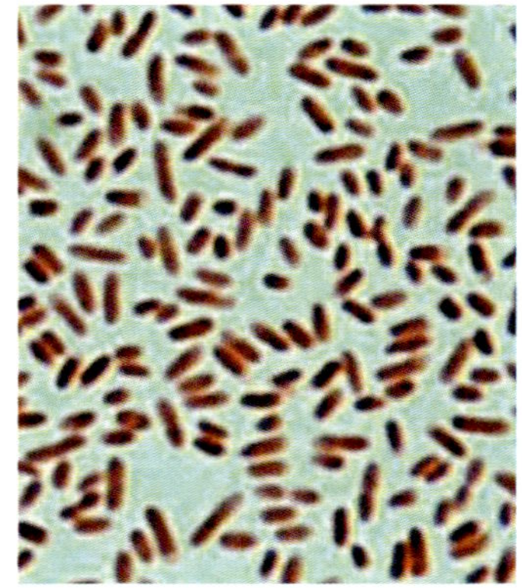

Fig. 11. *Aeromonas veronii* biovar Sobria, Gram stain.

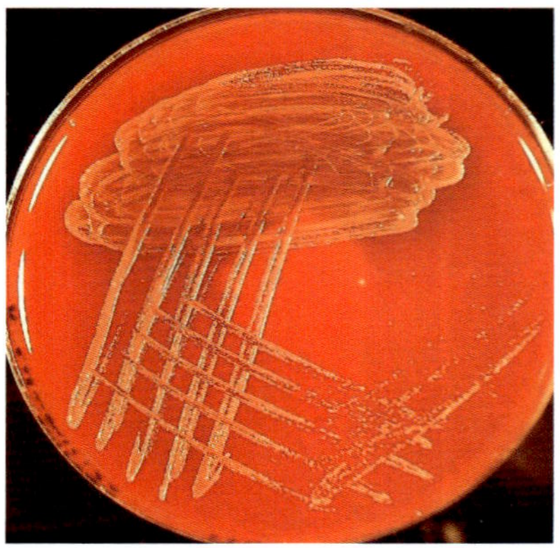

Fig. 12. Atypical *Aeromonas salmonicida* (Australian strain) on BA, 3 days.

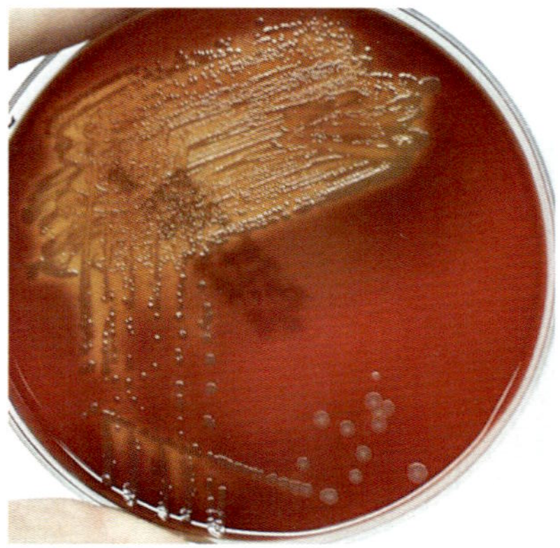

Fig. 13. Atypical *Aeromonas salmonicida* (Australian strain) on BA. 7 days.

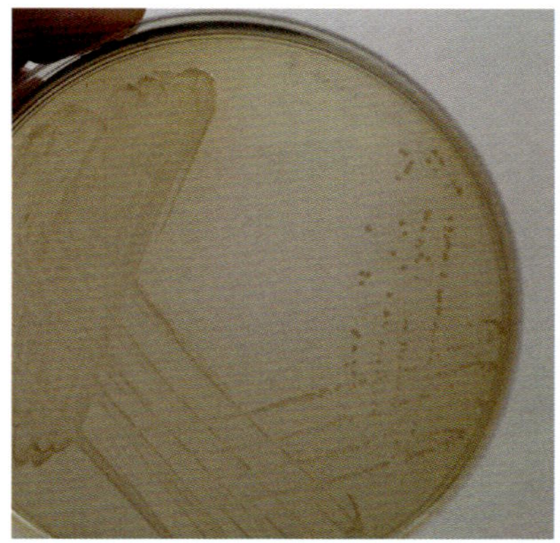

Fig. 14. Atypical *Aeromonas salmonicida* (Australian strain) showing pigment on NB agar.

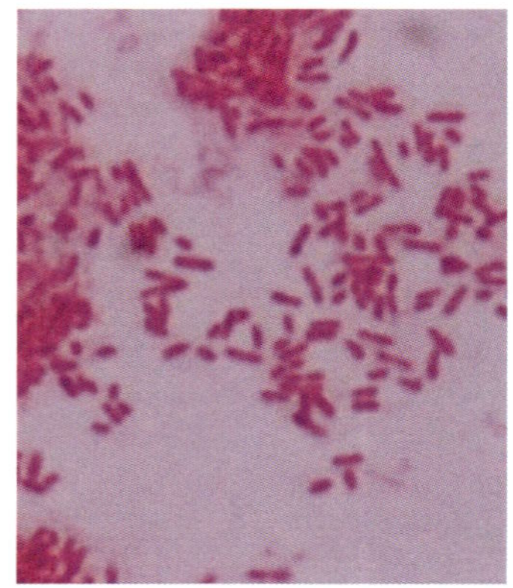

Fig. 15. Atypical *Aeromonas salmonicida* (Australian strain), Gram stain.

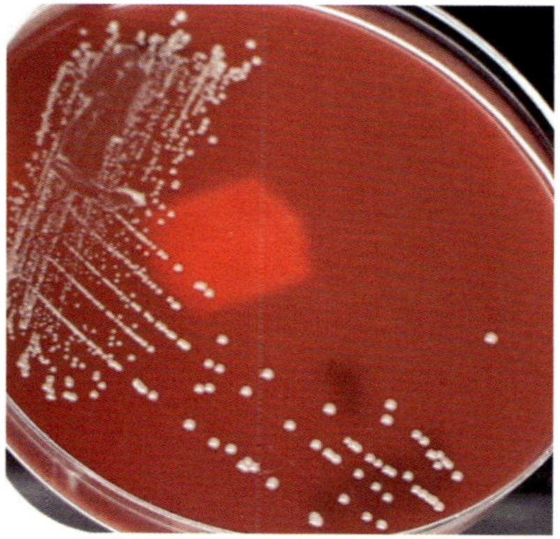

Fig. 16. *Cryptococcus neoformans* var. *gattii* subculture on BA, 3 days.

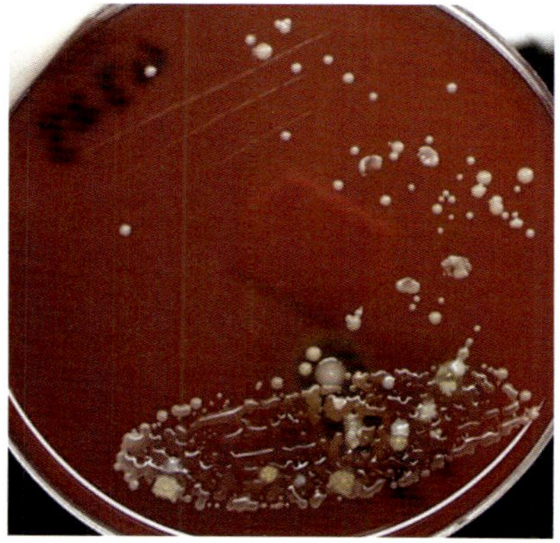

Fig. 17. *Cryptococcus*. Original isolation plate (Strep selective agar), 7 days.

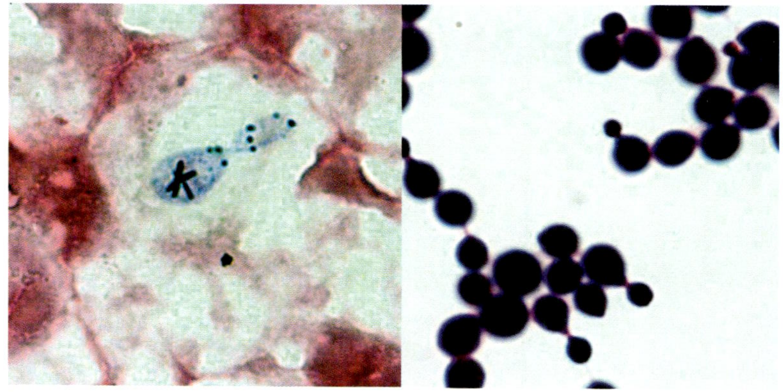

Fig. 18. *Cryptococcus*. Gram smear of culture, and within tissue.

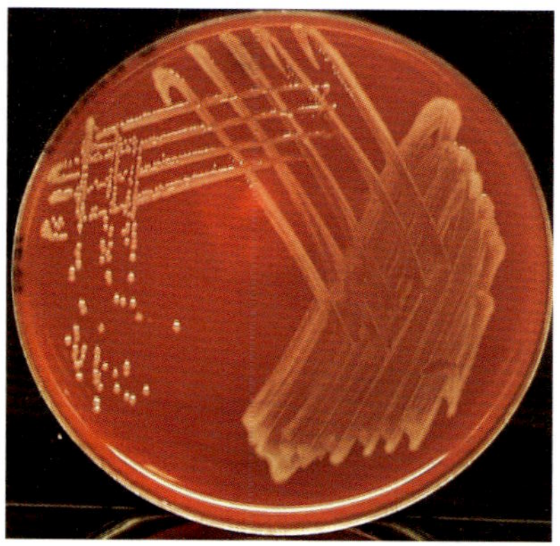

Fig. 19. *Edwardsiella hoshinae* on BA, 2 days, 25°C.

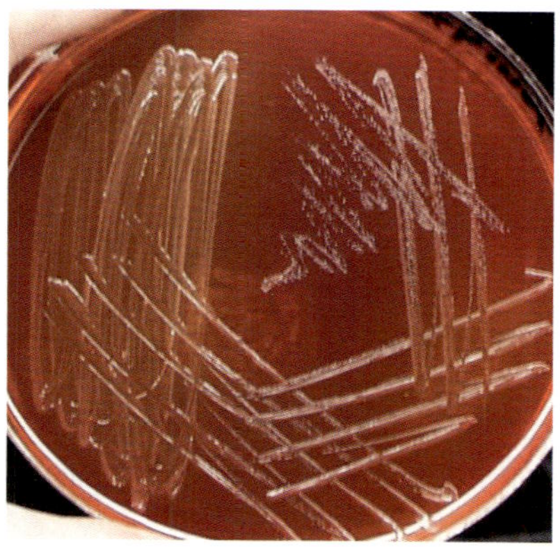

Fig. 20. *Edwardsiella hoshinae* on MCA, 24 h, 25°C.

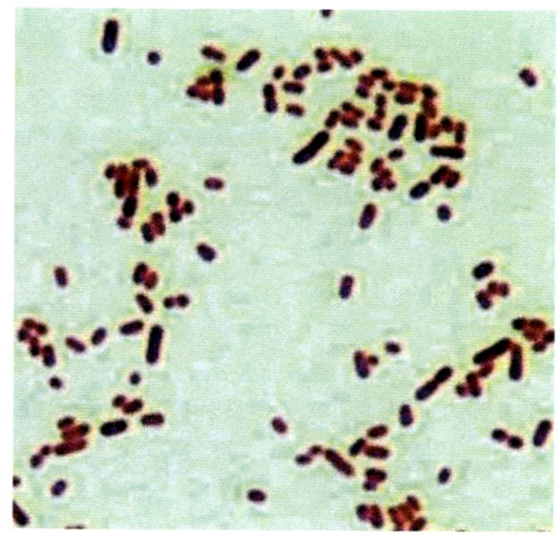

Fig. 21. *Edwardsiella hoshinae*, Gram stain.

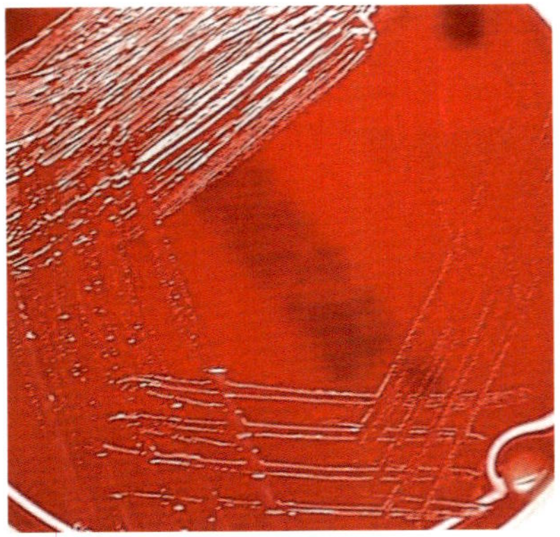

Fig. 22. *Edwardsiella tarda* on BA, 24 h, 25°C.

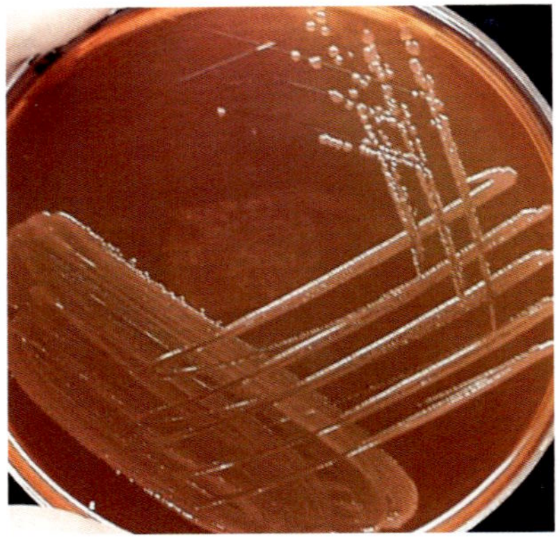

Fig. 23. *Edwardsiella tarda* on MCA, 24 h.

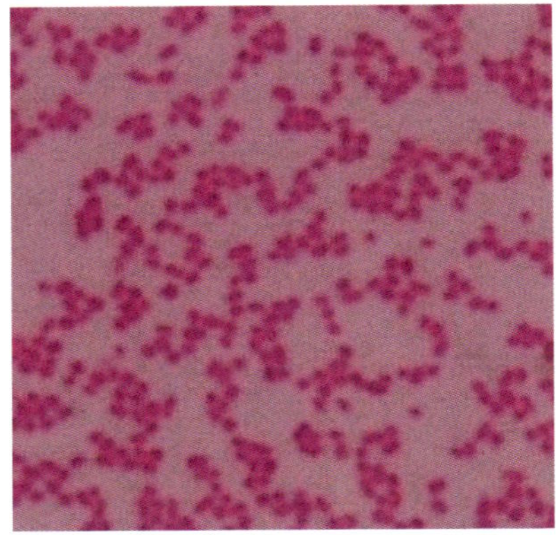

Fig. 24. *Edwardsiella tarda*, Gram stain.

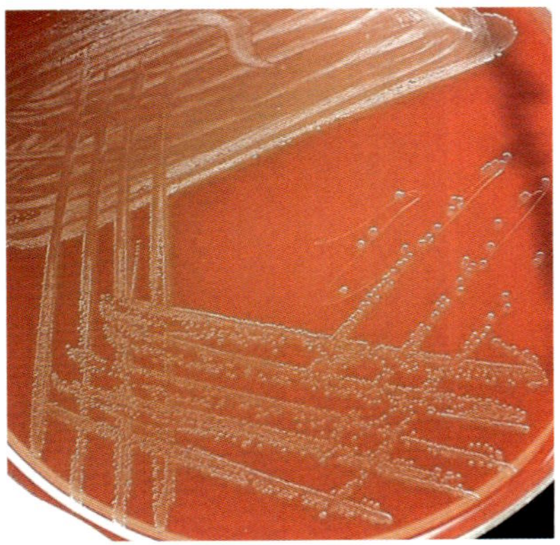

Fig. 25. *Erysipelothrix rhusiopathiae* on BA, 24 h, 25°C.

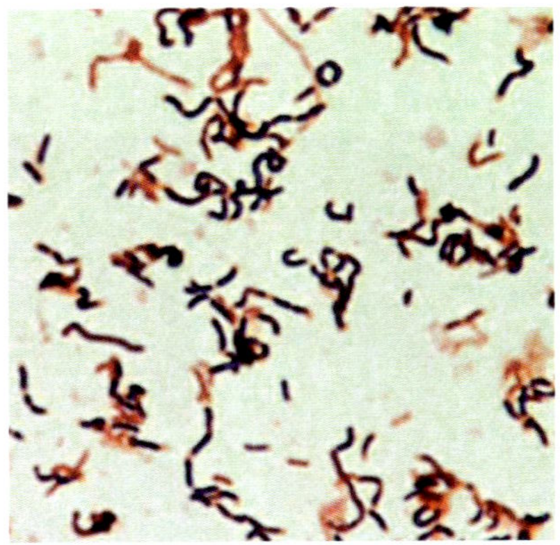

Fig. 26. *Erysipelothrix rhusiopathiae*, Gram stain.

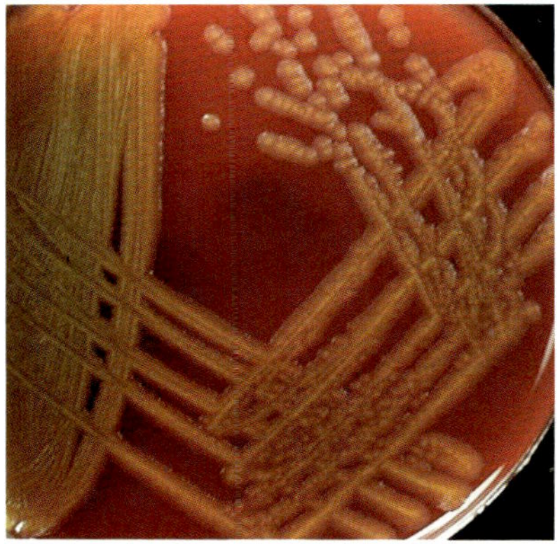

Fig. 27. *Flavobacterium johnsoniae* on BA, 48 h, 24°C.

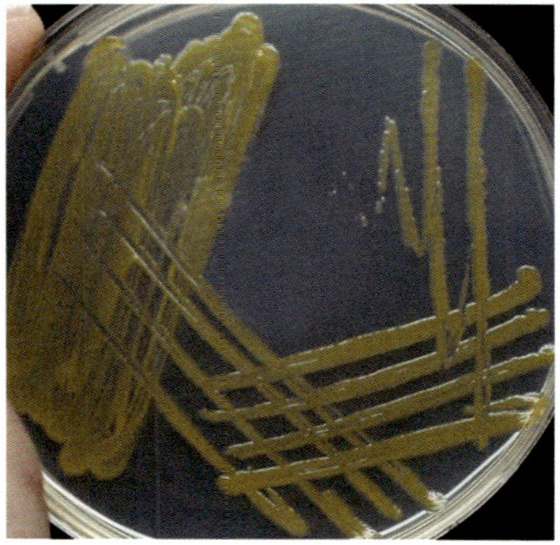

Fig. 28. *Flavobacterium johnsoniae* on NA, 48 h, 25°C.

Fig. 29. *Flavobacterium johnsoniae* on AO, 48 h, 25°C.

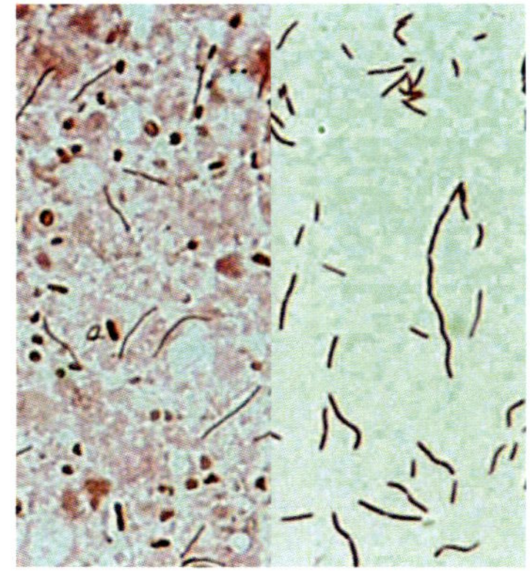

Figs 30 and 31. *F. johnsoniae*, Gram smear of gill tissue and cells from pure culture.

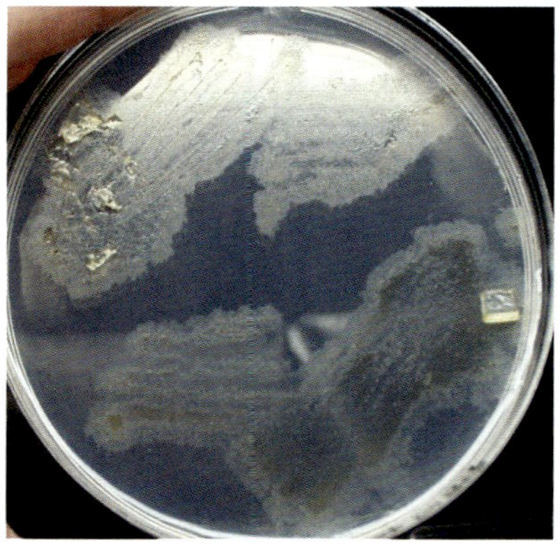

Fig. 32. *Flavobacterium columnare* subculture on AO, 5 days.

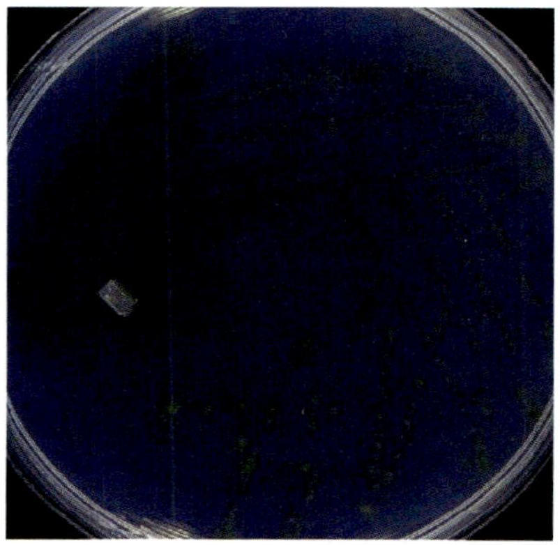

Fig. 33. *Flavobacterium columnare* subculture on AO, 5 days.

Fig. 34. *Flavobacterium columnare* magnified colonies.

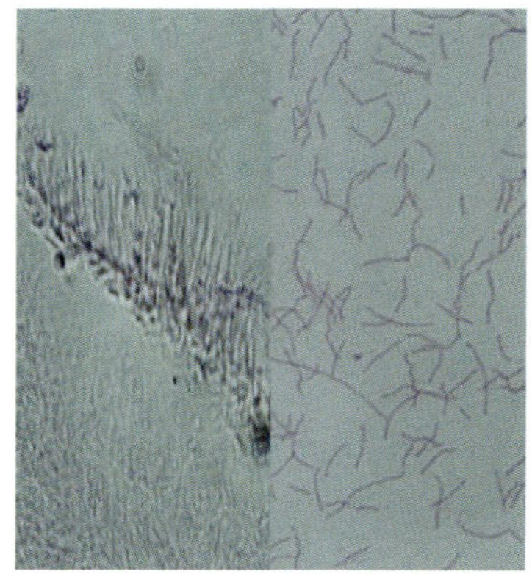

Figs 35 and 36. *Flavobacterium columnare*. HE section. Cells attached end-on to skin tissue. Gram smear of cells from culture.

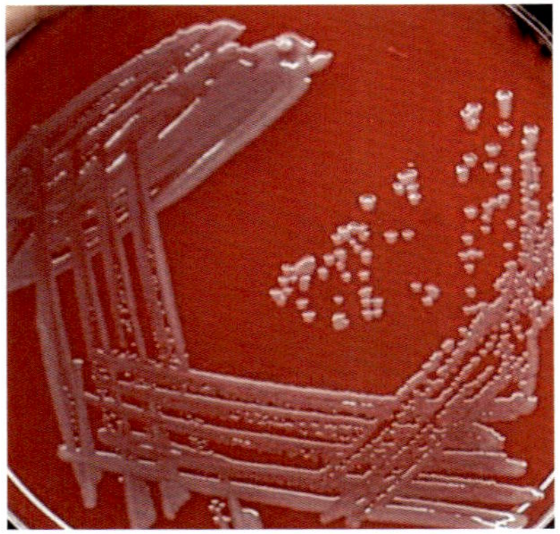
Fig. 37. *Hafnia alvei* on BA, 48 h, 25°C.

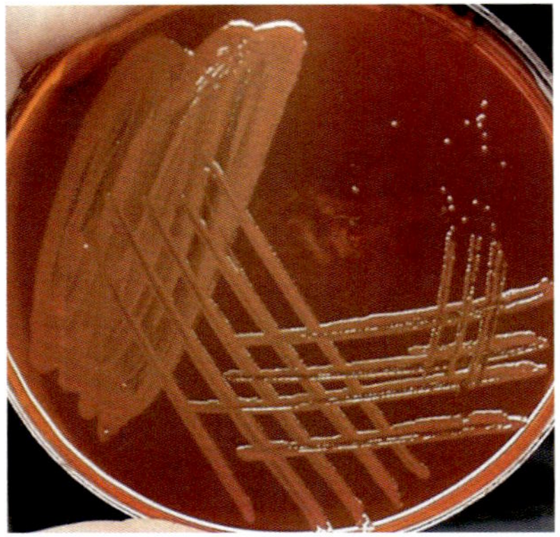
Fig. 38. *Hafnia alvei* on MCA, 48 h, 25°C.

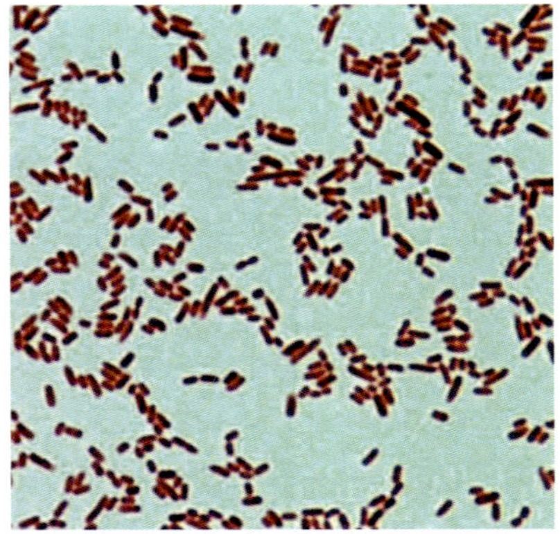
Fig. 39. *Hafnia alvei*, Gram stain.

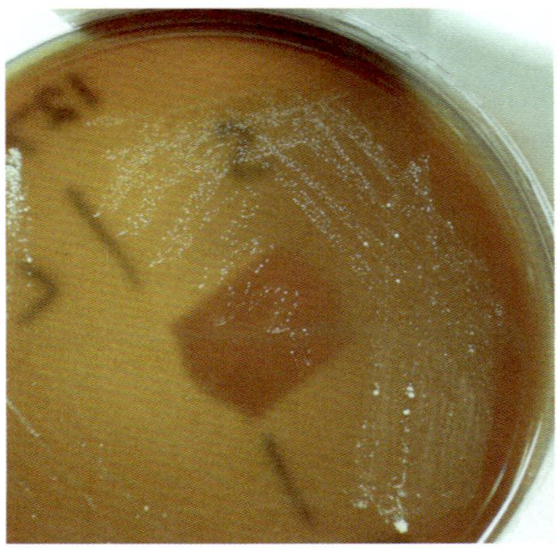
Fig. 40. *Mycobacterium marinum* on BA, 5 days.

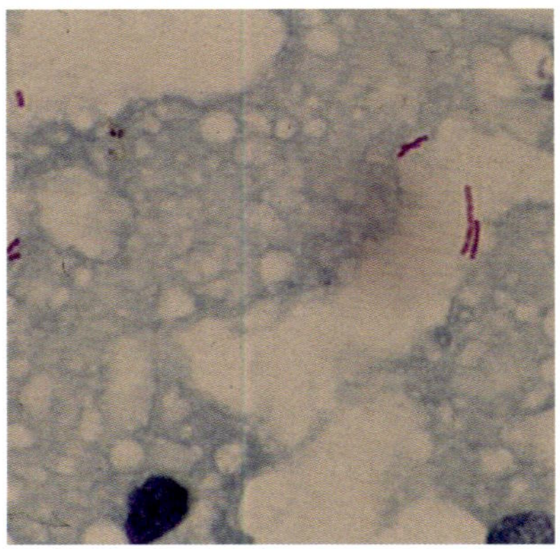
Fig. 41. *Mycobacterium marinum* in trout kidney. ZN stain.

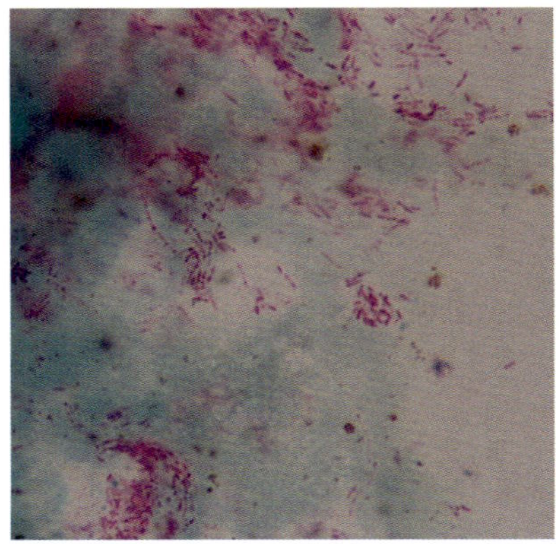
Fig. 42. *Mycobacterium marinum* in trout kidney. ZN stain.

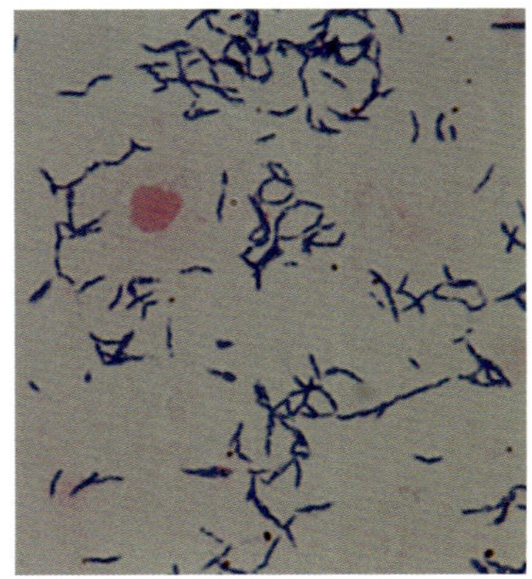
Fig. 43. *Nocardia*, Gram stain.

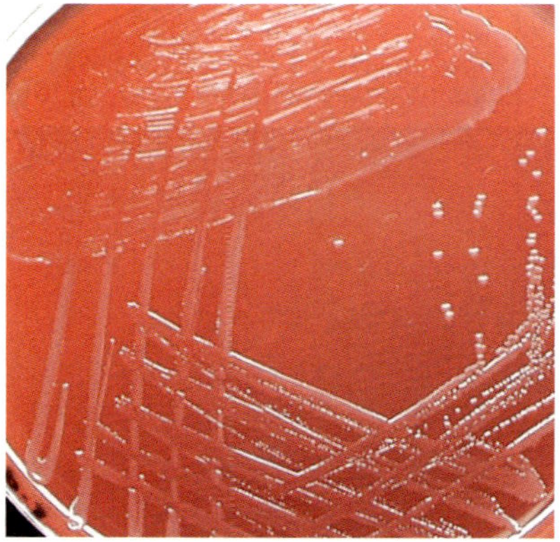

Fig. 44. *Photobacterium damselae* ssp. *damselae* on MSA-B, 2 days, 25°C.

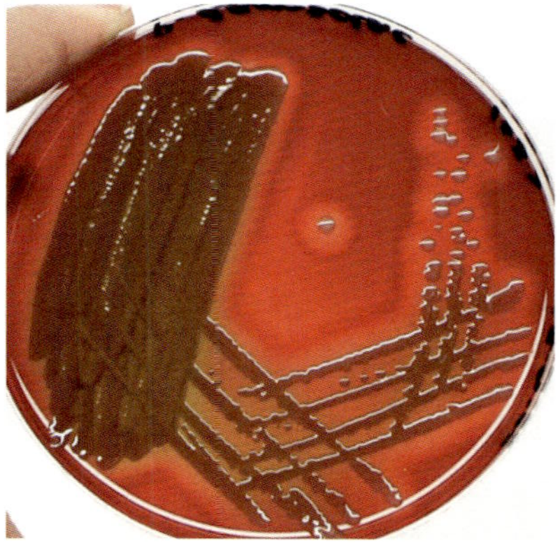

Fig. 45. *Photobacterium damselae* ssp. *damselae* on BA, 48 h.

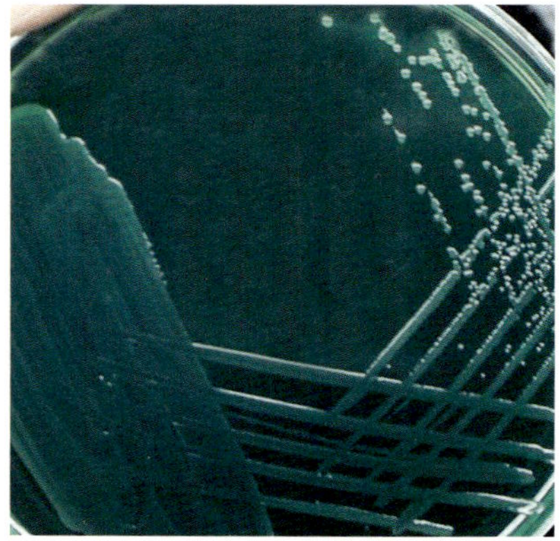

Fig. 46. *Photobacterium damselae* ssp. *damselae* on TCBS, 2 days, 25°C.

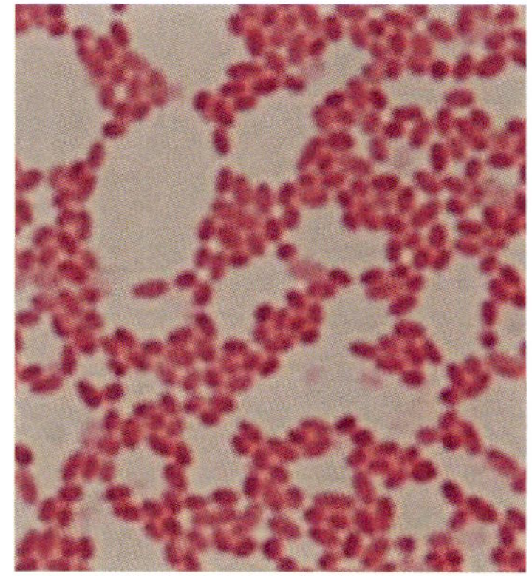

Fig. 47. *P. damselae* ssp. *damselae*, Gram stain.

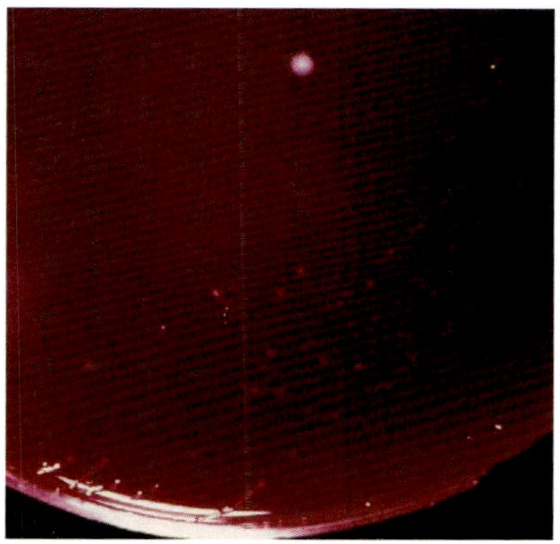

Fig. 48. *Photobacterium damselae* ssp. *piscicida*, 13 days.

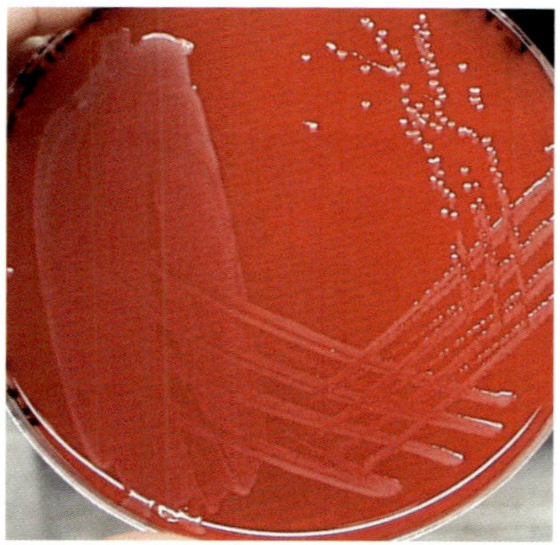

Fig. 49. *Plesiomonas shigelloides* on BA, 24 h, 25°C.

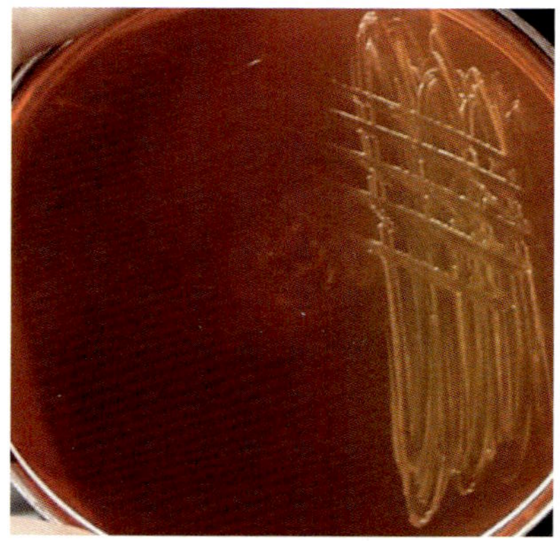

Fig. 50. *Plesiomonas shigelloides* on MCA, 24 h, 25°C.

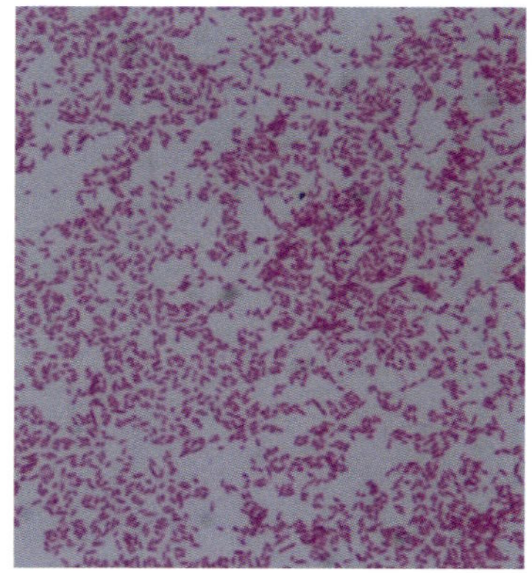

Fig. 51. *Plesiomonas shigelloides*, Gram stain.

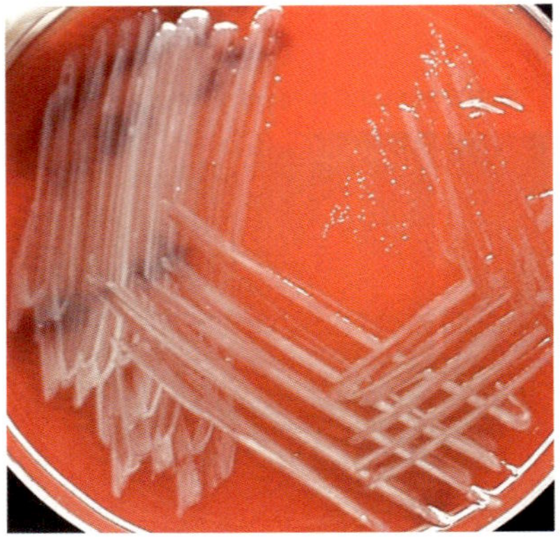

Fig. 52. *Pseudomonas fluorescens* on BA, 24 h.

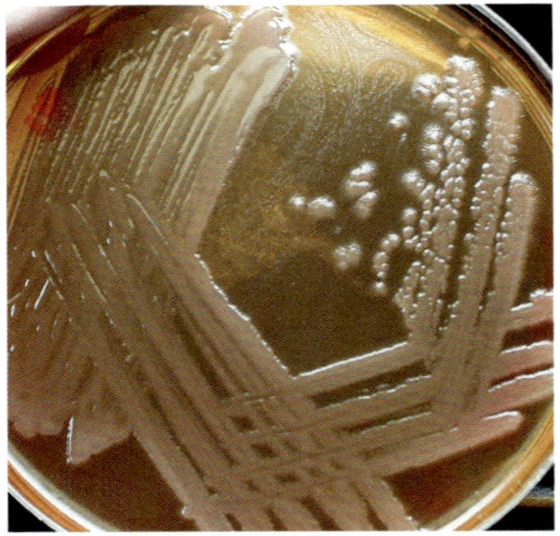

Fig. 53. *Pseudomonas fluorescens* on MCA, 2 days.

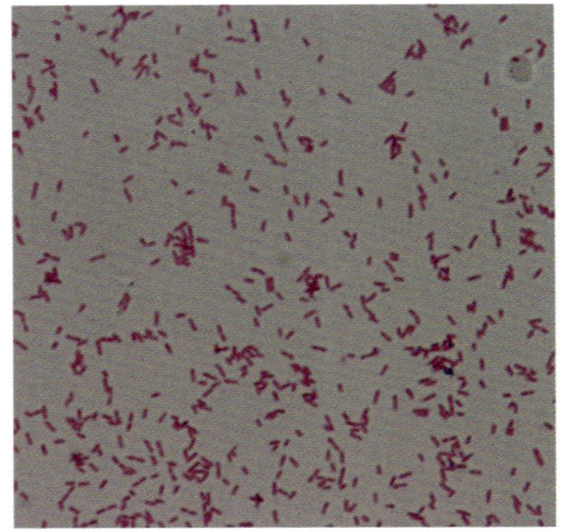

Fig. 54. *Pseudomonas fluorescens*, Gram stain.

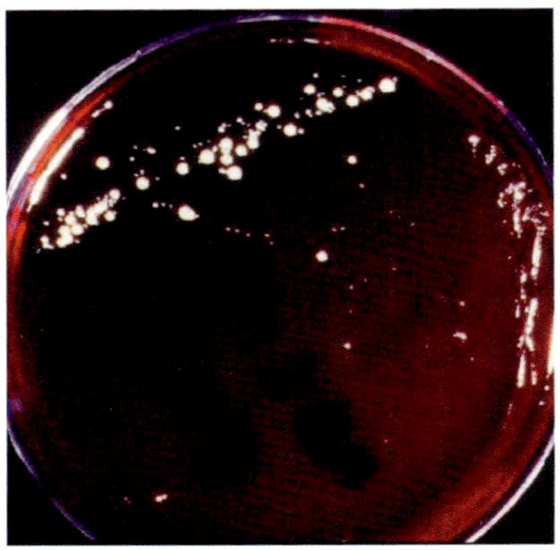

Fig. 55. *Renibacterium salmoninarum*, 2–3 weeks.

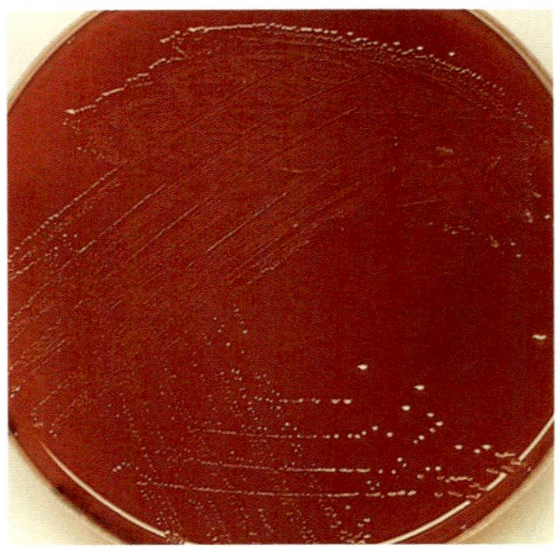

Fig. 56. *Streptococcus* (*difficile*) *agalactiae* Group B on BA, 3 days.

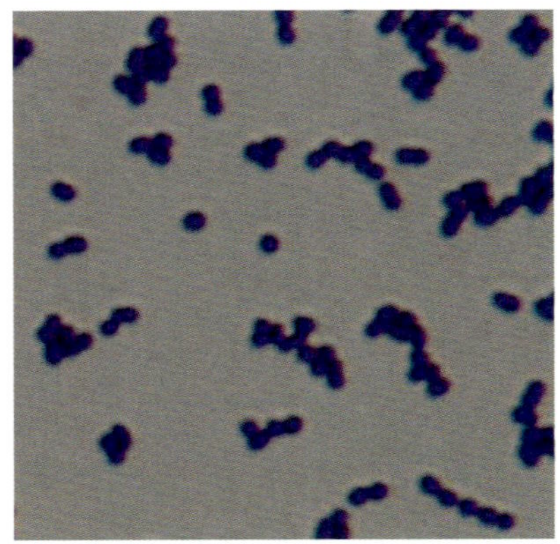

Fig. 57. *Streptococcus* (*difficile*) *agalactiae* Group B, Gram stain.

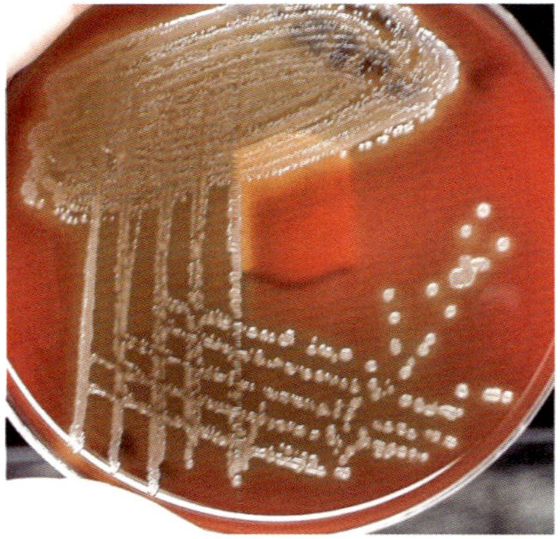

Fig. 58. Strain of *Streptococcus iniae* on BA, 2 days with weak β-haemolysis.

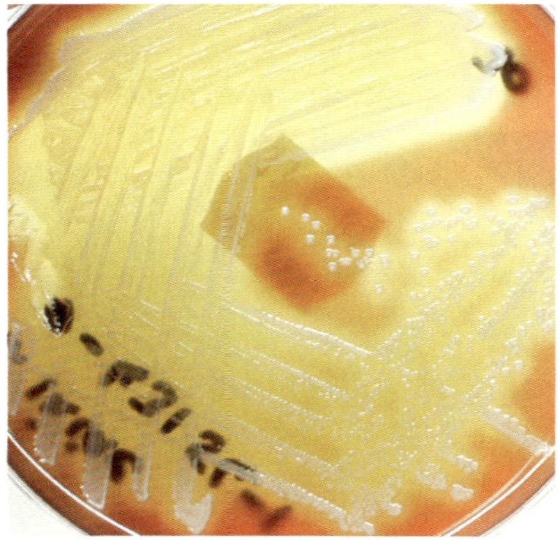

Fig. 59. Strain of *Streptococcus iniae* on BA, 2 days with stronger β-haemolysis.

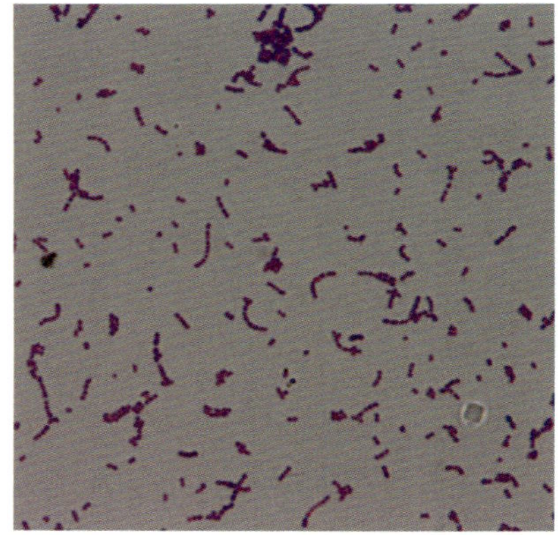

Fig. 60. *Streptococcus iniae*, Gram smear.

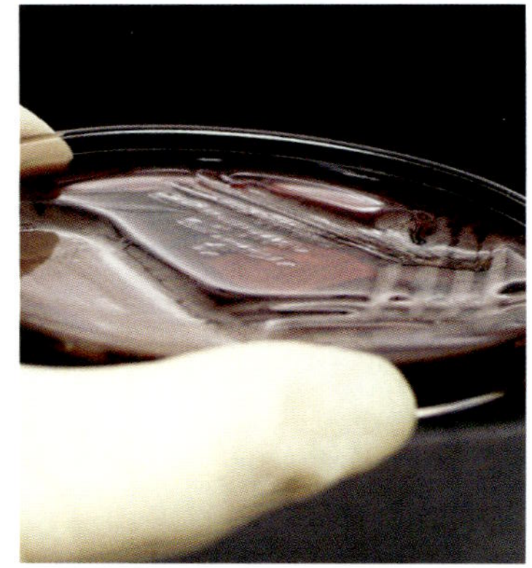

Fig. 61. *Vibrio agarivorans* on MSA-B, 7 days, 25°C.

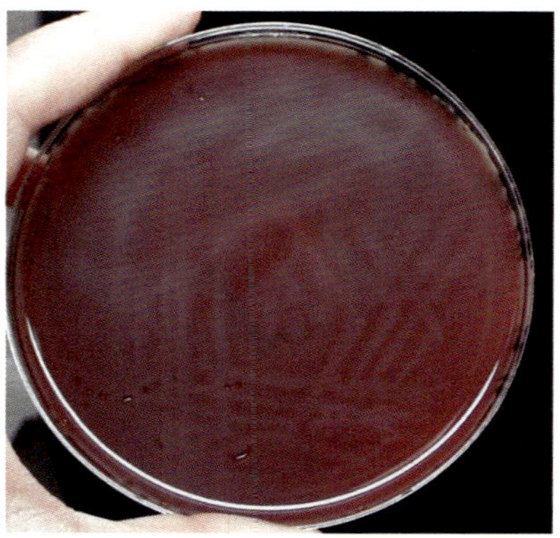

Fig. 62. *Vibrio alginolyticus* on MSA-B, 24 h. Growth has swarmed across plate.

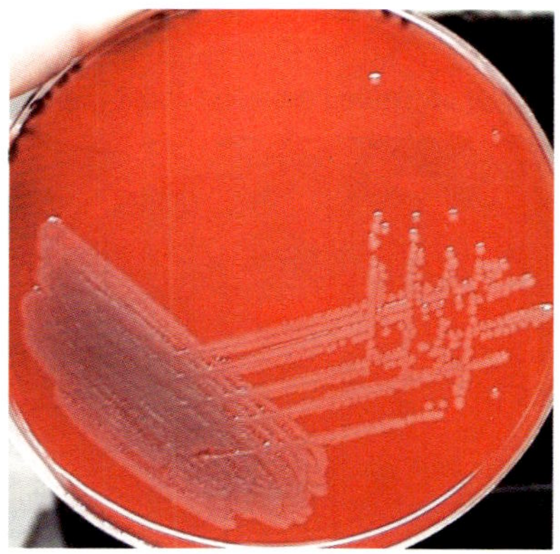

Fig. 63. *Vibrio alginolyticus* on BA, 24 h.

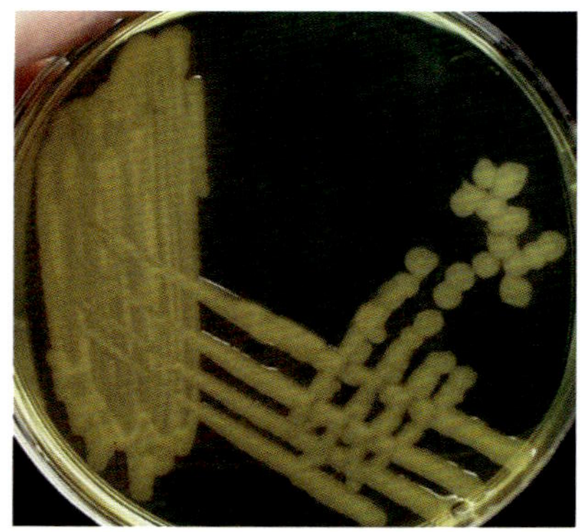

Fig. 64. *Vibrio alginolyticus* on TCBS, 24 h.

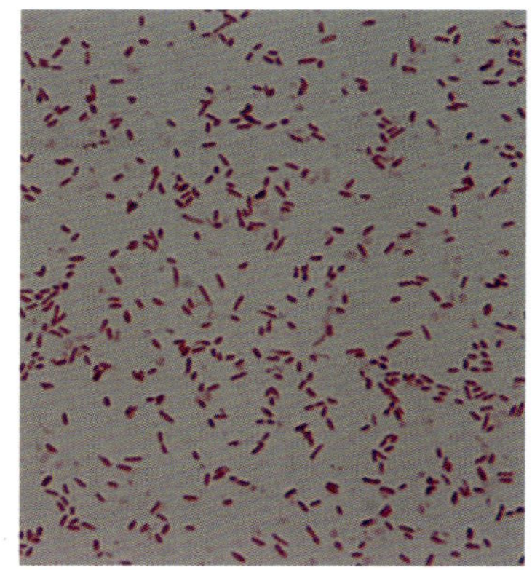

Fig. 65. *Vibrio alginolyticus*, Gram stain.

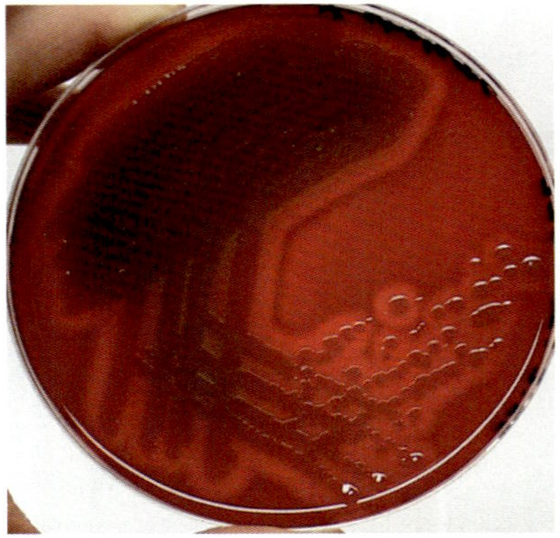

Fig. 66. *Vibrio cholerae* non-01 on BA, 24 h.

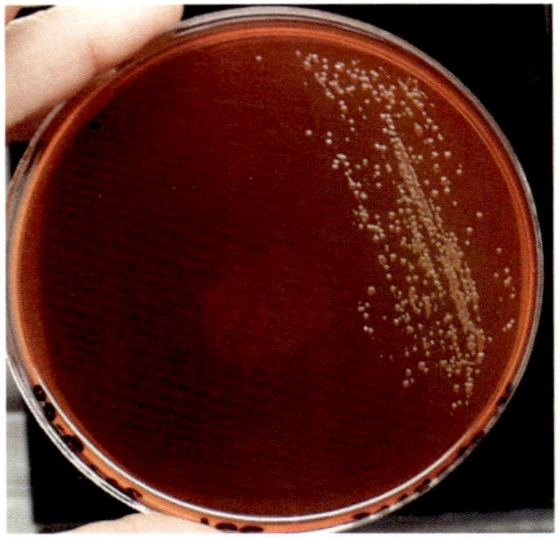

Fig. 67. *Vibrio cholerae* non-01 on MCA.

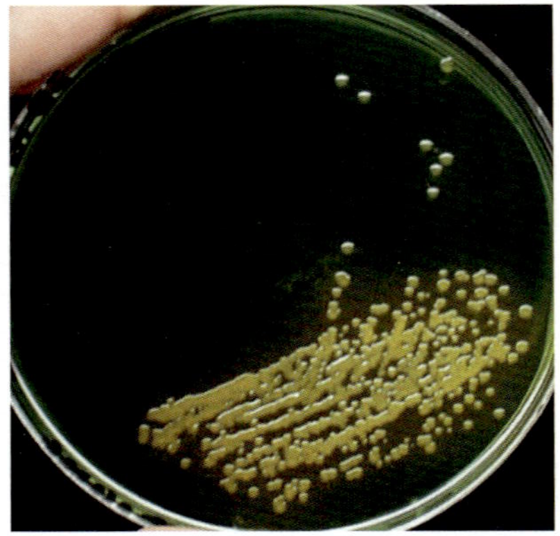

Fig. 68. *Vibrio cholerae* non-01 on TCBS.

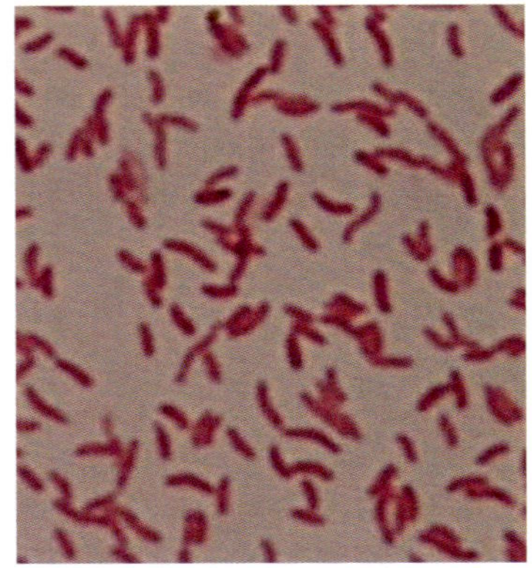

Fig. 69. *Vibrio cholerae*, Gram stain.

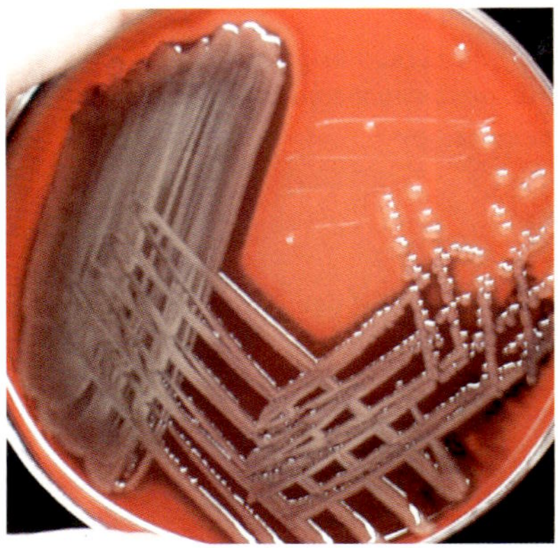

Fig. 70. *Vibrio fluvialis* on BA, 24 h, 25°C.

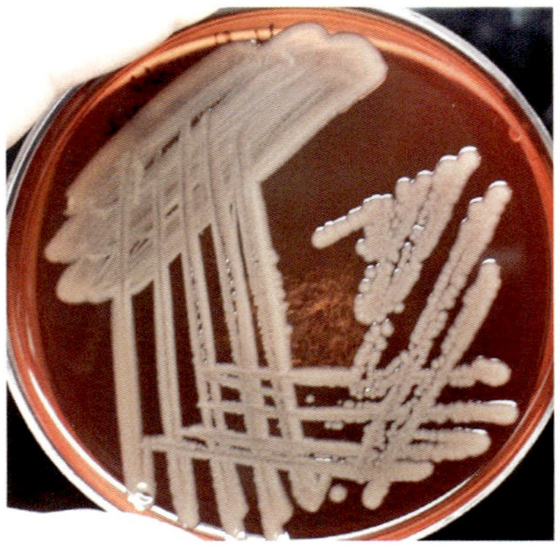

Fig. 71. *Vibrio fluvialis* on MSA-B, 48 h, 25°C.

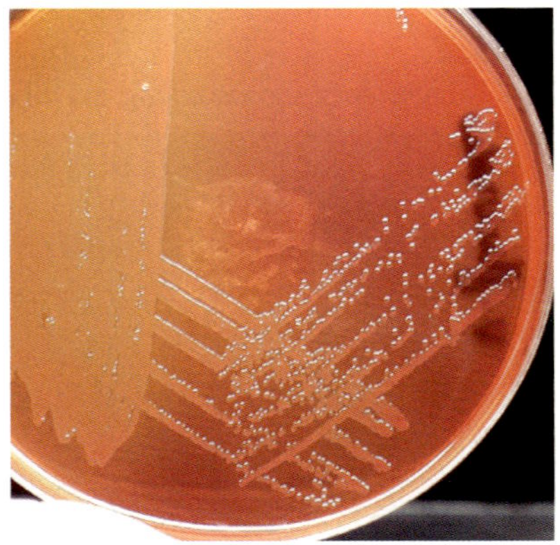

Fig. 72. *Vibrio fluvialis* on MCA, 24 h, 25°C.

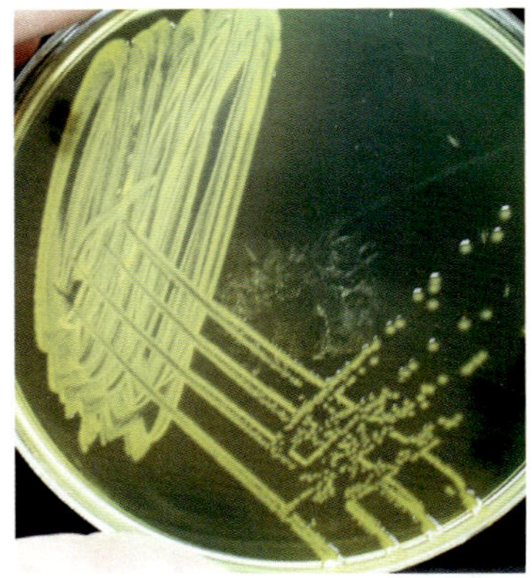

Fig. 73. *Vibrio fluvialis* on TCBS, 24 h, 25°C.

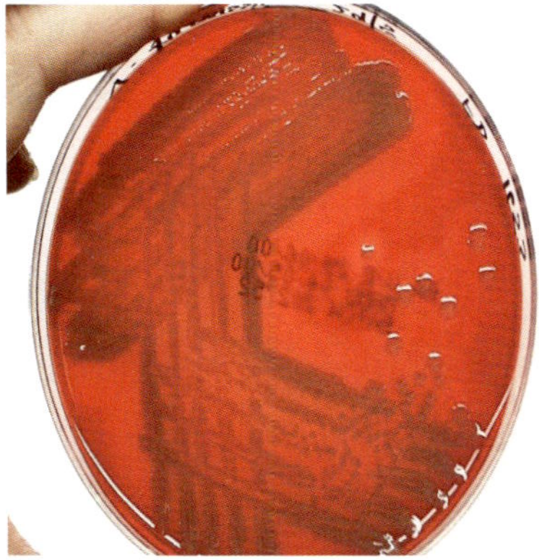

Fig. 74. *Vibrio furnissii* on BA, 48 h.

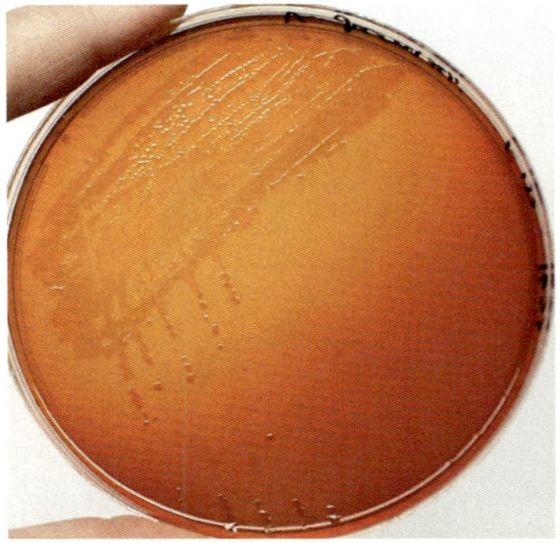

Fig. 75. *Vibrio furnissii* on MCA, 48 h.

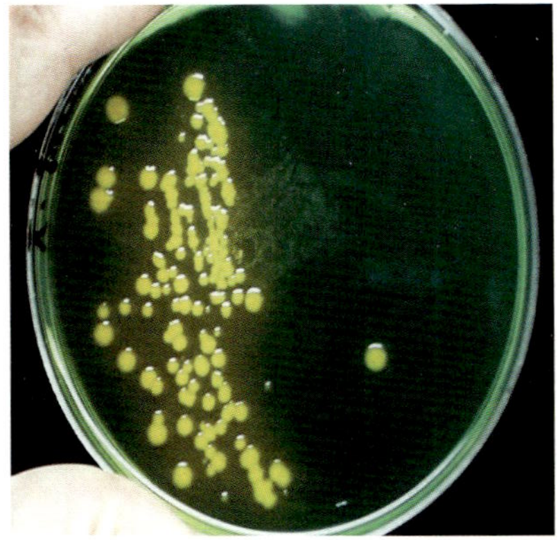

Fig. 76. *Vibrio furnissii* on TCBS, 48 h.

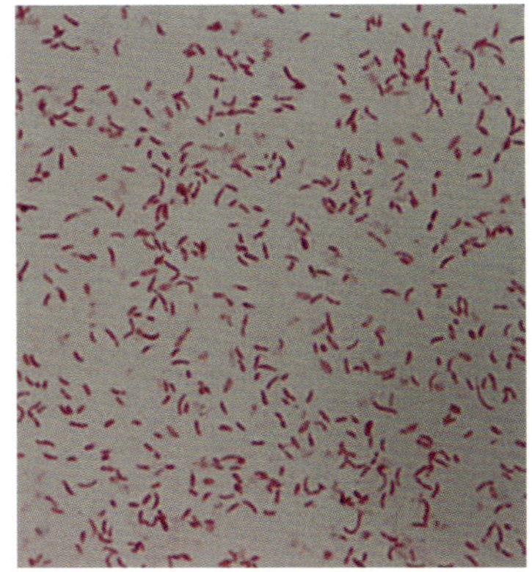

Fig. 77. *Vibrio fluvialis*, Gram stain.

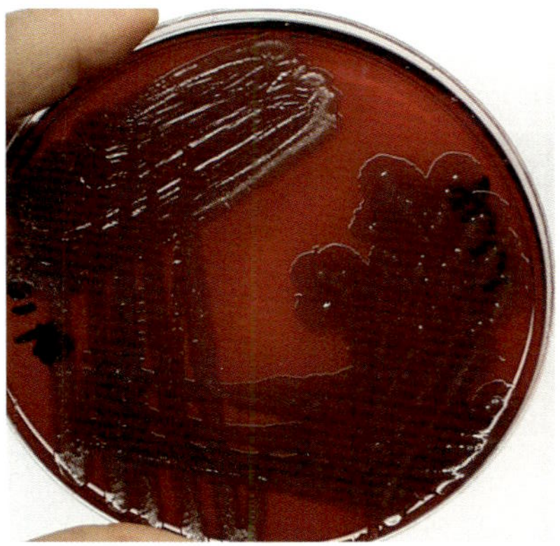

Fig. 78. *Vibrio harveyi* on MSA-B, 48 h.

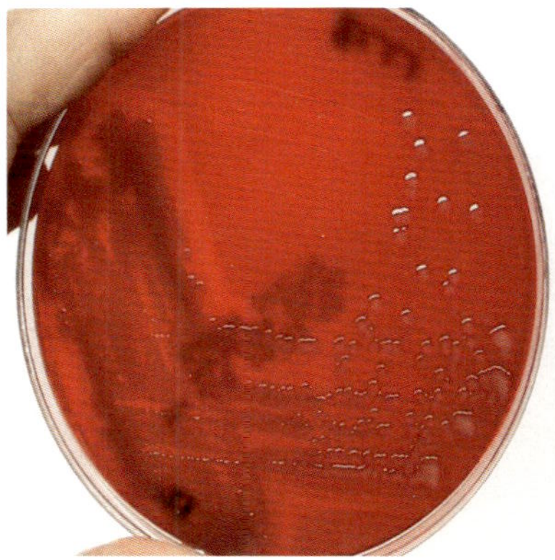

Fig. 79. *Vibrio harveyi* on BA, 48 h.

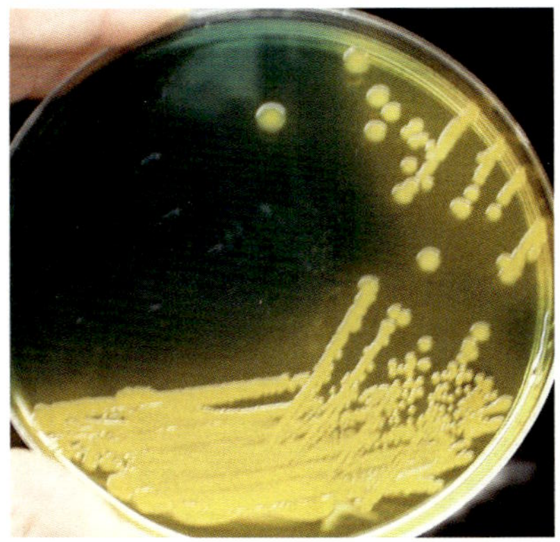

Fig. 80. *Vibrio harveyi* on TCBS, 48 h.

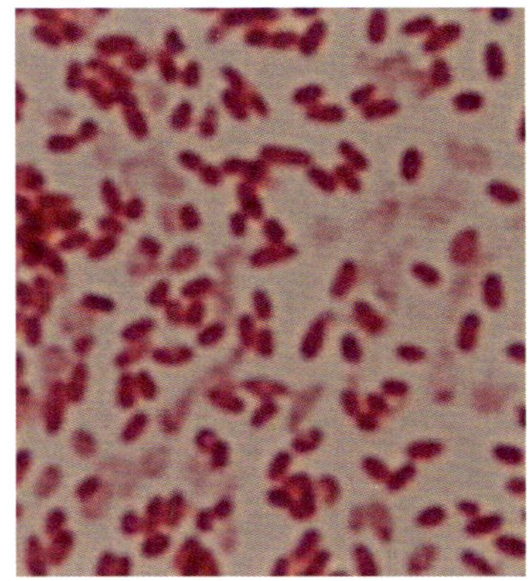

Fig. 81. *Vibrio harveyi*, Gram stain.

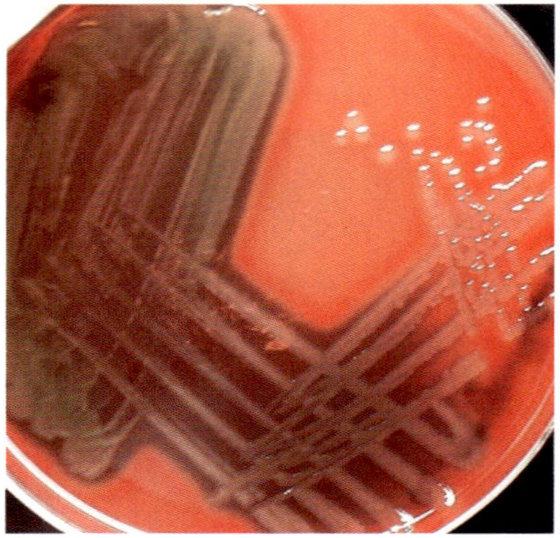

Fig. 82. *Vibrio* (*Listonella*) *anguillarum* on BA, 3 days, 25°C.

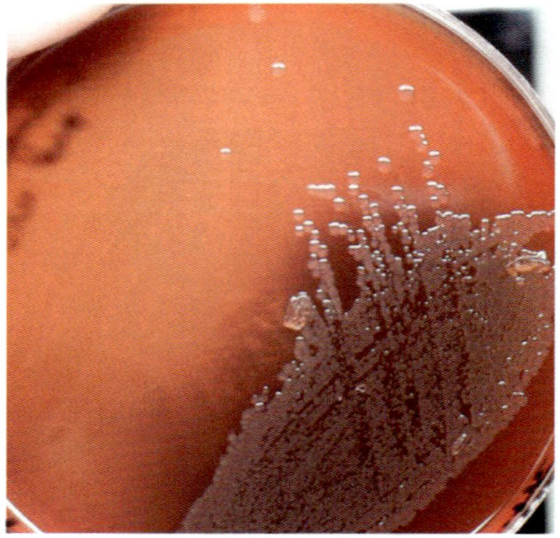

Fig. 83. *Vibrio* (*Listonella*) *anguillarum* on MSA-B, 24 h, 25°C.

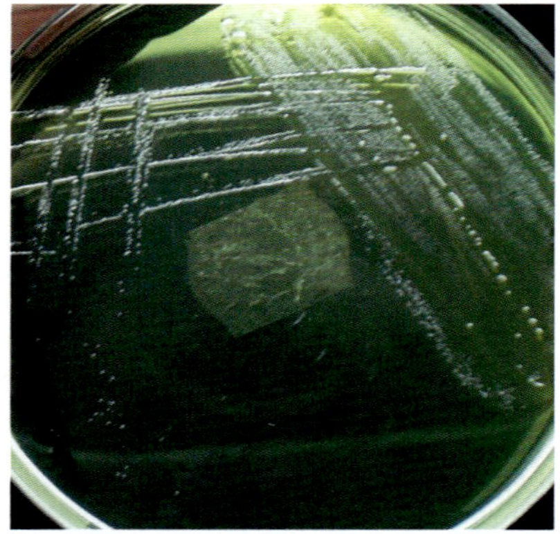

Fig. 84. *Vibrio* (*Listonella*) *anguillarum* on TCBS, 48 h, 25°C.

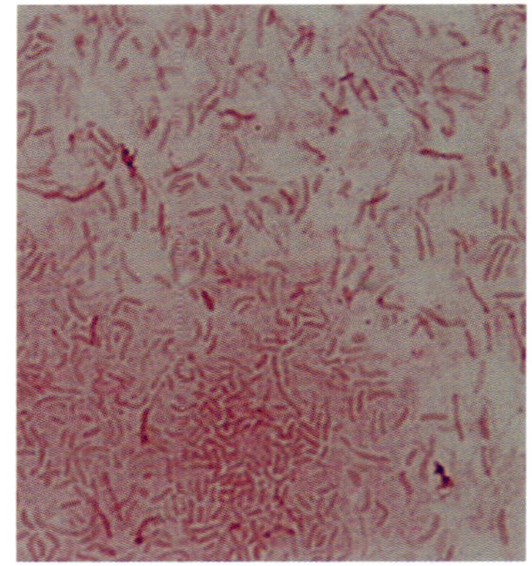

Fig. 85. *Vibrio* (*Listonella*) *anguillarum*, Gram stain.

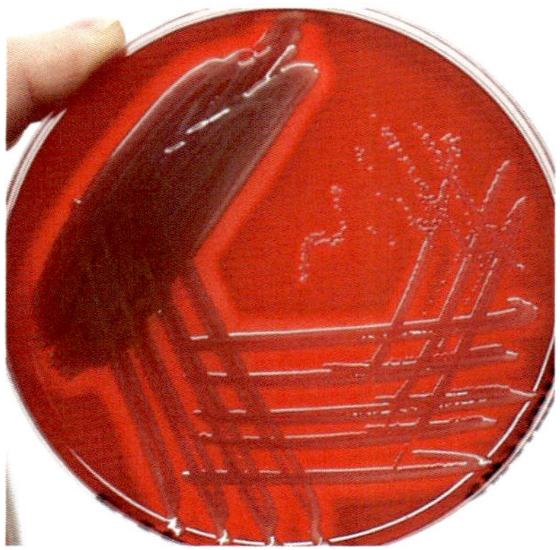

Fig. 86. *Vibrio mimicus* on BA, 24 h.

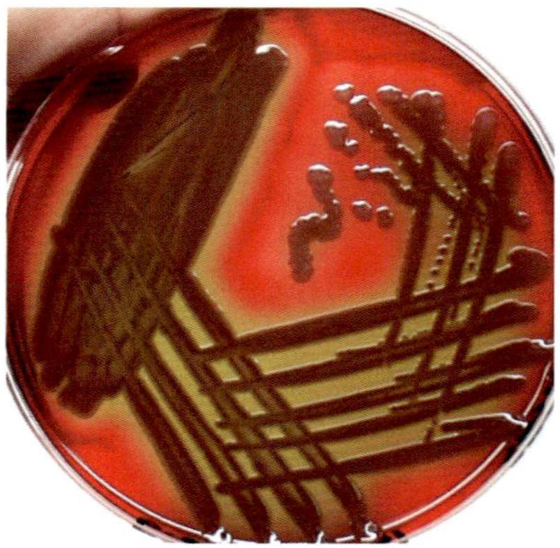

Fig. 87. *Vibrio mimicus* on BA, 48 h.

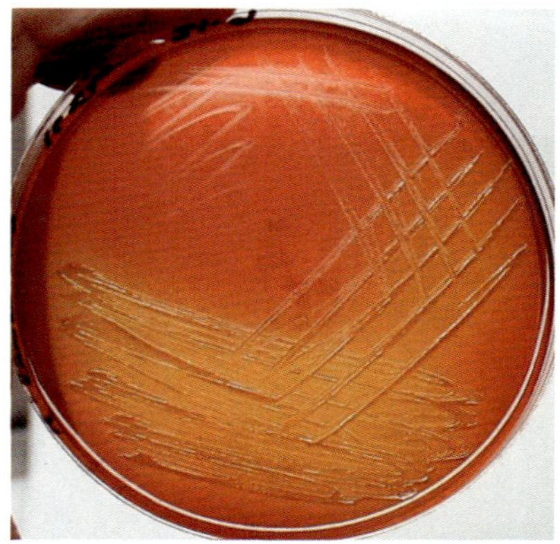

Fig. 88. *Vibrio mimicus* on MCA.

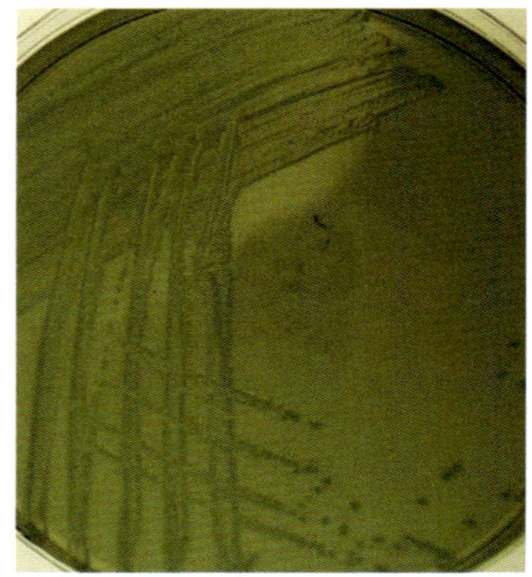

Fig. 89. *Vibrio mimicus* on TCBS.

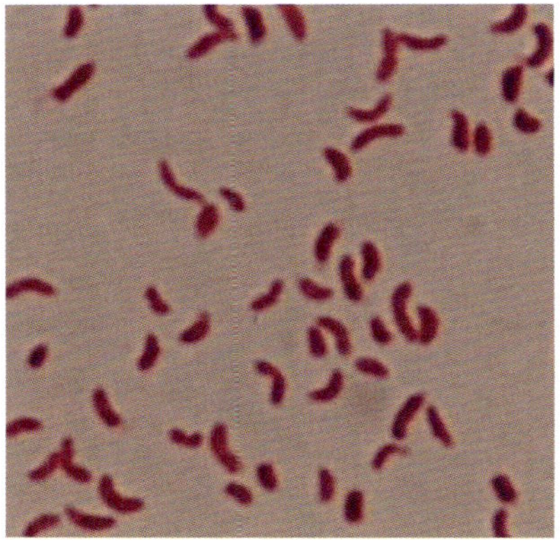

Fig. 90. *Vibrio mimicus*, Gram stain.

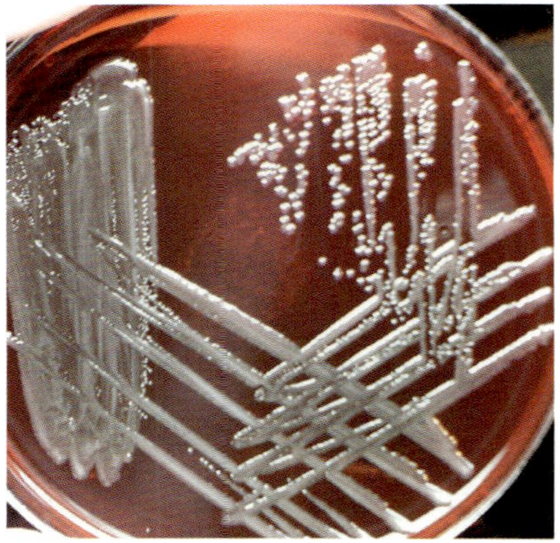

Fig. 91. *Vibrio ordalii* on MSA-B, 2 days.

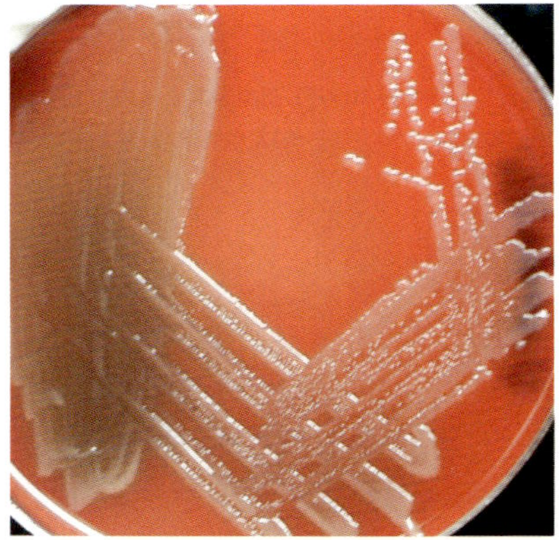

Fig. 92. *Vibrio ordalii* on BA, 3 days, 25°C. No growth on TCBS.

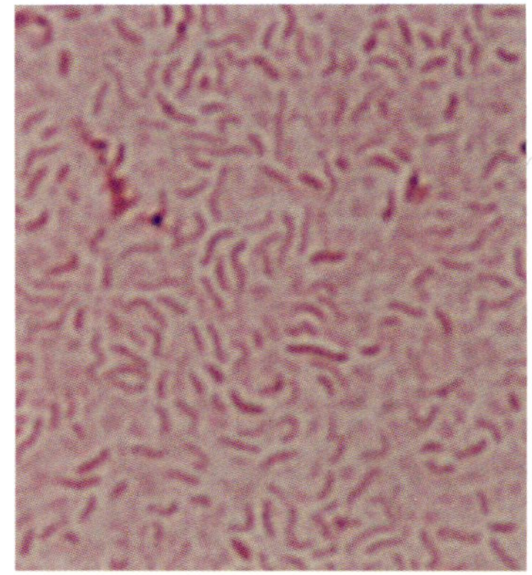

Fig. 93. *V. ordalii*, Gram stain.

Fig. 94. *Vibrio parahaemolyticus* on MSA-B, 48 h. Growth has swarmed across plate.

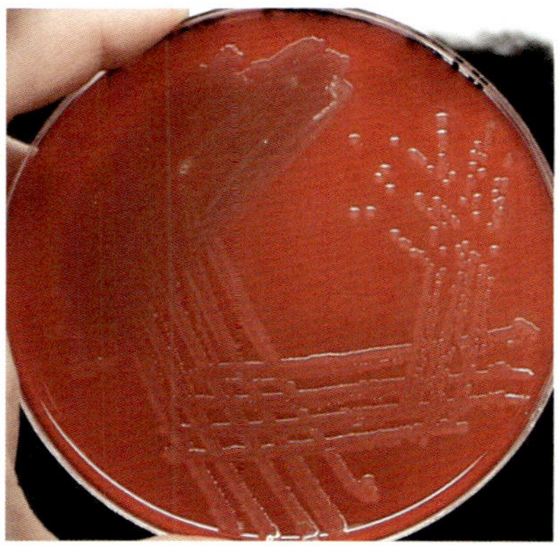

Fig. 95. *Vibrio parahaemolyticus* on BA, 24 h, 25°C.

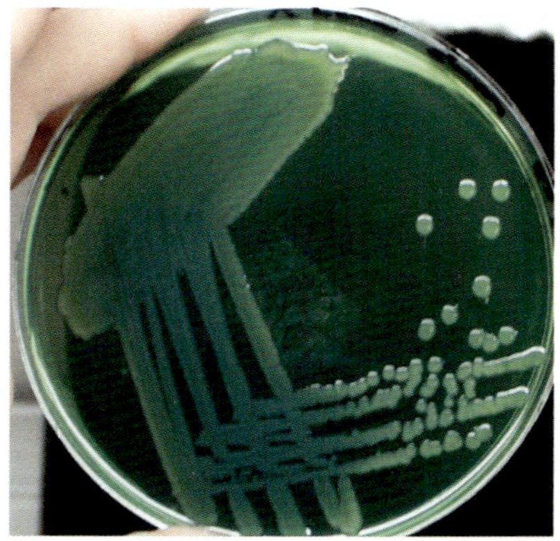

Fig. 96. *Vibrio parahaemolyticus* on TCBS, 24 h.

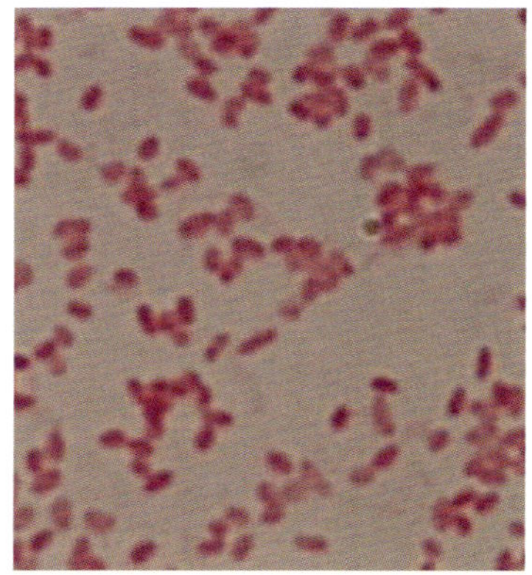

Fig. 97. *Vibrio parahaemolyticus*, Gram smear.

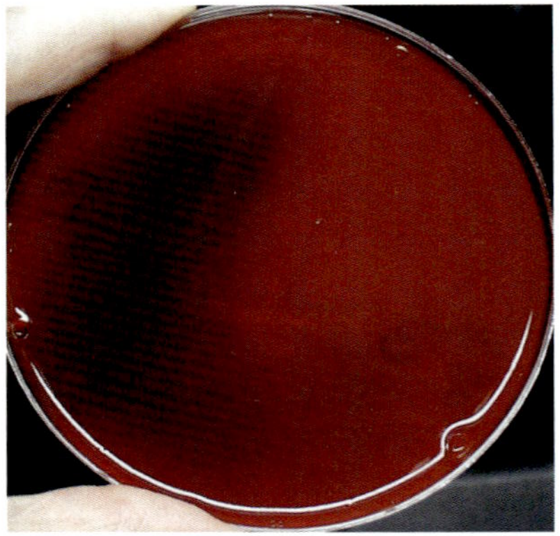

Fig. 98. *Vibrio proteolyticus* on BA, 24 h, 25°C. Showing swarming colonies.

Fig. 99. *Vibrio proteolyticus* completely covers an MSA-B plate at 24 h.

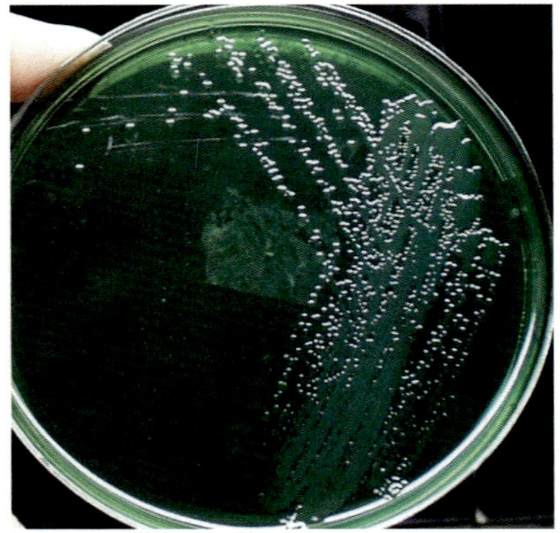

Fig. 100. *Vibrio proteolyticus* on TCBS, 24 h.

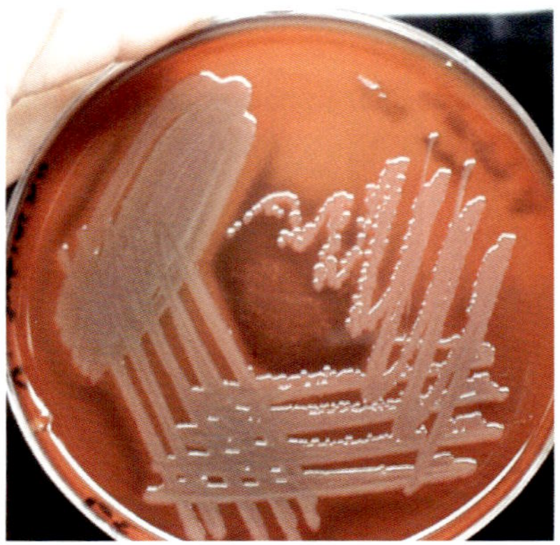

Fig. 101. *Vibrio tubiashii* on MSA-B, 2 days, 25°C.

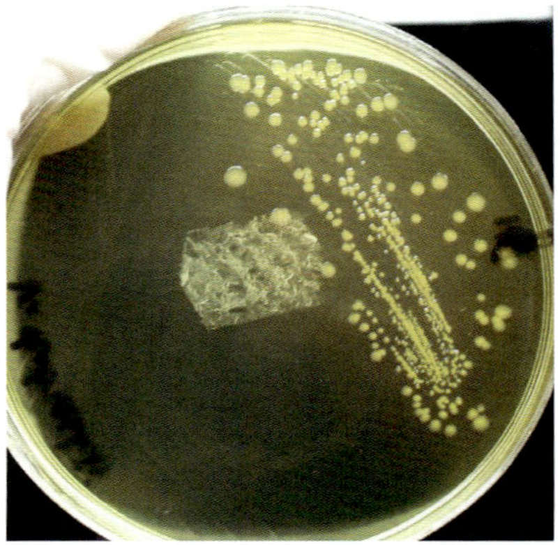

Fig. 102. *Vibrio tubiashii* on TCBS at 3 days, 25°C.

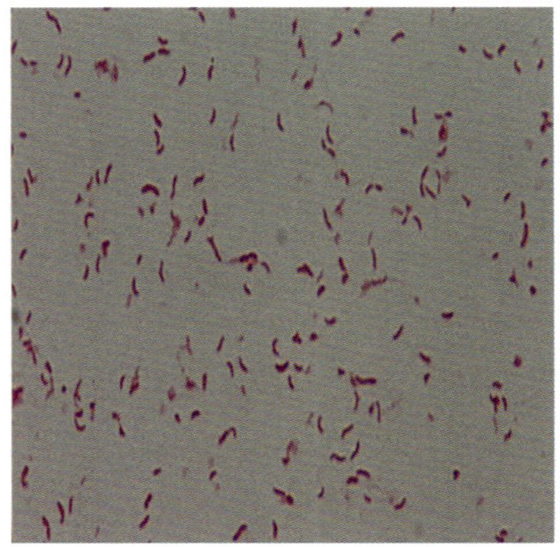

Fig. 103. *V. tubiashii*, Gram stain.

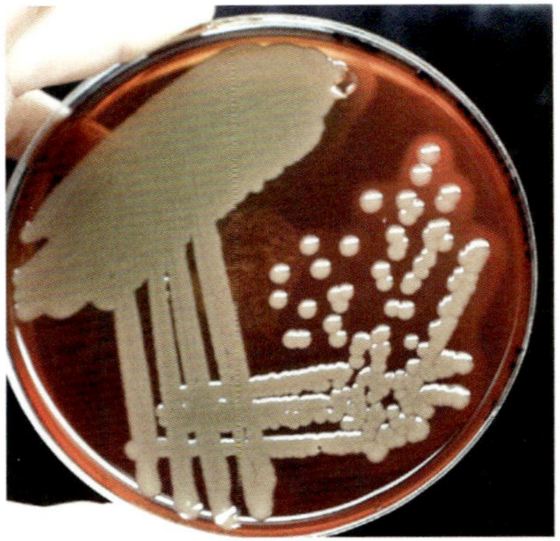

Fig. 104. *Vibrio vulnificus* on MSA-B, 2 days, 25°C.

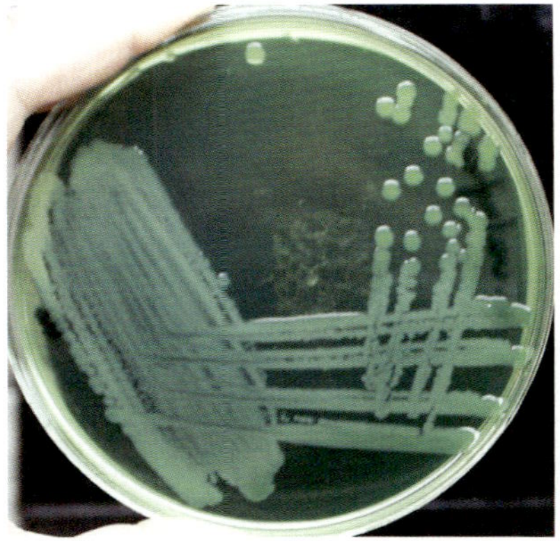

Fig. 105. *Vibrio vulnificus* on TCBS, 2 days, 25°C.

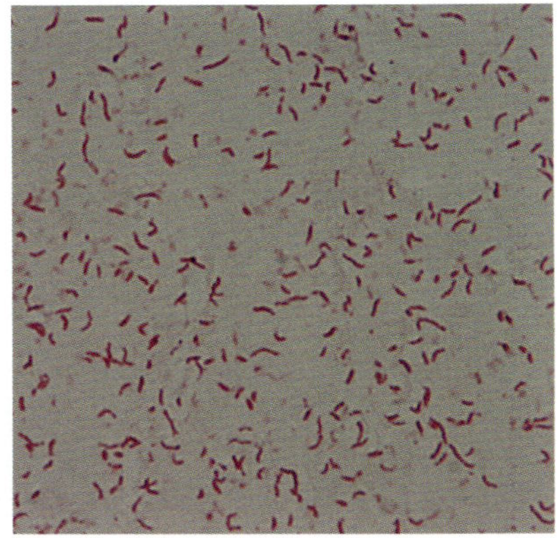

Fig. 106. *Vibrio vulnificus*, Gram stain.

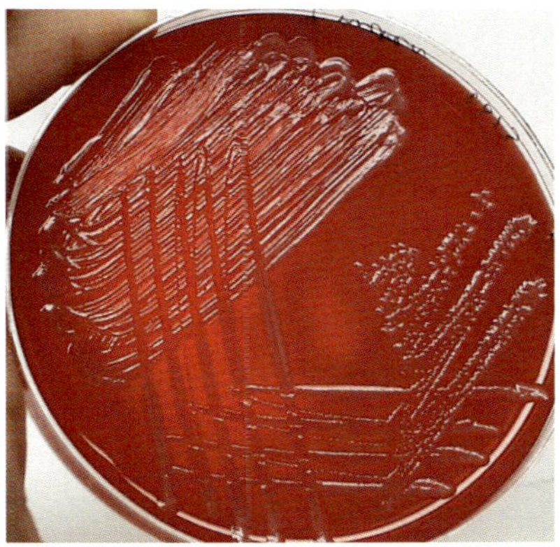

Fig. 107. *Yersinia ruckeri* on BA, 24 h, 25°C.

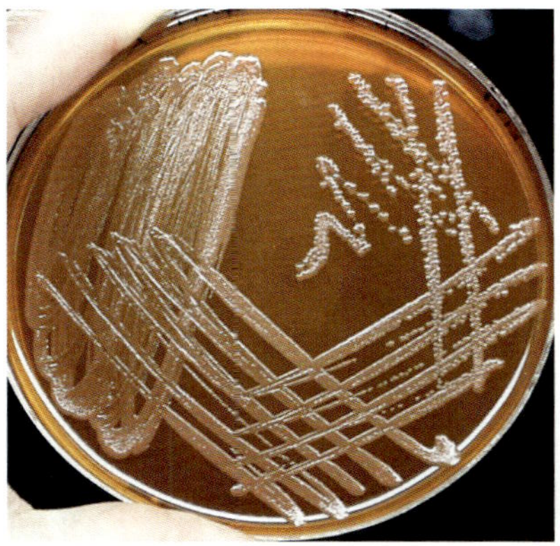

Fig. 108. *Yersinia ruckeri* on MCA, 24 h, 25°C.

Fig. 109. Vibrio identification discs. Top = 150 µg disc (Sensitive). Bottom = 10 µg disc (Resistant).

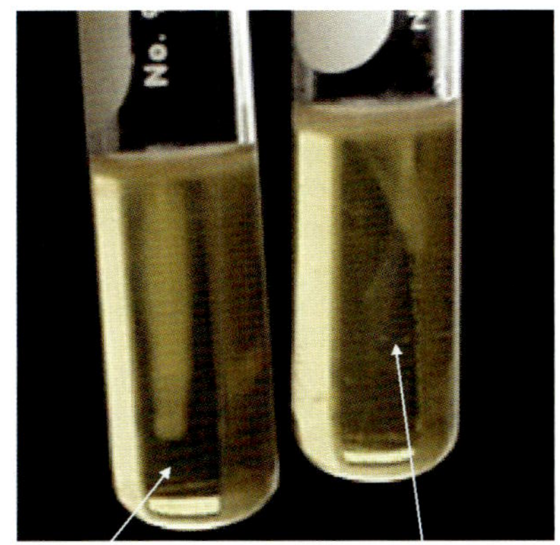

Fig. 110. Motility positive at 25°C (left), but negative at 37°C (right) for *Y. ruckeri.*

Fig. 111. Carbohydrate fermentation reaction: sucrose-positive (yellow), sucrose-negative (red) at 25°C, 24 h.

Fig. 112. Citrate test: *Yersinia ruckeri*, citrate-positive (blue) at 24°C, but citrate-negative (green) at 37°C.

Fig. 113. Decarboxylase reactions: Left to right, arginine dihydrolase (negative), lysine decarboxylase (positive), ornithine decarboxylase (positive), control tube (negative).

Fig. 114. Decarboxylase reactions: Left to right, arginine dihydrolase (positive), lysine decarboxylase (positive), ornithine decarboxylase (negative), control tube (negative).

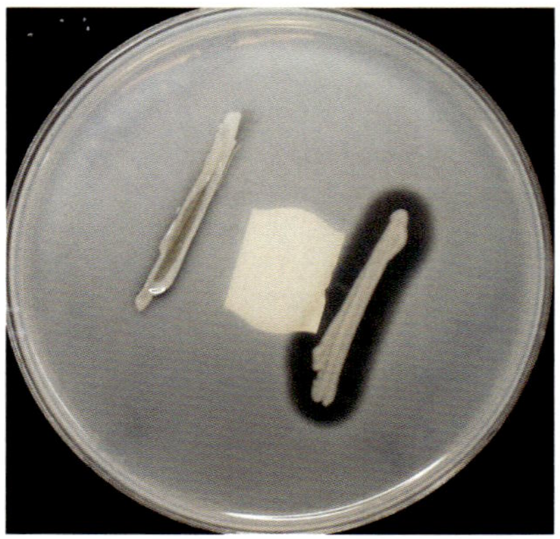

Fig. 115. DNase reaction: Negative reaction on left and positive reaction on right.

Fig. 116. Indole reaction: Negative on the left and positive on the right. *V. parahaemolyticus* with 0.85% NaCl and 2% NaCl, respectively.

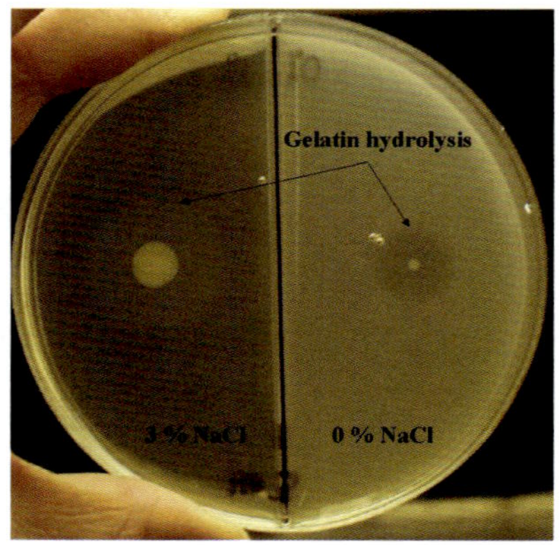

Fig. 117. Plate showing gelatin hydrolysis, and growth on 3% and 0% NaCl.

Fig. 118. Methyl Red reaction: positive.

1

Aquatic Animal Species and Organism Relationship

1.1 Host Species, Bacteria and Disease

This chapter deals with the relationship between the host species and the bacterial flora that may be either part of the normal flora of that host, or are pathogenic for that host. This information is presented in two formats. Table 1.1 lists the aquatic animal hosts in alphabetical order under their Class, followed by Order then Family and then listed in alphabetical order according to their common name. The scientific name is in parentheses. For example, to look for bacteria found in prawns (shrimps) look under the Phylum Arthropoda, Order Decapoda. To search for fish, look under the Phylum Chordata, Class Actinopterygii (ray-finned fish). These are then arranged under Order and Family. For anemones, corals and jellyfish look under the Phylum Cnidaria, Class Anthozoa.

Names listed in journal articles were checked against names according to Fishbase (http://www.fishbase.org), World Register of Marine Species (WoRMS) (http://www.marinespecies.org/index.php) and algaebase (http://www.algaebase.org). The adjacent columns in the table list the bacteria that have been reported to be either pathogens of the host or are considered part of the normal flora, the tissue site of infection, or the pathology presented and the disease state. Some organisms are considered to be opportunistic pathogens, and in a healthy host may be part of the normal flora. In a stressed animal, these same bacteria may overcome host defence mechanisms and cause morbidity or infection in the animal. Some organisms have been identified and isolated from a host but the pathogenicity of the organism is unknown, as virulence studies were not carried out.

In the second format in Table 1.2, the information is presented by listing the bacteria in alphabetical order with adjacent columns listing the name of the disease, the tissue site where the organisms may be found, the aquatic animals where the organism has been reported and the geographical location of the disease.

1.2 Bacteria and Relationship to Host

Table 1.2 lists the bacteria that may be pathogens or saprophytes of fish and other aquatic animals. The information is presented in tabular form summarizing the information in the text. Not all bacterial species listed are presented in the text.

The bacteria are listed in alphabetical order. The 'disease' column indicates the status of the organism as saprophyte, environmental organism or a pathogen of fish and other aquatic animals. Under 'disease signs' clinical information is summarized or the site of bacterial infection is listed. Under the 'aquatic species' and 'distribution' headings is information regarding the aquatic animals and the country where the organism was isolated according to the references cited. The final column lists the reference number.

Table 1.1. Host species and organism relationship.

Host	Pathogens	Normal flora	Tissue site	Disease status	Ref
KINGDOM ANIMALIA					
Abalone see under Phylum Mollusca, Class Bivalvia					
PHYLUM ANNELIDA					
CLASS Polychaeta					
Worm, Pompeii worm (deep-sea polychaete annelid, *Alvinella pompejana*)		*Vibrio diabolicus*		Environmental organism	1368
PHYLUM ARTHROPODA					
SUBPHYLUM CRUSTACEA					
ORDER Amphipoda					
Talorchestia sp.		*Vibrio harveyi*		Pathogenicity unknown	842
ORDER Anostraca					
Brine shrimp (*Artemia* species)	1. *Vibrio proteolyticus*	2. *Vibrio alginolyticus* 3. *Vibrio chagasii* 4. *Vibrio hispanicus*	1. Affects microvilli, disrupts gut epithelial cell junctions, devastates cells and tissues in the body cavity	1. Deaths 2. May be the predominant normal flora 3. Pathogenicity not known 4. No disease	244 636 1625 1635 1720
ORDER Decapoda					
Family Cambaridae					
Crawfish, American freshwater crayfish, Louisiana crawfish, red swamp crawfish (*Procambarus clarkii*) Family Cambaridae	*Vibrio cholerae* *Vibrio mimicus*			Mortalities	1073
Family Majidae					
Spider crab (*Maja brachydactyla* Balss)		*Photobacterium swingsii*, *Vibrio atypicus*, *Vibrio campbellii*, *Vibrio crassostreae*, *Vibrio cyclitrophicus*, *Vibrio fortis*, *Vibrio hepatarius*, *Vibrio nigripulchritudo*, *Vibrio rotiferianus*, *Vibrio splendidus*, *Vibrio tasmaniensis*	Haemolymph	Normal flora at less than 10^{3-4} CFU/ml	635 641 643
Family Nephropidae					

American lobster (*Homarus americanus*, *Homarus gammarus* L.)	1. *Aerococcus viridans* var *homari* 2. *Vibrio* (*Listonella*) *anguillarum*	3. *Vibrio metschnikovii*	1. Organism in haemolymph	1. Gaffkemia, high mortality 2. Vibriosis 3. Non-pathogenic	75 212 621 994 1547 1577 1583 1792 1793
Family Oregoniidae					
Tanner crab (*Chionecetes bairdi*)	*Aliivibrio logei*		Necrotic lesions	Pathogenicity unknown	112
Family Palaemonidae					
Giant freshwater prawn (*Macrobrachium rosenbergii*)	1. *Aeromonas caviae* 2. *A. veronii* biovar Veronii 3. *Lactococcus garvieae* 4. *L. lactis* ssp. *lactis* 5. *Vibrio alginolyticus*	6. *Flavobacterium macrobrachii*	3, 4. Yellowish-white spots in muscle, whitish muscle and swollen, yellow hepatopancreas, fluid accumulation between cuticle and muscle tissue 5. Anorexia, necrotic appendages, deaths 6. Pond water	1, 2. Pathogenic in virulence studies 3, 4. Mortality 4, 5. Pathogenic in experimental challenge 6. Environmental isolate	311 835 1514 1594 1763
Family Palinuridae					
Ornate spiny lobster (*Panulirus ornatus*)	*Vibrio owensii*			Pathogenic for cultured larvae in virulence studies	270
Southern rock lobster (*Jasus edwardsii*)	*Vibrio jasicida*		Phyllosoma larvae	Mortality. Pathogenicity confirmed by experimental challenge	450 1835
Spiny rock lobster or packhorse rock lobster (*Sagmariasus* [*Jasus*] *verreauxi*)	*Vibrio jasicida*		Phyllosoma larvae	Mortality. Pathogenicity confirmed by experimental challenge	450 1835
Family Parastacidae					
Australian freshwater crayfish yabby (*Cherax albidus*) red claw (*Cherax quadricarinatus*)	*Vibrio mimicus*		Haemolymph	Mortality, vibriosis	244 469 1812
Family Penaeidae					
Litopenaeus schmitti Burkenroad	1. *Vibrio harveyi*		1. Hepatopancreas	1. Haemorrhagic septicaemia, mortality	45

Continued

Table 1.1. Continued.

Host	Pathogens	Normal flora	Tissue site	Disease status	Ref
Litopenaeus (*Penaeus*) *stylirostris*	1. *Vibrio nigripulchritudo* 2. *Vibrio penaeicida* 3. *Vibrio harveyi* 4. *Vibrio sinaloensis*		1, 2. Septicaemia 3. Black spots on exoskeleton	1. Summer syndrome 2. Syndrome 93 3, 4. Mortality	45 371 543 625 626 627
Litopenaeus (*Penaeus*) *vannamei* Boone (Pacific white shrimp, whiteleg shrimp, penaeid shrimp or prawn)	1. *Mycobacterium peregrinum* 2. *Pseudomonas* species 3. Streptococcus Group B (non-haemolytic). Identified as *Streptococcus uberis* in API 20 Strep kit 4. *Vibrio alginolyticus* 5. *Vibrio campbellii* 6. *Vibrio harveyi* 7. *Vibrio penaeicida*	8. *Tenacibaculum litopenaei* 9. *Vibrio communis* 10. *Vibrio fortis* 11. *Vibrio hepatarius* 12. *Vibrio xuii*	1. Multifocal, melanized nodular lesions on carapace 2. Pink colour in larvae and tank water 3. Haemolymph, internal organs 4. Necrotic musculature 5. Haemolymph 6. Black spots on exoskeleton, red spots on abdomen. 8. Isolated from shrimp culture pond water 9. Isolated from digestive gland	1. Opportunistic infection 2. Mortality 3. Streptococcosis, 100% mortality in pathogenicity tests 4. Pathogenicity proven in experimental challenge 5. Co-infection with WSSV enhances mortality in experimental challenge 6. Mortality, bright-red syndrome 7. Virulent in challenge tests 8, 9. Possible normal flora 10, 12. Pathogenicity not known 11. Probiotic and health improvement	11 45 320 688 1034 1162 1334 1517 1557 1562 1632 1634
Black tiger prawn (*Penaeus monodon*)	1. *Photobacterium damselae* ssp. *damselae* 2. *Vibrio alginolyticus* 3. *Vibrio campbellii* 4. *Vibrio harveyi* 5. *Vibrio parahaemolyticus* 6. *Vibrio splendidus* II 7. *Vibrio owensii*		1. Pale yellow to black gills 2. White spots on carapace 3. Colonizes gut 4, 5, 6. Infects hepatopancreas with an inflammatory response in the intertubular sinuses 5. Highly virulent to prawn larvae in virulence studies	2. Mortality. Pathogenicity proven in experimental challenge 3, 4, 5, 6, 7. Mortalities 2. Pathogenicity proven with gnotobiotic *Artemia*	270 431 671 840 998 1434 1505 1705 1759

India prawn or shrimp (*Penaeus indicus*)	*Photobacterium damselae* ssp. *damselae*		Pale yellow to black gills	Mortality. Pathogenicity proven in experimental challenge	1705
Chinese prawn or shrimp see *Fenneropenaeus chinensis*					
Fairy shrimp (*Branchipus schaefferi* Fisher, *Chirocephalus diaphanus* Prévost, *Streptocephalus torvicornis* Waga)	*Aeromonas hydrophila*		Black nodules on thoracic appendages, cercopods and antennae	Black disease	449
Fenneropenaeus chinensis Referred to as Chinese shrimp, fleshy prawn or oriental shrimp. *Penaeus chinensis* is incorrect	1. *Vibrio alginolyticus* 2. *Vibrio harveyi* 3. *Vibrio vulnificus* serogroup E	4. *Vibrio pacinii* 5. *Vibrio kanaloae* 6. *Vibrio atypicus*	1, 2. Affects larval development from zoea stage onwards. Inactivity, anorexia, opaqueness and larvae settle to the bottom of the tank 6. Digestive tract	1, 2, 3. Vibriosis, mortalities 4. Isolated from healthy shrimp 5. Pathogenicity not determined 6. Part of digestive tract microflora	180 638 1698 1635 1767
Kuruma prawn see *Marsupenaeus japonicus*					
Marsupenaeus (*Penaeus*) *japonicus* (kuruma shrimp/prawn)	1. *Vibrio alginolyticus* 2. *Vibrio harveyi* 3. *Vibrio penaeicida*	4. *Acinetobacter* sp., *Alteromonas* sp., *Bacillus* sp., *Corynebacterium* sp., *Flavobacterium* sp., *Micrococcus* sp., *Moraxella* sp., *Pseudomonas* sp., *Staphylococcus* sp.	1. White spots on carapace 2. Black spots on exoskeleton 3. Septicaemia 4. Isolated from apparently healthy shrimp, but not from diseased shrimp	1. Mortality. Pathogenicity proven in experimental challenge 2, 3. Mortality 4. Non-pathogenic	45 803 998 1042
Prawn	1. *Aeromonas jandaei*	2. *Vibrio mimicus*		1. Mortality 2. Source of food poisoning in humans	276 412 497
Family Portunidae					
Crab		*Vibrio aestuarianus*		Present in apparently healthy crabs	1645
Blue crab (*Callinectes sapidus*)		*Vibrio cholerae*-like (2% of crabs) *Vibrio parahaemolyticus* (23% of crabs) *Vibrio vulnificus* (7% of crabs)	Haemolymph, digestive tract	Present in apparently healthy crabs	415
Mud crab	*Vibrio cincinnatiensis*			Mortality, disease	220

Continued

Table 1.1. Continued.

Host	Pathogens	Normal flora	Tissue site	Disease status	Ref
Swimming crab (*Portunus trituberculatus*), ORDER Decapoda, Family Portuninae	*Vibrio harveyi* (initially called *Vibrio* sp. zoea)			Mass mortality in zoeal larvae	805
ORDER Isopoda					
Wood-boring crustacean (*Limnoria tripunctata*)		*Vibrio proteolyticus*	Intestine	Possible normal flora	1133
PHYLUM CHORDATA					
CLASS Actinopterygii – ray-finned fish (a class or subclass of the bony fishes)					
ORDER Acipenseriformes (sturgeons and paddlefishes)					
Family Acipenseridae (sturgeons)					
Adriatic sturgeon (*Acipenser naccarii* Bonaparte)	*Lactococcus garvieae*		Inappetence, irregular swimming, some bilateral exophthalmia and abdominal ascites	Mortality	1466
Danube sturgeon (*Acipenser gueldens-taedtii*)	*Flavobacterium johnsoniae*		Skin lesions on ventral body and fins	Mortality	867
Siberian sturgeon (*Acipenser baerii baerii* Brandt)	*Yersinia ruckeri*		Can be infected with or without clinical signs	Yersiniosis	1737
White sturgeon (*Acipenser transmontanus*)		*Edwardsiella ictaluri*		Not susceptible to infection in bath immersion trials	131
ORDER Anguilliformes (eels and morays)					
Family Anguillidae					
African longfin eel (*Anguilla mossambica* Peters)	*Flavobacterium johnsoniae*		Skin ulcers	Mortality and disease	542
American eel (*Anguilla rostrata* Lesueur)	1. *Aeromonas salmonicida*		1. Focal lesions progress to large de-pigmented necrotic patches then detach at dermo-epidermal junction to form large ulcers that expose underlying muscle	1. Ulcerative skin disease, morbidity	1232

European eel (*Anguilla anguilla* Linnaeus)	1. *Aeromonas allosaccharophila* 2. *Aeromonas hydrophila*, *Aeromonas jandaei* 3. Atypical *Aeromonas salmonicida* 4. *Edwardsiella piscicida* 5. *Edwardsiella tarda* 6. *Flavobacterium psychrophilum* 7. *Pseudomonas anguilliseptica* 8. *Vibrio anguillarum* serotype 05 9. *Vibrio furnissii* 10. *Vibrio vulnificus* serovar O4 11. *Yersinia ruckeri*	12. *Aeromonas encheleia*	2. Kidney, liver 3. Skin lesions 5. Ulcers on body and internal lesions 6. Lesions on fins, peritoneum and intestine 7. Subcutaneous haemorrhages. Abdominal distension, organism in internal organs 8. Lesions 9. Haemorrhages in intestinal tract 10. Organism cultured from gills, intestinal contents, kidney, spleen 11. Can be infected with or without clinical signs	2, 8. Mortality. Virulent in experimental challenge 3. Infection 4. Pathogenic in experimental challege with zebrafish 6. Systemic infection 7. Septicaemia 8. Vibriosis 1, 10. Disease 11. Yersiniosis 12. Non-pathogenic	1 21 276 396 411 493 496 497 555 674 741 1004 1102 1136 1559
Japanese eel (*Anguilla japonica* Temminck and Schlegel)	1. Atypical *Aeromonas salmonicida* 2. *Edwardsiella piscicida* 3. *Edwardsiella tarda* 4. *Flavobacterium columnare* 5. *Lactococcus garvieae* 6. *Pseudomonas anguilliseptica* 7. *Vibrio vulnificus* biogroup 3, serovar E contains virulent and avirulent strains	Other biotypes of *V. vulnificus* are non-pathogenic for eels	3. Septicaemia, focal suppurative or granulomatous lesions, cutaneous ulcerations, abscessed or ulcerative lesions in kidney or liver 4. Gills 6. Petechiae on body. Haemorrhagic and ulcerative lesions. Lesions in mouth, operculum, brain, liver and kidney 7. Lesions	1. Pathogenic in experimental challenge with zebrafish 3. Edwardsiellosis, 'paracolon disease' 4. Columnaris disease 5. Mortality 6. 'Sekiten-byo' (red spot disease) 7. Vibriosis	1 50 414 757 954 1644 1659 1742 1743 1796
Marbled eel (*Anguilla marmorata*)	*Edwardsiella piscicida*			Pathogenic in experimental challege with zebrafish	1
Eel (*Anguilla reinhardtii*)	1. *Edwardsiella tarda* 2. *Photobacterium damselae* ssp. *damselae*		1. Muscle abscess and purulent meningitis	1. Australia	468 881
Eel	1. *Lactococcus garvieae* 2. *Vibrio parahaemolyticus*		2. Liver	2. Mortality in experimental infection	
Family Muraenidae (moray eels)					

Continued

Table 1.1. Continued.

Host	Pathogens	Normal flora	Tissue site	Disease status	Ref
Green moray eel (*Gymnothorax funebris* Ranzani) Spotted moray eel (*G. moringa* Cuvier)	*Mycobacterium montefiorense*		Florid skin nodules around the head and trunk that are soft, grey, gelatinous and tan-coloured	Proliferative skin disease	719 1013
ORDER Atheriniformes (silversides)					
Family Atherinopsinae (neotropical silversides)					
Argentinian silverside or pejerrey (*Odontesthes bonariensis*)	*Pseudomonas plecoglossicida*		Kidney	Bacterial haemorrhagic ascites (BHA)	1229
ORDER Aulopiformes (grinners)					
Family Synodontidae (lizardfishes)					
Variegated lizardfish (*Synodus variegatus* Lacepède), Red Sea fish, wild fish	*Streptococcus iniae*		Organism cultured from blood. No marked gross signs of disease	Morbidity	360
ORDER Beloniformes (needle fishes)					
Japanese medaka, Japanese rice fish (*Oryzias latipes* Temminck and Schlegel)	*Mycobacterium abscessus*			Mycobacteriosis	1620
Family Scomberesocidae (sauries)					
Pacific saury (*Cololabis saira* Brevoort)		*Shewanella sairae*	Intestine	Normal flora	1479
ORDER Beryciformes (sawbellies)					
Pineapple fish (*Cleidopus gloriamaris* De Vis), Family Monocentridae – pinecone fishes)	*Vibrio harveyi*		Haemorrhage around operculum and eyes	Mortality	1811
Japanese pinecone fish (*Monocentris japonica*), Family Monocentridae		*Aliivibrio fischeri*	Light organ	Symbiont	1435
ORDER Characiformes (characins)					
Family Characidae					
Black skirted tetra (*Hyphessobrycon* sp.)	*Edwardsiella tarda*		Septicaemia, focal suppurative or granulomatous lesions, cutaneous ulcerations	Edwardsiellosis	776 1156

Hyphessobrycon eques (previously *Serpae tetra*). Also known as red minor tetra	*Streptococcus agalactiae* Group B		Isolated in pure heavy growth	Deaths	244
Cardinal tetra (*Paracheirodon axelrodi*, Schultz)	*Mycobacterium gordonae*		Tissues	Mycobacteriosis	1012 1356
Neon tetra (*Paracheirodon innesi* Myers) (previously *Hyphessobrycon innesi*)	1. *Flavobacterium columnare* 2. *Mycobacterium fortuitum*		1. Skin discolouration, white areas of necrosis. Organism in skin, gills, muscle and inner surface of scales 2. Lesions	1. Muscle infections 2. Mycobacteriosis	216 1138 1428
Pacu (*Piaractus mesopotamicus*)	*Flavobacterium columnare*		Skin discolouration and ulceration	Columnaris disease	1343
Piracanjuba (*Brycon orbignyanus*)	*Flavobacterium columnare*		Skin discolouration and ulceration	Columnaris disease	1343
Tambaqui (*Colossoma macropomum*)	*Flavobacterium columnare*		Skin discolouration and ulceration	Columnaris disease	1343
ORDER Clupeiformes (herrings, shad)					
Family Clupeidae					
Herring, Atlantic herring (*Clupea harengus* Linnaeus). Also referred to as Baltic herring	1. *Pseudomonas anguilliseptica*	2. *Photobacterium iliopiscarium*	1. Haemorrhages in the eye 2. Intestine of oyster	1. Disease 2. Non-pathogenic	1050 1272 1682
Menhaden (*Brevoortia patronus* Goode)	*Streptococcus agalactiae*			Streptococcosis	501
Pacific herring (*Clupea pallasii pallasii*)	*Renibacterium salmoninarum*		Kidney	Bacterial kidney disease	504
Australian river gizzard shad or bony bream (*Nematalosa erebi*)	*Hafnia alvei*, *H. paralvei*		Intestine	Normal flora	1258
Shad, gizzard shad (*Dorosoma cepedianum* Lesueur)	*Aeromonas* species		Isolated in pure culture from kidney, liver, spleen. Isolates pathogenic for fingerling trout in virulence assay	Epizootic in spawning females	1652
South American pilchard (*Sardinops sagax* Jenyns). Also known as Australian pilchard, Californian pilchard, Pacific sardine etc. Synonym *Sardinops melanosticta*	1. *Tenacibaculum maritimum*	2. *Photobacterium* (*histaminum*) *damselae* ssp. *damselae*	1. Gliding bacteria seen as a tan-coloured pseudomembrane over the body	1. Disease 2. Skin microflora	309 904 1260
Family Engraulidae anchovy (*Engraulis mordax*)	*Tenacibaculum maritimum*		Haemorrhagic lesions on snout, eye, and midsection of the body	Disease	309
ORDER Cypriniformes					
Family Catostomidae					

Continued

Table 1.1. Continued.

Host	Pathogens	Normal flora	Tissue site	Disease status	Ref
Buffalo fish – *Ictiobus bubalus*, *I. cyprinellus*	*Flavobacterium columnare*		Skin lesions	Columnaris disease	414
Family Cyprinidae (minnows or carps)					
Bighead carp (*Hypophthalmichthys nobilis* Richardson) (*Aristichthys nobilis* older synonym)		*Edwardsiella ictaluri*		Not susceptible in experimental infection	1349
Common carp (*Cyprinus carpio carpio* Linnaeus). May be called koi carp and mirror carp	1. *Aeromonas bestiarum* 2. Atypical *Aeromonas salmonicida* 3. *Aeromonas veronii* biovar Veronii 4. *Citrobacter freundii* 5. *Flavobacterium columnare* 6. *Flavobacterium johnsoniae* 7. *Flavobacterium psychrophilum* 8. *Pseudomonas fluorescens* 9. *Mycobacterium marinum* 10. *Yersinia ruckeri*	11. *Aeromonas eucrenophila* 12. *Aeromonas sobria* 13. *Arcobacter cryaerophilus* 14. *Dietzia* (*Rhodococcus*) *maris*, *Rhodococcus fascians* 15. *Streptococcus iniae* and *Streptococcus agalactiae* (*Streptococcus difficile*) were non-pathogenic in virulence studies	1, 3. Haemorrhage, necrosis, ulcers 2. Ulcers, lesions 3. Also bloody ascites fluid 4. Internal organs in fingerlings 5. Gills, skin lesions 6. Skin lesions 7. Lesions on fins and internal organs 10. Can be infected with or without clinical signs 11. Ascites 14. Skin, intestine	1, 3. Pathogenic on virulence studies 2. Ulcerating dermal lesions 4. Heavy mortalities, septicaemia, opportunistic infection 5. Columnaris disease 6. Mortality 7, 8. Disease 9. Mycobacteriosis 10. Yersiniosis 11. Non-pathogenic 12. Non-pathogenic to carp in virulence studies 13. Non-pathogenic in experimental challenge 14. Part of normal flora	244 411 414 479 542 783 870 935 975 1004 1212 1356 1526 1662
Common shiner (*Luxilus* (*Notropis*) *cornutus* Mitchell)	*Renibacterium salmoninarum*		Kidney	Bacterial kidney disease	504
Crucian carp (*Carassius carassius* Linnaeus)	1. *Escherichia vulneris* 2. *Flavobacterium psychrophilum*		1. Haemorrhage in eyes, swollen, dark coloured abdomen, yellow liver, yellow fluid in intestine 2. Lesions on fins and internal organs	1. Mortality 2. Disease	85 1004
Grass carp (*Ctenopharyngodon idella* Valenciennes)	*Citrobacter freundii*		Intestine	Pathogenic in experimental challenge	1063

Freshwater bream (*Abramis brama* Linnaeus). Also called common bream, carp bream	*Aeromonas salmonicida* ssp. *achromogenes*		Large open lesions, areas of descalation. Typical 'furuncles' not seen	Mortalities	1117
White bream (*Blicca bjoerkna* Linnaeus)	*Aeromonas salmonicida* ssp. *achromogenes*		Large open skin lesions surrounded by areas of descalation. Typical 'furuncles' not seen	Furunculosis, mortalities	1117
Silver carp (*Hypophthalmichthys molitrix* Valenciennes)	1. *Providencia* (*Proteus*) *rettgeri* 2. *Staphylococcus aureus*		1. Large red ulcerative lesions on the abdomen, base of the pectoral fin and on the head. Organism isolated from internal organs. Pond had been fertilized with poultry faeces from which *P. rettgeri* was isolated 2. Red cornea becoming opaque. Degeneration of eye tissue	1. Mass mortality 2. Eye disease	149 1509
Chub, European chub (*Leuciscus cephalus* Linnaeus) synonym of *Squalius cephalus*	Atypical *Aeromonas salmonicida*		Skin ulcers and fin rot	Mortality	1808
Dace, common dace (*Leuciscus leuciscus* Linnaeus)	1. *Aeromonas salmonicida* ssp. *salmonicida* 2. Atypical *Aeromonas salmonicida*		1, 2. Skin ulcers	1. Furunculosis 2. Infection	674 733
Danio, sind danio (*Devario devario* Hamilton), previously *Danio devario*)	*Edwardsiella ictaluri*			Enteric septicaemia	
Flying fox (*Epalzeorhynchos kalopterus* Bleeker)	*Streptococcus iniae*		Isolated in pure growth	Deaths	244
Golden shiner (*Notemigonus crysoleucas* Mitchell)	1. *Streptococcus agalactiae* Group B	2. *Edwardsiella ictaluri*		1. Streptococcosis 2. Not susceptible in experimental infection	1349 1415
Goldfish (*Carassius auratus* Linnaeus)	1. Atypical *Aeromonas salmonicida* 2. *Aeromonas salmonicida* ssp. *nova* 3. *Flavobacterium branchiophilum* 4. *Mycobacterium abscessus* 5. *Mycobacterium chelonae* 6. *Mycobacterium fortuitum* 7. *Mycobacterium marinum* 8. *Mycobacterium smegmatis* 9. *Vibrio cholerae* (non-01)		1, 2. Ulcerative skin lesions, haemorrhages 3. Bacteria on gill surface, flared opercula, hyperaemic, swollen gill tissue, excess mucus production 6. Granulomas 8 Giant cells in peritoneum, granulomas 9. Septicaemia	1, 2. Goldfish ulcer disease (GUD) 3. Bacterial gill disease 4, 5, 6, 7. 8. Mycobacteriosis 9. Mortalities, opportunistic infection, stress-related	70 774 977 1281 1356 1382 1528 1607 1788 1789

Continued

Table 1.1. Continued.

Host	Pathogens	Normal flora	Tissue site	Disease status	Ref
Minnow	*Vibrio (Listonella) anguillarum*			Fish kill	244
Eurasian minnow (*Phoxinus phoxinus* Linnaeus)	Atypical *Aeromonas salmonicida*		Extensive haemorrhagic skin lesions	Mass mortality	690
Fathead minnow (American baitfish) (*Pimephales promelas* Rafinesque)	1. *Renibacterium salmoninarum* 2. *Yersinia ruckeri*		1. Kidney	1. Bacterial kidney disease 2. Yersiniosis, enteric redmouth (ERM)	411 504
Roach (*Rutilus rutilus* Linnaeus)	1. *Aeromonas salmonicida* ssp. *achromogenes* 2. Atypical *Aeromonas salmonicida* 3. *Yersinia ruckeri*		1. Large open lesions, areas of descalation. No typical 'furuncles' seen 2. Cutaneous ulcers 3. Disease is associated with poor environmental conditions	1. Mortalities 2. Ulcerative dermatitis 3. Yersiniosis	733 1117 1692 1788 1789
Rohu, roho labeo (*Labeo rohita* Hamilton)	1. *Edwardsiella tarda* 2. *Flavobacterium columnare*		1. Cutaneous lesions, necrotic abscesses, abdominal swelling, ascites fluid 2. Skin lesions and fin erosion	1. Mortality 2. Stress-related disease	946 1160
Odessa barb (*Puntius padamya* Kullander & Britz)	*Mycobacterium marinum*		Tissues	Mycobacteriosis	1356
Rosy barb (*Puntius conchonius* Hamilton)	1. *Edwardsiella ictaluri* 2, 3. *Mycobacterium marinum*, *Mycobacterium triplex*		1. Moribund, organism in internal organs	1. Mortality of 40% 2, 3. Mycobacteriosis	776 1356
Tench (*Tinca tinca* Linnaeus)	1. *Flavobacterium psychrophilum* 2. *Mycoplasma mobile*		1. Lesions in fins, musculature, intestine 2. Isolated from gills	1. Disease 2. Pathogenicity not stated	908 910 1004
Ya-fish (*Schizothorax prenanti*)	*Streptococcus agalactiae*		Exophthalmia, haemorrhage around eye, caudal abscess	Mortality	605
Zebrafish (*Danio rerio* Hamilton)	1. *Flavobacterium columnare* 2. *Flavobacterium johnsoniae* 3. *Mycobacterium chelonae*		1. Bacterium adheres to gills 2. Jaw and gill discolouration 3. Spleen, liver	1. Columnaris disease 2. Virulent by injection 3. Mycobacteriosis	1171 1262 1784
ORDER Cyprinodontiformes					
Family Cyprinodontidae (pupfishes)					

Charco Palma pupfish (*Cyprinodon veronicae* Lozano-Vilano & Contreras-Balderas)	*Mycobacterium gordonae*		Tissues	Mycobacteriosis	1356
Spanish toothcarp, Iberian toothcarp, (*Aphanius iberus* Valenciennes)	*Vibrio parahaemolyticus*		External haemorrhage, tail rot	Mortality	19
Family Fundulidae (topminnows and killifishes)					
Gulf killifish (*Fundulus grandis* Baird & Girard). Also called bullminnows	1. *Flavobacterium columnare* 2. *Streptococcus* sp. Non-haemolytic Group B		1. Skin lesions 2. Exophthalmia, petechial haemorrhage in abdomen, oedema in liver	1. Columnaris disease 2. Mortality, streptococcosis	414 1378
Family Goodeidae					
Golden skiffia (*Skiffia francesae* Kingston)	*Mycobacterium marinum*		Tissues	Mycobacteriosis	1356
Family Nothobranchiidae (African rivulines)					
Epiplatys bifasciatus Steindachner	*Mycobacterium ulcerans*		Intestine and gills	Pathogenicity not known	470
Family Poeciliidae (poeciliids)					
Aplocheilichthys species	*Mycobacterium ulcerans*		Intestine and gills	Pathogenicity not known	470
Guppy (*Poecilia reticulata* Peters, *Lebistes reticulatus* is a junior synonym)	*Mycobacterium fortuitum*		Fin lesions, emaciated, swim in circles	Mycobacteriosis	216
Balloon molly (*Poecilia* sp.)	*Escherichia vulneris*		Exophthalmia in eyes, pale gills, empty digestive tract, opened operculum, yellow liver	Mortality	85
Molly or black molly (*Poecilia sphenops* Valenciennes)	*Flavobacterium columnare*		White spots on the back, head and skin ulcers	Mortality	426
Silver molly (*Poecilia* sp.)	*Escherichia vulneris*		Pale liver, bent or curved body, thinness	Mortality	85
Platies, southern platyfish (*Xiphophorus* [*Platypoecilus*] *maculatus* Günther)	1. *Flavobacterium columnare* 2. *Mycobacterium marinum*		1. White spots on the back and head, and skin ulcers 2. Skin ulcer, bacteria in liver, spleen, gills	1, 2. Mortality	97 426
Pitch lake guppy	*Mycobacterium marinum*		Tissues	Mycobacteriosis	1356
ORDER Esociformes (pikes and mudminnows)					

Continued

Table 1.1. Continued.

Host	Pathogens	Normal flora	Tissue site	Disease status	Ref
Family Esocidae (pikes)					
Pike – northern pike (*Esox lucius* Linnaeus)	1. Atypical *Aeromonas salmonicida* 2. *Yersinia ruckeri*		1. Skin ulcers	1. Infection	411 674
ORDER Gadiformes					
Family Gadidae (cods and haddocks)					
Atlantic cod (*Gadus morhua* Linnaeus)	1. *Aliivibrio* (*Vibrio*) *logei* 2. *Aliivibrio* (*Vibrio*) *salmonicida* 3. Atypical *Aeromonas salmonicida* 4. *Francisella philomiragia* ssp. *noatunensis* 5. *Pseudomonas anguilliseptica* 6. *Vibrio anguillarum* serotype 02 7. *Vibrio splendidus* 8. *Vibrio xuii*	9. *Carnobacterium* species 10. *Photobacterium iliopiscarium* 11. *Shewanella algidipiscicola*, *Shewanella glacialipiscicola*, *Shewanella hafniensis*, *Shewanella morhuae* 12. *Vibrio pacinii*	1. Intestine 2. Septicaemia, haemorrhages 4. Granuloma 5. Kidney, brain 6. Lesions, Intestine 7, 9, 10. Intestine 11. Belly flap area 12. Intestine	1. Some strains are pathogenic 2. Coldwater vibriosis, Hitra disease 3. Skin ulcers 4. Granulomatous disease 5. Haemorrhagic septicaemia 6. Infection, vibriosis 7. Some strains pathogenic to larvae 8. Isolated from apparently healthy larvae. Pathogenicity not known 9. Non-pathogenic 10, 11. Normal flora 12. Intestinal flora when fed rotifers	100 241 370 473 525 852 1142 1266 1272 1395 1480 1481 1559 1682
Coalfish		*Photobacterium iliopiscarium*	Intestine	Non-pathogenic	1272 1682
Haddock (*Melanogrammus aeglefinus* Linnaeus)	Atypical *Aeromonas salmonicida*		Skin ulcers	Infection	674
Rockling, fourbeard rockling (*Enchelyopus cimbrius* Linnaeus)	Atypical *Aeromonas salmonicida*			Skin ulcers	1800

Fish – no details given		*Chryseobacterium piscium*		May be a food spoilage organism	421
ORDER Gonorynchiformes (milkfish)					
Family Chanidae (milkfish)					
Milkfish (*Chanos chanos* Forsskål)	1. *Mycobacterium abscessus* 2. *Vibrio harveyi*		1. Spleen, kidney, liver, gastrointestinal tract 2. Eye lesions, exophthalmia, opaqueness, haemorrhages in eyes	1. Mycobacteriosis 2. Eye disease and mortality	302 805
ORDER Gymnotiformes (knifefishes)					
Family Gymnotidae					
Banded knifefish (*Gymnotus carapo* Linnaeus)	*Edwardsiella tarda*			Septicaemia	1383
Family Sternopygidae					
Glass knifefish (*Eigenmannia virescens* Valenciennes), also called green knifefish	*Edwardsiella ictaluri*			Enteric septicaemia	871
ORDER Lophiiformes (anglerfishes)					
Family Antennariidae (frogfishes)					
Striated frogfish (*Antennarius striatus*)	*Mycobacterium marinum*		Acid fast rods in macrophages and interstitial infiltrates in liver, spleen, eyes, testes	Mycobacteriosis without typical granuloma formation	1826
ORDER Mugiliformes (mullet)					
Family Mugilidae					
Black mullet, striped mullet, grey mullet, flathead grey mullet (*Mugil cephalus* Linnaeus)	1. *Edwardsiella tarda* 2. *Eubacterium tarantellae* 3. *Lactococcus garvieae* 4. *Nocardia seriolae*		1. Septicaemia, focal suppurative or granulomatous lesions, cutaneous ulcerations 2. Organism recovered from brain, liver, kidney, blood	1. Edwardsiellosis 2. Mortality 3. Septicaemia, chronic meningitis 4. Nocardiosis	313 482 501 716 1506 1523

Continued

Table 1.1. Continued.

Host	Pathogens	Normal flora	Tissue site	Disease status	Ref
	5, 6. *Photobacterium damselae* ssp. *damselae* 6. *P. damselae* spp. *piscicida* 7. *Streptococcus agalactiae* Group B		3. Fish lethargic, exophthalmia, congestion and haemorrhage in spleen and kidney, white spots, ascites, granulomas, macrophages and bacteria in organs 4. Skin ulcers and tubercules in organs 5, 6. Spleen 7. Brain, eye, organs	5, 6. Carrier status and opportunistic infection 6. Fish pasteurellosis 7. Streptococcosis	1673
Golden grey mullet (*Liza aurata* Risso)	*Photobacterium damselae* ssp. *damselae*, *P. damselae* spp. *piscicida*		Spleen	Carrier status and opportunistic infection	1506
Borneo mullet, largescale mullet (*Liza macrolepis* Smith)	*Streptococcus iniae*		Susceptible to infection	Streptococcosis in experimental challenge	1840
White (silver) mullet (*Mugil curema* Valenciennes)	*Vibrio harveyi*		Organism isolated in pure culture	Haemorrhagic septicaemia	45
Wild mullet (*Liza klunzingeri* Day)	*Streptococcus agalactiae*		Haemorrhages on body especially in eye, mouth, operculum and fins. Organism cultured from brain, eye and blood	Streptococcosis, epizootic	501
ORDER Osmeriforms					
Family Plecoglossidae Ayu sweetfish (*Plecoglossus altivelis altivelis* Temminck and Schlegel)	1. *Flavobacterium psychrophilum* 2. *Pseudomonas anguilliseptica* 3. *Pseudomonas plecoglossicida* 4. *Renibacterium salmoninarum* 5. *Streptococcus iniae* 6. *Vibrio anguillarum* 01 and 02 (European designation) 7. *Vibrio cholerae* non-01 (negative for ornithine decarboxylase)		3. Haemorrhagic ascites 4. White nodules in kidney, abdomen swollen with fluid, exophthalmia 6. Haemorrhagic septicaemia	1. Cold water disease 2. Disease 3. Mortality, bacterial haemorrhagic ascites (BHA) 4. Bacterial kidney disease (BKD) 5. Mortality, streptococcosis 6. Vibriosis 7. Mass mortality	893 917 1182 1187 1194 1229 1559 1593 1659 1748 1749

ORDER Osteoglossiformes (bonytongues)					
Family Arapaimicae (bonytongues)					
Arapaima (*Arapaima gigas* Schinz), common name = pirarucu. Is a gigantic tropical freshwater fish (Synonym *Arapaima gigas* Cuvier)	*Salmonella enterica* ssp. *arizonae*		Organism isolated from liver, spleen, heart, kidney, bloody exudate in body cavity, corneal opacity	Septicaemia	923
Family Osteoglossidae (arowans)					
Bonyfish Spotted tongue (*Scleropages leichardti*)	*Edwardsiella tarda*			Septicaemia	1383
ORDER Perciformes (perch-like)					
Family Ammodytidae (sand lances)					
Eel (*Hyperoplus lanceolatus* Le Sauvage)	Atypical *Aeromonas salmonicida*		Haemorrhages on snout, caudal fin and tail. Typical umbonate furuncles on flank	Skin ulcers	398
Small sandeel (*Ammodytes tobianus*) Cuvier). *Ammodytes lancea* is a junior synonym	Atypical *Aeromonas salmonicida*		Haemorrhages on snout, caudal fin and tail. Typical umbonate furuncles on flank	Skin ulcers	398
Family Anarhichadidae (wolffishes)					
Atlantic wolffish (*Anarhichas lupus* Linnaeus), spotted wolffish (*A. minor* Olafsen)	1. Atypical *Aeromonas salmonicida*	2. *Carnobacterium divergens* 3. *Carnobacterium* sp.		1. Atypical furunculosis 2, 3. Part of intestinal microflora	1404
Family Apogonidae (cardinalfish)					
Banggai cardinal fish (*Pterapogon kauderni* Koumans, 1993)	*Mycobacterium marinum*		Tissues	Mycobacteriosis	1356
Family Carangidae (jacks and pompanos)	Order Perciformes				

Continued

Table 1.1. Continued.

Host	Pathogens	Normal flora	Tissue site	Disease status	Ref
Amberjack. Japanese amberjack (*Seriola quinqueradiata* Temminck and Schlegel) Also called yellowtail, rudderfish *Seriola* (*purpurascens*) *dumerili* Risso – greater amberjack	1. *Lactococcus garvieae* 2. *Mycobacterium marinum* 3. *Nocardia seriolae* 4. *Photobacterium damselae* ssp. *damselae* 5. *P. damselae* ssp. *piscicida* 6. *Pseudomonas putida* 7. *Streptococcus agalactiae* 8. *Streptococcus dysgalactiae* Group C 9. *Tenacibaculum maritimum* 10. *Vibrio anguillarum* 11. *Vibrio parahaemolyticus*		1. Erosion of tail fin, redness of anal fin, petechiae inside operculum, exophthalmia 2. Haemorrhagic ascites, tubercules in spleen and kidney 3. Abscesses in epidermis, formation of tubercules in gills, kidneys, spleen 4. Organism in spleen and kidney – toxin produced 5. Bacterial colonies of white-grey colour seen on spleen and kidney 6. Haemorrhagic abscesses on body 7. Organism isolated from brain 8. Lesions on caudal peduncle 9. Eroded mouth, frayed fins and tail 11. Liver	1, 4, 6. Mortality 2. Mycobacteriosis 3. Nocardiosis 5. Fish pasteurellosis, mortality 7. Septicaemia 8. Mortality 9. Tenacibaculosis 11. Mortality in experimental infection	19 171 479 482 868 905 944 955 957 959 1235 1461 1771 1828
Crevalle (*Caranx hippos* Linnaeus). Also known as trevally, crevalle jack, common jack, yellow jack	*Vibrio harveyi*		Dermal lesions	Infection	939
Pompanos, also called Florida pompano (*Trachinotus carolinus* L.), permit (*T. falcatus* L.)	*Vibrio harveyi*			Haemorrhagic septicaemia, mortality	45
Snubnose pompano (*Trachinotus blochii* Lacepède)	*Nocardia seriolae*		Skin ulcers and tubercules in organs	Nocardiosis	1523
Japanese jack mackerel (*Trachurus japonicus* Temminck and Schlegel)	1. *Lactococcus garvieae* 2. *Vibrio anguillarum* 3. *Vibrio harveyi* (reported as *Vibrio trachuri*)		3. Haemorrhages in organs and exophthalmia	3. Disease	818 1629

White trevally (*Pseudocaranx dentex* Bloch and Schneider). Also known as striped jack, silver bream, silver trevally, skippy etc.	1. *Lactococcus garvieae* 2. *Photobacterium damselae* ssp. *piscicida* 3. *Pseudomonas anguilliseptica*		1. Exophthalmia, swollen abdomen, haemorrhagic areas in mouth, base of fins and internal organs 2. Haemorrhagic septicaemia and white areas of granuloma in the kidney, liver, spleen 3. Haemorrhages in mouth, nose, operculum and brain. Organism in kidney	2. Fish pasteurellosis 1, 3 Mortalities	956 1088 1193
Family Centrarchidae (sunfishes)					
Largemouth bass (*Micropterus salmoides* Lacepède)	1. *Edwardsiella tarda* 2. *Flavobacterium columnare* 3. *Nocardia seriolae*	4. *Edwardsiella ictaluri*	1. Intestine, liver, spleen, dermal ulcers 2. Skin lesions 3. Skin ulcers and tubercules in organs	1. Internal organs pale and anaemic, blood watery, haemorrhagic nodules in stomach wall and serosa of intestine, cutaneous ulcers, kidney, liver 2. Columnaris disease 3. Nocardiosis 4. Not susceptible in experimental infection	414 562 1349 1523 1785
Bluegill (*Lepomis macrochirus* (Rafinesque) (also known as bream and coppernose), sunfish, bluegill sunfish green sunfish (*Lepomis cyanellus*)	*Streptococcus agalactiae* Group B		Brain, eye, organs	Streptococcosis in experimental infection	1415
Orange-spotted sunfish (*Lepomis humilis*)	*Flavobacterium columnare*		Skin lesions	Columnaris disease	414
Warmouth (*Lepomis gulosus*)	*Flavobacterium columnare*		Skin lesions	Columnaris disease	414
Crappie (*Pomoxis annularis*)	*Flavobacterium columnare*		Skin lesions	Columnaris disease	414
Family Centropomidae (snooks)					
Snook (*Centropomus undecimalis* Bloch)	*Vibrio harveyi*		Opaque cornea	Infection	939
Family Channidae (snakeheads)					
Snakehead (*Channa argus argus*). *Ophicephalus argus* is a senior synonym	1. *Aeromonas schubertii* 2. *Nocardia seriolae*		1. White nodules in kidney 2. Skin ulcers and tubercules in organs	1. Mortality 2. Nocardiosis	1036 1523

Continued

Table 1.1. Continued.

Host	Pathogens	Normal flora	Tissue site	Disease status	Ref
Blotched snakehead fish (*Channa maculata*)	*Aeromonas schubertii*		Skin ulcers, haemorrhage	Mortality	315
Striped snakehead fish (*Channa striata* Bloch)	1. *Mycobacterium fortuitum* (previously thought to be *Mycobacterium poriferae*) 2. *Mycobacterium marinum* 3. *Photobacterium damselae* ssp. *piscicida*		1, 2. Internal nodular lesions 3. White lesions in spleen and kidney	1, 2. Mycobacteriosis 3. Bacterial septi-caemia	1356 1365 1658 1671
Family Cichlidae (cichlids)					
Black acara (*Cichlasoma bimaculatum* Linnaeus)	*Mycobacterium chelonae abscessus*, *Mycobacterium fortuitum*, *Mycobacterium simiae*			Mycobacteriosis	977
Cockatoo cichlid (*Apistogramma cacatuoides* Hoedemann)	*Mycobacterium avium*, *Mycobacterium gordonae*, *Mycobacterium marinum*, *Mycobacterium triviale*		Organisms in skin, gill, muscle, liver, spleen, kidney	Mycobacteriosis	1012 1238
Electric blue hap (*Sciaenochromis ahli* Trewavas)	*Aeromonas jandaei*			Deaths	244
Firemouth cichlid (*Thorichthys meeki*, *Cichlasoma meeki* Brind)	*Mycobacterium chelonae abscessus*			Mycobacteriosis	977
Freshwater angelfish (*Pterophyllum scalare* Schultze)	*Mycobacterium gordonae*		Organisms in skin, gill, muscle, liver, spleen, kidney	Mycobacteriosis	1012 1238
Humphead cichlid (*Cyphotilapia frontosa* Boulenger)	*Mycobacterium fortuitum*		Organisms in skin	Mycobacteriosis	1238
Jewelfish (*Hemichromis bimaculatus* Gill)	*Mycobacterium ulcerans*		Intestine and gills	Pathogenicity not known	470
Oscar (*Astronotus ocellatus* Agassiz)	*Mycobacterium fortuitum*		Skin lesions on head	Mycobacteriosis	216
Ram (*Mikrogeophagus ramirezi*)	*Streptococcus agalactiae* Group B		Organism in pure growth in liver	Deaths	216
Mango tilapia (*Sarotherodon galilaeus galilaeus* Linnaeus)	*Mycobacterium ulcerans*		Intestine and gills	Pathogenicity not known	470

Nile tilapia (*Oreochromis* sp., *Oreochromis niloticus niloticus* Linnaeus) Also known as St Peter's fish	1. *Edwardsiella tarda* 2. *Flavobacterium columnare* 3. *Pseudomonas anguilliseptica* 4. *Pseudomonas fluorescens* 5. *Streptococcus agalactiae* 6. *Streptococcus dysgalactiae* Group C 7. *Streptococcus iniae*		1. Septicaemia, focal suppurative or granulomatous lesions, cutaneous ulcerations 2. Fin necrosis 3. Kidney, liver 4. Gonad 5. Brain, eye, organs, gills 7. Central nervous system involvement, lethargy, erratic swimming	1. Edwardsiellosis 2. Columnaris disease 3. Haemorrhagic septicaemia 4. Mortality 5, 6, 7. Streptococcosis	477 479 482 483 534 1135 1156 1213
Tilapia hybrid (*Oreochromis nilotica* × *Oreochromis aureus*)	*Streptococcus iniae*		Exophthalmia, petechiae around mouth, anus, fins, abnormality in liver, spleen, kidney	Streptococcosis	1329
Otjikoto tilapia (*Tilapia guinasana* Trewavas)	*Mycobacterium marinum*		Tissues	Mycobacteriosis	1356
Red devil (*Amphilophus labiatus* Günther)	*Mycobacterium fortuitum*		Organisms in skin and muscle	Mycobacteriosis	1238
Redbreast tilapia (*Tilapia rendalli*)		*Flavobacterium tilapiae*		Water from culture pond	314
Red tilapia (*Oreochromis* species)	*Streptococcus agalactiae*		Multiple necrotic foci in tissues	Streptococcosis	1183 1589
Striped barombi mbo cichlid (*Stomatepia mariae*)	*Mycobacterium stomatepiae*		Kidney, spleen	Mycobacteriosis	1355
Tilapia (*Oreochromis aureus* Steindachner), senior synonym to *Sarotherodon aureus*	1. *Edwardsiella ictaluri* 2. *Streptococcus agalactiae* 3. *Streptococcus iniae*		2. Gills, brain, organs 3. Loss of orientation, exophthalmia, petechiae around anus, mouth and pectoral fins, fluid in peritoneal cavity, enlarged organs	1. Slightly susceptible at high concentration 2. Virulence confirmed in bioassay 2, 3. Mortality	1135 1349
Family Coryphaenidae (dolphinfishes)					
Mahi-mahi or dolphin fish (*Coryphaena hippurus* Linnaeus)	*Pantoea* (*Enterobacter*) *agglomerans*		Haemorrhages in eye, and dorsal and lateral musculature	Mortality, opportunistic pathogen	676
Family Embiotocidae (surfishes)					
Shiner perch (*Cymatogaster aggregata* Gibbons)	*Renibacterium salmoninarum*		Kidney	Bacterial kidney disease	504
Family Ephippidae (spadefish, batfish, scats)					

Continued

Table 1.1. Continued.

Host	Pathogens	Normal flora	Tissue site	Disease status	Ref
Spadefish, Atlantic spadefish (*Chaetodipterus faber* Broussonet)	*Vibrio harveyi*		Bilateral exophthalmia, haemorrhages in and around eyes, corneal opacity	Mortality	45
Family Gobiidae (gobies)					
Longjaw mudsucker (*Gillichthys mirabilis* Cooper)		*Mycoplasma* ribotype	Gastrointestinal tract	Normal flora	116
Mud skipper (vernacular) *Boleophthalmus dussumieri* Valenciennes	*Vibrio* (*Listonella*) *anguillarum*			Fish kill	244
Family Haemulidae (grunts)					
Striped piggy (*Pomadasys stridens* Forsskål), Red Sea fish, wild fish	*Streptococcus iniae*		Organism cultured from blood. No marked gross signs of disease	Morbidity	360
Family Kyphosidae (sea chubs)					
Labracoglossa argenteiventris (Peters)		*Photobacterium* (*histaminum*) *damselae* ssp. *damselae*		Skin microflora	904 1260
Largescale blackfish (*Girella punctata*)	*Photobacterium damsela* ssp. *piscicida*		White granules in the spleen	Mortality	872
Family Labridae (wrasses)					
Clown coris (*Coris aygula*) or Red Sea wrasse	*Lactococcus garvieae*		Liver and kidney	Mortality	361
Wrasse, corkwing wrasse (*Symphodus melops*), cleaner fish	1. Atypical *Aeromonas salmonicida* 2. *Vibrio splendidus*, *Vibrio tapetis*		1. Haemorrhagic internal organs, bloody exudate 2. Kidney	1. Furunculosis 2. Mortality. Pathogenicity proven in experimental challenge	838 968
Family Lateolabracidae					
Seabass (*Lateolabrax japonicus* Cuvier), Japanese seaperch, Japanese seabass Perciformes, Family Lateolabracidae	1. *Mycobacterium marinum* 2. *Nocardia seriolae* 3. *Vibrio ponticus*		1. Tissues 2. White-yellow nodules in gill, heart, kidney, liver, spleen 3. Skin ulceration, liver	1. Mycobacteriosis 2. Nocardiosis 3. Vibriosis, deaths	312 1356 1523 1816
Family Latidae (lates perches)					

Barramundi (*Lates calcarifer* Bloch). Also known as barramundi perch or seabass	1. *Flavobacterium johnsoniae* 2. *Photobacterium damselae* ssp. *damselae* 3. *P. damselae* ssp. *piscicida* 4. *Streptococcus iniae* 5. *Tenacibaculum maritimum* 6. *Vibrio harveyi*		1. Superficial skin erosion on posterior flanks, pectoral fins and occasionally the lower jaw 3. Exophthalmia, haemorrhagic skin lesions, septicaemia 4. Organism isolated from brain 5. Skin and fin erosions 6. Isolated from lesions, liver, spleen, brain, eye	1. Disease in juveniles 2, 3, 4, 5, 6. Mortalities	228 284 378 1376 1396 1617
Family Latridae (trumpeters)					
Striped trumpeter (*Latris lineata* Forster)	*Tenacibaculum maritimum*		Dermal and fin erosions	Tenacibaculosis	673
Family Lutjanidae (snappers)					
Crimson snapper (*Lutjanus erythropterus*)	*Nocardia seriolae*		Skin ulcers and tubercules in organs	Nocardiosis	1523
Spotted rose snapper (*Lutjanus guttatus* Steindachner)		1. *Vibrio alfacsensis* 2. *Vibrio sinaloensis*	1. Liver, kidney 2. Liver, spleen, lesions	1, 2. Pathogenicity unknown	639 640
Family Moronidae (temperate basses)					
European seabass (*Dicentrarchus labrax* Linnaeus)	1. *Aeromonas hydrophila* 2. *Photobacterium damselae* ssp. *piscicida* 3. *Mycobacterium marinum* and *Mycobacterium* species 4. *Pseudomonas anguilliseptica* 5. *Streptococcus iniae* 6. *Tenacibaculum dicentrarchi* 7. *Tenacibaculum maritimum* 8. *Vibrio alginolyticus* 9. *Vibrio harveyi*	10. *Vibrio chagasii* 11. *Vibrio pacinii*	1. Enlargement of spleen, erythaema and swelling of anus 2. No obvious gross pathology, enlarged spleen with white nodules 4. Organism isolated from head kidney and spleen 3. Degeneration in eye, exophthalmia, skin lesions, necrotic areas in gills. 4. 5. Organisms in heart and spleen, extreme splenomegaly 5. Brain, kidney 6. Isolated from diseased fish 7. Pale skin zones with yellow edges, leading to necrotic lesions on body around fins, oral cavity, eyes and gills 8. Spleen, kidney, liver	1. Mortalities 2. Fish pasteurellosis 4. Haemorrhagic septicaemia 3. Mycobacteriosis 5. Exudative meningitis and panophthalmitis 6. Virulence not assessed 7. Dermal necrosis. Fingerlings very susceptible. Stress-related in older fish 10, 11. Pathogenicity not known 8, 9. Pathogenicity confirmed in challenge experiments	117 160 171 176 268 360 411 461 638 930 1346 1364 1855 1635

Continued

Table 1.1. Continued.

Host	Pathogens	Normal flora	Tissue site	Disease status	Ref
Striped bass (*Morone saxatilis* Walbaum) (*Roccus saxatilisi* is a synonym)	1. *Corynebacterium aquaticum* 2. *Carnobacterium (piscicola) maltaromaticum* 3. *Edwardsiella tarda* 4. *Mycobacterium marinum* 5. *Mycobacterium pseudoshottsii* 6. *Mycobacterium shottsii* 7. *Nocardia seriolae* 8. *Photobacterium damselae* ssp. *piscicida* 9. *Serratia marcescens* 10. *Streptococcus agalactiae* 11. *Streptococcus iniae*	12. *Edwardsiella ictaluri* 13. *Vibrio cholerae* non-01 and *Vibrio mimicus* Did not cause deaths in a virulence assay, however may be opportunistic pathogen in stressed fish	1. Organism in brain tissue, exophthalmia 2. Hyperaemia and haemorrhage in liver, kidney, spleen and brain 3. Haemorrhagic ulcers on body and fins, lesions kidney, liver, spleen 4. Nodular lesions in all organs 5, 6. skin lesions and granulomatous inflammation, spleen 7. Skin ulcers and tubercules in organs 8. Organism isolated from internal organs, enlarged kidney and spleen. Abnormal skin pigmentation 9. Necrosis of muscular tissues 10, 11. Brain, eye, organs	1. Disease 2. Mortality 3,4. Mortality 5, 6. Mycobacteriosis 7. Nocardiosis 8. Fish pasteurellosis 9. Mortality in fingerlings 10, 11. Streptococcosis 12. Not susceptible in immersion trials 13. Present in apparently healthy wild fish. May cause mortality in stressed farmed fish	133 135 136 137 479 703 711 977 1073 1231 1400 1401 1523 1550
White bass (*Morone chrysops*)	1. *Flavobacterium columnare* 2. *Photobacterium damselae* ssp. *piscicida*		1. Skin lesions 2. White tubercles 1–2 mm in spleen	1. Columnaris disease 2. Mortality	414 1231
White perch (*Morone americana*, Gmelin. Previously *Roccus americanus*)	1. *Mycobacterium pseudoshottsii*, *Mycobacterium shottsii* 2. *Photobacterium damselae* ssp. *piscicida*		1. Spleen, kidney 2. Organism isolated from internal organs	1. Mycobacteriosis 2. Fish pasteurellosis, massive mortality	832 1551 1585
Family Mullidae (goatfishes)					
Bensasi goatfish (*Upeneus japonicus* Houttuyn). *Upeneus bensasi* is a junior synonym		*Shewanella upenei*	Intestine	Normal flora	901
Family Oplegnathidae (knifejaws)					
Barred knifejaw (*Oplegnathus fasciatus*). Also know as rock bream	*Tenacibaculum maritimum*		Eroded mouth, frayed fins, tail rot	Mortality	1744
Family Osphronemidae (gouramies)					

Pearl gourami (*Trichogaster leerii* Bleeker)	*Mycobacterium marinum*		Organisms in skin, gill, liver, kidney, spleen	Mycobacteriosis	1012 1238
Siamese fighting fish (*Betta splendens* Regan)	1. *Edwardsiella tarda* 2. *Mycobacterium fortuitum* 3. *Mycobacterium marinum*		1. Moribund, single or multiple cutaneous ulcers of 1 mm diameter. Organism in internal organs 2. Lesions 3. Kidney	1. Mortality of 70% 2. Mycobacteriosis 2, 3. Mycobacteriosis but appeared clinically healthy	776 1190 1365
Three-spot gourami (*Trichopodus trichopterus* Pallas). Synonym *Trichogaster trichopterus*	1. *Mycobacterium gordonae* 2. *Mycobacterium* (*anabanti*) *marinum*		1. Organisms on skin 2. Necrotic lesions on head, body, muscle, fins, liver	1, 2. Mycobacteriosis	1012 1238 1475
Family Percichthyidae					
Golden perch (*Macquaria ambigua*)		*Hafnia alvei*, *Hafnia paralvei*	Intestine	Normal flora	1258
Family Percidae (perches)					
American yellow perch (*Perca flavescens* Mitchell)	1. *Flavobacterium columnare* 2. *Mycobacterium chelonae*		1. Skin lesions 2. Granulomatous peritonitis and hepatitis	1. Columnaris disease 2. Mycobacteriosis	403 414
European perch (*Perca fluviatilis* L.)	1. *Aeromonas salmonicida* ssp. *achromogenes* 2. *Flavobacterium psychrophilum* 3. *Yersinia ruckeri*	4. *Pseudomonas chlororaphis* ssp. *piscium*	1. Large open lesions, descalation. No typical 'furuncles' seen 2. Necrosis of mouth 3. Disease is associated with poor environmental conditions 4. Intestine	1. Mortality 2. Mortality 3. Yersiniosis 4. Normal flora	252 1051 1117 1692
Family Plesiopidae					
Comets (*Calloplesiops altivelis* Steindachner)	*Flavobacterium columnare*		Ulcers	Mortality	244
Family Pomacentridae (damselfishes)					
Damselfish (*Chromis chromis*)					
Sergeant-major (*Abudefduf saxatilis* L.)	*Mycobacterium marinum*		Spleen, liver	Mycobacteriosis	61
Staghorn damselfish, (*Amblyglyphidodon curacao* Bloch), yellowtail clownfish (*Amphiprion clarkii* Bennett)	*Pseudoalteromonas piscicida*	*Vibrio anguillarum* and *Vibrio parahaemolyticus* did not appear to be involved in disease in experimental infections	Diseased eggs, seen as white brain and spinal cord sunken irregular eyes, heart receded into yolk sac	Infection, mortality	1210

Continued

Table 1.1. Continued.

Host	Pathogens	Normal flora	Tissue site	Disease status	Ref
Blacksmith (*Chromis punctipinnis* Cooper)	*Photobacterium damselae* ssp. *damselae*		Skin ulcers	Fatal infection due to production of a cytolysin	1060
Family Pomatomidae (bluefishes)					
Bluefish (*Pomatomus saltatrix* L.). Known as tailor in Australia, shad and elf in South Africa. Other names include chopper, blue and anchoa	*Streptococcus agalactiae* Group B		Brain, eye, organs	Streptococcosis	134
Family Rachycentridae					
Marine cobia fish (*Rachycentron canadum* L.)	1. *Vibrio alginolyticus* 2. *Vibrio harveyi*		1. Kidney, ascites fluid 2. Gut	1. Pathogenicity confirmed in experimental challenge 2. Gastroenteritis	1043 1044
Family Scatophagidae (scats)					
Spotted scat (*Scatophagus argus* L.)	*Nocardia seriolae*		Skin ulcers or tubercules in organs	Nocardiosis	1523
Family Sciaenidae (drums or croakers)					
Croaker, Atlantic croaker (*Micropogonias undulatus* L.), spot croaker (*Leiostomus xanthurus* Lacepède – also known as spot)	1. *Mycobacterium marinum* 2. *Streptococcus agalactiae* Group B		1. Spleen, liver 2. Brain, eye, organs	1. Mycobacteriosis 2. Streptococcosis	61 134 501
Large yellow croaker (*Larimichthys crocea*, synonym *Pseudosciaena crocea*)	*Vibrio alginolyticus*		Intestine, particularly foregut	Pathogenicity proven in experimental challenge	310
Redfish, red drum (*Sciaenops ocellatus* L.)	1. *Eubacterium* species (tentative identification) 2. *Mycobacterium marinum* 3. *Streptococcus iniae* 4. *Vibrio harveyi*		1. Organism isolated from brain, liver, kidney and blood 2. Visceral and integumental lesions 3. Skin lesions, exophthalmia, eye degeneration, necrotic areas of gill rot Organism in heart and spleen 4. Gastroenteritis	1, 2, 3, 4. Mortality	360 440 716 1041

White weakfish or white seabass (*Atractoscion nobilis* Ayres)	*Tenacibaculum maritimum*		Lesions in eye, operculum. Lesions on body ranging from scale loss to deep ulcers in musculature.	Disease in juveniles	309
Weakfish, silver weakfish (*Cynoscion nothus* [Holbrook]), grey weakfish (*Cynoscion regalis*), weakfish (*Cynoscion regalis* Bloch & Schneider). Also known as silver sea trout	*Streptococcus agalactiae* Group B		Brain, eye, organs	Streptococcosis	134 501
Family Scombridae (mackerel, tunas, bonitos)					
Atlantic bluefin tuna (*Thunnus thynnus*)	*Photobacterium damselae* ssp. *piscicida*		No external clinical signs. Granulomas in kidney in chronic infection	Septicaemia	1159
Mackerel (*Scomber scombrus* Linnaeus)	*Mycobacterium* species		Grey-white nodules in spleen and visceral organs	Mycobacteriosis	1085
Chub mackerel (*Scomber japonicus* Houttuyn). May be known as Pacific mackerel (*Pneumatophorus japonicus* is a synonym)		*Shewanella pneumatophori*	Intestine	Normal flora	732
Family Scombropidae (gnomefishes)					
Gnomefish (*Scombrops boops*). Also called Japanese bluefish		*Photobacterium* (*histaminum*) *damselae* ssp. *damselae*		Skin microflora	904 1260
Family Serranidae (seabasses: groupers and fairy basslets)					
Black seabream (*Centropristis striata* L.)	*Mycobacterium marinum*		Liver, spleen	Mycobacteriosis	61
Grouper orange-spotted grouper (*Epinephelus coioides* Hamilton), greasy grouper, brown spotted grouper (*E. tauvina* Forsskål)	1. *Streptococcus iniae* 2. *Vibrio harveyi*		1. Moderately susceptible to infection 2. Tissues, liver, kidney, intestine	1. Streptococcosis in experimental challenge 2. Vibriosis. Pathogenicity proven in experimental challenge	1452 1830 1840
Grouper (*Epinephelus* species)	*Nocardia seriolae*		Skin ulcers and tubercules in organs	Nocardiosis	1523
Family Siganidae (rabbitfishes)					

Continued

Table 1.1. Continued.

Host	Pathogens	Normal flora	Tissue site	Disease status	Ref
Marbled spinefoot (*Siganus rivulatus* Forsskål)	1. *Mycobacterium marinum* 2. *Shewanella* (*Pseudomonas*) *putrefaciens* 3. *Streptococcus iniae*		1. White-yellow nodules in the spleen 2. Skin discolouration, and focal necrosis becoming haemorrhagic, abdominal ascites, exophthalmia, oedema and necrosis in kidneys 3. Systemic disease with diffuse visceral haemorrhages	1. Mycobacteriosis – infection spread from cage fish to wild rabbitfish 2, 3. Mortality	440 544 1453 1855
White-spotted spinefoot (*Siganus canaliculatus* Park)	*Streptococcus iniae*		Ascites, hepatomegaly, splenomegaly	Streptococcosis, mass mortality	1840
Family Sparidae (porgies)	Bream may be referred to as porgy or porgie in North America				
Black-head seabream, black seabream (*Acanthopagrus schlegelii schlegelii* [Bleeker])	1. *Vibrio alginolyticus* 2. *Halomonas cupida* 3. *Tenacibaculum maritimum*	4. *Shewanella schlegeliana* 5. *Vibrio nereis*	3. Eroded mouth, frayed fins, tail rot 4. Intestine	1, 2, 3. Mortality 4. Normal flora 5. Non-pathogenic	960 1479 1744
Crimson seabream (*Evynnis japonica*)	1. *Edwardsiella piscicida* 2. *Pseudomonas fluorescens*		2. Ulcerative and haemorrhagic skin lesions	1. Pathogenic in experimental challenge with zebrafish 2. Mortality	1 958
Dentex dentex	1. *Vibrio harveyi*	2. *Enterovibrio nigricans*	2. Isolated from head kidney along with *Vibrio harveyi*	1. Mortality 2. Pathogenicity undetermined	1314 1364
Gilthead seabream (*Sparus aurata* L.)	1. *Edwardsiella tarda* 2. *Photobacterium damselae* ssp. *damselae* 3. *P. damselae* ssp. *piscicida* 4. *Pseudomonas anguilliseptica* 5. *Streptococcus agalactiae* 6. *Streptococcus iniae* 7. *Tenacibaculum maritimum* 8. *Vibrio alginolyticus* 9. *Vibrio* (*Listonella*) *anguillarum*	12. *Hafnia alvei* 13. *Vibrio lentus* 14. *Vibrio ponticus* 15. *Enterovibrio nigricans*	1. Kidney, liver, spleen 2. Lethargy, distended abdomen, haemorrhages on fins and tail, pale liver 3. No external changes except anus red and protruded, abdomen distended, fluid in abdominal cavity, congested spleen, petechial haemorrhages on liver, granulomatous lesions in spleen and kidney, multifocal tissue necrosis	2, 9. Mortality 3. Fish pasteurellosis 4. Associated with 'winter disease', haemorrhagic septicaemia 5. Streptococcosis – Epizootic with 100% mortality 6. Exudative meningitis and panophthalmitis	99 108 109 117 135 176 186 455 478 501 521

	10. *Vibrio harveyi* 11. *Vibrio splendidus*		4. Erratic swimming at water surface, sink to the bottom of the cage and die. Ascites, renal haemorrhage 5. Haemorrhagic areas on body, mouth, eye, operculum and fins 6. Brain, kidney 7. Eroded mouth, frayed fins, rotten tail 8, 9, 10, 11. Ulcers, haemorrhages, exophthalmia 15. Isolated from head kidney along with *Vibrio harveyi*	1, 3, 7, 8, 9, 10, 11. Pathogenicity confirmed experimentally 9. Can also be resistant to disease in experimental challenge 12, 13, 14. Non-pathogenic in experimental infection 15. Pathogenicity undetermined	1080 1110 1295 1296 1314 1364 1649 1717 1855
Goldlined seabream *Rhabdosargus sarba* (*Sparus sarba* is a synonym). May be called silver seabream	*Vibrio alginolyticus*		Skin haemorrhages, ulcers, internal organs	Mortality proven in experimental challenge	1021
One-spot seabream (*Diplodus sargus kotschyi* Steindachner)				Not susceptible to infection with *Streptococcus iniae* in experimental challenge	1840
Pinfish (*Lagodon rhomboides* Linnaeus), a saltwater fish	*Streptococcus agalactiae* Group B		Brain, eye, organs	Streptococcosis	501 1348
Red porgy (*Pagrus Pagrus* Linnaeus)	*Streptococcus iniae*		Haemorrhagic septicaemia, exophthalmia	Mortality	478
Redbanded seabream (*Pagrus auriga*)	*Photobacterium damselae* ssp. *damselae*		Exophthalmia, dark skin pigmentation, ascites	Mortality	963
Red seabream (*Pagrus major* Temminck and Schlegel). May be called Japanese seabream	1. *Edwardsiella tarda* 2. *Vibrio anguillarum* 3. *Photobacterium damselae* ssp. *piscicida* 4. *Tenacibaculum maritimum*		1. Septicaemia, focal suppurative or granulomatous lesions, cutaneous ulcerations 3. Kidney, spleen 4. Eroded mouth, frayed fins, tail rot	1. Edwardsiellosis 2. Infection 3. Fish pasteurellosis 4. Mortality	1110 1744 1828
Seabass (*Diplodus* [*Puntazzo*] *puntazzo* Walbaum)	*Aeromonas hydrophila*		Enlargement of spleen, erythaema and swelling of anus	Mortalities	461
Silver seabream, seabream (*Pagrus aurata* Linnaeus)	*V. harveyi* and *Photobacterium damselae* ssp. *damselae*		Intestine and blood	Distended swimbladder (bloat), mortality	1579
Sobaity seabream (*Sparidentex hasta* [*Acanthopagrus cuvieri*]). May be called silver black porgy	*Vibrio harveyi*		Tissues	Mortality	1452

Continued

Table 1.1. Continued.

Host	Pathogens	Normal flora	Tissue site	Disease status	Ref
Yellowfin seabream (*Acanthopagrus latus* Houttuyn). May be known as black seabream	1. *Pseudomonas anguilliseptica*	*Streptococcus iniae*	1. Haemorrhage and ulcerative lesions	Not susceptible to infection with *Streptococcus iniae* in experimental challenge 1. Associated with 'winter disease'. Septicaemia	186 1195 1840
Family Symphysanodontidae					
Blue discus fish (*Symphysodon aequifasciatus* Pellegrin)	1. *Aeromonas hydrophila* 2. *Aeromonas jandaei* 3. *Mycobacterium fortuitum*		2. Organism in liver 3. Fin lesions	1, 2. Deaths 3. Mycobacteriosis	216 244
Family Terapontidae (grunters or tigerperches)					
Jarbua terapon (*Terapon jarbua* Forsskål)	*Nocardia seriolae*		Skin ulcers and tubercules in organs	Nocardiosis	1523
Rhynchopelates oxyrhynchus Temminck and Schlegel. Coastal fish – Japan (*Therapon oxyrhynchus* is a senior synonym)		*Grimontia* (*Vibrio*) *hollisae*	Intestinal contents	Non-pathogenic	1225
Silver perch (Bidyan perch) (*Bidyanus bidyanus* Mitchell)	Atypical *Aeromonas salmonicida*		Cutaneous ulcers	Skin disease, ulcerative dermatitis	1788 1789
Welch's grunter or silver bream (*Bidyanus welchi*)		*Hafnia paralvei*	Intestine	Normal flora	1258
Spangled perch (*Leiopotherapon unicolor*)		*Hafnia alvei*, *Hafnia paralvei*	Intestine	Normal flora	1258
Family Zoarcidae (eelpouts)					
Eelpout (*Zoarces viviparus* Linnaeus). May be called blenny, viviparous blenny	Atypical *Aeromonas salmonicida*		Red ulcers with white margins. Organism also isolated from internal organs	Skin ulcers, septicaemia	1800
Family Stromateidae (butterfishes)					
Pomfret, silver pomfret (*Pampus argenteus* Euphrasen)	*Streptococcus agalactiae*		Brain, kidney, ascites fluid	Streptococcosis, mortality	466
ORDER Pleuronectiformes (flatfishes)					
Family Cynoglossidae (tonguefishes)					

Tongue sole (*Cynoglossus semilaevis* Günther)	*Vibrio aestuarianus*		Haemorrhages on head, opercula, fins, liver necrosis	Mortality	1850
Family Paralichthyidae (large tooth flounders)					
Japanese flounder, also known as olive flounder, bastard halibut (*Paralichthys olivaceus* Temminck and Schlegel)	1. Atypical *Aeromonas salmonicida* 2. *Edwardsiella tarda* 3. *Lactococcus garvieae* 4. *Nocardia seriolae* 5. *Photobacterium damselae* ssp. *piscicida* 6. *Streptococcus iniae* 7. *Streptococcus parauberis* 8. *Tenacibaculum maritimum* 9. *Vibrio anguillarum* 10. *Vibrio harveyi*, *Edwardsiella tarda* 11. *Vibrio ichthyoenteri*	12. *Weissella hellenica* strain DS-12	2. Septicaemia, focal suppurative or granulomatous lesions, cutaneous ulcerations, abdominal swelling, ascites fluid 4. Tubercules in kidney, gills and spleen. Abscesses on epidermis 5. Haemorrhagic septicaemia and white areas of granuloma in the kidney, liver, spleen 7. Liver 10. Ascites, kidney, spleen 11. Intestinal necrosis, opaque intestines in larvae 12. Present in intestine	1. Infection 2. Edwardsiellosis 4. Nocardiosis 5. Fish pasteurellosis 3, 6, 7, 8, 9, 11. Mortality 10. Co-infection produced greater mortality in challenge experiments 12. Probiotic potential	93 171 260 572 796 804 899 917 944 1158 1193 1201 1215 1248 1524 1525
Summer flounder (*Paralichthys dentatus* L.)	1. *Mycobacterium* species with homology to *Mycobacterium marinum* and *Mycobacterium ulcerans* 2. *Vibrio harveyi*		1. Granulomas in the kidney, large red-tan coloured multilobulated mass in kidney 2. Reddening around anal area, distended abdomen filled with fluid, enteritis and necrosis	1. Mycobacteriosis 2. Flounder infectious necrotizing enteritis (FINE)	592 771 1555
Family Pleuronectidae (right-eye flounders)					
Common dab (*Limanda limanda* L.)	Atypical *Aeromonas salmonicida*		Round, red ulcers with white margin of necrotic tissue	Skin ulcers	1800
European flounder (*Platichthys flesus* L.)	Atypical *Aeromonas salmonicida* and oxidase-negative strains		Epidermal ulcerations – dark, red wounds usually round but some irregular forms seen	Skin ulcer disease	674 1795 1797 1800
Flounder		1. *Shewanella hafniensis* 2. *Shewanella morhuae*	1, 2. Belly flap area	Non-pathogenic, thought to be involved in food spoilage	1480

Continued

Table 1.1. Continued.

Host	Pathogens	Normal flora	Tissue site	Disease status	Ref
Greenback flounder (*Rhombosolea tapirina* Günther)	1. Atypical *Aeromonas salmonicida* 2. *Aeromonas salmonicida* biovar Flounder 3. *Tenacibaculum maritimum*		1, 2. Skin lesions 3. Skin erosion, lesions on dermis and fins	1, 2. Ulcerative dermal lesions 3. Tenacibaculosis	283 673 1788 1789
Halibut, Atlantic halibut (*Hippoglossus hippoglossus* L.)	1. Atypical *Aeromonas salmonicida* 2. *Tenacibaculum ovolyticum* 3. *Vibrio tapetis*, *Vibrio tapetis* ssp. *britannica*		1. Skin ulcers 2. Dissolves chorion and zona radiata of the egg shells 3. Head kidney	1. Infection 2. Opportunistic pathogen of eggs and larvae 3. Chronic mortality and co-infection with atypical *Aeromonas salmonicida*	103 674 675 1393
Greenland halibut (*Reinhardtius hippoglossoides* Walbaum)		*Arthrobacter rhombi*	Organism isolated from spleen and internal organs	Pathogenicity not determined	1276
Plaice, European plaice (*Pleuronectes platessa* Linnaeus)	1. *Moritella viscosa* 2. *Vibrio anguillarum* serotype 07	3. *Shewanella algidipiscicola*	1. Skin ulcers 2. Lesions 3. Flora from belly flap	1. Winter ulcer disease 2. Vibriosis 3. Normal flora	1071 1481 1559
Colistium nudipinnis Waite), also called turbct Brill (*C. guntheri*). Both species are also known as flatfish	*Vibrio campbellii*-like, *Vibrio splendidus* I		Organism in brain, kidney, liver, which showed lesions and haemorrhages	Acute mortality in juveniles – opportunistic infection	450
Family Scophthalmidae (turbots)					
Turbot (*Scophthalmus maximus* L.). Previously known as *Psetta maxima*	1. Atypical *Aeromonas salmonicida*. Also an oxidase-negative strain 2. *Chryseobacterium scophthalmum* 3. *Edwardsiella piscicida* 4. *Edwardsiella tarda* 5. *Mycobacterium chelonae* and *Mycobacterium marinum* 6. *Nocardia* sp. 7. *Photobacterium damselae* ssp. *damselae* 8. *P. damselae* ssp. *piscicida*	14. *Enterovibrio norvegicus* 15. *Vibrio chagasii* 16. *Vibrio cholerae* non-01 and *Vibrio mimicus* did not cause deaths in virulence assay 17. *Vibrio lentus* 18. *Vibrio neptunius* 19. *Vibrio pomeroyi* 20. *Vibrio scophthalmi*	1. Skin ulcers 2. Swollen intestines, haemorrhages in eye, skin and jaw, gill hyperplasia 4. Eyes, kidney, spleen, muscle 5. Granulomas in organs 7. Haemorrhage around anus, eyes, mouth 8. White tubercles in spleen 9. Organism in head kidney and spleen 10. Swollen kidney and spleen, yellow nodules, foci of liquefaction and necrosis	1. Mortality 2. Gill disease, haemorrhagic septicaemia 3. Pathogenic to zebrafish in experimental challenge 4. Edwardsiellosis. Pathogenicity confirmed in experimental challenge 5. Mycobacteriosis 7. Mortality 8. Fish pasteurellosis 9. Haemorrhagic septicaemia	1 55 100 176 286 293 456 459 521 556 590 971 979 1073

	9. *Pseudomonas anguilliseptica* 10. *Serratia liquefaciens* 11. *Streptococcus parauberis* 12 *Vibrio anguillarum* 01 and 02β, 02α 13. *Vibrio splendidus* biotype I		11. Lesions, haemorrhage in the anal and pectoral fins and petechiae on the abdomen, exophthalmia and pus in the eyes 13. Haemorrhages in mouth, abdominal distension, reddish fluid in peritoneal cavity 14, 15, 18, 20. Part of gut flora	10. Opportunist pathogen, mortality 11. Streptococcosis, Hepatomegaly, mucohaemorrhagic enteritis 12. Vibriosis 13. Epizootic in juveniles 14, 19. Normal flora 6, 15, 18, 20. Pathogenicity not known 17. Not pathogenic in experimental infection	1172 1173 1324 1559 1627 1632 1635 1639 1649 1650 1651 1653 1724 1800
Brill (*Scophthalmus rhombus* L.), also known as turbot	*Tenacibaculum soleae*		Ulcers on tail and body, liver, kidney	Mortality in adults	1056
Family Soleidae (soles)					
Common sole (*Solea solea* L.). Also known as Dover sole	1. *Tenacibaculum maritimum*	2. *Vagococcus fluvialis*	1. Blistering of the skin between the caudal and marginal fins, loss of dermal tissues, which progresses to necrotic ulcers	1. Black patch necrosis 2. Normal flora, probiotic potential	167 1127 1560 1725
Senegalese sole (*Solea senegalensis* Kaup)	1. *Edwardsiella tarda* 2. *Photobacterium damselae* ssp. *piscicida* 3. *Pseudomonas anguilliseptica* 4. *Tenacibaculum discolor* 5. *Tenacibaculum soleae*	6. *Vibrio alfacsensis*	1. Cutaneous lesions, organism in kidney, liver, spleen 2. Haemorrhagic septicaemia and white areas of granuloma in the kidney, liver, spleen 3. Kidney, brain 4. Kidney 5. Ulcers on tail and body, liver, kidney 6. Gills of healthy fish	1. Mortality 2. Fish pasteurellosis 3. Haemorrhagic septicaemia 4. Eroded mouth, fins and skin lesions similar to flexibacteriosis 5. Mortality in juveniles and adults 6. Pathogenicity unknown	100 286 640 1056 1344 1347 1857

Continued

Table 1.1. Continued.

Host	Pathogens	Normal flora	Tissue site	Disease status	Ref
Wedge sole (*Dicologlossa* [*Dicologoglossa*] *cuneata* [Moreau])	1. *Tenacibaculum soleae* 2. *Tenacibaculum maritimum* 3. *Vibrio harveyi* 4. *Vibrio tapetis*		1. Ulcers on tail and body, liver, kidney 2. Ulcers on tail and body, pale liver 3. Ulcers and internal organs 4. Ulcers on body and tail	1. Mortality in fry, juveniles and adults 2. Mortality in fry in challenge experiments 3. Mortality in fry and juveniles	1052 1053 1055 1056
ORDER Salmoniformes				Probably part of normal gut flora	
Family Salmonidae					
Arctic Charr (*Salvelinus alpinus* L.)	1. *Serratia liquefaciens*	2. *Carnobacterium divergens*	1. Haemorrhagic internal organs, ascites 2. Intestine	1. Mortality 2. Normal flora, potential probiotic	1405 1575
Atlantic salmon (*Salmo salar* L.)	1. *Aeromonas piscicola* 2. *Aeromonas salmonicida* ssp. *salmonicida* 3. *Alliivibrio* (*Vibrio*) *logei* 4. *Alliivibrio salmonicida* 5. Atypical *Aeromonas salmonicida* 6. *Aeromonas salmonicida* biovar Acheron 7. *Chryseobacterium chaponense* 8. *Chryseobacterium piscicola* 9. *Flavobacterium araucananum*, *Flavobacterium chilense* 10. *Flavobacterium branchiophilum* 11. *Flavobacterium psychrophilum* 12. *Moritella viscosa* 13. *Mycobacterium chelonae* 14. *Mycobacterium salmoniphilum* 15. *Nocardia seriolae* 16. *Pasteurella skyensis* 17. *Photobacterium damselae* ssp. *damselae*	34. *Carnobacterium inhibens* 35. *Carnobacterium* species 36. *Lactobacillus* sp. 37. Unidentified *Mycoplasma* phylotype 38. *Photobacterium iliopiscarium* 39. *Vibrio pacinii* 40. *Vibrio tasmaniensis* 41. *Vibrio fortis*	1. Fish moribund 2, 6. Skin ulcers 3. Skin lesions 4. Anaemia, haemorrhages 5. Organism from head kidney, cutaneous lesions 7. External lesions, gills, fins 8. Ulcerative lesions skin, muscle 9. Kidney, external lesions 10. Gills 11. Erosion of fins 12. Skin ulcers, internal dark-brown petechiae or ecchymotic haemor-rhage 13, 14. Granuloma-like nodules in tissues 15. Granulomatous lesions on body 16. Multifocal areas of coagulative necrosis in kidney, liver, spleen 18. Petechial haemorrhages on skin, mouth and anus, peritoneum and liver 19. Gills	1. Pathogenicity not known 2. Furunculosis 3. Possible pathogen 4. Coldwater vibriosis, Hitra disease 5, 6. Similar to Furunculosis 7. Pathogenicity not confirmed 8. Disease, virulence tested 9. Pathogenicity not confirmed 10. Bacterial gill disease (BGD) 11. Morbidity rather than mortality 12. Winter ulcer disease 13, 14. Mycobacteriosis 15. Nocardiosis 16. Mortality	71 83 141 153 154 181 199 217 235 236 249 282 397 420 473 614 638 673 689 733 743 798

	18. *Pseudomonas anguilliseptica* 19. *Pseudomonas fluorescens* 20. *Renibacterium salmoninarum* 21. *Rhodococcus erythropolis* 22. *Rhodococcus qingshengii* 23. *Serratia liquefaciens* 24. *Streptobacillus moniliformis*-like organism 25. *Streptococcus phocae* 26. *Tenacibaculum maritimum* 27. *Vagococcus salmoninarum* 28. *Varracalbmi* 29. *Vibrio anguillarum* 30. *Vibrio jasicida* 31. *Vibrio ordalii* 32. *Aliivibrio* (*Vibrio*) *wodanis* 33. *Yersinia ruckeri* serotype I		20. White nodules in kidney 21. 22 23. Swelling of kidneys, bleeding at the anus and intestine, gastroenteritis 24. Organism intracellular in tissues, endothelial cells of kidney glomeruli enlarged 25. Brain, eye, kidney, spleen, ulcers 26. Yellow-coloured plaques on teeth and oral cavity 27. Peritonitis, haemorrhagic ascites, retained eggs, engorged testes, languid swimming 28. Deep skin lesions, eye lesions, haemorrhagic and pyogranulomatous lesions in gills, kidney, liver and pseudobranch 30. Haemorrhagic septicaemia 33. Haemorrhagic musculature 34. Gastrointestinal tract 37. Gastrointestinal tract 38. Intestine	8, 17. Pathogenic in challenge experiments 18. Haemorrhagic disease 19. Atypical bacterial gill disease 20. Bacterial kidney disease 21. Septicaemia 22. Pseudomembranes on internal organs 21, 22. Associated with vaccination 23, 24. Mortality 25. Streptococcocis 26. Bacterial stomatitis (mouth rot) in smolts 27, 30. Mortality 28. Blindness 29. Vibriosis 31. Mortality 32. Opportunistic infection in 'winter ulcer' disease 33. Enteric redmouth (ERM) 34. Inhibits growth of *Vibrio anguillarum* and *Aeromonas salmonicida* 35. Potential probiotic 36, 38. Normal microflora 37. Non-pathogenic 40, 41. Pathogenicity not known	799 841 850 857 859 928 1071 1090 1123 1180 1187 1223 1265 1272 1282 1283 1302 1332 1404 1424 1472 1497 1498 1531 1559 1634 1637 1682 1689 1690 1783 1787 1794 1835 1845

Continued

Table 1.1. Continued.

Host	Pathogens	Normal flora	Tissue site	Disease status	Ref
Black sea salmon (*Salmo labrax*)	*Renibacterium salmoninarum*		Kidney	Bacterial kidney disease (BKD)	1484
Chinook salmon (*Oncorhynchus tshawytscha* Walbaum). *Oncorhynchus tschawytscha* is no longer valid	1. *Carnobacterium maltaromaticum* 2. *Edwardsiella ictaluri* 3. *Edwardsiella tarda* 4. *Flavobacterium branchiophilum* 5. *Flavobacterium columnare* 6. *Mycobacterium chelonae* 7. *Mycobacterium neoaurum* 8. *Mycobacterium salmoniphilum* 9. *Mycobacterium* species 10. *Renibacterium salmoninarum* 11. *Tenacibaculum maritimum* 12. *Vibrio anguillarum* 13. *Vibrio ordalii* 14. *Yersinia ruckeri*		1, 2. Kidney 3. Kidney 4. Bacteria attached to gill epithelium 5. Gill lesions 6. Kidney 7. Organism present in eyes, exophthalmia, and ocular lesions 8. Liver and kidney 10. Kidney lesions, gill lesions 12, 13. Haemorrhagic septicaemia 14. Kidney, liver, spleen, eye	1. Mortality associated with stress 2. Mortality in experimental infection 3. Edwardsiellosis. Pathogenicity confirmed by experimental challenge 4. Bacterial gill disease (BGD) 5. Columnaris disease 8. Mortality 7. Panophthalmitis 6, 8, 9. Mycobacteriosis 10. Bacterial kidney disease (BKD) 11, 12, 13. Mortality 14. Enteric redmouth (ERM)	48 57 62 90 115 131 168 249 309 411 505 683 736 1377 1427 1746 1783
Cutthroat trout (*Oncorhynchus clarkii clarkii* Richardson)	1. *Carnobacterium* (*piscicola*) *maltaromaticum* 2. *Mycobacterium chelonae*		1. Kidney	1. Mortality following stress at spawning 2. Mycobacteriosis	57 736
Chum salmon (*Oncorhynchus keta*)	1. *Flavobacterium psychrophilum* 2. *Renibacterium salmoninarum* 3. *Vibrio anguillarum* 4. *Vibrio ordalii*		1. Kidney lesions 2. Ovarian fluid, kidneys 3, 4. Muscle, skin, liver, spleen	1. Bacterial cold water disease 2. Bacterial kidney disease (BKD) 3. Haemorrhagic septicaemia	115 1151 1377 1471
Japanese amago trout (*Oncorhynchus rhodurus*)	*Pseudomonas chlororaphis*		Haemorrhages, increased ascites	Infection	696
Pacific salmon, coho salmon (*Oncorhynchus kisutch* Walbaum)	1. *Aeromonas salmonicida* (oxidase-negative strain) 2. *Flavobacterium columnare*		1. Fingerlings affected. Soft kidneys, occasional fish with haemorrhage on fin, otherwise no external signs	1. Furunculosis 3. Cold-water disease 4. Mycobacteriosis	57 115 303

	3. *Flavobacterium psychrophilum* 4. *Mycobacterium chelonae* 5. *Renibacterium salmoninarum* 6. *Vibrio anguillarum* 01 7. *Vibrio ordalii*		5. Kidney lesions 6. 7. Haemorrhagic septicaemia	5. Bacterial kidney disease (BKD) 6. 7. Haemorrhagic septicaemia	335 683 1377 1559 1650 1651 1659 1674 1749
Pink salmon (*Oncorhynchus gorbuscha*)	1. *Aeromonas salmonicida* ssp. *masoucida* 2. *Renibacterium salmoninarum*		1. Lesions 2. Lesions in kidney	1. Furunculosis 2. Bacterial kidney disease (BKD)	115 906
Masou or masu salmon, cherry salmon (*Oncorhynchus masou masou* Brevoort). Also known as Japanese salmon and yamame	1. *Aeromonas salmonicida* ssp. *masoucida* 2. *Flavobacterium branchiophilum* 3. *Hafnia alvei* 4. *Renibacterium salmoninarum*		1. Lesions 2, 3. Lesions in kidney	1. Furunculosis 2. Bacterial gill disease (BGD) 3. Mortality 4. Bacterial kidney disease	906 1187 1618 1746
Sockeye salmon (*Oncorhynchus nerka* Walbaum). Also called blueback salmon	1. *Flavobacterium branchiophilum* 2. *Nocardia salmonicida* 3. *Renibacterium salmoninarum* 4. *Yersinia ruckeri*		1. Bacterium attaches to gill epithelium 3. Lesions in kidney	1. Bacterial gill disease (BGD) 2. Nocardiosis 3. Bacterial kidney disease (BKD) 4. Enteric redmouth (ERM)	115 249 806 1436 1746
Brook trout (*Salvelinus fontinalis* Mitchell)	1. *Aeromonas salmonicida* spp. *salmonicida* 2. Atypical *Aeromonas salmonicida* 3. *Edwardsiella tarda* 4. *Flavobacterium branchiophilum* 5. *Flavobacterium columnare* 6. *Nocardia* species 7. *Renibacterium salmoninarum* 8. *Yersinia ruckeri*		1, 2. Skin ulcers 3. Lesions and haemorrhages on skin and organs, kidney, liver 4. Bacteria attach to gill epithelium 5. Skin lesions 6. Necrosis, thrombosis in kidney, spleen, gills 7. White nodules in kidney 8. Kidney, liver, spleen, eye	1. Furunculosis 2. Skin ulcers and septicaemia 3. Acute bacterial septicaemia 4. Bacterial gill disease (BGD) 5. Columnaris disease 6. Nocardiosis 7. Bacterial kidney disease 8. Enteric redmouth (ERM)	249 413 414 490 1280 1472 1674 1787

Continued

Table 1.1. Continued.

Host	Pathogens	Normal flora	Tissue site	Disease status	Ref
Sea trout, brown trout (*Salmo trutta trutta* L.)	1. *Aeromonas salmonicida* ssp. *salmonicida* 2. Atypical *Aeromonas salmonicida* 3. *Flavobacterium branchiophilum* 4. *Flavobacterium columnare* 5. *Hafnia alvei* 6. *Lactococcus lactis* spp. *tructae* 7. *Pseudomonas anguilliseptica* 8. *Renibacterium salmoninarum* 9. *Streptococcus agalactiae* 10. *Vagococcus salmoninarum* 11. *Vibrio anguillarum* 02		1, 2. Skin ulcers, internal organs 3. Skin around dorsal fin blanched, but no ulceration seen 5. Organism and lesions in the kidney 6. Intestinal mucus 7. Petechial haemorrhages on skin, mouth and anus, peritoneum and liver and at base of fins 8. White nodules in kidney 9. kidney, spleen, liver, brain 10. Spleen, liver 11. Lesions	1. Furunculosis 2. Skin ulcers and septicaemia 3. Bacterial gill disease 4. Columnaris disease, chronic mortality 5. Opportunistic pathogen 6. Normal flora 7. Haemorrhagic disease 8. Bacterial kidney disease 9. Streptococcosis, mortality 10. Mortality 11. Vibriosis	7 134 168 490 733 1330 1418 1472 1497 1559 1564 1746 1787 1794
Rainbow trout (*Oncorhynchus mykiss* Walbaum 1792) (*Salmo gairdneri* Richardson is a junior synonym). Also known as steelhead trout	1. *Aeromonas piscicola* 2. *Aeromonas salmonicida* ssp. *salmonicida* 3. Atypical *Aeromonas salmonicida* 4. Atypical *Aeromonas salmonicida* (growth at 37 °C) 5. *Aeromonas* sp. 6. *Arcobacter cryaerophilus* 7. *Carnobacterium (piscicola) maltaromaticum* 8. *Corynebacterium aquaticum* 9. *Chryseobacterium piscicola* 10. *Chryseobacterium tructae* 11. *Clostridium botulinum* 12. *Edwardsiella ictaluri* 13. *Edwardsiella tarda*	50. *Aeromonas tecta* 51. *Citrobacter freundii*	1. Fish moribund 2, 3, 4. Skin ulcers 5. Pathogenic for fingerlings in virulence assay 6. Kidney, liver, gills 7. Bilateral exophthalmia, periocular haemorrhages, ascites fluid and haemorrhages in liver, swimbladder, muscle and intestine, inflammation in the brain, kidney 8. Brain 9. Ulcerative lesions skin, muscle, anus peduncle 10. Isolated from liver 11. Toxin in serum and intestinal contents 12. Petechial haemorrhages on skin, liver, spleen, kidney	1, 34. Pathogenicity not known 2. Furunculosis 3, 4. Skin ulcers, septicaemia 5. Haemorrhagic septicaemia 6. Moderately virulent in experimental infection 7. Chronic disease with low level mortality or highly virulent 8. Exophthalmia in experimental infection 9. Lesions, virulence tested 10. Pathogenicity not known	29 39 48 55 57 70 73 76 77 79 80 85 86

	14. *Escherichia vulneris*		14. Haemorrhagic lesions on skin,	11. Botulism	115
	15. *Flavobacterium branchiophilum*		bloody exudate in digestive tract,	12, 13. Susceptibility	132
	16. *Flavobacterium oncorhynchi*		liver yellow and hyperaemic, gonads	confirmed	136
	17. *Flavobacterium oncorhynchi*		haemorrhagic, lesions in liver	experimentally	137
	18. *Flavobacterium psychrophilum*		13. Organism present in kidney	13. Edwardsiellosis	141
	19. *Hafnia alvei*		15. Bacteria attach to gill epithelium	14. Mortality	153
	20. *Janthinobacterium lividum*		16. External lesion	15. Bacterial gill	169
	21. *Klebsiella pneumoniae*		17. Liver	disease (BGD)	199
	22. *Lactococcus garvieae*		20. Exophthalmia, pale gills,	16, 17. Pathogenicity	244
	23. *Lactococcus lactis* spp. *tructae*		gastroenteritis	unknown	249
	24. *Lactococcus piscium*		21. Fin and tail	20. Mortality in fry,	253
	25. *Micrococcus luteus*		23. Intestinal mucus	possible opportunistic	269
	26. *Moritella viscosa*		25. Organism isolated from kidney,	infection	279
	27. *Mycobacterium chelonae*		spleen and ascites fluid	21. Fin and tail disease	389
	28. *Mycobacterium marinum*		26. Skin lesions, petechial	22. Mortality	406
	29. *Mycobacterium salmoniphilum*		haemorrhage in liver and peritoneal	23. Normal flora	413
	30. *Nocardia asteroides*		membranes	24. Pseudokidney	414
	31. *Photobacterium*		28. 31. Lesions in kidney	disease, lactobacillosis	435
	damselae ssp. *damselae*		29. Bacteria in liver and kidney	25. Vibriosis,	479
	32. *Planococcus* sp.		31. Organism present in	septicaemia	482
	33. *Plesiomonas shigelloides*		head kidney	25. Isolated from	533
	34. *Pseudomonas anguilliseptica*		33. Petechial haemorrhages in	moribund fish	542
	35. *Pseudomonas*		intestine	26. Mortality	556
	(*pseudoalcaligenes*) *oleovorans*		32. White spots on head, internal	27, 28, 29. Mortality,	604
	36. *Pseudomonas putida*		organs in natural and experimental	mycobacteriosis	612
	37. *Renibacterium salmoninarum*		infection, respectively	30. Nocardiosis	617
	01, 02, 03, 04		34. Petechial haemorrhages on skin,	31. Pathogenic in	673
	38. *Serratia marcescens*,		mouth and anus and base of fins,	challenge experiments	733
	Serratia plymuthica		peritoneum and liver	and natural infection	736
	39. *Staphylococcus warneri*		35. Haemorrhage internal tissues,	33. Possible opportunis-	799
	40. *Streptococcus agalactiae*		ascites	tic pathogen	857
	41. *Streptococcus iniae*		36. Ulcers dorsal surface	34. Haemorrhagic	880
	42. *Tenacibaculum maritimum*		37. White nodules in kidney	disease	921
	43. *Vagococcus salmoninarum*		38. Internal haemorrhaging, ascites	33, 35. Mortality in	1037
	44. *Vibrio anguillarum*		fluid, skin and muscle erosion	experimental challenge	1068

Continued

Table 1.1. Continued.

Host	Pathogens	Normal flora	Tissue site	Disease status	Ref
	45. *Vibrio brasiliensis*, *Vibrio coralliilyticus*, *Vibrio ezurae*, *Vibrio fortis*, *Vibrio kanaloae*, *Vibrio neptunius*, *Vibrio rotiferianus*, *Vibrio tubiashii* 46. *Vibrio splendidus* I 47. *Vibrio ordalii* 48. *Weissella* sp. 49. *Yersinia ruckeri*		39. Exophthalmia, ascites fluid, lesions on fins, organism in kidney and liver 40, 41. Organisms in brain and eye 42. Dermal and fin erosion 43. Loss of equilibrium, haemorrhage around eye and gill, lesions on body, congestion in spleen and liver. Peritonitis, haemorrhagic ascites, retained eggs, engorged testes, languid swimming 44. Lesions 45. Liver, kidney, muscle 48. Brain, kidney, liver 49. Haemorrhage around mouth and intestines 51. Kidney	36. Mortality 37. Bacterial kidney disease 38. Mortality in fingerlings 39. Diseased and dying trout – opportunistic infection 40, 41. Meningoenceph-alitis, septicaemia 42. Tenacibaculosis 43. Vagococcosis, chronic disease with mortality 44. Vibriosis, septicaemia 45. Pathogenic in experimental challenge 46. Mortalities in experimental infection 47. Vibriosis 48. Haemorrhagic septicamia 49. Enteric redmouth (ERM) 51. Opportunistic pathogen, carrier status, low pathogenicity in experimental challenge	1139 1187 1221 1222 1280 1322 1330 1383 1387 1429 1429 1468 1497 1551 1559 1564 1576 1650 1655 1659 1746 1749 1755 1787 1794 1798 1802 1845 1846
Grayling (*Thymallus thymallus* L.)	1. *Aeromonas salmonicida* ssp. *salmonicida* 2. Atypical *Aeromonas salmonicida* 3. *Yersinia ruckeri*		1, 2. Skin ulcers	1. Furunculosis 2, 3. Infection	411 674 733

Whitefish (*Coregonus* sp), lake cisco (*Coregonus artedi* Lesueur), lake whitefish (*Coregonus clupeaformis* Mitchell), peled (*Coregonus peled* Gmelin)	1. *Aeromonas salmonicida* ssp. *salmonicida* 2. Atypical *Aeromonas salmonicida* 3. *Carnobacterium maltaromaticum*-like 4. *Pseudomonas anguilliseptica* 5. *Yersinia ruckeri*		3. Kidney, swimbladder 4. Petechial haemorrhages on skin, mouth and anus, peritoneum and liver 5. Disease is associated with poor environmental conditions	1. Skin ulcers, furunculosis 2. Skin ulcers 3. Chronic mortality 4. Haemorrhagic disease 5. Yersiniosis	674 733 1047 1692 1794
Whitefish, mountain whitefish (*Prosopium williamsoni* Girard)	*Mycobacterium chelonae*			Mycobacteriosis	57
ORDER Scorpaeniformes (scorpion fish and flatheads)					
Family Agonidae (poachers)					
Percis (*Cottus*) *japonicus*	*Renibacterium salmoninarum*			May be carrier	1459
Family Anoplomatidae (sablefishes)					
Sablefish (*Anoplopoma fimbria* Pallas)	*Renibacterium salmoninarum*		Kidney	Bacterial kidney disease	504
Family Cottidae (sculpins)					
Goby (*Cottida* species)		*Aliivibrio logei*	Intestine	Part of normal flora	890
Pacific staghorn sculpin (*Leptocottus armatus* Girard)	*Mycobacterium scrofulaceum*		Liver white and friable	Mycobacteriosis	977
Family Cyclopteridae					
Lumpfish (*Cyclopterus lumpus* L.)	*Mycobacterium marinum*		Tissues	Mycobacteriosis	1356
Family Hexagrammidae					
Greenling – marine fish (*Hexagrammos otakii* Jordan & Starks)	Atypical *Aeromonas salmonicida*			Infection	796
Family Platycephalidae (flatheads)					
Flathead (*Platycephalus indicus* L.)	*Renibacterium salmoninarum*			May be carrier	1459
Family Sebastidae (rockfish, rockcod, thornyheads)					
Rockfish – Schlegel's black rockfish, Korean rockfish (*Sebastes schlegelii* Hilgendorf)	1. Atypical *Aeromonas salmonicida* 2. *Lactococcus garvieae*		1. Ulcer on the trunk. Organism isolated from the kidney and brain 2. Spleen and head kidney	1, 2. Mortality	796 824 863
ORDER Siluriformes					
Family Ariidae (Sea catfishes)					

Continued

Table 1.1. Continued.

Host	Pathogens	Normal flora	Tissue site	Disease status	Ref
Hardhead sea catfish (*Ariopsis* [*Arius*] *felis* L.). Family Ariidae (sea catfishes)	*Streptococcus agalactiae* Group B		Brain, eye, organs	Streptococcosis	501 1348
Family Bagridae (bagrid catfishes)					
Yellow catfish (*Tachysurus* [*Pelteobagrus*] *fulvidraco* Richardson)	*Edwardsiella ictaluri*		Brain, kidney, liver, spleen	Enteric septicaemia of catfish	1038 1829
Family Clariidae					
Philippine catfish, walking catfish (*Clarias batrachus* L.) (*Clarias gariepinus*)	1. *Aeromonas hydrophila* 2. *Edwardsiella ictaluri*			1. Ulcerative disease, mortality 2. Enteric septicaemia of catfish (ESC)	53 871
Family Ictaluridae (North American freshwater catfish)					
Black bullhead (*Ameiurus melas* – valid name) (*Ictalurus melas* Rafinesque – scientific name) Family Ictalurai	1. *Edwardsiella ictaluri* 2. *Flavobacterium columnare*		1. Organism in liver, spleen kidney 2. Yellow-edged skin erosions on head	1. Enteric septicaemia of catfish (ESC), Edwardsiellosis 2. Columnaris disease	166 414
Blue catfish (*Ictalurus furcatus* Valenciennes)	*Edwardsiella ictaluri*			Resistant to infection in immersion trials	1810
Brown bullhead (*Ameiurus nebulosus*) (synonym *is Ictalurus nebulosus*)	1. *Carnobacterium* (*piscicola*) *maltaromaticum* 2. *Edwardsiella ictaluri* 3. *Edwardsiella tarda*		1. Kidney 2. Infection in the brain, systemic dissemination and localization of the organism in the visceral organs and musculature and cutaneous ulcers 3. Septicaemia, focal suppurative or granulomatous lesions, cutaneous ulcerations	1. May be carriers 2. Enteric septicaemia of catfish (ESC), Edwardsiellosis 3. Edwardsiellosis. Opportunistic infection	136 702 819
Catfish species (*Ictalurus* sp. Rafinesque)	*Flavobacterium columnare*		Organism in kidney	Columnaris or saddleback disease	166

Channel catfish (*Ictalurus punctatus* Rafinesque)	1. *Aeromonas hydrophila* 2. *Bacillus mycoides* 3. *Carnobacterium* (*piscicola*) *maltaromaticum* 4. *Edwardsiella ictaluri*, anaerobic strains also isolated 5. *Edwardsiella tarda* 6. *Flavobacterium columnare* 7. *Stenotrophomonas maltophilia* 8. *Streptococcus ictaluri* 9. *Streptococcus iniae* 10. *Yersinia ruckeri*		2. Ulcerative skin lesions and focal necrosis of epaxial muscle 3. Hyperaemia and slight haemorrhage in liver, kidney, spleen and brain 4. Bacteria in brain and kidney 5. Bacteria isolated from lesions on skin and superficial muscle, necrosis in organs 6. Skin lesions, gills 7. Kidney, liver, ascites 8. Streptococcal arthritis, osteolysis, myositis, spinal meningitis 9. Petechiae on skin, necrotic inflammation in organs 10. Haemorrhagic rings around the eyes and frontal foramens	1, 2, 3. Infection and mortality 4. Enteric septicaemia of catfish (ESC) 5. Edwardsiellosis, enteric septicaemia, Opportunistic infection 6. Columnaris disease 7. Infectious intussusception syndrome 8. Bacterial disease 9. Mortality 10. Yersiniosis	136 166 267 307 402 606 648 702 1134 1152 1519 1714
Family Loricariidae (armoured catfishes)					
Suckermouth catfish or cascudo (*Hypostomus plecostomus*). Freshwater	*Flavobacterium columnare*		Ulceration and discolouration on skin and caudal fin	Columnaris disease	1343
White catfish (*Ameiurus catus* L.). Scientific name *Ictalurus catus*	*Edwardsiella ictaluri*			Enteric septicaemia of catfish (ESC)	702
Family Pangasiidae (shark catfishes)					
Striped catfish (*Pangasianodon hypophthalmus* Sauvage) previously *Pangasius*. Previously called sutchi catfish. Family Pangasiidae	*Edwardsiella ictaluri*		Mortality, multifocal irregular, white lesions on internal organs. Necrosis and pyogranulomatous inflammation	Bacillary necrosis of Pangasius	386
Family Siluridae (sheatfishes)					
Amur catfish (*Silurus asotus* L.)	1. *Edwardsiella piscicida* 2. *Edwardsiella tarda*	3. *Lactococcus lactis* ssp. *lactis*	2. Abdominal swelling, ascites fluid, abscess in kidney and spleen 3. Intestine	1. Pathogenic to zebrafish in experimental challenge (isolate previously identified as *Edwardsiella tarda*) 2. Edwardsiellosis 3. Normal flora	1 809 1838

Continued

Table 1.1. Continued.

Host	Pathogens	Normal flora	Tissue site	Disease status	Ref
Wels catfish (*Silurus glanis* L.). Also known as sheatfish, European catfish etc.	1. *Flavobacterium branchiophilum* 2. *Flavobacterium columnare*		1. Bacterium attaches to gill epithelium 2. Organism isolated from kidney, skin blanched and ulcerated	1. Bacterial gill disease (BGD) 2. Columnaris disease	166 516 1746
ORDER Syngnathiformes (seahorse and pipefish)					
Family Syngnathidae (seahorses and pipefish)					
Bluespotted cornetfish (*Fistularia commersonii*)		*Paracoccus fistulariae*	Intestine	Normal flora	903
Seadragon (leafy seadragon) (*Phycodurus eques*)	*Vibrio harveyi*		Isolated in pure culture from heart blood, liver	Death. Pathogenicity of the organism not determined	244
Spotted seahorse (*Hippocampus kuda* Bleeker) and *Hippocampus* species	*Vibrio harveyi*		External haemorrhages, haemorrhagic liver, ascites fluid	Mortality	20 244
Hippocampus guttulatus – long-snouted seahorse	1. *Vibrio alginolyticus* 2. *Vibrio splendidus*	3. *Bacillus galliciensis* 4. *Vibrio hippocampi*	1, 2. White spots on skin, lethargy 3, 4. Faeces	1, 2. Disease 3, 4. Normal flora	104 105 106
Hippocampus hippocampus – short-snouted seahorse	1. *Vibrio alginolyticus* 2. *Vibrio splendidus*		1, 2. White spots on skin, lethargy	1, 2. Disease	104
Yellow seahorse	*Mycobacterium chelonae*, *Mycobacterium marinum*		Tissues	Mycobacteriosis	1356
ORDER Tetraodontiformes (puffers and filefishes)					
Family Molidae (molas or ocean sunfishes)					
Ocean sunfish (*Mola mola* Linnaeus)	*Citrobacter freundii*		Kidney, spleen	Mortality	1478
Family Tetraodontidae (puffers)					
Pufferfish, white-spotted puffer (*Arothron hispidus*)		1. *Chryseobacterium (arothri) hominis* 2. *Ferrimonas senticii*	1. Kidney 2. Mucus, slime	1, 2. No details. Likely to be normal flora	265 266 860
Puffer (*Takifugu niphobles* Jordan & Snyder)		*Lactococcus lactis* ssp. *lactis*	Intestine	Normal flora	810

Puffer (*Takifugu obscures*)	*Vibrio harveyi* (ODC-negative strain)		Skin ulcer	Mortality. Pathogenicity confirmed experimentally	1811
Puffer (*Takifugu rubripes* Temminck & Schlegel)	*Vibrio harveyi*		Granulomas in liver, spleen, kidney, heart	Mortality. Pathogenicity confirmed experimentally	1161
PHYLUM CHORDATA					
CLASS Amphibia					
Family Bufonidae					
American toad (*Bufo americanus*)	*Streptococcus agalactiae* Group B		Brain, eye, organs	Streptococcosis in experimental infection	1415
Cane toad (*Bufo marinus*)	*Mycobacterium abscessus*		Liver, spleen	Natural infection. Pathogenicity confirmed experimentally	1163
Chaunus (*Bufo granulosus*)	*Mycobacterium abscessus*		Liver, spleen	Natural infection. Pathogenicity confirmed experimentally	1163
Family Pipidae					
South African clawed frog (*Xenopus laevis*)	*Mycobacterium chelonae*		Skin lesions, liver, spleen, heart blood	Mycobacteriosis	654
Family Ranidae					
American bullfrog (*Rana catesbeiana*). *Lithobates catesbeianus* may be the more correct name	*Aeromonas hydrophila* *Chryseobacterium indologenes* *Citrobacter freundii* *Edwardsiella tarda* *Elizabethkingia meningosepticum* *Pseudomonas* sp. *Streptococcus iniae*		Haemorrhage and lesions on legs, liver, spleen, kidney	Mortality	1112
Chinese bullfrog (*Rana rugulosa*)	*Aeromonas hydrophila* ssp. *ranae*		Liver, kidney	Septicaemia, mortality	791
Tiger frog (*Hoplobatrachus tigerinus*). *Rana tigerina rugulosa* and *Rana tigrina* are synonyms	*Elizabethkingia meningosepticum*		Brain	Cataract disease	1817
PHYLUM CHORDATA					
CLASS Aves (birds, penguins)					
ORDER Anseriformes					
Barnacle geese (*Branta leucopsis*). Anseriformes. Family Anatidae		*Helicobacter canadensis*	Faeces	Pathogenicity unknown, probable normal flora	1750

Continued

Table 1.1. Continued.

Host	Pathogens	Normal flora	Tissue site	Disease status	Ref
Canada geese (*Branta canadensis*). Family Anatidae		*Helicobacter anseris* *Helicobacter brantae* *Helicobacter canadensis*	Faeces	Pathogenicity unknown, probable normal flora	561
ORDER Accipitriformes					
Eagle. Bald eagle (*Haliaeetus leucocephalus*)	*Edwardsiella tarda*		Cloacal swab	Bird was sick but no other evidence of infection	1785
ORDER Charadriiformes					
Gull. Ring-billed gull (*Larus delawarensis*)		*Edwardsiella tarda*	Large intestine	No signs of disease	1785
Silver gull, seagull (*Chroicocephalus* [*Larus*] *novaehollandiae*). Family Laridae		*Helicobacter* spp.	Faeces	Asymptomatic	1289
ORDER Ciconiiformes					
Black stork (*Ciconia nigra*). Family Ciconiidae		*Corynebacterium ciconiae*	Trachea	Healthy animals	528
ORDER Gaviiformes					
Loon (common loon – *Gavia immer*)	*Edwardsiella tarda*		Intestinal content	haemorrhagic enteritis, intestinal content dark and tarry	1785
ORDER Gruiformes					
Crane (sandhill crane – *Grus canadensis*)		*Edwardsiella tarda Salmonella* Hartford, *Salmonella* Java	Large intestine	No signs of disease	1785
ORDER Pelecaniformes					
Heron (great blue heron – *Ardea herodias*)		*Edwardsiella tarda*	Large intestine	No signs of disease	1785
Pelican Brown pelican (*Pelecanus occidentalis carolinensis*)	*Edwardsiella tarda*		Intestinal content, lung, liver	haemorrhagic enteritis, intestinal content dark and tarry	1785
ORDER Procellariformes					
Albatross – grey-headed (*Diomedea chrysostoma*), black-browed albatross (*Diomedea melanophris*)		*Campylobacter subantarcticus*	Faeces	Pathogenicity not known	422
ORDER Sphenisciformes					
Family Spheniscidae					

King penguin (*Aptenodytes patagonicus*), southern rockhopper penguin (*Eudyptes chrysocome*) (*E. crestatus* is a synonym), gentoo penguin (*Pyoscelis papua*), African or black-footed penguin (*Spheniscus demersus*), Humboldt penguin (*Spheniscus humboldti*)		*Providencia (friedericiana) rustigianii*	Isolated from faeces of captive penguins	Part of normal flora	1175
Gentoo penguin (*Pygoscelis papua*) from subAntarctic region		1. *Campylobacter subantarcticus* 2. *Providencia (friedericiana) rustigianii* 3. *Salmonella* Enteritidis, *Salmonella* Havana, *Salmonella* Typhimurium	1. Faecal swab	1. Pathogenicity not known 2. Probable normal flora 3. Carrier status	422 1303 1175
Fairy or little blue penguin (*Eudyptula minor*)	1. *Erysipelothrix rhusiopathiae*	2. *Edwardsiella tarda*	1. Lung, liver, intestine 2. Faeces	1. Erysipelas, septicaemia 2. Isolated from healthy captive penguins	193 1383
Macaroni penguin (*Eudyptes chrysolophus*)	*Burkholderia (Pseudomonas) pseudomallei*		Pin-point, white lesions in liver and lung, fluid in airsacs, organism in spleen, liver, heart blood	Melioidosis	1086
Magellanic penguin (*Spheniscus magellanicus*)		*Corynebacterium sphenisci*, *Corynebacterium spheniscorum*, *Enterococcus faecalis*, *Staphylococcus sciuri*	Cloacae	Healthy penguins	652 653
PHYLUM CHORDATA					
CLASS Cephalaspidomorphi (lampreys)					
ORDER Petromyzontiformes (lampreys)					
Family Petromyzontidae (lampreys)					
Lamprey	*Aeromonas salmonicida* ssp. *salmonicida*		May be carriers of the disease. Organism found in kidney	Succumb to infection following stress in stress-test experiments	733
Pacific lamprey (*Lampetra tridentata*)	*Renibacterium salmoninarum*			No infection in experimental challenge	150

Continued

Table 1.1. Continued.

Host	Pathogens	Normal flora	Tissue site	Disease status	Ref
Sea lamprey (*Petromyzon marinus*)	1. *Aeromonas salmonicida* 2. *Flavobacterium psychrophilum* 3. *Renibacterium salmoninarum*		1. Furuncles on dorso-lateral musculature, and organism in the kidney 2. Skin ulcers and erosions on fins 3. Organisms isolated from kidney	1. Furunculosis 2. Cold water disease 3. Bacterial kidney disease	476 488 513
PHYLUM CHORDATA					
CLASS Chondrichthyes					
CLASS Elasmobranchii (sharks and rays)					
Brown shark (*Carcharhinus plumbeus*)	*Photobacterium damselae* ssp. *damselae*, *Vibrio harveyi*				659
Ornate wobbegong (*Orectolobus ornatus*). Also known as carpet or nurse shark. Animal held in captivity	*Photobacterium damselae* ssp. *damselae*		Organism isolated from organs	Mortality	1322
Dusky smooth-hound or smooth dogfish (*Mustelus canis* Mitchell) and spiny dogfish (*Squalus acanthias* L.) (mud shark)		*Alteromonas* sp., *Photobacterium* sp., *Pseudomonas* sp., *Shewanella putrefaciens*, *Vibrio* sp.	All isolated from lesions in the head kidney of healthy sharks	The role of the lesions, from which these organisms were cultured, in morbidity and mortality is unknown	203
Spiny dogfish or mud shark (*Squalus acanthias* L.)	*Photobacterium damselae* ssp. *damselae*			Mortality in experimental challenge	659
Blacktip shark (*Carcharhinus limbatus*), lemon shark (*Negaprion brevirostris* Poey), nurse shark (*Ginglymostoma cirratum*), tiger shark (*Galeocerdo cuvier*)	1. *Aeromonas salmonicida* *Alteromonas* species *Moraxella* species *Neisseria* species *Photobacterium damselae* ssp. *damselae* *P. damselae* ssp. *piscicida* *Plesiomonas shigelloides* *Vibrio alginolyticus* *Vibrio harveyi* 2. *Vibrio harveyi*	3. *Photobacterium damselae* ssp. *damselae*	1. Isolated from various tissue sites 2. Active splenic and hepatic disease on histological examination	1. Pathogenicity not assessed 2. Isolated from experimentally infected lemon sharks but did not show clinical disease. Stress-related 3. Not recovered from experimentally inoculated lemon sharks	658 660

Stingray, common stingray (*Dasyatis pastinaca* Linnaeus) Family Dasyatidae	1. *Photobacterium damselae* ssp. *damselae* 2. *Streptococcus agalactiae* Group B		1, 2. Organism isolated from organs	1. Mortality (animals held in captivity) 2. Streptococcosis	1322
PHYLUM CHORDATA					
CLASS Mammalia					
ORDER Carnivora					
Otter					
SUBORDER Caniformia					
Family Mustelidae					
European otter (*Lutra lutra*)	1. *Brucella* (*pinnipediae*) *pinnipedialis* 2. *Staphylococcus lutrae* 3. *Vagococcus lutrae*		1. Organism isolated from tissues 2. Organism in liver, spleen, lymph node 3. Organism found in blood, liver, lungs, spleen	1. Brucellosis 2. 3. Pathogenicity not known	545 546 551 988
Southern sea otter (*Enhydra lutris nereis*)	1. *Arcanobacterium phocae* 2. *Campylobacter* 3. *Clostridium difficile* 4. *Clostridium perfringens* 5. *Vibrio alginolyticus* 6. *Vibrio parahaemolyticus*	7. *Clostridium perfringens* 8. *Streptococcus phocae*	1. Tissues showing discharge or inflammation 2. Intestinal mural thickening 3. Faeces 4. Intestine 5, 6. Intestine and tissues 7. Intestine 8. Isolated from orogenital mucosa and lesions in live and dead stranded sea otters	1, 2, 3. Possible opportunistic pathogen 4, 7. Part of normal flora and opportunistic infection 5, 6. Lesions of enteritis and septicaemia 8. Pathogenicity unknown	800 845 1143
SUBORDER Pinnipedia					
Sea lion					
Australian sea lion (*Neophoca cinerea*). Family Otariidae	1. *Mycobacterium pinnipedii*	2. *Helicobacter* spp. 3. *Wolinella* spp., *Wolinella succinogenes*	1. Granulomatous lesions in lymph nodes, spleen, lungs, pleura, peritoneum 2. Faeces 3. Gastric mucosa	1. Tuberculosis 2. Asymptomatic 3. Gastrointestinal disease	374 1287 1288 1289

Continued

Table 1.1. Continued.

Host	Pathogens	Normal flora	Tissue site	Disease status	Ref
California sea lion (*Zalophus californianus*). Family Otariidae	1. *Arcanobacterium phocae* 2. *Burkholderia pseudomallei* 3. *Clostridium perfringens* 4. *Edwardsiella tarda* 5. *Escherichia coli* 6. *Leptospira interrogans* serovar pomona type kennewick 7. *Mycoplasma zalophi* 8. *Pasteurella multocida*	9. *Salmonella* Heidelberg, *Salmonella* Newport, *Salmonella* Oranienburg 10. *Helicobacter* spp.	1. Tissues with discharge or inflammation 2. Oedema, haemorrhage and nodules in lungs 3. Infection in muscle with gas and pus 4. Abscesses in lungs, sanguino-mucopurulent exudate in trachea and terminal bronchi or pathology associated with intestine. Organism in liver or lungs or mesenteric lymph nodes 5. Verrucous, grey-red lesions in atrioventricular valves 6. Kidney 7. Lesions in lung, neck abscess, carpal joint 8. Yellow, pus-filled fluid in pleural cavity 9. Recovered from healthy animals 10. Faeces	1. Possible opportunistic pathogen 2. Melioidosis 3. Clostridial myositis 4. Bacterial pneumonia, respiratory infection, peritonitis. Pathogenicity not conclusive 5. Endocarditis 6. Leptospirosis, renal disease 7. Pneumonia, or polyarthritis 8. Mortality 9. Non-pathogenic or carrier status 10. Asymptomatic	347 620 656 666 701 724 845 882 898 1289 1752
Southern sea lion (*Otaria flavescens*)	1. *Mycobacterium pinnipedii*		1. Granulomatous lesions in lymph nodes, spleen, lungs, pleura, peritoneum	1. Tuberculosis	374
Steller sea lion (*Eumetopias jubatus*)	*Edwardsiella tarda*		Peritoneal exudate	Opportunistic pathogen	347
Seal, INFRAORDER Pinnipedia					
Family Otariidae					
Australian fur seal (*Arctocephalus pusillus doriferus*), New Zealand fur seal (*Arctocephalus forsteri*)	1. *Helicobacter cetorum* 2. *Mycobacterium pinnipedii* 3. *Mycobacterium phocicerebrale*	4. *Helicobacter* spp. 5. *Mycoplasma phocidae*, *Mycoplasma zalophi*, *Mycoplasma* species	1. Gastric mucosa 2. Granulomatous lesions in lymph nodes, spleen, lungs, pleura, peritoneum 3. Aborted fetal tissue 4. Faeces 5. Nasal cavity	1. Unknown 2. Tuberculosis 3. Possible abortion cause 4. Asymptomatic 5. Probably upper respiratory tract commensal	374 630 1074 1288 1289

Northern fur seal (*Callorhinus ursinus* L.)	1. *Erysipelothrix rhusiopathiae* 2. *Leptospira interrogans* serovar pomona	3. *Salmonella* Adelaide, *Salmonella* Heidelberg, *Salmonella* Newport, *Salmonella* Oranienburg 4. *Acinetobacter calcoaceticus*, *Actinobacillus* sp., *Aeromonas eucrenophila*, *Alcaligenes faecalis*, *E. coli*, *Enterobacter* sp., *Klebsiella* sp., *Moraxella* sp., *Pseudomonas fluorescens*, *Pseudomonas* sp., *Staphylococcus epidermidis*, *Streptococcus* sp. 5. *Bacillus* sp., *Corynebacterium* sp., *E. coli*, *Listeria* sp., *Moraxella* sp., *Neisseria cuniculi*, *Proteus mirabilis*, *Staphylococcus epidermidis*, *Streptococcus* sp.	1. Teeth gum margin 2. Organism recovered from kidney 3. Recovered from healthy animals 4. Rectum 5. Oropharynx	1. Erysipelas 2. Leptospirosis, interstitial nephritis in adults, multiple haemorrhage syndrome in pups 3. Non-pathogenic or carrier status 4, 5. Normal flora	620 1540 1591 1707
South Georgian Antarctic fur seal (*Arctocephalus gazella*)		*Salmonella* Enteritidis, *Salmonella* Havana, *Salmonella* Newport, *Salmonella* Typhimurium		Carrier status	1303
Family Phocidae					
Caspian seal (*Pusa caspica*). *Phoca caspica* is a synonym		*Atopobacter phocae*, *Corynebacterium caspium*, *Streptococcus dysgalactiae* Group C	Penis	Pathogenicity unknown	357
Grey seal (*Halichoerus grypus*)	1. *Aeromonas hydrophila* 2. *Arcanobacterium phocae* 3. *Brucella* (*pinnipediae*) *pinnipedialis* 4. *Burkholderia pseudomallei* 5. *Mycoplasma phocicerebrale*, *Mycoplasma phocidae*, *Mycoplasma phocirhinis*, *Mycoplasma* species 6. *Streptococcus phocae* 7. *Streptococcus dysgalactiae* ssp. *dysgalactiae* serovar L	8. *Actinomyces marimammalium* 9. *Catellicoccus marimammalium* 10. *E. coli* 11. *Neisseria* sp. 12. *Pseudomonas* sp. 13. *Staphylococcus epidermidis* 14. *Streptococcus marimammalium* 15. *Streptococcus halichoeri*	1. Organism in lung and liver. Isolate positive for aerolysin gene, cytotoxin and haemolysin 3. Isolated from lung 4. Oedema, haemorrhage and nodules in lungs 5. Diseased lung 6, 7. Lesions in lung, exudate in bronchi 8, 9, 10. Intestine	1. Septicaemia 3. Brucellosis 4. Melioidosis 6, 7. Pneumonia – opportunistic infection in seal morbillivirus infection 10. Pathogenicity not clear. Isolated from seals with septicaemia and pneumonia 5, 8, 9, 14, 15. Pathogenicity unknown	88 545 546 554 724 763 942 984 986 987 1374 1538 1732

Continued

Table 1.1. Continued.

Host	Pathogens	Normal flora	Tissue site	Disease status	Ref
Harbour seal or common seal (*Phoca vitulina*), Pacific harbour seal (*P. vitulina richardii*)	1. *Arcanobacterium phocae* 2. *Bordetella bronchiseptica* 3. *Brucella* (*pinnipediae*) *pinnipedialis* 4. *Leptospira kirschneri grippotyphosa* 5. *Mycoplasma phocicerebrale* 6. *Mycoplasma phocirhinis* 7. *Streptococcus dysgalactiae* ssp. *dysgalactiae* serovar L 8. *Streptococcus equi* ssp. *zooepidemicus* 9. *Streptococcus phocae* 10. *Vagococcus fessus*	11. *Arcanobacterium phocisimile* 12. *Arthrobacter nasiphocae* 13. *Atopobacter phocae* 14. *Campylobacter insulaenigrae* 15. *Corynebacterium phocae* 16. *E. coli* 17. *Mycoplasma phocidae* 18. *Neisseria* spp. 19. *Pseudomonas* spp. 20. *Staphylococcus epidermidis* 21. *Streptococcus marimammalium*	1. Isolated from lung, peritoneal fluid 3. Gastric lymph node, spleen 4. Kidney 5. Isolated from brain, heart, lung, nose, throat 6. Isolated from pus in lung 7, 9. Lesions in lung, exudate in bronchi 8, 10. Organism isolated from liver and kidney in pure growth 11. Tissues with discharge or inflammation 13. Vagina and anus 14. Organism isolated from intestine and lymph nodes. The animal had lymphadenopathy and pulmonary haemorrhage 15. Isolated from nasal cavity 16. Isolated from intestine, lymph nodes, liver, spleen 18. Isolated from respiratory tract	3. Brucellosis 4. Leptospirosis 5, 6. Associated with respiratory disease 7, 8, 9. Pneumonia – opportunistic infection in seal morbillivirus infection 10. Possible opportunistic pathogen 1, 10, 11, 13, 14, 15, 21. Pathogenicity not known 17. Avirulent in experimental inoculation	15 88 358 545 546 552 554 587 616 727 762 827 845 926 986 989 1313 1374 1386 1430 1438 1538 1732
Harp seal (*Pagophilus groenlandicus*, previously *Phoca groenlandica*)	1. *Brucella pinnipedialis*	2. *Helicobacter* spp.	1. Organism in lymph nodes 2. Isolated from gastric tissues	1. Brucellosis 2. No clinical signs	545 554 681
Hooded seal (*Cystophora cristata*)	1. *Actinomyces marimammalium* 2. *Brucella* (*pinnipediae*) *pinnipedialis* 3. *Mycoplasma* sp. 4. *Nocardia asteroides*, *Nocardia brasiliensis*, *Nocardia farcinica*, *Nocardia otitidiscaviarum*		1, 3. Isolated from lung 4. Pyogranulomatous lesions in lung and organs	1, 3. Pathogenicity not determined 2. Brucellosis, stress related 4. Nocardiosis	88 546 547 763 1002 1667

Leopard seal (*Hydrurga leptonyx*)	*Nocardia* species		2. Pyogranulomatous lesions in lung, lymph node and brain	2. Nocardiosis	1002
Northern elephant seal (*Mirounga angustirostris*)	1. *Arcanobacterium phocae* 2. *Erysipelothrix rhusiopathiae* 3. *Leptospira interrogans* serovar pomona		1. Tissues with discharge or inflammation 2. Teeth gum margin 3. Kidney (renal cortex, renal tubules)	1. Possible opportunistic pathogen 2. Erysipelas 3. Renal disease, leptospirosis	345 845 1591
Ringed seal (*Pusa* [*Phoca*] *hispida*)	*Brucella pinnipedialis*		Organism in lymph nodes	Brucellosis	545 554
South American fur seal (*Arctocephalus australis*)	*Mycobacterium pinnipedii*		Granulomatous lesions in lymph nodes, spleen, lungs, pleura, peritoneum	Tuberculosis	374
Southern elephant seal (*Mirounga leonina*)		*Facklamia miroungae*	Nasal swab	Isolated from apparently healthy juvenile animal	763
Subantarctic fur seal (*Arctocephalus tropicalis*)	*Mycobacterium pinnipedii*		Granulomatous lesions in lymph nodes, spleen, lungs, pleura, peritoneum	Tuberculosis	374
Polar bear (*Ursus maritimus*). Family Ursidae	1. *Brucella* species	2. *Helicobacter* spp.	1. Serological diagnosis only 2. Gastric fluid PCR positive	1. Pathogenic potential unknown (bears healthy) 2. Pathogenicity unknown	1287 1665
Dolphin, ORDER Cetaceae					
Atlantic bottlenose dolphin (*Tursiops truncatus*). Family Delphinidae	1. *Brucella ceti* (or proposed *Brucella delphini*) 2. *Burkholderia pseudomallei* 3. *Clostridium perfringens* 4. *Edwardsiella tarda* 5. *Erysipelothrix rhusiopathiae* 6. *Helicobacter cetorum* 7. *Lactococcus garvieae* 8. *Mannheimia haemolytica* 9. *Nocardia asteroides*, *Nocardia levis*. *Nocardia* sp.	13. *Campylobacter* spp. 14. *Helicobacter* species	1. Abortion, abscess around vent 2. Oedema, haemorrhage and nodules in lungs 3. Abscess in dorsal muscle 4. Purulent mastitis 5. Rhomboid-shaped plaques, lesions or ulcers occur on skin, lungs congested and oedematous Organism in all organs 6. Organism cultured from faeces 7. Kidney	1. Brucellosis 2. Melioidosis 3. Clostridial myositis 4. Mastitis 5. Erysipelas, septicaemia 6. Ulcers in oesophagus and fore-stomach – gastritis 8. Septicaemia 9. Nocardiosis 10. Wound infection	417 500 507 570 607 632 633 656 678 724 1002 1128

Continued

Table 1.1. Continued.

Host	Pathogens	Normal flora	Tissue site	Disease status	Ref
	10. *Photobacterium damselae* ssp. *damselae* 11. *P. damselae* ssp. *piscicida* 12. *Streptococcus agalactiae*		8. Haemorrhagic tracheitis 9. Pyogranulomatous lesions in lung and organs 12. Subcutaneous fluid, liver, lung, skeletal muscle 13, 14. Present in dental plaque	11. Pasteurellosis 12. Necrotizing fasciitis, cellulitis 14. May act as a reservoir for gastric infection and gastric ulcers	1600 1847
Atlantic white-sided dolphin (*Lagenorhynchus acutus*). Family Delphinidae	1. *Brucella (cetaceae) ceti* 2. *Helicobacter cetorum*		1. Oesophageal ulceration, abortion, hepatic and splenic coagulative necrosis 2. Organism found in glandular mucosa of main stomach	1. Brucellosis 2. Gastric ulceration	546 547 679 678
Common dolphin (*Delphinus delphis*). Family Delphinidae	1. *Arcanobacterium phocae* 2. *Brucella (cetaceae) ceti* 3. *Helicobacter cetorum* 4. *Pasteurella multocida* 5. *Staphylococcus delphini*		1. Lung 2. Subcutaneous lesion 3. Organisms found in glandular mucosa of main stomach 4. Intestinal haemorrhage 5. Purulent skin lesions	1. Possible opportunistic pathogen 2. Brucellosis 3. Gastric ulceration 4. Enteritis 5. Skin infection	546 547 554 679 827 845 1430 1600 1704
Pacific white-sided dolphin (*Lagenorhynchus obliquidens*). Family Delphinidae	1. *Burkholderia pseudomallei* 2. *Helicobacter cetorum*		1. Oedema, haemorrhage and nodules in lungs 2. Organism cultured from faeces	1. Melioidosis 2. Ulcers in oesophagus and fore-stomach – Gastritis	678 724
Atlantic spotted dolphin (*Stenella frontalis*). Family Delphinidae	*Helicobacter* sp., possibly *Helicobacter cetorum*		Lesions in stomach and duodenal ampulla	Gastric ulceration	1590
Striped dolphin (*Stenella coeruleoalba*). Family Delphinidae	1. *Brucella (cetaceae) ceti* (proposed *Brucella delphini*) 2. *Cryptococcus neoformans* (yeast)	3. *Actinobacillus delphinicola*	1. Brain 2. Pulmonary infection 3. Isolated from various tissues	1. Meningitis, brucellosis 2. Pulmonary cryptococcosis 3. Pathogenicity not determined	416 546 547 554 580

Tursiops aduncus (Indian Pacific bottlenose dolphin). Family Delphinidae	*Pasteurella multocida*		Intestinal haemorrhage and bacteraemia. Source of infection was contamination from local bird rookery	Enteritis, mortality	1600
Amazon freshwater dolphin (*Inia geoffrensis*). Family Iniidae	*Streptococcus iniae*		Subcutaneous abscesses	Golf ball disease	1341
La Planta River dolphin (*Pontoporia blainvillei*). Family Pontoporiidae	1. *Helicobacter cetorum*		Gastric mucosa	Unknown	633
Family Phocoenidae					
Porpoise					
Harbour porpoise (*Phocoena phocoena*)	1. *Actinobacillus delphinicola* 2. *Actinobacillus scotiae* 3. *Actinomyces marimammalium* 4. *Brucella* (*cetaceae*) *ceti* 5. *Edwardsiella tarda* 6. *Erysipelothrix rhusiopathiae* 7. *Mycoplasma phocicerebrale* 8. *Streptococcus dysgalactiae* ssp. *dysgalactiae* Lancefield Group L 9. *Vagococcus fessus*	10. *Arcanobacterium pluranimalium* 11. *Campylobacter insulaenigrae* 12. *Catellicoccus marimammalium* 13. *Phocoenobacter uteri*	1. Lung, cervix, intestine 2. Liver, lung, brain and spleen of stranded porpoise 3. Lung 4. Lymph nodes 5. Liver and peritoneal exudate 6. Cutaneous lesions 7. Lung 8. Intestine, kidney, lung, spleen 9. Isolated from various tissues 10. Spleen 12. Kidney, intestine, pericardial fluid 13. Uterus	1, 2, 3, 7, 9, 10, 11, 12, 13. Pathogenicity not determined 4. Brucellosis 5. Peritonitis 6. Erysipelas 8. Septicaemia, broncho-pneumonia, myocarditis, pyelonephritis	347 546 548 550 547 548 549 552 553 763 763 827 984 985 1430 1503 1601
Yangtze finless porpoise (*Neophocaena phocaenoides asiaeorientalis*)	*Helicobacter cetorum*		Faeces	Unknown	1126
Whale ORDER Cetacea					
Family Balaenopteridae					
Minke whale (*Balaenoptera acutorostrata*)	1. *Brucella ceti* 2. *Granulicatella balaenopterae*		1. Isolated from liver and spleen 2. Isolated from pure growth from liver and kidney of beached whale	1. Disease status not known. Whale caught during commercial fishing 2. Pathogenicity not determined	340 352 990

Continued

Table 1.1. Continued.

Host	Pathogens	Normal flora	Tissue site	Disease status	Ref
Family Delphinidae					
False killer whale (*Pseudorca crassidens*)	1. *Burkholderia pseudomallei*	2. *Helicobacter* sp.	1. Oedema, haemorrhage and nodules in lungs 2. Gastric fluids	1. Melioidosis 2. Unknown	633 724
Killer whale (*Orcinus orca*)	1. *Burkholderia pseudomallei* 2. *Candida* sp. 3. *Clostridium perfringens* 4. *Helicobacter cetorum* 5. *Nocardia asteroides*, *Nocardia farcinica*		1. Oedema, haemorrhage and nodules in lungs 2. Necrotic skin lesions, blowhole lesions, can become systemic 3. Toxaemia, oedematous lymph nodes, liquefied muscle 4. Dental plaque 5. Pyogranulomatous lesions in lung and organs	1. Melioidosis 2. Infection 3. Clostridial myositis 4. Unknown 5. Nocardiosis	633 656 724 1002 1600
Pacific pilot whale (*Globicephala macrorhynchus*, previously *Globicephala scammoni*)	*Candida* sp.		Infection of nares usually as a secondary infection	Infection	1600
Family Monodontidae					
Beluga whale, white whale (*Delphinapterus leucas*)	1. *Helicobacter cetorum* 2. *Mycobacterium marinum* 3. *Nocardia brasiliensis*, *Nocardia cyriacigeorgica*, *Nocardia farcinica*, *Nocardia* sp.		1. Inappetence and lethargy. Esophageal and fore-stomach ulcers 2. Pyogranulomatous dermatitis and panniculitis 3. Pyogranulomatous lesions in lung and organs	1. Gastritis 2. Mycobacteriosis 3. Nocardiosis	207 678 679 680 1002
Family Ziphiinae					
Beaked whale (*Ziphius cavirostris* Cuvier)		1. *Flavobacterium ceti* 2. *Lactobacillus ceti*	1. 2. Liver, lung	1. 2. Isolated from stranded beached whale – no abnormalities	1709 1710
Sowerby's beaked whale (*Mesoplodon bidens*)	1. *Mycoplasma phocicerebrale*	2. *Actinobacillus delphinicola* 3. *Weissella ceti*	1. Kidney 2. Isolated from various organs 3. Brain, kidney, lymph nodes, spleen muscle	1, 2, 3. Pathogenicity not determined	548 553 1708
ORDER Monotremata					
Platypus (*Ornithorhynchus anatinus*), Family Ornithorhynchidae		*Hafnia paralvei*	Faeces	Normal flora	1258

Human, ORDER Primates					
Human (*Homo sapiens*)	1. *Aeromonas allosaccharophila*		1, 2. Faeces	1. Diarrhoea, renal failure	17
	2. *Aeromonas caviae*		4, 5. Faeces	in infant	220
	3. *Aeromonas diversa*		6. Wound infection, stool, blood	2, 4, 5, 9. Diarrhoea	222
	4. *Aeromonas hydrophila* ssp.		7. Wound, abscess, blood, pleural	6. Wound infection,	226
	dhakensis		fluid	diarrhoea	276
	5. *Aeromonas hydrophila*		9, 10, 11. Faeces	7, 9. Infection	277
	6. *Aeromonas jandaei*		12. Respiratory infection, abortion	8. Acute renal failure in	278
	7. *Aeromonas schubertii*		13. Headaches, lassitude, sinusitis	an infant	299
	8. *Aeromonas sobria* likely to be		14. Wound infections, pneumonia	10, 11, 18. Diarrhoea	339
	A. veronii biovar Sobria		and septicaemia	12, 13. Brucellosis	354
	9. *Aeromonas trota*		3, 11, 15. Wound infections	14. Melioidosis	388
	10. *Aeromonas veronii* biovar		16. Blood, dialysis fluid, pus, eye,	15, 16. Infection	393
	Sobria		aortic valve	17. Zoonosis, mortality	412
	11. *Aeromonas veronii* biovar		17. Cerebral spinal fluid	20. Erysipeloid	462
	Veronii		18. Intestine, faeces	21. Endocarditis	464
	12. *Brucella abortus*		19. Meningitis in neonates	3, 11, 23. Wound	510
	13. *Brucella* species from marine		20. Suppurating skin lesions	infection	540
	mammals		21. Endocarditis	24. Enteritis, bacter-	559
	14. *Burkholderia* (*Pseudomonas*)		23. Wound with watery discharge	aemia	575
	pseudomallei		following fish bite	25, 26. Infection	576
	15. *Chromobacterium violaceum*		24. Faeces, blood	28. Buruli ulcer	721
	16. *Chryseobacterium hominis*		25. blood, urine, skin	29. Wound infection	722
	17. *Cryptococcus neoformans*		26. Abscess	30. Sepsis. Also infection	723
	18. *Edwardsiella tarda*		27, 28. Skin lesions	in immunocompromised	747
	19. *Elizabethkingia*		30. Contaminated intravenous fluid	hosts	769
	meningosepticum		32. Leg ulcers, septicaemia, otitis	31. Septicaemia in	779
	20. *Erysipelothrix rhusiopathiae*		media, faeces	newborns	790
	21. *Granulicatella elegans*		33. Wounds following handling fish	32. Infection	807
	22. *Grimontia hollisae*		34. Bite-wound, blood culture	33. Cellulitis	932
	23. *Halomonas venusta*		isolates and from peritoneal fluid	33. Clinical infections	992
	24. *Helicobacter canadensis*		35, 36, 43. Faeces	36. Cholera	1031
	25. *Lactococcus garvieae*		37. Blood culture, faeces	37. Cellulitis,	1102
	Leptospira species		38. Blood and spinal fluid	meningoencephalitis,	1147
	26. *Mycobacterium chelonae*			septicaemia, diarrhoea	1155

Continued

Table 1.1. Continued.

Host	Pathogens	Normal flora	Tissue site	Disease status	Ref
	27. *Mycobacterium marinum* 28. *Mycobacterium ulcerans* 29. *Photobacterium damselae* ssp. *damselae* 30. *Rahnella aquatilis* 31. *Raoultella planticola* 32. *Shewanella algae* *Streptococcus agalactiae* 33. *Streptococcus iniae* 34. *Vagococcus fluvialis* 35. *Vibrio alginolyticus* 36. *Vibrio cholerae* 01 and 0139 37. *Vibrio cholerae* non-01 38. *Vibrio cincinnatiensis* 39. *Vibrio fluvialis* 40. *Vibrio furnissii* 41. *Vibrio metschnikovii* 42. *Vibrio mimicus* 43. *Vibrio parahaemolyticus* 44. *Vibrio vulnificus*		23, 39, 40. Faeces 42. Faeces, ear 43, 44. Wounds, faeces	38. Meningitis 23, 39, 40. Diarrhoea 41, 43. Wound infection, food poisoning 23, 35, 43 Gastroenteritis, food poisoning 42. Diarrhoea, ear infection 44. Infection, food poisoning	1170 1239 1420 1464 1613 1693 1700 1711 1731 1775
PHYLUM CHORDATA					
CLASS Reptilia					
Family Alligatoridae					
Caiman crocodilus	1. *Erysipelothrix rhusiopathiae* 2. *Mycoplasma alligatoris*		1. Necrotic foci under the scales of the abdomen and the jaw 2. Lung, liver, joints, brain	1. Septicaemia 2. Pathogenic in experimental inoculation	834 1367
Family Crocodylidae					
Alligator mississippiensis	1. *Edwardsiella tarda* 2. *Mycoplasma alligatoris* 3. *Pasteurella multocida* 4. *Staphylococcus*	5. *Edwardsiella tarda*, *Salmonella* Miami, *Salmonella* Java, *Salmonella* Hartford	1, 3. Large intestine, cloaca, congestion in kidneys with necrotic foci, peritonitis, ulcers in gastric mucosa 2. Pulmonary oedema, interstitial pneumonia, pericarditis, myocarditis, meningitis, synovitis 3, 4. Lungs	1. Nephritis, congested kidneys. Pathogenicity not conclusive 2. Acute multisystemic inflammatory disease 3, 4. Pneumonia 5. Apparently normal animals	229 1092 1752 1785

Crocodile		*Edwardsiella tarda*		Pathogenicity not known	
American crocodile (*Crocodylus acutus*)	*Erysipelothrix rhusiopathiae*		Blackened plaques on scales	Cutaneous lesions	834
Nile crocodile (*Crocodylus niloticus*)	*Mycoplasma crocodyli*		Swollen joints. Organism also found in lungs	Exudative polyarthritis	909
Saltwater crocodile (*Crocodylus porosus*)	1. *Providencia* (*Proteus*) *rettgeri* 2. *Streptococcus agalactiae*		1. Brain, liver 2. Necrotic skin lesions on ventral body wall or limbs	1. Septicaemia, meningitis 2. Necrotizing fasciitis	183 967
Siamese crocodile (*Crocodylus siamensis*)		*Mycoplasma alligatoris*		Non-pathogenic in experimental inoculation	1367
Snakes		*Edwardsiella tarda*	Isolated from faeces	Carrier status or part of the normal flora	817
ORDER Testudines (sea turtles)					
Tortoise (land turtle)	*Corynebacterium testudinoris*		Mouth lesions		351
Chinese soft-shelled turtle (*Pelodiscus* [*Trionyx*] *sinensis*)	*Mycobacterium marinum*		Granulomas in spleen, liver, lung, intestine, kidney	Mycobacteriosis	764
Turtles Family Geoemydidae: Caspian terrapin (*Mauremys caspica*), Travancore crowned turtle or cochin black turtle (*Melanochelys trijuga coronata*) Family Emydidae: painted turtle (*Chrysemys picta*), red-eared turtle (*Trachemys* [*Chrysemys*] *scripta elegans*), stinkpot or musk turtle (*Sternotherus odoratus*), eastern box turtle (*Terrapene carolina carolina*), Mississippi map turtle (*Graptemys* [*Malaclemys*] *pseudogeographica kohni*). The Family Emydidae are mostly freshwater species with some brackish water species Family Malaclemys: northern diamondback terrapin (*Malaclemys terrapin terrapin*)		*Edwardsiella tarda*, *Salmonella* Durham	Cloaca	Carrier status	1285

Continued

Table 1.1. Continued.

Host	Pathogens	Normal flora	Tissue site	Disease status	Ref
Green turtle (*Chelonia mydas*), loggerhead sea turtle (*Caretta caretta*), hawksbill sea turtle (*Eretmochelys imbricata*). All belong to Family Cheloniidae	1. *Aeromonas hydrophila* 2. *Austwickia* (*Dermatophilus*) *chelonae* 3. *Flavobacterium* sp. 4. *Mycobacterium* sp. 5. *Pseudomonas* sp. 6. *Vibrio alginolyticus*	7. *Acinetobacter calcoaceticus* 8. *Bacillus* sp. *Micrococcus* sp. *Moraxella* sp. *Proteus* sp.	2. Skin 4. Lung lesions 7. Oral cavity 8. Part of normal flora on skin, oral cavity and trachea	1. Associated with bronchopneumonia, ulcerative stomatitis, and traumatic ulcerative dermatitis caused by biting 2. Skin lesions, skin abscess, scabs 3. Associated with traumatic ulcerative dermatitis caused by biting, ulcerative stomatitis, bronchopneumonia and keratoconjunctivitis-ulcerative blepharitis 5. Associated with traumatic ulcerative dermatitis caused by biting, ulcerative stomatitis, adenitis, peritonitis, bronchopneumonia and keratoconjunctivitis-ulcerative blepharitis 6. Associated with traumatic ulcerative dermatitis caused by biting, bronchopneumonia and osteomyelitis	440 623 624 1109
Turtles (*Pseudemis scripta*)	*Aeromonas hydrophila*			Infection	1316
Yellow-spotted Amazon turtle (*Podocnemis unifilis*). Family Podocnemididae		*Salmonella* Durham	Faeces	Isolated from healthy zoo animals	1285

Hawaiian green turtle, also known as green sea turtle, Pacific sea turtle (*Chelonia mydas*)		*Citrobacter freundii* *Hafnia alvei* *Klebsiella oxytoca* *Photobacterium damselae* *Pseudomonas fluorescens* *Shewanella* (*Pseudomonas*) *putrefaciens* *Pseudomonas stutzeri* non-haemolytic *Streptococcus* sp. *Vibrio alginolyticus* *Vibrio fluvialis*	Nasal swabs and cloacal swabs from apparently healthy turtles.	Pathogenicity not determined. Most likely normal flora from these sites.	10
Leatherback turtle (*Dermochelys coriacea*)	*Photobacterium damselae* ssp. *damselae*		Fluid in body cavity, calcareous nodules in lung parenchyma, lungs congested	Valvular endocarditis and septicaemia	1245
Loggerhead sea turtle (*Caretta caretta*)	*Aerococcus viridans*		Gas in diverticulum, with multifocal granulomatous serositis. Green mucoid fluid and fibrinonecrotic membrane on mucosa	Oesophageal diverticulum	1657
Turtle	*Mycobacterium chelonae*				654
Ridley sea turtle (*Lepidochelys olivacea*)	*Vibrio mimicus*		Reduced viability of eggs	Cause of food poisoning in humans	8
PHYLUM CNIDARIA, contains the Classes for sea anemones (Anthozoa), corals (Anthozoa) and jellyfish (Medusozoa)					
CLASS Anthozoa, predatory animals of the order Actiniaria					
Anemone					
Sea anemone (*Aiptasia pulchella*)	*Tenacibaculum aiptasiae*			Disease in laboratory-cultured sea anemone	1761
Coral					
Acropora	1. *Vibrio harveyi*		1. Loss of tissue	1. White syndrome. Pathogenicity proved experimentally	1069
Barabattoia amicorum		*Photobacterium rosenbergii*	Water extract and tissue from bleached coral	Pathogenicity not determined	1636
Echinopora	1. *Vibrio harveyi*		1. Loss of tissue	1. White syndrome. Pathogenicity proved experimentally	1069

Continued

Table 1.1. Continued.

Host	Pathogens	Normal flora	Tissue site	Disease status	Ref
Eunicella verrucosa	*Vibrio tasmaniensis*		Isolated from necrotic tissue of diseased coral	Considered a pathogen of coral	1706
Favia sp. (Red Sea coral)		*Shewanella corallii*	Mucus layer	Normal flora. Has antibacterial properties	1527
Fungia granulosa (granular razor coral or mushroom coral)		*Corynebacterium maris*	Isolated from the mucus of healthy coral	Red Sea, Gulf of Eilat, Israel	152
Merulina ampliata		*Enterovibrio coralii* *Photobacterium rosenbergii*	Water extract from bleached coral	Pathogenicity not determined	1636
Montipora	1. *Vibrio harveyi*		1. Loss of tissue	1. White syndrome. Pathogenicity proved experimentally	1069
Mussismilia hispida		*Vibrio alfacsensis*, *Vibrio communis*	Isolated from the mucus of apparently healthy coral	May be part of the normal flora	320 640
Oculina patagonica	*Vibrio shilonii* (said to be a later subjective synonym of *Vibrio mediterranei*)		Binds to coral	Coral bleaching	114 952 1628
Pachyseris speciosa		*Photobacterium rosenbergii*	Tissue from bleached and healthy coral	Pathogenicity not determined	1636
Palythoa caribaeorum		*Vibrio maritimus* *Vibrio variabilis*	Isolated from the mucus of apparently healthy coral	May be part of the normal flora	321
Palythoa caribaeorum, *Palythoa variabilis*		*Vibrio communis*	Isolated from the mucus of apparently healthy coral	May be part of the normal flora	320
Phyllogorgia dilatata		*Vibrio communis*	Isolated from the mucus of apparently healthy coral	May be part of the normal flora	320
Pocillopora damicornis	1. *Vibrio coralliilyticus* 2. *Vibrio harveyi*		1. White spots seen at 3–5 days and complete tissue destruction after 2 weeks 2. Loss of tissue	1, 2. Tissue lysis and death 2. White syndrome. Pathogenicity proved experimentally	155 156 157 1069
Seriatopora histrix	1. *Vibrio harveyi*		1. Loss of tissue	1. White syndrome. Pathogenicity proved experimentally	1069

Stylophora pistilata	1. *Vibrio harveyi*		1. Loss of tissue	1. White Syndrome. Pathogenicity proved experimentally	1069
Stylophora species		*Vibrio stylophorae*		Apparently healthy coral	1516
Unidentified coral species		*Shewanella affinis*		No details	814
Clam – see Phylum Mollusca, Class Bivalvia					
Shrimp and Prawns – see Phylum Arthropoda, Order Decapoda					
Danio	See under ORNAMENTAL				
Discus fish	See under ORNAMENTAL				
Dolphin	See under SEA MAMMALS				
PHYLUM CHLOROPHYTA CLASS Florideophyceae Alga	Kingdom Plantae				
Marine alga (*Ulva lactuca* Linnaeus), CLASS Ulvales		*Pseudoalteromonas ulvae*		Has anti-fouling properties	471
Red alga (*Delesseria sanguinea*), CLASS Florideophyceae, PHYLUM RHODOPHYTA		*Zobellia galactanivorans*			119
Red algae (*Jania* species, Family Corallinaceae, KINGDOM ARCHAEPLASTIDA)		*Shewanella algae*	Surface	Produces neurotoxin, tetrodotoxin	1533
PHYLUM ECHINODERMATA (contains sea urchins, sea stars or starfish, and sea cucumbers)					
Sea urchin		*Vibrio diazotrophicus*	Part of gut flora		665
Sea urchin (*Strongylocentrotus intermedius*)		*Gramella marina*	Tissue	Normal flora	1205
Strongylocentrotus intermedius		*Echinicola pacifica*	Part of the microflora	Normal flora	1207
Sea cucumber (*Apostichopus japonicus*) (CLASS *Holothuroidea*)	1. *Aeromonas salmonicida* ssp. *masoucida* 2. *Pseudoalteromonas tetraodonis* 3. *Shewanella marisflavi* 4. *Shewanella* species 5. *Vibrio alginolyticus* 6. *Vibrio splendidus*		1. Ulcers 3. Skin ulceration, isolation from internal organs	1, 2. Pathogenicity proven in experimental challenge 3. Mortality	1020 1766 1823
Starfish		*Arcobacter marinus*	Tissue	Not known	897

Continued

Table 1.1. Continued.

Host	Pathogens	Normal flora	Tissue site	Disease status	Ref
SUBPHYLUM ASTEROZOA					
Crown of thorns starfish (*Acanthaster planci*). CLASS Asteroidea, ORDER Valvatida, Family Acanthasteridae	*Vibrio harveyi*, *Vibrio natriegens*, *Vibrio owensii*, *Vibrio rotiferianus*		Tissue	Isolated from diseased starfish but pathogencity not known	1411
PHYLUM MOLLUSCA					
CLASS Bivalvia					
Bivalve (*Nodipecten nodosus*)	1. *Vibrio fortis*	2. *Vibrio brasiliensis* 3. *Vibrio coralliilyticus* 4. *Vibrio fortis* 5. *Vibrio neptunius* 6. *Vibrio pomeroyi* 7. *Vibrio xuii*	1. Isolated from healthy and diseased larvae 5. Isolated from diseased and healthy animals 6. Isolated from healthy animals	1, 2, 3, 4, 5, 6, 7 Pathogenicity for bivalves not known	156 1634
Mollusc (*Bulinus sengalensis*)	*Mycobacterium ulcerans*		Tissue	Not known, possible carrier	932
CLASS Gastropoda					
Abalone (Family *Haliotidae*)	1. *Vibrio jasicida*	2. *Pseudoalteromonas* sp. 3. *Shewanella* sp. 4. *Vibrio agarivorans* (pathogenicity not determined) 4. *Vibrio mediterranei*	Tissue	1. Mortality. Confirmed experimentally 4, 5. Mortality, lesions. Pathogenicity not determined	244 1835
Haliotis discus discus		*Vibrio comitans* *Vibrio neonatus*	Gut	Normal microflora	1485 1486
Haliotis diversicolor aquatilis and *Haliotis diversicolor diversicolor*		*Vibrio ezurae*	Gut	Normal microflora	1486
Haliotis discus hannai		1. *Shewanella haliotis* 2. *Vibrio halioticoli*	1, 2. Gut	1, 2. Normal microflora	895 1490
Haliotis gigantea		*Vibrio comitans*	Gut	Normal microflora	1485
Haliotis laevigata and *Haliotis rubra*		*Vibrio superstes*	Gut	Normal microflora	704
Haliotis madaka		*Vibrio comitans*	Gut	Normal microflora	1485
(*Haliotis rufescens*) red abalone	1. *Vibrio alginolyticus*	2. *Vibrio rarus* 3. *Vibrio inusitatus*	1. Sick larvae are unable to swim and remain at the bottom of the pond 2, 3. Gut	1. Mortality in larvae, pathogenicity proven in experimental challenge 2, 3. Normal microflora	54 1485

Haliotis tuberculata	1. *Vibrio* (*carchariae*) *harveyi*	2. *Vibrio gallicus*	1. White pustules on foot 2. Gut flora	1. Mortality. Pathogenicity confirmed experimentally 2. Normal gut microflora	1219 1487
Small abalone (*Haliotis diversicolor supertexta*)	1. *Shewanella algae* 2. *Vibrio alginolyticus* 3. *Vibrio parahaemolyticus*		1. Shrunken muscle, whitening 2. Tissue in post-larvae 3. Organism in haemolymph	1, 2, 3. Pathogenicity confirmed in experimental challenge 3. Withering syndrome, mass mortality.	256 257 1040
(*Haliotis discus hannai* Ino, 1953) Japanese abalone	1. *Shewanella algae* 2. *Vibrio harveyi* (strain was non-luminescent, ODC-negative, urease-negative)		1. Whitening, shrunken muscle 2. White spots on foot, tissue lesions, necrotic degeneration in muscle fibres	1, 2. Mass mortality, lose ability to adhere. Pathogenicity confirmed experimentally	256 1228
Clam					
Carpet shell clam (*Venerupis* [*Ruditapes*] *decussata* L.)	1. *Vibrio alginolyticus* 2. *Vibrio splendidus* biovar II 3. *Vibrio tapetis*	4. *Vibrio atlanticus*, *Vibrio artabrorum*	1, 2. Velum, lesions and necrosis of tissue 3. Invasion of pallial (mantle) cavity and adherence to the periostracal lamina at the growing edge of the shell. Main characteristic feature is a brown conchiolin deposit on the inner surface of the shell	1, 2. Pathogenicity confirmed in experimental challenge 3. Brown ring disease (BRD) 4. Pathogenicity unknown	201 442 644
Geoduck clam (*Panope abrupta*)	*Vibrio tubiashii*		Abscess at margin of pallial cavity	Mortality. Pathogenicity confirmed experimentally	489
Manila clam (*Venerupis* [*Ruditapes*] *philippinarum* Adams & Reeve) and carpet shell clam (*Venerupis* [*Ruditapes*] *decussata* L.)	1. *Vibrio celticus* 2. *Vibrio tapetis*, *Vibrio tapetis* ssp. *britannicus*	3. *Aliivibrio finisterrensis* 4. *Vibrio artabrorum* *Vibrio atlanticus* 5. *Vibrio breoganii* sp. nov. 6. *Vibrio gallaecicus* sp. nov. 7. *Vibrio toranzoniae*	2. Invasion of pallial (mantle) cavity and adherence to the periostracal lamina at the growing edge of the shell. Main characteristic feature is a brown conchiolin deposit on the inner surface of the shell. 5. Normal flora in hepatopancreas, mantle, gonad, gills, extrapalial fluid. 1, 7. Tissue	1. Mortality in pathogenicity assays 2. Brown ring disease (BRD) 3, 4, 7. Pathogenicity unknown 5, 6. Healthy	33 103 139 140 142 143 144 201 285 442 463 980 1308

Continued

Table 1.1. Continued.

Host	Pathogens	Normal flora	Tissue site	Disease status	Ref
Clam (*Meretrix meretrix*)	1. *Vibrio* species MM5 2. *Vibrio parahaemolyticus*-like (strain MM21)		1. Lesion on foot 1, 2. organisms found in hepatopancreas, mantle and gill	1, 2. Mass mortality. Pathogenicity proven in experimental challenge	1841 1842
Clam (no details)		*Vibrio aestuarianus* *Vibrio metschnikovii*		Isolated from healthy clam	994 1645
Clam (*Mercenaria mercenaria*), bivalve molluscs	1. *Vibrio tubiashii*		1. Larvae cease swimming, digestion of tissue	1. Bacterial necrosis and systemic disease	668 1669
Clam (*Venerupis senegalensis* [Gmelin]) (*Venerupis pullastra* is an unaccepted name). Known as a pullet carpet shell	*Vibrio celticus*		Tissue	Disease. Mortality in pathogenicity tests	142
Cockles (*Cardium* sp.)		*Aeromonas bivalvium*, *Aeromonas molluscorum*		Pathogenicity unknown	1145 1146
Cockles (no details)		*Vibrio metschnikovii*		Isolated from healthy cockles	994
Razor shells (*Ensis* sp.)		*Aeromonas bivalvium*, *Aeromonas molluscorum*		Pathogenicity unknown	1145 1146
Mussel					
Far-eastern mussel (*Crenomytilus grayanus*)		*Pseudoalteromonas citrea*		Microflora	812
Giant mussel (no details given)		*Campylobacter lari* spp. *concheus*, *Campylobacter peloridis*		Pathogenicity unknown	424
Mytilus edulis (blue mussel)		*Vibrio mytili*	Tissue	Non-pathogenic	1363
Mytilus sp.	1. *Vibrio ponticus*	2. *Aeromonas molluscorum* 3. *Arcobacter ellisii* 4. *Arcobacter molluscorum* 5. *Arcobacter mytili*	1. Tissue	1. Mortality 2. Pathogenicity unknown 3, 4, 5. Isolated from healthy mussels	348 538 539 1080 1145
Mytilus galloprovincialis					
Perna canalicula (New Zealand mussel)	*Vibrio coralliilyticus*/*neptunius*-like		Digestive tract	Mortality proven in experimental challenge	879

Protothaca jedoensis Lischke (unaccepted synonym). *Leukoma* sp. accepted name		*Shewanella japonica*	Tissue		813
Oyster					
Eastern oyster (*Crassostrea virginica*) Family Ostreidae	1. *Roseovarius crassostreae* 2. *Vibrio tubiashii*	3. *Labrenzia* (*Stappia*) *alba* 4. *Shewanella colwelliana* 5. *Shewanella algae*, *Shewanella baltica*, *Shewanella hafniensis*, *Shewanella marisflavi*, *Shewanella putrefaciens*	1. Lesions, mantle retraction, conchiolin deposits inside shell 3. Probiotic properties 4. Promotes settlement of oyster larvae	1. Juvenile oyster disease (JOD) 3. Protects against JOD 4. Normal flora 5. Pathogenicity unknown	178 194 195 197 668 1402 1669 1773 1774
Flat oyster (*Ostrea edulis*) Family Ostreidae	1. *Vibrio anguillarum* 2. *Vibrio neptunius* 3. *Vibrio kanaloae* 4. *Vibrio tubiashii*		2. Reduced motility, larvae on bottom of tank, velum abnormalities, deaths 3. Diseased larvae 1, 4. Tissues	2. 98–100% mortality in pathogenicity tests 3. Pathogenicity not determined 1, 4. Mortality. Pathogenicity confirmed experimentally	453 668 1357 1632 1669
Kumamoto (*Crassostrea sikamea*)	*Vibrio tubiashii*		Abscess at margin of pallial cavity	Pathogenicity confirmed experimentally	489
Oyster (no details)		1. *Arcobacter molluscorum* 2. *Bacillus nanhaiensis*, *Bacillus zhanjiangensis* 3. *Grimontia hollisae* 4. *Vibrio metschnikovii* 5. *Vibrio mimicus*	2, 3, 4. Tissue	1, 2. Isolated from healthy oysters 3, 5. Pathogenic to humans 4. Non-pathogenic to oyster	316 317 412 538 721 994
Japanese oyster (*Crassostrea gigas*) Pacific oyster (*Crassostrea gigas*)	1. *Alteromonas* species (not speciated) 2. *Nocardia crassostreae* 3. *Photobacterium damselae* ssp. *damselae* 4. *Vibrio anguillarum* 5. *Vibrio aestuarianus* spp. *francensis*	14. *Aeromonas media* 15. *Aeromonas* species 16. *Alteromonas* species 17. *Photobacterium swingsii* 18. *Pseudomonas* species 19. *Tenacibaculum crassostreae* 20. *Vibrio pectenicida* 21. *Vibrio* species	1. Larval deaths 2. Focal areas of brown discolouration on the mantle, or green-yellow nodules on abductor muscle, gills, heart and mantle 4. Failure of veliger larvae to maintain themselves in the water column 4. Tissues	1, 3, 4, 5. Mortality 2. Nocardiosis 8. Mass mortality. Disease resembles bacillary necrosis 9. Summer mortality syndrome. Disease is stress-associated	50 89 156 453 489 491 524 566

Continued

Table 1.1. Continued.

Host	Pathogens	Normal flora	Tissue site	Disease status	Ref
	6. *Vibrio coralliilyticus* 7. *Vibrio fortis* 8. *Vibrio splendidus* 9. *Vibrio splendidus* biovar II 10. *Vibrio tubiashii* 11. *Vibrio vulnificus* serovar E (some strains are avirulent for Taiwanese eels) 12. *Vibrio gigantis* 13. *Vibrio crassostreae*	Total bacterial count in soft tissue = 2.9×10^4 colonies per gram, and in haemolymph = 2.6×10^4 colonies per ml.	5. Haemolymph 9. Organism isolated from gonads of broodstock. High bacterial load in larvae, bacteria in shell margin, loss of cilia and velum 10. Larvae cease swimming, digestion of tissue 19, 20. Haemolymph	10. Bacterial necrosis and systemic disease 14. Probiotic against *V. tubiashii* 16. Non-pathogenic 6, 7, 12, 13. Isolated from diseased oysters but pathogenicity not tested. 20. *Vibrio pectenicida* non-pathogenic in experimental infection	584 588 615 641 668 965 966 1001 1010 1220 1261 1592 1634 1669 1740
Pearl oyster		*Marinomonas ostreistagni*	Seawater from culture pond	Environmental organism	981
Mediterranean oysters		1. *Labrenzia* (*Stappia*) *alba* 2. *Marinomonas aquimarina* 3. *Vibrio lentus*		1. Probiotic potential 2. Environmental organism 3. Pathogenicity unknown	178 1079 1082 1362
Arctic scallop		*Aliivibrio logei*		Pathogenicity unknown	112
Scallop (*Argopecten purpuratus*) Family Pectinidae	*Aeromonas hydrophila* *Vibrio alginolyticus* *Vibrio splendidus*		Tissue	Larval deaths	1409
Scallop (*Pecten maximus*) Family Pectinidae	1. *Vibrio pectenicida* 2. *Vibrio splendidus*	3. *Phaeobacter* (*Roseobacter*) *gallaeciensis*	1, 2. Affects larvae	1, 2. Vibriosis. Pathogenicity confirmed 3. Normal flora	970 1099 1220 1441
Yesso or giant ezo scallop (*Patinopecten yessoensis*)		*Pseudoalteromonas citrea*		Microflora	812

Wedge-shells (*Donax trunculus*) Family Donacidae		*Aeromonas molluscorum*		Pathogenicity unknown	1145
CLASS Cephalopoda					
Octopus (*Octopus vulgaris*)	*Vibrio lentus*		Hard lesions on arm or head mantle, gill heart	Death and mortality. Pathogenicity confirmed	521
Squid					
Doryteuthis pealei (*Loligo pealei*), longfin inshore squid		1. *Shewanella pealeana* 2. *Aliivibrio* (*Vibrio*) *logei*	1. Associated with the reproductive organ, the accessory nidamental gland, in females 2. Symbiont in light organ	1, 2. Normal flora	531 1008
Hawaiian sepiolid or bobtail squid (*Euprymna scolopes*)		*Aliivibrio* (*Vibrio*) *fischeri*	Light organ	Light organ symbiont	196 531
Sepiola affinis, *Sepiola robusta*		*Aliivibrio* (*Vibrio*) *fischeri*, *Aliivibrio* (*Vibrio*) *logei*	Light organ	Light organ symbiont	531
PHYLUM PORIFERA					
Sponge – marine sponge (*Halichondria bowerbanki*). Also known as crumb-of-bread sponge		*Mycobacterium poriferae*			1294
Scleritoderma cyanea		*Vibrio caribbeanicus*		Pathogenicity unknown	738
ORNAMENTAL FISH					
Ornamental fish (no details given)		*Aeromonas aquariorum*	Associated with skin of diseased ornamental fish in aquaria	Pathogenicity unknown	1106
Porpoise	See under SEA MAMMALS				
PHYLUM ROTIFERA					
Rotifer (*Brachionus plicatilis*) CLASS Eurotatoria		1. *Vibrio chagasii* 2. *Vibrio rotiferianus* 3. *Vibrio neptunius* 4. *Vibrio anguillarum*, *Vibrio alginolyticus*, *Vibrio diazotrophicus*, *Vibrio mediterranei*, *Vibrio tubiashii*-like 5. *Vibrio xuii*	2, 3. Isolated from rotifer flow-through culture system 4, 5. Present in rotifer aquaculture systems	1, 2, 3. Pathogenicity not known 4. May be pathogenic to other aquatic species 5. Doesn't appear to be pathogenic to rotifer	241 637 1632 1635 1718
Nannochloropsis oculata		*Vibrio anguillarum*	Rotifer act as carrier for disease		1158

Continued

Table 1.1. Continued.

Host	Pathogens	Normal flora	Tissue site	Disease status	Ref
Seabream	See under bream				
Sea grass		*Vibrio aerogenes* *Vibrio diazotrophicus*	Present in sediment		665 1522
Seal. See under Sea Mammals					
Sea lion. See under Sea Mammals					
SEA MAMMALS (ORDER Cetacea) PHYLUM CHORDATA, CLASS Mammalia	Sea mammals include the three mammalian orders considered to be true aquatic mammals. These are *Cetacea* (whales, dolphins and porpoises), *Pinnipedia* (seals, sea lions and walruses) and Sirenians (sea cows) (Foster *et al.*, 2002)				
Shrimp – see under Prawn					
PHYLUM SIPUNCULA					
Benthic marine 'peanut worm' (*Golfingia margaritacea*). *Phascolosoma japonicum* is a junior synonym. CLASS Phascolosomatidea, ORDER Phascolosomatida, Family Phascolosoma		*Shewanella* (*affinis*) *colwelliana*		Normal flora	814
KINGDOM CHROMISTA					
Brown algae (kelp, seaweed) *Saccharina japonica*. Also known as makonbu (*Laminaria japonica* is a synonym). PHYLUM OCHROPHYTA, ORDER Laminariales	1. *Algicola* (*Pseudoalteromonas*) *bacteriolytica* 2. *Pseudoalteromonas elyakovii*		1. Produces a red pigment on the laminaria and causes damage to the seed supply 2. Degrades fronds	1. Red spot disease 2. Spot disease	1489 1491
KINGDOM PLANTAE					
PHYLUM CHLOROPHYTA CLASS Florideophyceae Alga					
Marine alga (*Ulva lactuca* Linnaeus), CLASS Ulvales		*Pseudoalteromonas ulvae*		Has anti-fouling properties	471
Red alga (*Delesseria sanguinea*), CLASS Florideophyceae, PHYLUM RHODOPHYTA		*Zobellia galactanivorans*			119

Red algae (*Jania* species, Family Corallinaceae, KINGDOM ARCHAEPLASTIDA)		*Shewanella algae*	Surface	Produces neurotoxin, tetrodotoxin	1533
WATER and ENVIRONMENT					
Aquaria water		*Aeromonas aquariorum*	Associated with skin of diseased ornamental fish in aquaria	Pathogenicity unknown	1106
Brackish water		*Arcobacter mytili*			348
Drinking water		*Aeromonas popoffii*			784
Fish-culture pond water		*Andreprevotia lacus*	Isolated in pond water of cultured fish	Part of pond water microflora	1513
Fish processing plant – drain pool		*Vibrio rumoiensis*		Drain pool at fish processing plant	1843
Freshwater	1. *Aeromonas veronii* biovar Veronii	1. *Aeromonas encheleia* 3. *Vibrio cincinnatiensis*	1. Potential cause of diarrhoea and wound infections in humans 2. Found in drinking water sources	1, 3. May cause infection in humans	220 723
Lake water		*Shewanella oneidensis*			1713
Mangroves		1. *Vibrio mangrovi* 2. *Vibrio plantisponsor* 3. *Vibrio porteresiae* 4. *Vibrio rhizosphaerae* 5. *Vibrio xiamenensis*	1, 2, 3, 4. Associated with rhizosphere of wild rice (*Porteresia coarctata*) plants in mangroves 5. Mangrove soil	1, 2. Shows nitrogen-fixing properties 4. Shows antibacterial properties against phytopathogens 5. Cellulase production	581 1371 1372 1373
Marsh mud (salt water)		*Vibrio gazogenes*			686
Plankton (used to include a range of drifting animals plants, bacteria, archaea)		*Vibrio mediterranei*			1361
Pond water		*Vogesella lacus*	Isolated from pond water used for culture of soft-shell turtles (*Pelodiscus sinensis*, previously *Trionyx sinensis*)	Part of pond water microflora	325
River water		1. *Aeromonas fluvialis* 2. *Aeromonas encheleia* 3. *Vibrio diazotrophicus*			37 493 665

Continued

Table 1.1. Continued.

Host	Pathogens	Normal flora	Tissue site	Disease status	Ref
Sand		*Vibrio areninigrae*		Environmental isolate	301
Seawater		1. *Aliivibrio sifiae* 2. *Arcobacter ellisii* 3. *Arcobacter marinus* 4. *Arcobacter molluscorum* 5. *Enterovibrio calviensis* 6. *Grimontia hollisae* 7. *Marinomonas aquimarina* 8. *Marinomonas ostreistagni* *Mesophilobacter marinus* 9. *Moritella japonica*, *Moritella marina* 10. *Shewanella aquimarina* 11. *Shewanella japonica* 12. *Shewanella marisflavi* 13. *Shewanella woodyi* 14. *Tenacibaculum gallaicum* 15. *Vibrio aestivus* 16. *Vibrio agarivorans* 17. *Vibrio azureus* 18. *Vibrio chagasii* 19. *Vibrio cyclitrophicus* 20. *Vibrio kanaloae* 21. *Vibrio lentus* 22. *Vibrio marisflavi* 23. *Vibrio mediterranei* 24. *Vibrio navarrensis* spp. *pommerensis* 25. *Vibrio nereis* 26. *Vibrio nigripulchritudo* 27. *Vibrio orientalis* 28. *Vibrio ponticus* 29. *Vibrio quintilis* 30. *Vibrio ruber* 31. *Vibrio sagamiensis*	1, 3, 5, 7, 8, 10, 11, 16, 19, 22, 23 Seawater 7. Oyster farm 8. Pearl oyster farm 6. Isolated from seawater from the coastal region of Keelung, Taiwan 16. Seawater surrounding mussel hatcheries 19. Harbour water 20. Isolated from seawater from holding tank for turbot (*Psetta maxima*)	Environmental isolate	126 127 348 363 436 538 539 709 721 813 851 897 981 1065 1079 1080 1082 1084 1093 1230 1233 1347 1361 1521 1635 1683 1760 1825 1832 1834 1836 1837

Sewage water		*Vibrio navarrensis*		Non-pathogenic	1687
Soil, sediment		*Vibrio hangzhouensis* *Vibrio diazotrophicus*			665 1818
Stream water		*Mycobacterium palustre*		Found in stream water from peatland run-off, and isolated from lymphadenitis in a child in Italy	1656
Yellow sea tidal flat		*Vibrio litoralis*			1202
Whale. See under Sea Mammals					

Table 1.2. Bacterial pathogens and saprophytes of fish and other aquatic animals.

Bacteria	Disease	Disease signs	Aquatic species	Distribution	Ref
Abiotrophia balaenopterae sp. nov.	See *Granulicatella balaenopterae* com. nov.				352
Abiotrophia elegans (nutritionally variant Streptococci)	See *Granulicatella elegans*				
Achromobacter denitrificans (previously *Achromobacter xylosoxidans* ssp. *denitrificans* and *Alcaligenes denitrificans*)			Found in marine and freshwater		1437
Acinetobacter baumannii Genospecies 2	Human infections		Isolated from human patients and environment		206
Acinetobacter calcoaceticus Genospecies 1	1. Environmental organism 2. Oral cavity flora		1. Isolated from soil 2. Turtles	2. Part of normal flora	206 623 624
Acinetobacter haemolyticus Genospecies 4	Pathogenicity not known for fish	Lesions in organs	Isolated from Atlantic salmon, channel catfish, environment and human clinical samples	Norway, USA	206
Actinobacillus scotiae	Pathogenicity not determined	Organisms in liver, lung, brain, spleen	Isolated from stranded porpoise	Scotland	550
Actinobacillus delphinicola	Pathogenicity not determined	Isolated from various tissues	Sea mammals – harbour porpoise (*Phocoena phocoena*), striped dolphin (*Stenella coeruleoalba*), Sowerby's beaked whale (*Mesoplodon bidens*)	Scotland	548
Actinomyces marimammalium	Pathogenicity not determined	Isolated from various tissues (lung, liver, spleen, intestine) with other organisms	Hooded seal (*Cystophara cristata*), grey seal (*Halichoerus grypus*), harbour porpoise (*Phocoena phocoena*)	UK	763
Aequorivita antarctica, Aequorivita lipolytica, Aequorivita crocea, Aequorivita sublithincola	Environmental isolates (member of *Flavobacteriaceae* family)		Isolated from marine environment – seawater, sea ice	Antarctica	211
Aerococcus viridans	Associated with infection in oesophageal diverticulum	Organism in oesophageal lesions	Loggerhead sea turtle	Spain	1657
Aerococcus viridans var. *homari*	High mortality, Gaffkemia disease. Fatal septicaemia	May see pink or red colouration in the haemolymph. Organisms multiply in the hepatopancreas, haemolymph and to a lesser extent in the heart and skeletal muscle	Lobster (*Homarus americanus, Homarus gammarus*), seawater and sediment in lobster ponds. Exist free living in marine benthos. Crabs and other crustaceans may act as reservoirs or carriers of the organism	Norway, Canada, England, Wales, UK	621 1552 1577 1583 1792 1793

Aeromonas allosaccharophila (HG15) (phenospecies and genomospecies) (previously HG 14 in 1995)	1. Diseased elvers 2. Diarrhoeic stools 3. Pathogencity unknown	1. Disease 2. Faeces	1. Elvers (*Anguilla anguilla*) on eel farm 2. Human 3. Pig carcass	1. Spain 2. South Carolina, USA 3. Spain	1102 1450
Aeromonas aquariorum (*Aeromonas hydrophila* ssp. *dhakensis* is a heterotypic synonym)	Pathogenicity unknown	Diseased imported fish	Skin and water of ornamental fish	Spain	1106
Aeromonas bestiarum (HG2) (genomospecies *A. bestiarum*, phenospecies *A hydrophila*-like)	Pathogenicity in doubt		Fish, river water, freshwater, lake water, seawater, shellfish, human faeces	Spain, US	30 873 1148
Aeromonas bivalvium	Pathogenicity unknown		Isolated from cockles (*Cardium* sp.) and razor shells (*Ensis* sp.) in a retail market	Barcelona, Spain	1146
Aeromonas caviae (HG4) (phenospecies and genomospecies) (*Aeromonas punctata* is a homotypic synonym)	1. Septicaemia, mortality when organism present in high numbers 2. Pathogenicity unknown 3. Infection, gastroenteritis	1. Dermal ulceration, present in hepatopancreas 2. Predominant microflora 3. Infection	1. Freshwater ornamental fish, Atlantic salmon, octopus, giant freshwater prawn turbot larvae 2. Bivalve molluscs 3. Human	Ubiquitous in the environment. Turkey, Kenya, Spain, Taiwan	42 1148 1594
Aeromonas culicicola	(Is a later subjective synonym of *Aeromonas veronii*)				537 788 1339
Aeromonas diversa (formerly Aeromonas Group 501, strain CDC 2478-85 (HG 13)			Isolated from patient with leg wound	USA	722 1147
Aeromonas encheleia (HG16) (phenospecies and genomospecies) (some strains previously HG11)	Non-pathogenic		Healthy eels, freshwater, drinking water sources, river water	Spain, Germany, Belgium, Finland, UK, USA	494 784 873
Enteric Group 5 (HG12)	See *Aeromonas schubertii*				722
Aeromonas enteropelogenes	See *Aeromonas trota*				789
Aeromonas eucrenophila (HG6) (phenospecies and genomospecies)	Non-pathogenic		Ascites of carp, drinking water, well water	Europe, Germany	784
Aeromonas fluvialis	Environmental isolate		River water	Spain	37
Aeromonas hydrophila ssp. *anaerogenes*	Species have been assigned to *Aeromonas caviae* and *Aeromonas eucrenophila*. Name is not in use				789

Continued

Table 1.2. Continued.

Bacteria	Disease	Disease signs	Aquatic species	Distribution	Ref
Aeromonas hydrophila ssp. *hydrophila* (HG1) (phenospecies and genomospecies (usually isolated in heavy reasonably pure growth for it to be considered the primary pathogen)	1. Haemorrhagic septicaemia, peritonitis, redsore disease, fin rot, red-fin disease, mortality. Associated with the fungal disease epizootic ulcerative syndrome, caused by *Aphanomyces invadans*, in catfish and snakehead fish, Thailand, Philippines. Opportunistic infection in seal morbillivirus infection 2. Opportunistic and primary pathogen 3. Mortality 4. Black disease. 5. Pathogenicity unclear. 6. Gastroenteritis.	1. Erosive or ulcerative dermal lesions, haemorrhage on fins and trunk, swelling of anus, erythaema, mortality 2. Red-leg frog disease 3. Deaths 4. Black nodules on thoracic appendages 5. Isolated in cases of abortion	1. Freshwater and ornamental fish, ayu, channel catfish, walking catfish, tilapia, trout, turtles, eels, reptiles, grey seal, occasionally marine fish – cultured Atlantic salmon, seabass. Increase in organic matter and temperatures above 18°C aid proliferation of *Aeromonas hydrophila* Found in fresh, brackish and coastal waters 2. Frogs, farmed-raised bullfrogs 3. Scallop larvae (*Argopecten purpuratus*) 4. Fairy shrimps 5. Cattle, horses, pigs 6. Humans	1, 2. Ubiquitous in the environment. Worldwide 3. Chile 4. Algeria, Germany, Spain.	42 53 449 461 493 623 624 942 945 1032 1112 1316 1409 1714
Aeromonas hydrophila ssp. *dhakensis* Group BD-2 (HG1). Identity in doubt. May be a heterotypic synonym of *Aeromonas aquariorum*	Diarrhoea	Cytotoxic and haemolytic properties	Isolated from diarrhoeal children	Bangladesh	790 1104
Aeromonas hydrophila ssp. *ranae* (HG1)	Septicaemia, mortalities	Isolated from liver and kidney. Cytotoxic and haemolytic properties (possesses *ASH1* gene)	Isolated from septicaemic Chinese bullfrogs (*Rana rugulosa*)	Thailand	791
Aeromonas ichthiosmia is a later heterotypic synonym of *Aeromonas veronii* biovar Sobria	See *Aeromonas veronii* biovar Sobria				353 785 1501
Aeromonas jandaei (HG9) (previously HG9 *Aeromonas sobria*) (phenospecies and genomospecies)	1. Pathogen 2. Pathogenicity unknown 3. Symbiont 4. Clinical significance	1. Liver, kidney 3. Intestine 4. Isolated from, blood, wound, diarrhoeal stools	1. Eel (*Anguilla anguilla*), electric blue hap (*Sciaenochromis ahli*), blue discus fish (*Symphysodon aequifasciatus*) 2. Prawn 3. Medicinal leech (*Macrobdella decora*) 4. Humans	1. Australia, Spain 2. Hawaii, 3, 4. USA	244 276 497 1530

Aeromonas media (HG5A) (genomospecies *Aeromonas caviae*, phenospecies *Aeromonas caviae*-like)	1. Environmental organism 2. Clinical, gastroenteritis		1. River water. Probiotic properties against *V. tubiashii*. Non-pathogenic for Pacific oyster (*Crassostrea gigas*) and eastern rainbow fish (*Melanotaenia splendida splendida*) 2. Humans	1. UK	35 615
Aeromonas molluscorum	Pathogenicity unknown		Isolated from cockles (*Cardium* sp.), mussels (*Mytilus* sp.), razor shells (*Ensis* sp.) and wedge-shells (*Donax trunculus*) in a retail market	Barcelona, Spain	1145
Aeromonas piscicola	Isolated from diseased fish	Moribund	Wild salmon (*Salmo salar*), trout (*Oncorhynchus mykiss*)	Spain	141
Aeromonas popoffii (HG17) (phenospecies and genomospecies)	Environmental organism		Drinking water reservoirs	Finland, Scotland	784
Aeromonas punctata	See *Aeromonas caviae*				
Aeromonas rivuli	Environmental organism		River water	Germany	536
Aeromonas salmonicida biovar Acheron	Skin lesions		Farmed Atlantic salmon (*Salmo salar*)	Australia (Tasmania)	283
Aeromonas salmonicida ssp. *achromogenes* (HG3) = phenomospecies genomospecies *Aeromonas salmonicida*	'Atypical' *Aeromonas salmonicida*. Variety of pathologies, carp erythrodermatitis, goldfish ulcer disease, ulcer disease of flounder	Large open skin lesions surrounded by areas of descalation with softened and haemorrhagic dermis	Salmonids and non-salmonids, Atlantic cod, silver bream (*Blicca bjoerkna*), common bream (*Abramis brama*), perch (*Perca fluviatilis*), roach (*Rutilus rutilus*) Found in fresh water, brackish water and marine environments	World-wide: Australia, England, Central and Northern Europe, Iceland, Japan, North America, South Africa	370 1117 1796
Aeromonas salmonicida biovar Flounder	Cutaneous ulcerative disease		Greenback flounder (*Rhombosolea tapirina*)	Australia	283 1788
Aeromonas salmonicida ssp. *masoucida* = phenospecies; genomospecies *Aeromonas salmonicida*	1. 'Atypical' *Aeromonas salmonicida* 2. Mortality	1. Superficial skin lesions 2. Ulcers	1. Salmonids, (sakuramasu – *Oncorhynchus masou* and pink salmon – *Oncorhynchus gorbuscha*) 2. Sea cucumber (*Apostichopus japonicus*)	Japan	906 1796 1823
Aeromonas salmonicida ssp. nova Proposed for all atypical isolates but not formally accepted	Goldfish ulcer disease, cutaneous ulcerative disease in goldfish	Cutaneous ulcers	Goldfish, salmonids/non-salmonids, eel, carp, marine fish	England, Japan, USA, Worldwide. Australian goldfish strains are thought to belong to this subspecies.	282 1117 1528 1664 1787 1789
Aeromonas salmonicida ssp. pectinolytica	Environmental organism		Polluted river	Argentina	1317

Continued

Table 1.2. Continued.

Bacteria	Disease	Disease signs	Aquatic species	Distribution	Ref
Aeromonas salmonicida ssp. *salmonicida* (HG3) = phenospecies; genomospecies *Aeromonas salmonicida* Known as 'Typical' *Aeromonas salmonicida*	1. Furunculosis 2. Goldfish ulcer disease (GUD)	1. Dermal ulceration showing typical umbonate furuncles. Organisms penetrate to underlying tissues, kidney, coelomic fluid, spleen, intestine	1. Salmonids, sea lamprey (*Petromyzon marinus*) 2. Many fish; goldfish, carp, silver perch, trout, Atlantic salmon, marine flounder, green-back flounder, eel, catfish, carp, cleaner fish. Frogs, after feeding from contaminated trout	Highly virulent strain from Canada, North America, Europe, UK, USA. Not present in Australia	490 513 968 1113 1232 1564
Aeromonas salmonicida ssp. *salmonicida*. Non-pigmented strains	Diseased salmon	Head kidney	Farmed Atlantic salmon. Injection experiments reveal the non-pigmented strain produced a higher mortality than pigmented strains	Norway	928
Aeromonas salmonicida ssp. *smithia* (HG3) = phenospecies; genomospecies *Aeromonas salmonicida*	'Atypical' *Aeromonas salmonicida*	Superficial skin lesions	Non-salmonids	England	81 1796
Aeromonas salmonicida Atypical strains	Mortality Variety of pathologies, carp erythrodermatitis, goldfish ulcer disease, ulcer disease of flounder, ulcer head disease of eels, ulcerative disease	Skin lesions, necrosis, haemorrhagic ulcers, umbonate furuncles and swelling on head of eels. Sometimes underlying tissues affected, particularly in farmed fish	Atlantic salmon (*Salmo salar*), blenny, carp, chub, cleaner fish, dab, eels, sand eels, flounder, greenback flounder (*Rhombosolea tapirina*), goldfish (*Carassius auratus*), minnow, rainbow trout (salt water), turbot (*Scophthalmus maximus*), silver perch (*Bidyanus bidyanus*), roach, rockfish	Australia, Baltic Sea, Denmark, England, Finland, Japan, Norway, South Africa, USA	199 282 398 690 733 774 796 824 968 979 1232 1788 1800 1808
Aeromonas salmonicida Atypical strains, oxidase-negative	Deaths, ulcerative disease	Lethargy, skin ulcers	Turbot (*Scophthalmus maxiumus*), and flounder from a salt water fish farm, coho salmon (*Oncorhynchus kisutch*)	Denmark, Baltic Sea, USA	304 1324 1800
Aeromonas salmonicida Atypical strains. Growth at 37°C	Death and morbidity	Skin ulcers	Carp, goldfish, roach. Garden ponds, aquaria, rivers.	England	70
Aeromonas sanarellii	Pathogenicity unknown	Trauma and abrasion wound	Human	Taiwan	36

Aeromonas schubertii (HG12) (previously called Enteric Group 501) (phenospecies and genomospecies)	1. Mortality 2. Environmental organism 3. Septicaemia, gastroenteritis, diarrhoea	3. Abscesses, wounds, pleural fluid, blood	1. Snakehead (*Channa maculata, Channa* [*Ophiocephalus*] *argus*) 3. Humans, often immunocompromised, leg injury–cellulitis	1. China 3. USA, Puerto Rico	5 278 315 722 1036
Aeromonas sharmana (the assignment of this organism to the *Aeromonas* genus is in doubt)	1. Reported from a warm spring			India	1103 1454
Aeromonas simiae	Isolated from healthy monkeys (*Macaca fascicularis*)			Mauritius	677
Aeromonas sobria (HG7) (phenospecies and genomospecies)	Pathogenicity unknown. Many strains erroneously identified		Fish, frogs	Ubiquitous in the environment	1352
Aeromonas taiwanensis	Pathogenicity unknown	Burn wound	Human	Taiwan	36
Aeromonas tecta (also known as *Aeromonas eucrenophila*-like)	Pathogenicity unknown		Isolated from surface of rainbow trout		435 677
Aeromonas trota (HG14) (junior heterotypic synonym of *Aeromonas enteropelogenes* and should be used instead of *Aeromonas trota*)	1. Non-pathogenic for fish 2. Diarrhoea	2. Stool samples	1. Found in marine, estuarine and freshwater environments 2. Humans	South-East Asia (Bangladesh, India, Indonesia, Thailand), worldwide	277 353 789
Aeromonas veronii biovar Sobria (HG8Y) = phenospecies; genomospecies *Aeromonas veronii* Phenospecies *Aeromonas sobria* (HG8X) = genomospecies *Aeromonas veronii*	1. Environmental organism 2. Humans	2. gastrointestinal disease, septicaemia, wound infection, meningitis	1. Ubiquitous in the environment 2. Humans – most pathogenic of *Aeromonas* taxa	Worldwide	540 1100
Aeromonas veronii biovar Veronii (HG10). Formerly known as Enteric Group 77 (an earlier heterotypic synonym of *Aeromonas ichthiosmia*) (phenospecies = genomospecies)	1. Mortality when organism present in high numbers 2. Diarrhoea, wounds, cholecystitis	1. Hepatopancreas 2. Faeces, wound site	1. Giant freshwater pawns 2. Humans Found in freshwater	1. Taiwan	5 353 723 785 1595
Aeromonas culicicola (a later subjective synonym of *Aeromonas veronii*)	Strains from ornamental fish cluster with *Aeromonas veronii*		Isolated from mosquito and drinking water supply	India, Spain	537 788 1339
Alcaligenes faecalis homari	See *Halomonas aquamarina*				14 75

Continued

Table 1.2. Continued.

Bacteria	Disease	Disease signs	Aquatic species	Distribution	Ref
Algicola (*Pseudoalteromonas*) *bacteriolytica*	Red-spot disease	Organism produces a red pigment on the Laminaria and induces damages to the seed supply	Culture beds of *Laminaria japonica*	Japan	1489
Aliivibrio finisterrensis sp. nov.	Pathogenicity not known		Manila clam (*Venerupis* [*Ruditapes*] *philippinarum*)	Spain	143
Aliivibrio (*Vibrio*) *fischeri* (synonymous with *Photobacterium fischeri*)	1. Pathogenicity in doubt 2. Symbiont	1. White nodules on skin, haemorrhagic ulceration, tumours in pancreas and bile duct 2. Present in light organ	1. Seabream, turbot 2. Sepiolid squid (*Euprymna scolopes* and *Sepiola species*), Japanese pinecone fish (*Monocentris japonica*)	1. Spain 2. Hawaii, Japan, USA	196 531 842 1435
Aliivibrio (*Vibrio*) *logei*	1. Possible pathogen 2. Some strains pathogenic 3. Pathogenicity unknown 4. Pathogenicity unknown 5. Pathogenicity not stated	1. Skin lesions 2. Mortality 3. Associated with moribund cod 4. Lesions 5. Isolated from intestine	1. Atlantic salmon 2, 3. Cod (*Gadus morhua*) larvae 4. Tanner crab, Arctic scallop 5. Goby fish (*Cottidae* species)	Arctic, Iceland, Norway, Russia, USA, UK, Western Pacific	112 241 890 1395
Aliivibrio (*Vibrio*) *salmonicida*	Coldwater vibriosis, hitra disease	Septicaemia, haemorrhages in the integument surrounding the organs. Organism found in blood and kidney	Atlantic salmon, rainbow trout, cod (*Gadus morhua*)	Canada, Faroe Islands (Denmark), Iceland, Norway, Scotland	237 397 473 852
Aliivibrio sifiae	Environmental isolate		Seawater	Japan	1834
Aliivibrio (*Vibrio*) *wodanis*	Non-pathogenic, may be an opportunist in winter ulcer disease		Atlantic salmon, rainbow trout, haddock	Cold water of Iceland, Finland, Norway, Sweden	153 1071
Allomonas enterica	See *Vibrio fluvialis*				855
Alteromonas species	Bacterial necrosis and septicaemia	Necrosis, septicaemia	Larval and juvenile molluscs (oyster)		
Alteromonas citrea	See *Pseudoalteromonas citrea*				
Alteromonas colwelliana	See *Shewanella colwelliana*				
Alteromonas communis	See *Marinomonas communis*				
Andreprevotia lacus sp. nov.	Environmental organism		Isolated from pond water of cultured fish	Taiwan	1513

Aquaspirillum sp.	Report of an association in Epizootic Ulcerative Syndrome – caused by the fungus, *Aphanomyces invadans*	*Aquaspirillum* induced slight dermomuscular necrotic lesions	Catfish – mild virulence only. Snakehead fish are not infected when challenged with *Aquaspirillum* species	Thailand	1032
Aquimarina latercula previously *Cytophaga latercula*	Environmental isolate		Marine	USA	328
Arcanobacterium phocae	Pathogenicity not determined, possible opportunistic pathogen	Tissues showing discharge or inflammation	Stranded animals: Californian sea lions (*Zalophus californianus*), Pacific harbour seals (*Phoca vitulina richardii*), northern elephant seal (*Mirounga angustirostris*), southern sea otter (*Enhydra lutris nereis*)	California coast, USA	845 1374
Arcanobacterium phocisimile	Pathogenicity not determined	No disease signs. Isolated from Vagina and anus	Harbour seal (*Phoca vitulina*)	North German Sea	727
Arcanobacterium pluranimalium	Pathogenicity not determined	Isolated from lung and spleen	Harbour porpoise (*Phocoena phocoena*), sallow deer	UK	985
Arcobacter bivalviorum					
Arcobacter cryaerophilus	Mortality	Loss of scales, organ damage	Trout (*Oncorhynchus mykiss*)	Turkey	86
Arcobacter ellisii	Possibly normal flora	None	Mussels (*Mytilus* species)	Spain	539
Arcobacter marinus	Environmental organism		Seawater containing seaweeds and a starfish	Korea	897
Arcobacter molluscorum	Isolated from healthy animals	None	Mussels (*Mytilus* species), oysters	Spain	538
Arcobacter mytili	Isolated from healthy animals	None	Mussels (*Mytilus* species), brackish water	Spain	348
Arthrobacter agilis	Environmental organism		Water, soil, human skin		
Arthrobacter nasiphocae	Possibly normal flora	Nasal cavity	Common seal (*Phoca vitulina*)		358
Arthrobacter rhombi	Pathogenicity not determined	Organism isolated from internal organs	Greenland halibut (*Reinhardtius hippoglossoides*)	Greenland	1276
Atopobacter phocae	Pathogenicity not determined	Intestine, lymph nodes, lung, penis	Seal (*Phoca vitulina*), Caspian seal (*Pusa* [*Phoca*] *caspica*)	Scotland, UK	989
Austwickia (*Dermatophilus*) *chelonae*	Dermatophilosis	Skin lesions, skin abscess, skin scabs	Testudines (Chelonians) – turtle and tortoise	Australia	672 1109
Bacillus cereus	Branchionecrosis	Found on necrotic gills of carp	Carp, striped bass	Russia, USA	132 1366
Bacillus galliciensis	Probable normal flora	Faeces	Long-snouted seahorse (*Hippocampus guttulatus*)	Spain	105

Continued

Table 1.2. Continued.

Bacteria	Disease	Disease signs	Aquatic species	Distribution	Ref
Bacillus hemicentroti					
Bacillus mycoides	Mortality	Ulcers on dorsum, focal necrosis of epaxial muscle	Channel catfish. Organism is ubiquitous in soil and has been implicated in disease in humans and parrots	Poland, USA	648
Bacillus nanhaiensis	Probable normal flora	Tissue	Oyster	China	317
Bacillus neizhouensis					
Bacillus subtilis	Part of bacterial flora in branchionecrosis		Carp	Poland	1366
Bacillus zhanjiangensis	Probable normal flora	Tissue	Oyster	China	316
Beneckea chitinovora (not on the list of recognized bacterial names. Beneckea transferred to *Vibrio*). Previously called *Bacillus chitinovorus*	Ulcerative shell disease, shell rot, spot disease, rust disease. Mildly contagious, chronic self-limiting disease	Affects the chitinous plates of the carapace (dorsal shell) and plastron (ventral shell). Shell becomes pitted and early lesions have blotchy dark colouration	Free ranging and captive turtles (spiny soft-shelled turtles, red-eared sliders, striped-necked musk, South American side-neck turtle, Eastern painted turtle)	USA	1754
Bordetella bronchiseptica	1. Bronchopneumonia, secondary pathogen in phocine morbillivirus (distemper) infection	Lung, trachea	1. Seals 2. Bronchopneumonia in dogs, laboratory animals, cats, rabbits, horses, turkeys, monkeys, humans, associated with atrophic rhinitis in pigs	Europe, Scotland, Denmark, UK	1386
Brevundimonas (*Pseudomonas*) *diminuta*	Environmental organism				1502
Brevundimonas (*Pseudomonas*) *vesicularis*	Environmental organism		Found in streams		1502
Brucella abortus	Brucellosis	Positive serology	Weddell seals (*Leptonychotes weddellii*)	Antarctica	1252
Brucella abortus, Brucella melitensis, Brucella suis and rarely *Brucella canis*	Brucellosis		Generally host specific: *Brucella abortus* (cattle), *Brucella canis* (dogs), *Brucella melitensis* (goats), *Brucella neotomae* (desert wood rats), *Brucella ovis* (sheep), *Brucella suis* (pigs, reindeer, hares) **Zoonotic.** Use Biological Safety Cabinet	Worldwide	369

Brucella (*cetaceae*) *ceti* (previously part of *Brucella maris* sp. nov. biovar I and II) Proposed names according to host; *Brucella phoecoenae* (porpoise), *Brucella delphini* (dolphin)	1. Brucellosis, abortion, infection, meningitis 2, 3. Brucellosis	1. Aborted foetus, subcutaneous lesions, spleen, lung, brain, vent lesion 2. Organism in liver and spleen of whale 3. Headaches, lassitude, severe sinusitis and positive blood culture	1. Isolated from cetaceans (dolphin, porpoise, whale). Atlantic white-sided dolphin (*Lagenorhynchus acutus*), common dolphin (*Delphinus delphis*), striped dolphin (*Stenella coeruleoalba*), porpoise (*Phocoena phocoena*), bottlenose dolphin (*Tursiops truncatus*), harbour porpoise (*Phocoena phocoena*) 2. Minke whale (*Balaenoptera acutorostrata*) 3. Human infection **Zoonotic.** Work with all suspect tissues and *Brucella* cultures in a Biological Safety Cabinet	Canada, England, Europe, Norway, Scotland, USA	226 340 342 416 417 545 546 554 827 1430
Brucella (*pinnipediae*) *pinnipedialis* (previously part of *Brucella maris* sp. nov. [biovar I and II]) Proposed name of *B. phocae* (seals)	1. Brucellosis, abortion, infection	1. Aborted foetus, subcutaneous lesions, spleen, lung. Organism in lymph nodes of seals	1. Isolated from pinnipeds (seal). Common seal (*Phoca vitulina*), grey seal (*Halichoerus grypus*), hooded seal (*Cystophora cristata*), harp seal (*Phoca groenlandicus*), ringed seal (Pusa [*Phoca*] *hispida*). European otter (*Lutra lutra*). Also found in lung worm (*Parafilaroides* species) infecting Pacific harbour seal **Possibly Zoonotic.** Work with all suspect tissues and *Brucella* cultures in a Biological Safety Cabinet	Canada, England, Europe, Scotland, USA	226 342 545 546 547 554 587 827 1430 1667
Brucella species (uncharacterized)	Brucellosis	No clinical sign of disease	Polar bear (*Ursus maritimus*)	Norway	1665
Budvicia aquatica	Environmental organism		Isolated from river and drinking water	Czech Republic, Sweden	205
Burkholderia (*Pseudomonas*) *cepacia*	Environmental organism		Fresh water, soil	Ubiquitous	1301
Burkholderia (*Pseudomonas*) *pseudomallei*	Melioidosis	Septicaemia, abscesses in lung, spinal column, liver, kidney	Cetaceans in oceanarium (false killer whale, bottlenose dolphin, white-sided dolphin, sea lion, grey seal), sheep, macaroni penguin (*Eudyptes chrysolophus*), goat, dog, galah, cockatoo, human Found in soil and water **Zoonotic.** Use Biological Safety Cabinet	Hong Kong. Disease of tropical and subtropical regions (Australia, South-East Asia)	724 1086
Campylobacter canadensis					

Continued

Table 1.2. Continued.

Bacteria	Disease	Disease signs	Aquatic species	Distribution	Ref
Campylobacter insulaenigrae	Possible normal flora	None	Common seal (*Phoca vitulina*), harbour porpoise (*Phocoena phocoena*)	UK	552
Campylobacter lari spp. *concheus*	Pathogenicity not known		Isolated from shellfish and from human faeces	Belgium, Canada, The Netherlands	424
Campylobacter lari spp. *lari*	Pathogen and non-pathogen		Isolated from faeces of herring gull (*Larus argentatus*), dogs, and chickens, horse intestine and cases of human diarrhoea and bacteraemia	Belgium, Canada, Sweden, UK	159 424
Campylobacter peloridis	Pathogenicity not known		Isolated from mussel and from human faeces and dialysis fluid	Belgium, The Netherlands	424
Campylobacter subantarcticus	Possible normal flora		Grey-headed albatross (*Diomedea chrysostoma*), black-crowned albatross (*D. melanophris*), gentoo penguin (*Pygoscelis papua*)	Brid Island, South Georgian archipelago, SubAntarctic	422
Campylobacter species	1. Enteritis 2. Unknown	1. Mild intestinal wall thickening 2. DNA from dental plaque	1. Southern sea otters (*Enhydra lutris nereis*) 2. Dolphin (*Tursiops truncatus*)	Argentina, USA	633 1143
Campylobacter volcris					
Candida sp. (yeast)	Opportunist infection		Cetaceans – particularly dolphins		1600
Carnobacterium alterfunditum	Environmental organism		Lake water	Antarctica	841
Carnobacterium divergens	Normal intestinal microflora of healthy fish	Intestine and digestive tract	Atlantic salmon juveniles, Atlantic cod, Arctic charr, saithe	France, Norway	355
Carnobacterium divergens Strain 6251	Potential probiotic against *Aeromonas salmonicida* ssp. *salmonicida*, *Vibrio anguillarum*, *Moritella viscosa*	Present in intestine	Arctic charr (*Salvelinus alpinus*)		1405
Carnobacterium funditum	Environmental organism		Lake water	Antarctica	841
Carnobacterium gallinarum	Environmental organism		Isolated from ice slush around chicken carcasses		355
Carnobacterium inhibens	Normal intestinal microflora of healthy fish	Found in digestive tract	Inhibits growth of *Vibrio anguillarum* and *Aeromonas salmonicida* in Atlantic salmon (*Salmo salar*)	Norway	841
Carnobacterium mobile	Environmental organism		Processed chicken meat		355

Carnobacterium maltaromaticum Previously *Carnobacterium piscicola* and *Lactobacillus piscicola*	1. Pseudokidney disease, Lactobacillosis. Post-stripping peritonitis. Seen in fish, 1 year or older, following stress such as handling and spawning. Most strains are opportunistic and possess low virulence, however other strains have higher virulence and cause high mortality	Epicarditis, peritonitis, visceral granulomas, distension of abdomen, ascites fluid, blood or blisters under the skin. Collect samples from kidney, spleen, swimbladder. Virulent strains cause bilateral exophthalmia, periocular and liver haemorrhages, and ascites fluid	Salmonid fish, cutthroat trout (*Oncorhynchus clarkii clarkii*), rainbow trout (*Oncorhynchus mykiss*), chinook salmon (*Oncorhynchus tshawytscha*), cultured striped bass (*Morone saxatilis*), channel catfish (*Ictalurus punctatus*), brown bullhead catfish (*Ictalurus nebulosus*). Less virulent in striped bass and channel catfish	Australia, Belgium, Canada, France, Spain, UK, USA	136 355 736 1655
Carnobacterium maltaromaticum-like	Chronic mortality	Diseased kidney and swimbladder	Lake whitefish (*Coregonus clupeaformis*)	USA	1047
Catellicoccus marimammalium	Pathogenicity not known	Intestine, kidney, peritoneal fluid, pericardial fluid	Porpoise (*Phocena phocoena*), grey seal (*Halichoerus grypus*)	Scotland	984
Catenibacterium species	See *Eubacterium* species				
Cellulophaga (*Cytophaga*) *lytica*	Environmental isolate		Marine, beach mud	Costa Rica	328
Chitinophaga (*Cytophaga*) *arvensicola*	Environmental isolate		Soil	Japan	168
Chitinophaga (*Flexibacter*) *sancti*	Environmental isolate			Argentina	168 1017
Chromobacterium violaceum Pigmented and non-pigmented strains	1. Environmental isolate 2. Wound infection, septicaemia, abscesses		1. Found in soil and water 2. Humans	1. Tropical and subtropical regions. 2. Australia, Malaysia, Senegal, Taiwan, USA, Vietnam	995
Chryseobacterium (*Flavobacterium*) *balustinum*	Flavobacteriosis		Marine fish	USA	1746
Chryseobacterium arothri	See *Chryseobacterium hominis*				266
Chryseobacterium (*Flavobacterium*) *gleum*	Non-pathogenic to fish		Found in human clinical specimens		751
Chryseobacterium chaponense	Pathogenicity not confirmed	External lesions, gills and fins	Atlantic salmon (*Salmo salar*)	Chile	859
Chryseobacterium (*arothri*) *hominis*	1. Probably non-pathogenic 2. Clinical infections	1. None reported 2. Isolated from blood, dialysis fluid, pus, aortic valve, eye	1. Pufferfish (*Arothron hispidus*) 2. Humans	1. Hawaii 2. Belgium, USA	266 860 1700

Continued

Table 1.2. Continued.

Bacteria	Disease	Disease signs	Aquatic species	Distribution	Ref
Chryseobacterium indologenes (previously *Sphingobacterium* and *Flavobacterium indologenes*)	1. Systemic infection 2. Clinical sources	1. Torticolis, gross lesions, granulomas, enlarged organs	1. Farmed bullfrogs (*Rana catesbeiana*) 2. Human specimens and hospital environment	USA	1112 1819
Chryseobacterium (*Flavobacterium*) *indoltheticum*					1173
Chryseobacterium meningosepticum	See *Elizabethkingia meningoseptica*				1693
Chryseobacterium oncorhynchi			Rainbow trout (*Oncorhynchus mykiss*)		1845
Chryseobacterium piscicola	Lesions	Ulcerative lesions on skin, muscle, flank, anus and peduncle	Atlantic salmon (*Salmo salar*), rainbow trout (*Oncorhynchus mykiss*)	Chile, Finland	798 799
Chryseobacterium piscium	None	None	Fish (no details given)	South Africa	421
Chryseobacterium (*Flavobacterium*) *scophthalmum*	Gill disease, haemorrhagic septicaemia. 100% mortality in juveniles	Swollen gill lamellae (hyperplasia), haemorrhagic septicaemia, swollen intestines containing yellow fluid, haemorrhages in eye, skin, jaw	Healthy and diseased turbot. Coastal waters	Scotland	1172 1173
Chryseobacterium tructae	Pathogenicity unknown	Liver	Isolated from diseased rainbow trout (*Oncorhynchus mykiss*)	Spain	1846
Citrobacter freundii	1. Opportunistic infection. Low virulence for trout 2. Systemic infection	1. Haemorrhagic spots on skin, eye and fins. Organism isolated from kidney, liver, spleen 2. Torticolis, gross lesions, granulomas, enlarged organs	1. Angel fish in aquaria, rainbow trout, sunfish, carp, grass carp (*Ctenopharyngodon idella*). Commonly isolated from soil, water, sewage, food, and from organs of diseased and healthy animals including mammals, birds, reptiles and amphibians 2. Farmed bullfrogs (*Rana catesbeiana*)	China, India, Japan, UK, USA, worldwide	870 1063 1112 1478 1654
Clostridium botulinum type E	Morbidity and mortality		Farmed trout, salmon, coho salmon. *Clostridium botulinum* can be a commensal in trout intestine and on gills. Toxin usually not produced in live fish, but is produced in dead fish as part of the decomposition process. Found in lake sediment. Bacterium produces toxin in an anaerobic environment	Britain, Canada, Denmark, Scotland, USA	251 269
Clostridium perfringens	Part of normal flora but can cause an opportunistic infection	Necrotizing enteritis, death	Southern sea otter (*Enhydra lutris nereis*)	USA	1143
Clostridium perfringens Type A	1. Enterotoxaemia 2. Myositis at injection site	1. Gut content 2. Abscess in muscle	Captive killer whale (*Orcinus orca*), dolphin (*Tursiops truncatus*), California sea lion (*Zalophus californianus*)	UK	656

Cobetia marina Previously *Halomonas marina, Pseudomonas marina* and *Deleya marina*	Environmental organism		Marine environment		124
Colwellia maris Previously *Vibrio* strain ABE-1	Environmental organism		Psychrophilic, found in seawater	Japan	1844
Corynebacterium aquaticum	1. Exophthalmia 2. Clinical infection	1. Organism seen in brain, haemorrhages in eyes	1. Striped bass, rainbow trout. Organism found in natural fresh and distilled water 2. Reported to cause infection in immunocompromised patients but may in fact be mis-identifed *Aureobacterium* species	1. USA 2. Worldwide	133
Corynebacterium caspium	Pathogenicity unknown	None	Caspian seal (*Pusa caspica*)	UK	357
Corynebacterium ciconiae	Probable normal flora	None	Black stork (*Ciconia nigra*)	Spain	528
Corynebacterium maris	Appears to be part of the normal flora in coral mucus		Coral – *Fungia granulosa*	Red Sea, Israel	152
Corynebacterium phocae	Pathogenicity not determined	None (nasal passage)	Common seal (*Phoca vitulina*)	Scotland, UK	1313
Corynebacterium sphenisci	Pathogenicity not determined	None (cloaca)	Magellanic penguin (*Spheniscus magellanicus*)	Argentina	652
Corynebacterium spheniscorum	Pathogenicity not determined	None (cloaca)	Magellanic penguin (*Spheniscus magellanicus*)	Argentina	653
Corynebacterium testudinoris	Associated with necrotic mouth lesions	Mouth lesions	Tortoise	Scotland	351
Cryptococcus lupi (yeast)	Environmental organism		Soil	Antarctica	94
Cryptococcus neoformans var. *gattii* (yeast)	Death, septicaemia	Organism isolated from lung, brain, lymph nodes	Dolphin Zoonotic organism Isolated from bat guano and associated with eucalyptus trees	Australia, Tropics and southern hemisphere	244 580
Cytophaga aquatilis	See *Flavobacterium hydatis*				172 1586
Cytophaga arvensicola	See *Chitinophaga arvensicola*				
Cytophaga aurantiaca	Environmental isolate		Swampy soil	Germany	168
Cytophaga columnaris	See *Flavobacterium columnare*				
Cytophaga fermentans	Environmental isolate		Marine mud	California	168 328
Cytophaga hutchinsonii	Environmental isolate		Soil		168
Cytophaga johnsonae	See *Flavobacterium johnsoniae*				
Cytophaga latercula	See *Aquimarina latercula*				

Continued

Table 1.2. Continued.

Bacteria	Disease	Disease signs	Aquatic species	Distribution	Ref
Cytophaga marinoflava	See *Leeuwenhoekiella marinoflava*				168
Cytophaga psychrophila	See *Flavobacterium psychrophilum*				169
Cytophaga species (*Flavobacterium* [*Sporocytophaga cauliformis*] *cauliformans*)	Environmental organism		Lake water		1115
Dermatophilus chelonae	See *Austwickia chelonae*				672 1109
Dermatophilus congolensis	Dermatophilosis (referred to as 'lumpy wool' and mycotic dermatitis in infected sheep)	Subcutaneous nodules and lesions containing caseous material	1. Aquatic species – crocodile, bearded dragon, polar bear, seal 2. Man, horse, sheep, blue tongue lizard, cat, cattle, goat, deer, monkey, pig, rabbit, owl, fox, giraffe, gazelle	Australia	650 856 1534
Dietzia (*Rhodococcus*) *maris*	Microflora	Skin and intestinal flora	Carp, soil	USSR	1212
Echinicola pacifica gen. nov., sp. nov.	Microflora		Sea urchin (*Strongylocentrotus intermedius*)	Gulf of Peter the great, East Sea (Sea of Japan)	1207
Edwardsiella hoshinae	Part of normal flora	Faeces	Birds (puffin, flamingo) and reptiles (monitor, lizard), water	Worldwide	662
Edwardsiella ictaluri Strains with a limited tolerance for oxygen have been isolated (Mitchell and Goodwin, 2000)	1. Enteric septicaemia of catfish 2. Environment	1. Initial infection in brain. Petechial haemorrhage under jaw and belly, head lesion, gills, kidney, liver, spleen infected	1. Freshwater ornamental fish (rosy barb, yoyo loach), cultured channel catfish (*Ictalurus punctatus*), *Danio*, white catfish, yellow catfish (*Tachysurus* [*Pelteobagrus*] *fulvidraco*) walking catfish (*Clarias batrachus*), freshwater catfish (*Pangasianodon hypophthalmus*), fork-tailed or high-fin catfish (*Neoarius berneyi*), narrowfront tandan (*Neosilurus ater*), toothless catfish (*Anodontiglanis dahli*), brown bullheads (*Ameiurus nebulosus*), green knifefish, bass, cyprinids and *Anguillidae*, rainbow trout. Catfish are the most susceptible species 2. Isolated from organically polluted waters, urine and faeces of man, and intestinal microflora of snakes	1. China, Thailand, Turkey, USA, Vietnam	386 702 718 776 819 871 880 1038 1152 1349 1829

Edwardsiella piscicida Some strains in fish originally identified as *Edwardsiella tarda*	Pathogenic to zebrafish in experimental challenge	Zebrafish show ulcerations on dorsal surface. Organism in liver and spleen	European eel (*Anguilla anguilla*), Japanese eel (*Anguilla japonica*), marbled eel (*Anguilla marmorata*), crimson seabream (*Evynnis japonicas*), amur or Korean catfish (*Silurus asotus*), turbot (*Scophthalmus maximus*)	China, Europe, Norway, Scotland, South Korea	1 971
Edwardsiella tarda (*Edwardsiella anguillimortifera* is the senior synonym, however *E. tarda* is conserved for use. Also named *Paracolobactrum anguillimortiferum*)	1. Edwardsiellosis, redpest, emphysematous putrefactive disease of catfish, fish gangrene 2. Systemic infection 3. Human clinical samples, diarrhoea 4. Non-pathogenic 5. Mortality (co-infection with *Vibrio harveyi*)	1. Septicaemia, and ulcerative dermatitis, intestinal infection. Lesions and abscesses in muscle, kidney, liver, spleen. Trout may have mucopurulent discharge from cloaca, congested spleen, enlarged liver, petechial haemorrhages on serosal fat and gills. Eels show abscessed or ulcerative lesions in kidney or liver or muscle 2. Torticolis, gross lesions, granulomas, enlarged organs 4. Faeces 5. Ascites fluid	1. Alligator (*Alligator mississippiensis*), angelfish, black mullet, bass, amur catfish (*Silurus asotus*), channel catfish (*Ictalurus punctatus*), chinook salmon (*Oncorhynchus tshawytscha*), ornamental fish (rosy barb), flounder (*Paralichthys olivaceous*), gilthead seabream (*Sparus aurata*), goldfish, largemouth bass (*Micropterus salmoides*), rainbow trout (*Oncorhynchus mykiss*), carp, Roho labeo (*Labeo rohita*), Senegalese sole (*Solea senegalensis*), ostrich, striped bass (*Morone saxatilis*), brook trout, turbot (*Scophthalmus maximus*), eels (*Anguilla anguilla, Anguilla japonica*), tilapia (*Oreochromis niloticus niloticus*), California sea lion (*Zalophus californianus*), harbour porpoise (*Phocoena phocoena*), Steller sea lion (*Eumetopias jubatus*), bullfrog (*Rana catesbeiana*), turtles. Also part of normal flora in many aquatic animals or animals associated with aquatic habitats (fish, frogs, amphibians, reptiles, snakes, mammals, captive little penguins, bald eagle (*Haliaeetus leucocephalus*), common loon (*Gavia immer*), great blue heron (*Ardea herodias*), pelican (*Pelecannus occidentalis*), ring-billed gull (*Larus delawarensis*), sandhill crane	Ubiquitous in environment, Asia, Australia, Canada, China, India, Japan, Korea, Northern and Southern Europe, South Korea, Spain, UK, USA	21 48 135 286 287 347 468 510 562 757 776 817 1112 1134 1156 1160 1201 1248 1285 1383 1387 1558 1674 1743 1752 1785 1815 1838

Continued

Table 1.2. Continued.

Bacteria	Disease	Disease signs	Aquatic species	Distribution	Ref
			(*Grus canadensis*), toads, turtles, and inhabitant of aquatic environment and surface water. Found in bile of healthy pigs 2. Banded knifefish, spotted bonytongue 3. Has been isolated from abscesses, meningitis, wounds, urine, blood, faeces, spinal fluid 4. Healthy captive fairy penguin, turtles, reptiles, human, birds, cattle 5. Olive flounder (*Paralichthys olivaceus*)		
Elizabethkingia meningoseptica Previously *Chryseobacterium* (*Flavobacterium*) *meningosepticum*	1. Non-pathogen for fish 2. Pericarditis, septicaemia 3. Systemic infection 4. Meningitis in infants	2. Isolated from pericardium, liver, eye lesion 3. Torticolis, gross lesions, granulomas, enlarged organs	1. Reported from fish blood, marine mud, soil, water 2. Birds (chicken, pigeon, finch) 3. Farmed bullfrogs (*Rana catesbeiana*), tiger frog (*Hoplobatrachus tigerinus*) 4. Human pathogen	China, USA, Worldwide	900 1112 1693 1817
Empedobacter brevis Previously *Flavobacterium breve*	1. Environmental 2. Clinical	2. Eye, urine, blood culture, bronchial secretion	1. Fish, canal water. Maybe pathogenic for laboratory animals 2. Humans	England, Ireland, Switzerland, Czech Republic	754 1695
Enterobacter agglomerans	See *Pantoea agglomerans*				
Enterococcus faecalis (reported as *E. faecalis* ssp. *liquefaciens*)	1. Possible pathogen. Identity in doubt 2. Systemic infection	1. Bacteria in liver and kidney, ulcers on fins	1. Rainbow trout, catfish, brown bullhead 2. Crabs Part of normal intestinal flora of man and animals. May cause nosocomial infections	1. Italy, Croatia. 2. French Mediterranean coast Worldwide	
Enterococcus faecium	Normal flora. May cause nosocomial infections		Part of normal intestinal flora of man and animals	Worldwide	
Enterococcus seriolicida	See *Lactococcus garvieae*				480 954 1614
Enterovibrio (*Vibrio*) *calviensis*	Environmental organism		Seawater	Western Mediterranean Sea, France	436
Enterovibrio coralii	Pathogenicity not determined		Isolated from water extract of bleached coral (*Merulina ampliata*)	Australia	1636
Enterovibrio nigricans	Pathogenicity not determined	Isolated from head kidney in association with *Vibrio harveyi*	Seabream (*Sparus aurata*) and *Dentex dentex*	Spain	1314

Enterovibrio norvegicus	Normal flora		Present in gut of turbot larvae	Norway	1627
Erysipelothrix rhusiopathiae	1. Non-pathogenic for fish 2. Erysipelas 3. Erysipeloid, skin disease, polyarthralgia, septic arthritis, renal failure, peritonitis	2. Systemic disease and skin disease 3. Skin disease, septicaemia	1. Parasitic on fish, lobster 2. Dolphins (*Tursiops truncatus*), northern fur seal (*Callorhinus ursinus*), northern elephant seal (*Mirounga angustirostus*), little blue penguin (*Eudyptula minor*), pig, kangaroo, emu, cattle, sheep, dog, horse, avian species, crocodile (*Caiman crocodilus, Crocodylus acutus*) 3. Human pathogen, occupational exposure	Worldwide	193 464 607 834 1128 1591
Escherichia albertii	Gastroenteritis	Diarrhoea	People		787
Escherichia coli	Endocarditis	Lesions in heart valves	Sea lion (*Zalophus californianus*)	Korea	898
Escherichia vulneris	1. Septicaemia, mortality 2. Wounds	1. Abnormalities and haemorrhages in gills, liver, kidney, spleen	1. Freshwater fish: rainbow trout (*Oncorhynchus mykiss*), balloon molly and silver molly (*Poecilia* sp.), common carp (*Cyprinus carpio*) 2. Humans, also isolated from faeces. Found in water of fish ponds and fish faeces	1. Turkey	85
Eubacterium species (Identification is tentative (Udey *et al*., 1977). Initially identified as *Catenabacterium* (Henley and Lewis, 1976)	Mortality	Organism recovered from brain, liver, kidney and blood	Grey mullet (*Mugil cephalus*), redfish (*Scianops ocellata*)	USA	716
Eubacterium tarantellae (original spelling *E. tarantellus*)	Meningitis. Possible secondary pathogen	Organism isolated from brain tissue	Striped mullet, black mullet (*Mugil cephalus*)	Florida	1673
Facklamia miroungae	Normal flora	Nasal cavity	Southern elephant seal	UK	763
Ferrimonas senticii	Probable normal flora	No clinical abnormalities seen	Pufferfish (*Arothron hispidus*)	Hawaii	265
Flavobacterium aquatile	Environmental organism		Deep well water	Kent, UK	172
Flavobacterium araucananum	Pathogenicity not confirmed	Kidney and external lesions	Atlantic salmon (*Salmo salar*)	Chile	857
Flavobacterium branchiophilum (*Flavobacterium branchiophila*)	Bacterial gill disease (BGD) High mortality	Anorexia, suffocation. Lesions and white-grey spots on gills. Organisms seen on gill surface	Freshwater fish: brook trout (*Salvelinus fontinalis*), rainbow trout (*Oncorhynchus mykiss*), Atlantic salmon (*Salmo salar*), chinook salmon (*Oncorhynchus tshawytscha*), sockeye salmon (*Oncorhynchus nerka*), yamame or masou salmon (*Oncorhynchus masou*), sheatfish or wels catfish (*Silurus glanis*), goldfish (*Carassias auratus*)	Canada, Hungary, Japan, Korea, USA	413 921 1280 1281 1282 1746

Continued

Table 1.2. Continued.

Bacteria	Disease	Disease signs	Aquatic species	Distribution	Ref
Flavobacterium ceti	Probable normal flora	No clinical abnormalities seen	Beaked whale (*Ziphius cavirostris*)	Canary Islands, Spain	1710
Flavobacterium chilense	Pathogenicity not confirmed	External lesions	Rainbow trout (*Oncorhynchus mykiss*)	Chile	857
Flavobacterium columnare (previously called *Cytophaga columnaris* and *Flexibacter columnaris*)	Columnaris disease, saddleback disease, systemic disease in severe cases	Yellow/brown lesions on gills. Ulcers and necrosis on body surface, 40% of infections eventually penetrate to internal tissues and organs	Freshwater fish: channel catfish, blue catfish, carp, white bass, large-mouth bass, barramundi, fathead minnow, black bullhead, salmonids (brown trout), nile tilapia, black comet, molly, eel, neon tetra, platies, tropical aquarium fish (cascudo, pacu, piracanjuba, tambaqui), zebrafish. Occurs where water temperature exceeds 14°C, particularly 25–32°C	Australia, Brazil, Finland, France, Hungary, India, Japan, Spain, USA, worldwide	166 168 244 414 426 534 946 1138 1171 1262 1343
Flavobacterium flevense (previously *Cytophaga flevensis*)	Environmental organism		Lake water	Netherlands	168 1115
Flavobacterium frigidarium	Environmental organism		Marine sediment	Antarctica	777
Flavobacterium gillisiae	Environmental organism		Environment	Antarctica	1115
Flavobacterium hibernum	Environmental organism		Freshwater lake	Antarctica	1115 1116
Flavobacterium hydatis (*Cytophaga aquatilis*)	Gill disease. Pathogenicity not determined	Isolated from gills of diseased salmon	Cultured salmonid	Europe, USA	1586
Flavobacterium johnsoniae Previously *Cytophaga johnsonae* (includes previously named *Flexibacter aurantiacus*)	False columnaris disease, gill disease, skin disease	Dermal and gill lesions	Barramundi, salmonids (rainbow trout), koi, longfin eel and other fish. Virulent by intraperitoneal injection in zebrafish (*Danio rerio*). Isolated from soil and mud	Australia, Europe, France, South Africa, UK, USA	168 284 542 867 1171 1573
Flavobacterium macrobrachii	Environmental organism		Water from prawn/shrimp (*Macrobrachium rosenbergii*) culture pond	Taiwan	1514
Flavobacterium meningosepticum	See *Elizabethkingia meningoseptica*				900
Flavobacterium oncorhynchi	Pathogenicity unknown	Isolated from liver	Diseased rainbow trout (*Oncorhynchus mykiss*)	Spain	1845
Flavobacterium (*Cytophaga*) *pectinovorum*	Environmental organism		Soil	England	172

Flavobacterium ponti	Environmental organism		Seawater		
Flavobacterium psychrophilum (previously *Flexibacter psychrophilus* and *Cytophaga psychrophila*)	Bacterial cold-water disease (BCWD), peduncle disease, rainbow trout fry syndrome (RTFS), fin rot	Erosions of the peduncle area, ulcers on scales, leads to penetration into tissues	Fish, especially fry and young fish, Atlantic salmon (*Salmo salar*), chum salmon (*Oncorhynchus keta*), coho salmon (*Oncorhynchus kisutch*), chinook salmon (*Oncorhynchus tshawytscha*), rainbow trout (*Oncorhynchus mykiss*), carp (*Cyprinus carpio, Carassius carassius*), eel (*Anguilla anguilla*), ayu (*Plecoglossus altivelis*), European perch (*Perca fluviatilis*), tench (*Tinca tinca*), sea lamprey (*Petromyzon marinus*). Occurs where water temperature is below 15°C	Australia, Canada, Chile, Denmark, Finland, France, Germany, Japan, Norway, Spain, UK, northern USA	168 169 253 335 488 1004 1051 1068 1151 1223 1224 1498 1648 1749 1798
Flavobacterium (*Cytophaga*) *saccharophilum*	Environmental organism		River water	England	1115
Flavobacterium scophthalmum	See *Chryseobacterium scophthalmum*				
Flavobacterium (*Cytophaga*) *succinicans*	Pathogenicity not confirmed	Isolated from superficial lesions on diseased fish	Salmon Found in freshwater	USA	172 328
Flavobacterium tegetincola	Environmental organism		Associated with cyanobacterial mats	Antarctica	1115
Flavobacterium tilapiae	Environmental organism		Pond water of tilapia (*Tilapia rendalli*)	Taiwan	314
Flavobacterium (*Cytophaga*) *xanthum*	Environmental organism		Mud pool	Antarctica	1115
Flexibacter canadensis	See *Solitalea canadensis*				327
Flexibacter columnaris	See *Flavobacterium columnare*				168
Flexithrix dorotheae Previously *Flexibacter aggregans*	Environmental organism		Marine environment, beach sand	Ghana	328
Flexibacter elegans	Environmental organism		Freshwater, hot spring		328
Flexibacter flexilis subspecies – *algavorum*, – *iolanthe*, – *pelliculosus*	Environmental organism		Found in freshwater, lily pond	Costa Rica	328
Flexibacter litoralis	Environmental organism		Marine and freshwater	California	328
Flexibacter maritimus	See *Tenacibaculum maritimum*				

Continued

Table 1.2. Continued.

Bacteria	Disease	Disease signs	Aquatic species	Distribution	Ref
Flexibacter ovolyticus	See *Tenacibaculum ovolyticum*				1597
Flexibacter polymorphus	Environmental isolate		Found in marine environment associated with a littoral ascidian (*Didemnum* sp.)	Mexico, USA	1016
Flexibacter psychrophilus	See *Flavobacterium psychrophilum*				168
Flexibacter roseolus	Environmental isolate		Hot spring	Costa Rica	327
Flexibacter ruber	Environmental isolate		Hot spring	Iceland	168
Flexibacter sancti	See *Chitinophaga sancti*				1017
Flexibacter tractuosus	See *Marivirga tractuosa*				1017
Francisella halioticida					
Francisella noatunensis					
Francisella philomiragia ssp. *noatunensis*	Granulomatous disease	Disease outbreaks	Farmed Atlantic cod (*Gadus morhua*)	Norway	1142 1266
Francisella tularensis	Tularamia in humans (Type A is highly virulent)		Found in mammals, rodents, rabbits and contaminated water		
Gramella echinicola	Normal flora		Sea urchin (*Strongylocentrotus intermedius*)	Sea of Japan (East Sea) Pacific Ocean	1206
Gramella marina	Normal flora		Sea urchin (*Strongylocentrotus intermedius*)	Sea of Japan (East Sea) Pacific Ocean	1205
Granulicatella adiacens and *Granulicatella elegans* (previously *Abiotrophia adiacens* and *Abiotrophia elegans*) Referred to as NVS – nutritionally variant Streptococci.	Clinical isolates. Streptococci that show satellite growth around other colonies	Normal flora of mouth, intestinal and urogenital tract. May cause endocarditis, conjunctivitis, otitis media	Humans. Require pyridoxal hydrochloride analogues (Vitamin B6) or L-cysteine HCl (*Granulicatella elegans*) for growth		352 861 1420
Granulicatella balaenopterae com. nov. (previously *Abiotrophia balaenopterae*)	Pathogenicity not determined	Isolated in pure growth from liver and kidney	Beached minke whale (*Balaenoptera acutorostrata*)	Scotland	352 990
Grimontia (*Vibrio*) *hollisae* Previously Group EF-13	1. Non-pathogenic for fish 2. Food poisoning, gastroenteritis, wound infection, bacteraemia	2. Diarrhoea, gastroenteritis, septicaemia	1. Present in fish intestine, oysters, clams, shrimp 2. Humans Has thermostable haemolysin (TDH) similar to *Vibrio parahaemolyticus*	1. Japan 2. USA	721 1170 1225

Haemophilus piscium	Re-classified as 'atypical' *Aeromonas salmonicida*. Most closely related to *Aeromonas salmonicida achromogenes*				69
Hafnia alvei	1. Haemorrhagic septicaemia. Mortalities 2. Non-virulent in experimental challenge 3. Normal flora 4. Intestinal disorders, pneumonia meningitis, abscesses and septicaemia	1. Kidney pathology	1. Cherry salmon (*Oncorhynchus masou*), brown trout (*Salmo trutta*), rainbow trout (*Oncorhynchus mykiss* 2. Gilthead seabream (*Sparus aurata*) 3. Bony bream (*Nematalosa erebi*), golden perch (*Macquaria ambigua*), spangled perch (*Leiopotherapon unicolor*) 4. Humans Ubiquitous in the environments of soil, sewage and water	1. Bulgaria, England, Japan, Spain 2. Australia	7 604 829 1258 1295 1418 1618
Hafnia paralvei	Normal flora		Bony bream (*Nematalosa erebi*), golden perch (*Macquaria ambigua*), spangled perch (*Leiopotherapon unicolor*), Welch's grunter (*Bidyanus welchi*), platypus (*Ornithorhynchus anatinus*)	Australia	786 1258
Halomonas aquamarina (synonymous with *Alcaligenes faecalis homari, Deleya aesta, Deleya aquamarina, Alcaligenes aquamarinus*. Transferred to genus *Halomonas* as *Halomonas aquamarina*)	Moribund	Softened shells, opaque areas on carapace. Organisms isolated from haemolymph	Lobsters	USA	14 75 1583
Halomonas cupida (previously *Alcaligenes cupidus* and *Deleya cupida*)	Mortalities		Blackhead seabream (*Acanthopagrus schlegelii*) fry	Japan	960
Halomonas elongata	Environmental organism		Hypersaline environments	The Netherlands	1735
Halomonas halodurans	Environmental organism		Hypersaline environments	USA, The Netherlands, Pacific Ocean	707
Halomonas marina (previously *Pseudomonas marina* and *Deleya marina*)	See *Cobetia marina*				124
Halomonas venusta (previously *Alcaligenes venustus*)	Clinical infection		Human – caused by fish bite. Present in marine waters	Maldive Islands	124 1731
Helicobacter anseris	Unknown	None	Canada geese (*Branta canadensis*)	USA	561
Helicobacter brantae	Unknown	None	Canada geese (*Branta canadensis*)	USA	561

Continued

Table 1.2. Continued.

Bacteria	Disease	Disease signs	Aquatic species	Distribution	Ref
Helicobacter canadensis	1. Unknown, probably normal flora 2. Possible enteritis and bacteraemia	1. None 2. Illness	1. Barnacle geese (*Branta leucopsis*), Canada geese (*Branta canadensis*) 2. Humans	Australia, Canada, Sweden, UK, USA	559 1750
Helicobacter cetorum	1. Organism found in dental placue, gastric fluid 2. Gastric ulcerations, gastritis 3. Faecal material PCR-positive	1. Potential reservoir for gastric infections 2. Stomach and oesophageal ulcers. Organism in glandular mucosa and the main stomach	1. Captive dolphins (*Tursiops truncatus*) 2. Dolphins (*Lagenorhynchus acutus, Stenella frontalis, Tursiops truncatus*), beluga whale (*Delphinapterus leucas*), killer whale (*Orcinus orca*), South American fur seals (*Arctocephalus australis*) 3. Yangtze finless porpoise (*Neophocaena phocaenoides asiaeorientalis*)	Argentina, China, USA, Venezuela	630 631 632 633 678 679 680 1126 1590
Helicobacter species	1, 2. Pathogenicity unknown	1. Faeces PCR-positive 2. Gastric mucosa and gastric fluid PCR-positive 3. Gastric tissues, culture and PCR-positive	1. Australian fur seal (*Arctocephalus pusillus doriferus*), New Zealand fur seals (*Arctocephalus forsteri*), Australian sea lions (*Neophoca cinerea*), California sea lion (*Zalophus californianus*), silver gull, seagull (*Chroicocephalus novaehollandiae*) 2. Australian sea lions (*Neophoca cinerea*), dolphin (*Tursiops truncatus*), false killer whale (*Pseudorca crassidens*), captive polar bear (*Ursus maritimus*) 3. Harp seal (*Pagophilus groenlandicus*)	Australia, Argentina, USA	633 681 1288 1289
Hydrogenophaga (*Pseudomonas*) *palleronii*	Environmental organism		Present in water	Germany, Russia	1801
Hydrogenophaga (*Pseudomonas*) *pseudoflava*	Environmental organism		Present in water, mud, soil	Germany	67 1801
Iodobacter fluviatilis (previously *Chromobacterium fluviatile*)	Environmental organism		Found in freshwater	Antarctic lakes, England, Scotland, Ubiquitous	1048
Janthinobacterium lividum	Anaemia	Exophthalmia, pale gills, internal symptoms	Rainbow trout Found in soil and spring water	Scotland	73

Klebsiella planticola *Klebsiella trevisanii*	See *Raoultella planticola*				462 529
Klebsiella pneumoniae	1. Fin and tail disease 2. Microflora		1. Rainbow trout 2. Mammalian tissue	UK Worldwide	406
Klebsiella ornithinolytica	See *Raoultella ornithinolytica*				462
Klebsiella oxytoca					462
Klebsiella terrigena	See *Raoultella terrigena*				462 820
Labrenzia (Stappia) alba Previously *Stappia stellulata*-like, Strain M1. Previously *Agrobacterium stellulatum*	Potential probiotic against juvenile oyster disease		Marine environment. May prevent colonization of juvenile oysters (*Crassostrea virginica* and Mediterranean oysters) by *Roseovarius crassostreae*	Spain, USA	178 194 1362
Lactobacillus ceti	Probable normal flora	None (stranded beached whale)	Beaked whale (*Ziphius cavirostris*)	Spain	1709
Lactobacillus species, especially a *Lactobacillus plantarum*-like isolate	Normal intestinal microflora of healthy fish	Intestine and digestive tract	Atlantic cod, Atlantic salmon, rainbow trout, wolffish, Arctic char	France, Norway	
Lactobacillus piscicola	See *Carnobacterium maltaromaticum*				736
Lactococcus garvieae biotypes 1–13. Previously *Enterococcus seriolicida, Streptococcus* type I and *Streptococcus garvieae*	1. Lactococcosis, Haemorrhagic septicaemia, haemorrhagic enteritis, meningoencephalitis 2. Subclinical mastitis 3. Infection, osteomyelitis	1. Bilateral exophthalmia, darkening of skin, congestion of intestine, liver, kidney, spleen, brain, distended abdomen, bloody ascites fluid in peritoneal cavity. Organism seen in heart, gills, skin, spleen, eyes, kidney 2. Milk 3. Blood, skin, urine, faeces	1. Adriatic sturgeon (*Acipenser naccarii*), black rockfish (*Sebastes schlegelii*), eel, grey mullet (*Mugil cephalus*), olive flounder (*Paralichthys olivaceus*), rainbow trout (*Oncorhynchus mykiss*), striped jack (*Pseudocaranx dentex*), yellowtail (*Seriola species*), giant freshwater prawn (*Macrobrachium rosenbergii*), turbot, bottlenose dolphin (*Tursiops truncatus*). Found in seawater, mud, intestines of wild fish e.g. Spanish mackerel, black scraper (*Thamnaconus* [*Navodon*] *modestus*). (Biotypes 1, 2, 3, 4, 5, 6, 10) 2. Cow, buffalo (Biotypes 4, 7, 8, 9) 3. Human (Biotypes 1, 2, 10, 11, 12, 13)	Australia, Europe, France, Italy, Israel, Japan, Korea, Kuwait, North America, Portugal, Saudi Arabia, Spain, South Africa, Taiwan, UK, USA	29 279 311 313 350 361 480 481 500 863 905 954 1088 1466 1524 1711 1614

Continued

Table 1.2. Continued.

Bacteria	Disease	Disease signs	Aquatic species	Distribution	Ref
Lactococcus lactis ssp. *lactis*	Normal flora	Present in intestine	Amur catfish (*Silurus asotus*), pufferfish (*Takifugu niphobles*), giant freshwater prawn (*Macrobrachium rosenbergii*)	Japan, Taiwan	810 1763
Lactococcus lactis ssp. *tructae*	Normal flora	Present in intestinal mucus	Rainbow trout (*Oncorhynchus mykiss*), brown trout (*Salmo trutta*)	Spain	1330
Lactococcus piscium	Lactobacillosis, pseudokidney disease		Rainbow trout	North America	1802
Leeuwenhoekiella (*Cytophaga*) *marinoflava*	Environmental isolate		Seawater	Scotland	172
Leptospira interrogans serovar pomona	1. Leptospirosis, renal failure 2. Leptospirosis, interstitial nephritis in adults, multiple haemorrhage syndrome in pups	1. Weight loss, dehydration, lethargy, swollen kidneys 2. Lethargy, dehydration, abortion	1. Northern elephant seal (*Mirounga angustirostris*) 2. Northern fur seal (*Callorhinus ursinus*)	Californian coast USA, Bering Sea	345 1540
Leptospira interrogans serovar pomona type kennewick	Leptospirosis, renal failure	Depression, anorexia, dehydration, swollen kidneys, loss of renule definition, renal lesions	Californian sea lion (*Zalophus californianus*)	Californian coast USA	666
Leptospira kirschneri grippotyphosa	Leptospirosis, renal failure		Harbour seal		
Listonella anguillarum	See *Vibrio anguillarum*				
Listonella ordalii	See *Vibrio ordalii*				1633
Listonella pelagia	See *Vibrio pelagius*				
Mannheimia haemolytica (previously *Pasteurella haemolytica*)	1. Ulcerative stomatitis 2. Disease 3. Haemorrhagic tracheitis		1. Reptiles 2. Sheep, goats, cattle 3. Dolphins	1. USA	1600
Marinilabilia salmonicolor biovar *agarovorans* (previously *Cytophaga salmonicolor* and *Cytophaga agarovorans*)	Environmental organism		Marine mud	California	172 328
Marinobacter hydrocarbonoclasticus (*Pseudomonas nautica*)			Found in marine environments		130
Marinobacter aquimarina					

Marinobacterium (Pseudomonas) stanieri	Environmental organism		Found in marine environment		130
Marinomonas aquimarina	Environmental organism		Oysters, seawater	Spain	1079
Marinomonas communis					
Marinomonas ostreistagni	Environmental organism		Seawater in pearl oyster pond	China	981
Mesophilobacter marinus	Environmental organism		Found in seawater	Japan, Indian Ocean	1230
Marivirga tractuosa previously *Flexibacter tractuosus*	Environmental organism		Marine and freshwater	Vietnam	327
Microbacterium (Flavobacterium) maritypicum	Environmental organism		Seawater		
Micrococcus luteus	Mortality	Pale gills, ascites fluid, gastroenteritis, internal haemorrhages	Rainbow trout (*Oncorhynchus mykiss*) fry	England	77
Moraxella species	Associated with mortalities		Striped bass	USA	138
Moritella japonica	Environmental organism		Sea bed sediment	Japan	1233
Moritella marina (Vibrio marinus)	Environmental organism		Seawater and sediment of Gulf of Naples, Italy, and North Pacific Ocean, USA	Italy, USA	153 363 1683
Moritella viscosa Previously *Vibrio viscosus*	Winter ulcer disease	Skin lesions, haemorrhages on internal organs	Atlantic salmon (*Salmo salar*), plaice (*Pleuronectes platessa*), lumpsucker, rainbow trout. Found in cold water	Cold water in Iceland, Norway, Scotland	153 154 236 1071
Mycobacterium species	Mycobacteriosis, systemic disease	Lesions on skin and kidneys. Nodules in organs, softening of spleen, kidney, liver. Granulomas in tissues	Many species of freshwater, saltwater and ornamental fish, yellowtail (*Seriola quinqueradiata*), freshwater snails, frogs, reptiles, turtles, Pacific green sea turtle, New Zealand fur seal, freshwater crocodile in Australia. Zoonotic	Worldwide	593 955 1252 1621
Mycobacterium abscessus	1. Granulomas, systemic disease 2. Mycobacteriosis 3. No clinical signs	1. No obvious clinical signs of piscine mycobacteriosis. Occasional external granuloma around buccal cavity and vent, internal granulomas 2. Nodules on internal organs 3. Organisms in liver and spleen	1. Japanese medaka (rice fish), freshwater tropical fish, black acara, goldfish, firemouth cichlid, oscar 2. Milkfish (*Chanos chanos*) 3. Cane toad (*Bufo marinus*), *Chaunus* (*Bufo*) *granulosus* **Zoonotic**	Brazil, Taiwan, USA	302 977 1163 1620

Continued

Table 1.2. Continued.

Bacteria	Disease	Disease signs	Aquatic species	Distribution	Ref
Mycobacterium chelonae	Mortality, granulomas, emaciation, exophthalmos, keratitis, skin ulcers, abnormal swimming behaviour	Multiple greyish-white miliary granuloma-like nodules in tissues, kidney, liver, spleen	Atlantic salmon (*Salmo salar*), chinook salmon (*Oncorhynchus tshawytscha*), cutthroat trout (*Oncorhynchus clarkii clarkii*), coho salmon (*Oncorhynchus kisutch*), goldfish, mountain whitefish (*Prosopium williamsoni*), yellow perch (*Perca flavescens*), yellow seahorse, snake, South African clawed frog (*Xenopus laevis*), turbot (*Scopthalmus maximus*), turtle, zebrafish (*Danio rerio*). **Zoonotic**	Australia, Canada, Portugal, Shetland Islands, Scotland, UK, USA, worldwide	57 62 235 388 403 459 654 655 775 1356 1621 1784
Mycobacterium chesapeaki	Thought to be *Mycobacterium shottsii*				593 708
Mycobacterium fortuitum	Septicaemia, mycobacteriosis	Granulomas on skin and in tissues. Seen as whitish spots on liver, kidney, heart, spleen. Emaciation, exophthalmia and inflammation of skin, internal nodular lesions	Aquarium/ornamental finfish (black acara, comet, red devil, discus fish, goldfish, gourami, guppy, humphead cichlid, neon tetra, oscar, Siamese fighting fish), Atlantic salmon, freshwater striped snakehead fish (*Channa striata*) **Zoonotic** – skin ulcers and diffuse pulmonary disease in humans	Australia, Czech Republic, Ghana, Malaysia, South Africa, Thailand, USA, worldwide	216 470 775 977 1238 1356 1365 1428 1607 1658
Mycobacterium gordonae	Mycobacteriosis	Granulomatous lesions	Charco palma pupfish (*Cyprinodon veronicae*), cardinal tetra, freshwater angelfish (*Pterophyllum scalare*), three-spot gourami (*Trichogaster trichopterus*), cockatoo cichlid (*Apistogramma cacatuoides*)	Czech Republic, UK	1012 1238 1356

Mycobacterium marinum	1, 4. Mycobacteriosis 2. Dermatitis and panniculitis in captive white whale 3. Skin lesions	1. Lesions in kidney and on skin. Nodular lesions may be systemic in all organs 1, 4. Granulomas in organs 2. Mycotic dermatitis	1. Freshwater trout, freshwater aquarium/ ornamental fish, banggai cardinal, cockatoo cichlid, crescent zoe, golden skiffia, goldfish, koi carp, pearl gourami, Odessa barb, otjikoto tilapia, Pitch lake guppy, rosy barb, Siamese fighting fish, three-spot gourami, Atlantic croaker (*Micropogonias undulatus*), hawksbill sea turtle, black seabass (*Centropristis striata*), lumpfish, rabbitfish (*Siganus rivulatus*), red drum (*Sciaenops ocellatus*), sergeant-major (*Abudefduf saxatilis*), southern platyfish (*Xiphophorus maculatus*), striated frogfish (*Antennarius striatus*), striped bass (*Morone saxatilis*), striped snakehead fish (*Channa striata*), summer flounder (*Paralichthys dentatus*), turbot (*Scopthalmus maximus*), yellowtail (*Seriola quinqueradiata*) 2. Captive white whale 3. Humans 4. Chinese soft-shelled turtle (*Pelodiscus sinensis*)	Australia, Columbia, Czech Republic, Israel, Japan, Malaysia, Portugal, Taiwan, Thailand, USA, Venezuela, worldwide, ubiquitous	61 97 207 244 440 459 711 764 771 977 1238 1365 1475 1771 1826
Mycobacterium montefiorense	Granulomatous dermatitis	Florid skin nodules – soft, gelatinous, grey- and tan-coloured around head and trunk	Green moray eels (*Gymnothorax funebris*), spotted moray eels (*Gymnothorax moringa*)	USA	719 1013
Mycobacterium neoaurum	Panophthalmitis	Ocular lesions, nodules in muscle and organs. Organism isolated with a *Rhodococcus* species	Atlantic salmon, chinook salmon **Zoonotic**	Canada	90
Mycobacterium palustre	Environmental isolate		Stream water from peatland run-off, and isolated from lymphadenitis in a child in Italy	Finland, Italy	1656
Mycobacterium peregrinum	Mycobacteriosis. Shrimp appeared healthy apart from black lesions on carapace	Multifocal, melanized nodular lesions in carapace	Pacific white shrimp (*Penaeus vannamei*). Causes skin infection in humans. Found in water and soil	USA	1162

Continued

Table 1.2. Continued.

Bacteria	Disease	Disease signs	Aquatic species	Distribution	Ref
Mycobacterium pinnipedii	Tuberculosis	Granulomatous lesions in lung, pleura, peritoneum, spleen and peripheral lymph nodes	Wild and captive Australian sea lion (*Neophoca cinerea*), southern sea lion (*Otaria flavescens*), Australian fur seal (*Arctocephalus pusillus doriferus*), New Zealand fur seal (*Arctocephalus forsteri*), South American fur seal (*Arctocephalus australis*), subantarctic fur seal (*Arctocephalus tropicalis*)	Australia, New Zealand, Uruguay, Argentina, UK	374
Mycobacterium poriferae			Marine sponge (*Halichondria bowerbanki*)		1294
Mycobacterium pseudoshottsii	Mycobacteriosis	External dermal ulcers and granulomas in spleen	Wild striped bass (*Morone saxatilis*), white perch (*Morone americana*)	USA (Chesapeake Bay and elsewhere)	708 1401 1585
Mycobacterium salmoniphilum	Mycobacteriosis	Mortalities, slow growth rate; swollen kidney and spleen	Atlantic salmon (*Salmo salar*), chinook salmon (*Oncorhynchus tshawytscha*), steelhead trout (*Oncorhynchus mykiss*)	Scotland, Norway, USA	235 1427 1783 1849
Mycobacterium simiae	Mycobacteriosis	Organism in liver and kidney	Black acara (*Cichlasoma bimaculatum*)	USA	977
Mycobacterium scrofulaceum	Mycobacteriosis	Lesions in kidney and liver. Liver white and friable	Pacific staghorn sculpin	USA	977
Mycobacterium shottsii	Mycobacteriosis	External dermal ulcers and granulomas in spleen	Wild striped bass (*Morone saxatilis*), white perch (*Morone americana*)	USA (Chesapeake Bay and elsewhere)	708 1400 1585
Mycobacterium simiae	Mycobacteriosis	Lesions in kidney and liver	Black acara Found in environmental water **Zoonotic**	USA	977
Mycobacterium smegmatis	Mycobacteriosis	Giant cells in peritoneum, granulomas	Goldfish (*Carassius auratus*)	USA	1607
Mycobacterium stomatepiae	Mycobacteriosis	Lesions in kidney and spleen	Striped barombi mbo cichlid	UK	1355
Mycobacterium triplex	Mycobacteriosis		Rosy barb	UK	1356
Mycobacterium triviale	Mycobacteriosis	Granulomatous lesions	Cockatoo cichlid (*Apistogramma cacatuoides*)	Czech Republic	1238
Mycobacterium ulcerans	1. Pathogenicity not known 2. Buruli ulcer	1. Intestine, gill positive by PCR 2. Chronic necrotic skin disease	1. *Epiplatys bifasciatus*, jewelfish (*Hemichromis bimaculatus*), mango tilapia (*Sarotherodon galilaeus galilaeus*), mollusc (*Bulinus sengalensis*), *Aplocheilichthys* sp., summer flounder (*Paralichthys dentatus*) 2. Humans	Benin, Ghana, USA	470 771 932

Mycoplasma alligatoris	1. Epizootic pneumonia, polyserositis and multifocal arthritis 2. Experimental inoculation	1. Organism found in trachea, lung, joint fluid, CSF	1. American alligator (*Alligator mississippiensis*), caiman crocodile (*Caiman latirostris*) 2. Non-pathogenic for Siamese crocodile (*Crocodylus siamensis*)	USA	229 1367
Mycoplasma crocodyli	Exudative polyarthritis	Swollen joints. Also found in lungs	Crocodiles (*Crocodylus niloticus*)	Zimbabwe	909
Mycoplasma mobile	Red disease	Gills	Tench (*Tinca tinca*)	USA	908 910
Mycoplasma phocicerebrale (previously *Mycoplasma phocacerebrale*)	1. Associated with respiratory disease. Avirulent in experimental inoculation but pathogenicity undecided 2. Pathogenicity unknown 3. Possible abortion	1. Isolated from, brain, nose, throat, lungs, heart 2. Kidney 3. Aborted fetal tissue	1. Common harbour seal (*Phoca vitulina*), grey seal (*Halichoerus grypus*) 2. Sowerby's whale (*Mesoplodon bidens*) 3. Australian fur seal (*Arctocephalus pusillus doriferus*)	Australia, North Sea, UK	88 553 616 926 1074
Mycoplasma phocidae (*Mycoplasma phocae* – name correction is not legitimate)	1. Avirulent in experimental inoculation but pathogenicity undecided 2. Possible commensal	1. Respiratory tract, flipper wound 2. Upper respiratory tract	1. Harbour seal (*Phoca vitulina*), grey seal (*Halichoerus grypus*) 3. Australian fur seal (*Arctocephalus pusillus doriferus*)	Australia, UK, USA	88 926 1074 1438
Mycoplasma phocirhinis (previously *Mycoplasma phocarhinis*)	Associated with respiratory disease	Isolated from nose, throat, trachea, lung, heart	Common harbour seal (*Phoca vitulina*)	UK, North Sea	88 616 926
Mycoplasma phylotype Mycoplasma ribotype	Possible normal gut flora		Farmed Atlantic salmon (*Salmo salar*) from Norway and Scotland and wild salmon, longjaw mudsucker (*Gillichthys mirabilis*)	Scotland, Norway, USA	116 743
Mycoplasma species	1. Unknown 2. Possible commensal	1. Diseased lung 2. Upper respiratory tract	1. Grey seal (*Halichoerus grypus*), hooded seal (*Cystophora cristata*) 2. Australian fur seal (*Arctocephalus pusillus doriferus*)	Australia, UK	88 1074
Mycoplasma testudinis	Non-pathogenic		Cloaca of tortoise	UK	728
Mycoplasma zalophi Name is yet to be validated	1. Death 2. Possible commensal	1. Pneumonia and/or arthritis 2. Upper respiratory tract	1. Californian sea lion (*Zalophus californianus*) 2. Australian fur seal (*Arctocephalus pusillus doriferus*)	Australia, USA	701 1074
Myroides (*Flavobacterium*) *odoratimimus*			Clinical specimens, hospital environments		1694

Continued

Table 1.2. Continued.

Bacteria	Disease	Disease signs	Aquatic species	Distribution	Ref
Myroides odoratus (previously *Flavobacterium odoratum*)			Clinical specimens (urine, wound swabs, leg ulcer), hospital environments.	UK, Czech Republic	753 1694
Nocardia asteroides	Nocardiosis	2. Pyogranulomatous lesions in lungs and organs	Neon tetra, rainbow trout (*Oncorhynchus mykiss*), largemouth bass, Formosa snakehead. Also causes infections in cats, cattle, dogs, fish, goats, humans 2. Hooded seal (*Cystophora cristata*), Atlantic bottlenose dolphin (*Tursiops truncatus*), killer whale (*Orcinus orca*), marine mammals	Argentina, Taiwan, USA	312 1002 1551
Nocardia brasiliensis	Nocardiosis	Pyogranulomatous lesions in lungs and organs	Beluga whale (*Delphinapterus leucas*), hooded seal (*Cystophora cristata*)	USA	1002
Nocardia brasiliensis and *Nocardia transvalensis*	Actinomycete mycetoma				1440
Nocardia crassostreae	Nocardiosis	Brown discolouration on mantle, green or yellow nodules on abductor muscle, gills, heart and mantle	Pacific oyster (*Crassostrea gigas*)	Canada, The Netherlands, USA	491 566
Nocardia cyriacigeorgica	Nocardiosis	Pyogranulomatous lesions in lungs and organs	Beluga whale (*Delphinapterus leucas*)	USA	1002
Nocardia farcinica	Nocardiosis	Pyogranulomatous lesions in lungs and organs	Hooded seal (*Cystophora cristata*), beluga whale (*Delphinapterus leucas*), killer whale (*Orcinus orca*)	USA	1002
Nocardia flavorosea	Environmental organism		Soil isolate	China	331
Nocardia levis	Nocardiosis	Pyogranulomatous lesions in lungs and organs	Atlantic bottlenose dolphin (*Tursiops truncatus*)	USA	1002
Nocardia nova			Human pathogen		1753
Nocardia otitidiscaviarum	Nocardiosis	Pyogranulomatous lesions in lungs and organs	Hooded seal (*Cystophora cristata*)	USA	1002
Nocardia salmonicida	Nocardiosis		Sockeye (blueback) salmon (*Oncorhynchus nerka*)		806 1436

Nocardia seriolae (previously *Nocardia kampachi*)	Nocardiosis	Abscesses and light-yellow coloured nodules in epidermis and tubercles and granulomas in gills, kidney, liver, heart and spleen	Grey mullet (*Mugil cephalus*), striped bass (*Morone saxatilis*), crimson snapper (*Lutjanus erythropterus*), seabass (*Lateolabrax japonicus*), largemouth bass (*Micropterus salmoides*), grouper (*Epinephelus* species), jarbua terapon (*Terapon jarbua*), snubnose pompano (*Trachinotus blochii*), spotted scaf (*Scatophagus argus*), rudderfish, yellowtail (*Seriola quinqueradiata*), Japanese flounder	China, Japan, Taiwan	312 868 944 1523
Nocardia sp.	1. Septicaemia, mortality 3. Nocardiosis	1. Small white spot lesions on dermis, muscle, gills and organs. Also found in internal organs 2. Co-isolated with *Mycobacterium marinum* and *Mycobacterium chelonae* 3. Pyogranulomatous lesions in organs	1. Atlantic salmon, farmed chinook salmon, marine fish, freshwater ornamental fish 2. Turbot (*Scopthalmus maximus*) 3. Atlantic bottlenose dolphin (*Tursiops truncatus*), beluga whale (*Delphinapterus leucas*), hooded seal (*Cystophora cristata*), leopard seal (*Hydrurga leptonyx*) Isolated from soil and plants	Worldwide, Australia, Canada, Japan, India, Portugal, Taiwan, USA	217 459 1002
Oceanimonas baumannii Previously *Oceanomonas baumannii*	Environmental organism	Degrades phenol	Estuary of River Wear	UK	232
Oceanimonas (*Pseudomonas*) *doudoroffii*	Environmental organism		Marine environment		130 232
Pantoea (*Enterobacter*) *agglomerans*	1. Pathogenicity in doubt – possible opportunist 2. Humans	1. Haemorrhages in eyes, dorsal musculature 2. Wounds, blood, urine	1. Dolphin fish (Mahi-mahi) 2. Humans. Also found in enteric tract 3. Found on plant surfaces, seeds, water. Also reported in enteric tract of deer without disease signs	USA, ubiquitous	509 602 676
Pantoea dispersa	Environmental organism		Plant surfaces, seeds, soil, environment	Ubiquitous	602
Paracoccus fistulariae	Normal flora		Bluespotted cornetfish (*Fistularia commersonii*)	Korea	903
Pasteurella multocida	1. Pneumonia, death, pleurisy (Fowl cholera) 2. Enteritis (contamination from nearby bird rookery)	1. Exudate in lungs, fluid in pleural cavity 2. Intestinal haemorrhage	1. Alligator, Californian sea lion (*Zalophus californianus*), penguins 2. Dolphins Respiratory disease in sheep, goats, cattle, rabbits	USA, worldwide	882 1092 1600
Pasteurella piscicida	See *Photobacterium damselae* ssp. *piscicida*				1641
Pasteurella skyensis	Mortality (low virulence)	Exhibit signs of loss of appetite, morbidity	Sea-farmed Atlantic salmon (*Salmo salar* L.)	Scotland	181 850

Continued

Table 1.2. Continued.

Bacteria	Disease	Disease signs	Aquatic species	Distribution	Ref
Pasteurella testudinis	Multifocal bronchopneumonia and commensal	Abscesses, lung lesions	Californian desert tortoise	USA	1553
Pedobacter heparinus (previously *Cytophaga* and *Sphingobacterium heparinum*) (*Cytophaga heparina*)	Environmental isolate		Soil. Degrades heparin		168 328
Pedobacter (*Sphingobacterium*) *piscium*	Environmental organism		Associated with frozen fish	Japan	1603
Phaeobacter gallaeciensis Previously *Roseobacter gallaeciensis*	Normal flora	Bacterial flora on larvae of scallop	Great scallop (*Pecten maximus*)	Europe	1099 1441
Phocoenobacter uteri	Pathogenicity unknown	Uterus	Harbour porpoise	UK	549
Photobacterium angustum	Environmental organism		Marine environment		128
Photobacterium damselae ssp. *damselae* (previously *Vibrio anguillarum* biovar II, *Vibrio damsela*, EF-5)	1. Vibriosis, systemic disease, granulomatous ulcerative dermatitis, deaths 2. Bloat 3. Carrier status and opportunistic infection 4. Wound infections	1. Skin ulcers in region of pectoral fin and caudal peduncle 2. Distended swimbladder 3. Organism in spleen 4. Soft tissue infection occurs due to production of cytolysin	1. Barramundi (*Lates calcarifer*), redbanded seabream (*Pagrus auriga*), seabream (*Sparus aurata*), damselfish (*Chromis punctipinnis*), eel (*Anguilla reinhardtii*), rainbow trout (*Oncorhynchus mykiss*), snapper or silver seabream (*Pagrus auratus*), turbot (*Scophthalmus maximus*), yellowtail (*Seriola quinqueradiata*), dolphin (*Tursiopsis truncatus*), ornate wobbegong (*Orectolobus ornatus*), brown shark (*Carcharhinus plumbeus*), stingray (*Dasyatis pastinaca*), tiger prawn (*Penaeus monodon*), Indian prawn (*Penaeus indicus*), oysters, leatherback turtle (*Dermochelys coriacea*), seahorses, octopus. May be part of microflora in healthy carcharhinid sharks and marine algae, Australian native and introduced fish 2. Silver seabream (*Pagrus auratus*) 3. Mullet (*Mugil cephalus* and *Liza aurata*) 4. Humans	Australia, Denmark, Europe, Hawaii, India, Italy, Japan, Spain, Tahiti, Taiwan, USA	128 339 556 570 659 881 963 1060 1170 1245 1322 1396 1461 1506 1547 1579 1641 1705 1717 1759

Photobacterium damselae ssp. *piscicida* Previousy *Pasteurella piscicida*	1. Pasteurellosis, fish pseudotuberculosis 2. Carrier status and opportunistic infection	1. Bacterial colonies and white nodules in spleen, kidney 2. Organism in spleen	1. Atlantic bluefin tuna (*Thunnus thynnus*), Japanese flounder or bastard halibut (*Paralichthys olivaceus*), gilthead seabream (*Sparus aurata*), red seabream (*Pagrus major*), seabass (*Dicentrarchus labrax*), striped bass (*Morone saxatilis*), white bass (*Morone chrysops*), sole (*Solea solea*, *Solea senegalensis*), blotched snakehead fish (*Channa maculata*), white perch (*Morone americana*), wild largescale blackfish (*Girella punctata*), yellowtail (*Seriola quinqueradiata*). Farmed and wild fish 2. Mullet (*Mugil cephalus* and *Liza aurata*)	Adriatic Sea (Croatia), France, Greece, Israel, Italy, Japan, Malta, Portugal, Scotland, Spain, Taiwan, Turkey, USA. Not in Australia or New Zealand	98 99 117 147 268 572 595 703 832 872 930 959 1089 1159 1231 1506 1551 1641 1649 1671 1828 1857
Photobacterium fischeri	See *Aliivibrio* (*Vibrio*) *fischeri* – homotypic synonym				
Photobacterium histaminum	Considered a later subjective synonym of *Photobacterium damselae* ssp. *damselae*				904 1260
Photobacterium iliopiscarium (previously *Vibrio iliopiscarius*)	Non-pathogenic	Intestine	Herring, coal fish, salmon and cod living in cold waters		1272 1682
Photobacterium leiognathi	Non-pathogenic	Light organ	Microflora of the luminous organ of a sea fish, *Leiognathus*		1388
Photobacterium logei	See *Aliivibrio* (*Vibrio*) *logei*				1685
Photobacterium phosphoreum	Environmental organism May cause spoilage of seafood		Marine environment. Symbiotic association with marine animals in light organs of teleost fishes		128
Photobacterium profundum	Environmental isolate		Isolated from deep sea sediment		1234

Continued

Table 1.2. Continued.

Bacteria	Disease	Disease signs	Aquatic species	Distribution	Ref
Photobacterium rosenbergii	Pathogenicity not determined		Bleached and apparently healthy corals (*Pachyseris speciosa*, *Merulina ampliata*, *Barabattoia amicorum*)	Australia	1636
Photobacterium swingsii	Part of normal microflora		Oyster (*Crassostrea gigas*) and haemolymph of spider crab (*Maja brachydactyla*)	Mexico	641
Planococcus citreus	Environmental isolate		Motile Gram-positive coccus isolated from seawater, marine clam and frozen boiled shrimp		1702
Planococcus kocurii	Environmental isolate		Skin of North Sea cod, fish curing brine, frozen boiled shrimp, frozen prawn	Japan	1702
Planococcus sp. Presumptive identification	Mortality	Pale gills, ascites fluid, gastroenteritis, internal haemorrhages	Rainbow trout (*Oncorhynchus mykiss*) fry	England	77 80
Planomicrobium okeanokoites (previously *Planococcus okeanokoites* and *Flavobacterium okeanokoites*)	Environmental organism		Isolated from marine mud	Japan	1192
Plesiomonas shigelloides (previously *Aeromonas shigelloides*)	Possible opportunist pathogen	Emaciation, petechial haemorrhages in intestine	African catfish, eel, gourami, rainbow trout, sturgeon. Captive penguin, aquatic reptiles. Ubiquitous in environment	Australia, Germany, Portugal	389 919
Providencia rettgeri (also known as *Proteus rettgeri*)	1, 2. Septicaemia	1. Organism isolated from internal organs, ulcerative external lesions 2. Septicaemia and meningitis	1. Silver carp (*Hypophthalmichthys molitrix*) 2. Crocodile (*Crocodylus porosus*) hatchlings Associated with poultry faeces	Australia, Israel	149 224 967
Providencia rustigianii (previously *Providencia friedericiana*)	Normal flora	Faeces	King penguin (*Aptenodytes patagonicus*, *Eudyptes chrysocome*), gentoo penguin (*Pyoscelis papua*), African or black-footed penguin (*Spheniscus demersus*), Humboldt penguin (*Spheniscus humboldti*)	Penguins in zoo in Germany	1175
Pseudoalteromonas (*Alteromonas*) *antarctica*	Environmental organism		Muddy soils and sediments in coastal areas	Antarctica	215
Pseudoalteromonas bacteriolytica	See *Algicola* (*Pseudoalteromonas*) *bacteriolytica*				1489
Pseudoalteromonas (*Alteromonas*) *citrea*	1. Environmental isolate 2. Microflora		1. Marine surface water 2. Far-eastern mussel (*Crenomytilus grayanus*) and yesso or giant ezo scallop (*Patinopecten yessoensis*), molluscs, ascidians, sponges	1. Mediterranean Sea, France 2. Sea of Japan, Bering Sea	598 812

Pseudoalteromonas (*Alteromonas*) *denitrificans*	Environmental organism			Norway Fiord coast	492
Pseudoalteromonas (*Alteromonas*) *distincta*	Microflora		Marine sponge	Komandorskie Islands, Russia	811
Pseudoalteromonas elyakovii (previously *Alteromonas elyakovii*)	1. Spot disease 2. Microflora	1. Fronds	1. Spot-wounded fronds of *Laminaria japonica* 2. Far-Eastern mussel (*Crenomytilus grayanus*)	1. Sea of Japan 2. Troitsa Bay (Russia), Sea of Japan	1491
Pseudoalteromonas (*Alteromonas*) *espejiana*	Environmental isolate		Marine environment	Californian coast	298
Pseudoalteromonas flavipulchra (previously *Pseudoalteromonas* and *Alteromonas aurantia*)	Environmental organism		Surface seawater	Mediterranean, France	600 815
Pseudoalteromonas haloplanktis haloplanktis	Environmental organism				598
Pseudoalteromonas haloplanktis tetraodonis	Mortality	Skin ulcers, viscera ejection, swollen mouth	Sea cucumber (*Apostichopus japonicus*)	China, Korea	1019
Pseudoalteromonas (*Alteromonas*) *luteoviolacea*	Environmental organism			Mediterranean, France	599
Pseudoalteromonas maricaloris	Environmental organism		Marine sponge (*Fascaplysinopsis reticulata*)	Coral Sea	815
Pseudoalteromonas (*Alteromonas*) *nigrifaciens*	Non-pathogenic		Isolated from seawater and mussels	Japan	816
Pseudoalteromonas piscicida (previously *Alteromonas piscicida*, *Pseudomonas piscicida*, *Flavobacterium piscicida*)	Diseased eggs, mortality	Whitening of egg cases	Damselfish (*Pomacentridae*) species: yellowtail (*Seriola quinqueradiata*), clownfish (*Amphiprion clarkii*) and staghorn damselfish (*Amblyglyphidodon curacao*)	Greece, Japan, Papua, New Guinea	147 242 595 1210
Pseudoalteromonas rubra (previously *Alteromonas rubra*)	Environmental organism		Marine water	Mediterranean, France	597
Pseudoalteromonas tetraodonis	Mortality		Sea cucumber (*Apostichopus japonicus*)	China, Korea	1019
Pseudoalteromonas ulvae	Environmental isolate	Inhibits germination of marine algal spores and settlement of invertebrate larvae	Marine environment. Found on the surface of marine alga *Ulva lactuca*	Australia	471
Pseudoalteromonas (*Alteromonas*) *undina*	Environmental isolate		Marine environment	Californian coast	298

Continued

Table 1.2. Continued.

Bacteria	Disease	Disease signs	Aquatic species	Distribution	Ref
Pseudomonas anguilliseptica	1. Red spot disease (Sekiten-byo) of eels 2. Haemorrhagic septicaemia 3. Associated with winter disease in seabream. Septicaemia 4. Eye lesions	1. Petechiae on body surface 2, 3. Haemorrhagic and ulcerative lesions, haemorrhages in eye, nose, operculum, brain, liver, kidney 4. Eye lesions	1. Eels (*Anguilla anguilla*, *Anguilla japonica*) 2. Atlantic salmon (*Salmo salar*), ayu (*Plecoglossus altivelis*), cod (*Gadus morhua*), rainbow trout (*Oncorhynchus mykiss*), sea trout (*Salmo trutta*), gilthead seabream (*Sparus aurata*), seabass, black spot seabream (*Pagellus bogaraveo*), striped jack (*Pseudocaranx dentex*), tilapia (*Oreochromis niloticus*), turbot (*Scophthalmus maximus*), whitefish (*Coregonus* sp.) 3. Seabream (*Sparus aurata*) 4. Baltic herring (*Clupea harengus*)	Canada, Denmark, Finland, France, Japan, Scotland, Spain, Taiwan. Not in Australia	100 176 455 477 525 956 1136 1194 1742 1794
Pseudomonas chlororaphis	1. Mortality 2. Opportunist pathogen	1. Haemorrhages, increased ascites fluid	1. Japanese amago trout (*Oncorhynchus rhodurus*) 2. Crayfish	1. Japan 2. UK	696
Pseudomonas chlororaphis ssp. *piscium*	Normal flora	Intestine	European perch (*Perca fluviatilis*)	Switzerland	252
Pseudomonas fluorescens	1. Mortalities, septicaemia. Opportunist pathogen. 2. Associated with 'atypical bacterial gill disease', water temperature <10°C	1. White nodules in spleen, abscesses in swim bladder. Fin or tail rot. 2. Bacteria on gills, bacterial cells adhere to lamellar epithelium	1. Freshwater ornamental fish – carp, red oscar, tilapia (*Oreochromis niloticus niloticus*), goldfish, crimson seabream (*Evynnis japonica*), seabream, rainbow trout 2. Cultured salmonids (rainbow trout, chinook salmon, Atlantic salmon)	1. Worldwide, Australia, Japan 2. Canada, Chile, Norway	958 1156 1280 1526
Pseudomonas species	Mortalities	Pink colour in larvae and tank water	Prawn (shrimp) (*Litopenaeus vannamei*)	Iran	1557
Pseudomonas (pseudoalcaligenes) oleovorans	Mortalities	Skin ulceration	Rainbow trout	Scotland	76
Pseudomonas plecoglossicida	Bacterial haemorrhagic ascites (BHA) – heavy mortalities	Bloody ascites	Cultured ayu (*Plecoglossus altivelis*), pejerrey or Argentinian silverside (*Odontesthes bonariensis*)	Japan	1229
Pseudomonas putida	1. Mortality 2. Mortality	1. Haemorrhagic abscess 2. Ulceration on dorsal surface	1. Yellowtail (*Seriola quinqueradiata*) 2. Rainbow trout (*Oncorhynchus mykiss*)	Japan, Turkey. Ubiquitous in environment	39 957
Pseudomonas putrefaciens	See *Shewanella algae*, *Shewanella baltica* and *Shewanella putrefaciens*				
Pseudomona stanieri	See *Marinobacterium stanieri*				130

Pseudomonas stutzeri (*Pseudomonas perfectomarina* is a junior synonym)	1. Lesions	1. Lesions	1. Octopus 2. Human, clinical material Found in wastewater, water and soil and marine environments	1. USA 2. Denmark Argentina, Spain, USA	130 746
Psychroflexus (*Flavobacterium*) *gondwanensis*	Environmental organism		Antarctic environment	Antarctica	1115
Rahnella aquatilis Previously known as *Enterobacter-cloacae-*like class H2	1. Environmental organism. 2. Clinical pathogen – rare. Bacteraemia, wound infection, urinary infection, respiratory infection		1. Isolated from water and snails 2. Human	1. France, ubiquitous, USA. 2. Korea and elsewhere	223 299
Raoultella ornithinolytica (previously known as ornithine-positive *Klebsiella oxytoca*)	Environmental organism				462
Raoultella planticola (previously *Klebsiella planticola* and *Klebsiella trevisanii*)	Environmental organism		Found in unpolluted water (drinking and surface water), soil, plants and occasionally mammalian tissue	Europe	462 529 820
Raoultella terrigena (previously *Klebsiella terrigena*)	Environmental organism		Found in unpolluted water (drinking and surface water), soil, plants and occasionally mammalian tissue. May be involved in nosocomial infections and neonatal infections	Europe	462
Renibacterium salmoninarum	1, 2. Bacterial kidney disease (BKD) 3. Carrier status, disease status unknown	1. Grey-white enlarged necrotic abscesses seen in kidney, blisters on flank, ulcers, abscesses, lesions in organs 2. Exophthalmia, ascites fluid, white nodules in kidney, liver	Salmonids: Atlantic salmon (*Salmo salar*), black sea salmon (*Salmo labrax*), brown trout (*Salmo trutta*), cutthroat trout (*Salmo gairdneri*), rainbow trout, brook trout (*Salvelinus fontinalis*), chinook salmon (*Oncorhynchus tshawytscha*), chum salmon (*Oncorhynchus keta*), coho salmon (*Oncorhynchus kisutch*), pink salmon (*Oncorhynchus gorbuscha*), sockeye salmon (*Oncorhynchus nerka*), sea lamprey (*Petromyzon marinus*) 2. Ayu (*Plecoglossus altivelis*), common shiner (*Luxilus* (*Notropis*) *cornutus*), flathead minnow (*Pimephales promelas*), North Pacific hake (*Merluccius productus*), Pacific herring (*Clupea pallasii pallasii*), sablefish (*Anoplopoma fimbria*), shiner perch (*Cymatogaster aggregata*) 3. Flathead (*Platycephalus indicus*), dragon poacher (*Percis japonicus*)	Canada, Chile, Denmark, France, Germany, Iceland, Japan, Spain, Turkey, USA, former Yugoslavia Not in Australia	115 476 504 1187 1459 1471 1472 1484

Continued

Table 1.2. Continued.

Bacteria	Disease	Disease signs	Aquatic species	Distribution	Ref
Rhodococcus erythropolis	Septicaemia	Severe peritonitis	Atlantic salmon (*Salmo salar*). Associated with vaccination	Chile, Norway, Scotland	1180 1265
Rhodococcus (luteus) fascians	Microflora	Skin and intestinal flora	Carp, soil	USSR	1212
Rhodococcus sp.	Panophthalmitis, ocular oedema. Pathogenicity not determined	Ocular lesions, nodules in muscle and organs. Found in association with *Mycobacterium neoaurum*	Atlantic salmon (*Salmo salar*), chinook salmon	Canada	90 1565
Rhodococcus qingshengii	Opportunistic pathogen	Pseudomembrane on internal organs	Atlantic salmon (*Salmo salar*). Associated with vaccination	Chile	83
Roseobacter gallaeciensis	See *Phaeobacter gallaeciensis*				1099 1441
Roseovarius crassostreae (previously known as *Roseobacter* species strain CVSP)	Juvenile oyster disease (JOD)	Reduced growth rate, fragile and uneven shell margins, cupping of left valve Internally – mantle retraction, lesions, proteinaceous deposits (conchiolin) inside shell	Hatchery reared oysters (*Crassostrea virginica*)	USA	194 195 197
Salegentibacter (*Flavobacterium*) *salegens*	Environmental organism		Hypersaline lake	Antarctica	1115
Salinivibrio (*Vibrio*) *costicola* ssp. *costicola*	Environmental organism		Hypersaline environments (salterns, cured and salted meats)	Australia, Canary Islands, Spain	582 766 1129
Salmonella enterica ssp. *arizonae*	Septicaemia	Organism isolated from organs	Piarucu (tropical freshwater fish) in zoo aquarium	Japan	923
Salmonella Durham	Carrier state	Cloacal swabs	African mud turtle, yellow-spotted Amazon turtle, helmeted terrapin, eastern box turtle, northern diamondback terrapin, Mississippi map turtle, Travancore crowned turtle, Caspian terrapin, painted turtle, red-eared turtle, stinkpot turtle	USA	1285
Salmonella enteritidis, *Salmonella* Havana, *Salmonella* Newport, *Salmonella* Typhimurium	Carrier state	Faeces	Gentoo penguin, macaroni penguin, grey-headed albatross, Antarctic fur seal	Bird Island South Georgian archipelago – Antarctica	1303

Salmonella Heidelberg, *Salmonella* Newport, *Salmonella* Oranienburg	Carrier state in healthy animals, or mild gastroenteritis	Rectal swabs	Californian sea lion (*Zalophus californianus*) pups, northern fur seal (*Callorhinus ursinus*) pups	USA	620
Salmonella group O type B	Death and illness		Sea turtles (*Chelonia mydas*)	Western Australia	
Serratia fonticola	Environmental organism		Water, spring water, soil, wild birds, gut content of slugs and snails	Ubiquitous	601 1176
Serratia liquefaciens	Mortalities, septicaemia. Opportunistic pathogen	Kidney, spleen, liver affected. Bloody ascites, haemorrhages in tissues. Redness and swelling around anus in Arctic char	Arctic char (*Salvelinus alpinus*), Atlantic salmon (*Salmo salar*), turbot (*Scophthalmus maximus*)	France, Scotland, USA (mid-Atlantic region)	1123 1575 1724
Serratia marcescens	Pathogenicity in doubt – opportunist	Kidney	Bass, trout, white perch	USA, polluted river	137
Serratia plymuthica	Pathogenicity in doubt – opportunist. Isolated from moribund trout	Organisms in internal organs	Rainbow trout. Association with pollution by domestic sewerage. Pathogenic in experimental challenge	Scotland, Spain	76 1222
Shewanella affinis Thought to be a later heterotypic synonym of *Shewanella colwelliana*	Probably normal flora		Benthic marine 'peanut' worm (*Golfingia margaritacea*), unidentified coral species	Sea of Japan, Pacific Sea	814 1481
Shewanella algae Previously identified as *Shewanella* (*Pseudomonas*) *putrefaciens* group IV, Gilardi biovar 2, CDC biotype 2	1. Pathogenic 2. Environmental isolate 3. Associated with human faeces, skin ulcers, otitis media, and found in bacteraemia	1. Whitening, shrunken muscle 2. Bacterial flora 3. Skin, blood	1. Post-larval abalone (*Haliotis diversicolor supertexta*) 2. Red algae, marine environment, oysters, seawater 3. Pathogenic to humans	1. Japan 2. Worldwide, Canada, China, Denmark, Sweden, USA	256 747 886 1239 1402 1533
Shewanella algidipiscicola	1. Normal flora, associated with food spoilage		1. Cod, plaice	Baltic Sea, Denmark	1481
Shewanella aquimarina	Environmental isolate		Seawater	Yellow Sea, Korea	1832
Shewanella baltica Previously *Shewanella* (*Pseudomonas*) *putrefaciens* Owen's group II	Environmental isolate		Oil brine, oysters, seawater	Baltic Sea, USA	1402 1853
Shewanella benthica			Intestine, holothurian (*Psychrobotes longicauda*)		209
Shewanella (*Alteromonas*) *colwelliana*	Autochthonous	Contributes to oyster larvae sediment. Adheres to surfaces such as oyster shell, glass, plastic	Estuarine oyster water. Eastern oyster (*Crassostrea virginica*)	UK	1773 1774

Continued

Table 1.2. Continued.

Bacteria	Disease	Disease signs	Aquatic species	Distribution	Ref
Shewanella corallii	Normal flora		Mucus layer from Red Sea coral (*Favia* sp.)	Israel	1527
Shewanella frigidimarina	Environmental isolate		Ice, ice algal biomass, cyanobacterial mat	Antarctica	209
Shewanella gelidimarina	Environmental isolate		Ice	Antarctica	209
Shewanella glacialipiscicola	Normal flora, associated with food spoilage		Cod	Baltic Sea, Denmark	1481
Shewanella hafniensis	1. Normal flora, associated with food spoilage 2. Possible normal flora		1. Cod, flounder 2. Oysters	Baltic Sea, Denmark, USA	1402 1480
Shewanella haliotis	Gut microflora of abalone		Abalone (*Haliotis discus*)	South Sea, Korea	895
Shewanella hanedai (*Alteromonas hanedai*)	Environmental isolate		Sediment, marine environment	Arctic	836
Shewanella japonica	Environmental isolate		Found in seawater and the mussel, (*Leukoma* sp. or *Protothaca jedoensis*). Digests agar	Coastal area of Sea of Japan	813
Shewanella marinintestina	1. Normal intestinal flora		1. Squid	Japan	1479
Shewanella marisflavi	1. Mortality 2. Possible normal flora	1. Skin ulceration, visceral ejection, mouth tumidity	1. Sea cucumber (*Apostichopus japonicus*), seawater 2. Oysters	Yellow Sea, Korea, USA	1020 1402 1832
Shewanella morhuae	Normal flora, associated with food spoilage		Cod, flounder	Baltic Sea, Denmark	1480
Shewanella oneidensis formerly *Shewanella* sp. MR-1	Environmental isolate		Lake water	Oneida Lake, New York; Lake Michigan, USA and Black Sea	1713
Shewanella pealeana	Normal flora, symbiotic	Accessory nidamental gland	Longfin inshore squid (*Doryteuthis pealeii*)		1008
Shewanella pneumatophori	Normal intestinal flora		Pacific mackerel (*Scomber japonicus*)		732
Shewanella putrefaciens Previously *Pseudomonas putrefaciens* Owens Group I, Gilardi biovar 1 and 3, CDC biotype 1	1. Septicaemia 2. Possible normal flora	1. Haemorrhagic necrosis on body. Frayed fins and exophthalmia	1. Rabbitfish 2. Oysters Found in aquatic environments including marine, sediments, oil fields, spoiling fish	Saudi Arabia, Antarctica, UK, USA, Worldwide	886 1402 1453
Shewanella sairae	Normal intestinal flora		Pacific saury (*Cololabis saira*)	Japan	1479
Shewanella schlegeliana	Normal intestinal flora		Black porgy (*Acanthopagrus schlegelii*)	Japan	1479
Shewanella upenei	Normal intestinal flora		Bensasi goatfish (*Upeneus japonicus*)	Korea	901

Shewanella woodyi	Non-pathogenic		Squid ink and seawater	Alboran Sea	1093
Solitalea (*Flexibacter*) *canadensis*	Environmental organism		Soil	Canada	327
Sphingobacterium (*mizutae*) *mizutaii* The name *Flavobacterium mizutaii* is not consistent with taxonomic opinion	Meningitis. Pathogenicity not determined	Spinal fluid	Premature infant. Natural habitat not known	Japan	1819
Sphingobacterium multivorum (previously *Flavobacterium multivorum* and Group Iik, biotype 2)	Clinical sources	Spleen	Human pathogen	UK, USA	750 752 1819
Sphingobacterium (*Flavobacterium*) *spiritivorum* *Flavobacterium yabuuchiae* is a junior synonym	Clinical isolates		Human		750 1819
Sphingomonas (*Pseudomonas*) *paucimobilis*	Environmental organism		Clinical, hospital environment, water		749
Staphylococcus aureus	Eye disease	Cornea red to opaque	Silver carp. Also found in marine environments	India	667 1509
Staphylococcus delphini	Skin lesions	Purulent skin lesions	Dolphins in captivity	Europe	1704
Staphylococcus epidermidis	1. Staphylococcosis 2. Skin microflora	1. Ulceration and haemorrhages on fins	1. Red seabream, yellow tail. Also found in marine and estuarine waters 2. Humans	1. Japan, Taiwan	667
Staphylococcus hominis	Not a known aquaculture pathogen		Reported from marine and estuarine environments and human sources		667 920
Staphylococcus lutrae	Pathogenicity not determined. May be pathogenic	Liver, spleen, lymph node	Isolated from European otters	UK	551
Staphylococcus species – *Staphylococcus capitis*, *Staphylococcus cohnii*, *Staphylococcus epidermidis*, *Staphylococcus haemolyticus*, *Staphylococcus saprophyticus*, *Staphylococcus simulans*, *Staphylococcus xylosus*	Microflora	Skin	Humans		667 920 1495
Staphylococcus warneri	1. Diseased and dying trout. Opportunistic infection 2. Associated with septicaemia, endocarditis, conjunctivitis, urinary tract and wound infections	1. Ulcerated lesions on fins and exophthalmia, ascitic fluid. Isolated from liver and kidney	1. Rainbow trout 2. Humans	1. Spain	617 667 920

Continued

Table 1.2. Continued.

Bacteria	Disease	Disease signs	Aquatic species	Distribution	Ref
Stappia stellulata-like	See *Labrenzia alba*				178
Streptobacillus moniliformis or possibly from the *Fusobacterium* group. Identification for this disease is not conclusive (Maher *et al*., 1995)	Bacterial disease	Intracellular organisms seen in tissues. Enlargement of endothelial cells of kidney glomeruli	Atlantic salmon Seen in pneumonia in koalas, cervical abscess in guinea pigs, polyarthritis in mice, tendon-sheath arthritis in turkey. Causative agent of rat-bite fever in humans	Ireland Worldwide	337 1090 1302
Stenotrophomonas maltophilia	Infectious intussusception syndrome	Enteritis, ascites, intussusception in lower intestine	Channel catfish (*Ictalurus punctatus*)	China	606
Streptococcus type I	Now recognized as *Lactococcus garvieae*				
Streptococcus agalactiae Group B Some strains were originally identified as *Streptococcus difficile* and later proved to be *Streptococcus agalactiae*	1. Mortality, exophthalmia, haemorrhage, septicaemia, meningoencephalitis 2. Mortality 3, 4. Necrotizing fasciitis	1. Haemorrhagic areas on body, mouth, fins. Organism in internal organs and brain 2. Swollen belly, lethargic, loss of appetite 3. Lethargy, necrotic skin lesions on ventral body wall and limbs 4. Cutaneous lesions, necrohaemorrhagic fasciitis, cellulitis	1. Bluefish (tailor, shad, elf), carp, cultured seabream, wild mullet, striped mullet, striped bass, pinfish (saltwater bream), Atlantic croaker, spot (croaker), weakfish, rainbow trout, sea-cultured yellowtails, stingray, sea trout, gulf menhaden, hardhead catfish, bullminnows, Nile tilapia, red tilapia, ya-fish (*Schizothorax prenanti*), aquarium fish (rams, golden shiner), frogs, mice 2. Silver pomfret (*Pampus argenteus*) 3. Captive saltwater crocodiles (*Crocodylus porosus*) 4. Captive bottlenose dolphin (*Tursiops truncatus*). Causes mastitis in cattle, infections in many other animals species, and neonatal meningitis in humans	Australia, Brazil, China, Israel, Italy, Japan, Kuwait, Malaysia, Thailand, USA	134 183 244 466 482 500 501 605 1183 1378 1589 1696 1847
Streptococcus species Group B (presumptive identification as *Streptococcus uberis* or *Streptococcus parauberis*)	Streptococcosis, bacteraemia, mortality	Deaths	Penaeid shrimp (*Litopenaeus vannamei*)	Latin America	688
Streptococcus difficile	Confirmed as *Streptococcus agalactiae* Group B, Type Ib capsular antigen				1696
Streptococcus dysgalactiae ssp. *dysgalactiae*, Group C	1. Streptococcosis 2. Pathogenicity unknown 3. Animal pathogen	2. Isolated from penis in mixed culture	1. Amerberjack (*Seriola dumerili*), yellowtail (*Seriola quinqueradiata*), Nile tilapia (*Oreochromis niloticus*) 2. Caspian seal (*Pusa* [*Phoca*] *caspica*) 2. Terrestrial animals – bovine mastitis	Brazil, UK, worldwide	357 1213 1235 1723

Streptococcus dysgalactiae ssp. *dysgalactiae* serovar L	1. Infections and septicaemia with resulting bronchopneumonia, myocarditis, osteomyelitis, pyelonephritis, abscesses 2. Isolated in conjunction with seal morbillivirus infection	1. Isolated in almost pure culture from lung, kidney, intestine, spleen	1. Isolated from harbour porpoises stranded or caught in fishing nets 2. Isolated from common or harbour seal (*Phoca vitulina*) and grey seal (*Halichoerus grypus*) with seal morbillivirus 3. Infections in cattle, dogs, pigs and other animals	Baltic Sea, North Sea, German North Sea	1601 1723 1732
Streptococcus dysgalactiae ssp. *equisimilis*, Group C, G or L	Animal and human pathogen		Animals and humans		1723
Streptococcus equi ssp. *zooepidemicus*, Group C	Possible secondary infection in phocine distemper virus outbreak	Respiratory problems	Harbour seal (*Phoca vitulina*)	German North Sea	15
Streptococcus halichoeri	Probable normal flora		Grey seal (*Halichoerus grypus*)	Scotland, Cornwall, UK	987
Streptococcus ictaluri	Bacterial infection	Streptococcal arthritis, osteolysis, myositis, spinal meningitis	Channel catfish (*Ictalurus punctatus*)	Mississippi Delta, USA	267 1519
Streptococcus iniae (*Streptococcus shiloi* is a junior synonym)	1. Deaths, meningoencephalitis 2. Golf ball disease in dolphins 3. Systemic infection 4. Wounds when cleaning fish. Deaths	1. Culture organism from the brain, kidney, liver 2. Subcutaneous abscess 3. Torticolis, gross lesions, granulomas, enlarged organs 4. Localized cellulitis, ulcers	1. Barramundi (*Lates calcarifer*), Borneo mullet (*Liza macrolepis*), channel catfish (*Ictalurus punctatus*), European seabream (*Dicentrarchus labrax*), gilthead seabream (*Sparus aurata*), pufferfish, rabbitfish (*Siganus canaliculatus*), red porgy (*Pagrus pagrus*), striped piggy (*Pomadasys stridens*), reef fish, rainbow trout (*Oncorhynchus mykiss*), stingray, tilapia (*Tilapia nilotica* × *Tilapia aurea*), cultured ayu, variegated lizard fish (*Synodus variegatus*), flying fox (*Epalzeorhynchos kalopterus*), aquarium fish, freshwater and saltwater fish 2. Freshwater Amazon dolphin (*Inia geoffrensis*) 3. Farmed bullfrogs (*Rana catesbeiana*) 4. Humans – especially elderly people and fish handlers Marine and brackish water	Australia, Bahrain, Barbados, Canada, China, Israel, Japan, South Africa, Spain, Taiwan, Thailand, USA	228 307 360 378 454 478 482 483 1112 1215 1329 1341 1342 1840 1855

Continued

Table 1.2. Continued.

Bacteria	Disease	Disease signs	Aquatic species	Distribution	Ref
Streptococcus marimammalium	1, 2. Pathogenicity unknown		1. Grey seal (*Halichoerus grypus*) 2. Common seal (*Phoca vitulina*)	Scotland	986
Streptococcus milleri (identification not certain. *Streptococcus milleri* not a recognized name)	Disease	Ulcers on flank and tail	Koi carp	UK	74 165
Streptococcus parauberis Previously *Streptococcus uberis* genotype II See also *Streptococcus* species Group B	1. Streptococcosis, mucohaemorrhagic enteritis 2. Mastitis	1. Haemorrhages in anal and pectoral fins, and eyes Organisms isolated from liver, kidney, spleen 2. Milk	1. Cultured turbot (*Scophthalmus maximus*), olive flounder (*Paralichthys olivaceus*) 2. Bovine	Korea, Spain	93 456 899 1653
Streptococcus phocae Lancefield Group, C, F or G	1. Opportunistic in viral infections 2. Isolated from stranded sea otter – pathogenicity unknown 3. Warm water Streptococcosis	1. Isolated in conjunction with seal morbillivirus; pneumonia 3. Mortality, exophthalmia, haemorrhage, pathological changes in liver, spleen, kidney	1. Harbour seal (*Phoca vitulina*), grey seal (*Halichoerus grypus*) 2. Sea otter (*Enhydra lutris nereis*) 3. Atlantic salmon (*Salmo salar*)	1. Norway, German North and Baltic seas 2. USA 3. Chile	614 1424 1538 1732
Streptococcus porcinus (previously *Streptococcus infrequens*, Lancefield Groups E, P, U, V)	Infections	Cervical lymph nodes and milk	Pigs	Worldwide	356
Streptococcus uberis see *Streptococcus* species Group B. Identification not confirmed	Streptococcosis, bacteraemia, mortality	Deaths	Penaeid shrimp (*Litopenaeus vannamei*)	Latin America	688
Streptomyces (*salmonis*) *hiroshimensis* (*Streptoverticillium salmonis* basonym)	Streptomycosis		Salmonids	USA	1436
Tenacibaculum aiptasiae	Diseased animals		Laboratory-cultured sea anemone (*Aiptasia pulchella*)	Taiwan	1761
Tenacibaculum crassostreae	Non-pathogenic	Healthy animals	Pacific oyster (*Crassostrea gigas*)	Korea	1001
Tenacibaculum dicentrarchi	Virulence not known	Isolated from diseased fish	European seabass (*Dicentrarchus labrax*)	Spain	1346
Tenacibaculum discolor	Similar to flexibacteriosis	Eroded mouth, rotten fins, shallow skin lesions and pale internal organs	Sole (*Solea senegalensis*)	Spain	1347
Tenacibaculum gallaicum	Environmental organism		Isolated from seawater from a holding tank for turbot (*Psetta maxima*) and from pond water of cultured shrimp (*Litopenaeus vannamei*)	Spain, Taiwan	1347 1517

Tenacibaculum litopenaei	Environmental organism		Isolated from water from a prawn (*Litopenaeus vannamei*) mariculture pond	Taiwan	1517
Tenacibaculum maritimum (previously *Cytophaga marina* and *Flexibacter maritimus*)	Tenacibaculosis, marine columnaris, erosive skin disease, gill lesions and ulcerative dermatitis, black patch necrosis (BPN), bacterial stomatitis (mouth rot)	Eroded mouth and fins, ulcerated skin lesions, tail rot	Marine fish, especially Atlantic salmon (*Salmo salar*), chinook salmon (*Oncorhynchus tshawytscha*), black bream (*Acanthopagrus schlegelii*), Dover sole (*Solea solea*), wedge sole (*Dicologoglossa cuneata*), flounder (*Paralichthys olivaceus*), greenback flounder (*Rhombosolea tapirina*), knifejaw or rock bream (*Oplegnathus fasciatus*), seabass (*Dicentrarchus labrax*), white seabass (*Atractoscion nobilis*), gilthead seabream (*Acanthopagrus schlegeli*), red seabream (*Pagrus major*), northern anchovy (*Engraulis mordax*), Pacific sardine (*Sardinops sagax*), striped trumpeter (*Latris lineata*), yellow-tail (*Seriola quinqueradiata*)	Atlantic, Australia, Canada, Europe, France, Japan, Scotland, Spain, USA	167 168 171 309 673 1055 1110 1127 1283 1725 1744
Tenacibaculum (*Flexibacter ovolyticus*) *ovolyticum*	Opportunist pathogen Larvae and egg mortalities	Dissolves chorion and zona radiata of the egg shells	Atlantic halibut eggs and larvae	Norway	675 1597
Tenacibaculum soleae	Mortality	Ulcers on tail and body, haemorrhagic or pale liver, petechial haemorrhages on fins	Cultured wedge sole (*Dicologoglossa cuneata*), brill (*Scophthalmus rhombus*), Senegalese sole (*Solea senegalensis*)	Spain	1056 1344
Trueperella (*Arcanobacterium*) *bernardiae*	Isolated from clinical sources especially blood and abscesses		Human-derived strains		573 1374 1827
Trueperella (*Arcanobacterium*) *pyogenes*	Associated with a variety of pyogenic conditions	Mucus membranes, tissues	Occurs in humans and animals	Worldwide	1374 1384 1827
Vagococcus fessus	Pathogenicity not determined	Organism isolated from liver, kidney	Isolated from a dead harbour seal and porpoise	Europe	762
Vagococcus fluvialis	1. Environmental organism 2. Normal flora, probiotic potential 3. Medical samples	3. blood, peritoneal fluid, wounds	1. River water, well water, chicken faeces, domestic animals (pig, cattle, cats, horse) 2. Sole (*Solea solea*), European seabass (*Dicentrarchus labrax*) 3. Humans	Spain, UK	354 1354 1560 1613

Continued

Table 1.2. Continued.

Bacteria	Disease	Disease signs	Aquatic species	Distribution	Ref
Vagococcus lutrae	Pathogenicity not determined	Organism found in blood, liver, lungs, spleen	Otter (*Lutra lutra*)	UK	988
Vagococcus salmoninarum	Vagococcosis, lactobacillosis, pseudokidney disease, peritonitis, septicaemia	Haemorrhages on gills and in ocular region, peritonitis, heart lesions, enlarged spleen, congested liver, listless behaviour	Atlantic salmon (*Salmo salar*), brown trout (*Salmo trutta*), rainbow trout (*Oncorhynchus mykiss*)	Australia, France, Italy, North America, Norway	612 1139 1442 1468 1497 1755
Varracalbmi sp. nov. (not in list of accepted names in nomenclature)	Eye lesions (bloody eye) and blindness	Lesions in kid, liver, gills, pseudobranch. Organism isolated from eyes. Skin ulcers	Atlantic salmon	Norway	1690
Vibrio aerogenes	Environmental organism		Associated with sea grass sediment, found in shallow coastal and oceanic seawaters	Taiwan	1522
Vibrio aestivus	Environmental organism		Seawater	Spain	1065
Vibrio aestuarianus	1. Mortality 2. Environmental organism	1. Haemorrhages on head, opercula, fins, necrosis of liver	1. Tongue sole (*Cynoglossus semilaevis*) 2. Found associated with shellfish (oyster, clam crab) and estuarine waters	China, USA	1645 1850
Vibrio aestuarianus spp. *francensis*	Mortality	Death, present in haemolymph	Oyster (*Crassostrea gigas*)	France	89 588
Vibrio agarivorans	Environmental organism		Seawater Abalone – pathogenicity not known	Australia, Mediterranean Sea (Spain)	244 1084
Vibrio alfacsensis	Probable normal flora		Isolated from gills, kidney and skin of healthy sole (*Solea senegalensis*); liver and kidney of spotted rose snapper (*Lutjanus guttatus*), and coral (*Mussismilia hispida*)	Brazil, Mexico, Spain	640

Vibrio alginolyticus	1. Mortality 2. Septicaemia 1. Systemic disease, ulcerative disease, necrosis, eye disease, vibriosis, scallop larval mortality 3. Disease 4. Predominate normal flora 5. Wound infections, external otitis, cellulitis, gastroenteritis	1, 2. Deaths 2. Darkened skin, haemorrhages 1. Organism isolated from organs and eye, scallop larvae 3. White spots on skin, lethargy 4. Non-pathogenic 5. Wounds on exposure to seawater or eating infected fish	1. Red abalone (*Haliotis rufescens*), small abalone (*Haliotis diversicolor supertexta*), carpet shell clam (*Venerupis* [*Ruditapes japonica*] *philippinarum*), scallop (*Argopecten purpuratus*), giant freshwater prawn (*Macrobrachium rosenbergii*), white shrimp (*Larimichthys vannamei*), kuruma prawn, tiger prawn, large yellow croaker (*Larimichthys crocea*), cobia (*Rachycentron canadum*) 2. Blackhead seabream (*Acanthopagrus schlegelii*), seabream (*Sparus auratus*), seabass (*Dicentrarchus labrax*), goldlined seabream (*Rhabdosargus sarba*), southern sea otter (*Enhydra lutris nereis*) 1. Australian native and introduced fish. Molluscs (larval and juvenile), South African abalone, eel, rainbow trout, seabream, turbot, turtles, seahorse, shrimps 3. Seahorse (*Hippocampus hippocampus*, *Hippocampus guttulatus*) 4. *Artemia* 5. Humans	Australia, Chile, China, Hong Kong, India, Japan, Mexico, worldwide Indo-Pacific and East Asia, Spain, Taiwan, Tunisia, USA	54 104 108 109 160 257 310 623 624 644 835 960 998 1021 1034 1044 1143 1155 1409 1462 1639
Vibrio (*Listonella*; *anguillarum* (serovars 01, 02, 08, 09) Previously *Vibrio anguillarum* biotype I. Most of the outbreaks are caused by serotypes 01 and 02	Vibriosis, systemic disease, ulcerative disease, necrosis	Red spots on ventral and lateral areas of fish, ulcerative skin lesions. Organisms in blood and haemopoietic tissues	1. Atlantic salmon (*Salmo salar*), chinook salmon (*Oncorhynchus tshawytscha*), coho salmon (*Oncorhynchus kisutch*), chum (*Oncorhynchus keta*), ayu (*Plecoglossus altivelis*), cod (*Gadus morhua*), eel (*Anguilla japonica*), halibut, plaice fry, sole, turbot (*Scophthalmus maximus*), Japanese flounder (*Paralichthys olivaceus*) winter flounder (*Pseudopleuronectes americanus*), lobster (*Homarus americanus*), oysters (*Crassostrea gigas*, *Ostrea edulis*), rainbow trout (*Oncorhynkiss mykiss*), seabass, red seabream (*Pagrus major*), gilthead seabream (*Sparus aurata*), octopus 2. Shrimps 3. Crabs	1. Worldwide 2. Indo-Pacific and East Asia, Spain 3. UK	109 127 212 453 714 1158 1182 1186 1323 1377 1395 1559

Continued

Table 1.2. Continued.

Bacteria	Disease	Disease signs	Aquatic species	Distribution	Ref
Vibrio areninigrae	Environmental organism		Sand	Korea	301
Vibrio artabrorum	Pathogenicity not known		Clam (*Venerupis philippinarum* and *Venerupis decussata*)	Spain	442
Vibrio atlanticus	Pathogenicity not known		Clam (*Venerupis philippinarum* and *Venerupis decussata*)	Spain	442
Vibrio atypicus	Part of microflora of digestive tract	None	Chinese prawn (*Penaeus chinensis*)	China	1767
Vibrio azureus	Environmental organism		Seawater	Japan	1836
Vibrio brasiliensis	1. Pathogenicity not known 2. Mortality	2. Deaths	1. Isolated from bivalve larvae (*Nodipecten nodosus*) 2. Pathogenic in experimental infection	Brazil	79 1632
Vibrio breoganii	Normal flora	Found in hepatopancreas, mantle, gills, gonad, extrapalial fluid	Isolated from clams (*Venerupis decussata*, *Venerupis philippinarum*)	Spain	140
Vibrio calviensis	See *Enterovibrio calviensis*				436
Vibrio campbellii	1. Pathogenic 2 Environmental organism	1. Organism in haemolymph	1. Prawns/shrimps (*Penaeus monodon*), experimental infection in dual challenge in *Litopenaeus vannamei* 2. Seawater	Florida, India, Philippines, Puerto Rico	431 671 1334 1780 1782
Vibrio campbellii-like	Mortalities in juveniles	Organism in brain, kidney, liver. Organs had lesions and haemorrhage	Hatchery-reared turbot and brill	New Zealand	450
Vibrio carchariae	Is a junior synonym of *Vibrio harveyi* Slight differences of phenotype between the strains are reported in the literature				1326
Vibrio caribbeanicus	Pathogenicity not known		*Scleritoderma cyanea*	Caribbean sea	739
Vibrio casei	Environmental (cheese)			France	190
Vibrio celticus	Mortality in pathogenicity assays	Mortality	Clams (*Venerupis philippinarum*, *Venerupis senegalensis*)	Spain	142
Vibrio chagasii	Pathogenicity not known		Found in water, rotifer (*Brachionus plicatilis*), *Artemia* sp. and gut of turbot larvae (*Scopthalmus maximus*)	Norway, Greece, Spain	1635

Vibrio cholerae (non-01)	1. Septicaemia 2. Gastroenteritis, septicaemia, cellulitis	1. Skin haemorrhages 2. Diarrhoea, vomiting	1. Ayu (*Plecoglossus altivelis*), goldfish (*Carassius auratus*), shellfish. Isolated from fresh and estuarine waters 2. Humans – consumption of raw shellfish	1. Australia, Japan. 2. Africa, Asian continent, Europe, USA, UK	393 575 769 893 1073 1382
Vibrio cholerae (01 and 0139)	Cholera	Severe, dehydrating diarrhoea and death if not treated	Humans Found in aquatic environment – surface waters	Serogroup 0139 in Bangladesh, Indian subcontinent only. Pandemic strain	17 575 769 807
Vibrio cholerae-like	Death		Lobsters	USA (organism doesn't grow at temperatures greater than 25°C)	
Vibrio cincinnatiensis	1. Environmental organism 2. Bacteraemia, meningitis 3. Mortality, disease		1. Environment, brackish, freshwater, mussels 2. Human pathogen 3. Mud crabs	1, 2. Italy, USA 3. China	220 1408
Vibrio comitans	Normal flora		Abalone (*Haliotis discus discus*, *Haliotis gigantea*, *Haliotis madaka*)	Japan	1485
Vibrio communis	See *Vibrio owensii*				269 320
Vibrio coralliilyticus Previously *Vibrio coralyticus* YB	1. Pathogen of coral 2. Mortality	1. Tissue lysis and death. Small white spots seen on coral at 3–5 days and complete destruction of tissue after 2 weeks 2. Deaths	1. Coral (*Pocillopora damicornis*). Also isolated from oyster larvae from the Atlantic Ocean, diseased oyster UK, and bivalve (*Nodipecten nodosus*) Brazil 2. Pathogenic in experimental infection of rainbow trout and *Artemia*	Atlantic Ocean, Indian Ocean and Red Sea	79 155 156 157
Vibrio coralliilyticus /*neptunius*-like	Mortality	Loss of velum, tissue deterioration	Greenshell™ mussel (*Perna canalicula*)	New Zealand	897
Vibrio crassostreae sp. nov.	Isolated from diseased oysters but pathogenicity unknown	Clinical signs not stated. Isolated from haemolymph	Cultured oysters (*Crassostrea gigas*)	France	524
Vibrio cyclitrophicus (originally spelt *Vibrio cyclotrophicus*)	Environmental organism		Eagle Harbor	Washington, USA	709
Vibrio diabolicus	Environmental organism		Isolated from polychaete annelid (*Alvinella pompejana*) in deep-sea vent	Pacific Ocean	1368

Continued

Table 1.2. Continued.

Bacteria	Disease	Disease signs	Aquatic species	Distribution	Ref
Vibrio diazotrophicus	Environmental organism		Gastrointestinal tract of sea urchins and marine animals, rotifers, surface of reeds, occurring in estuarine environments, seawater and sediments	Canada, Scotland, England	665 1718
Vibrio ezurae	1. Normal gut flora 2. Mortality	1. None 2. Death	1. Abalone (*Haliotis diversicolor aquatilis*, *Haliotis diversicolor diversicolor*) 2. Pathogenic in experimental infection of rainbow trout and *Artemia*	Japan	79 1486
Vibrio fischeri	See *Aliivibrio fischeri*				196
Vibrio fluvialis previously Group F (Furniss *et al.*, 1977) and Group EF6 (Huq *et al.*, 1980). *Allomonas enterica* shown to be *Vibrio fluvialis*	1. Mortality 2. Environment 3. Acute watery, cholera-like diarrhoea. Toxin produced	1. Organs 2. Non-pathogenic 3. Faeces	1. Plaice fry (*Pleuronectes platessa*) 2. Found in aquatic, particularly estuarine environment. Isolated from marine molluscs and crustacea 3. Humans	1. Denmark 2. Worldwide 3. Bangladesh, India, Kenya, Middle East, Philippines, Spain, Russia, Tanzania, Tunisia	576 778 855 996 1321 1504
Vibrio fortis	1. Pathogenicity unknown 2. Mortality	2. Death	1. Isolated from healthy shrimp larvae (*Litopenaeus vannamei*), healthy Atlantic salmon (*Salmo salar*), healthy and diseased bivalve larvae (*Nodipecten nodosus*) and diseased oyster larvae (*Crassostrea gigas*) 2. Pathogenic in experimental infection of rainbow trout and *Artemia*	Australia, Brazil, Mexico, UK	79 1634
Vibrio furnissii (formerly *Vibrio fluvialis* biovar 2 (aerogenic biogroup), Group F, group EF-6)	1. Pathogenicity in doubt 2. Environmental organism 3. Food poisoning, gastroenteritis	1. Haemorrhages in intestine 3. Diarrhoea	1. Eel 2. River water, animal faeces, marine molluscs and crustacean, tank water, amberjack (*Seriola dumerili*) 3. Humans Found in aquatic, particularly estuarine environment	1. Spain 2. UK, worldwide 3. Bahrain, Bangladesh, Indonesia, Peru, USA	18 222 496 576 996 1504
Vibrio gallaecicus	Healthy		Manila clam (*Venerupis* [*Ruditapes*] *philippinarum*)	North-west Spain	144
Vibrio gallicus	Non-pathogenic	Part of normal gut microflora	Abalone (*Haliotis tuberculata*)	France	1487
Vibrio gazogenes	Environmental organism		Marsh mud (saltwater)	USA	128 686
Vibrio gigantis	Isolated from diseased oysters but pathogenicity unknown	Clinical signs not stated. Isolated from haemolymph	Cultured oysters (*Crassostrea gigas*)	France	1010

Vibrio halioticoli	Non-pathogenic	Main gut microflora	Abalone	Japan	1490
Vibrio hangzhouensis	Environmental organism		Sediment	East China Sea	1818
Vibrio harveyi Organisms previcusly identified as *Vibrio carchariae* and *Vibrio trachuri* are junior synonyms of *Vibrio harveyi*	1. Mortality systemic disease, ulcerative disease, necrosis, vibriosis 2. Necrotizing enteritis 3. Gastroenteritis 4. Eye disease 5. Disease (organism originally called *Vibrio trachuri*) 6. Increased mortality in co-infection with *Edwardsiella tarda* 7. Pyogranulomatosis 8. Dermal lesions, ulcer, skin haemorrhages 10. Mortality 9. White syndrome 11. Mortality 12. Bloat 13. Pathogenicity not known 14. Wound (shark bite)	1. Lose ability to adhere. Necrotic degeneration and vacuolation in the lesion 2. Extended abdomen and red anus Fish – organism from kidney, eye. Prawns (shrimp) have black spots on exoskeleton or red spots on abdomen. Organism in lesions, eye, hepatopancreas 5. Haemorrhages in organs, exophthalmia 6. Ascites fluid 7. Granulomas in visceral organs 8. Lesions, skin ulcer 9. Tissue loss, exposure of exoskeleton 10. White pustules on foot of abalone 11. Haemorrhagic skin lesions 12. Distended swimbladder 13. Associated with diseased animals	1. Shark (brown shark, lemon shark), grouper, milkfish, common snook, crevalle jack, salmonids, barramundi, European seabass, seabream (porgy), gilthead seabream, common dentex, spadefish, wedge sole, molluscs, oysters (larval and juvenile), black tiger prawn, penaeid prawns and shrimps, swimming crab, seahorse, spiny rock lobster, sharks, octopus. Australian native and introduced fish 2. Flounder 3. Grouper, red drum, marine cobia, 4. Snook 5. Horse mackerel 6. Olive flounder (*Paralichthys olivaceus*) 7. Pufferfish (*Takifugu rubripes*) 8. Crevalle jack, pufferfish, pineapple fish 9. Corals 10. Abalone 11. Barramundi (seabass) 12. Silver seabream (*Pagrus aurata*) 12. Crown of thorns starfish (*Acanthaster planci*) 14. Humans	1. Australia, Europe, France, Indonesia, Indo-Pacific and East Asia, Indian Ocean, Japan, South Korea, Kuwait, Malaysia, New Zealand, Philippines, South America, South Korea, Spain, Taiwan, Thailand, Venezuela, USA, worldwide	20 45 109 431 450 591 592 658 659 805 818 840 939 1041 1042 1043 1053 1069 1161 1219 1228 1248 1326 1364 1376 1411 1452 1555 1562 1579 1617 1629 1811 1830

Continued

Table 1.2. Continued.

Bacteria	Disease	Disease signs	Aquatic species	Distribution	Ref
Vibrio hepatarius	Possible probiotic effect and health improvement for prawn		Hepatopancreas of wild adult prawn (*Penaeus vannamei*)	Ecuador	1634
Vibrio hippocampi	Normal flora	Faeces	Seahorse (*Hippocampus guttulatus*)	Spain	106
Vibrio hispanicus	Non-pathogenic		Isolated from *Artemia* sp. and culture water	Spain	636
Vibrio hollisae	See *Grimontia hollisae*				
Vibrio ichthyoenteri	Mortalities, intestinal necrosis	Opaque intestines	Japanese flounder larvae	Japan	804
Vibrio iliopiscarius	See *Photobacterium iliopiscarium*				
Vibrio inusitatus	Normal flora		Californian red abalone (*Haliotis rufescens*)	Japan	1485
Vibrio jasicida	1, 2. Mortality	1. Opaque body, red spots on body and pereiopods	1. Packhorse rock lobster or spiny lobster (*Sagmariasus* [*Jasus*] *verreauxi*), southern rock lobster (*Jasus edwardsii*) 2. Abalone (*Haliotis* sp.), Atlantic salmon (*Salmo salar*)	Australia, New Zealand	1835
Vibrio kanaloae	1. Pathogenicity not known 2. Mortality	2. Death	1. Seawater, shrimp (*Penaeus chinensis*) and diseased oyster larvae (*Ostrea edulis*) 2. Pathogenic in experimental infection of rainbow trout and *Artemia*	China, France, Hawaii	79 1635
Vibrio lentus	1. Disease and mortality 2. Pathogenicity not stated 3. Not pathogenic in experimental infection	1. Hard lesions in arm of head mantle	1. Octopus (*Octopus vulgaris*) 2. Associated with Mediterranean oysters 3. Turbot and seabream	Spanish Mediterranean coast	521 1082
Vibrio litoralis	Normal flora		Reported from Yellow Sea tidal flat	Korea	1202
Vibrio logei	See *Aliivibrio logei*				1685
Vibrio mangrovi	Environmental organism		Associated with the rhizosphere of wild rice (*Porteresia coarctata*) plants in mangrove	India	1373
Vibrio marinus	See *Moritella marina*				363 1683
Vibrio maritimus	Environmental organism		Found in mucus of coral (*Palythoa caribaeorum*)	Brazil	321
Vibrio marisflavi	Environmental organism		Reported from Yellow Sea	China	1760
Vibrio mediterranei	1. Non-pathogenic 2. Probiotic properties 3. Coral bleaching – Appears to be a heterogeneous species	3. Adheres and penetrates into the epidermis of the coral, *Oculina patagonica*. Culture from the coral mucus. The organism is intracellular and non-culturable	1. Plankton, sea sediments, seawater 2. Some strains are a probiotic for turbot larvae 3. Coral bleaching. Bacterium binds to the coral at elevated temperatures and disrupts the symbiotic process with the photosynthetic micro-algae endosymbionts (zooxanthellae). Heat-stable toxin produced	Spain, Mediterranean coast, Mediterranean Sea, Israel	114 952 1361 1628

Vibrio metschnikovii	1. Non-pathogenic for fish 2. Peritonitis, bacteraemia, wound infection	1. Isolated from fowl with choleraic disease	1. Found in marine and freshwater environment, especially rivers, estuaries and sewerage. Isolated from clams, cockles, oysters, lobster 2. Humans	Worldwide	994 1031
Vibrio mimicus Previously *Vibrio cholerae* Heiberg group V	1. Systemic disease and opportunist 2. Contaminant on sea turtle eggs 3. Food poisoning, gastroenteritis otitis media	1. Haemolymph, inflammation of pericardium 3. Diarrhoea, ear infection	1. Crustaceans – penaeid prawns, yabby, marron, freshwater crayfish. Barramundi 2. Reduced viability of turtle eggs and human pathogen following consumption of eggs 3. Humans – consumption of raw shellfish Found in fresh and brackish water	1, 3. Asia, Australia, Bangladesh, Canada, Guam, Mexico, New Zealand, USA, worldwide 2. Costa Rica	8 244 326 412 469 1073 1812
Vibrio species MM5	Pathogenic	High mortality, lesion on foot	Clam (*Meretrix meretrix*)	China	1841
Vibrio mytili	1, 2. Non-pathogenic	1. Normal flora	1. Mussels 2. Humans	Spain	1363
Vibrio natriegens	1. Environmental isolate 2. Pathogenicity not known	2. Associated with diseased animals	Salt marsh mud, water, oysters 2. Diseased crown of thorns starfish (*Acanthaster planci*)	Australia, UK, USA	78 1318 1411 1780
Vibrio navarrensis	Environmental isolate		Sewage and surface water	Spain	1687
Vibrio navarrensis spp. *pommerensis*	Environmental isolate		Seawater	German Baltic Sea	851
Vibrio neonatus	Normal gut flora	None	Abalone (*Haliotis discus discus*)	Japan	1486
Vibrio neptunius	1. Microflora 2, 3. Mortalities in pathogenicity tests	2. Reduced motility, sink to bottom of tank, velum abnormalities 3. Deaths	1. Dominant microflora in a recirculating system for rotifers. Isolated from healthy and diseased bivalve larvae (*Nodipecten nodosus*), gut of turbot larvae, rotifers 2. Flat oysters (*Ostrea edulis*) 3. Rainbow trout, *Artemia*	2. Spain	79 1357 1632
Vibrio nereis	1. Environmental isolate 2. Non-pathogenic		1. Seawater, 2. Black seabream (*Acanthopagrus schlegelii*)	Hawaii, Japan	126 960
Vibrio nigripulchritudo	1. Environmental isolate 2. Summer syndrome		1. Seawater, prawn pond water, sediment pore water, zooplankton from prawn ponds 2. Shrimp/prawn (*Litopenaeus stylirostris*)	USA, New Caledonia	129 625

Continued

Table 1.2. Continued.

Bacteria	Disease	Disease signs	Aquatic species	Distribution	Ref
Vibrio ordalii (previously *Vibrio anguillarum* biotype 2 and *Listonella ordalii*)	Vibriosis, bacterial necrosis and systemic disease	Organisms on muscle, skin, gills, digestive tract, heart, liver, kidneys, spleen. Necrosis and haemorrhage	Atlantic salmon (*Salmo salar*), coho salmon (*Oncorhynchus kisutch*), chum salmon (*Oncorhynchus keta*), chinook (*Oncorhynchus tshawytscha*), larval and juvenile molluscs	Australia, Chile, Japan, North-west Pacific, USA, worldwide	127 683 1377 1493 1531
Vibrio orientalis	1. Environmental isolate 2. Normal flora		1. Seawater 2. Haemolymph spider crab (*Maja brachydactyla*)	China, Spain, Canary Islands	635 1825
Vibrio owensii	1. Pathogenic to larvae in virulence studies 2. Pathogenicity not known 3. Non-pathogenic	1. Disease and mortality 2. Associated with diseased animals	1. Ornate spiny lobster (*Panulirus ornatus*) and prawns (*Penaeus monodon*) 2. Crown of thorns starfish (*Acanthaster planci*) 3. Coral (*Mussismilia hispida*, *Phyllogorgia dilatata*, *Palythoa caribaeorum*, *Palythoa variabilis*), prawn (*Litopenaeus vannamei*)	Australia, Brazil, Ecuador	270 320 1411
Vibrio pacinii	Pathogenicity not known		Isolated from healthy shrimp larvae (*Penaeus chinensis*), cod larvae (*Gadus morhua*), seabass (*Dicentrarchus*) and Atlantic salmon (*Salmo salar*)	Australia, China, Spain, UK	638 1395
Vibrio parahaemolyticus	1. Septicaemia, mortality 2. Withering syndrome 3. Food poisoning, gastroenteritis, wound infection, bacteraemia	1. External haemorrhage, tail rot, organism in internal organs, lesions of enteritis and septicaemia 2. Mass mortality, organisms found in haemolymph 3. Diarrhoea, vomiting	1. Crustaceans (shrimps, marron), mullet, octopus, tiger prawns (*Penaeus monodon*), Iberian toothcarp (*Aphanius iberus*), amberjack, eel, southern sea otter (*Enhydra lutris nereis*) 2. Cultured small abalone (*Haliotis diversicolor supertexta*) 3. Humans	1. Indo-Pacific and East Asia, Spain, Taiwan, Thailand, USA	19 244 569 853 1040 1143 1155 1434 1464
Vibrio parahaemolyticus-like	Mortality	Tissue oedema, increased mucus in gill, granules on mantle	Clam (*Meretrix meretrix*)	China	1842
Vibrio pectenicida	1. Pathogenic 2. Non-pathogenic in experimental infection	1. Moribund scallop	1. Scallop larvae (*Pecten maximus*), shrimps, prawn 2. Oyster (*Crassostrea gigas*) larvae	East Asia, Indo-Pacific, France	970 1220
Vibrio (*Listonella*) *pelagius* (previously *Vibrio pelagia* I and II, *Listonella pelagia*)	Mortalities	Erosion on fins and tail. Haemorrhages on fins and organs	Juvenile farmed turbot	Spain	1727

Vibrio penaeicida (previously *Vibrio* ᴾJ)	1. Vibriosis of prawns 2. Septicaemia. Causative agent of Syndrome 93	1. Brown spots (nodules) in lymphoid organs and gills 2. Haemolymph	1. Kuruma shrimp/prawn (*Marsupenaeus japonicus*) 2. Shrimp/prawn (*Litopenaeus stylirostris*, *Litopenaeus vannamei*)	1. Japan 2. New Caledonia, Mexico	11 371 626 803 1482
Vibrio plantisponsor	Environmental organism		Associated with the rhizosphere of wild rice (*Porteresia coarctata*) plants in mangrove	India	1372
Vibrio pomeroyi	Pathogenicity not known		Turbot (*Scopthalmus maximus*) and healthy bivalve larvae (*Nodipecten nodosus*)	Brazil, Spain	1635
Vibrio ponticus	1. Vibriosis, death 2. Pathogenicity in doubt 3. Environment	1. Skin ulceration, liver disease	1. Japanese seabass (*Lateolabrax japonicas*) 2. Gilthead seabream (*Sparus aurata*) 3. Seawater around mussel hatcheries	China, Spain	1080 1816
Vibrio porteresiae	Environmental organism		Associated with the rhizosphere of wild rice (*Porteresia coarctata*) plants in mangrove	India	1371
Vibrio proteolyticus	1. Pathogen 2. Normal flora	1. Affects microvilli, gut cells, devastates cells in body cavity	1. Juvenile *Artemia* species (brine shrimp), used as live diets for aquaculture species 2. Wood-boring crustacean (*Limnoria tripunctata*)	1. Australia, Europe 2. USA	1133 1720
Vibrio quintilis	Environmental organism		Seawater	Spain	1065
Vibrio rarus	Normal flora		Californian red abalone (*Haliotis rufescens*)	Japan	1485
Vibrio rhizosphaerae	Environmental organism		Associated with the rhizosphere of wild rice (*Porteresia coarctata*) plants in mangrove	India	947
Vibrio rotiferianus	1, 3. Pathogenicity not known 2. Mortality	2. Deaths 3. Associated with diseased animals	1. Isolated from rotifer flow-through culture system 2. Pathogenic for rainbow trout and *Artemia* in experimental infection 3. Diseased crown of thorns starfish (*Acanthaster planci*)	Australia, Belgium	79 637 1411
Vibrio ruber	Environmental isolate		Isolated from coastal seawater	Taiwan	1521
Vibrio rumoiensis	Environmental isolate		Isolated from drain pool of fish product processing plant	Japan	1843
Vibrio sagamiensis	Environmental isolate		Isolated from seawater	Japan and Equator in Pacific Ocean	1837
Vibrio salmonicida	See *Aliivibrio salmonicida*				397 473

Continued

Table 1.2. Continued.

Bacteria	Disease	Disease signs	Aquatic species	Distribution	Ref
Vibrio scophthalmi	Non-pathogenic	Part of gut microflora	Juvenile turbot	Spain	293 523
Vibrio shilonii (previously *Vibrio shiloi*, *Vibrio* species AK-1)	Considered a later synonym of *Vibrio mediterranei* (Thompson *et al.*, 2002)				114 952 1628
Vibrio sinaloensis	1. Pathogenicity unknown 2. Mortality	1. None	1. Spotted rose snapper (*Lutjanus guttatus*) 2. Prawn (*Litopenaeus vannamei*)	Mexico	543 639
Vibrio splendidus	1. Mortality in stressed oysters, summer mortality syndrome 2. Disease 3. Vibriosis 4, 5, 6. Mortality	2. Skin lesions, lethargy 3. Mortality, organism in kidney 4, 6. Mortality 5. Haemorrhages, ulcers, exophthalmia	1. Stressed oyster larvae (*Crassostrea gigas*) 2. Seahorse (*Hippocampus hippocampus*, *Hippocampus guttulatus*) 3. Corkwing wrasse (*Symphodus melops*) 4. Scallop (*Pecten maximus*), sea cucumber (*Apostichopus japonicus*) 5. Gilthead seabream (*Sparus aurata*) 6. Cod (*Gadus morhua*) larvae	China, France, Norway, Spain, UK	104 145 838 965 966 1395 1220 1766
1. *Vibrio splendidus* I 2. *Vibrio splendidus*-like	1. Septicaemia, mortalities in juveniles 2. Mortalities	1. Haemorrhages at mouth, anus and fins. Organisms in brain, kidney, liver 2. Skin haemorrhages, fin and tail rot erosion, decay of soft tissue between fins. Sample from kidney and spleen	1. Gilthead seabream, rainbow trout, turbot larvae (*Scophthalmus maximus*), shrimps (prawns). Juvenile hatchery-reared turbot and brill 2. Plaice (*Pleuronectes platessa*) fry	1. France, New Zealand, Scotland, Spain 2. Denmark	55 450 523 590 1321 1639
Vibrio splendidus II	1. Bacillary necrosis, mortality 2. Mortality	1. Larvae become inactive and settle to the bottom of the tank 2. Deaths, bacteria in velum and tissues	Japanese oyster larvae, cupped oyster (*Crassostrea gigas*) 2. Carpet shell clam (*Venerupis* [*Ruditapes*] *decussata*)	Japan, France, Spain	644 1592 1740
Vibrio stylophorae	Environmental organism		Coral, *Stylophora* species	Taiwan	1515
Vibrio superstes	Normal gut flora		Found in gut of Australian abalone *Haliotis laevigata* and *Haliotis rubra*	Australia	704

Vibrio tapetis (previously *Vibrio* P1 or VP1 or VTP)	1. Brown ring disease (BRD), high mortality 2. Vibriosis 3, 4. Mortality 5. Naturally resistant	1. Brown conchiolin deposit on inner surface of shell between pallial line and shell margins 2. Mortality, organism in kidney 3. Ulcers on tail and body	1. Clam (*Venerupis philippinarum*) 2. Corkwing wrasse (*Symphodus melops*) 3. Wedge sole (*Dicologlossa cuneata*) 4. Atlantic halibut (*Hippoglossus hippoglossus*) 5. European clam (*Venerupis decussata*)	England, France, Korea, Norway, Portugal, Scotland, Spain	33 34 201 463 838 1052 1300 1308 1393
Vibrio tapetis ssp. *britannicus*	Brown ring disease	Brown conchiolin deposit on inner surface of shell between pallial line and shell margins	Halibut (*Hippoglossus hippoglossus*), Manila clam seed (*Venerupis philippinarum*)	Ireland and UK	103
Vibrio tasmaniensis	1. Pathogenicity not known 2. Considered a pathogen	2. Disease outbreak in coral	1. Atlantic salmon (*Salmo salar*) 2. Coral (*Eunicella verrucosa*)	1. Tasmania, Australia 2. UK	1637 1706
Vibrio trachuri	See *Vibrio harveyi*				1629
Vibrio toranzoniae	Part of microflora	Pathogenicity unknown	Manila clam (*Venerupis philippinarum*) and carpet shell clam (*Venerupis decussata*)	Spain	980
Vibrio tubiashii	1. Bacterial necrosis and systemic disease 2. Mortality	1. Larvae cease swimming, tissue digestion 2. Deaths	1. Larval and juvenile molluscs (oyster, *Crassostrea gigas*, *Crassostrea virginica*, *Crassostrea sikamea*; and clam, *Mercenaria mercenaria*, *Panope abrupta*) 2. Pathogenic for rainbow trout and *Artemia* in experimental infection	Australia, UK, USA	79 489 668 1669
Vibrio variabilis	Environmental organism		Found in mucus of coral (*Palythoa caribaeorum*)	Brazil	321
Vibrio viscosus	See *Moritella viscosa*				153 1071
Vibrio vulnificus biotype I	1. Environmental 2. Infection, systemic disease, ulcerative disease, necrosis 3. Wound infection, septicaemia following consumption of raw seafood especially oysters. Septicaemia has a 50% fatality rate	1. Microflora of intestines or gills of fish, mussels, crabs and seabirds 3. Ingestion of contaminated seafood, damage to the intestinal wall. Wound infection	1. Reservoirs include clams, crabs, oysters, mussels, flounder, plankton, seabirds, marine and brackish waters 2. Cultured shrimps, molluscs – larval and juvenile, fish 3. Humans – people with high serum iron levels, immunocompromised, or with liver disease are particularly susceptible	1. Atlantic coastal regions, Denmark, Europe, Indo-Pacific and East Asia, Australia, Mexico 2. Australia, Indo-Pacific and East Asia Europe, Japan, USA 3. Australia, Belgium, Denmark, Germany, Japan, Holland, Sweden, USA	744 849 1263 1644

Continued

Table 1.2. Continued.

Bacteria	Disease	Disease signs	Aquatic species	Distribution	Ref
Vibrio vulnificus biotype 2 serovar E lipopolysaccharide-based O serogroup, 04	1. Vibriosis, haemorrhagic septicaemia 2. May cause illness and wound infection in humans following handling of eels	1. Organism isolated from gills, mucus, intestinal contents, spleen, and kidney	1. Eels – virulent and avirulent strains are found. Cultured shrimps. Found in brackish water, oysters and sediment 2. Humans	Denmark, Europe, Japan, Norway, Spain, Sweden, Taiwan	50 59 396 741 744 1263 1644
Vibrio vulnificus biotype 3	Severe wound infection and bacteraemia	Wounds and blood	Humans	Israel	182 1189
Aliivibrio (*Vibrio*) *wodanis*	See *Aliivibrio wodanis*				
Vibrio xiamenensis	Environmental organism		Found in mangrove soil	China	581
Vibrio xuii	1. Pathogenicity not known 2. Non-pathogenic		1. Isolated from shrimp culture water, white shrimp (*Litopenaeus vannamei*) and bivalve larvae (*Nodipecten nodosus*) and apparently healthy cod (*Gadus morhua*) larvae 2. Present in rotifer culture	China, Brazil, Ecuador, Norway	241 1632
Vogesella lacus	Environmental organism		Isolated from pond water of cultured soft-shelled turtle (*Pelodiscus sinensis*)	Taiwan	325
Weissella ceti	Pathogenicity unknown	Stranded whale – no lesions	Beaked whale (*Mesoplodon bidens*)	Spain	1708
Weissella hellenica strain DS-12	Potential probiotic	Non-pathogenic	Part of intestinal microflora of flounder. Species of *Weissella* are found in fermented foods such as fish and sausages	Korea	260
Weissella species	Haemorrhagic septicaemia	Haemorrhage on fins, eye	Rainbow trout (*Oncorhynchus mykiss*)	Brazil, China	533 1037
Wolinella species (putative *Candidatus Wolinella africanus*)	Pathogenicity unknown	Gastrointestinal disease	Australian sea lion (*Neophoca cinerea*)	Australia	1290
Wolinella succinogenes	Pathogenicity unknown	Gastrointestinal disease			
Yersinia aldovae Previously *Yersinia enterocolitica* Group X2	Environmental organism		Isolated from drinking water, river water, soil and fish		161
Yersinia bercovieri previously *Yersinia enterocolitica* biogroup 3B	Environmental organism		Isolated from freshwater and soil		1769
Yersinia enterocolitica ssp. *enterocolitica*	Generally non-pathogenic for mammals		Sucrose negative strains from small rodents. Found in terrestrial and aquatic habitats	Japan, Europe, Canada, USA	225 862

Yersinia frederiksenii	Environmental organism		Found in water. Fish, humans, mammals and birds may be healthy carriers		866
Yersinia intermedia	Environmental organism		Atlantic salmon. Found in water. Fish and mammals may be healthy carriers	Australia	866
Yersinia kristensenii	Environmental organism		Found in water, soil, animals. Fish, humans and mammals may be healthy carriers Sucrose-negative strains	Australia, Denmark, France, Germany, Greece, Japan, Norway, UK, USA	162 866
Yersinia mollaretii (previously *Yersinia enterocolitica* biogroup 3A)	Environmental organism		Isolated from freshwater and soil		1769
Yersinia rohdei	Environmental organism		Surface water, human faeces, dog faeces	California, Germany	28
Yersinia ruckeri	1. Enteric redmouth disease (ERM) – severe form of the disease Yersiniosis – milder form of the disease. Systemic disease 2. Salmonid blood spot 3. Non-pathogenic	1. Reddening of throat and mouth, haemorrhages on gills, fin. Isolated from kidney 2. Blood spots in eyes and haemorrhages in musculature of the caudal peduncle Organisms in liver and kidney. 3. Microflora of intestinal contents	1. Farmed and wild Atlantic salmon (*Salmo salar*), channel catfish (*Ictalurus punctatus*), goldfish, brown trout (*Salmo trutta*), rainbow trout (*Oncorhynchus mykiss*), perch (*Perca fluviatilis*), roach (*Rutilus rutilus*), whitefish (*Coregonus* sp.), eel, otter, freshwater ornamental and marine fish. Disease is usually associated with poor environmental conditions 2. Hatchery-reared trout and Atlantic salmon. Partially cross reacts with serovar I and was termed serovar I', however, proved to be serovar III 3. Seabirds	Alaska, Australia, Bulgaria, Canada, Denmark, Finland, France, Germany, Greece, Italy, New Zealand, Norway, South Africa, Scotland, Spain, Switzerland, Turkey, UK, USA 2. Australia	249 332 402 409 420 427 511 1045 1332 1429 1581 1692
Zobellia galactanivorans Previously *Zobellia galactanovorans*	Environmental organism	Associated with red alga (*Delesseria sanguinea*)	Marine environment	Brittany, France	119
Zobellia uliginosa (previously *Cytophaga* and *Flavobacterium uliginosum*)	Environmental organism		Marine environment		119

'Pathogenicity not determined' means that the organism was isolated from a dead or diseased animal, but that Koch's postulates for disease was not determined.

1.3 Taxonomy and Identification of Bacteria

Taxonomy is the classification, identification and naming (nomenclature) of organisms and the placement into a Kingdom, Phylum, Class, Order, Family, Genus and Species based on morphological and phenotypic characteristics and DNA–DNA relatedness. A Type strain is designated to represent the species and to compare new species. Identification to species level using biochemical identification tests can be difficult due to phenotypic diversity within a species or because of phenotypic similarity to other species. However, for many years morphological and phenotypic characteristics were the only methods available to the diagnostic laboratory for the identification of bacteria isolated from samples. Biochemical identification is an important tool for the diagnostic laboratory and new technologies such as matrix assisted laser desorption ionization time of flight (MALDI-TOF) mass spectrometry available commercially from companies such as Bruker Daltonics and bioMérieux offer exciting advances for the diagnostic laboratory (Dekker and Branda, 2011; Neville *et al.*, 2011). The technology is based on identifying bacteria and fungi by their protein profile. Biochemical methods rely on the fermentation, hydrolysis or carbon source utilization of a specific substrate, whereas protein profiling using mass spectrometry overcomes the shortcomings of traditional biochemical methods such as the length of time taken to complete the reaction, or false results due to incorrect growth conditions such as suboptimal NaCl concentration or growth temperature. Identification of a 24-h colony takes 45 s using the mass spectrometer. Current databases can identify more than 4000 species of bacteria and fungi including an extensive number of organisms commonly isolated in veterinary and aquatic microbiology laboratories. Unfortunately, at present the cost is inhibitory to the general diagnostic laboratory, however the technology allows the inoculum (1 μl from a bacterial colony mixed into matrix solution) to be prepared on specific cartridges off-site and sent through the post to a processing laboratory.

Molecular technologies such as polymerase chain reaction (PCR) and sequencing are more widely used now by diagnostic laboratories and are cheaper than a few years ago. PCR and sequencing can be used alone or in conjunction with biochemical methods to identify an organism.

Research laboratories are using phylogenetics, which examines the evolutionary relationship between organisms through molecular sequencing, and are applying this to relationships between species within a genus and attempting to improve upon biochemical identification. Even molecular identification is not straightforward and a PCR must be validated for specificity and sensitivity before it can be used in the diagnostic laboratory. Virulence genes and housekeeping genes may be present in a number of species within a genus and often these genes contain conserved and diverse regions. It is crucial that the optimum area of sequence is chosen for primer design. For interspecies identification, primers must be selected from a conserved region of the genome of the target species so as to detect and identify all strains within that target species, yet designed from a diverse region to allow differentiation from other species that harbour the same gene. For some genera the 16S rRNA gene is so similar between species that it offers little value for diagnostic identification unless used in conjunction with other identification methods. However, the 16S rRNA gene can assist or confirm biochemical identification results in placing an unknown bacterium into a genus and in some cases to species level and for a number of years has been used to assist taxonomy (Tindall *et al.*, 2010).

Strains belonging to a species have ≥70% DNA–DNA hybridization (DDH) relatedness with 5°C or less change in melt temperature (ΔT_m) (Wayne *et al.*, 1987), a related G+C (guanine cytosine) content and >97% similarity in the 16S rRNA sequence (Stackebrandt *et al.*, 2002). DDH measures hybridization efficiency between two sequences and not the sequence identity. A DDH of 70% corresponds to 95% average nucleotide identity (ANI); that is, the average nucleotide identity shared between two strains (Konstantinidis *et al.*, 2006). An ANI between two strains corresponds to >97% sequence identity in the 16S rRNA gene (Goris *et al.*, 2007).

Nomenclature of Bacteria

Certain rules govern the nomenclature of bacteria as stated in the International Code of Nomenclature of Bacteria (Lapage *et al.*, 1992). The Code was updated to replace the word Bacteria with Prokaryote and to include certain amendments as published in Labeda (2000). The date of validation of a new bacterium is considered to be the date of publication in the *International Journal of Systematic and Evolutionary Microbiology* (IJSEM). The International Code of Nomenclature of Prokaryotes is updated with each new volume of IJSEM and all organism names with relevant publications can be viewed online (http://www.bacterio.cict.fr). This List of Prokaryotic Names with Standing in Nomenclature should be consulted on a regular basis as in the aquatic world new bacteria are being discovered, identified and named at an increasing rate. Some are then reassigned to a different or newly described genus, while others change their species name as more research and study on the taxonomy follows.

Chapter Outline

The following section provides details on the current taxonomic status of a genus or species, a brief description of the disease, the aquatic species in which the organism has been identified and the phenotypic characteristics and identification tests, which are to be used in conjunction with the biochemical identification tables, and molecular identification. Other texts should be referred to for a comprehensive coverage of diseases. See Further Reading and Other Information Sources.

The bacterial species covered in this manual are part of the Domain *Bacteria*, Phylum BXII *Proteobacteria* (Class I *Alphaproteobacteria* – *Brucellaceae*, Class II – *Betaproteobacteria*) etc. Bacteria are grouped under the two main headings: Gram-negative rods and cocci, and Gram-positive rods and cocci. Within these two broad groups the bacteria are listed alphabetically under Phylum then Class, Family and Genus. Under Family, the genera and species are listed alphabetically.

A table at the beginning of each broad grouping (Tables 1.3 and 1.11, respectively) details the phylum, class, order, family and genus.

GRAM-NEGATIVE RODS AND COCCI

Phylum *Proteobacteria*
Class I Alphaproteobacteria
Order *Rhizobiales*

1.4 *Brucellaceae*

1.4.1 *Brucella*

Brucella are part of the Domain *Bacteria*, Class I *Alphaproteobacteria* (the alpha 2 subdivision of the *Proteobacteria*), Phylum BXII *Proteobacteria*, Order VI *Rhizobiales* and Family III *Brucellaceae*. Members of the *Brucella* genus are aerobic, Gram-negative cocci, coccobacilli or short rods, facultatively intracellular in the cells of mammals and have a DNA–DNA hybridization result of greater than 90%.

Brucella species cause abortion and reproductive disease and general infection in humans, terrestrial animals (camels, cattle, deer, desert wood rat, dogs, goats, hare, horses, pigs, reindeer, rodents, sheep and the common vole) and in marine mammals (cetaceans, pinnipeds, whales, otters, polar bears). *Brucella* species are generally host specific but can infect several animal species including humans. Currently, ten recognized species exist: *Brucella abortus* (bovine, human), *B. canis* (canine), *B. melitensis* (ovine, caprine), *B. neotomae* (desert wood rat), *B. ovis* (ovine), *B. suis* (porcine), *B. microti* (vole) and *B. inopinata* (human), and those described from marine mammals, *B. ceti* and *B. pinnipedialis* (Corbel and Banai, 2005). The *Brucella* species isolated from marine mammals are zoonotic (Brew *et al.*, 1999) and must be dealt with in a Class III Biological Safety Cabinet. The organism does not multiply outside the host, but can remain viable for years in frozen tissue (abortion material), or for months in moist conditions at 10–15°C, or for hours at 45–50°C. Transmission occurs between hosts by direct contact with infected material by ingestion,

Table 1.3. Taxonomy of Gram-negative rods and cocci listed in this manual.

Phylum	Class	Order	Family	Genera
Proteobacteria	I. *Alphaproteobacteria*	*Rhizobiales*	*Brucellaceae*	*Brucella*
		Rickettsiales	*Rickettsiaceae*	*Rickettsia*
		Rhodobacterales	*Rhodobacteraceae*	*Labrenzia*, *Paracoccus*, *Phaeobacter*, *Roseobacter*, *Roseovarius*, *Stappia*
	II. *Betaproteobacteria*		I. *Burkholderiacea* II. *Oxalobacteraceae* III. *Alcaligenaceae*	I. *Burkholderia* II. *Janthinobacterium* III. *Alcaligenes*, *Bordetella*
		IV. *Neisseriales*	*Neisseriaceae*	*Andreprevotia*, *Chromobacterium*, *Iodobacter*, *Kingella*, *Neisseria*, *Vogesella*
	III. *Gammaproteobacteria*	III. *Xanthomonadales*	*Xanthomonadaceae*	*Stenotrophomonas*
		V. *Thiotrichales*	III. *Francisellaceae*	*Francisella*
		VIII. *Oceanospirillales*	I. *Oceanospirillaceae* IV. *Halomonadaceae*	I. *Marinomonas* IV. *Halomonas*, *Deleya*
		IX. *Pseudomonadales*	I. *Pseudomonadaceae* II. *Moraxellaceae*	I. *Pseudomonas*, *Chryseomonas*, *Mesophilobacter* II. *Moraxella*, *Acinetobacter*, *Psychrobacter*
		X. *Alteromonadales*	*Alteromonadaceae*	*Alteromonas*, *Colwellia*, *Ferrimonas*, *Marinobacter*, *Marinobacterium*, *Moritella*, *Pseudoalteromonas*, *Psychromonas*, *Shewanella*
		XI. *Vibrionales*	*Vibrionaceae*	*Aliivibrio Vibrio*, *Enterovibrio*, *Grimontia*, *Listonella*, *Photobacterium*, *Salinivibrio*
		XII. *Aeromonadales*	*Aeromonadaceae*	*Aeromonas*
		XIII. *Enterobacteriales*	*Enterobacteriaceae*	*Escherichia*, *Citrobacter*, *Edwardsiella*, *Enterobacter*, *Hafnia*, *Klebsiella*, *Morganella*, *Plesiomonas*, *Proteus*, *Rahnella*, *Salmonella*, *Serratia*, *Shigella*, *Yersinia*
		XIV. *Pasteurellales*	*Pasteurellaceae*	*Pasteurella*, *Actinobacillus*, *Gallibacterium*, *Haemophilus*, *Mannheimia*, *Phocoenobacter*
	IV. *Epsilonproteobacteria*	I. *Campylobacterales*	I. *Campylobacteraceae* II. *Helicobacteraceae*	I. *Arcobacter*, *Campylobacter* II. *Helicobacter*, *Wolinella*
Bacteroidetes	I. *Bacteroidetes*	I. *Bacteroidales*	I. *Bacteroidaceae* II. *Rikenellaceae*	I. *Bacteroides* II. *Marinilabilia*
	II. *Flavobacteria*	I. *Flavobacteriales*	I. *Flavobacteriaceae* II. *Myroidaceae*	*Flavobacterium*, *Aequorivita*, *Arenibacter*, *Bergeyella*, *Capnocytophaga*, *Cellulophaga*, *Chryseobacterium*, *Coenonia*, *Croceibacter*, *Empedobacter*, *Gelidibacter*, *Myroides*, *Ornithobacterium*, *Polaribacter*, *Psychroflexus*, *Psychroserpens*, *Riemerella*, *Saligentibacter*, *Tenacibaculum*, *Weeksella*, *Zobellia* II. *Myroides*
	III. *Sphingobacteria*	I. *Sphingobacteriales*	I. *Sphingobacteriaceae*	I. *Sphingobacterium*
			III. *Flexibacteraceae*	I. *Flexibacter* III. *Cytophaga*
Fusobacteria	I. *Fusobacteria*	I. *Fusobacteriales*	I. *Fusobacteriaceae*	I. *Fusobacterium* VI. *Streptobacillus*

inhalation or contact with mucous membranes or damaged skin (Nymo *et al.*, 2011).

The taxonomy of *Brucella* has been controversial. Traditionally, new species were named according to host specificity with species identification and differentiation based on phenotypic tests such as metabolic activity, growth differences, phage type and antigenic reactions. However, Verger *et al.* (1985) proposed that the genus is monospecific according to the definition for species using DNA–DNA homology, which states that species within a genus are 50–70% homologous. The 90% DNA homology between the *Brucella* species means the genus is monospecific and only one species, *Brucella melitensis*, should be recognized in the genus with the other species recognized as biovars, thus *B. melitensis* biovar *abortus*, *B. melitensis* biovar *canis*, *B. melitensis* biovar *melitensis*, *B. melitensis* biovar *neotomae*, *B. melitensis* biovar *ovis* and *B. melitensis* biovar *suis* (Verger *et al.*, 1985). This is complicated because biovars already have been designated for several species, e.g. *B. suis* biovars 1, 2, 3, 4 and 5. Initially, the name *Brucella maris* comprising three biovars was proposed for isolates from seals, porpoises, dolphins, an otter and a whale (Jahans *et al.*, 1997), however the use of this name was disputed. Molecular investigation using PCR-restriction fragment length polymorphism of two outer membrane protein (OMP) genes, *omp2* and *omp2b* coding for porin genes, and IS*711*-based DNA fingerprinting together with phenotypic findings revealed the isolates from marine sources were a heterogeneous group (Bricker *et al.*, 2000; Cloeckaert *et al.*, 2001; Moreno *et al.*, 2002). Marine mammal isolates have a higher copy number of IS*711* compared to isolates from terrestrial animals and a specific marker for marine isolates exists for an IS*711* element downstream for the *bp26* gene (Clavareau *et al.*, 1998; Cloeckaert *et al.*, 2001). Polymorphism in the *omp2* locus divided the marine isolates into two groups, one group carrying one copy of each *omp2a* and *omp2b* genes (seal and otter isolates) and a second group (dolphin, porpoise, whale isolates) carrying two copies of the *omp2b* gene (Clavareau *et al.*, 1998; Cloeckaert *et al.*, 2001). Terrestrial isolates carry two copies of the *omp2a* gene. The subcommittee on the taxonomy of *Brucella* voted to return to traditional naming of *Brucella* based on host specificity (Osterman and Moriyon, 2006). Currently accepted names are *B. pinnipedialis* (originally *B. pinnipediae*) for strains isolated from pinnipeds (seals), and *B. ceti* (originally *B. cetaceae*) for strains isolated from cetaceans (whales, dolphins and porpoises) (Cloeckaert *et al.*, 2001; Foster *et al.*, 2002, 2007; Moreno *et al.*, 2002; Corbel and Banai, 2005). Further investigation using multilocus sequence analysis (MLSA), variable number tandem repeats (VNTRs) and multiple loci variable number tandem repeats analysis (MLVA), the latter termed MLVA-16 because of the 16 loci determined for the *Brucella* genus, confirmed the division of the marine isolates and the finding of three main groups with distinct host preferences (Groussaud *et al.*, 2007; Maquart *et al.*, 2009). Group one corresponds to *B. pinnipedialis* from seals and Group two comprises isolates of *B. ceti* from dolphins and porpoises, which is further divided into two distinct groups. The two groups within *B. ceti* are more similar to the group *B. pinnipedialis* than they are to each other. The naming of isolates according to the traditional naming based on host preferences has led to the suggestion of names such as *B. phocae* for seal isolates, *B. phoecoenae* for porpoise isolates and *B. delphini* for dolphin isolates, however these names are yet to be validated (Corbel and Banai, 2005; Groussaud *et al.*, 2007).

Brucella ceti

TAXONOMY AND SIGNIFICANCE. *Brucella ceti* (NCTC 12891^T=BCCN 94-74^T) was isolated from a skin lesion on a harbour porpoise (*Phocoena phocoena*) (Ross *et al.*, 1994; Foster *et al.*, 2007). The organism, like other members of the *Brucella* genus, is host specific and has been isolated from cetaceans: dolphins, porpoise and a whale (Foster *et al.*, 1996a , 2002; Clavareau *et al.*, 1998). The DNA–DNA relatedness to other *Brucella* species is >77%.

Molecular information suggests further division of isolates in the *B. ceti* group with subsequent delineation based upon host preferences such that isolates from dolphins would be classified as *B. delphini* and those from porpoises as *B. phoecoenae* (Corbel and Banai, 2005; Groussaud *et al.*, 2007).

An isolate isolated from the spleen and liver from a minke whale (*Balaenoptera acutorostrata*) fits the description of *B. ceti*, as the isolate possesses two copies of the *omp2b* gene and phenotypically resembles other *B. ceti* isolates from dolphins and porpoises except for a negative reaction for growth in the presence of safranin (Clavareau *et al.*, 1998). A *Brucella* species isolated from the brain of a European striped dolphin (*Stenella coeruleoalba*) from the coast of south-west England was identified as *B. ceti* and was positive for the IS*711* element common to marine mammal isolates but was type N(K) based on outer membrane (omp2) restriction digest, a common finding for dolphins. If the host-specific nomenclature was adopted then this isolate would be identified as *B. delphini* (Davison *et al.*, 2009). The organism was isolated in pure growth from the brain, and the animal exhibited non-suppurative meningoencephalitis. Cultures from heart blood, lung, liver and kidney were negative for *Brucella*. Similarly, an isolate, probably *B. ceti/delphini*, was isolated from a purulent abscess within blubber taken around the vent of a bottlenose dolphin (*Tursiops truncatus*) found dead on the south-west coast of England and was positive for IS*711* and omp type N(K) (Dawson *et al.*, 2006).

GROWTH CHARACTERISTICS AND PHENOTYPIC IDENTIFICATION. *Brucella ceti* is an aerobic, oxidase-positive, non-motile, Gram-negative coccus, coccobacillus or short rod (0.7–0.5×0.6–1.5 μm) occurring singly, in pairs, short chains or small groups. On Farrell's medium, Farrell's modified medium and Columbia sheep blood agar (CSBA) colonies are 0.5–1.0 mm raised, convex, circular and entire at 3–4 days at 37°C. On serum-glucose agar colonies are transparent, shiny, raised, convex, entire, and when viewed by transmitted light colonies are a pale honey colour. Non-haemolytic on CSBA. Growth in CO_2 is not required by the majority of strains. Growth is improved by the addition of serum or blood and occurs between 20 and 40°C with an optimum of 37°C. Catalase and urease are positive. L-glutamic acid, L-arabinose, D-galactose, D-ribose and D-xylose are oxidized, whereas L-alanine, L-asparagine, L-arginine, DL-ornithine and L-lysine are not oxidized. Results for *meso*-erythritol are variable between strains. Cultures are lysed by Iz and Wb phages but not by Tb and R/C phages (Foster *et al.*, 2007). A antigen is dominant. Both *B. ceti* and *B. pinnipedialis* give a numerical profile of 1200004 in the API 20NE. The manufacturer's database will incorrectly identify this as 'good identification for *Moraxella phenylpyruvica*' (Foster *et al.*, 1996a). These and other reactions are found in Tables 3.5 and 3.27–29.

Brucella cetorum

Brucella cetorum originally was proposed for *Brucella* species isolated from cetaceans, however this name was later changed to *B. ceti* (Foster *et al.*, 2002, 2007).

Brucella maris

The name *B. maris* and three biovars was proposed for isolates from marine mammals (Jahans *et al.*, 1997), however the name was not accepted due to the heterogeneous nature of the isolates from marine mammals (Cloeckaert *et al.*, 2001; Moreno *et al.*, 2002).

Brucella pinnipedialis

TAXONOMY AND SIGNIFICANCE. *Brucella pinnipedialis* (NCTC 12890^T=BCCN 94-73^T) was isolated from the spleen of a common seal (*Phoca vitulina*) off the coast of Scotland (Ross *et al.*, 1994; Foster *et al.*, 2007) and from lymph nodes of ringed seals (*Pusa* [*Phoca*] *hispida*) and a harp seal (*Phoca groenlandicus*) with brucellosis caught off the coast of Canada (Forbes *et al.*, 2000). Healthy hooded seals (*Cystophora cristata*) sampled in the ocean between Norway and Greenland were positive for *B. pinnipedialis* but ringed seals (*Pusa* [*Phoca*] *hispida*) sampled in the same area were negative (Tryland *et al.*, 2005). The organism was isolated from lung, lymph nodes and spleen.

A *Brucella* species, probably *B. pinnipedialis*, was isolated from a Pacific harbour seal (*Phoca vitulina richardsi*) in Puget Sound, Washington, USA. The seal was also infected with lungworm (*Parafilaroides* species) from which *Brucella* was also isolated (Garner *et al.*, 1997). The lungworms may be a source of transmission between animals.

Brucella pinnipedialis was isolated from the internal iliac lymph node from a European otter (*Lutra lutra*) found as road kill in Scotland (Foster *et al.*, 1996a).

The organism like other members of the *Brucella* genus is host specific and has been isolated from pinnipeds (seals) and an otter. The DNA–DNA relatedness to other *Brucella* species is >77%. An alternative name for seal isolates is *B. phocae*, however, to date this has not been adopted (Corbel and Banai, 2005; Groussaud *et al.*, 2007).

GROWTH CHARACTERISTICS AND PHENOTYPIC IDENTIFICATION. *Brucella pinnipedialis* is an aerobic, oxidase-positive, non-motile, Gram-negative coccus, coccobacillus or short rod (0.5–0.7×0.6–1.5 μm) occurring singly, and less frequently, in pairs, short chains or small groups. On Columbia sheep blood agar (CSBA) colonies are 0.5–1.0 mm raised, convex, circular and entire at 3–4 days at 37°C. Growth on Farrell's medium is slower occurring after 7–10 days, or may be absent (Foster *et al.*, 2007). Farrell's modified medium is recommended for marine mammal isolates, particularly those from seals (Foster *et al.*, 2002; Tryland *et al.*, 2005). On serum-glucose agar colonies are transparent, shiny, raised, convex, entire, and when viewed by transmitted light colonies are a pale honey colour. Growth is improved in the presence of CO_2 and is improved by the addition of serum or blood and occurs between 20 and 40°C with an optimum of 37°C. Catalase is positive and urease gives a rapid positive reaction. L-glutamic acid, D-ribose and *meso*-erythritol are oxidized, whereas L-alanine, L-arabinose, L-arginine, L-asparagine, D-galactose, L-lysine, DL-ornithine and D-xylose are not oxidized. Cultures are lysed by Iz and Wb (Weybridge) phages, but not by R/C phages. Lysis by Tb (Tbilisi) phage is positive for a small number of strains (Foster *et al.*, 2007). A antigen is dominant. The main differential characteristics between *B. ceti* and *B. pinnipedialis* are the lack of, or slower growth on Farrell's medium containing antibiotics, and CO_2 (10%) growth dependency for *B. pinnipedialis*, whereas *B. ceti* grows on Farrell's medium on primary culture and does not require CO_2 for growth (Foster *et al.*, 1996a).

Both *B. ceti* and *B. pinnipedialis* give a numerical profile of 1200004 in the API 20NE. The manufacturer's database will incorrectly identify this as 'good identification for *Psychrobacter phenylpyruvicus* (previously *Moraxella phenylpyruvica*)' (Foster *et al.*, 1996a). These and other reactions are found in Tables 3.5 and 3.27–29.

Brucella species

Serological detection of *Brucella* has been reported from clinically normal polar bears (*Ursus maritimus*) from Norway (Tryland *et al.*, 2001). Serological evidence of brucellosis has been reported in marine mammals from the northern Atlantic Ocean: harp seal (*Phoca groenlandicus*), hooded seal (*Cystophora cristata*), ringed seal (*Pusa* [*Phoca*] *hispida*), bearded seal (*Erignathus barbatus*), fin whale (*Balaenoptera physalus*), Sei whale (*B. borealis*) and minke whale (*B. acutorostrata*). *Brucella* was cultured from spleen, liver, kidney, lung, uterus, ovaries, testis and foetus of the minke whale (Tryland *et al.*, 1999).

PHENOTYPIC TESTING METHODS FOR *BRUCELLA* SPECIES. A modified Farrell's medium is recommended for isolates from marine mammals because bacitracin and nalidixic acid may be inhibitory to *Brucella* species from these sources (Foster *et al.*, 2002; Tryland *et al.*, 2005). See Farrell's modified medium in the media section.

Brucella species do not liquefy gelatin, indole is not produced and methyl red and Voges–Proskaüer tests are negative. Cells stain weakly positive in the modified acid fast stain. Carbohydrates in conventional tests are not fermented by *Brucella* species except for *B. neotomae*. Peptone in carbohydrate basal media may inhibit the growth of *Brucella* (Pickett and Nelson, 1955). Carbohydrate and amino acid may be oxidized and, therefore, an assay of metabolic activity using oxidation of carbohydrates is used for phenotypic differentiation (Jahans *et al.*, 1997). A commercially available biotyping system, Taxa Profile™ from Merlin Diagnostika (Al Dahouk *et al.*, 2010) that tests metabolization of various substrates provides differentiation between *B. ceti* and *B. pinnipedialis*. Media and details for oxidation of carbohydrates are found in the media section under *Brucella* – Assay of Metabolic Activity and *Brucella* – Growth on Substrate Media. Recipes for media are found in the media section.

MOLECULAR IDENTIFICATION. A multiplex PCR (Bruce-ladder) has been described and refined for the differentiation of all currently recognized *Brucella* species (López-Goñi *et al.*, 2008; Mayer-Scholl *et al.*, 2010; Kang *et al.*, 2011; Table 1.4). Primer pairs amplify gene regions that allow differentiation between the *Brucella* species based on insertions, deletions and point mutation differences between the species (López-Goñi *et al.*, 2008). Eight primer pairs in a multiplex PCR reaction result in amplification of a combination of amplicons specific to each *Brucella* species. Primers BMEII436f

Table 1.4. PCR primers for differentiation of *Brucella* species (López-Goñi *et al*., 2008; Kang *et al*., 2011).

Primer	Sequence 5′–3′	Amplicon size bp
BMEI0998f	ATC CTA TTG CCC CGA TAA GG	1682
BMEI0997r	GCT TCG CAT TTT CAC TGT AGC	
BMEI0535f	GCG CAT TCT TCG GTT ATG AA	450 (1320b)
BMEI0536r	CGC AGG CGA AAA CAG CTA TAA	
BMEII0843f	TTT ACA CAG GCA ATC CAG CA	1071
BMEII0844r	GCG TCC AGT TGT TGT TGA TG	
BMEI1436f	CCA ACC GTA TGT CCT CTC T	766
BMEI1435r	TGC GGG AAC TGG TGT TCG ACG	
BMEII0428f	GCC GCT ATT ATG TGG ACT GG	587
BMEII0428r	AAT GAC TTC ACG GTC GTT CG	
BR0953f	GGA ACA CTA CGC CAC CTT GT	272
BR0953r	GAT GGA GCA AAC GCT GAA G	
BMEI0752f	CAG GCA AAC CCT CAG AAG C	218
BMEI0752r	GAT GTG GTA ACG CAC ACC AA	
BMEII0987f	CGC AGA CAG TGA CCA TCA AA	152
BMEII0987r	GTA TTC AGC CCC CGT TAC CT	
BME1r02f	CTA CTC AAG GAC AAC AGG TG	344
BME1r02r	TGT GTC GTT TAA GGC AAT AGG	

and 435r, which resulted in a 794-bp amplicon (López-Goñi *et al*., 2008), were redesigned to produce a 766-bp amplicon that differentiates between *B. pinnipedialis* and *B. ceti* due to a deletion site in *B. ceti* (Kang *et al*., 2011). This primer pair produces four amplicons from *B. ceti* and five amplicons from *B. pinnipedialis*. The specific patterns are published in the respective papers.

Brucella cells are inactivated by heating at 80°C for 2 h. DNA is extracted with the QIAamp DNA mini kit (Mayer-Scholl *et al*., 2010). Heat lysis is also effective for DNA template (López-Goñi *et al*., 2008). A 25-µl reaction volume contains 12.5 µl 2× Qiagen Multiplex PCR master mix (Qiagen), 2.5 µl primer mix (each primer 2 pmol/µl) and 1 µl of template. Thermocycling conditions consist of an initial denaturation step at 95°C for 15 min followed by 25 cycles of denaturation at 94°C for 30 s, primer annealing at 58°C for 90 s and primer extension at 72°C for 3 min. A final extension step is undertaken at 72°C for 10 min (Mayer-Scholl *et al*., 2010).

PCR primers that amplify the *bp26* gene, a gene that encodes for a diagnostic protein antigen for brucellosis, enable differentiation between *Brucella* species from terrestrial animal and human isolates, and those from marine mammals (Cloeckaert *et al*., 2000). The primers do not differentiate between *Brucella* species isolated from marine mammals. The primers 26Af (5′-GCC CCT GAC ATA ACC CGC TT-3′) and 26Br (5′-GAG CGT GAC ATT TGC CGA TA-3′) produce amplicon of 1900 bp from *Brucella* species isolated from marine mammals and 1029 bp from all other *Brucella* species.

Order *Rhodobacterales*

1.5 *Rhodobacteraceae*

1.5.1 *Labrenzia*

Labrenzia aggregans

Stappia aggregata, *S. alba* and *S. marina* have transferred to a new genus *Labrenzia*.

Stappia (stellulata-like) aggregans

The *Stappia stellulata*-like isolate (M1), identified as a potential probiotic organism protective against juvenile oyster disease (JOD) affecting oysters *Crassostrea virginica*, is suggested to be *Stappia aggregata* based on 16S rRBNA sequence similarity (Pujalte *et al*., 2005). *Stappia stellulata*-like (M1) is a Gram-negative, catalase and oxidase positive, motile rod. The organism grows on SWT (see media section) and MA 2216 at 23–25°C in 2–7 days. Growth on MA 2216 and SWT at 23°C is seen as mucoid colonies with light-brown pigment at 3 days. Growth is poor under anaerobic conditions. Star-shaped aggregates are formed when grown in liquid media. These and other reactions are presented in Tables 3.11 and 3.27–29.

Labrenzia alba

TAXONOMY AND SIGNIFICANCE. *Labrenzia (Stappia) alba* (5OM6^T=CECT 5095^T=CIP 108402^T) was isolated as part of the normal flora in Mediterranean oysters and originally named *Stappia alba* as a new species (Pujalte *et al.*, 2005), however it was transferred to a new genus *Labrenzia* as *Labrenzia alba* along with *Stappia aggregata* and *S. marina* (Biebl *et al.*, 2007).

GROWTH CHARACTERISTICS AND PHENOTYPIC IDENTIFICATION. *Labrenzia alba* is a motile, Gram-negative straight rod (2–4 × 1 μm). On MA 2216 colonies are non-pigmented, white, opaque, mucoid and non-luminescent after 6 days at 25°C. Growth also occurs on yeast extract medium of Baumann (Pujalte *et al.*, 2005). Catalase, oxidase and nitrate are positive. Negative in OF medium. Growth occurs at 1–8% NaCl and at 13–28°C but not at 4 or 37°C. Negative for Thornley's arginine dihydrolase and Moeller's ODC and LDC. Negative hydrolysis of gelatin, casein, starch, alginate and agar. DNAse is positive when NaCl is incorporated into the medium. These and other reactions are presented in Table 3.11.

1.5.2 *Paracoccus*

Paracoccus fistulariae

TAXONOMY AND SIGNIFICANCE. *Paracoccus fistulariae* (22-5^T=KCTC 22803^T=CCUG 58401^T) was isolated from the intestine of blue spotted cornetfish (*Fistularia commersonii*) (Kim *et al.*, 2010b). *Paracoccus* belongs to the class *Alphaproteobacteria* with 26 species reported from terrestrial environments and marine environments such as intertidal sediment and seawater. According to 16S rRNA sequencing, *P. fistulariae* is most closely related to *P. homiensis* (97.4%) and *P. zeaxanthinifaciens* (96.9%).

GROWTH CHARACTERISTICS AND PHENOTYPIC IDENTIFICATION. *Paracoccus fistulariae* is an oxidase-positive, Gram-negative, short rod or rod-shaped bacterium (0.3–0.6×0.7–2.5 μm), non-motile and non-spore forming, which grows at 30°C on MA 2216 medium. Colonies are circular, slightly convex, smooth, glistening, grey-yellow and 2–3 mm in diameter after 3 days on MA 2216. Growth occurs at 10–40°C with an optimum of 30°C, and no growth at 4 or 45°C. Growth occurs at 0–14.0% (w/v) NaCl with an optimum of 2–5% NaCl. Mg^{2+} is required for growth. Growth does not occur under anaerobic conditions. Tests are positive for nitrate reduction, catalase, lipase, hydrolysis of Tween 60, Tween 80, gelatin and aesculin and fermentation of cellobiose, D-galactose and D-ribose. These and other reactions are presented in Tables 3.11, 3.24–26 and 3.38. The inoculum is prepared with artificial seawater.

1.5.3 *Phaeobacter*

Phaeobacter gallaeciensis

TAXONOMY AND SIGNIFICANCE. *Phaeobacter gallaeciensis* (BS107^T=CIP 105210^T=ATCC 700781^T=NBRC 16654^T=DSM 17395^T), previously known as *Roseobacter gallaeciensis*, was isolated from seawater, larval cultures and collectors used in the settlement of the larvae of the giant scallop (*Pecten maximus*) cultured in Spain (Ruiz-Ponte *et al.*, 1998; Martens *et al.*, 2006). *Phaeobacter gallaeciensis* has antibacterial activity against a number of *Vibrio* species.

GROWTH CHARACTERISTICS AND PHENOTYPIC IDENTIFICATION. *Phaeobacter gallaeciensis* is a non-fermentative, strictly aerobic, motile, Gram-negative ovoid rod (1.7–2.5×0.7–1.0 μm). Oxidase and catalase are positive. On MA 2216, colonies are 0.5 mm, smooth, convex and brownish in young cultures and brown after 72–96 h at 23°C. After 7 days colonies are 2 mm and brown with irregular edges. A diffusible pigment is produced. Bacteriochlorophyll *a* is not produced. Growth occurs at 15–37°C with an optimum of 23–27°C and no growth at 4°C. Growth occurs at pH 7.0–10.0 with an optimum of 7.0. Growth occurs in 0.1–2 M NaCl with an optimum of 0.2 M NaCl. Thiamine is required for growth. Nitrate is not reduced. Gelatin, DNA and urea are not hydrolysed and reactions are negative for ADH, LDC, ODC and Simmons' citrate. Growth does not occur on TCBS. The isolate is sensitive to vibriostatic agent 0/129150 µg (Ruiz-Ponte *et al.*, 1998; Martens *et al.*, 2006). These and other reactions are presented in Table 3.11.

1.5.4 *Roseobacter*

Roseobacter gallaeciensis. See *Phaeobacter gallaeciensis*.

1.5.5 *Roseovarius*

Roseovarius crassostreae

TAXONOMY AND SIGNIFICANCE. *Roseovarius crassostreae* (CV919-312^T=ATCC BAA-1102^T=DSM 16950^T) is the causative agent of juvenile oyster disease, a seasonal disease affecting hatchery-reared juvenile *Crassostrea virginica* in the north-eastern USA (Boettcher *et al.*, 1999, 2005). Previously, it was referred to as *Roseobacter* species CSVP (Boettcher *et al.*, 1999, 2000). The disease occurs in late summer when water temperature is greater than 20°C and salinity ranges from 27 to 33 ppt. Oysters less than 25 mm in shell height are the most affected. Disease signs include reduced growth rate, fragile or uneven shell margins, cupping of the left valve, mantle retraction and a thick ring of conchiolin (thick proteinaceous material indicating a host response) around the periphery of the mantle (Boettcher *et al.*, 1999).

GROWTH CHARACTERISTICS AND PHENOTYPIC IDENTIFICATION. *Roseovarius crassostreae* is a non-fermentative, aerobic, motile, Gram-negative, oxidase-positive ovoid to rod-shaped cell (0.28×0.9 µm). Catalase is positive. Sodium ions are required for growth and growth occurs on a medium containing seawater-based salts. On seawater-tryptone (SWT) agar (see media section) colonies are 1 mm, pinkish-beige, chalky in consistency, umbonate, round, non-mucoid and semi-translucent after 5 days at 23°C. Some strains are smaller, growing to 0.5 mm after 5 days and have a greenish-yellow appearance. The colony leaves an impression in the agar when removed. Optimal growth occurs at 34–37°C with no growth at 4 or 42°C. Optimal pH is 6.5–8.0 with no growth at less than pH 6.0. Optimal salinity ranges from 1.0 to 1.5% tested on SWT agar containing 0.5% tryptone, 0.3% yeast extract, 0.3% glycerol and a range of NaCl test concentrations (Boettcher *et al.*, 1999, 2005). The pink pigment is not due to bacteriochlorophyll *a* production. Reactions for ADH, urease, gelatinase and aesculin are negative. Nitrate is reduced. These and other reactions are presented in Tables 3.11, 3.27–29 and 3.38.

MOLECULAR IDENTIFICATION. PCR primers based on the 16S-23S rDNA region have been designed to detect *R. crassostreae* direct in oyster tissue (Maloy *et al.*, 2005). Amplicon identity is confirmed by restriction enzyme analysis.

DNA is extracted from oyster tissue using the QIAamp DNA Micro kit (Qiagen) using the manufacturer's protocol for Isolation of Genomic DNA from Swabs beginning at step 2. Oysters are rinsed in 70% filter-sterilized seawater and then aseptically dissected for removal of the soft tissue. The inner valve surfaces are rinsed in sterilized seawater and a moistened swab is used to collect material from the inner valves. Collected material is suspended in 1 ml of 70% sterilized seawater. The suspension is pelleted by centrifugation at 19,000 × *g* for 4 min at 4°C. DNA is extracted from the pelleted material according to the QIAamp kit. A 25-µl master mix contains 0.75 µM forward primer ROS-109F (5′-GTT GGT TCT ACC TGA CGG-3′), 0.75 µM reverse primer ROS+955R (5′-CTC TTT GAA TGT CCT ACC GC-3′), 2.0 mM $MgCl_2$, 200 µM each dNTP, 1.25U *Taq* polymerase (Invitrogen) and 2.5 µl 10× PCR buffer, 2 µl DNA and water to 25 µl. A product of 1071 or 1100 bp is amplified. DNA templates of 1 µl of 10^5 bacterial cells or 50 ng of bacterial DNA can also be used. When using whole cells as DNA template, a $MgCl_2$ concentration of 1.5 mM is used. Thermocycling is conducted in a Controller PTC-100™ (MJ Research) using an initial denaturation step of 95°C for 5 min followed by 40 cycles of denaturation at 94°C for 30 s, annealing at 60°C for 30 s, elongation at 72°C for 1 min with a final elongation step at 72°C for 7 min. The sensitivity is 10 cells per suspension or 1.2 bacterial cells when DNA is extracted from inner valve tissue, and 100 cells per suspension or 12.5 bacterial cells when surface material from host tissue is used as template. The PCR is specific for *R. crassostreae* and product is not amplified from *Roseovarius nubinhibens* (ISM^T),

Roseovarius denitrificans (OCH114[T]), *Roseobacter litoralis* (ATCC49566[T]), *Roseovarius tolerans* (EL-172[T]) or *Marinovum* (*Ruegeria*) *algicola* (ATCC51440[T]).

Confirmation of amplicon specificity is determined by restriction of the amplicon with *Ava*I. A 10-μl restriction mix contains 5 μl amplicon, 1 μl NEB buffer #4, 5 U *Ava*I (New England Bio Labs) made up to 10 μl with nuclease-free water. The reaction is incubated at 37°C overnight and restricted fragments are analysed in a 1.5% gel. Depending upon the strain, a 1100-bp amplicon will be restricted into 430-, 360- and 310-bp fragments; a 1071-bp amplicon will be restricted into 631- and 430-bp fragments. Strains are identified as genetic signature I (GSI) or GSII, respectively.

1.5.6 *Stappia*

Stappia alba, *Stappia stellulata*-like. See *Labrenzia alba*.

Class II Betaproteobacteria

1.6 *Burkholderiaceae*

1.6.1 *Burkholderia*

Burkholderia pseudomallei

TAXONOMY AND SIGNIFICANCE. *Burkholderia* (*Pseudomonas*) *pseudomallei* causes melioidosis, an often fatal disease of humans and animals. Melioidosis occurs in mainly tropical areas in northern Australia and south-eastern Asia including Hong Kong. *Burkholderia pseudomallei* is present in soil and water in endemic areas. Melioidosis resulted in the death of a macaroni penguin (*Eudyptes chrysolophus*) acquired from Chile and housed in a zoo in Hong Kong (MacKnight *et al.*, 1990). The penguin displayed lethargy, anorexia and produced a light brown diarrhoea. The organism was cultured from lung, liver, kidney, spleen, proventriculus, abdominal air sacs and intestine. A study of marine mammals that died over a 25-year period at an oceanarium in Hong Kong found meliodidosis in false killer whales (*Pseudorca crassidens*), killer whale (*Orcinus orca*), bottlenosed dolphins (*Tursiops struncatus*), Pacific white-sided dolphins (*Lagenorhynchus obliquidens*), Californian sea lion (*Zalophus californianus*) and grey seal (*Halichoerus grypus*). The animals had been sourced mainly from Japan and Taiwan. The animals presented with a variety of symptoms including septicaemia, lethargy, inappetence, anorexia, respiratory distress and mammary swelling. *Burkholderia pseudomallei* was isolated from lung, heart blood, liver and in mixed culture from intestine and stomach (Hicks *et al.*, 2000).

GROWTH CHARACTERISTICS AND PHENOTYPIC IDENTIFICATION. *Burkholderia pseudomallei* is an aerobic, Gram-negative rod with oval to round cells that have bipolar staining seen after 3–4 days of culture. The cells may be mistaken for spores as only the periphery stains. Positive for oxidase and catalase. On BA, glycerol plates or Ashdown's medium, colonies are 0.5–1 mm at 24 h and 3–4 mm at 3–4 days, are white, convex and smooth with a white sheen and become uneven, umbonate and wrinkled after 1–4 days at 37°C. Broth culture is turbid with wrinkled pellicle. See media section for media recipes. *Burkholderia pseudomallei* is zoonotic and all samples and plates must be processed and examined in a Class II Biological Safety Cabinet. Reactions are presented in Table 3.11.

1.7 *Oxalobacteraceae*

1.7.1 *Janthinobacterium*

Janthinobacterium lividum

Janthinobacterium lividum was associated with mortality of rainbow trout (*Oncorhynchus mykiss*) cultured in Scotland and Northern Ireland (Austin *et al.*, 1992). Infected fish had exophthalmia, pale gills and pale internal organs and swollen abdomen. In experimental challenge using 5×10^2 cells/fish injected intramuscularly or intraperitoneally, 100% mortality occurred within 14 days. At an inoculum of 5×10^6 cells/fish 50% mortality occurred within 5 h.

Janthinobacterium lividum is a strictly aerobic, motile, oxidase positive, Gram-negative rod (1.8–6 × 0.8–1.5 μm). Catalase is positive. On BA, TSA, AO, KDM2 colonies are purple-pigmented, round and low convex after 24–48 h at 22°C.

Growth occurs at 4–30°C but not at 37°C and in 0–2% NaCl but not in 3% NaCl. Nitrate is reduced. VP is positive and gelatin is hydrolysed. These and other reactions are presented in Table 3.11.

Order *Neisseriales*

1.8 *Neisseriaceae*

1.8.1 *Andreprevotia*

Andreprevotia lacus

The genus *Andreprevotia* contains one other species, *A. chitinilytica*, and belongs to the family *Neisseriaceae*. *Andreprevotia lacus* (GFC-1^T=BCRC 17832^T= LMG 24502^T) was isolated from pond water from cultured fish in Taiwan (Sheu *et al.*, 2009). Cells are Gram-negative, aerobic, non-sporing, rod-shaped and motile. Colonies are cream, 1.0–2.0 mm, round, entire and convex on R2A agar at 25°C after 48 h. Growth occurs at 15–37°C, 0–1% (w/v) NaCl and with an optimum of 30°C and 0.5% NaCl. Oxidase, catalase and chitin hydrolysis are positive. Reactions for hydrolysis of lipase (corn oil), Tween 20, 40 60 and 80, starch, cellulose and casein are negative. Reactions are presented in Tables 3.11, 3.27–30 and 3.38.

1.8.2 *Vogesella*

Vogesella lacus sp. nov.

TAXONOMY AND SIGNIFICANCE. *Vogesella lacus* (GR13^T=BCRC 17836^T=LMG 24504^T) was isolated from a freshwater pond used for the culture of the soft-shell turtle *Pelodiscus sinensis* (previously *Trionyx sinensis*) in Taiwan (Chou *et al.*, 2009). *Vogesella lacus* is a Gram-negative, aerobic, motile, non-spore forming rod (0.8 × 1.5–2.3 µm). Growth on nutrient agar occurs as transparent or colourless colonies and on R2A agar it grows as 1–2 mm round, entire, convex, colourless colonies after 48 h at 25°C. Growth occurs at a temperature range of 10–42°C and 0–3% (w/v) NaCl with optimal growth at 25–30°C and 0.5% NaCl. Catalase, oxidase, hydrolyis of casein, Tween 60 and 80 are positive, whereas DNase is weak. These and other reactions are presented in Tables 3.11, 3.27–29 and 3.38.

Class III Gammaproteobacteria

Order *Aeromonadales*

1.9 *Aeromonadaceae*

The family *Aeromonadaceae* belongs to the *Gammaproteobacteria* (Class III) and the Order *Aeromonadales* and is the only family in the order.

1.9.1 *Aeromonas*

TAXONOMY. The genus *Aeromonas* belongs to the class *Gammaproteobacteria* and the family *Aeromonadaceae* and contains Gram-negative, non-spore-forming bacilli or coccobacilli that are facultatively anaerobic, oxidase-positive, catalase-positive, nitrate-positive, do not require NaCl for growth and are resistant to vibriostatic agent 0/129 (2,4-diamino-6,7-diisopropylpteridine). Aeromonads are widespread in the environment, especially aquatic habitats, and can be isolated from foods. Some species may be primary or opportunistic pathogens of invertebrates and vertebrates including humans.

The taxonomy of *Aeromonas* species is in a continual state of flux as new species are described and the descriptions of the existing taxa are refined. Assignment of new species to the genus *Aeromonas* is based on DNA G+C (guanine+cytosine) content of 57–63 mol%. Species delineation is based on the 16S rRNA gene, which has a high conservation rate of 97.8–100% corresponding to 0–32 nucleotide differences (Martínez-Murcia *et al.*, 1992). Delineation of species and subspecies is based on a DNA–DNA relatedness of equal or greater than 70% and with a 5°C or less ΔT_m (Wayne *et al.*, 1987).

A phenotypic and genotypic classification system exists where phenotype (phenospecies) is determined according to key biochemical tests and genotypic (genomospecies) classification is based on DNA–DNA hybridization. To date 20 DNA hybridization groups (HG) are recognized. Discrepancies in DNA–DNA hybridization results have been reported (Huys *et al.*, 1997a, 2001, 2005; Martínez-Murcia, 1999; Esteve *et al.*, 2003). The designation of the genospecies is dependent on the type strain within the HG. Problems exist for the diagnostic microbiologist in the phenotypic identification of strains. Phenotypic diversity of strains occurs within a DNA HG and strains from different HGs may be biochemically similar. Biochemical reactions are presented in Tables 3.1, 3.2 and 3.3 (Abbott *et al.*, 1992; Kaznowski, 1998).

All strains of motile *Aeromonas* species are positive for the fermentation of glucose and maltose, and ONPG. All strains are negative for urease and fermentation of inositol and xylose and are resistant to the vibriostatic reagent 0/129.

More recently, in addition to biochemical tests, sequencing of housekeeping genes has been recommended for species delineation as these regions are highly conserved. Sequence comparison of up to five housekeeping genes is recommended when describing a new species (Stackebrandt *et al.*, 2002), although for the *Aeromonas* genus seven genes may be more appropriate for species delineation (Martínez-Murcia *et al.*, 2011). The products (often proteins) expressed from housekeeping genes are needed for cell maintenance or metabolic functions and are transcribed at a relatively constant level making these regions of the genome ideal for species comparison. The housekeeping genes reported for species delineation of the *Aeromonadaceae* include: the *gyrB* gene, which encodes the B-subunit of DNA gyrase, a type II DNA topoisomerase; the genes that encode the different subunits of the multisubunit enzyme, DNA-dependent RNA polymerase (*rpoD*), which encodes the sigma^{70} factor, which confers promoter-specific transcription initiation on RNA polymerase; *rpoB*, the gene encoding the beta-subunit of the RNA polymerase, *rpoA*, which encodes the two alpha subunits, and *rpoC*, which encodes one B′ subunit); *dnaJ*, which encodes heat-shock protein 40; and *cpn60*, a type I chaperonin (Yáñez *et al.*, 2003; Soler *et al.*, 2004; Küpfer *et al.*, 2006; Nhung *et al.*, 2007; Miñana-Galbis *et al.*, 2009). Gene sequences of *gyrB* and *rpoD* are good molecular chronometers for phylogenetic inference in the *Aeromonas* genus as they show similar substitution rates. Sequence similarity of the *gyrB* gene for all *Aeromonas* species is 86.3–100% corresponding to 0–131 nucleotide differences with a rate of nucleotide substitution of 0–2.3% at the intraspecies level (Yáñez *et al.*, 2003; Soler *et al.*, 2004) and a maximum interspecies divergence of 15.2% (mean=7.76%) (Küpfer *et al.*, 2006). The *rpoB* gene also has a high resolving power, although it is more conserved than the *gyrB* gene, and has a similarity of 81.7–100% (1–148 nucleotides) for all *Aeromonas* species with an intraspecies rate of substitutions of 0–2.6% (Soler *et al.*, 2004) and an interspecies divergence of 13% (mean=6.07%) (Küpfer *et al.*, 2006). The sequence divergence of the *gyrB* and *rpoD* genes is greater than the *cpn60* gene. All housekeeping genes give better discrimination between species of *Aeromonas* compared to the 16S rRNA gene with the *gyrB* gene having a mean substitution rate six times that of the 16S rRNA (Yáñez *et al.*, 2003).

PHYLOGENY AND IDENTIFICATION OF *AEROMONAS* USING HOUSEKEEPING GENES.

gyrB gene

The housekeeping gene *gyrB*, which encodes the B-subunit of DNA gyrase, a type II DNA topoisomerase is suitable for species delineation of the *Aeromonadaceae* (Yáñez *et al.*, 2003).

A 50-μl master mix contains 20 pmol of primer gyrB3F (5′-TCC GGC GGT CTG CAC GGC GT-3′), 20 pmol primer gyrB14R (T5′-TGT CCG GGT TGT ACT CGT C-3′), 50 mM KCl, 10 mM Tris/HCl (pH 9.0), 1.5 mM $MgCl_2$, 0.1% Triton X-100, 0.2 mM each dNTP (Roche Diagnostics), 1 U *Taq* DNA polymerase (Amersham Biosciences) and 1 μl DNA. Thermocycling is carried out with an initial denaturation at 95°C for 5 min followed by 35 cycles of denaturation at 94°C for 30 s, annealing at 55°C for 30 s, extension at 72°C for 1 min. A product of approximately 1100 bp is amplified. Product is purified and sent for sequencing.

Cpn60 gene

Sequencing of the universal target (UT) of the housekeeping gene *cpn60*, a type I chaperonin, can assist in the identification and phylogeny of *Aeromonas* species. Intraspecies divergence rates are less than or equal to 3.5% (<19 nucleotide differences) with interspecific divergence rates of 3.7–16.9% (20–84 nucleotide differences) (Miñana-Galbis *et al.*, 2009). *Aeromonas sharmana* falls outside this range and its assignment to the *Aeromons* genus is in contention (Saavedra *et al.*, 2006; Martínez-Murcia *et al.*, 2007). The range of sequence divergence for the *cpn60* UT gene is less than that found for the *gyrB* and *rpoD* genes, but greater than that of the 16S rRNA gene, thus giving better discrimination between species of *Aeromonas* compared to the 16S rRNA gene (Miñana-Galbis *et al.*, 2009). A web-based taxonomic database for the *cpn60* gene is available (http://cpndb.cbr.nrc.ca).

DNA is extracted using the REALPURE kit (RBMEG03, Durviz).

Two primers (C175 and C938) corresponding to nucleotides 175–194 and 938–920, respectively, are used to amplify a 555-bp fragment corresponding to positions 274–828 of the complete *cpn60* gene. The same primers are used for DNA and for sequencing amplified product.

A total master mix volume of 50 µl contains 50 mM KCl, 15 mM Tris/HCl, pH 8.0, 1.5 mM $MgCl_2$, 0.25 mM dNTPs (Amersham Biosciences), 2.5 U Ampli*Taq* Gold DNA polymerase (Applied Biosystems) and 50 pmol of each primer: C175 (5′-GAA ATY GAA CTG GAA GAC AA-3′) and C938 (5′-GTY GCT TTT TCC AGC TCC A-3′).

Thermocycling is performed on a 2720 thermal cycler (Applied Biosystems) using a protocol of denaturation at 95°C for 10 min, 35 cycles of denaturation at 94°C for 1 min, annealing at 52°C for 1 min and extension at 72°C for 1 min; and a final extension step at 72°C for 10 min. The amplified product is purified in a MSB Spin PCRapace kit (Initek). Sequencing is performed using the ABI PRISM BigDye Terminator v3.1 Cycle Sequencing kit (Applied Biosystems) using thermal cycling conditions at 96°C for 1 min; 25 cycles of 96°C for 10 s, 52°C for 5 s and 60°C for 4 min. Nucleotide sequence is determined using a ABI PRISM 3730 DNA analyser.

Sequence alignments are performed using Clustal W software (Thompson *et al.*, 1994). DNA polymorphism data are conducted with DnaSP software (Rozas and Rozas, 1999). Phylogeny calculations, including synonymous and non-synonymous substitutions, the Z-test for neutrality (dS=dN) and neighbour-joining and maximum-parsimony trees of DNA sequence alignments are performed using MEGA software version 3.1 (Kumar *et al.*, 2004).

PATHOGENICITY. A number of virulence factors contribute to the pathogenicity of *Aeromonas* species and these include the aerolysin/haemolysin group of genes, the cytotonic enterotoxins (Ast, Alt), and the cytotoxin encoded by the *act* gene. A type III secretion system delivers toxins such as AexT directly to the host cell and induces apoptosis.

Aeromonas allosaccharophila

TAXONOMY AND SIGNIFICANCE. *Aeromonas allosaccharophila* (CECT 4199^T=289=ATCC 51208 =CIP 104616=DSM 11576=LMG 14059) was isolated from diseased elvers (*Anguilla anguilla*) on an eel farm in Spain, and from a female patient with diarrhoea in the USA (Martínez-Murcia *et al.*, 1992). *Aeromonas allosaccharophila* also has been isolated from pig carcasses and from equipment at an abattoir (Saavedra *et al.*, 2007). *Aeromonas allosaccharophila* was originally proposed as hybridization group 14 (Esteve *et al.*, 1995b) but later proposed to be HG 15. The status of *A. allosaccharophila* is controversial as studies using amplified fragment length polymorphism (AFLP) typing and DNA–DNA hybridization suggest *A. allosaccharophila* strains LMG 14059^T and LMG 14021 are closely related to *A. veronii* biogroup Sobria HG 8 (strains LMG 13068, LMG 13693 and LMG 13071-LMG 13074) with values of 78–90% homology (Huys *et al.*, 2001). Other studies found 0% homology between the two species (Esteve *et al.*, 1995b).

Analysis of housekeeping genes gives mixed results with some confirming species status, whereas others do not. Phylogenetic analysis of the *gyrB* gene supports the species status of *A. allosaccharophila*, recognizing a close relationship between *A. allosaccharophila* and the strains of *A. veronii* biovar Sobria separating them in different clusters (Yáñez *et al.*, 2003; Soler *et al.*, 2004; Saavedra *et al.*, 2006). These analyses have confirmed strains CECT 4199^T, CECT 4220=LMG 14021, LMG 13071=CECT 4909, LMG 13073=CECT 4911, LMG 13074=CECT 4912, LMG 14058=CECT 4200, MDC12, MDC45, MDC234, MCD380, MDC561 as being *A. allosaccharophila*, yet some of them were received as *A. veronii* biovar Sobria (strains LMG 13071 =CECT 4909, LMG 13073=CECT 4911, LMG 13074=CECT 4912) (Saavedra *et al.*, 2006; MDC=Molecular Diagnostics Centre, Spain). Sequence analysis of the *cpn60* gene, which reveals a 20–24 nucleotide difference (3.7–4.5% divergence) between *A. allosaccharophila* and *A. veronii* group, also supports separate species (Miñana-Galbis *et al.*, 2009).

The allocation of strain CECT 4911, previously assigned to *A. veronii* by Huys *et al.* (1996), has been confirmed as *A. allosaccharophila* using *gyrB* analysis by Yáñez *et al.* (2003), Soler *et al.*

(2004), Küpfer *et al.* (2006), Saavedra *et al.* (2006) and Martínez-Murcia *et al.* (2011).

The *rpoD* gene sequences indicate a borderline relationship between the strains (Soler *et al.*, 2004; Saavedra *et al.*, 2006), whereas the *dnaJ* gene suggests *A. allosaccharophila* is a later heterotypic synonym of *A. veronii* (Nhung *et al.*, 2007); however, multilocus phylogenetic analysis using seven genes (*gyrB*, *rpoD*, *recA*, *dnaJ*, *gyrA*, *dnaX* and *atpD*) supports retention of *A. allosaccharophila* (Martínez-Murcia *et al.*, 2011).

Aeromonas veronii biogroup Sobria strain LMG 13072 has been identified as belonging to the *A. jandaei* group (Esteve *et al.*, 2003).

GROWTH CHARACTERISTICS AND PHENOTYPIC IDENTIFICATION. *Aeromonas allosaccharophila* is a motile, Gram-negative, straight rod. Reactions are positive for oxidase, catalase and nitrate. Growth occurs in 0–3% NaCl and at 4–42°C. Casein is hydrolysed but not elastin (Martínez-Murcia *et al.*, 1992). These and other reactions are listed in Tables 3.3, 3.24–26 and 3.30.

Aeromonas aquariorum

TAXONOMY AND SIGNIFICANCE. *Aeromonas aquariorum* ($MDC47^T$=$CECT\ 7289^T$=$DSM\ 18362^T$) was the name given to *Aeromonas* species isolated from the skin of ornamental fish and their water and imported to Portugal from Asian countries, and from patients (blood and faeces) in Taiwan (Martínez-Murcia *et al.*, 2008; Figueras *et al.*, 2009). Phylogenetic analysis of the *gyrB* gene and 16S rRNA gene sequences shows that *A. aquariorum* is closely related to *A. trota*, yet information from the *rpoD* gene indicates similarity to *A. hydrophila*, with *A. trota* being distantly related. DNA–DNA hybridization indicates levels of 20.4–55.7% to all strains of *Aeromonas* suggesting it is a species in its own right with close levels of relatedness to *A. trota* and *A. hydrophila* being 44.4 and 46.0%, respectively. Further research indicates that *A. aquariorum* and *A. hydrophila* ssp. *dhakensis* are heterotrophic synonyms, and although Martínez-Murcia *et al.* (2009) suggested that *A. hydrophila* ssp. *dhakensis* belongs to the species *A. aquariorum*, this is not valid according to the rules of nomenclature (Euzéby: http://www.bacterio.cict.fr/a/aeromonas.html). Multilocus phylogenetic analysis using seven genes (*gyrB*, *rpoD*, *recA*, *dnaJ*, *gyrA*, *dnaX* and *atpD*) supports *A. aquariorum* being separate from *A. hydrophila*, but confirms evidence from Martínez-Murcia *et al.* (2009) that the type strain of *A. hydrophila* ssp. *dhakensis* and two other strains belong to *A. aquariorum* (Martínez-Murcia *et al.*, 2011).

Strains from ornamental fish and Taiwanese patients possessed the genes encoding the pore-forming toxin aerolysin/haemolysin with 84% and 28% of strains possessing the two cytotoxic genes, the heat-stable *ast* and the heat-labile *alt*, respectively (Figueras *et al.*, 2009).

GROWTH CHARACTERISTICS AND PHENOTYPIC IDENTIFICATION. *Aeromonas aquariorum* is a Gram-negative, motile, straight rod. Growth occurs as yellow colonies at 28°C (optimum 30–37°C) in 24 h on glutamate starch penicillin medium (Oxoid) and as cream colonies on TSA. Growth occurs on MacConkey agar and at 0–3% NaCl. Brown pigment is not produced. Nitrate is reduced, oxidase and catalase are positive and it is resistant to vibriostatic agent 0/129 150 µg. VP is reported as negative in the species description but positive in the table for key tests (Martínez-Murcia *et al.*, 2008), however according to Aravena-Román *et al.* (2011), *A. aquariorum* is positive for VP (95% of strains). Positive for LDC (95%), and gas from glucose (90%), H_2S from cysteine and negative for fermentation of L-arabinose. These and other reactions are listed in Tables 3.3, 3.24–26 and 3.30.

Aeromonas bestiarum

TAXONOMY AND SIGNIFICANCE. *Aeromonas bestiarum* was originally identified as hybridization group 2 (HG 2) and strains have been isolated from freshwater, lake water and common carp (Ali *et al.*, 1996; Kaznowski, 1998; Kozińska *et al.*, 2002; Miñana-Galbis *et al.*, 2002). In experimental challenge, *A. bestiarum* is virulent to carp (*Carpio carpio*). Infected fish display ulcers, haemorrhage and necrosis (Kozińska *et al.*, 2002).

Aeromonas bestiarum is highly similar to *A. salmonicida* ssp. *salmonicida* and the two species are almost impossible to separate using biochemical tests, 16S rRNA sequencing or cluster analysis of the 16S–23S intergenic spacer region RFLP. However, they can be separated by phylogenetic analysis using housekeeping genes *gyrB* and *rpoD* (Martínez-Murcia *et al.*, 2005). Multilocus phylogenetic analysis using seven genes (*gyrB*, *rpoD*, *recA*, *dnaJ*, *gyrA*, *dnaX* and *atpD*) demonstrates a main branch containing four species clusters, which are *A. bestiarum*, *A. salmonicida*, *A. popoffii* and *A. piscicola*, confirming their species status. In this analysis *A. bestiarum* and *A. piscicola* were the most closely related (Martínez-Murcia *et al.*, 2011). Reactions are presented in Tables 3.3, 3.24–26 and 3.30.

Aeromonas bivalvium

TAXONOMY AND SIGNIFICANCE. *Aeromonas bivalvium* (868^T=CECT 7113^T=LMG 23376^T) was isolated from cockles (*Cardium* species) and razor shells (*Ensis* sp.) found in a retail market in Barcelona, Spain (Miñana-Galbis *et al.*, 2007). Sequence similarity to other *Aeromonas* species is 96.75–99.71% based on the 16S rRNA gene. Two strains of *A. bivalvium* are closely related to *A. popoffii* with 99.71% similarity and four to six nucleotide differences. According to fluorescent amplified fragment length polymorphism (FAFLP) banding patterns, *A. bivalvium* does not cluster significantly with other *Aeromonas* species. Analysis of the *cpn60* gene sequence confirms its identity as a separate species (Miñana-Galbis *et al.*, 2009).

GROWTH CHARACTERISTICS AND PHENOTYPIC IDENTIFICATION. *Aeromonas bivalvium* is a motile, Gram-negative straight coccoid/rod (0.5–2.0×0.3–1.0 µm). On TSA colonies are 3–4 mm in diameter, beige, non-swarming and circular at 48 h at 25°C. Growth is optimal at 30–37°C but occurs at a range of 4–37°C with strain $868E^T$ growing at 40.5°C. Oxidase and catalase are positive, nitrate is reduced to nitrite, brown pigment is not produced and it is resistant to vibriostatic agent 0/129. Growth occurs on MacConkey agar and m-Aeromonas selective base Havelaar and at 0–3% NaCl, and weakly at 6%. Starch is hydrolysed but not erythrocytes or elastin. These and other reactions are presented in Tables 3.3 and 3.24–26.

Aeromonas caviae

Aeromonas caviae is a homotypic synonym of *A. punctata*. *A. caviae* is ubiquitous in the environment. Infection has resulted in septicaemia and mortality in freshwater ornamental fish, Atlantic salmon, octopus, giant freshwater prawn and turbot larvae, and gastroenteritis in humans (Altwegg *et al.*, 1990; Sung *et al.*, 2000).

In a study of *Aeromonas* species isolated from environmental samples in Spain, *A. caviae* predominated the microflora of bivalve molluscs (Miñana-Galbis *et al.*, 2002). In experimental challenge, *A. caviae* caused 100% mortality in giant freshwater prawn (*Macrobrachium rosenbergii*) when injected at 3.8×10^6 cells/g body weight, which resulted in an LD_{50} of 51.2×10^3 cells/g (Sung *et al.*, 2000).

PCR primers have been designed for *A. caviae*. Primers AER8 (5′-CTG CTG GCT GTG ACG TTA CTC GCA G-3′) and AER9 (5′-TTC GCC ACC GGT ATT CCT CCA GAT C-3′) produce amplicon of 260 bp. A thermocycling programme consists of an annealing temperature at 62°C and 30 cycles. The product is digested with *Alu*I, 37°C for 1 h. Fragments of 180 and 80 bp are produced (Khan and Cerniglia, 1997).

Aeromonas culicicola

Aeromonas culicicola is a later heterotypic synonym of *A. veronii* biovar Veronii (Huys *et al.*, 2005; Nhung *et al.*, 2007; Martínez-Murcia *et al.*, 2011).

Aeromonas diversa

TAXONOMY AND SIGNIFICANCE. A strain (CDC 2478-85^T= CECT 4254^T and CDC 2555-87=CECT 5178) was isolated in 1988 from a leg wound of a patient in the USA and, although showing less similarity according to phenotypic tests and DNA–DNA hybridization, was placed into Enteric Group 501. A second strain (CDC 2555-87), isolated from a bone fracture wound in 1991, was found to be similar to strain CDC 2478-85, and both strains were given the name *Aeromonas* group 501, HG 13 (Hickman-Brenner *et al.*, 1988). These strains have been re-characterized and named *Aeromonas diversa* (Miñana-Galbis *et al.*, 2010a). *Aeromonas diversa* has high sequence similarity of 97.83–99.13% (31 to 13 nucleotide differences, respectively) with other *Aeromonas* species except with *A. schubertii* (previously Enteric Group 501), which has a 99.67% similarity (only five nucleotide differences).

GROWTH CHARACTERISTICS AND PHENOTYPIC IDENTIFICATION. *Aeromonas diversa* is a facultatively anaerobic, Gram-negative straight rod, oxidase and catalase positive with optimal growth between 30 and 37°C and occuring up to 40.5°C but not at 4°C. The organism grows in 0–3% NaCl but not at 6% NaCl. Growth occurs on BA, TSA, MacConkey agar and on m-Aeromonas selective agar base Havelaar, but not on TCBS. Beta-haemolysis is produced, brown pigment is not produced, nitrate is reduced to nitrite, and is resistant to vibriostatic agent 0/129 150 µg. Strain CDC 2478-85^T is positive for VP, whereas strain CDC 2555-87 is negative. These and other reactions are presented in Table 3.3.

Aeromonas encheleia

TAXONOMY AND SIGNIFICANCE. Four strains of *A. encheleia* (S181^T=CECT 4342^T) were isolated from healthy European eels (*Anguilla anguilla*) in Spain (Esteve *et al.*, 1995a). In experimental challenge using intraperitoneal inoculation the LD_{50} was 10^8 cells/fish for three strains, indicating the strains were avirulent, and one strain was weakly virulent with an LD_{50} of 1.7×10^7 cells/fish.

A group of strains assigned to *A. eucrenophila* subgroup II consisted of strain CCUG 30345=LMG 13061 found in an artesian well in the UK, strain CCUG 30346=LMG 13062 from an infiltration well in Germany, strain CCUG 30364=LMG 13075 from an ankle fracture in New Zealand and strain CCUG 30365=LMG 13076 from the Mohawk River, New York, belonged to *A. encheleia* based on DNA homology of 74–105% (Huys *et al.*, 1997a). However, phenotypic tests for LMG 13075 and 13076 show differences from *A. encheleia* in key biochemical tests such as arginine dihydrolase, ornithine and lysine decarboxylases. Other strains isolated from drinking water in Finland, Belgium and Scotland and previously identified as *A. encheleia*-like were confirmed to be *A. encheleia* based on DNA homology.

Sequence analysis of the *cpn60* gene reveals that strain CECT 4253, which represents HG 11, belongs to *A. encheleia* (Miñana-Galbis *et al.*, 2009). According to a dendrogram constructed from *cpn60* sequence information, *A. encheleia*, *A. caviae*, *A. media* and *A. eucrenophila* display related but different phylogenetic lines.

GROWTH CHARACTERISTICS AND PHENOTYPIC IDENTIFICATION. *Aeromonas encheleia* is a facultatively anaerobic, Gram-negative, motile, straight rod, with rounded ends that occur singly, in pairs or in short chains. Oxidase, catalase and nitrate are positive and the organism is resistant to vibriostatic agent 0/129 150 µg. On TSA and BA colonies are non-pigmented, haemolytic (on BA) after 24 h at 28°C. Growth occurs at 4–37°C. Brown, water-soluble pigment is not produced but old cultures of 10–15 days' growth may have colonies with light brown centres. NaCl is not required for growth but 0–3% NaCl is tolerated. Growth occurs on MacConkey agar, but not TCBS agar. Gas is produced from glucose under anaerobic conditions with the exception of one strain (S177). Acid is produced from cellobiose; arbutin, casein, collagen and chitin are hydrolysed but not elastin. Arginine dihydrolase is positive by the methods of both Moeller and Thornley (Esteve *et al.*, 1995a). These and other reactions are presented in Tables 3.3, 3.24–26 and 3.30.

Aeromonas enteropelogenes

Aeromonas enteropelogenes is an earlier heterotypic synonym of *A. trota*. *Aeromonas enteropelogenes* was the name given to strains that were similar to *A. sobria* but were negative for hydrolysis of arbutin, aesculin and salicin, negative for the fermentation of arabinose and sucrose and negative for Voges–Proskaüer. The strains were isolated in India from faeces (Schubert *et al.*, 1990a). See further information under *Aeromonas trota*.

Aeromonas fluvialis

TAXONOMY AND SIGNIFICANCE. *Aeromonas fluvialis* (717^T= CECT 7401^T=LMG 224681^T) was isolated from water collected from the Muga River, Girona, north-eastern Spain (Alperi *et al.*, 2010b). Sequencing of the 16S rRNA gene demonstrates sequence similarity with the *Aeromonas* genus at a range of 97.6–99.5% resulting in 36–7 base pair differences. *Aeromonas fluvialis* is most closely related to *A. veronii* and *A. allosaccharophila* based on *rpoD* sequence analysis. DNA–DNA hybridization levels are 54–66% and below the level of 70% required for a new species.

GROWTH CHARACTERISTICS AND PHENOTYPIC IDENTIFICATION. *Aeromonas fluvialis* is a Gram-negative, motile, straight rod (3–3.5×0.6–0.7 μm). Positive for oxidase, catalase and nitrate, and resistant to vibriostatic agent 0/129 150 μg. Colonies on TSA with blood are 4–5 mm at 48 h at 30°C, beige, non-haemolytic and circular. Brown diffusible pigment is not produced. Growth occurs at 30–37°C but not at 4 or 40°C. Grows occurs without NaCl but not in 6% NaCl. Colonies are lactose positive on MacConkey agar and in nutrient broth containing 1% lactose and phenol red indicator, but lactose is negative using the API 50CH. These and other reactions are presented in Tables 3.3 and 3.24–30.

Aeromonas hydrophila ssp. *hydrophila*

TAXONOMY AND SIGNIFICANCE. *Aeromonas hydrophila* ssp. *hydrophila* is ubiquitous in the aquatic environment worldwide, and is an opportunistic pathogen following a primary stressor for many aquatic animal species including ornamental fish, ayu, channel catfish, walking catfish, tilapia, trout, eels, reptiles, grey seal, Atlantic salmon, seabream, turtles, scallops and prawns (Ventura and Grizzle, 1988; Pasquale *et al.*, 1994; Angka *et al.*, 1995; Riquelme *et al.*, 1996; Dierckens *et al.*, 1998; Doukas *et al.*, 1998; Mauel *et al.*, 2002). Infection may result in haemorrhagic septicaemia, peritonitis or fin rot. It is a cause of infection and diarrhoea in humans. The pathogenesis and virulence of this organism remains unclear. A number of toxins are produced including haemolysins, cytotoxins, enterotoxin, a cytolytic haemolysin and aerolysin. Aerolysin is a channel-forming toxin that kills cells, is lethal to mice and possesses haemolytic and cytotoxic activities. It is the main contributor to pathogenicity (Chopra *et al.*, 1993). The aerolysin gene is not found in non-haemolytic strains of *A. hydrophila*, non-haemolytic strains of *A. caviae* or strains of *A. sobria* (Pollard *et al.*, 1990). Different phenotypes (lysine decarboxylase, cellobiose fermentation, indole, agglutination of 0.2% acriflavine) have been examined for an association with disease outbreaks, however conflicting results have been reported (Lallier and Higgins, 1988; Nielsen *et al.*, 2001).

Strain ATCC 7965=NCTC 7812=CIP 57.50, originally identified as *A. hydrophila*, is in fact *A. salmonicida* based on results of phenotypic characteristics and sequence analysis using *cpn60*, *dnaJ*, *gyrB* and *rpoD* genes (Miñana-Galbis *et al.*, 2010).

Virulence studies in strains isolated from walking catfish suggest strains range in virulence from avirulent, weakly virulent to strongly virulent. All virulent strains possessed haemolysin genes, and were able to lyse red blood cells from a wide variety of animal species. The majority of avirulent or weakly virulent strains did not possess all three haemolysin genes tested (*AHH*$_1$, *AHH*$_5$, *ASA*) or had combinations of only one or two of the genes. These strains generally were unable to lyse the red blood cells from a number of animal species. The virulent strains were able to cause infection at a dose rate of 10^4 CFU/ml, whereas avirulent strains did not cause disease at this dose (Angka *et al.*, 1995). Disease in channel catfish (*Ictalurus punctatus*) presented as either a septicaemic disease with a diffuse necrosis in internal organs, or as a skin infection with cutaneous lesions (Ventura and Grizzle, 1988).

Aeromonas hydrophila caused mortality in scallop (*Argopecten purpuratus*) larvae cultured in Chile and was the sole pathogen isolated from dead larvae on the bottom of the tank. Other pathogens cultured included *Vibrio alginolyticus*, which was isolated from swimming larvae, seawater of larval tanks, a reservoir tank of filtered (50 μm) seawater and microalgae (*Chaetoceros* sp. and *Isochrysis galvana*), and *Vibrio splendidus*, which was isolated from water pipe walls and reservoir tank 10 μm filtered seawater (Riquelme *et al.*, 1996). All isolates were pathogenic at concentrations of 5.5×10^4 to 5.5×10^2 cells/ml in experimental challenge with some strains of *A. hydrophila* causing the highest mortality at 5.5×10^2 CFU/ml. Infection in fairy shrimp (*Branchipus schaefferi*) is referred to as black disease due to the black nodules on thoracic appendages, the cercopods and antennae (Dierckens *et al.*, 1998).

Although *A. hydrophila* is mainly associated with freshwater environments, it has also been reported from mortalities of marine fish and cultured seabream (*Dicentrarchus labrax* and *Puntazzo puntazzo*) in Greece (Doukas *et al.*, 1998).

Aeromonas hydrophila is a pathogen of cultured American bullfrogs (*Rana castesbeiana* – note *Lithobates catesbeianus* is thought to be the more correct name). Infected frogs display torticollis and stupor. Focal areas of haemorrhage occur in the muscle, legs are swollen and liver, spleen and kidney may be enlarged (Mauel *et al.*, 2002).

GROWTH CHARACTERISTICS AND PHENOTYPIC IDENTIFICATION. *Aeromonas hydrophila* ssp. *hydrophila* is a motile, facultatively anaerobic, Gram-negative rod. On BA colonies are 2 mm, cream and beta-haemolytic after 24 h at 25–37°C. Growth occurs on MacConkey agar. Weak growth may occur on TCBS.

Variable results exist in the literature for lysine decarboxylase (LDC) for strains including the type strain ATCC 7966 (Abbott *et al.*, 1992; Kaznowski 1998; Nielsen *et al.*, 2001; Huys *et al.*, 2002b). This may be due to misinterpretation of the test, or the use of different methods. In the conventional tube test using Moeller's medium, LDC is a pale purple colour, particularly in comparison to the ADH tube. Also, a stronger result or a positive result may be obtained after 48 h incubation rather than at 24 h incubation. LDC should be positive (Martin-Carnahan and Joseph, 2005). Reports also vary for utilization of citrate and reaction for methyl red. Citrate is positive for 60% of strains of *A. hydrophila*, and MR is positive for 53% of strains (Abbott *et al.*, 1992). These and other reactions are presented in Tables 3.3, 3.24–30 and 3.38.

MOLECULAR DETECTION. A multiplex PCR has been developed for the simultaneous detection of *Flavobacterium columnare*, *Edwardsiella ictaluri* and *Aeromonas hydrophila* (Panangala *et al.*, 2007). Primers designed to the aerolysin gene are AerOFd (5′-CCA AGG GGT CTG TGG CGA CA-3′) and reverse primer AeroRs (5′-TTT CAC CGG TAA CAG GAT TG-3′). For the PCR reaction see under *E. ictaluri*.

Aeromonas hydrophila ssp. dhakensis

TAXONOMY AND SIGNIFICANCE. *Aeromonas hydrophila* ssp. *dhakensis* (group BD-2, LMG 19562^T=CCUG 45377^T) was isolated from the faeces of children with diarrhoea in Bangladesh in 1997 (Huys *et al.*, 2002b). DNA–DNA hybridization showed a 77–94% homology within the BD-2 isolates and a 78–92% homology to *A. hydrophia*, and these findings together with phenotypic differences, and FAFLP and ERIC-PCR fingerprinting suggested a subspecies. Sequence analysis of the *cpn60* gene indicates 37–40 nucleotide differences (7–7.6% divergence) between *A. hydrophila* ssp. *dhakensis* and other *A. hydrophila* subspecies suggesting that it is a novel *Aeromonas* species (Miñana-Galbis *et al.*, 2009). Other phylogenetic analysis using sequence information from housekeeping genes suggests that *A. hydrophila* ssp. *dhakensis* cannot be separated from the newly described *A. aquariorum* and that they are heterotypic synonyms (Martínez-Murcia *et al.*, 2009). According to the rules of nomenclature *A. hydrophila* ssp. *dhakensis* cannot be reclassified as *A. aquariorum* as suggested by Martínez-Murcia *et al.* (2009) (Euzéby, 1997; http://www.bacterio.cict.fr/a/aeromonas.html).

Of the ten strains tested, nine were haemolytic against sheep red blood cells and eight had high cytotoxin titres (Huys *et al.*, 2002b).

GROWTH CHARACTERISTICS AND PHENOTYPIC IDENTIFICATION. *Aeromonas hydrophila* ssp. *dhakensis* is a Gram-negative, motile, straight rod. Positive for oxidase, catalase and nitrate and resistant to vibriostatic agent 0/129 150 μg. Optimal growth on TSA occurs at 28°C after 24 h but growth can occur at 42°C after 24–48 h. Brown pigment is not produced.

Phenotypically *A. hydrophila* ssp. *dhakensis* can be differentiated from *A. hydrophila* ssp. *hydrophila* as the former is negative for acid production from L-arabinose, and negative for utilization of methyl α-D-mannoside, L-fucose and L-arabinose, whereas *A. hydrophila* ssp. *hydrophila* is positive for acid production from L-arabinose and positive for utilization of methyl α-D-mannoside, L-fucose and L-arabinose (Huys *et al.*, 2002b).

Aeromonas hydrophila ssp. proteolytica

See *Vibrio proteolyticus*.

Aeromonas hydrophila ssp. ranae

TAXONOMY AND SIGNIFICANCE. Seven sucrose-negative strains of *Aeromonas* sp. (Au-1D12^T= LMG 19707^T=CCUG 46211^T) showing haemolytic activity against frog erythrocytes and designated as Group Au were isolated from the liver and kidney of septicaemic, farmed Chinese bullfrogs (*Rana rugulosa*) in Thailand (Pearson *et al.*, 2000). These strains were subsequently identified as *A. hydrophila* ssp. *ranae* (Huys *et al.*, 2003b). The isolates belong to *A. hydrophila* HG 1 and have >99.0% 16S rRNA sequence similarity with *A. hydrophila* ssp. *hydrophila*, but are a specific subgroup based on FAFLP, which reveals a genotypic relatedness of 51% to *A. hydrophila* spp. *hydrophila* and a 33% correlation to *A. hydrophila* spp. *dhakensis* (Huys *et al.*, 2003b). DNA–DNA hybridization indicates a relatedness of 80–93% to *A. hydrophila* ssp. *hydrophila*, 75–87% to *A. hydrophila* ssp. *dhakensis* and 46–64% to all other *Aeromonas* species.

Aeromonas hydrophila ssp. *ranae* has strong haemolytic activity against frog erythrocytes due to the presence of the *ASH1* gene, a putative virulence gene. The organism also produces elastase and is cytotoxic to rainbow trout cells, but not to a mammalian cell line (Pearson *et al.*, 2000). A challenge trial with *A. hydrophila* ssp. *ranae* resulted in septicaemia and mortality of frogs.

GROWTH CHARACTERISTICS AND PHENOTYPIC IDENTIFICATION. *Aeromonas hydrophila* ssp. *ranae* is a motile, Gram-negative, straight rod, oxidase and catalase positive with optimal growth at 28°C after 24 h. Brown soluble pigment is not produced on TSA. The type strain is negative for VP, but three of the other six strains were positive (Huys *et al.*, 2003b). These and other reactions are presented in Tables 3.3 and 3.24–26.

Aeromonas ichthiosmia

TAXONOMY AND SIGNIFICANCE. *Aeromonas ichthiosmia* (Type strain LMG 12645^{T}) is a later synonym of *A. veronii* (Type strain 9075^{T}, HG 10) as shown by 16S rRNA sequence studies, AFLP and DNA–DNA hybridization (Collins *et al.*, 1993; Huys *et al.*, 2001).

Aeromonas jandaei

TAXONOMY AND SIGNIFICANCE. Originally, nine strains of sucrose-negative *Aeromonas* species were isolated from clinical sources (wound infections, diarrhoea), water and from a prawn and were found to cluster with *A. sobria* (CDC 0787-80=ATCC 49568=CECT 4228), the definition strain for DNA HG 9. These strains were identified as a new species, *A. jandaei* (HG 9) (Carnahan *et al.*, 1991a). This taxon has been amended after a further polyphasic study that included 14 sucrose-positive strains from freshwater and from reared European eels (*Anguilla anguilla*). *Aeromonas jandaei* now includes sucrose-positive strains from clinical and environmental sources. Sucrose-positive strains from clinical sources were often identified as *A. veronii* biogroup Sobria and some, such as CECT 4910 (CCUG 30361=LMG 13072), are now re-classified as *A. jandaei*. Other strains of the *A. veronii* biovar Sobria do not cluster with *A. jandaei* and the *A. veronii* biovar Sobria needs further revision (Esteve *et al.*, 2003). According to DNA–DNA hybridization, isolates within the *A. jandaei* cluster, identified as phenon 1, have a similarity of 81.6% with a range of 75–87%.

Sucrose-negative strains of *A. jandaei* have been isolated from the North American medicinal leech, *Macrobdella decora*, where the bacterium is the predominant digestive tract symbiont (Siddall *et al.*, 2007). Other strains have been detected in diseased aquarium fish: electric blue hap (*Sciaenochromis ahli* Trewavas) and blue discus fish (*Symphysodon aequifasciatus* Pellegrin) (Buller, unpublished).

Pathogenicity trials conducted on European elvers (*Anguilla anguilla*) showed that *Aeromonas jandaei* was a primary pathogen for eels. Intraperitoneal inoculation produced a LD_{50} of $10^{5.4}$–10^{7} CFU/ml with mortalities beginning at 18 h post-challenge. In bath challenge experiments mortalities occurred at 9–13 days post-challenge with 17–50% mortalities for challenge doses of 1.2×10^{7} to 3.0×10^{8} CFU/ml (Esteve *et al.*, 1993). Ulcerations occurred on the tail and purulent fluid was present around the gills, and the bacterium was isolated in low numbers and pure growth from the liver and kidney. In the same set of experiments, *A. jandaei* was non-pathogenic for trout (*Oncorhynchus mykiss*) held at temperatures of 13–15°C. Further pathogenicity trials on juvenile eels established *A. jandaei* isolates as highly virulent (LD_{50} $<10^{7.0}$ CFU) or moderately virulent (LD_{50} $10^{7.0}$–$10^{7.5}$ CFU) (Esteve *et al.*, 2003).

Most strains of *A. jandaei* produce elastase when tested by modified Sharman's medium (Hsu *et al.*, 1981) and elastase is thought to be a virulence factor for many *Aeromonas* species (Esteve *et al.*, 2003).

GROWTH CHARACTERISTICS AND PHENOTYPIC IDENTIFICATION. *Aeromonas jandaei* is a motile, Gram-negative, straight rod. Growth occurs on TSA, MacConkey agar and TCBS after 24 h at 28°C. Brown pigment is not produced. Growth occurs in the presence of 0–3% NaCl and at temperatures of 4–42°C. Oxidase, catalase and nitrate are positive and the organism is resistant to vibriostatic agent O/129. Fermentation of sucrose is variable but the type strain is sucrose-negative. The organism is positive for Moeller's and Thornley's arginine dihydrolase and positive for hydrolysis of elastin, casein, starch and red blood cells but is negative for hydrolysis of arbutin (Esteve *et al.*, 2003). These and other reactions are presented in Tables 3.3, 3.24–26 and 3.30.

Aeromonas media

TAXONOMY AND SIGNIFICANCE. *Aeromonas media* (R_m=ATCC 33907^T) was isolated from river water in England (Allen *et al.*, 1983). In the ATCC information for this strain it is listed as being isolated from fish-farm effluent. Fifteen strains were examined in the initial isolation. *Aeromonas media* has been reported as a potential probiotic organism with activity against *Vibrio tubiashii* in challenge experiments with Pacific oyster (*Crassostrea gigas*) (Gibson *et al.*, 1998).

GROWTH CHARACTERISTICS AND PHENOTYPIC IDENTIFICATION. *Aeromonas media* is a non-motile, facultatively anaerobic, Gram-negative rod (1×2 μm). Positive for catalase and oxidase. On TSA and BA colonies are 2 mm, cream, raised, round, shiny and entire after 48 h at 22°C. A brown diffusible, non-fluorescent pigment is produced. Growth occurs at 4–37°C but not at 42°C. The type strain is negative for growth at 3% NaCl but other strains are positive. Growth does not occur at 5% NaCl. Nitrate is reduced. Cellulose, elastin and urea are not degraded. Reactions are variable for hydrolysis of DNA, chitin and aesculin, but the type strain is negative for DNAse and aesculin and positive for hydrolysis of chitin. Sodium citrate is not utilized by the type strain but is variable for other strains. H_2S is not produced on Klinger iron agar (Oxoid). These and other reactions are presented in Tables 3.3 and 3.30.

Aeromonas molluscorum

TAXONOMY AND SIGNIFICANCE. Five strains isolated from cockles (*Cardium* sp.), mussels (*Mytilus* sp.), razor shells (*Ensis* sp.) and wedge-shells (*Donax trunculus*) collected at retail markets in Barcelona, Spain, were identified as *A. molluscorum* (848^T= CECT 5864^T=LMG 22214^T) (Miñana-Galbis *et al.*, 2004). DNA–DNA hybridization showed that *A. molluscorum* was <70% (27–45%) similar to the type strains of the *Aeromonas* HGs and had a G+C content of 59.0–59.4%. Sequence similarity of the 16S rRNA gene indicates similarity of <97% to other *Aeromonas* species with the highest similarity being to *A. encheleia*. Sequence analysis using the *gyrB* and *rpoD* genes demonstrates that *A. molluscorum* strains are different from each other, but form a tight group that branches close to *A. encheleia* confirming 16S rRNA gene sequence information and the designation of this species (Saavedra *et al.*, 2006). Sequence analysis using the *cpn60* gene also confirms the designation of the species *A. molluscorum* and clearly separates it from *A. bivalvium* (Miñana-Galbis *et al.*, 2009). According to FAFLP banding patterns, *A. bivalvium* does not cluster significantly with other *Aeromonas* species.

GROWTH CHARACTERISTICS AND PHENOTYPIC IDENTIFICATION. *Aeromonas molluscorum* is a motile, Gram-negative rod 1.2–2.2×0.5–1.0 μm, oxidase and catalase positive with optimal growth between 25 and 30°C and a temperature range of 4–37°C. Growth does not occur at 40.5°C. Growth occurs at 0–3% NaCl but not at 6% NaCl. Colonies on TSA supplemented with 0.5% (w/v) NaCl are 3–4 mm in diameter, non-swarming, circular and beige at 48 h and 25°C. Brown pigment is not seen, growth occurs on MacConkey agar and the organism is resistant to vibriostatic agent 0/129 150 μg. Arbutin is hydrolysed but not starch or elastin and acid is produced from D-cellobiose (Miñana-Galbis *et al.*, 2004). These and other reactions are presented in Table 3.3.

Aeromonas piscicola

TAXONOMY AND SIGNIFICANCE. *Aeromonas piscicola* (S1.2^T=CECT 7443^T=LMG 24783^T) was isolated from moribund fish, salmon (*Salmo salar*) and from trout (*Oncorhynchus mykiss*) in Spain (Beaz-Hidalgo *et al.*, 2009b). Sequence analysis of the five strains using the 16S rRNA gene revealed an almost identical sequence to *A. bestiarum*, *A. salmonicida* ssp. *masoucida* and to *A. salmonicida* ssp. *achromogenes* for four strains whereas a fifth strain clustered with *A. salmonicida* ssp. *salmonicida*. Phylogenetic analysis using *rpoD* and *gyrB* genes indicated *A. piscicola* formed a separate cluster and was a new species. Data using *recA* and *dnaJ* sequences confirmed a separate species, however they did show close similarity with *A. bestiarum*.

DNA–DNA hybridization confirmed species delineation of values below 70%.

GROWTH CHARACTERISTICS AND PHENOTYPIC IDENTIFICATION. *Aeromonas piscicola* is a motile, Gram-negative rod (1.5–2.0×0.6–1.0 µm). Colonies do not produce diffusible brown pigment, but haemolysis is seen on BA. Oxidase and catalase are positive; growth is optimal at 25–30°C with a range of 4 to 37°C and growth occurs at 0–3% NaCl. Nitrate is reduced, elastin, starch and Tween 80 are hydrolysed and strains are resistant to vibriostatic agent 0/129 150 µg. Fermentation of salicin and hydrolysis of arbutin are positive in the API 50CH when incubated at 25°C for 24 h, but are negative when tested at 37°C. Fermentation of salicin at 37°C is positive when tested in nutrient broth (Difco) containing a final concentration of salicin of 1% (Beaz-Hidalgo *et al.*, 2009b). These and other reactions are presented in Tables 3.3, 3.24–26 and 3.30.

Aeromonas popoffii

TAXONOMY AND SIGNIFICANCE. Seven isolates named *A. popoffii* (LMG 17541^T) were isolated from drinking water production plants in Belgium and Scotland. The strains are closely related to *A. bestiarum* with a DNA–DNA relatedness of 51–63%, followed by *A. hydrophila* at 50–60% and *A. salmonicida* at 39–55%. Results from ribotyping using AFLP and from cellular fatty acids confirmed the relatedness with *A. bestiarum* (Huys *et al.*, 1997b). Similarity using the 16S rRNA gene showed a 99.71% similarity with a four to six nucleotide difference to *A. bivalvium* although FALP showed that *A. popoffii* did not cluster significantly with *A. bivalvium* (Miñana-Galbis *et al.*, 2007).

GROWTH CHARACTERISTICS AND PHENOTYPIC IDENTIFICATION. *Aeromonas popoffii* is a facultatively anaerobic, Gram-negative, motile, straight rod. Oxidase, catalase and nitrate are positive and the organism is resistant to vibriostatic agent 0/129 150 µg. Growth occurs on TSA in 24 h at an optimal temperature of 28°C but growth also occurs at 37°C. Brown water-soluble pigment is not produced. Growth occurs in 0% NaCl but not in 3–10% NaCl. Casein, gelatin, starch, Tween 80 and lethicin are hydrolysed, but not aesculin or chitin. The type strain LMG 17541 is indole-negative, but other strains are indole-positive (Huys *et al.*, 1997b). These and other reactions are presented in Tables 3.3, 3.24–26 and 3.30.

Aeromonas rivuli

TAXONOMY AND SIGNIFICANCE. *Aeromonas rivuli* (WB4.1-19^T=CECT 7518^T= DSM 22539^T=MDC 2511^T) was isolated from a karst water rivulet in northern Germany (Figueras *et al.*, 2011a). Based on 16S rRNA sequence analysis *A. rivuli* is 99.7% similar to *A. sobria*, however multilocus phylogenetic analysis using sequence information from housekeeping genes *gyrB*, *rpoD*, *recA*, *dnaJ* and *gyrA* indicates *A. rivuli* clusters with *A. molluscorum* and *A. bivalvium*.

GROWTH CHARACTERISTICS AND PHENOTYPIC IDENTIFICATION. *Aeromonas rivuli* is a motile, facultatively anaerobic, Gram-negative rod (2.5–2.0×0.7–0.5 µm). Oxidase and catalase are positive. Nitrate is reduced. Resistant to vibriostatic agent 0/129 150 µg. On TSA and BA colonies are non-haemolytic, 2–2.5 mm, beige and opaque at 30°C after 48 h, and 1–2 mm after 48 h at 37°C. Brown diffusible pigment is not produced. Growth occurs at 7–37°C with an optimum at 30°C and at 0–3% NaCl. Utilization of citrate is variable (Figueras *et al.*, 2011a). These and other reactions are presented in Tables 3.3 and 3.24–30.

Aeromonas salmonicida

In contrast to all other *Aeromonas* species apart from *A. sobria*, *A. salmonicida* is non-motile. *A. salmonicida* spp. *salmonicida* belongs to *Aeromonas* DNA HG 3, which contains *A. hydrophila*. Although the use of hybridization group is now considered outdated for taxonomic purposes (Janda and Abbott, 2010), noting these designations assists in maintaining the history of particular species. Five subspecies are recognized, *salmonicida*, *achromogenes*, *masoucida*, *smithia* and *pectinolytica* (Smith, 1963; Schubert, 1967; Kimura, 1969; Austin *et al.*, 1989; Pavan *et al.*, 2000). In addition to the recognized subspecies, a number of strains exhibiting phenotypic differences to the species has been reported. These differences include variation in growth temperature, oxidase reaction, delayed or absent brown pigment production, and slow or fastidious growth. The subspecies *nova* was proposed to accommodate all atypical strains (McCarthy, 1980), however this has not been formally accepted and it is virtually impossible with the great variation in phenotypic tests to designate these strains to one subspecies.

Aeromonas salmonicida ssp. *salmonicida* ('typical') strains are a homogeneous genetic group regardless of geographical origin as shown by DNA homology (Belland and Trust, 1988). *Aeromonas salmonicida* is related to *A. hydrophila* at the genus level but is distinct at the species level. *Aeromonas salmonicida* ssp. *achromogenes* and *A. salmonicida* ssp. *masoucida* are closer to the motile aeromonads based on DNA:DNA homology. Belland and Trust (1988) proposed that these two subspecies should be combined and called *A. salmonicida* ssp. *achromogenes*, however this has not been formally accepted. DNA homology of 15 atypical strains from non-salmonids demonstrated a diverse heterogeneous group containing two distinct biotypes, one from goldfish and one from European carp, with the remaining isolates not conforming to a group. It has been suggested that the subspecies designations, with the exception of *smithia*, are invalid and should be replaced by biovar classifications; however, this is yet to be adopted (Holmes, 1998; Farmer and Holmes, 2009).

Strain ATCC7965=NCTC 7812=CIP 57.50, originally identified as *A. hydrophila*, is in fact *A. salmonicida* based on results of phenotypic characteristics and sequence analysis using *cpn60*, *dnaJ*, *gyrB* and *rpoD* genes (Miñana-Galbis *et al.*, 2010b).

Aeromonas salmonicida ssp. salmonicida

TAXONOMY AND SIGNIFICANCE. *Aeromonas salmonicida* ssp. *salmonicida* (previously *Bacillus salmonicida*) was first described by Emmerich and Weibel (1894) as causing disease in cultured brown trout (*Salmo trutta fario*) and brook trout (*Salmo fontinalis*) in Germany in 1888. The disease, furunculosis, affects mainly salmonid fish and is typified by characteristic umbonate ulcers or abscess-like cavities termed furuncles, in the dermis, which develop internally in the musculature and can discharge pus or extend deeply to expose bones. Infection leads to septicaemia and acute mortality. The organism penetrates into the underlying tissues and organs and can be cultured from the blood, muscles and internal organs (Arkwright, 1912; Cipriano and Bullock, 2001). *Aeromonas salmonicida* ssp. *salmonicida* has been reported from salmon, brown trout (*Salmo trutta*), rainbow trout (*Oncorhynchus mykiss*), grayling, bream, eels and sea lamprey (*Petromyzon marinus*), from countries including Germany, Austria, Finland, France, Portugal, Switzerland, the UK and the USA (Arkwright, 1912; Noga and Berkhoff, 1990; Hirvelä-Koski *et al.*, 1994; Sousa *et al.*, 1997; Cipriano and Bullock, 2001; Faisal *et al.*, 2007a). The Smith strain, strain 20, NCMB 1102 was isolated from *Salmo salar* in Welsh rivers and stored in the Public Health Laboratory, Carmarthen (Smith, 1963). Schubert (1967) incorrectly refers to this strain as *A. salmonicida* ssp. *achromogenes*.

Production of a brown water-soluble pigment is often one of the distinguishing features between *Aeromonas salmonicida* ssp. *salmonicida* and the atypical strains, which produce either no pigment, or have a delayed pigment production. However, this is not absolute as a non-pigment-producing strain of *A. salmonicida* ssp. *salmonicida* caused disease outbreaks in Atlantic salmon (*S. salar*), and in experimental infection the non-pigment-producing strain caused significantly higher mortalities than the reference strain (Koppang *et al.*, 2000).

Aeromonas bestiarum is highly similar to *A. salmonicida* ssp. *salmonicida* and they are almost impossible to separate using biochemical tests, 16S rRNA sequencing or cluster analysis of the 16S-23S intergenic spacer region RFLP. However, they can be separated by phylogenetic analysis using housekeeping genes *gyrB* and *rpoD* (Martínez-Murcia *et al.*, 2005).

GROWTH CHARACTERISTICS AND PHENOTYPIC IDENTIFICATION. *Aeromonas salmonicida* ssp. *salmonicida* (NCMB 1102) is a gram-negative, evenly staining coccobacillus occurring in pairs, chains and clumps when grown in nutrient broth. On nutrient agar colonies are raised, round, yellowish-grey, translucent and friable and produce a brown water-soluble pigment (Smith, 1963). Production of brown diffusible pigment is examined over 5–7 days. Pigment production is detected using TSA plates supplemented with l-tyrosine (0.1%) and l-phenylalanine (0.1%) and incubated at 5, 15 and 22°C for 7 days (Griffin *et al.*, 1953) and on Columbia Base agar with and without 0.1% D-glucose at 22°C (Altmann *et al.*, 1992).

Furunculosis Agar and TSA containing 0.5% NaCl may be used for growth of the organism and production of pigment. Pigment production occurs under aerobic conditions but not under anaerobic conditions (Donlon *et al.*, 1983). Pigment production may be reduced in the presence of D-glucose. Addition of 0.1% w/v glucose delays pigment production, and 0.15% (w/v) glucose completely inhibits pigment production (Altmann *et al.*, 1992). Atypical *A. salmonicida* strains from goldfish in Australia have intense pigment production after 3 days on Columbia agar, becoming light brown at 6 days on TSA.

Incubation temperature is critical, as differences in biochemical reactions have been noted at different temperatures such as 11, 18 and 28°C (Hahnel and Gould, 1982). For consistency, an incubation temperature of 22°C over 3–5 days is recommended and was used in the study by Koppang *et al.* (2000). Some strains are capable of growth at 37°C (Austin, 1993).

The A-protein layer, a major virulence factor for *A. salmonicida* ssp. *salmonicida*, can be detected using TSA supplemented with Coomassie brilliant blue (CBB agar) where bacterial colonies containing the A-protein layer are seen as blue coloured colonies, and A-protein-negative colonies appear white (Evenberg *et al.*, 1985). Autoagglutination in distilled water also indicates the presence of the A-protein layer. However, these tests are not reliable indicators of the *in vivo* virulence of *A. salmonicida* (Bernoth, 1990; Olivier, 1990). These and other reactions are presented in Tables 3.1 and 3.24–30.

MOLECULAR IDENTIFICATION. Self-quenched fluorogenic PCR primers for the detection of *Aeromonas salmonicida* have been developed (Balcázar *et al.*, 2007). DNA is extracted from bacterial cultures by homogenizing 1.0 ml of broth culture in 200 µl of TE buffer (1 0mM Tris/HCl, pH 8.0, 1 mM EDTA) and centrifuging at 1200 ×*g* for 1 min. The pellet is treated with InstaGene Matrix (Bio-Rad) according to the manufacturer's instructions. The LUX fluorogenic primers are 5′-CGG AAC GTA ATC TGA ATT GTT CTT TTC(FAM) G-3′ and primer 5′-ATT GCT TAT CGA GGC AGC CAA C-3′. A 50-µl master mix contains 25 µl Platinum Quantitative PCR SuperMix-UDG (uracil DNA glycosylase) 60 U Platinum *Taq* DNA polymerase/ml, 40 mM Tris/HCl (pH 8.4), 100 mM KCl, 6 mM $MgCl_2$, 400 µM each dGTP, dATP and dCTP, 800 µl dUTP, 40 U UDG/ml and stabilizers, FAM-labelled LUX primer and unlabelled primer at 200 nM each as a final concentration, ROX reference dye and 10 µl DNA template. Thermocycling is carried out on an ABI Prism 7000 sequence detection thermocycler using an initial step at 50°C for 2 min and denaturation at 95°C for 2 min, followed by 40 cycles of denaturation at 95°C for 15 s, annealing at 55°C for 30 s, and extension at 72°C for 30 s.

Atypical Aeromonas salmonicida

TAXONOMY AND SIGNIFICANCE. The term 'atypical' is used to describe strains not identified as *A. salmonicida* ssp. *salmonicida* and comprises strains that are slow growing, have slow or no pigment production, and biochemical characteristics different from *A. salmonicida* ssp. *salmonicida* and the recognized subspecies. The atypical strains are a diverse heterogeneous group both genetically and phenotypically. Studies of the strains using RAPD, ribotyping, presence of genes for the extracellular A-layer protein, serine protease, glycerophospholipid:cholesterol acetyltransferase gene, the 16S rRNA gene and biochemical methods have not been able to categorize the strains (Austin, 1993; Hirvelä-Koski *et al.*, 1994; Wiklund *et al.*, 1994; Hänninen and Hirvelä-Koski, 1997; Austin *et al.*, 1998). Recognized subspecies include *A. salmonicida* ssp. *achromogenes*, *A. salmonicida* ssp. *masoucida*, *A. salmonicida* ssp. *pectinolytica* and *A. salmonicida* ssp. *smithia*. Other phenotypic variants have been reported such as an oxidase-negative strain, a strain capable of growth at 37°C and biovars *A. salmonicida* biovar Acheron and *A. salmonicida* biovar Flounder (Smith, 1963; Austin, 1993; Pedersen *et al.*, 1994; Larsen and Pedersen, 1996; Iida *et al.*, 1997a; Wiklund and Dalsgaard, 1998; Carson and Gudkovs, 2001).

The pathogenicity of atypical strains shows great variability, and virulence mechanisms differ from those described for 'typical' *Aeromonas salmonicida*. An extracellular A-layer, an extracellular metallo-protease and a different iron utilization mechanism have been described for atypical strains compared to typical strains (Wiklund and Dalsgaard, 1998). Current thought suggests that the skin ulceration seen in the disease may be due to only one well-defined biotype of atypical *A. salmonicida*, yet these organisms cannot be isolated consistently. The organisms are most commonly isolated in the healing stage of the ulcer, suggesting that the infection is not lethal (Wiklund *et al.*, 1999).

The atypical strains infect a wider variety of fish species than does *A. salmonicida* ssp. *salmonicida* and have been reported from a number of non-salmonid species from freshwater, estuarine and marine environments often with a diverse pathology and variation in virulence and phenotypic tests (Wiklund and Dalsgaard, 1998). Affected species have included marine fish, sable fish (*Anoplopoma fimbria*), moribund sockeye salmon (*Oncorhynchus nerka*) and chum salmon (*O. keta*) (Evelyn, 1971), from ulcerated dab (*Limanda limanda*), flounder (*Platichthys flesus*) and turbot (*Scopthalmus maximus*) caught in the Baltic Sea off Finland (Wiklund *et al.*, 1999) and wrasse (*Labridae*) used as cleaner fish of farmed Atlantic salmon (*Salmo salar*) in Scotland (Laidler *et al.*, 1999).

Disease in non-salmonids with the atypical strains is often characterized by cutaneous ulceration without septicaemia. In goldfish, infection with *A. salmonicida* is referred to as goldfish ulcer disease (GUD) and causes carp erythrodermatitis in cultured and wild stocks of carp (Elliott and Schotts, 1980).

Infected sand-eels (*Ammodytes lancea* and *Hyperoplus lanceolatus*) caught off the coast of Denmark and held in recirculating tanks containing seawater, developed small haemorrhages on the caudal fin and flank, necrosis of the tail, and umbonate furuncles on the flank (Dalsgaard and Paulsen, 1986). Herring held in the same tank were not affected. The strain was phenotypically different to strains from rainbow trout farmed in Denmark and in experimental infection was pathogenic to rainbow trout.

Atypical *A. salmonicida* was reported from rainbow trout (*Oncorhynchus mykiss*) farmed in seawater in South Africa. The lesions resembled furunculosis, however, bacterial colonies rarely occurred in internal organs and few gross changes were seen in the kidney and spleen (Boomker *et al.*, 1984). The organism could not be identified to any known subspecies.

In Australia, an atypical biovar was established in wild goldfish in south-eastern Australia due to infected carrier fish imported from Japan in 1974 (Trust *et al.*, 1980; Humphrey and Ashburner, 1993). Infected fish developed dermal ulcers typical of GUD. Since the initial spread of the disease it has been reported from koi carp (*Cyprinus carpio*) (Langdon 1988), roach (*Rutilus rutilis*), silver perch (*Bidyanus bidyanus*) in Victoria, New South Wales and Western Australia (Humphrey and Ashburner, 1993). Restriction enzyme analysis of Australian strains of atypical *Aeromonas salmonicida* from goldfish and silver perch demonstrated genetic similarity and were similar to strains from the USA, but slightly different to a strain found in infected fish imported to Tasmania from Singapore. Strains from greenback flounder formed a separate genetic group (Whittington *et al.*, 1995). The atypical goldfish strains were highly virulent to Atlantic salmon with an LD_{50} of 3 CFU/fish in experimental challenge using bath immersion and intraperitoneal inoculation. The organism was recovered from kidney and from cutaneous hyperaemic lesions with marked inflammatory processes associated with muscle and dermal necrosis (Carson and Handlinger, 1988). The goldfish strain inoculated into brown trout (*Salmo trutta*), rainbow trout (*O. mykiss*) and brook trout (*Salvelinus fontinalis*) was highly virulent with an LD_{50} of 10^2–10^3 CFU. Clinical and histopathological signs consistent with furunculosis included sudden death, cutaneous melanization, bacterial colonization of gills and internal organs and lack of inflammatory cell infiltration. Necrotizing skin lesions consistent with subacute furunculosis occurred in some fish.

Two biovars have been reported from Tasmania, Australia. In 1993, *A. salmonicida* biovar Flounder was isolated from kidney and caused ulcerated dermal lesions in juvenile hatchery-reared and wild-caught greenback flounder (*Rhombosolea tapirina*) held in shore-based tanks during grow-out trials. In 2000, biovar Acheron was isolated from a limited disease outbreak in Atlantic salmon. Whole cell protein profiles exhibited differences between atypical *A. salmonicida* from goldfish, and biovars Acheron and Flounder (Carson and Gudkovs, 2001).

GROWTH CHARACTERISTICS AND PHENOTYPIC IDENTIFICATION. Isolation of 'atypical' strains is improved using BA or TSA-B as the primary isolation medium. Identification of *A. salmonicida* is based on the following characteristics: Gram-negative rod, negative motility, catalase positive, oxidase positive, production of acid from glucose, resistance to vibriostatic agent 0/129 and usually lack of growth at 37°C (Shotts *et al.*, 1980). Atypical strains display variations in biochemical reactions, may have slow growth, or are slow to produce pigment, or are negative for oxidase, and may produce a variety of extracellular proteases (Wiklund and Dalsgaard, 1998). In an inter-laboratory study to evaluate biochemical methods for identification of the species, even after standardization of tests laboratories reported discrepancies for a number of tests (Dalsgaard *et al.*, 1998). Studies involving ribotyping, presence of genes and RAPD have not resolved the many strains into any particular groups.

Pigment production occurs under aerobic conditions but not under anaerobic conditions (Donlon *et al.*, 1983). Pigment production may be reduced in the presence of D-glucose. Addition of 0.1% w/v glucose to TSA delays pigment production, and 0.15% (w/v) glucose completely inhibits pigment production (Altmann *et al.*, 1992). Atypical *A. salmonicida* strains from goldfish in Australia have intense pigment production after 3 days on Columbia agar becoming light brown at 6 days on TSA. Pigment production can vary and may depend upon the media used. In a study of atypical strains by Hänninen and Hirvelä-Koski (1997), furunculosis agar detected the highest number of strains with pigment (100%), followed by BHIA (86%) and TSA (74%). *Aeromonas salmonicida* ssp. *achromogenes* produces pigment on furunculosis agar, TSA and BHIA but not on nutrient agar. The addition of L-tyrosine to TSA or BHIA leads to an increase in pigment detection, however pigment is still detected earlier on furunculosis agar within 3 to 7 days at 20°C. TSA, BHIA and furunculosis agar are not recommended for primary isolation, as the number of organisms detected is not as high as when blood agar is used. However, they are suitable for subculture (Bernoth and Artz, 1989; Austin, 1993).

Atypical *A. salmonicida* strains causing GUD grow as small friable colonies on trypticase soy agar at 48–72 h incubation at 25°C. Optimal growth temperature is 18–25°C with no growth at 37°C. Brown water-soluble pigment is not seen until prolonged incubation after day 8 at which time colonies are 4–8 mm in diameter and are convex, umbonate with entire edges. Granular growth is seen in liquid culture. Cells autoagglutinate in 0.9% saline. Cells are non-motile, Gram-negative with some bipolar staining. Cells are coccobacilli 0.9–5 μm, but some filamentous forms may be present. Atypical strains can be differentiated from classical *A. salmonicida* ssp. *salmonicida* by fermentation of sucrose, arabinose, salicin and trehalose, production of indole, aesculin hydrolysis, β-galactosidase activity and alkaline reaction in Moeller's ADH. Phenotypic variations are seen with Australian goldfish strains. A FAT for the Australian strain of *Aeromonas salmonicida* has been noted to cross-react with *Aeromonas sobria*. These and other reactions are presented in Tables 3.1, 3.24–30 and 3.38.

Aeromonas salmonicida ssp. *achromogenes*

Aeromonas salmonicida ssp. *achromogenes* (NCTC 1102) was first isolated from seawater off the coast of Scotland (Smith, 1963). The organism caused subacute furunculosis in wild silver bream (*Blicca bjoerkna*), perch (*Perca fluviatilis*), roach (*Rutilus rutilus*) and common bream (*Abramis brama*) found in a drainage channel in England (McCarthy, 1975). The strain did not produce brown pigment but showed cross-reaction to *A. salmonicida* antisera. In experimental challenge the isolates had low virulence for rainbow trout with an LD_{50} of 1.7×10^9 causing death in 10 days. Signs of furunculosis were seen with the organism cultured from heart blood, kidney and spleen. Naturally infected fish had large open lesions on the skin with extensive areas of descalation on the body. Lesions were not typical of furunculosis but had softened and haemorrhagic centres with occasional exposure of underlying musculature. *Aeromonas salmonicida achromogenes* caused mass mortalities in minnows (*Phoxinus phoxinus*) in Lake Tveitevatn, Norway in 1970 (Håstein *et al.*, 1978). As well as skin ulcers some fish had erosion of the bones of the jaws. In both minnows and perch secondary invasion by *Saprolegnia* species was seen. *Aeromonas salmonicida* ssp. *achromogenes* was reported from Atlantic cod (*Gadus morhua*) collected from Terrence Bay, Nova Scotia and held in seawater tanks (Cornick *et al.*, 1984).

Aeromonas salmonicida ssp. *masoucida*

TAXONOMY AND SIGNIFICANCE. *Aeromonas salmonicida* ssp. *masoucida* was first identified as causing furunculosis in 'Sakuramasu' (*Oncorhynchus masou*) and pink salmon (*O. gorbuscha*) cultured in Japan (Kimura, 1969).

Virulent and avirulent strains of *A. salmonicida* ssp. *masoucida* have been isolated from sea cucumber cultured in China (Yang *et al.*, 2008). Virulent strains have an LD_{50} of $1.0\times10^{5.5}$ CFU/animal in immersion trials and an LD_{50} of $1.0\times10^{5.25}$/animal when injected with the organism. Virulence is affected by temperature, as the strains are more virulent at low temperatures (12°C).

Atypical strains caused infection in cultured marine fish, greenling (*Hexagrammos otakii*), Japanese flounder (*Paralichthys olivaceus*), Schlegel's black rockfish (*Sebastes schlegeli*) cultured in Oita Prefecture, Japan (Iida *et al.*, 1997a), and in rockfish cultured in the Okayama Prefecture (Izumikawa and Ueki, 1997). An atypical strain was isolated from ulcerated chub (*Leuciscus cephalis*) held in a spring-fed body of water in Lancashire, England (Wilson and Holliman, 1994). Wild and cultured eel (*Anguilla rostrata*) in North America and Japanese eel (*Anguilla japonica*) infected with atypical strains present as localized depigmented loci progressing to large ulcers that expose the underlying muscle (Kitao *et al.*, 1985; Noga and Berkhoff, 1990).

Aeromonas salmonicida ssp. *masoucida* is a Gram-negative straight rod with rounded ends (1.2×0.5 μm). No flagellum is present (Yang *et al.*, 2008).

Aeromonas salmonicida ssp. pectinolytica

Aeromonas salmonicida ssp. *pectinolytica* ($34mel^T$= DSM 12609^T) was isolated from a heavily polluted river in Argentina (Pavan *et al.*, 2000). It is able to degrade polypectate. Cells are Gram-negative, straight rods that occur singly, in pairs or short chains. Oxidase and catalase are positive and nitrate is reduced. On TSA colonies are seen after 24 h at 35°C. A brown diffusible pigment is produced. These and other reactions are presented in Table 3.1.

Aeromonas salmonicida ssp. smithia

TAXONOMY AND SIGNIFICANCE. *Aeromonas salmonicida* ssp. *smithia* was isolated from diseased fish. Clinical signs include haemorrhages on fins and vent and gastroenteritis (Austin *et al.*, 1989). On TSA colonies are rough or smooth. Non-haemolytic and non-motile. Brown diffusible pigment is not produced. Growth occurs at 4–25°C but not at 30°C. Growth occurs at 0–2% NaCl but not at 3% NaCl. DNA, starch and gelatin are hydrolysed, but aesculin, chitin and elastin are not. These and other reactions are presented in Table 3.1 and some reactions are presented in Tables 3.24–26.

Oxidase-negative strains of atypical Aeromonas salmonicida

Oxidase-negative strains have been reported from turbot (*Scophthalmus maximus*) cultured from saltwater fish farms in Denmark and Norway (Pedersen *et al.*, 1994; Larsen and Pedersen, 1996). Skin lesions on the body start as small erosions with a white centre and a thin haemorrhagic zone and develop into 0.5–3 mm diameter ulcers with an umbonate appearance and a haemorrhagic centre. Bacteria are isolated from the lesions and kidney. The slide agglutination test using rabbit antisera raised to *A. salmonicida* ssp. *salmonicida* is positive for the oxidase-negative strain and Coomassie blue and Congo red tests are positive. The strain has no or low pathogenicity for rainbow trout when tested in experimental challenge. Oxidase-negative strains were reported from flounder (*Platichthys flesus*) in the Baltic Sea (Wiklund and Bylund, 1993) and from coho salmon (*Oncorhynchus kisutch*; strain ATCC 49385) from a hatchery near Belfair, Washington, USA (Chapman *et al.*, 1991). *Aeromonas salmonicida* spp. *salmonicida* was co-isolated, which had a different antibiotic profile to the oxidase-negative strain.

Atypical strains capable of growth at 37°C

Strains capable of growth at 37°C were isolated from goldfish (*Carassius auratus*), carp (*Cyprinus carpio*) and roach (*Rutilis rutilis*) collected from aquaria, garden ponds and rivers in England in 1991 and 1992. Infection resulted in surface ulcers and septicaemia in experimental challenge with rainbow trout and goldfish (Austin, 1993).

Haemophilus piscium

Strains originally identified as *H. piscium* have been included with strains of *Aeromonas salmonicida*. Phenotypic evidence suggests the strain is synonymous with *A. salmonicida* ssp. *achromogenes* yet molecular studies involving ribotyping, RAPD and PCR grouping do not support this (Austin *et al.*, 1998).

Aeromonas sanarellii

TAXONOMY AND SIGNIFICANCE. *Aeromonas sanarellii* (A2-67^T=CECT 7402^T =LMG 24682^T) was isolated from a trauma and abrasion wound in a patient in Taiwan (Alperi *et al.*, 2010a). Based on the 16S rRNA gene, *A. sanarellii* has 99.6% similarity to *A. trota*, 99.7% to *A. caviae* and 99.6% to *A. aquariorum*, which corresponds to 5, 4 and 5 bp differences, respectively. In contrast, multilocus phylogenetic analysis using *gyrB*, *rpoD*, *recA*, *dnaJ* and *gyrA* suggests a greater similarity to *A. media* and *A. caviae*. A phylogenetic tree based on sequence information from the *rpoD* gene indicates *A. sanarellii* has similarity of 92.7% (56 bp differences) to *A. taiwanensis* and 92.2% (63 bp differences) to *A. caviae*. Intraspecies similarity between *Aeromonas* species in the *rpoD* gene is 97% and the similarities of 92.7 and 92.2% are below this threshold, confirming species status (Soler *et al.*, 2004; Alperi *et al.*, 2010a).

GROWTH CHARACTERISTICS AND PHENOTYPIC IDENTIFICATION. *Aeromonas sanarellii* is a facultatively anaerobic, motile, Gram-negative straight rod (2.5×0.7 μm). Oxidase and catalase are positive. Nitrate is reduced. It is resistant to vibriostatic agent 0/129 150 μg. On TSA and BA colonies are non-haemolytic, 4–5 mm after 48 h at 30°C with two different morphologies. One colony type appears beige, opaque and circular, and the other colony type is translucent. Brown diffusible pigment is not produced. Growth occurs at 30–45°C but not at 4 or 50°C after 7 days. Growth occurs at 0% NaCl but not at 6% NaCl (Alperi *et al.* 2010a). These and other reactions are presented in Tables 3.3 and 3.24–3.30.

Aeromonas schubertii

TAXONOMY AND SIGNIFICANCE. *Aeromonas schubertii* (ATCC 43700^T = CDC 2446-81) is the name given to seven strains previously known as Enteric Group 501 (Hickman-Brenner *et al.*, 1988). The strains were isolated from human sources including abscess, leg wound with cellulitis, pleural fluid and blood from patients in Puerto Rico and from the southern coastal states of the USA and Chesapeake Bay. A number of patients appeared to have an underlying disease. Strains were tested for virulence factors and found to be negative for cytotoxins, cell-free haemolysin, haemagglutinins, pectin and chitinase. However, two other clinical strains from patients were positive for cytotoxin, protease, haemolysin and DNAse (Carnahan *et al.*, 1989).

Until recently *A. schubertii* had not been reported as a pathogen for aquatic animals. Mortality due to *A. schubertii* was reported from two species of snakehead cultured in China, *Channa maculata* and *C.* (*Ophiocephalus*) *argus*. The disease symptoms varied as *C. maculata* displayed ulcers and haemorrhages on the ventral body and head, necrosis in liver, intestine and spleen, whereas *C. argus* had white nodules in the kidney (Chen *et al.*, 2012b; Liu and Li, 2012). The identity of the organism was confirmed to be *A. schubertii* through biochemical results, by 16S rRNA sequencing (99% with *A. schubertii*) and the *C. argus* isolate was a 100% match with *A. schubertii* using concatenated sequences from the *gyrB*, *rpoD* and *dnaJ* genes. Pathogenicity was confirmed for both isolates in challenge experiments using intraperitoneal injection. Mortality of 100, 87.5 and 100% was achieved with 5.78×10^7, 5.78×10^6 and 5.78×10^5 CFU/ml, respectively, for *C. argus*, and for *C. maculata* the LD_{50} was 1.4×10^4, 4.5×10^4 and 6.4×10^6 CFU/g, respectively (Chen *et al.*, 2012b; Liu and Li, 2012). Three isolates from *C. maculata* showed phenotypic differences between themselves and with the type strain (Chen *et al.*, 2012b).

GROWTH CHARACTERISTICS AND PHENOTYPIC IDENTIFICATION. *Aeromonas schubertii* is oxidase and nitrate positive and is resistant to vibriostatic agent 0/129 150 μg. The organism grows without NaCl but does not grow in 6% NaCl. It grows at 37°C. Gelatin is hydrolysed at 22°C, DNAse is positive at 25°C and 37°C. The Voges–Proskaüer reaction is positive for the type strain when tested using 1% NaCl added to the MRVP medium. Two clinical strains were later reported as being negative for VP (Carnahan *et al.*, 1989). The strains from snakehead were generally negative for VP, with one strain being positive. Strains from *C. maculata* are variable for ODC and urease, and are different from the type strain in reactions for LDC, urea and DNase (Chen *et al.*, 2012b; Liu and Li, 2012). Using conventional methods the strains are positive for acid from D-galactose and negative for acid from adonitol, D-arabitol, cellobiose, dulcitol, erythritol, melibiose, raffinose and glycerol. The strains are positive for lipase using corn oil (Hickman-Brenner *et al.*, 1988). Haemolysis is weak when tested using TSA and sheep red blood cells, but strong when tested using TSA and horse red blood cells at 22 and 37°C at 24 h (Carnahan *et al.*, 1989). These and other reactions are presented in Tables 3.3 and 3.24–26.

Aeromonas sharmana

TAXONOMY AND SIGNIFICANCE. An organism (GPTSA^T-6=MTCC 7090^T and DSMZ 1744^T) isolated from a warm-temperature spring in India was assigned to the family *Aeromonadaceae* and, based on 16S rRNA gene sequence similarity, to the genus *Aeromonas* (Saha and Chakrabarti, 2006). However, the placement of this organism in the genus *Aeromonas* is in doubt as the 16S rRNA sequence similarity of 94.8–95.1% is too low, as the *Aeromonas* species show a phylogenetic similarity of 96.7–100%. The proposed *A. sharmana* has nucleotide differences in the 16S rRNA regions 86–106 and 584–604 that are conserved in all other *Aeromonas* species identified to date and are used as genus-specific targets (Martínez-Murcia *et al.*, 2007). Results from *cpn60* gene sequence analysis support these findings as 86–111 nucleotide differences (17.4–23.3% divergence) were found, which is significantly higher than the mean of 54.7 nucleotides and 10.6% divergence (Miñana-Galbis *et al.*, 2009). As discussed by Martínez-Murcia *et al.* (2007), sequence information initially desposited in Genbank was first submitted as *Manjusharmella aquatica* gen. nov. sp. nov. and their recommendation is to retain this name.

GROWTH CHARACTERISTICS AND PHENOTYPIC IDENTIFICATION. Isolate GPTSA^T-6 shows phenotypic differences that are not consistent with other members of the *Aeromonas* genus such as a negative reaction for nitrate and a weak or negative catalase reaction (Martínez-Murcia *et al.*, 2007). The organism is a Gram-negative facultative anaerobic, motile, oxidase-positive, short rod occurring singly or in pairs. It grows on TSA with 1.5% NaCl as opaque, whitish, round, convex colonies with slightly irregular margins, and at a temperature range of 15–42°C. Growth occurs up to 2% NaCl but not at 3%. It is resistant to vibriostatic agent 0/129 and susceptible to 10 µg ampicillin (Saha and Chakrabarti, 2006), hydrolyses starch but not Tween 80 and ferments cellobiose (Martínez-Murcia *et al.*, 2007).

Aeromonas simiae

TAXONOMY AND SIGNIFICANCE. Two strains of *A. simiae* (IBS S6874^T=IP 107798^T=CCUG 47378^T) were isolated from the faeces of two healthy monkeys (*Macaca fascicularis*) obtained from Mauritius and housed in a quarantine facility in France (Harf-Monteil *et al.*, 2004). Sequencing analysis of the 16S rRNA gene revealed high similarity to *Aeromonas* at 2.0–3.5% nucleotide substitution and 1.7% with *A. schubertii*. DNA-DNA hybridization revealed 18 and 16% similarity to *A. trota* and *A. schubertii* indicating *A. simiae* formed a new species. In agreement with 16S rRNA information, *gyrB* and *rpoD* gene sequence information also found that *A. simiae* formed a distinct species that was most closely related to *A. schubertii* (Saavedra *et al.*, 2006).

GROWTH CHARACTERISTICS AND PHENOTYPIC IDENTIFICATION. *Aeromonas simiae* is a facultatively anaerobic, motile, Gram-negative, straight rod that is positive for catalase and oxidase, reduces nitrate and is resistant to vibriostatic agent 0/129 150 µg. Growth occurs on TSA and colonies are non-haemolytic on sheep blood agar. Growth occurs at 0–3% NaCl but not at 6% NaCl and at a range of 30 to 42°C with an optimal temperature of 30°C. Hydrolysis of aesculin and fermentation of arbutin are variable for strains but the type strain is negative for aesculin and positive for arbutin (Harf-Monteil *et al.*, 2004). These and other reactions are presented in Tables 3.3 and 3.30 and some reactions are in Tables 3.24–26.

Aeromonas sobria

TAXONOMY AND SIGNIFICANCE. *Aeromonas sobria* (208=CIP 7433^T=NCMB 12065^T) was isolated from a fish and other strains were isolated from fish, frogs and one human isolate (Popoff and Véron, 1976). According to Martin-Carnahan and Joseph (2005), *A. sobria* has not been reported from human or animal sources although there are incorrect references to *A. sobria* from human clinical sources. This arises from the difficulty in phenotypic identification to species level and the confusion caused by *A. veronii* biovar Sobria.

Aeromonas sobria is genomospecies *A. sobria* and phenospecies *A. sobria* and HG 7.

Isolates identified as *A. sobria* have been reported as causing disease in farmed perch (*Perca fluviatilis*) and adult gizzard shad (*Dorosoma cepedianum*) (Toranzo *et al.*, 1989; Wahli *et al.*, 2005), however, the reported biochemical results do not suggest *A. sobria*. The isolates were haemolytic and were positive for ADH, two tests that are negative for *A. sobria*.

GROWTH CHARACTERISTICS AND PHENOTYPIC IDENTIFICATION. *Aeromonas sobria* is a facultatively anaerobic, Gram-negative, straight rod. Positive for oxidase and negative for catalase. Nitrate is reduced. Resistant to vibriostatic agent 0/129 150 µg. Growth occurs on TSA and BA after 24 h at 25°C. Haemolysis is not produced on TSA containing 5% sheep blood. Maximum temperature for growth is 28°C for 65% of strains and 37°C for 34% of strains. VP is produced from 57% of strains according to Popoff and Véron (1976), but is listed as negative by Martin-Carnahan and Joseph (2005) and Abbot *et al.* (2003). Aesculin, urease and elastin are not hydrolysed. Gas is produced from fermentation of glucose. Salicin is not fermented. H_2S is produced from cysteine. Motility is listed as positive by Popoff and Véron (1976) but negative by Martin-Carnahan and Joseph (2005) and Abbot *et al.* (2003). *A. sobria* does not hydrolyse gelatin and is negative for ADH, whereas *A. veronii* biovar Sobria is positive for gelatin hydrolysis and ADH (Popoff and Véron, 1976; Martin-Carnahan and Joseph, 2005). ADH was tested using Moeller's medium. Tests are conducted at an incubation temperature of 25°C, and tests for VP are read at 3 days and urea, citrate, ODC, LDC and ADH are read at 4 days (Abbot *et al.*, 2003). Tests to differentiate *A. sobria* from *A. veronii* biovar Sobria are haemolysis, ADH and gelatin. These and other reactions are presented in Tables 3.3 and 3.24–26.

Aeromonas taiwanensis

TAXONOMY AND SIGNIFICANCE. *Aeromonas taiwanensis* (A2-50^T=CECT 7403^T=LMG 24683^T) was isolated from a burn wound from a patient in Taiwan (Alperi *et al.*, 2010a). Based on the 16S rRNA gene, *A. taiwanensis* has 99.7% similarity to *A. trota*, 99.6% to *A. caviae* and 99.8% to *A. aquariorum*, which corresponds to 4, 5 and 2 bp differences. In contrast, multilocus phylogenetic analysis using *gyrB*, *rpoD*, *recA*, *dnaJ* and *gyrA* suggests a greater similarity to *A. media* and *A. caviae*. A phylogenetic tree based on sequence information from the *rpoD* gene indicates *A. taiwanensis* has similarity of 92.7% (56 bp differences) to *A. sanarellii* and 92.2% (63 bp differences) to *A. caviae*. Intraspecies similarity between *Aeromonas* species in the *rpoD* gene is 97% and the similarities of 92.7 and 92.2% are below this threshold, confirming species status (Soler *et al.*, 2004; Alperi *et al.*, 2010a).

GROWTH CHARACTERISTICS AND PHENOTYPIC IDENTIFICATION. *Aeromonas taiwanensis* is a motile, facultatively anaerobic, Gram-negative rod (2.5–1.8×1–0.8 µm). Catalase and oxidase are positive. Nitrate is reduced. It is resistant to vibriostatic agent 0/129 150 µg. On TSA and BA colonies are non-haemolytic, 2–5 mm, beige, opaque, circular after 48 h at 30°C. Brown diffusible pigment is not produced. Growth occurs at 30–36°C but not at 4 or 40°C after 24 h. Growth occurs at 0% NaCl but not in 6% NaCl (Alperi *et al.*, 2010a). Reactions in the API 20E are the same as *Vibrio fluvialis* NCTC 11327. They can be differentiated using the vibriostatic discs and growth on TCBS. The reactions in the API 20NE are the same as those for *A. sanarellii*. These two species can be separated by use of citrate. These and other reactions are presented in Tables 3.3 and 3.24–30.

Aeromonas tecta

TAXONOMY AND SIGNIFICANCE. Fourteen strains of unidentified *Aeromonas* species (F518^T=CECT 7082^T=DSM 17300^T=MDC 91^T) from clinical and environmental sources including two strains isolated from rainbow trout were identified as *A. tecta* (Demarta *et al.*, 2008). *Aeromonas tecta* is closely related to a number of other *Aeromonas* species based on 16S rRNA gene sequence, particularly *A. eucrenophila* and *A. encheleia*, but cluster in a separate group according to FAFLP. The designation of a separate species was confirmed by *gyrB* and *rpoD* gene sequencing and by DNA–DNA hybridization values of 5–45%.

GROWTH CHARACTERISTICS AND PHENOTYPIC IDENTIFICATION. *Aeromonas tecta* is a Gram-negative, motile, straight rod, positive for catalase and oxidase, reduces nitrate and is resistant to vibriostatic agent 0/129 150 µg. Beta-haemolytic, non-pigmented colonies occur on sheep blood agar after 24 h at 30°C. Growth does not occur at 4 or 42°C. The type strain is positive for VP and negative for hydrolysis of aesculin but other strains are variable and positive, respectively (Demarta *et al.*, 2008). These and other reactions are presented in Tables 3.3 and 3.30.

Aeromonas trota

TAXONOMY AND SIGNIFICANCE. *Aeromonas trota* (AH2^{T}=ATCC 49657^{T}) was proposed for 13 *Aeromonas* species isolated from human faecal samples from southern and south-eastern Asia (Carnahan *et al.*, 1991b). *Aeromonas enteropelogenes* was the name given to a group of psychrophilic *A. sobria*-like strains isolated from patients with diarrhoea in India (Schubert *et al.*, 1990a; Schubert *et al.*, 1990b). Later, sequence analysis of the 16S rRNA gene revealed 100% identity between *A. trota* and *A. enteropelogenes* (Collins *et al.*, 1993) and a ten nucleotide difference (1.8% divergence) in *cpn60* sequence (Miñana-Galbis *et al.*, 2009). Biochemical reactions in the API 20E and API 50E are identical and DNA–DNA hybridization shows a relatedness of 81–99% confirming that the two strains are the same. *Aeromonas enteropelogenes* was published first making *A. trota* a junior synonym, but because the name *A. trota* is more widely used this name has been retained (Huys *et al.*, 2002a).

Sterile filtered culture supernatant has a cytotoxic effect on Vero cells, haemolyses human red blood cells and shows proteolysis on TSA containing skim milk (Carnahan *et al.*, 1991b).

GROWTH CHARACTERISTICS AND PHENOTYPIC IDENTIFICATION. *Aeromonas trota* is a motile, facultatively anaerobic, Gram-negative, straight rod with rounded ends. Positive reactions are obtained for oxidase, catalase, nitrate, haemolysis of sheep and human red blood cells. It is resistant to vibriostatic agent 0/129 150 µg (Schubert *et al.*, 1990a; Carnahan *et al.*, 1991b). H_2S in the Triple Sugar Iron (TSI) tube is negative but is positive on modified GCF media. Elastase is negative whereas casein, gelatin and DNA are hydrolysed. These and other reactions are presented in Tables 3.3, 3.24–26 and 3.30.

Aeromonas veronii

Aeromonas veronii (ATCC 35604=CDC 1169-83) was originally identified as Enteric group 77 (Hickman-Brenner *et al.*, 1987) and HG 10. Genetically, strains belonging to HG 10 were identical to strains belonging to HG 8, which contained strains called 'clinical *A. sobria*' (as opposed to environmental *A. sobria* belonging to HG7) (Martin-Carnahan and Joseph, 2005). Strains in HG 8 and HG 10 are phenotypically different and so were divided into biovars. Currently, the genomospecies *A. veronii* belongs to hybridization groups HG 8X (CDC 0437-84) and 8Y (ATCC 9071). The phenospecies of HG 8X (CDC 0437-84) is *A. sobria* and the phenospecies of HG 8Y is *A. veronii* biovar Sobria (Martin-Carnahan and Joseph, 2005).

Aeromonas veronii biovar Sobria

TAXONOMY AND SIGNIFICANCE. *Aeromomas veronii* biovar Sobria (ATCC 9071) HG 8Y causes gastrointestinal disease, septicaemia, wound infections and meningitis in humans (Martin-Carnahan and Joseph, 2005).

A close relationship exists between *A. veronii* biovar Sobria and *A. allosaccharophila* and some species have been mis-identified. Strains LMG 13071=CECT 4909, LMG 13073=CECT 4911, LMG 13074=CECT 4912 reported as *A. veronii* biovar Sobria are in fact *A. allosaccharophila* based on phylogenetic analysis of *gyrB* gene (Saavedra *et al.*, 2006).

Aeromonas veronii biogroup Sobria strain LMG 13072 has since been identified as belonging to the *A. jandaei* group (Esteve *et al.*, 2003).

Analysis of housekeeping genes gives mixed results with some confirming species status whereas others do not. Phylogenetic analysis of the *gyrB* gene supports the species status of *A. allosaccharophila*, recognizing a close relationship between *A. allosaccharophila* and the strains of *A. veronii* biovar Sobria separating them in different clusters (Yáñez *et al.*, 2003; Soler *et al.*, 2004; Saavedra *et al.*, 2006). These analyses have confirmed strains CECT 4199^{T}, CECT 4220=LMG 14021, LMG 13071=CECT 4909, LMG 13073=CECT 4911, LMG 13074=CECT 4912, LMG 14058=CECT 4200, MDC12, MDC45, MDC234, MCD380, MDC561 (MDC=Molecular Diagnostics Centre, Spain) as being *A. allosaccharophila* yet some of them were received as *A. veronii* biovar Sobria (strains LMG 13071=CECT 4909, LMG13073=CECT4911, LMG13074=CECT4912) (Saavedra *et al.*, 2006). Sequence analysis of the *cpn60* gene, which reveals a 20–24 nucleotide difference (3.7–4.5% divergence) between *A. allosaccharophila* and *A. veronii* group, also supports separate species (Miñana-Galbis *et al.*, 2009).

GROWTH CHARACTERISTICS AND PHENOTYPIC IDENTIFICATION. *Aeromonas veronii* biovar Sobria is a facultatively anaerobic, motile, Gram-negative rod. Growth occurs on BA and TSA and at 0–3% NaCl. Colonies are haemolytic. Aesculin is not hydrolysed, VP and indole are positive. Gelatin is hydrolysed. LDC and ADH are positive, but ODC is negative. These and other reactions are presented in Tables 3.3 and 3.24–26.

Aeromonas veronii biovar Veronii

TAXONOMY AND SIGNIFICANCE. *Aeromonas veronii* (ATCC 35604=CDC 1169-83), originally classified as Enteric Group 77, was isolated from clinical sources including infected wounds from patients exposed to freshwater, respiratory secretions from victims of near-drowning in freshwater and from the faeces of patients with diarrhoea (Hickman-Brenner *et al.*, 1987).

Two *Aeromonas* species, *A. ichthiosmia* and *A. culicicola* are now identified as *A. veronii* according to phylogenetic studies. *Aeromonas ichthiosmia*, an *A. sobria*-like *Aeromonas* species isolated from surface water (Schubert *et al.*, 1990b), is a later heterotypic synonym of *A. veronii* according to 16S rRNA gene sequence analysis (Collins *et al.*, 1993), AFLP typing (Huys *et al.*, 1996) and DNA–DNA hybridization (Huys *et al.*, 2001). Analysis using the *cpn60* gene showed 14–20 nucleotide differences (2.6–3.7% divergence) between *A. veronii* biovar Veronii, *A. veronii* biovar Sobria, *Aeromonas culicicola* and *A. ichthiosmia* (Miñana-Galbis *et al.*, 2009). *Aeromonas culicicola* (MTCC 3249=NCIM 5147=CECT 5761) isolated from mosquitoes *Culex quinquefasciatus* and *Aedes aegypti* in India (Pidiyar *et al.*, 2002), from drinking water in Spain (Figueras *et al.*, 2005) and from ornamental fish (Saavedra *et al.*, 2006) is a later heterotypic synonym of *A. veronii* biovar Veronii based on a DNA–DNA hybridization result of 79–88%, which was different to the original report of 44% (Huys *et al.*, 2005). Phylogenetic analysis using the *gyrB* and *rpoD* gene sequences found that the strains from drinking water and from *A. aegypti* clustered with *A. veronii*, whereas the type strain from *C. quinquefasciatus* and the strains from ornamental fish formed a subcluster that was borderline to *A. veronii*. This study concluded that the relationship between *A. culicicola* and *A. veronii* remains unclear (Saavedra *et al.*, 2006). However, analysis of the *cpn60*, *rpoD* and *dnaJ* genes supports the proposal that *A. culicicola* is a later synonym of *A. veronii* (Huys *et al.*, 2005; Nhung *et al.*, 2007). Strains originally identified as *A. culicicola* produce cytolytic enterotoxin (AHCYTOEN gene) (Pidiyar *et al.*, 2003), a significant virulence factor in human isolates (Kingombe *et al.*, 1999).

The allocation of strain CECT 4911 previously assigned to *A. veronii* by Huys *et al.* (1996) has been confirmed as *A. allosaccharophila* using *gyrB* analysis by Yáñez *et al.* (2003), Soler *et al.* (2004), Küpfer *et al.* (2006), Saavedra *et al.* (2006) and Martínez-Murcia, *et al.* (2011).

Aeromonas veronii biovar Veronii is pathogenic to giant freshwater prawn (*Macrobrachium rosenbergii*) in experimental challenge with 3.7×10^5 cells/g body weight with 100% mortality. The LD_{50} is 2.0×10^3 cells/g (Sung *et al.*, 2000).

GROWTH CHARACTERISTICS AND PHENOTYPIC IDENTIFICATION. *Aeromonas veronii* biovar Veronii is a Gram-negative, motile, straight rod, which is oxidase-, catalase- and nitrate-positive and is resistant to vibriostatic agent 0/129 150 µg. On TSA and BA, colonies are beta-haemolytic, 2–3 mm, grey and may have a metallic sheen after 24 h at 30°C. Brown soluble pigment is not produced. The organism grows as 2 mm non-lactose fermenting colonies on MacConkey agar and generally does not grow on TCBS, although one of nine strains produces yellow colonies in light growth (plating efficiency less than 10^{-3}). Optimal growth occurs at 18 h at 30°C. NaCl is not required for growth although the number of strains with positive reactions is increased for ornithine and methyl red (89% compared to 100% positive at 48 h incubation for both tests) when 1% NaCl is added to the test medium. Growth occurs in 0–5% NaCl but not at concentrations of 6% and higher. *Aeromonas veronii* biovar Veronii differs from the majority of *Aeromonas* species by being positive for ornithine decarboxylase and negative for arginine dihydrolase. According to the API 20E reactions for lysine decarboxylase (LDC) are variable with five of nine (55.5%) strains positive for LDC (Hickman-Brenner *et al.*, 1987). Due to positive reactions for ODC and LDC, the API 20E profiles may produce an identification of *V. cholerae*. Aesculin is hydrolysed, VP and indole are positive. *Aeromonas veronii* is positive for the production of gas in the glucose fermentation test and negative for the string test (see under 'string' test for notes about false-positive reactions), with weak or no growth on TCBS and resistance to vibriostatic agent, whereas *V. cholerae* is negative for gas production, positive for the string test, grows as 2–3 mm colonies on TCBS and is sensitive to vibriostatic agent at 10 and 150 μg (Hickman-Brenner *et al.*, 1987). Lipase (corn oil) is positive. DNAse is positive for 100% of strains when tested at 25°C, whereas only 75% of strains are positive after 48 h incubation at 37°C. These and other reactions are presented in Tables 3.3, 3.24–26 and 3.30.

Order *Alteromonadales*

1.10 *Alteromonadaceae*

The family *Alteromonadaceae* belongs to the order *Alteromonadales*, part of the *Gammaproteobacteria* (Class III). *Alteromonadaceae* is the only family in this order and consists of eight genera that include *Alteromonas*, *Colwellia*, *Ferrimonas*, *Marinobacter*, *Moritella*, *Pseudoalteromonas* and *Shewanella*. Genera in this family form a phylogenetic clade in the *Gammaproteobacteria* between the family *Pseudomonadaceae* and the family *Enterobacteriaceae* (Gauthier *et al.*, 1995a; Urakawa *et al.*, 1998; Bowman and McMeekin, 2005). They are Gram-negative, motile, straight or curved rods.

1.10.1 *Alteromonas*

Phylogenetic analysis resulted in the division of the genus *Alteromonas* into two genera: *Alteromonas*, of which *A. macleodii* is the only species, and *Pseudoalteromonas*, which contains 11 species (Gauthier *et al.*, 1995a). Growth occurs on marine agar 2216 (Difco) (Gauthier *et al.*, 1995a).

1.10.2 *Colwellia*

Colwellia maris

Colwellia maris (*Vibrio* species strain ABE-1, JCM 10085^T) was isolated from the Hokkaido coast, Japan (Yumoto *et al.*, 1998). The organism is psychrophilic, with growth occurring at 0–24°C but not greater than 25°C, with an optimum growth range of 15–22°C. Cells are Gram-negative, curved rods, reduce nitrate and hydrolyse gelatin and DNA. Growth occurs on MA 2216. It is sensitive to vibriostatic agent 0/129 10 and 150 μg. These and other reactions are presented in Table 3.11.

1.10.3 *Ferrimonas*

Ferrimonas senticii

TAXONOMY AND SIGNIFICANCE. *Ferrimonas senticii* (P2S11^T=ATCC BAA-1480^T=DSM 18821^T) was isolated from the mucus or slime of pufferfish (*Arothron hispidus*) found in the waters around the Hawaiian islands (Campbell *et al.*, 2007). The genus *Ferrimonas* also contains *F. balearica*, *F. marina*, *F. futtsuensis* and *F. kyonanensis*. Based on 16S rRNA sequencing *F. senticii* is most closely related to *F. marina* (93.5%) and *F. balearica* (82.9%).

GROWTH CHARACTERISTICS AND PHENOTYPIC IDENTIFICATION. *Ferrimonas senticii* is an aerobic, facultative anaerobic, non-motile (using hanging drop and stab agar methods), Gram-negative rod. Colonies on half strength MA 2216 agar (Difco) are 1–2 mm, off-white and circular at 1–2 days at 30°C. Optimum temperature range is 20–37°C with slower growth at 20°C and no growth at 4°C. The optimal NaCl concentration is 2.0% with a range of 1.5–7.0%. H_2S is produced at 24 h in TSI agar slants when the inoculum is prepared with 2% NaCl (Campbell *et al.*, 2007). These and other reactions are presented in Tables 3.27–29 and 3.38.

1.10.4 *Moritella*

The genus *Moritella* belongs to the class *Gammaproteobacteria*, order *Alteromonadales* and the family *Alteromonadaceae*. There are seven species in the genus (Euzéby, http://www.bacterio.cict.fr/m/moritella.html). They are psychrophilic and found in marine environments particularly water and sediments at depths of 500–8600 m and temperatures of 1–11°C (Urakawa *et al.*, 1998).

Moritella marina

TAXONOMY AND SIGNIFICANCE. *Moritella marina* (MP-1^T=ATCC 15381^T) was isolated as *Spirillum marinum* originally in 1891 by H. Russell from the Gulf of Naples, but the culture was lost. Strains were re-isolated from seawater sampled at a depth of 1200 m off the coast of Oregan, USA and identified as the 1891 *Vibrio marinus* isolate based on phenotype (Colwell and Morita, 1964). *Vibrio marinus* was assigned to the newly created genus *Moritella* (Urakawa *et al.*, 1998), which is closely related to *Shewanella* and *Pseudoalteromonas* in the family *Alteromonadaceae* (Gauthier *et al.*, 1995a). *Moritella marina* has 99.1% rRNA sequence similarity to *M. viscosa* (Benediktsdóttir *et al.*, 2000).

GROWTH CHARACTERISTICS AND PHENOTYPIC IDENTIFICATION. *Moritella marina* is a motile, oxidase positive, catalase positive, facultatively anaerobic, Gram-negative curved or straight rod (1.2–0.6×0.6–1.0 μm) with rounded ends and occurs singly or in pairs. The nutritional requirements are restricted but not complex. NaCl is required for growth. Growth occurs on MA 2216 but not on TCBS. On seawater peptone agar (see media section), colonies are 2–5 mm, greyish-cream, translucent, circular, convex and smooth (Colwell and Morita, 1964). Growth occurs at a maximum of 20°C and only some strains are able to grow at 24°C. Fermentation of D-mannose using 0.7% mannose was reported positive by Lunder *et al.* (2000). It is negative for Simmons' citrate. H_2S tested by production of H_2S from cysteine and sodium thiosulfate is negative. Casein, chitin and gelatin are hydrolysed; indole and cellulose are negative (Colwell and Morita, 1964). *N*-acetyl glucosamine is utilized (Urakawa *et al.*, 1998). Results are variable for LDC. There is a marked sensitivity to the vibriostatic agent 0/129. These and other reactions are presented in Tables 3.19, 3.24–26 and 3.38.

Moritella viscosa

TAXONOMY AND SIGNIFICANCE. *Moritella* (*Vibrio*) *viscosa* (NVI 88/478^T=NCIMB 13584^T) is the causative agent of 'winter ulcer disease', an economically significant disease for the sea-farmed salmon industry in Norway, Iceland and Scotland (Bruno *et al.*, 1998b; Lunder *et al.*, 2000). It has been isolated from Atlantic salmon (*Salmo salar*), plaice (*Pleuronectes platessa*) and rainbow trout with winter ulcer disease, and from healthy lumpsucker from Iceland and Norway (Benediktsdóttir *et al.*, 2000; Lunder *et al.*, 2000). Pathogenicity was established by experimental challenge. The disease occurs at water temperatures below 10°C. Infected fish develop ulcers on the skin that extend into the underlying musculature leading to necrosis in the tissues. The organism can be cultured from skin, muscle, liver, spleen, head, kidney and intestine (Bruno *et al.*, 1998b; Løvoll *et al.*, 2009).

Moritella viscosa has a sequence similarity of 99.1% to *M. marina* based on 16S rDNA sequencing and clusters within the *Pseudoalteromonas, Shewanella* group (Benediktsdóttir *et al.*, 2000).

GROWTH CHARACTERISTICS AND PHENOTYPIC IDENTIFICATION. *Moritella viscosa* is a motile, oxidase positive, catalase positive, facultatively anaerobic, Gram-negative rod that exhibits short or elongated curved rods on solid medium and long cells when grown in liquid media. On MSA-B, BA with 1.5% NaCl, TSA with 2% NaCl and MA 2216, colonies are 0.5 mm, grey, round, translucent, non-swarming and non-luminescent after 24 h at 15–22°C. Colonies are viscous and adhere to the medium and form long threads when lifted from the plate. This phenomenon is more evident on BA with 1.5% NaCl than MA 2216. Some strains may show strong haemolysis, whereas with other strains a light haemolysis is seen underneath the colony only. There is some biotype variation for strains isolated from Norway, Scotland, south-west Iceland and northern Iceland (Benediktsdóttir *et al.*, 1998). Growth does not occur on TCBS, although some strains may show a faint growth. Growth occurs at 4–21°C but not at 30°C. Growth occurs at 1–4 % NaCl but not at 0% NaCl. Of the strains, 95% are negative for methyl red. Tween 80, casein, lecithin and gelatin are hydrolysed, whereas alginate is not. Fermentation of cellobiose, glycerol, melibiose, raffinose and rhamnose is negative, whereas dextrin is positive (Lunder *et al.*, 2000). *Moritella viscosus* often needs prolonged incubation times for growth in test media. Most strains require 1% peptone for optimal growth. Growth does not occur in 0.5% peptone. Media used for fermentation of *Vibrio* species do not always support the growth of *M. viscosa* (Benediktsdóttir *et al.*, 2000). Results for salt requirements and temperature growth limits are inconsistent in the literature. One study suggests 1–4% NaCl and a temperature of 25°C (Lunder *et al.*, 2000). A second study suggests 2–3% NaCl with no growth at 4% NaCl, and a temperature range of 4–21°C with no growth at 25°C (Benediktsdóttir *et al.*, 2000). The reason for these differences may be the use of different growth media. Benediktsdóttir *et al.* (2000) used a base consisting of MA 2216 (Difco) and Lunder *et al.* (2000) used a proteose peptone base with mineral salts for carbohydrate fermentation (see media section), but also tested the strains using commercially available media (API Zym) following the manufacturer's instructions. Bruno *et al.* (1998b) successfully grew *M. viscosa* on TSA containing 2% NaCl and tested biochemical fermentation reactions using commercially available kits, API 20E and API 50CH using incubation at15°C and 72 h. When using identification systems it is important to compare the results with those obtained for that identification system. These and other reactions are presented in Tables 3.19, 3.24–26, 3.30 and 3.38.

MOLECULAR IDENTIFICATION. A real-time PCR enables detection of *M. viscosa* from infected muscle, skin, head kidney, spleen and liver (Løvoll *et al.*, 2009). DNA is extracted from RNA*later*-stored tissue samples using TissueLyser (Qiagen) and the NucliSens easyMag platform (Biomerieux). The master mix consists of 300 nM each of primer Fw (5′-CGT TGC GAA TGC AGA GGT-3′) and Rev (5′-AGG CAT TGC TTG CTG GTT A-3′), 150 nM probe (FAM-5′-TGC AGG CAA GCC AAC TTC GAC A-3′), 2× TaqMan Universal PCR master mix and DNA in a 25 μl volume. Thermocycling is carried out on a Stratagene thermocycler at 50°C for 2 min, 95°C for 10 min and 45 cycles of 95°C for 15 s and 60°C for 60 s.

1.10.5 *Pseudoalteromonas*

The genus *Pseudoalteromonas* was formed to include most of the previously named *Alteromonas* species with the exception of *A. macleodii*. The majority of *Pseudoalteromonas* are associated with eukaryotic hosts and are frequently isolated from marine waters. The majority grow on MA 2216 and show optimal growth at 23°C.

Pseudoalteromonas citrea

TAXONOMY AND SIGNIFICANCE. *Pseudoalteromonas citrea* (NCMB 1889^T=ATCC 29719^T), a yellow-pigmented, Gram-negative rod isolated from seawater, was first described by Gauthier (1977) as *Alteromonas citrea*. *Pseudoalteromonas citrea* has been isolated from the far-eastern mussel (*Crenomytilis grayanus*) and the yesso or giant ezo scallop (*Patinopectin yessoensis*) collected from Troitza Bay, Gulf of Peter the Great, Sea of Japan (Ivanova *et al.*, 1998).

GROWTH CHARACTERISTICS AND PHENOTYPIC IDENTIFICATION. *Pseudoalteromonas citrea* is a strictly aerobic, motile, Gram-negative straight rod (0.7–1.5×1.5–4 μm) that is catalase and oxidase positive. On MA 2216 colonies are 0.5 mm bright and whitish after 24 h at 23°C becoming 4 mm and lemon-yellow in colour after 4 days. On MSSA, colonies are 6–7 mm bright, mucoid, whitish colonies that turn black around the colony after 2–3 days. The development of haemolysis is slow and a ring of haemolysis is seen after 5 days. Sea salts are required for growth although some strains can grow with only NaCl (Gauthier, 1977). Phenotypic variation was found between the type strain and between the strains isolated from mussels and scallops and this is thought to be due to the different ecological niches. Some mussel and scallop strains hydrolysed agar and some had a slightly different temperature range. All strains grew between 10°C and 28°C but growth was variable at 4°C and 35°C (Ivanova *et al.*, 1998). All strains hydrolyse gelatin but not chitin. The type strain hydrolyses casein but not chitin or agar. Results for H_2S production, methyl red, Voges–Proskaüer and hydrolysis of aesculin are negative. Brown pigment is produced on tyrosine-containing media. The phenotypic reactions described in this manual are those of the type strain. These and other reactions are presented in Table 3.11 and Table 3.38.

Pseudoalteromonas haloplanktis ssp. *tetraodonis*

See *Pseudoalteromonas tetraodonis.*

Pseudoalteromonas piscicida

TAXONOMY AND SIGNIFICANCE. *Pseudoalteromonas piscicida* (RTY) was isolated from a fish kill involving perch (*Gerres*), snapper (*Lutjanus*), snook (*Centropomus*), red fish (*Sciaenops*), needlefish (*Strongylura*), catfish (*Felichthys*) and mullet (*Mugil*) on the west coast of Florida, Gulf of Mexico, USA in 1946 and in subsequent years, and named *Flavobacterium piscicida* (Bein, 1954). Pathogenicity was proven in experimental challenge. The bacterium was reclassified as *Pseudomonas piscicida* (MB-1^T=ATCC 15057^T) (Buck *et al.*, 1963) and then *Pseudoalteromonas* (Gauthier *et al.*, 1995).

Pseudoalteromonas piscicida causes mortality of eggs of damselfish (*Pomacentridae*) species, yellow clownfish (*Amphiprion clarkii*) and staghorn damselfish (*Amblyglyphidodon curacao*), popular marine aquarium fish (Nelson and Ghiorse, 1999). Eggs collected from dead coral substrate off the coast of Papau New Guinea died within 24–36 h of collection. Experimental challenge demonstrated pathogenicity of *P. piscicida* for damselfish eggs. Infected eggs changed from being transluscent with orange pigments in the yolk sac to a whitening of the egg cases.

GROWTH CHARACTERISTICS AND PHENOTYPIC IDENTIFICATION. *Pseudoalteromonas piscicida* is a motile, Gram-negative, ovoid, straight rod (0.6–0.8×1.2–1.8 μm) occurring singly and in chains. Colonies are spreading, irregular and glistening with undulate to lobate edges. Peptone (0.1%) is required for growth (Bein, 1954). Larger more uniform cells are seen in the logarithmic phase of growth, and in the stationary phase of growth cells are smaller and coccoid. Cells occur singly and in occasional clumps. Growth occurs on high-nutrient marine agar containing peptone, yeast extract and glycerol (Jensen *et al.*, 1996) (see media section). On this medium colonies are 3–6 mm, light orange to white in the centre and a darker orange at the edge after 24 h at 30°C. A water soluble, yellow-orange diffusible pigment is produced. After 48 h colonies are circular, convex and smooth with rough edges. It is resistant to penicillin G. Pigment production is inhibited around the pencillin disc. Growth occurs at 22–40°C. The strain from damselfish eggs is referred to as strain Cura-d and deposited as ATCC 700661. Using MOF medium of Leifson (1963) glucose, maltose and sucrose are oxidized but not fermented. Lactose, xylose and D-mannitol are not fermented (Buck *et al.*, 1963). These and other reactions are presented in Tables 3.11 and 3.30.

Pseudoalteromonas tetraodonis

Pseudoalteromonas haloplanktis ssp. *tetraodonis* has been assigned to *P. tetraodonis*. *Pseudoalteromonas tetraodonis* (*P. haloplanktis* ssp. *tetraodonis*) was isolated along with a number of other bacteria from sea cucumber (*Apostichopus japonicus*), a predominant aquaculture species in China (Li *et al.*, 2010a). In experimental challenge the LD_{50} was 7.24×10^7 CFU/g body weight, compared to *Vibrio splendidus* (LD_{50} 1.74×10^7 CFU/g) and *Shewanella marisflavi* (7.76×10^6 CFU/g).

Pseudoalteromonas tetraodonis was identified using biochemical methods and 16S rRNA sequencing (Li *et al.*, 2010a). The strain from sea cucumber grew at 4°C, whereas *P. haloplanktis* ssp. *tetraodonis* strain IAM 14160 was reported as negative. Strain IAM is negative for hydrolysis of agar and chitin. These and other reactions are presented in Table 3.11.

1.10.6 *Shewanella*

The *Shewanella* genus contains over 35 species that at various times have been classified as *Pseudomonas*, *Achromobacter* or *Alteromonas*. *Shewanella* are found in the environment, particularly in freshwater and marine water, and are important in the breakdown of organic matter. Species have been reported to cause infection in some aquatic animals and some cause soft tissue and skin infection in humans. Other species are a cause of food spoilage of seafood products. *Shewanella* are facultatively anaerobic, Gram-negative rods, and are generally oxidase positive and indole negative. Most strains produce H_2S in TSI (triple sugar iron agar). Acid may be produced from glucose and *N*-acetylglucosamine and some other carbohydrates. Arginine dihydrolase (ADH) is usually negative. Colonies on solid media may produce a tan or pink-orange pigment. Species grow on MA 2216 and nutrient agar (NA). Ideally, the medium of Leifson (1963) (see media section) containing 0.5% of carbohydrate is the best method for determining carbohydrate fermentation of *Shewanella* species, however in many diagnostic laboratories in-house preparation of these biochemicals is not feasible.

Shewanella affinis

TAXONOMY AND SIGNIFICANCE. *Shewanella affinis* (KMM 3587^T=CIP 107703^T=ATCC BAA-642^T) was isolated from the benthic marine 'peanut worm' *Golfingia margaritacea* (*Phascolosoma japonicum*, as reported in the orginal paper, is an unaccepted synonym) from the Gulf of Peter the Great in the Sea of Japan region, and from an unidentified hydrocoral species collected in the Kuril Islands, north-west Pacific Ocean (Ivanova *et al.*, 2004). *Shewanella affinis* has 99% 16S rRNA sequence similarity to *Shewanella colwelliana*. DNA–DNA hybridization of 45–52% demonstrates the strain is a new species, however further studies based on *gyrB* sequences suggest *S. affinis* is a later heterotypic synonym of *S. colwelliana* (Satomi *et al.*, 2007).

GROWTH CHARACTERISTICS AND PHENOTYPIC IDENTIFICATION. *Shewanella affinis* is a motile, oxidase positive, catalase positive, facultatively anaerobic, Gram-negative bacillus (1.0–2.0×0.6–0.8 μm). On MA 2216 colonies are slightly pink in colour, circular, smooth, convex with an entire edge. NaCl is required for growth and growth occurs at 0.5–6.0% NaCl with some strains growing at 8% NaCl. Growth occurs at 10–34°C with an optimum of 20–25°C. Growth does not occur at 37°C. These and other reactions are presented in Tables 3.11 and 3.27–29.

Shewanella algae

TAXONOMY AND SIGNIFICANCE. *Shewanella algae* (OK-1^T=IAM 14159^T) was isolated from the surface of red algae, *Jania* species, a branching segmented seaweed (Simidu *et al.*, 1990). *Shewanella algae* produces tetrodotoxin, a neurotoxin. The spelling of the original name *S. alga* was corrected to *S. algae* (Nozue *et al.*, 1992). *Shewanella putrefaciens* genomic group IV, Gilardi biovar 2, Centre for Disease Control and prevention (CDC) biotype 2 was reclassified as *S. algae* (Khashe and Janda, 1998).

Shewanella algae caused mass mortality in 10- to 20-day-old post-larval abalone (*Haliotis diversicolor supertexta*) reared in hatchery ponds in southern China (Cai *et al.*, 2006a). Abalone exhibited shrunken tissue and whitening and slower movement. Infected post-larvae became detached from diatom films. The bacterium was identified by API 20E and by 16S rRNA sequencing. Pathogenicity was confirmed in challenge experiments and infected animals demonstrated the same clinical signs as naturally infected animals. The LD_{50} was 1.8×10^4 CFU/ml. A disease outbreak in rabbitfish (*Signaus rivulatus*) at an aquaculture facility in the Red Sea, Saudia Arabia was said to be due to *Shewanella* (*Pseudomonas*) *putrefaciens*, which was identified using the API 20E system (Saeed *et al.*, 1987). The isolate grew at 42°C and in up to 9% NaCl, which suggests the isolate may have been *S. algae* rather than *S. putrefaciens*. Infected fish had haemorrhagic areas on the body and mouth, frayed fins and exophthalmia. The organism was cultured from liver, kidney and spleen. Pathogenicity was confirmed by experimental challenge.

Shewanella algae, along with other *Shewanella* species (*S. putrefaciens*, *S. baltica*, *S. hafniensis*, *S. marisflavi*), is part of the bacterial flora found in oysters in Delaware Bay, USA (Richards *et al.*, 2008).

Shewanella algae is a human pathogen causing ear infection, wound and soft tissue infections (Holmes *et al.*, 1975; Nozue *et al.*, 1992; Khashe and Janda, 1998; Richards *et al.*, 2008).

GROWTH CHARACTERISTICS AND PHENOTYPIC IDENTIFICATION. *Shewanella algae* is a motile, oxidase positive, short, Gram-negative, straight rod (1.6×0.8 μm). On MA 2216 colonies are pale tan to salmon pink or yellow-orange to brown, convex, entire and have a butyrous consistency after 2 days at 37°C. Growth occurs at 10–42°C and at 6% NaCl but not at 4°C (Nozue *et al.*, 1992). Growth occurs on Salmonella-Shigella (SS) agar. H_2S is strongly positive in TSI and blackening occurs within several hours. Beta-haemolysis is produced on sheep blood agar after 48 h at 37°C. Haemolysis may not be seen at 24 h. Colonies of strain ATCC 51192 are alpha-haemolytic at 25°C on TSA containing 5% sheep blood. Isolates from oysters produce beta-haemolysis and not all isolates grow at 6% NaCl (Richards *et al.*, 2008). It is non-luminescent. D-ribose and D-glucose are positive oxidatively in Leifson's medium (1963) (see media section) after 4–12 days. Strains vary in reactions for urea and Simmons' citrate. Positive for lipase and negative for amylase, chitinase and alginase (Simidu *et al.*, 1990). Fermentation is negative for D- and L-arabinose, D-mannose, mannitol, *iso*-inositol, sucrose, maltose, lactose and rhamnose. The API 20E may misidentify *S. algae* as *S. putrefaciens*. *Shewanella algae* can be differentiated from *S. putrefaciens* by being positive for growth on SS agar, at 42°C and 6% NaCl and beta-haemolysis on BA, and negative for growth at 4°C and negative for fermentation of maltose and arabinose. *Shewanella putrefaciens* is negative for growth on SS agar, at 42°C and at 6% NaCl and beta-haemolysis, but is positive for growth at 4°C and for fermentation of L-arabinose (Nozue *et al.*, 1992). These and other reactions are presented in Tables 3.11, 3.24–29 and 3.38.

Shewanella algidipiscicola

TAXONOMY AND SIGNIFICANCE. *Shewanella algidipiscicola* ($S13^T$=LMG 23746^T=NBRC 102032^T) was isolated from the belly flap area from plaice and cod caught in the Baltic Sea, Denmark (Satomi *et al.*, 2007). It is thought to be involved in spoilage of stored seafood. *Shewanella algidipiscicola* has 16S rRNA gene sequence similarity to *S. colwelliana* and *S. affinis* of >99.0% but has only 85% similarity based on the sequence of the *gyrB* gene.

GROWTH CHARACTERISTICS AND PHENOTYPIC IDENTIFICATION. *Shewanella algidipiscicola* is a motile, oxidase positive, Gram-negative rod (1.0–1.2×0.5–0.7 µm). On TSA containing 5% sheep blood colonies are non-haemolytic, dull white, round, undulate with irregular margins and non-luminescent at 24 h at 25°C (Satomi *et al.*, 2007). Growth occurs at 4–30°C with 70% of strains able to grow at 37°C, although the type strain does not grow at 37°C (Richards *et al.*, 2008). Optimal growth is at 25°C. Growth occurs on 50% MA 2216. Growth can occur in 6% NaCl but not for the type strain. H_2S is produced in TSI, and aesculin is hydrolysed. Gelatin and DNase are not hydrolysed. The API 20E database may erroneously identify *S. algidipiscicola* as *S. putrefaciens* (Richards *et al.*, 2008). These and other reactions are presented in Tables 3.11 and 3.24–29.

Shewanella aquimarina

TAXONOMY AND SIGNIFICANCE. *Shewanella aquimarina* (SW-120^T=KCCM 41821^T=JCM 12193^T) was isolated from seawater from the Yellow Sea, Korea (Yoon *et al.*, 2004). The isolate has 16S rRNA gene sequence similarity to *S. marisflavi* (97.4%), *S. affinis* (98.5%), *S. colwelliana* (98.5%) and *S. waksmanii* (97.0%).

GROWTH CHARACTERISTICS AND PHENOTYPIC IDENTIFICATION. *Shewanella aquimarina* is a motile, oxidase positive, Gram-negative rod (2.0–4.0×0.6–0.9 µm). On MA 2216 colonies are 2–4 mm, light brown, smooth, glistening, circular, flat to slightly raised after 3 days at 30°C (Yoon *et al.*, 2004). Growth occurs at 10–42°C, but not at 4°C or above 43°C. Optimal growth temperature is 30–37°C. Growth occurs at 2–8% NaCl but not at 0 or 9% NaCl. Growth occurs under anaerobic conditions on MA 2216. Casein, and Tween 80 are hydrolysed but aesculin and urea are not. D-ribose is fermented but L-arabinose, D-cellobiose, D-melezitose, D-raffinose and L-rhamnose are not. These and other reactions are presented in Tables 3.11 and 3.38.

Shewanella baltica

Shewanella baltica was reported from oysters and seawater in Delaware Bay, USA (Richards *et al.*, 2008). Growth occurs at 25°C and at lower temperatures, however some isolates from oysters and seawater grow at 37°C. Growth does not occur at 6% NaCl or at 42°C and colonies are non-haemolytic, although one strain from seawater was haemolytic. The API 20E database may erroneously identify S. *baltica* as *S. putrefaciens* (Richards *et al.*, 2008). These and other reactions are presented in Tables 3.11, 3.24–29 and 3.38.

Shewanella colwelliana

TAXONOMY AND SIGNIFICANCE. *Shewanella colwelliana* (LST^T=ATCC 39565^T) was isolated from oysters and holding tanks for oyster spat at a facility in the USA (Weiner *et al.*, 1985, 1988). The organism adheres to glass, plastic, teflon, copper and polycarbonate surfaces and to oyster shell, and is thought to be involved in larval settlement. It has been proposed that *S. colwelliana* is an earlier heterotypic synonym of *S. affinis* as the two species have an almost identical *gyrB* sequence (Satomi *et al.*, 2007).

GROWTH CHARACTERISTICS AND PHENOTYPIC IDENTIFICATION. *Shewanella colwelliana* is a motile, Gram-negative rod (1–2×0.5 μm). The cells become elongated and then helical during late stationary and declining phases of growth in broth culture, and on agar media. On solid media in the late growth phase cells become helical and filamentous and grow up to 20 μm (Weiner *et al.*, 1985). A brown-black pigment is produced during the late growth phase in liquid media. Growth occurs on MA 2216 and in media supplemented with 3% NaCl. On MA 2216 colonies are 1 mm circular, convex with undulate margins. After 7 days colonies become irregular and 5 mm in diameter (Weiner *et al.*, 1988). In liquid media a uniform turbidity is produced without a pellicle. In 2216 broth a red-brown pigment is produced. In brain heart infusion broth containing 3% NaCl mucoid floculates form. Cells adhere to the surface of the container. Growth does not occur at 0% NaCl or 6% NaCl, but occurs at 1–5% NaCl with optimal growth at 2–4%. Growth occurs at 8–30°C with an optimum at 25°C, however growth does not occur at 42°C. It is positive for amylase, gelatinase and nitrate and negative for luminescence, urease and chitinase. These and other reactions are presented in Tables 3.11 and 3.27–29.

Shewanella corallii

TAXONOMY AND SIGNIFICANCE. *Shewanella corallii* (fav-2-10-05^T=LMG 24563^T=DSM 21332^T) is a bacterium with antibacterial properties isolated from the mucus layer of Red Sea coral (*Favia* species) in the Gulf of Eilat, Israel (Shnit-Orland *et al.*, 2010). *Shewanella corallii* has 16S rRNA gene sequence similarity to *S. marisflavi* (96.0%) and *S. haliotis* (95.9%).

GROWTH CHARACTERISTICS AND PHENOTYPIC IDENTIFICATION. *Shewanella corallii* is a motile, oxidase positive, Gram-negative rod (2×0.8–1.0 μm). Growth occurs on 50% MA 2216 as 1 mm yellowish-brown colonies at 24 h and 26°C. Growth occurs aerobically whereas slow growth occurs in an anaerobic atmosphere (Shnit-Orland *et al.*, 2010). Growth occurs in Luria–Bertani broth and marine broth but it grows slowly in nutrient broth. Growth occurs at 0.5–6% NaCl with weak growth at 8% NaCl and no growth at 10% NaCl. On media containing NaCl only, at either 0.5% or 8%, pigment is a yellow colour compared to yellow-orange in media containing artificial seawater salts. Growth occurs at 10–37°C, with weak growth at 37°C and optimal growth at 26°C. Growth does not occur at 4 or 42°C. Beta-haemolysis is seen at 4 days but not at 2 days' incubation. These and other reactions are presented in Tables 3.11, 3.27–29 and 3.38.

Shewanella frigidimarina

Shewanella frigidimarina (ACAM 591^T) was isolated from sea ice, sea ice algal biomass and cyanobacterial mats in the Antarctic (Bowman *et al.*, 1997). Strain NCIMB 400 isolated from the North Sea, Scotland, and originally identified as *S. putrefaciens*, was reclassified as *S. frigidimarina* (Reid and Gordon, 1999). Cells are curved or straight rods (1–2.5×0.5–0.8 μm) and motile. On MA 2216, colonies are tan-coloured at 3–5 days at 10°C and develop mucoid colonies after 10 days. Optimal temperature for growth is 20–22°C with a range of 0–27°C. NaCl is not required for growth but can tolerate up to 9% NaCl. Negative for chitinase. These and other reactions are presented in Tables 3.11 and 3.24–26.

Shewanella gelidimarina

Shewanella gelidimarina (ACAM 456^T) was isolated from Antarctic Sea Ice (Bowman *et al.*, 1997). It is a Gram-negative rod (1.5–2.5 μm), straight or curved and occurs singly or in pairs. On MA 2216 colonies are tan coloured at 3–5 days at 10°C. Colonies become mucoid after 10 days. Optimal temperature for growth is 15–17°C. Growth does not occur above 23°C or at 8% NaCl. Seawater is required for growth and growth does not occur on media containing less than 1% NaCl. Growth occurs on MA 2216 prepared with up to 2× artificial seawater or on NA containing 6% NaCl. It is positive for chitinase and does not ferment D-glucose. These and other reactions are presented in Tables 3.11 and 3.24–26.

Shewanella glacialipiscicola

TAXONOMY AND SIGNIFICANCE. *Shewanella glacialipiscicola* (T147^T=LMG 23744^T=NBRC 102030^T)

was isolated from the belly flap area of cod caught in the Baltic Sea, Denmark (Satomi *et al.*, 2007). It is thought to be involved in spoilage of stored seafood. *Shewanella glacialipiscicola* has 16S rRNA gene sequence similarity to *S. morhuae* of 99.9%, and a similarity of 95.6% based on the *gyrB* sequence.

GROWTH CHARACTERISTICS AND PHENOTYPIC IDENTIFICATION. *Shewanella glacialipiscicola* is a motile, oxidase positive, Gram-negative rod (1.0–1.2×0.5–0.7 μm). On TSA colonies are beige, round, undulate with irregular margins and are non-luminescent after 24 h at 25°C (Satomi *et al.*, 2007). Growth occurs at 4–37°C. Growth occurs on 50% MA 2216. H_2S is produced, and DNase and aesculin are hydrolysed. These and other reactions are presented in Tables 3.11 and 3.27–29.

Shewanella hafniensis

TAXONOMY AND SIGNIFICANCE. *Shewanella hafniensis* ($P010^T$=ATCC BAA-1207^T=NBRC 100975^T), thought to be involved in food spoilage, was isolated from cod and flounder caught in the Baltic Sea off the coast of Denmark (Satomi *et al.*, 2006). *Shewanella hafniensis* along with other *Shewanella* species is part of the bacterial flora found in oysters and seawater in Delaware Bay, USA (Richards *et al.*, 2008). Based on 16S rRNA sequencing it is most closely related to *S. putrefaciens* (99.0% similarity) and *S. baltica* (98.3%), however *gyrB* gene sequence analysis indicates only 90.0% and 93.9% similarity, respectively.

GROWTH CHARACTERISTICS AND PHENOTYPIC IDENTIFICATION. *Shewanella hafniensis* is a motile, Gram-negative rod (0.3–0.7×1.0–1.2 μm), which is oxidase and catalase positive. On TSA colonies are dull white, round, undulate with irregular margins, and non-luminescent after 24 h at 25°C (Satomi *et al.*, 2006). Growth occurs at 0–6% NaCl and 4–25°C but is negative at >37°C. It is positive for production of H_2S, hydrolysis of gelatin, utlization of citrate and production of DNase. Glucose is not fermented. These and other reactions are presented in Tables 3.11 and 3.24–29.

Shewanella haliotis

TAXONOMY AND SIGNIFICANCE. *Shewanella haliotis* ($DW01^T$=KCTC 12896^T=JCM 14758^T) was isolated from the gut microflora of abalone (*Haliotis discus hannai*) found in the South Sea, Republic of Korea (Kim *et al.*, 2007). Based on 16S rRNA sequence information it has similarity to *S. algae* (98.3%), *S. profunda* (94.9%), *S. kaireitica* (94.4%) and *S. schlegeliana* (94.1%).

GROWTH CHARACTERISTICS AND PHENOTYPIC IDENTIFICATION. *Shewanella haliotis* is a motile, facultatively anaerobic, Gram-negative rod (2.0–4.3×0.5–0.7 μm), which is oxidase and catalase positive. Growth occurs on MA 2216, TSA and nutrient agar. On TSA colonies are 5 mm, pink-orange coloured, opaque, circular, low convex, entire with a smooth edge after 5 days at 30°C (Kim *et al.*, 2007). Growth occurs at 0–10% NaCl with an optimum of 4%, and at 10–42°C with an optimum of 37°C. It does not ferment glucose or produce ADH or LDC. It produces H_2S and is positive for ODC. These and other reactions are presented in Tables 3.11, 3.24–30 and 3.38.

Shewanella hanedai

Shewanella hanedai (281^T =ATCC 33224^T) was originally named *Alteromonas hanedai*. It is a non-fermentative, motile Gram-negative straight rod found in Arctic sediments and seawater (Jensen *et al.*, 1980). It is luminescent on MA 2216 agar.

Shewanella japonica

TAXONOMY AND SIGNIFICANCE. *Shewanella japonica* (KMM 3299^T=LMG 19691^T=CIP 106860^T) was isolated from seawater and the mussel *Protothaca jedoensis* found in the Sea of Japan, the Gulf of Peter the Great and the Yellow Sea (Ivanova *et al.*, 2001). (According to the World Register of Marine Species, *Protothaca* is an unaccepted synonym of *Leukoma* species.) Based on 16S rRNA sequence information it has similarity to *S. frigidimarina* (95%) and *S. baltica* (93%).

GROWTH CHARACTERISTICS AND PHENOTYPIC IDENTIFICATION. *Shewanella japonica* is a motile, facultatively anaerobic, Gram-negative rod (1.0–2.0×0.6–0.8 μm), which is oxidase and catalase positive. On MA 2216 and MSA-B colonies are 3–5 mm, slightly pink in colour, smooth, circular, convex and with an entire edge when grown at 25°C for 2–4 days. It is beta-haemolytic on TSA containing

5% sheep blood. Seawater or NaCl ions are not required for growth. It grows at 1–3% NaCl but not at 6% NaCl. It grows at 10–37°C with an optimum of 20–25°C. There is no growth at 42°C. It is cytotoxic for Ehrlich cells. Digests agar. It is negative for chitinase and elastase. It hydrolyses gelatin and DNase and utilizes galactose, fructose, maltose, *N*-acetyl-glucosamine, xylose, glucose, acetate, D-glucosamine, pyruvate and cellobiose as sole sources of carbon. Sucrose, D-mannose, lactose, melibiose, D-gluconate, fumarate, citrate, aconitate, meso-erythritol, D-mannitol, glycerol, D-sorbitol, DL-malate and trehalose are not utilized. These and other reactions are presented in Table 3.11.

Shewanella marinintestina

TAXONOMY AND SIGNIFICANCE. *Shewanella marinintestina* (JCM 11558^T=LMG 21403^T) was isolated from the intestine of a squid from Japan (Satomi *et al.*, 2003). *Shewanella marinintestina* has >99% 16S rRNA sequence similarity to *S. schlegeliana* and *S. sairae*, and a 97% sequence similarity to *S. pealeana*.

GROWTH CHARACTERISTICS AND PHENOTYPIC IDENTIFICATION. *Shewanella marinintestina* is a facultatively anaerobic, motile, Gram-negative rod (2.0–3.0 μm long), which is oxidase and catalase positive. On MA 2216 colonies are ivory in colour, opaque and, circular after 48 h at 20°C. Growth occurs at 4–30°C with an optimum of 20–25°C. Growth occurs at 1–5% (w/v) NaCl with an optimal range of 2.0–3.0%. Hydrolysis of gelatin, lipase, casein and hippurate is positive. Nitrate is reduced and H_2S is produced (Satomi *et al.*, 2003). These and other reactions are presented in Tables 3.11, 3.24–26, 3.30 and 3.38.

Shewanella marisflavi

TAXONOMY AND SIGNIFICANCE. *Shewanella marisflavi* (SW-117^T=KCCM 41822^T=JCM 12192^T), present in seawater in the Yellow Sea of Korea (Yoon *et al.*, 2004), is a pathogen of sea cucumber (*Apostichopus japonicus*), a marine animal used as a food source in some countries such as China (Li *et al.*, 2010b). Clinical signs in juveniles and adults include skin ulceration, mouth tumidity, visceral ejection and death. Pathogenicity studies using intraperitoneal injection resulted in mortality at an inoculum of 5.6×10^8 CFU/ml but not at 5.6×10^4 CFU/ml. The LD_{50} at 14 days was 7.76×10^6 CFU/g body weight. *Shewanella marisflavi* was more virulent for mice and swordtail fish with an LD_{50} of 6.8×10^4 and 4.85×10^4 CFU/g body weight, respectively (Li *et al.*, 2010b).

Shewanella marisflavi, *S. putrefaciens*, *S. baltica* and *S. algae* are part of the bacterial flora found in oysters and seawater in Delaware Bay, USA (Richards *et al.*, 2008).

Shewanella marisflavi has 16S rRNA gene sequence similarity to *S. aquimarina* (97.4%), *S. affinis* (98.5%), *S. colwelliana* (98.5%) and *S. waksmanii* (97.0%) (Yoon *et al.*, 2004).

GROWTH CHARACTERISTICS AND PHENOTYPIC IDENTIFICATION. *Shewanella marisflavi* is a motile, Gram-negative rod (0.8–1.1×2.0–3.5 μm), which is catalase and oxidase positive. Colonies are 2.0–4.0 mm, circular, smooth, glistening, flat to slightly raised, sticky and are an apricot to pale orange colour after 3 days on MA 2216 at 25°C (see colour photographs in Li *et al.*, 2010b). Growth also occurs on tryptic soy agar with 2% final salt concentration and nutrient agar with 1.5% NaCl concentration. Colonies are green on TCBS and red on MacConkey agar. Beta-haemolysis is seen on sheep blood agar. Growth occurs at 4–35°C but not at 40°C for the sea cucumber isolate, whereas the type strain from seawater grows at 42°C but not at 43°C. Growth occurs at 0–6% NaCl but not at 8% NaCl. The organism is resistant to vibriostatic agent 0/129. *Shewanella marisflavi* produces H_2S, and ferments glycerol and D-fructose. Fermentation is negative for raffinose and rhamnose. Casein, tyrosine and Tween 80 are hydrolysed. Citrate is weak for the sea cucumber isolate and negative for the type strain. Fermentation of sucrose using conventional methods (Cowan and Steele, 1970) was reported negative for the type strain (Yoon *et al.*, 2004) and positive for the sea cucumber strain (Li *et al.*, 2010b), although the sea cucumber strain was negative according to the colour reaction on TCBS. These and other reactions are presented in Tables 3.11, 3.24–26 and 3.38.

Shewanella morhuae

TAXONOMY AND SIGNIFICANCE. *Shewanella morhuae* (U1417^T=ATCC BAA-1205^T=NBRC 100978^T) was isolated from cod and flounder caught in the Baltic Sea off the coast of Denmark and is thought to be involved in food spoilage (Satomi *et al.*, 2006). The isolate is less than 97% similar to other *Shewanella* species but is approximately 98% similar to *S. hafniensis* using 16S rRNA sequencing and approximately 89% similar to *S. hafniensis* based on *gyrB* gene sequence information.

GROWTH CHARACTERISTICS AND PHENOTYPIC IDENTIFICATION. *Shewanella morhuae* is an aerobic, motile, Gram-negative rod (0.5–0.7×1.0–1.2 μm), which is oxidase and catalase positive. On TSA colonies are dull white, round, undulate with irregular margins, and non-luminescent after 24 h at 25°C (Satomi *et al.*, 2006). Growth occurs at 0–6% NaCl and 4–25°C but is negative at 37°C. It is positive for hydrolysis of gelatin and production of DNase. Production of H_2S was reported positive by Satomi *et al.* (2006) but negative in TSI by Richards *et al.* (2008). It is non-haemolytic. Glucose is not fermented and utilization of citrate is negative. These and other reactions are presented in Tables 3.11 and 3.24–29.

Shewanella oneidensis

Shewanella oneidensis (ATCC 700550^T, previously *S. putrefaciens* MR-1) was isolated from water in Oneidensis Lake, USA (Venkateswaran *et al.*, 1999). *Shewanella oneidensis* is closely related to *S. putrefaciens* (97%), *S. frigidimarina* (96.2%) and *S. benthica* (91%) based on the 16S rRNA gene sequence, and difference from *S. putrefaciens* was confirmed by *gyrB* sequence. Cells are 2–3×0.4–0.7 μm, Gram-negative and facultatively anaerobic. Growth occurs on MA 2216 and MSA-B as pale pink to beige coloured colonies, 1–4 mm, circular, smooth, entire, convex and non-luminescent. Colour deepens with age. These and other reactions are presented in Tables 3.11 and 3.27–29.

Shewanella pealeana

TAXONOMY AND SIGNIFICANCE. *Shewanella pealeana* (ANG-SQ1^T=ATCC 700345^T) is part of the microbial flora of the accessory nidamental gland, an organ in the reproductive system of the female longfin inshore squid (*Doryteuthis pealeii*, previously *Loligo pealei*) (Leonardo *et al.*, 1999).

GROWTH CHARACTERISTICS AND PHENOTYPIC IDENTIFICATION. *Shewanella pealeana* is a facultatively anaerobic, motile, Gram-negative rod (2.0–3.0×0.4–0.6 μm). On MA 2216 colonies are salmon-coloured, opaque, with a mucoid surface after 48 h at 25°C and 2 mm in diameter after several days of growth. It is non-haemolytic. Growth occurs at 4–30°C, with optimal growth occurring at 25°C. Growth occurs at 0.125–0.75 M NaCl with an optimum of 0.5 M NaCl. Cells appear filamentous when grown under suboptimal conditions. It is positive for catalase, oxidase and lipase and negative for hydrolysis of gelatin and starch. 'Sugars' are not fermented (Leonardo *et al.*, 1999). These and other reactions are presented in Tables 3.11 and 3.27–29.

Shewanella pneumatophori

TAXONOMY AND SIGNIFICANCE. *Shewanella pneumatophori* (SCRC-2738^T=JCM 13187^T=NCIMB 14060^T), a bacterium that produces eicosapentaenoic acid, was isolated from the intestines of the Pacific mackerel (*Pneumatophorus japonicus*) (Hirota *et al.*, 2005). Based on 16S rRNA sequencing it is most closely related to *S. marinintestina* (99.3% similarity), *S. sairae* (99.3%) and *S. schlegeliana* (99.2%).

GROWTH CHARACTERISTICS AND PHENOTYPIC IDENTIFICATION. *Shewanella pneumatophori* is a motile, Gram-negative rod (0.4–0.6×1.4–4.6 μm) that is oxidase and catalase positive. On MA 2216 colonies have a slight pink colour, are circular and round after 24 h at 25°C (Hirota *et al.*, 2005). Growth occurs at 1–8% NaCl but not at 0% NaCl and 4–32°C, but is negative at 37°C. It is positive for production of H_2S, hydrolysis of gelatin, casein, DNA and Tween 24, 40, 60 and 80. Results are negative for hydrolysis of chitin, starch and hippurate. Acid is produced from D-glucose, D-fructose and D-galactose under anaerobic conditions but acid is not produced oxidatively from L-arabinose, D-fructose, melibiose, raffinose, D-galactose, L-rhamnose or glycerol. These and other reactions are presented in Table 3.11.

Shewanella putrefaciens

TAXONOMY AND SIGNIFICANCE. *Shewanella putrefaciens* was divided into four genomic groups based on DNA–DNA hybridization, and Owen's genomic group II was reclassified as *S. baltica* (Ziemke *et al.*, 1998). *Shewanella putrefaciens* genomic group IV, Gilardi biovar 1 and 3, Centre for Disease Control and prevention (CDC) biotype 1 was reclassified as *S. algae* (Khashe and Janda, 1998).

Strain NCIMB 400 has been reclassified as *Shewanella frigidimarina* (Reid and Gordon, 1999).

Shewanella putrefaciens, along with other *Shewanella* species (*S. abalonesis*, *S. algae*, *S. baltica*, *S. hafniensis*, *S. marisflavi*), is part of the bacterial flora found in oysters in Delaware Bay, USA (Richards *et al.*, 2008). A bacterium identified using the API 20E system as *Shewanella* (*Pseudomoas*) *putrefaciens* was isolated from a disease outbreak in rabbitfish (*Signaus rivulatus*) at an aquaculture facility in the Red Sea, Saudi Arabia (Saeed *et al.*, 1987). The isolate grew at 42°C and up to 9% NaCl, which suggests the isolate may have been *S. algae*.

GROWTH CHARACTERISTICS AND PHENOTYPIC IDENTIFICATION. Colonies produce a reddish-brown or pink pigment. Growth occurs at 25 and 37°C but not at 42°C. H_2S is negative on TSI. Growth does not occur at 6% NaCl. Colonies produce alpha-haemolysis on TSA with 5% sheep blood (Richards *et al.*, 2008). It is non-luminescent. Differentiation of *S. putrefaciens* from *S. algae* is based on growth occurring at 4°C but not at 42°C or 6% NaCl, fermentation of L-arabinose, and lack of beta-haemolysis on BA, and lack of growth on SS agar (Nozue *et al.*, 1992).

Shewanella putrefaciens is listed in the API 20E database, however other *Shewanella* species may be misidentified as *S. putrefaciens* because of the biochemical non-reactivity of the *Shewanella* genus (Richards *et al.*, 2008). These and other reactions are presented in Tables 3.11, 3.24–29 and 3.38.

Shewanella sairae

TAXONOMY AND SIGNIFICANCE. *Shewanella sairae* (SM2-1T=JCM 11563^T=LMG 21408^T) was isolated from the intestine of a Pacific saury (*Cololabis saira*) from Japan (Satomi *et al.*, 2003). *Shewanella sairae* has >99% 16S rRNA sequence similarity to *S. schlegeliana* and *S. marinintestina*, and a 97% sequence similarity to *S. pealeana*. It clusters within the *S. pealeana*/*S. gelidimarina* cluster. Sequencing of the *gyrB* gene provides improved differentiation between *S. sairae*, *S. schlegeliana* and *S. marinintestina*.

GROWTH CHARACTERISTICS AND PHENOTYPIC IDENTIFICATION. *Shewanella sairae* is a facultatively anaerobic, motile, Gram-negative rod (2.0–3.0 µm), which is oxidase and catalase positive. On MA 2216 colonies are ivory in colour, opaque, circular after 48 h at 20°C. Growth occurs at 4–27°C with an optimum of 20–25°C. Growth occurs at 1–5% (w/v) NaCl with an optimal range of 2.0–3.0%. Hydrolysis of gelatin, lipase, casein and hippurate is positive. Nitrate is reduced and H_2S is produced. It is negative for hydrolysis of chitin, agarose and alginase. These and other reactions are presented in Tables 3.11, 3.27–30 and 3.38.

Shewanella schlegeliana

TAXONOMY AND SIGNIFICANCE. *Shewanella schlegeliana* (HRKA1^T=JCM 11561^T=LMG 21406^T) was isolated from the intestine of a black porgy (*Acanthopagrus schlegelii*) from Japan (Satomi *et al.*, 2003). *Shewanella schlegeliana* has >99% 16S rRNA sequence similarity to *S. sairae* and *S. marinintestina*, and a 97% sequence similarity to *S. pealeana*. It clusters within the *S. pealeana*/*S. gelidimarina* cluster. Sequencing of the *gyrB* gene provides improved differentiation between *S. sairae*, *S. schlegeliana* and *S. marinintestina*.

GROWTH CHARACTERISTICS AND PHENOTYPIC IDENTIFICATION. *Shewanella schlegeliana* is a facultatively anaerobic, motile, Gram-negative rod (2.0–3.0 µm), which is oxidase and catalase positive. On MA 2216 colonies are ivory in colour, opaque, circular after 48 h at 20°C. Growth occurs at 4–32°C with an optimum of 20–25°C. Growth occurs at 1–5% (w/v) NaCl with an optimal range of 2.0–3.0%. Hydrolysis of gelatin, urease, aesculin and hippurate is positive. Nitrate is reduced and H_2S is produced. It is negative for hydrolysis of chitin, lipase, agarose and alginase. These and other reactions are presented in Tables 3.11, 3.27–30 and 3.38.

Shewanella upenei

TAXONOMY AND SIGNIFICANCE. *Shewanella upenei* (20-23R^T=KCTC 22806^T=CCUG 58400^T) was isolated from the intestine of bensasi goatfish (*Upeneus japonicus*) collected from the South Sea in Korea (Kim *et al.*, 2011). *Shewanella upenei* has 99.5% 16S rRNA sequence similarity to *S. algae*, 99.2% to *S. haliotis* and 97.5% to *S. chilikensis*.

GROWTH CHARACTERISTICS AND PHENOTYPIC IDENTIFICATION. *Shewanella upenei* is a motile, Gram-negative rod (1.0–4.5×0.3–0.7 µm), which is oxidase and catalase positive. On MA 2216 colonies are 3–5 mm, orange-yellow, circular, raised, smooth, glistening and slightly irregular after

3 days at 30°C. Growth occurs at 10–40°C with an optimum of 30°C. Growth does not occur at 4 or 45°C. Growth occurs at 0–11% (w/v) NaCl with an optimal range of 2.0–5.0%. Casein, DNAse, gelatin, Tween 20, 40, 60 and 80 are hydrolysed when tested on MA 2216 with added substrate. Agar is not hydrolysed. Nitrate is reduced (Kim *et al.*, 2011). These and other reactions are presented in Tables 3.11, 3.24–26 and 3.38.

Shewanella woodyi

Shewanella woodyi (MS32T=ATCC 51908T) was isolated from squid ink and water from the Alboran Sea (Makemson *et al.*, 1997). It is a non-sporulating rod (0.5–1.0×1.4–2 µm). It is luminescent on MA 2216. NaCl is required for production of luminescence. Growth occurs at 4°C but not at 30°C with an optimum of 25°C. No 'sugars' are fermented. On MA 2216 colonies are pink-orange at 20°C. No growth on TCBS. Negative for chitinase. These and other reactions are presented in Tables 3.11 and 3.24–26.

Order *Enterobacteriales*

1.11 *Enterobacteriaceae*

The family *Enterobacteriaceae* belongs to the *Gammaproteobacteria* (Class III), and the Order '*Enterobacteriales*'. The family *Enterobacteriaceae* contains 39 genera including *Escherichia*, *Citrobacter*, *Edwardsiella*, *Enterobacter*, *Erwinia*, *Ewingella*, *Hafnia*, *Klebsiella*, *Morganella*, *Pantoea*, *Plesiomonas*, *Proteus*, *Providencia*, *Rahnella*, *Salmonella*, *Serratia*, *Shigella* and *Yersinia*.

1.11.1 *Citrobacter*

Citrobacter freundii

Citrobacter freundii belongs to the family *Enterobacteriaceae* and is a common inhabitant of soil, water and sewerage and has been isolated from food. It is part of the normal intestinal flora in a number of animals including mammals, reptiles, birds and amphibians, but can exist in a carrier status, however, and is considered an opportunistic pathogen in situations of stress. Experimental challenge to assess pathogenicity indicates the organism has a low to moderate virulence ranging from LD_{50} of 10^6 cells for rainbow trout, 10^5–10^6 for common carp (*Cyprinus carpio*) and a 66.7% mortality tested by immersion challenge using 10^8 CFU/ml in grass carp (*Ctenopharyngodon idella*) (Karunasagar *et al.*, 1992; Toranzo *et al.*, 1994; Lü *et al.*, 2011). In experimental challenge using rainbow trout the organism was recovered from the internal organs of survivors indicating its carrier status (Toranzo *et al.*, 1994).

Grass carp are intensively farmed in China and Citrobacter gastroenteritis is an increasing problem. Other *Citrobacter* species, namely *C. gillenii* and *C. werkmanii*, are also isolated from the intestinal tract but only *C. freundii* results in pathogenicity in experimental challenge (Lü *et al.*, 2011). Naturally infected common carp fingerlings had high mortalities due to septicaemia and displayed haemorrhages on the skin, eye and base of the fins and in the peritoneum. The organism was isolated from kidneys, liver, spleen, heart, and peritoneal fluid (Karunasagar *et al.*, 1992). Although isolates varied in their production of haemolysis on laboratory media no correlation with virulence was found.

An epizootic in sunfish (*Mola mola*) held in an aquarium in Japan was due to *C. freundii* and likely caused by poor water quality. Infected fish displayed haemorrhagic lesions on the skin and around the anus, multiple lipoid masses in the kidney and the organism was isolated in pure growth from the kidney and spleen (Sato *et al.*, 1982).

Citrobacter freundii is a pathogen of cultured American bullfrogs (*Rana catesbeiana* – note *Lithobates catesbeianus* is thought to be the more correct name) causing the disease red-leg syndrome or bacterial dermatosepticaemia. Infected frogs display torticollis and stupor. Focal areas of haemorrhage occur in the muscle, legs are swollen and liver, spleen and kidney may be enlarged (Mauel *et al.*, 2002).

Strains of *C. freundii* show variation in reactions for ADH, ODC, fermentation of sucrose, melibiose, amygdalin and salicin (API 20E) regardless of source or geographical location (Toranzo *et al.*, 1994). Isolates from naturally infected common carp varied from reactions reported in Bergey's manual by being positive for indole production and negative for fermentation of D-sorbitol (Karunasagar *et al.*, 1992). Urea may be positive by Christensen's method but usually is negative when tested for the preformed enzyme, and is often negative in the API 20E. These and other reactions are presented in Tables 1.3 and 3.24–26.

1.11.2 *Edwardsiella*

The *Edwardsiella* genus belongs to the Family *Enterobacteriaceae*. There are four species within the genus *Edwardsiella*: *E. tarda* (type species of the genus), *E. hoshinae*, *E. ictaluri* and the newly described *E. piscicida*, which contains strains previously identified as *E. tarda*.

Edwardsiella hoshinae

Edwardsiella hoshinae may smell like *Plesiomonas shigelloides*, which has a strong, pungent, sweetish smell. *Vibrio (carchariae) harveyi* ATCC 35084 has a similar pungent smell, but not as strong as *P. shigelloides*. *Edwardsiella hoshinae* was reported as indole positive by Grimont *et al.* (1980), but negative or weak results were found by Farmer and McWhorter (1984). Likewise for TSI, Farmer and McWhorter (1984) report a negative result, whereas a positive result was recorded by Grimont *et al.* (1980).

Edwardsiella ictaluri

TAXONOMY AND SIGNIFICANCE. *Edwardsiella ictaluri* causes enteric septicaemia of catfish (ESC), and was first reported from cultured catfish (*Ictalurus punctatus*) in the USA (Hawke *et al.*, 1981). ESC is a major disease problem for the commercial channel catfish industry. Disease resistance is variable among channel catfish species with blue catfish (*Ictalurus furcatus*) and Red River strain showing the most disease resistance (Wolters and Johnson, 1994). Ornamental species and salmonids are susceptible and it has also been reported in seabream (*Dicentrarchus labrax*), white catfish (*Ameiurus catus*), green knife fish and brown bullhead (*Ameiurus nebulosus*) (Hawke *et al.*, 1981; Iwanowicz *et al.*, 2006). It has been reported from a number of countries in species such as walking catfish (*Clarias batrachus*) in Thailand (Kasornchandra *et al.*, 1987), yellow catfish (*Tachysurus* [*Pelteobagrus*] *fulvidraco*) in China (Ye *et al.*, 2009; Liu *et al.*, 2010b) and rainbow trout (*Oncorhynchus mykiss*) cultured in Turkey (Keskin *et al.*, 2004). In Vietnam, *E. ictaluri* was isolated from cultured striped catfish (*Pangasianodon hypophthalmus*) and is referred to as bacillary necrosis (Crumlish *et al.*, 2002). Infection in *Pangasianodon* presents as multifocal, white lesions irregular in size and found in internal organs, predominantly the kidney, liver and spleen. In histology the lesions appear as areas of necrosis and pyogranulomatous inflammation.

Edwardsiella ictaluri is exotic to Australia but has been detected in imported fish including rosy barb (*Puntius conchonius*), almorha or yoyo loach (*Botia almorhae*), while held in quarantine facilities in Northern Territory, Queensland, Victoria and Western Australia and hatchery reared salmonids in Australia (Humphrey *et al.*, 1986; Herbert, 2012). Australian native catfish are susceptible and include eel-tailed or toothless catfish (*Anodontiglanis dahli*), narrowfront tandan (*Neosilurus ater*) and fork-tailed or highfin catfish (*Neoarius berneyi*) (Herbert, 2012).

Outbreaks of the disease occur in the temperature range of 18 to 28°C, whereas a low-level mortality and carrier status may be seen at temperatures outside this range. In the acute form of the disease, the disease expresses as an acute septicaemia. Petechial haemorrhages occur around the mouth, the throat and the fins, and internally in the liver and other organs. The organism crosses the intestinal mucosa into the internal organs. A chronic form of the disease occurs as a meningoencephalitis with behavioural changes, and ulceration or 'hole-in-the-head' (Hawke *et al.*, 1981). The organism can be isolated from the peritoneal fluid, liver, kidney and spleen.

In an outbreak among tank-reared wild caught brown bullheads (*Ameiurus nebulosus*) the main clinical signs were a swollen and reddened vent, external petechial haemorrhages on the ventral surface, haemorrhagic lesions around the eyes and raised circular reddish lesions on the lateral trunk (Iwanowicz *et al.*, 2006). Internal examination showed petechial haemorrhages along the oesophagus and lower intestine and reddening around the mouth, and enlarged spleen and trunk kidney. The 'hole in the head' seen in classic ESC was only seen in fish that died 8 weeks post-outbreak.

In experimental infection *E. ictaluri* caused 100% mortality in channel catfish fingerlings injected with 0.1 ml of 1.5×10^3 cells. Tilapia, *Oreochromis* (*Sarotherodon*) *aureus*, were slightly susceptible with 70% deaths at an inoculum of 1.5×10^8. Golden shiner (*Notemigonus crysoleucas*), bighead carp (*Hypophthalmichthys nobilis*) and largemouth bass (*Micropterus salmoides*) were not susceptible (Plumb and Sanchez, 1983). Rainbow trout (*Oncorhynchus mykiss*) were susceptible to infection when inoculated with 0.5 ml of 10^6 cells/ml (Keskin *et al.*, 2004). In immersion challenge experiments, channel catfish and chinook salmon (*Oncorhynchus tshawytscha*) were susceptible with 32% and 75% mortality, respectively, compared to striped bass (*Morone saxatilis*) with 5% mortality (Baxa *et al.*, 1990). The LD_{50} for chinook salmon was 3.4×10^7 CFU/ml. White sturgeon (*Acipenser transmontanus*) and blue catfish (*I. furcatus*) were not susceptible in bath immersion experiments (Baxa *et al.*, 1990; Wolters and Johnson, 1994).

GROWTH CHARACTERISTICS AND PHENOTYPIC IDENTIFICATION. *Edwardsiella ictaluri* is a facultatively anaerobic, oxidase-negative, Gram-negative rod to coccobacillus (0.75×1.5–2.5 μm) and catalase positive. After 48 h at 28–30°C colonies appear on BA as 1–2 mm round, circular to slightly irregular, low convex with a smooth surface, non-pigmented, pale grey with a butyrous consistency. A greenish tinge may be seen around the colony and slight beta-haemolysis under the colony. Growth also occurs on TSA, BHIA (Brain Heart Infusion Agar), nutrient agar and MacConkey agar. strains are motile at 25°C but non-motile at 37°C (Hawke *et al.*, 1981; Humphrey *et al.*, 1986). Growth was reported to occur on brilliant green agar, however a Vietnamese strain grew poorly on this medium but gave good growth on XLD agar (Buller, unpublished data). Growth occurs on SS agar (BGA, SS and XLD are used for selective growth of *Salmonella* species). A selective medium (EIM) has been reported for *E. ictaluri* EIM (Shotts and Waltman, 1990) and is detailed in the media section.

MR and VP are positive and negative, respectively, at both 37°C and 20°C, although strains from brown bullhead were negative for MR. Growth may be negative or slow at 37°C and 22°C but there is no growth at 42°C. The organism grows in 0–1.5% NaCl, but not 2% NaCl. H_2S is negative in TSI medium but variable when lead acetate paper is used. Results are negative for chitin and collagen but positive for chondroitin sulfate, which is a major component of cartilage and may explain the 'hole in the head' lesion (Waltman *et al.*, 1986). The strains from rainbow trout cultured in Turkey were positive for mannitol and melibiose when tested using conventional biochemical methods and the API 20E kit (Keskin *et al.*, 2004), whereas *E. ictaluri* from other fish species have been reported negative for these two tests. The API 20E number is 410414057 for the rainbow trout isolates, which gives a 56.7% probability of *Salmonella* species. A number of 410400057 is obtained when these two tests are negative and the database gives a 99.2% probability of *E. tarda* with tests against this organism being H_2S and indole. Isolates from Vietnamese catfish were only positive for lysine and glucose using the API 20E kit, profile 400400057, when incubated at 28°C for 48 h (Crumlish *et al.*, 2010). This profile gives a probability of *E. tarda* of 52.3%. Microscopically the bacterial cells show a greater pleomorphism in length and size than normally seen with other strains of *E. ictaluri* (Crumlish *et al.*, 2002). *Edwardsiella ictaluri* is not in the API 20E database, therefore it is important to interpret the reactions according to these results listed here. Tests should be read at 24 and 48 h.

Edwardsiella ictaluri is differentiated from *E. tarda* by indole and hydrogen sulfide as *E. ictaluri* is negative for both tests whereas *E. tarda* is positive for both tests. Other similar organisms such as *Yersinia ruckeri*, *E. tarda*, *E. hoshinae* and *Hafnia alvei* grow well at 37°C whereas *E. ictaluri* grows slowly or poorly at 37°C. *Edwardsiella ictaluri* can be differentiated from *Y. ruckeri* as *E. ictaluri* is negative for citrate, trehalose and gelatin whereas *Y. ruckeri* is positive for these tests.

Strains with a limited tolerance to oxygen were isolated from kidney of channel catfish with deep ulcerative lesions (Mitchell and Goodwin, 2000). The bacteria were cultured on TSA containing blood at 25°C for 2 days and under anaerobic conditions. Cells are short Gram-negative rods and are negative for methyl red at 25°C, and negative for hanging drop motility at 22°C and 25°C, but are positive in motility indole ornithine agar medium at 28°C and 96 h (Mitchell and Goodwin, 2000). Biochemical tests should be read at 48 h. These and other reactions are presented in Tables 3.10, 3.24–26 and 3.38.

MOLECULAR IDENTIFICATION. A PCR specific for *E. ictaluri* was reported using primers that amplify the region between the intervening sequence (IVS) and the inter-ribosomal space (IRS) (Williams and Lawrence, 2010). The IRS, a region between the *rrs* and *rrl* genes and also referred to as the intergenic transcribed spacers (ITS), is found in the *Edwardsiella* genus but is absent from the *Enterobacteriaceae*. The IVS sequence in the 23S rRNA gene is only found in *E. ictaluri* and not in *E. tarda*. Williams and Lawrence (2010) suggest that the combination of the IVS and IRS primers is specific for *E. ictaluri* and does not produce a product for *E. tarda* whereas a PCR reported by Panangala *et al.* (2007) requires high stringency for cycling parameters and the PCR mix because the forward primer (EiFd-1) has 100% identity with *E. tarda* and the reverse primer (EiRs-1) only has a 2 bp difference to *E. tarda*.

Another PCR based on a putative transposon, a mobile element between the *ser*C and *aro*A genes, has a 4–10 bp difference for each primer compared to other bacterial genomes, however using primer sites on a mobile element may lead to false negatives due to the instability of the transposase (Williams and Lawrence, 2010).

DNA is extracted from colonies using the DNeasy Blood and Tissue kit (Qiagen) or the following manual method where a single colony is resuspended in 50 µl sterile saline and incubated at 95°C for 5 min. The suspension is centrifuged and 2 µl of supernatant used in the PCR reaction. Primer IVS (5′-TTA AAG TCG AGT TGG CTT AGG G-3′) and IRS (5′-TAC GCT TTC CTC AGT GAG TGT C-3′) are used in the PCR reaction mixture. Thermocycler conditions consist of an initial denaturation step at 95°C for 4 min followed by 30 cycles at 95°C for 30 s; 60°C for 30 s; 72°C for 2 min with a final cycle at 72°C for 10 min. Expected size of the amplicon is 2000 bp.

The *Edwardsiella* genus can be detected with a 1300-bp amplicon produced from the IVS region using primer 16S flank (5′-TAT CTA ATC CTG TTT GCT CCC C-3′) and primer 23S-F (5′-GAC GTT GAT AGG CTG GGT GT-3′) with the same PCR conditions (Williams and Lawrence, 2010).

A multiplex PCR was developed for the simultaneous detection of *E. ictaluri*, *Flavobacterium columnare* and *Aeromonas hydrophila* (Panangala *et al.*, 2007). Primers for *E. ictaluri* were designed to the 16S rRNA gene, and it is these primers that require stringent conditions for specificity as mentioned above by Williams and Lawrence (2010). Primers for *F. columnare* were from Welker *et al.* (2005), primers for *A. hydrophila* were from Pollard *et al.* (1990), and could detect these organisms from cultured cells, and 30 to 50 mg of tissue from blood, gill and kidney using the High Pure PCR template preparation kit (Roche Diagnostics) as per manufacturer's instructions. Tissues are macerated using micro-pestles and 10 µl buffer (10 mM Tris HCl, 100 mM EDTA). The reaction mix consists of primers for *F. columnare* (FCISRFL 5′-TGC GGC TGG ATC ACC TCC TTT CTA GAG ACA-3′ and FCISRR1 5′-TAA TYR CTA AAG ATG TTC TTT CTA CTT GTT TG-3′), which produces an amplicon of 504 bp; primers for *A. hydrophila* based on the aerolysin gene (Aero1a 5′-CCA AGG GGT CTG TGG CGA CA-3′ and reverse primer Aero1b 5′-TTT CAC CGG TAA CAG GAT TG-3′) producing an amplicon of 209 bp; primers for *E. ictaluri* (EiFd-1 5′-GTA GCA GGG AGA AAG CTT GC-3′ and EiRs-1 5′-GAA CGC TAT TAA CGC TCA CAC C-3′), which produce an amplicon of 407 bp. (Note: primer designations from the original papers are used here.) Conditions were optimized to produce an optimal analytical sensitivity in 1 g of tissue of 3.4×10^2 to 2.5×10^5 cells/g tissue.

The 25-µl master mix consists of 12.5 µl of 1× Epicentre Fail Safe PreMix Buffer-D, 1 µl (5U) *Taq* DNA polymerase (Promega), 1 µg (20 ng) sample DNA, 3 µl sterile distilled water and 0.6 µM each primer FCISRFL and FCISRR1, 0.4 µM each primer EiFd-1 and EiRs-1, and 0.5 µM each primer AeroFd and AeroRs. Cycling is carried out on a Tgradient thermocycler (Whatman Biometra) using an initial denaturation at 95°C for 4 min with 30 cycles consisting of 95°C for 30s, with annealing at 63°C to 54°C (decreased by 0.3 for each cycle) for 45 s and extension at 72°C for 30 s and one final cycle with extension at 72°C for 10 min. A no template control and positive controls for each organism is included.

Amplified products are visualized on a 3:1 Nu-Sieve agarose gel (Cambrex Bio Science) under electrophoresis of 5 volts/cm and stained in the dark with 1× Sybr Green-1. The stepwise reduction in annealing temperature improves the band intensities for the three products.

The multiplex PCR is specific when tested against *A. sobria*, *A. caviae*, *E. tarda*, *E. hoshinae*, *F. psychrophilum*, *V. anguillarum*, *Y. ruckeri*, *E. coli*, *Enterobacter sakazakii*, *Streptococcus iniae* and *S. agalactiae*. Cross-reaction is obtained from *A. salmonicida* ssp. *salmonicida* as an amplicon of 209 bp is produced from the primers designed to detect the aerolysin gene (Panangala *et al.*, 2007).

Edwardsiella piscicida

TAXONOMY AND SIGNIFICANCE. A number of strains isolated from fish and eels previously identified as *E. tarda* have been named as a new species, *Edwardsiella piscicida* (ET883^T=CCUG 62929) (Abayneh *et al.*, 2013). DNA–DNA hybridization of *E. piscicida* strains with *E. tarda* (ATCC 15947^T) is 68.2–93.9%, which is below the minimum level of 70% and indicates a new species. Sequence analysis using 16S rRNA and housekeeping genes *gyrB*, *mdh*, *adk*, *dnaK*, *phoR*, *metG*, *pyrG* and *aroE2* place the majority of *E. piscicda* strains in a cluster separate from *E. tarda*, whereas other strains cluster with *E. ictaluri*. *Edwardsiella tarda* strain NCIMB 2034, from an unknown fish species, clusters with *Edwardsiella tarda*. The *gyrB* gene sequence similarity for isolate LTB-4 (*E. piscicida* strain from turbot cultured in China) is 97% with NCIMB 2034 (Lan *et al.*, 2008).

Strains previously identified as *E. tarda* and now identified as *E. piscicida* are strains (ET883^T, ET2640) from European eel (*Anguilla anguilla*) from Norway isolated in 1989, strains (RM 298.1, HI 9.1) from turbot (*Scophthalmus maximus*) in northern and southern Europe isolated in 2006, strains (ETA1, ETB1) from turbot in Scotland in 2007, strains (LTB4, WY 18) from turbot in China in 2006 (Lan *et al.*, 2008) and strain ETK01 from Korean catfish (*Silurus asotus*) in Korea in 2008. Isolates previously identified as *E. tarda* and now identified as *E. piscicida* but belonging to a separate phylogenetic cluster are isolate ET080813 from marbled eel (*A. marmorata*) and isolate ET080814 from Japanese eel (*A. japonica*), both from China in 2008, and NCIMB 2056 from crimson seabream (*Evynnis japonica*) in the NCIMB culture collection (Abayneh *et al.*, 2013). *Edwardsiella piscicida* (strain LTB-4) from cultured turbot in China caused high mortality of 300 fish/day with infected fish displaying eye tumefaction, haemorrhages on the body, fins, head and mouth, abdominal swelling, ascites and small white lesions on the liver (Lan *et al.*, 2008). The organism was isolated from the liver, spleen, blood and brain. The LD_{50} in experimental challenge of turbot was 2.2×10^4 CFU/fish.

Edwardsiella piscicida is pathogenic to zebrafish in experimental challenge whereas *E. tarda* (ATCC 15947^T) is not. Infected zebrafish display erratic swimming, low feeding rates and ulcerations on dorsal surfaces. The organism was re-isolated from liver and spleen. Challenge with the type strain ET883 resulted in cumulative mortalities of 100% by day 7 compared to three other strains that resulted in 95%, 87.5% and 11.1% cumulative mortalities during the same period (Abayneh *et al.*, 2013).

GROWTH CHARACTERISTICS AND PHENOTYPIC IDENTIFICATION. *Edwardsiella piscicida* is a facultatively anaerobic, oxidase-negative, motile, Gram-negative small rod (0.5–0.9×1.0–2.0 μm) that is positive for catalase. The isolate (LTB-4) from turbot cultured in China was initially reported as devoid of flagella and negative for motility tests (Lan *et al.*, 2008), however this was later reported to be motile but has one or two flagella only (He *et al.*, 2011). Other isolates are motile (Abayneh *et al.*, 2013). On BA, colonies are pinpoint, circular, slightly convex, smooth and glistening. Beta-haemolysis is visible underneath the colony. Growth occurs at 25–37°C, with an optimum of 28–30°C, and at 1–5% NaCl. Growth does not occur at 12 or 42°C or at 6% NaCl. *Edwardsiella piscicida* is negative for growth at 42°C whereas *E. tarda* is positive. Variation exists for fermentation of D-mannitol and L-arabinose with the two (25%) strains from marbled eel and Japanese eel being positive for D-mannitol and the one (8.3%) strain from marbled eel being positive for L-arabinose. Nine (75%) strains are negative for Simmons' citrate, whereas strains from marbled and Japanese eel and crimson seabream are positive. The majority (75%) of strains of *E. piscicida* give a number of 454400077 in the API 20E, which is the same profile given by the majority of *E. tarda* strains. *Edwardsiella piscicida* has different reactions for β-methyl-D-glucoside, citric acid and L-proline compared to *E. tarda* when tested using the Biolog GN2 MicroPlates (Biolog Inc, USA). *Edwardsiella piscicida* is negative for these reactions whereas *E. tarda* is positive (Abayneh *et al.*, 2013). It is not known whether these tests using other identification methods would give the same result. These and other reactions are presented in Tables 3.10 and 3.24–26.

Edwardsiella tarda

TAXONOMY AND SIGNIFICANCE. *Edwardsiella tarda* (CDC 1483-59^T=ATCC 15947^T) was used first to describe a group of isolates cultured from faeces and wounds in people, and faeces from cattle and snakes (Ewing *et al*., 1965). A bacterium isolated from red disease of eel in Japan was named *Paracolobactrum anguillimortiferum* (Hoshina, 1962), but was subsequently found to be identical to *E. tarda* (Wakabayashi and Egusa, 1973).

Edwardsiella tarda can be found in the environment in pond water, lakes, rivers, soil and in the cloacae of crocodiles, lizards, seagulls, snakes, tortoises and waterfowl and may be part of the normal flora or exist in a carrier status (Ewing *et al*., 1965; Iveson, 1971; White *et al*., 1973; Wyatt *et al*., 1979). *Edwardsiella tarda* is a pathogen of many aquatic species and a cause of enteritis in humans. Disease signs and pathogencity are variable according to different host species, and variation in phenotype and virulence exists within *E. tarda* strains.

Some strains previously described as *E. tarda* have been re-named *E. piscicida*. These are strains isolated from European eel (*Anguilla anguilla*) in Norway in 1989, from marbled eel (*A. marmorata*) and Japanese eel (*A. japonica*) from China in 2008, from Korean catfish in South Korea in 2008 and from turbot (*Scophthalmus maximus*) in Europe, Scotland and China in 2006 and 2007 (Abayneh *et al*., 2013). It is not known whether both *E. tarda* and *E. piscicida* were isolated from these sources.

Edwardsiella tarda infection in channel catfish (*Ictalurus punctatus*) in the USA has been referred to as emphysematous putrefactive disease of catfish (EPDC) to describe the gross appearance of infected fish due to the 3–5 mm cutaneous lesions on the postero-lateral areas of the body and the development of abscesses within the muscles of the flank and caudal peduncle (Meyer and Bullock, 1973). The lesions become large gas-filled cavities appearing as convex swollen areas. The organism is isolated from the kidney and lesions. Disease signs are similar in carp Roho labeo (*Labeo rohita*) cultured in India (Mohanty and Sahoo, 2010).

Edwardsiella tarda infection has been reported from brook trout (*Salvelinus fontinalis*) in Canada (Uhland *et al*., 2000), in wild striped bass (*Moroxone saxatilis*) in Chesapeake Bay and in pre-spawning chinook salmon (*Oncorhynchus tshawytscha*) in the Rogue River in Oregon, and largemouth bass (*Micropterus salmoides*) in Florida, USA (White *et al*., 1973; Amandi *et al*., 1982; Baya *et al*., 1997). Infected striped bass and largemouth bass present with malodorous haemorrhagic ulcers on the body and fins, small lesions in the kidney, liver and spleen, empty intestines and yellowish or sanginous mucoid fluid in the body cavity. The organism is isolated from kidney and intestine in chinook salmon. Experimental challenge indicates *E. tarda* is pathogenic for striped bass (LD_{50} 4×10^5 cells), gilthead seabream (LD_{50} 7×10^5) and turbot (LD_{50} 3×10^5 cells) when injected intramuscularly, and for chinook salmon (LD_{50} 4.1×10^6), steelhead trout (LD_{50} 5.6×10^6) and channel catfish (LD_{50} 4.0×10^5) when injected intraperitoneally and maintained at 18°C. At a lower temperature of 12°C the LD_{50} for chinook salmon and rainbow trout is 6.4×10^7 and 1.7×10^6, respectively (Amandi *et al*., 1982; Baya *et al*., 1997). The disease is usually associated with poor water quality and elevated water temperature, often above 30°C (Meyer and Bullock, 1973), but this is not always the case as disease in tilapia occurred in winter and spring when temperatures were 20–30°C (Miyashita, 1984) and at 17°C for epizootics in Senegalese sole in Spain (Castro *et al*., 2012). Tilapia (*Oreochromis niloticus niloticus*) cultured in Japan had haemorrhagic lesions in the gonad, particularly the ovary. Virulence was proven through intramuscular injection (Miyashita, 1984). Disease signs in Senegalese sole (*Solea senegalensis*) cultured in a marine farm off the coast of Spain included cutaneous lesions on the dorsal surface, haemorrhagic areas on the ventral surface and tumefactions around the eye (Castro *et al*., 2012).

Infection has occurred in rainbow trout (*Oncorhynchus mykiss*) cultured around the world in the Czech Republic, in Australia and in Brazil (Reddacliff *et al*., 1996; Řehulka *et al*., 2012). In experimental infection *E. tarda* did not cause disease in common carp (*Cyprinus carpio*), Prussian carp and tench (*Tinca tinca*) when inoculated intraperitoneally with 8×10^6 cells (Řehulka *et al*., 2012).

Edwardsiella tarda has been reported from epizootics in turbot (*Scophthalmus maxiumus*) in northern and southern Europe. Infected fish display small haemorrhages in the musculature, mainly in the head region, discolouration of the skin, increased ascites fluid in the abdominal cavity and haemorrhagic kidney (Castro *et al*., 2006). The organism was cultured from cutaneous lesions, spleen, liver and kidney. In amur catfish (*Silurus asotus*) also known as Japanese or Korean catfish, cultured in Korea, infection results in abdominal swelling and enteritis and small abscess-like lesions in the spleen and kidney (Yu *et al*., 2009). It is likely that some of the strains isolated from turbot and catfish were the newly described *E. piscicida* (Abayneh *et al.*, 2013).

Edwardsiella tarda infects eels (*Anguilla japonica* and *A. anguilla*) and the disease is termed edwardsiellosis, and has been reported from Japan and Spain (Hoshina, 1962; Wakabayashi and Egusa, 1973; Alcaide *et al.*, 2006). It is associated with pond-cultured eel disease in Japan and infected animals display abscessed or ulcerative lesions in the kidney and liver. Infected eels caught from a Mediterranean freshwater coastal lagoon, Spain, had an ulcer on the body particularly the opercula region, overproduction of mucus, petechiae on the belly and haemorrhagic fins. Experimental infection with intraperitoneal or intramuscular injection of 0.1–0.4 mg of cells resulted in mortalities in 3–7 days for Japanese eels whereas an LD_{50} of 7.4×10^4 to 6.6×10^6 CFU/ml was recorded for the strain isolated from Spanish eels (Hoshina, 1962; Wakabayashi and Egusa, 1973; Alcaide *et al.*, 2006). Experimental challenge with the type strain CECT 849^T did not result in mortality of eels; however, there is the possibility this strain has been attenuated due to high levels of subculturing (Alcaide *et al.*, 2006), although Abayneh *et al.* (2013) also reported the type strain non-pathogenic in experimental challenge with zebrafish.

Two phenotypes have been reported in Japan: a typical strain which is motile and an atypical non-motile strain. Motile and non-motile strains have been reported from eels (*A. japonica*), flounder (*Paralichthys olivaceous*), red seabream (*Pagrus major*) and yellowtail (*Seriola quinqueradiata*) (Nakatsugawa, 1983; Tomomasa *et al.*, 2005). When tested for pathogenicity, differences in virulence occurred as flounder and yellowtail were susceptible to both strains but only red seabream was susceptible to the atypical strain (Tomomasa *et al.*, 2005). Strain LTB-4 from turbot cultured in China was non-motile and devoid of flagella and was later identified as *E. piscicida* (Lan *et al.*, 2008; Abayneh *et al.*, 2013), although a later study found that the bacterium possessed one or two flagella and exhibited motility (He *et al.*, 2011). It is not known whether these other non-motile strains are in fact *E. piscicida*. The 12 strains of *E. piscicida* identified to date are motile (Abayneh *et al.*, 2013).

Edwardsiella tarda infects aquarium fish and was reported from fighting fish (*Betta splendens*) imported from Singapore and held in a quarantine facility in Australia. Moribund fish had single or multiple 1 mm cutaneous ulcers and the organism was isolated in pure growth from the kidney, liver and spleen (Humphrey *et al.*, 1986).

Other reports of *E. tarda* infections have included spotted bonytongue (called saratoga in Australia) (*Scleropages leichardti*), banded knife fish (*Gymnotus carapo*) at a zoo, and captive little penguin or fairy penguin (*Eudyptula minor*) and it has been found in subdermal abscesses in eels in Australia. Cutaneous ulcerations or haemorrhages may be seen on the skin and organs (Eaves *et al.*, 1990; Reddacliff *et al.*, 1996). *Edwardsiella tarda* was isolated from several species of healthy turtle (Table 1.1) at the New York Zoological Park and appeared to harbour the organism with no ill health (Otis and Behler, 1973).

A number of aquatic animals can be carriers of *E. tarda* and it may be part of their normal intestinal flora, however there are reports of *E. tarda* being associated with disease. In a study investigating a number of aquatic animals *E. tarda* was found to be associated with enteric disease and intestinal lesions in brown pelican (*Pelecanus occidentalis carolinensis*), common or great northern loon (*Gavia immer*) and bald eagle (*Haliaeetus leucocephalus*) but was considered part of the normal intestinal flora of American alligator (*Alligator mississippiensis*), sandhill crane (*Grus canadensis*), great blue heron (*Ardea herodias*) and ring-billed gull (*Larus delawarensis*), which all appeared normal with no signs of disease (White *et al.*, 1973). However, others have reported *E. tarda* in association with a captive alligator (*A. mississippiensis*) that died showing fibrinous peritonitis, congested kidneys from which *E. tarda* was isolated and gastric ulcers (Wallace *et al.*, 1966).

Edwardsiella tarda may also be an opportunistic pathogen in sick or injured marine mammals. The organism was isolated in pure growth from peritoneal exudate from a Steller sea lion (*Eumetopias jubatus*) presenting with a perforated ulcer in the posterior colon. A pure culture was also obtained from the liver of a pregnant harbour porpoise (*Phocena phocena*) with purulent metritis and fibrinopurulent peritonitis. *Edwardsiella tarda* was co-isolated with *Staphylococcus aureus* from the liver of a California sea lion (*Zalophus californiensis*) that died with fibrinopurulent peritonitis and abscessation of the lumbar lymph nodes. A pure growth of *S. aureus* was isolated from the peritoneal exudate (Coles *et al.*, 1978). A sea lion (*Z. californianus*) that died showed extensive pathological changes along the intestine and enlarged mesenteric lymph nodes from which *E. tarda* was isolated (Wallace *et al.*, 1966).

Edwardsiella tarda has been isolated from frogs, turtles and crayfish in catfish ponds and from pond water and mud (Wyatt *et al.*, 1979).

The *sodB* gene (coding for iron-cofactored superoxide dismutase) from *Edwardsiella* species was amplified (454 bp for all species) and sequenced. Differences in nucleotide sequence divided the species of *Edwardsiella* into a pathogenic and non-pathogenic cluster. Cluster I comprises pathogenic strains of *E. tarda* from Japanese eel, Japanese flounder (*Paralichthys olivaceus*), Nile tilapia (*Oreochromis niloticus*) and ayu (*Plecoglossus altivelis*), atypical *E. tarda* from red seabream, *Edwardsiella* species from Japanese eel and *E. ictaluri*. Non-pathogenic *E. tarda* and *E. hoshinae* were found in cluster II (Yamada and Wakabayashi, 1999).

A serotyping scheme combining schemes previously developed at the National Institute of Health in Japan and Centers for Disease Control, USA, resulted in the designation of 61 O group antigens and 45 H antigens (Tamura *et al.*, 1988). *Edwardsiella tarda* has four serotypes, A, B, C, D with serotype A being reported as the most virulent serotype for eels, tilapia and loach in experimental infection (Park *et al.*, 1983) and from an epizootic in wild striped bass in Chesapeake Bay, USA (Baya *et al.*, 1997).

Because it is widespread in the environment, the assessment of its pathogenicity from some aquatic animals may be difficult. However, site of isolation, clinical signs, disease status and histopathological examination should enable a correct diagnosis (Wallace *et al.*, 1966; Otis and Behler, 1973; White *et al.*, 1973; Miyashita, 1984; Humphrey *et al.*, 1986; Reddacliff *et al.*, 1996).

Edwardsiella tarda can be cultured from ulcers, eyes, kidney, spleen and muscle using trypticase soya agar containing 1% NaCl. An improved isolation rate may be obtained by culturing kidney tissue in thioglycollate broth at 22°C for 24–96 h before subculturing to brain heart infusion agar (Amandi *et al.*, 1982).

GROWTH CHARACTERISTICS AND PHENOTYPIC IDENTIFICATION. *Edwardsiella tarda* is a facultatively anaerobic, oxidase-negative, motile, Gram-negative small rod (2×0.6 μm). Catalase is positive. On nutrient agar or TSA, colonies are 0.5–1 mm, greyish-white, transparent, round, entire, glistening, convex and moist after 24 h at 25°C. On BA colonies are whitish, raised, glistening, smooth and round (Hoshina, 1962; Wakabayashi and Egusa, 1973). The type strain ATCC 15947^T produces haemolysis underneath the colony and barely beyond the colony perimeter when grown on BA containing bovine red blood cells at 30°C (Abayneh *et al.*, 2013). Growth occurs on MacConkey agar as non-lactose fermenting colonies and on Salmonella–Shigella (SS) agar as 1–3 mm colonies either predominantly black or with a black centre. On SS agar it is similar in appearance to *Salmonella* species but has a more rapid production of H_2S (the black appearance) and the colony is raised whereas *Salmonella* colonies are flatter and opaque with black centres. On xylose-lysine-deoxycholate (XLD) agar colonies are clear with a black centre surrounded by a reddish colour in the medium due to the decarboxylation of lysine. On bismuth-sulfite agar (BSA) colonies are 1 mm and black. Growth does not occur on brilliant green agar (BGA) (Wyatt *et al.*, 1979). A selective medium (ET medium) based on MacConkey agar and the addition of colistin increases the recovery rate of *E. tarda* from mixed cultures compared to SS agar. On ET medium colonies are clear to whitish with a black centre (Castro *et al.*, 2011). A selective medium, strontium chloride B broth, followed by subculture to solid media is effective for the isolation of *E. tarda* and *Salmonella* species (Iveson, 1971). The media are detailed in the media section.

Growth occurs at 0–4% NaCl, pH 4.5–9.5 and 15–42°C. It is motile at 25 and 37°C. Phenotypic differences have been reported. It is positive for Christensen's citrate but negative for Simmons' citrate (Ewing *et al.*, 1965; Alcaide *et al.*, 2006). Citrate reaction in the API 20E may be variable for strains isolated from eels (Alcaide *et al.*, 2006). Methyl red is positive at 37°C and strains are motile at 25 and 37°C although non-motile strains have been reported. Hydrolysis of casein, elastin, gelatin and starch is negative (Baya *et al.*, 1997). Turbot isolates from Europe have an API 20E profile of 4744000057 when incubated at 25°C and are citrate positive whereas isolates from other species have been reported as negative for citrate (Castro *et al.*, 2006). Isolates from infected wild eels in Spain have API 20E profiles of 474400057, 414400057 and 434400057 and show variations in reactions for citrate, and

H_2S (Alcaide *et al.*, 2006). The type strain CECT 849^T has a profile of 474400057. These and other reactions are presented in Tables 3.10, 3.24–26 and 3.38.

MOLECULAR IDENTIFICATION. A study that investigated four reported primer pairs for the amplification of *E. tarda* found mixed results and only the primer pair amplifying the *etfD* fimbrial unit reported by Sakai *et al.* (2007) detects all strains of *E. tarda* (Castro *et al.*, 2010). The primer pair combination gyrBF1/gyrBR1, which amplifies a 415-bp product (Lan *et al.*, 2008), did not amplify any of the strains tested in the study. Primer pairs detecting the *etfA* major fimbrial subunit reported by Sakai *et al.* (2007) amplify a 415-bp product and primer pairs tardaF/tardaR amplify 1109 bp of the ORFII-ORFIII haemolysin gene (Chen and Lai, 1998), but do not detect all strains of *E. tarda*. As a number of the strains used in the study were from turbot and eels it is possible that the study contained some of the newly described *E. piscicida* strains, but this requires investigation.

DNA from cultured cells is extracted using Insta-Gene Matrix (BIO-RAD) and eluted in a final volume of 200 µl of Matrix. The concentration of DNA is adjusted to 20 ng/µl. A 25-µl master mix is prepared with Ready-To-Go PCR beads (Amersham Pharmacia Biotech), 1 µl of 100 µM primer etfD-F (5′-GGT AAC CTG ATT TGG CGT TC-3′), 1 µl of 100 µM primer etfD-R (5′-GGA TCA CCT GGA TCT TAT CC-3′) and 1 µl DNA (Sakai *et al.*, 2007; Castro *et al.*, 2010). Amplification is conducted with an initial denaturation step at 94°C for 2 min followed by 30 cycles of denaturation at 95°C for 20 s, annealing at 55°C for 20 s, extension at 72°C for 1 min and a final elongation step at 72°C for 7 min (Sakai *et al.*, 2007). An amplicon of 445 bp is obtained. The primers detect *E. tarda* CECT 849 and *E. tarda* NCIMB 2034, and strains isolated from red seabream, gilthead seabream, Japanese flounder, tilapia, angel fish, Pacific salmon, striped bass, catfish and Japanese eel. The primers are specific against *E. ictaluri* (ATCC 33202), *E. hoshinae* (DSM 13771), *Escherichia coli* (CECT 433), *Enterobacter cloacae* (TW 83/03), *E. aerogenes* (RPM 799.1), *Yersinia ruckeri* (1651), *Hafnia alvei* (15/1403), *Serratia marcescens* (PC 955.1), *Aeromonas salmonicida* ssp. *salmonicida* (ACR 218.1), *Vibrio anguillarum* (R82), *Photobacterium damselae* ssp. *piscicida* (ATCC 29690), *Pseudomonas anguilliseptica* (CECT 899), *P. fluorescens* (07/1139), *P. aeruginosa* (CECT 110), *Lactococcus garvieae* (SRG 1.1) and *Streptococcus parauberis* (ACC 33.1) (Castro *et al.*, 2010).

EDWARDSIELLA TARDA DETECTION BY LAMP. Detection of *E. tarda* is based on amplification of the haemolysin gene, *eth*A. The Backward Inner Primer (BIP) consists of primer B1, a linker and primer B2 (5′-TTG GTA CCA TCG GCA AGC CGT TTT GGT ATC GCT GCT GCT CTG C-3′). The Forward Inner Primer (FIP) is composed of primer F1c, a linker and a complementary sequence F2c (5′-GCC TTT CTT CAC CGC CCC TTT TTT GGC GTT AGC GTC GAC TAC AG-3′). Primer B3 is 5′-TGG ATC TGG GTG GTC GTC-3′. Primer F3 is 5′-AGC CAA CGT ACC CAG GTC-3′. Amplification is carried out at 65°C for 45 min. Amplification does not occur at 60°C or 63°C. Sensitivity is 10 to 10^9 CFU.

1.11.3 *Escherichia*

Escherichia vulneris

Escherichia vulneris caused mortalities in naturally infected fish including rainbow trout (*Oncorhynchus mykiss*), balloon and silver molly (*Poecilia* species) and common carp (*Cyprinus carpio*) at a fish production unit in Turkey (Aydin *et al.*, 1997). Pathogenicity was confirmed by experimental challenge using 10^6 cells/ml injected into 1.5-year-old rainbow trout. Clinical signs varied between species of fish but included haemorrhagic lesions on the skin, exophthalmia, bloody exudate in the intestinal tract or empty digestive tract and abnormal liver. Bacteria were isolated from kidney, liver, spleen and muscle. Growth was obtained on TSA incubated at 37°C for 24 h. The identification was confirmed by biochemical reactions in the API 20E. These and other reactions are presented in Tables 3.10 and 3.24–26.

1.11.4 *Hafnia*

Hafnia alvei and Hafnia paralvei

TAXONOMY AND SIGNIFICANCE. *Hafnia alvei* belongs to the family *Enterobacteriaceae* and has been reported from soil, water and sewerage and is a common inhabitant of the gastrointestinal tract of humans and animals including reptiles, insects, mammals, frogs, fish and birds, and has been isolated from foodstuffs including fish products (Okada and Gordon, 2003; Janda and Abbott, 2006). Biochemical and genetic variation exists within the genus, however this is being clarified with the recognition that two genetic or DNA hybridization groups are in fact two species (*H. alvei* and *H. paralvei*), and that strains isolated previously from cases of gastroenteritis and possessing the *eae* gene are not *H. alvei* but a newly described species, *Escherichia albertii* (Huys *et al.*, 2003a, 2010). True *H. alvei* strains are negative for the *eae* gene. DNA HG 1 remains as *H. alvei* (ATCC 13337^T) and DNA HG 2 has now been given the name *H. paralvei* (ATCC 29927^T). Sequence similarity of the 16S rRNA gene is 98.8 to 99.1% between the two species (Huys *et al.*, 2010). A high percentage (89%) of *H. alvei* strains tends to be associated with water, frogs, birds and invertebrates, whereas 83% of freshwater fish strains have been identified as *H. paralvei*. Strains from mammals and reptiles are evenly distributed between the two species (Okada and Gordon, 2003).

Hafnia alvei is an opportunistic pathogen and has been associated with infections and gastroenteritis in people and infection in some animals and also fish: cherry salmon (*Oncorhynchus masou*), brown trout (*Salmo trutta*) and rainbow trout in Japan and Spain (Gelev *et al.*, 1990; Teshima *et al.*, 1992; Rodríguez *et al.*, 1998; Acosta *et al.*, 2002; Janda and Abbott, 2006). Very little is known about virulence and pathogenicity of *H. alvei* or *H. paralvei* and it is not known whether the strains previously reported from fish are in fact *H. paralvei* although in some reports strains are bona fide *H. alvei* as their identity was confirmed by a bacteriophage specific to the strain 1672 (ATCC 51873) (Acosta *et al.*, 2002). Recently, strains of *H. alvei* that produce a Shiga-like toxin have been described from cases of human gastroenteritis (Abbott *et al.*, 2011). This cytoxin is not found in *H. paralvei*. A study investigating adherence to a fibroblastic-like cell line from rainbow trout gonad (RTG-2) indicated differences in the ability of strains of *H. alvei* to adhere to cells, as *H. alvei* strain 1967–82, a human strain, was more adherent than strain C-34 recovered from rainbow trout (Padilla *et al.*, 2010).

In juvenile cherry salmon (*Oncorhynchus masou*) cultured in Japan, infection with *H. alvei* caused a disease similar to bacterial kidney disease. Infected fish swam slowly and had swollen abdomens and darkened skin. The kidney had grey-white furuncle-like lesions. Intraperitoneal injections reproduced the disease after 3 months (Teshima *et al.*, 1992). In wild brown trout (*Salmo trutta*) in Spain, infection with *H. alvei* caused haemorrhagic septicaemia (Acosta *et al.*, 2002). Infected fish swam abnormally, had swollen abdomens, darkening of the eyes and dark areas on the head and dorsal area. The kidney was enlarged and had areas of blackening. In experimental challenge using intraperitoneal injection with strain C-34 and strain 2446 (isolated from rainbow trout) the symptoms and mortalities were reproduced and the organism was re-isolated from the kidney. The identification of *H. alvei* was confirmed by biochemical tests and reaction to a specific bacteriophage. Further studies involving inoculation of brown trout (*Salmo trutta*) with strain C-34 resulted in an LD_{50} of 1.3×10^6 indicating the strain is weakly virulent to virulent, whereas a strain of *H. alvei* isolated from a human food source was highly virulent as it had an LD_{50} of $<10^5$ (Rodríguez *et al.*, 1998). These same two strains of *H. alvei* when inoculated into gilthead seabream (*Sparus aurata*) were not virulent, had an LD_{50} of $>10^8$ and no mortalities occurred (Padilla *et al.*, 2010).

GROWTH CHARACTERISTICS AND PHENOTYPIC IDENTIFICATION. *Hafnia alvei* and *H. paralvei* are facultatively anaerobic, non-spore-forming, oxidase-negative, straight Gram-negative rods (1.0×2.0–5.0 μm). Growth occurs on BA, BHIA, TSA, MacConkey agar, Hektoen enteric, eosin-methylene blue, and xylose-lysine-deoxycholate agar (XLD). Colonies on XLD are red or pink. Growth on desoxycholate-citrate agar (DCA) is variable, only 25–60% of strains grow on Salmonella-Shigella (SS) agar and 75–100% of isolates grow on Brilliant Green agar (BGA). Lactose is not fermented, therefore colonies appear as non-lactose fermenting colonies on MCA. On many media colonies are 2–3 mm in diameter, smooth, convex and translucent. On some media, colonies may have an irregular edge similar to *Salmonella* species on MCA

(Abbott *et al.*, 2011). Growth occurs at 20–35°C but results from biochemical reactions can vary at an incubation temperature of 35°C compared to 22 or 25°C. They are motile at 25–30°C but may be non-motile at 35°C. Methyl red is negative at 22°C but positive at 35°C. VP is positive at 22–28°C but negative at 35°C. Simmons' citrate is positive for 60% of strains at 25–30°C after 3–4 days but negative at 35°C (Sakazaki, 2005). *Hafnia alvei* and *H. paralvei* can be differentiated by the following tests. *H. alvei* is positive for β-glucosidase and malonate utilization, negative for D-arabinose fermentation and variable for salicin fermentation and aesculin hydrolysis. *H. paralvei* is negative for β-glucosidase and malonate utilization, variable for D-arabinose fermentation and negative for salicin fermentation and aesculin hydrolysis. β-glucosidase is tested using 4-methylumbelliferone fluorescent discs (Key Scientific) and read at 24 h according to the manufacturer's instructions. Malonate, aesculin and salicin are read at 48 h and D-arabinose is read at 96 h. *Hafnia alvei* and *H. paralvei* are positive for L-arabinose fermentation. Strains from clinical sources are incubated at 37°C and strains from fish are incubated at 28°C (Huys *et al.*, 2010). Some discrepancies exist: *Hafnia alvei* ATCC 13337^T is atypical being negative for malonate, whereas *H. paralvei* ATCC 29927T is only weakly positive for β-glucosidase (Abbott *et al.*, 2011).

Some clinical strains previously called *H. alvei* have now been identified as *Escherichia albertii*. The API 20E does not distinguish between *H. alvei* and *E. albertii* (Huys *et al.*, 2003a). *Hafnia alvei* is positive for chitinase, whereas *E. albertii* is negative. *Hafnia alvei* is positive for phage 1672, whereas *E. albertii* is negative (Janda and Abbott, 2006).

Hafnia alvei is known to cross-react with *Yersinia ruckeri* antisera (Stevenson and Airdrie, 1984; and in the author's experience).

Hafnia alvei may be differentiated from *Yersinia ruckeri* as *H. alvei* is positive for xylose fermentation, whereas *Y. ruckeri* is negative.

These and other reactions are presented in Table 3.10.

1.11.5 *Proteus*

Proteus rettgeri

See under *Providencia rettgeri*.

1.11.6 *Providencia*

Providencia belongs to the family *Enterobacteriaceae*.

Providencia rettgeri and Proteus rettgeri

Providencia rettgeri and *Proteus rettgeri* are homotypic synonyms, which means they share the same type strain and both names are valid (Brenner *et al.*, 1978).

Providencia rettgeri is a cause of septicaemia and meningitis in hatchling croccodiles held in captivity and results from poor husbandry such as overcrowding, stress due to suboptimal temperatures or incorrect feeding regimes (Ladds *et al.*, 1996). Abnormal behaviour, nervous signs and mortality are seen in infected animals. The organism can be isolated from the liver and brain.

Providencia rettgeri (*Proteus rettgeri*) was associated with mass mortality of farmed silver carp (*Hypophthalmichthys molitrix*) in Israel. The disease was prevalent in winter months and exacerbated by the stress caused by netting and handling (Bejerano *et al.*, 1979). Fish had large and often deep ulcerative lesions on the base of pectoral fin, around the head and on the abdomen. A *Saprolegnia* species was isolated from a number of lesions but mortality was not associated with this fungus. *Providencia rettgeri* was cultured on BA, NA and TSA at 35°C from internal organs: liver, kidney, heart and lesions. In experimental challenge with 5×10^2 cells injected intramuscularly or by immersion following scarring, 50% or more mortality was obtained. Biochemical results are presented in Table 3.10.

Providencia (*friedericiana*) *rustigianii* is part of the faecal flora in penguins held captive in a zoo (Müller, 1983). *Providencia rustigianii* was isolated from Humboldt penguin (*Spheniscus humboldti*), African or black-footed penguin (*S. demersus*), gentoo penguin (*Pygoscelis papua*), southern rockhopper penguin (*Eudyptes chrysocome*; *E. crestatus* is a synonym) and king penguin (*Aptenodytes patagonicus*).

Providencia species are facultatively anaerobic, motile, oxidase-negative, Gram-negative rods. Most strains are non-lactose fermenting. *Providencia rustigianii* is a Gram-negative rod (3.0×1.0 μm). Strains are flagellated at 25°C but non-flagellated or sparsely flagellated when grown at 37°C.

On BA colonies are 1–2 mm, non-haemolytic, smooth, glossy and semitransluscent after 24 h at 36°C. A brown pigment is produced similar to other members of the *Providencia* genus. Growth occurs poorly at 4 and 44°C, whereas good growth occurs between 10 and 40°C. Nitrate is reduced whereas ADH, LDC, ODC, H_2S and urease are negative. These and other reactions are presented in Table 3.10.

1.11.7 *Salmonella*

The nomenclature of *Salmonella* has been regulated by two systems and the background to this and the current nomenclature is outlined online by J.P. Euzéby (http://www.bacterio.cict.fr/s/salmonella.html).

Salmonella species are fermentative, oxidase-negative, Gram-negative rods and may exist in a carrier status in animals or humans, or are pathogenic causing diarrhoea in humans and a variety of conditions in animals.

Salmonella arizonae and *Salmonella choleraesuis* ssp. *arizonae* are homotypic synonyms.

Salmonella enterica ssp. *arizonae* caused septicaemia in a giant tropical freshwater fish, arapaima or pirarucu (*Arapaima gigas*) exhibited in an aquarium in Japan (Kodama *et al.*, 1987). The organism was isolated from gills and internal organs: liver, kidney, spleen, heart, head kidney and kidney. The eyes were opaque and bloody fluid was present in the body cavity, and catarrhal lesions and congestion was observed in the mucous membranes of the stomach and intestines.

Salmonella Newport, *S.* Oranienburg and *S.* Heidelberg were isolated from 40% of apparently healthy California sea lion (*Zalophus californianus*) pups and 33% of apparently healthy northern fur seal (*Callorhinus ursinus*) pups from San Miguel Island, and in this instance the isolation probably represents a carrier status (Gilmartin *et al.*, 1979). It is likely that human visitors to Antarctica have introduced *Salmonella* species to animals in the region as *S.* Enteritidis, *S.* Havana and *S.* Typhimurium have been reported from fur seals (*Arctocephalus gazella*) and gentoo penguins (*Pygoscelis papua*) and *S.* Newport from fur seals. At the time of that sampling *Salmonella* species were not found in macaroni penguin (*Eudyptes chrysolophus*) or black-browed albatross (*Diomedea melanophris*) or grey-headed albatross (*D. chrysostoma*) (Palmgren *et al.*, 2000).

GROWTH CHARACTERISTICS AND PHENOTYPIC IDENTIFICATION. *Salmonella* species are facultatively anaerobic, motile, oxidase negative, Gram-negative rods. On BA, colonies are 2 mm, grey, non-haemolytic after 24 h at 37°C. On MacConkey agar colonies are non-lactose fermenting and often have a ragged edge. A number of selective agars and selective and enrichment broths are available commercially for their isolation. These include brilliant green agar (BGA), bismuth sulfite agar (BSA), deoxycholate agar (DCA), Salmonella-Shigella (SS) agar, xylose lysine deoxycholate (XLD) agar and broths such as strontium chloride B (Iveson, 1971) and rapport medium.

Salmonella species ferment glucose with the production of gas, and ferment mannitol, maltose and sorbitol, but do not ferment lactose, sucrose or salicin. VP and urea are negative. MR and citrate are positive. H_2S is produced in triple sugar iron (TSI) and glissuda media (Iveson, 1971). *Salmonella* species are usually identified using Salmonella O and Salmonella H antiserum, available commercially. Identification to serovar is done by a specialized laboratory. These and other reactions are presented in Tables 3.10 and 3.24–26.

1.11.8 *Serratia*

Serratia are members of the *Enterobacteriaceae* and occur in the environment in soil and water. They do not produce H_2S and are positive for β-galactosidase and fermentation of maltose, D-mannitol, D-mannose and trehalose. Fermentation of L-sorbose is negative. All species of *Serratia*, except *S. fonticola*, are positive for gelatin hydrolysis, VP and DNase (Grimont and Grimont, 2005).

Serratia liquefaciens

Serratia liquefaciens is an opportunistic pathogen and has been reported in disease outbreaks in turbot (*Scophthalmus maximus*), brook trout (*Salvelinus fontinalis*), Arctic char (*S. alpinus*) and Atlantic salmon (*Salmo salar*) from France, Scotland and the USA (McIntosh and Austin, 1990; Vigneulle and Laurencin, 1995; Starliper, 2001). In Atlantic salmon clinical signs included swollen kidney and nodules on the kidney and spleen and some ascites fluid (McIntosh and Austin, 1990). The disease was reproduced in experimental challenge and resulted in muscle liquefaction, swelling of kidneys, white nodules on kidney and spleen, ascites fluid and gastroenteritis. Challenge with 10^3 cells resulted in 100% mortality in 72 h and in 24 h with 10^4 cells/ml. Challenge with type strains of *S. liquefaciens* (CCM 2715 and CCM 2717) required a higher dose (10^7 cells/ml) to reproduce muscle liquefaction and haemorrhagic septicaemia. In turbot, infection caused low-level mortality and clinical signs were similar to those seen in Atlantic salmon. Ascites fluid was present in the peritoneal cavity, the kidney and spleen were swollen and had yellowish nodules and foci of necrotic liquefaction, and petechial haemorrhages were observed in the liver. In turbot the disease could not be replicated in experimental challenge with 1.3×10^6 cells (Vigneulle and Laurencin, 1995). In Arctic char the liver, spleen and kidney were severely haemorrhagic and ascites fluid was present in the body cavity. The disease was replicated in experimental challenge. Intramuscular inoculation resulted in a lower LD_{50} than intraperitoneal injection: 1.5×10^3 to 4.26×10^4 CFU/fish and 1.56×10^6 to 3.78×10^6 CFU/fish, respectively. The LD_{50} for Atlantic salmon, brook trout and rainbow trout was 1.11×10^6, 6.79×10^5 and 7.33×10^5 CFU/fish, respectively (Starliper, 2001).

Serratia liquefaciens is a motile, Gram-negative rod (2–3 μm), strongly catalase positive and strongly proteolytic. Growth occurs on BA, brain–heart infusion agar (BHIA), TSA and MCA but not MA 2216. It is non-haemolytic. Growth occurs at 5–37°C but not at 40°C. The strain (K2) from Atlantic salmon grows better at 25°C than at 37°C and is positive for motility, DNase, citrate and fermentation of lactose, maltose and melibiose at 25°C but not at 37°C. Although all *Serratia* species, with the exception of *S. fonticola*, are positive for gelatin hydrolysis, the strain from Atlantic salmon is negative for gelatin in the API 20E (McIntosh and Austin, 1990). The strains from turbot are non-motile at 25°C and do not grow at 37°C (Vigneulle and Laurencin, 1995), whereas the strains from Arctic char are motile at 25 and 37°C (Starliper, 2001). The Arctic char isolates are negative for fermentation of *myo*-inositol, whereas Atlantic salmon and turbot isolates are positive. These and other reactions are presented in Tables 3.10 and 3.24–26.

Serratia marcescens

Serratia marcescens was isolated from the head kidney of apparently healthy natural populations of white perch (*Morone americanus*) during a survey of the Black River, Maryland, USA. The isolate was pathogenic in experimental challenge of striped bass (*Morone saxatilis*) and rainbow trout (*Oncorhynchus mykiss*) with a mean LD_{50} of 1×10^5 and 2.1×10^4, respectively, by intraperitoneal and intramuscular injection. Clinical signs include necrosis of muscular tissues and a reddish colour on the head, probably due to production of prodigiosin pigment by the bacterium (Baya *et al.*, 1992c).

Serratia marcescens is a motile, Gram-negative rod that is oxidase negative and catalase positive. Growth occurs at 4–45°C and 0–8% NaCl, and on MacConkey agar and brilliant green agar. The red non-diffusible pigment prodigiosin or the diffusible rose-coloured pigment pyrimine is produced. The strain from white perch is positive for gelatin hydrolysis whereas strain ATCC 8100^T is negative. It is negative for haemolysis of trout red blood cells but a weak or delayed β-haemolysis occurs with sheep blood cells. *Serratia marcescens* is negative for fermentation of arabinose, melibiose, xylose and raffinose and can be differentiated from *S. plymuthica* based on these reactions. LDC may give a weak or delayed reaction. In conventional tube media the fermentation of inositol and arabinose is positive and negative, respectively, whereas the API 20E gives the opposite results (Baya *et al.*, 1992c). These and other reactions are presented in Tables 3.10 and 3.24–26.

Serratia plymuthica

Infection with *Serratia plymuthica* resulted in a low level chronic mortality of rainbow trout (*Oncorhynchus mykiss*) fingerlings cultured in Spain, and in rainbow trout in the UK it was co-isolated with *Yersinia ruckeri* (Nieto *et al.*, 1990; Austin and Stobie, 1992a). Experimental challenge resulted in a LD_{50} of 10^5 cells and fish displayed abdominal swelling and discolouration of the skin (Nieto *et al.*, 1990). Intraperitoneal and intramuscular injection with 10^4 cells/fish resulted in 100% mortality in 7 days and produced internal haemorrhages, ascites fluid in the peritoneal cavity and severe erosion of muscle (Austin and Stobie, 1992a).

Serratia plymuthica is a motile, Gram-negative rod that is oxidase negative. The red, non-diffusible, pigment prodigiosin is produced by most strains. On TSA colonies are 1–2 mm, red, round after 24–48 h at 22°C. Growth occurs at 4–30°C and at 37°C growth is variable amongst strains. Growth occurs at 0–5% NaCl. The strain from rainbow trout is variable for motility. Methyl red is positive at 37°C only. These reactions can be variable amongst strains. The rainbow trout strains are negative for β-haemolysis on sheep blood agar (Nieto *et al.*, 1990). Citrate is positive by Simmons' tube media (Nieto *et al.*, 1990), but was reported negative by the API 20E (Austin and Stobie, 1992a). The API 20E database reports citrate as being positive for 65% of strains. These and other reactions are presented in Tables 3.10 and 3.24–26.

1.11.9 *Yersinia*

Yersinia enterocolitica

Yersinia enterocolitica was divided into the subspecies *Y. enterocolitica* subspecies *enterocolitica* and *Y. enterocolitica* subspecies *palearctica*. *Yersinia mollaretii* and *Y. bercovieri* were formerly *Y. enterocolitica* biogroups 3A and 3B, respectively (Wauters *et al.*, 1988).

Yersinia frederiksenii

Yersinia frederiksenii are rhamnose-positive strains, formerly called atypical *Y. enterocolitica*. *Yersinia intermedia* comprises rhamnose-positive, melibiose-positive, raffinose-positive strains, formerly called *Y. enterocolitica* or *Y. enterocolitica*-like. *Yersinia kristensenii* refers to sucrose-negative strains.

Yersinia ruckeri

TAXONOMY AND SIGNIFICANCE. Infection with the Gram-negative bacterium *Yersinia ruckeri* causes enteric redmouth disease (ERM) and affects nearly all species of salmonid (Ross *et al.*, 1966; Ewing *et al.*, 1978). ERM is a serious economic disease in many countries. Disease outbreaks generally occur as a result of poor environmental conditions that lead to low oxygen levels, high water temperature and poor water quality, which stresses the fish, making them more susceptible to disease. Disease signs include haemorrhagic areas around the mouth, in the intestines and other organs. A milder form of the disease is termed Yersiniosis (Tobback *et al.*, 2007). There are a number of modes of transmission of the disease, including birds, wild fish and carrier fish (Willumsen, 1989). The organism also forms biofilms on fish tanks, which is a source of re-infection (Coquet *et al.*, 2002).

ERM had been described in rainbow trout (*Oncorhynchus mykiss*) for many years before the causative organism was named (Ewing *et al.*, 1978). *Yersinia ruckeri* has been reported from wild and farmed Atlantic salmon (*Salmo salar*), channel catfish (*Ictalurus punctataus*), perch (*Perca fluviatilis*), brown trout (*S. trutta*), rainbow trout, roach (*Rutilus rutilis*), steelhead trout, sockeye salmon, chinook salmon, whitefish (*Coregonus* sp.) from Alaska, Canada, Denmark, Finland, Scotland and the USA (Ross *et al.*, 1966; Ewing *et al.*, 1978; Dear, 1988; Valtonen *et al.*, 1992; Petrie *et al.*, 1996; Danley *et al.*, 1999).

Yersinia ruckeri can be characterized by serotype, biotype and outer membrane proteins (OMPs) and the division into clonal groups. An original serotyping scheme based on whole-cell reactions identified six serovars, known as the Hagerman serovars. Most naturally occurring outbreaks in diseased fish and rainbow trout are caused by serovar I, known as the Hagerman strain, and this is the most virulent strain (McCarthy and Johnson, 1982). Serovar II may be significant when it is associated with ERM in chinook salmon (Cipriano *et al.*, 1986). Serovar II and serovar V isolates are sorbitol-positive (de Grandis *et al.*, 1988). Serovar III is found in Australia. Some Australian isolates of *Y. ruckeri* that cross-react with serovar I are termed serovar I′.

A further serotyping scheme based on heat-stable O-antigens identified serotypes 01, 02, 05, 06 and 07 in Europe, North America, Australia and South Africa (Davies, 1990). All serotypes occur in Europe and North America, whereas in Australia and South Africa only serotype 01 has been identified. This study suggested that the Australian isolate and other isolates described as Hagerman serovar III had been incorrectly typed and were in fact serotype O1 by their scheme. Five OMP profiles (1–5) were identified that enabled differentiation of isolates within a serotype. Serotype O1 consists of six clonal groups. Clone 2 was found in disease outbreaks in the UK and clone 5 was found in disease outbreaks in mainland Europe, North America and South Africa. Clones 3, 4 and 6 do not appear to be associated with disease outbreaks. Clones 1 and 3 were found in Australian isolates with clone 3 also being found in Finland, France, West Germany and the USA. Clone 4 was from Norwegian isolates, and clone 6 from isolates in Finland, France, Norway and Canada (Davies, 1991).

Two biotypes have been described based on hydrolysis of Tween 20 and Tween 80, and motility. Biotype 2 strains are non-motile, and do not hydrolyse Tween 80 (Davies and Frerichs, 1989). Division of strains into motile and non-motile types tends to be geographically biased. Strains from the UK tend to be non-motile, with occasional non-motile strains reported from Canada and Norway.

Initially it was suggested that sorbitol fermentation was associated with pathogenicity, however this is not a reliable indicator of virulence (Stevenson and Airdrie 1984, Cipriano *et al.*, 1986).

GROWTH CHARACTERISTICS AND PHENOTYPIC IDENTIFICATION. *Yersinia ruckeri* is a facultatively anaerobic, oxidase negative, Gram-negative rod (1–3 μm) (Ewing *et al.*, 1978). Motility is variable. Beta-haemolysis is negative. On BA and TSA, colonies are 2–3 mm, off-white, grey, shiny, raised, entire and appear as a typical *Enterobacteriaciae*-like colony after 24–48 h at 25°C. Australian strains tend to be 3–4 mm with irregular edges after 48 h incubation at 25°C. The colony has a dark centre at 48 h. Growth occurs on MCA and XLD but not on Yersinia selective agar (YSA; Oxoid) or TCBS. Growth occurs at 22–37°C with 22–25°C being optimal. Motility is variable between strains and according to incubation temperature. Motility and citrate are negative at 37°C but positive at 25°C.

Non-motile strains lack lipolytic activity when tested using the Tween 20 and Tween 80 tests. Therefore, Shotts–Waltman medium (Waltman and Shotts, 1984) is inappropriate for the differential isolation of non-motile strains of *Yersinia ruckeri* as it contains Tween 80 (Davies and Frerichs, 1989).

Results between conventional tube tests and the API 20E system may give variable results. These include citrate utilization, gelatin hydrolysis, VP and nitrate. The nitrate test in the API 20E system may give unreliable results for *Y. ruckeri*, thus the conventional tube nitrate test is recommended. After 24 h incubation at 25°C citrate and gelatin may be falsely negative, therefore 48 h incubation is recommended. Motility and citrate are negative at 37°C but positive at 25°C. The API 20E may show more VP-positive results than the conventional tube test (Davies and Frerichs, 1989). In the API 20E LDC and ODC may be negative at 24 h but positive at 48 h.

Yersinia ruckeri may be differentiated from *Hafnia alvei* by xylose fermentation. *Yersinia ruckeri* is negative, whereas *H. alvei* is positive for xylose fermentation. *Hafnia alvei* is known to cross-react with *Yersinia ruckeri* antisera (Stevenson and Daly, 1982; Stevenson and Airdrie 1984; and in the author's experience).

These and other reactions are presented in Tables 3.10, 3.24–26 and 3.38. Tests should be incubated at 25°C.

MOLECULAR IDENTIFICATION. A number of PCRs have been developed for the detection of *Y. ruckeri*. One PCR is based on primers designed within the 16S rRNA gene (Carson and Wilson, 2009). DNA is extracted from blood or fish tissue using the QIAamp (Qiagen) or PureGene (Gentra Systems) tissue digestion and extraction kits. DNA concentration is adjusted to 10–20 ng/μl. A 20-μl master mix is prepared using 4 μl of 10 μM (2.0 μM final concentration) for each primer, forward primer Yrf (5′-AAC CCA GAT GGG ATT AGC TAG TAA-3′) and reverse primer Yrr (5′-GTT CAG TGC TAT TAA CAC TTA ACC C-3′), 2 μl 10× PCR buffer, 1.6 μl of 2.5 mM each dNTP (200 μM final concentration), 0.8 μl of 50 mM $MgCl_2$ (2.0 mM final), 0.1 μl of 5U/μl *Taq* (0.5 U/20 μl), 6.5 μl purified water and 1 μl of 10–20 ng/μl DNA. Controls include a DNA-positive control, a DNA-free control and a DNA-spiked tissue control when testing tissue. Thermocycling is carried out using an initial denaturation at 94°C for 3 min followed by 35 cycles of denaturation at 94°C for 30 s, annealing at 60°C for 30 s, extension at 72°C for 30 s and a final extension step at 72°C for 4 min. An amplicon of 247 bp is obtained. The lower limit of detection for purified DNA is 5–10 fg, which equates to 1–2 bacterial genome equivalents. For tissue the lower limit of detection is 10^3–10^4 cells/g.

Another PCR is based on the quorum sensing system, termed *yruR/yruI* (Temprano *et al.*, 2001). Two sets of primers were designed with one set flanking the two primers internal to the flanking primers. Both sets of primers are specific against other *Yersinia* species: *Y. aldovae*, *Y. enterocolitica*, *Y. frederiksenii*, *Y. intermedia*, *Y. kristensenii*, and other *Enterobacteriaceae*: *Aeromonas hydrophila*, *Enterobacter cloacae*, *Escherichia coli*, *Klebsiella pneumoniae*, *Proteus vulgaris*, *Pseudomonas putida*, *Serratia marcescens* and *Vibrio alginolyticus*. The flanking primers of FF-1 (5′-TCC CCT GGT GTT GGC TAA TAT TGC-3′) and FR1 (5′-GCG CTA GTC ACT ACT GGC TAT CCT-3′) generate an amplicon of 1900 bp. The second set of primers IF2 (5′-GAG CGC TAC GAC AGT CCC AGA TAT-3′) and IR2 (5′-CAT ACC TTT AAC GCT CAG TTC GAC-3′) generate an amplicon of 1000 bp. A 50-μl master mix is composed of 1 μM of each primer, 0.4 mM dNTP, 5 μl of 10× PCR buffer (100 mM Tris-HCl, 20 mM $MgCl_2$, 500 mM KCl (pH 8.3)), 1.25 U *Taq* DNA polymerase, 5 μl DNA and purified water. The concentration of $MgCl_2$ is crucial to this PCR as no product is obtained when a concentration of 1.5 mM is used and non-specific amplification occurs at concentrations above the optimum of 2 mM.

Thermocycling is carried out with an initial denaturation at 94°C for 2 min, followed by 40 cycles of denaturation at 92°C for 1 min, annealing at 65°C for 1 min, extension at 72°C for 1.5 min and a final elongation step at 72°C for 5 min. The sensitivity of the PCR is 12 CFU. No product is obtained when the number of cells exceeds 10^6 CFU.

A loop-mediated isothermal PCR was designed using primers to the quorum sensing gene, *yruI/yruR* (Saleh *et al.*, 2008). The reaction mix contains 40 pmol of inner primers FIP (5′-TGT TCG TTT ATT GAA CTT CAC CGA TTT TCG TCG AAC TGA GCG TTA A-3′) and BIP (5′-AAG CTG ATT TCC ATA AAT TCC GAG TTT TTA ATG ACA TGG AGT TTG ATG AG-3′), 5 pmol of each outer primer F3 (5′-TCG ATA TAG TTA CCT TCC GG-3′) and B3 (5′-ATG GGC AGT GAA CTG TAG-3′), 20 pmol of loop forward primer LF (5′-AGG TAT CGT GTG TTA GGA TTA TCG T-3′), 1.4 mM dNTP mix, 1.6 M betaine (Sigma-Aldrich), 4.5 mM SO_4, 8 U *Bst* DNA polymerase (New England Biolabs) and 1× of the supplied Thermopol buffer and DNA template. The reaction is carried out in a Loopamp realtime turbimeter (LA-200 Teramecs Co) at 60°C, 63°C and 65°C for 30, 45 and 60 min followed by 80°C for 2 min to terminate the reaction. Amplified product is detected by either measuring the change in absorbance at 650 nm in the Loopamp realtime turbimeter where samples with an optical density of less than 0.1 are negative for *Y. ruckeri*, or by visual detection using 1 μl of fluorescent detection reagent (Eiken Chemical Co.) added before incubation at 63°C, or by the addition of 1 μl of 1:10 SYBR Green I nucleic acid gel stain (10,000× concentration in DSMO; Cambrex BioScience) to the LAMP product tube after termination of the reaction. The LAMP PCR is specific when tested against *Y. aldovae*, *Y. enterocolitica*, *Y. frederiksenii*, *Y. intermedia*, *Y. kristensenii*, *Aeromonas salmonicida*, *A. sobria*, *Renibacterium salmoninarum*, *Flavobacterium columnare* and *Pseudomonas aeruginosa*. The sensitivity is 10^{-6}, which is equivalent to 1 pg of genomic DNA.

Yersinia ruckeri can be detected using primers YER 8 (5′-GCG AGG AGG AAG GGT TAA GTG-3′) and YER 10 (5′-GAA GGC ACC AAG GCA TCT CTG-3′), which produce an amplicon of 575 bp. The thermocycling programme consists of 35 cycles and an annealing temperature of 60°C (Gibello *et al.*, 1999). Primers designed to the 16S rDNA are Ruck1 (5′-CAG CGG AAA GTA GCT TG-3′) and Ruck2 (5′-TGT TCA GTG CTA TTA ACA CTT AA-3′), which produce an amplicon of 409 bp. The thermocycling programme consists of 30 cycles and an annealing temperature of 55°C (LeJeune and Rurangirwa, 2000).

Order *Thiotrichales*

1.12 *Francisellaceae*

Francisellaceae is part of the Class III *Gammaproteobacteria* and belongs to the Order *Thiotrichales*.

1.12.1 *Francisella*

The genus *Francisella* belongs to the *Gammaproteobacteria* (Class III) and the Order *Thiotrichales*, which also contains the family *Piscirickettsiaceae*. *Francisella* species are aerobic, Gram-negative, short rod-shaped or coccoid cells. Cysteine is required for growth or enhances growth.

Francisella philomiragia ssp. *noatunensis*

TAXONOMY AND SIGNIFICANCE. *Francisella philomiragia* ssp. *noatunensis* (2005/50/F292-6C^T=NCIMB 14265^T=LMG 23800^T) was isolated from farmed Atlantic cod (*Gadus morhua*) during disease outbreaks along the Norwegian coast. Clinical signs were few but some fish showed weight loss, raised haemorrhagic nodules on the skin or unilateral or bilateral ocular damage, opacity and corneal perforation (Olsen *et al.*, 2006b; Mikalsen *et al.*, 2007). Granulatomous lesions were found in the heart, liver, spleen and kidney; the latter organs were swollen and the intestine showed a thickened mucosa. The bacteria were present in the cytoplasmic vacuoles of the macrophages.

Francisella philomiragia ssp. *noatunensis* is most closely related to *F. philomiragia*, which is considered to be an opportunistic pathogen in humans, particularly the immunocompromised, and is often associated with water-borne transmission. The two species share 99.3% 16S rRNA gene sequence similarity. Housekeeping genes (*groEL*, *pgm*, *shdA*, *rpoB*, *rpoA*) were analysed individually and as concatenated sequences and both analyses resulted in a level of similarity greater than 95% (with the exception of *rpoA*), and together with a DNA–DNA re-association of 68% with *F. philomiragia* suggests that the designation of *F. philomiragia* ssp. *noatunensis* as a separate species is inconclusive (Mikalsen *et al.*, 2007).

GROWTH CHARACTERISTICS AND PHENOTYPIC IDENTIFICATION. *Francisella philomiragia* ssp. *noatunensis* is an intracellular, strictly aerobic, non-motile, Gram-negative coccobacillus. Cells may vary in size and the Gram stain may be weak. Catalase is weakly positive. The oxidase test is negative for *F. philomiragia* ssp. *noatunensis*, as is *F. tularensis*, yet *F. philomiragia* is oxidase positive. Growth occurs after 3 days incubation at 22°C on cysteine heart agar containing 5% sheep blood (CHAB) as pinpoint, <1 mm, slightly opaque, whitish, mucoid, non-haemolytic colonies and at 4 days are 1 mm and will increase to 2–3 mm in size with further incubation. Greening of the agar may be seen (Olsen *et al.*, 2006b; Mikalsen *et al.*, 2007). On cysteine heart agar (CHA, Difco) without blood, growth is poor and although the inoculum may be heavy, individual colonies are not seen. It does not grow in media without cysteine. Growth does not occur on blood agar (with or without 1.5% NaCl) and does not occur, or is not sustained, on media such as buffered charcoal yeast agar (BCYA, Difco) and bacto chocolate agar enriched (CA, Difco), which are media that support growth of other *Francisella* species. Growth occurs at a range of 10–30°C but not at 37°C, whereas *F. philomiragia* grows well at 37°C on CHAB. Optimal growth is at 22°C on CHAB agar. H_2S is produced from cysteine-supplemented media. Conventional biochemical tests using phenol red broth base (Difco) and supplemented with 1% (w/v) of one of the following carbohydrates, glucose, ribose, maltose, mannitol, sorbitol, trehalose, sucrose, arabinose, are negative. These and other reactions are presented in Table 3.38. Incubation is carried out at 22°C and 37°C for 4 h for the API Zym.

Order *Oceanospirillales*

1.13 *Oceanospirillaceae*

The Order *Oceanospirillales* and the Family *Oceanospirillaceae* belong to the Class *Gammaproteobacteria* under the phylum *Proteobacteria*.

1.13.1 *Marinomonas*

Marinomonas aquimarina (11SM4^T=CECT 5080^T=CIP 108405^T=CCUG 49439^T) was isolated from seawater and water surrounding oysters cultured in Spain (Macián *et al.*, 2005). The name *M. aquamarina* was corrected to *M. aquimarina* (J.P. Euzéby: http://www.bacterio.cict.fr/m/marinomonas.html). *Marinomonas ostreistagni* (UST010306-043^T=-JCM 13672^T=NRRL B-41433^T) was isolated from seawater from a pearl-oyster pond in China (Lau *et al.*, 2006). Both strains grow on MA 2216 agar at 25°C. *Marinomonas aquimarina* is a strictly aerobic, oxidase positive, motile, Gram-negative, straight rod (1–4×0.5–1 μm). On MA 2216 colonies are 2 mm, non-swarming, non-luminescent, non-pigmented, smooth, convex with regular edges after 24 h at 25°C. NaCl is required for growth. Growth occurs in 1–10% NaCl. Growth occurs at 13–40°C but not at 4°C. Nitrate is not reduced, and gelatin, casein, starch, alginate, Tween-80, DNA and agar are not hydrolysed. Results are negative for ADH by Moeller and Thornley's methods (Macián *et al.*, 2005).

Marinomonas ostreistagni is a strictly aerobic, non-fermentative, motile, oxidase positive, Gram-negative rod (0.6–3.6×0.4–0.6 μm). On MA 2216 colonies are 1–2 mm, non-pigmented, opaque, circular, convex with an entire margin after 24 h at 30°C. Growth occurs in 1–10% but not in 15% NaCl. The optimum for growth is 3% NaCl. Growth occurs at 16–40°C but not at 4°C or 42°C, with an optimum of 33–37°C. Glucose is oxidized but not fermented (Lau *et al.*, 2006). These and other reactions are presented in Table 3.11.

Order *Pasteurellales*

1.14 *Pasteurellaceae*

The family *Pasteurellaceae* belongs to the order *Pasteurellales* under the class *Gammaproteobacteria* and the phylum *Proteobacteria*. It contains the genera *Actinobacillus*, *Gallibacterium*, *Haemophilus*, *Mannheimia*, *Pasteurella* and *Phocoenobacter*.

1.14.1 *Actinobacillus*

Actinobacillus delphinicola

TAXONOMY AND SIGNIFICANCE. *Actinobacillus delphinicola* (NCTC 12870^T) was isolated from lung, stomach, cervix, uterus, intestine and intestinal lymph node of stranded sea mammals: harbour porpoise (*Phocoena phocoena*), striped dolphin (*Stenella coeruleoalba*) and Sowerby's beaked whale (*Mesoplodon bidens*) found around the coast of Scotland (Foster *et al.*, 1996b). Sequence similarity of the 16S rRNA gene varied from 90.2 to 95.1% for members of the *Haemophilus–Pasteurella* species complex.

Pathogenicity of the isolate for cetaceans is unknown. The organism was not isolated from common seal (*Phoca vitulina*), grey seal (*Halichoerus grypus*), hooded seal (*Cystophora cristata*) or harp seal (*Phoca groenlandicus*) investigated at the same time (Foster *et al.*, 1996b).

GROWTH CHARACTERISTICS AND PHENOTYPIC IDENTIFICATION. *Actinobacillus delphinicola* is a facultatively anaerobic, non-motile, oxidase-positive, Gram-negative pleomorphic rod. Catalase is negative. Growth occurs anaerobically and in air with added CO_2, which is required for growth. Colonies on Columbia sheep blood agar (CSBA) are 0.75–1 mm, circular, entire, low, convex and smooth after 24 h at 37°C and in 10% CO_2 (Foster *et al.*, 1996b). Haemolysis is negative or weak. Growth is enhanced by blood or serum and colonies are smaller if grown on CBA without added blood. Growth does not occur on MacConkey agar. Growth occurs at 37–42°C but not at 22°C. Nitrate is reduced. Glucose and mannose are fermented but not adonitol, lactose, galactose, sucrose or xylose. VP is positive. ADH is variable. X and V factor are not required. Growth does not occur in some biochemical media such as MRVP and tryptone water medium. A heavy inoculum is required for some biochemical reactions such as fermentation of carbohydrates (sugars). Variable reactions occur for ADH, ODC and LDC and are based on animal species. These and other reactions are presented in the Tables 3.11, 3.24–29 and 3.38.

Actinobacillus scotiae

TAXONOMY AND SIGNIFICANCE. *Actinobacillus scotiae* (NCTC 12922^T) was isolated from three harbour porpoises (*Phocoena phocoena*) stranded off the coast of Scotland (Foster *et al.*, 1998). The isolates were recovered from brain, lung, spleen, liver, kidney, small intestine and mesenteric lymph node. One animal had septicaemia. The pathogenicity is unknown. *Actinobacillus scotiae* has 16S rRNA sequence similarity of between 91.2 to 94.0% to members of the *Haemophilus–Pasteurella* complex that includes *Actinobacillus*. *Actinobacillus scotiae* has a 16S rRNA sequence similarity of 96.3% with *A. delphinicola*.

GROWTH CHARACTERISTICS AND PHENOTYPIC IDENTIFICATION. *Actinobacillus scotiae* is a non-motile, facultatively anaerobic, Gram-negative pleomorphic, oxidase-positive rod. Catalase is negative. On CSBA colonies are 0.5 mm after 24 h at 37°C. Growth occurs anaerobically and requires the presence of 10% CO_2 for growth in air. Haemolysis is weak. The presence of blood or serum enhances growth. Growth does not occur on MacConkey agar. Growth occurs at 25–37°C but not at 42°C. X and V factors are not required. Nitrate is reduced. Results are negative for ADH and LDC. The type strain is positive for ODC and fermentation of galactose whereas other strains are variable for ODC and galactose (Foster *et al.*, 1998).

These and other reactions are presented in Tables 3.11 and 3.24–29.

1.14.2 *Pasteurella*

Pasteurella species are found in tissues and mucous membranes of mammals, birds and turtles.

Pasteurella multocida

TAXONOMY AND SIGNIFICANCE. *Pasteurella multocida* was isolated from a Californian sea lion (*Zalophus californianus*) housed in a marine aquarium, and which had died after a prolonged spell of fainting attacks. A turbid, yellowish fluid filled the pleural cavity and stomach, and pleura were inflamed and brachial lymph nodes were large and inflamed. *Pasteurella multocida* was isolated from samples (Keyes *et al.*, 1968). *Pasteurella multocida* causes respiratory disease in a number of terrestrial mammals and birds.

GROWTH CHARACTERISTICS AND PHENOTYPIC IDENTIFICATION. *Pasteurella multocida* is a facultatively anaerobic, non-motile, Gram-negative coccobacillus, and includes cocci and short rods. Bipolar staining may be seen in the Gram stain. Older cultures may exhibit pleomorphic forms; pleomorphic rods and short filaments. Most strains produce a capsule. On BA colonies are 1–2 mm, non-haemolytic, light grey, round after 24 h at 37°C. Most strains produce a distinctive sweetish smell. Some strains are mucoid and rough and smooth colonies also may be produced. The strain from sea lion produced 0.5 mm colonies that became rough with age and small 'daughter' colonies developed around the original colony. Nitrate is reduced, VP, urease and gelatin are negative. ODC and indole are usually positive. These and other reactions are presented in Tables 3.11, 3.24–26 and 3.38.

Pasteurella skyensis

TAXONOMY AND SIGNIFICANCE. *Pasteurella skyensis* (95A1^T=NCTC 13204^T =NCIMB 13593^T) was isolated from Atlantic salmon (*Salmo salar*) farmed off the coast of Scotland (Jones and Cox, 1999; Birkbeck *et al.*, 2002). Infected salmon presented with no feed in the stomach, petechiae on peritoneal surfaces and white lesions on liver, kidney and spleen. Advanced infection presented with a generalized peritonitis. Sequencing of the 16S rRNA gene demonstrated the closest organism is *Phocenobacter uteri* (NCTC 12872) with 97.1% similarity. It was 92.3% similar to both *Actinobacillus delphinicola* and *Bibersteinia trehalosi*. Experimental challenge by intraperitoneal injection of 1–4×10^6 CFU confirmed pathogencity of the isolate.

GROWTH CHARACTERISTICS AND PHENOTYPIC IDENTIFICATION. *Pasteurella skyensis* is a non-motile, facultatively anaerobic, pleomorphic, Gram-negative rod that is catalase negative (Birkbeck *et al.*, 2002). Blood is required for growth. On trypticase soy agar containing 1.5% w/v sea salts and 5% defibrinated horse blood, colonies are 0.5 mm, grey, circular, smooth, low, convex when incubated in air at 22°C for 48 h. Haemolysis is weak or negative. Growth does not occur on MA 2216 or TCBS. Oxidase is weak. Nitrate, ONPG and VP are negative. Indole is positive. A small volume and heavy inoculum is needed for conventional biochemical tests. These and other reactions are presented in Tables 3.11 and 3.38.

1.14.3 *Phocoenobacter*

Phocoenobacter uteri

Phocoenobacter uteri (NCTC 12872^T) was isolated from the uterus of harbour porpoise (*Phocoena phocoena*) as part of an ongoing investigation into microflora (Foster *et al.*, 2000). Disease signs were not stated and the pathogenicity is not known. *Phocoenobacter uteri* belongs to the family *Pasteurellaceae* and is related to other members of the family, *Actinobacillus* and *Pasteurella*, at only 92–93% based on 16S rRNA sequence analysis.

GROWTH CHARACTERISTICS AND PHENOTYPIC IDENTIFICATION. *Phocoenobacter uteri* is a facultatively anaerobic, oxidase positive, non-motile, Gram-negative pleomorphic rod. Catalase is negative. On BA, colonies are 0.5 mm, non-haemolytic, grey, circular, entire, low, convex after growth in air supplemented with 10% CO_2 after 48 h at 37°C, although CO_2 is not required for growth. Growth is absent or poor on media without blood or serum. Growth does not occur on MacConkey agar. Nitrate is reduced. VP and ONPG are positive.

Glucose is fermented but not inositol, maltose, mannose, mannitol, sucrose or xylose. Urea and indole are negative. A heavy inoculum is needed in the identification sets (Foster *et al.*, 2000). These and other reactions are presented in Tables 3.11, 3.24–29 and 3.38.

Order *Pseudomonadales*

1.15 *Pseudomonadaceae*

1.15.1 *Mesophilobacter*

Mesophilobacter marinus

Mesophilobacter marinus describes marine coccobacilli that morphologically resemble the *Acinetobacter–Moraxella* group.

1.15.2 *Pseudomonas*

Pseudomonas anguilliseptica

TAXONOMY AND SIGNIFICANCE. *Pseudomonas anguilliseptica* (NCMB 1949^T=ATCC 33660^T=CCUG 35503^T=CIP 106711^T) was first isolated in 1971 from pond-cultured eel (*Anguilla japonica*) in Japan (Wakabayashi and Egusa, 1972). The infection, which resulted in mass mortality, was named 'Sekiten-byo' or red spot disease. Clinical signs include petechiae around the mouth, opercula and ventral side of the body. Occasional petechiae are observed in the liver and peritoneum. Pathogenicity was proven through experimental challenge. Since then *P. anguilliseptica* has been isolated from European eel (*A. anguilla*), ayu (*Plecoglossus altivelis*), Atlantic salmon (*Salmo salar*), Baltic herring (*Clupea harengus*), cod (*Gadus morhua*), rainbow trout (*Oncorhynchus mykiss*), seabream (*Dicentrarchus labrax*), seabream (*Sparus aurata*), sea trout (*Salmo trutta*), sole (*Solea sengalensis*), striped jack (*Pseudocaranx dentex*), tilapia (*Oreochromis niloticus*), turbot (*Scophthalmus maximus*) and whitefish (*Coregonus* species) from Canada, Egypt, Finland, France, Japan, Spain and the UK (Nakai *et al.*, 1985; Wiklund and Bylund, 1990; Michel *et al.*, 1992; Lönnström *et al.*, 1994; Berthe *et al.*, 1995; Kusuda *et al.*, 1995; Doménech *et al.*, 1999; Blanco *et al.*, 2002; Park *et al.*, 2002; López-Romalde *et al.*, 2003; Ferguson *et al.*, 2004; Balboa *et al.*, 2007; Eissa *et al.*, 2010).

In Baltic herring infection results in haemorrhagic areas in the eye and sometimes a punctured cornea. Haemorrhage may occur on the head and fins. The organism can be isolated from the eyes and kidney but not from the liver or spleen. Herring may act as carriers. In challenge experiments the herring isolate had low pathogenicity for rainbow trout (Lönnström *et al.*, 1994). Infected cod also have lesions in the cornea and in the connective tissues associated with the head and eye region, and the bacterium can be isolated from the eye, liver and kidney (Ferguson *et al.*, 2004). Clinical signs in Atlantic salmon, rainbow trout, sea trout and whitefish are similar to those seen in eels. Petechiae occur on the skin, peritoneum and liver, and the bacterium can be isolated from kidney, liver and spleen, and occasionally from skin, intestine, eye and base of fins. In challenge experiments 100% of sea trout injected with 0.6×10^9 cells/fish died within 7 to 9 days, and 88% died within 6 to 10 days when injected with 0.6×10^7 cells/fish (Wiklund and Bylund, 1990). A similar range of virulence was found in experimentally challenged turbot and sole where the LD_{50} ranged from 9×10^4 to 5×10^7 cells/fish for turbot and 7×10^4 to 3.8×10^7 for sole. The organism was re-isolated from the kidney and brain (Balboa *et al.*, 2007). Infection with *P. anguilliseptica* was seen in association with 'winter disease', a mortality syndrome that occurs in farmed gilthead seabream (*Sparus aurata*) along the coast of Spain (Doménech *et al.*, 1999). Clinical signs include slow swimming at the water surface and on one side only, with death occurring in 1–2 days. Fish display abdominal distension due to ascites fluid. Cutaneous lesions are not seen. The kidney is haemorrhagic, liver is pale and the intestine is congested with a fibrinous yellow exudate. The bacterium is isolated from kidney and occasionally from liver and brain.

GROWTH CHARACTERISTICS AND PHENOTYPIC IDENTIFICATION. *Pseudomonas anguilliseptica* is a Gram-negative non-sporing rod (2×0.4 μm). Long filamentous cells may occur on media containing blood. Metachromatic granules are seen in the cells stained with Loeffler's methylene blue or Neisser's solution. Oxidase and catalase are positive. On nutrient agar growth is slow. On NA (containing 0.5% NaCl), BA, TSA and BHIA colonies are <1 mm, pale-grey, round, shiny at 4–7 days after 72 h at 25°C. Growth is enhanced by the addition of 10% horse blood to the medium. It is non-haemolytic. On TSA colonies are viscous. Growth occurs on MacConkey agar, but not on Pseudomonas isolation agar (Difco) or TCBS. It is motile at 15°C but not at 25 or 30°C. Growth in liquid medium is poor but can be enhanced by the addition of bovine serum. Growth occurs optimally at 15–20°C, and from 5–30°C, but does not occur at 37°C. Growth occurs at 0–3% NaCl but not at 4% NaCl (Wakabayashi and Egusa, 1972; Lönnström *et al.*, 1994). Isolates from ayu in Japan are negative for gelatin hydrolysis, whereas eel isolates, Finnish isolates and the type strain are positive (Nakai *et al.*, 1985). The ability to hydrolyse gelatin may be lost after subculture. Motility is seen only when the organism is grown below 20°C (Michel *et al.*, 1992). Isolates from salmonid species in Europe are negative for hydrolysis of gelatin (López-Romalde *et al.*, 2003). The type strain is positive for utilization of citrate, but isolates from salmon and Baltic herring are variable (Lönnström *et al.*, 1994; López-Romalde *et al.*, 2003). Citrate is positive for all isolates using Simmons' tube citrate. Reactions in the API 20NE are variable (Balboa *et al.*, 2007). Hydrolysis of Tween 80 is variable between strains, as eel isolates are positive whereas salmonid species from Europe are negative (Wiklund and Bylund, 1990; López-Romalde *et al.*, 2003). ADH reaction in Moeller's medium is negative, whereas the reaction is positive in Thornley's medium. ADH is positive in the API 20NE in only 50% of isolates that are positive by Thornley's method. *Pseudomonas anguilliseptica* can be identified using the API 20NE although it is not in the commercial database. To obtain consistent results Columbia sheep blood agar (CBA) must be used instead of TSA as the growth medium and the inoculum must be prepared in sterile distilled water and the concentration of the inoculum must be obtained spectrophotometrically to absorbance of 1 at a wavelength of 580. Incubation is carried out at 25°C. Four profiles were obtained from 49 European strains and the eel type strain (López-Romalde *et al.*, 2003).

There are three serotypes of *P. anguilliseptica* based on a heat-labile surface K antigen. Serotypes from eels in Japan are K– and K+, ayu have K+ serotype as do the Finnish isolates (Wiklund and Bylund, 1990). Antisera against *P. anguilliseptica* may show cross-reaction with *Pseudomonas putida* and *Vibrio* (*Listonella*) *anguillarum*, however used diluted at 1:5 to 1:10 may overcome the cross-reaction (Toranzo *et al.*, 1987). These and other reactions are presented in Tables 3.11, 3.24–29 and 3.38.

MOLECULAR IDENTIFICATION. Primers PAF (5′-GAC CTC GCG CCA TTA-3′) and PAR (5′-CTC AGC AGT TTT GAA AG-3′) for use in a conventional PCR were designed within the non-conserved regions of the 16S rRNA gene. The master mix is prepared in a 100 µl volume containing 1 mM of each primer (PAF and PAR), 1× PCR buffer (Applied Biosystems), 2 mM of each dNTP, 2.5 U Ampli*Taq* gold (Applied Biosystems) and ultrapure water to 100 µl. DNA from bacterial cells (50–70 ng), or 20 µl boiled cells or 10 µl of DNA extracted from fish tissue is added to each tube of reaction mixture. The thermocycling protocol on a PT-100 thermal cycler (MJ Research) is 95°C for 7 min, followed by 35 cycles at 95°C for 1 min, 46°C for 1 min and 72°C for 1.5 min with a final extension step at 72°C for 5 min. A 439-bp amplicon is produced. The limit of detection is 1×10^4 cells/ml equating to 200 cells per reaction tube, from tissue is 170 cells per reaction tube, which equates to 3.4×10^4 cells/g tissue. At an annealing temperature of 43°C, amplicon (size not stated) is obtained from *P. fluorescens* (CECT 378^T) and *P. aeruginosa* (CECT 110^T). The primers are specific against *P. alcaligenes* (CECT 929^T), *P. mendocina* (CECT 320^T), *P. oleovorans* (CECT 4079^T), *P. pseudoalcaligenes* (CECT 318^T), *P. putida* (CECT 324^T), *P. stutzeri* (CECT 930^T), *P. tolaasii* (CRCT 4472^T), *Brevundimonas diminuta* (CECT 317^T), *Burkholderia cepacia* (CECT 322), *B. caryophylli* (CECT 117), *Delftia* (*Comamonas*) *acidovorans* (CECT 311^T), *Ralstonia pickettii* (CECT 330^T) and *Serratia marcescens* (CECT 846) (Blanco *et al.*, 2002).

Pseudomonas chlororaphis ssp. *chlororaphis*

Pseudomonas chlororaphis ssp. *chlororaphis* caused mortality in Japanese amago trout (*Oncorhynchus rhodurus*) fry cultured in Japan. Clinical signs included haemorrhagic areas on the body and a distended abdomen due to ascites fluid. The organism was virulent for eel in experimental challenge and the same clinical signs were reproduced (Hatai *et al*, 1975).

The bacterium produces a green pigment seen as crystals in the colony. Biochemical reactions are presented in Table 3.11.

Pseudomonas chlororaphis ssp. *piscium*

TAXONOMY AND SIGNIFICANCE. *Pseudomonas chlororaphis* spp. *piscium* (JF 3835^T=NCIMB 14478^T=DSM 21509^T) was isolated as part of the normal flora in the intestine of European perch (*Perca fluviatilis*) caught in Lake Neuchâtel, Switzerland. The organism has 16S rRNA gene similarity to *P. chlororaphis* ssp. *chlororaphis* (99.84%), *P. chlororaphis* ssp. *aurantiaca* (99.75%) and *P. chlororaphis* ssp. *aureofaciens* (99.4%) (Burr *et al.*, 2010).

GROWTH CHARACTERISTICS AND PHENOTYPIC IDENTIFICATION. *Pseudomonas chlororaphis* spp. *piscium* is a motile, aerobic, Gram-negative rod, positive for catalase and oxidase. A yellow-green fluorescent pigment is produced on Pseudomonas agar F (Difco), but there is no pyocyanin production on Pseudomonas agar P (Difco). On Luria–Bertani agar a non-diffusible pigment is produced after 4–5 days at 25 or 30°C. Growth occurs at 18–30°C but is weak at 37°C. Growth does not occur under anaerobic conditions. These and other reactions are presented in Tables 3.27–29.

Pseudomonas fluorescens

TAXONOMY AND SIGNIFICANCE. Large ulcerative lesions with haemorrhagic borders occurred on the body of crimson seabream (*Evynnis japonica*) cultured in Japan, and was due to infection with *Pseudomonas fluorescens*. The disease outbreak occurred in late autumn and spring. The bacterium was isolated from the blood and kidney. Virulence was proven through experimental challenge (Kusuda *et al.*, 1974). Infection in tilapia (*Oreochromis niloticus niloticus*) occurred in winter and spring and clinical features presented as white nodules in the spleen and abscesses in the swim bladder. Pathogenicity was proven through intramuscular injection and increased mortalities were related to a water temperature range of 15–20°C (Miyashita, 1984). *Pseudomonas fluorescens* caused disease in cultured carp (Shiose *et al.*, 1974). *P. fluorescens* was thought to be responsible for a disease outbreak in intensively reared salmon (rainbow trout, chinook salmon, brown trout and coho salmon in Canada, Norway and Chile) and was referred to as atypical bacterial gill disease. Clinical presentations were different to classical bacterial gill disease, which is caused by *Flavobacterium branchiophilum*. The affected fish did not show respiratory distress and did not orientate at the surface of the water column or at the margin of the tanks as is typical for bacterial gill disease. Although *F. psychrophilum* was isolated from the gills of salmon cultured in Canada, the bacterium common to all outbreaks and that was predominant on affected gills was *P. fluorescens* (Ostland *et al.*, 1999a).

GROWTH CHARACTERISTICS AND PHENOTYPIC IDENTIFICATION. *Pseudomonas fluorescens* is a Gram-negative, oxidase-positive rod (1.8×0.4 μm) and grows on MacConkey agar, Salmonella-Shigella agar and is beta-haemolytic on BA. On nutrient agar colonies grow within 24 h. A diffusible fluorescein pigment is produced. Growth occurs at 25°C but not at 37°C and at 0–3% NaCl. Positive for ADH but not ODC or LDC. It is resistant to vibriostatic agent 0/129 (Kusuda *et al.*, 1974). These and other reactions are presented in Table 3.11. *Pseudomonas fluorescens* is in the API 20NE commercial database (Biomerieux).

Pseudomonas (*pseudoalcaligenes*) *oleovorans*

TAXONOMY AND SIGNIFICANCE. *Pseudomonas* (*pseudoalcaligenes*) *oleovorans* was co-isolated with *Serratia plymuthica* from rainbow trout infected with *Yersinia ruckeri* in the UK. In experimental challenge injection of 10^5 cells/fish resulted in 100% mortality over 7 days. Fish had internal haemorrhaging and ascites fluid in the peritoneal cavity and some haemorrhaging at the vent (Austin and Stobie, 1992a). *Pseudomonas pseudoalcaligenes* (ATCC 17440^T) is a later synonym of *P. oleovorans* (ATCC 8062^T) (Saha *et al.*, 2010).

GROWTH CHARACTERISTICS AND PHENOTYPIC IDENTIFICATION. The isolate of *P.* (*pseudoalcaligenes*) *oleovorans* recovered from rainbow trout is a short Gram-negative rod that is positive for catalase and oxidase. Colonies are cream-coloured on TSA. Tests are positive for ADH and ODC but negative for ONPG, H_2S, indole, LDC, TDA and hydrolysis of DNA, starch and urea. Hydrolysis of gelatin is positive. Glucose and arabinose are fermented but not amygdalin, inositol, mannose, melibiose, rhamnose or sorbitol (Austin and Stobie, 1992a). These and other reactions are presented in Tables 3.11 and 3.24–26.

Pseudomonas plecoglossicida

TAXONOMY AND SIGNIFICANCE. *Pseudomonas plecoglossicida* (FPC 951^T=ATCC 700383^T) causes bacterial haemorrhagic ascites in ayu (*Plecoglossus altivelis*) and Argentinian silverside or pejerrey (*Odontesthes bonariensis*) cultured in Japan (Nishimori *et al.*, 2000). The bacterium can be cultured from the ascites, blood, liver, spleen and kidney. In experimental infection using bath immersion it was demonstrated that the organism invades the dermal layer of the skin and also enters the fish via the gills. At 6 h post-challenge the organism was detected in the liver, spleen and kidney (Sukenda and Wakabayashi, 2000). Based on 16S rRNA sequence analysis it is similar to *P. stutzeri* (99.6%), *P. putida* (98.9%), *P. asplenii* (98.4%), *P. fluorescens* (98.25%), *P. flavescens* (98.2%), *P. mendocina* (97.4%), *P. chlororaphis* ssp. *aureofaciens*, *P. chlororaphis* ssp. *chlororaphis* (97.1%) and *P. aeruginosa* (95.9%) (Nishimori *et al.*, 2000).

GROWTH CHARACTERISTICS AND PHENOTYPIC IDENTIFICATION. *Pseudomonas plecoglossicida* is a motile Gram-negative rod (0.5–1×2.5–4.5 µm). Growth occurs on TSA, BA, MSA-B at 10–30°C and at a range of 0–5% NaCl but not at 4 or 41°C. Positive for oxidase and nitrate reduction. Beta-haemolysis is seen on media containing blood. It is similar to *P. putida* biovar A, however *P. putida* is negative for reduction of nitrate. These and other reactions are presented in Tables 3.11, 3.24–26 (some reactions), 3.27–29 and 3.38.

MOLECULAR IDENTIFICATION. *Pseudomonas plecoglossicida* can be detected in kidney tissue using a real time quantitative PCR with primers designed to the *gyrB* coding region. Primer GBPA-F 5′-CCT GCT GAA GGA CGA GCG TTC G-3′, primer GBPA-R 5′-AAC CAG GTG AGT ACC ACC GTC G-3′ and a dual labelled fluorogenic hybridization probe TGT-P 5′-(FAM)AGA TGG CGT GGG CGT TGA AGT AGC GC(TAMRA)-3′ are used in one reaction tube and the same primers and a dual labelled fluorogenic hybridization internal control probe ISD-P 5′-(VIC)CCT TCA CCA CCA CGG CCG AGC GTG AG(TAMRA)-3′ are used in a second tube. The master mix is prepared in a 20-µl volume containing 1 µM of each primer (GBPA-F and GBPA-R), 250 nM of either the detection probe (TGT-P) or the internal control probe (ISD-P), 2 µl 10× PCR buffer (*Taq*Man PCR Core reagent kit, Applied Biosystems), 5 mM $MgCl_2$, 200 µM of each dATP, dCTP, dGTP, 400 µM dUTP, 0.5 U Ampli*Taq* gold (Applied Biosystems), 0.2 U AmpErase uracil N-glycosylase, 1 µl of DNA and ultrapure water to make up to 20 µl. The thermocycling protocol is 50°C for 2 min, 95°C for 10 min, followed by 50 cycles of 95°C for 50 s, 68°C for 1 min. The qPCR detected 10 copy numbers of target DNA per µg.

The primers are specific against *Aeromonas hydrophila* (IAM 12337), *A. salmonicida* (FPC 365), *Escherichia coli* (FPC 350), *Edwardsiella tarda* (FPC 801), *E. ictaluri* (FPC 447), *Flavobacterium columnare* (FPC 970), *F. psychrophilum* (FPC 828), *Photobacterium damselae* ssp. *piscicida* (FPC 851), *Pseudomonas aeruginosa* (IAM 1514), *P. anguilliseptica* (FPC 52), *P. chlororaphis* ssp. *chlororaphis* (FPC 308), *P. fluorescens* (FPC 96), *P. putida* (ATCC 12633) and *Vibrio anguillarum* (FPC 346) (Sukenda and Wakabayashi, 2000).

A nested PCR using two primer sets designed to detect the *gyrB* gene has been tested for sensitivity and specificity. For the first round PCR, the master mix consists of 10 pmol of each primer PL-G1F (5′-CAC AGG CTC CAA TGG CTG TA-3′) and primer PL-G1R (5′-AGC CCA GTG CAG TGA TCA GC-3′), 0.1 nmol of each dNTP, 0.5 U *Taq* polymerase (Bioneer, Daejeon, Korea), 2 µl of DNA supernatant in a 10-µl reaction volume. Thermocycling is undertaken in a Techgene Thermal Cycler (Techne, Stone, UK) using a preheating step of 94°C for 5 min followed by 35 cycles of denaturation at 94°C for 15 s, annealing at 66°C for 20 s and extension at 72°C for 60 s, with a final extension step of 72°C for 5 min. For the second round PCR, the internal primers are PL-G2F (5′-TGC TGA AGG ACG AGC GTT CG-3′) and PL-G2R (5′-ATC ATC TTG CCG ACA ACA GC-3′). The thermocycling protocol is the same as the first round PCR. An amplicon of 978 bp is obtained from the first round primers. Non-specific amplicon at a molecular weight and lower intensity is obtained from *P. anguilliseptica* (FPC 48) and *Aeromonas salmonicida* ssp.

salmonicida (FPC 367). The second round PCR amplifies a product of 520 bp. This is specific against *Pseudomonas aeruginosa*, *P. anguilliseptica*, *P. fluorescens*, *P. putida*, *Aeromonas salmonicida* ssp. *masoucida*, *A. salmonicida* ssp. *salmonicida*, *Vibrio* (*Listonella*) *anguillarum*, *Flavobacterium branchiophilum*, *F. columnare*, *F. psychrophilum*, *Edwardsiella tarda* and *Escherichia coli* (Izumi *et al*., 2007). The limit of detection is 0.86 CFU per reaction tube.

DNA is extracted from either kidney or intestinal tissue homogenate or bacterial cells by adding 10 µl of source material to 300 µl of 5% Chelex100 (Sigma) and incubating at 55°C for 30 min. The mixture is vortexed at high speed for 5–10 s and then boiled for 20 min. The mixture is centrifuged for 10 min at 10,000 × *g* and the supernatant used in the PCR (Izumi *et al*., 2007).

Pseudomonas pseudoalcaligenes

See *Pseudomonas oleovorans*.

Pseudomonas putida

In Japan, an epizootic in cultured yellowtail (*Seriola quinqueradiata*) was caused by *Pseudomonas putida*. On the body surface infected fish had large abscesses with a haemorrhagic border. The bacterium was isolated from the spleen and kidney (Kusuda and Toyoshima, 1976). *Pseudomonas putida* caused a disease outbreak in rainbow trout (*Oncorhynchus mykiss*) in grow-out tanks in Turkey. Ulcerations occurred almost exclusively on the dorsal surface of infected fish. In advanced cases the integument and musculature degenerated to the extent that the underlying vertebral column was exposed. Other signs were exophthalmia, dark pigmentation and frayed dorsal fin. Internal organs appeared normal. The stomach was empty and the intestine contained a yellow exudate. Identification was confirmed by 16S rRNA sequencing and API 20NE results (Altinok *et al*., 2006). Pathogenicity was confirmed by experimental challenge. The bacterium was isolated from skin lesion, liver, spleen and kidney.

Pseudomonas putida is a motile, Gram-negative rod (1.5–4.4×0.5–1.0 µm). On nutrient agar, colonies grow within 24 h. Growth occurs at 25–30°C but not at 41°C. Growth at 4°C is variable, with biovar A being variable and biovar B being positive. Nitrate is not reduced. These and other reactions are presented in Tables 3.11 and 3.27–29.

Pseudomonas species

An unidentified *Pseudomonas* species that produces a red pigment in the water column of hatchery tanks of zoeal and post-larvae shrimp (prawn) (*Litopenaeus vannamei*) cultured in Iran caused mass mortality in these animals. The disease was replicated in experimental challenge. Mortality of 100% occurred in 24–48 h after challenge with 1.5–2.0×10^5 CFU/ml. The bacterium is closely related to *P. mesophilica* and *P. anguilliseptica* based on phenotypic characterization (Soltani *et al*., 2010).

The bacterium is a motile, Gram-negative coccobacilli. Colonies produce a red pigment on marine and nutrient agar. Growth does not occur on TCBS. Growth does not occur on media containing less than 5% seawater (Soltani *et al*., 2010). Growth does not occur in NaCl only. The biochemical reactions are presented in Table 3.11.

Order *Vibrionales*

1.16 *Vibrionaceae*

The family *Vibrionaceae* belongs to the class *Gammaproteobacteria* and is the only family in the order *Vibrionales*. Bergey's manual (*Bergy's Manual of Systematic Bacteriology*, Vol II Part B Gamma Proteobacteria) lists three genera within the family *Vibrionaceae*; *Vibrio*, *Photobacterium* and *Salinivibrio*, however the number of genera has been expanded to include *Enterovibrio*, *Grimontia* and the recently created *Aliivibrio* genus.

To date 122 species have been identified within the family *Vibrionaceae*. In general, those isolated from fish grow well on ordinary plate media to which either NaCl or sea salts have been added if the isolates are of marine origin.

1.16.1 *Aliivibrio*

The genus *Aliivibrio* was created to encompass closely related species *Vibrio fischeri*, *V. logei*, *V. salmonicida* and *V. wodanis*, which form a distinct phylogenetic group within the *Vibrionaceae* based on phenotype, 16S rRNA sequencing and multilocus sequence analysis (MLSA) of the genes *recA*, *rpoA*, *pyrH* and *gyrB* (Urbanczyk *et al.*, 2007). Further phylogenetic analysis using *gapA*, *gyrB*, *pyrH*, *recA*, *rpoA*, the *luxABE* region and the 16S rRNA gene confirmed the reclassification and also demonstrated two clades within the *A. fischeri* group. A strain ATCC 33715 (Y-1) previously identified as *A. fischeri* was shown to be *A. sifiae* (Ast *et al.*, 2009). Biochemical characteristics that are common to all *Aliivibrio* species are sensitivity to vibriostatic agent 0/129 10 μg, negative for ADH, ODC, VP, indole and hydrolysis of gelatin and sensitive to novobiocin 5 μg (Urbanczyk *et al.*, 2007; Beaz-Hidalgo *et al.*, 2010b). All species except *A. salmonicida* have yellow-orange pigmented colonies.

Aliivibrio finisterrensis

TAXONOMY AND SIGNIFICANCE. Four strains of *A. finisterrensis* (CMJ 11.1^T=CECT 7228^T=LMG 23869^T) were isolated from Manila clams (*Venerupis* [*Ruditapes*] *philippinarum*) on the north-western coast of Spain (Beaz-Hidalgo *et al.*, 2010b). *Aliivibrio finisterrensis* is closely related to *A. wodanis*, *A. salmonicida*, *A. fischeri* and *A. logei* based on 16S rRNA gene sequence similarities of 98.1–96.0%. The isolates form a different genomic group from other species of *Aliivibrio* based on *rpoA* (RNA polymerase alpha chain), *recA* (recA protein), *atpA* (α-subunit of bacterial ATP synthase), UMP (uridine monophosphate kinase) and *pyrH* genes.

GROWTH CHARACTERISTICS AND PHENOTYPIC IDENTIFICATION. *Aliivibrio finisterrensis* is a motile, facultatively anaerobic, Gram-negative, oxidase-positive, large rod (1.36–2.15×0.8–1.05 μm). It is susceptible to vibriostatic agent 0/129 150 μg disc and grows in media supplemented with 1–6% NaCl (optimum 1–3%) and at 4–30°C. It does not grow at 0 or 0.5% or 8% NaCl or at a temperature of 35°C. It is non-luminescent, does not hydrolyse starch or alginate, but is positive for hydrolysis of Tween 80. The Type strain and two other strains produce 1.0–1.5 mm green colonies on TCBS after 24 h incubation whereas one strain (C 17.13) produces yellow colonies. Biochemical identification can be done using standard biochemical tests, API 20E, API Zym according to the manufacturer's instructions with the exception of using 0.85% saline as an inoculum, and the API 50CH using an inoculum of 0.85% saline adjusted to optical density of 1.0 at 580 nm and mixed (1:90 v/v) with ZOF medium (Beaz-Hidalgo *et al.*, 2010b). Two of the four strains produce acid from arbutin and melibiose. Variations occur for fermentation of arbutin, D-mannitol and sucrose. These and other reactions are found in Tables 3.21, 3.24–26, 3.30 and 3.38.

Aliivibrio fischeri

TAXONOMY AND SIGNIFICANCE. *Aliivibrio fischeri* was isolated from the Baltic Sea and described and named by Professor Bernhard Fischer as *Photobacterium fischeri* (Bacteriol. Centralblatt, vol. 3, p. 105, 1888, as referenced by Beijerinck, 1889; the pdf was available at https://myweb-space.wisc.edu/wollenberg/web/science/Beijerinck1889.pdf). The type strain is ATCC 7744^T (=NCIMB 1281), and is given the number 398 in Reichelt *et al.*, (1976) and was deposited by Beijerinck as *Achromobacter fischeri* following assignment to this genus according to Bergey's manual, 1930. In the 1957 manual (Breed *et al.*, 1957) it was listed as *Photobacterium fischeri*. Strain ATCC 25918 (=61) was isolated from the sea and deposited as *P. fischeri* by Baumann *et al.* (1971a). There are some differences in biochemical reactions between the two species, which might be attributable to different methods or different concentrations of NaCl used for the inoculum. *A Vibrio* species named as *V. marinus* in 1891 by H. Russell and re-isolated by Colwell and Morita (1964) was later found to be a later subjective synonym of *A. fischeri* (Hendrie *et al.*, 1971b). Strain ATCC 15382 isolated from skin of Pacific cod and originally identified as *A. fischeri* (previous designations *Photobacterium fischeri*, *Vibrio fischeri*) (Hendrie *et al.*, 1971b) was shown to be *A. logei* (Bang *et al.*, 1978).

Aliivibrio fischeri is a symbiont in the light organ of the Japanese pinecone fish (*Monocentris japonica*) (Ruby and Nealson, 1976), and the sepiolid squid, *Euprymna scolopes*, also known as the Hawaiian bobtail squid (Boettcher and Ruby, 1990). Along with *A. logei* it is a symbiont in the light organ of the analogous bobtail squid (*Sepiola affinis*) and the robust bobtail squid (*S. robusta*) found in the Mediterranean sea and the Atlantic

Ocean (Fidopiastis *et al.*, 1998). It was isolated from a dead squid at Woods Hole, USA (Johnson and Shunk, 1936). *A. fischeri* has been isolated from water surrounding cultured amberjack (*Seriola dumerili*) in Spain (Alcaide, 2003).

MLSA identified *A. fischeri* as belonging to the Fischeri clade, which also contains *A. logei*, *A. salmonicida* and *A. wodanis* (Sawabe *et al.*, 2007b). Members of this clade have >36% DDH and 39–42 GC mol%, 89.8–94.4% similarity of concatenated MLSA sequence and AAI of 95.2%. Phylogenetic analysis using *gapA*, *gyrB*, *pyrH*, *recA*, *rpoA*, the *luxABE* region and the 16S rRNA gene demonstrated two clades within the *A. fischeri* group: clade one containing ATCC 7744^T and clade two containing ATCC 25918 (Ast *et al.*, 2009). A strain ATCC 33715 (Y-1) previously identified as *A. fischeri* was shown to be *A. sifiae*.

GROWTH CHARACTERISTICS AND PHENOTYPIC IDENTIFICATION. *Aliivibrio fischeri* is a motile, facultatively anaerobic, Gram-negative, oxidase-positive, short thick rod (1.0–2.5×0.4–0.8 µm) and occurs singly and in pairs. Occasional rods are slightly curved with pointed ends. It has tufts of polar flagella. On nutrient agar containing seawater, colonies are 1–2 mm, circular, smooth, slightly raised and iridescent, off-white initially, becoming greyish to yellowish and may be viscous. Older colonies become yellowish with slightly serrated margins (Johnson and Shunk, 1936; Breed *et al.*, 1957; Hendrie *et al.*, 1971b). On MSA-B colonies are non-haemolytic, 1 mm light grey to cream and 0.5 mm on MA 2216 after 48 h at 25°C. On BA colonies are non-haemolytic and 0.5 mm after 24 h. Bacterial cells lyse the red cells in the medium after 48 h. Growth does not occur on TCBS or MacConkey agar. Colonies are golden yellow on seawater tryptone yeast extract agar (SWT). Alginate is hydrolysed (Fidopiastis *et al.*, 1998). Luminescence is produced and a higher level of luminescence is produced when cells are grown at 25°C compared to lower temperatures. Growth does not occur in 10% NaCl or at 4°C or 37°C. Growth occurs at 30°C. Most strains grow at 5°C but not the type strain ATCC 7744^T (=NCMB 1281^T) (Hendrie *et al.*, 1971b). Results are positive for urease and fermentation of gentiobiose. Agar is not digested. Fermentation is negative for arabinose, erythritol, fucose, lactose, raffinose, sucrose, trehalose and xylose (Johnson and Shunk, 1936). Strain ATCC 7744^T is negative for gelatin using the plate method (this manual) but reported positive by Hendrie *et al.* (1971b) using the tube method. The type strain (ATCC 7744) is positive for fermentation of salicin but other strains are variable in their reactions. In 'sugar' tubes reactions for fermentation of inositol, lactose, sorbitol, sucrose and trehalose may be difficult to read due to a pale yellow colour at the top of the tubes. These should be read as negative. Indole gives a strong reaction in the tube method. Methyl red is positive using the MRVP medium but needs five drops of methyl red reagent (usually one or two drops is sufficient to show a positive reaction for most bacteria). In the API 20E VP, citrate, indole and urea are negative when 0.85% NaCl is used as an inoculum, whereas an inoculum of 2% NaCl results in a weak VP reaction, and a positive reaction for citrate, indole and urea. VP with 2% NaCl inoculum is negative in Difco MRVP medium (this manual). A positive VP in the API 20E but negative in MRVP medium using an inoculum containing 1.5% NaCl was also reported for ATCC 25918 by Thyssen *et al.* (1998). Reports for fermentation from mannitol are variable in the literature. Fermentation of 'sugars' in the API 20E may show a stronger result with an inoculum of 0.85% NaCl, but a weak result with a 2% NaCl inoculum. ADH may be positive with 2% NaCl but is negative for 0.85% NaCl. Moeller's and Thornley's ADH are negative. Aesculin was reported positive by Urbanczyk *et al.* (2007), but found to be negative for ATCC ATCC 7744^T. These and other reactions are found in Tables 3.21, 3.24–26 and 3.38.

MOLECULAR IDENTIFICATION. *Aliivibrio fischeri* can be detected using primers developed to the *luxA* gene (Lee and Ruby, 1995). Forward primer LuxA-F (5′-GTT CTT AGT TGG ATT ATT GG-3′) and reverse primer LuxA-R (5′-TCA GTT CCA TTA GCT TCA AAT CC-3′) at 100 pmol each are used in a reaction mixture containing PCR buffer, 2 mM $MgCl_2$, 0.2 mM each dNTP, 2.5 U *Taq* DNA polymerase in a volume of 95 µl. Thermocycling is carried out at 94°C for 5 min, followed by 40 cycles of at 94°C for 1.5 min, 40°C for 2 min, 72°C for 2 min and a final extension at 72°C for 7 min. A product of 428 bp is produced. Lee and Ruby (1995) noted some non-specific products were produced. The sensitivity was 1 CFU. Specificity results were not recorded.

Aliivibrio logei

TAXONOMY AND SIGNIFICANCE. *Aliivibrio logei* (584^T=ATCC 29985^T), originally named *Photobacterium logei* and later *Vibrio logei*, was first isolated from the intestine of scallop from the Arctic and other strains were isolated from Arctic sediment, and necrotic lesions on tanner crabs found off the Oregon coast, USA (Bang *et al.*, 1978). An isolate (PS 207=ATCC 15382=NCMB 1143) from the skin of Pacific cod caught off Port Orchard, USA was identified as *forma specialis* of *Vibrio marinus* (Colwell and Morita, 1964) but was later found to be *A. logei* (Bang *et al.*, 1978). *Aliivibrio logei* strain KCh1 (Kamchatka isolate) was isolated from the intestine of a goby (*Cottida* species) found in the Okhotsk Sea, western Pacific Ocean, Russia (Khrulnova *et al.*, 2010). Isolates thought to be *A. logei* and obtained from the light organ of the bobtail squid (*Sepiola affinis*) and the robust bobtail squid (*S. robusta*) found in the Mediterranean Sea and the Atlantic Ocean (Fidopiastis *et al.*, 1998) were found to cluster in a new clade named 'thorii' based on MLSA (Ast *et al.*, 2009).

Aliivibrio logei is generally considered to be non-pathogenic, however some recent studies have described pathogenic strains of *A. logei* confirmed by experimental challenge. During a study undertaken on cod (*Gadus morhua*) larvae in the UK to investigate changes in gut flora according to diet, one tank of larvae experienced heavy mortality due to infection with *V. anguillarum* (strain 91079) and *A. logei* (strain n8c1-8) in equal densities of approximately 1×10^4 CFU/ml. Both *V. anguillarum* and *A. logei* caused mortalities in challenge experiments, with *V. anguillarum* causing greater mortalities than *A. logei* (Reid *et al.*, 2009). *Aliivibrio logei* was also associated with moribund fry (*Gadus morhua*) cultured in Norway. The isolate has been detected in gut of cod larvae in populations with high mortality, however experimental challenge experiments did not indicate it was a primary pathogen (Brunvold *et al.*, 2007).

According to 16S rRNA sequence information *A. logei* is closely related to *A. salmonicida* and less so to *A. fischeri*. MLSA identified *A. logei* as belonging to the Fischeri clade, which also contains *A. fischeri*, *A. salmonicida* and *A. wodanis* (Sawabe *et al.*, 2007b). Members of this clade have >36% DDH and 39–42 GC mol%, 89.8–94.4% similarity in concatenated MLSA sequence and AAI of 95.2%. Phylogenetic analysis using *gapA*, *gyrB*, *pyrH*, *recA*, *rpoA*, the *luxABE* region and the 16S rRNA gene found that some strains (SA12 and SR6) from the light organs of squid, *Seriola affinis* and *S. robusta* identified by Fidopiastis *et al.* (1998) as *A. logei*, did not group with the type strain (ATCC 29985^T) but grouped with *A. wodanis* (ATCC BAA-104^T) (Ast *et al.*, 2009). A clade containing unidentified *Aliivibrio* species was proposed as *A. sifiae*, however, these strains appear to be different to the *A. sifiae* (NBRC 105001^T) proposed by Yoshizawa *et al.* (2010a).

GROWTH CHARACTERISTICS AND PHENOTYPIC IDENTIFICATION. *Aliivibrio logei* is a motile, facultatively anaerobic, Gram-negative, oxidase-positive, straight rod with tufts of two to eight polar flagella (Bang *et al.*, 1978). Colonies are golden-yellow on seawater tryptone yeast extract agar (SWT) (Fidopiastis *et al.*, 1998). Most strains of *A. logei* require nutritional supplementation with 0.05 g/l (final concentration) yeast extract and 0.05 g/l tryptone to assist growth in biochemical identification media (Bang *et al.*, 1978). On MSA-B and MSSA colonies are slightly sticky, grey and pinpoint at 2–3 days and become orange with age. On MA 2216, colonies are 0.2 mm, light yellow at 3 days and 25°C. Colonies are pinpoint and green on TCBS at 48–72 h. Growth does not occur on BA or MacConkey agar. Large zones of growth inhibition are seen around 10 µg and 150 µg vibriostatic discs. Growth occurs after 4 days at 4°C and 20°C and slower growth is observed at 25°C. Growth does not occur at 30°C. Luminescence is produced and is greater when grown at 20°C compared to 25°C. Results are positive for lipase, chitinase and acid from D-gluconate. ADH and VP are negative (Bang *et al.*, 1978). The strain KCh1 produces a bright luminescence at 20°C but not at temperatures above 20°C or at temperatures of 8–16°C (Khrulnova *et al.*, 2010). *Aliivibrio logei* is very similar phenotypically to *A. fischeri* and the two species are mainly differentiated using growth at 4°C and 30°C. *Aliivibrio logei* is able to grow at 4°C but not at 30°C, whereas *A. fischeri* grows at 30°C but not at 4°C. All conventional biochemical tests in Table 3.21 and the API 20E strip may be negative or have no growth without the addition of 0.05% yeast extract (final concentration). Hydrogen sulfide is produced when grown on a lead–acetate–seawater medium. H_2S is produced from cysteine but not from sodium thiosulfate (Colwell and Morita, 1964). Although strains ATCC 15382 and NCMB 1143 are the same isolate, different biochemical reactions for methyl red, lysine decarboxylase and fermentation of salicin have been recorded in the literature by the same authors (Lunder *et al.*, 2000). The fermentation of salicin may be a weak delayed reaction (Hendrie *et al.*, 1971b). Chitin and aesculin are hydrolysed, cellulose and gelatin are not degraded and citrate is not used as a sole carbon source in Simmons' medium (Colwell and Morita, 1964). Colwell and Morita (1964) performed biochemical tests for gelatin, nitrate, citrate, indole, methyl red and Voges–Proskaüer using standard tests but with artificial seawater as the inoculum. Fermentation tests were performed using the method of Hugh and Leifson (1953). Lunder *et al.* (2000) used a medium containing proteose peptone, a mineral salt solution, phenol red indicator and 0.7% of the required carbohydrate for the fermentation tests (see media section). These and other reactions are presented in Tables 3.21 and 3.24–3.26, with some reactions also in Tables 3.30 and 3.38.

Aliivibrio salmonicida

TAXONOMY AND SIGNIFICANCE. *Aliivibrio* (*Vibrio*) *salmonicida* (H1 7751^T=NCMB 2262^T) is the causative agent of coldwater vibriosis, also known as Hitra disease, and was first isolated from Atlantic salmon (*Salmo salar*) cultured around the island of Hitra south of Trondheim, Norway (Egidius *et al.*, 1986). It has been isolated from Atlantic salmon farmed off the coast of the Shetland Islands, Scotland (Bruno *et al.*, 1986) and from cod (*Gadus morhua*) in Norway (Jørgensen *et al.*, 1989). A less severe infection occurs in rainbow trout (Egidius *et al.*, 1984). Infected fish become anaemic and septicaemic, and extensive haemorrhages occur in the integument surrounding the internal organs and on the skin and gills. The disease occurs in late autumn winter and early spring particularly when water temperature is below 10°C (Egidius *et al.*, 1984, 1986).

MLSA identified *A. salmonicida* as belonging to the Fischeri clade, which also contains *A. logei*, *A. fischeri* and *A. wodanis* (Sawabe *et al.*, 2007b). Members of this clade have >36% DDH and 39–42 GC mol%, 89.8–94.4% similarity of concatenated MLSA sequence and AAI of 95.2%.

GROWTH CHARACTERISTICS AND PHENOTYPIC IDENTIFICATION. *Aliivibrio salmonicida* is a motile, facultatively anaerobic Gram-negative curved rod (2–3×0.5 μm) that may be pleomorphic. Oxidase is positive and the organism is strongly susceptible to vibriostatic agent 0/129 with a zone diameter of 30 mm (Egidius *et al.*, 1986). Growth on primary isolation media tryptone soy agar (TSA) supplemented with 1.5% NaCl, or nutrient agar containing 5% human blood is slow and may take 3–5 days for colonies to appear when incubated at 15°C, but subcultured growth occurs within 48 h. Colonies are 0.5 mm, greyish, smooth with entire edges, non-haemolytic (NCMB 2262^T) and non-swarming after 3 days at 15°C. Haemolysis of horse red cells by NCMB 2245 was reported positive by Bruno *et al.* (1986). Growth occurs at 0.5–4.0% NaCl with an optimum of 1.5%. Growth occurs at 1–20°C and may or may not occur at 22°C but does not occur at 25°C. Optimal growth occurs at 12–15°C. Results are negative for nitrate, citrate, indole, ADH, H_2S, VP, gelatinase, lipase, urease and chitinase (Egidius *et al.*, 1986). Simmons' citrate was reported positive for NCMB 2245 by Bruno *et al.* (1986). Visible luminescence is produced only under certain inducing conditions (Fidopiastis *et al.*, 1999). Different reactions in the API Zym for NCMB 2245 are recorded depending upon the temperature and time of incubation and the inoculum. For the API Zym, more positive results are obtained when 2% marine salts are used as an inoculum compared to 0.85% NaCl as used by Bruno *et al.* (1986) and Lunder *et al.* (2000). These and other reactions are presented in Tables 3.21, 3.24–26, 3.30 and 3.38.

Aliivibrio sifiae

TAXONOMY AND SIGNIFICANCE. *Aliivibrio sifiae* (H1-1^T=NBRC 105001^T) was isolated from surface water in Tokyo Bay, Japan (Yoshizawa *et al.*, 2010a). According to 16S rRNA sequence information *A. sifiae* has similarity to *A. wodanis* (99.3%), *A. logei* (99.0%), *A. salmonicida* (98.4%) and *A. finisterrensis* (98.5%). MLSA analysis of 16S rRNA, *pyrH*, *recA*, *rpoA*, *gapA* and *gyrB* constructed from 4195 bp demonstrated similarity to *A. wodanis* (96.4%), *A. logei* (95.3%), *A. salmonicida* (95.1%) and *A. finisterrensis* (95.1%). Phylogenetic analysis using the *gapA* gene did not provide sufficient discrimination between *Aliivibrio* species.

GROWTH CHARACTERISTICS AND PHENOTYPIC IDENTIFICATION. *Aliivibrio sifiae* is a motile, oxidase positive, Gram-negative coccoid-rod (1.7–2.5×1.0 μm). On MA 2216 colonies are non-pigmented, round, entire and translucent and show a weakly swarming growth. A yellow luminescence is produced. The type strain grows as 5–7 mm yellow colonies on TCBS whereas another strain (H1-2) is green on TCBS. Growth occurs at 0.5–5% NaCl with an optimum of 3–4%. Growth does not occur at 8%. Growth occurs at 4–30°C with an optimum of 10–25°C but not at 37°C. H_2S is negative in sodium thiosulfate. Tween 20, 40, 60 and 80 are hydrolysed but starch and gelatin are not. An inoculum containing 2% NaCl is used and sets are incubated at 20°C. These and other reactions are presented in Tables 3.21, 3.24–26 and 3.38.

Aliivibrio wodanis

TAXONOMY AND SIGNIFICANCE. *Aliivibrio* (*Vibrio*) *wodanis* (NVI 88/441^T=NCIMB 13582^T) was isolated from cases of winter ulcer in Atlantic salmon (*Salmo salar*), an economically significant disease for the sea-farmed salmon industry in Norway, Iceland and Scotland, and is thought to be an opportunistic bacterium in the disease (Lunder *et al.*, 1995, 2000; Benediktsdóttir *et al.*, 2000). The primary causative agent of winter ulcer is *Moritella viscosa*. Phenotypically, *A. wodanis* is most closely related to *Moritella* (*Vibrio*) *viscosa*, *A. fisheri* (ATCC 15381), *A. logei* and *A. salmonicida*. *Aliivibrio wodanis* has 98.2% similarity to *A. salmonicida*, 98.1 and 98.8% to *A. logei* and 96.0% similarity to *A. fischeri* based on 16S rRNA gene sequence (Benediktsdóttir *et al.*, 2000; Lunder *et al.*, 2000). MLSA identified *A. wodanis* as belonging to the Fischeri clade, which also contains *A. logei*, *A. salmonicida* and *A. fischeri* (Sawabe *et al.*, 2007b). Members of this clade have >36% DDH and 39–42 GC mol%, 89.8–94.4% similarity of concatenated MLSA sequence and AAI of 95.2%. *A. wodanis* has a sequence similarity of 98.8% to *Aliivibrio* (*Vibrio*) *logei* strain CIP 103204 based on 16S rRNA sequencing (Benediktsdóttir *et al.*, 2000).

GROWTH CHARACTERISTICS AND PHENOTYPIC IDENTIFICATION. *Aliivibrio wodanis* is a motile, oxidase positive, Gram-negative short or elongated rod. On MSA-B colonies are yellow, round, opaque, non-swarming and non-luminescent after 48 h at 15°C (Lunder *et al.*, 2000). Growth occurs at 4–25°C but not at 30°C. NaCl is required for growth, which occurs at 1–4% NaCl. Reactions for indole (majority are positive) and DNase are variable amongst strains. Starch, gelatin and Tween 80 are hydrolysed, and a minority of strains hydrolyse alginate (23% positive), lethicin (3%), aesculin (3%) and urea (20%). Casein is not hydrolysed. Fermentation of cellobiose, melibiose, raffinose and L-rhamnose is negative, but results are positive for dextrin (Lunder *et al.*, 2000). Requirements for NaCl are variable amongst strains and range from 0.5 to 5% (Lunder *et al.*, 2000). A study found that 1 of 16 isolates showed growth at 0.5% NaCl, and 16 of 23 isolates showed growth at 4% (Benediktsdóttir *et al.*, 2000). These and other reactions are presented in Tables 3.21, 3.24–26, 3.30 and 3.38.

1.16.2 *Enterovibrio*

Enterovibrio species are part of the *Vibrionaceae* and are closely related to the genus *Grimontia*, sharing 94% 16S rRNA sequence similarity. *Enterovibrio norvegicus*, the type species, shares only 91.6% 16S rRNA gene sequence similarity with *V. cholerae* (Thompson *et al.*, 2002a). Species within the genus are motile, positive for oxidase, catalase, β-galactosidase, negative for the Voges–Proskaüer (VP) reaction and show variable results for arginine dihydrolase (ADH), indole production, nitrate reduction and susceptibility to vibriostatic agent 0/129 (Thompson *et al.*, 2002a; Pascual *et al.*, 2009).

Enterovibrio calviensis

TAXONOMY AND SIGNIFICANCE. *Enterovibrio calviensis* (RE35F/12^T=DSM 14347^T=CIP 107077^T), originally identified as *Vibrio calviensis*, was isolated from seawater in the western Mediterranean Sea, Bay of Calvi, France and is most closely related to *Vibrio scophthalmi* with 94.3% similarity for the 16S rRNA gene (Denner *et al.*, 2002). Further studies using the 16S rRNA, *rpoD* and *recA* genes suggest the bacterium belongs to the *Enterovibrio* genus (Pascual *et al.*, 2009).

GROWTH CHARACTERISTICS AND PHENOTYPIC IDENTIFICATION. *Enterovibrio calviensis* is a facultatively anaerobic, motile, oxidase-positive, Gram-negative, straight to slightly curved rod (0.25–1.0×0.75–2.5μm). On MA 2216 colonies are brownish, translucent, circular, smooth, with an entire edge (Denner *et al.*, 2002). It produces green colonies on TCBS. Growth occurs at 20–30°C and at 4°C within 10 days, but not at 37°C. Na^+ is required for growth with growth occurring between 1.5 and 12% NaCl with an optimum of 2.5–3.5% but no growth at 0 or 1.0% NaCl. Luminescence is not seen. It is susceptible to 0/129 vibriostatic agent. It hydrolyses Tween 80 but not casein, starch or agar. Nitrate is positive in the conventional tube method but negative in the API 20E. These and other reactions are presented in Tables 3.23, 3.24–26 and 3.38. An inoculum of 2% NaCl is used for the tests.

Enterovibrio coralii

TAXONOMY AND SIGNIFICANCE. *Enterovibrio coralii* (CC17^T=CBMAI 623^T=LMG 22228^T) was isolated from water extract of bleached coral (*Merulina ampliata*) growing around Magnetic Island, Australia (Thompson *et al.*, 2005). *Enterovibrio coralii* was isolated from head kidney of *Dentex dentex* (Pascual *et al.*, 2009). The pathogenicity status of the bacterium was not determined for either isolation.

In experimental challenge using rainbow trout (*Oncorhynchus mykiss*) and *Artemia* nauplii inoculated with 10^6 CFU/fish or 10^6 CFU/ml, respectively, *E. coralii* (LMG 22228^T) was rated as non-virulent or low virulence. Although there was some evidence of exophthalmia, kidney necrosis, ascites and liquefaction of internal organs, clinical disease was not seen, and at the end of the experiment the organism was recovered from tissue at either a low rate or not at all (Austin *et al.*, 2005).

Enterovibrio coralii has 95.8% and 95.5% 16S rRNA gene sequence similarity to *E. calviensis* and *E. norvegicus*, respectively, and 96.1% sequence similarity to *G. hollisae* based on the *rpoA* gene. Analysis of the *recA* gene revealed 88.2 and 85% similarity to *E. norvegicus* and *G. hollisae*, respectively.

GROWTH CHARACTERISTICS AND PHENOTYPIC IDENTIFICATION. *Enterovibrio coralii* is a motile, oxidase-positive, Gram-negative rod. On TSA containing 2% NaCl colonies are 5 mm, beige and transparent, round, smooth umbonate, with an entire edge after 48 h at 28°C (Thompson *et al.*, 2005). On TCBS colonies are 2 mm, green with raised margins. Growth occurs at 20–30°C but not at 4 or 40°C, and occurs at 2–6.0% NaCl but not at 0 or 8% NaCl. Glucose and D-mannitol are fermented. It is resistant to 150 μg 0/129 vibriostatic agent. ADH was reported positive for *E. coralii* from coral (Thompson *et al.*, 2005), whereas strains from *Dentex* species had variable reactions with six of eight strains positive for ADH using Thornley's method with marine broth as an inoculum and after prolonged incubation for more than 48 h (Pascual *et al.*, 2009). The original description reported indole as negative for the type strain, whereas it was found to be positive by Pascual *et al.* (2009). These and other reactions are presented in Tables 3.23, 3.24–26 and 3.38.

Enterovibrio nigricans

TAXONOMY AND SIGNIFICANCE. *Enterovibrio nigricans* (DAl 1-1-5^T=CECT 7320^T=CAIM 661^T) was isolated from the head kidney of seabream (*Sparus aurata*) and *Dentex dentex* cultured on the coast of the Mediterranean, Spain (Pascual *et al.*, 2009). The pathogenicity of this organism is unknown as other organisms were isolated from the head kidney in both fish species, including a number of *Vibrio* species with *V. harveyi* predominating (Company *et al.*, 1999; Pujalte *et al.*, 2003). In the *Dentex* similar species were seen including the unicellular parasite *Ichthyophonus* (Company *et al.*, 1999).

GROWTH CHARACTERISTICS AND PHENOTYPIC IDENTIFICATION. *Enterovibrio nigricans* is a facultatively anaerobic, motile, Gram-negative bacillus or coccobacillus (0.4–0.7×1.0–1.5 μm), which is oxidase positive. On MA 2216 agar some strains produce 1–2 mm, black to deep brown pigmented colonies with a slightly diffusible pigment at 20–24°C after 5 days. Colonies are round, entire and have a bright surface. A picture is available in the original article (Pascual *et al.*, 2009). Growth also occurs on TSA with 1% NaCl. Slow, sparse growth of green (sucrose-negative) colonies is seen on TCBS. Growth occurs at 15–28°C but not at 4 or 37°C, and occurs at 0.9–6.0% NaCl but not at 0.4 or 7% NaCl. Glucose is fermented under anaerobic conditions without gas production and some strains reduce nitrate to nitrite. Hydrolysis is positive for Tween 80, weak for DNA and negative for casein, gelatin, starch, alginate and lecithin. Luminescence is not produced on MA 2216 (note this medium may not be optimal for luminescence production). Other reactions are presented in Table 3.23.

Enterovibrio norvegicus

TAXONOMY AND SIGNIFICANCE. *Enterovibrio norvegicus* was isolated from the gut of healthy turbot (*Scophthalmus maximus*) larvae cultured in Norway (Thompson *et al.*, 2002a). In experimental challenge using rainbow trout and *Artemia* nauplii inoculated with 10^6 CFU/fish or 10^6 CFU/ml, respectively, *E. norvegicus* (LMG 19839^T, LMG 19840, but not LMG 19842) was rated as non-virulent or low virulence. Although there was some evidence of exophthalmia, kidney necrosis, ascites and liquefaction of internal organs, clinical disease was not seen, and at the end of the experiment the organism was recovered from tissue at either a low rate or not at all (Austin *et al.*, 2005).

GROWTH CHARACTERISTICS AND PHENOTYPIC IDENTIFICATION. *Enterovibrio norvegicus* is a facultatively anaerobic, motile, Gram-negative rod (1.0–1.2×0.8 μm) that is catalase and oxidase positive. On MA 2216 agar colonies are 1 mm, beige, round, smooth with raised margins after 2 days at 28°C (Thompson *et al.*, 2002a). Growth also occurs on TSA and BHI agar supplemented with 1.5% NaCl. Green colonies grow slowly on TCBS after 3 days. Growth does not occur at 0 or 8% NaCl or at 4 or 35°C. Optimal growth occurs at 2% NaCl and 20–28°C. Luminescence is negative (method not stated) and it is resistant to 0/129 10 μg and 150 μg vibriostatic agent. These and other reactions are presented in Tables 3.23, 3.24–26 and 3.38. An inoculum of 1.5% NaCl, with MacFarland 3 opacity is used for the API 20E. Incubation is at 25°C and tests are read at 48 h (Thompson *et al.*, 2002a).

1.16.3 *Grimontia*

Grimontia hollisae

TAXONOMY AND SIGNIFICANCE. *Grimontia hollisae* (LMG 17719^T=LMG 21416=LMG 21538), originally named *Vibrio hollisae* (special bacteriology group EF-13=ATCC 33564^T), causes gastroenteritis and septicaemia in humans (Hickman *et al.*, 1982). It has been isolated from raw oysters, clams, shrimps, intestine of coastal fish (*Rhynchopelates* [*Therapon*] *oxyrhynchus*) in Japan and seawater (Hickman *et al.*, 1982; Nishibuchi *et al.*, 1988). *Vibrio hollisae* was reclassified as *Grimontia hollisae* based on lack of similarity to *V. cholerae* (type species for *Vibrio*) as it has only 90.8% similarity based on 16S rRNA gene sequence. *Grimontia hollisae* has 16S rRNA gene sequence similarity of 93% to *Photobacterium* and 91.2% to *Salinivibrio* and is most closely related to *Enterovibrio norvegicus* (94.6%) (Thompson *et al.*, 2003a). *Grimontia hollisae* produces a haemolysin similar to thermostable direct haemolysin (TDH) produced by clinical isolates of *V. parahaemolyticus*. TDH is rarely produced by environmental isolates of *V. parahaemolyticus* and is a major virulence factor in human infections (Nishibuchi *et al.*, 1988).

GROWTH CHARACTERISTICS AND PHENOTYPIC IDENTIFICATION. *Grimontia hollisae* is a facultatively anaerobic, oxidase-positive, small Gram-negative rod with some slightly curved cells (1.5–2.0×0.5 μm). On BA (sheep blood) colonies are 1–2 mm at 24 h with a weak zone of haemolysis seen around individual colonies. On MA 2216 agar, colonies are 1–2 mm at 24 h at 36°C (Hickman *et al.*, 1982). Growth also occurs on TSA containing 1.5% NaCl after 24 h at 28°C (Thompson *et al.*, 2003a). Growth does not occur on TCBS or on MCA after 48 h, although some strains may show light growth in the heaviest area of streaking (the pool area). It was noted that growth may be seen with 58% of isolates after 48 h when MCA is inoculated from a heart infusion broth culture and with 75% of isolates after 7 days' incubation (Hickman *et al.*, 1982). Two colony types may be seen on MA 2216 and BA with one type more opaque, however the two types give identical biochemical results. In semi-solid motility medium none of the strains are motile at 36°C after 48 h, but 88% are motile after 7 days. Growth on MSSA and BA (horse blood) is seen as 0.5 mm grey colonies at 24 h and 2 mm grey colonies at 48 h. Colonies are an orange-tan colour when collected on a white cotton swab. Growth does not occur without NaCl and conventional media (MR, VP, Moeller's decarboxylase tests, aesculin, nitrate) must be supplemented with NaCl (1%). Indole using peptone water contains 0.5% NaCl, however it also must be supplemented with NaCl. A strong indole reaction is seen in the tube test. In the conventional tube media fermentation of arabinose and mannose are positive, however glucose fermentation is negative or weak. In the API 20E, reactions for glucose, sucrose and arabinose are very weak. Growth occurs in 1–6% NaCl. MR was reported negative, but MR-positive strains occur. Simmons' tube citrate is negative at 48 h but becomes positive after a prolonged incubation of 11 days. Citrate is negative in the API 20E. Growth occurs on DNase medium, but the result is negative. An incomplete or partial zone of inhibition of growth is seen around the 0/129 vibriostatic agent but complete inhibition is not seen. The string test using sodium deoxycholate (see biochemical tests section) is positive. Bioluminescence on MA 2216 is negative at 25°C after 24 h. These and other reactions are presented in Tables 3.23 and 3.24–26.

MOLECULAR IDENTIFICATION. Two conventional PCRs based on the *toxR* and *gyrB* genes have been reported (Vuddhakul *et al.*, 2000). If using colony material then a simple boiling method can be used, however if using direct detection from tissue then a commercial method such as silica-coated magnetic beads is recommended to overcome PCR inhibitors present in tissue material. For the *gyrB* PCR, 2 μl DNA is used in the PCR reaction containing 1 μl of 2.5 mM dNTP, 5 μl of each primer (2 μM) HG-F1 (5′-GCT CTG TCG GAA AAA CTT GA-3′) and HG-R2 (5′-ATG CTC AAA ATG GAA CAC AG-3′), 0.1 μl of *Taq* polymerase (5 U/μl), 2 μl of 10× buffer, 1.2 μl of 25 mM $MgCl_2$ and 3.7 μl ultrapure water. The DNA is amplified in 30 cycles consisting of denaturation at 94°C for 1 min, annealing at 55°C for 1 min and extension at 72°C for 1 min with a final cycle of 72°C for 7 min using a Genius thermal cycler, model FGENO5TY. A product size of 363 bp is visualized after electrophoresis and staining of a 2% gel.

The *toxR* gene PCR consists of 10 μl DNA, 4 μl of 2.5 mM dNTP, 1 μl of each primer (20 μM) HT-F3 (5′-CTG CCC AGA CAC TCC CTC TTC-3′) and HT-R2 (5′-CTC TTT CCT TAC CAT AGA AAC CG-3′), 0.25 μl of *Taq* polymerase (ExTaq; Takara), 5 μl of 10× buffer containing 20 mM $MgCl_2$ (ExTaq buffer; Takara) and 28.75 μl ultrapure water. The DNA is amplified in 24 cycles consisting of denaturation at 95°C for 1 min, annealing at 62°C for 1.5 min and extension at 72°C for 1.5 min with a final cycle of 72°C for 5 min using a Perkin-Elmer 480 thermal cycler. A product size of 306 bp is visualized after electrophoresis and staining of a 2% gel. The sensitivity for direct detection is 10–100 CFU/ml.

1.16.4 *Listonella*

The genus *Listonella* was proposed to accommodate *Vibrio anguillarum* and *V. pelagius* as *Listonella anguillarum* and *L. pelagia* based on 5S rRNA gene sequence information (MacDonell and Colwell, 1985). However, recent phylogenetic analysis using 16S rRNA sequence analysis, average aminoacid identity (AAI) and multilocus sequence analysis (MLSA) has suggested *Listonella* is a later heterotypic synonym of *Vibrio* and recommends the return of these two species to the genus *Vibrio* (Thompson *et al.*, 2011).

Listonella anguillarum

See *Vibrio anguillarum*.

Listonella damsela

See *Photobacterium damselae* ssp. *damselae*.

Listonella pelagia

See *Vibrio pelagius*.

1.16.5 *Photobacterium*

Photobacterium belong to the class *Gammaproteobacteria* and the family *Vibrionaceae*.

Sequence information from the 16S rRNA gene groups the genera of the *Vibrionaceae* into five clades; Vibrio, Photobacterium, Salinivibrio, Enterovibrio and Grimontia (Thompson *et al.*, 2001b). MLSA using concatenated sequences from nine genes (*ftsZ*, *gapA*, *gyrB*, *mreB*, *pyrH*, *recA*, *rpoA*, *topA* and 16S rRNA) identifies 14 clades using the SplitsTree decomposition method (Sawabe *et al.*, 2007b). Some of the clades correspond to families (*Salinivibrio*, *Photobacterium*) and others correspond to species within the *Vibrio* genus. Ten *Photobacterium* strains analysed by eight housekeeping genes indicate that species within the genus *Photobacterium* share 87.5–95.8% similarity according to concatenated MLSA sequence data and 98.6–99.4% amino acid identity (AAI). The data suggested that *Salinivibrio* was the ancestor of the vibrios and this clade contains *Enterovibrio* and *Grimontia*. *Photobacterium* formed a clade and the species were divided into one of five subclades.

Photobacterium angustum

TAXONOMY AND SIGNIFICANCE. MLSA identified *P. angustum* as belonging to the Photobacterium subclade, Leiognathi clade, which also contains *P. leiognathi* (Sawabe *et al.*, 2007b). Members of this clade have >44% DDH, 40–44 GC mol%, 94.0% similarity of concatenated MLSA sequences and 98.6% AAI.

Photobacterium damselae ssp. damselae

TAXONOMY AND SIGNIFICANCE. *Photobacterium damselae* ssp. *damselae* causes disease in aquatic species inhabiting marine, brackish and freshwater environments and has been reported worldwide. Originally, *P. damselae* ssp. *damselae* (2588-80^T=ATCC 33539^T) was isolated from dermal ulcers in damselfish (*Chromis punctipinnis*) in the USA and named *Vibrio damsela* (Love *et al.*, 1981). Based on 5S rRNA sequence data it was placed into a new genus termed *Listonella* (MacDonell and Colwell, 1985) but it was found that 5S rRNA sequence data is unreliable for assignment to genus for the *Vibrionaceae* and that in fact the organism belongs to the genus *Photobacterium* (Smith *et al.*, 1991). MLSA identified *P. damselae* ssp. *damselae* as belonging to the Photobacterium subclade, Damselae (Sawabe *et al.*, 2007b). Members of this clade have 42% GC mol%.

Photobacterium damselae ssp. *damselae* has been associated with disease in rainbow trout (*Oncorhynchus mykiss*), barramundi or seabass (*Lates calcarifer*), redbanded seabream (*Pagrus auriga*), seabream (*Sparus aurata*), snapper or silver seabream (*P. auratus*), mullet (*Mugil cephalus*, *Liza aurata*), turbot (*Scophthalmus maximus*), yellowtail (*Seriola quinqueradiata*), eel (*Anguilla reinhardtii*), stingray (*Dasyatis pastinaca*) and ornate wobbegong (*Orectolobus ornatuis*) also known as nurse or carpet shark, brown shark (*Carcharhinus plumbeus*), spiny dogfish or mud shark (*Squalus acanthias*), dolphin (*Tursiops truncatus*), tiger prawn or shrimp (*Penaeus monodon*) and leatherback turtle (*Dermochelys coriacea*) in Australia, Denmark, Hawaii, India, Italy, Japan, Spain, Tahiti and the USA (Grimes *et al.*, 1984; Obendorf *et al.*, 1987; Fujioka *et al.*, 1988; Sakata *et al.*, 1989; Vera *et al.*, 1991; Fouz *et al.*, 1992; Renault *et al.*, 1994; Pedersen *et al.*, 1997; Labella *et al.*, 2006; Stephens *et al.*, 2006; Vaseeharan *et al.*, 2007; Serracca *et al.*, 2011). *Photobacterium damselae* ssp. *damselae* is part of the bacterial flora found in oysters and seawater in Delaware Bay, USA (Richards *et al.*, 2008).

Photobacterium damselae ssp. *damselae* is part of the aquatic environment. It is found in seawater and the skin of fish: labracoglossid fish (*Labracoglossa argenteiventris*), gnomefish or Japanese bluefish (*Scombrops boops*), and sardine (*Sardinops* [*melanostictus*] *sagax*) collected from fish markets in Japan. These isolates were originally identified as a new species and named *P. histaminum* (Okuzumi *et al.*, 1994), but were found to be *P. damselae* ssp. *damselae* (Kimura *et al.*, 2000).

Photobacterium damselae is also a human pathogen; usually wound infections occur after exposure to marine or brackish water, which can develop into a nectrotizing fascitis that may be fatal (Morris *et al.*, 1982; Clarridge and Zighelboim-Daum, 1985).

The most common signs of disease in finfish are dermal ulcers, areas of haemorrhage especially on the fins, and septicaemia. In the original isolation from damsel fish ulcers were 0.5–2 cm in diameter and occurred near the pectoral fin and caudal peduncle. The disease was reproduced in experimental challenge using fish scarified before application of 10^7 to 10^8 cells to the damaged site. Ulcers were also reproduced without scarification and deaths resulted in other fish challenged with *P. damselae*: *C. caerulea*, garibaldi (*Hypsypops rubicundus*) and *Dascyllus trimaculatus*.

Photobacterium damselae ssp. *damselae* is found in water and can act as a primary and opportunistic infection. Factors that stress the fish and impact on their immune system such as a change in temperature, a change in salinity or other adverse environmental conditons can lead to infection. A survey of wild mullet (*Mugil cephalus*, *Liza aurata*) in the rivers of Italy suggested fish can harbour the organism in internal organs such as the spleen and exist in a carrier status, but cause disease when suboptimal environmental conditions of high water temperature and higher salinity exist (Serracca *et al.*, 2011). Stress due to suboptimal water conditions was a factor for ongoing mortality in cultured snapper (*P. auratus*) caused by co-infection with *P. damsela* ssp. *damselae* and *V. harveyi*. Infected fish showed either no obvious clinical signs or darkened skin pigmentation, over-inflated swimbladders and distended intestines containing yellowish fluid. *Photobacterium damselae* was isolated from intestines and *V. harveyi* from the blood (Stephens *et al.*, 2006). In cultured tiger prawn (*P. monodon*) a change in temperature, either increased to 34°C or decreased to 22°C from a control temperature of 26°C, resulted in greater mortalities of prawns challenged experimentally with 8.48×10^4 CFU/prawn than those held at 26°C without a change in temperature (Wang and Chen, 2006).

Virulence varies depending on species of aquatic animal and strain of the organism. *Photobacterium damselae* is highly virulent for rainbow trout and turbot with an LD_{50} of 1×10^3 to 5×10^5 cells/fish (Fouz *et al.*, 1992). In contrast, the type strain (ATCC 35083) isolated from brown shark was not virulent in the same assay. In tiger prawn (*P. monodon* and *P. indicus*) a similar highly virulent range for the LD_{50} was obtained, of 2×10^3 to 5×10^5 CFU and 4×10^2 to 2×10^4 CFU, respectively (Vaseeharan *et al.*, 2007).

Virulence is strongly related to the production of haemolysin, seen as a clear zone on a blood agar plate. In experimental challenge using rainbow trout, isolates producing a narrow zone of haemolysis are less virulent (LD_{50} 1.5×10^8 CFU) compared to isolates producing a wide zone of haemolysis (LD_{50} 3.9×10^3 CFU) (Pedersen *et al.*, 2009). The strongly haemolytic isolates produce a heat-labile cytotoxin (Kreger, 1984) named damselysin (Dly), which is encoded by the *dly* gene found within a 150–170 kb conjugative plasmid pPHDD1. The plasmid also encodes a homologue of the pore-forming toxin HlyA and both the pPHDD1 encoded *dly* and *hlyA* genes are required for virulence (Rivas *et al.*, 2011). Production of cytotoxin is optimal during mid- to late logarithmic phase of growth in broth culture. In contrast to optimal cell growth in medium containing 0.8% to 2% sodium ions, production of cytolysin is optimal at 0.5% Na^+ and is reduced at ≥0.8% (Kreger, 1984).

GROWTH CHARACTERISTICS AND PHENOTYPIC IDENTIFICATION. *Photobacterium damselae* is a facultatively anaerobic, oxidase positive, Gram-negative rod that is relatively pleomorphic and includes coccoid forms and long rods. On MSA-B colonies are beta-haemolytic, 2–3 mm, greyish-white, translucent, smooth and entire after 24–48 h at 25°C. Colonies are 2–3 mm, green on TCBS. Growth occurs at 18 and 37°C but for initial isolation, growth is best at 25°C. It is non-luminescent. Motility is weak. Growth occurs in 1–6% NaCl. Gas is produced from fermentation of carbohydrates. Glucose, mannose and maltose are fermented. Gelatin is not hydrolysed (Love *et al.*, 1981).

Some biochemical differences are recorded according to the biochemical method used and the amount of NaCl in the inoculum. The tube urease is positive, however in the API 20E the urease result may depend on the salt concentration in the inoculum. When an inoculum of 0.85% NaCl is used a positive urea result is obtained, but at an inoculum concentration of 2% NaCl a negative urea may be obtained. Urease was also found to be negative at 3% NaCl (Clarridge and Zighelboim-Daum, 1985). Time of incubation is important also, as urea may be weak or negative at 24 h incubation, but there is a strong positive at 48 h incubation. *Photobacterium damselae* ssp. *damselae* has been grouped into biotypes (Pedersen *et al.*, 1997) and many of these are listed in the conventional biochemical table for *Photobacterium* (Table 3.20). The following

Additional tests for differentiation of biotypes. (Pedersen *et al.*, 1997.)

Biotypes	1	2	3	4	5	6	7	8	9
Lipase	+	−	+	+	+	+	+	+	+
Cellobiose	+	+	+	+	−	−	+	+	+

table lists other biochemical tests not listed in Table 3.20 (conventional biochemical tests). These and other reactions are presented in Tables 3.20, 3.24–30 and 3.38.

MOLECULAR IDENTIFICATION. See under *Photobacterium damselae* ssp. *piscicida* for a multiplex PCR for the detection of both subspecies of *P. damselae* as described by Amagliani *et al.* (2009).

Photobacterium damselae ssp. piscicida

TAXONOMY AND SIGNIFICANCE. *Photobacterium damselae* ssp. *piscicida* (ATCC 17911^{T}) was originally identified as a *Pasteurella* species when it was isolated as the causative organism of an epizootic of white perch (Morone [*Roccus*] *americana*) in Chesapeake Bay, USA in 1963 (Snieszko *et al.*, 1964b). It was later identified as *Pasteurella piscicida* (Janssen and Surgalla, 1968) but based on rRNA cistron hybridization the bacterium was identified as belonging to the *Vibrionaceae* family (De Ley *et al.*, 1990) before being identified as *Photobacterium damselae* ssp. *piscicida* using rRNA sequence information and DNA–DNA hybrization results (Gauthier *et al.*, 1995b) and phenotypic studies (Thyssen *et al.*, 1998). Recently, phylogenetic analysis using concatenated sequences from the 16S rRNA, *gapA*, *gyrB*, *pyrH*, *recA* and *rpoA* genes confirmed the placement of *P. damselae* ssp. *piscicida* in the genus *Photobacterium* and its close similarity to *P. damselae* ssp. *damselae* (Urbanczyk *et al.*, 2011).

Photobacterium damselae ssp. *piscicida* has been reported from big scale sand smelt (*Atherina buoyeri*), Atlantic bluefin tuna (*Thunnus thynnus*), blackfish (*Girella punctata*), gilthead seabream (*Sparus aurata*), European seabass (*Dicentrarchus labrax*), Japanese flounder or bastard halibut (*Paralichthys olivaceus*), mullet (*Mugil cephalus*), red seabream (*Pagrus major*), Senegalese sole (*Solea senegalensis*), sole (*S. solea*), snakehead fish (*Channa maculata*), striped bass (*Morone saxatilis*), white bass (*M. chrysops*), white perch (*Morone americanus*) and yellowtail (*Seriola quinqueradiata*) cultured or found wild in the Adriatic Sea (Croatia), France, Greece, Israel, Italy, Japan, Malta, Portugal, Spain, Taiwan, Turkey and the USA (Snieszko *et al.*, 1964b; Kusuda and Yamaoka, 1972; Yasunaga *et al.*, 1983; Tung *et al.*, 1985; Hawke *et al.*, 1987; Toranzo *et al.*, 1991; Bakopoulos *et al.*, 1995; Baptista *et al.*, 1996; Candan *et al.*, 1996; Fukuda *et al.*, 1996; Bakopoulos *et al.*, 1997; Kawakami *et al.*, 1999; Zorrilla *et al.*, 1999; Nitzan *et al.*, 2001; Korun and Timur, 2005; Mladineo *et al.*, 2006). It has not been reported from Australia or New Zealand.

The disease has been called fish Pasteurellosis due to the initial naming as *Pasteurella piscicida*, and has also been referred to as bacterial pseudotuberculosis in infections in yellowtail due to the white nodules seen in the kidney and spleen (Kusuda and Yamaoka, 1972; Hawke *et al.*, 1987; Nagano *et al.*, 2011). In the acute form mass mortality occurs and few pathological or clinical changes are seen, but abnormal skin pigmentation may occur. In the chronic form there is low-level mortality and whitish-grey bacterial colonies 0.5–1.0 mm in size occur in the spleen and kidney (Kusuda and Yamaoka, 1972; Hawke *et al.*, 1987; Mladineo *et al.*, 2006). Fingerlings are particularly susceptible but older fish also succumb to the disease (Baptista *et al.*, 1996).

The bacterium adheres to the skin, gills and external areas of the lower jaw and pectoral fin, and enters the blood stream and the spleen and kidney where white nodules or colonies of bacteria are formed (Nagano *et al.*, 2011). The disease tends to occur in summer when water temperatures are over 23°C, however mortalities have also been reported at lower temperatures of 18–19°C (Toranzo *et al.*, 1991; Korun and Timur, 2005). In experimental challenge of bass, the bacterium was re-isolated from fish held at 24°C but not from fish held at 15–16°C or at 30°C (Nitzan *et al.*, 2001). The LD_{50} for yellowtail is $<3.8\times10^2$ CFU/fish and 3.8×10^3 CFU/fish for largescale blackfish (Kawakami *et al.*, 1999) and for gilthead seabream and turbot the LD_{50} is 1.6×10^5 to 9.5×10^5 and $\leq1.6\times10^4$, respectively (Toranzo *et al.*, 1991). The LD_{50} for red seabream and yellowtail is 0.001 mg to 0.01 mg per fish (Yasunaga *et al.*, 1983).

GROWTH CHARACTERISTICS AND PHENOTYPIC IDENTIFICATION. *Photobacterium damselae* ssp. *piscicida* is a non-motile, facultatively anerobic, Gram-negative rod (1.5×1.0 μm). The rods stain darker at each end (bipolar staining). Depending on media composition considerable pleomorphism may be seen and is dependent on NaCl concentration (Snieszko *et al.*, 1964b). Shape can range from coccoid to rods and filamentous forms (Bakopoulos *et al.*, 1995).

On agar medium containing 0.5–3% NaCl, nutrient agar, BA, MA 2216 and MSA-B colonies are 1 mm, regular, round, convex and slightly viscous after 48 h at 25°C. Colonies have a soft consistency and may become a greyish-yellow colour. It is non-haemolytic. Growth on nutrient agar and choc agar is less than that on blood agar or brain heart infusion agar. There is no growth on MacConkey agar or TCBS. Growth occurs at 20–30°C but not at 10, 35 or 37°C. There is no growth at 0% or 5% NaCl. It is sensitive to vibriostatic agent 0/129 (Snieszko *et al.*, 1964b; Kusuda and Yamaoka, 1972; Hawke *et al.*, 1987; Magariños *et al.*, 1992).

Strains from striped bass, ATCC 17911 and ATCC 29687, are strongly positive for fermentation of glucose, mannose, galactose and fructose and weakly positive for maltose, and negative for arabinose, inositol, lactose, mannitol and sucrose using conventional sugars (prepared with semi-solid agar and phenol red indicator, 1.5% NaCl and 1% of the desired carbohydrate) (Hawke *et al.*, 1987). The fermentation reactions may be improved under anaerobic conditions. This is easily achieved by overlaying the medium with sterile paraffin oil. The glucose reaction in the API 20E kit may be very weak or negative. The VP reaction is variable, and research into the biochemical properties of isolates from France, Greece, Italy and Japan found that all isolates tested were negative for VP in the API 20E system (Bakopoulos *et al.*, 1995), however tuna isolates from the Adriatic coast are VP positive in the API 20E (Mladineo *et al.*, 2006). Hawke *et al.* (1987) reported striped bass strains, ATCC 17911^T and ATCC 29687, as negative for VP using Clark and Lubs MRVP medium supplied by Difco. Kusuda and Yamaoka (1972) reported yellowtail strains as weakly positive for VP and positive for MR, whereas weak MR and positive VP was reported for gilthead seabream strains (Baptista *et al.*, 1996). A weak MR (Difco) was reported by Magariños *et al.* (1992) for European, Japanese and US strains. A latex agglutination test, Mono-Pp, is available from Bionor.

These and other reactions are presented in Tables 3.20, 3.24–30 and 3.38.

MOLECULAR IDENTIFICATION. The following PCRs reported by Osorio *et al.* (2000b) and Amagliani *et al.* (2009) are used to detect and differentiate both subspecies of *P. damselae*.

A conventional multiplex PCR using primers designed to detect the 16S rRNA gene from both species and the *ureC* from *P. damselae* ssp. *damselae* only, detects and differentiates the two subspecies (Osorio *et al.*, 2000b). A 100-µl master mix comprises 160 pmol of each primer, being forward primer Car1 (5′-GCT TGA AGA GAT TCG AGT-3′), reverse primer Car2 (5′-CAC CTC GCG GTC TTG CTG-3′) for the 16S rRNA gene, forward primer Ure-5 (5′-TCC GGA ATA GGT AAA GCG GG-3′) and reverse primer Ure-3 (5′-CTT GAA TAT CCA TCT CAT CTG C-3′) for the urease gene, and 2 U *Taq* polymerase (Perkin Elmer), 1× *Taq* buffer, 2 mM $MgCl_2$, 200 µM of each dNTP and 100 ng DNA. Thermocycling is carried out using an initial denaturation at 95°C for 4 min followed by 30 cycles of denaturation at 95°C for 1 min, annealing at 60°C for 1 min, extension at 72°C for 40 s and a final extension step at 72°C for 5 min. *Photobacterium damselae* ssp. *damselae* produces amplicons at 267 and 448 bp corresponding to the 16S rRNA genes and *ureC* genes, respectively. *Photobacterium damselae* ssp. *piscicida* produces product of 267 bp only and does not produce amplicon to the *ureC* gene. Specificity and sensitivity were not provided.

A conventional multiplex PCR was developed for the detection of *P. damselae* ssp. *piscicida* and *P. damselae* ssp. *damselae* together with an internal amplification control for use on fish tissue based on primers designed to part of the gene coding for a putative penicillin-binding protein (Pbp) 1A, and primers Ure5 and Ure-3 detecting the urease gene in *P. damselae* ssp. *damselae* as reported by Osorio *et al.* (2000b) (Amagliani *et al.*, 2009). DNA is extracted from fish tissue using the DNeasy Blood and Tissue Kit (Qiagen) according to the manufacturer's protocol. An internal amplification control is prepared using the pDEF recombinant plasmid and a pCR 2.1 vector (Invitrogen) into which has been inserted a 186-bp sequence of murine leukaemia virus *gag* gene that is detected using the primers SDR3 (5′-CCG GTC TGA CCT TTA CAC TG-3′) and SDR4 (5′-GGG GTT CTC TGT GAG AAG GT-3′) that amplify 112 bp. The 50-µl reaction mix consists of 0.08 µmol/l each *P. damselae* ssp. *piscicida* primers 76a (5′-CCG ACT CAA CTA CAG ATC ACC CAG TC-3′) and 76b (5′-GTG CGG CCT AAA TTT CGA CGA-3′), 1 µmol/l of each *P. damselae* ssp. *damselae* primers Ure-5 (5′-TCC GGA ATA

GGT AAA GCG GG-3′) and Ure-3 (5′-CTT GAA TAT CCA TCT CAT CTG C-3′), 0.6 μmol/l of each internal control primer SDR3 and primer SDR4, 10^4 copies of the internal control template, 200 μmol/l each dNTP and 1 U Hot-Rescue DNA polymerase (Diatheva), 1× PCR buffer with final $MgCl_2$ concentration adjusted to 7 mmol/l and 0.2 mg/ml BSA. The thermocycling protocol begins with a denaturation step of 95°C for 10 min, followed by 50 cycles of 95°C for 30 s, 67°C for 30 s and 72°C for 1 min with a final extension step at 72°C for 10 min. A 297-bp amplicon is produced for *P. damselae* ssp. *piscicida*, however 22% (n=9) of *P. damselae* ssp. *damselae* isolates also produce amplicon of the same size. The Ure primers are used to differentiate true *P. damselae* ssp. *piscicida* from *P. damselae* ssp. *damselae* and produce amplicon of 448 bp. The sensitivity is 500 pg–50 fg of genomic DNA per reaction tube or about 100 bacterial cells (Amagliani *et al.*, 2009). Some non-specific amplification was noted from tissue samples.

Photobacterium fischeri

See *Aliivibrio fischeri*.

Photobacterium frigidiphilum

MLSA identified *P. frigidiphilum* as belonging to the *Photobacterium* subclade, Phosphoreum, which also contains *P. phosphoreum* and *P. iliopiscarius* (Sawabe *et al.*, 2007b). Members of this clade have 38–43.8 GC mol%, 94.24–95.8% similarity of concatenated MLSA sequences and 99.4% AAI.

Photobacterium histaminum

Photobacterium histaminum is a later heterotypic synonym of *P. damselae* ssp. *damselae* (Okuzumi *et al.*, 1994; Kimura *et al.*, 2000).

Photobacterium iliopiscarium

TAXONOMY AND SIGNIFICANCE. *Photobacterium iliopiscarium* (PS1^T=ATCC 51760^T) is part of the intestinal microflora of herring, coalfish, cod and salmon found in cold seawater. The bacterium was initially named *Vibrio iliopiscarius* (Onarheim *et al.*, 1994) but later transferred to the genus *Photobacterium* (Urakawa *et al.*, 1999). It is closely related to *P. phosphoreum* based on 16S rRNA gene sequence information. MLSA identified *P. iliopiscarium* as belonging to the *Photobacterium* subclade, Phosphoreum, which also contains *P. phosphoreum* and *P. frigidiphilum* (Sawabe *et al.*, 2007b). Members of this clade have 38–43.8 GC mol%, 94.24–95.8% similarity of concatenated MLSA sequences and 99.4% AAI.

GROWTH CHARACTERISTICS AND PHENOTYPIC IDENTIFICATION. *Photobacterium iliopiscarium* is a Gram-negative straight or curved rod with some pleomorphism. It is non-haemolytic. On MSA-B and MA 2216 colonies are small colourless to greyish, opaque with an entire edge. Colonies are slow growing and may take up to 14 days for growth to become visible. It is catalase and oxidase positive. Other reactions are presented in Tables 3.20, 3.24–26 and 3.30.

Photobacterium indicum

TAXONOMY AND SIGNIFICANCE. MLSA identified *P. indicum* as belonging to the *Photobacterium* subclade, Profundum, which also contains *P. profundum* and *P. lipolyticum* (Sawabe *et al.*, 2007b). Members of this clade have 40–42 GC mol%, 87.5–92.3% similarity of concatenated MLSA sequences.

Photobacterium leiognathi

MLSA identified *P. leiognathi* as belonging to the *Photobacterium* subclade, Leiognathi clade, which also contains *P. angustum* (Sawabe *et al.*, 2007b). Members of this clade have >44% DDH, 40–44 GC mol%, 94.0% similarity of concatenated MLSA sequences and 98.6% AAI. Some biochemical reactions are presented in Table 3.20.

Photobacterium lipolyticum

MLSA identified *P. lipolyticum* as belonging to the *Photobacterium* subclade, Profundum, which also contains *P. profundum* and *P. indicum* (Sawabe *et al.*, 2007b). Members of this clade have 40–42 GC mol%, 87.5–92.3% similarity of concatenated MLSA sequences.

Photobacterium logei

See *Aliivibrio logei*.

Photobacterium phosphoreum

MLSA identified *P. phosphoreum* as belonging to the *Photobacterium* subclade, Phosphoreum, which also contains *P. frigidiphilum* and *P. iliopiscarium* (Sawabe *et al.*, 2007b). Members of this clade have 38–43.8 GC mol%, 94.24–95.8% similarity of concatenated MLSA sequences and 99.4% AAI. Some biochemical reactions are presented in Tables 3.20 and 3.24.

Photobacterium profundum

MLSA identified *P. profundum* as belonging to the *Photobacterium* subclade, Profundum, which also contains *P. indicum* and *P. lipolyticum* (Sawabe *et al.*, 2007b). Members of this clade have 40–42 GC mol%, 87.5–92.3% similarity of concatenated MLSA sequences. Some biochemical reactions are presented in Table 3.20.

Photobacterium rosenbergii

TAXONOMY AND SIGNIFICANCE. *Photobacterium rosenbergii* (CC1^T=CBMAI 622^T=LMG 22223^T) was isolated from corals (bleached and apparently healthy *Pachyseris speciosa*, bleached *Merulina ampliata*, bleached *Barabattoia amicorum*) in the water around Magnetic Island, Australia (Thompson *et al.*, 2005b). The pathogenicity was not determined.

In experimental challenge using rainbow trout (*Oncorhynchus mykiss*) and *Artemia* nauplii inoculated with 10^6 CFU/fish or 10^6 CFU/ml, respectively, strains of *P. rosenbergii* (LMG 22223^T, LMG 22224, LMG 22225, LMG 22227) were rated as non-virulent or low virulence. Although there was some evidence of exophthalmia, kidney necrosis, ascites and liquefaction of internal organs, clinical disease was not seen, and at the end of the experiment the organism was recovered from tissue at either a low rate or not at all (Austin *et al.*, 2005).

Photobacterium rosenbergii has 95.6% 16S rRNA gene sequence similarity to *P. leiognathi* and 91.2% sequence similarity to *P. phosphoreum* based on the *rpoA* gene. Analysis of the *recA* gene revealed 83.9% similarity to *P. leiognathi*. MLSA identified *P. rosenbergii* as belonging to the *Photobacterium* subclade, Rosenbergii, currently the only species in this clade (Sawabe *et al.*, 2007b). The GC mol% is 47.6–47.9.

GROWTH CHARACTERISTICS AND PHENOTYPIC IDENTIFICATION. *Photobacterium rosenbergii* is a motile, oxidase-positive, Gram-negative rod (2–4×1–2 μm). On TSA containing 2% NaCl, colonies are beige and opaque, round, smooth with an entire edge after 48 h at 28°C (Thompson *et al.*, 2005b). On TCBS colonies are yellow with raised margins. Growth occurs at 20–30°C but not at 4 or 40°C, and occurs at 2–6.0% NaCl but not at 0 or 8% NaCl. Glucose and D-mannitol are fermented. It is resistant to 150 μg 0/129 vibriostatic agent. Some reactions are variable within the species compared to the type strain. These and other reactions are presented in Tables 3.20, 3.24–26 and 3.38.

Photobacterium swingsii

TAXONOMY AND SIGNIFICANCE. *Photobacterium swingsii* (CAIM 1393^T=CECT 7576^T) was isolated as part of the bacterial microflora in cultured oysters (*Crassostrea gigas*) in Mexico and the haemolymph of wild spider crab (*Maja brachydactyla*) in Spain (Gomez-Gil *et al.*, 2011). The most closely related bacteria based on 16S rRNA gene sequence analysis are *P. aplysiae* and *P. frigidiphilum* (44.0–53.85%).

GROWTH CHARACTERISTICS AND PHENOTYPIC IDENTIFICATION. *Photobacterium swingsii* is a facultatively anaerobic, Gram-negative coccobacillus. Oxidase is positive and five of six strains are positive for catalase. On MA 2216 colonies are 2–3 mm, white, smooth and non-luminescent after 24 h at 25°C. On TCBS colonies are 2–3 mm, green and round. Growth occurs at 3–6% NaCl but not at 0 or 10% NaCl. Growth occurs at 4 to 30°C but not at 35°C (except for the type strain and one other strain). No growth at 40°C. Sensitive to vibriostatic agent at 10 and 150 μg (Gomez-Gil *et al.*, 2011). These and other reactions are presented in Tables 3.20, 3.24–26 and 3.38. An inoculum of 2.5% NaCl is used.

1.16.6 *Salinivibrio*

Salinivibrio (*Vibrio*) *costicola*

TAXONOMY AND SIGNIFICANCE. *Salinivibrio* species belong to the class *Gammaproteobacteria* and the family *Vibrionaceae*. Phylogenetic information from MLSA using concatenated sequences from nine genes (*ftsZ*, *gapA*, *gyrB*, *mreB*, *pyrH*, *recA*, *rpoA*, *topA* and 16S rRNA) suggests *Salinivibrio* is the ancestor of the vibrios (Sawabe *et al.*, 2007b). The *Salinivibrio* clade is one of 14 clades and contains *Enterovibrio* and *Grimontia*.

Salinivibrio costicola (NCMB 701^T) was isolated from Australian cured bacon (Smith, 1938, as referenced in Garcia *et al.*, 1987) and also from salterns in Spain and the Canary Islands and originally named *Vibrio costicola* (Garcia *et al.*, 1987). It is isolated also from salted meats. 16S rRNA gene sequence information revealed the organism formed a separate phylogenetic branch from the vibrios and it was placed in a new genus *Salinivibrio* as *Salinivibrio costicola* (Mellado *et al.*, 1996).

GROWTH CHARACTERISTICS AND PHENOTYPIC IDENTIFICATION. *Salinivibrio costicola* is a motile, facultatively anaerobic, Gram-negative curved rod (0.5×1.5–3.0 μm). On marine salt medium (10% w/v marine salts) colonies are 2–3 mm, cream-coloured after 48 h at 37°C (Garcia *et al.*, 1987). Growth occurs at 0.5–20% NaCl but not at 0% NaCl. Optimal NaCl concentration for growth is 5–10%. Growth occurs at 5–45°C. It is oxidase and catalase positive and negative for nitrate reduction, ODC, ONPG and indole. ADH and VP are positive (Garcia *et al.*, 1987; Mellado *et al.*, 1996). These and other reactions are presented in Table 3.23.

MOLECULAR IDENTIFICATION. *Salinivibrio costicola* can be detected using IGS^{Glu} (cosB)-specific primers VCOS-F (5′-CTG ACG CTA TTC TTG CGA-3′) and VCOS-R (5′-GTA ATC ACA TTC GTA AAT GC-3′) that produce an amplicon of 186 bp. In the thermocycling programme an annealing temperature of 55°C is used and 35 cycles (Lee *et al.*, 2002).

1.16.7 *Vibrio*

TAXONOMY. The *Vibrio* genus consists of facultative anaerobic, fermentative, curved Gram-negative rods that belong to the class *gammaproteobacteria* and the family *Vibrionaceae*. *Vibrio* species from the marine environment grow on agar medium containing salt (NaCl or sea salts), whereas *Vibrio* species isolated from the freshwater environment do not have a salt requirement. Vibrios isolated from aquatic animals from brackish water may or may not have a salt requirement for some biochemical and enzyme identification tests. The majority of *Vibrio* species grow on thiosulfate citrate bile salt agar (TCBS), are oxidase positive and nitrate positive although there are exceptions to these tests. To date the genus consists of 106 species with more being regularly identified. *Vibrio* species are predominantly found in the marine environment but also in freshwater and are either part of the normal flora of aquatic animals and organisms or are pathogenic and cause disease. Some species also cause disease in humans and are generally food-borne or water-borne infections.

Identification of a *Vibrio* species is based on biochemical tests that involve carbohydrate fermentation, carbon source utilization or enzymatic hydrolysis of a specific substrate. Biochemical tests may be 'in-house', that is, prepared in the laboratory, or obtained commercially such as the API strips from bioMérieux, or the Biolog. Problems occur due to phenotypic diversity within a species and this may be due to isolation of the bacterium from different animal (aquatic) species or from different geographical areas although phenotypic diversity of an organism also occurs within the same ecological site (unpublished data). Some biochemical tests are less reliable because the test outcome is dependent upon certain growth conditions such as the optimal NaCl concentration or the requirement for other 'salts' such as MgCl and KCl found in seawater, or optimal temperature. Some bacteria become adapted to their hosts' environment to the extent that a test such as urea may be negative when NaCl is added to the inoculum when the bacterium is isolated from brackish water, yet positive when NaCl is not used in the inoculm. Variation in test results with and without NaCl can occur, for example, with *Photobacterium damselae* ssp. *damselae*.

Molecular methods are being used to establish evolutionary relationships among vibrios, to determine epidemiology and to assist in identification to species level. For the *Vibrio* genus the 16S rRNA is so similar between species that it offers little value for diagnostic identification unless used in conjunction with other identification methods. For example, the 16S rRNA gene positions 455–477 (*Escherichia coli* numbering system) are conserved across *V. alginolyticus*, *V. campbellii*, *V. harveyi* and *V. natriegens*, whereas positions 73–100 and 180–196 have some diversity (Dorsch *et al.*, 1992). However, sequencing the 16S rRNA alone is not recommended for differentiation of *Vibrio* species. Even many of the housekeeping genes are conserved across species and must be carefully selected if sequence information is to be used to design primers for a specific PCR for diagnostic use. For example, the *toxR* gene is found in a number of *Vibrio* species, and a PCR based on this gene gives cross-reactions with a number of species. Diversity exists within this gene and primer design in an area conserved for a particular species yet diverse for other species is crucial for specificity and use as a diagnostic identification tool.

The genetic diversity of the *Vibrio* species was investigated using amplified fragment length polymorphisms (AFLPs). In this technique, DNA is digested with two restriction enzymes, and a subset of restriction fragments is selected based on ligation of adaptors specific to the restriction sites at each end of the cut fragments. The selected fragments are amplified in a PCR reaction using fluorescent labelled primers (fluorescent amplified fragment length polymorphisms; FALP). A dendrogram of FALP patterns from 506 strains identifies 69 clusters (Thompson *et al.*, 2001b). The same data analysed using the software BinClass, which is based on a minimization of stochastic complexity (SC), identified 64 classes that correspond to the clusters obtained by FALP. The SC classification for *Vibrio fortis*, *V. diazotrophicus*, *V. campbellii* and *Enterovibrio norvegicus* is in agreement with 16S rRNA sequence information and DNA–DNA hybridization, which shows that these species are not as distantly related to the other *Vibrio* clusters as suggested by the FALP method (Dawyndt *et al.*, 2005).

Molecular studies undertaken on the *Vibrio* species have investigated the phylogenetics of the genus, that is, determining the evolutionary relationships through the sequencing of genes, particularly housekeeping genes, with a view to confirming the allocation to species of new isolates and to determine the true identity of other isolates that are phenotypically similar. In their report, the ad hoc committee for re-evaluating the species definition recommended the use of multilocus sequence typing (MLST) (or multilocus sequence analysis, MLSA) in addition to the current recommendation of 16S rRNA sequencing, DNA–DNA hybridization and phenotypic tests when describing a new species (Stackebrandt *et al.*, 2002). MLST uses sequence information from 450–500 bp of a housekeeping gene to infer genetic relatedness within or between species (Maiden *et al.*, 1998). In some cases the conserved genes may not offer sufficient discriminatory power to differentiate species. Sequence information from between five and nine housekeeping genes may be needed for delineation of some species of *Vibrio*. As well as investigating evolutionary history and epidemiology, MLST and sequencing of other genes such as virulence genes (MLVST) is being investigated as an aid to species identification in the diagnostic laboratory. MLST information can be obtained at two main sites (http://www.mlst.net and http://pubmlst.org) although very little sequence information exists for the *Vibrio* species. MLSA information for *Vibrio* species can be found at http://www.taxvibrio.lncc.br (Sawabe *et al.*, 2007b).

Housekeeping genes that have been investigated for *Vibrio* species include *ftsZ* (cell division protein), *gapA* (glyceraldehyde-3-phosphate dehydrogenase), *gyrB* (gyrase B), *mreB* (rod shape-determining protein), *pyrH* (encodes UMP kinase), *recA* (protein involved in DNA repair), *rpoA* (RNA polymerase, sigma factor 70), *rpoD* (RNA polymerase, sigma 70 factor), *rctB* (replication initiator protein), *topA* (topoisomerase I) and *toxR* (virulence regulatory protein) (Table 1.5).

Sequence information from the 16S rRNA gene groups the genera of the *Vibrionaceae* into five clades; Vibrio, Photobacterium, Salinivibrio, Enterovibrio and Grimontia (Thompson *et al.*, 2001b). MLSA using concatenated sequences from nine genes (*ftsZ*, *gapA*, *gyrB*, *mreB*, *pyrH*, *recA*, *rpoA*, *topA* and 16S rRNA) from 78 isolates identifies 14 clades using the SplitsTree decomposition method (Sawabe *et al.*, 2007b). Some of the clades correspond to families (*Salinivibrio*, *Photobacterium*) and others correspond to species within the *Vibrio* genus (Anguillarum clade, Cholerae clade, Coralliilyticus clade, Diazotrophicus clade, Fischeri clade, Halioticoli clade, Harveyi clade, Nereis clade, Orientalis clade, Scophthalmi clade, Splendidus clade and Vulnificus clade). Based on the eight housekeeping genes, *Vibrio* species share >95% gene sequence similarity and >99.4% amino acid identity (AAI). The species within a clade have <5% GC variation, >20% DNA–DNA hybridization, >85% MLSA sequence similarity and >89% average amino acid identity. The data suggest that *Salinivibrio* is the ancestor of the vibrios. The Salinivibrio clade contains *Enterovibrio* and *Grimontia*. The Vibrio core group, a term previously suggested through 16S rRNA sequence information that identified *V. alginolyticus*, *V. campbellii*, *V. harveyi*, *V. natriegens*, *V. proteolyticus*, *V. parahaemolyticus* and *V. vulnificus* as closely related (Dorsch *et al.*, 1992; Kita-Tsukamoto, 1993), was identified in the Harveyi clade using MLSA. However, MLSA identified *V. vulnificus* as belonging to its own clade with *V. proteolyticus* lying between the Harveyi and Vulnificus clades.

Other housekeeping genes have been investigated in order to test their usefulness for phylogenetic studies. Analysis using the *rpoA*, *recA* and *pyrH* genes is in general agreement with information obtained from the 16S rRNA gene. Strains within a species have a similarity of 98% for the *rpoA* gene, and 94% each for the *recA* and *pyrH* genes (Thompson *et al.*, 2005a). The *rpoA* gene is less discriminatory than the *recA* and *pyrH* genes for differentiation between *V. splendidus* and *V. tubiashii*. The *recA* and *atpA* genes are also more discriminatory than the 16S rRNA gene and strains within a species have at least 94% *recA* gene sequence similarity and 99% similarity for the *atpA* gene (Thompson *et al.*, 2004, 2007a). The *rpoB* gene also shows suitable discriminatory ability (Ki *et al.*, 2009) and in clinical laboratories has been used to identify *Vibrio* species that are not detected using a multiplex PCR for *V. cholerae*, *V. mimicus*, *V. parahaemolyticus* and *V. vulnificus* (Tarr *et al.*, 2007) (see under *V. parahaemolyticus* molecular section).

Concatenated sequences of *rpoD*, *rctB* and *ToxR* genes may be useful in discriminating between strains belonging to the Vibrio core group (*V. alginolyticus*, *V. campbellii*, *V. harveyi*, *V. natriegens*, *V. parahaemolyticus* and *V. rotiferianus*). The analysis of these three genes appears to be more useful than using results from a total of seven genes that also included 16S rRNA, *recA*, *pyrH* and *gyrB* genes (Pascual *et al.*, 2010).

The housekeeping genes (*gapA*, *gyrB*, *recA*, *rpoA*, *pyrH*, *atpA* and *dnaJ*) have been used as phylogenetic markers for delineating the *Vibrio* species. The *rpoA* gene has been proposed as a phylogenetic marker for bacterial classification because it is ubiquitous in bacteria, is present as a single copy, is resistant to lateral gene transfer and evolves slowly like the 16S rRNA gene (Zeigler, 2003).

TopA has highest resolution followed by *mreB*, *ftsZ* and *pyrB* for both species and strain differentiation (Thompson *et al.*, 2007b). *V. harveyi* LMG 19643, LMG 19714 and 823WBZ7 were identical in all loci except *topA* where 823WBZ7 accumulated three silent point mutations.

RecA gene sequence analysis shows *V. campbellii* forms a cluster nested within *V. harveyi* with strains sharing 94% sequence similarity. *GapA* reveals *V. harveyi* in a tight cluster within *V. campbellii* with 99.5% sequence similarity. *Vibrio campbellii* is heterogeneous within the *gapA* gene with 97.7% similarity between strains. A gene showing the lowest resolution value has the most discriminatory power for species identification in the *Vibrionaceae* (Thompson *et al.*, 2007b).

Gene sequences are aligned and a phylogenetic tree is constructed using concatenated sequences by neighbour-joining (NJ) analysis using MEGA4 software (Kumar *et al.*, 2008). Appropriate type or reference strains are used in the construction of the phylogenetic tree. The robustness of each topology is checked using 500 bootstrap replications.

Genes to be used as phylogenetic markers should fulfil several criteria. They should be present in a majority of genomes and be present as a single copy within a genome. Useful sequence information should be contained in segments short enough for sequencing (900–2250 nucleotides). The sequence information must be capable of predicting whole genome relationships with accuracy and precision, and should correlate with data from the 16S rRNA gene and DNA–DNA hybridization information (Zeigler, 2003).

Table 1.5. Comparison of genes for sequencing and identification of *Vibrio* species.

Gene	Useful for identification	Does not resolve	Species resolution/ resolving power	Strain resolution	Sequence similarity	Reference
atpA					99.00%	Thompson *et al.*, 2007
ftsZ	Good differentiation for *V. harveyi* and *V. campbellii*	Low sequence variation	81%		94–96%	Thompson *et al.*, 2007
gapA		Low sequence variation	86.00%			Thompson *et al.*, 2007
gyrB		*V. campbellii* and *V. harveyi*	99.5% sequence similarity *V. campbellii* and *V. harveyi*. 77%			
mreB	Good differentiation for *V. harveyi* and *V. campbellii*		83%		94–96%	
pyrH	Good differentiation for *V. harveyi* and *V. campbellii*	Low sequence variation	Yes, resolving power of 83%	yes	94–96%	Thompson *et al.*, 2005, 2007
recA			81%		94%	Thompson *et al.*, 2005
rrn			No, 96%			
rpoA			89%		98%	Thompson *et al.*, 2005
topA	Good differentiation for *V. harveyi* and *V. campbellii*		Resolving power of 74%		90% sequence similarity	

*atp*A (encodes the α-subunit of bacterial ATP synthase), *fts*Z (cell division protein), *gap*A (glyceraldehyde-3-phosphate dehydrogenase), *gyr*B (partial gyrase subunit), *mre*B (rod shape-determining protein), *pyr*H (uridylate kinase), *rec*A (encodes RecA protein), *rrn* (16S rRNA), *rpo*A (encodes the α-chain of RNA polymerase), *top*A (topoisomerase I).

DNA–DNA hybridization is still a necessary technique for discriminating species when describing new species, especially those that have >97% 16S rRNA gene sequence similarity with known species (Stackebrandt *et al.*, 2002). This technique is laborious and not available to all laboratories. A real-time PCR technique has been described to measure GC content and DNA–DNA similarity in an attempt to provide a simpler, reliable alternative to current techniques (Moreira *et al.*, 2011). DNA–DNA hybridization involves the measurement of the thermal stability of hybrid complexes versus homologous complexes that result when DNA from two strains is heated to denaturation and then cooled to allow the hybridization of similar sequences. The method of de Ley *et al.* (1970) based on thermal denaturation is the most commonly used method. The melting temperature or the thermal denaturation midpoint (ΔT_m) is the temperature at which 50% of the initial double-stranded DNA has denatured to single strands. DNA–DNA relatedness is measured by the difference between T_m of the reference strain and T_m of the hybrid DNA. Strains with >70% DDH similarity will have <5°C ΔT_m. The DDH obtained using the real-time methodology indicated that strains of the same species will have <4°C ΔT_m (Moreira *et al.*, 2011). The ΔT_m of *Vibrio harveyi* and *V. campbellii* was 5.2°C and ΔT_m of *V. coralliilyticus* and *V. neptunius* was 8.75°C.

Vibrio species contain two chromosomes: chromosome one is more conserved and contains genes for growth and viability, whereas chromosome two is more variable and contains genes for adaption to the environment (Okada *et al.*, 2005).

There are over 94 recognized species of *Vibrio*. For the diagnostic laboratory, identification of *Vibrio* species isolated from diagnostic samples is a challenge. Phenotypic identification using biochemical tests based on either fermentation of sugars (carbohydrates and alcohols), utilization of sugars or sole carbon source sugars, and detection of enzymes by hydrolysis of a specifc substrate is becoming more and more of a challenge as new species of *Vibrio* are being discovered. Many *Vibrio* species are closely related. Dorsch *et al.* (1992) identified a Vibrio core group (*V. harveyi, V. campbellii, V. rotiferianus, V. parahaemolyticus, V. alginolyticus* and *V. natriegens*), which has greater than 97.6% similarity according to 16S rRNA sequence information. Currently, genomics methods are being explored to try and overcome the shortfalls of phenotypic identification, which is complicated by the phenotypic diversity within a species as it constantly evolves to adapt to its environment.

PRIMERS FOR SEQUENCING THE 16S RDNA GENE FOR THE IDENTIFICATION OF VIBRIO SPECIES. High sequence homology exists within the 16S rRNA for all *Vibrio* species, and therefore sequencing part of the 16S rDNA is of little use when trying to identify to species level. The entire 16S rDNA must be sequenced for identification to species level. This can be achieved by amplifying the entire 16S rDNA with universal primers (Table 1.6). A standard master mix is used for the Universal PCR with standard cycling parameters and an annealing temperature of 55°C. A product of approximately 1.5 kilo base pairs (Kbp) is amplified. This amplified product is used as a template for eight sequencing primers (Table 1.7) (Thompson *et al.*, 2001b).

SEQUENCING THE HOUSEKEEPING GENES FROM VIBRIO SPECIES FOR TAXONOMIC IDENTIFICATION. The housekeeping genes in Table 1.8 have been used for amplification of *Vibrio* species followed

Table 1.6. Universal primers for amplification of 16S rDNA from *Vibrio* species.

Forward primer 5′–3′	Reverse primer 5′–3′	Reference
EUBB (7-26) AGAGTTTGATCMTGGCTCAG	EUB-A AAGGAGGTGATCCANCCRCA	Suzuki and Giovannomi, 1996; Weisburg *et al.*, 1991; Thompson *et al.*, 2001
MH1AGTTTGATCMTGGCTCAG	MH2 TACCTTGTTACGACTFCACCCCA	Thompson *et al.*, 2001

Table 1.7. Sequencing primers for the 16S rDNA from *Vibrio* species.

Primer name	E. coli position	Sequence 5′–3′	Reference
16F358	339–358	CTC CTA CGG GAG GCA GT	Modified from primer 357r from Lane, 1991
16F536	519–536	CAG CAG CCG CGG TAA TAC	Thompson *et al.*, 2001
16F926	908–926	AAC TCA AAG GAA TTG ACG G	Modified from primer 926f from Lane, 1991
16F1112	1093–1112	AGT CCC GCA ACG AGC GCA AC	Thompson *et al.*, 2001
16F1241	1222–1241	GCT ACA CAC GTG CTA CAA TG	Thompson *et al.*, 2001
16R339	358–339	ACT GCT GCC TCC CGT AGG AG	Modified from primer 342r from Lane, 1991
16R519	536–519	GTA TTA CCG CGG CTG CTG	Modified from primer 519r from Lane, 1991
16R1093	1112–1093	GTT GCG CTC GTT GCG GGA CT	Thompson *et al.*, 2001. Similar to 1100r Lane, 1991

F, forward primer; R, reverse primer.

Table 1.8. Sequencing primers for housekeeping genes from *Vibrio* species.

Gene target	Forward 5′–3′	Reverse 5′–3′	Reference
atpA	atpA-01-F=CTD AAT TCH CAN GAA ATY AGY G	atpA-04-R=TTA CCA RGW YTG GGT TGC-3'	Thompson *et al*., 2007
gyrB	gyrB_Vfmod.for=CGT TTY TGG CCR AGT G	gyrB.rev=TCM CCY TCC ACW ATG TA	Preheim *et al*., 2011
pyrH	pyrh-04-F=ATG ASN ACB AAY CCW AAA CC	pyrh-02-R=GTR AAB GCN GMY ARR TCC A	Pascual *et al*., 2010
recA	recA.for=TGG ACG AGA ATA AAC AGA AGG C	recA.rev=CCG TTA TAG CTG TAC CAA GCG CCC	Preheim *et al*., 2011
recA	recA-01-F=TGA RAA RCA RTT YGG TAA AGG position 222	recA-02-R=TCR CCN TTR TAG CTR TAC C position 1040	Thompson *et al*., 2005
recA	recA-03-F=TYG GBG TGA TGT TYG GTA ACC position 767	recA-04-R=GGG TTA CCR AAC ATC ACV CC position 769	Thompson *et al*., 2005
topA	topA_F=CAR CAA GCW CGT CGM TTC AT	topA_R=AGT TRG TYG AGT CRG TAC GC	Sawabe *et al*., 2007

recA-01-F and recA-02-R are used to amplify *recA* fragments from all *Vibrio* species (Thompson *et al*., 2005).
gyrB, DNA gyrase B subunit; *pyrH*, uridylate kinase; *recA*, recombination repair protein; *topA*, topoisomerase I.

by sequencing and analysis for identification to species level. See molecular chapter for further information.

Specific amplification and thermocycling conditions are given for some housekeeping genes as follows.

PCR type	Conventional PCR for amplification of partial *gyrB* gene and for amplification and identification of *Vibrio* species (Preheim *et al*., 2011)		
DNA extraction	From overnight plate culture		
PCR master mix			
Primer *gyrB*_Vfmod.for	CGT TTY TGG CCR AGT G		
Primer *gyrB*.rev	TCM CCY TCC ACW ATG TA		
Reagent	**Stock**	**Final concentration**	**Volume × 1**
Promega master mix	× 2	× 1	12.5 µl
Primer *gyrB*_Vfmod.for	50 µM	0.3 µM	0.15 µl
Primer *gyrB*.rev	50 µM	0.3 µM	0.15 µl
Water			7.2 µl
DNA 1 µg			5 µl
Total volume			**25 µl**

Thermocycling

Step 1	Step 2	Step 3
95°C, 3 min	95°C, 30 s	95°C, 30 s
	37–45°C, 30 s	37–45°C, 30 s
	72°C, 60 s	72°C, 5 min
× 1	× 30	×1

Specific annealing temperature: *gyrB*, 40°C; *recA*, 45°C

The *rpoB* gene from *Vibrio* species was amplified using the primers CM32b and 1110F followed by sequencing using degenerate primers 1661F and 1783R (Tarr *et al*., 2007).

PCR type	Amplification of the *rpoB* gene for *Vibrio* species followed by sequencing for species identification (Tarr *et al.*, 2007)		
DNA extraction	A colony from overnight growth is suspended in 200 µl 1 × Tris-EDTA, pH 8.0 and heated at 95°C for 10 min. The supernatant is used in the PCR reaction		
PCR master mix			
Primer 1110F (*rpoB*)	5′-GTA GAA ATC TAC CGC ATG ATG-3′		
Primer CM32b (*rpoB*)	5′-CGG AAC GGC CTG ACG TTG CAT-3′		
Sequencing primer 1661F	5′-TTY ATG GAY CAR AAC AAC CC-3′		
Sequencing primer 1783R	5′-GGA CCT TYA GGN GTT TCG AT-3′		
Reagent	**Stock**	**Final concentration**	**Volume × 1**
Buffer	× 10	× 1	5.0 µl
$MgCl_2$	25 mM	1.5 mM	3.0 µl
dNTP	10 mM	0.2 mM	1 µl
Betaine (Sigma-Aldrich)		1 mM	5 µl
Primer	10 µM	0.5 µM	2.5 µl
Primer	10 µM	0.5 µM	2.5 µl
Ampli*Taq* Gold (Applied Biosystems)	250 U/µl	2.5 U	0.5 µl
Water			29.0 µl
DNA			1.5 µl
Total volume			50 µl
Thermocycling	Touch down PCR		
Step 1	**Step 2**	**Step 3**	**Step 4**
93°C, 15 min	92°C, 40 s	92°C, 40 s	92°C, 40 s
	60–50°C, 1 min decreasing 1°C each cycle	50°C, 1 min	50°C, 1 min
	72°C, 1.5 min	72°C, 1.5 min	72°C, 7 min
× 1	(i.e. 11 cycles)	× 30	× 1
Product	984 bp		

The product is viewed after gel electrophoresis. Amplified product is purified (QIAquick, Qiagen) and then used in the sequencing reaction.

PCR type	Sequencing reaction for *Vibrio* species identification (Tarr *et al.*, 2007)		
DNA	Purified amplicon from *rpoB* gene amplification DTCS Quickstart sequencing kit (Beckman Coulter)		
PCR master mix			
Sequencing primer 1661F	5′-TTY ATG GAY CAR AAC AAC CC-3′		
Sequencing primer 1783R	5′-GGA CCT TYA GGN GTT TCG AT-3′		
Reagent	**Stock**	**Final Concentration**	**Volume ×1**
DTCS sequencing master mix (Beckman Coulter)			8 µl
Betaine (Sigma-Aldrich)	5 M	1 mM	4 µl
Primer	20 µM	0.1 µM	1 µl
Primer	20 µM	0.1 µM	1 µl
Water			1 µl
DNA (50 fmol)			5 µl
Total volume			20 µl
Thermocycling	**Step 1**	**Step 2**	
	96°C, 20 s	96°C, 20 s	

Continued

Continued.

	50°C, 20 s	50°C, 1 20 s
	60°C, 4 min	60°C, 4 min
	× 30	4°C hold
Send off for sequencing		

Twelve *Vibrio* species that were analysed by the *rpoB* gene could be distinguished from each other, however only two nucleotides were different between *V. harveyi* and *V. campbellii* (0.23%). Genetic distances for other species were *V. fluvialis* and *V. furnissii* 2.1%, *V. alginolyticus* and *V. parahaemolyticus* 3.2% and *V. cholerae* and *V. mimicus* 4.2%.

Vibrio aerogenes

TAXONOMY AND SIGNIFICANCE. *Vibrio aerogenes* (FG1^T=ATCC 700797^T=CCRC 17041^T) was isolated from seagrass sediment collected from Nanwan Bay, Taiwan (Shieh *et al.*, 2000). The highest level of similarity based on the 16S rRNA gene was *V. mytili* (96.2%). MLSA identified *V. aerogenes* as belonging to the Gazogenes clade, which also contains *V. gazogenes* and *V. ruber* (Sawabe *et al.*, 2007b). Members of this clade have >32% DDH and 46–47 GC mol%. *Vibrio mytili* belongs to the Harveyi clade.

GROWTH CHARACTERISTICS AND PHENOTYPIC IDENTIFICATION. *Vibrio aerogenes* is a facultatively anaerobic, motile, Gram-negative, slightly curved rod (2–3×0.6–0.8 μm) (Shieh *et al.*, 2000). Growth occurs on media containing NaCl but does not grow on BA. Growth occurs on MSA-B, MSSA, MA 2216 and TSA containing 2% NaCl. Beta-haemolysis is seen on MSA-B and MSSA (both are blood-containing media) with pinpoint colonies at 24 h, and light grey 1 mm colonies at 48 h. Colonies are green on TCBS. ONPG in the tube is a strong yellow colour. Motility is negative using a stab culture. Indole using tryptone water (see media section) is positive at 48 h and may need up to 18 drops of Kovács' reagent added to the tube to obtain a positive result. In conventional tube media fermentation of mannose and glucose are positive but the reaction is only seen in the top 1 cm of the tube. Indole is positive in the API 20E. On polypeptone yeast agar (used by Shieh), colonies are off-white, flat, circular and non-swarming after 1–2 days at 30°C. Luminescence is negative. Growth occurs at 20–35°C with weak growth seen at 15 and 38°C and no growth at 4 or 42°C. Growth does not occur in the absence of NaCl but occurs at 1–6% NaCl and is optimal at 4%. Na ions rather than Cl ions are required for growth. Results are positive for Thornley's ADH (Shieh *et al.*, 2000) but negative with Moeller's ADH. Fermentation is positive for cellobiose, and hydrolysis is positive for DNase (needs 2% NaCl added to the medium to obtain growth) and lipase and negative for agarase. Shieh *et al.* (2000) used a base of polypeptone yeast broth to prepare the tests for carbohydrate fermentation. Weak or negative results are obtained using a conventional base containing phenol red pH indicator (see results in Table 3.23, and recipe in the media section). Shieh *et al.* (2000) reported resistance to vibriostatic agent for the type strain (ATCC 700797, FG1) and a negative oxidase reaction, however strain DSMZ 14438 (FG1) is sensitive to vibriostatic agent at 10 and 150 μg and is strongly positive to oxidase using oxidase reagent (see interpretation of biochemical tests section). The catalase reaction is strongly positive. For strain DSMZ 14438 the nitrate reaction is positive in the tube test but may be negative in the API 20E even after the addition of zinc. These and other reactions are presented in Tables 3.23–26 and 3.38. An inoculum containing 2% NaCl is used for all identification sets and results are read at 48 h.

Vibrio aestivus

TAXONOMY AND SIGNIFICANCE. *Vibrio aestivus* (M22^T= CECT 7558^T=CAIM 1861^T and M61=CECT 7559) was isolated from seawater off the coast of Valencia, Spain (Lucena *et al.*, 2012). Based on 16S rRNA gene sequence *V. aestivus* is most closely related to *V. marisflavi* (97.1–97.2%),

V. variabilis (97.0–97.1%), *V. maritimus* (96.6–96.8%) and *V. stylophorae* (96.2%), but based on phylogenetic methods and other tree methods, it belongs to the Marisflavi clade and has the greatest similarity to *V. marisflavi* and *V. stylophorae*.

GROWTH CHARACTERISTICS AND PHENOTYPIC IDENTIFICATION. *Vibrio aestivus* is a facultatively anaerobic Gram-negative rod (1.0–2.0×0.5 μm) that is positive for oxidase and catalase. On MA 2216 colonies are 0.5–1.0 mm, beige, bright, translucent, round after 48 h at 25°C and after 5 days become tan-brown in colour. Colonies are green on TCBS. Growth occurs at 15–37°C but not at 4 or 40°C. Growth occurs in 0.85–7% NaCl, but not at 0% NaCl. Gelatin, casein, starch, Tween 80 and alginate are not hydrolysed (Lucena *et al.*, 2012). These and other reactions are presented in Tables 3.23, 3.30 and 3.38.

Vibrio aestuarianus

TAXONOMY AND SIGNIFICANCE. *Vibrio aestuarianus* (OY-0-002^T=ATCC 35048^T) was isolated from estuarine waters and shellfish (oyster, clam, crab) found in Oregon Bay, USA (Tison and Seidler, 1983). MLSA identified *V. aestuarianus* as belonging to the Anguillarum clade, which also contains *V. ordalii* and *V. anguillarum* (Sawabe *et al.*, 2007b). Members of this clade have >30% DDH, 43–46 GC mol% and a MLSA concatenated similarity of 88.9–98.6% and AAI of 95.7–99.6%.

Summer outbreaks of mortalities in oyster (*Crassostrea gigas*) in France were proven, under experimental conditions, to be due to phenotypically different strains of *V. aestuarianus*. It was proposed that the type strain be referred to as *V. aestuarianus* ssp. *aestuarianus* and the strain causing mortalities in oyster be referred to as *V. aestuarianus* ssp. *francensis* (02/041^T=LMG 24517=CIP 109791) (Garnier *et al.*, 2008). In *in vivo* pathogenicity tests virulent strains caused greater than 50% mortality in oysters when injected into the abductor muscle with 0.1 ml of bacterial inoculum resulting in 10^8 CFU per oyster (the concentration found in septicaemic oysters in the wild). Some strains gave a moderate mortality of 15–35%, and other strains including the type strain ATCC 35048 gave mortalities of <15% and were considered non-pathogenic. The major virulence factor appears to be a zinc metalloprotease (Vam) that has the ability to avoid the host's immune system through immunosuppression of the oyster haemocytes. A non-pathogenic strain of *V. splendidus* transformed with the *vam* gene became toxic to oysters. The deduced protein sequence of the *vam* gene has high homology to other metalloproteases involved in *Vibrio* pathogenicity (Labreuche *et al.*, 2010).

Mortalities of tongue sole (*Cynoglossus semilaevis* Günther) cultured in China were caused by *V. aestuarianus* ssp. *aestuarianus* (Zhang *et al.*, 2011). Clinical signs of disease included erratic swimming and loss of appetite, haemorrhages on the head, opercula and the fins. The abdomen was filled with ascitic fluid and focal areas of necrosis with haemorrhages were present in the liver. The organism was cultured from the liver, kidney and spleen. The strain was identified using the characteristics reported by Garnier *et al.* (2008). An isolate from the diseased fish was shown experimentally to cause infection with 100% mortality at 4–7 days using an inoculum of 10^6 CFU/ml per fish injected intraperitoneally.

Vibrio aestuarianus was co-isolated with *Nocardia crassostreae* from diseased Pacific oysters (*Crassostrea gigas*) in Lake Grevelingen in the Netherlands (Engelsma *et al.*, 2008). Disease signs were more typical of infection with *N. crassostreae* and infection with the vibrio was suspected to be secondary. The onset of mortalities was thought to be due to poor environmental conditions such as low oxygen and high water temperature.

A strain of *V. aestuarianus* identified as 15 was not pathogenic with an LD_{50} of $>10^8$ CFU/g body weight when cultured gilthead seabream (*Sparus auratus*) were challenged intraperitoneally (Balebona *et al.*, 1998b).

GROWTH CHARACTERISTICS AND PHENOTYPIC IDENTIFICATION. *Vibrio aestuarianus* is a facultatively anaerobic, motile, oxidase-positive, Gram-negative straight or slightly curved rod (1.5–2.0×0.5 μm). On BA and MSA-B *V. aestuarianus* ssp. *aestuarianus* colonies are 2 mm pale grey, non-swarming and with a large zone of beta-haemolysis at 24 h. At 48 and 72 h the zone of haemolysis is about 3 mm and a double zone of haemolysis is seen in areas of heavy growth. Colony size is up to 3.8 mm at 72 h. On MSSA colonies are pinpoint (<0.5 mm) at 24 h and 2 mm, shiny grey at 48 h.

On MA 2216, colonies are 1 mm cream at 24 h and 1.3 mm at 48 h and 1.6 mm at 72 h. On TCBS colonies are yellow, 1 mm at 24 h and 2 mm at 48 h. On MCA colonies are not seen at 24 h but are pinpoint and lactose positive at 48 h. Luminescence is not detected on BA, MSA-B, MSSA or MA 2216. Swarming does not occur on any of these media. Growth does not occur at 0 or 7% NaCl. Optimal growth occurs at 20–25°C and at 5 and 37°C but not at 42°C (Tison and Seidler, 1983). Differences in reactions were recorded for *V. aestuarianus* spp. *aestuarianus* depending on the method used. The original description by Tison and Seidler (1983) recorded positive reactions for ADH, gelatinase, fermentation of melibiose (77.8% of strains) and glycerol (88.9% of strains) using conventional media containing 1.5% NaCl. These reactions were reported as negative using the API 20E and API 50CH (Garnier *et al.*, 2008). Fermentation of glucose in the API 20E can be negative or weak. *Vibrio aestuarianus* spp. *aestuarianus* strain NCIMB 2236 is positive for ADH, gelatin and glucose at 24 h. Motility is negative by the tube method and in a wet preparation cells show a slow movement. ADH is negative at 24 h but positive at 48 h using Moeller's tube method. The reaction in Simmons' citrate is negative at 24 h but positive at 48 h. Citrate in the API 20E may show a weak reaction at 48 h. Gelatin shows a very weak hazy zone on the 3% NaCl agar plate. Aesculin using the tube method gives a weak reaction at 48 h but the colour does not darken over 72 h (this manual).

Cells of *V. aestuarianus* spp. *francensis* are motile, Gram-negative curved rods (0.4–0.5 × 1–1.5 μm). Colonies on MA 2216 are less than 1 mm, circular, slightly white and translucent after 24 h at 20°C (Garnier *et al.*, 2008). The strains from oysters do not grow on TCBS or at 6% NaCl (except strain 02/041) or at 35°C. Growth occurs at 4–30°C. Fermentation is positive for D-amygdalin and cellobiose (77.9%), and negative for dulcitol, raffinose and L-rhamnose. Some reactions are variable for *V. aestuarianus* spp. *francensis*. Many of the strains were oxidase-negative using oxidase discs from Sanofi-Pasteur. These and other reactions are presented in Tables 3.22, 3.24–26, 3.30 and 3.38. The commercial kits are inoculated using opacity of McFarland tube 1 and 2% NaCl and read at 24 and 48 h. The Zym is inoculated with a 2% NaCl to McFarland opacity tube 5 or 6 and read at 24 h after incubation at 25°C.

MOLECULAR IDENTIFICATION. *Vibrio aestuarianus* can be identified by sequencing the *gyrB* gene sequence using the universal primers for *gyrB* (Zhang *et al.*, 2011) or using a specific real-time PCR that targets a region in the *dnaJ* gene specific to *V. aestuarianus* that has been used to amplify DNA of the organism in cultures, seawater and oyster homogenate (Saulnier *et al.*, 2009).

DNA is extracted from colonies by the boiling method where one bacterial colony is emulsified in 100 μl of ultrapure water and heated for 10 min at 98°C then cooled to 4°C and 5 μl is used in the PCR reaction mixture.

For direct detection from seawater and oyster tissue (can be stored at −20°C), oyster tissue is crushed in sterile artificial seawater (2.3% NaCl, 20 mM KCl, 5 mM $MgSO_4$, 2 mM $CaCl_2$) and DNA is extracted from 200 μl using the QIAamp DNA minikit (Qiagen). DNA is resuspended in 10 ml of ultrapure water for oyster homogenate, and 200 μl for seawater samples. Re-suspending the oyster homogenate DNA at a more concentrated dilution causes inhibition of the PCR likely to be due to PCR inhibitors in the tissue.

The Brilliant QPCR Core Reagent kit (Stratagene) is used. A 25-μl reaction mixture contains 5 μl DNA, 5 mM $MgCl_2$, 20 μM of each dNTP, 300 nM of each primer, 200 nM of probe and 1.25 units of hot start Sure *Taq* polymerase.

The probe consists of an oligonucleotide dually labelled with a 5′-reporter dye Texas Red and a downstream 3′-quencher dye BHQ-2 (Eurogentec).

PCR type	Dual labelled hydrolysis probe PCR using a black hole quencher (Saulnier *et al.*, 2009)
DNA extraction	See text
PCR master mix	Brilliant QPCR Core Reagent kit (Stratagene)
Primer *DNAj* aes F1	5′-GTA TGA AAT TTT AAC TGA CCC ACA A-3′ 58°C T_m

Continued

Continued.

Primer *DNAj* aes R1	5′-CAA TTT CTT TCG AAC AAC CAC-3′ 59°C T_m		
DNAj probe	5′-TEXAS RED TGG TAG CGC AGA CTT CGG CGA C BHQ-2-3′ 71°C T_m		
Reagent	**Stock**	**Final concentration**	**Volume ×1**
Core PCR buffer	×10	×1	2.5 µl
$MgCl_2$	50 mM	5 mM	2.5 µl
dNTP mix	20 mM	20 µM	1.0 µl
SureStart *Taq* DNA polymerase	5 U/µl		0.25 µl
Primer *DNAj* aes F1	50 µM	300 nM	0.15 µl
Primer *DNAj* aes R1	50 µM	300 nM	0.15 µl
DNAj probe	10 µM	200 nM	0.5 µl
Water			12.95 µl
DNA			5 µl
Total volume			25 µl
Thermocycling	MX3000 Stratagene thermocycler		
Step 1	**Step 2**		
95°C, 10 min	95°C, 15 s		
	60°C, 90 s		
×1	×40		
Ct (or Cq) value	The threshold is set using the algorithm from the thermocycler software or set manually based on the Ct at which the increase above the base line is detected		
Sensitivity	1.6×10^2 cells/ml for seawater and 1.6×10^2 cells/mg for oyster tissue		
PCR specific against:	*Vibrio brasiliensis* (DSMZ 17184ᵀ), *V. chagasii* (LMG 21353ᵀ), *V. coralliilyticus* (CIP 107925ᵀ), *V. crassostreae* (LGP 7ᵀ), *V. cyclitrophicus* (LMG 21359ᵀ), *V. ezurae* (DSMZ 17533ᵀ), *V. fortis* (LMG 20547ᵀ), *V. gallicus* (DSMZ 16639ᵀ), *V. gigantis* (LGP 13ᵀ), *V. kanaloe* (LMG 20539ᵀ), *V. lentus* (CIP 107166ᵀ), *V. mediterranei* (DSMZ 19502ᵀ), *V. pacinii* (LMG 19999ᵀ), *V. pelagius* (LMG 3897ᵀ), *V. penaeicida* (DSMZ 14398ᵀ), *V. pomeroyi* (LMG 20537ᵀ), *V. rotiferianus* (DSMZ 17186ᵀ), *V. shilonii* (DSMZ 13774ᵀ), *V. splendidus* (LMG 4042ᵀ), *V. superstes* (DSMZ 16383ᵀ), *V. tapetis* (CECT 4600ᵀ), *V. tasmaniensis* (LMG 20012ᵀ), *V. xuii* (DSMZ 17185ᵀ)		

CAL Fluor Red 610 can be used instead of Texas Red.

Amplification and sequencing of the *gyrB* gene sequence using universal primers for *gyrB* are presented as follows (Zhang *et al.*, 2011).

PCR type	Amplification of *gyrB* gene using universal primers followed by sequencing (Zhang *et al.*, 2011)
DNA extraction	
PCR master mix	
Universal primer *gyrB* UP1	5′-GAA GTC ATC ATG ACC GTT CTG CAY GCN GGN GGN AAR TTY GA-3′[a]
Universal primer *gyrB* UP2r	5′-AGC AGG GTA CGG ATG TGC GAG CCR TCN ACR TCN GCR TCN GTC AT-3′[a]

Continued

Continued.

Reagent	Stock	Final concentration	Volume ×1
PCR buffer	×10	×1	2.0 µl
$MgCl_2$	20 mM	1.5 mM	1.6 µl
dNTP mix	20 mM	1.5 mM	0.4 µl
Taq DNA polymerase	2.5 U/µl		0.2 µl
Primer UP1	50 µM	0.5 µM	0.2 µl
Primer UP2r	50 µM	0.5 µM	0.2 µl
Water			14.4 µl
DNA			2 µl
Total volume			20 µl
Thermocycling	Thermocycler not stated		
Step 1	**Step 2**		**Step 3**
95°C, 5 min	94°C, 1 min 57°C, 1 min 72°C, 2 min		94°C, 1 min 57°C, 1 min 72°C, 7 min
×1	×30		×1
Product	1200 bp		
Similarity	98–99% similarity to *V. aestuarianus*. 82% similarity to *gyrB* gene of *Photobacterium kishitanii*, *V. anguillarum*		

[a]For degeneracy codes see molecular section

PCR type	Conventional PCR targeting 16S rRNA gene specific to *V. aestuarianus* (Azandégbé *et al*., 2010)		
DNA extraction	SDS (sodium dodecyl sulfate) lysis method		
PCR master mix			
Primer Vaest 64-99 dir	5′-GAG CGG TTA CAT TTC AAA ACG TTG CTT TTG AAG ATG-3′		
Primer Vaest 664-683rev	5′-CCT CTA CAG TAC TCT AGT CC-3′		
Reagent	**Stock**	**Final concentration**	**Volume ×1**
PCR buffer	×10	×1	2.0 µl
$MgCl_2$	25 mM	1.5 mM	1.5 µl
dNTP mix	10 mM	4.0 mM	10 µl
Ampli*Taq* DNA polymerase	250 U	1 U	0.1 µl
Primer Vaest 64-99 dir	50 µM	2.0 µM	1.0 µl
Primer Vaest 664-683rev	50 µM	2.0 µM	1.0 µl
Water			6.9 µl
DNA (10 ng)			2 µl
Total volume			25 µl
Thermocycling	BioRad thermocycler		
Step 1	**Step 2**		**Step 3**
95°C, 5 min	95°C, 30 s 65°C, 60 s 72°C, 1 min 30 s		95°C, 30 s 65°C, 60 s 72°C, 7 min
×1	×30		×1
Product	Not reported		
Sensitivity	Limit of detection = 10^3 CFU/g		
Specificity	Not reported		

PCR products are purified using the Wizard PCR Preps kit (Promega) before sending for sequencing.

Vibrio agarivorans

TAXONOMY AND SIGNIFICANCE. *Vibrio agarivorans* (289^T=CECT 5085^T=DSM 13756^T) was isolated from seawater in the Mediterranean, Spain (Macián *et al.*, 2001b). It is most closely related to *V. fischeri* (96.5%), *Aliivibrio logei* (95.2%) and *V. salmonicida* (94.8%) based on 16S rRNA sequencing. *Vibrio agarivorans* was isolated from abalone cultured off the south-west coast of Western Australia. The pathogenicity is not known.

GROWTH CHARACTERISTICS AND PHENOTYPIC IDENTIFICATION. *Vibrio agarivorans* is a facultatively anaerobic, motile, oxidase-positive, Gram-negative rod (2–4×0.4–0.6 μm). On MA 2216 and TSA containing 2% NaCl colonies are unpigmented and regular in shape and are green on TCBS (Macián *et al.*, 2001b). Luminescence is not seen. Growth occurs at 20–37°C. Growth does not occur in the absence of NaCl and up to 2% NaCl must be added to culture media and to identification tests. Results are negative for both Moeller's and Thornley's ADH. Agar is hydrolysed (see colour plate section) and on MSA-B and MSSA small craters develop under the colony and become larger over time. Agar digestion is slower on MA 2216. Initially, agar hydrolysis is seen as small depressions on an agar plate after 3 days incubation at 25°C. This effect increases over time, and by about 7 days the effect is very marked. This was particularly the case with isolates from abalone in Western Australia. The carbohydrate fermentation reactions may be difficult to read and this effect has been noticed with all strains tested. All carbohydrate fermentation tubes should be compared alongside the glucose reaction, as this will be a bright yellow colour. The positive reactions are easy to determine, but those that appear as a weak reaction are in fact negative reactions. These may appear as a pale yellow colour in the middle of the tube. However, after 48 and 72 h incubation the reactions are clearly negative. *Vibrio agarivorans* strains first isolated and reported from the Mediterranean Sea are positive for aesculin, and a negative reaction for gelatin (Macián *et al.*, 2001b). Strains from Western Australia are negative for aesculin using the conventional tube test, and positive when the plate method is used. However, the gelatin result in the API 20E is negative after 48 h incubation at 25°C. These and other reactions are presented in Tables 3.23–26 and 3.38.

Vibrio albensis

Vibrio albensis is a synonym of *V. cholerae*.

Vibrio alfacsensis

TAXONOMY AND SIGNIFICANCE. *Vibrio alfacsensis* (CAIM 1831^T=DSM 24595^T) was isolated from gills, kidney and skin of healthy sole (*Solea senegalensis*) cultured in Alfacs Bay, Spain, from liver and kidney of spotted rose snapper (*Lutjanus guttatus*) caught off the coast of Mexico, and from coral (*Mussismilia hispida*) off the coast of Brazil (Gomez-Gil *et al.*, 2012). The strains of *V. alfacsensis* are most closely related to *V. ponticus* (98.2–98.3%), *V. furnissii* (97.2–97.3%) and *V. fluvialis* (96.9–97.1%), *V. orientalis* (97.04%) and *V. hepatarius* (97.0%). MLSA using concatenated sequences from *gyrB*, *pyrH*, *recA*, *topA* and the 16S rRNA genes demonstrated similarity to *V. orientalis* and *V. hepatarius* rather than *V. ponticus*.

GROWTH CHARACTERISTICS AND PHENOTYPIC IDENTIFICATION. *Vibrio alfacsensis* is a facultatively anaerobic, Gram-negative coccobacillus (2.09×0.86 μm) that is catalase positive and oxidase positive. Growth occurs at 0.5–7.0% NaCl with an optimum of 2.5–5.0%. Growth occurs at 20–37°C with variable growth between strains at 8°C, and no growth at 40°C. Optimal temperature for growth is 30–37°C. The five strains from sole are 1–5 mm, yellow, smooth and umbonate on TCBS, whereas the two strains from spotted rose snapper and the one strain from coral are 3–5 mm, green, raised, circular with undulated margins on TCBS. Growth occurs on MA 2216 agar as small circular translucent colonies. ADH is negative for all strains using Moeller's method, and is positive in the API 20E. These and other reactions are presented in Tables 3.23, 3.24–26 and 3.38.

Vibrio alginolyticus

TAXONOMY AND SIGNIFICANCE. *Vibrio alginolyticus* (XII-53^T=ATCC 17749^T) was isolated in 1955 by Dr I. Takikawa from faeces of patients admitted to the National Yokohama Hospital, Japan, with acute gastroenteritis. The organism was named *Pseudomonas enteritidis* and later renamed *Oceanomonas alginolytica* synonym *P. enteritis* Takikawa's serotype XII (Miyamoto *et al.*, 1961), although the proposal for the genus *Oceanomonas* was never accepted (Sakazaki *et al.*, 1963). A study of further food-poisoning outbreaks in Japan identified two biotypes of *V. parahaemolyticus* (Sakazaki *et al.*, 1963), but subsequently biotype 2 was named *V. alginolyticus* (Sakazaki, 1968). Baumann *et al.* (1971a) placed it in the genus *Beneckea*, however with the abolition of the genus *Beneckea* the name reverted to *Vibrio alginolyticus* (Baumann *et al.*, 1980).

MLSA identified *V. alginolyticus* as belonging to the Harveyi clade, which also contains *V. harveyi*, *V. campbellii*, *V. mytili*, *V. natriegens*, *V. parahaemolyticus* and *V. rotiferianus* (Sawabe *et al.*, 2007b). Members of this clade have >25% DDH and 42–48 GC mol%, a MLSA concatenated sequence similarity of 90.1–96.2% and AAI of 97.2–99.4%. Strains within the Harveyi clade (*V. alginolyticus*, *V. campbellii*, *V. harveyi*, *V. natriegens*, *V. parahaemolyticus*, *V. rotiferianus*) can be discriminated using concatenated sequences from three housekeeping genes, *rctB*, *rpoD* and *toxR*, with *toxR* and *rpoD* providing the most discrimination (Pascual *et al.*, 2010) (see under *Vibrio harveyi*). *Vibrio alginolyticus* has 99.7% homology with the 16S rRNA gene of *V. parahaemolyticus* (Ruimy *et al.*, 1994) and for a 1451-bp sequence there are only 5 bp differences between the two organisms compared to 166 bp differences for a 1258-bp sequence of the *gyrB* gene relating to 86.8% homology (Venkateswaran *et al.*, 1998).

Vibrio alginolyticus (strain Swy) caused septicaemia in kuruma prawns and its pathogenic status to both kuruma and tiger prawns was proven in experimental challenge with an LD_{50} of 4.43×10^4 and 1.57×10^5 CFU/g body weight, respectively (Lee *et al.*, 1996). The disease was referred to as white spotted syndrome due to white spots on the carapace, and has occurred in kuruma prawns cultured in Taiwan since 1992. The organism was cultured from the hepatopancreas and haemolymph. Prawns (*Penaeus monodon*) carrying the white spot syndrome virus (WSSV) may be more susceptible to bacterial infection. In India, cultured *P. monodon* had a secondary bacterial infection caused by *V. alginolyticus* (strain QS7) in which infected prawns had a reddish body colouration due to the bacterial infection plus white spots in the cuticle due to WSSV (Selvin and Lipton, 2003). The LD_{50} in experimental challenge was 5×10^6 CFU/shrimp. At 10^8 CFU/shrimp 100% mortality occurred at 24 h, whereas at 10^5 CFU/shrimp no mortalities were seen within 7 days. In stage 9 larvae of the giant freshwater prawn *Macrobrachium rosenbergii* (de Man), infection with *V. alginolyticus* caused anorexia, with resulting poor growth and inactivity, and necrotic appendages. In experimental infection using strain MRNL-3, 80% mortality occurred with a challenge of 10^6 CFU/ml, 87% with 10^7 CFU/ml and 100% mortality with a challenge of 10^8 CFU/ml (Jayaprakash *et al.*, 2006). *Vibrio alginolyticus* caused infection outbreaks in white shrimp, *Litopenaeus vannamei*, cultured in Taiwan (Liu *et al.*, 2004). Animals showed anorexia, poor growth and inactivity and had reddish pleural borders on antennae, uropods and telson and musculature was white or pale. In experimental challenge the LD_{50} at 96 h was 3.0×10^5 CFU/animal. Identification of *V. alginolyticus* (strain CH003) was confirmed by 16S rRNA sequencing with 99.9% sequence similarity to ATCC 17749. Biochemical tests using API 20E was the same for strain CH003 and ATCC 17749.

Vibrio alginolyticus (ATCC 17749) caused mortalities in larvae and 4-day old post-larvae of red abalone (*Haliotis rufescens*) (Anguiano-Beltrán *et al.*, 1998). In experimental challenge, deaths occurred at 24 h using a concentration of 10^5 cells/ml for larvae and at a concentration of 10^6 cells/ml for post-larvae. Clinical signs included inability to swim with larvae remaining on the bottom of the container. Post-larvae showed weak foot attachment. Disease outbreaks in post-larval abalone (*Haliotis diversicolor supertexta*) cultured in southern China were due to *V. alginolyticus*, which had an LD_{50} of 1×10^4 CFU/ml (Cai *et al.*, 2006b).

In one outbreak, *V. alginolyticus* caused 62% mortality in carpet shell clam (*Venerupis* [*Ruditapes*] *decussata*) cultured in Spain and 73% mortality in a second outbreak in which *V. splendidus* biovar II was also cultured (Gómez-León *et al.*, 2005). Pathogenicity was proven in experimental challenge. Carpet shell clam (6 mm in length) were inoculated by immersion challenge and by intravalvar injection using 1×10^6 CFU/ml and 20 μl of 5×10^6 CFU/ml, respectively. *V. alginolyticus* (strain TA15) was slightly more virulent than

V. splendidus II (strain TA2) with 60 and 48% mortality, respectively. *Vibrio alginolyticus* caused mortality in scallop (*Argopecten purpuratus*) larvae cultured in Chile and was isolated from swimming larvae, seawater of larval tanks, a reservoir tank of filtered (50 μm) seawater and microalgae (*Chaetoceros* sp. and *Isochrysis galvana*). Other pathogens cultured included *Aeromonas hydrophila* and *V. splendidus* (Riquelme *et al.*, 1996). Only *A. hydrophila* was cultured from the dead larvae on the bottom of the tank. All were pathogenic at concentrations of 5.5×10^4 to 5.5×10^2 cells/ml in experimental challenge. Strains of *V. alginolyticus* cultured from swimming larvae were not pathogenic in experimental challenge. *Vibrio splendidus* was cultured from the water pipe walls and a reservoir tank of 10 μm-filtered seawater.

In captive seahorses (*Hippocampus guttulatus* and *H. hippocampus*) in Spain, *V. splendidus* and *V. alginolyticus* were the predominant flora on mucosal surfaces (Balcázar *et al.*, 2010a). Seahorses had white spots on the skin and were lethargic and some had necrotic tail lesions.

A number of disease outbreaks in cultured large yellow croaker (*Larimichthys crocea*, synonym *Pseudosciaena crocea*) in China have been attributed to *V. alginolyticus*, strain ND-01 (Chen *et al.*, 2008). The route of infection is through the intestine, particularly the foregut where the organism adheres to intestinal mucus. *Vibrio alginolyticus* causes disease outbreaks in cultured gilthead seabream (*Sparus aurata*) in Spain with clinical signs of dark skin and in some cases skin ulcers, septicaemia, haemorrhages, accumulation of fluid in the peritoneal cavity and haemorrhagic liver may be seen (Balebona *et al.*, 1998a; Balebona *et al.*, 1998b). The organism was isolated from spleen, liver and kidney. In experimental challenge the LD_{50} for fish challenged intraperitoneally was 5.4×10^4 to 1.0×10^6 CFU/ml. Fish with skin damage were more susceptible to infection. In mass mortalities of cage-cultured goldlined seabream (*Rhabdosargus sarba*, also known as *Sparus sarba* and silver seabream) in Hong Kong, infection with *V. alginolyticus* resulted in tail and fin rot, extensive haemorrhages and ulcerations on the body (Li *et al.*, 2003). Pathogenicity was confirmed in experimental challenge using different inoculation routes. Intraperitoneally injected fish died within 2–5 days post-injection, whereas mortality in skin-injured fish challenged by immersion was delayed by 3–4 days. The extracellular product from *V. alginolyticus* was pathogenic when injected into juvenile silver seabream with a resulting 14-day LD_{50} of 0.92 μg/g fish. Disease is also seen in gilthead seabream and seabream (*Dicentrarchus labrax*) cultured on the Tunisian Mediterranean (Ben Kahla-Nakbi *et al.*, 2009). *Vibrio alginolyticus* (strain E333) caused deaths in yellow grouper (*Epinephelus awoara*) inoculated with 100 μl of 10^5 CFU/ml. Clinical signs included fading of skin colour, petechiae or ecchymoses on the skin, some ulceration, and swelling of liver and kidney (Jing-jing *et al.*, 2011). Mortality outbreaks in juvenile cobia (*Rachycentron canadum*) cultured in Taiwan have been due to *V. alginolyticus* (Liu *et al.*, 2004a). Clinical signs include lethargy, eye damage in some fish, dark skin and ascites fluid. Strain C3c01 had an LD_{50} of 3.28×10^4 CFU/g body weight. The organism was isolated from ascites fluid and kidney.

Vibrio alginolyticus may form the predominant flora in *Artemia* reared for fish food but usually does not cause mortality in *Artemia* (Thomson *et al.*, 2005).

The pathogenesis of *V. alginolyticus* is not well understood. *Vibrio alginolyticus* may carry several potential virulence factors such as ability to adhere to mucus, outer membrane proteins (OMPs), production of extracellular products (ECPs), thermostable and thermolabile haemolysin, and may carry virulence genes such as those found in *V. cholerae*. In a study using *V. alginolyticus* strains isolated from diseased gilthead seabream and seabream cultured in Tunisia, strains varied in the type of virulence factors present. Of 34 isolates tested, 50% were positive for *toxS*, 47% positive for *toxR*, 35% positive for the *V. cholerae* virulence pathogenicity island (VPI) and 20% were positive for the *ace* gene (Ben Kahla-Nakbi *et al.*, 2009). Only two (6%) strains were positive for all four virulence genes. Four strains tested in experimental challenge had one or more of these genes, however two strains without these virulence factors were pathogenic in experimental challenge. These results were similar to the findings for *V. harveyi* where no correlation was found between the presence of toxin genes and pathogenicity in experimental challenge (Bai *et al.*, 2008; Ruwandeepika *et al.*, 2010). A study using live organisms and extracellular protein (ECP) indicated up-regulation of heat shock protein hsp70 in seabream (*Rhabdosargus sarba*, also known as *Sparus sarba*) challenged with both components, but up-regulation of hsp90 with ECP only, and no change in transcription levels of hsp60 (Deane *et al.*, 2004). A multiplex PCR for the detection of virulence genes collagenase and *toxR* differentiated between virulent strains from diseased fish and avirulent isolates from seawater (Cai *et al.*, 2009). Virulent isolates had both genes present, whereas avirulent isolates had none or one of either gene (see under molecular identification). The *ompK* gene was also tested but it did not appear to correlate to virulence. In the study the presence of both the collagenase gene and the *toxR* gene and the relationship to virulence was not confirmed by experimental challenge in fish, although challenged fish were used to assess the detection limit. Further work needs to be done to confirm these findings. Sequence differences in a 470-bp region of the *V. alginolyticus* collagenase gene was analysed using single strand conformation polymorphism (SSCP). PCR-SSCP identified 13 genotypes (A1, A2, A3, A4, B, C, D, E, F, G1, G2, H, I) from 78 isolates. Genotype A1 was predominant (48.1%). In experimental challenge in red snapper (*Lutjanus sanguineus*) genotypes A4, D, E, G2 and H were highly virulent and genotypes A1, A2, A3, B, C, F, G1 and I were of low virulence when injected at a concentration of 1×10^7 CFU/ml (Yishan *et al.*, 2011). PCR-SSCP analysis of the *gyrB* gene identified four genotypes from the 78 strains. Sequence information from a 1167-bp region of the *gyrB* gene from 11 strains of *V. alginolyticus* identified 54 bp variations (Luo and Hu, 2008).

GROWTH CHARACTERISTICS AND PHENOTYPIC IDENTIFICATION. *Vibrio alginolyticus* is a facultatively anaerobic, Gram-negative rod. On BA individual colonies (ATCC 17749^T) may not show swarming at 24 h but swarming develops over time. Some swarming may be seen in the pool. At 24 h and 25°C growth swarms over the plate, completely covering the agar surface. Swarming is seen on MSA-B and MSSA plates as grey, shiny growth. Swarming is not seen on MA 2216 plates, and at 24 h colonies are 1–1.5 mm and slightly mucoid in parts especially in the pool and at 48 h are 2 mm and cream. Colonies are 1–2 mm at 24 h, and 3 mm and yellow after 48 h on TCBS. The sucrose in the media may be rapidly fermented and colonies may appear green due to a pH reversal after 48 h or prolonged incubation, however even at 24 h some greenish colonies may be seen, therefore it is important to record the colour reaction on TCBS within 24 h. Growth does not occur on MCA. A selective medium (VAL agar) using 9% NaCl and incubation at 37°C was developed to detect *V. alginolyticus* and to retard growth of many other *Vibrio* species. Some strains of *V. proteolyticus* will grow on this agar as blue colonies and *Staphylococcus epidermidis* produces green-yellowish colonies similar to *V. alginolyticus* (Chang *et al.*, 2011). When the efficiency of detection of VAL for *V. alginolyticus* was compared to MA 2216 and TCBS, VAL detected fewer *V. alginolyticus* colonies compared to the other two media. The number of colonies on MA 2216 was 2.83×10^8 CFU/ml, 2.59×10^7 CFU/ml on TCBS and 1.27×10^7 CFU/ml on VAL media. These findings are consistent for use of selective media where a reduced number of colonies are seen on a selective medium compared to a general purpose medium. The method for VAL can be found in the media section. A selective medium should always be used in conjunction with a general purpose medium. Growth occurs in media containing 0.85% NaCl to 10% NaCl and at a temperature range of 22–42°C.

Strain ATCC 17749^T is positive in ODC and LDC using Moeller's medium. Strain ATCC 17749^T and the strain CH003 from white shrimp cultured in Taiwan were both negative for ODC in the API 20E (Liu *et al.*, 2004a; Cai *et al.*, 2006b). Different reactions have been reported for ATCC 17749^T using the API 20E, and this may be due to different inoculum concentrations of NaCl, incubation temperature and time for incubation. *Vibrio alginolyticus* strains (DPI-HE-4, CAN, AO35) from seabream were negative in the ODC using conventional media (Balebona *et al.*, 1998a). *Vibrio alginolyticus* is positive for hydrolysis of alginate and chitin and is able to grow on deoxycholate agar (Miyamoto *et al.*, 1961).

Vibrio alginolyticus is in the database of the Vitek 2 as it may cause disease in humans. Some isolates from aquatic species may not be identified as *V. alginolyticus* as the reactions are affected by reduced saline, 0.085% as used by the Vitek and by incubation at 37°C. Reactions that may give a false result are ODC, LDC, glucose and others. When compared with conventional tube media the colour change for ODC and LDC is less defined when incubated at 37°C compared to incubation at 25°C. Hydrolysis is positive for gelatin, casein, starch, DNA, aesculin and Tween 40, 60 and 80 but negative for Tween 20 and chitin. These and other reactions are presented in Tables 3.22, 3.24–26 and 3.38.

MOLECULAR IDENTIFICATION. A PCR using the *rpoD* and *toxR* housekeeping genes followed by sequencing and analysis using concatenated sequences can be used to enable differentiation between strains comprising the Vibrio core group (Pascual *et al.*, 2010) (see method under *Vibrio harveyi*). Amplification and analysis of a partial sequence of the *toxR* gene enables differentiation between *V. alginolyticus* and *V. parahaemolyticus* as these two species share 99.8% sequence similarity in the 16S rRNA sequence but only 61.7% similarity in a partial *toxR* sequence (Montieri *et al.*, 2010).

PCR type	PCR and sequence analysis based on the *toxR* gene for differentiation of *V. alginolyticus* and *V. parahaemolyticus* (Montieri *et al.*, 2010)		
DNA extraction	Cultured cells: mix 1 colony in 200 µl sterile distilled water. Boil for 3 min. Centrifuge at 12,000 ×*g* for 5 min		
PCR master mix			
Primer VaVp-ToxRf	5′-GAG CAR GGK TTY GAG GTG G-3′		
Primer VaVp-ToxRr	5′-CRC CWG TGG CRA TYA CYT C-3		
Reagent	**Stock**	**Final concentration**	**Volume ×1**
Buffer (MBI Fermentas)	×10	×1	5.0 µl
dNTP mix	10 mM	2.0 mM	1.0 µl
$MgCl_2$	25 mM	2.0 mM	4.0 µl
Primer VaVp-ToxRf	50 µM	0.10 µM	0.1 µl
Primer VaVp-ToxRr	50 µM	0.10 µM	0.1 µl
Taq DNA polymerase (MBI Fermentas)	250 U	1.0 U	0.2 µl
Bovine serum albumin	20 mg/ml	0.1 mg/ml	0.25 µl
Water			34.35 µl
DNA (50 ng final)			5 µl
Total volume			50 µl
Thermocycling			
Step 1	**Step 2**		**Step 3**
94°C, 2 min	95°C, 1 min		95°C, 1 min
	56°C, 1 min		56°C, 1 min
	72°C, 1.5 min		72°C, 5 min
×1	×35		×1
Product	570 bp		
Sensitivity	Not reported		
Specific against	Not reported		

A number of conventional, real-time and LAMP PCR methods have been reported for the detection of *V. alginolyticus*.

A conventional PCR designed for the *gyrB* gene was used and compared for sensitivity using the LAMP PCR. The LAMP PCR (see further below) was 10 times more sensitive than the conventional PCR (Cai *et al.*, 2010). The conventional PCR detected *V. alginolyticus* in tissues from experimentally challenged humphead snapper (*Lutjanus sanguineus*) and gave the same detection rates as the LAMP PCR.

PCR type	Conventional PCR based on the *gyrB* gene for detection of *V. alginolyticus* (Cai *et al.*, 2010)
DNA extraction	Cultured cells: 1 colony mixed in 200 µl sterile distilled water, boiled for 3 min and centrifuged at 12,000 ×*g* for 5 min
PCR master mix	Fish tissue: QIAamp DNA Mini kit (Qiagen)

Continued

Continued.

Primer Va-PF	5′-CTG TTG TGG GTG ATA CGG-3′		
Primer Va-PR	5′-TTC GCT TCT GTC GGG TTC-3		
Reagent	**Stock**	**Final concentration**	**Volume ×1**
Buffer for Ex*Taq* (TaKaRa) containing $MgCl_2$	×10	×1	2.5 µl
dNTP mix	25 mM	2.5 mM	2.5 µl
Primer Va-PF	50 µM	0.10 µM	0.05 µl
Primer Va-PR	50 µM	0.10 µM	0.05 µl
Ex*Taq* DNA polymerase (TaKaRa)	250 U	1.25 U	2.5 µl
Water			15.4 µl
DNA			2 µl
Total volume			25 µl
Thermocycling	DNA thermocycler 9700 (Applied Biosystems)		
Step 1	Step 2		Step 3
94°C, 4 min	94°C, 60 s		94°C, 60 s
	55°C, 60 s		55°C, 60 s
	72°C, 90 s		72°C, 4 min
×1	×30		×1
Product	Not given, but based on gel photo in paper, approximately 600 bp		
Sensitivity	3.7×10^3 CFU/ml (37 CFU/reaction)		
Specific against	Not stated		

PCR type	Conventional PCR based on the *gyrB* gene for detection of *V. alginolyticus* (Luo and Hu, 2008)		
DNA extraction	Overnight culture in Alkaline Peptone Water (APW, see media section) at 30°C followed by extraction using the MiniBest Bacterial genomic DNA extraction kit (Takara Biotechnology)		
PCR master mix			
Primer AlgF1	5′-TCA GAG AAA GTT GAG CTA ACG ATT-3′		
Primer AlgR1	5′-CAT CGT CGC CTG AAG TCG CTG T-3		
Reagent	**Stock**	**Final concentration**	**Volume ×1**
Buffer for *Taq* (TaKaRa)	×10	×1	2.5 µl
dNTP mix	10 mM	0.2 mM	0.5 µl
$MgCl_2$	25 mM	2 mM	2 µl
Primer AlgF1	50 µM	0.4 µM	0.2 µl
Primer AlgR1	50 µM	0.4 µM	0.2 µl
Taq DNA polymerase (TaKaRa)	250 U	1 U	0.1 µl
Water			18.5 µl
DNA			1 µl
Total volume			25 µl
Thermocycling			
Step 1	**Step 2**		**Step 3**
94°C, 4 min	94°C, 30 s		94°C, 30 s
	64°C, 30 s		64°C, 30 s
	72°C, 60 s		72°C, 8 min
×1	×32		×1
Product	568 bp		
Sensitivity	Not given		

Continued

Continued.

Specific against	*Vibrio* (*Listonella*) *pelagius* (CGMCC 1.1588), *Photobacterium damselae* (A0508), *V. anguillarum* (APCCC Z0512), *V. campbellii* (CGMCC 1.1597), *V. cholerae* non-01 (EN SC0312), *V. fluvialis* (CGMCC 1.1609), *V. furnissii* (ATCC 33841), *V. harveyi* (EN SD0613, CGMCC 1.1593), *V. mimicus* (CGMCC 1.1969), *V. natriegens* (CGMCC 1.1594), *V. nereis* (CGMCC 1.1623), *V. parahaemolyticus* (ATCC 17802), *V. proteolyticus* (CGMCC 1.1826), *V. splendidus* (CGMCC 1.1606), *V. vulnificus* (CGMCC 1.1758), *Bacillus subtilis*, *Aeromonas hydrophila* (APCCC 0388), *A. sobria* (APCCC 0398), *Edwardsiella tarda* (APCCC S1), *Lactobacillus plantarum* (ATCC 8014), *Proteus mirabilis* (E06119), *Pseudoalteromonas* sp. (EN GS0432), *Pseudomonas aeruginosa* (APCCC 1.50), *Pseudomonas fluorescens* (ATCC 10646), *Rhodococcus* sp., *Salmonella* Typhimurium, *Shewanella algae* (EN YJ06114), *Staphylococcus aureus* (ATCC 6538)

A multiplex PCR designed to detect the collagenase and the *toxR* gene differentiated between virulent and avirulent isolates of *V. alginolyticus* sourced from diseased fish and prawns, and from seawater off the coast of China (Cai *et al.*, 2009). Virulent strains possessed both genes (collagenase and *toxR*) whereas non-virulent strains had one only or none of these genes. Virulence did not appear to be confirmed in experimental challenge and the authors suggest that because the isolates were sourced from one geographical location the findings may not correlate with other geographical areas. The test requires a pre-enrichment step in alkaline peptone water (APW) to improve the sensitivity of detection (West and Colwell, 1984). The test is not suitable as a diagnostic test for the identification of *V. alginolyticus* because not all strains possessed the genes.

PCR type	Multiplex PCR for detection of virulent strains of *V. alginolyticus* (Cai *et al.*, 2009)		
DNA extraction	Cultured cells: 1 colony mixed in 200 µl sterile distilled water, boiled for 3 min and centrifuged at 12,000 ×*g* for 5 min.		
	Fish tissue: QIAamp DNA Mini kit (Qiagen).		
Pre-enrichment	Homogenize 1 g of fish tissue in 10 ml APW and incubate at 28°C for 6–12 h. Centrifuge the broth and re-suspend the pellet in 1 ml of saline. Use 10 µl as template DNA in the multiplex PCR		
PCR master mix			
Primer col-F	5′-TGG TGA ACA GCC AGT AAA-3′		
Primer col-R	5′-CAA ACC CAT CAT AAG TAG TC-3		
Primer toxR-F	5′-CAG AAG AAT CGG AAG AAC A-3′		
Primer toxR-R	5′-TAG AAT GAC GCA CAA AGG-3′		
Reagent	**Stock**	**Final concentration**	**Volume ×1**
Buffer for Ex*Taq* (TaKaRa)	×10	×1	5.0 µl
dNTP mix	2.5 mM	250 µM	5.0 µl
Primer col-F	50 µM	0.5 µM	0.5 µl
Primer col-R	50 µM	0.5 µM	0.5 µl
Primer toxR-F	50 µM	0.5 µM	0.5 µl
Primer toxR-R	50 µM	0.5 µM	0.5 µl
Ex*Taq* DNA polymerase (TaKaRa) 5U/µl	250 U	2 U/50 µl	0.4 µl
Water			35.6 µl
DNA			2 µl

Continued

Continued.

Total volume		50 µl
Thermocycling	DNA thermocycler 9700 (Applied Biosystems)	
Step 1	**Step 2**	**Step 3**
94°C, 5 min	94°C, 60 s	94°C, 60 s
	65°C, 60 s	65°C, 60 s
	72°C, 90 s	72°C, 10 min
×1	×30	×1
Product	collagenase gene = 424 bp *toxR* gene = 173 bp	
Interpretation	Virulent strains positive for both genes Avirulent strains positive for one only or none of the genes	
Sensitivity	8.8×10^2 CFU/ml for bacterial cells	
Specific against	The following strains were negative for collagenase gene and *toxR* gene: *Grimontia hollisae* (CAS), *Photobacterium damselae* (Huang01), *V. anguillarum* (CGMCC Va101), *V. campbellii* (CGMCC 1.1597), *V. cholerae* non-01 (CGMCC), *V. fluvialis* (CGMCC 1.1609), *V. furnissii* (CGMCC 1.7450), *V. harveyi* (CGMCC 1.1593), *V. metschnikovii* (CGMMC 1.7350), *V. mimicus* (CGMCC 1.7300), *V. natriegens* (CGMCC 1.1594), *V. parahaemolyticus* (ATCC 17803), *V. splendidus* (CGMCC 1.1606), *V. vulnificus* (CGMCC 1.7500), *Pseudomonas aeruginosa*, *Aeromonas hydrophila*, *Escherichia coli*	

Conventional PCR with only one primer set consisted of a 50-µl master mix prepared with the following at the final concentration for a single reaction using 1× buffer (5 µl), 1 µM of each primer (1 µl), 200 mM dNTP (4 µl), 1 U *Taq* polymerase (0.2 µl) and 2 µl DNA.

The following multiplex PCR was developed for the simultaneous detection of *V. alginolyticus*, *V. parahaemolyticus* and *V. cholerae* O1 and non-O1 from seafood samples following enrichment in alkaline peptone water (APW), and may be suitable for detection of these organisms in aquatic species.

PCR type	Multiplex PCR based on collagenase gene for detection of *V. alginolyticus*, *V. parahaemolyticus* and *V. cholerae* (Di Pinto *et al*., 2005, 2006)		
Pre-enrichment	Add 25 g of tissue sample to 225 ml APW containing 2% NaCl final concentration. Incubate at 37°C overnight. Culture loopful of broth to TCBS. Incubate 37°C overnight. Pick suspect colonies and extract DNA using QIAamp DNA minikit (Qiagen)		
DNA extraction	QIAamp DNA Mini kit (Qiagen)		
PCR master mix			
Primer VA-F	5′-CGA GTA CAG TCA CTT GAA AGC C-3′		
Primer VA-R	5′-CAC AAC AGA ACT CGC GTT ACC-3		
Primer VP-F	5′-GAA AGT TGA ACA TCA TCA GCA CGA-3′		
Primer VP-R	5′-GGT CAG AAT CAA ACG CCG-3′		
Primer VC-F	5′-CGG CGT GGC TGG ATA CAT TG-3′		
Primer VC-R	5′-GTC ACA CTT AAA TAG TAG CGT CC-3′		
Reagent	**Stock**	**Final concentration**	**Volume ×1**
HotStar*Taq* Master Mix (Qiagen)	×2	×1	12.5 µl
Primer VA-F	50 µM	0.5 µM	0.25 µl
Primer VA-R	50 µM	0.5 µM	0.25 µl
Primer VP-F	50 µM	0.5 µM	0.25 µl
Primer VP-R	50 µM	0.5 µM	0.25 µl
Primer VC-F	50 µM	0.5 µM	0.25 µl
Primer VC-R	50 µM	0.5 µM	0.25 µl
Water			9.0µl
DNA			2 µl
Total volume			**25 µl**

Continued

Continued.

Thermocycling	Mastercycler (Eppendorf)	
Step 1	**Step 2**	**Step 3**
95°C, 15 min	94°C, 30 s	94°C, 30 s
	57°C, 30 s	57°C, 30 s
	72°C, 60 s	72°C, 5 min
×1	×35	×1
Product	*V. alginolyticus* = 737 bp; *V. parahaemolyticus* = 271 bp; *V. cholerae* = 389 bp	
Sensitivity	Not given	
Specific against	*Grimontia hollisae* (ATCC 33565), *Photobacterium damselae* (ATCC 33536), *V. anguillarum* (ATCC 43306), *V. fluvialis* (ATCC 33810), *V. harveyi* (ATCC 35084), *V. metschnikovii* (ATCC 7708), *V. mimicus* (ATCC 33653), *V. vulnificus* (ATCC 27562), *Aeromonas hydrophila* (ATCC 23213), *E. coli* (ATCC 35421), *Listeria* sp. (ATCC 7646), *Salmonella* sp. (ATCC 35664)	

A PCR using the SYBR Green chemistry was designed using the *rpoX* gene, which is a *rpoS*-like sigma factor involved in stress response and biofilm formation. This gene was chosen because the *gyrB* gene in *V. alginolyticus* has greater than 95% homology to *V. campbellii* (Jing-Jing *et al.*, 2011). The PCR was designed using sequence information entered into GenBank for *V. alginolyticus* strains 12G01 and Zj-51.

PCR type	SYBR Green I PCR targeting the rpoX gene (Jing-Jing *et al.*, 2011)		
DNA extraction	Axyprep Bacterial Genomic DNA Miniprep kit (Axygen Biosciences)		
PCR master mix			
Primer rpoX-F	5′-ACG CTG CTC AGG GGT GAA AGC-3′		
Primer rpoX-R	5′-CGG AGA CGT TTC TGC ACG CGT-3′		
Reagent	**Stock**	**Final concentration**	**Volume ×1**
Platinum SYBR Green qPCR SuperMix-UDG (Invitrogen)	×2	×1	25 µl
Primer rpoX-F	10 µM	0.2 µM	1.0 µl
Primer rpoX-R	10 µM	0.2 µM	1.0 µl
ROX Reference dye (optional) (volume may depend on thermocycler)		0.5 µM	1.0 µl
Water			17 µl
DNA			5 µl
Total volume			**50 µl**
Thermocycling			
Step 1	**Step 2**		
94°C, 60 s	94°C, 20 s		
	64°C, 20 s		
	72°C, 30 s		
×1	×40		
Product	354 bp		
R^2 (linear regression coefficient)	0.9959		
Melt temperature	85°C		
Sensitivity	0.1 pg with a Ct value of 31.11		
Limit of detection	Pure culture = 10^3 CFU/ml with a Ct value of 24.99 Cells cultured in seawater = 10^2 CFU/ml		

Continued

Continued.

Specific against	*Vibrio anguillarum* (wild strain), *V. parahaemolyticus* (ATCC 17802), *V. harveyi* (wild strains), *V. campbellii* (wild strains), *V. proteolyticus* (CGMCC 1.1826), *V. furnissii* (ATCC 33813), *V. natriegens* (ATCC 33788), *V. ichthyoenteri* (MCCC 1A00057), *V. aestuarianus* (MCCC 1H00002), *V. gazogenes* (MCCC 1H00014), *V. diazotrophicus* (MCCC 1H00024), *V. metschnikovii* (MCCC 1H00048), *V. vulnificus* (ATCC 27562), *V. mimicus* (ATCC 33653), *V. fluvialis* (ATCC 33810), *Escherichia coli*

Notes: See Platinum qPCR SYBR Green manual available online. The use of ROX dye will depend on the type of real-time thermocycler, and the volume will depend on the model. These are detailed in the manual. The manual suggests a hot start protocol for thermocycling, which is different to the protocol reported by Jing-jing *et al.* (2011).

The PCR was tested on yellow grouper (*Epinephelus awoara*) inoculated with 100 µl of 10^5 CFU/ml of *V. alginolyticus*. Tissue (1 g) from skin ulcers, liver and kidney was homogenized in 1 ml of 0.9% NaCl and centrifuged at 4000 rpm for 1 min. The supernatant is boiled for 10 min and 2 µl is used in the PCR reaction.

A SYBR green PCR was designed to detect the *gyrB* gene in *V. alginolyticus* and was used to detect the organism in seawater, fish, prawn and oyster tissue (Zhou *et al.*, 2007). A 3-h pre-enrichment step improved the sensitivity of detection.

PCR type	SYBR Green I PCR targeting the *gyrB* gene (Zhou *et al.*, 2007)		
DNA extraction	Centrifuge 1 ml of seafood tissue homogenate grown in SWP-20 at 12,000 ×*g* for 10 min. Resuspend cell pellet in 100 µl TE buffer. Boil for 10 min. Centrifuge 3000 ×*g* for 3 min and use 2 µl supernatant for DNA template		
PCR master mix			
Primer gyrB-F	5′-ATT GAG AAC CCG ACA GAA GCG-3′		
Primer gyrB-R	5′-AAT GCG GTG ATC AGT GTT ACT-3′		
Reagent	**Stock**	**Final concentration**	**Volume ×1**
SYBR Green real-time PCR mix (Toyobo Co)	×2	×1	12.5 µl
Primer gyrB-F	10 µM	0.1 µM	0.05 µl
Primer gyrB-R	10 µM	0.1 µM	0.05 µl
Water			10.4 µl
DNA			2 µl
Total volume			25 µl
Thermocycling	ABI PRISM 7900 thermocycler		
Step 1	**Step 2**		
95°C, 60 s	95°C, 15 s 60°C, 15 s 72°C, 45 s		
×1	×45		
Product	340 bp		
R^2 (linear regression coefficient)	0.993		
Melt temperature	88.6°C		
Sensitivity	0.4 pg DNA (= 72 bacterial cells) with a Ct value of 31.56 102 CFU/ml from seawater and tissue homogenate with a Ct value of 30.18 1 CFU/ml after pre-enrichment		

Continued

Continued.

Specific against	*Vibrio anguillarum*, *Photobacterium damselae* (A01), *V. cholerae* non-01 (20050166), *V. harveyi*, *V. mimicus* (1.1969), *V. para-haemolyticus* (1.1615), *V. vulnificus* (1.1758), *Cardiobacterium hominis* (C20), *Chryseomonas luteola* (A10), *Enterobacter aerogenes* (C22), *E. cloacae* (C23), *Escherichia coli* (A03), *Kingella denitrificans* (A02), *Listeria innocua* (20041662), *L. monocytogenes* (54003), *Moellerella wisconsensis* (A12), *Moraxella* sp. (B14), *Pseudomonas fluorescens* (A09), *Salmonella* sp., *Shewanella putrefaciens* (A07), *Staphylococcus aureus* (IAM 1011)

SWP-20: autoclaved seawater (20 ppt salinity) supplemented with 0.2% (w/v) Bacto-peptone. Homogenized tissue is added to SWP-20 (1:10) and incubated for 3 h at 30°C with shaking (180 rev/min) as a pre-enrichment step. The time is based on an optimal doubling time of 20 min for *V. alginolyticus*. The sensitivity is 1 CFU/ml for a 1-ml sample using the pre-enrichment step.

A loop-mediated isothermal PCR was designed targeting the *gyrB* gene and amplification confirmed using *V. alginolyticus* ATCC 17749^{T} (Cai *et al.*, 2010). The PCR successfully amplified *V. alginolyticus* from experimentally challenged humphead snapper (*Lutjanus sanguineus*).

PCR type	Loop-mediated isothermal PCR targeting the *gyrB* gene (Cai *et al.*, 2010)		
DNA extraction	Cultured cells: 1 colony mixed in 200 µl sterile distilled water, boiled for 3 min and centrifuged at 12,000 ×*g* for 5 min Fish tissue: QIAamp DNA Mini kit (Qiagen)		
LAMP PCR master mix			
Primer Va-FIP (F1c-TTTT-F2)	5′-GAG CAC GTG CTG CAT CGA TGA T-TTTT-TCC TGA TTG AGA ACC CGA CA-3′		
Primer Va-BIPB1c-TTTT-B2)	5′-GAA GCA GCG CGT AAA GCT CGT-TTTT-TGG AAG GCC AGC TAG GTC-3′		
Primer Va-F3	5′-TTG AGT CTG CAA TGG GTG AA-3′		
Primer Va-B3	5′-GCC GGA TCT TTT TCC TGA CA-3′		
Reagent	**Stock**	**Final concentration**	**Volume ×1**
Polymerase buffer	×1	×1	12.5 µl
dNTP 10 mM mix		200 µM	0.5 µl
Primer Va-FIP	50 µM	0.4 µM	0.02 µl
Primer Va-BIP	50 µM	0.4 µM	0.02 µl
Primer Va-F3	50 µM	0.05 µM	0.025 µl
Primer Va-B3	50 µM	0.05 µM	0.025 µl
Bst DNA polymerase	8U		1 µl
Water			8.91 µl
DNA			2 µl
Total volume			25 µl
Reaction conditions	**Amplification**	**Termination**	
	64°C for 60 min	80°C for 5 min	
Interpretation	Add 1 µl (1:10) of SYBR Green I. Positive = green colour. Negative = no colour change		
Sensitivity	3.7×10^{2} CFU/ml (3.7 CFU/reaction)		
Specific against	*Aeromonas hydrophila*, *Grimontia hollisae* (SCSIO GZH01), *Photobacterium damselae* (GDOU Huang01), *V. campbellii* (IMCAS 1.1597), *V. cholera* non-01 (SCSIO GZ06), *V. fluvialis* (IMCAS 1.1609), *V. harveyi* (ATCC 14126), *V. metschnikovii* (IMCAS 1.7350), *V. mimicus* (IMCAS 1.7300), *V. natriegens* (IMCAS 1.1594), *V. parahaemolyticus* (ATCC 17802, ATCC 17803), *V. splendidus* (IMCAS 1.1606), *V. vulnificus* (IMCAS 1.7500), *V. anguillarum* (IMCAS Va101), *Escherichia coli*, *Streptococcus agalactiae* from diseased fish (ZQ0901), *S. iniae* (ZJ0801), *Pseudomonas aeruginosa*		

The sequencing of the 16S rRNA gene is not reliable for definitive identification of *Vibrio* species because of the degree of sequence similarity. The following PCR using two primer sets was used to confirm biochemical identification of *V. alginolyticus*. Specificity testing in the original paper was not performed so the degree of sequence similarity to other *Vibrio* species or in particular the Harveyi clade to which *V. alginolyticus* belongs based on MLSA typing, was not determined. Sequence analysis of the 16S rRNA gene suggested 99.7% homology between *V. alginolyticus* and *V. parahaemolyticus* (Ruimy *et al.*, 1994).

PCR type	Conventional PCR for detection of *V. alginolyticus* based on 16S rRNA gene (Liu *et al.*, 2004a)		
DNA extraction	Cultured cells. DNA extracted using the Genomic DNA purification kit from Promega		
PCR master mix			
Primer Va-16F1	5′-ATT GAA GAG TTT GAT CAT GGC TCA GA-3′		
Primer Va-16R1	5′-CAG CTA TTA ACT ACA CTA-3′		
Primer Va-16F2	5′-CCT TCG GGT TGT AAA GCA CT-3′		
Primer Va-16R2	5′-TCC TCC CGT AGT TGA AAC TAC CTA CT-3′		
Reagent	**Stock**	**Final concentration**	**Volume ×1**
Buffer for *Taq*	×10	×1	2.5 µl
dNTP mix	25 mM	0.25 mM	0.25 µl
$MgCl_2$	25,000	2.5 mM	2.5 µl
Primer Va-PF	50 µM	1.0 µM	0.5 µl
Primer Va-PR	50 µM	1.0 µM	0.5 µl
Taq DNA polymerase	5 U/µl	5 U	0.5 µl
Water			13.25 µl
DNA			5 µl
Total volume			25 µl

Thermocycling	DNA thermocycler 9700 (Applied Biosystems)	
Step 1	**Step 2**	**Step 3**
94°C, 4 min	94°C, 1 min	94°C, 1 min
	50°C, 1 min	50°C, 1 min
	72°C, 2 min	72°C, 10 min
×1	×30	×1
Product	1486 bp	
Sensitivity	Not stated	
Specific against	Not stated	

Vibrio anguillarum

TAXONOMY AND SIGNIFICANCE. *Vibrio anguillarum* (NCMB 6^T=NCIMB 6^T=ATCC 19264^T), was first isolated and identified in 1909 by A. Bergman as a cause of vibriosis in eels. Various isolates (*Achromobacter ichthyodermis*, *Vibrio ichthyodermis*, *V. piscium* var. *japonicus*, *V. piscium*, *Pseudomonas ichthyodermis*) isolated from carp, cod, finnock and rainbow trout have since been shown to be synonyms of *V. anguillarum* (Hendrie *et al.*, 1971a). Two biotypes (I and II) were recognized (Baumann *et al.*, 1978), but later shown to be two separate species, *V. anguillarum* previously biotype I, and *V. ordalii* previously biotype II (Schiewe *et al.*, 1981). *Vibrio anguillarum* was placed into the genus *Beneckea* (Baumann *et al.*, 1978) and then into the genus *Vibrio* following the abolition of the genus *Beneckea* (Baumann *et al.*, 1980). A new genus *Listonella* was proposed by MacDonell and Colwell (1985) for *L. anguillarum* and *L. pelagia* based on 5S rRNA gene sequence analysis. However, *Listonella* is suggested as a later heterotypic synonym of *Vibrio* and it is recommended that these species be returned to the genus *Vibrio* as *V. anguillarum* and *V. pelagius*, respectively (Thompson *et al.*, 2011). MLSA identified *V. anguillarum* as belonging to the Anguillarum clade, which also contains *V. aestuarianus* and *V. ordalii* (Sawabe *et al.*, 2007b). Members of this clade have >30% DDH, 43–46 GC mol% and a MLSA concatenated similarity of 88.9–98.6% and AAI of 95.7–99.6%.

Vibrio anguillarum is ubiquitous in freshwater, brackish and marine environments. It is of economic importance to the salmonid industry as salmonids are particularly susceptible to disease when being transferred from freshwater to salt water. Infection with *V. anguillarum* results in a fatal haemorrhagic septicaemia and has been reported worldwide from numerous fish species and molluscs and include Atlantic salmon (*Salmo salar*), Pacific salmon or coho salmon (*Oncorhynchus kisutch*), chum salmon (*O. keta*), sockeye salmon (*O. nerka*), pink salmon (*O. gorbuscha*), masou or cherry salmon (*O. masou*), chinook salmon (*O. tshawytscha*), rainbow trout (*O. mykiss*), brown trout (*Salmo trutta*), ayu (*Plecoglossus altivelis*), cod (*Gadus morhua*), plaice, saithe (*Pollachius virens*), gilthead seabream (*Sparus aurata*), red seabream (*Pagrus major*), seabream, sea mullet (*Mugil cephalus*), turbot (*Scophthalmus maximus*), Japanese flounder (*Paralichthys olivaceus*), winter flounder (*Pseudopleuronectes americus*), eels (*Anguilla anguilla, A. japonica*), feral fish and oysters (*Crassostrea gigas* and *Ostrea edulis*) (DiSalvo *et al.*, 1978; Muroga *et al.*, 1984; Ransom *et al.*, 1984; Sørensen and Larsen, 1986; Toranzo and Barja, 1990; Myhr *et al.*, 1991; Balebona *et al.*, 1998b; Pedersen *et al.*, 1999a; Mizuki *et al.*, 2006).

Clinical signs of infection with *V. anguillarum* include ulcerative skin lesions or boils with red putrefying and liquefying flesh and red spots on ventral and lateral areas of fish. The bacterium can be isolated from the blood and haemopoietic tissues. In three species of salmon (chum, coho and chinook salmon), *V. anguillarum* infects the vascular areas and haemopoeitic tissues and bacterial cells are observed in the blood vessel and areas of haemorrhage. Heavy infestations occur in the liver, kidney, spleen, visceral peritoneum, cardiac muscle and loose connective tissue (Ransom *et al.*, 1984).

Vibrio anguillarum was isolated from wild fingerlings of ayu (*Plecoglossus altivelis*) in Japan and is a potential source of infection for farmed fish (Muroga *et al.*, 1984). In a Japanese flounder (*Paralichthys olivaceus*) hatchery, *V. anguillarum* was found in 61% of larval and juvenile flounder at densities of 5.0×10^1 to 7.4×10^5 CFU/ml, in 100% of rotifer (*Nannochloropsis oculata*) and 25% of *Artemia* (*Artemia nauplii*) live diets at 2.5×10^3 to 4.6×10^6 CFU/ml and was transient in the intestinal microflora of flounder (Mizuki *et al.*, 2006). The link between live diets and intestinal flora was also found in a seabream hatchery where *V. anguillarum* serotype O20 was found in rotifers (live feed), fish tank water and intestinal microflora of seabream larvae (Pedersen *et al.*, 1999b). In a study undertaken on cod (*Gadus morhua*) larvae in the UK, a live diet of rotifer led to digestive tract microbiota in the larvae being dominated by *Ruegeria*/*Roseobacter*, *Pseudoalteromonas* and *Microbacterium* species, whereas a diet of *Artemia* resulted in greater than 90% *Vibrio* species in the digestive tract with *V. alginolyticus* dominating (Reid *et al.*, 2009). When changed to dry feed, *V. splendidus* dominated. One tank of larvae experienced heavy mortality due to infection with *V. anguillarum* (strain 91079) that was negative for serotypes O1 and O2, and an equal density of *Aliivibrio logei* (strain n8c1-8). Both *V. anguillarum* and *A. logei* caused mortalities in challenge experiments with *V. anguillarum* causing greater mortalities than *A. logei*.

Vibrio anguillarum has an LD_{50} of 5.0×10^4 to 9.2×10^5 CFU/g body weight in experimental intraperitoneal challenge of gilthead seabream (*Sparus aurata*). The pathogen has been involved in disease outbreaks in gilthead seabream cultured in Spain (Balebona *et al.*, 1998b).

Vibrio anguillarum may be found as part of the bacterial flora occurring in oysters and seawater in Delaware Bay, USA (Richards *et al.*, 2008). It was also a cause of mass mortality in oysters (*Crassostrea gigas* and *Ostrea edulis*) cultured off the Californian coastline, USA (DiSalvo *et al.*, 1978).

Very little is known about its pathogenesis, but the organism produces a variety of virulence factors that include haemolysin *vah1*, a metalloprotease encoded by the *empA* gene, lipopolysaccharide, exopolysaccharides involved in penetration of skin mucus and formation of a biofilm, and outer membrane proteins involved in osmoregulation and adaption to different salinities (Kao *et al.*, 2009; Naka and Crosa, 2011). Strains of *V. anguillarum* that possess the *rtxA* gene, which encodes an exotoxin and is part of the *rtx* operon and posses a second haemolysin gene cluster, are virulent, whereas strains without a functional *rtxA* gene are avirulent (Li *et al.*, 2008a).

Initially, different serotyping systems existed but a serological study of strains from Europe (Denmark), America and Japan coordinated the systems and divided isolates into ten serotypes (European serotype designation retained) based on the detection of the heat-stable somatic O antigen (lipopolysaccharide). Serotype 01 corresponds to serotype I (USA) and serotype C or J-0-3 (Japan). Serotype 02 corresponds to serotype II (USA) and serotype A or J-0-1 (Japan). Serotype 03 corresponds to serotype B or J-0-2 (Japan). The Japanese serotypes of D, E and F were non-typable by the coordinated system. Serotypes 01 and 02 are predominant worldwide and have been isolated from diseased fish involving outbreaks at a distribution rate of 70% and 15%, respectively. In 75% of vibriosis infections in feral fish, serotype 02 is isolated. The other serotypes have only been isolated from the environment (Sørensen and Larsen, 1986; Myhr *et al.*, 1991). Additional serotypes have been added to total 23 serotypes (01–023), however a number of isolates from the environment remain untypable (Pedersen *et al.*, 1999b). Serotype 01 strains tend to be arabinose-positive. Isolates from serotype 01 are a homogeneous group, whereas isolates from serotype 02 can be further divided into groups 02α and 02β based on differences detected by double immunodiffusion, dot-blot assay and enzyme-linked immunoabsorbant assay (Toranzo and Barja, 1990). *Vibrio ordalii*, previously classified as *V. anguillarum* biotype II, cross-reacts with serotype 02 antiserum (Toranzo *et al.*, 1987). In Japan, virulent serotypes tend to be J-0-1, J-0-2, J-0-3, with J-0-1 predominant in infected ayu. Other unknown serotypes tend to be avirulent (Muroga *et al.*, 1984).

GROWTH CHARACTERISTICS AND PHENOTYPIC IDENTIFICATION. *Vibrio anguillarum* is a facultatively anaerobic motile, short, straight or curved Gram-negative rod (0.5–0.7×1–2 μm) with rounded ends occurring singly and in pairs, and is pleomorphic. It exhibits a rapid motility. On nutrient agar colonies are 1–2 mm, off-white to buff coloured, translucent or opaque, circular, convex, entire, smooth and shiny after 24 h. On horse blood agar, colonies are greyish and usually beta-haemolytic (Hendrie *et al.*, 1971a). On MSA-B, colonies are 2 mm, glistening, cream-coloured in young colonies and become greenish-pale brown in older colonies. Haemolysis is underneath the colony. It is non-luminescent. On MSA-B, *V. cholerae* and *V. mimicus* have a similar appearance but grow more quickly. Growth occurs at 5–30°C with no growth at 37°C with an optimum of 20–25°C. Growth occurs at 0–5% NaCl. Results are positive for Moeller's ADH. Some strains are negative for citrate in both the conventional tube test and in the API 20E. The Simmons' tube citrate method may be slow and take 2–3 days. The MR tube may show weak growth and a weak positive reaction after 48 h incubation and some strains may be negative. Growth on TCBS tends to be slow and there is substantially less growth when compared to the amount of growth on an MSA-B or BA plate. Strains isolated from an estuarine environment may grow better in 0.85% NaCl, whereas those strains from a marine environment show a preferred growth with 2–3% NaCl, and this can be seen clearly on a gelatin-salt plate. Use growth on this plate or growth on a BA as opposed to growth on MSA-B, as a guide to the salt requirement of the organism when preparing the NaCl for the inoculating fluid for either the conventional biochemical set or the API 20E. Strain NCIMB 2129 is reported to be negative for fermentation of sorbitol and trehalose (Benediktsdóttir *et al.*, 1998). Differences in citrate, indole and MR are reported for NCIMB 2129, NCIMB 6 and ATCC 14181 (Myhr *et al.*, 1991; Benediktsdóttir *et al.*, 1998; Lunder *et al.*, 2000). Serotype 01 strains tend to be positive for L-arabinose (Toranzo and Barja, 1990).

A selective medium (VAM) for the presumptive identification of *V. anguillarum* contains bile salts, NaCl and sorbitol for differentiation of *V. anguillarum* from other *Vibro* species (Alsina *et al.*, 1994) (see media section). These and other reactions are presented in Tables 3.22, 3.24–26 and 3.38.

MOLECULAR IDENTIFICATION. The following conventional PCR detects the haemolysin gene, *vah1* of *V. anguillarum*, which is found in serotypes A to I (Hirono *et al.*, 1996). However, *vah1* may not be present in all strains of *V. anguillarum* as the PCR detected only 17 of 20 isolates belonging to serotype A. DNA is extracted from bacterial growth from 20 colonies suspended in 40 μl of distilled water to which is added 10 μl of proteinase K (1 mg/ml) and 50 μl of 40 mmol/l Tris buffer (pH 8.0) containing 1% Tween 20, 0.5% Nonidet p-40 and 1 mmol/l EDTA. The solution is incubated at 60°C for 20 min and then boiled for 5 min to inactivate the proteinase K.

The solution is centrifuged at 7740 ×*g* for 10 min and 12.5 µl is used in the PCR reaction (Mizuki *et al*., 2006). The master mix contains 0.06 µmol/l each of primers VAH1-P1 (5′-ACC GAT GCC ATC GCT CAA GA-3′) and VAH-P2 (5′-GGA TAT TGA CCG AAG AGT CA-3′), 5 µl of 10× PCR reaction buffer, 2 mmol/l $MgCl_2$, 200 µmol/l of each dNTP and 2.5 U *Taq* polymerase (Promega) in a final volume of 50 µl (Hirono *et al*., 1996; Mizuki *et al*., 2006). Thermocyling is undertaken in an iCycler (BioRad) using an initial denaturation at 94°C for 60 s followed by 30 cycles at 94°C for 30 s, 55°C for 30 s and 72°C for 45 s with a final extension at 72°C for 10 min (Mizuki *et al*., 2006). A 490-bp amplicon is produced. The sensitivity of the PCR is 10 pg template DNA (Hirono *et al*., 1996). The PCR detected strains belonging to serotypes B to I, but only detected 17 of 20 isolates belonging to serotype A. The PCR was also used to detect *V. anguillarum* direct from fish tissue using DNA extracted from tissue with the Instagene matrix (BioRad). The primers are specific against *Aeromonas hydrophila*, *A. salmonicida*, *Aliivibrio fischeri*, *Escherichia coli*, *Grimontia hollisae*, *Photobacterium leiognathi*, *Salinivibrio costicola*, *Vibrio alginolyticus*, *V. campbellii*, *V. cholerae*, *V. diazotrophicus*, *V. gazogenes*, *V. harveyi*, *V. natriegens*, *V. nereis*, *V. ordalii*, *V. parahaemolyticus*, *V. pelagius*, *V. splendidus* and *V. vulnificus* (Hirono *et al*., 1996).

Another conventional PCR based on the *toxR* gene was developed for use on cultured cells (Okuda *et al*., 2001). DNA is extracted by boiling for 5 min a 1 ml aliquot from an overnight broth culture. The boiled solution is centrifuged at 15,000 ×*g* and the supernatant diluted 1:10 in distilled water. The master mix consists of 1 µl of 20 pmol/µl primer VA-U2 (5′-CAC TTC GCA ACC CGA AGA GAC A-3′) and 1 µl of 20 pmol/µl primer VA-D1 (5′-CTG CTT AGG TGC CAG TTC TCC A-3′), 5 µl 1× PCR buffer (ExTaq Takara), 4 µl of 2.5 mM dNTPs, 0.25 µl *Taq* polymerase (ExTaq Takara), 28.75 µl distilled water and 10 µl of DNA supernatant. The thermocycling protocol is 20 cycles at 95°C for 1 min, 62°C for 1.5 min and 72°C for 1.5 min with a final extension step at 72°C for 5 min. A 307-bp amplicon is produced. The primers are specific against *Aeromonas hydrophila*, *A. salmonicida* (NCMB 1102, NCMB 2020), *Aliivibrio fischeri* (ATCC 7744[T]), *A. logei* (ATCC 15382[T]), *Escherichia coli*, *Edwardsiella tarda* (NUF251, ET-82021), *Enterococcus seriolicida* (ATCC 49156), *Grimontia hollisae* (525-82), *Photobacterium damselae* ssp. *damselae* (ATCC 33539), *P. damselae* ssp. *piscicida* (SJ-9107, OT-8447), *P. iliopiscarium* (ATCC 51760), *Pseudomonas anguilliseptica* (NCMB 1950), *Streptococcus iniae* (HS95-06) *Vibrio alginolyticus* (NCMB 1903), *V. campbellii* (ATCC 25920), *V. cholerae* 01 and 0139, *V. cincinnatiensis* (ATCC 35912), *V. diazotrophicus* (ATCC 33466), *V. fluvialis* (NCTC 11327), *V. furnissii* (RIMD 2223001), *V. gazogenes* (ATCC 29988), *V. harveyi* (ATCC 35084, ATCC 14126), *V. ichthyoenteri* (IFO 15847), *V. mediterranei* (ATCC 43341), *V. metschnikovii* (IAM 1039), *V. mimicus* (RIMD 2218002), *V. mytili* (ATCC 51288), *V. navarrensis* (ATCC 51183), *V. nereis* (ATCC 25917), *V. nigripulchritudo* (ATCC 27043), *V. ordalii* (ATCC 33509), *V. orientalis* (ATCC 33934), *V. parahaemolyticus*, *V. pelagius* (ATCC 25916), *V. penaeicida* (IFO 15640), *V. proteolyticus* (NCMB 1326), *V. splendidus* (ATCC 33125), *V. tubiashii* (ATCC 19109) and *V. vulnificus* (RIMD 2219022) (Okuda *et al*., 2001).

A conventional PCR using primers VARPO1 5′-AGA CCA AGA GAT CAT GGA TT-3′ and VARPO2 5′-AGT TGT TCG TAT CTG GGA TG-3′ designed to detect the *rpoS* gene was specific against 25 *Vibrio* species and against some *Enterobacteriaceae* species (*Aeromonas hydrophila*, *Aliivibrio fischeri*, *A. salmonicida*, *Escherichia coli*, *Enterobacter cloacae*, *Edwardsiella tarda*, *Grimontia hollisae*, *Klebsiella oxytoca*, *K. pneumoniae*, *Photobacterium damselae*, *Salmonella typhi*, *Shigella flexneri*, *S. sonnei*, *V. campbellii*, *V. cholerae*, *V. cincinnatiensis*, *V. diazotrophicus*, *V. fluvialis*, *V. furnissii*, *V. harveyi*, *V. logei*, *V. mediterranei*, *V. metschnikovii*, *V. mimicus*, *V. natriegens*, *V. navarrensis*, *V. nereis*, *V. ordalii*, *V. orientalis*, *V. parahaemolyticus*, *V. proteolyticus*, *V. tubiashii* and *V. vulnificus*) (Kim *et al*., 2008). Details of the master mix reagents and concentrations were not provided. The thermocycling protocol included an initial denaturation step at 94°C for 10 min followed by 25 cycles of denaturation at 94°C for 30 s, annealing at 51°C for 30 s, extension at 72°C for 30 s with a final elongation step at 72°C for 7 min. An amplicon of 689 bp is produced. The PCR was tested on bacterial cells and was sensitive to 3 pg of purified DNA or 6 CFU/ml and 50 ng of tissue from experimentally infected flounder and 10 ng from experimentally infected prawn (*Penaeus monodon*) tissue (Kim *et al*., 2008).

Primers designed to the *toxR* gene and the 16S rRNA gene were compared by real-time PCR for sensitivity and specificity of detection (Crisafi *et al.*, 2011). One primer set uses 50 nM forward primer 16SrDNA F (5′-CCA CGC CGT AAC GAT GTC TA-3′) and 100 nM reverse primer 16SrDNA R (5′-CCA GGC GGT CTA CTT AAC GCG T-3′). An amplicon of 81 bp is produced and the melting temperature is 82°C. The second primer set uses 100 nM toxR F (5′-ACA CCA CCA ACG AGC CTG A-3′) and 50 nM toxR R (5′-TTG TCT CTT CGG GTT GCG A-3′). The amplicon is 93 bp and the melting temperature is 80°C. A 25-μl master mix volume is composed of 12.5 μl SYBR® Green PCR master mix (Applied Biosystems), 0.5 μl forward and reverse primers, 1 μl DNA containing 20 ng from pure culture samples or 50 ng from tissue samples. The thermocycling protocol included an initial denaturation step at 95°C for 10 min followed by 45 cycles of denaturation at 95°C for 15 s and annealing/elongation at 60°C for 60 s. A dissociation step was used to check for primer–dimer formation.

Vibrio anguillarum serotypes 01, 02 and 03 are detected by both primer sets. The primers are specific against *Aeromonas hydrophila* (ATCC 7966^T), *A. salmonicida* ssp. *salmonicida* (ATCC 33658^T), *Aliivibrio fischeri* (ATCC 7744^T), *A. logei* (ATCC 15382^T), *Escherichia coli* (ATCC 8739^T), *Lactococcus garvieae* (ATCC 43921^T), *Photobacterium damselae* ssp. *damselae* (ATCC 33539), *P. damselae* ssp. *piscicida* (ATCC 17911^T), *Rhodococcus erythropolis* (ATCC 25544^T), *Salinivibrio costicola* (ATCC 33508^T), *Vibrio alginolyticus* (ATCC 14582^T, 17749^T, 19108^T), *V. furnissii* (CGMCC 1.1613), *V. mediterranei* (ATCC 43341^T), *V. ordalii* (ATCC 33509^T), *V. pelagius* (ATCC 25916^T), *V. ponticus* (ATCC 14391^T), *V. splendidus* (ATCC 33125^T) and *V. vulnificus* (ATCC 27562^T, 33147) (Crisafi *et al.*, 2011).

The specificity of a PCR designed to the *rpoN* gene (codes cellular sigma factor σ^{54}) is compromised if the annealing temperature of 62°C is not adhered to, as at an annealing temperature of 58°C weak amplicon is obtained from *V. ordalii* (Gonzalez *et al.*, 2003).

Vibrio areninigrae

TAXONOMY AND SIGNIFICANCE. *Vibrio areninigrae* (J74^T=KCTC 22122^T=JCM 14949^T) was isolated from black sand collected from Jeju Island, Korea (Chang *et al.*, 2008). Sequencing of the 16S rRNA gene identified *V. areninigrae* as most closely related to *V. hispanicus* (98.7%) with DNA–DNA hybridization of 37.7% indicating separate species.

GROWTH CHARACTERISTICS AND PHENOTYPIC IDENTIFICATION. *Vibrio areninigrae* is a facultatively anaerobic motile, oxidase-positive, Gram-negative slightly curved rod (0.5–0.8×2–3 μm). On MA 2216 colonies are 2–3.5 mm, pale yellow, circular, slightly irregular, smooth, low convex after 72 h at 30°C (Chang *et al.*, 2008). Growth occurs at 4–45°C and at 1–7.5% NaCl but not below 1% or with more than 10% NaCl. Growth occurs at pH 5–10 but not below pH 4.5 or above pH 11. It is sensitive to 150 μg O/129 vibriostatic agent. These and other reactions are presented in Tables 3.23–26 and 3.38.

Vibrio artabrorum

TAXONOMY AND SIGNIFICANCE. *Vibrio artabrorum* (Vb 11.8^T=CECT 7226^T=LMG 23865^T) was isolated from clam (*Venerupis* [*Ruditapes*] *philippinarum*) cultured in Galicia in north-west Spain (Diéguez *et al.*, 2011). *Vibrio atabrorum* was originally identified as belonging to cluster 70 in a study using amplified fragment length polymorphisms (AFLPs) to examine *Vibrio* species isolated from clam (Beaz-Hidalgo *et al.*, 2008). A neighbour-joining tree composed of concatenated sequences of the housekeeping genes *recA*, *rpoA*, *rpoD*, *atpA* and the 16S rRNA sequence, showed *V. artabrorum* was related to the Splendidus clade with sequence similarity to *V. gigantis* (99.9%), *V. pomeroyi* (99.4%), *V. crassostreae* (99.5%), *V. tasmaniensis* (99.2%) and *V. splendidus* (99%).

GROWTH CHARACTERISTICS AND PHENOTYPIC IDENTIFICATION. *Vibrio artabrorum* is a facultatively anaerobic, oxidase-positive, Gram-negative rod (0.8–1.0×1.4–1.8 μm). Motility is positive in 37.5% of strains (three of eight strains). On MA 2216 colonies are non-swarming, non-luminescent, beige, round, smooth and 2–6 mm at 24 h at 25°C (Diéguez *et al.*, 2011). One isolate produced a brown diffusible pigment. Growth occurs at 4–25°C but not at 35°C. Growth occurs at 3–6% NaCl. It is sensitive to 150 μg O/129 vibriostatic agent. These and other reactions are presented in Tables 3.23, 3.24–26, 3.30 and 3.38.

Vibrio atlanticus

TAXONOMY AND SIGNIFICANCE. *Vibrio atlanticus* (Vb 11.11ᵀ=CECT 7223ᵀ=LMG 24300ᵀ) was isolated from clam (*Venerupis* [*Ruditapes*] *philippinarum*) reared at Galicia in north-west Spain (Diéguez *et al.*, 2011). *Vibrio atlanticus* was originally identified as belonging to cluster 5 in a study using AFLP (amplified fragment length polymorphisms) to examine *Vibrio* species isolated from clam (Beaz-Hidalgo *et al.*, 2008). Using a neighbour-joining tree composed of concatenated sequences of the housekeeping genes *recA*, *rpoA*, *rpoD*, *atpA* and the 16S rRNA sequence, *V. atlanticus* is related to the Splendidus clade with sequence similarity to *V. tasmaniensis* (99.3%), *V. lentus* (99.1%) and *V. splendidus* (99%). Sequence similarity using *recA* and *rpoA* indicates *V. atlanticus* is most closely related to *V. tasmaniensis* with sequence similarities higher than those for interspecies delineation as proposed by Thompson *et al.* (2005a), however other authors (Pascual *et al.*, 2010) state these limits are inaccurate for the central clade of the vibrios.

GROWTH CHARACTERISTICS AND PHENOTYPIC IDENTIFICATION. *Vibrio atlanticus* is a facultatively anaerobic motile, oxidase-positive, Gram-negative rod (0.7–0.8×1.1–1.4 μm). On MA 2216 colonies are non-swarming, non-luminescent, beige, round, smooth and 2–6 mm after 24 h at 25°C. After 72 h at 30°C they are 2–3.5 mm, pale yellow, circular, slightly irregular, smooth, low convex (Diéguez *et al.*, 2011). The result for catalase is positive. Growth occurs at 4°C but not at 35°C. Growth occurs at 3–6% NaCl but not at 8% NaCl. The indole reaction is described as usually positive. Gelatin is negative for the type strain but positive for other strains. Starch and Tween 80 are hydrolysed. These and other reactions are presented in Tables 3.23, 3.30 and 3.38.

Vibrio atypicus

TAXONOMY AND SIGNIFICANCE. *Vibrio atypicus* (HHSO2ᵀ=CGMCC 18461ᵀ=LMG 24781ᵀ) was isolated from the digestive tract of healthy Chinese prawn (*Penaeus chinensis*) (Wang *et al.*, 2010b). *Vibrio atypicus* is closely related to *V. harveyi*, *V. nereis* and *V. orientalis* with 98% similarity of the 16S rRNA gene, and although it clusters with the *V. harveyi* clade it forms a separate branch using neighbour-joining and maximum-parsimony trees. *Vibrio atypicus* clusters with the *V. orientalis* clade based on concatenated sequences of housekeeping genes (*gapA*, *mreB*, *pyrH*, *recA*, *topA*, *gyrB*).

Vibrio atypicus was identified along with a number of other *Vibrio* species as being part of the haemolymph of healthy spider crab (*Maja brachydactyla*) caught off the coast of Spain and the Canary Islands (Gomez-Gil *et al.*, 2010). The numbers of bacterial flora in healthy haemolymph is usually in the order of 10^3–10^4 CFU/ml (Gomez-Gil *et al.*, 1998). Under experimental conditions *V. atypicus* was non-pathogenic to rainbow trout (21 g) when inoculated intraperitoneally with 100 μl of bacteria at 8.8×10^7 CFU/ml.

GROWTH CHARACTERISTICS AND PHENOTYPIC IDENTIFICATION. *Vibrio atypicus* is a facultatively anaerobic, motile, oxidase-positive, Gram-negative slightly curved rod (0.5–0.6×1–1.5 μm). It is catalase negative, which is an unusual finding for *Vibrio* species. On MA 2216 colonies are 2 mm, non-pigmented, circular, smooth, convex with entire margins after 1 day at 28°C, and grow as bright yellow colonies on TCBS (Wang *et al.*, 2010b). It is non-luminescent on MA 2216. Growth occurs at 16–30°C, with an optimum of 20°C, and in 0.5–7% NaCl with an optimum of 1–3% (w/v) and at pH 7–10 with an optimum of pH 8–9. Growth does not occur at 4°C or 37°C. D-cellobiose is positive when tested using the Vitek 2 GN card and negative results are obtained for fermentation of L-arabitol, adonitol, maltose, D-mannose and D-tagatose. These and other reactions are presented in Tables 3.23 and 3.24–26. Identification sets are incubated according to the manufacturer's instructions with the exception that an inoculum is used of 1.5% NaCl and an incubation temperature of 28°C.

Vibrio azureus

TAXONOMY AND SIGNIFICANCE. *Vibrio azureus* (LC2-005ᵀ=NBRC 104587ᵀ=KCTC 22352ᵀ) was isolated from seawater off the coast of Japan (Yoshizawa *et al.*, 2009). *Vibrio azureus* clusters in the *V. harveyi* clade.

GROWTH CHARACTERISTICS AND PHENOTYPIC IDENTIFICATION. *Vibrio azureus* is a facultatively anaerobic, motile, oxidase-positive, Gram-negative bacillus. On MA 2216 colonies are non-pigmented, translucent and show slight swarming and are luminescent. On TCBS colonies are 5–7 mm, and round green after 48 h at 20°C (Yoshizawa *et al.*, 2009). Growth occurs at 0.5–6% NaCl and 10–37°C, but not at 8–10% NaCl or 4°C. Nitrate is not reduced. The indole reaction is positive or weak. These and other reactions are presented in Tables 3.23, 3.24–26 and 3.38. An inoculum of 2% NaCl is used with incubation temperature of 20°C, time of 48 h for the API 20E and 6 h for the API Zym.

Vibrio brasiliensis

TAXONOMY AND SIGNIFICANCE. *Vibrio brasiliensis* (LMG 20546[T]) was isolated from larval bivalve molluscs, *Nodipecten nodosus*, cultured in Brazil (Thompson *et al.*, 2003b). In experimental challenge, *V. brasiliensis* caused 100% mortality when inoculated into rainbow trout (*Oncorhynchus mykiss*) at a dose of 10^6 cells/fish for strain LMG 20010 and 40% mortality for strain LMG 20546[T] with an LD_{50} of 10^4 for both strains (Austin *et al.*, 2005). The two strains caused 75% mortality in *Artemia* nauplii exposed to 10^6 CFU/ml. Extracellular product (0.1 ml) from the two strains caused 100% and 20% mortalities, respectively, when inoculated into trout, and 75% and 85%, respectively, in *Artemia*. In rainbow trout, infection resulted in softening of the organs, soft kidney and presence of the organism in the kidney, and distention of the abdomen and anus resulting in gastroenteritis and muscle necrosis.

Vibrio brasiliensis, initially identified as FAFLP group A8 (Thompson *et al.*, 2001b) together with *V. neptunius* and *V. xuii*, was related to *V. tubiashii* (98–98.8%), *V. nereis* (97.6–98.8%), *V. coralliilyticus* (96.8–98.5%), *V. mytili* (96.8–98.2%) and *V. diabolicus* (97.1–98.1%) based on 16S rRNA sequencing. MLSA identified *V. brasiliensis* as belonging to the Orientalis clade, which also contains *V. orientalis* and *V. hepatarius*, and *V. tubiashii* (Sawabe *et al.*, 2007). Members of this clade have >24% DDH and 43–46 GC mol%, 91.2–94.2% similarity based on concatenated MLSA sequence data and AAI of 97–97.9%. *V. neptunius* and *V. coralliilyticus* belong to the Coralliilyticus clade, *V. xuii* and *V. nereis* to the Nereis clade, and *V. mytili* to the Harveyi clade. *V. diabolicus* is not available for study.

GROWTH CHARACTERISTICS AND PHENOTYPIC IDENTIFICATION. *Vibrio brasiliensis* is a facultatively anaerobic, motile, oxidase-positive, Gram-negative slightly curved rod (2.5–3 μm). On TSA containing 2% NaCl colonies are 2.5–3 mm, beige, smooth, round with entire edges after 48 h at 28°C (Thompson *et al.*, 2003b). On TCBS colonies are 3 mm, yellow, umbonate, wavy, shiny and round with scalloped edges after 24 h at 28°C. Growth does not occur without NaCl, or at 8% NaCl, or at temperatures of 4 or 45°C. These and other reactions are presented in Tables 3.23, 3.24–26 and 3.38. An inoculum is used of 2% NaCl and opacity McFarland 3 and sets are incubated at 25°C for 48 h.

Vibrio breoganii

TAXONOMY AND SIGNIFICANCE. *Vibrio breoganii* sp. nov. (CECT 7222[T] = LMG 23858[T]) was isolated from the hepatopancreas, mantle, gonad, gills and extrapalial fluid from cultured clams, *Venerupis* [*Ruditapes*] *philippinarum* and *V. decussata*, on the north-west coast of Spain (Galicia). *Vibrio breoganii* belongs to the *V. halioticoli* species group according to phylogenetic analysis based on the housekeeping genes *rpoA*, *recA* and *atpA*, and is closely related to *V. comitans*, *V. rarus* and *V. inusitatus* with ≥99.0% sequence similarity according to 16S rRNA gene sequence information (Beaz Hidalgo *et al.*, 2009a).

GROWTH CHARACTERISTICS AND PHENOTYPIC IDENTIFICATION. *Vibrio breoganii* is a facultatively anaerobic, Gram-negative rod. On MA 2216 agar colonies are 0.3–0.7 mm diameter, non-pigmented, translucent, non-swarming and smooth after 24 h at 25°C. Green colonies of 0.7–1.0 mm are produced on TCBS. Growth does not occur in the absence of NaCl, but growth occurs in 1.5–6.0% NaCl but not above 8% NaCl. Growth occurs at 4°C, but not above 37°C. Non-luminescent, does not hydrolyse agar, and produces alginase. These and other reactions are presented in Tables 3.23, 3.24–26, 3.30 and 3.38. An inoculum of 0.85% NACl is used for the API 20E and API Zym and performed according to the manufacturer's instructions. For the API 50CH, the inoculum is prepared in 0.85% NaCl and adjusted to an optical density of 1.0 at a wavelength of 580 and mixed with ZOF medium at 1:10 v/v and the results read at 24 h, 48 h, 96 h, 120 h and 6 days (Beaz Hidalgo *et al.*, 2009a).

Vibrio calviensis

See *Enterovibrio calviensis*.

Vibrio campbellii

TAXONOMY AND SIGNIFICANCE. *Vibrio campbellii* (ATCC 25920[T]) was isolated from ocean water off the coast of Hawaii and originally designated *Beneckea campbellii* (Baumann *et al.*, 1971a) and reassigned to the genus *Vibrio* with the abolition of *Beneckea* (Baumann *et al.*, 1980). The G+C content is 45.9–47.2 mol% (Baumann *et al.*, 1971a). The DNA–DNA hybridization with *V. harveyi* is 65% (Reichelt *et al.*, 1976). *Vibrio campbellii* is closely related to *V. harveyi* both phenotypically and genetically. MLSA using four housekeeping genes, *topA* (topoisomerase I), *pyrH* (urydilate kinase), *ftsZ* (cell division protein) and *mreB* (actin-like cytoskeleton protein) separates *V. campbellii* and *V. harveyi* into different clusters (Thompson *et al.*, 2007a). Gene sequence similarity for *ftsZ*, *mreB* and *pyrH* is 94–96% with *topA* being the most resolving resulting in 90% gene similarity. Gene sequence variation within each cluster is 1.5%. Other housekeeping genes, *recA* and *gapA*, show that both strains are related with 94% and 99.5% similarity, respectively. The gene *gyrB* does not distinguish the two species. A concatenated sequence similarity of at least 95% using all seven genes indicates strains are the same species.

MLSA identified *V. campbellii* as belonging to the Harveyi clade, which also contains *V. harveyi*, *V. alginolyticus*, *V. mytili*, *V. natriegens*, *V. parahaemolyticus* and *V. rotiferianus* (Sawabe *et al.*, 2007b). Members of this clade have >25% DDH and 42–48 GC mol%, an MLSA concatenated sequence similarity of 90.1–96.2% and AAI of 97.2–99.4%. Strains within the Harveyi clade (*V. alginolyticus*, *V. campbellii*, *V. harveyi*, *V. natriegens*, *V. parahaemolyticus*, *V. rotiferianus*) can be discriminated using concatenated sequences from three housekeeping genes, *rctB*, *rpoD* and *toxR*, with *toxR* and *rpoD* providing the most discrimination (Pascual *et al.*, 2010) (see under *Vibrio harveyi*).

High expression levels of *luxR*, *toxR* and *vhp* genes were found using reverse transcriptase PCR in two strains PN9801=LMG 21363 and BB120=ATCC BAA-1116, originally identified as *V. harveyi* but identified as *V. campbellii* by Gomez-Gil *et al.* (2004b) and Lin *et al.* (2010), respectively (Ruwandeepika *et al.*, 2011). Lower levels of expression occurred for *vhh* (haemolysin gene) and *srp* (serine protease gene). The two isolates are moderately virulent (ATCC BAA1116) and highly virulent (LMG 21363) in experimental challenge in gnotobiotic *Artemia*. A significant correlation was found between expression levels of *toxR* and *luxR* together with the protease genes *srp* and *vhp*, whereas no correlation was found between expression levels of *toxR* and *luxR* with the haemolysin gene, *vhh*. There was a significant correlation between the expression of the serine protease gene *srp* and the haemolysin gene *vhh*. Contrary to the finding of Ruwandeepika *et al.* (2011), strain PN9801=LMG 21363 was defined as either non-virulent or of low virulence when used in experimental challenge with rainbow trout (*Oncorhynchus mykiss*) and *Artemia* nauplii, as clinical signs of disease were only seen at the highest inoculated dose of 10^6 CFU/fish and 10^6 CFU/ml/*Artemia*, and at the end of the experiment the organism was not recovered from tissues (Austin *et al.*, 2005).

Vibrio campbellii can be difficult to differentiate from *V. harveyi* as it has many phenotypic characteristics in common, however a main differentiating test is ornithine decarboxylase (ODC). *Vibrio harveyi* is positive for ODC and *V. campbellii* is negative. The ornithine decarboxylase (*odc*) gene from *V. harveyi* has sequence similarity to the *odc* gene in *V. vulnificus* of 94%, 88% to *V. parahaemolyticus* and 81% to *V. cholerae*. The *odc* gene is not detected in *V. campbellii* (Hedreyda and Rañoa, 2007).

Isolate ATCC BAA-116 (BB120) is a model for quorum sensing research and was initially identified as *V. harveyi*, but has been identified as *V. campbellii* based on comparative genomic hybridization and MLSA using *ftsZ*, *mreB* and *topA* to construct a phylogenetic tree, with these results confirmed in a second tree using *rpoD*, *rctB* and *toxR* genes (Lin *et al.*, 2010). Isolate HYO1 isolated from moribund shrimp in Thailand initially identified as *V. harveyi* (Rattanama *et al.*, 2009) was also shown to be *V. campbellii* using the same techniques. The following strains have been confirmed as *V. campbellii* by Gomez-Gil *et al.* (2004b): LMG 11216[T] (=ATCC 25920[T]= CAIM 519[T]), LMG 11256 (=ATCC 25921=CAIM 512), LMG 13241, LMG 16828 (=VIB 403),

LMG 16830 (=VIB 406), LMG 20369 (=8P), CAIM 94 (=HL115), CAIM 113 (=HL135), CAIM 114 (=HL136), CAIM 126 (=HL148), CAIM 128 (=HL150=LMG 21365), CAIM 129 (=HL151), CAIM 130 (=HL152), CAIM 133 (=HL155), CAIM 135 (=HL157), CAIM 136 (=HL158), CAIM 137 (=HL159), CAIM 149 (=HL34), CAIM 222 (=HP130), CAIM 284 (=IPL8), CAIM 333 (=M_1=LMG 21362), CAIM 371 (=PL96-11-6), CAIM 372 (=PN-9801=LMG 21363), CAIM 392 (=STD3-1002), CAIM 395 (=STD3-131), CAIM 401 (=SW-9702), CAIM 415 (=Z_1=LMG 21461), CAIM 416 (=Z_2), CAIM 417 (=Z_3), CAIM 679 (=1A-1), CAIM 684 (=Na-1), R-14899 (VIB 629) and R-14905 (VIB 394). The following strains identified as *V. harveyi* were shown to be *V. campbellii* using concatenated sequences from housekeeping genes (*topA*, *ftsZ*, *toxR*, *rctB*, *rpoD*, *mreB*): LMG 11658, LMG 22893, LMG 16862, LMG 22891, LMG 16863, LMG 16829 and LMG 1660 (Hoffman *et al.*, 2012a). It is important to check the identity of species, particularly if certain strains are being used for pathogenicity or virulence investigations, design of primers or molecular and phenotypic investigations.

Vibrio campbellii is found in seawater and was considered to be non-pathogenic, however due to the close association between *V. campbellii* and *V. harveyi* a number of reports indicate mis-identification of the two species with the result that *V. campbellii* has been responsible for a number of disease outbreaks in aquatic animals.

Pathogenic and non-pathogenic strains of *V. campbellii* were isolated from diseased prawns (shrimps) and identification and pathogenicity confirmed using sequencing of the 16S rRNA and *pyrH* (uridylate kinase) genes, specific PCR based on the haemolysin gene and experimental challenge using gnotobiotic *Artemia* (Haldar *et al.*, 2011) (see molecular identification for PCRs for *pyrH* and specific PCR). Of the 26 strains 30.8% were highly virulent in experimental challenge, 11.5% moderately virulent and 57.7% were non-virulent. All highly virulent isolates came from diseased prawns (shrimp), the moderately virulent isolates came from diseased prawn and estuary water and the non-virulent isolates came from diseased prawns and pond soil. Two *V. harveyi* isolates from pond water were not virulent. The virulent isolates colonized the gut wall, whereas the non-virulent isolates did not. The 50% lethal concentration (LC_{50}) was $<10^3$ CFU/ml for pathogenic strains, 10^3–10^7 CFU/ml for moderately pathogenic strains and $>10^7$ CFU/ml for non-pathogenic strains.

Pathogenicity of *V. campbellii* may be enhanced in polymicrobial infection. In experimental studies, the *in vivo* multiplication and pathogenicity of *V. campbellii* was enhanced when specific pathogen-free prawns, *Litopenaeus vannamei*, were pre-infected with white spot syndrome virus (WSSV) prior to challenge with *V. campbellii* (Phuoc *et al.*, 2008). In experimental challenge with 10^6 CFU/animal using the bacterium only, mortalities were 16.7% compared to 100% by 72–96 h when challenged with both organisms. No mortalities were observed with $<10^6$ CFU/animal of *V. campbellii* only. Mortality was not seen in prawns injected with supernatant from *V. campbellii* cultures or with sonicated bacterial cells. In animals with dual infection, the bacterial load of *V. campbellii* was 1000 times higher in the haemolymph compared to single challenge with *V. campbellii* only. It was hypothesized that healthy prawns can eliminate low bacterial densities due to the clearing mechanism of the haemocytes but when infected with WSSV the animal's defence mechanisms are compromised allowing *V. campbellii* to multiply within the host.

Factors occurring in the environment, such as low oxygen (hypoxia) and elevated carbon dioxide (hypercapnia), also compromise the immune system in Pacific white shrimp (*L. vannamei*), the Atlantic blue crab (*Callinectes sapidus*) and the eastern oyster (*Crassostrea virginica*) and prevent the haemocytes from clearing *V. campbellii* from the haemolymph (Holman *et al.*, 2004; Burgents *et al.*, 2005; Macey *et al.*, 2008).

Vibrio campbellii caused mortalities in pond-cultured *Penaeus monodon* in the Philippines and pathogenicity was proven in experimental challenge (de la Peña *et al.*, 2001). Two isolates of *V. campbellii*, PN-9801 (originally identified as *V. harveyi* but later proven to be *V. campbellii* by Soto-Rodriguez *et al.*, 2003) and PIZ-9806, resulted in 100% mortality at 7 days when injected at 10^8 CFU/animal. Challenge with 10^7 CFU/animal resulted in 70–90% mortality. *V. harveyi* was also isolated from the diseased prawns and 100% mortality occurred with 10^7 CFU/animal.

Vibrio campbellii was identified along with a number of other *Vibrio* species as being part of the haemolymph of healthy spider crab (*Maja brachydactyla*) caught off the coast of Spain and the Canary Islands (Gomez-Gil *et al.*, 2010). The numbers of bacterial flora in healthy haemolymph is usually in the order of 10^3–10^4 CFU/ml (Gomez-Gil *et al.*, 1998).

GROWTH CHARACTERISTICS AND PHENOTYPIC IDENTIFICATION. On BA, colonies are pinpoint, grey, translucent with weak lysis seen around the pool, with slight greening of the agar in this area. Colonies are 2 mm, pale grey and shiny with a zone of lysis at 48 h. A double zone of lysis is seen at 72 h. On MSA-B, colonies are 1 mm, shiny and slightly mucoid at 24 h, and 1.8 mm light grey at 48 h with a double zone of beta-haemolysis. Colonies on MA 2216 are 0.8–1.3 mm, cream, slightly mucoid and at 48 h colonies are either round or elongated 1.6–2.5 mm. On MSSA, colonies are <0.5 mm, grey, shiny and slightly mucoid at 24 h, and 2.5 mm in size and mucoid at 48 h. On TCBS, colonies are green with a slightly darker green centre and are <0.5 mm at 24 h and 2.7 mm at 48 h. Colonies are non-swarming and non-luminescent on these media. Motility is negative in the tube test but positive in a wet preparation. These and other reactions are presented in Tables 3.22, 3.24–26 and 3.38.

MOLECULAR IDENTIFICATION. See under *Vibrio harveyi* for a PCR using the *rpoD* and *toxR* housekeeping genes followed by sequencing and analysis using concatenated sequences to enable differentiation between strains comprising the Vibrio core group.

A multiplex PCR based on the haemolysin gene (*hly*) for detection and differentiation of *V. campbellii*, *V. harveyi* and *V. parahaemolyticus* (Haldar *et al.*, 2010) and the *gyrB* gene for *V. alginolyticus* (Luo and Hu, 2008) is detailed under *V. harveyi*.

Vibrio carchariae

See *Vibrio harveyi*.

Vibrio caribbeanicus

TAXONOMY AND SIGNIFICANCE. *Vibrio caribbeanicus* (N384^T=DSM 23640^T=ATCC BAA-2122^T) was isolated from a marine sponge *Scleritoderma cyanea*, phylum Porifera, collected at a depth of 795 feet (242m)off the west coast of the island of Curaçao situated in the Caribbean Sea off the coast of Venezuela (Hoffman *et al.*, 2012b). *Vibrio caribbeanicus* is most closely related to *V. brasiliensis* (98.8%), *V. nigripulchritudo* (98.5%), *V. tubiashii* (98.6%) and *V. sinaloensis* (98.2%) based on 16S rRNA sequencing. The G+C content is 41.6 mol%. MLSA concatenated sequence analysis using *ftsZ*, *gyrB*, *mreB*, *pyrH* and *rpoA* identifies *V. caribbeanicus* as belonging to the Orientalis clade. Trees constructed from the genes *gapA* and *recA* group *V. caribbeanicus* with *V. alginolyticus* whereas analysis using *topA* clusters the isolate with *Aliivibrio fischeri*. Sequence comparisons indicate high similarity of 99.9% with both *V. alginolyticus* and *A. fischeri*. The incongruence between the gene data sets suggests genetic homogeneity for *gapA*, *recA* and *topA* between *V. caribbeanicus*, *V. alginolyticus* and *A. fischeri* and is probably the result of horizontal gene transfer within these alleles (Hoffman *et al.*, 2012b).

GROWTH CHARACTERISTICS AND PHENOTYPIC IDENTIFICATION. *Vibrio caribbeanicus* is a facultatively anaerobic, motile, oxidase-positive, Gram-negative, slightly curved rod (2.5–3.1×1.0 μm). Colonies are 1–2 mm, cream, translucent, round, smooth and non-swarming on MA 2216, and on TSA containing 1% NaCl (w/v) after 24 h at 28°C. Green, round, convex colonies 1 mm in size are produced on TCBS after 48 h. Luminescence is negative. Negative for catalase. Growth occurs at 10–35°C but not at 37°C, and at 0.5–8.0% NaCl and pH 6.0–10.0. Optimum growth is at 28°C, 1.5% NaCl and pH 7. DNase is positive using TSA containing 1.5% NaCl, DNA and methyl green (Hoffman *et al.*, 2012b). Results are negative for hydrolysis of starch, Tween 80 (lipase) and casein. These and other reactions are presented in Tables 3.23–30 and 3.38. Sets are inoculated with an inoculum in 1.5% NaCl and incubated at 28°C for 48 h.

Vibrio casei

TAXONOMY AND SIGNIFICANCE. *Vibrio casei* (WS 4539^T=DSM 22364^T=LMG 25240^T) was isolated from the surfaces of two smear-ripened French cheeses (Munster and Livarot) (Bleicher *et al.*, 2010). *Vibrio casei* is most closely related to *V. rumoiensis* (96.3%) and *V. litoralis* (95.9%) based on 16S rRNA sequencing with DNA–DNA hybridization of 38–42% and 28–37%, respectively, delineating the new species.

GROWTH CHARACTERISTICS AND PHENOTYPIC IDENTIFICATION. *Vibrio casei* is a facultatively anaerobic, motile, oxidase-positive, Gram-negative straight rod (0.6–1.2×1.5–1.8 μm). On MA 2216, colonies are 1–2 mm, cream, round or slightly irregular in shape, non-swarming after 3 days at 30°C (Bleicher *et al.*, 2010). Luminescence is negative on MA 2216. Growth occurs at 2–30°C, a pH of 6–8.5 and at 2–10% NaCl but not at 37°C or 0 or 12% NaCl. These and other reactions are presented in Table 3.23, limited reactions in Tables 3.24–29, and Table 3.38. An inoculum of 1.5% NaCl is used with a cell opacity of McFarland 3, and incubation temperature of 30°C with time of 24 h for the API 20NE and 6 h for the API Zym.

Vibrio celticus

TAXONOMY AND SIGNIFICANCE. *Vibrio celticus* (Rd 8.15^T=CECT 7224^T=LMG 23850^T) is a newly described species isolated from clams (*Venerupis* [*Ruditapes*] *philippinarum* and *V. senegalensis*) cultured in north-western Spain (Beaz-Hidalgo *et al.*, 2010a). *Vibrio celticus* belongs to the Splendidus clade based on sequence information from the 16S rRNA gene (98.6% similarity to eight species within the clade) and the housekeeping genes, *rpoA*, *rpoD*, *recA* and *atpA* (98.7–99.9% similarity). Pathogenicity tests demonstrate virulence of *V. celticus* for both species of clams. Cumulative mortalities of 100% were obtained when clams were challenged with 10^4 or 10^6 CFU/ml.

GROWTH CHARACTERISTICS AND PHENOTYPIC IDENTIFICATION. *Vibrio celticus* is a motile, facultatively anaerobic, large, Gram-negative rod of variable size (1.3–1.5×0.78–0.81 μm) that is oxidase positive, non-luminescent and susceptible to vibriostatic agent 0/129 150 μg. On MA agar colonies are round, smooth, beige and non-swarming, and yellow on TCBS after 24 h at 24°C. NaCl is required for growth with optimal concentration of 1–3% (w/v), with growth at 6% NaCl but not at 8%. Growth is optimal at 25°C and occurs at 4°C, but growth does not occur at 35°C. Glucose is fermented without gas production; starch and Tween 80 are hydrolysed (Beaz-Hidalgo *et al.*, 2010a). These and other reactions are presented in Tables 3.22, 3.24–26 and 3.38.

Vibrio chagasii

TAXONOMY AND SIGNIFICANCE. *Vibrio chagasii* (LMG 21353^T=CAIM 431^T) was isolated from the gut of turbot larvae (*Scophthalmus maximus*) from Norway, rotifer (*Brachionus plicatilis*) from Greece, water from Greece, *Artemia* species from Spain and seabass (*Dicentrarchus labrax*) from Greece (Thompson *et al.*, 2003d). The pathogenicity was not determined.

In experimental challenge using rainbow trout (*Oncorhynchus mykiss*) and *Artemia* nauplii inoculated with 10^6 CFU/fish or 10^6 CFU/ml, respectively, *V. chagasii* (LMG 21353^T) was rated as non-virulent or low virulence. Although there was some evidence of exophthalmia, kidney necrosis, ascites and liquefaction of internal organs, clinical disease was not seen, and at the end of the experiment the organism was recovered from tissue at either a low rate or not at all. Strains LMG 13219 and LMG 21354 did not cause any disease signs (Austin *et al.*, 2005).

Vibrio chagasii together with *V. kanaloae* and *V. pomeroyi* are phylogenetically related to *V. splendidus*. MLSA identifies *V. chagasii* as belonging to the Splendidus clade, which contains *V. splendidus*, *V. crassostreae*, *V. cyclitrophicus*, *V. fortis*, *V. gigantis*, *V. kanaloae*, *V. lentus*, *V. pelagius*, *V. pomeroyi* and *V. tasmaniensis* (Sawabe *et al.*, 2007b). Members of this clade have >30% DDH and 39–47 GC mol%, 90.6–99.8% similarity based on concatenated MLSA sequence data and AAI of 96.5–99.8%.

GROWTH CHARACTERISTICS AND PHENOTYPIC IDENTIFICATION. *Vibrio chagasii* is a facultatively anaerobic, motile Gram-negative slightly curved rod (1.0×2–3 μm), oxidase and catalase positive. On

TSA supplemented with 2% NaCl, colonies are 3–4 mm, non-swarming, opaque, dark beige, smooth, round, with an entire edge after 48 h at 28°C. The type strain produces yellow colonies on TCBS, whereas 12 other strains produce green colonies. NaCl is required for growth with growth occurring up to 6% NaCl, but not at 0 or 10% NaCl. Growth occurs at 15°C but not at 35°C. Optimal conditions are 28°C and 2% NaCl. Luminescence is not produced. ADH is positive for the type strain whereas 58% of strains are negative (Thompson *et al.*, 2003d). Other reactions are presented in Tables 3.23, 3.24–26 and 3.38.

Vibrio cholerae

TAXONOMY AND SIGNIFICANCE. *Vibrio cholerae* (ATCC 14035ᵀ, NCTC 8021ᵀ), serogroup 01 and 0139, is the cause of cholera in humans and was first described in Italy in 1854 by Dr Filippo Pacini as the cause of the Asiatic Cholera Pandemic of 1846–63 (Hugh, 1964).

A biotyping scheme based on the fermentation of sucrose, D-mannose and L-arabinose categorizes the strains into biotypes known as Heiberg types I to VIII (Heiberg, 1936). A further subdivision places strains into serogroups based on the somatic O antigen, with over 200 serogroups being recognized. Two biotypes designated El Tor and Classical cannot be differentiated serologically and belong to serovar 01. A classification scheme was suggested where *V. cholerae* isolated from cholera patients was classified as 01, and all other strains were classified as non-01. Non-01 strains cause gastroenteritis and septicaemia and other infections in humans, and also have been reported to cause infections in terrestrial and aquatic animals. Serovar 01 causes the classical cholera epidemics seen in humans in Asia, Africa and South America. Serotype 0139 Bengal, also an epidemic strain, arose in Bangladesh. Antisera is commercially available for the detection of 01 and 0139 strains. The non-01 serogroups can still cause disease in humans in the form of diarrhoea, wound infections and sepsis (Dakin *et al.*, 1974; Furniss *et al.*, 1978; Albert *et al.*, 1993; Sack *et al.*, 2004).

Vibrio cholerae non-01 is widespread in aquatic environments (Lee *et al.*, 1982). *Vibrio cholerae* has been reported as a disease in some aquatic animals and in others can exist as a carrier status.

Some laboratories may isolate *V. cholerae* from imported fish held in quarantine and it is important to have the diagnostic capacity to differentiate *V. cholerae* non-01 from *V. cholerae* 01 and 0139, which are responsible for cholera in humans, especially in those countries which are free of the disease.

Vibrio cholerae non-01 was isolated from apparently healthy goldfish (*Carassius auratus*) in an aquarium and from mortalities of farmed tilapia (*Oreochromis niloticus*) from Chesapeake Bay in the USA (Lupiani *et al.*, 1993). Experimental challenge using intraperitoneal injection demonstrated that *V. cholerae* non-01 was not pathogenic to tilapia as the LD_{50} was $>10^7$. An opportunistic infection with *V. cholerae* non-01 caused mortalities in goldfish (*C. auratus*) held in a quarantine facility in Australia (Reddacliff *et al.*, 1993). Infected fish had small 1 mm lesions on the body and mucus and hyperplasia in the gills. *Vibrio cholerae* non-01 has caused epizootics in ayu (*Plecoglossus altivelis*) in the rivers of Japan (Kiiyukia *et al.*, 1992). The isolate did not agglutinate with non-01 antisera but possessed other somatic antigens. It was biochemically different and was negative for ODC.

Phylogenetic analysis using sequences from the 16S rRNA gene, lux operon (*luxRI*) and glyceraldehyde phosphate dehydrogenase (*gapA*) gene indicate substantial similarity with *V. mimicus*, which is thought to be an atypical strain of *V. cholerae* (Nishiguchi and Nair, 2003). The 16S rRNA gene has sequence similarity of 99.06–100% to *V. mimicus* and 70% DDH similarity (Kita-Tsukamoto *et al.*, 1993).

MLSA identified *V. cholerae* as belonging to the Cholerae clade, which also contains *V. cincinnatiensis*, *V. furnissii*, *V. fluvialis*, *V. metschnikovii* and *V. mimicus* (Sawabe *et al.*, 2007b). Members of this clade have 44–50 GC mol%, a MLSA concatenated similarity of 85.4–94.7% and AAI of 92.8–99.6%.

GROWTH CHARACTERISTICS AND PHENOTYPIC IDENTIFICATION. *Vibrio cholerae* is a facultatively anaerobic, motile, oxidase positive, asporogenous Gram-negative rod (0.6–0.3 μm) that is straight or slightly curved (Hugh, 1964). On BA and MSA-B, colonies are round, smooth, 2–3 mm and greenish-grey. The production of beta-haemolysis is variable. Traditionally, the classical and El Tor biotypes were differentiated by haemolysis, as the classical strains are non-haemolytic and the El Tor biotype is haemolytic, however this has changed and now there are beta-haemolytic and non-haemolytic El Tor strains. On plates prepared with sheep red blood cells, non-haemolytic strains may produce a greenish colour, which occurs in the 'pool' where there is heavy growth, but is not seen around isolated colonies. This is due to the production of metabolic by-products and the effect occurs with prolonged incubation (>24 h). Colonies are yellow on TCBS.

Vibrio cholerae 0139 is a Gram-negative curved rod that is beta-haemolytic. On BA and MSA-B, colonies are greyish-opaque with darker centres. *Vibrio cholerae* 0139 is resistant to vibriostatic agent pteridine 0/129 (Albert *et al.*, 1993). Resistance to the vibriostatic agent pteridine 0/129 at 150 µg has been noted in some developing countries. Vibriostatic agent pteridine, available as commercial discs at concentrations of 0/129 50 µg and 150 µg, is used as one of the means of differentiating *Vibrio* species from *Aeromonas* species, but caution may be required (Huq *et al.*, 1992; Nair and Holmes, 1999).

A zone of 9 mm is classified as susceptible for the 0/129 500 µg disc for *Vibrio* species (Bernardet and Grimont, 1989). *Vibrio cholerae* strain 0139 is resistant to 0/129 500 µg disc (Albert *et al.*, 1993; Islam *et al.*, 1994).

Growth does not occur on deoxycholate agar. Growth occurs at 4–42°C and at 0–6% NaCl. Fermentation is positive for galactose, maltose, mannitol, mannose and sucrose. Gas is not produced from glucose. Fermentation is positive for lactose and trehalose but is delayed from 2 to 4 days. Fermentation is negative for arabinose, cellobiose, inositol, raffinose, salicin and xylose. Indole, citrate and gelatin are positive. Urea is negative. After the addition of methyl red reagent the MR test colour is orange rather than red (positive) or yellow (negative). VP is weakly positive. Casein, DNase, elastase and lipase are hydrolysed (Hugh, 1964; Lupiani *et al.*, 1993). Lupiani *et al.* (1993) reported strains of *V. cholerae* 01 as negative for VP, whereas a majority of *V. cholerae* non-01 are positive.

Vibrio cholerae is negative for aesculin, however some strains cause blackening of the medium due to the production of melanin (Coyne and Al-Harthi, 1992). To determine true hydrolysis as opposed to melanin production, the aesculin tube must be tested for loss of fluorescence (Choopun *et al.*, 2002). Aesculin solution will fluoresce under long-wave ultraviolet (UV) light (354 nm), therefore a loss of fluorescence indicates hydrolysis of aesculin and a positive test result (MacFaddin, 1980). Some plastic tubes are UV opaque, so to detect if fluorescence is present or not, the tube of medium must be examined under the UV light with the top removed, or the test medium is poured into a Petri dish and examined with a Wood's UV light in a darkened room. Usually fluorescence is seen without the need to examine in a darkened room. An uninoculated control and a negative result will show a ring of fluorescence around the edge of the liquid, whereas a true positive for hydrolysis of aesculin will not fluoresce. The operator's exposed skin must be protected with laboratory coat and gloves and the eyes protected with UV opaque glasses.

Vibrio cholerae non-01 strains isolated from ayu in Japan were found to be negative for ornithine decarboxylase (ODC) (Kiiyukia *et al.*, 1992). Normally, *V. cholerae* isolates are positive for ODC.

These and other reactions are presented in Tables 3.22, 3.24–29 and 3.38.

MOLECULAR IDENTIFICATION. A multiplex PCR for detection of *V. cholerae*, *V. mimicus*, *V. parahaemolyticus* and *V. vulnificus* (Tarr *et al.*, 2007) is detailed under *V. parahaemolyticus*. A multiplex PCR based on collagenase gene for detection of *V. cholerae*, *V. alginolyticus* and *V. parahaemolyticus* (Di Pinto *et al.*, 2005, 2006) is detailed under *V. alginolyticus*. A multiplex PCR was designed for the detection of the *atpA* gene (encodes the ATP synthase subunit A) in *V. cholerae*, *V. parahaemolyticus* and *V. vulnificus* (Izumiya *et al.*, 2011) (see under *V. parahaemolyticus*). The PCR was specific and gave comparable results to the PCR that detects *V. cholerae* based on the *ompW* gene (Nandi *et al.*, 2000).

Vibrio cincinnatiensis

TAXONOMY AND SIGNIFICANCE. *Vibrio cincinnatiensis* (ATCC 35912[T]) was isolated in the USA from a patient with meningitis (Brayton *et al.*, 1986) and it has been found in brackish and freshwater, and in the estuarine and coastal waters of Chesapeake Bay, USA. In the Adriatic Sea, Italy, it has been reported from mussels (*Mytilus galloprovincialis*) and may be a source of contamination for human infections. In China it has been reported as a pathogen for mud crab. *Vibrio cincinnatiensis* was found to be resistant to complement-mediated lysis when tested using the humoral fluids of the Japanese lancelot or amphioxus (*Branchiostoma belcheri*) (Ripabelli *et al.*, 1999; Li *et al.*, 2008b).

MLSA identified *V. cincinnatiensis* as belonging to the Cholerae clade, which also contains *V. cholerae*, *V. furnissii*, *V. fluvialis*, *V. metschnikovii* and *V. mimicus* (Sawabe *et al.*, 2007b). Members of this clade have 44–50 GC mol%, a MLSA concatenated similarity of 85.4–94.7% and AAI of 92.8–99.6%. Based on 5S rRNA sequence analysis *V. cincinnatiensis* is most closely related to *V. gazogenes* (98%), and also related to *V. mimicus*, *V. fluvialis* and *V. metschnikovii* (Brayton *et al.*, 1986). *Vibrio gazogenes* belongs to the Gazogenes clade according to MLSA.

GROWTH CHARACTERISTICS AND PHENOTYPIC IDENTIFICATION. *Vibrio cincinnatiensis* is a facultatively anaerobic, motile, oxidase-positive, Gram-negative bacillus (0.7×2.0 μm). On NA (nutrient agar) containing 1% NaCl colonies are 1–2 mm, cream, round, smooth, glossy after 24 h at 25°C, and grow as yellow colonies on TCBS (Brayton *et al.*, 1986). Luminescence is not seen on a nutrient agar medium containing NaCl, $MgCl_2$ and KCl reported by West and Colwell (1984). Growth occurs at 25–35°C but not at 4 or 42°C. Growth does not occur in the absence of NaCl or at 8% NaCl but occurs in the range 1–6%. Chitinase and amylase are produced but not alginase, caseinase or elastase. Hydrolysis is negative for Tween 20, 40, 60 and 80. Results are negative for Thornley's ADH. These and other reactions are presented in Tables 3.23 and 3.24–26. Sets are incubated according to the manufacturer's instructions with the exception that an inoculum is used of 1% NaCl.

Vibrio comitans

TAXONOMY AND SIGNIFICANCE. *Vibrio comitans* (GHG2-1[T]=LMG 23416[T]=NBRC 102076[T]) was isolated from the intestine of wild-caught abalone (*Haliotis discus discus*, *H. gigantea*, *H. madaka*) collected off the coast of Japan (Sawabe *et al.*, 2007a). Strains of *V. comitans*, together with *V. inusitatus* and *V. rarus*, are closely related to *V. superstes* (99.3–99.5%) based on 16S rRNA sequencing. *Vibrio comitans* has 94.6–94.8% similarity to *V. inusitatus* based on the *gapA* gene.

GROWTH CHARACTERISTICS AND PHENOTYPIC IDENTIFICATION. *Vibrio comitans* is a facultatively anaerobic, non-motile, oxidase-positive, Gram-negative rod (0.5–1.0×1.2–2.0 μm). On MA 2216 agar colonies are beige, circular, round and convex with an entire edge. Green colonies are produced on TCBS (Sawabe *et al.*, 2007a). Luminescence is not seen. Growth occurs at 4°C (for 83% of strains) and at 30°C but not at 37°C. Growth does not occur in the absence of NaCl but occurs at 1–6% NaCl. Hydrolysis is positive for alginate, and negative for starch, chitin, Tween 80 and agar. These and other reactions are presented in Table 3.23; some reactions are reported in Tables 3.24–29.

Vibrio communis

TAXONOMY AND SIGNIFICANCE. *Vibrio communis* (R-40496[T]=LMG 25430[T]=CAIM 1816[T]) was isolated from mucus of apparently healthy coral (*Mussismilia hispida*, *Phyllogorgia dilatata*, *Palythoa caribaeorum*, *P. variabilis*) endemic to the coast of Brazil, and prawn or white shrimp (*Litopenaeus vannamei*) cultured in Ecuador (Chimetto *et al.*, 2011a). The eight isolates described from these marine animals belonged to AFLP cluster A31 (Thompson *et al.*, 2001b) and are closely related to the *Vibrio harveyi* clade based on 16S rRNA sequencing (more than 99% similarity), *gyrB* (83.9–91.7%), *recA* (89.0–98.3%), *rpoA* (95.8–98.3%), *topA* (80.7–90.7%), *pyrH* (87.5–96.3%) and *mreB* (86.4–96.5%) (Chimetto *et al.*, 2011a). The closest phylogenetic neighbours based on 16S rRNA sequencing are *V. rotiferianus* (99.5%) and *V. mytili* (98.8%). MLSA using concatenated sequence information from protein-coding genes, *rpoA*, *pyrH*, *topA*, *ftsZ* and *mreB*, suggests *V. communis* is the same species as *V. owensii*

(Cano-Gomez *et al.*, 2011). The evidence that *V. communis* is identical to *V. owensii* was confirmed using concatenated sequence information from *topA*, *ftsZ*, *toxR*, *rctB*, *rpoD* and *mreB* (Hoffman *et al.*, 2012a. The biochemical reactions are the same for both organisms. Hoffman *et al.* (2012a also identified that some strains (LMG 11659, LMG 10948, LMG 4043) previously assigned to *V. harveyi* are in fact *V. communis*.

Vibrio communis was isolated from apparently healthy corals and it is hypothesized that it may be involved in a protective effect against pathogens or assist in nitrogen fixation.

GROWTH CHARACTERISTICS AND PHENOTYPIC IDENTIFICATION. *Vibrio communis* is a facultatively anaerobic, motile, oxidase-positive, Gram-negative rod (2.5–3.5×1.0 μm). On TSA containing 2% NaCl, colonies are 1 mm, beige, translucent, smooth and convex with an entire edge after 24 h at 28°C (Chimetto *et al.*, 2011a). Yellow colonies are seen on TCBS. Prolific growth occurs at 15–37°C and at 0.5–6.0 NaCl (w/v) but growth is weak at 8.0% NaCl. Growth does not occur in the absence of NaCl or above 9% NaCl and at 7°C or below, or at 42°C. Some biochemical diversity was noted amongst the eight isolates with 87.5% of strains positive for hydrolysis of gelatin and 87.5% of strains negative for citrate utilization and one of eight strains was negative for ODC. These and other reactions are presented in Tables 3.23, 3.24–26 and 3.38. Sets are incubated according to the manufacturer's instructions with the exception that an inoculum is used of 2.0% NaCl.

MOLECULAR IDENTIFICATION. Currently no PCR test is available for *V. communis*. The following sequence information is available in Genbank, EMBL and DDBJ: 16S rRNA (GU078670-GU078676, AJ345066), *gyrB* (GU078680-GU078684), *recA* (GU078693-GU078696, EU717055, EF596446), *rpoA* (GU078697-GU078703, AJ842625), *topA* (GU078704-GU078710), *pyrH* (GU078691, GU078692, EU251617, EU251620, EU251630, EU251638, EU716913, EF596375), *mreB* (GU078685-GU078690) (Chimetto *et al.*, 2011a).

Vibrio coralliilyticus

TAXONOMY AND SIGNIFICANCE. *Vibrio coralliilyticus* ($YB1^T$=ATCC BAA-450^T=LMG 20984^T) causes tissue lysis of coral (*Pocillopora damicornis*) found in the Indian Ocean and the Red Sea. *Vibrio coralliilyticus* has also been reported from diseased oyster larvae in Kent, UK and from bivalve larvae in Brazil (Ben-Haim *et al.*, 2003a). All strains from these sources are pathogenic to coral (*Pocillopora damicornis*) causing tissue lysis and death when tested in pathogenicity experiments (Ben-Haim and Rosenberg, 2002). In the same experiment *V. shilonii* (considered to be *V. mediterranei*) did not cause tissue lysis of coral. Pathogenicity is related to water temperature as tissue damage occurs only at temperatures greater than 25°C and at a temperature of 29°C tissue lysis occurs within 2 weeks. A large increase in the production of an extracellular protease produced by *V. coralliilyticus* occurs in the temperature range 24–28°C, which suggests lysis and death of the coral is due to the protease. In contrast, coral bleaching appears to be due to bacterial infection of the photosynthetic microalgal endosymbionts (zooxanthellae), which exist intracellularly in a symbiotic relationship with the coral animals. *Vibrio shilonii* (*V. mediterranei*) does not cause bleaching in corals lacking zooxanthellae. It is thought that coral bleaching by *V. coralliilyticus* is also a result of infection of the micro-algae (Ben-Haim *et al.*, 2003b).

In experimental challenge, *V. coralliilyticus* (LMG 20984^T and LMG 21349) caused 100% mortality at 24 h when inoculated into rainbow trout (*Oncorhynchus mykiss*) at a dose of 10^4 cells/fish with an LD_{50} of 10^1 and 10^2 for each strain, respectively (Austin *et al.*, 2005). The two strains caused 90 and 95% mortality, respectively, in *Artemia* nauplii exposed to 10^6 CFU/ml. Extracellular product (ECP) (0.1 ml) from LMG 20984^T caused 100% mortality in rainbow trout whereas ECP from strain LMG 21349 was less virulent, causing 20% mortality. In rainbow trout, infection resulted in neurological signs, liquefaction of kidneys, extensive muscle necrosis and haemorrhaging, and ascites fluid. Both strains produce caseinase, gelatinase, elastase, phospholipase, lipase (Tween 80) and elastase and are beta-haemolytic on sheep blood agar.

A bacterium identified as *V. coralliilyticus*/*neptunius*-like was isolated from Greenshell™ mussel larvae (*Perna canalicula*) in New Zealand (Kesarcodi-Watson *et al.*, 2009). In experimental challenge the isolate caused 75% mortality of larvae at a concentration of 10^2 CFU/ml. Clinical signs included erratic swimming, loss of lipids, deformed velum with aggregation of bacteria around the velum, and deterioration and loss of soft tissue. Infection occurred in the digestive system. 16S rRNA sequencing suggested similarity to either *V. coralliilyticus* (99.3–99.5%) or *V. neptunius* (99.5%).

Vibrio coralliilyticus is most closely related to *V. tubiashii* (97.2%), *V. nereis* (96.8%) and *V. shilonii* (96.6%) based on 16S rRNA sequencing (Ben-Haim *et al.*, 2003a). MLSA identified *V. coralliilyticus* as belonging to the Coralliilyticus clade, which also contains *V. neptunius* (Sawabe *et al.*, 2007b). Members of this clade have >64% DDH, 45–46 mol% GC, a MLSA concatenated similarity of 95.6% and AAI of 99.6%. MLSA identified *V. nereis* as belonging to the Nereis clade and *V. tubiashii* as belonging to the Orientalis clade. *Vibrio shilonii* is a junior synonym of *V. mediterranei* and this isolate does not belong to a known clade at this time.

GROWTH CHARACTERISTICS AND PHENOTYPIC IDENTIFICATION. *Vibrio coralliilyticus* is a motile, facultatively anaerobic, Gram-negative rod (1.2–1.5×0.8 μm), catalase positive and oxidase positive. On MA 2216 agar and TSA supplemented with 2% NaCl, colonies are cream-coloured with smooth edges at 30°C after 48 h and 3 mm after 3 days (Ben-Haim *et al.*, 2003a). Colonies are yellow on TCBS and 3 mm at 3 days. Luminescence is negative. Growth occurs at 30°C and in 1–7% NaCl but not at 0 or 8% NaCl. The *V. coralliilyticus/neptunius*-like isolate from mussels grew on TSA supplemented with 2% NaCl and produced yellow colonies on TCBS. There is some biochemical variation between strains with two of the six strains positive for ADH as detected in the API 20NE, and four strains are negative (Ben-Haim *et al.*, 2003a). These and other reactions including those for the *V. coralliilyticus/neptunius*-like isolate from mussels are presented in Tables 3.22 and 3.27–29. An inoculum of 3% NaCl is used for the API 20NE.

MOLECULAR IDENTIFICATION. *Vibrio coralliilyticus* can be detected using a real-time PCR and hybridization probes directed to the *dnaJ* gene, which codes for the heat shock protein (Pollock *et al.*, 2010). The nucleotide sequences from ten *V. coralliilyticus* strains were aligned and primers were designed to detect a 128-bp region with single nucleotide polymorphisms conserved for these strains but different from other *Vibrio* species.

PCR type	qPCR with hybridization probe to the *dnaJ* gene of *V. coralliilyticus* (Pollock *et al.*, 2010)		
DNA extraction	Promega Wizard Prep DNA Purification Kit (Promega)		
PCR master mix			
Primer Vc_dnaJ_F1	5′-CGG TTC GYG GTG TTT CAA AA-3′		
Primer Vc_dnaJ_R1	5′-AAC CTG ACC ATG ACC GTG ACA-3′		
Probe Vc_dnaJ_TMP (Taqman)	5′-6-FAM-CAG TGG CGC GAA G-MGBNFQ-3′(6-FAM = 6-carboxyfluorescein, MGBNFQ is molecular groove binding non-fluorescent quencher)		
Reagent	**Stock**	**Final concentration**	**Volume ×1**
*Taq*Man buffer A	×10	×1	2 μl
AmpliTaq Gold DNA polymerase	5 U/μl	0.5 U	2 μl
dATP	10 mM	0.2 μM	0.4 μl
dGTP	10 mM	0.2 μM	0.4 μl
dCTP	10 mM	0.2 μM	0.4 μl
dUTP		0.4 μM	0.4 μl
AmpErase uracil N-glycosylase (UNG)	1 U/μl	0.2 U	4 μl
$MgCl_2$	25 mM	3 mM	2.4 μl
Primer Vc_dnaJ_F1	50 μM	0.6 μM	0.24 μl
Primer Vc_dnaJ_R1	50 μM	0.6 μM	0.24 μl
Probe Vc_dnaJ_TMP	10 μM	0.2 μM	0.4 μl
Water			6.12 μl
DNA			1 μl
Total volume			**20 μl**
Thermocycling	RotoGene 300		

Continued

Continued.

Step 1	**Step 2**
50°C, 120 s(UNG activation)	95°C, 15 s
95°C, 10 min	60°C, 60 s
×1	×40
Fluorescence measurement	FAM channel 470 nm excitation and 510 nm detection during annealing/ extension
Product	128 bp
Ct (positive)	9–14.5
Ct (negative)	>25.4
R^2 (linear regression coefficient)	≥0.998 for DNA and 0.953 for cells
Sensitivity	One Caribbean strain was not detected. In seawater 1 CFU/ml was detected, and in coral tissue 10^4 CFU/ml, and 0.1 pg of purified DNA
Specific against	*V. alginolyticus*, *V. brasiliensis*, *V. campbellii*, *V. fortis*, *V. furnissii*, *V. harveyi*, *V. natriegens*, *V. neptunius*, *V. ordalii*, *V. parahaemolyticus*, *V. proteolyticus*, *V. rotiferianus*, *V. splendidus*, *V. tubiashii*, *V. xuii*, *Enterovibrio calviensis*, *Escherichia coli*, *Psychrobacter* sp., *Shewanella* sp.

See Molecular section for degeneracy codes.

A loop-mediated isothermal PCR is available, which targets the *rpoA* gene in *V. coralliilyticus* (Liu *et al.*, 2010a).

A conventional PCR is also described using the F3 and B3 primers from the LAMP PCR (Liu *et al.*, 2010a). This PCR is 10-fold less sensitive than the LAMP PCR.

PCR type	LAMP PCR for detection of *V. coralliilyticus* strains (Liu *et al.*, 2010a)		
DNA extraction	Manual method – boiling. A loopful of bacterial cells is selected from a pure growth and emulsified in 200 µl distilled water and the cell solution boiled for 10 min. Boiled cells are frozen for 2 min and then centrifuged at 12,000 ×*g* for 5 min. The supernatant is used in the PCR reaction		
LAMP PCR master mix			
Primer Vc-FIP	5′-CAT WGC GAT CTC AGC GTT GTC TTT TGA TCG CTA ACC CAG AAC ACG-3′		
Primer Vc-BIP	5′-TGG TTA CGT TCC AGC TTC AGC TTT CAA CCA GTA GGC GAC CRA T-3′		
Primer Vc-F3	5′-TGG TTG CAG GTG ACA TCA C-3′		
Primer Vc-B3	5′-TCT ACT GGG CTG TAC GTA GC-3		
Reagent	**Stock**	**Final concentration**	**Volume ×1**
Buffer*	×2	×1	12.5 µl
Primer Vc-FIP	50 µM	0.4 µM	0.18 µl
Primer Vc-BIP	50 µM	0.4 µM	0.18 µl
Primer Vc-F3	50 µM	0.05 µM	0.02 µl
Primer Vc-B3	50 µM	0.05 µM	0.02 µl
Bst DNA polymerase	8 U		1 µl
Water			7.09 µl
DNA			2 µl
Total volume			**25 µl**

* 40 mM Tris-HCl (pH 8.8), 20 mM KCl, 16 mM $MgSO_4$, 20 mM $(NH_4)_2SO_4$, 0.2% Triton X-100, 1.6 M betaine, 2.8 mM dNTPs each

Reaction conditions	**Amplification**	**Termination**
	65°C for 70 min	80°C for 5 min
Interpretation	Cool to 4°C and observe white turbidity for positive result	

Continued

Continued.

Sensitivity	3.6×10³ CFU/ml (8 CFU/reaction)
Specific against:	*V. alginolyticus*, *V. anguillarum*, *V. campbellii*, *V. fluvialis*, *V. harveyi*, *V. mimicus*, *V. parahaemolyticus*, *V. splendidus*, *V. tubiashii*, *V. vulnificus*, *Escherichia coli*, *Staphylococcus aureus*, *Pseudomonas aeruginosa*, *Bacillus subtilis*

PCR type	Conventional PCR for detection of *V. coralliilyticus* strains (Liu *et al.*, 2010a)		
DNA extraction	Manual method – boiling. A loopful of bacterial cells is selected from a pure growth and emulsified in 200 µl distilled water and the cell solution boiled for 10 min. Boiled cells are frozen for 2 min and then centrifuged at 12,000 ×*g* for 5 min. The supernatant is used in the PCR reaction		
PCR master mix			
Primer Vc-F3	5′-TGG TTG CAG GTG ACA TCA C-3′		
Primer Vc-B3	5′-TCT ACT GGG CTG TAC GTA GC-3		
Reagent	**Stock**	**Final concentration**	**Volume ×1**
Buffer for ExTaq (TaKaRa)	×10	×1	2.5 µl
dNTP mix	25 mM	2.5 mM	2.5 µl
Primer Vc-F3	50 µM	0.10 µM	0.05 µl
Primer Vc-B3	50 µM	0.10 µM	0.05 µl
ExTaq DNA polymerase (TaKaRa)	5 U/µl		2.5 µl
Water			15.4 µl
DNA			2 µl
Total volume			**25 µl**

Thermocycling		
Step 1	**Step 2**	**Step 3**
94°C, 4 min	94°C, 60 s	94°C, 60 s
	55°C, 60 s	55°C, 60 s
	72°C, 90 s	72°C, 4 min
×1	×30	×1
Product bp	213 bp	
Sensitivity	3.6×10⁴ CFU/ml (72 CFU/reaction)	
Specific against:	*Vibrio alginolyticus*, *V. anguillarum* *V. campbellii*, *V. fluvialis*, *V. harveyi*, *V. mimicus*, *V. parahaemolyticus*, *V. splendidus*, *V. tubiashii*, *V. vulnificus*, *Escherichia coli*, *Staphylococcus aureus*, *Pseudomonas aeruginosa*, *Bacillus subtilis*	

Vibrio costicola

See *Salinivibrio costicola*.

Vibrio crassostreae

TAXONOMY AND SIGNIFICANCE. Five strains of *V. crassostreae* (LGP 7T ^{T}=LMG 22240^{T}=CIP 108327^{T}) were isolated from the haemolymph of diseased cultured oysters (*Crassostrea gigas*) in France. *Vibrio crassostreae* and *V. gigantis* (also isolated from haemolymph of diseased oyster (Le Roux *et al.*, 2005) are closely related to *V. splendidus* with a 98% similarity based on 16S rRNA gene sequence similarity (Faury *et al.*, 2004). It is thought that some of the strains isolated from outbreaks of summer mortality in oysters in France and identified as *V. splendidus* may be in fact *V. crassostreae*.

Vibrio crassostreae was part of the normal flora in which many *Vibrio* species were identified in haemolymph of healthy spider crab (*Maja brachydactyla*) caught off the coast of Spain and the Canary Islands (Gomez-Gil *et al.*, 2010). The numbers of bacterial flora in healthy haemolymph is usually in the order of 10^3–10^4 CFU/ml (Gomez-Gil *et al.*, 1998).

MLSA identified *V. crassostreae* as belonging to the Splendidus clade, which contains *V. splendidus*, *V. chagasii*, *V. cyclitrophicus*, *V. fortis*, *V. gigantis*, *V. kanaloae*, *V. lentus*, *V. pelagius*, *V. pomeroyi* and *V. tasmaniensis* (Sawabe *et al.*, 2007b). Members of this clade have >30% DDH and 39–47 GC mol%, 90.6–99.8% similarity based on concatenated MLSA sequence data and AAI of 96.5–99.8%.

GROWTH CHARACTERISTICS AND PHENOTYPIC IDENTIFICATION. *Vibrio crassostreae* is isolated on tryptone soy agar (TSA, Oxoid) containing 2% NaCl and incubated at 20°C for 48 h. Growth occurs between 4 and 25°C but not at 35°C. Growth is optimal at 2–4% NaCl, with variable growth at 6% NaCl and no growth at 8 or 10% NaCl (Faury *et al.*, 2004). Yellow, 5 mm colonies are formed on TCBS. Commercial kits API 20E and API 50CH are inoculated with an inoculum of 2% NaCl. These and other reactions are presented in Tables 3.22, 3.24–26 and 3.30.

Vibrio cyclitrophicus

TAXONOMY AND SIGNIFICANCE. *Vibrio cyclitrophicus* (P-2P44^T=ATCC 700982^T=PICC 106644^T) (originally named *V. cyclotrophicus*) is a phenanthrene-degrading bacterium isolated from a creosote-contaminated site in Eagle Harbor, USA (Hedlund and Staley, 2001). It is closely related to *V. splendidus* (98.3%) according to 16S rRNA sequencing. MLSA identified *V. cyclitrophicus* as belonging to the Splendidus clade, which contains *V. splendidus*, *V. chagasii*, *V. crassostreae*, *V. fortis*, *V. gigantis*, *V. kanaloae*, *V. lentus*, *V. pelagius*, *V. pomeroyi* and *V. tasmaniensis* (Sawabe *et al.*, 2007b). Members of this clade have >30% DDH and 39–47 GC mol%, 90.6–99.8% similarity based on concatenated MLSA sequence data and AAI of 96.5–99.8%.

The haemolymph of healthy spider crab (*Maja brachydactyla*) caught off the coast of Spain and the Canary Islands contained *V. cyclitrophicus* along with a number of other *Vibrio* species (Gomez-Gil *et al.*, 2010). The numbers of bacterial flora in healthy haemolymph is usually in the order of 10^3–10^4 CFU/ml (Gomez-Gil *et al.*, 1998).

In experimental challenge using rainbow trout (*Oncorhynchus mykiss*) and *Artemia* nauplii inoculated with 10^6 CFU/fish or 10^6 CFU/ml, respectively, *V. cyclitrophicus* (LMG 21354) was rated as non-virulent or low virulence. Although there was some evidence of exophthalmia, kidney necrosis, ascites and liquefaction of internal organs, clinical disease was not seen, and at the end of the experiment the organism was recovered from tissue at either a low rate or not at all (Austin *et al.*, 2005).

GROWTH CHARACTERISTICS AND PHENOTYPIC IDENTIFICATION. *Vibrio cyclitrophicus* is a facultatively anaerobic, motile, oxidase-positive, Gram-negative rod (1.5–2.5×0.6 μm). On MA 2216, colonies are cream, circular, flat and grow to 4 mm. Luminescence is not seen. Growth occurs at 4–37°C but not at 42°C. Growth does not occur in the absence of NaCl. Results are positive for casein hydrolysis, lecithinase and lipase (Tween 20 and Tween 80) and negative for alginase and chitinase. ADH is weak. These and other reactions are presented in Tables 3.23–26 and 3.38. Sets are inoculated with an inoculum in 2% NaCl and incubated at 25°C for 48 h.

Vibrio damsela

See *Photobacterium damselae* spp. *damselae*.

Vibrio diabolicus

TAXONOMY AND SIGNIFICANCE. *Vibrio diabolicus* (HE800^T=CNCM I-1629^T) was isolated from the dorsal integument of the Pompeii worm, a Polychaete Annelid, *Alvinella pompejana*, found in a deep-sea hydrothermal vent in the East Pacific rise (Raguénès *et al.*, 1997). Based on DNA–DNA re-association, *V. diabolicus* is most closely related to *V. mytili*, *V. nereis* and *V. tubiashii*.

GROWTH CHARACTERISTICS AND PHENOTYPIC IDENTIFICATION. *Vibrio diabolicus* is a facultatively anaerobic, motile, oxidase-positive, Gram-negative straight rod (0.8×2.2 μm). On MA 2216, colonies are 2 mm and non-pigmented, but are 9 mm, mucoid and swarming after 3 days at 25°C when MA 2216 agar is supplemented with 30 g/l glucose (Raguénès *et al.*, 1997). Luminescence is not seen. Growth occurs at 20–45°C (optimum 30–45°C) but not at 4°C. A seawater base is required for growth. Chitin is hydrolysed. These and other reactions are presented in Tables 3.23–26 and 3.38. The inoculum is prepared with 2% NaCl and the sets read after 48 h at 25°C.

Vibrio diazotrophicus

TAXONOMY AND SIGNIFICANCE. *Vibrio diazotrophicus* (strain 1^T=NS1^T=ATCC 33466^T), a nitrogen-fixing bacterium, was isolated from the intestinal tract of sea urchins collected in Nova Scotia, Canada, and from the surface of reeds, ditch water and ditch sediment in a drainage ditch, Kent, UK (Guerinot *et al.*, 1982). It has been reported from a rotifer (*Brachionus plicatilis*) aquaculture system (Verdonck *et al.*, 1997). *Vibrio diazotrophicus* is closely related to *V. alginolyticus*, *V. campbellii*, *V. hispanicus* and *V. proteolyticus* (96.8–97.9%) based on 16S rRNA gene sequence similarity (Gomez-Gil *et al.*, 2004b). MLSA identified *V. diazotrophicus* as belonging to the Diazotrophicus clade, which also contains *V. hispanicus* (Sawabe *et al.*, 2007b). Members of this clade have 43–47 mol% GC, a MLSA concatenated similarity of 91.2% and AAI of 97.0%.

GROWTH CHARACTERISTICS AND PHENOTYPIC IDENTIFICATION. *Vibrio diazotrophicus* is a facultatively anaerobic, motile, oxidase-positive, Gram-negative short, straight or curved rod (1.5–2.0×0.5 μm). Cells may occur singly, or joined as S or spiral shapes (Guerinot *et al.*, 1982). Growth occurs on BA as 0.5 mm light grey colonies and as 1 mm light grey-cream colonies with weak lysis of the red cells after 24 h at 25°C. On MA 2216, colonies are 0.5 mm, off-white to cream after 48 h. Colonies are yellow on TCBS and show weak lactose fermentation on MCA. Luminescence is negative on an NA medium containing NaCl, KCl and $MgCl_2$ (Lee *et al.*, 1978). Hydrolysis is negative for chitin, casein, chondroitinase, elastin, Tween 80 and lecithinase. The oxidase reaction is strong. Aesculin is weak at 48 h and there is a deep black colour at 4 days. Salicin fermentation is difficult to read as the tube (5 ml) is yellow with a red band at the top of the tube. Growth occurs at 20–37°C but not at 42°C. Growth occurs in 0.5–10% NaCl but not at 0% NaCl. ADH is positive by Thornley's method and positive at 3–4 days using Moeller's method. Glucose in the API 20E may give a weak reaction at 48 h and nitrate may be negative even after the addition of zinc dust. These and other reactions are presented in Tables 3.23–26 and 3.38. Sets are inoculated with an inoculum in 2% NaCl and incubated at 25°C for 48 h.

MOLECULAR IDENTIFICATION. *Vibrio diazotrophicus* can be detected using primers VDIA-F (5′-AGA TTC TCT TGA TGA GTG CC-3′) and VDIA-R (5′-TAC CTA CAT CTC TAA GAG ACA TAG-3′), which produce an amplicon of 300 bp. The thermocycling programme consists of 35 cycles and an annealing temperature of 55°C (Lee *et al.*, 2002).

Vibrio ezurae

TAXONOMY AND SIGNIFICANCE. *Vibrio ezurae* (HDS1-1^T=IAM 15061^T=LMG 19970^T) was isolated from the gut of abalone (*Haliotis diversicolor aquatilis* and *H. diversicolor diversicolor*) in Japan (Sawabe *et al.*, 2004a). In experimental challenge, *V. ezurae* (LMG 19970^T) caused 70% mortality at 7 days when inoculated into rainbow trout (*Oncorhynchus mykiss*) at a dose of 10^6 cells/fish and 60% mortality with 10^4 cells/fish. The LD_{50} was 10^3 (Austin *et al.*, 2005). No mortalities occurred at an inoculation of 10^2 cells/fish. A mortality of 85% occurred for *Artemia* nauplii exposed to 10^6 CFU/ml. Extracellular product (ECP) (4 μg/ml) did not cause mortality in rainbow trout, but caused 90% mortality in *Artemia*. In rainbow trout, infection resulted in softening of muscles and presence of ascites fluid. LMG 19970^T was negative for caseinase, gelatinase, elastase, phospholipase, lipase (Tween 80), elastase and non-haemolytic on sheep blood agar.

Vibrio ezurae forms a strong clade with *V. neonatus*, *V. halioticoli*, *V. gallicus* and *V. superstes* using phylogenetic data based on the 16S rRNA and the *gap* (glyceraldehyde 3-phosphate dehydrogenase) gene and is most closely related to *V. neonatus* and *V. halioticoli* based on phenotypic tests and both *V. neonatus* and *V. ezurae* differ from *V. halioticoli* in only two of 72 phenotypic tests. MLSA identified *V. ezurae* as belonging to the Halioticoli clade, which also contains *V. halioticoli*, *V. gallicus*, *V. neonatus* and *V. superstes* (Sawabe *et al.*, 2007b). Members of this clade have >22% DDH, 39–42 mol% GC, a MLSA concatenated similarity of 88.0–97.7% and AAI of 94.7–99.5%.

GROWTH CHARACTERISTICS AND PHENOTYPIC IDENTIFICATION. *Vibrio ezurae* is a facultatively anaerobic, non-motile, Gram-negative rod (0.5–1.0×0.8–3.0 μm). It is oxidase and catalase positive. On MA 2216 agar, colonies are beige, circular, smooth and convex with an entire edge (Sawabe *et al*., 2004a. Growth occurs at 15–30°C with no growth at 4 or 37°C. Growth does not occur without Na^+ ions and is negative at 1 and 6% NaCl with growth occurring at 2 and 3% NaCl. Luminescence is negative on MA 2216 medium and results are negative for hydrolysis of gelatin, chitin, agar, lipid, starch and Tween 80. Hydrolysis of alginate is positive. *Vibrio ezurae* differs from *V. neonatus* with positive results for utilization of D-mannose and D-gluconate, whereas *V. neonatus* is negative for these tests. *Vibrio ezurae* differs from *V. halioticoli* with a positive reaction for utilization of D-gluconate, which is negative in *V. halioticoli* and *V. neonatus*. These and other reactions are presented in Table 3.23. Note that the carbohydrates ('sugars') in this table are the results from fermentation, not utilization. Some reactions are presented in Tables 3.24–26.

Vibrio fischeri

See *Aliivibrio fischeri*.

Vibrio fluvialis

TAXONOMY AND SIGNIFICANCE. *Vibrio fluvialis* (ATCC 33809^T, 606=NCTC 11327) was originally identified as Group F, a group of strains isolated from diarrhoea in humans from Bahrain and Bangladesh, and from shellfish and estuarine waters in England (Furniss *et al*., 1977; Lee *et al*., 1978). The strains were divided into aerogenic strains – those that produced gas from fermentation of glucose, and those that were anaerogenic – gas not produced (Lee *et al*., 1978; Seidler *et al*., 1980). The anaerogenic strains were named *V. fluvialis* (Lee *et al*., 1981), whereas the aerogenic strains were named *V. furnissii* (Brenner *et al*., 1983). *Vibrio fluvialis* is a cause of a bloody diarrhoea in humans and has been isolated from water and seafood product. It was isolated from diseased plaice (*Pleuronectes platessa*) fry cultured in Denmark, however in challenge experiments using salmon and turbot, the LD_{50} was $>10^6$ indicating low virulence (Pedersen *et al*., 1999a).

MLSA identified *V. fluvialis* as belonging to the Cholerae clade, which also contains *V. cholerae*, *V. cincinnatiensis*, *V. furnissii*, *V. metschnikovii* and *V. mimicus* (Sawabe *et al*., 2007b). Members of this clade have 44–50 GC mol%, a MLSA concatenated similarity of 85.4–94.7% and AAI of 92.8–99.6%.

An isolate identified as *Vibrio fluvialis*-like has been implicated in limp lobster disease (Tall *et al*., 2003). The organism is recovered from heart, mid-gut and haemolymph. Strains harbouring a plasmid were more virulent in challenge experiments than those strains without the plasmid.

GROWTH CHARACTERISTICS AND PHENOTYPIC IDENTIFICATION. *Vibrio fluvialis* is a facultatively anaerobic, short, straight or curved Gram-negative rod that occurs singly or in pairs and pleomorphism may be seen. It is oxidase and catalase positive. On BA, colonies are 2.5–3.5 mm, light grey, flat, shiny after 24 h at 25°C and 3.8–4.6 mm with colour becoming grey-green at 48 h. Haemolysis is strong and defined at 24 and 48 h. Growth on MSA-B and MSSA appears as 1.7 mm colonies, light grey with strong beta-haemolysis on MSA-B, and on MSSA colonies appear shiny and slightly mucoid. On MA2216 colonies are 3.3–4.7 mm, cream and mucoid and at 48 h are up to 5.6 mm in size and mucoid colonies can drip from the plate (similar to *Klebsiella pneumoniae*). Growth on MCA occurs as 0.3 mm pinpoint colonies at 24 h and 25°C and increase to 1.7 mm, shiny and pale pink at 48 h with good growth across the plate. Colonies on TCBS are 1 mm yellow at 24 h and 2.9 mm and a strong yellow colour, raised, round and shiny at 48 h. Strong positive reactions are seen at 48 h for gelatin using the gelatin plate method and DNase plate, and conventional tests for citrate, ONPG and aesculin (media section). Fermentation of glucose is positive at 25, 37 and 42°C. ADH is negative after 3 days in Moeller's medium but is positive in the API 20E after 24 h at 25°C incubation. Motility may appear negative in the tube test, but it is strongly motile in a wet preparation or hanging drop. Detection of colour change after the addition of up to 10 drops of Kovács' reagent is slow in the conventional tube test for indole but is rapidly positive in the API 20E (this manual). The API 20E database may identify these species as *Aeromonas* species (Brenner *et al*., 1983).

V. fluvialis is negative for L-rhamnose, and 63% of *V. fluvialis* are positive for cellobiose (Lee *et al.*, 1981; Brenner *et al.*, 1983). These and other reactions are presented in Tables 3.22, 3.24–26 and 3.38.

Vibrio fortis

TAXONOMY AND SIGNIFICANCE. *Vibrio fortis* (LMG 21557^T=CAIM 629^T) was isolated from healthy prawn (*Penaeus vannamei*) in Mexico, from Atlantic salmon in Tasmania, Australia, from healthy and diseased bivalve larvae (*Nodipecten nodosus*) from Brazil and from diseased oyster larvae (*Crassostrea gigas*) from Brazil (Thompson *et al.*, 2003c). It was originally identified as FALP Group A9 (Thompson *et al.*, 2001b). *Vibrio fortis* was identified with other *Vibrio* species as being part of the haemolymph of healthy spider crab (*Maja brachydactyla*) caught off the coast of Spain and the Canary Islands (Gomez-Gil *et al.*, 2010). The numbers of bacterial flora in healthy haemolymph is usually in the order of 10^3–10^4 CFU/ml (Gomez-Gil *et al.*, 1998).

In experimental challenge, *V. fortis* (LMG 21557^T and LMG 20547) caused 100% mortality at 2–3 days when inoculated into rainbow trout (*Oncorhynchus mykiss*) at a dose of 10^4 cells/fish. At an inoculation concentration of 10^2 cells/ fish strain LMG 21557^T caused 100% mortalities compared to 40% mortalities for strain LMG 20547. The LD_{50} was <10^2 and 10^2, respectively (Austin *et al.*, 2005). A mortality of 100% and 90%, respectively, occurred for *Artemia* nauplii exposed to 10^6 CFU/ml. Extracellular product (ECP) (19–28 µg/ml) caused 100% mortality in rainbow trout, and 95% and 80%, respectively, in *Artemia*. In rainbow trout, clinical signs included erratic swimming and spinning, exophthalmia, liquefaction of internal organs and muscles. The strains were positive for caseinase and gelatinase but only one strain (LMG 20547) was haemolytic to horse blood and trout blood.

Vibrio fortis is closely related to *Vibrio* (*Listonella*) *pelagius* (98.8%) according to 16S rRNA sequencing. MLSA identified *V. fortis* as belonging to the Splendidus clade, which contains *V. splendidus*, *V. chagasii*, *V. crassostreae*, *V. cyclitrophicus*, *V. gigantis*, *V. kanaloae*, *V. lentus*, *V. pelagius*, *V. pomeroyi* and *V. tasmaniensis* (Sawabe *et al.*, 2007b). Members of this clade have >30% DDH and 39–47 GC mol%, 90.6–99.8% similarity based on concatenated MLSA sequence data and AAI of 96.5–99.8%.

GROWTH CHARACTERISTICS AND PHENOTYPIC IDENTIFICATION. *Vibrio fortis* is a facultatively anaerobic, motile, oxidase-positive, Gram-negative slightly curved rod (3×1 µm). On TSA containing 2% NaCl, colonies are 4 mm, translucent to opaque, low, convex and non-swarming after 48 h at 28°C and yellow 4–5 mm colonies are produced on TCBS after 48 h. Growth occurs at 4–35°C and at 1–8% NaCl with an optimum salt content of 2.5% in media. Results for the type stain are presented in the following tables. Some wild-type strains give variable results for gelatin (19% positive), sucrose (37%), melibiose (62%), amygdalin (56%) and arabinose (0.6%) (Thompson *et al.*, 2003c). The type strain is positive for caseinase, phospholipase (egg yolk) and gelatinase and negative for lipase (Tween 80) and elastase, and negative for haemolysis of sheep, horse and rainbow trout red cells (Austin *et al.*, 2005). Strain LMG 20547 is haemolytic to horse blood and trout blood and positive for caseinase and gelatinase. These and other reactions are presented in Table 3.23, Tables 3.24–26 and 3.38. Sets are inoculated with an inoculum in 2% NaCl and incubated at 25°C for 48 h.

Vibrio furnissii

TAXONOMY AND SIGNIFICANCE. *Vibrio furnissii* (ATCC 35016^T=CDC B3215^T) was originally named Group F for a group of isolates from humans suffering from diarrhoea in Bahrain and Bangladesh, and from shellfish and estuarine waters in England (Furniss *et al.*, 1977; Lee *et al.*, 1978). Within this group some strains were aerogenic, that is, produced gas from fermentation of glucose, whereas other strains were anaerogenic with no gas production (Lee *et al.*, 1978; Seidler *et al.*, 1980). Aerogenic strains were named *Vibrio furnissii* and the anaerogenic strains *V. fluvialis* (Brenner *et al.*, 1983; Lee *et al.*, 1981). *Vibrio furnissii* has been isolated from human faeces, river water in England, pig and rabbit faeces in England and from water surrounding cultured amberjack (*Seriola dumerili*) in Spain (Lee *et al.*, 1978; Brenner *et al.*, 1983; Alcaide, 2003). *Vibrio furnissii* is pathogenic for European eels (*Anguilla anguilla*) at an LD_{50} of 10^6 CFU/fish when challenged intraperitoneally and the extracellular products are pathogenic to elvers (Esteve *et al.*, 1995c).

MLSA identified *V. furnissii* as belonging to the Cholerae clade, which also contains *V. cholerae*, *V. furnissii*, *V. fluvialis*, *V. metschnikovii* and *V. mimicus* (Sawabe *et al.*, 2007b). Members of this clade have 44–50 GC mol%, a MLSA concatenated similarity of 85.4–94.7% and AAI of 92.8–99.6%.

GROWTH CHARACTERISTICS AND PHENOTYPIC IDENTIFICATION. *Vibrio furnissii* is a facultatively anaerobic, motile, straight to slightly curved, Gram-negative rod. Oxidase positive. On BA, MSA-B and BHIA, colonies are opaque, shiny, smooth, round, domed, may be mucoid and are 2–3 mm after 24 h at 30°C. Yellow colonies are produced on TCBS. Results for ONPG may be variable. Fermentation of cellobiose is negative and 57% of strains are positive for fermentation of L-rhamnose. Indole is negative when tested in peptone water containing 1% NaCl. When indole is tested using heart infusion broth, 14% of *V. furnissii* are positive. Indole is positive using the API 20E. The API 20E may identify *V. furnissii* as *Aeromonas* species (Lee *et al.*, 1981; Brenner *et al.*, 1983). In the indole test using tryptone water, a weak reaction is seen even with the addition of approximately 12 drops of Kovác's reagent to a 48 h tryptone water (5 ml). The colour disappears after 1 min and appears as a dirty orange-brown colour. *Vibrio furnissii* is negative for ADH reaction in Moeller's medium if no NaCl is added to the reagent tube. ADH is positive when the final salt concentration is 2%. VP is negative in MRVP medium but may show a weak reaction in the API 20E (this manual). Gas is seen in the glucose fermentation tube. Simmons' citrate is positive. The main differentiating test from *V. fluvialis* is production of gas in fermentation tests, especially glucose. These and other reactions are presented in Tables 3.22, 3.24–26 and 3.38. The inoculum must be prepared with 1–2% NaCl.

Vibrio gallaecicus sp. nov.

TAXONOMY AND SIGNIFICANCE. *Vibrio gallaecicus* sp. nov. (VB 8.9^T=CECT 7244^T=LMG 24045^T) was isolated from healthy, cultured Manila clams (*Venerupis* [*Ruditapes*] *philippinarum*) from north-western Spain. Molecular and phenotypic profiling suggests *V. gallaecicus* belongs to the *V. splendidus* clade (Beaz-Hidalgo *et al.*, 2009c). Three isolates were first described from the clams and subsequently proposed as a new species, *V. gallaecicus*, within the genus *Vibrio*. The type strain is VB 8.9^T, and the other two isolates are identified as VB 5.12 and C 8.10. The three strains were 99.9% genetically similar to each other based on the 16S rRNA sequence, and the type strain was 97.6% similar to *V. splendidus* and 96.0% similar to *V. chagasii* based on 16S rRNA gene similarity (Beaz-Hidalgo *et al.*, 2009c).

GROWTH CHARACTERISTICS AND PHENOTYPIC IDENTIFICATION. *Vibrio gallaecicus* is a motile, facultative anaerobe, Gram-negative rod of variable size (1.45–1.82×0.76–0.91 μm), oxidase positive, nitrate positive and sensitive to vibriostatic agent 0/129 150 μg. Colonies are round, smooth, beige colour and non-swarming on MA plates at 25°C. On TCBS the type strain VB 8.9^T and C 8.10 produce 1–1.2 mm green-coloured colonies after 24 h, whereas isolate VB 5.12 produces 1–1.5 mm yellow-coloured colonies. All strains grow at 4°C and require NaCl for growth with an optimal range of 1–3% and no growth above 6% NaCl. Luminescence is not produced. Identification can be performed using traditional biochemical identification media, API 20E and API Zym using 0.85% NaCl as the inoculum. Tests using the API 50CH are performed with a modification to the manufacturer's specifications for inoculation. A bacterial suspension was prepared in 0.85% saline, adjusted to an optical density of 1.0 at 580 nm and mixed 1:90 v/v with ZOF medium (Beaz-Hidalgo *et al.*, 2009c). These and other reactions are presented in Tables 3.23–26, 3.30 and 3.38.

Vibrio gallicus

TAXONOMY AND SIGNIFICANCE. *Vibrio gallicus* (HT2-1^T=CIP 107863^T=LMG 21878^T) was isolated from the gut of healthy abalone (*Haliotis tuberculata*) collected from the coast of Brest (Brittany), France (Sawabe *et al.*, 2004b).

In experimental challenge using rainbow trout (*Oncorhynchus mykiss*) and *Artemia* nauplii inoculated with 10^6 CFU/fish or 10^6 CFU/ml, respectively, *V. gallicus* (LMG 21354) was rated as non-virulent. Clinical disease was not seen, and at the end of the experiment the organism was not recovered from tissue (Austin *et al.*, 2005).

Phylogenetic analysis indicated that the five strains of *V. gallicus* were related at <97% similarity to *Aliivibrio* (*Vibrio*) *fischeri*, *A. logei*, *A. salmonicida* and *A. wodanis* based on 16S rRNA gene sequence similarity and did not form a clade with any known *Vibrio* species (Sawabe *et al.*, 2004b). MLSA identified *V. gallicus* as belonging to the Halioticoli clade, which also contains *V. halioticoli*, *V. ezurae*, *V. neonatus* and *V. superstes* (Sawabe *et al.*, 2007b). Members of this clade have >22% DDH, 39–42 mol% GC, a MLSA concatenated similarity of 88.0–97.7% and AAI of 94.7–99.5%.

GROWTH CHARACTERISTICS AND PHENOTYPIC IDENTIFICATION. *Vibrio gallicus* is a non-motile, non-pigmented rod and requires the presence of NaCl for growth. Colonies on MA 2216 agar are beige, circular, smooth and convex. Growth occurs between 15 and 30°C but not at 37°C. Alginate is hydrolysed, but not chitin or agar. It is non-luminescent. It is phenotypically similar to *V. halioticoli* (Sawabe *et al.*, 2004b). These and other reactions are presented in Table 3.23.

Vibrio gazogenes

TAXONOMY AND SIGNIFICANCE. *Vibrio gazogenes* ($PB1^T$=ATCC 29988^T) (formerly *Beneckea gazogenes*) was isolated from mud of a saltwater marsh located near Woods Hole, Massachusetts, USA (Harwood, 1978). *Vibrio gazogenes* has 98% similarity to the 5S rRNA gene of *V. cincinnatiensis* (Ripabelli *et al.*, 1999), and forms a cluster with *V. mangrovi*, *V. ruber* and *V. rhizosphaerae* based on the 16S rRNA gene (Rameshkumar *et al.*, 2010). MLSA identified *V. gazogenes* as belonging to the Gazogenes clade, which also contains *V. aerogenes* and *V. ruber* (Sawabe *et al.*, 2007b). Members of this clade have >32% DDH and 46–47 GC mol%. *Vibrio cincinnatiensis* belongs to the Cholerae clade, and *V. mangrovi* and *V. rhizosphaerae* were not identified at the time of the MLSA report.

GROWTH CHARACTERISTICS AND PHENOTYPIC IDENTIFICATION. *Vibrio gazogenes* is a facultatively anaerobic, motile, oxidase-negative, nitrate-negative, slightly curved Gram-negative rod (1.3×2.5 μm). On MA 2216, colonies are 0.3–0.9 mm dry or rough, and red-pigmented after 24 h at either 25 or 36°C. White colonies may be present (Harwood, 1978; Farmer *et al.*, 1988). The pigment produced is prodigiosin (the same as produced by *Serratia marcescens*). *Vibrio gazogenes* was originally reported as oxidase positive but has been corrected as being negative for oxidase. Growth is poor on media containing 0.1–1% NaCl. Growth occurs at 1–10% NaCl and at 15–42°C with an optimum of 25–30°C. Results are negative for hydrolysis of alginate, cellulose and chitin and positive for lipase and positive for fermentation of cellobiose. ADH, LDC and ODC are negative by Moeller's methods. Luminescence and swarming are negative. These and other reactions are presented in Tables 3.23–26. An inoculum containing 2% NaCl is used and sets are incubated at 25°C.

Vibrio gigantis

TAXONOMY AND SIGNIFICANCE. Four strains of *V. gigantis* (LGP 13^T=LMG 22741^T=CIP 108656^T) were isolated from the haemolymph of diseased cultured oysters (*Crassostrea gigas*). The organism is closely related to *V. splendidus* with a 98% similarity based on 16S rRNA gene sequence similarity (Le Roux *et al.*, 2005). MLSA identified *V. gigantis* as belonging to the Splendidus clade, which contains *V. splendidus*, *V. chagasii*, *V. crassostreae*, *V. cyclitrophicus*, *V. fortis*, *V. kanaloae*, *V. lentus*, *V. pelagius*, *V. pomeroyi* and *V. tasmaniensis* (Sawabe *et al.*, 2007b). Members of this clade have >30% DDH and 39–47 GC mol%, 90.6–99.8% similarity based on concatenated MLSA sequence data and AAI of 96.5–99.8%.

GROWTH CHARACTERISTICS AND PHENOTYPIC IDENTIFICATION. *Vibrio gigantis* was isolated on tryptone soy agar (TSA, Oxoid) containing 2% NaCl and incubated at 20°C for 48 h. Colonies are translucent, non-swarming, round and entire. On TCBS, colonies are 5 mm and yellow. Commercial kits API 20E and API 50CH were inoculated with an inoculum of 2% NaCl (Le Roux *et al.*, 2005). Growth is optimum at 2–4% NaCl, with no growth at 6% NaCl, and growth at temperatures of 4°C and 25°C but not at 35°C. These and other reactions are presented in Tables 3.22, 3.24–26 and 3.30.

Vibrio halioticoli

TAXONOMY AND SIGNIFICANCE. *Vibrio halioticoli* (IAM 14596^T), an alginolytic bacterium, was isolated from the gut of abalone (*Haliotis discus hannai*). It is phenotypically close to other alginolytic bacteria, *Aliivibrio fischeri* and *Vibrio* (*Listonella pelagia*) *pelagius* (Sawabe *et al.*, 1998b). *Vibrio halioticoli* has 98.3% 16S rRNA gene similarity to *V. ezurae* and *V. neonatus* and together these bacteria form a clade that includes *V. gallicus* and *V. superstes* based on the 16S rRNA gene and the *gap* (glyceraldehyde 3-phosphate dehydrogenase) gene (Sawabe *et al.*, 2004b). *Vibrio halioticoli* is different from *V. neonatus* by only two of 72 phenotypic tests. MLSA identified *V. halioticoli* as belonging to the Halioticoli clade, which also contains *V. gallicus*, *V. ezurae*, *V. neonatus* and *V. superstes* (Sawabe *et al.*, 2007b). Members of this clade have >22% DDH, 39–42 mol% GC, a MLSA concatenated similarity of 88.0–97.7% and AAI of 94.7–99.5%.

GROWTH CHARACTERISTICS AND PHENOTYPIC IDENTIFICATION. *Vibrio halioticoli* is a facultatively anaerobic, non-motile, oxidase-positive, Gram-negative rod (1.7–2.0×0.6–0.8 μm). On MSA-B, colonies are pinpoint after 24 h at 25°C and show good growth at 48 h. On MA 2216, colonies are beige, circular, smooth and convex with an entire edge after 48 h. Growth occurs at 15–30°C with no growth at 37 or 40°C or below, or at 0% NaCl. Alginate is hydrolysed but not starch, chitin, Tween 80 or agar. Luminescence and swarming are negative (Sawabe *et al.*, 1998b). The nitrate test may be negative in the API 20E. These and other reactions are presented in Tables 3.23, 3.24–26 and 3.38. An inoculum containing 2% NaCl is used and sets are incubated at 25°C.

Vibrio hangzhouensis

TAXONOMY AND SIGNIFICANCE. *Vibrio hangzhouensis* (CN83^T=CGMCC 1.7062^T=JCM 15146^T) was isolated from sediment collected from the East China Sea (Xu *et al.*, 2009). Based on 16S rRNA sequencing the closest sequence similarity is with *V. agarivorans* (98.4%) and *V. campbellii* (97.8%).

GROWTH CHARACTERISTICS AND PHENOTYPIC IDENTIFICATION. *Vibrio hangzhouensis* is a facultatively anaerobic, motile, oxidase-positive, Gram-negative, straight or slightly curved rod (1–2×0.5–0.8 μm). On MA 2216, colonies are 1–2 mm, non-pigmented, smooth, circular with slightly irregular borders after 48 h at 25°C. Growth occurs on MacConkey agar as pink (lactose-fermenting) colonies (Xu *et al.*, 2009). Growth occurs at 20–37°C with no growth at 15°C or below, or at 42°C or above. Growth occurs at 0.5–7.5% with an optimum of 3% NaCl. Starch, Tween 20 and tyrosine are hydrolysed, but agar, casein and Tween 80 are not hydrolysed. These and other reactions are presented in Tables 3.23–29 and 3.38. Sets are inoculated according to the manufacturer's instructions and incubated at 30°C.

Vibrio harveyi

TAXONOMY AND SIGNIFICANCE. *Vibrio harveyi* (ATCC 14126^T) was isolated from a dead amphipod (*Talorchestia* species) at Woods Hole, Massachusetts, USA in 1935 and named *Achromobacter harveyi* (Johnson and Shunk, 1936). Later, it was assigned to the genus *Lucibacterium* and then *Beneckea* and with the abolition of the genus *Beneckea* it was assigned to the genus *Vibrio* (Baumann *et al.*, 1980).

Vibrio harveyi has been found in a wide range of aquatic habitats and may be part of intestinal flora or cause disease in a variety of aquatic animals. It is a major pathogen for many finfish and prawns (shrimp).

A bacterium isolated from a shark and named *V. carchariae* (Grimes *et al.*, 1984) was later found to be a junior synonym of *V. harveyi*, as the two organisms are indistinguishable by AFLP, DNA–DNA hybridization, ribotyping (Pedersen *et al.*, 1998) and 16S ribosomal DNA sequencing (Gauger and Gómez-Chiarri, 2002). A bacterium named *V. trachuri*, isolated from cultured Japanese horse mackerel (Iwamoto *et al.*, 1995), was found to be a junior synonym of *V. harveyi* (Thompson *et al.*, 2002b).

Vibrio harveyi (originally named *V. carchariae*) caused disease in captive sandbar sharks (*Carcharhinus plumbeus*) (Grimes *et al.*, 1984), and in physiologically compromised lemon sharks (*Negaprion brevirostris*) caused infection in experimental challenge (Grimes *et al.*, 1985). *Vibrio harveyi* was the agent responsible for an opportunistic infection in the eyes of common snook (*Centropomus undecimalis* Bloch) thought to be damaged during netting of broodstock from the wild (Kraxberger-Beatty and McGarey, 1990) in the USA. The organism was also reported from deep dermal lesions in a crevally jack (*Caranx hippos*) captured nearby. In summer flounder (*Paralichthys dentatus*) cultured in the USA, *V. harveyi* caused an outbreak of disease termed flounder infectious necrotizing enteritis (FINE) that was characterized by opaque, blood-stained fluid in the abdomen, distended abdomen, reddening around the anal area, necrosis and enteritis of the posterior intestine (Soffientino *et al.*, 1999). *Vibrio harveyi* was the predominant organism in the intestine and kidney. Other potential pathogens, *Photobacterium damselae* spp. *damselae*, *V. scophthalmi* and *V. ichthyoenteri* (from the kidneys), were isolated in low numbers, and in contrast to *V. harveyi*, did not cause mortality in experimental challenge when inoculated into fish at 10^7 CFU/fish. Differences in virulence of *V. harveyi* strains was noted with some strains causing no mortality whereas other *V. harveyi* strains caused mortality and classic signs of FINE (Gauger *et al.*, 2006). In South Korea, mass mortality of olive flounder (*Paralichthys olivaceus*) occurred following a co-infection of *V. harveyi*, *Edwardsiella tarda* and marine birnavirus. In experimental challenge via bath immersion of 1×10^5 cells/ml for 30 min, greater than 90% mortality occurred when fish were exposed to both bacteria (and virus) compared to 70% mortality when inoculated with each individual bacterium (and without virus). Virus alone did not cause mortality (Oh *et al.*, 2006). *Vibrio harveyi* was responsible for eye disease of milkfish (*Chanos chanos*) in the Philippines and mass mortality in larvae of swimming crab (*Portunus trituberculatus*) cultured in Japan (Ishimaru and Muroga, 1997), and in Kuwait caused mass mortality in cultured silver black porgy (*Sparidentex hasta* [*Acanthopagrus cuvieri*]) and greasy grouper (brown-spotted grouper) (*Epinephelus tauvina*) (Saeed, 1995). Infection in grouper (*E. coioides*) cultured in Taiwan resulted in mortality with clinical signs of swollen intestine containing yellow fluid, and isolation of the organism from liver, kidney and intestinal fluid (Yii *et al.*, 1997). Pathogenicity of the isolated strain (EmI82KL) was proven in experimental challenge resulting in an LD_{50} of 2.53×10^7 CFU/ml. Infection in Atlantic spadefish (*Chaetodipterus faber*) in Venezuela resulted in bilateral exophthalmia, corneal opacity in severe cases and haemorrhages around the eyes and occasionally on the fins, with a pure growth of the organism obtained from these sites. White mullet (*Mugil curema*), pompanos, also called Florida pompano (*Trachinotus carolinus*), and permit (*T. falcatus*) from the same region also suffered infection and showed haemorrhagic septicaemia (Alvarez *et al.*, 1998). Strains of *V. harveyi* were luminescent. In experimental infection mortalities occurred at 10^6 CFU/fish but not at 10^4 CFU/fish and there were no disease signs at the lower challenge rate. Exophthalmia and mortality were seen in cage-cultured barramundi (seabass) (*Lates calcarifer*) in the Philippines (Tendencia, 2002). Non-luminous *V. harveyi* was cultured from haemorrhagic skin lesions, spleen, liver, kidney, eyes and brain. In seabass (*Lates calcarifer*) cultured in Malaysia infected fish showed deep skin lesions, and haemorrhagic areas on the base of the fins and anus, and had fin and tail rot (Ransangan and Mustafa, 2009). Isolates (n=21) cultured from kidney, heart, liver and spleen were all identified as *V. harveyi* using 16S rRNA sequencing, however phenotypic results identified *V. harveyi* (n=4), *V. parahaemolyticus* (n=16) and *V. alginolyticus* (n=1). Pathogenicity for European seabass (*Dicentrarchus labrax*) and common dentex (*Dentex dentex*) cultured along the Spanish Mediterranean coast was confirmed by experimental challenge with an LD_{50} of 10^5–10^6 CFU/fish, but was less virulent for cultured gilthead seabream (*Sparus aurata*) (Pujalte *et al.*, 2003). However, others reported virulence for gilthead bream at an LD_{50} of 10^5 CFU/g body weight (Balebona *et al.*, 1998a; Balebona *et al.*, 1998b). Vibriosis in wedge sole (*Dicologlossa cuneata*) cultured in Spain was confirmed by experimental challenge to be due to a non-luminescent strain of *V. harveyi* (López *et al.*, 2009a). Mortality of 90–100% was found with intraperitoneal injection of 10^6 CFU/fish but mortality did not occur with bath immersion of 10^7 cells/ml. Identification of *V. harveyi* was confirmed by PCR directed to the *toxR* gene of Pang *et al.* (2006) (see under molecular identification). In red drum (*Sciaenops ocellatus*) cultured in Taiwan, *V. harveyi* was responsible for gastroenteritis characterized by a swollen intestine containing yellow fluid. The organism was cultured from the kidney, liver and intestine and in experimental challenge the LD_{50} was 2.9×10^7 CFU/fish for the bacterium and 3.85 μg protein/g fish body weight for an extracellular product produced by the organism (Liu *et al.*, 2003). Gastroenteritis was also found in cultured marine cobia fish (*Rachycentron canadum*) from Taiwan (Liu *et al.*, 2004b).

In snapper (*Pagrus auratus*) a combination of infection with *V. harveyi* and *Photobacterium damselae* ssp. *damselae* caused bloat resulting in low mortality at an aquaculture facility in Western Australia (Stephens *et al.*, 2006). Fish were noted to have gas-distended swim bladders and swam at the water surface prior to death. *Photobacterium damselae* was isolated from the intestine and *V. harveyi* from the blood of affected fish.

An isolate negative for ornithine decarboxylase and identified by conventional biochemical tests and 16S rRNA sequencing as being *V. harveyi* caused mortality in pufferfish (*Takifugu obscures*) and pineapple fish (*Cleidopus gloriamaris*) in an aquarium in Korea (Won *et al.*, 2009). The pufferfish exhibited a single deep ulcer on the lateral or ventral side of the skin and haemorrhagic areas at the base of the pectoral, pelvic and anal fins, whereas the pineapple fish showed haemorrhage around the operculum and eyes but did not have ulcers. In pufferfish (*T. rubripes*) cultured in Japan, infection with *V. harveyi* resulted in a pyogranulomatosis with granulation tissue in the liver, kidney, spleen and heart. Moderate mortality occurred in natural infection, and in experimental infection via intramuscular injection of 1.0×10^8 CFU/fish, 20% mortality occurred (Mohi *et al.*, 2010).

Infection in seahorses (*Hippocampus* species and *H. kuda*) held in an aquarium in Spain resulted in 90% mortality. Clinical signs included external haemorrhages, ascites fluid and haemorrhagic liver. The organism was isolated in pure growth from these sites, including the mouth (Alcaide *et al.*, 2001).

Vibrio harveyi is a major pathogen of prawns (shrimp) and has been reported from a number of countries including the Philippines, Taiwan, Thailand, Mexico and Venezuela. Infection occurred in the heptopancreas in black tiger prawns (*Penaeus monodon*) cultured in Thailand (Jiravanichpaisal *et al.*, 1994). Clinical signs included brittle shells, black or brown spots on the body, pink or brown gills and whitish muscle and an abnormal hepatopancreas that showed discolouration and was hard or swollen. Pathogenicity was confirmed in challenge experiments with 80–100% mortality in 24 h when injected at a concentration of 10^5 CFU/prawn. A luminous strain of *V. harveyi* was the cause of disease outbreaks in tiger prawn and kuruma prawn (*Marsupenaeus japonicus*, previously *Penaeus japonicus*) cultured in Taiwan (Liu *et al.*, 1996). Juvenile prawn (shrimp) *Litopenaeus vannamei* cultured in Mexico have been affected by *V. harveyi* causing a syndrome known as 'bright-red syndrome' characterized by red spots of discolouration on the abdomen, lethargy, anorexia and opacity of the abdominal muscle (Soto-Rodriguez *et al.*, 2010). The organism was isolated from hepatopancreas, lymphoid organ, heart, muscle, antennal glandule and stomach epithelium. In challenge experiments, 100 µl of 10^7 CFU of *V. harveyi* injected into the third abdominal segment caused death within 2.5 h of injection, and at a concentration of 10^6 CFU produced muscle opacity at 24 h post-injection. Identification was confirmed using REP-PCR (Cabanillas-Beltrán *et al.*, 2006). Farmed and wild-caught prawns (*Penaeus schmitti*, *Litopenaeus vannamei*, *L. stylirostris*) in Venezuela displayed haemorrhagic septicaemia caused by a luminescent strain of *V. harveyi*. Mortalities in experimental challenge reached 100% with a dose of 10^6 CFU/prawn resulting in an LD_{50} of 10^5–10^6 (Alvarez *et al.*, 1998). Luminescent *V. harveyi* strains caused mortalities in *Penaeus monodon* pond-cultured in the Philippines, and in experimental challenge, 100% mortalities occurred at 7 days at an inoculum of 10^7 CFU/animal (de la Peña *et al.*, 2001).

Mass mortalities of cultured Japanese abalone (*Haliotis discus hannai* [*Sulculus diversicolor supratexta*]) occurred with a sucrose-negative, ornithine decarboxylase-negative strain of *V. harveyi* (TK1091). White spots occurred on the foot and the abalone lost the ability to adhere (Nishimori *et al.*, 1998). Sequencing by 16S rRNA indicated 99.9% similarity to *V. harveyi* ATCC 35084^T and 99.4% similarity to ATCC 14129^T. Pathogenicity was confirmed experimentally with abalone deaths occurring 3 days post-injection with 1.5×10^8 CFU/animal. White pustules on the foot also were a feature of infected abalone (*Haliotis tuberculata*) grown naturally or farmed along the French coast. In this outbreak, a sucrose-negative, ornithine-positive strain of *V. harveyi* (ORM4) caused 80% mortality (Nicolas *et al.*, 2002). The organism was isolated from haemolymph, cardiac sinus, muscle and mucus gland, and identified using the API 20E, Biolog and 16S rRNA sequencing. Sequencing indicated 99.6% similarity (5 bases difference) to ATCC 35084^T. The disease was replicated under experimental challenge with injection of 0.5 ml of

5×10^8 CFU. Mortality was associated with temperature, as deaths occurred when the temperature increased above 17°C. Abalone susceptibility to the infection is also associated with immune status and spawning (Travers *et al.*, 2008).

Vibrio harveyi was found to be associated with white syndrome in tropical stony corals (*Acropora*, *Echinopora*, *Montipora*, *Pocillopora damicornis*, *Seriatopora histrix*) sampled from either natural reefs in the Indian Ocean, Indonesia, or in aquaria. *Vibrio harveyi* predominated the *Vibrio* species isolated from diseased corals and was not found in healthy corals (Luna *et al.*, 2010). *Vibrio coralliilyticus* was reported as an agent of coral bleaching in *Pocillopora damicornis* (Ben-Haim *et al.*, 2003a) but was not isolated in the corals sampled from natural reef. Pathogenicity was proven experimentally with coral exposed to 1×10^6 CFU/ml. Infected coral showed tissue loss, observed as white bands, spots or patches and exposure of exoskeleton, signs typical of white syndrome.

Vibrio harveyi, along with *V. natriegens*, *V. owensii* and *V. rotiferianus*, was reported from diseased crown of thorns starfish (*Acanthaster planci*) sampled from the Great Barrier Reef, Australia. Currently, the pathogenicity of *V. harveyi* and these *Vibrio* species for the starfish is not known (Rivera-Posada *et al.*, 2011).

A number of papers report variability in virulence of strains of *V. harveyi*. A range of virulence factors including exotoxin (*toxR* gene), chitinase, metalloprotease, serine protease and haemolysin *vhh*/*tlh* genes have been noted for different strains of *V. harveyi*. The gene *toxS*, which acts with *toxR* to activate expression of cholera toxin, *ctx*, in *V. cholerae* is found in only 16% of *V. harveyi* strains. Two virulence genes found in *V. cholerae*, *zot*, which produces zonula occludens toxin responsible for increasing intestinal mucosa permeability, and *tcpA*, which encodes a subunit of the toxin-regulated pilus involved in colonization, are not found in *V. harveyi* (Bai *et al.*, 2008). Variation exists in the presence of the flagellum genes, *flaB* (flagellin B) and *flaC* (flagella C subunit of *V. anguillarum*), with 81.25% of *V. harveyi* strains being positive for *flaB* and 37.5% of strains positive for *flaC* (Bai *et al.*, 2008). Under experimental challenge, no correlation was found between pathogenicity, as measured by mortalities, and the presence of these genes in strains inoculated into the zebrafish (Bai *et al.*, 2008).

Virulence genes *vhh* (haemolysin of *V. vulnificus*), *chiA* (chitinase), *vhpA* (metalloprotease), $toxR_{vh}$ (specific primers for *toxR* in *V. harveyi*), *luxR* (produces LuxR a quorum sensing regulator) and serine protease are all found in *V. harveyi*, but the presence of these genes does not correlate with pathogenicity when tested in gnotobiotic *Artemia* nauplii (Ruwandeepika *et al.*, 2010). The relative expression levels of the genes *luxR*, *toxR*, *vhh*, *srp* (serine protease gene) and *vhp* were measured by reverse transcriptase PCR using eight strains of *V. harveyi* and two strains PN9801=LMG 21363 and BB120=ATCC BAA-1116, originally identified as *V. harveyi*, but identified as *V. campbellii* by Gomez-Gil *et al.* (2004b) and Lin *et al.* (2010), respectively. The two *V. harveyi*/*V. campbellii* isolates had the highest expression levels of *luxR*, *toxR* and *vhp* compared to the other *V. harveyi* strains (Ruwandeepika *et al.*, 2011). In experimental challenge in gnotobiotic *Artemia*, *V. harveyi*/*V. campbellii* strain ATCC BAA1116 was moderately virulent and *V. harveyi*/*V. campbellii* strain LMG 21363 was highly virulent and both showed 160-fold higher *in vivo* expression levels of *luxR*, *toxR*, *vhh* (haemolysin), *srp* (serine protease) and *vhp* (metalloprotease) genes compared to an avirulent *V. harveyi* isolate LMG 11226 (confirmed as *V. harveyi* by Gomez-Gil *et al.* 2004b). A significant correlation was found between expression levels of *toxR* and *luxR* with the protease genes *srp* and *vhp*, whereas there was no correlation between expression levels of *toxR* and *luxR* with the haemolysin gene, *vhh*. There was a significant correlation between the expression of the serine protease gene *srp* and the haemolysin gene *vhh* (Ruwandeepika *et al.*, 2011). The *luxR* gene was expressed by all eight *V. harveyi* isolates and the two *V. harveyi*/*V. campbellii* isolates with relatively low variation in expression levels between the strains. Two of the eight *V. harveyi* isolates (VIB 571, VIB 645) had high levels of *vhh* and *srp* and moderately high expression levels of *toxR*, whereas another *V. harveyi* strain (ACMM 642) had high expression levels of *srp* and *vhp* and lower levels of *toxR*. These strains were not tested for pathogenicity in *Artemia*.

A major virulence factor is an exotoxin that is a transmembrane transcription regulator that controls the expression of virulence genes and has been found in a number of *Vibrio* species including *V. cholerae*, *V. alginolyticus*, *V. fluvialis* and *V. mimicus* and *G. hollisae*. In *V. harveyi* sequence variation has been found in the *toxR* gene between different strains and has impacted on some PCRs designed to amplify the *toxR* gene (Conejero and Hedreyda, 2003; Franco and Hedreyda, 2006; Pang *et al.*, 2006).

Vibrio harveyi can enter a viable but non-culturable (VBNC) state, which may enable it to respond to stressors such as reduced nutrients or changes in temperature. Virulence potential can be retained and studies have demonstrated that dormant cells recovered when grown under the right conditions. Cells were resuscitated at 26°C in sterile seawater containing yeast extract (0.025%) with the addition of either Tween 20 (final concentration of 6% v/v) or compound vitamin B (final concentration 1 mg/ml). Resuscitated cells caused mortalities when injected into zebrafish with the LD_{50} being 2.85×10^{4} CFU/ml (Sun *et al.*, 2008).

A number of molecular methods have been investigated to improve taxonomy and the identification of new species, and to improve the diagnostic ability to identify strains to species level. These techniques include sequencing genes, multilocus sequence analysis and PCR for detection of specific genes.

The *toxR* gene produces the protein ToxR, a transmembrane transcription regulator involved in the expression of virulence genes. The gene consists of a periplasmic domain, plus a cytoplasmic DNA-binding-transcriptional activation domain and a transmembrane domain separated by a region thought to be responsible for the tethering of the transcriptional activation domain to the cytoplasmic membrane (Osorio and Klose, 2000). The tether region shows variation amongst the *Vibrio* species and has been used for PCR primers (Conejero and Hedreyda, 2003; Franco and Hedreyda, 2006; Pang *et al.*, 2006). The gene has been found in *Aliivibrio fischeri*, *Grimontia hollisae*, *Photobacterium damselae* spp. *damselae*, *P. damselae* ssp. *piscicida*, *P. profundum*, *V. alginolyticus*, *V. cholerae*, *V. fluvialis*, *V. mimicus* and *V. vulnificus* (Osorio and Klose, 2000). The *toxR* gene from *V. harveyi* shows similarity to the *toxR* gene of *V. parahaemolyticus* (86%), *V. fluvialis* (84%) and *V. vulnificus* (83%) and similarity of 85% to the partial sequence of the *toxR* gene from *V. campbellii* (Franco and Hedreyda, 2006).

A Vibrio Core group was initially described based on close similarity of the 16S rRNA gene sequence for *V. harveyi*, *V. alginolyticus*, *V. campbellii*, *V. natriegens*, *V. parahaemolyticus*, *V. proteolyticus* and *V. vulnificus* (Dorsch *et al.*, 1992). The term Harveyi clade was formed based on MLSA information from nine genes, *ftsZ*, *gapA*, *gyrB*, *mreB*, *pyrH*, *recA*, *rpoA*, *topA* and 16S rRNA that grouped *V. harveyi*, *V. campbellii*, *V. alginolyticus*, *V. mytili*, *V. natriegens*, *V. parahaemolyticus* and *V. rotiferianus* as being closely related (Sawabe *et al.*, 2007b). Members of the Harveyi clade have >25% DDH (about 70% for most species pairs), 42–48 GC mol%, a MLSA concatenated sequence similarity of 90.1–96.2% and AAI of 97.2–99.4%. Concatenated sequences from three housekeeping genes, *rctB*, *rpoD* and *toxR*, enable discrimination between *V. alginolyticus*, *V. campbellii*, *V. harveyi*, *V. natriegens*, *V. parahaemolyticus* and *V. rotiferianus*. The inclusion of sequence information from another four genes (16S rRNA, *recA*, *pyrH* and *gyrB*) does not improve discrimination. Two genes in particular are the most resolving: *toxR* and *rpoD* (Pascual *et al.*, 2010) (see under molecular identification). The intraspecies sequence similarity for the three most differentiating genes is *toxR* (77.2–100%), *rctB* (85.6–100%) and *rpoD* (95.6–100%). The intraspecies sequence similarity for the other four genes is *gyrB* (86.8–100%), *recA* (92.7–100%), *pyrH* (93.7–100%) and 16S rRNA (98.8–100%). Interspecies similarity is *toxR* (33.8–72.5%), *rctB* (74.3–92.7%), *rpoD* (79.1–96.0%) followed by *gyrB* (83.1–99.5%), *pyrH* (86.4–97.8%), *recA* (87.9–99.9%) and 16S rRNA (97.6–99.9%).

Vibrio campbellii can be difficult to differentiate from *V. harveyi* as it has many phenotypic characteristics in common, however a main differentiating test is ornithine decarboxylase (ODC). *V. harveyi* is positive for ODC and *V. campbellii* is negative. The ornithine decarboxylase (*odc*) gene from *V. harveyi* has sequence similarity to the *odc* gene in *V. vulnificus* of 94%, 88% to *V. parahaemolyticus* and 81% to *V. cholerae*. The *odc* gene is not detected in *V. campbellii* (Hedreyda and Rañoa, 2007). The two species are not differentiated by sequence analysis of the *gyrB* (DNA gyrase B subunit) gene. The housekeeping genes *recA* (encodes recombinant repair protein) and *gapA* (glyceraldehyde 3-phosphate dehydrogenase) show that both strains are related with 94% and 99.5% similarity, respectively. *Vibrio harveyi* and *V. campbellii* are separated into different clusters using concatenated sequence analysis of four housekeeping genes, *topA* (topoisomerase I), *pyrH* (urydilate kinase), *ftsZ* (cell division protein) and *mreB* (actin-like cytoskeleton protein) (Thompson *et al.*, 2007). Gene sequence similarities for *ftsZ*, *mreB* and *pyrH* is 94–96% with *topA* being the most resolving resulting in 90% gene similarity. Gene sequence variation within each cluster is 1.5%.

The following strains have been confirmed as *V. harveyi* by Gomez-Gil *et al.* (2004b): LMG 4044^T (=ATCC 14126^T=CAIM 326^T), LMG 7890 (ATCC 35084=CAIM 517), LMG 11226 (=ATCC 33843=CAIM 520), LMG 19643 (CAIM 529), LMG 20977 (CAIM 541=AK2), CAIM 2 (=11 MZ), CAIM 79, CAIM 462 (=R-14950=VIB 23), CAIM 463 (=R-14951=VIB 22), CAIM 464 (R-14952=VIB 568) and CAIM 689 (=10 MZ).

The following strains identified as *V. harveyi* were shown to be *V. campbellii* using concatenated sequences from housekeeping genes (*topA*, *ftsZ*, *toxR*, *rctB*, *rpoD*, *mreB*): LMG 11658, LMG 22893, LMG 16862, LMG 22891, LMG 16863, LMG 16829, LMG 1660 (Hoffman *et al.*, 2012a.

The following strains have been confirmed as *V. campbellii* by Gomez-Gil *et al.* (2004b): LMG 11216^T (=ATCC 25920^T=CAIM 519^T), LMG 11256 (=ATCC 25921=CAIM 512), LMG 13241, LMG 16828 (=VIB 403), LMG 16830 (=VIB 406), LMG 20369 (=8P), CAIM 94 (=HL115), CAIM 113 (=HL135), CAIM 114 (=HL136), CAIM 126 (=HL148), CAIM 128 (=HL150=LMG 21365), CAIM 129 (=HL151), CAIM 130 (=HL152), CAIM 133 (=HL155), CAIM 135 (=HL157), CAIM 136 (=HL158), CAIM 137 (=HL159), CAIM 149 (=HL34), CAIM 222 (=HP130), CAIM 284 (=IPL8), CAIM 333 (=M_1=LMG 21362), CAIM 371 (=PL96-11-6), CAIM 372 (=PN-9801=LMG 21363), CAIM 392 (=STD3-1002), CAIM 395 (=STD3-131), CAIM 401 (=SW-9702), CAIM 415 (=Z_1=LMG 21461), CAIM 416 (=Z_2), CAIM 417 (=Z_3), CAIM 679 (=1A-1), CAIM 684 (=Na-1), R-14899 (VIB 629) and R-14905 (VIB 394).

GROWTH CHARACTERISTICS AND PHENOTYPIC IDENTIFICATION. *Vibrio harveyi* is a facultatively anaerobic, motile, oxidase-positive, Gram-negative rod. Cells swarm on some media such as MSA-B and MSSA. Swarming does not occur on MA 2216 or TCBS.

Fermentation is positive for β-gentiobiose and negative for arbutin. Hydrolysis of casein is negative (Yoshizawa *et al.*, 2012).

Vibrio (*carchariae*) *harveyi* ATCC 35084 strain is negative for citrate and gelatin (plate method) at 24 h and needs to be read after 48 h or longer. The culture has a pungent smell similar to *Plesiomonas shigelloides*. Strain ATCC 14126 is positive for luminescence and negative for urease, whereas strain ATCC 35084 is negative for luminescence and positive for urease (Alcaide *et al.*, 2001). Strains NCIMB 1280, ATCC 14126 and ATCC 14129 are reported to be gelatin positive (Benediktsdóttir *et al.*, 1998). Strains ATCC 14126 and ATCC 14129 are reported to be negative for fermentation of sorbitol, whereas strain ATCC 35084 is positive after 2 days (Alcaide *et al.*, 2001; Buller, unpublished). Haemolysis is variable, and reported positive against sheep red blood cells (Alcaide *et al.*, 2001).

Vibrio alginolyticus and *Vibrio harveyi* may be difficult to differentiate. *V. harveyi* is negative for the VP test, whereas *V. alginolyticus* is positive. *V. harveyi* is positive for fermentation of D-glucuronate, whereas *V. alginolyticus* is negative. *Vibrio alginolyticus* swarms and completely covers an MSA-B plate in 24 h at 25°C, whereas *V. harveyi* has a slow or spreading type of growth. *Vibrio alginolyticus* is urease-negative, whereas *V. harveyi* is usually urease-positive (50% reported as positive).

Differentiation of *V. alginolyticus* and *V. harveyi.*

Test	VP	10% NaCl	Growth at 42°C	Urease	D-Glucuronate
Organism					
Vibrio alginolyticus	+	+	+	–	–
Vibrio harveyi	–	–	–	V	+

V=variable reaction

These and other reactions are presented in Tables 3.22, 3.24–29 and 3.38. Sets are inoculated with an inoculum in 2% NaCl and incubated at 25°C or 30°C for 48 h, or as indicated in the tables.

MOLECULAR IDENTIFICATION. A multiplex PCR based on the haemolysin gene (*hly*) was designed for detection and differentiation of *V. harveyi*, *V. campbellii* and *V. parahaemolyticus* (Haldar *et al.*, 2010). The *hly* gene was found to be stable in the genome as no mobile genetic elements were detected in the flanking regions. The gene is well conserved in the *Vibrio* species with high intraspecies sequence identity, yet has considerable interspecies sequence diversity making it suitable for identification of the three *Vibrio* species. The *hly* gene sequence of *V. campbellii* had sequence similarity of 79.7% with *V. harveyi* and 35.7% with *V. mimicus*.

PCR type	Multiplex PCR for detection of the haemolysin gene (*hly*) in *V. harveyi*, *V. campbellii* and *V. parahaemolyticus* (Haldar *et al.*, 2010)		
DNA extraction	Suspend bacterial cells in 500 µl TE buffer and boil for 10 min. Cool. Centrifuge for 5 min at 12,800 ×*g*. Store supernatant at –20°C. For direct detection use a pre-enrichment in APW containing 2% NaCl and incubate for 12 h at 37°C		
PCR master mix			
Primer Vh-hly 1F	5′-GAG TTC GGT TTC TTT CAA G-3′		
Primer Vh-hly1R	5′-TGT AGT TTT TCG CTA ATT TC-3′		
Primer Vca-hly5	5′-CTA TTG GTG GAA CGC AC-3′		
Primer Vca-hly3	5′-GTA TTC TGT CCA TAC AAA C-3′		
Primer Vp-tlh1	5′-GAT TTG GCG AAC GAG AAC-3′		
Primer Vp-tlh2	5′-CGT CTC GAA CAA GGC G-3′		
Reagent	**Stock**	**Final concentration**	**Volume ×1**
Taq buffer containing $MgCl_2$	×10	×1	3.0 µl
dNTP	10 mM	200 µM	0.6 µl
Primer Vh-hly 1F	50 µM	0.25 µM	0.15 µl
Primer Vh-hly1R	50 µM	0.25 µM	0.15 µl
Primer Vca-hly5	50 µM	0.25 µM	0.15 µl
Primer Vca-hly3	50 µM	0.25 µM	0.15 µl
Primer Vp-tlh1	50 µM	0.5 µM	0.30 µl
Primer Vp-tlh2	50 µM	0.5 µM	0.30 µl
r*Taq* DNA polymerase (Takara Bio)	250 U	0.75 U	0.09 µl
Water			23.11 µl
DNA			2 µl
Total volume			**30 µl**
Thermocycling			
Step 1	**Step 2**	**Step 3**	
94°C, 5 min	94°C, 30 s 54°C, 30 s 72°C, 30 s	94°C, 30 s 54°C, 30 s 72°C, 5 min	
×1	×30	×1	
Product	*V. harveyi* (Vh-hly), 454 bp; *V. campbellii* (Vca-hly), 328 bp; *V. parahaemolyticus* (Vp-thl), 695 bp		
Sensitivity	10 bacterial cells/tube for *V. campbellii* and 100 cells/tube for *V. harveyi* and *V. parahaemolyticus*		
Specific against	*Aliivibrio fischeri* (NCIMB 1281^T), *A. logei* (NCIMB 2252^T), *A. salmonicida* (ATCC 43839^T), *Photobacterium damselae* (ATCC 33539^T), *V. alginolyticus* (ATCC 17749^T), *V. anguillarum* (IFO 13266^T), *V. campbellii* (ATCC 25920^T), *V. fluvialis* (ATCC 33809^T), *V. harveyi* (ATCC 35804^T and 14126^T), *V. mediterranei* (UM 5172), *V. metschnikovii* (IAM 1039), *V. mimicus* (ATCC 33653^T), *V. natriegens* (ATCC 14048^T), *V. nigripulchritudo* (ATCC 27043^T), *V. ordalii* (CIP 103205^T), *V. orientalis* (ATCC 33934^T), *V. parahaemolyticus* (NBRC 12711^T), *V. pelagius* (ATCC 25916^T), *V. proteolyticus* (ATCC 15338^T), *V. splendidus* II (IAM 14411), *V. vulnificus* (IFO 15645^T). Tested with no adverse results or non-specific amplicon (apart from the specific amplicon expected) against 27 strains of *V. campbellii*, 69 strains of *V. parahaemolyticus* and 16 strains of *V. harveyi*. Other strains tested included *Campylobacter jejuni*, *C. coli*, *C. fetus*, *Escherichia coli*, *Salmonella enterica*, *Shigella dysenteriae*, *Shigella flexneri*, *Shigella sonnei* and *Yersinia enterocolitica*. Strains are wild types unless otherwise indicated		

Amplification and sequencing of the *pyrH* gene is suitable for differentiation between *V. campbellii* and *V. harveyi* (Haldar *et al.*, 2011). Most *Vibrio* species will be detected with these primers. The *pyrH* gene has a resolving power among the *Vibrio* species of approximately 83% (Thompson *et al.*, 2007b).

PCR type	PCR and sequencing of the *pyrH* gene for differentiation of *V. harveyi* and *V. campbellii* (Haldar *et al.*, 2010). Will detect most *Vibrio* species			
DNA extraction	Extracted from overnight growth			
PCR master mix				
Primer VpyrH80F	5′-GAT CGT ATG GCT CAA GAA G-3′			
Primer VpyrH530R	5′-TAG GCA TTT TGT GGT CAC G-3′			
Reagent	**Stock**	**Final concentration**	**Volume ×1**	
Taq buffer containing $MgCl_2$	×10	×1	5.0 µl	
dNTP	10 mM	200 µM	1.0 µl	
Primer VpyrH80F	10 µM	0.5 µM	2.5 µl	
Primer VpyrH530R	10 µM	0.5 µM	2.5 µl	
Ampli*Taq* DNA polymerase	250 U	1 U	0.2 µl	
Water			33.8 µl	
DNA (0.01 µg/µl)			5.0 µl	
Total volume			**50 µl**	
Thermocycling				
Step 1	**Step 2**		**Step 3**	**Step 4**
95°C, 5 min	95°C, 1 min		95°C, 35 s	95°C, 35 s
	50°C, 2 min 15 s		50°C, 1 min 15 s	50°C, 1 min 15 s
	72°C, 1 min 15 s		72°C, 1 min 15 s	72°C, 7 min
×1	×3		×30	×1
Product	450–467 bp			

A phylogenetic tree constructed of concatenated sequences of partial (496 bp) 16S rRNA gene and *pyrH* gene sequences enabled differentiation of *V. campbellii* and *V. harveyi*.

Two PCRs have been described for the detection of the *toxR* gene in *V. harveyi*. The *toxR* gene consists of conserved areas and variable region. Designing primers specific to *V. harveyi* has been problematic as the *toxR* gene is found in a number of *Vibrio* species.

PCR type	Conventional PCR for amplification and sequencing of the *toxR* gene (Pang *et al.*, 2006)		
DNA extraction	1.5 ml from overnight culture, centrifuged 5000 *xg* for 5 min, washed in saline. Pellet resuspended in 100 µl distilled water and boiled for 10 min. Supernatant used in PCR		
PCR master mix			
Primer toxRF1	5′-GAAGCAGCACTCACCGAT-3′		
Primer toxRR1	5′-GGTGAAGACTCATCAGCA-3′		
Reagent	**Stock**	**Final concentration**	**Volume ×1**
PCR buffer	×10	×1	5.0 µl
$MgCl_2$	25 mM	1.5 mM	3.0 µl
dNTPs	25 mM	0.5 mM	1 µl
Taq	5 U/µl	0.03 U	0.25 µl
Primer F	50 µM	1.0 µM	1.0 µl
Primer R	50 µM	1.0 µM	1.0 µl
Water			38.25 µl

Continued

Continued.

DNA		0.5 µl
Total volume		**50 µl**
Thermocycling	ABR 2400 (Advanced Biotechnologies, USA) thermocycler	
Step 1	**Step 2**	**Step 4**
94°C, 5 min	94°C, 1 min	94°C, 1 min
	55°C, 1 min	55°C, 1 min
	72°C, 1 min	72°C, 10 min
×1	×30	×1
Product for *toxR*	382 bp	
Sensitivity	4.9 ×10^3 CFU/ml	
Specific against	*Aliivibrio* (*Vibrio*) *fischeri*, *Aliivibrio* (*Vibrio*) *logei*, *Grimontia hollisae*, *Photobacterium damselae* ssp. *damselae*, *Salinivibrio costicola* ssp. *costicola*, *Vibrio* (*Listonella*) *anguillarum*, *V. aestuarianus*, *V. alginolyticus*, *V. campbellii*, *V. cincinnatiensis*, *V. diazotrophicus*, *V. fluvialis*, *V. furnissii*, *V. gazogenes*, *V. mediterranei*, *V. mimicus*, *V. natriegens*, *V. nereis*, *V. parahaemolyticus*, *V.* (*Listonella*) *pelagius*, *V. proteolyticus*, *V. tubiashii*, *V. vulnificus*	

The following primers are used to amplify segments for the *toxR* and *rpoD* genes.

PCR type	MLSA, sequencing and concatenated sequence analysis for differentiation of *Vibrio* core group or the Harveyi Clade (Pascual *et al.*, 2010).		
DNA extraction	From bacterial cells		
PCR master mix			
Primers for *rpoD* gene (factor σ70 RNA polymerase)	70F = 5′-ACGACTGACCCGGTACGCATGTAYATGMGNGARATGGGNACNGT-3′ 70R = 5′-ATAGAAATAACCAGACGTAAGTTNGCYTCNACCATYTCYTTYT-3′		
Primers for *toxR* gene (transmembrane regulatory protein)	ToxRs = 5′-GANCARGGNTTYGARGTNGAYGAYTC-3′ ToxRas = 5′-TTDKKTTGNCCNCYNGTVGCDATNAC-3′		
Reagent	**Stock**	**Final concentration**	**Volume ×1**
PCR buffer	×10	×1	5.0 µl
$MgCl_2$	100 mM	1.5 mM	0.75 µl
dATP	10 mM	200 µM	1 µl
dCTP	10 mM	200 µM	1 µl
dGTP	10 mM	200 µM	1 µl
dTTP	10 mM	200 µM	1 µl
Taq (Genotaq)	250 U/µl	2.5 U	0.5 µl
Primer F	50 µM	1.0 µM	1.0 µl
Primer R	50 µM	1.0 µM	1.0 µl
Water			32.75 µl
DNA (50 ng/µl)			5 µl
Total volume			**50 µl**
Thermocycling	PTC-100 thermocycler (MJ Research)		
Step 1	Step 2	Step 3	Step 4
95°C, 5 min	95°C, 1 min	95°C, 35 s	95°C, 35 s
	55°C, 2 min 15 s	55°C, 2 min 15 s	55°C, 2 min 15 s
	72°C, 1 min 15 s	72°C, 1 min 15 s	72°C, 10 min
×1	×3	×30	×1
Product for *rpoD*	780 bp		
Product for *toxR*	477 bp		

The amplified products are examined using a 1.2% agarose gel. Product is purified and then used in the sequencing reaction, which uses the same primers diluted ten-fold to 5 pmol (Pascual *et al.*, 2010).

Sequencing of the *mreB*, *topA* and 16S rRNA genes can identify and differentiate vibrios in the Harveyi clade (Rivera-Posada *et al.*, 2011). See under *V. natriegens*.

Vibrio hepatarius

TAXONOMY AND SIGNIFICANCE. *Vibrio hepatarius* (LMG 20362^T=CAIM 693^T) was isolated from the hepatopancreas of prawn (*Litopenaeus vannamei*) and was a predominant colonizer that outcompeted *Vibrio harveyi* and appeared to offer health enhancement of prawn. It was originally identified as FALP Group A26 (Thompson *et al.*, 2001b, 2003c). It was also isolated from the haemolymph of healthy spider crab (*Maja brachydactyla*) caught off the coast of Spain and the Canary Islands (Gomez-Gil *et al.*, 2010). The numbers of bacterial flora in healthy haemolymph is usually in the order of 10^3–10^4 CFU/ml (Gomez-Gil *et al.*, 1998).

In experimental challenge using rainbow trout (*Oncorhynchus mykiss*) and *Artemia* nauplii inoculated with 10^6 CFU/fish or 10^6 CFU/ml, respectively, *V. hepatarius* (LMG 20362^T) was rated as non-virulent. Clinical disease was not seen, and at the end of the experiment the organism was not recovered from tissue (Austin *et al.*, 2005).

Vibrio hepatarius is most closely related to *V. xuii* (99.1%), *V. tubiashii* (98.6%) and *V. nereis* (99.0%) based on 16S rRNA sequence similarity. MLSA identified *V. hepatarius* as belonging to the Orientalis clade, which also contains *V. orientalis* and *V. brasiliensis*, and *V. tubiashii* (Sawabe *et al.*, 2007b). Members of this clade have >24% DDH and 43–46 GC mol%, 91.2–94.2% similarity based on concatenated MLSA sequence data and AAI of 97–97.9%. *V. xuii* and *V. nereis* belong to the Nereis clade.

GROWTH CHARACTERISTICS AND PHENOTYPIC IDENTIFICATION. *Vibrio hepatarius* is a facultatively anaerobic, motile, oxidase-positive, Gram-negative slightly curved rod (2–3×1.0 μm). On TSA containing 2% NaCl colonies are 6 mm beige, convex, smooth and rounded after 48 h at 28°C, and grow as 6 mm yellow, round, smooth colonies on TCBS after 48 h at 28°C. Growth occurs at 4–35°C and at 0–8% NaCl (Thompson *et al.*, 2003c). These and other reactions are presented in Tables 3.23–26 and 3.38. Sets are inoculated with an inoculum in 2% NaCl and incubated at 25°C for 48 h.

Vibrio hippocampi

TAXONOMY AND SIGNIFICANCE. *Vibrio hippocampi* (BFLP-4^T=DSM 22717^T=LMG 25354^T) was isolated from the faeces of wild seahorses (*Hippocampus guttulatus*) caught off the coast of north-west Spain (Balcázar *et al.*, 2010c). It is most closely related to *V. ichthyoenteri* (97.1%), *V. mediterranei* (96.7%), *V. scophthalmi* (96.7%) and *V. sinaloensis* (96.6%) based on 16S rRNA sequencing.

GROWTH CHARACTERISTICS AND PHENOTYPIC IDENTIFICATION. *Vibrio hippocampi* is a motile, facultatively anaerobic, Gram-negative, slightly curved rod (1.0×2.0–2.5 μm), catalase- and oxidase positive. On TSA agar containing 1.5% w/v NaCl, colonies are 1.5–2.0 mm, cream coloured and circular after 3 days at 20°C (Balcázar *et al.*, 2010c). Growth does not occur below 4°C or above 35°C and is optimal at 20°C. Growth occurs between 0 and 7% NaCl, but not at 8% NaCl, and at pH 5.5–9 but not below pH 5.0 or above pH 9.0. These and other reactions are presented in Tables 3.23, 3.27–30 and 3.38.

Vibrio hispanicus

TAXONOMY AND SIGNIFICANCE. Three strains of *V. hispanicus* (VIB 213^T=LMG 13240^T=CAIM 525^T) were isolated from *Artemia* species and seawater from Barcelona, Spain (Gomez-Gil *et al.*, 2004a).

In experimental challenge using rainbow trout (*Oncorhynchus mykiss*) and *Artemia* nauplii inoculated with 10^6 CFU/fish or 10^6 CFU/ml, respectively, *V. hispanicus* (LMG 13213) was rated as non-virulent or low virulence. Although there was some evidence of exophthalmia, kidney necrosis, ascites and liquefaction of internal organs, clinical disease was not seen, and at the end of the experiment the organism was recovered from tissue at either a low rate or not at all (Austin *et al.*, 2005).

The organism is closely related to *V. alginolyticus*, *V. campbellii*, *V. diazotrophicus*, *V. proteolyticus* and others with a 96.8–97.9% similarity based on 16S rRNA gene sequence similarity (Gomez-Gil *et al.*, 2004a). MLSA identified *V. hispanicus* as belonging to the Diazotrophicus clade, which also contains *V. diazotrophicus* (Sawabe *et al.*, 2007b). Members of this clade have 43–47 mol% GC, a MLSA concatenated similarity of 91.2% and AAI of 97.0%. MLSA identified *V. alginolyticus* and *V. campbellii* as belonging to the Harveyi clade. *Vibrio proteolyticus* does not belong to a clade at the present time.

GROWTH CHARACTERISTICS AND PHENOTYPIC IDENTIFICATION. *Vibrio hispanicus* was isolated on tryptic soy agar (TSA, Oxoid) containing 1.5% NaCl and incubated at 28°C for 48 h. Growth occurs in 0–10% NaCl (w/v) but not at 12.0% NaCl, and at temperatures of 4–40°C (Gomez-Gil *et al.*, 2004a). These and other reactions are presented in Tables 3.23, 3.24–3.26 and 3.38. Commercial kits API 20E and API Zym are inoculated using 2% NaCl.

Vibrio hollisae

See *Grimontia hollisae*.

Vibrio ichthyoenteri

TAXONOMY AND SIGNIFICANCE. *Vibrio ichthyoenteri* (F-2^T=IFO 15847^T) was isolated from the intestine of Japanese flounder (*Paralichthys olivaceus*) and causes intestinal necrosis of flounder larvae (Ishimaru *et al.*, 1996). It has also been isolated from rose-spotted snapper (Gomez-Gil *et al.*, 2007). 16S rRNA sequencing indicates 97.1% similarity to *V. hippocampi* (Balcázar *et al.*, 2010c). MLSA identified *V. ichthyoenteri* as belonging to the Scophthalmi clade, which also contains *V. scophthalmi* (Sawabe *et al.*, 2007b). Members of this clade have >32% DDH and 43–44 GC mol%, 95.5% similarity based on concatenated MLSA sequence data and AAI of 99.4%.

GROWTH CHARACTERISTICS AND PHENOTYPIC IDENTIFICATION. *Vibrio ichthyoenteri* is a facultatively anaerobic, motile, oxidase-positive, Gram-negative straight or slightly curved rod (1.6–2.5×0.6–0.8 µm). Colonies are pinpoint after 24 h at 25°C on MA 2216 and BA, and 0.5 mm on MSA-B. Colonies of the type strain are 0.1 mm green (due to weak fermentation of sucrose) at 24 h on TCBS and grow to 1 mm after 48 h. Other strains produce yellow colonies on TCBS. Growth occurs between 15 and 30°C but not at 4 or 35°C. Growth occurs at 1–6% NaCl but not at 0 or 8% NaCl. The oxidase reaction is strong. Luminescence and swarming are not seen. Results are negative for both Moeller's and Thornley's ADH and negative for chitinase, alginase, lipase and amylase. Ishimaru *et al.* (1996) reported negative results for fermentation of D-mannitol and positive results for sucrose for strain IFO 15847, whereas strain DSMZ 14397 is positive for mannitol and negative for sucrose using the conventional tube method with phenol red pH indicator, and in the API 20E (this manual). These and other reactions are presented in Tables 3.22, 3.24–26 and 3.38. Sets are inoculated with an inoculum in 2% NaCl and incubated at 25°C for 48 h.

Vibrio iliopiscarius

See *Photobacterium iliopiscarium*.

Vibrio inusitatus

TAXONOMY AND SIGNIFICANCE. *Vibrio inusitatus* (RW14^T=LMG 23434^T=NBRC 102082^T) was isolated from the intestine of Californian red abalone (*Haliotis rufescens*) (Sawabe *et al.*, 2007a). *Vibrio inusitatus*, together with *V. comitans* and *V. rarus*, is closely related to *V. superstes* (99.3–99.5%) based on 16S rRNA sequencing with *V. inusitatus* having 98.7% similarity to *V. superstes*. *Vibrio inusitatus* has 94.6–94.8% similarity to *V. comitans* based on the *gapA* gene.

GROWTH CHARACTERISTICS AND PHENOTYPIC IDENTIFICATION. *Vibrio inusitatus* is a facultatively anaerobic, non-motile, oxidase-positive, Gram-negative rod (1–2×0.5–1.0 µm). On MA 2216, colonies are beige, circular, smooth, convex with an entire edge. Colonies are green on TCBS (Sawabe *et al.*, 2007a). Luminescence is not seen. Growth occurs at 4–30°C but not at 37°C. Growth does not occur in the absence of NaCl but occurs at 1–6% NaCl. Starch, Tween 80, agar and chitin are not hydrolysed. These and other reactions are presented in Table 3.23 and some reactions are reported in Tables 3.24–3.29.

Vibrio jasicida

TAXONOMY AND SIGNIFICANCE. *Vibrio jasicida* (TCFB 0772^T=LMG 25398^T=JCM 16453^T) was isolated from packhorse lobster, also known as spiny lobster, *Sagmariasus* (*Jasus*) *verreauxi*, from New Zealand. Other strains were isolated from southern rock lobster (*Jasus edwardsii*), abalone (*Haliotis* sp.) and Atlantic salmon (*Salmo salar*) in Australia (Yoshizawa *et al.*, 2012). Deaths occurred in packhorse rock lobster phyllosoma larvae held in an experimental facility and the causative organism was originally identified as a luminous *Vibrio harveyi* (Diggles *et al.*, 2000). The larvae exhibited small red spots throughout the body and pereiopods. The body appeared opaque and faintly luminous. Bacteria were isolated from the gut and hepatopancreas. Pathogenicity was confirmed experimentally using immersion baths containing a concentration of greater than 10^4 CFU/ml.

Vibrio jasicida is closely related to *V. communis* (99.8%), *V. owensii* (99.8%), *V. natriegens* (99.4%), *V. parahaemolyticus* (99.4%), *V. rotiferianus* (99.2%), *V. alginolyticus* (99.1%) and *V. campbellii* (99.1%) based on 16S rRNA sequencing. MLSA using housekeeping genes *gyrB*, *pyrH*, *ftsZ*, *mreB* and *gapA* (a total of 2443 bp) indicates similarity to *V. communis* (94.6%), *V. rotiferianus* (94.4%), *V. owensii* (94.1%), *V. campbellii* (92.3%), *V. parahaemolyticus* (90.3%), *V. alginolyticus* (89.3%) and *V. natriegens* (88.8%).

GROWTH CHARACTERISTICS AND PHENOTYPIC IDENTIFICATION. *Vibrio jasicida* is a motile, facultatively anaerobic, oxidase-positive, Gram-negative rod (1.0–1.5×0.5 µm). On MA 2216, colonies are 5–7 mm, non-pigmented and non-swarming. Colonies are green and translucent on TCBS (Yoshizawa *et al.*, 2012). Luminescence is positive. Growth occurs at 10–37°C, with an optimum of 25–30°C. Growth does not occur in the absence of NaCl, but occurs at 0.5–8% NaCl with an optimum of 2.0–6.0%. ODC and LDC are positive and ADH is negative. Fermentation is positive for β-gentiobiose, amygdalin, arbutin, cellobiose, D-galactose and negative for L-rhamnose and raffinose. Aesculin, casein and starch are hydrolysed. Tests are positive for alkaline phosphatase, acid phosphatase and α-D-galactosidase, β-D-galactosidase and γ-glutamyl transpeptidase. These and other reactions are presented in Table 3.22.

Vibrio kanaloae

TAXONOMY AND SIGNIFICANCE. *Vibrio kanaloae* (LMG 20539^T=CAIM 485^T) was isolated from diseased oyster larvae (*Ostrea edulis*) from France, seawater from Hawaii and shrimp (*Penaeus chinensis*) from China (Thompson *et al.*, 2003d). In experimental challenge, *V. kanaloae* (LMG 20539^T) caused 100% mortality within 7 days when inoculated into rainbow trout (*Oncorhynchus mykiss*) at a dose of 10^2 cells/fish. The LD_{50} was <10^2 (Austin *et al.*, 2005). Mortality of *Artemia* nauplii exposed to 10^6 CFU/ml was 90%. Extracellular product (ECP) (4 µg/ml) caused 100% mortality in rainbow trout 1 h post-injection, and 70% in *Artemia*. In rainbow trout, clinical signs included haemorrhages in the eye and exophthalmia, liquefaction of internal organs and muscles, gastroenteritis and protruded anus. The strain is positive for caseinase, gelatinase, lipase and phospholipase and is haemolytic on sheep blood agar but produces no haemolysis against horse red blood cells or trout red cells.

Vibrio kanaloae together with *V. pomeroyi* and *V. chagasii* are phylogenetically related to *V. splendidus*, *V. lentus*, *V. cyclitrophicus*, *V. mediterranei* and *V. orientalis*. MLSA identified *V. kanaloae* as belonging to the Splendidus clade, which contains *V. splendidus*, *V. chagasii*, *V. crassostreae*, *V. cyclitrophicus*, *V. fortis*, *V. gigantis*, *V. lentus*, *V. pelagius*, *V. pomeroyi* and *V. tasmaniensis* (Sawabe *et al.*, 2007b). Members of this clade have >30% DDH and 39–47 GC mol%, 90.6–99.8% similarity based on concatenated MLSA sequence data and AAI of 96.5–99.8%. *Vibrio mediterranei* did not cluster in a clade and *V. orientalis* belonged to the Orientalis clade.

GROWTH CHARACTERISTICS AND PHENOTYPIC IDENTIFICATION. *Vibrio kanaloae* is a facultatively anaerobic, motile, Gram-negative, slightly curved rod (1.0×2–3 µm), catalase and oxidase positive. On TSA agar containing 2% NaCl colonies are 5 mm beige, round, smooth with entire margins after 2 days at 28°C (Thompson *et al.*, 2003d). Yellow, 5–10 mm colonies are produced on TCBS. Growth does not occur at 0% NaCl or at 10% NaCl. Growth occurs at 4°C, but not at temperatures ≥35°C. The type strain does not grow at 8% NaCl, whereas three of five strains do grow. ADH is variable (two of five strains positive), but positive for the type strain. Other reactions are presented in Tables 3.23–26 and 3.38.

Vibrio lentus

TAXONOMY AND SIGNIFICANCE. *Vibrio lentus* (40M4^T=CECT 5110^T=DSM 13757^T) was isolated from seawater and Mediterranean oysters on the Spanish coast. Pathogenicity was not stated (Macián *et al.*, 2001a). *Vibrio lentus* causes disease and mortality in wild octopus (*Octopus vulgaris*) (Farto *et al.*, 2003). Clinical signs include round, hard lesions on the arms or head mantle with advanced stages of the disease showing loss of skin and exposure of underlying muscle. The organism can be isolated from the skin lesion and gill heart. Pathogenicity was proven in infectivity assays using a bath challenge with an inoculum of 2×10^8 CFU/ml over 6 h. Clinical signs and mortality were recorded at 72 h. *Vibrio lentus* did not cause infection in seabream (2–3 g) or turbot (5 g) inoculated intraperitoneally with 10^7 CFU/fish. Further studies indicate that the octopus strains produces an extracellular product of 39 kDa that has similar proteolytic properties to the 39 kDa protease produced by *Vibrio* (*Listonella*) *pelagius*. The extracellular product causes mortality and vibriosis in turbot when inoculated intraperitoneally with 4 ng/g fish (Farto *et al.*, 2006).

The type strain from Mediterranean oysters has 16S rRNA gene sequence similarity of 98.6% to *V. splendidus* biotype I, and 98.0% to *V. splendidus* biotype 2 and similarities of 97.4% and 97.2%, respectively, based on the 23S rDNA gene (Macián *et al.*, 2001a). The levels of similarity obtained by Macián *et al.*, (2001a) are above the 97% level set for species delineation (Stackebrandt and Goebel, 1994), however based on DNA–DNA hybridization of 59.0 and 48.0% with both biotypes of *V. splendidus*, and phenotypic variations, *V. lentus* can be delineated as a separate species. The octopus strain of *V. lentus* has 89% 16S rRNA gene sequence similarity to *V. splendidus* and 98.6% sequence similarity to the Mediterranean oyster strain (Farto *et al.*, 2003).

In experimental challenge using rainbow trout (*Oncorhynchus mykiss*) and *Artemia* nauplii inoculated with 10^6 CFU/fish or 10^{6}CFU/ml, respectively, *V. lentus* (LMG 21356 and LMG 21034) were rated as non-virulent or low virulence. Although there was some evidence of exophthalmia, kidney necrosis, ascites and liquefaction of internal organs, clinical disease was not seen, and at the end of the experiment the organism was recovered from tissue at either a low rate or not at all. Strain LMG 21355 was non-virulent and did not show signs of infection and the organism was not recovered from tissues (Austin *et al.*, 2005).

MLSA identified *V. lentus* as belonging to the Splendidus clade, which contains *V. splendidus*, *V. chagasii*, *V. crassostreae*, *V. cyclitrophicus*, *V. fortis*, *V. gigantis*, *V. kanaloae*, *V. pelagius*, *V. pomeroyi* and *V. tasmaniensis* (Sawabe *et al.*, 2007b). Members of this clade have >30% DDH and 39–47 GC mol%, 90.6–99.8% similarity based on concatenated MLSA sequence data and AAI of 96.5–99.8%.

GROWTH CHARACTERISTICS AND PHENOTYPIC IDENTIFICATION. *Vibrio lentus* is a facultatively anaerobic, oxidase-positive, motile, Gram-negative rod. Growth occurs on MSA-B and MA 2216 as pinpoint colonies at 24 h and 0.5 mm at 48 h. No growth on BA. On TSA containing 2% NaCl, colonies are 0.2–0.4 mm, non-pigmented, round, non-swarming after 48 h at 22°C (Farto *et al.*, 2003). Green colonies are produced on TCBS. It is non-luminescent on MA 2216 and resistant to 0/129 150 µg vibriostatic agent. Growth occurs at 4–30°C for the type strain whereas some of the octopus strains do not grow at 4°C. All strains grow at 10–22°C but not at 37°C or at 10% NaCl. Results are positive for ADH by Thornley's method but negative using Moeller's method (Macián *et al.*, 2001a). Indole is positive for the type strain, mostly positive for the octopus strains (75%) and variable among oyster and seawater strains. Starch, Tween 80 and gelatin are hydrolysed. Hydrolysis of casein and starch is positive for the type strain and positive for 75% and 50%, respectively, for octopus strains. Chitin is hydrolysed by the type strain but is negative for the octopus strains. DNase production is negative for the type strain and positive for 50% of octopus strains. The DNase plate needs to be prepared with 2% NaCl for growth to occur on the plate. Acid production is negative from D-melibiose, D-raffinose, rhamnose and D-amygdalin. Fermentation using conventional tube tests may show negative reactions for maltose and mannitol and very weak positive results for mannose and trehalose, and a weak reaction for glucose at 5 days. In the API 20E the glucose reaction may be negative or very weak. ADH in the API 20E may be weak. Methyl red is positive for all octopus strains but negative for the type strain. An inoculum of 2% NaCl is required for identification sets. These and other reactions are presented in Tables 3.22, 3.24–26 and 3.38.

Vibrio litoralis

TAXONOMY AND SIGNIFICANCE. Two strains of *V. litoralis* (MANO22D^{T}=KCTC 12520^{T}=DSM 17657^{T}) were isolated from a tidal flat area near Dae-Chun, Chung-Nam area of the Yellow Sea in Korea (Nam *et al.*, 2007). *Vibrio litoralis* is related to *V. rumoiensis* with 98.65% 16S rRNA gene sequence similarity. These two strains are phylogenetically distant from the main *Vibrio* branch.

GROWTH CHARACTERISTICS AND PHENOTYPIC IDENTIFICATION. *Vibrio litoralis* grows as pale yellow colonies 1.5–2.0 mm, smooth, circular with a slightly irregular shape after 3 days on MA 2216 agar, are rod-shaped (0.6–0.6×2.0–3.0 μm), motile, grow in 1–12% NaCl (but not outside this range), 4–45°C, pH 4.1–8.8, oxidase positive and non-luminescent (Nam *et al.*, 2007). These and other reactions are presented in Tables 3.23, 3.27–29 and 3.38.

Vibrio logei

See *Aliivibrio logei*.

Vibrio mangrovi

TAXONOMY AND SIGNIFICANCE. *Vibrio mangrovi* (MSSRF38^{T}=DSM 19641^{T}=LMG 24290^{T}) is a nitrogen-fixing bacterium isolated from the rhizosphere of wild rice (*Porteresia coarctata*) found in mangroves in India (Rameshkumar *et al.*, 2010). Based on 16S rRNA sequencing it is most closely related to *V. ruber* (98.3%), *V. rhizosphaerae* (98.2%) and less than 96% to other *Vibrio* species. Multilocus sequencing indicates low levels of similarity (82–90%) to all *Vibrio* species using the *ftsZ*, *gapA*, *gyrB* and *mreB* genes, and similarities of 85.3–91.6% to *V. rhizosphaerae*, *V. ruber* and *V. gazogenes*.

GROWTH CHARACTERISTICS AND PHENOTYPIC IDENTIFICATION. *Vibrio mangrovi* is a facultatively anaerobic, motile, Gram-negative slightly curved rod (0.8×1.5–5.0 μm), occurring singly or in pairs. It is catalase positive but oxidase negative. Growth occurs on TSA containing 2% NaCl. A red non-diffusible prodigiosin pigment is produced. Growth is optimal at 28–30°C and 0.1–8% NaCl with no growth at 0 or 10% NaCl. Growth does not occur on TCBS and it is resistant to 0/129 vibriostatic agent (Rameshkumar *et al.*, 2010). It is phenotypically similar to *V. rhizosphere* and *V. ruber*. Hydrolysis is positive for casein (skim milk). Other reactions are presented in Tables 3.23–29 and 3.38.

Vibrio marinus (ATCC 15381^{T})

See *Moritella marina*.

Vibrio marisflavi

TAXONOMY AND SIGNIFICANCE. *Vibrio marisflavi* (VH134^{T}=CGMCC1.89941^{T}=LMG25284^{T}=DSM 23086^{T}) was isolated from seawater collected at a depth of 10 m near the Yellow Sea Cold Water Mass, China (Wang *et al.*, 2011). Based on 16S rRNA sequencing it is most closely related to *V. rumoiensis* (96.6%). Analysis of *gyrB*, *topA*, *recA*, *ftsZ*, *mreB*, *gapA* and *rpoA* genes confirms similarity with *V. rumoiensis*.

GROWTH CHARACTERISTICS AND PHENOTYPIC IDENTIFICATION. *Vibrio marisflavi* is a facultatively anaerobic, motile, oxidase-positive, Gram-negative slightly curved rod (0.8–1×0.4–0.6 μm). On MA 2216, colonies are 1–1.5 mm, light yellow, smooth, circular and non-swarming. Colonies are yellow and translucent on TCBS (Wang *et al.*, 2011). Luminescence is not produced. Catalase and oxidase are weakly positive. Growth occurs at 16–37°C but not at 4 or 40°C. Growth occurs at 1–6% NaCl but not at 0% NaCl. Caseinase and lipase are positive and amylase is negative. These and other reactions are presented in Tables 3.23–30 and 3.38. Sets are incubated according to the manufacturer's instructions with the exception that an inoculum is used of 1.5% NaCl.

Vibrio maritimus

TAXONOMY AND SIGNIFICANCE. *Vibrio maritimus* (R-40493^T=LMG 25439^T=CAIM 1455^T) was isolated from coral, a zoanthid called *Palythoa caribaeorum* from a reef off the coast of Brazil (Chimetto *et al.*, 2011b). *Vibrio variabilis* was isolated at the same time. Both species share 99% homology based on 16S rRNA sequencing and are 55–57% similar according to DNA–DNA hybridization. According to 16S rRNA gene sequence *V. maritimus* has similarity to *V. neptunius* (97.1%), *V. coralliilyticus* (97.3%), *V. nigripulchritudo* (98.5%), *V. sinaloensis* (98.3%) and *V. brasiliensis* (98.4%). Multilocus sequence analysis concatenated sequences of *ftsZ*, *gyrB*, *recA*, *rpoA*, *topA*, *pyrH*, *mreB* and 16S rRNA genes indicates the two species form a separate branch within the *Vibrio* genus with less than 92% similarity.

GROWTH CHARACTERISTICS AND PHENOTYPIC IDENTIFICATION. *Vibrio maritimus* is a facultatively anaerobic, motile, oxidase-positive, Gram-negative rod (1.5–4.0×1.0 μm). On TSA containing 2% NaCl, colonies are 1 mm, beige, convex, round, and smooth after 24 h at 28°C. Colonies are yellow on TCBS (Chimetto *et al.*, 2011b). Optimal temperature is 28–37°C and growth occurs at 15–37°C, but not at 40°C. Optimal NaCl concentration is 1.5–6.0% with growth occurring at a range of 0.5–8.0% and weak growth at 8%. These and other reactions are presented in Tables 3.23–26 and 3.38.

Vibrio mediterranei

TAXONOMY AND SIGNIFICANCE. *Vibrio mediterranei* (50^T=CECT 621^T) was isolated from seawater, seawater sediment and sea plankton collected from a coastal area south of Valencia, Spain (Pujalte and Garay 1986). *Vibrio shilonii*, identified as the causative agent of coral bleaching of *Oculina patagonica* (Banin *et al.*, 2000; Kushmaro *et al.*, 2001), was found to be a later synonym of *V. mediterranei* (Thompson *et al.*, 2001a). Strain AK2 isolated from bleached coral has been identified as *V. harveyi* (Gomez-Gil *et al.*, 2004b). In challenge experiments *V. mediterranei* (*V. shilonii*) did not cause tissue lysis of coral (Ben-Haim *et al.*, 2003b). *V. coralliilyticus* has been identified as the aetiological agent of tissue lysis in the coral *Pocillopora damicornis* (Ben-Haim *et al.*, 2003a).

In experimental challenge using rainbow trout (*Oncorhynchus mykiss*) and *Artemia* nauplii inoculated with 10^6 CFU/fish or 10^6CFU/ml, respectively, *V. mediterranei* (strain *V. shilonii* LMG 20977 from *Oculina patagonica*) was rated as non-virulent or low virulence. Although there was some evidence of exophthalmia, kidney necrosis, ascites and liquefaction of internal organs, clinical disease was not seen, and at the end of the experiment the organism was recovered from tissue at either a low rate or not at all (Austin *et al.*, 2005).

GROWTH CHARACTERISTICS AND PHENOTYPIC IDENTIFICATION. *Vibrio mediterranei* is a facultatively anaerobic, motile, oxidase-positive, Gram-negative rod (1–2×0.5 μm). On MA 2216, colonies are 0.7 mm after 24 h at 25°C and 1.3 mm after 48 h, cream, circular and non-swarming. On MSA-B, colonies are 2 mm and dull grey with beta-haemolysis seen in the pool only at 24 h, but haemolysis is seen around colonies after 48 h. Colonies increase in size to 3.7 mm, and are a dirty grey colour with a slight poached egg morphology at 48 h. On MSSA, colonies are 3–5 mm and a dirty cream-grey colour after 48 h. Growth occurs on BA as 0.5 mm colonies at 24 h increasing to 2.2 mm and non-haemolytic, pale grey and raised after 48 h. On TCBS, colonies are 5 mm yellow at 48 h (this manual). Luminescence is not seen in the type strain but 25% of strains were reported to be luminescent (Pujalte and Garay, 1986). Growth does not occur at 40°C or at 10% NaCl. Motility is seen as a slow motion in a wet preparation but is negative in the tube test after 48 h at 25°C. The oxidase reaction is strong occurring within a few seconds using the filter paper method (see chapter on interpretation of biochemical identification tests and media section). Starch and Tween 80 are hydrolysed. *Vibrio mediterranei* may show a positive ADH result using Thornley's method and in the API 20E, but negative in Moeller's ADH. In the API 20E, LDC, citrate, inositol and sorbitol are negative at 24 h but positive at 48 h for strain BAA-91, originally identified as *V. shilonii*. Motility is positive in the tube test.

These and other reactions are presented in Tables 3.22, 3.24–26 and 3.38. Sets are inoculated with an inoculum in 2% NaCl and incubated at 25°C for 48 h.

Vibrio metschnikovii

TAXONOMY AND SIGNIFICANCE. The original isolate of *V. metschnikovii* (NCTC 8443^{T}) was isolated from a fowl that died of choleraic disease. The result of a numerical taxonomic study found that this strain, which had been in culture collections for many years, had some characteristics different from fresh isolates. A fresh strain (NCTC 11170) isolated from cockles was recommended as the working strain (Lee *et al.*, 1978). *Vibrio metschnikovii* is widely distributed in the aquatic environment including rivers and sewerage. It has been isolated from shellfish such as cockles, clams, oysters and lobster and also from animal intestines. Reports from humans are rare, but have been reported from intestine, wound infections and food poisoning. It produces a cytotoxin (Farmer *et al.*, 1988; Linde *et al.*, 2004). MLSA identified *V. metschnikovii* as belonging to the Cholerae clade, which also contains *V. cholerae*, *V. cincinnatiensis*, *V. fluvialis*, *V. furnissii*, *V. metschnikovii* and *V. mimicus* (Sawabe *et al.*, 2007b). Members of this clade have 44–50 GC mol%, a MLSA concatenated similarity of 85.4–94.7% and AAI of 92.8–99.6%.

GROWTH CHARACTERISTICS AND PHENOTYPIC IDENTIFICATION. *Vibrio metschnikovii* is a facultatively anaerobic, motile, oxidase-negative, nitrate-negative, Gram-negative, straight or slightly curved rod (1.5–2.5×0.5 μm) occurring singly, in pairs or short chains of three to four cells (Lee *et al.*, 1978). On Columbia sheep blood agar, colonies are non-swarming, 2–3 mm, greyish, opaque with complete haemolysis after 24 h at 37°C. Growth on MCA is poor (or negative) and colonies are lactose-fermenting (Linde *et al.*, 2004). Colonies are 0.5 mm and light grey after 24 h on BA and MSA-B, and are 1 mm at 48 h on MA 2216 and become a tan/grey colour after 5 days. Luminescence is not seen on nutrient agar containing NaCl. Growth on TCBS is negative or variable amongst strains. Growth occurs at 0.5–7% NaCl but not at 0 or 10% NaCl. Growth occurs at 15–37°C but not at 42°C. Pinpoint colonies grow after 10 days at 4°C. Although reported as negative growth at 42°C, growth at this temperature has been noted. In the conventional tube method the type strain gives a strong indole reaction after 48 h incubation. Simmons' citrate is positive at 72 h and positive at 48 h in the API 20E. ADH and LDC are positive for the type strain by Moeller's method. LDC was reported negative by Farmer *et al.* (1988), as was citrate using Simmons' and Christensen's methods. The organism can be identified in the Vitek 2 GN card and the API 32E (Biomerieux). These and other reactions are presented in Tables 3.23–26 and 3.38. Sets are incubated according to the manufacturer's instructions with the exception that an inoculum is used of 1.5% NaCl.

MOLECULAR IDENTIFICATION. A conventional and SYBR Green PCR based on the *infC* (initiation factor 3) gene are available for *V. metschnikovii* (Cao *et al.*, 2010). Both PCRs were specific when tested against *V. cholerae* serotypes 01, 0139, non-01, non-0139, *V. parahaemolyticus*, *V. alginolyticus*, *V. vulnificus*, *V. mimicus*, *V. fluvialis*, *V. proteolyticus*, *Aeromonas hydrophila*, *Plesiomonas shigelloides*, *Listeria monocytogenes*, *L. ivanovii*, *L. innocua*, *Salmonella enterica*, *S. enteritidis*, *Pseudomonas aeruginosa*, *Enterobacter sakazakii*, *Escherichia coli*, *Citrobacter freundii*, *Shigella flexneri*, *Staphylococcus aureus*, *Yersinia enterocolitica*, *Bacillus cereus* and *Clostridium perfringens*.

PCR type	Conventional	
DNA extraction	TaKaRa MiniBEST bacterial genomic DNA extraction kit	
PCR master mix		
Primer 1	5′-AGCATGACCAAGCTGCTCTT-3′	
Primer 2	5′-GCAAATATCCGACAGCACCATT-3	
Reagent	**Stock**	**Volume ×1**
Buffer	×10	2.5 µl
dNTPs	10 mmol/l	2 µl
P 1	10 µmol/l	1 µl
P 2	10 µmol/l	1 µl
Taq DNA polymerase	5 U/ul	0.2 µl
Water		16.3 µl
DNA		2 µl
Total volume		**25 µl**
Thermocycling	**Thermocyler (Applied Biosystems)**	
Step 1	**Step 2**	**Step 3**
94°C, 3 min	94°C, 1 min	94°C, 1 min
	60°C, 1 min	60°C, 1 min
	72°C, 1 min	72°C, 7 min
×1	×35	×1
Product	260 bp	
Sensitivity	1790 CFU/ml	

PCR type	SYBR Green	
DNA extraction	TaKaRa MiniBEST bacterial genomic DNA extraction kit	
PCR master mix		
Primer 1	5′-AGCATGACCAAGCTGCTCTT-3′	
Primer 2	5′-GCAAATATCCGACAGCACCATT-3	
Reagent	**Stock**	**Volume ×1**
SYBR Premix Ex *Taq*		10 µl
P 1	10 µmol/l	0.5 µl
P 2	10 µmol/l	0.5 µl
Water		7.0 µl
DNA		2 µl
Total volume		**20 µl**
Thermocycling	LightCycler Real-Time PCR System (Roche)	
Step 1	Step 2	
95°C, 10 s	95°C, 5 s	
	60°C, 20 s	
×1	×40	
Product	260 bp	
Melt temperature	79°C	
Sensitivity	180 CFU/ml	

Vibrio mimicus

TAXONOMY AND SIGNIFICANCE. *Vibrio mimicus* (1721-77^{T}=ATCC 33653^{T}) was isolated from shellfish (oysters and prawn as a cause of food poisoning), water mite, water and from cases of diarrhoea and ear infections in humans. It is phenotypically related to *V. cholerae* but is negative for the fermentation of sucrose and was originally identified as *V. cholerae* Heiberg group 5 (Davis *et al.*, 1981). Phylogenetic analysis using sequences from the 16S rRNA gene, lux operon (*luxRI*) and glyceraldehyde phosphate dehydrogenase (*gapA*) gene indicates that *V. mimicus* is an atypical strain of *V. cholerae* and has a substantially smaller 16S rRNA gene, thought to be due to a reduced genome component from its ancestral parent, *V. cholera* (Nishiguchi and Nair, 2003). The 16S rRNA gene has sequence similarity of 99.06–100% with *V. cholerae* and about 70% DDH similarity (Kita-Tsukamoto *et al.*, 1993).

MLSA identified *V. mimicus* as belonging to the Cholerae clade, which also contains *V. cholerae*, *V. cincinnatiensis*, *V. fluvialis*, *V. furnissii* and *V. metschnikovii* (Sawabe *et al.*, 2007b). Members of this clade have 44–50 GC mol%, a MLSA concatenated similarity of 85.4–94.7% and AAI of 92.8–99.6%.

In natural infection, *V. mimicus* was isolated in mixed bacterial flora at a rate of 14% from the haemolymph of Australian freshwater crayfish, yabby (*Cherax albidus*) and red claw (*C. quadricarinatus*) (Eaves and Ketterer, 1994; Wong *et al.*, 1995). Three of seven strains tested were virulent and resulted in 100% mortality when tested in intramuscular challenge tests using an inoculum of 10^5 CFU/ml. Avirulent strains did not cause mortalities. Infection and mass mortality was related to postharvest purging before storage and shipment, or stress due to high stocking density and poor water quality including low dissolved oxygen and temperatures above 28°C.

Vibrio mimicus was isolated from apparently healthy brown bullhead (*Ictalurus nebulosus*) and striped bass (*Morone saxatilis*), and from mortalities of farmed striped bass from Chesapeake Bay in the USA (Lupiani *et al.*, 1993). The results of experimental challenge using intraperitoneal injection indicated that *V. mimicus* was not pathogenic to striped bass as LD_{50} was $>10^7$, however the strains were virulent to mice at an LD_{50} of 10^6.

GROWTH CHARACTERISTICS AND PHENOTYPIC IDENTIFICATION. *Vibrio mimicus* is a facultatively anaerobic, motile, oxidase-positive, Gram-negative, curved rod. On BA and MSA-B, colonies are 1.5 mm, grey to translucent, round and with a strong zone of beta-haemolysis after 24 h at 25°C. Areas of heavy growth appear grey-green after 72 h. On MSSA, colonies are 2 mm grey at 48 h, but haemolysis is not easily seen on this medium. On MA 2216, colonies are 0.5 mm at 24 h and 1 mm and cream after 48 h at 25°C. Colonies are pinpoint at 24 h and 1–2 mm on MCA after 48 h, and 1–2 mm green on TCBS (see colour plate section). Growth occurs at 0–3% NaCl and 50% of strains grow at 6% NaCl, but not at 10% NaCl (Davis *et al.*, 1981 and this manual). Lethicin, Tween 20 and Tween 80 are hydrolysed but starch is negative (Eaves and Ketterer, 1994). *Vibrio mimicus* is negative in the VP test whereas 65% of *V. cholerae* strains are VP positive. *Vibrio mimicus* is negative for fermentation of sucrose whereas *V. cholerae* is positive. In the API 20E citrate may be weak positive when an inoculum containing a final concentration of 2% NaCl is used, but is clearly positive with an inoculum of 0.85% NaCl. *Vibrio mimicus* can be identified using the Vitek 2 GN card and the API 32E (Biomerieux). These and other reactions are presented in Tables 3.22, 3.24–26 and 3.38. Some reactions are listed for the API 20NE. Sets are incubated according to the manufacturer's instructions with the exception that an inoculum is used of 1–2% NaCl or as indicated in the relevant table.

MOLECULAR IDENTIFICATION. A multiplex PCR for detection of *V. mimicus*, *V. cholerae*, *V. parahaemolyticus* and *V. vulnificus* (Tarr *et al.*, 2007) is detailed under *V. parahaemolyticus*.

Vibrio species MM5

TAXONOMY AND SIGNIFICANCE. An unidentified *Vibrio* named *Vibrio* species MM5 caused high mortality in clam (*Meretrix meretrix*) cultured in China (Yue *et al.*, 2011). Based on 16S rRNA sequencing the highest similarity is with *V. furnissii* (98%). A tree constructed using the neighbour-joining method and composed of concatenated sequences of *gapA*, *ftsZ*, *mreB* and *topA* showed close association with *V. furnissii* and *V. fluvialis*, which lie in the Cholerae clade identified by Sawabe *et al.* (2007b). In infected clam the organism was found in the hepatocytes, hepatopancreas, mantle and gill, and lesions occurred on the foot. Pathogenicity was proven in experimental challenge in which cumulative mortalities reached 70% at 132 h.

GROWTH CHARACTERISTICS AND PHENOTYPIC IDENTIFICATION. *Vibrio* species MM5 is a facultatively anaerobic, Gram-negative rod. On Tryptic soy agar containing 0.6% yeast extract, colonies are white changing to translucent cyan colour on prolonged incubation (Yue *et al.*, 2011). Isolate MM5 grew at up to 8% NaCl, but not at 12% NaCl.

It was resistant to vibriostatic compound 0/129, but no concentration was given. Reactions in the Vitek (bioMerieux) gram-negative card did not identify the isolate to any of the strains capable of growth at 37°C, and which are in the Vitek database. The fact that reactions were obtained by the Vitek system suggests the organism can grow in 0.85% saline and at 37°C. Apart from the Vitek, reactions were also carried out using conventional methods according to Bergey's Manual. These and other reactions are presented in Table 3.22.

Vibrio mytili

TAXONOMY AND SIGNIFICANCE. *Vibrio mytili* (165^{T}= CECT 632^{T}) was isolated from mussels (*Mytilus edulis*) harvested off the Atlantic Spanish coast. It is not known to be pathogenic (Pujalte *et al.*, 1993). MLSA identified *V. mytili* as belonging to the Harveyi clade, which also contains *V. campbellii*, *V. alginolyticus*, *V. harveyi*, *V. natriegens*, *V. parahaemolyticus* and *V. rotiferianus* (Sawabe *et al.*, 2007b). Members of this clade have >25% DDH and 42–48 GC mol%, a MLSA concatenated sequence similarity of 90.1–96.2% and AAI of 97.2–99.4%. Strains within the Harveyi clade (*V. alginolyticus*, *V. campbellii*, *V. harveyi*, *V. natriegens*, *V. parahaemolyticus*, *V. rotiferianus*) can be discriminated using concatenated sequences from three housekeeping genes, *rctB*, *rpoD* and *toxR*, with *toxR* and *rpoD* providing the most discrimination (Pascual *et al.*, 2010) (see under *Vibrio harveyi*).

GROWTH CHARACTERISTICS AND PHENOTYPIC IDENTIFICATION. *Vibrio mytili* is a facultatively anaerobic, motile, oxidase-positive, Gram-negative coccobacillus. Growth occurs on BA as 0.5 mm, light grey colonies at 24 h and 25°C. At 48 h, colonies are 3–4 mm on BA, 1–2 mm on MA 2216 and 4 mm on MSA-B. Haemolysis and swarming are negative. Yellow colonies grow on TCBS. Growth does not occur on MCA (this manual). Growth was reported from TSA and yeast extract agar containing 2% NaCl (Pujalte *et al.*, 1993). Growth occurs at 1–10% NaCl but not at 20% NaCl. Growth occurs at 10–37°C, but not at 4 or 40°C for CECT 632^{T}. Strain CIP 103929 grew at 42°C. Luminescence is not seen. Results are positive for Thornley's ADH. Moeller's ADH is delayed and a faint purple is seen at 5 days with a strong positive at 9 days. The catalase reaction is slow. The oxidase reaction is strongly positive using the oxidase reagent (see 'Interpretation of Biochemical Tests'). Motility is slow in the tube method. Casein, starch and alginate are not hydrolysed, but Tween 80 and tyrosine are degraded. Arbutin, D-cellobiose and D-galactose are fermented but not D-melibiose or adonitol. These and other reactions are presented in Tables 3.23–26 and 3.38. An inoculum of 2% NaCl is used with an incubation temperature of 25°C.

Vibrio natriegens

TAXONOMY AND SIGNIFICANCE. *Vibrio natriegens* (ATCC 14048^{T}) was isolated from salt marsh mud on Sapelo Island, Georgia, USA and originally named *Pseudomonas natriegens* (Payne *et al.*, 1961), but was placed into the genus *Beneckea* (Baumann *et al.*, 1971a) and later into the genus *Vibrio* based on DNA base composition and phenotypic characteristics (Austin *et al.*, 1978). It has been reported from water, sediment and oysters (West *et al.*, 1985), and from diseased crown of thorns starfish (*Acanthaster planci*), Great Barrier Reef, Australia (Rivera-Posada *et al.*, 2011). Other *Vibrio* species (*V. harveyi*, *V. owensii* and *V. rotiferianus*) were found in the diseased crown of thorns starfish, but currently, the pathogenicity of *V. natriegens* and these *Vibrio* species for the starfish is not known (Rivera-Posada *et al.*, 2011).

Phylogenetic analysis of strain NCIMB 1900 (ATCC 25916) and CIP 102762, originally identified as *V. pelagius*, indicated that these strains are more likely to be *Vibrio natriegens*, and that the original strain deposited at NCIMB may have been lost. Phenotypic tests for strain NCIMB 1900 corresponded to those expected for *V. natriegens* being positive for acid production from L-arabinose, melibiose and L-rhamnose, with a negative result for lactose fermentation and alginate degradation (Macián *et al.*, 2000). The phenotypic results recorded as *V. pelagia* NCIMB 1900 by Lunder *et al.* (2000) have been recorded as *V. natriegens* in this book.

MLSA identified *V. natriegens* as belonging to the Harveyi clade, which also contains *V. campbellii*, *V. alginolyticus*, *V. harveyi*, *V. mytili*, *V. parahaemolyticus* and *V. rotiferianus* (Sawabe *et al.*, 2007b). Members of this clade have >25% DDH and 42–48 GC mol%, a MLSA concatenated sequence similarity of 90.1–96.2% and AAI of 97.2–99.4%. Strains within the Harveyi clade (*V. alginolyticus*, *V. campbellii*, *V. harveyi*, *V. natriegens*, *V. parahaemolyticus*, *V. rotiferianus*) can be discriminated using concatenated sequences from three housekeeping genes, *rctB*, *rpoD* and *toxR*, with *toxR* and *rpoD* providing the most discrimination (Pascual *et al.*, 2010) (see under *Vibrio harveyi*).

GROWTH CHARACTERISTICS AND PHENOTYPIC IDENTIFICATION. *Vibrio natriegens* is a facultatively anaerobic, motile, Gram-negative rod, positive for catalase and oxidase. On MSSA, colonies are 2 mm, light grey to cream, shiny, round, non-swarming after 24 h at 25°C, and up to 5 mm after 3 days. It is non-haemolytic. Colonies are 1–2 mm dark yellow, round, shiny on TCBS after 2 days at 25°C. Reports vary for sensitivity to vibriostatic agent 0/129. *Vibrio natriegens* is resistant to 10 and 150 μg after 24 h on MSSA (this manual). It is negative for luminescence. Growth occurs at 15–37°C but not at 4°C or 42°C. Growth occurs at 2–6% NaCl, but not at 0 or 8% NaCl (West *et al.*, 1985). Hydrolysis of gelatin using the gelatin/NaCl plate (see media section) is more easily seen if the plate (after 48 h incubation at 25°C) is incubated at 4°C for 20 min before reading and interpretation. The aesculin tube method will only show a weak reaction if insufficient NaCl is added as an inoculum. Using 2% NaCl in the inoculum will produce a positive reaction in 24 h. In the API 20E citrate may give a weak positive reaction at 48 h. VP is positive when read at 48 h, and gelatin is negative at 24 h but positive at 48 h (this manual). These and other reactions are presented in Tables 3.23–29 and 3.38. An inoculum of 2% NaCl is used and the sets are incubated at 25°C or 30°C for 48 h or as indicated in the tables.

MOLECULAR IDENTIFICATION. See under *Vibrio harveyi* for a PCR using the *rpoD* and *toxR* housekeeping genes followed by sequencing and analysis using concatenated sequences to enable differentiation between strains comprising the Vibrio Core group.

A polyphasic approach using sequence information from the 16S rRNA gene, *topA* (topoisomerase I) and *mreB* (rod shaping protein) enables differentiation of closely related species within the Harveyi clade (Rivera-Posada *et al.*, 2011). For 16S rRNA sequencing, primer 27F and 1492R (Lane, 1991) are used (see under molecular section). Primers for amplification of the *topA* gene are VtopA400F (5′-GAG ATC ATC GGT GGT GAT G-3′) and VtopA1200R (5′-GAA GGA CGA ATC GCT TCG TG-3′) and for the *mreB* gene VmreB12F (5′-ACT TCG TGG CAT GTT TTC-3′) and VmreB999R (5′-CCG TGC ATA TCG ATC ATT TC-3′). Amplification and sequencing follows protocols described by Sawabe *et al.* (2007b). A 20-μl master mix consists of 0.5 μM each primer, 1× PCR buffer (Qiagen), 200 μM dNTPs, 0.5 μl *Taq* polymerase (Qiagen) and 20 ng DNA. An annealing temperature of 50°C is used for both primer sets (Rivera-Posada *et al.*, 2011).

Vibrio navarrensis

TAXONOMY AND SIGNIFICANCE. *Vibrio navarrensis* (1397-6^{T}=CIP 103381^{T}) was isolated from sewage and surface water in Navarra, Spain (Urdaci *et al.*, 1991). *Vibrio navarrensis* ssp. *pommerensis* was isolated from the German Baltic Sea and identified as a biotype (Jores *et al.*, 2007). It is closely related to *V. vulnificus*. MLSA identified *V. navarrensis* as belonging to the Vulnificus clade, which contains *V. vulnificus* (Sawabe *et al.*, 2007b). Members of this clade have >30% DDH and 45–48 GC mol%, 88.6% similarity based on concatenated MLSA sequence data and AAI of 96.4%.

GROWTH CHARACTERISTICS AND PHENOTYPIC IDENTIFICATION. *Vibrio navarrensis* is a facultatively anaerobic, motile, oxidase-positive, Gram-negative rod (1–2×0.8–1.0 μm) susceptible to vibriostatic agent. After 24 h at 25°C colonies on BA are 0.5 mm and 0.1 mm on MA 2216. At 48 h colonies are 0.5 mm cream on MA 2216 and 2 mm grey on MSA-B and 1 mm on BA. Haemolysis and swarming are not seen. On NA containing 2% NaCl colonies are 2–3 mm, round, opaque and non-pigmented after 24 h at 30°C. Colonies are yellow on TCBS (Urdaci *et al.*, 1991; and this manual). Luminescence is not seen on NA with NaCl. Growth occurs at 10–40°C with 82% of strains positive for growth at 42°C. Growth occurs at 1–7% NaCl, but not at 10% NaCl. Results are negative for both Moeller's and Thornley's ADH. Starch is hydrolysed. Gelatin hydrolysis was reported positive (Urdaci *et al.*, 1991), but is negative in the API 20E, and on gelatin plate media (see media section) a zone of gelatin hydrolysis is only seen after 2–3' days incubation followed by placement of the 3% NaCl gelatin plate at 4°C for 30 min. The type strain was reported positive for Simmons' citrate, whereas another five strains are negative (Urdaci *et al.*, 1991). Simmons' citrate becomes positive after prolonged incubation of 3–4 days, but is negative in the API 20E after 48 h. Growth in aesculin medium (see media section) is poor and a positive result is only recorded after 4 days.

Vibrio navarrensis ssp. *pommerensis* is a motile, Gram-negative curved rod (3×1 µm), oxidase-positive, and is susceptible to vibriostatic agent (Jores *et al.*, 2007). On NA containing 2% NaCl colonies are 2–3 mm, round, entire, non-swarming after 24 h at 30°C and produce a greenish-brown pigment after 2 days. Luminescence is negative. Growth is optimal at 28–37°C with no growth at 4 or >40°C. Haemolysis is seen on sheep blood agar after 12–24 h. Gelatin is hydrolysed. Simmons' citrate is negative. These and other reactions are presented in Tables 3.23–26 and 3.38. Sets are incubated at 25°C with an inoculum of 2% NaCl.

Vibrio neonatus

TAXONOMY AND SIGNIFICANCE. *Vibrio neonatus* (HDD3-1^T=IAM 15060^T=LMG 19973^T) was isolated from the gut of abalone (*Haliotis discus discus*) in Japan (Sawabe *et al.*, 2004a). The isolate is closely related to *V. ezurae* and *V. halioticoli*, with *V. neonatus* and *V. ezurae* having 98.3% 16S rRNA gene sequence similarity to *V. halioticoli*. A strong clade is formed with *V. neonatus*, *V. ezurae*, *V. halioticoli*, *V. gallicus* and *V. superstes* based on phylogenetic analysis using the 16S rRNA gene and the *gap* (glyceraldehyde 3-phosphate dehydrogenase) gene. Sawabe *et al.* (2004a) noted that six undetermined positions in the *V. halioticoli* 16S rRNA gene sequence means that the similarity could be as close as 99%. *Vibrio neonatus* was different to *V. halioticoli* in 2 of 72 phenotypic tests. MLSA identified *V. neonatus* as belonging to the Halioticoli clade, which also contains *V. gallicus*, *V. ezurae*, *V. halioticoli* and *V. superstes* (Sawabe *et al.*, 2007b). Members of this clade have >22% DDH, 39–42 mol% GC, a MLSA concatenated similarity of 88.0–97.7% and AAI of 94.7–99.5%.

GROWTH CHARACTERISTICS AND PHENOTYPIC IDENTIFICATION. *Vibrio neonatus* is a facultatively anaerobic, Gram-negative rod (0.5–0.6×1.2–3.0 µm), non-motile, and oxidase and catalase positive. On MA 2216 agar, colonies are beige, circular, smooth, convex, with an entire edge at 25°C. Colonies are green on TCBS. Growth occurs at 15–30°C but not at 4 or 37°C. Na^+ ions are required for growth and growth does not occur at 1% or 6% NaCl. Results are positive for hydrolysis of alginate and negative for luminescence, hydrolysis of starch, gelatin, chitin, Tween 80 and agar. *Vibrio neonatus* differs from *V. ezurae* being negative for utilisation of D-mannose and D-gluconate, which are both positive for *V. ezurae*. These and other reactions are presented in Table 3.23. Note that these reactions report the results of fermentation of carbohydrates ('sugars') not utilization. Some results are presented in Table 3.24.

Vibrio neptunius

TAXONOMY AND SIGNIFICANCE. *Vibrio neptunius* (LMG 20536^T=CAIM 532^T), initially identified as FAFLP group A5, was isolated from healthy and diseased bivalve larvae (*Nodipecten nodosus*), from the gut of turbot larvae (*Scophthalmus maximus*) and from healthy rotifers (*Brachionus plicatilis*) and the culture water of rotifers (Thompson *et al.*, 2001b, 2003b). *Vibrio neptunius* caused mortalities in flat oyster, *Ostrea edulis*, cultured along the coast of Spain (Prado *et al.*, 2005). Clinical signs included growth depression, abnormal swimming, reduction in motility, deformation of the velum in the larvae, and mortalities. Diseased spat had 10^3 CFU/spat compared to healthy spat with levels of <10 CFU/spat. Pathogenicity tests conducted on the larvae resulted in 98–100% mortality in 72–96 h.

In experimental challenge, *V. neptunius* (LMG 20536^T and LMG 20610) caused 100% and 80% mortality, respectively, within 7 days when inoculated into rainbow trout (*Oncorhynchus mykiss*) at a dose of 10^6 cells/fish. At an inoculum of 10^4 CFU/fish, 100% and 60% mortality occurred. The LD_{50} was $<10^2$ for the type strain and 10^2 for LMG 20619 (Austin *et al.*, 2005). Mortality of *Artemia* nauplii exposed to 10^6 CFU/ml was 100 and 90%, respectively. Injection with extracellular product caused 20% mortality in rainbow trout using the type strains and 100% mortality with strain LMG 20610 and 55–65% in *Artemia*. In rainbow trout, clinical signs included exophthalmia and liquefaction of internal organs and muscles. The strains were positive for caseinase, elastase, lipase, phospholipase and were haemolytic on sheep blood agar. The type strain was also positive for gelatinase.

In Greenshell™ mussels cultured in New Zealand, an isolate tentatively identified as *V. coralliilyticus/neptunius*-like caused 75% larval mortality in experimental challenge with 10^2 CFU/ml (Kesarcodi-Watson *et al.*, 2009). Larvae showed erratic swimming, bacterial accumulation around velum, loss of velum and deterioration and loss of soft tissue. The infection occurred in the digestive system. Sequencing of the 16S rRNA gene indicated similarity to *V. coralliilyticus* (99.3–99.5%) and similarity to *V. neptunius* (99.5%).

Vibrio neptunius together with *V. brasiliensis* and *V. xuii* is related to *V. tubiashii* (98–98.8%), *V. nereis* (97.6–98.8%), *V. coralliilyticus* (96.8–98.5%), *V. mytili* (96.8–98.2%) and *V. diabolicus* (97.1–98.1%) based on 16S rRNA sequencing (Thompson *et al.*, 2003b). MLSA identifies *V. neptunius* as belonging to the Coralliilyticus clade, which also contains *V. coralliilyticus* (Sawabe *et al.*, 2007b). Members of this clade have >64% DDH, 45–46 mol% GC, a MLSA concatenated similarity of 95.6% and AAI of 99.6%. MLSA identifies *V. nereis* and *V. xuii* as belonging to the Nereis clade, *V. brasiliensis* and *V. tubiashii* as belonging to the Orientalis clade and *V. mytili* as belonging to the Harveyi clade. *Vibrio diabolicus* is not accessible and was not included in the analysis.

GROWTH CHARACTERISTICS AND PHENOTYPIC IDENTIFICATION. *Vibrio neptunius* is a facultatively anaerobic, motile, oxidase-positive, Gram-negative, slightly curved rod (2.3–3×1 μm). On TSA containing 2% NaCl colonies are non-luminescent, non-swarming, 3 mm, beige, translucent, smooth, round at 28°C after 48 h and grow as 2–3 mm yellow colonies on TCBS after 24 h (Thompson *et al.*, 2003b). Growth does not occur at 4 or 40°C or at 8% NaCl. Growth does not occur in the absence of NaCl, and up to 2% NaCl is added to culture media and to identification tests. The type strain and the oyster strain are positive for ADH in the API 20E but 14.2% of strains are negative. The type strain is positive for gelatin but strain LMG 20610 is negative. Oyster strains were reported as negative for citrate, indole, VP and nitrate, whereas these tests were all positive for bivalve larvae strains (Thompson *et al.*, 2003b; Prado *et al.*, 2005). The *V. coralliilyticus/neptunius*-like isolate was positive for ADH in the API 20NE and produced yellow colonies on TCBS and grew on TSA supplemented with 2% NaCl. These and other reactions are presented in Tables 3.22, 3.24–29 and 3.38. An inoculum of 2% NaCl is used and opacity McFarland 3 and sets are incubated at 25°C for 48 h.

Vibrio nereis

TAXONOMY AND SIGNIFICANCE. *Vibrio nereis* (80^T= ATCC 25917^T) was isolated from seawater off the coast of Hawaii and originally named *Beneckea nereida* (Baumann *et al.*, 1971a), but later changed to *V. nereis* following the abolition of the genus *Beneckea* (Baumann *et al.*, 1980). *Vibrio nereis* was isolated along with *V. alginolyticus* and *Deleya* (*Alcaligenes*) *cupida* from mass mortality in larval black seabream (*Acanthopagrus schlegeli*) cultured in Japan. In experimental challenge *V. nereis* was not pathogenic, however *V. alginolyticus* and *Deleya* (*Alcaligenes*) *cupida* produced high mortality (Kusuda *et al.*, 1986). Non-pathogenicity of *V. nereis* was confirmed also in experimental challenge using rainbow trout (*Oncorhynchus mykiss*) and *Artemia* nauplii inoculated with 10^6 CFU/fish or 10^6CFU/ml, respectively, with results rating *V. nereis* (LMG 3895^T) as non-virulent. Signs of infection were not seen and the organism was not recovered from tissues (Austin *et al.*, 2005).

MLSA identified *V. nereis* as belonging to the Nereis clade, which also contains *V. xuii* (Sawabe *et al.*, 2007b). Members of this clade have >30% DDH, 39–47 GC mol%, a MLSA concatenated sequence similarity of 91.2% and AAI of 96.8%.

GROWTH CHARACTERISTICS AND PHENOTYPIC IDENTIFICATION. *Vibrio nereis* is a facultatively anaerobic, motile, straight, Gram-negative rod with occasional curved rods. On MSA-B and MSSA, colonies are 1.5 mm, non-swarming, non-haemolytic, light grey, shiny and raised after 24 h at 25°C. At 3 days colonies are 4 mm dark cream in colour and shiny. On BA colonies are 3 mm flat, grey and non-haemolytic after 3 days at 25°C. Growth on TCBS is negative at 24 h and may be weak at 48 h (this manual). Growth occurs at 4–40C° and at 2–6% NaCl (Baumann *et al.*, 1971a; Kusuda *et al.*, 1986). In the API 20E the reaction for glucose may show a pale green colour at 24 and 48 h rather than a bright yellow. Fermentation of glucose using the conventional tube method is positive. An inoculum is prepared using 2% NaCl. ADH is positive at 48 h using Moeller's and Thornley's methods. Indole may be weak in the conventional tube method. Luminescence is negative. Reactions are presented in Tables 3.23–26 and 3.38.

Vibrio nigripulchritudo

TAXONOMY AND SIGNIFICANCE. *Vibrio nigripulchritudo* (164^T=ATCC 27043^T), originally named *Beneckea nigrapulchrituda*, was isolated from seawater (Baumann *et al.*, 1971b, 1980). It is the aetiological agent of summer syndrome, a disease affecting grow-out in prawns/shrimps (*Litopenaeus stylirostris*) in New Caledonia (Costa *et al.*, 1998; Goarant *et al.*, 2006a). The disease occurs at high temperatures in contrast to Syndrome 93 caused by *V. penaeicida*, which occurs when temperatures are below 25°C. High mortality is seen, but low mortality associated with chronic disease occurs and the organisms can be isolated from healthy carriers before, during and after the disease. *Vibrio nigripulchritudo* can be isolated from prawn pond water and sediment pore water and is associated with crustaceans and zooplankton in the pond water. Highly virulent, moderately virulent and non-virulent strains exist and were delineated by experimental challenge (Goarant *et al.*, 2006a). The virulent strains are clonal as determined by molecular epidemiology using arbitrarily primed PCR (AP-PCR) and multi-locus sequence typing (MLST) (Goarant *et al.*, 2006b). Two plasmids appear to be responsible for virulence, a small plasmid pB1067 (11.2 kb) and a large plasmid pA1066 (247 kb); the latter is essential for full virulence. Both plasmids are present in highly virulent strains (represented by strain pSFn1) and their presence is essential for virulence expression as proven by challenge experiments. The plasmids are rarely found in non-pathogenic isolates. A heat-sensitive extracellular factor, toxic when injected into prawns, is produced by highly virulent strains, but not by moderately virulent strains. The production and/or activity of this unknown factor, thought to be a metalloprotease, is dependent upon the presence of the large plasmid pA1066 (Le Roux *et al.*, 2011). Strain CIP 103195 is not virulent for *L. stylirostris* (Goarant *et al.*, 2007).

Vibrio nigripulchritudo was isolated amongst other *Vibrio* species in the haemolymph of healthy spider crab (*Maja brachydactyla*) caught off the coast of Spain and the Canary Islands (Gomez-Gil *et al.*, 2010).

MLSA identifies *V. nigripulchritudo* and *V. penaeicida* as belonging to the Nigripulchritudo clade (Sawabe *et al.*, 2007b). Members of this clade have >36% DDH, 46–47 GC mol% and a MLSA concatenated sequence similarity of 89.0%.

GROWTH CHARACTERISTICS AND PHENOTYPIC IDENTIFICATION. *Vibrio nigripulchritudo* is a facultatively anaerobic, motile, oxidase-positive, Gram-negative rod. Intense blue-black colonies are produced after 2–4 days at 25°C on media containing glycerol: basal medium agar (BMA: 50 mM tris(hydroxymethyl)aminomethane (tris)-hydrochloride (pH 7.5), 190 mM NH_4C1, 0.33 mM $K_2HPO_4.3H_2O$, 0.1 mM $FeSO_4$ $7H_2O$, 20 g agar, 0.2% glycerol (v/v) and 1 L half-strength artificial seawater) (Baumann *et al.*, 1971b), and MA 2216 containing 2% (w/v) glycerol (Goarant *et al.*, 2006a). The surrounding medium clears after 2–4 days at 25°C. Under a dissecting microscope colonies can be seen to contain crystals of pigment. The pigment is not soluble in water, or solvents such as methanol, ethanol, propanol, butanol, acetone, ethyl ether, ethyl acetate, xylene, benzene or chloroform. Pigmentation is slower to develop and less intense on BMA containing yeast extract at 5g/l (YEA). In artificial seawater cells clump and are difficult to disperse. Cells begin as straight rods in young cultures in the exponential phase and become curved in the stationary phase (Baumann *et al.*, 1971b). On BA colonies are pinpoint at 24 h, beta-haemolytic and opaque, and on MSA-B are 0.5 mm grey but haemolysis is not seen. Colonies are green on TCBS (this manual). Growth occurs at 25–30°C but not at 4 or 35°C. Luminescence is not seen on Farghaly's medium supplemented with 5 g of Difco yeast extract, 5 g of Difco tryptone, 10 g of $CaCO_3$(powder) and 20 g of Difco agar in 1 L adjusted to pH 7.2 (Baumann *et al.*, 1971a). Results are positive for lipase, chitinase and negative for alginase. Citrate was positive according to Baumann *et al*., (1971a) for strain 164[T] but strain CIP 103195 (which is three generations from the type strain) is negative in Simmons' tube citrate after 48 h, but positive in the API 20E. Nitrate is positive in the tube test and a weak positive in the API 20E. The ONPG reaction in tube media is strong and the oxidase reaction is strong. Reactions in tube media (see media section) for fermentation of maltose and trehalose were weak at 48 h. The catalase reaction is slow (this manual). These and other reactions are presented in Tables 3.22, 3.24–26 and 3.38. Sets are inoculated with an inoculum in 2% NaCl and incubated at 25°C for 48 h.

MOLECULAR IDENTIFICATION. *Vibrio nigripulchritudo* can be detected using a conventional PCR based on the 16S-23S rRNA intergenic spacer region (Lee *et al*., 2002), a conventional PCR based on 16S rRNA primers (Goarant *et al*., 2006a), a conventional PCR specific for highly virulent strains (Walling *et al*., 2010), a SYBR Green real-time PCR designed for the genus based on the *gyrB* gene (Goarant *et al*., 2007), the emerging virulence clone using a specific hybridization probe PCR designed for virulent strains (Goarant *et al*., 2007) and a loop-mediated isothermal amplification (LAMP) PCR for the genus (Fall *et al*., 2008).

PCR type	SYBR Green PCR to detect all strains of *V. nigripulchritudo* (Goarant *et al*., 2007)		
DNA extraction	High Pure™ PCR Template Preparation kit (Roche Diagnostics) or Chelex™ 100 (Sigma)		
PCR master mix			
Primer VngF2	5′-CCCGAACGAAGCGAAA-3′		
Primer VngR2	5′-ACCTTTCAGTGGCAAGATG-3		
Reagent	**Stock**	**Final concentration**	**Volume ×1**
LightCycler FastStart DNA Master PLUS SYBR Green 1 kit (Roche)			12.5 µl
Primer VngF2	50 µM	0.5 µM	0.15 µl
Primer VngR2	50 µM	0.5 µM	0.15 µl
Water			2.2 µl
DNA			5 µl
Total volume			**15 µl**
Thermocycling	LightCycler Real-Time PCR System (Roche)		
Step 1	**Step 2**		
95°C, 10 min	95°C, 5 s 62°C, 3 s 72°C, 12 s		
×1	×45		
Product	258 bp		
Melt temperature	87.5–87.8°C for the virulent clonal strains 88.2°C for the type strain 88.4–89.1°C for all other *V. nigripulchritudo* strains		
Sensitivity	20 gene copies		
Specific against:	*V. penaeicida* (KH-1T from Japan), *V. penaeicida* (AM 101) from New Caledonia, *V. alginolyticus*, *V. harveyi*, *V. aestuarianus*, *V. tubiashii*, *Photobacterium damselae*, *V. orientalis*, *V. splendidus*, *V. fluvialis* and *V. nereis*		

The SYBR Green PCR does not allow differentiation between the emerging pathogenic strain and all other strains of *V. nigripulchritudo*.

PCR type	Hybridization (FRET) probe/s to detect the emerging virulent clone of *V. nigripulchritudo* (Goarant *et al*., 2007) using primers and FRET probes to detect point mutations in the *gyrB* gene specific to the emerging virulent clone		
DNA extraction	High Pure™ PCR Template Preparation kit (Roche Diagnostics) or Chelex™ 100 (Sigma)		
PCR master mix			
Primer VngF2	5′-CCCGAACGAAGCGAAA-3′		
Primer VngR2	5′-ACCTTTCAGTGGCAAGATG-3		
Probe VngA1	5′-TGTGGAGGGTGACTCTGC-fluorescein-3′		
Probe VngS1	5′-Red705-CCCAGCACTCTCTGAACTCTAT-phosphate		
Reagent	**Stock**	**Final concentration**	**Volume ×1**
LightCycler FastStart DNA Master PLUS Hybridization Probe kit (Roche)		×1	12.5 µl
Primer VngF2	50 µM	0.5 µM	0.2 µl
Primer VngR2	50 µM	0.5 µM	0.2 µl
Probe VngA1	10 µM	0.2 µM	0.4 µl
Probe VngS1	10 µM	0.2 µM	0.4 µl
Water			1.3 µl
DNA			5 µl
Total volume			**20 µl**
Thermocycling	LightCycler Real-Time PCR System (Roche)		
Step 1	**Step 2**		
95°C, 10 s	95°C, 5 s 60°C, 3 s 72°C, 12 s		
×1	×50		
Product	258 bp		
Sensitivity	5 gene copies		
Specific against:	*V. penaeicida* (KH-1T from Japan), *V. penaeicida* (AM 101) from New Caledonia, *V. alginolyticus*, *V. harveyi*, *V. aestuarianus*, *V. tubiashii*, *Photobacterium damselae*, *V. orientalis*, *V. splendidus*, *V. fluvialis* and *V. nereis*		

The 258 bp generated using the primers and the hybridization probe can not be checked by gel electrophoresis because sequence polymorphism prevents hybridization of the sensor VngS1 probe.

PCR type	Conventional PCR to detect the virulence plasmid pSFn1 (Walling *et al*., 2010) found in highly virulent strains only		
DNA extraction	High Pure™ PCR Template Preparation kit (Roche Diagnostics) or Chelex™ 100 (Sigma)		
PCR master mix			
Primer pSFn1-6020	5′-TGTCTTCTGGATCGCTTCGCC-3′		
Primer pSFn1-6974	5′-CGTCGTAAGGAGCGATAAGCC-3		
Reagent	**Stock**	**Final concentration**	**Volume ×1**
Qiagen Fast Cycling PCR Kit		×1	10.0 µl
Primer pSFn1-6020	50 µM	0.5 µM	0.2 µl
Primer pSFn1-6974	50 µM	0.5 µM	0.2 µl
Water			9.6 µl
DNA			2 µl
Total volume			**20 µl**
Thermocycling	No details given for thermocycler		
Step 1	**Step 2**		**Step 3**
95°C, 5 min (hot start)	96°C, 5 s		96°C, 5 s

Continued

Continued.

	55°C, 5 s	55°C, 5 s
	68°C, 3 s	72°C, 60 s
×1	×35	×1
Product	954 bp	
Sensitivity	100 CFU/g sediment	

No specific details were given for the reagent concentrations apart from the use of Qiagen polymerase and buffers. The concentrations listed here are those recommended by Qiagen for the Fast Cycling PCR Kit. This PCR is specific for highly pathogenic strains which by definition carry both plasmids. Strains rated by experimental challenge as moderately virulent or non-virulent do not amplify using this PCR.

PCR type	Conventional PCR (Fall *et al.*, 2008)		
DNA extraction	DNeasy Tissue kit (Qiagen)		
PCR master mix			
Primer Fw	5′-GGTTCTTAACGAATCACGCTCT-3′		
Primer Rv	5′-CCACCGTGTACGCTTAGTCA-3		
Reagent	**Stock**	**Final concentration**	**Volume ×1**
dNTP mix (2.5 mM each dNTP)	25 mM	2.5 mM	5.0 µl
Buffer, Gene *Taq* universal	10×	×1	5.0 µl
Taq polymerase (Nippon Gene, Japan)	5U/uL		0.5 µl
Primer Fw	10 µM	1 µM	5.0 µl
Primer Rv	10 µM	1 µM	5.0 µl
Water			28.5 µl
DNA			1.0 µl
Total volume			**50 µl**
Thermocycling	No details given for thermocycler		
Step 1	**Step 2**		**Step 3**
94°C, 5 min	94°C, 1 min		94°C, 1 min
	60°C, 1 min		60°C, 1 min
	72°C, 1 min		72°C, 3 min
×1	×28		×1
Product	278 bp		
Sensitivity	1000 CFU/ml		

The following LAMP PCR for the detection of *V. nigripulchritudo* strains is based on two inner and two outer primers designed to detect the intergenic spacer region between the 16S and 23S rRNA genes (Fall *et al.*, 2008).

PCR type	LAMP PCR for detection of *V. nigripulchritudo* strains (Fall *et al.*, 2008)		
DNA extraction	DNeasy Tissue Kit (Qiagen)		
LAMP PCR master mix			
Primer Vn-FIP	5′-CCCACCAAGGTTCAGGTGAGTTTTTCAGGCTGGGTTAGAGGG-3′		
Primer Vn-BIP	5′-ACATTGAATCTTTATGGGGCTTTTCCTCCAGTCGTCAAGCTAGG-3		
Primer Vn-F3	5′-AGAAGCCAAATGTTGTTG-3′		
Primer Vn-B3	5′-TCTGTGGTGGAGCTAAGC-3		
Reagent	**Stock**	**Final concentration**	**Volume ×1**
Master mix[a] (LAMP kit)	×2	×1	12.5 µl
Primer Vn-FIP	50 µM	0.4 µM	0.18 µl
Primer Vn-BIP	50 µM	0.4 µM	0.18 µl
Primer Vn-F3	50 µM	0.05 µM	0.02 µl

Continued

Continued.

Primer Vn-B3	50 µM	0.05 µM	0.02 µl
Bst DNA polymerase	8 U		1 µl
Water			9.09 µl
DNA			1 µl
Total volume			**23 µl**
Reaction temperature	63°C		
Reaction time	60 min		
Sensitivity	100 CFU/ml		
Specific against	*V. nigripulchritudo* (ATCC 27043), *V. penaeicida* (NBRC 15640), *V. tapetis*, *V. harveyi*, *V. parahaemolyticus*, *Listonella anguillarum* (V06-25, V06-28), *V. alginolyticus*, *Pseudoalteromonas haloplanktis* and *Photobacterium damselae*		

[a]40 mM Tris-HCl, 20 mM KCl, 16 mM $MgSO_4$, 20 mM $(NH_4)_2SO_4$, 0.2% Tween-20, 1.6 M betaine, 2.8 mM dNTPs each.

DNA is extracted from artificially infected prawn tissues such as heart, lymphoid organ and muscle using the DNeasy Tissue Kit (Qiagen) according to the manufacturer's instructions. The LAMP PCR has a 10-fold sensitivity of detection compared to the conventional PCR.

Vibrio ordalii

TAXONOMY AND SIGNIFICANCE. *Vibrio ordalii* (DF_3K^T= $DomF^3$ kid^T=ATCC 33509^T) was isolated from kidney of diseased coho salmon (*Oncorhynchus kisutch*) in the USA. The name *V. ordalii* was proposed for those strains previously identified as *Vibrio anguillarum* biotype 2 (Harrell *et al.*, 1976; Schiewe *et al.*, 1981). *Vibrio anguillarum* biotype 2 was placed into the genus *Beneckea* (Baumann *et al.*, 1971a) and then returned to the genus *Vibrio* following the abolition of the genus *Beneckea* (Baumann *et al.*, 1980).

Vibrio ordalii has been isolated from diseased salmonid fish in the USA and Japan and from ayu (*Plecoglossus altivelis*) and rockfish (*Sebastes schlegeli*) in Japan (Muroga *et al.*, 1984). Infection with *V. ordalii* in chum (*Oncorhynchus keta*), coho (*O. kisutch*) and chinook salmon (*O. tshawytscha*) results in a haemorrhagic septicaemia with the organism observed as aggregates or colonies in muscle, skin and gills and in loose connective tissue throughout the gastrointestinal tract, and occasionally in spleen and liver (Ransom *et al.*, 1984). Strains of *V. ordalii* have been isolated from Atlantic salmon (*Salmo salar*) in Chile that show some biochemical differences to ATCC 33509^T, being positive for fermentation of trehalose and negative for mannitol (Colquhoun *et al.*, 2004; Silva-Rubio *et al.*, 2008). The Chilean strains agglutinate with *V. anguillarum* serotype 02 antisera, which is a feature of *V. ordalii* as shown by Toranzo *et al.* (1987).

MLSA identifies *V. ordalii* as belonging to the Anguillarum clade, which also contains *V. aestuarianus* and *V. anguillarum* (Sawabe *et al.*, 2007b). Members of this clade have >30% DDH, 43–46 GC mol%, a MLSA concatenated similarity of 88.9–98.6% and AAI of 95.7–99.6%.

GROWTH CHARACTERISTICS AND PHENOTYPIC IDENTIFICATION. *Vibrio ordalii* is a motile, facultatively anaerobic, curved, Gram-negative rod (2.5–3×1 µm) (Harrell *et al.*, 1976; Schiewe *et al.*, 1981). On MSA-B colonies are 1–2 mm, circular, convex, off-white to grey after 4–6 days at 22°C. Growth is enhanced by the addition of NaCl. No growth at 37°C. It is negative for ADH in Moeller's and Thornley's media and negative for Simmons' and Christensen's citrate. These and other reactions are presented in Tables 3.22, 3.24–26 and 3.38. Add NaCl to sets.

Vibrio orientalis

TAXONOMY AND SIGNIFICANCE. *Vibrio orientalis* (717^T=ATCC 33934^T) is a luminescent bacterium isolated from seawater off the coast of China (Yang *et al.*, 1983). It is also found in the haemolymph of the spider crab (*Maja brachydactyla*) caught off the Galician coast, Spain, and the Canary Islands (Gomez-Gil *et al.*, 2010). *Vibrio orientalis* is closely related (98% similarity) to *V. atypicus*, *V. harveyi* and *V. nereis* based on the

16S rRNA gene, and clusters with *V. atypicus* based on concatenated sequences of housekeeping genes (*gapA*, *mreB*, *pyrH*, *recA*, *topA* and *gyrB*) (Wang *et al.*, 2010). MLSA identifies *V. orientalis* as belonging to the Orientalis clade, which also contains *V. hepatarius*, *V. brasiliensis* and *V. tubiashii* (Sawabe *et al.*, 2007b). Members of this clade have >24% DDH and 43–46 GC mol%, 91.2–94.2% similarity based on concatenated MLSA sequence data and AAI of 97–97.9%. *Vibrio nereis* belongs to the Nereis clade and *V. harveyi* belongs to the Harveyi clade.

GROWTH CHARACTERISTICS AND PHENOTYPIC IDENTIFICATION. *Vibrio orientalis* is a facultatively anaerobic, motile, oxidase-positive, Gram-negative rod (Yang *et al.*, 1983). On MSSA, colonies are 1 mm and dark grey after 24 h at 25°C and increase in size to 4–5 mm with a yellow-grey colour after 4 days. Growth occurs on MA 2216, and on TCBS individual colonies are a faint yellow colour. Luminescence is positive. Growth occurs at 4–35°C but not at 40°C. ADH is positive in Thornley's method and in the API 20E, and negative in Moeller's ADH. Growth does not occur in the conventional MRVP medium, however VP was reported negative by Yang *et al.* (1983) and is negative in the API 20E (a pale pink colour may be seen, which is a negative result). Results are positive for amylase, chitinase and lipase and negative for alginase. These and other reactions are presented in Tables 3.23–26 and 3.38. Sets are inoculated with an inoculum in 2% NaCl and incubated at 25°C for 48 h.

Vibrio owensii

TAXONOMY AND SIGNIFICANCE. *Vibrio owensii* (DYO5^T=JCM 16517^T=ACM 5300^T) was isolated from diseased larvae of the ornate spiny lobster (*Panulirus ornatus*), and strain 47666-1 was isolated from diseased prawns (*Penaeus monodon*) grown in aquaculture systems in Queensland, Australia (Cano-Gómez *et al.*, 2010). The epithet *owensii* is incorrect according to the Bacteriological Code (Euzéby, http://www.bacterio.cict.fr). Virulence studies demonstrated both strains were highly virulent. *Vibrio owensii* along with *V. harveyi*, *V. natriegens* and *V. rotiferianus* have been reported from diseased crown of thorns starfish (*Acanthapaster planci*) on the Great Barrier Reef, Australia, however at this stage the pathogenicity of *V. owensii* and these *Vibrio* species for the starfish is not known (Rivera-Posada *et al.*, 2011).

The two strains (DYO5^T and 47666-1) belong to the *V. harveyi* clade based on 16S rRNA sequence analysis with 99.2–99.5% sequence identity to other species in the clade. The multilocus sequence analysis of five concatenated protein-coding loci (*rpoA* [RNA polymerase α-subunit], *pyrH* [uridylate kinase], *topA* [topoisomerase I], *ftsZ* [cell division protein FtsZ], *mreB* [rod shaping protein MreB]) indicates that *V. owensii* is highly related to *V. harveyi*, *V. campbellii* and *V. rotiferianus* (Cano-Gómez *et al.*, 2010). Concatenated sequence information from protein-coding genes, *rpoA*, *pyrH*, *topA*, *ftsZ* and *mreB*, suggests *V. owensii* is the same species as *V. communis* (Cano-Gómez *et al.*, 2011). The evidence that *V. owensii* and *V. communis* are identical was confirmed using concatenated sequence information from *topA*, *ftsZ*, *toxR*, *rctB*, *rpoD* and *mreB* (Hoffman *et al.*, 2012a). The biochemical reactions are the same for both organisms. Hoffman *et al.* (2012a found that some strains (LMG 11659, LMG 10948, LMG 4043) originally identified as *V. harveyi* were found to be *V. communis*, and so now these would be strains of *V. owensii*.

GROWTH CHARACTERISTICS AND PHENOTYPIC IDENTIFICATION. *Vibrio owensii* is a Gram-negative slightly curved rod (1.0–3.1 μm). On MA 2216 colonies are 2–3 mm, translucent or opaque smooth, and round, non-swarming colonies at 48 h and 28°C and bright yellow 2–3mm colonies on TCBS. Strain DYO5 is 1 mm grey, shiny and round with a weak beta-haemolysis on MSA-B after 24 h at 25°C, and becomes dry with a slightly wrinkled appearance after 48 h. Strain 47666-1 is 2–4 mm, shiny, round and slightly elongated in the direction of the streak line on MSA-B after 24 h at 25°C, and remains mostly smooth and cream-coloured with an occasional wrinkled colony. Strains of *V. harveyi* can also obtain a wrinkled appearance and tend to show a slow swarming-type of growth on MSA-B (this manual). Growth of *V. owensii* occurs in 1–8% NaCl, but not at 0 or 10% NaCl and has a growth temperature range of 12–15°C to 35–37°C and no growth at 4°C. None of the strains are luminescent, although strain 47666-1 was initially reported as luminescent (Cano-Gómez *et al.*, 2010).

The aesculin tube turns black after 2–3 days, however when examined under fluorescent light the medium fluoresces indicating a negative result for aesculin hydrolysis (strain DY05), yet strain 47666-1 shows a weak fluorescence indicating some aesculin hydrolysis (see media section for test interpretation). Citrate is negative in Simmons' tube medium, but positive at 48 h in the API 20E. Nitrate is positive in the tube test, but is negative in the API 20E (also confirmed by a red colour after adding zinc dust). The reactions for glucose, sucrose and amygdalin are weak in the API 20E for strain DY05. Strains DY05 and 47666-1 differ in reactions for ONPG and gelatin (this manual). These and other reactions are presented in Tables 3.22, 3.24–29 and 3.38. An inoculum containing 2% NaCl is used and sets are incubated at 25°C or 30°C for 48 h as indicated in the tables.

MOLECULAR IDENTIFICATION. Sequencing of the *mreB*, *topA* and 16S rRNA genes can identify and differentiate vibrios in the Harveyi clade (Rivera-Posada *et al.*, 2011). See under *V. natriegens*.

Vibrio pacinii

TAXONOMY AND SIGNIFICANCE. *Vibrio pacinii* (LMG 19999^T=CAIM 530^T=STD3-1057^T) was isolated from healthy shrimp larvae (*Penaeus chinensis*) cultured in China, from seabream (*Dicentrarchus labrax*) in Spain and from Atlantic salmon (*Salmo salar*) from Tasmania, Australia (Gomez-Gil *et al.*, 2003b).

In experimental challenge using rainbow trout (*Oncorhynchus mykiss*) and *Artemia* nauplii inoculated with 10^6 CFU/fish or 10^6 CFU/ml, respectively, *V. pacinii* (LMG 19999^T and LMG 21514) was rated as non-virulent. Signs of infection were not seen and the organism was not recovered from tissues (Austin *et al.*, 2005). *Vibrio pacinii* was found as part of the normal intestinal flora when cod (*Gadus morhua*) larvae were fed rotifers (Reid *et al.*, 2009).

The organism is closely related to *V. metschnikovii*, *V. kanaloae*, *V. pomeroyi*, *V. aestuarianus*, *V. tapetis* and *V. furnissii*.

GROWTH CHARACTERISTICS AND PHENOTYPIC IDENTIFICATION. *Vibrio pacinii* is a facultatively anaerobic, motile, oxidase-positive, Gram-negative rod. On MA 2216 and TSA containing 2% NaCl, colonies are translucent, non-pigmented, non-swarming and non-luminescent. Colonies are 1.5–2.8 mm, yellow and round on TCBS (Gomez-Gil *et al.*, 2003b). Growth occurs at 4–35°C but not at 40°C. Growth occurs at 1.5–8.0% NaCl, but not in the absence of NaCl or at 10% NaCl. Catalase is positive. Some strain variation is seen with biochemical tests, with the seabream strain being negative for VP, whereas the other two strains are positive. The seabream strain is strongly positive for citrate, whereas the other two strains produce a weak reaction. In some identification schemes, *V. pacinii* may be identified as *V. splendidus* II or *V. furnissii* (Gomez-Gil *et al.*, 2003b). API Zym test reactions may vary in colour development for some strains. Strains are negative for α-chymotrypsin, but the prawn strain gives a weak positive reaction. These and other reactions are presented in Tables 3.23–26 and 3.38. Sets are incubated according to the manufacturer's instructions with the exception that an inoculum is used of 1.5% NaCl.

Vibrio parahaemolyticus

TAXONOMY AND SIGNIFICANCE. *Vibrio parahaemolyticus* (EB 101^T=ATCC 17802^T), originally named *Pasteurella parahaemolytica*, was isolated in 1950 from Japanese patients with 'shirasu' food poisoning after eating boiled and semi-dried sardines (Fujino *et al.*, 1974). The organism was renamed *Oceanomonas parahaemolytica* (Miyamoto *et al.*, 1961), although the proposal of the genus *Oceanomonas* was never accepted and it was placed into the genus *Vibrio* as *Vibrio parahaemolyticus* (Sakazaki *et al.*, 1963). Baumann *et al.* (1971a) placed it in the genus *Beneckea*, however with the abolition of this genus it reverted to the genus *Vibrio* (Baumann *et al.*, 1980).

MLSA identifies *V. parahaemolyticus* as belonging to the Harveyi clade, which also contains *V. campbellii*, *V. alginolyticus*, *V. harveyi*, *V. mytili*, *V. natriegens* and *V. rotiferianus* (Sawabe *et al.*, 2007b). Members of this clade have >25% DDH and 42–48 GC mol%, a MLSA concatenated sequence similarity of 90.1–96.2% and AAI of 97.2–99.4%. Strains within the Harveyi clade (*V. alginolyticus*, *V. campbellii*, *V. harveyi*, *V. natriegens*, *V. parahaemolyticus* and *V. rotiferianus*) can be discriminated using concatenated sequences from three housekeeping genes, *rctB*, *rpoD* and *toxR*, with *toxR* and *rpoD* providing the most discrimination (Pascual *et al.*, 2010) (see under *Vibrio harveyi*). *Vibrio parahaemolyticus* has 99.7% homology with the 16S rRNA gene of *V. alginolyticus* (Ruimy *et al.*, 1994) and for a 1451-bp sequence only 5 bp differences exists between the two organisms compared to 166 bp differences for a 1258-bp sequence of the *gyrB* gene, relating to 86.8% homology (Venkateswaran *et al.*, 1998).

In humans, *V. parahaemolyticus* causes gastroenteritis usually associated with eating shellfish or raw fish, and can cause wound infections and septicaemia. It is also a pathogen for some aquatic animals. The main virulence factor in clinical isolates causing gastroenteritis in people is thermostable direct haemolysin (TDH), which is not inactivated by heating at 100°C for 10 min. A particular beta-haemolysis is detected using Wagatsuma agar containing 5% NaCl, mannitol and either human or rabbit defibrinated blood (see media section). This haemolytic phenomenon is called the Kanagawa phenomenon (KP). Environmental strains tend to be KP negative. Beta-haemolysis is not seen with horse blood, or if the medium is heat-sterilized (autoclaved), or if the NaCl concentration is less than 5%. If these conditions are not met, strains that are non-haemolytic on Wagatsuma agar may be haemolytic under other conditions (Miyamoto *et al.*, 1969). As well as haemolytic properties, the TDH has enterotoxic, cardiotoxic and cytotoxic properties. It is encoded by the *tdh* gene, which has well-conserved coding regions, but sequence variants occur and are nominated *tdh1* to *tdh5* and have 96–98% sequence similarity. High transcription levels of the *tdh2* gene variant are responsible for the Kanagawa phenomenon. Two copies of *tdh* are found in KP-positive strains and it is present in only a few environmental strains. It is found in all strains of *G. hollisae*, some strains of *V. mimicus* and some strains of *V. cholerae* non-01. Some strains of *V. parahaemolyticus* produce a weak haemolysis on Wagatsuma agar (KP-intermediate) and these strains including 16% of KP-negative strains possess one copy only of *tdh* and a low level of expression may be the cause of the KP-intermediate and KP-negative phenotype. The expression of the *tdh* gene appears to be influenced by the involvement of ToxRS in transcriptional regulation. The *toxRS* operon in *V. parahaemolyticus* has functional and structural similarities to the *V. cholerae toxRS* operon, which is responsible for encoding a transcriptional activator of the cholera enterotoxin. *Vibrio parahaemolyticus toxR* and *toxS* are 52 and 62% similar in their nucleotide sequence to the equivalent genes in *V. cholerae* (Nishibuchi and Kaper, 1995). A TDH-related haemolysin, produced by the *trh* gene, has similarities to TDH and may have evolved from a common ancestor due to a single base change. The *trh* strains are also associated with gastroenteritis and clinical isolates carry both genes, whereas environmental isolates do not. The *trh* genes vary in their nucleotide sequence and are clustered into two main groups designated *trh1* and *trh2* with 84% identity. Only some strains of *V. parahaemolyticus* produce urease, however there is a strong correlation between production of urease and the presence of the *trh* gene (Iida *et al.*, 1997b). A thermolabile haemolysin is produced by all KP-negative strains and most, but not all, KP-positive strains (Sakurai *et al.*, 1974). The thermolabile haemolysin does not appear to be related to virulence in humans. For clinical (human) strains a serotyping scheme based on the somatic (O) antigens and capsular (K) antigens is used, and serotype 03:K6 is a major pandemic strain of *V. parahaemolyticus*.

Very little is known about the pathogenesis of *V. parahaemolyticus* in aquatic animals. The bacterium may be carried by shellfish and fish without signs of infection, or it can be pathogenic and cause mortalities. For example, clams can eliminate the organism without causing mortality. Strains of *V. parahaemolyticus* pathogenic (positive for the *tdh* gene) and non-pathogenic (negative for *tdh* and *trh* genes) to humans were studied in clams, *Venerupis decussata* and *V. philippinarum*, to determine their uptake and retention in the clams. Both species of clam retained the bacteria for up to 72 h before eliminating them (Lopez-Joven *et al.*, 2011).

Mass mortality of small abalone (*Haliotis diversicolor supertexta*) cultured in Taiwan was due to *V. parahaemolyticus* (strains CH-1 and B4) isolated from the haemolymph (Liu *et al.*, 2000). Due to the appearance of the abalone the disease was called withering syndrome. Strain B4 and its extracellular products were pathogenic to abalone in experimental challenge with LD_{50} of 1.6×10^5 CFU/g and 7.58 μg protein/g body weight, respectively. Mass mortality in clam

(*Meretrix meretrix*) in China was caused by a *V. parahaemolyticus*-like organism that was LDC-negative using the GN card on the Vitek 2, and also identified by 16S rRNA sequencing (Yue *et al.*, 2010). As the Vitek has a limited number of *Vibrio* species in the database and incubates at 37°C, and the 16S rRNA gene is unreliable for species identification because of the high similarity within the *Vibrio* genus, it is not known whether other *Vibrio* species may have been involved or identified. The organism was positive for the *toxR* and *tlh* genes and negative for the *tdh* gene. Infected clams had increased lipid droplets in the hepatocytes, excessive secretions in the gills and granules in the mantle. Pathogenicity was proven in experimental challenge with an LD_{50} of 6×10^6 CFU/ml.

Vibrio parahaemolyticus was reported from diseased black tiger prawn (shrimp) (*Penaeus monodon*) cultured in Thailand (Ruangpan and Kitao, 1991). Infected animals showed weak response to stimuli, loss of appetite, unbalanced swimming and had a tendency to lie on the slope of the pond before dying. The hepatopancreas and musculature was inflamed and a milky fluid was seen in the intestine. *Vibrio parahaemolyticus* was the predominant species (46.8%) followed by *V. vulnificus* (36.1%). *Vibrio vulnificus* was thought to be an opportunist pathogen following results of experimental challenge.

Pathogenicity was related to salinity in outbreaks of mortality in laboratory-cultured Iberian toothcarp (*Aphanius iberus*), an endangered fish found in the salt mashes and coastal lagoons of Spain (Alcaide *et al.*, 1999). *Vibrio parahaemolyticus* caused high mortality and infection resulted in tail rot, red spots on the head and an intestine swollen with ascites fluid. The organism was isolated in pure culture from internal organs, particularly the liver. Pathogenicity was confirmed in experimental challenge through intraperitoneal injection with a resulting LD_{50} of 10^6 CFU/fish when fish were kept in 5–30% salinity. The isolate was non-virulent for fish (LD_{50} >10^9 CFU/fish) kept at 45% salinity. The isolate was pathogenic to amberjack and eel at 5×10^3 CFU/fish and 6.2×10^5 CFU/fish, respectively, at 5–30% salinity.

Southern sea otters (*Enhydra lutris nereis*) found along the coast of California, USA, may be susceptible to enteric pathogens such as *V. parahaemolyticus*, which are associated with surface runoff from urban areas (Miller *et al.*, 2010). The organism was detected in dead otters that had lesions of enteritis and septicaemia.

GROWTH CHARACTERISTICS AND PHENOTYPIC IDENTIFICATION. *Vibrio parahaemolyticus* is a facultatively anaerobic, motile, Gram-negative rod that is straight or occasionally slightly curved with some pleomorphism (Fujino *et al.*, 1974). On MSA-B and MSSA, colonies swarm as a light grey film over the entire plate at 24 h at 25°C (see colour plate section). On MA 2216 agar, colonies are non-swarming, non-mucoid, 1 mm cream at 24 h, and after 48 h are 1.5 mm round, slightly elongated colonies following the streak lines. Colonies on TCBS are 1–2 mm green at 24 h and 3.9 mm at 48 h. Growth is not seen on MCA at 24 h, but after 48 h colonies are 0.5 mm and lactose positive. Luminescence is negative on these media (this manual). Growth occurs at 22–42°C and in 0.85–7% NaCl. Results are positive for hydrolysis of chitin, starch, casein and negative for alginate. Fermentation of arabinose and hydrolysis of aesculin is variable. Fermentation of cellobiose is negative at 24 h, but may be positive after several days of incubation (Fujino *et al.*, 1974). Growth does not occur in 10% NaCl.

As with many of the *Vibrio* species different results may be obtained when different incubation times, temperature and NaCl inoculum concentrations are used. In the API 20E, a clear positive reaction in the arabinose test is obtained for strain ATCC 43996 when incubated for 48 h and prepared with a 2% NaCl inoculum. However, a weak false-positive reaction is seen for ADH under these conditions. In the conventional tube tests, a false-negative result for indole will be recorded if normal saline (0.85%) is used as the inoculum. A positive indole result is recorded with 1.5% and 2% NaCl as a final concentration (this manual).

In an assessment of the API 20E and API 20NE for accuracy in identifying *V. parahaemolyticus*, the API 20E had a sensitivity (percentage of true positives) of 59% and a specificity (percentage of true negative identifications) of 100% and a concordance (reproducibility in different laboratories) of 84%. The sensitivity of the API 20NE was 74%, with a specificity of 73% and a concordance of 65% (Croci *et al.*, 2007a). The API 20E did not detect as many *V. parahaemolyticus* strains compared to the API 20NE, but was more accurate as it did not detect any false positives. The API 20NE incorrectly identified some strains of *V. harveyi* (ATCC 14126), *V. alginolyticus* and *V. campbellii* as *V. parahaemolyticus*, and did not identify *V. parahaemolyticus* strain ATCC 43996. The API 20E correctly identified *V. parahaemolyticus* strain ATCC 43996, but did not identify *V. parahaemolyticus* strains ATCC 17802 and ATCC 27519 according to the API database. A reliable identification for *V. parahaemolyticus* should be based on biochemical tests and the *toxR* gene PCR.

These and other reactions are presented in Tables 3.22, 3.24–26 and 3.38. Sets are incubated according to the manufacturer's instructions with the exception that an inoculum is used of 2% NaCl and incubation temperature of 25°C.

MOLECULAR IDENTIFICATION. In the diagnostic laboratory molecular techniques can be used to either confirm the biochemical identification of an isolate by using amplification and sequencing of housekeeping genes, or amplify DNA to detect and identify an isolate using virulence genes or housekeeping genes. When biochemical tests may not provide definitive differentiation between species, then amplification of a housekeeping gene and sequence comparison can assist in identification. The *toxR* gene can assist in providing differentiation between *V. parahaemolyticus* and *V. alginolyticus*, two species that are closely related by the 16S rRNA gene, but have only 61.7% identical nucleotides within a partial sequence of the *toxR* gene (Montieri *et al.*, 2010). See method under *V. alginolyticus*.

A PCR based on the *rpoD* and *toxR* housekeeping genes followed by sequencing and analysis using concatenated sequences may enable differentiation between strains comprising the Vibrio Core group (Pascual *et al.*, 2010). See method under *Vibrio harveyi*.

Many PCRs have been designed for the detection of *V. parahaemolyticus*, particularly from seafoods because of the importance of this bacterium in causing food poisoning. These PCRs usually target the virulence genes for clinical isolates and may not always be appropriate for use as a diagnostic test for identification of *V. parahaemolyticus* from aquatic animals. Virulence genes such as the *tdh* gene may not be present in environmental isolates of *V. parahaemolyticus*. To date, there does not appear to be any study that investigates the relationship of virulence genes of significance to clinical isolates and their presence in strains causing infection in aquatic animals. PCRs based on housekeeping genes or conserved regions should be used in addition to those based on virulence factors. A number of PCRs are detailed here but this list is not extensive and the PCRs may not have been tested on isolates pathogenic to aquatic animals. The list has been selected to try and encompass validated PCRs and those that might be of value to laboratories in the detection of *V. parahaemolyticus* from aquatic animals.

Validation of the more commonly used PCR methods for detection of *V. parahaemolyticus* in seafoods indicate that the PCR for detection of the *toxR* gene (Kim *et al.*, 1999b) has the highest performance (100% inclusivity and exclusivity) compared to the *gyrB* gene PCR (Venkateswaran *et al.*, 1998), the *tlh* gene tested as a single gene PCR instead of the multiplex as designed by Bej *et al.* (1999) and the PCR for fragment pR72H (Lee *et al.*, 1995) (Croci *et al.*, 2007b). The *gyrB* gene PCR and the pR72H fragment PCR were the least reliable and most difficult to assess. Although the *gyrB* gene PCR produced the correct amplicon in *V. parahaemolyticus* it also produced amplicons of the same molecular weight in a number of *Vibrio* species, and in *V. alginolyticus* ATCC 17749 two amplicons were produced, one at the same molecular weight as *V. parahaemolyticus*. A non-specific amplicon was obtained from *V. harveyi* ATCC 14126 and from *V. campbellii* ATCC 25920. The pR72H fragment PCR gave an amplicon at the expected molecular weight but a number of *V. parahaemolyticus* isolates gave an amplicon at a similar molecular weight indicating insertion and deletion within the gene region and some strains produced no amplicon. Although the *toxR* gene gave the best results, non-specific amplicons occurred in *V. vulnificus* ATCC 27562 (=LMG 13545^{T}) and ATCC 33149. The PCRs were assessed in an intra-laboratory study and only one strain was not identified by the *toxR* PCR (Croci *et al.*, 2007b). This study tested the *tlh* PCR by Bej *et al.* (1999) as a single PCR rather than in the multiplex, and tested two different reaction mixes with different concentrations of $MgCl_2$, primers and *Taq* DNA polymerase. The *tlh* PCR using the two primers and the protocols designated A and B gave false-positives from some strains of *V. alginolyticus* (ATCC 17749) and *V. harveyi* (wild strain). In the same study a PCR designed on the pR72H (a cloned *Hind*III DNA fragment) by Lee *et al.* (1995) and Robert-Pillot *et al.* (2002) did not correctly identify all *V. parahaemolyticus* strains and also gave false-positives for *V. mimicus* (ATCC 33653) and *V. alginolyticus* (wild strain). These results stress the importance of optimizing and validating specificity and sensitivity for all PCRs adopted into a laboratory. Two validated PCRs may need to be conducted for definitive identification of *V. parahaemolyticus* from closely related species.

In another comparison of PCR methods for *V. parahaemolyticus*, primers designed to the metalloprotease gene (*vpm*) were compared to the *gyrB* PCR (Venkateswaran *et al.*, 1998), the *tlh, tdh, trh* multiplex PCR (Bej *et al.*, 1999) and the *toxR* gene PCR (Kim *et al.*, 1999b) and tested for specificity and sensitivity and used to detect *V. parahaemolyticus* from environmental sources (Luan *et al.*, 2007). The comparison used identical primer concentrations and conditions for all primer pairs apart from recommended annealing temperatures. The *gyrB* gene (encodes the B subunit protein of DNA gyrase, topoisomerase type II) detected all reference and wild-type strains of *V. parahaemolyticus*, however the DNA template concentration was critical and required a minimum of 0.8 ng in a 50-μl PCR mixture. The *tlh* gene was present in all isolates making it a suitable gene for detection of *V. parahaemolyticus*, however the *tdh* was present in only 75% of the 85 *V. parahaemolyticus* isolates and none of the isolates were positive for the *trh* gene. The *tdh* and *trh* are suitable for the detection of pathogenic isolates of clinical importance, but their significance in aquatic animals is not known. The primers targeting the *toxR* gene (Kim *et al.*, 1999b) demonstrated amplicon from all *V. parahaemolyticus* but gave non-specific product from *V. vulnificus* (LMG 13545^T) as also found by Croci *et al.* (2007b).

A multiplex PCR based on the haemolysin gene (*hly*) for detection and differentiation of *V. campbellii, V. harveyi* and *V. parahaemolyticus* (Haldar *et al.*, 2010) and the *gyrB* gene for *V. alginolyticus* (Luo and Hu, 2008) is detailed under *V. harveyi*.

PCR type	Conventional PCR based on the metalloprotease (*vpm*) gene for detection of *V. parahaemolyticus* (Luan *et al.*, 2007)		
DNA extraction	Phenol-chloroform extraction from bacterial cells		
PCR master mix			
Primer VPM1	5′-CAG CTA CCG AAA CAG ACG CTA-3′		
Primer VPM2	5′-TCC TAT CGA GGA CTC TCT CAA C-3		
Reagent	**Stock**	**Final concentration**	**Volume ×1**
Buffer for *Taq*	×10	×1	5.0 μl
dNTP mix	10 mM	0.2 mM	1.0 μl
$MgCl_2$	25 mM	1.5 mM	3.0 μl
Primer VPM1	50 μM	0.5 μM	0.5 μl
Primer VPM2	50 μM	0.5 μM	0.5 μl
Taq DNA polymerase (Shanghai Sangon)	250 U	1.5 U	0.3 μl
Water			38.7 μl
DNA			1 μl
Total volume			**50 μl**

Thermocycling	Applied Biosystems 2720 Thermal cycler	
Step 1	**Step 2**	**Step 3**
94°C, 5 min	94°C, 60 s	94°C, 60 s
	58°C, 60 s	58°C, 60 s
	72°C, 60 s	72°C, 7 min
×1	×30	×1
Product	675 bp	
Sensitivity	4 pg DNA, 10^2 CFU per reaction tube	

Continued

Continued

Specific against	*Vibrio aestuarianus* (LMG 7909^T), *V. alginolyticus* (LMG 4408T, LMG 4409T), *V. anguillarum* (CW4 from fish), *V. cincinnatiensis* (LMG 7891T), *V. furnissii* (LMG 7910T), *V. (Listonella) pelagius* (LMG 3897T), *V. mimicus* (LMG 7896T), *V. tubiashii* (LMG 10936T), *V. vulnificus* (LMG 13545T), *Aeromonas caviae* (CCCCM 1.1960), *A. hydrophila* (PR156, Herriot-Watt Uni). *A. hydrophila* (CCCM 1.2017), *Edwardsiella tarda* (CW7 from fish), *Pseudomonas alcaligenes* (CCCM 1.1805)

The following conventional PCR based on the *toxR* gene, which encodes a transmembrane regulatory protein ToxR, has been used in validation tests. Non-specific amplicons were reported from five *V. vulnificus* strains including the type strain LGM 13545 (Luan *et al.*, 2007), a finding also reported by Croci *et al.* (2007b).

PCR type	Conventional PCR based on the *toxR* gene for detection of *V. parahaemolyticus* (Kim *et al.*, 1999b)		
DNA extraction	1 ml of overnight culture. Boil for 5 min. Centrifuge at 13,000 rpm for 5 min. Dilute 1:10 in distilled water		
PCR master mix			
Primer ToxR1	5′-GTC TTC TGA CGC AAT CGT TG-3′		
Primer ToxR2	5′-ATA CGA GTG GTT GCT GTC ATG-3′		
Reagent	**Stock**	**Final concentration**	**Volume ×1**
Buffer for Ex *Taq* (with $MgCl_2$)	×10	×1	5.0 µl
dNTP mix	10 mM	0.2 mM	1.0 µl
Primer VPM1	50 µM	0.4 µM	0.4 µl
Primer VPM2	50 µM	0.4 µM	0.4 µl
Ex *Taq* DNA polymerase (Takara)	250 U	1.25 U	0.25 µl
Water			39.95 µl
DNA			3 µl
Total volume			**50 µl**

Thermocycler	Applied Biosystems 2720 Thermal Cycler	
Step 1	**Step 2**	**Step 3**
94°C, 5 min	94°C, 1 min	94°C, 1 min
	63°C, 1.5 min	63°C, 1.5 min
	72°C, 1.5 min	72°C, 7 min
×1	×20	×1
Product	368 bp	
Sensitivity	Not stated	
Specific against	*Grimontia hollisae* (525-82), *Photobacterium damselae* (RIMD2222001), *Vibrio alginolyticus* (219), *V. anguillarum* (PT-87050), *V. cholerae* 01 (NIH41), *V. cholerae* 0139 (MO45), *V. cholerae* non-01 (AM2), *V. fluvialis* (RIMD2220002), *V. furnissii* (RIMD2223001), *V. metschnikovii* (RIMD2208006), *V. mimicus* (RIMD2218002), *V. vulnificus* (RIMD2219022). Non-specific amplicon from 5 of 11 strains of *V. vulnificus*. Also see specificity undertaken for this PCR in validation tests (Luan *et al.*, 2007)	

Primers to detect the *gyrB* gene were used to detect and identify *V. parahaemolyticus* in contaminated prawns (Venkateswaran *et al.*, 1998). The *gyrB* gene codes the B subunit protein of DNA gyrase (topoisomerase type II) and is a single-copy gene. Substantial specificity testing was conducted using type strains. Specificity testing was also conducted in the comparison analysis by Luan *et al.* (2007). They found DNA concentration was critical to success of the PCR. An annealing temperature of 58°C was used (Venkateswaran *et al.*, 1998), however a temperature of 60°C improved specificity against *V. alginolyticus* as the lower annealing temperature produced an amplicon of the expected size from some strains of *V. alginolyticus* (Kim *et al.*, 1999b). However, specificity of the *gyrB* PCR was also in doubt as reported by Croci *et al.* (2007b), where two amplicons were obtained from *V. alginolyticus* ATCC 17749, one at the expected size for *V. parahaemolyticus* and one non-specific amplicon, with a non-specific amplicon also obtained from *V. campbellii* ATCC 25920 and *V. harveyi* ATCC 14126.

PCR type	Conventional PCR based on the *gyrB* gene for detection of *V. parahaemolyticus* (Venkateswaran *et al.*, 1998)		
DNA extraction	Phenol-chloroform extraction from bacterial cells		
PCR master mix			
Primer VP-1	5′-CGG CGT GGG TGT TTC GGT AGT-3′		
Primer VP-2r	5′-TCC GCT TCG CGC TCA TCA ATA-3′		
Reagent	**Stock**	**Final concentration**	**Volume ×1**
Buffer for *Taq*	×10	×1	5.0 µl
dNTP mix	10 mM	0.2 mM	1.0 µl
$MgCl_2$	25 mM	1.5 mM	3.0 µl
Primer VP-1	50 µM	1.0 µM	1.0 µl
Primer VP-2r	50 µM	1.0 µM	1.0 µl
Taq DNA polymerase	250 U	2.5 U	0.5 µl
Water			37.5 µl
DNA 1 µg			1 µl
Total volume			**50 µl**

Thermocycling	Applied Biosystems 2720 Thermal Cycler	
Step 1	**Step 2**	**Step 3**
94°C, 5 min	94°C, 60 s	94°C, 60 s
	60°C, 60 s	60°C, 60 s
	72°C, 60 s	72°C, 7 min
×1	×30	×1
Product	285 bp	
Sensitivity	Pre-enrichment is critical. 1.5 CFU bacterial cells/g shrimp homogenate detected following pre-enrichment in APW for 18 h at 37°C but 1.5×10^5 CFU/g homogenate was negative at time zero before enrichment incubation. 5 CFU or 4 pg DNA per 100 µl reaction tube using bacterial cells	
Specific against	*Aliivibrio fischeri* (ATCC 7744^T), *A. logei* (ATCC 29985^T), *A. salmonicida* (ATCC 43839^T), *Grimontia hollisae* (CDC 75-80), *Moritella marinus* (ATCC 15381^T), *Photobacterium damselae* (ATCC 33539^T), *Salinivibrio costicola* (ATCC 33508^T), *Vibrio aestuarianus* (ATCC 35048^T), *V. alginolyticus* (ATCC 17749^T), *V. anguillarum* (ATCC 19264^T), *V. campbellii* (ATCC 25920^T), *V. cholerae* 01 (*V. albensis* ATCC 14547^T), *V. cholerae* non-01 (unknown), *V. cincinnatiensis* (ATCC 35912^T), *V. diazotrophicus* (ATCC 33466^T), *V. fluvialis* (JCM 3752), *V. furnissii* (ATCC 35016^T), *V. gazogenes* (ATCC 29988^T), *V. harveyi* (ATCC 35084^T, ATCC 14126^{Tc}), *V. ichthyoenteri* (IFO 15847), *Vibrio* (*Listonella*) *pelagius* (ATCC 25916^T), *V. mediterranei* (ATCC 43341^T), *V. metschnikovii* (ATCC 7708^T), *V. mimicus* (unknown), *V. mytili* (NCIMB 13275), *V. natriegens* (ATCC 14048^T), *V. navarrensis* (NCIMB 13120), *V. nereis* (ATCC 25917^T), *V. nigripulchritudo* (ATCC 27043^T), *V. ordalii* (ATCC 33509^T), *V. orientalis*	

Continued

Continued

	(ATCC 33934^T), *V. proteolyticus* (ATCC 15338^T), *V. splendidus* (ATCC 33125^T), *V. tubiashii* (ATCC 19109^T), *V. vulnificus* (ATCC 2046^T), *Aeromonas hydrophila* (ATCC 19570), *Alteromonas macleodii* (ATCC 27126^T), *Deleya aquamarina* (JCM 27128^T), *D. cupida* (JCM 27124^T), *D. halophila* (unknown), *D. marina* (JCM 25374^T), *D. pacifica* (JCM 27122^T), *D. venusta* (JCM 27125^T), *Escherichia coli* (ATCC 25922^T), *Marinomonas communis* (ATCC 27118^T), *M. vaga* (ATCC 27119^T), *Pseudoalteromonas atlantica* (ATCC 19262^T), *P. aurantia* (ATCC 33046^T), *P. carrogeenovara* (ATCC 43555^T), *P. citrea* (ATCC 29719^T), *P. espejiana* (ATCC 29659^T), *P. haloplanktis* (ATCC 14393^T), *P. luteoviolacea* (ATCC 33492^T), *P. nigrifaciens* (ATCC 19375^T), *P. rubra* (ATCC 29570^T), *P. tetraodonis* (unknown), *P. undina* (ATCC 29660^T), *Pseudomonas aeruginosa* (IFO 12689), *P. fluorescens* (IFO 14160), *Salmonella typhimurium* (ATCC 13311^T), *Shewanella algae* (ATCC 51192^Tc), *S. benthica* (ATCC 43992^T), *S. hanedai* (ATCC 33224^T), *S. putrefaciens* (ATCC 8071^Tc), *Shigella dysenteriae* (ATCC 13313^T), *S. sonnei* (ATCC 29930^T), *Staphylococcus aureus* (ATCC 12600^T). Also see under VPM PCR for further specificity isolates (Luan *et al*., 2007)

A multiplex PCR was developed for the detection of *V. cholerae*, *V. mimicus*, *V. parahaemolyticus* and *V. vulnificus* using the *sodB* gene for *V. cholerae* and *V. mimicus*, the *flaE* gene for *V. parahaemolyticus* and the *hsp60* gene for *V. vulnificus* (Tarr *et al*., 2007). An internal positive control was designed from the 16S rRNA sequence from all *Vibrio* and *Enterobacteriaceae*. The amplicon generated by the 16S rRNA primers was usually outcompeted by three of the specific target species but was usually seen when co-amplified with *V. parahaemolyticus* (Tarr *et al*., 2007). Strains can also be identified by amplification of the *rpoB* gene followed by sequencing (see under *Vibrio* genus).

PCR type	Multiplex PCR for detection of *V. parahaemolyticus*, *V. cholerae*, *V. mimicus* and *V. vulnificus* (Tarr *et al*., 2007)		
DNA extraction	A colony from overnight growth is suspended in 200 µl 1 × Tris-EDTA, pH 8.0 and heated at 95°C for 10 min. The supernatant is used in the PCR reaction		
PCR master mix			
V. cholerae	Primer *Vc.sodB*-F = 5′-AAG ACC TCA ACT GGC GGT A-3′		
	Primer *Vc.sodB*-R = 5′-GAA GTG TTA GTG ATC GCC AGA GT-3′		
V. mimicus	Primer *Vm.sodB*-F = 5′-CAT TCG GTT CTT TCG CTG AT-3′		
	Primer *Vm.sodB*-R2 = 5′-GAA GTG TTA GTG ATT GCT AGA GAT-3′		
V. parahaemolyticus	Primer *Vp.flaE*-79F = 5′-GCA GCT GAT CAA AAC GTT GAG T-3′		
	Primer *Vp.flaE*-934R = 5′-ATT ATC GAT CGT GCC ACT CAC-3′		
V. vulnificus	Primer *Vv.hsp*-326F = 5′-GTC TTA AAG CGG TTG CTG C-3′		
	Primer *Vv.hsp*-697R = 5′-CGC TTC AAG TGC TGG TAG AAG-3′		
Internal control for all *Vibrio* species	Primer *V.*16S-700F = 5′-CGG TGA AAT GCG TAG AGA T-3′		
	Primer *V.*16S-1325R = 5′-TTA CTA GCG ATT CCG AGT TC-3′		
Reagent	**Stock**	**Final concentration**	**Volume ×1**
Ampli*Taq* Gold (Applied Biosystems)	5 U/µl	2 U	0.2 µl
Buffer	×10	×1	2 µl
$MgCl_2$	25 mM	1.5 mM	1.2 µl
dNTP	20 mM	0.2 mM	2 µl
Primer *Vc.sodB*-F	10 µM	0.5 µM	1 µl
Primer *Vc.sodB*-R	10 µM	0.5 µM	1 µl
Primer *Vm.sodB*-F	10 µM	0.75 µM	1.5 µl
Primer *Vm.sodB*-R2	10 µM	0.75 µM	1.5 µl
Primer *Vp.flaE*-79F	10 µM	1.0 µM	2 µl

Continued

Continued.

Primer *Vp.flaE*-934R	10 µM	1.0 µM	2 µl
Primer *Vv.hsp*-326F	10 µM	0.25 µM	0.5 µl
Primer *Vv.hsp*-697R	10 µM	0.25 µM	0.5 µl
Primer *V.*16S-700F	10 µM	0.05 µM	0.1 µl
Primer *V.*16S-1325R	10 µM	0.05 µM	0.1 µl
Water			2.9 µl
DNA			1.5 µl
Total volume			**20 µl**
Thermocycling			
Step 1	**Step 2**		**Step 3**
93°C, 15 min	92°C, 40 s		92°C, 40 s
	57°C, 1 min		57°C, 1 min
	72°C, 1.5 min		72°C, 7 min
×1	×35		×1
Product	*Vibrio cholerae,* 248 bp; *V. mimicus*, 121 bp; *V. parahaemolyticus*, 897 bp; *V. vulnificus*, 410 bp; 16S rRNA (internal control), 663 bp		
Sensitivity	Not determined		
Specific against	No details given		

A multiplex PCR based on collagenase gene for detection of *V. cholerae, V. alginolyticus* and *V. parahaemolyticus* (Di Pinto *et al.*, 2005, 2006) is detailed under *V. alginolyticus.*

A multiplex PCR was designed for the detection of the *atpA* gene (encodes the ATP synthase subunit A) in *V. parahaemolyticus, V. cholerae* and *V. vulnificus* (Izumiya *et al.*, 2011). The PCR was specific and gave comparable results to the PCR that detects *V. cholerae* based on the *ompW* gene (Nandi *et al.*, 2000) and the PCR for *V. parahaemolyticus* based on the *toxR* gene (Kim *et al.*, 1999b). The *atpA* gene for *V. vulnificus* detected two strains that were not detected using the *vvhA* gene for *V. vulnificus* (Hill *et al.*, 1991).

PCR type	Multiplex PCR for detection of *atpA* gene in *V. parahaemolyticus*, *V. cholerae* and *V. vulnificus* (Izumiya *et al.*, 2011)		
DNA extraction	Genomic DNA extracted using Genomic DNA isolation kit for bacteria (Toho KK)		
Pre-enrichment	1:10 dilution in 10 ml APW and incubate at 28°C or 37°C overnight. Use 1 µl as template DNA in the multiplex PCR		
PCR master mix			
Primer *atpA*-VP-F	5′-TAC TAG GCC GCG TAG TA-3′		
Primer *atpA*-VP-R	5′-CGC TGG ACG TAC ACC T-3′		
Primer *atpA*-VC-F	5′-AAT GGG TCC ATA CGC GGA T-3′		
Primer *atpA*-VC-R	5′-TGG TGA AGT YTG TTT TGC ACC-3′		
Primer *atpA*-VV-F	5′-ACG GTG CTC TGA AGA ACA CA-3′		
Primer *atpA*-VV-R	5′-GCA GAA ACG TCA CCA GCC TGA-3′		
Reagent	**Stock**	**Final concentration**	**Volume ×1**
EmeraldAmp PCR Master Mix (TaKaRa)	×2	×1	12.5 µl
Primer *atpA*-VP-F	50 µM	0.5 µM	0.25 µl
Primer *atpA*-VP-R	50 µM	0.5 µM	0.25 µl
Primer *atpA*-VC-F	50 µM	0.5 µM	0.25 µl
Primer *atpA*-VC-R	50 µM	0.5 µM	0.25 µl
Primer *atpA*-VV-F	50 µM	0.5 µM	0.25 µl
Primer *atpA*-VV-R	50 µM	0.5 µM	0.25 µl

Continued

Continued.

Water			9.0 µl
DNA <500 ng			2 µl
Total volume			**25 µl**
Thermocycling	DNA Thermocycler 9700 (Applied Biosystems)		
Step 1	**Step 2**	**Step 3**	
95°C, 2 min	95°C, 20 s	95°C, 20 s	
	65°C, 20 s	65°C, 20 s	
	72°C, 60 s	72°C, 7 min	
×1	×25	×1	
Product	*Vibrio parahaemolyticus*, 794 bp; *V. cholerae*, 160 bp; *V. vulnificus*, 373 bp		
Sensitivity	10 bacterial cells with or without pre-enrichment		
Specific against	*V. alginolyticus*, *V. brasiliensis*, *V. diabolicus*, *V. fluvialis* (ATCC 33809), *V. furnissii* (NCTC 11328), *V. harveyi*, *V. mediterranei*, *V. mimicus* (ATCC 33653), *V. natriegens*, *Aeromonas* sp., *Listonella* sp., *Shewanella* sp., *Shigella* sp., *Salmonella* sp.		

A multiplex PCR designed to detect the *tdh*, *trh* and *tlh* haemolysin genes enables detection and differentiation of total *V. parahaemolyticus* and clinically virulent strains in shellfish (Bej *et al.*, 1999). The *tlh* gene PCR was tested as a single PCR using two protocols with different concentrations of $MgCl_2$, *Taq* and primers and a number of false-positive results occurred (Croci *et al.*, 2007a, 2007b). The multiplex PCR was validated against other methods by Luan *et al.* (2007) and no false positives or false negatives were detected as a multiplex or when using the same reagent concentrations as the *vpm* PCR.

PCR type	Multiplex PCR for detection of total and clinically virulent isolates of *V. parahaemolyticus* based on *tdh*, *trh* and *tlh* genes (Bej *et al.*, 1999)		
DNA extraction	CTAB method (Ausubel, 1987). See molecular section		
Pre-enrichment	1:10 dilution in 10 ml APW and incubate at 28°C or 37°C for 8 h or overnight. Use 2 µl as template DNA in the multiplex PCR		
PCR master mix			
Primer L-tl	5′-AAA GCG GAT TAT GCA GAA GCA CTG-3′		
Primer R-tl	5′-GCT ACT TTC TAG CAT TTT CTC TGC-3′		
Primer L-tdh	5′-GTA AAG GTC TCT GAC TTT TGG AC-3′		
Primer R-tdh	5′-TGG AAT AGA ACC TTC ATC TTC ACC-3′		
Primer L-trh	5′-TTG GCT TCG ATA TTT TCA GTA TCT-3′		
Primer R-trh	5′-CAT AAC AAA CAT ATG CCC ATT TCC G-3′		
Reagent	**Stock**	**Final concentration**	**Volume ×1**
Buffer containing $MgCl_2$	×10	×1	5.0 µl
dNTP	10 mM	200 µM	1.0 µl
Primer L-tl	50 µM	1 µM	1.0 µl
Primer R-tl	50 µM	1 µM	1.0 µl
Primer L-tdh	50 µM	1 µM	1.0 µl
Primer R-tdh	50 µM	1 µM	1.0 µl
Primer L-trh	50 µM	1 µM	1.0 µl
Primer R-trh	50 µM	1 µM	1.0 µl
Taq DNA polymerase (Perkin Elmer)	250 U	2.5 U	0.5 ul
Water			35.5 µl
DNA (0.5 µg final conc)			2 µl
Total volume			**50 µl**
Thermocycling	**DNA Thermocycler 480 (Perkin Elmer)**		
Step 1	**Step 2**	**Step 3**	

Continued

Continued.

94°C, 3 min	94°C, 1 min 55°C, 1 min 72°C, 1 min	94°C, 1 min 55°C, 1 min 72°C, 5 min
×1	×30	×1
Product	Thermolabile haemolysin (*tlh*), 450 bp; thermostable direct haemolysin gene (*tdh*), 269 bp; thermostable related haemolysin gene (*trh*), 500 bp	
Sensitivity	*tdh* gene ≥100 cells, and *trh* and *tlh* ≥10 cells per 10 g pre-enriched oyster tissue homogenate	
Specific against	*Photobacterium damselae*, *V. alginolyticus*, *V. fluvialis*, *V. furnissii*, *V. metschnikovii*, *V. mimicus*, *V. proteolyticus*, *V. vulnificus*, *Plesiomonas* sp. Also see under VPM PCR for further specificity isolates (Luan *et al*., 2007)	

A multiplexed real-time PCR allows simultaneous detection of the *tlh*, *tdh* and *trh* genes and the open reading frame 8 of phage f237 (ORF8) for detection of total *V. parahaemolyticus* and strains associated with clinical virulence and the human pandemic serotype O3:K6 as detected by ORF8 (Ward and Bej, 2006).

PCR type	Multiplex qPCR with hybridization probes for detection of *tlh*, *tdh* and *trh* genes and ORF8 for *V. parahaemolyticus* (Ward and Bej, 2006)		
DNA extraction	Kit for extraction from tissue or CTAB method (Ausubel, 1987). See molecular section		
PCR master mix			
Primer F-tl (*tlh* gene)	5′-AAA GCG GAT TAT GCA GAA GCA CTG-3′		
Primer R-tl (*tlh* gene)	5′-GCT ACT TTC TAG CAT TTT CTC TGC-3′		
P-tl952 (*tlh* probe)	5′-TexR-AAG AAC TTC ATG TTG ATG ACA CT-BHQ2-3′		
F-tdh170DG	5′-GTA RAG GTC TCT GAC TTT TGG AC-3′		
R-tdh403	5′-CTA CAG AAT YAT AGG AAT GTT GAA G-3′		
P-tdh-341R (*tdh* probe)	5′-Cy5-ATT TTA CGA ACA CAG CAG AAT GA-Iowa Black-RQ-3′		
F-trh82	5′-CCA TCM ATA CCT TTT CCT TCT CC-3′		
R-trh287	5′-ACY GTC ATA TAG GCG CTT AAC-3′		
P-trh275 (*trh* probe)	5′-TET-TAT TTG TYG TTA GAA ATA CAA CAA T-BHQ1-3′		
F-O3MM824 (ORF8)	5′-AGG ACG CAG TTA CGC TTG ATG-3′		
R-O3MM1192 (ORF8)	5′-CTA ACG CAT TGT CCC TTT GTA G-3′		
P-ORF8-853 (ORF8 probe)	5′-6-FAM-AAG CCA TTA ACA GTT GAA GGC GTT GAC T-BHQ1-3′		
Reagent	**Stock**	**Final concentration**	**Volume ×1**
Taq buffer	×10	×1	2.5 µl
dNTP	10 mM	400 µM	1.0 µl
$MgCl_2$	25 mM	5 mM	5.0 µl
BSA	20 mg/ml	100 ng	0.13 ul
Primer F-tl (*tlh* gene)	50 µM	0.4 µM	0.2 µl
Primer R-tl (*tlh* gene)	50 µM	0.4 µM	0.2 µl
P-tl952 (*tlh* probe)	10 µM	0.2 µM	0.5 µl
F-tdh170DG	50 µM	0.4 µM	0.2 µl
R-tdh403	50 µM	0.4 µM	0.2 µl
P-tdh-341R (*tdh* probe)	10 µM	0.2 µM	0.5 µl
F-trh82	50 µM	0.4 µM	0.2 µl
R-trh287	50 µM	0.4 µM	0.2 µl
P-trh275 (*trh* probe)	10 µM	0.2 µM	0.5 µl
F-O3MM824 (ORF8)	50 µM	0.4 µM	0.2 µl
R-O3MM1192 (ORF8)	50 µM	0.4 µM	0.2 µl
P-ORF8-853 (ORF8 probe)	10 µM	0.2 µM	0.5 µl
Platinum Taq DNA polymerase (Invitrogen)	5 U/µl	2 U	0.2 µl

Continued

Continued.

Water			7.58 µl
DNA			5 µl
Total volume			**25 µl**
Thermocycling	Cepheid Smart Cycler II (Cepheid)		
Step 1	**Step 2**		
95°C, 120 s	95°C, 20 s		
	56°C, 20 s		
	72°C, 30 s		
×1	×45		
Fluorescence measurement			
Product bp	*tlh*, 450 bp; *tdh*, 229 bp; *trh*, 207 bp; ORFF8, 369 bp		
Ct	15 fluorescent units		
Ct (positive)	At 200 pg DNA 26.98–28.53; at 2 µg DNA 14.05–15.54 for all four primers. Background non-specific DNA or bacteria increased the Ct value by one cycle		
Ct (negative)	Not stated		
R^2 (linear regression coefficient)	0.988–0.993 for purified DNA and 0.984–0.999 for bacterial cells		
Sensitivity	200 pg DNA or 10^4 CFU. Reaction inconsistencies occurred above 7 µg background DNA and inhibition of the reaction occurred at 10 µg background DNA		
Specific against	Used BLAST search. Quoted results obtained using previous primers (Bej *et al.*, 1999)		
Optimization	All components in the PCR mixture were optimized against a range of concentrations		

Tex-R, sulforhodamine 101 (Texas Red) fluorescent dye; BHQ2, Black Hole-2 quencher dye; 6-FAM, 6-fluorescein fluorescent dye; Cy5, carbocyanine fluorescent dye; Iowa Black-RQ, Iowa Black quencher dye; TET, tetrachloro-6-carboxyfluorescein dye.

A multiplex PCR was reported using newly designed primers for the *toxR* genes of *V. parahaemolyticus* and *V. cholerae* and and the *vvh* gene of *V. vulnificus* for the detection of these three *Vibrio* species and was tested for specificity using type strains (Neogi *et al.*, 2010).

PCR type	Multiplex PCR for detection of *toxR* gene for *V. parahaemolyticus* and *V. cholerae* and *vvh* gene for *V. vulnificus* (Neogi *et al.*, 2010)		
DNA extraction	Suspend bacterial cells in 500 µl TE buffer and boil for 10 min. Cool. Centrifuge for 5 min at 12,800 ×*g*. Store supernatant at −20°C. For direct detection use a pre-enrichment in APW containing 2% NaCl and incubate for 12 h at 37°C		
PCR master mix			
Primer VP toxR 325F	5′-TGT ACT GTT GAA CGC CTA A-3′		
Primer VP toxR 828R	5′-CAC GTT CTC ATA CGA GTG-3′		
Primer VC toxR 403F	5′-GAA GCT GCT CAT GAC ATC-3′		
Primer VC toxR 678R	5′-AAG ATC AGG GTG GTT ATT C-3′		
Primer vvhA 870F	5′-ACT CAA CTA TCG TGC ACG-3′		
Primer vvhA 1236R	5′-ACA CTG TTC GAC TGT GAG-3′		
Reagent	**Stock**	**Final concentration**	**Volume ×1**
Taq buffer containing $MgCl_2$	×10	×1	2.0 µl
dNTP	10 mM	200 µM	0.4 µl
Primer VP toxR 325F	50 µM	0.15 µM	0.06 µl
Primer VP toxR 828R	50 µM	0.15 µM	0.06 µl
Primer VC toxR 403F	50 µM	0.5 µM	0.2 µl
Primer VC toxR 678R	50 µM	0.5 µM	0.2 µl
Primer vvhA 870F	50 µM	0.3 µM	0.12 µl
Primer vvhA 1236R	50 µM	0.3 µM	0.12 µl

Continued

Continued.

rTaq DNA polymerase (Takara Bio)	5 U/µl	0.625 U	0.05 µl
Water			15.79 µl
DNA			1 µl
Total volume			**20 µl**
Thermocycling			
Step 1	**Step 2**		**Step 2**
94°C, 5 min	94°C, 30 s 55°C, 30 s 72°C, 30 s		94°C, 30 s 55°C, 30 s 72°C, 5 min
×1	×30		×1
Product	*Vibrio parahaemolyticus* (VP toxR), 503 bp; *V. cholerae* (VC toxR), 275 bp; *V. vulnificus* (vvhA), 366 bp		
Sensitivity	10 CFU per tube (multiplex)		
Specific against	*Aliivibrio fischeri* (NCIMB 1281^T), *A. logei* (NCIMB 2252^T), *A. salmonicida* (ATCC 43839^T), *Photobacterium damselae* (ATCC 33539^T), *P. phosphoreum*, *V. alginolyticus* (ATCC 17749^T), *V. anguillarum* (IFO 13266^T), *V. campbellii* (ATCC 25920^T), *V. fluvialis* (ATCC 33809^T), *V. harveyi* (ATCC 35804^T and 14126^T), *V. mediterranei* (UM 5172), *V. metschnikovii* (IAM 1039), *V. mimicus* (ATCC 33653^T), *V. natriegens* (ATCC 14048^T), *V. nigripulchritudo* (ATCC 27043^T), *V. ordalii* (CIP 103205^T), *V. orientalis* (ATCC 33934^T), *V. parahaemolyticus* (NBRC 12711^T), *V. pelagius* (ATCC 25916^T), *V. proteolyticus* (ATCC 15338^T), *V. splendidus* I (ATCC 33125^T), *V. splendidus* II (IAM 14411), *V. vulnificus* (IFO 15645^T). Tested with no adverse results or non-specific amplicon (apart from the specific amplicon expected) against 322 strains of *V. cholerae* (01 classical, 01 El Tor, 0139 non-01 and non-0139), 82 strains of *V. parahaemolyticus* and 12 strains of *V. vulnificus*. Other strains tested included *Aeromonas* sp., *Bacillus* sp., *Campylobacter jejuni*, *C. coli*, *C. fetus*, *Cronobacter sakazakii*, *Enterobacter cloacae*, *Escherichia coli*, *Klebsiella pneumoniae*, *Klebsiella* sp., *Salmonella enterica*, *Salmonella* sp., *Shigella dysenteriae*, *S. flexneri* and *S. sonnei*. Strains are wild types unless otherwise indicated.		
Optimization	The primer concentrations were optimized		
Validation	Tested using pond water spiked with the target organisms and no non-specific amplicons were detected. Amplicons from all three species were still detected in the multiplex when a 100-fold difference in cell concentration existed		

A loop-mediated isothermal PCR was developed based on the *gyrB* gene for the detection of *V. parahaemolyticus* (Cai *et al.*, 2010). DNA is extracted from a bacterial colony and homogenized in 200 µl sterile distilled water and boiled for 2–3 min. The mixture is centrifuged for 5 min at 12,000 ×*g* and the supernatant used for template. The primers are Va-FIP (Forward inner (F1c-TTTT-F2) (GAG CAC GTG CTG CAT CGA TGA T-TTTT-TCC TGA TTG AGA ACC CGA CA), Va-BIP Backward inner (B1c-TTTT-B2) (GAA GCA GCG CGT AAA GCT CGT-TTTT-TGG AAG GCC AGC TAG GTC), Va-F3 Forward outer (TTG AGT CTG CAA TGG GTG AA), Va-B3 Backward outer (GCC GGA TCT TTT TCC TGA CA), PF Forward (CTG TTG TGG GTG ATA CGG), PR Reverse (TTC GCT TCT GTC GGG TTC). The 25-µl reaction mixture contains 2 µl genomic DNA, 40 pmol of each primer FIP and BIP, and 5 pmol of each primer F3 and B3, 1.6 pmol/l dNTP and 8 U *Bst* DNA polymerase and buffer (New England labs). The reaction mixture is heated at 64°C for 60 min before terminating the reaction by heating at 80°C for 5 min. The LAMP products are detected visually by ethidium bromide staining of an electrophoresis gel, and by the addition of 1 µl (1:10) of SYBR Green I (Molecular Probes). The sensitivity is 3.7×10^2 CFU/ml (3.7 CFU per reaction). The LAMP is specific and does not amplify DNA from *V. parahaemolyticus* (ATCC 17802), *V. harveyi* (ATCC 14126), *V. vulnificus* (1.7500), *V. anguillarum* (Va101), *Grimontia hollisae* (GZH01), *V. metschnikovii* (1.7350), *V. natriegens* (1.1594), *V. campbellii* (1.1597), *V. splendidus* (1.1606) or *V. fluvialis* (1.1609).

Vibrio pectenicida

TAXONOMY AND SIGNIFICANCE. *Vibrio pectenicida* (A365^T=CIP 105190^T) was isolated from diseased scallop (*Pecten maximus*) larvae cultured in a hatchery in Brittany, France (Lambert *et al.*, 1998). Phylogenetic investigation showed a close relationship with *V. tapetis* and *V. splendidus*. Pathogenicity was confirmed for scallop larvae through experimental infection with an inoculum of 10^4 cells/ml in water and a resulting mortality of 100% in 4 days (Nicolas *et al.*, 1996). *Vibrio pectenicida* was not pathogenic for oyster (*Crassostrea gigas*) larvae. In bath challenge experiments with great scallop (*Pecten maximus*) larvae, *V. pectenicida* caused mortalities within 48 h (Sandlund *et al.*, 2006).

GROWTH CHARACTERISTICS AND PHENOTYPIC IDENTIFICATION. *Vibrio pectenicida* is a facultatively anaerobic, motile, oxidase-positive, Gram-negative rod. Colonies on MSA-B are pinpoint and light grey after 24 h at 25°C and show faint lysis of the red cells. Growth does not occur on BA after 24–72 h. On MA 2216, colonies are 1 mm and cream at 48 h. Growth on TCBS is negative at 24 h with weak growth at 48 h and 1 mm green colonies by day 4. Swarming growth on MA 2216 was reported for the type strain A365, CIP 105190 (Lambert *et al.*, 1998), but this was not found for strain ATCC 700783 (cultured from CIP 105190) on MA 2216 and the media reported here (this manual). Gelatin was negative by the plate method (see media section) and in the API 20E, however was reported positive by Lambert *et al.* (1998). Luminescence is not seen. Growth occurs at 4–30°C but not at 35°C. Growth occurs at 1–6% but not at 0% or 8% NaCl. It is negative for ADH in the API 20E and by Moeller's method. Starch and Tween 80 are hydrolysed. Citrate is negative in the API 20E after 48 h, but becomes positive in Simmons' tube citrate after 5 days. The catalase reaction is strong. These and other reactions are presented in Tables 3.22, 3.24–26 and 3.38. Sets are inoculated with an inoculum in 2% NaCl and incubated at 25°C for 48 h.

Vibrio pelagius

TAXONOMY AND SIGNIFICANCE. *Vibrio* (*Listonella pelagia*) *pelagius* (CECT 4202^T=ATCC 25916^T) was isolated from seawater off the coast of Hawaii and originally named *Beneckea pelagia* (Baumann *et al.*, 1971a), but renamed *Vibrio pelagius* following the abolition of the genus *Beneckea* (Baumann *et al.*, 1980). It was later placed into a new genus as *Listonella pelagia* (MacDonell and Colwell, 1985). A recent study has recommended that *Listonella* is a later heterotypic synonym of the genus *Vibrio* and that *L. pelagia* revert to *V. pelagius* (Thompson *et al.*, 2011).

Phylogenetic analysis of strain NCIMB 1900 and CIP 102762, originally identified as *V. pelagius*, has indicated that these strains are more likely to be *Vibrio natriegens*, and that the original strain deposited at NCIMB may have been lost. Phenotypic tests for strain NCIMB 1900 corresponded to the results expected for *V. natriegens*, being positive for acid production from L-arabinose, melibiose and L-rhamnose, with a negative result for lactose fermentation and alginate degradation (Macián *et al.*, 2000). The phenotypic results recorded as *L. pelagia* NCIMB 1900 by Lunder *et al.* (2000) have been recorded as *V. natriegens* in this manual.

MLSA identified *V. pelagius* as belonging to the Splendidus clade, which contains *V. splendidus*, *V. chagasii*, *V. crassostreae*, *V. cyclitrophicus*, *V. fortis*, *V. gigantis*, *V. kanaloae*, *V. lentus*, *V. pomeroyi* and *V. tasmaniensis* (Sawabe *et al.*, 2007b). Members of this clade have >30% DDH and 39–47 GC mol%, 90.6–99.8% similarity based on concatenated MLSA sequence data and AAI of 96.5–99.8%.

Vibrio pelagius may be part of the microflora of larval, juvenile and adult turbot (*Scophthalmus maximus*), and the organism was thought to provide nutritional benefits through the production of a polyunsaturated fatty acid, eicosapentaenoic acid (Ringø *et al.*, 1992). However, highly pathogenic strains causing mass mortality of turbot larvae cultured in Spain have been reported (Villamil *et al.*, 2003).

GROWTH CHARACTERISTICS AND PHENOTYPIC IDENTIFICATION. *Vibrio pelagius* is a facultatively anaerobic, motile, oxidase-positive, Gram-negative rod. *Vibrio pelagius* I (ATCC 25916) grows on BA as 0.6 mm, non-haemolytic, light grey colonies, shiny, round and slightly mucoid in the pool at 24 h, and 2 mm, non-haemolytic and whitish, raised, round and entire, shiny, glistening and slightly mucoid at 48 h and 25°C. Colonies on MSA-B and MSSA are similar but are larger in size being 1.3 mm at 24 h and 2.2 mm at 48 h. On MA 2216 colonies are 1–3.6 mm, cream and mucoid at 24 h and 2.1–3.7 mm and up to 5.2 mm raised shiny, cream and mucoid at 48 h. A combination of large and small colonies is seen on MA 2216. Swarming growth is not seen. On TCBS colonies are 2 mm and yellow at 24 h and 3.6 mm and yellow with a darker centre after 48 h. Growth does not occur on MacConkey agar. The margins of the zones of no growth around the vibriostatic discs are very defined. Growth does not occur at 37°C or 42°C. Motility is negative in the tube test, but positive when examined microscopically using a wet preparation. The tube ONPG is a strong yellow reaction. Growth does not occur in the MRVP tube medium. Supplementation with NaCl is required in the biochemical test media. In the API 20E kit, nitrate is negative, but is positive using the tube method. In the API 20E the glucose gives a pale green colour at 24 and 48 h and mannitol is yellow indicating a positive reaction. Both tests are positive in the conventional tube method. Reactions in OF media are weak (this manual).

Vibrio pelagius II (ATCC 33784) grows on BA as 1–2 mm grey, raised, shiny colonies at 3 days. On MSA-B and MSSA colonies are 3 mm, dark cream with a slight apricot colour, shiny and slightly mucoid after 3 days at 25°C. On TCBS colonies are 1.5–2 mm and yellow. On MA 2216 and TSA containing 2% NaCl, colonies are unpigmented and regular in shape and grow as green colonies on TCBS (Macián *et al.*, 2001). Luminescence is not seen. Growth occurs at 20–37°C. Growth does not occur in the absence of NaCl and up to 2% NaCl must be added to culture media and to identification tests. Results are negative for both Moeller's and Thornley's ADH. Agar is hydrolysed (see colour plate section). Motility is negative in the tube method. The plate gelatin is slow and a positive reaction may not be seen until 4 days of incubation (this manual). These and other reactions are presented in Tables 3.22, 3.24–26 and 3.38. An inoculum containing 2% NaCl is used.

Vibrio penaeicida

TAXONOMY AND SIGNIFICANCE. *Vibrio penaeicida* (KH-1^T=IFO 15640^T=JCM 9123^T), a pathogen for prawn, was isolated in Japan from diseased cultured kuruma prawn (*Fenneropenaeus japonicus*, misnamed *Penaeus japonicus*), apparently healthy prawn and from prawn pond water (Ishimaru *et al.*, 1995). In New Caledonia it is the causative agent of Syndrome 93, which results in acute and chronic mortalities of *Litopenaeus stylirostris* especially in winter (Costa *et al.*, 1998). Infected prawns are weak and sensitive to stress and swim close to the water surface and pond edges. The shell is soft and dark and may moult. The muscle is opaque. Lesions containing bacteria are seen in the hepatopancreas intertubular spaces, lumen of the antennal gland and gills. Haemolymph is slow to clot and in moribund animals is heavily contaminated with a number of species of bacteria, whereas haemolymph in healthy animals is relatively free of bacteria. *Vibrio penaeicida* was confirmed as a pathogen in experimental challenge resulting in a median lethal dose (LD_{50}) of 1.3×10^4 CFU/ml in immersion challenge and <5 CFU/prawn in intramuscular challenge (Saulnier *et al.*, 2000). In experimental challenge, cysteine protease-like exotoxin extracted from culture supernatants of *V. penaeicida* strains KH-1 (Japan) and AM101 (Syndrome 93 New Caledonia) proved pathogenic to juvenile white shrimp (*Litopenaeus vannamei*) cultured in Mexico. *Vibrio penaeicida* has not been reported from Mexico (Aguirre-Guzman *et al.*, 2005). Strain AM101 was highly virulent and resulted in greater mortality compared to strain KH-1 with a LD_{50} of 5.61 μg protein per prawn.

MLSA identified *V. penaeicida* and *V. nigripulchritudo* as belonging to the Nigripulchritudo clade (Sawabe *et al.*, 2007b). Members of this clade have >36% DDH, 46–47 GC mol%

and a MLSA concatenated sequence similarity of 89.0%.

GROWTH CHARACTERISTICS AND PHENOTYPIC IDENTIFICATION. *Vibrio penaeicida* is a facultatively anaerobic, motile, oxidase-positive, Gram-negative short, straight or slightly curved rod (1.5–2×0.5–0.8 μm). Colonies on MSSA are pinpoint at 24 h and 0.5 mm at 48 h, and on MSA-B colonies are 0.2 mm and 1.5 mm pale grey at 24 and 48 h, respectively. There is no growth on BA after 48 h at 25°C. On MA 2216 colonies are pinpoint at 24 h and 0.5 mm, cream, circular and low convex at 48 h. Swarming and luminescence are not seen. Growth occurs at 20–30°C and 1–3% NaCl, but not at 4 or 35°C, or 0.5 or 6% NaCl. The catalase reaction is slow. Results are positive for alginase, amylase, chitinase and lipase. ADH is negative in Moeller's tube method but positive in the API 20E. Simmons' tube citrate was reported positive for KH-1T (Ishimaru *et al.*, 1995), but strain DSMZ 14398 reported in this manual is negative for Simmons' citrate but positive in the API 20E. Nitrate was reported positive (Ishimaru *et al.*, 1995; Costa *et al.*, 1998), but was negative at 48 h in the nitrate tube and API 20E for strain DSMZ 14398. Costa *et al.* (1998) reported green colonies on TCBS, however no growth occurred for DSMZ 14398. The tube VP test (see media section) is negative (for KH-1 and DSMZ 14398) but the API 20E is positive (for DSMZ 14398). In the API 20E the indole needs three to four drops of Kovács' reagent to show the positive reaction. The glucose reaction is seen as a pale green. The oxidase reaction needs a heavy inoculum and is slow using the filter paper method, taking 30–60 s for a heavy inoculum and 2 min for a light streak of growth (see interpretation of biochemical tests and media section). Motility is positive using a wet preparation, but is negative using the tube method. Sensitivity to vibriostatic disc 10 μg was reported, however, strain DSMZ 14398 was resistant to 10 μg and partially sensitive (small zone) to 150 μg on MSA-B medium. The tube aesculin (see media section) becomes positive after 4–5 days' incubation. These and other reactions are presented in Tables 3.22, 3.24–26 and 3.38. Sets are inoculated with an inoculum in 2% NaCl and incubated at 25°C for 48 h.

MOLECULAR IDENTIFICATION. *Vibrio penaeicida* can be detected using either a conventional PCR or a SYBR Green PCR, both of which use the same primers to detect 310 bp of the 16S rRNA gene (Saulnier *et al.*, 2000; Goarant and Merien 2006). DNA extraction was optimized for samples that included haemolymph, cultured cells, pond water, seawater and sediment pore water. High Pure™ PCR Template Preparation Kit (Roche Diagnostics) or Chelex™ 100 (Sigma) produced increased sensitivity of detection compared to a manual phenol-chloroform-isoamyl-alcohol procedure, or a method involving lysis with sodium dodecyl sulfate (SDS) and digestion with proteinase K. The High Pure method detected more positives from water compared to the Chelex method. The best results were obtained using 200 μl of original sample and using the pellet in the DNA extraction procedure. For haemolymph samples a 10-μl volume used in the DNA extraction procedure gave the best results compared to volumes of 50, 100 and 200 μl. Haemolymph clotting appeared to be a major factor resulting in low yields.

PCR type	Conventional (Saulnier *et al.*, 2000)		
DNA extraction	High Pure™ PCR Template Preparation Kit (Roche Diagnostics) or Chelex™ 100 (Sigma)		
PCR master mix			
Primer VpF	5′-GTGTGAAGTTAATAGCTTCATATC-3′		
Primer VpR	5′-CGCATCTGAGTGTCAGTATCT-3′		
Reagent	**Stock**	**Final concentration**	**Volume ×1**
Buffer[a]	×10	×1	2.0 μl
dNTPs	20 mM	200 μM	0.2 μl
Primer VpF	50 μM	0.2 μM	0.1 μl

Continued

Continued.

Primer VpR	50 μM	0.2 μM	0.1 μl
Goldstar DNA polymerase	5 U/μl		0.4 μl
Water			7.2 μl
DNA			10 μl
Total volume			**20 μl**
Thermocycling	Thermocylcer (MJ Research)		
Step 1	**Step 2**		**Step 3**
94°C, 1 min	94°C, 30 s		94°C, 30 s
	62°C, 30 s		62°C, 30 s
	72°C, 30 s		72°C, 5 min
×1	×35		×1
Product	310 bp		
Sensitivity	20 CFU/ml		
Specific against	See SYBR Green PCR results		

[a]Buffer = 10mM Tris-HCl, pH 9.0, 50 mM KCl, 2.5 mM $MgCl_2$, 0.1% Triton X100.

PCR type	SYBR Green (Goarant and Merien, 2006)		
DNA extraction	Bacterial culture, haemolymph, seawater, sediment pore water. High Pure™ PCR Template Preparation Kit (Roche Diagnostics) or Chelex™ 100 (Sigma)		
PCR master mix			
Primer VpF	5′-GTGTGAAGTTAATAGCTTCATATC-3′		
Primer VR	5′-CGCATCTGAGTGTCAGTATCT-3′		
Reagent	**Stock**	**Final concentration**	**Volume ×1**
LightCycler FastStart DNA Master SYBR Green I kit (Roche)[a]	×10	×1	2 μl
Primer VpF	50 μM	0.5 μM	0.2 μl
Primer VR	50 μM	0.5 μM	0.2 μl
Water			17.6 μl
DNA			2 μl
Total volume			**20 μl**
Thermocycling	LightCycler Real-Time PCR System (Roche)		
Step 1	**Step 2**		
95°C, 10 min (hot start)	95°C, 8 s		
	60°C, 5 s		
	72°C, 13 s		
×1	×50		
Data acquisition	Continuous fluorescence acquisition		
Product	310 bp		
Melt temperature	87.1°C		
Sensitivity	660 CFU/ml for pumping water, 330 CFU/ml for pond water, 52 CFU/ml for sediment pore water, 50 CFU/μl for 10 μl haemolymph		
Specific against	*Vibrio aestuarianus*, *V. alginolyticus*, *V. campbellii*, *V. harveyi*, *V. fluvialis*, *V. gazogenes*, *V. navarrensis*, *V. nereis*, *V. nigripulchritudo*, *V. orientalis*, *V. parahaemolyticus*, *V. proteolyticus*, *V. splendidus*, *V. tubiashii*, *Grimontia hollisae*, *Photobacterium damselae* spp. *damselae*, *Proteus mirabilis*, *Escherichia coli*, *Staphylococcus aureus*, *Streptococcus agalactiae*, *Acinetobacter baumannii*, *Shigella flexneri*, *Klebsiella pneumoniae* and *Salmonella* species.		

The LightCycler FastStart DNA Master SYBR Green I kit (Roche) master mix contains dUTP instead of dTTP and FastStart *Taq* polymerase, a hot start polymerase.

Vibrio plantisponsor

TAXONOMY AND SIGNIFICANCE. *Vibrio plantisponsor* (MSSRF60[T]=DSM 21026[T]=LMG 24470[T]=CAIM 1392[T]) was isolated from the roots of wild rice collected from the Pichavaram mangroves, India (Rameshkumar *et al.*, 2011). It is closely related to the Diazotrophicus clade using sequence analysis of five housekeeping genes, *recA*, *gyrB*, *pyrH*, *mreB* and *gapA*, and based on 16S rRNA sequencing it is most closely related to *V. diazotrophicus* (99.7%), *V. areninigrae* (98.2%) and *V. hispanicus* (97.8%).

GROWTH CHARACTERISTICS AND PHENOTYPIC IDENTIFICATION. *Vibrio plantisponsor* is a facultatively anaerobic, motile Gram-negative slightly curved rod, and is oxidase and catalase positive. On MA 2216, colonies are 2–4 mm, opaque, circular and smooth after 48 h at 28°C. Yellow colonies are produced on TCBS. Growth occurs at 0.5–8% (w/v) NaCl, but not at 0% NaCl. Optimal temperature is 28°C with growth occurring at 10–40°C. Tween 80 and alginate are not hydrolysed whereas casein is hydrolysed (Rameshkumar *et al.*, 2011). These and other reactions are presented in Tables 3.23–26 and 3.30. An inoculum of 2.5% NaCl was used and results read at 48 h after incubation at 28°C.

Vibrio pomeroyi

TAXONOMY AND SIGNIFICANCE. *Vibrio pomeroyi* (LMG 20537[T]=CAIM 578[T]) was isolated from healthy bivalve larvae (*Nodipecten nodosus*) from Brazil and turbot (*Scophthalmus maximus*) from Spain (Thompson *et al.*, 2003d). The pathogenicity was not determined.

In experimental challenge using rainbow trout (*Oncorhynchus mykiss*) and *Artemia* nauplii inoculated with 10^6 CFU/fish or 10^6CFU/ml, respectively, *V. pomeroyi* (LMG 21352[T]) was rated as non-virulent or low virulence. Although there was some evidence of exophthalmia, kidney necrosis, ascites and liquefaction of internal organs, clinical disease was not seen, and at the end of the experiment the organism was recovered from tissue at either a low rate or not at all. Strain LMG 20537[T] did not cause any disease signs (Austin *et al.*, 2005).

Vibrio pomeroyi together with *V. kanaloae* and *V. chagasii* are phylogenetically related to *V. splendidus*. MLSA identified *V. pomeroyi* as belonging to the Splendidus clade, which contains *V. splendidus*, *V. chagasii*, *V. crassostreae*, *V. cyclitrophicus*, *V. fortis*, *V. gigantis*, *V. kanaloae*, *V. lentus*, *V. pelagius* and *V. tasmaniensis* (Sawabe *et al.*, 2007b). Members of this clade have >30% DDH and 39–47 GC mol%, 90.6–99.8% similarity based on concatenated MLSA sequence data and AAI of 96.5–99.8%.

GROWTH CHARACTERISTICS AND PHENOTYPIC IDENTIFICATION. *Vibrio pomeroyi* is a facultatively anaerobic, motile Gram-negative slightly curved rod (1.0×2–3 µm), oxidase and catalase positive. On TSA supplemented with 2% NaCl, colonies are non-swarming, beige, circular, smooth, round, with an entire edge after 48 h at 28°C. The type strain produces green colonies on TCBS, whereas six other strains produce yellow colonies. NaCl is required for growth with growth occurring up to 8% NaCl, but not at 0 or 10% NaCl. Growth occurs at 4°C but not at 35°C. Optimal conditions are 28°C and 2% NaCl. Luminescence is not produced. ADH is positive for the type strain, whereas 50% of strains are negative. *Vibrio pomeroyi* was reported nitrate positive (Thompson *et al.*, 2003d), but others have reported nitrate negative using the API 20E (Faury *et al.*, 2004; Le Roux *et al.*, 2005). These and other reactions are presented in Tables 3.23–26, 3.30 and 3.38.

Vibrio ponticus

TAXONOMY AND SIGNIFICANCE. *Vibrio ponticus* (369[T]= DSM 16217[T]=CECT 5869[T] from mussels) was isolated from diseased and healthy gilthead seabream (*Sparus aurata*), mussels and seawater from mussels from the Mediterranean coast, Spain (Macián *et al.*, 2004) and from diseased Japanese seabass (*Lateolabrax japonicus*), China (Xie *et al.*, 2007). The mass death of Japanese seabream was related to a rise in water temperature. Haemorrhagic areas occurred on the liver and on the skin, the latter developing into pale-coloured ulcers, some with deep penetration of the musculature. The seabass isolate was highly virulent with an LD_{50} of 3.2×10^3 CFU/fish when injected intramuscularly and 2.5×10^3 when injected intraperitoneally. The strain from gilthead seabream did not cause deaths in healthy fish when injected intraperitoneally (Macián *et al.*, 2004). Based on 1338 bp, strain

RP30 from Japanese seabream had 99.3% 16S rRNA gene sequence similarity to *V. ponticus* from mussels and seabream (Xie *et al.*, 2007). *Vibrio ponticus* is most similar to *V. fluvialis* (97.1%) and to *V. furnissii* (97.3%) according to 16S rDNA sequencing.

GROWTH CHARACTERISTICS AND PHENOTYPIC IDENTIFICATION. *Vibrio ponticus* is a motile, Gram-negative rod (3.0–1.5 μm). The strain from Japanese seabream was reported to be slightly curved, whereas the strain from mussels and seabream was reported as a straight rod. Colonies on MA 2216 are 2–3 mm, white, non-swarming, non-pigmented, shiny, round with regular edges after 48 h at 25°C (Macián *et al.*, 2004). Colonies are yellow on TCBS. Na^{+} ions are required for growth with growth occurring in up to 8% NaCl, but not at 10% NaCl. Growth does not occur at 4 or 40°C. Luminescence is not seen after 24 h growth on seawater agar. Reactions are negative for the hydrolysis of casein, starch, agar and Tween 80. A strong indole reaction is seen. ADH was negative by the methods of Thornley and Moeller, and LDC (Moeller's method) was positive according to Macián *et al.* (2004), but ADH was positive and LDC negative in the API 20E according to Xie *et al.* (2007). These and other reactions are presented in Tables 3.22 and 3.24–26. An inoculum is used of 2% NaCl.

Vibrio porteresiae

TAXONOMY AND SIGNIFICANCE. *Vibrio porteresiae* (MSSRF30^{T}=LMG 24061^{T}=DSM 19223^{T}) was isolated from the roots of wild rice (*Porteresia coarctata* Tateoka) in mangroves in Pichavaram, India and has nitrogen-fixing properties (Rameshkumar *et al.*, 2008). Two strains (MSSRF30^{T} and MSSRF31) are related to *V. fluvialis* (LMG 7894^{T}) at 96.8% similarity, to *V. furnissii* (LMG 7910T) at 96.8% and to *V. tubiashii* (CIP 102760^{T}) at 96.7% similarity based on 16S rRNA gene sequencing, and occupy a distinct phylogenetic branch based on 16S rRNA, *rpoA*, *recA* and *pyrH* genes. Low levels of sequence similarity of 83–93% are found to all *Vibrio* species using MLSA of *recA*, *pyrH*, *rpoA* and *nifH*. The level of similarity for 16S rRNA sequence is below the recommended level of 97%, although MLSA using the housekeeping genes suggests the strain is a member of the *Vibrio* genus.

GROWTH CHARACTERISTICS AND PHENOTYPIC IDENTIFICATION. *Vibrio porteresiae* grows at 28°C on TSA with 1.5% NaCl as non-swarming, white, round, smooth colonies and produces exopolysaccharide within 2 days. Cells are straight to slightly curved (0.9×1.0–2.0 μm), occurring singly or in pairs. The organism is motile, positive for oxidase and catalase, and does not reduce nitrate. It has some uncommon phenotypic characteristics for *Vibrio* species including no growth on TCBS and resistance to 10 μg and 150 μg vibriostatic agent 0/129. It grows in 1–5% NaCl, but not in 8% NaCl or in the absence of NaCl, and has an optimal NaCl requirement of 1–2%. Growth occurs between 20 and 37°C whereas growth does not occur at 4°C or 45°C. *Vibrio porteresiae* ferments glucose, fructose, cellobiose, galactose and rhamnose, but does not ferment raffinose, melibiose, dulcitol, sorbose, melezitose, inulin, glycerol or adonitol. These and other reactions are presented in Table 3.23.

Vibrio proteolyticus

TAXONOMY AND SIGNIFICANCE. *Vibrio proteolyticus*, originally named *Aeromonas proteolytica* (ATCC 33125), was isolated from the intestine of a marine wood-boring crustacean (*Limnoria tripunctata*) found on the South Carolina and Florida coasts, USA (Merkel *et al.*, 1964). The name was later changed to *Aeromonas hydrophila* ssp. *proteolytica*, *Beneckea proteolytica* and finally placed in the genus *Vibrio* (Baumann *et al.*, 1980). Strain CW8T2 is a pathogen for *Artemia* (brine shrimp), which are used as a source of live feed for some cultured aquatic species. Infection with *V. proteolyticus* leads to disruption of the gut epithelial cells and microvilli and penetration into body cells (Verschuere *et al.*, 2000). Under experimental conditions 100% mortality of *Artemia* occurred after 48 h infection with 1×10^{2}–5×10^{6} CFU/ml of the organism. *Vibrio proteolyticus* has 16S rRNA gene sequence similarity to *V. alginolyticus*, *V. campbellii*, *V. diazotrophicus* and *V. hispanicus* (Gomez-Gil *et al.*, 2004b).

GROWTH CHARACTERISTICS AND PHENOTYPIC IDENTIFICATION. *Vibrio proteolyticus* is a facultatively anaerobic, motile, oxidase-positive, Gram-negative rod (1.1–1.8×0.6–0.8 μm) (Merkel *et al.*, 1964). On MSA-B and MSSA colonies swarm and completely cover the plate after 24 h at 25°C and could be mistaken for *V. alginolyticus*. The colony colour darkens with age. On MA 2216, colonies are 1 mm, cream, and show spreading growth 3–5 mm from the edge of the colony and up to 5–10 mm from the edge of the colony at 48 h. Colonies do not swarm on BA and are beta-haemolytic, 0.5 mm pale grey, round, circular at 24 h and 2.5 mm at 48 h with a slight greenish colour in the pool. Colonies are green, non-swarming on TCBS and 1 mm at 24 h and 3 mm at 48 h. Luminescence is not seen on any of these media. The oxidase reaction is rapid. Conventional tube media and API 20E are positive for LDC and ADH. LDC in the API rapid ID 32E is negative at 48 h (this manual). Growth occurs at 25°C and 37°C but not at 42°C. Extracellular aminopeptidase is produced (Merkel *et al.*, 1964). These and other reactions are presented in Tables 3.22, 3.24–26 and 3.38.

MOLECULAR IDENTIFICATION. The following primers have been reported. IGS[IA] (proC)-specific primers: VPRO-CGC ATT CTT ACG AGT GTG and VPRO-R ATT AGT TGT ATT CAA ATA. Annealing temperature is 55°C using 35 cycles. An amplicon of 133 bp is produced (Lee *et al.*, 2002).

Vibrio quintilis

TAXONOMY AND SIGNIFICANCE. *Vibrio quintilis* (M62^T=CECT 7734^T=CAIM 1863^T) was isolated from seawater off the coast of Valencia, Spain (Lucena *et al.*, 2012). Based on the 16S rRNA gene sequence *V. quintilis* is most similar to *V. aerogenes* (97.6%) and is related at 95.1–96.0% to other species in the Gazogenes clade including *V. gazogenes*, *V. ruber*, *V. rhizosphaerae* and *V. mangrovi*.

GROWTH CHARACTERISTICS AND PHENOTYPIC IDENTIFICATION. *Vibrio quintilis* is a motile, Gram-negative rod (0.7–2.0×0.6 μm). Oxidase is negative. On MA 2216 colonies are 0.5 mm, non-pigmented and transparent after 48 h at 25°C. Growth does not occur on TCBS. Growth occurs at 28–37°C but not at 4 or 40°C. Na^+ is required for growth and growth occurs at 1.7–6% NaCl but not at 0 or 8% NaCl. Results are positive for Thornley's ADH. Nitrate is not reduced. DNA is hydrolysed but hydrolysis is weak for casein, and negative for Tween 80 and alginate. Gelatin is negative in the API 20E, but is weakly positive in a tube method containing marine broth in semi-solid agar and 12% gelatin (Lucena *et al.*, 2012). These and other reactions are presented in Tables 3.23, 3.30 and 3.38.

Vibrio rarus

TAXONOMY AND SIGNIFICANCE. *Vibrio rarus* (RW22^T-=LMG 23674^T=NBRC 102084^T) was isolated from the intestine of Californian red abalone (*Haliotis rufescens*) cultured in the USA (Sawabe *et al.*, 2007a). Based on 16S rRNA gene sequence similarity, *V. rarus* has 99.6–99.7% similarity to *V. comitans*, 98.6–99.0% similarity to strains of *V. inusitatus* and 98.6% similarity to *V. superstes*. Sequence similarity of the *gapA* gene reveals similarity to *V. neonatus* (93.3%), *V. ezurae* (93.3%) and *V. halioticoli* (93.4%).

GROWTH CHARACTERISTICS AND PHENOTYPIC IDENTIFICATION. *Vibrio rarus* is a facultatively anaerobic, non-motile, oxidase-positive, Gram-negative rod (1–2×0.5–1.0 μm). On MA 2216, colonies are beige, circular, smooth, convex with an entire edge. Colonies are green on TCBS (Sawabe *et al.*, 2007a). Luminescence is not seen. Growth occurs at 15–30°C, but not at 37°C. Growth does not occur at 0 or 1% NaCl. Alginate is hydrolysed but not starch, chitin, Tween 80, agar or DNA. These and other reactions are presented in Table 3.23, and some reactions are reported in Tables 3.24–29.

Vibrio rhizosphaerae

TAXONOMY AND SIGNIFICANCE. *Vibrio rhizosphaerae* (MSSRF3^T=LMG 23790^T=DSM 18581^T) was isolated from the rhizosphere of wild rice (*Porteresia coarctata* Tateoka) in mangroves in Pichavaram, India and has antibacterial activity against phytopathogens (Kumar and Nair, 2007). The two strains of *V. rhizosphaerae* are closely related to *V. ruber* (JCM 11486^T) based on 16S rRNA (98.3–98.5%), *rpoA* (98.3–99.7%), *recA* (90.2–99.8%) and *pyrH* (91.3–99.4%) gene sequences. The strains are related to *V. gazogenes* LMG 19540^T at 95.7 and 95.5% similarity, and below 95.7% similarity to other *Vibrio* species based on 16S rRNA gene sequence.

GROWTH CHARACTERISTICS AND PHENOTYPIC IDENTIFICATION. *Vibrio rhizosphaerae* grows on TSA with 2% NaCl at 28°C and produces a bright pink-red (rose-pink), non-diffusible pigment. The pigment is prodigiosin. Cells are 1.8–3.2×0.4–0.5 μm. The organism is motile, catalase-positive, oxidase-negative, nitrate-negative, produces caseinase and lipase and is non-luminescent (Kumar and Nair, 2007). Growth occurs between 20 and 42°C with an optimum of 25–30°C and 0.1–10% NaCl. Growth does not occur at 4 or 45°C, or in the absence of NaCl. These and other reactions are presented in Table 3.23.

Vibrio rotiferianus

TAXONOMY AND SIGNIFICANCE. *Vibrio rotiferianus* (LMG 21460^T=CAIM 577^T) was isolated from a rotifer (*Brachionus plicatilis*) culture system in Ghent, Belgium (Gomez-Gil *et al.*, 2003a). It was also found in the haemolymph of healthy spider crab (*Maja brachydactyla*) caught off the coast of Spain and the Canary Islands (Gomez-Gil *et al.*, 2010). The numbers of bacterial flora in healthy haemolymph is usually in the order of 10^3–10^4 CFU/ml (Gomez-Gil *et al.*, 1998).

Different strains are variable in their pathogenicity. In experimental challenge, *V. rotiferianus* (LMG 21460^T and LMG 21456) caused 30% and 100% mortality, respectively, within 4 days when inoculated into rainbow trout (*Oncorhynchus mykiss*) at a dose of 10^6 cells/fish. At an inoculum of 10^4 CFU/fish, 20% and 60% mortality occurred. The LD_{50} was 10^5 for the type strain and $<10^2$ for LMG 21456 (Austin *et al.*, 2005). Mortality of *Artemia* nauplii exposed to 10^6 CFU/ml was 85% for both strains. Injection with extracellular product caused 100% mortality in rainbow trout using both strains and 65% mortality using both strains in challenges with *Artemia*. In rainbow trout, clinical signs included liquefaction of internal organs, protrusion of the anus and haemorrhage and necrosis in muscles. The strains were positive for caseinase, gelatinase and phospholipase, negative for elastase and lipase (Tween 80). The strains were non-haemolytic to red cells from sheep, horse and trout.

Vibrio rotiferianus has been isolated from diseased crown of thorns starfish (*Acanthapaster planci*) from the Great Barrier Reef, Australia, in association with other *Vibrio* species: *V. harveyi*, *V. owensii* and *V. natriegens* (Rivera-Posada *et al.*, 2011). Their role in pathogenicity is not known at this stage.

Vibrio rotiferianus is most closely related to *V. campbellii* (99.86%) and *V. harveyi* (99.11%) based on 16S rRNA sequencing. MLSA identifies *V. rotiferianus* as belonging to the Harveyi clade, which also contains *V. campbellii*, *V. alginolyticus*, *V. harveyi*, *V. mytili*, *V. natriegens* and *V. parahaemolyticus* (Sawabe *et al.*, 2007b). Members of this clade have >25% DDH and 42–48 GC mol%, a MLSA concatenated sequence similarity of 90.1–96.2%, and AAI of 97.2–99.4%. Strains within the Harveyi clade (*V. alginolyticus*, *V. campbellii*, *V. harveyi*, *V. natriegens*, *V. parahaemolyticus* and *V. rotiferianus*) can be discriminated using concatenated sequences from three housekeeping genes, *rctB*, *rpoD* and *toxR*, with *toxR* and *rpoD* providing the most discrimination (Pascual *et al.*, 2010) (see under *Vibrio harveyi*).

GROWTH CHARACTERISTICS AND PHENOTYPIC IDENTIFICATION. *Vibrio rotiferianus* is a facultatively anaerobic, motile, oxidase-positive, Gram-negative curved rod (2.0–3.5×0.8–1.2 μm). On MA 2216, colonies are non-pigmented, non-swarming, non-luminescent, and on TCBS are 2–3 mm, bright yellow after 24 h at 25°C (Gomez-Gil *et al.*, 2003a). Growth occurs at 28–40°C but not at 4°C. Growth occurs at 1.5–6% NaCl but not at 0 or 8% NaCl. These and other reactions are presented in Tables 3.23–29 and 3.38.

MOLECULAR IDENTIFICATION. See under *Vibrio harveyi* for a PCR using the *rpoD* and *toxR* housekeeping genes followed by sequencing and analysis using concatenated sequences to enable differentiation between strains comprising the *Vibrio* core group. Sequencing of the *mreB*, *topA* and 16S rRNA genes can identify and differentiate vibrios in the Harveyi clade (Rivera-Posada *et al.*, 2011) (see under *V. natriegens*).

Vibrio ruber

TAXONOMY AND SIGNIFICANCE. *Vibrio ruber* (VR1^T=JCM 11486^T=CCRC 17186^T) was isolated from shallow seawater from the coastal region of Keelung, Taiwan (Shieh *et al.*, 2003). 16S rRNA gene sequencing revealed <96% similarity to *Vibrio* species, with the greatest similarity being to *V. gazogenes* (ATCC 29988^T) at 95.8%. MLSA identifies *V. ruber* as belonging to the Gazogenes clade, which also contains *V. aerogenes* and *V. gazogenes* (Sawabe *et al.*, 2007b). Members of this clade have >32% DDH and 46–47 GC mol%.

GROWTH CHARACTERISTICS AND PHENOTYPIC IDENTIFICATION. *Vibrio ruber* is a Gram-negative, motile, curved rod (1.8–2.4×0.7–0.8 µm). On peptone yeast (PY) agar (6 g Bacto peptone (Difco), 2 g Bacto yeast extract (Difco), 25 g NaCl, 3 g $MgSO_4.7H_20$, 0.01g $CaCl_2$, 15 g Bacto agar (Difco), 1 l deionized water) colonies are flat, circular, non-swarming with a red pigment (Shieh *et al.*, 2003). The pigment is prodigiosin. Growth does not occur in the absence of NaCl, but occurs at 1–10% with an optimum of 2% NaCl. Growth occurs at 20–40°C with the most rapid growth occurring at 25–30°C. Growth does not occur at 4°C or at 45°C. *Vibrio ruber* is catalase-positive and oxidase-negative, non-luminescent and resistant to vibriostatic agent at 10 and 150 µg. Fermentation reactions are positive for cellobiose, and galactose. Amylase and lipase are produced, but reactions are negative for agarase and caseinase. These and other reactions are presented in Table 3.23.

Vibrio rumoiensis

TAXONOMY AND SIGNIFICANCE. *Vibrio rumoiensis* (S-1^T=FERM P-14531^T) was isolated from the drain pool of a fish processing plant, Japan and expresses high catalase activity (Yumoto *et al.*, 1999). Based on 16S rRNA sequencing the organism has similarity to the *Aliivibrio fischeri-logei* group (93.3–93.8%), *Photobacterium* (93.0–94.3%) and to *Salinivibrio costicola* (89.5%).

GROWTH CHARACTERISTICS AND PHENOTYPIC IDENTIFICATION. *Vibrio rumoiensis* is a facultatively psychrophilic, non-motile, oxidase-positive, Gram-negative rod (0.7–2.1×0.5–0.9 µm). On PYS-2 agar (8.0 g polypeptone, 3.0 g yeast extract, 5.0 g NaCl, 15 g agar/L at pH 7.5) colonies are white to colourless, and circular at 27°C (Yumoto *et al.*, 1999). Growth occurs at 2–34°C and at 3–6% NaCl, but not at 0% NaCl. Chitin, starch and Tween 20, 40, 60 and 80 are hydrolysed, but not casein or alginic acid. These and other reactions are presented in Table 3.23. An inoculum is used of 2–3% NaCl.

Vibrio sagamiensis

TAXONOMY AND SIGNIFICANCE. *Vibrio sagamiensis* (LC2-047^T=NBRC 104589^T=KCTC 22354^T) was isolated from seawater in Sagami Bay, Japan and a second strain LC1-908 from the equator region in the Pacific Ocean (Yoshizawa *et al.*, 2010b). According to 16S rRNA sequence information the strains are closely related to *V. azureus* (99.4%), *V. parahaemolyticus* (99.3%), *V. campbellii* (99.1%), *V. natriegens* (99.1%) and *V. alginolyticus* (99.1%). Multilocus sequence analysis using six genes (*gyrB*, *pyrH*, *gapA*, *ftsZ*, *mreB* and *recA*) indicates *V. sagamiensis* belongs to the Harveyi clade.

GROWTH CHARACTERISTICS AND PHENOTYPIC IDENTIFICATION. *Vibrio sagamiensis* is a facultatively anaerobic, motile, oxidase-positive, Gram-negative short rod or coccoid-rod (1.7–2.0×1.0 µm). On MA 2216, colonies are translucent, circular with entire margins and show weak swarming and positive luminescence. Growth occurs at 10–37°C with an optimum of 25–30°C, but no growth at 4°C (Yoshizawa *et al.*, 2010b). Growth occurs at 0.5–6% NaCl with an optimum of 3–5%, but growth does not occur at 8 or 10%. Gelatin, Tween 20, 40, 60 and 80 are hydrolysed but not starch. These and other reactions are presented in Tables 3.23–26 and 3.38. Sets are incubated at 20°C with an inoculum of 2% NaCl and read at the time indicated in the appropriate table.

Vibrio salmonicida

See *Aliivibrio salmonicida*.

Vibrio scophthalmi

TAXONOMY AND SIGNIFICANCE. *Vibrio scophthalmi* (A089^T=CECT 4638^T) was isolated from the intestines of juvenile turbot (*Scophthalmus maximus*) cultured in a fish hatchery in Spain (Cerdà-Cuéllar *et al.*, 1997) and was part of the dominant gastrointestinal flora in this fish species. *Vibrio scophthalmi* is most closely related to *V. aestuarianus* (97.8%) based on 16S rRNA sequencing. MLSA identifies *V. scophthalmi* as belonging to the Scophthalmi clade, which also contains *V. ichthyoenteri* (Sawabe *et al.*, 2007b). Members of this clade have >32% DDH and 43–44 GC mol%, 95.5% similarity based on concatenated MLSA sequence data and AAI of 99.4%. *Vibrio aestuarianus* belongs to the Anguillarum clade.

GROWTH CHARACTERISTICS AND PHENOTYPIC IDENTIFICATION. *Vibrio scophthalmi* is a facultatively anaerobic, motile, oxidase-positive, Gram-negative short rod. On MA 2216 colonies are non-swarming, non-luminescent, regular and non-pigmented. Colonies are yellow on TCBS. Growth occurs at 22–30°C but not at 4, 37 or 40°C. Grows occurs at 1–3% NaCl but not at 0% or 6% NaCl. Of the strains, 90% are positive for ADH (Moeller's method). The ADH in the API 20E is weak at 24 and 48 h. Aesculin was reported positive (Cerdà-Cuéllar *et al.*, 1997) but is negative in the tube method. Motility using a stab culture is negative, but was reported as positive using the method of Hugh and Leifson (1953) (see media section) and Leifson's oxidation-fermentation basal medium. *Vibrio scophthalmi* and *V. splendidus* biovar I may be difficult to differentiate based on phenotypic tests and this is compounded due to the variation in phenotypic test results between strains in each species. The type strain is positive for chitin hydrolysis, whereas other strains are negative (Farto *et al.*, 1999). Indole, ONPG and fermentation of mannitol appear to be the best differentiating phenotypic tests. *Vibrio scophthalmi* is negative for all three, whereas *V. splendidus* biovar I is positive for all three. These and other reactions are presented in Tables 3.23–26 and 3.38. Sets are incubated at 25°C with an inoculum of 2% NaCl.

Vibrio shilonii

See *Vibrio mediterranei*.

Vibrio sinaloensis

TAXONOMY AND SIGNIFICANCE. *Vibrio sinaloensis* (CAIM 797^T=CECT 7298^T) was isolated from liver, spleen and lesions of cultured spotted rose snapper (*Lutjanus guttatus* Steindachner) in Mexico (Gomez-Gil *et al.*, 2008). A number of *Vibrio* species were isolated from cultured and wild-caught snapper but the pathogenesis of *V. sinaloensis* is unknown. Strains of *V. sinaloensis* caused mortality of prawns (*Litopenaeus vannamei*) when used in challenge experiments (Flores-Miranda *et al.*, 2011). Based on 16S rRNA sequencing the most closely related *Vibrio* species are *V. brasiliensis* (98.17%) and *V. hepatarius* (97.72%). DNA fingerprinting using rep-PCR indicates that *V. fortis* is also closely related, whereas 16S rRNA sequencing showed a similarity of 96.7%.

GROWTH CHARACTERISTICS AND PHENOTYPIC IDENTIFICATION. *Vibrio sinaloensis* is a facultatively anaerobic, motile, oxidase-positive, Gram-negative curved rod. Growth occurs on MA 2216 agar, and TSA with 2% NaCl. Colonies are bright yellow on TCBS (Gomez-Gil *et al.*, 2008). Growth occurs at 15–37°C and in 0.35–9% NaCl. Growth does not occur at 0 or 10% NaCl or at 4°C. Growth at 40°C is variable with three of five strains showing no growth at 40°C. Hydrolysis is positive for casein, starch and Tween 80 and negative for alginate and agar. These and other reactions are presented in Tables 3.23–26. Sets are incubated according to the manufacturer's instructions with the exception that an inoculum is used of 2.5% NaCl.

Vibrio splendidus

TAXONOMY AND SIGNIFICANCE. *Vibrio splendidus* (NCMB1^T=ATCC 33125^T), originally identified around 1899 by M.W. Beijerinck as *Photobacterium splendidum*, and isolated from seawater off the Netherlands coast, was not formally reported but was mentioned in several of his papers including 1889 and 1901 (Beijerinck, 1889, 1901). *Photobacterum splendidum* was reassigned to the genus *Beneckea* as *B. splendida* using strain NCMB1 from the Delft Culture Collection, which had been isolated by H.J. Van Zutphen, as the original Beijerinck strain submitted in 1922 was deemed 'no longer available for study' (Reichelt *et al.*, 1976). In later work the genus *Beneckea* was abolished and the isolate was re-named *Vibrio splendidus* biovar I (Baumann *et al.*, 1980). Several strains of *V. splendidus* exist in the Delft Culture Collection (now the Netherlands Culture Collection of Bacteria following the 1998 merger with the Delft University of Technology Bacterial Collection founded by Beijerinck) and they are referred to by their originator (de Graaf NCCB 38017; Ylstra NCCB 48036; and van Zutphen NCCB 53037). In 1979, a *P. splendidum* culture stored in 1925 (and thought to be from Beijerinck) was successfully revived and all strains of *V. splendidus* were compared using physiological tests and the results suggested the neotype strain of van Zutphen NCCB 53037 was different from the other *V. splendidus* strains (Robertson *et al.*, 2011). Information on strain source is available at http://www.straininfo.net/ strains then click on 'strain passport' for history of subculture information. Currently, the strains of van Zutphen, Kluyver, Spencer, NCMB1, Verdonck VIB 308, DSM 19640 and some others do not have *V. splendidus* listed against them and are awaiting confirmation of identification. Strain ATCC 33125, NCIMB1, NCCB 53037, LMG 4042, NBRC 15643 and others are listed as being *V. splendidus*.

'Verzamelde Geschriften' (The Collected Writings) of Martinus Beijerinck written in Dutch, French or English is available at http://www.archive.org. An English translation of his 1889 paper on *Photobacterium liminosum* is available as a pdf and is fascinating to read (https://mywebspace.wisc.edu/wollenberg/web/science/Beijerinck1889.pdf).

Vibrio splendidus is found in marine and brackish water such as estuaries and can cause disease in a number of marine animals including shellfish and finfish. Two biogroups are described based on DNA–DNA hybridization and phenotypic characteristics. Biogroup I is luminescent whereas biogroup II is not. Confirming the identity of *V. splendidus* is difficult because of its similarity to a number of *Vibrio* species now referred to as the *V. splendidus* group. In this group, sequencing of the 16S rRNA gene identifies two clades: *V. chagasii* in one clade and *V. lentus*, *V. splendidus*, *V. tasmaniensis*, *V. kanaloae* and *V. pomeroyi* in the second clade (Le Roux *et al.*, 2004). Further discrimination of species is not obtained and because of the sequence similarity they are considered to be of the same rRNA species complex (Fox *et al.*, 1992). MLSA identifies *V. splendidus* as belonging to the Splendidus clade, which contains *V. chagasii*, *V. crassostreae*, *V. cyclitrophicus*, *V. fortis*, *V. gigantis*, *V. kanaloae*, *V. lentus*, *V. pelagius*, *V. pomeroyi* and *V. tasmaniensis* (Sawabe *et al.*, 2007b). Members of this clade have >30% DDH and 39–47 GC mol%, 90.6–99.8% similarity based on concatenated MLSA sequence data and AAI of 96.5–99.8%. A phylogenetic tree of the partial gyrase subunit (the *gyrB* gene) identifies five clusters: *V. chagasii* in cluster 1, *V. kanaloae*, *V. pomeroyi* and *V. tasmaniensis* in cluster 2, *V. lentus* in cluster 3, *V. splendidus* in cluster 4 and as yet three unidentified strains in cluster 5 (Le Roux *et al.*, 2004). The housekeeping genes *recA* and *pyrH* have better resolving power than *rpoA* for discriminating within the *V. splendidus* group (Thompson *et al.*, 2005a).

Strains of *V. splendidus* vary in their virulence and this may be due to the secretion by some strains of a 35–37 kDa metalloprotease (Vsm), which together with other unknown extracellular proteins has a cytopathic effect on mollusc cells. A second metalloprotease (101 kDa) may be involved in inhibition of host immunity as it has significant identity to an immune inhibitor produced by *Bacillus thuringiensis* (Binesse *et al.*, 2008). The outer membrane protein OmpU is required for virulence and exerts its pathogenic effect by binding to fibronectin resulting in a competitive advantage in colonization of oyster tissue (Duperthuy *et al.*, 2010). OmpU is resistant to the antimicrobial peptides and proteins produced by the oyster (*Crassostrea gigas*) as part of its innate immune system.

A lethal 39 kDa extracellular protease from *Vibrio* (*Listonella*) *pelagius* pathogenic to cultured turbot (*Psetta maximus*) in Spain was found in strains identified as *V. splendidus*-*V. lentus* group isolated from diseased turbot (*Scopthalmus maximus*) and gill heart of octopus (*Octopus vulgaris*) also cultured in Spain (Farto *et al.*, 2006).

Vibrio splendidus I caused disease outbreaks in juvenile turbot raised in Spain with clinical signs including erratic swimming, abdominal extension, haemorrhagic areas in the mouth and accumulation of red-coloured fluid in the abdomen (Angulo *et al.*, 1994). In experimental challenge an LD_{50} of 1.2×10^4 CFU/fish for turbot and an LD_{50} of 2.2×10^4 CFU/fish was obtained for rainbow trout. Further work indicated variation in virulence of *V. splendidus* I isolates with four (DMC-1, DTC-5, LTS-3, HNF-8) of 11 isolates showing high pathogenicity in experimental challenge of 1×10^5 CFU/ml against turbot larvae (Thompson *et al.*, 2005). *Artemia* used as the food source did not harbour *V. splendidus*. The bacterial flora of the *Artemia* was *V. alginolyticus*, but this was not pathogenic to turbot larvae. Differences in pathogenicity of *V. splendidus* isolates were found in turbot larvae (*S. maximus*) in France and virulence appears to be correlated with production of valine aminopeptidase using the API Zym kit (Gatesoupe *et al.*, 1999). *Vibrio splendidus* (strain A515) caused mortality in scallop (*Pecten maximus*) larvae cultured off the coast of Brittany, France, and pathogenicity was confirmed by experimental challenge (Nicolas *et al.*, 1996).

Vibrio splendidus II caused mass mortality in oyster (*Crassostrea gigas*) larvae in Japan. Clinical signs included accumulation of bacteria around the shell margin, loss of cilia and velum. In experimental challenge 100% mortality occurred at a concentration of 10^5 CFU/ml with LD_{50} values of 5.2×10^4 CFU/ml. Not all strains of *V. splendidus* II were virulent. Virulent and non-virulent strains were detected in larvae, whereas strains from the breeder tank and seawater were either less virulent, or non-virulent (Sugumar *et al.*, 1998). *Vibrio harveyi* was also isolated from these samples, but did not cause mortality when tested at 10^7 CFU/ml in the challenge experiments. The experiments were conducted at 25°C, with the observation that this temperature may have favoured the multiplication of the bacteria. Temperature was also a factor in the investigation of 'summer mortality syndrome' using challenge experiments with juvenile *C. gigas* cultured in France. Mortality caused by *V. splendidus* increased with temperature when oysters were injected into the abductor muscle of healthy juveniles at a concentration of 10^4, 10^6 and 10^7 CFU/oyster (Lacoste *et al.*, 2001a). Mortality was <10% at temperatures of 7°C. However, mortality was associated with oysters showing neuroendocrine system dysfunction through a lower than normal concentration of noradrenaline following stress. Juveniles that were not 'pre-stressed' did not succumb to infection. Oysters that had low noradrenaline levels (<5000 pg/ml haemolymph) after being stressed by shaking for 1 min, succumbed to infection. Stressed oysters with high levels of noradrenaline (>5000 pg/ml haemolymph) following stress did not succumb to infection. Two-year old oysters did not show significant mortality. *Vibrio splendidus* was identified by API 20E and by 16S rRNA sequencing (Lacoste *et al.*, 2001b). *Vibrio splendidus* (strain 0529) caused mortality in Greenshell™ mussel larvae cultured in New Zealand (Kesarcodi-Watson *et al.*, 2009). Larvae showed irregular swimming, detached cilia and aggregation of bacteria around the velum and deterioration of tissue. Infection occurred in the digestive system.

A mortality of 73% occurred in carpet shell clam (*Venerupis decussata*) cultured in Spain. *Vibrio alginolyticus* and *V. splendidus* biovar II were cultured (Gómez-León *et al.*, 2005). Pathogenicity was proven in experimental challenge using 6 mm carpet shell clam inoculated by immersion challenge with 1×10^6 CFU/ml and by intravalvar injection using 20 µl of 5×10^6 CFU/ml. *Vibrio alginolyticus* (strain TA15) was slightly more virulent than *V. splendidus* II (strain TA2) with 60 and 48% mortality, respectively.

A *V. splendidus*-like strain caused mortalities within 48 h in great scallop (*Pecten maximus*) larvae in bath challenge experiments (Sandlund *et al.*, 2006).

In Spain, in diseased captive seahorses (*Hippocampus guttulatus* and *H. hippocampus*), *V. splendidus* and *V. alginolyticus* were the predominant flora on mucosal surfaces (Balcázar *et al.*, 2010a). Seahorses were lethargic and had white spots on the skin and some animals had necrotic tail lesions.

In corkwing wrasse (*Symphodus melops*), cleaner fish used to reduce the infestation of ectoparasitic sea lice on Atlantic salmon farmed in Norway, *V. splendidus* caused mortality that was confirmed by experimental challenge (Jensen *et al.*, 2003). Fish displayed loss of appetite, reduced swimming and skin ulcerations and the organism was isolated from the kidneys.

In a survey of pathogens in gilthead seabream (*Sparus aurata*) cultured in south-western Spain, *V. splendidus* (strains AO28, 25/900, DC10-R-3) was one of a number of pathogens proven to be virulent with an LD_{50} of 1.7×10^4 to 1.2×10^5 CFU/g body weight in experimental challenge (Balebona *et al.*, 1998b).

Vibrio splendidus has been isolated from sea cucumber (*Apostichopus japonicus*) from aquaculture farms in China and together with *Shewanella* species and *Pseudoalteromonas tetraodonis* was shown to be pathogenic in experimental challenge with a LD_{50} of 1.74×10^7 CFU/g body weight (Li *et al.*, 2010a). Clinical signs include high mortality, swollen mouth, skin ulceration and viscera ejection. Bacteria were isolated from skin ulcerations, coelomic fluid and viscera.

Vibrio splendidus was identified along with a number of other *Vibrio* species as being part of the haemolymph of healthy spider crab (*Maja brachydactyla*) caught off the coast of Spain and the Canary Islands (Gomez-Gil *et al.*, 2010). The numbers of bacterial flora in healthy haemolymph is usually in the order of 10^3–10^4 CFU/ml (Gomez-Gil *et al.*, 1998).

GROWTH CHARACTERISTICS AND PHENOTYPIC IDENTIFICATION. *Vibrio splendidus* is a facultatively anaerobic, motile, oxidase-positive, Gram-negative rod. *V. splendidus* has been differentiated into two biovars based on phenotypic differences such as luminescence and colony appearance on TCBS, where biovar I is luminescent and produces yellow colonies on TCBS, and biovar II is non-luminescent and produces green colonies on TCBS. However, these characteristics may be unreliable. Strain CRL 40, *V. splendidus* biovar I, produces green colonies on TCBS that are 1 mm at 24 h and 1.5 mm at 48 h. It is weakly luminescent on MA 2216. *Vibrio splendidus* ATCC 33789, biovar II shows no growth at 24 h and poor growth and pinpoint to 1 mm green colonies on TCBS at 48 h. Growth is only seen in the 'pool'. On BA strain ATCC 33789 produces no growth at 24 h, but at 48 h produces 1 mm grey colonies and at 72 h colonies are 2.7 mm with a small zone of beta-haemolysis and a greening of the agar in areas of heavy growth. Strain CRL 40 does not grow on BA after 72 h. On MSA-B ATCC 33789 produces 2 mm grey colonies with a slight pink tinge with beta-haemolysis underneath the colony and at 48 h colonies are 4.5 mm, cream and mucoid. On MA 2216, strain ATCC 33789 produces 1 mm, cream, non-luminescent colonies increasing in size to 1.5 mm at 48 h. Colonies are non-swarming and non-mucoid on MA 2216. Swarming and luminescence is not seen on BA, MSA-B, MSSA or MA 2216 for ATCC 33789. Strain CRL 40 produces 1.5–2 mm colonies on MA 2216 at 48 h, which have a dirty-grey colour with a yellow tinge. Both these strains are negative for fermentation of sucrose in conventional tube media and in the API 20E. Most strains of *V. splendidus* I show a positive ADH result using Thornley's method, but negative using Moeller's method (see notes under ADH in 'Interpretation of Biochemical Identification Tests and Sets'). *Vibrio splendidus* I strain CRL 40 is negative for ADH in both the conventional method Moeller's method (see Table 3.22) and the API 20E (Table 3.24–3.26) when examined up to 72 and 48 h, respectively. *Vibrio splendidus* II, strain ATCC 33789, shows a delayed reaction for ADH in the API 20E and may be negative at 48 h but positive at 72 h. A recording of a negative result for the ADH will give the same API 20E number as *V. campbellii*. Strain CRL 40 is positive for the ONPG in the API 20E but negative in the conventional tube method. The oxidase shows a strong reaction. The tube motility is negative and cells in a wet preparation show a slow motility or a tumbling action. Growth does not occur on MCA or the DNase plate (this manual).

Both biotypes hydrolyse chitin, amylase and lipase (Furniss *et al.*, 1978) and biotype I, strain ATCC 33125 hydrolyses starch, lecithin and Tween 20 and is negative for casein, cellulose, Tween 40 and Tween 80 (Farto *et al.*, 1999). Strain LMG 4042^{T} grows at 4–35°C but not at 40°C and is negative for nitrate in the API 20E (Faury *et al.*, 2004). Strains CRL 40 and ATCC 33789 are positive for nitrate in both the conventional tube test and the API 20E. *Vibrio scophthalmi* and *V. splendidus* biovar I may be difficult to differentiate based on phenotypic tests and this is compounded due to the variation of phenotypic test results between strains in each species (Farto *et al.*, 1999). Indole, ONPG and fermentation of mannitol appear to be the best differentiating phenotypic tests. *Vibrio scophthalmi* is negative for all three, whereas *V. splendidus* biovar I is positive for all three. These and other reactions are presented in Tables 3.22, 3.24–26 and 3.38.

MOLECULAR IDENTIFICATION. A quantitative PCR for *V. splendidus* was designed using primers VSF (5′-CAC AGC GAT AAC CGC TAC AA-3′) and VSR (5′-GTA CGT TGG CGA AGA CAT GA-3′) (Prol *et al.*, 2009). DNA is extracted from a bacterial colony using the NucleoSpin Tissue kit and 10 μl is used in the PCR reaction. The Power SYBR® Green master mix (Applied Biosystems) is used and contains AmpliTaq Gold® DNA polymerase, SYBR Green fluorescent dye, and reference dye ROX that is required by some real-time thermocycling machines. Primers are used at a final concentration of 4.3 μmol/l. Thermocycling is carried out on a 7500 Fast Real-Time machine (Applied Biosystems) using an initial step of 95°C for 10 min followed by 40 cycles of 95°C for 30 s, 62°C for 60 s and 72°C for 30 s (R^2= 0.97–0.99).

The primers amplify product from *V. splendidus* NCIMB 1 (=ATCC 33125=LMG 4042) but not from *Aeromonas hydrophila*, *V. vulnificus* or *V. (Listonella) anguillarum*.

Degenerate primers designed to amplify the *gyrB* gene were used to detect *V. splendidus* in sea cucumber (Wang *et al.*, 2010a). Some specificity testing was conducted, but not all *Vibrio* species within the Splendidus clade were used for specificity testing. The primers amplify product of approximately 254 bp from *V. mediterranei* CECT 615 and *V. (Listonella) pelagius* ATCC 25916 (Buller, unpublished data). A product of 252 bp is produced from *V. splendidus* ATCC 33789.

The master mix consists of 2 μl of 10× PCR buffer containing 1.5 mM $MgCl_2$, 2.0 μM primer VspF-*gyrB* (5′-ACC AAC AAA ACA CCR ATY ATY C-3′), 2.0 μM primer VspR-*gyrB* (5′-ATC ACC YGA TGT TGM WGT CTT C-3′), 100 μM dNTP mix, 5 U *Taq* DNA polymerase, 20 ng DNA in a 20 μl volume. Thermocycling is carried out using an initial denaturation step at 95°C for 5 min followed by 31 cycles of denaturation at 94°C for 30 s, annealing at 60°C for 30 s, extension at 72°C for 30 s and a final step using an extension at 72°C for 6 min. Amplified product is 252 bp. The sensitivity is 0.34 pg/μl DNA, 10^3 CFU/ml bacterial cells. The primers were reported by Wang *et al.*, (2010a) to be specific against *V. aestuarianus*, *V. alginolyticus*, *V. campbellii*, *V. cyclitrophicus*, *V. harveyi*, *V. parahaemolyticus*, *V. tasmaniensis*, *Listonella anguillarum*, *Aeromonas hydrophila*, *A. sobria*, *Bacillus subtilis*, *E. coli*, *Photobacterium phosphoreum* and *Pseudoalteromonas* sp., however, as previously mentioned, product of the same molecular weight is amplified from *V. mediterranei* CECT 615 and *Vibrio (Listonella) pelagius* ATCC 25916.

Primers (VSPN-F 5′-GAT TTA GTT AAA GCC AGA GC-3′ and VSPN-R 5′-CCT GAT AAC TGT TTG CCG-3′) targeting the 16S-23S rDNA intergenic spacer region were designed to amplify and detect *V. splendidus* and produce an amplicon of either 240 or 294 bp from *V. splendidus* ATCC 33125 biotype I (Lee *et al.*, 2002). These primers were used in addition to biochemical testing and sequencing of the 16S rRNA gene to confirm identification of *V. splendidus* isolated from sea cucumber (Li *et al.*, 2010a). The primers do not amplify DNA from *V. splendidus* ATCC 33789 biotype II and produce amplicons of various sizes from *V. aestuarianus* NCIMB 2236 and *Escherichia coli* and a number of unidentified *Vibrio* species (Buller, unpublished data).

Vibrio stylophorae

TAXONOMY AND SIGNIFICANCE. *Vibrio stylophorae* (KTW-12^T=BCRC 80105^T=LMG 25357^T) was isolated from the reef-building coral, *Stylophora* species, off the coast of Taiwan (Sheu *et al.*, 2011b). Based on 16S rRNA sequence the isolate is closely related to *V. porteresiae* with 94.8% similarity and together forms a clade with *V. cholerae* (93.9%) and *V. mimicus* (93.6%). Low levels (74.6–85.0%) of similarity are seen for other *Vibrio* species using multilocus sequence analysis of *rpoA*, *recA* and *pyrH* genes. For the *rpoA* gene the closest similarity is to *A. fischeri* (85.0%), *V. mimicus* (83.9%), *V. cholerae* (83.8%) and *V. porteresiae* (82.0%). The *recA* gene shows similarity to *V. diazotrophicus* (85.0%), *V. cholerae* (84.1%), *V. mimicus* (84.0%) and *V. porteresiae* (82.5%). The *pyrH* gene sequence has similarity to *V. porteresiae* (75.9%) and 77.7–74.6% to all other *Vibrio* species. *Vibrio stylophorae* does not cluster with any known *Vibrio* species when all four genes (16S rRNA, *pyrH*, *recA* and *rpoA*) are analysed using a neighbour-joining phylogenetic tree based on concatenated sequences of the genes.

GROWTH CHARACTERISTICS AND PHENOTYPIC IDENTIFICATION. *Vibrio stylophorae* is a facultative anaerobe, non-motile, oxidase-positive, slightly curved rod. On MA 2216, colonies are 0.3–0.5 mm in diameter and 1.0–3.1 mm in length, semi-transparent, round, smooth and convex with entire margins at 48 h at 30°C. Colonies are 0.5–0.8 mm at 72 h. Growth does not occur on TCBS. Optimal growth occurs at 30°C with a range of 15–35°C. Optimal NaCl is 2% with a range of 0.5–6%. Chitin and DNA are hydro lysed, whereas hydrolysis is negative for lipase, starch, casein, alginate, lecithin CM-cellulose, Tween 20, 40, 60 and 80. Hydrolysis of gelatin was reported as positive in the API 20E but negative in the API 20NE (Sheu *et al.*, 2011b). These and other reactions are presented in Tables 3.23–29 and 3.38.

Vibrio succinogenes

See *Wolinella succinogenes*.

Vibrio superstes

TAXONOMY AND SIGNIFICANCE. *Vibrio superstes* (LMG 21323^T) was isolated from the gut of healthy abalone *Haliotis laevigata* and *H. rubra* collected off the Victorian coast at Clifton Springs, Australia. *Vibrio superstes* in considered part of the normal flora of the gut of Australian abalone occurring at between 0–20% of the gut microflora (Hayashi *et al.*, 2003).

In experimental challenge using rainbow trout (*Oncorhynchus mykiss*) and *Artemia* nauplii inoculated with 10^6 CFU/fish or 10^6CFU/ml, respectively, *V. superstes* (LMG 21323^T and LMG 21322) was rated as non-virulent. Clinical disease was not seen, and at the end of the experiment the organism was not recovered from tissue (Austin *et al.*, 2005).

The strain is closely related to *V. halioticoli* at 98% based on 16S rDNA sequence similarity and also based on phenotypic tests (Hayashi *et al.*, 2003). MLSA identifies *V. superstes* as belonging to the Halioticoli clade, which also contains *V. gallicus*, *V. ezurae*, *V. halioticoli* and *V. neonatus* (Sawabe *et al.*, 2007b). Members of this clade have >22% DDH, 39–42 mol% GC, a MLSA concatenated similarity of 88.0–97.7% and AAI of 94.7–99.5%.

GROWTH CHARACTERISTICS AND PHENOTYPIC IDENTIFICATION. *Vibrio superstes* is a facultative anaerobe, non-motile, alginolytic, non-pigmented, non-swarming. It is non-luminescent, does not produce chitinase and has a salt requirement for growth, but does not grow at 1% or 6% NaCl. Growth occurs at 15–30°C, but not at 37°C. On MA 2216 agar, colonies are beige, circular, smooth, convex with entire edges. Traditional biochemical tests were reported by Hayashi *et al.* (2003). *Vibrio superstes* is negative for amylase, chitinase, agarose and fermentation of L-rhamnose and is positive for alginase (Hayashi *et al.*, 2003). These and other reactions are presented in Table 3.23.

Vibrio tapetis

TAXONOMY AND SIGNIFICANCE. *Vibrio tapetis* (P1^T =B1090^T=CECT 4600^T) is the causative agent of brown ring disease in Manila clam (*Venerupis* [*Ruditapes*] *philippinarum*) and was first reported in France in 1987 (Borrego *et al.*, 1996). Infection causes mass mortality in clams. The disease is characterized by a brown conchiolin deposit of several layers inside the shell between the pallial line and the edge of the shell. A shell repair process occurs and this can be seen as white calcified areas that cover the brown deposit (Paillard, 2004). The disease is seasonal, occurring at cooler temperatures; water temperature and salinity are related to pathogenicity of the disease. The disease occurs at temperatures of 14°C, but at temperatures of 21°C recovery takes place. High mortalities are related to salinity of 2.0% (20 parts per 1000) but are reduced at 4.0% (40 parts per 1000) (Reid *et al.*, 2003b; Paillard *et al.*, 2004). The bacterium has been reported from cross-cut carpet shell clam (*Venerupis* [*Ruditapes*] *decussata*), venus clam (*Polititapes* [*Venerupis aurea*] *aureus*), common cockle (*Cerastoderma edule*), and fish species corkwing wrasse (*Symphodus melops*), wedge sole (*Dicologlossa cuneata*) and Atlantic halibut (*Hippoglossus hippoglossus*), in France, Italy, Ireland, Korea, Portugal, Japan, Norway, Spain and the UK (Castro *et al.*, 1992; Paillard *et al.*, 1994, 2004; Allam *et al.*, 2000; Jensen *et al.*, 2003; Reid *et al.*, 2003a; Park *et al.*, 2006; Rodríguez *et al.*, 2006; Drummond *et al.*, 2007; López *et al.*, 2011). Intraspecies biochemical and genetic variability were noted and recently multilocus sequence analysis resulted in two clusters based on host origin and geographical source, with the result that isolates in cluster 2 have been named *V. tapetis* ssp. *britannicus* (Balboa and Romalde, 2013).

Clams are the most susceptible species, whereas cockles seem to be carriers of the organism. Experimental challenge did not result in disease when tested using cockle (*C. edule*), hard clam (*Mercenaria mercenaria*) and oysters (*Crassostrea gigas*, *C. virginica*, *Ostrea edulis*) (Paillard, 2004). The native European clam (*V. decussata*) rapidly eliminates the bacterium from the soft tissues and haemolymph and is resistant to brown ring disease (Allam *et al.*, 2002).

Infection in fry and juvenile wedge sole results in ulcers on the body and tail. The bacterium is found in the ulcers but not in internal organs. Mortality rates are 5–10% over 1 week. Identification was confirmed by PCR, MLSA phylogenetic analysis and phenotypic tests. Experimental challenge using intraperitoneal injection of 10^6 CFU/fish with one of the four strains of *V. tapetis* isolated from infected fish did not reproduce the disease (López *et al.*, 2011a). In corkwing wrasse (*Symphodus melops*), cleaner fish used to reduce the infestation of ectoparasitic sea lice on Atlantic salmon farmed in Norway, *V. tapetis* caused a high daily mortality that was confirmed by experimental challenge (Jensen *et al.*, 2003). Fish displayed loss of appetite, reduced swimming and skin ulcerations and the organism was isolated from the kidneys. *Vibrio tapetis* was co-isolated with an atypical strain of *Aeromonas salmonicida* from head kidney of Atlantic halibut. Re-infection with *V. tapetis* did not occur in experimental challenge and the significance of the isolation could not be resolved (Reid *et al.*, 2003a).

Based on the 16S rRNA sequence *V. tapetis* has similarity to *V. shilonii* (98.4%), *V. mediterranei* (98.3%), *V. penaeicida* LMG 19663T (97.5%), *V. splendidus* (97.9%) and *V. tasmaniensis* LMG 21574T (98.1%). Phylogenetic analysis using concatenated sequences of the 16S rRNA gene and the *atpA*, *pyrH* and *rpoA* genes can separate strains from other *Vibrio* species (López *et al.*, 2011a).

GROWTH CHARACTERISTICS AND PHENOTYPIC IDENTIFICATION. *Vibrio tapetis* is a motile, facultatively anaerobic, Gram-negative coccobacillus (1.0–1.5×0.5 μm) that is positive for catalase and oxidase. Growth is affected by incubation temperature and type of media. The outcome of an evaluation of different media and incubation temperatures recommends the use of MA 2216 (Difco) or agar seawater (ASWT) and incubation at 15°C as growth reaches log phase at 28 h at this temperature compared to 36 h at 25°C. Mannitol marine agar (MMA) assists in the primary isolation of *V. tapetis* from bivalve samples, as the mannitol appears to improve the growth of the bacterium from these samples, although overall growth is slightly less than that on MA 2216 or ASWT. Although mannitol may improve the growth, the addition of mannitol to ASWT resulted in slightly less growth than that obtained from ASWT. Growth was less on TSA containing added NaCl, and is poor on TCBS (Balboa *et al.*, 2012). On MA 2216 and TSA containing 2% NaCl, colonies are non-swarming, non-pigmented, circular and translucent after 48 h at 20°C. On TCBS colonies are green in colour. Growth does not occur in 0% NaCl or >5% NaCl, but occurs at 1–3% NaCl. Growth occurs at 4–22°C but not above 27°C. Gelatin, casein, starch and Tween 80 are hydrolysed but not urea or alginate (Borrego *et al.*, 1996). On MSA-B and MA 2216 colonies of strain CIP 104856 are 0.5 to 1 mm after 48 h at 25°C. Colonies of different sizes are seen on both media. Growth is not seen at 24 h on either medium. On MSA-B colonies are non-haemolytic, a light grey colour and become a cream colour with age (this manual). Strains from clams and wedge sole are negative for fermentation of sucrose and mannitol, whereas strains from corkwing wrasse and halibut are positive when tested using the API 20E (Jensen *et al.*, 2003; Reid *et al.*, 2003a; Lopez *et al.*, 2011a). These and other reactions are presented in Tables 3.22, 3.24–26 and 3.38. Results for isolates obtained from wedge sole were incubated at 20°C and the inoculum was prepared with sterile seawater (López *et al.*, 2011a).

MOLECULAR IDENTIFICATION. An evaluation was conducted of the three PCRs that have been developed to date for the detection and identification of the 16S rRNA gene of *V. tapetis* and tested against strains from molluscs and fish (Balboa *et al.*, 2011). The PCR primer pairs were VtF-VtR (Paillard *et al.*, 2006), VtKf-VtKr (Park *et al.*, 2006) and Jvt1-Jvt2 (Romalde *et al.*, 2007 as referenced in Balboa *et al.*, 2011). The primer pairs Jvt1-Jvt2 were the most sensitive (2–20 bacterial cells) and specific (no cross-reactions). The VtF-VtR primers produced cross-reaction with *V. ezurae*, *V. mytili*, *V. nigripulchritudo*, *V. pomeroyi*, *V. proteolyticus* and *V. splendidus*, whereas primers VtKf-VtKr produced cross-reaction with *V. mytili*, *V. pectenicida* and *V. pomeroyi* (Balboa *et al.*, 2011).

DNA is extracted using cultured bacterial cells suspended in 1 ml of sterile ultrapure water and centrifuged for 1 min at 12,000 rpm. The pellet is resuspended in 200 μl Instagene Matrix (BioRad) and incubated at 56°C for 30 min. The mixture is vortexed and then boiled for 8 min. After another vortex the mixture is centrifuged for 3 min at 12,000 rpm and the supernatant used in the PCR reaction. The concentration of the DNA was measured spectrophotometrically at 580 nm and adjusted to 1000 ng/μl and 100 ng used in the PCR reaction (Balboa *et al.*, 2011).

Primers Jvt1 (5′-TAC GTT AAT AGC GTG CAT CC-3′) and Jvt2 (5′-TCA CCA TCG CTG GTT GGC TG-3′) are used in the PCR reaction mixture prepared with Ready-To-Go PCR beads (Amersham Pharmacia Biotech, UK) with 1.5 mM $MgCl_2$. The thermocycling protocol consists of denaturation at 94°C for 3 min followed by 35 cycles at 94°C for 1 min, 64°C for 1 min and

72°C for 1.5 min and a final elongation step at 72°C for 10 min. An amplicon of 816 bp is produced.

Vibrio tapetis ssp. britannicus

TAXONOMY AND SIGNIFICANCE. Multilocus sequence analysis revealed two clusters: cluster 1 containing the majority of strains, and cluster 2 containing three strains from Ireland and the UK and isolated from halibut (*Hippoglossus hippoglossus*) and Manila clam (*Venerupis philippinarum*) seed. Isolates in cluster 2 are identified as *Vibrio tapetis* ssp. *britannicus* (Balboa and Romalde, 2013). Phenotypically *V. tapetis* ssp. *britannicus* is different from *V. tapetis* in terms of fermentation of mannitol and arabinose, and utilization of citrate. These and other reactions are presented in Tables 3.22 and 3.24–26. The results are read at 24, 48 and 72 h with 72 h being the final reading. Incubation temperature is 25°C and sea salts or 1% NaCl are used to prepare the inoculum (Balboa and Romalde, 2013).

Vibrio tasmaniensis

TAXONOMY AND SIGNIFICANCE. *Vibrio tasmaniensis* (VIB 836^T=LMG 20012^T), FAFLP cluster A45, was isolated from Atlantic salmon (*Salmo salar*) cultured in Tasmania, Australia (Thompson *et al.*, 2001b, 2003e). A strain (CC056) of *V. tasmaniensis* was isolated from necrotic tissue of *Eunicella verrucosa*, a gorgonian octocoral, growing off Lundy Island, Devon, UK (Vattakaven *et al.*, 2006). Identification was confirmed by F. Thompson using FAFLP.

In experimental challenge using rainbow trout (*Oncorhynchus mykiss*) and *Artemia* nauplii inoculated with 10^6 CFU/fish or 10^6 CFU/ml, respectively, *V. tasmaniensis* (LMG 20012^T and LMG 21574) was rated as non-virulent. Clinical disease was not seen, and at the end of the experiment the organism was not recovered from tissue (Austin *et al.*, 2005).

Vibrio tasmaniensis is closely related to *V. cyclitrophicus*, *V. lentus* and *V. splendidus* (98.2–98.9%) based on 16S rRNA sequencing. MLSA identified *V. tasmaniensis* as belonging to the Splendidus clade, which contains *V. splendidus*, *V. chagasii*, *V. crassostreae*, *V. cyclitrophicus*, *V. fortis*, *V. gigantis*, *V. kanaloae*, *V. lentus*, *V. pelagius* and *V. pomeroyi* (Sawabe *et al.*, 2007b). Members of this clade have >30% DDH and 39–47 GC mol%, 90.6–99.8% similarity based on concatenated MLSA sequence data and AAI of 96.5–99.8%.

GROWTH CHARACTERISTICS AND PHENOTYPIC IDENTIFICATION. *Vibrio tasmaniensis* is a facultatively anaerobic, motile, oxidase-positive, Gram-negative, slightly curved rod (2–3×1.0 μm). On TSA containing 2% NaCl colonies are 4 mm, non-swarming, beige, translucent, smooth, round convex after 48 h at 28°C and 4–5 mm green colonies on TCBS (Thompson *et al.*, 2003e). Growth occurs at 4–35°C. Growth does not occur in the absence of NaCl or at 8% NaCl. Up to 2% NaCl must be added to culture media and to identification tests. These and other reactions are presented in the Environmental *Vibrio* table (Table 3.23) and API 20E (Tables 3.24–26). 2% NaCl is used for the inoculum and sets are incubated for 48 h at 25–28°C.

Vibrio toranzoniae

TAXONOMY AND SIGNIFICANCE. *Vibrio toranzoniae* (Vb 10.8^T=CECT 7225^T=CAIM 1869^T) was isolated from cultured Manila clam (*Venerupis philippinarum*) in Galicia, Spain. Three other strains were isolated, two from cross-cut carpet shell clam (*V. decussata*) (Lasa *et al.*, 2013). The isolates were identified originally as cluster 26 from AFLP analysis investigating *Vibrio* species diversity in cultured clams. The pathogenicity is unknown. The 16S rRNA gene sequence is 99.6% similar to *V. kanaloae* and 96.4% to *V. gallaecicus*, and has greater than 98.5% similarity to species within the Splendidus clade: *V. kanaloae*, *V. artabrorum*, *V. gigantis*, *V. celticus*, *V. pomeroyi*, *V. crassostreae* and *V. splendidus*. Sequence analysis of housekeeping genes *atpA*, *recA*, *pyrH*, *rpoA* and *rpoD* indicates similarity to different *Vibrio* species; for example, *atpA* indicates the closest neighbour is *V. splendidus* and *pyrH* indicates similarity to *V. kanaloae*, however concatenated sequence information confirms the strains are a monophyletic group.

GROWTH CHARACTERISTICS AND PHENOTYPIC IDENTIFICATION. *Vibrio toranzoniae* is a facultatively anaerobic, motile, oxidase-positive, Gram-negative rod (0.7–0.8×1.5 μm). On MA 2216, colonies are round, beige, smooth, non-swarming and non-luminescent after 24 h at 25°C. On TCBS colonies are yellow. Catalase is positive. Glucose is fermented without gas production. The type strain (Vb 10.8) is sensitive to vibriostatic agent 0/129 at 150 µg per disc, but the other three strains are resistant. Indole is positive for the type strain Vb 10.8 and strain Cmf 13.9, but the other two strains are negative. Growth occurs at 4°C to 25°C, however an upper limit was not established. Growth occurs at 3 to 10% NaCl. Identification media is supplemented with 1% NaCl. Gelatin and Tween 80 are hydrolysed but negative for starch (Lasa *et al.*, 2013). These and other reactions are presented in Tables 3.23, 3.27–30 and 3.38. Conventional media in the biochemical set (as listed in Table 3.23) are supplemented with 1% NaCl and for the commercial sets 0.85% NaCl is used for the inoculum. Sets are incubated for 48 h at 25°C.

Vibrio trachuri

See *Vibrio harveyi*.

Originally isolated from Japanese horse mackerel (*Trachurus japonicus*) and identified as a new species of *Vibrio* (Iwamoto *et al.*, 1995), it was later identified as a junior synonym of *Vibrio harveyi* (Thompson *et al.*, 2002b).

Vibrio tubiashii

TAXONOMY AND SIGNIFICANCE. *Vibrio tubiashii* (ATCC 19109^{T}) was isolated from diseased larvae of clams and oysters in Japan and is the cause of bacillary necrosis and vibriosis in larval and juvenile hard clam (*Mercenaria mercenaria*) and eastern oyster (*Crassostrea virginica*) (Tubiash *et al.*, 1965; Hada *et al.*, 1984). It was also reported from the haemolymph of healthy spider crab (*Maja brachydactyla*) caught off the coast of Spain and the Canary Islands (Gomez-Gil *et al.*, 2010). The aquaculture industry in North America has seen a re-emergence of vibriosis caused by *V. tubiashii*, which has been reported from Pacific oysters (*C. gigas*), Kumamoto oysters (*C. sikamea*) and geoduck clam (*Panope abrupta*) and episodes appear to be related to mixing of unusually warm surface seawater with cooler seawater containing a high concentration (1.6×10^{5} CFU/ml) of *Vibrio* species (Elston *et al.*, 2008). Pathogenicity for these species was confirmed by experimental challenge. Manila clams and mussel (*Mytilus galloprovincialis*) larvae appear to be more resistant to the disease.

In experimental challenge, *V. tubiashii* (LMG 10936^{T}) caused 100% mortality within 2 days when inoculated into rainbow trout (*Oncorhynchus mykiss*) at a dose of 10^{6} cells/fish, and 80% mortality when inoculated with a dose of 10^{4} CFU/fish. The LD_{50} was 10^{2} (Austin *et al.*, 2005). *Artemia* nauplii exposed to 10^{6} CFU/ml resulted in 85% mortality. Injection with extracellular product (18 µg/ml) resulted in 100% mortality in rainbow trout and 40% mortality in *Artemia*. In rainbow trout, clinical signs included liquefaction of internal organs, discharge from and protrusion of the anus and necrosis of the muscle. The strain was positive for caseinase, gelatinase, phospholipase (egg yolk), lipase (Tween 80) and negative for elastase. Beta-haemolysis was seen for sheep, horse and trout red blood cells.

Vibrio tubiashii produces metalloprotease coded by the *vtpA* gene, which has homology with the metalloprotease genes of other *Vibrio* species that include *V. alginolyticus*, *V. cholerae*, *V. fluvialis*, *V. mimicus*, *V. parahaemolyticus*, *V. splendidus* and *V. vulnificus*. A cytolysin/haemolysin is produced and the two genes (*vthA* and *vthB*) have homology to *vvhA* and *vvhB* genes from *V. cholerae*. The secreted metalloprotease appears to be the major virulence factor for *V. tubiashii* and inhibits the extracellular haemolysin VthA (Hasegawa and Häse, 2009).

Vibrio tubiashii is closely related to *V. brasiliensis*, *V. neptunius* and *V. xuii* (98–98.8%) (Thompson *et al.*, 2003b) and to *V. coralliilyticus* (97.2%) (Ben-Haim *et al.*, 2003a) based on 16S rRNA sequencing. MLSA identified *V. tubiashii* as belonging to the Orientalis clade, which also contains *V. hepatarius*, *V. brasiliensis* and *V. orientalis* (Sawabe *et al.*, 2007b). Members of this clade have >24% DDH and 43–46 GC mol%, 91.2–94.2% similarity based on concatenated MLSA sequence data and AAI of 97–97.9%. *Vibrio coralliilyticus* and *V. neptunius* belong to the Coralliilyticus clade and *V. xuii* belongs to the Nereis clade. The housekeeping genes *recA* and *pyrH* have better resolving power than *rpoA* for discriminating within the *V. tubiashii* group (Thompson *et al.*, 2005a).

GROWTH CHARACTERISTICS AND PHENOTYPIC IDENTIFICATION. *Vibrio tubiashii* is a facultatively anaerobic, motile, oxidase-positive, Gram-negative straight or slightly curved rod (0.5×1.5 µm). On MA 2216, colonies are 1 mm, smooth, circular and off-white and may be mucoid at 48 h. Luminescence is negative (Hada *et al.*, 1984). On BA colonies are pinpoint at 24 h, grey, opaque becoming 2 mm lighter grey and show greening in the pool at 48 h. On MSA-B colonies are 1 mm grey, beta-haemolytic at 24 h and 3 mm and light grey at 48 h. Growth occurs as 2 mm yellow colonies at 24 h on TCBS. The optimum concentration is 1–3% NaCl but growth does not occur at 0 or 8% NaCl. Growth occurs at 25°C but not at 37 or 42°C. The tube indole reaction is strong. Chitin, DNA (strong reaction), gelatin, lecithin, starch, Tween 80, tyrosine and xanthine are hydrolysed, but elastin is negative. The type strain ATCC 19109 was reported as positive using Moeller's arginine, with 83% of strains positive by this method (Hada *et al.*, 1984). Strain NCIMB 1340 does not show as positive using Moeller's medium until the fifth day of incubation, whereas ADH is positive in the API 20E at 24 h (this manual). The glucose in Moeller's medium appears to inhibit the reaction due to catabolite repression of the inducible ADH system (Macián *et al.*, 1996). See notes under ADH in 'Interpretation of Biochemical Identification Tests and Sets'. These and other reactions are presented in Tables 3.22, 3.24–26 and 3.38. Two per cent NaCl is used for the inoculum and sets are incubated for 48 h at 25–28°C.

MOLECULAR IDENTIFICATION. The following primers have been published: *Vibrio tubiashii* IGS[IA] (tubA)-specific primers, VTUB-F TGG GTC TTT CAG GCC CG and VTUB-R CGA CGA ATG ACC GTT GTC. An annealing temperature of 55°C and 35 cycles are used, and an amplicon of 394 bp is produced (Lee *et al.*, 2002).

A specific SYBR Green PCR based on the *vtpA* encoding a metalloprotease has been designed for *V. tubiashii* (Gharaibeh *et al.*, 2009) and a conventional PCR was designed for the protease gene and the haemolysin gene (Elston *et al.*, 2008).

PCR type	SYBR Green (Gharaibeh *et al.*, 2009)		
DNA extraction	200 µl of boiled sample applied to Qiagen DNeasy Blood and Tissue DNA extraction kit		
PCR master mix			
Primer *vtpA*-F	5′-CAA ATG CTT TGG CTG ATT GCT-3′		
Primer *vtpA*-R	5′-CCA TCT CTG CGG CTG TAA CTG-3′		
Reagent	**Stock**	**Final concentration**	**Volume ×1**
Platinum SYBR Green qPCR Super Mix-UDG (Invitrogen)		×1	12.5 µl
Primer *vtpA*-F	9 µM	0.9 µM	2.5 µl
Primer *vtpA*-R	9 µM	0.9 µM	2.5 µl
ROX reference dye (Invitrogen)	500 nM		1 µl
Water			1.5 µl
DNA			5 µl
Total volume			**25 µl**
Thermocycling	ABI PRISM 7500 Sequence Detection System (Applied Biosystems)		
Step 1	**Step 2**		
50°C, 2 min	95°C, 15 s		
95°C, 2 min	60°C, 31 s		
×1	×40		
Product	63 bp		
Ct (positive)	12–32 (10^7 cells–10 cells)		
Ct (negative)	>33		
R^2 (linear regression coefficient)	≥0.995		
Melt temperature	Not stated		
Sensitivity	10 cells. 0.1 ng of sample gave a Ct value of 23, which corresponded to 8154 cells		
Specific against	*Vibrio alginolyticus*, *V. cholerae*, *V. fluvialis*, *V. mimicus*, *V. parahaemolyticus*, *V. splendidus* and *Escherichia coli*		

A high concentration of DNA (650 ng) oversaturated the qPCR and erroneous results were obtained.

Primers were designed to detect the protease gene and the haemolysin gene from *V. tubiashii*, however no details were given for the master mix components, the thermocycling conditions or the sensitivity and specificity.

PCR type	Conventional PCR (Elston *et al.*, 2008)
DNA extraction	DNeasy Blood and Tissue DNA extraction kit (Qiagen)
Primers	
Primer protease-F	5′-ATG AAA CAA CGT CAA ATG CTT TG-3′
Primer protease-R	5′-GAA CGG AGC ATC AGA AGC CCC AAG AGA-3′
Primer haemolysin-F	5′-TCA GCT ATA AGA ATG AAA AAA CAC-3′
Primer haemolysin-R	5′-GTA CGT GGA ATT GAC CAT CCT GTC-3′

The protease primers correspond to the putative translational start site of 1–417 and the primers for the haemolysin gene correspond to the putative translation start site 575–971.

Vibrio variabilis

TAXONOMY AND SIGNIFICANCE. *Vibrio variabilis* (R-40492^T=LMG 25438^T=CAIM 1454^T) was isolated from the mucus of apparently healthy coral, a zoanthid called *Palythoa caribaeorum*, from a reef off the coast of Brazil (Chimetto *et al.*, 2011b). *Vibrio maritimus* was isolated at the same time. The two species have 99% homology based on 16S rRNA sequencing and are 55–57% similar according to DNA–DNA hybridization. Based on 16S rRNA gene sequence *V. variabilis* has similarity to *V. neptunius* (98.5%), *V. coralliilyticus* (98.2%), *V. nigripulchritudo* (98.2%), *V. sinaloensis* (98.2%) and *V. brasiliensis* (98.1%). Multilocus sequence analysis concatenated sequences of *ftsZ*, *gyrB*, *recA*, *rpoA*, *topA*, *pyrH*, *mreB* and 16S rRNA genes indicates the two species form a separate branch within the *Vibrio* genus with less than 92% similarity.

GROWTH CHARACTERISTICS AND PHENOTYPIC IDENTIFICATION. *Vibrio variabilis* is a facultatively anaerobic, motile, oxidase-positive, Gram-negative rod (1.5–2.5×0.9 μm). On TSA containing 2% NaCl colonies are 1 mm, beige, convex, round and smooth after 24 h at 28°C. Colonies develop a black colour over time and under limited light conditions. Colonies are yellow on TCBS (Chimetto *et al.*, 2011b). Optimal temperature is 28–35°C and growth occurs at 15–40°C but not at 42°C. Optimal NaCl concentration is 1.5–5.0% with growth occurring at a range of 0.5–7.0% with no growth at 8%. These and other reactions are presented in Tables 3.20–26 and 3.38.

Vibrio viscosus

See *Moritella viscosa*.

Vibrio vulnificus

TAXONOMY AND SIGNIFICANCE. *Vibrio vulnificus* is a serious human pathogen and may cause wound infection or seafood-related deaths. It is found in aquatic environments; in water, sediment, fish, crabs, oysters, mussels, prawns and seabirds, and may be part of the microflora of intestines or gills of fish (Oliver *et al.*, 1983; Jones and Oliver, 2009). It was first named *Beneckea vulnifica* (Reichelt *et al.*, 1976).

Vibrio vulnificus was identified as a pathogen of eels in Japan in 1977 and named *V. anguillarum* type B, but this did not have valid taxonomic standing and the strains were classified as *V. vulnificus* biogroup 2 (Tison *et al.*, 1982). MLSA identifies *V. vulnificus* as belonging to the Vulnificus clade, which also contains *V. navarrensis*. Members of this clade have >30% DDH and 45–48 GC mol%, 88.6% similarity based on concatenated MLSA sequence data and AAI of 96.4% (Sawabe *et al.*, 2007b).

Vibrio vulnificus is grouped into three biotypes; biotypes 1, 2 and 3. Biotype 1 includes strains from seawater, oysters and human infection. Biotype 2 is a pathogen of eels and occurs worldwide (Tison *et al.*, 1982; Fouz *et al.*, 2006). Biotype 2 comprises four serovars based on two different serotyping systems. Serovar E is equivalent to serovar 04 and is the most virulent serovar for eels and is a cause of infection in humans. Humans are not affected by the other serovars. All known serovar E strains have been isolated from eels, however serovar E strains have been reported that are avirulent for eels (Tison *et al.*, 1982; Biosca *et al.*, 1997; Amaro *et al.*, 1999; Bisharat *et al.*, 1999). In eels clinical signs include haemorrhage on the fins, and petechiae and ulcers occasionally occur on the head. Serovar A has been reported in eels in Denmark and Spain with differing clinical signs. Danish cultured eels displayed areas of haemorrhage on the head and gills and necrotic areas in soft tissues. Serovar 2 causes occasional outbreaks in eels cultured in Spain and in these infections haemorrhage around the vent is a common sign together with haemorrhagic areas in the intestine and blood in the faeces (Høi *et al.*, 1998; Dalsgaard *et al.*, 1999). Biotypes 1 and 2 can infect cultured shrimps/prawns.

Biotype 3 has been isolated from human infection in Israel and is associated with people handling tilapia fish. It may be difficult to associate a phenotype to a biotype (Bisharat *et al.*, 1999).

Vibrio vulnificus caused disease in cultured ovate pompano (*Trachinotus ovatus*) in the Gulf of Yangjiang, China. Clinical signs included ulcers, and haemorrhagic areas on the body, the gills and on internal organs, particularly the liver and intestine (Li *et al.*, 2006).

GROWTH CHARACTERISTICS AND PHENOTYPIC IDENTIFICATION. In the original classification by Tison *et al.* (1982), biovar 1 isolates were positive for indole and ornithine decarboxylase and were mainly isolated from human clinical sources, whereas biovar 2 strains were isolated mainly from diseased eels and were indole- and ornithine-negative. However, in Denmark and Sweden 85% of biotype 2 isolates showing pathogenicity for eels were indole-positive (Høi *et al.*, 1998). Biotype 2 strains are serovar E and these can be further divided into 03 and 04 based on LPS-O antigen. Some biotype 2, serovar E, strains from Taiwan have been found to be avirulent for eels. These strains are positive for mannitol fermentation (Amaro *et al.*, 1999). *Vibrio vulnificus* biotype 3 has been put forward by the Centers for Disease Control, USA, as a possible identification for *Vibrio* species isolated from wound infections in humans in Israel (Bisharat *et al.*, 1999). It is different from currently recognized *V. vulnificus* species because of negative reactions in cellobiose, citrate, lactose, salicin and ONPG. It is positive by PCR for the *V. vulnificus* haemolysin gene (Bisharat *et al.*, 1999; Nair and Holmes, 1999).

Strains of *V. vulnificus* are mostly sucrose negative although some sucrose positive strains have been reported. Isolates from pompano are positive for sucrose and identification was confirmed by 16S rRNA sequencing and sequencing of the haemolysin gene. These isolates did not fit with biotype 1, 2 or 3. The organism was isolated from the jaw and kidney (Li *et al.*, 2006).

Note that sucrose media that is autoclaved for sterility may give false-positive reactions. Filter-sterilized sucrose media is preferred for more accurate fermentation results.

In the conventional biochemical identification tables (Table 3.23) and API 20E tables (Tables 3.24–3.26), the various *V. vulnificus* strains and their different reactions are listed so as to assist the microbiologist when making an identification. These and other reactions are presented in Tables 3.22 and 3.24–29.

MOLECULAR IDENTIFICATION. A multiplex PCR for detection of *V. vulnificus*, *V. cholerae*, *V. mimicus* and *V. parahaemolyticus* (Tarr *et al.*, 2007) is detailed under *V. parahaemolyticus*.

A multiplex PCR designed for the detection of the *atpA* gene (encodes the ATP synthase subunit A) in *V. vulnificus*, *V. parahaemolyticus* and *V. cholerae* (Izumiya *et al.*, 2011) is detailed under *V. parahaemolyticus*. The *atpA* gene for *V. vulnificus* detected two strains that were not detected using the *vvhA* gene for *V. vulnificus* (Hill *et al.*, 1991).

Vibrio wodanis

See *Aliivibrio wodanis*.

Vibrio xiamenensis

TAXONOMY AND SIGNIFICANCE. *Vibrio xiamenensis* (G21^T=DSM 22851^T=CGMCC 1.10228^T) was isolated from mangrove soil in China (Gao *et al.*, 2012). Based on 16S rRNA sequence information the organism has the closest similarity to *V. furnissii* (97.4%), *V. fluvialis* (97.1%) and *V. ponticus* (96.1%). MLSA information based on the *rpoA*, *recA*, *mreB*, *gapA*, *gyrB* and *pyrH* genes demonstrated similarities to *V. furnissii* of

92.4%, 86.0%, 84.7, 87.1%, 79.5% and 84.4%, respectively. Similarity to *V. fluvialis* was 92.6%, 86.9%, 84.6%, 92.3%, 79.9% and 83.2%, respectively.

GROWTH CHARACTERISTICS AND PHENOTYPIC IDENTIFICATION. *Vibrio xiamenensis* is a facultatively anaerobic, motile, oxidase-positive, Gram-negative, slightly curved rod (1.4–1.55×0.7–0.8 μm). On MA 2216, colonies are white, circular to fusiform after incubation at 37°C for 24 h (Gao *et al.*, 2012). Growth occurs in 0.5–10% NaCl (w/v) with an optimum of 2–4%. Growth occurs at 15–40°C with optimal growth at 30–37°C. Growth does not occur below 10°C or above 45°C. It does not grow on TCBS. Starch and Tween 80 are hydrolysed, and gelatin, casein, DNase and alginate are not hydrolysed. These and other reactions are presented in Tables 3.23–26 and 3.38. An inoculum is used of 2% NaCl and sets are incubated at 30°C for 24 h and 8 h for API 20E and API Zym, respectively.

Vibrio xuii

TAXONOMY AND SIGNIFICANCE. *Vibrio xuii* (LMG 21346^T=CAIM 146^T) was isolated from shrimp culture water in China, white leg shrimp (*Litopenaeus* [*Penaeus*] *vannamei*) in China and Ecuador and bivalve larvae (*Nodipecten nodosus*) in Brazil (Thompson *et al.*, 2003b).

In experimental challenge using rainbow trout (*Oncorhynchus mykiss*) and *Artemia* nauplii inoculated with 10^6 CFU/fish or 10^6 CFU/ml, respectively, *V. xuii* (LMG 20001 from *N. nodosus* and strain LMG 21347 from *L. vannamei*) was rated as non-virulent or low virulence. Although there was some evidence of exophthalmia, kidney necrosis, ascites and liquefaction of internal organs, clinical disease was not seen, and at the end of the experiment the organism was recovered from tissue at either a low rate or not at all. Strain LMG 21346^T from prawn culture water did not cause any disease signs (Austin *et al.*, 2005). *Vibrio xuii* was isolated from rotifer culture and from apparently healthy cod (*Gadus morhua*) larvae in Norway (Brunvold *et al.*, 2007).

Vibrio xuii together with *V. neptunius* and *V. brasiliensis* is related to *V. tubiashii* (98–98.8%), *V. nereis* (97.6–98.8%), *V. coralliilyticus* (96.8–98.5%), *V. mytili* (96.8–98.2%) and *V. diabolicus* (97.1–98.1%) based on 16S rRNA sequencing. MLSA identified *V. xuii* as belonging to the Nereis clade, which also contains *V. nereis* (Sawabe *et al.*, 2007b). Members of this clade have >30% DDH, 39–47 GC mol%, a MLSA concatenated sequence similarity of 91.2% and AAI of 96.8%. *V. neptunius* belongs to the Coralliilyticus clade, *V. brasiliensis* and *V. tubiashii* belong to the Orientalis clade, and *V. mytili* belongs to the Harveyi clade. *Vibrio diabolicus* is not available for study.

GROWTH CHARACTERISTICS AND PHENOTYPIC IDENTIFICATION. *Vibrio xuii* is a facultatively anaerobic, motile, oxidase-positive, Gram-negative, slightly curved rod (2–3×1 μm). On TSA containing 2% NaCl colonies are non-luminescent, non-swarming, 3–4 mm, beige, translucent, smooth, round at 28°C after 48 h and grow as 2 mm yellow, round, convex, shiny colonies on TCBS after 24 h (Thompson *et al.*, 2003b). Growth does not occur at 4 or 45°C or at 10% NaCl. Growth does not occur in the absence of NaCl, and up to 2% NaCl is added to culture media and to identification tests. These and other reactions are presented in Tables 3.23–26 and 3.38. An inoculum is used of 2% NaCl and opacity McFarland 3 and sets are incubated at 25°C for 48 h.

1.16.8 *Varracalbmi*

Varracalbmi

TAXONOMY AND SIGNIFICANCE. *Varracalbmi* was isolated from diseased farmed Atlantic salmon (*Salmo salar*) that presented with a haemorrhagic and nectotizing pyogranulomatous inflammation targeting the eye and resulting in panophthalmitis and blindness (Valheim *et al.*, 2000). Infected fish had lesions in the kidney, liver, gills and pseudobranch and deep dermal lesions in severely affected fish. Cases were sporadic with low morbidity and mortality and occurred under subarctic conditions with high salinity. The bacterium was isolated from eyes, dermal vesicles, dermal ulcers and lesions in the kidney and liver. In experimental challenge 100% mortality was obtained when salmon smolts were injected intraperitoneally with 4×10^7 CFU and 51% mortality after 21 days from an injection of 4×10^4 CFU.

The genus *Varracalbmi* is not recognized officially and is not listed in the list of prokaryotic names with standing in nomenclature (Euzéby) or in Bergey's Manual. Valheim *et al.* (2000) suggest the bacterium belongs to the *Vibrionaceae/Pasteurellaceae*.

GROWTH CHARACTERISTICS AND PHENOTYPIC IDENTIFICATION. *Varracalbmi* is a facultatively anaerobic, oxidase-positive, non-motile, Gram-negative, slender rod (1.7–3.5×0.3 μm), straight or slightly curved that occurs in short chains when grown in liquid medium. Catalase is negative. On BA containing 2% NaCl and on MA2216 (Difco) colonies are 1 mm, shiny grey, round, even, convex, opaque and adhere to the agar after 48 h at 22°C. An indent of the colony in the agar is seen when the colony is removed. Colonies are initially non-haemolytic but develop alpha-haemolysis after 1 week. Growth occurs at 4–22°C with an optimum at 15°C. Growth is enhanced by supplementation with 1% mineral salts solution (0.2 g $MgCl_2$, 0.05 g $CaCl_2$, 0.005 g $FeCl_2$ $(6H_2O)$, 0.005 g $MnSO_4$ (H_2O), 0.0025 g $CoCl_2$ $(6H_2O)$, 0.0025 g $CuSO_{24}$ $(5H_2O)$ and 0.0025 g $NaMoO_4$ in 1000 ml H_2O). Growth does not occur without NaCl. Fermentation is slow and occurs after 7 days. It is sensitive to vibriostatic compound O/129 150 μg. Nitrate is not reduced. Reactions are weak for amylase and lecithinase. These and other reactions are presented in Table 3.11.

Order *Xanthomonadales*

1.17 *Xanthomonadaceae*

1.17.1 *Stenotrophomonas*

Stenotrophomonas maltophilia

TAXONOMY AND SIGNIFICANCE. *Stenotrophomonas maltophilia* is an opportunistic pathogen causing infectious intussusception syndrome in channel catfish (*Ictalurus punctatus*) cultured in China (Geng *et al.*, 2010). Infected fish are lethargic, swim near the surface and have entertitis, ascites and intussusception in the lower intestine. Cutaneous haemorrhages may be seen. The organism is cultured from the liver, kidney and ascites. Pathogenicity was confirmed by experimental challenge using intraperitoneal injection with 1.5×10^6, 1.5×10^5, 1.5×10^4 and 1.5×10^3 CFU/fish, which resulted in mortalities of 90–100%, 65–70%, 40–50% and 0–5%, respectively. Identification was confirmed by API 20NE and 16S rRNA sequencing.

Stenotrophomonas maltophilia was previously named *Pseudomonas maltophilia* and *Xanthomonas maltophilia*.

GROWTH CHARACTERISTICS AND PHENOTYPIC IDENTIFICATION. *Stenotrophomonas maltophilia* is a strictly aerobic, non-fermentative, motile, oxidase negative, Gram-negative straight or slightly curved rod (1.5×0.5 μm) occurring singly or in pairs. Catalase is positive. On TSA colonies are 0.6–1.4 mm, white, round after 24–48 h at 28°C. Strains may have yellow, greenish or grey coloured colonies and may go dark brown with age. Strains from channel catfish are colourless initially and became grey-white with a zone of beta-haemolysis at 48 h on TSA with sheep red blood cells. The API 20NE bionumber is 1472341. These and other reactions are presented in Tables 3.10 and 3.27–29.

Class IV Epsilonproteobacteria
Order *Campylobacterales*

The Order *Campylobacterales* comprises the family *Campylobacteraceae* containing the genera *Arcobacter* and *Campylobacter*, and the family *Helicobacteraceae* containing the genera *Helicobacter* and *Wolinella*.

1.18 *Campylobacteraceae*

Members of the Family *Campylobacteraceae* include *Arcobacter*, *Campylobacter* and *Sulfurospirillum* and are characterized as being aerotolerant. *Arcobacter* and *Campylobacter* are found in a variety of animals including humans and some are pathogenic.

1.18.1 *Arcobacter*

Arcobacter species have been isolated from animals, humans and aquatic animals, and were originally classified as *Campylobacter* species before the genus *Arcobacter* was described to encompass aerotolerant strains. Currently

13 species are recognized. *Arcobacter* are classified as Gram-negative rods that are usually curved, S-shaped or helical with a darting, corkscrew motion, non-spore-forming with growth occurring at the range from 15 to 37°C but not at 4 or 42°C. Optimal growth usually occurs in a microaerobic atmosphere (Vandamme *et al.*, 1991). *Arcobacter* species are differentiated from *Campylobacter* as *Arcobacter* will grow in air and at lower temperatures of 15–30°C. Isolation of *Arcobacter* is improved using a selective medium and an enrichment step, although it has been noted that some methods favour faster-growing species (Collado and Figueras, 2011). In a study investigating the isolation and recovery rate of *Arcobacter*, an enrichment step (1:10 dilution of sample to broth) using H broth as described by Houf *et al.* (2001) followed by subculture to CC plates (modified Campylobacter charcoal deoxycholate agar, mCCDA, medium supplemented with cefoperazone, amphotericin B, teicoplanin, known as CAT) as described by Kemp *et al.* (2005), was the most sensitive and specific of all five methods tested (Merga *et al.*, 2011). The study did not include those species isolated from aquatic animals. Methods and media for isolation and identification tests are detailed in the media section. Phenotypic tests used to identify the species are listed under the relevant species and in Table 3.12.

MOLECULAR IDENTIFICATION. Amplification and sequencing of housekeeping genes *rpoB*, *gyrA*, *gyrB* and *hsp60* have a high discriminatory power and has given similar results to 16S rRNA gene phylogeny information (Figueras *et al.*, 2011).

PCR type	Conventional PCR for amplification of *rpoB* gene and sequencing for identification of *Arcobacter* species (Korczak *et al.*, 2006; Collado *et al.*, 2009)		
DNA extraction	From 2–3 day plate culture		
PCR master mix			
Primer CamrpoB-L	5′-CCA ATT TAT GGA TCA AAC-3′		
Primer RPOB-R	5′-GTT GCA TGT TNG NAC CCA T-3′		
Reagent	**Stock**	**Final concentration**	**Volume ×1**
Buffer	×10	×1	5.0 µl
$MgCl_2$	25 mM	2.5 mM	5.0 µl
dNTP mix	10 mM	100 µM	0.5 µl
Primer CamrpoB-L	50 µM	0.2 µM	0.2 µl
Primer RPOB-R	50 µM	0.2 µM	0.2 µl
Taq FIREPol DNA polymerase (Solis BioDyne)	250 U	2.5 U	0.5 µl
Water			33.6 µl
DNA 100 ng			5.0 µl
Total volume			**50 µl**

Thermocycling		
Step 1	**Step 2**	**Step 3**
94°C, 3 min	94°C, 30 s	94°C, 30 s
	54°C, 30 s	54°C, 30 s
	72°C, 30 s	72°C, 7 min
×1	×35	×1
Product	524 bp	

Note: the same primers were used for the sequence reaction.

The *hsp60* gene is also known as *cpn60* or *groEL* and encodes for a 60-kDa chaperonin protein that is found in almost all members of the *Bacteria*. The following primers for the *hsp60* gene were also used for analysis of *Campylobacter* and *Helicobacter* species (Hill *et al.*, 2006). The results can be compared to a database of *hsp60* gene sequences (http://cpndb.cbr.nrc.ca) that have been prepared using the following primers.

PCR type	Conventional PCR for amplification of *hsp60* gene and sequencing for identification of *Arcobacter* species (Hill *et al.*, 2006; Debruyne *et al.*, 2010b)		
DNA extraction	From 2–3 day plate culture using Instagene (BioRad)		
PCR master mix			
Primer H729	5′-CGC CAG GGT TTT CCC AGT CAC GAC GAI III GCI GGI GAY GGI ACI ACI AC-3′		
Primer H730	5′-AGC GGA TAA CAA TTT CAC ACA GGA YKI YKI TCI CCR AAI CCI GGI GCY TT-3′		
Reagent	**Stock**	**Final concentration**	**Volume ×1**
Buffer	×10	×1	5.0 µl
$MgCl_2$	25 mM	2.5 mM	5.0 µl
dNTP mix	10 mM	250 µM	1.25 µl
Primer H729	50 µM	0.2 µM	0.2 µl
Primer H730	50 µM	0.2 µM	0.2 µl
Taq DNA polymerase (Invitrogen)	250 U	2.5 U	0.5 µl
Water			32.85 µl
DNA 100 ng			5.0 µl
Total volume			**50 µl**
Thermocycling	iCycler BioRad)		
Step 1	**Step 2**	**Step 3**	
95°C, 2 min	95°C, 30 s	95°C, 30 s	
	46°C, 30 s	46°C, 30 s	
	72°C, 30 s	72°C, 5 min	
×1	×40	×1	
Product	555 bp		

Note: Reactions on a Stratagene Robocycler were done at times of 1 min at 94°C, annealing of 36°C for 2 min and final extension of 72°C for 10 min. The primers contain landing sites (underlined) for sequencing primers M13 (-40) universal forward 5′-GTT TTC CCA GTC ACG AC-3′ and M13(48) universal reverse 5′-AGC GGA TAA CAA TTT CAC ACA GGA-3′ (also called M13/pUC) for the sequence reaction (Hill *et al.*, 2006).

PCR type	Conventional PCR for amplification of *gyrB* gene and sequencing for identification of *Arcobacter* species (Collado and Figueras, 2011)		
DNA extraction	From 2–3 day plate culture (InstaGene Matrix, BioRad)		
PCR master mix			
Primer gyrB-Arc-7F	5′-GTT TAY CAY TTT GAA GGT GG-3′		
Primer gyrB-Arc-14R	5′-CTA GAT TTT TCA ACA TTT AAA AT-3′		
Reagent	**Stock**	**Final concentration**	**Volume ×1**
Buffer	×10	×1	5.0 µl
$MgCl_2$	25 mM	2.0 mM	4.0 µl
dNTP mix	10 mM	200 µM	0.5 µl
Primer gyrB-Arc-7F	50 µM	0.2 µM	0.2 µl
Primer gyrB-Arc-14R	50 µM	0.2 µM	0.2 µl
Taq DNA polymerase (Invitrogen)	250 U	1.0 U	0.2 µl
Water			38.9 µl
DNA 100 ng			1.0 µl
Total volume			**50 µl**
Thermocycling			
Step 1	**Step 2**	**Step 3**	
94°C, 3 min	94°C, 15 s	94°C, 15 s	
	50°C, 30 s	50°C, 30 s	
	72°C, 45 s	72°C, 5 min	
×1	×35	×1	
Product	722 bp		

Note: the same primers were used for the sequence reaction.

Arcobacter bivalviorum

TAXONOMY AND SIGNIFICANCE. *Arcobacter bivalviorum* (F4^T=CECT 7835^T=LMG 26154^T) was isolated from healthy mussels from the Ebro Delta, Spain (Levican *et al.*, 2012). Similarity based on the 16S rRNA ranged from 91.1% to 94.8% to other *Arcobacter* species with the most similar being *A. defluvii* (97.1%) and *A. ellisii* (97.0%). Further analysis using sequences from *rpoB*, *gyrB* and *hsp60* genes and phenotypic information and MALDI-TOF profiles demonstrated the identification of a new species. The organism shows cross-reaction with the multiplex PCR for *A. cryaerophilus* described by Houf *et al.* (2000) and either no amplicon or amplicon of molecular weight for *A. butzleri* for the multiplex PCR described by Douidah *et al.* (2010).

GROWTH CHARACTERISTICS AND PHENOTYPIC IDENTIFICATION. *Arcobacter bivalviorum* is a motile, oxidase-positive, slightly curved Gram-negative rod (0.9–2.0×0.3–0.5 μm). On blood agar colonies are 2–4 mm, beige to off-white, circular with entire margins, convex and non-swarming at 30°C at 48 h in an aerobic atmosphere (Levican *et al.*, 2012). Growth occurs at 18–37°C, but not at 42°C under both aerobic and microaerobic atmosphere. Growth is weak under an anaerobic atmosphere. Beta-haemolysis is not seen on TSA containing 5% sheep blood. Growth occurs on MA 2216, on NA containing 0.5% NaCl and 2% and 4% NaCl (w/v). Growth does not occur on minimal medium, Campylobacter charcoal deoxycholate agar (CCDA) or MacConkey agar. Growth does not occur on nutrient agar supplemented with 5% sheep blood and containing the following compounds: 1% glycine; 0.1% sodium deoxycholate; 1% oxgall; 0.01%, 0.04% or 0.1% 2,3,5 triphenyl tetrazolium chloride; 0.001% brilliant green; 0.05% safranin; 0.0005% crystal violet; 0.005% basic fuchsin; and medium with 64 mg/l cefoperazone. These and other reactions are presented in Table 3.12.

Arcobacter cryaerophilus

TAXONOMY AND SIGNIFICANCE. *Arcobacter cryaerophilus* was isolated from naturally infected fingerling and juvenile rainbow trout (*Oncorhynchus mykiss*) on trout farms in Turkey and pathogenicity was proven by intramuscular injection. The LD_{50} was 7.8×10^5 viable cells when the trout were kept at 4°C and 2.25×10^4 viable cells at 16°C (Aydin *et al.*, 2000). Clinical signs included exophthalmia, oedema and haemorrhage at the base of fins and gill filaments, loss of scales, darkening of skin, necrosis of kidney, pale liver, swollen gallbladder and haemorrhagic spots on heart. The organism was isolated from kidney, liver, spleen and gills.

Arcobacter cryaerophilus has been isolated from milk of cows with mastitis, and the reproductive tracts and aborted foetuses of farm animals and from animal faeces (Neill *et al.*, 1985).

Arcobacter cryaerophilus is heterogeneous and two subgroups are recognized based on DNA–DNA hybridization studies. Analysis of the *hsp60* gene confirmed the heterogeneity and a rooted neighbour-joining tree shows different clusters of strains (Debruyne *et al.*, 2010).

GROWTH CHARACTERISTICS AND PHENOTYPIC IDENTIFICATION. *Arcobacter cryaerophilus* is a motile, oxidase-positive, Gram-negative rod with curved S-shaped or helical rods (1.8×0.4 μm). Unusually long cells, >20 μm, may occur. Darting motility may be seen. Growth occurs at 30°C with some strains growing between 5 and 40°C. All strains grow at 15°C. Colonies are 1 mm, smooth, convex with entire edges after 48–72 h at 30°C in a microaerobic atmosphere. Colonies may become flattened and irregular in size. Subcultured strains grow in air, 10% CO_2 or anaerobically. Catalase is positive, nitrate is reduced for 98.6% of strains. NaCl at 2% is tolerated and growth occurs on 0.1% glycine and 4% glucose, but not on media containing 1% glycine or 1 mg/ml triphenyltetrazolium chloride (TTC). Methyl red, Voges–Proskaüer, indole, ODC, LDC, ADH, urease, DNAse and citrate are negative. Hydrolysis of casein, gelatin, starch, hippurate and aesculin are negative. Carbohydrates are not fermented (Neill *et al.*, 1985). Colonies may swarm on moist media (Vandamme *et al.*, 1991). Isolates from infected trout grew on tryptic soya agar (Merck) as small white colonies, and on Campylobacter selective media

(Merck) at 25°C after 48 h in a microaerobic atmosphere (Aydin *et al.*, 2000). Growth was positive at 15–42°C. No growth at 3 or 6% NaCl. Growth was negative on MacConkey agar and TCBS. Growth occurs on Campylobacter charcoal deoxycholate agar (Figueras *et al.*, 2011c). These and other reactions are presented in Table 3.12.

Arcobacter ellisii

TAXONOMY AND SIGNIFICANCE. *Arcobacter ellisii* (F79-6^T=CECT 7837^T=LMG 26155^T) was isolated from mussels collected from the Ebro Delta, Spain (Figueras *et al.*, 2011c). It is closely related to *A. defluvii* (from sewerage water) with 99.1% similarity based on the 16S rRNA gene sequence. Pathogenicity is not known and animals were healthy at the time of collection.

GROWTH CHARACTERISTICS AND PHENOTYPIC IDENTIFICATION. *Arcobacter ellisii* is a motile, oxidase-positive, Gram-negative slightly curved rod (1–1.8×0.3–0.9 μm). Some filamentous forms up to 7 μm long may be seen. On BA, colonies are 2–4 mm non-haemolytic, beige to off-white, circular entire margins, convex and non-swarming when incubated aerobically at 30°C for 48 h. Growth occurs at 18–37°C in aerobic and microaerobic atmospheres. Growth is weak when grown anaerobically at 30°C and aerobically at 42°C (Figueras *et al.*, 2011c). Growth also occurs on MA 2216, media containing 2% NaCl or 0.1% deoxycholate, and there is weak growth after 5 days on CCDA. Growth does not occur at 4% NaCl or with 1% glycine, 1% oxgall or 0.01%, 0.04% or 0.1% TTC. Casein, starch and lecithin are not hydrolysed. Two of the three strains grew on MCA. These and other reactions are presented in Table 3.12.

MOLECULAR IDENTIFICATION. No specific PCR tests are available, however amplification and sequencing of the *rpoB*, *gyrB*, *hsp60* and 16S rRNA genes can be used for identification (Figueras *et al.*, 2011c). See under *Arcobacter*.

Arcobacter marinus

TAXONOMY AND SIGNIFICANCE. *Arcobacter marinus* (CL-S1^T=KCCM 90072^T=JCM 15502^T) was isolated from the sea near Dokdo Island, East Sea, Korea (Kim *et al.*, 2010a). Seaweeds and starfish were added to the seawater in the laboratory and maintained for 15 days before the organism was cultured from the homogenate. Based on 16S rRNA and *gyrA* gene sequence analysis *A. marinus* is closely related to *A. halophilus* with 96.1% and 88.2% similarity, respectively. Similarity to other *Arcobacter* species based on the *gyrA* gene sequence is 81.2–84.0% and for the 16S rRNA is 92.4–94.3%.

GROWTH CHARACTERISTICS AND PHENOTYPIC IDENTIFICATION. *Arcobacter marinus* is a motile, oxidase-positive, Gram-negative slightly curved rod (1.7–0.5×0.3–0.1 μm). On MA 2216 colonies are 1 mm, circular , convex and creamy white after 3 days at 37°C in either an aerobic or microaerobic atmosphere. On sheep blood agar colonies are non-haemolytic, 4 mm, grey, irregular, and umbonate after 3 days at 37°C. Growth occurs at 10–40°C with an optimum 30–37°C, and at pH 5.9–8.5 (optimum pH 7.0). Growth occurs at 1–9% NaCl (optimum 3–5%). Results are negative for catalase, urea, hippurate and H_2S. Nitrate is positive. Indoxyl acetate is hydrolysed when cells are grown in a microaerobic atmosphere but negative when cells are grown aerobically (Kim *et al.*, 2010a). Growth is negative on CCDA (Figueras *et al.*, 2011c). These and other reactions are presented in Table 3.12.

MOLECULAR IDENTIFICATION. No specific PCR has been described, however primers for the *gyrA* sequence (encodes the subunit A of DNA gyrase) from all currently known *Arcobacter* were designed for amplification followed by sequencing for differentiation of *Arcobacter* species.

Primers 27F and 1492R (Lane, 1991) (see Molecular section) were used for amplification and sequencing of the 16S rRNA gene (Kim *et al.*, 2010a).

Arcobacter molluscorum

TAXONOMY AND SIGNIFICANCE. *Arcobacter molluscorum* (F98-3^T=CECT 7696^T=LMG 25693^T) was isolated from healthy mussels (phylum *Mollusca*) and oysters on a mussel farm in the Ebro Delta, north-eastern Spain (Figueras *et al.*, 2011b). Based on 16S rRNA sequence analysis *A. molluscorum* is related to *A. marinus* (97.6%), *A. halophilus*

PCR type	Conventional PCR for amplification of partial *gyrA* gene and for amplification and identification of *Arcobacter* species (Kim *et al.*, 2010a)		
DNA extraction	From 2–3 day plate culture		
PCR master mix			
Primer gyrA-F	5′-TGG ACG TGC ATT ACC AGA TG-3′		
Primer gyrA-R	5′-GCA ACT TTT CCT TTT CCA CCT-3′		
Reagent	**Stock**	**Final concentration**	**Volume ×1**
Buffer (containing 1.5 mM $MgCl_2$)	×10	×1	2.0 µl
dNTP mix	10 mM	125 µM	0.25 µl
Primer gyrA-F	50 µM	0.5 µM	0.2 µl
Primer gyrA-R	50 µM	0.5 µM	0.2 µl
Taq DNA polymerase (Bioneer)	250 U	1.5 U	0.12 µl
Water			15.23 µl
DNA 30 ng			2.0 µl
Total volume			**20 µl**

Thermocycling		
Step 1	**Step 2**	**Step 3**
95°C, 5 min	95°C, 30 s	95°C, 30 s
	55°C, 30 s	55°C, 30 s
	72°C, 3 min	72°C, 5 min
×1	×30	×1

Note: the same primers were used for the sequence reaction.

(95.6%) and *A. mytili* (94.7%) and with the remaining *Arcobacter* species down to 92.4%. DNA–DNA hybridization with *A. marinus* is 54.8% ±1.05. Phylogenetic trees constructed from individual sequence analysis and concatenated sequence analysis of housekeeping genes, *rpoB*, *gyrB* and *hsp60*, confirmed the identification of the new species.

GROWTH CHARACTERISTICS AND PHENOTYPIC IDENTIFICATION. *Arcobacter molluscorum* is a motile, oxidase-positive, Gram-negative slightly curved rod (1–3.4×0.3–0.5 µm). Some long (7 µm) filamentous forms may occur. On BA colonies are non-swarming, non-haemolytic, 2–4 mm, beige to off-white, circular with entire margins, convex after 48 h at 30°C in an aerobic atmosphere. Growth occurs in aerobic and microaerobic atmospheres without visual differences, and growth is enhanced with the addition of 2% NaCl (final concentration) to the medium. Growth occurs at 18–37°C but is weak at 30°C under anaerobic conditions and weak at 42°C under aerobic conditions. It is oxidase and catalase positive. Growth is positive on MacConkey agar and on media containing 0.1% sodium deoxycholate, 1% oxgall, 0.05% safranin, 0.0005% crystal violet and media with 64 mg cefoperazone/l. Growth is enhanced on these media when supplemented with 2% NaCl. Growth occurs on media containing 0.005% basic fuchsin or 0.001% brilliant green only when supplemented with 2% NaCl. Growth is negative on minimal media, and on media containing 1% glycine, 0.04% 2,3,5-triphenyl tetrazolium chloride (TTC), or on CCDA. Hydrolysis is negative for indoxyl acetate, urease, casein, lecithin and starch. Nitrate is reduced after 72 h in a microaerobic atmosphere and after 5 days aerobically. H_2S is negative in triple sugar iron agar medium (Figueras *et al.*, 2011b). These and other reactions are presented in Table 3.12.

MOLECULAR IDENTIFICATION. Figueras *et al.* (2011b) noted that *A. molluscorum* cross-reacts and produces amplicon at the same molecular weight as for *A. cryaerophilus* and *A. skirrowii* in a multiplex PCR designed on the 16S rRNA gene for *A. skirrowii* and the 23S rRNA for *A. cryaerophilus* (Houf *et al.*, 2000). Another PCR (Douidah *et al.*, 2010) designed for these and other *Arcobacter* species was specific and did not cross-react with *A. molluscorum* (Figueras *et al.*, 2011b). Currently no specific PCR has been described for *A. molluscorum*.

Arcobacter mytili

TAXONOMY AND SIGNIFICANCE. *Arcobacter mytili* (F2075^T=CECT 7386^T=LMG 24559^T) was isolated from mussels (*Mytili* species) and brackish water from Catalonia, north-eastern Spain (Collado *et al.*, 2009).

GROWTH CHARACTERISTICS AND PHENOTYPIC IDENTIFICATION. *Arcobacter mytili* is a motile, oxidase-positive, Gram-negative slightly curved rod (1–3×0.4–0.6 μm). On BA colonies are non-swarming, non-haemolytic, 2–4 mm, beige to off-white, circular, entire, convex after 48 h at 30°C in an aerobic atmosphere (Collado *et al.*, 2009). Growth occurs at 18–37°C with no differences in aerobic or microaerobic atmospheres, but is negative at 4°C. Growth is weak when grown anaerobically at 30°C and aerobically at 42°C. Growth occurs on MacConkey agar. Growth is negative on CCDA. These and other reactions are presented in Table 3.12.

Arcobacter venerupis

TAXONOMY AND SIGNIFICANCE. *Arcobacter venerupis* (F67^T=CECT 7836^T=LMG 26156^T) was isolated from healthy clams from Galicia, Spain (Levican *et al.*, 2012).

Similarity based on the 16S rRNA ranged from 91.1% to 94.8% to other *Arcobacter* species with the most similar being *A. defluvii* (97.1%) and *A. ellisii* (97.0%). Further analysis using sequences from *rpoB*, *gyrB* and *hsp60* genes and phenotypic information and MALDI-TOF profiles demonstrated the identification of a new species. The organism shows cross-reaction in the multiplex PCR for *A. cryaerophilus* described by Houf *et al.* (2000), and either no amplicon or amplicon of the same molecular weight as *A. butzleri* in the multiplex PCR described by Douidah *et al.* (2010).

1.18.2 *Campylobacter*

Campylobacter insulaenigrae

TAXONOMY AND SIGNIFICANCE. *Campylobacter insulaenigrae* (NCTC 12927^T=CCUG 48653^T) was isolated from the rectum of three common seals (*Phoca vitulina*) and from the small intestine of a stranded harbour porpoise (*Phocoena phocoena*) in Scotland (Foster *et al.*, 2004) and from two elephant seals in the USA (Debruyne *et al.*, 2009). Based on 16S rRNA sequence analysis it is closely related to *Campylobacter jejuni* ssp. *jejuni* and *C. jejuni* ssp. *doylei* (98.8%), *C. lari* (98.3%) and *C. coli* (97.6%).

GROWTH CHARACTERISTICS AND PHENOTYPIC IDENTIFICATION. *Campylobacter insulaenigrae* is a motile, oxidase-positive, Gram-negative comma-shaped rod. On Columbia Base agar (CBA) supplemented with Blaser-Wang selective supplements (Oxoid) and 5% (w/v) citrated sheep blood agar (CSBA) (Oxoid) colonies are alpha-haemolytic, 0.75–1.0 mm, circular, entire, low-convex, smooth, shiny, grey, translucent and butyrous after 48 h at 37°C. Growth occurs on nutrient agar. Colonies are easily emulsified. Growth does not occur at 25°C or 42°C or in aerobic or anaerobic atmospheres (Foster *et al.*, 2004). Does not grow at 2 or 3.5% NaCl. These and other reactions are presented in Table 3.12.

MOLECULAR IDENTIFICATION. Identification of *C. insulaenigrae* can be confirmed by amplification and sequence analysis of the 16S rRNA gene using universal primers pA (5′-AGA GTT TGA TCC TGG CTC AG-3′), position 8–28 in *Escherichia coli* numbering, and primer pH (5′-AAG GAG GTG ATC CAG CCG CA-3′) *E. coli* position 1542–1522 (Foster *et al.*, 2004).

Campylobacter lari

TAXONOMY AND SIGNIFICANCE. *Campylobacter lari* consists of two subspecies, *C. lari* ssp. *lari* and *C. lari* ssp. *concheus*. *Campylobacter lari* ssp. *lari* (NCTC 11352^T=LMG 8846^T) was isolated from the faeces of a gull in the UK in 1976 and strains have been isolated from faeces of wild birds, dogs, chickens, horse intestine and from cases of human diarrhoea and bacteraemia (Benjamin *et al.*, 1983; Debruyne *et al.*, 2009). These strains are nalidixic acid-resistant and urease-negative.

Campylobacter lari ssp. *concheus* (2897R^T=CCUG 55786^T=LMG 21009^T) was isolated from shellfish from the Netherlands in 1993 and has also been isolated from human faeces in Canada and the Netherlands (Debruyne *et al.*, 2009).

Based on the 16S rRNA gene sequence it has 97.6% similarity to *C. peloridis* and >97% sequence similarity to *C. jejuni*, *C. coli* and *C. insulaenigrae*. Sequence analysis of 555 bp of the *hsp60* (also known as *cpn60* and *groEL*) gene demonstrated similarity to *C. lari* ssp. *concheus* to *C. jejuni* and *C. coli* of <89%. Pathogenicity for shellfish is not known.

GROWTH CHARACTERISTICS AND PHENOTYPIC IDENTIFICATION. Strains of *Campylobacter lari* are motile, oxidase-positive, Gram-negative spiral to curved rod (1.7–2.4×0.3 µm). On BA, colonies are alpha-haemolytic, 1–1.5 mm, colourless, round, entire and convex after 72 h at 37°C in a microaerobic atmosphere (Debruyne *et al*., 2009). Growth occurs on nutrient agar in microaerobic conditions. Cells have a rapid darting motility. No growth at 25°C but growth occurs at 37 and 42°C when grown in a microaerobic atmosphere. Growth does not occur in aerobic conditions at 25 or 37°C. Growth occurs in a microaerobic atmosphere on unsupplemented CCDA. Negative results for glucose fermentation. Strains of *C. lari* ssp. *lari* show alpha-haemolysis on BA. Growth occurs on nutrient agar containing 0.02 or 0.05% safranin or 0.1% sodium deoxycholate in an microaerobic atmosphere. Results are negative for urease and DNase. Of the strains, 80–94% grow on BA supplemented with 32 mg/l nalidixic acid and all strains grow on BA containing 32 mg/l cephalothin. *Campylobacter lari* ssp. *concheus* does not grow on nutrient agar supplemented with 0.05% safranin and growth on 0.1% sodium deoxycholate is variable for strains. Growth is positive on media containing 32 mg/l cephalothin and negative for nalidixic acid. These and other reactions are presented in Table 3.12.

MOLECULAR IDENTIFICATION. The following PCR for amplification and sequencing of the *hsp60* gene for discrimination of *Arcobacter*, *Campylobacter* and *Helicobacter* (Hill *et al*., 2006) (see under *Arcobacter*) was optimized for *Campylobacter* species (Debruyne *et al*., 2009).

PCR type	Conventional PCR for amplification of *hsp60* gene and sequencing for identification of *Campylobacter* species (Hill *et al*., 2006; Debruyne *et al*., 2009)		
DNA extraction	From 2–3 day plate culture using Instagene (BioRad)		
PCR master mix			
Primer H729	5′-CGC CAG GGT TTT CCC AGT CAC GAC GAI III GCI GGI GAY GGI ACI ACI AC-3′		
Primer H730	5′-AGC GGA TAA CAA TTT CAC ACA GGA YKI YKI TCI CCR AAI CCI GGI GCY TT-3′		
Reagent	**Stock**	**Final concentration**	**Volume ×1**
Buffer	×10	×1	2.5 µl
$MgCl_2$	25 mM	2.5 mM	2.5 µl
dNTP mix	10 mM	200 µM	0.5 µl
Primer H729	50 µM	0.5 µM	0.25 µl
Primer H730	50 µM	0.5 µM	0.25 µl
Taq DNA polymerase	250 U	0.5 U	0.25 µl
Water			17.75 µl
DNA 25 ng			1.0 µl
Total volume			**25 µl**

Thermocycling		
Step 1	**Step 2**	**Step 3**
95°C, 2 min	95°C, 30 s	95°C, 30 s
	46°C, 30 s	46°C, 30 s
	72°C, 30 s	72°C, 5 min
×1	×30	× l
Product	555 bp	

The primers contain landing sites (underlined) for sequencing primers M13 (-40) universal forward 5′-GTT TTC CCA GTC ACG AC-3′ and M13(48) universal reverse 5′-AGC GGA TAA CAA TTT CAC ACA GGA-3′ (also called M13/pUC) for the sequence reaction (Hill *et al*., 2006).

Campylobacter peloridis

TAXONOMY AND SIGNIFICANCE. *Campylobacter peloridis* (2314BVA^T=CCUG 55787^T=LMG 23910^T) was isolated from shellfish, possibly a giant mussel, in 1993 from the Netherlands. It also has been isolated from human faeces and dialysis fluid (Debruyne *et al.*, 2009). *Campylobacter peloridis* has 97.6% 16S rRNA gene sequence similarity to *C. lari* ssp. *concheus* and >97% similarity to *C. lari*, *C. jejuni*, *C. coli* and *C. insulaenigrae*. Based on the *hsp60* gene it is 88–90% similar to *C. lari*. Pathogenicity is unknown.

GROWTH CHARACTERISTICS AND PHENOTYPIC IDENTIFICATION. *Campylobacter peloridis* is a motile, oxidase-positive, Gram-negative slightly curved rod. On BA (5% blood) colonies are 1–1.5 mm, colourless, circular, entire, convex after 72 h at 37°C in a microaerobic atmosphere (Debruyne *et al.*, 2009). There is no growth on nutrient agar containing 0.05% safranin but growth occurs on nutrient agar containing 1% glycine and 4 mg metronidazole/l. Of the strains, 80–94% grow on media containing 32 mg/l nalidixic acid and 5–24% of strains grow on media containing 32 mg/l cephalothin. These and other reactions are presented in Table 3.12.

For molecular identification using amplification and sequencing of the *hsp60* gene, see under *C. lari*.

Campylobacter species

TAXONOMY AND SIGNIFICANCE. DNA sequences from non-*Helicobacter* species from the order *Campylobacterales* were detected in dental plaque from dolphins (*Tursiops truncatus*) held captive in an oceanarium in Argentina (Goldman *et al.*, 2011). Primers F1, F0 and R1 were used to differentiate between *Helicobacter* and non-*Helicobacter* species (see under *Helicobacter*). The sequences clustered with *Campylobacter* but had less than 90.1% similarity to *C. concisus* and 91.2% with *C. insulaenigrae* and the bacteria appears to be a new species.

Campylobacter subantarcticus

TAXONOMY AND SIGNIFICANCE. *Campylobacter subantarcticus* (R-3023^T=CCUG 38513^T=LMG 24377^T) was isolated from grey-headed albatross (*Diomedea chrysostoma*) chicks in the subAntarctic region in 1996 (Debruyne *et al.*, 2010a). Other strains were isolated from black-browed albatross (*Diomedea melanophris*) chicks and from a gentoo penguin (*Pygoscelis papua*). The organism was not isolated from faecal samples taken from 10 adult and 40 female Antarctic fur seal pups (*Arctocephalus gazella*) tested at the same time. *Campylobacter subantarcticus* is closely related to *C. lari* ssp. *concheus*, *C. lari* ssp. *lari*, *C. jejuni*, *C. coli*, *C. insulaenigrae* and *C. peloridis* with >97% sequence similarity in the 16S rRNA gene. It could not be distinguished from *C. lari* ssp. *concheus*, which had 99.4–99.9% sequence identity. Sequence similarity with the *hsp60* gene was 93.3–93.9% for these species. Pathogenicity is unknown.

GROWTH CHARACTERISTICS AND PHENOTYPIC IDENTIFICATION. *Campylobacter subantarcticus* is a motile, oxidase-positive, Gram-negative slightly curved rod. On BA (5% blood) colonies are alpha-haemolytic, 1–1.5 mm, colourless, circular, entire, convex after 72 h at 37°C in a microaerobic atmosphere (Debruyne *et al.*, 2010a). There is no growth on nutrient agar containing 0.02% or 0.05% safranin, 4% NaCl, 0.1% sodium deoxycholate, 32 mg/ml cephalothin. Under a microaerobic atmosphere growth occurs at 37 and 42°C but not at 18–25°C. Growth does not occur under aerobic conditions at 25 or 37°C but growth occurs at 37°C under anaerobic conditions on BA and BA supplemented with 0.1% trimethylamine *N*-oxide. Most strains do not grow on MacConkey agar or on media containing 4 mg/ml metronidazole. Growth does not occur on unsupplemented nutrient agar, casein agar, lecithin agar or tyrosine agar. These and other reactions are presented in Table 3.12.

For molecular identification using amplification and sequencing of the *hsp60* gene, see under *C. lari*.

1.19 *Helicobacteraceae*

1.19.1 *Helicobacter*

The genus *Helicobacter* comprises about 32 species, and belongs to the family *Helicobacteraceae* and the order *Campylobacterales*. *Helicobacter* species have been isolated from the gastrointestinal tract of humans, marine mammals and a range of animals including rodents, cats, dogs, horses, monkeys, birds, pigs and cattle. The organisms are microaerophilic and the cells are fusiform, curved or spiral and possess flagella in different numbers and locations according to the species. *Helicobacter* species are grown on TSA+5% Blood, BA, Brucella agar or Skirrow's medium (VPT). To improve the isolation rate stomach tissue is homogenized in 1 ml Brucella broth (Difco) containing 5% faecal calf serum. A volume of 100 µl is inoculated to plates and/or filtered through a 0.45 or 0.8 µm filter before inoculating to plates (Butzler *et al.*, 1973; Harper *et al.*, 2002a). The plates are incubated at 37°C in a microaerophilic atmosphere of N_2, H_2, CO_2 (80:10:10) for 2–4 weeks. Commercial systems are available for generation of correct atmosphere, e.g. MGC Anaero Pak™ or Campylo from Mitsubishi Gas Chemical Company.

The results of biochemical identification tests are found in Table 3.13 and methods for preparation of the identification media and reagents are in the media section under *Arcobacter*, *Campylobacter* or *Helicobacter*.

MOLECULAR IDENTIFICATION. *Helicobacter* species-specific primers C97 and C05 (Fox *et al.*, 2000) and H276 and H676 (Germani *et al.*, 1997) based on the 16S rRNA gene are available for direct detection from tissue or confirmation of phenotypic identification through sequencing of the 16S rRNA using universal bacterial primers F24 and F25 (Dewhirst *et al.*, 1999). Primers are also available to differentiate between *Helicobacter* species (primer F1 and R1, pair CG1) and non-*Helicobacter* species (primer F0 and R1, pair CG2) within the order *Campylobacterales* (Goldman *et al.*, 2009, 2011). A positive result with both primer pairs indicates the presence of both *Helicobacter* species and non-*Helicobacter* species within the order *Campylobacterales*, whereas a negative result for pair CG1 and a positive result for pair CG2 indicates DNA from non-*Helicobacter* within the order *Campylobacterales*.

PCR type	Conventional PCR for amplification of *Helicobacter* species-specific region of 16S rRNA gene amplification and sequencing (Fox *et al.*, 2000; Harper *et al.*, 2002a)		
DNA extraction	Cells from plate culture extracted using High-Pure template preparation kit (Roche)		
PCR master mix			
Primer C97	5′-GCT ATG ACG GGT ATC C-3′		
Primer C05	5′-ACT TCA CCC CAG TCG CTG-3′		
Reagent	**Stock**	**Final concentration**	**Volume ×1**
Taq 2000 buffer with 15 mM $MgCl_2$ (Stratagene)	×10	×1	5.0 µl
dNTP 100 mM mix containing 25 mM each dNTP	100 mM	400 µM	0.2 µl
Primer C97	50 µM	0.2 µM	0.2 µl
Primer C05	50 µM	0.2 µM	0.2 µl
Taq 2000 polymerase (Stratagene)	250 U	1U	0.2 µl
Water			43.2 µl
DNA			1 µl
Total volume			**100 µl**

Thermocycling		
Step 1	**Step 2**	**Step 3**
95°C, 8 min	95°C, 45 s 60°C, 45 s 72°C, 90 s (+5 s for each additional cycle)	95°C, 45 s 60°C, 45 s 72°C, 10 min
×1	×30	×1
Product)	1200 bp	
Sensitivity	50 fg for pure DNA and 500 fg for DNA from faeces	

PCR type	Conventional PCR for amplification of *Helicobacter* species, using degenerate primers for 16S rRNA gene amplification and sequencing (Fox *et al.*, 1995, 2006)
PCR master mix	As for PCR using primers C97 and C05
Primer C70	5′-AGA GTT TGA TYM TGG C-3′
Primer B37	5′-TAC GGY TAC CTT GTT ACG A-3′
Thermocycling	As for PCR using primers C97 and C05
Product	1500 bp

PCR type	Conventional PCR for amplification of *Helicobacter* species-specific region of 16S rRNA gene amplification and sequencing (Germani *et al.*, 1997; Oxley and McKay, 2005)		
DNA extraction	Cells from plate culture extracted using High-Pure template preparation kit (Roche). Tissue samples extracted using QIAamp DNA minikit (Qiagen)		
PCR master mix			
Primer H276 forward	5′-TAT GAC GGG TAT CCG GC-3′		
Primer H676 reverse	5′-ATT CCA CCT ACC TCT CCC A-3′		
Reagent	**Stock**	**Final concentration**	**Volume ×1**
Taq buffer	×10	×1	3.0 μl
dNTP mix	10 mM	170 μM	0.51 μl
$MgCl_2$	25 mM	2.5 mM	3 μl
Primer H276 forward	50 μM	0.25 μM	0.15 μl
Primer H676 reverse	50 μM	0.25 μM	0.15 μl
Taq polymerase (Biotech Int.)	250 U	1U	0.12 μl
Water			21.07 μl
DNA			2 μl
Total volume			**30 μl**
Thermocycling	Eppendorf Mastercycler gradient (Eppendorf)		
Step 1	**Step 2**	**Step 3**	
95°C, 2 min	94°C, 30 s	94°C, 30 s	
	66°C, 30 s	66°C, 30 s	
	72°C, 30 s	72°C, 10 min	
×1	×30	×1	
Product	400 bp		

PCR type	Conventional PCR for amplification of 16S rRNA gene using universal primers F24 and F25 (Dewhirst *et al.*, 1999; Harper *et al.*, 2002a)
DNA extraction	Cells from plate culture extracted using High-Pure template preparation kit (Roche)
PCR master mix	As for PCR using primers C97 and C05
Primer F24	5′-GAG TTT GAT YMT GGC TCA G-3′
Primer F25	5′-AAG GAG GTG WTC CAR CC-3′
Thermocycling	As for PCR using primers C97 and C05

PCR type	Conventional PCR for amplification and differentiation of *Helicobacter* species and non-*Helicobacter* species within the order *Campylobacterales* using the 16S rRNA gene (Goldman *et al.*, 2009b, 2011)
DNA extraction	Tissue extracted with QIAamp mini kit (Qiagen)
PCR master mix	
Primer F1	5′-GTA TCC GGC CTG AGA-3′

Continued

Continued.

Primer R1	5′-ATT TTA CCC CTA CAC CAA-3′		
F0	5′-GAG TTT GAT CCT GGC TCA GAG-3′		
Reagent	**Stock**	**Final concentration**	**Volume ×1**
Taq buffer	×10	×1	5.0 μl
dNTP 10 mM	10 mM	200 μM	1.0 μl
$MgCl_2$	25 mM	1.5 mM	3.0 μl
Primer F1	50 μM	0.3 μM	0.3 μl
Primer R1	50 μM	0.3 μM	0.3 μl
Platinum *Taq* polymerase (Invitrogen)	250 U	1U	0.2 μl
Water			30.2 μl
DNA			10 μl
Total volume			**50 μl**
Thermocycling			
Step 1	**Step 2**	**Step 3**	
91°C, 1 min	94°C, 30 s	94°C, 30 s	
	50°C, 30 s[a]	50°C, 30 s	
	72°C, 30 s	72°C, 5 min	
×1	×30	×1	
Product	F1 and R1 (pair CG1), 387 bp; F0 and R1 (pair CG2), 650 bp		
Interpretation	A positive result with both primer pairs indicates the presence of both *Helicobacter* species and non-*Helicobacter* species within the order *Campylobacterales*, whereas a negative result for pair CG1 and a positive result for pair CG2 indicates DNA from non-*Helicobacter* within the order *Campylobacterales*		

[a]Conditions are the same for primers F1 and R1 with the exception of an annealing temperature of 51°C.

Helicobacter anseris

TAXONOMY AND SIGNIFICANCE. *Helicobacter anseris* (MIT-04-9362[T]=ATCC BAA-1299[T]) was isolated from the faeces of 20.6% of migratory Canada geese (*Branta canadensis*) in Boston, USA (Fox *et al.*, 2006). Pathogenicity is unknown. *Helicobacter anseris* has 96.4% 16S rRNA sequence similarity to *H. brantae*, also isolated from geese. Both organisms were co-isolated in 30% of geese with *H. anseris*.

GROWTH CHARACTERISTICS AND PHENOTYPIC IDENTIFICATION. *Helicobacter anseris* is a motile, oxidase-positive, Gram-negative, slender slightly curved to rod-shaped rod (0.5×1.8 μm). For isolation, faeces is homogenized in 1 ml phosphate-buffered saline and the slurry passed through a 0.45 μm filter on to a trypticase soy agar plate (TSA) containing 5% sheep blood. Unfiltered homogenate is cultured to TSA plates containing either trimethoprim, vancomycin and polymyxin (VPT plates or Skirrow's medium; see media section) or TSA containing cefoperazone, vancomycin and amphotericin B (see CVA medium in media section). All plates are incubated at 37°C under microaerobic atmosphere (80% N_2, 10% H_2, 10% CO_2). On BA (5% blood) colonies grow as a spreading layer. Growth occurs at 37 and 42°C. Growth does not occur under aerobic or anaerobic conditions. *Helicobacter anseris* is urease positive. These and other reactions are presented in Table 3.13.

MOLECULAR IDENTIFICATION. See under *Helicobacter* genus and protocol using primers C70 and B37 for amplification of the 16S rRNA gene.

Helicobacter brantae

TAXONOMY AND SIGNIFICANCE. *Helicobacter brantae* (MIT-04-9366[T]=ATCC BAA-1298[T]) was isolated from the faeces of 7.2% of migratory Canada geese (*Branta canadensis*) in Boston, USA (Fox *et al.*, 2006). Pathogenicity is unknown. *Helicobacter brantae* has 98.5% 16S rRNA sequence similarity to *H. cholecystus*. Of the geese that had *H. anseris*, 30% also had *H. brantae*.

GROWTH CHARACTERISTICS AND PHENOTYPIC IDENTIFICATION. *Helicobacter brantae* is a motile, oxidase-positive, Gram-negative, slender slightly curved to rod-shaped rod (0.5×1.8 μm). See under *H. anseris* for isolation techniques and colony morphology. *Helicobacter brantae* is urease negative. These and other reactions are presented in Table 3.13.

Helicobacter canadensis

TAXONOMY AND SIGNIFICANCE. *Helicobacter canadensis* (NLEP-16143^T=MIT 98-5491^T=ATCC 700968^T) is an emerging human pathogen thought to be responsible for cases of enteritis and bacteraemia in patients in Canada and Australia (Fox *et al.*, 2000). It has also been isolated from barnacle geese (*Branta leucopsis*) in Sweden and Canada geese (*B. canadensis*) in the UK (Waldenström *et al.*, 2003). *Helicobacter canadensis* is closely related to *H. pullorum* based on 16S rRNA sequence information and biochemical reactions.

GROWTH CHARACTERISTICS AND PHENOTYPIC IDENTIFICATION. *Helicobacter canadensis* is a motile, oxidase-positive, Gram-negative, slender, curved to spiral shaped rod (0.3×1.5–4 μm) with one to three spirals. Growth on BA is seen as a spreading film at 37°C and 42°C. Growth occurs in microaerobic atmosphere but not under aerobic or anaerobic conditions. Urease is negative (Fox *et al.*, 2000). *Helicobacter canadensis* is tolerant to growth on media containing 0.5% sodium fluoride, whereas *H. pullorum* is negative (Waldenström *et al.*, 2003). The avian strains of *H. canadensis* showed some phenotypic variation to strains from humans, as all seven avian strains were positive for nitrate reduction, whereas only two of four human strains were reported positive for nitrate. The avian isolates were positive for alkaline phosphatase and the human strains were negative. These and other reactions are presented in Table 3.13. DNA can be amplified using primers C97 and C05 and the amplicon sequenced (see under *Helicobacter* genus).

Helicobacter cetorum

TAXONOMY AND SIGNIFICANCE. *Helicobacter cetorum* (MIT 99-5656^T=ATCC BAA-540^T) was isolated from the main stomach of wild stranded Atlantic white-sided dolphin (*Lagenorhynchus acutus*), and from the faeces of captive beluga whale (*Delphinapterus leucas*) (isolate MIT 00-7128=ATCC BAA-429), captive Pacific white-sided dolphin (*Lagenorhynchus obliquidens*) and an Atlantic bottle-nose dolphin (*Tursiops truncatus*) (Harper *et al.*, 2000, 2002a). Clinical signs in the beluga whale included inappetance and lethargy. Endoscopic examination of the captive beluga whale and captive Atlantic bottlenose dolphin revealed ulcers in the oesophagus and fore-stomach. The dolphin had chronic regurgitation. The stranded Pacific white-sided dolphin had a history of chronic regurgitation and weight loss and linear erosions were seen in the oesophagus. The prevalence of *H. cetorum* in the faeces of Atlantic bottlenose is approximately 50% (Harper *et al.*, 2003b). Sequencing of the 16S rRNA gene revealed 98% sequence similarity to *H. pylori*. A *Helicobacter* species with 98–99% homology to the 16S rRNA gene of *H. cetorum* was isolated from an Atlantic spotted dolphin (*Stenella frontalis*) stranded on the Caribbean coast of Venezuela (Suárez *et al.*, 2010). Lesions were present in the duodenal ampulla and ulcerations and leucocyte infiltration occurred in the glandular main and the lower pyloric stomach. PCR was positive for stomach tissue, and for oesophageal and intestinal contents. DNA sequences with 99.5–100% homology to *H. cetorum* were isolated from gastric fluid, dental plaque and saliva from captive dolphins (*Tursiops truncatus*, *T. gephyreus*) and captive killer whale (*Orcinus orca*) in Argentina, wild La Plata river dolphins (*Pontoporia blainvillei*) in Argentina, and stranded South American fur seals (*Arctocephalus australis*) (Goldman *et al.*, 2002, 2009a,b, 2011). *Helicobacter cetorum* was identified by PCR in faecal samples collected from wild and captive Yangtze finless porpoises in China (McLaughlin *et al.*, 2011). Disease status of the porpoises was not known.

GROWTH CHARACTERISTICS AND PHENOTYPIC IDENTIFICATION. *Helicobacter cetorum* is a motile, oxidase-positive, Gram-negative fusiform (wider at the middle and tapers at the ends), slightly spiral rod 0.6–4 μm. For isolation, 0.5 g of faeces or gastric mucosa is homogenized in 1 ml brucella broth containing 5% fetal calf serum. 100 μl of homogenate is cultured to Skirrow's VPT medium, CVA and AMB (see media section under *Helicobacter* medium) and the remainder is filtered through a 0.45 μm filter on to a BA plate. All plates are incubated at 37°C under microaerobic atmosphere (80% N_2, 10% H_2, 10% CO_2). On isolation media after 2–4 weeks' incubation, colonies grow as a thin spreading film. Pure subcultured growth grows on BA after 2–5 days (Harper *et al.*, 2002a). Growth occurs at 37 and 42°C but not at 25°C. All cetacean isolates are susceptible to cephalothin. Isolates from the wild stranded Atlantic white-sided dolphins were sensitive or intermediate to nalidixic acid, whereas the isolates from the captive animals were resistant to nalidixic acid. These and other reactions are presented in Table 3.13.

MOLECULAR IDENTIFICATION. See under *Helicobacter* genus for primers specific to the genus.

Primers Hce F and Hce R were developed for the direct detection of *H. cetorum* from faeces (Harper *et al.*, 2003b).

PCR type	Conventional PCR for amplification of 16S rRNA gene from *Helicobacter cetorum* (Harper *et al.*, 2003)		
DNA extraction	Faeces: QIAamp mini DNA kit (Qiagen)		
PCR master mix			
Primer Hce F	5′-GTT ATG TGC CTC TTA GTT TG-3′		
Primer Hce R	5′-AGA GTT CTC AGC ATA ACC T-3′		
Reagent	**Stock**	**Final concentration**	**Volume ×1**
High Fidelity buffer with $MgCl_2$ (Roche)	×10	×1	5.0 µl
dNTP mix	10 mM	200 µM	1.0 µl
Primer Hce F	50 µM	0.5 µM	0.5 µl
Primer Hce R	50 µM	0.5 µM	0.5 µl
BSA 20 mg/ml	20 mg/ml	200 µg/ml	0.5 µl
Taq High Fidelity polymerase (Roche)[a]	250 U	2.5 U	0.5 µl
Water			32.0 µl
DNA			10 µl
Total volume			**50 µl**

Thermocycling		
Step 1	**Step 2**	**Step 3**
94°C, 4 min	94°C, 1 min	94°C, 1 min
Cool to 58°C. Add 2.5 U *Taq* to samples	58°C, 2 min	58°C, 2 min
	72°C, 3 min	72°C, 8 min
×1	×35	×1
Product	1022 bp	
Sensitivity	5 fg for pure DNA and 50 fg for DNA from faeces	
Specific against	*Helicobacter pylori*, *Escherichia coli*, *Enterococcus faecalis*, *Proteus mirabilis*, and other *Helicobacter* species isolated from harp seals, sea otters, sea lions	

Helicobacter species

Helicobacter species not specifically identified to any known species have been isolated from wild and captive Australian fur seals (*Arctocephalus pusillus doriferus*), wild New Zealand fur seals (*A. forsteri*), captive Australian sea lion (*Neophoca cinerea*), captive California sea lion (*Zalophus californianus*) and wild silver gulls also known as the seagull (*Chroicocephalus* [*Larus*] *novaehollandiae*) (Oxley *et al.*, 2004; Oxley and McKay, 2005). The unknown *Helicobacter* species were divided into three groups with 10–20 bp, or 94.9–97.5% difference in a 400-bp region of the 16S rRNA gene amplified using *Helicobacter*-specific primers designed by Germani *et al.* (1997). Group 1 isolated from captive seals clustered with *H. cetorum* with a 6–8 bp or 98–98.6% difference. Group 2 isolated from wild New Zealand fur seals did not cluster with any previously reported sequences and may represent a novel species. Group 3 contained isolates from captive and wild seals and silver gulls and clustered with *H. canis*.

DNA sequences which had less than 97.0% similarity to *H. cetorum* and which may represent a new *Helicobacter* species were isolated from dolphins (*Tursiops truncatus*) and a false killer whale (*Pseudorca crassidens*) held captive in an oceanarium in Argentina (Goldman *et al.*, 2011). Gastric fluid from captive polar bears (*Ursus maritimus*) held in an aquarium, Queensland, Australia, was positive using *Helicobacter* species-specific primers and the sequence was different by 11 bp (97.3%) to other *Helicobacter* species (Oxley *et al.*, 2005). *Helicobacter* species were isolated from the gastric mucosa of harp seals, *Pagophilus* (*Phoca*) *groenlandicus*, stranded on the New England coast, USA (Harper *et al.*, 2003a). No clinical abnormalities were seen in the animals. Two different isolates (01-5529-A and 01-5529-B) were cultured from one animal. Both isolates were microaerophilic and grew as a thin spreading film on trypticase soy agar containing 5% sheep blood, VPT and CVA plates (see media section under *Helicobacter* media) after 2–4 weeks at 37°C and in a microaerobic atmosphere. Subcultures grew within 3–5 days. One DNA sequence had 98% similarity to *H. canis* but had different reactions for phenotypic tests. DNA sequence from the isolate clustered with sequence information from a sea lion and sea otter and had phenotypic similarities to the other harp seal isolate except for the urease reaction. Reactions are presented in Table 3.13.

1.19.2 *Wolinella*

DNA sequences similar (98.7%) to *Wolinella succinogenes* and a putative *Wolinella* species (*Candidatus Wolinella africanus*) were recovered using *Helicobacter*-specific primers (H276 forward and H676 reverse) from gastric mucosa biopsies taken from an Australian sea lion (*Neophoca cinerea*) named Duran (Oxley *et al.*, 2005). Erosion, ulceration, fibrosis of the lamina propria and spiral organisms 5–12 μm in size were seen in the gastric mucosa. The animal had a 10-year history of gastrointestinal disease including abdominal pain, loss of appetite and lethargy.

Phylum *Bacteroidetes*

This phylum contains Gram-negative anaerobic rods that are found in the environment of soil, sediments, seawater and intestine and skin of animals. The classes *Bacteroidetes*, *Flavobacteria* and *Sphingobacteria* are in this phylum.

Class Flavobacteria

The *Cytophaga-Flavobacterium-Bacteroides* phylum contains the family *Flavobacteriaceae*, which consists of 26 genera. Two previously unaffiliated organisms misclassified with the *Cytophaga* have been reclassified as *Aquamarina* (*Cytophaga*) *latercula* and *Leeuwenhoekiella* (*Cytophaga*) *marinoflava* (Bernardet and Nakagawa, 2006). At various times this group has been known as the yellow-pigmented rods, *Flavobacterium*-like, or *Cytophaga* group or *Flexibacter* group, and the genera and species within this group have undergone numerous name changes and reclassification in recent years. *Cytophaga* species belong to the family *Cytophagaceae* with *Cytophaga* as the type genus. The genus *Flexibacter* contains *Flexibacter flexilis*, which is the type species. Many species previously assigned to the genus *Flexibacter* have been reclassified to other genera (see List of Names with Standing in Nomenclature at http://www.bacterio.cict.fr). The type genus of the family *Flavobacteriaceae* is *Flavobacterium*.

Order *Flavobacteriales*

1.20 *Flavobacteriaceae*

Many samples from the aquatic environment, both freshwater and marine, may produce yellow-pigmented colonies on culture. It is important to identify whether these colonies are pathogens or saprophytes. Clinical information is extremely useful in knowing how much time and effort to put into their identification. Thus, if bacterial cells, usually long and thin, are seen adhering to the surface or epithelium of aquatic animals, then genera of the *Flavobacteriaceae* family may be suspected. To aid in their identification it is important for the microbiologist to understand the complexities of this family. The following information will assist in this regard, and much of this information has been taken from the paper by J.-F. Bernardet, Y. Nakagawa and B. Holmes in their proposed guidelines for describing members of the family *Flavobacteriaceae* (Bernardet *et al.*, 2002).

Genera of the family *Flavobacteriaceae* consist of a group of halophilic organisms (tolerates saline conditions), some of which are psychrophilic (tolerate cold temperatures) (Tables 1.9, 1.10). They are described as having cells that are

Table 1.9. Differential characteristics of the genera within the family *Flavobacteriaceae*.

Genus	Host	Pigment type[a]	Seawater requirement	Gliding motility	Atmosphere	Growth on MCA
Aequorivita	Seawater, sea ice, green alga	Carotenoid	Variable	Negative	Aerobic	Not tested
Arenibacter	Marine sediment, sea urchin, green alga		Variable	Variable	Aerobic	
Bergeyella	Human, cat, dog	None	Negative	Negative	Aerobic	No growth
Capnocytophaga	Human and dog	Flexirubin	Negative	Positive	Microaerophilic	No growth
Cellulophaga	Seawater, marine alga and beach mud	Yes	Variable	Positive	Aerobic	Not tested
Chryseobacterium	Fish, marine mud, human, cow's milk, soil	Flexirubin	Negative	Negative	Aerobic	Variable
Coenonia	Peking duck, ducks, geese	Not tested	Negative	Negative	Microaerophilic	No growth
Croceibacter	Seawater	Carotenoid			Aerobic	
Elizabethkingia meningoseptica	Human, bullfrogs	None	Negative		Aerobic	Variable
Empedobacter	Human	Flexirubin	Negative	Negative	Aerobic	Positive
Flavobacterium	Fish, water, sea ice, soil, mud, marine lake, Antarctica	Yes (carotenoid)	Negative	11 species, positive	Aerobic	Negative
Gelidibacter	Sea ice	Carotenoid	Positive	Positive	Aerobic	Negative
Myroides	Human	Flexirubin	Negative	Negative	Aerobic	Positive
Ornithobacterium	Turkey	None	Negative	Negative	Microaerophilic	Negative
Polaribacter	Seawater, sea ice, marine lake	Carotenoid	Positive	Negative	Aerobic	Not tested
Psychroflexus	Antarctica	Carotenoid	Variable	Variable	Aerobic	Negative
Psychroserpens	Antarctica	Carotenoid	Positive	Negative	Aerobic	Negative
Riemerella	Duck, pigeon	Variable	Negative	Negative	Microaerophilic	Negative
Salegentibacter	Antarctica	Carotenoid	Negative	Negative	Aerobic	Not tested
Tenacibaculum	Marine algae, marine sponge, fish	Carotenoid, or weak reaction	Variable	Positive	Aerobic	Not tested
Weeksella	Human	None	Negative	Negative	Aerobic	Positive
Zobellia	Red marine alga, marine sediment	Flexirubin	Positive	Positive	Aerobic	Not tested

[a]Some genera only produce one type of pigment, others none, whereas other genera contain species that produce either a flexirubin-type pigment, a carotenoid-type pigment or both. A 'yes' in this column indicates that both or one type of pigment may be produced.
MCA = MacConkey agar.

Table 1.10. Further differential characteristics for the genera in the family *Flavobacteriaceae*. (Data from Vandamme *et al.*, 1994.)

Genus	Aesculin	Catalase	DNase	Gelatin	Glucose acid	Indole	ONPG	Nitrate	Sucrose	Urease
Aequorivita	V	+	80% Neg	+	Neg	NT	NT	Neg	Neg	V
Arenibacter										
Bergeyella	Neg	+	Neg	+	Neg	+	Neg	Neg	Neg	+
Capnocytophaga	V	V	NT	V	+	Neg	V	V	+	V
Cellulophaga	NT	+	V	V	V	NT	NT	V	V	V
Chryseobacterium	+	+	+v	+	+	+	V	V	V, neg	V
Coenonia	+	+	NT	Neg	+	Neg	+	Neg	Neg	Neg
Elizabethkingia meningoseptica	+	+	+	+	Neg	+	+	Neg	Neg	V (82% neg)
Empedobacter	Neg	+	+	+	+	+	Neg	Neg	Neg	V
Flavobacterium	10/14 =+	+ or weak	V	11/14 =+	V	Neg	V	V	V	V
	neg			Neg	+				Pos	Neg
Gelidibacter	+	+	+	V	+	Neg	Neg	Neg	Neg	Neg
Myroides	Neg	+	+	+	Neg	Neg	Neg	Neg	Neg	+
Ornithobacterium	Neg	Neg	Neg	Neg	V	Neg	+	Neg	Neg	+
Polaribacter	V	+ or weak	NT	V	+	Neg	V	Neg	V	Neg
Psychroflexus	V	+	+	V	V	Neg	Neg	Neg	Neg	V
Psychroserpens	Neg	+	Neg	V	Neg	Neg	V	Neg	Neg	Neg
Riemerella	V	+	NT	+	+	V	Neg	Neg	Neg	V
Salegentibacter	+	+	+	+	V	NT	+	+	V	V
Tenacibaculum	Neg	+	+	+	Neg	NT	NT	V	NT	NT
Weeksella	Neg	+	Neg	+	Neg	+	Neg	Neg	Neg	Neg
Zobellia	+	+	+	+	+	+	+	+	+	Neg

V, variable; NT, not tested; Neg, negative; +, positive.
10/14 =+ indicates the number of the 14 species in the genus that are positive.

Gram-negative, 1–10 µm long and 0.3–0.6 µm wide with some species forming filamentous flexible cells and others coiled or helical cells. Some members exhibit a gliding motility and the rest are non-motile. The optimum temperature for growth for all genera is 25–35°C with some species being psychrophilic or psychrotolerant. Growth is aerobic for most of the genera, however microaerophilic or anaerobic conditions are required for some genera.

Sphingolipids are absent and this feature differentiates the family *Flavobacteriaceae* from the family *Sphingobacteriaceae*. None of the genera in the family *Flavobacteriaceae* digest crystalline cellulose when tested with filter paper and this characteristic distinguishes them from the genus *Cytophaga*, which only contains species that digest crystalline cellulose. It is important to differentiate the presence of the enzyme cellulase from other enzymes that can degrade cellulose derivatives such as carboxymethylcellulose or hydroxyethylcellulose, which may be present in some species of the *Flavobacteriaceae* family. Only cellulase is capable of degrading crystalline cellulose, which is tested using filter paper (see under media section, and tests and their interpretation).

A number of other families are contained under the *Cytophaga-Flavobacterium-Bacteroides* phylum. Like the family *Flavobacteriaceae* these include genera that produce yellow-pigmented colonies that may be cultured from samples received from the aquatic environment. Some of these species include *Cyclobacterium marinum*, *Cytophaga hutchinsonii*, *Flexibacter flexilis*, *Marinilabilia salmonicolor*, *Pedobacter heparinus*, *Sphingobacterium spiritivorum* and others. Most of the members of the *Flavobacteriaceae* family are aerobic except for *Capnocytophaga*, *Coenonia*, *Ornithobacterium* and *Riemerella*. An appropriate isolation technique is culture on to blood agar and incubation in a CO_2-enriched atmosphere with 5–10% CO_2 or a commercial gas generated atmosphere that achieves 5% O_2, 10% CO_2 and 85% N_2. Growth is either poor or absent under anaerobic conditions (Bernardet *et al.*, 2002).

Agar digestion is positive for *Cellulophaga*, *Zobellia* and variable for *Flavobacterium*. None of the other genera in the family *Flavobacteriaceae* digest agarose.

Researchers describing new species in the family *Flavobacteriaceae* are recommended to follow the guidelines for description and identification of new taxa and species in Bernardet *et al.* (2002).

1.20.1 *Aequorivita*

The genus *Aequorivita* belongs to the family *Flavobacteriaceae* with five recognized species, which are characterized as psychrotolerant, non-agarolytic, and are yellow-orange on MA 2216 agar due to the production of carotenoid pigment (Bowman and Nichols, 2002; Park *et al.*, 2009). Species are aerobic, oxidase-negative and catalase-positive and non-motile, with growth occurring between 4 and 40°C with an optimum of 25°C.

1.20.2 *Chryseobacterium*

Chryseobacterium arothri

See *Chryseobacterium hominis*.

Chryseobacterium chaponense

TAXONOMY AND SIGNIFICANCE. *Chryseobacterium chaponense* (Sa 1147-06^T=DSM 23145^T=CCM 7737^T) was isolated in mixed culture with *Flavobacterium psychrophilum* from external lesions, gills and fins of diseased Atlantic salmon (*Salmo salar*) farmed in Chile (Kämpfer *et al.*, 2011). Pathogenicity is not known. According to 16S rRNA sequence *C. chaponense* is similar to *C. jeonii* (98.9%) and *C. antarcticum* (97.5%).

GROWTH CHARACTERISTICS AND PHENOTYPIC IDENTIFICATION. *Chryseobacterium chaponense* is an aerobic, non-motile, Gram-negative rod (1–2 µm) that is positive for oxidase and catalase. On nutrient agar colonies are 2 mm, yellowish, smooth, round, translucent, shiny and entire after 48 h at 30°C. On prolonged incubation the colonies do not appear as single colonies but appear to merge together due to production of extracellular substances. A yellow, non-luminescent, non-diffusible pigment is produced. Flexirubin pigment (tested by KOH) is negative. Growth occurs on R2A agar and tryptic soya agar (Oxoid) but not on MacConkey agar. Growth occurs at 11–36°C but does not occur below 4°C or above 37°C. Growth occurs in 0–1% NaCl. Glucose, L-arabinose,

D-mannitol and trehalose are not fermented. Gelatin (plate method), casein, starch, DNA (Oxoid) and Christensen's urease are not hydrolysed and Kovac's and Ehrlich's indole are negative (Kämpfer *et al.*, 2011). These and other reactions are presented in the Pathogenic *Flavobacteriaceae* table (Table 3.7).

Chryseobacterium hominis

TAXONOMY AND SIGNIFICANCE. *Chryseobacterium hominis* (P2K6^T=CIP 109575^T= DSM 19326^T) is a newly described species isolated from pufferfish (*Arothron hispidus*) caught off the coast of Hawaii (Campbell *et al.*, 2008). The isolate was recovered from the kidneys, but no details were given regarding the disease status of the fish. The isolate was originally named *C. arothri* but was later confirmed to be a later heterotypic synonym of *C. hominis* (Kämpfer *et al.*, 2009), which was described a few months earlier for clinical isolates, CDC groups 11-c and 11-h (Vaneechoutte *et al.*, 2007). Based on 16S rRNA sequencing the isolate is most closely related to *C. formosense* (96.6%) and *C. hispanicum* and *C. caeni* (both 96.1%).

GROWTH CHARACTERISTICS AND PHENOTYPIC IDENTIFICATION. *Chryseobacterium hominis* is an aerobic, facultatively anaerobic, Gram-negative rod (1.0×0.5 μm), catalase positive and oxidase positive. It is non-motile by both hanging drop and stab tube methods. Colonies on TSA are yellowish and circular after 24 h at 30°C and yellowish on half-strength MA 2216. It grows at a range of 0.5–3.0% (w/v) NaCl with an optimum of 2.0%, and an optimum range of 20–37°C with slow growth at 20°C and no growth at 42°C. Hydrolysis of Tween 80 was reported positive by Campbell *et al.* (2008) but negative by Kämpfer *et al.* (2009). H_2S is positive in triple sugar iron agar containing 2% NaCl (Campbell *et al.*, 2008) but was reported negative using Kligler agar (Kämpfer *et al.*, 2009). Nitrate reduction is negative in the isolates from the pufferfish but 8 of 11 human strains were positive for nitrate (Vaneechoutte *et al.*, 2007). Flexirubin pigment was not detected. These and other reactions are presented in Tables 3.8, 3.27–29 and 3.38. Rosco diagnostic tablets can be used for some of the biochemical tests (Kämpfer *et al.*, 2009).

Chryseobacterium indologenes

TAXONOMY AND SIGNIFICANCE. *Chryseobacterium indologenes* is a pathogen of cultured American bullfrogs (*Rana castesbeiana* – note *Lithobates catesbeianus* is thought to be the more correct name). Infected frogs display torticollis and stupor. Focal areas of haemorrhage occur in the muscle, legs are swollen and liver, spleen and kidney may be enlarged (Mauel *et al.*, 2002).

Chryseobacterium piscicola

TAXONOMY AND SIGNIFICANCE. *Chryseobacterium piscicola* (VQ-6316s^T=CECT 7357^T=DSM 21068^T) is a newly described species isolated from diseased rainbow trout (*Oncorhynchus mykiss*) and Atlantic salmon (*Salmo salar*) in Chile (Ilardi *et al.*, 2009). The organism was also isolated from farmed diseased Atlantic salmon in Finland and found to belong to the same serological group as the Chilean isolates (Ilardi *et al.*, 2010). Clinical symptoms were similar to infection with *Flavobacterium psychrophilum* with ulcerative lesions on the flank, muscle, skin, anus and peduncle areas. The isolates were differentiated from *F. psychrophilum* by 16S rRNA sequencing. Sequence information revealed the closest phylogenetic organisms were *C. soldanellicola* (96.9%) and *C. soli* (97.1%).

An experimental infection study using Finnish isolates suggested a moderate virulence for salmon (Ilardi *et al.*, 2010).

GROWTH CHARACTERISTICS AND PHENOTYPIC IDENTIFICATION. *Chryseobacterium piscicola* is a Gram-negative rod (0.8–1.2×1.6–3.5μm) and is non-motile and non-gliding. Colonies are yellow, smooth, shiny, circular and alpha-haemolytic on blood agar and growth also occurs on R2A agar, nutrient agar and TSA, but growth is negative on MacConkey agar and MA 2216 (Ilardi *et al.*, 2009). Hydrolysis is negative for Tween 80, tyrosine, agar, starch and casein. DNase is weakly positive. Flexirubin pigment is detected using 20% KOH, but carotenoid pigment tested by Congo red is negative. These and other reactions are presented in Tables 3.7 and 3.38.

Chryseobacterium piscium

TAXONOMY AND SIGNIFICANCE. *Chryseobacterium piscium* (LMG 23089^T=CCUG 51923^T) is a newly described species isolated from freshly caught fish from the South Atlantic Ocean off the coast of South Africa (de Beer *et al.*, 2006). It is suggested that the organism may be a spoilage organism for raw fish. Based on 16S rRNA sequencing the isolate is related to *C. balustinum* (99.3%) and *C. scophthalmum* (98.9%) with DNA–DNA hybridization levels for these two species below 57%.

GROWTH CHARACTERISTICS AND PHENOTYPIC IDENTIFICATION. *Chryseobacterium piscium* is a Gram-negative rod (1.0×<0.5 μm), catalase positive and oxidase positive. Colonies on TSA are yellow, shiny, translucent with entire edges at 24 h. Growth occurs weakly on nutrient agar but not on MacConkey agar (no. 3, Oxoid CM115). Growth occurs in the range 4–32°C but not at 37 or 42°C. Growth occurs in 5% NaCl in nutrient broth (de Beer *et al.*, 2006). Hydrolysis is negative for tyrosine and starch. DNase is weakly positive. Aesculin is positive in 24 h, and tests are positive for hydrolysis of gelatin, casein, lecithinase, phosphatase and urea. H_2S is negative in triple sugar iron agar and SIM medium (Oxoid CM435). Flexirubin pigment is detected using 20% KOH. An alkaline reaction is seen in O-F medium. These and other reactions are presented in Table 3.8.

Chryseobacterium scophthalmum

TAXONOMY AND SIGNIFICANCE. *Chryseobacterium (Flavobacterium) scophthalmum* (CCM 4109^T=LMG 13028^T) was isolated from gills of diseased turbot (*Scophthalmus maximus*) and from seawater collected off the coast of Scotland (Mudarris and Austin, 1989; Mudarris *et al.*, 1994; Vandamme *et al.*, 1994). Fish had gill hyperplasia and haemorrhagic septicaemia with haemorrhages in the eyes, skin, jaw, brain, stomach, liver and kidney. Based on 16S rRNA sequencing *C. scophthalmum* is related to *C. piscium* (98.9%) and *C. balustinum* (99.3%) (de Beer *et al.*, 2006).

GROWTH CHARACTERISTICS AND PHENOTYPIC IDENTIFICATION. *Chryseobacterium scopthalmum* is a non-motile, Gram-negative rod (2.0×0.5 μm) with rounded ends. Rods become shorter in late exponential phase of growth. Gliding may be present in fresh colonies but may be lost on storage. Growth occurs on medium K (see media section), TSA and MA 2216 and colonies are 5–6 mm, shiny, smooth, round, raised and entire after 48 h at 25°C. An orange, non-diffusible, non-fluorescent pigment is produced. Growth occurs at 15–25°C and weak growth occurs at 4°C but growth is negative at 35°C. Growth occurs at 0–4% NaCl but not at 5% NaCl. It does not grow on MacConkey agar. Results are negative for nitrate, H_2S, indole, ADH, LDC and ODC. Agar, chitin, cellulose and starch are not hydrolysed (Mudarris *et al.*, 1994). These and other reactions are presented in Tables 3.7 and 3.38.

Chryseobacterium tructae

TAXONOMY AND SIGNIFICANCE. *Chryseobacterium tructae* (1084-08^T=CECT 7798^T=CCUG 60111^T) was isolated from the liver and gills of diseased rainbow trout (*Oncorhynchus mykiss*) presenting with septicaemia and sampled from Spain (Zamora *et al.*, 2012b). Virulence has not been assessed by experimental challenge. Three isolates of *C. tructae* were 98.5% similar to *C. indologenes* CCUG 14556^T based on 16S rRNA gene sequence, and 2–39% similar based on DNA–DNA hybridization.

GROWTH CHARACTERISTICS AND PHENOTYPIC IDENTIFICATION. *Chryseobacterium tructae* is a non-gliding, Gram-negative rod (2.6×0.7 μm). It is catalase and oxidase positive. On tryptone glucose extract agar (TGE, Difco) and nutrient agar colonies are pale orange, circular, smooth and entire after 72 h at 25°C and non-haemolytic on Columbia agar. A diffusible flexirubin-type pigment is produced. A brown pigment is produced on tyrosine agar. Congo red is negative. Growth is best under aerobic conditions and weak under microaerobic conditions. Growth occurs at 5–30°C with an optimum of 25°C and no growth at 37°C. Results are negative for nitrate reduction. It is non-agarolytic. Results are positive for hydrolysis of gelatin, starch, tyrosine and casein (Zamora *et al.*, 2012b). These and other reactions are presented in Tables 3.7, 3.27–29 and 3.38.

1.20.3 *Cytophaga*

This genus contains *C. hutchinsonii* the type species, and *C. aurantiaca*, which display gliding motility and are cellulolytic. Their closest relative is the genus *Sporocytophaga*, containing *S. myxococcoides*, which is cellulolytic and produces microcysts (Bernardet *et al.*, 1996). *Cytophaga hutchinsonii* grows on Dubos medium supplemented with 30% (w/v) cellulose.

1.20.4 *Elizabethkingia*

Elizabethkingia meningosepticum

TAXONOMY AND SIGNIFICANCE. *Elizabethkingia meningosepticum* is a Gram-negative rod found in the environment. Previously, it was known as *Flavobacterium meningosepticum* and *Chryseobacterium meningosepticum* (Kim *et al.*, 2005).

Elizabethkingia meningosepticum is a pathogen of cultured American bullfrogs (*Rana catesbeiana* – note *Lithobates catesbeianus* is thought to be the more correct name) and cultured tiger frog (*Hoplobatrachus tigerinus* – note *Rana tigerina* and *R. tigrina* are synonyms) (Mauel *et al.*, 2002; Xie *et al.*, 2009). Infected American bullfrogs display torticollis and stupor. Focal areas of haemorrhage occur in the muscle, legs are swollen and liver, spleen and kidney may be enlarged. In cultured tiger frog, *E. meningosepticum* causes cataract disease and the main symptoms are an opaque eye lens, torticollis and ascites. The organism is cultured from the brain.

Elizabethkingia meningosepticum is an aerobic, non-motile, oxidase positive, Gram-negative rod. Glucose is fermented but not sucrose, arabinose or xylose. ONPG is positive and aesculin and gelatin are hydrolysed. Indole and urea can be variable but the strain from tiger frogs was positive. These and other reactions are presented in Tables 3.7, 3.27–29 and 3.38.

1.20.5 *Flavobacterium*

The *Flavobacterium* genus represents predominantly gliding, yellow-pigmented bacteria that belong to the *Flexibacter-Bacteroides-Flavobacterium* phylum. *Flavobacterium* is the type genus of the family *Flavobacteriaceae* that includes the genera of *Bergeyella*, *Capnocytophaga*, *Chryseobacterium*, *Empedobacter*, *Gelidibacter*, *Myroides*, *Ornithobacterium*, *Polaribacter*, *Psychroflexus*, *Psychroserpens*, *Riemerella*, *Weeksella* and misclassified *Cytophaga* and *Flexibacter* species. Members of *Flavobacterium* genus are aerobic, Gram-negative rods, produce yellow-pigmented colonies, are motile by gliding, do not hydrolyse cellulose and are widely distributed in soil and freshwater habitats. Some are pathogenic for fish (Bernardet *et al.*, 1996). Currently there are 111 species of *Flavobacterium* (Euzéby http://www.bacterio.cict.fr/f/flavobacterium.html)

Flavobacterium araucananum

TAXONOMY AND SIGNIFICANCE. *Flavobacterium araucananum* (LM-19-Fp^T=LMG 26359^T=CCM 7939^T) was isolated from kidney and external lesions on diseased Atlantic salmon (*Salmo salar*) cultured in Chile (Kämpfer *et al.*, 2012). Pathogenicity has not been confirmed.

GROWTH CHARACTERISTICS AND PHENOTYPIC IDENTIFICATION. *Flavobacterium araucananum* is a gliding, aerobic, Gram-negative rod (1.0–1.5×0.8–1.0 μm) that is oxidase positive. On TSA colonies are yellow, round and convex. Growth occurs on NA, TSA, Columbia sheep blood agar, Luria-Bertani, R2, brain heart infusion agar and Mueller-Hinton agar but not on MA 2216 or TCBS. Growth occurs at 4–30°C with an optimum of 25–30°C, but does not occur at 2°C or 36°C. Growth occurs at 0–3% NaCl. Glucose, lactose, maltose and D-mannose are fermented but L-arabinose, sucrose and D-mannitol are not. Gelatin, casein and starch are hydrolysed and H_2S is produced (Kämpfer *et al.*, 2012). These and other reactions are presented in Table 3.7.

Flavobacterium branchiophilum

TAXONOMY AND SIGNIFICANCE. *Flavobacterium branchiophilum* (*Flavobacterium branchiophila*) (BGD-7721^T=ATCC 35035^T) causes bacterial gill disease (BGD) in freshwater fish with resulting high mortality (Davis, 1926; Wakabayashi *et al.*, 1989). Infection is characterized by numerous filamentous, Gram-negative bacteria on the gills of fish and clubbing of gill filaments with proliferation of gill epithelium but not necrosis. Gill tissue is usually swollen and there is excess mucus production. Infected fish display laboured ventilation and usually die within 24 h of infection. The disease is usually associated with adverse environmental conditions. Acute outbreaks occur in juvenile fish (Snieszko, 1981). Bacterial gill disease has been reported from rainbow or steelhead trout (*Oncorhynchus mykiss*), brook trout (*Salvelinus fontinalis*), black-spotted trout, Atlantic salmon (*Salmo salar*), chinook salmon (*O. tshawytscha*), yamame or masou salmon (*O. masou*), sockeye salmon (*O. nerka*), sheatfish or wels catfish (*Silurus glanis*) and goldfish (*Carassius auratus*) from Canada, Japan, Hungary, Korea and the USA (Davis, 1926; Ostland *et al.*, 1989, 1994; Wakabayashi *et al.*, 1989; Ko and Heo, 1997).

GROWTH CHARACTERISTICS AND PHENOTYPIC IDENTIFICATION. *Flavobacterium branchiophilum* is a non-motile, non-gliding, Gram-negative rod (5–8×0.5 μm) occurring as short chains of two or three cells. Catalase and oxidase are positive. Samples from gill arches are cultured to minimal nutrient media such as Anacker Ordal's medium. On Anacker Ordal's medium colonies are 0.5–1 mm, light yellow, transparent, round and smooth after 5 days at 18°C. Colonies do not spread. Results are negative for flexirubin pigment (KOH test). It requires minimal nutrients for growth and growth only occurs on trypticase soya agar (TSA) if diluted 1:20. Good growth occurs at 10–25°C, and some strains can grow at 5°C and 30°C. Growth occurs at 0 and 0.025% NaCl and some strains can tolerate 0.1% NaCl. It does not grow anaerobically. It does not grow on DNase medium, TSA, nutrient agar or MacConkey agar. Gelatin, casein and starch are hydrolysed but not cellulose, aesculin, chitin or agar. Nitrate is not reduced. Glucose, sucrose, maltose and trehalose are fermented (Wakabayashi *et al.*, 1989; Ko and Heo, 1997). These and other reactions are presented in Tables 3.7 and 3.38.

MOLECULAR IDENTIFICATION. *Flavobacterium branchiophilum* can be detected using primer BRA1 (5′-ACT TCT ICG AGT AGA AG-3′) where I equals inosine, and universal primer 1500R (5′-GGT TAC CTT GTT ACG ACT T-3′) of Weisberg *et al.* (1991). A reverse primer BRA2 (5′-CGC TCC TAC TCA CGT AG-3′) did not amplify *F. branchiophilum* (Toyama *et al.*, 1996). The master mix is composed of 10 pmol of each primer (BRA1 and 1500R), 1.5 mM $MgCl_2$, 0.5 nmol of each dNTP, 2.5 U *Taq* DNA polymerase and 5 μl of template DNA in a 50-μl reaction volume. Thermocycling is carried out at an initial denaturation at 94°C for 2 min, followed by 30–35 cycles of denaturation at 94°C for 2 min, 45°C for 1.5 min, 72°C for 2 min and a final elongation step at 72°C for 5 min (Toyama *et al.*, 1996).

Flavobacterium ceti

TAXONOMY AND SIGNIFICANCE. *Flavobacterium ceti* ($454\text{-}2^T$=CECT 7184^T=CCUG 52969^T) is a newly described species isolated from stranded beached, beaked whales (*Ziphius cavirostris*) found on the coast of the Canary Islands, Spain (Vela *et al.*, 2007). Sequencing of the 16S rRNA gene identified the new species with closely related species being *F. johnsoniae* (93.7%), *F. frigidimaris* (93.4%), *F. aquidurense* (93.4%), *F. hibernum* (93.4%) and *F. degerlachei* (93.4%).

Members of the *Flavobacterium* genus comprise saprophytic and pathogenic species for fish species and humans and are widely distributed in nature, particularly in aquatic environments. The whale isolates were detected in liver and lung in animals that showed no abnormal clinical signs.

GROWTH CHARACTERISTICS AND PHENOTYPIC IDENTIFICATION. *Flavobacterium ceti* is a Gram-negative rod (0.3×2.0 μm) and is non-motile and non-gliding. Growth occurs aerobically with weak growth in microaerobic conditions and poor growth in an anaerobic atmosphere. On Columbia blood agar, colonies are orange-pigmented, circular, smooth, entire and non-haemolytic at 37°C after 24 h. It grows on TSA agar, weakly on marine agar (MA 2216, Difco) and nutrient agar after 48 h, but does not grow on MacConkey agar. Growth occurs at 22–37°C with an optimum of 35°C with no growth at 4, 15 or 45°C. In brain–heart infusion broth growth occurs in 3% NaCl but not at 4.5 or 6.5% NaCl. Results are positive for catalase and oxidase, and for hydrolysis of gelatin, casein and starch and negative for degradation of CM-cellulose, alginate, pectin and agar. A brown pigment is seen on tyrosine agar but degradation of tyrosine is not detected. The test for carotenoid pigment using Congo red is negative, whereas flexirubin pigment is detected using 20% KOH. These and other reactions are presented in Tables 3.8, 3.27–29 and 3.38.

Flavobacterium chilense

TAXONOMY AND SIGNIFICANCE. *Flavobacterium chilense* (LM-09-fp^T=LMG 26360^T=CCM 7940^T) was isolated from an external lesion on rainbow trout (*Oncorhynchus mykiss*) farmed in Chile (Kämpfer *et al.*, 2012). Pathogenicity has not been confirmed by experimental challenge.

GROWTH CHARACTERISTICS AND PHENOTYPIC IDENTIFICATION. *Flavobacterium chilense* is an aerobic, gliding, Gram-negative long rod (2.0–3.0×0.8–1.0 μm). The result for oxidase is positive. On TSA colonies are yellowish, round and convex. Growth occurs on TSA, nutrient agar, Columbia sheep blood agar, Luria-Bertani, R2A, brain–heart infusion agar and Mueller-Hinton agar but not on MA 2216 or TCBS. Growth occurs at 4–30°C with an optimum of 25–28°C. Growth does not occur at 2 or 36°C. Growth occurs at 0–3% NaCl. L-arabinose, glucose, lactose, maltose and D-mannose are fermented but not sucrose (Kämpfer *et al.*, 2012). These and other reactions are presented in Table 3.7.

Flavobacterium columnare

TAXONOMY AND SIGNIFICANCE. *Flavobacterium columnare* (1-S-2cl^T =ATCC 23463^T) was isolated from warm-water fish in the Mississippi, USA, and although the organism could not be cultured at that time, it was named *Bacillus columnaris* to reflect the columnar masses of bacterial cells attached to fish tissue when examined microscopically (Davis, 1922). The bacterium was later cultured from sockeye salmon (*Oncorhynchus nerka*) and named *Chondrococcus columnaris* (Ordal and Rucker, 1944). Other names include *Cytophaga columnaris* and *Flexibacter columnaris* (Bernardet and Grimont, 1989; Bernardet *et al.*, 1996).

Flavobacterium columnare has been isolated from freshwater fish, tropical fish and eels including the following: Arctic char (*Salvelinus alpinus*), fingerling black bass (*Micropterus salmoides*, *M. dolomieu*), black molly (*Poecilia sphenops*), brook trout (*Salvelinus fontinalis*), buffalo fish (*Ictiobus bubalus*, *I. cyprinellus*), bullhead (*Ameiurus nebulosus*, *A. melas*), carp (*Cyprinus carpio*), cascudo or suckermouth catfish (*Hypostomus plecostomus*), channel catfish (*Ictalurus punctatus*), chinook salmon (*O. tshawytscha*), crappie (*Pomoxis annularis*), killifish (*Fundulus heteroclitus*), minnow (*Pimephales notatus*), neon tetra (*Paracheirodon innesi*), Nile tilapia (*Oreochromis niloticus*), pacu (*Piaractus mesopotamicus*), piracanjuba (*Brycon orbignyanus*), platies or southern platyfish (*Xiphophorus* [*Platypoecilus*] *maculatus*), rohu labeo (*Labeo rohita*), sheatfish or wels catfish (*Silurus glanis*), sunfish (*Lepomis humilis*), tambaqui (*Colossoma macropomum*), warmouth (*Lepomis gulosus*), white bass (*Morone chrysops*), yellow perch (*Perca flavescens*), zebrafish (*Danio rerio*) and Japanese eel (*Anguilla japonica*) from Australia, Belgium, Brazil, Japan, Finland, France, Hungary, India, Spain and the USA (Davis, 1922; Ordal and Rucker, 1944; Wakabayashi *et al.*, 1970; Kumar *et al.*, 1986; Bernardet, 1989; Koski *et al.*, 1993; Decostere *et al.*, 1998; Michel *et al.*, 2002; Figueiredo *et al.*, 2005; Pilarski *et al.*, 2008).

Infection with *F. columnare* is a disease of freshwater fish and is generally referred to as columnaris disease or saddleback disease, and occurs worldwide. It is of particular economic importance in the aquaculture of channel catfish in the USA. The disease is characterized by lesions on the body surface and gills. The appearance of the lesions varies according to fish species. In scaleless fish such as catfish, the infection begins as small circular lesions with a blue-grey necrotic centre and a margin of red inflamed skin due to the infiltration of red cells. Columns of bacteria can be seen at the centre and the edge of the lesion, however the edge of the lesion is almost pure *F. columnare* cells, whereas the centre contains secondary invaders (Davis, 1922; Bullock *et al.*, 1986). In scaled fish the lesions begin at the outer margins of the fins and spread inward over the body giving an appearance of a saddle, hence the name saddleback disease. The scales loosen and slough off as the integument disintegrates. Severe lesions result from penetration of the dermis and necrosis of the underlying muscles. Microscopically the bacterial cells can be seen attached to fish tissue and scales in column-like masses (Davis, 1922).

In experimental challenge of zebrafish (*Danio rerio*), disease onset is rapid with clinical signs of a saddleback lesion occurring within 24 h of bath immersion of abraded fish and in fish injected either intraperitoneally or intramuscularly. The LD_{50} ranges from 1.1×10^6 to 7.2×10^7 CFU according to the challenge method (Moyer and Hunnicutt, 2007).

Virulence is related to strain type and water temperature (Davis, 1922; Wakabayashi *et al.*, 1970; Bullock *et al.*, 1986). Disease outbreaks occur more commonly in the summer months when water temperatures are high as the disease is not seen at tempertures of 10–15°C.

Genetic diversity has been reported amongst strains of *F. columnare* that enable division into three groups or genomovars. However, the phenotypic characteristics are identical for the three genomovars and there is no evidence to classify the groups as new species or subspecies (Triyanto and Wakabayashi, 1999). A genomovar is defined as phenotypically similar but genetically distinct (Ursing *et al.*, 1995).

GROWTH CHARACTERISTICS AND PHENOTYPIC IDENTIFICATION. *Flavobacterium columnare* is a Gram-negative, slender, flexible rod (5–12×0.5 μm) (Davis, 1922). On Anacker Ordal's medium, Hsu-Shott's selective medium, Shieh's medium, modified Shieh's medium or TYES medium (see media section), colonies are pale yellow, with a convoluted centre and rhizoid edges with a tendancy to adhere to the medium (Anacker and Ordal,, 1955; Anacker and Ordal, 1959; Shieh, 1980; Bullock *et al.*, 1986; Triyanto and Wakabayashi, 1999; see colour plate section). Growth occurs at 18–30°C but not below 14°C or above 33°C. Spheroplasts are seen at suboptimal temperatures (Michel *et al.*, 2002).

Gelatin and casein are hydrolysed, whereas chitin, agar, starch and cellulose are not hydrolysed (Wakabayashi *et al.*, 1970; Bernardet, 1989). Congo red is absorbed and flexirubin pigment is produced (as detected using KOH).

Different reactions for the API Zym are reported and this may be due to the different incubation time and temperatures used, and different inoculum preparation. These are noted in the relevant tables.

A selective medium (Shieh, 1980) is detailed in the media section. Tobramycin was added to improve selectivity and reduce the growth and colony size of *Aeromonas hydrophila* and to prevent the growth of *A. salmonicida*, *S. putrefaciens* and *P. fluorescens* (Decostere *et al.*, 1997).

In genomic group I, which contains the type strain IAM 14301^T, 50% of strains are positive for the nitrate test. Nitrate reductase is also variable for genomic group II, whereas strains in genomic group III are negative for nitrate. Strains for genomic group II and III are able to grow at 37°C but not at 15°C, whereas strains of genomic group I are variable with 85% showing growth at 15°C, and only 75% at 37°C (Triyanto and Wakabayashi, 1999).

MOLECULAR IDENTIFICATION. A number of PCRs have been reported for the detection of *F. columnare* but problems have risen with some primer sets that do not detect all strains of *F. columnare* due to the genetic diversity within the species (Triyanto *et al.*, 1999; Bader *et al.*, 2003). In addition, strain ATCC 43622 (RR-1a) identified as *F. columnare* (Bernardet and Grimont, 1989) and used to test some PCRs (Bader and Shotts, 1998) has been shown to be *F. johnsoniae* (Darwish *et al.*, 2004). Primers ColF and ColR were designed to detect all species of *F. columnare* (Darwish *et al.*, 2004). The reaction mixture contains the *Taq* PCR Core kit (Qiagen) plus 0.2 μM of primer ColF (5′-CAG TGG TGA AAT CTG GT-3′), 0.2 μM primer ColR (5′-GCT CCT ACT TGC GTA GT-3′), 20 μM dNTPS, 1× Q solution, 2.5 U *Taq* polymerase, 2.5 mM $MgCl_2$ and 100–500 ng DNA template in a 100-μl reaction volume. Thermocycling consists of initial denaturation at 95°C for 10 min followed by 30 cycles of denaturation at 94°C for 30 s, annealing at 45°C for 30 s, elongation at 72°C for 2 min and a final extension at 72°C for 8 min. An amplicon of 675 bp is produced. The PCR is specific against *F. aquatile* (ATCC 11947), *F. johnsoniae* (ATCC 29584), *F. psychrophilum* (ATCC 49511) and *Tenacibaculum maritimum* (ATCC 43398) (Darwish *et al.*, 2004).

A PCR for detecting and differentiating the three genomovars is based on amplification using primer sets designed to the 16S rRNA gene (Triyanto *et al.*, 1999). All strains of *F. columnare* are amplified using primer Col-72F (5′-GAA GGA GCT TGT TCC TTT-3′) and primer Col-1260R (5′-GCC TAC TTG CGT AGT G-3′). Specific primer sets are then used in a second-round PCR using the amplified product as template to differentiate the genomovars. Primers for genomovar I are Col-Ta (5′-TTC AGA TGG CTT CAT TTG-3′) and primer Col-Tb (5′-CCG TTT ACG GGC GTT GGA ATA CAG-3′); for genomovar II, primer Col-T1 (5′-ATT AAA TGG CAT CAT TTA-3′) and primer Col-T2 (5′-TCG TTT ACG GCG TGG ACT ACC A-3′); for genomovar III, primer Col-T11 (5′-GAT GTG GCC TCA CAT TGT G-3′) and primer Col-Tb (5′-CCG TTT ACG GGC GTT GGA ATA CAG-3′). Concentrations of reagents in the master mix were not given in the original paper, however Michel *et al.* (2002) reported using 0.4 µl *Taq* polymerase (Gibco), 2 µl 10 mM dNTPs (Invitrogen), 2 µl 20 mM primers and 5 µl DNA in a 50-µl reaction volume. Thermocycling is conducted using an initial denaturation at 94°C for 1 min followed by 35 cycles of denaturation at 94°C for 1 min, annealing temperature (60°C for primers Col-72F-Col-1260R; 54°C for genomovar I primers; 52°C for genomovar II primers; and 56°C for genomovar III primers) for 1 min and extension at 72°C for 1.5 min, with a final step of elongation at 72°C for 7 min. Amplified product of 1200 bp is produced from primers Col-72F-Col-1260R. Primers for amplifying genomovar II produce a product band at 621 bp, but also produce bands at 800 and 1000 bp in all *Flavobacterium* species (Triyanto *et al.*, 1999). Triyanto *et al.* (1999) extracted DNA from cultured cells using Chelex® 100, however Michel *et al.* (2002) reported that this method did not result in amplicon using primers Col-72F and Col-1260R, and only a weak band was obtained with boiled cells. DNA extraction was successful using Tris-EDTA and proteinase K. Fish tissue was extracted using the same method but diluted 1:10 prior to use.

A PCR designed to the 16S-23S rDNA region was developed using primers FCISRFL (5′-TGC GGC TGG ATC ACC TCC TTT CTA GAG ACA-3′) and FCISRR1 (5′-TAA TYR CTA AAG ATG TTC TTT CTA CTT GTT TG-3′). DNA is extracted from fish tissue using the DNeasy tissue kit (Qiagen) following the manufacturer's instructions and eluting in 50 µl of nuclease-free water (Welker *et al.*, 2005). A 3-µl reaction volume contains 1× PCR premix (buffer H, Epicentre), 20 pmol of each primer, 3 U *Taq* DNA polymerase (Promega) and 50 ng template DNA. Thermocycling protocol is denaturation at 94°C for 5 min followed by 40 cycles of 94°C for 30 s, 55°C for 45 s, 72°C for 60 s with a final extension step at 72°C for 7 min and at 4°C for infinity. Amplified product varied from 500 to 550 bp for isolates from channel catfish, black bass, carp, brown trout, rainbow trout and chinook salmon from Israel, France and the USA, whereas isolates from Nile tilapia from Brazil had products at 450 and 475 bp. The primers are specific and did not amplify DNA from *F. aquatile* (ATCC 11947), *F. branchiophilum* (ATCC 35035), *F. johnsoniae* (ATCC 43622, ATCC 17061), *F. hydatis* (ATCC 29551), *F. psychrophilum* (ATCC 49418), *Tenacibaculum maritimum* (ATCC 43397), *Aeromonas hydrophila* (MN-01-01), *Edwardsiella ictaluri* (AL-93-75), *E. tarda* (AL-93-68), *Streptococcus iniae* (ARS-60) and *S. agalactiae klunzingeri* (35KU). The sensitivity is 7 CFU from cultured cells, 29.7 CFU/g from liver, 40.0 CFU/g from blood, 40.2 CFU/g from skin and 58.5 CFU/g from gill. The PCR is more sensitive than culture when testing gill and tank water, whereas culture is more sensitive than PCR when testing skin. Gill and skin are the best samples for detecting *F. columnare* compared to liver and blood, as these sites are usually only involved in chronic infections and may not be involved in acute infection (Welker *et al.*, 2005). The primers designed to the 16S-23S rDNA region gave the same sensitivity on cultured cells as the nested PCR of Bader *et al.* (2003).

The Col1 and Col2 primers of Bader and Shotts (1998) and the Col1-1500R and Cola-Colb primers of Toyama *et al.* (1996) do not amplify all strains of *F. columnare* due to the genetic variability within the species (Toyama *et al.*, 1996; Triyanto and Wakayashi, 1999; Bader *et al.*, 2003). In addition, the Col1 and Col2 primers are prone to primer dimer formation and a prolonged preheating step (94°C for 10 min) is needed to overcome this. The Col1 and Col2 primers were redesigned and named FvpF1 and FvpR1 to overcome these issues (Bader *et al.*, 2003). The primers were tested in a singleplex but were not reported to have been retested in the multiplex PCR. An annealing temperature of 59°C is optimal as below this non-specific amplification occurs. The PCR mixture contains 50 pmol of primer FvpF1 (5′-GCC CAG AGA AAT TTG GAT-3′), 50 pmol of primer FvpR1 (5′-TGC GAT TAC TAG CGA ATC C-3′), 200 µM of each dNTP, 2.5 U *Taq* DNA polymerase (Promega), PCR buffer, 1 ng DNA per 50 µl reaction volume and water in a 50 µl volume. Thermocyling is carried out with an initial denaturation at 95°C for 5 min followed by 25 cycles of 95°C for 30 s, 59°C for 30 s, 72°C for 1 min with a final extension step at 72°C for 6 min and 4°C for infinity. Amplicon of 1193 bp is produced. The sensitivity is 100 CFU. To improve sensitivity a first-round PCR can be done with universal 16S rRNA primers 20F (Bader *et al.*, 2003). The primers are specific against *F. aquatile* (ATCC 11947), *F. branchiophilum* (ATCC 35035), *F. johnsoniae* (ATCC 17061), *F. psychrophilum*

(ATCC 49512), *Tenacibaculum maritimum* (ATCC 43397), *A. hydrophila* (ATCC 7966), *Escherichia coli* (J-5), *Edwardsiella ictaluri* (AAHRU AL-93-75), *E. tarda* (AAHRU FL6-60) and *Streptococcus iniae* (AAHRU 98-60).

A multiplex PCR was developed for the simultaneous detection of *F. columnare*, *E. ictaluri* and *Aeromonas hydrophila* (Panangala *et al.*, 2007). Primers for *F. columnare* were based on Welker *et al.* (2005). For the PCR reaction see under *Edwardsiella ictaluri*.

Flavobacterium johnsoniae

TAXONOMY AND SIGNIFICANCE. *Flavobacterium johnsoniae* was isolated from soil and mud and originally named *Cytophaga johnsonae* (Stanier, 1947) but later transferred to the genus *Flavobacterium* as *F. johnsoniae* (Bernardet *et al.*, 1996). *Flexibacter* (*Cytophaga aurantiaca*) *aurantiacus*, strain NCIMB 1382 (DWO) described by Lewin and Lounsbery (1969), is thought to be an isolate of *Flavobacterium johnsoniae* (Bernardet *et al.*, 1996).

Flavobacterium johnsoniae causes skin lesions on juvenile barramundi (*Lates calcarifer*) cultured in hatcheries in Australia (Carson *et al.*, 1993). The bacteria attach to the scales and cause thinning and loss of scales, loss of pigmentation and loss of epithelium. In experimental challenge *F. johnsoniae* causes infection in zebrafish (*Danio rerio*) when injected at a high dose, and led to inflammation and loss of scales at the injection site, a yellow discolouration of the body and a red colour around the jaw and gills, and lesions between the caudal and pelvic fins. The LD_{50} was $3.2–4.2\times10^{10}$ CFU for intramuscular and intraperitoneal injection. Bath challenge did not result in infection (Moyer and Hunnicutt, 2007). *Flavobacterium johnsoniae* causes losses in cultured Danube sturgeon (*Acipenser gueldenstaedtii*) in Turkey (Karatas *et al.*, 2010). Clinical signs included pale grey skin lesions with excessive mucus. Lesions and ulcerations occurred on the pelvic and pectoral fins and on the ventral side of the body.

GROWTH CHARACTERISTICS AND PHENOTYPIC IDENTIFICATION. *Flavobacterium johnsoniae* is a Gram-negative rod (1.5–15×0.2–0.4 μm). Long rods predominate in young cultures in minimal medium, and branched cells may be seen in young cultures. Cells become shorter and coccoid forms can be seen in older cultures. Catalase and oxidase are positive. Growth occurs on Anacker-Ordal's (AO) medium, BA, nutrient agar, TSA, agar containing 10% milk, and on peptone agar. Colonies are yellow. Different colony morphology is seen on peptone agar depending upon the concentration of the agar and peptone. When peptone concentration is 1.0% and greater, colonies are bright yellow, round, raised, compact with an entire edge. When the peptone concentration is less than 1.0% colonies show swarming and at 0.5% peptone colonies are broad and flat with a slightly raised centre with a spreading edge. At less than 0.25% peptone colonies rapidly swarm and may almost become invisible. At an agar concentration of 1.0% and peptone 0.5% colonies sink into the agar (Stanier, 1947). Flexirubin pigment (KOH test) is produced and Congo red is absorbed (see media section, and identification tests and their interpretation). Chitin, gelatin and starch are hydrolysed. Carbohydrates are not fermented. Casein is hydrolysed on agar containing 10% milk. Results are negative for indole. Optimum temperature is 25–30°C but it grows at 10–30°C but not at 37°C. It grows at 0.5% to 1.0% NaCl (Stanier, 1947; Carson *et al.*, 1993). These and other reactions are presented in Tables 3.7, 3.24–26 and 3.38.

Flavobacterium macrobrachii

TAXONOMY AND SIGNIFICANCE. *Flavobacterium macrobrachii* (an-8^T=BCRC 17965^T=LMG 25203^T) was isolated from water in a freshwater aquaculture pond for giant shrimp/prawn (*Macrobrachium rosenbergii*) in China (Sheu *et al.*, 2011a). According to the 16S rRNA sequence, *F. macrobrachii* is most closely related to *F. terrigena* (95.1%) and *F. terrae* (94.9%).

GROWTH CHARACTERISTICS AND PHENOTYPIC IDENTIFICATION. *Flavobacterium macrobrachii* is an aerobic, non-motile, Gram-negative rod (0.8–2.0×0.3–0.5 μm) that is positive for oxidase and catalase. On R2A agar colonies are 0.8–1.5 mm, yellow, round, smooth, entire and convex after 72 h at 25°C. Growth is weak on nutrient agar and does not occur on trypticase soy agar. Growth occurs at 15–30°C with an optimum of 25°C, but does not grow at 10°C or 37°C. Growth occurs at 0–1% NaCl with an optimum of 0%

NaCl. Gelatin, starch, casein, DNA and Tween 40, 60 and 80 are hydrolysed, but not chitin or agar. Flexirubin pigment (KOH) is negative and Congo red is not absorbed (Sheu *et al.*, 2011a). These and other reactions are presented in Tables 3.24–29 and 3.38.

Flavobacterium oncorhynchi

TAXONOMY AND SIGNIFICANCE. *Flavobacterium oncorhynchi* (63-081^T=CECT 7678^T=CCUG 59446^T) was isolated from the liver of diseased rainbow trout (*Oncorhynchus mykiss*) cultured in Spain (Zamora *et al.*, 2012a). The fish displayed clinical signs similar to infection with *F. branchiophilum*. Pathogenicity has not been established. Based on sequence similarity of the 16S rRNA gene, *F. oncorhynchi* is similar to *F. chungangense* (98.6%), *F. frigidimaris* (98.1%), *F. hercynium* (97.9%) and *F. aquidurense* (97.8%).

GROWTH CHARACTERISTICS AND PHENOTYPIC IDENTIFICATION. *Flavobacterium oncorhynchi* is a non-gliding, Gram-negative rod (2–3×0.5 μm) that is oxidase and catalase positive. On TSA, sheep blood agar and nutrient agar colonies are pale yellow, smooth, round and entire at 25°C after 72 h. It is non-haemolytic. A pale brown pigment is produced on tyrosine agar. Growth does not occur on MA 2216, MacConkey agar or nutrient agar containing 3, 4.5 or 6.5% NaCl. Growth occurs aerobically, but is weak under microaerophilic conditions. Growth occurs at 15–30°C with an optimum at 25°C and no growth at 37°C. Flexirubin pigment (detected by KOH) is positive, whereas Congo red is not absorbed. Nitrate is reduced. Gelatin and agar are not hydrolysed. Starch is hydrolysed. Results for hydrolysis of casein are variable amongst strains (Zamora *et al.*, 2012a). These and other reactions are presented in Tables 3.7 and 3.27–29.

Flavobacterium psychrophilum

TAXONOMY AND SIGNIFICANCE. *Flavobacterium psychrophilum* (*Cytophaga psychrophila*, *Flexibacter psychrophilus*) was first described in the USA, and isolated from open lesions on the peduncle region of juvenile coho salmon (*Oncorhynchus kisutch*) (Borg, 1948). The disease has been referred to as bacterial cold water disease (BCWD), low temperature disease, peduncle disease and rainbow trout fry syndrome. *Flavobacterium psychrophilum* mainly causes disease in salmonid species but also has been associated with disease outbreaks in non-salmonids.

Flavobacterium psychrophilum has caused disease outbreaks in Atlantic salmon (*Salmo salar*) in Australia (Schmidtke and Carson, 1995), in ayu (*Plecoglossus altivelis*), coho salmon, chinook salmon, silver salmon and rainbow trout in Japan and the USA (Borg, 1948; Pacha, 1968; Wakabayashi *et al.*, 1994; Cipriano *et al.*, 1996), wild and farmed rainbow trout in Canada, Finland, France and Spain (Bernardet and Kerouault, 1989; Toranzo and Barja, 1992; Wiklund *et al.*, 1994b; Lumsden *et al.*, 1996), rainbow trout and Atlantic salmon in Chile and Norway (Bustos *et al.*, 1995; Nilsen *et al.*, 2011a,b), farmed perch (*Perca fluviatilis*) in Finland (Lönnström *et al.*, 2008), in eel and cyprinids in Germany (Lehmann *et al.*, 1991) and in sea lamprey (*Petromyzon marinus*) in Lake Ontario, USA (Elsayed *et al.*, 2006). In Japan, *F. psychrophilum* also has been isolated from wild-caught pike gudgeon (*Pseudogobio esocinus*), pale chub (*Zacco platypus*) and reddish bullhead (*Liobagrus reini*) (Suzuki *et al.*, 2008).

Disease outbreak in all species is associated with low temperatures, usually up to 15°C, although the upper temperature varies between fish species. Clinical signs also vary between species. In rainbow trout disease occurs at a water temperature of 15–16°C. Fish have external ulcerations that extend into the musculature, an enlarged spleen, and in some fish the kidney may be swollen (Wiklund *et al.*, 1994b). Fry are anorexic with a distended abdomen and show whirling before death. Larger fish show ulcerations of the skin and dark colouration of the caudal peduncle. The spleen is swollen and organs are pale. Yellow ascites fluid may be present (Nilsen *et al.*, 2011b). The bacterium may be recovered from the spleen, kidney and liver, but this is not always the case, and isolation from external ulcers may be the only positive sample (Wiklund *et al.*, 1994b). Infected Atlantic salmon in Australia had eroded fins (pectoral and pelvic and occasionally the dorsal fin) and tails, and in some cases complete loss of the fin. In infrequent cases the caudal fin was involved including erosion of the peduncle musculature. At the time of infection the water temperature was 2–9°C. The organism was isolated from eroded fins but not from the kidney or blood (Schmidtke and Carson, 1995). In cage-cultured rainbow trout in Canada, the infection presented as a necrotic myositis characterized by large bullae within the musculature of the body wall, including the caudal fin and caudal peduncle (Lumsden *et al.*, 1996). The bullae often protruded from the body and eventually ruptured. The bacterium was isolated from the lesions and from kidney. Infected Atlantic salmon in Norway also presented with a necrotic myositis (Nilsen *et al.*, 2011a).

Flavobacterium psychrophilum infection in perch (*Perca fluviatilis*) farmed in Finland resulted in tissue necrosis of the mouth region (Lönnström *et al.*, 2008). In wild eels (*Anguilla anguilla*) in rivers in Germany, the main clinical signs were haemorrhages on the unpaired fins and internal organs such as the swim bladder, peritoneum and dorsal musculature system, and inflammation in the intestine (Lehmann *et al.*, 1991). Infection in carp (*Cyprinus carpio*), tench (*Tinca tinca*) and crucian carp (*Carassius carassius*) resulted in similar clinical symptoms as seen in infected eels (Lehmann *et al.*, 1991). The water temperature was 8–12°C. Infected sea lamprey had shallow skin lesions and ulcers on the dorsal and caudal fins, and some fish had darkening of the skin and the inner wall of the nostrils and eyes were covered in a white slimy film (Elsayed *et al.*, 2006).

GROWTH CHARACTERISTICS AND PHENOTYPIC IDENTIFICATION. *Flavobacterium psychrophilum* is a Gram-negative, slender, flexible rod (1–7×0.5 μm). Cells become shorter as the culture ages. Catalase is positive, although weak reactions have been reported (Schmidtke and Carson, 1995). Oxidase reaction may be weak or negative, and variable reactions have been reported. Growth occurs on Anacker and Ordal's (AO) medium (Anacker and Ordal, 1955), Shieh's medium (Shieh, 1980), Hsu-Shott's medium (Bullock *et al.*, 1986), AO plus skimmed milk and carbohydrates (Daskalov *et al.*, 1999) and TYES agar (Triyanto and Wakabayashi, 1999) (see media section), but not trypticase soy agar (TSA) or Mueller-Hinton agar (Pacha, 1968; Toranzo and Barja, 1992; Schmidtke and Carson, 1995; Elsayed *et al.*, 2006). Colonies are 1–2 mm, bright yellow, smooth, glossy and have thin spreading edges after 5 days at 15–25°C and do not adhere to the agar. Some colonies may have an entire edge, whereas others have a flat rhizoid growth and often both colony types are seen on the plate (Bernardet and Kerouault, 1989). *Flavobacterium psychrophilum* does not grow on nutrient-rich media and a nutrient-limited medium is required. A study investigating different media and supplements found that the addition of activated charcoal improved the number of cells recovered, and it is hypothesized that activated charcoal, which has a high porosity but no nutritional value, adsorbs any toxic factors in the medium such as hydrogen peroxide and superoxide ions. Six media were tested: nutrient agar (NA), nutrient agar with activated charcoal (NAC), enriched Anacker Ordal serum supplemented with activated charcoal and aromatic compounds (EAOCa) and other combinations based on Anacker Ordal medium, which included enriched Anacker Ordal serum, enriched Anacker Ordal serum supplemented with aromatic compounds and enriched Anacker Ordal serum supplemented with activated charcoal. NAC and EAOCa performed the best with the greatest number of cells recovered, however results were more variable with NAC and therefore EAOCa is recommended (Álvarez and Guijarro, 2007). (Recipes for NAC and EAOCa are in the media section.) A second study investigated the ingredients in the most commonly used media (Tryptone-enriched AO [EAO], TYES and TYES-maltose) for *F. psychrophilum* and found that sodium acetate, calcium chloride and magnesium sulfate inhibit growth, whereas skimmed milk (0.2%) and horse serum (1%) slightly improve the growth (Oplinger and Wagner, 2012). Maltose has no effect on growth. EAO medium gave the best growth results compared to TYES and TYES-maltose. The best medium was EAO without sodium acetate and with the addition of 0.2% skimmed milk and 1% horse serum (EAO-SH) (see media section). *Flavobacterium psychrophilum*, therefore, requires protein supplied by tryptone a pancreatic digest of casein, and casein from skimmed milk. Albumin in horse serum provides an additional protein source. Higher concentrations of serum did not improve growth (Oplinger and Wagner, 2012).

Samples are best taken from lesions, mucus and kidney. In some cases culturing serial dilutions of tissue may improve the isolation rate. It should be incubated at 14–18°C for 5–7 days. Growth occurs at 4–20°C, is weak or slow at 6, 22 and 25°C and does not grow at 30°C. Growth occurs at 0.8% NaCl and only 60% of isolates are capable of growth at 1% NaCl, and 2% NaCl is inhibitory for growth (Pacha, 1968; Bernardet and Kerouault, 1989). Gliding motility may be difficult to observe under microscopic examination as the movement is unusually slow (Lehmann *et al.*, 1991; Schmidtke and Carson, 1995).

Casein and tyrosine are hydrolysed, but agar, starch, aesculin, chitin, cellulose and CM-cellulose are not hydrolysed. Albumin is digested. Autoclaved cells of *Escherichia coli* are digested. Glucose is not fermented (Pacha, 1968; Bernardet and Kerouault, 1989). Hydrolysis of gelatin is variable between strains and is suggested as a virulence factor as a positive gelatin result has been linked to muscle necrosis. Hydrolysis of elastase as detected using AO medium containing 0.05% elastin (derived from bovine neck ligament, Sigma) is variable between strains (Nilsen *et al.*, 2011a). Flexirubin (KOH) is produced and Congo red is not absorbed (Lehmann *et al.*, 1991; Schmidtke and Carson, 1995). Schmidtke and Carson (1995) noted that isolates from Atlantic salmon cultured in Australia produced brown pigment on media containing tyrosine, which has not been found with other isolates (Lehmann *et al.*, 1991).

These and other reactions are presented in Tables 3.7, 3.30 and 3.38.

MOLECULAR IDENTIFICATION. A number of PCRs have been developed to target the 16S rRNA, the *gyrB*, *gyrA* and *ppiC* genes, and use nested PCR, single PCR, real-time PCR or LAMP techniques. In a study to investigate the specificity and sensitivity of primers targeting the 16 rRNA (Toyama *et al.*, 1994), *gyrB* (Izumi and Wakabayashi, 2000), *gyrA* (Izumi and Aranishi, 2004) and *ppiC* (Yoshiura *et al.*, 2006) genes, primers that targeted the 16S rRNA, *gyrB* and *ppiC* genes had the same sensitivity, however the *gyrA* gene target gave some false positives. The 16S rRNA target gene was ten times more sensitive than the other PCRs due to the six copies of the 16S rRNA gene in *F. psychrophilum* compared to one copy of the other genes, however it is less specific and gave some occasional false positives when tested against gill washings, and therefore the *gyrB* and *ppiC* PCRs are recommended (Suzuki *et al.*, 2008). Primers PSY1 and PSY2 (Toyama *et al.*, 1994) were also reported to produce some non-specific amplicons (Urdaci *et al.*, 1998).

The PCR using primers targeting the *gyrB* gene that encodes the subunit B protein of DNA gyrase consists of 10 pmol primer PSY-G1F (5′-TGC AGG AAA TCT TAC ACT CG-3′), 10 pmol primer PSY-G1R (5′-GTT GCA ATT ACA ATG TTG T-3′), 2 nmol of each dNTP, 1 µl 10× PCR buffer, 0.25 U *Taq* DNA polymerase (Takara) and 1 µl template DNA in a 10-µl reaction volume (Izumi and Wakabayashi, 2000; Suzuki *et al.*, 2008). Thermocycling is carried out with an initial denaturation at 94°C for 5 min, followed by 35 cycles of denaturation at 94°C for 60 s, annealing at 56°C for 60 s, elongation at 72°C for 120 s and a final cycle using an extension at 72°C for 5 min. A 1017-bp amplicon is produced. The primers are specific against *F. breve* (ATCC14234), *F. columnare* (JIP P06/90), *F. flevensis* (IAM 12375), *F. flavescens* (ATCC 14231), *F. johnsoniae* (ATCC 21123), *F. sulfureum* (ATCC 14232), *Aeromonas salmonicida* (JCM 7874), *A. hydrophila* (IAM 12460), *Pseudomonas aeruginosa* (IAM 1514), *P. anguilliseptica* (NCINB 1950), *P. ayucida* (FPC 941), *P. chlororaphis* (FPC 308), *P. fluorescens* (FPC 96), *P. putida* (ATCC 12633), *Edwardsiella tarda* (SU 226), *Photobacterium damselae* ssp. *piscicida* (KG 8702), *Vibrio anguillarum* (FPC 346), *V. pelagius* (ATCC 25916), *V. mimicus* (ATCC 33653), *V. harveyi* (ATCC 35084^{T}), *V. tubiashi* (ATCC 19109) and *Yersinia ruckeri* (FPC 442). Primers PSY-G1F and PSY-G1R identified the same number of positive samples as primers PSY1 and PSY2 (Toyama *et al.*, 1994) designed to the 16S rRNA gene (Izumi and Wakabayashi, 2000).

The other PCR found to be specific by Suzuki *et al.* (2008) is based on primers targeting the *ppiC* gene (encodes peptidyl-prolyl *cis-trans* isomerase C) as described by Yoshiura *et al.* (2006). The PCR reaction mixture contains primer fp-PPIC1F (5′-GTA CCA TGA TAC AGT CAG GTT TTT ATA CCA-3′) and primer fp PPIC1R (5′-GCG TTT TTA AAT CCA ACT CTT GCT TCG-3′) with other reagents as used by Suzuki *et al.* (2008) for the *gyrB* PCR. Thermocycling was performed using initial denaturation at 94°C for 5 min followed by 35 cycles of denaturation at 95°C for 15 s, annealing at 60°C for 30 s, elongation at 72°C for 30 s with a final elongation at 72°C for 5 min (Yoshiura *et al.*, 2006; Suzuki *et al.*, 2008). An amplicon of 346 bp is produced.

Primers PSY1 and PSY2 designed to the 16S rRNA gene (Toyama *et al.*, 1994) have been reported extensively (Bader and Shotts, 1998; Wiklund *et al.*, 2000), but as previously mentioned some non-specific amplicons and false-positives have been reported with these primers (Urdaci *et al.*, 1998; Suzuki *et al.*, 2008). Primers PSY1 (5′-CGA TCC TAC TTG CGT AG-3′) and PSY2 (5′-GTT GGC ATC AAC ACA CT-3′) produce amplicon of 1089 bp after 30 cycles using an annealing temperature of 45°C for 1.5 min (Toyama *et al.*, 1994). Suzuki *et al.* (2008) tested this primer pair using the protocol reported for the *gyrB* primers which has an annealing temperature of 56°C. Annealing temperatures of 50°C (Bader and Shotts, 1998) or 57°C (Wiklund *et al.*, 2000) increased the amount of amplicon from primers PSY1 and PSY2.

Other primers (FP1 and FP2) designed to the 16S rRNA gene were specific when tested against over 40 species from the *Flavobacterium*, *Cytophaga*, *Chryseobacterium* and *Sphingobacterium* genera (Urdaci *et al.*, 1998). The reaction mixture consists of 10 pM primer FP1 (5′-GTT AGT TGG CAT CAA CAC-3′) and 10 pM primer FP2 (5′-TCG ATC CTA CTT GCG TAG-3′), 50 µM each dNTP, 1 U *Taq* DNA polymerase, 2.5 µl 10× PCR buffer and 25–50 ng DNA template in a 25-µl volume. Thermocyling is carried out with an initial denaturation at 95°C for 5 min, followed by 35 cycles of 94°C for 30 s, 54°C for 1 min, 72°C for 1 min and a final elongation step of 72°C for 10 min. An amplicon of 1088 bp is produced. If an annealing temperature of less than 54°C (e.g. 47°C) is used, non-specific amplicons occur with *F. columnare*, *F. johnsoniae* and *Tenacibaculum maritimum* (Urdaci *et al.*, 1998). The primers can be used in a nested PCR with universal primers of Weisburg *et al.* (1991), where forward primer is 5′-AGA GTT TGA TCA TGG CTC AG-3′ and reverse primer is 5′-CGG TTA CCT TGT TAC GAC TT-3′, which are used in the first-round reaction.

A qPCR with primers and probe designed to the 16S rRNA gene (del Cerro *et al.*, 2002) had a sensitivity of 11 copies per reaction or 1.1 pg DNA per reaction and a detection limit of 4.7 CFU (Suzuki *et al.*, 2008). Liver from experimentally infected fish is collected and homogenized as a 10% (w/v) suspension in TE buffer (1 mmol/l Tris-HCl and 0.5 mmol/l EDTA, pH 8). An aliquot of 300 µl is centrifuged at 12,000 rpm for 5 min. The supernatant is discarded and the pellet resuspended in 200 µl ultrapure water and boiled for 5 min. The solution is placed on ice before centrifuging for 30 s to remove cell debris. DNA from *F. psychrophilum* in liver tissue can also be extracted using the InstaGene Matrix (BioRad). The reaction mixture consists of 1 µmol/l of forward primer FP1 (5′-GTT AGT TGG CAT CAA CAC-3′) (Urdaci *et al.*, 1998) and 1 µmol/l of newly designed primer FP3 (5′-ACA CTG GCA GTC TTG CTA-3′), 200 nmol/l probe FPP1 (5′-FAM-TGA CGA CAA CCA TGC AGC ACC TTG-TAMRA-3′), 200 µmol/l each dNTP, 1 U Dynazyme II DNA polymerase and buffer (Finnymes OY, Espoo, Finland), 3 mmol/l $MgCl_2$ final concentration, 2–5 µl template DNA in a 50-µl volume. Thermocycling is carried out on a GeneAmp 9700 using an initial denaturation of 94°C for 2 min followed by 45 cycles of denaturation at 94°C for 40 s, 60°C for 40 s, extension at 72°C for 60 s and a final extension step at 72°C for 5 min. An amplicon of 971 bp is produced. The qPCR was specific when tested against *F. columnare* (LMG 13035), *F. johnsoniae* (UW101), *Aeromonas hydrophila* (CECT 4221), *Lactococcus garvieae* (CECT 5073) and *Yersinia ruckeri* (147) (del Cerro *et al.*, 2002).

Another qPCR using SYBR Green and primers designed to the 16S rRNA gene comprises a reaction mixture of 1 µl primer Fp_16S1_fw (5′-GAG TTG GCA TCA ACA CAC-3′) 2 µmol/l, 1 µl primer Fp_16Sint1_rev (5′-TCC GTG TCT CAG TAC CAG-3′) 2 µmol/l, 12.5 µl Brilliant SYBR Green qPCR master mix 2× (Stratagene), 5 µl bacterial culture and water in a-25 µl reaction volume. Thermocycling is carried out with an initial denaturation at 96°C for 10 min followed by 40 cycles of denaturation at 96°C for 30 s, annealing at 56°C for 30 s and extension at 72°C for 30 s. A melt curve analysis is performed from 65°C to 95°C with temperature increasing at a rate of 2°C/min. Fluorescence is recorded using the SYBR Green I optic channel set at wavelength 495 nm on a MX3000p Stratagene thermocycler. A 146-bp amplicon is produced. Positive samples are detected at Ct values of 18.9±2.2, whereas negatives are recorded at 28.5–38.4. The melt temperature of the amplicon is 80.5°C±0.6°C. Weak cross-reaction with *F. branchiophilum* is seen, and although Ct values are high (Ct 35) for *F. columnare*, *Shewanella putrefaciens* and *Bacillus megaterium*, the melt temperatures are 81.03°C±0.04°C (Orieux *et al.*, 2011).

A loop-mediated PCR using primers designed to the DNA topoisomerase IV subunit B gene, *parE*, has a sensitivity of 2.0×10^1 copies per reaction. DNA from cultured cells is extracted using the Generation Capture Column kit (Gentra Systems). The reaction mixture consists of 8 U *Bst* DNA polymerase large fragment (New England Biolabs), 0.8 mM dNTPs (GE Healthcare UK), 1.6 µM primer FIP-17 (5′-CTG AGG CTT CGA ATG GTT TGT GAG GTA CAC ATT TAG GCG C-3′), 1.6 µM primer BIP-17 (5′-CTG CCA TTG TAG TTC GTG TCG GGT CCC ATA TCT GTA GAT CC-3′), 0.2 µM primer F3-17 (5′-ACT CCT TTG TAA ACG GGC-3′), 0.2 µM primer B3-17 (5′-ATG TTC TCA CCG AAG GC-3′), 0.8 M betaine, 20 mM Tris-HCl (pH 8.8), 10 mM KCl, 10 mM $(NH_4)_2SO_4$, 8 mM $MgSO_4$, 0.1% Tween 20 and 2 µl DNA template in a 25-µl reaction volume. The reaction proceeded at 65°C for 80 min followed by inactivation at 80°C for 5 min. Turbidity, due to the production of magnesium phosphate, a by-product of the reaction, was read with a real-time turbidimeter (Realoop-30, Moritex Corp.). The LAMP PCR is specific when tested against *F. aquatile* (NBRC 15052^T), *F. branchiophilum* (NBRC 15030^T), *F. columnare* (NBRC 100251^T), *F. flevense* (NBRC 14960^T), *F. hydatis* (NBRC 14958^T), *F. johnsoniae* (NBRC 14942^T), *Chryseobacterium balustinum* (NBRC 15053^T), *Tenacibaculum maritimum* (NBRC 15946^T) and *Vibrio anguillarum* (M93) (Fujiwara-Nagata and Eguchi, 2009).

Flavobacterium tilapiae

TAXONOMY AND SIGNIFICANCE. *Flavobacterium tilapiae* (Ruye-71^T=BCRC 80262^T=KCTC 23312^T) was isolated from freshwater from the culture pond of apparently healthy tilapia (*Tilapia rendalli*) in Taiwan (Chen *et al.*, 2012a). Based on 16S rRNA sequence *F. tilapiae* is most closely related to *F. defluvii* (97.7%), *F. johnsoniae* (96.4%) and *Flexibacter aurantiacus* (96.3%).

GROWTH CHARACTERISTICS AND PHENOTYPIC IDENTIFICATION. *Flavobacterium tilapiae* is an aerobic, non-motile, Gram-negative rod (2.0–7.7×0.6–0.3 µm). Gliding motility is observed. Results are positive for catalase and oxidase. On R2A agar colonies are 1.1–3.2 mm, bright yellow, slightly sticky, convex, circular with irregular curled margins after 72 h at 25°C. Growth occurs at 10–30°C with an optimum of 15–30°C. Growth occurs at 0–2% NaCl with an optimum of 0–1%. Casein, starch, chitin, DNA, gelatin and CM-cellulose are hydrolysed. Flexirubin (using KOH) is positive but Congo red is not absorbed (Chen *et al.*, 2012a). These and other reactions are presented in Tables 3.8, 3.27–29 and 3.38.

1.20.6 *Flexibacter*

Most of the *Flexibacter* species have been reclassified into other genera (Nedashkovskaya *et al.*, 2010).

Flexibacter aurantiacus

Bernardet *et al.* (1996) proposed that *Flexibacter aurantiacus* (NCIMB 1382^T and NCIMB 1455) was a later homotypic synonym of *Flavobacterium johnsoniae*, however this has not been formally changed (http://www.bacterio.cict.fr/f/flexibacter.html).

1.20.7 *Gramella*

Gramella marina

TAXONOMY AND SIGNIFICANCE. *Gramella marina* (KMM 6048^T=KCTC 12366^T=25418^T) was isolated from the sea urchin *Strongylocentrotus intermedius* collected from Troitsa Bay, Gulf of Peter the Great, Sea of Japan (also known as East Sea), Pacific Ocean (Nedashkovskaya *et al.*, 2010). It is likely to be part of the normal flora. *Gramella* belongs to the family *Flavobacteriaceae*. *Gramella marina* has 16S rRNA sequence similarity with *G. echinicola* (97.5%) and *G. portivictoriae* (98.4%).

Gramella marina is a strictly aerobic, oxidase positive, Gram-negative gliding rod (1.7–2.5×0.4–0.6 µm). Catalase is positive. On MA 2216 colonies are 2–3 mm, yellow-orange, circular, shiny, with entire edges afer 24 h at 28°C. Seawater salts are required for growth. Growth occurs at 4–37°C with an optimum of 28–30°C and at 1–15% NaCl with an optimum of 2–5%. Aesculin, casein, gelatin, starch and Tween 80 are hydrolysed. These and other reactions are presented in Tables 3.8, 3.24–29 and 3.38.

1.20.8 *Tenacibaculum*

The genus *Tenacibaculum* is part of the family *Flavobacteriaceae* and currently consists of 15 species. They are found as part of the environment in water, tidal flats and mud. Some species are pathogens of fish, some have been isolated from diseased fish and sea anemone, but virulence capacity is yet to be established, and some are part of the microflora of oysters. *Tenacibaculum* species are strictly aerobic, Gram-negative rods that do not have flagella but are motile by gliding. They may or may not adhere to agar. A yellow pigment is produced that is mainly zeaxanthin. Flexirubin-type pigment is not produced (Suzuki *et al.*, 2001).

Tenacibaculum aiptasiae

TAXONOMY AND SIGNIFICANCE. *Tenacibaculum aiptasiae* (a4^{T}=BCRC 17655^{T}=LMG 24004^{T}) was isolated from diseased laboratory-cultured sea anemone (*Aiptasia pulchella*) in Taiwan (Wang *et al.*, 2008a). It was most closely related to *T. lutimaris* (97.6%) and *T. aestuarii* (97.7%) based on 16S rRNA sequencing.

GROWTH CHARACTERISTICS AND PHENOTYPIC IDENTIFICATION. *Tenacibaculum aiptasiae* is an aerobic, Gram-negative rod (0.4–0.6×2.0–7.0 μm), catalase positive and oxidase positive and exhibits gliding motility. Cells are straight but occasional degenerative spherical cells may be seen in ageing MA 2216 broth cultures. On MA 2216 agar colonies are pale yellow, circular and convex with spreading margins, 10–20 mm in diameter after 5 days at 25°C (Wang *et al.*, 2008a). Growth occurs at 8–40°C and in 1–10% NaCl at pH 7–9. Optimal conditions are 30–35°C, 3–4% NaCl and pH 8.0. Flexirubin pigment is not produced but carotenoid pigments using Congo red are positive. Hydrolysis is positive for gelatin, casein, starch, DNA, aesculin and Tween, 40, 60 and 80 but negative for Tween 20 and chitin. These and other reactions are presented in Tables 3.7, 3.27–29 and 3.38.

Tenacibaculum crassostreae

TAXONOMY AND SIGNIFICANCE. *Tenacibaculum crassostreae* (JO^{T}=KCTC 22329^{T}=JCM 15428^{T}) was isolated from apparently healthy Pacific oyster (*Crassostrea gigas*) collected from Wan Island in Korea (Lee *et al.*, 2009). The organism was identified as a new species based on significant difference within the 16S rRNA gene giving a similarity of 94.6–97.8%, and by DNA–DNA hybridization with less than 28% relatedness with major *Tenacibaculum* species.

GROWTH CHARACTERISTICS AND PHENOTYPIC IDENTIFICATION. *Tenacibaculum crassostreae* is an aerobic, Gram-negative straight rod (0.3×1.5–5.0 μm), catalase positive, oxidase positive and exhibiting gliding motility. On MA 2216 agar colonies are translucent yellow, circular, low-convex with regular edges, 5 mm in diameter after 5 days at 30°C (Lee *et al.*, 2009). Growth occurs at 15–37°C, at pH 6.0–7.0, and in 3–7% (w/v) sea salts but not in NaCl alone. No growth at 40°C. Flexirubin pigment is not produced. Hydrolysis is positive for casein, starch and Tween 80 but negative for chitin. These and other reactions are presented in Tables 3.8, 3.24–29 and 3.38. Each kit is used according to the manufacturer's instructions except that the inoculum is prepared with artificial seawater or 2% sea salts (Lee *et al.*, 2009).

Tenacibaculum dicentrarchi

TAXONOMY AND SIGNIFICANCE. *Tenacibaculum dicentrarchi* (35/09^{T}=NCIMB 14598^{T}=CECT 7612^{T}) was isolated from diseased European seabream (*Dicentrarchus labrax*) cultured in Spain (Piñeiro-Vidal *et al.*, 2012). Based on 16S rRNA gene sequence *T. dicentrarchi* is similar to *T. ovolyticum* (97.3%), *T. solea* (96.4%), *T. aiptasiae* (96.3%) and *T. gallaicum* (95.3%).

GROWTH CHARACTERISTICS AND PHENOTYPIC IDENTIFICATION. *Tenacibaculum dicentrarchi* is a strictly aerobic, gliding, Gram-negative straight rod (2–40× 0.3–0.5 μm). Long filamentous cells up to 150 μm may be seen, and spherical cells may occur in old cultures. Positive for catalase and oxidase. On Flexibacter maritimus medium (FMM) (see media section) and MA 2216 (Difco) colonies are pale yellow, non-adherent, flat, circular with uneven edges after 48 h at 25°C. Flexirubin pigment is not produced. Growth occurs in media containing 30–100% strength seawater but not in media with NaCl only. Growth occurs at 4–30°C with an optimum of 22–25°C. Gelatin and casein are hydrolysed but not Tween 80 or starch. Carbohydrates are not fermented. Negative for nitrate, H_2S, indole and VP. These and other reactions are presented in Tables 3.7 and 3.38. The API kit is used according to the manufacturer's instructions with the exception that sterile seawater or artificial seawater salts is used as the inoculum.

Tenacibaculum discolor

TAXONOMY AND SIGNIFICANCE. *Tenacibaculum discolor* (LL04 11.1.1^T=NCIMB 14278^T=DSM 18842^T) was isolated from the kidney of diseased sole (*Solea senegalensis*) in north-western Spain (Piñeiro-Vidal *et al.*, 2008b). Sequence similarity of the 16S rRNA gene occurs with *T. litoreum* (99.4%), *T. mesophilum* (97.4%), *T. aestuarii* (97.3%), *T. lutimaris* (97.3%), *T. skagerrakense* (96.4%) and *T. amylolyticum* (96.0%). Signs typical of flexibacteriosis, as seen in infection with *T. maritimum*, occur also in sole, which display an eroded mouth, rotten fins, shallow lesions on the skin and pale-coloured internal organs. Pathogenicity testing confimed *T. discolor* is virulent for turbot and sole (Piñeiro-Vidal *et al.*, 2007).

GROWTH CHARACTERISTICS AND PHENOTYPIC IDENTIFICATION. *Tenacibaculum discolor* is a strictly aerobic, Gram-negative rod (0.5×2.0–30 μm), catalase positive and oxidase positive and exhibits gliding motility. Growth occurs on FMM (Flexibacter maritimus medium) agar as flat, bright yellow colonies with uneven edges. It does not adhere to the agar. On MA 2216 colonies are a bright yellow viewed straight on, but appear bright green in colour when viewed at a 30–45° angle (Piñeiro-Vidal *et al.*, 2008b). Growth occurs in media supplemented with seawater at concentrations of 30–100%, but growth does not occur in media supplemented with NaCl only. Growth temperature ranges from 14 to 38°C with an optimum of 25–30°C, and pH tolerance ranges from 6.0 to 8.0. Flexirubin pigment is not produced. Hydrolysis of gelatin and casein is positive, but results are negative for starch and Tween 80. Acid is not produced from carbohydrates. These and other reactions are presented in Tables 3.7 and 3.38.

Tenacibaculum gallaicum

TAXONOMY AND SIGNIFICANCE. *Tenacibaculum gallaicum* (A37.1^T=NCIMB 1417^T=DSM 18841^T) was isolated from the seawater in a holding tank containing turbot (*Psetta maxima*) in north-western Spain (Piñeiro-Vidal *et al.*, 2008b). Sequence analysis of the 16S rRNA gene identifies the closest species as *T. litoreum* (98.4%), *T. aestuarii* (97.3%), *T. lutimaris* (96.8%), *T. mesophilum* (96.7%), *T. amylolyticum* (96.3%) and *T. skagerrakense* (95.9%). Diseased fish show signs typical of flexibacteriosis, which includes an eroded mouth, rotten fins, shallow skin lesions and pale internal organs. Pathogenicity testing confims *T. gallaicum* is virulent for turbot and sole (Piñeiro-Vidal *et al.*, 2007).

GROWTH CHARACTERISTICS AND PHENOTYPIC IDENTIFICATION. *Tenacibaculum gallaicum* is a strictly aerobic, Gram-negative rod (0.5×2.0–30 μm), catalase positive and oxidase positive and exhibits gliding motility. Spherical degenerative cells rarely occur in ageing 2216 broth cultures. Growth occurs on FMM (Flexibacter maritimus medium, see media section) agar and MA 2216 agar as flat, bright yellow colonies with uneven edges. It does not adhere to the agar (Piñeiro-Vidal *et al.*, 2008b). Growth occurs in media supplemented with seawater at concentrations of 30–100%, but growth does not occur in media supplemented with NaCl only. The optimal growth temperature is 25–30°C but occurs over a range of 14–38°C and a pH range of 6.0–8.0. Hydrolysis of gelatin and casein is positive, but negative for starch and Tween 80. Nitrate is positive but VP, indole and H_2S are negative. Acid is not produced from carbohydrates. These and other reactions are presented in Tables 3.7 and 3.38. The kits are used according to the manufacturer's instructions with the exception that sterile seawater or artificial seawater salts is used as the inoculum.

Tenacibaculum litopenaei

TAXONOMY AND SIGNIFICANCE. *Tenacibaculum litopenaei* (B-1^T=BCRC 17590^T=LMG 23706^T) was isolated from pond water of cultured prawn (shrimp) (*Litopenaeus vannamei*) in Taiwan (Sheu *et al.*, 2007).

GROWTH CHARACTERISTICS AND PHENOTYPIC IDENTIFICATION. *Tenacibaculum litopenaei* is an aerobic,

Gram-negative rod (0.3–0.5×2.0–10.0 μm in length), catalase negative and oxidase positive and exhibits gliding motility. Growth occurs on MA 2216 at 28°C after 48 h as 5–10 mm yellow colonies with spreading edges (Sheu *et al.*, 2007). Growth occurs at 10–39°C and in 2–10% NaCl at pH 5–10. Optimal conditions are 28–37°C, 3% NaCl and pH 7.0–8.0. The occasional spherical cell is seen in ageing broth culture (MA 2216). Flexirubin pigment is not produced, but carotenoid pigments are present when detected using Congo red. It hydrolyses gelatin, chitin and starch, but reactions are negative for nitrate, indole, aesculin, urease, Tween 80 and arginine dihydrolase. These and other reactions are presented in Tables 3.8, 3.27–29 and 3.38.

Tenacibaculum maritimum

TAXONOMY AND SIGNIFICANCE. *Tenacibaculum maritimum* (R2^T=NBCM 2514^T) was originally named *Flexibacter maritimus* (Wakabayashi *et al.*, 1986). Phylogenetic analysis based on 16S rRNA gene sequence and the DNA gyrase B subunit gene (*gyrB*) identified the organism as belonging to the *Flavobacteriaceae* family and a new genus *Tenacibaculum* was established (Suzuki *et al.*, 2001). The name *Cytophaga marina* (NCMB 2153) is synonymous with *T. maritimum* (Holmes, 1992).

Tenacibaculum maritimum is the causative agent of tenacibaculosis, an ulcerative disease that affects a variety of marine fish species. Over the years a number of names have been used for the disease including salt water columnaris disease, bacterial stomatitis, black patch necrosis and eroded mouth syndrome. Clinical signs typically include ulcers and necrosis on the body surface, eroded mouth, frayed fins and tail rot, and the disease may progress to septicaemia (Avendaño-Herrera *et al.*, 2006). Initial signs may be pale areas of skin with yellow edges that develop rapidly into necrotic lesions. Microscopic examination of the periphery of lesions may reveal slender Gram-negative rods in parallel (Bernardet *et al.*, 1994).

The first isolation was in Japan from diseased cultured red seabream (*Pagrus major*) fry 15 to 60 mm in length, gilthead or black seabream (*Acanthopagrus schlegeli*) fry and rock bream (*Oplegnathus fasciatus*) (Masumura and Wakabayashi, 1977; Wakabayashi *et al.*, 1986). Infected fish displayed eroded mouths, frayed fins, tail rot, and skin lesions that progressed to shallow ulcers. Mortality rates were 20–30%. Under microscopic examination long slender bacilli are seen attached to lesions and affected surfaces. Strain R2 (NCMB 2514^T) and B2 (NCMB 2513) are from red bream. Since then *T. maritimum* has been reported from Atlantic salmon (*Salmo salar*), barramundi (*Lates calcarifer*), flounder (*Paralichthys olivaceus*), greenback flounder (*Rhombosolea tapirina*), rainbow trout (*Oncorhynchus mykiss*), sole (*Solea senegalensis*), Dover sole (*S. solea*), wedge sole (*Dicologoglossa cuneata*), seabream (*Dicentrarchus labrax*), striped trumpeter (*Latris lineata*), white weak fish or white seabream (*Atractoscion nobilis*), Pacific sardine (*Sardinops sagax*), northern anchovy (*Engraulis mordax*) and yellowtail (*Seriola quinqueradiata*) cultured or found wild in Australia, Canada, France, Japan, Scotland, Spain and the USA (McVicar and White, 1979; Bernardet *et al.*, 1990, 1994; Chen *et al.*, 1995; Handlinger *et al.*, 1997; Ostland *et al.*, 1999a; Vilar *et al.*, 2012). *Tenacibaculum maritimum* has also been detected in jellyfish (*Phialella quadrata* and *Muggiaea atlantica*), which are thought to act as carriers of the disease (Fringuelli *et al.*, 2012).

Infection in Dover sole (*S. solea*) cultured in Scotland in the 1970s was referred to as black patch necrosis due to the slight blistering of the skin surface and darkening of the skin between the caudal and marginal fins, which led to loss of the epithelium and exposure of the underlying haemorrhagic dermal tissues. Necrotic ulcers developed due to secondary invading organisms (McVicar and White, 1979; Bernardet *et al.*, 1990). The disease was related to stress.

Tenacibaculum maritimum causes classic signs in wedge sole (*Dicologoglossa cuneata* Moreau) cultured in south-western Span (López *et al.*, 2009b). Mortality rates of up to 18% over 1–3 weeks occurred in juvenile fish of 7–21 g. Pathogenicity was confirmed by experimental bath challenge with strain a443. The LD_{50} was 2.2×10^5 CFU/ml. At an inoculum of 10^6–10^7 CFU/ml, mortalities occurred within the first 7 days of exposure but lower inoculum doses resulted in delayed mortality rates of 6 days post-infection. The identity was confirmed using the API 20E, API ZYM, 16S rRNA sequencing and two specific PCRs with primers MAR1-MAR2 (Toyama *et al.*, 1996), and primers Mar1 and Mar2 (Bader and Schotts, 1998).

Strains of *T. maritimum* may vary in their virulence, as in experimental challenge experiments using Atlantic salmon challenge with adherent strains resulted in a greater mortality than non-adherent strains (van Gelderen *et al.*, 2010).

In Canada, the disease in Atlantic salmon (*Salmo salar*) smolts cultured in seawater in the first year was referred to as bacterial stomatitis or mouth rot due to the yellow plaques on the teeth and in the oral cavity (Ostland *et al.*, 1999b). Jellyfish (*Phialella quadrata*), which can be a problem for cage-reared fish, may be vectors for the disease as the bacterium is present on the lips of the jellyfish (Ferguson *et al.*, 2010).

GROWTH CHARACTERISTICS AND PHENOTYPIC IDENTIFICATION. *Tenacibaculum maritimum* is an aerobic, oxidase positive, Gram-negative, slender, flexible rod (0.5×2–30 μm) with an occasional cell of 100 μm in length (Wakabayashi *et al.*, 1986). Cells from older cultures are shorter in length and spherical cells may be seen. The latter are not viable. Flagella are not present but cells have a gliding motility that can be seen in a wet preparation (hanging drop) made with seawater. Growth occurs on cytophaga medium, also known as Anacker-Ordal's (AO) medium containing sea salts (AO-M) (see media section under Anacker-Ordal agar). Colonies are slow growing, flat, thin, light yellow in colour and have ragged edges (rhizoid) and grow to about 5 mm after 4–5 days at 25°C. Colonies are never larger than 5 mm. In liquid medium growth is seen as a thin layer of growth (pellicle) at the surface. Growth does not occur anaerobically. Growth occurs at 15–34°C with an optimum of 28°C. Growth does not occur when AO agar is supplemented with NaCl. At least 30% seawater or 38 g/l artificial seawater salts (ASW) (Sigma) is required for growth. KCl as well as NaCl is required for growth and Ca^{2+} and MG^{2+} enhance growth, whereas SO_4^{2-} is mildly inhibitory (Wakabayashi *et al.*, 1986). Growth also occurs on Flexibacter maritimus medium (FMM, see media section) (Pazos *et al.*, 1996) as light yellow, flat colonies with irregular edges. This medium contains the same ingredients as MA 2216 but without the chloride, phosphate and other ions contained in MA 2216. It also contains sodium acetate. Good growth occurs on MA 2216 (although some reports state growth is variable) but colonies are yellow and round and do not display the irregular edges seen on other media (Bernardet *et al.*, 1994; Pazos *et al.*, 1996). Some colonies may adhere strongly to the agar (López *et al.*, 2009b). Brown pigment is produced on AO-M containing 0.5% tyrosine.

Glucose, galactose, fructose, cellobiose and rhamnose are not fermented. Agar, chitin and starch are not degraded but casein is degraded. Casamino acids, yeast extract and tryptone are used as sources of carbon and nitrogen for growth (Wakabayashi *et al.*, 1986). Colonies absorb Congo red stain (see media section) but do not produce flexirubin-type pigments (tested with KOH – see media section and identification tests and their interpretation). For Canadian strains isolated from Atlantic salmon smolts some differences were noted between these strains and the reference strains. The salmon strains had an optimum temperature of 18–25°C, which is slightly lower than the reference strains, and also had an optimum for media containing 70% seawater. Many of the strains had α- and β-glucosidase activity, an unusual finding for *T. maritimum* isolates (Ostland *et al.*, 1999b).

These and other reactions are presented in Tables 3.7 and 3.38. The API ZYM should be incubated at 22–25°C with overnight incubation.

MOLECULAR IDENTIFICATION. Conventional, nested and a real-time PCR for the detection of *T. maritimum* have been reported.

Primers MAR1 (5′-AAT GGC ATC GTT TTA AA-3′) and MAR2 (5′-CGC TCT CTG TTG CCA GA-3′) are based on the 16S rRNA gene (Toyama *et al.*, 1996). The master mix is composed of 10 pmol of each primer (MAR1 and MAR2), 1.5 mM $MgCl_2$, 0.5 nmol of each dNTP, 2.5 U *Taq* DNA polymerase, and 5 μl template DNA in a 50-μl reaction volume. Thermocycling is carried out at an initial denaturation at 94°C for 2 min, followed by 30–35 cycles of denaturation at 94°C for 2 min, 45°C for 1.5 min, 72°C for 2 min and a final elongation step at 72°C for 5 min (Toyama *et al.*, 1996).

Other primers Mar1 (5′-TGT AGC TTG CTA CAG ATG A-3′) and Mar2 (5′-AAA TAC CTA CTC GTA GGT ACG-3′), also designed to the 16S rRNA gene, are specific for *T. maritimum* (Bader and Shotts, 1998). The reaction mix contains 10 pmol of each primer (Mar1 and Mar2). Thermocycling is carried out using a preheating cycle of 94°C for 10 min to reduce primer-dimer formation followed by 39 cycles of denaturation at 94°C for 2 min, annealing at 58°C for 2 min, extension at 72°C for 2 min and a final extension step at 72°C for 4 min and an indefinite hold at 4°C. An amplicon of 400 bp is produced. The primers are specific against *F. aquatile* (ATCC 11947), *F. breve* (ATCC 14234), *F. psychrophilum* (ATCC 49418), *F. johnsoniae* (ATCC 17061) and strain ATCC 43622, which has since been identified as *F. johnsoniae* (Darwish *et al.*, 2004) and not *F. columnare* as used for specificity testing by Bader and Shotts (1998). A nested PCR using universal primers of 20F and 1500R (Weisberg *et al.*, 1991; see molecular section) in the first PCR reaction followed by a nested PCR using primers Mar1 and Mar2 in the nested reaction allows detection of this organism direct from fish tissue (Cepeda *et al.*, 2003). Ready-to-Go PCR beads (Amersham Pharmacia Biotech) were used, with 1 pmol of each primer in a 25-μl reaction. Cycling conditions for the first PCR reaction used a preheating step at 95°C for 5 min followed by 30 cycles of denaturation at 95°C for 30 s, annealing at 57°C for 30 s, extension at 72°C for 60 s with a final cycle of extension at 72°C for 5 min. The second PCR reaction used a preheating step at 94°C for 2 min, followed by 40 cycles of 94°C for 2 s, 54°C for 2 s, 72°C for 10 s with a final cycle that used an extension step of 4 min. Amplification using singleplex PCR with Mar and Mar2 primers did not amplifiy DNA from seeded fish tissue, but DNA was detected using the nested PCR. The nested PCR has a sensitivity of 75 CFU/g fish tissue, which equates to 0.2 cells per reaction tube. In addition to strains tested for specificity by Bader and Shotts (1998), the Mar1 and Mar2 primers were specific against *F. columnare* (NCIMB 2248), *F. flevensis* (NCIMB 12056), *F. succinicans* (NCIMB 2277), *Vibrio anguillarum* (ATCC 14181), *V. splendidus* (ATCC 25914) and *Aeromonas salmonicida* ssp. *salmonicida* (ATCC 35658) (Cepeda *et al.*, 2003).

A TaqMan MGB probe qPCR using primers designed to the 16S rRNA gene has been applied to formalin-fixed paraffin-embedded (FFPE) fish gill tissue and jellyfish tissue for the detection of *T. maritimum* (Fringuelli *et al.*, 2012). FFPE tissue is cut to a thickness of 40 μm using a new microtome blade for each block and the microtome is cleaned with DNA zap (Ambion) after each block. DNA is extracted using the QIAamp DNA FFPE tissue kit (Qiagen) according to the manufacturer's instructions. A slice from a wax block without tissue is used as a DNA extraction negative control after every ten samples. DNA from jellyfish tissue is extracted using the QIAamp DNA mini kit (Qiagen). The reaction mixture consists of 10 μl Quantitect Multiplex PCR NoRox Master mix kit (Qiagen), 0.5 μM primer MAR 4 F (5′-TGC CTT CTA CAG AGG GAT AGC C-3′), 0.5 μM primer MAR reverse (5′-CTA TCG TTG CCA TGG TAA GCC G-3′), 0.05 μM probe MAR probe (5′-6-FAM-CAC TTT GGA ATG GCA TCG-MGB quencher-3′) (Applied Biosystems), and 2 μl DNA template in a total volume of 20 μl. Thermocycling is carried out on a Rotor-Gene-Q (Qiagen) real-time PCR thermocycler with an initial denaturation at 95°C for 15 min followed by 45 cycles of denaturation at 95°C for 20 s and combined annealing/extension at 52°C for 20 s. Fluorescence from FAM is acquired during the annealing extension stage. An amplicon of 155 bp is produced. The sensitivity of the assay is 4.8 copy number/μl with a detection limit of 1–5100 cells per reaction (Fringuelli *et al.*, 2012).

Tenacibaculum ovolyticum

TAXONOMY AND SIGNIFICANCE. *Tenacibaculum ovolyticum* (EKD002^T=NCIMB 13127^T) was isolated from the eggs of Atlantic halibut (*Hippoglossus hippoglossus*) and originally named *Flexibacter ovolyticus* (Hansen *et al.*, 1992). It was later transferred to the new genus *Tenacibaculum* (Suzuki *et al.*, 2001).

Tenacibaculum ovolyticum adheres to Atlantic halibut eggs and dissolves the chorion and zona radiata of the shells. It causes high mortality a few days before and after hatching.

GROWTH CHARACTERISTICS AND PHENOTYPIC IDENTIFICATION. *Tenacibaculum ovolyticum* is a strictly aerobic, Gram-negative long, slender rod (0.4×2–20 μm) that is motile by gliding. Long filaments up to 70–100 μm may be seen. Oxidase and catalase are positive. On MA 2216 colonies are light brownish yellow, flattened and elevated with regular edges. A rapid loss of viability occurs, which is seen as an area of lysis in the centre of a 5-day-old colony with viable cells at the edge. It is non-haemolytic. Growth occurs on Anacker Ordal's medium (AO) prepared with 70% seawater. Acid (fermentation) is not produced from carbohydrates. Gelatin, tyrosine, DNA and Tween 80 are hydrolysed. Chitin, starch and cellulose are not hydrolysed. Nitrate is reduced. Negative for Congo red and flexirubin pigment (tested by KOH – see media section, and identification tests and their interpretation). It does not grow under anaerobic conditions (Hansen *et al.*, 1992). These and other reactions are presented in Tables 3.7 and 3.38.

Tenacibaculum soleae

TAXONOMY AND SIGNIFICANCE. *Tenacibaculum soleae* (LL04 12.1.7^T=CECT 7292^T=NCIMB 14368^T) was isolated from diseased cultured sole (*Solea senegalensis*), cultured wedge sole (*Dicologoglossa cuneata*) and brill (*Scophthalmus rhombus*) in Spain (Piñeiro-Vidal *et al.*, 2008a; López *et al.*, 2010). Based on 16S rRNA sequencing it is most closely related to *T. ovolyticum* and *T. aestuarii* (96.7%), *T. lutimaris* and *T. mesophilum* (96.4%), *T. gallaicum* (96.2%), *T. amylolyticum* (96.0%), *T. litoreum* (95.9%), *T. discolor* (95.8%), *T. skagerrakense* (95.7%), *T. litopenaei* (95.0%) and *T. maritimum* (94.8%). In disease outbreaks, mortalities occurred in fry, juveniles and adults of wedge sole, adult brill and juvenile and adult Senegalese sole with the main clinical signs being ulcers on the body and tail, and with some fish showing petechial haemorrhages on fins, jaw and surface of the ventral body, haemorrhagic or pale liver and erratic swimming behaviour. The organism was cultured from the liver, kidney and ulcers often in mixed infection with *Vibrio harveyi* and *T. maritimum* (López *et al.*, 2010). Virulence testing using bath immersion and intraperitoneal injection demonstrated the organism was pathogenic for fingerlings of sole and turbot but not for mice. The LD_{50} was 7.8×10^5 CFU/ml.

GROWTH CHARACTERISTICS AND PHENOTYPIC IDENTIFICATION. *Tenacibaculum soleae* is a strictly aerobic, Gram-negative rod (0.5×2.0–25.0 μm), catalase positive and oxidase positive and exhibits gliding motility. Cells are straight but degenerative spherical cells may be seen in ageing MA 2216 broth cultures. On MA 2216 agar and FMM agar, colonies are flat, bright yellow, with spreading margins and uneven edges and do not adhere to the agar. Growth occurs at 15–30°C but not at 4, 35 or 40°C and has an optimal range of 22–25°C. It grows at pH 6.0–8.0, and in media containing 50–100% seawater. It does not grow on media containing NaCl alone. Carotenoid pigment is detected with Congo red stain, but flexirubin pigment is not produced. Hydrolysis is positive for gelatin, DNA and casein but negative for starch, Tween 20, Tween 80 and haemolysin. Carbohydrates are not fermented, H_2S and indole are negative whereas VP and nitrate are positive. Tests were conducted using artificial seawater salts as an inoculum and an incubation temperature of 20°C (López *et al.*, 2010). Some reactions were variable between the type strain and strains isolated from other sole, and brill. These and other reactions are presented in Tables 3.7, 3.24–26 and 3.38.

MOLECULAR IDENTIFICATION. Two conventional PCRs have been published for the detection of *T. soleae*. One is based on the 16S rRNA gene (García-González *et al.*, 2011) and the other is based on the 16-23S region (López *et al.*, 2011b).

For the PCR based on the 16S-23S intergenic spacer region (López *et al.*, 2011b) DNA extraction is carried out as follows. For cultured cells: boil for 10 min. Centrifuge at 12,400 × *g*. For fish tissue: homogenize 100 mg tissue in TE buffer (Sigma). Add 1% SDS and 100 mg/ml Proteinase K and incubate 3 h or overnight at 56°C. Add 20 μg/ml pancreatic RNase and incubate for 1 h at 37°C. Transfer solution to a phase-lock gel (Eppendorf) and purify the DNA using the phenol/chloroform/isoamyl alcohol procedure. Precipitate the DNA with ethanol and dissolve in distilled water. The PCR mixture contained RedTaq ReadyMix (Sigma), 200 μmol/l of each dNTP, 200 nmol/l primer G47F (5′-ATG CTA ATA TGT GGC ATC AC-3′) and primer Primer G47R (5′-CGT AAT TCG TAA TTA ACT TTG T-3′), 3 U Taq DNA polymerase, 2 μg–100 ng template DNA and distilled water in a 50-μl volume. Thermocycling is conducted using a Mastercycler gradient (Eppendorf) with denaturation at 94°C for 5 min followed by 45 cycles of 94°C for 1 min, 57°C for 45 s, 72°C for 1 min and a final elongation at 72°C for 5 min. The amplified product is 1555 bp. In some cases a faint band at 750 bp is also seen. The sensitivity was 1 pg with purified DNA. The limit of detection from seeded fish tissue was 10 pg of DNA in 1 µg of fish tissue. The primers are specific against *Tenacibaculum discolor* (DSM 18842^{T}), *T. gallaicum* (DSM 18841^{T}), *T. litoreum* (JCM 13039^{T}), *T. maritimum* (CECT 4276), *T. ovolyticum* (LMG 13025), *Flavobacterium johnsoniae* (ATCC 17061^{T}), *F. marinotypicum* (CECT 578^{T}), *F. psychrophilum* (NCIMB 1947^{T}), *Aeromonas salmonicida* ssp. *salmonicida* (CECT894^{T}), *Aliivibrio fischeri* (CECT 524^{T}), *Vibrio alginolyticus* (CECT 521^{T}), *V. campbellii* (CECT 523^{T}), *V. harveyi* (CECT 525^{T}), *V. natriegens* (CECT 526^{T}), *V. parahaemolyticus* (CECT 511^{T}), *V. tubiashii* (CECT 4196^{T}), *V. splendidus* (CECT 528^{T}), *Photobacterium angustum* (CECT 5690^{T}), *P. damselae* ssp. *damselae* (CECT 626^{T}), *P. damselae* ssp. *piscicida* (CECT 5895), *P. leiognathi* (CECT 4191^{T}), *P. phosphoreum* (CECT 4192^{T}), *Polaribacter* sp. (a500), *Pseudomonas aeruginosa* (ATCC 27853), *P. anguilliseptica* (CECT 899^{T}), *P. baetica* (a390^{T}), *P. fluorescens* (CECT 378^{T}), *P. putida* (CECT 385) and *Escherichia coli* (ATCC 25922) (López *et al.*, 2011b).

The conventional PCR based on the 16S rRNA gene is detailed in the following table.

Class Sphingobacteria

Order Sphingobacteriales

The order *Sphingobacteriales* contains the family *Cyclobacteriaceae* and the family *Sphingobacteriaceae*.

1.21 *Cyclobacteriaceae*

1.21.1 *Echinicola*

Echinicola pacifica

Three strains of *Echinicola pacifica* (KMM 6172^{T}=KCTC 12368^{T}=LMG 23350^{T}) were isolated from the sea urchin *Strongylocentrotus intermedius* found in the Troitsa Bay, Gulf of Peter the Great, East Sea, which is also known as the Sea of Japan (Nedashkovskaya *et al.*, 2006). The genus *Echinicola* was created due to significant molecular divergence from other taxa. *Echinicola pacifica* is a Gram-negative rod (0.3–0.4×1.2–1.9 µm), motile by gliding, and agarolytic. On MA 2216

PCR type	Conventional PCR based on the 16S rRNA gene (García-González *et al.*, 2011)		
DNA extraction	Cultured cells and from mixed cultures on MA 2216 using Ultraclean® Microbial DNA isolation kit (Mo Bio Labs)		
PCR master mix			
Primer Sol-Fw	5′-TGC TAA TAT GTG GCA TCA CAA-3′		
Primer Sol-Rv	5′-CAA CCC ATA GGG CAG TCA TC-3′		
Reagent	**Stock**	**Final concentration**	**Volume ×1**
Premix Taq (Takara Biotech)	×2	×1	25.0 µl
Primer Sol-Fw	50 µM	0.2 µM	0.2 µl
Primer Sol-Rv	50 µM	0.2 µM	0.2 µl
Water			23.6 µl
DNA			1 µl
Total volume			**50 µl**
Thermocycling	Mastercycler Epigradient thermal cycler (Eppendorf)		
Step 1	**Step 2**		**Step 3**
95°C, 5 min	95°C, 30 s		95°C, 30 s
	60.8°C, 20 s		60.8°C, 20 s
	72°C, 20 s		72°C, 5 min
×1	×30		×1
Product bp	248 bp		
Sensitivity	1–10 bacterial cells per reaction		
Specific against	*Aeromonas salmonicida, A. hydrophila, Aliivibrio fischeri, Listonella anguillarum, L. pelagia, Photobacterium damselae* spp. *piscicida, Tenacibaculum maritimum, T. discolor, T. gallaicum, T. ovolyticum, T. litoreum, V. fluvialis, V. harveyi, V. scophthalmi*		

colonies are 2–3 mm, circular, convex, shiny, pink-coloured and sunken into the agar. NaCl is not required for growth and growth occurs at 0–12% NaCl and at a temperature of 6–41°C with an optimum of 25–28°C. Non-diffusible carotenoid pigment is produced but not flexirubin-type pigment. Results are positive for catalase and oxidase. Agar is degraded and gelatin is weakly hydrolysed. Results are positive for hydrolysis of Tween 40, starch, aesculin and production of hydrogen sulfide, and negative for casein, DNAse, cellulose and chitin. It does not grow under strictly anaerobic conditions. These and other reactions are presented in Tables 3.24–30 and 3.38.

1.22 *Sphingobacteriaceae*

1.22.1 *Sphingobacterium*

Sphingobacterium species

Members of the *Sphingobacterium* genus have a low G+C content (39–42 mol%) and contain sphingolipids. They are found in soil and compost.

Phylum *Fusobacteria*
Class Fusobacteria
Order *Fusobacteriales*

1.23 *Fusobacteriaceae*

1.23.1 *Streptobacillus*

Streptobacillus moniliformis-like organism

TAXONOMY AND SIGNIFICANCE. A *Streptobacillus moniliformis*-like organism (AVG2115) was isolated from Atlantic salmon (*Salmo salar*) cultured in Ireland. Infected fish displayed abnormal swimming behaviour, which included being lethargic, a loss of balance and swimming close to the surface. Petechiae were evident on the abdominal wall, muscle, swimbladder and pyloric caeca and white mucus was present in the gut (Palmer *et al.*, 1994). A small cocco-bacillus was present extracellularly and intracellularly in tissue macrophages and glomerular endothelial cells but not found in the endothelial cells of other tissues. The organism was cultured from the kidney, liver and spleen.

Sequencing of the 16S rRNA was unable to assign a definitive genus to the organism but it was closely related to *Streptobacillus moniliformis*, *Sebaldella termitidis*, *Leptotrichia buccalis*, *Propionigenium modestum* and the *Fusobacterium* group (Maher *et al.*, 1995). At the time of writing a Blast search revealed only a 90–91% match with *Streptobacillus moniliformis* DSM 12112. The closest match is 93% with *Leptotrichia* species ES2714, indicating isolate AVG2115 is a new genus as proposed by Palmer *et al.* (1994) and Maher *et al.* (1995).

GROWTH CHARACTERISTICS AND PHENOTYPIC IDENTIFICATION. The *Streptobacillus moniliformis*-like organism is a slow-growing, moderately halophilic facultative or aerotolerant anaerobic, non-motile, Gram-negative rod that shows filamentous cells (0.4×<5 μm) with possible branching in older cultures and in broth culture. On blood agar containing 7% horse red blood cells and 1% NaCl (BA-NaCl), and brain–heart infusion agar containing 10% fetal calf serum and 1% NaCl (BHIA-FCS-NaCl) colonies are off-white, friable and convex with a granular 'breadcrumb' morphology but become concave with 'molar tooth' morphology. Colony size is 0.1 mm after 4–8 days and up to 0.3 mm after 10–14 days at 15 to 22°C when incubated aerobically. Growth is not improved under aerobic conditions supplemented with 5% CO_2 but a colony size of up to 0.6 mm is produced under anaerobic conditions. A small zone of beta-haemolysis is seen after 7–14 days on HBA-NaCl. Growth is slightly improved by the addition of 0.5 mg/ml L-cysteine HCL but not by the addition of 2 μg/ml cocarboxylase. These and other reactions are presented in Table 3.10.

GRAM-POSITIVE RODS and COCCI

The following information on phylum, class, order, family and genus is based on the information in *Bergey's Manual of Systematic Bacteriology*, 2nd edition (Table 1.11). The information is not as complete as in Bergey's Manual, but is subsectioned for listing of the bacteria that appear in this manual. The information here is

Table 1.11. Taxonomy of Gram-positive rods and cocci listed in this manual.

Phylum	Class	Order	Family	Genera
Actinobacteria	I. *Actinobacteria*	I. *Actinomycetales*	I. *Actinomycetaceae* II. *Micrococcaceae* VIII. *Dermatophilaceae* IV. *Mycobacteriaceae* V. *Nocardiaceae* *Corynebacteriaceae*	I. *Actinomyces*, *Arcanobacterium*, *Trueperella* II. *Micrococcus*, *Arthrobacter*, *Renibacterium* VIII. *Dermatophilus* IV. *Mycobacterium* V. *Nocardia*, *Rhodococcus* *Corynebacterium*
Firmicutes	I. *Clostridia*	I. *Clostridiales*	*Clostridiaceae* *Eubacteriaceae*	*Clostridium* *Eubacterium*
	II. *Mollicutes*	I. *Mycoplasmatales*	*Mycoplasmataceae*	*Mycoplasma*, *Ureaplasma*
		V. *Incertae sedis*	*Erysipelotrichaceae*	*Erysipelothrix*
	III. *Bacilli*	I. *Bacillales*	I. *Bacillaceae* IV. *Listeriaceae* VI. *Planococcaceae* VII. *Sporolactobacillaceae* VIII. *Staphylococcaceae*	I. *Bacillus* IV. *Listeria* VI. *Planococcus* VII. *Marinococcus* VIII. *Staphylococcus*, *Gemella*
		II. *Lactobacillales*	I. *Lactobacillaceae* II. *Aerococcaceae* III. *Carnobacteriaceae* IV. *Enterococcaceae* V. *Leuconostocaceae* VI. *Streptococcaceae*	I. *Lactobacillus* II. *Aerococcus*, *Abiotrophia*, *Facklamia* III. *Carnobacterium*, *Granulicatella* IV. *Enterococcus*, *Vagococcus* V. *Weissella* VI. *Streptococcus*, *Lactococcus* *Catellicoccus*
Spirochaetes	I. *Spirochaetes*	I. *Spirochaetales*	I. *Spirochaetaceae* III. *Leptospiraceae*	I. *Spirochaeta* III. *Leptospira*

listed alphabetically based on the phylum, class, family name, and within the family listed alphabetically based on the genus name.

Phylum *Actinobacteria*
Class Actinobacteria
Order *Actinomycetales*

1.24 *Actinomycetaceae*

1.24.1 *Actinomyces*

Actinomyces marimammalium

TAXONOMY AND SIGNIFICANCE. *Actinomyces marimmalium* (CCUG 41710T) was isolated in mixed culture (*Escherichia coli*, *Streptococcus* species, *Actinobacillus* species) from the lung, liver, spleen, kidney and mesenteric lymph node of a dead hooded seal (*Crystophora cristata*) with pneumonia (Hoyles *et al.*, 2001b). The organism was also isolated from the small intestine of a grey seal (*Halichoerus grypus*) that had been shot, and from the lung of a dead harbour porpoise (*Phocoena phocena*). Pathogenicity is not known. The strains have 99.5% 16S rRNA sequence similarity to each other and 93% to other *Actinomyces* species, the highest being *A. hordeovulneris*.

GROWTH CHARACTERISTICS AND PHENOTYPIC IDENTIFICATION. *Actinomyces marimmalium* is a facultatively anaerobic, non-motile, Gram-positive straight to slightly curved rod with some branching. It is non-acid fast. On Columbia sheep blood agar (CSBA) colonies are pinpoint to 0.5 mm, grey, non-haemolytic, entire, circular, convex after 48 h at 37°C and cultured in air with 5% CO_2 (Hoyles *et al.*, 2001b). Catalase is negative. Aesculin, gelatin and hippurate are not hydrolysed. Nitrate is not reduced, urease and VP are negative. Fermentation of glycogen is variable. These and other reactions are presented in Table 3.15, 3.31 and 3.36

1.24.2 *Arcanobacterium*

According to chemotaxonomic and phylogenetic studies, the genus *Arcanobacterium* is not monophyletic but consists of two lines, which have now been divided into two genera: *Arcanobacterium* consisting of *A. haemolyticum* (type species of the genus), *A. phocae*, *A. pluranimalium* and others; and the genus *Trueperella* consisting of *T. pyogenes*, *T. bernardiae* and others (Yassin *et al.*, 2011). Many of the species in *Arcanobacterium* and *Trueperella* were originally assigned to the genus *Corynebacterium*.

Members of the *Arcanobacterium* genus are facultatively anaerobic, non-motile, non-acid fast and Gram-positive. Cells grown on blood agar are slender, irregular and predominantly rod-shaped or arranged at an angle to produce V-formations during the first 18 h of growth, and over time, become granular and segmented and resemble small, irregular cocci. Growth is enhanced when grown in air supplemented with CO_2, and enhanced when grown on media containing blood or serum. The CAMP test is positive. Mycolic acids are not present (Yassin *et al.*, 2011).

Arcanobacterium phocae

TAXONOMY AND SIGNIFICANCE. *Arcanobacterium phocae* (M1590^T=DSM 10002^T) was isolated from the lung of a common seal (*Phoca vitulina*) (Ramos *et al.*, 1997). Other strains have been isolated from abscesses, wounds, lung, spleen, intestine, mesenteric and submaxillary lymph nodes, mouth lesion, nasal swabs, vaginal swabs and peritoneal fluid of common seals, grey seals (*Halichoerus grypus*), northern elephant seal (*Mirounga angustirostris*), Pacific harbour seals (*Phoca vitulina richardii*), California sea lion (*Zalophus californianus*), common dolphin (*Delphinus delphis*) and southern sea otter (*Enhydra lutris nereis*) (Ramos *et al.*, 1997; Johnson *et al.*, 2003). A study of animals stranded between 1994 and 2000 suggests the prevalence is 8%. It is thought that *A. phocae* may originally have been reported as *Listeria ivanovii* due to the phenotypic similarity between the two bacteria. Motility and aesculin tests differentiate the two bacteria. The significance of *A. phocae* is not known, however it would appear to be an opportunistic pathogen in mammals, as it has been isolated from superficial pyogenic wounds and exudates in mixed populations and from deep-seated and systemic infections such as brain abscesses and osteomyelitis and in mixed flora from cases of pneumoniae and septicaemia (Ramos *et al.*, 1997; Johnson *et al.*, 2003).

GROWTH CHARACTERISTICS AND PHENOTYPIC IDENTIFICATION. *Arcanobacterium phocae* is a facultatively anaerobic, non-motile, non-spore-forming Gram-positive short rod that occurs singly, in pairs or in clusters. Short diphtheroid forms may occur. Primary branching is not observed. It is non-acid fast. Catalase is variable, although the type strain is positive. On BA colonies are tiny, white, circular, low, convex and have a strong zone of beta-haemolysis that is two to three times the diameter of the colony after 24 h in air containing 5% CO_2 at 37°C. Growth is not enhanced by increased CO_2 concentration to 10%. Glucose is fermented without gas. Nitrate is not reduced. Gelatin, aesculin and urea are not hydrolysed (Ramos *et al.*, 1997). These and other reactions are presented in Tables 3.15, 3.30–35 and 3.38.

Arcanobacterium phocisimile

TAXONOMY AND SIGNIFICANCE. *Arcanobacterium phocisimile* (2698^T=LMG 27073^T=CCM 8430^T) was isolated from the vagina of a harbour seal during a monitoring programme of free-living seals in the German North Sea (Hijazin *et al.*, 2012). A second strain (strain 4112) was isolated from the anus of a male seal. Pathogenicity is unknown.

The isolates are 95.5–98.4% similar to *Arcanobacterium* species based on the 16S rRNA sequence and most closely related to *A. haemolyticum*. The strains are related to *Trueperella* species at 93.2–94.1% based on the 16S rRNA sequence.

GROWTH CHARACTERISTICS AND PHENOTYPIC IDENTIFICATION. *Arcanobacterium phocisimile* is a non-motile, non-spore-forming, oxidase negative, Gram-positive rod (1–2×0.5 μm). On BA containing sheep blood cells colonies display a weak zone of haemolysis when incubated under a microaerobic atmosphere. Haemolysis is less pronounced when incubated under aerobic and anaerobic atmospheres. A CAMP-like reaction occurs with indicator strain *Staphylococcus aureus*. DNA is hydrolysed but not urea, hippurate, gelatin or aesculin. Nitrate is not reduced. These and other reactions are presented in Tables 3.15 and 3.31–33.

Arcanobacterium pluranimalium

TAXONOMY AND SIGNIFICANCE. *Arcanobacterium pluranimalium* (M430/94/2^T=CCUG 42575^T) was isolated from the spleen of a dead harbour porpoise (*Phocoena phocoena*) and strain CCUG 41729 was isolated from an abscess in the lung of a dead fallow deer (Lawson *et al.*, 2001). Pathogenicity is unknown.

GROWTH CHARACTERISTICS AND PHENOTYPIC IDENTIFICATION. *Arcanobacterium pluranimalium* is a facultatively anaerobic, non-motile, non-spore-forming, Gram-positive straight to slightly curved, non-branching, slender rod. It is non-acid fast. Catalase is positive. On Columbia base agar containing 5% horse blood colonies are alpha-haemolytic after 24 h at 37°C and in air plus 5% CO_2. Maltose is fermented using the API Coryne kit but not with the API rapid ID 32 Strep. Aesculin is weakly hydrolysed. Hydrolysis is positive for gelatin and hippurate. Nitrate is not reduced. VP is negative (Lawson *et al.*, 2001). These and other reactions are presented in Tables 3.15, 3.31–33 and 3.36.

1.24.3 *Trueperella*

Some species originally assigned to the genus *Arcanobacterium* have been re-assigned to a new genus, *Trueperella*. Chemotaxonomic and phylogenetic studies suggests the genus *Arcanobacterium* is not monophyletic but consists of two lines, one of these being *Trueperella*, which consists of *T. pyogenes*, *T. bernardiae* and others. Other species are retained in the genus *Arcanobacterium*, these being *A. haemolyticum* (type species of the genus), *A. phocae*, *A. pluranimalium* and others (Yassin *et al.*, 2011). Many of the species in *Arcanobacterium* and *Trueperella* were originally assigned to the genus *Corynebacterium*.

Members of the *Trueperella* genus are aerobic and facultatively anaerobic, non-motile, non-acid fast and Gram-positive. Cells vary in shape and size according to different growth media from coccobacilli to short rods (0.2–0.9×0.3–2.5 μm) that occur singly, in pairs, in V, T and palisade formations or in clusters. Cells from 24 h broth cultures are Gram-positive but may become Gram-variable in older cultures. On BA, colonies are pinpoint and beta-haemolytic after 24 h at 37°C, and 0.5–1.5 mm after 48–72 h incubation. Mycolic acids are not present (Yassin *et al.*, 2011). Biochemical reactions for *T. pyogenes* are presented in Table 3.15 and for *T. bernardiae* in Table 3.31.

1.25 *Corynebacteriaceae*

1.25.1 *Corynebacterium*

Corynebacterium aquaticum

TAXONOMY AND SIGNIFICANCE. *Corynebacterium aquaticum* (strain RB 968BA) was isolated from the brain and eye of diseased striped bass (*Morone saxatilis*) suffering from bilateral exophthalmia whilst being held in an experimental aquaculture facility in the USA (Baya *et al.*, 1992b). In infected fish the haemorrhages occurred in the eye and brain and blood occurred in the brain cavity. The organism was recovered from the brain, but not from the liver or kidney. Recovery of the organism only occurred in liquid medium: TSB and BHIB. Experimental challenge confirmed pathogenicity with an LD_{50} of 1.0×10^5 cells for striped bass and 5.8×10^4 cells for rainbow trout. Extracellular products caused mortality when injected into striped bass. In addition to the injected *C. aquaticum* being re-isolated from dead fish it was recovered also from surviving fish, indicating a carrier status. The type strain (ATCC 14665) was not pathogenic. *Corynebacterium aquaticum* lost virulence after being stored frozen for 6 months.

Corynebacterium aquaticum (ATCC 14665^T) was originally isolated from distilled and stored freshwater (Leifson, 1962). The name has not been officially listed in the list of Prokaryotic names with standing in nomenclature (Euzéby) but is recognized in the commercial database API Coryne (bioMérieux). *Corynebacterium aquaticum* has been reported from cases of infection and septicaemia in patients, usually immunocompromised, but is difficult to differentiate from *Aureobacterium* species (Funke *et al.*, 1994).

GROWTH CHARACTERISTICS AND PHENOTYPIC IDENTIFICATION. *Corynebacterium aquaticum* is a facultatively anaerobic, non-spore-forming, non-acid fast, oxidase-negative, Gram-positive rod ($1–3\times0.5–0.8$ μm) that exhibits typical coryneform morphology with club-shaped forms and angular arrangements of cells. Catalase is positive and oxidase is negative. On BA and TSA colonies are 1–3 mm, yellow, round, raised, entire, opaque and slightly viscid after 48 h at 25°C. The yellow pigment is non-diffusible. Beta-haemolysis only occurs at 37°C. Growth occurs at 4–42°C and 0–5% NaCl. Carbohydrates are not fermented, citrate is not utilized but the VP reaction is positive and aesculin is hydrolysed. The strain from striped bass is negative for nitrate hydrolysis whereas the type strain is positive in the API Coryne kit. Gelatin and casein are hydrolysed for the strain from striped bass but not from the type strain (Baya *et al.*, 1992b). Funke *et al.* (1994) reported that *Aureobacterium* strains hydrolysed gelatin and casein whereas strains of *C. aquaticum* from humans did not. *Corynebacterium aquaticum* strains from human sources are positive for reduction of nitrate (Funke *et al.*, 1994). It cannot be determined from the data whether the strains from striped bass belong to the genus *Aureobacterium*. In the analysis of strains of *Aureobacterium* from human isolates they could not be identified to species level using the API Coryne, API Zym or API 50CH. These and other reactions are presented in Tables 3.15 and 3.31–33.

Corynebacterium caspium

TAXONOMY AND SIGNIFICANCE. *Corynebacterium caspium* (M/106/00/5^T=CCUG 44566^T=CIP 107965^T) was isolated from the penis of a Caspium seal (*Pusa [Phoca] caspium*) in mixed culture with *Atopobacter phocae* and *Streptococcus dysgalactiae* Group C during the study of the microflora of marine mammals (Collins *et al.*, 2004). Pathogenicity is unknown. Sequence analysis demonstrates the organism is most closely related to *C. mastitidis* and *C. renale*.

GROWTH CHARACTERISTICS AND PHENOTYPIC IDENTIFICATION. *Corynebacterium caspium* is a facultatively anaerobic, oxidase-negative, non-spore-forming, Gram-positive rod consisting of irregularly tapered rods and club forms. Catalase is positive. On BA colonies are 0.5 mm, non-haemolytic, circular, convex, opaque and dull at 24 h, and 2 mm at 72 h after incubation at 37°C. The colonies resemble some *Streptococcus* species in that the colony can be moved in its entirety across the plate. Nitrate is negative. These and other reactions are presented in Tables 3.15, 3.31–33 and 3.38.

Corynebacterium maris

Corynebacterium maris (Coryn-1^T=DSM 45190^T =LMG 24561^T) was isolated from the mucus of healthy coral, *Fungia granulosa*, in the northern

Red Sea, Gulf of Eilat, Israel (Ben-Dov *et al.*, 2009). *C. maris* is a Gram-positive, aerobic, non-sporing, non-motile coccobacilli (0.5–0.8×0.8–1.5 µm). Growth occurs at 30°C on marine agar 2216 as 1 mm colonies at 48 h, and on nutrient agar and LB agar as 2.5 mm colonies at 72 h, which are yellow or yellowish, circular, convex, smooth and non-haemolytic. Growth occurs at 26–37°C with an optimum of 35°C, and no growth at 4°C. Optimal NaCl concentration is 0.5–4.0% with a range of 0–10% NaCl. No growth occurs under anaerobic conditions. These and other reactions are presented in Tables 3.15, 3.31–33 and 3.38.

Corynebacterium phocae

TAXONOMY AND SIGNIFICANCE. *Corynebacterium phocae* (M408/89/1^T=38205^T) was isolated from the nasal cavity of four common seals (*Phoca vitulina*) (Pascual *et al.*, 1998). Pathogenicity is unknown.

GROWTH CHARACTERISTICS AND PHENOTYPIC IDENTIFICATION. *Corynebacterium phocae* is a facultatively anaerobic, non-motile, Gram-positive, short, asporogenous rod with coryneform morphology. It is non-acid fast. Catalase is positive. On BA colonies are 1 mm, circular, convex, slightly glistening after 24 h at 37°C. Glucose, D-mannose and maltose are fermented whereas D-arabinose, L-arabinose, inositol and mannitol are not. Nitrate is not reduced. Aesculin and gelatin are not hydrolysed. Hydrolysis of urea is variable. Mycolic acids are present. These and other reactions are presented in Tables 3.15, 3.30–33 and 3.38.

Corynebacterium sphenisci

TAXONOMY AND SIGNIFICANCE. *Corynebacterium sphenisci* (CECT 5990^T=CCUG 46398^T) was isolated together with *Staphylococcus sciuri* from the cloacae of apparently healthy wild magellanic penguins (*Spheniscus magellanicus*) (Goyache *et al.*, 2003a). Pathogenicity is unknown. The sequence similarity of the 16S rRNA gene is 96.2, 97.0, 96.3% to *C. amycolatum*, *C. xerosis* and *C. freneyi*, respectively.

GROWTH CHARACTERISTICS AND PHENOTYPIC IDENTIFICATION. *Corynebacterium sphenisci* is a facultatively anaerobic, non-motile, Gram-positive rod displaying coryneform morphology. On BA colonies are 1–2 mm, non-haemolytic, whitish, low convex, dry and rough after 48 h at 37°C. Catalase is positive. The CAMP test is negative. Nitrate is reduced. Gelatin, urea and aesculin are not hydrolysed (Goyache *et al.*, 2003a). These and other reactions are presented in Tables 3.15, 3.30–33 and 3.38.

Corynebacterium spheniscorum

TAXONOMY AND SIGNIFICANCE. *Corynebacterium spheniscorum* (PG 39^T=CECT 5986^T=CCUG 45512^T) was isolated from the cloacae of apparently healthy wild magellanic penguins (*Spheniscus magellanicus*). Some 30% of penguins also harboured *Staphylococcus sciuri* and *Enterococcus faecalis* in the cloacae suggesting these isolates are normal flora (Goyache *et al.*, 2003b). Pathogenicity is unknown. The sequence similarity of the 16S rRNA gene is 95.7, 95.6, 95.3% to *C. argentoratense*, *C. felinum* and *C. ulcerans*.

GROWTH CHARACTERISTICS AND PHENOTYPIC IDENTIFICATION. *Corynebacterium spheniscorum* is a facultatively anaerobic, non-motile, non-spore-forming, Gram-positive short rod. Catalase is positive. On BA colonies are 1–2 mm, non-haemolytic, whitish, circular, smooth and entire after 24 h at 37°C. Growth is slight under anaerobic conditions. The CAMP reaction with *Staphyloccoccus aureus* is positive. Urea, gelatin and aesculin are not hydrolysed. Nitrate is not reduced (Goyache *et al.*, 2003b). These and other reactions are presented in Tables 3.15, 3.30–33 and 3.38.

1.26 *Micrococcaceae*

1.26.1 *Micrococcus*

Micrococcus luteus

TAXONOMY AND SIGNIFICANCE. *Micrococcus luteus* was isolated from diseased rainbow trout fry farmed in the UK that displayed signs of rainbow trout fry syndrome (Austin and Stobie, 1992b). Infected fish had exophthalmia, pigmentation of the skin and pale gills and spleen. The kidney was swollen and ascities fluid was in the abdomen. Experimental challenge using intraperitoneal and intramuscular injections with 10^5–10^7

cells/ml resulted in 54% mortality. The organism was recovered from spleen, kidney and ascites fluid but not from the liver.

GROWTH CHARACTERISTICS AND PHENOTYPIC IDENTIFICATION. *Micrococcus luteus* is a large, non-motile Gram-positive coccus (2 μm) that occurs in tetrads. Catalase and oxidase are positive. On BHIA and BA colonies are yellow-pigmented after 24 h at 37°C. Reactions for the isolate from fish were variable for urease and growth in 2% NaCl (Austin and Stobie, 1992b).

Micrococcus luteus does not grow in 15% NaCl and growth occurs at 45°C but may be weak. *Micrococcus luteus* can be divided into three biovars. Casein is hydrolysed by biovars II and III. Tween 80 is not hydrolysed by biovar I or III and is variable for biovar II (Wieser *et al.*, 2002). These and other reactions are presented in Table 3.14.

1.26.2 *Arthrobacter*

Arthrobacter species are asporogenous rod-shaped bacteria that have a coryneform morphology, are catalase positive and strictly aerobic.

Arthrobacter nasiphocae

TAXONOMY AND SIGNIFICANCE. *Arthrobacter nasiphocae* (CCUG 42953^T) was isolated in mixed culture (*Corynebacterium phocae*, *Pseudomonas aeruginosa*, *Staphylococcus aureus* and *Staphylococcus* species) from the nasal cavities of two common seals (*Phoca vitulina*) housed in a rehabilitation centre (Collins *et al.*, 2002a). The pathogenicity is unknown.

GROWTH CHARACTERISTICS AND PHENOTYPIC IDENTIFICATION. *Arthrobacter nasiphocae* is a strictly aerobic, non-spore-forming Gram-positive short rod that displays coccoid forms and irregular-shaped rods. Catalase is positive. On CBA containing 5% horse blood, colonies are 1 mm, grey-white, non-haemolytic, circular and convex after 24 h at 37°C. Blood or serum is not required for growth. Growth occurs at 25–42°C and in 5% NaCl but not in 10% NaCl. Fermentation of glucose, mannitol, maltose and sucrose is negative. Hippurate and gelatin are hydrolysed but aesculin and starch are not. Nitrate is not reduced (Collins *et al.*, 2002a). These and other reactions are presented in Tables 3.15, 3.31–33 and 3.38.

Arthrobacter rhombi

TAXONOMY AND SIGNIFICANCE. *Arthrobacter rhombi* (CCUG 38813^T) was isolated from the spleen of Greenland halibut (*Reinhardtius hippoglossoides*) captured by deep trawl off the coast of Newfoundland (Osorio *et al.*, 1999). Pathogenicity is unknown. *Arthrobacter rhombi* is most closely related to *A. sulfureus* (96.7%) and *A. nicotianae* (96.6%) based on 16S rRNA sequence.

GROWTH CHARACTERISTICS AND PHENOTYPIC IDENTIFICATION. *Arthrobacter rhombi* is a strictly aerobic, non-spore-forming, non-motile, Gram-positive short rod to ovoid cell and occurs singly. It is positive for catalase and oxidase. On brain–heart infusion agar (BHIA) colonies are 1 mm, yellowish or whitish, smooth convex after 48 h at 25°C. Growth occurs at 1–10% NaCl and 4–30°C. Aesculin hydrolysis and beta-galactosidase are positive in the API CORYNE kit. These and other reactions are presented in Tables 3.15 and 3.30–33.

1.26.3 *Renibacterium*

Renibacterium salmoninarum

TAXONOMY AND SIGNIFICANCE. *Renibacterium salmoninarum* (Lea-1-74=ATCC 33209) is a small Gram-positive diplococcus with fastidious growth requirements and the causative agent of bacterial kidney disease (BKD). The disease also has been referred to as corynebacterium kidney disease and salmonid kidney disease. It was first referred to as Dee disease with the initial isolation from Atlantic salmon (*Salmo salar*) in Scotland in 1930. Later, it was found in brook trout (*Salvelinus fontinalis*) and brown trout (*Salmo trutta*) in the USA. The disease has been reported from parts of Asia, Canada, Chile, England, France, Japan, Iceland, Italy, North America, Scotland, Spain, Turkey and the former Yugoslavia. It is exotic to Australia and New Zealand. *Renibacterium* originally was included in the genus *Corynebacterium*, however it does not contain mycolic acids in the cell wall and possesses a different peptidoglycan from *Corynebacteria* found in animals and humans. *Renibacterium* is related to *Arthrobacter* and *Micrococcus*. The G+C content is 53.0 mol% (Sanders and Fryer, 1980; Fryer and Sanders, 1981; Sanders and Barros, 1986). Whole genome sequencing reveals that *R. salmoninarum* evolved from non-pathogenic *Arthrobacter* species by genomic reduction and horizontal gene transfer (Wiens *et al.*, 2008).

BKD begins as a chronic form that develops full expression when the fish are 1 year old. Systemic granulomatous lesions are seen in all organs, but particularly in the kidney where grey to grey-white necrotic abscesses occur throughout the tissue causing enlargement and necrosis of the kidney. Clinical signs may not be seen, but generally include a distended abdomen, exophthalmos, petechiae and haemorrhagic areas around the base of the fins and swollen, purulent or granular kidney (Fryer and Sanders, 1981). It is a disease of wild and farmed fish from the family *Salmonidae* (salmon, trout, char), however infection from non-salmonids has been reported. Mortality rate often is high. BKD was listed in the 2000 OIE manual of diagnostic tests for aquatic animals but has since been delisted.

Renibacterium salmoninarum has been cultured from chinook trout (*Oncorhynchus tshawytscha*), coho (*O. kisutch*), chum (*O. keta*), pink (*O. gorbuscha*), and sockeye salmon (*O. nerka*), and rainbow trout (*Oncorhynchus mykiss*) in the USA (Banner *et al.*, 1986), chum salmon (*O. keta*), Chile, and from cultured black salmon (*S. labrax*) in the Black Sea region of Turkey (Savas *et al.*, 2006). Infection occurs in cultured ayu (*Plecoglossus altivelis*), a non-salmonid fish, cultured in Japan (Nagai and Iida, 2002). Clinical signs include white nodules in the kidney, swollen abdomen and ascites fluid. Pathogenicity was confirmed through experimental challenge in ayu and masou salmon (*O. masou*) resulting in mortality of 100% and 60%, respectively, when challenged with 10^3 cells/fish.

Renibacterium salmoninarum has also been reported from non-salmonid fish including moribund North Pacific hake (*Merluccius productus*) (Kent *et al.*, 1998). In experimental challenge, *R. salmoninarum* produced disease in non-salmonid fish including Pacific herring (*Clupea pallasii pallasii*), sablefish (*Anoplopoma fimbria*), shiner perch (*Cymatogaster aggregata*), common shiner (*Luxilus* [*Notropis*] *cornutus*) and flathead minnow (*Pimephales promelas*) (Evelyn, 1993; Nagai and Iida, 2002).

Infection was reported from sea lamprey (*Petromyzon marinus*) found in the Great Lakes, USA and Canada (Eissa *et al.*, 2006). The organism was cultured from kidney but not blood or internal organs from 4 to 16% of lamprey. In an earlier study investigating the carrier status of Pacific lamprey (*Lampetra tridentata*) in the Stamp River, Vancouver Island, Canada, experimental challenge with 1.3×10^6 CFU injected intraperitoneally did not result in infection when tested 3 months post-inoculation (Bell and Traxler, 1986). An eel (*Anguilla anguilla*) from the River Itchen, southern England, was positive for *R. salmoninarum* when tested by PCR (Chambers *et al*, 2009).

Some fish such as bartail flathead (*Platycephalus indicus*) and Japanese sculpin (referred to as *Cottus japonicus*, but accepted name is *Percis japonicus*) may be carriers of the infection as they were positive for antigen (57-kDa protein) when tested by indirect fluorescent antibody test, although the organism was not cultured (Sakai and Kobayashi, 1992). Studies on the blue mussel (*Mytilus edulis*), commonly a cause of biofouling of sea cages, indicates that although cells of *R. salmoninarum* are taken up by the bivalve, a bactericidal effect is exerted in the digestive tract and it is unlikely the mussel is a source of re-infection in salmon farms (Paclibare *et al.*, 1994).

A major virulence factor is a 57-kD protein, which is heat stable at 100°C for 30 min and is present on the cell surface and in culture supernatant (Getchell *et al.*, 1985). Protein p57 is encoded by the major soluble antigen (*msa*) gene (Chien *et al.*, 1992).

GROWTH CHARACTERISTICS AND PHENOTYPIC IDENTIFICATION. *Renibacterium salmoninarum* is a slow-growing, aerobic, non-motile, short, strongly staining, Gram-positive rod or diplococcus (0.3–1.0×1.0–1.5 μm) occurring singly and in pairs. Cells are not acid-fast and non-encapsulated. Cysteine is required for growth and most strains also require serum or whole blood for growth. The addition of albumin, charcoal or Tween 60 to the medium may enhance growth (Sanders and Fryer, 1980; Goodfellow *et al.*, 1985).

A number of media have been reported for the culture of *R. salmoninarum*: Mueller Hinton agar containing 0.1% cysteine (MHC or MHCA);

kidney disease medium-2 (KDM2); selective kidney disease medium (S-KDM), which consists of KDM2 plus antimicrobials; KDM-C which is KDM2 medium with 0.1% w/v activated charcoal replacing the serum content; spent KDM2 with serum and used as a nurse culture or included as a 1.5% v/v supplement; and KDM2 containing 5% or 10% of the unknown supplement from the spent medium (Five-M or Ten-M) (Bullock *et al.*, 1974; Austin *et al.*, 1983; Daly and Stevenson, 1985, 1993; Evelyn *et al.*, 1990; Teska, 1994). In a study to compare these media, medium Ten-M resulted in the greatest number of positive cultures (80%) from infected rainbow trout (*O. mykiss*) compared to S-KDM (42.5%), KDM-C (17.5%), KDM (10%) and MHC (5%) (Chambers and Barker, 2006). Samples are collected from lesions especially those seen in the kidney. Coelomic fluid is suitable from mature females (OIE, 2000b). Sampling of the head kidney resulted in a greater number of positives compared to the anterior kidney. Although *R. salmoninarum* is said to be a strict aerobe, growth on KDM2 agar and incubated under a microaerobic atmosphere produced a larger colony size of 3–4 mm compared to 1–2 mm under aerobic conditions at 12 weeks of incubation (Hirvelä-Koski, 2008). A microaerobic atmosphere of 5–8% oxygen and 12–15% CO_2 was achieved using CampyGen™ packs and equipment (Oxoid). Inconsistent growth performance may occur on KDM2 medium due to batch-to-batch variation in the composition of peptone, however these inconsistencies can be overcome by using a nurse culture technique or supplementation with 'spent' culture broth (Evelyn *et al.*, 1990). The nurse culture techniques involve placing a 25-μl drop of nurse culture (prepared from cells suspended in saline or peptone) into the centre of a plate inoculated with the test sample. Once the drop is dry the plates are inverted, sealed to prevent drying, and incubated at 15°C for 21 days. The supplementation with spent culture broth method can be achieved by preparing new KDM2 medium with 1.5% (v/v) of spent culture medium. The use of spent culture medium as a supplement is not possible in countries where *Renibacterium* is exotic and a range of media such as KDM2 (with the addition of serum and not spent culture medium) and S-KDM should be used.

Renibacterium salmoninarum grows at 15–18°C with maximum cell growth occurring after 20–30 days of incubation. Growth is even slower at 5 and 22°C, and does not occur at 37°C. Colonies are white to cream-yellow, circular, convex and of varying size on cysteine serum agar, and on Loeffler's coagulated serum medium colonies are creamy with a matt surface. Colonies are raised, shiny, smooth and yellow on Dorset egg medium. In cysteine serum broth growth occurs at the surface only. Growth is seen at 15 days on semi-defined medium (SDM) with raised, smooth, shiny cream-yellow colonies (Sanders and Fryer, 1980; Goodfellow *et al.*, 1985). On KDM-2 and KDMC media pinpoint colonies are visible between 2 to 8 weeks at 15°C. Growth does not occur on blood agar or on trypticase soya agar.

Biochemical properties of 26 strains of *R. salmoninarum* from different geographical locations (America, Canada and the UK) were found to be reasonably consistent (Bruno and Munro, 1986). Some differences in reactions have been reported between Bruno and Munro (1986) and Goodfellow *et al.* (1985). Hydrolysis of DNA was reported positive for 26 strains, whereas reactions were negative for 44 strains; gelatin hydrolysis was positive for 26 strains but quoted as negative by Goodfellow *et al.* (1985). Conventional methods were used with the exception that MHC was used as a base medium for hydrolysis of gelatin, DNA and detection of beta-haemolysis (Bruno and Munro, 1986).

It is catalase positive and oxidase negative. Litmus milk is hydrolysed without a change in pH. There is no cross-reaction with Lancefield group G antigen. Glucose is not fermented and there is no apparent acid production from sugars. Chitin is not hydrolysed but reactions are positive for Tween 20, 40 and 60 but negative for Tween 80 (Sanders and Fryer, 1980; Goodfellow *et al.*, 1985).

Bacterial cells are seen in smears prepared from kidney tissue and they may appear in pairs, short chains or pleomorphic forms giving the appearance of 'Chinese letters'.

Biochemical reactions are presented in Tables 3.15 and 3.38.

Commercial enzyme-linked immunosorbent assay (ELISA) kits are available for serological testing. An immunogold chromatographic strip test for the detection of the p57 antigen is available commercially from Cer*Test* Biotec, Spain.

MOLECULAR IDENTIFICATION. A number of methods are available to detect *R. salmoninarum* in kidney tissue, however no single assay is recommended, but rather sensitivity of detection is improved by use of multiple diagnostic tests (Sandell and Jacobson, 2011). Three real-time PCRs (qPCR) and ELISA were compared for analytical sensitivity and specificity using samples testing positive with a nested PCR (nPCR) reported by Chase and Pascho (1998). A PCR targeting the multiple copy major soluble antigen with a non-fluorescent quencher probe (*msa*/NFQ) (Chase *et al.*, 2006) detected 53.2% of samples positive by nPCR compared to 21.8% positive by the single copy *abc* gene with a non-fluorescent quencher probe (*abc*/NFQ), and 18.2% positive with an *abc*/TAMRA assay (Rhodes *et al.*, 2006). The ELISA detected 16.4% of the positive samples detected by nested PCR. However, although the *msa*/NFQ PCR detected a higher number of the nPCR positives, the *abc*/NFQ PCR detected a small number of positives that were not detected by the *msa*/NFQ PCR (Sandell and Jacobson, 2011). The authors could not verify whether this was due to cross reaction or not. Although the *msa*/NFQ PCR detected more positives it had an average threshold cycle (Ct; also known as quantification cycle, Cq) of 32.0 compared to 27.4 and 24.6 for the *abc*/TAMRA and *abc*/NFQ PCRs, respectively, indicating the *msa*/NFQ PCR was detecting at the near limit of detection, which may account for the failure to detect some positive samples. The PCRs were compared using the same concentration of DNA. The analysis of the qPCRs was complicated by the fact that qPCR performance appeared to be related to infection intensity. The *msa*/NFQ PCR performed better than the *abc*/NFQ PCR on tissue from hatchery salmon with BKD compared to samples from salmon naturally exposed to the infection, whereas the *abc*/NFQ PCR rated slightly higher for the samples naturally exposed to infection. All PCRs were specific when tested against *Arthrobacter globiformis*, *Corynebacterium aquaticum*, *Kocuria (Micrococcus) varians*, *Micrococcus luteus*, *Aeromonas hydrophila*, *A. salmonicida*, *Edwardsiella tarda*, *Flavobacterium columnare*, *F. psychrophilum*, *Pseudomonas fluorescens*, *Vibrio anguillarum*, *V. ordalii* and *Yersinia ruckeri*.

The nPCR of Chase and Pascho (1998) amplifies a 320-bp fragment of the p57 protein gene and detects 10 bacterial cells per reaction compared to 1000 cells per reaction for a conventional PCR. DNA extraction from kidney tissue is improved by the use of the QIAamp tissue kit (Qiagen) compared to a phenol/chloroform extraction method. The first round consists of a forward primer P3 (75–93) (5′-AGC TTC GCA AGG TGA AGG G-3′) and a reverse primer M21 (438–458) (5′-GCA ACA GGT TTA TTT GCC GGG-3′). The primers for the second round amplification reaction are forward primer P4 (95–119) (5′-ATT CTT CCA CTT CAA CAG TAC AAG G-3′) and reverse M38 (394–415) (5′-CAT TAT CGT TAC ACC CGA AAC C-3′). The reaction mix consists of 5 μl DNA, 0.2 mM of each dNTP, 50 mM KCl, 10 mM Tris-HCl (pH 8.3), 1.5 mM $MgCl_2$, 0.4 μM of each primer P3 and M21 and 2 U *Taq* polymerase in a 50-μl reaction mix. Thermocycling consists of 30 cycles of 94°C for 30 s, annealing at 60°C for 30 s and extension at 72°C for 1 min. The second PCR round is performed using 1 μl of amplified product and primers P4 and M38 with 20 cycles of the thermocycling protocol. A 320-bp product is positive. The sensitivity is 10 cells per reaction (Chase *et al.*, 2006). The first set of primers can be used in a conventional PCR using 35 cycles. A 383-bp product is amplified. Precautions must be observed when performing nested PCR due to the high potential for false positives due to contamination. The addition of amplified template for the second-round PCR must be performed in a biological safety-cabinet (or equivalent) different to the one used for the addition of DNA template to the first-round PCR. PCR master mix must always be prepared in a 'clean' room and in a Biological Safety Cabinet or equivalent. The nPCR is specific when tested against *C. aquaticum*, *P. fluorescens*, *Carnobacterium piscicola*, *A. hydrophila*, *A. salmonicida*, *Y. ruckeri*, *L. anguillarum* and *Rothia dentocariosa* (Chase and Pascho, 1998). An optimized thermocycling protocol for a MJ Research PJ100 thermocycler was used by Sandell and Jacobson (2011) and consists of 88°C for 3 min, 94°C for 2 min, followed by 30 cycles of 94°C for 30 s, 60°C for 1 min and 72°C for 1 min.

The *msa*/NFQ PCR (Chase *et al.*, 2006) consists of forward primer RS1238 (5′-GTG ACC AAC ACC CAG ATA TCC A-3′), reverse primer RS1307 (5′-TCG CCA GAC CAC CAT TTA CC-3′) and probe RS1262.NFQ (5′-FAM-CAC CAG ATG GAG CAA C-MGB/NFQ) as modified by Sandell and Jacobson (2011). The 25-μl PCR mixture contains 12.5 μl 2× *Taq*Man universal PCR master mix, 900 nM of each primer, 250 nM probe and 5 μl DNA template. Thermocycling is carried out using 50°C for 2 min, 95°C for 10 min, followed by 40 cycles of 95°C for 15 s and 60°C for 1 min (Chase *et al.*, 2006). The sensitivity is 5 cells/reaction (Rhodes *et al.*, 2006). Regression value exceeds 0.99, the coefficient of variation (CV) is 28.7 and the average Ct is 32.0 (Sandell and Jacobson 2011). A Ct of <39 is positive (Chase *et al.*, 2006). The *msa*/NFQ PCR had the same specificity as the nPCR.

The *abc*/NFQ PCR consists of a forward primer ABCtransfor2 (5′-CTA AAC GAT TTC CCG GTC AA-3′), reverse primer ABC transrev2 (5′-GAT TTT GCC TGC TGG TAT TTC C-3′) and a probe ABC.NFQ (5′-FAM-AAG CGC CAG CAG TCG ACG GC-MGB/NFQ-3′) as modified by Sandell and Jacobson (2011). The PCR master mix contains 900 nM primers, 200 nM probe and 100 ng template DNA. Thermocycling consists of 50°C for 2 min, 95°C for 10 min, followed by 45 cycles of 95°C for 15 s and 60°C for 1 min. The sensitivity is 2–4 cells/mg tissue (Rhodes *et al.*, 2006). Regression value exceeds 0.99, CV=20.1 and the average Ct is 24.6 (Sandell and Jacobson, 2011).

A qPCR using primers designed to the 57-kDa cell surface protein precursor *msa2* gene amplifies *R. salmoninarum* from whole blood (Bruno *et al.*, 2007). An internal control was designed to quantify the total DNA in the sample plus indicate the presence of PCR inhibitors. Blood is collected into heparin or EDTA, stored at 4°C and tested within 24 h. DNA is extracted using the QIAamp kit (Qiagen) in a procedure optimized for whole blood. Blood (10 μl) is added to 190 μl phosphate-buffered saline (PBS) to which 20 μl of Proteinase K and 200 μl lysis buffer are added. The mixture is vortexed for 30 s and incubated at 70°C for 2 h. The remaining steps of binding, washing and elution are carried out according to the manufacturer's instructions. The forward primer RS-F (5′-GGA GCA ACT CCG GTT ACT GGT A-3′), reverse primer RS-R (5′-TGG CCG TCC TTG AAC CAT-3′) and probe RS-P (5′-6-FAM-TGG TCT GGC GAC AAC AAC ACG TAT GGT-MGB-3′) amplify a 74-bp region of the *msa2* gene from *R. salmoninarum* from the blood. An internal control probe, ELF, was designed to the rainbow trout elongation factor α1, a housekeeping gene. Forward primer ELFSTD-F (5′-GGC CAG ATC TCC CAG GGC TAT-3′), reverse primer ELFSTD-R (5′-TGA ACT TGC AGG CGA TGT GA-3′) and probe ELDSTD-P (5′-CCT GTG CTG GAT TGC CAT ACT G-3′) amplify a 66-bp fragment of the Atlantic salmon and rainbow trout elongation factor α1 gene. The 20-μl PCR reaction mixture contains 1 μl of first strand cDNA synthesis reaction, 10 μl *Taq*Man 2× PCR Mix (Applied Biosystems), 900 nM each of forward and reverse primers, and 250 nM probe. Thermocycling consists of 50°C for 2 min, 95°C for 10 min followed by 45 cycles of 95°C for 15 s, and 60°C for 1 min. The fluorescence is recorded at the completion of the entire run. A standard curve is prepared using 10-fold serial dilutions of genomic DNA from *R. salmoninarum* or from Atlantic salmon liver tissue using the QIAamp kit (Qiagen) and used for qPCR with the RS or ELF primer/probe sets. The efficiency of each PCR is estimated from the Ct value and the -log (dilution). Ct values from samples are converted to Equivalent Target Amount (ETA) using the standard curve. The level of gene expression is expressed relative to the ELF control by the ratio ETA RS by ETA ELF (Bruno *et al.*, 2007). The ratio RS/ELF gives an estimation of the amount of *R. salmoninarum* in the initial 10 μl of blood.

Primers designed to detect a section of the *p57* gene (McIntosh *et al.*, 1996) cross-reacted with *Yersinia ruckeri* and were re-designed for specificity (Chambers *et al.*, 2009). The new primers are DN1 (5′-GAA AAT AAA AAA AAT TTT AGC GCT G-3′) and DN2 (5′-C TT GGC AGG ACC ATC TTT GT-3′). A 100-μl master mix contains 10 μl buffer IV (Abgene), 1.5 mM $MgCl_2$, 0.25 mM each dNTP, 100 pmol of each primer, and 10 μl DNA template. Thermocycling consisted of 35 cycles of 95°C for 1 min, 57°C for 1 min, 72°C for 2 min followed by one round of 94°C for 2 min, 55°C for 2 min and 72°C for 10 min. The sensitivity of the PCR was less than 100 cells per reaction or 0.2–82 CFU per reaction. A DNA extraction method using DNAzol (Invitrogen) was used but it was noted that the presence of kidney tissue appeared to slightly inhibit the PCR reaction (Chambers *et al.*, 2009). The sensitivity of the test is not compromised when 50 mg of head kidney tissue is pooled from up to five fish. The redesigned primers are specific against type strains of *Aerococcus viridans*,

Aeromonas caviae, *A. hydrophila*, *A. salmonicida*, *A. salmonicida* ssp. *achromogenes*, *Arthrobacter globiformis*, *Bacillus cereus*, *B. polymyxa*, *B. sphaericus*, *B. subtilis*, *Brevibacterium linens*, *Corynebacterium aquaticum*, *Carnobacterium piscicola*, *Escherichia coli*, *Flavobacterium psychrophilum*, *Klebsiella aerogenes*, *Lactobacillus brevis*, *Micrococcus luteus*, *Pseudomonas aeruginosa*, *P. anguilliseptica*, *P. fluorescens*, *Rothia dentocariosa*, *Salmonella typhimurium*, *Serratia marcescens*, *Staphylococcus aureus*, *S. epidermidis*, *Stenotrophomonas* (*Pseudomonas*) *maltophilia*, *Vibrio anguillarum*, *V. ordalii*, *V. salmonicida*, *Tenacibaculum maritimum* and *Yersinia ruckeri*.

A PCR using primers designed to the 16S rRNA gene and an internal control (Jansson *et al*., 2008) was slightly less sensitive than the *msa*/NFQ PCR (Chase *et al*., 2006) according to results from spiked samples. A 107-bp amplicon is produced from forward primer 5′-TGG ATA CGA CCT ATC ACC GCA T-3′ and reverse primer 5′-TCG CCT TGG TTA GCT ATT ACC-3′. A probe (5′-TTT TTG CGG TTT TGG ATG GAC TCG-3′) was labelled with the fluorescent reporter dye, 6-carboxyfluorescein at the 5′ end and with the black hole quencher (BHQ-1) at the 3′ end. For an internal control a mimic molecule was constructed by amplification of a segment of the human amelogenin gene with target sequences of the *Renibacterium salmoninarum* forward and reverse primers as flanking segments. An amplicon of 123 bp was generated. It was cloned into the pCR®2.1 vector using the Original TA Cloning® kit (Invitrogen). After overnight growth the plasmid DNA was purified and the concentration adjusted to a minimal amount that did not interfere with the assay (actual concentration not stated). The internal control is detected using an IC probe (5′-Cy5-CAA CCA ATG ATG CCC GTT CCT-BHQ-2-3′). The PCR master mix contains 2 U Ampli*Taq* Gold DNA polymerase, 5 µl GeneAmp PCR buffer II, 2 mM $MgCl_2$, 0.8 mM GeneAmp dNTP mixture (Applied Biosystems), 8% glycerol (Sigma Aldrich), 0.2 µM probe, 0.25 µM of each primer, 2.0 µl internal control, 0.2 µM IC probe in a total volume of 50 µl. Thermocycling consists of 95°C for 8 min, followed by 45 cycles of 95°C for 30 s, 60°C for 60 s and a final round of 72°C for 6 min. A Ct value of equal to or less than 38 is positive. The sensitivity is 1–10 genomes of pure DNA per reaction, and detection of 40–400 cells/µl in spiked ovarian fluid. DNA extraction using the DNeasy tissue mini kit (Qiagen) gave a sensitivity of 4–20 cells/mg kidney tissue. It is ten times more sensitive than ELISA (Jansson *et al*., 2008). The internal control did not inhibit the reaction. The PCR is specific against *Aeromonas hydrophila*, *A. salmonicida* ssp. *achromogenes*, *Arthrobacter globiformis*, *A. oxydans*, *Erysipelothrix rhusiopathiae*, *Flavobacterium columnare*, *F. psychrophilum*, *Leucobacter albus*, *L. komagatae*, *Vibrio anguillarum* and *Yersinia ruckeri*.

A loop-mediated isothermal amplification PCR using primers designed from the *p57* gene is reported to be more sensitive than qPCR (Bruno *et al*., 2007) when tested using diluted DNA extracted from kidney tissue using the DNeasy animal tissue kit from Qiagen (Gahlawat *et al*., 2009). The forward inner primer (FIP) consists of complementary sequences of F1c (22 nt) and F2 (18 nt): 5′-GGA GTT GCT CCA TCT GGT GCA TCC GCA ACA GCA ACT GAC A-3′. The backward inner primer (BIP) consists of complementary sequences of B1c (22 nt) and B2 (19 nt): 5′-CTG GTA AAT GGT GGT CTG GCG ACC ATT GGC CGT CCT TGA AC-3′. Two outer primers are F3 (20 nt): 5′-GCC CGG TAG AGG TTA AAG TC-3′; and B3 (19 nt): 5′-CGG AAC CAG CAT TTG GCT T-3′. A 25-µl master mix contains 60 mM Tris HCl (pH 8.8), 30 mM KCl, 6 mM $MgSO_4$, 30 mM $(NH4)_2SO_4$, 0.1% Triton X-100, 4.0 µl 5 M betaine (Sigma), 1.0 µl 25 mM dNTPs (Invitrogen), 4.0 µl each of FIP and BIP (10 µM), 0.5 µl each of F3 and B3 primers (10 µM), 0.5 µl water and 2 µl DNA template. The mixture is heated at 95°C for 5 min, then chilled on ice and 1.0 µl of 8 U Bst DNA polymerase (New England BioLabs) was added. Tubes are incubated at 65°C for 1 h before the reaction is terminated by heating at 80°C for 5 min. The LAMP PCR is 10-fold more sensitive compared to a qPCR based on serial dilutions of DNA. However, it appears the LAMP PCR is affected by PCR inhibitors in kidney tissue, as samples were positive by qPCR, but not by LAMP PCR unless the extracted DNA was diluted 1000-fold. The LAMP PCR is specific against *Nocardia kampachi*, *Carnobacterium piscicola*, *Pseudomonas* species and *Photobacterium* species.

PCR inhibitors were reported in kidney tissue and it was found that dilution overcomes the inhibitors.

1.27 *Mycobacteriaceae*

1.27.1 *Mycobacterium*

Mycobacteria are Gram-positive, acid fast, alcohol fast, aerobic, pleomorphic, non-sporing rods 0.2–0.6 μm to 1–10 μm in length and belong to the genus *Mycobacterium*, the order *Actinomycetales*, suborder *Corynebacterineae* and the family *Mycobacteriaceae*. Mycobacteria have a unique cell wall composed of long chains of 60–90 carbons of 3-hydroxymycolic acids. These high molecular weight mycolic acids, which precipitate from cold ether-ethanol, confer acid-fastness on the bacterial cell. Other genera such as *Caseobacter*, *Corynebacterium*, *Nocardia* and *Rhodococcus* contain mycolic acids of the cell wall chemotype IV in their cell walls (Hartmans *et al.*, 2006).

Over 156 species and subspecies of *Mycobacterium* have been described from a variety of habitats including soil, and a number of species cause disease in humans, terrestrial animals, fish and other aquatic animals. Some Mycobacteria isolated from fish are known human pathogens. An example is *M. marinum*, which causes infection referred to as fish fancier's finger, fish tank or swimming pool granuloma, and affects fishermen and aquaculturists. Infection leads to granulomatous lesions in skin and peripheral deep tissues. *Mycobacterium peregrinum* and *M. scrofulaceum* also cause granulomas, and *M. fortuitum* and *M. chelonae* are opportunistic pathogens infecting immunocompromised people. *Mycobacterium ulcerans* causes Buruli ulcer but there is potential for other mycolactone-producing mycobacteria isolated from fish to cause opportunistic infection in people. *Mycobacterium marinum* is a biosafety level 2 organism and is not transmitted to humans via aerosols, whereas *M. tuberculosis*, the cause of tuberculosis in humans, is a biosafety level 3 organism and is transmitted through aerosols.

Some Mycobacteria infecting fish belong to the *Mycobacterium tuberculosis* clade and these include *M. marinum*, *M. shottsii*, *M. pseudoshottsii* and *M. ulcerans*. Other species included in the *Mycobacterium tuberculosis* clade include those known as the *M. tuberculosis* complex (*M. tuberculosis*, *M. bovis*, *M. africanum*, *M. caprae*, *M. canettii*, *M. pinnipedii* and *M. microti*). Most species within the *M. tuberculosis* clade are slow growers with the exception of *M. marinum* whose colonies appear in 7–14 days compared to 2–3 months for other species. Non-*Mycobacterium* tuberculosis clade, often referred to as atypical Mycobacteria, are usually rapid growers.

In fish, infection with a *Mycobacterium* species normally occurs as a chronic, systemic disease characterized by granulomas in multiple organs or tissues. Aquatic animals may be infected for 2–4 years before the infection is detected via low mortality rate and failure to thrive (Ashburner, 1977; Gauthier and Rhodes, 2009). Typical gross pathology may include skin ulcers, exophthalmia, keratinitis, with clinical signs including abnormal swimming behaviour and emaciation. The route of infection is usually by consumption of contaminated food and water.

Mycobacterium abscessus (ATCC 19977) initially proposed as a subspecies of *M. chelonae* ssp. *chelonae*, as *M. chelonae* ssp. *abscessus*, was elevated to species level based on DNA–DNA hybridization (Kusunoki and Ezaki, 1992). *Mycobacterium abscessus* was cultured from internal granulomas found in Japanese medaka (*Oryzias latipes*) held captive in an experimental facility in the USA (Teska *et al.*, 1997). It also has been reported from infected goldfish (*Carassius auratus*), firemouth cichlid (*Cichlasoma meeki*) and black acara (*C. bimaculatum*) (Lansdell *et al.*, 1993). The organism caused an outbreak of mycobacteriosis with cumulative mortalities of up to 66.7% over 12 months in milkfish (*Chanos chanos*) cultured in Taiwan (Chang *et al.*, 2006). Infection resulted in gross reddish or grey-white nodules on the peritoneal surface and in spleen, liver, kidney and gastrointestinal tract. *Mycobacterium abscessus* was identified by PCR using degenerate primers targeted to the heat shock protein gene. It has been found in wild cane toad (*Bufo marinus*) and *Chaunus* (*Bufo*) *granulosa* from Brazil. The toads appeared clinically normal and no lesions were seen in the viscera, but acid-fast bacilli were cultured from liver and/or spleen (Mok and Carvalho, 1984). Experimental challenge confirmed the infection, and again toads appeared clinically normal for the first few month post-inoculation, but some toads became emaciated, yet no visible lesions were observed. The organism was recovered from all organs.

Mycobacterium avium was isolated from granulomatous lesions in a cockatoo cichlid

(*Apistogramma cacatuoides*) as part of a study of aquarium fish in the Czech Republic (Lescenko *et al.*, 2003).

Mycobacterium chelonae belongs to Runyon's Group IV, a group of fast-growing bacteria. *Mycobacterium chelonae* is found in soil and water and has been reported from infected fish and other aquatic animals with granulomas, and from cutaneous abscesses in humans following injection, trauma or surgery, and from immunosuppressed individuals. The type strain for *M. chelonae* (ATCC 6841^T) was isolated from a patient in Rio de Janeiro (Cruz, 1938) but was first isolated from a turtle in 1903 (referred to by the discoverer Freidmann as the turtle tubercle bacillus) and later found in chinook salmon (*Oncorhynchus tshawytscha*) and other Pacific salmonids (Ashburner, 1977; Grange, 1981). Severe granulomatous peritonitis and hepatitis occurred in yellow perch (*Perca flavescens*) in a lake in Canada (Daoust *et al.*, 1989). The organism has been isolated from skin lesions and in some fish, the spleen and/or kidney. A report of a chronic infection in farmed Atlantic salmon (*Salmo salar*) in the Shetland Isles, Scotland originally identified as *M. chelonae* was later proven to be *M. salmoniphilum* (Bruno *et al.*, 1998a; Whipps *et al.*, 2007). A mycobacterial species isolated from salmonids (cutthroat salmon, coho salmon, fall chinook salmon, mountain whitefish) in the USA and fall chinook salmon in Australia and proposed as a new species, *M. chelonae* ssp. *piscarium*, was later demonstrated to be *M. chelonae* (Arakawa *et al.*, 1986). *Mycobacterium chelonae* was isolated from paraffin-embedded tissue from yellow seahorse held at the London Zoo, UK and from goldfish in aquaria in Scotland (Pourahmad *et al.*, 2009). The organism was also reported from zebrafish (*Danio rerio*) held at a research facility in the USA (Whipps *et al.*, 2008). *Mycobacterium chelonae* together with *M. phocaicum* and *M. neoaurum* were identified by PCR from biofilm in the tanks. It has been reported from the aquatic frog, the South African clawed frog (*Xenopus laevis*) (Green *et al.*, 2000).

Mycobacterium fortuitum belongs to Runyon's Group IV, fast-growing mycobacteria and contains two subspecies: *M. fortuitum* ssp. *fortuitum* and *M. fortuitum* ssp. *acetamidolyticum*. *Mycobacterium fortuitum* ssp. *fortuitum* (ATCC 6842^T) was first reported from a person with an abscess (Cruz, 1938) and *M. fortuitum* ssp. *acetamidolyticum* was isolated from a human lung infection (Tsukamura *et al.*, 1986). *Mycobacterium fortuitum* was isolated from neon tetra (*Paracheirodon innesi*) in the USA (Ross and Brancato, 1959), black acara (*Cichlasoma bimaculatum*) (Lansdell *et al.*, 1993), Siamese fighting fish (*Betta splendens*) in Thailand (Puttinaowarat *et al.*, 2002) and in Malaysia (Najiah *et al.*, 2011), and in ornamental aquarium fish in South Africa: guppy (*Poecilia reticulata*), oscar (*Astronotus ocellatus*) and discus fish (*Symphysodon aequifasciatus*) (Bragg *et al.*, 1990). It has been reported from red devil (*Amphilophus labiatus*) and humphead cichlid (*Cyphotilapia frontosa*) held in aquaria in the Czech Republic (Novotny *et al.*, 2010), and isolated from paraffin-embedded tissue from goldfish held in aquaria in Scotland (Pourahmad *et al.*, 2009). Experimental challenge proved its pathogenicity to goldfish (*Carassius auratus*) (Talaat *et al.*, 1999). An isolate originally identified as *M. poriferae* and isolated from mycobacteriosis in cultured striped snakehead fish (*Channa striata*) in Thailand was later proven to be *M. fortuitum* using PCR (Tortoli *et al.*, 1996; Puttinaowarat *et al.*, 2002).

Mycobacterium gordonae was isolated from formalin-fixed paraffin-embedded tissue from Charco Palma pupfish and cardinal tetra held at the London Zoo Aquarium, UK, and from aquaria in the Czech Republic (Lescenko *et al.*, 2003; Pourahmad *et al.*, 2009). The organism also was detected in freshwater angelfish (*Pterophyllum scalare*) and cockatoo cichlid (*Apistogramma cacatuoides*) in the Czech Republic (Lescenko *et al.*, 2003; Novotny *et al.*, 2010).

Mycobacterium marinum has been reported from a wide range of aquatic animals and can cause skin infections in people, particularly fish handlers and those in the aquarium trade. *Mycobacterium marinum* (ATCC 927^T) was first named and reported from marine fish that died at the Philadelphia Aquarium, USA (Aronson, 1926). Fish with mycobacteriosis included sergeant-major (*Abudefduf saxatilis*), Atlantic croaker (*Micropogonias undulatus*) and black seabream (*Centropristis striatus*) (Aronson, 1926). Tubercules or greyish-white, elevated, firm nodules 1–8 mm in diameter were seen in gills, kidney, roe, pericardium and eye of some or all of the fish and the organism was isolated from the liver and spleen. *Mycobacterium marinum* has been reported from a variety of aquarium fish. In the Czech Republic in a study of freshwater ornamental fish, *M. marinum* was isolated from skin, gill, muscle, liver, spleen and kidney of cockatoo cichlid (*Apistogramma cacatuoides* Hoedemann) and pearl gourami (*Trichogaster leerii* Bleeker) (Novotny *et al.*, 2010). Siamese fighting fish (*Betta splendens*), a popular aquarium fish, were infected with both *M. marinum* and *M. fortuitum* causing outbreaks of mycobacteriosis in Thailand and Malaysia (Puttinaowarat *et al.*, 2002; Najiah *et al.*, 2011). Infection occurred in three-spot gouramies (*Trichopodus trichopterus*) imported into Venezuela from Columbia (Santacana *et al.*, 1982). The organism was originally called *M. anabanti*, which had been reported from roundtail paradise fish (*Macropodus opercularis*) and pumpkinseed sunfish (*Lepomis gibbosus*), however based on published phenotypic and pathogenesis data *M. anabanti* is thought to be *M. marinum* and the name was not reported in the list of Type and Reference strains (Sneath and Skerman, 1966; Gauthier and Rhodes, 2009). Likewise, an isolate recovered from infected southern (Mexican) platyfish (*Xiphophorus* [*Platypoecilus*] *maculatus*), which caused emaciation, skin ulcer and a friable liver, was named *M. platypoe cilus* (Baker and Hagan, 1942). This name was not accepted and again is considered to be *M. marinum*. In a study using PCR amplification of formalin-fixed paraffin-embedded tissue *M. marinum* was detected in goldfish, Koi carp, rosy barb and Odessa barb from samples from Scotland, UK; from striped barombi cichlid, lumpsucker, Banggai cardinal fish, Otjikoto tilapia, yellow seahorse, golden skiffia, Pitch Lake guppy and crescent zoe from London Zoo, UK, and from seabream in Turkey (Pourahmad *et al.*, 2009). Mycobacteriosis due to *M. marinum* was found in wild rabbitfish (*Siganus rivulatus*) associated with sea-cage farming in the Gulf of Eilat in the Red Sea (Diamant *et al.*, 2000). Although the rabbitfish appeared clinically normal white-yellow nodules were seen in a slightly enlarged spleen. In heavily infected fish, nodules containing acid-fast bacilli occurred in the liver, kidney, alimentary tract and mesenteries. *Mycobacterium marinum* has been reported from a variety of fish in the Red Sea including European seabream (*Dicentrarchus labrax*) and red drum (*Sciaenops ocellatus*) (Diamant *et al.*, 2000) and from striped bass (*Morone saxatilis*) farmed in the USA (Hedrick *et al.*, 1987; Lansdell *et al.*, 1993). Clinical signs were absent from most fish apart from emaciation, however some had small blisters and shallow ulcerations on the skin containing acid-fast bacilli, and numerous nodular lesions in the kidney, liver and spleen. A *Mycobacterium* species identified as belonging to *Mycobacterium* Group I photochromogenic based on Runyon's classification was isolated from epizootics in yellowtail (*Seriola quinqueradiata*) cultured in Sukumo Bay, Japan in 1986 (Kusuda *et al.*, 1987). Classical signs were hypertrophy of the kidney and spleen with tubercules and visceral adhesions, and haemorrhagic ascites. More recent isolations have identified *M. marinum* as causing infections in yellowtail (Weerakhun *et al.*, 2007). Experimental challenge experiments confirmed the pathogenicity of the isolate. The *Mycobacterium* species reported in 1987 had some phenotypic differences to the *M. marinum* isolate. The *M. marinum* isolate was identical to *M. pseudoshottsii* and had a one nucleotide difference with the *M. marinum* type strain according to partial 16S rRNA sequencing, however biochemical results confirmed identification as *M. marinum*. An atypical presentation of mycobacteriosis appeared in striated frogfish (*Antennarius striatus*) held in a marine laboratory aquarium, USA after being imported from Brazil (Yanong *et al.*, 2003). Severe systemic mycobacteriosis occurred without the presence of typical granulomas. Fish had external lesions, reproductive problems or abnormal behaviour and necropsy findings included pale yellow or tan livers, ocular opacity, enlarged coelomic cavities with ascites and raised nodules. Culture of tissue was generally negative except for liver from one fish, which grew *M. marinum* when incubated at 23–25°C but not at 30°C. Acid-fast bacilli were present within macrophages in blood vessels and in interstitial infiltrates in liver, spleen, heart, testes and other tissue. Mycobacteriosis due to *M. marinum* occurred in snakehead fish (*Channa striata*) cultured in Thailand (Puttinaowarat *et al.*, 2002; Pourahmad *et al.*, 2009). The organism caused mycobacteriosis in pond-cultured soft-shell turtles (*Trionyx sinensis*) in Taiwan (Hsieh *et al.*, 2006). Partial sequencing of the heat shock protein gene identified two groups of *M. marinum* in the turtles. Infection with Group one strains led to heterophilic granulomas. This group had high sequence similarity of 99.8–100% with *M. marinum* (AF456471) and *M. pseudoshottsii* (AY5550226), whereas infection with Group two strains led to histiocytic granulomas and had sequence similarity of 99.8–100% to *M. marinum* (ATCC 927, AF56470) and *M. shottsii* (AY550225).

In a captive white whale (*Delphinapterus leucas*) *M. marinum* caused dermatitis, panniculitis and chronic pleuritis and was cultured from skin and pleura (Bowenkamp *et al*., 2001). Infection with *M. chelonae* and *M. marinum* occurred in turbot (*Scophthalmus maximus*) farmed in Portugal (dos Santos *et al*., 2002). *M. chelonae* was isolated in all cases from the kidney, whereas *M. marinum* was isolated from the spleen and *Nocardia* was isolated from some fish but did not appear to be the dominant cause of the infection. Granulomas were seen in organs of both moribund fish and fish that appeared clinically normal. A mycobacterium that caused multiple granulomas in kidney, heart, stomach, intestine and mesentery of summer flounder (*Paralichthys dentatus*) held captive at a facility in the USA was identified by culture and 16S rRNA sequencing as having close homology to *M. marinum* and *M. ulcerans* (Hughes *et al*., 2002). Sequence similarity for the 16S rRNA gene is 99.8% representing a two nucleotide difference between *M. marinum* and *M. ulcerans* (Portaels *et al*., 1996).

Mycobacterium montefiorense (originally identified as *Mycobacterium triplex*-like) was isolated from captive green moray eel (*Gymnothorax funebris*) and spotted moray eel (*G. moringa*) (Herbst *et al*., 2001; Levi *et al*., 2003). Infected eels had granulomatous dermatitis and fasciitis. It is closely related to *M. triplex*, *M. simiae* and *M. avium* complex based on biochemical tests. It does not grow at temperatures above 30°C. *Mycobacterium montefiorense* can be identified by sequencing the internal spacer rRNA using primers Sp1 and Sp2, and sequencing the *hsp65* gene using primers Tb11 and Tb12. With regard to the *hsp65* gene *M. montefiorense* differs from *M. triplex* by 10 of 441 bp, and from *M. genavense* by 14 of 441 bp (Levi *et al*., 2003).

Mycobacterium neoaurum was isolated in Canada from farmed chinook salmon (*Oncorhynchus tshawytscha*) with a progressive eye disease characterized by a pyogranulomatous and necrotizing panophthalmitis (Backman *et al*., 1990). Acid-fast bacilli were mostly confined to the eye but were also found in low numbers from kidney and spleen but only in severely affected fish. A *Rhodococcus* species was co-isolated from all samples.

Mycobacterium palustre (E846^T=ATCC BAA-377^T=DSM 44572^T) was isolated from stream water from peatland run-off, from an Italian child with cervical lymphadenitis and from lymph nodes of two Finnish pigs (629=DSMZ 44571) sampled at an abattoir (Torkko *et al*., 2002). Based on 16S rRNA sequence *M. palustre* is most closely related to *M. kubicae*.

Mycobacterium peregrinum (ATCC 14467^T) is closely related to *M. fortuitum* based on mycolic acids and was proposed as a subjective synonym of *M. fortuitum*, however DNA–DNA hybridization and phenotypic tests demonstrated *M. peregrinum* as a separate species (Kusunoki and Ezaki, 1992). *Mycobacterium peregrinum* was isolated from infected Pacific white shrimp (prawn) (*Penaeus vannamei*) in the USA (Mohney *et al*., 1998). Multifocal black or melanized nodular lesions were evident on the carapace of cultured animals. *Mycobacterium peregrinum* is found free-living in water and soil and can cause skin infections in people (Mohney *et al*., 1998).

A *Mycobacterium* species initially called the seal bacillus and identified as *M. pinnipedii* (6482^T=ATCC BAA-688^T=NCTC 13288^T) was isolated from cases of tuberculosis in Australian sea lion (*Neophoca cinerea*), southern sea lion (*Otaria flavescens*), Australian fur seal (*Arctocephalus pusillus doriferus*), New Zealand fur seal (*A. forsteri*), South American fur seal (*A. australis*) and subantarctic fur seal (*A. tropicalis*) found wild or captive in Australia, New Zealand, the UK, Argentina and Uruguay (Cousins *et al*., 2003). Signs of infection included granulomatous lesions in lung, pleura, peritoneum, spleen and peripheral lymph nodes. The organism can cause infection in humans, cattle, guinea pigs and rabbits. *Mycobacterium pinnipedii* belongs to the *M. tuberculosis* complex based on spoligotyping, and gene sequences for *gyrA* and *katG* and phenotypic tests.

Mycobacterium poriferae (47^T=ATCC 35087^T), a scotochromogenic, rapidly growing species, was isolated from a marine sponge (*Halichondria bowerbanki*) (Padgitt and Moshier, 1987).

For many years epizootics in striped bass (*Morone saxatilis*) in Chesapeake Bay, USA caused by a mycobacterial species have been known to be due to two recently described species, *M. pseudoshottsii* (L15^T=ATCC BAA-883^T=NCTC 13318^T) and *M. shottsii*, however a variety of Mycobacteria have been isolated during the epizootics including *M. interjectum*, *M. marinum*, *M. scrofulaceum*, *M. szulgai* and *M. triplex* (Heckert *et al.*, 2001; Rhodes *et al.*, 2003, 2004, 2005). Mycobacteriosis in striped bass is characterized by skin lesions and granulomas containing non-branching acid-fast bacilli and isolation of the organism from the spleen. Analysis of 16S rRNA, exported repeated protein (*erp*) and the 65-kD heat-shock protein (*hsp65*) gene sequences demonstrated an unknown species which was closely related to *M. shottsii*, *M. ulcerans* and *M. marinum*. *Mycobacterium shottsii* (M175^T=ATCC 700981^T=NCTC 13215^T) was also isolated from infected striped bass in Chesapeake Bay (Rhodes *et al.*, 2003). *Mycobacterium shottsii* is closely related to *M. ulcerans* and *M. marinum* having 99.2% similarity to the 16S rRNA gene. *Mycobacterium pseudoshottsii* and *M. shottsii* have been isolated from infected striped bass and white perch (*Morone americana*) found in other parts of the USA (Stine *et al.*, 2009).

Mycobacterium salmoniphilum (ATCC 13758^T) was initially proposed for a *Mycobacterium* species isolated from Chinook salmon (*Oncorhynchus tshawytscha*) and rainbow trout (*O. mykiss*) in 1960 in the USA (Ross, 1960), however the name was not accepted formally because of the difficulty in distinguishing the isolate from *M. fortuitum*. Other isolates from salmon have been attributed to *M. chelonae* or called *M. chelonae*-like, which was proposed as *M. chelonae* ssp. *piscarium*, however this name was not accepted either because of the difficulty in distinguishing it from *M. chelonae*. Further analysis of isolates from salmon using internal transcribed spacer region, *hsp65* and RNA polymerase β-subunit (*rpoB*) determined the validity of *M. salmoniphilum* (Whipps *et al.*, 2007). It is closely related to the *M. chelonae* complex, which includes *M. chelonae*, *M. abscessus*, *M. immunogenum* and *M. bolletii* (*M. massiliense*) based on genotypic and phenotypic tests. More recently, *M. salmoniphilum* has been isolated from mortalities of Atlantic salmon (*Salmo salar*) cultured in Norway (Zerihun *et al.*, 2011). Infected fish displayed low body weight, swollen spleen and kidney and some fish had greyish-white granulomas in or on the liver and spleen, however overall, fish displayed few clinical signs. A chronic infection with *M. salmoniphilum*, originally identified as *M. chelonae*, caused mortalities in farmed Atlantic salmon (*Salmo salar*) in the Shetland Isles, Scotland (Bruno *et al.*, 1998a).

The sequence deposited in Genbank as *M. 'chesapeaki'* (Heckert *et al.*, 2001) contains high variability at the 3′-end, which Rhodes *et al.* (2005) suggested represents contamination and the organism should be re-sequenced before claiming status as a new species. Sequences from the 16S rRNA and internal transcribed spacer regions of *M. 'chesapeaki'* are identical to *M. shottsii* (Gauthier and Rhodes, 2009).

Mycobacterium scrofulaceum was reported from infected Pacific staghorn sculpin (*Leptocottus armatus*) (Lansdell *et al.*, 1993). The liver was white and friable with gross fatty degenerative changes. Lipid droplets were seen in the liver.

Mycobacterium simiae is commonly found in environmental water sources and has been reported from black acara (*Cichlasoma bimaculatum*) (Lansdell *et al.*, 1993).

Mycobacterium smegmatis produced mycobacteriosis when injected into goldfish (*Carassius auratus*). Differences in virulence and pathology were noted for the two strains tested. *Mycobacterium smegmatis* ATCC 19420 caused profuse giant cell involvement in the peritoneum whereas *M. smegmatis* mc^{2}155 infection led to granuloma formation (Talaat *et al.*, 1999).

Mycobacterium stomatepiae, a slow-growing non-chromogenic mycobacteria, was isolated from striped barombi mbo cichlids (*Stomatepia mariae*) held at the London Zoo Aquarium, UK (Pourahmad *et al.*, 2008, 2009). The organism was isolated from granulomas in the kidney and spleen. Based on 16S rRNA sequencing *M. stomatepiae* is closely related to *M. florentinum*, whereas sequence information from the *rpoB* gene indicated similarity to *M. mageritense*.

Mycobacterium triplex was isolated from formalin-fixed paraffin-embedded tissue from rosy barb from Scotland, UK (Pourahmad *et al.*, 2009).

Mycobacterium ulcerans, the cause of Buruli ulcer in humans, was isolated in fish sampled in Ghana and Benin in locations where Buruli ulcer is endemic (Eddyani *et al.*, 2004). Intestines and gills of jewelfish (*Hemichromis bimaculatus*), mango tilapia (*Sarotherodon galilaeus galilaeus*), *Epiplatys bifasciatus* and *Aplocheilichthys* species were positive by PCR. *Mycobacterium ulcerans* was also found in samples of gill and intestine from unidentified fish, tilapia and tissue from mollusc (*Bulinus senegalensis*) (Kotlowski *et al.*, 2004). The viability of *M. ulcerans* is adversely affected by commonly used decontamination treatments such as oxalic acid. A treatment using mild HCl is recommended for *M. ulcerans* followed by culture using egg-yolk-based media.

GROWTH CHARACTERISTICS AND PHENOTYPIC IDENTIFICATION. Mycobacteria do not stain or may stain weakly with Gram's stain due to the inability of stain to penetrate the cell wall because of the mycolic acids (long chain fatty acids). Cells are visualized in tissues and culture material using the Ziehl-Neelsen stain, which has a heating step to allow the stain to penetrate the cell wall. A lack of staining can occur with some tissue decalcification procedures but may be overcome or improved through the use of Fite's acid fast stain or pretreatment of sections with periodic acid (Daoust *et al.*, 1989) (see media section, Stains). It is important to use both phenotypical and genotypic methods to improve accuracy of identification.

DECONTAMINATION PROCEDURES FOR CULTURE. Because mycobacteria are often slow growing, they are outgrown by contaminating bacteria such as post-mortem invaders, or samples selected for culture that have a high bacterial load such as faeces or gastrointestinal tract. Ideally, aseptically excised tissues from euthanased animals should be used for culture, however this is not always possible. A number of decontamination methods are available to reduce the numbers of contaminating bacteria, but these are used at the risk of adversely affecting the viability of mycobacteria. Not all decontamination procedures used for mycobacteria are appropriate for every *Mycobacterium* species, as their effect on viability varies between species. For example *M. ulcerans*, although not a common pathogen for aquatic animals, is sensitive to commonly used decontamination methods such as: (i) oxalic acid, where samples are treated with 1.1 N oxalic acid (final concentration) for 20 min; (ii) the Petroff method, which uses 0.5 N NaOH (final concentration) for 20 min followed by neutralization with 1 N HCl; and (iii) the reverse Petroff method, which treats samples with 0.5 N HCl (final concentration) for 20 min followed by neutralization with 1 N NaOH. The best recovery of *M. ulcerans* is obtained using a mild HCl treatment of 0.03 N HCl (final concentration) for 20 min (Palomino and Portaels, 1998). The use of egg yolk in media improves the recovery of *M. ulcerans* from all decontamination methods. Rhodes *et al.* (2004) reported the use of Middlebrook 7H10 agar modified with the addition of 0.025% malachite green, 0.04% cycloheximide and 0.0035% nalidixic acid as used by Teska *et al.* (1997) for the recovery of *M. abscessus* resulted in poor growth of *M. avium*, *M. flavescens*, *M. gordonae*, *M. interjectum*, *M. intracellulare*, *M. marinum*, *M. nonchromogenicum*, *M. peregrinum*, *M. phlei*, *M. scrofulaceum*, *M. simiae*, *M. terrae* and *M. szulgai*. Recovery of *M. kansasii* did not occur on this medium, whereas *M. chelonae*, *M. fortuitum* and some strains of *M. scrofulaceum* and *M. simiae* were recovered with good growth on this medium.

Decontamination methods can reduce recovery numbers by 1–3 logs. In a comparison of NaOH, HCl and Zephiran™ treatments, the best results were found with HCl or NaOH compared to Zephiran™. The latter treatment detected only half the number of positive samples, whereas the use of 2% NaOH or 2% HCl produced similar results (44% positive) for tissue. More than three times the number of positive samples were obtained when treated material was cultured on to the egg yolk-containing medium Lowenstein-Jensen (L-J) compared to Blood agar, or Middlebrook 7H10 agar containing 0.5% glycerol and oleic acid, albumin, dextrose and catalase (OADC) or containing albumin, dextrose and catalase (ADC) (Rhodes *et al.*, 2004). Overgrowth by fungi and other bacteria were a problem. In that study *M. marinum* and *M. terrae* were least affected by HCl and NaOH compared to *M. interjectum*, *M. peregrinum* and *M. scrofulaceum*.

Tissues should be excised aseptically to reduce the possibility of contamination, and immersed in 70% alcohol and flamed quickly. Small pieces of tissue can be immersed in 70% alcohol for 5 min, drained and air dried and/or rinsed in three to five changes of sterile water or saline. Rinsing only should be used for skin samples to avoid affecting viability as a result of alcohol immersion or flaming. Samples are plated to culture media. Some methods recommend plating of aliquots from serial dilutions of tissue homogenate to aid in recovery and differentiation from non-mycobacterial species and recommend research into the use of growth enhancers such as mycobactin (Gauthier and Rhodes, 2009).

Contaminated tissue samples are homogenized in 2–3 ml Butterfields buffer using a tissue grinder. For safety this should be done in a Biological Safety Cabinet to prevent the risk from aerosols. A 1-ml aliquot of homogenate is centrifuged at 4000 ×*g* for 30 min at 4°C and the pellet resuspended in either 2% NaOH or 2% HCl and incubated at room temperature for 15 min. Suspensions are re-centrifuged and pellets resuspended in buffer containing bromocresol purple pH indicator. Neutralization is performed by adding either 1% HCl or 1% NaOH until a blue colour persists (pH 6.8). Samples are centrifuged and resuspended in 0.5–1 ml buffer. For skin samples neutralization is performed before centrifugation to reduce the time of exposure to either acid or base (Rhodes *et al.*, 2004). Aliquots are plated to solid media and incubated at 23–25°C and at 30°C. The use of egg-yolk-containing media such as LJ medium results in a greater number of positive results. For samples such as spleen collected aseptically these authors found good recovery rates were obtained when samples were plated direct to Middlebrook 7H10 agar and incubated at 23°C for 3 months. The detection of *M. marinum*, *M. simiae*, *M. scrofulaceum*, *M. chelonae* and *M. fortuitum* was achieved by homogenizing excised tissue in 10 ml sterile distilled water to which an equal volume of 2–3% NaOH was added and vortexed for 45 s. The mixture was allowed to stand at room temperature for 15 min before centrifugation for 15 min at 3000 ×*g* to achieve a 95% sedimentation rate of bacteria (Lansdell *et al.*, 1993). The solution was neutralized with 2 N HCl before an aliquot was used to inoculate media such as L-J. Media were incubated at 25 and 30°C in 3–5% CO_2. For the detection of *M. marinum*, *M. chelonae*, *M. fortuitum*, *M. gordonae* and *M. stomatepiae* from a variety of fish, decontamination involved exposing tissue homogenate to 2% NaOH (w/v) for 15 min at 21°C followed by centrifugation at 8000 ×*g* for 15 min. The supernatant was removed and the pellet washed twice with 1 ml phosphate-buffered saline (PBS) and then resuspended in 150 µl PBS. An aliquot (25 µl) was used to inoculate a Middlebrook 7H10 OADC (0.5% glycerol, 10% oleic acid-albumin-dextrose-catalase) plate and incubated at 22°C and 30°C for 3 months (Pourahmad *et al.*, 2009).

Incubation temperature and time also influence the successful recovery of mycobacteria. Some species may take 3 months to grow and some may not grow at temperatures greater than 25°C. Replicate incubations at 23–25°C, 30°C and 37°C is recommended.

Decontamination procedures used for direct PCR also may affect the viability of the bacterial cell for culture.

DECONTAMINATION PROCEDURE FOR DIRECT PCR. A one tube decontamination and DNA extraction method was used to extract DNA of *M. ulcerans* from tissue samples from molluscs and gills and intestines from fish (Kotlowski *et al.*, 2004). Samples of 200 mg are homogenized in 4 ml TE buffer. A 1-ml aliquot is centrifuged for 4000 ×*g* for 15 min at 4°C. The pellet is resuspended in 150 µl lysis buffer (10 mM Tris/HCl, pH 8.0; 5 mM EDTA, pH 8.0; 4 M guanidinium isothiocyanate, pH 7.5; 50 g/l Sarcosyl, 2.5 g/l SDS, 5 g/l sodium citrate and 5 g/l TritonX-100). Chloroform (300 µl) and Tris-saturated phenol, pH 6.9 (300 µl) are added to the mixture and incubated at −20°C for 1 h. Ratio of lysis buffer, chloroform and tris-saturated phenol is 1:2:2. The tubes are centrifuged for 20 min at 10,000 ×*g*. The supernatant is transferred to a fresh tube and a ¼ volume of 2-propanol is added. The mixture is loaded to a silica-cellulose membrane in columns (A & A Biotechnology, Poland). The sample is filtered through the column by gravity. Absolute ethanol (300 µl) is passed through the column twice to wash the membrane then the DNA is eluted with 400 µl hot (75°C) TE buffer. The collected DNA is precipitated using two portions of absolute ethanol and the resulting pellet is suspended in 25 µl 0.5× TE buffer (pH 8.0). Guanidium isothiocyanate denatures protein and inhibits nucleases and the thiocyanate anion combined with the guanidinum cation is more effective at denaturing protein than guanidinium chloride. The use of the chloroform/phenol mixture together with the guanidinum isothiocyanate and sarcosyl improved the yield of DNA compared to lysis buffer in the absence of phenol and chloroform. Phenol dissolves proteins and lipids leaving a water-soluble solution, whereas the chloroform diffuses into the cell leading to cell lysis and inactivation of DNAses. The use of sodium citrate (an anticoagulant) and TritonX-100 (a detergent) prevents clumping of the mycobacterial cells. The mycolic acid composition of the cell wall can make it difficult to lyse using lysozyme and proteinase K. This method was not suitable for extraction of DNA from soil.

CULTURE MEDIA AND CULTURE PROCEDURES. The use of Middlebrook 7H10 agar (Middlebrook *et al.*, 1960) with added oleic acid-albumin-dextrose-catalase (OADC) is a mildly selective media and is considered to improve the isolation rate of mycobacteria from aquatic sources compared to the use of Lowenstein-Jensen (LJ) medium, as it has a lower concentration of malachite green, an inhibitory substance used to decrease contamination from other bacteria, and the transparency of the medium improves visibility for establishing growth of colonies (Lansdell *et al.*, 1993).

The use of LJ medium supplemented with glycerol may improve the isolation rate of *M. ulcerans*. The use of both Middlebrook 7H10 agar and LJ medium containing glycerol is recommended for culture of mycobacteria from aquatic sources.

A commonly used classification system based on phenotypic characteristics, called the Runyon classification (Runyon, 1959), groups culturable Mycobacteria (except *M. tuberculosis*) according to phenotypic characteristics such as pigment production on exposure to light and rate of growth, and can group distantly related mycobacteria but can also split closely related species. Groups I, II and III are slow growers. Group I consists of photochromogens, which produce a yellow-orange pigment when exposed to light. *Mycobacterium marinum* although a rapid grower is in Runyon Group I. Group II organisms are scotochromogens and produce a yellow-orange pigment in light or dark. *Mycobacterium scrofulaceum* is in this group. Group III are non-chromogenic and do not produce pigment, although some species may produce a very pale yellow or tan pigment but this does not intensify on exposure to light. Group IV are rapid growers producing visible colonies in 5–7 days. Pigment is not produced. *Mycobacterium fortuitum*, *M. peregrinum*, *M. abscessus* and *M. chelonae* belong to this group. The Runyon grouping is a convenient descriptive tool to group isolates, however identification cannot be assigned based on phenotype alone due to limited discrimination and variability in performance and interpretation (Springer *et al.*, 1996).

Many strains produce pigments and these can be classified according to whether pigment production occurs in the presence or absence of light. Photochromogenic strains are those that produce pigment in the presence of light and scotochromogenic are those strains that produce pigment in the dark or light (Runyon, 1959).

When isolating mycobacteria from fish and other aquatic animals it is important to distinguish between the pathogen and other mycobacteria that may be contaminants from the environment or skin surface or part of ingested material. In this respect culture is important as mycobacteria detected through direct amplification of DNA may be contaminants. Slow-growing mycobacteria may be outcompeted by fast-growing contaminating bacteria or secondary invaders and decontamination procedures can assist in isolating mycobacteria. Certain substances such as benzalkonium chloride and hypochlorite are used in decontamination procedures. Mycobacteria are hydrophobic and therefore decontamination procedures using acids or bases or compounds such benzalkonium chloride or hypochlorite are used to pre-treat samples prior to culture. However, concentrations and exposure or treatment times must be validated to ensure against adverse recovery.

Minimal standards for identification of new species of *Mycobacterium* proposed by Lévy-Frébault and Portaels (1992) include growth on Lowenstein Jensen medium, usually a modified formula, or for strains that do not grow on LJ medium, growth on Ogawa egg yolk medium or Middlebrook 7H10 agar or a modified Middlebrook. A division is based on slow or rapid growth with rapid growers being classified as those that produce visible colonies in 7 days. Rapid growers can usually grow on standard bacteriological agars such as BA, whereas slow growers cannot. A number of biochemical tests have been used but the ones required for descriptions of new species are: growth at 25, 30, 33, 37, 42 and 45°C; pigmentation; resistance to isoiacid (1 and 10 µg/ml); thiophene-2-carboxylic hydrazide (2 pg/ml), hydroxylamine (500 pg/ml), *p*-nitrobenzoic acid (500 pg/ml), sodium chloride (5%), thiacetazone (10 pg/ml), picrate (0.2%) and oleate (250 pg/ml); two catalase tests (semiquantitative test and heat tolerance test); Tween hydrolysis; urease activity; niacin production; and nitrate reductase, acid phosphatase, arylsulfatase, pyrazinamidase and α-esterase activities. For tests not described in this manual the reader is referred to

Lévy-Frébault and Portaels (1992) for a full description of the tests and original references. A kit for phenotypic identification of 16 species of *Mycobacteria* includes tests for Tween 80 hydrolysis, pigmentation, arylsulfatase reduction, nitrate reduction and urease, and can be purchased from Kyokuto Seiyaku industry.

Mycobacterium abscessus is an acid-fast rod (1.0–2.5×0.5 µm). Growth occurs after 7 days at 28°C and 37°C but not at 43°C on egg medium. Colonies are non-photochromogenic, white to grey and are intermediate between rough and smooth. Growth occurs on MCA at 28 and 37°C (Kusunoki and Ezaki, 1992). Recovery of *M. abscessus* from Japanese medaka was improved with the use of Middlebrook 7H10 agar with albumin, dextrose and catalase enrichment and modified by the addition of malachite green and antibiotics, which assisted in suppression of *Pseudomonas* species (Teska *et al.*, 1997). The formulation is as follows: modified 7H10-ADC plus 0.05% NH_4SO_4, 0.15% KH_2PO_4, 0.15% Na_2HPO_4, 0.04% sodium citrate, 0.0025% $MgSO_4$, 0.00005% $CaCl_2$, 0.0001% $ZnSO_4$, 0.0001% $CuSO_4$, 0.05% L-glutamic acid, 0.004% ferric ammonium citrate, 0.0001% pyridoxine, 0.00005% biotin, 0.025% malachite green, 0.5% glycerol, 0.0035% nalidixic acid and 0.04% cycloheximide. The ADC consisted of 0.5% bovine albumin V, 0.2% D-(+)-glucose and 0.0003% catalase (Sigma). The medium was made as a broth or as plates with the addition of 1.5% agar. This modification of Middlebrook 7H10 was not suitable for isolation of mycobacteria from striped bass (Rhodes *et al.*, 2004). Samples from medaka were placed into modified Middlebrook 7H10-ADC broth 1:10 (w/v) and incubated for 1 h at room temperature and then homogenized. The homogenate was serially diluted and 25 µl volumes dropped on to the surface of a modified Middlebrook 7H10-ADC plate. The drop was allowed to absorb into the agar before incubating at 25°C for up to 14 days (Teska *et al.*, 1997). Colonies had a light-green blue tint due to the dye in the medium but with age the colonies became off-white to tan in colour.

Mycobacterium chelonae is classified as a rapid growing mycobacterium although primary isolation may take weeks or months (Grange, 1981). Colonies may be colourless, smooth, and dome-shaped or rough and spreading and develop a grey-purple colour with age. Growth does not occur at 42°C.

Mycobacterium fortuitum (ATCC 6841^T) is non-photochromogenic, grows rapidly with colonies appearing in 5–7 days and grows at 28 and 37°C but not at 42°C. Growth occurs on BA, brain–heart infusion (BHI), NA, TSA, Ogawa egg medium, LJ agar, Middlebrook 7H10 and Sauton agar (Najiah *et al.*, 2011).

Mycobacterium marinum can be isolated from spleen and liver after decontamination using 2% solution of NaOH for 30–60 min at 37°C and neutralizing the sediment with 5% HCl (Aronson, 1926). Tissues collected aseptically from euthanased fish do not require decontamination. *Mycobacterium marinum* grows on BA and BHI agar in 3–5 days as rough, yellow-coloured colonies when incubated at 25°C and in 7 days when incubated at 37°C (Hedrick *et al.*, 1987), although Najiah *et al.* (2011) reported that the type strain ATCC 972 did not grow on BHI agar. Pigment may become deep orange (see colour plate section). Cells from tissue smears or culture material are 0.3–0.5×1.0–1.5 µm and acid-fast with a banded or beaded appearance. Tween 80 hydrolysis was reported positive by Bowenkamp *et al.* (2001), but negative after 7 and 14 days by Najiah *et al.* (2011). Similarly, arylsulfatase was reported negative by Bowenkamp *et al.* (2001) but negative at 3 days and positive at 14 days by Najiah *et al.* (2011). Both sets of reactions are reported in Tables 3.16 and 3.17.

Mycobacterium montefiorense is a slow growing mycobacterium and does not grow at temperatures above 30°C. It is non-chromogenic and acid-fast rods show beading, and short rods and coccoid forms occur. On Middlebrook agar at 25°C colonies are small and transparent (Levi *et al.*, 2003). Growth occurs on BA at room temperature and growth of alpha-haemolytic colonies can be seen after 1 week using a dissecting microscope. Visible colonies at 12 weeks are rough, dry with raised centres and flat borders (fried egg appearance) (Herbst *et al.*, 2001). BA plates may need to be incubated for up to 20 weeks.

The *M. neoaurum* grown from eye of chinook salmon are acid-fast rods (3–4×0.6 µm). Growth occurs on blood agar as colonies of variable size, 0.3–0.4 mm, smooth, round and begin to develop a light yellow colour in 5–8 days at room temperature. No growth occurs at 37°C or on agar containing 5% salt. Arylsulfatase is positive after 14 days.

Mycobacterium palustre was isolated from water by filtering 125 ml of water through a 0.2 μm filter to collect the bacterial cells. The cells are eluted from the membrane filter by placing filter fragments into 5 ml of sample water and adding glass beads and shaking the tube in a rotary mixer. The sample concentrate (5 ml) is subject to decontamination using 5 ml 2 M H_2SO_4 and 0.7 ml cycloheximide (8 mg/ml) for 30 min. The sample is centrifuged and neutralized using 30 ml of sterile distilled water, re-centrifuged and the deposit resuspended in 0.7 ml sterile distilled water. Decontamination using 1 M NaOH and 5% oxalic acid was not successful (Torkko *et al*, 2002). *Mycobacterium palustre* is acid and alcohol-fast. Colonies on Middlebrook 7H11 agar and egg-based media grow after 2–4 weeks at 36°C, and are 1–1.5 mm (4 weeks growth), smooth, yellow to pale yellow, and scotochromogenic at 36°C and photochromogenic or scotochromogenic at 42°C. Optimal growth temperature is 36°C and colonies are slightly larger on Middlebrook agar compared to egg-based media. Acid phosphatase tested using the API Zym after overnight incubation at 36°C was positive whereas 85% of strains were positive for alkaline phosphatase and all strains were negative for α-galactosidase and β-galactosidase. Other tests from the API Zym were not reported.

Mycobacterium peregrinum is an acid-fast rod (1.5–4×0.5 μm). Growth on egg medium occurs after 7 days at 28 and 37°C but not at 43°C. Colonies are non-photochromogenic, white to slightly yellowish and are intermediate between rough and smooth. On Middlebrook 7H11 agar colonies are buff-coloured and growth occurs at 4–7 days at 37°C in CO_2. On LJ medium colonies are buff-coloured, rough or filamentous. Growth occurs on MCA at 28 but not at 37°C (Kusunoki and Ezaki, 1992; Mohney *et al*., 1998). Strains from shrimp were reported as negative for Tween 80 hydrolysis whereas the type strain was positive (Kusunoki and Ezaki, 1992; Mohney *et al*., 1998).

Mycobacterium pinnipedii grows on egg-based media in 3–6 weeks at 36–37°C as rough, flat, non-photochromogenic colonies. Growth is enhanced by sodium pyruvate. Loose cord formation is seen in a ZN smear (Cousins *et al*., 2003). Reaction is generally negative for niacin, however some isolates give a low or medium reaction to niacin.

Mycobacterium poriferae is a rapidly growing mycobacterium, producing smooth, moist, shiny, dome-shaped colonies after 4 days at 28°C on Middlebrook 7H-11 agar or tryptic soy agar with 5% (v/v) glycerol (Padgitt and Moshier, 1987). Optimal growth occurs at 28–30°C. Growth does not occur at 42°C. *Mycobacterium poriferae* is scotochromogenic producing orange colonies. Cells are 0.7–2.5 wide × 1.1–4.9 μm in length and can appear coccoid. Non-dividing cells are 2.2–2.3×2.7–3.1 μm long. Branching and Y-shaped cells are not seen, but cells can become enlarged at one end with an ovoid non-acid-fast body.

Mycobacterium pseudoshottsii is a slow-growing photochromogenic acid-fast bacillus (0.4–0.6×0.8–1 μm) and forms cell aggregates in culture (Rhodes *et al*., 2005). On Middlebrook 7H10 agar colonies are 1–3 mm, rough, flat, with a slightly irregular margin at 23°C after 2 months with growth becoming visible at 4–6 weeks. Colonies become slightly umbonate on further incubation. Little or no growth occurs at 30°C and growth does not occur at 37°C. On exposure to light, colonies become pale yellow to gold-coloured with the colour intensifying over time.

Mycobacterium salmoniphilum grows in 4–6 days at 20–30°C as cream-coloured, smooth, shiny colonies on Middlebrook 7H10 agar, blood agar, MacConkey agar and LJ media. After 10 days colonies appear waxy with a fried egg appearance and irregular edges. Weak or delayed growth occurs at 10°C whereas growth does not occur at 37°C. On Tryptone soya agar (TSA) containing 2% NaCl colonies are pale-cream, circular, smooth in 7 days at 15°C. Growth also occurs on BHI, Anacker and Ordal's medium using 0.5% tryptone and Mueller Hinton agar (Ross, 1960; Bruno *et al*, 1998a; Whipps *et al*., 2007). Colonies on serum dextrose agar are 1.5–2.0 mm, light cream, glistening and smooth at 4 days at 25°C. On ageing, colonies become flatter and edges spread with the centre appearing heaped or papilliform. Colonies on Sabouraud agar are off-white and sightly dull. Some flat, dry, granular colonies are seen (Ross, 1960). Two of the three strains identified by Ross (1960) developed a water-soluble purple pigment on Sabouraud dextrose agar when incubated at 19–25°C for 2–3 weeks in the absence and presence of light. An amber-coloured diffusible pigment was noted in the medium. At 5–6 weeks the colonial growth and the diffusible pigment were black-purple in colour. Cells are strongly acid-fast, stain evenly without beading except for some cells which contain terminal or subterminal granules, are non-motile, slender, straight or slightly curved (1–7×0.25–0.6 μm). Urease was reported positive by Zerihun *et al.* (2011) but negative by Ross (1960).

Mycobacterium shottsii is a slow-growing non-pigmented mycobacterium that does not grow at 37°C.

Mycobacterium stomatepiae grows on Middlebrook 7H10 media supplemented with OADC and 10% glycerol as smooth, off-white and non-chromogenic colonies, and LJ media after 4 weeks' incubation at 30°C. Growth does not occur at 15, 37 or 42°C when tested at 12 weeks' incubation.

The recovery of *M. stomatepia* is obtained with Middlebrook 7H10 medium supplemented with 10% oleic acid-albumin-glucose-catalase (OADC) (Becton Dickinson) and 0.5% glycerol (Sigma) for 4 weeks at 30°C following decontamination with 2% NaOH (w/v) for 15 min. Middlebrook 7H10 agar with OADC can be purchased from Becton Dickinson.

Mycobacterium ulcerans may require up to 19 weeks of culture before visible growth occurs (Eddyani *et al.*, 2004).

Biochemical reactions for *Mycobacteria* species are presented in Tables 3.16 and 3.17. Methods for media and stains are presented in the media section.

MOLECULAR IDENTIFICATION

DNA extraction methods

1. DNA extraction from tissue: see molecular chapter for a simple boiling method.

2. DNA extraction from bacterial cells: DNA is extracted using a commercial method for mycobacteria or by a manual DNA extraction method, but whichever method is used it must be suitable for mycobacteria, as the waxy bacterial wall can be difficult or impossible to disrupt and many of the DNA extraction methods used for other bacteria are not suitable for mycobacteria. Also, the thickness of the cell wall in mycobacteria increases its resistance to simple disruption methods. The following manual method has been used for mycobacteria. A loop of bacterial cells grown on solid media or a pellet from 250 μl of broth culture is suspended in 1 ml TE buffer (10 mM Tris, 1 mM EDTA, pH 8.0) and inactivated by heating at 80°C for 10 min. Cells are pelleted by centrifugation at 10,000 ×*g* for 10 min and the supernatant discarded. A loopful or 100 μl of acid-washed glass beads diameter 100 μm (Sigma) and 100 μl of TE buffer are added to the pellet. Cells are disintegrated by placing the tube in a tissue disintegrator (Mickle Laboratories) for 2 min at maximum speed. After centrifugation (10,000 ×*g* for 10 min) the supernatant is transferred to a new tube (Telenti *et al.*, 1993). A 5-μl volume of DNA is used in a 50 μl PCR mixture (Telenti *et al.*, 1993), or diluted 1/100 and 10 μl used in a 100-μl PCR master mix (Ringuet *et al.*, 1999).

DNA from bacterial cultures has been extracted using the QIAamp DNA mini kit (Qiagen), DNeasy kit (Qiagen) and the UltraClean microbial DNA kit (MoBio Laboratories) (Whipps *et al.*, 2007).

3. DNA extraction from formalin fixed paraffin-embedded tissue (FFPE): the mycobacterial cell can be difficult to disrupt because of the waxy mycolic acids plus the thickness of the cell wall, however fixation in formalin also presents problems. Unbuffered and neutral buffered formalin can inhibit PCR efficiency, although fixation for up to 7 days in 10% neutral buffered formalin usually does not affect sensitivity. Fixation of tissues using formaldehyde (HCHO) leads to DNA-protein and RNA-protein crosslinkages. Unbuffered fixative, due to the low pH often <1, fragments DNA. These effects lead to a reduction in both the quantity of amplifiable DNA and the length (Gilbert *et al.*, 2007). A number of methods for improving the quantity and length of DNA amplified from FFPE tissue are explained in the molecular chapter.

The following method has been used for amplification of mycobacteria from FFPE tissues. Sections are cut to 1 μm thickness and placed into a 1.5-ml micro-centrifuge tube. A new microtome blade is used for each sample to avoid cross-contamination. The paraffin is removed by adding 1 ml of xylene and mixing on a rocker. The sections are centrifuged at 12,000 ×*g* for 5 min and the supernatant removed. The pellet is washed twice with 1 ml of absolute ethanol and centrifuged at 12,000 ×*g* for 4 min. The pellet is air dried then 25 µl of proteinase K (20 mg/l) and 180 µl of lysis buffer (from NucleoSpin® kit, Macherey-Nagel GmbH, Germany) are added and the mixture is incubated at 56°C overnight. The proteinase K is inactivated by treatment at 70°C for 10 min. The DNA is extracted using the NucleoSpin kit (Pourahmad *et al.*, 2009). PCR is conducted using Tb11 and Tb12 primers for the *hsp65* gene.

PCR. A number of genes have been targeted for species identification and differentiation of mycobacteria and include the 16S-23S internal transcribed spacer, heat-shock protein 65 (*hsp65*), exported repeated protein (*erp*), RNA polymerase B subunit (*rpoB*) genes, and insertion sequences MPM-associated IS2404 and IS2606, and IS90I/IS1245 used to identify *M. avium* in a tropical fish (Lescenko *et al.*, 2003). Sequencing the 16S rRNA gene followed by restriction enzyme fragment length polymorphisms (RFLPs) analysis can assist in identification of *M. chelonae*, *M. fortuitum* and *M. marinum* (Talaat *et al.*, 1997), but has limitations for differentiation of some mycobacteria due to similarity in this region of closely related mycobacteria such as *M. pseudoshottsii* and *M. shottsii* as they differ by 8 of the 1454 nucleotides.

Amplification and sequencing of *hsp65* enables the differentiation of *Mycobacterium* species and is used for clinical and environmental isolates. The gene is present in all mycobacteria and is more variable than the 16S rRNA gene. DNA is extracted using the method of Telenti *et al.* (1993). A 50-µl master mix is prepared from 50 mM KCl, 10 mM Tris-HCl (pH 8.3), 1.5 mM $MgCl_2$, 10% glycerol, 200 µM of each dNTP, 0.5 µM of each primer (Tb11 and Tb12), 1.25 U *Taq* polymerase and 5 µl DNA (Telenti *et al.*, 1993). The same primers for *hsp65* were designed to include M13 sequencing primer sites and are called 21M13-Tb11 and M13-Tb1 (Ringuet *et al.*, 1999). DNA is diluted 1/100 and 10 µl used in a 100-µl PCR master mix containing 0.25 µM of each primer (21M13-Tb11 and M13-Tb12), 200 µM each dNTP, ×1 *Taq* buffer and 0.5 U *Taq* polymerase (ATGC; Biotechnologies, France). Amplification is also successful using the Promega Master Mix (15 µl), 1.8 µl of each primer (Tb11 and Tb12, final concentration 1.2 µM), 6.4 µl water, 5 µl DNA in a 30-µl total volume master mix (Sam Hair, pers. comm.). Amplification is performed in a thermocycler (9600 Perkin-Elmer) for 35 cycles of 94°C for 20 s, 60°C for 20 s and 72°C for 45 s. Amplified products (441 bp) are checked on a gel before being sent for sequencing (Ringuet *et al.*, 1999). Sequences are checked against Genbank. *Mycobacterium chelonae* and *M. abscessus* have only four nucleotide differences in the 16S rRNA gene but have a difference of 30 nucleotides in the *hsp65* gene, making sequencing of the *hsp65* gene a more suitable measure for differentiation of mycobacteria. Telenti *et al.* (1993) restricted the amplified product with *Bst*EII and *Hae*III enzymes to obtain unique restriction patterns (PRA) that differentiated many *Mycobacterium* species, although this does not differentiate members of the *M. tuberculosis* complex. A website is available to determine the species identity from the resulting restriction fragments (http://app.chuv.ch/prasite/index.html).

The exported repetitive protein (*erp*) gene is found in all mycobacteria and due to differences between species sequencing of the gene can assist in species identification. Primers P1 and P2 (see Table 1.12) were used in a master mix containing *Pfu* Turbo *Taq* polymerase (Stratagene), an annealing temperature of 55°C and 30 cycles (Rhodes *et al.*, 2005).

1.28 *Nocardiaceae*

1.28.1 *Nocardia*

TAXONOMY AND SIGNIFICANCE. *Nocardia* are aerobic, Gram-positive actinomycetes that contain mycolic acids in the cell wall. They produce branched substrate hyphae that fragment *in situ* into rod-shaped to coccoid non-motile elements and have sparse to abundant aerial hyphae that are always present but some may only be seen microscopically, and are acid-alcohol fast at some stages of the growth cycle. Hyphae on a 3-day-old colony may only be 1 mm in length. *Nocardia* belong to the *Corynebacteriaceae* suborder of the *Actinomycetales* order (Brown-Elliott *et al.*, 2006). Over 83 species have been reported (J.P. Euzéby, http://www.bacterio.cict.fr). *Nocardia* species are found in the environment, particularly soil. They cause disease in human and animals, and have been

Table 1.12. Primers for the amplification of *Mycobacterium* species.

Organism	Forward primer, 5′–3′	Reverse primer, 5′–3′	AT°C	C	Bp	Reference
Mycobacterium species *hsp65*	Tb11 ACC AAC GAT GGT GTG TCC AT	Tb12 CTT GTC GAA CCG CAT ACC CT	60	45	439	Telenti *et al.*, 1993
Mycobacterium species *hsp65* with sequencing primers[a]	21M13-Tb11 TGT AAA ACG ACG GCC AGT ACC AAC GAT GGT GTG TCC AT	M13-Tb12 CAG GAA ACA GCT ATG ACC CTT GTC GAA CCG CAT ACC CT	60	35	441	Ringuet *et al.*, 1999
Mycobacterium species *erp* gene	P1 ATG ACC GCG CTG TCG CAG GGT	P2 GGG GTC ACC CAG GAT CGG ATA	55	30		Rhodes *et al.*, 2005
Mycobacterium species internal transcribed spacer rRNA genes	Sp1 ACC TCC TTT CTA AGG AGC ACC	Sp2 GAT GCT CGC AAC CAC TAT CCA	59	38		Levi *et al.*, 2003
Mycobacterium genus	T39 GCG AAC GGG TGA GTA ACA CG	T13 TGC ACA CAG GCC ACA AGG GA	50	30	924	Talaat *et al.*, 1997
Mycobacterium chelonae	Restrict with *Apa*I to give fragments at 812, 112 bp	Restrict with *Ban*I to give fragments at 562 and 362 bp				Talaat *et al.*, 1997
Mycobacterium fortuitum	Restrict with *Apa*I to give fragments at 677, 132, 115 bp	Restrict with *Ban*I to give fragments at 562 and 362 bp				Talaat *et al.*, 1997
Mycobacterium marinum	Restrict with *Apa*I to give fragments at 677, 132, 115 bp	No restriction with *Ban*I				Talaat *et al.*, 1997
Mycobacterium montefiorense Nested PCR	For1 CGAAAGCGTGGGG-AGCGAACA	Rev1 AGACCCCGATCCGA-ACTGAGACC	55	38		Herbst *et al.*, 2001
	For2 GGTGTGGGTTTCCTTCCTT	Rev2 ACGGGCCATTGTAGCAT	55	55	409	

[a]Underline indicates sequencing primer.

reported from diseased aquatic animals such as fish, oysters and marine mammals and in many cases have not been identified to species level due to the difficulty with phenotypic identification for this genus. The disease is often chronic but leads to necrosis and death.

Unidentified *Nocardia* species have been reported from Atlantic bottlenose dolphin (*Tursiops truncatus*), beluga whale (*Delphinapterus leucas*), hooded seal (*Cystophora cristata*) and leopard seal (*Hydrurga leptonyx*) (Leger *et al.*, 2009). These *Nocardia* species were not able to be identified to known species using phenotypic or sequencing methods. The disease was systemic with pyogranulomatous lesions in organs particularly the lung. *Nocardia* species that were identified included *N. asteroides*, *N. brasiliensis*, *N. cyriacigeorgica*, *N. farcinica*, *N. levis* and *N. otitidiscaviarum* and were isolated from hooded seal (*Cystophora cristata*), Atlantic bottlenose dolphin (*Tursiops truncatus*), beluga whale (*Delphinapterus leucas*) and killer whale (*Orcinus orca*).

Nocardia asteroides

TAXONOMY AND SIGNIFICANCE. *Nocardia asteroides* was responsible for disease outbreaks in rainbow trout at a hatchery in the USA (Snieszko *et al.*, 1964a). The mortality rate was low and infected fish had a distended abdomen and displayed atypical swimming behaviour. The organism was seen in smears of brain capsule in some infected fish. *Nocardia asteroides* also has been reported causing Nocardiosis in hooded seal (*Cystophora cristata*) and Atlantic bottlenose dolphin (*Tursiops truncatus*)

(Leger *et al.*, 2009). The disease is systemic with pyogranulomatous lesions in lung and other organs.

Nocardia crassostreae

TAXONOMY AND SIGNIFICANCE. *Nocardia crassostreae* (NB4H=ATCC 700418^{T}) was isolated from Pacific oysters (*Crassostrea gigas*) in British Columbia, Canada (Friedman *et al.*, 1998). Green or yellow nodules composed of Gram-positive, acid-fast filamentous bacteria surrounded by haemocytes occur in the abductor muscle, gills, heart and mantle, and areas of brown discolouration are seen in the mantle. The organism has also been reported from yellow-green nodules on gill and mantle of *C. gigas* in Lake Grevelingen in the Netherlands (Engelsma *et al.*, 2008). Identity was confirmed by sequencing the 16S rRNA gene and by specific PCR. *Vibrio aestuarianus* also was isolated from Pacific oyster in the same area.

GROWTH CHARACTERISTICS AND PHENOTYPIC IDENTIFICATION. *Nocardia crassostreae* is an aerobic, non-motile, acid-fast, Gram-positive actinomycete that forms a branched substrate mycelium which fragments into irregularly sized rods. Aerial hyphae are not formed (Friedman *et al.*, 1998). On brain–heart infusion agar (BHI) colonies are pale yellow, dry, waxy, wrinkled with slight appearance of filamentous margins. Soluble pigment is not produced. Cells are beaded and filamentous and easily fragment into rods and cocci. Growth occurs in 2% NaCl at 22°C. Growth does not occur at 5°C. Results are positive for catalase, urease and aesculin hydrolysis and negative for oxidase, and hydrolysis of starch and gelatin. It ferments glucose and sucrose but not D- or L-arabinose, galactose, lactose, maltose, rhamnose or xylose and does not use L-arginine and lysine as sole carbon and nitrogen sources (Engelsma *et al.*, 2008). These and other reactions are presented in Table 3.16.

MOLECULAR IDENTIFICATION. *Nocardia crassostreae* can be identified by amplification and sequencing of the 16S rRNA gene or by a specific PCR (Bower *et al.*, 2005).

Nocardia salmonicida

TAXONOMY AND SIGNIFICANCE. *Nocardia salmonicida* (JCM 4826^{T}), isolated from sockeye (blueback) salmon (*Oncorhynchus nerka*), was originally identified as *Streptomyces salmonicida* (Rucker, 1949; Isik *et al.*, 1999). Nodules in the spleen and kidney showed a mass of mycelia. Pathogenicity was proven in experimental challenge with fingerlings, which were injected either intraperitoneally or intramuscularly. Both naturally infected and experimentally infected fish developed a secondary infection with *Saprolegnia parasitica*.

Nocardia salmonicida forms a monophyletic clade with *N. asteroides* with a similarity of 97% (Isik *et al.*, 1999).

GROWTH CHARACTERISTICS AND PHENOTYPIC IDENTIFICATION. *Nocardia salmonicida* is an aerobic, Gram-positive, partially acid-fast actinomycete, which produces extensively branched substrate mycelium that fragments into rod and coccoid forms (Rucker, 1949; Isik *et al.*, 1999). On glucose yeast extract agar (GYEA) plates at 14 days at 30°C, colonies are convex and irregular with filamentous colony margins, and diameter size varies from 1 to 3 mm. Colonies may show marked radial foldings. The surface of the colony takes on a powdery appearance due to the development of conidia and white to pink aerial hyphae are seen on an orange substrate mycelium. Diffusible pigment is not seen, although Rucker (1949) reported a brown water-soluble pigment on 1% agar plates containing 0.5% tryptone. Growth occurs between 20 and 30°C with an optimum at 30°C, but does not occur at 10 or 35°C. Growth occurs in 3% NaCl but not at 10% NaCl. Tween 20 and Tween 80 are degraded, aesculin and urea are hydrolysed. Hydrolysis is negative for casein, starch and chitin. H_2S is positive using a lead acetate soaked filter paper over tryptone broth.

These and other reactions are presented in Table 3.16.

Nocardia seriolae

TAXONOMY AND SIGNIFICANCE. *Nocardia seriolae* (JCM 3360^{T}) was isolated from yellowtail (*Seriola quinqueradiata*) in Japan (Kudo *et al.*, 1988). *Nocardia seriolae* was originally reported as *N. kampachi*, however the name was never validated. Infection is referred to as Nocardiosis and is characterized by abscesses in the epidermis and tubercules in the kidney, spleen and gills. *Nocardia seriolae* is similar in phenotypic results to *N. asteroides*. The organism has been reported from a variety of fish in Japan, Taiwan and China including grey mullet (*Mugil cephalus*), striped bass (*Morone saxatilis*), crimson snapper (*Lutjanus erythropterus*), seabream (*Lateolabrax japonicus*), largemouth bass (*Micropterus salmoides*), grouper (*Epinephelus* sp.), snubnose pompano (*Trachinotus* blochii), spotted scat (*Scatophagus argus*), Japanese flounder (*Paralichthys olivaceus*), large yellow croaker (*Larimichthys crocea*), snakehead (*Channa* [*Ophicephalus*] *argus argus*) and three-striped tigerfish (*Terapon jarbua*) (Kudo *et al.*, 1988; Shimahara *et al.*, 2008; Wang *et al.*, 2009). Infection in pond-cultured tigerfish was characterized by haemorrhages and ulcers on the skin with ascites, enlarged spleen, kidney and liver with light yellow nodules on these and other internal organs (Wang *et al.*, 2009). Cumulative mortality was 2.5% over 60 days. Virulence was confirmed by experimental infection and re-isolation of the organism. Identification of *N. seriolae* was undertaken using API Zym, and sequencing of the 16S rRNA gene, the heat shock protein gene and RNA polymerase gene, *rpoB*. The tigerfish isolates formed a monophyletic clade of 99.9% similarity to *N. seriolae* BRC 13745. Partial sequence of the 16S rRNA gene showed similarities of 100% to *N. seriolae*, 97.3% to *N. nova*, 95.8% to *N. asteroides*, 94% to *N. salmonicida* and 93.6% to *N. otitidiscaviarum*, with similar results for the other genes.

GROWTH CHARACTERISTICS AND PHENOTYPIC IDENTIFICATION. *Nocardia seriolae* is Gram-positive and acid fast. On glucose yeast extract agar (GYEA) and yeast extract-malt extract agar (ISP medium no. 2) colonies are yellowish orange with no macroscopic aerial mycelia. Growth also occurs on TSA, BA, BHIA and Lowenstein Jensen medium. Growth occurs at 15–30°C but not at 10 or 37°C. Growth at 50°C for 8 h is negative. Fermentation is positive for glucose and glycerol but is negative for adonitol, galactose and rhamnose. Citrate is utilized as a sole carbon source. It does not require NaCl for growth. Nitrate has been reported positive but was reported negative for isolates from three-striped tigerfish (Wang *et al.*, 2009). These and other reactions are presented in Tables 3.16 and 3.38. The API Zym is inoculated according to the manufacturer's instructions except that the inoculum is prepared by sonicating cells in 2 ml sterile saline (0.85%) for 10–20 s (Shimahara *et al.*, 2008) or suspending cells in 3 ml 0.85% sterile saline and shaking with glass beads (2 mm diameter) for 10 min (Wang *et al.*, 2009). The turbidity is adjusted to No. 6 McFarland as per instructions. The test is incubated at 25°C for 4 or 5 h before adding the reagents and reading the results. Incubation times and temperatures are detailed in Table 3.38.

MOLECULAR IDENTIFICATION. Sequencing of the 16S rRNA gene, the heat shock protein gene and the RNA polymerase gene, *rpoB*, can assist in the identification of *Nocardia* species. A specific primer set based on the 16S rRNA gene also has been designed using primer NS1 and primer NG1 (Laurent *et al.*, 2000).

Nocardia seriolae can be identified by loop-mediated isothermal amplification using sequence information based on the 16S-23S ribosomal

PCR type	Conventional PCR for identification of *N. seriolae* based on the 16S rRNA gene (Laurent *et al.*, 2000; Wang *et al.*, 2009)
DNA extraction	Suspend a bacterial colony in 500 µl sterile water and 150 µl InstaGene Matrix (BioRad). Vortex for 2 min. Heat at 100°C for 30 min. Centrifuge at 9000 ×*g* for 8 min. Dilute supernatant 1:50 and use 5 µl in PCR reaction.
PCR master mix	
Primer NS1	5′-ACT CAC AGC TCA ACT GTG G-3′
Primer NG1[a]	5′-ACC GAC CAC AAG GGG G-3′

Continued

Continued.

Reagent	Stock	Final concentration	Volume ×1
Polymerase buffer containing $MgCl_2$	×10	×1	5.0 µl
dNTP 10 mM mix	10 mM	200 µM	1.0 µl
Primer NS1	50 µM	0.25 µM	0.25 µl
Primer NG1	50 µM	0.25 µM	0.25 µl
*Taq*DNA polymerase		2.5 U	0.5 µl
Water			38.0 µl
DNA			5 µl
Total volume			**50 µl**

Thermocycling		
Step 1	**Step 2**	**Step 3**
94°C, 3 min	94°C, 1 min	94°C, 1 min
	57°C, 20 s	57°C, 20 s
	72°C, 1 min	72°C, 10 min
×1	×30	×1
Product bp	432 bp	
Sensitivity	100 cells	
Specific against	*Nocardia asteroides* BCRC 13364, *N. brasiliensis*, *N. farcinica* BRCRC 13364 & BCR 13380, *N. nova*, *N. otitidiscaviarum*, *N. salmonicida* BCRC 12441, *Lactococcus garvieae*, *Mycobacterium fortuitum* BCRC 15320, *Vibrio alginolyticus*, *Vibrio anguillarum*, *Escherichia coli*	

[a]Primer NG1 was reported incorrectly as GGG G-3′ by Laurent *et al*. (1999; erratum 2000, 38:2805) but correctly as GGG GG-3′ by Wang *et al*. (2009).

PCR type	Partial sequencing of the 16S rRNA gene (Laurent *et al*., 1999, 2000; Wang *et al*., 2009)
DNA extraction	Prepare lysis solution of 750 µl ultrapure water and 200 µl chelex solution (15% (w/v) chelex and 1% Tween (v/v)). Suspend bacterial colonies in 950 µl lysis solution and stir for 2 min. Centrifuge for 20 s at 12,000 ×*g*. Dilute supernatant 1:20 and use 15 µl in PCR mix
PCR master mix	
Primer NG1	5′-ACC GAC CAC AAG GGG G-3′
Primer NG2	5′-GGT TGT AAA CCT CTT TCG A-3′

Reagent	Stock	Final conentration	Volume ×1
Polymerase buffer	×1	×1	5.0 µl
dNTP 10 mM mix		200 µM	1.0
Primer NG1	50 µM	1.0 µM	1.0 µl
Primer NG2	50 µM	1.0 µM	1.0 µl
*Taq*DNA polymerase		1 U	0.2 µl
Water			36.8 µl
DNA			5 µl
Total volume			**50 µl**

Thermocycling		
Step 1	**Step 2**	**Step 3**
94°C, 11 min	94°C, 1 min	94°C, 1 min
	55°C, 20 s	55°C, 20 s
	72°C, 1 min	72°C, 10 min
×1	×30	×1
Product	596 bp	

Continued

Continued.

PCR type	Amplification for sequencing of the heat shock protein gene *hsp* 65 (Telenti *et al*., 1993; Wang *et al*., 2009)		
DNA extraction			
PCR master mix			
Primer Tb11	5′-ACC AAC GAT GGT GTG TCC AT-3′		
Primer Tb12	5′-CTT GTC GAA CCG CAT ACC CT-3′		
Reagent	**Stock**	**Final concentration**	**Volume ×1**
Polymerase buffer	×10	×1	5.0 µl
dNTP 10 mM mix	10 mM	200 µM	1.0 µl
$MgCl_2$	25 mM	1.5 mM	3.0 µl
Primer Tb11	50 µM	0.5 µM	0.5 µl
Primer Tb12	50 µM	0.5 µM	0.5 µl
*Taq*DNA polymerase		1.25 U	0.25 µl
Water			34.8 µl
DNA			5 µl
Total volume			**50 µl**
Thermocycling			
Step 1	**Step 2**	**Step 3**	
95°C, 1 min	94°C, 1 min	94°C, 1 min	
	60°C, 1 min	60°C, 1 min	
	72°C, 1 min	72°C, 10 min	
×1	×45	×1	
Product bp	439 bp		
Similarities	*Nocardia asteroides* AY756513 95.4%, *N. brevicatena* Z36930 93.2%, *N. farcinica* AY756523 93.8%, *N. nova* AY756527 92.7%, *N. otitidiscaviarum* AY756528 94.9%		

PCR type	Amplification for sequencing of the RNA polymerase gene *rpoB* (Kim *et al*., 1999; Wang *et al*., 2009)		
PCR master mix			
Primer MF59-39	5′-CGA CCA CTT CGG CAA CCG-3′		
Primer MR59-39	5′-TCG ATC GGG CAC ATC CGG-3′		
Reagent	**Stock**	**Final concentration**	**Volume ×1**
Polymerase buffer	×10	×1	2.0 µl
dNTP 10 mM mix	10 mM	250 µM	0.5 µl
$MgCl_2$	25 mM	1.5 mM	1.2 µl
Primer MF59-39	50 µM	1.0 µM	0.4 µl
Primer MR59-39	50 µM	1.0 µM	0.4 µl
*Taq*DNA polymerase		1.0 U	0.2 µl
Water			13.3 µl
DNA			2 µl
Total volume			**20 µl**
Thermocycling			
Step 1	**Step 2**	**Step 3**	
95°C, 1 min	95°C, 30 s	95°C, 30 s	
	60°C, 30 s	60°C, 30 s	
	72°C, 45 s	72°C, 5 min	
×1	×30	×1	
Product bp	342 bp		
Similarities	*Mycobacterium fortuitum* AY147156=92.9%, *M. mucogenicum* AY147174=92.2%		

RNA transcribed spacer region (Itano *et al.*, 2006). Specificity testing was conducted using wild-type strains and the use of type strains was not reported.

Phylum *Firmicutes*
Class Clostridia
Order *Clostridiales*

1.29 *Clostridiaceae*

1.29.1 *Clostridium*

Clostridium botulinum

TAXONOMY AND SIGNIFICANCE. *Clostridium botulinum* caused mortalitites in farmed rainbow trout (*Oncorhynchus mykiss*) in Scotland (Burns and Williams, 1975; Cann and Taylor, 1982). The toxin was detected in the serum and intestinal contents of the infected fish. Experimental challenge confirmed the source of the infection.

The organism may be a commensal on the gills and in the intestine and is found in fish farm sediments, fish and fish products. In the food industry contamination is a source of fatal food poisoning (Burns and Williams, 1975; Hielm *et al.*, 1998; Hyytiä *et al.*, 1998). The toxin is not usually produced in live fish, but is produced in dead fish as part of the decomposition process. The toxin is produced in an anaerobic environment.

Clostridium perfringens

TAXONOMY AND SIGNIFICANCE. Infection with *Clostridium perfringens* causes enterotoxaemia, peritonitis or clostridial myositis in marine mammals including pinnipeds, killer whale (*Orcinus orca*), seal, sea lion and dolphin. A number of cases of clostridial myositis have resulted from insufficient disinfection of the fur and skin before intramuscular injections (such as multivitamin injections) in captive animals as has been observed in killer whale, bottlenosed dolphin (*Tursiops truncatus*) and Californian sea lion (*Zalophus californianus*) (Greenwood and Taylor, 1978). Some of these infections have been due to toxin type A.

Class Mollicutes
Order *Mycoplasmatales*

1.30 *Mycoplasmataceae*

1.30.1 *Mycoplasma*

Mycoplasma belongs to the Phylum XVI *Tenericutes*, Class *Mollicutes*, Order I *Mycoplasmatales* and the Family *Mycoplasmataceae*. Previously, the class *Mollicutes* was classified in the Phylum XIII *Firmicutes*. The family contains two genera, *Mycoplasma* and *Ureaplasma* (Ludwig *et al.*, 2009).

Mycoplasma are bacteria that lack a cell wall but are enclosed by a plasma membrane and are the smallest self-replicating prokaryotes devoid of a cell wall. Based on16S rRNA sequence information they are phylogenetically related to Gram-positive bacteria, in particular *Clostridia*, but if they can be seen in a Gram stain they tend to stain Gram-negative (Krieg, 2010). They are found as common inhabitants of the mucous membranes of animals and humans and have been reported from fish, marine mammals, insects and plants and may be saprophytes, parasites or pathogens. Currently 123 species are listed (J.P. Euzéby; http://www.bacterio.cict).

Two genera *Eperythrozoon* and *Haemobartonella*, recently placed into the family *Mycoplasmataceae*, are currently of uncertain taxonomic position and now assigned as Family II *incertae sedis* (Ludwig *et al.*, 2009). *Haemobartonella* and *Eperythrozoon* species are non-culturable organisms and are referred to as haemotrophic unculturable *Mycoplasma* species or hemoplasmas. Some of these species have been proposed as *Candidatus Mycoplasma* species. *Candidatus* is proposed as a temporary category for unculturable bacteria (Neimark *et al.*, 2001).

Because of their nutritional requirements and the need to use specialized metabolic tests (which are not available commercially) for identification to species level, many laboratories are unable to speciate Mycoplasmas. Sequencing of the 16S rRNA gene together with sequencing the *rpoB* gene (RNA polymerase beta-subunit) and the 16-23S rRNA ITS (intergenic transcribed spacer region) have been suggested as suitable molecular markers for possible speciation or to give an indication of appropriate phenotypic tests before final species identification (Volokhov *et al.*, 2012). Sequence similarity within the phylogenetic units within the family *Mycoplasmataceae* ranges from 78 to 95% for the 16S rRNA gene, 69 to 94% for the 16S-23S ITS and 67 to 91% for the *rpoB* gene. *Mycoplasmataceae* belong to a polyphyletic taxon, that is, the family consists of unrelated species descended from more than one ancestor. The 16S rRNA phylogenetic tree divides the species into eight phylogenetic clusters, with groups within these clusters. The clusters are combined into a Mycoplasma clade and a Mycoplasma-Ureaplasma clade. Within the Mycoplasma clade several *Mycoplasma* species are above the 97% sequence similarity value for differentiation of species based on the 16S rRNA gene, however phenotypic differences based on serology, metabolic activity and ecological site continues to define them as separate species. Information from 16S rRNA sequence needs to be considered in conjunction with sequence information from the ITS and *rpoB* genes, although even some pairs such as *M. cottewii*/*M. yeatsii* and *M. capricolum*/*M. leachii* have similarity values of 99.7%, 99.7%, 97.7% and 99.2%, 98.2% and 97% for the 16S, ITS and *rpoB* genes, respectively (Volokhov *et al.*, 2012).

All *Mycoplasma* species require cholesterol for growth, a distinguishing factor that separates them from *Acholeplasma*, which do not have a cholesterol requirement. Likewise, *Mycoplasma* species are urease negative, which separates them from *Ureaplasma*, which do hydrolyse urea. A lack of a cell wall means that *Mycoplasma* species are resistant to penicillin.

Descriptions of new species must follow the required minimal standards for description of new species of Mollicutes (Brown *et al.*, 2007).

MOLECULAR IDENTIFICATION. Amplification of the 16S rRNA sequence can assist in identification of *Mycoplasma* species. The following primers have been used (Kirchhoff *et al.*, 1997). The sense strand nucleotides 11 to 30, 5′-AGA GTT TGA TCC TGG CTC AGG A-3′ and the antisense strand nucleotides 1512 to 1493 5′-GGT AGG GAT ACC TTG TTA CGA CT-3′. Primer concentration was 1 μM in a standard PCR mix and the annealing temperature was 59°C.

Mycoplasma alligatoris

TAXONOMY AND SIGNIFICANCE. *Mycoplasma alligatoris* (A21JP2^T=ATCC 700619^T) was isolated from trachea, lung, joint synovial fluid and cerebrospinal fluid from captive American alligators (*Alligator mississippiensis*) that had died during an epizootic episode in 1995 (Brown *et al.*, 2001a). The disease was characterized by anorexia, lethargy, bilateral white ocular discharge and periocular, facial and limb oedema with signs of pneumonia and arthritis leading to death. Pathogenicity was confirmed by experimental inoculation. Experimental inoculation of broad-snouted (or broad-nosed) caiman crocodiles (*Caiman latirostris*) and siamese crocodiles (*Crocodylus siamensis*) with 10^6 CFU/ml instilled through the glottis demonstrated pathogenicity for caiman crocodiles but not for siamese crocodiles (Pye *et al.*, 2001).

Partial 16S rRNA sequencing revealed a 98% similarity to *M. crocodyli* and 91% similar to *M. sturni* and that it belonged to the *M. fermentans* clade (Brown *et al.*, 2001a), although more recent phylogenetic analysis now puts *M. alligatoris* and *M. crocodyli* along with *M. sturni* in the Synoviae cluster with *M. fermentans* in the Bovis-lipophilum cluster (Volokhov *et al.*, 2012).

GROWTH CHARACTERISTICS AND PHENOTYPIC IDENTIFICATION. *Mycoplasma alligatoris* is non-motile and lacks a cell wall. Electron micrographs show non-helical cells. Using a stereomicroscope, colonies are seen on ATCC 988 medium after 24 h at 30°C and 5% CO_2. Colonies exhibit a typical fried egg appearance on medium containing 0.8% agar. Sterol or serum is required for growth. Growth is slow at 22–34°C but occurs rapidly at 30–34°C with acidification of broth media occurring after 8 h at 30°C (Brown *et al.*, 2001a). Growth does not occur at 4, 37 or 42°C. Growth occurs with 5% CO_2 and under anaerobic atmosphere. It is resistant to penicillin and sensitive to digitonin with growth inhibition zones of 6–9 mm wide. Growth occurred in broth containing 20 and 2% fetal calf serum but not in broth containing 0.2% serum. Sucrose was fermented. Chicken red blood cells and guinea pig red blood cells are not agglutinated. The organism passed through a 0.22-μm filter. These and other reactions are found in Table 3.18. Details of biochemical media are found in the media section.

Mycoplasma crocodyli

TAXONOMY AND SIGNIFICANCE. *Mycoplasma crocodyli* (MP145^T=ATCC 51981^T) was isolated from joints and lungs from crocodiles (*Crocodylus niloticus*) with exudative polyarthritis, which were farmed in Zimbabwe (Kirchhoff *et al.*, 1997).

Pathogenicity was confirmed by experimental inoculation of yearling crocodiles, whereas goats were not susceptible. Sequencing of the 16S rRNA gene demonstrates relationship to the *M. fermentans* phylogenetic clade and 98% similarity to *M. alligatoris* followed by 91% similarity to *M. corogypsi* and 90% to *M. sturni* (Kirchoff *et al.*, 1997). More recent phylogenetic analysis now puts *M. crocodyli* and *M. alligatoris* along with *M. sturni* and *M. corogypsi* in the Synoviae cluster with *M. fermentans* in the Bovis-lipophilum cluster (Volokhov *et al.*, 2012).

GROWTH CHARACTERISTICS AND PHENOTYPIC IDENTIFICATION. *Mycoplasma crocodyli* is non-motile when examined by a wet preparation, and electron micrographs indicate coccoid or round cells. Colonies are visible after 2–3 days on agar medium containing serum and incubated at 37°C with 10% CO_2 and show typical fried egg morphology (Kirchoff *et al.*, 1997). Cells pass through 0.45 and 0.22 μm filters but overall numbers of cells are reduced by about 100-fold after each filtration. Growth occurs between 25 and 42°C with an optimum of 37°C. Cholesterol or serum is required for growth with serum concentrations of 20% horse serum or 27% porcine serum producing rapid heavy growth. It is sensitive to digitonin with a zone of growth inhibition of 5–10 mm. Glucose and maltose are fermented, arginine, aesculin and urea are not hydrolysed. Tetrazolium chloride (TTC) is reduced aerobically and anaerobically and phosphatase activity is positive. Bovine, ovine and rabbit red blood cells are lysed.

Mycoplasma mobile

TAXONOMY AND SIGNIFICANCE. *Mycoplasma mobile* (163K^T=ATCC 43663^T) was isolated from the gills of tench (*Tinca tinca*) with symptoms of red disease (Kirchhoff and Rosengarten, 1984; Kirchhoff *et al.*, 1987). Pathogenicity is not known. *Mycoplasma mobile* clusters in the Moatsii-sualvi-mobile group on the 16S rRNA phylogenetic tree and forms a unique group between the Hominis and Neurolyticum-hyopneumoniae clusters (Volokhov *et al.*, 2012)

GROWTH CHARACTERISTICS AND PHENOTYPIC IDENTIFICATION. *Mycoplasma mobile* lacks a cell wall and can attach to glass, plastic and living cells, and shows a gliding motion on glass and plastics at a speed of 2–4 μm/s. Electron micrographs show elongated, ovoid or flask-shaped cells consisting of a thicker body and a slender part with a terminal hemispherical structure, which can also be seen under dark ground illumination (Kirchhoff *et al.*, 1987). When grown in broth cultures in a Petri dish, cells attach to the bottom of the plate and cannot be removed by washing. Typical fried egg-shaped colonies are produced on modified Hayflick medium containing horse or bovine serum or on modified Friis medium. Colonies are 10–500 μm after 2–6 days at 25°C. Growth occurs between 17 and 30°C with an optimum of 25°C and weak growth at 4°C. Cells do not survive longer than 10 h at 37°C. Growth occurs under aerobic and anaerobic atmosphere. Catalase reaction is weak and tends to occur only with young cultures. Oxidase is positive for 50% of cultures. Cells pass through 0.45 and 0.22 μm filters but with reductions in numbers. Sensitive to digitonin with a growth inhibition zone of 12–15 mm. Serum is required for growth and growth is enhanced by the addition of 1–20 μg/ml cholesterol. Film and spots occurs on modified Hayflick medium containing horse serum but is negative on media containing bovine or porcine serum. Reduction of TTC occurs under aerobic and anaerobic atmospheres. Coagulated serum is not digested. Human, equine, bovine, ovine, porcine, rabbit, guinea pig and fish red blood cells are haemolysed.

Mycoplasma phocicerebrale

TAXONOMY AND SIGNIFICANCE. *Mycoplasma (phocacerebrale) phocicerebrale* (1049^T=ATCC 49640^T) was isolated from the brain, nose, throat, lung and heart of harbour seals (*Phoca vitulina*), found after mass mortality of 18,000 animals in the Baltic and North Sea in 1988 (Giebel *et al.*, 1991). Phocine distemper virus and Influenza A virus were isolated and were the main cause of the epidemic. The acute disease was characterized by pneumonia, skin lesions, diarrhoea, poly-arthritis, nervous signs, and abortion in females. *Mycoplasma phocicerebrale* was co-isolated with *M. phocirhinis*. Despite exhibiting strong cytopathic activity when tested against trachea cultures with resulting inhibition of ciliary action and damage to the underlying epithelium, the role of *M. phocicerebrale* in pathogenicity as a primary or secondary pathogen is not known.

Mycoplasma phocicerebrale was reported in the UK from a grey seal (*Halichoerus grypus*) with an infected flipper resulting from bite wounds (Ayling *et al.*, 2011). Other possible pathogens also were isolated and included *Arcanobacterium phocae* and *Streptococcus halichoeri*. Culture of wound infections, mouth ulcer, eye, lung of grey seals and tooth of a common seal (*P. vitulina*) resulted in identification of *M. phocicerebrale*. *Mycoplasma phocicerebrale* and *M. phocirhinis* frequently are present on teeth of healthy seals.

Mycoplasma phocicerebrale was reported from the lungs of harbour porpoises (*Phocoena phocoena*) and the liver from one animal that stranded on the coast of Scotland (Foster *et al.*, 2011). It is thought to be a common inhabitant of the upper respiratory tract of Australian fur seals (*Arctocephalus pusillus doriferus*) as 72% of seals sampled in the Bass Strait in south-eastern Australia were apparently healthy with no clinical or pathological signs of infection (Lynch *et al.*, 2011). The organism also was isolated from the lung of one adult Australian fur seal but pneumonia was not detected. In the same investigation *M. phocicerebrale* was isolated in the thymus of an aborted fetus and other *Mycoplasma* species were detected by PCR. The fetal tissue showed inflammatory lesions suggesting *Mycoplasma* infection may lead to gestational failure. *Mycoplasma phocicerebrale* is the causative agent of seal finger in humans (Hartley and Pitcher, 2002). It is hypothesized that *M. phocicerebrale* and *M. phocirhinis* are opportunistic pathogens when they penetrate the skin and cause severe wound infections (Ayling *et al.*, 2011).

Mycoplasma phocicerebrale belongs to the Hominis cluster on the 16S rRNA phylogenetic tree along with *M. phocidae* with a 97.9–98.5% similarity, whereas *M. zalophi* forms a separate group within the Hominis cluster (Königsson *et al.*, 2001; Volokhov *et al.*, 2012).

GROWTH CHARACTERISTICS AND PHENOTYPIC IDENTIFICATION. *Mycoplasma phocicerebrale* lacks a cell wall and is coccoid, round or dumbbell shaped. On media containing 0.8% agar, colonies exhibit a typical fried egg appearance when viewed with a stereomicroscope. Optimum growth temperature is 37°C. Cholesterol or serum are required for growth. Glucose is not fermented, urea is not hydrolysed. Results are positive for arginine hydrolysis. TTC and potassium tellurite are not reduced under aerobic or anaerobic conditions, and film and spots are negative (Giebel *et al.*, 1991). Sheep red cells are lysed. Coagulated serum is not liquefied. These and other reactions are found in Table 3.18. Details of biochemical media are found in the media section.

Mycoplasma phocidae

TAXONOMY AND SIGNIFICANCE. *Mycoplasma phocidae* (*M. phocae*) (105^T=ATCC 33657^T) was isolated from the respiratory tract of harbour seals (*Phoca vitulina*) that died of an epizootic pneumonia where phocine distemper virus and influenza A virus were thought to be the primary pathogens (Ruhnke and Madoff, 1992). In experimental challenge *M. phocidae* appeared to be avirulent for grey and harp seals but it was suggested the organism may contribute to respiratory diseases in association with environmental or other infectious challenges. *Mycoplasma phocidae* was isolated from a puncture wound of a grey seal pup (*Halichoerus grypus*) (Ayling *et al.*, 2011). Other bacteria isolated included *Arcanobacterium phocae* and *Atopobacter phocae*. It also has been isolated from diseased lung in a common seal in co-isolation with a novel *Pasteurellaceae*, a novel *Streptococcus* species and phocine distemper virus. Apparently healthy juvenile and one adult Australian fur seal samples from Bass Strait, south-eastern Australia, harboured *M. phocidae* along with *M. zalophi*, and an unidentified *Mycoplasma* species in the nasal cavity (Lynch *et al.*, 2011).

Mycoplasma phocidae along with *M. phocicerebrale* belongs to the Hominis cluster and the *M. alkalescens* subcluster on the 16S rRNA phylogenetic tree with species within the subcluster having a similarity of 97.9–98.5%, whereas *M. zalophi* forms a separate group within the Hominis cluster (Königsson *et al.*, 2001; Volokhov *et al.*, 2012).

There is contention regarding the naming of the organism. The original spelling of the epithet *phocidae* was corrected to *phocae* (Königsson *et al.*, 2001), however the correction is illegitimate according to the rules of nomenclature (De Vos and Trüper, 2000), therefore *M. phocidae* continues to be used.

GROWTH CHARACTERISTICS AND PHENOTYPIC IDENTIFICATION. *Mycoplasma phocidae* appears as mainly coccoid forms with a clearly defined three-layer membrane but no cell wall when viewed using ultra-thin sections observed by electron microscopy (Ruhnke and Madoff, 1992). Under phase contrast microscopy mainly coccoid forms are seen with some pleomorphism and filaments of 0.1–0.4 µm in diameter. Helical forms are not seen. Colonies have a typical fried egg appearance on agar media when viewed with a stereomicroscope, and produce an alpha-haemolysis or greening of the agar on horse blood agar. Growth is enhanced by incubation in CO_2 or in an anaerobic atmosphere. Optimum growth temperature is 37°C and occurs under microaerobic and anaerobic conditions. Serum is required for growth as growth does not occur on media without the addition of serum. Glucose is not fermented, urea is not hydrolysed. Arginine is hydrolysed. TTC is not reduced. Film and spots are positive on 1.1% agar containing 20% horse serum. Coagulated serum is not liquefied and casein is not digested. The organism causes agglutination of red blood cells from sheep, chickens, guinea pigs and seals. Susceptible to 1.5% digitonin with a zone of inhibition of 8 mm. Resistant to penicillin and thallium acetate. The cells of *M. phocidae* pass through a 0.45-µm filter but not through a 0.25-µm filter. These and other reactions are found in Table 3.18. Details of biochemical media are found in the media section.

Mycoplasma phocirhinis

TAXONOMY AND SIGNIFICANCE. *Mycoplasma phocirhinis* (*phocarhinis*) (852^T=ATCC 49639^T) was isolated from the nose, throat, trachea, lung and heart of harbour seals (*Phoca vitulina*) found after mass mortality in the Baltic and North Sea in 1988 (Giebel *et al.*, 1991). The organism was co-isolated with *M. phocicerebrale.* The role in pathogenicity as a primary or secondary pathogen is not known, however the organisms were isolated in high numbers from diseased organs and exhibited strong cytopathic activity when tested against trachea cultures with resulting inhibition of ciliary action and damage to the underlying epithelium.

Mycoplasma phocirhinis, similar to *M. phocicerebrale*, is frequently isolated from teeth of grey seals (*Halichoerus grypus*) and common seals (*Phoca vitulina*) (Ayling *et al.*, 2011). It also has been isolated from diseased lung of a common seal. Other possible pathogens such as *Bordetella bronchiseptica*, *Streptococcus phocae* and *Corynebacterium phocae* also were isolated. It is hypothesized that *M. phocicerebrale* and *M. phocirhinis* are opportunistic pathogens when they penetrate the skin and cause severe wound infections.

Mycoplasma phocirhinis occurs in the Bovis-lipophilum cluster on the 16S rRNA phylogenetic tree (Königsson *et al.*, 2001; Volokhov *et al.*, 2012).

The original spelling of the epithet *phocarhinis* was corrected to *phocirhinis* (Königsson *et al.*, 2001).

GROWTH CHARACTERISTICS AND PHENOTYPIC IDENTIFICATION. *Mycoplasma phocirhinis* lacks a cell wall and appears as coccoid or round in shape. On media containing 0.8% agar, colonies usually exhibit a typical fried egg appearance when viewed with a stereomicroscope. Optimum growth temperature is 37°C. Cholesterol or serum are required for growth. Glucose is not fermented, urea and arginine are not hydrolysed. Tetrazolium chloride is reduced under aerobic and anaerobic conditions, whereas potassium tellurite is reduced under anaerobic conditions only. Film and spots are positive on 1.1% agar containing 20% horse serum (Giebel *et al.*, 1991). Sheep red cells are lysed. Coagulated serum is not liquefied. These and other reactions are found in Table 3.18. Details of biochemical media are found in the media section.

Mycoplasma species

TAXONOMY AND SIGNIFICANCE. Uncharacterized *Mycoplasma* species were isolated from diseased lung of grey seal (*Halichoerus grypus*) and hooded seal (*Cystophora cristata*) (Ayling *et al.*, 2011).

Novel *Mycoplasma* species were detected from the lungs of harbour porpoises (*Phocoena phocoena*) and the kidney of one animal that had stranded on the coast of Scotland (Foster *et al.*, 2011). A *Mycoplasma* closely related to *Mycoplasma* species 13CL from a pelican was isolated from the kidney of a Sowerby's beaked whale (*Mesoplodon bidens*). The significance of these isolations in terms of pathogenicity or relationship to the strandings is not known. Some animals had a parasitic infection.

An unidentified *Mycoplasma* phylotype, detected by partial 16S rRNA amplification and sequencing of DNA extracted from the gastrointestinal tract, was found as a predominant gut microflora in Atlantic salmon (*Salmo salar*) farmed in Norway and Scotland and also in wild salmon (Holben *et al.*, 2002). The sequence was most similar to *M. penetrans* and *M. volis* both belonging to the *M. iowae* subgroup. A *Mycoplasma* ribotype identified by PCR and denaturing gradient gel electrophoresis of gut microflora of the longjaw mudsucker (*Gillichthys mirabilis*) from the USA identified an unknown *Mycoplasma* species that was most similar (95%) to *M. mobile* based on genetic information (Bano *et al.*, 2007).

A *Mycoplasma* species identified by 16S rRNA sequencing most closely matching (99%) GenBank number EU714238.1 (*Mycoplasma* sp. CSL7503-Lung) was isolated from nasal swabs of apparently healthy Australian fur seals (*Arctocephalus pusillus doriferus*) and from lymph nodes of juveniles found in northern Bass Strait in south-eastern Australia (Lynch *et al.*, 2011). Lymphocytic and plasmacytic tracheitis was observed in some juveniles. The isolate was positive for glucose and digestion of casein and negative for phosphate, arginine, TTC and film and spots. *Mycoplasma zalophi* and *M. phocae* also were isolated. *Mycoplasma* species CSL7503-Lung was isolated from Californian sea lions (*Zalophus californianus*) undergoing rehabilitation (Volokhov *et al.*, 2012).

Mycoplasma zalophi

TAXONOMY AND SIGNIFICANCE. *Mycoplasma zalophi* was isolated from subdermal abscesses in the majority of affected animals and from pleural fluid, lung tissue, joint material and a muscle abscess in other Californian sea lions (*Zalophus californianus*) undergoing rehabilitation in the USA (Haulena *et al.*, 2006). Some animals presenting with pneumonia or polyarthritis subsequently died, whereas other animals presenting with abscesses were successfully treated and later released. The organism was isolated in pure growth in a number of animals. Results of biochemical testing using methods for mycoplasma and partial sequencing of the 16S rRNA suggested a new *Mycoplasma* species. Based on culture and identification using 16S rRNA primers (>98% similarity), *M. zalophi* was detected in nasal swabs of apparently healthy Australian fur seals (*Arctocephalus pusillus doriferus*) found in northern Bass Strait in south-eastern Australia (Lynch *et al.*, 2011). The organism was also isolated from lymph nodes of juveniles exhibiting lymphocytic and plasmacytic tracheitis.

Mycoplasma zalophi is most similar to *M. gypis* in the *M. hominis* clade, which has been confirmed by a more recent phylogenetic tree suggesting *M. zalophi* and *M. gypis* form the Zalophi-gypis group within the Hominis cluster (Volokhov *et al.*, 2012). The validity of the proposed name *M. zalophi* has yet to be accepted.

GROWTH CHARACTERISTICS AND PHENOTYPIC IDENTIFICATION. *Mycoplasma zalophi* grows on blood agar and chocolate agar as pinpoint colonies embedded into the agar at 3–6 days at 35°C in CO_2, and grows on SP4 mycoplasma medium as pinpoint to 2 mm colonies in 24 h in air at 37°C (Haulena *et al.*, 2006). Tests are negative for glucose, phosphate, casein digestion and film and spots, and positive for arginine and TTC. The culture of *M. zalophi* obtained from Australian fur seals was positive for film and spots (Lynch *et al.*, 2011). These and other reactions are found in Table 3.18. Details of biochemical media are found in the media section.

Class Erysipelotrichia
Order *Erysipelotrichales*

1.31 *Erysipelotrichaceae*

1.31.1 *Erysipelothrix*

Erysipelothrix are Gram-positive rods that are straight, slightly curved and slender with a tendency to form long filaments.

Erysipelothrix rhusiopathiae

TAXONOMY AND SIGNIFICANCE. *Erysipelothrix rhusiopathiae* is the infective agent of erysipelas, a disease that affects a variety of domestic and farmed animals, humans and aquatic animals and aquatic mammals. *Erysipelothrix rhusiopathiae* causes erysipelas in bottle-nosed dolphin (*Tursiops truncatus*), Atlantic spotted dolphin (*Stenella plagiodon*), Risso's dolphin (*Grampus griseus*), Pacific white-sided dolphin (*Lagenorhynchus obliquidens*) either captive or free-ranging in the Mediterranean Sea or the USA, and has been isolated from the teeth and gum margins in northern fur seal (*Callorhinus ursinus*) and northern elephant seal (*Mirounga angustirostris*) (Geraci *et al.*, 1966; Suer and Vedros, 1988; Dunbar and Clarridge, 2000; Melero *et al.*, 2011). Infection resulting in septicaemia and death occurred in a little blue penguin (*Eudyptula minot*) held in an aquarium. The organism was isolated from lung, liver and intestine (Boerner *et al.*, 2004). The organism has also been isolated from herring fish meal and from bites and wound abrasions from marine mammal handlers (Suer and Vedros, 1988). In patients infection has lead to polyarthralgia, renal failure, joint swelling and pain, localized lesions on fingers and is a hazard for fish handlers, butchers, abattoir workers and people working with swine and other susceptible farm animals (Dunbar and Clarridge, 2000). It has been reported from a caiman crocodile (*Caiman crocodilus*) and an American crocodile (*Crocodilus acutus*) housed in a zoo in the USA (Jasmin and Baucom, 1967). Irregular diagonal brownish-black plaques or lesions occurred on the dorsal surface of the tail and back.

Infected animals display ulcers or lesions on the skin on the dorsal aspect of the tail and fin that may be square, rectangular or often described as rhomboid-shaped, but also may be poorly delineated. Ulcers can bleed and vary in size from 4 to 12 cm and 0.4–0.9 cm deep. The skin may be wrinkled, scaly or raised with oedema. All organs may be affected and show associated pathology including pulmonary congestion, brain congestion, foam through blowholes and rales (Geraci *et al.*, 1966; Melero *et al.*, 2011).

GROWTH CHARACTERISTICS AND PHENOTYPIC IDENTIFICATION. *Erysipelothrix rhusiopathiae* is a Gram-positive to Gram-variable rod (2–5 μm) that is pleomorphic and displays tangled forms. On BA, it forms 0.5 mm grey-green colonies with a small zone of alpha-haemolysis after 24–48 h at 37°C. *Erysipelothrix rhusiopathiae* is easily overgrown by contaminants and may not be predominant in cultured material. To improve the isolation rate, tissue or swab material is placed into Woods broth, a selective and enrichment medium (see media section), and incubated at 37°C for 24 and 48 h. Subcultures are prepared at 24 and 48 h to BA and again incubated at 37°C for 24 and 48 h. A selective medium, Packer's plate (see media section) is also available for culture of the organism, however the use of Woods broth as a selective and enrichment medium followed by subculture to BA is a sensitive method for detection and is recommended in addition to direct sampling to solid medium.

A few drops of sterile serum added to the inoculating medium improves reactions when using conventional biochemical media. Nitrate reduction is negative. Urea and esculin are negative. Glucose, lactose and maltose are fermented. Reactions for L-arabinose, mannose, sucrose and trehalose are variable (Geraci *et al.*, 1966; Suer and Vedros, 1988). These and other reactions are presented in Tables 3.15, 3.31–33 and 3.36.

Class Bacilli
Order *Bacillales*

1.32 *Bacillaceae*

1.32.1 *Bacillus*

Bacillus mycoides

TAXONOMY AND SIGNIFICANCE. *Bacillus mycoides* caused high mortality, lesions and ulcers on the dorsum and necrosis of the expaxial muscle of channel catfish (*Ictalurus punctatus*) cultured in the USA (Goodwin *et al.*, 1994). In experimental challenge using intraperitoneal injection, intramuscular injection, subcutaneous injection and abrasion, the underlying musculature became necrotic in fish receiving subcutaneous and intramuscular injections, whereas infection did not develop in the other treatments. The bacterium was cultured from muscle, liver and trunk kidney of all treated fish except the abraded skin-treated fish. In the epizootic, green sunfish (*Lepomis cyanellus*), threadfin shad (*Dorosoma petenense*) and golden shiner (*Notemigonus crysoleucas*) in the same culture pond did not show symptoms or mortalities.

GROWTH CHARACTERISTICS AND PHENOTYPIC IDENTIFICATION. *Bacillus mycoides* is a non-motile, Gram-positive rod occurring in chains. On BA and Mueller–Hinton agar, colonies are rhizoid with a filamentous, counterclockwise swirling growth pattern. Catalase is positive. Glucose is fermented,

VP is positive and lethinase is produced. Parasporal crystals are negative. Growth does not occur at 45°C (Goodwin *et al.*, 1994). These and other reactions are presented in Table 3.15.

Bacillus zhanjiangensis

TAXONOMY AND SIGNIFICANCE. *Bacillus zhanjiangensis* (JSM 099021^T=DSM 23010^T=KCTC 13713^T) was isolated from an oyster from the South China Sea (Chen *et al.*, 2011b). Based on 16S rRNA sequence analysis closely related strains are *B. halmapalus* (99.0%), *B. horikoshii* (98.4%) and *B. cohnii* (98.0%).

GROWTH CHARACTERISTICS AND PHENOTYPIC IDENTIFICATION. *Bacillus zhanjiangensis* is an aerobic, Gram-positive endospore-forming rod (3.0–6.0×0.6–1.0 µm), oxidase-positive, catalase-positive and motile. Spores are ellipsoidal and subterminal with slightly swollen sporangia. Chains of four to eight cells tend to occur. On MA colonies are 2–3 mm, cream to white, flat, semi-translucent, smooth, glistening with circular to slightly irregular margins (Chen *et al.*, 2011b). Growth occurs at 0–15% NaCl (optimum 2–4%), 10–45°C (optimum 30–35°C). Casein, hippurate, starch, Tween 40, 60 and 80 are hydrolysed, but results are negative for Tween 20 and cellulose. These and other reactions are presented in Table 3.15 and 3.38.

1.33 *Planococcaceae*

1.33.1 *Planococcus*

Planococcus species are aerobic, motile, coccoid cells occurring singly, in pairs, in groups of three or in tetrads and are Gram-positive to Gram-variable. Catalase is positive and oxidase is generally negative. Colonies are yellow-orange in colour and growth occurs on nutrient agar containing 12% NaCl. Urea is not hydrolysed (Nakagawa *et al.*, 1996). They occur in a range of habitats including seawater, marine clams, fish, prawns, cyanobacterial mats and freshwater ponds.

Planococcus species

TAXONOMY AND SIGNIFICANCE. *Planococcus* species was isolated from spots on the head of juvenile rainbow trout (*Oncorhynchus mykiss*) held in a recirculation freshwater facility in the UK (Austin *et al.*, 1988). The spots were recessed and off-white, 2–4 mm on the heads of three of 2000 trout. In experimental challenge, the *Planococcus* species was pathogenic when challenged intraperitoneally with a high concentration (10^7 cells/ml) of cells and a cumulative mortality of 40% occurred from 48 h post-infection. Infected fish displayed swollen and haemorrhagic intestine, and slight liquefaction of the kidney. The organism was recovered from the gastrointestinal tract, peritoneal cavity and kidney. Fish challenged by bathing following scarification did not become infected. The organism most closely resembled *P. citreus* based on biochemical analysis.

Another *Planococcus* species was isolated from moribund and dead trout fry displaying symptoms of rainbow trout fry syndrome on fish farms in the UK (Austin and Stobie, 1992b). *Micrococcus luteus* was isolated also from the fish. Infected fish had exophthalmia, pale gills and swollen abdomens and the organism was isolated from kidney, spleen and ascites fluid. Experimental challenge with moderate to high concentration of *Planococcus* cells (10^5–10^7 cells/ml) injected intramuscularly or intraperitoneally resulted in 32% mortalities in 14 days. The gills, liver and spleen of infected fish were pale, and there were internal haemorrhages, gastroenteritis and ascites fluid. The *Planococcus* species was biochemically different to that previously reported in trout (Austin *et al.*, 1988).

GROWTH CHARACTERISTICS AND PHENOTYPIC IDENTIFICATION. *Planococcus* species reported in 1988 and presumptively identified as *P. citreus* was an aerobic, motile, Gram-positive coccus (1–2 µm) occurring in pairs. Catalase, citrate, urease, ADH, ODC are positive, whereas LDC, gelatinase and oxidase are negative (Austin *et al.*, 1988). A positive reaction for urease is in contrast to the description of the genus (Nakagawa *et al.*, 1996). Growth occurs at 37°C and in 0–15% NaCl (Austin *et al.*, 1988). The *Planococcus* species isolated in 1992 from trout was a motile Gram-positive coccus occurring in clumps and appearing as shiny yellow colonies on BHIA. Reactions are negative for ADH, ODC and urease (Austin and Stobie, 1992b). These and other reactions are presented in Table 3.14.

Order *Lactobacillales*

1.34 *Aerococcaceae*

1.34.1 *Abiotrophia*

See under *Granulicatella* species.

1.34.2 *Aerococcus*

Aerococcus viridans

TAXONOMY AND SIGNIFICANCE. *Aerococcus viridans* var. *homari* (previously *Gaffkya homari*) causes a disease in lobsters known as Gaffkemia or red tail disease (Hitchner and Snieszko, 1947; Snieszko and Taylor, 1947; Williams *et al.*, 1953). Outbreaks of disease tend to occur in holding facilities when fighting between live lobsters results in breaks in the integment, which allows entry of the pathogen. The disease has been reported in lobsters (*Homarus gammarus*) in England, Wales and Norway (Gjerde *et al.*, 1984; Wiik *et al.*, 1987; Stebbing *et al.*, 2012) and in *H. americanus* in the USA and Canada (Snieszko and Taylor, 1947; Stewart, 1972; Stebbing *et al.*, 2012). Molecular typing indicates that strains from *H. americanus* introduced the disease to the UK. Virulent strains from Canada, the USA, England, Wales and Norway show very little genetic variation (Stebbing *et al.*, 2012). Mortality is usually high and in the order of 20–50% and infected lobsters display a pink colouration on the ventral side of the abdomen. The blood has a prolonged clotting time, is less viscous and is pink in colour and bacteria can be seen in the bloodstream in advanced cases (Snieszko and Taylor, 1947).

In lobsters, virulent and avirulent strains are reported. Virulent strains have antigen b, possess a polysaccharide capsule and are not agglutinated by lobster haemolymph serum, whereas avirulent strains do not possess antigen b, do not have a capsule, are agglutinated by haemolymph serum and do not grow well in lobster haemolymph serum (Steenbergen *et al.*, 1977; Stewart *et al.*, 2004). The chaperonin 60 protein is expressed in virulent strains but not in avirulent strains (Clark and Greenwood, 2011).

Aerococcus viridans was a cause of oesophageal diverticulum in a loggerhead sea turtle (*Caretta caretta*) stranded on the coast of Spain (Torrent *et al.*, 2002).

Aerococcus viridans occurs in soil and a wide range of environmental sources and is generally saprophytic.

GROWTH CHARACTERISTICS AND PHENOTYPIC IDENTIFICATION. *Aerococcus viridans* is non-motile, facultatively anaerobic, Gram-positive coccus in tetrads with a wide pseudocapsule when seen in blood (0.5×2.7 μm) and is non-motile, catalase negative, non-spore forming and occurs singly, in pairs or short chains. On BA or TSA colonies are 1 mm, alpha-haemolytic, greyish-white, circular and convex (Snieszko and Taylor, 1947). On media the cells form irregular conglomerations without capsules. A selective medium, SEIM (Stewart, 1972; Gjerde, 1984), is detailed in the media section. On SIEM colonies are small, whitish with entire transparent edges and surrounded by a yellow zone (Wiik *et al.*, 1987).

Litmus milk is not changed and gelatin is not hydrolysed. VP is negative. Mannitol, dextrose, sucrose, maltose and lactose are fermented without gas production (Snieszko and Taylor, 1947). Most strains hydrolyse hippurate.

The presence of a capsule is detected using a smear prepared from a bacterial culture and air-dried. The polysaccharide capsule is stained by applying alcian blue (1% in 3% acetic acid) to the slide for 15 min. The smear is washed in distilled water and drained. A counterstain of 1% aqueous neutral red is applied for 1 min before the slide is washed with distilled water. The slide is air-dried and blotted dry and examined under oil immersion (×100). Capsules appear as a blue-green band around the tetrads of cells, which stain red (Stewart *et al.*, 2004). Biochemical reactions are presented in Tables 3.14, 3.30, 3.36, 3.38 and API STAPH (see apiweb™ database, Biomerieux).

1.34.3 *Facklamia*

Facklamia miroungae

TAXONOMY AND SIGNIFICANCE. *Facklamia miroungae* (A/G13/99/2^{T}=CCUG42728^{T}=CIP 106764^{T}) was isolated from the nasal cavity of a juvenile southern elephant seal (*Mirounga leonina*) as part of a study of the microflora of animal species in the South Orkneys, Southern Ocean, near the north-eastern tip of Antarctica (Hoyles *et al.*, 2001a).

GROWTH CHARACTERISTICS AND PHENOTYPIC IDENTIFICATION. *Facklamia miroungae* is a facultatively anaerobic, non-motile, Gram-positive ovoid coccus (0.8–0.9 μm) that occurs in pairs or short chains. Catalase is negative. On BA colonies are <0.5 mm, alpha-haemolytic, grey-coloured, round, entire and shiny after 24 h at 37°C and in air supplemented with 5% CO_2. Growth occurs at 25–42°C and in 5% NaCl. Nitrate is not reduced and VP is negative. Aesculin, gelatin and hippurate are not hydrolysed. Glucose is fermented but not mannitol, maltose, lactose or sorbitol (Hoyles *et al.*, 2001a). These and other reactions are presented in Tables 3.14, 3.36 and 3.38.

1.35 *Carnobacteriaceae*

The family *Carnobacteriaceae* belongs to the order *Lactobacillales*. Within the family *Carnobacteriaceae* are the genera *Atopobacter*, *Carnobacterium* and *Granulicatella* (*Bergey's Manual of Systematic Bacteriology*, Vol. 3, The Firmicutes, 2009).

1.35.1 *Atopobacter*

Atopobacter phocae

TAXONOMY AND SIGNIFICANCE. *Atopobacter phocae* (M1590/94/2^T=CCUG 42358^T=CIP 106392^T) was isolated from the small intestine and lymph nodes (mesenteric, external iliac, precapsular pancreatic) from a common seal (*Phoca vitulina*) pup with lymphadenopathy, congested lungs and pulmonary haemorrhage (Lawson *et al.*, 2000). *Arcanobacterium phocae* was co-isolated from the lungs. A second isolate (M955/98/1) was isolated in mixed culture from liver, spleen and blood of a common seal that died of gunshot wounds. The pathogenicity is unknown. Sequencing of the 16S rRNA gene revealed a phylogenetic association with lactic acid bacteria, in particular, *Granulicatella* species.

GROWTH CHARACTERISTICS AND PHENOTYPIC IDENTIFICATION. *Atopobacter phocae* is a facultatively anaerobic, Gram-positive short irregular rod. Catalase is negative. On BA colonies are pinpoint, grey, smooth and non-haemolytic after 24 h at 37°C in air containing 5% CO_2. Growth occurs at 25–37°C but not at 42°C. Blood and serum are not necessary for growth. Glucose is fermented without gas production. Hydrolysis is negative for aesculin, gelatin, hippurate and urease. Nitrate is not reduced. These and other reactions are in Tables 3.15, 3.36 and 3.38.

1.35.2 *Carnobacterium*

The genus *Carnobacterium* belongs to the family *Carnobacteriaceae*, the order *Lactobacillales* within the class *Bacilli*. The genus *Carnobacterium* is a member of the lactic acid group of bacteria and was created to group the heterofermentative, rod-shaped lactic acid bacteria that produce L-lactic acid from glucose (Collins *et al.*, 1987). Cells are Gram-positive short to slender rods, which occur singly, in pairs or short chains. It does not grow on acetate agar and nitrate is not reduced. Species may or may not be motile. Growth occurs at pH 9. It is psychrotolerant as most strains can grow at 0°C but not at 45°C. It is usually catalase negative and is oxidase negative. The *Carnobacterium* species are phylogenetically closer to the genera *Enterococcus* and *Vagococcus* than to members of the genus *Lactobacillus*. *Carnobacterium* is distinguished from *Lactobacillus* by the ability to grow at pH 9.0, but not at pH 4.5 and failure to grow on acetate agar (pH 5.4), whereas *Lactobacillus* grow on this medium (Rogosa medium, Oxoid; Rogosa *et al.*, 1951).

Carnobacterium species are implicated as spoilage organisms in chilled meat and seafoods, but a number of strains of different species also have probiotic potential in fish diseases or as protective cultures in chilled foods.

Discrepancies have been reported between conventional tube tests and the API 50CH system with results for inulin, lactose, sorbitol and galactose (Michel *et al.*, 1986; Baya *et al.*, 1991).

Table 1.14 provides a guide for differentiation between similar genera: *Carnobacterium*, *Lactobacillus*, *Streptococcus iniae*, *Vagococcus* and *Renibacterium*.

Carnobacterium divergens

TAXONOMY AND SIGNIFICANCE. *Carnobacterium divergens* strain 6251, with probiotic potential against *Aeromonas salmonicida* ssp. *salmonicida*, *Vibrio anguillarum* and *Moritella viscosa*, was isolated from the small intestine of Arctic char (*Salvelinus alpinus*) (Ringø *et al.*, 2002). Based on 16S rRNA sequence analysis strain 6251 has 96.2% similarity to the type strain based on 16S rRNA sequence information.

GROWTH CHARACTERISTICS AND PHENOTYPIC IDENTIFICATION. *Carnobacterium divergens* strain 6251 is a non-motile, Gram-positive asporogenous rod (0.2×0.3–0.7 µm) occurring singly or in pairs in the exponential phase in broth culture, and in the stationary phase cells are 0.2×0.5–1.2 µm in pairs or small chains. It is negative for catalase and oxidase. On tryptic soy agar colonies are 2–3 mm, whitish, raised, entire and circular when incubated in air at 23°C for 48 h. There is no growth on TCBS (Ringø *et al.*, 2002). Growth occurs at 0–6% NaCl, 0–35°C and 5.5–9.1 pH. Growth does not occur at 37°C or on acetate agar. Nitrate is not reduced, H_2S is not produced and gelatin and urea are not hydrolysed. These and other reactions are presented in Tables 3.15 and 3.38.

Carnobacterium inhibens

TAXONOMY AND SIGNIFICANCE. *Carnobacterium inhibens* ($K1^T$=CCUG 31728^T) was isolated from the intestine of Atlantic salmon (*Salmo salar*). It has probiotic potential as *in vitro* tests showed growth inhibition of *Vibrio anguillarum* and *Aeromonas salmonicida* (Jöborn *et al.*, 1999). Based on 16S rRNA sequence analysis *C. inhibens* is most closely related to *C. alterfunditum*.

GROWTH CHARACTERISTICS AND PHENOTYPIC IDENTIFICATION. *Carnobacterium inhibens* is a motile, Gram-positive, asporogenous rod (0.2×0.3–0.7 µm), occurring singly or in pairs at the exponential phase of growth in broth, and 0.2×0.5–1.2 µm at the stationary phase. V forms may be present. Growth does not occur on TCBS or MRS (Lactobacillus MRS agar from Difco based on the formulations of deMan, Rogosa and Sharpe). It is oxidase and catalase negative. On tryptic soy agar containing 2% NaCl colonies are 1–2 mm, white, entire and circular when grown aerobically, and buff-coloured when grown anaerobically. On blood agar colonies show alpha-haemolysis (Jöborn *et al.*, 1999). Growth occurs at 0–6% NaCl, pH 5.5–9.0 and 0–30°C. Reactions are presented in Tables 1.11, 3.34–35, 3.36 and 3.38.

Carnobacterium (*piscicola*) *maltaromaticum*

TAXONOMY AND SIGNIFICANCE. *Carnobacterium* (*Lactobacillus*) *piscicola* and *Lactobacillus maltaromicus* have been reclassified as *Carnobacterium maltaromaticum* based on 16S-23S rRNA sequence information, randomly amplified polymorphic DNA (RAPD) and restriction digest of amplified 16S rRNA (Mora *et al.*, 2003). *Carnobacterium maltaromaticum*, previously named *Lactobacillus piscicola* ($B270^T$=ATCC 35586^T), was isolated from diseased rainbow trout (*Oncorhynchus mykiss*), spring chinook salmon (*O. tshawytscha*) and cutthroat trout (*Salmo clarkii clarkii*) in Oregon, USA in 1970 (Hiu *et al.*, 1984). The disease, which caused septicaemia in fish, was termed pseudo-kidney disease and caused a low level mortality associated with stress such as spawning. Pathological signs vary and include septicaemia, splenomegaly, distended abdomen, abscesses in the musculature, ascites fluid, internal haemorrhages and blood-filled blisters under the skin. Chronic mortality was also observed in post-spawned rainbow trout and fish of spawning age in the USA (Starliper *et al.*, 1992). Clinical signs included lethargy, loss of appetite, darkening of the body surface and an occasional blister-like lesion. Kidney, liver and spleen were swollen, ascites fluid increased and haemorrhage was seen in liver, swim bladder and lower intestine, and bacteria were isolated from these tissues.

Challenge experiments demonstrated variable virulence between strains and for host species. Some strains were not pathogenic for rainbow trout, whereas other strains were. Challenge in striped bass (*Morone saxatilis*) did not result in mortalities yet the organism was cultured from kidney, suggesting a carrier status. In wild and cultured species, mortalities were found in cultured striped bass, channel catfish (*Ictalurus punctatus*) and cutthroat trout, and a carrier status was found in brown bullhead (*Ameiurus nebulosus*) (Baya *et al.*, 1991).

A highly virulent strain (PT-31) was responsible for epizootics in market-size rainbow trout (*Oncorhynchus mykiss*) farmed in Spain (Toranzo *et al.*, 1993). Infected fish displayed bilateral exophthalmia, periocular haemorrhages, ascites fluid and haemorrhages on liver, muscle and intestine. The LD_{50} was <5–6×10^4 cells.

A bacterium identified as *C. maltaromaticum*-like was isolated from kidney and swim bladder of lake whitefish (*Coregonus clupeaformis*) from Lakes Michigan and Huron (Loch *et al.*, 2008). Infected fish had splenomegaly, renal and splenic congestion, and swim bladders had hyperplasia, fibrin deposits and neovascularization indicating chronic irritation. The prevalence of infection is low at 1.8% but reached 33% in one seasonal sample. Spawning stress may be a disease trigger. Partial sequencing 600 bp of the 16S-23D rRNA gene demonstrated 97% similarity with the type strain of *C. maltaromaticum*, and sequencing 300 bp of the *pisA* precursor gene for the piscicolin 126 protein yielded 98% similarity indicating a four nucleotide difference.

Based on 16S rRNA sequence analysis, closely related strains are *Bacillus halmapalus* (99.0%), *B. horikoshii* (98.4%) and *B. cohnii* (98.0%).

Carnobacterium (*Lactobacillus*) *piscicola* may have been reported as *Corynebacterium* species, *Lactobacillus* species, *Listeria* species or *Vagococcus salmoninarum* in earlier reports in the literature.

GROWTH CHARACTERISTICS AND PHENOTYPIC IDENTIFICATION. *Carnobacterium maltaromaticum* is a facultatively anaerobic, non-motile, Gram-positive, non-spore-forming rod (1.1–1.4×0.5–0.6 µm), occurring singly, in pairs or short chains. Cells can become Gram-variable within 24 h. In Gram-stained smears of kidney, cells appear smaller and occur as single cells more often compared to cells grown in broth or on agar plates (Hiu *et al.*, 1984). It is oxidase and catalase negative. On TSA colonies are pinpoint at 24 h, and at 48 h are less than 2 mm, white, round, entire at 25°C. Growth may be optimal at 30°C. Growth rate decreases at temperatures below 15°C and growth does not occur above 40°C. Folic acid, riboflavin, pantothenate and niacin are required for growth. *Carnobacterium maltaromaticum* is similar to *Lactobacillus acidophilus* in its growth requirements but growth occurs on many standard laboratory media including BA, TSA, brain–heart infusion agar and Rogosa acetate agar and thioglycolate broth. Optimum pH is 6.0–7.0. H_2S is not reduced on a TSI slant.

Carnobacterium maltaromaticum-like organisms isolated from white lakefish produce 1–2 mm, whitish, convex, entire, semi-translucent colonies on trypticase soy agar after 24–48 h at 22°C. Fermentation of carbohydrates was performed using a phenol red base containing a final concentration of 1% of the desired carbohydrate and other biochemical reactions were tested using traditional methods. Fermentation was negative for adonitol, inositol, raffinose and rhamnose. Fermentation was positive for cellobiose, galactose, lactose, and variable for glycerol, inulin and melibiose. Growth does not occur on MacConkey agar or Rogosa acetate agar.

Discrepancies between results for some carbohydrates fermented by *C. maltaromaticum* have been reported in the literature. In particular, a weak reaction for a positive fermentation of a carbohydrate tested using phenol red broth base has been reported as negative when purple broth basal medium is used in the test. Sorbitol and lactose may be positive when phenol red broth base is used, but negative when purple broth basal medium is used (Toranzo *et al.*, 1993).

These and other reactions are reported in Tables 3.15 and 3.38.

MOLECULAR IDENTIFICATION. *Carnobacterium maltaromaticum*-like organisms isolated from white lakefish were compared to *C. maltaromaticum* by partial sequencing of the *pisA* precursor gene for the piscicolin 126 protein (Loch *et al.*, 2008). Forward primers were PisA forward (5′-GTC ACA GCA TTG ATG CGT ATC-3′) and PisA reverse (5′-GAT GTG ATA CAG TCA GCA TGT-3′). The master mix comprised 4 µl DNA template, 2.5 µl 10× PCR buffer containing 1.5 mM $MgCl_2$ (Invitrogen), 0.5 µl dNTP (Invitrogen), 1 µl of each primer (5 µmol/l), 17.5 µl water and 0.5 µl *Taq* polymerase. Thermocycling was carried out with an initial cycle of 94°C for 10 min, followed by 35 cycles of 94°C for 1 min, 55°C for 1 min, and 72°C for 1 min with a final step of 72°C for 10 min. The 300-bp product was cloned and sequenced. The sequence shows a four nucleotide difference between the two species.

1.35.3 *Granulicatella and Abiotrophia*

Granulicatella balaenopterae

TAXONOMY AND SIGNIFICANCE. *Granulicatella balaenopterae* belongs to a group of organisms that were known as nutritionally variant streptococci (NVS), and characterized by satellite growth around colonies of other bacteria. NVS were separated from the genus *Streptococcus* based on satellite growth, nutritional requirements for growth, pyrrolidonyl arylamidase production and 16S rRNA sequencing. The NVS were divided into the genus of *Abiotrophia* containing the species *A. defectiva*, and the genus *Granulicatella* comprising *G. adiacens*, *G. elegans* and *G. balaenopterae* (Collins and Lawson, 2000). *Abiotrophia defectiva*, *G. adiacens* and *G. elegans* are reported from medical samples and are part of the normal flora of the human upper respiratory tract, however they may cause sepsis, bacteraemia and infective endocarditis. *Granulicatella* (*Abiotrophia*) *balaenopterae* (M1975/96/1^T=CCUG 37380^T) was reported from a minke whale (*Balaenoptera acutorostrata*) found beached on the north-west coast of Scotland (Lawson *et al.*, 1999b). The isolate was found in heavy mixed growth from the lung and the spleen and in pure growth from the liver and kidney. The pathogenicity of the organism remains unknown. The organism was shown to be a new species from *G. adiacens* and *G. elegans* by a 56 and 72 bp difference, respectively, over 1370 bp of the 16S rRNA gene, resulting in a >3% divergence.

GROWTH CHARACTERISTICS AND PHENOTYPIC IDENTIFICATION. *Granulicatella balaenopterae* is a non-motile, Gram-positive coccus occurring as single cells, pairs or short chains. It is catalase negative and facultatively anaerobic. Growth occurs at 37°C in an atmosphere of 5% CO_2 on Columbia base agar containing 5% horse blood. Colonies are non-haemolytic, 0.2 mm at 24 h, do not require pyridoxal hydrochloride or *Staphylococcus* feeder streak to support growth and do not show satellitism around other bacterial colonies.

Other species of *Abiotrophia* and *Granulicatella* exhibit satellite growth on blood agar. Colonies are 0.2 mm, greyish-white at 24 h, and 1–3 mm at 48 h on blood agar at 37°C. Growth does not occur at 10°C or at 45°C. They may fail to grow on subculture unless cultured on to blood agar containing a Staph streak or media containing pyridoxal to support growth. Commercial blood-culture media (for the culture of blood) usually contains pyridoxal and will support the growth of these organisms, however this type of media may not be routine in a veterinary microbiology laboratory. The organisms can grow on Choc agar and brucella media as these contain pyridoxal or a disc impregnated with pyridoxal can be placed on to a blood agar plate to show satellitism. Growth can also be assisted by flooding a blood agar plate with pyridoxal HCl (10 mg/l) or subculture on to a Choc agar plate and incubation in an anaerobic atmosphere, or five drops of pyridoxal HCl added to a 7 ml Todd Hewitt broth, or five drops in 5 ml of thioglycolate broth. *Granulicatella* may be pleomorphic and occur as Gram-positive cocci mostly in pairs, but cellular shapes can be bizarre including bacilli and elongated, swollen, and are easily over-decolourized in the Gram stain. Growth in a Todd Hewitt broth containing pyridoxal HCl with show typical streptococcal forms in a Gram stain.

It is important to use the recommended concentration of inoculum for all tests, but especially for LAP (production of leucine aminopeptidase) and PYR (pyrolidonyl arylamidase), which often produce a weak reaction for these species, and therefore subculturing a number of plates at the same time will assist in obtaining sufficient growth for inoculum preparation. Even with the correct inoculum weak positive reactions may be seen as a pink colour rather than the recommended red colour (when using Rosco diagnostic discs). Many other tests can give atypical reactions. The key tests for *Abiotrophia* and *Granulicatella* are LAP, PYR, arginine, bile aesculin, growth in 6.5% NaCl, pyridoxal HCl dependence, satellitism around a *Staphylococcus* streak, beta-glucuronidase (βGur), VP, catalase and susceptibility to vancomycin (Table 1.13).

All strains of *Abiotrophia* and *Granulicatella* are positive for pyrrolidonylarylamidase and leucine aminopeptidase, and sensitive to vancomycin. They do not grow in 6.5% NaCl or at 10°C or 45°C.

Granulicatella adiacens shows satellite growth and alpha-haemolysis on blood agar. Growth occurs in Todd Hewitt broth supplemented with 0.01% L-cysteine hydrochloride or when supplemented with 0.001% pyridoxal HCl. A red

Table 1.13. Key tests for identification of *Abiotrophia* and *Granulicatella* species.

	G. balaenopterae	*G. adiacens*	*G. elegans*	*A. defectiva*
Satellite growth	–	+	+	+
βGur	–	+	–	–
PYR	–	w	w	w
LAP	+	w	w	w
NaCl 6.5%	–	–	–	–
aesculin	+	–	–	–
ADH	+	–	+	–
Sucrose	–	+	+	+
Hip	–	–	+	–
Pul	+	–	–	+
Tag	–	+	–	v
Tre	+	–	–	+

ADH, arginine dihydrolase; βGur, beta-glucuronidase; Hip, hippurate; LAP, leucine aminopeptidase; Pul, pullulan fermentation; PYR, pyrolidonyl arylamidase; Tag, tagatose fermentation; –, negative; +, positive; W, weak.

chromophore is produced when the bacterium is boiled for 5 min at pH 2. *Granulicatella elegans* colonies are alpha-haemolytic, 0.2 mm at 48 h on sheep blood agar with minimal growth on Choc agar. Growth occurs at 27–37°C but not at 20°C or 42°C. Growth occurs in Todd Hewitt broth supplemented with 0.01% L-cysteine hydrochloride but not when supplemented with 0.001% pyridoxal HCl. A red chromophore is produced after boiling (Collins and Lawson, 2000).

The *Abiotrophia* and *Granulicatella* biochemical results are reported in Tables 3.14, 3.34–35, 3.36 and 3.38. Rosco Diagnostic Tablets can be used to test for PYR, LAP, α-galactosidase, β-galactosidase, β-glucuronidase, alkaline phosphatase and α-fucosidase, hippurate.

1.36 *Enterococcaceae*

1.36.1 *Enterococcus*

Enterococcus faecalis and *Enterococcus faecium*

Enterococcus faecalis was previously known as *Streptococcus liquefaciens* and *Streptococcus faecalis*.

Enterococcus faecalis and *E. faecium* are positive for Lancefield Group D antigen, grow at 45°C and in the presence of 6.5% NaCl. Reliable tests for differentiation between the two species are acid production from L-arabinose (*E. faecalis* is negative and *E. faecium* is positive) and pyruvate utilization (*E. faecalis* is positive and *E. faecium* is negative).

Biochemical results are reported in Tables 3.14, 3.34–35 and 3.36.

Enterococcus seriolicida

Enterococcus seriolicida (YT-3=ATCC 49156[T]) was isolated from diseased yellowtail (*Seriola quinqueradiata*) and eels (*Anguilla japonica*) and thought to be a new species of *Enterococcus* (Kusuda *et al.*, 1991), but later was shown to be a junior synonym of *Lactococcus garvieae* (Eldar *et al.*, 1994; Teixeira *et al.*, 1996).

Biochemical results are reported in Tables 3.14 and 3.34–35.

1.36.2 *Vagococcus*

The genus *Vagococcus* was proposed to accommodate Gram-positive cocci that are motile and react with Lancefield Group N antisera (Collins *et al.*, 1989). However, some species are non-motile and *Vagococcus salmoninarum* is negative for Lancefield Group N (Schmidtke and Carson, 1994).

Vagococcus fessus

TAXONOMY AND SIGNIFICANCE. *Vagococcus fessus* (M2661/98/1[T]=CUCG 41755[T]) was isolated in pure growth from the liver and kidney from a dead seal, and strain M520/99/2 (CCUG 42419) was isolated in heavy growth from brain, kidney, liver, lung, spleen, placenta, peritoneum and small intestine from a harbour porpoise (Hoyles *et al.*, 2000). The pathogenicity was not determined but the only other isolations from the harbour porpoise were scanty growth of *Salmonella* species from the lung and *Proteus* species from the brain. The two isolates are identical based on sequence information of 1470 nucleotides of the 16S rRNA gene and have 95.2–95.6% sequence similarity to other *Vagococcus* species.

GROWTH CHARACTERISTICS AND PHENOTYPIC IDENTIFICATION. *Vagococcus fessus* is a facultatively anaerobic, Gram-positive coccus occurring singly, in pairs or in chains with cells elongated in the direction of the chain. It is catalase-negative. Colonies are alpha-haemolytic when cultured on Columbia blood agar and incubated at 37°C in air plus 5% CO_2. Glucose is fermented but gas is not produced. Cyclodextrin is variable. Fermentation is negative for L-arabinose, lactose, mannitol, sucrose, trehalose and D-xylose. Results are negative for urea, Voges–Proskaüer and nitrate. Hippurate is not hydrolysed (Hoyles *et al.*, 2000). These and other reactions are presented in Tables 3.14, 3.36 and 3.38.

Vagococcus fluvialis

TAXONOMY AND SIGNIFICANCE. *Vagococcus fluvialis* (NCDO 2497^T) was isolated from river water and chicken faeces and the name was proposed to accommodate Lancefield Group N strains that are motile (Collins *et al.*, 1989). Strains of *Vagococcus fluvialis* have been isolated from well water, humans and pigs (Teixeira *et al.*, 1997) and are part of the normal flora in the intestine of sole (*Solea solea*), and one strain showed probiotic potential in cultured European seabream (*Dicentrarchus labrax*) (Sorroza *et al.*, 2012).

GROWTH CHARACTERISTICS AND PHENOTYPIC IDENTIFICATION. *Vagococcus fluvialis* is a facultatively anaerobic, Gram-positive coccus that is ovoid and occurs singly, in pairs or in chains. Motility is variable. Catalase is negative. Growth occurs at 10–37°C but is variable for growth at 45°C and in 6.5% NaCl. The type strain is negative for VP but some strains are positive. All strains are negative for fermentation of raffinose and positive for fermentation of mannitol and sorbitol, but variable for fermentation of sucrose (Collins *et al.*, 1989; Teixeira *et al.*, 1997). These and other reactions are presented in Tables 3.14, 3.30, 3.34–35, 3.36, and 3.38.

Vagococcus lutrae

TAXONOMY AND SIGNIFICANCE. *Vagococcus lutrae* (M1134/97/1^T=CCUG 39187^T) was isolated from the blood, liver, lungs and spleen from an otter (*Lutra lutra*) in the UK (Lawson *et al.*, 1999a). The pathogenicity is unknown. Based on 16S rRNA sequence analysis the most closely related strains are *V. fluvialis* and *V. salmoninarum*.

GROWTH CHARACTERISTICS AND PHENOTYPIC IDENTIFICATION. *Vagococcus lutrae* is a facultatively anaerobic, motile, Gram-positive coccus occurring singly, in pairs or in chains with elongation of the cells occurring in the direction of the chain. On blood agar colonies are smooth, 0.1–0.2 mm in diameter when incubated at 37°C in air containing 5% CO_2 (Lawson *et al.*, 1999a). Catalase is negative. Acid is produced from glucose. Hydrolysis of aesculin is positive and VP is positive, whereas reduction of nitrate is negative. Gelatin and hippurate are not hydrolysed. Other reactions are presented in Tables 3.14, 3.36 and 3.38.

Vagococcus penaei

TAXONOMY AND SIGNIFICANCE. Five strains of *Vagococcus penaei* (CD276^T=LMG 24833^T=CIP 109914^T) were isolated from the microbiota of spoiled cooked shrimp (*Penaeus vannamei*) harvested from the Pacific Ocean (Jaffrès *et al.*, 2010). The five strains are 100% homologous based on 16S rRNA sequence but have some biochemical differences.

GROWTH CHARACTERISTICS AND PHENOTYPIC IDENTIFICATION. *Vagococcus penaei* is a facultatively anaerobic, non-motile, Gram-positive coccus (0.5–1 μm) occurring singly, in pairs or in chains with elongation of the cells occurring in the direction of the chain. On brain–heart infusion agar colonies are white, smooth, 0.5–1.0 mm in diameter when incubated at 30°C in air containing 5% CO_2. Negative for oxidase, catalase and beta-haemolysis. Growth occurs at 10°C but not at 45°C or in 6.5% NaCl (Jaffrès *et al.*,

2010). Other reactions are presented in Tables 3.14, 3.36 and 3.38.

Vagococcus salmoninarum

TAXONOMY AND SIGNIFICANCE. *Vagococcus salmoninarum* (OS1-68^T=NFCB 2777^T) was isolated from an adult rainbow trout (*Oncorhynchus mykiss*) in the USA in 1968 (Wallbanks *et al.*, 1990). Vagococcosis is an emerging disease in the trout industry in Europe and has affected rainbow trout in Spain (Ruiz-Zarzuela *et al.*, 2005), juvenile and adult rainbow trout in France (Michel *et al.*, 1997) and rainbow trout broodstock and those over 50 g in Italy (Ghittino *et al.*, 2004; Salogni *et al.*, 2007). It has been isolated from rainbow trout, brown trout (*Salmo trutta*) and Atlantic salmon (*Salmo salar*) in Australia and Norway (Schmidtke and Carson, 1994). Disease outbreaks are associated with water temperatures of 10–12°C and post-spawning stress (Ghittino *et al.*, 2004; Salogni *et al.*, 2007). Clinical signs include chronic mortality, languid swimming and anorexia, haemorrhage in gills and marked bilateral exophthalmos which may lead to disruption of the eyeball. Necropsy findings include peritonitis, enlarged and congested liver and spleen, brain congestion, haemorrhagic ascites, retained eggs, engorged testes and inflamed swim bladder (Schmidtke and Carson, 1994; Michel *et al.*, 1997; Ghittino *et al.*, 2004; Salogni *et al.*, 2007). Samples for culture are best taken from the eye and brain and can result in 100% positive cultures compared to 10% when sampled from the kidney (Ghittino *et al.*, 2004).

In experimental challenge when inoculated intraperitoneally with 10^9 CFU/fish, the LD_{50} occurs at day 40 post-challenge (Ghittino *et al.*, 2004).

Vagococcus salmoninarum has 96% 16S rRNA sequence similarity with *V. fluvialis*, and is 92–94% similar to *Carnobacterium* species and *Enterococcus* species (Wallbanks *et al.*, 1990).

GROWTH CHARACTERISTICS AND PHENOTYPIC IDENTIFICATION. *Vagococcus salmoninarum* is a facultatively anaerobic, non-motile, Gram-positive coccus occurring singly, in pairs or in chains. Cells are ovoid or short rods depending upon the culture conditions. Catalase is negative. On blood agar colonies are 0.1–1 mm, alpha-haemolytic at 48–72 h when incubated at 20–30°C in air or air with 5% CO_2. Growth occurs at 5 and 30°C but not at 40°C (Wallbanks *et al.*, 1990). Growth at 37°C and hydrolysis of hippurate is variable between strains (Michel *et al.*, 1997; Ruiz-Zarzuela *et al.*, 2005). Growth occurs at pH 9.6 but not in 6.5% NaCl. Growth on MacConkey agar and TCBS is variable. Aesculin is hydrolysed and H_2S is produced. Nitrate, ADH and urea are negative. The genus *Vagococcus* is positive for Lancefield Group N, however *V. salmoninarum* is negative for the N antigen (Schmidtke and Carson, 1994). These and other reactions are presented in Tables 3.14, 3.30, 3.34–35, 3.36 and 3.38.

MOLECULAR IDENTIFICATION. Primers are designed to detect the 16S rRNA gene of *V. salmoninarum* and produce a product of 300 bp. The forward primer is pSal-1 (5′-GTT TTA GCC GCA TGG CTG AGA TAT-3′) and reverse primer pSal-2 (5′-AGG TGG GAA CAG TTA CTC TCC CA-3′) (Ruiz-Zarzuela *et al.*, 2005). Concentrations for the master mix were not detailed. Thermocycling consists of 94°C for 3 min, followed by 35 cycles of 94°C for 60 s, 55°C for 60 s, 72°C for 60 s and a final cycle with an extension of 72°C for 10 min. Sensitivity and specificity of the PCR were not provided.

1.36.3 *Catellicoccus*

The genus *Catellicoccus* is a 'sister' group to the family *Enterococcaceae* and has not yet been assigned to a family.

Catellicoccus marimammalium

TAXONOMY AND SIGNIFICANCE. *Catellicoccus marimammalium* (M35/04/03^T=CCUG 49459^T=CIP 108575^T) was isolated from kidney, small intestine, mesentery, peritoneal fluid and pericardial fluid from a porpoise (*Phocoena phocoena*) found on the coast of Scotland that had died of severe enteritis and peritonitis (Lawson *et al.*, 2006). A second strain (M271/94/1) was isolated from the small intestine of a grey seal (*Halichoerius grypus*) that had died from a gunshot. Phylogenetic analysis suggests the species is related to enterococci, vagococci, tetragenococci and *Melissococcus plutonius*.

GROWTH CHARACTERISTICS AND PHENOTYPIC IDENTIFICATION. *Catellicoccus marimammalium* is a facultatively anaerobic, non-spore forming, Gram-positive coccus occurring in pairs or chains. Catalase is negative. It does not grow in conventional liquid media (brain–heart infusion broth, Oxoid; Todd-Hewitt broth, Oxoid; nutrient broth, Oxoid; fastidious anaerobe broth, LabM) with or without the addition of serum. On BA colonies are 0.25–0.55 mm, non-haemolytic, semi-translucent, round and entire after 24 h at 37°C. Growth is enhanced by incubation in air supplemented with CO_2. Growth does not occur at 10°C. Growth occurs in 10% bile. Nitrate is not reduced. Aesculin and hippurate are not hydrolysed. There is a positive reaction with Lancefield Group D antisera (Lawson *et al.*, 2006). These and other reactions are presented in Tables 3.14, 3.36 and 3.34–35.

1.37 *Lactobacillaceae*

1.37.1 *Lactobacillus*

Lactobacilli are generally non-pathogenic and are found as part of the normal intestinal and mucus membrane flora of animals and humans, and some species are used in fermentation processes in the food industry. The whale isolates were detected in liver and lung in animals that showed no abnormal clinical signs.

Lactobacillus ceti

TAXONOMY AND SIGNIFICANCE. *Lactobacillus ceti* (142-2^T=CECT 7185^T=CCUG 53626^T) is a newly described species isolated from stranded beached, beaked whales (*Ziphius cavirostris*) (Vela *et al.*, 2008). Sequencing of the 16S rRNA gene identified the new species with closely related species being *L. satsumensis* (94.2%), *L. salivarius* (94.0%), *L. nagelii* (94.0%) and *L. saerimneri* (93.8%).

GROWTH CHARACTERISTICS AND PHENOTYPIC IDENTIFICATION. *Lactobacillus ceti* is a facultatively anaerobic, Gram-positive rod (0.5×2.7 µm in length) and is non-motile, catalase negative, non-spore forming and occurs singly, in pairs or short chains. After incubation at 37°C for 48 h, colonies on Columbia blood agar are non-haemolytic, 1 mm in diameter, circular, smooth and entire. Growth occurs at 22 to 37°C after 48 h and at 15°C after 5 days and in 3% NaCl. Growth does not occur at 4 or 42°C or on MRS agar or at 6.5% NaCl. L-Lactate is produced as an end product from hexoses and pentoses. These and other reactions are presented in Tables 3.15, 3.30–33 and 3.38.

1.38 *Leuconostocaceae*

1.38.1 *Weissella*

Weissella are Gram-positive cocci or short rods that belong to the group of bacteria known as lactic acid bacteria. They occur in a variety of habitats including soil, vegetables, fermented food, and some fish, animal and human sources.

Weissella ceti

TAXONOMY AND SIGNIFICANCE. *Weissella ceti* (1119-1A-09^T=CECT 7719^T=CCUG 59653^T) was isolated from the spleen of a beaked whale (*Mesoplodon bidens*) stranded off the Spanish coast (Vela *et al.*, 2011). Other strains were isolated from spleen, muscle, brain, kidney and lymph nodes of another three animals. All nine strains have 100% 16S rRNA sequence similarity to each other and 96.3% and 96.0% sequence similarity to *W. viridescens* and *W. minor*, respectively. The strains have 99.2–99.5% sequence similarity to an unidentified *Weissella* species reported by Liu *et al.* (2009) from diseased trout cultured in China (Vela *et al.*, 2011).

GROWTH CHARACTERISTICS AND PHENOTYPIC IDENTIFICATION. *Weissella ceti* is a non-motile, non-spore forming, facultatively anaerobic, Gram-positive coccoid rod to short rod (1.5×0.2 µm). On blood agar colonies are 0.75–1.0 mm, alpha-haemolytic, circular, non-pigmented after 24 h at 37°C. Growth occurs at 22–37°C but not at 15 or 42°C. Growth occurs at 3–6.5% NaCl, but not at pH 3.9. Glucose is fermented but gas is not produced. Nitrate is not reduced. VP reaction is positive. Hydrolysis is positive for aesculin but not gelatin or hippurate. Hydrolysis of ADH is variable for strains. These and other reactions are presented in Tables 1.11, 3.30–33 and 3.38.

Table 1.14. Differentiation of *Carnobacterium*, *Lactobacillus*, *Lactococcus*, *Vagococcus* and *Renibacterium*.

Test	*Lactobacillus sp.*	*Carnobacterium piscicola*	*Carnobacterium divergens*	*Lactococcus garvieae*	*Lactococcus piscium*	*Vagococcus fluvialis*	*Vagococcus lutrae*	*Vagococcus salmoninarum*	*Renibacterium salmoninarum*
Growth on acetate agar (RAA)	+	–	–	–	–	–	–	–	
ADH	V	+	+	+	–	–	–	–	
Acid from mannitol	V	+	–	+	+	+	–	–	–
H_2S	–	–	–	–	–	–	ND	+	
Lancefield Group D reaction		Positive		Negative		Weak, delayed reaction	ND	Weak, delayed reaction	
Lancefield Group N reaction				Positive[b]	Positive[b]	Positive	ND	Negative	
Streptococcal grouping kit Groups A–G (Oxoid)		D		Negative	Negative	D weak reaction	ND	D weak reaction	Negative

[a]Eldar *et al*. (1999) reported a positive result. Teixeira *et al*. (1996) reports as negative. [b]Schmidtke and Carson, (1994) reported a negative result.
V = Variable result
D = Streptococcus Group D

Table 1.15. Differential tests for some non-fastidious fish pathogenic Gram-positive cocci and rods. (Table kindly supplied by Dr Jeremy Carson, Department of Primary Industries, Water and Environment, Tasmania.)

Test	*L. garvieae**	*C. piscicola*	*V. salmoninarum*[5]	*L. piscium*	*S. iniae*
Gram	+	+	+	+	+
Cell shape	ec	sr	sr/cb	c (sr/cb)[233]	c
Haemolysis	α	–	α	–	α[816], β[816,797]
Catalase	–	–	–	–	–
VP (plate)	89%+	+	–	+	–
H_2S	–	–	+	–	–
Bile-aesculin	+	+	62%+	–	–
PYR	+	+	+	–	95%+[223]
ADH	+	+	–	–	71%+[223]
Aesculin	+	+	+	+	+
Glucose	+	+	+	+	+
Galactose	+	65%+	–	+	+
Lactose	–	35–60%+	–	+	–
Maltose	90%+	94%+	17%+	+	+
Mannitol	90%+	88%+	4%+	+	+
Raffinose	–	29%+	–	+	–
Salicin	98%+	+	87%+	+	+
Sucrose	7%+	+	52%+	+	+
Sorbitol	–	–	4%+	–	–
Trehalose	+	+	87%+	+	+
Glycerol	–	40–100%+	–	–	–
Inulin	–	0–100%+	–	–	–
L-Arabinose	–	–	–	– (+)[233]	–
Dulcitol	–	–	–	–	–
Fructose	+	+	+	+	+
Starch	–	0–100%+	13%+	–	+
Xylose	–	–	–	–	–
Adonitol	–	–	–	–	–
Melibiose	–	30%+	–	+ (–)[233]	–

All reactions determined by miniaturized microtitre tray tests (Schmidtke and Carson, 1994) or conventional tests. [#]Type strain is positive; [233]original description reaction in brackets; [816]α on bovine blood, β on sheep blood. **Lactococcus garvieae* (syn. *Enterococcus seriolicida*), *Carnobacterium piscicola*, *Vagococcus salmoninarum*, *Lactococcus piscium*, *Streptococcus iniae*.

Weissella hellenica

TAXONOMY AND SIGNIFICANCE. *Weissella hellenica* strain DS-12 was isolated from the intestine of flounder (*Paralichthys olivaceous*) cultured in South Korea. It has antibacterial activity against *Edwardsiella tarda*, *Photobacterium damsela* ssp. *piscicida*, *Aeromonas hydrophila* and *Vibrio anguillarum* and has probiotic potential whereas the type strain (JCM 10103T) has no antibacterial activity against these species (Cai *et al.*, 1998).

GROWTH CHARACTERISTICS AND PHENOTYPIC IDENTIFICATION. *Weissella hellenica* strain DS-12 is a non-motile, non-spore forming, Gram-positive coccoid rod (1.4–1.7×0.8–1.0 μm). Growth occurs at 15–40°C with weak growth at 45°C. Growth occurs in MRS broth containing 3–10% NaCl, or 10% bile, and growth occurs for 90 min at 30°C in MRS broth at a pH of 3.5. Glucose is fermented and gas is produced. The type strain (JCM 10103^T) is a coccus (0.9–1×0.9–1.0 μm) and growth occurs at 20–30°C and 3–5% NaCl but not at 40°C, or in 10% bile, or at pH 3.5. Catalase is negative for both strains. Growth occurs on Man Rogosa and Sharpe (MRS; Sigma-Aldrich) agar. Biochemical reactions are presented in Tables 3.14 and 3.30.

Weissella species

TAXONOMY AND SIGNIFICANCE. Un-named *Weissella* species have been isolated from diseased rainbow trout in China (isolates JZ-1L, RT-2L) and from haemorrhagic septicaemia in rainbow trout (*Oncorhynchus mykiss*) (isolates WS-08, WS-71 and others) cultured in Brazil (Liu *et al.*, 2009; Figueiredo *et al.*, 2012). A 1500-bp product from the 16S rRNA gene showed 98% sequence similarity to the *Weissella* species isolated from rainbow trout in China and in a phylogenetic analysis using a neighbour-joining tree the boostrap percentages were 100% for all strains (Figueiredo *et al.*, 2012). This analysis did not include *W. ceti*, which has a 99.2–99.5% sequence similarity to the *Weissella* species strains from China (Vela *et al.*, 2011). Further work needs to be done to determine if the strains from rainbow trout are *W. ceti*.

Pathogenicity for rainbow trout was confirmed in experimental challenge using intraperitoneal injection (3.4×10^4 CFU), immersion (3×10^8 CFU) and co-habitation (1.6×10^8 CFU). The *Weissella* species was pathogenic for tilapia (*Oreochromis niloticus*) at 1.2×10^8 CFU injected intraperitoneally (Figueiredo *et al.*, 2012). Infected fish displayed anorexia and lethargy, ascites, exophthalmia, rectal prolapse and haemorrhagic areas on the lateral fin, pectoral fin, anal region and eye.

GROWTH CHARACTERISTICS AND PHENOTYPIC IDENTIFICATION. *Weissella* species are Gram-positive cocci, and negative for catalase and oxidase. Alpha-haemolysis is variable. Growth occurs on MRS and BA but not on MacConkey agar. Growth occurs at 20–45°C.

Biochemical results are presented in Tables 3.14 and 3.36.

MOLECULAR IDENTIFICATION. Genus specific primers are available for *Weissella* species.

1.39 *Streptococcaceae*

1.39.1 *Lactococcus*

Lactococcus is a genus of the family *Streptococcaceae*, which is part of the order *Lactobacillales*. The *Lactococcus* genus was created to accommodate 'lactic' or group N Streptococci. There are currently seven species, including *L. garvieae*, *L. lactis*, *L. piscium*, *L. plantarum* and *L. raffinolactis*. They have been isolated from milk, activated sludge foam, vegetable matter and fish (http://www.bacterio.cict.fr/l/lactococcus.html). Lactococci are mesophilic and grow at a temperature range of 10–40°C but some may grow at lower temperatures. There are varying reports in the literature for Group N results for *L. garvieae* and *L. piscium*.

Lactococcus garvieae

TAXONOMY AND SIGNIFICANCE. *Lactococcus garvieae* (NCDO 2155^T=ATCC 43921^T) was originally identified as *Streptococcus garvieae* and isolated from bovine mastitis (Collins *et al.*, 1983). A bacterium causing infection in yellowtail (*Seriola quinqueradiata*) in Japan was identified as a new species, *Enterococcus seriolicida* YT-3=ATCC 49156 (Kusuda *et al.*, 1991), but later shown to be a junior synonym of *L. garvieae* (Teixeira *et al.*, 1996; Eldar *et al.*, 1996). *Lactococcus garvieae* has been responsible for disease in a number of aquatic species including farmed rainbow trout (*Oncorhynchus mykiss*) in Australia, South Africa and the UK (Carson *et al.*, 1993; Algöet *et al.*, 2009), Adriatic sturgeon (*Acipenser naccarii*) farmed in Italy (Salati *et al.*, 1996), wild Red Sea wrasse or clown wrasse (*Coris aygula*) (Colorni *et al.*, 2003), giant freshwater prawn (*Macrobrachium rosenbergii*) and mullet (*Mugil cephalus*) in Taiwan (Chen *et al.*, 2001, 2002), black rockfish (*Sebastes schlegelii*) and olive flounder (*Paralichthys olivaceus*) in Korea (Kang *et al.*, 2004; Shin *et al.*, 2009), striped jack (*Pseudocaranx dentex*) in Japan (Madinabeitia *et al.*, 2009) and yellowtail (*Seriola* sp.) in Japan (Kimura and Kusuda, 1982; Kusuda *et al.*, 1991). In experimental challenge, grayling (*Thymallus thymallus*) were highly susceptible whereas Atlantic salmon (*Salmo salar*), brown trout (*S. trutta*) and cyprinid species (carp [*Cyprinus carpio*], tench [*Tinca tinca*], rudd [*Scardinius erythrophthalmus*], barbel [*Barbus barbus*], chub [*Leuciscus cephalus*], dace [*L. leuciscus*] and roach [*Rutilus rutilus*]) were less susceptible (Algöet *et al.*, 2009). It has also been detected in marine mammals and was isolated from the kidney of a freshly dead wild bottlenose dolphin (*Tursiops truncatus*) found in Kuwait Bay (Evans *et al.*, 2006).

Lactococcus garvieae produces an exotoxin that when injected intramuscularly prior to percutaneous challenge with 10^6–10^7 cells leads to death of yellowtail within 4 days (Kusuda *et al.*, 1991). Injection with cells alone did not lead to infection of yellowtail, however intramuscular injection of grey mullet resulted in an LD_{50} of 1.2×10^6 CFU/fish (Chen *et al.*, 2002). Clinical signs include congestion in the internal organs, haemorrhage, exophthalmos, corneal opacity, redness on tail and ventral fins and petechiae inside the operculum (Kusuda *et al.*, 1991; Chen *et al.*, 2002).

In the giant freshwater prawn (*Macrobrachium rosenbergii*) infected prawns have whitish muscles with marked oedema. Necrotic lesions occur in the muscles and hepatopancreas (Chen *et al.*, 2001). The LD_{50} is 6.6×10^5 CFU/prawn.

GROWTH CHARACTERISTICS AND PHENOTYPIC IDENTIFICATION. *Lactococcus garvieae* is a facultatively anaerobic, non-motile, non-spore forming, Gram-positive ovoid coccus (1.4×0.7 µm), which forms short chains. Cells are elongated in the direction of the chain. Catalase is negative. Growth occurs on BA, nutrient agar (NA), brain–heart infusion agar, 40% bile agar and in 0.1% methylene blue-milk, but it does not grow on Salmonella-Shigella agar (Kusuda *et al*, 1991). Growth on MacConkey agar and TCBS may be negative or weak. On BA or NA colonies are circular, smooth and entire, non-pigmented (Collins *et al.*, 1983; Salati *et al.*, 1996). There is alpha-haemolysis on BA. Growth occurs at 10–40°C and weak or delayed growth may be seen at 45°C. Growth occurs at 0–6% NaCl and weak growth may be seen at 6.5% NaCl. Results are negative for Lancefield groups A, B, C, D, E, F, G, H, K, L, M, N and O (Kusuda *et al*, 1991). It does not reduce potassium tellurite or tetrazolium. Different results have been noted for ribose, depending on the identification system used. Ribose is positive by the API 50CH system, but negative with the API Rapid ID 32 Strep system (Vela *et al.*, 2000). *Lactococcus garvieae* may be biochemically indistinguishable from *Streptococcus parauberis* (Doménech *et al.*, 1996). *Lactococcus garvieae* (*Enterococcus seriolicida*) is phenotypically similar to *L. lactis*, however they can be differentiated by sensitivity testing to clindamycin. *L. garvieae* is resistant to clindamycin, whereas *L. lactis* is sensitive (Elliott and Facklam, 1996). *Lactococcus lactis* is now known as *L. delbrueckii* ssp. *lactis*.

Lactococcus garvieae was divided into biotypes (Vela *et al.*, 2000), however these divisions have been disputed. Further research indicates that *L. garvieae* strains are relatively homogeneous regardless of geographical location or aquatic host. What is important in biochemical identification is that the density of the inoculum be standardized when using the API Rapid ID 32 Strep system and that the bacterial cells to be used for the inoculum are taken from blood agar plates. Reliable, repetitive results were only obtained when the inoculum density was adjusted using a spectrophotometer to an optical density of 0.8 at a wavelength of 580 nm. Also, cells grown on blood agar produced reliable results for biochemical tests whereas growth taken from tryptone soya agar (Oxoid) produced variability in biochemical tests. However, slight variability was still observed for biochemical tests such as β-galactosidase, hippurate, β-mannosidase, acid from melezitose, N-acetyl-β-glucosaminidase and acid from pululan (Ravelo *et al.*, 2001).

MOLECULAR IDENTIFICATION. PCRs for the detection of *L. garvieae* have been designed based on amplification of the 16S rRNA gene, the 16S-23S rRNA intergenic spacer region and a multiplex PCR that differentiates seven species of *Lactococcus* including *L. garvieae* and *L. piscium* (Zlotkin *et al.*, 1998a; Jung *et al.*, 2010; Odamaki *et al.*, 2011; Dang *et al.*, 2012).

A conventional PCR with primers designed to detect the 16S rRNA and 23S rRNA spacer region was tested for specificity against a comprehensive range of closely related species and genera and compared to previously published primers (Dang *et al.*, 2012). The master mix consists of 1 mM of each primer ITSLg30F (5′-ACT TTA TTC AGT TTT GAG GGG TCT-3′) and primer ITSLg 319R (5′-TTT AAA AGA ATT CGC AGC TTT ACA-3′), 2.5 µl reaction buffer, 0.2 mM dNTPs, 1.5 mM $MgCl_2$, 2.5 U *Taq* polymerase (Roche Diagnostics) and 50 ng DNA in a 25-µl reaction volume. Thermocycling is carried out using an initial denaturation step at 94°C for 5 min followed by 30 cycles of denaturation at 94°C for 30 s, annealing at 58°C for 30 s, extension at 72°C for 40 s and a final step using an extension at 72°C for 7 min. Amplified product is 290 bp. The limit of detection is 2.63 pg DNA.

The primers are specific when tested against the genera and species listed for the CAU12F and CAU15R primers plus *Streptococcus anginosus* ATCC 33397^T, *S. australis* KCOM 1439, *S. infantis* KCOM 1375, *S. pseudopneumoniae* and *S. sinensis* KCOM 1017. Dang *et al.* (2012) tested previously published primers and found that they were not specific. The primers CAU12F and CAU15R (Jung *et al.*, 2010) were not specific for one of six species of *L. lactis* ssp. *lactis* (KCTC 2013) tested; primers SA1B10-1-F and SA1B10-1-R (Aoki *et al.*, 2000) are not specific against *Enterococcus casseliflavis* KCTC 3638^T and *T. solitarius* KCTC 3923^T; primers pLG-1 and pLG-2 (Zlotkin *et al.*, 1998a) are not specific against *T. solitarius* KCTC 3923^T; primers LcG-F and Lc-R (Odamaki *et al.*, 2011) are not specific against *L. lactis* ssp. *lactis* KCTC 2013 and KCTC 3191, *Streptococcus anginosus* ATCC 33397^T and *S. pyogenes* KCTC 3984^T.

The master mix for a conventional PCR based on primers for detection of the 16S rRNA gene comprises 8 ng/μl primer pLG-1 (5′-CAT AAC AAT GAG AAT CGC-3′), 8 ng/μl primer pLG-2 (5′-GCA CCC TCG CGG GTT G-3′), 1 U Vent DNA polymerase (New England Biolabs), 1× buffer (New England Biolabs), a final concentration of 0.25 mmol/l for each dNTP, 0.1 mg/μl bovine serum albumin in a 50-μl volume. DNA is obtained by touching a pure colony and placing into the reaction tube. Thermocycling is carried out using an initial denaturation step at 94°C for 3 min followed by 35 cycles of denaturation at 94°C for 1 min, annealing at 55°C for 1 min, extension at 72°C for 1.5 min and a final step using an extension at 72°C for 10 min. The amplified product is 1100 bp. The primers detect *L. garvieae* strains from Australia, Asia and Europe, and are specific against *L. lactis* (NCFB 604^T), *L. piscium* (NCFB 2778^T), *Vagococcus salmoninarum* (NCFB 2777^T), *Streptococcus iniae* (ATCC 29178^T), *S. difficile* (=*S. agalactiae* CIP 103769^T) and *Aeromonas salmonicida* (3173/86). The sensitivity is 4 CFU per reaction tube (Zlotkin *et al.*, 1998a). Jung *et al.* (2010) suggest the pLG1 and pLG2 primers (Zlotkin *et al.*, 1998a) or the primers SA1B10-1-F (5′CAT TTT ACG ATG GCG CAG-3′) and SA1B10-1-R (5′-CGT CGT GTT GCT GCA ACA-3′) (Aoki *et al.*, 2000) are not specific against *Tetragenococcus solitarius.*

A real-time PCR based on the detection of the 16S rRNA gene comprises a master mix containing 10 pmol primer CAU12F (5′-ACT CGT GCT ATC CTT-3′), 10 pmol primer CAU15R (5′-TGG GTA CTC CCA ACT TCC-3′), 10 μl SYBR Green master mix (Finnzymes Inc., Woburn, Massachusetts, USA), 1 μl template DNA and 7 μl nuclease-free water in a 20-μl reaction volume. Thermocycling is carried out using an initial denaturation step at 95°C for 15 min followed by 60 cycles of denaturation at 94°C for 20 s, annealing at 58°C for 30 s and extension at 72°C for 45 s. A melt curve analysis is conducted from 95 to 65°C at a rate of 0.1°C/s with continuous acquisition of fluorescent data. The melting temperature of the amplicon is 84.69°C. A standard curve is generated from serial dilutions of purified *L. garvieae* cells (1 to 10^7 CFU/ml) against the log input cells/ml. Cell concentrations are determined by viable count. These data are used to obtain the limit of detection, which is 32 fg with a mean Ct value of 36.75 (Jung *et al.*, 2010).

The same primers were used in a conventional PCR, which consists of a 20-μl master mix containing 10 pmol of each primer, Tris–HCl (pH 9.0), 40 mmol/l KCl, 250 μmol/l dNTPs, 1 U *Taq* polymerase, 1.5 mmol/l $MgCl_2$ and 1 μl (10 ng) template DNA in a 20-μl reaction volume. Thermocycling is carried out using an initial denaturation step at 95°C for 5 min followed by 35 cycles of denaturation at 95°C for 1 min, annealing at 58°C for 1 min, extension at 72°C for 1 min and a final step using an extension at 72°C for 7 min. The amplified product is 415 bp. The sensitivity of the conventional PCR is 1 ng DNA. Both PCRs are specific against *Lactococcus piscium* KCTC 3639^T, *Lactococcus lactis* ssp. *lactis* KCTC 3769^T, *Lactococcus lactis* ssp. *hordniae* KCTC 3768^T, *Lactococcus lactis* ssp. *cremoris* DSM 20069^T, *Lactococcus raffinolactis* KCTC 3982^T, *Lactococcus plantarum* DSM 20686^T, *Lactococcus chungangensis* KCTC 13185^T, *Streptococcus oralis* DSM 20066, *Streptococcus gordonii* KCTC 3286^T, *Streptococcus mitis* KCTC 3556^T, *Streptococcus pyogenes* KCTC 3984^T, *Streptococcus intermedius* KCTC 3268^T, *Streptocococcus parasanguinis* KCTC 13046^T, *Streptococcus sanguinis* KCTC 3284^T, *Streptococcus pneumoniae* KCTC 5080^T, *Tetragenococcus solitarius* KCTC 3553^T, *Enterococcus hirae* KCTC 3616^T, *Enterococcus mundtii* KCTC 3630^T, *Enterococcus casseliflavus* KCTC 3638^T, *Enterococcus malodoratus* KCTC 3641^T, *Enterococcus cecorum* KCTC 3642^T, *Enterococcus saccharolyticus* KCTC 3643^T, *Enterococcus villorum* KCTC 13904^T, *Enterococcus haemoperoxidus* KCTC 13910[T], *Enterococcus moraviensis* KCTC 13911[T], *Enterococcus phoeniculicola* KCTC 3818[T], *Tetragenococcus solitarius* KCTC 3923[T] *Enterococcus raffinosus* KCTC 5189[T], *Enterococcus avium* KCTC 5190[T], *Enterococcus faecalis* KCTC 3206[T], *Vagococcus salmoninarum* LMG 11491[T], *Vagococcus lutrae* LMG 19537[T] and *Vagococcus fluvialis* LMG 9464[T] (Jung *et al.*, 2010). These primers were developed for analysis of food and seafood products.

Lactococcus lactis

TAXONOMY AND SIGNIFICANCE. *Lactococcus lactis* ssp. *lactis* is found in various environments and is used as a starter culture for fermented dairy products and in cheese making. It is part of the normal intestinal flora of pufferfish (*Takifugu niphobles*) and Amur catfish (*Silurus asotus*) in Japan (Itoi *et al.*, 2008). *Lactococcus lactis* ssp. *lactis* causes white muscle disease in giant freshwater prawn (*Macrobrachium rosenbergii*) farmed in Taiwan (Wang *et al.*, 2008b). Prawns had poor growth rates and a white colouration to the body. Muscles appeared white and swollen and the hepatopancreas was yellow. Fluid accumulated between the cuticle and the muscle, and the subcuticular area, epidermis and dermis was infiltrated with haemocytes. Tissue smears revealed numerous Gram-positive cocci. Necrotic foci were seen in the hepatopancreas. Identification was confirmed using 16S rRNA sequencing, and biochemical result from conventional tests, API 20 Strep and API Rapid ID 32 Strep (Biomerieux). In experimental challenge with an injection of 1.2×10^8 CFU, all prawns died within 3 days. The LD_{50} was 10^6–10^7 CFU/g body weight.

Lactococcus lactis ssp. *tructae* (L105[T]=LMG 24662[T]=DSM 21502[T]) is part of the normal flora of intestinal mucus of rainbow trout (*Oncorhynchus mykiss*) and brown trout (*Salmo trutta*) found in Spain (Pérez *et al.*, 2011). Sequencing of the 16S rRNA gene demonstrated similarity to *L. lactis* ssp. *lactis* and *L. lactis* ssp. *hordniae* of 99.4%, and a similarity of 99.9% to *L. lactis* ssp. *cremoris*. The housekeeping genes *rpoB* and *recA* show sequence similarity of 99.3% and 99.7%, respectively, to *L. lactis* ssp. *cremoris*.

GROWTH CHARACTERISTICS AND PHENOTYPIC IDENTIFICATION. *Lactococcus lactis* is a facultatively anaerobic, non-motile, Gram-positive coccus with ovoid or spherical cells. Cells are often elongated in the direction of the chain. Cells of *L. lactis* ssp. *lactis* from giant freshwater prawn are 0.1–0.3 µm. Colonies on TSA, BA and brain-heart infusion agar are 1 mm and non-haemolytic at 48 h and 25°C. Growth occurs at 6.0% NaCl but not at 6.5% or higher. Growth occurs at 4–40°C with 45% of strains able to grow at 45°C (Wang *et al.*, 2008b).

Lactococcus lactis ssp. *tructae* is a facultatively anaerobic, non-motile, Gram-positive coccus with ovoid or spherical cells. Cells are often elongated in the direction of the chain. Catalase is negative. On Man–Rogosa–Sharpe (MRS) agar (Difco), colonies are white-cream, opaque, round, convex at 22°C for 24 h (Pérez *et al.*, 2011). Growth occurs on BA and TSA. Growth occurs in 4% NaCl.

Reactions are presented in Tables 3.15, 3.30, 3.34–35 and 3.36.

MOLECULAR IDENTIFICATION. *Lactococcus lactis* ssp. *lactis* can be identified using specific primers or through amplification of the 16S-23S rRNA region (Blaiotta *et al.*, 2002; Wang *et al.*, 2008b). Primers LLF (5′-GCA ATT GCA TCA CTC AAA GA-3′) and LLR (5′-ACA GAG AAC TTA TAG CTC CC-3′) are used at a concentration of 150 ng of each primer, together with 1 mM of each dNTP, 1 U *Taq* DNA polymerase (Takara), 1× PCR buffer and 25 ng template in a 100-µl reaction volume. Thermocycling was carried out with an initial cycle of 94°C for 2 min, followed by 30 cycles of 92°C for 1 min, 50°C for 1.5 min, and 72°C for 2 min with a final step of 72°C for 5 min. A 650-bp product is indicative of *L. lactis* ssp. *lactis*. Amplification of the 16S-23S rRNA region is carried out using forward primer G1 (5′-GAA GTC GTA ACA AGG-3′) and reverse primer L1 (5′-CAA GGC ATC CAC CGT-3′). Thermocycling was carried out with an initial cycle of 94°C for 1 min, followed by 35 cycles of 94°C for 1 min, 55°C for 2 min, and 72°C for 3 min with a final step of 72°C for 10 min. *Lactococcus lactis* ssp. *lactis* produces a product of 380 bp, *L. garvieae* 430 bp and *L. raffinolactis* 450 bp. Other bacteria also produce one or more amplicons including *Enterococcus faecalis*, *E. casseliflavus* and *E. durans* (two bands at 300 and 400 bp), *Streptococcus salivarius* ssp. *thermophilus* and *S. salivarius* ssp. *salivarius* (one band at 360 bp) (Blaiotta *et al.*, 2002).

1.39.2 *Streptococcus*

The *Streptococcus* genus contains over 94 species and 17 subspecies, occurring predominantly from clinical and veterinary sources. They were originally known by four broad categories nominated by Sherman as pyogenic, viridans, lactic and enterococci (Facklam, 2002). The pyogenic division includes β-haemolytic strains of *Streptococcus*, which are classified into Lancefield groups according to the carbohydrate composition of the bacterial antigen on their cell wall (Lancefield, 1933; Table 1.16).

The Lancefield grouping system is useful but cannot be used for accurate identification of specific β-haemolytic Streptococci (Facklam, 2002). *Streptococcus agalactiae* is the only species with the Group B antigen, however cross-reactions in the commercial Strep grouping kits have been reported for species such as *S. porcinus*. Recently, a Group B *Streptococcus* was reported to be virulent for penaeid shrimp (*Litopenaeus vannamei*) but was identified as being *S. uberis* using the API 20 Strep and as *S. parauberis* using the Biolog system (Hasson *et al.*, 2009). Further identification needs to be done to determine the true identity of this organism.

The viridans group are non-haemolytic, or show a green colouration around the colony on blood agar (referred to as alpha-haemolysis), do not tolerate high pH, salt, or temperatures at or below 10°C. The lactic division is now known as the genus *Lactococcus* and the enterococci as the genus *Enterococcus*. Most strains are non-haemolytic (often referred to as gamma-haemolysis) but some are β-haemolytic. They are capable of growth at high pH and high salt concentrations, and grow at a temperature range of 10–45°C.

The green zone of haemolysis or α-haemolysis is a result of partial lysis of the red cells due to the production of hydrogen peroxide by the bacterial colony. In the absence of oxygen in the growth atmosphere then peroxide is not produced and greening of the agar does not occur (Facklam, 2002).

Streptococcus agalactiae

TAXONOMY AND SIGNIFICANCE. *Streptococcus agalactiae* is a Lancefield Group B *Streptococcus* comprising non-haemolytic and β-haemolytic strains. Nine serogroups of Group B *Streptococcus* are recognized based on their capsular polysaccharide.

Streptococcus agalactiae Group B causes disease in many animal species including neonatal meningitis in humans and mastitis in cows. In fish the disease has been referred to as Streptococcosis, although this term has been applied to a number of *Streptococcus* species causing disease. Clinical signs in fish include abnormal behaviour such as erratic swimming, whirling on the surface, and C-shaped curvature of the body while swimming at the surface. The eye may be opaque, exophthalmia seen and haemorrhages present. Haemorrhagic areas occur on the head and body, in particular around the mouth, snout, operculum and fins. Haemorrhagic enteritis may be present.

The disease affects many aquatic animals and has caused epizootics in hatchery reared freshwater golden shiner (*Notemigonus crysoleucas*), wild gulf menhaden (*Brevoortia patronus*),

Table 1.16. *Streptococcus* species and Lancefield group.

Lancefield group	*Streptococcus species*
Group A	*S. pyogenes* (bacitracin sensitive)
Group B	*S. agalactiae* (bacitracin resistant), (β-haemolytic and non-haemolytic strains)
Group C	*S. dysgalactiae* ssp. *dysgalactiae* (non-haemolytic), *S. dysgalactiae* ssp. *equisimilis*, *S. equi* ssp. *equi*, *S. equi* ssp. *zooepidemicus*
Group D	*Enterococcus* species, *S. bovis*
Group E	*S. mutans*
Group F	*S. anginosus*
Group G	*S. canis*
Group H	*S. sanguinis*
Group L	*S. dysgalactiae*
Group N	*Lactococcus lactis*
Group R, S	*S. suis*

hardhead sea catfish (*Arius felis*), striped mullet (*Mugil cephalus*), pinfish (*Lagodon rhomboides*), Atlantic croaker (*Micropogonias undulatus*), spot (*Leiostomus xanthurus*), stingray (*Dasyatis* sp.) and silver weakfish (*Cynoscion nothus*). Disease outbreaks caused by *S. agalactiae* have been reported from cultured bullminnows (*Fundulus grandis*), bluefish (*Pomatomus saltatrix*), striped bass (*Morone saxatilis*), grey weakfish (*C. regalis*) and weakfish (*Cynoscion regalis*, known as sea trout in some countries) in the USA, and exhibited high virulence in infection trials in Nile tilapia (*Oreochromis niloticus*) and *O. aureus* (Rasheed *et al*, 1985; Baya *et al.*, 1990a; Evans *et al.*, 2002). *Streptococcus agalactiae* strains from infected ya-fish (*Schizothorax prenanti*) cultured in China were also virulent in experimental challenge of Nile tilapia and crucian carp (*Carassius carassuis*) (Geng *et al.*, 2012). *Streptococcus agalactiae* has become a major problem in farmed Nile tilapia in Brazil causing meningoencephalitis and septicaemia. Virulence studies indicated an LD_{50} of 90 bacterial cells, and predisposing factors of high stocking density and water temperatures above 26°C (Mian *et al.*, 2009). Considerable phenotypic variation was seen with 11 different phenotypes from 29 isolates as tested using the API 20 Strep (see Tables 3.34 and 3.35). It has been reported from red hybrid tilapia (*O. niloticus*) and red tilapia (*Oreochromis* sp.) on commercial farms in Malaysia and Thailand (Suanyuk *et al.*, 2008; Musa *et al.*, 2009).

Experimental infection using intraperitoneal and intramuscular injection confirmed pathogenicity with 100% mortality within 4 to 5 days for golden shiners, bluegill (*Lepomis macrochirus*), green sunfish (*L. cyanellus*) and American toads (*Bufo americanus*) (Robinson and Meyer, 1966). Infection was not established in channel catfish (*Ictalurus punctatus*), largemouth bass (*Micropterus salmoides*), bigmouth buffalo (*Ictiobus cyprinellus*), goldfish (*Carassius auratus*) and black crappie (*Pomoxis nigromaculatus*), although other research indicated channel catfish susceptibility using intramuscular injections (Plumb *et al.*, 1974). Chickens, white mice and hamsters were not susceptible.

Cultured gilthead seabream (*Sparus aurata*) and wild mullet (*Liza klunzingeri*) in Kuwait Bay, Arabian Gulf, infected with *S. agalactiae* showed abnormal behaviour by erratic swimming and circling, with clinical signs of peri-orbital and intraocular haemorrhage, exophthalmia with haemorrhages in the integumental and musculoskeletal systems, cranial area and body surface particularly around the mouth, operculum and fins. The organism was cultured from the brain, eye, kidney, blood and intestine with the brain yielding the highest number of organisms. Virulence testing of these isolates performed on tilapia (*Oreochromis niloticus*) with an inoculum of 1×10^7 CFU/fish resulted in 90–100% mortality in 7 days. The LD_{50} was $1.9\times10^{3.3}$ CFU (Evans *et al.*, 2002). Silver pomfret (*Pampus argenteus*) undergoing aquaculture in Kuwait were also affected by a β-haemolytic *S. agalactiae* Group B. Fish were lethargic, had loss of appetite and swollen intestines containing thick fluid. The organism was cultured from ascites fluid, brain, kidney, spleen and liver (Duremdez *et al.*, 2004).

Streptococcosis due to a Group B *Streptococcus* species was reported from penaeid shrimp (*Litopenaeus vannamei*), however identification using the API 20 Strep kit identified the organism as *S. uberis* and the Biolog gave an identification of *S. parauberis*. The two species are almost impossible to differentiate biochemically and further tests need to be conducted to confirm the identify of the organism (Hasson *et al.*, 2009).

A streptococcal species reported from St Peter's fish (Nile tilapia) and rainbow trout (*Oncorhynchus mykiss*) initially thought to be a new species and named *S. difficile* (Eldar *et al.*, 1994), was later identified as non-haemolytic *S. agalactiae* Group B (Vandamme *et al.*, 1997).

Necrotizing fasciitis was reported from captive juvenile saltwater crocodiles (*Crocodylus porosus*) in the Northern Territory, Australia (Bishop *et al.*, 2007), and from a captured common bottlenose dolphin (*Tursiops truncatus*) in Italy (Zappulli *et al.*, 2005). The dolphin had severe necrohaemorrhagic fasciitis and cellulitis and a necrotizing myosis in the head and dorsocranial thorax. *Streptococcus agalactiae* was isolated from affected areas and tissues such as lung, liver and skeletal muscle. Streptococcosis was reported also from a wild dolphin that had eaten *S. agalactiae* infected mullet during a epizootic in Kuwait Bay, Kuwait (Evans *et al.*, 2006). In crocodiles clinical symptoms included lethargy, and necrotic lesions and sloughing of soft tissue on the ventral body wall, ventral neck, thorax with swollen limbs in some animal's limbs. Affected animals died within a few days. The isolate was serotype Ia and multi-locus sequence type ST23, suggesting a highly virulent strain of human origin. Other organisms, considered secondary invaders, were also isolated and these included *Salmonella* species, *Providencia rettgeri*, *Pseudomonas aeruginosa* and *Edwardsiella tarda*.

Phylogenetic studies of *S. agalactiae* Group B from aquatic species indicated two divergent populations based on capsular serotype and multi-locus sequence typing (Evans *et al.*, 2008). Isolates from mullet, seabream and a dolphin from Kuwait, hybrid striped bass from Israel and Nile tilapia from USA, Brazil and Honduras were capsular serotype Ia. Three of the fish isolates were capsular type Ib. *Streptococcus agalactiae* isolated from tilapia in Thailand were capsular type Ia and III (Suanyuk *et al.*, 2008).

GROWTH CHARACTERISTICS AND PHENOTYPIC IDENTIFICATION. *Streptococcus agalactiae* is a catalase-negative, Gram-positive coccus occurring in pairs or short chains and is positive for Lancefield group B. On media containing 5% blood, colonies are 0.5 mm in diameter, white, grey, round and may be β-haemolytic or non-haemolytic at 24 h at 37°C. *Streptococcus agalactiae* lyses red cells weakly and the zone of haemolysis is typically small. Strains of *S. agalactiae* from mammalian sources are alpha-haemolytic, however, most of the strains isolated from fish are reported to be non-haemolytic and initially were identified as *S. difficile*. They are now identified to be *S. agalactiae* Group B, capsular type Ib (Vandamme *et al.*, 1997). These isolates are slow growing and adhere strongly to the agar plate, which makes it difficult to resuspend the cells in an inoculating fluid. Conventional biochemical tests may need incubating for more than 48 h for sufficient growth to occur in the tubes. MR and VP reactions may be weak, and poor growth is seen in the tryptone water (indole test). The ADH test using conventional media may take more than 3 days' incubation before a slight colour change is seen.

Different reactions are reported for *S. agalactiae* isolated from human, fish and animal sources and this may in part be due to the optimum incubation temperature for enzyme reactions. For example, hippurate may be positive at an incubation temperature of 25°C but negative at 37°C for isolates from fish. The VP reaction by the conventional tube method is negative, but positive in the API rapid ID 32 (Vandamme *et al.*, 1997). This difference was also reported for strains from mullet, bream and type strains ATCC 13813 (non-haemolytic) and ATCC 27956 (β-haemolytic) (Evans *et al.*, 2002). *Streptococcus (difficile) agalactiae* strain LMG15977^{T} has an optimal growth temperature of 30°C and does not grow at 37°C in air but growth occurs at 37°C when grown in an anaerobic or microaerophilic atmosphere (Vandamme *et al.*, 1997). These reactions are listed separately in Table 3.36.

Other biochemical differences have been reported between fish isolates and type strains. In the API Rapid ID 32 Strep the test for β-galactosidase was negative for the fish isolates but positive for the ATCC strains (13813, 27956) whereas β-glucuronidase was positive for the fish isolates and negative for the ATCC strains.

Phenotypic differences have been reported from different fish species. In the fish isolates variable reactions occurred for hippurate, α-galactosidase, alkaline phosphatase, glycogen, pullulan, melezitose and tagatose (Evans *et al.*, 2002). The isolates from bluefish, striped bass and sea trout were sensitive to bacitracin disc (0.4 IU) (Baya *et al.*, 1990a). Silver pomfret isolates grew at 20–42°C but not at 10, 15 or 45°C. They grew at 0.5–6.0% NaCl but not at 6.5% NaCl, nor at a pH of 9.6 (Duremdez *et al.*, 2004). Isolates from bluefish, striped bass and sea trout in Chesapeake Bay, USA grew at 20–30°C but not at 15 or 40°C (Baya *et al.*, 1990). *Streptococcus agalactiae* isolated from infected Nile tilapia resulted in 11 phenotypes from 29 isolates when tested using the API 20 Strep (Mian *et al.*, 2009). These are listed in Tables 3.34 and 3.35.

The strains from bream, mullet and crocodiles gave typical reactions in the API Rapid ID32 Strep, and the strain from silver pomfret gave typical reactions in the API 20 Strep. Capsular serotype (Ia, Ib, II, III, IV, V) can be performed using a kit (Denka Seiken, Japan). These and other reactions are presented in Tables 3.14, 3.34–35, 3.36 and 3.38. The manufacturer's recommendations are followed for incubation time and temperature.

Streptococcus difficile

See *Streptococcus agalactiae*. *Streptococcus difficile* has now been identified as *Streptococcus agalactiae* type Ib (Vandamme *et al.*, 1997).

Streptococcus dysgalactiae

TAXONOMY AND SIGNIFICANCE. *Streptococcus dysgalactiae* consists of two subspecies: *Streptococcus dysgalactiae* ssp. *dysgalactiae* strains, which are alpha- or non-haemolytic and Lancefield Group C, and *S. dysgalactiae* ssp. *equisimilis*, which are beta-haemolytic and may have the Lancefield Group C, G or L antigen and some have been shown to have the A antigen (Vieira *et al.*, 1998; Facklam, 2002).

Streptococcus dysgalactiae ssp. *dysgalactiae* Lancefield group C is an emerging disease in aquaculture and has been reported from amberjack (*Seriola dumerili*) and yellowtail (*S. quinqueradiata*) cultured in Japan (Nomoto *et al.*, 2004) and Nile tilapia (*Oreochromis niloticus*) cultured in Brazil (Netto *et al.*, 2011). In these species infection presents as severe necrosis of the caudal peduncle and septicaemia.

Streptococcus dysgalactiae ssp. *dysgalactiae* Lancefield group L producing beta-haemolysis was isolated from various organs from harbour porpoise (*Phocena phocena*) either stranded or caught in fishing nets in the North and Baltic seas (Swenshon *et al.*, 1998).

GROWTH CHARACTERISTICS AND PHENOTYPIC IDENTIFICATION. *Streptococcus dysgalactiae* is a catalase negative, Gram-positive coccus occurring in chains. On BA colonies are 1 mm, light grey and have a zone of haemolysis after 24 h at 37°C.

Streptococcus dysgalactiae ssp. *dysgalactiae* Lancefield group C isolated from amberjack and yellowtail grows as white colonies on Todd-Hewitt agar, is umbonate and auto-agglutinates when emulsified in saline on a glass slide. Colonies are alpha-haemolytic at 24–72 h but become beta-haemolytic after prolonged incubation. Long chains are formed in Todd-Hewitt broth. In the API 20Strep variable results are obtained for ADH, fermentation of ribose, trehalose, starch (AMD) and glycogen depending upon the incubation temperature being 25°C or 37°C (Nomoto *et al.*, 2004).

Streptococcus dysgalactiae Lancefield group L from porpoises is beta-haemolytic on BA after 24–48 h at 37°C. Strep grouping using a commercial kit is negative for groups A to G. It is sensitive to bacitracin disc and 40% bile. Growth does not occur at 10 or 45°C, 6.5% NaCl or pH 9.6 (Nomoto *et al.*, 2004).

These and other reactions are presented in Tables 3.14, 3.34–35, 3.36 and 3.38.

Streptococcus equi ssp. zooepidemicus

TAXONOMY AND SIGNIFICANCE. *Streptococcus equi* ssp. *zooepidemicus* is a Group C *Streptococcus*. It causes infections in animals and is a major cause of bovine mastitis, and has been reported in human infections (Facklam, 2002). A Group C *Streptococcus* found in harbour seals (*Phoca vitulina*) infected with phocine distemper in an outbreak reported in 2002 from the German North Sea was identified as *Streptococcus equi* ssp. *zooepidemicus* (Akineden *et al.*, 2005). The organism was isolated from kidney, liver, lung, intestine, spleen, vagina and anus of 28 seals. Other bacteria including *S. phocae*, *Bordetella bronchiseptica*, *Arcanobacterium phocae*, *Brucella* species and *Erysipelothrix rhusiopathiae* were also isolated during the same outbreak (Skaar *et al.*, 1994; Vossen *et al.*, 2004). The virulence of *S. equi* ssp. *equisimilis* in this situation is not known.

GROWTH CHARACTERISTICS AND PHENOTYPIC IDENTIFICATION. *Streptococcus equi* ssp. *zooepidemicus* isolated from seals grew on sheep blood agar plates as mucoid colonies with a zone of complete haemolysis and was positive for Lancefield Group C (Oxoid Strep Grouping kit). Biochemical identification was performed using conventional tests (results not given) and confirmed using the *sodA*-*seeI* species-specific PCR described by Alber *et al.*, 2004 (Akineden *et al.*, 2005).

Streptococcus equi ssp. *zooepidemicus* can be identified using Tables 3.34–35 and 3.36. See molecular section for the PCR.

Streptococcus halichoeri

TAXONOMY AND SIGNIFICANCE. *Streptococcus halichoeri* sp. nov. (CCUG 48324^T=CIP 108195^T) was isolated from three grey seals (*Halichoerus grypus*) found in Scotland, and from three seals sampled at a rehabilitation centre in Cornwall, UK (Lawson *et al.*, 2004). The sampling sites and disease status of the animals were not given. The organism has a high sequence similarity with *S. phocae* (96.9%) based on 16S rRNA sequencing of over 1400 bp, and with results from whole-cell protein profiling and phenotypic testing was identified as a new species. The G+C content is 39 mol%.

GROWTH CHARACTERISTICS AND PHENOTYPIC IDENTIFICATION. *Streptococcus halichoeri* is a facultatively anaerobic, catalase-negative, Gram-positive coccus occurring in pairs or short chains. On Columbia base agar containing 5% sheep blood colonies are 0.5 mm in diameter, white, umbonate and non-haemolytic at 24 h at 37°C (Lawson *et al.*, 2004). It is positive for Lancefield Group B. The organism can be identified using the API Rapid ID 32 Strep, and API 20 Strep (bioMerieux) using the recommended incubation time and temperature. The profile numbers are distinctive with a profile number of 1430(1)3141001 for the API Rapid ID32 Strep, and 1163100(1) for the API 20 Strep. See Tables 3.14, 3.34–35 and 3.36.

Streptococcus ictaluri

TAXONOMY AND SIGNIFICANCE. *Streptococcus ictaluri* sp. nov. (707-05^T=SO2-1108^T=ATCC BAA-1300^T=CCUG 52536^T) was isolated from channel catfish (*Ictalurus punctatus*) broodstock held in commercial aquaculture facilities in the Mississippi Delta, USA (Shewmaker *et al.*, 2007). Infection results in spinal meningitis, myositis, arthritis and osteolysis in broodstock resulting in low-grade mortalities and poor reproduction. Lesions on the jaw and fin bases developed into haemorrhagic ulcers leading to ulceration of the articular surfaces and joints of the lower jaw and cranium and exudate-filled fistula tracts. The organism was isolated from lesions (Camus *et al.*, 2008).

Based on biochemical identification, 16S rRNA sequencing and DNA–DNA hybridization studies the organism was identified as a new species. Sequencing revealed phylogenetic closeness to *S. iniae*, *S. uberis*, *S. parauberis* (2.0–2.3% divergence) with greater divergence between *S. pyogenes*, *S. urinalis*, *S. dysgalactiae* ssp. *dysgalactiae* and *S. canis* (2.5–3.2%). The G+C content is 38.5 mol% (Shewmaker *et al.*, 2007).

GROWTH CHARACTERISTICS AND PHENOTYPIC IDENTIFICATION. *Streptococcus ictaluri* is a catalase-negative, Gram-positive coccus occurring in pairs or short chains. It is non-haemolytic and non-pigmented on tryptic soy agar with 5% defibrinated sheep blood. The optimal growth temperature is 30°C and growth is negative at 45°C and at 6.5% NaCl. The organism can be identified using the API Rapid ID 32 Strep (bioMerieux) using the recommended incubation time and temperature. The profile numbers give good identification at 96% confidence for *Gemella haemolysans* because both organisms are relatively inert apart from a positive reaction for pyroglutamylaminopeptidase (PYR), however they can be differentiated by morphology using a Gram stain (Shewmaker *et al.*, 2007). *Gemella haemolysans* is a facultative anaerobic Gram-positive coccus and is negative for catalase and oxidase. It may be easily decolourized in a Gram stain. It closely resembles *Streptococcus viridans* and *Neisseria* sp. *Gemella* species occur in pairs, tetrads, clusters and sometimes short chains of cells. *Streptococcus ictaluri* is positive for fermentation of ribose in the API 20 Strep but negative in the API Rapid ID 32 Strep (Shewmaker *et al.*, 2007; Camus *et al.*, 2008). These and other reactions are presented in Tables 3.14, the 3.34–35 and 3.36. The API 20 Strep is incubated at 29°C and read at 24 h, although the reaction for ADH takes 2–3 days.

Streptococcus iniae

TAXONOMY AND SIGNIFICANCE. *Streptococcus iniae* was first isolated from Amazon freshwater dolphins (*Inia geoffrensis*) held at an aquarium in California and in New York, USA (Pier and Madin 1976; Pier *et al.*, 1978). Both dolphins had numerous discrete, raised subcutaneous abscesses 2–5 cm in diameter on the thoracic and abdominal cavities. *Streptococcus iniae* was cultured from the wall of the abscesses but few organisms were cultured from the purulent exudate. The disease is referred to as golf ball disease due to the appearance of the abscess foci, and can be fatal for dolphins unless treated with antibiotics. An isolate originally named *S. shiloi* and isolated from tilapia (St Peter's fish) and rainbow trout (*Oncorhynchus mykiss*) from Israel (Eldar *et al.*, 1994) was later found to be *S. iniae* (Eldar *et al.*, 1995b).

Streptococcus iniae has been reported worldwide from numerous fish species including barramundi (*Lates calcarifer*), channel catfsh (*Ictalurus punctatus*), European seabream (*Dicentrachus labrax*), gilthead seabream (*Sparus aurata*), seabream (*Dicentrarchus labrax*), rainbow trout (*Oncorhynchus mykiss*), Japanese flounder (*Paralichthys olivaceus*), Nile tilapia (*Oreochromis niloticus*), hybrid tilapia (*O. niloticus* × *O. aurata*), striped piggy (*Pomadasys stridens*), rabbit fish (*Siganus canaliculatus*), red porgy (*Pagrus pagrus*), marbled spinefoot (*Siganus rivulatus*), variegated lizardfish (*Syndodus variegatus*), ornamental fish, flying fox (*Epalzeorhynchos kalopterus*) in Australia, Bahrain, China, Israel, Spain and the USA (Perera *et al.*, 1994; Zlotkin *et al.*, 1998b; Bromage *et al.*, 1999; Nguyen and Kanai, 1999; Yuasa *et al.*, 1999; Colorni *et al.*, 2002; Creeper and Buller, 2006; El Aamri *et al.*, 2010; Chen *et al.*, 2011a).

Clinical signs of the disease may vary according to the species of fish infected. Generally there is a meningoencephalitis, panophthalmis and exudative meningitis, haemorrhages or petechiae on pectoral fins, mouth and anus, visceral haemorrhages and necrotic inflammation in internal organs (Perera *et al.*, 1994; Zlotkin *et al.*, 1998b; Yuasa *et al.*, 1999; Creeper and Buller, 2006; Chen *et al.*, 2011a). The organism is isolated from sites where the organism occurs and these include the brain, eye, spleen, liver and kidney.

Susceptibility to infection varies with species of fish. In experimental challenge, seabream (*Acanthopagrus latus*) and one-spot bream (*Diplodus sargus*) were not susceptible to *S. iniae* infection when injected intramuscularly with 10^8 CFU/ml. In contrast, Borneo mullet (*Liza macrolepis*) was susceptible with mortality rates of greater than 50% when 10^6 and 10^8 CFU/ml were injected intramuscularly, and for tests with brown-spot grouper (*Epinephelus coioides*) a high dose of 10^8 CFU/ml was required before high mortality occurred. *Streptococcus iniae* is highly virulent for rabbitfish and high mortality occurred in experimental challenge when injected intramuscularly with 10^4 CFU/ml (Yuasa *et al.*, 1999). Red porgy are highly susceptible (LD_{50} 1.7×10^4 CFU/fish) whereas gilthead seabream are slightly less susceptible (LD_{50} 1.32×10^5 CFU /fish) (El Aamri *et al.*, 2010). Chronic infection commonly occurs at a temperature of 25°C, whereas a more acute form of the disease occurs when the temperature ranges from 28 to 32°C (Yuasa *et al.*, 1999).

Streptococcus iniae is a pathogen of cultured American bullfrogs (*Rana castesbeiana*; note *Lithobates catesbeianus* is thought to be the more correct name). Infected frogs display torticollis and stupor. Focal areas of haemorrhage occur in the muscle, legs are swollen and liver, spleen and kidney may be enlarged (Mauel *et al.*, 2002).

GROWTH CHARACTERISTICS AND PHENOTYPIC IDENTIFICATION. *Streptococcus iniae* is a non-motile, Gram-positive coccus (0.3–0.5 μm) occurring in long chains when grown in broth culture. Colonies are pinpoint at 24 h and 25°C on BA, NA, BHIA and TSA, and grow to 1 mm, white, round, umbonate, with an opaque centre spot after 48 h, with a zone of β-haemolysis. β-Haemolysis may be surrounded by a diffuse ring of alpha-haemolysis. Complete haemolysis is seen on sheep and horse blood agar but only a partial haemolysis is observed when the medium is supplemented with human or bovine blood. Variations in haemolysis may be seen between strains. Growth occurs on *Streptococcus* selective medium (Oxoid) (Pier and Madin, 1976; Pier *et al.*, 1978; Creeper and Buller, 2006). Growth in Todd Hewitt broth occurs with a coarse white granular material at the base of the tube and speckled granular material in the upper part of the tube when grown overnight at 37°C. When conventional tests are used acid is produced from dextran, fructose and galactose, and growth occurs on 0.1% tetrazolium medium and in 4.0% NaCl. Results are negative for fermentation of dulcitol, inulin, melibiose, raffinose and rhamnose and negative for growth at 6.5% NaCl, 10% and 40% bile broth and growth at 45°C (Pier and Madin, 1976). Reports vary on growth at 10°C. Some reports state that growth at 37°C is better than growth at incubation temperatures of 10, 25 or 40°C (Perera *et al.*, 1994). Other test results are presented in Table 3.14.

Strains isolated from either fish or humans have a slightly different biochemical profile. *Streptococcus iniae* strains isolated from cultured and wild fish in the Red Sea are similar in biochemical profile to other *S. iniae* strains, except for a negative arginine dihydrolase (ADH) in the majority of strains, and late fermentation of galactose and amygdalin as detected in the API 50CH, which was incubated for 72 h before results were recorded (Colorni *et al.*, 2002). The biochemical profile obtained from the API 20 Strep agrees with other results recorded for fish isolates where 30% of strains are negative for ADH, and variation is seen in the fermentation of mannitol (Dodson *et al.*, 1999). Isolates from fish are variable in some biochemical reactions as detailed in the identification tables. ADH using Moeller's medium requires a heavy inoculum and an incubation time of greater than 48 h at 25°C. Reactions in the API rapid ID 32 Strep for β-glucuronidase, pyroglutamic acid arylamidase and β-mannoside are positive when incubated at 37°C but are negative or weak or difficult to interpret after incubation at 25°C. Reactions are negative for *Streptococcus* groups A to G (Prolex™ latex agglutination system, Pro-Lab Diagnsotics) (Creeper and Buller, 2006). These and other reactions are presented in Tables 3.14, 3.30 and 3.34–35 and 3.36. Reactions at 25°C are slower than at 37°C.

MOLECULAR IDENTIFICATION. PCRs have been reported for the detection of *S. iniae* based on primers designed to the lactate oxidase-encoding (*lctO*) gene, the 16S rRNA and the 16S-23S intergenic spacer region (Zlotkin *et al.*, 1998b; Mata *et al.*, 2004; Zhou *et al.*, 2011).

Primers for detection of the lactate oxidase-encoding (*lctO*) gene are LOX-1 (5′-AAG GGG AAA TCG CAA GTG CC-3′) and LOX-2 (5′-ATA TCT GAT TGG GCC GTC TAA-3′). A 50-μl reaction mix contains 1 μM of each primer, 0.2 mM of each dNTP, 2 U DNA polymerase (Biotools B & M Labs), 1× reaction buffer and 20 μl boiled cells or 10 μl DNA extracted from tissue. Thermocycling is carried out using an initial denaturation step at 95°C for 1 min followed by 35 cycles of denaturation at 92°C for 1 min, annealing at 60°C (55–60°C can be used) for 1 min, extension at 72°C for 2 min and a final step using an extension at 72°C for 5 min. The amplified product is 870 bp. The detection limit is 62–31 cells per reaction and 25 pg of DNA. The primers are specific when tested against *S. agalactiae* (STR 34), *S. agalactiae* (previously identified as *S. difficile* CIP 103768^T), *S. suis* (CECT 958^T), *S. uberis* (CECT 994^T), *S. parauberis* 94/16, *Aerococcus viridans* (NCDO 1225), *Carnobacterium piscicola* (01/5423), *Lactococcus garvieae* (ATCC 43921^T), *L. piscium* (CECT 4493^T), *Vagococcus fluvialis* (NCDO 2497) and *V. salmoninarum* (NCDO 2777) (Mata *et al.*, 2004).

The master mix using primers designed to the 16S rRNA genes contains 8 ng/ml of each primer Sin-1 (CTA GAG TAC ACA TGT ACT (AGC T)AAG) and Sin-2 (GGA TTT TCC ACT CCC ATT AC), I U Vent DNA polymerase (New England Biolabs), 1× buffer 0.25 mmol/l, final concentration of each dNTP, 0.1 mg/ml bovine albumin in a 50-μl volume (Zlotkin *et al.*, 1998). Thermocycling is carried out using an initial denaturation step at 94°C for 3 min followed by 35 cycles of denaturation at 94°C for 1 min, annealing at 55°C for 1 min, extension at 72°C for 1.5 min and a final step using an extension at 72°C for 10 min. The amplified product is 300 bp. Sensitivity was not reported. The primers are specific when tested against *S. agalactiae* (ATCC 13813), *S. parauberis*, *Aeromonas salmonicida*, *Lactococcus garvieae*, *L. lactis* and *Vagococcus salmoninarum* (Zlotkin *et al.*, 1998), however Mata *et al.* (2004) found primers Sin-1 and Sin-2 were not specific against *S. agalactiae* (*S. difficile* CIP 103768^T and CIP 103853) as amplicon was produced at the same molecular weight as *S. iniae*.

The master mix using primers designed to the 16S-23S intergenic spacer region contains 0.1 μM of each primer SP-1 (5′-GAA AAT AGG AAA GAG ACG CAG TGT C-3′) and SP-2 (5′-CCT TAT TTC CAG TCT TTC GAC CTT C-3′), 200 μM of each dNTP, PCR buffer (10 mm Tris HCl, pH 8.8, 50 mm KCl, 2.5 mm $MgCl_2$, 0.08% Nonidet P40), 1.0 U *Taq* polymerase (Fermentas™) and DNA in a 50-μl reaction volume (Zhou *et al.*, 2011). Thermocycling is carried out using an initial denaturation step at 94°C for 3 min followed by 35 cycles of denaturation at 94°C for 1 min, annealing at 60°C for 1 min, extension at 72°C for 1 min and a final step using an extension at 72°C for 10 min. The amplified product is 377 bp. The limit of detection is 20 pg of DNA and 10 cells. The primers are specific at annealing temperatures of 55 and 60°C when tested against *Streptococcus agalactiae*, *S. dysgalactiae*, *S. parauberis*, *Lactococcus garvieae*, *Photobacterium damselae* ssp. *piscicida*, *Nocardia seriolae*, *Lactococcus lactis*, *Staphylococcus aureus*, *Vibrio anguillarum* and *Aeromonas hydrophila*.

Streptococcus marimammalium

TAXONOMY AND SIGNIFICANCE. *Streptococcus marimammalium* sp. nov. (M54/01/1^T=CCUG

48494[T]=CIP 108309[T]) was isolated from grey seal (*Halichoerus grypus*) and from the lung of the common seal (*Phoca vitulina*) found in Inverness, Scotland (Lawson *et al.*, 2005) and was identified as a new species of *Streptococcus* based on sequence information from 1400 bp of the 16S rRNA gene. Based on this information the isolate is most closely related to *S. hyovaginalis*, *S. pluranimalium*, *S. thoraltensis*, *S. ovis*, *S. minor*, *S. suis* and *S. acidominimus* (93.7–94.3%). The G+C content is 38 mol%.

GROWTH CHARACTERISTICS AND PHENOTYPIC IDENTIFICATION. *Streptococcus marimammalium* is a facultatively anaerobic, catalase-negative, Gram-positive coccus occurring in pairs or short chains. On sheep blood agar, colonies are 0.2–0.3 mm in diameter, grey-white and non-haemolytic at 24 h incubation but produce beta-haemolysis after 3 days. Growth does not occur on MacConkey agar. It is positive for Lancefield Group C as tested by the Streptex system (Remell, Europe). These and other reactions are presented in Tables 3.34–35, 3.36 and 3.38.

Streptococcus parauberis and *Streptococcus uberis*

TAXONOMY AND SIGNIFICANCE. *Streptococcus parauberis* has been isolated from infected fish as a cause of Streptococcosis, and in terrestrial animals it is a cause of mastitis in cows. *Streptococcus parauberis* and *S. uberis* are unable to be differentiated based on phenotypic tests and *Streptococcus parauberis* may be biochemically indistinguishable from *Lactococcus garvieae* (Doménech *et al.*, 1996; Facklam, 2002).

A *Streptococcus* species positive for Group B Lancefield group was isolated from penaeid shrimp (*Litopenaeus vannamei*) and identified as *S. uberis* using the API 20 Strep and as *S. parauberis* using the Biolog system (Hasson *et al.*, 2009). A bioassay using juvenile shrimp confirmed pathogenicity of the *Streptococcus* isolate when injected intramuscularly at a concentration of 3–5×10^5 CFU/shrimp and in water-borne exposure assays.

Streptococcus parauberis was reported from cultured olive flounder (*Paralichthys olivaceous*) in Korea. Pathogenicity was confirmed through experimental challenge. Signs of infection include haemorrhagic areas around the mouth, eyes and pectoral fins. Ascites fluid was present (Baeck *et al.*, 2006; Kim *et al.*, 2006).

GROWTH CHARACTERISTICS AND PHENOTYPIC IDENTIFICATION. *Streptococcus parauberis* is a Gram-positive short rod to coccobacilli in pairs or short chains. Catalase and oxidase are negative. On BA and TSA colonies are 1.5–2 mm at 24 h, round, whitish and slightly alpha-haemolytic after 24–48 h at 22–37°C.

The *S. uberis/parauberis* isolate from shrimp (prawns) produced 0.3–0.5 mm white non-haemolytic colonies.

There are different reports for the reaction of *S. uberis* to pyrrolidonylarylamidase. *Bergey's Manual of Determinative Bacteriology* 1994, reports a negative reaction, whereas Collins *et al.* (1984) and Doménech *et al.* (1996) report a positive reaction (Doménech *et al.*, 1996).

The two species *S. uberis* and *S. parauberis* are placed together in the following Table 1.17 because differentiation of the two species by conventional phenotypic tests is not possible. Species-specific probes have been developed for identification of both species (Facklam, 2002).

These and other reactions are presented in Tables 3.14, 3.30, 3.34–35, 3.36 and 3.38. Kits are incubated at 37°C for 24 h.

Table 1.17. Identification of unusual *Streptococcus* species and other genera of Gram-positive cocci in chains.

Species	LAP	PYR	NaCl	BE	Aes	ADH	VP	Hip	Man	Sor
S. acidominimis	+	+	–	–	+	–	–	–	+	–
S. pluranimalium	+	–	+	+	+	–	–	–	–	–
S. uberis	+	+	+	–	+	V	V	+	+	+
S. parauberis										
Facklamia species	+	+	+	–	V	V	–	V	–	V

LAP, leucine aminopeptidase; PYR, pyrrolidonylarylamidase; NaCl, growth in 6.5% NaCl broth; BE bile–aesculin reaction; Aes, aesculin hydrolysis; ADH, arginine hydrolysis; VP, Voges–Proskaüer; Hip, hippurate hydrolysis; Man, mannitol fermentation; Sor, sorbitol fermentation. V = Variable reaction.

Streptococcus phocae

TAXONOMY AND SIGNIFICANCE. *Streptococcus phocae* (8399 HI[T]=NCTC 12719[T]) was isolated from common harbour seal (*Phoca vitulina*) and grey seal (*Halichoerus grypus*) during an outbreak of morbillivirus among seals from northern Europe near Norway, and was thought to cause an opportunistic infection that in many cases led to the death of the animals (Skaar *et al.*, 1994). *Streptococcus phocae* was isolated from lung, liver, spleen and kidney of infected animals. Based on DNA–DNA hybridization studies and biochemical tests the isolate was identified as a new species.

Since then *S. phocae* has been reported from harbour seals and grey seals from the German North and Baltic seas (Vossen *et al.*, 2004), from Cape fur seal (*Arctocephalus pusillus pusillus*), ringed seal (*Pusa* [*Phoca*] *hispida*), harbour porpoise (*Phocena phocena*) and cetaceans, and from southern sea otters (*Enhydra lutris nereis*) stranded on the Californian coast, USA (Imai *et al.*, 2009).

Streptococcus phocae caused disease outbreaks during the summer months at temperatures above 15°C in caged-farmed Atlantic salmon (*Salmo salar*) in the Reloncaví estuary in Chile (Gibello *et al.*, 2005; Romalde *et al.*, 2008). Smolts and adult fish cultured in floating cages in both marine and estuarine waters were affected with 3 to 25% cumulative mortality and showed signs of exophthalmia with haemorrhagic and purulent fluid around the eye, ventral petechiae, skin abscesses, haemorrhage in abdominal fat, pericarditis, enlarged, yellow liver and pathological changes in the kidney and spleen. Some fish showed muscle liquefaction and deep ulcerations. Gram-positive cocci were seen in these sites, including the brain. The strains were pathogenic for Atlantic salmon (20 g) with LD_{50} values of 1.9×10^5 to 5.6×10^5 bacteria per fish. Mortalities of 80 and 100% for the low and high dose, respectively, began at day 4. Neither the salmon strains nor the type strain (DSMZ 15635[T] from a seal in Norway) were pathogenic for rainbow trout (15 g) using intraperitoneal inoculation, or for BALB/C mice using an inoculum of 0.1 µl of 10^7 cells per mouse (Romalde *et al.*, 2008).

GROWTH CHARACTERISTICS AND PHENOTYPIC IDENTIFICATION. *Streptococcus phocae* is a facultatively anaerobic, catalase-negative, non-motile, Gram-positive coccus (0.6–0.9 µm diameter) occurring singly, in pairs or in chains. On Columbia base agar containing sheep blood, colonies are white, pinpoint (0.5 mm in diameter), smooth and glistening, and beta-haemolytic after 24 h incubation at 22–25°C and are up to 0.8 mm in diameter at 37°C at 24 h under aerobic conditions (Skaar *et al.*, 1994; Romalde *et al.*, 2008). Aesculin hydrolysis is negative, growth occurs at 25 and 37°C and at pH 9.6 but not at 6.5% NaCl or in 40% bile salts. Growth is negative at 10°C and 45°C. *Streptococcus phocae* is sensitive to bacitracin discs (0.04 IU, Neo-Sensitabs) with ≥30 mm zones of growth inhibition, and resistant to optochin discs (Neo-Sensitabs, Rosco). It is differentiated from *S. pyogenes* Group A by Lancefield group, and biochemical reactions such as lactose, trehalose and ADH (all positive for *S. pyogenes* and negative for *S. phocae*). Most Group F strains belong to *S. milleri*, and *S. phocae* is differentiated from this species by negative reactions for ADH and VP. The majority of seal isolates (90%) are positive for Lancefield Group F including the type strain (DSMZ 15635[T]) from a seal from Norway, whereas other seal isolates are positive for Group C or may be untypeable (Skaar *et al.*, 1994; Vossen *et al.*, 2004). Isolates from sea otters are Lancefield Group G, F or untypeable (Imai *et al.*, 2009). Isolates from salmon are Group G but also show a weak reaction to Group A (Romalde *et al.*, 2008). The seal Group C isolates were positive using both the standard Lancefield system and the Streptex kit (Wellcome Diagnostics) (Skaar *et al.*, 1994). In the API Rapid ID 32 Strep the profile may give an identification of *Gemella* or *Gardnerella vaginalis* because *S. phocae* is not in the API database. *Gemella* species tend to be alpha-haemolytic, Gram-variable cocci occurring in pairs or tetrads. The API Rapid ID 32Strep profiles reported for salmon isolates are 04102001000 (Gibello *et al.*, 2005), 04000000000 and 04000001000 (Romalde *et al.*, 2008). The API 20 Strep profile for *S. phocae* is 0062007 and will result in an identification of S. *dysgalactiae* ssp. *equisimilis* in the API database. These and other reactions are presented in Tables 3.30, 3.34, 3.36 and 3.38.

Phylum *Spirochaetes*
Class Spirochaetes

1.40 *Leptospiraceae*

The leptospiraceae family belongs to the class *Spirochaetes* and the phylum *Spirochaetes*.

1.40.1 *Leptospira*

Leptospira are thin spirochaetes 6–20 μm in length with one or both ends ending in a hook shape and are obligate aerobes. The *Leptospira* genus consists of at least 16 species based on DNA hybridization studies of which 12 are pathogenic and four are saprophytic. The *Leptospira* are divided into 260 serovars based on their lipopolysaccharide antigens. They are found worldwide except for Antarctica, and are associated with high humidity and neutral pH (6.9–7.4) and can be isolated from water such as ponds, lakes and reservoirs (Adler and de la Peña Moctezuma, 2010). *Leptospira* are zoonotic and have been reported from almost all mammals including marine mammals. *Leptospira* can penetrate broken skin and mucosal surfaces and tend to colonize the renal tubules of hosts. The organisms are shed into the environment via the urine.

Growth occurs in media containing vitamin B1, B12, long chain fatty acids and ammonium salts at an optimum temperature of 28–30°C. The most commonly used medium is EMJH, which contains oleic acid, bovine serum albumin and Tween. Leptospirosis is diagnosed by culture of infected material (fresh kidney, blood or urine) into EMJH medium or testing of the serum against live cultures using the microscopic agglutination test (MAT). Detection by culture may take up to 13 weeks with weekly examination of culture material using dark field microscopy. Culture and identification of *Leptospira* is best done by a laboratory that specializes in this organism.

Leptospira interrogans

TAXONOMY AND SIGNIFICANCE. *Leptospira interrogans* serovar pomona has been isolated from northern elephant seals (*Mirounga angustirostris*), Californian sea lions (*Zalophus californianus*) and northern fur seals (*Callorhinus ursinus*) (Smith *et al.*, 1977; Gulland *et al.*, 1996; Colegrove *et al.*, 2005) and has been associated with live strandings, epizootics and enzootic disease.

Clinical signs in northern elephant seals include lethargy, weight loss and dehydration and abnormal blood biochemistry reflects renal failure. Kidneys are swollen and histology reveals tubulointerstitial nephritis with tubule degeneration. Spirochaetes may be seen in the renal cortex and lumen of renal tubules using a Warthin–Starry stain or immunohistochemistry (Colegrove *et al.*, 2005). Infection in northern fur seals caused interstitial nephritis in adults and a multiple haemorrhagic syndrome and stillbirth in pups. Leptospires were isolated from the kidney in adults and from the placenta and occasionally the liver and kidney in stillborn pups (Smith *et al.*, 1977). Renal disease was a common finding in infected Californian sea lions and premature births and reproductive failure has also been reported (Gulland *et al.*, 1996). Clinical signs were similar to northern elephant seals with lethargy, anorexia, dehydration and reluctance to use hind flippers. Swelling of the kidneys was seen along with loss of differentiation between the renal medullae and cortices, renal lesions and friable livers. Leptospires were seen by dark ground illumination in samples of kidney and urine.

Abortion occurred in Californian sea lions (Smith *et al.*, 1974).

Leptospira kirshneri grippotyphosa

TAXONOMY AND SIGNIFICANCE. *Leptospira kirshneri grippotyphosa* has been isolated from Pacific harbour seals (*Phoca vitulina richardii*) rehabilitated in a facility in California, USA and in the wild (Stamper *et al.*, 1998; Stevens *et al.*, 1999). The seals had signs consistent with interstitial nephritis; tubular necrosis and organisms were present in the kidney parenchyma.

GROWTH CHARACTERISTICS AND PHENOTYPIC IDENTIFICATION OF *LEPTOSPIRA* SPECIES. Using a sterile Pasteur pipette a plug of kidney or a few drops of urine are inoculated into a 5-ml vial of EMJH medium. It is important the material is fresh so that viable organisms are cultured. To assist in reducing contamination with other bacteria, samples can be serially diluted and dilutions used to inoculate the EMJH medium. 5-Fluorouracil (0.1 g/l) and actidione (0.1 g/l) can be added to the medium to assist in reduction of bacterial contamination. The use of semi-solid agar also assists in culturing *Leptospira*. The inoculated culture is incubated at 28–30°C for up to 13 weeks before calling a negative result. The culture is examined each week by making a wet preparation of the culture and examining using dark field illumination. The slide is scanned in a grid pattern using a magnification of ×200 (i.e. ×10 eye piece and ×20 objective lens). Close examination to confirm the presence of leptospires is done using the ×40 objective lens.

The identification to species level is done using the microscopic agglutination test (MAT) performed on serum. This test is best done by a specialist laboratory as the laboratory needs to hold and maintain a range of type strains for use in the test.

2

Bacteriological Culture Techniques: Microscopy, Culture and Identification

Successful isolation and accurate identification of a suspected pathogen depends on the use of standardized bacteriological culture methods. It is important to work through these methods in a precise and logical step-by-step manner. The bacteria are grown on culture media, then inoculated into biochemical identification tests and those results are recorded and compared with standard results. The following table sets out the steps for the isolation and identification of an organism and also directs the reader to the location of the technique or method in this manual.

2.1 Specimen Collection and Submission

Samples should be collected from live sick (moribund) animals or from recently dead animals. Animals that have been dead for more than 6 h usually have grossly contaminated organs because of the overgrowth of post-mortem invading bacteria. It is very difficult for the laboratory to work with these samples. Unfortunately, it is not always possible to get the samples to the laboratory within the ideal time, because of large distances between some laboratories and the aquaculture farm.

Ideally all samples for bacteriological examination should be collected before the start of antimicrobial treatment. Samples collected from aquatic animal species after commencement of treatment are unlikely to grow any pathogenic bacteria.

Samples may be submitted as swabs, tissues or whole fish. The most appropriate samples for identification of a systemic disease may be kidney, haemolymph or blood. For skin disease, swabs or samples of skin are appropriate. Refer to Tables 1.1 and 1.2 for tissue sites where the bacteria are likely to be located.

2.1.1 Transport medium and transport to the laboratory

If swabs have been collected, then they must be placed into a transport medium to prevent the swab drying out and for the organisms to remain viable. Amies transport medium is the medium of choice for transport of swabs to the laboratory.

Fish, tissues and swabs in Amies transport medium should be transported to the laboratory as soon as possible after collection, and preferably on ice or at 4°C.

Stuart's transport medium may be unsuitable for transport and survival of *Vibrio* sp. Numbers of viable cells of *Vibrio alginolyticus*, *Vibrio harveyi* and *Photobacterium damselae* ssp. *damselae* were greatly reduced after 24 h and were almost non-viable at 48 and 72 h when tested from Stuart's media kept at 4°C (McLetchie and Buller, unpublished). However, there was no

Table 2.1. Outline of steps for culture and identification.

Day	Activity	Method or technique
Day 1	**1.** Sample collection and preparation	Section 2.1
	2. Inoculation of sample to primary isolation plate media (or broths where appropriate)	Section 2.2, Tables 2.2 and 2.3, text
	3. Incubate at appropriate temperature and atmosphere	Section 2.2, Tables 2.2 and 2.3, text
Day 2 (24 h)	**1.** Examine culture plates	Section 2.3
	2. Select suspect colonies and subculture to BA or MSA-B to obtain pure growth (secondary plates)	Table 2.4, text for cultural and microscopic appearance
	3. Re-incubate primary plates	Section 2.2, Tables 2.2 and 2.3
	4. Incubate all plates at appropriate temperature and atmosphere as before	
Day 3 (48 h)	**1.** Re-examine primary plates for slow-growing pathogens	Section 2.3
	2. Check that subcultures on secondary plates are pure	
	3. Perform primary identification tests	Section 2.4 and media section
	4. Inoculate appropriate biochemical identification set	Sections 2.5 and 2.6
	5. Re-incubate primary plates as before	Sections 2.2, Tables 2.2 and 2.3
Day 4 (72 h)	**1.** Re-examine primary plates for growth of slow-growing pathogens. Re-incubate if disease suggests a pathogen that requires more than 3 days for growth	
	2. Examine biochemical identification set/s and record results at 24 h incubation	Table 2.5
Day 5 (96 h)	**1.** Examine biochemical identification set/s and record results for 48 h incubation. Add reagents for tests for indole, methyl red, nitrate, Voges–Proskaüer	Table 2.5 Chapter 3 and Tables 3.1–3.23 for biochemical results (conventional biochemicals), and Tables 3.24–3.39 for results for API kits
	2. Interpret results from appropriate identification table	

BA = blood agar; MSA-B = marine salt agar with blood.

reduction of viable cells when all three organisms were tested from Amies transport media held at 4°C for 24, 48 and 72 h.

2.1.2 Sample preparation

Swabs

Swabs require no sample preparation, and are inoculated directly to the appropriate media.

Tissues

Aseptically remove a piece of infected tissue from the fish and place into a McCartney bottle. Macerate the tissue with flame-sterilized scissors, and use a sterile, cotton-tipped swab to inoculate to appropriate media. Tissues that have not been aseptically removed from the fish need to be surface-sterilized. This can be done by either rinsing the piece of tissue in three or four washes of sterile water or saline, or for larger pieces of tissue (walnut-sized), dipping in 70% alcohol and flaming briefly for 3–5 s. The tissue is then placed into a sterile McCartney bottle or appropriate container and macerated with flame-sterilized scissors. A sterile, cotton-tipped swab is used to sample the tissue and inoculate to appropriate agar plates and make smears for Gram stain.

Lesions

Always collect material from the edge of the lesion, including some of the immediate tissue to ensure sampling for true invading bacteria. Sampling from the middle of the lesion may culture secondary invading bacteria that will confuse the results. Also, the invading bacterium may no longer be viable in the centre of the lesion.

Whole fish

For small fish, the whole fish is surface-sterilized by rinsing the fish in three or four changes of sterile distilled water or normal saline. For larger fish, surface-sterilize by either flaming with 70% alcohol or washing the area with three or four washes of sterile distilled water or normal saline. Using flamed-sterilized forceps and scissors, make an incision from behind the gills and to the mid-line. Next cut along the mid-line towards the anus. Lift back the flap of flesh to expose the internal organs. Select the required organs for analysis and remove with flame-sterilized forceps and scissors. Chop tissues as before and inoculate to plates.

2.2 Culture and Incubation

2.2.1 Culture media

All specimens should be cultured to a general-purpose medium. For freshwater specimens use blood agar (BA) and MacConkey agar (MCA). Tryptone soya agar (TSA) can also be used; however, the addition of blood can improve the isolation of some organisms such as *Aeromonas salmonicida*, *Carnobacterium* and *Streptococci*. For marine specimens, use a medium containing Na ions, BA and MCA. Marine salt agar (MSA-B), which is prepared from TSA with 5% horse blood and 2% NaCl (see Media section), is a good general-purpose media for marine organisms. Marine 2216, commercially available from Difco and based on ZoBell's (1941) original formulae, is also suitable. MSA-B may give an improved growth, particularly on primary culture with commonly isolated *Vibrio* sp. MA 2216 is more suitable when a complex mixture of electrolytes is required by some marine organisms.

2.2.2 Isolation and growth conditions: media, temperature and time of incubation

All specimens should be inoculated to the general-purpose medium. This medium is then incubated at 25°C for 2–5 days as indicated in Table 2.2 for general culture conditions. However, specimens from a particular environment such as a cold water environment or a marine environment may need a particular incubation temperature or NaCl may be required in the medium. These variations are listed either in the text under the specific organism or in Table 2.3 for those organisms not detailed in the text.

Some organisms have special growth requirements or may be better detected with the use of a selective and/or enrichment medium if it is available for that organism. Such requirements for specific organisms are detailed either in the text or in Table 2.3 for those organisms not detailed in the text.

The preparation of all media mentioned in Tables 2.2 and 2.3 is listed in the section on media preparation.

Table 2.2. General culture.

Fish, aquatic animals	Media: freshwater animals	Media: salt water animals	Temperature: optimum (range)
All specimens, tissues	BA, MCA	MSA-B, BA, MCA	20–25°C for 2–5 days
Coldwater	BA, MCA	MSA-B, BA, MCA	15–22°C for 3–5 days
Tropical	BA, MCA	MSA-B, BA, MCA	25°C for 25 days

BA = blood agar; MCA = MacConkey agar; MSA-B = marine salt agar containing blood.

Table 2.3. Specific culture requirements of organisms not detailed in the text.

Bacterium	Media for freshwater animals	Media for salt water animals	Temperature, atmosphere, time
Aequorivita sp.		MA 2216 NA, TSA, AO with either 2.5% NaCl or 35 g/l seawater salts added	20°C, 3–7 days
Aerococcus viridans var. *homari*		NA+5% blood, TSA, BA, BHIB SIEM selective media	30°C, 24–48 h
Aeromonas sp., unless indicated	TSA, BA		25–37°C, 24–48 h
Alcaligenes faecalis var. *homari*		Marine 2216	18°C for 7 days
Alteromonas		Marine 2216 Growth is poor on TSA and NA even with added salt	Strict aerobes. 25°C, 24–48 h
Aquaspirillum sp.	TSA, BA		25°C, 2–4 days
Aquimarina (*Cytophaga*) *latercula*		Marine Flexibacter medium (Lewin and Lounsbery, 1969)	
Austwickia (*Dermatophilus*) *chelonae*	BA, poly plate		25°C, 2–5 days
Bordetella bronchiseptica	BA, CFPA, MCA		37°C, 48–72 h
Clostridium botulinum	Anaerobe. ANA plates Identified by testing for toxin production after inoculation into Robertson's cooked meat broth		Incubate broths for 6 days at 30°C prior to testing for toxicity
Corynebacterium ciconiae	BA		24 h 37°C
Cytophaga hutchinsonii	Dubos medium supplemented with 30% (wt/vol) D-cellobiose		22°C, 48–72 h
Dermatophilus congolensis	BA, poly plate		37°C, 2–3 days
Empedobacter (*Flavobacterium*) *brevis*	TSA		30°C, 2–3 days
Enterococcus sp.	BA, TSA, BHIA		25°C, 48–72 h
Eubacterium tarantellae	BHIB or BA (anaerobe), ANA (Medium 110 plus vitamins is recommended by DSMZ)		20–25°C, 7 days. Anaerobic conditions
Facklamia miroungae	BA	MSA-B	37°C, 24–48 h
Flavobacterium gillisiae, *F. tegetincola*, *F. xanthum*	NA, TSA, R2A	MA 2216	20°C, 2–7 days

Flexibacter sp. – marine origin		AO-M (A0 medium prepared with artificial seawater (Sigma) at 38 g/l)	25°C, 2–7 days
Flexibacter polymorphus	Flexibacter medium cf Lewin (1974)		30°C, 2–7 days
Flexibacter roseolus	NA		25°C, 2–7 days
Flexibacter ruber	NA		25°C, 2–7 days
Halomonas elongata		TSA+8% salt	30°C, 24–48 h
Myroides (Flavobacterium) odoratus	TSA		30°C, 48–72 h
Pasteurella skyensis		TSA+blood+1.5% NaCl. MA 2216+blood (no growth without blood or 1.5% NaCl)	22–30°C, in air, 48 h
Pedobacter heparinus, *P. piscium*	NA, PY		28°C, 48 h
Pseudoalteromonas sp.		MA 2216	15–30°C, 1–5 days
Pseudoalteromonas antarctica	TSA, TSB	MA 2216, MSA-B	15°C, 5 days
Pseudoalteromonas elyakovii		MA 2216	10–37°C, optimum 25°C
Pseudoalteromonas maricaloris	TSA	MA 2216	25–35°C, 5 days
Pseudoalteromonas tetraodonis		MA 2216, TSA	25°C, 2–3 days
Pseudomonas fluorescens	BA, Pseudomonas selective agar (Becton, Dickinson Co.)		10–25°C, 2–5 days
Salegentibacter salegens		MA 2216, NA, TSA, R2A	20°C, 2–7 days
Shewanella, *Alteromonas*, *hanedai*		TSA+0.05 g/l yeast extract, MA 2216	15°C, 48–72 h
Sphingobacterium, *Flavobacterium*, *multivorum*, *S. spiritivorum*	NA, TSA, PY		28–30°C, 48 h
Staphylococcus lutrae	BA		37°C, 24 h
Staphylococcus warneri	BA, TSA		22–25°C, 48–72 h
Streptobacillus moniliformis	BA. Requires addition of 20% serum to broth for growth		25–37°C, 5–7 days

2.3 Examination of Culture Plates

Many bacteria from fish and aquatic animals need 48 h before colonies appear, or before they are of a suitable size for examination and subculture. At 24-h incubation, the colony appearance of different *Vibrio* sp. may not be distinguishable; however, at 48-h incubation, the physical appearance between different species may be more apparent. The addition of blood into a medium, such as MSA-B, for *Vibrio* sp., improves the ability to differentiate between the colonies of different species, and to determine haemolysis. Generally, a bacterium that is present in a moderate to heavy growth or as the predominant organism will be significant. Bear in mind the age of the samples at the time they were received at the laboratory. Specimens older than 24 h from the collection date may be overgrown with commensal bacteria.

Select appropriate colonies for subculture. These may be given an individual number (e.g. #1, #2). For example, selected colonies are circled and numbered on the underside of the Petri dish using a felt pen, and then picked off and subcultured to fresh plates. Freshwater samples are subcultured to BA, MCA and thiosulfate–citrate–bile salts–sucrose agar (TCBS) (the latter for suspect *Vibrio* sp.). Samples from marine sources are subcultured to BA, MCA, MSA-B and TCBS. Organisms that require special media for growth are subcultured to the appropriate media (Table 2.3). Some organisms from marine environments may need to be cultured to MA 2216 media. Incubate for a further 24 h or as required for sufficient growth. Pure subculture growths are used for the inoculation of biochemical identification tests.

2.3.1 Microscopic and cultural characteristics

Table 2.4 details microscopic and colony appearance, plus the results of the primary tests of Gram reaction, oxidase and catalase for those organisms not detailed in the text. The 'Identification sets' column lists the identification sets that can be performed for identification of that organism.

2.4 Biochemical Identification Tests

Biochemical identification of a bacterium proceeds through a number of steps. A pure subculture of the organism is used to perform primary identification tests and to inoculate biochemical identification sets (secondary identification tests), either in-house prepared media or commercial identification sets such as API available from bioMérieux. The identification sets are incubated at the appropriate temperature and for the appropriate time for reactions to occur. The results are recorded on a laboratory work sheet, or in the case of the commercial identification kits on the supplied recording sheets. See Section 2.7 for the interpretation of biochemical identification sets. The media and reagents used in the biochemical identification are detailed in the Media section. Growth characteristics or reagent reactions are also detailed in the Media section.

2.4.1 Primary tests

The primary tests include microscopic examination of smears, in particular the Gram stain, catalase, oxidase, presence of haemolysis, motility and growth on MacConkey agar. More information on the performance of these tests and their interpretation is found later in this section and in the media and reagent preparation section.

2.4.2 Secondary tests: biochemical identification sets

Biochemical identification is achieved by secondary identification tests, which are the main tests used to identify an organism to species level.

Tubed media

CARBOHYDRATE FERMENTATION. L-arabinose, glucose, inositol, lactose, maltose, mannitol, mannose, salicin, sorbitol, sucrose, trehalose, xylose. The carbohydrates are commonly called 'sugars'.

DECARBOXYLASES. Arginine dihydrolase (ADH), lysine decarboxylase (LDC), ornithine decarboxylase (ODC), decarboxylase control tube.

Table 2.4. Microscopic and cultural characteristics.

Bacterium	Gram	Morphology	βH	TCBS	Colony characteristics	Cat	Ox	Identification sets
Acinetobacter sp	Neg	Predominantly diplococcal forms 1.0×0.7 µm from plate growth	–		Colonies circular, smooth, translucent to slightly opaque, butyrous to mucoid non-pigmented. 0.5–2 mm at 24 h at 30°C	+		Gram-negative coccus table
Acinetobacter haemolyticus	Neg	Predominantly diplococcal forms 1.0×0.7 µm from plate growth	βH		Colonies on TSA are circular, convex, smooth, slightly opaque and may have a sticky consistency	+	–	Gram-negative coccus table
Actinobacillus delphinicola	Neg	Pleomorphic rods	–w	NG	Colonies on BA in 10% CO_2 are 0.75–1 mm at 24 h, round, smooth grey. Blood or serum enhances growth. Growth at 42 but not at 22°C	–	+	Gram-negative oxidase-positive table, API 20E, API 20NE, API-ZYM. Use a heavy inoculum
Actinobacillus scotiae	Neg	Pleomorphic rods	–	NG	Colonies 0.5 mm round, grey, on BA at 24 h. Requires 10% CO_2. May be weakly haemolytic on sheep blood	–	+	Gram-negative oxidase-positive table, API 20E
Actinomyces marimammalium	Pos	Straight or slightly curved rods, some branching	–		On BA, colonies are 0.5 mm grey, entire, convex at 48 h. Growth at 37°C in air with 5% CO_2	–		API Coryne, API rapid ID 32 Strep
Aequorivita sp.	Neg	Rods 0.5–20×0.2–0.3 µm			On MA 2216 yellow or orange colonies, compact, circular, convex, smooth with an entire edge, non-spreading, butyrous consistency. Growth on NA+2.5% NaCl			Environmental *Flavobacteriaceae* table, API-ZYM (some reactions)
Aerococcus viridans var. *homari*	Pos	Tetrad-forming coccus	α	NG	αH on BA. Examine with India ink for encapsulated cocci	+		Gram-positive coccus table, API rapid ID 32 Strep
Algicola (*Pseudoaltero-monas*) *bacteriolytica*	Neg	Rods with rounded ends, 0.6–0.9×1.9–2.5 µm			Growth on MA 2216, red pigment may or may not be present	+w	+	Gram-positive oxidase-negative rod table, API 20NE
Alteromonas sp.	Neg	Rods		NG	Growth on MA 2216 as small pale yellow colonies	+	+, v	Gram-negative oxidase-positive table
Arcanobacterium phocae	Pos	Cocco-bacilli and short rods, singly, pairs and clusters. Non-acid fast	βH		Colonies on BA at 24 h, white, tiny, round, with large zone of βH	v		Gram-positive rod table, API 50CH, API-ZYM
Arcanobacterium pluranimalium	Pos	Straight to slightly curved, non-branching, slender rods	αH		Alpha haemolysis on BA, small colonies	+		API Coryne, API rapid ID32 Strep

Continued

Table 2.4. Continued.

Bacterium	Gram	Morphology	βH	TCBS	Colony characteristics	Cat	Ox	Identification sets
Austwickia (*Dermatophilus*) *chelonae*	Pos	Branching filaments containing rows of cocci (zoospores). Zoospores are motile	βH		Colonies on BA in CO_2 are grey-white to grey-yellow, adherent and pit the agar. At 48 h, colonies are tiny, rough, granular, raised. Further incubation produces colonies up to 2–3 mm, umbonate, molar tooth crater forms and become mucoid	+		Gram-positive rod table. Make impression smears from underside of freshly removed scabs. Stain with Giemsa stain. Filaments are best stained with Giemsa rather than Gram stain
Bacillus cereus	Pos	Rods	βH		On BA colonies are white-grey, ground-glass appearance, 2 mm at 24 h. Become slightly yellow with age	+		Gram-positive rod table
Bordetella bronchiseptica	Neg	Thin rods with tapered ends, singly and pairs. May have long filamentous forms	NH		On BA and CFPA colonies are 1 mm at 48 h and may be haemolytic or non-haemolytic. Depending on phase variation rough and smooth colonies occur and may be rough, translucent with a raised centre and undulating outer margin, or smooth, opaque and pearl-like	+	+	Gram-negative oxidase-positive rod table. See media section for selective media
Brevundimonas diminuta	Neg	Short rods 1–4×0.5 µm	αH		Growth BA, NA, TSA, 30–37°C, 2 days. Pantothenate, biotin, cyanocobalamin are required as growth factors	+	+	Gram-negative oxidase-positive rod table, API 50CH API 20NE, API-ZYM
Brevundimonas vesicularis	Neg		αH		Yellow colonies on BA, TSA, NA	+	+	Gram-negative oxidase-positive rod table
Budvicia aquatica	Neg	Straight rods			Small colonies on NA, 0.5 mm 24 h at 30°C. Translucent with smooth entire edges	+	–	Gram-negative rod table
Chromobacterium violaceum	Neg		NH or βH		Pigmented strains are deep purple, round, slightly raised on BA. Non-pigmented strains show β-haemolysis on BA	+	+	Gram-negative oxidase-positive rod table, API 20E, API 20NE
Chryseobacterium balustinum	Neg	Rods, 0.5×1.0–3.0 µm	–	NG	Mucoid, yellow colonies on AO	+	+	Pathogenic *Flavobacteriaceae* table, API-ZYM, API 20NE
Chryseobacterium (*Flavobacterium*) *gleum*	Neg	Non-sporing rods with rounded ends	NH	NG	On NA circular, entire, viscid become mucoid and translucent after 5 days. Bright yellow pigment	+	+	Environmental *Flavobacteriaceae* table, API-ZYM, API 20NE

Chryseobacterium indologenes	Neg	Straight rods 0.5×1.3–2.5 μm			Colonies on heart infusion agar are 1 mm at 24 h at 30°C. On 0.3% agar growth spreads out in a flower-like growth pattern	+	+	Pathogenic *Flavobacteriaceae* table, API 20NE, API-ZYM
Citrobacter freundii	Neg	Rods		NG	2 mm grey colonies at 24 h on BA	+	–	Gram-negative rod table, API 20E
Clostridium botulinum	Pos	Oval, subterminal spores, 3.4–7.5×0.7 μm	βH		Semi-opaque to translucent, matt colony, 1–3 mm, irregular lobate margin and raised centre	–		Gram-positive rod table, API rapid 32 A
Colwellia maris	Neg	Curved rods, 0.6–1×2–4 μm			Marine 2216 agar, optimal growth at 15°C. Growth range 0–22°C, but no growth at >25°C	+	+	Gram-negative oxidase-positive table
Corynebacterium ciconiae	Pos	Short rods, non-motile	NH		ON BA colonies are 1–2 mm, circular, convex, dry after 24 h at 37°C	+	–	Gram-positive rod table, API Coryne, API ZYM, API 50CH
Corynebacterium testudinoris	Pos	Diphtheroid rods			Colonies on BA are yellow-pigmented	+		API Coryne, API-ZYM
Cryptococcus neoformans var. *gattii*	Pos	Budding yeasts	NH		Slow growing. Colonies 0.5–1 mm cream dull, 3 days on BA. Colonies change from cream to dirty yellow, light tan colour after 7 days on BA			India ink for capsule stain, urease-positive. Growth on Strep-selective media (Oxoid), and SAB plates. Use Biological Safety Cabinet
Dermatophilus	Pos	Branching filaments containing rows of cocci (zoospores). Zoospores are motile	βH		Colonies on BA in CO_2 are grey-white to grey-yellow, adherent and pit the agar. At 48 h, colonies are tiny, rough, granular, raised. Further incubation produces colonies up to 2–3 mm, umbonate, molar tooth crater forms and become mucoid	+		GPR Set. Make impression smears from underside of freshly removed scabs. Stain with Giemsa stain. Filaments are best stained with Giemsa rather than Gram stain
Edwardsiella hoshinae		Straight rods			Growth on NA, BA, MCA. Colonies flat or slightly convex, 1–2 mm at 24–48 h, 30 and 37°C	+	–	Gram-negative rod table, API 20E
Empedobacter brevis	Neg	Rods			Yellow colonies on NA and BA, 0.2–2.5 mm at 24 h and 30°C, low convex, entire edge. At day 7 on BA may see slight α-haemolysis	+	+	Environmental *Flavobacteriaceae* table

Continued

Table 2.4. Continued.

Bacterium	Gram	Morphology	βH	TCBS	Colony characteristics	Cat	Ox	Identification sets
Eubacterium sp.	Pos	Long chains of pleomorphic rods. No spores. In older cultures cells may have 'ovoid' bodies	βH		On BA under anaerobic conditions, colonies are haemolytic, translucent, slowly spreading, flat, and contoured with filamentous edges. Anaerobic atmosphere required	–	–	Anaerobe table
Eubacterium limosum	Pos	0.6–0.9×1.6–4.8 μm. Rods may have swollen ends	NH		Punctate, circular, convex, entire colonies that are translucent to slightly opaque. Less than 1 mm at 48 h. Anaerobic	–		Anaerobe table, API rapid 32 A
Eubacterium tarantellus	Pos	Very long filamentous, unbranched rods (10 μm)	βH	NG	2–5 mm, translucent colonies, slightly rhizoid and mucoid. Anaerobic	–		Anaerobe table
Flavobacterium aquatile	Neg				Colonies on AO media low, convex, round	+	+	Environmental *Flavobacteriaceae* table, API-ZYM
Flavobacterium flevense	Neg				On AO media, low convex, round colonies sunken into agar	+	+	Environmental *Flavobacteriaceae* table, API-ZYM
Flavobacterium frigidarium	Neg	Rods 0.8–2×0.5–0.7 μm, singly and in pairs			Growth on AO, flat round yellow colonies with entire margins. Grows on NA, TSA and MA 2216	+	+	Environmental *Flavobacteriaceae*, API-ZYM
Flavobacterium gillisiae	Neg	Rods 2–5×0.4–0.5 μm			Orange-pigmented colonies, butyrous, circular and convex with entire edge. Growth on MA 2216, NA, TSA, R2A	+	–	Environmental *Flavobacteriaceae*, API 20E
Flavobacterium hibernum	Neg	Rods 0.7×1.8–13 μm			Growth on TSA. Yellow, mucoid colonies at 25°C, gelatinous at 4°C	+	–	Environmental *Flavobacteriaceae* table, API 20E, API 20NE
Flavobacterium hydatis (*Cytophaga aquatilis*)	Neg	Rods, 8.0×0.5 μm	NH	NG	Mucoid, yellow-orange colonies with flat spreading filamentous margins. Incubate at 14°C for 14 days	+	–	API-ZYM, API 20E
Flavobacterium pectinovorum	Neg				On AO media, colonies low round with entire margins		+	Environmental *Flavobacteriaceae*, API-ZYM
Flavobacterium saccharophilum	Neg				On AO media, colonies flat, spreading, sunken into agar	+	–	Environmental *Flavobacteriaceae*, API-ZYM
Flavobacterium succinicans	Neg				On AO media, colonies flat, spreading, with filamentous margins			Environmental *Flavobacteriaceae*, API-ZYM

Flavobacterium tegetincola	Neg	Rods 2–5×0.4–0.5 µm			Yellow-pigmented colonies, butyrous, circular, convex, entire. Growth on MA 2216, NA, TSA, R2A	+	–	Environmental *Flavobacteriaceae*, API 20E
Flavobacterium xanthum	Neg				Growth on MA 2216, NA, TSA, R2A	+	+	Environmental *Flavobacteriaceae*, API 20E
Flexibacter polymorphus	Neg	Filaments multicellular. At end of each cell is a refractile granule of lipid material			Marine agar. Cobalamin required for growth. Peach-coloured pigment	–	+	Environmental *Flavobacteriaceae*, API 50CH, API-ZYM
Halomonas aquamarina (previously *Alcaligenes faecalis homari*)	Neg	Straight or curved rods 1.5 µm, bipolar staining			Colonies on MA 2216, off-white translucent, raised smooth 2–3 mm at 24 h at 18°C and 37°C. Slight tendency to spread	+	+	Gram-negative oxidase-positive table
Halomonas elongata	Neg	Rods singly or paired			Colonies 2 mm at 24 h, smooth, glistening, opaque white. After 24 h colonies spread due to flexuous filaments, but are no larger than 4 mm. Requires 8% NaCl for growth	+	+V	Gram-negative rod table. Add 8% salt to medium
Halomonas halodurans	Neg	Rods single, pairs or chains			On MA 2216 colonies are smooth, glistening, translucent white, convex, 1–2 mm in 24 h at 20 or 30°C			Gram-negative rod table
Halomonas venusta	Neg	Rod			At 48 h colourless, mucoid colonies on BA, MCA	+	+	Gram-negative oxidase-positive table, API 20NE
Hydrogenophaga (Pseudomonas) pseudo-flava	Neg	Ovoid cells. Rods of 2.5 µm in older cultures		NG	On NB agar yellow-pigmented colonies 2–4 mm at 3 days, slightly irregular shape with an undulating margin	W	+	Gram-negative oxidase-positive table
Iodobacter fluviatile	Neg	Small rod, 0.7×3.0–3.5 µm		NG	Violet-coloured colonies, thin spreading margin on low nutrient agar such as ¼ strength NA. Non-spreading on NA	+	+	Gram-negative oxidase-positive table
Mannheimia haemolytica	Neg	Pleomorphic rods, long and short forms	βH		Grey colonies 1–2 mm on BA at 24 h. β-Haemolysis seen	+	+	Gram-negative oxidase-positive table, API-ZYM
Marinilabilia salmonicolor	Neg				Gliding, yellow to pale pink-pigmented colonies	+	+	Environmental *Flavobacteriaceae*, API-ZYM

Continued

Table 2.4. Continued.

Bacterium	Gram	Morphology	βH	TCBS	Colony characteristics	Cat	Ox	Identification sets
Mesophilobacter marinus	Neg	Coccobacilli, pleomorphic	NH	NG	Growth on MA 2216 and NA. Colonies circular sometimes irregular, convex, glistening, opaque, pale yellow-brown. Describes marine coccobacilli that morphologically resemble the Acinetobacter–Moraxella group	+	+	Gram-negative coccus table
Mycoplasma mobile					Cells glide in wet preparation as seen by dark ground illumination. Cells are elongated. Optimum temperature is 25°C			Mycoplasma table
Myroides (*Flavobacterium*) *odoratimimus*	Neg	Rods 0.5×1–4 µm	NH		Yellow-pigmented colonies. Growth on MCA, NA and TSA. No gliding or swarming. Fruity odour	+	+	Gram-negative oxidase-positive rod table, API 20NE, API-ZYM
Myroides (*Flavobacterium*) *odoratus*	Neg	Rods 0.5×1–4 µm	NH		Yellow-pigmented colonies, 3–4 mm on MCA, NA and TSA at 24 h. No gliding or swarming. Fruity odour	+	+	Gram-negative oxidase-positive rod table, API 20NE, API-ZYM
Pantoea (*Enterobacter*) *agglomerans*	Neg	Non-encapsulated, non-spore forming, broad, straight rod, 0.5–1.0× 1.0–3.0 µm	NH	NG	Colonies on NA are smooth, convex, entire, and translucent and may have a yellow pigment. Growth on BA, MSA-B and MA 2216. Growth at 37°C but not at 44°C	+	–	Gram-negative rod table, API 20E
Pasteurella multocida	Neg	Cocco-bacilli	NH		Growth on BA, 1–2 mm at 24–48 h. Colony size and appearance vary from different animal species. Colonies may be smooth or mucoid, dark grey with greenish appearance. Distinctive musty sweetish smell	+	+	Gram-negative oxidase-positive rod table, API-ZYM, API 20E
Pasteurella testudinis	N	Pleomorphic rod 0.2×1.5–2 µm	βH		BA white, mucoid, 0.5–1 mm 24–48 h, 20–37°C	+	+	Gram-negative oxidase-positive rod table
Pedobacter heparinus	Neg	Non-sporing rods, 0.4–0.5×0.5–1.0 µm	βH slow		Growth on NA, 1–3 mm at 48 h, circular, low convex, smooth, opaque. A yellow or creamy white non-fluorescent pigment is produced. Creamy white colonies on PY	+	+	Environmental *Flavobacteriaceae*, API 50CH, API-ZYM

Pedobacter piscium	Neg	Non-sporing rods 0.4–0.5×0.5–1.0 µm			Colonies on NA are circular, entire, low convex at 2 days. A yellow or creamy white, non-fluorescent pigment is produced	+	+	Environmental *Flavobacteriaceae*, API 50CH, API-ZYM
Planococcus kocurii	Pos	Spherical cells 1.0–1.2 µm, singly, pairs and tetrads. Motile			Colonies on peptone yeast extract agar are circular, smooth, convex and yellow-orange pigmented	+	–	Gram-positive coccus table
Planococcus okeanokoites	Neg, v	Rods 0.4–0.8×1–20 µm			Cells bright orange to yellow. Requires 3% NaCl	+	W	Gram-positive coccus table. Add NaCl (3%)
Plesiomonas shigelloides	Neg	Straight, variable length rod	NH	W, G	Growth on BA and MCA 1–2 mm 24 h at 25–37°C. Pungent smell	+	+	Gram-negative oxidase-positive rod table, API-ZYM, API 20E
Providencia friedericiana	Neg	Non-sporing rods 0.5×1–3 µm	NH		On BA, colonies 1–2 mm at 24 h, glossy, semitranslucent, smooth. Orange-red on MCA	+	–	API 20E
Providencia rettgeri	Neg	Rod 0.6×1.0–1.5 µm	NH		Colonies on BA, TSA circular, discrete, convex, glistening, creamy-white 24 h at 37°C	+	–	Gram-negative rod table, API 20E
Pseudoalteromonas genus	Neg	Straight or curved, non-spore forming rods 0.2–1.5×1.8–3 µm. Single flagella			Strict aerobes, growth on MA 2216 at 20°C. Negative for ADH. All positive for gelatin and DNase	+, w, v	+	Gram-negative oxidase-positive rod table
Pseudoalteromonas antarctica	Neg	Rods, 0.9×1–3 µm, become larger and filamentous with age (10 µm)			Growth at 4–30°C on TSA, MSA-B, MA 2216. Colonies round, smooth, beige, convex, mucoid, 1–2 mm in 5 days at 15°C. Growth factors not required. Aerobic	+	+	Gram-negative oxidase-positive rod table, API 20NE, API-ZYM
Pseudoalteromonas flavipulchra, *Pseudoalteromonas aurantia*	Neg	Straight rods 0.7–1.5×1.5–4 µm			On MA 2216, colonies 1 mm, bright pale yellow 24 h, 23°C, At 5 days, orange-brown to green-brown 4 mm On MSA-B, bright, mucoid, dark-brown colonies 2 days, then medium turns black, and haemolysis is seen. Poor growth on TSA and NA even with 2% NaCl	+	+	Gram-negative oxidase-positive rod table
Pseudoalteromonas denitrificans	Neg	Rods 2–4×0.5–0.7 µm			On MA 2216, colonies have pink to red pigment	–	+	Gram-negative oxidase-positive rod table

Continued

Table 2.4. Continued.

Bacterium	Gram	Morphology	βH	TCBS	Colony characteristics	Cat	Ox	Identification sets
Pseudoalteromonas elyakovii	Neg	Rods with rounded ends, 0.5–0.8×1.8–4.0 μm			Growth on MA 2216 as beige coloured colonies, round, circular, smooth, convex	+	+	Gram-negative oxidase-positive rod table
Pseudoalteromonas espejiana	Neg	Straight rods, 0.2–1.0× 2.0–3.5 μm			Growth on MA 2216	+	+	Gram-negative oxidase-positive rod table
Pseudoaltermonas flavipulchra	Neg	Rods, single cells, 0.5–1.5 μm			Growth MA 2216 with orange colonies	+	+	Gram-negative oxidase-positive rod table
Pseudoalteromonas luteoviolacea	Neg				Growth on MA 2216. Colonies 3–5 mm at 4 days at 25°C, regular, convex, opaque, violet-coloured	–	+	Gram-negative oxidase-positive rod table
Pseudoaltermonas maricaloris	Neg	Rods, single cells, 0.7–0.9×1.0–1.2 μm			Growth on MA 2216, as round colonies 2–3 mm, circular, regular, convex, translucent, smooth, lemon-yellow pigment at 48 h. Optimum 25–35°C, 0.5–10% NaCl	+	+	Gram-negative oxidase-positive rod table
Pseudoalteromonas rubra	Neg	Straight or slightly curved rods, 2–4×0.8–1.5 μm	βH		Growth on MA 2216 bright pinkish-white at 24 h at 23°C. Centre turns red and sometimes blue at 4 days, size 6–7 mm with two or three concentric rings. With blood added to MA 2216, colonies are bright, mucoid, red almost black with β-haemolysis. Colonies produce a diffusible brown pigment with an odour of hydrogen cyanide	+	+	Gram-negative oxidase-positive rod table, API-ZYM
Pseudoalteromonas ulvae	Neg	Rods 1.75–2.5×1–1.5 μm			Growth on TSA with 2% NaCl, MSA-B, MA 2216. Colonies on TSA are white, whereas on MA 2216 medium colonies are purple at 48 h, 23°C	+	+	Gram-negative oxidase-positive rod table, API 20E
Pseudoalteromonas undina	Neg	Curved rods 0.7–0.9× 1.8–3.0 μm			Growth on MA 2216	+	+	Gram-negative oxidase-positive rod table
Pseudomonas stutzeri	Neg	Straight and slightly curved rods, some bipolar staining	NH	NG	There may be rough and smooth colonies. Colonies 0.5 mm, grey, wrinkled, dry and buff-coloured. May be slightly yellow in colour	+	+	Gram-negative oxidase-positive rod table
Rhodococcus erythropolis	Pos	Short ovoid rods	NH		On BA colonies are round, shiny, beige. Strictly aerobic	+	–	Gram-positive rod table, API 50CH

Rhodococcus (luteus) fascians	Pos	Straight or slightly curved rods 0.6–1×3–6 µm in angular or parallel arrangement			Weak growth on NA colonies yellow, raised, glistening, smooth. Growth on Lowenstein–Jensen medium is abundant and yellow-orange	+		Gram-positive rod table
Rhodococcus maris	Pos	Short ovoid rods 0.6–1×1–2 µm			Weak growth on NA, colonies yellow, raised, glistening, smooth	+		Gram-positive rod table
Rhodococcus qingshengii	Pos	Short ovoid rods				+	–	Gram-positive rod table, API 20Strep, API ZYM
Rhodococcus sp.	Pos	Rods, slightly club-shaped, 2–3x0.6 µm			Growth on BA and MSA-B in 3–4 days at 25°C. No growth at 37°C (except for *R. equi*). At 8 days colonies slightly domed, round, smooth, dry, deep creamy yellow	–	–	Gram-positive rod table
Salegentibacter salegens	Neg	Rod-shaped cells, singly, pairs, occasional chains			Growth on MA 2216, NA, TSA	+	+	Environmental *Flavobacteriaceae*
Serratia fonticola	Neg	Rods 0.5×3.0 µm			Growth on NA at 30 and 37°C	+	–	Gram-negative rod table, API 20E
Sphingobacterium multivorum	Neg	Non-sporing rods 1 µm, singly or pairs			Yellow colonies on NA, circular, low, convex	+	+	API-ZYM
Sphingobacterium spiritivorum	Neg	Non-sporing rods 1 µm	NH		Colonies on NA, yellow, low convex, smooth 30–37°C	+	+	API-ZYM
Sphingomonas (Pseudo-monas) paucimobilis	Neg	Rods, 0.7×1.4 µm	NH		Growth on NA and BA. Yellow-pigmented colonies circular, low, convex at 2 days at 22°C	+	+	Environmental *Flavobacteriaceae*
Staphylococcus delphini	Pos	Cocci 0.8–1 µm, mostly in clusters but also singly and paired	βH	NG	On NA, colonies 5–7 mm circular, smooth, opaque to translucent with incubation	+	–	Gram-positive coccus table, coagulase test, API-ZYM
Staphylococcus hominis	Pos	Cocci in tetrads, occasionally pairs	NH		Colonies 3–5 mm, smooth, dull, opaque, slightly umbonate with bevelled edge. Pigment pale yellow to grey-white	+		Gram-positive coccus table, coagulase, DNase
Staphylococcus lutrae	Pos	Cocci singly, pairs and clusters	βH		Colonies 1.5–2 mm, smooth, round 24 h, 37°C	+	–	Gram-positive coccus table, coagulase, DNase
Staphylococcus warneri	Pos	Cocci 0.5–1.2 µm diameter in pairs and singly, occasionally tetrads	NH		Colony growth in 24 h at 37°C, 3–5 mm, smooth, round, sticky. Most strains are bright yellow-orange or with a yellow ring around the edge of the colony. 20% of strains are non-pigmented	+		Gram-positive coccus table, coagulase, DNase

Continued

Table 2.4. Continued.

Bacterium	Gram	Morphology	βH	TCBS	Colony characteristics	Cat	Ox	Identification sets
Streptobacillus moniliformis	Neg	Pleomorphic rods frequently in chains, tangled filaments with bulbous swellings	NH		Clear, non-haemolytic colonies, 0.5 mm at 48 h on BA in CO_2 and 37°C. In broth culture produces discrete fluff-ball-like colonies	–	–	Gram-negative rod table
Streptobacillus monili-formis-like organism ID not conclusive (Maher *et al.*, 1995)	Neg	Coccobacilli 0.4–0.6 µm in tissues. Older cultures show filamentous cells with possible branching and some terminal swellings	βH		Growth on BHIA containing fetal calf serum and NaCl. Colonies 0.1 mm at 4–8 days, aerobically at 15–22°C. Colonies off-white, friable, convex, granular, 'breadcrumb' appearance. Older colonies were concave 'molar tooth' appearance. βH after 7–14 days	–	–	Gram-negative rod table. Weakly acid-fast
Streptococcus porcinus	Pos	Spherical to ovoid cells in small chains	βH		On BA colonies are small, elevated, entire and haemolytic	–		Gram-positive coccus table, API 20 Strep, API 50CH
Streptomyces salmonis	Pos	Mycelia		ND	Brick-red to orange substrate mycelia and white to pink and yellow shades of aerial mycelia			Gram-positive rod table
Yersinia kristensenii	Neg	Rods			Growth on NA, BA in 24 h at 28°C 1–2 mm. Cultures may have strong, musty or cabbage-like odour	+	–	Gram-negative rod table, API 20E
Zobellia galactanovorans	Neg	Rods 0.3–0.4×3–8 µm with rounded ends			Growth on MA 2216 as yellow spreading colonies at 30°C. Hydrolyse agar	+	+	Environmental *Flavobacteriaceae*, API 20E, API 20NE, API-ZYM. Addition of NaCl, optional
Zobellia uliginosa	Neg				On MA 2216 agar colonies are orange with tenacious slime. Spreading activity is weak or negative. Hydrolyse agar	+	+	Environmental *Flavobacteriaceae*, API 20E, API 20NE, API-ZYM. Addition of NaCl, optional

Where NaCl is required for optimal growth, add NaCl to the set at a final concentration of 2% NaCl unless stated otherwise.

OTHER BIOCHEMICAL TESTS. Aesculin, motility, methyl red Voges–Proskaüer (MRVP), nitrate, oxidative fermentative tubes, *o*-nitrophenyl β-D-galactopyranoside (ONPG), triple sugar iron (TSI) for H_2S, indole, urea. See next section for inoculation methods.

GROWTH AT DIFFERENT TEMPERATURES. Use either a tryptone soy broth (TSB) or glucose tube and incubate at 37°C, 40°C or desired test temperature.

GROWTH IN 10% NACL. To test an organism's ability to grow in the presence of 10% NaCl, dispense equal volumes of TSB and 20% NaCl into a sterile McCartney bottle or bijou bottle to give a final concentration of 10% NaCl. Inoculate with the organism to a concentration of McFarland opacity tube 1. Incubate at 25°C or optimum temperature and after 24 h incubation observe for evidence of growth as seen by an increase in the turbidity of the broth.

Plate media

DNase plate, gelatin/salt plate, MCA, TCBS. Use an MSA-B or BA plate for purity check of the inoculum.

Discs

Ampicillin 10 µg, 'Vibrio discs' 0129 10 µg and 150 µg. The discs are placed on a lawn inoculum of the bacterium on either a BA or MSA-B plate according to the organism's growth requirements. The inoculum should be prepared to a density of McFarland tube 1. Use for *Listonella*, *Moritella*, *Photobacterium*, *Vibrio* and *Aeromonas* sp. The Vibrio discs (vibrio static agent pteridine 0/129) differentiate between *Vibrio* sp. and *Aeromonas* sp., the latter being resistant to both concentrations. A zone of 9 mm is classified as susceptible for the 0/129 150 µg disc (Bernardet and Grimont, 1989). Also see section for media and reagent preparation.

2.5 Inoculation of Biochemical Identification Sets

Examine the subculture plate and, if the growth is pure, use to inoculate biochemical identification sets. For marine samples, NaCl needs to be added to the set at a final concentration of 2%. Many enzymes from a variety of *Vibrio* sp. will not be active at lower NaCl concentrations and false negative results will be obtained. This may apply to indole, Voges–Proskaüer (VP), ADH, ODC and LDC reactions. Some tests need a heavy inoculum. These are the decarboxylases and arginine dihydrolase, urea and citrate. Incubate the set at the appropriate temperature for a minimum of 48 h. ODC, LDC, ADH, methyl red (MR), VP and indole usually require 48 h incubation as a minimum for a positive reaction even if reasonable growth is seen in the tubes.

To inoculate the tubes plunge a sterile Pasteur pipette three-quarters into the tubed media, releasing three or four drops of inoculum as the pipette is withdrawn.

After inoculating all the tubes, place one drop of the inoculum onto a purity plate and streak out for isolated colonies. After suitable incubation, check that the growth is pure.

2.5.1 Freshwater isolates

For bacteria isolated from freshwater sources, inoculate bacterium into sterile normal saline or sterile distilled water (usually 10 ml) to opacity McFarland 3. Add three to five drops of inoculum to each tube. Some media require a heavy inoculum (opacity approx. McFarland 6). These are aesculin, ADH, LDC, ODC, the decarboxylase control, MRVP, citrate and urea.

2.5.2 Marine isolates and addition of NaCl

Fish organs are at physiological saline level. Therefore, be aware that some bacteria may adapt to this NaCl concentration. By subculturing an organism to BA and MSA-B, or culturing to a 0% and 3% salt plate, the salt requirement or preference can be determined. However, many of these organisms produce more accurate results in biochemical tests when the NaCl is added to the media, usually at a final concentration of 2%. Therefore check the optimal NaCl concentration range of the suspect organism from the NaCl column in the tables for biochemical results or in the text under the relevant organism.

For bacteria isolated from the marine environment, the optimal final salt concentration in the medium is all-important. Insufficient salt concentration will lead to false negative results even though good growth may be seen in the tube. An example of this is *Vibrio parahaemolyticus*, which will give a negative indole result when an inoculum of physiological saline is used, but a positive result when a final concentration of 2% NaCl is achieved in the tryptone water (indole test) medium (*See photographic section*).

Tubed media can be prepared with or without added NaCl. However, a laboratory may want to keep media preparation to a minimum and prepare all media without additional NaCl. At the time of inoculation, 500 μl of a 20% sterile NaCl stock solution is added to all liquid media (5 ml) (aesculin, indole test, ADH, ODC, LDC, nitrate, MRVP). A sterile 20% salt stock solution can be prepared in a Schott bottle to which an autoclavable 2-ml volumetric dispenser is added, and set to 500 μl. An example of a volumetric dispenser is a Socorex Calibrex 520, 2 ml with 0.05-ml divisions.

2.5.3 Paraffin oil overlay

ADH, LDC, ODC and control and one tube of the oxidative fermentative (OF) media is overlaid with sterile paraffin oil, approximately 5 mm depth. For ease of use, paraffin oil can be sterilized in a Schott bottle to which a 2-ml volumetric dispenser is added (Socorex, Calibrex 520; 2 ml with 0.05-ml divisions).

2.5.4 Incubation

Plate and tubed media are incubated at the appropriate temperature and duration for the appropriate bacterium. A temperature of 25°C and 2–5 days is used as a general rule. Refer to incubation guidelines in Table 2.3 or the text for the relevant organism.

2.6 API Identification Systems

bioMérieux, France, produces a number of kits (listed below) for the identification of bacteria. Information on the kits, their inoculation and interpretation can be found at http://www.biomerieux-diagnostics.com under Customer Service and apiweb®.

API 20E
API 20NE
API 50CH
API rapid A
API rapid ID 32 STREP
API 20 Strep
API Coryne
API ZYM

These kits have been established for the identification of bacteria from medical laboratories. However, their use for the identification of bacteria from the aquatic environment has been reported in the literature. Specific information on how to inoculate and interpret the results for bacteria from aquatic animals is detailed below.

2.6.1 Identification using API 20E

There have been a number of reports in the literature about the failure of the API 20E system to identify many *Aeromonas* and *Vibrio* sp. (Santos *et al*., 1993). Some of this is due in part to the lack of such information in the API database. However, it is well noted that differences in reactions occur between the API system and conventional biochemical tests. This is especially true of the decarboxylases, citrate, urea, indole and VP. Where such differences have been reported in the literature, these have been indicated in the conventional tables (Table for pathogenic *Vibrio* sp. etc.). When using the biochemical identification tables in this book, make sure the correct database is referred to, i.e. the conventional database or the API 20E database.

MacDonell *et al*. (1982) recommend using a diluent containing 20% salt. They used marine salts mix (Instant Ocean) purchased from Aquarium Systems, Mentor, Ohio, which is composed of marine salts with salinities adjusted to 20%. Kent (1982) recommended suspending bacterial cells in 50% sterile artificial seawater for inoculating the API 20E strips. Artificial seawater salts can also be purchased from Sigma (see Media section). Alternatively, an inoculum with 2% sterile NaCl works well for the commonly isolated *Vibrio*, *Photobacterium* and *Listonella* sp. The optimal NaCl concentration is extremely important for tests such as citrate, urea, MR, VP and indole. For example, *V. parahaemolyticus* gives a negative result for indole when an inoculum of 0.85% NaCl is used, but a positive indole result is achieved when 2% NaCl is used.

Adjust the suspension of cells to an opacity of McFarland number 1 and inoculate the set. It is recommended that the API 20E system is incubated at 25°C for 48 h. The sugars are read at 24 h and the remaining tests read at 48 h. The decarboxylases (LDC, ODC) and arginine dihydrolase (ADH) reactions may not be detected until 48 h. As recommended by the manufacturer, a negative nitrate reaction should always have zinc dust added to the reaction cupule to determine a true negative result (see Media section and Table 2.5). See later in this section for interpretation and results from the API kits.

2.6.2 Identification using the Vitek (bioMérieux)

The Vitek is an automated system for the identification of bacteria, yeasts and fungi. It contains some *Vibrio* sp. in its database, predominantly those *Vibrio* sp. that cause infection in humans. However, as the machine is programmed for incubation at 37°C, only *Vibrio* sp. that grow at 37°C can produce reactions.

2.6.3 Identification using the MALDI-TOF

Matrix-assisted laser desorption ionization time of flight (MALDI-TOF) mass spectrometry is gaining prominence in medical and veterinary laboratories (Dekker and Branda, 2011; Neville *et al.*, 2011). Bacteria, yeasts and fungi are identified using the abundant ribosomal proteins predominantly in the 2–20 kDA range and the resulting mass spectrograph is compared against a database. A number of *Vibrio* sp. are in the database; however, at this stage, some results need to be treated with caution because with the majority of vibrios only one representative from each species is present in the database and often a significant identification level is obtained simultaneously for two or more species. Once the database is strengthened using more representative isolates within each species, this technology will come to the fore for microbial identification. Because of similarities between *Aeromonas* sp., the database may not be able to distinguish between species. In these cases further biochemical tests or sequencing may assist in a more definitive identification.

2.7 Interpretation of Biochemical Identification Tests and Sets

2.7.1 Conventional media

Positive sugar fermentation results may be recorded at 24 h. All liquid media must be incubated for at least 48 h, and sometimes longer, for positive results to be achieved. This applies in particular to ADH, LDC, ODC, aesculin, citrate and MRVP.

Other tests: For a description of other tests, such as hippurate hydrolysis etc., refer to the Media section.

2.7.2 Identification tests and their interpretation

This section contains notes on the interpretation of identification tests and any problems that may occur. A description of each test is also found in the chapter containing the test media recipes.

Aesculin

A positive result is recorded when the colour is pitch-black and when half or more of the tube is blackened (MacFaddin, 1980). Shades of grey should not be recorded as positive. Results should be read up to 3 days. Some organisms will not grow in the aesculin medium despite the optimal salt concentration. These are *Vibrio* (*Listonella*) *anguillarum*, *Vibrio* (*Listonella*) *pelagia*, *Vibrio ordalii*. These organisms are all negative for aesculin production.

Some bacteria, especially some *Vibrio* sp., produce melanin (Coyne and Al-Harthi, 1992) and in the aesculin medium will cause blackening of the medium and can appear as a false-positive result (Choopun *et al.*, 2002). To determine a true positive aesculin result, the aesculin test must be examined for loss of fluorescence under long-wave ultraviolet (UV) light at 354 nm. Aesculin will fluoresce and therefore the presence of fluorescence will indicate that the aesculin has not been hydrolysed. A positive result, that is, the hydrolysis of aesculin, will be negative for fluorescence (MacFaddin, 1980). Some plastic tubes are UV opaque; therefore, test for fluorescence by holding a Wood's UV lamp over the top of an opened tube, or pour the contents into a Petri dish to observe presence or absence of fluorescence. When using UV light, protect

Table 2.5. Interpretation of conventional tests as reported in the tables for Gram-positive, Gram-negative, etc.

Test	Reagents added	Result
Carbohydrate fermentation	Nil	Colour change from red to yellow (if using phenol red pH indicator). *See photographic section*
Aesculin (method from Cowan and Steel, 1974)	Nil	A positive is a dark black colour. All positive reactions should be checked for true hydrolysis, because a colour change may be due to melanin production from some organisms such as *Vibrio cholerae*. Aesculin is naturally fluorescent, therefore a loss of fluorescence indicates hydrolysis of aesculin and a true positive result. Test for loss of fluorescence using a UV light such as a Wood's lamp. (Wear appropriate protective clothing and eyewear)
ADH (Arginine dihydrolase) Moeller's method	Nil	A positive result is a change of colour from green-brown to bright purple colour (Moeller's method). Thornley's arginine method is recommended for some *Vibrio* sp. *See photographic section* In the Vibrio, Aliivibrio, Moritella and Photobacterium tables (4.19, 4.20, 4.21, 4.22) Moeller's ADH is reported unless indicated
Citrate (Simmons')	Nil	A positive is a change of colour from green to blue. *See photographic section*
DNase plate	Flood plate with 1 M HCl. Leave 1–2 min and pour off	Zone of hydrolysis. Examine plate over a dark-coloured tile and look for zones of clearing (hydrolysis) around bacterial growth. *See photographic section*
ODC, LDC	Nil	A positive reaction is a colour change from green-brown to purple/grey. *See photographic section*
Gelatin/salt plate (tests for gelatin hydrolysis and NaCl requirement)	Nil	The growth or lack of growth on 0% or 3% NaCl is recorded for the NaCl requirement. Gelatin hydrolysis is observed by holding the plate against a dark background and with transmitted light examining for a zone of hydrolysis around the bacterial colony. The hydrolysis of gelatin is seen as either a zone of clearing on the medium with 0% NaCl or as a zone of opacity on the 3% NaCl side. The appearance of the zones may be improved by chilling the plate at 4°C for 20 min, or flooding with ammonium sulfate. *See photographic section*
H_2S (TSI medium)	Nil	Blackening of medium especially along the inoculum line
Indole (tryptone water)	Add three to seven drops of Kovács' reagent	A positive reaction is the formation of an upper pink layer in tube. *See photographic section* The test solution must contain NaCl (2% final concentration) if the organism requires NaCl for growth. NaCl can be added at the time of preparation of the batch of media or added to single tubes at the time of inoculation
1. Motility – tube method 2. Motility – hanging drop method	1. Nil 2. Nil	**1.** Tube method. Motile bacteria grow and spread through the semisolid gel. Spreading is seen as diffuse growth emanating from the line of inoculation. Non-motile bacteria do not migrate from the line of inoculation **2.** Hanging drop method. Place a drop of saline onto a coverslip. Emulsify some organisms from a plate or broth culture in logarithmic phase of growth. Invert the coverslip onto a glass slide that has been prepared with four lots of 3–4 mm spots of plasticine or equivalent, so that the drop containing the organisms is suspended from the coverslip. The plasticine spots are placed according to the corners of the coverslip. A hanging drop is the optimal method for observing motility

		The coverslip may be placed directly onto the microscope slide, but must be examined before the slide has dried. Also, care must be taken that the liquid is not 'pulled' from the slide if it comes into contact with the microscope stage apparatus, because the appearance of bacterial cells being drawn to the side of the coverslip may give the appearance of motility Examine under ×40 or ×100 objective and inspect bacterial cells for definite individual motion that causes the cell to traverse the glass slide. Do not mistake motility for Brownian motion, which is seen as movement of the cell within a small confined area, usually the diameter of the cell
MR (methyl red)	Add three to five drops of methyl red reagent after 48 h incubation (test for VP reaction first)	Persistence of red colour The reaction should not be tested in less than 48 h incubation because a false-negative reaction may be obtained. The optimal time is 3 days' incubation. Aliquots can be taken and tested daily. A negative reaction should be incubated for 3 days. *See photographic section*
VP (Voges–Proskaüer)	Remove 200 µl of medium from MRVP tube and place into a 0.6-ml microfuge tube Add one drop each of VP I and VP II reagents (can use the reagents from the API 20E kit)	A positive reaction is indicated by a pink colour development after 10 to 20 min If an aliquot is negative at 24 h, incubate the original tube for a further 24 h and repeat the addition of reagents The temperature of incubation is important for VP because some organisms may be negative at 37°C, but positive at 25°C
Nitrate	Add five drops of each reagent, nitrate A and nitrate B Can use in-house prepared reagents or reagents from API 20E kit	Formation of a red colour is positive For negative reactions add a match-head size amount of zinc dust. Formation of a pink colour confirms a negative nitrate result, whereas no further colour with Zn indicates a positive reaction
OF medium (oxidative fermentative)	Nil	A yellow colour is recorded as positive, and indicates the fermentation of glucose. Fermentation of glucose in the tube without paraffin oil (open tube), but not in the tube with oil indicates an oxidative organism. Growth in both tubes indicates the organism is a facultative anaerobe and ferments glucose under aerobic and anaerobic conditions
ONPG	Nil	Yellow colour is positive. Clear or no colour is recorded as negative
TCBS (thiosulfate–citrate–bile salts–sucrose agar)	Nil	Record growth of yellow or green colonies, or record 'no-growth'. *See photographic section*
Urea (Christensen's)	Nil	A bright pink colour is positive
'Vibrio discs' 0/129 10 µg and 150 µg	Nil	Record a zone around the disc as 'sensitive', and no zone as 'resistant'. A zone of 9 mm is classified as susceptible for the 0/129 150 µg disc

Some organisms will not grow in the aesculin medium despite the optimal salt concentration. These are *Vibrio* (*Listonella*) *anguillarum*, *V. pelagius* and *V. ordalii*. These organisms are all negative for aesculin production.

the eyes with UV opaque glasses, and exposed skin with laboratory clothing and gloves.

Arginine dihydrolase (ADH)

Some strains of *Vibrio* may show contradictory results for ADH when using Moeller's or Thornley's medium. These include *V. mediterranei*, *V. mytili*, *V. orientalis*, most strains of *V. splendidus* I, and some strains of *V. tubiashii*. They are all positive according to Thornley's method, but negative in Moeller's ADH. The glucose in Moeller's medium appears to inhibit the reaction due to catabolite repression of the inducible ADH system (Macián *et al.*, 1996). *Photobacterium* sp. were found to produce alkaline products in Thornley's medium, yet none possessed a constitutive ADH system when tested with more sensitive analytical methods (Baumann *et al.*, 1971; West and Colwell, 1984). Other *Vibrio* sp. such as *Vibrio* (*Listonella*) *anguillarum*, *Photobacterium* (*Vibrio*) *damselae* ssp. *damselae*, *V. fluvialis* and *V. furnissii* are positive for ADH by both methods. Thornley's method is the recommended method for *Vibrio* sp. (Macián *et al.*, 1996). However, the use of Thornley's medium may not be an option for some laboratories, whereas Moeller's medium is easily obtainable commercially (e.g. Difco).

Carbohydrate fermentation

In some reports in the literature, the terms 'utilization' and 'fermentation' have been used interchangeably and it has been difficult to ascertain which result was reported from which test method. Fermentation should be used to describe a result from the breakdown of a carbohydrate or 'sugar' reaction, such as used in conventional biochemical media and the API 20E. A pH indicator (usually phenol red) in the medium detects the acid change caused by the breakdown products. Utilization should only refer to tests that determine an organism's ability to use a substance as a sole carbon source. There is usually no pH indicator in the medium and growth is observed as an increase in opacity or turbidity of the medium. The exception is Simmons' citrate medium, which determines an organism's ability to use citrate as a sole carbon source and ammonium ions as a sole nitrogen source. There are no other nutrients in this medium. It contains the pH indicator bromothymol blue, which is green at pH below 6.9 and turns blue at a pH of 7.6 or more (see photographic section). The agar medium is prepared as a slant in a tube (see Media section). On the other hand, Christensen's citrate method contains other nutrients and is not a utilization test for citrate as a sole carbon source (Cowan and Steel, 1970; MacFaddin, 1980).

Carotenoid pigment detection

A few drops of 0.01% aqueous Congo red (CR) are placed on a few isolated colonies growing on a medium such as Anacker–Ordal (AO) agar. After 2 min the colonies are rinsed with water. The colonies develop a red colour if the test is positive and the colour can last for a number of hours. CR detects the presence of extracellular galactosamine glycan (Johnson and Chilton, 1966, citing, E.J. Ordal, personal communication).

Catalase

A loopful of bacterial growth is removed from the culture plate and smeared onto a glass slide. A drop of 30% hydrogen peroxide is placed on to the bacterial cells and the appearance of bubbles indicates a positive test. When removing cells from an agar medium that contains blood, care must be taken not to transfer any of the blood-containing medium with the bacterial growth, because the presence of blood may lead to a false-positive result.

Cellulose digestion

The production of the enzyme cellulase can be tested by the digestion of cellulose. A tube containing Anacker–Ordal broth (also called Cytophaga broth, Anacker and Ordal, 1955) and a strip of Toyo no. 2 filter paper is inoculated with bacterial growth. The tube is incubated at an appropriate temperature for up to 7 days, and examined for disintegration of the filter paper, which indicates a positive result (Wakabayashi *et al.*, 1989).

Congo red (CR)

Used for the detection of carotenoid pigment. See under carotenoid pigment detection, above.

Decarboxylases

ADH, ODC and LDC should be inoculated with a heavy inoculum of the organism. These media should be incubated for a minimum of 2 days and maximum of 14 days. For marine organisms, NaCl is added to a final concentration of 2%. Without NaCl many tests may produce a false-negative reaction. Sterile paraffin oil must be added to these media (about 0.5 ml for a 5-ml tube), otherwise false-positive reactions may occur.

Flexirubin pigment

Flexirubin pigment may be produced by some of the members of the *Flavobacteriaceae* family. For its detection, growth is taken from a culture plate (usually Anacker–Ordal medium) and smeared onto a glass slide, which is placed on a white background such as a piece of paper. The colony mass is flooded with 20% KOH and examined for an immediate colour change to reddish purple or brown. It may be helpful to place two bacterial masses on the glass slide, one that is flooded with KOH and the other to act as a control for colour differentiation. The KOH can also be used directly on the colony growth on a culture plate; however, the colour change may not be noticeable if there is only a thin layer of growth (Bernardet *et al.*, 2002). A reversion to the original colour occurs when the bacterial cells are flooded with an acidic solution (Reichenbach *et al.*, 1974, 1981).

Gliding motility

The demonstration of gliding motility of the *Cytophaga–Flavobacterium–Bacteroides* group may be difficult due to the choice of culture technique and the definition of gliding motility. The organism needs to be grown on a low-nutrient concentration agar such as AO, which contains 1.5% w/v agar. Gliding is affected by the amount of surface moisture and best results are obtained with freshly poured plates and incubation in a humid atmosphere. Gliding is observed by direct microscopic examination of the swarming edge of bacterial growth on a thinly poured plate after overnight incubation. A high-powered dry lens is used. Gliding is defined as 'movement that is continuous and regularly follows the long axis of the cells that are predominantly organised in bundles during movement'. It should be noted that spreading colonies are not necessarily an indication of gliding motility, because spreading may result from other surface translocation mechanisms (Henrichsen, 1972). Gliding can usually be suspected if a colony has a rhizoid shape on AO agar; however, this will not be seen if the agar surface is too dry. Nor will gliding motility be seen if bacterial growth is suspended in saline. Gliding motility can also be observed in a hanging drop prepared from a liquid culture (Bernardet *et al.*, 2002).

Haemolysis

Haemolysis is observed on BA. A clear zone of lysis of the red cells around a bacterial colony is referred to as beta-haemolysis (βH). A greenish tinge seen around some colonies, such as some of the *Streptococci* sp., is referred to as alpha-haemolysis (αH). *Streptococcus iniae* shows complete βH on sheep blood agar, but only partial haemolysis when the medium is supplemented with human or bovine blood.

Some *Vibrio* sp. are haemolytic on blood agar without salt, but non-haemolytic on salt-containing media such as MSA-B, even though their preferred growth medium is with Na ions. It is suggested that haemolysins may be produced when the organism is under stress.

Indole

A false-negative result may be recorded for salt-requiring organisms if there is an insufficient final concentration of NaCl in the test medium. Growth may be seen in the test medium; however, the organism may not express the enzyme unless the NaCl concentration is optimal. To achieve a 2% NaCl final concentration that is optimal for the majority of marine organisms, add 500 µl of 20% NaCl stock solution to 5 ml of test medium. See under Set Inoculation.

KOH

Used for the detection of flexirubin pigment. See under Flexirubin pigment.

Luminescence

Luminescence may be detected by growing the organism on an appropriate growth medium such as MSA-B or nutrient broth No. 2 (Oxoid) (25 g), NaCl (17.5 g), KCl (1.0 g), $MgCl_2.6H_2O$ (4.0 g), agar (12.0 g), distilled water (1000 ml) (Furniss *et al.*, 1978). MA 2216 is also a suitable medium for production of luminescence for a number of *Vibrio* sp. Not all luminescent bacteria will luminesce on all of these media and luminescence may be seen on MSA-B but not MA 2216. Reliance on luminescence needs to be treated with caution, because expression of luminescence appears to be dependent on a number of factors, including media, and ideally should be measured with a luminometer rather than by eye (J. Carson, Department of Primary Industries, Water and Environment, Tasmania, 2003, personal communication). Luminescence is optimal after 18–24 h of incubation at 25°C. It is also dependent on aerobic conditions, so that broth cultures need to be shaken to aerate the medium before luminescence can be detected. Luminescence is detected by observing the plates or broth culture in a dark room and allowing the eyes to adjust to the dark for 5 min. However, if possible luminescence should be measured from a broth culture using an instrument such as a Wallac Microbeta Plus liquid scintillation counter, which measures relative light units (Manefield *et al.*, 2000).

Organisms that are positive for luminescence include: *Aliivibrio* (*Vibrio*) *fischeri*, *A.* (*Vibrio*) *logei*, *A.* (*Vibrio*) *salmonicida*, *Photobacterium leiognathi*, *P. phosphoreum*, *V. orientalis*, *V. splendidus* biovar I (Furniss *et al.*, 1978; Lunder *et al.*, 2000).

Organisms that are negative for luminescence include: *Aeromonas salmonicida*, *Aliivibrio* (*Vibrio*) *salmonicida*, *A.* (*Vibrio*) *wodanis*, *Enterovibrio norvegicus*, *Vibrio* (*Listonella*) *anguillarum*, *V.* (*Listonella*) *pelagius*, *Moritella marina*, *M. viscosa*, *Photobacterium angustum*, *P. damselae* ssp. *damselae*, *Plesiomonas shigelloides*, *V. agarivorans*, *V. alginolyticus*, *V. brasiliensis*, *V. calviensis*, *V. campbellii*, *V. coralliilyticus*, *V. diazotrophicus*, *V. fluvialis*, *V. furnissii*, *V. gazogenes*, *V. halioticoli*, *V. ichthyoenteri*, *V. lentus*, *V. metschnikovii*, *V. natriegens*, *V. navarrensis*, *V. neptunius*, *V. nereis*, *V. nigripulchritudo*, *V. ordalii*, *V. pacinii*, *V. parahaemolyticus*, *V. penaeicida*, *V. proteolyticus*, *V. rotiferianus*, *V. splendidus* biovar II, *V. tapetis*, *V. tubiashii*, *V. vulnificus* and *V. xuii* (Furniss *et al.*, 1978; Lunder *et al.*, 2000; Gomez-Gil *et al.*, 2003). *Aliivibrio* (*Vibrio*) *fischeri* NCMB 1281T was reported to be negative for luminescence (Lunder *et al.*, 2000).

Strains of *A.* (*Vibrio*) *logei*, *V. cholerae* and *V. harveyi* are variable for luminescence (Furniss *et al.*, 1978; Lunder *et al.*, 2000).

Methyl red (MR)

The MRVP medium (Difco) should be inoculated with a heavy inoculum of the organism. Salt needs to be added if required as a growth factor by the organism under identification. Incubate at either 25°C or 37°C according to the organism's requirements (see Tables 2.2–2.4, or in the text under the appropriate bacterium). Incubate for 48 h, and then detect the presence of acetylmethylcarbinol by adding three to five drops of methyl red. Continuation of a red colour indicates a positive result. For a negative reaction the red colour disappears as the methyl red reagent is added to the reaction tube. The success of this test for a positive result is dependent on incubation time, not the amount of growth, although obviously growth must be seen in the tube to indicate that the organism has grown in the medium. The test can be carried out at 24 h if desired, by removing an aliquot (200 μl) to a microfuge tube and testing for MR by the addition of one drop of MR reagent. A negative reaction must be incubated for a further 24 h and re-tested after this time before being called negative. MR may take 3 days for a positive reaction.

Oxidase test

Oxidase test strips are available commercially, and these are recommended because of standardization of the reagent. The BBL™ DrySlide™ Oxidase test from Becton Dickinson Company is a reliable kit method. Some methods have been known to produce false-positive reactions with *Escherichia coli* (data from IFM proficiency program). For the manual oxidase test, use the oxidase reagent tetramethyl-*p*-phenylenediamine. Prepare a 1% solution in distilled water and in a light-protecting bottle and store at 4°C. For use, place a few drops of the reagent onto some filter paper. Using a wooden matchstick such as an orange stick or a platinum loop, smear bacterial growth onto the moistened filter paper. Development of a purple colour within 10–30 s denotes a positive. The 1% solution usually lasts 1 h and should be discarded once a deep purple colour develops.

A nichrome loop should not be used, because it causes false-positive reactions. This test should not be performed from organisms grown on media that contain sucrose or nitrates. Therefore TCBS is an unsuitable medium from which to determine an oxidase reaction (Furniss and Donovan, 1974; Jones, 1981).

TCBS

This medium is used to identify an organism as a *Vibrio* sp. However, there are some *Vibrio* sp. that do not grow on TCBS. These are *Moritella* sp., *Photobacterium leiognathi*, *Grimontia* (*Vibrio*) *hollisae* and *Vibrio ordalii*. *Vibrio* (*Listonella*) *anguillarum* may show slow growth on TCBS. Strains of *Aliivibrio* (*Vibrio*) *fischeri* may show weak or negative growth.

TCBS is not strictly selective and some organisms will grow, although at a reduced colony size. Species of *Aeromonas* and *Enterococcus* will grow as small (1 mm) yellow colonies on TCBS plates. *Proteus* sp. may grow as 1-mm yellow/green colonies on TCBS. *Plesiomonas* does not grow well on TCBS. Some yellow colonies of *Vibrio* sp. may revert to green colonies after a few days' growth as they use up the sucrose in the medium (Oxoid manual).

VP

The MRVP medium (Difco) should be inoculated with a heavy inoculum of bacterium. NaCl needs to be added if required by the organism. Incubate at either 25°C or 37°C, according to the organism's requirements. As a general rule, use an incubation temperature of 25°C for all bacteria isolated from aquatic environments and aquatic animals. Incubate for 48 h, and then detect acetoin production using the reagents. VP can be tested at 24 h if there is sufficient growth in the test medium. A negative result, however, should be re-incubated for 48–72 h. Test an aliquot by placing 200 μl of the incubated medium into a 0.6-ml microfuge tube. Add one drop each of VP reagent A and VP reagent B. Commercial reagents from the API 20E kit work well with the Difco MRVP medium. Examine for red colourization up to 20 min. Although the two tests can be performed in the same tube, it is preferable to test the reactions at 24, 48 and 72 h. The incubation time influences the production of acetoin.

Vibriostatic agent

Resistance to the vibriostatic agent pteridine (0/129) has been noted in some developing countries. Microbiologists are advised to continue using 0/129 as a means of differentiating *Vibrio* sp. from *Aeromonas* sp., but caution may be required in some countries (Huq *et al.*, 1992; Nair and Holmes, 1999). A zone of 9 mm is classified as susceptible for the 0/129 500 μg disc for *Vibrio* sp. (Bernardet and Grimont, 1989). A zone size of 22 mm is considered sensitive for *Photobacterium damselae* ssp. *damselae* (Love *et al.*, 1981). *Vibrio cholerae* strain 0139 is resistant to 0/129 500 μg disc (Albert *et al.*, 1993; Islam *et al.*, 1994).

2.7.3 Using the Biochemical Identification Tables

The following sections are devoted to the identification of an unknown organism using phenotypic or biochemical tests. The variations reported in the literature for different strains are also recorded, so that the microbiologist is aware of the difficulties encountered when identifying some particular species. Hopefully this system should not prove too tedious for the user of the manual; it is intended to give users more confidence in naming an organism and to alert them to potential difficulties.

In some cases, species are relatively newly described and different researchers have obtained different biochemical reactions. Rather than presenting a consensus result, both results are listed. It must be stressed that difficulties arise when assessing results of some phenotypic tests when two different versions of a biochemical test have been used. Where differences were noted from different journal articles, these are either stated in the tables, or are detailed under the notes on interpretation and the problems encountered, or listed in the text under the relevant organism. Other notes are from the author's experience with particular species. Phenotypic tests are still the primary identification system, and therefore microbiologists must be aware of the difficulties encountered with some tests for particular organisms. Some species have been newly described on one strain only and as more information is gathered about this species it may become evident that the type strain is not in actual fact representative of the species (Janda and Abbott, 2002). Subsequent strains isolated and identified as belonging to this species may show slightly different results in the biochemical results. Reporting these differences may assist the user of the manual with identification. For example, the type strain of *Vibrio lentus* is negative for mannitol fermentation, yet all other isolates reported thus far are positive. Likewise with *Vibrio penaeicida* where the type strain is negative for indole and 50% of other strains are positive. The atypical *Aeromonas salmonicida* group is another example. In this case, different biochemical results are reported from strains isolated from different fish species. For the diagnostic laboratory trying to identify an unknown isolate, it may be difficult to say for certain that the isolate is an atypical *A. salmonicida*. However, with the phenotypic details listed of all isolates from different fish species, it may enable a more definitive identification, or at least an identification can be made with slightly more confidence.

There are limitations to all phenotypic (and genotypic) identification systems (Janda and Abbott, 2002). By detailing the biochemical variations it is hoped that the user of the manual will be assisted in making a more informed identification.

2.7.4 Interpretation and identification of genera and species

In the identification tables, where authors have recorded a different result for the same type strain, then these results are listed separately. Rather than present a consensus from the literature, listing the different results provides an indication to the reader of this manual of those tests that are likely to produce disparate results for the individual organisms. However, for some results a 'v' or variable reaction is recorded. Variable results may have been reported in the literature or the organism itself may be variable in its fermentation of the biochemical.

2.7.5 Commercially available test kits

Brucella abortus antiserum (Difco).
Listonella anguillarum, specific monoclonal antibody for serotypes 01, 02, 03 and 04, 05 and 07 for environmental serotypes (BIONOR Mono-Va, BIONOR Aqua, Skien, Norway).
Photobacterium damselae ssp. *piscicida*, specific monoclonal antibody (BIONOR Mono-Pp, BIONOR Aqua, Skien, Norway).
Staphylase test for identification of *Staphylococcus aureus* (Oxoid).
Streptococcus Group A, B, C antiserum (Difco). Antiserum for Groups A, B, C, D, F, G (Oxoid).
Vibrio cholerae. Anti-V. cholerae 01 serum (Denka Seiken). Bacto Vibrio Cholerae antiserum for detecting three serotypes in group 01. These are serotype Ogawa (AB O antigen factors), Inaba (AC O antigen factors) and Hikojima (ABC O antigen factors) (Difco).

BIONOR™ Mono-Aqua rapid diagnostic agglutination test kits are available for *Aeromonas salmoninicida* (MONO-As), *Photobacterium damselae* ssp. *piscicida* (MONO-Pp), *Vibrio anguillarum* (MONO-Va), *V. salmonicida* (MONO-Vs) and *Yersinia ruckeri* (MONO-Yr) available from Tamar Laboratory supplies Ltd. The MONO Va kit detects serotypes 01, 02 and 03 of *Listonella anguillarum* and the environmental serotypes of 04, 05 and 07. However, non-specific agglutination occurs against strains of *V. splendidus* and motile *Aeromonas* sp. A cellular concentration of 10^8 ml^{-1} is recommended for positive agglutination (Romalde *et al.*, 1995). Strains of *Photobacterium damselae* ssp. *piscicida* are detected by the MONO-Pp kit, with no cross-reaction against *Aeromonas salmonicida*. However, non-specific agglutination occurs with strains of *Actinobacillus pleuropneumoniae*, *Haemophilus parasuis*, *Mannheimia haemolytica* and *Pasteurella multocida* (Romalde *et al.*, 1995). MONO-Yr kit detects classical serotypes 01 and 03 of *Yersinia ruckeri*. For the detection of 02 serotypes the test needs to be performed at either 48 h culture growth or from subcultures. The kit is unable to detect *Y. ruckeri* serotypes 05 and 06 because of the composition of the serotypes included in the kit. A cell concentration of 10^8 ml^{-1} is needed for positive agglutination. No cross-reaction with other bacterial pathogens was found (Romalde *et al.*, 1995).

3

Biochemical Identification Tables

3.1. Conventional Biochemical Tests

The following tables are to be used for interpreting the results obtained from in-house or conventional biochemical media. The tables are listed in alphabetical order of the table heading. The organisms in the tables are grouped into those that are considered to be either pathogens in aquatic animals or environmental and saprophytic organisms. The saprophytic, environmental and other species of bacteria are included under 'environmental' so as to assist in differentiating a pathogen from a closely related species and thus obtaining a definitive identification of the unknown organism.

In a number of cases, different or variable biochemical tests were recorded in the literature by different laboratories for the same type strain. Also, differences in a biochemical were recorded for different strains. These results are included here so that the reader is aware that some test results may not be reliable and that different results may be recorded by different laboratories. In the cases reported in this manual, the tests that gave variable results appeared to be performed using the same method. It is known that different methods can give different results for the same organism, and therefore it is important to be consistent with the methods used. Where a difference between the conventional biochemical test and the API 20E was reported, this is recorded in the tables, so that the reader is aware of variations for different test methods for that particular organism. The numbers in the tables report the percentage of positives as reported in the literature when a number of strains were tested.

With the exception of the *Vibrio* tables, all other tables list the organisms alphabetically within the groupings of 'pathogenic' or 'environmental' organisms.

The species in the *Vibrio* tables are grouped according to their ODC–LDC–ADH reaction. Therefore all species with a ODC+, LDC+, ADH− result are grouped together. Likewise other ODC–LDC–ADH reaction combinations are grouped together. Use these groupings as a starting point to identifying an unknown.

Use the tables in conjunction with the section on interpreting test results and specific reactions noted for some genera and species.

An example of a Laboratory Worksheet for recording results is included.

Use the following schematic (Fig. 3.1) as a guide to which table or tables (conventional media) to use when identifying an unknown. Start with the result for the Gram stain, then cell morphology and, for the Gram-negatives, a further division based on oxidase result.

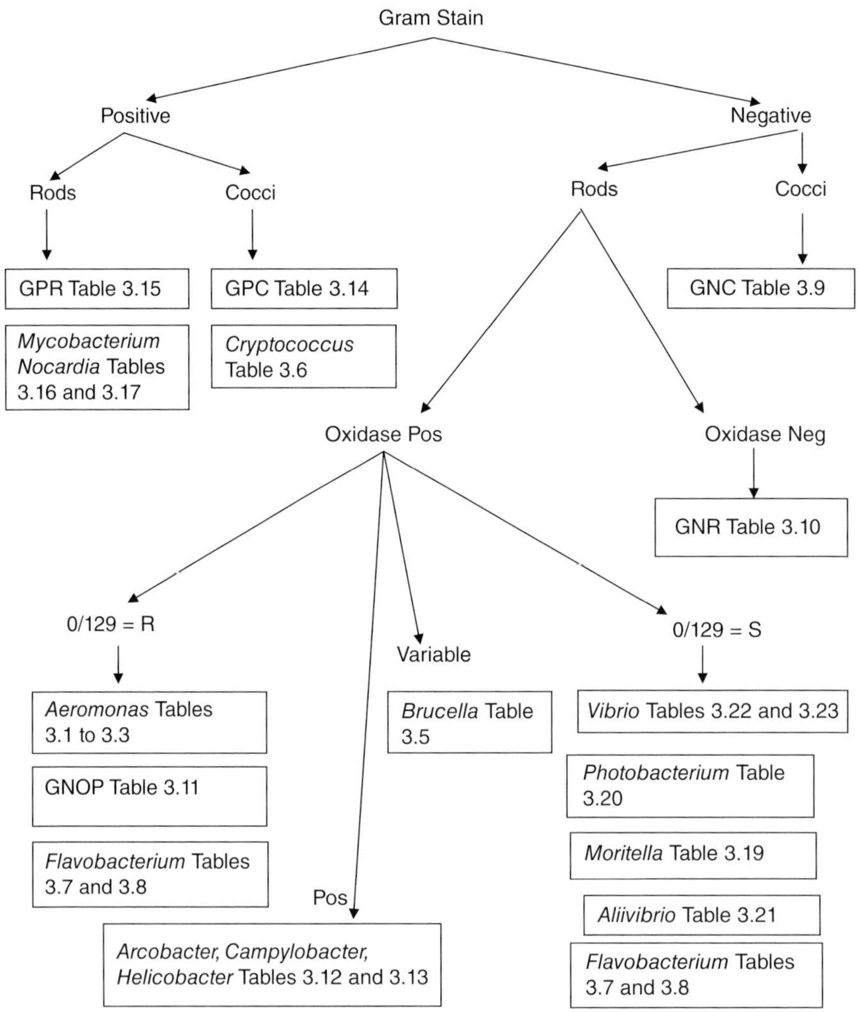

Fig. 3.1. Schematic for using the biochemical identification tables.

3.2. Results for API Kits

3.2.1 Interpretation of API 20E profile

Use the profile index provided by the manufacturer as a guide only. Kent (1982) does not recommend using the profile index provided with the kit because of misidentifications, especially when attempting to identify an organism not found in the commercial database. The API 20E system will even misidentify organisms when the identification is given at the 99% confidence level. For example, Dalsgaard *et al.* (1996) found that the API 20E system misidentified *V. vulnificus*. The system gave a greater than 98% confidence level as *Aeromonas hydrophila*, *Flavobacterium meningosepticum*, *Burkholderia cepacia*, and between 90 and 95% confidence levels as *V. alginolyticus* and *V. parahaemolyticus*. All these 'isolates' were in fact *V. vulnificus*, which they proved using a species-specific DNA probe.

It is recommended that the API 20E system be used with caution. Some additional tests may be required to give a definitive answer. This is particularly so for the identification of species from the genus *Aeromonas*. Adding conventional media of aesculin, MR and salicin in addition to the API 20E system will improve the likelihood of the correct identification.

The API 20E database and the conventional media database for *Vibrio* species listed in this book are collations from the literature, and the results are from type strains tested at the Animal Health Laboratory at the Department of Agriculture and Food, Western Australia. For some organisms, such as *Vibrio vulnificus*, many profile numbers were obtained from the literature, indicating the phenotypic strain variation among species.

The reactions for the API 20E are listed in three different formats. The first format indicates the reactions obtained for the different tests in the same order as the API 20E test strip. The organisms are listed in alphabetical order. The second format lists all organisms in alphabetical order and then the corresponding API 20E profile number. The third format lists the API 20E profile number in numerical order. Once the profile number is obtained and checked for organism identity, the reactions can be checked in the corresponding tables to assess any similarity with close profile numbers and to determine whether additional tests should be done to confirm this initial identification. In some cases the complete nine-digit API 20E profile number was not reported in the journal article, because the tests for growth on MacConkey agar and oxidative-fermentative were not reported. In these cases the API 20E profile number is recorded in the manual as a seven- or eight-digit profile number.

The numbers in the cells of the tables are the percentage of positive strains recorded with that test result.

3.2.2 Identification using API ZYM

The API ZYM tables are included because many newly described species are tested for the production of enzymes using this kit. It is not generally considered a routine diagnostic kit; however, it is useful for some bacteria from veterinary isolations such as the species from *Pasteurella*, *Actinobacillus* and *Histophilus ovis* and is able to differentiate between species when phenotypic tests may be in doubt (Cousins and Lloyd, 1988). For these organisms the intensity of the colour reactions is consistent for a species and allows species identification. Therefore, the correct inoculum concentration is important. It can also be used to confirm or back up phenotypic tests when identifying an unknown bacterium isolated from an aquatic animal species.

Species within a genus give very similar results for the API ZYM; however, there are often differences with one or two enzymes between species that can give an indication of the identification of the species. The intensity of the colour reaction is important. Thus it is important to be consistent with the inoculum density. The manufacturer recommends a density of McFarland tube 6.

For *Flavobacterium*, *Flexibacter* and *Cytophaga* species, an incubation time of 12 h is recommended when incubating at a temperature of 18–30°C (Bernardet and Grimont, 1989). For an incubation temperature of 37°C, use an incubation time of 4 h, as recommended by the manufacturer. In the author's experience, isolates from an aquatic environment give better reactions when incubated for 24 h at a temperature of 25°C. However, it must be stressed that if the incubation time and temperature is stated in the tables, then these parameters must be used because the results are reported on these. Some references did not state the time and temperature of incubation that was used, and therefore it is assumed that the manufacturer's recommendations were used.

Improved results may also be obtained if ammonium salt sugar (ASS) broth is used as the inoculating fluid for the *Flavobacterium*/*Cytophaga* group from marine and freshwater sources (Bernardet and Grimont, 1989).

3.2.3. Miscellaneous tests

Chitinase activity (spot filter paper method)

A quick spot filter paper method is available to test for chitinase ability in bacteria (O'Brien and Colwell, 1987).

A stock solution is prepared by dissolving 50 µmol of 4-methylumbelliferyl-*N*-acetyl-β-D-glucosaminide (Sigma Chemicals) in 2 ml of dimethylformamide (4-MUF.GlcNAc) (Sigma). A buffered working solution is prepared by diluting

0.6 ml of stock solution of 4-MUF.GlcNAc in 9.4 ml phosphate buffer (0.1 M, pH 7.4). The stock solution and buffered working solutions are used immediately or stored without sterilization at −20°C. Stability is several months. For the test 1–5 colonies are selected from 24–48 h growth and rubbed onto Whatman no. 1 filter paper. Twenty microlitres 4-MUF.GlcNac-buffered working solution is added to the smeared bacterial cells. The filter paper is incubated at 25°C (for *Vibrio* sp.) for 10 min. Add one drop of saturated sodium bicarbonate solution and expose to ultraviolet light at a wavelength of 366 nm. Positive reactions produce a strong light-blue fluorescence.

String test

The string test is a presumptive test for the identification of *Vibrio cholera*, but is not specific to *V. cholerae* (Smith, 1970). *Vibrio cholerae*, classical and El Tor strains, give a strong positive result, whereas non-cholera vibrios give a result varying from positive to strong positive. Other members of the *Enterobacteriaceae* (*Escherichia*, *Aerobacter*, *Alcaligenes*, *Klebsiella*, *Proteus*, *Pseudomonas*, *Salmonella* and *Shigella*) are negative. The string test detects the sensitivity to 2,4-diamino-6,7,-diisopropylpteridine phosphate (vibriostatic compound 0/129 ([Sigma Chemical Co.]). The test is conducted by suspending colonies, from 24 h at 25°C growth from a trypticase soy agar plate, in one drop of 0.5% sodium deoxycholate in water. After 1 min the loop is touched to the drop and slowly raised. If a 'mucoid string' forms between the drop and loop as it is raised, the strain is considered to be positive for the string test. The sodium deoxycholate lyses the bacterial cells and releases the DNA, which causes the mixture to become viscous. A negative control is performed using a drop of saline instead of the deoxycholate. Non-cholerae vibrios produce a finer shorter string compared with *V. cholerae*. Note that *Aeromonas* can also be positive when grown on phosphate-containing media (Smith, 1970). This test is not definitive and other appropriate tests must be performed.

Table 3.1. *Aeromonas salmonicida* (non-motile *Aeromonas* species).

Test	Gm	Ox	cat	βH	mot	Pig	ODC	LDC	ADH	Nit	Ind	Cit	urea	mr	vp	aes	G	onpg	OF	arab	glu	inos	lac	malt	man	mano	sal	sor	suc	tre	Xyl	H_2S	MCA	TCBS	Dnase	temp	NaCl	0129 10	Kf	Amp	Ref
Organism																																									
Pathogenic																																									
Typical																																									
A. salmonicida ssp. *salmonicida* NCMB 1102 HG3	–	+	+	+	–	+	–	+	+	+	–	–	–	+	–	+	+	+	F	+	$+_{g+}$	–	–	+	+	+	+	–	–	+	–	–	+	–	+	4–30	0–3	R	S	S	395, 669, 928, 979
	no growth 37°C, API 20E LDC may be neg																																								
A. salmonicida ssp. salmonicida NCMB 1102[T]	–cb	+	+	+	–	+[ft]				+	–		–	–	–	+	+		F	+	$+_{g-}$		–	+	+			–	–	–		–				5–22	0–3	R	R	R	1546
	No growth 37°C																																								
A. salmonicida ssp. *salmonicida* ATCC 14174	–	+			–	+	–	+		+	–	–			–		+		F	–	+		–	+	+	–	–		–	–		–									370
A. salmonicida ssp. *salmonicida*	–r	+	+	+	–	–*	–	–	+	+	–	–			–	+	+	–	F	+	$+_{g+}$		–	+	+		+	–	–	+	–	–	+	–	+	4–30	0.5–3	R			928
	Non-pigmented strain																																								
'Atypical'																																									
A. salmonicida ssp. *achromogenes* NCMB 1110[T]	–r	+	+	v	–	+[ft]	–	–	v	+	+	–	–	–	+	–	v	+	F	–	$+_{g-}$	–	–	v	+	–	–	–	+	–	–	v	+	–	+	4–28	0–4	R	R	R	395, 669, 928, 1117
	No growth 37°C. API Ind may be neg																																								
A. salmonicida ssp. *masoucida*	–	+	+	+	–	–	–	+	+	+	+	–	–	+	+	+	+	+	F	+	$+_{g+}$	–	–	+	+	+	–	–	+	+	–	+	+	–	+	30	0–2	R		R	669, 906, 928
	No growth 37°C. In API 20E LDC, VP & H_2S may be neg																																								
A. salmonicida ssp. *nova*	–	+	+	+	–	–	–	–	–	+	+	v	–		–	–	–	–	F	–	$+_{g-}$	–	–	+	+	v	–	–	+	–	–	–				18–25		R		R	1529
	Haemin requirement																																								
A. salmonicida ssp. *smithia*	–	+	+	25	–	–	–	–	–		–	v	–	–	–	–	+	+	F		+		–	–	–v			–	v	–		+	–	–	+	5–20	0–2	R		R	81,928
	No growth 37°C																																								
American eel	–	+	–			–[b]	–	–	v	+		–	–		–		+	v	F	–	+	–		+		–		–	v			–	–								1232

Continued

Table 3.1. Continued.

Test	Gm	Ox	cat	βH	mot	Pig	ODC	LDC	ADH	Nit	Ind	Cit	urea	mr	vp	aes	G	onpg	OF	arab	glu	inos	lac	malt	man	mano	sal	sor	suc	tre	Xyl	H_2S	MCA	TCBS	Dnase	temp	NaCl	0129 10	Kf	Amp	Ref
Australian goldfish	−r	+	+	−	−	$+^{b}$	−	−	−	+	+	−	−	−	−	−	−	17	F	−	$+_{g-}$	−	−	83	+	−	−	−	+	−	−	−				22	0.5–3	R			282, 1789
Australian goldfish FD1334	−	+	+	−	−	$+^{bt}$	−	−	−	+s	+s	−	−	+s	−	+s	w	w	F	−	+	−	−	+	+	−	−	−	+	−	−	−			+s						244
Baltic sea turbot	−r	−	+	+	−	$-^{b}$	−	−	v	−	−	−			v	v	+	−	F		$+_{g-}$				−		−		−						+			R	R	S	395, 1800
Baltic sea flounder	−r	−	+	+	−	$-^{b}$	−	−	v	v	−	−			+	+	+	v	F		$+_{g-}$				+		+v		+						+			R	S	S	395, 1800
Baltic sea (Finnish) flounder	−	−	+	+		$-^{h}$	−	−	+	v	−	−			−	+	+	+	F	+	$+_{g-}$	−	−	+				−			−	−	+	−	+			R	S	S	1797
Coho salmon (USA)	−	−	+		−	$+^{t}$	−	−	−	+	−	−	−	+	−	+	+			−	+	−	−	+	+	+	+	−	−	−	−										304
Baltic sea dab	−r	+	+	−	−	$-^{b}$	−	−	−	−	+	−			+	+	+	−	F		$+_{g-}$				−		+		+						+w			R	S	S	1800
Baltic sea blenny	−r	+	+	−	−	$+^{b}$	−	−	+	+	+	−			+	−	+	−	F		$+_{g-}$				+		−		+						+			R	R	R	1800
Canadian Atlantic cod	−	+			−	$+^{s}$				+	+	−			−				F	−	+		−	+	+				+												
Canadian Atlantic salmon Keij strain	−	+			−	$-^{b}$				−	+	−			−				F	−	+		−	+	+				+												*,370
Canadian sablefish AS_2	−	+			−	$+^{b}$				+	+	−			−				F	−	+		−	+	−				+												*,370
Danish cod	−	+			−	+	−	v	+		−				−	+			F		$+_{gv}$				+				−							20	0–2				398
Danish sand eels	−	+			−	$+^{bs}$	−	−	−		+				+	−			F		$+_{g-}$				+				+							20	0–2				398
Danish turbot	−	−	+	−	−	−	−	−	+		−				+	+	+		F	−	$+_{g-}$				−		+		−	−								R	R		1324
English non-salmonid	−	+	+	+	−	$+^{t}$	−	−	+		−	v		−	−	−	v	+	F		+	−	v	v	+			+	+	−		+	+		+	4–37	0–3			R	70
Norway minnows	−	+			−	$-^{b}$				+			−	+		−			F	−		−	−	+w	+	+w	−	−	+	−	−	−						R		S	690
South African rainbow trout	−r	+	+w	−		$+^{b}$				+	−					+	+		F	+	+	−		+	+				+	−			+			22					199
pigment positive		+																		−	+	−			+				+		−								6	3	674

pigment negative																																						56	72	674
Environmental																																								
A. salmonicida pectinolytica	–	+	+	+	–	+	–	–	–v	+	+	+	–		+	–	+	+	F	+	$+_{g+}$	–	+		+		–	+	+	+	–	–			+	35	0	R	R		1317
	VP at 25°C. Brown pigment																																							

Furunculosis agar (FA) is more sensitive to pigment production than TSA, TSA+t
P=pigment produced
b=Pigment tested on BA
f=pigment on FA
h=pigment on BHIA
s=pigment production slow
t=Pigment tested on TSA
+s=slow positive reaction
*=Pigment tested on TSA containing L-tyrosine (TSA+t)

The following methods are recommended so as to standardize testing and reduce discrepancies from lab to lab. Carbohydrate fermentation is performed in phenol red broth (Difco) with 1% carbohydrate. Haemolysis is tested using BA containing horse blood, other methods are from Cowan and Steele (and in this manual). For marine isolates the addition of NaCl to a final concentration of 1.5% in 'liquid' tubes such as MRVP, aesculin, nitrate, ADH, LDC, ODC may improve consistency of results. Zones of 20 mm and greater are considered sensitive for Amp (ampicillin) and Kf (cephalothin). Incubate at 20°C and read at 14 days. Dalgaard *et al.* (1998); Austin and Stobie (1992a); Kalina (1984); Kampfer (1992); and Khashe and Janda (1998) also use these methods.

Table 3.2. *Aeromonas*. Phenotypic tests according to DNA Hybridization Groups.

Test	Gm	Ox	cat	βH	mot	pig	ODC	LDC	ADH	Nit	Ind	Cit	urea	vp	aes	G	OF	arab	glu	inos	lac	malt	man	mano	sal	sor	suc	tre	Xyl	Dnase	40°C	0129 10	0129 150	Amp
DNA Hybridization Group																																		
HG 1	–	+	+	+	+	–	–	96	+	+	+	–	–	91	96	+	F	87	$+_{g+}$	–	–	+	96	+	+	–	+	+	–	+	+	R	R	R
HG 2	–	+	+	+	+	–	–	59	94	+	+	+	–	83	+	+	F	+	$+_{g+}$	–	67	+	+	+	+	+	+	+	–	+	–	R	R	R
HG 3	–	+	+	92	+	–	–	55	90	+	+	46	–	90	+	+	F	+	$+_{g+}$	–	54	67	+	+	+	+	+	+	–	+	–	R	R	R
HG 4	–	+	+	10	+	–	–	–	+	+	97	70	–	–	97	+	F	+	+	–	87	+	+	31	+	7	+	+	–	+	87	R	R	R
HG 5	–	+	+	–	–	50	–	–	92	+	+	+	–	–	92	+	F	+	+	–	+	+	+	+	+	–	+	+	–	+	–	R	R	R
HG 6	–	+	+	–	+	–	–	–	87	+	+	+	–	–	+	+	F	87	+	–	+	+	+	+	+	–	+	+	–	+	–	R	R	R
HG 8/10	–	+	+	90	+	–	–	57	+	+	+	9[illegible]	–	90	10	+	F	14	$+_{g+}$	–	10	+	+	+	–	–	+	+	–	+	+	R	R	R

Gm=Gram reaction, Ox=oxidase, cat=catalase, βH=beta haemolysis, mot=motility, pig=brown pigment, ODC=ornithine decarboxylase, LDC=lysine decarboxylase, ADH=arginine dihydrolase, Nit=nitrate reduction, Ind=indole, Cit=citrate, Urea=urea hydrolysis, vp=Voges–Proskaüer, aes=aesculin hydrolysis, G=gelatin hydrolysis, OF=oxidative fermentative, arab=L-arabinose fermentation, glu=glucose fermentation, inos=inositol fermentation, lac=lactose fermentation, malt=maltose fermentation, man=manitol fermentation, mano=mannose fermentation, sal=salicin fermentation, sor=sorbitol fermentation, suc=sucrose fermentation, tre=trehalose fermentation, xylose fermentation, Dnase=hydrolysis of DNA, 40°C=growth at 40°C, 0129/10=sensitivity to vibriostatic agent 0129 at 10 μg concentration, 0129/150=sensitivity to vibriostatic agent 0129 at 150 μg concentation, Amp=sensitivity to ampicillin. Numbers refer to percentage of positive strains, –=negative, +=positive reaction. HG=Hybridization Group.
Data from Abbott *et al.*, 1992; Kaznowski, 1998.

Table 3.3. *Aeromonas* species – motile.

Test	Gm	Ox	cat	βH	mot	SW	ODC	LDC	ADH	Nit	Ind	Cit	urea	mr	vp	aes	G	onpg	OF	arab	glu	inos	lac	malt	man	mano	sal	sor	suc	tre	Xyl	H_2S	MCA	TCBS	Dnase	temp	NaCl	0129 10	0129 150	Amp	Ref
Organism																																									
Pathogenic																																									
A. allosaccharophila HG15	–	+	+	+	+	–	33	+	67	+	+	67	–		–	+	+	+	F	67	$+_{g+}$	–	–	+	+	+	–	–	+	+	–	+	+		+	4–42	0–3	R	R	R	1102
A. bestiarum HG2	–	+	+	+	+	–	–	+	+	+	+	+	–	+	+	+	+	+	F	+	$+_{gv}$	–	+	+	+	+	v	–	+	+	–	+			+	4–42		R	R		873
A. caviae HG4	–	+	+	10	+	–	–	–	+	+	+	+	–	+	–	+	+	+	F	+	$+_{g-}$	–	87	+	+	31	+	–	+	+	–	–	+		+	4–37		R	R	R	42, 277, 497
A. aquariorum	–	+	+		+	–	–	+	+	+	+	93	–		+	+	+	+	F	–	$+_{g+}$	–	–	+	+	+	+	–	+	+	–	+	+			25–37	0–3	R	R	R	535,1106
A. hydrophila dhakensis	–	+	+	+	+	–	–	+	+	+	+	+	–	–	+	+	+	+	F	–	$+_{g+}$	–	–	+	+	+	+	–	+	+	–	–	+	–w		25–42		R	R	R	790
A. hydrophila LMG 2844[T] HG1	–	+	+	+	+	–	–	+	+	+	+	60	–	53	+	+	+	+	F	+	$+_{g+}$	–	–	+	+	+	+	–	+	+	–	+	+	–w	+	4–42	0–2	R	R	R	2, 790, 873
A. hydrophila ranae	–	+	+		+	–	–	+	+		+				v	–	+	+	F	–								–	–			–									791
A. jandaei HG9	–	+	+	+	+	–	–	+	+	+	+	75	–	+	+	–	+	+	F	–	$+_{g+}$	–	–	+	+	+	–	–	v	+	–	+	+	+w	+	4–42	0–3	R	R	R	244, 276, 497
A. piscicola	–	+	+	+	+	–	–	+	+	+	+			–	+	+	+	+	F	–	$+_{g+}$	–	–	+	+	+	+		+	+	–	+			+	4–37	0–3	R	R	R	141
A. schubertii HG12	–	+	+	+	+	–	–	+	+	+	–	+	–	+	+	–	+	+	F	–	$+_{g-}$	–	–	+	–	+	–	–	–	+	–	–		–	+	4–42	0–3	R	R	R	278, 722
	gel at 22°C, βH – weak on sheep blood, pos on horse blood agar																																								
A. veronii biovar Veronii HG10	–	+	+	+	+	–	+	55	–	+	+	+	–	+	+	+	67	89	F	–	$+_{g+}$	–	–	+	+	+	+	–	+	+	–	v	+	–	+	4–42	0–5	R	R	R	276, 723
Environmental																																									
A. sobria HG7	–	+	50	–	v	–	–	+	–	+	+	+	–	45	57	–	–	+	F	–	$+_{gv}$	–	–	+	+	+	–	–	+	+	–	+	+	–	–	15–28	0–1	R	R	R	42, 858, 1100
	88% VP pos in API 20E																																								
A. simiae	–	+	+	–	+	–	–	+	+	+	–				–	v			F	–	$+_{g-}$	–	–	+	–	+	–	–	+	+	–	–	+	–		4–42	0–3	R	R	R	677
A. tecta CECT 7082[T]	–	+	+	+	+	–	–	+	+	+	–	–			+	–			F	–	$+_{g+}$	–	–	+	+	+	+	–	–	+	–					25–30	0–3	R	R	R	435
A. tecta from trout	–	+	+	+	+	–	–	+	+	+	–	+			–	+			F	–	$+_{g+}$	–	–	+	+	+	+	–	–	+	–					25–30	0–3	R	R	R	435
A. trota HG14	–	+	+	+	+	–	–	+	+	+	+	+	–	+	–	–	+	+	F	–	$+_{g+}$	–	–	+	+	+	–	–	–	+	–	+	+	–		4–42		R	R	S	277, 789
A. veronii biovar Sobria HG8	–	+	+	+	+	–	–	+	+	+	+	50	–	+	+	–	+	+	F	v	$+_{g+}$	–	5	+	+	+	–	35	+	+	–	+	+		+	4–42		R	R	R	42, 276, 497, 873, 945
A. diversa group 501 HG13	–	+	+		+		–	+	+	+					–	–		+	F	–	$+_{g+}$	–			+				23	+	–		+				0–3	R	R	R	722
A. diversa CDC 2478–85[T]	–	+	+	+	+	–	–	–	+	+	+	–		+	v	–	+	–	F	–	$+_{g-}$	–	–		–	–	–	–	–	+	–	–	+	–	+		0–3	R	R	R	1147
A. bivalvium 868[T]	–	+	+	–	+	–	–	+	–	+	+	–	–	+	–	+	+	+	F	+	$+_{g-}$	–	–		+	–	+	–	+	+	–	–	+	–	+	4–37	0–3	R	R	R	1146
A. encheleia HG16	–	+	+	+	+	–	–	–	+	+	+	–	–	v	25	+	+	+	F	–	$+_{g+}$	–	–	+	+	+	+	–	75	+	–	+	+	–	+	4–37	0–3	R	R	R	494, 783

A. eucrenophila HG6	–	+	+	50	+	–	–	–	+	+	+	v	–	+	–	+	+	+	F	+	$+_{g+}$	–	+	+	+	+	+	–	v	+	–	+			+	25–37	0–2	R	R	R	2, 42, 783, 858, 873
A. media HG5A/5B	–	+	+	25	–	–	–	–	+	+	+	55	–	+	–	v	+	+	F	+	$+_{g-}$	–	+	+	+	+	94	–	+	+	–	–	+	–	v	4–37	0–2	R	R	R	35, 615
A. popoffii	–	+	+	–w	+	–	–	–	+	+	v	70	–		+	–	+	+	F	50	$+_{g+}$	–	–	85	+	+	–	–	–	+	–	50		–	+	28–37	0–2	R	R	R	784
	The type strain is indole-negative, other strains are indole-positive																																								
A. molluscorum 848^T	–	+	+	+	+	–	–	–	+	+	–	–	–	+	–	+	+	+	F	80	$+_{g-}$	–	–		+	+	v	–	+	+	–	–	+	–	+	4–37	0–3	R	R	R	1145
A. taiwanensis	–	+	+	–	+	–	–	–	+	+	+	+	–		–	+	+	+	F	+	$+_{g+}$	–	–	+	+	–	+	–	+	+	–	–	+	–	+	30–36	0–3	R	R		36
A. sanarellii	–	+	+	–	+	–	–		+	+	+	–	–		–	+	+	+	F	+	$+_{g-}$	–	–	+	+	–	+	–	+	+	–	–	+	–	+	30–45	0–3	R	R		36
Aeromonas rivuli	–	+	+	–	+	–	–	–	+	+	–	v			–	+	+	+	F	–	$+_{g-}$	–	–	+	+	+	+		+	+	–	–		–	+	Jul-37	0–3	R	R		536
A. fluvialis 717^T	–	+	+	–	+	–	–	–	–	+	+	+	–		–	–	–	+	F	–	$+_{g+}$	–	+	+	–	+	+	–	+	+	–	–	+	–	–	25–37	0–3	R	R	R	37

Table 3.4. Anaerobes.

Test	Gm	Ox	cat	βH	mot	SW	ODC	LDC	ADH	Nit	Ind	Cit	urea	mr	vp	aes	G	onpg	OF	arab	glu	inos	lac	malt	man	mann	sal	sor	suc	tre	Xyl	H_2S	MCA	TCBS	Dnase	temp	NaCl	0129 10	0129 150	Amp	Ref
Organism																																									
Pathogenic																																									
Clostridium botulinum E	+r		–	+	+	–				–	–		–		–	–	–				+				–		–		–		–	–									269
	Rods with ova subterminal endospores with appendages and exosporia																																								
Clostridium perfringens	+r				–					v	–					v				–		v	+	+	–	+	–	v	+	v	–										
Edwardsiella ictaluri oxygen intolerant strain	–	–	+		–	–	+	+	–	+	–	–	–	–	–	–	–	–	F	–	$+_{g-}$	–	+	+	–	+	–	–	–	–	–	–	+			25–37					1152
	Non-motile at 37°C																																								
Eubacterium species 841	+r	–	–		–	–				–							+				$+_{g+}$		+	+			–		–		–	+				20	0–1				716
Eubacterium species 1065	+r	–	–		–	–				+							+				$+_{g+}$		+	+			–		–		–	–				20	0–1				716
Eubacterium tarantellae	+r		–	+	–	–				–	–					–	–			–	+		+	–	–	–	–		–	–	-	–		+		25–37	0–1				1673
	Long unbranched filamentous rods																																								

Table 3.5. *Brucella.*

Test	Gm	Ox	cat	βH	mot	CO_2	Nit	cit	ind	MR	VP	H_2S	urea	BF	SO	TH	Ala	Ara	Arg	Asp	Glut	Orn	Lys	Gal	Rib	Xyl	Ery	Uro	A	M	R	Ref	Host
Organism																													Surface antigens				
Pathogenic																																	
Brucella ceti	−	+	+		−	−	+					−	+	+	+	+	−	+	−	−	+	−	−	+	+	+	+V	−	+	V	−	547,554, 827	UK dolphin, porpoise
Brucella ceti	−	+	+			−						−	+	+	−	+													+			340	Minke whale
Brucella pinnipedialis	−	+	+		−	+	+					−	+	+	+	+	−	−	−	−	+	−	−	−	+	−	+	−	+	V	−	545,547, 554,827	Seal, otter
Brucella abortus	−		+	V	−		+	−	−	−	−	V	+	+			+	+	−	+	+	−	−	++	+++	++	+++	+	+	V	−	369,827	Cattle
Brucella canis	−		+				+	−	−	−	−	−	+	−			−	−	+	−	+	++	++	−	+++	−	−	−	−	−	+	369,827	Dogs
Brucella inopinata						−						−V	+	+		+																	
Brucella melitensis	−		+				+	−	−	−	−	−	+	+			++	−	−	++	+++	−	−	+	+	+	+++	+	−	+		369,827	Sheep, goat
Brucella microti						−						−	+	+		+													−	+			
Brucella neotomae	−	−	+				+	−	−	−	−	+	+	−			−	+	−	+	++	−	+	++	−	+	+++	+	−	−		369,827	Desert wood rat
Brucella ovis	−	−	+				−	−	−	−	−	+	+	+			−	−	−	+	+	−	−	−	−	+	−	+	−	−	+	369,827	Sheep
Brucella suis 1	−		+					−	−	−	−	+	+	−			−	+	++	−	−	++	++	++	+++	+++	+++	++	+	+	−	369,827	Pig
Brucella suis 2	−		+					−	−	−	−	−	+	−			−	+	+	−	++	+	+	++	+++	+++	++	+++	+	+	−	369,827	Pig, hare
Brucella suis 3	−		+					−	−	−	−	−	+	+			−	−	++	−	++	++	++	−	+++	+	+++	++	+	+	−	369,827	Pig
Brucella suis 4	−		+					−	−	−	−	−	+	+			−	−	++	−	++	++	++	−	+++	−	+++	−	+	+	−	369,827	Reindeer
Brucella suis 5	−		+					−	−	−	−		+				−	−	−	+	++	−	+	−	+++	−	+++	+	+	+	−	827	

Ala=L-alanine, Ara=L-arabinose, Arg=L-arginine, Asp=L-asparagine, BF=growth on media containing basic fuchsin 20 µg/ml (1/50,000). βH=beta haemolysis, CO_2=carbon dioxide requirement, cat=catalase, cit=citrate, Ery=*meso*-erythritol, Gal=D-galactose, Glut=L-glutamic acid, Gm=Gram reaction, H_2S=hydrogen sulfide, ind=indole, Lys=L-lysine, MR=methyl red, Nit=nitrate reduction, Orn=DL-ornithine, Ox=oxidase, Rib=D-ribose, urea=urea, SO=growth on medium containing safranin O 100 µg/ml (1/10,000), TH=growth on medium containing thionin 20 µg/ml (1/50,000), Uro=urocanic acid, VP=Voges–Proskaüer, Xyl=D-xylose, Surface antigens, A, M & R = agglutination with monospecific antisera.

Table 3.6. *Cryptococcus.*

Test	Gm	Ox	cat	βH	mot	Cap	ODC	LDC	ADH	Nit	Ind	Cit	urea	mr	vp	aes	G	onpg	OF	arab	glu	inos	lac	malt	man	mano	sal	sor	suc	tre	Xyl	GT	temp	NaCl
Organism																																		
Pathogenic																																		
Cryptococcus neoformans var gattii	+			−	−	+	+	−	−	+			+					+			+	−	−	−					+	−		−	20–42	
Environmental																																		
C. lupi	+											+	−				−			+	+	−		+	+		+	+		+	+		4–25	0–2
	No fermentation, only assimilation. Antarctic habitat. Growth on glucose peptone yeast extract																																	
Candida species	+					−							−																					

GT=Germ tube, cap=capsule

Table 3.7. *Cytophaga–Flavobateriaceae – Bacteroides* Group – pathogenic.

Test	Gm	Ox	cat	βH	mot	Glid	ODC	LDC	ADH	Nit	Ind	Cit	urea	mr	vɔ	aes	G	onpg	OF	arab	glu	inos	lac	malt	man	mann	sal	sor	suc	tre	Xyl	H_2S	MCA	CR	KOH	Dnase	temp	NaCl	0129 150	NA	TSA	Ref
Organism																																										
Pathogenic																																										
Chryseobacterium balustinum	–	+	+	–	–v	–	–	–	–	+v	+	–	–v			+	+	–v	Ov	–	+	–	–	–	–		–	–	–v	–v	–	–	+	–	+	+v	15–35	0–2		+	+	168,1746
	Previously *Flavobacterium balustinum*. Yellow pigment.																																									
Chryseobacterium chaponense	–	+	+		–		–	–	–		–	–	–				–	–	O	–	–	–	–	–	–		–	–	–	–	–	–	–		–	–	5–36	0–1		+	+	859
Chryseobacterium indologenes	–	+	+		–	+	–	–	–	v	+		–			+	+	v	O		+		–	+	+		–		–	+	–		+		+	+	36					1173
Chryseobacterium piscicola	–	+	+	α	–	–	–	–	–	–		–		–	–	+	+		O														–	–	+	w	4–28	0–3		+	+	799
Chryseobacterium scophthalmum	–	+	+	+	–	v	–	–	–	–	–	–	+	–	–	+	+		F	–	+	–	–	+	–		–	–	–		–	–	–		+	+	15–25	0–4			+	1172,1173
	Previously *Flavobacterium scophthalmum*. Orange pigment																																									
Chryseobacterium tructae	–	+	+	–		–			–	–	+	+	+			+	+	–	O		–											–	–	–	+	–	5–30	0–3		+		1846
Elizabethkingia meningoseptica	–	+	+	–	–	–	–	–	–	–	v	v	v	–	–	+	+	+	O	–	+	–	v	+	v	+	–	–	–	v	–v	–	w	–	–	+	37–42	0–2		+	+	168,1746, 1817
	previously *Chryseobacterium* (*Flavobacterium*) *meningosepticum*																																									
Flavobacterium araucananum	–	+				+	–	–	–		+	–	–			+	+		O	–	+	–	+	+	–	+	–	–	–	+	+	+		–	+		4–30	0–3		+	+	857
Flavobacterium branchiophilum ATCC 35035	–	+	+		–	–				–	–					–	+	+	O	–w	$+_{g-}$	–	–	+	–	–v	+v	–	+	+	–w	–	–	–	–	–	10–25	0–0.1	S	–	–	172,1282, 1746
	Yellow pigment, no agar hydrolysis. Growth on 1/20 TSA																																									
Flavobacterium chilense	–	+				+	–	–	–		–	–	–			+	+		O	+	+	–	+	+	–	+	–	–	–	+	+	–		–	+		4–30	0–3		+	+	857
Flavobacterium columnare IAM 14301^T	–	+	+	–	–	+	–	–	–	80	–	–	–	–	–	–	+	–	O	–	–	–	–	–	–	–	–		–	–	v	+		+	+	+	10–37	0–0.5	S	–	–	168,244, 284,1282
	Adheres strongly to agar. H_2S from lead acetate medium.																																								(other references 111, 139, 298, 299)	1663
Flavobacterium hydatis	–	–v	+	+	–	+	–	–	–	+	–	–	–	–	–	+	+	+	F	+	+	–	+	+	+	+	–	–	+	–	+	–	–	–	+	+	20–30	0–1	R	+	+	168,172, 1586
	Previously *Cytophaga aquatilis*. Mucoid, orange-yellow colonies.																																									
F. johnsoniae AHLDA 1714	–	+	+	–		+	–	–	–	+	–	+	+s	–	–	+	+	+	F	–	w	–	–	–	–	–	–	–	–	–	–	–	–		+	–			S	+	+	244,867
F. johnsoniae ATCC 17061	–	+	+	–		+	–	–	–	+	–		+			+	+	+	O	+	+		+	+	–	+	+		–	+	+w	–		+	+	+	5–30	0–1	S	+	+	284,1282
F. johnsoniae ATCC 17061	–	+	+			+	–	–	–	–	–					+	+	–	O		+											+		–	–		5–30	0–1		+		534,1068

Continued

Table 3.7. Continued.

Test	Gm	Ox	cat	βH	mot	Glid	ODC	LDC	ADH	Nit	Ind	Cit	urea	mr	vp	aes	G	onpg	OF	arab	glu	inos	lac	malt	man	mann	sal	sor	suc	tre	Xyl	H_2S	MCA	CR	KOH	Dnase	temp	NaCl	0129 150	NA	TSA	Ref
F. johnsoniae DSM 2064	–	+	+			+	–	–	–	–	–		+				+	+	O		+											–		–	–	+	10–30	0–2	S		+	168
F. johnsoniae DSM 29585*	–	+	+			+	–	–	–	+	–		+				+	+	O		+											–		–	+	+	10–30	0–1	S	+	+	168
Flavobacterium oncorhynchi	–	+	+	–		–			–	+	–	–	–			+	–	+	O		–											–	–	–	+	–	15–30	0–1		+	+	1845
Flavobacterium psychrophilum	–	+v	+	70	–	w	–	–	–	–	–	–	–	–	–	–	+	–	O	–	–	–	–	–	–	–	–	–	–v	–	–	–	–	–	+	+w	4–25	0–0.8	S		–	167,168, 172
	Yellow colonies. Growth enhanced in AO media with 0.5% tryptone																																									335,1004
Tenacibaculum aiptasiae	–	+	+			+			–	–	–	–	–			+	+	–	O		–													+	–	+	8–40	1–10				1761
Tenacibaculum dicentrarchi	–	+	+			+				–	–				–		+		O		–											–			–		4–30	ASW				1346
Tenacibaculum discolor	–	+	+			+				+	–				–		+		O		–											–			–		14–38	ASW				1347
Tenacibaculum gallaicum	–	+	+			+				+	–				–		+		O		–											–					14–38	ASW		–		1347
Tenacibaculum maritimum	–	+	+		–	+				+	–		–	–	–	–	+	–	I	–	–	–	–	–	–	–	–	–	–	–	–	–		+	–	+w	15–34	ASW	S		–	167, 168, 171
	Pale yellow colonies. No agar hydrolysis. Prepare media with ASW base																																									1744
Tenacibaculum (Flexibacter) ovolyticum	–	+	+	–		+	–	–	–	+	–	–	–				+	–	O	–	–		–		–		–		–		–	–		–	–	+	4–25	3+			–	675
	Pale yellow colonies. NG on TCBS																																									
Tenacibaculum soleae	–	+	+			+	–	–	–	+	–	–	–		+	–	+	–	O	–	–	–			–			–	–			–		+	–	+	14–30	3	R		–	1056,1344
	No growth using NaCl alone. Supplement tests with FMM or MA 2216 broth																																									

*This organism is no longer in the DSMZ database

1/20 TSA=TSA prepared at a 1:20 dilution, +s=slow reaction, –w=negative or weak reaction, Ov=Oxidative variable results, v=variable reaction, NG=no growth.

Table 3.8. *Cytophaga – Flavobacteriaceae – Bacteroides* Group – Environmental.

Test	Gm	Ox	cat	βH	mot	Glid	ODC	LDC	ADH	Nit	Ind	Cit	urea	mr	vp	aes	G	onpg	OF	arab	glu	inos	lac	malt	man	mano	sal	sor	suc	tre	Xyl	H_2S	MCA	CR	KOH	Dnase	temp	NaCl	0129 150	NA	TSA	Ref
Organism																																										
Aequorivita antarctica	–		+		–	–	–	–	–	–			–v			+	+		O	–	–	–	–	–	–	–		–	–	–		–			–	–	0–25	0.5–6				211
	Orange pigmented colonies																																									
Aequorivita crocea	–		+		–	–	–	–	–	–			–			+	+		O	–	–	–	–	–	–	–		–	–	–		–			–	+v	0–25	0.5–6				211
	Yellow pigmented colonies																																									
Aequorivita lipolytica	–		+		–	–	–	–	–	–			+v			+	+		O	–	–	–	–	–	–	–		–	–	–		–			–	–	0–25	0.5–6				211
	Yellow pigmented colonies																																									
Aequorivita sublithincola	–		+		–	–	–	–	–	–			+			–	+		O	–	–	–	–	–	–	–		–	–	–		–			–	–	0–25	1–6				211
	Orange pigmented colonies																																									
Aquimarina (Cytophaga) latercula	–	+	–		–		–	–	–	+							+	–	O		+								+			+			+		15–30	1–2				168,1016
	Bright red pigment																																									
Cellulophaga lytica	–	+	+		–	+	–	–	–	–							–	+	O		–								+			–		–	–		15–37	1–2				168,
Chitinophaga (Cytophaga) arvensicola	–	+	+		–	–	–	–	–	+							+	+	O													–		–	+		15–37	0–2			+	168
Chitinophaga (Flexibacter) sancti	–	+	+		–	–				–							+	+	O													–		+	+		15–35	0–1			+	168, 309
Chryseobacterium (Flavobacterium) gleum	–	+	+	–	–	–	–	–	–	v	–	–	v			+		+	I		+	–	–		–		–	–	–			–	+		+	+	25–37			+		751
	Indole pos with Ehrlich's but negative by Kovács'																																									
Chryseobacterium hominis	–	+	+		–		–	–	–	–	+	–	–			+	+	–	F	–	+		–	+	–		–		+	–	–	v	–		–	+	20–37	0.5–3			+	266, 860
Chryseobacterium indoltheticum	–									–	+		–			+	+	–			+												+				36					1173
Cytophaga allerginae	–	+	+		–		–	–	–	–							+	+	O													–		–	+		15–25	0–1			+	168
Cytophaga fermentans	–	+	v		–	–	–	–	–	–							+	+	F		+								+			–		–	–		15–25	1–2				168, 328
Cytophaga hutchinsonii	–	+	+		–	+	–	–	–	–	–		+				v	–	O		+											–		–	+	–	15–25	0–1				168
	pigmented colonies. Digests cellulose.																																									
Empedobacter (Flavobacterium) brevis	–	+	+	–	–	–	–	–	–	–	+	–	–			–	+	–	O	–	–v	–	–	+	–	–	–	–	–	–	–	–	+	–	+	+	28–37	0–2		+	+	168,754, 1746
	Pigmented colonies, fruity odour																																									
Flavobacterium aquatile	–	+	+		–	+	–	–	–	+v	–	–	–			+v	+v	+v	O	–	+	–	+	+	–		–	–	+	–	–	–	–	–	–	–	5–30	0–0.5	R	–v	+	168,172, 1282 1746

Continued

Table 3.8. Continued.

Test	Gm	Ox	cat	βH	mot	Glid	ODC	LDC	ADH	Nit	Ind	Cit	urea	mr	vp	aes	G	onpg	OF	arab	glu	inos	lac	malt	man	mano	sal	sor	suc	tre	Xyl	H_2S	MCA	CR	KOH	Dnase	temp	NaCl	0129 150	NA	TSA	Ref
Flavobacterium aquatile	–	–	+				–	–	–	–						+	+				+											+		–	–							534
Flavobacterium aquatile	–	+	+			–				–	–					+	+	+	O													+			–		5–30	0.2–0.5		v		1068
Flavobacterium ceti	–	+	+	–	–	–	–	–	–	–	–		–		–	–	+		O	–	–	–		–	–	–		–	–			–	–	–	+	–	22–37	0–3		w	+	1710
Flavobacterium flevense	–	+	+		–	+	–	–	–	–v						+	–	+	O		+											–		–	–	–	15–22	0–2	S	+	+	168,172, 1115
	Hydrolyses agar																																									
Flavobacterium frigidarium	–	+	+		–	–			–	–	–	–	–		–	+	+	–	O		–			–	–						–	–	–	+	–	–	15–28	0–5		+	+	777
Flavobacterium gillisiae	–	–	+			–	–	–	+	–	–	–	–		–	+	–	–	O	–	+	–	–	+	+	+		–	+	+	–	–		–	–	–	0–27	0–5	R	+	+	1115
Flavobacterium hibernum	–	–	+			+	–	–	–	+	–	–	–		–	+	+	+	F	+	+	–				–		–	+			–		–	+	+	0–30	0–2		+	+	1115,1116
Flavobacterium macrobrachii	–	+			–		–	–	–	–	–	–	–		–	–		–	O	–	–	–	–	–	–	–	–	–	–	–	–	–		–	–		15–30	0–1		w		1514
Flavobacterium mizutaii	–	+	+			–				–v	–	+v	–			+	–		O		+		+	+	–	+	+		+						–	–	30–37			+	+	1819
Flavobacterium pectinovorum	–	+				+				+						+	+	+			+											v		–	+	+	25	0–1	S	+	+	172,1115
Flavobacterium saccharophilum	–	–	+			+				+						+	+	+			–											+		–	+	–	4–30	0–2	S	+	+	172
	Hydrolyses agar																																									
Flavobacterium succinicans	–	+	+			G				v						+	+	+	F		+								w			–		–	–	+	25	0	S	+	+	172,328, 1115
Flavobacterium tegetincola	–	–	+			+	–	–	–	–	–	–	–		–	–	–	–	O	–	+	–		–	+			–	–	–	–	–		–	–	–	0–27	1–5	R	+	+	1115
Flavobacterium tilapiae	–	+	+		–	+			–	–	–		–			+	+	+	O		–													–	+	+	10–30	0–2				314
Flavobacterium xanthum	–	+	+			–	–	–	–	+	–	–	–		–	+	+	–	O	–	+	–	–	+	+	+		–	+	+	–	+		–	–	–	0–20	2	R	+	+	1115
Flexithrix (Flexibacter aggregans) dorotheae	–	+	–		–	+	–	–	–	–							–	+	O		+								+			+		–	–		22–37	1–2				168, 1017
	Hydrolyses agar																																									
Flexibacter elegans	–	+	–		–					–							+		O		–								–			–					22–25	0–2				327
Flexibacter flexilis	–	+	–		–	+	–	–	–	–	–						+	–	O	–	+		–	+		–	–		+	–	–	–		–	–		5–30	0–0.2		–	–	168, 1282
	Cells 15–20 μm, Orange pigmented colonies																																									
Flexibacter flexilis	–	+	+			+				–	–					+	+	–	O													v			+		5–30			–		1068

Flexibacter litoralis	−	+	−		−		−	−	−	−							+	−	O		−								−			+		−	−		15–25	2–5				168, 327
	Pink pigment																																									
Flexibacter polymorphus	−	+	−		−	+	−	−	−	+							v	+	O													−		−	−		22–32	2				168, 1016
	Cells >100 µm. Hydrolyses agar. Peach-coloured pigment																																									
Flexibacter roseolus	−	+	−		−		−	−	−	−							+	−	O		−								−			−		−	−		22–37	0–2		+	+	168, 327
	Cells 50–100 µm. Bright red orange pigment																																									
Flexibacter ruber	−	+	−		−		−	−	−	−							+	−	O		+								+			−		−	−		22–37	0–2		+	+	168, 328
	Bright red orange pigment. Cells 50–100 µm																																									
Gramella marina	−	+	+		+	+	−	−	−			+			+	+	+		O	+				+		−			+	+						−	4–37	1–15				1205
Leeuwenhoekiella (Cytophaga) marinoflava	−		+							+							+		F		+								+			−										328
Marivirga (Flexibacter) tractuosa	−	+	+		−		−	−	−	−							+	−	O													−/+		−	−		22–37	0–2				168, 328
	Orange pigment																																									
Marinilabilia salmonicolor	−	+	+		−		−	−	−	−							+	+	F		+								+			+		−	−		15–37	1–2				168, 328
	salmon-pink colour																																									
Marinobacter hydrocarbonoclasticus	−	+			+				−	+		+					−				−																40–45	2				124
	Previously *Pseudomonas nautica*																																									
Myroides odoratimimus	−	+	+	−	−	−			−	−	−	−	+			−	+	−	O		−												+				18–37					168, 753
Myroides odoratus	−	+	+	−	−	−	−	−	−	−	−	−	+			−	+	−	alk	−	−	−	−	−	−		−	−	−	−	−	−	+	−	−	+	18–37	0–2		+	+	168, 753, 1746
	previously *Flavobacterium odoratum,* Fruity odour																																									
Pedobacter heparinus	−	+	+	+s	−	−	−	−	−	v	−	−	v	−	−	+	−	+	O	−	+s		+	+	−		−	+	+	−	+	−	−	−	−	+	5–37	0–3		+	+	172, 328, 1603
Pedobacter piscium	−	+	+		−	−				−	−		−			+	−		O	−	−			−			+	−	−	+	−					+	5–28			+	+	1603
Salegentibacter salegens	−	+	+		−	−													O	−		−						−			−			−	−		0–30	0–20	R			1115
Solitalea (Flexibacter) canadensis	−	+	+	−	−	+	−	−	−	+	−	−		−	−		+	+	F		+		+		−				−			−v	−	−	−	+	10–40	0–1			+	168,
Sphingobacterium multivorum	−	+	+		−	−	−	−	−	−	−	−	+			+	14	+	O	+	+v	−	+	+	−	+	+	−	+	+	+	−	+	−	−	60	15–37	0–2		+	+	168,
Sphingobacterium spiritivorum	−	+	+	−	−	−	−	−	−	−	−	−	+			+	+	+	O	27	+v	−	+v	+	+	+	+	−	+	+	+	−	+	−	−	+	15–37	0–2		+	+	168, 750, 1746
Tenacibaculum litopenaei	−	+	−		−	+			−	−	−	−	−			−	+		O		−													+	−		10–39	2–10				1517
Zobellia galactanivorans	−	+	+		−	+			−	+	−	−	−		−	+	+	+	O	+	+			+	+	+		−	+			−			+	+	13–45	0.5–6				119
Zobellia uliginosa	−	+	+		−	+			−	+	−		−			+	+	+	O		+			+	−				+						+	+	13–30	0.5–2				119

alk = alkaline reaction, CR = congo red, Glid = gliding motility, KOH = potassium hydroxide, NA = growth on nutrient agar, TSA = growth on tryptone soya agar.

Table 3.9. Gram-negative coccobacilli and cocci.

Test	Gm	Ox	cat	βH	mot	SW	ODC	LDC	ADH	Nit	Ind	Cit	urea	mr	vp	aes	G	onpg	OF	arab	glu	inos	lac	malt	man	mann	sal	sor	suc	tre	Xyl	H_2S	MCA	TCBS	Dnase	temp	NaCl	0129 10	0129 150	Amp	Ref
Organism																																									
Pathogenic																																									
Acinetobacter sp.	–cr	+	+	+	–		–	–	–	–	–		–	–	–		–	–	F	–	–	–	–	+	–	+	–	–			–	–									68
	Note: Acinetobacter do not usually produce oxidase																																								
Moraxella sp.	cb	+	+	+			–	–	–	–	–	–	–		–		–	–		–	–	–	–	–	–	+w	–	–	–			–									138
	Haemolysis tested on sheep blood.																																								
	Taxonomic position of both these organisms is in doubt																																								
Environmental																																									
Acinetobacter baumannii	–	–	+	–		–				–	–	v				–	–	–	O		+											–			–	15–37					206
Acinetobacter calcoaceticus	–	–	+	–		–				–	–	+				–	–	–	O		+											–			–	15–37					206
Acinetobacter haemolyticus	–	–	+	+		–				–	–	+				–	+	–	O		52											–			–	15–37					206
Mesophilobacter marinus	–cb	+	+	–	–	–				+	v	+	v	+	–		v		O		+				+							–				5–37	1–6				1230
	Growth on NA, MA 2216																																								

Table 3.10. Gram-negative, oxidase-negative rods.

Test	Gm	Ox	cat	βH	mot	SW	ODC	LDC	ADH	Nit	Ind	Cit	urea	mr	vp	aes	G	onpg	OF	arab	glu	inos	lac	malt	man	mano	sal	sor	suc	tre	Xyl	H_2S	MCA	TCBS	Dnase	temp	NaCl	0129 10	0129 150	Amp	Ref
Organism																																									
Pathogenic																																									
Citrobacter freundii	–	–	+	v	+	–	v	–	v	+	v	+	–	+	–	–	–	+	F	+	$+_{g+}$	–	+	+	+		–	v	10	+	+	+	+		–	20–37	0–2				870
Citrobacter freundii	–	–	+	v	+	–	12	–	v	+	–	+	–	+	–	–	–	+	F	+	$+_{g+}$	–	+	+	+		–	+	10	+	+	+	+		–	20–37	0–2				844, 870, 1063
Citrobacter freundii	–	–	+			–	–	–	–		–	–	–		–		–		F	+	+	–			+			+	+			+									1478
Edwardsiella hoshinae	–	–	+	–	+	–	+	+	–	+	w+	–	–	+	–	–	–	–	F	30	$+_{gv}$	–	–	+	+	+	75	–	+	+	–	+	+		–	25–40	0–1.5	S			662
	Motile at 25°C and 35°C																																								
Edwardsiella ictaluri	–	–	+	w	+	–	+	+	–	+	–	–	–	+	–	–	–	–	F	–	$+_{g+}$	–	–	+	–	+	–	–	–	–	–	–	+		–	20–37	0–1.5				702, 776, 1758
	Weakly motile at 25°C but not at 35°C. Read MR at 48 h																																								
Edwardsiella ictaluri oxygen intolerant strain	–	–	+		–	–	+	+	–	+	–	–	–	–	–	–	–	–	F	–	$+_{g-}$	–	+	+	–	+	–	–	–	–	–	–	+			25–37					1152
	Culture anaerobically. Non-motile at 37°C																																								
Edwardsiella piscicida	–	–	+	w	v	–	+	+	–	+	+	–	–	+	–		–	–	F	8	$+_{g+}$	–	–		17			–	–	–		+	+			25–37	1–5				1, 971
Edwardsiella tarda	–	–	+	w	+	–	+	+	–	+	+	–S	–	+	–	–	–	–	F	–	$+_{g+}$	–	–	+	–	+	–	–	–	–	–	+	+		–	10–42	0–2				135, 776, 1383,
	Motile at 25°C and 35°C																																								1387, 1815
Edwardsiella tarda CECT 849^{T}	–	–	+	w	+	–	+	+	–	+	+	+C	–	+	–	–	–	–	F	–	$+_{g+}$	–	–	+	–	+	–	–	–	–	–	+	+		–	15–42	0–4				21, 286, 510
Edwardsiella tarda eels	–	–	+		+	–	+	+	–	+	+	v	–	+	–	–	–	–	F	–	$+_{g+}$	–	–	+	–	+	–	–	–	–	–	v	+			28	0–2				21,1743
E. tarda non-motile strain	–	–	+		–														F	+					+	+															1647
Escherichia vulneris	–	–	+	+	+		–	+	+	+	–	–	–	+	–	+	–		F	+	$+_{g+}$	–	+	+	+		+	–	–	+	+	–	+	–	–	20–37		R	R		85
Hafnia alvei	–sr	–	+	–	td	–	+	+	–	+	–	td	–	td	td	v	–	+	F	–*	$+_{g+}$	–	–	+	+	+	v	–	–	+	+	–	+	–	–	4–40	0–4	R	PS		786, 829, 1458, 1418
Hafnia paralvei	–sr	–	+	–	td	–	+	+	–	+	–	td	–	td	td	–	–	–	F	v*	$+_{g+}$	–	–	+	+	+	–	–	–	+	+	–	+	–	–						786
Halomonas cupida	–	–	+	–	+		+	+	+	+	–			–	–				I				+	+	+		+	+	$+_{w}$	+		–				10–25		S			
Klebsiella pneumoniae	–	–	+	–	–	–	–	+v	–	+	–	+	+	60	40	+	–	+	F	+	$+_{g+}$	+	+	+	+		+	+	+	+	+	–	+		–	35	0–5	R	R		844
	Mucoid																																								
Pantoea agglomerans	–	–	+	–	+	–	–	–	–	+	–	+	–	–	+	+	+	+	F	+	$+_{g-}$	–	–	+	+	+	+	–	+	+	+	–	+	–	–	4–37	0–6				509, 676
	Opportunist, yellow pigment, MR +ve with 2% NaCl																																								

Continued

Table 3.10. Continued.

Test	Gm	Ox	cat	βH	mot	SW	ODC	LDC	ADH	Nit	Ind	Cit	urea	mr	vp	aes	G	onpg	OF	arab	glu	inos	lac	malt	man	mano	sal	sor	suc	tre	Xyl	H_2S	MCA	TCBS	Dnase	temp	NaCl	0129 10	0129 150	Amp	Ref
Providencia (Proteus) rettgeri	–	–	+		+	–	–	–	–	+	+	+	+	+	–	+	–	–	F	–	+	+	–	–	+	+	v	–	–	–	–	–	+		+	18–37	0–3				149
Salmonella arizonae	–	–	+	–	+	–		+		+	–	+	–	+	–	–	–	–	F	–	$+_{g+}$		+	+	+		–	+	+	+	+	+	+			15–41	0–6	R	R		923
Salmonella species	–	–			–v	–	+					+	–	+	–		–	–	F		$+_{g+}$		–	+	+		–		–			+	+	–							
Serratia liquefaciens	–	–	+	+	+	–	+	+	–	–	–	+	–	–	+	+	+	+	F	+	$+_{g+}$	+	–	+	+	+	+	+	+	+	+	–	+	–	+	4–37	0–5	R	R		511, 601
	Ref 192 is MR +86% at 37°C, 17% at 30°C																																								1575
Serratia liquefaciens	–	–	+	+	+	–	+	+	–	+	–	+	–	–	+	+	+	+	F	+	$+_{g+}$	–	–	+	+	+	+	+	+	+	+	–	+	–	+	4–37	0–5	R	R		1575
Serratia liquefaciens	–	–	+	+	+	–	+	+	–	+	–	+	–	–	+	+	+	+	F	+	$+_{g+}$	+	+	+	+	+		+	+	+	+	–	+	–	+	25	0–5	R	R	R	1123
	Atlantic salmon. Incubation at 25°C																																								
Serratia marcescens	–	–	+	v	+	–	+	+	–	+	–	+	–	–	+	+	v	+	F	–	$+_{g+}$	+	–	+	+	+	+	+	+	+	–	–	+	–	+	4–45	0–8	R	R	R	137
	Red pigment																																								
Serratia plymuthica	–	–	+	–	–	–	–	–	–	+	–	+	–	v	+	+	+	+	F	+	$+_{g-}$	+	–	+	+	+	+	v	+	+	+	–	–	–	+	4–37	0–8	R	R	R	137, 1222
	Red pigment. MR +ve at 37°C only																																								
Stenotrophomonas maltophilia	–	–	+	β	+	–	–	+	–	–	–	+	–v			v	+	+	O	–	+w	–	+v	+	–	+	+	–	+v	+	54	+	+		+	35	0–5				606
Streptobacillus moniliformis-like	–	–	–		–	–	–	–	+	–	v	–	–		–	–	–	–	F		+		–	+	–	–			–		–	v				10–25	1–4				1090, 1302
	ID of organism was not conclusive																																								
Streptobacillus moniliformis	–	–	–	–	–	–	–	–	–	–	–	–	–			–	–		F	–	w		–	w	–			–	–		–	–	–			25–35					337
	Reactions of type strain. Requires addition of 20% serum for growth																																								
Yersinia intermedia	–	–	+		+		+	–	–	+	+	+	+	+	+w	+	–	+	F	+	+	+	–	+	+	+	+	+	+	+	50	–			–						820
	Motile at 25°C but not at 35°C. MR, VP, cit +ve at 35°C, –ve at 25°C																																								
Yersinia ruckeri	–	–	+	+	+	–	+	+	–	85	–	+	–	+	–	–	52	+	F	–	$+_{g-}$	–	–	+	+	97	–	–	–	97	–	–	+		–	37	0–3				249, 511
	Motile at 25°C but not at 35°C. Citrate +ve at 25°C but –ve at 37°C																																								1045
Y. ruckeri	–	–	+	–	+	–	+	+	–	+	+	+	–		+	–	+	+	F	–	+	–	–	+	+	+	–	+	–	+	–		+		–		0–5				1045
	Australian strain reported by Llewellyn																																								
Y. ruckeri serovar I																												–													
Y. ruckeri serovar II																												+													
Y. ruckeri serovar III																												–													
Y. ruckeri serovar IV	These are not strains of *Y. ruckeri*																			+											+										
Y. ruckeri serovar V																												+													
Y. ruckeri serovar VI																																									
	Fish isolates that ferment arab and rhamnose are not *Y. ruckeri*, but may be *Hafnia alvei*																																								

Environmental																																									
Budvicia aquatica	–	–	+		+	–	–	–	–	+	–	–	+	v	–	–	–w		F	+	+	–		–	$+_{s}$	–	–	–	–	–	+	+	+		–	4–37	0–4				
Citrobacter diversus	–	–	+		+	–	+	–	60	+	+	+	+	+	–	–	–	+	F	+	$+_{g+}$	–	33	+	+		10	+	17	+	+	–						R	R		508
Enterobacter aerogenes	–	–	+		+	–	+	+	–		–		–	–	+	+	–		F	+	$+_{g+}$	+	+	+			40	+	+	+	+	–			–		0–5				844
Enterobacter cloacae	–	–	+	–	+	–	+	–	+	+	–	+	–	–	+	+	+	+	F	+	$+_{g+}$	70	+	+	+		+	+	+	+	+	–	+	–	–		0–5	R	R		844
Enterobacter liquefaciens	–	–	+		+	–	+	65	–		–		–	25	+	+	+		F	+	$+_{g+}$	+	60	+			+	+	+	+	+	–			+		0–5				844
Escherichia albertii	–	–	+		–	–	+	+	–	+	–	–	–	+	–		–		F	+	$+_{g+}$	–	–	v	+		–	–	–	v	–				–						787, 829
Escherichia coli	–	–	+	v	50	–	+v	+	–	+	+	–	–	+	–	40	–		F	+	$+_{gv}$	–	55	+	+		30	+	55	+	+	–			–	25–44	0–5				844
Halomonas elongata	–	+v	+		+	–	+	+		+	22		66	11	1ˇ	56	44	33	F		+		67			+			+			–				30	0.1–20		R		1735
Halomonas halodurans	–	–	+			–	+	+		–	–		+		–	+		–	F	+	+		–			–					+	–				22–37	0.1–20				707
Halomonas marina						–																																			
Klebsiella oxytoca	–	–	+			–	–				+		+		+			+	F																	10–41					462
Klebsiella ornithinolytica	–	–	+			–	+				+				v				F																	10–37					462
Klebsiella pneumoniae ozaenae	–	–	+		–	–	–	–	–	+	–	+	–	+	–	+	–			+	$+_{g-}$	–	+	+			+	+	–	+	+	–	+		–	25–37	0–5				844
Klebsiella rhinoscleromatis	–	–	+	–	–	–	–	–	–	+	–	+	–	+	–	–	–		F	+	$+_{g-}$	+	–	+			+	+	+	+	+	–	+		–	25–37	0–5				844
Kluyvera ascorbata	–	–	+		+	–	+	+	–	+	+	+	–	+	–		–	+	F		$+_{g+}$				+			41	+v			–				5–					
Kluyvera cryocrescens	–	–	+		+	–	+	+	–	+	+	+	–	+	–		–	+	F		$+_{g+}$				+			41	+v			–				5+					
Pantoea dispersa	–	–			+	–	10	–	–	+	–	+	–		+	–	$+_{s}$	+	F	+	$+_{g-}$	+	–	+	+	+	–	–	+	+	+	–			–	30–41	0	R	R		602
	Yellow pigment. Type strain is ODC +ve																																								
Providencia (friedericiana) rustigianii	–	–	+	–	+rt	–	–	–	–	+	+	w	–	+	–	–	–	–	F	–	$+_{g+}$	–	–	–	–	+	–	–	$+_{s}$	–	–	–	+		–	10–40	0				1175
	Motile at 25°C but neg at 37°C																																								
Rahnella aquatilis	–	–			+	–	–	–	–	+	–	+	–	89	+	+	–	+	F	+	$+_{g+}$	–	+	+	+	+	+	+	+	+	+	–	+	–	–	4–37	0	R	R		223
	Motile at 25°C but not at 35°C. Man = –ve by API 20E																																								
R. aquatilis Genospecies 2	–	–			31	–	–	–	–	+	–	79	–	+	+	+	34	+	F	+	$+_{g+}$	–	+	+	+	+	+	+	+	+	+	–	+	–	–						223
R. aquatilis Genospecies 3	–	–			–	–	–	–	–	+	–	–	–	52	72	+	–	+	F	+	$+_{g+}$	–	+	+	+	+	+	+	+	+	+	–	+	–	–						223

Continued

Table 3.10. Continued.

Test	Gm	Ox	cat	βH	mot	SW	ODC	LDC	ADH	Nit	Ind	Cit	urea	mr	vp	aes	G	onpg	OF	arab	glu	inos	lac	malt	man	mano	sal	sor	suc	tre	Xyl	H_2S	MCA	TCBS	Dnase	temp	NaCl	0129 10	0129 150	Amp	Ref
Raoultella planticola	–	–	+	–	–	–	–	+	–	+	+	+	+	+	+	+	–	+	F	+	$+_{g+}$	+	+	+	+	+	+	+	+	+	+	–	+	–	–	4–41	0				462
Raoultella terrigena	–	–	+	–	–	–	–	+	–	+	–	+	+	+	+	+	–	+	F	+	$+_{g+}$	+	+	+	+	+	+	+	+	+	+	–	+		–	4–35	0	R	R		462, 820
	Similar morphologically to *K. pneumoniae*																																								
Serratia fonticola	–	–	+		+	–	+	+	–	+	–	+	–	+	–	+	–	+	F	+	$+_{g+}$	+	+	+	+	+	+	+	$+_{s}$	+	70	–	+		–	4–37					601, 1176
Serratia marcescens	–	–	+	–	+	–	+	+				+	40	18	+	73	70		F	27	–	77	17				+	+	+		–				+						601
Serratia liquefaciens	–	–	+	–	+	–	$-_{v}$	+	–	+	–	+	v	37	+	+	+		F	+	$+_{g+}$	64	16	+	+	+	+	+	+	+	+	–	+	–	+						1045
Serratia rubidaea	–	–	+		+	–	–	+	–			+		24	+	+	+		F	+	–	35	+				+	–	+		+				+						511, 601
Yersinia aldovae	–	–	+	–	$+_{v}$	–	+	–	–	+	–	+	+	+	+	+	–	+	F	+	+	+		$-_{v}$	+	+		+	–	+	+	+	+		–						162
Yersinia bercovieri	–	–	+	–	+	–	+	–	–	+	–	–	+	+	–	$+_{s}$	–	+	F	+	+	–	–	+	+	+	$+_{s}$	+	+	+	+	–	+		–						1769
Yersinia enterocolitica	–	–	+	–	–	–	V	–		+	V	–	+	+	+	15	–	+	F	+		–					–	+	+	+	36										511
Yersinia mollaretii	–	–	+	–	+	–	+	–	–	+	–	–	+	+	–	$+_{s}$	–		F	+	+	+S	–	+	+	+	$+_{s}$	+	+	+	+	–	+		–						1769
Yersinia frederiksenii	–	–	+	–	+	–	+	–	–	+	+	–V	+	+	+	+	–	+	F	+	+	44	22	+	+	+	+	+	+	+	20	–			–	25–37	0				866
	Motility and VP positive at 25°C, but negative at 37°C																																								
Yersinia kristensenii	–	–	+	–	+	–	+	–	–	+	43	–	+	+	–	–	–	+	F	+	$+_{g-}$	27	60	+	+	+	–	+	–	+	+	–			–	4–41	0				162, 820
	Motile at 25°C but not at 35°C. VP –ve at 25°C																																								
Yersinia rohdei	–	–	+	–	+	–	$+_{s}$	–	–	+	–	+	$+_{s}$	+	–	–	–		F	+	+	–	+	+	+	+	–	+	+	+	+	–	+	–	–	25	0	R	R		28
	All *Yersinia* and *Hafnia* species should be incubated at 25 to 28°C																																								
Yersinia pseudotuberculosis	–	–	+		+	–	–	–	–	+	–	–	+	+	–	+	–		F	+	+	–	–	+	+	+	+	–	–	+	+	–			V						225

td=temperature dependent, –*, +*=reaction for D-arabinose; '–S=negative for Simmons' citrate; +C=positive for Christensen's citrate.

Table 3.11. Gram-negative, oxidase-positive rods.

Test	Gm	Ox	cat	βH	mot	SW	ODC	LDC	ADH	Nit	Ind	Cit	urea	mr	vp	aes	G	onpg	OF	arab	glu	inos	lac	malt	man	mano	sal	sor	suc	tre	Xyl	H_2S	MCA	TCBS	Dnase	temp	NaCl	0129 10	0129 150	Amp	Ref
Organism																																									
Pathogenic																																									
Actinobacillus delphinicola	–	+	–	–w	–	–	+	+	–	+	–		–	–	+		–	–	F		+	–	–	–	–	+	–	–	–	–	–		–			30–42					548
	Strain from *Mesoplodon bidens*. 10% CO_2 required for growth																																								
Actinobacillus delphinicola	–	+	–	–w	–	–	70	+	21	+	–		–	–	+		–	–	F		+	–	–	–	–	+	–	–	–	–	–		–			30–42					548
	Strains from *Phocoena phocoena*. 10% CO_2 required for growth																																								
Actinobacillus delphinicola	–	+	–	–w	–	–	–	–	+	+	–		–	–	+		–	–	F		+	–	–	–	–	+	–	–	–	–	–		–			30–42					548
	Strains from *Stenella coeruleoalba*. CO_2 required for growth																																								
Actinobacillus scotiae	–	+	–	–w	–	–	70	–	–	+	–	–	+		+		–	+	F	–	+	–	+	–	–	+	–	–	–	–	–		–	–		25–37	0				550
	10% CO_2 required for growth. Nit = neg using Rosco tablets																																								
Beneckea chitinovora	–			+	+	–		+		+	+	+	–				+		F		$+_{g+}$								+			+				20–25	1–4				1754
Bordetella bronchiseptica	–	+	+	+	+	–	–	–	–	+	–	+	+	–		–	–	–	O	–	–		–	–	–		–	–	–	–	–	–	+			25–37	0–4				1386
Brucella abortus	–	+	+	v	–					+	–	–	+s			–	–		O		v		–	–	–				–		+	–	v			35	0				337
Burkholderia pseudomallei	–	+	+	v	+	–	–	–	+	+	–	v	–	–	–	+	v	–	O	+	+	+	+	+	+	+	–	+	–	+	+	–	+			20–42					64, 1337
	API reactions may be neg for urea and nitrogen, 18% pos for cit																																								
Deleya aquamarinus	–	+	+	–	+	–	+		–	+	–	–	+	–	–	–	–	–	O						+							+s			–	4–42	1–10			S	14,75
	Previously *Alcaligenes faecalis homari*																																								
Janthinobacterium lividum	–	+	+		+	–			+	+	–	+	–		+	+	+	–	O	+s	–	+	–	+	+	+	+	+	+	–	+s					4–30	0–2				73
	Purple p gment																																								
Pasteurella multocida	–	+	+	–	–	–	+	–	–	+	+	–	–	–	–	–	–	–	F	v	+		–	v	+		–	v	+	v	v	–	–			25–37	0				337, 1185
Pasteurella skyensis	–	+w	–	–w	–	–	+	+	–	–	+		–		–			–	F		+	–	+	+	+	+	–	–	–	+	–			–		14–32	1.5–2				181
Pasteurella testudinis	–	+	+	+	–	–	–	–	–	+	+	–	–			+		+	F	v	$+_{g-}$	+s	–	85	+	–	–	v	+	v	+		v								1443
Plesiomonas shigelloides	–	+	+	90	+	–	+	+	95	+	50	–c	–	v	–	–	–	+c	F	–	$+_{gv}$	50	+	+	50	+	–	50	–	+	–	–	+	–G	50	10–42	0–3	R	S	S	337
Pseudoalteromonas piscicida	–	+	+		+	–			–	+		–					+		O		–		–	–	–				–		–				+	22–40	5–10				242, 1210

Continued

Table 3.11. Continued.

Test	Gm	Ox	cat	βH	mot	SW	ODC	LDC	ADH	Nit	Ind	Cit	urea	mr	vp	aes	G	onpg	OF	arab	glu	inos	lac	malt	man	mano	sal	sor	suc	tre	Xyl	H_2S	MCA	TCBS	Dnase	temp	NaCl	0129 10	0129 150	Amp	Ref
Pseudomonas anguilliseptica	–	+	+	–	+	–	–	–	–	–	–	+	–	–	–	–	v	–	I	–	–	–	–	–	–	–	–	–	–		–	–	+	–	–	5–30	0–3	S	R	S	176, 1050, 1136, 1194
	Finnish isolates are gel pos, Japanese isolates are gel neg																																								1742, 1794
Pseudomonas chlororaphis	–	+	+		+	–	v	v	+	+	–	+	–				+	–	O		+		–		+			–	+	+						4–35					252
	Green colonies. 5 days for growth																																								
Ps fluorescens biovar I	–	+	+		+	–		–	+	–		+					+	–	O	+	+		–	+	+	+	–	+	+		+		+			4–37					
Ps fluorescens biovar II	–	+	+		+	–		–	+	+		+					+	–	O	+	+		–	+	+	+	–	+	+		+		+			4–37					
Ps fluorescens biovar III	–	+	+		+	–		–	+	+		+					+	–	O	+	+		–	+	v	+	–	v	–		+		+			4–37					
Ps fluorescens biovar IV	–	+	+		+	–		–	+	+		+					+	–	O	+	+		–	+	+	+	–	+	+		+		+			4–37					
Ps fluorescens biovar V	–	+	+		+	–		–	+	–		+					+	–	O	+	+		–	+	v	+	–	v	v		+		+			4v–37					
Pseudomonas fluorescens	–	+	+	+	+	–	–	–	+	+	–	+	+	–	–	–	+	–	O	+	+	+	–	–	+	+	–	+	–	v	+	–	+			4–30		R	R		957, 958, 1229
Pseudomonas plecoglossicida	–	+	+	+	+	–	–	–	+	+	–	+	–			–	–		O		–															10–30	0–5	R	R		1229
	Closely related to *Ps putida*																																								
P. (pseudoalcali-genes) oleovorans	–	+	+	–	+	–	v	–	+	+	–	+	–		–	–	–	–	OA	+	+	–	–	14	–	–		–	–		11	–	+		–	15–41	0				618
Pseudomonas putida	–	+	+	+	+	–	–	–	+	–	–	+	44	–	–	–	–	–	O	+	+	–	–	21	19	+	–	–v	9	–	+	–	+			4–35	0–5				39, 618, 1337
Pseudomonas species	–	+	+		+	–		–		–		–		–	–		–			–	–		–	–		–	–	–				–		–		0–40					1557
Pseudomonas stutzeri	–	+	+	–	+	–	–	–	–	+	–	+	14			–	–	–	O	v	+	–v	–	+	68	+	–	–	v	–	+	–v	+		–	25–41	0–6				618, 746, 1337
	Wrinkled colonies, slightly yellow																																								
Roseovarius crassostreae	–	+	+		+	–			–	+			–			–	–		I																	23–34	1–1.5				194, 197
Shewanella algae IAM 14159[T]	–	+	+	+	+	–	+	–	–	+	–	v	v	–	–	–	+	–	I	–	–	–	–	–	–	–		–	–	–	–	+	+	–	+	10–42	0–10				886, 1239, 1533, 1853

Shewanella marisflavi	–	+	+	+	+	–	–	–	–	+	+	w	–		+	–	+	–	F	–	+	–	v	+	–	–	–	–	–	–	–	+	+	G		4–40	0–12	R	R		1020, 1832
Varracalbimi	–	+	–	–α	–	–	–	–	+	–	–		–			–	–		F	–	+g–		+	+	+	+		+								4–22	2	S	S		1690
Environmental																																									
Achromobacter (xylosoxidans) denitrificans	–	+	+	v	+	–	–	–	–	+	–	+	–	–		–	–	–	O	–	–		–	–	–				–		–	–	+		–	10–37	0–4				618, 1437
	Previously *Alcaligenes denitrificans*																																								
Algicola bacteriolytica	–	+	+w		+	–				–		–					+		O		+															15–35	3				1489
	Red pigment																																								
Alteromonas macleodii	–	+	+		+	–			–	–		–					+		O		+		+	+	–	–	+	–	+	+	20					35–40	2	R	R		600, 1773
Andreprevotia lacus	–	+	+		+				+	+	–	+	+			–					+														–	15–37	0–1				1513
Aquaspirillum sp.	–	+	+		+		–	–	–	+	–	–	+		–	–	–		I	–	–		–	+	–	–			+	–	–	–				25	0–2	R	R		1032
Brevundimonas diminuta	–	+	+	α	+	–	–	–	–	–	–	–	–			–	–	–	O	–	–v	–	–	–	–	–	–	–	–	–	–	–	+		+14	30–37	0–5				1502
Brevundimonas (Pseudomonas) vesicularis	–	+	+	α	+	–	–	–	–	–	–	–	–			+	58	52	O		+		–	46	–	–	–	–	–	–	16	–	20		–	25	0–5				618, 1502
	Yellow colonies																																								
Burkholderia (Pseudomonas) cepacia	–	v+	+	–	+	–	66	+	–	–	–	+	42			28	73	79	O		+v		+	+	+	+	–		81		+	–	+		–	25–40	0	R	R		618
	Freshwater, yellow pigment																																								
Chromobacterium violaceum	–	+	+	+v	+	–	–	–	+	+	20	+	–	37	–	–	86	–	F	–	+g–		–	–	–				–	+	–	–	+	–	–	25–37	0	R	R		992
	Violet pigment. Some non-pigmented strains																																								
Colwellia maris	–	+	+		+	–				+	–			–	–		+		O	–	+			–		–			–						+	0–22	3–4	S	S		1844
Myroides odoratimimus	–	+	+	–		–			–	–	–	–	+			–	+	–	O		–								–				+			18–37	0–1				1694
Myroides odoratus	–	+	+	–		–			–	–	–	–	+			–	+	–	O		–								–				+			18–37	0–1				1694
Phocoenobacter uteri	–	+	–	–	–	–	–	–	–	+	–		–		+			+	F		+	–	–	–	–	–	–	–	–	–	–		–			22–37					549
Pseudoalteromonas haloplanktis tetraodonis	–	+	+		+					–		–	–				+		O		+		+	+		–			+							4–35	2				598
Halomonas aquamarina	–	+	+	–	+	–	–	–	–	–	–	+	–	–	–	–	–	–	O		–		–	–	–				–		–	–	+		–	25–37	0–4				337, 618
	Previously *Alcaligenes faecalis*																																								
Halomonas elongata	–	+v	+		+	–	+	+		+	22		66	11	11	56	44	33	F		+		67			+			+			–				30	0.1–20		R		1735

Continued

Table 3.11. Continued.

Test	Gm	Ox	cat	βH	mot	SW	ODC	LDC	ADH	Nit	Ind	Cit	urea	mr	vp	aes	G	onpg	OF	arab	glu	inos	lac	malt	man	mano	sal	sor	suc	tre	Xyl	H_2S	MCA	TCBS	Dnase	temp	NaCl	0129 10	0129 150	Amp	Ref
Halomonas venusta	−	+	+	−	+	−	−	−	−	+	−	+	+			+	−		O	−	+	+	−	v		−	−	v	v	v	−		+	−		4–37	0–6.5		R	S	123, 1731
Hydrogenophaga (*Pseudomonas*) *flava*	−	+	w							−	−	−	−			−	−		O	+	$+_{g-}$			+	+	+		+	+	+	−	−									67
Hydrogenophaga (*Pseudomonas*) *palleronii*	−	+	w		+		−	−	−	−	−	+	−				−		O	−	−		−	−	−	−	−	−	−	−	−	+				30					67, 1801
	Yellow pigment																																								
Hydrogenophaga pseudoflava	−	+	w		+		+	+		−	−	+	−			−	−		O	+	$+_{g-}$		−	+	+	+	−		+	+	+	+				35–41					67, 1801
	Yellow pigment. Nitrate pos with yeast extract																																								
Iodobacter fluviatilis	−	+	+	+w	+		−	−	−	+		85	−			−	+		F	−	$+_{g-}$	−	−	+	−	+	−	−	−	+	−	−	+			4–25	0–1	R	R	R	1048
	Violet on MCA. Spreading colonies on 1/4 NA																																								
Labrenzia alba	−	+	+		+	−	−	−	−	+	−	+					−		I		−											−			+	13–28	1–8				1362
Labrenzia (*Stappia*) *aggregata*	−	+	+			−			−	+	−	+	+			+	−	+			−														−	13–40	1–10				1362
Stappia (*stellulata*) *aggregata* M1	−	+	+	−	+	−	−	+		+	+							+	F																	25	1.5				1362
Mannheimia haemolytica	−	+	+	+		−	−	−	−	+		−	−			−		+	F		+												+	−							
Marinomonas aquimarina	−	+	+		+	−	−	−	−	−	−	+			−		−		O													−			−	13–40	1–10				1079
Marinomonas communis	−	+							−			+					−		O		+		−	+	+	+	−	+	−	25	+					35–40					600
Marinimonas ostreistagni	−	+	+									+							O		−				−				−			−			−	16–40	1–10				981
Marinomonas vaga	−	−							−	−		+					−				+		−	40	+	+	−	+	−	20	60					35					600
Oceanimonas baumannii	−	+			+	−			−	+	−	+	−			−	−	−	O		−															10–41	1–7				232
Oceanimonas doudoroffii	−	+			+	−			−	+	−	+	−			−	−	−	O		−															10–41	2				130, 232
	Nitrate may be neg by tube test, pos by API 20NE																																								
Paracoccus fistulariae	−	+	+		−		−	−	−	+	−	+	−			+	+	−	O	+	+	−	+	+	−	+		−	+	−	+	−				10–40	0–14				903
Phaeobacter (*Roseobacter*) *gallaeciensis*	−	+	+		+		−	−	−	−		−	−				−	−	O		+				+			+	+	+						15–37	2				1099, 1441
	Ovoid rods, brown diffusible pigment																																								
Phocoenobacter uteri	−	+	−	−	−	−	−	−	−	+	−	−	−		+		−	+	F	−	+	−	−	−	−	−	−	−	−	−	−		−			22–37					549

Pseudoalteromonas antarctica	–	+	–		+	–			–	–	–	–	–		+	–	+	–	I	–	+	–		+	+	+	–	–	–			–	+		–	4–30	0.1–9				215
Pseudoalteromonas citrea	–	+	+	+	+	–	–	–	–	–		–	–	–	–	–	+	–	O		+		–	–	–	+	–	–	–	+		–			+	10–30	1–10	R	R		598
Pseudoalteromonas denitrificans	–	+	–		+				–	+		–					+		O																+	4–22	1.5–5	R	R		492
	Red pigment																																								
Pseudoalteromonas distincta	–	+	+																																						
Pseudoalteromonas elyakovii	–	+	+		+	–				–		+					+		O		+															10–37	2				1489
Pseudoalteromonas espejiana	–	+	+							–		–					+																			35	2				298
Pseudoalteromonas flavipulchra	–	+	+	+	+	–	–	–	–	–		–	–	–	–	–	+	–	O		+		–		–	+	–	–	–	+		–			+	10–44	0.5–10	R	R		600, 815
Pseudoalteromonas haloplanktis haloplanktis	–	+	–w		+	–			–	–		+	+				+		O		+		–	+	50	+	–	–	+	+	–				+	4–15	2				598
Pseudoalteromonas luteoviolacea	–	+	–					–	–	–	–	–	–			+	+	–	O		+v		–	+	–	–			–	+	–	–			+	10–30	2–4	R	R		599
	Produces a violet pigment																																								
Pseudoalteromonas maricaloris	–	+	+		+	–			–	–							+		O																	10–37	0.5–10	R	R		815
Pseudoalteromonas nigrifaciens	–	+			+	–			–	–		+					+		O																	4–28	2				816
	Produces black pigment																																								
Pseudoalteromonas rubra	–	+	+	+	+	–	–	–	–	–		–	–			+	+		O		+		–	–	–	+	–	–	–	+					+	25–35	2	R	R		597
	Red pigment																																								
Pseudoalteromonas ulvae	–	+	+		+	–	–	–	–			+	–				+	–	I		–															4–25	1–2	S	S	S	471
Pseudoalteromonas undina	–	+								–		–					+		O				–		–	–	–	–	+							25–35	2				298
Pseudomonas acidovorans	–	+	+	–v	+	–	–	–	–	+	–	+	–			–	–v	–	I		–		–	–	+w	–			–		–	v	+		–	25–35	0				618
Pseudomonas aeruginosa	–	+	+	–	+	–	–	–	+	+	–	+	+v			–	+	–	O	+	+v	–	–	–	+	80	–	–	–	v	83	–	+		–	25–41	0–5	R	R		1229, 1337
	Green pigment. May have brick-red pigment																																								
Pseudomonas alcaligenes	–	+	+	14	+	–	–	–	+	+	–	v	–			–	–		O	–	–		–	–	–	–	–	–	–	–	–	–	+		–	25–41	0–5				337
	Rod or filament																																								
Pseudomonas aureofaciens	–	+	+		+	–			+	–		+					+				v		–		+			–	v	v						4–37					
	Orange pigment																																								
P. chlororaphis ssp. *piscium*	–	+	+		+	–			–	+	–		–				+	–	O		–															18–30					252

Table 3.11. Continued.

Test	Gm	Ox	cat	βH	mot	SW	ODC	LDC	ADH	Nit	Ind	Cit	urea	mr	vp	aes	G	onpg	OF	arab	glu	inos	lac	malt	man	mano	sal	sor	suc	tre	Xyl	H_2S	MCA	TCBS	Dnase	temp	NaCl	0129 10	0129 150	Amp	Ref
Pseudomonas mendocina	–	+	+	+	+	–	–	–	+	+	–	+	50			–	–	–	O		+		–	–	–	+	–	–	–	–	+	+	+		–	20–41	0–6				618
	Yellow pigment																																								
Pseudomonas mesophilica	–	+	+	–	+	–				v	–	–	v			–	–		O		+w		–	–	–				–		–	–	v			25–35	0				
	Fat cells, pleomorphic, vacuolated. Coral-pink pigment																																								
P. perfectomarina	–	+			–				–	+		+					–				+															40	2				
Pseudomonas stanieri	–	+			+				–	–		+					–				–															40	2				
Salinivibrio costicola ssp. *costicola*	–cv	+	+	+	+	–	–	–	+	12	–	20	–	53	+	+	+	–	F	–	$+_{g-}$	–	–	+	+	50	20	–	+	+	–	80			+	5–45	0.5–20	R	S	S	
	Previously *Vibrio costicola*. Found in hypersaline habitats.																																								
S. costicola NCMB 701[T]	–cv	+	+	+	+	–	+	–		–	–	–	–	+	+	–	–	–	F	–	$+_{g-}$	–	–	+	+			–	+	+	–	–			+	5–45	0.5–20	R	S	S	
Shewanella affinis	–	+	+	+	+	–		–	–	+	–	–			–		+		F		–											+				10–34	0.5–6				814
Shewanella algidipiscicola	–	+	+	–	+	–	–		–	+	–	–	–		–	+	–	–	O		–											+			–	4–37	1–6				1481
Shewanella amazonensis	–	+	+	+	+	–	–	–	–	+	–	+			–		+		O		–											+				4–40	0–3				1713
	Antarctic marine organism																																								
Shewanella aquimarina	–	+	+	+	+	–				+		–	–			–	+		F	–	–	–	–	–	–	–		–	–	–	–	+				10–42	2–3				1832
Shewanella baltica	–	+	+	–	+	–	+	–	–	+	–	+	–	–	–	–	–	–	O										+			+	+		+	4–30					1853
Shewanella benthica	–	+	+		+	–	–	–	–	+	–				–		+		O	–	+	–	–	–	–	–		–	–	–	–	+				4–15	0–2				1234, 1713
Shewanella colwelliana	–	+	+	–	+	–	–	–	–	+	+		–		–		+	–	O	–	–	–		–	–	–	–	–	–	–		–			+	8–30	1–5	R	R		209, 1773, 1774
	Adheres to surfaces. Sugars go alkaline. Older broth cultures have red-brown pigment																																								
Sherwanella corallii	–	+	+	s	+	–			+	+	–		–			+	+	–	F		+															10–37	0.5–8				1527
Shewanella frigidimarina	–	+	+	–	+	–	–v	–	–v	+	–	+v	–		–	v	+	–v	O	–	+	–	–	+	+	–v		–	+	+v	–v	+v			+	0–27	0–8				209
	Antarctic marine organism																																								
Shewanella gelidimarina	–	+	+		+	–	–	–	–	+	–	–	–		–	–	+	–	O	–	–	–	–	–	–	–		–	–	–	–	+			+v	0–23	1–6				209
	Antarctic marine organism																																								
Shewanella glacialipiscicola	–	+	+		+		+		–	+	–	–	–			+	+	–	O		–											+			+	4–30	0–3				1481

Shewanella hafniensis	–	+	+		+	–	+		–	+	–	+	–			+	+	–	O		–											+			+	4–25	0–6				1480
Shewanella haliotis	–	+	+		+	–	+	–	–	–	–	–	–		–		+	–	F		–											+				10–42	0–10				895
Shewanella hanedai	–	+	+		+	–	–	–	–	+	–	–			–		+		O	–	–	–	–	–	–	–	–	–	–	–	–	+				4–25	2	S	S		836,1234
Shewanella japonica	–	+	+	+	+	–				+		–					+		F		+														+	10–37	0–3				813
	Degrades agar																																								
Shewanella marinintestina	–	+	+	–	+	–	–	–	–	+	–		–		–	–	+	+	F	–	–	–	–	–	–			–	–		–	+		+	+	4–30	1–5	R	R		1479
Shewanella morhuae	–	+	+	–	+	–	+		–	+	–	–	–			+	91	–	O		–											v			+	4–25	0–4				1402, 1480
Shewanella oneidensis	–	+	+	–	+	–	–	–	–	+	–	–			–		+		F		+		–	–	–			–	–			+				15–40	0–3				1713
Shewanella pealeana	–	+	+	–	+	–	–	–	–	+	–	+	–		–	–	–	–	O	–	–			–	–	–		–	–	–	–	+			–	4–30	0.1–0.7				1008, 1479
Shewanella pneumatophori	–	+	+		+	–				–	–			–	–	–	+	–	O	–	–	–	–	–	–			–	–	–	–	+			+	4–32	1–8				732
Shewanella putrefaciens	–	+	+	–	+	–	+	–	–	+	–	v	–	–	–	–	+	–	O	+	s	–	–	70	–	9	–	–	–	–	–	+	+	–	+	4–40	0–5	S	S		886, 1234, 1402
Shewanella sairae	–	+	+	–	+	–	–	–	–	+	–		–		–	–	+	–	F	–	–	–	–	–	–			–	–		–	+		+	–	4–27	1–5	R	R		1479
Shewanella schlegeliana	–	+	+	–	+	–	–	–	–	+	–		+		–	+	+	+	F	–	–	–	–	–	–			–	–		–	+		+	+	4–32	1–5	R	R		1479
Shewanella upenei	–	+	+		+					+		–	–			–	+			–		–	–	–	–	–		–	–	–	–				+	10–40	0–11				901
Shewanella woodyi	–	+	+		+	–	–	–	–	+	–	–			–		+		O	–	–				–			–	–		–	+				4–25	2				209, 1093
Sphingomonas paucimobilis	–	+	+	–	+	–	–	–	–	–	–	–	–			+	–	+	O		–		+	+	–	+			+		+	–	–		+	30–37					618
	Yellow pigment																																								
Stappia stellulata IAM 12621[T]	–	+	+						–	+	–	–	–			–	–	–			–															5–37	2–8				1362
Vogesella lacus	–	+	+		+	–			–	+	+	+	–			–	–	–			–														w	25–30	0.5				195,325

Also see for *Brucella* species.

Table 3.12. *Arcobacter*.

Test	Gm	Ox	cat	mot	Na	Kf	Gly	IA	1.5 NaCl	2 NaCl	4 NaCl	25°C	37°C O_2	37°C mO_2	42°C	urea	Nit	hip	GGT	AP	TTC	MCA	Saf	B	Cef	H_2S	Ref
Organism																											
Arcobacter																											
Arcobacter bivalviorum	–	+	+	+			–	+	+	+	+	+	+	+	–	–	–				–	–	–	–	S		1014
Arcobacter butzleri	–		v	+			–	+			–		+	+	v	–	+				+	+	+	v	R		349
Arcobacter cibarius	–		v	+			–	+			–		–	+	–	–	–				+	+	+	+	R		349
Arcobacter cryaerophilus	–	+	+	+	S,v	R	–	99	+		–	+	v	v	$-^{b}$	–	98	–		–	–	v	+	+	R	–	349, 1208
Arcobacter defluvii	–	+	w	+			–	+	+		–	+	+	+	+	+	+				–	+	+	+	v	–	349
Arcobacter ellisii	–	+	+	+	I	S	–	+	+		–	+	+	+	w	33	+		+	–	–	66	–	–	S	–	539
Arcobacter halophilus	–		–	+			$+^{a}$	+			+		+	+	–	–	+				–	–	–	–	S		349
Arcobacter marinus	–	+	–	+			$+^{a}$	+			+	+	+	+	–	–	+	–		–	–	–	+	–	S		349
Arcobacter molluscorum	–	+	+	+			–	–	+		+	+	+	+	+	–	+s				–	+	+	+	R	–	538
Arcobacter mytili	–	+	w	+			+	–	+		+	+	+	+	w	–	+				–	+	–	+	S	–	348, 349
Arcobacter nitrofigilis	–		+	+	S	S	–	+	+		+		v	–	–	+	+				–	–	–	–	S	+	349
Arcobacter skirrowii	–		+	+			–	+	+		+		+	+	–	–	+				+	–	+	+	R		349
Arcobacter thereius	–		+	+			+	+			–		–	–	–	–	+				+	v	+	–	R		349
Arcobacter trophiarum	–		+	+			v	+			–		–	–	–	–	–				+	v	v	+	R		349
Arcobacter venerupis	–	+	+	+			–	+			–		–	+	–	+	+				–	+	–	–	S		1014
Campylobacter																											
Campylobacter coli	–	+	+				+	+		–		–		+	+		+	–								v	552
C. insulaenigrae	–	+	+	+	R	R	+	–		–	–	–	–	+	–	–	+	–				–	–			–	552
C. jejuni jejuni	–	+	+				+	+		–		–		+		+	+	+								–	552
C. jejuni doylei	–	+	v				v	+		–		–		+	–		–	+								–	552
C. lari spp. *concheus*	–	+	+	+	S	R	+		+	85		–	–	+	+	–	+	–				+	–		+		424
C. lari spp. *lari*	–	+	+	+	85	R	+		+	+		–	–	+	+	–	+	–				85	+		+	–	424
C. peloridis	–	+	+	+	85	15	+		+	80		–	–	+	+	–	+	–				+	–		–		424
C. subantarcticus	–	+	+	+	R	S	85	–	+	+	–	–	–	+	+			–				15	–	+	–	–	422
Wolinella succinogenes	–	+	–		R	R			–			–			v	–	+	–	–	–						–	

AP=alkaline phosphatase hydrolysis, B=bile or oxgall 1%, cat=catalase, Cef=growth on media containing 64 mg/l cefoperazone, GGT=gammaglutamyl transpeptidase activity, Gly=growth in 1% glycine, Gm=Gram reaction, Hip=hippurate hydrolysis, H_2S=production of hydrogen sulfide, I=intermediate susceptibility, IA=indoxyl acetate hydrolysis, Kf=sensitivity to cephalothin 30 μg, MCA=growth on MacConkey agar, Mot=motility, Na=susceptibility to naladixic acid 30 μg, 1.5% NaCl=growth in the presence of 1.5% NaCl, 4 NaCl=growth in the presence of 4% NaCl, Nit=nitrate reduction, Ox=oxidase, Saf=safranin 0.05%, S=sensitive, R=resistant, TTC=triphenyltetrazolium chloride, Urea=urease activity, v=variable reaction, w=weak reaction, 25°C, 37°C, 42°C=growth at these temperatures, 37°C O_2=growth at 37°C in air, 37°C mO_2=growth at 37°C in microaerobic atmosphere, $+^{a}$=test done using 2% NaCl, $-^{b}$= trout isolates grew at 42°C, +s=positive after 72 h at mO_2 and 5 days at O_2, numbers indicate percentage of positive strains.
Tests for alkaline phosphatase and hydrolysis of indoxyl acetate are done using the disc method (media section), GGT, AP, hip, Nit are done using the API Campy identification system (bioMerieux). H_2S is tested in triple sugar iron agar. Urea hydrolysis was tested using a rapid urea test (media section) from Remel. Growth on 1% glycine can be tested on a growth medium such as BA, containing 1% glycine. Tests for nitrate, glycine, 1.5% NaCl, 4% NaCl and growth at different temperatures are done using nutrient agar base media or BA plates (see media section). Sensitivity to Na and Kf are done on BA using standard antibiotic sensitivity test methods.

Table 3.13. *Helicobacter* species.

Test	Gm	Ox	cat	mot	Na	Kf	Gly	IA	1.5 NaCl	25°C	37°C	42°C	urea	Nit	hip	GGT	AP	TTC	H_2S	NaF	Ref
Organism																					
H. acinonychis	–		+	+	R	S	–	–		–	+	–	+	–		+	+				679, 680, 1130
H. anseris	–	+	+	+	S	R	w+	+			+	+	+	–		–	–				561
H. bilis	–	+	+		R	R	+	–	–	–	+	+	+	+					+		1130
H. brantae	–	+	+	+	S	R	+	+			+	+	–	–		–	–				561
H. canadensis Avian	–	+	+	+		R		+			+	+	–	+			+			+	1750
H. canadensis Human	–	+	+	+	R	R	+	+			+	+	–	50		–	–			+	559
H. canis	–		–		S	I		+				+	–	–			+			–	1130
H. cetorum	–	+	+	+	I/R	S	+	–		–	+	+	+	–		+	–				678, 679, 680
H. cholecystus	–	+	+	+	I	R	+	–	–	–	+	+	–	+	–	–	+		–	–	1130
H. cinaedi	–		+		S	I	+	–	–	–	+	–	–	+	–	–	–		–	–	1130
H. felis	–	+	+	+	R	S	–	–	–	–	+	+	+	+	–	+	+		–		679, 1130
H. fennelliae	–		+		S	S	+	+				–	–	–		–	+			v	1130
H. hepaticus	–	+	+		R	R	+	+	+	–	+	–	+	+	–				+	–	1130
H. muridarum	–	+	+		R	R	–	+	–	–	+	–	+	–	–	+	+		+		1130
H. mustelae	–	+	+		S	R	–	+	–	–	+	+	+	+	–	+	+		–		679, 1130
H. nemestrinae	–		+		R	S	–	–				+	+	–			+				679, 1130
H. pametensis	–	+	+		S	S	+	–	–		+	+	–	+	–	–	+		–	+	1130
H. pullorum	–		+		S	R		–				+	–	+			–			–	1130
H. pylori	–	+	+	+	R	S	–	–	–	–	+	–	+	–	–	+	+		–		679, 1130
H. rodentium	–	+	+		R	R	+	–	+	–	+	+	–	+		–	–			+	559
H. trogontum	–	+	+	+	R	R	–	–	–	–	+	+	+	+	–	+	33		–	+	1130
Helicobacter spp. MIT 01-5529-A	–	+	+	+	I	R		–		–	+	+	+	–			–				681
Helicobacter spp. MIT 01-5529-B	–	+	+	+	I	R		–		–	+	+	–	–			–				681

AP=alkaline phosphatase hydrolysis, cat=catalase, GGT=gammaglutamyl transpeptidase activity, Gly=growth in 1% glycine, Gm=Gram reaction, Hip=hippurate hydrolysis, H_2S=production of hydrogen sulfide, I=intermediate susceptibility, IA=indoxyl acetate hydrolysis, Kf=sensitivity to cephalothin 30 μg, Mot=motility, Na=susceptibility to nalidixic acid 30 μg, 1.5% NaCl=growth in the presence of 1.5% NaCl, Nit=nitrate reduction, Ox=oxidase, S=sensitive, R=resistant, TTC=triphenyltetrazolium chloride, Urea=urease activity, 25°C, 37°C, 42°C=growth at these temperatures.

Tests for alkaline phosphatase and hydrolysis of indoxyl acetate were done using ANI-Ident discs from bioMerieux, GGT, AP, hip, Nit, H_2S are done using the API Campy identification system (bioMerieux). Urea hydrolysis was tested using a rapid urea test from Remel. Growth on 1% glycine can be tested on a growth medium such as BA, containing 1% glycine. Sensitivity to Na and Kf were done on BA using standard antibiotic sensitivity test methods.

Table 3.14. Gram-positive cocci.

Test	Gm	Ox	cat	βH	mot	SW	ODC	LDC	ADH	Nit	Ind	Cit	urea	mr	vp	aes	G	onpg	OF	arab	glu	inos	lac	malt	man	mano	sal	sor	suc	tre	Xyl	H_2S	MCA	Coag	Dnase	temp	NaCl	Man An	RAA	Hip	Ref
Organism																																									
Pathogenic																																									
Aerococcus viridans	+		–	α	–	–				–					–						+																				1657
Aerococcus viridans var *homari*	+	–	w	α	–	–			–	–					–	+	–		F	50	+		+	+	50	+	60	50	+	+				–		10–37	0–10			+	621
Enterococcus faecalis var *liquefaciens*	+cb	–	–			–				+	–	+		–	–		+															–			+						
Granulicatella balaenopterae	+c		–	–	–	–			+	–			w+		–	+	–	–	F	–	+		–	+	–			–	–	+	–									–	352, 990
Lactococcus garvieae	+	–	–	α	–	–	–	–	+	–	–	–	–	+	+	+	–	–	F	–	+	–	–	+	+	+	+	–	v	+	–	–	v			10–45	0–6.0		–	–	954, 1466, 1614
	Growth in 6.0% NaCl, optimum 0% NaCl, 37°C. Resistant to clindamycin																																								1380, 1711
biotype 1, 2, 12	+	–		α	–				+						+	+				–			+	+	+			–	+	+			–			10–45	0–6.5				1711
biotype 3, 4, 5, 6, 7, 10, 11, 13	+	–		α	–				+						+	+				–			+	+	+			–	–	+			–			10–45	0–6.5				1711
biotype 8, 9	+	–		α	–				+						+	+				–			+	+	–			–	–	+			–			10–45	0–6.5				1711
Lactococcus garvieae	+		–	–	–	–			+	–	–			+	+	+	–	–	F	–	+	–	v	+	v	+	+	–	v	+	–	–	–			10–40	0–4			–	350, 954, 1614
Lactococcus piscium	+cb	–	–	–	–	–			–	–			–		+	+			F	+	+	–	+	+	+	+	+	–	+	+	+	–	+			5–30			–	–	1802
Micrococcus luteus	+	+	+		–	–	–	–	v	–	–	–	v		–		+	–	O	–	–	–			v			–	v			–				10–45	0–10				77
Planococcus species	+	–	+		+	–	–	–	–	+	–		–		+		–	+	O	–	+	–			–			–	–			–				37	2				77
Planococcus species	+	–	+		+		+	–	+			+	+				–	–	O		+															37	0–15				80
Staphylococcus aureus	+	–	+	+	–	–			+	+			+		+				F	–	+		+	+	+	+			+	+	–			+	+			+			617
Staphylococcus delphini	+	–	+	+	–	–			+	+			+		–		–		F	–	+		+	+	+s	+			+	–	–			+	+w	37–45	0–15	–			551, 1704
Staphylococcus epidermidis	+	–	+	–	–	–			+	80			+		+				F	–	+	–	+	+	–	+	–	–	+	–	–			–	–w	25–45	0–7				617

Staphylococcus lutrae	+c	–	+	+	–	–			–	+			+		–			+	F		+	+	+	+	v	+		+		+	+			+	+w	25–42	0–10				551
Staphylococcus warneri	+	–	+	–w	–	–			+	–			+		+				F	–	+	–	–	$+_s$	63	–	–	–	+	+	–			–	–	25–40	0–10				617
Streptococcus agalactiae	+c	–	–	v	–	–	–	–	+	–	–	–		+	v	–	–	–	Fw	–	+	–	v	+	–	+	–	–	+	+	–	–	–		+	25–37	0–3			+	134, 501
	Group B, No growth at 10°C or 45°C. VP may be neg in conventional test and pos in API Rapid ID 32 Strep																																								
Streptococcus agalactiae (*S.difficile*)	+	–	–	–	–	–	–	–	–	–	–	–	–	+	+	–	–	–	F	–	+	–	–	+	–	+	–	–	v	–	–		–			30	0–1			+	244, 1696
	Group B, Type Ib. Hip +ve at 30, –ve at 37°C																																								
Streptococcus dysgalactiae	+	+	–	+	–	–			–		–			+		–	–		F	+			+					–				–								+	1824
	Isolates from sturgeon. βH at 4°C but not at 37°C																																								
Streptococcus dysgalactiae ssp. *dysgalactiae*	+	–	–	+	–	–			+				–		–	v			F	–	+		+		–		v	–		+							0–4			–	1601, 1726
	Positive to Lancefield Group L antisera. Bacitracin = S																																								
Streptococcus dysgalactiae ssp. *dysgalactiae*	+	–	–	v	–	–			v						–	–			F	–	+		v		–		–	–	+	+			–				0–4			–	244, 1235, 1696
	Group C																																								
Streptococcus ictaluri	+		–	–	–	–			–s				–		–	–								+	–				–	–						25–35	0–2			–	1519
Streptococcus iniae ATCC 29177	+	–	–	+	–	–	–	–	+	–	–		–	+	–	–	–		F	–	+	–	+	+	+	+	–	–	+	+	–	–	–w		+	10–35	2–4			–	1215, 1342
	No growth at 6.5% NaCl, or 10°C or 45°C. Isolates from fish																																								
Streptococcus iniae ATCC 29178	+	–	–	+	–	–			+	–				–	–	+	–	–	F	–	+	–	–	+	+	+	+	–	+	+	–		–w			10–37	2–4			–	454, 1341
	No growth at 6.5% NaCl, or 10°C or 45°C. Isolates from fish																																								
S. iniae barramundi WA	+		–	+	–	–	–	–	+s	–	–	–	–	–	–	+	–	+	F	–	+	–	–	+	+	+	+	–	+	+	–	–	–		–w		0–3				378
Streptococcus iniae	+	–	–	+	–	–			–	–					–	+	–	–	F	–	+	–	–	+	+	+	+	–	+	+	–					37	2–4			–	454
	Strains from humans																																								
Streptococcus milleri	+	–	–	+w	–	–			+	–	–	+	–	–		+	–	+		–	+	+	–	+	–	+	+	–	+	+	+	–	+		–						165
Streptococcus parauberis	+cb	–	–	α	–	–			+		–				+	+		–	F	–	+	–	+	+	+	+	+	+	+	+	–		v			10–37	0–4			v	356, 456
	No growth at 6.5% NaCl, or 4°C or 45°C. Neg for Strep Group D																																								
Streptococcus phocae	+	+	–	–	–				–						–	–			F	–	+	–	–	+	–	+	–	–	–	–	–					25–37	0			–	1424, 1538
	Sensitive to bacitracin disc, Strains may react with Strep Group C, F or G antisera or are untypeable																																								

Continued

Table 3.14. Continued.

Test	Gm	Ox	cat	βH	mot	SW	ODC	LDC	ADH	Nit	Ind	Cit	urea	mr	vp	aes	G	onpg	OF	arab	glu	inos	lac	malt	man	mano	sal	sor	suc	tre	Xyl	H_2S	MCA	Coag	Dnase	temp	NaCl	Man An	RAA	Hip	Ref
Streptococcus uberis	+	–	–	–	–	–			+						+	+			F	–			+		+			+		+										+	688
	An isolate from prawns was positive for Lancefield Group B																																								
Vagococcus fessus	+cr		–	α	+	–			–	–			–		–				F	–	$+_{g-}$		–	–	–			–	–	–	–					37				–	762
Vagococcus lutrae	+cr		–	α	+	–			–	–			–		–	+	–	+	F	–	+			+	–			+	+	+	–					37	0			–	988
Vagococcus salmoninarum	+cb	–	–	α	–	–			–	–	–	–	–		+	+	–	–	F	–	$+_{g-}$	–	–	+	–	+	+	v	+	+	–w	+	v		–	5–30	0–3		–	v	1139, 1497, 1755
	May grow on TCBS, weak delayed reaction with Strep Group D antisera. Hip & VP pos with API 20 STREP																																								1613
	Vagococcus describes motile Lancefield Group N. *Vagococcus salmoninarum* has been reported as negative for Lancefield Group N																																								
Environmental																																									
Abiotrophia defectiva	+	–	–	α	–	–			–				–		–			+	F				+						+	+						20–42				–	861
Abiotrophia para-adiacens	+	–	–	α	–	–			–									–	F										+	–											861
Catellicoccus marimammalium	+		–	–		–			+	–			v		–	–			F	–			–	–	–			–	–	+						37				–	984
Enterococcus faecalis	+	–	–	α	–	–	–	–	+		–	–	–		+	+	v	–	F	–	+	–	+	+	+	+	+	v	+	+	–		+			10–45	0–6.5			+	1380
Enterococcus faecium	+		–	α	v	–			+			–			+	+	–	+	F	+	+	–	+	+	+	+	+	–	+	+	–					10–50	0–6.5			+	
Facklamia miroungae	+		–	α	–	–			+	–			+		–	–	–		F	–			–	–	–			–	–	+						25–42	0–5			–	761
Gemella haemolysans		–	–			–																																			
Granulicatella adiacens	+	–	–	α	–	–			–				–		–			–	F				–						+	–						20–42				–	990
Granulicatella elegans	+	–	–	α	–	–			+				v		–			–	F				–						+	–						27–37				+	1420
Lactococcus lactis. Serogroup N	+	–	–		–				+						+	+			F	–			+	+	–			–	–	+						10–45	0–2			–	1614
	Phenotypically similar to *E. seriolicida*. Sensitive to clindamycin																																								
Micrococcus luteus	+	+	+		–	–						–	+																							37–45	0–10				1791
Micrococcus luteus	+	+	+		–	–						v	v																							37–45	0–10				1791

Micrococcus luteus	+	+	+		−	−						−	+																							37–45	0–10				1791
Planococcus citreus	+	−	+		+	−	−	−	−	−	−	v	−	−	−	−	+			−	+	−	−	−	−	−	−	−	−	−	−	−			+	5–30	0–15			v	1688
Planococcus kocurii	+	−	+		+	−	−	−	−	−	−	v	−	−	−	−	v			−	v	−	−	−	−	−	−	−	−	−	−	−			+	5–30	0–10			v	1688
Planococcus okeanokoites	−v	w	+		+	−	+	+	+	−	−		−	−	−	−	+		O	−	−		−	−	−				−		−					20–37	3–5				1192
Staphylococcus aureus anaerobius				+											−									+						−	−			+	+						551
	Anaerobic growth only																																								
Staphylococcus capitis	+		+w	−w	−					+					v				F	−	+	−	−	−	+	+	−	−	+s	−	−			−	w	30–45	0–10				920
Staphylococcus carnosus	+	−		−	−					+					+				F		+		v	−	+	+			−	v	−			−		15–45	0–15				
Staphylococcus chromogenes									+				v		−				F					v						+				−	−			−			
Staphylococcus cohnii	+		+	−w	−					−					+w				F	−	+	−	+	80	80	80	−	−	−	+	−			−	20	15–45	0–10				1495
Staphylococcus cohnii urealyticum	+	−	+		−		−		28	15			+		25	28		+	F	−	+		−	22	+	+	−		−	+	−			−				−			
Staphylococcus haemolyticus	+		+	+v	−					+					+				F	−	+	−	50	+	50	−	−	−	+	+	−			−	−	20–45	0–10				1495
Staphylococcus hominus	+		+	−	−					+					w				F	−	+	−	60	+	−	−	−	−	+	+	−			−	w	20–45	0–7				920
Staphylococcus hyicus	+		+	−					+	−									F					−		+				+	−			v	+			−			551
Staphylococcus intermedius	+		+	+						+					−				F					−w		+				+	−			+	+			−			551
Staphylococcus saprophyticus	+		+	−	−					−					+				F	−	+	−	+	+	+	−	−	−	+	+	−			−	−	15–40	0–15				1495
S. schleiferi coagulans	+			+											+				F					−						−	−			+	+						551
S. schleiferi schleiferi	+			−											+				F					−						v	−			−	+						551
Staphylococcus simulans	+		+	−w	−					+					−				F		+	−	+	−	+	77	−	−	+	77	−			−	v	15–45	0–10	+			920
Staphylococcus warneri	+		+w	−	−					−					+				F	−	+	−	−	+s	+	−	−	−	+	+	−			−	−w	25–40	0–10				920

Continued

Table 3.14. Continued.

Test	Gm	Ox	cat	βH	mot	SW	ODC	LDC	ADH	Nit	Ind	Cit	urea	mr	vp	aes	G	onpg	OF	arab	glu	inos	lac	malt	man	mano	sal	sor	suc	tre	Xyl	H_2S	MCA	Coag	Dnase	temp	NaCl	Man An	RAA	Hip	Ref
Staphylococcus xylosus	+		+	–	–					80					–				F	80	+	–	80	+	+	+	–	–	80	80	+			–	–	15–40	0–10				1495
Streptococcus dysgalactiae equisimilis	+		–	+	–	–			+						–				F	–								–		+										–	1723
Streptococcus halichoeri	+		–	–	–	–			+						+	–				–					+	–										25–37				–	987
Streptococcus marimammalium Group C	+		–	s	–	–			v				v		–	–			F	–			+	v	–			–	–	–			–	–		25				–	986
	Beta haemolysis delayed up to 3 days																																								
Streptococcus porcinus	+			+					+						+	+		–	F	–	+	–	+	+	+	+	+	+	+	+	–					10–37	0–6.5			–	356, 1341
Streptococcus uberis	+		–	–	–	–			+						+	+		–	F	–	+	–	+	+	+	+	+	+	+	+	–					10–37	0–6.5			+	356, 1341
Vagococcus fluvialis ATCC 49515	+cr		–	α	+	–			–	+			–		–	+	–	v	F	–	$+_{g-}$	–	–	+	+	+	+	–	–	+	–	–				5–40	0–6.5		–	–	354, 1354, 1613
	Weak delayed reaction with Strep Group D antisera																																								
Vagococcus penaei	+	–	–	–	–				+	–			–		+			–	F	–	$+_{g-}$		–	+	–			–	+	+						10–30				–	826
Weissella ceti CECT 7719[T]	+cb		–	α	–	–			v	–					+	+	–	–	F	–	$+_{g-}$	–	–	+	–	–	–	–	–	+	–					22–37	3–6.5			–	1708
Weissella hellenica DS–12	+cr		–		–	–														+	$+_{g-}$	–	–	+	–	+	–	–	+	+	–					15–40	3–10				260
Weissella hellenica JCM 10103[T]	+c		–		–	–														+	+	–	–	+	–	+	–	–	+	+	–					20–30	3–5				260
Weissella species	+c	–	–	–	–	–			v						+	+								+	–			–	–	+			–			20–45					533

The list of Streptococci and Staphylococci is not comprehensive, and readers are urged to use a text such as Bergey's manual, if a *Streptococcus* species does not fit with the species listed here.
Coag=coagulase; Hip=hippurate hydrolysis; Man An=fermentation of mannitol under anaerobic conditions; RAA=growth on Rogosa Acetate Agar

Table 3.15. Gram-positive rods.

Test	Gm	Ox	cat	βH	mot	SW	ODC	LDC	ADH	Nit	Ind	Cit	urea	mr	vp	aes	G	onpg	OF	arab	glu	inos	lac	malt	man	mano	sal	sor	suc	tre	Xyl	H_2S	MCA	Dnase	temp	NaCl	ZN	MAF	RAA	Hip	Ref
Organism																																									
Pathogenic																																									
Actinomyces marimammalium	+		–		–	–			–	–			–		–	–	–		F	–	+		+	+	–			–	–	–	–						–			–	763
Arcanobacterium phocae	+		10	+	–	–			–	–			–			–	–		F	–	$+_{g-}$	+	68	+	40	60	–	–	+	+	10						–	–		–	845, 1374
Austwickia chelonae	+		+	+	+	–				+	–		–	–	–		+		F		+		–	v	–		–	–	–	–	–				25		–				1109
Bacillus cereus	+		+	+	+						–				+		+		O		+														25–45		–	–			132, 648
Bacillus mycoides	+		+	+	–						–				+		+		O		+														25–37		–	–			648
Carnobacterium (piscicola) maltaromaticum	+sr	–	–	–α	–	–	–	–	v	–	–	–	–	+	v[w]	+	–	59	F	–	$+_{g-}$	–	40*	+	88	+	+	–*	+	+	–	–	–	–	0–35	0–6	–	–	–	–	136, 355, 736, 1047, 1167, 1576
	* Weak reactions in phenol red medium, negative in purple base medium																																								1655
Carnobacterium maltaromaticum-like	+sr	–	–		–	–	–	–	+	–	–	–		+	v	+		+	F	–	$+_{g-}$	–	+	+	v	+	+	–	+	+	–	–	–			0–6			–		1047
Carnobacterium sp.	+r	–	–	–	–	–	–	–	+	–	–	–	–	+	–	+	–	v	F		$+_{g-}$		+v	+	+	+	+	v	+	+		–	–	–	10–37	0–6	–	–	–		136
Corynebacterium aquaticum	+	–	+	+	+	–	–	–	–	–	–	–	–		–	+	+	+	I		–		–	–	–				–		–	–	–		4–42	0–5					133
Corynebacterium aquaticum	+	–	+	+	+	–	–	–	–	+	–	–			–	+	–		I		+			+		+		+	+		+	–	–		25–42	0–3					132, 1005
C. pseudotuberculosis	+r	–	+	+	–		–	–	–	–	–	–	+		–	–	–	+	F		$+_{g-}$		–	–	–				–		–	–			20–42	0–3					235
Corynebacterium testudinoris	+		+		–	–				+			–			+	–		F		+		–	+	–				+		–				37	0					351
Dermatophilus congolensis	+		+	+	+	–				–	–		+	–	–		+		F		+		–	+v	–		–	–	–	–	–				37		–				649
Erysipelothrix rhusiopathiae	+r	–	–	α	–	–				–	–		–		–	–	–		F	v	+	–	+	+	–	v	–	–	v	v	–w	+			5–37						607, 1591
Renibacterium salmoninarum	+	–	+	+	–	–			–	–	–		–	–	–	–	+	–	I	–	–		–	–	–		–	–	–	–	–			v	15–18		–				238, 647, 1472

Continued

Table 3.15. Continued.

Test	Gm	Ox	cat	βH	mot	SW	ODC	LDC	ADH	Nit	Ind	Cit	urea	mr	vp	aes	G	onpg	OF	arab	glu	inos	lac	malt	man	mano	sal	sor	suc	tre	Xyl	H_2S	MCA	Dnase	temp	NaCl	ZN	MAF	RAA	Hip	Ref
Rhodococcus erythropolis	+	–	+	–	–	–						+							O																15–30	0–1					1265
Rhodococcus qingshengii	+	–	+		–	–	–	–	–	+		–	+	+	–	+	–		O														–	–	15–37	0–3					83
Rhodococcus sp.	+r	–	–			–							+						F		–	+	–	–	–			–	–		+				25	0–5	–				90
	No growth 37°C. Growth in 5% NaCl																																								
Streptomyces salmonaris	+																+		O	–	+	+			–				w	+	–	–		+	12–37						68
	Brick-red to orange pigmented mycelia																																								
Environmental																																									
Arcanobacterium haemolyticum	+																																								
Arcanobacterium phocisimile	+	–	+	w		–				–			–			–	–	+			+		+	+	–				+		–			+						–	727
Arcanobacterium pluranimalium	+		+	α	–	–			–	–			–		–	+w	+		F	–	+		–	v	–			–	–	–	–									+	985
Arthrobacter agilis	+c	+	+	–	+v	–	–	–	–	–	–	–		–	–	+	+	+	O		–		–			–						–			v	20–30	0–1				
Arthrobacter rhombi	+cb	+	+		–	–				–			–			+	–	+	O	–	+	–	+	+	+	+	+	–	+	+	–				4–30	1–10					1276
Arthrobacter nasiphocae	+		+	–						–			–		–	–	+		O		–		–	–	–				–		–				25–40	5				+	358
Atopobacter phocae	+		–	–					+	–			–		–	–	–		F	–	+		v	+	–			–	v	v	–				25					–	989
Bacillus zhanjiangensis	+r	+	+		+	–	–	–	–	–	–	+	–	–	–	+	+		I		–											–		–		0–15				+	316
Carnobacterium alterfunditum	+r	–	–	–	+	–				–			nt			–	–		F	–	+	–	–	w	–	w	nt	–	w	–	–	nt			0–20	1–6	–	–		–	841
Carnobacterium divergens	+r	–	–		–	–	–	–	+	–	–		–		+	+	–	+	F	–	+	–	–	+	–	+	+	–	+	+	–	–			0–40	0–7	–	–	–		355, 1405
Carnobacterium divergens 6251	+r	–	–		–	–				–	–		–				–		F		$+_{g-}$		+		+							–			0–35	0–6			–		1405
Carnobacterium funditum	+r	–	–	–	+	–				–			nt			–	–		F	–	+	–	–	+	+	+	+	–	+	+	–	nt			0–20	1–6	–	–		–	841
Carnobacterium gallinarum	+r	–	–	–	–	–	–	–	+	–	–		–		+	+		+	F	–	$+_{g-}$	–	+	+	–	–	+	–	+	+	+	–			0–35				–		355
Carnobacterium inhibens	+r	–	–		+	–			–	–	–	–	–		–	+	–	–	F	+	$+_{g-}$	–	w	+	+	–	+	–	+	+	–	–			0–30	1–6	–	–	–	+	841
Carnobacterium mobile	+r	–	–		+	–	–	–	+	–	–		–		–	+		+	F	–	$+_{g+}$	–	–	+	–	+	+	–	+	+		–			0–35				–		355
Carnobacterium maris	+cb	+	+	–	–	–				–			–				+		O		–		–	–	–				–		–				26–37	0–10					152

Corynebacterium caspium	+r	–	+	–						–			+			–	–		F		+		–	–	–				–		–				37						357
Corynebacterium ciconiae	+r		+	–	–	–				–			v			–	–		F	–	+	–	+	v		+	–	–	+	v	+				37						528
Corynebacterium phocae	+r		+	–	–	–				–			v			–	–		F	–	+	–	v	+	–	+	–	–	v	–	–				37		–				1313
C. pseudodiphtheriticum	+r	–	+	+	–					+	–	–	+		–	–	–	–	I		–		–	–	–				–		–	–			20–42	0–5					235
Corynebacterium sphenisci	+r		+	–	–	–				+			–			–	–		F		+		–	+	–	+			–	+	–				37						652
Corynebacterium spheniscorum	+r		+	–	–	–				–			–			–	–		F	–	+	–	–	+	–	+	–	–	–	+	–				37						653
C. xerosis	+r	–	+	+	–		–	–	–	+	–	–	–		–	–	–	–	F		+g–		–	–	–				–		–	–			20–35	0–3	–	–			235
Lactobacillus ceti	+r		–	–	–	–				–			–			–	+		F	–	–	–	–	–	–	–	–	–	–	–	–				22–37	0–3					1709
L. lactis sp. *tructae*	+cb		–		–				+										F				+	+	+				+		–					0–4					1330
L. piscium																																									
Rhodococcus equi	+									+		–	+				–			–		–			–	+	+	–	+		–	–				0–7	–	–			1212
Rhodococcus (luteus) fascians	+		+		–	–				15	–	+	+	–	–		36		O	+	+	–	–	–	+	+	–	+	+		+	40			25	0–5	–	–			1212
Rhodococcus maris	+		+		–	–				+	–	–	30	–	–		–		O	–	+	–	–	–	–	–	–	–	–		–	–			25	0–7	–	–			1212
Trueperella bernardiae	+		–	v	–					–			–			–			F		+			+	–				–		–										1374
Trueperella pyogenes	+		–	+	–	–	–	–	–	–	–		–			–	+	+	F	–	+g–	+	+	+	v–	+	–	v+	v–	+	+				20–40		–	–		+,–	1374
	Acid, clot, reduction in litmus milk																																								

Carnobacterium species are very similar to *Lactobacillus* sp. *Vagococcus* describes motile Lancefield Group N strains, but there are exceptions.
Also see tables for *Nocardia* and *Mycobacteria*.

Table 3.16. *Mycobacterium* and *Nocardia* species.

Test	Gm	Ox	cat	βH	mot	ODC	LDC	ADH	Nit	Ind	Cit	urea	mr	vp	aes	G	onpg	arab	glu	inos	lac	malt	man	mano	sal	sor	suc	tre	Xyl	H_2S	temp	NaCl	NaCl 4	50°C	Hipp	Ref
Organism																																				
Mycobacterium abscessus	AFB		+						–		–	+						–	–	–			–	–		–	–	–	–			6.5	+			961, 1620
Mycobacterium avium	AFB		+		–				+			–																			37–45	5				1621
Mycobacterium chelonae	AFB	–	+		–				–		v	+						–	v	–			–	–		–	–	–	–		15–25	0–3	–		–	235, 403, 961
M. fortuitum ATCC 6841[T]	AFB		+		–				+		–	+						–	+	–			–	+		–	–	–	–		28–37	0–5	+			207, 961, 1621
M. fortuitum ATCC 6841[T]	AFB		+		–				+			–						+	+	–			+	+		–		–	–		28–37	0–5	+	–		1190
M. fortuitum ATCC 6842[T]	AFB																	–		–	–	–	–	+		–	–	+	–							1427
M. fortuitum ssp. *acetamidolyticum* ATCC 35931[T]	AFB								+		–	+						–	–	–			–	–		–	–	–	–		28–43					961
M. marinum ATCC 927[T]	AFB		+		–				–			+						–		–	–	–	–	–		–	–	–	–		28–30	0		–		207, 1427
M. marinum ATCC 927[T]	AFB		+		–				–			–						+	+	–			–	–		+		–	–		28–30	5	+	–		1190
M. marinum, yellowtail	AFB		+		–				–		–	+						–	–	–	–	–	–	–		–		–	–		15–37		–			1771
M. montefiorense	AFB				–				–			–																			25–30			–		719, 1013
Mycobacterium neoaurum	AFB																														25			–		90
M. palustre E846[T]	AFB		–		–				54			+																			30–42					1656
Mycobacterium peregrinum ATCC 14467[T]	AFB		+						+		–							–	+	–			–	+		–	–	+	+		25–37	+				961, 1162
M. pinnipedii 6482[T]	AFB				–				–																						25–37					374
M. poriferae ATCC 35087[T]	AFB		+						–		+	+						–	+	–	–		+	+		–	+	–	+		25–37	5	+			1294
M. pseudoshottsii ATCC BAA-883[T]	AFB		–						–			+																			22–25	0		–		1401
M. salmoniphilum ATCC 13578[T]	AFB	–	+	–					–	–	v	v	–	–		–		–	+	–	–	–	–	+		–	–	+	–	–	20–30	0–2		–		1427, 1783
Mycobacterium scrofulaceum	AFB		+						–			+																								1621
M. shottsii M175[T]	AFB		–						–			+																			22–25					1400

M. simiae	AFB								–			+																								1621
M. smegmatis	AFB								+									+														5	+			1621
Mycobacterium species	AFB		–		–				–			+																			28					708
M. stomatepiae DSM 45059^{T}	AFB								–			–																			30			–		1355
Mycobacterium triplex	AFB		–																												30–37					1013
Nocardia asteroides	+	–	+		–				+	–	+	+w	–	–	+	–	+	–	+	–	–	–	–	–	–	–	–	+	–	–	20–37	0–5	+	+	v	566, 944, 951
N. brasiliensis	+		+						+	–	+	+	–	–	+	+	+	–	+	+	–	–	+	–		–	–	–	–		30–45	0–5	+	–	–	566, 944, 951
N. brevicatena	+								–		–	–			+			–	–	–			–			–		+								331
N. carnea	+								+		–	–			+			–	+	v			+			+		+								331
N. crassostreae	+	–	+		–						–	+			+	–		–	+		–	–	–				+	+	–		22–30	0–2		–		566
N. farcinica	+	–	+						+		–	+			+	–		–	+	–	–	v	–	–		–		v	–		20–45	0–4		+		944
N. flavorosea	+				–				–		–	–			–	–		–	–	–	–	–	–	–	–	–	–	–	–					+	–	331
N. nova	+								+		–	–			+			–	+	–	–	–	–	–		–	–	–	–		20–40	0–4	–	+		944, 1753
N. (caviae) otitidiscaviarum	+	–	+		–				+	–	v	+w	–	–	+	–	+	–	+	+	–	–	+	–	–	–	–	v	–			0–4		+	v	951
N. otitidiscaviarum	+								+		–	–			+			v	+	+			v			–		v								331
N. pseudobrasiliensis	+		+		–				–		+	+			+			–	+	+	–	–	+	+				+	–		20–37					331, 1440
N. seriolae	+	–	+		–				+	–	+	–			+	–		–	+	–	–	–	–	40	–	–		25	–	+	15–30	0–5	+	–		566, 944
N. salmonicida	+	–	+		–				+	–	+	+			+	–		–	+	–	–	–	–	–		–	–	–	–	+	10–30	0–4	–	+		806, 944, 1436
N. transvalensis									+			+			+			–	+	v			v			–		v								331
N. vaccinii									+		–	+			+			+	+	v			+			–		v								331
Nocardia species (Aust)	+									–	+							–	+	+		–	–													217
	AFB = acid-fast bacilli in ZN stain																																			

Table 3.17. *Mycobacterium* species – additional tests.

Test	Growth 22–25°C	Growth 28°C	Growth 37°C	Growth 5–7 days	Pigment in light	niacin	Nitrate	Tween 80	cat	Cat heat stable	aryl	urease	pyr	Fe	thio	5% NaCl 28°C	5% NaCl 37°C	MCA 28°C	MCA 37°C	Ref
Organism																				
Mycobacterium abscessus		+	+	+	–	–	–	v	+		+	+	+	–	+	+	+	+	+	961
Mycobacterium avium	+	+	+			–	–	–	–	v	–	–	+		+		–			
Mycobacterium chelonae		+	+	+	–	v,–	–	–	+		+3 d	+	+	–	+	+	–	+	–	235, 403, 961, 1621
Mycobacterium fortuitum ATCC 6841ᵀ	+	+	+	+	–	–	+	–7d	+	+	+3 d	+	+	+	+	+	+	+	+	207, 216, 961, 1621
M. fortuitum ssp. *acetamidolyticum*		+	+	+	–		+	–			+	+	+	+		–	–	–	–	961
Mycobacterium marinum ATCC 927ᵀ		+	+	+	+	–	–	+	–		–3 d	+	+	–	+	+	–	–	–	207, 1621
Mycobacterium marinum ATCC 927ᵀ		+	+	+	+		–	–	+	+	+14 d	–		–		+	–	–	–	1190
Mycobacterium marinum	+	+	15–49	+	+	–	–	+	15–49	+	15–49	+	50–80		+	–	–			711, 1401
M. montefiorense ATCC BAA–256ᵀ	+	+	–	–	–	–	–	–	–	w	–	–				–				1013
Mycobacterium neoaurum					+		+				+			+		+	–		–	90, 1621
Mycobacterium palustre E846ᵀ			+	–	85		54	+	–		45	+	+							1656
Mycobacterium peregrinum ATCC 14467ᵀ	–	+	+	+	–	–	+	v	+	+	+	+	+	+			+	+	–	961, 1162
M. pinnipedii 6482ᵀ			+	–	–	v	–													374
M. poriferae ATCC 35087ᵀ		+	+	+	+		–	+	+	+	–	+	80	+		+				1294
M. pseudoshottsii ATCC BAA–883ᵀ	+	+	–	–	+	+	–	–	–	50–80	–3 d	+	–			–	–	–	–	1401
M. salmoniphilum ATCC 13578ᵀ	+	+	–	+	–		–		+		+	v					–	+	–	1427, 1783, 1849

Mycobacterium scrofulaceum	+	+	+		+	–	–	–	+	+	–	+	+		+					1621
M. shottsii M175[T]	+	15–49	–	–	–	+	–	–	–	15–49	–3 d	+	–							1400
Mycobacterium simiae	+	+	+		50–80	15–49	–	–	+	+	+10 d	+	+		+		–			1401, 1621
M. smegmatis		+	+	+	–	–	+	+			+14 d			+	+	+		–		1621
M. stomatepiae DSM 45059[T]	+	+	–	–	–		–	–	–	+	+	–		–		–	–	–	–	1355
Mycobacterium species (new)							–	–	–		–	+	+				–			708
M. triplex	w	+	+			–	–	–	+	+	+10 d	+					–			1013
Mycobacterium ulcerans	+	+	15–49		–	–	–	–	–			15–49								1401

aryl=arylsulfatase after 3 days; cat=catalase pH 7.0 at 68°C, Fe=iron uptake; Growth 28°C=growth on Ogawa egg medium at 28°C; Growth 37°C=growth on Ogawa egg medium at 37°C; MCA=growth on MacConkey agar without crystal violet at 28°C and 37°C; niacin=niacin accumulation; nitrate=nitrate reduction; Tween 80=tween 80 hydrolysis; pyr=pyrazinamidase after 5 days; thio=growth on 2-thiophenecarboxylic acid hydrazide 0, 1, 5 µg/ml; urease=urease hydrolysis; 5% NaCl=growth in 5% NaCl; numbers indicate percentage of positive strains; 10 d=positive result at 10 days.

Table 3.18. *Mycoplasma* species.

Test	Gm	Ox	cat	βH	mot	PO_4	ODC	LDC	ADH	Nit	Ind	Cit	urea	mr	vp	aes	G	TTC	OF	arab	glu	inos	lac	malt	man	mano	sal	C	S	tre	Xyl	F&S	temp	NaCl	Pen	Ref
Organism																																				
Pathogenic																																				
Mycoplasma alligatoris					–	+			–				–			–					+		+			+		+	+			–	30–34			232
Mycoplasma crocodyli				+		+			–				–			–		+			+					+		+	+			–	25–42			909
Mycoplasma mobile		v	w	+	+g	+V			–				–				–	+		+	+		+	+		+			+			+	4–30			908
	Cells adhere strongly to glass or plastic. Cells glide. Weak growth on BA. F&S with horse serum, not bovine serum																																			
Mycoplasma phocirhinis				+		–			–				–					+			–							+	+			+	37			616
Mycoplasma phocicerebrale				+		–			+				–					–			–							+	+			–	37			616
Mycoplasma phocidae				α		+			+				–					–			–								+			+	37			1438
Mycoplasma species (fur seals)						–			–									–			+											–				1074
Mycoplasma testudineum				+		–			–				–			–		–		–	+	–	–	+	+	+	–			–	–	F				
Mycoplasma testudinis				+		–			–				–			–		–		–	+	–	–	+	+	+	–			–	–	F				728
Mycoplasma zalophi						–			+									+			–											v	37			701

See Media Preparation for specific biochemical tests for Mycoplasma.

C=cholesterol requirement for growth, F&S=film and spots, Glu=glucose fermentation, Gm=Gram, Ox=oxidase, βH=lysis of sheep blood cells, mot=motility, PO_4=reduction of potassium tellurite, S=serum required for growth, temp=temperature range or optimum temperature, TTC=reduction of tetrazolium chloride under aerobic conditions, urea=hydrolysis of urea, +g=gliding motility.

Table 3.19. *Moritella* species.

Test	Gm	Ox	cat	βH	mot	SW	ODC	LDC	ADH	Nit	Ind	Cit	urea	mr	vp	aes	G	onpg	OF	arab	glu	inos	lac	malt	man	mano	sal	sor	suc	tre	Xyl	H_2S	MCA	TCBS	Dnase	temp	NaCl	0129 10	0129 150	Amp	Ref
Organism																																									
Pathogenic																																									
Moritella viscosa NCIMB 13584[T]	–	+	+	+	+	–	–	+	–	+	–	–	+	5	–	–	+	–	I	–	$+_{g-}$	–	–	–	–	–	–	–	–	–	–			–	+	4–25	1–4	R	S	S	1071
	Isolates from Scotland, Norway. May need 1% peptone added to media for growth. No growth at 30°C																																								
Moritella viscosa	–	+	+	+	+	–	–	+	–								+		I	–	+	–	–	+	–	+	–	–	–	–				–	+	4–21	2–3	R	S		153
	Isolates from south-west Iceland only. No growth at 25°C																																								
Moritella viscosa	–	+	+	w	+	–	–	–	–								+		I	–	+	–	–	–	–	–	–	–	–	–				–	+	4–21	2–3	R	S		153
	Isolates from North Iceland																																								
Environmental																																									
Moritella japonica	–	+	+		+	–	–	+	–	+	–						+		F	–	+	–	–	–	–	–		–	–	–	–	–				4–15		R	R		1233
Moritella marina ATCC 15381[T]	–	+	+	+	+	–	–	–	–	+	–	–	+	–	–	+	+		F	–	$+_{g-}$	–	–		–	+	–	–	–	–	–			–	–	0–20	3–5	R	S	S	1071
Moritella marina ATCC 15381[T]	–	+	+	–	+	–	–	+	–	+	–	–		+	–	–	+		F	–	$+_{g-}$	–	–	w,–	v	–	–	–	–	–	–	–		–	+	4–20	3	PS	S		153, 154, 363, 715, 1683

Also check tables for *Aliivibrio* species, *Photobacterium* species, *Vibrio* – Pathogenic species, *Vibrio* – Environmental species.

Table 3.20. *Photobacterium* species.

Test	Gm	Ox	cat	βH	mot	SW	ODC	LDC	ADH	Nit	Ind	Cit	urea	mr	vp	aes	G	onpg	OF	arab	glu	inos	lac	malt	man	mano	sal	sor	suc	tre	Xyl	H_2S	MCA	TCBS	Dnase	temp	NaCl	0129 10	0129 150	Amp	Ref
Organism																																									
Pathogenic																																									
P. damselae ssp. *damselae* ATCC NCIMB 2184^{T}	–	+	w	+	+w	–	–	–	+	+	–	–	+	+	+	–	–	–	F	–	$+_{g+}$	–	–	+	–	+	–	–	–	–	–	–	+	G	+	10–37	0.5–5	S	S	S	154, 556, 1071
	80% of *P. damselae* strains may be positive for trehalose (Ref 745)																																								595, 1322
biotype 7	–	+	+	+	+	–	–	+	+	+	–		+	+	+	–	+	–	F	–	$+_{g+}$	–		+			–	–				–		G			0.5–5	R	R		1322
biotype 1	–	+	+	+	+	–	–	–	+	+	–		+	+	+	–	–	–	F	–	$+_{g+}$	–		+			–	–				–		G			0.5–7	S	S		1322
biotype 2	–	+	+	+	+	–	–	–	+	+	–		+	+	+	–	–	–	F	–	$+_{g+}$	–		+			–	–				–		G			0.5–7	S	S		1322
biotype 3	–	+	+	+	+	–	–	–	+	+	–		+	+	+	–	–	–	F	–	$+_{g+}$	–		+			–	–				–		G			0.5–5	S	S		1322
biotype 4	–	+	+	+	+	–	–	–	+	+	–		+	+	–	–	–	–	F	–	$+_{g+}$	–		+			–	–				–		G			0.5–7	S	S		1322
biotype 5	–	+	+	+	+	–	–	+	+	+	–		+	+	+	–	–	–	F	–	$+_{g+}$	–		+			–	–				–		G			0.5–7	S	S		1322
biotype 6	–	+	+	+	+	–	–	–	+	+	–		+	+	+	–	–	–	F	–	$+_{g+}$	–		+			–	–				–		G			0.5–7	S	S		1322
biotype 7	–	+	+	+	+	–	–	+	+	+	–		+	+	+	–	+	–	F	–	$+_{g+}$	–		+			–	–				–		G			0.5–5	R	R		1322
biotype 8	–	+	+	+	+	–	–	–	+	+	–		+	+	+	–	+	–	F	–	$+_{g+}$	–		+			–	–				–		G			0.5–7	R	R		1322
biotype 9	–	+	+	+	+	–	–	–	+	+	–		+	+	+	–	+	–	F	–	$+_{g+}$	–		+			–	–				–		G			0.5–7	S	S		1322
	Haemolysis may be more apparent on BA than MSA-B																																								
Photobacterium damselae ssp. *piscicida* ATCC 17911	–cb	+	+	–	–	–	–	–	+	–	–	–	–	w	v	–	–	–	F	–	$+_{g-}$	–	–	–	–	+	–	–	–	–	–	–	–	–	73	15–30	0.5–3	S	S	S	595, 702, 959
	Short rods, bipolar staining. VP results variable or neg by API 20E																																								1089, 1641
Photobacterium rosenbergii sp. nov	–	+			+		–	–	+	+	–	+	–		–		–	+		–	+	v			+			–	+					Y		20–30	2–6				1636
Environmental																																									
P. angustum NCIMB 1895	–	+	+	–	–	–	–	–	+	+	–	–	–	+	+	+	+	+	F	–	$+_{g-}$	–	v	+	–	+	–	–	+	v	+	–	–	–	+	4–37	0.5–6	S	S	S	1641
P. iliopiscarium ATCC 51760	–	+	+	–	+		–	+	+	+	–	–	–	+	+		–		F	–	$+_{g+}$	–	–	+	–	+	–	–	–	–		–		G	+	4–25	0.5–2	S	S		154
P. iliopiscarium ATCC 51760	–		–	–	–	–	–	+	+	+	–	–	–	+	+	–	–	–	F	–	$+_{g-}$	–	–	+	–	+	–	–	–	–	–	–	–	–	–	25	0–, 3w	S	S		244
P. leiognathi ATCC 25521	–	–	+		+		–	+	–	v	–			+	+		+s		F	–	$+_{g-}$	–	–	–		+	–		–		–					20–30	0.5–5				713, 1641
P. leiognathi LMG 4228	–	–	–	–	–	–	–	–	+	–	–	–	–	–	+	–	–	+	F	–	$+_{g-}$	–	v	–	–	+	–	–	–	–	–	–	+	–	–	20–35	0.5–6	S	S	S	1641

P. phosphoreum NCIMB 844	–	–	+	–	v	–	–	+	+	+	–	–	–	+	+	v	–	+	F	–	+$_{g+}$	–	–	+	+	+	–	–	–	–	–	–			+	4–30	0.5–8	R	S		154
P. phosphoreum NCIMB 1282	–	–	+	–	v	–	–	–	+	+	–	–	–	+	+	v	–	+	F	–	+$_{g+}$	–	–	+	–	+	–	–	–	–	–	–			–	4–21	1–8	S	S		154
P. profundum	–	–					–	+	+		–				+		–		F		+$_{g+}$			+	–	+	–	–	–	–						5–20					1234
P. rosenbergii LMG 22223^T	–	+			+	–	–	–	+	+	–		–		–		–		F	–	+	+			+			–	+					Y		20–30	2–6		R		1636
P. rosenbergii	–	+			+	–	–	–	+	+	–		–		–		–		F	–	+	v			+			–	+					Y		20–30	2–6		R		1636
P. swingsii CAIM 1393^T	–	+	+				–	–	+	+	16	–	–		+		16	50	F	+	+	–		+	33	+			–	+		–		G		4–30	2.5–6	S	S		641

Also check tables for *Aliivibrio* species, *Moritella* species, *Photobacterium* species, *Vibrio* – pathogenic species, *Vibrio* – environmental species.

Table 3.21. *Aliivibrio.*

Test	Gm	Ox	cat	βH	mot	SW	ODC	LDC	ADH	Nit	Ind	Cit	urea	mr	vp	aes	G	onpg	OF	arab	glu	inos	lac	malt	man	mano	sal	sor	suc	tre	Xyl	H_2S	MCA	TCBS	Dnase	temp	NaCl	0129 10	0129 150	Amp	Ref
Organism																																									
Pathogenic																																									
A. finisterrensis CMJ 11.1[T]	–	+	+		+	–	–	+	+	+	–	–	+	+	–	+	–	+	F	–	$+_{g-}$	–	+	+	–	+	+	–	–	–	–			G	+	4–30	1–6	S	S	R	143
A. finisterrensis C 17.13	–	+	+		+	–	–	+	+	+	–	–	+	+	–	+	–	+	F	–	$+_{g-}$	–	+	+	+	+	+	–	+	–	–			Y	+	4–30	1–6	S	S	R	143
A. fischeri ATCC 7744[T]	–	+	+	w	+	–	–	+	–	+	–	–	+	+	–	–	–	–	F	–	$+_{g-}$	–	–	+	+	+	+	–	–	–	–	–	–	–	+	25	1–3	S	S	R	244
A. fischeri ATCC 7744[T]	–	+	+		+	–	–	+	–	+	–			+	–		+		F	–	$+_{g-}$	–	–	+	–	+	+	–	–		–					10–30	1–5	S	S	R	715
A. fischeri ATCC 7744	–	+	+	–v	+	–	–	+	–	+	–	v	+	+	–	–	–	20	F	–	$+_{g-}$	–	–	+	+	–	€0	–	40	–	–			–, G	+	10–30	0.5–6	S	S	S	
	Yellow-orange pigment. West description=mano+ve																																								
A. fischeri NCMB 1281, ATCC 7744	–	+	+		+	–	–	+	–	+	–			+	–		+		F	–	$+_{g-}$	–	–	–	+	+	–	–	–	–	–			v	+	21–30	0.5–5	S	S		154
A. fischeri NCMB 1281	–	+	+	+	+	–	–	+	–	+	–		–	–	+	+	–		F	–	$+_{g-}$	–	–		+	–	–	–	+	–	–			v	+	10–35	0.5–7	S	S	S	143, 1071
A. fischeri ATCC 7744	–			–	–	–	–	+	–	+	+	–	–	+	–	–	–	w	F	+	$+_{g-}$	–	–	+	+	+	–	+	–	+	+	–	w	–	–		1–3	R	PS		244
A. fischeri ATCC 25918	–	+	+	–	+	–	–	+	–	+	–	–	–	–	–	–	–	–	F	–	+	–	–	+	–	–	–	–	–	–	–	–	–	–	+	10–30	1–6	R	S	R	1641
	VP +ve in API 20E, –ve in MRVP medium. Indole strong reaction. MR needs five drops																																								
A. logei ATCC 35077								+		+																															531, 1234
A. logei ATCC 29985[T]	–	+	+	–	+	–	–	+	–	+	–	–	–	+	–	+	–	–	F	–	$+_{g-}$	–	–	+	+	+	–	–	–	+	–				+	4–22	0.5–5	S	S	S	112, 1071
A. logei ATCC 15382	–	+	+	–	+	–	–	–	–	+	–	–	+	+	–	+	–		F	–	$+_{g-}$	–	+s	+	–	+	+s	–	+	+	–	v			+	4–25	0.5–5	S	S	S	112, 143, 363, 715, 1071
A. logei NCMB 1143	–	+	+	–	+	–	–	+	–	+	–		+	–	–	+	–		F	–	$+_{g-}$	–	+	+	–	+	+	–	+	+	–				+	4–30	0.5–5	S	S	S	154, 1071, 1641
A. logei NCIMB 2252	–	+	+	–	+	–	+	+	–	+	–			+	–		+		F	–	$+_{g-}$	–	–	+	–	+	–	–	–	–					+	4–22	0.5–5	S			154

A. salmonicida NCMB 2262[T]	−	+	+	−	+	−	−	−	−	−	−	−	−	−	−	−	−	−	F	−	$+_{gv}$	−	−	+	v	v	−	−	−	+	−	−			+	1–20	0.5–5	S	S	S	473, 1071
A. salmonicida NCIMB 2262	−	+	+	−	+	−	−	−	−	−	−		−	−	−	−	+		F	−	$+_{g-}$	−	−	+	+	+	−	−	−	−	−				+	1–21	0.5–4	S	S	S	154, 1071
A. salmonicida NCMB 2245	−	+	+	+	+	−	−	−	−	−	−	+	−		−	−	−	−	w	−	+	−	−	+	+	+	−	w	−	+	−	−			+	5–20		S	S		236
A. sifiae H1-1[T]	−	+	+		+	+		+	−	+	+		−				−	+	F	−	w				+				+			−		Y		4–30	0.5–5				1834
A. sifiae H1-2	−	+	+		+	+		+	−	+	−		−				−	+	F	−	w				+				−			−		G		4–30	0.5–5				1834
A. wodanis ATCC BAA-104[T]	−	+	+	+	+	−	−	−	−		+		−	+	−	−	+	−	w	−	v	−	−	+	43	96	−	−	78	+				−,w	+	4–25	2–4	S	S		143, 153, 1834
A. wodanis NCIMB 13582[T]	−	+	+	94	+	−	−	−	−		89		20	+	−	3	+			−	$+_{g-}$	−	−		34	+	−	6	83	+	−				89	4–20	0.5–5		S	S	1071

Also check tables for *Moritella* species, *Photobacterium* species, *Vibrio* – pathogenic species, *Vibrio* – environmental species.

Table 3.22. *Vibrio* – pathogenic species.

Test	Gm	Ox	cat	βH	mot	SW	ODC	LDC	ADH	Nit	Ind	Cit	urea	mr	vp	aes	G	onpg	OF	arab	glu	inos	lac	malt	man	mano	sal	sor	suc	tre	Xyl	H_2S	MCA	TCBS	Dnase	temp	NaCl	0129 10	0129 150	Amp	Ref
Organism																																									
V. alginolyticus	–	+	+	+v	+	+	53	+	–	+	+	60	–	+	83	–	+	–	F	–	+g–	–	–	+	+	+	33	–	+	+	–	–	+	Y	+	15–42	1–10	R	S	R	1164, 1779
V. alginolyticus ATCC 17749[T]	–	+	+	–	+	+	+	+	–	+	+	w	–	+	+	–	w	–	F	–	+g–	–	–	+	+	s	–	–	+	+	–	–	w	Y	+	25–43	1–10	PS	S		244
V. alginolyticus ATCC 17749[T]	–	+	+		+	+	–	+	–	+	+	–	–		–		+	–	F	–	+	–			+			–	+			–		Y							998
V. alginolyticus NCIMB 1903	–	+					–	+	–	+	+			+	+		+		F	–	+g–	–	–	+	+	–	–	–	+	–					+	15–42	1–10	S	S		108, 154
V. alginolyticus	–	+	+			+	–	+	–	+	+	+	–		+	–	+	–	F	–	+g–	–	–	+	+	+		–	+					Y		18–40	1–10				108
Vibrio cholerae 01	–	+	+	+	+	–	+	+	–	+	+	+v	–	–	–	–[a]	+	+	F	–	+g–	–	–s	+	+	80	+v	–	+	+	–	–	+	Y	+	4–42	0–6v	S	S	R	769, 1073, 1779
Vibrio cholerae non 01	–	+	+	+	+	–	+	+	–	+	+	+	–	–	+	–[a]	+	+	F	–	+g–	–	–s	+	+	70	–	–	+	+	–	–	+	Y	+	10–42	0–3	S	S	Sv	1073
Vibrio cholerae 0139	–	+		+	+	–	+	+	–	+	+		–		75		+		F	–	+g–	–	–	+	+	+			+			–		Y			0–3	R	R	S	17, 807
V. mimicus	–	+	+	+	+	–	+	+	–	+	88	95	–	15	–	–	90	+	F	–	+g–	–	–	+	+	+	–	–	–	+	–	–	+	G	+	4–42	0–6	S	S	R	412, 496, 1073
V. mimicus ATCC 33653[T]	–	+	+	+	+	–	+	+	–	+	+	+	–	+	–	–	+	+	F	–	+g–	–	–	+	+	+	–	–	–	+	–	–	+	G	+	25–43	0–3	S	S	R	244
V. parahaemolyticus	–	+	+	+	+	+	+	+	–	62	+	63	–	+	–	–	+	–	F	80	+g–	–	–	+	+	+	53	–	–	+	–	–	+	G	+	20–40	3–8	R	S	V	569, 1164
V. parahaemolyticus ATCC 43996	–	+	+	+	+	+	+	+	–	+	+	–	+s	+	–	–	+	–	F	+	+g–	–	–	+	+	+	–	–	–	+	–	–	+	G	w	20–40	0–3	R	PS		244
V. parahaemolyticus ATCC 27969	–	+	+		+		+	+	–	+	–	+	–		–		+	–	F	–	+g–	–		+	+			–	–			–		G					S		1040
V. parahaemolyticus ATCC 17802[T]	–	+	+	–	+	–	+	+	–	+	+	+	+	+	–	–	+	–	F	+	+g–	–	–	+	+	+	–	–	–	+	–	–	+	G	+	20–40	1–3	S	S	R	244
V. (*carachiae*) *harveyi* ATCC 35084	–	+	+	–	+	+s	+	+	–	+	+	+s	+	+	–	–	v	–	F	–	+g–	–	–	+	+	+	–	+s	+	+	–	–	+	Y/G	+	10–40	0.5–8	R	S		20, 154, 244,
	Colonies of *V. harveyi* slowly spread across the plate. Cit may take 3–5 days																																								
ATCC 35048		+	+	–	+	+	+	+	–	+	+	s	+	+	–	s	–	–	F	–	+g–	–	–	+	+	+	–	–	+	+	–		–	Y	+	25–37	1–3				244
V. harveyi ATCC 14126, ATCC 14129	–	+	+	w	+	–	+	+	–	+	+	+w	–		–		+	–	F	–	+g–	–			+	+	–	–	+			–		Y		12–40	3–6	R	S		20
	Haemolysis against sheep RBC																																								
V. harveyi ATCC 6		+	+		+	–	+	+	–	+	+	–	–	+	–	w	s	–	F	–	+g–	–	–	+	+	+	–	–	+	+	–	–	+	Y	–	25–37	1.5–3	S	S	R	244
V. (*trachuri*) *harveyi*	–	+			+		–	+	–	+	+		–		–			–	F	–	+	–	+	+				+	+			–		Y		30	3–7	S	S		818

V. chagasii	–	+	+	–	+	+	–	–	–	+	+	+s	–	+	–	–	w	–	F	–	$+_{g-}$	–	–	+	+	+	–	–	–	+	–	–	–	G/v	+		2–8	R	S		1635
V. jasicida JCM 16453[T]	–	+	+		+		+	+	–	+	+	+	+	+	–	+	+	+	F	–	+	–			+	+	+	v	–			–		G	+	10–37	0.5–8	S	V	R	1835
V. owensii DY05[T]	–	+	+	v	+	–	+	+	–	+	+	–	–	+	–	–	+	–	F	–	+	–	–	+	+	+	–	–	+	+	–	–		Y	+	15–37	1–8	S	S	R	244, 270
	Aesculin goes black but fluorescence indicates neg result																																								
V. owensii 47666-1	–	+	+	w	+	–	+	+	–	+	+	–	–	+	–	w	w	+	F	–	+	–	–	+	+	+	–	–	+	+	–			Y	+		1–3	S	S		244, 270
	Aesculin goes black but weak fluorescence indicates weak pos result																																								
V. (*communis*) *owensii*	–	+	+		+	–	+	+	–	+	+	–	–		–		+	w	F	–	+	–			+			–	+			–		Y		15–37	0.5–6				270, 320
V. rotiferianus LMG 21460[T]	–	+		–	+	–	+	+	–		+	–	83		–		+	+	F	+	$+_{g-}$	–			+			–	+			–		Y		28–40	2–6	S	S		79, 637
V. aestuarianus ATCC 35048	–	+	+	+	+	–	–	–	+	+	+	+	–		–	+	+	+	F	–	$+_{g-}$	–	+		+	+	–	v+	+	+	–	–	+	Y	+	4–37	0.5–6	S	S	S	1645
V. aestuarianus ATCC 35048	–	+	+		+	–	–	–	+	+	+	–					–	+	F	–	$+_{g-}$	–	–	+	+	+	–	–	+	+	–			Y		4–37	0.5–6			S	588
V. aestuarianus NCIMB 2236	–	+	+	+	–t	–	–	–	+	+	+	+	–	+	–	w	w	+	F	–	$+_{g-}$	–	–	+	+	+	–	–	+	+	–	–	s	Y	–	4–37	0.5–6	S	S		244
V. aestuarianus spp. *francensis*	–	+	+		+	–	–	–	–	+	+	86					57	+	F	–	$+_{g-}$	–	–	+	–	+	–	–	+	+	–			NG		4–30	0–3				588
Vibrio species MM5							–	–	+		–	–	–	+	–			+	F	–	+	–		+	+				+			–				25–37	1–8	R			1841
V. campbellii ATCC 2590	–	+	+	+	–t	–	–	+	–	+	+	–	–	+	–	s	–	–	F	–	$+_{g-}$	–	–	+	+	+	–	–	–	+	–	–	–	G	+	4–37	0.5–6	S	S		244
V. campbellii	–	+	+	–	+	–	–	–	–	+	+	58	–	50	–	50	+	–	F	–	$+_{g-}$	–	–	+	48	50	82	–	–	+	–	–		G	+	20–35	3–6	R	S	R	1780
V. vulnificus ATCC 27562 biotype 1, serovar non-E	–	+	+	+	+	–	+	+	–	+	+	+	–	+	–	–	+	+	F	–	$+_{g-}$	–	+	+	–	–	–	–	–v	+	–	–	+	G	–	20–42	3–6	S	S		180
	Human strain, USA – non-virulent for eels. Cit=neg in API 20E but pos in Simmons' tube citrate																																								
V. vulnificus ATCC 27562 biotype 1	–	+	+	+	+	–	89	+	–	+	+	+	–	+	–		+	+	F	–	$+_{g-}$	–	+	+	44	+	+	44	–	+	–	–		G		37	0.5–7	S	S		1644
	As originally described by Tison (1982) using conventional media																																								
V. vulnificus ATCC 33184 biotype 1, serovar non-E	–	+	+	+	+	–	+	+	–	+	+	+	–	+	–	–	+	+	F	–	$+_{g-}$	–	+	+	+	+		–	–	+	–	–	+	G		37	0.5–6	S	S	S	180
	Human strain – USA																																								
V. vulnificus biotype 1, serovar non-E	–						+				+														v																396
	Isolated from diseased eels. Belgium (man=neg), Sweden (man=pos)																																								
V. vulnificus ATCC 33187 biotype 2, serovar E	–	+	+	+	+	–	+	+	–	+	–	+	–	+	–	–	+	+	F	–	$+_{g-}$	–	+	+	–	–	–	–	–	+	–	–	+	G	+	20–35	3–6	S	S		180
	Human strain, USA – virulent for eels																																								
V. vulnificus ATCC 33149 biotype 2, serovar E, 04	–	+	+	+	+	–	–	+	–	+	–	+	–	+	–	–	+	+	F	–	$+_{g-}$	–	+	+	–	–	–	–	–	+	–	–	+	G	+	20–35	3–6	S	S		180
	Virulent for eels – Japanese strain																																								

Continued

Table 3.22. Continued.

Test	Gm	Ox	cat	βH	mot	SW	ODC	LDC	ADH	Nit	Ind	Cit	urea	mr	vp	aes	G	onpg	OF	arab	glu	inos	lac	malt	man	mano	sal	sor	suc	tre	Xyl	H_2S	MCA	TCBS	Dnase	temp	NaCl	0129 10	0129 150	Amp	Ref
V. vulnificus biotype 2 serovar E	−	+	+	+	+	−	20	+	−	+	−	+	−	+	−	−	+	+	F	−	$+_{g-}$	−	+	+	+	−	−	−	−	+	−	−	+	G	+	20–35	3–6	S	S		50
	Taiwanese strains – avirulent for eels																																								
V. vulnificus ATCC 33148 biotype 2	−	+	+	+	+	−	−	+	−	+	−	+	−	+	−		+	+	F	−	$+_{g-}$	−	+	+	−	+	+	−	−	+	−	−		G		37	0.5–7	S	S		1644
	As originally described by Tison (1982) using conventional media																																								
V. vulnificus biogroup 3	−	+	+	+	+	−	+	+	−	+	+	−	−	+	−	−	+	−s	F	−	$+_{g-}$	−	−	+	−		−	−	−	+	−	−	−	G				S	S		182, 1189
	Human strain – Israel																																								
V. mediterranei	−	+	+	−	+	−	−	+	−	+	+	−	−	+	−	75	−	+	F	−	$+_{g-}$	−	−	+	+	+	25	+	+	+	−	−	−	Y	+	20–30	3–6	S	S	S	1361
V. mediterranei CECT 615	−	+	+	+	+	−	−	+	−	+	+	−	−	+	−	−	−	+	F	−	$+_{g-}$	+	−	+	+	+	+	+	+	+	−	−	−	Y	−	20–30	0.5–6	S	S		244
V. (*shilonii*) *mediterranei*	−	+	+		+	+	−	+	−	+	+		−				+	+	F		$+_{g+}$								+					Y		16–37	2–4	S	S		114, 952
V. (*shilonii*) *mediterranei* ATCC 1337		+	+s	−	+	−	−	+	−	+	+	−	−	−	−	−	s	+	F	−	$+_{g-}$	−	−	+	+	+	−	−	+	+	−	−	−	Y	+	25–37	1–3	S	S	R	244
V. anguillarum NCIMB 6	−	+	+	+	+	−	−	−	+	+	+	−	−	+	+	−	+	+	F	+	$+_{g-}$	−	+	+	+	+	−	+	+	+	−	−	−	wG	+	4–30	0.5–8	S	S		154, 1186
	MR may be negative or weak. Citrate is positive or negative																																								
V. anguillarum NCIMB 2129	−	+	+	+	+	−	−	−	+	+	+	−	−	+	+	−	+	+	F	+	$+_{g-}$	−	−	+	+	+	−	−	+	−	−	−	−	wG	+	4–37	0.5–8	S	S		154, 1186
V. anguillarum CRL 11	−	+		+	+	−	−	−	+	+	+	s	−	+	+	−	+	+	F	+s	$+_{g-}$	−	−	+	+	+	−	+	+	+	−	−	+	Y	+	25–37	1–3	S	S	R	244
V. anguillarum ATCC 19264	−	+	+s	+	+	−	−	−	+	+	+	+	−	−	+	−	+	+	F	+	$+_{g-}$	−	−	+	+	+	−	+	−	+	−	−	−	G	+w	25	1–3	S	S	R	244
V. anguillarum ATCC 14181	−	+	+	+	+	−	−	−	+	+	−	+	−	−	+	−	+	+	F	−	$+_{g-}$	−	−	+	+	+	−	+	+	+	−	−	−	Y	+	10–37	0.5–7	S	S	R	1071
V. ponticus	−	+	+		+	−	−	+	−	+	+	+	−		−		−	−	F	+	$+_{g-}$	−			−			−	+			−	−	Y	−	25–37	2–8				1080
	LDC neg, ADH pos by API 20E																																								1816
V. ponticus								−	+										F																						1816
V. proteolyticus	−	+	+		+	+				+	+	+		+	+		+		F	−	$+_{g-}$	−	−	+	+	+	−	+	−	+	−	−				20–40	1–10	R	S	S	1133, 1720
V. proteolyticus	−	+	+	+	+	+	−	+	+	+	+	+	33	+	+	−	+	−	F	−	$+_{g-}$	−	−	+	+	−	−	50	76	+	−	−		Y/G	+	20–40	1–10	R	S	S	1720
V. proteolyticus AHLDA 1735	−	+	+	+	+	+	−	+	+	+	+	+	+	+	+	−	+	−	F	−	$+_{g-}$	−	−	+	+	w	−	+	−	+	−	−	w	G	−	42–	1–10	PS	S		244
V. proteolyticus CIP 102892	−	+	+	+	+	+	−	+	+	+	+	+	−	+	+	−	−	−	F	−	$+_{g-}$	−	−	+	+	+	−	+	−	+	−	−	−	G	+	20–37	0.5	R	S		244
V. neptunius LMG 20536[T]	−cr	+	+	+	+	−	−	−	+	+	+	+	−	+	+		+	−	F	−	+	−			−			−	+			−		Y		20–35	1.5–6	S	S		79, 1632

V. neptunius PP–145.98	–cr	+	+		+	–	–	–	+	–	–	–	–		–	–	+	–	F	–	$+_{g-}$	–			–			–	+			–		Y		15–35	1.5–6	S	S		1632
V. neptunius LMG 20610	–cr	+	+	+													–																								79
V. fluvialis NCTC 11327	–	+		+	+	–	–	–	–	+	+	+	–	+	–	+	+	+	F	+	$+_{g-}$	–	–	+	+	+	+	–	+	+	–	–	+	Y	+	20–42	0–3	PS	S		244
V. fluvialis ATCC 33809	–	+	+	v	+	–	–	–	+	+	+	+	–	+	–	72	+	+	F	+	$+_{g-}$	–	–	+	+	+	–	–	+	+	–	–	+	Y	+	10–37	0–8	S	S	V	154, 1321
V. fluvialis NCTC 11327	–	+	+	v	+	–	–	–	+	+	–	+	–	+	–	–	+	+s	F	+	$+_{g-}$	–	–	+	+	+	–	–	+	+	–	–	+	Y	+	10–35	1–6	R	S	V	222, 996
	API 20 E pos for VP and indole																																								
V. furnissii ATCC 35016^{T}	–	+	+	v	+	–	–	–	+	+	v	+	–	+	–	–	+	v	F	+	$+_{g+}$	–	–	+	+	+	–	–	+	+	–	–	+	Y	+	20–37	1–8	R	S	R	222, 996
	API 20 E pos for VP and indole																																								
V. furnissii ATCC 35016^{T}	–	+	+	+	+	–	–	–	+	+	–	+	–	+	–	–	+	+	F	+	$+_{g+}$	–	–	+	+	+	–	–	+	+	–	–	+	Y	+	20–43	1–10	R	S	R	244
V. furnissii ATCC 11218	–	+	+	+	+	–	–	–	+	+	–	+	–	+	–	–	+	+	F	+	$+_{g+}$	–	–	+	+	+	–	–	+	+	–	–	+	Y	w	20–37	1–8	R	S		244
	ADH = neg at 0 % NaCl. Pos ADH at 2% NaCl																																								
V. splendidus I ATCC 33125^{T}	–	+	+	v	+	–	–	–	+	+	+	43	–	+	–	+	+	+	F	–	$+_{g-}$	–	–	+	+	+	–	–	–	+	–	–	–	G	–	4–37	0.5–6	S	S	R	154, 523, 1071
	Type strain is citrate negative																																								1321
V. splendidus I ATCC 33125	–	+	+	–	+	–	–	–	+T	+	+	+	–	+	–	+	+	+	F	–	+	–	–	+	+	+	–	–	+	+		–		Y	+	4–28	1–8	S	S		55, 154, 523
V. splendidus	–	+	+		–	–	–	–	89	+	+	11	–		22	+	78	+	F	–	+	–	22	+	+	+	–	–	–	+	–	–	–	G							590
V. splendidus, oysters	–	+	+		–		–	–	–	+	+	–	–		–		–	–	F	–	+	–			+			–	+			–		Y							965
V. splendidus LGP32	–	+	+	–		–	–	–	+	+		–	–			–	+	–	F		+		–		–	+			–					G							603
V. splendidus I CRL 40	–	+	+	–	–	–	–	–	–M	+	+	–	–	+	–	–	+s	–	F	–	$+_{g-}$	–	–	+	+	+	–	–	–	+	–	–	–	G	NG		1–3	R	R		244
V. splendidus II NCMB 2251	–	+	+	+	+	–	–	–	+	+	+	+	–	+	–	–	+	+	F	–	$+_{g-}$	–	–	+	+	–	–	–	–	+	–			G	+	10–30	0.5–7	S	S	R	154, 1071, 1782
V. splendidus II ATCC 33789	–	+		w	+	–	–	–	–M	+	+	–	–	+	–	+	–	–w	F	–	+	–	–	+	+	–	–	–	–	+	–	–	–	wG	NG	25	0–3	S	S		244
V. celticus	–	+	+		+	–	–	–	+	+	+	–	–	+	+	+	+	–	F	25	$+_{g-}$	–	+	+	+	+	–	–	+	+	+			Y		25	1–6	S	S		142
V. brasiliensis LMG 20546^{T}	–cr	+	+		+	–	–	–	+	+	+		–	+	+		+	+	F	–	+	–			+			–	+			–		Y		20–40	2–6	S	S		79, 1632
V. tubiashii ATCC 19109^{T}	–	+	+		+	–	–	–	83	+	+	v	30	+	–	–	+	+	F	–	$+_{g-}$	–	–		+	+	–	–	+	+	–	–		Y	+	10–30	1–3	S	S	S	201, 668

Continued

Table 3.22. Continued.

Test	Gm	Ox	cat	βH	mot	SW	ODC	LDC	ADH	Nit	Ind	Cit	urea	mr	vp	aes	G	onpg	OF	arab	glu	inos	lac	malt	man	mano	sal	sor	suc	tre	Xyl	H_2S	MCA	TCBS	Dnase	temp	NaCl	0129 10	0129 150	Amp	Ref
V. tubiashii A TCC 19109		+	+	+	–	–	–	–	–	+	+	–	–	+	–	–	+	–	F	–	$+_{g-}$	–	–	+	+	+	–	–	+	+	–	–	–	Y	+	25	1–3	S	S		244
V. tubiashii NCMB 1340^{T}	–	+	+	+	–	–	–	–	+s	+	+	–	–	+	–	–	+	–	F	–	$+_{g-}$	–	–	+	+	+	–	–	+	+	–	–	–	Y	+	25	1–3	S	S		244
V. tubiashii NCMB 1340^{T}	–	+	+	w	+	–	–	–	+s	+	+	–	–	+	–	–	+	+	F	–	$+_{g-}$	–	–	+	+	+	–	–	+	+	–	–	–	Yw	+	10–30	0.5–7	S	S	S	244, 1071
Enterovibrio coralii sp.nov.	–	+			+		–	–	+	+	–		–		–		–	+	F	–	+	–			+			–	–					G		20–30	2–6		R		1636
V. crassostreae LMG 22240^{T}	–	+	+		+	–	–	–	+	+	+	–	–		–	+	+	–	F	–	$+_{g}$	–	–	v	+	+	–	–	+	+	–	–		Y		4–25	2–4	S	S		524
V. gigantis LMG 22741^{T}	–	+	+		+	–	–	–	+	+	+	–	–		–	+	+	–	F	–	$+_{g}$	–	–	+	+	+	–	–	–	+	–	–		G		4–25	2–4	S	S		1010
V. kanaloae LMG 20539^{T}	–	+	+	v	+	–	–	–	+	+	+		–	–	+		+		F	+	+	–			+			–	+			–		Y		4–30	1–6	S	V		1635
V. kanaloae	–	+	+		+	–	–	–	+	–	+	–	–		–	+	+	+	F	–	+	–	–	+	+	+	–	–	+	+	–	–		Y		20–28	2–4				524
V. coralliilyticus LMG 20984^{T}	–	+	+	+	+	–			+	+	+	+	–				+	+	F		+								+					Y		25–30	1–7	S	S	R	79, 155, 156
V. coralliilyticus	–	+	+		+	–			20	+	+	80	–				+	+	F		+								+					Y		25–30	1–7	S	S	R	155, 156
V. coralliilyticus/ neptunius-like	–	+	+		+	–			+	+	–	–	–			–	+	–	F		+								+					Y		17–37	0.5–5				879
V. lentus CECT 5110^{T}	–	+	+	–	+	–	–	–	+	+	+	+		–	–	–	+		F	–	$+_{g-}$	–		+	–	+	–	–	–	+	–	–		G	+	4–30	2–6	R	R		521, 1082
	Thornley's ADH=pos, Moeller ADH=neg. Type strain=man neg																																								
V. lentus CECT 5110	–	+	+	+	–	–	–	–	+	–	+	–	–	–	–	–	+	+	F	–	$+_{g-}$	–	–	–	+	vw	–	–	–	vw	–	–	–	wG	NG	25	1–3	R	R	S	244
V. lentus octopus strains	–	+	+	75	+	–	–	–	75	–	+	–		+	–	25	+		F	–	$+_{g-}$	–	–		+	+	–	–	–	+		–		G	–	10–22	2–6	R	R		521
V. nigripulchritudo 164^{T}	–	+				–	–	–	–	+	+	+	–		–	50	+	+	F	–	$+_{g-}$	+	+	+	+	–	–	v+	–	+	–			G	+	20–30	3–5	R	S	S	577
	colonies blue-black on MA 2216 with 2% glycerol																																								
V. nigripulchritudo CIP 103195	–	+	s	+	w	–	–	–	–	+	+	–	–	+	–	–	+	+	F	–	$+_{g-}$	–	–	w	–	+	–	–	–	w	–	–	–	G	–	25	1–3	S	S		244
V. pelagius I ATCC 25916^{T}	–	+		–	+	–	–	–	–	+	–	–	–	–	–	–	–	+	I	–	$+_{g-}$	–	–	+	+	–	–	–	+	+	–	–	–	Y	–	25	1–3	S	S		244

V. pelagius II ATCC 33784	–	+	+	–	–	–	–	–	–	+	+	–	–	+	–	–	s	+	Fw	–	$+_{g-}$	–	–	+	+	+	–	–	+	+	–	–		Y	–	25	1–3	PS	S		244
V. (Listonella) pelagius NCMB 2253*	–	+	+	–	+	–	–	–	–	+	+	+	–	+	–	–	+	+	F	–	$+_{g-}$	–	–	+	+	+	–	–	+	+	–	–		v	–	10–30	0.5–7	S	S	S	1071, 1186
	Previously *Vibrio pelagius* II																																								
V. ezurae LMG 19979	–	+	+	–	–	–	–	–	–	+	+	–		+	–		–	–	F	–	$+_{g-}$	–		+	+			–	–					G		15–30	2–5	S	S		79, 1486
V. fortis LMG 21557^{T}	–	+	+	–	+	–	–	–	–	+	+	–	–		+		v	+	F	–	+	–			+			–	+			–		Y		4–35	1–8	S	S	S	79, 1634
V. fortis LMG 20547	–			v													+																								79
V. ichthyoenteri IFO 15847^{T}	–	+	+		+	–	–	–	–	+	–	–			–		–	–	F	–	$+_{g-}$	–	–	+	–	+	–	–	+	+	–	–		Yw		15–30	1–6	S	S		804
V. ichthyoenteri DSMZ 14397	–	+	+	–	+	–	–	–	–	+	–	–	–	+	–	–	–	–	F	–	$+_{g-}$	–	–	+	+	+	–	–	–	+	–	–	–	G	–	25	1–3	S	S	S	244
V. tapetis CECT 4600^{T}	–cb	+	+		+	–	–	–	–	+	+	–	–		–	–	+	+	F	–	$+_{g-}$	–	–	+	–	+		–	–	+	–	–		G		4–22	1–3	S	S	S	201, 285, 1237
	Aesculin reported positive by Paillard 1994																																								
V. tapetis CECT 4600^{T}						–	–	–	–		+				–		+			–	$+_{g-}$		–		–	+			–							4–22	2			S	285
V. tapetis P1									–			–			–		+	+		–	$+_{g-}$	–			–	+		–	–												285
V. tapetis CIP 104856	–cb	+	+	–	–	–	–	–	–	+	–	–	–	+w	–	–	+s	–	I	–	–	–	–	–	–	–	–	–	–	–	–	–	–	Gw	–		0–3	S	S		244
V. tapetis wedge sole	–	+	+	–	–	–	–	–	–		+		–	+		–	+				$+_{g-}$		–		–	+	–	–						G		4–20	2–3				1052
V. tapetis ssp. *britannicus*	–	+	+	–	–	–	–	–	–			+							I	+					+																103
V. ordalii ATCC 33934	–cv	+	+	+	+	–	–	–	–	–	–	–v	–	–	–	–	+	–	F	–	$+_{g-}$	–	–	+	+	–	–	–	+	–	–	–		–	+	4–30	0.5–7	S	S	Rv	683, 1493, 1782
V. ordalii DF–3K	–cv	+	+	+	+	–	–	–	–	–	–	–	–	–	–	–	+	–	F	–	$+_{g-}$	–	–	+	–	–	–	–	+	–	–	–		–	+	4–30	0.5–7	S	S	Rv	397, 1493
V. ordalii NCIMB 2167^{T}	–	+	+	+	+	–	–	–	–	–	–		–	–	–	–	v		F	–	$+_{g-}$	–	–	+	+	–	–	–	+	v	–				+	21–37	0.5–5	S	S	S	154, 1071
V. pectenicida A365^{T}	–	+	+		+	+	–	–	–	+	–	+			–		+	–	F	–	$+_{g-}$	–		+		–		–	–		–			–	+	4–30	1–6	S	S		970
V. pectenicida ATCC 700783	–	+	+	L	–t	–	–	–	–	+	–	s	–	+	–	–	–	–	F	–	$+_{g-}$	–	–	+w	–	–	–	–	–	+w	–	–	–	Gw	NG	25	1–3	S	S	R	244
V. penaeicida	–	+	+		+	–	–	–	–	+	50	67	–	+	–		+	+	F	–	$+_{g-}$	–	50	+	–	+	–	–	–	+	–	–		G		20–30	1–3	S			371, 803
	Type strain=indole neg																																								
V. penaeicida DSMZ 14398	–	s	s	–	+	–	–	–	–	–	–	–	–	w	–	–	s	w	F	–	–	–	–	–	–	–	–	–	–	–	–	–	–	–	NG	25	1–3	S	S		244

[a]*Vibrio cholerae* is negative for aesculin, but some strains cause blackening of the medium due to production of melanin. To determine a true negative the loss of fluorescence must be measured at 354 nm. See under 'aesculin' in interpretation of tests.

In this table, Vibrios are grouped according to their ODC, LDC, ADH reactions. Use this as a starting point in identifications. 42–=negative growth at 42°C; t=tube method.

Also check tables for *Aliivibrio* species, *Moritella* species, *Photobacterium* species, *Vibrio* – environmental species.

Table 3.23. *Vibrio* – environmental species.

Test	Gm	Ox	cat	βH	mot	SW	ODC	LDC	ADH	Nit	Ind	Cit	urea	mr	vp	aes	G	onpg	OF	arab	glu	inos	lac	malt	man	mano	sal	sor	suc	tre	Xyl	H_2S	MCA	TCBS	Dnase	temp	NaCl	0129 10	0129 150	Amp	Ref
Organism																																									
Environmental																																									
V. diabolicus	–	+	+		+	+	+	+	–	+	+	–	–		–	–	+	–	F	–	+	–	–	+	+	+	–	–	+	+	–	–				20–45	2–5	S	S		1368
V. cincinnatiensis	–	+	+		+	–	–	+	–	+	–	+	–		+		–	+	F	+	$+_{g-}$	+	–		–	+	+	–	+	+		–		Y	+	22–37	1–6	R	S	S	220
V. alfacsensis CAIM 1831^T	–	+	+		+		v	v	–M	v	+	–	–		v		–	+	F	–	+	–			v			–	v			–							S		640
V. alfacsensis CAIM 1836	–	+	+		+		v	v	–M	v	+	–	+		v		–	+	F	–	+	–			v			–	v			–							S		640
V. alfacsensis CAIM 404	–	+	+		+		v	v	–M	v	+	+	–		v		–	+	F	–	+	–			v			–	v			–							S		640
V. marisflavi WH134^T	–	w	w		+	–	–	+	+	+	–	–	–		–	+	+	+	F	–	+	–		+	–	+	+	–	+	+	+	–		Y	+	16–37	1–6		S		1760
V. metschnikovii NCTC 8443	–	–	+	+	+	–	–	+	+	–	+	+	–	+	+	–	+	+	F	–	$+_{g-}$	+	–	+	+	+	–	–	+	+	–	–	–	NG	w+	10–42	1–3	S	S	S	244
V. metschnikovii	–cr	–	+	+	+	–	–	38	81	–	+	86	–	+	+	62	73	54	F	–	$+_{g-}$	36	59	+	+	+	–	46	+	+	–	–	v	Y, –	76	4–40	0.5–7	S	S	V	518, 994, 1782
V. xiamenensis G21^T	–cr	+	+		+		–	–	–	+	–	–	–	+	–		–	+	F	–	+	–			+			–	+			–		NG	–	15–40	0.5–10				581
V. pacinii LMG 19999^T	–	+	+		+	–	–	–	+	+	–	w	–	+	66		v		F		+		+		–							–		Y		4–35	1.5–6	S	S		638
V. xuii LMG 21346^T	–cr	+	+		+	–	–	–	+	+	+	–	–	+	+		–	–	F	+	+	–			+			–	+			–		Y		20–40	2–8	S	R		1632
V. orientalis NCMB 2195^T	–	+	+	+	+t	–	–	w	–	+	+	–	–	ng	ng	–	+	–	F	–	$+_{g-}$	–	–	+	+	+	–	–	+	+	–	–	–	Y	+	4–35	0–3	S	S		244
V. orientalis NCMB 2195^T	–	+	+	–	+	–	–	–	+	+	+	+	–	+	–	–	+		F	–	$+_{g-}$	–	–		+	+	–	–	+	+	–			Y	+	4–35	0.5–8	S	S	S	1071
	Some LDC positive strains																																								
V. aerogenes (ATCC 700797T)	–	–	+		+	–	–	–	+T	+	+	+			–		+		F	–	$+_{g+}$	+	–		+	+		–	+	–	+				+	20–35	1–7	R	R	S	1522
V. aerogenes DSMZ 1314	–	+	+	+	–t	–	–	–	–M	+	+	–	–	+	–	–	+	+	F	–	$+_{g-}$	–	–	w	–	w	–	–	–	–	–	–	–	G	NG	25	1–3	S	S	S	244
V. diazotrophicus	–	+	+	–	+	–	–	–	+	+	+	+	16	+	–	+	–	+	F	+	$+_{g-}$	–	+	+	+	–	50	–	+	+	+	–		Y	–	10–35	0.5–6	R	S	S	665, 1780, 1782
V. diazotrophicus ATCC 33466	–	+	+	L	+t	–	–	–	+s	+	+	+	–	+	–	s	–	+	F	+	$+_{g-}$	–	–	+	+	–	w	–	+	+	+	–	+	Y	–	10–37	0.5–10	R	S	S	244

V. sinaloensis	−	+	+	+	+	−	−	−	+	+	−	?	−	+	−		+	+	F		$+_{g+}$								+			−		Y	+	15–37	0.5–9	S	S		639
V. chagasii LMG 21353^T	−	+	+		+	−	−	−	+	+	+		−	+	+		+		F	+	+	−			+			−	−			−		G		15–30	1–6	S	S		1635
V. chagasii	−	+	+		+	−	−	−	58	+	+		−	+	+		+		F	+	+	−			+			−	v−			−		v		15–30	1–6	S	S		1635
V. chagasii LMG 21353^T	−	+	+		+	−	−	−	+	+	+	−	−		−		+	−	F	−	+	−		+	+	+	−	−	+	+	−	−				20–30	1–6				524
V. pomeroyi LMG 20537^T	−	+	+		+	−	−	−	+	+	+		−	+	+		+		F	−	+	−			+			−	−			−		G		4–30	1–8	S	S		1635
V. pomeroyi	−	+	+		+	−	−	−	50	+	+		−	+	83		83		F	−	+	−			+			33	+			−		Y		4–30	1–8	S	S		1635
V. pomeroyi LMG 20537^T	−	+	+		+	−	−	−	+	−	+	−	−		−	+	+	+	F	−	+	−	−	+	+	+	−	−	+	+	−	−		G		20–28	2–4				524
V. nereis	−	+		80	+	−	−	−	+	+	+	+	−	+	−	−	80	−	F	−	$+_{g-}$	−	−	+	−	+	−	−	+	+	−	−		−	+	4–40	2–6	R	S	S	1781, 1782
V. nereis ATCC 25917	−	+		−	w	−	−	−	+s	+	+	−	−	+	−	−	−	−	F	−	$+_{g-}$	−	−	+	−	+w	−	−	+	+	−	−		−,w	+	20–35	3–6	S	S		244
V. mytili CECT 632^T	−	+	+		+	−	−	−	+	+	−	+v	−		−	+	−		F	+	$+_{g+}$	−		+	+	−	+	−	+	+	+	−	−	Y	−	10–37	1–10	R	S		1363
V. mytili CIP 103929	−	+	+s	−	s	−	−	−	+s	+	−	+	−	+	−	+	−	+	F	+	+	−	+	+	+	−	+	−	+	+	+	−	−	Y	−	25–43	1–10	R	S	S	244
V. hispanicus	−	+	+		+	−	−	−	+	+	+	+	−	+	−		−		F		$+_{g-}$											−		Y				S	S		636
V. plantisponsor MSSRF60^T	−	+	+		+	−	−	−	+	+	+	+	−		−	+	−	+	F	−	$+_{g-}$	−	+	+	+	+	+	−	+	+	+	−		Y	−	10–40	0.5–8	S	S		1372
V. maritimus R-40493^T	−	+	+		+	−	−	−	+	+	+	−	−		−		+	+	F	−	+	+			+			−	+			−		Y		15–37	0.5–8				321
V. variabilis R-40492^T	−	+	+		+	−	−	−	+	+	+	−	−		−		+	+	F	−	+	−			+			−	+			−		Y		15–35	0.5–7				321
V. stylophorae KTW-12^T	−	+	+		−	−	−	−	+	+	−	−	−		+	−	v	−	F	−	+	−						+	−			−		NG	+	15–35	0.5–6				1516
V. scophthalmi	−	+	+	−	+	−	−	−	90	+	−	−	−	+	−	+	−	−	F	−	$+_{g-}$	−	−	+	−	+	−	−	+	−	−	−	−	Y	−	22–35	0.5–3	S	S		293, 523
V. scophthalmi CECT 4638	−	+	+	−			−	−	−T	+	−	−	−	+	−	+	−		F	−	$+_{g-}$	−	−		−	+	−	−	+	−		−		Y	−	4–28	0.5–3	S	S		523
V. scophthalmi CECT 4638	−	−	+	−	−t	−	−	−	−M	+	−	−	−	+	−	−	−	−	F	−	$+_{g-}$	−	−	+	−	+	−	−	+	−	−	−	−	Y	−	25	1–3	S	S		244
Vibrio quintilis CECT 7734^T	−	−			+		−	−	+T	−	−	+			+		w	−	F	−	+	−	−	+	+	+	+	−	+		+			NG	+	28–37	1.7–6				1065
Salinivibrio costicola	−	−	−		−	−	−	−	+	−	−				+		+	−	F	−	+	−	−	+	+	+			+	+	−					5–45	0.5–20				582, 766
Enterovibrio coralii	−	+	+		+	−	−	−	+	+	−		−				−	+	F	−	+	−			+			−	−					G		20–30	2–6	R	R		1636

Continued

Table 3.23. Continued.

Test	Gm	Ox	cat	βH	mot	SW	ODC	LDC	ADH	Nit	Ind	Cit	urea	mr	vp	aes	G	onpg	OF	arab	glu	inos	lac	malt	man	mano	sal	sor	suc	tre	Xyl	H_2S	MCA	TCBS	Dnase	temp	NaCl	0129 10	0129 150	Amp	Ref
Enterovibrio nigricans	–	+	+		+	–	–	–	v	v		–			–		–	+	F	–	$+_{g-}$	–	+	–	+		–	+	–	–	–			wG	w+	15–28	0.9–6				1314
Enterovibrio norvegicus	–	+	+		+	–	–	–	+	–	+	–	–	–	–		–	+	F	–	+	–	–	+	–			–	–	+		–		G	–	20–28	1.5–6	R	R		1627
Enterovibrio calviensis	–	+	+		+	–			–	+	–	+	+	+	–		+	+	F		$+_{g-}$													G		4–30	1.5–12	S	S	S	436
Enterovibrio calviensis DSMZ14347	–	+	+	–	–	–	–	+	–	+	–	–	–	ng	–	–	s	+	F	–	$+_{g-}$	–	–	–	–	+	–	–	–	–	–	–	–	G	–	25	0–3	S	S	S	244
Enterovibrio nigricans DA1 1-1-5^T	–	+	+		+	–	–	–	–	+		–			–		–	+	F	–	$+_{g-}$	–	–	–	–		–	–	–	–	–			wG	w+	15–28	0.9–6				1314
	The type strain produces black pigment																																								
V. cyclitrophicus LMG 21359^T	–	+	+		+	–	–	–	–	+	+	–	–		–		+	+	F	–	+	–		+	+	+	–	–	+	+	–	–									524
V. cyclitrophicus	–	+	+		+		–	–	+w	–	–		–		–		+	–	F	–	$+_{g-}$	–			+	–	–	–	+	+	+	–				4–37	2–10			S	709
V. rarus	–	+	+		–	–	–	–	–	+	+	–		+	–		–	–	F	–	$+_{g-}$	–		+	+			–	–					G	–	15–30	2–6	S	S		1485
V. tasmaniensis	–	+	+		+	–	–	–	–	+	+		–		+		–	–	F	–	+	–			+			–	–			–		G		4–35	1–6	S	S	R	1637
V. tasmaniensis LMG 20012^T	–	+	+		+	–	–	–	–	+	+	–	–		–	+	–	–	F	–	+	–	–	+	+	+	–	–	–	+	–	–		G		20–28	2–4				524
V. atlanticus LMG 24300^T	–	+	+		+	–	–	–	–	+	v				+	+	v		F	+	+	–		+	+	+	–	–	+	+	+			Y		4–25	3–6	S	S		442
V. gallaecicus VB 8.9^T	–	+	+		+	–	–	–	–	+	+	–	–	+	–	+	+	–	F	–	$+_{g-}$	–	–	+	+	+	–	–	v–	+	–			v	+	4–25	1–3	S	S	R	144
	Results from API 20E & API CH50. Type strain is green on TCBS. One of the three strains was yellow on TCBS																																								
V. gallaecicus LMG 24045^T	–	+	+		+	–	–	–	–	+	+	–	–	+	–	+	+	–	F	–	$+_{g-}$	–	–	+	+	+	–	–	+v	+	–	–		Yv	+	4–30	1–3	S	S	R	144
V. halioticoli IAM 14596^T	–	+	+	–	–	–	–	–	–	+	+	–	–	+	–	–	–	–	F	–	$+_{g-}$	–	–	+	+	–	–	–	–	–	–	–	–	G	–	15–30	2–3	S	S		1490
V. halioticoli IAM 14596^T	–	+	+	–	–	–	–	–	–	+	–	–	–	+	–	–	–	–	F	–	w	–	–	+	+	–	–	–	–	–	–	–	–	wG	–	15–30	2–3	S	S		244
V. hangzhouensis CN83^T	–	+	+		+		–	–	–	+	+	+	–		–	+	+		F		+		+	+	+	+			+	+	–	+	+		–	20–37	0.5–7.5	S	S		1818
Grimontia hollisae	–cv	+		w	+w	–	–	–	–	+	+	–	–	–	–	–	–	–	F	+	$+_{g-}$	–	–	–	–	+	–	–	–	–	–	–	–	–	–	25–37	1–6	R	S	S	721, 1225
	Haemolysis weak at 24 h, positive after 7 days																																								
G. hollisae CIP 101886	–	+	s	–	+	–	–	–	–	+	+	–	–	+	–	–	–	–	F	+	–	–	–	–	–	+	–	–	–	–	–	–	–	NG	–		1–3	R	S		244

V. hepatarius LMG 20362[T]	–	+	+		+	–	–	–	–	+	+		–		+		–	–	F	–	+	–			+			–	+			–		Y		4–35	0–8	S	S	R	1634
V. sagamiensis LC2–047[T]	–	+	+		+	w	–	–	–	+	w	–	–				+	–	F	–	–				–				–			–		G		0.5–6	10–37				1837
V. navarrensis	–	+	+		+	–	–	–	–	+	+	+	–		–	66	+		F	–	$+_{g-}$		–	+	+	+		–	+			–		Y		10–40	0.5–7	S	S		1687
V. navarrensis CIP 103381	–	+	s	–	+	–	–	–	–	+	+	s	–	+	–	s	s	s	F	–	$+_{g-}$	–	–	+	+	s	–	–	+	+	–	–	+	Y	+	25–43	1–3	S	S	R	244
V. navarrensis ssp *pommerensis*	–	+	+	+	+	–	–	–	–	+	+	–	–	+	–		+		F		+	–	–	+	+			–	+	+	–	–		Y		28–37	0.5–5	S	S		851
	green-brown pigment, colonies opaque																																								
V. neonatus	–	+	+		–	–	–	–	–	+	+	–		+	–		–	–	F	–	$+_{g-}$	–		+	+			–	–					G		15–30	2–5	S	S		1486
V. areninigrae	–	+	+		+	–			–	+	+		–						F		$+_{g-}$															4–45	1–7	S	S		301
V. toranzoniae Vb10.8T	–	+	+		+	–	–	–	–	+	+	–	–		–	–	+	–	F	+	$+_{g-}$	–		+	+	+			+	+				Y		4–25	1–10	S	S	R	980
V. toranzoniae Cmf 13.9	–	+	+		+	–	–	–	–	+	+	–	–		–	+	+	–	F	+	$+_{g-}$	–		+	+	–			+	+				Y		4–25	1–10	S	R	R	980
V. toranzoniae CMJ 9.4	–	+	+		+	–	–	–	–	+	–	–	–		–	–	+	–	F	+	$+_{g-}$	+		+	+	–			+	+				Y		4–25	1–10	S	R	R	980
V. toranzoniae CMJ 9.11	–	+	+		+	–	–	–	–	+	–	–	–		–	–	+	–	F	+	$+_{g-}$	–		+	+	+			+	–				Y		4–25	1–10	S	R	R	980
V. gallicus LMG 21878[T]	–	+	+		–	–	–	–	–	+	+v	–	–	+	–		–	–	F	–	$+_{g-}$	–		+	+			–	–					G		15–30	2–3	S	S		1487
V. gallicus CIP 107863[T]	–	+	+		–		–	–	–	+		–			–		–	–	F	–	$+_{g-}$	–		+	+			–	–							20–30					1487
V. artabrorum LMG 23865[T]	–	+	+		37	–	–	–	–	+	–				+		–		F	+	+	–		+	+	+	–	–	+	+	+			Y		4–25	3–6	S	S		442
V. caribbeanicus DSM 23640[T]	–	+	–		+	–	–	–	+	–	–	–	–	–	–	–	–	–		–	–	–		–	–	–		–	+			–		Y	+	10–35	0.5–8	S	S		739
V. breoganii LMG 23858[T]	–	+	+		–	–	–	–	–	+	–	–	–	+	–	+	–	+	F	–	$+_{g-}$	–	–	+	+		+	–	–	–	+	–		G	–	4–30	2–3	S	S	R	140
V. comitans	–	+	+		–	–	–	–	–	+	–	33		+	–		–	83	F	–	$+_{g-}$	–		+	+			–	–					G	67	4–30	1–6	S	S		1485
V. inusitatus	–	+	+		–	–	–	–	–	+	–	–		+	–		–	–	F	–	$+_{g-}$	–		+	+			–	–					G	50	4–30	1–6	S	S		1485
V. agarivorans	–	+	+	–	+	–	–	–	–	+	–	–	–	+	–	–	+	+	F	–	$+_{g-}$	–	–	+	w	–	–	–	–	–	+	–	–	G	–	20–37	1–6	S	S		244, 1084
	Hydrolyses agar. Gelatin neg in API 20E																																								
V. aestivus CECT 7558[T]	–	+	+		+		–	–	–	+	–				–		–	+	F		+	–			–			–	–					G		15–37	0.8–7				1065
V. atypicus	–	+	–		+	–	–	–	–	+	–	–	+		–		+	+	F	–	+	–		–	+	–		–	+			–		Y	+	16–30	0.5–7	S	S		1767

Continued

Table 3.23. Continued.

Test	Gm	Ox	cat	βH	mot	SW	ODC	LDC	ADH	Nit	Ind	Cit	urea	mr	vp	aes	G	onpg	OF	arab	glu	inos	lac	malt	man	mano	sal	sor	suc	tre	Xyl	H_2S	MCA	TCBS	Dnase	temp	NaCl	0129 10	0129 150	Amp	Ref
V. gazogenes	–cr	–	+		+	–	–	–	–	–	–	+	–		–		+		F	+	$+_{g+}$	–	–	+	+	+	+	+	+	+	+	–		NG	+	20–42	3–6	S	S		518,686
	Red to orange colonies																																								
V. natriegens	–	+	+	–	+	–	–	–	–	+	–	+	62		–	+	+	67	F	+	$+_{g-}$	60	–	+	+	30	+	–	+	+	–	–		–Y	+	10–37	2–6	R	S	S	78,1780
V. natriegens NCMB 1900	–	+	+	–	+	–	–	–	–	+	–	+	–	–	–	+	–	+	F	+	$+_{g-}$	–	–	+	+	+	+	–	+	+	–	–		Y	+	4–37	0.5–7	S	S	S	1071, 1186
V. natriegens ATCC 14048^T	–	+	–	–	+	–	–	–	–	+	–	+	–	+	–	+	+	–	F	+	$+_{g-}$	–	–	+	+	–	+	–	+	+	–	–		Y	–	25	2	R	R		244
V. litoralis MANO22^T	–	+	+		+				–	+	–	+	–			–	–		F		$+_{g-}$															4–44	1–12	S	S		1202
V. hippocampi	–	+	+		+				–	+	–	–	–			–	–		F	–	+	–	–	+	+	–	–	–	–	+	–			G		10–35	0–7				106
V. mangrovi MSSRF38^T	–	–	+				–	–	–	+	–		–		+	+			F	+	$+_{g+}$	–	+	+	+	+	+	–	+	–	+	–		NG		28–30	0.1–8	R	R		1373
	Red pigment																																								
V. casei	–	+	+		+	–				+		+				+		+	F		+															2–30	1–10	S	S		190
V. azureus	–	+	+		+	+	–	–		–	w	–							F	–	w	–			+			–	–					G		10–37	0.5–6				1836
V. porteresiae MSSRF30^T	–	+	+		+	–	–	–	–	–	–	+			+	+		–	F		$+_{g+}$	–	+	+	+		+	–	+	–	+	–		NG		20–37	1–5	R	R		1371
V. rhizosphaerae MSSRF3^T	–	–	+		+	–	–	–	–	–	–				+		+		F		$+_{g+}$	–	+		+	+	+	–	+	–	+					20–42	0.1–10				947
V. ruber VR1^T	–	–	+		+	–	–	–	–	+	–						+		F		$+_{g+}$	–	+		+	+	+	–	+	–	+					20–40	1–10	R	R		1521
	Red colonies																																								
V. rumoiensis	–	+	+		–	–			–	+	–	+		+	–		–		F	+	+	–		+	+	+			+		+	–			+	2–34	3–6	S	S		1843
V. superstes LMG 21323^T	–	+	+		–	–	–	–	–	+	–	–		+	–		–	–	F	–	$+_{g-}$	–		+	+			–	–					G		15–30	2–3	S	S		704

+s=slow reaction, w=weak reaction

Vibrio natriegens strain NCMB 1900 was initially recorded as the type strain for *Listonella pelagia* I (Macián *et al*., 2000).

In this table, Vibrios are grouped according to their ODC, LDC, ADH reactions. Use this as a starting point in identifications. M=Moeller's ADH; T=Thornley's ADH.

Also check tables for *Aliivibrio* species, *Moritella* species, *Photobacterium* species, *Vibrio* – pathogenic species; t=tube motility test.

Laboratory Worksheet

Case Number																																								
Test	Gm	Ox	cat	βH	mot	SW	ODC	LDC	ADH	Nit	Ind	Cit	urea	mr	vp	aes	G	onpg	OF	arab	glu	inos	lac	malt	man	mano	sal	sor	suc	tre	Xyl	H_2S	MCA	TCBS	Dnase	temp	NaCl 0/3	0129 10	0129 150	Amp	**Identification**
Isolated No.																																								

Gm=Gram, Ox=oxidase, cat=catalase, βH=beta haemolysis, mot=motility, SW=swarming, ODC=ornithine decarboxylase, LDC=lysine decarboxylase, ADH=arginine dihydrolase, Nit=nitrate, Ind=indole, Cit=citrate, urea=urea, mr=methyl red, vp=Vogas–Proskaüer, aes=aesculin, G=gelatin, onpg=*o*-nitrophenyl-β-galactopyranoside (β-galactosidase test), OF=oxidative fermentative, arab=arabinose, glu=glucose, inos=inositol, lac=lactose, malt=maltose, man=mannitol, mano=mannose, sal=salicin, sor=sorbitol, suc=sucrose, tre=trehalose, xyl=xylose, H_2S=hydrogen sulfide, MCA=MacConkey agar, TCBS=thiosulfate–citrate–bile–sucrose agar, Dnase=deoxyribonuclease agar, temp=temperature, NaCl 0/3=growth in 0% and 3% salt, 0129 10=sensitivity to vibrio static disc 10 μg, 0129 150=sensitivity to vibrio static disc at 150 μg, Amp=sensitivity to ampicillin 10 μg

Table 3.24. API 20E database biochemical results.

Tests	onpg	ADH	LDC	ODC	cit	H_2S	ure	TDA	ind	vp	G	glu	man	inos	sor	rha	suc	mel	amy	arab	ox	NO_2	N_2	mot	MCA	O	F	Temp	Time	Inoc	Ref	Strain
Organism																														%NaCl		
Acinetobacter/Moraxella	–	–	–	–	–	–	–	–	–	–	–	–	–	–	–	–	–	–	–	–	+										68	
Actinobacillus delphinicola	–	+	–	–	–	–	–	–	–	+	–	+	–	–	–	–	–	–	–	–	+	+		–	–	+	+	37	48	0.85	548	Dolphin
Actinobacillus delphinicola	–	21	+	70	–	–	–	–	–	+	–	+	–	–	–	–	–	–	–	–	+	+		–	–	+	+	37	48	0.85	548	Porpoise
Actinobacillus delphinicola	–	+	–	–	–	–	–	–	–	+	–	+	–	–	–	–	–	–	–	–	+	+		–	–	+	+	37	48	0.85	548	Whale
Actinobacillus scotiae	+	–	–	+	–	–	+	–	–	+	–	+	–	–	–	–	–	–	–	–	+	+		–	–	+	+	37	48	0.85	550	NCTC 12922[T]
Actinobacillus scotiae	+	–	–	–	–	–	+	–	–	+	–	+	–	–	–	–	–	–	–	–	+	+		–	–	+	+	37	48	0.85	550	Strains
Aeromonas allosaccharophila	+	67	+	33	v	–	–	–	+	–	+	+	+	–	–	67	+	33	–	67	+	+		+	+	+	+	37	24	0.85	1102	Various
Aeromonas aquariorum	+	+	+	–		+	–	–	+	–	+	+	+	–	–	–	+	–	–	–	+	+		+	+	+	+	30	24	0.85	1106	CECT 7289[T]
Aeromonas bestiarum	+	+	+	–	–	–	–	–	+	+	+	+	+	–	–	+	+	–	+	+	+	+		+	+	+	+	27	24	0.85	935	Carp
Aeromonas bivalvium	+	–	+	–	+	–	–	–	+	–	+	+	+	–	–	–	+	–		+	+	+		+	+	+	+	25–30	24	1	1146	868E[T]
Aeromonas caviae	+	+	–	–	10	–	–	–	+	8	80	+	+	–	–	–	+	–	–	+	+	+		+	+	+	+				42	
Aeromonas caviae	+	+	–	–	+	–	–	–	+	–	+	+	+	–	–	–	+	–	–	+	+	+		+	+	+	+	25	48	0.85	494	ATCC 15468[T]
Aeromonas diversa	–	+	–	–	–	–	–	–	+	v	+	+	–	–	–	–	–	–		–	+	+		+	+	+	+	25–30	24	1	1147	CDC 2478-85[T]
Aeromonas encheleia	+	+	–	–	–	–	–	–	+	25	+	+	+	–	–	75	75	–	+	–	+	+		+	+	+	+	27	24	0.85	494,935	Esteve 1995
Aeromonas eucrenophila	+	+	–	–	–	–	–	–	+	–	+	+	+	–	–		–				+	+		+	+	+	+					
Aeromonas fluvialis	+	–	–	–	+	–	–	–	+	–	–	–	–	–	–	–	+	–	–	–	+	+		+	+	+	+	30	48	0.85	37	CECT 7401[T]
Aeromonas hydrophila	+	+	+	–	–	–	–	–	+	+	+	+	+	–	–	–	+	–	–	+	+	+		+	+	+	+	25	48	0.85	493	ATCC 7966[T]
Aeromonas hydrophila	+	+	–	–	–	–	–	–	+	+	–	+	+	–	–	–	+	–	–	+	+	+		+	+	+	+				1476	ATCC 7966[T]
Aeromonas hydrophila	+	+	–	–	v	–	–	–	+	+	+	+	+	–	–	+	+	–	+	+	+	+		+	+	+	+				669	ATCC 14715
Aeromonas hydrophila	+	+	–	–	+	–	–	–	+	–	+	+	+	–	–	–	+	–	+	+	+	+		+	+	+	+			2	878	WFM 504
Aeromonas hydrophila	+	+	–	–	–	–	–	–	+	+	+	+	+	–	+	–	+	–		+	+	+		+	+	+	+	23	48	2	461	
Aeromonas jandaei	+	+	+	–	+	–	–	–	+	+	+	+	+	–	–	–	–	50	–	–	+	+		+	+	+	+				276	
Aeromonas jandaei	+	+	+	–	+	–	–	–	+	+	+	+	w	–	–	–	–	+	–	–	+	+		+	+	+	+	25	48	0.85	244	AHLDA 1718
Aeromonas molluscorum	+	+	–	–	+	–	–	–	–	–	+	+	+		–	–	+	–		+	+	+		+	+	+	+	25–30	24	1	1145	848[T]
Aeromonas piscicola	+	+	+	–		+			+	+	+	+	+	–		–	+	–	–	–	+	+		+		+	+	25	24	0.85	141	S1.2[T]
Aeromonas popoffii	+	+	–	–	+	50	–	–	v	+	+	+	+	–	–	–	–	–			+	+		+	+	+	+	25	24	0.85	784	LMG 17541[T]
Aeromonas rivuli	+	+	–	–	–	–	–	–	–	–	+	+	+	–	–	–	+	+	+	+	+	+		+	+	+	+	30	48	0.85	536	CECT 7518[T]
Aeromonas salmonicida	–	+	–	–	–	–	–	–	–	–	+	+	+	–	–	–	–	–	–	–	+	+		–	+	+	+				928	

A. salmonicida achromogenes	–	–	–	–	–	–	–	–	–v	–	–	+	+	–	–	–	+	–	–	–	+	+		–	+	+	+				669	
A. salmonicida achromogenes	–	–	–	–	–	–	–	–	+	+	–	+	+	–	–	–	+	–	–	–	+	+		–	+	+	+				968	NCIMB 1110
A. salmonicida masoucida	+	+	–	–	–	–	–	–	+	+v	–	+	+	–	–	–	+	–	+	–	+	+		–	+	+	+				669	
A. salmonicida salmonicida	+	+	–	–	–	–	–	–	–	–	+	+	+	–	–	–	–	–	+	–	+	+		–	+	+	+				669	
A. salmonicida salmonicida	–	+	–	–	–	–	–	–	–	–	+	+	+	–	–	–	–	–	–	–	+	+		–	+	+	+				1476	
A. salmonicida salmonicida	–	+	+	–	–	–	–	–	–	–	+	+	+	–	–	–	–	–	–	–	+	+		–	+	+	+				1476	
A. salmonicida salmonicida	–	–	–	–	–	–	–	–	–	+	–	+	–	–	–	–	–	–	–	–	+	+		–	+	+	+	15	72	0.85	513	
A. salmonicida salmonicida	+	+	+	–	–	–	–	–	–	–	+	+	+	–	–	–	–	–	+	–	+	+		–	+	+	+				968	NCIMB 1102
A. salmonicida atypical									+w		–		–							–											1664	Australian strain
A. salmonicida atypical	–	+w	–	–	–	–	–	–	+	–	–	+	–	–	–	–	w	–	–	–	+	+		–		w	w	25	72	0.85	244	AHLDA 1334
Aeromonas sanarellii	+	+	–	–	–	–	–	+	+	–	+	+	+	–	–	–	+	–	+	+	+	+		+	+	+	+	30	48	0.85	36	CECT 7402^T
Aeromonas schubertii	+	+	+	–	+	–	–	–	–	–	+	+	–	–	–	–	–	–	–	–	+	+		+	+	+	+	25	48	0.85	278	ATCC 43700^T
Aeromonas schubertii	+	+	+	–	+	–	–	–	–	+	+	+	–	–	–	–	–	–	–	–	+	+		+	+	+	+	25	48	0.85	278	AMC 4396-87
Aeromonas simiae		+	+	–					–	–		+	–	–	–	–	+	–	–	–	+	+		+	+	+	+	25–37	24	1	677	IBS S 6874^T
Aeromonas sobria	+	–	+	–	+	+	–	–	+	v	–	+	+	–	–	–	+	+	–	–	+	+		v	+	+	+	25	72	0.85	4	
Aeromonas sobria	+	–	+	–	+	+	–	–	+	–	–	+	+	–	–	–	+	+	–	–	+	+		v	+	+	+	25	72	0.85	4	
Aeromonas taiwanensis	+	+	–	–	+	–	–	+	+	–	+	+	+	–	–	–	+	–	+	+	+	+		+	+	+	+	30	48	0.85	36	CECT 7403^T
Aeromonas trota	+	+	+	–	+	–	–	–	+	–	+	+	+	–	–	–	–	–	+	–	+	+		+	+	+	+	30	24	0.85	277	ATCC 49657^T
A. veronii biovar Sobria	+	+	+	–	v	v	–	–	+	v	+	+	+	–	–	–	+	–	–	v	+	+		+	+	+	+	25	72	0.85	4	
A. veronii biovar Veronii	+	–	–	+	–	–	–	–	+	+	+	+	+	–	–	–	+	–	–	–	+	+		+	+	+	+	27	24	0.85	5,723	ATCC 35604^T
A. veronii biovar Veronii	+	–	–	+	–	–	–	–	+	–	+	+	+	–	–	–	+	–	+	–	+	+		+	+	+	+	27	24	0.85	5,723	ATCC 35604^T
A. veronii biovar Veronii	+	–	–	+	+	–	–	–	+	–	+	+	+	–	–	–	+	–	+	–	+	+		+	+	+	+	27	24	0.85	5,723	ATCC 35604^T
A. veronii biovar Veronii	+	–	–	+	+	–	–	–	+	+	+	+	+	–	–	–	+	–	+	–	+	+		+	+	+	+	27	24	0.85	5,723	ATCC 35604^T
A. veronii biovar Veronii	+	–	+	+	–	–	–	–	+	+	+	+	+	–	–	–	+	–	+	–	+	+		+	+	+	+	27	24	0.85	5,723	ATCC 35604^T
A. veronii biovar Veronii	+	–	+	+	+	–	–	–	+	–	+	+	+	–	–	–	+	–	+	–	+	+		+	+	+	+	27	24	0.85	5,723	ATCC 35604^T
A. veronii biovar Veronii	+	–	+	+	+	–	–	+	+	+	+	+	+	–	–	–	+	–	+	–	+	+		+	+	+	+	27	24	0.85	5,723	ATCC 35604^T

Continued

Table 3.24. Continued.

Tests	onpg	ADH	LDC	ODC	cit	H_2S	ure	TDA	ind	vp	G	glu	man	inos	sor	rha	suc	mel	amy	arab	ox	NO_2	N_2	mot	MCA	O	F	Temp	Time	Inoc	Ref	Strain
Aeromonas culicicola, a heterotypic synonym of *A. veronii*	+	+	+	–	+		–	–	+	+	+	+	+	–	–	–	+	+	–	–	+	+		+		+	+	30	24	0.85	1339	NCIM 5147[T]
Aliivibrio finisterrensis	+	+	+	–	–	–	+	–	–	–	–	+	–	–	–	–	–	+	–	–	+	+		+				25	48	0.85	143	CMJ 11.1[T]
Aliivibrio finisterrensis	+	+	+	–	–	–	+	–	–	–	–	+	+	–	–	–	+	–	–	–	+	+		+				25	48	0.85	143	C 17.13
Aliivibrio finisterrensis	+	+	+	–	–	–	+	–	–	–	–	+	–	–	–	–	–	–	–	–	+	+		+				25	48	0.85	143	C 13.19
Aliivibrio finisterrensis	+	+	+	–	–	–	+	–	–	–	–	+	–	–	–	–	–	–	–	–	+	+		+				25	48	0.85	143	V 14.7
Aliivibrio fischeri	–	–	+	–	–	–	–	–	–	+	–	+	–	–	–	–	–	–	+	–	+	+		+	–	+	+	26	72	1.5	1641	ATCC 25918
Aliivibrio fischeri ATCC 7744[T]	–	+	+	–	+	–	+	–	–	w	–	w	w	–	–	–	–	–	w	–	+	+		+	–	+	+	25	48	2	244	ATCC 7744[T]
Aliivibrio fischeri ATCC 7744[T]	–	–	+	–	–	–	–	–	–	–	–	+	+	–	–	w	–	w	–	–	+	+		+	–	+	+	25	48	0.85	244	ATCC 7744[T]
Alivibrio logei	–	–	–	–	–	–	–	–	–	–	–	–	–	–	–	–	–	–	–	–	+	–		+	–	–	–	25	48	2	244	ATCC 29985
Aliivibrio salmonicida	–	–	–	–	–	–	–	–	–	–	–	+	–	–	–	–	–	–	–	–	+	–		+		+	+	15	48	1.5	68	
Aliivibrio salmonicida	–	–	–	–	+	–																										
Aliivibrio sifiae H1-1[T]	+	–	+			–	–		+		–	w	+				+		+	–	+	+		+				20	48	2	1834	H1-1T
Aliivibrio sifiae H1-2	+	–	+			–	–		–		–	w	+				–		+	–	+	+		+				20	48	2	1834	H1-2
Aliivibrio wodanis	–	–	–	–	–		–	–	–		+	–	–	–	–	–	–	–	–	w	+			+				20	48	2	1834	ATCC BAA-104[T]
Aliivibrio wodanis	–	–	–	–	–	–	–	–	+	–	+	+	+	–	–	–	+	–	–	–	+			+				20	48	0.85	143	ATCC BAA-104[T]
Aliivibrio wodanis	–	–	–	–	–	–	20	–	89	–	+	+	34	–	–	–	83	–	–	–	+			+		+	+	22	48	1.5	153,1071	
Allomonas sp.	+	+	–	–	–	–	–	–	+	–	+	+	+	–	–	–	+	–	+	–								20	48	2	1834	WFM401
Burkholderia pseudomallei	–	+	–	–	63	–	–	–	–	–	+	+	+	+	+	–	43	–	60	+	+	+		+	+	+	–	37	48	0.85	64	
Carnobacterium divergens	+	+	–	–		–	–	–	–	+			–																			
Citrobacter freundii	+	v	–	–	–	+	–	–	–	–	–	–	–	–	+	+	+	+	–	+	–										68	
Citrobacter freundii	+	–	–	–	–	+	–	–	–	–	–	+	+	–	+	+	+	+	–	+	–	+		+	+	+	+	37	48	0.85	1478	
Citrobacter freundii	90	24	–	–	75	75	–	–	–	–	–	+	+	25	+	+	+	82	40	+	–	+		+	+	+	+	37	48	0.85	*	
Echinicola pacifica	+	v	–	–	–	+	–	–	–	–	w	–	–	–	+	+	+	–	–	+	–	–		+				25			1207	
Edwardsiella ictaluri	–	–	+	+	–	–	–	–	–	–	–	+	–	–	–	–	–	–	–	–	–	+		+	+	+	+	28	48	0.85	68,702	
Edwardsiella ictaluri	–	–	+	–	–	–	–	–	–	–	–	+	–	–	–	–	–	–	–	–	–	+		+	+	+	+	28	48	0.85	386	
Edwardsiella ictaluri	–	–	+	+	–	–	–	–	–	–	–	+	+	–	–	–	–	+	–	–	–	+		+	+	+	+	28	48	0.85	880	
Edwardsiella piscicida	–	–	+	+	–	+	–	–	+	–	–	+	–	–	–	–	–	–	–	–	–	+		+	+	+	+	28	48	0.85	1	

Edwardsiella piscicida	–	–	+	+	+	+	–	–	+	–	–	+	+	–	–	–	–	–	–	+	–	+		+	+	+	+	28	48	0.85	1	
Edwardsiella piscicida	–	–	+	+	+	+	–	–	+	–	–	+	+	–	–	–	–	–	–	–	–	+		+	+	+	+	28	48	0.85	1	
Edwardsiella piscicida	–	–	+	+	+	+	–	–	+	–	–	+	–	–	–	–	–	–	–	–	–	+		+	+	+	+	28	48	0.85	143	
Edwardsiella tarda	–	–	+	w	+	+	–	–	+	–	–	+	–	–	–	–	–	–	–	–	–	+			+	+	+	25	48	0.85	244	AHLDA 135
Edwardsiella tarda	–	–	+	+	–	+	–	–	+	–	–	+	–	–	–	–	–	–	–	–	–	+		+	+	+	+	22	48	0.85	68,1674	
Edwardsiella tarda	–	–	+	+	+	+	–	–	+	–	–	+	–	–	–	–	–	–	–	–	–	+		+	+	+	+	25	48	0.85	68,286	
Edwardsiella tarda	–	–	+	+	+	+	–	–	+	–	–	+	–	–	–	–	–	–	–	–	–	+		+	+	+	+	28	48	0.85	21	CECT 849^T
Edwardsiella tarda	–	–	+	+	–	–	–	–	+	–	–	+	–	–	–	–	–	–	–	–	–	+		+	+	+	+	28	48	0.85	21	eels
Edwardsiella tarda	–	–	+	+	+	–	–	–	+	–	–	+	–	–	–	–	–	–	–	–	–	+		+	+	+	+	28	48	0.85	21	eels
Edwardsiella tarda – atypical strains	–	–	+	–	+	+	–	–	+	–	–	+	+	–	–	–	–	–	–	–	–							25	48	0.85	1820	
Enterovibrio calviensis DSMZ 14347	+	–	–	–	–	–	–	–	+	–	–	w	w	–	–	–	–	–	w	–	+	–		+	–	+	+	25	48	2	244	DSMZ 1337
Enterovibrio coralii	+	+	–	–			–	–	–	–	–	+	+	–	–	–	–	–	–	–	+	+		+		+	+	28	48	2	1636	LMG 22228^T
Enterovibrio coralii		75s	–	–					+	–	–										+	+		+		+	+	28	48	2	1314	
Enterovibrio norvegicus	+	+	–	–	–	–	–	–	+	–	–	+	–	–	–	–	–	–	–	–	+	–		+		+	+	25	48	2	1627	LMG 19839^T
Escherichia vulneris	+	30	50	–	–	–	–	–	–	–	–	+	+	–	–	+	25	–	+	+	–	+		+	+	+	+	37	24	0.85	*	
Flavobacterium branchiophilum	–	–	–	–	–	–	–	–	–	–	+	+	–	–	–	–	+	+	–	–	–										68	
Flavobacterium columnare	–	–	–	–	–	+	–	–	–	–	+	–	–	–	–	–	–	–	–	–	+										68	
Flavobacterium columnare	+	–	–	–		+	–	–	–	–	+	–	–	–	–	–	–	–	–	–	+	+		–	–	–	–	25	48	0.5	1511	
Flavobacterium gillisiae	–	+	–	–	–	–	–	–	–	–	–	+	+	–	–	–	+	–		–	–	–						20	48	1.5	1115	
Flavobacterium hibernum	+	–	–	–	–	–	–	–	–	–	+	+	–	–	–	+	+	–	–	+	–	+						25	48	1.5	1116	ATCC 51468^T
Flavobacterium hydatis	–	–	–	–	–	–	–	–	–	–	+	+	+	–	–	–	+	–	–	+	–										68	
Flavobacterium johnsoniae	+	–	–	–	+	–	–	–	–	+	+	–	–	–	–	–	–	–	–	–	+	+		–		–	–	25	72	0.85	244	AHLDA 1714
Flavobacterium macrobrachii	–	–	–	–	–	–	–	–	–	–	+	–	–	–	–	–	–	–	–	–	+	–		–	–	–	–	25	48	0.85	1514	LMG 25203^T
Flavobacterium psychrophilum	–	–	–	–	–	–	–	–	–	–	+	–	–	–	–	–	–	–	–	–	+										68	
Flavobacterium psychrophilum	–	–	–	–	–	–	–	–	–	+	+	–	–	–	–	–	–	–	–	–								18	48	0	1745	
Flavobacterium tegetincola	–	–	–	–	–	–	–	–	–	–	–	+	+	–	–	–	–	–	–	–	–	–						20	48	1.5	1115	
Flavobacterium xantham	–	–	–	–	–	+	–		–	–	+	+	+	–	–	–	+	–	–	–	+	+						20	48	1.5	1115	
Gramella marina	+	–	–	–	+	–	–	–	–	+	+	+	–	–	–	–	+	+	+	+	+	–		+	–	+	–	28	48	2	1205	KMM 6048^T

Continued

Table 3.24. Continued.

Tests	onpg	ADH	LDC	ODC	cit	H_2S	ure	TDA	ind	vp	G	glu	man	inos	sor	rha	suc	mel	amy	arab	ox	NO_2	N_2	mot	MCA	O	F	Temp	Time	Inoc	Ref	Strain
Grimontia hollisae		–	–	–	–				+	–											+	+		+	–	+	+	25	48	2	1631	
Grimontia hollisae	–	–	–	–	–	–	–	–	+	–	–	w	–	–	–	–	w	–	–	w	+	+		+	–	+	+	25	48	2	244	CIP 101886
Hafnia alvei	–	–	+	+	+	–	–	–	–	–	–	+	+	–	–	–	–	–	–	+	–	+									68	
Hafnia alvei	+	–	+	+	–	–	–	–	–	–	–	+	+	–	–	+	–	–	–	+	–	+		+	+	+	+	37	48	0.85	1418	
Hafnia alvei	+	–	+	+	+	–	–	–	–	+	–	+	+	–	–	+	–	–	+	+	–	+		+	+	+	+	25	48	0.85	244	AHLDA 1729
Hafnia alvei	+	–	+	+	+	–	–	–	–	–	–	+	+	–	–	+	–	–	–	+	–	+		+	+	+	+	25	48	0.85	402	ATCC 51873
Halomonas cupida	–	+	+	+	–	–	–	–	–	–	–	–	+	–	+	+	+	–	–	–	–										68	
Janthinobacterium lividum	–	+	–	–	+	–	–	–	–	+	+	+	+	–	–	–	–	–	–	–	+	+		+		+	–	22	48	1	73	
Klebsiella pneumoniae	+	–	v	–	+	–	v	–	–	+	–	+	+	+	+	+	+	+	+	+	–	+		+	+	+	+				68	
Moritella marina	–	–	–	–	–	–	–	–	–	–	+	+	–	–	–	–	–	–	–	–	+	+		+	–	+	+	15	72	2	1683	ATCC 15381^T
Moritella viscosa	–	–	+	–	–	–	–	–	–	–	v	+	–	–	–	–	–	–	–	–	+	+		+				15	72	2	236	NCIMB 13484
Moritella viscosa	–	–	+	–	–	–	–	–	–	–	v	+	–	–	–	–	–	–	–	–	+	+		+		+						
Pantoea agglomerans	+	–	–	–	+	–	–	–	–	+	+s	+	+	–	–	+	+	–	–	+	–	+		+		+	+	37	48	0.85	602	
Pantoea agglomerans	+	–	–	–	+	–	–	–	–	+	+	+	+	–	–	–	+	–	–	–	–										68	
Pantoea dispersa	+	–	–	–	+	–	–	–	–	+	+s	+	+	–	–	+	+	v	+	+	–	+		+		+	+				602	
Paracoccus fistulariae	–	–	–	–	+	–	–	–	–		+	+	–	–	–	+	+	+	–	+	+	+		–		+	–	30	48	ASW	903	KCTC 22803^T
Pasteurella multocida	–	–	–	+v	–	–	–	–	+	–	–	+	+	–	+	–	+	–	–	–	+	+		–	–	+	+	37	48	0.85	878,882	
Pasteurella testudinis	+	–	–	–	–	–	–	–	+	–	–	+	+	70	–	60	+	30	–	30	+	+		–	+	+	+	37	48	0.85	1553	
Phocoenobacter uteri	–	–	–	–	–	–	–	–	–	+	–	+w	–	–	–	–	–	–	–	–	+	+		–	–	+	+	37	48	0.85	549	NCTC 12872^T
Photobacterium angustum	–	+	–	–	–	–	–	–	–	+	+	+	–	–	–	–	+	–	–	–	+	–		–	–	+	+	26	72	1.5	1641	NCIMB 1895
P. damselae ssp. *damselae*	–	+	–	–	–	–	+	–	–	–	–	+	–	–	–	–	–	–	–	–	+	+		+	+	+	+	25	48	2	339	
P. damselae ssp. *damselae*	–	+	–	–	–	–	+	–	–	+	–	+	–	–	–	–	–	–	–	–	+	+		+	+	+	+	25	48	0.85	244	12-1859-1
P. damselae ssp. *damselae*	–	+	–	–	–	–	–	–	–	+	–	+	–	–	–	–	–	–	–	–	+	+		+	+	+	+	25	48	2	244	12-1859-1
P. damselae ssp. *damselae*	–	+	–	–	–	–	+	–	–	+	–	+	–	–	–	–	–	–	–	–	+	+		+	+	+	+	26	72	1.5	1641	ATCC 33539^T
P. damselae ssp. *damselae*	–	+	+	–	–	–	–	–	–	–	–	+	–	–	–	–	–	–	–	–	+	+		+	+	+	+	25	48	0.85	496	ATCC 33539^T
P. damselae ssp. *damselae*	–	+	–	–	+	–	–	–	–	–	–	+	–	–	–	–	–	–	+	–	–	+		+	+	+	+	26	72	1.5	1641	NCIMB 2184
P. damselae ssp. *damselae*	–	+	55	–	15	–	85	–	–	+	20	+	–	–	–	–	–	–	15	–	75	+		+	+	+	+	26	72	1.5	1641	Strains

P. damselae ssp. *damselae*	−	+	+	−	−	−	+	−	−	+	−	+	−	−	−	−	−	−	−	−	+							22	48	0.85	963	
P. damselae ssp. *piscicida*	−	+	−	−	18	−	−	−	−	96	−	97	−	−	−	−	−	−	−	−	+	−		−	46	+	+	26	72	1.5	1641	Strains
P. damselae ssp. *piscicida*	−	+	−	−	−	−	−	−	−	−	−	+	−	−	−	−	−	−	−	−	+	−		−	−	+	+	25	48	1.5	98,703, 1089	ATCC 17911, 29687
P. damselae ssp. *piscicida*	−	+	−	−	−	−	−	−	−	+	−	+	−	−	−	−	−	−	−	−	+	−		−	−	+	+	25	48	2	1159	Tuna
P. damselae ssp. *piscicida*	−	+	−	−	−	−	−	−	−	−	−	+w	−	−	−	−	−	−	−	−	+	−		−		+	+				878	ATCC 17911, 29687
P. damselae ssp. *piscicida*	−	+	−	−	−	−	−	−	−	+	−	+	−	−	−	−	−	−	−	−	+	−		−		+	+	22	48	1.5	#	ATCC 29690, 17911
Photobacterium iliopiscarium	−	+	+	−	−	−	−	−	−	+	−	+	−	−	−	−	−	−	−	−	+	w		+		+	+	25	48		244	ATCC 51760
Photobacterium iliopiscarium	−	+	+	−	−	−	−	−	−	−	−	+	−	−	−	−	−	−	−	−	+	+		+		+	+				1272	PS1
Photobacterium leiognathi	−	+	−	−	−	−	−	−	−	+	−	+	−	−	−	−	−	−	−	−	−	+		−	+	+	+	26	72	1.5	1641	LMG 4228
Photobacterium phosphoreum	−	+	w	−	−	−	−	−	+	−	+	+	−	−	−	−	+	−	+	−										2	1076	IB39
Photobacterium rosenbergii	+	+	−	−	+	−	−	−	−	−	−	+	+	+	−	+	+	+	+	−	+	+		+		+	+	28	48	2	1636	LMG 22228^T
Photobacterium rosenbergii	+	+	−	−	+	−	−	−	−	−	−	+	+	v	−	v	+	+	+	−	+	+		+		+	+	28	48	2	1636	
Photobacterium swingsii	50	+	−	−	−	−	−	16	−	+	16	+	33	−	−	−	−	83	83	+	+	+		+		+	+	25	48	2.5	641	CAIM 1393^T
Photobacterium swingsii	+	+	−	−	−	−	−	−	−	+	−	+	−	−	−	−	−	+	+	+	+	+		+		+	+	25	48	2.5	641	CAIM 1393^T
Photobacterium swingsii	−	+	−	−	−	−	−	−	−	+	−	+	−	−	−	−	−	+	+	+	+	+		+		+	+	25	48	2.5	641	CAIM 1393^T
Plesiomonas shigelloides	+	+	+	+	−	−	−	−	+	−	−	+	−	+	−	−	−	−	−	−	+										68	
# References: Forbes 2000, Aiso 1968, Aluotto 1970, Baptista 1996, Dalsgaard 1998, Dienes 1939																																
Providencia rustigianii	−	−	−	−	+w	−	−	+	+	−	−	+	−	−	−	−	+w	−	−	−	−	+		+	+	+	+	37	48	0.85	1175	
Providencia (Proteus) rettgeri	−	−	−	−	−	−	+	+	+	−	−	+	+	+	−	+	−	−	−	−	−										68	
Pseudoalteromonas ulvae	−	−	−	−	+	−	−	+			+	−									+			+		−	−	23	48	1.5	471	
Pseudomonas anguilliseptica	−	−	−	−	−	−	−	−	−	−	+	−	−	−	−	−	−	−	−	−	+										68	
Pseudomonas fluorescens	−	+	−	−	+	−	−	−	−	−	+	+	+	+	+	−	+	−	−	+	+	+		+	+	−	−				68	
Pseudomonas fluorescens/putida	−	75	−	−	75	−	−	−	−	10	27	25	−	−	−	−	−	25	1	20	+	26	−	+	+	+	−				*	

Continued

Table 3.24. Continued.

Tests	onpg	ADH	LDC	ODC	cit	H_2S	ure	TDA	ind	vp	G	glu	man	inos	sor	rha	suc	mel	amy	arab	ox	NO_2	N_2	mot	MCA	O	F	Temp	Time	Inoc	Ref	Strain
P. (Pseudoalcaligenes) oleovorans	–	+	–	+	+	–	–	–	–	–	+	+	–	–	–	–	–	–	–	+	+							25	48	0.85	76	
Pseudomonas plecoglossicida		–	–	–					–	–	–	+	–				+	–		–	+	–									878	
Pseudomonas putida							–					–v																				
Rahnella aquatilis	+	–	–	–	+	–	–	–	–	+	–	+	–	–	+	+	+	+		+	–	+		+	+	+	+				299	
Salegentibacter salegens																								–				20	48	1.5	1115	
Salmonella cholerasuis arizonae	+	–	+	–	+	+	–	–	–	–	+	+	–	+	–	+	–	–	–	–	–										68	
Serratia fonticola	+	–	+	+	+	–	–	–	–	+	–	+	+	–	+	+	–	+		+	–	+		+	+	+	+				1176	
Serratia fonticola	+	–	+	+	+	–	–	–	–	–	–	+	+	+	+	+	20	+		+	–	+		+	+	+	+	37	48	0.85	601	
Serratia liquefaciens	+	–	+	+	+	–	v	–	–	v	+	+	+	v	+	–	+	v	+	–	–	+									68	
Serratia liquefaciens	+	–	+	+	+	–	–	–	–	+	–	+	+	+	–	–	+	+	–	–	–	+		+	+	+	+	25	48	0.85	1123	K2
Serratia liquefaciens	+	–	+	+	+	–	–	–	–	+	–	+	+	+	–	–	+	+	+	+	–	+		+	+	+	+	25	48	0.85	1123	K2
Serratia liquefaciens	+	–	+	+	+	–	–	–	–	+	+	+	+	+	+	–	+	+	–	+	–	+		+	+	+	+	25	48	0.85	1724	Turbot
Serratia marcescens	+	–	+	+	+	–	–	–	–	+	+	+	+	–	+	–	+	–	+	+	–	+		+	+	+	+	37	48	0.85	137	Perch
Serratia marcescens	94	–	95	95	96	–	25	–	–	70	87	+	+	85	+	–	+	68	97	25	–	+		+	+	+	+	37	48	0.85	*	
Serratia plymuthica	+	–	–	–	–	–	–	–	–	+	+	+	+	+	–	–	+	+	+	+	–	+		+	+	+	+	22	48	0.85	76	Trout
Serratia plymuthica	+	–	–	–	+	–	–	–	–	+	+	+	+	+	–	–	+	+	+	+	–	+		+	+	+	+	22	48	0.85	1222	Trout
Serratia plymuthica	+	–	–	–	65	–	–	–	–	65	50	+	+	70	–	–	+	85	+	+	–	+		+	+	+	+	37	48	0.85	*	
Shewanella algae	–	–	–	+	–	+	–	–	–	–	+	+	–	–	–	–	–	–	–	–	+	+		+	+	+	–	35	48	0.85	886, 1239	
Shewanella algae	–	–	–	–	+	+	–	–	–	–	–	–	–	–	–	–	–	–	–	–	+	+		+				25	48	2	256	ATCC 55627^T
Shewanella algae	–	–	–	–	–	+	–	–	–	–	+	–	–	–	–	–	–	–	–	–	+	+		+				37	48	2	1402	ATCC 51192^T
Shewanella algidipiscicola	–	–	–	–	–	+	–	–	–	–	–	–	–	–	–	–	–	–	–	–	+	+		+				25	48	2	1402	LMG 23746^T
Shewanella baltica	–	–	–	+	–	+	–	–	–	–	+	–	–	–	–	–	–	–	–	–	+	+		+				25	48	2	1402	NCTC 10735^T
Shewanella baltica	–	–	–	–	–	+	–	–	–	–	+	–	–	–	–	–	–	–	–	–	+	+		+				37	48	2	1402	Oysters
Shewanella baltica	–	–	–	–	–	–	–	–	–	–	+	–	–	–	–	–	–	–	–	–	+	+		+				37	48	2	1402	Seawater
Shewanella frigidimarina	–v	–v	–	–v	+v	+	–	–	–	–	+	+	+	–	–	–	+	–	–	–	+	+		+		+	+	20	48	2	209	ACAM 591^T
Shewanella frigidimarina	–	–	–	–	–	+	–	–	–	–	–	–	–	–	–	–	–	–	–	–	+	+		+				25	48	2	1402	ACAM 591^T
Shewanella gelidimarina	–	–	–	–	–	+	–	–	–	–	+	–	–	–	–	–	–	–	–	–	+	+		+				15	48	2	209	ACAM 456^T
Shewanella hafniensis	–	–	–	–	–	–	–	–	–	–	–	–	–	–	–	–	–	–	–	–	+	+		+				37	48	2	1402	Oysters
Shewanella haliotis	–	–	–	+	–	+	–	–	–	–	+	–	–	–	–	–	–	–	–	–	+	+		+		+	+	25	48	2	895	KCTC 12896^T
Shewanella hanedai		–			–						+	–	–	–	–	–	–	–	–	–	+	+		+		+	–	25	48	2	1093	ATCC 33224^T

Shewanella marisflavi	–	–	–	–	–	+	–	–	–	–	–	–	–	–	–	–	–	–	–	–	+	+		+				37	48	2	1402	Oysters
Shewanella morhuae	–	–	–	–	–	+	–	–	–	–	+	–	–	–	–	–	–	–	–	–	+	+		+				37	48	2	1402	ATCC U1417^T
Shewanella putrefaciens	–	–	–	+	+	+	–	–	–	–	+	w	–	–	–	–	35	–	–	49	+	+		+	+	+	–	35	48	0.85	886	
Shewanella putrefaciens	–	–	–	–	–	–	–	–	–	–	–	–	–	–	–	–	–	–	–	–	+	+		+				37	48	2	1402	ATCC BAA-1097^T
Shewanella putrefaciens	–	–	–	–	–	–	–	–	–	–	+	–	–	–	–	–	–	–	–	–	+	+		+				37	48	2	1402	Oysters
Shewanella upenei							–				+		–	–	–	–	–	–		–	+	+		+				30	48	2	901	KCTC 22806^T
Shewanella woodyi	–	–	–	–	–	–	–	–	–	–	+	–	–	–	–	–	–	–	–	–	+	+		+		+	–	25	48	2	1093	ATCC 51908^T
Tenacibaculum crassostreae		–				–	–		–		+	–									+	–					–	37	24	2	1004	
Tenacibaculum maritimum	–	–	–	–	–	–	–	–	–	–	+	–	–	–	–	–	–	–	–	–	+										68	
Tenacibaculum maritimum	–	–	–	–	–	–	–	–	–	–	+	–	–	–	–	–	–	–	–	–	+	+		–	–	–	–	20	48	ASW	1054	
Tenacibaculum soleae	–	–	–	–	–	–	–	–	–	+	+	–	–	–	–	–	–	–	–	–	+	+		+	–	–	–	20	24	ASW	1055, 1344	
Vibrio aerogenes	+	–	–	–	–	–	–	–	+	–	–	w	–	–	–	–	–	–	+	–	+	+		w	–	+	+	25	48	2	244	DSMZ 14438
Vibrio aestivus	+	–	–	–					–	–	–	+	–	–	–	–	–	–			+	+		+		+	+	25	48	0.85	1065	CECT 7558^T
Vibrio aestuarianus	+	+	–	–	–	–	–	w	+	–	+	+	+	–	–	–	+	–	+	–	+	+		w	s	+	+	25	48	2	244	NCIMB 2236
Vibrio aestuarianus spp. *aestuarianus*	+	–	–	–	–	–	–	+	+	–	–	–	+	–	–	–	+	–	–	–	–	+		+		+	+	20	48	2	588	ATCC 35048^T
Vibrio aestuarianus spp. *aestuarianus*	+	–	–	–	25	–	–	75	+	–	–	25	–	–	–	–	25	–	–	–	25	+		+		+	+	20	48	2	588	Oysters
Vibrio aestuarianus spp. *francensis*	+	29	–	–	86	–	–	71	+	–	57	14	29	14	14	14	71	–	29	–	14	+		+		+	+	20	48	2	588	Oysters
Vibrio agarivorans	+	–	–	–	–	–	–	–	–	–	–	+	+	–	–	–	–	–	+	–	+	+		+	–	+	+	25	48	2	244	AHLDA 1732
Vibrio agarivorans	+	–	–	–	–	–	–	–	–	–	–	+	+	–	–	–	–	–	+	–	+	+		+	–	+	+	25	48	2	244	CECT 5085^T
Vibrio alfacsensis	+	+	+	+	–	–	–	w	+	+	–	+	+	–	–	–	+	v	+	–	+	+		+		+	+	30	48	2	640	CAIM 1831^T
Vibrio alfacsensis	+	+	+	+	–	–	+	w	+	+	–	+	+	–	–	–	+	v	+	–	+	+		+		+	+	30	48	2	640	CAIM 1836
Vibrio alfacsensis	+	+	+	+	–	–	–	w	+	+	–	+	+	–	–	–	–	v	+	–	+	+		+		+	+	30	48	2	640	CAIM 1835
Vibrio alfacsensis	+	+	+	+	+	–	–	w	+	+	–	+	+	–	–	–	–	v	+	–	+	+		+		+	+	30	48	2	640	CAIM 404
Vibrio alginolyticus	–	–	+	53	v	–	–	–	+	83	√	+	+	–	–	–	+	–	67	–	+	+		+		+	+				1164	
Vibrio alginolyticus	–	–	+	–	–	–	–	–	+	w	+	+	+	–	–	–	+	–	–	–	+	+		+	–	+	+	25	48	2	244	
Vibrio alginolyticus	–	–	+	+	–	–	–	–	+	+	+	+	+	–	–	–	+	v	v	–	+	+		+		+	+				1409	
Vibrio alginolyticus	–	–	+	+	+	–	+	–	+	–	+	+	+	–	–	–	+	–	–	–	+	+		+		+	+	25	48	2	244	ATCC 17749^T
Vibrio alginolyticus	–	–	+	–	–	–	–	–	+	–	+	+	+	–	+	–	+	–	–	–	+	+		+		+	+	28	48	2.5	1044	ATCC 17749^T
Vibrio alginolyticus	–	–	+	–	–	–	–	–	+	–	+	+	+	–	–	–	+	–	–	–	+	+		+		+	+	25	48	1	1001	ATCC 17749^T
Vibrio alginolyticus	–	–	+	–	–	–	–	–	+	+	+	+	+	–	+	–	+	–	–	–	+	+		+		+	+	30	24	1	257	ATCC 17749^T

Continued

Table 3.24. Continued.

Tests	onpg	ADH	LDC	ODC	cit	H_2S	ure	TDA	ind	vp	G	glu	man	inos	sor	rha	suc	mel	amy	arab	ox	NO_2	N_2	mot	MCA	O	F	Temp	Time	Inoc	Ref	Strain
Vibrio alginolyticus	–	–	+	+	–	–	–	–	+	–	+	+	+	–	–	–	+	–	–	–	+	+		+				25	48	1	1001	Swy
Vibrio anguillarum	+	+	–	–	+	–	–	–	+	+	+	+	+	–	+	–	+	–	–	–	+	+		+	v	+	+	25	48	0.85	496	NCMB 6
Vibrio anguillarum	+	+v	–	–	+	–	–	–	+	+	+	+	+	–	+	–	+	–	+	+v	+	+				+	+				68, 878	
Vibrio anguillarum	+	+	–	–	–	–	–	–	+	+	+	+	w	–	+	–	w	–	–	–	+	+			–	+	+	25	48	2		
Vibrio anguillarum	+	+	–	–	+	–	–	–	+	+	+	+	+	–	+	–	+	–	–	+	+	+			–	+	+	25	48	2	244	AHLDA 1730
Vibrio anguillarum ATCC 19264	+	+	–	–	+	–	–	–	+	+	+	+	+	–	+	–	–	–	w	–		+		+	–	+	+	25	48	2	244	ATCC 19264
Vibrio anguillarum CRL 11	+	+	–	–	–	–	–	–	+	+	+	+	+	–	+	–	+	–	+	–		+			+	+	+	25	48	2	244	CRL 11
Vibrio artabrorum		–	–	–																	+	+				+	+	25	48	2	442	LMG 23865[T]
Vibrio artabrorum		–	–	–																	+	+				+	+	25	48	2	442	Strains
Vibrio atlanticus	+	–	–	–					+	+	–	+	+	–	–	–	+	–		+	+	+		+		+	+	25	48	2	442	LMG 24300[T]
Vibrio atlanticus	75	–	–	–						+	+	+		–	–	–	+	–		25	+	+		+		+	+	25	48	2	442	Strains
Vibrio atypicus	+	–	–	–	–	–	+	–	–	–	+	+	+	–	–	–	+	–	–	–	+	+		+		+	+	28	48	1.5	1767	HHSO2[T]
Vibrio azureus			–	–	–			–	+w			w	–	–	–	–	–	–	+	–	+	–		+		+	+	20	48	2	1836	NBRC 104587[T]
Vibrio brasiliensis	+	+	–	–	+	–	–	–	+	+	+	+	+	–	–	–	+	–	+	–	+	+		+	nt	+	+	25	48	2	1632	LMG 20546[T]
Vibrio brasiliensis		+	–	–				–	+	–	+	+	+	–		–	+	–	+	–	+	+		+		+	+	25	48	2	320, 739	LMG 20546[T]
Vibrio breoganii	+	–	–	–	–	–	–		–	–	–	+	+	–	–	–	–	–	+	–	+	+		–	nt	+	+	25	48	0.85	140	LMG 23858[T]
Vibrio campbellii	–	–	–	–	–	–	–	w	+	–	–	+	+	–	–	–	–	–	+	–	+	+		+	–	+	+	25	48	2	244	ATCC 2590
Vibrio caribbeanicus	–	+	–	–	–	–	–	–	–	–	–	–	–	–	–	–	–	–	–	–	+	–		+			–	28	248	1.5	739	DSM 23640[T]
Vibrio casei	+				+																+	+				+	+	30	24	1.5	190	
Vibrio celticus	–	+	–	–	–	–	–	–	+	+	+	+	+	–	–	+	+	+	+	–	+	+		+		+	+	25	48	0.85	142	Rd 8.15[T]
Vibrio celticus	–	+	–	–	–	–	–	–	+	+	+	+	+	–	–	+	+	–	+	–	+	+		+		+	+	25	48	0.85	142	RD 6.8
Vibrio celticus	–	+	–	–	–	–	–	–	+	+	+	+	+	–	–	+	+	–	+	+	+	+		+		+	+	25	48	0.85	142	Rd2L5
Vibrio chagasii	–		–	–		–	–	–	+			+	+	–		–	–	–	–	–	+	+		+		+	+	25	48	2	1635	LMG 21353[T]
Vibrio chagasii	–		–	–		–	–	–	+			+	+	–		–	–	–	–	–	+	+		+		+	+	25	48	2	1635	12 strains
Vibrio chagasii	–	+	–	–	–	–	–	+	+	–	+	+	+	–	–	–	+	–	–	–	+	–		+		+	+	25	48	2	524	LMG 21353[T]
Vibrio cholerae non-01	+	–	+	+	v	–	–	–	+	+	+	+	+	–	–	–	+	–	–	–	+			+	+	+	+					
Vibrio cholerae 01	+	–	+	+	v	–	–	–	+	–	+	+	+	–	–	–	+	–	–	–	+			+	+	+	+					
Vibrio cholerae	+	–	+	+	w	–	–	–	+	+	w	+	+	–	–	–	+	–	+	–	+	+		+	+	+	+			2	1076, 1164	WF110r
Vibrio cincinnatiensis	+	–	+	–	+	–	–		–	+	–	+	–	+	–		+			+	+	+		+	+	+	+	25	48	1	220	ATCC 35912
Vibrio comitans	–	–	–	–	33				–	–	–	+	+	–	–	–	–			–	+	+		–		+	+	25	48	2	1485	
Vibrio communis	+	–	+	+	–	–	–	–	+	+	+	+	+	–	–	–	+	–	+	–	+	+		+		+	+	25	48	2	320	LMG 25430[T]
Vibrio communis	+	–	+	+	–	–	–	–	+	+	–	+	+	–	–	–	+	–	+	–	+	+		+		+	+	25	48	2	320	R-40504

Vibrio communis	+	−	+	+	+	−	−	−	+	+	+	+	+	−	−	−	+	−	+	−	+	+		+		+	+	25	48	2	320	R-40498
Vibrio communis	+	−	+	−	−	−	−	−	+	+	+	+	+	−	−	−	+	−	+	−	+	+		+		+	+	25	48	2	320	R-40900
Vibrio coralliilyticus		+	−	−				+	+	+	+	+	+	−		−	+	−	−	+		+						25	48	2	321, 738	LMG 20984^T
Vibrio crassostreae	−	+	−	−	−	−	−	v	+	−	+	+	+	−	−	−	+	+	+	−	+	+		+		+	+	20	48	2	524	LMG 22240^T
Vibrio cyclitrophicus	−	w	−	−	−	−	−	−	−	−	+	+	+	−	−	−	+	−	+	−	+	−		+		+	+	25	48	2	709	P-2P44^T
Vibrio cyclitrophicus	+	−	−	−	−	−	−	−	+	−	−	+	+	−	−	−	+	−	+	−	+	+		+		+	+	25	48	2	524	LMG 21359^T
Vibrio diabolicus	−	−	+	+	−	−	−	+	+	−	+	+	+	−	−	−	+	−	+	−	+	+		+		+	+	25	48	2	1368	CNCM I-1629^T
Vibrio diazotrophicus	+	+	−	−	+	−	−	−	+	−	−	w	+	−	−	w	+	−	+	+	+	w		+	+	+	+	25	48	2	244	ATCC 33466^T
Vibrio ezurae	−	−	−	−	−				+	−	−	+	+	−	−	+	−		−	−	+	+		−		+	+	25	48	2		LMG 20536^T
Vibrio fluvialis	+	+	−	−	+	−	−	w	+	−	+	+	+	−	−	−	+	−	+	+	+	+		+	+	+	+	25	48	2	244	NCTC 11327
Vibrio fluvialis	+	+	−	−	+	−	−	−	+	−	+	+	+	−	−	−	+	−	−	+	+	+		+	+	+	+	25	48	0.85	496	MEJ 311
Vibrio fluvialis	+	+	−	−	81	−	−	−	+	56	+	+	+	−	−	−	+	−	−	+	+	+		+	+	+	+	37	48	0.85	996	Lee 1981
Vibrio fortis	+	−	−	−	−	−	−	+	+	+	−	+	+	−	−	−	+	+	+	−	+	+		+	nt	+	+	25	48	2	1634	LMG 21557^T
Vibrio fortis	+	−	−	−	−	−	−	+	+	v	19	+	+	−	−	−	+	62	56	1	+	+		+	nt	+	+	25	48	2	1634	Strains
Vibrio furnissii	+	+	−	−	63	−	−	−	88	75	+	+	+	−	−	−	+	−	−	+	+	+		+	+	+	+	37	48	0.85	1504	
Vibrio furnissii	+	+	−	−	w	−	−	−	+	w	−	+	+	−	−	−	+	−	−	+	+	+		+	+	+	+	25	48	0.85	244	ATCC 11218
Vibrio furnissii	+	+	−	−	+	−	−	−	+	−	+	+	+	−	−	−	+	−	−	+	+	+		+	+	+	+	25	48	0.85	496	ATCC 35016^T
Vibrio furnissii ATCC 35016^T	+	+	−	−	+	−	−	−	−	−	−	+	+	−	−	−	+	−	−	+	+	+		+	+	+	+	25	48	2	244	ATCC 35016^T
Vibrio gallaecicus	−	−	−	−	−	−	−	−	+	−	+	+	+	−	−	−	33	−	66	−	+	+		+	−	+	+	25	48	0.85	144	LMG 24045^T
Vibrio gigantis	−	+	−	−	−	−	−	−	+	−	+	+	+	−	−	−	−	+	+	−	+	+		+		+	+	20	48	2	1010	LMG 22741^T
Vibrio gazogenes		−	−	−	+	−	−		−	−	+	+	+	−	+	−	+	v		+	−	−		+		+	+	25	48	2	518, 686	ATCC 29988^T
Vibrio harveyi	+	−	+	+	−	−	−	−	+	−	+	+	+	−	−	−	−	−	+	−	+	+		+		+	+				1042	ATCC 14126
Vibrio harveyi	+	−	+	+	+	−	−	−	+	−	+	+	+	−	−	+	−	−	+	+	+	+		+		+	+				1042	ATCC 25919
Vibrio harveyi	−	−	+	+	+w	−	−	−	+	−	−	+	+	−	−	−	+	−	+	−	+	+		+		+	+	25	48	2	1228	ATCC 14129
Vibrio harveyi	−	−	+	+	+	−	+	−	+	−	v	+	+	−	−	−	+	−	+	−	+	+		+		+	+	25	48	2	1219	ATCC 35084^T
Vibrio harveyi	−	−	+	+	w	−	−	−	+	−	−	+	+	−	−	−	−	−	+	−	+	+		+		+	+	25	48	2	1219	Abalone
Vibrio (carchariae) harveyi	−	−	+	+	+	−	+	−	+	−	−	+	+	−	−	−	+	−	+	−	+	+		+	+	+	+	25	48	2	244,1830	ATCC 35084
Vibrio harveyi	−	−	+	+	w	−	+	−	+	−	−	+	+	−	−	−	+	−	+	−	+	+		+		+	+	25	48	2	1228	ATCC 35084^T
Vibrio harveyi	−	−	+	+	w	−	−	−	+	−	−	+	+	−	−	−	+	−	+	−	+	+		+		+	+	25	48	2	1228	ATCC 14129^T
Vibrio harveyi	−	−	+	−	w	−	−	−	+	w	−	+	+	−	−	−	−	−	+	−	+	+		+		+	+	25	48	2	1228	Abalone
Vibrio harveyi	v	−	+	+	+	−	v	−	+	−	v	+	+	−	+	−	+	−	+	−	+										1043	ATCC 35084
Vibrio harveyi	−	−	+	+	+	−	+	−	+	−	−	+	+	−	+	−	+	−	+	−	+	+		+	−	+	+	25	48	2	592	DN01
Vibrio halioticoli	−	−	−	−	−	−	−	−	−	−	−	w	+	−	−	−	−	−	−	−	+	+		−	−	w	w	25	48	2	244	IAM 14596^T

Continued

Table 3.24. Continued.

Tests	onpg	ADH	LDC	ODC	cit	H_2S	ure	TDA	ind	vp	G	glu	man	inos	sor	rha	suc	mel	amy	arab	ox	NO_2	N_2	mot	MCA	O	F	Temp	Time	Inoc	Ref	Strain
Vibrio halioticoli	–	–	–	–	–	–	–	–	–	–	–	w	+	–	–	–	–	–	–	–	+	+		–	–	w	w	25	48	2	244	AHLDA 1734
Vibrio hangzhouensis		–	–	–	+	+	–	–	+	–	+	+	+			–	+				+	+		+	+	+	+	30	48	1	1818	CN83^T
Vibrio hepatarius	–	–	–	–	–	–	–	+	+	+	–	+	+	–	–	–	+	–	+	–	+	+		+	nt	+	+	25	48	2	1634	LMG 20362^T
Vibrio hepatarius		+	–	–				–	+		+		+				+	–	+	–		+									738	LMG 20362^T
Vibrio hispanicus		+	–	–	+	–	–	–	+	–	–	+	+								+	+		+		+	+	28	48	2	636	LMG 13240^T
Vibrio ichthyoenteri	–	–	–	–	–	–	–	–	–	–	–	+	–	–	–	–	+w	–	–	–	+	+		+		+	+	28	48	2	804	IFO 15847
Vibrio ichthyoenteri	–	–	–	–	–	–	–	–	–	–	–	+	+	–	–	–	–	–	–	–	+	+		+	–	+	+	25	48	2	244	DSMZ 14397
Vibrio inusitatus	–	–	–	–	–				–	–	–	+	+	–	–	–	–			–	+	+		–		+	+	25	48	2	1485	
Vibrio kanaloae	+	+	–	–	–	–	–	+	+	+	+	+	+	–	–	–	+	–	–	+	+	+		+		+	+	25	48	2	1635	LMG 20539^T
Vibrio kanaloae	+	+	–	–	–	–	–	+	+	–	+	+	+	–	–	–	+	–	+	–	+	–		+		+	+	25	48	2	524	LMG 20539^T
Vibrio lentus CECT 5110	+	w	–	–	–	–	–	–	+	–	+	w	+	–	–	–	–	–	–	–	+	–		–	–	+	+					CECT 5110
Vibrio lentus	+	–	–	–	–	–	–	–	+	–	+	+	+	–	–	–	–	–	–	–	+	–		w	–	+	+	25	48	2	524	CIP 107166^T
Vibrio mangrovi		–	–	–		–	–		–	+		+	25	48	2		+	+	+	+	–	+		+		+	+				1373	
Vibrio maritimus	+	+	–	–	–	–	–	–	+	–	+	+	+	+	–	–	+	–	+	–	+	+		+		+	+	25	48	2	321	R-40493^T
Vibrio marisflavi	+	+	+	–	–	–	–	–	–	–	+	+	28	48	2		+	+	+	–	+	+		+		+	+	28	48	1.5	1760	WH134^T
Vibrio mediterranei	+	+	+	–	–	–	–	–	+	–	+	+	+	+	+	–	+	–	+	–	+	+		+	–	+	+	24	48	2	244	AHLDA 1733
Vibrio mediterranei	w	+	–	–	–	–	–	–	+	–	s	+	+	+	+	–	+	w	+	–	+	+		+	–	+	+	24	48	2	244	CECT 615
Vibrio (shilonii) mediterranei	+	+	+	–	+	–	–	–	+	–	–	+	+	+	+	+	+	+	+	–	+	+		–	–	+	+	25	48	2	244	ATCC BAA-91
Vibrio metschnikovii	+	+	+	–	+	–	–	w	+	+	+	w	w	+	–	–	+	–	–	–	–	–		+	–	+	+	25	48	2	244	NCTC 8443^T
Vibrio mimicus	+	–	+	+	+	–	–	–	+	–	+	+	+	–	–	–	–	–	–	–	+	+		+	+	+	+				326, 412	
Vibrio mimicus	+	–	+	+	–	–	–	–	+	–	+	+	+	–	–	–	–	–	–	–	+	+		+	+	+	+				326	
Vibrio mimicus	+	–	+	+	w	–	–	–	+	–	+	+	+	–	–	–	–	–	–	–	+	+		+	+	+	+	25	48	2	244	ATCC 33653^T
Vibrio mimicus	+	–	+	+	+	–	–	–	+	–	+	+	+	–	–	–	–	–	–	–	+	+		+	+	+	+	25	48	0.85	244	ATCC 33653^T
Vibrio mytili	+	–	–	–	–	–	–	+	+	–	–	+	+	–	–	–	+	–	+	–	+	+		+	–	+	+	25	48	2	1368	CECT 632^T
Vibrio mytili	+	+	–	–	–	–	–	–	–	–	–	w	+	–	–	–	+	–	+	+	+	+		s	–	+	+	25	48	2	244	CIP 103929
Vibrio natriegens	–	–	–	–	w	–	–	–	–	–	w	+	+	–	–	+	+	–	+	+	+	+		+		+	+	24	48	2	244	ATCC 14048
Vibrio navarrensis		–	–	–	s	–	–	–	+	–	+	+	+	–	–	–	+	–	+	–	+	+		+		+	+	30	48	2	1687	CIP 103381
Vibrio navarrensis	–	–	–	–	–	–	–	–	+	–	–	w	+	–	–	–	+	–	+	–	+	+		+	+	+	+	25	48	2	244	CIP 103381
V. navarrensis spp. *pommerensis*		–	–	–	–	–	–	–	+	–	+	+	+	–	–		+				+	+		+		+	+	25	48	2	851	CH-291^T
Vibrio neonatus	–	–	–	–	–				+	–	–	+	+	–	–	+	–		–	–	+	+		–		+	+	25	48	2		LMG 20536^T
Vibrio neptunius	–	86	–	–	+	–	–	+	+	+	+	+	–	–	–	–	+	–	–	–	+	+		+		+	+	25	48	2	1632	LMG 20536^T
Vibrio neptunius	–	+	–	–	–	–	–		–	–	+	+	–	–	–	–	+	–	–	–	+	–		+		+	+	25	48	0.85	1357	PP-145.98

Vibrio neresis	–	–	–	–	–	–	–	+	+	–	–	+	–	–	–	–	+	–	–	–	+	+		+		+	+	25	48	2	1368	LMG 3895^T
Vibrio nereis	–	+	–	–	+	–	–	–	+	–	–	w	–	–	–	–	+	–	–	–	+	+		+		+	+	25	48	2	244	ATCC 25917
Vibrio nigripulchritudo	+	–	–	–	+	–	–	–	+	–	+	+	–	–	–	–	–	+	+	–	+	w		w	–	+	+	25	48	2	244	CIP 103195
Vibrio ordalii	–	–	–	–	–	–	–	–	–	–	+	+	+	–	–	–	+	–	–	–	+	–		+	–	+	+				878	DF 3K
Vibrio ordalii	–	–	–	–	–	–	–	–	–	–	–	+	w	–	–	–	+	–	–	–	+	–		+	–	+	+	25	48	0.85	244	ATCC 33509
Vibrio ordalii	–	–	–	–	–	–	–	–	–	–	+	+	+	–	–	–	+	–	–	–	+	–		+	–	+	+	25	48	2	1531	Type strain
Vibrio ordalii	–	–	–	–	–	–	–	–	–	–	–	+	–	–	–	–	+	–	–	–	+	–		+	–	+	+	25	48	2	1531	Chilean strains
Vibrio ordalii	–	–	–	–	–	–	–	–	–	–	–	–	–	–	–	–	+	–	–	–	+	–		+	–	+	+	25	48	2	1531	Chilean strains
Vibrio orientalis	–	+	+	–	–	–	–	–	+	–	+	+	+	–	–	–	+	–	+	–	+	+		+	–	w	w	25	48	2	244	NCMB 2195^T
Vibrio orientalis	–	+	+	–	–	–	–	w	+	–	+	+	+	–	–	–	+	+	+	–	+	+		+	–	+	+	25	48	2	244	ATCC 33935
Vibrio orientalis		–	+	–				–			+	'	+				+	–	+	–								28	48	1.5	738	ATCC 33935
Vibrio owensii	–	–	+	+	+	–	–	+	+	–	+	w	+	–	–	–	w	–	w	–	+	–		+		+	+	28	48	2	244,270	DY05^T
Vibrio owensii	+	–	+	+	+	–	–	+	+	–	–	+	+	–	–	–	+	–	+	–	+	–		+		+	+	25	48	2	244,270	47666-1
Vibrio pacinii	+	+	–	–	w	–	–	+	–	+	\	+	–	–	–	–	+	33	+	–	+	+		+		+	+	25	48	1.5	638	LMG 1999^T
Vibrio pacinii	–	+	–	–	w	–	–	+	–	+	\	+	–	–	–	–	+	33	+	–	+	+		+		+	+	28	48	1.5	638	Salmon
Vibrio pacinii	+	+	–	–	+	–	–	+	–	–	\	+	–	–	–	–	+	33	+	–	+	+		+		+	+	28	48	1.5	638	Seabass
Vibrio parahaemolyticus	21	–	+	+		–	–	–	–	–	–	+	+	–	–	–	–		+	77	+	+				+	+				*,1164	
Vibrio parahaemolyticus	–	–	+	+	w	–	–	–	+	–	–	+	+	–	–	–	–	–	–	w	+	+		+	+w	+	+	25	48	2	244,	ATCC 43996
Vibrio parahaemolyticus	–	–	+	+	+	–	–	–	+	–	–	+	+	–	–	–	–	–	–	+	+	+			+w	+	+				878	
Vibrio parahaemolyticus	–	–	+	+	–	–	+	–	+	–	–	+	+	–	–	–	–	–	–	+	+	+		+	+	+	+	25	48	2	19	ATCC 17802^T
Vibrio parahaemolyticus	–	–	+	+	+	–	–	w	+	–	–	+	+	–	–	–	–	–	+	+	+	+		+	+	+	+	25	48	2	244	ATCC 17802^T
Vibrio pectenicida	–	–	–	–	+	–	–	–	–	–	–	+	–	–	–	–	–	–	–	–	+	+		+	nt	+	+			2	970	CIP 105190^T
Vibrio pectenicida	–	–	–	–	–	–	–	–	–	w	–	w	–	–	–	–	–	–	–	–	+	+		–	–	w	w	25	48	2	244	ATCC 700783
Vibrio (Listonella) pelagius I	+	–	–	–	–	–	–	–	–	–	–	w	+	–	–	–	+	–	–	–	+	–		+	–	w	w	24	48	2	244	ATCC 25916^T
Vibrio (Listonella) pelagius II	+	–	–	–	–	–	–	–	+	–	+	+	+	–	–	–	+	–	–	–	+	+		+	–	w	w	24	48	2	244	ATCC 33784
Vibrio penaeicida	+	–	–	–	+	–	nt	–	–	–	+	+	–	–	–	–	–	–	–	–	+	+		+	nt	+	+	25			803	KH-1^T
Vibrio penaeicida	+	–	–	–	+	–	nt	–	50	–	+	+	–	–	–	–	–	–	–	–	+	+		+	nt	+	+	25			803	Strains
Vibrio penaeicida	+	–	–	–	+	–	–	–	+	–	+	+	–	–	–	–	–	–	–	–	+	+		+	nt	+	+	25			371	Noumea strain
Vibrio penaeicida	+	+	–	–	+	–	–	–	+	+	+	–	–	–	–	–	–	–	–	–	–	–		–t	–	+	+	25	48	2	244	DSMZ 14398
Vibrio penaeicida	+	+	–	–	+	–	–	–	+	–	+	+	+	+	–	–	+	–	+	–	–	–		+	+	+	+	25	48	2	244	DSMZ 14398
Vibrio plantisponsor	+	+	–	–	+	–	–	–	+	–	–	+	+	–	–	+	+	–	+	+	+	+		+	nt	+	+	28	48	2.5	1372	MSSRF60^T

Continued

Table 3.24. Continued.

Tests	onpg	ADH	LDC	ODC	cit	H_2S	ure	TDA	ind	vp	G	glu	man	inos	sor	rha	suc	mel	amy	arab	ox	NO_2	N_2	mot	MCA	O	F	Temp	Time	Inoc	Ref	Strain
Vibrio pomeroyi	+	+	–	–		–	–	–	+	+	+	+	+	–	–	–	+	–	+	–	+	+		+		+	+	25	48	2	1635	LMG 20537^T
Vibrio pomeroyi	+	50	–	–		–	–	–	+	83	83	+	+	–	33	–	+	33	+	–	+	+		+		+	+	25	48	2	1635	6 strains
Vibrio pomeroyi	+	+	–	–	–	–	–	+	+	–	+	+	+	–	–	–	–	–	+	–	+	–		+	–	+	+	25	48	2	524	LMG 20537^T
Vibrio ponticus	–	+	–	–	+	–	–	–	+	–	–	+	–	–	–	–	+	–	–	+	+	+		+	–	+	+	25	48	2	1816	RP30
Vibrio ponticus		–	+	–	+				+	–	–	+					+				+	+		+							1080	CECT 5869^T
Vibrio proteolyticus	–	+	+	–	+	–	–	–	+	+	+	+	+	–	+	–	–	–	–	–	+	+		+	w	+	+	25	48	2	244	AHLDA 1735
Vibrio proteolyticus	–	+	+	–	+	–	–	+	+	+	+	+	+	–	+	–	–	–	+	–	+	+s		+	–	+	+	25	48	2	244	CIP 102892
Vibrio quintilis	–	+	–	–					–		–			–	–	–		–		–	–	–						25	48	2	1065	CECT 7734^T
Vibrio rarus	–	–	–	–	–				+	–	–	+	+	–	–	–	–			–	+	+		–		+	+	25	48	2	1485	
V. rumoiensis		–			+	–			–	–	–	+		–		–					+	+		–		+	+			2		
Vibrio rotiferianus	+	–	+	+	–	–	83	+	+	–	+	+	+	–	–	–	+	+	+	+	+	nr		+		+	+	25	48	2	637	LMG 21460^T
Vibrio sagamiensis	–	–	–	–	–	–	–	–	w		+	–	–	–	–		–	–	–	–	+	+		+	nt	+	+	20	48	2	1837	LC2-047^T
Vibrio scophthalmi	–	90	–	–	–	–	–	–	–	–	–	+	–	–	–	–	+	–	–	–	+	+		+	nt	+	+	25	48	1.5	293	CECT 4638
Vibrio scophthalmi	–	–	–	–	–	–	–	–	–	–	–	+	–	–	–	–	+	–	–	–	+	+		+	–	+	+	25	48	1.5	523	CECT 4638
Vibrio scophthalmi	–	w	–	–	–	–	–	–	–	–	–	+	–	–	–	–	+	–	–	–	+	+		+	–	+	+	25	48	2	244	CECT 4638
Vibrio shilonii ATCC 1272	+	+	–	–	–	–	–	–	nt	nt	–	+	+	–	–	+	+	+	+	–								25	24	2	244	ATCC 1272
Vibrio sinaloensis	+	+	–	–	–	–	–	+	–	–	+	+	–	–		–	+	–	–		+	+		+		+	+	25	48	2.5	639	CAIM 797^T
Vibrio sinaloensis	+	+	–	–				+	+		+		+				+	+	+	–		+						28	48	1.5	738	DSM 21326^T
Vibrio splendidus biovar I	+	+	–	–	–	–	–	–	+	–	+	+	+	–	–	–	+	–	–	–	+	+		+	–	+	+	22	48	0.85	523	ATCC 33125^T
Vibrio splendidus	+	88	–	–	22	–	–	–	+	25	75	+	+	–	–	–	–	38	+	–	+	+		w	–	+	+	25	48	2	590	Turbot
Vibrio splendidus	+	+	–	–	–	–	–	–	+	–	+	+	+	–	–	–	–	+	+	–	+	+		w	–	+	+	25	48	2	590	AH5, turbot
Vibrio splendidus	+	+	–	–	–	–	–	–	+	–	+	+	+	–	–	–	–	+	+	–	+	+		w	–	+	+	25	48	2	590	A515, scallop
Vibrio splendidus	+	+	–	–	+	–	–	–	+	–	+	+	+	–	–	–	–	+	+	–	+	+		w	–	+	+	25	48	2	590	VS1, turbot
Vibrio splendidus	+	+	–	–	–	–	–	–	+	+	+	+	+	–	–	–	–	+	+	–	+	+		w	–	+	+	25	48	2	590	VS4, turbot
Vibrio splendidus	+	+	–	–	–	–	–	–	+	–	–	+	+	–	–	–	–	–	+	–	+	+		w	–	+	+	25	48	2	590	VS5, turbot
Vibrio splendidus	+	+	–	–	–	–	–	–	+	–	+	+	+	–	–	–	–	–	+	–	+	+		w	–	+	+	25	48	2	590	VS6, VS11, turbot
Vibrio splendidus	+	+	–	–	–	–	–	–	+	+	–	+	+	–	–	–	–	–	+	–	+	+		w	–	+	+	25	48	2	590	VS10, turbot
Vibrio splendidus	+	–	–	–	–	–	–	–	+	–	+	+	+	–	–	–	–	–	+	–	+	+		w	–	+	+	25	48	2	590	VS9 turbot
Vibrio splendidus	+	–	–	–	–	–	–	–	+	–	+	+	+	–	–	–	–	+	+	–	+	+		w	–	+	+	25	48	2	524,1009	LMG 4042^T
Vibrio splendidus	+	–						–			+													w	–	+	+	25	48	2	1010	LMG 19031^T
Vibrio splendidus	–	–	–	–	–	–	–	–	+	–	–	+	+	–	–	–	+	+	+	–	+	+	–	–	–	+	+	25	48	2	965	Oyster

Vibrio splendidus	+	–			–		–		–		+	–					–				+	+		+		+	+	25	48	2	879	Mussel
Vibrio splendidus I	+	–	–	–	–	–	–	–	+	–	+	+	+	–	–	–	–	+	+	–	+	+	–	–	–	+	+	25	48	2	244	CRL 40
Vibrio splendidus II	–	s	–	–	–	–	–	w	+	–	–	+	+	–	–	–	–	–	+	–	+	+	–	w	–	+	+	25	48	2	244	ATCC 33789
Vibrio splendidus II	–	–	–	–		–			+	–	+	+					–		+		+	+		+		+	+				644	TA2
Vibrio tapetis	+	–	–	–	–	–	–	–	+	–	+	+	–	–	–	–	–	–	+	–	+	+		+		+	+	25		2	201, 285	
Vibrio tapetis CIP 104856	+	–	–	–	w	–	–	–	–	–	–	w	–	–	–	–	–	–	–	–	+	+		–	–			25	48	2	244	CIP 104856
Vibrio tapetis	+	–	–	–	–	–	–	–	+	–	+	+	–	–	–	–	–	–	–	–	+	+		+				20	48	SW	1052	Wedge sole
Vibrio tapetis	+	–	–	–	–	–	–	–	+	–	+	+	+	–	–	–	–	–	–	–	+	+		+				25	48	0.85	1393	Halibut
V. tapetis ssp. *britannicus*	+	–	–	–	+	–	–	–	+	–	+	+	+	–	–	–	–	–	–	+	+	+		+	–	–	–	25	72	1	103	
Vibrio tasmaniensis	–	–	–	–	–	–	–	+	+	+	–	+	+	–	–	–	–	–	+	–	+	+		+		+	+	25	48	2	1637	LMG 20012^{T}
Vibrio tasmaniensis	–	–	–	–	–	–	–	–	+	–	–	+	+	–	–	–	–	–	–	–	+	+		+		+	+	25	48	2	524, 1009	LMG 20012^{T}
Vibrio tubiashii	+	–	–	–	–	–	–	+	+	–	+	+	+	–	–	–	+	–	+	–	+	+		+	–	+	+	25	48	2	668, 1368	LMG 10936^{T}
Vibrio tubiashii	+	+s	–	–	–	–	–	–	+	–	–	+	+	–	–	–	+	–	+	–	+	+		+	–	+	+	25	48	2	244	
Vibrio tubiashii	+	+	–	–	–	–	–	–	+	–	+	+	+	–	–	–	+	+	+	–	+	+		+	–	+	+	25	48	2	244	NCIMB 1340
Vibrio tubiashii	+	+	–	–				+	+		+		+				+	+	+	–		+						28	48	1.5	738	ATCC 19109^{T}
Vibrio variabilis	+	+	–	–	–	–	–	–	+	–	+	+	+	–	–	+	+	–	+	–	+	+		+		+	+	25	48	2	321	R-40492^{T}
V. vulnificus biotype 1	+	–	+	+	–	–	–	–	+	–	+	+	+	–	–	–	–	–	+	–	+	+								f	50	
V. vulnificus biotype II	+	–	+	–	+	–	–	–	–	–	+	+	–	–	–	–	–	–	+	–	+										50	
V. vulnificus biotype II	+	–	+	+	+	–	–	–	–	–	+	+	–	–	–	–	–	–	+	–	+										50	
V. vulnificus biotype 2 sero E	+	–	+	+	+	–	–	–	–	–	+	+	+	–	–	–	–	–	+	–	+	+						25	48	g	50	
V. vulnificus biotype 2 sero E	+	–	+	+	+	–	–	–	–	–	+	+	–	–	–	–	–	–	+	–	+	+								a	180	ATCC 33187
V. vulnificus biotype 2 sero E	+	–	+	–	–	–	–	–	–	–	+	+	–	–	–	–	–	–	+	–	+	+								b	180	ATCC 33149
V. vulnificus biotype 2 sero E	+	–	+	–	+	–	–	–	–	–	+	+	–	–	–	–	–	–	+	–	+	+								c	180	
V. vulnificus biotype 2 sero E	+	–	+	–	–	–	–	–	–	–	+	+	–	–	–	–	–	–	–	–	+	+								d	180	
V. vulnificus biotype 2 sero E	–	–	+	–	–	–	–	–	–	–	+	+	–	–	–	–	–	–	+	–	+	+								e	180	
V. vulnificus	+	–	+	+	+	–	–	–	+	–	+	+	+	–	–	–	–	–	w	+									48		1075	WF8A1110
V. xiamenensis	+	–	–	–	–	–	–	–	–	–	–	+	+	–	–	+	+	–	+	–	+	+		+		+	+	30	24	2	581	G21^{T}
V. xuii	–	+	–	–	–	–	–	+	+	+	–	+	+	–	–	–	+	–	+	+	+	+		+		+	+	25	48	2	1632	LMG 21346^{T}
Yersinia intermedia	+	–	–	+	–	–	+	–	+	–	–	+	+	v	+	+	+	+	+	v	–										68	
Yersinia ruckeri	+	–	–	+	–	–	–	–	–	–	–	+	+	–	–	–	–	–	–	–	–										1425	
Yersinia ruckeri	+	–	–	+	–	–	–	–	–	–	–	+	+	–	+	–	–	–	–	–	–										1582	

Continued

Table 3.24. Continued.

Tests	onpg	ADH	LDC	ODC	cit	H_2S	ure	TDA	ind	vp	G	glu	man	inos	sor	rha	suc	mel	amy	arab	ox	NO_2	N_2	mot	MCA	O	F	Temp	Time	Inoc	Ref	Strain
Yersinia ruckeri	+	–	–	+	–	–	–	–	–	–	–	+	+	–	+	–	–	–	+	–	–										1425	
Yersinia ruckeri	+	–	–	+	–	–	–	–	–	–	–	+	+	–	–	–	–	–	–	–	–										1476, 1582	
Yersinia ruckeri	+	–	–	+	–	–	–	–	–	–	–	+	+	+	+	–	–	–	–	–	–										1425	
Yersinia ruckeri	+	–	–	+	–	–	–	–	–	+	–	+	+	–	–	–	–	–	–	–	–										1425	
Yersinia ruckeri	+	–	–	+	–	–	–	–	–	+	–	+	+	–	+	–	–	–	–	–	–										1425, 1582	
Yersinia ruckeri	+	–	–	+	–	–	–	–	–	+	+	+	+	–	–	–	–	–	–	–	–										1425	
Yersinia ruckeri	+	–	–	+	–	–	–	–	–	+	+	+	+	–	+	–	–	–	–	–	–										1425	
Yersinia ruckeri	+	–	+	+	–	–	–	–	–	–	–	+	+	–	–	–	–	–	–	–	–	+		+	+	+	+	25	48	0.85	367, 1476	
Yersinia ruckeri	+	–	+	+	–	–	–	–	–	–	+	+	+	–	+	–	–	–	–	–	–	+		–	+	+	+	25	48	0.85	1643	
Yersinia ruckeri	+	–	+	+	–	–	–	–	–	–	+	+	+	–	–	–	–	–	–	–	–	+		–	+	+	+	25	48	0.85	1643	
Yersinia ruckeri	+	–	+	+	–	–	–	–	–	+	–	+	+	–	–	–	–	–	–	–	–	+		+	+	+	+	25	48	0.85	244, 402	AHDA 1313
Yersinia ruckeri	+	–	+	+	–	–	–	–	–	+	–	+	+	–	+	–	–	–	–	–	–	+		+	+	+	+	25	48	0.85	367	
Yersinia ruckeri	+	–	+	+	–	–	–	–	–	+	+	+	+	–	+	–	–	–	–	–	–	+		+	+	+	+	22	48	0.85	367, 420	
Yersinia ruckeri	+	–	+	+	–	–	–	–	–	+	+	+	+	–	–	–	–	–	–	–	–	+		–	+	+	+	25	48	0.85	1643	
Yersinia ruckeri	+	–	+	+	+	–	–	–	–	–	+	+	+	–	–	–	–	–	–	–	–										1425	
Yersinia ruckeri	+	–	+	+	+	–	–	–	–	+	+	+	+	–	V	–	–	–	–	–	–	+		+	+	+	+	25	48	0.85	411	
Zobellia galactanivorans	+	–	–	–	–	–	–	–	–	–	+	+	+	–	–	+	+	–	–	+	+	+		–		+	–	30	7d	1.5	119	
Zobellia uliginosa	+	–	–	–	–	–	–	–	–		+	+	–	–	–		+				+	+		–		+	–	30	7d	1.5	119	

onpg=β-galactosidase, ADH=arginine dihydrolase, LDC=lysine decarboxylase, ODC=ornithine decarboxylase, cit=citrate, H_2S=production of hydrcgen sulfide, ure=urease, TDA=tryptophane deaminase, ind=indole, vp=Voges–Proskaüer, G=gelatin, glu=glucose, man=mannitol, inos=inositol, sor=sorbitol, rha=rhamnose, suc=sucrose, mel=melibiose, amy=amygdalin, arab=arabinose, ox=oxidase, NO_2=nitrate reduction, N_2=reduction to nitrite gas, mot=motility, MCA=growth on MacConkey, O=oxidative fermentation, F=fermentation, Temp=temperature of incubation, Time=time of incubation, h=hours, d=days, Inoc=Inoculum, %NaCl=final NaCl concentration in inoculum, w=weak, +=positive reaction, –=negative reaction, s=slow reaction, numbers indicate percentage of positive strains. a=note that *A. hydrophila* strains are considered to be negative for sorbitol, b=The incubation time was not stated for this result. It is important to incubate for 48 h to ensure correct LDC result. s=slow reaction, –t=negative tube reaction, V=variable reaction, *from API database, Ref=reference number, s=slow reaction.

Table 3.25. API 20E database numbers (organisms listed alphabetically).

Organism	Strain	API 20E number	No. of strains	Notes	Ref
		0000004	1		68
Actinobacillus delphinicola		200500416	2	Dolphin, whale	548
Actinobacillus scotiae	NCTC 12922^T	111500416	1	Porpoises	550
Actinobacillus scotiae	strains	101500416	2	Porpoises	550
Aeromonas allosaccharophila		724613657	1		68
Aeromonas allosaccharophila		724616657	1		68
Aeromonas allosaccharophila		724617657	1		68
Aeromonas bestiarum		704713757	5	Carp	935
Aeromonas caviae	ATCC 15468^T	324612657			42, 495
Aeromonas encheleia		304613457			494
Aeromonas fluvialis	CECT 7401^T	124002457	1		37
Aeromonas hydrophila	ATCC 7966^T	704712657	1		495
Aeromonas hydrophila	ATCC 7966^T	304512657	1	Important to incubate for 48 h for correct LDC result	1476
Aeromonas hydrophila	ATCC 14715	324713757	1		669
Aeromonas jandaei		724714457	2		244, 276
Aeromonas rivuli	CECT 7518^T	300616757	2	Creek water	536
Aeromonas salmonicida		200610417			928
Aeromonas salmonicida	AHLDA 1334	204402417	1	'Atypical' strain, imported goldfish (Western Australia)	244
Aeromonas salmonicida achromogenes		004412417			669
Aeromonas salmonicida achromogenes	NCIMB 1110	004512417			968
Aeromonas salmonicida masoucida		304512517			669
Aeromonas salmonicida salmonicida		200610417			1476
Aeromonas salmonicida salmonicida		300610517			669
Aeromonas salmonicida salmonicida		600610417			1476
Aeromonas salmonicida salmonicida	NCIMB 1102	700610517			968
Aeromonas salmonicida salmonicida		000500417		Lamprey	513
Aeromonas sanarellii	CECT 7402^T	306612757	1	Human strain	36
Aeromonas schubertii	ATCC 43700^T	720600457	2		278
Aeromonas schubertii	AMC 4396-87	720700457	1		278
Aeromonas sobria	CIP 7433^T	264416417			4
Aeromonas sobria	CIP 7433^T	264516417			4
Aeromonas sobria	CIP 7433^T	264416457			4
Aeromonas sobria	CIP 7433^T	264516457			4
Aeromonas taiwanensis	CECT 7403^T	326612757		Human strain	36

Continued

Table 3.25. Continued.

Organism	Strain	API 20E number	No. of strains	Notes	Ref
Aeromonas trota	ATCC 49657^T	724610557	1		277, 789
A. veronii biovar Sobria		704612457			4
A. veronii biovar Sobria		704612657			4
A. veronii biovar Sobria		704712457			4
A. veronii biovar Sobria		704712657			4
A. veronii biovar Sobria		724612457			4
A. veronii biovar Sobria		724612657			4
A. veronii biovar Sobria		724712457			4
A. veronii biovar Sobria		724712657			4
A. veronii biovar Sobria		744612457			4
A. veronii biovar Sobria		744612657			4
A. veronii biovar Sobria		744712457			4
A. veronii biovar Sobria		744712657			4
A. veronii biovar Sobria		764612457			4
A. veronii biovar Sobria		764612657			4
A. veronii biovar Sobria		764712457			4
A. veronii biovar Sobria		764712657			4
A. veronii biovar Veronii	ATCC 35623	114612557	1	Faeces, diarrhoea, USA	723
A. veronii biovar Veronii	ATCC 35604^T	114712457	1	Sputum, drowning victim, USA	723
A. veronii biovar Veronii	ATCC 35606	134612557	1	Maxillary sinus, USA	723
A. veronii biovar Veronii	ATCC 35605	134712557	1	Wound isolate, human, USA	723
A. veronii biovar Veronii	ATCC 35622, 1067–84	514712557	2	Faeces, diarrhoea, USA	723
A. veronii biovar Veronii	0935–84, 1068–84	534612557	2	Faeces, diarrhoea, USA	723
A. veronii biovar Veronii	0265–84	534712557	1	Endotracheal tube, drowning victim, USA	723
A. veronii biovar Veronii		534712557	12	Various strains	5, 723, 935
Aliivibrio finisterrensis	C 13.19	70140045.	1		143
Aliivibrio finisterrensis	CMJ 11.1^T	70140445.	1		143
Aliivibrio finisterrensis	C 17.13	70141245.	1		143
Aliivibrio fischeri	ATCC 25918	400500556	1	1.5% NaCl	1641
Aliivibrio fischeri	ATCC 7744^T	404450656	1	0.85% NaCl	244
Aliivibrio fischeri	ATCC 7744^T	404455656	1	0.85% NaCl	244
Aliivibrio fischeri	ATCC 7744^T	421400456	1	2 % NaCl	244
Aliivibrio fischeri	ATCC 7744^T	421400556	1	2 % NaCl	244
Aliivibrio fischeri	ATCC 7744^T	421500456	1	2 % NaCl	244
Aliivibrio fischeri	ATCC 7744^T	421500556	1	2 % NaCl	244

Aliivibrio fischeri	ATCC 7744^T	421410456	1	2 % NaCl	244
Aliivibrio fischeri	ATCC 7744^T	421410556	1	2 % NaCl	244
Aliivibrio fischeri	ATCC 7744^T	421510456	1	2 % NaCl	244
Aliivibrio fischeri	ATCC 7744^T	421510556	1	2 % NaCl	244
Alivibrio logei	ATCC 29985^T	000000440	1	2 % NaCl	244
Aliivibrio wodanis	ATCC BAA-104^T	0046124..	1		143
Aliivibrio wodanis	ATCC BAA-104^T	0002004..	1		1834
Aliivibrio wodanis	ATCC BAA-104^T	0002006..	1		1834
Aliivibrio fischeri	ATCC 7744^T	621510556	1	2 % NaCl	244
Aliivibrio fischeri	ATCC 7744^T	621410556	1	2 % NaCl	244
Aliivibrio fischeri	ATCC 7744^T	621000456	1	2 % NaCl	244
Burkholderia cepacia		0004004	3/10		64
Burkholderia cepacia		0206006	1/10		64
Burkholderia cepacia		4304004	4/10		64
Burkholderia cepacia		5304004	1/10		64
Burkholderia pseudomallei		0006727	1/91		64
Burkholderia pseudomallei		0206706	1/91		64
Burkholderia pseudomallei		2006704	1/91		64
Burkholderia pseudomallei		2006706	4/91		64
Burkholderia pseudomallei	NCTC8016	2206707	1/91		64
Burkholderia pseudomallei		2006707	5/91		64
Burkholderia pseudomallei		2006726	1/91		64
Burkholderia pseudomallei		2006727	22/91		64
Burkholderia pseudomallei		2202704	1/91		64
Burkholderia pseudomallei		2202706	3/91		64
Burkholderia pseudomallei		2206704	4/91		64
Burkholderia pseudomallei		2206706	21/91		64
Burkholderia pseudomallei		2206707	12/91		64
Burkholderia pseudomallei		2206727	15/91		64
Citrobacter freundii		3644152			870
Citrobacter freundii		3644153			870
Citrobacter freundii		3644552			870
Citrobacter freundii		3644553			870
Citrobacter freundii		3644572			870
Citrobacter freundii		3644573			870
Citrobacter freundii		3644172			870
Citrobacter freundii		3744152			870
Citrobacter freundii		3744153			870
Citrobacter freundii		3744172			870

Continued

Table 3.25. Continued.

Organism	Strain	API 20E number	No. of strains	Notes	Ref
Citrobacter freundii		3744173			870
Citrobacter freundii		3744552			870
Citrobacter freundii		3744553			870
Citrobacter freundii		3744573			870
Citrobacter freundii		3744573			870
Edwardsiella hoshinae		454412057			*
Edwardsiella hoshinae		474412057			*
Edwardsiella ictaluri		410400057		Channel catfish	702
Edwardsiella ictaluri		400400057		Vietnamese catfish	386
Edwardsiella ictaluri		410414057		Turkish rainbow trout	880
Edwardsiella piscicida		454400057		European eel, turbot, Korean catfish	1
Edwardsiella piscicida		474400057		NCIMB 2056 (crimson seabream)	1
Edwardsiella piscicida		474410057		Japanese eel, China	1
Edwardsiella piscicida		474410257		Marbled eel, China	1
Edwardsiella tarda		454400057		Brook trout	68, 1674
Edwardsiella tarda		474400057		Turbot (*Psetta maxima*), Europe, CECT 849^T, eel Spain, AHLDA 135	21, 244, 286
Edwardsiella tarda		414400057		Eels Spain	21
Edwardsiella tarda		434400057		Eels Spain	21
Edwardsiella tarda	AHLDA 135	474400057			244
Edwardsiella tarda – atypical		4644100..	4	Atypical strains from red seabream, Japan	1820
Enterobacter agglomerans		120712057			68
Enterovibrio calviensis	DSMZ 1337	1044105046			244
Enterovibrio norvegicus	LMG 19839^T	304400446			1627
Escherichia vulneris		10041137			*
Escherichia vulneris		10041337			*
Escherichia vulneris		30041137			*
Escherichia vulneris		30041337			*
Escherichia vulneris		50041137			*
Escherichia vulneris		50041337			*
Escherichia vulneris		70041137			*
Escherichia vulneris		70041337			*
Flavobacterium branchiophilum		0006060..			68
Flavobacterium columnare		0402004..			68
Flavobacterium columnare		140200410			1511

Flavobacterium gillisiae		200412000			1115
Flavobacterium hibernum	ATCC 51468ᵀ	100603210			1116
Flavobacterium hydatis		000612210			68
Flavobacterium johnsoniae		120300410			244
Flavobacterium macrobrachii	LMG 25203ᵀ	000200400			1514
Flavobacterium psychrophilum		0002004..			68
Flavobacterium tegetincola		00041000.			1115
Flavobacterium xanthum		04061241.			1115
Gramella marina	KMM 6048ᵀ	120706742			1205
Grimontia hollisae	CIP 101886	004000456			244
Grimontia hollisae	CIP 101886	004000656			244
Grimontia hollisae	CIP 101886	004002456			244
Grimontia hollisae	CIP 101886	004002656			244
Grimontia hollisae	CIP 101886	004400456			244
Grimontia hollisae	CIP 101886	004400656			244
Grimontia hollisae	CIP 101886	004402456			244
Grimontia hollisae	CIP 101886	004402656			244
Hafnia alvei		430410257			68
Hafnia alvei		510411257			1418
Hafnia alvei		530511357			244
Hafnia alvei	ATCC 51873	5304112..	1		402
Halomonas cupida		6100530..			68
Janthinobacterium lividum		220710452			73
Klebsiella pneumoniae		521577357			68
Moritella marina	ATCC 15381ᵀ	00060045.		Pacific Ocean	1683
Moritella viscosa	NCIMB 13485	40440045.		Atlantic salmon, Scotland	236
Moritella viscosa	NCIMB 13484	40460045.		Atlantic salmon, Scotland	236
Pantoea agglomerans	CDC 1429–71	100577357			602
Pantoea agglomerans	ATCC 14589	120516357			602
Pantoea agglomerans	NCPPB 2285	120517257			602
Pantoea agglomerans	DNA HG 14589	120517357			602
Pantoea agglomerans		120713257			602
Pantoea dispersa		120713357			602
Pantoea dispersa		120717357			602
Pasteurella multocida		014452416			878
Phocoenobacter uteri	NCTC 12872ᵀ	000100416			549
Phocoenobacter uteri	NCTC 12872ᵀ	000500416			549
Photobacterium angustum	NCIMB 1895	200702406			1641
Photobacterium damselae damselae		201400457		25°C and 35°C incubation	339

Continued

Table 3.25. Continued.

Organism	Strain	API 20E number	No. of strains	Notes	Ref
Photobacterium damselae damselae		201500457		0.85% NaCl inoculum	244
Photobacterium damselae damselae		200500457		2% NaCl inoculum	244
Photobacterium damselae damselae	LMG 7892	200500457			71
Photobacterium damselae damselae	ATCC 33539^T	201500457			1641
Photobacterium damselae damselae	NCIMB 2184	220400157			1641
Photobacterium damselae damselae	ATCC 33539^T	600400457			496
Photobacterium damselae damselae	LMG 13639	600500457	6	Various sources	71
Photobacterium damselae damselae		6015004		Redbanded seabream	963
Photobacterium damselae piscicida	P90029	001100407	1	Atypical strain	1641
Photobacterium damselae piscicida	ATCC 17911	200400406		ATCC 29687	878
Photobacterium damselae piscicida	ATCC 17911^T, 29687	200400406		Type strains, European, Japanese, USA strains	98,702,1089
Photobacterium damselae piscicida		200500406		Strains from tuna	1159
Photobacterium damselae piscicida	ATCC 17911	200500406		ATCC 29690, and other strains. Refs: 7, 22, 60, 200	595,1476,1641
Photobacterium damselae piscicida	ATCC 17911	200500407	majority	NCIMB 2058	1641
Photobacterium damselae piscicida		220500407			1641
Photobacterium iliopiscarium	PS1^T	600400456	1		1272
Photobacterium iliopiscarium	ATCC 51760	600500456			244
Photobacterium leiognathi	LMG 4228	200500017			1641
Photobacterium phosporeum	IB39	6046021..			1076
Photobacterium swingsii	CAIM 1393^T	200504756		2.5% inoculum	641
Photobacterium swingsii	CAIM 1393^T	300504756		2.5% inoculum	641
Plesiomonas shigelloides		714420457			68
Providencia rustigianii	DSM 2620	026402057			1175
Providencia rettgeri		007431057			71
Pseudomonas aeruginosa		0206006	1/18		64
Pseudomonas aeruginosa		2002004	1/18		64
Pseudomonas aeruginosa		2006004	5/18		64
Pseudomonas aeruginosa		2206004	9/18		64
Pseudomonas aeruginosa		2206006	1/18		64
Pseudomonas anguilliseptica		000200440			68
Pseudomonas anguilliseptica		020000440			176,1136,1794
Pseudomonas anguilliseptica		020200440			176,1136,1794
Pseudomonas fluorescens		0204004..	2/8		64
Pseudomonas fluorescens		2000004..	5/8		64
Pseudomonas fluorescens		220000443	1/8		64
Pseudomonas fluorescens		220000453			

P. (Pseudoalcaligenes) oleovorans		2306006..		Strain from rainbow trout	76
Pseudomonas stutzeri		0000004..	1/9		64
Pseudomonas stutzeri		0004004..	5/9		64
Pseudomonas stutzeri		0004104..	3/9		64
Rahnella aquatilis		100557317			
Rahnella aquatilis		120547257			299
Rahnella aquatilis		120557317			
Salmonella cholerasuis arizonae		560621057			68
Serratia fonticola		530475257			601
Serratia fonticola		530477257			601
Serratia fonticola		530555257			1176
Serratia liquefaciens	K2	530536057		From Atlantic salmon	1123
Serratia liquefaciens	K2	530536357		From Atlantic salmon	1123
Serratia liquefaciens	Turbot	530776217		From turbot	1724
Serratia marcescens	Perch	530752157		White perch	137
Serratia marcescens	Perch	530772357		White perch	137
Serratia plymuthica		120732257			68
Serratia plymuthica	Trout	100736357		Rainbow trout	76
Serratia plymuthica	Trout	120736357		Rainbow trout	1222
Shewanella algae	IAM 14159^T	050200453			886
Shewanella algae	IAM 14159^T	050600453			1239
Shewanella algae	ATCC 55627^T	06000045.			256
Shewanella algae	ATCC 51192^T	04020045.			1402
Shewanella algidipiscicola	LMG 23746^T	04000045.			1402
Shewanella baltica	NCTC 10735^T	05020045.			1402
Shewanella baltica		04020045.		Oysters	1402
Shewanella baltica		00020045.		Seawater	1402
Shewanella frigidimarina	ACAM 591^T	04061245.			209
Shewanella frigidimarina	ACAM 591^T	06061245.			209
Shewanella frigidimarina	ACAM 591^T	24061245.			209
Shewanella frigidimarina	ACAM 591^T	26061245.			209
Shewanella frigidimarina	ACAM 591^T	04000045.			1402
Shewanella gelidimarina	ACAM 456^T	04020045.			209
Shewanella hafniensis		00000045.		Oysters	1402
Shewanella haliotis	KCTC 12896^T	05020045.			895
Shewanella marisflavi		04000045.		Oysters	1402
Shewanella morhuae	ATCC U1417^T	04020045.			1402
Shewanella putrefaciens		070200453			886
Shewanella putrefaciens		070200653			886

Continued

Table 3.25. Continued.

Organism	Strain	API 20E number	No. of strains	Notes	Ref
Shewanella putrefaciens		070202653			886
Shewanella putrefaciens	ATCC BAA-1097^T	00000045.			1402
Shewanella putrefaciens		00020045.		Oysters	1402
Shewanella woodyi		000200452			209
Tenacibaculum maritimum		000200410			68
Tenacibaculum maritimum	a442	000200410		Wedge sole	1055
Tenacibaculum soleae	NCIMB 14368^T	000300450		9	1056,1344
Vibrio aerogenes	DSMZ 14438	104400556	1	Strain FG1, seagrass sediment	244
Vibrio aerogenes	DSMZ 14438	104400546	1	Strain FG1, seagrass sediment	244
Vibrio aestuarianus	ATCC 35048^T	106012057	1		588,1645
Vibrio aestuarianus	ATCC 35048^T	106012056	1		588,1645
Vibrio aestuarianus	ATCC 35048^T	106012457	1		588,1645
Vibrio aestuarianus	ATCC 35048^T	106012456	1		588,1645
Vibrio aestuarianus	NCIMB 2236	304612556	1		244
Vibrio aestuarianus	NCIMB 2236	306612556	1		244
Vibrio agarivorans	AHLDA 1732	100410556	2	From abalone. Pathogenicity unknown	244
Vibrio agarivorans	CECT 5085^T	100410556	1	From seawater	244
Vibrio alginolyticus		404712456			244
Vibrio alginolyticus		404712457			244
Vibrio alginolyticus		414712457			1409
Vibrio alginolyticus		414712557			1164
Vibrio alginolyticus		414716557			1409
Vibrio alginolyticus		434712457			1164
Vibrio alginolyticus	ATCC 17749^T	435612456	1		244
Vibrio alginolyticus	ATCC 17749^T	404652456	1		1044
Vibrio alginolyticus	ATCC 17749^T	404612456	1		998
Vibrio alginolyticus	Swy	414612456	1	Kuruma prawn	998
Vibrio alginolyticus	ATCC 17749^T	404752456	1		257
Vibrio anguillarum	CRL 11	304752557			244
Vibrio anguillarum	ATCC 19264	324750456			244
Vibrio anguillarum	ATCC 19264	324750556			244
Vibrio anguillarum	V10	304452456	1	Japan	71
Vibrio anguillarum	V239	304572557	1	Japan	71
Vibrio anguillarum	PT-87050	304652456	2	Japan	71
Vibrio anguillarum		304752456	1		

Vibrio anguillarum	V244 to V246	304752476	5	Japan	71
Vibrio anguillarum	HT-77003	304752657	1	Japan	71
Vibrio anguillarum	V240	324472757	1	Japan	71
Vibrio anguillarum	V241	324562757	1	Japan	71
Vibrio anguillarum	ET-78063	324632657	1	Japan	71
Vibrio anguillarum	UB 4346, 434	324712677	2	Spain	71
Vibrio anguillarum	NCMB 6	3247524..	1		496
Vibrio anguillarum	UB (ET-1)	324752557	1	Japan	71
Vibrio anguillarum	V318, AHLDA 1730	324752656	2	Japan, Australia	71, 244
Vibrio anguillarum	RH-8101, AVL	324752657	6	France, Japan	71, 244
Vibrio anguillarum		3247527..			68
Vibrio anguillarum		324752756			878
Vibrio anguillarum		324752756			71
Vibrio anguillarum	LMG 3347	324752757	4	Japan, Norway, Spain	71
Vibrio anguillarum	UB A078	324752777	1	Spain	71
Vibrio anguillarum	V320	324772656	1	Spain	71
Vibrio anguillarum	UB A054-56	324772657	4	Japan, Spain	71
Vibrio atypicus	HHSO2T	101612456	1	healthy Chinese prawn	1767
Vibrio brasiliensis	LMG 20546[T]	32471255.		bivalve larvae	1632
Vibrio breoganii	LMG 22858[T]	100410516	7	Clams, Spain	140
Vibrio campbellii	ATCC 2590	004410556	1		244
Vibrio campbellii	ATCC 2590	006410556	1		244
Vibrio caribbeanicus	DSM 23640[T]	20000044	1	Sponge, *Scleritoderma* cyanea	739
Vibrio celticus	Rd 8.15[T]	204717556	1	Clams, Spain	142
Vibrio celticus	RD 6.8	204713556	1	Clams, Spain	142
Vibrio celticus	Rd2L5	204713756	1	Clams, Spain	142
Vibrio chagasii	LMG 21353[T]	206612446	1		524
Vibrio cholerae	LMG 16741	204612457	1	Shrimp, Thailand	71
Vibrio cholerae	LMG 16742	324602557	1	*P. orientalis*, PRC, Thailand	71
Vibrio cholerae 0⁻		514612457			
Vibrio cholerae non 01		514712457			
Vibrio cholerae	PS-7701	524712457	1	*P. altivelis*, Japan	71
Vibrio cholerae	PS-7705	524712476	1	*P. altivelis*, Japan	71
Vibrio cholerae	91/1198	534612457	1	*Carassius auratus*, Australia	71

Continued

Table 3.25. Continued.

Organism	Strain	API 20E number	No. of strains	Notes	Ref
Vibrio cholerae 01		534612457			
Vibrio cholerae non 01	LMG 16743	534712457	1		71
Vibrio cholerae	WF 110r	534712557			1076,1164
Vibrio (communis) owensii	LMG 25430^T	514712556	6	Coral (*M. hispida*, *P. dilatata*, *P. caribaeorum*), prawn	320
Vibrio (communis) owensii	R-40504	514512556	1	Coral (*P. variabilis*)	320
Vibrio (communis) owensii	R-40498	534712556	1	Coral (*M. hispida*)	320
Vibrio (communis) owensii	R-40900	504712556	1	Coral (*M. hispida*)	320
Vibrio crassostreae	LMG 22240^T	204616556	5	Diseased oysters, pathogenicity not tested	524
Vibrio crassostreae	LMG 22240^T	206616556	5	Diseased oysters, pathogenicity not tested	524
Vibrio cyclitrophicus	P-2P44^T	000612546	1		709
Vibrio cyclitrophicus	P-2P44^T	200612546	1		709
Vibrio cyclitrophicus	LMG 21359^T	104412556	1		524
Vibrio diabolicus	CNCM I-1629^T	416612556	1	The Pompeii worm, deep sea vent	1368
Vibrio diazotrophicus	ATCC 33466^T	324412757	1	Gut of sea urchin and rotifers, surface of reeds, seagrass	244
Vibrio diazotrophicus	ATCC 33466^T	324413757	1	Gut of sea urchin and rotifers, surface of reeds, seagrass	244
Vibrio diazotrophicus	ATCC 33466^T	324013457	1	Gut of sea urchin and rotifers, surface of reeds, seagrass	244
Vibrio diazotrophicus	ATCC 33466^T	324012457	1	Gut of sea urchin and rotifers, surface of reeds, seagrass	244
Vibrio fischeri	ATCC 25918	400500556			1641
Vibrio fluvialis	NCTC 11327	324612757	1		244
Vibrio fluvialis	NCTC 11327	326612757	1		244
Vibrio fluvialis		304612657			996,1504
Vibrio fluvialis		304712657			996,1504
Vibrio fluvialis		324412757			996,1504
Vibrio fluvialis	MEJ 311	324612657	2		496,996,1504
Vibrio fluvialis		324712657			996,1504
Vibrio fortis	LMG 21557^T	106516556	1	Salmon, seawater, diseased and healthy bivalve larvae	1634
Vibrio furnissii	ATCC 35016^T	324612657			496
Vibrio furnissii		304712657			1504
Vibrio furnissii		324712657			1504
Vibrio furnissii	ATCC 11218	304412657		Citrate, indole, VP weak reaction	244
Vibrio furnissii	ATCC 11218	304512657		Citrate, indole, VP weak reaction	244
Vibrio furnissii	ATCC 11218	324412657		Citrate, indole, VP weak reaction	244
Vibrio furnissii	ATCC 11218	324512657		Citrate, indole, VP weak reaction	244

Vibrio furnissii	ATCC 35016^T	320412657	1		244
Vibrio gallaecicus	LMG 24045^T, VB 8.9^T	004610556	1	Negative for sucrose and amygdalin	144
Vibrio gallaecicus	VB 5.12	004612556	1	Positive for sucrose and amygdalin	144
Vibrio gigantis	LMG 22741T	204614556	4	Diseased oysters, pathogenicity not tested	1010
Vibrio (carchariae) harveyi		415412557			244,1228
Vibrio (carchariae) harveyi		415612557			244,1228
Vibrio (carchariae) harveyi		435412557			244,1228
Vibrio (carchariae) harveyi	ATCC 35084	435612557			244,1228
Vibrio harveyi	DN01	435652556		Summer flounder disease, FINE	592
Vibrio harveyi	ATCC 14126	514610556			1042
Vibrio harveyi	ATCC 14129^T	414612556			1228
Vibrio harveyi	ATCC 14129^T	434612556			1228
Vibrio harveyi	ATCC 25919	534611756			1042
Vibrio harveyi	ATCC 35084^T	415412556	1		1219,1228
Vibrio harveyi	ATCC 35084^T	415612556	1		1219
Vibrio harveyi	ATCC 35084^T	435412556	1		1219,1228
Vibrio harveyi	ATCC 35084^T	435612556	1		1219
Vibrio harveyi	Abalone	414610556	1	Abalone	1219
Vibrio harveyi	Abalone	434610556	1	Abalone	1219
Vibrio harveyi	Abalone	404410556	1	Japanese abalone	1228
Vibrio harveyi	Abalone	404510556	1	Japanese abalone	1228
Vibrio harveyi	Abalone	424410556	1	Japanese abalone	1228
Vibrio harveyi	Abalone	424510556	1	Japanese abalone	1228
Vibrio halioticoli	IAM 14596^T	000410456	1	Abalone	244
Vibrio halioticoli	IAM 14596^T	000010456	1	Abalone	244
Vibrio halioticoli	AHLDA 1734	000410456	1	Abalone	244
Vibrio hepatarius	LMG 20362^T	106512556	1	Hepatopancreas of prawn	1634
Vibrio ichthyoenteri	IFO 15847^T	000402456	1	Intestinal necrosis of Japanese flounder	804
Vibrio ichthyoenteri	DSMZ 14397	000410456	1	Intestinal necrosis of Japanese flounder	244
Vibrio kanaloae	LMG 20539^T	30671656	1		1635
Vibrio kanaloae	LMG 20539^T	306612546	1		524
Vibrio lentus	CECT 5110	104210406	1	Oysters	244
Vibrio lentus	CECT 5110	104210446	1	Oysters	244

Continued

Table 3.25. Continued.

Organism	Strain	API 20E number	No. of strains	Notes	Ref
Vibrio lentus	CECT 5110	104610406	1	Oysters	244
Vibrio lentus	CECT 5110	104610446	1	Oysters	244
Vibrio lentus	CECT 5110	304210406	1	Oysters	244
Vibrio lentus	CECT 5110	304210446	1	Oysters	244
Vibrio lentus	CECT 5110	304610406	1	Oysters	244
Vibrio lentus	CECT 5110	304610446	1	Oysters	244
Vibrio lentus	CIP 107166^T	104610406	1		524
Vibrio lentus	CIP 107166^T	104610446	1		524
Vibrio maritimus	R-40493^T	304632556	1	Coral, *Palythoa caribaeorum*	321
Vibrio marisflavi	WH134^T	700606556	1	Seawater, China	1760
Vibrio mediterranei	AHLDA 1733	704672556	1	Abalone	244
Vibrio mediterranei	CECT 615	304472556	1	Seawater	244
Vibrio mediterranei	CECT 615	304476556	1	Seawater	244
Vibrio mediterranei	CECT 615	304576556	1	Seawater	244
Vibrio mediterranei	CECT 615	304572556	1	Seawater	244
Vibrio (shilonii) mediterranei	ATCC BAA-91	724477556	1	Bleached coral	244
Vibrio metschnikovii	NCTC 8443^T	724732046	1	Isolated from cockles	244
Vibrio metschnikovii	NCTC 8443^T	726732046	1	Isolated from cockles	244
Vibrio mimicus		514610457	1		326
Vibrio mimicus		534610457	1		326, 412
Vibrio mimicus	ATCC 33653^T	514610457	1	Human ear infection	244
Vibrio mimicus	ATCC 33653^T	534610457	1	Human ear infection	244
Vibrio mytili	CECT 632^T	106412556	1		1368
Vibrio mytili	CIP 103929	300411756	1		244
Vibrio natriegens	ATCC 14048	000413756	1		244
Vibrio natriegens	ATCC 14048	000613756	1		244
Vibrio natriegens	ATCC 14048	020413756	1		244
Vibrio natriegens	ATCC 14048	020613756	1		244
Vibrio navarrensis	CIP 103381^T	004412557	1	Sewage and surface water	244
Vibrio navarrensis	CIP 103381^T	024412557	1	Sewage and surface water	1687
Vibrio neptunius	LMG 20536^T	226702456	9	Bivalve larvae, rotifers, turbot larvae	1632
Vibrio neptunius	PP-145.98	20 0/2 602446	1	Oyster larvae	1357
Vibrio nereis	LMG 3895^T	006402456	1		1368

Vibrio nereis	ATCC 25917	224402456	1		244
Vibrio nereis	ATCC 25917	224002456	1		244
Vibrio nigripulchritudo	CIP 103195	124604556	1	Seawater	244
Vibrio ordalii	LMG 13544	000402446	1		71
Vibrio ordalii	LMG 10951	000402476	3	Salmon	71
Vibrio ordalii	V-306	000402556	1	Amago trout, Japan	71
Vibrio ordalii	RF-2, PT-81025	000402576	8	*O. mykiss*, *Pl. altivelis*, Japan	71
Vibrio ordalii	V-11	000412446	1	*O. mykiss*, Japan	71
Vibrio ordalii	F378, F380, V-250	000602546	4	Salmon, Australia; *O. mykiss*, Japan	71
Vibrio ordalii	F379, F381	000612446	2	Salmon, Australia	71
Vibrio ordalii	DF 3K	000612446	1	Denmark	878
Vibrio ordalii	V-302	304752476	4	*O. mykiss*, Japan	71
Vibrio ordalii	Chilean strains	0004024..			1531
Vibrio ordalii	DF 3K	000612446		Coho salmon	878
Vibrio ordalii	ATCC 33509^T	000402546		Coho salmon	244
Vibrio ordalii	ATCC 33509^T	000412546		Coho salmon	244
Vibrio ordalii	ATCC 33509^T	000612446			1531
Vibrio ordalii	Chilean strains	000402446		Atlantic salmon	1531
Vibrio ordalii	Chilean strains	000602446		Atlantic salmon	1531
Vibrio orientalis	NCMB 2195^T	604612556	1	Seawater, spider crab	635,1825
Vibrio orientalis	ATCC 33935	604616556	1		244
Vibrio orientalis	ATCC 33935	606616556	1		244
Vibrio owensii	DY05^T	416612556	1	Ornate spiny lobster (*P. ornatus*), Australia	270
Vibrio owensii	47666-1	516612556	1	Tiger prawn (*P. monodon*), Australia	270
Vibrio owensii	DY05^T	434112546	1	Ornate spiny lobster (*P. ornatus*), Australia	244
Vibrio owensii	DY05+	434612546	1	Ornate spiny lobster (*P. ornatus*), Australia	244
Vibrio owensii	DY05^T	436112546	1	Ornate spiny lobster (*P. ornatus*), Australia	244
Vibrio owensii	DY05^T	436612546	1	Ornate spiny lobster (*P. ornatus*), Australia	244
Vibrio owensii	47666-1	536412546	1	Tiger prawn (*P. monodon*), Australia	244
Vibrio pacinii	LMG 1999^T	322 6/7 0 2/6 556	1	Prawn	638
Vibrio pacinii	LMG 21514	222 6/7 0 2/6 556	1	Salmon	638
Vibrio pacinii	LMG 13245	322 4/6 0 2/6 556	1	Seabass	638
Vibrio parahaemolyticus	ATCC 17802^T	434610757	1		244
Vibrio parahaemolyticus	ATCC 17802^T	436610757	1		244
Vibrio parahaemolyticus		410610657			1164

Continued

Table 3.25. Continued.

Organism	Strain	API 20E number	No. of strains	Notes	Ref
Vibrio parahaemolyticus		410610757			1164
Vibrio parahaemolyticus	ATCC 43996	414610657	1		244
Vibrio parahaemolyticus	ATCC 43996	434610657	1		244
Vibrio parahaemolyticus		430610657			1164
Vibrio parahaemolyticus		434610657			878
Vibrio parahaemolyticus	ATCC 17802	415610657	1		18
Vibrio pectenicida	CIP 105190^T	020600456	1	Pathogenic to scallop larvae	970
Vibrio pectenicida	ATCC 700783	000000416	1	Pathogenic to scallop larvae	244
Vibrio pectenicida	ATCC 700783	000400416	1	Pathogenic to scallop larvae	244
Vibrio pelagius I	ATCC 25916^T	100412446	1		244
Vibrio pelagius I	ATCC 25916^T	100012446	1		244
Vibrio pelagius II	ATCC 33784	104612456	1		244
Vibrio pelagius		100412456			
Vibrio pelagius		104612456			
Vibrio pelagius		124612456			
Vibrio penaeicida	KH-1^T	120600456		Pathogen of prawns (*Penaeus japonicus* & *P. stylirostris*)	803
Vibrio penaeicida	Noumea strain	124600456		Pathogen of prawns (*Penaeus japonicus* & *P. stylirostris*)	371
Vibrio penaeicida	DSMZ 14398	324300006	1	Pathogen of prawns (*Penaeus japonicus* & *P. stylirostris*)	244
Vibrio plantisponsor	MSSRF60^T	324413756	2	Roots of wild rice in mangroves	1372
Vibrio pomeroyi	LMG 20537^T	306610546	1		524
Vibrio ponticus	RP30	224402656		Japanese seabass	1816
Vibrio proteolyticus	AHLDA 1735	624750456	1	Artemia	244
Vibrio proteolyticus	CIP 102892	626750556	1		244
Vibrio rotiferianus	LMG 21460^T	5176167..	1	Rotifer, Ghent Belgium	637
Vibrio sagamiensis	LC2-047^T	000200456	2	Seawater	1837
Vibrio sagamiensis	LC2-047^T	004200456	2	Seawater	1837
Vibrio salmonicida	RVAU 890206-1/12	000410040	1	*Salmo salar*, Faroe Islands	71
Vibrio salmonicida	RVAU 881129-1/2	000410440	5	*Salmo salar*, Faroe Islands	71
Vibrio salmonicida	RVAU 881129-1/1	000410446	1	*Salmo salar*, Faroe Islands	71
Vibrio salmonicida	RVAU 890206-1/10	000412640	1	*Salmo salar*, Faroe Islands	71
Vibrio salmonicida	RVAU 890206-1/7	520600556	1	*Salmo salar*, Faroe Islands	71
Vibrio scophthalmi	CECT 4638	200402456	1	Intestine of juvenile turbot	293

Vibrio scophthalmi	CECT 4638	000402456	1	Intestine of juvenile turbot	244
Vibrio scophthalmi	CECT 4638	200402456	1	Intestine of juvenile turbot	244
Vibrio scophthalmi	CECT 4638	000402456	1	Intestine of juvenile turbot	523
Vibrio splendidus		004416516		Juvenile oyster (*C. gigas*)	965
Vibrio splendidus	UB S292	204610456	1	Seawater, Sweden	71
Vibrio splendidus	90-0652	224614446	1	Striped trumpeter, Australia	71
Vibrio splendidus	RVAU 88-12-686	304610456	3	*Salmo salar,* Norway	71
Vibrio splendidus	UB S236	304610556	1	Seawater, Sweden	71
Vibrio splendidus	RVAU 88-12-711	304612456	1	*Sp. auratus*, Norway	71
Vibrio splendidus	89-1638	304612556	1	*O. mykiss*, Australia	71
Vibrio splendidus	A515, AH5	304614556	1	Scallop strain A515, turbot strain AH5	590
Vibrio splendidus	A515, AH5	304614516	1	Scallop strain A515, turbot strain AH5	590
Vibrio splendidus	VS4	304714516	1	Turbot strain VS4	590
Vibrio splendidus	VS4	304714556	1	Turbot strain VS4	590
Vibrio splendidus	VS1	324614516	1	Turbot strain VS1	590
Vibrio splendidus	VS1	324614556	1	Turbot strain VS1	590
Vibrio splendidus	VS5	304410516	1	Turbot strain VS5	590
Vibrio splendidus	VS5	304410556	1	Turbot strain VS5	590
Vibrio splendidus	VS6, VS11	304610516	1	VS6, VS11, turbot	590
Vibrio splendidus	VS6, VS11	304610556	1	VS6, VS11, turbot	590
Vibrio splendidus	VS10	304510516	1	VS10, turbot	590
Vibrio splendidus	VS10	304510556	1	VS10, turbot	590
Vibrio splendidus	VS9	104610516	1	VS9, turbot	590
Vibrio splendidus	VS9	104610556	1	VS9, turbot	590
Vibrio splendidus		304616556			590
Vibrio splendidus	LMG 4042^T	104614516	1		524,1009
Vibrio splendidus	LMG 4042^T	104614556	1		524,1009
Vibrio splendidus	RVAU 88-12-717	324410556	1	*S. maximus*, Norway	71
Vibrio splendidus	LMG 16752	324414557	1	Oyster, Spain	71
Vibrio splendidus	RVAU 88-12-712	324610456	1	*Solea solea*, Norway	71
Vibrio splendidus	UB S308, LMG 16747	324610556	2	Seawater, Sweden; oyster, Spain	71
Vibrio splendidus	LMG 16745	324610557	1	*S. auratus*, Greece	71
Vibrio splendidus	LMG 16744, 16750	324612557	2	*S. auratus*, Greece; oyster, Spain	71
Vibrio splendidus	LMG 16748	324614556	1	Oyster, Spain	71

Continued

Table 3.25. Continued.

Organism	Strain	API 20E number	No. of strains	Notes	Ref
Vibrio splendidus	LMG 16749	324616556	1	Oyster, Spain	71
Vibrio splendidus	LMG 16746	324712556	1	Oyster, Spain	71
Vibrio splendidus I	ATCC 33125^{T}	304612456	1		523
Vibrio splendidus I	CRL 40	104614556	1	Mar ne fish	244
Vibrio splendidus II	ATCC 33789	006410556	1		244
Vibrio splendidus II	ATCC 33789	006410516	1		244
Vibrio splendidus II	ATCC 33789	206410556	1		244
Vibrio splendidus II	ATCC 33789	206410516	1		244
Vibrio tapetis		10460055.			201,285
Vibrio tapetis	Wedge sole	104600456	4	Isolates from wedge sole	1052
Vibrio tapetis	Halibut	104610456	1	Isolates from Halibut	1393
V. tapetis ssp. *britannicus*		124610656	3	Halibut and clam seed, UK	103
Vibrio tasmaniensis	LMG 20012^{T}	004410456	1		524,1009
Vibrio tubiashii	LMG 10936^{T}	106612556	1		668,1368
Vibrio tubiashii		304412556			668,1368
Vibrio tubiashii	NCIMB 1340	304616556	1	Hard clam, USA	244
Vibrio variabilis	R-40492^{T}	304613556	1	Mucus of apparently healthy coral, *Palythoa caribaeorum*	321
V. vulnificus biovar 1 non-serovar E	M06-24	434610557	1	Human blood, USA	180
V. vulnificus biovar 1 non-serovar E	532	504600557	2	Diseased eel, Belgium, avirulent for eels	59,180
V. vulnificus biovar 1 non-serovar E	628	514412757	1	Paguara fish, Venezuela	180
V. vulnificus biovar 1 non-serovar E	530	514602557	7	Diseased eel, Belgium, virulent	59,180
V. vulnificus biovar 1 non-serovar E	ATCC 27562	514610557	1	Human strain, USA	50,180
V. vulnificus biovar 1 non-serovar E		514610557	23	Human (USA), eels (Spain, Sweden), shrimp (Thailand)	59,180
V. vulnificus biovar 1 non-serovar E	B9629, C7184	514610557		Human wound infection, USA	59,180
V. vulnificus biovar 1 non-serovar E	UMH1, 374	514610557		Human wound infection, USA	59,180
V. vulnificus biovar 1 non-serovar E	ATCC 27562	534600557		Human strain, USA	59,180
V. vulnificus biovar 1 non-serovar E	E109	534600557		Hea thy eel, Spain, avirulent for eels	180
V. vulnificus biovar 1 non-serovar E	C7184	534610557	1	Human blood, USA, avirulent for eels	180
V. vulnificus biovar 1 serovar E	534	514410557	2	Diseased eel, Sweden, avirulent for eels	59,180
V. vulnificus biovar 1 serovar E	ATCC 33186	514610557		Human blood, USA, avirulent for eels	180
V. vulnificus biovar 1 serovar E	521	534600557		Unknown, Australia, avirulent for eels	180
V. vulnificus biovar 1	LMG 12092	104602557	1	Eel, Belgium	59

V. vulnificus biovar 1	169	104610557	1	Eel, Belgium	59
V. vulnificus biovar 1	E109	114000557	1	Eel, Spain	59
V. vulnificus biovar 1	160	114610557	1	Eel, Belgium	59
V. vulnificus biovar 1	M626	410610457	1	Eel, Spain	59
V. vulnificus biovar 1	Vv1	504600557		Human wound infection, USA	59
V. vulnificus biovar 1	UMH1	504610557	1	Human wound infection, USA	59
V. vulnificus biovar 1	M631	510610557	1	Eel, Spain	59
V. vulnificus biovar 1	167	514650557	1	Eel, Belgium	59
V. vulnificus biovar 1	VIB 521	516600557	1	Unknown	59
V. vulnificus biovar 1		524610557			180
V. vulnificus biovar 2 serovar E	171	100600557	1	Diseased eel, Sweden	180
V. vulnificus biovar 2	171	104600557	1	Eel, Belgium	59
V. vulnificus biovar 2 serovar E	NCIMB 2136	400600557	1	Diseased eel, Japan	180
V. vulnificus biovar 2		420600557			180
V. vulnificus biovar 2 serovar E	520	500400557	1	Shrimp, Taiwan, virulent for eels	180
V. vulnificus biovar 2 serovar E	NCIMB 2137	500600457	1	Diseased eel, Japan	59,180
V. vulnificus biovar 2 serovar E	ATCC 33149	500600557	1	Diseased eel, Japan	59,180,741
V. vulnificus biovar 2	NCIMB 2138	500600557	1	Diseased eel, Japan	180
V. vulnificus biovar 2 serovar E	121	510200457	1	Diseased eel, Sweden	180
V. vulnificus biovar 2 serovar E	526	510200557	1	Diseased eel, Sweden	180
V. vulnificus biovar 2 serovar E	524	510600557		Diseased eel, Norway	180
V. vulnificus biovar 2	NCIMB 2136	510600557	1	Diseased eel, Japan	59
V. vulnificus biovar 2		510600557	14	Eels (Japan, Norway – serovar E, Spain, Sweden), shrimp, human	59,180
V. vulnificus biovar 2 serovar E	523	514600557	5	Eels (Belgium – serovar E, Spain, Sweden), seabream (Spain)	59,180
V. vulnificus biovar 2 serovar E	NCIMB 2138	520600557	1	Diseased eel, Japan	180
V. vulnificus biovar 2 serovar E	E86	520600557	2	Diseased eel, (Japan, Spain)	59,180
V. vulnificus biovar 2 serovar E	525	530600457	1	Diseased eel, Sweden	180
V. vulnificus biovar 2 serovar E	ATCC 33187	530600557	1	Human strain, USA	180
V. vulnificus biovar 2 serovar E	E105	530600557		Diseased eel, Spain	50,180
V. vulnificus biovar 2 serovar E		530600557	2	Human (USA), eel (Spain)	50,180
V. vulnificus biovar 2 serovar E		530610557	1	Taiwanese strains, avirulent for eels	50
V. vulnificus	818	414412557	1	Unknown, France	180
V. vulnificus	822	504410557	1	Shrimp, Senegal	180

Continued

Table 3.25. Continued.

Organism	Strain	API 20E number	No. of strains	Notes	Ref
V. vulnificus	WF8A1110	534610357			1076
V. xiamenensis	G21^T	100413756	1	Mangrove soil	581
Vibrio xuii	LMG 21346^T	206512756	3	Shrimp culture water, shrimp, bivalve larvae	1632
Yersinia intermedia		115457157			68
Yersinia intermedia		115457357			68
Yersinia intermedia		115477357			68
Yersinia intermedia		115477357			68
Yersinia ruckeri		110410057			1425,1476
Yersinia ruckeri		110450057			1582
Yersinia ruckeri		1104501			1425
Yersinia ruckeri		1104700			1425
Yersinia ruckeri		1105100			1425
Yersinia ruckeri		110550057			1425
Yersinia ruckeri		1107100			1425
Yersinia ruckeri		1107500			1425
Yersinia ruckeri	ATCC 29473	510410057	2		367,402,1425,1476
Yersinia ruckeri	AHLDA 1313	510510057	1		244,1425,1643
Yersinia ruckeri		5105500	20	Environmental strains	367,1425
Yersinia ruckeri		510610017			1643
Yersinia ruckeri		510650017			1643
Yersinia ruckeri		510710017			1425,1643
Yersinia ruckeri		510750017	1	Environmental strains	367,1643
Yersinia ruckeri		530710057			411,1425
Yersinia ruckeri		530750057			411
Zobellia galactanivorans		10061361.			119
Zobellia uliginosa		100.0..5.			119

* from API database.

To be used in conjunction with the API Database supplied by bioMerieux.

Table 3.26. AFI 20E database numbers (numbers in ascending order).

API 20E number	Organism	Strain	No. of strains	Notes	Ref
0000004..	*Acinetobacter/Moraxella*		1		68
0000004..	*Pseudomonas stutzeri*		1/9		64
000000440	*Alivibrio logei*	ATCC 29985^T			244
00000045.	*Shewanella hafniensis*			Oysters	1402
00000045.	*Shewanella putrefaciens*	ATCC BAA-1097^T			1402
000000416	*Vibrio pectenicida*	ATCC 700783	1	Pathogenic to scallop larvae	244
000010456	*Vibrio halioticoli*	IAM 14596^T	1	Abalone	244
000100416	*Phocoenobacter uteri*	NCTC 12872^T			549
0002004..	*Flavobacterium psychrophilum*				68
0002004..	*Aliivibrio wodanis*	ATCC BAA-104^T			1834
000200400	*Flavobacterium macrobrachii*	LMG 25203+			1514
000200410	*Tenacibaculum maritimum*	a443			68,1055
000200440	*Pseudomonas anguilliseptica*				68
00020045.	*Shewanella baltica*			Seawater, oysters	1402
00020045.	*Shewanella putrefaciens*			Oysters	1402
000200452	*Shewanella woodyi*				209
000200456	*Vibrio sagamiensis*	LC2-047^T	2	Seawater	1837
0002006..	*Aliivibrio wodanis*	ATCC BAA-104^T			1834
000300450	*Tenacibaculum soleae*	NCIMB 14368^T	9		1056,1344
0004004..	*Pseudomonas stutzeri*		5/9		64
0004004..	*Burkholderia cepacia*		3/10		64
000400416	*Vibrio pectenicida*	ATCC 700783	1	Pathogenic to scallop larvae	244
000402546	*Vibrio ordalii*	ATCC 33509^T		Coho salmon	244
000402446	*Vibrio ordalii*	Chilean strains		Atlantic salmon	1531
000402446	*Vibrio ordalii*	LMG 13544			71
000402456	*Vibrio ichthyoenteri*	IFO 15847^T	1	Intestinal necrosis of Japanese flounder	804
000402456	*Vibrio scophthalmi*	CECT 4638	1	Intestine of juvenile turbot	244,523
000402476	*Vibrio ordalii*	LMG 10951		Salmon	71
000402576	*Vibrio ordalii*	RF-2, PT-81025		*O, mykiss*, *P. altivelis*, Japan	71
00041000.	*Flavobacterium tegetincola*				1115
000410040	*Aliivibrio salmonicida*	RVAU 890206-1/12		*Salmo salar*, Faroe Islands	71
0004104..	*Pseudomonas stutzeri*		3/9		64
000410456	*Vibrio ichthyoenteri*	DSMZ 14397	1	Intestinal necrosis of Japanese flounder	244

Continued

Table 3.26. Continued.

API 20E number	Organism	Strain	No. of strains	Notes	Ref
000410440	*Aliivibrio salmonicida*	RVAU 881129-1/2		*Salmo salar*, Faroe Islands	71
000410446	*Aliivibrio salmonicida*	RVAU 881129-1/1		*Salmo salar*, Faroe Islands	71
000410456	*Vibrio halioticoli*	AHLDA 1734	1	Abalone	244
000410456	*Vibrio halioticoli*	IAM 14596^T	1	Abalone	244
000412546	*Vibrio ordalii*	ATCC 33509		Coho salmon	244
000412640	*Aliivibrio salmonicida*	RVAU 890206-1/10		*Salmo salar*, Faroe Islands	71
000413756	*Vibrio natriegens*	ATCC 14048			244
000500416	*Phocoenobacter uteri*	NCTC 12872^T			549
000500417	*Aeromonas salmonicida salmonicida*			Lamprey	513
000602446	*Vibrio ordalii*	Chilean strains		Atlantic salmon	1531
00060045.	*Moritella marina*	ATCC 15381^T		Pacific Ocean	1683
000602546	*Vibrio ordalii*	F378, F380		Salmon, Australia	71
0006060..	*Flavobacterium branchiophilum*				68
000613756	*Vibrio natriegens*	ATCC 14048			244
000612210	*Flavobacterium hydatis*				68
000612446	*Vibrio ordalii*	F379, F381		Salmon, Australia	71
000612446	*Vibrio ordalii*	DF 3K			878
000612446	*Vibrio ordalii*	ATCC 33509^T			1531
000612546	*Vibrio cyclitrophicus*	P-2P44^T	1		709
0006727..	*Burkholderia pseudomallei*		1/91		64
001100407	*Photobacterium damselae piscicida*	P90029	1	Atypical strain	1641
001412556	*Vibrio navarrensis*	CIP 103381	1		244
004000456	*Grimontia hollisae*	CIP 101886	1		244
004000656	*Grimontia hollisae*	CIP 101886	1		244
004002456	*Grimontia hollisae*	CIP 101886	1		244
004002656	*Grimontia hollisae*	CIP 101886	1		244
004200456	*Vibrio sagamiensis*	LC2-047^T	2	Seawater	1837
004400456	*Grimontia hollisae*	CIP 101886	1		244
004400656	*Grimontia hollisae*	CIP 101886	1		244
004402456	*Grimontia hollisae*	CIP 101886	1		244
004402656	*Grimontia hollisae*	CIP 101886	1		244
004410456	*Vibrio tasmaniensis*	LMG 20012^T	1		524,1009
004410556	*Vibrio campbellii*	ATCC 2590	1		244

004412417	*Aeromonas salmonicida achromogenes*				669
004412557	*Vibrio navarrensis*	CIP 103381^T	1	Sewage and surface water	244
004416516	*Vibrio splendidus*			Juvenile oyster (*C. gigas*)	965
004512417	*Aeromonas salmonicida achromogenes*	NCIMB 1110			968
004610556	*Vibrio gallaecicus*	LMG 24045^T, VB 8.9^T	2	Negative for sucrose and amygdalin	144
004612556	*Vibrio gallaecicus*	VB 5.12	1	Positive for sucrose and amygdalin	144
0046124..	*Aliivibrio wodanis*	ATCC BAA-104^T	1		143
006402456	*Vibrio nereis*	LMG 3895^T	1		1368
006410516	*Vibrio splendidus* II	ATCC 33789	1		244
006410556	*Vibrio splendidus* II	ATCC 33789	1		244
006410556	*Vibrio campbellii*	ATCC 2590	1		244
006510556	*Vibrio tasmaniensis*	LMG 20012^T	1	Atlantic salmon, Tasmania, Australia	1637
007431057	*Providencia rettgeri*				68
014452456	*Pasteurella multocida*				878
020000440	*Pseudomonas anguilliseptica*				176,1136,1794
020200440	*Pseudomonas anguilliseptica*				176,1136,1794
0204004..	*Pseudomonas fluorescens*		2/8		64
020413756	*Vibrio natriegens*	ATCC 14048			244
020600456	*Vibrio pectenicida*	CIP 105190^T	1	Pathogenic to scallop larvae	970
0206006..	*Burkholderia cepacia*		1/10		64
0206006..	*Pseudomonas aeruginosa*		1/18		64
020613756	*Vibrio natriegens*	ATCC 14048			244
0206706..	*Burkholderia pseudomallei*		1/91		64
024412557	*Vibrio navarrensis*	CIP 103381^T	1	Sewage and surface water	1687
026402057	*Providencia rustigianii*	DSM 2620			1175
04000045.	*Shewanella algidipiscicola*	LMG 23746^T			1402
04000045.	*Shewanella frigidimarina*	ACAM 591^T			1402
04000045.	*Shewanella marisflavi*			Oysters	1402
0402004..	*Flavobacterium columnare*				68
04020045.	*Shewanella gelidimarina*				209
04020045.	*Shewanella algae*	ATCC 51192^T			1402
04020045.	*Shewanella baltica*			Oysters	1402
04020045.	*Shewanella morhuae*	ATCC U1417^T			1402
04061241.	*Flavobacterium xantham*				1115
04061245.	*Shewanella frigidimarina*	ACAM 591^T			209

Continued

Table 3.26. Continued.

API 20E number	Organism	Strain	No. of strains	Notes	Ref
05020045.	*Shewanella baltica*	NCTC 10735^T			1402
05020045.	*Shewanella haliotis*	KCTC 12896^T			895
050200453	*Shewanella algae*	IAM 14159^T			886
050600453	*Shewanella algae*	IAM 14159^T			1239
06000045.	*Shewanella algae*	ATCC 55627^T			256
06061245.	*Shewanella frigidimarina*	ACAM 591^T			209
070200453	*Shewanella putrefaciens*				886
070200653	*Shewanella putrefaciens*				886
070202653	*Shewanella putrefaciens*				886
100.0..5.	*Zobellia uliginosa*				119
100012446	*Vibrio pelagius* I	ATCC 25916^T			244
10041137	*Escherichia vulneris*				*
10041337	*Escherichia vulneris*				*
100410516	*Vibrio breoganii*	LMG 23858^T	7	Normal flora from cultured clams, Spain	140
100410556	*Vibrio agarivorans*	CECT 5085^T, AHL 1732	2	From abalone, pathogenicity unknown	244
100412446	*Vibrio pelagius* I	ATCC 25916^T			244
100412456	*Vibrio pelagius*				
100413756	*Vibrio xiamenensis*	G21^T	1	Mangrove soil	581
100577357	*Pantoea agglomerans*	CDC 1429-71			602
100600557	*V. vulnificus* biovar 2 serovar E	171		Diseased eel, Sweden	180
100603210	*Flavobacterium hibernum*	ATCC 51468^T			1116
10061361.	*Zobellia galactanivorans*				119
100736357	*Serratia plymuthica*	Trout		Rainbow trout	76
101500416	*Actinobacillus scotiae*		2	Porpoises	550
101612456	*Vibrio atypicus*	HHSO2^T	1	Healthy Chinese prawn	1767
104210406	*Vibrio lentus*	CECT 5110	1	Oysters	244
104210446	*Vibrio lentus*	CECT 5110	1	Oysters	244
104400546	*Vibrio aerogenes*	DSMZ 14438	1	Strain FG1, seagrass sediment	244
104400556	*Vibrio aerogenes*	DSMZ 14438	1	Strain FG1, seagrass sediment	244
1044105046	*Enterovibrio calviensis*	DSMZ 1337	1		244
104412556	*Vibrio cyclitrophicus*	LMG 21359^T	1		524
104600456	*Vibrio tapetis*	Wedge sole	4	Isolates from wedge sole	1052
104610406	*Vibrio lentus*	CIP 107166^T	1		524

104610446	*Vibrio lentus*	CIP 107166^{T}	1		524
104610456	*Vibrio tapetis*	Halibut		Isolates from halibut	1393
10460055.	*Vibrio tapetis*				201,285
104600557	*V. vulnificus* biovar 2	171		Eel, belgium	59
104602557	*V. vulnificus* biovar 1	LMG 12092		Eel, Belgium	59
104610406	*Vibrio lentus*	CECT 5110	1	Oysters	244
104610446	*Vibrio lentus*	CECT 5110	1	Oysters	244
104610516	*Vibrio splendidus*	VS9	1	VS9, turbot	590
104610556	*Vibrio splendidus*	VS9	1	VS9, turbot	590
104610557	*V. vulnificus* biovar 1	169		Eel, Belgium	59
104612456	*Vibrio pelagius*				
104612456	*Vibrio pelagius* II	ATCC 33784			244
104614516	*Vibrio splendidus* I	LMG 4042^{T}	1		524,1009
104614556	*Vibrio splendidus* I	CRL 40	1	Marine fish	244
104614556	*Vibrio splendidus* I	LMG 4042^{T}	1		524
106012056	*Vibrio aestuarianus*	ATCC 35048^{T}	1		588,1645
106012057	*Vibrio aestuarianus*	ATCC 35048^{T}	1		588,1645
106012456	*Vibrio aestuarianus*	ATCC 35048^{T}	1		588,1645
106012457	*Vibrio aestuarianus*	ATCC 35048^{T}	1		588,1645
106412556	*Vibrio mytili*	CECT 632^{T}	1		1368
106512556	*Vibrio hepatarius*	LMG 20362^{T}	1	Hepatopancreas of prawn	1634
106516556	*Vibrio fortis*	LMG 21557^{T}	1	Salmon, seawater, diseased and healthy bivalve larvae	1634
106612556	*Vibrio tubiashii*	LMG 10936^{T}	1		1368,1582
110410057	*Yersinia ruckeri*				1425,1476
110450057	*Yersinia ruckeri*				1582
1104501	*Yersinia ruckeri*				1425
1104700	*Yersinia ruckeri*				1425
1105100	*Yersinia ruckeri*				1425
110550057	*Yersinia ruckeri*				1425,1582
1107100	*Yersinia ruckeri*				1425
1107500	*Yersinia ruckeri*				1425
111500416	*Actinobacillus scotiae*	NCTC 12922^{T}	1	Porpoises	550
114000557	*V. vulnificus* biovar 1	E109		Eel, Spain	59
114610557	*V. vulnificus* biovar 1	160		Eel, Belgium	59
114612557	*Aeromonas veronii* biovar Veronii	ATCC 35623	1	Faeces, diarrhoea, USA	723
114712457	*Aeromonas veronii* biovar Veronii	ATCC 35604	1	Sputum, drowning victim, USA	723

Continued

Table 3.26. Continued.

API 20E number	Organism	Strain	No. of strains	Notes	Ref
115457157	*Yersinia intermedia*				68
115457357	*Yersinia intermedia*				68
115477357	*Yersinia intermedia*				68
120300410	*Flavobacterium johnsoniae*				244
120516357	*Pantoea agglomerans*	ATCC 14589			602
120517257	*Pantoea agglomerans*	NCPPB 2285			602
120517357	*Pantoea agglomerans*	DNA HG 14589			602
120547257	*Rahnella aquatilis*				223
120600456	*Vibrio penaeicida*	KH-1^{T}	1	Pathogen of prawns (*P. japonicus* & *P. stylirostris*)	803
120706742	*Gramella marina*	KMM 6048^{T}			1205
120712057	*Enterobacter agglomerans*				68
120713257	*Pantoea agglomerans*				602
120713357	*Pantoea dispersa*				602
120717357	*Pantoea dispersa*				602
120732257	*Serratia plymuthica*				68
120736357	*Serratia plymuthica*	Trout		Rainbow trout	1222
124002457	*Aeromonas fluvialis*	CECT 7401^{T}	1	Riverwater	37
124600456	*Vibrio penaeicida*	Noumea strain		Pathogen of prawns (*P. japonicus* & *P. stylirostris*)	371
124604556	*Vibrio nigripulchritudo*	CIP 103195	1	Pathogen of prawns (*P. japonicus* & *P. stylirostris*)	244
124610656	*V. tapetis* ssp. *britannicus*		3	Halibut and clam seed, UK	103
124612456	*Vibrio pelagius*				
134612557	*Aeromonas veronii* biovar Veronii	ATCC 35606	1	Maxillary sinus, USA	723
134712557	*Aeromonas veronii* biovar Veronii	ATCC 35605	1	Wound isolate, human, USA	723
140200410	*Flavobacterium columnare*				1511
2000004..	*Pseudomonas fluorescens*		5/8		64
20000044.	*Vibrio caribbeanicus*	DSM 23640^{T}	1	Sponge, *Scleritoderma cyanea*	739
2002004..	*Pseudomonas aeuriginosa*		1/18		64
200400406	*Photobacterium damselae piscicida*	ATCC 17911		ATCC 29687	878
200400406	*Photobacterium damselae piscicida*	ATCC 17911^{T}		1.5 & 2.0% NaCl inoculum	98,703,1089
200402456	*Vibrio scophthalmi*	CECT 4638	1	Intestine of juvenile turbot	244,293
200412000	*Flavobacterium gillisiae*				1115

200500017	*Photobacterium leiognathi*	LMG 4228			1641
200500406	*Photobacterium damselae piscicida*				595,1649,1857
200500406	*Photobacterium damselae piscicida*	ATCC 17911		ATCC 29690	1476,1641
200500407	*Photobacterium damselae piscicida*	ATCC 17911	Majority	NCIMB 2058	1641
200500406	*Photobacterium damselae piscicida*			Strains from tuna	1159
200500416	*Actinobacillus delphinicola*		2	Dolphin, whale	548
200500457	*Photobacterium damselae damselae*	LMG 7892			71
200500457	*Photobacterium damselae damselae*			2% NaCl inoculum	71
200504756	*Photobacterium swingsii*	CAIM 1393^T		2.5% inoculum	641
2006004..	*Pseudomonas aeuriginosa*		5/18		64
200602446	*Vibrio neptunius*	PP-145.98	1	Oyster larvae	1357
200610417	*Aeromonas salmonicida*				928
200610417	*Aeromonas salmonicida salmonicida*				1476
200612546	*Vibrio cyclitrophicus*	P-2P44^T	1		709
2006704..	*Burkholderia pseudomallei*		1/91		64
2006706..	*Burkholderia pseudomallei*		4/91		64
2006707..	*Burkholderia pseudomallei*		5/91		64
2006726..	*Burkholderia pseudomallei*		1/91		64
2006727..	*Burkholderia pseudomallei*		22/91		64
200702406	*Photobacterium angustum*	NCIMB 1895			1641
201400457	*Photobacterium damselae damselae*			25°C and 35°C incubation	339
201500457	*Photobacterium damselae damselae*			0.85% NaCl inoculum	244
201500457	*Photobacterium damselae damselae*	ATCC 33539			1641
202602446	*Vibrio neptunius*	PP-145.98	1	oyster larvae	1357
204402417	*Aeromonas salmonicida*	AHLDA 1334	1	'Atypical' strain, goldfish	244
204610456	*Vibrio splendidus*	UB S292		Seawater, Sweden	71
204612457	*Vibrio cholerae*	LMG 16741		Shrimp, Thailand	71
204614556	*Vibrio gigantis*	LMG 22741^T	4	Diseased oysters	1010
204616556	*Vibrio crassostreae*	LMG 22240^T	5	Diseased oysters	524
204713556	*Vibrio celticus*	RD 6.8	2	Clams, Spain	142
204713756	*Vibrio celticus*	Rd2L5	1	Clams, Spain	142
204717556	*Vibrio celticus*	Rd 8.15T	1	Clams, Spain	142
206410516	*Vibrio splendidus* II	ATCC 33789	1		244
206410556	*Vibrio splendidus* II	ATCC 33789	1		244

Continued

Table 3.26. Continued.

API 20E number	Organism	Strain	No. of strains	Notes	Ref
206512756	*Vibrio xuii*	LMG 21346^T	3	Shrimp culture water, shrimp, bivalve larvae	1632
206612446	*Vibrio chagasii*	LMG 21353^T	1		524
206616556	*Vibrio crassostreae*	LMG 22240^T	5	Diseased oysters	524
2200004..	*Pseudomonas fluorescens*		1/8		64
2202704..	*Burkholderia pseudomallei*		1/91		64
2202706..	*Burkholderia pseudomallei*		3/91		64
220400157	*Photobacterium damselae damselae*	NCIMB 2184			1641
220500406	*Photobacterium damselae piscicida*				1641
2206004..	*Pseudomonas aeruginosa*		9/18		64
2206006..	*Pseudomonas aeruginosa*		1/18		64
2206704..	*Burkholderia pseudomallei*		4/91		46
2206706..	*Burkholderia pseudomallei*		21/91		64
2206707..	*Burkholderia pseudomallei*	NCTC8016	1/91		64
2206707..	*Burkholderia pseudomallei*		12/91		64
2206727..	*Burkholderia pseudomallei*		15/91		64
220710452	*Janthinobacterium lividum*				73
222 6/7 0 2/6 556	*Vibrio pacinii*	LMG 21514	1	Salmon	638
224002456	*Vibrio nereis*	ATCC 25917			244
224402456	*Vibrio nereis*	ATCC 25917			244
224402656	*Vibrio ponticus*	RP30	1	Japanese seabass	1816
224614446	*Vibrio splendidus*	90-0652		Striped trumpeter, Australia	71
226702456	*Vibrio neptunius*	LMG 20536^T	9	Rotifers, bivalve larvae	1632
2306006..	*P. (Pseudoalcaligenes) oleovorans*			Strain from rainbow trout	76
24061245.	*Shewanella frigidimarina*	ACAM 591^T			209
26061245.	*Shewanella frigidimarina*	ACAM 591^T			209
264416417	*Aeromonas sobria*	CIP 7433^T			4
264516417	*Aeromonas sobria*	CIP 7433^T			4
264416457	*Aeromonas sobria*	CIP 7433^T			4
264516457	*Aeromonas sobria*	CIP 7433^T			4
30041137	*Escherichia vulneris*				*
30041337	*Escherichia vulneris*				*
300610517	*Aeromonas salmonicida salmonicida*				669
300616757	*Aeromonas rivuli*	CECT 7518^T		Creek water	536

324612657	*Aeromonas caviae*	ATCC 15468^T			494
300411756	*Vibrio mytili*	CIP 103929	1		244
300504756	*Photobacterium swingsii*	CAIM 1393^T		2.5% inoculum	641
304210406	*Vibrio lentus*	CECT 5110	1	Oysters	244
304210446	*Vibrio lentus*	CECT 5110	1	Oysters	244
304400446	*Enterovibrio norvegicus*	LMG 19839^T	1	Turbot	1627
304410516	*Vibrio splendidus*	VS5	1	Turbot strain VS5	590
304410556	*Vibrio splendidus*	VS5	1	Turbot strain VS5	590
304412556	*Vibrio tubiashii*				668,1368
304412657	*Vibrio furnissii*	ATCC 11218		Citrate, indole, VP weak reaction	244
304452456	*Vibrio anguillarum*				71
304472556	*Vibrio mediterranei*	CECT 615	1	Seawater	244
304476556	*Vibrio mediterranei*	CECT 615	1	Seawater	244
304510516	*Vibrio splendidus*	VS10	1	VS10, turbot	590
304510556	*Vibrio splendidus*	VS10	1	VS10, Turbot	590
304512517	*Aeromonas salmonicida masoucida*				669
304512657	*Aeromonas hydrophila*	ATCC 7966	1		1476
304512657	*Vibrio furnissii*	ATCC 11218	1	Citrate, indole, VP weak reaction	244
304572556	*Vibrio mediterranei*	CECT 615	1	Seawater	244
304572557	*Vibrio anguillarum*				71
304576556	*Vibrio mediterranei*	CECT 615	1	Seawater	244
304610406	*Vibrio lentus*	CECT 5110	1	Oysters	244
304610446	*Vibrio lentus*	CECT 5110	1	Oysters	244
304610456	*Vibrio splendidus*	RVAU 88-12-686		*S. salar*, Norway	71
304610516	*Vibrio splendidus*	VS6, VS11	1	VS6, VS11, turbot	590
304610556	*Vibrio splendidus*	VS6, VS11	1	VS6, VS11, turbot	590
304610556	*Vibrio splendidus*	UB S236		Seawater, Sweden	71
304612456	*Vibrio splendidus*	RVAU 88-12-711		*S. auratus*, Norway	71
304612456	*Vibrio splendidus* I	ATCC 33125^T	1		523
304612556	*Vibrio splendidus*	89-1638		*O. mykiss*, Australia	71
304612556	*Vibrio aestuarianus*	NCIMB 2236	1	Citrate should be positive	244
304612657	*Vibrio fluvialis*				996,1504
304613457	*Aeromonas encheleia*				494
304613556	*Vibrio variabilis*	R-40492^T	1	Mucus of apparently healthy coral, *P. caribaeorum*	321

Continued

Table 3.26. Continued.

API 20E number	Organism	Strain	No. of strains	Notes	Ref
304614516	*Vibrio splendidus*	A515, AH5	1	Scallop strain A515, turbot strain AH5	590
304614556	*Vibrio splendidus*	A515, AH5	1	Scallop strain A515, turbot strain AH5	590
304616556	*Vibrio tubiashii*	NCIMB 1340	1	Hard clam, USA	244
304616556	*Vibrio splendidus*				590
304632556	*Vibrio maritimus*	R-40493^T	1	Coral, *Palythoa caribaeorum*	321
304652456	*Vibrio anguillarum*				71
304672556	*Vibrio mediterranei*	CECT 615	1		244
304712657	*Vibrio fluvialis*				996
304712657	*Vibrio furnissii*				1504
304714516	*Vibrio splendidus*	VS4	1	Turbot strain VS4	590
304714556	*Vibrio splendidus*	VS4	1	Turbot strain VS4	590
304752476	*Vibrio anguillarum*				
304752476	*Vibrio anguillarum*				71
304752557	*Vibrio anguillarum*	CRL 11			244
304752657	*Vibrio anguillarum*				71
306610546	*Vibrio pomeroyi*	LMG 20537^T	1		524
306612546	*Vibrio kanaloae*	LMG 20539^T	1		524
306612556	*Vibrio aestuarianus*	NCIMB 2236			244
306612757	*Aeromonas sanarellii*	CECT 7402^T		Human strain	36
30671656	*Vibrio kanaloae*	LMG 20539^T	1		1635
320412657	*Vibrio furnissii*	ATCC 35016^T	1		244
322 4/6 0 2/6 556	*Vibrio pacinii*	LMG 13245		Seabass	638
322 6/7 0 2/6 556	*Vibrio pacinii*	LMG 1999^T		Prawn	638
324012457	*Vibrio diazotrophicus*	ATCC 33466^T	1	Reeds, seagrass, gut of sea urchin, rotifers	244
324013457	*Vibrio diazotrophicus*	ATCC 33466^T	1	Reeds, seagrass, gut of sea urchin, rotifers	244
324300006	*Vibrio penaeicida*	DSMZ 14398	1	Pathogen of prawns (*Penaeus japonicus*)	244
324410556	*Vibrio splendidus*	RVAU 88-12-717		*S. maximus*, Norway	71
324412757	*Vibrio fluvialis*				996,1504
324412757	*Vibrio diazotrophicus*	ATCC 33466^T	1	Reeds, seagrass, gut of sea urchin, rotifers	244
324413756	*Vibrio plantisponsor*	MSSRF60^T	2	Roots of wild rice in mangroves	1372
324413757	*Vibrio diazotrophicus*	ATCC 33466^T	1	Reeds, seagrass, gut of sea urchin, rotifers	244
324414557	*Vibrio splendidus*	LMG 16752		Oyster, Spain	71
324472757	*Vibrio anguillarum*				71

324412657	*Vibrio furnissii*	ATCC 11218	1	Citrate, indole, weak reaction	244
324512657	*Vibrio furnissii*	ATCC 11218	1	Citrate, indole, weak reaction	244
324562757	*Vibrio anguillarum*				71
324614516	*Vibrio splendidus*	VS1	1	Turbot strain VS1	590
324614556	*Vibrio splendidus*	VS1	1	Turbot strain VS1	590
324632657	*Vibrio anguillarum*				71
324602557	*Vibrio cholerae*	LMG 16742		*P. orientalis*, PRC	71
324610456	*Vibrio splendidus*	RVAU 88-12-712		*Solea solea*, Norway	71
324610556	*Vibrio splendidus*	UB S308		Seawater, Sweden	71
324610557	*Vibrio splendidus*	LMG 16745		*S. auratus*, Greece	71
324612557	*Vibrio splendidus*	LMG 16744		*S. auratus*, Greece	71
324612657	*Aeromonas caviae*				42, 494
324612757	*Vibrio fluvialis*	NCTC 11327	1		244
326612757	*Vibrio fluvialis*	NCTC 11327	1		244
324612657	*Vibrio fluvialis*	MEJ 311	3		496, 996,1504
324612657	*Vibrio furnissii*	ATCC 35016^T			496
324614556	*Vibrio splendidus*	LMG 16748		Oyster, Spain	71
324616556	*Vibrio splendidus*	LMG 16749		Oyster, Spain	71
32471255.	*Vibrio brasiliensis*	LMG 20546^T		Bivalve larvae	1632
324712556	*Vibrio splendidus*	LMG 16746		Oyster, Spain	71
324712657	*Vibrio fluvialis*				996,1504
324712657	*Vibrio furnissii*	ATCC 11218			244,1504
324712677	*Vibrio anguillarum*				71
324713757	*Aeromonas hydrophila*	ATCC 14715	1		669
324750456	*Vibrio anguillarum*	ATCC 19264			244
324750556	*Vibrio anguillarum*	ATCC 19264			244
324752456	*Vibrio anguillarum*	NCMB 6^T			496
324752557	*Vibrio anguillarum*				71
324752656	*Vibrio anguillarum*	AHLDA 1730	2	Japan, Australia (freshwater fish)	71, 244
324752756	*Vibrio anguillarum*				68
324752756	*Vibrio anguillarum*				878
324752756	*Vibrio anguillarum*				71
324752757	*Vibrio anguillarum*				71
324752777	*Vibrio anguillarum*				71
324772656	*Vibrio anguillarum*				71
324772657	*Vibrio anguillarum*				71

Continued

Table 3.26. Continued.

API 20E number	Organism	Strain	No. of strains	Notes	Ref
326612757	*Vibrio fluvialis*	NCTC 11327	1		244
326612757	*Aeromonas taiwanensis*	CECT 7403^T		Human strain	36
3644152	*Citrobacter freundii*				870,1654
3644153	*Citrobacter freundii*				870,1654
3644552	*Citrobacter freundii*				870,1654
3644553	*Citrobacter freundii*				870,1654
3644572	*Citrobacter freundii*				870,1654
3644573	*Citrobacter freundii*				870,1654
3644172	*Citrobacter freundii*				870,1654
3744152	*Citrobacter freundii*				870,1654
3744153	*Citrobacter freundii*				870,1654
3744172	*Citrobacter freundii*				870,1654
3744173	*Citrobacter freundii*				870,1654
3744552	*Citrobacter freundii*				870,1654
3744553	*Citrobacter freundii*				870,1654
3744573	*Citrobacter freundii*				870,1654
3744573	*Citrobacter freundii*				870,1654
400400057	*Edwardsiella ictaluri*			Vietnamese catfish	386
400500556	*Aliivibrio fischeri*	ATCC 25918		1.5% NaCl	1641
400600557	*V. vulnificus* biovar 2 serovar E	NCIMB 2136		Diseased eel, Japan	180
40440045.	*Moritella viscosa*	NCIMB 13484		Atlantic salmon, Scotland	236
404410556	*Vibrio harveyi*	Abalone		Japanese abalone	1228
404450656	*Aliivibrio fischeri*	ATCC 7744^T		0.85% NaCl	244
404455656	*Aliivibrio fischeri*	ATCC 7744^T		0.85% NaCl	244
404510556	*Vibrio harveyi*	Abalone		Japanese abalone	1228
40460045.	*Moritella viscosa*	NCIMB 13485		Atlantic salmon, Scotland	236
404612456	*Vibrio alginolyticus*	ATCC 17749^T			998
404652456	*Vibrio alginolyticus*	ATCC 17749^T			1034
404712456	*Vibrio alginolyticus*				244
404712457	*Vibrio alginolyticus*				244
404752456	*Vibrio alginolyticus*	ATCC 17749^T			257
410400057	*Edwardsiella ictaluri*			Channel catfish	68,702
410414057	*Edwardsiella ictaluri*			Turkish rainbow trout	880

410610457	*V. vulnificus* biovar 1	M626		Eel, Spain	59
410610657	*Vibrio parahaemolyticus*				1164
410610757	*Vibrio parahaemolyticus*				1164
414400057	*Edwardsiella tarda*	Eels		Spain	21
414412557	*V. vulnificus*	818		Unknown, France	180
414610556	*Vibrio harveyi*	Abalone		France	1219
414610657	*Vibrio parahaemolyticus*	ATCC 43996			244
414612456	*Vibrio alginolyticus*	Swy		Kuruma prawn	998
414612556	*Vibrio harveyi*	ATCC 14129^T			1228
414712457	*Vibrio alginolyticus*				1409
414712557	*Vibrio alginolyticus*				1164
414716557	*Vibrio alginolyticus*				1409
415412557	*Vibrio (carchariae) harveyi*				244,1228,1830
415412556	*Vibrio harveyi*				1219,1228
415610657	*Vibrio parahaemolyticus*	ATCC 17802^T	1		19
415612556	*Vibrio harveyi*				
415612557	*Vibrio (carchariae) harveyi*				244,1228,1830
416612556	*Vibrio owensii*	DY05^T	1	Diseased ornate spiny lobster, Australia	270
416612556	*Vibrio diabolicus*	CNCM I-1629^T	1	The Pompeii worm, deep sea vent	1368
420600557	*V. vulnificus* biovar 2				180
421400456	*Aliivibrio fischeri*	ATCC 7744^T		2% NaCl	244
421400556	*Aliivibrio fischeri*	ATCC 7744^T		2% NaCl	244
421410456	*Aliivibrio fischeri*	ATCC 7744^T		2% NaCl	244
421410556	*Aliivibrio fischeri*	ATCC 7744^T		2% NaCl	244
421500456	*Aliivibrio fischeri*	ATCC 7744^T		2% NaCl	244
421500556	*Aliivibrio fischeri*	ATCC 7744^T		2% NaCl	244
421510456	*Aliivibrio fischeri*	ATCC 7744^T		2% NaCl	244
421510556	*Aliivibrio fischeri*	ATCC 7744^T		2% NaCl	244
424410556	*Vibrio harveyi*	Abalone		Japanese abalone	1228
424510556	*Vibrio harveyi*	Abalone		Japanese abalone	1228
4304004..	*Burkholderia cepacia*		4/10		64
430410257	*Hafnia alvei*				68
430610657	*Vibrio parahaemolyticus*				1164
434112546	*Vibrio owensii*	DY05^T	1	Ornate spiny lobster (*P. ornatus*), Australia	244
434400057	*Edwardsiella tarda*	Eels		Spain	21

Continued

Table 3.26. Continued.

API 20E number	Organism	Strain	No. of strains	Notes	Ref
434610556	*Vibrio harveyi*	Abalone		Abalone, France	1219
434610557	*V. vulnificus* biovar 1 non-serovar E	M06-24		Human blood, USA	180
434610657	*Vibrio parahaemolyticus*				878
434610757	*Vibrio parahaemolyticus*	ATCC 17802^T	1		244
434610657	*Vibrio parahaemolyticus*	ATCC 43996	1		244
434612546	*Vibrio owensii*	DY05^T	1	Ornate spiny lobster (*P. ornatus*), Australia	244
434612556	*Vibrio harveyi*	ATCC 14129^T	1		1228
434612557	*Vibrio harveyi*	ATCC 14129			1228
434712457	*Vibrio alginolyticus*				1164
435412556	*Vibrio harveyi*	ATCC 35084	1		1219,1228
435412557	*Vibrio (carchariae) harveyi*				244,1228,1830
435612556	*Vibrio harveyi*	ATCC 35084	1		1219
435652556	*Vibrio harveyi*	DN01		Summer flounder disease, FINE	592
435612456	*Vibrio alginolyticus*	ATCC 1507	1		244
435612557	*Vibrio (carchariae) harveyi*	ATCC 35084			244,1228,1830
436112546	*Vibrio owensii*	DY05^T	1	Ornate spiny lobster (*P. ornatus*), Australia	244
436610757	*Vibrio parahaemolyticus*	ATCC 17802	1		244
436612546	*Vibrio owensii*	DY05^T	1	Ornate spiny lobster (*P. ornatus*), Australia	244
454400057	*Edwardsiella tarda*	ATCC 15947^T			68,1674
454400057	*Edwardsiella piscicida*	NCIMB 14824^T	9	European eel, turbot, Korean catfish	1
4644100..	Atypical *Edwardsiella tarda*		4	Atypical strains from red seabream, Japan	1820
474400057	*Edwardsiella piscicida*	NCIMB 2056	1	crimson seabream	1
474400057	*Edwardsiella tarda*			Turbot (*Psetta maxima*) Europe, Eels Spain	21, 286
474400057	*Edwardsiella tarda*	AHLDA 135		ATCC 15947^T, CECT 849^T	244,1820
474410057	*Edwardsiella piscicida*	ET080814		Japanese eel, China	1
474410257	*Edwardsiella piscicida*	ET080813		Marbled eel, China	1
500400557	*V. vulnificus* biovar 2 serovar E	520		Shrimp, Taiwan, virulent for eels	180
50041137	*Escherichia vulneris*				
50041337	*Escherichia vulneris*				
500600457	*V. vulnificus* biovar 2 serovar E	NCIMB 2137		Diseased eel, Japan	59,180
500600557	*V. vulnificus* biovar 2 serovar E	ATCC 33149		Diseased eel, Japan	59,180,741
500600557	*V. vulnificus* biovar 2	NCIMB 2138		Diseased eel, Japan	180
504410557	*V. vulnificus*	822		Shrimp, Senegal	180

504600557	*V. vulnificus* biovar 1 non-serovar E	532		Diseased eel, Belgium, avirulent for eels	180
504600557	*V. vulnificus* biovar 1	Vv1		Human wound infection, USA	59
504610557	*V. vulnificus* biovar 1	UMH1		Human wound infection, USA	59
504712556	*Vibrio (communis) owensii*	R-40900	1	Coral (*M. hispida*)	320
510200457	*V. vulnificus* biovar 2 serovar E	121		Diseased eel, Sweden	180
510200557	*V. vulnificus* biovar 2 serovar E	526		Diseased eel, Sweden	180
510410057	*Yersinia ruckeri*	ATCC 29473			367, 402, 1425, 1476
510411257	*Hafnia alvei*				1418
510510057	*Yersinia ruckeri*	AHLDA 1313	2	Channel catfish	244, 402, 1425, 1643
510550057	*Yersinia ruckeri*		20	Environmental isolates	367,1425
510600557	*V. vulnificus* biovar 2	NCIMB 2136		Diseased eel, Japan	59
510600557	*V. vulnificus* biovar 2 serovar E	524		Diseased eel, Norway	180
510610017	*Yersinia ruckeri*				1643
510610557	*V. vulnificus* biovar 1	M631		Eel, Spain	59
510650017	*Yersinia ruckeri*				1643
510710017	*Yersinia ruckeri*				1425,1643
510710057	*Yersinia ruckeri*	YR55, YR80	2	Environmental isolates	367
510750057	*Yersinia ruckeri*		1	Environmental isolates	367,1643
514410557	*V. vulnificus* biovar 1 serovar E	534		Diseased eel, Sweden, avirulent for eels	180
514412757	*V. vulnificus* biovar 1 non-serovar E	628		Paguara fish, Venezuela	180
514512556	*Vibrio (communis) owensii*	R-40504	1	Coral (*P. variabilis*)	320
514600557	*V. vulnificus* biovar 2 serovar E	523		Diseased eel	180
514602557	*V. vulnificus* biovar 1 non-serovar E	530		Diseased eel, Belgium, virulent	180
514610457	*Vibrio mimicus*	ATCC 33653^{T}			244, 326
514610556	*Vibrio harveyi*	ATCC 14126			1042
514610556	*Vibrio harveyi*	ATCC 25919			1042
514610557	*V. vulnificus* biovar 1 non-serovar E	ATCC 27562		Human strain, USA	50,180
514610557	*V. vulnificus* biovar 1 non-serovar E	B9629, C7184		Human wound infection, USA	59,180
514610557	*V. vulnificus* biovar 1 non-serovar E	UMH1, 374		Human wound infection, USA	59,180
514610557	*V. vulnificus* biovar 1 serovar E	ATCC 33186		Human blood, USA, avirulent for eels	180
514612457	*Vibrio cholerae* 01				
514650557	*V. vulnificus* biovar 1	167		Eel, Belgium	59
514712556	*Vibrio (communis) owensii*	LMG 25430^{T}	6	Coral (*M. hispida*, *P. dilatata*, *P. caribaeorum*), prawn	320

Continued

Table 3.26. Continued.

API 20E number	Organism	Strain	No. of strains	Notes	Ref
514712457	*Vibrio cholerae* non 01				
514712557	*Aeromonas veronii veronii*	ATCC 35622	1	Faeces, diarrhoea, USA	723
516600557	*V. vulnificus* biovar 1	VIB 521		Unknown	59
516612556	*Vibrio owensii*	47666-1	1	Diseased tiger prawn, Australia	270
5176167..	*Vibrio rotiferianus*	LMG 21460^T		Rotifer	637
520600556	*Aliivibrio salmonicida*	RVAU 890206-1/7		*Salmo salar*, Faroe Islands	71
520600557	*V. vulnificus* biovar 2 serovar E	NCIMB 2138		Diseased eel, Japan	180
520600557	*V. vulnificus* biovar 2 serovar E	E86		Diseased eel, Spain	50,180
521577357	*Klebsiella pneumoniae*				68
524610557	*V. vulnificus* biovar 1				180
524712457	*Vibrio cholerae*	PS-7701		*P. altivelis*, Japan	71
524712476	*Vibrio cholerae*	PS-7705		*P. altivelis*, Japan	71
530710057	*Yersinia ruckeri*				411,1425
530750057	*Yersinia ruckeri*				411
5304004..	*Burkholderia cepacia*		1/10		64
5304112..	*Hafnia alvei*	ATCC 51873	1		402
530475257	*Serratia fonticola*				601
530477257	*Serratia fonticola*				601
530510057	*Yersinia ruckeri*	AHLDA 1313			244
530511357	*Hafnia alvei*	AHLDA 1729			244
530536057	*Serratia liquefaciens*	K2		From Atlantic salmon	1123
530536357	*Serratia liquefaciens*	K2		From Atlantic salmon	1123
530555257	*Serratia fonticola*				1176
530600457	*V. vulnificus* biovar 2 serovar E	525		Diseased eel, Sweden	180
530600557	*V. vulnificus* biovar 2 serovar E	ATCC 33187		Human strain, USA	180
530600557	*V. vulnificus* biovar 2 serovar E	E105		Diseased eel, Spain	50,180
530610057	*Yersinia ruckeri*				68
530610557	*V. vulnificus* biovar 2 serovar E			Taiwanese strains, avirulent for eels	50
530710057	*Yersinia ruckeri*				411
530750057	*Yersinia ruckeri*				411
530752157	*Serratia marcescens*	White perch		White perch	137
530772357	*Serratia marcescens*	White perch		White perch	137

530776217	*Serratia liquefaciens*	Turbot		From turbot	1724
534600557	*V. vulnificus* biovar 1 non-serovar E	ATCC 27562		Human strain, USA	59
534600557	*V. vulnificus* biovar 1 non-serovar E	E109		Healthy eel, Spain, avirulent for eels	180
534600557	*V. vulnificus* biovar 1 serovar E	521		Unknown, Australia, avirulent for eels	180
534610357	*V. vulnificus*	WF8A1110			1076
534610457	*Vibrio mimicus*	ATCC 33653^T		Human ear infection	244, 326, 412
534610557	*V. vulnificus* biovar 1 non-serovar E	C7184		Human blood, USA, avirulent for eels	180
534612457	*Vibrio cholerae*	91/1198		*Carassius auratus*, Australia	71
534612457	*Vibrio cholerae* 01				
534612557	*Aeromonas veronii veronii*	0935-84, 1068-84	2	Faeces, diarrhoea, USA	723
534712457	*Vibrio cholerae* non 01				
534712556	*Vibrio (communis) owensii*	R-40498		Coral (*M. hispida*)	320
534712557	*Aeromonas veronii veronii*	0265-84	1	Endotracheal tube, drowning victim, USA	59,723
534712557	*Vibrio cholerae*	WF 110r			1076,1164
536412546	*Vibrio owensii*	47666-1	1	Tiger prawn (*P. monodon*), Australia	244
560621057	*Salmonella cholerasuis arizonae*				68
600400456	*Photobacterium iliopiscarium*	PS1^T			1272
600400457	*Photobacterium damselae damselae*	ATCC 33539^T		Usually VP positive	496
600500456	*Photobacterium iliopiscarium*	ATCC 51760			244
600500457	*Photobacterium damselae damselae*	LMG 13639	6	Various sources	71
600610417	*Aeromonas salmonicida salmonicida*				1476
6015004	*Photobacterium damselae damselae*			Redbanded seabream	963
604612556	*Vibrio orientalis*	NCMB 2195^T	1	Seawater, spider crab	244
604616556	*Vibrio orientalis*	ATCC 33935	1		244
6046021..	*Photobacterium phosphoreum*	IB39			1076
606616556	*Vibrio orientalis*	ATCC 33935	1		244
6100530..	*Halomonas cupida*				68
624750456	*Vibrio proteolyticus*	AHLDA 1735	1	Artemia	244
626750556	*Vibrio proteolyticus*	CIP 102892			244
700606556	*Vibrio marisflavi*	WH134^T	1	Seawater, China	1760
700610517	*Aeromonas salmonicida salmonicida*	NCIMB 1102			968
70140045.	*Aliivibrio finisterrensis*	C 13.19	1		143
70140445.	*Aliivibrio finisterrensis*	CMJ 11.1^T	1		143
70141245.	*Aliivibrio finisterrensis*	C 17.13	1		143
70041137	*Escherichia vulneris*				*

Continued

Table 3.26. Continued.

API 20E number	Organism	Strain	No. of strains	Notes	Ref
70041337	*Escherichia vulneris*				*
704672556	*Vibrio mediterranei*	AHLDA 1733	1	Abalone	244
704612457	*A. veronii* biovar Sobria				4
704612657	*A. veronii* biovar Sobria				4
704712457	*A. veronii* biovar Sobria				4
704712457	*Aeromonas sobria*	CIP 74.33^{T}			493
704712657	*Aeromonas sobria*				
704712657	*Aeromonas hydrophila*	ATCC 7966^{T}	1		493
704713757	*Aeromonas bestiarum*		5	Carp	935
714420457	*Plesiomonas shigelloides*				68
720600457	*Aeromonas schubertii*	ATCC 43700^{T}	2	Clinical strains	278
720700457	*Aeromonas schubertii*	AMC 4396-87	1	Clinical strains	278
724477556	*Vibrio (shilonii) mediterranei*	ATCC BAA-91	1	Bleached coral	244
724610557	*Aeromonas trota*	ATCC 49657^{T}	1		277, 789
724612457	*A. veronii* biovar Sobria				4
724612657	*A. veronii* biovar Sobria				4
724613657	*Aeromonas allosaccharophila*		1		68
724616657	*Aeromonas allosaccharophila*		1		68
724617657	*Aeromonas allosaccharophila*		1		68
724712457	*A. veronii* biovar Sobria				4
724712657	*A. veronii* biovar Sobria				4
724714457	*Aeromonas jandaei*		2		244, 276
724732046	*Vibrio metschnikovii*	NCTC 8443^{T}	1	Isolated from cockles	244
726732046	*Vibrio metschnikovii*	NCTC 8443^{T}	1	Isolated from cockles	244
744612457	*A. veronii* biovar Sobria				4
744612657	*A. veronii* biovar Sobria				4
744712457	*A. veronii* biovar Sobria				4
744712657	*A. veronii* biovar Sobria				4
764612457	*A. veronii* biovar Sobria				4
764612657	*A. veronii* biovar Sobria				4
764712457	*A. veronii* biovar Sobria				4
764712657	*A. veronii* biovar Sobria				4

Not all references reported the nine-digit profile number, hence some species only have a seven- or eight-digit profile number with no results for growth on MCA or OF results. ID=identification.

Table 3.27. API 20NE database biochemical results.

	NO_3	TRP	Glu	ADH	Ure	Esc	Gel	Png	Glu	Ara	Mne	Man	Nag	Mal	Gnt	Cap	Adi	Mlt	Cit	Pac	Ox	Temp	Time	Inoc	Ref	Strain
Organism																										
Acinetobacter calcoaceticus	–	–	–	–	–	–	–	–	–	–	–	–	–	–	–	+	–	–	+	+	–		24		68	
Actinobacillus delphinicola	+	–	–	–	–	–	–	–	–	–	–	–	–	-	-	-	-	-	-	-	+	37	48	0.85	548	Dolphin
Actinobacillus delphinicola	+	–	–	+	–	–	–	–	–	–	–	–	–	–	–	–	–	–	–	–	+	37	48	0.85	548	Porpoise
Actinobacillus delphinicola	+	–	–	–	–	–	–	–	–	–	–	–	–	–	–	–	–	–	–	–	+	37	48	0.85	548	Whale
Aeromonas fluvialis	+	+	+	–	–	–	–	+	+	–	+	–	+	+	+	–	–	+	–	–	+	30	48	0.85	37	CECT 7401^T
Aeromonas hydrophila	+	v	+	+	–	+	+	+	+	v	+	+	+	+	+	+	–	+	v	–	+				68	
Aeromonas rivuli	+	–	+	+	–	+	+	+	+	–	+	+	+	+	–	+	–	+	–	–	+	30	48	0.85	536	CECT 7518^T
A. salmonicida achromogenes/masoucida	+	v	–	–	–	v	v	–	v	–	v	v	–	–	–	–	–	–	–	–	+				68	
A. salmonicida ssp. *achromogenes*		78	21	14		7			78		7	21	–	–				50							69	
A. salmonicida ssp. *masoucida*		60	+	80		+			+		20	+	80	10				–							69	
A. salmonicida ssp. *salmonicida*	+	v	v	v	–	+	+	–	+	–	v	+	+	+	+	–	–	+	–	–	+				68	
A. salmonicida ssp. *salmonicida*	+	–	+	–	–	+	+	–	–	–	–	–	–	–	–	–	–	–	–	–	+	15	72	0.85	513	
A. salmonicida ssp. *smithia*		50	25	25		25			+		75	75	25	75				50							69	
Aeromonas sanarellii	+	+	+	+	–	+	+	+	+	+	–	+	+	+	+	+	–	+	–	–	+	30	48	0.85	36	CECT 7402^T
Aeromonas taiwanensis	+	+	+	+	–	+	+	+	+	+	–	+	+	+	+	+	–	+	–	–	+	30	48	0.85	36	CECT 7403^T
Algicola (*Pseudoalteromonas*) *bacteriolytica*	–		+				+		+		+	–	–	–	–			–	–	+	+				1489	
Alteromonas nigrifaciens	–			–			+		+		+	–		+	+	+			+	+	+				816	
Andreprevotia lacus	+	–	+	+	+	–			+	+	+	+	+	+	–	–	+	+	+	–	+	30	48	0.85	1513	Type strain
Bordetella bronchiseptica	–	–	–	–	+	–	–	–	–	–	–	–	–	–	–	–	+	–	+	+	+	37		0.85	184	Type strain
Bordetella bronchiseptica	+	–	–	–	+	–	–	–	–	–	–	–	–	–	–	–	+	–	+	+	+	37		0.85	184	Type strain
Bordetella bronchiseptica	+	–	–	–	+	–	–	–	–	–	–	–	–	–	–	–	+	–	+	–	+	37		0.85	184	Type strain
Brevundimonas diminuta	–	–	–	–	–	–	–	–	–	–	–	–	–	–	–	–	–	–	–	–	+	28	48	0.85	1502	ATCC 11568^T
Brucella ceti and *B. pinnipedialis*	+	–	–	–	+	–	–	–	–	–	–	–	–	–	–	–	–	–	–	–	+	37	48	0.85	546	
Chromobacterium violaceum	–	–	+	+	–	–	+	–	+	–	–	–	+	–	+	+	–	+	+	–	+				992	Pigmented
Chromobacterium violaceum	+	–	+	+	–	–	+	–	+	–	–	–	+	–	+	+	–	+	+	–	+				992	Non-pigmented
Chryseobacterium balustinum	+	+	–	–	+	+	–		–	–	w	–	–	w	–	–	–	–	w	–	+	37	48	0.85	1693	
Chryseobacterium hominis	–										+			+	+						+				266	
Chryseobacterium hominis	+	+			–		–	–		–	+	–		+			–		–		+	25	48	0.85	1846	CCUG 52711^T
Chryseobacterium indologenes	-	+	-	-	+	+	–		w	-	w	-	-	w	-	-	-	-	w	-	+	37	48	0.85	1693	
Chryseobacterium indologenes	–	+			+		–	+		–	+	–		+			–		+		+	25	48	0.85	1846	CCUG 14556^T

Continued

Table 3.27. Continued.

Organism	NO_3	TRP	Glu	ADH	Ure	Esc	Gel	Png	Glu	Ara	Mne	Man	Nag	Mal	Gnt	Cap	Adi	Mlt	Cit	Pac	Ox	Temp	Time	Inoc	Ref	Strain
Chryseobacterium indoltheticum	–	+	–	–	–	+	+		w	–	w	–	–	w	–	–	–	–	–	w	+	37	48	0.85	1693	
Chryseobacterium oncorhynchi	–	+			+		+	–		–	+	–		+			–		+		+	25	48	0.85	1846	701B–08[T]
Chryseobacterium scophthalmum	–	–	w	–	+	+	+		+	–		–		+					–		+	22	48	0.85	1173	
Chryseobacterium tructae	–	+	–	–	+	+	+	–	+	–	+	–	–	+	–	–	–	–	+	–	+	25	48	0.85	1846	CECT 7798[T]
Elizabethkingia meningoseptica	–	+	–	–	–	+	+	+	+	–	+	+	+	+	–	–	–	–	–	–	+	37	48	0.85	1693	LMG 12279[T]
Elizabethkingia meningoseptica	–	+	–	–	+	+	+	+	+	–	+	+	+	w	–	–	–	–	+	–	+	37	48	0.85	1693	LMG 13004
Elizabethkingia meningoseptica	–	+	–	–	+	+	+	+	+	–	+	w	w	w	–	–	–	–	+	–	+	37	48	0.85	1693	
Ferrimonas senticii					–																				265	P2S11[T]
Flavobacterium ceti	–	–	–	–	–	–	+	–	–	–	–	–	–	–	–	–	–	–	–	–	+	37	48	0.85	1710	
Flavobacterium gleum	+	v	–	–	+	+	+		+	+	w	–	–	+	–	–	–	–	+	–	+	37	48	0.85	1693	
Flavobacteriun hibernum	+	–	+	–	–	+	+	+	+	+	+	–	+	+	–	–	–	–	–	–	–	25	48	0.85	1116	ATCC 51468[T]
Flavobacterium johnsoniae	+	–	–	–	–	+	+	+	+	–	+	–	+	+	–	–	–	–	+	–	+	25	48	0.85	867	Sturgeon
Flavobacterium macrobrachii	–	–	–	–	–	–	+	–	–	–	–	–	–	–	–	–	–	–	–	–	+	25	48	0.85	1514	LMG 25203[T]
Flavobacterium oncorhynchi	+	–	–	–	–	+	–	+	–	+	+	–	+	+	–	–	–	–	–	–	+	25	48	0.85	1845	CECT 7678[T]
Flavobacterium tilapiae	–	–	–	–	–	+	+	+	+	+	+	–	+	+	–	–	–	–	–	–	+	25	48	0.85	314	BCRC 80262[T]
Gramella marina	–	–	+	–	–	+	+	+	+	+	+	–	–	+	+	–	+	+	+	+	+	28	48	2	1205	KMM 6048[T]
Halomonas cupida	+									+	+		+	+	+	+					–				124	
Halomonas venusta	+	–	–	–	+	+	–	+	+	–	–	–	+	+	+	+	–	+	+	+	+	37	48	0.85	1731	Human
Janthinobacterium lividum	+	–	–	–	–	+	–	–	+	+	+	+	v	+	–	v	v	+	+	+	+				1030	
Vibrio anguillarum	+	+	+	+	–	–	+		+	+	+	+	+	+	+			+	+		+				210	
Myroides odoratimimus	–	–	–	–	+	–	+	–	–	–	–	–	–	–	–	–	–	+w	–	–	+	28	48	0.85	1694	NCTC 11180[T]
Myroides odoratus	–	–	–	–	+	–	+	–	–	–	–	–	–	–	–	–	–	+w	–	–	+	28	48	0.85	1694	NCTC 11036[T]
Oceanimonas baumannii	+	–	–	–	–	–	–	–	–	–	–	–	–	–	–	+	–	+	+	–	+	25	48	1.5	232	ATCC 700832[T]
Oceanimonas doudoroffii	+	–	–	–	–	–	–	–	–	–	–	–	–	–	–	+	–	+	+	–	+	25	48	1.5	232	ATCC 27123[T]
Phocoenobacter uteri	+	–	+w	–	–	–	–	–	–	–	–	–	–	–	–	–	–	–	–	–	+	37	48	0.85	549	NCTC 12872[T]
Photobacterium eurosenbergii	+	+	+	–	–	+	+	+	+	–	v	–	w	–	–	–	–	v	–	w	+	30	48	2	1411	Wild types
Photobacterium damselae	+	–	+	+	+	–	–	–	–	–	–	–	–	–	–	–	–	v	–	–	+				68	
P. damselae damselae	+	–	+	+	+	v	–	–	+	–	+	–	+	+	–	–	+	–	–	+	+				595	
P. damselae piscicida	–	–	+	+	–	–	–	–	+	–	+	–	+	–	–	–	+	–	–	–	+				595	
Plesiomonas shigelloides	+	+	+	+	–	–	–	+	v	–	–	–	v	v	v	v	–	v	–	–	+				68	
Pseudoalteromonas antarctica	–	–	+	–	–	–	+	–	+	–	+	+	+	+	+	–	–	–	+	–	+	5	5d		215	
Pseudoalteromonas haloplanktis					+		+		+				–		–			–							594	

Pseudomonas anguilliseptica	–w	–	–	–	–	–	+	–	–	–	–	–	v	+	–	–	+	–	–	–	+	22	48	0.85	1136	
Pseudomonas anguilliseptica	–	–	–	+	–	–	–	–	–	–	–	–	–	–	–	+	–	+	+	–	+	25	48	DW	100, 1059	
Pseudomonas anguilliseptica	–	–	–	–	–	–	–	–	–	–	–	–	–	–	–	+	–	+	+	–	+	25	48	DW	1059	
Pseudomonas anguilliseptica	–	–	–	+	–	–	–	–	–	–	–	–	–	–	–	–	–	+	+	–	+	25	48	DW	100, 1059	
Pseudomonas anguilliseptica	–	–	–	–	–	–	–	–	–	–	–	–	–	–	–	–	–	+	+	–	+	25	48	DW	1059	
Pseudomonas anguilliseptica	–	–	–	+	–	–	–	–	–	–	–	–	–	–	–	+	–	+	–	–	+	25	48	DW	100	
Pseudomonas chlororaphis ssp *aurantiaca*	–	–	–	–	–	–	+	–	+	+	+	+	+	–	+	+	–	+	+	+	+	25	48	0.85	252	NCIMB 10068[T]
Pseudomonas chlororaphis ssp *aureofaciens*	+	–	–	–	–	–	+	–	+	+	+	+	+	–	+	+	–	+	+	+	+	25	48	0.85	252	DSM 6698[T]
Pseudomonas chlororaphis ssp *chlororaphis*	–	–	–	–	–	v	–	–	+	+	+	+	–	–	+	v	–	+	+	–	v				68	
Pseudomonas chlororaphis ssp *chlororaphis*	+	–	–	+	–	–	+	–	+	–	+	+	+	–	+	+	–	+	+	–	+	25	48	0.85	252	DSM 50083[T]
P. chlororaphis ssp. *piscium*	+	–	–	–	–	–	+	–	+	–	+	+	+	–	+	+	–	+	+	+	+	25	48	0.85	252	JF 3835T
Pseudomonas fluorescens	–	–	–	+	–	–	–	–	+	+	+	+	+	–	+	+	–	+	+	–	+	25	48	0.85	252	DSM 50090[T]
Pseudomonas fluorescens	v	–	–	v	–	–	v	–	+	v	+	+	v	–	+	+	–	+	+	–	+				68	
Pseudomonas fluorescens	27	–	–	80	1	1	39	1	99	71	97	89	85	1	99	99	10	99	99	16	99	25	48	0.85	*	
Pseudomonas plecoglossicida	+	–	–	+	–	–	–	–	+	–	–	–	–	–	+	+	–	+	+	+	+	25	48	0.85	1229	ATCC 700383[T]
Pseudomonas putida	+	–	–	+	–	–	–	–	+	–	–	–	–	–	+	+	–	+	+	+	+	25	48	0.85	252	DSM 291[T]
Pseudomonas putida	–	–	–	+	–	–	–	–	+	–	–	–	–	–	+	+	–	+	+	–	+	25	48	0.85	39	Trout
Pseudomonas putida	+			+			–			v	v	v						v		+					1229	
Roseovarius crassostreae	+			–	–	–	–														+				197	ATCC BAA-1102[T]
Shewanella affinis	+	–	–	–	–	+	–	–	–	–	–	–	–	–	–	–	–	–	–	–	+	25	48	2	1480	ATCC BAA-642[T]
Shewanella algae LMG 2265	+	–	–	–	–	–	+	–	–	–	–	–	+	–	–	+	–	+	–	–	+				1853	
Shewanella algae	+	–	–	–	–	+	+	–	+	–	–	–	+	–	–	+	–	+	–	–	+				1480	IAM 14159[T]
Shewanella algidipiscicola	+	–	–	–	–	+	–	–	–	–	–	–	+	–	–	3	–	+	–	–	+	25	48	2	1481	LMG 23746[T]
Shewanella baltica	+	–	+	–	–	–	+	–	+	–	–	–	+	+	+	–	–	+	+	–	+	25	48	2	1480, 1853	NCTC 10735[T]
Shewanella baltica	+	–	–	–	–	+	+	–	+	–	–	–	+	+	+	+	–	+	+	–	+	25	48	2	1480	NCTC 10337[T]
Shewanella colwelliana	+	–	–	–	–	+	+	–	–	–	–	–	–	–	+	–	–	–	–	–	+	25	48	2	1480	ATCC 39565[T]
Shewanella corallii	+	–	+	+	–	+	+	–	–	–	–	–	+	+	–	–	–	–	–	–	+	25	48	2	1527	LMG 24563[T]
Shewanella glacialipiscicola	+	–	–	–	–	+	+	–	–	57	–	–	+	–	+	–	–	+	–	–	+	25	48	2	1481	LMG 23744[T]
Shewanella hafniensis	+	–	–	–	–	+	+	–	+	21	–	–	+	+	+	71	–	+	+	–	+	25	48	2	1480	ATCC BAA-1207[T]

Continued

Table 3.27. Continued.

Organism	NO_3	TRP	Glu	ADH	Ure	Esc	Gel	Png	Glu	Ara	Mne	Man	Nag	Mal	Gnt	Cap	Adi	Mlt	Cit	Pac	Ox	Temp	Time	Inoc	Ref	Strain
Shewanella haliotis	+	–	–	–	–	–	+	–	–	–	–	–	+	–	–	+	–	+	–	–	+	25	48	2	895	KCTC 12896[T]
Shewanella marinintestina	+	–	–	–	–	–	+		+		–	–	+	–	–						+	25	48	2	1479	
Shewanella morhuae	+	–	–	–	–	+	91	–	–	61	–	–	+	–	+	–	–	+	–	–	+	25	48	2	1480	
Shewanella oneidensis	+	–	–	–	–	+	+	–	–	–	–	–	+	–	–	–	–	–	–	–	+	25	48	2	1480	ATCC 700550[T]
Shewanella pealeana	+	–					–	–	+	–									+		+	25	48	2	1008	
Shewanella putrefaciens	+	–	+	–	–	–	91	–	–	+	–	–	+	+	–	+	–	+	17	–	+	35	48	0.85	886	
Shewanella putrefaciens	+	–	–	–	–	+	–	–	–	+	–	–	+	–	–	–	–	+	–	–	+	25	48	2	1480	ATCC 8071[T]
Shewanella sairae		–	–	–	–	–					–	–	+	–	–						+	25	48	2	1479	
Shewanella schlegeliana		–	–	–	+	+					–	–	+	–	–						+	25	48	2	1479	
Sphingomonas subarctica	–	–	–	–	–	+	–	+	+	+	–	–	+	+	–	–	–	+	+	–						
Stappia stellulata-like strain M1	+			–	+	+	–		+	+	+	+	+	+	+	–	+	+	+	–	+	25	48	1.5	195	
Stenotrophomonas maltophilia	+	–	–	–	–	+	+	+	+	–	+	–	+	+	–	–	–	+	+	–	–	28	48	0.85	606	
Tenacibaculum aiptasiae	–	–	–	–	–	+	+	–	–	–	–	–	–	–	–	–	–	–	–	–	+	25	72	1.5	1761	
Tenacibaculum crassostreae	–	–	–	–	–	–	+												–		+	37	24	2	1761	
Tenacibaculum aiptasiae	–	–	–	–	–	+	+	–	–	–	–	–	–	–	–	–	–	–	–	–	+	25	72	1.5	1761	
Tenacibaculum litopenaei	–	–	–	–	–	–	+		–										–		+	25	72	1.5	1517	
Vagococcus salmoninarum	–	–	+	–	–	+	–	–													–	25	48	0.85	1139, 1442	
Vibrio alginolyticus	+	+	+	–	–	v	+	–	v	–	v	v	v	v	v	–	–	+	–	–	+				68	
Vibrio alginolyticus	+	+	+	–	–	–	+		+	–	+	+	+	+	+	nt	nt	+	+	nt	+				201	
Vibrio alginolyticus	+	+	+	–	–	–	+	+	+	+	+	+	+	+	+	–	–	+	–	–	+					
Vibrio areninigrae	+	+	+	–	–				+	–	–	+		+					–		+	25	24	2	301	KCTC 22122[T]
Vibrio caribbeanicus	–	–	–	+	–	–	–	–	+	–	–	–	–	–	–	–	–	–	–	–	+	28	48	1.5	739	DSM 23640[T]
Vibrio casei	+					+			+	+	+	+	+	+	+			+	+		+	30	24	1.5	190	
Vibrio comitans	+	–	+	–			–		+		–	+	+	+	+			–	33		+	25	48	2	1485	
Vibrio coralliilyticus	+	+	+	+	–	+	+	+	+	–	+	+	+	+	+				+		+	30	48	3	155,156	YB1[T]
Vibrio coralliilyticus	+	–	+	+	–	–	+	–	–	–	–	–	–	–	–	–	–	–	–	–	+	28	48	3	879	DOI
Vibrio cholerae	+	+	+	–	–	–	+	+	+	–	v	v	v	+	+	–	–	+	+	–	+				68	
Vibrio diabolicus	+	+	+	–	–	–	+	+	+	–	+	+	+	+	+	+	–	+	+	–	+	25	48	2	1368	CNCM I-1629[T]
Vibrio fortis	+	+	+	–	–	+	+	+	+	–	w	w	w	w	+	–	–	+	–	–	+	30	48	2	1411	Wild types
Vibrio hangzhouensis	+	+	+	–	–	+	+	+	+	–	+	+		+				+	+		+	30	48	1	1818	CN83[T]
Vibrio harveyi	+	+	+	–	–	+	+	+	+	+	+	+	+	+	+	–	–	+	+	–	+				20	ATCC 14126
Vibrio harveyi	+	+	+	–	–	+	+	+	+	–	–	+	–	–	+	–	–	+	+	–	+				20	

Vibrio (carchariae) harveyi	+	+	+	–	+	+	+	+	+	–	+	+	–	+	+	–	–	+	+	–	+	24	48	2	20	ATCC 35084T
Vibrio harveyi	+	+	+	–	–	+	+	+	+	+	+	+	+	+	+	–	–	+	+	–	+	24	48	2	20	ATCC 14126T
Vibrio harveyi	+	+	+	–	–	+	+	+	+	–	+	+	–	+	+	–	–	+	+	–	+	24	48	2	20	Seahorse strain
Vibrio harveyi	+	+	+	–	–	+	+	+	+	–	+	+	+	v	+	–	–	+	+	v	+	30	48	2	1411	Wild types
Vibrio hippocampi	+	–	+	–	–	–	–	–	+	–	–	–	+	+	–	–	–	–	–	–	+	20	48	1.5	106	
Vibrio ichthyoenteri	+	–	+	–			–					33			50			–			+				804	
Vibrio inusitatus	+	–	+	–			–		+		+	+	+	+	+			–	–		+	25	48	2	1485	
Vibrio litoralis	+	–	+	–	–	–	–	–	+	+	+	+	+	+	+	–	–	+	+	–	+	25–30		3	1202	MANO22T
Vibrio mangrovi	+	–	+	–	–	+															–	28	48	2	1373	
Vibrio marisflavi	+	–	+	+	–	+	–												–		+	28	48	1.5	1760	WH134T
Vibrio mimicus	+	+	+	–	–		–		+	–	–	+	+	+	+			–			+	24	48	0.85	469	Yabby
Vibrio mytili	+	+	+	+	–	–	–	+	+	+	+	+	+	+	+	–	–	+	+	–	+	25	48	2	1368	CECT 632T
Vibrio natriegens	+	+	+	–	–	+	+	v	+	–	–	+	+	+	+	v	–	+	+	–	+	30	48	2	1411	Wild types
Vibrio navarrensis	+	+	+	–	–	+	+		+	–		+	+	+	+	–		+	60	–	+				1687	
Vibrio nereis	+	+	+	–	–	–	v	–	+	–	+	–	+	+	+	–	–	+	+	–	+	25	48	2	1368	LMG 3895T
Vibrio owensii	+	+	+	–	–	+	+		+	–	+	+			+			+	–		+	28	48	2	270	DY05T
Vibrio owensii	+	+	+	–	–	+	v	+	+	–	+	w	w	w	+	–	–	+	–	–	+	30	48	2	1411	Wild types
Vibrio penaeicida	+	–	+	–			+					+			+			+	+		+				803	
Vibrio rarus	+	–	+	–			–		+		+	+	+	+	+			–	–		+	25	48	2	1485	
Vibrio rotiferianus	+	+	+	–	–	+	+	+	+	–	+	v	+	+	+	–	–	+	v	v	+	30	48	2	1411	Wild types
Vibrio scophthalmi	+	–	+	–	–	+	–	+											+		+	25	48	2	523	CECT 4638
Vibrio shiloi	+	+	+	–	–		+				+	+	+								+				952	
Vibrio splendidus	+	–	–	–	–	–	+	+	–	–	–	–	–	–	–	–	–	–	–	–	+	25	48	2	879	Mussel
Vibrio splendidus	+	+	+	+	–	+	+	+											+		+	25	48	2	523	ATCC 33125T
Vibrio tapetis	+	+		–	–	–	+		+	–	+	–	+	–	–			–	–						201	
V. toranzoniae Vb10.8T	+	+	+	–	–	–	+	–	+	+	+	–	+	+	–						–	25	48	2	980	Vb10.8T
V. toranzoniae Cmf 13.9	+	+	+	–	–	+	+	–	+	+	–	+	+	+	+						–	25	48	2	980	Cmf 13.9
V. toranzoniae CMJ 9.4	+	–	+	–	–	–	+	–	+	+	–	+	+	+	–						–	25	48	2	980	CMJ 9.4
V. toranzoniae CMJ 9.11	+	–	+	–	–	–	+	–	+	+	+	+	+	+	+						–	25	48	2	980	CMJ 9.11
Vibrio tubiashii	+	+	+	+	–	–	+	+	+	–	+	+	+	+	+	–	–	+	–	–	+	25	48	2	1368	LMG 10936T
Vibrio vulnificus	+	+	+	–	–	+	+	+	–	–	–	+	–	–	+	–	–	+	+	–	+	25	48	2	*	99.8%
Vibrio vulnificus	+	+	+	–	–	+	+	+	+	–	–	+	–	–	+	–	–	+	+	–	+	25	48	2	*	95.1 %
Vogesella lacus	+	+	–	+	–	–	–		+	–	+	+	–	–	–	–	–	+	+	–	+	25–30	24	0.5	325	
Zobellia galactanivorans	+	–	+	–	–	+	+	+	+	w	+	+	+	+	–	–	–	–	–	–	+	30	7d	2.5	119	DSM 12802T
Zobellia uliginosa	+	–	+	–	–	+	+	+	+	+	+	+	+	+	w	–	–	–	–	–	+	30	7d	2.5	119	

NO_3=nitrate, TRP=indole, GLU=glucose fermentation, ADH=arginine dihydrolase, Ure=urease, Esc=aesculin, Gel=gelatin hydrolysis, Png=p-nitrophenyl-β-D-galactopyranoside, Glu=glucose assimilation, Ara=L-arabinose assimilation, Mne=mannose assimilation, Man=mannitol assimilation, Nag=*N*-acetylglucosamine, Mal=maltose assimilation, Gnt=gluconate assimilation, Cap=caprate assimilation, Adi=adipate assimilation, Mlt=malate assimilation, Cit=citrate assimilation, Pac=phenylacetate, Ox=oxidase, *=API database.

Table 3.28. API 20NE database numbers (organism listed alphabetically).

Organism	Strain	API 20NE number	No. of strains	Notes	Ref
Actinobacillus delphinicola	Dolphin	1000004			548
Actinobacillus delphinicola	Porpoise	1100004			548
Actinobacillus delphinicola	Whale	1000004			548
Aeromonas fluvialis	CECT 7401ᵀ	7062744		Riverwater	37
Aeromonas rivuli	CECT 7518ᵀ	5576354		Creek water	536
A. salmonicida ssp. *salmonicida*		5410004		Lamprey	513
Aeromonas sanarellii	CECT 7402ᵀ	7575754		Human strain	36
Aeromonas taiwanensis	CECT 7403ᵀ	7575754		Human strain	36
Brevundimonas diminuta	ATCC 11568ᵀ	0000004			1502
Brucella ceti and *B. pinnipedialis*		1200004	2		546
Chryseobacterium tructae	CECT 7798ᵀ	2652205	3		1846
Elizabethkingia meningoseptica	LMG 12279ᵀ	2476304	1	Must be read at 48 h	1693
Elizabethkingia meningoseptica	LMG 13004	2676105	2	Must be read at 48 h	1693
Elizabethkingia meningoseptica	LMG 13004	2676305	2	Must be read at 48 h	1693
Flavobacterium hibernum	ATCC 51468ᵀ	5473300			1116
Flavobacterium johnsoniae	Sturgeon	1472305			867
Flavobacterium macrobrachii	LMG 25203ᵀ	0010004			1514
Flavobacterium oncorhynchi	CECT 7678ᵀ	1423304			1845
Flavobacterium tilapiae	BCRC 80262ᵀ	0473304			314
Gramella marina	KMM 6048ᵀ	4473667			1205
Halomonas venusta	Human	1660757			1731
Myroides odoratimimus	NCTC 11180ᵀ	0210004	5		1694
Myroides odoratimimus	NCTC 11180ᵀ	0210044	15		1694
Myroides odoratus	NCTC 11036ᵀ	0210004			1694
Myroides odoratus	NCTC 11036ᵀ	0210044			1694
Oceanimonas baumannii	ATCC 700832ᵀ	1000055			232
Oceanimonas doudoroffii	ATCC 27123ᵀ	1000055			232
Phocoenobacter uteri	NCTC 12872ᵀ	5000004	1		549
Pseudomonas anguilliseptica	Strains	0000045	8	Culture on CBA	1059
Pseudomonas anguilliseptica	Strains	0000055	18	Culture on CBA	1059
Pseudomonas anguilliseptica	Strains	0100045	2	Culture on CBA	1059
Pseudomonas anguilliseptica	Cod	0100054		Culture on CBA	100
Pseudomonas anguilliseptica	Strains	0100055	22	Culture on CBA	1059
Pseudomonas chlororaphis ssp. *aurantiaca*	NCIMB 10068ᵀ	0057557			252
Pseudomonas chlororaphis ssp. *aureofaciens*	DSM 6698ᵀ	1057557			252
Pseudomonas chlororaphis ssp. *chlororaphis*	DSM 50083ᵀ	1157655			252
P. chlororaphis ssp. *piscium*	JF 3835ᵀ	1056557			252
Pseudomonas fluorescens	DSM 50090ᵀ	0147555			252
Pseudomonas plecoglossicida	ATCC 700383ᵀ	1140457	1		1229
Pseudomonas putida	DSM 291ᵀ	1140457			252

Continued

Table 3.28. Continued.

Organism	Strain	API 20NE number	No. of strains	Notes	Ref
Pseudomonas putida	Trout	0140455			39
Shewanella affinis	ATCC BAA-642[T]	1400004	1		1480
Shewanella algae	IAM 14159[T]	1450154	1		1480
Shewanella algidipiscicola	LMG 23746[T]	1400144	30		1481
Shewanella algidipiscicola	LMG 23746[T]	1400154	3		1481
Shewanella baltica	NCTC 10337[T]	1450755	1		1480
Shewanella colwelliana	ATCC 39565[T]	1410404	1		1480
Shewanella corallii	LMG 24563[T]	5510304	1		1527
Shewanella glacialipiscicola	LMG 23744[T]	1410544	4		1481
Shewanella glacialipiscicola	LMG 23744[T]	1411544	3		1481
Shewanella hafniensis		1450745	14		1480
Shewanella hafniensis		1450755	14		1480
Shewanella hafniensis		1451745	14		1480
Shewanella hafniensis		1451755	14		1480
Shewanella haliotis	KCTC 12896[T]	1010154	1		895
Shewanella morhuae		1400544	33		1480
Shewanella morhuae		1401544	33		1480
Shewanella morhuae		1410544	33		1480
Shewanella morhuae		1411544	33		1480
Shewanella oneidensis	ATCC 700550[T]	1410104	1		1480
Shewanella putrefaciens	ATCC 8071[T]	1401144	1		1480
Stenotrophomonas maltophilia		1472341	4	Channel catfish	606
Vibrio caribbeanicus	DSM 23640[T]	0140004	1		739
Vibrio harveyi	ATCC 14126[T]	7477745	1	Must be read at 48 h	20
Vibrio harveyi	ATCC 35084[T]	7676645		Must be read at 48 h	20
Vibrio splendidus	0529			Greenshell mussel	879
V. toranzoniae Vb10.8T	Vb10.8[T]	70533..			980
V. toranzoniae Cmf 13.9	Cmf 13.9	74557..			980
V. toranzoniae CMJ 9.4	CMJ 9.4	50557..			980
V. toranzoniae CMJ 9.11	CMJ 9.11	50553..			980
Vibrio vulnificus		7434445		99.8% probability	
Vibrio vulnificus		7474445		95.1% probability	
Zobellia galactanivorans	DSM 12802[T]	5476304			119
Zobellia galactanivorans	DSM 12802[T]	5477304			119
Zobellia uliginosa		5477304			119
Zobellia uliginosa		5477704			119

Table 3.29. API 20NE database numbers (numbers in ascending order).

API 20NE number	Organism	Strain	No. of strains	Notes	Ref
0000004	*Brevundimonas diminuta*	ATCC 11568^T			1502
0000045	*Pseudomonas anguilliseptica*	Strains	8		1059
0000055	*Pseudomonas anguilliseptica*	Strains	18		1059
0010004	*Flavobacterium macrobrachii*	LMG 25203^T			1514
0057557	*Pseudomonas chlororaphis* ssp. *aurantiaca*	NCIMB 10068^T			252
0100045	*Pseudomonas anguilliseptica*	Strains	2		1059
0100054	*Pseudomonas anguilliseptica*	Cod			100
0100055	*Pseudomonas anguilliseptica*	Strains	22		1059
	Pseudomonas putida	Trout			39
0147555	*Pseudomonas fluorescens*	DSM 50090^T			252
0210004	*Myroides odoratimimus*	NCTC 11180^T	5		1694
0210044	*Myroides odoratimimus*	NCTC 11180^T	15		1694
0210004	*Myroides odoratus*	NCTC 11036^T			1694
0210044	*Myroides odoratus*	NCTC 11036^T			1694
0473304	*Flavobacterium tilapiae*	BCRC 80262^T			314
1000004	*Actinobacillus delphinicola*			Dolphin, whale	548
1000055	*Oceanimonas baumannii*	ATCC 700832^T			232
1000055	*Oceanimonas doudoroffii*	ATCC 27123^T			232
1010154	*Shewanella haliotis*	KCTC 12896^T			895
1056557	*Pseudomonas chlororaphis* ssp. *piscium*	JF 3835^T			252
1057557	*Pseudomonas chlororaphis* ssp. *aureofaciens*	DSM 6698^T			
1100004	*Actinobacillus delphinicola*			Porpoise	548
1140457	*Pseudomonas plecoglossicida*	ATCC 700383^T			1229
1140457	*Pseudomonas putida*	DSM 291^T			252
1157655	*Pseudomonas chlororaphis* ssp. *chlororaphis*	DSM 50083^T			252
1200004	*Brucella ceti* and *B. pinnipedialis*		2		546
1400004	*Shewanella affinis*	ATCC BAA-642^T	1		1480
1400144	*Shewanella algidipiscicola*	LMG 23746^T	30		1481
1400154	*Shewanella algidipiscicola*	LMG 23746^T	3		1481
1400544	*Shewanella morhuae*		33		1480
1401144	*Shewanella putrefaciens*	ATCC 8071^T	1		1480
1401544	*Shewanella morhuae*		33		1480
1410104	*Shewanella oneidensis*	ATCC 700550^T	1		1480
1410404	*Shewanella colwelliana*	ATCC 39565^T	1		1480
1410544	*Shewanella glacialipiscicola*	LMG 23744^T	4		1481
1410544	*Shewanella morhuae*		33		1480
1411544	*Shewanella morhuae*		33		1480
1411544	*Shewanella glacialipiscicola*	LMG 23744^T	3		1481
1423304	*Flavobacterium oncorhynchi*	CECT 7678^T			1845
1450154	*Shewanella algae*	IAM 14159^T	1		1480

Continued

Table 3.29. Continued.

API 20NE number	Organism	Strain	No. of strains	Notes	Ref
1450745	*Shewanella hafniensis*		14		1480
1450755	*Shewanella hafniensis*		14		1480
1450755	*Shewanella baltica*	NCTC 10337^T	1		1480
1451745	*Shewanella hafniensis*		14		1480
1451755	*Shewanella hafniensis*		14		1480
1472305	*Flavobacterium johnsoniae*	Sturgeon			867
1472341	*Stenotrophomonas maltophilia*	Channel catfish	4		606
1660757	*Halomonas venusta*	Human			1731
2476304	*Elizabethkingia meningoseptica*	LMG 12279^T	1		1693
2652205	*Chryseobacterium tructae*	CECT 7798^T			1846
2676105	*Elizabethkingia meningoseptica*	LMG 13004	1		1693
2676305	*Elizabethkingia meningoseptica*	LMG 13004	1		1693
4473667	*Gramella marina*	KMM 6048^T			1205
5000004	*Phocoenobacter uteri*	NCTC 12872^T	1		549
50553..	*V. toranzoniae* CMJ 9.11	CMJ 9.11	1		980
50557..	*V. toranzoniae* CMJ 9.4	CMJ 9.4	1		980
5410004	*A. salmonicida* ssp. *salmonicida*			Lamprey	513
5473300	*Flavobacterium hibernum*	ATCC 51468^T			1116
5476304	*Zobellia galactanivorans*	DSM 12802^T			119
5477304	*Zobellia galactanivorans*	DSM 12802^T			119
5477304	*Zobellia uliginosa*				119
5477704	*Zobellia uliginosa*				119
5510304	*Shewanella corallii*	LMG 24563^T	1		1527
5576354	*Aeromonas rivuli*	CECT 7518^T	2		536
70533..	*V. toranzoniae* Vb10.8T	Vb10.8^T	1		980
7062744	*Aeromonas fluvialis*	CECT 7401^T			37
7434445	*Vibrio vulnificus*			99.8% probability	
74557..	*V. toranzoniae* Cmf 13.9	Cmf 13.9	1		980
7477745	*Vibrio harveyi*	ATCC 14126^T		Must be read at 48 h	20
7575754	*Aeromonas sanarellii*	CECT 7402^T			36
7575754	*Aeromonas taiwanensis*	CECT 7403^T			36
7676645	*Vibrio harveyi*	ATCC 35084^T		Must be read at 48 h	20

*=API database.

Table 3.30. API 50 CH database biochemical results.

	1	2	3	4	5	6	7	8	9	10	11	12	13	14	15	16	17	18	19	20	21	22	23	24	25	26	27	28	29	30	31	32	33	34	35	36	37	38	39	40	41	42	43	44	45	46	47	48	49					
Test	Gly	Ery	Dara	Lara	rib	dXyl	lXyl	ado	mdx	gal	glu	fru	mne	Lsor	rha	dul	ino	man	sor	mdm	mdg	nag	amy	arb	esc	sal	cel	mal	lac	mel	suc	tre	inu	mlz	raf	amd	glyg	xylt	gen	tur	lyx	tag	Dfuc	Lfuc	Darl	Larl	gnt	2ket	5ket	Temp	Time	Inoc	Ref	Strain
Organism																																																						
Aerococcus viridans var *homari*					w					+	+	+	+					+	v		v	+	+	+	+		+	+	+	v	+	+	+		+				+	v							+		v				1792	
Aeromonas allosaccharophila	+	–	–	67	+	–	–	–	–	+	+	+	+	–	50	–	–	+	–	–	+	+	–	–	–	–	+	+	–	50	+	+	–	–	50	+	+	–	–		–	–	–	–	–	–		–	–	30	48 h	0.85	873	CDC 9533-76
Aeromonas aquariorum	+	–	–	–	+	–	–	–	–	+	+	+	+	–	–	–	–	+	–	–	+	+	–	+	+	+	–	+	–	–	+	+	–	–	–	+	+	–	–	–	–	–	–	–	–	–	+	–	–	30	48 h	0.85	1106	CECT 7289^T
Aeromonas bestiarum HG 2	+	–	–	+	+	–	–	–	–	+	+	+	+	–	–	–	–	+	+	–	+	+	–	+	+	+	67	67	67	–	+	+	–	–	–	+	+	–	–		–	–	–	–	–	–		–	–	30	48 h	0.85	873	CDC 9533-76
Aeromonas caviae HG 4	+	–	–	+	+	–	–	–	–	+	+	+	31	–	–	–	–	+	7	–	–	+	–	+	+	+	90	+	87	–	+	+	–	–	–	+	+	–	36	–	–	–	–	–	–	–	+	–	–	30	48 h	0.85	42, 873	ATCC 15468^T
Aeromonas encheleia	+	–				–		–		+	+		+			–	–	+	–		–				+	+	–	+	–	–	+	+			–										–					30	48 h	0.85	783	
Aeromonas eucrenophila HG6	+	–	–	+	+	–	–	–	–	+	+	+	+	–	–	–	–	+	–	–	–	+	–	+	+	+	+	+	+	–	+	+	–	–	–	+	+	–	–		–	–	–	–	–	–		–	–	30	48 h	0.85	873	ATCC 23309^T
Aeromonas fluvialis	+	–	–	–	+	–	–	–	–	+	+	+	+	–	–	–	–	–	–	–	–	+	–	–	–	+	+	+	+	–	+	+	–	–	–	+	+	–	+	–	–	–	–	–	–	–	+	–	–	30	48 h	0.85	37	CECT 7401^T
Aeromonas jandaei HG9	+		–	–	+	–	–			+	+	+	+	–	–	–	–	+	–				–	–	–	–	–	+	–	–	v	+			–			–			–									28	48 h	0.85	497	ATCC 49568^T
Aeromonas hydrophila HG1	+	–	–	+	+	–	–	–	–	+	+	+	+	–	–	–	–	+	–	–	–	+	–	+	+	+	–	+	–	–	+	+	–	–	–	+	+	–	–	–	–	–	–	–	–	–	+	–	–	30	48 h	0.85	873	ATCC 7966^T
Aeromonas hydrophila HG3	+	–	–	+	+	–	–	–	–	+	+	+	+	–	–	–	–	+	+	–	+	+	–	+	+	+	62	+	54	–	+	+	–	–	–	+	+	–	–		–	–	–	–	–	–		–	–	30	48 h	0.85	873	CDC0 434-84
Aeromonas media HG 5	+	–	–	+	+	–	–	–	–	+	+	+	+	–	–	–	–	+	–	–	–	+	–	+	+	+	+	+	+	–	+	+	–	–	–	+	+	–	–		–	–	–	–	–	–		–	–	30	48 h	0.85	873	CDC 0862-83, CDC 9072-83, CDC 9072-83
Aeromonas piscicola	+	–	–	–	+	–	–	–	–	+	+	+	+	–	v	–	–	+	–	–	+	+	–	+	+	+	–	+	–	–	+	+	–	–	–	+	+	–	–	–	–	–	–	–	–	–		–	–	25	24 h	0.85	141	S1.2^T
Aeromonas popoffii	+	–				–		–		+	+	v	+		–	–	–	+	–		+	+			–	–	–	+	–	–	–	+			–										–		+			25	24 h	0.85	783	LMG 17541^T
Aeromonas rivuli	+	–	–	–	+	–	–	–	–	+	+	+	+	–	–	–	–	+	–	–	–	+	–	+	+	+	v	+	–	–	+	+	–	–	–	+	+	–	–	–	–	–	–	–	–	–	–	–	–	30	48 h	0.85	536	CECT 7518^T
Aeromonas sanarellii	+	–	–	+	+	–	–	–	–	+	+	+	–	–	–	–	–	+	–	–	–	+	+		+	+	–	+	–	–	+	+	–	–	–		+	–	–	–	–	–	–	–	–	–		–	–	30	48 h	0.85	36	CECT 7402^T

Aeromonas simiae	+	–	–	–	+	–	–	–	–	+	+	+	+	–	–	–	–	–	–	–	–	+	–	v	v	–	+	+	–	–	+	+	–	–	–	+	+	–	–	–	–	–	–	–	–	–	–	–	–	30	48 h	0.85	677	IBS S6874^T, IBS S6652
***A. salmonicida* 'atypical'**																																																						
A. salmonicida achromogenes	+			–	+					+	+		+					+			–	–		–		–	–				+	+		–		+	+			–							–		+s	20	7 d	0.5	1797	
Baltic sea turbot	–	–	–	–	–	–	–	–	–	–	+	+	–	–	–	–	–	–	–	–	–	–	–	–	v	–	–	–	–	–	–	–	–	–	–	–	–	–	–	–	–	–	–	–	–	–	–	–	+	20	14 d		1800	
Baltic sea flounder	–	–	–	–	v	–	–	–	–	–	+	+	v	–	–	–	–	v	–	–	–	–	–	–	+	v	–	+	–	–	v	+	–	–	–	v	+	–	–	–	–	–	–	–	–	–	–	–	+	20	14 d		1800	
Baltic sea dab	+	–	–	–	–	–	–	–	–	v	+	+	v	–	–	–	–	–	–	–	–	+w	–	–	+	+	–	–	–	–	+	–	–	–	–	–	–	–	–	–	–	–	–	–	–	–	–	–	+	20	14 d		1800	
Baltic sea blenny	+	–	–	–	+	–	–	–	–	+	+	+	+	–	–	–	–	+	–	–	–	+	–	–	–	–	–	+	–	–	+	+	–	–	–	+	+	–	–	–	–	–	–	–	–	–	–	–	+w	20	14 d		1800	
'atypical' Pigment pos	+	–	–	–	+	–	–	–	–	+	+	+		–	–	–	–	+		–	–	17	–								+	68	–	–		100	3		–		–	–	–		–			–		20	7 d		674, 733	
'atypical' Pigment neg	44	–	–	–	+	–	–	–	–	44	+	+		–	–	–	–	+		–	–	88	–								+	27	–	–		63	75		–		–	–	–		–			–		20	7 d		674	
'atypical' Pigment neg	44	–	–	–	+	–	–	–	–	44	+	+		–	–	–	–	–		–	–	88	–								+		–	–		63	75		–		–	–	–		–			–		20	7 d		674	
Australian strain	–			–						–								–			–	–		–	–	–					+	–															–						1664	
***A. salmonicida* typical**	+			+	+					+		+	+					+	–		+	+		+	+	+	–	+			–	–			–	+	+				–						+		–	20–22	7 d		733	
A. salmonicida salmonicida	+			–	+					+	+		–					+			+	+		+		+	–				–	–		–		+	+			–							+		+s	20	7 d	0.5	1797	
Aeromonas sobria	+	–	–	28	+	–	–	–	–	+	+	+	+	–	–	–	–	+	–	–	24	+	–	–	16	–	52	+	16	12	+	+	–	–	–	+	+	–	–	–	–	–	–	–	–	–	+	–	–	29	48 h		42	
Aeromonas taiwanensis	+	–	–	+	+	–	–	–	–	+	+	+	–	–	–	–	–	+	–	–	–	+	+		+	+	–	+	–	–	+	+	–	–	+			–	–	–	–	–	–	–	–	–	+	–	–	30	48 h	0.85	36	CECT 7403^T
Aeromonas tecta F518^T	+	–	–	–	+	–	–	–	–	+	+	+	+	–	–	–	–	+	–	–	–	+	–	+		+	–	+	–	–	–	+	–	–	–	+	+	–	+	–	–	–	–	–	–	–	+	–	–	30	24 h	0.85	435	CECT 7082^T
Aeromonas tecta from trout	+	–	–	–	+	–	–	–	–	+	+	+	+	–	–	–	–	+	–	–	–	+	–	+	+	+	–	+	–	–	–	+	–	–	–	+	+	–	+	–	–	–	–	–	–	–	+	–	–	30	24 h	0.85	435	CECT 7081
Aeromonas trota	+	–	–	–	+	–	–	–	–	+	+	+	+	–	–	–	–	+	–	–	–	+	+	–	–	–	+	+	–	–	–	+	–	–	–	+	+	–	–	–	–	–	–	–	–	–	+	–	–	37	24 h	0.85	789	ATCC 49657^T
Aeromonas veronii sobria HG8/10	+	–	–	14	+	–	–	–	–	+	+	+	+	–	–	–	–	+	–	–	29	+	–	–	10	–	57	+	10	–	+	+	–	–	–	+	+	–	–		–	–	–	–	–	–		–	–	30	48 h	0.85	873	CDC 0437-84
Aeromonas veronii veronii	+	–		–		–		–		+	+		+		–	–	–	+	–		89		89		+	+	78	+	–	–	+	+			–										–					25	48 h	0.85	723	
Aeromonas culicicola, a synonym of *A. veronii*	+	–	–	–	+	–	–	–	–	+	+	+	+	–	–	–	–	+	–	–	–	+	–	–	–	–	–	+	–	+	+		–	–	–	+	+	–	–	–	–	–	–	–	–	–		–	–	30	24 h	0.85	1339	NCIMB 5147^T

Continued

Table 3.30. Continued.

	1	2	3	4	5	6	7	8	9	10	11	12	13	14	15	16	17	18	19	20	21	22	23	24	25	26	27	28	29	30	31	32	33	34	35	36	37	38	39	40	41	42	43	44	45	46	47	48	49					
Test	Gly	Ery	Dara	Lara	rib	dXyl	lXyl	ado	mdx	gal	glu	fru	mne	Lsor	rha	dul	ino	man	sor	mdm	mdg	nag	amy	arb	esc	sal	cel	mal	lac	mel	suc	tre	inu	mlz	raf	amd	glyg	xylt	gen	tur	lyx	tag	Dfuc	Lfuc	Darl	Larl	gnt	2ket	5ket	Temp	Time	Inoc	Ref	Strain
Organism																																																						
Aliivibrio finisterrensis	+	–	–	–	+	–	–	–	–	+	+	+	+	–	–	–	–	–	–	–	–	+	–	+	+	+	+	+	+	+	–	–	–	–	–	–	–	–	+	–	–	–	–	–	–	–		+	–	25	48 h	0.85	143	CMJ 11.1^T
Aliivibrio finisterrensis	+	–	–	–	+	–	–	–	–	+	+	+	+	–	–	–	–	+	–	–	–	+	–	–	+	+	+	+	+	–	+	–	–	–	–	–	–	–	+	–	–	–	–	–	–	–		+	–	25	48 h	0.85	143	C 17.13
Aliivibrio logei	+			–		–							+		–		–	+	–							+	+			–	–	+			–															20	48 h	0.85	143	LMG 19806^T
Aliivibrio logei	+			–		–					+		+		–		–	–	–							–	+		+	–	+	+			–																		1071	ATCC 15382
Aliivibrio salmonicida	+sl	–	–	–	+	–	–	–	–	+	+	+	–	–	–	–	–	+	–	–	–	+	–	–	–	–	–	+	–	–	–	+	–	–	–	–	–	–	–	–	–	–	–	–	–	–	+	–	–	15	48 h	1.5	473	NCMB 2262^T
Aliivibrio salmonicida	w	–	–	–	+	–	–	–	–	+	+	+	+	–	–	–	–	+	w	–	–	+	–	–	–	–	–	+	–	–	–	+	–	–	–	–	–	–	–	–	–	–	–	–	–	–	+	–	–	15	48 h	2	237	NCMB 2245
Aliivibrio wodanis	+			–		–				+	+		+				–	v	–						–	–	–		–	–	+	+			–	+														20	48 h	0.85	143	ATCC BAA-104^T
Andreprevotia lacus	–	–	–	–	+	–	–	–	–	+	+	+	+	–	–	–	–	–	–	–	–	+	–	–	–	–	–	–	–	–	–	–	–	–	–	–	–	–	–	–	+	–	–	–	–	–	–	–	+	30	24 h	0.85	1513	LMG 24502^T
Arcanobacterium phocae	+	–	–	–	+	–	–	–	–	+	+	+	70	–	–	–	+	30	–	–	–	+	–	–	–	–	–	+	+	–	+	+	–	70	–	+	+	–	–	+	–	30	–	–	–	–	–	–	+	37	24 h	0.85	1374	DSM 10002^T
Arthrobacter rhombi	+	–	–	–	–	–	–	–	–	+	+	+	+	–	v	–	–	+	–	–	–	–	+	+	+	+	+	+	+	+	+	+	–	–	–	–	–	–	+	+	–	–	–	–	+	–	+	–	–	25	48 h		1276	CCUG 38813^T
Brevundimonas diminuta	–	–	–	–	–	–	–	–	–	–	–	–	–	–	–	–	–	–	–	–	–	–	–	–	–	–	–	–	–	–	–	–	–	–	–	–	–	–	–	–	–	–	–	–	–	–	–	–	–	28	4 h	0.85	1502	ATCC 11568^T
Brevundimonas vesicularis	–	–	–	–	–	+	–	–	–	–	+	–	–	–	–	–	–	–	–	–	–	–	–	–	–	–	–	+	–	–	–	–	–	–	–	–	–	–	–	–	–	–	–	–	–	–	–	–	–	28	4 h	0.85	1502	ATCC 11426^T
Carnobacterium alterfunditum					+					–								–		–	–		+				–	+					–	–						+		+								25	7 d	0.85	1405	
Carnobacterium divergens		–	–	–	–	–	–	–	–	–	+	+	+	–	–	–	–	–	–	–	–	+	+			+	–	+	–	–	+	+	–	+	–	–	–	–	+	–	–	–	–	–	–	–	+	–	–	25	7 d	0.85	355	
Carnobacterium divergens					+					+								+		+	+		+				+	+		+			+	+						–		–										0.85	1405	6251
Carnobacterium funditum					+					+								+		+	+		+				+	+					–	+						+		+								25	7 d	0.85	1405	
Carnobacterium gallinarum		–	–	–	+	+	–	–	–	+	+	+	+	–	–	–	–	–	–	–	+	+	+			+	+	+			+	+	–	+	–		–	–		+	–	+	–	–	–	–	+		–				355	
Carnobacterium inhibens					+					+								+		–	–		+				+	–					+	–						+		–								25	7 d	0.85	1405	

Carnobacterium inhibens	–	–	–	–	+	–	–	–	–	+	+	+	+	–	–	–	–	+	–	–	–	+	+	+	+	+	+	+	w	–	+	+	w	–	–	–	–	–	+	–	–	–	–	–	–	–	–	–	–				841	
Carnobacterium maltaromaticum	+	–	–	–	+	–	–	–	–	64	+	+	+	–	–	–	–	88	v	v	+	+	+	+	+	+	+	+	35	29	+	+	v	12	29	v	–	–	+	+w	–	–	–	–	–	–	+	–	–	25	72 h	0.85	136, 355, 736	
Carnobacterium mobile		–	–	–	+	–	–	–	–		+	+	+	–	–	–	–	–	–	–	–	+	–			+	+	+			+	+	+	–	–		–	–		–	–	–v	–	–	–	–	–		–					
Carnobacterium mobile					+					–								–		–	–		+				–	–					+	–						–		–												
Chryseobacterium scophthalmum				–							–							–								–	–		–		–				–																			
Corynebacterium ciconiae	+	–	–	–	+	–	–	–	–	–	+	+	+	–	–	–	–	–	–	–	–	+	–	–	–	–	–	+	–	–	–	–	–	–	–	–	–	–	–	–	–	–	–	–	–	–		–	–	37	72 h	0.85	528	CECT 5779^{T}
Corynebacterium ciconiae	+	–	+	–	+	–	–	–	–	–	+	+	+	–	–	–	–	–	–	–	–	+	–	–	–	–	–	–	–	–	–	–	–	–	–	–	–	–	–	–	–	–	–	+	–	–	+	–	–	37	72 h	0.85	528	
Corynebacterium ciconiae	+	–	–	–	+	–	–	–	–	–	+	+	+	–	–	–	–	–	–	–	–	+	–	–	–	–	–	–	–	–	–	–	–	–	–	+	–	–	–	–	–	–	–	–	–	–	–	–	–	37	72 h	0.85	528	
Corynebacterium ciconiae	+	–	–	–	+	–	–	–	–	–	+	+	+	–	–	–	–	–	–	–	–	+	–	–	–	–	–	–	–	–	–	–	–	–	–	–	–	–	–	–	–	–	–	–	–	–	–	–	–	37	72 h	0.85	528	
Corynebacterium ciconiae	+	–	–	–	+	–	–	–	–	–	+	+	+	–	–	–	–	–	–	–	–	+	–	–	–	–	–	–	–	–	–	+	–	–	–	–	–	–	–	–	–	–	–	–	–	–	–	–	–	37	72 h	0.85	528	
Corynebacterium phocae	–	–	–	–	–	–	–	–	–	v	+	–	+	–	–	–	–	–	–	–	–	+	–	–	–	–	–	+	v	–	v	–	–	–	v	–	–	–	–	–	–	–	–	–	–	–	–	–	–	37	48 h	0.85	1313	
Corynebacterium sphenisci	–	–	–	–	–	–	–	–		–	+	+	+	–	–	–	–	–	–	–	–	–	–	–	–	–	–	+	–	–	–	+	–	–	–		–	–	–	–	–	–	–	–	–	–		–	–	37	72 h	0.85	652	
Corynebacterium spheniscorum	–	–	–	–	+	–	–	–	–	–	+	+	+	–	–	v	–	–	–	–	–	+	–	–	–	–	–	+	–	–	–	+	–	–	–		–	–	–	–	–	–	–	–	–	–	–	–	–	37	48 h	0.85	653	
Enterococcus faecium	v			+		–									–			+	–	–	–									v	v		–		–	–	–			–		–					v							
Enterococcus seriolicida	–	–	–	–	+	–	–	–	–	+	+	+	+	–	–	–	–	+	–	–	–	+	+	+	+	+	+	+	–	–	–	+	–	–	–	–	–	–	+	–	–	–	–	–	–	–	–	–	–	28	24–96 h		350, 483	ATCC 49156
Flavobacterium columnare	–	–	–	–	–	–	–	–	–	–	–	–	–	–	–	–	–	–	–	–	–	–	–	–	–	–	–	–	–	–	–	–	–	–	–	–	–	–	–	–	–	–	–	–	–	–	–	–	–	22	12 h	ASW	166	
Flavobacterium psychrophilum	–	–	–	–	–	–	–	–	–	–	–	–	–	–	–	–	–	–	–	–	–	–	–	–	–	–	–	–	–	–	–	–	–	–	–	–	–	–	–	–	–	–	–	–	–	–	–	–	–	22	12 h	0.85	168	
Flexibacter litoralis	–	–	–	–	–	–	–	–	–	–	–	–	–	–	–	–	–	–	–	–	–	–	–	–	–	–	–	–	–	–	–	–	–	–	–	–	–	–	–	–	–	–	–	–	–	–	–	–	–				168	
Flexibacter polymorphus	–	–	–	–	–	–	–	–	–	–	–	–	–	–	–	–	–	–	–	–	–	–	–	–	–	–	–	–	–	–	–	–	–	–	–	–	–	–	–	–	–	–	–	–	–	–	–	–	–				168	
Flexibacter roseolus	–	–	–	–	–	–	–	–	–	–	–	–	–	–	–	–	–	–	–	–	–	–	–	–	–	–	–	–	–	–	–	–	–	–	–	–	–	–	–	–	–	–	–	–	–	–	–	–	–				168	
Flexibacter ruber	–	–	–	–	–	–	–	–	–	–	–	–	–	–	–	–	–	–	–	–	–	–	–	–	–	–	–	–	–	–	–	–	–	–	–	–	–	–	–	–	–	–	–	–	–	–	–	–	–				168	
Janthinobacterium lividum	+	–	+	+	+	+	–	–	–	+	+	+	+	–	v	–	+	+	+	–	–	v	–	+	–	+	+	+	–	–	+	–	v	–	v	–	–	+	–	–	+	–	–	+	+	–	–	+	–				1030	

Continued

Table 3.30. Continued.

	1	2	3	4	5	6	7	8	9	10	11	12	13	14	15	16	17	18	19	20	21	22	23	24	25	26	27	28	29	30	31	32	33	34	35	36	37	38	39	40	41	42	43	44	45	46	47	48	49					
Test	**Gly**	**Ery**	**Dara**	**Lara**	**rib**	**dXyl**	**lXyl**	**ado**	**mdx**	**gal**	**glu**	**fru**	**mne**	**Lsor**	**rha**	**dul**	**ino**	**man**	**sor**	**mdm**	**mdg**	**nag**	**amy**	**arb**	**esc**	**sal**	**cel**	**mal**	**lac**	**mel**	**suc**	**tre**	**inu**	**mlz**	**raf**	**amd**	**glyg**	**xylt**	**gen**	**tur**	**lyx**	**tag**	**Dfuc**	**Lfuc**	**Darl**	**Larl**	**gnt**	**2ket**	**5ket**	**Temp**	**Time**	**Inoc**	**Ref**	**Strain**
Organism																																																						
Klebsiella planticola	+	–		+	+	+		+		+	+	+	+	+	+	–	+	+	+		+				+	+	+	+	+	+	+	+		–	+	–	–																	
Lactococcus garvieae	–	–	–	–	+	–	–	–	–	+	+	+	+	–	–	–	–	75	–	–	–	+	+	+	+	+	+	75	–	–	50	+	–	–	–	–	–	–	+	–	–	50	–	–	–	–	50	–	–	28	24–96 h	0.85	350, 481, 1711	ATCC 49156
Biotypes 1, 2, 12																		+													+											+											1711	
Biotypes 3, 7, 11, 13																		+													–											+											1711	
Biotypes 4, 5, 6, 10																		+													–											–											1711	
Biotypes 8, 9																		–													–											–											1711	
Lactococcus lactis ssp. *cremoris*	–	–	–	–	–	–	–	–	–	–	+	+	+	–	–	–	–	+	+	–	–	+	w	+	+	+	+	+	–	–	+	+	–	+	–	–	–	–	67	67	–	–	–	–	–	–	–	–	–	25	24 h	0.85	350	NCDO 1869
Lactococcus lactis ssp. *lactis*	–	–	–	+	+	+	–	–	–	+	+	+	+	–	–	–	–	+	–	–	–	+	+	+	+	+	+	+	+	–	+	–	+	–	–	+	–	–	+	–	–	–	–	–	–	–	+	–	–	25	48 h	0.85	809	Catfish
Lactococcus lactis ssp. *lactis*	–	–	–	+	+	+	–	–	–	+	+	+	+	–	–	–	–	+	–	–	–	+	+	+	+	+	+	+	+	–	+	+	–	–	–	+	–	–	+	–	–	–	–	–	–	–	+	–	–	25	48 h	0.85	809	Pufferfish
Lactococcus lactis spp. *tructae*					+	–												+					+					+	+	+	+				+												+			25	48 h	0.85	1330	L105^T
Lactococcus piscium	–	–	–	+	+	+	–	–	–	+	+	+	+	–	–	–	–	+	–	–	–	+	+	+	+	+	+	+	+	+	+	+	–	+	+	+w	–	–	+	+	–	–	–	–	–	–	+	–	–	25	48 h	0.85	1802	NCFB 2778
Lactococcus raffinolactis	–	–	–	33	17	83	–	–	–	+	+	+	+	17	–	–	–	67	–	–	17	+	33	83	+	+	+	+	+	+	+	+	50	50	+	+	–	–	67	50	–	–	–	–	–	–	–	–	–	25	24 h	0.85	350	NCDO 617
Lactobacillus ceti	–	–	–	–	+	–	–	–	–	–	–	–	–	–	–	–	–	–	–	–	–	–	–	–	–	–	–	–	–	–	–	–	–	–	–	–	–	–	–	–	–	–	–	–	–	–	–	–	+	37	7 d	0.85	1709	
Moritella viscosa	–	–	–	–	+	–	–	–	–	–	+	+	–	–	–	–	–	–	–	–	–	+	–	–	–	–	–	+	–	–	–	–	–	–	–	+	+	–	–	–	–	–	–	–	–	–	–	+	–	15	72 h	2	236	NCIMB 13484
Pedobacter heparinus	–	–	–	+	–	+	–	–	–	+	+	+	+	–	–	–	–	–	+	–	–	–	–	–	+	–	+	+	+	–	+	–	–	–	–	+	–	–	–	+	–	–		+	–	–	–	–	+	28	48 h	0.5	1603	
Pedobacter piscium	–	–	–	–		v	–	–		+	+	+	+		–	–	–		–	+	+		+	+	+	+	+	+	+		+	+		–	+			–	+	+	–	–	–	–	+	–	–	–	–	28	48 h	0.5	1603	
Photobacterium angustum	–	–	–	–	+	–	–	–	–	+	+	+	+	–	–	–	–	–	–	–	–	+	–	–	–	–	–	+	–	–	–	–	–	–	–	–	–	–	–	–	–	–	–	–	–	–	–	–	–	26	72 h	1.5	1641	NCIMB 1895
P. damselae damselae	+	–	–	–	+	–	–	–	–	+	+	+	+	–	–	–	–	–	–	–	–	+	–	–	–	–	+	+	–	–	–	+	–	–	–		–	–	–	–	–	–	–	–	–	–	–	–	–	26	72 h	1.5	1641	ATCC 35083, ATCC 33539

P. damselae piscicida	–	–	–	–	+	–	–	–	–	+w	+	+	+	–	–	–	–	–	–	–	–	+	–	–	–	–	–	–	–	–	+w	–	–	–	+w	v	–	–	–	–	–	–	–	–	–	–	–	–	–	26	72 h	1.5	595, 1641	NCIMB 2058, NCIMB 25918, ATCC 17911
Photobacterium iliopiscarium	v	–	–	–	+	–	–	–	–	+	+	+	+	–	–	–	–	–	–	–	–	+	–	–		–	–	+	–	–	–	v	–	–	–	–	–	–	–	–	–	–	–	–	–	–	v	–	–	22	72 h	2	1272	
Photobacterium leiognathi	+	–	–	–	+	–	–	–	–	+	+	+	+	–	–	–	–	–	–	–	–	+	–	–	–	–	–	–	–	–	–	–	–	–	–	–	–	–	–	–	–	–	–	–	–	–	+	–	–	26	72 h	1.5	1641	LMG 4228
Rhodococcus erythropolis	+	v	–	–	v	–	–	v	–	–	+		–	–	–	–	+	+	+	–	–	+	–	v	–	–	–	–	–	–	+	+	–	–	–	–	–	v	–	–	–	–	–	–	+	v		–	v	30	72 h	0.85	1265	
Shewanella haliotis	–		–	–	–	–		–		–	–		–	–		–	–	–	–			+			–	–		–	–	–	–		–										–	–						25	48 h	2	895	KCTC 12896ᵀ
Shewanella marinintestina	–	–	–	–	+	–	–	–	–	–	–	–	–	–	–	–	–	–	–	–		+	–	–	–	–	–	–	–	–	–	–	–	–	–	–		–		–	–	–	–		–	–		–	–	20	48 h	2	1479	JCM 11558ᵀ
Shewanella sairae	–	–	–	–	+	–	–	–	–	–	–	–	–	–	–	–	–	–	–	–		+	–	–	–	–	–	–	–	–	–	–	–	–	–	–		–		–	–	–	–		–	–		–	–	20	48 h	2	1479	JCM 11563ᵀ
Shewanella schlegeliana	–	–	–	–	+	–	–	–	–	–	–	–	–	–	–	–	–	–	–	–		+	–	–	+	–	–	–	–	–	–	–	–	–	–	–		–		–	–	–	–		–	–		–	–	20	48 h	2	1479	JCM 11561ᵀ
Streptococcus agalactiae	–	–	–	–	w	–	–	–	–	–	+	+	+	–	–	–	–	–	–	–	–	+	–	–	–	–	–	+	–	–	+	–	–	–	–	–	–	–	–	–	–	–	–	–	–	–	–	–	–	30	24 h	0.85	1696	
Streptococcus agalactiae	–	–	–	–	w	–	–	–	–	–	+	+	+	–	–	–	–	–	–	–	–	+	–	–	–	–	–	+	–	–	+	–	–	–	–	–	–	–	–	–	–	–	–	–	–	–	–	–	–	37	24 h	0.85	1696	
Streptococcus agalactiae	–	–	–	–	+	–	–	–	–	–	+	+	+	–	–	–	–	–	–	–	–	+	–	–	–	–	–	+	–	–	+	–	–	–	–	–	–	–	–	–	–	–	–	–	–	–	–	–	–	30	72 h	0.85	482	ND 2-22
Streptococcus agalactiae	–	–	–	–	+	–	–	–	–	–	+	+	+	–	–	–	–	–	–	–	–	+	–	–	–	–	–	+	–	–	+	–	–	–	–	–	–	–	–	–	–	–	–	–	–	–	–	–	–	30	72 h	0.85	482	ND 2-22
S. dysgalactiae ssp. *dysgalactiae* Gp L	–	–	–	–	+	–	–	–	–	+	+	+	+	–	–	–	–	–	–	–	–	+	–	–	–	–	–	+	–	–	+	+	–	–	–	+	+	–	–	–	–	–	–	–	–	–	–	–	–	37	24 h	0.85	1732	
Streptococcus iniae	–	–	–	–	+	–	–	–	–	?	+	+	+	–	–	–	–	+	–	+	–	+	–	+	+	+	+	+	–		+	+	–	+	–	+	+	–	+	–	–	–	–	–	–	–	–	–	–					
Streptococcus iniae	–	–	–	–	+	–	–	–	–	–	+	+	+	–	–	–	–	+	–	–	–	+	–	+	+	+	+	+	–	–	+	+	–	+	–	+	+	–	+	–	–	–	–	–	–	–	–	–	–	30	72 h		483	ND 2-16
Streptococcus iniae	–	–	–	–	+	–	–	–	–	–	+	+	+	–	–	–	–	+	–	+	–	+	–	+	+	+	+	+	–	–	+	+	–	+	–	+	+	–	+	–	–	–	–	–	–	–	–	–	–	30	72 h		483, 479	ND 2-16
Streptococcus iniae	–	–	–	–	+	–	–	–	–	+	+	+	+	–	–	–	–	+	–	–	–	+	+	+	+	+	+	+	–	–	+	+	–	+	–	+	+	–	+	–	–	–	–	–	–	–	–	–	–	24	72 h	0.85	360	
Streptococcus parauberis	–	–	–	–	+	–	–	–	–	+	+	+	+	–	–	v	–	+	+	–	–	+	+	+	+	+	+	+	+	–	+	+	v	–	–v	–v	–	–	+	–	–	+	–	–	–	–	–	–	–	37	24 h		456	
Streptococcus phocae	–	–	–	–	+	–	–	–	–	–	+	+	+	–	–	–	–	–	–	–	–	+	–	–	–	–	–	+	–	–	–	–	–	–	–	–	–	–	–	–	–	–	–	–	–	–	–	–	–	37	24 h	0.85	1538, 1732	
Streptococcus porcinus	+	–	–	–	+	–	–	–	–	+	+	+	+	–	–	–	–	+	+	–	–	+	v		+	+	+	+	v	–	+	+	–		–	v	–	–		–	–	–	–	–	–	–	–	–	–	37	24–48 h	0.5	356	

Continued

Table 3.30. Continued.

	1	2	3	4	5	6	7	8	9	10	11	12	13	14	15	16	17	18	19	20	21	22	23	24	25	26	27	28	29	30	31	32	33	34	35	36	37	38	39	40	41	42	43	44	45	46	47	48	49					
Test	Gly	Ery	Dara	Lara	rib	dXyl	lXyl	ado	mdx	gal	glu	fru	mne	Lsor	rha	dul	ino	man	sor	mdm	mdg	nag	amy	arb	esc	sal	cel	mal	lac	mel	suc	tre	inu	mlz	raf	amd	glyg	xylt	gen	tur	lyx	tag	Dfuc	Lfuc	Darl	Larl	gnt	2ket	5ket	Temp	Time	Inoc	Ref	Strain
Organism																																																						
Streptococcus uberis	–	–	–	–	+	–	–	–	–	+	+	+	+	–	–	v	–	+	+	–	–	+	+	+	+	+	+	+	+	–	+	+	+	–	–	–	–	–	+	–	–	+	–	–	–	–	–	–	–	37	24 h		456	
Tenacibaculum maritimum	–	–	–	–	–	–	–	–	–	–	–	–	–	–	–	–	–	–	–	–	–	–	–	–	–	–	–	–	–	–	–	–	–	–	–	–	–	–	–	–	–	–	–	–	–	–	–	–	–	22	12 h	ASW	168, 413	
Trueperella bernardiae	+	v			v	–		+			+	+						–		–			–	–	–			+			–					+	v	+							+	v		–	+	37	48 h		573	
Trueperella pyogenes	–	+		–	+	+		+		+	+	+	+		–		+	–	+				–		–	–	+	+	+	–	–	+	–	+	–	+	+													37	48 h		1384	
Vagococcus fluvialis	–	–	–	–	+	–	–	–	–	–	+	+	+	–	–	–	–	+	+	–	+	+	+	+	+	+	+	+	–	–	+	+	–			–	–	–	+	+	–	–	–	–	–	–		–	–				354	Species
Vagococcus fluvialis	–	–	–	–	+	–	–	–	–	–	+	+	+	–	–	–	–	+	–	–	+	–	+	+	+	+	+	+	–	–	–	+	–	–	–	–	–	–	+	v	–	–	–	–	–	–	–	–	–				354, 1033	NCDO 2497
Vagococcus salmoninarum	–	–	–	–	60	v	–	–	–	–	+	+	+	90	–	–	–	v	70	–	–	+	+	+	+	+	60	70	–	–	+	+	–	–	–	–	–	–	50	–	–	+	–	60	–	–	–	+w	–	25	2–7 d	0.85	1139, 1755	NCFB 2777^T
Vibrio aestivus	–	–	–	–	–	–	–	–	–	+	+	+	–	–	–	–	–	–	–	–	–	–	–	–	–	–	+	+	–	–	–	+	–	–	–	–	–	–	–	–	–	–	–	–	–	–	–	–	–	25	48 h	0.85	1065	CECT 7558^T
V. aestuarianus ssp. *aestuarianus*	–	–	–	–	+	–	–	–	–	+	+	+	+	–	–	–	–	–	–	–	–	+	–	–	–	–	+	+	–	–	+	+	–	–	–	+	+	–	–	–	–	–	–	–	–	–	+	–	–	24	48 h	2	588	ATCC 35048^T
V. aestuarianus ssp. *francensis*	–	–	–	–	43	–	–	–	–	86	86	86	86	–	–	–	–	–	–	–	–	+	–	–	–	–	29	43	–	–	+	+	–	–	–	43	43	–	–	–	–	–	–	–	–	–	57	–	–	24	48 h	2	588	Oysters
Vibrio artabrorum	+	–		+	+	+	–	–	–	+		+	+	–	+		–	+	–	–	–	+	–			–	+	+			+	+	–	–	–	+	+	–	+				–			–			–	25	48 h	2	442	LMG 23865^T
Vibrio artabrorum		–			+	57	–	–	–	+		+	+	–			–	+	–	–	–	+	–			–	+	+			+	+	–	–	–	+	+	–	+				–			–			–	25	48 h	2	442	Strains
Vibrio atlanticus	+	–	–	+	+		–	–	–	+	+	+	+	–	–	–	–	+	–	–	–	+	–	–	+	–	+	+		–	+	+	–	–	–	+	+	–	+	–	–		–	+	–	–	+	–	–	25	48 h	2	442	LMG 24300^T
Vibrio atlanticus	+	–	–		+	+	–	–	–		+	+	+	–	–	–	–		–	–	–	+	–	–	+	–	+	+		–	+	+	–	–	–	+	+	–		–	–		–		–	–		–	–	25	48 h	2	442	Strains
Vibrio breoganii	–		–	–	28	+	14			+	+		14		57		–	+					+	71	+	+	+		14	28	14	–					71		+											24	48 h	0.85	140	LMG 23858^T
Vibrio caribbeanicus				–							–		–				–	–	–			–			–			–		–	–					–														28	48 h	1.5	739	DSM 23640^T
Vibrio celticus	+	–	–	–	+	+	–	–	–	+	+	+	+	–	+	–	–	+	–	–	–	+	–	–	–	–	+	+	+	+	+	+	–	–	–	+	+	–	–	–	–	–	–	–	–	–	–	–	–	24	24 h	0.85	142	Rd 8.15^T
Vibrio celticus	+	–	–	–	+	+	–	–	–	+	+	+	+	–	+	–	–	+	–	–	–	+	–	–	–	–	+	+	+	–	+	+	–	–	–	+	+	–	–	–	–	–	–	–	–	–	–	–	–	24	24 h	0.85	142	RD 6.8
Vibrio celticus	+	–	–	+	+	+	–	–	–	+	+	+	+	–	+	–	–	+	–	–	–	+	–	–	–	–	+	+	+	–	+	+	–	–	–	+	+	–	–	–	–	–	–	–	–	–	+	–	–	24	24 h	0.85	142	Rd2L5
Vibrio celticus	+	–	–	–	+	+	–	–	–	+	+	+	+	–	+	–	–	+	–	–	–	+	–	–	–	–	+	+	+	–	+	+	–	–	–	+	+	–	–	–	–	–	–	–	–	–	–	–	–	24	24 h	0.85	142	Rd 16.13
Vibrio chagasii	–	–	–	–	+	–	–	–	–	–	+	+	+	–	–	–	–	+	–	–	–	+	–	–	+	–	+	+	–	–	+	+	–	–	–	+	+	–	–	–	–	–	–	+	–	–	–	–	–	25	48 h	2	524	LMG 21353^T

Vibrio crassostreae	+	–	–	–	+	–	–	–	–	+	+	+	+	–	–	–	–	+	–	–	–	+	+	–	+	–	+	–	–	+	+	+	–	–	–	+	+	–	–	–	–	–	–	+	–	–	–	–	–	24	48 h	2	524	LMG 22240^{T}
Vibrio cyclitrophicus	–	–	–	–	+	–	–	–	–	+	+	+	+	–	–	–	–	+	–	–	–	+	+	–	+	–	+	+	–	–	+	+	–	–	–	–	–	–	–	–	–	–	–	+	–	–	–	–	–	25	48 h	2	524	LMG 21359^{T}
Vibrio diabolicus	+	–	–	–	+	–	–	–	–	+	+	+	+	–	–	–	–	+	–	–	–	+	+	–	–	–	–	+	–	–	+	+	–	–	–	+	+	–	–	–	–	–	–	–	–	–	+	–	–	25	48 h	2	1368	CNCM I-1629^{T}
Vibrio fischeri	–	–	–	–	+	–	–	–	–	+	+	+	+	–	–	–	–	–	–	–	–	+	–	–	–	+	+	+	–	–	–	–	–	–	–	+	–	–	–	–	–	–	–	–	–	–	–	–	–	26	72 h	1.5	1641	ATCC 25918
Vibrio gallaecicus	+	–	–	–	+	–	–	–	–	–	+	+	+	–	–	–	–	+	–	–	–	+		–	+	–	+	+	–	–	33	+	–	–	–	+	+		–	–	–	–	–	–	–	–	–	+	–	24	48 h	0.85	144	LMG 24045^{T}
Vibrio gigantis	+	–	–	–	+	–	–	–	–	+	+	+	+	–	–	–	–	+	–	–	–	+	+	–	+	–	+	+	–	+	–	+	–	–	–	+	+	–	–	–	–	–	–	–	–	–		–	–	24	48 h	2	1010	LMG 22741^{T}
Vibrio harveyi ATCC 14129^{T}	–	–	–	–	+	–	–	–	–	+	+	+	+	–	–	–	–	+	–	–	–	+	–	–	+w	–	+	+	–	–	+	+	–	–	–	+	+	–	–	–	–	–	–	–	–	–	+	–	–	25	24 h	2	1228	ATCC 14129^{T}
Vibrio harveyi ATCC 35084^{T}	–	–	–	–	+	–	–	–	–	+	+	+	+	–	–	–	–	+	–	–	–	+	–	–	+w	–	+	+	–	–	+	+	–	–	–	+	+	–	–	–	–	–	–	–	–	–	+	–	–	25	24 h	2	1228	ATCC 35084^{T}
Vibrio hippocampi	–	–	–	–	–	–	–	–	–	–	+	–	–	–	–	–	–	+	–	–	–	+	–	–	–	–	+	+	–	–	–	+	–	–	–	–	–	–	–	–	–	–	–	–	–	–	–			20	48 h	1.5	106	
Vibrio hispanicus										+	+		+		+			+	–		+						+		+			+			+				+				+							28	48 h	2	636	LMG 13240^{T}
Vibrio kanaloae	+	–	–	–	+	–	–	–	–	+	+	+	+	–	–	–	–	+	–	–	–	+	+	–	+	–	+	+	–	–	+	+	–	–	–	+	+	–	–	–	–	–	–	+	–	–		–	–	25	48 h	2	524	LMG 20539^{T}
Vibrio lentus	–	–	–	–	+	–	–	–	–	–	+	+	+	–	–	–	–	+	–	–	–	+	–	–	+	–	+	–	–	–	–	+	–	–	–	–	+	–	–	–	–	–	–	+	–	–		–	–	25	48 h	2	524	CIP 107166^{T}
Vibrio mangrovi	+	–	–	+		+	–	–	–	+	+	+	+	–	+	–	–	+	–	–	–	+	+	+	+	+	+	+	+	+	+	–	–	–	–	+	+	–	+	–	–	–	–	–	–	–	+	–	–	28	48 h	2	1373	
Vibrio marisflavi	+	–	–	–	+	+	–	–	–	+	+	+	+	–	–	–	–	–	–	–	+	+	+	+	+	+	+	+	–	+	+	+	–	–	–	–	–	–	+	–	–	–	–	–	–	–	+	–	–	28	48 h	1.5	1760	WH134^{T}
Vibrio mytili	+	–	–	+	+	+	–	–	–	+	+	+	+	–	–	–	–	+	–	–	–	+	+	+	+	+	+	+	–	–	+	+	–	+	–	+	+	–	+	+	–	–	–	–	–	–	+	–	–	25	48 h	2	1368	CECT 632T
Vibrio nereis	–	–	–	–	+	–	–	–	–	–	+	+	+	–	–	–	–	–	–	–	–	+	–	–	–	–	–	+	–	–	+	+	–	–	–	+	–	–	–	+	–	–	–	–	–	–	+	–	–	25	48 h	2	1368	LMG 3895^{T}
Vibrio penaeicida	–			–		–		–		–	+	+	+		–	–	–	–	–							–	+	+	–	–	–	+			–		+													25	7 d		803	JMC 9123
Vibrio plantisponsor	–	–	–	+	+	+	–	–	–	+	+	+	+	–	+	–	–	+	–	–	–	+	+	+	+	+	+	+	+	–	+	+	–	–	+	+	+	–	+	–	–	–	–	–	–	–	–	–	–	28	48 h	2.5	1372	MSS-RF60^{T}
Vibrio pomeroyi	–	–	–	–	+	–	–	–	–	+	+	+	+	–	–	–	–	+	–	–	–	+	+	–	+	–	+	+	–	–	–	+	–	–	–	+	+	–	–	–	–	–	–	+	–	–	–	–	–	25	48 h	2	524	LMG 20537^{T}
Vibrio quintilis	–	–	–	–	+	+	–	–	–	–	+	+	+	–	–	–	–	+	–	–	–	+	–	+	+	+	+	+	–	–	+	–	–	–	–	+	+	–	–	–	–	–	–	–	–	–	–	–	–	25	48 h	2	1065	CECT 7734^{T}
Vibrio salmonicida	+sl	–	–	–	+	–	–	–	–	+	+	+	–	–	–	–	–	+	–	–	–	+	–	–	–	–	–	+	–	–	–	+	–	–	–	–	–	–	–	–	–	–	–	–	–	–	+	–	–				473	
Vibrio splendidus	+	–	–	–	+	–	–	–	–	+	+	+	+	–	–	–	–	+	–	–	–	+	–	–	+	–	+	+	+	+	–	+	–	–	–	+	+	–	–	–	–	–	–	+	–	–	+	–	–	25	48 h	2	590	A515, scallop
Vibrio splendidus	+	–	–	–	+	–	–	–	–	+	+	+	+	–	–	–	–	+	–	–	–	+	–	–	+	–	+	+	+	+	–	+	–	–	–	+	+	–	–	–	–	–	–	–	–	–	+	–	–	25	48 h	2	590	AH5, turbot

Continued

Table 3.30. Continued.

	1	2	3	4	5	6	7	8	9	10	11	12	13	14	15	16	17	18	19	20	21	22	23	24	25	26	27	28	29	30	31	32	33	34	35	36	37	38	39	40	41	42	43	44	45	46	47	48	49					
Test	**Gly**	**Ery**	**Dara**	**Lara**	**rib**	**dXyl**	**lXyl**	**ado**	**mdx**	**gal**	**glu**	**fru**	**mne**	**Lsor**	**rha**	**dul**	**ino**	**man**	**sor**	**mdm**	**mdg**	**nag**	**amy**	**arb**	**esc**	**sal**	**cel**	**mal**	**lac**	**mel**	**suc**	**tre**	**inu**	**mlz**	**raf**	**amd**	**glyg**	**xylt**	**gen**	**tur**	**lyx**	**tag**	**Dfuc**	**Lfuc**	**Darl**	**Larl**	**gnt**	**2ket**	**5ket**	**Temp**	**Time**	**Inoc**	**Ref**	**Strain**
Organism																																																						
Vibrio splendidus	+	–	–	–	+	–	–	–	–	+	+	+	+	–	–	–	–	+	–	–	–	+	–	–	+	–	+	+	–	+	–	+	–	–	–	+	+	–	–	–	–	–	–	–	–	–	+	–	–	25	48 h	2	590	VS1, 4, 6 turbot
Vibrio splendidus	–	–	–	–	+	–	–	–	–	+	+	+	+	–	–	–	–	+	–	–	–	+	–	–	+	–	+	+	–	+	–	+	–	–	–	+	+	–	–	–	–	–	–	–	–	–	+	–	–	25	48 h	2	590	VS5, 9, 10, 11, turbot
Vibrio splendidus	+	–	–	–	+	–	–	–	–	+	+	+	+	–	–	–	–	+	–	–	–	+	+	–	+	–	+	+	–	+	–	+	–	–	–	+	+	–	–	–	–	–	–	+	–	–	–	–	–	25	48 h	2	524	LMG 4042^T
Vibrio tasmaniensis	–	–	–	–	+	–	–	–	–	–	+	+	+	–	–	–	–	+	–	–	–	+	–	–	+	–	+	+	–	–	–	+	–	–	–	–	–	–	–	–	–	–	–	+	–	–	–	–	–	25	48 h	2	524	LMG 20012^T
V. toranzoniae Vb10.8T	+		+		+					+	+	+	+				–	+				+					+	+			+	+				+	+	–										+		25	48 h	2	980	Vb10.8^T
V. toranzoniae Cmf 13.9	+		+		+					+	+	+	–				–	+				+					+	+			+	+				+	+	+										+		25	48 h	2	980	Cmf 13.9
V. toranzoniae CMJ 9.4	+		+		+					+	+	+	–				+	+				+					+	+			+	+				+	+	+										+		25	48 h	2	980	CMJ 9.4
V. toranzoniae CMJ 9.11	+		+		+					+	+	+	+				–	+				+					+	+			+	–				–	+	–										+		25	48 h	2	980	CMJ 9.11
Vibrio tubiashii	–	–	–	–	+	–	–	–	–	+	+	+	+	–	–	–	–	+	–	–	–	+	+	–	–	–	+	+	–	–	+	+	–	–	–	+	+	–	–	–	–	–	–	–	–	–	+	–	–	25	48 h	2	1368	LMG 10936^T
Weissella ceti CECT 7719^T	–	–	–	–	+	–	–	–	–	–	+	–	–	–	–	–	–	–	–	–	–	+	–	–	–	–	–	+	–	–	–	+	–	–	–	–	–	–	–	–	–	–	–	–	–	–	–	–	–	37	7 d	0.85	1708	CECT 7719^T
Weissella hellenica DS-12	–	–	–	+	–	–	–	–	–	+	+	+	+	–	–	–	–	–	–	–			–	–	+	–	+	+	–	+	+	+	–	–	+	–	–	–	–	+	–	–	–	–			+	–	–	30	48 h	0.85	260	DS-12
Weissella hellenica JCM 9890^T	–	–	–	+	–	–	–	–	–	–	+	+	+	–	–	–	–	–	–	–			–	–	–	–	–	+	–	–	+	+	–	–	–	–	–	–	–	–	–	–	–	–	–	–	+	–	–	30	48 h	0.85	260	DS-12
Yersinia kristensenii	+	–	–	+	+	+	–	–	–	+	+	+	+	+	–	–	–	+	+	–	–	+	V	+	–	–	+	+	+	–	–	+	–	–	–	+	–													28	48 h	0.85	162	
Yersinia ruckeri	+	–	–	–	+	–	–	–	–	+	+	+	+	–	–	–	–	+	V	–	–	+	–	–	–	–	–	+	–	–	–	+	–	–	–	–	–	–	–	–	–	–	–	–	–	–	+	–	–	22	48 h	0.85	420	
Zobellia uliginosa	–										+				–			–							+			+			+										w	+						+		30	7 d	2.5	119	DSM 2061^T
Zobellia galactanovorans	–			+						+	+	+	+		+			+	–				+		+			+			+										–	–						–		30	7 d	2.5	45	DSM 12802^T

Gly=glycerol, Ery=erythritol, Dara=D-arabinose, Lara=L-arabinose, rib=ribose, dXy=D-xylose, ado=adonitol, mdx=b-methyl-D-xyloside, gal=galactose, glu=glucose, fru=fructose, mne=mannose, Lsor=L-sorbose, rha=rhamnose, dul=dulcitol, ino=inositol, man=mannitol, sor=sorbitol, mdm=α-methyl-D-mannoside, mdg=α-methyl-D-glucoside, nag=*N*-acetylglucosamine, amy=amygdalin, arb=arbutin, esc=esculin, sal=salicin, cel=cellobiose, mal=maltose, lac=lactose, mel=melibiose, suc=sucrose, tre=trehalose, inu=inulin, mlz=melizitose, raf=raffinose, amd=starch, glyg=glycogen, xylt=xylitol, gen=gentiobiose, tur=D-turanose, lyx=D-lyxose, tag=D-tagatose, Dfuc=D-fucose, Lfuc=L-fucose, Darl=D-arabitol, Larl=L-arabitol, gnt=gluconate, 2ket=2-ketogluconate, 5ket=5-ketogluconate. Numbers indicate the percentage of strains with a positive result.

Tables 3.31. API Coryne database biochemical results.

	Nit	Pyz	Pyra	Pal	β gur	β gal	α glu	BNAG	esc	ure	gel	Cont	glu	rib	xyl	man	mal	lac	sac	gly	cat	Ref
Actinomyces marimammalium	–	–	–	v	–	–		+	–	–	–	–	+				+					985
Arcanobacterium haemolyticum	–	+	–	+	+	–	+	+	–	–	–	–	+	–	–	–	+	+		–	+	985
Arcanobacterium haemolyticum	–	+	–	+	–	+	+	+	–	–	–	–	+	+	–	–	+	+	–	–	–	727
Arcanobacterium phocae	–	+	–	+	–	+	+	–	–	–	–	–	+	+	v	30	+	v	+	+	v	989
Arcanobacterium phocae	–	+	–	+	–	+	+	+	–	–	–	–	+	+	+	+	+	+	+	+	+	1374
Arcanobacterium phocae	–	+	–	+	–	+	+	+	–	–	–	–	+	+	v	v	+	+	+	+	v	1374
Arcanobacterium phocae	–	–	–	+	–	+	+	+	–	–	–	–	+	+	w	–	+	+	+	+	+	727
Arcanobacterium phocisimile	–	+	–	w	–	+	+	–	–	–	–	–	+	+	–	–	+	+	+	+	+	727
Arcanobacterium pluranimalium	–	+	+	–	+	–	–	–	+	–	+	–	+	+	–	–	+	–	–	–	+	727
Arcanobacterium pluranimalium	–	v	+	–	+	–	–	–	w+	–	+	–	+	+	–	–	v	–	–	–	+	985
Arthrobacter nasiphocae	–	–	+	+	–	–	+	–	–	–	+	–	–	–	–	–	–	–	–	–	+	358
Arthrobacter rhombi	–	–	–	–	–	+	–	–	+	–	–	–	+	–	–	+	+	+	+	–	+	1276
Corynebacterium aquaticum	–	+	+	+	–	+	+	–	+	–	+v	–					–					235
Corynebacterium aquaticum	–	+	+	+	–	+	+	–	+	–	w	–	–	–	–	–	–	–	–	–	+	133
Corynebacterium aquaticum	+	+	–	+	–	+	+	+	+	–	–	–	–	–	–	–	–	–	–	–	+	
Corynebacterium caspium	–	+	–	–	–	–	–	–	–	+	–	–	+	+	–	–	–	–	–	–	+	357
Corynebacterium ciconiae	–	+	–	+	–	–	–	–	–	+	–	–	+	+	–	–	–	–	–	–	+	528
Corynebacterium ciconiae	–	+	–	+	–	–	–	–	–	–	–	–	+	+	–	–	+	–	–	–	+	528
Corynebacterium maris	–	+	+	+	–	–	+	–	–	–	+	–	–	–	–	–	–	–	–	–	+	152
Corynebacterium phocae	–	–	+	+	–	–	+	–	–	+	–	–	+	–	–	–	+	–	–	–	+	1313
Corynebacterium phocae	–	–	+	+	–	–	+	–	–	–	–	–	+	–	–	–	+	+	+	–	+	1313
C. pseudodiphtheriticum	+	+	+	+	–	–	–	–	–	+	–	–					-					235
C. pseudodiphtheriticum	+	+	+	+	–	–	–	–	–	+	–	–	–	–	–	–	–	–	–	–	+	133
C. pseudotuberculosis	–	–	–	+	–	+	–	–	–	+	–	–					–					235
C. pseudotuberculosis	–	–	–	+	–	+	–	–	–	+	–	–	+	+	–	–	–	–	–	–	+	133
Corynebacterium sphenisc.	+	+	–	–	–	–	–	–	–	–	–	–	+	–	–	–	+	–	–	–	+	652
Corynebacterium spheniscorum	–	+	–	–	–	–	–	–	–	–	–	–	+	+	–	–	+	–	–	–	+	653

Continued

Table 3.31. Continued.

	Nit	Pyz	Pyra	Pal	β gur	β gal	α glu	BNAG	esc	ure	gel	Cont	glu	rib	xyl	man	mal	lac	sac	gly	cat	Ref
Corynebacterium testudinoris	+	–	–	–	–	–	v	–	+	–	–	–	+	+	–	–	+	–	+	–	+	351
Corynebacterium xerosis	+	–	+	+	–	–	–	–	–	–	–	–					–					235
Erysipelothrix rhusiopathiae	–	+	–	–	–	–	–	+	–	–	–	–	+	+	–	–	–	+	+	–	–	464
Erysipelothrix rhusiopathiae	–	–	+	+	–	–	–	+	–	–	–	–	+	+	–	–	–	+	–	–	–	464
Erysipelothrix rhusiopathiae	–	–	–	–	–	–	–	+	–	–	–	–	+	–	–	–	–	+	–	–	–	464
Erysipelothrix rhusiopathiae	–	–	+	–	–	–	–	+	–	–	–	–	+	+	–	–	–	+	–	–	–	464
Lactobacillus ceti	–	–	–	+	–	–	–	–	–	–	+	–	–	+	–	–	–	–	–	–	–	1709
Trueperella bernardiae					–	–			–		–	–		+	–			–		+	–	985
Weissella ceti	–	+	–	+	–	–	–	–	+	–	–	–	+	+	–	–	+	–	–	–	–	1708

Nit=nitrate, Pyz=pyrrolidonyl acrylamidase, Pyra=pyrazinamidase, Pal=alkaline phosphatase, β gur=β glucuronidase, β gal=β galactosidase, α glu=α glucosidase, βNAG=*N*-acetyl-beta glucosaminidase, esc=esculin, ure=urease, gel=gelatin, glu=glucose, rib=ribose, xyl=xylose, man=mannitol, mal=maltose, lac=lactose, sac=sucrose, gly=glycogen, cat=catalase. Ref=reference, +=positive reaction, –=negative reaction, numbers indicate percentage of positive strains, w=weak reaction, cont=negative control tube.

Table 3.32. API Coryne database numbers (organism listed alphabetically).

Organism	Strain	API Coryne number	No. of strains	Notes	Ref
Actinomyces marimammalium					763
Arcanobacterium haemolyticum		2330164			985
Arcanobacterium haemolyticum		2530360			727
Arcanobacterium phocae	DSM 10002^T	2530777	1		1374
Arcanobacterium phocae	Strains	2530363	5		1374
Arcanobacterium phocae		0530767			727
Arcanobacterium phocisimile		2510367			727
Arcanobacterium pluranimalium		6242367			727
Arcanobacterium pluranimalium		4242304			985
Arcanobacterium pluranimalium		4242324			985
Arcanobacterium pluranimalium		6242304			985
Arcanobacterium pluranimalium		6242324			985
Arthrobacter nasiphocae		4112004			358
Arthrobacter rhombi		0440175			1276
Corynebacterium aquaticum	RB 965BA	6552004			133
Corynebacterium aquaticum	ATCC 14665	3570004			133
Corynebacterium caspium	CCUG 44566^T	2001304	1		357
Corynebacterium ciconiae		2101304	5		528
Corynebacterium ciconiae		2100324	3		528
Corynebacterium maris		6112004			152
Corynebacterium phocae	CCUG 38205^T	4111124	1		1313
Corynebacterium phocae	Others	4110165	3		1313
C. pseudodiphtheriticum	ATCC 10700	7101004			133
C. pseudotuberculosis	ATCC 19410	0501304			133
Corynebacterium sphenisci	CECT 5990^T	3000124	6		652
Corynebacterium spheniscorum	CECT 45512^T	2000324	20		653
Corynebacterium testudinoris	CCUG 41823^T	1040325	1		351
Corynebacterium testudinoris	CCUG 41823^T	1050325	1		351
Erysipelothrix rhusiopathiae		2020341			464
Erysipelothrix rhusiopathiae		4120340			464
Erysipelothrix rhusiopathiae		0020140			464
Erysipelothrix rhusiopathiae		4020340			464
Erysipelothrix rhusiopathiae		4020340			464
Lactobacillus ceti					1709
Weissella ceti		2140320			1708

Table 3.33. API Coryne database numbers (numbers in ascending order).

API 20NE number	Organism	Strain	No. of strains	Notes	Ref
0020140	*Erysipelothrix rhusiopathiae*				464
0440175	*Arthrobacter rhombi*				1276
0501304	*C. pseudotuberculosis*	ATCC 19410			133
0530767	*Arcanobacterium phocae*				727
1040325	*Corynebacterium testudinoris*	CCUG 41823^T	1		351
1050325	*Corynebacterium testudinoris*	CCUG 41823^T	1		351
2000324	*Corynebacterium spheniscorum*	CECT 45512^T			653
2001304	*Corynebacterium caspium*	CCUG 44566^T			357
2020341	*Erysipelothrix rhusiopathiae*				464
2100324	*Corynebacterium ciconiae*		3		528
2101304	*Corynebacterium ciconiae*		5		528
2330164	*Arcanobacterium haemolyticum*				985
2510367	*Arcanobacterium phocisimile*				727
2530360	*Arcanobacterium haemolyticum*				727
2530363	*Arcanobacterium phocae*	Strains	6		1374
2530777	*Arcanobacterium phocae*	DSM 10002^T	1		1374
3000124	*Corynebacterium sphenisci*	CECT 5990^T	6		652
3570004	*Corynebacterium aquaticum*	ATCC 14665			133
4020340	*Erysipelothrix rhusiopathiae*				464
4110165	*Corynebacterium phocae*	Others	3		1313
4111124	*Corynebacterium phocae*	CCUG 38205^T			1313
4112004	*Arthrobacter nasiphocae*				358
4120340	*Erysipelothrix rhusiopathiae*				464
4242304	*Arcanobacterium pluranimalium*				985
4242324	*Arcanobacterium pluranimalium*				985
6112004	*Corynebacterium maris*				152
6242304	*Arcanobacterium pluranimalium*				985
6242367	*Arcanobacterium pluranimalium*				727
6242324	*Arcanobacterium pluranimalium*				985
6552004	*Corynebacterium aquaticum*	RB 965BA			133
7101004	*C. pseudodiphtheriticum*	ATCC 10700			133

Table 3.34. API 20 Strep database biochemical results.

	vp	hip	aes	pyra	α-gal	β gur	β-gal	Pal	Lap	ADH	rib	ara	man	sor	lac	tre	inu	raf	amd	glyg	Hem	Ref	Strain
Abiotrophia defectiva	–	–	–	+	+	–	+	–	+	–	–	–	–	–	+	V	–	+	+	–			
Arcanobacterium phocae	–	–	–	52	+	–	+	+	+	–	+	–	–	–	68	–	–	–	+	+	+	845	
Carnobacterium inhibens	–	+	+	–	–	–	–	–	–	–	+	–	+	–	+W	+	+W	–	–	–	α	841	
Carnobacterium maltaromaticum	+	–	+	+				nt		+	+	–	+	+	V	+	+	V	–	+		1139	
Catellicoccus marimammalium	–	–	–	–	–	–	–	–	+	+	–	–	–	–	–	+	–	–	–	–	–	984	CCUG 49459T
Enterococcus avium	+	–	+	+	–	–	–	–	–	–	+	+	+	+	+	+	+	+	+	–	α	311,1763	
Enterococcus durans	+	–	+	+	+	–	+	–	–	+	+	–	–	–	+	–	–	–	+	–	α	311,1763	
Enterococcus faecalis	+	–	+	+	–	–	+	–	+	+	+	–	+	+	+	+	–	–	+	–	α	311,1763	
Enterococcus faecium		+			–	–	+			+		+	+	–			–	–	–	V			
Granulicatella adiacens	–	–	–	+	–	V	–	–	+	–	–	–	–	–	–	–	V	–	–	–	γ		
Granulicatella balaenopterae	–	–	+	w	–	–	–	–	+	+	–	–	–	–	–	+	–	–	–	–		352,990,1693	
Granulicatella elegans	–	V	–	+	–	–	–	–	+	+	–	–	–	–	–	–	–	V	–	–	α	352,1420	
Lactococcus garvieae	+	–	+	+	–	–	–	–	+	+	+	–	+	–	–	+	–	–	–	–	α	311,481	ATCC 43921T
Lactococcus garvieae	+	–	+	+	–	–	–	–	+	+	+	–	+	–	–	+	–	–	+	–	α	311	Rainbow trout
Lactococcus garvieae	+	–	33	66	–	–	–	–	33	33	–	–	–	–	–	–	–	–	–	–	α	311	Prawn
Lactococcus garvieae	+	–	+	+	–	–	–	–	–	–	+	–	+	+	–	+	+	–	–	–	–	93	
Lactococcus lactis ssp. *cremoris*	+	–	+	–	–	–	–	–	+	+	+	–	–	–	–	+	–	–	–	–	γ	311	
Lactococcus lactis ssp. *lactis*	+	–	+	+	–	–	–	–	+	+	+	–	–	–	–	+	–	–	+	–	γ	1763	
L. lactis ssp. *lactis*	+	–	+	+	–	–	–	–	36	82	36	–	27	–	27	73	64	–	36	–	γ	1763	
Lactococcus lactis ssp. *tructae*										+	+		+		+			V				1330	
Lactococcus raffinolactis	+	–	+	–	+	–	–	–	+	–	–	–	–	–	+	+	–	+	+	–	γ	311,1763	
Rhodococcus qingshengii	–		+				+	+	+	–												83	
Streptococcus agalactiae	+	–	–	–	–	–	–	+	+	+	+	–	–	–	–	–	–	–	–	–	–	482	
Streptococcus agalactiae	+	+	–	–	–	+	–	+	+	–	+S	–	–	–	–	–	–	–	–	+		1696	
Streptococcus agalactiae	+	+	–	–	+	+	–	+	+	+	+	–	–	–	–	+	–	–	–	–	β	1589	
Streptococcus agalactiae	+	–	–	–	+	+	–	+	+	+	+	–	–	–	–	+	–	–	–	–	β	1589	
Streptococcus agalactiae	+	+	–	–	–	–	–	+	+	+	+	–	–	–	–	+	–	–	–	–	β	466	
Streptococcus agalactiae Biotype 1	+	+	–	–	–	–	–	+	+	+	+	–	–	–	–	–	–	–	–	–	–	1135	
	Biotypes from tilapia																						
Biotype 2	+	+	–	–	–	–	–	+	+	–	+	–	+	–	–	+	–	+	–	–	β	1135	
Biotype 3	+	+	–	–	–	–	–	+	+	+	+	–	–	–	+	–	–	–	–	–	–	1135	
Biotype 4	+	+	–	–	–	+	–	+	+	+	+	–	–	–	–	+	–	–	–	–	β	1135	
Biotype 5	+	+	–	–	–	+	–	+	+	+	+	–	–	–	–	+	–	–	+	–	β	1135	

Continued

Table 3.34. Continued.

	vp	hip	aes	pyra	α-gal	β gur	β-gal	Pal	Lap	ADH	rib	ara	man	sor	lac	tre	inu	raf	amd	glyg	Hem	Ref	Strain
Biotype 6	+	+	–	–	–	–	–	+	+	+	+	–	–	–	–	–	–	–	+	–	–	1135	
Biotype 7	+	+	–	–	–	–	–	+	+	–	+	–	–	–	–	–	–	–	+	–	–	1135	
Biotype 8	+	–	–	–	–	–	–	+	+	+	+	–	–	–	–	–	–	–	+	–	–	1135	
Biotype 9	+	+	–	–	–	–	–	+	+	+	+	–	–	–	–	+	–	–	–	–	–	1135	
Biotype 10	+	+	–	–	–	–	–	+	+	–	+	–	–	–	–	–	–	–	–	–	–	1135	
Biotype 11	+	+	–	–	–	–	–	+	–	–	+	–	–	–	–	–	–	–	–	–	–	1135	
Streptococcus dysgalactiae	–	v	v	–	–	+	–	+	+	v	–	–	–	–	–	+	–	–	–	v	+	1213	
S. dysgalactiae ssp. *dysgalactiae* Gp C	–	–	–	–	–	+	–	+	+	10	10	–	–	–	–	60	–	–	70	40	–	1235	25°C
S. dysgalactiae ssp. *dysgalactiae* Gp C	–	–	–	–	–	+	–	+	+	33	80	–	–	–	–	–	–	–	+	50	–	1235	37°C
S. dysgalactiae ssp. *dysgalactiae* Gp C	–	–	–	–	–	+	–	+	+	–	+	–	–	+	+	+	–	–	+	–	–	1235	ATCC 43078
S. equi ssp. *zooepidemicus*	–	–	–	–	–	+	–	+	+	+	+	–	–	+	+	–	–	–	+	+		*	
Streptococcus ictaluri	w	–	–	+	–	+	–	+	+	+s	+	–	–	–	–	–	–	–	–	–	–	267	
Streptococcus iniae Human strains	–	–	+	–	–	–	–	+	+	–	–	–	+	–	–	+	–	–	+	–	αβ	454	
Streptococcus iniae Fish strains	–	–	+	+	–	+	–	+	+	70	+	–	46	–	–	+	–	–	+	+	αβ	360,454	
Streptococcus iniae	–	–	+	+	–	+	–	+	+	+	+	–	+	–	–	+	–	–	+	+	β	483,1840	
Streptococcus iniae	–	–	+	+	–	+	–	+	+	+	+	–	+	–	–	+	–	–	–	+	β	478	
Streptococcus iniae	–	–	+	+	–	+	–	+	+	–	+	–	+	–	–	+	–	–	+	+	+	228,454	
Streptococcus halichoeri	+	–	–	+	–	–	–	+	+	+	+	–	+	–	v	–	–	–	v	–		987	
Streptococcus marimammalium	–	–	–	–	–	–	–	+	+	v	–	–	–	–	+	–	–	–	–	–	–	986	
Streptococcus parauberis	+	+	–	+	–	–	–	+	+	+	–	–	+	–	+	+	–	–	–	–	–	93	
Streptococcus phocae	–	–	–	–	–	–	–	+	–	–	+	–	–	–	–	–	–	–	–	–	β	1538	
Streptococcus phocae Group G	–	v	–	–	–	–	–	+	+	v	v	–	–	–	v	v	–	–	+	+	β	800	
Streptococcus phocae Group F	–	–	–	–	–	–	–	+	+	–	v	–	–	–	v	v	–	–	v	v	β	800	
Streptococcus phocae ungroupable	–	–	–	–	–	–	–	+	+	–	+	–	–	–	–	–	–	–	+	+	β	800	
Streptococcus porcinus	+	–	+	–	–	+	–	+	+	+	+	–	+	+		+	–	–	v	–	β	356	
Streptococcus uberis	+	+	+	–	–	+	–	+	+	+	+	–	+	+	+	+	–	–	+	+	–	688	
Streptococcus uberis	+	+	+	v	v	+	–	–	+	+	+	–	+	+	+	+	+	+	–	–	–	*	
Streptococcus uberis	+	+	+	+	+	+	–	–	+	+	+	–	+	+	+	+	+	+	–	–	–	*	
Vagococcus fluvialis	+v	–	+	+	–	–	–	–v	+v	–	+	–	+	+	–	+	–	+	+	+	β	1354	
Vagococcus salmoninarum	+	v	+	+	–	–	–	+	+	–	70	–	–	70	–	+	–	–	–	–	β	1139,1442	

VP=Voges–Proskaüer, hip=hippurate, aes=esculin, pyra=pyrrolidonylarylamidase, α-gal=a galactosidase, β-gur=β glucuronidase, β-gal=β galactosidase, Pal=alkaline phosphatase, Lap=leucine arylamidase, ADH=arginine dihydrolase, rib=ribose, ara=arabinose, man=mannitol, sor=sorbitol, lac=lactose, tre=trehalose, inu=inulin, raf=raffinose, amd=amidon (starch), glyg=glycogen, Hem=Haemolysis, Ref=reference, nt=not tested. Numbers show percentage of positive strains. +=positive results, –=negative result, w=weak reaction, s=slow reaction. α=alpha haemolysis, β=beta haemolysis, γ=no haemolysis, *=API database.

Table 3.35. API 20 Strep database numbers (numbers in ascending order).

API 20E number	Organism	Strain	Notes	Ref
0022004	*Streptococcus phocae*		API database identifies as *Gemella haemolysans* 88.6%	1538
0041010	*Catellicoccus marimammalium*	CCUG 49459[T]		984
0060004	*Streptococcus phocae* Group F		API database identifies as *Gemella* sp.	800
0060005	*Streptococcus phocae* Group F		API database identifies as *Gemella* sp.	800
0060006	*Streptococcus phocae* Group F		API database identifies as *Gemella* sp.	800
0060007	*Streptococcus phocae* Group G		API database identifies as *S. equi* ssp. *equi* 95.8%	800
0060014	*Streptococcus phocae* Group F		API identifies as a number of Streptococcal or *Gemella* species	800
0060015	*Streptococcus phocae* Group F		API identifies as a number of Streptococcal or *Gemella* species	800
0060016	*Streptococcus phocae* Group F		API identifies as a number of Streptococcal or *Gemella* species	800
0060017	*Streptococcus phocae* Group G		API identifies as a number of Streptococcal species	800
0060400	*Streptococcus marimammalium*		API database identifies as *Streptococcus mitis* 1 82.2%	986
0060404	*Streptococcus phocae* Group F		API identifies as a number of Streptococcal or *Gemella* species	
0060405	*Streptococcus phocae* Group F		API identifies as a number of Streptococcal species	
0060406	*Streptococcus phocae* Group F		API identifies as a number of Streptococcal or *Gemella* species	
0060407	*Streptococcus phocae* Group G		API identifies as a number of Streptococcal species	800
0060417	*Streptococcus phocae* Group G		API identities as *Streptococcus pyogenes* 93.2%	800
0061007	*Streptococcus phocae* Group G		API database identifies as *S. equi* ssp. *equi* 99.9%	800
0061017	*Streptococcus phocae* Group G		API identifies as a number of Streptococcal species	800
0061400	*Streptococcus marimammalium*		API database identifies as *Streptococcus mitis* 1 78.5%	986
0061407	*Streptococcus phocae* Group G		API identifies as a number of Streptococcal species	800
0061417	*Streptococcus phocae* Group G		API identities as *Streptococcus pyogenes* 96.2%	800
0062004	*Streptococcus phocae* Group F		API database identifies as *Gemella* sp.	800
0062005	*Streptococcus phocae* Group F		API identifies as a number of Streptococcal or *Gemella* species	800
0062006	*Streptococcus phocae* Group F		API identifies as a number of Streptococcal or *Gemella* species	800
0062007	*Streptococcus phocae*	from seals	API database identifies as *S. dysgalactiae* ssp. *equisimilis* 87.1%	800

Continued

Table 3.35. Continued.

API 20E number	Organism	Strain	Notes	Ref
0062014	*Streptococcus phocae* Group F		API database identifies as *S. dysgalactiae* ssp. *equisimilis* 82.0%	800
0062015	*Streptococcus phocae* Group F		API database identifies as *S. dysgalactiae* ssp. *equisimilis* 99.5%	800
0062016	*Streptococcus phocae* Group F		API database identifies as *S. dysgalactiae* ssp. *equisimilis* 96.5%	800
0062017	*Streptococcus phocae* Group F		API database identifies as *S. dysgalactiae* ssp. *equisimilis* 94.5%	800
0063007	*Streptococcus phocae* Group G		API identifies as a number of Streptococcal species	800
0063017	*Streptococcus phocae* Group G		API database identifies *S. dysgalactiae* ssp. *equisimilis* 81.1%	800
0063407	*Streptococcus phocae* Group G		API identifies as a number of Streptococcal species	800
0063417	*Streptococcus phocae* Group G		API identifies as a number of Streptococcal species	800
0062404	*Streptococcus phocae* Group F		API identifies as a number of Streptococcal species	800
0062405	*Streptococcus phocae* Group F		API database identifies as *S. canis* 87.5%	800
0062406	*Streptococcus phocae* Group F		API identifies as a number of Streptococcal species	800
0062407	*Streptococcus phocae* Group F		API identifies as a number of Streptococcal species	800
0062414	*Streptococcus phocae* Group F		API database identifies as *S. dysgalactiae* ssp. *equisimilis* 83.6%	800
0062415	*Streptococcus phocae* Group F		API database identifies as *S. dysgalactiae* ssp. *equisimilis* 97.1%	800
0062416	*Streptococcus phocae* Group F		API database identifies as *S. dysgalactiae* ssp. *equisimilis* 91.4%	800
0062417	*Streptococcus phocae* Group F		API database identifies as *S. dysgalactiae* ssp. *equisimilis* 82.7%	800
0140000	*Granulicatella adiacens*		*Granulicatella adiacens* 80.4%. Read at 48 h	1420
0140020	*Granulicatella adiacens*		API database identifies as *Gemella morbillorum* 85.6%	1420
0141000	*Granulicatella elegans*		API database identifies as *Gemella morbillorum* 85.2%	
0141040	*Granulicatella elegans*		API database identifies as *Gemella morbillorum* 90.1%	
0272007	*Arcanobacterium phocae*			845
0350451	*Abiotrophia defectiva*			*
0372407	*Arcanobacterium phocae*			845
0460001	*S. dysgalactiae* ssp. *dysgalactiae* Gp C		25°C	1235
0460003	*S. dysgalactiae* ssp. *dysgalactiae* Gp C		25°C	1235

Continued

Table 3.35. Continued.

API 20E number	Organism	Strain	Notes	Ref
0460013	*S. dysgalactiae* ssp. *dysgalactiae* Gp C		25°C	1235
0462001	*S. dysgalactiae* ssp. *dysgalactiae* Gp C		37°C	1235
0462003	*S. dysgalactiae* ssp. *dysgalactiae* Gp C		37°C	1235
0463001	*S. dysgalactiae* ssp. *dysgalactiae* Gp C		25°C, 37°C	1235
0463003	*S. dysgalactiae* ssp. *dysgalactiae* Gp C		25°C, 37°C	1235
0463011	*S. dysgalactiae* ssp. *dysgalactiae* Gp C		25°C	1235
0460013	*S. dysgalactiae* ssp. *dysgalactiae* Gp C		25°C	1235
0462611	*S. dysgalactiae* ssp. *dysgalactiae* Gp C		ATCC 43078	1235
0540000	*Granulicatella adiacens*		API database identifies as *Granulicatella adiacens* 99.8%	1420
0540020	*Granulicatella adiacens*		API database identifies as *Granulicatella adiacens* 97.9%	1420
0463607	*Streptococcus equi* ssp. *zooepidemicus*		API database identifies as *S. equi zooepidemicus* 99.9%	*
1063000	*Streptococcus agalactiae*		API database identifies a number of taxa	482
1063001	*Streptococcus agalactiae*		API database identifies as *Streptococcus agalactiae* 61.1%	1135
1163100	*Streptococcus halichoeri*		API database identifies a number of taxa	987
1163101	*Streptococcus halichoeri*		API database identifies a number of taxa	987
1163500	*Streptococcus halichoeri*		API database identifies a number of taxa	987
1163501	*Streptococcus halichoeri*		API database identifies as *L. lactis* ssp. *lactis* 87.4%	987
1563000	*Streptococcus ictaluri*			267
1663014	*Streptococcus agalactiae*	Thailand, Tilapia	API database identifies as *Streptococcus agalactiae* 95.9%	1589
2060007	*Streptococcus phocae* Group G		API database identifies a number of taxa	800
2060017	*Streptococcus phocae* Group G		API identifies as a number of Streptococcal species	800
2060407	*Streptococcus phocae* Group G		API identifies as a number of Streptococcal species	800
2060417	*Streptococcus phocae* Group G		API identifies as a number of Streptococcal species	800
2061007	*Streptococcus phocae* Group G		API database identifies as *Streptococcus equi* ssp. *equi* 97.3%	800
2061017	*Streptococcus phocae* Group G		API identifies as *Streptococcus* Group L 88.9%	800

Continued

Table 3.35. Continued.

API 20E number	Organism	Strain	Notes	Ref
2061407	*Streptococcus phocae* Group G		API identifies as a number of Streptococcal species	800
2061417	*Streptococcus phocae* Group G		API identifies as a number of Streptococcal species	800
2063007	*Streptococcus phocae* Group G		API identifies as a number of Streptococcal species	800
2063017	*Streptococcus phocae* Group G		API identifies as *Streptococcus* Group L 97.8%	
2063407	*Streptococcus phocae* Group G		API identifies as *Streptococcus* Group L 93.7%	
2063417	*Streptococcus phocae* Group G		API identifies as *Streptococcus* Group L 99.3%	
2141000	*Granulicatella elegans*		API database identifies a number of taxa	
2141040	*Granulicatella elegans*		API database identifies as *Aerococcus urinae* 92.5%	
3022000	*Streptococcus agalactiae*		API database identifies a number of taxa	1135
3062000	*Streptococcus agalactiae*		API database identifies a number of taxa	1135
3062001	*Streptococcus agalactiae*		API database identifies a number of taxa	1135
3062154	*Streptococcus agalactiae*		API database identifies low discrimination for *S. agalactiae*	1135
3063000	*Streptococcus agalactiae*		API database identifies as *Streptococcus agalactiae* 99.3%	1135
3063001	*Streptococcus agalactiae*		API database identifies as *Streptococcus agalactiae* 99.1%	1135
3063010	*Streptococcus agalactiae*		API database identifies as *Streptococcus agalactiae* 99.8%	1135
3063014	*Streptococcus agalactiae*		API database identifies as *Streptococcus agalactiae* 99.9%	466
3063400	*Streptococcus agalactiae*		API database identifies as *Streptococcus agalactiae* 99.1%	1135
3161510	*Streptococcus parauberis*			93
3462002	*Streptococcus (difficile) agalactiae*		API database identifies as *Streptococcus agalactiae* 99.4%	1696
3463014	*Streptococcus agalactiae*		API database identifies as *Streptococcus agalactiae* 99.9%	1135
3463015	*Streptococcus agalactiae*		API database identifies as *Streptococcus agalactiae* 99.9%	1135
3663014	*Streptococcus agalactiae*	Thailand, Tilapia	API database identifies as *Streptococcus agalactiae* 99.9%	1589
4060111	*Streptococcus iniae*	Human strains	API database identifies a number of taxa	454
4060115	*Streptococcus iniae*	Human strains	number of Streptococcal species identified in the API database	454
4141010	*Granulicatella balaenopterae*		API database identifies a number of taxa	990,1693
4142353	*Vagococcus fluvialis*		identifies as *Aerococcus viridans* in API database	1354

Continued

Table 3.35. Continued.

API 20E number	Organism	Strain	Notes	Ref
4162353	*Vagococcus fluvialis*		Identifies as *Aerococcus viridans* in API database	1354
4562013	*Streptococcus iniae*	Fish strains	Identifies *S. dysgalactiae* ssp. *equisimilis* 97.8% in the API database	360,454
4562017	*Streptococcus iniae*	Fish strains	Identifies *S. dysgalactiae* ssp. *equisimilis* 99.9% in the API database	360,454
4562113	*Streptococcus iniae*	Fish strains	Identifies *S. dysgalactiae* ssp. *equisimilis* in the API database	360,454
4562117	*Streptococcus iniae*	Fish strains	Identifies *S. dysgalactiae* ssp. *equisimilis* in the API database	228,360, 454
4563013	*Streptococcus iniae*	Fish strains	Identifies *S. dysgalactiae* ssp. *equisimilis* 93.3% in the API database	360,454
4563017	*Streptococcus iniae*	Fish strains	Identifies *S. dysgalactiae* ssp. *equisimilis* 99.8% in the API database	360,454
4563113	*Streptococcus iniae*	Fish strains	Identifies *S. dysgalactiae* ssp. *equisimilis* in the API database	360,454
4563116	*Streptococcus iniae*	Red porgy		478
4563117	*Streptococcus iniae*	Fish strains	Identifies *S. dysgalactiae* ssp. *equisimilis* 99.9% in the API database	360,454, 483,1840
5053010	*Lactococcus lactis* ssp. *cremoris*		Identifies as *Enterococcus* and *Lactococcus* sp. in API database	311,1763
5102330	*Lactococcus garvieae*			93
5103332	*Carnobacterium piscicola*		Identifies as *Aerococcus viridans* in API database	1137
5103372	*Carnobacterium piscicola*		Identifies as possibility of *Enterococcus* in API database	1137
5103732	*Carnobacterium piscicola*		Identifies as possibility of *Enterococcus* in API database	1137
5103772	*Carnobacterium piscicola*		Identifies as *Aerococcus viridans* in API database	1137
5106771	*Enterococcus avium*		Identifies as *Aerococcus viridans* in API database	311,1763
5142310	*Enterococcus seriolicida*		Identifies as *Enterococcus avium* in API database	311
5142353	*Vagococcus fluvialis*		Identifies as *Aerococcus viridans* in API database	1354
5143010	*Lactococcus lactis* ssp. *lactis*		Identifies as *Enterococcus* and *Lactococcus* sp. in API database	311
5143011	*Lactococcus lactis* ssp. *lactis*			1763
5143110	*Lactococcus garvieae*		API database identifies a number of taxa	481,1763
5153711	*Enterococcus faecalis*		Enterococcus *faecalis* in API database	311,1763
5162353	*Vagococcus fluvialis*		Identifies as *Aerococcus viridans* in API database	1354
5240451	*Lactococcus raffinolactis*		Identifies as Streptococcal species in API database	311,1763
5313401	*Enterococcus durans*		*Enterococcus durans* in API database	311,1763

Continued

Table 3.35. Continued.

API 20E number	Organism	Strain	Notes	Ref
5463314	*Streptococcus porcinus*			356
5463315	*Streptococcus porcinus*			356
6002530	*Carnobacterium inhibens*		Identifies as *Aerococcus viridans* in API database	841
67157510	*Enterococcus faecium*			*
7160010	*Vagococcus salmoninarum*		API database identifies as possibility of *S. acidominimus* or *S. pluranimalium*	1139
7160211	*Vagococcus salmoninarum*			1139
7160210	*Vagococcus salmoninarum*		API database identifies as possibility of *S. acidominimus* or *S. pluranimalium*	1139
7162010	*Vagococcus salmoninarum*		API database identifies as possibility of *S. acidominimus* or *S. pluranimalium*	1139
7162210	*Vagococcus salmoninarum*		API database identifies as possibility of *S. acidominimus* or *S. pluranimalium*	1139
7463713	*Streptococcus uberis*	From penaeid shrimp	API database identifies as 99.9% *S. uberis*	688
7463733	*Streptococcus uberis*			*
7543660	*Streptococcus uberis*			*
7643660	*Streptococcus uberis*			*
7743660	*Streptococcus uberis*			*

*=API database.

Table 3.36. API Rapid ID 32 STREP database biochemical results.

	1	1.1	1.2	1.3	1.4	2	2	1.7	2	2	1A	1B	0	0.1	0.2	0.3	0.4	1	1	0.7	1	0.9	0A	0B	1C	1D	1E	0C	0D	0E	1F	0F	Ref	Strains
	ADH	βglu	βgar	βgur	αgal	pal	rib	man	sor	lac	tre	raf	vp	appa	βgal	pyrA	βnag	gta	hip	glyg	pul	mal	mel	mlz	suc	lara	darl	mbdg	tag	βman	cdex	ure		
Organism																																		
Abiotrophia defectiva	–	–	50	–	+	–	–	–	–	50	+	50	–	50	+	75	–	–	–	–	+	+	–	–	+	–	–	–	v	–	–	–	*, 352	
Abiotrophia para-adiacens	–	+	–	50	–						–				–		50				–				+				–	–			352	
Actinomyces marimammalium	–	–	+	–	–	v	–	–	–	+	–	–	–	+	+	–	+	+	–	v	–	+	–	–	–	–	–	–	–	–	–	–	763	
Aerococcus viridans	–	70	3	30	60	–	28	75	25	79	91	42	–	–	10	83	–	–	92	10	10	95	–	–	+	–	–	65	–	–	–	–	*	
Arcanobacterium haemolyticum			–	v	–		–			+					–	v			–	–										–			985	
Arcanobacterium pluranimalium	–	–	–	+	–	–	+	–	–	–	–	–	–	+	–	+	–	v	+	–	–	–	–	–	–	–	–	–	–	–	–	–	985	
Arcanobacterium phocae			+	–	+w		+			v					+	–			–	+										–			989	
Atopobacter phocae	+	–	–	–	–	+	+	–	–	v	v	–	–	w–	–	w	–	–	–	+	+	+	–	–	v	–	–	–	–	–	+	–	989	
Carnobacterium alterfunditum	–						–	–					–			–	–	–				+			+					–			359	
Carnobacterium divergens	+						+	–					+			+	+	+				+			+					–			359	
Carnobacterium funditum	–						–	–					–			–	–	+				+			+					–			359	
Carnobacterium gallinarum	+						+	–					+			+	+	+				+			+					–			359	
Carnobacterium mobile	+						–	–					–			+	–	–				–			+					–			359	
Carnobacterium maltaromaticum	+						+	–					+			+	+	+							+					–			359	
Carnobacterium maltaromaticum	+	–	–	+		–	+	+	v	–	+	w	+			+	+	+	–	–	–	+	w	w	+	–		w		–	+	–	985	
Carnobacterium inhibens	–	+	–	–	–	–	+	+	–	+w	+	–	–	–	–	–	–	–	+	–	–	+	–	–	+	–	–	+	–	–	–	–	841	
Catellicoccus marimammalium	+	–	–	–	–	–	–	–	–	–	+	–	–	–	–	–	–	–	–	–	–	–	–	–	–	–	–	–	–	–	–	–	984	CCUG 49459^T
Catellicoccus marimammalium	+	–	–	–	–	–	–	–	–	–	+	–	–	–	–	–	–	+	–	–	–	–	–	–	–	–	–	–	–	–	–	+	984	Strains
Enterococcus faecalis	+	+	–	–	–	–	+	+	–	+	+	–	+	+	–	+	+	–	–	–	–	+	–	–	+	–	–		+	–	+	–	1380	
Erysipelothrix rhusiopathiae	42	–	+	–	–	28	–	–	–	75	–	–	–	64	–	+	80	85	–	–	–	–	–	–	–	–	–	–	–	–	–	–	*	
Facklamia miroungae	+	–	–	–	–	–w	nr	–	–	–	+	–	–	+	–	+	–w	w	–	–	–	–	–	–	–	–	–	–	–	–	–	+	761	
Granulicatella adiacens	–	+	–	30	–	–	–	–	–	–	–	–	–	–	7	w	25	–	–	–	–	+	–	–	89	–	–	–	+	–	–	–	*, 861	
Granulicatella balaenopterae	+	–	–	–	–	–	–	–	–	–	+	–	–	–	–	+w	+	–	–	–	+	+	–	–	–	–	–	–	–	–	–	+w	352, 985, 989, 990	
Granulicatella elegans	+	–	–	–	–	–	–	–	–	–	–	v	–	–	–	w	–	–	v	–	–	–	–	–	v	–	–	–	–	–	–	–v	352, 861	
Lactococcus garvieae	+	+	–	–	–	–	–	–	–	–	+	–	+	+	–	–	–	–	–	–	–	+	–	–	+	–	–	+	+	–	–	–	1763	
Lactococcus garvieae	+	+	–	–	–	–	+	+	–	–	+	–	+	+	–	+	+	–	–	–	–	+	–	–	+	–	–		+	–	+	–	1380	
Lactococcus garvieae dolphin isolate	+	+	–	–	–	–	+	–	–	–	+	–	+	+	–	–	–	–	–	–	–	–	–	–	–	–	–	+	–	–	–	–	500	
	conventional VP is negative, non-haemolytic																																	

Continued

Table 3.36. Continued.

	1	1.1	1.2	1.3	1.4	2	2	1.7	2	2	1A	1B	0	0.1	0.2	0.3	0.4	1	1	0.7	1	0.9	0A	0B	1C	1D	1E	0C	0D	0E	1F	0F	Ref	Strains
	ADH	βglu	βgar	βgur	αgal	pal	rib	man	sor	lac	tre	raf	vp	appa	βgal	pyrA	βnag	gta	hip	glyg	pul	mal	mel	mlz	suc	lara	darl	mbdg	tag	βman	cdex	ure		
Lactococcus garvieae	+	+	–	–	–	–	35	75	–	50	+	–	+	+	–	74	10	–	–	–	–	75	–	–	50	–	–	85	50	–	50	–	*	
Biotype 1								+								+	–								+				+		–		*, 1711	
Biotype 2								+								+	–								+				+		+		*, 1711	
Biotype 3								+								+	–								–				+		–		*, 1711	
Biotype 4								+								–	–								–				–		–		*, 1711	
Biotype 5								+								–	+								–				–		–		*, 1711	
Biotype 6								+								–	+								–				–		+		*, 1711	
Biotype 7								+								+	–								–				+		–		*, 1711	
Biotype 8								–								+	–								–				–		–		*, 1711	
Biotype 9								–								+	+								–				–		–		*, 1711	
Biotype 10								+								–	–								–				–		+		*, 1711	
Biotype 11								+								+	–								–				+		+		*, 1711	
Biotype 12								+								–	–								+				+		–		*, 1711	
Biotype 13								+								–	–								–				+		–		*, 1711	
Lactococcus lactis ssp. *cremoris*	+	+	–	–	–	–	–	+	–	+	+	–	–	–	+	–	–	+	–	–	+	+	–	+	+	+	+	–	+	+	+	–	1763	BCRC 11067
Lactococcus lactis lactis	+	+	5	–	–	–	95	26	–	50	75	–	74	+	26	–	50	–	26	–	–	+	–	–	26	–	–	85	3	50	95	–		
Lactococcus lactis lactis	+	+	–	–	–	–	+	–	–	+	+	–	+	+	–	–	–	–	+	–	–	+	–	–	+	–	–	–	–	–	+	–	1763	BCRC 10791
Lactococcus lactis lactis	+	+	–	–	–	–	–	+	–	+	+	–	27	27	54	–	–	82	54	–	–	+	–	–	+	+	54	–	+	+	+	–	1763	Prawns
Streptococcus agalactiae	+	–	–	+	–	+	–	–	–	–	–	–	+	+	–	–	–	–	+	–	+	+	–	–	+	–	–	+	–	–	–	–	*	
S. agalactiae ATCC 13813/27956	+	–	–	–	+	+	+	–	–	+	+	+	+	+	–	–	–	–	+	–	–	+	–	–	+	–	–	+	+	–	–	–	500	
Streptococcus agalactiae	+	–	–	50	10	+	85	–	–	30	74	–	90	+	–	–	–	–	+	4	+	+	–	–	+	–	–	90	26	–	–	–	*	
Streptococcus agalactiae	+	–	–	+	31	45	+	–	–	–	+	–	+	+	–	–	–	–	72	27	45	+	–	14	+	–	–	+	27	–	–	–	500	
	Haemolytic strains from mullet																																	
Streptococcus agalactiae	+	–	–	–	–	–	+	–	–	+	+	+	+	–	–	–	–	–	+	–	+	+	–	–	+	–	–	+	+	–	–	+	500	
	Non-haemolytic strain from dolphin																																	
Streptococcus agalactiae (*difficile*)	+	–	–	–	–	–	–	–	–	–	–	–	+	–	–	–	–	–	–	–	–	+	–	–	–	–	–	–	–	–	–	–	244	
Streptococcus difficilis	+			–	–			–	–	–	–	–	+			–	+			–			–		+					–			614	

Streptococcus dysgalactiae ssp. *dysgalactiae*	+	–	–	+	–	+	+	–	–	+	+	–	–	+	–	–	–	–	–	+	+	+	–	–	+	–	–	–	–	–	40	–	1601	
	Lancefield Group L=positive. Streptex Groups A–G = negative																																	
Streptococcus equi ssp. *equisimilis*	Get from API database																																	
Streptococcus halichoeri	+	–	–	–	–	+	+	+	–	v	–	–	+	+	–	+	–	–	–	–	+	+	–	–	–	–	–	–	–	–	+	–	987	
Streptococcus ictaluri	–	–	–	–	–	+	–	–	–	–	–	–	+	+	–	+	–	–	–	–	+	+	–	–	–	–	–	–	–	–	+	–	1519	
Streptococcus iniae – fish	67			+	–		83	+	–	–	+	–	–			+			–	+						–								
Streptococcus iniae – human	–			–	–		–		–	+	–	–			–			–	–						–									
Streptococcus iniae – fish	+s	+	–	+	–	w	–	+	–	–	+	–	–	+	–	+	–	–	–	–	–	–	–	–	–	–	–	–	–	–	–	–	5σ, 244	
Streptococcus iniae – fish	+	+	–	–	–	–	–	+	–	–	+	–	–	+	–	+	–	–	–	–	–	+	–	–	+	–	–	–	–	–	–	–	25°C, Buller	
Streptococcus iniae – fish	+	+	–	+	–	–	+	+	–	–	+	–	–	+	–	+	–	–	–	+	–	+	–	–	+	–	–	–	–	–	–	–	37°C, Buller	
Streptococcus iniae – fish	+	+	–	w	–	+	+	+	–	–	+	–	–	+	–	w	–	–	–	+	+	+	–	+	+	–	–	v	–	w	–	–	25°C, 378	
Streptococcus iniae – fish	+	+	v	+	–	+	+	+	–	–	+	–	–	+	–	+	–	–	–	+	+	+	–	+	+	–	–	v	–	+	–	–	37°C, 378	
Streptococcus iniae – fish	+	+	–	+	–	+	+	+	–	–	+	–	–	–	–	+	–	+	–	+	+	+	–	+	+	–	–	–	–	v	–	–	378	
Streptococcus marimammalium	v	–	+	–	–	+	–	–	–	+	–	–	–	+	–	–	–	v	–	–	–	+	–	–	–	–	–	–	–	–	–	v	986	
Streptococcus parauberis	+	+	–	–	15	–	+	+	+	+	+	10	+	+	–	10	–	–	50	–	–	+	–	10	+	–	–	+	30	30	–	–	*, 456	
Streptococcus phocae	–	–	–	–	–	+	–	–	–	–	–	–	–	–	–	–	–	–	–	–	–	+	–	–	–	–	–	–	–	–	–	–	1424	
DSMZ 15635[T]	4 of 34 salmon isolates had this profile																																	
Streptococcus phocae	–	–	–	–	–	+	–	–	–	–	–	–	–	–	–	–	–	–	–	–	–	–	–	–	–	–	–	–	–	–	–	–	1424	
Streptococcus phocae	–	–	–	–	–	+	+	–	–	–	–	–	–	+	–	–	–	–	–	–	–	+	–	–	–	–	–	–	–	–	–	–	614	
Streptococcus uberis	+	+	30	+	15	5	+	+	+	+	+	20	+	+	–	30	–	–	90	5	–	+	–	–	+	–	–	+	5	–	–	–	*	
Trueperella bernardiae			–	–	–		+			–					–	–			–	+													985	
Trueperella pyogenes			+	+	–		+			+					+	–			v	v													985	
Vagococcus fessus	–	v	v	–	–	v	–	–	–	–	–	–	–	–	–	+	–	v	–	–	–	–	–	–	–	–	–	–	–	–	v	–	762	
Vagococcus fluvialis	–	+	–	–	–	–	+	+	+w	–	+	–	–	–	–	+w	+	+	–	–	–	+	–	–	+	–	–	–	–	v+	+	–	354, 762, 826, 1354	
Vagococcus lutrae	–	+	–w	–	+	–w	+	–	+	–	+	–	–	–	–w	+	+	+	–	–	–	+	–	–	+	–	–	+	–	+	+	–	988	
Vagococcus penaei CD276[T]	+	–	–	–	–	–	+	–	–	–	+	+	+	–	–	+	–	+	–	–	–	+	–	+	+	–	–	–	–	–	+	–	826	
Vagococcus penaei CD279	–	–	–	–	–	–	+	–	–	–	+	+	+	–	–	+	–	–	–	–	–	+	–	+	+	–	–	–	–	–	–	–	826	
Vagococcus penaei CD285	–	–	–	–	–	–	+	–	–	–	+	+	+	–	–	+	–	–	–	–	–	+	–	+	+	–	–	–	–	–	–	–	826	
Vagococcus penaei CD380	–	–	–	–	–	–	+	–	–	–	+	+	+	–	–	+	–	–	–	–	–	+	–	+	+	–	–	–	–	–	–	–	826	
Vagococcus penaei CD310	+	+	–	–	–	–	+	–	+	–	+	–	+	–	–	+	–	+	–	–	–	+	–	+	+	–	–	–	–	+	–	–	826	

Continued

Table 3.36. Continued.

	1	1.1	1.2	1.3	1.4	2	2	1.7	2	2	1A	1B	0	0.1	0.2	0.3	0.4	1	1	0.7	1	0.9	0A	0B	1C	1D	1E	0C	0D	0E	1F	0F	Ref	Strains
	ADH	βglu	βgar	βgur	αgal	pal	rib	man	sor	lac	tre	raf	vp	appa	βgal	pyrA	βnag	gta	hip	glyg	pul	mal	mel	mlz	suc	lara	darl	mbdg	tag	βman	cdex	ure		
Vagococcus salmoninarum	–						–	+	+	–	+	–	–						–			+	–		+	–	–				+	–	1497, 1614	
Vagococcus salmoninarum	–	w			–	w	–		–		+		–			+		+				–		–	–			–		–	–		826	CCUG 33394[T]
Vagococcus salmoninarum	–						+	+	+	–	+	–	–			+			–			+	–	–	+	–	–				+	–	1497, 1614	
Weissella species	v				–	v	+	–	–	–	+		+						v	–	+	+	–	–	–		–	–	–		–		533	

ADH=arginine dihydrolase, βglu=β-glucosidase, βgar=β-galactosidase, βgur=β-glucuronidase, αgal=α-galactosidase, pal=alkaline phosphatase, rib=ribose, man=mannitol, sor=sorbitol, lac=lactose, tre=trehalose, raf=raffinose, vp=acetoin production (Voges–Proskaüer), appa=alanine-phenylalanine-proline arylamidase, βgal=β-galactosidase, pyrA=pyroglutamic acid arylamidase, βnag=*N*-acetyl-beta glucosaminidase, gta=glycyl-tryptophane arylamidase, hip=hippurate, glyg=glycogen, pul=pullulan, mal=maltose, mel=melibiose, mlz=melezitose, suc=sucrose, lara=L-arabinose, darl=D-arabinose, mbdg=methyl-B-D-glucopyranoside, tag=tagatose, βman=β-mannosidase, cdex=cyclodextrin, ure=urease. Ref=reference, w=weak reaction, +=positive reaction, –=negative reaction, s=slow reaction. Numbers indicate percentage of positive strains. *=from API database, #=AHLDA 1722 strain from imported aquarium fish (a=results at 37°C incubation, b=results from 25°C incubation), s=strain from Queensland, g=strains from mullet and seabream from Kuwait.

Table 3.37. API Rapid ID 32A database biochemical results.

	ure	adh	αgal	βgal	βgp	αglu	αglu	βara	βgur	βnag	mne	raf	gdc	αfuc	nit	ind	pal	argA	proA	lga	pheA	leuA	pyrA	tyrA	alaA	glyA	hisA	gga	serA	Temp	Time	Inoc	Ref	Strain
Organism																																		
Actinomyces viscosus	–	–	+	+	–	+	+	–	–	–	+	+	–	–	+	–	–	–	+	–	+	+	–	+	v	–	–	–	–	37	4.5	0.85	*	ATCC 15987
Bacteroides fragilis	–	–	+	+	–	+	+	–	–	+	+	+	+	+	–	–	+	v	–	+	v	+	–	v	+	–	–	+	–	37	4.5	0.85	*	ATCC 23745
Capnocytophaga sputigena	–	v	v	+	+	+	+	–	–	+	v	+	–	–	–	–	+	+	+	+	+	+	–	+	+	+	+	+	+	37	4.5	0.85	*	ATCC 33612
Clostridium sordellii	+	–	–	v	–	v	–	–	–	–	–	–	–	–	–	+	–	–	+	–	–	–	–	–	–	–	–	–	–	37	4.5	0.85	*	ATCC 9714
Clostridium sporogenes	–	+	–	–	–	v	v	–	–	–	–	–	–	–	–	–	–	–	v	–	–	–	+	–	–	–	–	–	–	37	4.5	0.85	*	ATCC 19404

Ure=urease, ADH=arginine dihydrolase,βgal=β-galactosidase, βgal=β-galactosidase, βgp=β-galactosidase 6 phosphate, αglu=α-glucosidase, βglu=β-glucosidase, αara=α-arabinosidase, βgur=β-glucuronidase, βnag=*N*-acetyl-β-glucosaminidase, mne=mannose fermentation, raf=raffinose fermentation, gdc=glutamic acid decarboxylase, αfuc=α-fucosidase, nit=nitrate reduction, ind=indole production, pal=alkaline phosphatase, argA=arginine arylamidase, proA=proline arylamidase, lga=leucyl glycine arylamidase, pheA=phenylalanine arylamidase, leuA=leucine arylamidase, pyrA=pyroglutamic acid arylamidase, tyrA=tyrosine arylamidase, alaA=alanine arylamidase, glyA=glycine arylamidase, hisA=histidine arylamidase, gga=glutamyl glutamic acid arylamidase, serA=serine arylamidase, *=API database

Table 3.38. API Zym database results.

Well	1	2	3	4	5	6	7	8	9	10	11	12	13	14	15	16	17	18	19	20	Temp	Time	Inoc	Strain	Ref
Enzyme	con	alk	C4	C8	C14	aryl	val	cys	try	chr	acp	np	αgal	βgal	βglucr	αglu	βglu	Naβglu	man	fuc	°C	h	% NaCl		
Organism			est	est lip	lip																				
Actinobacillus delphinicola	–	5	1	3	–	3	–	–	–	–	5	0–2	–	–	–	–	–	–	–	–	37	4	0.85		548
Actinomyces viscosus	–	–	–	+	–	+	+	–	–	–	+	+	+	+	–	–	+	–	–	+					68
Aequorivita antarctica	–	+	+																	–	20				211
Aerococcus viridans var *homari*	–	+	+	+	–	+	–	–	–	–	–	+	–	–	–	–	–	–	–	–					68
Aeromonas caviae	–	+	+	+	+	+	–	–	+	–	+	+	–	+	–	–	–	+	–	–	37	4	0.85	ATCC 15468	1824
Aeromonas hydrophila	–	5	–	4	2	4	1	–	3	–	5	1	–	3	–	1	–	4	–	–	24	4	0.85	NCTC 7810	244
Aeromonas jandaei	–	1	2	3	–	2	w	–	–	–	2	2	–	2	–	w	–	2	–	–	24	4	0.85	AHLDA 1718	244
A. salmonicida ssp. *achromogenes*	–		93				7			57				71		–	–	78							69
A. salmonicida ssp. *masoucida*	–	–	60				40			+				80		20	60	+							69
Aeromonas salmonicida atypical	–	3	–	1	–	2	w	–	1	–	5	1	–	–	–	–	–	2	–	–	24	20	0.85	AHLDA 1334	244
A. salmonicida ssp. *smithia*	–		75				25			–				+		+	50	25							69
Aeromonas veronii sobria	–	5	–	4	1	3	w	–	3	–	5	1	1	4	–	1	–	5	–	–	24	4	0.85	AHLDA 1684	244
Aliivibrio finisterrensis	–	+	–	+		+	–	–		–	–	–			–	–	–	+	–	–	24	4	0.85	CMJ 11.1^{T}	143
Aliivibrio finisterrensis	–	+	+	+		+	+	–		–	–	–			–	–	–	+	–	–	24	4	0.85	C 17.13	143
Aliivibrio finisterrensis	–	+	–	+		+	–	–		–	–	–			–	–	–	+	–	–	24	4	0.85	C 13.19	143
Aliivibrio finisterrensis	–	+	–	+		+	–	–		–	–	–			–	–	–	+	–	–	24	4	0.85	V 14.7	143
Aliivibrio (*Vibrio*) *fischeri*	–	+	–	–		w	–	–		–	+	+		–	–	–	–	–	–	–	20	6	2	NCIMB 1281^{T}	143,1834
Aliivibrio (*Vibrio*) *fischeri*	–	–	–	–	–	–	–	–	–	–	w	–	–	–	–	–	–	–	–	–	24	24	2	ATCC 7744^{T}	244
Aliivibrio (*Vibrio*) *fischeri*	–	+	–	+	–	+	–	–	–	–	+	+	–	–	–	–	–	+	–	–	22	24	0.85	NCMB 1281^{T}	1071
Aliivibrio (*Vibrio*) *fischeri*	–	–	–	–	–	–	–	–	–	–	w	–	–	–	–	–	–	–	–	–	25	24	2	ATCC 7744	244
Aliivibrio (*Vibrio*) *fischeri*	–	–	–	–	–	–	–	–	–	–	w	–	–	–	–	–	–	–	–	–	25	24	2	ATCC 7744	244
Aliivibrio (*Vibrio*) *logei*	–	5	w	w	–	–	–	–	–	–	w	w	–	–	–	–	–	–	–	–	24	24	2	ATCC 29985^{T}	244
Aliivibrio (*Vibrio*) *logei*	–		–	w		w	–				+			–		–		–			20	6	2	ATCC 29985^{T}	1834
Aliivibrio (*Vibrio*) *logei*	–	+	+	+	–	+	–	–	–	–	+	+	–	–	–	–	–	–	–	–	22	24	0.85	ATCC 15382	1071
Aliivibrio (*Vibrio*) *logei*	–	+	+	+	–	+	–	–	–	–	+	+	+	+	–	–	–	–	–	–	22	24	0.85	NCMB 1143	1071
Aliivibrio (*Vibrio*) *logei*	–	+	+	+	–	+	–	–	–	–	+	+	–	–	–	–	–	–	–	–	22	24	0.85	ATCC 29985^{T}	1071
Aliivibrio sifiae	–	+	–	–	–	+	w	–	–	–	+	+	–	+	–	–	–	w	–	–	20	6	2	H1-1^{T}	1834
Aliivibrio sifiae	–	+	–	–	–	+	–	–	–	–	+	+	–	+	–	–	–	w	–	–	20	6	2	H1-2	1834
Aliivibrio salmonicida	–	+	w	w		v	w	–	–	–	+	+		+	–	+	–	+	–	–	20	6	2	NCMB 2262^{T}	143,1834
Aliivibrio salmonicida	–	+	+	+	–	–	–	–	–	–	+	+	–	–	–	–	–	–	–	–	20	24	0.85	NCMB 2262^{T}	1071

Table 3.38. Continued.

Well	1	2	3	4	5	6	7	8	9	10	11	12	13	14	15	16	17	18	19	20	Temp	Time	Inoc	Strain	Ref
Enzyme	con	alk	C4	C8	C14	aryl	val	cys	try	chr	acp	np	αgal	βgal	βglucr	αglu	βglu	Naβglu	man	fuc	°C	h	% NaCl		
Organism			est	est lip	lip																				
Aliivibrio salmonicida	–	+	–	+	–	–	–	–	–	–	+	+	–	–	–	–	–	–	–	–	20	24	0.85	NCMB 2245	1071
Aliivibrio salmonicida	–	+	+	+	+	+	+	–	–	–	+	+	–	–	–	–	–	–	+	–	22	24	2	NCMB 2245	237
Aliivibrio salmonicida	–	+	+	+	–	–	–	–	–	–	+	+	–	–	–	–	–	–	–	–	20	24	0.85	90/1667-10c	1071
Aliivibrio wodanis	–	+	v	w		v	–	–		–	v	+		–	–	–	–	–	–	–	20	6	2	ATCC BAA-104^{T}	143,1834
Aliivibrio wodanis	–	+	83	+	–	–	86	–	–	–	69	+	–	–	–	–	–	77	–	–	22	24	0.85	NCIMB 13582^{T}	1071
Alteromonas macleodii	–	4	–	2	–	2	2	–	4	–	4	1	–	–	–	–	–	–	–	–					594
Andreprevotia lacus	–	+	+	+	–	+	–	–	–	–	+	+	–	–	–	+	–	+	–	–	25	4	0.85	LMG 24502^{T}	1513
Arcanobacterium phocae	–	4	2	2	–	3	–	1	v	–	5	–	3	1	–	5	–	–	–	–	37	4	0.85	DSM 10002^{T}	1374
Arthrobacter aurescens	–	–	–	+	–	+	–	+	+	–	+	+	–	+	+	+	+	–	+	–					68
Arthrobacter nasiphocae	–	+	1	–	–	+	–	–	–	–	+		–	–	–	+	–	–	–	–	37	4	0.85	CCUG 42953	358
Atopobacter phocae	–	+	1	1	–	–w	–	–	–	–	+	–	–	v	–	–	–	–	–	–	37	4	0.85	CCUG 42358^{T}	989
Bacillus zhanjiangensis	–	+	–	+	+	+	–	–	+	+	–	+	–	–	–	–	–	+	–	–	37	4	3	DSM 23010^{T}	316
Bordetella bronchiseptica	–	w–1	1	1	–	3–4	w	–	–	–	2	w–2	–	–	–	–	–	–	–	–	37	4	0.85	Various	244
Brevundimonas diminuta	–	+		+	–	+	–	–	+	+	+		–	–		–	–	–	–	–	28	4	0.85	ATCC 11568^{T}	1502
Brevundimonas vesicularis	–	+		+	–	+	v	–	+	–	+		–	–		+	–	–	–	–	28	4	0.85	ATCC 11426^{T}	1502
Carnobacterium alterfunditum	–	–	+	+	–	–	–	–	–	–	+	+	–	–	–	–	+	–	–	–	25	4	0.85	CCUG 34643	1405
Carnobacterium divergens	–	–	+	+	–	–	–	–	–	–	w	+	–	–	–	–	–	–	–	–	25	4	0.85	CCUG 30094	1405
Carnobacterium divergens 6251	–	–	+	w	–	–	–	–	–	–	–	w	–	–	+	–	–	–	–	–	25	4	0.85	6251	1405
Carnobacterium funditum	–	–	+	+	–	–	–	–	–	–	+	+	–	–	–	–	+	–	–	–	25	4	0.85	CCUG 34644	1405
Carnobacterium gallinarum	–	–	+	+	–	–	–	–	–	–	–	w	–	–	–	–	–	–	–	–	25	4	0.85	CCUG 30095	1405
Carnobacterium inhibens	–	–		–	–	–	–	–	–	–	–	–	–	–	–	–	+	–	–	–	25	4	0.85	CCUG 31728^{T}	841
Carnobacterium inhibens	–	–	–	–	–	–	–	–	–	–	–	+	–	–	–	–	–	–	–	–	25	4	0.85	CCUG 31728	1405
Carnobacterium mobile	–	–	+	+	–	–	–	–	–	–	–	w	–	–	–	–	–	–	–	–	25	4	0.85	CCUG 30096	1405
Carnobacterium (piscicola) maltaromaticum	–	–	+	+	–	–	–	–	–	–	+	+	–	–	–	–	–	–	–	–	25	4	0.85	CCUG 34645	1405
Cellulophaga lytica	–	5	3	4	1	5	4	2	3	1	5	5	1	1	–	3	2	5	2	–	22	12	ASS		168
Chryseobacterium balustinum	–	5	1	2	2	5	5	3	2	1	5	5	–	–	–	3	3	2	–	–	25	12	ASS		168,1173
Chryseobacterium gleum	–	5	2	3	1	5	5	2	2	2	5	5	–	–	–	4	5	4	–	–	22	12			1173
Chryseobacterium hominis	–		–	+			+		–					–		–	–	–			25	4	0.85	CCUG 52711^{T}	266,1846
Chryseobacterium indologenes	–	5	2	3	1	5	4	2	2	2	5	5	–	–	–	5	–	4	–	–	22	12			1173
Chryseobacterium indoltheticum	–	5	2	3	1	5	4	1	–	–	4	4	–	–	–	4	–	3	–	–	22	12			1173

Chryseobacterium oncorhynchi	–		–	+			+		+					–		+	–	–			25	4	0.85	701B-08^T	1845
Chryseobacterium piscicola	–	+	+	+	+	+	+	+	–	+	+	+	–	–	+	+	+	+	–	–	15	4	0.85	Type strain	799
Chryseobacterium piscium																									
Chryseobacterium scophthalmum	–	5	3	4	1	5	4	2	4	1	5	5	–	–	–	3	1	2	–	–	22	12	0.85	CCM 4109^T	1173
Chryseobacterium tructae	–	+	–	+	–	+	+	–	–	–	+	+	–	–	–	–	–	–	–	–	25	4	0.85	CECT 7798^T	1846
Corynebacterium caspium	–	–	+	+	–	–	–	–	+	–	–	–	–	–	–	–	–	–	–	–	37	4	0.85	CCUG 44566^T	357
Corynebacterium ciconiae	–	+	+	+	+	+	–	–	–	–	+	+	–	–	–	–	+	–	–	–	37	4	0.85	CECT 5779^T	528
Corynebacterium maris	–	+	+	+	+	+	–	–	–	–	–	–	–	–	–	+	–	–	–	–	37	4	0.85	LMG 24561^T	152
Corynebacterium phocae	–	+	–	+	–	v	–	–	–	–	+	–	–	–	–	+	–	–	–	–	37	4	0.85	CCUG 38205^T	1313
Corynebacterium sphenisc	–	–	+	+	+	+	+	+	–	+	+	+	–	–	–	–	–	–	–	–	37	4	0.85	CECT 5990^T	652
Corynebacterium spheniscorum	–	–	+	+	–	w	–	–	–	–	–	–	–	–	–	–	–	–	–	–	37	4	0.85	CECT 45512^T	653
Corynebacterium testudinoris	–	–	w	w	–	+	–	–	–	–	+w	–	–	–	–	v	+	–	–	–	37	4	0.85	CCUG 41823^T	351
Cytophaga allerginae	–	5	2	4	1	5	5	2	1	3	5	5	3	3	2	5	5	4	–	–	22	12	ASS		168
Cytophaga arvensicola	–	5	2	3	1	5	5	2	4	1	4	4	3	4	–	4	3	5	2	3	22	12	ASS		168
Cytophaga fermentans	–	5	1	3	–	5	2	–	2	–	5	5	–	5	1	2	5	5	–	–	22	12	ASS		168
Cytophaga hutchinsonii	–	4	2	4	1	4	4	3	–	–	3	2	–	–	–	–	–	–	–	–	22	12	ASS		168
Cytophaga latercula	–	5	2	4	1	5	5	3	5	5	4	4	–	–	–	–	–	–	–	–	22	12	ASS		168
Edwardsiella tarda	–	5	–	–	w	5	1	–	–	–	4	2	–	–	–	–	–	3	–	–	24	20	0.85	AHLDA 135	244
Edwardsiella ictaluri	–	+	+	+	–	+	+	–	–	+	+	+	–	–	–	–	–	+	–	–	30	4	0.85	Yellow catfish	1038
Elizabethkingia meningoseptica	–	5	4	4	3	5	5	5	3	4	4	5	3	–	–	4	2	5	2	2	37	4	ASS		168
Empedobacter brevis	–	5	3	4	–	5	5	3	3	2	5	5	–	–	–	4	–	–	–	–	22	12	ASS		168
Enterovibrio (Vibrio) calviensis	–	+	+	+	+	+	–	–	–	–	+	+	–	+		–			–					DSM 14347^T	436
Enterovibrio (Vibrio) calviensis	–	–	–	–	–	–	–	–	–	–	w	–	–	–	–	–	–	–	–	–	24	24	2	DSMZ 14347	244
Enterovibrio (Vibrio) calviensis	–	+	+	+	+	+	–	–	–	–	+	+	–	+		–			–					DSM 14347^T	436
Enterovibrio coralii	–	+	+	+	–	+	+	–	–	–	–	+	+	+	–	+	–	+	–	–	28	48	2	LMG 22228^T	1636
Enterovibrio norvegicus	–	+	+	+	+	+	–	–	–	–	+	+	–	+	–	nr	nr	nr	nr	nr	25	48	2	LMG 19839^T	1627
Facklamia miroungae	–	–w	–	–	–	+	w	w	–	w	w	w	–	–	–	–	–	–w	–	–	37	4	0.85	CCUG 42728^T	761
Flavobacterium aquatile	–	4	2	4	1	5	5	2	–	–	1	2	–	–	–	5	–	–	–	–	22	12	ASS		168,172
Flavobacterium aquatile	–	+	+	+	–	+	+	+	+	–	+	+	–	–	–	+	–	+	–	–	18	48	PBS	ATCC 11947	1282
Flavobacterium aquatile	–	+	+	+	<2	+	+	+	+	<2	+	+	<2	<2	<2	+	<2	+	<2	<2	18	7 d	0	ATCC 11947	1068
Flavobacterium aquatile	–	+	+	+	–	+	+	+	+	–	+	+	–	+	–	+	+	+	+	–	25	4	0.85	ATCC 11947	534
Flavobacterium branchiophilum	–	5	2	3	–	5	4	2	2	–	4	4	–	–	–	1	–	–	–	–	25	12	ASS		172
Flavobacterium branchiophilum	–	+	+	+	–	+	+	+	+v	–	+	+	v	–	–	+	+	–	–	–	18	48	PBS		1282
Flavobacterium branchiophilum	–	+	+	+	<2	+	+	+	+	<2	+	+	<2	<2	<2	+	+	<2	<2	<2	18	7 d	0	ATCC 35035	1068
Flavobacterium branchiophilum	–	+	+	+	–	+	+	+	–	–	+	+	–	–	–	–	–	–	–	–	25	4	0		1746
Flavobacterium ceti	–	w	–	w	–	+	+	–	–	–	w	w	–	–	–	–	–	–	–	–	37	4	0.85	CECT 7184^T	1710
Flavobacterium columnare	–	5	2	3	–	4	4	-	3	1	3	3	–	–	–	–	–	–	–	–	25	12	ASS		168,172

Continued

Table 3.38. Continued.

Well	1	2	3	4	5	6	7	8	9	10	11	12	13	14	15	16	17	18	19	20	Temp	Time	Inoc	Strain	Ref
Enzyme	con	alk	C4	C8	C14	aryl	val	cys	try	chr	acp	np	αgal	βgal	βglucr	αglu	βglu	Naβglu	man	fuc	°C	h	% NaCl		
Organism			est	est lip	lip																				
Flavobacterium columnare	–	5	–	2	–	5	5	3	–	–	5	4	–	–	–	3	–	1	–	–					777
Flavobacterium columnare	–	+	+	+	–	+	+	+	–	–	+	+	–	–	–	–	–	–	–	–	18	48	PBS	NCMB 2248	1282
Flavobacterium columnare	–	+	+	+	–	+	+	–	+	–	+	+	–	–	–	–	–	–	–	–	25	4	0.85	ATCC 23463	534
Flavobacterium columnare	–	+	<2	+	<2	+	+	<2	<2	<2	+	<2	<2	<2	<2	<2	<2	<2	<2	<2	18	7 d	0	ATCC 49513	1068
Flavobacterium columnare	–	5	–	2	–	5	2	2	–	–	1	1	–	–	–	–	–	–	–	–	24	4	0.85	AHLDA 1468	244
Flavobacterium flevense	–	5	1	2	1	5	1	1	–	–	3	3	1	5	–	3	–	4	–	–	22	12	ASS		168, 172
Flavobacterium frigidarium	–	5	–	2	–	5	5	1	–	1	5	5	–	–	–	–	–	–	–	–					777
Flavobacterium hydatis	–	5	2	4	1	4	5	2	–	–	4	5	–	–	–	4	–	2	–	–	22	12	ASS		168, 172
Flavobacterium hydatis	–	5	–	1	–	5	5	3	–	–	5	5	–	–	–	5	–	5	–	–					777
Flavobacterium johnsoniae	–	5	1	3	1	5	5	2	1	1	4	5	–	3	–	4	1	4	–	–	25	12	ASS	DSM 2064^T	168, 172
Flavobacterium johnsoniae	–	5	3	4	1	4	5	3	1	2	5	4	5	5	1	5	2	5	–	–	25	12	ASS	ATCC 29585	168
Flavobacterium johnsoniae	–	5	2	3	1	5	5	3	1	2	5	4	3	3	1	5	2	5	–	–	25	12	ASS	ATCC 29586	168
Flavobacterium johnsoniae	–	5	–	1	–	3	1	1	–	–	4	3	2	1	–	3	3	2	–	–	25				777
Flavobacterium johnsoniae	–	+	+	+	+	+	+	+	+	+	+	+	+	+	–	+	+	+	–	–	18	48	PBS	ATCC 17061	1282
Flavobacterium johnsoniae	–	+	+	+	+	+	+	+	+	+	+	+	+	+	<2	+	+	+	<2	<2	18	7 d	0	ATCC 17061	1068
Flavobacterium johnsoniae	–	+	+	+	+	+	+	+	–	–	+	+	–	+	–	+	+	+	–	–	25	4	0.85	ATCC 17061	534
Flavobacterium macrobrachii	–	+	–	+	–	+	+	–	–	–	–	+	–	–	–	–	–	–	–	–	25	4	0.85	LMG 25203^T	1514
Flavobacterium pectinovorum	–	5	3	3	–	4	4	2	–	–	4	4	–	2	1	4	5	3	–	–	22	12	0.85	NCIMB 9059	172
Flavobacterium psychrophilum	–	5	2	3	1	5	1	–	–	–	3	3	–	–	–	–	–	–	–	–	20	12	0.85	NCIMB 1947	168, 172
Flavobacterium psychrophilum	–	+	–	+	–	+	+	–	–	–	+	+	–	–	–	–	–	–	–	–	18	48	PBS	NCMB 1947	1282
Flavobacterium psychrophilum	–	+	<2	+	<2	+	+	+	<2	<2	+	+	<2	<2	<2	<2	<2	<2	<2	<2	18	7 d	0	ATCC 49510	1068
Flavobacterium saccharophilum	–	5	3	4	–	4	4	2	–	–	5	3	2	4	–	5	–	4	–	–	22	12	ASS	NCIMB 2072	172
Flavobacterium saccharophilum	–	5	–	1	–	2	1	–	–	–	5	2	–	1	–	1	–	1	–	–				NCIMB 2072	777
Flavobacterium succinicans	–	5	3	3	–	4	4	2	1	–	5	5	–	–	–	4	2	4	–	–	22	12	ASS	NCIMB 2277	172
Flavobacterium tilapiae	–	+	+	+	+	+	+	+	–	–	+	+	–	–	–	+	–	+	–	–	25	4	0.85	BCRC 80262^T	314
Flexibacter aurantiacus	–	5	2	4	1	5	5	3	–	–	5	3	–	1	–	5	3	3	–	–	22	12	ASS	NCMB 1382	168
Flexibacter aurantiacus	–	5	2	4	1	5	5	3	–	–	5	4	–	–	–	5	3	4	–	–	22	12	ASS	NCMB 1455	168
Flexibacter canadensis	–	5	2	3	1	5	5	3	4	1	5	5	4	1	–	5	3	4	–	3	22	12	ASS	ATCC 29591	168
Flexibacter flexilis	–	5	3	4	–	5	4	3	1	2	3	1	–	–	–	3	–	–	–	–	22	12	ASS	NCMB 1377	168
Flexibacter flexilis	–	+	–	+	–	+	+	+	–	+	+	+	–	–	–	+	+	–	–	–	18	48	PBS	ATCC 23079	1282
Flexibacter flexilis	–	+	<2	+	<2	+	+	+	<2	+	+	+	<2	<2	<2	+	+	<2	<2	<2	18	7 d	0	ATCC 23079	1068
Flexibacter litoralis	–	5	2	4	1	5	5	3	5	1	5	5	–	–	–	–	–	–	–	–	22	12	ASS	NCMB 1366	168
Flexibacter polymorphus	–	5	2	3	1	5	5	3	4	1	2	2	–	–	–	2	–	–	–	–	30	12	ASS	ATCC 27820	168
Flexibacter roseolus	–	4	2	3	1	3	3	2	1	3	2	2	–	–	–	–	–	–	–	–	25	12	ASS	NCMB 1433	168

Flexibacter ruber	–	5	3	4	2	3	3	2	1	3	3	3	–	–	–	–	–	–	–	–	25	12	ASS	NCMB 1436	168
Flexibacter sancti	–	5	–	2	–	4	1	–	4	–	4	5	4	4	–	4	3	4	–	2	22	12	ASS	NCMB 1379	168
Flexibacter tractuosus	–	5	2	3	1	5	4	3	1	4	5	4	–	–	–	5	1	–	–	–	22	12	ASS	NCMB 1408	168
Flexithrix (*Flexibacter aggregans*) *dorotheae*	–	5	3	4	1	5	4	3	3	–	5	5	1	3	–	4	2	5	3	3	22	12	ASS	NCMB 1443	168
Francisella philomiragia	–	+	+	+	–	–	–	–	–	–	+	+	–	–	–	–	–	+	–	–	37	4	0.85	ATCC 25015[T]	1142
Francisella philomiragia	–	+	+	+	–	–	–	–	–	–	+	+	–	–	–	–	–	+	–	–	37	4	0.85	CCUG 12603	1142
Francisella philomiragia ssp. *noatunensis*	–	+	+	+	–	–	–	–	–	–	+	+	–	–	–	–	–	+	–	–	22	4	0.85	2005/50/ F292-6C[T]	1142
Gramella marina	–	+	+	+	+	+	+	+	+	+	+	+	+	+	–	+	+	–	+	–	28	4	2	KMM 6048[T]	1205
Granulicatella balaenopterae	–	–	–	+	–	+	–	–	–	–	–	–	–	–	–	–	–	+	–	–	25	4	0.85	CCUG 37380[T]	352,990
Hafnia alvei	–	5	–	w	1	3	2	–	w	–	4	1	–	2	–	2	1	w	–	–	24	4	0.85	AHLDA 1729	244
Lactobacillus ceti	–	+	w	w	–	w	–	–	–	–	+	+	–	–	–	–	–	–	–	–	25	4	0.85	CECT 7185[T]	1709
Lactococcus garvieae	–	–	+	+	–	+	–	–	–	–	+	+	–	–	–	+	–	+	–	–	25	4	0.85		93
Mannheimia haemolytica	–	5	1	1–w	–	1–3	0–1	–	–	–	5	1–2	–	1–2	–	–	–	–	–	0–2	37	4	0.85	Various	244
Marinilabilia salmonicolor	–	5	1	2	–	–	–	–	–	4	2	1	3	–	1	5	5	5	–	–	25	12	ASS	NCMB 2216	168
Moritella marina	–	+	+	+	–	+	–	–	–	–	+	+	–	–	–	–	–	–	–	–	22	24	2	NCMB 1144[T]	1071
Moritella viscosa	–	+	+	+	+	+	–	–	–	–	+	+	–	–	–	–	–	–	–	–	22	24	2	NCMB 13584[T]	1071
Myroides odoratimimus	–	+	+	+	–w	3	–w	–w	–	–w	+	+	–	–	–	–	–	–	–	–	28	4	0.85	NCTC 11180[T]	1694
Myroides odoratus	–	+	+	+	–w	1	–w	–w	–	–w	+	+	–	–	–	–	–	–	–	–	28	4	0.85	NCTC 11036[T]	1694
Myroides odoratus	–	5	2	3	–	2	–	–	1	–	5	5	–	–	–	–	–	–	–	–	30	12		NCTC 11036	168,172
Nocardia asteroides	–	+	–	+	–	+	–	+	–	–	+	+	–	–	–	+	+	–	–	–	25	4	0.85	BCRC 13364	1762
Nocardia farcinica	–	+	–	–	–	+	+	–	+	–	+	+	–	–	–	+	+	–	–	–	35	4	0.85	BCRC 13380	1762
Nocardia salmonicida	–	+	+	+	+	+	–	–	+	–	+	+	–	–	–	+	+	–	–	–	25	4	0.85	BCRC 12441	1762
Nocardia seriolae	–	+	–	+	–	+	–	–	+		+	+	–	–	–	+	+	–	–	–	25	4	0.85	BCRC 13745	1762
Nocardia seriolae	–	+	+	+	–	+	+	–	+		+	+	–	–	–	+	+	–	–	–	25	4	0.85	NS127	1762
Nocardia seriolae	–	26	88	+	–	+	–	–	89	–	+	+	–	–	–	98	+	–	–	–	25	5	0.85		1523
Paracoccus fistulariae	–	w	–	–	–	w	–	–	–	–	w	w	–	–	–	+	–	–	–	–	30		ASW	KCTC 22803[T]	903
Pasteurella multocida	–	5	1–w	1–2	–	2	0–w	–	–	–	5	1	–	–	–	–	–	–	–	–	37	4	0.85	Various	244
Pasteurella skyensis	–	5	1	2	–	2	–	–	–	–	5	5	–	–	–	–	–	–	–	–	25	72	1.5	NCTC 13204[T]	181
Pedobacter heparinus	–	+	+	+	–	+	–	–	–	–	+	+	+	+	–	+	+	+	+	–	28	4	0.85	NCIB 9290	1603
Pedobacter heparinus	–	5	2	4	–	4	2	1	–	–	4	2	–	–	–	1	–	4	1	–	22	12		NCIB 9290	168
Pedobacter piscium	–	+	+	+		+	+	+	+	+	+	+	–	+	–	+	+	+	+	–	28	4	0.85		1603
Phocoenobacter uteri	–	5	1	5	–	1	–	–	–	–	5	5	–	–	–	–	–	–	–	–	37	4	0.85	NCTC 12872[T]	549
Photobacterium damselae	–	+	+	+	–	+	–	–	–	–	+	+	–	–	–	–	–	+	–	–	22	24	0.85	NCMB 2184	1071
Photobacterium damselae	–	4–5	2–3	3–4	–	1–2	–	–	–	1–2	5	1–2	–	–	–	–	–	3–4	–	–	22	4	0.85		963
Photobacterium damselae	–	5	–	2	–	4	1	1	–	w	5	1	–	1	–	–	–	4	–	–	24	20	2	AHLDA 1683	244
Photobacterium piscicida	–	+	+	+	–	+	–	–	–	–	+	+	–	–	–	–	–	+	–	–					1857
Photobacterium piscicida	–	3–4	2–3	3–4	–	1–2	–	–	–	–	3–4	2–3	–	–	–	–	–	3–4	–	–	22	4	0.85	ATCC 17911	963

Continued

Table 3.38. Continued.

Well	1	2	3	4	5	6	7	8	9	10	11	12	13	14	15	16	17	18	19	20	Temp	Time	Inoc	Strain	Ref
Enzyme	con	alk	C4	C8	C14	aryl	val	cys	try	chr	acp	np	αgal	βgal	βglucr	αglu	βglu	Naβglu	man	fuc	°C	h	% NaCl		
Organism			est	est lip	lip																				
Photobacterium rosenbergii	–	+	+	+	+	+	+	–	–	–	+	+	+	+	–	+	–	+	–	–	28	48	2	LMG 22228^T	1636
Photobacterium rosenbergii	–	+	+	+	v	+	+	–	–	–	+	+	v	+	–	v	–	+	–	–	28	48	2		1636
Photobacterium swingsii	–	+	w	+	+		67	–,w		–	+	+	–	–	–	–	–	67	–	–	25	4	2.5	CAIM 1393^T	641
Plesiomonas shigelloides	–	4	–	1	w	4	1	–	–	–	5	1	–	w	–	2	–	5	–	–	24	4		AHLDA 192	244
Pseudoalteromonas antarctica	–	+	+	+	–	+	+	+	+	+	+	+	–	–	–	+	–	+	+	–					216
Pseudoalteromonas citrea	–								+	+	+			–	–		–	w	–	–					598
Pseudoalteromonas rubra	–	4	–	2	1	2	1	–	4	1	4	2	–	–	–	–	–	2	–	–					596
Pseudomonas anguilliseptica	–	+	–	+	+	+	–	–	–	–	+	+	–	–	–	–	–	–	–	–				NCIMB 1949^T	176
Pseudomonas anguilliseptica	–	–	+	+	–	+	–	–	–	–	+	+	–	–	–	–	–	–	–	–	22	4		Cod	100
Pseudomonas anguilliseptica	–	–	+	+	+	–	+	–	–	–	–	+	+	–	–	–	–	–	–	–	25	4	0.85	Strains	1059
Pseudomonas plecoglossicida	–	+	–	+	–	+	+	–	+	–	+	+	–	–	–	–	–	–	–	–	25	4	0.85	ATCC 700383^T	1229
Renibacterium salmoninarum	–	+	+	+	–	+	–	–	+	–	+	+	–	–	–	+	–	–	+	–	15	6	DW		68, 647
Rhodococcus qingshengii	–	+	–	+	–	+	+	–	–	–	–	+	–	–	–	+	+	–	–	–	30	4	0.85		83
Roseovarius crassostreae	–	+	+	+	w	+	w	–	–	–	+	–	–	–	–	–	–	–	–	–	37	4	SW	ATCC BAA-1102^T	197
Shewanella algae	–	+	–	+	–	+	13	–	+	+	+	+	–	–	–	–	–	+	–	–	35	4	0.85		886
Shewanella aquimarina	–	+	+	+	–	w	–	–	–	+	+	+	–	–	–	–	–	+	–	–	30		2	KCCM 41821^T	1832
Shewanella baltica	–	–	–	–	–	–	–	–	–	w	–	–	–	–	–	–	–	+	–	–	30	4	2	NCTC 10735^T	1853
Shewanella corallii	–	+	w	w	–	–	–	–	–	+	–	w	–	–	–	–	–	w	–	–	25	48	2	LMG 24563^T	1527
Shewanella haliotis	–	+			–					+											25	24	2	KCTC 12896^T	985
Shewanella marinintestina	–	+	+	+	+	+	–		–	+	+		–	–	–	–	–	+	–	–	20	48	2	JCM 11558^T	1479
Shewanella marisflavi	–	+	+	+	–	w	–	–	–	+	+	+	–	–	–	–	–	+	–	–	30		2	KCCM 41821^T	1832
Shewanella putrefaciens	–	+	–	+	–	+	67	–	78	89	+	+	–	–	–	–	–	+	–	–	35	4	0.85	ATCC 8071^T	886
Shewanella sairae	–	+	+	+	+	+	–		–	+	+		–	–	–	–	–	–	–	–	20	48	2	JCM 11563^T	1479
Shewanella schlegeliana	–	+	+	–	–	+	–		–	+	–		–	–	–	–	–	+	–	–	20	48	2	JCM 11561^T	1479
Shewanella upenei	–	+	+	+	–	w	–	–	–	+			–	–	–	–	–		–	–	30	4	2	KCTC 22806^T	901
Sphingobacterium multivorum	–	+	–	+	–	+	+	–	–	–	+	–	–	+	–	+	–	+	–	–					752
Sphingobacterium multivorum	–	5	4	4	1	5	1	1	–	–	5	5	2	4	–	4	3	5	3	–	37	4	ASS	NCTC 11343	168
Sphingobacterium spiritivorum	–	+	–	–	–	–	–	–	+	–	+	1	–	1	–	+	–	+	–	–				NCTC 11386	750
Sphingobacterium spiritivorum	–	5	2	4	–	5	1	1	5	2	5	5	2	4	1	3	2	5	3	4	30	12	ASS	NCTC 11386	168
Staphylococcus lutrae	–	5	5	5	–	–	–	–	–	–	5	5	–	1	–	–	–	–	–	1					551

Streptococcus dysgalactiae	–	+	–	–	–	+	–	–	–	–	+	+	–	–	+	+	–	–	–	–	37	4	0.85	Sturgeon	1824
Streptococcus dysgalactiae ssp. *dysgalactiae* ATCC 43078^T	–	+	–	–	–	+	–	–	–	–	+	–	–	–	+	+	–	–	–	–	37	4	0.85	ATCC 43078^T	1824
Streptococcus dysgalactiae ssp. *equisimilis*	–	+	–	–	–	+	–	–	–	–	+	+	–	–	+	+	–	–	–	–	37	4	0.85	ATCC 35666	1824
Streptococcus marimammalium	–	+	–	–	–	+	+	–	–	+	+	w	–	–	–	–	–	–	–	–	25	4	0.85	CCUG 48494^T	986
Streptococcus parauberis	–	+	–	+	–	+	–	–	–	+	+	+	–	–	–	+	–	–	–	–	37	4	0.85		93
Streptococcus phocae	–	+	–	–	–	+	–	–	–	–	–	–	–	–	–	–	–	–	–	–	25	4	0.85	All strains	1424
Tenacibaculum aiptasiae	–	+	+	+	+	+	+	+	+	+	+	+	–	–	–	–	–	–	–	–	37	4	1.5		1761
Tenacibaculum crassostreae	–	–	+	+	–	+	+	–	+	–	+	+	–	–	–	–	–	–	–	–	37	4	2		1001
Tenacibaculum dicentrarchi	–	+	+	+		+	+	+	–	–						–					25	4	ASW	CECT 7612^T	1346
Tenacibaculum discolor	–	+	+	+	+	+	+	+	+	+	+	+	–	–	–	–	–	–	–	–	37	4	ASW		1347
Tenacibaculum gallaicum	–	+	+	+	+	+	+	+	+	+	+	+	–	–	–	–	–	–	–	–	37	4	ASW		1347
Tenacibaculum litopenaei	–	+	+	+	+	+	+	+	–	–	+	+	+		–	+	+	+	+	+	37	4	1.5		1517
Tenacibaculum maritimum	–	5	1	3–4	2–5	5	5	5	5	5	5	5	–	–	–	0–1	–	–	–	–	18	7 d	2	Salmon smolts	1283
Tenacibaculum maritimum	–	5	3	4	1	5	5	3	1	2	5	5	–	–	–	–	–	–	–	–	25	12	2		168
Tenacibaculum maritimum	–	5	2	3	1	4	5	2	3	1	5	5	–	–	–	–	–	–	–	–	22	12	0.85	ATCC 43398	
Tenacibaculum maritimum	–	+	+	+	+	+	+	+	+	+	+	+	–	–	–	–	–	–	–	–	20	4	ASW	a443	1055
Tenacibaculum maritimum	–	5	2	4	2	5	5	3	1	2	5	5	–	–	–	–	–	–	–	–	22	24	ASW	NCIMB 2154^T	171, 309
Tenacibaculum maritimum	–	5	3	4	1	5	5	3	1	3	5	5	–	–	–	–	–	–	–	–	22	24	ASW	NCIMB 2153	171
Tenacibaculum maritimum	–	5	3	4	1	4	5	3	1	2	5	5	–	–	–	–	–	–	–	–	22	24	ASW	NCIMB 2158	171
Tenacibaculum maritimum	–	5	2	3	–	5	5	3	1	–	4	3	–	–	–	–	–	–	–	–	22	24	ASW	Seabass	171
Tenacibaculum ovolyticum	–	+	+	+	–	+	+	–	–	–	+	+	–	–	–	–	–	–	–	–	20	12	ASW	NCIMB 13127^T	675
Tenacibaculum soleae	–	+	+	+	+	+	+	+	–	–	–	–	–	–	–	–	–	–	–	–	20	4	ASS	NCIMB 14368^T	1344
Tenacibaculum soleae	–	+	+	v	–	+	+	+	–	–	+	+	–	–	–	–	–	v	–	–	20	4	ASS	Sole, brill	1056
Vagococcus fessus	–	v	+w	+w	–	+	–	v	–	+	–	w	–	v	–	–	v	–	–	–	37	4	0.85	CCUG 41755^T	762
Vagococcus fluvialis	–	–	w	+		–				w	+	+	–			w	+								826
Vagococcus lutrae	–	–w	w	+	–	+	–	–	–	+	+	w	+	–w	–	+	+	+	–	–	37	4	0.85	CCUG 39187^T	988
Vagococcus penaei	–	–	+	+	–	–	–	–	–	–	+	+	–	–	–	–	–	+	–	–	37	4	0.85	CD276^T	826
Vagococcus penaei	–	–	+	+	–	+	–	–	–	+	+	+	–	–	–	+	–	+	–	–	37	4	0.85	CD279	826
Vagococcus penaei	–	–	+	+	–	+	–	–	–	+	+	+	–	–	–	+	–	+	–	–	37	4	0.85	CD285	826
Vagococcus penaei	–	–	+	+	–	+	–	–	–	+	+	+	–	–	–	+	–	+	–	–	37	4	0.85	CD380	826
Vagococcus penaei	–	–	+	+	–	+	–	–	–	+	+	+	–	–	–	+	+	+	–	–	37	4	0.85	CD310	826
Vagococcus salmoninarum	–	w	w	w		+				+	+	–	–			–	w				37	4	0.85		826
Vibrio aerogenes	–	–	–	–	–	–	–	–	–	–	–	–	–	–	–	–	–	–	–	–	25	24	2	DSMZ 14438	244
Vibrio aestivus	–	+	–	–	–	+	–	–	–	–	–	–	–	–	–	–	–	–	–	–	25	4	0.85	CECT 7558^T	1065

Continued

Table 3.38. Continued.

Well	1	2	3	4	5	6	7	8	9	10	11	12	13	14	15	16	17	18	19	20	Temp	Time	Inoc	Strain	Ref
Enzyme	con	alk	C4	C8	C14	aryl	val	cys	try	chr	acp	np	αgal	βgal	βglucr	αglu	βglu	Naβglu	man	fuc	°C	h	% NaCl		
Organism			est	est lip	lip																				
Vibrio aestuarianus	–	2	vw	1	w	4	w	–	–	–	–	vw	–	–	–	–	–	1	–	–	25	24	2	NCIMB 2236	244
Vibrio agarivorans	–	1	w	1	–	1	–	–	–	–	–	w	–	2	–	–	–	–	–	–	25	24	2	AHLDA 1732	244
Vibrio alfacsensis	–		+		–	+	–	–	+	–	+	–	–	–	–	–	–	–	–	–	30	24	2.5	CAIM 1831^{T}	640
Vibrio alfacsensis	–		–		–	+	–	–	+	–	+	–	–	–	–	–	–	–	–	–	30	24	2.5		640
Vibrio alfacsensis	–		–		–	+	–	–	+	+	+	–	–	–	–	–	–	–	–	–	30	24	2.5	CAIM 1835	640
Vibrio alfacsensis	–		–		–	+	–	–	+	+	+	–	–	+	–	–	–	–	–	–	30	24	2.5	CAIM 404	640
Vibrio alfacsensis	–		–		–	+	–	–	+	–	+	+	–	–	–	–	–	–	–	–	30	24	2.5	CAIM 1834	640
Vibrio alginolyticus	–	5	–	2	w	4	1	–	2	–	5	1	–	–	–	–	–	1	–	–	25	4			244
Vibrio alginolyticus	–	+	+	+	+	+			+	+	+	+									25	24	2	ATCC 17749^{T}	
Vibrio alginolyticus	–	+	+	+	+	+	+		+	+	+	+									25	24	2	ATCC 33787	
Vibrio alginolyticus	–	+	+	+	+	+	+	–	+	–	+	+	–	–	–	+	–	+	–	–	25	24	1	Swy	998
Vibrio alginolyticus	–	5	3	3		5	3	3	5	1	5							5			22	24	1	TA15	644
Vibrio (Listonella) anguillarum	–	+	+	+	–	+	+	+	–	–	+	+	–	–	–	–	–	+	–	–	22	24	0.85	ATCC 14181	1071
Vibrio (Listonella) anguillarum	–	4	–	2	–w	1	–	–	–	–	–	–	–	–	–	–	–	–	–	–	25	24	2	NCMB6	154
Vibrio (Listonella) anguillarum	–	3	1	2	2	2	1	1	–	–	–	w	–	w	–	–	2	–	–	–	24	24	2	AHLDA 1730	244
Vibrio (Listonella) anguillarum	–	4	–	2	–w	1	–	–	–	–	–	–	–	–	–	–	–	–	–	–	25	24	2	ATCC 19264	244
Vibrio (Listonella) anguillarum	–	4	–	2	–w	1	–	–	–	–	–	–	–	–	–	–	–	–	–	–	25	24	2	CRL 11	244
Vibrio areninigrae	–	+	–	+	–	+	+	–	+	–	+	+	–	–	–	–	–	–	+	–	25	24	2	KCTC 22122^{T}	301
Vibrio azureus	–	+	–	+		+				–	–	+				–		+			20	48	2	NBRC 104587^{T}	1836
Vibrio brasiliensis	–	+	+	+	+	+	+	–	–	–	–	–	–	+	–	–	–	–	–	–	25	24	1.5	LMG 20546^{T}	1632
Vibrio breoganii	–		+	+		+					+	+						+			24	24	0.85	LMG 23858^{T}	140
Vibrio campbellii	–	5	w	2	w	2	–	–	–	5	–	1	–	–	–	–	–	–	–	–	25	24	2	ATCC 2590	244
Vibrio caribbeanicus	–	+	+	+	–	+	–	–	+	–			–	–	–	–	–	–	–	–	28	4	1.5	DSM 23640^{T}	738
Vibrio casei	–	+	+	+		+					+	+		+				+			30	6	1.5	DSM 22364^{T}	190
Vibrio celticus	–	+	+	+		+	+		+		+	–									24	24	0.85	Rd 8.15^{T}	142
Vibrio celticus	–	+	+	+		+	+		+		+	+									24	24	0.85	RD 6.8	142
Vibrio celticus	–	+	+	+		+	–		+		+	–									24	24	0.85	Rd2L5	142
Vibrio celticus	–	+	+	+	+	+	+		+		+	+									24	24	0.85	Rd 16.13	142
Vibrio chagasii	–	+	+	+	–	–	–		+		–	+	–		–	–	–	–	–	–				LMG 21353^{T}	1635
Vibrio chagasii	–	+	+	+	17	–	8		83		67	+	–		–	–	–	42	–	–				12 strains	1635
Vibrio cholerae non-01	–	3	–	3	1	2	1	w	–	–	1	1	–	2	–	1	–	4	–	–	24	24	0.85	AHLDA 996	244
Vibrio (communis) owensii	–	+	+	+	+	+	+	–	+	+	+	+	–	+	–	+	–	+	–	–	24	24	2	LMG 25430^{T}	320
Vibrio (communis) owensii	–	+	+	+	+	+	+	–	+	+	+	+	–	+	–	–	–	+	–	–	24	24	2	R-40498	320

Vibrio (communis) owensii	–	+	+	+	+	+	+	–	+	+	+	+	–	+	–	+	w	+	–	–	24	24	2	R-40501	320
Vibrio diabolicus	–	3	2	3	1	3	–	–	2	2	3	–	–	–	–	–	–	–	–	–	25	48	2	CNCM I-1629^T	1368
Vibrio diazotrophicus	–	–	–	–	–	–	–	–	–	–	–	–	–	–	–	–	–	–	–	–	24	24	2	ATCC 33466^T	244
Vibrio fluvialis	–	4	1	2	2	5	1	–	–	–	5	w	–	2	–	w	w	1	–	–	25	24	2	NCTC 11327	244
Vibrio furnissii	–	3	–	2	1	4	2	2	–	–	1	1	–	1	–	w	–	4	–	–	25	24	0.85	ATCC 11218	244
Vibrio furnissii	–	3	–	2	1	4	2	2	–	–	1	1	–	1	–	w	–	4	–	–	25	24	0.85	ATCC 35016	244
Vibrio gallaecicus	–	+	+	+		+			–		–	–									25	24	0.85	LMG 24045^T	144
Vibrio halioticoli	–	1	w	1	–	1	w	–	–	–	–	1	–	–	–	–	–	3	–	–	24	24	2	AHLDA 1734	244
Vibrio halioticoli	–	1	w	1	–	1	w	–	–	–	–	1	–	–	–	–	–	3	–	–	25	24	2	ATCC 700680	244
Vibrio hangzhouensis	–	+	+	+	–	–	–	–	–	–	+	+	–	–	–	–	–	–	–	–	30	4	1	CN83^T	1818
Vibrio harveyi	–	3	2	2	–	1	w	–	–	1	5	1	–	–	–	–	–	1	–	–	25	4	0.85	ATCC 35084	244
Vibrio harveyi	–	4	2	2	–	1	w	–	–	2	3	1	–	–	–	–	–	2	–	–	25	4	2	ATCC 35084	244
Vibrio harveyi	–	+	+	+	–	–	–	–	–	–	+	+	–	–	–	–	–	–	–	–				ATCC 35084	1830
Vibrio harveyi	–	+	+	+	–	–	–	–	–	–	+	+	–	–	–	–	–	–	–	–				ATCC 14126	244
Vibrio harveyi	–	5	3	4	–	5	2	–	–	2	5	1	–	–	–	–	–	–	–	–	25	24	2	ATCC 6	244
Vibrio hippocampi	–	–	+	–	–	+	–	–	–	–	+	–	–	–	–	–	–		–	–	20	48	1.5	DSM 22717^T	106
Vibrio hispanicus	–	+	+	+	–	+	–	–	–	–	+	+		–	–	–		–	–	–	28	24	2	LMG 13240^T	636
Vibrio ichthyoenteri	–	–	–	–	–	–	–	–	–	–	–	–	–	–	–	–	–	–	–	–	25	24	2	DSMZ 14397	244
Vibrio kanaloae	–	+	+	+		+	+	–	+	–		+	–	–	–	–	–	+	–	–				LMG 20539^T	1635
Vibrio kanaloae	–	+	+	+		+	60	–	+	–		+	–	–	–	–	–	60	–	–				LMG 20539^T	1635
Vibrio lentus	–	–	–	–	–	–	–	–	–	–	–	–	–	–	–	–	–	–	–	–	25	24	2	CECT 5110	244
Vibrio litoralis	–	+	+	+	–	+	–	–	–	–	+	+	–	–	–	–	–	+	–	–	25–30	24	2	MANO22D^T	1202
Vibrio maritimus	–	+	+	+	+	+	w	–	–	–	+	+	+	+	–	+	–	w	–	–	25	24	2	R-40493^T	321
Vibrio marisflavi	–	+	+	+	–	+	–	–	–	–	+	+	–	+	–	–	–	–	–	–	28	24	1.5	WH134^T	1760
Vibrio mediterranei	–	5	2	2	2	1	–	–	–	–	4	2	–	–	–	–	–	2	–	–	25	24	2	AHLDA 1733	244
Vibrio shilonii	–	5	2	2	2	1	–	–	–	–	4	2	–	–	–	–	–	2	–	–	25	24	2	ATCC BAA-91	244
Vibrio mediterranei	–	5	1	1	–	1	–	–	–	–	–	–	–	–	–	1	1	–	–	–	25	24	2	CECT 615	244
Vibrio metschnikovii	–	–	–	–	–	–	–	–	–	–	–	–	–	–	–	–	–	+	–	–	24	24	2	NCTC 8443	244
Vibrio mimicus	–	5	–	2	–	4	1	–	–	w	5	1	–	1	–	–	–	3	–	–	24	4	2	AHLDA 1654	244
Vibrio mimicus	–	5	1	2	–	3	w	–	–	–	5	2	–	–	–	–	–	w	–	–	25	24	2	ATCC 33653^T	244
Vibrio mimicus	–	5	–	2	–	4	1	–	–	w	5	1	–	1	–	–	–	3	–	–	24	4	2	ATCC 33653^T	244
Vibrio mytili	–	–	2	3	1	3	–	–	1	–	2	–	–	–	–	–	–	2	–	–	25	48	2	CECT 632^T	1368
Vibrio mytili	–	–	–	–	–	–	–	–	–	–	–	–	–	–	–	–	–	2	–	–	24	24	2	CIP 103929	244
Vibrio natriegenes	–	+	+	+	–	–	–	–	–	–	+	+	–	–	–	–	–	–	–	–	22	24	0.85	NCMB 1900	1071
Vibrio navarrensis	–	5	2	2	1	2	–	–	–	–	1	1	–	–	–	w	–	–	–	–	25	24	2	CIP 103381	244
Vibrio neptunius	–	+	+	+	+	+	+	–	+	–	–	–	–	–	–	–	–	+	–	–	22	24	1.5	LMG 20536^T	1632
Vibrio neptunius	–	+	+	+	20	+	20	–	20	–	–	–	–	–	–	–	–	+	–	–	22	24	1.5	Others	1632
Vibrio nereis	–	–	1	1	–	2	–	–	–	–	–	–	–	–	–	–	–	–	–	–	25	48	2	LMG 3895	1368
Vibrio nigripulchritudo	–	–	–	–	–	–	–	–	–	–	–	–	–	–	–	–	–	–	–	–	24	24	2	CIP 103195	244
Vibrio ordalii	–	+	+	+	–	+	–	–	–	–	+	+	–	–	–	–	–	+	–	–	22	24	0.85	NCMB 2167	1071

Continued

Table 3.38. Continued.

Well	1	2	3	4	5	6	7	8	9	10	11	12	13	14	15	16	17	18	19	20	Temp	Time	Inoc	Strain	Ref
Enzyme	con	alk	C4	C8	C14	aryl	val	cys	try	chr	acp	np	αgal	βgal	βglucr	αglu	βglu	Naβglu	man	fuc	°C	h	% NaCl		
Organism			est	est lip	lip																				
Vibrio ordalii	–	5	–	w	–	4	w	–	–	–	w	w	–	–	–	–	–	–	–	–	25	4	2	DF 3K	878
Vibrio orientalis	–	+	+	+	+	+	+	–	–	–	+	+	–	–	–	–	–	+	–	–	22	24	0.85	NCMB2195	1071
Vibrio orientalis	–	5	1	1	w	3	1	1	–	–	–	w	–	–	–	–	–	w	–	–	25	24	2	ATCC 33935	244
Vibrio owensii	–	+	w	+		+					+	+	–	–	–	–	–	–	–	–	28	48	2	DY05^T	270
Vibrio owensii	–	5	2	3	–	4	w	w	2	5	5	w	–	–	–	–	–	1	–	–	25	24	2	DY05^T	244
Vibrio owensii	–	5	1	1	w	4	1	1	2	4	5	1	–	–	–	–	–	w	–	–	25	24	2	47666-1	244
Vibrio pacinii	–	+	+	+	–	+	+	+	–	w	+	+	–	+	–	+	–	–	–	–	28	4	1.5	LMG 1999^T	638
Vibrio pacinii	–	+	+	+	–	w	w	w	–	–	+	w	–	–	–	w	–	–	–	–	28	4	1.5	LMG 21514	638
Vibrio pacinii	–	w	w	w	–	w	w	–	–	–	+	w	–	+	–	+	–	–	–	–	28	4	1.5	LMG 13245	638
Vibrio parahaemolyticus	–	5	–	3	w	5	3	–	1	–	5	1	–	–	–	–	–	–	–	–	25	20	2	ATCC 43996	244
Vibrio parahaemolyticus	–	+	–	–	–	–	–	–	+	–	+	+	–	–	–	–	–	+	–	–	30	4	2	ATCC 17802	1040
Vibrio parahaemolyticus	–	+	–	+	–	–	–	–	+	–	+	+	–	–	–	–	–	+	–	–	30	4	2	ATCC 27969	1040
Vibrio parahaemolyticus	–	–	–	–	–	–	–	–	–	–	–	–	–	–	–	–	–	+	–	–	24	24	2	ATCC 17802	244
Vibrio pectenicida	–	–	–	–	–	–	–	–	–	–	–	–	–	–	–	–	–	5	–	–	24	24	2	ATCC 700783	244
Vibrio (Listonella) pelagius I	–	5	1	2	–	2	w	1	–	–	w	w	–	–	–	–	–	w	–	–	24	24	2	ATCC 25916^T	244
Vibrio (Listonella) pelagius II	–	+	+	+	–	+	–	–	+	–	+	+	–	–	–	–	–	–	–	–	22	24	0.85	NCMB 2253	1071
Vibrio penaeicida	–	3	2	1	1	4	2	–	–	–	5	2	–	–	–	–	–	2	–	–	25	24	2	DSMZ 14398	244
Vibrio pomeroyi	–	+	+	+	–		–	–			+	+	–	+	–	–	–		–	–				LMG 21353T	1635
Vibrio proteolyticus	–	2	1	2	–	2	–	–	–	–	–	w	–	–	–	–	–	w	–	–	25	24	2	AHLDA 1735	244
Vibrio proteolyticus	–	5	2	2	w	5	1	–	–	–	2	1	–	–	–	–	–	2	–	–	25	24	2	CIP 102892	244
Vibrio quintilis	–	+	–	–	–	+	+	–	–	–	+	–	–	–	+	–	–	–	–	–	25	4	2	CECT 7734^T	1065
Vibrio rotiferianus	–	+	+	+	–	+	+	–	+	+	+	+	–	–	–	–	–	–	–	–	25	24	2	LMG 21460^T	637
Vibrio sagamiensis	–	+	w	w	w	+	–	–	–	–	w	+	–	–	–	–	–	–	–	–	20	6	2	LC2-047^T	1837
Vibrio salmonicida	–	+	+v	+	–	–	–	–	–	–	+	+	–	–	–	–	–	–	–	–	22	24	0.85	NCMB 2262	1071
Vibrio scophthalmi	–	5	2	w	–	4	w	–	–	–	3	1	–	–	–	–	–	+	–	–	25	24	2	CECT 4638	244
Vibrio splendidus I	–	+	+	+	+	+	–	–	+	–	+	+	–	–	–	–	–	+	–	–	22	24	0.85	NCMB 1^T	1071
Vibrio splendidus I	–	+	+	+	50	+	63	–	w	–	88	50	–	63	–	50	–	–	–	–	25	4	2	Turbot	590
Vibrio splendidus I	–	+	+	+	+	+	–	–	+	–	+	+	–	+	–	+	–	–	–	–	25	4	2	Scallop A515	590
Vibrio splendidus I	–	+	+	+	+	w	–	–	w	–	–	–	–	–	–	–	–	–	–	–	25	4	2	Turbot AH5	590
Vibrio splendidus I	–	+	+	+	+	w	–	–	w	–	+	–	–	–	–	–	–	–	–	–	25	4	2	Turbot VS1, 4	590
Vibrio splendidus I	–	+	+	+	–	+	+	–	w	–	+	+	–	+	–	+	–	–	–	–	25	4	2	Turbot VS5	590

Vibrio splendidus I	–	+	+	+	+	+	+	–	w	–	+	+	–	+	–	–	–	–	–	–	25	4	2	Turbot VS6	590
Vibrio splendidus I	–	+	+	+	+	+	+	–	w	–	+	+	–	+	–	+	–	–	–	–	25	4	2	Turbot 10,11	590
Vibrio splendidus II	–	+	+	+	–	+	+	–	+	–	+	+	–	–	–	–	–	–	–	–	22	24	0.85	NCMB 2251	1071
Vibrio splendidus II	–	5	4	3		3	1	w	3	w	5							3			22	24	1	TA2	644
Vibrio splendidus II	–	4	1	2	–	2	w	–	w	–	–	vw	–	–	–	–	–	–	–	–	25	24	2	ATCC 33789	244
Vibrio stylophorae	–	+	–	+	–	+	+	+	–	–	+	+	–	–	–	–	–	+	–	–	30	4	0.85	KTW-12^T	1515
Vibrio tapetis	–	+	+	+	–	+	+	–	–	–	+	–	–	–	–	–	–	+	–	–	22	18	2	B1090T	1237
Vibrio tapetis	–	–	–	–	–	–	–	–	–	–	–	–	–	–	–	–	–	–	–	–	24	24	2	CIP 104856	244
Vibrio tapetis	–	+	+	+	+	+	+	+	–	–	w	w	–	–	–	–	–	–	–	–	20	4	SW	Wedge sole	1052
Vibrio tubiashii	–	2	1	3	1	2	–	–	–	–	–	–	–	–	–	–	–	–	–	–	25	48	2	LMG 10936^T	1368
Vibrio tubiashii	–	+	+	+	–	+	+	+	–	–	–	+	–	–	–	–	–	+	–	–	22	24	0.85	NCMB 1340^T	1071
Vibrio tubiashii	–	–	–	–	–	–	–	–	–	–	–	–	–	–	–	–	–	–	–	–	24	24	2	NCMB 1340^T	244
Vibrio variabilis	–	+	+	+	–	+	+	+	+	–	+	+	+	+	–	+	–	+	–	–	25	24	2	R-40492^T	321
Vibrio vulnificus biovar I	–	5	w	1	–	1	w	–	–	–	w	1	–	–	–	–	–	–	–	–	25	4	0.85	AHLDA 1716	244
Vibrio vulnificus biovar I	–	5	1	1	–	1	w	w	–	–	–	w	–	–	–	–	–	–	–	–	25	4	2	AHLDA 1716	244
Vibrio wodanis	–	+	83	+	–	+	86	–	–	–	69	+	–	–	–	–	–	77	–	–	22	24		NCIMB 13582	1071
Vibrio xiamenensis	–	+	–	+	–	+	–	–	–	–	+	+	–	–	–	–	+	–	–	–	30	8	2	G21^T	581
Vibrio xuii	–	+	+	+	+	+	+	–	–	–	+	+	–	–	–	–	–	–	–	–	25	24	2	LMG 21346^T	1632
Vogesella lacus	–	+	+	+	–	+	+	+	–	–	–	+	–	–	–	–	–	+	–	–	25	24	0.5	LMG 24504^T	325
Weissella ceti	–	+	–	–	–	–	–	–	–	–	+	+	–	–	–	–	–	–	–	–	37	4	0.85	CECT 7719^T	1708
Yersinia ruckeri	–	1	–	w	w	4	1	–	–	–	2	1	–	2	–	1	–	3	–	–	25	20	0.85	AHLDA 1313	244
Zobellia galactanivorans	–											3	3	1						3	30	7 d	2.5		119
Zobellia uliginosa	–											1	1	3						1	30	7 d	2.5		119

1=control well, 2=2-naphthyl-phosphate, 3=2-naphthyl-butyrate, 4=2-naphthyl-caprylate, 5=2-naphthyl-myristate, 6=L-leucyl-2-naphthylamide, 7=L-valyl-2-naphthylamide, 8=L-cystyl-2-naphthylamide, 9=*N*-benzoyl-DL-arginine-2-naphthylamide, 10=*N*-glutaryl-phenylalanine-2-naphthylamide, 11=2-naphthyl-phosphate, 12=naphthol-AS-BI-phosphate, 13=6-Br-2-naphthyl-α-D-galactopyranoside, 14=2-naphthyl-β-D-galactopyranoside, 15=naphthol-AS-BI-β-D-glucuronide, 16=2-naphthyl-α-D-glucopyranoside, 17=6-Br-2-naphthyl-β-D-glucopyranoside, 18=1-naphthyl-*N*-acetyl-β-D-glucosaminide, 19=6-Br-2-naphthyl-β-D-mannopyranoside, 20=2-naphthyl-β-L-fucopyranoside. DW=Distilled Water, Temp=temperature of incubation, Time=time of incubation, h=hours, d=days, Inoc=inoculum, %NaCl indicates final NaCl concentration in the inoculum. MI=Manufacturer's Instructions. nr=not reported, Ref=reference number, SW=inoculum prepared with seawater. Numbers 0–5 refer to strength of colour reaction. Numbers 10–90 indicate percent of positive strains. Ostland 1994 called a reaction of 2 or less as negative.

Table 3.39. Strains quoted in References.

Organism	Strain number	Reference number
Actinobacillus delphinicola	NCTC 12870^T	548
Actinobacillus scotiae	NCTC 12922^T	550
Actinomyces marimammalium	CCUG 41710^T=CIP 106509^T	763
Aequorivita antarctica	ACAM 640^T, DSM 14231^T	211
Aequorivita lipolytica	ACAM 641^T, DSM 14236^T	211
Aequorivita crocea	ACAM 642^T, DSM 14239^T	211
Aequorivita sublithincola	ACAM 643^T, DSM 14238^T	211
Aeromonas allosaccharophila HG15	CECT 4199^T	1102
Aeromonas aquariorum	MDC 47^T=CECT 7289^T=DSM 18362^T	1106
Aeromonas bestiarum HG2	CDC 9533-76	2, 42, 873
Aeromonas bivalvium	868E^T=CECT 7113^T=LMG 23376^T	1146
Aeromonas caviae HG4	ATCC 15468	42, 873
Aeromonas diversa HG 13	CDC 2478-85^T=CECT 4254^T=ATCC 43946^T=LMG 17321^T	1147
Aeromonas encheleia HG16	CECT 4342=LMG 16330	494, 783
Aeromonas eucrenophila HG6	ATCC 23309, CDC 0859-83	42, 873
Aeromonas eucrenophila	LMG 3774=NCMB 74	783
Aeromonas eucrenophila	LMG 13057	783
Aeromonas fluvialis	717^T=CECT 77401^T=LMG 24681^T	37
Aeromonas hydrophila dhakensis	P21^T=LMG 19562^T=CCUG 45377^T	790
Aeromonas hydrophila hydrophila	ATCC 7966^T=LMG 2844^T	790
Aeromonas hydrophila HG1	ATCC 7966^T, CDC 9079-79	42, 873, 1781
Aeromonas hydrophila HG3	CDC 0434-84	873
A hydrophila anaerogenes = A. caviae	ATCC 15468	1781
A hydrophila anaerogenes	ATCC 15467	
Aeromonas jandaei HG 9	CDC 0787-80, ATCC 49568	276, 873
Aeromonas media HG 5B	CDC 9072-83, CDC 0435-84	42, 873
Aeromonas media HG 5A	CDC 9072-83, CDC 0862-83	42, 873
Aeromonas media	ATCC 33907^T=LMG 7907=CCM 3653=DSM4881	35
Aeromonas molluscorum	848^T=CECT 5864^T=LMG 22214^T	1145
Aeromonas piscicola	S1.2^T= CECT 7443^T=LMG 24783^T	141
Aeromonas popoffii	LMG 17541^T	784
Aeromonas rivuli	CECT 7518^T	536
Aeromonas salmonicida achromogenes	NCMB 1110	928, 979
Aeromonas salmonicida pectinolytica	DSM 12609^T=34 mel	1317
A. salmonicida salmonicida	ATCC 14174	928, 1781
A. salmonicida salmonicida	SVLT -1, -2, -5, -6, non-pigmented strains	928
A. salmonicida salmonicida	NCMB 1102	979
Aeromonas salmonicida HG3	CDC 0434-84	42
Aeromonas sanarellii	CECT 7402^T	36
Aeromonas schubertii HG12	CDC 2446-81, ATCC 43700^T, formerly Enteric Group 501	278, 873
Aeromonas simiae	IBS S6874^T=CIP 107798^T=CCUG 47378^T	677

Continued

Table 3.39. Continued.

Organism	Strain number	Reference number
Aeromonas sobria HG7	CIP 7433, CDC 9538-76	42, 873, 1781
Aeromonas veronii sobria HG8/10	CDC 0437-84	873
Aeromonas species HG11	CDC 1306-83	873
Aeromonas taiwanensis	CECT 7403^T	36
Aeromonas tecta	F5187^T=CECT 7082^T=DSM 17300^T=MDC 91^T	435
Aeromonas trota	ATCC 49657^T=LMG 12223^T HG 14	277
Aeromonas veronii	HG 10, ATCC 35604^T=CDC 1169-83	42, 723
Aeromonas veronii	HG 8, CDC-0437-84	42
Algicola bacteriolytica	IAM 14595^T	1489
Aliivibrio fischeri	ATCC 7744^T=NCIMB 1281^T	244, 715
Aliivibrio fischeri	ATCC 25918	1641
A. finisterrensis	CMJ 11.1^T and C 17.13	143
Aliivibrio logei	584^T=ATCC 29985^T	112
Aliivibrio logei	PS 207=ATCC 15382=NCMB 1143	112, 363, 715
Aliivibrio salmonicida	HI 7751^T=NCMB 2262^T	473, 1071
Aliivibrio salmonicida	NCMB 2245	1071
Aliivibrio salmonicida	90/1667-10c	1071
Aliivibrio sifiae	H1-1^T=NBRC 15001^T	1834
Aliivibrio (Vibrio) wodanis	NCIMB 13582^T=NVI 88/441^T	1071
Alteromonas aurantia	ATCC 33046, NCMB 2052	600
Alteromonas citrea	NCMB 1889	598
Alteromonas rubra	NCMB 1890	596
Andreprevotia lacus	GFC-1^T=BCRC 17832^T=LMG 24502^T	1513
Arcanobacterium phocae	DSM 10002^T, M1590/94/3^T	1374
Arcanobacterium phocisimile	2698^T= LMG 27073^T=CCM 8430^T	727
Arcanobacterium pluranimalium	CCUG 42575^T=CIP 106442	985
Arcobacter ellisii	F79-6^T=CECT7837^T=LMG 26155^T	539
Arcobacter marinus	CL-S1^T=KCCM 90072^T=JCM 15502^T	897
Arthrobacter nasiphocae	CCUG 42953^T	358
Arthrobacter rhombi	CCUG 38813^T	1276
Atopobacter phocae	CCUG 42358^T=CIP 106392	989
Bacillus galliciensis	BFLP-1^T=DSM 21539^T=LMG 24668^T	105
Bacillus hemicentroti	JSM 076093^T=DSM 23007^T=KCTC 13710^T	
Bacillus nanhaiensis	JSM 082006^T=DSM 23009^T=KCTC 13712^T	317
Bacillus neizhouensis	JSM 071004^T=CCTCC AB 207161^T=DSM 19794^T=KCTC 13187^T	
Bacillus zhanjiangensis	JSM 099021^T=DSM 23010^T=KCTC 13713^T	316
Bordetella bronchiseptica	ATCC 19395, ATCC 4617, NCTC 8344	184
Brevundimonas diminuta	ATCC 11568^T, LMG 2089^T=CCUG 1427^T	1502
Brevundimonas vesicularis	ATCC 11426^T=CCUG 2032^T=LMG 2350^T	1502
Brucella maris biovar I	NCTC 12890, 2/94	827
Brucella maris biovar II	NCTC 12891, 1/94	827

Continued

Table 3.39. Continued.

Organism	Strain number	Reference number
Brucella maris biovar III		507, 827
Campylobacter insulaenigrae	NCTC 12927^T=CCUG 48653^T	552
Campylobacter lari spp. *concheus*	2897R^T=LMG 21009^T=CCUG 55786^T	424
Campylobacter lari spp. *lari*	LMG 8846^T=NCTC 11352^T	159, 424
Campylobacter peloridis	2314BVA^T=LMG 23910^T=CCUG 55787^T	424
Campylobacter subantarcticus	R-3023^T=LMG 24377^T=CCUG 38513^T	422
Carnobacterium divergens	6251	1405
Carnobacterium inhibens	CCUG 31728^T, strain K1	841
Carnobacterium (*piscicola*) *maltaromaticum*	B270^T=ATCC 35586^T=DSM 20730^T=DSM 20722	136, 736, 1167, 1497
Catellicoccus marimammalium	M35/04/3^T=CCUG 49459^T=CIP 108575^T	984
Chromobacterium violaceum		992
Chryseobacterium chaponense	Sa 1147-06^T=DSM 23145^T=CCM 7737^T	859
Chryseobacterium (*arothri*) *hominis*	P2K6^T=CIP 109575^T=DSM 19326^T	266
Chryseobacterium piscicola	VQ-6316s^T=CECT 7357^T=DSM 21068^T	799
Chryseobacterium piscium	LMG 23089^T=CCUG 51923^T	421
Chryseobacterium scophthalmum	CCM 4109^T=LMG 13028^T	1173
Chryseobacterium tructae	1084-08^T=CECT 7798^T=CCUG 60111^T	1846
Corynebacterium aquaticum	RB 968 BA	133, 235
Corynebacterium aquaticum	ATCC 14665^T	133
Corynebacterium caspium	M/106/00/5T=CCUG 44566^T=CIP 107965^T	357
Corynebacterium ciconiae	BS13^T=CECT 5779^T=CCUG 47525^T	528
Corynebacterium maris	Coryn-1^T=DSM 45190^T=LMG 24561^T	152
Corynebacterium phocae	M408/98/1^T=CCUG 38205^T	1313
Corynebacterium sphenisci	CECT 5990^T=CCUG 46398^T	652
Corynebacterium spheniscorum	CECT 45512^T=CECT 5986^T	653
Corynebacterium testudinoris	CCUG 41823^T	351
Edwardsiella ictaluri	CDC 1976-78, ACC 33202	702
Edwardsiella piscicida	ET883^T=NCIMB 14824^T=CCUG 62929	1
Edwardsiella tarda	ATCC 15947^T	776, 1383
Edwardsiella tarda	AHLDA 135	244
Enterococcus faecalis	ATCC 19433	1380
Vibrio calviensis	RE35/F12^T=CIP 107077^T=DSM 14347^T	436
Enterovibrio coralii	CC17^T=LMG 22228^T=CBMAI 623^T	1636
Enterovibrio nigricans	DA1 1-1-5^T=CECT 7320^T=CAIM 661^T	1314
Enterovibrio norvegicus	LMG 19839^T, CAIM 430^T	1627
Facklamia miroungae	A/G13/99/2^T=CCUG 42728^T=CIP 106764^T	761
Ferrimonas senticii	P2S11^T=ATCC BAA-1480^T=DSM 18821^T	265
Flavobacterium aquatile	NCIB 8694^T, LMG 4008^T	168, 172
Flavobacterium araucananum	LM-19-Fp^T=LMG 26359^T=CCM 7939^T	857
Flavobacterium branchiophilum	ATCC 35035^T, BGD-7721^T	1746
Flavobacterium ceti	454-2^T=CECT 7184^T=CCUG 52969^T	1710

Continued

Table 3.39. Continued.

Organism	Strain number	Reference number
Flavobacterium chilense	LM-09-Fp^{T}=LMG 26360^{T}=CCM 7940^{T}	857
Flavobacterium frigidarium	ATCC 700810^{T}=NCIMB 13737^{T}	777
Flavobacterium gillisiae	ACAM 601^{T}	1115
Flavobacterium hibernum	ATCC 51468^{T}=ACAM 376^{T}	1116
Flavobacterium hydatis (*C. aquatilis*)	ATCC 29551	1586
Flavobacterium johnsoniae	ATCC 17061^{T}	284
Flavobacterium macrobrachii	an-8^{T}=BCRC 17965^{T}=LMG 25203^{T}	1514
Flavobacterium oncorhynchi	631-08^{T}=CECT 7678^{T}=CCUG 59446^{T}	1845
Flavobacterium tegetincola	ACAM 602^{T}	1115
Flavobacterium tilapiae	Ruye-71^{T}=BCRC 80262^{T}=KCTC 23312^{T}	314
Flexibacter polymorphus	ATCC 27820	1016
Francisella philomiragia	ATCC 25015^{T}, CCUG 12603	1142
Francisella philomiragia ssp. *noatunensis*	2005/50/F292-6C^{T}	1142
Gramella marina	KMM 6048^{T}=KCTC 12366^{T}=LMG 25418^{T}	1205
Granulicatella balaenopterae	CCUG 37380^{T}, M1975/96/1^{T}	990
Grimontia (*Vibrio*) *hollisae*	KUMA871, ATCC 33564^{T}	721, 1225
Helicobacter anseris	MIT 04-9362^{T}=ATCC BAA-1299^{T}	561
Helicobacter brantae	MIT 04-9366^{T}=ATCC BAA-1298^{T}	561
Helicobacter canadensis	NLEP-16143^{T}=MIT 98-5491^{T}=ATCC 700968^{T}	559
Helicobacter cetorum	MIT 99-5656^{T}=ATCC BAA-540^{T}	678
Iodobacter fluviatile	NCTC 11159^{T}	1048
Janthinobacterium lividum	NCIMB 9230, NCIMB 9414, DSM 1522	1030
Labrenzia alba	5OM6^{T}=CECT 5095^{T}=CIP 108402^{T}	1362
Lactococcus garvieae	NCDO 2155^{T}=ATCC 43921^{T}	350, 456, 1380, 1614
Lactococcus garvieae	ATCC 49156^{T}	480, 954, 1497
Lactococcus lactis ssp. *lactis*	ATCC 19435	1614
Lactococcus lactis spp. *tructae*	L105^{T}=LMG 24662^{T}=DSM 21502^{T}	1330
Lactococcus piscium	NCFB 2778	1802
Lactobacillus ceti	142-2^{T}=CECT 7185^{T}=CCUG 53626^{T}	1709
Listonella anguillarum	ATCC 14181	1071
Listonella anguillarum	NCMB 6=ATCC 19264 (Bagge and Bagge strain)	714
Listonella anguillarum	NCMB 407=PL 1	714
Listonella anguillarum	NCMB 571 (Hoshina)	714
Listonella anguillarum	NCMB 828=ATCC 14181 (strain 4063, Smith)	714
Listonella pelagia I (said to be *V. natriegenes*)	NCMB 1900^{T}	1071
Listonella pelagia II	NCMB 2253	1071
Marinimonas aquimarina	11SM4^{T}=CECT 5080^{T}=CIP 108405^{T}	1079
Marinimonas ostreistagni	UST010306-043^{T}=JCM 13672^{T}	981
Mesophilobacter marinus	IAM 13185	1230
Moritella japonica	DSK1^{T}=JCM 10249^{T}	1233

Continued

Table 3.39. Continued.

Organism	Strain number	Reference number
Moritella (Vibrio) marina	MP-1^T=ATCC 15381^T=NCMB 1144^T	363, 594, 1683
Moritella marina	S1.2T= CECT 7443T=LMG 24783^T	153
Moritella viscosa	NCIMB 13584^T=NVI 88/478^T	153, 1071
Moritella viscosa	MT 1887=NCIMB 13484, 1896=NCIMB 13485	236
Mycobacterium abscessus	ATCC 19977^T	961
Mycobacterium chelonae	ATCC 6841^T, NCTC 946^T	388, 961
Mycobacterium fortuitum	ATCC 6841^T	961
Mycobacterium fortuitum	ATCC 6842^T	388
Mycobacterium fortuitum ssp. *acetamidolyticum*	ATCC 35931^T	961
Mycobacterium marinum	ATCC 927^T	61
Mycobacterium montefiorense	ATCC BAA-256^T=DSM 44602^T	1013
Mycobacterium palustre	E846^T=ATCC BAA-377^T=DSMZ 44572^T	1656
Mycobacterium peregrinum	ATCC 14467^T	961
Mycobacterium pinnipedii	6482^T=ATCC BAA-688^T=NCTC 13288^T	374
Mycobacterium poriferae	47^T=ATCC 35087^T	1294
Mycobacterium pseudoshottsii	L15^T=ATCC BAA-883^T=NCTC 13318^T	1401
Mycobacterium salmoniphilum	SC^T=ATCC 13578^T=DSM 43276^T	1427, 1783
Mycobacterium shottsii	M175^T=ATCC 700981^T=NCTC 13215^T	1400
Mycobacterium stomatepiae	T11^T=DSM 45059^T=CIP 109275^T	1355
Mycoplasma alligatoris	A21JP2^T=ATCC 700619^T	229
Mycoplasma crocodyli	MP145^T=ATCC 51981^T	909
Mycoplasma mobile	163K^T=ATCC 43663^T	909
Mycoplasma phocicerebrale	1049^T=ATCC 49640^T	616
Mycoplasma phocidae	105^T=ATCC 33657^T	1438
Mycoplasma phocirhinis	852^T=ATCC 49639^T	616
Myroides odoratimimus	NCTC 11180^T, LMG 4029^T	1694
Myroides odoratus	NCTC 11036^T, LMG 1233^T	1694
Nocardia crassostreae	NB4H^T=ATCC 700418^T	566
Nocardia salmonicida	JCM 4826^T=ATCC 27463^T	806
Nocardia seriolae	JCM 3360^T	944
Nocardia sp. Australian strain	98/1655	217
Oceanimonas baumannii	ATCC 700832^T	232
Oceanimonas doudoroffii	ATCC 27123^T	232
Pantoea agglomerans	ATCC 27155^T, NCTC 9381	509, 602
Pantoea agglomerans	ATCC 27155^T ATCC 12287	676
Pantoea dispersa	ATCC 14589^T	602
Paracoccus fistulariae	22-5^T, CCUG 58401^T, KCTC 22803^T	903
Pasteurella skyensis	95A1^T=NCTC 13204^T, NCIMB 13593^T	181
Phaeobacter (Roseobacter) gallaeciensis	BS107^T=CIP 105210^T=ATCC 700781^T=DSM 17395^T	1099, 1441
Phocoenobacter uteri	NCTC 12872^T	549
Photobacterium angustum	NCIMB 1895	1641

Continued

Table 3.39. Continued.

Organism	Strain number	Reference number
Photobacterium damselae ssp. *damselae*	ATCC 33539=NCIMB 2184T	595, 1060, 1071, 1547, 1641
Photobacterium damselae ssp. *damselae*	ATCC 35083	556, 1641
Photobacterium damselae ssp. *piscicida*	ATCC 17911 & NCIMB 2058	595, 1641
Photobacterium damselae ssp. *piscicida*	ATCC 29690, ATCC 17911^{T}	703, 1089
Photobacterium fischeri	NCMB 1281^{T}=ATCC 7744	1071
Photobacterium fischeri	ATCC 25918	1641
Photobacterium fischeri	ATCC 7744	715, 1781
Photobacterium iliopiscarium	PS1^{T}=ATCC 51760^{T}	1272, 1682
Photobacterium leiognathi	LMG 4228, NCIMB 1895	1641
Photobacterium logei	ATCC 15382=NCMB 1143=PS 207	1071
Photobacterium logei	NCMB 1143	1071
Photobacterium logei	ATCC 29985^{T}	1071
Photobacterium logei	ATCC 15382	153
Photobacterium logei	NCIMB 2252, ATCC 29985	1071
Photobacterium rosenbergii	CC1^{T}=LMG 22223^{T}=CBMAI 622^{T}	1636
Providencia rustigianii	DSM 2620	1175
Pseudoalteromonas antarctica	CECT 4664^{T}, NF3	215
Pseudoalteromonas citrea	ATCC 29719^{T}, NCMB 1889	598
Pseudoalteromonas elyakovii	ATCC 700519^{T}, KMM 162^{T}	1491
Pseudoalteromonas piscicida	MB-1^{T}=ATCC 15057^{T}	242
Pseudomonas anguilliseptica	ATCC 33660^{T}	1742
Pseudomonas chlororaphis ssp. *piscium*	JF 3835^{T}=NCIMB 14478^{T}=DSM 21509^{T}	252
Pseudomonas plecoglossicida	FPC 951^{T}=ATCC 700383^{T}	1229
Raoultella planticola	ATCC 33531	462
Raoultella terrigena	ATCC 33257	462
Renibacterium salmoninarum	Lea-1-74=ATCC 33209^{T}	1472
Roseovarius crassostreae	CV919-312^{T}=ATCC BAA-1102^{T}=DSM 16950^{T}	197
Salinivibrio costicola	NCMB 701^{T}	582
Serratia fonticola	ATCC 29844	601
Shewanella affinis	KMM 3587^{T}=CIP 107703^{T}=ATCC BAA-642^{T}	814
Shewanella algae	OK-1^{T}=IAM 14159T	886, 1239, 1853
Shewanella algidipiscicola	S13^{T}=LMG 23746^{T}=NBRC 102032^{T}	1481
Shewanella aquimarina	SW-120^{T}=KCCM 41821^{T}=JCM 12193^{T}	1832
Shewanella baltica	NCTC 10735^{T}=DSM 9439^{T}= CECT 323^{T}= IAM 1477^{T}= LMG 2250^{T}	1853
Shewanella colwelliana	LST^{T}=ATCC 39565	1773, 1774
Shewanella corallii	fav-2-10-05^{T}=LMG 24563^{T}=DSM 21332^{T}	1527
Shewanella frigidimarina	ACAM 591T	209
Shewanella glacialipiscicola	T147^{T}=LMG 23744^{T}=NBRC 102030^{T}	1481
Shewanella gelidimarina	ACAM 456^{T}	209
Shewanella hafniensis	P010^{T}=ATCC BAA-11207^{T}=NBRC 100975^{T}	1480
Shewanella haliotis	DWO1^{T}=KCTC 12896^{T}=JCM 14758^{T}	895

Continued

Table 3.39. Continued.

Organism	Strain number	Reference number
Shewanella (Alteromonas) hanedai	ATCC 33224^{T}	836
Shewanella japonica	KMM 3299^{T}=LMG 19691^{T}=CIP 106860^{T}	813
Shewanella marinintestina	IK-1^{T}=JCM 11558^{T}=LMG 21403^{T}	1479
Shewanella marisflavi	SW-117^{T}=KCCM 41822^{T}=JCM 12192^{T}, AP629	1020, 1832
Shewanella morhuae	U1417^{T}=ATCC BAA-1205^{T}=NBRC 100978^{T}	1480
Shewanella oneidensis	ATCC 700550^{T}	1713
Shewanella pealeana	ANG-SQ1^{T}=ATCC 700345^{T}	1008
Shewanella pneumatophori	SCRC-2738^{T}=JCM 13187^{T}=NCIMB 14060^{T}	732
Shewanella putrefaciens	ATCC 8071^{T}	1239
Shewanella sairae	SM2-1^{T}=JCM 11563^{T}=LMG 21408^{T}	1479
Shewanella schlegeliana	HRKA1^{T}=JCM 11561^{T}=LMG 21406^{T}	1479
Shewanella upenei	20-23R^{T}=KCTC 22806^{T}=CCUG 58400T	901
Shewanella woodyi	ATCC 51908^{T}, MS32	1093
Sphingobacterium spiritivorum	NCTC 11386	750
Staphylococcus delphini	DSM 20771^{T}	1704
Staphylococcus lutrae	DSM 10244, M340/94/1	551
Stappia (stellulata–like) aggregans	M1	1362
Streptococcus (difficile) agalactiae	ND 2-22, CIP 103768	482
Streptococcus (difficile) agalactiae	LMG 15977^{T}	1696
Streptococcus halichoeri	CCUG 48324^{T}=CIP 108195^{T}	987
Streptococcus ictaluri	707-05^{T}=SO2-1108^{T}=ATCC BAA-1300^{T}=CCUG 52536^{T}	1519
Streptococcus iniae	ND 2-16, CIP 103769	1342
Streptococcus iniae	BU=ATCC 29177, PW=ATCC 29178	1341, 1342
Streptococcus marimammalium	M54/01/1^{T}, CCUG 48494^{T}, CIP 108309^{T}	986
Streptococcus parauberis	NCDO 2020	456
Streptococcus phocae	NCTC 12719^{T}, 8399 HI^{T}	1538
Streptococcus uberis	NCDO 2038	456
Tenacibaculum aiptasiae	a4^{T}=BCRC 17655^{T}=LMG 24004^{T}	1761
Tenacibaculum crassostreae	JO-1^{T}=KCTC 22329^{T}=JCM 15428^{T}	1001
Tenacibaculum dicentrarchi	35/09^{T}=CECT 7612^{T}=BCIMB 14598^{T}	1346
Tenacibaculum discolor	LL04 11.1.1^{T}=NCIMB 14278^{T}=DSM 18842^{T}	1347
Tenacibaculum gallaicum	A37.1^{T}=NCIMB 14147^{T}=DSM 18841^{T}	1347
Tenacibaculum litopenaei	B-1^{T}=BCRC 17590^{T}=LMG 23706^{T}	1517
Tenacibaculum (Flexibacter) maritimum	R2^{T}=NCMB 2154^{T}	1744
Tenacibaculum (Flexibacter) ovolyticum	NCIMB 13127^{T}=EKD002^{T}	675
Tenacibaculum soleae	LL04 12.1.7^{T}=CECT 7292^{T}=NCIMB 14368^{T}	1344
Vagococcus fessus	CCUG 41755^{T}	762
Vagococcus fluvialis	NCDO 2497, NCFB 2497	354, 1354
Vagococcus lutrae	CCUG 39187^{T}	988
Vagococcus penaei	CD276^{T}=LMG 24833^{T}=CIP 109914^{T}	826
Vagococcus salmoninarum	NCFB 2777^{T}	1497, 1755
Varracalbmi		1690

Continued

Table 3.39. Continued.

Organism	Strain number	Reference number
Vibrio aerogenes	ATCC 700797^{T}=CCRC 17041^{T}=FG1^{T}= DSMZ 14438	1522
Vibrio aestivus	M22^{T}=CECT 7558^{T}=CAIM 1861^{T}	1065
Vibrio aestuarianus	ATCC 35048^{T}, LMG 7909	1645
Vibrio aestuarianus spp. *francensis*	02/041^{T}=LMG 24517^{T}	588
Vibrio agarivorans	CECT 5084, CECT 5085^{T}=DSM 13756	1084
Vibrio agarivorans	AHLDA 1732	244
Vibrio alfacsensis	CAIM 1831^{T}	640
Vibrio alginolyticus	ATCC 17749	244
Vibrio areninigrae	J74^{T}=KCTC 22122^{T}=JCM 14949^{T}	301
Vibrio artabrorum	Vb 11.8T=LMG 23865^{T}=CECT 7226^{T}	442
Vibrio atlanticus	Vb 11.11T=LMG 24300^{T}=CECT 7223^{T}	442
Vibrio atypicus	HHSO2^{T}=CGMCC 1.8461^{T}=LMG 24781^{T}	1767
Vibrio azureus	LC2-005^{T}=NBRC 104587^{T}=KCTC 22352^{T}	1836
Vibrio brasiliensis	LMG 20546^{T}	1632
Vibrio breoganii	LMG 23858^{T}, CECT 7222^{T}	140
Vibrio caribbeanicus	DSM 23640^{T}	739
Vibrio casei	WS 4539^{T}=DSM 22364^{T}=LMG 25240^{T}	190
Vibrio celticus	Rd 8.15^{T}=CECT 7224^{T}=LMG 23850^{T}	142
Vibrio chagasii	LMG 21353^{T}=CAIM 431^{T}	1635
Vibrio cincinnatiensis	ATCC 35912	220
Vibrio comitans	LMG 23416^{T}	1485
Vibrio communis	LMG 25430^{T}	320
Vibrio coralliilyticus	YB1^{T}=ATCC BAA-450^{T}=LMG 20984^{T}	155, 156, 157
Vibrio coralliilyticus/neptunius-like	DOI	879
Vibrio crassostreae	LMG 22240^{T}=CIP 108327^{T}=LGP 7^{T}	524
Vibrio cyclitrophicus	P-2P44^{T}=ATCC 700982^{T}	709
Vibrio diabolicus	CNCM I-1629=HE800^{T}	1368
Vibrio diazotrophicus	ATCC 33466^{T}	665
Vibrio ezurae	HDS1-1^{T}=IAM 15061^{T}=LMG 19970^{T}	1486
Vibrio fluvialis	ATCC 49515, NCDO 2497=NCFB 2497	354, 1613
Vibrio fluvialis	NCTC 11327	222, 996
Vibrio fortis	LMG 21557^{T}	1634
Vibrio furnissii	ATCC 35016^{T}=CDC B3215^{T}	222
Vibrio furnissii	ATCC 11218	244
Vibrio gazogenes	ATCC 29988^{T}	518, 686
Vibrio gallaecicus	VB 8.9^{T}=LMG 24045T=CECT 7244^{T}, VB 5.12	144
Vibrio gallicus	LMG 21878^{T}=CIP 107863^{T}=HT2-1^{T}	1487
Vibrio gigantis	LMG 22741^{T}=CIP 108656^{T}=LGP 13^{T}	1010
Vibrio halioticoli	IAM 14596^{T}	1490
Vibrio hangzhouensis	CN83^{T}	1818
Vibrio harveyi	ATCC 14126^{T}	20, 1781
Vibrio (carchariae) harveyi	ATCC 35084^{T}	20, 244

Continued

Table 3.39. Continued.

Organism	Strain number	Reference number
Vibrio hepatarius	LMG 20362^{T}	1634
Vibrio hippocampi	BFLP-4^{T}=DSM 22717^{T}=LMG 25354^{T}	106
Vibrio hispanicus	LMG 13240^{T}=CAIM 525^{T}=VIB 213^{T}	636
Vibrio ichthyoenteri	IFO 15847^{T}=DSMZ 14397^{T}	804
Vibrio inusitatus	LMG 23434^{T}=NBRC 102082^{T}	1485
Vibrio jasicida	TCFB 0772^{T}=JCM 16453^{T}=LMG 25398^{T}	1835
Vibrio kanaloae	LMG 20539^{T}=CAIM 485^{T}	1635
Vibrio lentus	CECT 5110^{T}=DSM 13757=40MA	1082
Vibrio litoralis	MANO22D^{T}=KCTC 12520^{T}=DSM 17657^{T}	1202
Vibrio mangrovi	MSSRF38^{T}=LMG 24290^{T}=DSM 19641^{T}	1373
Vibrio maritimus	R-40493^{T}=LMG 25439^{T}=CAIM 1455^{T}	321
Vibrio marisflavi	WH134^{T}=CGMCC 1.8994^{T}=LMG 25284^{T}=DSM 23086^{T}	1760
Vibrio mediterranei	CECT 621^{T}, LMG11258	244, 293, 1361
Vibrio metschnikovii	NCTC 8563	1782
Vibrio metschnikovii	NCTC 8443^{T}	994
Vibrio mimicus	ATCC 33653^{T}	244, 412, 1073, 1781
Vibrio species MM5	MM5	1841
Vibrio mytili	CECT 632T, CIP 103929	244, 1363
Vibrio natriegens	ATCC 14048T	78, 126, 1318
Vibrio navarrensis	CIP 1397-6=CIP 103381^{T}	1682, 1687
Vibrio navarrensis spp. *pommerensis*	CH-291^{T}	851
Vibrio neonatus	HDD3-1^{T}=LMG 19973^{T}=IAM 15060^{T}	1486
Vibrio neptunius	LMG 20536T^{T}	1632
Vibrio neptunius	PP-145.98	1357
Vibrio nereis	LMG 3895^{T}	1368
Vibrio nigripulchritudo	164^{T}=ATCC 27043^{T}, CIP 103195 (3 generations)	129, 244
Vibrio ordalii	NCMB 2167^{T}, ATCC 33509=DF$_3$K=Dom F$_3$	1071, 1493
Vibrio orientalis	NCMB 2195^{T}; ATCC 33934^{T}	244, 1071
Vibrio owensii	DY05^{T}=JCM 16517^{T}=ACM 5300^{T}	270
Vibrio owensii	47666-1	270
Vibrio pacinii	LMG 1999^{T}	638
Vibrio parahaemolyticus	ATCC 17802	244
Vibrio parahaemolyticus	ATCC 17802, ATCC 27969	1040
Vibrio pectenicida	CIP 105190^{T}=A365^{T}	970
Vibrio pectenicida	ATCC 700783	244
Vibrio penaeicida	JCM 9123^{T}=KH-1^{T}= IFO 15640^{T}	803
Vibrio penaeicida	DSMZ 14398	244
Vibrio plantisponsor	MSSRF60^{T}=DSM 21026^{T}=LMG 24470^{T}=CAIM 1392^{T}	1372
Vibrio pomeroyi	LMG 20537^{T}=CAIM 578^{T}	1635
Vibrio ponticus	369^{T}=CECT 5869^{T}=DSM 16217^{T}	1080
Vibrio ponticus	RF30	1816

Continued

Table 3.39. Continued.

Organism	Strain number	Reference number
Vibrio porteresiae	MSSRF30^{T}=LMG 24061^{T}=DSM 19223^{T}	1371
Vibrio proteolyticus	AHLDA 1735	244
Vibrio proteolyticus	ATCC 33125^{T}, CW8T2	1133, 1720
Vibrio quintilis	M62^{T}=CECT 7734^{T}=CAIM 1863^{T}	1065
Vibrio rarus	LMG 23674^{T}=NBRC 102084T	1485
Vibrio rhizosphaerae	MSSRF3^{T}=LMG 23790^{T}=DSM 18581^{T}	947
Vibrio rotiferianus	LMG 21460^{T}	637
Vibrio ruber	VR1^{T}=JCM 11486^{T}=CCRC 17186^{T}	1521
Vibrio rumoiensis	S-1^{T}=FERM P-14531^{T}	1843
Vibrio sagamiensis	LC2-047^{T}=NBRC 104589^{T}=KCTC 22354^{T}	1837
Vibrio scophthalmi	A089, CECT 4638^{T}	293, 523
Vibrio shilonii (possibily *V. mediterranei*)	ATCC BAA-91^{T}=DSM 13774=AK-1	114, 952
Vibrio sinaloensis	CAIM 797^{T}=CECT 7298^{T}	639
Vibrio splendidus I	NCMB 1^{T} (=ATCC 33125^{T})	154, 293, 1071
Vibrio splendidus I	ATCC 33125	590, 1321
Vibrio splendidus II	NCMB 2251	293, 1071
Vibrio stylophorae	KTW-12^{T}=BCRC 80105^{T}=LMG 25357^{T}	1515
Vibrio tapetis	B1090^{T}, CECT 4600^{T}	201, 1237
Vibrio tasmaniensis	LMG 20012^{T}	1637
Vibrio tubiashii	NCMB 1340^{T}	244, 1071
Vibrio tubiashii	ATCC 19109^{T}	668
Vibrio tubiashii	LMG 10936^{T}	1368
Vibrio tubiashii	NCIMB 706	244
Vibrio variabilis	R-40492^{T}=LMG 25438^{T}=CAIM 1454^{T}	321
Vibrio vulnificus		293
V. vulnificus biovar I non-serovar E f	ATCC 27562	50, 1644
V. vulnificus biovar 2, serovar E a	ATCC 33187	180
V. vulnificus biovar 2, serovar E b	ATCC 33149	180
V. vulnificus biovar 2, serovar 04 b	ATCC 33149	741
V. vulnificus biovar2, serovar E c	NCIMB 2138	180
V. vulnificus biovar 2 serovar E d	NCIMB 2137	180
V. vulnificus biovar 2, serovar E e	NCIMB 2136	180
V. vulnificus biovar 2, serovar E g	Taiwanese strains	50
V. vulnificus biovar 2	ATCC 33148	1644
V. xiamenensis	G21^{T}	581
Vibrio xuii	LMG 21346^{T}	1632
Vogesella lacus	GR13^{T}=BCRC 17836^{T}=LMG 24504^{T}	325
Weissella ceti	1119-1A-09^{T}=CECT 7710^{T}=CCUG 59653^{T}	1708
Weissella hellenica	DS-12	260
Weissella hellenica	JCM 10103^{T}	260
Yersinia rohdei	ATCC 43380, CDC 3022-85	28
Yersinia aldovae	ATCC 35236, CDC 669-83	161
Zobellia galactanivorans	Dsij^{T}=DSM 12802^{T}	119

4

Technical Methods

4.1 Total Bacterial Count

Diagnostic, research or industry-based laboratories often monitor the pond or tank water for bacterial load as an indicator of water quality and hence as a disease indicator. This is done by a test known as the total bacterial count (TBC).

There are a number of methods for total bacterial counts and these may be found in most microbiology textbooks. A basic method for total bacterial counts is described here.

Generally, total bacterial counts are performed using a good general-purpose medium that will support the growth of the majority of the organisms that are anticipated to be cultured. A selective medium for *Vibrio* sp. such as thiosulfate-citrate-bile-salts-sucrose (TCBS) is often used as a culture medium for total bacterial counts. However, this medium has a tendency to produce a lower count than an equivalent sample inoculated to marine salt agar (MSA-B) plate (or blood agar (BA) in the case of freshwater samples) even when the majority of bacteria cultured are Vibrios. Therefore, it is not the best primary culture medium, however it is useful when doing a total vibrio count.

The time between collection of the water sample and testing is important. Bacteria in the water sample will multiply within a few hours at room temperature. Thus, a delay of more than 1–2 h between collection and testing may result in an inaccurate assessment of the bacterial load of the original water sample. The water sample must be stored at 4°C either during transport or while waiting to be tested in the laboratory. There is no effect on bacterial count due to the size of the collection container (Simon and Oppenheimer, 1968).

There is no predetermined optimal bacterial load of pond or tank water, and this will vary anyway according to the number of fish in the water, that is, the stocking density. It is suggested that laboratories regularly monitor a particular tank or set of tanks and keep a record of their results. The bacterial loads or TBC is then related back to the health of the fish. If a daily log or graph is kept of the TBC against the health of the fish, then the laboratory can eventually determine the optimal bacterial load for healthy fish, as opposed to the load seen when the fish appear diseased.

Suggested sites for monitoring bacterial load are the intake water supply, the filtered water supply, fish tanks and the feed supply. The sediment at the bottom of the tank will have a greater bacterial count than the water body, particularly when there are dead or diseased larvae or fish.

Research in an oyster hatchery indicated that four main genera were involved in the bacterial load, and these were *Vibrio* sp., *Alteromonas* sp., *Pseudomonas* sp. and *Flavobacterium* sp. *Vibrio* sp. were the predominant organism. Virulence testing showed that a third of these *Vibrio* sp., which included *Listonella anguillarum*, and two of the ten *Alteromonas* sp. were pathogenic. The isolates were not identified to species level. Infection was not seen in the larvae when they were dosed with the non-pathogenic bacteria. However, fatal infection was produced when the bacterial load of the pathogenic bacteria was 1×10^7–5×10^7. At a lower dose of 1×10^5–5×10^5 infection still occurred, but took 2–3 days longer to develop and for clinical signs to be seen. Thus, infection rates are related to the bacterial load of pathogenic bacteria (Garland *et al.*, 1983).

Bacteria are part of the diet of marine filter feeders. In another study on healthy oysters (*Crassostrea gigas*) and horse mussel (*Modiolus modiolus*), the total bacterial load of normal flora in the haemolymph and soft tissue was 2.6×10^4 colonies/ml and 2.9×10^4 colonies/g, respectively. The predominant flora and percentage of the total flora were *Pseudomonas* sp. (61.3%), *Vibrio* sp. (27%) and *Aeromonas* sp. (11.7%) in the haemolymph, with predominating bacteria in the soft tissues being *Vibrio* sp. (38.5%), *Pseudomonas* sp. (33%) and *Aeromonas* sp. (28.5%). When the oysters and horse mussel were challenged in virulence studies using *Vibrio salmonicida*, the total bacterial counts increased to 10^5 colonies/ml in the haemolymph and 6×10^7 colonies/g in the soft tissue (Olafsen *et al.*, 1993).

The following table indicates the bacterial counts obtained in both studies.

The microflora in the intestines of healthy jack mackerel (*Trachurus japonicus*) was found to be 4.6×10^6 colonies/g and consisted of *Vibrio* sp. The total bacterial load in the stomach was 2.6×10^5 colonies/g with the predominant flora being *Vibrio* sp., and *Achromobacter* sp., with smaller counts of *Pseudomonas* sp., *Flavobacterium* sp., *Corynebacterium* sp., *Bacillus* sp. and *Sarcina* sp. (Aiso *et al.*, 1968).

4.1.1 Equipment

A pipette or pipettes capable of measuring 100 µl to 1000 µl.
1.5 ml microfuge tubes.
An agar medium that supports the growth of the target bacteria, such as BA for freshwater or (MSA-B or MA 2216) for salt water.
Inoculating loop.
Bunsen burner.
Incubator set to 25°C.

4.1.2 Water

Collect a volume of water.
Do dilutions at neat, 10^{-1}, 10^{-2}, 10^{-3}.

4.1.3 Dilutions

Neat:=100 µl to a plate and lawn inoculate (spread evenly over the plate).
10^{-1}:=900 µl of sterile distilled water+100 µl of sample.

Table 4.1. Total bacterial counts at sites in oyster hatcheries: indication of healthy and diseased states.

Sample site	Total bacterial count, cells/ml
Seawater (filtered, no larvae)[a]	10^3–10^{3}[a]
Tank water[a]	10^4–10^{5}[a]
Sediment[a]	10^7–10^8 cells/g[a]
Haemolymph healthy oyster[b]	2.6×10^{4}[b]
Soft tissue healthy oyster[b]	2.9×10^{4}[b]
Load of pathogenic bacteria at which mortalities are seen. Rapid onset of infection[a]	1×10^7–5×10^{7}[a]
Load of pathogenic bacteria at which mortalities are seen. Slower onset of infection[a]	1×10^5–5×10^{5}[a]
Haemolymph in diseased oyster[b]	10^{5}[b]
Soft tissue in diseased oyster[b]	6×10^{7}[b]

Data from [a]Garland *et al.* (1983); [b]Olafsen *et al.* (1993).

10^{-2}: =900 µl of sterile distilled water+100 µl of dilution 10^{-1}.
10^{-3}: =900 µl of sterile distilled water+100 µl of dilution 10^{-2}.

4.1.4 Incubation

Place plates in sealed plastic container and incubate at room temperature or an incubator set to 25°C.

4.1.5 Counting the bacterial colonies

Examine the plates at 24 and 48 h.
Count the number of colonies on a plate. Using a felt pen, mark each colony from the back of the Petri dish to assist with counting. Count=*N*.

4.1.6 Calculation to obtain colony forming units per ml (CFU/ml)

If 100 µl is placed onto a plate and then lawn inoculated, this is a plate dilution of 10^{-1}. If 10 µl were placed on a plate, this would be a plate dilution of 10^{-2}.

CFU/ml=*N* × dilution × plate dilution.

4.2 Microscopy

Most laboratory staff will be familiar with the use of a light microscope. However, for those who are new to the use of a microscope, some basics are explained here.

Before use, a microscope should be set up for Koehler illumination. This ensures that the light entering the microscope is focused so as to produce an evenly illuminated field.

- Place a stained smear onto the stage. Focus using the ×10 objective.
- Close the field diaphragm – this is the diaphragm at the base of the microscope.
- Close the aperture of the iris diaphragm. This is the aperture just under the stage that the glass slide rests on.
- There will be a small circle of light. Centre this circle of light using the condenser centring screws.
- Next, adjust the height of the condenser under the stage so that the edge of the circle of light is in sharp focus.
- Now open the field diaphragm so that it just disappears from view.
- Open the aperture diaphragm to suit the contrast required.

Staff who are new to the use of a microscope should remember that examination of a Gram-stained smear of bacterial cells is best observed using the oil immersion lens, which has a magnification of 1000×. The aperture on the sub-stage condenser should be open to half its setting, with the light intensity set so that it is comfortable for the eyes. For the examination of wet preparations, this aperture will need to be almost closed so that a good contrast is obtained and cellular structures are defined.

Only use lens tissue paper for cleaning the objective lens. Always remove the oil from the lens after use, as the oil can damage the structures within the objective. If oil does get onto any of the lenses, it can be removed by soaking some lens tissue with petroleum spirit and gently wiping across the lens until the oil has been removed.

4.3 Storage of Isolates

Isolates may be stored at −80°C in Lab Lemco broth supplemented with 10% glycerol. For marine organisms, use marine broth 2216 supplemented with 30% glycerol (Bowman and Nichols, 2002).

There are a number of different media for freeze-drying of bacteria. One method is to suspend organisms in 1 ml of inositol horse serum in Wheaton serum bottles or appropriate container depending on freeze-drying equipment. Snap freeze in liquid nitrogen and follow instructions for freeze-drying equipment (see Chapter 7).

For *Vibrio* sp. the use of *meso*-inositol in nutrient broth followed by reconstitution using *Vibrio* recovery medium is recommended (Carson *et al.*, 2009). See Chapter 7 for recipes.

An analysis of different cryoprotectants tested for *Flavobacterium columnare*, *F. psychrophilum* and *Tenacibaculum maritimum* recommends the use of a medium containing two-thirds Brucella broth (Difco) and one-third horse or fetal calf serum (Desolme and Bernardet, 1996).

5

Fungi, Yeasts and Oomycetes from Fish and Other Aquatic Organisms

5.1 Introduction

A number of different fungi are pathogens or saprophytes on fish and other aquatic animals. These fungi come from a number of different Families.

Taxonomic overview

Within the system of life there are six kingdoms: Animalia, Bacteria, Fungi, Chromista, Protozoa and Plantae (Çavalier-Smith, 1998). Fungi are eukaryotic organisms. The kingdom Fungi contains one subkingdom, seven phyla and ten subphyla. Within the kingdom Fungi are the Eumycota or True fungi. Slime moulds and water moulds (oomycetes) used to be classified in the kingdom Fungi but now oomycetes belong to the kingdom Chromista and the phylum *Oomycota*. Slime moulds are grouped with the *Amoebozoa*. Oomycetes have cellulose in their cell walls as do plants, whereas True fungi have chitin and glucans in their cell walls.

The kingdom Fungi contains the following phyla: *Ascomycota, Basidiomycota, Chytridiomycota, Glomeromycota, Incertae sedis, Zygomycota* and others.

The kingdom Chromista contains the following phyla: *Oomycota, Bigyra* and others.

Phyla may also be referred to as divisions.

The naming of fungi is governed by the International Code of Nomenclature for Algae, Fungi and Plants (ICN) that came into effect in January 2012 and all names are scrutinized by the Nomenclature Committee for Fungi of the International Mycological Association (http://www.ima-mycology.org). All fungi must be registered with a recognized repository such as MycoBank (http://www.mycobank.org) (Karthick and Williams, 2012). MycoBank provides pictures and illustrations and descriptions of growth (Crous *et al.*, 2004).

Table 5.2 lists fungi in alphabetical order and details the disease status, disease signs and the affected aquatic animal species and geographical distribution reported from references cited in this manual.

KINGDOM FUNGI

In this chapter the fungi are grouped according to their phylum and listed in alphabetical order under phylum. The phyla are also listed in alphabetical order. Apart from listing according to phylum, no other taxonomic groupings of fungi are made in this chapter.

Phylum *Ascomycota*

The phylum *Ascomycota* contains filamentous fungi such as *Aspergillus, Fusarium* and *Penicillium*, and yeasts such as *Saccharomyces* and *Candida*.

Table 5.1. Host species and organism relationship.

Host	Pathogens	Normal flora	Tissue site	Disease status	Ref
KINGDOM ANIMALIA PHYLUM *ARTHROPODA* SUBPHYLUM *CRUSTACEA* ORDER *Anostraca*					
Brine shrimp (*Artemia salina*)		*Metschnikowia bicuspidate* var. *bicuspidata*		Non-pathogenic in experimental infection	1567
ORDER *Decapoda* Family *Astacidae*					
Freshwater crayfish, European crayfish, noble crayfish, broadfingered crayfish (*Astacus astacus*)	1. *Aphanomyces astaci* 2. *Saprolegnia ferax* 3. *Trichosporon beigelii*		1. Melanized areas on cuticle of abdomen and limbs 3. Cuticle	1. Crayfish plague 3. Infection	445 1554 1677
Narrow-clawed or Turkish crayfish (*A. leptodactylus*)	*Aphanomyces astaci*		Melanized areas on cuticle of abdomen and limbs	Crayfish plague	118 1369
Signal crayfish (*Pacifastacus leniusculus*)	1. *Aphanomyces astaci* 2. *Aphanomyces repetans*		1. Soft cuticle of abdomen and joints 2. Isolated from soft cuticle	1. Carrier status. Crayfish plague under stressful conditions 2. Non-pathogenic in experimental challenge	264 443 1432 1677
White-clawed crayfish or Atlantic stream crayfish (*Austropotamobius pallipes*)	1. *Aphanomyces astaci* 2. *Aphanomyces frigidophilus* 3. *Fusarium tabacinum*		1. Soft cuticle of abdomen and joints 2. Subabdominal cuticle 3. Gills	1. Crayfish plague Non-pathogenic in experimental challenge but associated with a mass mortality event in the wild 3. Black gill disease	25 110 111 264 390 447
Family *Cambaridae*					
Freshwater crayfish (*Orconectes propinquus*)	*Saprolegnia australis*		Ulcerative lesions on cuticle	Ulcerative disease	943
Red swamp crawfish, Louisiana swamp crayfish (*Procambarus clarkii*)	1. *Aphanomyces astaci* 2. *Aphanomyces repetans*		1. Soft cuticle of abdomen and joints 2. Isolated from soft cuticle	1. Carrier status and crayfish plague under stressful conditions 2. Non-pathogenic in experimental challenge	264 443 1432

Zarigani or Japanese crayfish (*Cambaroides japonicus*)	*Aphanomyces astaci*		1. Soft cuticle of abdomen and joints	Crayfish plague in experimental infection	1680
Spinycheek or mud crayfish (*Orconectes limosus*)	*Aphanomyces astaci*		1. Soft cuticle of abdomen and joints	1. Carrier status and crayfish plague under stressful conditions	937
Family *Nephropidae*					
American lobster (*Homarus americanus*)	*Haliphthoros milfordensis*		Gill and thoracic segments	Mycosis	541
Lobster (*Homarus gammarus*). *H. vulgaris* is an unaccepted synonym	*Fusarium solani*		Exoskeleton and occasionally stomach	Infection	23
Family *Palinuridae*					
Japanese spiny lobster (*Panulirus japonicus*)	*Atkinsiella panulirata*		Phyllosoma	Japan	914
Family *Pandalidae*					
Coonstripe shrimp *Pandalus prensor* (*P. hypsinotus* is an unaccepted name)	*Lagenidium myophilum*		Muscle	Mortality	1200
Family *Parastacidae*					
Gippsland crayfish (*Euastacus kershawi*), *E. clydensis*, *E. crassus*	*Aphanomyces astaci*		1. Soft cuticle of abdomen and joints	Crayfish plague in experimental infection	1680 1681
Tasmanian giant freshwater lobster (*Astacopsis gouldi*), *A. fluviatilis*	*Aphanomyces astaci*		1. Soft cuticle of abdomen and joints	Crayfish plague in experimental infection	1681
Yabby (*Cherax destructor*), *C. quinquecarinatus*, *C. papuanus*	*Aphanomyces astaci*		1. Soft cuticle of abdomen and joints	Crayfish plague in experimental infection	1432 1681
Geocharax gracilis	*Aphanomyces astaci*		1. Soft cuticle of abdomen and joints	Crayfish plague in experimental infection	1681
Family *Penaeidae*					
Kuruma prawn (*Penaeus japonica*)	*Fusarium solani*, *F. oxysporon*		Black gills	Infection	887 888
Tiger prawn (*P. monodon*)	1. *Haliphthoros milfordensis* 2. *Lagenidium callinectes* 3. *L. thermophilum*		1. Zoeae 2. Thorax, abdomen, eyestalks 3. Eggs and larvae	1, 2, 3. Mycosis	330 1181 1370

Continued

Table 5.1. Continued.

Host	**Pathogens**	**Normal flora**	**Tissue site**	**Disease status**	**Ref**
Family *Pinnotheroidea*					
Pea crab (*Pinnotheres pisum*)	*Leptolegnia marina*		Eggs, gills	Mortality	66
Family *Portunidae*					
Blue crab (*Callinectes sapidus*)	*Lagenidium callinectes*		Eggs	Mycosis	1419 1474
Blue swimmer crab (*Portunus pelagicus*)	*Atkinsiella okinawaensis*, *Haliphthoros milfordensis*, *Lagenidium callinectes*		Eggs and zoeae	Mycosis	1197
Swimming crab, Japanese blue crab, gazami crab, horse crab (*Portunus trituberculatus*)	1. *Atkinsiella dubia* 2. *Halodaphnea okinawaensis* 3. *Metschnikowia bicuspidata*		1, 2. Mantle 3. Muscle, haemolymph, heart	1, 2. Fungal infection 3. Milky crab disease	1197 1765
Mangrove or mud crab (*Scylla serrata*)	1. *Haliphthoros milfordensis* 2. *Halocrusticida baliensis* 3. *Halodaphnea hamanaensis* 4. *Lagenidium thermophilum*		1, 2, 4. Zoeae and eggs 3. Eggs	1, 2, 3, 4. Fungal infection, mortality	177 700 1199
Family *Ucididae*					
Mangrove crab (*Ucides cordatus*)	*Exophiala cancerae* and *Fonsecaea brasiliensis*		Heart, gill, thoracic ganglion, hemocele	Lethargic crab disease	192 1340 1722
Family *Varunidae*					
Chinese mitten crab (*Eriocheir sinensis* Milne Edwards)	*Aphanomyces astaci*		Synovial membrane in limbs	Crayfish plague under experimental conditions	158
ORDER *Mysidae*					
Plankton crustacean (*Mysis relicta*)	*Aphanomyces astaci*			Resistant in experimental infection	1680
ORDER *Sessilia*					
Barnacle (*Chthamalus fragilis* var. *denticulate*	*Lagenidium chthamalophilum*		Ova	Mycosis	846
PHYLUM *CHORDATA*					
CLASS *ACTINOPTERYGII* Ray-finned fish (a class or subclass of the bony fishes)					

ORDER *Acipenseriformes* (sturgeons and paddlefishes)					
Family *Acipenseridae* (sturgeons)					
Persian sturgeon (*Acipenser persicus* Borodin)	*Saprolegnia parasitica* and *S. diclina*		Eggs	Saprolegniosis	611
ORDER *Anguilliformes* (eels and morays)					
Family *Anguillidae*					
European eel (*Anguilla anguilla*)	*Saprolegnia* sp. (most likely *S. parasitica*)		Loss of epithelium and ulceration	Mortality	366
Japanese eel (*A. japonica*)	*Branchiomyces sanguinis*		Gill	Gill rot	318 475
ORDER *Characiformes* (Characins)					
Family *Characidae*					
Tambaqui or cachama (*Colossoma macropomum*)	*Branchiomyces* sp.		Gill lesions	Gill rot, respiratory disease	1328
ORDER *Clupeiformes* (herrings)					
Atlantic menhaden (*Brevoortia tyrannus*)	*Aphanomyces invadans*		Dermal ulcers	Epizootic ulcerative syndrome (EUS)	188 189 467 1561
Bony fish or Australian river gizzard shad (*Nematalosa erebi*)	1. *Aphanomyces invadans* 2. *Saprolegnia ferax*		1. Dermal ulcers 2. Lesions	1. Epizootic ulcerative syndrome 2. Saprolegniosis	214 445
ORDER *Cypriniformes* (carps)					
Family *Cyprinidae* (Minnows or carps)					
Bunnei (*Mesopotamichthys [Barbus] sharpeyi*)	*Saprolegnia parasitica*		Gills, fins, skin	Saprolegniosis	27
Common carp (*Cyprinus carpio*)	*Branchiomyces sanguinis*		Gills	Gill rot	1039

Continued

Table 5.1. Continued.

Host	Pathogens	Normal flora	Tissue site	Disease status	Ref
Goldfish (*Carassius auratus*)	*Aphanomyces invadans*		Skin	Epizootic ulcerative syndrome (EUS)	1333
Roho labeo (*Labeo rohita*)	*Aphanomyces invadans*		Skin	EUS	1413
Silver barb (*Barbonymus gonionotus*)	*Aphanomyces invadans*		Skin	EUS	1413
Tench (*Tinca tinca*)	*Branchiomyces demigrans*		Gills	Gill rot	1039
Zebra danio, also called zebrafish (*Danio* [*Brachydanio*] *rerio*)	*Saprolegnia brachydanis*		Tissues	Fungal disease	875
ORDER *Cyprinodontiformes* (rivulines, killifishes and live bearers)					
Family *Aplocheilidae* (killifishes)					
Blue panchax (*Aplocheilus panchax*)	*Aphanomyces laevis*		Body and gills	Mycosis	1165
ORDER *Esociformes* (pikes and mudminnows)					
Family *Esocidae* (pikes)					
Northern pike (*Esox lucius*)	*Branchiomyces demigrans*		Gills	Gill rot	1039
ORDER *Gadiformes*					
Family *Gadidae* (cods and haddocks)					
Atlantic cod (*Gadus morhua* Linnaeus)	*Exophiala angulospora*		Kidney	Systemic mycosis	622
ORDER *Mugiliformes* (mullets)					
Family *Mugilidae* (mullets)					
Flathead grey mullet, or sea mullet (*Mugil cephalus*)	*Aphanomyces invadans*		Ulcerated skin lesions, may be red-coloured	Epizootic ulcerative syndrome (EUS)	262 564 1561
ORDER *Perciformes* (Perch-like)					

Family *Ammodytidae* (Sand lances)					
American sand lance (*Ammodytes americanus*)	*Exophiala aquamarina*			Mycosis	428
Family *Carangidae* (jacks and pompanos)					
White Trevally (*Pseudocaranx dentex* Bloch and Schneider). Also known as Striped Jack, silver bream, silver trevally, skippy etc.	*Exophiala xenobiotica*		Kidney	Mycosis	1178
Family *Centrarchidae*					
Smallmouth bass (*Micropterus dolomieu*)	*Saprolegnia ferax*			Resistant to infection	392
Largemouth bass (*Micropterus salmoides*)	*Aphanomyces invadans*		Ulcerative skin lesions	Epizootic ulcerative syndrome or ulcerated mycosis	1561
Family *Channidae* (Snakeheads)					
Striped snakehead (*Channa striata*)	*Aphanomyces invadans*		Ulcerative skin lesions	Epizootic ulcerative syndrome	263
Family *Cichlidae*					
Blue tilapia (*Oreochromis aureus* Steindachner), senior synonym to *Sarotherodon aureus*)	*Paecilomyces lilacinus*		Spleen, kidney	Wasting disease	1375
Hybrid tilapia, red tilapia (*O. niloticus* × *O. mossambicus*) and green tilapia (*O. niloticus* × *O. aureus*)	*Branchiomyces* sp.		Microscopic lesions in gills	Mortality	1307
Mozambique tilapia (*O. mossambicus*)	*Paecilomyces lilacinus*		Spleen, kidney	Wasting disease	1375
Mozambique tilapia hybrid (*O. mossambicus* × *O. hornorum*)	*Paecilomyces marquandii*		Kidney	Renal mycosis	1024

Continued

Table 5.1. Continued.

Host	Pathogens	Normal flora	Tissue site	Disease status	Ref
Nile tilapia (*O. niloticus*)	1. *Aspergillus flavus* 2. *Achlya klebsiana*, *Aphanomyces laevis*, *Phoma herbarum*, *Purpureocillium lilacinum*		1. Abdomen, eye, liver, kidney 2. Grey to black lesions on body	1. Aspergillomycosis 2. Fungal infection	31 1267 1269
Banded tilapia (*Tilapia sparramanii*)	*Aphanomyces invadans*		Ulcerative skin lesions	Epizootic ulcerative syndrome	768
Nembwe (*Serranochromis robustus*)	*Aphanomyces invadans*		Ulcerative skin lesions	Epizootic ulcerative syndrome	768
Family *Lutjanidae* (snappers)					
Northern red snapper (*Lutjanus campechanus*)	1. *Cladosporium sphaerospermum* 2. *Penicillium corylophilum*		1. Posterior kidney 2. Swim bladder	1, 2. Mycoses	187
Family *Gobiidae* (gobies)					
Tank toby (*Glossogobius giuris*)	*Aphanomyces invadans*		Ulcerative skin lesions	Epizootic ulcerative syndrome	262
Family *Osphronemidae* (gouramies)					
Three spot gourami (*Trichopodus trichopterus*)	*Aphanomyces invadans*		Ulcerative skin lesions	Epizootic ulcerative syndrome	262
Family *Percichthyidae*					
Golden perch (*Macquaria ambigua*)	*Aphanomyces invadans*		Ulcerative skin lesions	Epizootic ulcerative syndrome	214
Murray cod (*Maccullochella peelii*)	*Aphanomyces invadans*		Ulcerative skin lesions	Epizootic ulcerative syndrome	214
Family *Scombridae* (mackerels, tunas, bonitos)					
Little tunny fish (*Euthynnus alletteratus*)	*Exophiala aquamarina*			Mycosis	428
King George whiting (*Sillaginodes punctatus*)	*Exophiala salmonis*		Skin ulceration	Mortality	1397
Family *Sparidae* (porgies)					

Southern black bream (*Acanthopagrus butcheri*), yellowfin bream or surf bream (*A. australis*)	*Aphanomyces invadans*		Ulcerated skin lesions	Epizootic ulcerative syndrome (EUS)	244 564
Family *Sillaginidae* (smelt-whiting)					
Sand sillago or sand whiting (*Sillago ciliata*)	*Aphanomyces invadans*		Ulcerated skin lesions	Epizootic ulcerative syndrome (EUS)	564
Family *Terapontidae* (Grunters or tigerperches)					
Silver perch (*Bidyanus bidyanus*)	1. *Aphanomyces invadans* 2. *Saprolegnia parasitica*		1. Ulcerated skin lesions 2. Skin lesions	1. Epizootic ulcerative syndrome (EUS) 2. Winter saprolegniosis	1431 1561
Spangled perch (*Leiopotherapon unicolor*)	*Aphanomyces invadans*		Ulcerated skin lesions	Epizootic ulcerative syndrome (EUS)	214
ORDER *Salmoniformes*					
Family *Salmonidae*					
Atlantic salmon (*Salmo salar* L.)	1. *Exophiala pisciphila* 2. *E. psychrophila* 3. *E. salmonis* 4. *Isaria farinosa* (previously *Paecilomyces farinosus*) 5. *Phialophora* sp. 6. *Saprolegnia diclina*		1. Fistulas on caudal-lateral surface of head 2. Kidney granuloma 3. Granulomas kidney, liver, spleen 4. Swim bladder 5. Fungal masses in swimbladder and attaching to visceral organs 6. Gills	1, 2, 3, 4, 5, 6. Mycoses, mortality	234 239 487 976 1003 1284 1327 1403
Brown trout (*Salmo trutta trutta*)	1. *Saprolegnia parasitica* 2. *Saprolegnia diclina*, *S. australis*, *S. ferax*, *S. furcata*, *S. hypogyna*, *S. unispora*, *S. parasitica* 3. *S. diclina*, *S. ferax*, *S. australis*, *S. parasitica* 4. *S. ferax*		1. Cutaneous lesions on broodstock 2. Eggs 3. Alevins 4. Skin and fins	1. Pathogenic 2, 3. Saprophytic (colonization) 4. Mortality	392 565

Continued

Table 5.1. Continued.

Host	Pathogens	Normal flora	Tissue site	Disease status	Ref
Brook trout (*Salvelinus fontinalis*)	*Saprolegnia ferax*		White areas on skin, and ragged fins	Mortality	392
Japanese char (*Salvelinus japonicus*). *S. leucomaensis* is a synonym	*Aphanomyces frigidophilus*		Eggs	Mortality	915
Lake trout (*Salvelinus namaycush*)	*Saprolegnia ferax*		White areas on skin, and ragged fins	Mortality	392
Chinook salmon (*Oncorhynchus tshawytscha*)	*Phoma herbarum*		Swim bladder, kidney	Systemic mycosis	514
Chum salmon (*O. keta*)	*Saprolegnia australis*, *S. parasitica*, *S. salmonis*		Infected eggs and lesions on head, peduncle, caudal fin	Mortality	781
Coho salmon (*O. kisutch*)	*Saprolegnia parasitica*		Lesions on head, adipose fin and caudal fin	Mortality	691
Cutthroat trout (*O. clarkii*)	*Exophiala salmonis*		Brain	Mycetoma	428
Masou salmon (*O. masou*)	1. *Ochroconis* sp. 2. *Saprolegnia parasitica*, *S. salmonis*		1. Kidney 2. Infected eggs and lesions on head, peduncle, caudal fin	1. Mortality	692 781
Rainbow trout (*O. mykiss*)	1. *Saprolegnia australis* 2. *S. parasitica*		1. Affected the dorsal fin 2. Ventral lesions	1. Saprolegniosis 2. Mortality	697 760
Sockeye salmon (*O. nerka*)	*Saprolegnia parasitica*, *S. salmonis*		Lesions on head, peduncle, caudal fin	Mortality	780 781
ORDER *Pleuronectiformes*					
Family *Pleuronectidae* (right eye flounders)					
American plaice (*Hippoglossoides platessoides* Fabricius)	*Hormoconis resinae*		Skin lesions	Mycosis	1587
Winter flounder (*Pseudopleuronectes americanus*)	*Exophiala aquamarina*			Mycosis	428

ORDER *Scorpaeniformes* (scorpion fish and flatheads)					
Family *Cyclopteridae* (lumpfishes)					
Lumpfish (*Cyclopterus lumpus*)	*Exophiala aquamarina*			Mycosis	428
Family *Synanceiidae* (stonefishes)					
Devil stinger (*Inimicus japonicus*)	*Ochroconis humicola*		Skin ulcers	Fungal infection	1739
ORDER *Siluriformes* (Catfish)					
Family *Ariidae* (sea catfishes)					
Fork-tailed or Highfin catfish (*Neoarius berneyi*)	*Edwardsiella tarda*			Mortality	718
Family *Bagridae* (bagrid catfishes)					
Yellow catfish (*Tachysurus* [*Pelteobagrus*] *fulvidraco* Richardson)	1. *Saprolegnia ferax* 2. *Achlya klebsiana*		Eggs	Saprolegniosis	272 273
Family *Clariidae*					
Philippine catfish (*Clarias batrachus*)	*Aphanomyces invadans*		Ulcerative skin lesions	Epizootic ulcerative syndrome (EUS)	262
African catfish (*Clarias gariepinus*)	1. *Achlya klebsiana*, *Aphanomyces laevis*, *Phoma herbarum*, *Purpureocillium lilacinum* 2. *Aphanomyces invadans*		1. Grey to black lesions on body 2. Ulcerative skin lesions	1. Fungal infection 2. Epizootic ulcerative syndrome (EUS)	31 768
Family *Ictaluridae* (North American freshwater catfish)					
Channel catfish (*Ictalurus punctatus* Rafinesque)	1. *Branchiomyces sanguinis* 2. *Exophiala pisciphila* 3. *Saprolegnia parasitica*		1. Gills 2, 3. Skin lesions	1. Gill rot 2. Systemic mycosis 3. Winter saprolegniosis	191 889 1122
Family *Loricariidae* (armoured catfishes)					

Continued

Table 5.1. Continued.

Host	Pathogens	Normal flora	Tissue site	Disease status	Ref
Amazonian loricariid (*Baryancistrus* sp.)	*Branchiomyces*-like		Microscopic lesions on gills	Mortality	1307
Family *Mochokidae* (squeakers)					
Family *Plotosidae* (eeltail catfish)					
Narrowfront tandan (*Neosilurus ater*) and Toothless catfish (*Anodontiglanis dahli*)	*Edwardsiella ictaluri*			Mortality	718
ORDER *Synbranchiformes* (spiny eels)					
Zig-zag or tyre track eel (*Mastacembelus armatus*)	*Achlya debaryana*			Mycosis	891
Squeakers (*Synodontis* sp.)	*Aphanomyces invadans*		Ulcerative skin lesions	Epizootic ulcerative syndrome (EUS)	768
ORDER *Syngnathiformes* (seahorses and pipefish)					
Family *Syngnathidae* (seahorses and pipefish)					
Potbelly seahorse (*Hippocampus bleekeri*)	*Exophiala pisciphila*				428
Seadragon (Leafy seadragon) (*Phycodurus eques*), weedy seadragon (*Phyllopteryx taeniolatus*)	*Exophiala angulospora*, *E. aquamarina*		Skin lesions, gill, skeletal muscle, kidney	Disseminated phaeohyphomycosis	428 1240
ORDER *Osmeriforms*					

Family *Plecoglossidae* Ayu sweetfish (*Plecoglossus altivelis altivelis* Temminck and Schlegel)	1. *Phoma* sp. 2. *Pythium flevoense* 3. *Saprolegnia parasitica*		1. Swim bladder, kidney, liver 2. Swim bladder, kidney, intestine, pancreas, spleen, abdominal cavity, musculature, spinal cord 3. Stomach, spleen, kidney, liver, swim bladder	1, 2. Visceral mycosis 3. Mycotic gastritis	693 1738
Family *Salangidae* (icefishes or noodlefishes) Japanese ice fish (*Salangichthys microdon*)	*Aphanomyces salsuginosus*		Fungi on head, fins, dermal layers and muscle layers	Mycosis	1604
CLASS *AMPHIBIA* ORDER *Anura*					
Family *Bufonidae*					
Green toad (*Pseudepidalea viridis*)	*Exophiala cancerae*		Liver	Mycosis	428
Natterjack toad (*Bufo calamita*)	*Saprolegnia diclina*		Embryos	Mortality	527
Chiriqui Harlequin frog (*Atelopus chiriquiensis*), Costa Rican Variable Harlequin toad or clown frog (*A. varius*)	*Batrachochytrium dendrobatidis*			Mortality, cutaneous chytridiomycosis	164 1822
Cane toad (*B. marinus*), *Bufo haematiticus*, Asian toad (*Bufo gargarizans*)	1. *Batrachochytrium dendrobatidis* 2. *Mucor amphibiorum*		2. Lesions in liver and bladder	1. Mortality, cutaneous chytridiomycosis 2. Mucormycosis	164 1566
Bleeding toad or fire toad (*Leptophyrne cruentata*)	*Batrachochytrium dendrobatidis*			Mortality, cutaneous chytridiomycosis	953
Beaked or Rio Viejo toads (*Rhinella quechua*)	*Batrachochytrium dendrobatidis*		Keratinized mouthparts of tadpole	Mortality, cutaneous chytridiomycosis	120
Family *Centrolenidae*					
Yellow-flecked glass frog (*Cochranella albomaculata*), *C. prosoblepon*	*Batrachochytrium dendrobatidis*			Mortality, cutaneous chytridiomycosis	164

Continued

Table 5.1. Continued.

Host	Pathogens	Normal flora	Tissue site	Disease status	Ref
Family *Ceratophryniae*					
Lesser chini frog or Chacoan burrowing frog (*Chacophrys pierotti*)	*Batrachochytrium dendrobatidis*		Mucus and sloughed skin on ventral abdomen, drink patch and digits	Mortality, cutaneous chytridiomycosis	1675
Family *Dendrobates*					
Poison dart frogs (*Dendrobates* sp.), *D. auratus*, *D. azureus*, *D. pumilio*	*Batrachochytrium dendrobatidis*			Mortality, cutaneous chytridiomycosis	1184 1331
Family *Discoglossidae*					
Iberian painted frog (*Discoglossus galganoi*)	*Batrachochytrium dendrobatidis*			Apparently healthy	725
Family *Hylidae*					
Duellmanohyla soralia	*Batrachochytrium dendrobatidis*			Mortality, cutaneous chytridiomycosis	924
Japanese tree frog (*Hyla japonica*)	*Batrachochytrium dendrobatidis*			Mortality, cutaneous chytridiomycosis	1822
Mediterranean tree frog (*Hyla meridionalis*)	*Batrachochytrium dendrobatidis*			Apparently healthy	725
Montevideo tree frog (*Hypsiboas pulchellus*), *Scinax squalirostris*	*Batrachochytrium dendrobatidis*		Mouthparts of tadpoles	Mortality, cutaneous chytridiomycosis	202
Slender tree frog (*Litoria adelaidensis*), white-lipped tree frog (*L. infrafrenata*), green tree frog (*L. caerulea*)	*Mucor amphibiorum*		Granulomas in skin, intestine, lung, nasal passages, brain	Mortality, mucormycosis	163 379
Spikethumb frog (*Plectrohyla dasypus*)	*Batrachochytrium dendrobatidis*			Mortality, cutaneous chytridiomycosis	924
Bullfrog – spring peeper (*Pseudacris crucifer*)	*Leptolegnia* sp.		Eggs	Mortality	1444
Pacific tree frog (*Pseudacris regilla*)	*Saprolegnia ferax*		Larvae	Mortality in experimental challenge	1426
Copan stream frog (*Ptychohyla hypomykter*)	*Batrachochytrium dendrobatidis*			Mortality, cutaneous chytridiomycosis	924

Family *Leiuperidae*					
Chilean four-eyed frog (*Pleurodema thaul*)	*Batrachochytrium dendrobatidis*			Mortality, cutaneous chytridiomycosis	204
Family *Leptodactylidae*					
Grey wood frog (*Batrachyla leptopus*), Cranwell's horned frog (*Ceratophrys cranwelli*), Surinam horned frog (*C. cornuta*), Argentine horned frog (*C. ornata*), Venezuelan horned frog (*C. calcarata*), *Craugastor rostralis*, Budgett's frog (*Lepidobatrachus laevis*), smokey jungle frog (*Leptodactylus pentadactylus*)	*Batrachochytrium dendrobatidis*		Mucus and sloughed skin on ventral abdomen, drink patch and digits	Mortality, cutaneous chytridiomycosis	204 924 1675
Eleutherodactylus cruentus, *E. emcelae*	*Batrachochytrium dendrobatidis*			Mortality, cutaneous chytridiomycosis	164
Odontophrynus maisuma, Hensel's dwarf frog (*Physalaemus henselii*)	*Batrachochytrium dendrobatidis*		Mouthparts of tadpoles	Mortality, cutaneous chytridiomycosis	202
Telmatobius atacamensis and *T. pisanoi*	*Batrachochytrium dendrobatidis*		Sloughing of the *stratum corneum*	Mortality, cutaneous chytridiomycosis	121
Family *Megophyridae*					
Hasselt's toad (*Leptobrachium hasseltii*)	*Batrachochytrium dendrobatidis*			Mortality, cutaneous chytridiomycosis	953
Family *Microhylidae*					
Plethodontohyla tuberata	*Batrachochytrium dendrobatidis*		Mucus and sloughed skin on ventral abdomen, drink patch and digits	Mortality, cutaneous chytridiomycosis	1675

Continued

Table 5.1. Continued.

Host	Pathogens	Normal flora	Tissue site	Disease status	Ref
Family *Myobatrachidae*					
Pobblebonk or banjo frog (*Limnodynastes dumerilii*), spotted marsh frog (*L. tasmaniensis*), green tree frog (*L. caerulea*), Lesueur's tree frog (*L. lesueuri*), Torrent tree frog (*L. nannotis*), creek frog (*L. rheocola*), spotted tree frog (*L. spenceri*), great barred frog (*Mixophyes fasciolatus*), sharp-snouted day frog (*Taudactylus acutirostris*), Eungella Torrent frog (*T. eungellensis*)	*Batrachochytrium dendrobatidis*			Mortality, cutaneous chytridiomycosis	164
Family *Pelobatidae*					
Western spadefoot (*Pelobates cultripes*)	*Batrachochytrium dendrobatidis*			Apparently healthy	725
Family *Ranidae*					
Yunnan pond frog (*Babina pleuraden* synonym *R. pleuraden*)	*Batrachochytrium dendrobatidis*		Toe-clip	PCR-positive	95
Southern bell or growling grass frog (*Litoria raniformis*), green tree frog (*L. caerulea*)	*Batrachochytrium dendrobatidis*			Mortality, cutaneous chytridiomycosis	1733 1751
Perez's tree frog (*Pelophylax perezi*)	*Batrachochytrium dendrobatidis*			Apparently healthy	725
Edible frog (*Pelophylax* kl. *esculentus*), synonym *R. esculenta*	*Batrachochytrium dendrobatidis*		Fungus in abdominal, pelvic and femoral skin areas	No mortality	1537

American bullfrog (*Rana catesbeiana*)	1. *Batrachochytrium dendrobatidis* 2. *Saprolegnia* sp.		1. Toe clips, femoral skin of juveniles 2. Eggs	1. PCR-positive 2. Mortality	391 1444 1822
Hydrophylax chalconotus, previously *Rana chalconota*	*Batrachochytrium dendrobatidis*			Mortality, cutaneous chytridiomycosis	953
Chaochiao frog (*R. chaochiaoensis*)	*Batrachochytrium dendrobatidis*		Toe-clip	PCR-positive	95
Limnonectes microdiscus	*Batrachochytrium dendrobatidis*			Mortality, cutaneous chytridiomycosis	953
Green puddle frog (*Occidozyga lima*)	*Batrachochytrium dendrobatidis*		Mucus and sloughed skin on ventral abdomen, drink patch and digits	Mortality, cutaneous chytridiomycosis	1675
Odorrana andersonii	*Batrachochytrium dendrobatidis*		Toe-clip	PCR-positive	95
Rana maculata (synonym *Lithobates maculata*)	*Batrachochytrium dendrobatidis*			Mortality, cutaneous chytridiomycosis	924
Cascades frog (*Rana cascadae*)	*Saprolegnia ferax*		Larvae	Unaffected in experimental challenge	1426
Red-legged frog (*R. aurora*)	*Saprolegnia ferax*		Larvae	Mortality in experimental challenge	1426
Family *Rhacophoridae*					
Java tree frog (*Rhacophorus javanus*)	*Batrachochytrium dendrobatidis*			Mortality, cutaneous chytridiomycosis	953
Family *Rhinodermatidae*					
Darwin's frog (*Rhinoderma darwinii*)	*Batrachochytrium dendrobatidis*			Mortality, cutaneous chytridiomycosis	204
ORDER *Caudata*					
Family *Ambystomatidae*					
Long-toed salamander (*Ambystoma macrodactylum*)	*Saprolegnia ferax*		Larvae	Unaffected in experimental challenge	1426

Continued

Table 5.1. Continued.

Host	Pathogens	Normal flora	Tissue site	Disease status	Ref
Family *Salamandridae*					
Bosca's newt (*Lissotriton boscai*)	*Batrachochytrium dendrobatidis*			Apparently healthy	725
Southern or pygmy marbled newt (*Triturus pygmaeus*)	*Batrachochytrium dendrobatidis*			Apparently healthy	725
ORDER *Urodela*					
Family *Salamandridae*					
Spanish ribbed newt (*Pleurodeles waltl*)	*Batrachochytrium dendrobatidis*			Apparently healthy	725
CLASS *AVES* (birds, penguins) Family *Spheniscidae*					
Magellanic penguin (*Spheniscus magellanicus*), Humboldt penguin (*S. humboldti*), African or Jackass penguin (*S. demersus*)	*Aspergillus fumigatus*		Kidney, air sac, liver, spleen	Invasive aspergillosis	46
CLASS *CHONDRICHTHYES* ORDER *Caracharhiniformes*					
Bonnethead shark (*Sphyrna tiburo*), scalloped hammerhead shark (*Sphyrna lewini*)	*Fusarium solani*		Ulcerative skin lesions	Hawaii, USA	385 1174
Smooth dogfish (*Mustelus canis* Mitchill) and spiny dogfish (*Squalus acanthias* L.) (mud shark)	*Exophiala pisciphila*		Brain and skin lesions	Mycosis	589
CLASS *MAMMALIA* ORDER *Cetacea* Family *Delphinidae*					
Indo-Pacific bottlenose dolphin (*Tursiops aduncus*)	*Lacazia loboi*		Multiple grey nodules on body	Lobomycosis	375 912

Common bottlenose dolphin (*T. truncatus*)	1. *Lacazia loboi* 2. *Rhodotorula* sp.		1. Multiple grey nodules on body 2. Superficial subcutis	1. Lobomycosis 2. Pathogenicity unknown, likely to be secondary invader in necrotizing fasciitis caused by *Streptococcus agalactiae*	261 407 912 1141 1847
Family *Ornithorhynchidae*					
Platypus (*Ornithorhynchus anatinus*)	*Mucor amphibiorum*		Skin ulcerations on bill, tail and body	Mucormycosis	1179 1246
CLASS *REPTILIA*					
Family *Cheloniidae* turtle					
Loggerhead sea turtle (*Caretta caretta*)	*Fusarium solani*		Eggs, embryos	Mortality	1477
CLASS *SAUROPSIDA*					
ORDER *Testudines*					
Chinese soft-shelled turtle (*Pelodiscus* [*Trionyx*] *sinensis*)	*Aphanomyces sinensis*		Small whitish maculae on carapace and hyphae in epidermis but not dermis	Mortality. Pathogenicity confirmed by experimental challenge	1605
PHYLUM *CNIDARIA* CLASS *ANTHOZOA*					
Venus sea fan (*Gorgonia flabellum*) and purple sea fan (*G. ventalina*)	*Aspergillus sydowii*		Tissues	Mortality	32
PHYLUM *MOLLUSCA* CLASS *BIVALVIA*					
Family *Cardiidae* *Acanthocardia* [*Cardium*] *echinata*	*Leptolegnia marina*		Mantle, foot	Mortality	66
Family *Pholadidae* *Barnea candida*	*Leptolegnia marina*		Mantle, foot	Mortality	66
CLASS *GASTROPODA*					
Abalone (*Haliotis* [*sieboldii*] *gigantea*)	1. *Halodaphnea* (*Atkinsiella*) *awabi* 2. *Atkinsiella dubia* 3. *Haliphthoros milfordensis*		1. Mantle and peduncle 2. Mantle 3. Mantle, epipode or foot	1, 2, 3. Lesions	694 916 1197
PHYLUM *ROTIFERA*					
Rotifer (*Brachionus plicatilis*)	*Halodaphnea* (*Atkinsiella*) *parasitica*		Eggs and body	Fungal disease	1196

Table 5.2. Fungal pathogens and saprophytes of fish and other aquatic animals.

Organism	Disease	Disease signs	Aquatic species	Distribution	Ref
Achlya debaryana	Mycosis		Zig-zag or tyre-track eel (*Mastacembelus armatus*)	India	891
Achlya klebsiana	1, 2. Fungal infection	2. Infected eggs	1. Nile tilapia (*Oreochromis niloticus niloticus*) 2. Yellow catfish (*Tachysurus* [*Pelteobagrus*] *fulvidraco*)	China, Egypt	31 272
Aphanomyces astaci	1. Crayfish plague 2. Carrier status and susceptible 3. Crayfish plague under experimental conditions 4. Resistant in experimental challenge	1. Mortality, slow response 3. Death at time of moulting, incomplete moulting, fungal infestation of synovial membranes	1. Atlantic stream crayfish or white-clawed freshwater crayfish (*Austropotamobius pallipes*), European or noble crayfish (*Astacus astacus*), narrow-clawed or Turkish crayfish (*A. leptodactylus*) 2. North American freshwater crayfish Louisiana red swamp crayfish (*Procambarus clarkii*), signal crayfish (*Pacifastacus leniusculus*), mud crayfish or spinycheek crayfish (*Orconectes limosus*) 3. Chinese mitten crab (*Eriocheir sinensis*), Gippsland crayfish (*Euastacus kershawi*), *E. clydensis*, *E. crassus*, Japanese crayfish or Zarigani (*Cambaroides japonicus*), yabby (*Cherax destructor*), *C. quinquecarinatus*, *C. papuanus*, *Geocharax gracilis*, Tasmanian giant freshwater crayfish (*Astacopsis gouldi*), *A. fluviatilis* 4. Plankton crustacean (*Mysis relicta*)	Czech Republic, Finland, France, Germany, Greece, Hungary, Ireland, Italy, Norway, Spain, Sweden, Turkey, UK, USA	118 158 264 390 443 447 936 937 1369 1432 1677 1680 1681
Aphanomyces frigidophilus	1. Mycosis 2. Associated with mass mortality event but non-pathogenic in experimental challenge	1. Fungus on eggs 2. Subabdominal cuticle	1. Japanese char (*Salvelinus japonicus*) 2. Atlantic stream crayfish or white-clawed freshwater crayfish (*Austropotamobius pallipes*)	Japan, Spain	110 111 915
Aphanomyces invadans	Epizootic ulcerative syndrome (EUS)	Dermal ulcers	American shad, Atlantic menhaden (*Brevoortia tyrannus*), *Barbus* sp., goldfish (*Carassius auratus*), largemouth bass, bony herring (*Nematalosa erebi*), catfish species (sharptooth catfish – *Clarias gariepinus*, Philippine catfish – *C. batrachus*), flathead grey mullet (*Mugil cephalus*), golden perch (*Macquaria ambigua*), *Labeo rohita*, Murray cod (*Maccullochella peelii*), silver mullet, striped mullet, Nembwe	Australia, Bangladesh, Botswana, Burma, Japan, Namibia, New Guinea, Philippines, Thailand, USA	188 189 214 262 263 467 564 768 1125

			(*Serranochromis robustus*), pike species, *Sargochromis* sp., *Serranochromis* sp., sand whiting (*Sillago ciliata*), silver perch (*Bidyanus bidyanus*), snakehead (*Channa striatus*), southern black bream (*Acanthopagrus butcheri*), yellowfin or surf bream (*A. australis*), tank goby (*Glossogobius giuris*), three-spot gourami (*Trichogaster trichopterus*), *Tilapia* sp., squeakers (*Synodontis* sp.)		1333 1413 1561
Aphanomyces laevis	Fungal infection	White cottony growth on body	Nile tilapia (*Oreochromis niloticus niloticus*), African catfish (*Clarias gariepinus*), blue panchax (*Aplocheilus panchax*)	Egypt, India	31 1165
Aphanomyces repetans	Non-pathogenic in experimental challenge	Isolated from cuticle	Louisiana red swamp crayfish (*Procambarus clarkii*), signal crayfish (*Pacifastacus leniusculus*)	Spain	264 1432
Aphanomyces salsuginosus	Fungal infection	White cottony growth on head. Fungi in fins, muscle and dermal layers	Japanese ice fish (*Salangichthys microdon*)	Japan	1604
Aphanomyces sinensis	Fungal infection	Whitish maculae on carapace and hyphae in epidermis but not dermis	Juvenile soft-shelled turtle (*Pedodiscus sinensis*)	Japan	1605
Aspergillus flavus	Aspergillomycosis	Abdominal distension and darkening, lethargy	Tilapia (*Oreochromis niloticus*)	Kenya	1267 1269
Aspergillus fumigatus	Aspergillosis		Magellanic penguin (*Spheniscus magellanicus*), Humboldt penguin (*S. humboldti*), African or jackass penguin (*S. demersus*)	Captive Spain	46
Aspergillus sydowii	Aspergillosis	Purpling of tissue, lesions, galling	Sea fan corals (*Gorgonia ventalina*, *G. flabellum*)	Caribbean	32
Atkinsiella awabi	See *Halodaphnea awabi*				916
Atkinsiella dubia	Fungal disease	1. Isolated from mantle 2. Isolated from gills	1. Abalone (*Haliotis gigantea*) 2. Swimming crab (*Portunus trituberculatus*)	Japan	1197
Atkinsiella okinawaensis	See *Halodaphnea okinawaensis*				1198
Atkinsiella panulirata	Fungal disease	Phyllosoma	Japanese spiny lobster (*Panulirus japonicus*)	Japan	914
Atkinsiella parasitica	See *Halodaphnea parasitica*				1196

Continued

Table 5.2. Continued.

Organism	Disease	Disease signs	Aquatic species	Distribution	Ref
Batrachochytrium dendrobatidis	Cutaneous chytridiomycosis	Areas of abnormal epidermal sloughing in frogs and abnormality of the keratinized mouthparts in tadpoles	Chiriqui Harlequin frog (*Atelopus chiriquiensis*), Costa Rican variable harlequin toad or clown frog (*A. varius*), Yunnan pond frog (*Babina pleuraden*), grey wood frog (*Batrachyla leptopus*), Asian toad (*Bufo gargarizans*), *Bufo haematiticus*, cane toad (*Bufo marinus*), yellow-flecked glass frog (*Cochranella albomaculata*), *C. prosoblepon*, *Dendrobates* sp., *D. auratus*, *D. azureus*, *D. pumilio*, Iberian painted frog (*Discoglossus galganoi*), *Eleutherodactylus cruentus*, *E. emcelae*, *Craugastor rostralis*, *Duellmanohyla soralia*, Java tree frog (*Hydrophylax chalconotus*, previously *Rana chalconota*), Mediterranean tree frog (*Hyla meridionalis*), Montevideo tree frog (*Hypsiboas pulchellus*), Hasselt's toad (*Leptobrachium hasseltii*), bleeding toad or fire toad (*Leptophyrne cruentata*), the pobblebonk or banjo frog (*Limnodynastes dumerilii*), spotted marsh frog (*L. tasmaniensis*), green tree frog (*L. caerulea*), Lesueur's tree frog (*L. lesueuri*), southern bell or growling grass frog (*Litoria raniformis*), Torrent tree frog (*L. nannotis*), creek frog (*L. rheocola*), spotted tree frog (*L. spenceri*), green tree frog (*L. caerulea*), great barred frog (*Mixophyes fasciolatus*), western spadefoot (*Pelobates cultripes*), edible frog (*Pelophylax* kl. *esculentus*), Perez's tree frog (*P. perezi*), Chilean four-eyed frog (*Pleurodema thaul*), Copan stream frog (*Ptychohyla hypomykter*), *Rhacophorus javanus*, sharp-snouted day frog (*Taudactylus acutirostris*), Eungella Torrent frog (*T. eungellensis*), *Telmatobius atacamensis*, Chaochiao frog (*R. chaochiaoensis*), *Rana maculata* and *T. pisanoi*, beaked or Rio Viejo toads (*Rhinella quechua*), *Odotophrynus maisuma*, spikethumb frog (*Plectrohyla dasypus*), Hensel's dwarf frog (*Physalaemus henselii*), *Scinax squalirostris*, Budgett's or wide-mouth frog (*Lepidobatrachus laevis*), Surinam horned frog (*Ceratophrys cornuta*), Cranwell's horned frog (*C. cranwelli*), Argentine horned frog (*C. ornata*), Venezuelan horned frog (*C. calcarata*), lesser chini frog or Chacoan burrowing frog (*Chacophrys pierotti*), green puddle	Argentina, Australia, Bolivia, Britain, Chile, China, Germany, Honduras, Indonesia, Italy, Japan, New Zealand, Panama, South Korea, Spain, Uruguay, USA	95 120 121 164 202 204 391 725 924 953 1184 1331 1675 1733 1751 1822

			frog (*Occidozyga lima*), smoky jungle frog (*Leptodactylus pentadactylus*), Japanese tree frog (*Hyla japonica*), *Odorrana andersonii*, American bullfrog (*Rana catesbeiana*), Darwin's frog (*Rhinoderma darwinii*), Bosca's newt (*Lissotriton boscai*), Spanish ribbed newt (*Pleurodeles waltl*), southern or pygmy marbled newt (*Triturus pygmaeus*		
Branchiomyces demigrans	Gill rot	Lesions in gills	Northern pike (*Esox lucius*), tench (*Tinca tinca*)	Germany	1039
Branchiomyces sanguinis	Gill rot	Lesions in gills	Carp (*Cyprinus carpio*), catfish (*Ictalurus punctatus*), Japanese eel (*Anguilla japonica*)	Germany, Japan, Taiwan, USA	318 475 889 1039
Branchiomyces-like	Gill rot	Microscopic lesions in gills	Amazonian loricariid (*Baryancistrus* sp.), hybrid tilapia – green tilapia (*Oreochromis niloticus* × *O. aureus*), red tilapia (*O. niloticus* × *O. mossambicus*), tambaqui or chacama (*Colossoma macropomum*)	Brazil, Israel	1307 1328
Cladosporium sphaerospermum	Infection in posterior kidney	Erratic swimming	Red snapper (*Lutjanus campechanus*)	USA	187
Exophiala angulospora	1. Systemic mycosis 2. Disseminated phaeohyphomycosis	1. Erratic swimming, darkened skin 2. Skin lesions and internal necrotic foci	1. Atlantic cod (*Gadus morhua*) 2. Weedy seadragon (*Phyllopteryx taeniolatus*) and leafy seadragon (*Phycodurus eques*)	Norway, USA	622 1240
Exophiala aquamarina	Mycosis	Skin, head, necrotic tissue, bone	Winter flounder (*Pseudopleuronectes americanus*), leafy seadragon (*Phycodurus eques*), weedy seadragon (*Phyllopteryx taeniolatus*), lumpfish (*Cyclopterus lumpus*), little tunny fish (*Euthynnus alletteratus*), sandlance (*Ammodytes americanus*)	Canada, USA	428 1240
Exophiala cancerae	1. Lethargic crab disease 2. Mycosis	1. Weak motor control, poor balance	1. Mangrove crab (*Ucides cordatus*) 2. Green toad (*Pseudepidalea viridis*), water, fruit drink, corn, human	Australia, Brazil Germany, Japan, Netherlands, USA	192 428 1340 1722
Exophiala pisciphila	1, 2, 3, 4. Systemic mycosis	1. Skin lesions 2. Erratic swimming, skin darkening and surface fistulas 4. Black lesions on body	1. Smooth dogfish (*Mustelus canis*), potbelly seahorse (*Hippocampus bleekeri*) 2. Atlantic salmon (*Salmo salar*) 3. Channel catfish (*Ictalurus punctatus*) 4. American plaice (*Hippoglossoides platessoides*)	Australia, Canada, USA	428 589 976 1122 1587

Continued

Table 5.2. Continued.

Organism	Disease	Disease signs	Aquatic species	Distribution	Ref
Exophiala psychrophila	Mycosis	Mortality, kidney granuloma	Atlantic salmon (*Salmo salar*) smolt	Ireland, Norway	428 1327
Exophiala salmonis	1. Mycosis 2. Mycetoma 3. Mortality	1. Abdominal swelling, granulomas internal organs 2. Brain 3. Skin ulcerations	1. Atlantic salmon (*Salmo salar*) 2. Cutthroat trout (*Oncorhynchus clarkii*) fingerlings 3. King George whiting (*Sillaginodes punctatus*)	Australia, Canada, Scotland, UK, USA	428 1284 1397 1403
Exophiala xenobiotica	Mycosis	Swollen abdomen, distended kidney	Striped jack or white trevally (*Pseudocaranx dentex*)	Japan	1178
Fonsecaea brasiliensis	Involved in lethargic crab disease	Weak motor control, poor balance	Mangrove crab (*Ucides cordatus*)	Brazil	1722
Fusarium oxysporon	Black gill disease	Black coloration of gills	Kuruma prawn (*Penaeus japonica*)	Japan	887
Fusarium solani	Infection, mortality	1. Unusual coloured spots on egg shells, grey mycelium covering eggs 2. Erosive lesions on exoskeleton 3. Black gills 4. Ulcerative skin lesions	1. Loggerhead turtle (*Caretta caretta*) 2. Lobster (*Homarus gammarus*) 3. Kuruma prawn (*Penaeus japonicus*) 4. Bonnethead shark (*Sphyrna tiburo*), scalloped hammerhead shark (*S. lewini*)	Cape Verde Africa, Hawaii, Japan, Wales, UK, USA	23 385 888 1174 1477
Fusarium tabacinum	Black gill disease	Black coloration of gills	White-clawed crayfish or Atlantic stream crayfish (*Austropotamobius pallipes*)	England, UK	25
Haliphthoros milfordensis	1. Fungal infection 2, 3. Mycosis	1. Zoea and eggs 2. Lesions on mantle, epipode and foot 3. Gill and thoracic segments	1. Mangrove crab (*Scylla serrata*), blue swimmer crab (*Portunus pelagicus*), black tiger prawn (*Penaeus monodon*), Atlantic oyster drill (*Urosalpinx cinerea*) 2. Abalone (*Haliotis gigantea*) 3. American lobster (*Homarus americanus*)	Indonesia, Japan, USA, Vietnam	330 541 694 700 1197 1728
Halocrusticida baliensis	Fungal infection	Zoea and eggs	Mangrove crab (*Scylla serrata*)	Indonesia	700
Halodaphnea (Atkinsiella) awabi	Fungal disease	Tubercule-like swelling on mantle and lesions in peduncle	Abalone (*Haliotis gigantea*)	Japan	916
Halodaphnea hamanaensis	Fungal disease	Eggs	Mangrove crab (*Scylla serrata*)	Japan	177

Halodaphnea okinawaensis. Synonyms are *Atkinsiella okinawaensis* and *Halocrusticida okinawaensis*	Fungal disease	Brown eggs and whitish zoeae	Blue swimmer crab (*Portunus pelagicus*)	Japan	1197
Halodaphnea (Atkinsiella) parasitica	Fungal disease	Diseased eggs and lesions on body	Rotifer (*Brachionus plicatilis*)	Japan	1196
Isaria farinosa (previously *Paecilomyces farinosus*)	Swim bladder infection and low mortality	Swollen swim bladder, reddening of vent area	Atlantic salmon (*Salmo salar*)	Germany, Scotland	234 1003
Lacazia loboi	Lobomycosis	Grey nodules on body	Indo-Pacific bottlenose dolphin (*Tursiops aduncus*), Atlantic bottlenose dolphin (*T. truncatus*)	Brazil, Madagascar, Mozambique, USA	261 375 407 912 1141
Lagenidium callinectes	Mycosis	1. Thorax, abdomen, eyestalks 2. Eggs	1. Tiger prawn (*Penaeus monodon*) 2. Blue crab (*Callinectes sapidius*), blue swimmer crab (*Portunus pelagicus*) 3. Mangrove crab (*Scylla serrata*)	India, Indonesia, Japan, USA	700 1197 1370 1419 1474
Lagenidium chthamalophilum	Mycosis	Mortality of eggs	Barnacle (*Chthamalus fragilis* var *denticulata*)	USA	846
Lagenidium myophilum	Mortality	Mortality and whitish or black lesion in muscle	Juvenile coonstripe shrimp (*Pandalus prensor*)	Japan	1200
Lagenidium thermophilum	Mycosis	Eggs, larvae or zoeae	Mangrove or mud crab (*Scylla serrata*), tiger prawn (*Penaeus monodon*)	Indonesia, Thailand	1181 1197
Leptolegnia marina	Mortality	Death, hyphae in eggs	Pea crab (*Pinnotheres pisum*), bivalves *Barnea candida* and *Acanthocardia echinata*	UK	66
Leptolegnia sp.	Mortality of eggs	Death, presence of white hyphae on eggs	Bullfrog (*Pseudacris crucifer*)	USA	1444
Metschnikowia bicuspidata var *bicuspidata*	1. Milky disease 2. Non-pathogenic	Yeast in muscle, heart, hepatopancreas	1. Swimming crab (*Portunus trituberculatus*) 2. Brine shrimp (*Artemia salina*)	China, USA	1567 1765
Metschnikowia kamienskii	See *Metschnikowia bicuspidata* var. *bicuspidata*				

Continued

Table 5.2. Continued.

Organism	Disease	Disease signs	Aquatic species	Distribution	Ref
Mucor amphibiorum	1. Ulcerative skin disease 2. Mucormycosis	1. Ulcers and thickened lesions on body, tail, bill and lung 2. Lesions in liver, bladder, lungs, brain, skin, intestine	1. Platypus (*Ornithorhynchus anatinus*) 2. Cane toad (*Bufo marinus*), slender tree frog (*Litoria adelaidensis*), white-lipped tree frog (*L. infrafrenata*), green tree frog (*L. caerulea*)	Australia	163 379 1179 1246 1566
Ochroconis sp.	Visceral mycosis	Abdominal and anal swelling	Masou salmon (*Oncorhynchus masou*)	Japan	692
Ochroconis humicola	Fungal infection	Lesions on body	Devil stinger (*Inimicus japonicus*)	Japan	1739
Paecilomyces lilacinus	See *Purpureocillium lilacinum*				
Paecilomyces marquandii	Renal mycosis	Granulomatous kidney	Hybrid tilapia (*Oreochromis mossambica* × *O. hornorum*)	USA	1024
Penicillium corylophilum	Swim bladder infection	Erratic swimming	Red snapper (*Lutjanus campechanus*)	USA	187
Phialophora sp.	Systemic mycosis	Mortality, haemorrhagic fins	Atlantic salmon (*Salmo salar*)	Scotland	487
Phoma herbarum	Systemic mycosis	Erratic swimming, swollen abdomen, protruded vent	Chinook salmon (*Oncorhynchus tshawytscha*), Nile tilapia (*Oreochromis niloticus niloticus*), African catfish (*Clarias gariepinus*)	Egypt, USA	31 514
Phoma sp.	Visceral mycosis	Fungal hyphae in swim bladder, kidney and liver	Ayu (*Plecoglossus altivelis*)	Japan	693
Purpureocillium (*Paecilomyces*) *lilacinum*	Wasting disease	Emaciation, sunken eyes, eroded fins	Blue tilapia (*Oreochromis* [*Tilapia*] *aureus*), Mozambique tilapia (*O. mossambica*), Nile tilapia (*Oreochromis niloticus niloticus*), African catfish (*Clarias gariepinus*)	Egypt, Puerto Rico	31 1375
Pythium flevoense	Visceral mycosis	Opaque abdomen	Ayu (*Plecoglossus altivelis*)	Japan	1153
Rhodotorula sp.	Necrotizing fasciitis caused by *Streptococcus agalactiae*	Pathogenicity unknown, likely to be secondary invader	Common bottlenose dolphin (*Tursiops truncatus*)	Italy	1847
Saprolegnia australis	1. Pathogenicity unknown 2. Saprolegniosis 3. Ulcerative disease	1. Co-colonization with other *Saprolegnia* sp. 2. Infected eggs, lesions on head, peduncle and caudal fin 3. Ulcerated lesions on cuticle	1. Brown trout (*Salmo trutta trutta*) eggs, and alevins (yolk-sac fry) 2. Chum salmon (*Oncorhynchus keta*), rainbow trout (*O. mykiss*) fingerlings 3. Freshwater crayfish (*Orconectes propinquus*)	Japan, Spain, USA	565 697 781 943

Saprolegnia brachydanis	Fungal infection	Infected tissues	Zebra danio (*Danio* [*Brachydanio*] *rerio*)	China	875
Saprolegnia diclina	1. Pathogenicity unknown 2, 3. Mortality 4. Saprolegniosis	1. Co-colonization with other *Saprolegnia* sp. 2. Gill pathology 4. Infected eggs	1. Brown trout (*Salmo trutta trutta*) eggs, and alevins (yolk-sac fry) 2. Atlantic salmon (*Salmo salar*) fry 3. Natterjack toad (*Bufo calamita*) 4. Persian sturgeon (*Acipenser persicus*)	Iran, Scotland, Spain	239 527 565 611
Saprolegnia ferax	1. Pathogenic 2. Pathogenic 3. Pathogenicity unknown 4. Resistant to infection 5. Pathogenic in experimental challenge 6. Unaffected in experimental challenge	1. White areas on skin, frayed tail and fins 2. Fungal growth on eggs 3. Co-colonization with other *Saprolegnia* sp.	1. Brown trout (*Salmo trutta trutta*), brook trout (*Salvelinus fontinalis*), green tench (*Tinca tinca*), bony bream (*Nematalosa erebi*), crayfish (*Astacus astacus*) 2. Yellow catfish (*Tachysurus* [*Pelteobagrus*] *fulvidraco*) eggs 3. Brown trout eggs, and alevins (yolk-sac fry) 4. Smallmouth bass (*Micropterus dolomieu*) 5. Pacific tree frog (*Pseudacris regilla*), red-legged frog (*R. aurora*) 6. Cascades frog (*Rana cascadae*), long-toed salamander (*Ambystoma macrodactylum*)	Australia, China, Finland, Spain, USA	273 392 445 565 1426
Saprolegnia furcata	Pathogenicity unknown	Co-colonization with other *Saprolegnia* sp.	Brown trout (*Salmo trutta trutta*) eggs	Spain	565
Saprolegnia hypogyna	Pathogenicity unknown	Co-colonization with other *Saprolegnia* sp.	Brown trout (*Salmo trutta trutta*) eggs	Spain	565
Saprolegnia parasitica	1, 4. Pathogenic, saprolegniosis 2. Mortality 3. Mortality in experimental infection following stress 5. Mycotic gastritis	1. Cutaneous lesions, infected eggs 2, 3. Loss of epithelium and ulceration 4. Infected eggs 5. Stomach and other organs	1. Brown trout (*Salmo trutta trutta*) broodstock, coho salmon (*Oncorhynchus kisutch*), chum salmon (*O. keta*), masou salmon (*O. masou*), sockeye salmon (*O. nerka*) 2. European eel (*Anguilla anguilla*), silver perch (*Bidyanus bidyanus*), channel catfish (*Ictalurus punctatus*), bunnei (*Mesopotamichthys* [*Barbus*] *sharpeyi*) 3. Rainbow trout (*O. mykiss*) 4. Persian sturgeon (*Acipenser persicus*) 5. Ayu (*Plecoglossus altivelis*)	Australia, Iran, Iraq, Japan, Scotland, Spain, USA	27 191 366 565 611 691 760 781 1431 1738
Saprolegnia salmonis	Mortality	Lesions on head, peduncle and caudal fin	Sockeye salmon (*Oncorhynchus nerka*), chum salmon (*O. keta*), masou salmon (*O. masou*),	Japan	780 781
Saprolegnia sp.	Mortality of eggs	Death, presence of white hyphae on eggs	American bullfrog (*Rana catesbeiana*)	USA	1444
Trichosporon bergelii	Infection	Melanization of cuticle	Freshwater crayfish (*Astacus astacus*)	Sweden	1554

Ascomycetes fungi reproduce sexually by forming ascospores in an ascus, a sac-like cell.

5.1.1 *Aspergillus* (Genus)

Aspergillus belongs to the phylum *Ascomycota*, class *Eurotiomycetes*, order *Eurotiales* and family *Trichocomaceae*.

For isolation of *Aspergillus* species from aquatic animals, suitable media include Sabouraud's dextrose agar (SAB), corn meal agar (CMA), malt extract agar (MEA) and Czapek-Dox agar (CDA). Media are incubated at 25–30°C. Growth is more rapid at 30°C, but slower growth at incubation of 25°C may be better for identifying culture characteristics and other morphology for identification purposes. *Aspergillus* species usually grow within 48–72 h. Growth on SAB is usually profuse, but may not be useful for identifying colony characteristics. Growth on CMA is sparse, but allows the subculture of individual colonies, whereas culture on CMA and CDA is usually the best for identifying morphological features for species identification (Olufemi and Roberts, 1983).

Aspergillus flavus

Aspergillus flavus caused mortality in cultured tilapia (*Oreochromis niloticus*) in Kenya (Olufemi *et al*., 1983). Infected fish displayed lethargy, distended abdomen, darkening of the skin and exophthalmia. Abdominal fluid was abundant and either clear or bloodstained. Some cases had necrotic liver with liquefaction. Lesions occurred in the liver, spleen and heart, and the fungus was cultured from affected sites including the eye. The source of the infection was contaminated feed, which was proven experimentally (Olufemi and Roberts, 1986). *Aspergillus flavus* produces aflatoxin.

On SAB, colonies grow within 48–72 h. On SAB and Czapek's agar (CZA), colonies are yellow-green at 25°C. Growth on CMA is sparse.

Aspergillus fumigatus

Aspergillus fumigatus is responsible for an invasive aspergillosis in captive penguins, Magellanic penguin (*Spheniscus magellanicus*), Humboldt penguin (*S. humboldti*) and African or Jackass penguin (*S. demersus*) held captive at a zoo-aquarium in Spain (Alvarez-Perez *et al*., 2010). Whitish nodules occurred on the lungs and air sacs and the fungus was isolated from these sites and from kidney, liver and lung. The fungus is ubiquitous in the environment and an opportunist. In the penguins, all strains were positive for alkaline protease using skim milk, but differences were observed for elastase, indicating a potential difference in virulence.

Aspergillus fumigatus grows rapidly on SAB, CDA, MEA and blood agar (BA) as blue-green colonies with a white rim. Colonies are 7 cm in 10 days at 24–26°C. The colony is composed of a dense mat of conidiophores that produce conidia 2–3 μm. Conidiophores are pigmented and have clavate vesicles arising from clearly differentiated thick-walled foot cells. Conidia form long chains and are globose to subglobose. Conidial heads are columnar and conidiogenous cells are uniseriate. Growth occurs up to 55°C and conidia can survive at 70°C.

Aspergillus sydowii

Aspergillus sydowii is responsible for causing epizootics in sea fan corals: venus sea fan (*Gorgonia flabellum*) and purple sea fan (*G. ventalina*) found in the Caribbean. Signs of aspergillosis in sea fan corals include purpling of the tissue, lesions and galling. The fungus is not generally part of the marine environment. Growth occurs on PYG medium (0.1% peptone, 0.1% yeast extract, 0.3% glucose, 3% Instant Ocean [Aquarium systems, USA]). Colonies are blue-green with reddish-brown shades. Growth occurs at 25 and 30°C with higher growth at 30°C, which is thought to allow the fungus to overcome the defence systems of the coral (Alker *et al*., 2001). Conidiophores are smooth-walled, hyaline and 500 μm in length. Conidia are 2.5–4.0 μm, green in mass, spherical to subspherical.

5.1.2 *Cladosporium* (Genus)

Cladosporium belongs to the phylum *Ascomycota*, class *Dothideomycetes*, order *Capnodiales* and family *Davidiellaceae*.

Cladosporium sphaerospermum

Cladosporium sphaerospermum infected the posterior kidney of two of 15 wild-caught, tank-held red snapper (*Lutjanus campechanus*) in the USA (Blaylock *et al.*, 2001). *Penicillium corylophilum* was isolated from the swim bladder at the same time. Fish showed erratic swimming behaviour. The kidney and swim bladder were distended. The kidney was infiltrated by masses of hyphae and appeared brownish-black in colour and showed areas of haemorrhage, necrotic areas and multiple granulomas. *Cladosporium sphaerospermum* grows on TSA plates containing NaCl and on potato dextrose agar (PDA) and colonies are brown-black with a velvety appearance. Conidia are spherical to subspherical and 4 μm in diameter and conidia chains branch laterally in an irregular pattern. Conidiophores are relatively unbranched and are not nodose. In wet preparations of the kidney tissue hyphae, conidiophores and conidial chains are seen (Blaylock *et al.*, 2001).

5.1.3 *Exophiala* (Genus)

Exophiala belongs to the phylum *Ascomycota*, class *Ascomycetes*, order *Chaetothyriales* and family *Herpotrichiellaceae*. It is one of the genera described as demataciaceous *Hyphomycetes* because they produce conidia that result in dark green, olive green, green-black or black colonies. *Exophiala* is also described as a black yeast or black-yeast-like fungi. *Exophiala* is an anamorph (asexual) genus, which is an annellide meaning that the conidiogenous cell gives rise to successive conidia in a basipetal manner producing slimy heads of conidia. It is phylogenetically related to the order *Chaetothyriales*, which is characterized by production of budding cells with the yeast/hyphae transition mainly occurring via torulose hyphae (de Hoog *et al.*, 2011). Phylogenetic analysis using the small subunit ribosomal RNA (SSU rRNA) gene sequence information indicates the majority of mesophilic waterborne black yeast species belong to a single clade and many species cause cutaneous or disseminated infection in cold-blooded aquatic animals including fish, crabs, frogs, toads and turtles. Animals with smooth, moist or mucous skins are most susceptible, including immunocompromised humans (de Hoog *et al.*, 2011).

Exophiala angulospora

Exophiala angulospora caused systemic inflammation and moderate mortality in Atlantic cod (*Gadus morhua*) cultured in indoor tanks in Norway (Gjessing *et al.*, 2011). Infected fish showed abnormal swimming behaviour, dark colouration of the skin, particularly the caudal region. Internal organs, particularly kidney, and in some fish liver and brain, were affected and contained yellow-brown, branched, septate hyphae. Disseminated phaeohyphomycosis caused by *E. angulospora* and a novel *Exophiala* species occurred in leafy sea dragons (*Phycodurus eques*) and weedy seadragons (*Phyllopteryx taeniolatus*) held in an aquarium in the USA but which had been imported from Australia (Nyaoke *et al.*, 2009). The seadragons had lesions or ulcerations on the head, trunk, dorsum of tail, or at the base of the fins that were well demarcated, often with raised black margins. The fungus had invaded the dermis and extended into the underlying hypodermis, fascia and skeletal muscle. Black necrotic foci occurred in the kidney, swim bladder, gill and intestinal wall.

Exophiala angulospora is a slow-growing fungus that produces floccose, dark-grey to olive grey-brown or black coloured colonies that grow on SAB containing chloramphenicol, and PDA. Colony diameter is 33 mm after 30 days. Young colonies are yeast-like, but become velvety with age. On MEA after 2 weeks at 24°C colonies are mucoid in the centre and velvety towards the edge and greyish-green to olivaceous black. Hyphae are 1.5–3.0 μm wide, smooth-walled and pale olivaceous. Budding cells are seen. Conidiophores are darkly pigmented and may be branched or unbranched and bear conidiogenous cells that are hyaline to pale olive, and flask-shaped to cylindrical 6–16×2.5–2.0 μm. Conidia are produced in masses or aggregate in slimy heads and are smooth, thin-walled, subhyaline or pale olivaceous, single-celled, angular or may be elliptic to subglobose and are also described as being mostly triangular with rounded ends 2.5–4.0×2–3 μm (de Hoog *et al.*, 2011; Gjessing *et al.*, 2011). Growth does not occur at 35°C. Brown diffusible pigment is not produced (Nyaoke *et al.*, 2009). Identification can be obtained by amplification using universal primers ITS1 and ITS4 that amplify the internal transcribed spacer region between the 18S and the 28S rRNA genes, or by amplifying the D1/D2 regions of the 28S rRNA gene using primers NL-1 and NL-4 followed by sequencing the amplicon (Nyaoke *et al.*, 2009).

Exophiala aquamarina

Exophiala aquamarina caused a disseminated infection and was isolated from skin, skull, bone and necrotic tissue of leafy (*Phycodurus eques*) and weedy (*Phyllopteryx taeniolatus*) seadragons held at an aquarium in the USA, winter flounder (*Pseudopleuronectes americanus*) and little tunny fish (*Euthynnus alletteratus*), sand lance (*Ammodytes americanus*) from the USA, and lumpfish (*Cyclopterus lumpus*) from Canada. Skin lesions were necrotic and the fungus invaded the blood vessels, tissues, skull and bone, but not the brain (Nyaoke *et al.*, 2009; de Hoog *et al.*, 2011).

Growth occurs on MEA and after 2 weeks at 27°C. Colonies are restricted, olivaceous black, velvety with aerial mycelia in the centre. The reverse of the colony is olivaceous black. Diffusible pigment is not produced. Conidiogenous cells are flask-shaped with short annellated zones and sometimes with sympodial conidiogenesis. Hyphae that are spirally twisted are seen. Conidia are 6.7–19.2×4.0–4.8 μm, ellipsoidal to cylindrical. Yeast cells are rarely present. Optimal growth occurs at 24–30°C with a maximum of 33–36°C, and no growth at 37°C and a minimum of ≤4°C (de Hoog *et al.*, 2011).

Exophiala cancerae

Exophiala cancerae is the causative organism of lethargic crab disease (LCD), a disease causing mortality in mangrove crab (*Ucides cordatus*) found along the Atlantic coast of Brazil (Boeger *et al.*, 2005; Pie *et al.*, 2011). Crabs lose motor skills, leading to lethargy and eventually death. Fungal elements are seen in the gills, heart, thoracic ganglion and hemocele and stain pink using Grocott's methionine stain and periodic acid-Schiff (PAS). *Exophiala cancerae* has been isolated from the liver of a green toad (*Pseudepidalea viridis*) in Israel, from water in Germany, from clean tank water in the Netherlands, from a fruit juice drink in Australia and from skin infection in a patient. Some GenBank accessions submitted as *E. salmonis* have been identified as *E. cancerae* and these include a strain from a patient, from a bathroom in Japan and from corn in Australia (de Hoog *et al.*, 2011).

Conidiophores and somatic hyphae are seen in the crab cardiac tissue and thoracic ganglion, and sometimes in the intestine and musculature of the claw. Somatic hyphae are branched, hyaline and septate, and the cell wall is composed of two layers. Conidiophores are non-specific and are not specialized for the production of conidia. Conidia occur as single cells or in masses in nodules in the tissue. Conidia are 6.52×2.68 μm, unicellular, ovate and hyaline. The wall of the conidiophore forms the conidium and chains may form (Boeger *et al.*, 2005). Growth occurs on MEA after 2 weeks at 24°C with a temperature range of 24–27°C and a maximum of 30–33°C, and no growth at 37°C. Colonies are circular and initially flat, olivaceous black, slimy with a velvety dark olivaceous grey centre. The reverse is olivaceous black. There is no diffusible pigment. Conidiophores are short, erect, brown, cylindrical, multi-celled and poorly differentiated. Conidia are septate, subhyaline and pale brown, obovoid to cylindrical.

In experimental challenge of crabs, mortalities began at day 11 post-inoculation and rose to 60% 33 days post-inoculation (Vicente *et al.*, 2012). The organism has not been isolated from the environment, only from crabs.

Specific primers have been designed for the amplification and detection of *E. cancerae* (Pie *et al.*, 2011). Components such as chitin in the cell wall of the fungal spores (conidia) are highly resistant to extraction reagents but the following method was successful in obtaining fungal DNA from crab haemolymph. Haemolymph (400 μl) is centrifuged at 14,000 rpm for 2 min. The precipitate is suspended in 98 μl of digestion buffer consisting of TE-SDS buffer (50 mM Tris, pH 8.0, 100 mM EDTA [Ethylenediaminetetraacetic acid], 0.5% SDS), 2 μl proteinase K and the mixture is incubated at 50°C for 3 h. The host DNA is removed using the initial part of the method for solubilization of the DNA as per the kit EZ-DNA (Biosystems, Brazil). The sample is centrifuged at 14,000 rpm for 1 min and the supernatant containing the host DNA is discarded. The precipitate is suspended in 500 μl of sterile distilled water and centrifuged. The supernatant is discarded. Washing is repeated three times to ensure removal of host DNA. The precipitate containing fungal material is suspended in 500 μl of sterile distilled water, cells are ruptured using an ultrasound homogenizer (Glues Parmer, Chicago) at 5×50 KHz for 30 s. Extraction of DNA is completed using the animal tissue protocol from the DNeasy® kit (Qiagen) (Boeger *et al.*,

2005). Fungal DNA can be amplified for sequencing using universal ITS primers ITS1 and ITS4 (White *et al.*, 1990). Specific primers LCDF5 (5′-CCC TCT GGC CCG CGC TCG TCG ACA GC-3′) and LCDR3 (5′-GAG AAA GAT TCT CCC TGA GGG GAG GCA-3′) can detect *E. cancerae* DNA. A 25-μl reaction mix consists of 3 mM $MgCl_2$, 0.4 mM dNTPs, 1×reaction buffer, 2.5 U Platinum *Taq* DNA polymerase (Invitrogen), 2 μM of each primer and 0.05–5 ng DNA template. Thermocycling is carried out with an initial denaturation at 95°C for 5 min followed by 25 cycles of denaturation at 95°C for 40 s, annealing at 70.5°C for 30 s, elongation at 72°C for 30 s and a final extension step at 72°C for 2 min (Pie *et al.*, 2011). Although not stated, the amplicon is approximately 450 bp. The primers are specific when tested against *E. jeanselmei*, *E. spinifera*, *Cladophialophora immunda*, *C. saturnica*, *Ramichloridium atrovirens* and *Fonsecaea pedrosoi*.

Exophiala pisciphila

Exophiala pisciphila caused systemic mycosis in channel catfish (*Ictalurus punctatus*) in the USA (McGinnis and Ajello, 1974) and was reported in Atlantic salmon (*Salmo salar*) held at an experimental facility in Australia (Langdon and MacDonald, 1987). The fungus was isolated from the brain and from skin lesions. Infection resulted in erratic swimming behaviour or 'whirling', depression and darkening of the skin. One or more fistulas occurred on the caudal–lateral surface of the head near the eye and extended into underlying connective tissue and brain. Dark red gelatinous material surrounded eroded cranial cartilage. Fungal hyphae were cultured from these lesions, but not from kidney, liver or spleen (Langdon and MacDonald, 1987). It has also been reported in smooth dogfish (*Mustelus canis*) held at the New York Aquarium shark tank in the USA (Gaskins and Cheung, 1986) and in potbelly seahorse. An isolate originally identified as *Hormoconis resinae* and isolated from American plaice (*Hippoglossoides platessoides*) held at an aquarium in Canada was identified later as *E. pisciphila* (Strongman *et al.*, 1997; de Hoog *et al.*, 2011). The fungus caused round black lesions on the ventral side of the body and caudal peduncles of the plaice. The epidermis remained intact in early lesions, but advanced lesions had a central ulcer. Necrosis and fungal hyphae penetrated through the dermis to the underlying musculature. Invasion of fungal hyphae did not occur in the liver, kidney and gonads.

Growth occurs on BA and MEA and it is slow growing and compact with brownish-green aerial hyphae. The reverse of the colony is black. Hyphae are 1.6–2.1 μm, septate and branched. Conidiophores are rare. Sporulation occurs after 2 months in either distilled water or 80% seawater (Strongman *et al.*, 1997).

Exophiala pisciphila grows on a number of fungal isolation and growth media, including PDA, SAB, CMA, Sabouraud's maltose agar (SMA) and CDA at 25°C, however growth varies and some media are better for identifying morphological features. Growth is best on PDA and malt agar (MA). Colonies are 25–28 mm, humped, blackish mouse grey to iron grey or olive black at the centre after 14 days. Aerial hyphae are velvety at the edge and floccose at the centre. The reverse of the colony is olivaceous black. On CZA, SDA and SMA, very few aerial hyphae are produced, whereas on PDA and MA aerial hyphae are raised and floccose. Colonies on BA and nutrient agar (NA) do not increase in size after 15 days. Sporulation occurs on PDA, but is poor on SAB. CMA has the advantage of easier visualization of conidia, because sparser mycelia are produced. As part of the growing process, hyphae branch and produce conidia, which bud from the hyphae. Hyphae become melanized. Aerial hyphae are produced over time. Sporogenous cells are monophialidic of variable length and 1.8–3.6 μm in diameter and are integrated, terminal, discrete, determinate or percurrent, lageniform of ampulliform and without a collarette. Germination of conidia occurs 24–36 h after inoculation and group in a slimy mass at the apex of a phialide. Lipid globules are seen within cells. Conidia produce germ tubes that become septate. As well as being produced via hyphae, conidia are also produced via annelids and via pre-existing conidia. *Exophiala pisciphila* produces smaller conidia than *E. salmonis* or *E. brunnea* and they are aseptate and subglobose to obovoid conidia (McGinnis and Ajello, 1974; Gaskins and Cheung, 1986).

Exophiala psychrophila

Exophiala psychrophila caused high mortality in Atlantic salmon (*Salmo salar*) smolts cultured in Norway (Pedersen and Langvad, 1989).

The fungus was isolated from granulomas in the kidney. It has also been isolated from salmon in Ireland (de Hoog *et al.*, 2011).

E. psychrophilum is slow growing and grows on MEA and CMA, with slightly better growth on MEA. On MEA, colonies are 7–10 mm after 2 weeks, 14–17 mm after 3 weeks and 20–24 mm after 4 weeks at 18°C. Initially colonies are black and yeast-like and develop into a dome shape with a dark centre that contains a mass of mycelia and a zone of mouse-grey mycelia. Growth occurs at 0°C after 6 months, but optimal temperature range is 17–21°C and a maximum of 23°C. There is no growth at 37°C. Hyphae are 1–3 μm wide, pale brown, septate and branch sparingly. Moniliform cells are common and are 3–5×5–15 μm and occur in chains commonly of 4–12 cells but can be from two to hundreds of cells. Conidia are aseptate and vary in shape from globose to oblong or may be tapered, 1.5–2.5×3–6 μm, and tend to accumulate in a slimy mass at the apex of the conidiogenous cells. Lipid globules are present and may falsely indicate that the conidia are septate. Conidia may be produced from conidiogenous cells, from hyphae, from moniliform cells or from conidia. Differentiating the *Exophiala* species is difficult due to the few morphological differences; however, a major criterion for *E. psychrophila* is growth at 17–21°C with a maximum of 23°C (Pedersen and Langvad, 1989).

Exophiala salmonis

Exophiala salmonis caused mortality in Atlantic salmon (*Salmo salar*) in a hatchery in Scotland (Richards *et al.*, 1978) and lesions in Atlantic salmon held at a facility in the USA (Otis and Wolke, 1985). It has been reported from mycetoma in the brain of cutthroat trout (*Oncorhynchus clarkii*) fingerlings at a hatchery in Canada and drinking water in the Netherlands.

Table 5.3. Differentiation of *Exophiala* species.

Species	Septation of conidia	Temperature range (°C)	Max. temp (°C)
E. jeanselmei	1	20–30	36
E. pisciphila	0	23–29	32
E. psychrophila	20	17–21	23
E. salmonis	1	20–30	36

The infections appeared to relate to a reduced immune system caused by some sort of stress, such as unsuccessful spawning and reduced feed intake (de Hoog *et al.*, 2011). Infected fish had abdominal swelling and enlarged kidney exhibiting white nodules 3–5 mm in size. Nodules were also seen in the liver, spleen and heart. Straw-coloured fluid was found underneath the kidney capsule and adhesions occurred between the kidney and abdominal wall. Necrotic lesions in the musculature and fungal hyphae were seen in all affected sites (Richards *et al.*, 1978). In the fish exhibiting lesions, masses of fungal hyphae were seen in the posterior kidney, which was swollen and had raised areas on its surface, and a white opaque fluid internally. Intestinal mucosa was sloughed and had haemorrhagic necrotic areas (Otis and Wolke, 1985). An organism, most likely *E. salmonis*, was isolated from King George whiting (*Sillaginodes punctatus*) captured off the coast of South Australia and held as bloodstock (Reuter *et al.*, 2003). Fish had ulcerations on the body that extended to the underlying musculature, and on the head with resulting necrotic cranial bones. The kidney had white foci of necrosis and the swim bladder was thickened.

E. salmonis is 7–10 mm, grey to dark grey on CMA after 10 days at 25°C and 25 mm, grey, velvety and domed after 2 months at 22°C on MEA and PDA (Richards *et al.*, 1978; Otis and Wolke, 1985). The reverse of the colony is dark brown. The isolates from whiting grew on SAB containing chloramphenicol and gentamycin and was black to dull olive-green with a sharp margin and grey aerial mycelia forming with age (Reuter *et al.*, 2003). On MEA after 2 weeks at 24°C, colonies are dry, depressed, hairy and olivaceous black. Yeast cells are rare. Hyphae are septate and branching and 2–3 μm in diameter. Conidiogenous cells are poorly differentiated, intercalary or flask-shaped, and annellated zones are short and inconspicuous. Conidia are hyaline, subhyaline to pale brown, smooth, cylindrical to ellipsoidal, 8–17 μm long by 2.5–3 μm wide, with none, one or three septae. They form from an annellide and accumulate in balls at the tip of the annellide. Optimal temperature range is 18–24°C with a maximum of 30–33°C, a minimum of ≤4°C. It does not grow at 37°C (Richards *et al.*, 1978; Otis and Wolke, 1985; de Hoog *et al.*, 2011).

Exophiala xenobiotica

Exophiala xenobiotica caused mortality in striped jack or white trevally (*Pseudocaranx dentex*) cultured in Japan (Munchan *et al.*, 2009). Clinical signs included swelling of the abdomen and kidney distension. Fungal hyphae and necrosis occurred in the kidney, gills and heart. *Exophiala xenobiotica* causes cutaneous infection in humans and is found in the environment, particularly those that are rich in monoaromatic hydrocarbons and alkanes (de Hoog *et al.*, 2006).

On PDA and MEA, colonies are black and yeast-like, slimy at 1 week, and become olive brown, velvety and woolly with an umbonate centre and flat margin over 4 weeks. The reverse of the colony is black. There is no diffusible pigment. Optimal growth is at 30°C and up to 36°C. There is no growth on PDA containing 11% NaCl (Munchan *et al.*, 2009). Hyphae are pale brown to olive, 1.3–3.0 μm wide and irregularly septate every 7–28 μm. Anastomoses are abundant. Conidiophores are one to seven celled and are at right angles or acute to the hyphae. Conidia are single-celled and occur in small groups and are obovoid 3.3–4.0×1.6–2.0 μm. Chlamydospores may be present. Conidiogenous cells are lemon-shaped or fusiform and inserted laterally on the hyphae with a single, terminal annellated zone. Morphologically, *E. xenobiotica* is indistinguishable from *E. oligosperma* and from immature colonies of *E. jeanselmei* (de Hoog *et al.*, 2006).

E. xenobiotica can be identified through sequencing the ITS1-5.8S-ITS2 regions. DNA is extracted using the DEXAPAT® kit (TaKaRa), which is recommended for extraction of DNA from paraffin-embedded sections, but was found to be convenient for fungi (Munchan *et al.*, 2009). Fungal hyphae (100 μl) are placed into a 1.5-ml microfuge tube and 0.5 ml DEXAPAT® solution is added. Hyphae are homogenized with a micropestle. The solution is incubated at 100°C for 10 min then centrifuged at 13,201 × *g* for 10 min. The supernatant is used for DNA template. The master mix is composed of Ready-to-Go beads (Amersham Pharmacia), 2.5 μl of 10 pM of primer ITS-5 (5′-GGA AGT AAA AGT CGT AAC AAG G-3′), 2.5 μl of 10 pM of primer ITS-4 (5′-TCC TCC GCT TAT TGA TAT GC-3′), 17.5 μl distilled water and 2.5 μl of DNA template. Thermocycling is carried out in a thermal cycler MP (TaKaRa) with an initial denaturation at 95°C for 4 min followed by 30 cycles of amplification at a denaturation temperature of 94°C for 1 min, annealing at 50°C for 1 min, extension at 72°C for 2 min with a final extension at 72°C for 10 min (Munchan *et al.*, 2009). Sequencing primers are ITS-5, ITS-4, ITS-2 (5′-GCT GCG TTC TTC ATC GAT GC-3′) and ITS3 (5′-GCA TCG ATGAAG AAC GCA GC-3′) from White *et al.* (1990).

5.1.4 *Fonsecaea* (Genus)

Fonsecaea belongs to the phylum *Ascomycota*, subdivision *Pezizomycotina*, class *Eurotiomycetes*, subclass *Chaetothyriomycetidae*, order *Chaetothyriales* and family *Herpotrichiellaceae*.

Fonsecaea brasiliensis

Fonsecaea brasiliensis is a newly described black yeast-like fungus that together with *Exophiala cancerae* is involved in lethargic crab disease (LCD) of mangrove crab (*Ucides cordatus*) found along the Brazilian coast (Boeger *et al.*, 2005; Pie *et al.*, 2011; Vicente *et al.*, 2012). LCD is characterized by weak motor control of the pereiopods and chelae with resulting poor balance and lethargy leading to death (Boeger *et al.*, 2005). In experimental challenge, mortality due to *E. cancerae* began on day 11 post-inoculation and increased to 60% over time, whereas mortality due to *F. brasiliensis* began on day 29 post-inoculation and rose to 30%. The evidence suggests *F. brasiliensis* is a secondary invader with a preferred temperature of 37°C, in contrast to *E. cancerae*, which is a primary pathogen but has a preferred temperature range of 24–27°C and a maximum of 33°C. It could not be ascertained whether infection with *Fonsecaea brasiliensis* contributed to increased severity of the disease. *Fonsecaea brasiliensis* was isolated from infected crabs as well as the environment, but *E. cancerae* was only isolated from infected crabs (Vicente *et al.*, 2012).

Fonsecaea brasiliensis was isolated from heart, hepatopancreas, epidermis and connective tissue. On MEA and CMA incubated at 24°C, colonies are olivaceous brown at day 3 and develop a velvety greyish-black outer with a grey centre by day 14. The margin is straight and entire. The reverse is greyish to olivaceous black. Diffusible pigment is not produced. Hyphae and conidia are olivaceous brown. Conidia form in densely branched, short acropetal chains of a maximum of five conidia, often located on distinct denticles. Conidia are 6.5–10.0×2.0–3.5 µm, olivaceous brown, smooth-walled, ellipsoidal to lemon-shaped and become narrow at each end. Conidia can occur as a sympodial cluster at the end of a hypha. Growth occurs optimally at 27–30°C and up to 37°C (Vicente *et al.*, 2012).

5.1.5 *Fusarium* (Genus)

Fusarium belongs to the phylum *Ascomycota*, class *Sordariomycetes*, order *Hypocreales* and family *Nectriceae* and is one of the hyaline moulds. Colonies are usually fast growing and may have coloured mycelia depending on the species. They produce fusiform macroconidia.

Fusarium oxysporon

Fusarium oxysporon infection in Kuruma prawns (*Penaeus japonica*) cultured in Japan led to black gill disease (Khoa and Hatai, 2005). Pathogenicity was confirmed through experimental infection. Other *Fusarium* species have also been reported from black gill disease (Khoa *et al.*, 2005).

Growth occurs on PDA after 4 days with white felt-like aerial mycelium. Pigment changes over time from white to pale beige in younger colonies and purple or dark violet in older cultures. Conidiophores are short and single, and later arranged in densely branched clusters. Macroconidia are fusiform and slightly curved with three to five septae. The phialides are short and mostly non-septate in contrast to *F. solani*.

Fusarium solani

Fusarium solani (Mart.) Saccardo 1881 belongs to the Ascomycetes and is widely distributed in the environment, particularly in soil and plant debris. It is a common plant pathogen. *Fusarium solani* caused mass egg mortality in loggerhead turtles (*Caretta caretta*) on Boavista island, Cape Verde, Africa (Sarmiento-Ramírez *et al.*, 2010). Disease signs included unusual coloured spots of yellow, blue or grey egg shells or grey mycelium completely covering the shell, and dead embryos. Challenge experiments conducted using non-infected eggs exposed to pieces of infected shell resulted in 83.3% mortality. In lobster (*Homarus gammarus*) cultured in North Wales, UK, infection produced dark-brown erosive patches on the exoskeleton and in one animal dark brown lesions occurred in the stomach (Alderman, 1981). Infection in kuruma prawn (*Penaeus japonicus*) cultured in Japan resulted in black gills due to melanization of the gills (Khoa *et al.*, 2005). Microscopically, hyphae and conidia were seen on the gills. Pathogenicity was confirmed experimentally. *Fusarium solani* caused a necrotizing mycotic infection in newborn Bonnethead sharks (*Sphyrna tiburo*) at an aquarium in the USA (Muhvich *et al.*, 1989). Ulcerative skin lesions were present on the dorsal tip of the caudal fin, the mandible and the eye from which *F. solani* was cultured. Extensive necrosis occurred in the muscle and dermis and the cartilaginous backbone and fungal hyphae infiltrated the cartilaginous matrix. In scalloped hammerhead shark (*S. lewini*) held at an aquarium in Hawaii, infection of the lateral canal line occurred and a whitish fluid was discharged from the cephalic canal pores (Crow *et al.*, 1995).

Fusarium solani grows on Sabouraud's glucose agar (SGA) supplemented with chloramphenicol, PDA and peptone glucose agar containing penicillin (100 mg/l). Some researchers use plates without antibiotic supplementation but apply a small volume of antibiotic solution (ampicillin and streptomycin) direct to the plates (Khoa *et al.*, 2005). Culture plates are incubated at room temperature (25°C). On PDA, colonies are blue or bluish-green. *Fusarium solani* produces septate microconidia (oval 9–5×2–4 µm), characteristic fusiform macroconidia (spindle or banana-shaped) and chlamydospores. The monophialides are elongated, 50–70×2–3 µm and usually have a distinct collarette, and bear microconida. The macroconidia are slightly curved with pointed ends (fusiform) and have three to five septa and are produced after 4–7 days from short multi-branched conidiophores.

Identification can be carried out by amplifying the DNA using universal fungal primers based on the internally transcribed spacers ITS1 and ITS2 followed by sequencing using primer pair ITS5 and ITS4 according to White *et al.* (1990; Sarmiento-Ramírez *et al.*, 2010), or amplification using primer pair ITS5 and NL4 followed by direct sequencing (Khoa *et al.*, 2005). See section 5.3, Molecular Identification.

Fusarium tabacinum

Fusarium tabacinum caused black gill disease in white-clawed or Atlantic stream freshwater crayfish (*Austropotamobius pallipes*) held in a laboratory tank after being collected 12 weeks before from a river in England, UK (Alderman and Polglase, 1985). Fungal hyphae were evident in gill tissue and throughout parts of the exoskeleton, particularly the walking legs. Melanization occurred where hyphae were present.

Fusarium tabacinum grows slowly, is whitish to beige and floccose. Aerial mycelia are sparse. Conidiogenous cells are monophialidic, macroconidia are cylindrical and slightly curved and have 13 septae. Microconidia are absent. Growth occurs on RGY agar, also known as IM medium (Alderman and Polglase, 1986).

Photographs and description of growth can be found at http://www.mycobank.org/MycoTaxo.aspx?Link=T&Rec=156563.

5.1.6 *Hormoconis* (Genus)

Hormoconis belongs to the phylum *Ascomycota*, class *Eurotiomycetes*, subclass *Eurotiomycetidae* and family *Amorphothecaceae*.

Hormoconis resinae see *Amorphotheca resinae*

An isolate originally identified as *Hormoconis resinae* and isolated from American plaice was found to be *Exophiala pisciphila* (Strongman *et al*., 1997; de Hoog *et al*., 2011).

5.1.7 *Isaria* (Genus)

Isaria belongs to the phylum *Ascomycota*, class *Sordariomycetes*, order *Hypocreales* and family *Cordycipitaceae*.

Isaria farinosa

Isaria farinosa (formerly *Paecilomyces farinosus*) is a nematophagous fungus that is carnivorous for nematodes and has been used as a biological control of some insect species. *Isaria farinosa* caused a swim bladder infection resulting in low mortality in Atlantic salmon (*Salmo salar*) farmed in Scotland and Germany. Clinical signs include reddening of the vent and a swollen abdomen. The swim bladder was swollen with some haemorrhage seen around the swim bladder, which was filled with a white mass that on microscopic examination showed a mass of branching, septate hyphae (Bruno, 1989; Lehmann *et al*., 1999). The fungus is found worldwide.

Isaria farinosa grows on SMA, MA and dextrose agar at room temperature (25°C) to 4–6 cm in diameter in 14 days. Colonies are white with a dense mycelium that produces conidiophores and conidia and the colony becomes slightly granular and tufted. After prolonged growth the colony may develop a yellowish basal felt with aerial hyphae and conidia becoming pale yellow or orange.

Photographs and description of growth can be found at http://www.mycobank.org/MycoTaxo.aspx?Link=T&Rec=156563.

5.1.8 *Lacazia* (Genus)

Lacazia belongs to the kingdom Fungi, phylum *Ascomycota*, class *Eurotiomycetes* and order *Onygenales*; the family is undefined (*Incertae sedis*) at present. The genus *Lacazia* contains a single species, *L. loboi*. To date, it has not been cultured.

Lacazia loboi

Lacazia loboi (formerly *Loboa loboi*) is an uncultured dimorphic fungal pathogen belonging to the order *Onygenales*. It was first described in a patient in the 1930s by a Brazilian dermatologist Jorge Lôbo. Infection results in a chronic mycotic skin disease called lobomycosis or lacaziosis that presents as a chronic proliferative and ulcerative dermatitis and cellulitis and tends to be associated with older or immunocompromised animals. It was thought to belong to the genus *Paracoccidioides* as *Paracoccidioides loboi* but has now been given a genus in its own right (Taborda *et al*., 1999). Further phylogenetic studies using sequence information from 18S rDNA, internal transcribed spacers 1 and 2, partial coding sequences of chitin synthase 4 (*CH4*) and chitin synthase 2, ADP-ribosylation factor (*ADP-rf*) and the antigenic protein gene *gp43* confirmed that *L. loboi* forms a monophyletic clade distinct from *Paracoccidioides brasiliensis* and other members of the *Ajellomycetaceae* such as *Ajellomyces capsulatus* and *A. dermatitidis* (Herr *et al.*, 2001; Mendoza *et al.*, 2001; Vilela *et al.*, 2009).

To date lobomycosis has been reported from dolphins and humans. In dolphins the disease has been reported from Indo-Pacific bottlenose dolphins (*Tursiops aduncus*) from the Indian Ocean between Mozambique and Madagascar (Kiszka *et al.*, 2009), from inshore and estuarine *Delphinidae* dolphins, the common bottlenose dolphin (*Tursiops truncatus*) and Guiana dolphins (*Sotalia guianensis*) found on the US Atlantic coast, Brazil, the Bay of Biscay in Europe and Surinam (Migaki *et al.*, 1971; Caldwell *et al.*, 1975; Cowan, 1993; Durken *et al.*, 2009; Daura-Jorge and Simões-Lopes, 2011). The disease presents as a chronic proliferative, ulcerated dermatitis and cellulitis and is characterized by grey-white to pale pink, verrucous lesions, which may be raised, ulcerate and form plaques that can be up to 30 cm in diameter. The head, flippers, flukes, dorsal fin and peduncle are most affected (Kiszka *et al.*, 2009). In humans it infects cutaneous and subcutaneous tissue presenting as nodular parakeloidal lesions occurring on the ears and the extremities. The disease tends to occur in the tropics and people such as aquaculture farmers, aquarium workers or fishermen are most at risk.

Lacazia loboi appears as thick-walled yeast cells (6–12 μm) (globose or subglobose cells) occurring singly with small budding cells or in chains connected by thin tubular bridges called connector cells (7.6–7.9 μm), or connected by catenulate cells that are variable in branching (Taborda *et al.*, 1999). In a skin biopsy or histology section, the cells are stained using a Grocott methamine silver stain. *In vivo*, blastoconidia are produced from globose or subglobose cells. Blastoconidia can be singular or occur as branching chains and are connected by narrow connector cells 7.6–7.9 μm or by catenulate cells that are variable in branching (Cowan, 1993; Taborda *et al.*, 1999).

MOLECULAR IDENTIFICATION. Recovering DNA from tissue may be difficult because the organism produces proteases and endonucleases that destroy the cells within minutes of collection of samples, and aliquots of tissue may need to be placed into a DNA extraction reagent immediately after collection (Mendoza *et al.*, 2001). Small pieces of tissue can be ground under liquid nitrogen before being subjected to treatment with sodium dodecyl sulfate and proteinase K digestion and extraction using phenol chloroform (Vilela *et al.*, 2009). *Lacazia loboi* is identified by amplification of the ITS region (using primers ITS1 and ITS4), the chitin synthase 4 (*CHS4*) gene, ADP-ribosylation factor (*ADP-rf*) and/or the antigenic protein gene *gp43* followed by sequencing. Primers are as follows: *L. loboi* chitin synthase detected by primer 4 LlCHS4-1 (5′-CAC CAC CTG TCT AAA GCT-3′) and primer LlCHS4-2 (5′-CGA TTT CAA TGT CAG AAT A-3′); *L. loboi ADP-ribosylation factor* (*ADP-rf*) amplified by primer LlRibosyl-1 (5′-GYC TCG ATG CTG CCG GAA-3′) and primer LlRibo-syl-2 (5′-ACG ACA CGG TCA CGA TCG-3′); and the *gp43*-like gene amplified by primer NL2 (5′-AAC GGC TTC GAC AAC AGC-3′) and primer NL4 (5′-TAG ATA CAT GGC GCA GTC-3′) (Vilela *et al.*, 2009). Primers ITS1 and ITS4 according to White *et al.* (1990) are detailed in the molecular section. The master mix is prepared with AmpliTaq Gold DNA polymerase (Applied Biosystems) using the manufacturer's recommendations. The thermocycling protocol is conducted using an initial denaturation at 95°C for 10 min followed by 40 cycles of denaturation at 95°C for 1 min, annealing at 60°C for 2 min and extension at 72°C for 3 min with a final extension step at 72°C for 10 min. Primers NL2 and NL4 amplify two fragments at 431 and 486 bp.

5.1.9 *Metschnikowia* (Genus)

Metschnikowia is a yeast that belongs to the kingdom Fungi, phylum *Ascomycota*, subdivision *Saccharomycotina*, subclass *Saccharomycetidae*, order *Saccharomycetales* and family *Metschnikowiaceae*. Cells may be spheroidal to ellipsoidal, pyriform or cylindroid, and vegetative reproduction occurs through multilateral budding. Pseudomycelia are simple. Asci are elongate, clavate, sphaeropedunculate or ellipsoidopedunculate. Ascospores are needle-shaped at one or both ends and do not have a whip-like appendage.

Metschnikowia bicuspidata

Metschnikowia bicuspidata (strain WCY) is the causative agent of milky disease in swimming crab (*Portunus trituberculatus*) (also known as Japanese blue crab, horse crab and gazami crab) cultured in China (Wang *et al.*, 2007). The yeast is found in the muscle, hepatopancreas and the heart, and tissues have a milky appearance. Pathogenicity was confirmed by experimental challenge. Growth occurs on YPD medium (glucose 2.0%, polypeptone 2.0% and yeast extract 1.0%) at 28°C. Chloramphenicol 0.05% can be added to inhibit bacterial growth. Colonies are white to cream and butyrous. Strain WCY has budding cells and pseudohyphae. Glucose is fermented, but not galactose, lactose, maltose, raffinose, sucrose or trehalose. L-Arabinose, D-arabinose, cellobiose, glucose, galactose, inulin, lactose, maltose, melibiose, L-sorbose, raffinose, sucrose, trehalose are assimilated. Gelatin is not liquefied. Vitamins are required for growth (Wang *et al.*, 2007).

The yeast can be identified by amplification of the DNA using universal primers for the 18S rDNA, primer P1 (5′-ATC TGG TTG ATC CTG CCA GT-3′) and reverse primer P2 (5′-GAT CCT TCC GCA GGT TCA CC-3′) or primers for the internal transcribed spacer region, primer 11 (5′-TCC GTA GGT GAA CCT GCG G-3′) and primer P21 (3′-TCC TCC GCT TAT TGA TAT GC-3′). The master mix consists of 2.5 µl of 10× buffer, 0.8 µmol/l dNTPs, 1.5 mmol/l $MgCl_2$ and 0.5 µmol/l of each primer, 1.25 U *Taq* DNA polymerase, 1 µl of DNA in a 25-µl reaction mix. Thermocycling is performed using an initial denaturation at 94°C for 10 min followed by 32 cycles of denaturation at 94°C for 1 min, annealing at 53°C for 1 min, extension at 72°C for 2 min and a final extension at 72°C for 10 min (Wang *et al.*, 2007). The amplified product is checked on a gel before following the process for sequencing and submission to a sequencing facility.

An isolate originally named *Metschnikowia kamienskii* was isolated from brine shrimp (*Artemia salina*) found in a lake in Canada (Spencer *et al.*, 1964). *Metschnikowia kamienskii* is a synonym of *Metschnikowia bicuspidata* var. *bicuspidata* and the latter name is the currently accepted name (MycoBank). The isolate was not pathogenic when fed to brine shrimp.

Growth occurs slowly on MEA and after 1 month colonies are cream-white, semi-dull, smooth and have fine transverse striations. The colony edge is entire but has occasional tufts of primitive pseudomycelium. In broth culture, after 3 days cells are 3.5–6.6×5.3–10 µm with some cells up to 16 µm. Cells occur singly, in pairs or in small clusters or chains, are ovoid to globose, sometime pear-shaped or slightly elongated. Cells are produced by budding. Sporulation occurs on yeast extract malt extract peptone ager (YM), vegetable agar plus yeast and MA, and occurs in fresh cultures at 10–12°C. During sporulation vegetative cells become elongated and club-shaped, 4–7×25–42 µm. A single needle-shaped spore that is pointed at each end is produced from each ascus mother cell. The spores are the same length as the asci. Glucose is fermented slowly, whereas fermentation is negative for sucrose, lactose, maltose and raffinose. It does not grow at 30°C or 37°C. Glucose, L-sorbose, maltose, sucrose, cellobiose and trehalose are assimilated but lactose, melibiose, inulin, L-arabinose are not assimilated (Spencer *et al.*, 1964).

5.1.10 *Mucor* (Genus)

Mucor amphibiorum

Mucor amphibiorum belongs to the kingdom Fungi, phylum *Ascomycota*, subdivision *Mucoromycotina*, order *Mucorales* and family *Mucoraeae*. It is a dimorphic fungus occurring only in a spherical form in tissues, but displaying fungal reproductive structures in culture (Speare *et al.*, 1994).

Mucor amphibiorum has been reported from infected platypus, cane toad, green tree frog (*Litoria caerulea*), slender tree frog (*L. adelensis*) and white-lipped tree frogs (*L. infrafrenata*) in Australia (Munday and Peel, 1983; Obendorf *et al.*, 1993; Speare *et al.*, 1994; Berger *et al.*, 1998; Creeper *et al.*, 1998). It causes a fatal, severe ulcerative skin disease in platypus (*Ornithorhynchus anatinus*) found in Tasmania, Australia (Munday and Peel, 1983; Obendorf *et al.*, 1993). To date *M. amphibiorum* has not been found in platypus on the mainland. In platypus, lesions occur on the bill, tail and body and can be up to 50 mm or more and penetrate 10 mm into the musculature. Lesions present as ulcers or thickened skin lesions that may be bleeding and have a serous exudate. The fungus can penetrate to the lungs. Histologically a severe, suppurating caseous granulomatous dermatitis and myositis is present.

On mainland Australia, *M. amphibiborum* has been reported from granulomas in tissues including liver and bladder in the cane toad (*Bufo marinus*) (Speare *et al.*, 1994). The fungus was isolated from the soil of one site of infected toads and also from the faeces of toads. It caused disseminated granulomas in green tree frog (*Litoria caerulea*), slender tree frog (*L. adelensis*) and white-lipped tree frogs (*L. infrafrenata*). In the green tree frog the only signs were sluggishness and a solid, cream coloured mass protruding from the snout.

In squash preparations of tissues or lesions, refractile spherules containing daughter spherules are seen. These appear as large round yeast-like organisms of 13–23 μm with a thick cell wall. The spherules are tissue forms of sporangia and sporangiophores, and contain 2–11 daughter spherules, which are the sporangiospores or chlamydospores. The spherules may or may not contain daughter spherules or signs of developing spherules. Large and small spherules are present. In a Gram stain the wall stains pink or red and the contents are pink, and in a Giemsa stain the wall is dark blue or purple and the contents blue or purple. In a haematoxylin and eosin stain the wall is red and the contents blue-red. Spherules have not been reported on SAB, although spherical-like structures were reported when grown on BA at 35°C. No mycelia or hyphal development is seen *in vivo* (Munday and Peel, 1983; Obendorf *et al.*, 1993; Speare *et al.*, 1994).

Mucor amphibiorum grows on SAB agar at 28°C. Colonies are 50–56 mm in diameter, suede-like, flat and cream to off-white after 6 days. There is little or no aerial mycelia and a ring of substrate mycelium. Growth fills the Petri dish after 21 days. The centre of the colony has slightly raised white mycelia. Colonies become grey as sporangia develop. The reverse of the colony is bright yellow in the centre and a pale yellow at the edge of the colony. On PDA (Oxoid) colonies are pale smoky grey to yellowish and pale yellow on the reverse. Hyphae are non-septate. Sporangiophores may be branched and form terminal, globose to spherical sporangia containing numerous sporangiospores. Sporangiospores are hyaline, grey to brown, globose, to elliptoidal. After dispersal of the sporangiospores a collarette is visible at the base of the columella. *Mucor amphibiorum* is dimorphic and heterothallic. The fungus has positive and negative mating types and only the positive mating types have been isolated from infected platypus. Positive and negative mating types were isolated from cane toad (Obendorf *et al.*, 1993; Speare *et al.*, 1994).

Assessment of fungal morphology and mating reactions are performed on PDA (Oxoid) at 25°C. Mating is done with a mating-positive and a mating-negative strain. The test (unknown mating type) and control cultures (a known mating-positive strain and a known mating-negative strain) are embedded into the agar about 3 cm apart. If the strain under test is a positive mating strain, a black line due to the production of zygospores occurs when mated with the known mating-negative strain. The black line, if present, occurs where the two cultures meet (Speare *et al.*, 1994).

5.1.11 *Ochroconis* (Genus)

Ochroconis belongs to the phylum *Ascomycota*.

Ochroconis humicola

Infection with *Ochroconis humicola* caused ulcers on the dorsal surface of devil stinger (*Inimicus japonicus*) cultured in Japan (Wada *et al.*, 1995). Fish showed inappetance but no mortality. Ulcers became necrotic and developed into the underlying musculature from which fungal hyphae were isolated.

On PYGS plates containing streptomycin and ampicillin, colonies are pale brown, velvety to floccose and slightly domed when grown at 25°C. Hyphae are 1–2 μm in width and septate. Conidia are sparse, pale brown, cylindrical with round ends, 1.8–2.2×1–10 μm, two-celled and smooth-walled (Wada *et al.*, 1995).

Ochroconis species

An unidentified *Ochroconis* species causing chronic mortalities was isolated from kidney of masou salmon (*Oncorhynchus masou*) cultured in Japan (Hatai and Kubota, 1989). Infected fish had a swollen abdomen and swelling of the anal area, anaemic gills and an accumulation of pale red ascites fluid. The posterior kidney was discoloured and contained white nodular foci containing fungal hyphae. Hyphae were seen in liver and spleen of some fish. On PYGS, SDA, NA, PDA and glucose yeast extract agar (GY), colonies are slow growing, slightly domed, velvety and a pale dark brown colour and 14–18 mm after 30 days at 25°C. Conidia are sparse and either two- or four-walled, verrucosa, cylindrical with rounded ends and a pale brown.

5.1.12 *Paecilomyces* (Genus)

Paecilomyces farinosus see Isaria farinosa

Paecilomyces lilacinus see Purpureocillum lilacinum

Paecilomyces marquandii

A renal mycosis was caused by infection with *Paecilomyces marquandii* infecting hybrid tilapia (*Oreochromis mossambicus* × *O. hornorum*) in an aquarium in the USA (Lightner *et al.*, 1988). The infected fish had a distended kidney with patches of hyphae on the peritoneum covering the kidney. The kidney had multiple granulomas that contained masses of hyphae.

Paecilomyces marquandii is a moderately fast growing fungus with a dense matt of hyphae and a floccose overgrowth of aerial mycelium. Colonies change colour from white to brown. The reverse of the colony is bright yellow to orange-yellow with a yellow-brown exudate diffusing into the agar. Hyphae are hyaline, smooth-walled and 2.5–3.2 μm wide. Conidiophores are hyaline and smooth-walled and arise from submerged hyphae or are formed as side branches on the aerial hyphae and consist of verticillate branches with whorls of two to four phialides. Conidia occur in chains and are ellipsoidal or fusiform. Chlamydospores are single, thin-walled, globose to ellipsoidal and 3.5 μm in diameter.

5.1.13 *Penicillium* (Genus)

Penicillium belongs to the phylum *Ascomycota*, class *Eurotiomycetes*, order *Eurotiales* and family *Trichocomaceae*.

Penicillium corylophilum

Penicillium corylophilum infected the swim bladder of two of 15 red snapper (*Lutjanus campechanus*), which were caught and held in tanks for broodstock in the USA (Blaylock *et al.*, 2001). Infection was not systemic and was confined to the swim bladder. Co-infection with *Cladosporium sphaerospermum* occurred at the same time in the posterior kidney. *Penicillium corylophilum* was isolated on TSA containing NaCl and was subcultured to PDA. Predominantly biverticillate penicillia with some monoverticillate and terverticillate penicillia are produced with ampulliform phialides 9–11 μm in length. Metulae are variable in length on the same penicillus, but generally 12–20 μm. Conidia are smooth-walled, spherical to subspherical and 2.5–3.0 μm in diameter. Growth does not occur on Czapek Yeast Extract Agar (CYA) at 37°C, but occurs at 25°C after 1 week as 27–29 mm in diameter with grey-green conidia and white mycelia with a slightly yellow colour on the reverse of the colony. Pigment is not soluble. On 25% glycerol nitrate agar, colonies are 9–12 mm and on MEA, colonies are 28–30 mm at 25°C after 1 week (Blaylock *et al.*, 2001).

5.1.14 *Phialophora* (Genus)

Phialophora belongs to the phylum *Ascomycota*, class *Dothideomycetes*, order *Chaetothyriales* and family *Herpotrichiellaceae*. *Phialophora* is one of the *Hyphomycetes* or dematiaceous moulds.

A fungus identified as *Phialophora* species caused systemic fungal disease in Atlantic salmon (*Salmo salar*) farmed in Scotland (Ellis *et al.*, 1983). Infected fish were emaciated with swollen pelvic girdle, haemorrhages on the base of the fins and petechiae on the ventral surface with an excess of skin mucus. A whitish mucoid material was present in the swim bladder, which was pale and deflated. Firm white-pink masses consisting of fungal hyphae were adherent to visceral organs. Swim bladder was cultured to SAB plates at 22°C. Fungal growth appeared in 2–3 days and was subcultured to MA, PDA and SDA. Colonies on SDA are slow growing, 25 mm after 2 months, beige in colour, becoming olive then dark brown with a convoluted surface. Colonies are initially slimy. Aerial mycelia are not seen, but a paler submerged edge 3 mm wide has some sparse grey aerial mycelia. The reverse of the colony is dark brown with a paler edge. Growth on PDA is more rapid and 25 mm in size in 3 weeks. Colonies are dull, olive-grey with aerial mycelia. Mycelia are hyaline, thin walled, smooth and septate 1–3 μm wide, anastomosing, and are branched at right angles and occasionally at an acute angle. Hyphae may become slightly pigmented and wider or irregularly swollen and form moniliform chains of cells. Conidiophores are variable and may consist of only one cell attached at a right angle to the hyphae or may be composed of several conidiogenous cells attached at right angles or acute angles to one or more terminal intercalary cells on short hyphal branches. Conidiogenous cells are variable, being either true conidiogenous cells, anatomically intercalary hyphal cells or anatomically conidia but functioning as conidiogenous cells. Conidiogenous cells may or may not have collarettes. Conidia are hyaline, aseptate, thin walled and smooth.

5.1.15 *Phoma* (Genus)

Phoma belongs to the phylum *Ascomycota*, class *Dothideomycetes*, order *Pleosporales*; family is undefined (*Incertae sedis*) at present. The fungus occurs in soil, water and food, and is a plant pathogen.

Phoma herbarum

Phoma herbarum caused systemic mycosis in hatchery-reared fingerlings of chinook salmon (*Oncorhynchus tshawytscha*) in USA (Faisal *et al.*, 2007b). Fish swam abnormally, had swollen abdomens, exophthalmia, protruding vents and areas of muscle softening. A whitish, creamy fungal mass was in the swim bladder, swim bladder walls were dark red as was the kidney and musculature. Fungal hyphae occurred in the swim bladder, stomach and kidney. On Sabouraud's medium, colonies grow profusely in 3 days and are a dark brown woolly mass with a white-brown powdery surface and the reverse of the colony is dark black. A dark diffusible pigment is produced as the colony develops. Hyphae are 5–8 μm in diameter, brown, filamentous, hyaline and septate. Fruiting bodies are not produced. The formation of pycnidia (dark brown spherical bodies) occurred after 10–20 days on CMA, a finding consistent for *Phoma* species. Pycnidia are 50–200 μm, with some measuring 300 μm in diameter and containing one or more pores or ostioles. Pycnidia are globose to elongate and solitary or confluent, as is typical for the genus. Identification was confirmed using sequencing of the 18S subunit rDNA and ITS1, 5.8S and ITS2 regions using universal primers followed by sequencing of the amplicons.

In a study to investigate the identification of fungal disease in Nile tilapia (*Oreochrmoms niloticus niloticus*) and African catfish (*Clarias gariepinus*), two popular freshwater fish species farmed in Egypt, *Phoma herbarum* along with *Purpureocillium lilacinum* (*Paecilomyces lilacinus*) were the two most commonly isolated soil fungi, and *Aphanomyces laevis* and *Achlya klebsiana* were the most commonly isolated Oomycetes fungi (Ali *et al.*, 2011). In challenge experiments, intramuscular injection led to higher mortality than bath exposure for *Phoma herbarum*.

An unidentified *Phoma* species caused a visceral mycosis in fry of ayu (*Plecoglossus altivelis*) farmed in Japan (Hatai *et al.*, 1986). Clinical signs were similar to those found in infected chinook salmon, with opaque abdomen, and fungal hyphae present in the swim bladder, kidney, intestine and liver. The fungus was similar to *P. herbarum* with the exception of being a dark green colour and the production of chlamydospores, which were not found in *P. herbarum* ATCC 24909 cultured at the same time. Pycnidia were thin-walled, brown, globose and 100–390 μm in diameter.

5.1.16 *Purpureocillium* (Genus)

Purpureocillium belongs to the phylum *Ascomycota*, class *Sordariomycetes*, order *Hypocreales* and family *Ophiocordycipitaceae*. It contains one species. Purple conidia are produced.

Purpureocillium lilacinum

Paecilomyces lilacinus is a synonym, and *Penicillum lilacinum* is a basionym of *Purpureocillium lilacinum*.

Purpureocillium lilacinum (originally reported as *Paecilomyces lilacinus*) is a hyphomycete fungus that causes wasting disease in blue tilapia (*Tilapia aurea*) and feral Mozambique tilapia (*Tilapia mossambica*) in Puerto Rico (Rand *et al.*, 2000). The fungus is found in soil and infects insects and occasionally vertebrates and fish. Infected fish become emaciated, have sunken eyes, eroded fins, and some have haemorrhagic and scaleless lesions on their flanks. A clear amber fluid is seen in the body cavity and gastrointestinal tract. Spleen, kidney, heart and liver have numerous golden to reddish brown granulomas. Fungal elements are seen in necrotic areas. Poor water quality appears to be a predisposing factor. Samples of infected tissues are cultured to trypticase soya agar (TSA) and fungal growth is subcultured to CYA, Blakeskee MEA and PDA and incubated at 20°C for 7–14 days. On CYA, colonies are 23–27 mm after 7 days, pinkish-white with white mycelium, which are usually inconspicuous, the colony is radially sulcate, and a brown to reddish brown exudate is produced. The reverse of the colony is dark brown with a creamy brown centre. On MEA, colonies are 21–23 mm after 7 days with conspicuous white mycelia. The surface texture is finely granular to floccular and a creamy pink exudate is produced. The reverse of the colony is pale to dark brown. On PDA, colonies are 25–28 mm after 7 days with a finely granular texture to the surface of the colony and granular to flocculose centrally, with white inconspicuous mycelia and buff-pink conidia. The reverse of the colony is orange to dark brown centrally. Growth does not occur at 5°C and slower growth occurs at 37°C. Hyphae are septate with smooth walls. Conidiophores have simple to complex heads composed of whorls of vertically arranged branches, each bearing 2–11 phialides. Phialades consist of a swollen basal part and taper to a thin neck. On MEA, strains of *P. lilacinus* usually develop a white mycelium initially, which changes to a purple colour, vinaceous, when sporulating.

In a study to investigate the identification of fungal disease in Nile tilapia (*Oreochromis niloticus niloticus*) and African catfish (*Clarias gariepinus*), two popular freshwater fish species farmed in Egypt, *Purpureocillium lilacinum* along with *Phoma herbarum* and two Oomycetes fungi, *Aphanomyces laevis* and *Achlya klebsiana* were the most commonly isolated fungi (Ali *et al.*, 2011). In challenge experiments using *P. lilacinum*, intramuscular injection led to higher mortality than bath exposure.

Phylum *Basidiomycota*

Members of the phylum *Basidiomycota* include *Malassezia* and *Cryptococcus*.

5.1.17 *Cryptococcus* (Genus) (yeasts)

All *Cryptococcus* species are positive for urease, whereas *Candida* species are negative.

5.1.18 *Trichosporon* (Genus)

Trichosporon belongs to the phylum *Basidiomycota*, class *Tremellomycetes*, order *Tremellales* and family *Trichosporon*.

Trichosporon beigelii

Trichosporon beigelii grew within the cuticle of freshwater crayfish (*Astacus astacus*) held in aquaria at an experimental facility in Sweden. The aquaria were filled with lake water and supplemented with magnesium chloride as part of an experiment investigating infectivity of the crayfish plaque fungus *Aphanomyces astaci* (Söderhäll *et al.*, 1993). Hyphae were seen in the cuticle and a strong melanin response was seen. *Trichosporon beigelii* is a saprophyte in soil and water, but is known to be parasitic for man and animals and, in the case of infection, in *A. astacus* appears to be opportunistic when crayfish were stressed resulting in a reduced immune system.

Infected cuticle is examined microscopically for fungal hyphae, which are excised from the cuticle and washed in 70% ethanol and rinsed in sterile distilled water. Hyphae are placed inside a sterile glass ring, which is placed onto the surface of a peptone glucose agar plate and incubated at 20°C. A 5-µl volume of 0.05% potassium tellurite is placed into the ring to inhibit the growth of bacteria. The ring forces the fungus to grow through the agar and away from the glass ring and contaminating bacteria. Fungus is subcultured to CMA and 2% MEA. Fungal colonies are flat, with an ochre colour, entire to fimbriate margins and 23–27 mm after 3 days and 45–65 mm after 10 days. On malt yeast extract peptone agar or MEA, colonies are more yeast-like and are wrinkled and a greyish-orange colour. Mycelia are mostly submerged. Advancing colony hyphae are straight or flexuous and 4 µm in diameter and initially non-septate, but develop septa. Branching hyphae are smaller (3 µm) and more densely septate. The main and secondary hyphae fragment readily into cylindrical arthroconidia 8–35 × 3–4 μm in size and by the third day numerous globose blastoconidia of 4 μm diameter form on the arthroconidia and the ends of the hyphae fragments.

Phylum *Chytridiomycota*

5.1.18 *Batrachochytrium* (Genus)

Batrachochytrium belongs to the kingdom Fungi, phylum *Chytridiomycota*, class *Chytridiomycetes* and order *Chytridiales*. There is one species in the genus.

Fungi belonging to the phylum *Chytridiomycota* have a simple, non-mycelial morphology and cell walls that contain chitin. The fungal body, the thallus, does not form mycelia but has rhizoids that are hair-like protuberances. The cytoplasm of the thallus divides to form zoospores that are motile by one flagellum. The thallus containing the zoospores is referred to as the zoosporangium (Pessier *et al.*, 1999).

Batrachochytrium dendrobatidis

Batrachochytrium dendrobatidis was first identified in 1998 as being responsible for the disease cutaneous chytridiomycosis, causing mortality of amphibian (the order *Anuran*) populations in Queensland (Australia) and Panama (Berger *et al.*, 1998). *Batrachochytrium dendrobatidis* was proven to be the causative organism through experimental challenge of blue and yellow poison dart frogs (*Dendrobates tinctorius*) (Longcore *et al.*, 1999). The disease has become a pandemic and is responsible for the decline and loss of many amphibian species and has been reported from six continents (Hyatt *et al.*, 2010). It is listed by the World Organization for Animal Health (OIE) as notifiable (Schloegel *et al.*, 2010). A retrospective survey of archived material from 1879 to 1999 found the first recording of cutaneous chytridiomycosis was from an African clawed frog (*Xenopus laevis*) in 1938 in Southern Africa. Results suggest the international trade of the African clawed frog that began in the mid-1930s is the origin of the world-wide dissemination of the fungus (Weldon *et al.*, 2004). Another retrospective study conducted on more than 3000 archived samples, held in the USA and collected from frogs brought in from 25 countries between the years 1895 and 2001, found that positive samples came from five Canadian provinces and seven American states, whereas all other samples (440) from 23 other countries were negative (Ouellet *et al.*, 2005).

In 1998, the species of frogs affected in Australia included the cane toad (*Bufo marinus*), the pobblebonk or banjo frog (*Limnodynastes dumerilii*), spotted marsh frog (*L. tasmaniensis*), green tree frog (*L. caerulea*), Lesueur's tree frog (*L. lesueuri*), Torrent tree frog (*L. nannotis*), creek frog (*L. rheocola*), spotted tree frog (*L. spenceri*), great barred frog (*Mixophyes fasciolatus*), sharp-snouted day frog (*Taudactylus acutirostris*), Eungella Torrent frog (*T. eungellensis*), and those from Panama included the Chiriqui harlequin frog (*Atelopus chiriquiensis*), Costa Rican variable harlequin toad or clown frog (*A. varius*, *Bufo haematiticus*), yellow-flecked glass frog (*Cochranella albomaculata*, *C. prosoblepon*, *Eleutherodactylus cruentus*, *E. emcelae*) (Berger *et al.*, 1998). Since then, *B. dendrobatidis* has been reported from poison dart frogs (*Dendrobates azureus*, *D. auratus*) native to South America and White's or Australian green tree frogs (*Litoria caerulea*) held at a zoo in the USA (Pessier *et al.*, 1999; Longcore *et al.*, 1999), and in southern bell frogs (*Litoria raniformis*) in a pond in New Zealand (Waldman *et al.*, 2001).

In South America, cutaneous chytridiomycosis has been found in *Telmatobius atacamensis* and *T. pisanoi* from Argentina (Barrionuevo and Mangione, 2006), from keratinized mouthparts of tadpoles of beaked or Rio Viejo toads (*Rhinella quechua*) in Bolivia (Barrionuevo *et al.*, 2008), from tadpole mouth parts of Montevideo tree frog (*Hypsiboas pulchellus*), *Odontophrynus maisuma*, Hensel's dwarf frog (*Physalaemus henselii*), *Scinax squalirostris* found in Uruguay (Borteiro *et al.*, 2009), from grey wood frog (*Batrachyla leptopus*), Chilean four-eyed frog (*Pleurodema thaul*) and Darwin's frog (*Rhinoderma darwinii*) in Chile (Bourke *et al.*, 2011) and *Craugastor rostralis*, *Duellmanohyla soralia*, *Rana maculata* (synonym *Lithobates maculata*), spikethumb frog (*Plectrohyla dasypus*) and Copan stream frog (*Ptychohyla hypomykter*) in Honduras (Kolby *et al.*, 2010).

In Asia the disease has been reported from Japan at a private breeding collection housing imported and wild-caught frogs. Cutaneous chytridiomycosis was confirmed by PCR in Budgett's or wide mouth frog (*Lepidobatrachus laevis*), Surinam horned frog (*Ceratophrys cornuta*), Cranwell's horned frog (*C. cranwelli*), Argentine horned frog (*C. ornata*), Venezuelan horned frog (*C. calcarata*), lesser chini frog or Chacoan burrowing frog (*Chacophrys pierotti*), green puddle frog (*Occidozyga lima*), smoky jungle frog (*Leptodactylus pentadactylus*) and *Plethodontohyla tuberata* (Une *et al.*, 2008). At a national park in Indonesia the following frogs were found to have the disease: Java tree frog (*Rhacophorus javanus*), *Hydrophylax chalconotus*, previously *Rana chalconota*, Hasselt's toad (*Leptobrachium hasseltii*), *Limnonectes microdiscus*, bleeding toad or fire toad (*Leptophyrne cruentata*) (Kusrini *et al.*, 2008). A survey of wild frogs in South Korea indicated the presence of *B. dendrobatidis* in three species: Asian toad (*Bufo gargarizans*), Japanese tree frog (*Hyla japonica*) and American bullfrog (*Rana catesbeiana*) (Yang *et al.*, 2009). In a 2010 survey of wild Dybowski's frog (*Rana dybowskii*), the disease was not found in China (Wei *et al.*, 2010). *Batrachochytrium dendrobatidis* has been detected by PCR but not by histological examination of four species of frog in China: American bullfrog (*Rana catesbeiana*), Chaochiao frog (*R. chaochiaoensis*), Yunnan pond frog (*Babina pleuraden* synonym *R. pleuraden*) and *Odorrana andersonii* collected from the wild or from markets (Bai *et al.*, 2010).

The disease was first reported in Europe in green and black poison dart frogs (*Dendrobates auratus*) and *D. pumilo* imported into Germany and was also found in captive-bred frogs in Belgium and Germany (Mutschmann *et al.*, 2009). *Batrachochytrium odorrana batidis* has been reported from wild frogs, Edible frog (*Rana esculenta*) in Italy, but no evidence was seen of mortality (Simoncelli *et al.*, 2005). In Spain, seven species of frog and newt were found positive by PCR for *B. dendrobatidis*, but were apparently healthy and no mortalities were detected (Hidalgo-Vila *et al.*, 2012). Amphibian species included Iberian painted frog (*Discoglossus galganoi*), Mediterranean tree frog (*Hyla meridionalis*), tadpoles of Western Spadefoot (*Pelobates cultripes*), Perez's tree frog (*Pelophylax perezi*), Bosca's newt (*Lissotriton boscai*), adults of Iberian ribbed newt or Spanish ribbed newt (*Pleurodeles waltl*) and southern marbled newt or pygmy marbled newt (*Triturus pygmaeus*). In Britain, introduced juvenile North American bullfrogs in a pond in Kent tested PCR-positive to the fungus (Cunningham *et al.*, 2005).

Infected animals may show no obvious signs of disease but behavioural changes may be noted such as slow movement. Generally, sick or dying animals display areas of abnormal epidermal sloughing particularly on the ventral abdomen, the hypervascularized pelvic patch or 'drink patch' and digits. Spherical or non-spherical sporangia may be seen microscopically in the keratinized layers, the stratum corneum and stratum granulosum of the digits, and ventral body, in particular the drink patch. The fungus only invades the stratum corneum and stratum granulosum. Tadpoles of some species appear normal, but on microscopic examination zoosporangia may be seen in the keratinized mouth parts (Berger *et al.*, 1998; Une *et al.*, 2008), whereas others show obvious deformity such as loss of keratinization of the jaw sheaths, loss of teeth, erosion of the cutting edge and absence of serrations (Borteiro *et al.*, 2009). Common histological findings are the presence of mature zoosporangia 7.5–10 μm in diameter that are round, flask-shaped or pot-shaped with a thin wall. Mature zoosporangia contain basophilic staining zoospores 1 μm in diameter. Empty post-discharge zoosporangia can be seen in the stratum corneum and sometimes show discharge papillae and chytrid thalli, and some may be septate (Barrionuevo *et al.*, 2008; Une *et al.*, 2008; Borteiro *et al.*, 2009). Sloughed skin may be examined under the microscope for sporangia, which are round or oval, 5–13 μm and occur in clumps in the host cells, and might be empty or contain zoospores. The use of lactophenol blue or Diff-Quick can can help to distinguish features.

The fungus is thought to kill infected amphibians by interfering with the osmoregulation ability of the epidermis, and the resulting loss of electrolytes and osmotic balance leads to death (Voyles *et al.*, 2007). Some amphibian species such as the African clawed toad (*Xenopus laevis*) appear to be carriers of the disease because they show no clinical signs of the disease and sudden mortality events have not been noted (Weldon *et al.*, 2004).

Isolating the fungus in pure culture free from bacteria and other contaminating fungi is difficult (Longcore *et al.*, 1999). Remove sloughed skin and place in sterile distilled water and examine under a microscope for the presence of fungal elements – zoosporangia, which appear as round, pot or flask-shaped. If present, culture this piece of skin and embed slightly into PmTg, TGhL or mTGh agar (Longcore *et al.*, 1999; OIE 2012c). Skin of size 1 mm^2 may be 'cleaned' by using a needle to drag the skin through or across the agar to remove surface bacteria. Place cleaned skin onto a new agar plate. Incubate at 20–23°C. To increase the chance of growth, culture up to six pieces of skin over two plates. Examine plates under a compound microscope (×100) after 1–3 days for the presence of motile 3–4 μm zoospores near the cultured skin or the presence of spherical thalli. These may take up to 2 weeks to develop. Remove and examine micoscopically in a drop of water on a glass slide. Compare with photographs of sporangia and zoospores (http://www.umaine.edu/chytrids/Batrachochytrium/Photographs.html). Transfer to 1% tryptone agar and incubate for up to 2 weeks. It is important to transfer groups of thalli or developing zoosporangia because they do not grow singly but as groups. Colonies of zoosporangia can also be grown in broth culture. In culture, the zoospores are ovate to spherical and vary in size from 3 to 5 μm. They may become elongate. Many cytoplasmic extensions may be seen projecting and retracting from the zoospore. Zoospores contain dark bodies or lipid globules and a dark shadowy area that consists of ribosomes. They have a posteria flagellum of 19–20 μm. Zoospores do not encyst on PmTG or 2% tryptone NA but lyse and do not develop into zoosporangia; however, on TGhL agar some zoospores form germlings with fine thread-like rhizoids. These thalli or developing zoosporangia continue to grow over 4–5 days and can have several rhizoids. In amphibian epidermis, thalli are single or aggregated. Each segment of the thallus forms a flask-shaped zoosporangium up to 40 μm in diameter. The wall of the zoosporangium is smooth and of uniform thickness. Large groups of zoosporangia grow, but small groups and single zoospores do not grow in culture, and when transferring material from skin it is important to select areas containing large groups of zoosporangia or aggregated thalli. The mature zoosporangium contains 4–10 zoospores and has one or more discharge papillae through which the zoospores are released. The empty zoosporangium retains its structure and thin septate may be seen dividing the zoosporangium into compartments but later the empty zoosporanigium collapses into an irregular shape (Longcore *et al.*, 1999).

Liquid cultures can be stored at 4°C for up to 4 months. Broth cultures with very active zoospores can be stored by centrifugation of the broth at 1700×*g* for 10 min. The pellet is resuspended in 1 ml 10% DMSO, 10% fetal calf serum in 1% tryptone broth and frozen at −80°C overnight and the next day in liquid nitrogen for permanent storage (OIE, 2012c).

MOLECULAR IDENTIFICATION. Diagnosis of cutaneous chytridiomycosis is conducted by histology, and the organism is detected by culture and PCR. A number of invasive and non-invasive protocols have been reported for sampling: toe clips, bathing of animals, filtering water and swabbing. The diagnostic assays and sampling protocols were assessed for specificity and sensitivity using experimentally infected animals, and recommendations made (Hyatt *et al.*, 2007). The sampling protocols target the sloughed epidermal skin from the limbs, feet and ventral surfaces of the abdomen. Toe clipping of live animals is not recommended due to the stress caused to the animal and there are no significant advantages in using this method, even though toe clips detect a higher number of zoospore equivalents early in the infection (at day 4 post-infection) compared with other methods. The swab method had diagnostic sensitivity of 100% at day 31 post-infection. Bathing frogs in water was not practical in the field because of the constraints of the method; it was found that zoospores adhere to the sides of the container and therefore lead to decreased sensitivity on storage of the water. Water collected in the field must be centrifuged immediately and the pellet stored at 4°C. Swabbing of the ventral surfaces (underside of legs, feet and drink patch) and the digits for adults, and swabbing the mouth parts for tadpoles, was the most sensitive sampling method. The swab is passed across the skin three to five times. Recommended swabs are MW 100-100 from Medical Wire and Equipment Co. Inoculated swabs are stored dry in 1.5 ml microfuge tubes at −20°C, 4°C, or 23°C for 18 months with no loss in number of zoospores at any temperature. The use of 70% alcohol as a storage medium does not confer advantage over dry swabs.

DNA EXTRACTION. A Medical Wire and Equipment swab MW 100-100 (available from bioMérieux) is used to swab vigorously two to three times on the underside of the feet, legs, digits and drink patch. The swab is broken off into a 1.5-ml screw-cap tube containing 30–40 mm of 0.5 mm zirconium/silica beads (Biospec. Products cat # 11079105z) and 50 µl PrepMan Ultra reagent (Applied Biosystems). The solution is homogenized using a beadbeater (Minibeadbeater 8, BioSpec Products) for 2×45 s and centrifuged at 13,000×*g* for 30 s between each homogenization. Samples are heated at 100°C for 10 min, cooled to room temperature and centrifuged at 13,000×*g* for 3 min. Supernatant can be stored at 4°C for a week or at −20°C for longer periods. Samples of toe clips, filters and tadpole oral discs are extracted with PrepMan Ultra reagent 10% w/v in the same manner.

PCR. A few PCRs have been developed for *B. dendrobatidis*, but the one that has been validated (Hyatt *et al*., 2007) and recommended by the OIE (2012c) is the qPCR of Boyle *et al*. (2004), which uses a hydrolysis probe (TaqMan®, Applied Biosystems) and primers designed to the internal transcribed spacer region ITS1 and 5.8S junction. An internal control is used to detect PCR inhibitors, such as phenolic compounds of humic and tannic acid that occur in the soil, contaminating some swabs. This is a synthetic amplicon available from Applied Biosystems, TaqMan® exogenous internal positive control reagents (VIC probe), cat. # 4308323 (Hyatt *et al*., 2007). Adding 400 ng/µl to the reaction mix was found to reduce inhibition caused by phenolic compounds. In a comparison of the PCR of Boyle *et al*. (2004) with and without the addition of BSA, 72.5% of DNA extracts gave a false-negative reaction without the use of BSA, but all these extracts were detected as positive with the addition of BSA (Garland *et al*., 2010). Other DNA extraction techniques should be tested to determine capability of removing PCR inhibitors.

The master mix is prepared in a 25-µl reaction volume containing 12.5 µl 2× TaqMan® Master Mix (Applied Biosystems), 900 nM primer ITS1-3 Bd (5′-CCT TGA TAT AAT ACA GTG TGC CAT ATG TC-3′), 900 nM primer 5.8S Bd (5′-AGC CAA GAG ATC CGT TGT CAA A-3′), 250 nN probe Chytr MGB2 (5′-6FAM-CGA GTC GAA CAA AAT-MGBNFQ-3′) and 5 µl DNA previously diluted 1:10 in ultra pure water. To improve sensitivity of samples containing a low load of zoospores, the tests are done in triplicate (Boyle *et al*., 2004; Hyatt *et al*., 2007). Well (or tube) 1 contains the master mix as well as internal control (1 µl 10×Exo IPC mix and 0.5 µl 50×Exo IPC DNA). Thermocycling is conducted on a ABI Prism 7700 (Applied Biosystems) thermocycler with an uracil *N*-deglycosylase digest at 50°C for 2 min followed by activation of the Taq Gold thermostable DNA polymerase at 95°C for 10 min. Denaturation is conducted at 95°C for 15 s followed by 45 cycles of annealing and amplification at 60°C for 1 min (Boyle *et al*., 2004; OIE, 2012c). An amplicon of approximately 146 bp is produced. According to the OIE (2012c), the results should be analysed by performing outliers, baseline and threshold (OBT) by setting the baseline range at 3, 15 and the threshold bar to a Delta R_N of 0.1 (*B. dendrobatidis* and the internal control). If positive samples have a Ct value <18, then the upper baseline value is reduced from 15 to at least 3 lower than the sample Ct. As stated in the OIE method (2012c), the assay is determined to be valid by visual assessment to see if the amplification curves have the characteristic shape, that the exogenous control is not more than twofold higher than the determined value of the negative control, because this indicates inhibition and the test should be repeated using DNA template diluted 1:100. The non-template control and the negative control extraction should have a Ct value ≥39. The standard curve must be performed at concentrations of 100, 10, 1 and 0.1 zoospores and be within the specified range of the reference standard and an R^2 greater than 0.98. A positive result will have an amplification curve similar to the control and have a Ct value of ≤39 (in all three wells); a negative result will have no characteristic amplification curve and a Ct value of >41 (in all three wells); an indeterminate result is one that has a characteristic amplification curve similar to the positive control but with a Ct value of 39–41 and the test should be repeated.

A negative result for the internal control and a positive result for the sample mean PCR inhibitors are not present in the sample. A negative result for both internal control and sample means that PCR inhibitors are present in the sample and the template DNA should be diluted 10-fold and re-tested (Hyatt *et al*., 2007).

The analytical sensitivity of the hydrolysis probe PCR for the detection of the fungus is

0.1 zoospore equivalents (Boyle *et al.*, 2004). An animal that has just shed infected skin may give a negative PCR result (Hyatt *et al.*, 2007). The hydrolysis probe PCR is specific and does not detect DNA from other fungi of the order *Chytridiales*, *Rhizophydium* spp., *Rhizophydium haynaldii*, *Rhizophlyctis-Rhizophydium*-like, or *Gonapodya* spp., a fungus of the order *Monoblepharidales* or *Allomyces macrogynus* from the order *Blastocladiales* (Boyle *et al.*, 2004).

The hydrolysis probe was designed from the ITS1 region for the forward primer and from the adjacent 5.8S region for the reverse primer. G-rich and an AT-rich stem-loop structure within the ITS and 5.8S sequence were problematic for primer and probe design and these areas were avoided because they caused failure of the TaqMan® assay. The sequences chosen were conserved for isolates from Australia, Ecuador, Panama and the USA, because some regions have either four base-pair deletions or changes among the isolates and again these areas were avoided for probe and primer design (Boyle *et al.*, 2004).

The primers designed by Boyle *et al.* (2004) were used in a conventional PCR using different polymerase and reaction mix and achieved the same level of sensitivity as the TaqMan® hydrolysis probe PCR and offer an alternative for laboratories that do not have a real-time thermocycler (Garland *et al.*, 2011). A 20-μl reaction mix contains 0.8 U FastStart Taq DNA polymerase (Roche), 0.25 mM of each dNTP (Invitrogen 10297-018), 900 nM of each primer (ITS1-3 Bd and 5.8S Bd), 3 mM $MgCl_2$, 400 ng/μl BSA (Sigma A4161) and 4 μl of DNA template diluted 1:10. Thermocycling consists of an initial denaturation at 95°C for 4 min followed by 50 cycles of denaturation at 95°C for 30 s, annealing at 55°C for 30 s, and extension at 72°C for 45 s in 0.2 ml tubes on a Mastercycler Gradient (Eppendorf) PCR machine (Garland *et al.*, 2011). Detection is performed by agarose gel electrophoresis and ethidium bromide staining, with the suggestion that SYBR Gold staining (Invitrogen) could also be used.

Methods for collection of samples, DNA extraction and qPCR are documented in *Infection with Batrachochytrium dendrobatidis* (Chapter 2.1.1, OIE, 2012; http://www.oie.int/fileadmin/Home/eng/Health_standards/aahm/2010/2.01.01_BATRACHO.pdf).

KINGDOM CHROMISTA

Within the kingdom Chromista are the phyla *Oomycota*, *Bigyra* and others. Phyla may also be referred to as divisions.

Phylum *Oomycota*

Organisms within the phylum *Oomycota* have cellulose in their cell walls like plants, as opposed to True fungi, which have chitin in their cell walls.

5.1.19 *Achyla* (Genus)

Achlya belongs to the kingdom Chromista, phylum *Oomycota*, class *Oomycetes*, order *Saprolegniales* and family *Saprolegniaceae*.

Achlya resemble *Saprolegia* in the size, growth and appearance of some vegetative parts. In *Achlya* species, the spores are released and immediately come to rest, and for a short period exhibit a slow rocking motion before swimming. Sporangia are extremely rare and sometimes can be induced in small numbers in acidified culture water. The oogonia may be intercalary and the dolioform wall is densely ornamented or smooth, and if smooth they generally have eight or more oospores; no hyaline envelope around the oogonia; and hypogynous antheridial cells sometimes present. The sporangium is not filamentous but is cylindrical, fusiform, clavate or naviculate. Spores occur in more than one row at least in the apical portion of the sporangium (Coker, 1923; Johnson and Seymour 2002 online).

Achlya debaryana

Achlya debaryana was reported to cause a mycosis epizootic in catfish (*Mastacembelus armatus*) at a fish production centre in India (Khulbe *et al.*, 1994). According to fishbase, *M. armatus* is a zig-zag or tyre track eel. The outbreak was related to stress caused to the fish by the suboptimal water conditions of low water temperature (22–24°C) and high dissolved oxygen content (9.5 mg/l).

Achlya debaryana has large chlamydospores that are seen as a white fringe at the edge of the colony. The chlamydospores have been described as being arranged into groups resembling the branching horns of deer. The hyphae are sectioned, each section being a chlamydospore and containing dense protoplasm. The chlamydospores develop to become either sporangia or thin hyphae. Sporangia vary in size and may be large and stout and rounded at the end, or may be long and slender with a drawn-out narrow point that may be bent. The spores are bound by a matrix and form a sphere at the tip of the sporangium but some form detached groups. An emergence papilla is formed just before spores are released. Some spores do not emerge from the sporangium but develop into hyphae. When first released, the spores are inactive apart from a slight rocking movement for 5–10 min, and then become active and swim away. Just before emergence a vacuole can be seen in the spore and this becomes smaller at emergence. The walls of oogonia are unpitted. Stalks of the oogonia are one to three times as long as the oogonial diameter. Antheridia are diclinous and occur on all oogonia and arise only from the main hyphae (Coker, 1923).

Achlya klebsiana

In a study to investigate the identification of fungal disease in Nile tilapia (*Oreochrmoms niloticus niloticus*) and African catfish (*Clarias gariepinus*), two popular freshwater fish species farmed in Egypt, *Achlya klebsiana* and *Aphanomyces laevis* along with *Phoma herbarum* and *Purpureocillium lilacinum* (*Paecilomyces lilacinus*), two soil fungi, were the most commonly isolated oomycetes fungi (Ali *et al.*, 2011). In challenge experiments, bath exposure led to higher mortality than intramuscular injection for the two oomycete fungi. *Achlya klebsiana* causes mycosis in cultured yellow catfish (*Tachysurus* [*Pelteobagrus*] *fulvidraco*) eggs in China (Cao *et al.*, 2013). Pathogenicity was confirmed by experimental challenge. Identification was confirmed by typical morphology and by amplification and sequencing of the ITS region using universal primers ITS1 and ITS4, as designed by White *et al.* (1990).

Hyphae are aseptate. Zoosporangia are cyclindrical, clavate or fusiform, and renewal is basipetal and infrequently in sympodial arrangement. Primary spores are discharged from the tip of the zoosporangium and form a ball of primary cysts before being released as secondary zoospores. The oogonia are spherical, obpyriform and contain 1–15 oospores. The oogonial wall is nearly always unpitted. Oospores are eccentric and produce a laterally placed oil droplet. Antheridial branches are frequently diclinous, sometimes monoclinous. Gemmae are globular or pyriform and mainly form in chains (Coker, 1923; Cao *et al.*, 2013).

5.1.20 *Aphanomyces* (Genus)

Aphanomyces belongs to the kingdom Chromista, phylum *Oomycota*, class *Oomycetes*, order *Saprolegniales* and family *Leptolegniaceae*.

The genus *Aphanomyces* is defined by the sporangia not being thicker than the vegetative hyphae, the sporangium is filamentous and spores are borne in a single row, emerge apically in elongated form, then round up and encyst in a clump at the end of the sporangium, re-emerge and swim again. Hyphae are delicate, long and sparingly branched. Specialized gemmae are absent. The oogonium is terminal on long or short branches, smooth or warted, and the wall is thin and unpitted. Antheridia are not always present and are diclinous or androgenous (Coker, 1923).

Aphanomyces astaci

Aphanomyces astaci is the causative agent of crayfish plague (Schikora, 1906), a fungal disease affecting freshwater crayfish (*Decapoda*: *Astacidae*, *Cambaridae*). Crayfish plague is listed as significant by the World Organization for Animal Health (OIE, previously Office International des Epizooties).

Aphanomyces astaci is highly virulent and, in countries where it occurs, outbreaks may be associated with 100% mortality. Susceptible species include European or noble crayfish (*Astacus astacus*), Turkish, Danube or narrow-clawed crayfish (*A. leptodactylus*), European or white-clawed crayfish (*Austropotamobius pallipes*) and stone crayfish (*A. torrentium*). Other freshwater crayfish species, which occur in countries where crayfish plague has not been detected, are susceptible in challenge experiments and include: from Australia, Gippsland crayfish (*Euastacus kershawi*), *E. clydensis*, Alpine spiny crayfish (*E. crassus*), common yabby (*Cherax destructor*), gilgie or jilgi (*C. quinquecarinatus*), *Geocharax gracilis*, Tasmanian giant freshwater lobster (*Astacopsis gouldi*), *A. fluviatilis*; from New Guinea (*Cherax papuanus*); and from Japan, Japanese crayfish or zarigani (*Cambaroides japonicus*) (Unestam and Weiss, 1970; Unestam, 1972; Unestam 1975, Unestam and Ajaxon, 1976; Alderman, 1996). The majority of the crayfish such as *Euastacus* spp. and *Astacopsis fluviatilis* are highly susceptible, like the European crayfish species *A. astacus*. *Cherax destructor*, although susceptible, is considered slightly less susceptible than the European crayfish species because increased melanin deposits suggests some immune response and mean time to death was 16–28 days compared with 7–14 days for *A. astacus*, but *C. destructor* is not resistant like the North American species (Unestam, 1975). Schikora (1906) noted an acute and chronic phase of the disease because some apparently healthy crayfish from Poland only succumbed to infection following stress due to transportation. Some resistance has been noted also in *A. leptodactylus* tested from the Romanian stretch of the Danube because these apparently healthy crayfish were positive according to TaqMan® minor grove binder real-time PCR (Pârvulescu *et al.*, 2012). Under experimental challenge, the plankton crustacean (*Mysis relicta*) is resistant to infection (Unestam, 1972).

In transmission experiments using infected brook crayfish (*Astacus astacus*), the Chinese mitten crab (*Eriocheir sinensis*), a pest in Europe, was found to be susceptible to infection with *A. astaci*. The disease progressed slowly compared with more susceptible crayfish species and some crabs were resistant to infection. In infected crabs fungal hyphae were present on the inner surface of the synovial membranes of the limbs, and in some cases incomplete moulting of limbs occurred (Benisch, 1940).

North American freshwater crayfish, such as signal crayfish (*Pacificus leniusculus*), mud or spinycheek crayfish (*Orconectes limosus*) and Louisiana swamp crayfish (*Procambarus clarkii*) may be resistant or asymptomatic carriers of the infection under normal conditions (Unestam and Weiss, 1970; Unestam, 1972; Vey *et al.*, 1983; Diéguez-Uribeondo and Söderhäll, 1993). However, they are susceptible to infection and may succumb to the disease when stressed or exposed to a sufficient concentration of the pathogen, as proven in experimental bath challenge using 50,000 spores/ml of *A. astaci*. This resulted in 80% mortality of *P. clarkii* and a mean time to death of 25 days (Diéguez-Uribeondo and Söderhäll, 1993). The LD_{50} for *P. leniusculus* is 2×10^6 spores compared with 5×10^4 spores for *Astacus astacus*, a highly susceptible crayfish species (Unestam and Weiss, 1970). North American freshwater crayfish are capable of harbouring the fungus and transmitting infection to highly susceptible species. They have been introduced into Europe at various times to stock waterways after crayfish plague decimated native freshwater crayfish, but unfortunately re-introduced the disease.

The disease was first reported in Italy in 1859, and the causative agent *A. astaci* named in 1906 (Schikora, 1906) and cultured by Nybelin (Nybelin, 1936). By then the disease had been reported in France, Germany, Austria, Slovenia, Russia, Finland and Sweden (Alderman, 1996). The disease has now been reported in the Baltic region, Czech Republic, UK, Greece, Hungary (species, *Orconectes limosus*), Ireland, Norway, Romania (species, *Astacus leptodactylus*, *Orconectes limosus*), Spain (species, *A. leptodactylus*, *Austropotamobius pallipes*) and Turkey (species, *A. leptodactylus*) (Cuellar and Coll, 1983; Baran and Soylu, 1989; Rahe and Soylu, 1989; Kozubíková *et al.*, 2010). Recently, it was re-confirmed in Italy, where *A. astaci* was detected in dead *A. pallipes* in 2009 (Cammà *et al.*, 2010). It was co-isolated with *A. repetans*. *Aphanomyces repetans*, a newly described species, was isolated from crayfish (*Procambarus clarkii* and *Pacifastacus leniusculus*) in Spain and Italy in 2004. *Aphanomyces repetans* is not pathogenic to noble crayfish (*A. astacus*) or the yabby (*Cherax destructor*), an Australian freshwater crayfish, in experimental challenge. In contrast to *A. astaci*, it has a high growth rate, does not produce chitinase constitutively, and its zoospores germinate in response to PG1 (Royo *et al.*, 2004). *Aphanomyces invadans*, a fungal species that infects freshwater and brackish fish, does not infect freshwater crayfish (*Astacus astacus*) (Lilley and Roberts, 1997).

To date five genetic types of *A. astaci* have been identified (Huang *et al.*, 1994; Diéguez-Uribeondo *et al.*, 1995; Kozubíková *et al.*, 2011a). Three types originally identified using a dendrogram constructed from molecular types generated from random amplified polymorphic DNA (RAPD) showed a similarity index of 0.3 from 15 strains of *A. astaci* isolated between 1962 and 1988 from noble crayfish (*A. astacus*), *A. leptodactylus*, signal crayfish (*P. leniusculus*) and *P. clarkii* in Sweden, Turkey and the USA: group A (Ra, Vå, A1, J1, Uv, L1, Vl, Da, Sv), group B (P1, SA, Si) and group C (Ti, Yx, Kv) (Huang *et al.*, 1994). Group A contains isolate A1 from Turkish crayfish (*A. leptodactylus*) in Turkey, and Swedish isolates (Ra, Vå) from noble crayfish, and suggests isolates in this cluster are the descendants from a very early introduction of the fungus into Europe and are genetically different to the strains introduced from 1969 onwards (Huang *et al.*, 1994). Group B contains an isolate (P1) from signal crayfish from Lake Tahoe in the USA and some of the strains from noble crayfish and signal crayfish in Sweden, suggesting crayfish plague detected in Sweden originated from signal crayfish imported from California, USA, in 1969. Also in group B is isolate Pa from white-clawed or European crayfish (*Austropotamobius pallipes*) in Spain, and indicates transmission of crayfish plague from introduced signal crayfish to the native crayfish (Diéguez-Uribeondo and Söderhäll, 1999). Group C is a unique cluster containing isolate Kv from signal crayfish imported into Sweden from Canada. A fourth group, Group D, has 22% similarity to groups A, B and C and is composed of isolate (Pc) isolated in Spain from red swamp crayfish (*Procambarus clarkii*), a warm water species (Diéguez-Uribeondo *et al.*, 1995). This strain is physiologically adapted to higher temperatures than those strains isolated from cold water freshwater crayfish found in Sweden and the USA as the growth rate is faster at temperatures above 25°C, a higher number of spores are produced above 20°C and a greater percentage of motile spores is produced above 18°C compared with the cold water strains at these temperatures. A fifth group, Group E, was isolated in the Czech Republic from spiny-cheek crayfish (*Ornectes limosus*), a North American species introduced into Europe (Kozubíková *et al.*, 2011a).

Massive stock losses of up to 100% mortality are often the first sign of the disease in highly susceptible species. In carrier animals, infection may not become evident unless the animals suffer stress through transportation or unusual environmental conditions. The fungus infiltrates the exoskeleton particularly at the soft cuticle in the joints of the legs, antennae and pereiopods, the tips of the pleopods and the soft cuticle of the abdomen. Death occurs within 1–3 weeks (Schikora, 1906; Nybelin, 1936; Unestam, 1972; Bangyeekhun *et al.*, 2001). Hyphae can protrude from the joints and melanin deposits, a result of the immune response, and occur on the body, although in acute infection and with highly susceptible species melanin deposits may not be seen (Alderman, 1987). Hyphae become encapsulated with melanin, more so in chronic infections and in the more resistant crayfish species (Schikora, 1906; Unestam and Weiss, 1970).

The most suitable specimens for culture are live or moribund crayfish showing signs of infection, such as slow or lack of tail flip response, melanin deposits, whitening of the musculature beneath the thin areas of the cuticle especially the ventral abdomen and limb (period) joints, and unusual behaviour such as being seen in daylight hours, although such signs may not be observed until the terminal stages of the infection. Whitening of the musculature should not be confused with porcelain disease of Thelohaniasis, a disease caused by the microsporidium *Thelohania contejeani* (Alderman, 1996). Dead crayfish should be collected within 24 h of death, as post-mortem invaders will predominate and often kill *A. astaci* (Unestam, 1965).

Aphanomyces astaci is present in and can be isolated from uncalcified cuticle, nerve cord, muscle tissue, connective tissue, gut and eye (Schikora, 1906; Unestam, 1965; Alderman and Polglase, 1986). Samples of skin or cuticle are washed alternately in sterile distilled water and 70% alcohol before being embedded slightly into agar medium such as peptone-glucose medium, PG1 (Unestam, 1965), or isolation medium IM (Alderman and Polglase, 1986), the latter containing oxolinic acid and penicillin to reduce bacterial contamination. *Aphanomyces astaci* does not grow on SDA, CMA or MEA (Lilley and Roberts, 1997). Magnesium, potassium, iron and calcium are important for growth of the fungus and these are ingredients in PG1 medium. Some agars contain some minerals such as magnesium and calcium. Hyphae of *A. astaci* can produce substances that become toxic by lowering the pH of the medium, therefore it is important to have the agar in the plates at a minimum depth of 4 mm because this assists in diluting the toxic substances, as does the buffering effect of the phosphate buffer. If broth cultures are used, shaking at 20–25°C improves growth (Unestam, 1965).

Kozubíková *et al.* (2011) found the use of the whole animal cuticle and all the walking legs improved the rate of isolation of the fungus. Alderman and Polglase (1986) suggest that if contamination cannot be overcome, then excised infected limbs or the entire juvenile animal placed into sterile distilled water initiates sporulation of *A. astaci* zoospores. The earliest formed zoospores tend to be those from the pathogen, rather than contaminants. Aliquots of zoospores are plated to IM agar to produce germlings, which in turn are subcultured to fresh medium. Oxolinic acid and penicillin do not impact on zoospore production. Another method for isolation of *A. astaci* is to place small pieces of infected material inside a sterile glass, stainless steel or a heat-resistant plastic ring about 1 cm in diameter that is heated and placed onto the agar medium and melted into the agar surface by about 1–2 mm. Adding 0.5 ml of sterile 0.05% potassium-tellurite solution reduces the bacterial contamination. *Aphanomyces astaci* can grow through the agar and under the ring and away from contaminants that are contained within the ring (Cerenius *et al.*, 1988).

On IM agar colonies are colourless, flat and grow within the surface of the agar and are 10 mm in diameter by 4–7 days when incubated at 16°C (Alderman and Polglase, 1986). At incubation below 7°C, some growth occurs at the surface. Some strains, such as those from Sweden, produce colonies that are ropey on the surface and have a slightly spiral growth. Optimum growth temperature is 22–24°C and growth occurs between 4 and 25°C, but not at 0 or 30°C. Plate cultures are viable for up to 30 days at 16°C, but do not survive drying (Alderman and Polglase, 1986). In the initial stages of isolation, new fungal growth should be subcultured regularly to fresh medium to ensure pure subcultures from which to study zoospore production and encystment to aid identification.

Hyphae are branched, aseptate and 7–10 µm in width (Schikora, 1906; Alderman and Polglase, 1986). Vegetative branches tend to be narrower for the first 20–30 µm of growth. Young actively growing hyphae have densely packed coarsely granular cytoplasm with many refractile globules. Older hyphae are vacuolated with cytoplasm on the periphery and thin strands of protoplasm bridging the central vacuole. Old hyphae may be empty of contents. Coiled hyphae tend to occur when grown on agar medium.

Sporangia (zoosporangia) form in 20–30 h at 16°C and 10–25 h at 20°C when actively growing thalli from agar or broth culture are transferred to distilled water (Alderman and Polglase, 1986). As well as distilled water, lake water or diluted salt solution also trigger sporulation. Sporulation is limited or does not occur in growth medium such as PG1. Sporulation can be carried out *in vitro* by the use of drop cultures, that is, placing 170–250 µl volumes of PG1 into a Petri dish and adding mycelia to the drops and incubating at 13–20°C (Unestam, 1969). Sporulation is induced by removing PG1 medium with a pipette, washing once with sterile lake water and replacing with the same volume of lake water. The number of spores germinated can be increased by the addition of different compounds including $CaCl_2$, NaCl or mannitol at 150 mOsm/kg water. Lake water that has a concentration of 0.3 mM Ca^{2+} is optimal for both zoospore motility and mycelial growth. Germination induced by NaCl occurs at temperatures of 15–23°C and an optimum of 20°C, whereas cultures not induced by additives are less affected by temperature. Sporulation is inhibited at 30°C. The microcultures are examined with the aid of an inverted microscope (Unestam, 1969; Svensson and Unestam, 1975). Growth in drop cultures of PG1 medium, washing three times in distilled water and incubating in 750 µl of 1 mM $CaCl_2$ for 14 h at 13°C will also stimulate sporulation (Cerenius *et al.*, 1988). Sporangia are myceloid, terminal or intercalary and develop from undifferentiated hyphae. Terminal sporangia are delineated by a single basal septum, and intercalary sporangia are delineated by a septa at either end of the sporangial segment. These septae are thicker than the hyphal wall, are highly refractile and are related to the sporangium, and are not part of the vegetative hyphae, which is aseptate. During the formation of sporangia the cytoplasm cleaves into elongated units 16–25×8 µm with rounded ends. Discharge of spores is achlyoid, a process characterized by an aplanospore that encysts at the sporangial orifice. Intercalary sporangia are complex and develop from new lateral extramatrical hyphae, which form the discharge tube of the sporangium (Alderman and Polglase, 1986). The primary spores extrude through the hyphal tip, a process that takes less than 5 min, and accumulate at the sporangial orifice. The spores begin the encystment process by a change from an elongated shape to a round shape and the development of a cyst wall. The time from release to encystment is 2–5 min. The encysted spore mass adheres to the sporangial tip and is not easily disrupted and remains encysted for 8–12 h. Each sporangium contains 15–30 primary spores of 9–11 μm. Sporangial development and spore discharge is optimal at temperatures of 16–24°C. Not all sporangia discharge. In some cases the primary spores develop as spheres and encyst within the sporangium, a process that is observed *in vivo* and *in vitro*. These spores often germinate to produce hyphae (Alderman and Polglase, 1986).

Secondary spores develop by the emergence of spore cytoplasm through a narrow pore at the tip of a papilla. The extruded cytoplasm rounds up and begins a gentle rocking motion as the spore changes to a reniform shape and flagella emerge. Secondary biflagellated zoospores are 8×12 μm. Formation of secondary spores is optimal at 20°C, but does not occur at 24°C. Active motility takes 5–20 min (Alderman and Polglase, 1986) and is affected by temperature. At a temperature range of 12–23°C, 30–50% of spores are motile after 1–5 h, whereas motility stops at 24–25°C (Unestam, 1975). Cysts either germinate to produce new hyphae or undergo repeated zoospore emergence (Cerenius and Söderhäll, 1984).

To induce secondary zoospores, primary cysts are produced, as described for induction of sporulation, and collected using a pipette, filtered through double-layer cheese cloth and concentrated by centrifugation at 400 × *g* for 15 min. The cysts are washed twice in distilled water and once with 1 mM $CaCl_2$ and suspended in 1 mM $CaCl_2$ and incubated overnight at 13°C in drop cultures. Zoospores are collected by pipette and placed into 1 ml drop cultures. To induce encystment, an equal volume of PG1 medium or 50 mM $CaCl_2$ is added to the drop cultures in Petri dishes and incubated for 6 h at 13°C. Encystment can also be induced by placing the 1-ml zoospore suspension containing 1 mM $CaCl_2$ in a test tube and vortexing for 20 s before placing back into an equal volume of PG1 medium in drop cultures in Petri dishes and incubating for 6 h at 13°C. The cysts are collected, centrifuged for 15 min at 400 × g and washed twice in 1 mM $CaCl_2$ to induce secondary zoospore emergence, although secondary zoospore emergence can be achieved from vortexed solutions without the washing steps (Cerenius and Söderhäll, 1984).

Other *Aphanomyces* species can be induced to produce sexual structures by growth on CMA, however this is not the case with *A. astaci* and sexual structures cannot be produced (Cerenius *et al.*, 1988).

Aphanomyces astaci does not survive freeze-drying or −80°C and so is difficult to maintain for a culture collection. Cultures may survive for up to 11 months at room temperature when grown on PG1 medium on sloped cultures in MacCartney bottles and covered with mineral or autoclaved, dried paraffin oil before the colony reaches the sides of the container. The use of paraffin oil appears to lower the metabolic rate of the organism. The use of small pieces (10 mm^3) taken from stock cultures for propagation is more successful for culturing than larger pieces (Unestam, 1965). Cultures prepared by the same manner using IM medium are viable for 6 months when stored in the dark at 7°C, or 3–4 months if stored at 15–18°C (D. Alderman, personal communication, 2003). Cultures then need to be subcultured to fresh storage medium to maintain the collection.

MOLECULAR IDENTIFICATION. PCRs to detect *A. astaci* have been developed; a conventional PCR (Oidtmann *et al.*, 2006) and a TaqMan® minor grove binder (MGB) real-time PCR (Vrålstad *et al.*, 2009) use primers designed to the ITS region and both PCRs are reported in the 2012 OIE manual (OIE, 2012b), and a third PCR utilizes real-time PCR with primers designed to the chitinase genes (Hochwimmer *et al.*, 2009). In a study to compare the sensitivity and specificity of these PCRs, the TaqMan® MGB real-time PCR of Vrålstad *et al.* (2009) was the most sensitive for detecting both mycelium (50 fg) and zoospores (one zoospore). In contrast, the other two PCRs are less sensitive because both detect a minimum of 500 fg of mycelium, or a minimum of 10 zoospores for the conventional PCR and 100 zoospores for the qPCR (Tuffs and Oidtmann, 2011). The conventional PCR of Oidtmann *et al.* (2006) lacks specificity and produces cross-reaction with bacteria that may be present in aquatic samples, these being *Citrobacter freundii* and *Hafnia alvei*, a common bacterium of aquatic habitats. All three PCRs detected genotypes A, B, C and D of *A. astaci*. All three PCRs are specific when tested against *Aphanomyces invadans*, *A. frigidophilus*, *A. laevis*, *A. cladogamus*, *Saprolegnia parasitica*, *S. diclina*, *Pythium flevoense*, *Aeromonas hydrophila* and *Variovirax* species (Tuffs and Oidtmann, 2011). The recently described *Aphanomyces repetans* co-isolated with *A. astaci* in Spain and Italy (Royo *et al.*, 2004; Cammà *et al.*, 2010) was not tested for specificity. The conventional PCR of Oidtmann *et al.* (2006) based on primers designated 42f and 640r that produce an amplicon of 569 bp, and reported in the 2012 OIE method, was modified from a previous conventional PCR using primers 525f and 640r (Oidtman *et al.*, 2004). These primers (525f and 640r) lack specificity because an amplicon of the same size as that generated from *A. astaci* is also generated from *Aphanomyces frigidophilus*, *A. repetans*, *Saprolegnia parasitica* and *S. australis* when tested using the conventional PCR protocol and when used with SYBR® Green chemistry in a real-time PCR (Ballesteros *et al.*, 2007). Semi-nested PCRs using primers 42f and 640r and primers 525f and 640r were not specific against *A. invadans* or *A. frigidophilus*, but a singleplex using 42f and 640r was specific (Oidtmann *et al.*, 2006). *Saprolegnia* spp. were not tested. The sequence of primer 525f differs from *A. frigidophilus* at two single nucleotide sites, whereas primer 640r differs by three nucleotides (two nucleotides and a single nucleotide site) (Ballesteros *et al.*, 2007). For the SYBR® Green real-time PCR, the melt curve of the amplicon for *A. frigidophilus* is only 0.5°C different from *A. astaci* (Ballesteros *et al.*, 2007). These results highlight the need to test and report sensitivities and specificities of PCRs as new fungal species are identified that might be present in the same ecological site as the pathogen of interest.

To date, studies suggest that the TaqMan® MGB real-time PCR (Vrålstad *et al.*, 2009) offers the best sensitivity and specificity for the detection of *A. astaci* from crayfish material, zoospores and contaminated waters (Tuffs and Oidtmann, 2011). This PCR targets a 59-bp sequence from the ITS1 region and uses a probe that contains an insertion-deletion motif that is unique to the four genotypes of *A. astaci*. It is located in the most variable part of the ITS. The master mix is prepared in a 25-μl volume containing 500 nM (0.5 μM) forward primer AphAstITS-39F (5′-AAG GCT TGT GCT GGG ATG TT-3′), 500 nM (0.5 μM) reverse primer AphAstITS-97R (5′-CTT CTT GCG AAA CCT TCT GCT A-3′) and 200 nM (0.2 μM) TaqMan® MGB probe AphAstITS-60T (5′-6-FAM-TTC GGG ACG ACC C-MGBNFQ-3′) labelled with the fluorescent reporter dye FAM at the 5′-end and a non-fluorescent quencher MGBNFQ at the 3′-end, 12.5 μl TaqMan® Environmental Master Mix (Applied Biosystems), 1.5 μl sterile Milli-Q water and 5 μl template DNA (undiluted and diluted 10-fold). The thermocycling protocol has been tested on the Mx3005P qPCR thermocycler supplied by Stratagene, the ABI Prism 7900HT Sequence Detection System qPCR thermocycler from Applied Biosystems and Mastercycler ep realplex S from Eppendorf and on a iQ5 real-time PCR detection system from BioRad (Vrålstad *et al.*, 2009; Strand *et al.*, 2011; Pârvulescu *et al.*, 2012). An initial decontamination step is conducted at 50°C for 2 min for optimal uracil-DNA glycosylase (UNG) enzymatic activity followed by 95°C for 10 min to simultaneously deactivate the UNG, activate the DNA polymerase and denature template DNA, and continued with 50 cycles of denaturation at 95°C for 15 s and annealing at 60°C for 60 s (Vrålstad *et al.*, 2009; OIE, 2012b). All runs should include a no template control every 10 samples, an extraction blank control and an environmental control (see DNA extraction, below). On machines where ROX is selected as a passive reference, the software-defined baseline is automatically set at 2.0. The limit of detection is five PCR forming units (target template copies), which equates to less than one *A. astaci* genome.

The results from the qPCR are quantified in terms of the PCR-forming units (PFUs), limit of quantification (LOQ) and the level of PCR inhibition (Vrålstad *et al.*, 2009). When the qPCR is initially validated in each laboratory, a calibration or standard curve is calculated by plotting concentration or mass (ng/μl) of DNA of fourfold serially diluted DNA from a pure culture of *A. astaci* (as measured using a NanoDrop ND-1000 spectrometer [NanoDrop Technologies], or other appropriate measure) versus cycle threshold (Ct). Thirteen dilutions (calibrant or standard points) are prepared and each dilution is run in 30 replicates to establish the repeatability of the quantification. To establish the PFU, the three most diluted concentrations are run in 40 replicates. This establishes the number of amplifiable ITS copies per calibrant or standard and is based on single molecule quantification or SIMQUANT as detailed by Berdal *et al.* (2008). In this process the DNA is quantified based on PFU, where one PFU corresponds to one amplifiable copy of target DNA. PFU describes the performance and distribution of the amplifiable target copies in template DNA and does not include damaged DNA that is non-amplifiable or copies of target DNA that are aggregated and not uniformly distributed, such as tandem repeat ribosomal RNA genes. The PFU in each PCR replicate is calculated from the mass (or concentration ng/μl) of DNA in each calibrant or dilution, which is multiplied by the volume of DNA template (5 μl) used in the qPCR reaction. The LOQ is 50 PFU.

PCR inhibition is detected by calculating the cycle threshold (Ct) difference (ΔCt) between undiluted and 10-fold diluted DNA replicates, i.e. ΔCt=Ct diluted 10-fold – Ct undiluted. Under ideal conditions the theoretical ΔCt value is 3.32; however, in practice an acceptable level of variance due to inaccuracies arising from pipetting and amplification efficiency is 15%. This results in an acceptable ΔCt of 3.32±0.5 or a range of 2.82–3.82 indicating no inhibition. A difference of <2.82 in ΔCt value between the undiluted and diluted DNA is interpreted as evidence of inhibition in the undiluted DNA sample (Kozubíková *et al.*, 2011), although Strand *et al.* (2011) took this value to be <3.

In the original report (Vrålstad *et al.*, 2009), Universal PCR Master Mix (Applied Biosystems) with passive reference dye ROX was used; however, TaqMan® Environmental Master Mix reduces PCR inhibition caused by the humic substances present in lake water; the result of decaying organic matter and arising from soil or humus, peat, coal in natural waterways of which humic acid is the major component. TaqMan® Environmental Master Mix is almost 100% efficient at removing inhibition and is more efficient in reducing inhibition than the addition of bovine serum albumin (BSA; 5 mg in 0.25 ml, Fermentas) used at a final concentration of between 0.1 and 0.5 μg/μl, depending on the water source (Strand *et al.*, 2011).

The TaqMan® MGB real-time PCR is specific against *A. frigidophilus* (SAP172/VI04838), *A. laevis* (CBS105.52/VI03559), *A. repetans* (Se/VI03560), *A. stellus* (CBS568.67/VI03561), *Saprolegnia australis* (VI03661, VI03840, VI03841), *S. ferax* (VI04022), *S. diclina* (ATCC 90215) and *S. parasitica* (ATCC 90213) (Vrålstad *et al.*, 2009).

Although the conventional PCR using primers 42f and 640r can give false-negatives because it is 100-fold less sensitive than qPCR and is prone to false-positives due to cross-reaction with bacterial species and other *Aphanomyces* species, it can be used in certain circumstances and for sequencing to confirm the presence of *A. astaci* (Kozubíková *et al.*, 2011). Primers 42f and 640r are referred to as BO 42 and BO 640 in the 2012 OIE manual. The master mix is prepared in a 50-μl reaction volume with 0.5 μM primer BO 42 (5′-GCT TGT GCT GAG GAT GTT CT-3′), 0.5 μM primer BO 640 (5′-CTA TCC GAC TCC GCA TTC TG-3′), 1 × PCR buffer (75 mM Tris HCl, pH 8.8, 20 mM $[NH_4]_2SO_4$, 0.01% v/v Tween 20), 1.5 mM $MgCl_2$, 0.2 mM of each dATP, dCTP, dTTP, dGTP, 1.25 units DNA polymerase (e.g. Thermoprime Plus DNA Polymerase, AB Gene) and 2 μl template. Alternatively, the same concentration of primers and volume of DNA template can be used with Reddy Mix PCR Mastermix (ABgene with 1.5 mM $MgCl_2$). The thermocycling protocol is conducted using an initial denaturation at 96°C for 5 min followed by 40 cycles of denaturation at 96°C for 1 min, annealing at 59°C for 1 min and elongation at 72°C for 1 min with a final extension step at 72°C for 7 min. The amplified product of 569 bp is examined using gel electrophoresis. The sensitivity is 500 fg of target DNA and 10 zoospores. Due to the issues with specificity with these primers, the amplicon should be sequenced to confirm identity of *A. astaci*. If sequencing using primers BO 42 and BO 640 does not produce 100% similarity with *A. astaci*, then the template DNA should be sequenced using universal primers ITS1 and ITS4 from White *et al.* (1990) (Oidtmann *et al.*, 2006; Tuffs and Oidtmann, 2011; OIE, 2012b). Universal primers are listed in the fungal molecular section.

DNA is extracted from soft abdominal cuticle. Not all DNA extraction methods are suitable for oomycetes because like plants and algae, they contain cellulose, a complex polysaccharide, in their cell walls, therefore DNA extraction methods must be capable of breaking down cellulose. DNA extraction can be undertaken using a commercial kit, preferably with a proteinase K digestion step or the CTAB method (Ausubel, 1987) used by Vrålstad *et al.* (2009). Pre-treatment with liquid nitrogen is also recommended (Oidtmann *et al.*, 2006) because this is a mechanical means of breaking down the cell wall. It involves grinding 30–50 mg of excised soft abdominal wall pieces under liquid nitrogen using a mortar and pestle until a fine powder is achieved.

DNA extraction using the CTAB method (Ausubel, 1987) is performed according to Vrålstad *et al.* (2009). Surface contamination from crayfish tissue is removed by wiping with a towel and sterile water. Mycelium or crayfish tissue (10–40 mg) is placed into a 1.5-ml microfuge tube and frozen at −80°C for at least 15 min. To the frozen tube material is added 700 µl of pre-heated (65°C) CTAB buffer (20 g/l CTAB [Hexadecyl trimethyl-ammonium bromide] 1.4 M NaCl, 0.1 M Tris-HCl, 20 mM Na_2EDTA) and the material is ground in the buffer using a micropestle (Eppendorf). Alternatively grinding can be done by automation using a Precellys® 24 tissue homogenizer (Bertin Technologies, France) and disposable 2-ml tubes with steel beads (Kit Precellys MK28). Tissue samples are ground at 6500 rpm for 1 min three times, with 2-min intervals between centrifugation. Ground samples are frozen at −80°C for a minimum of 10 min before being heated to 65°C for 5–10 min. Samples are centrifuged for 5 min to remove potential bubbles or vapours. RNase solution (10 µl of 10 mg/ml) is added, mixed and heated at 65°C for 30 min. Proteinase K (10 µl of 20 mg/ml) is added, mixed and incubated at 65°C for 30 min. The tubes are vortexed and centrifuged at 12,000 × *g* for 5 min. A 600-µl aliquot of supernatant is transferred to a new 1.5-ml microfuge tube and 600 µl of chloroform is added. The tubes are centrifuged at 18,000 × *g* for 15 min before 400 µl of the aqueous phase is transferred to a new tube and 300 µl of ice cold isopropanol is added. The liquids are mixed carefully by immersion. The tubes are incubated at room temperature for 20 min before the DNA is pelleted by centrifugation at 18,000 × *g* for 10 min. The DNA is washed with 300 µl of ice-cold ethanol and centrifuged at 18,000 × *g* for 5 min. The supernatant is discarded and the DNA is dried in a vacuum dryer or heating block at 50°C for 15–30 min. The DNA is re-suspended in TE buffer and incubated at room temperature for at least an hour, then mixed carefully and centrifuged for 30 s before use in the qPCR. Each extraction run includes (i) an environmental control, which consists of a microfuge tube containing 200 µl of Milli-Q water that is left open on the bench during the DNA extraction process, and (ii) an extraction blank control, which consists of a microfuge tube containing 700 µl of CTAB-buffer that is taken through the extraction process. Each DNA is tested neat and at 1:10 dilution.

Aphanomyces invadans

Aphanomyces invadans is the causative organism of epizootic ulcerative syndrome (EUS) and is listed as a significant disease by the World Organization for Animal Health (OIE, Office Internationale des Epizooties). It is an ulcerative skin disease affecting wild and farmed freshwater and estuarine finfish in Australia, Bangladesh, Indonesia, India, Japan, Pakistan, Papua New Guinea, Philippines, Malaysia, Sri Lanka, Thailand, the USA and recently in Africa, in Botswana and Namibia. The disease has not been reported from Europe.

The disease and the fungus have been known by different names. In Japan, where the disease was first reported in 1971 from farmed ayu (*Plecoglossus altivelis*), it was called mycotic granulomatoses and the fungus, *Aphanomyces piscicida* (Egusa and Masuda, 1971). The disease was reported a year later in grey mullet (*Mugil cephalus*) in eastern Australia, where it was called 'red spot disease' because of the distinctive red-coloured lesions (McKenzie and Hall, 1976). The disease continued to spread and was reported from Papua New Guinea, south-east and south Asia and then Pakistan (Lilley *et al.*, 1998). In North America the disease was known as ulcerative mycosis when first reported from Atlantic menhaden (*Brevoortia tyrannus*) from Chesapeake Bay (Dykstra *et al.*, 1986). These strains were genetically different to the later isolates from menhaden (Blazer *et al.*, 1999) and there is some thought that the first isolation might have been *A. laevis*. The later isolates are genetically similar to the strains from Australia and Asia. *Aphanomyces invadans* was confirmed as the pathogenic organism causing ulcerative mycosis in menhaden through experimental challenge with zoospores. Subcutaneous injection of zoospores from *Achlya bisexualis* and *Phialemonium dimorphosporum* did not reproduce the disease (Sosa *et al.*, 2007). PCR detection confirmed the disease in largemouth bass and American shad, black drum, sheepshead, silver mullet, striped mullet and silver perch.

The organism was first named *Aphanomyces invadaris* but this has been corrected to

A. invadans (Willoughby *et al.*, 1995; David and Kirk, 1997).

In countries where EUS occurs, over 100 species of fish have been reported to be susceptible, but some species are resistant or show a milder response in experimental challenge. Tilapia are mostly resistant, stickleback are resistant, resistance is variable for roach, and rainbow trout show signs of infection and development of lesions but this occurs at a slower rate compared with highly susceptible fish. Milkfish and Chinese carp were found to be resistant to infection (Khan *et al.*, 1998; Lilley *et al.*, 1998). In Australia, EUS and the fungus have been reported from a variety of fish occurring in natural river systems in Western Australia, New South Wales, Northern Territory, Queensland and South Australia; southern black bream (*Acanthopagrus butcheri*), bony bream (*Nematalosa erebi*), Murray cod (*Maccullochella peelii*), sea mullet (*Mugil cephalus*), Yellowfin bream (*Acanthopagrus australis*), sand whiting (*Sillago ciliate*), silver perch (*Bidyanus bidyanus*), spangled perch (*Leiopotherapon unicolor*) and golden perch (*Macquaria ambigua*) (McKenzie and Hall, 1976; Fraser *et al.*, 1992; Buller 2004, 2007; Boys *et al.*, 2012; Rupert Mathwin personal communication). It has been reported from silver barb (*Barbonymus* [*Puntius*] *gonionotus*) in Myanmar (Burma) and Roho labeo (*Labeo rohita*) in Bangladesh (Roberts *et al.*, 1993).

In the USA, *A. invadans* has been confirmed in menhaden (*Brevoortia tyrannus*), largemouth bass and American shad, black drum, sheepshead, silver mullet, striped mullet and silver perch (Blazer *et al.*, 2002; Sosa *et al.*, 2007). EUS was reported from captive bullseye snakehead (*Channa marulius*) collected from waterways in Florida, USA (Saylor *et al.*, 2010). In Namibia and Botswana in Africa, *A. invadans* was reported from 24 species of fish found in the Zambezi river (Huchzermeyer and Van der Waal, 2012). Fish with the highest prevalence of EUS included squeakers (*Synodontis* spp.), sharptooth catfish (*Clarias gariepinus*) and Nembwe (*Serranochromis robustus*). In Egypt, *Aphanomyces* spp. was isolated from ulcerated lesions present on striped mullet (*Mugil cephalus*) and thin lip grey mullet (*Liza ramada*), but due to lack of sexual structures could not be speciated (Shaheen *et al.*, 1999).

Although the disease has not been reported from Europe, in experimental challenge, European catfish (*Silurus glanis*) are highly susceptible. Rainbow trout (*Oncorhynchus mykiss*) have a low or moderate susceptibility and eels are resistant (Oidtmann *et al.*, 2008).

EUS is a seasonal disease and occurs during periods of low temperatures and after periods of heavy rainfall (Bondad-Reantaso *et al.*, 1992), conditions that favour sporulation of *A. invadans* (Lumanlan-Mayo *et al.*, 1997). Low temperatures inhibit the inflammatory response of the fish (Chinabut *et al.*, 1995). The fungus can only invade and cause infection once the dermis is exposed (Lilley *et al.*, 1998). Events such as heavy rainfall lead to acid sulfate run-off into the waterways, causing damage to the skin and making it vulnerable to infection (Callinan *et al.*, 1995). Other factors such as viruses and bacteria may be involved in damage to the dermis and Rhabidoviruses, *Aeromonas* spp., *Vibrio* spp. and other Gram-negative organisms have been implicated in complicating the aetiology.

Clinical signs in infected animals include large red oval or circular ulcerated lesions or dermal ulcers up to 4 cm in diameter that extend to varying degrees into the underlying musculature. The lesions begin as an erythematous dermatitis; a mild to severe focal, chronic, active dermatitis usually seen after an initial assault such as acid sulfate runoff following extensive rain or flooding. Lesions are usually less than 1 cm, involve one or two scales and epidermis is present at the edge of the lesion and irregularly present over the lesion. The epidermis may be hyperplastic and oedematous with mononuclear cells present. At this stage there is no fungal involvement and the dermatitis may or may not resolve. Lesions that don't resolve progress into a necrotizing dermatitis that develops into dermal ulcers associated with a severe necrotizing granulomatous myositis. Lesions tend to occur on the dorsal and posterior areas of the body, but may occur on the ventral and anterior areas. Dermal ulcers have a sharply defined margin that may be reddened, and extend and expose the underlying musculature. Fungi are present in the lesions, particularly at the margins, but rarely penetrate the musculature. The hyphae can extend throughout the stratum spongiosum, stratum compactum and the underlying subcutaneous tissue and skeletal muscle. Epidermis may be absent from affected areas. The disease can cause high mortality and up to 80% of fish may be infected. The large red ulcerated lesions make the fish unsaleable (Callinan *et al.*, 1989).

Based on protein profiles from SDS-PAGE, *A. invadans* isolates from Australia and from striped snakehead (*Channa striatus*), mullet (*Mugil* spp.), bar-eyed goby or tank goby (*Glossogobius giuris*), walking catfish (*Clarias batrachus*) and three-spot gourami (*Trichogaster trichopterus*) from the Philippines were very similar (Callinan *et al*., 1995). Analysis using random amplification of polymorphic DNA indicated a single clonal type was responsible for the rapid spread of the disease throughout Australia and South-East Asia. Restriction fragment length polymorphism showed that *A. invadans* is more closely related to *A. astaci*, the agent of crayfish plague, than to other *Aphanomyces* species (Lilley *et al*., 1997). In experimental challenge *A. invadans* did not cause infection in freshwater crayfish (Lilley and Roberts, 1997).

Aphanomyces invadans is difficult to isolate in pure culture because it is slow growing and is rapidly overgrown by contaminating or secondary invaders such as *Saprolegnia* spp., *Aspergillus* spp. or bacteria. The fungus is killed by drying or temperatures below 7°C, therefore special consideration needs to be given to appropriate packaging and transport of specimens to the laboratory to maintain viability of the fungus. Ideally, moribund whole fish or fish collected within 12 h of death should be transported to the laboratory at temperatures between 7 and 25°C. The fungus will no longer be viable in autolysed tissue, making it unsuitable for culture. Tissue at the leading edge of the lesion and just beyond the leading edge and leading into the immediate musculature underlying the lesion is removed for culture or PCR. Hyphae in the centre of the lesion are likely to be dead and devoid of internal components, and therefore DNA, so several samples should be taken from a lesion to optimize the chances of obtaining viable fungal material (Roberts *et al*., 1993). Fungal isolation is most successful from muscle that appears yellow and soft (Fraser *et al*., 1992).

Tissue is removed and washed with three to four changes of sterile distilled water or saline or seared with a hot spatula or scalpel blade to remove surface bacteria. The seared skin is removed and small (2×2×2 mm) pieces of tissue are removed and pressed 2–3 mm into the surface of a culture plate. Bits of tissue should be surrounded by agar to protect emerging hyphae from drying. Plates are best used without pre-drying the surface because the fungus does not tolerate drying.

Plates are inverted and incubated at 20–25°C for 7–10 days. Petri dishes can be taped with parafilm or plates kept in a sealed box in the refrigerated incubator to prevent drying of the agar surface and tissue inoculum. Other pieces of tissue are placed onto a glass slide and flattened under a cover slip and examined for the presence of non-septate, branching hyphae. Hyphae are strong and difficult to break apart. For histology, fungal elements are visualized using Grocott's modification of Gomori's hexamine silver method (Grocott, 1955; Carleton *et al*., 1980), which stains the tissue green and fungal elements black. Haematoxylin and eosin with a silver counterstain can also be used.

Plates are incubated at 15–25°C and examined daily for up to 10 days. Use of a stereomicroscope can assist in detecting and removing emerging hyphae and subculturing to a fresh plate. *Aphanomyces invadans* does not produce aerial hyphae but grows within the agar. One way to separate *A. invadans* from contaminating fungi is to surround the tissue with a sterile stainless steel washer or glass ring. The contaminating fungi are kept within the ring, allowing *A. invadans* to grow under and away from the ring.

A procedure for culture and isolation of *A. invadans* using broth culture followed by plate culture and the use of antibiotics may give a higher isolation rate than culturing direct to an agar plate (Willoughby and Roberts, 1994). Thin slices of muscle tissue from the edge of the ulcer are cut from successively deeper muscle and each slice is placed into a Petri dish containing GP-POX broth (see under Fungal media). Media supplemented with yeast extract is not used at this stage because it encourages growth of bacteria. Liquid cultures are incubated for 6 h at 22°C. The tissues are examined under the microscope for newly formed hyphae. Hyphae should have a clear continuum with the hyphae in the muscle tissue and should be narrower than the parent hyphae. Hyphae that are extending from the tissue should be selected and some of the tissue included and transferred to fresh GP-POX medium (Roberts *et al*., 1993). Slices should be incubated for a further 24 h. Colonies should grow to about 4 mm in size. Large colonies may be saprophytic *Aphanomyces* species. Colonies of potential *A. invadans* need to reach this size so that when transferred to solid media they are able to survive. If transferred too soon they may not survive the drying effect on plate media. However, colonies need to be transferred before too much bacterial overgrowth happens. Colonies in broth culture are transferred to GPY-PS agar (glucose peptone yeast extract with penicillin and streptomycin). Colonies are examined hourly for overgrowth of bacterial contaminants. Bacteria-free margins of fungal growth are removed to fresh media. A bacterial-free margin of fungal growth is transferred to GPY broth and incubated at 10°C and subcultured every 4–5 weeks.

Optimal growth for strains from Australia, Bangladesh, Indonesia, Japan, Philippines, Thailand and the USA occurs on GYP agar (Roberts *et al.*, 1993) at 26–30°C. IM agar (Alderman and Polglase, 1986), a medium used for *A. astaci*, also supports the growth of *A. invadans* (Buller, 2004). On IM agar, colonies are flat without aerial hyphae and opaque with a white surface. At the leading edge of the colony, individual hyphae can be seen by eye. Lilley and Roberts (1997) reported that strains from Australia and Asia do not grow on SAB, CMA, MEA or PDA, and had poor growth on CDA compared with growth on GPY agar. Hatai *et al.* (1994) reported growth on PDA, but absence of growth on CDA agar and MA. Fraser *et al.* (1992) reported good growth for three Australian strains on glucose yeast extract agar (GYEA), and Griffin's GY agar (5 g glucose, 2.5 g yeast extract and agar), as cited by Dykstra *et al.* (1986), with colony size reduced by half on CDA, fish meat extract agar, Czapek yeast autolysate agar and CMA supplemented with 2% glucose and 1% yeast extract. However, it was suggested that even though growth was slow on CDA, it was a suitable medium as water from syneresis occurred around pieces of inoculum and it was hypothesized that this might prevent drying of the hyphae as it left the tissue and began growing within the agar (Fraser *et al.*, 1992).

The *Aphanomyces* strain (84-1240=ATCC 62427) isolated from menhaden in 1984 (Dykstra *et al.*, 1986) does not appear to be *A. invadans* because the isolate has a preference for slightly colder temperatures than *A. invadans* isolated from Australia and Asia and is able to grow on CMA and MEA, and slowly on SDA (Lilley and Roberts, 1997). The isolates have genetic differences and are thought to be *A. laevis* (Blazer *et al.*, 1999). Since then, *A. invadans* has been confirmed from menhaden and other freshwater fish in the USA (Blazer *et al.*, 2002; Sosa *et al.*, 2007).

Aphanomyces invadans grows slowly at a rate of 4 mm per day on glucose-peptone agar. Growth does not occur at 6°C or 37°C and is slow at 10–18°C, with optimal growth between 22 and 30°C and decreasing slightly at 34°C (Lilley and Roberts, 1997). Saprophytic fungi tend to grow at 37°C. The hyphae of *A. invadans* are non-septate, sparsely branched and 11.7–16.7 μm in width *in situ* (Roberts *et al.*, 1993; Willoughby *et al.*, 1995). In culture, young actively growing hyphae are narrower, 8.3 μm. In infected fish or in culture, hyphae become wavy or undulate as they die. At almost all stages of the disease the hyphae may have autolysed cell contents with no reproductive structures. Sporangia are filamentous, aseptate and indistinguishable from hyphae, except for a septum that separates the hyphae and sporangium. Sporangia are terminal on the hyphae and often complex, and up to 930 μm long and 10 μm wide. Along the sporangium one or four spore balls containing primary zoospores are produced laterally. Primary zoospores are 6.7–10 μm in diameter and release the secondary zoospores containing two flagella that encyst immediately (Roberts *et al.*, 1993; Willoughby *et al.*, 1995). It has an 'achlyoid' manner of primary spore discharge and encystment. After zoospore release, mature zoospores that are not released are viable and germinate readily, often as a large unit.

Australian and Philippine isolates have one or two lateral evacuation tubes per sporangium, whereas isolates from Thailand have four (Callinan *et al.*, 1995).

It may be difficult to produce zoospore and cyst formation *in vitro* and a number of different media have been reported. Before inducing sporulation a slice of pure growth from a culture plate is inoculated into a solution containing 30 ml of GY broth and 11 ml of distilled water. The culture medium is shaken gently at one oscillation per second for 7 days at 20–25°C. The mycelial mat is washed three times in distilled water before placing in a sporulation medium. In a study comparing sporulation media, salinity and temperature, it was found that temperature and salinity are critical to sporulation with sporulation occurring only at 15–25°C and at salinities of 0, 1 and 2% (Kiryu *et al.*, 2005). Use of natural water sources induced sporulation, whereas it did not occur in artificial media using artificial sea salts at salinity of 1‰ or SM medium (0.25 mM each of $CaCl_2$ and KCl) from Griffin (1978, as cited by Kiryu *et al.*, 2005). However, Fraser *et al.* (1992) reported spore formation in water and in Griffin's SM medium with and without 2% NaCl. All media contained two sterilized hemp seed halves. Secondary zoospores appear 18–24 h after sporulation and peak at 1–2 days. Encystment occurs immediately when physical shock is used, that is, when the zoospore solution is vortexed for 10 s at maximum speed. Willoughby and Roberts (1994) produced fungal spores by selecting a 4-mm plug of agar containing the leading edge of a colony and placing into GP broth at 22°C for 48 h. The plug is washed in autoclaved pond water and left overnight. An improved production of zoospores was achieved by growing *A. invadans* mycelium on sterile hemp or buckwheat seeds and once the mycelium is growing placing the seeds and fungus into V8 broth (5% Campbell's V8 juice, 0.2% $CaCO_3$, pH 6.1) for 48 h (Khan *et al.*, 1998). (Note, some countries may need Government permits to use hemp seeds, but buckwheat [*Fagopyrum esculentum*] seeds can also be used as bait for *Saprolegniaceae* [Kiziewicz, 2005]. Sporulation is induced by washing in autoclaved pond water and leaving overnight at 15–25°C in autoclaved pond or river water containing 0–2 ppt. Sporulation occurs 1–3 days after induction and zoospores can remain motile for 12–18 h. The process begins with the transformation of a long piece of hyphae into a sporangium, which is delineated by a highly refractive septum. Primary spores (cytoplasmic units) are formed from the sporangium's contents and line up as elongated spores linked by thin strands of cytoplasm (Fraser *et al.*, 1992). Spores are released from the sporangial discharge tubes either terminally or laterally. Encystment of spores begins after 12–18 h, but begins immediately with vigorous shaking or vortexing. In the encystment phase the spores become round and a cell wall develops to enclose a spherical cluster of 30–50 spores. Secondary zoospores form from the cysts. Secondary zoospores are seen within 12 h of sporangial development and are subspherical and biflagellate. A stereomicroscope or light microscope with a stage that can accommodate a Petri dish are used to view the sporulation process.

MOLECULAR IDENTIFICATION. *Aphanomyces invadans* is difficult to culture because it is slow growing and is quickly overtaken by rapidly growing saprophytic fungi and bacteria. A higher number of positive samples are detected using PCR compared with culture (Buller, 2007). PCR has been successful from fish tissue transported for 5 days in 70% non-denatured alcohol and after 3 days' transportation from tissue containing the lesion but without any solute (Buller, unpublished results). Careful selection of tissue material is crucial to successful amplification of DNA. Tissue from the leading edge and the immediate underlying tissue is selected for PCR. Hyphae in the centre of the lesion are likely to be dead and devoid of internal components (Roberts *et al.*, 1993) and some of the hyphal tips at the leading edge of the lesion may contain minimal DNA, therefore several samples should be taken from the leading edge of the lesion to optimize the chances of obtaining viable fungal material and hence DNA for PCR.

Choice of DNA extraction method is also important for successful detection by PCR. Hyphae of *A. invadans* and other members of the *Saprolegniaceae* contain cellulose in their cell walls and therefore DNA extraction methods suitable for plants, algae and fungi should be tested and validated. DNA extraction methods that are suitable for *A. invadans* are the DNeasy® mini plant tissue kit (Qiagen), and DNAzol reagent (Invitrogen) are suitable. These commercially available kits are suitable for the extraction of *A. invadans* from fish tissue or from fungal growth from broth culture and used according to the manufacturer's instructions. Some methods advocate a pre-extraction step of grinding the fungus under liquid nitrogen before proceeding to DNA extraction. This method can be difficult because the liquid nitrogen 'spits' on contact with the fungus. Mycelial mats from broth cultures need to be blotted between filter paper to remove excess moisture in an attempt to reduce the splattering of material when exposed to liquid nitrogen. In the three commercial kits mentioned here, pre-grinding under liquid nitrogen is not necessary. For PCR, fungal hyphae must first be grown in a broth culture such as GP broth because it is not possible to obtain hyphae without agar contamination when grown on agar plates since *A. invadans* grows within the agar and does not produce aerial hyphae. Agar contains PCR inhibitors.

A number of PCRs have been reported (Blaser *et al.*, 2002; Phadee *et al.*, 2004; Buller, 2004, 2007; Vandersea *et al.*, 2006; Oidtmann *et al.*, 2008).

Primers designed to the internal transcribed spacer regions are forward primer AIFP14 (5′-CTG ACT CAC ACT CGG CTA GC-3′) and reverse primer AIRP10 (5′-ATT ACA CTA TCT CAC TCC GC-3′) (Buller, 2004). The 25-μl master mix contains 1.5 μl of each primer (20 pmol working solution), 12.5 μl of 2× Promega master mix (Promega) and 2 μl of DNA. Thermocycling is conducted with an initial denaturation at 94°C for 5 min followed by 30 cycles of denaturation at 95°C for 1 min, annealing at 55°C for 1 min and extension at 72°C for 1 min. A final step consists of elongation at 72°C for 10 min. A product of 554 bp is obtained. The primers can be used at a range of working concentrations (20–100 pmol) with no loss of amplification. The primers detect *A. invadans* strains from Australia (several strains), Japan (NJ9701), Thailand (T99G2, IMI 836084) and the USA (WIC, WIC2). The primers are specific and do not amplify product from fungal species: *Aphanomyces astaci* (Ti, Pc), *A. frigidophilus* (ATCC 204464), *Achlya diffusa* (ATCC 16111), *Aspergillus niger*, *A. fumigatus*, *Pythium sulcatum* (WAC 8470), *Penicillium* spp. (AS-01-4050), *Saprolegnia diclina* (ATCC 36144), *S. parasitica* (AS-03-3827-355), *S. salmonis* (AS-03-3827-354), *Saprolegnia* spp., *Trichoderma* spp., or bacterial species: *Aeromonas hydrophila* (ATCC 7966), atypical *Aeromonas salmonicida* (AHLDA 1334), *Photobacterium damselae* ssp. *damselae* (AHLDA 1683), *Streptococcus dysgalactiae* (AHLDA 1616), *S. iniae* (AHLDA), *Vibrio alginolyticus* (01-618), *V. anguillarum* (AHLDA 1730), *V. cholerae* non-01 (AHLDA 996), *V. mimicus* (AHLDA 1654) (Buller, 2004, 2007). Using serially diluted DNA, the sensitivity of primers AIFP14 and AIRP10 was 19.5 pg/ml DNA when prepared with the Promega master mix and 19.5 ng/ml DNA when prepared in an in-house PCR mix (i.e. individual components added). In a comparison of primers AIFP14 and AIRP10 against other primers using the same master mix and DNA concentration, primers AIFP14 and AIRP10 (Buller, 2004) produced amplicon at a concentration of greater than 100 ng/band as did primers ITS11 and ITS23 (Phadee *et al.*, 2004). Primers 2F and R1 (Vandersea *et al.*, 2006) and primers P1 and P2 (Blazer *et al.*, 2002) produced a reduced concentration of amplicon (234 bp and 98 bp, respectively) at 70–90 ng/band, which was observed at half the intensity on the gel. Primers 2F and R1 were not specific and amplified product of the same molecular weight from *Penicillium* spp., *Aspergillus niger*, *Saprolegnia* spp. and *S. salmonis*. Primers P1 and P2 are specific at their recommended annealing temperature of 56°C, but at an annealing temperature of 57.5°C cross-reaction occurs with *Penicillium* spp.

Primers reported by Phadee *et al.* (2004) were designed to amplify the ITS1 and ITS 2 regions of *A. invadans* and are forward primer ITS11 (5′-GCC GAA GTT TCG CAA GAA AC-3′) and reverse primer ITS23 (5′-CGT ATA GAC ACA AGC ACA CCA-3′). Although these primers were originally reported as specific against *A. frigidophilus*, Oidtmann *et al.* (2008) reported cross reaction at concentrations of 10 ng or more and that primer ITS11 is fully homologous to *A. frigidophilus* and primer ITS23 deviates by 1 nucleotide only.

A semi-nested PCR was designed (Oidtmann *et al.*, 2008) and evaluated against primers of Phadee *et al.* (2004) and Vandersea *et al.* (2006). However, the second set of primers were not specific against *A. frigidophilus* and therefore only the single round primers are recommended. These are BO73 (5′-CTT GTG CTG AGC TCA CAC TC-3′) and BO639 (5′-ACA CAA GAT TAC ACT ATC TC-3′) using a 50-μl reaction volume, 0.6 μM of each primer, 2.5 μl DNA template and Reddy-Mix PCR Master Mix (ABgene AB-0575). Thermocycling was carried out at an initial denaturation at 96°C for 5 min followed by 20 cycles of denaturation at 96°C for 1 min, annealing at 58°C for 1 min, extension at 72°C for 1 min and a final extension step at 72°C for 5 min (Oidtmann *et al.*, 2008). Amplicon is 564 bp. The primers are specific against *Aphanomyces astaci*, *A. brassicae*, *A. cladogamus*, *A. cochlioides*, *A. euteiches*, *A. frigidophilus*, *A. helicoides*, *A. iridis*, *A. irregulare*, *A. laevis*, *Achlya racemosa*, *Pythium flevoense*, *Saprolegnia diclina*, *S. ferax*, *S. furcata*, *S. litoralis*, *S. parasitica*, *S. terrestris*, *Aspergillus* spp., *Fusarium solani*, *Candida albicans*, *Aeromonas hydrophila*, *A. salmonicida* ssp. *salmonicida*, *Citrobacter freundii*, *Edwardsiella tarda*, *Escherichia coli*,

Flavobacterium psychrophilum, *Vibrio anguillarum*, *Mycobacterium fortuitum*, *Mycobacterium marinum*, *Pseudomonas aeruginosa* and *Staphylococcus aureus*. The sensitivity for the single round PCR was 25 fg (0.1–1 genomic units). Oidtmann *et al.* (2008) found that the primers (2F and R1) of Vandersea *et al.* (2006) had a limit of detection of 2.5 pg corresponding to 10–100 genomic units and the primers (ITS11 and ITS23) of Phadee *et al.* (2004) detected 2 ng corresponding to 8000 and 8000 units.

Aphanomyces frigidophilus

Aphanomyces frigidophilus (NJM 9500) was reported from eggs of Japanese char (*Salvelinus japonicus*, reported as *S. leucomaensis*, which is a synonym) in Japan (Kitancharoen and Hatai, 1997) and has been reported from crayfish (*Austropotamobius pallipes*) in rivers in Spain (Ballesteros *et al.*, 2006). Strain SAP233 did not cause mortality in crayfish in experimental challenge, but was originally isolated from subabdominal cuticle of crayfish that died in a mass mortality event.

A. frigidophilus grows on PG1 agar and GY agar to which Streptomycin sulfate (150 μg/ml) and ampicillin (100 μg/ml) can be added to inhibit bacterial growth (Kitancharoen and Hatai, 1997; Ballesteros *et al.*, 2006). Growth occurs at 5–30°C with an optimum of 25°C. Growth occurs at a pH range of 5.0–11.0 with an optimum of 7.0–9.0. Zoospores germinate at a pH range of 5.0–11.0 in GY broth and zoospore production is more active at low temperatures such as 10°C and does not occur at 30°C. Hyphae are aseptate, 7–10 μm in diameter, smooth, slightly wavy and moderately branched. Zoosporangia are isodiametric, long, with cross walls seldom observed. Zoospores form as a single row of elongated tapering spores and are usually connected by a thin strand of protoplasm and emerge from an orifice. Zoospores emerging later than the primary zoospores become elongated with rounded ends as they move through the zoosporangium. At emergence spores are irregularly shaped but soon round up into a spherical shape and encyst as a cluster at the orifice. Secondary zoospores are produced and are reniform and laterally biflagellate. Oogonia are 16–25 μm in diameter, abundant, have a short oogonial stalk, are usually pyriform, spherical or subspherical, have a short papillate and are crenulated or have an irregular or roughened outer contour without spines or tubercules. Roughened contours also occur in young oogonia. Oogonia are occasionally elongate and intercalary in the hyphae. A single oospore is 14–22 μm in diameter, mostly fills the oogonium and is mainly spherical but occasionally obovate or irregular. Oospores are predominantly subcentric with a large shiny vesicle surrounded by fine granules. Antheridium are not detected in the sexual stage (Kitancharoen and Hatai, 1997). The *A. frigidophilus* strain isolated from freshwater crayfish does not produce chitinase constitutively and no oogonia were observed when grown on CMA, hemp seed or snake skin. It has a 95% sequence similarity to *A. astaci* in the ITS 1 region and differs by 10 of 614 bases in the ITS2 region.

Aphanomyces laevis

In a study to investigate the identification of fungal disease in Nile tilapia (*Oreochromis niloticus niloticus*) and African catfish (*Clarias gariepinus*), two popular freshwater fish species farmed in Egypt, *Aphanomyces laevis* was one of the most commonly isolated oomycetes fungi, along with *Achlya klebsiana* (Ali *et al.*, 2011). In experimental challenge experiments, bath exposure led to higher mortality than intramuscular injection. Also commonly isolated were two soil fungi, *Phoma herbarum* and *Purpureocillium lilacinum*. *Aphanomyces laevis* was isolated from a fatal mycosis affecting blue panchax (*Aplocheilus panchax*) native to ponds, lakes and other freshwater bodies in India and utilized for mosquito control. A white cottony growth occurred on the body and gills and patches of epithelium loss occurred (Mondal and De, 2002). Hyphae penetrated to the underlying musculature.

In *A. laevis* the oogonial walls are smooth, not warted and do not have spines. The eggs are 16.5–25 μm thick. Hyphae are slender and branched and 5–7.5 μl thick. Spores are rod-shaped in the sporangium. Sporangium are filamentous and indistinguishable from somatic hyphae and contain a single row of aplanospores. The spores are released singly at the apex of the sporangium and when released the spores aggregate and form a hollow ball-like structure. The spores become spherical and encyst. Reniform, biflagellated spores emerge after encystment and swim before encysting again. Encysted spores are 8–10 μm in diameter. Spores eventually germinate to form hyphae (Coker, 1923; Mondal and De, 2002).

Aphanomyces piscicida

The name *Aphanomyces piscicida* is invalid and the organism is *A. invadans* (MycoBank; www.mycobank.com).

Aphanomyces repetans

Aphanomyces repetans was isolated from *Procambarus clarkii* and *Pacificastacus leniusculus* in Italy and Spain (Royo *et al.*, 2004), and from white-clawed freshwater crayfish (*Austropotamobius pallipes*) in watercourses in central Italy along with *A. astaci*, the causative agent of crayfish plague (Cammà *et al.*, 2010). *Aphanomyces repetans* is considered a saprophyte because it does not cause mortality in noble crayfish (*Astacus astacus*) or the yabby (*Cherax destructor*) from Australia when tested in experimental challenge. *Aphanomyces repetans* does not express chitinase constitutively, has a rapid growth and zoospores germinate in PG1 medium, all factors that are different to *A. astaci* (Royo *et al.*, 2004). *Aphanomyces repetans* is not listed in MycoBank.

Aphanomyces salsuginosus

Aphanomyces salsuginosus (NJM 0801^{T}=NBRC 106578) was isolated from ice fish (*Salangichthys microdon*), a brackish water fish that had been captured in a lake in Japan and held in an aquarium for a few weeks (Takuma *et al.*, 2010). Fish had white cotton wool growth on the mouth, fins and head. Histologically, fungal hyphae were observed in the oral cavity and hyphae had penetrated the cartilage tissue of the mandible and maxilla. The hyphae penetrated the dermis and musculature.

Growth occurs on GY agar containing 0.5% NaCl. Initial isolation is conducted on GY plus 0.5% with the addition of 500 mg/ml of both ampicillin and streptomycin. On GY agar with 0.5% NaCl, colonies are white with flat mycelia. Growth occurs after 4–6 days at 15°C and optimal growth occurs at 20°C with a range of 10–25°C. Growth does not occur at 30°C. Growth occurs slowly at NaCl concentrations of 1 or 2% and generally declines at 3 and 5% according to strain variation. Hyphae are aseptate, moderately branched, 5–10 μm in width, smooth and slightly wavy. Zoosporangia are slender and the same diameter as the hyphae, i.e. isodiametric. Primary zoospores are 8–11 μm in diameter and occur in a single row in the zoosporangium and cluster at the orifice of the zoosporangium. Sexual reproductive structures are produced when hyphae are incubated in sterile tap water containing 0.5% NaCl for 3 days. Oogonia are 21–33 μm in diameter, abundant, usually spherical or subspherical and very rarely pyriform and have irregular short papillae in old and young oogonia. Pits are absent from the oogonial wall. Single spores 19–27 μm in diameter are spherical, contain a subcentric refractive globule in the cytoplasm and fill the oogonium. Antheridial branches, if present, are predominantly androgenous, but monoclinous and diclinous branches occur infrequently. Oogonia with single antheridium predominate, but a few oogonia may be present with two or three antheridia. Oogonial stalks are unbranched (Takuma *et al.*, 2010).

Morphologically *A. salsuginosus* is similar to *A. laevis* and *A. stellatus* as similar sexual structures are produced. *Aphanomyces salsuginosus* can be differentiated from *A. laevis*, which has a smooth ooginial wall in comparison with *A. salsuginosus* and *A. stellatus*, which have irregular short papillae. The basal portion of the oogonium is broad in both *A. salsuginosus* and *A. laevis*, but is tapered in *A. stellatus*. Atheridial branch formation is monoclinous and diclinous in *A. laevis* and *A. stellatus*, whereas *A. salsuginosus* is androgenous (Takuma *et al.*, 2010).

Sequence analysis of the ITS over 659 bp indicates 93.5% similarity to *A. astaci* (Takuma *et al.*, 2010).

Aphanomyces sinensis

Aphanomyces sinensis (NJM 0719^{T}=NBRC 106581, MB 518613) was isolated from juvenile soft-shelled turtle (*Pelodiscus sinensis*) cultured in Japan (Takuma *et al.*, 2011). Clinical signs of infection include small white maculae on the carapace. Histologically hyphae were observed in the epidermis but did not penetrate to the dermis or internal organs. *Aphanomyces sinensis* was pathogenic in experimental challenge. It is likely that infection with *A. sinensis* leads to hypoxia by preventing cutaneous respiration, as in turtles, 70% of oxygen uptake is achieved through the shell.

Colonies on GY agar are white with flat mycelial growth and almost cover the plate in 5 days at 20°C. Optimal growth occurs at 25–30°C and does not occur at 35°C. Optimal pH for growth is 6–9. Hyphae are 3–12 µm in width, aseptate, moderately branched, smooth and slightly wavy. Zoosporangia are isodiametric, this is, the same diameter as the hyphae. Primary zoospores are 8–12 µm in diameter and occur as a single row in the zoosporangium and cluster and encyst at the orifice. Secondary zoospores are produced and are reniform and laterally biflagellated. Sexual structures are produced in hemp seed culture. Oogonia are 15–22 µm in diameter, smooth, spherical or subspherical and rarely pyriform. Oospores in the oogonia are mainly centric to subcentric, with a central nuclear spot and a large shiny vesicle, and usually fill the oogonium. Oogonial stalks are unbranched. Antheridia are usually present and are predominantly diclinous, and infrequently monoclinous and androgynous. Oogonia have either two or three anteridia. Antheridial cells are simple and vermiform and often irregular (Takuma *et al.*, 2011).

Morphologically, *A. sinensis* is similar to *A. laevis* as both have spherical oogonia. The antheridial branches of *A. laevis* are predominantly diclinous and monoclinous and rarely androgynous. Sequence analysis of 662 bp of the ITS region demonstrates 80.3% similarity to *A. frigidophilus* as the closest species (Takuma *et al.*, 2011).

Aphanomyces stellatus

Aphanomyces stellatus was isolated from ice fish (*Salangichthys microdon*) from brackish water in Japan (Shah *et al.*, 1977).

5.1.21 *Atkinsiella* (Genus)

Atkinsiella belongs to the kingdom Chromista, phylum *Oomycota*, class *Oomycetes* and family *Haliphthoraceae*.

Atkinsiella are characterized by a saccate lobed thallus, vacuoles at early stages of zoosporogenesis and diplanetic biflagellate zoospores. Zoospores encyst within the thallus and emerge in the second swimming stage (Vishniac, 1958).

Atkinsiella awabi see *Halodaphnea awabi*

Atkinsiella dubia

Atkinsiella dubia was isolated from the mantle of abalone (*Haliotis sieboldii*) and from the gills of the swimming crab (*Portunus trituberculatus*) cultured in Japan (Nakamura and Hatai, 1995b).

Growth occurs on PYGS and colonies are 30 mm, moist and moderately heaped in the centre, crystalline and tuberculate. In PYGS broth hyphae are 250 µm in diameter, non-septate, radially branched, stout and have clusters of shiny spherical granules but no oil droplets or vacuoles. Pigment is not produced. A 3-day-old culture is strongly saccate. In seawater, clusters of granules are evenly distributed inside the mycelia and mycelia develop narrow branches or discharge tubes through which zoospores emerge. Zoospores are produced after 35 h at 20–25°C and there are two stages to zoospore production. In the first stage the clusters or protoplasmic masses, which are supported at the centre of the zoosporangia by protoplasmic threads, differentiate into loose networks of zoospores and finally individual zoospores. The zoosporangium is the same size and shape as the mycelium. Zoospores encyst within the zoosporangium and discharge tubes and are 8–11 µm, biflagellate and subglobose to globose, and swim sluggishly. In the second stage zoospores are released individually and swim freely. They are 4–7×8–12 µm, dimorphic, diplanetic, laterally biflagellated, pyriform, slipper-shaped and isokont.

Atkinsiella hamanaensis see *Halodaphnea hamanaensis*

Atkinsiella okinawaensis see *Halodaphnea okinawaensis*

Atkinsiella panulirata see *Halodaphnea panulirata*

Atkinsiella parasitica see *Halodaphnea parasitica*

5.1.22 *Branchiomyces* (Genus)

Branchiomyces belongs to the kingdom Chromista, phylum *Oomycota*, class *Oomycetes* and order *Saprolegniales*.

Branchiomyces demigrans and *B. sanguinis*

Branchiomyces demigrans and *B. sanguinis* cause gill rot, a respiratory disease reported from a number of fish species including eel, carp, tench, pike, perch, bass, pumpkinseed, bluegill, burbot, stickleback, salmonids and tambaqui in Europe including Germany, and in Brazil, Japan, Taiwan and the USA. *Branchiomyces sanguinis* has been reported from cultured eels (*Anguilla japonica*) in Japan and Taiwan, cultured channel catfish (*Ictalurus punctatus*) fry in the USA and common carp (*Cyprinus carpio*) in Germany. *Branchiomyces demigrans* has been reported from northern pike (*Esox lucius*) and tench (*Tinca tinca*) in Europe (Egusa and Ohiwa, 1972; Chien *et al.*, 1979; Liu *et al.*, 1985; Khoo *et al.*, 1998). In channel catfish fry the disease presented when temperatures were 28–32°C, with varying numbers of mortality and number of ponds affected (Khoo *et al.*, 1998). *Branchiomyces sanguinis* occurs intravascularly and *B. demigrans* tends to occur outside the vessels. Gills are striated or marbled with pale areas of infection or dying tissue. Macroscopic evidence of infection may not be seen, but fungal elements are seen histologically (Paperna and Smirnova, 1997; Khoo *et al.*, 1998). The fungus infiltrates filamental arteries and lamella capillaries of the gills and appears as strings or clusters of bead-like structures that are clusters of zoosporangia contained within a thick hyphal wall. Metastatic lesions containing hyphae and spores occur in the epicardium and spores but no hyphae may be seen in the spleen (Chien *et al.*, 1979; Liu *et al.*, 1985). The spores within the zoosporangium are released into the host tissue when the zoosporangium ruptures. Hyphae measure 10–30 µm in width, and spores are 4.5–7 µm in diameter. In channel catfish isolates, the hyphae in the tissues are light brown, refractile, branching, non-septate with granular material, spores 8–15 µm, occurring in the centre of the hypha (Khoo *et al.*, 1998). In an H&E stain the hyphae have eosinophilic walls and the central granular material is eosinophilic to basophilic.

The isolate from channel catfish was cultured on CMA at 16–37°C, with the highest growth rate occurring at 37°C at a rate of 0.7 mm/h compared with 0.1 mm/h at 16°C (Khoo *et al.*, 1998). Hyphae measured 7 µm. No sexual or asexual reproductive structures were produced on hemp seed water culture or CMA.

A *Branchiomyces*-like infection was reported from red tilapia (*Oreochromis niloticus* × *O. mossambicus*) and green tilapia (*O. niloticus* × *O. aureus*) cultured in Israel, and from Amazonian loricariid (*Baryancistrus* spp.) (Paperna and Smirnova, 1997) and tambaqui or Chacama (*Colossoma macropomum*) (Pereira *et al.*, 2012). On some farms fish showed subclinical infection with little damage to the gills, whereas on other farms 85% mortality occurred. Undivided and sporulating stages of the fungus are observed histologically in gill tissue. Hyphae of up to 11 µm may contain a single multinucleated plasmodium with a foamy cytoplasm. These divide into uninuclear single cells 6–7×5–6 µm. Daughter plasmodia containing single and multinucleated cells and a slightly eosinophilic cytoplasm extend through the hypha. Hyphae become filled with soft and hard-walled sporonts that may be single or binucleated. The final stage is a sporont filled with spores that are then released into the surrounding gill tissue and the environment (Paperna and Smirnova, 1997).

5.1.23 *Haliphthoros* (Genus)

Haliphthoros belongs to the kingdom Chromista, subkingdom Harosa, Infrakingdom Heterokonta, phylum *Oomycota*, class *Peronosporea*, subclass *Saprolegniidae*, order *Saprolegniales* and family *Haliphthoraceae*. Members of the family *Haliphthoraceae* are filamentous, branching, holocarpic and parasitize diatoms or marine invertebrates. Large monoor diplanetic zoospores are produced within the thallus (Vishniac, 1958). Members of the genus have zoospores that are laterally biflagellate and monoplanetic.

Haliphthoros milfordensis

Haliphthoros milfordensis was first described from eggs of eastern or Atlantic oyster drill (*Urosalpinx cinerea*), a small predatory sea snail (Vishniac, 1958). The fungus caused mortality in larvae of the mangrove or mud crab (*Scylla serrata*) at a fisheries research station in Indonesia (Hatai *et al.*, 2000), infected zoeae of black tiger prawn (*Penaeus monodon*) cultured in Vietnam (Chukanhom *et al.*, 2003), eggs and zoeae of the blue swimmer crab (*Portunus pelagicus*) cultured in

Japan (Nakamura and Hatai, 1995a), and gill and thoracic segments of juvenile American lobster (*Homraus americanus*) in the USA (Fisher *et al.*, 1975). Infection in abalone (*Haliotis gigantea*) held in an aquarium in Japan resulted in lesions that were flat or had tubercule-like swellings on the mantle, epipode and dorsal surface of the foot and contained fungal hyphae (Hatai, 1982).

On PYGS agar, colonies are 20–25 mm, whitish with damp centres after 5 days at 25°C. Hyphae are 7.5–30 μm in width, aseptate, branched and have numerous shiny granules and at times concentrated masses of protoplasm. The masses of protoplasm are also seen when grown in artificial seawater. Zoosporangia with discharge tubes arise from fragments that are tuberculate, saccate or irregular and variable in size and shape. Zoospores are produced after 8–12 h when mycelia are cultured into artificial seawater. Sporulation only occurs with 100–60% actual seawater and does not occur with more dilute seawater (Vishniac, 1958). A discharge tube 5–10 μm×15–300 μm, straight or slightly curved, forms on the lateral side of the zoosporangium. The protoplasm begins to divide in the sporangia and continues in the discharge tubes just before zoospores are liberated. Zoospores are 6–7.5×7–12 μm and 6.5×8.5 μm, kidney or slipper-shaped, elongated, reniform and laterally biflagellate. After a swarming period of 48 h, spores encyst. Encysted zoospores are 3–7 μm, globose or subglobose and a birefringent granule of 2 μm in diameter is seen near the cell wall. After encystment, spores germinate with a hair-like filament 15–150 μm in length (Vishniac, 1958; Hatai *et al.*, 2000). Growth occurs at 15–40°C with an optimum of 25–30°C and no growth at 5–10°C. Growth occurs at 50–100% seawater with poor growth at 25% seawater. Growth does not occur on PYG agar without seawater, and growth is poor on PYG agar prepared with 5% NaCl and 1–5% KCl (Chukanhom *et al.*, 2003).

5.1.24 *Halocrusticida* (Genus)

Halocrusticida belongs to the kingdom Chromista, phylum *Oomycota*, class *Oomycetes* and family *Haliphthoraceae*.

Halocrusticida baliensis

Halocrusticida baliensis was isolated from the zoea of a mangrove crab (*Scylla serrata*) cultured in Indonesia (Hatai *et al.*, 2000).

On PYGS agar containing streptomycin and ampicillin, colonies are 17 mm, yellowish after 4 weeks at 25°C. Hyphae are 15–40 μm, aseptate, irregularly branched, saccate-lobed. The thallus is aseptate and becomes septate as it divides into subthalli that are cylindrical, saccate, irregular, tuberculate and variable in size and shape. During zoosporogenesis vacuoles and numerous shiny granules are seen. Gemmae 20–60 μm develop from zoosporangia in seawater and are saccate-lobed, thick-walled and have shiny granules. Zoosporangia are the same size as the subthalli. One to several discharge tubes containing zoospores in lines of two are extended from a sporangium. The discharge tubes are 7.6–25 μm in width to 30–450 μm in length, have a broad base and may or may not taper. Zoospores develop 22–24 h after inoculation into artificial seawater. Zoospores are 7.2×6.1 μm, biflagellated laterally, monoplanetic, pyriform, slipper-shaped, oblong and spherical. Encysted zoospores develop a hair-like filament 7.5–210 μm in length that is enlarged at the tip to form a hyphal bud (Hatai *et al.*, 2000).

5.1.25 *Halodaphnea* (Genus)

Halodaphnea spp. belong to the kingdom Chromista, phylum *Oomycota*, class *Oomycetes* and order *Myzocytiopsidales*.

Halodaphnea awabi

Halodaphnea awabi (synonyms are *Atkinsiella awabi* and *Halocrusticida awabi*) caused disease in abalone (*Haliotis* [*sieboldii*] *gigantea*) cultured in Japan (Kitancharoen *et al.*, 1994). Infected abalone had a tubercule-like swelling on the mantle with melanized lesions on the peduncle and fungal hyphae were isolated from infected areas.

Growth occurs on PYGS agar at 20°C. Hyphae are 16–140 μm in diameter, stout, branched and irregular in size. Wavy or coiled lateral or terminal discharge tubes form and produce sporangia. Zoospores are pyriform, biflagellated and diplanetic. A hair-like filament with a globular enlarged tip is produced from an encysted spore, although direct germination without filament formation also occurs.

Halodaphnea hamanaensis

Halodaphnea hamanaensis is the accepted name and *Atkinsiella hamanaensis* and *Halocrusticida hamanaensis* are synonyms (Global Biodiversity Information Facility and MycoBank, www. gbif. org and www.mycobank.org).

Halodaphnea hamanaensis was isolated from the ova of the mangrove crab (*Scylla serrata*) cultured in Japan (Bian and Egusa, 1980). The eggs become filled with fungal hyphae, which emerge from the eggs forming discharge tubes. Growth occurs on PYGS agar containing NaCl at a final concentration of 3%, on NA+3% NaCl, and YpSs +NaCl (powdered yeast extract 4 g, soluble starch 15 g, K_2HPO_4 1 g, $MgSO_4.7H_2O$ 0.5 g, 30 g NaCl, 1 l water). Colonies are cream-white, 10–13 mm after 7 days at 29–32°C (optimum temperature), flat and filamentous and folded in the centre. The margin of the colony becomes irregular over time due to the swollen hyphal tips. Over time the colour changes to grey to light brown and a light brown pigment is produced that diffuses into the agar. Pigment is produced in 2 days when grown on brain–heart infusion agar, but may take up to a month on YpSs, PYGS and NA agars. Pigment is not produced on CMA. Growth occurs at 15–32°C in nutrient broth containing NaCl and at 17–32°C in PYGS broth. Maximum growth occurs at 2–3% NaCl. The fungus develops a saccate-lobed thallus, discharge tubes are elongated, vacuoles occur during zoosporogenesis and zoospores are diplanetic, features that place it in the genus *Halodaphnea*. Hyphae are 12–40 µm tubular or lobed and sparingly septate, and contain large oil droplets, vacuoles and shiny granules that begin as small dots but become rod-shaped up to 13 µm in length. Discharge tubes are 40–1150×10–22 µm, simple, non-septate equal in diameter but taper distally. The fungus is holocarpic, meaning the thallus is entirely converted into sporangia. Zoospores are formed and cleaved within the discharge tubes and the sporangium, and after a short motile stage are released and form a dense mass near the tips of the discharge tubes, but may remain encysted within the sporangia and discharge tubes. Zoospores are 6.3×4.5 µm, pyriform, slipper-shaped and have two lateral flagella. Spores swim and then encyst. Encysted spores may have a hair-like filament. No morphological differences are seen between primary and secondary zoospores. In shaker-grown broth culture, colonies may have inflated, spherical, obovate, naviculate hyphal tips (Bian and Egusa, 1980).

Halodaphnea okinawaensis

Halodaphnea okinawaensis was isolated from the zoeae and eggs of the blue swimmer crab (*Portunus pelagius*) cultured in Japan (Nakamura and Hatai, 1995a). *Atkinsiella okinawaensis* and *Halocrusticida okinawaensis* are synonyms. Infected eggs were brown in colour, whereas infected Zoeae become whitish due to the infiltration of hyphae.

Growth occurs on PYGS agar at 25°C. Zoosporangia are the same size and shape as the subthalli and have one to several discharge tubes from the sporangium.

Halodaphnea panulirata

Halodaphnea panulirata infected the phyllosoma of Japanese spiny lobster (*Panulirus japonicus*) cultured in Japan (Kitancharoen and Hatai, 1995). *Atkinsiella panulirata* and *Halocrusticida panuliri* are synonyms.

Growth occurs on PYGS agar at 25°C. Growth is slow and colonies are creamy white, raised and moist with some growth submerged in the agar. Hyphae are 10–22 µm, arranged in a radiating pattern, irregularly branched and occasionally separated by cross walls into subthalli. Sporangia form from the subthalli. Sporangia may have discharge tubes at the terminal or subterminal area that are wavy or partly coiled. Zoospores are biflagellated, siokont, diplanetic, pyriform and reniform. Zoospores are produced in artificial seawater. Encysted spores are connected by a hair-like filament with a globular enlarged tip. Encysted spores also germinate directly as stout hyphae. Growth occurs in 1.5–6.0% seawater and in 2.0–4.0% NaCl.

Halodaphnea parasitica

Halodaphnea parasitica was isolated from the eggs and body lesions of rotifer (*Brachionus plicatilis*) cultured in Japan (Nakamura and Hatai, 1994). The fungus was initially named *Atkinsiella parasitica* but this is now a synonym along with *Halocrusticida parasitica* and *Halodaphnea parasitica* is the accepted name (Global Biodiversity Information Facility).

Growth occurs on PYGS agar incubated at 25°C. Monoplanetic biflagellate zoospores are produced when cultured in artificial seawater. Discharge tubes are infrequently branched.

5.1.26 *Lagenidium* (Genus)

Lagenidium spp. belong to the kingdom Chromista, phylum *Oomycota*, class *Peronospores*, subclass *Saprolegniidae*, order *Pythiales* and family *Pythiaeceae*.

Lagenidium callinectes

The name *Salilagenidium callinectes* is an obligate synonym. *Lagenidium callinectes* caused mycosis in tiger prawn (*Penaeus monodon*) larvae cultured in India (Ramasamy *et al.*, 1996). Fungal hyphae were recovered from the thorax, abdomen, swimming appendages and eyestalks, and the musculature. The fungus infects eggs of the blue crab (*Callinectes sapidus*) found in Chesapeake Bay, and rivers of Virginia in the USA, and affects production of crab larvae (Sandoz *et al.*, 1944; Rogers-Talbert, 1948). It has also been isolated from eggs of the blue swimmer crab (*Portunus pelagicus*) cultured in Japan (Nakamura and Hatai, 1995a). The zoospores of the fungus are capable of penetrating the egg membrane. In experimental infection, eggs of other crabs species, oyster crab (*Pinnotheres ostreum*) and mud crab (*Neopanope texiana*) became infected, but eggs of *Sesarmas cinereum* did not (Rogers-Talbert, 1948).

Lagenidium callinectes grows on PDA and on SAB with dextrose and supplemented with 7 μg/ml streptomycin to reduce bacterial contamination. On PYGS agar colonies are 35–40 mm, whitish with damp centres after 5 days at 25°C. Hyphae are 5–30 μm in width, contorted, irregularly branched and sparsely septate or aseptate and contain numerous shiny-rod-shaped granules. Growth occurs at 15–25°C. Zoospores form after 12 h of inoculation of mycelia into artificial seawater. Zoospore formation begins with a flow of protoplasm into the tip of the discharge tube and vesicles appear. The protoplasmic masses are connected in a chain by a protoplasmic thread. Once all the protoplasm has entered the vesicles, flagellae appear within 5 min and by 10 min zoospores form and swim freely within the vesicle for 25 min before being released after 40 min. Mature vesicles are 10×80 μm, globose to subglobose and gelatinous. Discharge tubes are broad at the orifice and 5–17×18–200 μm. Zoospores may be released singly through a hole in the vesicle wall or by complete rupture of the vesicle. Zoospores are 5–15×8–15 μm, laterally biflagellated, reniform to elongate, isokont, monoplenetic and swim for up to 20 h, but not beyond 24 h. Encysted zoospores are 6–13 μm, globose to subglobose and without flagella. A range of salinity from 5 to 30 ppt is tolerated by the fungus and growth of hyphae and spore production occurs at a rapid rate at this salinity range (Rogers-Talbert, 1948; Hatai *et al.*, 2000).

Lagenidium chthamalophilum

Lagenidium chthamalophilum was isolated from the ova of the barnacle, *Chthamalus fragilis* var. *denticulata* collected in the USA and found at the waterline attached to pilings, rocks, sea walls and other structures (Johnson, 1958). Growth occurs on sea water agar and on sea water agar to which is added 0.1% glucose and 0.05% yeast extract. Slender, sparingly branched and vacuolate hyphae with a contorted appearance are produced on culture. In the infected ovum, hyphae are 10–18 μm, stout, highly vacuolated, branched and contain numerous cytoplasmic oil droplets. Occasionally, lateral globose swellings of up to 39 μm in diameter are produced. Sporangia are produced by segmentation of intramatrical hyphae. Antheridia are not observed.

Lagenidium myophilum

Lagenidium myophilum caused a fungal infection resulting in 100% mortality of juvenile coonstripe prawn or shrimp (*Pandalus* [*hypsinotus*] *prensor*) cultured in Japan (Nakamura *et al.*, 1994). Whitish and sometimes black lesions occurred on the muscle.

Growth occurs on PYGS agar at 10–15°C.

Lagenidium thermophilum

Lagenidium thermophilum caused mycosis in eggs and larvae of black tiger shrimp or prawn

(*Penaeus monodon*) cultured in Thailand (Muraosa *et al.*, 2006), and eggs and zoeae of mangrove or mud crab (*Scylla serrata*) cultured in Indonesia (Nakamura *et al.*, 1995). Infected eggs appear white. In the infected crabs, fungal hyphae grow throughout the crab. In experimental challenge, *L. thermophilum* infected the swimming crab (*Portunus tribuerculatus*).

Samples are cultured to PYGS plates, onto which streptomycin sulfate and ampicillin are added to reduce bacterial growth. Some methods add the antibiotics to the medium at the time of preparation, whereas other methods rinse the surface of the plate with antibiotic solution before inoculation of the medium.

Lagenidium thermophilum grows on PYGS plates at 25°C as whitish, flat and filamentous colonies in 4 days, and reaches 30 mm diameter in 7 days. Vegetative hyphae are 15–30 µm in width, aseptate, irregularly branched and have numerous protoplasmic oil droplets. Some hyphae have swollen ends 40 µm thick. Growth occurs at 15–35°C, with an optimum of 25°C. The type strain of *L. thermophilum* grows at 20–40°C, with an optimum of 30°C. *Lagenidium thermophilum* from crabs grew at 15–45°C with an optimum of 30–40°C. Growth occurs on PYGS agar supplemented with 0 or 5% NaCl. Zoospores are produced 2–36 h following transfer of mycelia to sterile artificial seawater. Discharge tubes 7–10 µm, straight or irregularly curved, develop gradually from the hyphae and vesicles form at the tip of the discharge tubes. Protoplasmic masses with oil droplets move into the vesicles and laterally biflagellated zoospores form within 15 min, and swimming zoospores are seen at 25 min. The vesicle separates and then the zoospores are discharged at 30 min. Zoospores are pyriform to subglobose, monoplanetic and 13×9 µm. The zoospores encyst after several hours and lose their flagella and become globose (Nakamura *et al.*, 1995; Muraosa *et al.*, 2006).

5.1.27 *Pythium* (Genus)

Pythium belongs to the kingdom Chromista, phylum *Oomycota*, class *Oomycetes*, order *Pythiales* and family *Pythiaceae*.

Pythium flevoense

Pythium flevoense caused a visceral mycosis in ayu (*Plecoglossus altivelis*) larvae cultured in Japan (Miura *et al.*, 2010). The ayu had opaque abdomens from which fungal hyphae were seen microscopically. Hyphae were present in the musculature, spinal cord, swim bladder, kidney, spleen, pancreas, intestine and abdominal cavity. Growth occurs on GY agar (1% glucose and 0.25% yeast extract) to which a solution of 10 µg/plate of streptomycin and ampicillin was spread onto the surface of the plate to inhibit bacterial contaminants. Colonies are 31.2 mm, white after 5 days at 25°C. Growth occurs at 5–30°C, but not at 36.5°C. Growth occurs at 0–3.5% NaCl, but not at 4% NaCl, and growth is optimal at 0.5% NaCl. On CMA, colonies are submerged. Hyphae are 5 µm in diameter and non-septate. Sporangia are filamentous. Oogonia (22.0–31.0 µm in diameter) have several antheridia, which entwine around the oogonia. Oospores are 18.0–27.0 µm and encysted zoospores are 7.0–10.0 µm in diameter (Miura *et al.*, 2010). Identification was confirmed through amplification and sequencing of the ITS and 5.8 rDNA regions using primers ITS5 and ITS4 (White *et al.*, 1990), which produce an amplicon of about 800 bp. DNA extraction was performed using the Puregene Yeast and Gram-Positive Bacteria Kit (Gentra System). The thermocycling protocol consists of initial denaturation at 95°C for 5 min followed by 35 cycles of denaturation at 95°C for 30 s, annealing at 55°C for 30 s and extension at 72°C for 1 min with a final extension at 72°C for 7 min. After purification of the amplicon using QIAquick PCR purification kit (Qiagen), sequencing is performed using primers ITS1 and ITS3 or ITS4 (White *et al.*, 1990).

5.2 *Saprolegniaceae* (Family)

The family *Saprolegniaceae* (order *Saprolegniales*) belongs to the class *Oomycota*. They are not 'true fungi' but belong to the phylum *Heterokontophyta* under the kingdom Chromalveolata and as such are members of the *Stramenopila*, which includes diatoms and algae.

Phyla *Oomycota*, kingdom Chromista contains a number of aquatic fungi that infect aquatic animals, and also includes pathogens and saprophytes of plants. The main species affecting fish and aquatic animals as either saprophytes or pathogens include *Achlya*, *Pythiopsis*, *Saprolegnia*, *Leptolegnia*, *Aphanomyces*, *Thraustotheca*, *Dictyuchus* (Johnson *et al.*, 2003; http://dl.uncw.edu/digilib/biology/fungi/taxonomy%20and%20systematics/padgett%20book).

All *Saprolegniaceae* are filamentous with non-septate hyphae, although some have cross-walls that delimit reproductive cells. As with all water moulds, the hyphae of the *Saprolegniaceae* elongate by apical growth, which involves the wall vesicle formation. A primary characteristic of this family is the ability to produce motile spores (zoospores). They are differentiated on the shape and flagellar insertion, two features that are more stable than the number of times the spores encyst or re-emerge and swim. Two morphological types of spores occur in the *Saprolegniaceae*, the pyriform, anteriorly biflagellate primary spore and the reniform, laterally biflagellate secondary spore. Species may be monomorphic, which means one spore type, and dimorphic for those species that produce both primary and secondary spores.

The cell wall of hyphae in the family *Saprolegniaceae* is mainly composed of glucan, a polysaccharide (http://uncclc.coast.uncwil.edu).

5.2.1 *Dictyuchus* (Genus)

Sporangia are cylindrical, fusiform or clavate, but not filamentous, and spores occur in more than one row in the sporangium. Oogonia are smooth walled and contain only one oospore each.

5.2.2 *Leptolegnia* (Genus)

Leptolegnia belong to the kingdom Chromista, phylum *Oomycota*, class *Oomycetes*, order *Saprolegniales* and family *Leptolegniaceae*. Members of the *Leptolegniaceae* are parasitic or saprobic on animals, plants or fungi. Hyphae are non-septate, moderately narrow, 15–25 μm in diameter, and have granular protoplasm. Mycelium are filamentous and slightly sinuous. Asexual reproduction occurs by zoosporangia and the zoosporangium is almost undifferentiated from the hypha and may be branched. Zoosporogenesis is intrasporangial and may not be completed before discharge of zoospores. During discharge zoospores may be linked by protoplasmic strands. Usually zoospores are in a single row in the zoosporangium, which elongate on emergence then changes shape to pyriform. Sexual reproduction occurs by differentiated oogonia. Oogonia may be smooth or papillate and often without pits. Each mature oogonium usually contains only one oospore. The oospore may appear semi-solid with a few granules that are confined to the periphery. Inclusions are rarely in Brownian motion. Antheridium may be present or absent (Dick *et al.*, 1999).

Leptolegnia marina

Leptolegnia marina is a pathogen for the pea crab (*Pinnotheres pisum*). The crabs inhabit bivalves such as oysters, clams and mussels. The fungus infects the eggs of the female crabs and also infects the gills and causes death of mature crabs. The fungus also caused the death of the bivalve molluscs *Barnea candida* and *Acanthocardia* (*Cardium*) *echinata* held in a laboratory aquarium. The fungus occurred in the gills of *B. candida* and invaded the mantle of *A. echinata* and caused lesions on the mantle and the foot (Atkins, 1954).

Hyphae are 7.5–20 μm and sometimes up to 40 μm in diameter and branched. The rows of zoospores within the sporangium varies from one to several rows. Primary zoospores are 8–14 μm in length, pyriform and biflagellated when swimming. Once released the zoospores swim for 5–50 min before encysting. Encysted spores are 6–11 μm. Sexual reproduction is rare, but when it occurs oogonia are abundant in the host tissue. Oogonia are mostly spherical but may be pyriform or oval, thin-walled and smooth, and are borne on short side branches. Antheridia are mostly hypogynous. Oospores are smooth-walled, colourless, transparent, thick, two layered and occur singly and completely fill the oogonium, which is 17.5–30 μm and up to 37 μm (Atkins, 1954).

An unidentified *Leptolegnia* spp. caused mortality in experimental challenge of the spring peeper frog (*Pseudacris crucifer*) found in the USA and Canada (Ruthig, 2009).

5.2.3 *Saprolegnia* (Genus)

Saprolegnia belong to the kingdom Chromista, phylum *Oomycota*, class *Oomycetes*, order *Saprolegniales* and family *Saprolegniaceae*.

The identification of species within the genus *Saprolegnia* is difficult and has traditionally been based on morphological features of sexual structures; oogonia, oospores and antheridia (Coker, 1923; Seymour, 1970; Johnson *et al.*, 2002 online). A revision of the systematics conducted by Johnson *et al.* (2002 online) proposed the following characteristics to delineate *Saprolegnia* spp. Sporangia may be fusiform, clavate, cylindrical or irregular. Spores are dimorphic and sometimes polyplanetic. Primary spores swim away immediately on release. Spores occur in more than one row in sporangium and are not elongated on emergence. Oogonia are lateral, terminal intercalary or sessile and predominately of spherical or obpyriform shape. The oogonia usually produce more than one oospore, but don't produce only a single oospore in every oogonium throughout the colony. Oogonial walls may be pitted, unpitted, smooth or crenated. Oospores may be centric, subcentric, subeccentric or eccentric. Antheridial branches may or may not be present, and are androgynous, monoclinous, diclinous or hypogynous.

Many species overlap with their morphological traits and some species don't produce sexual structures *in vitro*, making identification to species level difficult. Recent publications have investigated the relationship between sequence information and traditional morphological characteristics as a means of improving the identification to species level.

Combined analysis of the partial 28S rRNA gene and the complete internal transcribed spacer (ITS) region differentiated 55 species of *Saprolegnia* into 10 clades. Analysis of DNA sequence information according to clades against morphological characteristics plus a number of morphological features not normally considered for species delineation suggested a further revision of the systematics of the *Saprolegnia* spp. is needed (Hulvey *et al.*, 2007).

Saprolegnia parasitica is one of the species that rarely produces sexual structures *in vitro* and all asexual isolates from fish were named *S. parasitica* by Coker (1923). *Saprolegnia parasitica* is difficult to differentiate from *S. diclina* because the structure of the oospore is the main morphological feature between the two species, where *S. parasitica* has subcentric oospores and *S. diclina* has centric oospores. Because of the difficulty in distinguishing between the two species, it was suggested any isolations be referred to as the *S. parasitica–diclina* complex (Seymour, 1970). A further definition of the complex included the species *S. australis* Elliott, *S. diclina* Humphrey, *S. parasitica* Coker, *S. shikotsuensis* Hatai, Egusa and Awakura, *S. salmonis* and *S. multispora* based on oogonia with predominantly diclinous antheridial branches and oogonial walls that were thin-walled, unpitted or inconspicuously pitted. Salmonid isolates were divided into *S diclina* Type 1, 2, and 3. To clarify the taxonomy, the analysis of the internal transcribed spacer rDNA sequences of all species in this complex sourced from a range of aquatic fish and animals resulted in five (Diéguez-Uribeondo *et al.*, 2007) and four phylogenetic clades (Ke *et al.*, 2009). Both studies found *S. parasitica*, *S. diclina* Type I, *S. salmonis*, *S. hypogyna* and unnamed *Saprolegnia* spp. clustered together in a clade, and the name *S. parasitica* should be applied to these isolates. Despite a wide variety of geographical locations for the isolates and variations in morphological characteristics and pathogenicity, their sequence information was almost identical. The morphological characteristics of species in this clade are bundles of long-hooked hairs on the secondary cysts, a high frequency of retracted germination and, if it occurs, oogonia production at 7°C (Diéguez-Uribeondo *et al.*, 2007). Both studies agreed on clustering of *S. ferax* and *S. diclina* Type 2 together with *S. anomalies*, *S. bulbosa*, *S. litoralis*, *S. longicaulis*, *S. mixta* and *S. oliviae* (Diéguez-Uribeondo *et al.*, 2007; Ke *et al.*, 2009). *Saprolegnia diclina* Type 3 (also referred to as *S. diclina* s. str.) was found in two clades and was closely related to *S. australia* isolates from Chile, and *S. multispora* (Diéguez-Uribeondo *et al.*, 2007). Ke *et al.* (2009) found a cluster composed of *S. monilifera*, *S. terrestris* and *S. torulosa*), and a cluster composed of *S. australis*, *S. diclina* and *S. plymorpha*.

Saprolegnia colonize immunocompromised or injured hosts or digest organic matter. Infection by Saprolegnia of a host is known as mycoses or saprolegniosis. A wide temperature range of 3–33°C is tolerated and Saprolegnia are most commonly found in freshwater and also brackish water. On laboratory media and on infected hosts, Saprolegnia grows as white fluffy colonies.

Saprolegnia australis

Saprolegnia australis has been reported from infections of finfish and crayfish. The fungus was isolated from eggs of chum salmon (*Oncorhynchus keta*) (Hussein *et al.*, 2001) and from fin rot in rainbow trout (*Oncorhynchus mykiss*) fingerlings cultured in Japan (Hatai *et al.*, 1977). *Saprolegnia australis* was one of seven *Saprolegnia* species found to colonize brown trout (*Salmo trutta trutta*) eggs in a hatchery in Spain (Fregeneda-Grandes *et al.*, 2007). It was isolated at a rate of 14.9% on eggs, and at a rate of 11.1% along with three other *Saprolegnia* spp. on alevins. *Saprolegnia australis* has been isolated from cuticle ulcerations in native crayfish (*Orconectes propinquus*) in a lake in the USA (Krugner-Higby *et al.*, 2010). The normal habitat is freshwater streams and lakes (Elliott, 1968).

Saprolegnia australis grows on IM and GP agar (glucose peptone) as fluffy white colonies. Hyphae are 30 μm in diameter and sparingly branched. Oogonial formation is abundant with a spherical or pyriform shape 39–99 μm, and usually terminal, sometimes lateral or intercalary, pyriform to obovate and sometimes with a long neck. The oogonium is usually 48–110 μm in diameter. Spores are subcentric and 10.5 μm. Each oogonium contains 10–14 oospores. The oogonial wall is smooth with conspicuous pits or unpitted or with a few conspicuous pits. Oogonial stalks are long and two to many times the diameter of the oogonium. Antheridial branches are diclinous and occasionally androgynous, tubular to clavate, sometimes branched, often intercalary. Usually many anteridia encircle the oogonium. Bundles of long hairs are absent. Encysted zoospores are 10.5 μm in diameter. Gemmae may be abundant, are filiform, clavate, spherical or irregular (Elliott, 1968; Fregeneda-Grandes *et al.*, 2007).

Saprolegnia brachydanis

Saprolegnia brachydanis was isolated from the tropical freshwater zebra fish (*Brachydanio rerio*) held at a research facility in China (Ke *et al.*, 2009). The name *S. brachydanis* was used in the original paper but is listed in MycoBank as *S. brachydanionis* with a comment that the name is invalid.

The isolate is closely related to *S. australis* based on morphology and sequence information using internal transcribed spacer primers ITS1 and ITS4. Growth occurs on CMA at 18–23°C. Hyphae are 18–55 μm, slender, sparingly branched. Morphological features are observed after culture in sterile lake water. Zoosporangia are 115–455×24–55 μm, cylindrical, clavate and fusiform, and are straight or bent. Zoospores are discharged from the tip of the zoosporangium. Primary zoospores are pyriform. Encysted zoospores are 10–12 μm, and globose. Secondary zoospores are reniform. Gemmae are spherical, oviform and terminal or intercalary. Oogonia are 31–90 μm in diameter, glomerulate, spherical or obovate, but mostly pyriform and spherical, and walls are smooth, and unpitted or pitted under the point of atheridial attachment. Oogonial stalks are short, slender, straight or curved and may be twisted. Atheridial cells are simple or branched, curved or twisted, mostly monoclinous, but some are diclinous and up to eight may be attached to an oogonium. The oogonia may be formed between two and three hyphae (Ke *et al.*, 2009).

Saprolegnia diclina

Mortalities due to *S. diclina* occurred in Atlantic salmon (*Salmo salar*) fry cultured in Scotland (Bruno and Stamps, 1987). Fish were thin, swam slowly and irregularly, fins were reddened with increased mucus around the gill filaments and a cottony mycelium was seen between the gill filaments and the operculum. The fungus did not penetrate to the internal organs. *Saprolegnia diclina* was co-isolated with *S. parasitica* from coho salmon cultured in Japan, but the isolation of *S. diclina* was thought to be secondary to a primary infection by *Saprolegnia parasitica* (Hatai and Hoshiai, 1992). In experimental infection of Atlantic salmon, *S. diclina* caused no or very low mortality, which is in line with other findings (Stueland *et al.*, 2005). *Saprolegnia diclina* has also been reported from fungal infections of Persian sturgeon (*Acipenser persicus*) eggs (Ghiasi *et al.*, 2010).

Over half (51.4%) of the seven *Saprolegnia* spp. found to colonize eggs of brown trout

(*Salmo trutta trutta*) in a hatchery in Spain and was attributed to *S. diclina* (Fregeneda-Grandes *et al.*, 2007). It was isolated at rates of 44.4% on alevins and dominated the three other *Saprolegnia* spp. isolated from alevins.

Saprolegnia diclina causes infection in embryos of the natterjack toad (*Bufo calamita*) found in the mountains in Spain (Fernández-Benéitez *et al.*, 2008). Pathogenicity was confirmed by experimental challenge.

Saprolegnia diclina grows on IM and GP agar (glucose peptone) as fluffy white colonies. Growth occurs on CMA and PDA at 25°C (Fernández-Benéitez *et al.*, 2008). On SDA and tryptic soya agar at 15 and 25°C, colonies are 1 cm, white with a slightly brown centre after 24–48 h. Hyphae are 20–40 μm, non-septate, multinucleated and branched. Growth occurs at 15–30°C (Bruno and Stamps, 1987). Oogonial formation is abundant and it has a spherical or pyriform shape 36–90 μm, and each oogonium may contain up to 30 centric or subcentric oospores 24 μm in diameter. Oogonia develop laterally on hyphae. The oogonial wall is unpitted or with a few inconspicuous pits only under the points of antheridial attachment. Antheridial cells are often branched, applied laterally over the oogonium and are diclinous. Bundles of long hairs are absent, or long hooked hairs on secondary zoospores are absent (Fregeneda-Grandes *et al.*, 2007). Up to three successive generations of zoospores are produced. Direct germination from an encysted spore occurs (Hatai and Hoshiai, 1992). See under *S. parasitica* for spore germination method to differentiate *S. diclina* from *S. parasitica*.

Identification can be confirmed by amplification of the internal transcribed spacer regions using universal primers ITS5 and ITS4 (White *et al.*, 1990) and sequenced. DNA extraction is successful using the EZNA-fungi DNA miniprep kit (Omega Biotek) (Fernández-Benéitez *et al.*, 2008).

Saprolegnia ferax

Saprolegnia ferax is found in cold water streams and rivers and is considered generally as being non-pathogenic for fish; however, this may depend on the fish species and any predisposing conditions such as stress or compromised immune status. Earlier studies indicated that *S. ferax* is pathogenic to fish when their resistance is compromised and the fungus has been responsible for disease outbreaks in the wild and in hatcheries. *S. ferax* caused death in brown trout (*Salmo trutta*), brook trout (*Salvelinus fontinalis*) and green tench (*Tinca tinca*) in experiments conducted in the USA, whereas lake trout (*Salvelinus namaycush*) was moderately susceptible and smallmouth bass (*Micropterus dolomieu*) was resistant to infection (Cunningham, 1899). Infected fish showed patches of fungal growth on the head, dorsal fins and body; the fungus generally attacks those parts not covered in scales. The membranous parts of the fins and tails are attacked and the tail appears ragged and frayed. The initial pale areas of infection on the skin may develop into ulcers due to abrasion as the fish rub against the sides of the tank to ease the irritation caused by the infection, and also due to secondary bacterial invaders (Cunningham, 1899). *Saprolegnia ferax* was isolated in low numbers (5.4%) compared with a total of seven *Saprolegnia* spp. found to colonize brown trout (*Salmo trutta trutta*) eggs in a hatchery in Spain (Fregeneda-Grandes *et al.*, 2007). It was isolated at rates of 22.2% from three other *Saprolegnia* species on alevins. An isolate identified as *S. diclina* type 2 from bony bream (*Nematalosa erebi*) from Australia was found to be *Saprolegnia ferax* based on cluster analysis (Diéguez-Uribeondo *et al.*, 2007). *Saprolegnia* spp. isolated from fungal infection in silver crucian carp or Prussian carp (*Carassius gibelio*) and zebrafish (*Danio rerio*) from fish farms in China were identified as belonging to the *S. ferax* clade (Ke *et al.*, 2009). Also in China, *S. ferax* caused saprolegniosis of yellow catfish (*Tachysurus* [*Pelteobagrus*] *fulvidraco*) eggs and was proven to be the primary pathogen through experimental challenge (Cao *et al.*, 2012). In experimental infection of Atlantic salmon, *S. ferax* caused no or very low mortality, which is in line with other findings for Atlantic salmon (Stueland *et al.*, 2005).

Saprolegnia ferax may be implicated in the decline of amphibian populations. In experimental challenge, it was found that *S. ferax* is pathogenic for different life stages of amphibians. An exposure time of 1 week killed mid-to-late-stage larvae of Pacific tree frog (*Pseudacris regilla*), but not those of Cascades frog (*Rana cascadae*), whereas after a 2-week exposure red-legged frog (*R. aurora*) larvae were killed, but larvae of long-toed salamander (*Ambystoma macrodactylum*) were not (Romansic *et al.*, 2009).

Saprolegnia ferax grows on PDA, IM and GP agar (glucose peptone) as fluffy white colonies. Hyphae are non-septate and sparingly branched. Sporangia are abundant, clavate, filiform or spherical. Zoospores usually discharge singly and encyst shortly after discharge. Cysts are 9–12 μm. Oogonial formation is very abundant, terminal or lateral and they have a spherical or pyriform shape 37–97 μm and contain 1–20 centric or subcentric oospores. The oogonial wall is conspicuously pitted. Antheridial branches are androgynous, monoclinous or absent. Rarely diclinous. Bundles of long hairs are absent (Johnson *et al.*, 2003; Fregeneda-Grandes *et al.*, 2007). Discharge of zoospores was examined by inoculating fungal elements to PDA and placing an autoclaved canola seed (*Brassica napus*) at the edge of the colony. When covered in hyphae, the seed was transferred to sterile filtered river water to observe zoospore discharge (Cao *et al.*, 2012).

Saprolegnia furcata

Saprolegnia furcata colonized brown trout eggs (*Salmo trutta trutta*) at rates of 5.4% compared with six other *Saprolegnia* spp. isolated in a study conducted on a hatchery in Spain (Fregeneda-Grandes *et al.*, 2007).

Saprolegnia furcata grows on IM and GP agar (glucose peptone) as fluffy white colonies. Oogonial formation is abundant with a spherical or subspherical shape and centric oospores. The oogonial wall is conspicuously pitted. Antheridial branches are androgynous. Bundles of long hairs are absent (Fregeneda-Grandes *et al.*, 2007).

Saprolegnia hypogyna

Saprolegnia hypogyna grows on IM and GP agar (glucose peptone) as fluffy white colonies. Oogonial formation is very abundant with a spherical or subspherical shape and centric oospores. The oogonial wall is conspicuously pitted. Antheridial branches are hypogynous. Bundles of long hairs are absent (Fregeneda-Grandes *et al.*, 2007).

Saprolegnia parasitica

Saprolegnia parasitica is found worldwide in fresh water and is responsible for mortalities in wild populations of salmonids, and in the aquaculture of salmon, trout, catfish, coarse fish, crayfish, and found in water and soil in many countries including Australia, Canada, Chile, Finland, Iran, Iraq, Japan, Scandinavia, Scotland, Spain, Sweden, Turkey and the USA (Van West, 2006; Diéguez-Uribeondo *et al.*, 2007).

Due to the difficulty in producing sexual structures in *S. parasitica* using artificial media, lack of sexual structures can lead to mis-identification and some species of *Saprolegnia* may inadvertently be assigned to *S. parasitica*. A *Saprolegnia* sp., thought to be *S. parasitica* because of the absence of sexual structures, caused mortality in European eel (*Anguilla anguilla*) cultured in Scotland. Infection resulted in lesions that led to a loss of epithelium and progressed to ulceration (Copland and Willoughby, 1982). In Australia, a study investigating winter saprolegniosis in silver perch (*Bidyanus bidyanus*), farmed in New South Wales, found the causative organism to be *S. parasitica* (Rowland *et al.*, 2007). Winter saprolegniosis occurs when the temperature drops to below 16°C and is associated with rapid drops in temperature of 4–5°C over 5–7 days. Fungal lesions are seen on the gills and pale lesions occur on the body and the skin is abnormally dark. Winter saprolegniosis in channel catfish (*Ictalurus punctatus*) also occurs in the USA. Again, rapid decrease in temperature is a factor, in this case 22°C to 10°C. This rapid decrease in temperature impacts on the fish defence mechanisms, leaving them immunocompromised. Experimental infection confirmed a *Saprolegnia* sp. (isolate CF1=ATCC 200048) as the primary pathogen. The *Saprolegnia* sp. from channel catfish has not been speciated. In experimental challenge using immunocompromised catfish, *S. parasitica* ATCC 22284 and *S. diclina* ATCC 42062 did not infect or kill catfish, whereas isolate CF1 killed catfish within 1 week. Other *Saprolegnia* spp. isolates from winter saprolegniosis showed carrying degrees of pathogenicity (Bly *et al.*, 1992). In experimental infection of rainbow trout (*Oncorhynchus mykiss*), challenge with *S. parasitica* ATCC 22284 resulted in 37% mortality after 7 days and 70% mortality after 14 days when fish were stressed by abrasion and temperature increase. The two stressors of abrasion and non-optimal water temperature were necessary to induce infection and one stressor alone did not result in infection (Howe and Stehly, 1998).

In a study investigating colonization of brown trout eggs by *Saprolegnia* spp., *S. parasitica* was found at rates of 2.7% with six other *Saprolegnia* spp. isolated from the eggs, whereas in cutaneous lesions of brood stock from the same hatchery, *S. parasitica* was the sole *Saprolegnia* sp. isolated (Fregeneda-Grandes *et al.*, 2007). It was isolated at rates of 22.2% from three other *Saprolegnia* spp. on alevins. In Japan, *S. parasitica* caused mass mortality in cultured coho salmon (*O. kitsuch*), chum salmon (*O. keta*) and sockeye salmon (*O. nerka*) (Hatai and Hoshiai, 1992; Hussein *et al.*, 2001) and a mycotic gastritis in juvenile ayu (*Plecoglossus altivelis*) (Wada *et al.*, 1993). Lesions containing mycelia occurred on the caudal peduncle, caudal fin, adipose fin and head. Fungal hyphae occurred in the dermis and musculature, but did not penetrate to the internal organs and no internal gross signs were seen. In infected ayu the pyloric region of the stomach was red and fungal hyphae were present in this area and had penetrated the abdominal adipose tissue, the pancreas, spleen, kidney, liver and swim bladder.

Saprolegnia parasitica and *S. diclina* have both been isolated from cases of saprolegniosis in Persian sturgeon (*Acipenser persicus*) eggs cultured in Iran (Ghiasi *et al.*, 2010). In Iraq, the fungus caused saprolegniosis in bunnei fish (*Mesopotamichthys* [*Barbus*] *sharpeyi*). Fungal growth occurred on the skin, gills and fins of fish (Al-Duboon *et al.*, 2006).

Strains of *S. parasitica* appear to have varying levels of pathogenicity. One factor that appears to correlate with virulence when tested by experimental challenge is that pathogenic strains have a significantly higher initial growth rate of germinating cysts in sterile tap water and this together with the presence of long hooked hairs on zoospores can differentiate pathogenic and non-pathogenic strains of *S. parasitica* and other *Saprolegnia* spp. (Stueland *et al.*, 2005).

The identification of *S. parasitica* is difficult because it rarely produces sexual structures under laboratory conditions. Also, *S. parasitica* is difficult to differentiate from *S. diclina* as the structure of the oospore is the main morphological feature between the two species, where *S. parasitica* has subcentric oospores and *S. diclina* has centric oospores. Because of the difficulty in distinguishing between the two species, it was suggested any isolations be referred to as the *S. parasitica–diclina* complex (Seymour, 1970); however, most isolations are now reported to species level.

Saprolegnia parasitica grows on IM and GP agar (glucose peptone) as fluffy white colonies. Indirect germination occurs from an encysted spore resulting in a long empty, septate, germ tube from the spore. Long-hooked hairs occur on empty zoospores (Hatai and Hoshiai, 1992) and their presence is best examined using phase contrast microscopy or electron microscopy. It is this structure that infects fish. In asexual reproduction, zoosporangia fill with zoospores that are discharged through the zoosporangium tip and primary zoospores are released. Zoosporangia are variable, usually bent and irregular, and may be up to 70 μm long. Zoospores are 9–11.5 μm. Primary zoospores swim for a short time before encysting and releasing secondary zoospores. Secondary zoospores swim for a longer time than primary zoospores. The zoospores can encyst again and release a new zoospore and this process of repeated encystment and release is referred to as polyplanetism. Sexual structures such as oogonia and antheridia are not normally seen or may be seen on special media (not stated, Coker and Mathews, 1937). Oogonia, if present, are borne on the tips of hypha or intercalary, are thin-walled and smooth, unpitted, and 65–95 μm in diameter, spherical, subspherical, pyriform or clavate. Oospores are 18–22 μm in diameter, subcentric and 3–25 are contained in an oogonium. Antheridial branches are long and slender, diclinous or androgynous, and antheridia are clavate or subcylindric and usually one to five are attached to an oogonium (Coker and Mathews, 1937; Van West, 2006).

A major difference between *S. parasitica* and *S. diclina* is the difference in germination of spores. To facilitate the production of spores, a culture is grown in GP broth at 20°C for 24 h. Hyphae are collected from the broth and rinsed three times in sterile tap water and then incubated in 30 ml of sterile tap water and incubated at 20°C for 20 h. Zoospores (300 μl) are collected close to the water surface to avoid collecting encysted or germinating spores and inoculated into 29.7 ml of one quarter dilution of GY broth in a Petri dish. After 3 h at 20°C, the culture is observed for germination of spores. *Saprolegnia diclina* does not show any indirect germination of spores in one-quarter GY broth, whereas at other concentrations of GY broth low numbers of indirect germination may be observed. *Saprolegnia parasitica* shows indirect germination when grown in one-quarter GY broth. Neither species shows spore germination in distilled water after 24 h (Yuasa *et al.*, 1997). The use of one-quarter strength GY broth for germination gave better results than sporulation using hemp seed cultures.

Saprolegnia salmonis

Saprolegnia salmonis caused mass mortality in sockeye salmon (*Oncorhynchus nerka*), chum salmon (*O. keta*) and masou salmon (*O. masou*) cultured in Japan (Hussein and Hatai, 1999; Hussein *et al.*, 2001). Lesions occurred on the caudal fin, peduncle region and the head, and the fungus was isolated from the muscle. Growth occurs on GY agar at 15–20°C, but growth ceases after 3 days at 30°C. Growth occurs at 3% NaCl, but not at 3.5% NaCl. Most species within the genus *Saprolegnia* are freshwater isolates. Colonies are 82 mm at 5 days. Hyphae are 22 μm, non-septate, delicate and moderately branched. Zoosporangia are 178–186×24–36 μm, clavate (club-shaped) or straight and cylindrical and sometimes curved. Primary zoospores are 11 μm, biflagellated and pyriform and encyst 30 min to several hours after discharge. Secondary zoospores are reniform. Oogonia may be spherical, pyriform or elongated. Elongated oogonia contain 16–20 oospores and spherical oogonia contain 4–12 oospores. Oospores are 20 μm and spherical, and internally are mostly subcentric but also centric. Antheridial cells may or may not be attached to the Oogonium cells. The wall of the oogonium is smooth and conspicuously pitted under the attachment of the antheridial cell, but may be pitted without antheridial attachment. The oogonial stalks are slightly bent or straight and not branched. Anteridial cells are 8 μm, unbranched, simple, tubular and diclinous. Gemmae (or chlamydospores) are not observed.

Saprolegnia species

An unidentified *Saprolegnia* sp. caused mortality in experimental challenge of the American bullfrog (*Rana catesbeiana*) (Ruthig, 2009).

Saprolegnia unispora

Saprolegnia unispora colonized brown trout eggs (*Salmo trutta trutta*) at rates of 1.6% compared with six other *Saprolegnia* spp. isolated in a study conducted on a hatchery in Spain (Fregeneda-Grandes *et al.*, 2007).

Saprolegnia unispora grows on IM and GP agar (glucose peptone) as fluffy white colonies. Oogonial formation is scarce and occurs only after growth at 7°C and prolonged incubation of 20–30 days. Oogonia have a spherical shape and the walls are unpitted with subcentric oospores. Antheridial branches are absent. Bundles of long hairs are absent (Fregeneda-Grandes *et al.*, 2007).

5.2.4 Fungi basics

The following terminology is used to describe microscopic or macroscopic portions of fungi or sexual or asexual reproductive cells of fungi.

Anamorph: an asexual reproductive stage or form of a fungus.
Annellide: a type of conidiogenous cell that produces conidia in a basipetal manner. The tip of the annellide becomes longer and narrower as each conidium is formed and released.
Antheridia (pleural): the male reproductive structures such as gametangia.
Antheridium (singular): male gametangium.
Ascus (singular) or **Asci** (plural): the cell that contains the ascospores in fungi belonging to the Ascomycota.
Asexual reproduction: reproduction without meiosis, ploidy reduction or fertilization and the mechanism by which fungi propagate or reproduce. In Oomycetes, asexual reproduction is through the zoosporangium and the formation and discharge of zoospores.
Biflagellate: those zoospores having two flagella.
Chlamydospore: a reproductive unit, thick-walled, asexual spore that is derived from the hyphal cell and can function as a resting spore, which can survive unfavourable conditions.
Chlamydospore: a resting spore that develops from a hypha, and is a thick-walled melanized thallic conidium. May also be called a gemmae.
Clavate: club-shaped.
Collarette: a small collar.
Conidiogenous: a cell that gives rise to conidia.
Conidiophore: a hypha that gives rise to conidia (also called conidiospores).
Conidiospores or **conidia**: produced asexually and form on a conidiophore.

Conidium: an asexual spore supported on a hypha. It is not contained in a sporangium.
Dematiaceous: a dark brown, green-grey or black colour and a term used for fungi that produce melanin in their cell walls. The brown colour comes from the melanin. Dematiaceous fungi have septate hyphae with irregular branching.
Dermatophyte: a fungus belonging to one of the three genera *Epidermophyton*, *Microsporum* or *Trichophyton* that infect nail, skin and hair, and have the ability to degrade keratin.
Dimorphic: fungi that exist in a filamentous form as a fungus or as a yeast, depending on the environmental conditions such as temperature or whether they are inside host tissue or not.
Diplanetic: fungi that produce two different types of zoospores, such as some species of *Saprolegnia*.
Encystment: the formation of a thick wall around a zoospore after it settles and loses its flagellum.
Gametangium: a structure that produces gametes.
Gemmae: also referred to as a chlamydospore.
Holocarpic: the thallus is entirely converted into sporangia.
Hyphae: the long microscopic filaments (that contain zoospores). The hyphae or filament is the basic structure of a fungus. Some hyphae grow above the medium and are called aerial hyphae or aerial mycelia. Vegetative hyphae grow within the medium to absorb nutrients.
Monoplanetic: fungi that produce one type of spore as opposed to those that are diplanetic and produce two types of zoospores.
Mycelium: the mass of hyphae growing on a plate (or host) that is seen macroscopically.
Oogonia (plural): the large round egg-shaped structures that contain one or more female gametes.
Oogonium (singular): a female gametangium.
Oosphere: a female gamete within an oogonium.
Oospore: a diploid spore that is produced by the *Oomycota* fungi.
Phialiade: an open-ended, tubular or flask-like conidiophore that produces conidia.
Polyplanetism: repeated cycles of encystment and zoospore release.
Pseudohyphae: an elongated blastoconidia formed in some yeasts and resembles hyphae. Pseudohyphae divide by budding, whereas true hyphae elongate from the tip. Pseudohyphae form due to incomplete budding where the cells remain attached after division.
Reniform: kidney-shaped.
Repeated zoospore emergence: repeated zoospore emergence is the ability of an encysted spore to release a new zoospore generation ingested of germinating.
Saccate: sac-like or having a sac or forming a sac.
Septum (singular), **septa** (pleural): the internal cross wall seen in some hyphae, referred to as septate hyphae, that divides hyphal cells.
Sexual reproduction: through fertilization and in the oomycetes occurs when the antheridium penetrates the oogonium.
Sporangiophore: a hypha that bears the sporangium.
Sporangium (singular) or **sporangia** (plural): a sac-like structure that produces and holds the asexual spores.
Spores (or conidia): are the reproductive cell and develop from the hyphae.
Teleomorph: a sexual form or stage of a fungus.
Uniseriate: arranged in a single row.
Zoosporangium: a sexual structure where zoospores develop.
Zoospore: a motile spore (sporangiospore), which propels itself by one or more flagella. It is asexual and used for dispersal of the fungus. Oomycetes produce biflagellated spores.

A number of terms are used in the process of identifying fungi and most of these describe the morphological characteristics of the fungi, which include zoosporangial discharge; sporangium length and width; diameter of the zoospore cyst; renewal; appression; oogonial origin, shape, diameter, stalk length, inner wall appearance, pitting; oospore number and oospore position. A major morphological characteristic is zoosporangial discharge and a number of states are used to describe the process or method of zoosporangial discharge. Achlyoid discharge is used to describe a process whereby primary zoospores emerge and encyst in a hollow space at the sporangial orifice. In saprolegnoid discharge, the primary zoospores emerge and immediately swim away. Dictyuchoid discharge describes a process whereby the primary zoospores encyst within the zoosporangium before releasing motile secondary spores through a small hole in the sporangial wall, leaving behind empty primary spore cysts in the sporangium. Renewal describes the process for production of secondary sporangia. Appression describes the attachment of the antheridial cell to the oogonium such that it is appressed laterally or is attached as short projections. Oogonial shape may be spherical, pyriform, fusiform, oval or angular, and oogonial ornaments may be papillate, tuberculate, spiny, mammiform, truncate, crenated, wavy or absent.

5.2.5 Method to detect repeated zoospore emergence

Mycelia are grown in PG-1 drop cultures for 3 days at 20°C. Mycelia are washed three times with sterile water and incubated in river water in Petri dishes for 14 h at 20°C. The swimming zoospores are transferred to test tubes and agitated using a vortex mixer for 45 s. The zoospores are incubated for 150 min at 20°C. The release of secondary zoospores is examined under a light microscope (Diéguez-Uribeondo *et al.*, 1994).

5.2.6 Further information

Kirk, P.M., Cannon, P.F., Minter, D.W. and Stalpers, J.A. (eds) (2008) *Dictionary of the Fungi*, 10th edn CAB International, Wallingford, UK.
http://dl.uncw.edu/digilib/biology/fungi/taxonomy and systematics
http://www.doctorfungus.org/thefungi/nomenclature.htm
http://tomvolkfungi.net/
http://www.mycolog.com/CHAP1.htm

5.3 Molecular Identification of Fungi

Fungi can be identified by amplification of the internal transcribed spacer (ITS) regions, the 18S or 28S rRNA regions using universal primers (White *et al.*, 1990).

5.3.1 Combinations of primers

Primers ITS1 and ITS2 amplify a 300-base pair region, the ITS-1, located between the 18S and 5.8S rRNA. Primers ITS1 and ITS4 amplify an 800-base pair region from the 18S rRNA to the 28S rRNA (large subunit) that includes the spacer regions 1 and 2 and the 5.8 S rRNA. Primers C1 and D2 amplify the 28S rRNA of all fungal species. An annealing temperature of 55°C is used for 35 cycles (White *et al.*, 1990).

ITS5 (forward) and ITS4 (reverse) (Munchan *et al.*, 2009) primers are used at a concentration of 10 pmol and an annealing temperature of 50°C at 1 min. ITS1 (forward) and ITS4 (reverse) amplify the intergenic spacer region located between 18S and 28S rRNA genes and have been used to amplify DNA from *Exophiala angulospora* prepared in the master mix at concentrations recommended by the manufacturer of the master mix reagent (Nyaoke *et al.*, 2009). Primers NL-1 and NL-4 amplify DNA from the D1/D2 region of the 28S rRNA gene. Primers are used at concentrations recommended by the manufacturer of the master mix reagent. Thermocycling protocols for ITS1/ITS4 and NL-1/NL-4 consist of an initial denaturation at 95°C for 10 min followed by 30 cycles of denaturation at 95°C for 1 min, annealing at 58°C

Table 5.4. Universal primers for the amplification of fungi.

Primer	Primer sequence 5′–3′	References
ITS1	TCC GTA GGT GAA CCT GCG G	White *et al.* (1990)
ITS2	GCT GCG TTC TTC ATC GAT GC	White *et al.* (1990)
ITS3	GCA TCG ATG AAG AAC GCA GC	White *et al.* (1990)
ITS-4	TTC TCC GCT TAT TGA TAT GC	White *et al.* (1990)
ITS-5	GGA AGT AAA AGT CGT AAC AAG G	White *et al.* (1990)
ITS-4	TCC TCC GCT TAT TGA TAT GC	White *et al.* (1990)
C1	ACC CGC TGA TTT AAG CAT	White *et al.* (1990)
D2	TCC GTG TTT CAA GAC GG	White *et al.* (1990)
NL-1	GCA TAT CAA TAA GCG GAG GAA AAG	Nyaoke (2009)
NL-4	GGT CCG TGT TTC AAG ACG G	Nyaoke (2009)

for 1 min, extension at 72°C for 1 min with a final extension at 72°C for 10 min.

5.4 Media for Fungi

The following media have been used to grow various fungi and the recipes are detailed in the media section. BA=blood agar, CMA=cornmeal agar, CZA=Czapek solution agar, MA=malt agar, MEA=malt extract agar, NA=nutrient agar, PDA=potato dextrose agar, RGY agar (Alderman and Polglase, 1986), SAB=Sabouraud's agar and may or may not contain chloramphenicol SDA=Sabouraud's dextrose agar. PDA and MA are enriched media, whereas CMA is a low nutrient medium.

5.4.1 GY agar (Hatai and Egusa, 1979)

Dissolve 10 g glucose, 2.5 g yeast extract and 15 g agar in 1 l of distilled water and autoclave at 121°C for 15 min. Cool to 50°C and dispense 18–21 ml into Petri dishes. Place into plastic bags, seal and store at 4°C (Hatai and Egusa, 1979). A solution of 10 μg/plate of streptomycin and ampicillin can be spread onto the surface of the plate to inhibit bacterial contaminants.

5.4.2 GP agar (Roberts *et al.*, 1993)

Dissolve 3 g glucose, 1 g peptone, 0.128 g $MgSO_4.7H_2O$, 0.014 g KH_2PO_4, 8 mg Ca, 0.5 mg Fe, 0.5 mg Mn, 0.1 mg Cu, 0.1 mg Zn and 15 g agar in 1000 ml distilled water.

Prepare trace elements as a stock solution. Add 1 ml of trace element solution to 1000 ml distilled water. Add other reagents to the distilled water. Mix to dissolve and autoclave at 121°C for 15 min. Cool to 50°C and dispense 18–21 ml into Petri dishes. Place into plastic bags, seal and store at 4°C.

GP agar is used as primary growth medium for isolating *Aphanomyces invadans*. GP agar supports the growth of *Aphanomyces* spp., *Saprolegnia* spp., *Achlya* spp. and other oomycetes fungi.

Preparation of trace elements stock solution

Prepare a 100× stock solution (100 ml) of the trace element solution using the amounts in the last column of the table below.

Element	Molecular weight
Cu	63.546
Zn	65.38
Mn	54.94
Fe	55.85
Ca	40.08

See inline table at bottom of the page.
Example of calculation:

$FeCl_3.6H_2O$. Formula weight = 270.29

Formula weight obtained from:

[MW of $FeCl_3$=162.20]

MW of $6H_2O = 6\times[(2\times1.008)+(1\times15.99)] = 108.09$.

FW of $FeCl_3.6H_2O$ = 270.29 (=162.20+108.09).

Prepared from the following:

Calculations for amount of trace element to weigh for 100 × stock solution of trace elements

Compound	Formulae weight	Molecular weight of trace element	Final concentration required	Amount to weigh for 100 × stock solution
$FeCl_3.6H_2O$	270.29	55.85	0.5 mg	242.00 mg
$CaCl_2.2H_2O$	110.98	40.08	8.0 mg	2.215 g
$MnCl_2.4H_2O$	197.91	54.94	0.5 mg	180.00 mg
$ZnSO_4.7H_2O$	161.45	65.38	0.1 mg	43.97 mg
$CuSO_4.5H_2O$	249.686	63.546	0.1 mg	39.29 mg

Want Fe at a final concentration of 0.5 mg (in 1 l).

Therefore, 270.30/55.85×0.5=2.42mg. For 100×stock solution (100 ml) weigh out 242 mg into stock solution.

Note: Compounds with water attached are usually easier to dissolve than the anhydrous solutions.

5.4.3 GP broth (Roberts *et al.*, 1993)

As for GP agar, but without the addition of agar. Mix to dissolve and autoclave at 121°C for 15 min. Cool to 50°C and dispense 10 ml aliquots into McCartney bottles. Store at 4°C.

GP broth is used as primary growth medium for isolating *Aphanomyces invadans*. GP broth supports the growth of *Aphanomyces* spp., *Saprolegnia* spp., *Achlya* spp. and other oomycetes fungi.

5.4.4 GP-POX broth (Roberts *et al.*, 1993)

Prepare GP broth in 1000 ml distilled water. Autoclave at 121°C for 15 min. Cool to 50°C and aseptically add filter-sterilized antibiotics; 10 mg/l penicillin and 10 mg/l oxolinic acid. The original concentration of antibiotics was 100 mg/l (Roberts *et al.*, 1993), but this was reduced to 10 mg/l (Lilley and Roberts, 1997).

5.4.5 GPY-PS agar (Roberts *et al.*, 1993)

Prepare the same reagents and concentrations as for GP agar (Roberts *et al.*, 1993), but add 0.5 g yeast extract. Autoclave at 121°C for 15 min. Cool to 50°C and aseptically add filter-sterilized antibiotics; 10 mg/l penicillin and 10 mg/l streptomycin sulfate. The original concentration of antibiotics was 100 mg/l (Roberts *et al.*, 1993), but this was reduced to 10 mg/l (Lilley and Roberts, 1997).

5.4.6 GPY-PS broth (Roberts *et al.*, 1993)

Prepare as for GYP-PS agar, but without the agar.

5.4.7 Isolation medium (IM) (Alderman and Polglase, 1986)

This medium was designed for the growth of *Aphanomyces astaci* (Alderman and Polglase, 1986), but also is suitable for the growth of *Oomycetes* fungi, including *A. invadans* (Buller, 2004).

Dissolve 2.4 g of agar (Oxoid agar no. 1), 0.2 g yeast extract (Dicfo), 1.0 g glucose, in 200 ml of distilled water. Autoclave at 121°C for 15 min. Cool to 50°C. Mix 2.0 mg oxolinic acid and 200 mg penicillin G and filter sterilize. Aseptically add to cooled mixture. Dispense 18–21 ml into Petri dishes. Place into plastic bags, seal and store at 4°C.

5.4.8 Malt extract agar (MEA)

Dissolve malt extract (20 g), peptone (1 g), dextrose (20 g) and agar (15 g) in 1 l of distilled water and autoclave at 121°C for 15 min. Cool to 50°C and dispense 18–21 ml into Petri dishes. Place into plastic bags, seal and store at 4°C.

5.4.9 mTGh agar

This medium is used for the growth of *Batrachochytrium dendrobatidis* (OIE, 2012c). Dissolve 8 g tryptone, 2 g gelatin hydrolysate and 10 g agar in 1 l of distilled water. Autoclave at 121°C for 15 min. Cool to 50°C and aseptically add filter-sterilized 200 mg/l penicillin G and 200–500 mg/l streptomycin sulfate. Dispense into Petri dishes. Place into plastic bags, seal and store at 4°C. If bacterial contamination persists, 1 mg/l ciprofloxacin can be added to the medium.

5.4.10 Peptone-glucose agar (PG1) (Unestam, 1965)

Peptone glucose (PG1) prepared as a broth or agar is used to grow *Aphanomyces astaci* (Unestam, 1965). Combine 0.165 g $MgCl_2.6H_2O$, 0.147 g $CaCl_2.2H_2O$, 0.373 g KCl, 0.02 g $FeCl_3.6H_2O$, 1.5 ml 0.2 M EDTA (neutralized with solid NaOH), 0.3% peptone, 0.6% glucose, 1/75 mol phosphate buffer ($NaH_2PO_4.H_2O$+$Na_2HPO_4.2H_2O$), 1 ml trace element stock solution. To make solid medium, add 15 g/l agar. The use of peptone derived from enzyme hydrolysis supports a greater amount of growth of the fungus than peptone derived from acid hydrolysis (Unestam, 1965).

The phosphate buffer (5/3×final concentration) is prepared in a Schott bottle and autoclaved at 121°C for 15 min. The rest of the basal medium (10×concentration) is autoclaved separately before adding aseptically to the Schott bottle. pH is 6.0–7.0. Cool to 50°C and dispense into Petri dishes. Place into plastic bags, seal and store at 4°C.

The plates should be poured to a depth of 4 mm, as the agar assists in diluting toxic substances produced by the fungus (Unestam, 1965).

The stock trace element solution is prepared by combining 0.25 g $CuSO_4.5H_2O$, 0.286 g H_3BO_3, 0.181 g $MnCl_2.4H_2O$, 0.022 g $ZnSO_4.7H_2O$ and 0.009 g $H_2MoO_4.4H_2O$ in 1000 ml distilled water.

5.4.11 Potato dextrose agar (PDA)

Dissolve potato infusion (4.0 g), dextrose (20 g) and agar (15 g) in 1 l of distilled water and autoclave at 121°C for 15 min. Cool to 50°C and dispense into Petri dishes. Place into plastic bags, seal and store at 4°C.

5.4.12 PYGS agar (Nakamura *et al.*, 1995)

Peptone (1.25 g), yeast extract (1.25 g), glucose (3.0 g), agar (12 g) and 37.6 g artificial seawater (Aqua-Ocean®, Japan Pet Drugs), or 1 l of sterile seawater (Nakamura *et al.*, 1995; Muraosa *et al.*, 2006). A selective media can be prepared by adding 500 μg/ml of streptomycin sulfate and 500 μg/ml ampicillin (Nakamura *et al.*, 1995). Mix ingredients and autoclave at 121°C for 15 min. Cool to 50°C and add filter-sterilized antibiotic solution. Dispense into Petri dishes. Place into plastic bags, seal and store at 4°C.

5.4.13 PmTG agar (Longcore *et al.*, 1999)

PmTG nutrient agar is used for culturing *Batrachochytrium dendrobatidis* (Longcore *et al.*, 1999). One gram of peptonized milk, 1 g tryptone, 5 g glucose and 10 g agar are added to 1 l of distilled water. Mix ingredients and autoclave at 121°C for 15 min. Cool to 50°C and add filter-sterilized antibiotic solution of 400 mg streptomycin sulfate and 200 mg penicillin G. Dispense into Petri dishes. Place into plastic bags, seal and store at 4°C.

5.4.14 Sabouraud agar

Sabouraud agar, first described by Sabouraud in 1892, is used widely for culturing true fungi and yeasts and has a pH of 5.6 that inhibits growth of bacteria. The media can be purchased commercially. It is composed of 40 g dextrose, 10 g peptone and 20 g agar in 1 l of distilled water. Autoclave at 121°C for 15 min. Cool to 50°C and dispense into Petri dishes. Place into plastic bags, seal and store at 4°C. Chloramphenicol can be added to the medium and aids in the recovery of dermatophytes from skin specimens. The addition of cycloheximide aids in the recovery of slow growing fungi because it inhibits the growth of fast-growing fungi such as *Aspergillus* spp., *Trichosporon* spp. and Zygomycete fungi. The growth of some *Candida* spp. and *Cryptococcus neoformans* are also inhibited by cycloheximide.

5.4.15 TGhL

This medium is used to culture *Batrachochytrium dendrobatidis* for zoospore production (Longcore *et al.*, 1999). Add 16 g tryptone, 4 g gelatin hydrolysate, 2 g lactose and 10 g agar to 1 l of distilled water. Autoclave at 121°C for 15 min. Cool to 50°C and dispense into Petri dishes. Place into plastic bags, seal and store at 4°C.

5.4.16 Tryptone agar

Add 10 g tryptone and 10 g agar to 1 l of distilled water. Autoclave at 121°C for 15 min. Cool to 50°C and pour into Petri dishes. Place into plastic bags, seal and store at 4°C.

5.5 Stains

5.5.1 Grocott's modification of Gomori's hexamine silver method (as detailed in Carletons' Histological Technique (Carleton *et al.*, 1980)

This stain results in fungal elements being outlined in black. Mucin stains grey and the background stains a pale green. Aldehydes are formed from the carbohydrate compounds in the fungi by the oxidation of chromic acid and in turn the aldehydes reduce the hexamine-silver nitrate solution to a black metallic silver compound.

Reagents

1. sodium tetraborate (borax) in distilled water (5% solution);
2. silver nitrate in distilled water (5% solution);
3. hexamine (methenamine) in distilled water (3% solution);
4. chromic acid (5%).

Prepare a stock solution of hexamine (methenamine) silver solution by adding 5 ml of 5% silver nitrate to 100 ml of 3% hexamine. A white precipitate forms, but dissolves when shaken. The stock solution will keep for several months at 4°C.

Prepare a working solution by mixing 2 ml of sodium tetraborate with 25 ml of distilled water. Add 25 ml of the stock hexamine silver solution. Prepare fresh each time.

To stain paraffin-embedded sections, take sections to water. Oxidize in 5% chromic acid for 1 h. Wash in running water. Rinse with 1% sodium bisulfite to remove any residual chromic acid. Wash in tap water for 5 min. Wash in a minimum of three changes of distilled water. Place sections in preheated working hexamine silver nitrate solution and incubate at 56°C for 30–60 min until the sections are yellowish-brown. The time taken depends on the tissue fixative used. Rinse in six changes of distilled water. Tone in 0.1% gold chloride solution for 2–5 min. Rinse in distilled water. Place in 2% sodium thiosulfate ('hypo') solution for 2–5 min to remove the unreduced silver. Wash thoroughly in water. Counterstain with a light green solution (0.2 g light green SF, 0.2 ml glacial acetic acid, 500 ml distilled water) for 30 s. Dehydrate, clear and mount.

Safety precautions: Silver nitrate and chromic acid are toxic. Avoid contact with skin. Methenamine is a flammable solid and an irritant. Wear protective clothing (laboratory coat), gloves and protective glasses.

6

Techniques for the Molecular Identification of Bacteria

The polymerase chain reaction is an *in vitro* method that exponentially amplifies a specific region of DNA (Saiki *et al.*, 1985; Mullis and Faloona, 1987) and has wide applications in both diagnostic and research laboratories. It is routinely being used in the diagnosis and identification of bacteria. Likewise, 16S rDNA sequencing is being used more widely and is a useful tool for the identification of bacteria that are difficult to identify by biochemical tests.

DNA extraction techniques, basic PCR protocols and preparation of reagents are covered in this chapter. Specific PCRs for the identification of different species are detailed either under the relevant organism or in this chapter.

6.1 Molecular Identification by PCR Using Specific Primers

6.1.1 *Aeromonas* spp. PCR

The primers used for *Aeromonas hydrophila* (Nielsen *et al.*, 2001) also produce a band at 685 base pairs (bp) with an Australian strain of atypical *A. salmonicida* (AHLDA 1334). *Aeromonas hydrophila* strains that are positive using these primers are negative for the phenotypic tests of lysine decarboxylase and cellobiose fermentation. This PCR may assist in determining that isolates are *A. hydrophila*, as phenotypic tests are not always reliable (Nielsen *et al.*, 2001). However, phenotypic tests would need to be carried out to differentiate from atypical *A. salmonicida*. *Aeromonas hydrophila*, ATCC 7810, which is positive for lysine decarboxylase, does not produce an amplified product by this PCR.

A PCR that detected the presence of the aerolysin gene detected DNA from all haemolytic strains of *A. hydrophila* that were cytotoxic to vero and CHO cells, and produced enterotoxin as detected by suckling-mouse assays. No aerolysin gene was detected in non-haemolytic strains of *A. hydrophila*, non-haemolytic strains of *A. caviae* or strains of *A. sobria*. Extraneous bands were found with some strains of *A. caviae* and *Plesiomonas shigelloides*, but not at the correct molecular weight (Pollard *et al.*, 1990). The aerolysin gene is considered to be a useful virulence marker for detecting virulent pathogenic *Aeromonas* spp. (Kong *et al.*, 2002). Performing PCRs using the primers of Nielsen *et al.* (2001) and Pollard *et al.* (1990) may be useful in identifying virulent strains of *A. hydrophila*.

6.1.2 Multiplex PCR

A multiplex PCR enabled simultaneous detection of the human pathogens *Escherichia coli*, *Salmonella typhimurium*, *V. cholerae*, *V. parahaemolyticus* and *V. vulnificus* from shellfish samples. An optimized PCR was used with 2.5 mM $MgCl_2$, and an annealing temperature of 55°C (Brasher *et al.*, 1998).

6.1.3 Nested PCR

Any of the primers in Table 6.1 that are designed within the 16S rRNA can be used in a nested PCR reaction to improve detection sensitivity. The first PCR reaction is done with eubacterial primers that amplify the entire 16S rDNA, and the specific primers are then used for a nested PCR reaction. The detection of *Flavobacterium columnare* was improved in tissue samples when specific primers were used in a nested PCR with eubacterial universal primers (Triyanto *et al.*, 1999; Bader *et al.*, 2003).

An optimized nested PCR was reported for *Aeromonas salmonicida*, *Flavobacterium psychrophilum* and *Yersinia ruckeri*. The first round PCR used universal primers for the 16S rDNA gene and the second round of primers were species-specific for the three organisms tested. The PCR conditions were optimized so that the same conditions could be used for all primers (Taylor and Winton, 2002). The universal primers of Weisburg *et al.* (1991) with forward primer fD2 5′-AGA GTT TGA TCA TGG CTC AG-3′ and reverse primer rP2 5′-GTT TAC CTT GTT ACG ACT T-3′ were used to generate a 1500-bp fragment of the 16S rDNA gene. This template was used in a second round PCR with primers for *Aeromonas salmonicida* AS1 and AS2 (Høie *et al.*, 1997), *Flavobacterium psychrophilum* PSY1 and PSY2 (Toyama *et al.*, 1994) and primers YER8 and YER10 for *Yersinia ruckeri* (Gibello *et al.*, 1999). In the PCR reaction mix the final concentration of $MgCl_2$ was 2.0 mM, 200 μM dNTP, with 100 pmol final concentration of each primer and 1.25 U/50 μl of *Taq*. Thermocycling conditions were also standardized with an initial denaturation step of 95°C for 4 min, followed by 30 cycles at 95°C for 45 s, annealing at 45°C for universal primers and at 55°C for specific primers, with an extension at 72°C for 90 s. A final cycle was performed with an extension step at 72°C for 4 min. A final cycle of 4°C infinity was used as a holding cycle (Taylor and Winton, 2002).

Outline of steps for PCR using specific primers

Step	Method
1	Extract DNA from bacterial colonies
2	Amplify DNA using specific primers
3	Visualize DNA on agarose gel

6.1.4 DNA extraction from bacterial cells

Many methods for the extraction of DNA from bacterial cells are available and these include manual methods and commercially available kits. Such kits include InstaGene™ (Bio-Rad), AquaPure™ genomic DNA kits (Bio-Rad), Chelex®-based resin (Bio-Rad), Puregene® (Gentra Systems), PrepMan™ (Applied Biosystems), MasterPure™ (Astral Scientific – Epicentre), Wizard® Genomic (Promega), and the DNeasy® Tissue system and QiAamp® System from Qiagen. These methods differ in their degree of purity of the resulting DNA. Simple boiling methods will provide less pure DNA than kits that use resins or membranes. Boiling methods using pure bacterial cells from a culture plate provide suitable DNA for PCR methods, however tissue material contains many PCR inhibitors and this material will require more advanced methods of DNA extraction to obtain high-quality DNA.

The quality of the DNA and absence of PCR inhibitors affects the sensitivity and specificity of the PCR, and testing different DNA extraction methods is one of the steps for validation of a PCR method. Different bacteria and fungi may require specific methods. For example, Gram-positive organisms require a step using lysozyme to open the cell wall. Oomycete fungi belong to the Chromista and the cell wall may be difficult to lyse.

Manual method 1

Suspend bacterial cells in 100 μl of a solution of 1 mM EDTA-0.5% Triton-X-100. Boil in a microwave on HIGH setting for 5 min (Lee *et al.*, 1998).

An alternative method is as follows: Place 50 μg of tissue into a 1.5-ml microfuge tube. Add 100 μl of 1% Triton-X100, 20 mM Tris-HCl pH 8 and 2 mM EDTA. Immerse the tube up to two-thirds depth in boiling water and boil for 12–15 min. Use supernatant in the PCR reaction (Valsecchi *et al.*, 1998). This has been used successfully for amplification of mycobacteria from tissue.

Manual method 2

Make a suspension of bacterial cells to 0.5 McFarland density. Centrifuge at 13,000 rpm for 5 min. Suspend pellet of cells in a

Table 6.1. List of specific primers available for PCR detection of aquatic organisms.

Organism	Forward primer 5′–3′	Reverse primer 5′–3′	AT	C	Bp	Reference
Aeromonas hydrophila Doesn't differentiate from *A. encheleia**	AH-F GAAAGGTTGATGCCTAATACGTA	AH-R CGTGCTGGCAACAAAGGACAG	60	28	685	Nielsen *et al.* (2001)
Aeromonas hydrophila Aerolysin gene	Aero1a CCAAGGGGTCTGTGGCGACA	Aero1b TTTCACCGGTAACAGGATTG	55	60	209	Pollard *et al.* (1990)
Aeromonas hydrophila (Specific for HG1) Produces primer dimer bands at base of gel	7e5-F6 GCTCTCATTGCAGAGGCCGTTACTAGG	7e5-R78 TCTGCACGAAACTTCAAGCAGCTTTG	55	35	242	Oakey *et al.* (1999)
Aeromonas salmonicida ssp. *salmonicida*	ASS-F AGCCTCCACGCGCTCACAGC	ASS-R AAGAGGCCCCATAGTGTGGG	60	30	512	Miyata *et al.* (1996)
Aeromonas salmonicida ssp. *salmonicida* Use in nested PCR with Universal 16S rDNA primers	AS1 GGCCTTTCGCGATTGGATGA	AS2 TCACAGTTGACACGTATTAGGCGC	55	30	271	Høie *et al.* (1997) Taylor and Winton (2002)
Aeromonas sobria Detects cytotoxin gene	SOBF GCG ACC AAC TAC ACC GAC CTG	SOBB GGACTTGTAGAGGGCAAC CCG				Filler *et al.* (2000)
Aeromonas trota	AL1 TTGCCGCCCAGGCCGGTGCTG	AL2 ACCACTGTGTGGACCAGGGTA	66	30	622	Khan *et al.* (1999)
Edwardsiella sp. Amplifies sodB gene from all *Edwardsiella* sp.	E1F ATGTCRTTCGAATTACCTGC	497R TCGATGTARTARGCGTGTTCCCA	42	35	454	Yamada and Wakabayashi (1999)
Lactococcus garvieae Based on the dihydropteroate synthase gene	SA1B10-1-F CATTTTACGATGGCGCAG	SA1B10-1-R CGTCGTGTTGCTGCAACA	58	30	709	Aoki *et al.* (2000)
Lactococcus garvieae Based on 16S rDNA	pLG-1 CATAACAATGAGAATCGC	pLG-2 GCACCCTCGCGGGTTG	55	35	1100	Zlotkin *et al.* (1998a)
Lactococcus garvieae	IRL TTTGAGAGTTTGATCCTGG	LgR AAGTAATTTTCCACTCTACTT	45	35	482	Pu *et al.* (2002)
Lactococcus piscium Doesn't differentiate from *L. plantarum*, *L. raffinolactis*	IRL TTTGAGAGTTTGATCCTGG	PiplraR CGTCACTGAGGGCTGGAT	45	35	863	Pu *et al.* (2002)
Nocardia genus	NG1 ACCGACCACAAGGGG	NG2 GGTTGTAACCTCTTCGA	55	30	596	Laurent *et al.* (1999)

Continued

Table 6.1. Continued.

Organism	Forward primer 5′–3′	Reverse primer 5′–3′	AT	C	Bp	Reference
Photobacterium damselae ssp.	Ure-5	Ure-3	60	30	448	Osorio *et al.* (2000)
damselae and *piscicida*	TCCGGAATAGGTAAAGCGGG	CTTGAATATCCATCTCATCTGC				
Utilizes presence or absence of	Car1	Car2	60	30	267	
urease gene	GCTTGAAGAGATTCGAGT	CACCTCGCGGTCTTGCTG				
Streptococcus agalactiae	DSF1 TGCTAGGTGTTAGGCCCTTT	DSR1 CTTGCGACTCGTTGTACCAA	67	30	450	Ahmet *et al.* (1999)
(β-haemolytic, Group B).	DSF2 GGCCTAGAGATAGGCTTTCT	DSR1	67	30	265	
Semi-nested PCR.						
Cross-reaction with *S. porcinus*						
Streptococcus iniae	Sin-1	Sin-2	55	35	300	Zlotkin *et al.* (1998b)
16S rRNA	(CTAGAGTACACATGTACT(AGCT)AAG)	GGATTTTCCACTCCCATTAC				
Vibrio nigripulchritudo	VNIG-F	VNIG-R	55	35	114	Lee *et al.* (2002)
IGS^{Glu} (nigA)-specific primer	CATTTCTTTGAAACAGAAAGT	TAGATAAGGGGATTGTTGCTA				
Vibrio penaeicida	VpF	VR	62	35	310	Saulnier *et al.* (2000)
16S rDNA	GTGTGAAGTTAATAGCTTCATATC	CGCATCTGAGTGTCAGTATCT				
Vibrio salmonicida	VSAM-F	VSAM-R	55	35	275	Lee *et al.* (2002)
IGS^{0} (salA)-specific primer	TGCGATTTATGAGTGTTCA	ACTCTTCATTGAGAGTTCTG				
Vibrio splendidus	VSPN-F	VSPN-R	55	35	240,	Lee *et al.* (2002)
IGS^{0} (spnA; spnD)-specific primer	GATTTAGTTAAAGCCAGAGC	CCTGATAACTGTTTGCCG			294	
Vibrio trachuri	*Pst*I-1a	*Pst*I-1b	60	35	417	Iwamoto *et al.* (1995)
	TGCGCTGACGTGTCTGAATT	AAGCAGCGATGACAAGCAGT				
Vibrio vulnificus	P1	P2				Lee *et al.* (1998)
Nested PCR	GACTATCGCATCAACAACCG	AGGTAGCGAGTATTACTGCC	57	50	704	
	P3	P4				
	GCTATTTCACCGCCGCTCAC	CCGCAGAGCCGTAAACCGAA	59	50	222	
Vibrio vulnificus	F	R	64.5	30–50		Coleman *et al.* (1996)
Cytolysin-haemolysin gene	CGCCGCTCACTGGGGCAGTGGCTG	GCGGGTGGTTCGGTTAACGGCTGG				
Multiplex PCR*						
Virulent *Aeromonas* sp.	*Aero*-F	*Aero*-R	62	35	720	Kong *et al.* (2002)
	TGTCGGSGATGACATGGAYGTG	CCAGTTCCAGTCCCACCACTTCA				
Vibrio cholerae	Esp-F	Esp-R				
	GAATTATTGGCTCCTGTGCAGG	ATCGCTTGGCGCATCACTGCCC				
Vibrio parahaemolyticus	V*para*-F	V*para*-R				
	GCTGACAAAACAACAATTTATTGTT	GGAGTTTCGAGTTGATGAAC				

For multiplex reaction, primers were used at 0.2 µM for Aero, 0.1 µM for Esp and 1.0 µM for V*para*.						Kong *et al.* (2002)
The *Aero* primer pairs detect virulent species of *A. hydrophila*, *A. enteropelogenes*, *A. eucrenophila*, *A. popoffii*, *A. sobria* and *A. trota*.						
E. coli (*uidA* gene)	L-UIDA	R-UIDA	55–60	30	147	Brasher
	TGGTAATTACCGACGAAAACGGC	ACGCGTGGTTACAGTCTTGCG				*et al.* (1998)
Vibrio cholerae (*ctx* gene)	L-CTX	R-CTX				
	CTCAGACGGGATTTGTTAGGCACG	TCTATCTCTGTAGCCGGTATTACG			302	
Vibro parahaemolyticus	L-TL	R-TL				
(*tl* gene)	AAAGCGGATTATGCAGAAGCACTG	GCTACTTTCTAGCATTTTCTCTGC			450	
(may amplify *V. alginolyticus*)	L-CTH	R-CTH				
Vibrio vulnificus (*cth* gene)	TTCCAACTTCAAACCGAACTATGAC	GCTACTTTCTAGCATTTTCTCTGC			205	
	L-INVA	R-INVA				
Salmonella typhimurium (*invA*)	CTCTACTTAACAGTGCTCGTTTAC	TTGATAAACTTCATCGCACCGTCA			273	
Universal protocol						
Vibrio cholerae (toxin gene)	VC-1	VC-2			563	Wang *et al.* (1997)
	GGCAGATTCTAGACCTCCT	TCGATGATCTTGGAGCATTC				(quotes Fields *et al.* [1992])
Vibrio vulnificus (cytolysin gene)	VV-1	VV-2			383	Wang *et al.* (1997)
	CTCACTGGGGCAGTGGCT	CCAGCCGTTAACCGAACCA				(quotes Brauns *et al.* [1991])
Vibrio parahaemolyticus	VIP-1	VIP-2			381	Wang *et al.* (1997)
(genomic DNA)	GAATTCGATAGGGTGTTAACC	ATCCTTGAACATACGCAGC				(quotes Lee *et al.* [1995])

AT=annealing temperature, C=number of cycles, Bp=product size in base pairs, IGS=Intergenic spacers.

digestion buffer of 50 mM Tris-HCl (pH 8.5), 1mM EDTA, 0.5% SDS and 200 μg/ml of proteinase K. Incubate for 3 h at 55°C with agitation. Heat inactivate the proteinase K for 10 min at 95°C. Cool to 4°C, then centrifuge 13,000 rpm for 10 min. Use supernatant in the PCR reaction. However, the PCR master mix must contain a final concentration of Tween 20 to neutralize the effect of SDS, which is inhibitory to *Taq* polymerase at concentrations as low as 0.01% (Goldenberger *et al.*, 1995).

This method is also suitable for tissue samples with an overnight incubation in digestion buffer followed by sonication (Goldenberger *et al.*, 1995).

CTAB method (Ausubel et al., 1987)

Wash cells in TE (10 mM Tris-Cl, pH 8.0, 1 mM EDTA) buffer. A grain of rice-sized cell deposit is required. To the cell deposit add 567 μl TE, 30 μl 10% sodium dodecyl sulfate (SDS) and 3 μl 20 mg/ml Proteinase K. Incubate 37°C for 1 h. Add 100 μl 5 M NaCl and 80 μl hexadecyltrimethyl ammonium bromide (CTAB reagent). Vortex to mix. Incubate 65°C for 10 min. Add 600 μl chloroform/isoamyl alcohol (24:1) or phenol chloroform-isoamyl alcohol (25:24:2) v/v. Vortex gently, 20 s. Centrifuge 5 min at 10,000 rpm. Repeat chloroform/isoamyl alcohol. Add 400 μl ice-cold isopropanol to aqueous part (top). Centrifuge for 5 min at 6500 rpm. Wash precipitate with 400 μl 70% alcohol, twice. Centrifuge at 6500 rpm for 5 min. A final wash should be done with redistilled 95% alcohol. Dry in incubator, lids open with tubes on their side. Reconstitute with 20 μl sterile milliQ water.

Reagents:

CTAB/NaCl solution

NaCl, 4.1 g; distilled water, 80 ml; CTAB, 10 g. Dissolve NaCl in water, slowly add CTAB and heat to 65°C. Adjust final volume to 100 ml.

Sodium dodecyl sulfate (10%)

SDS, 100 g; distilled water, 900 ml.

Heat to 68°C to dissolve. Adjust pH to 7.2 using HCl and adjust volume to 1 l.

5M NaCl

NaCl 29.22 g.

Distilled water 100 ml.

Manual method 2. For Gram-positive bacteria

For Gram-positive bacteria treat cells with lysozyme (1 mg lysozyme/ml TE, pH 8.0).

6.1.5 Extraction of DNA from tissue

As for extraction of DNA from cultured bacterial cells, bacterial DNA can also be extracted from tissue samples. There are many methods available for both manual and commercially available methods. The following are some suggestions.

Manual method 1

To 100 μl of homogenized tissue add 100 μl of Chelex-100 resin (Bio-Rad Laboratories, or Sigma-Aldrich). Heat at 56°C for 10 min. Add 200 μl of 0.1% Triton-X-100 and boil for 10 min. Cool on ice, then centrifuge at 12,000 rpm for 3 min. Use 5 μl in PCR reaction (Khan and Cerniglia, 1997).

Manual method 2

Culture 10 g of tissue into 90 ml of TSBYE medium (30 g tryptic soy broth powder with dextrose, Difco; 6 g yeast extract; 1 l of water). Incubate at 25°C overnight with shaking. Take 0.5 ml of the upper phase of the sample and mix with 1 ml of sterile phosphate buffered saline (PBS, 0.05 mol/l, pH 7.4). Centrifuge at 9000×*g* for 3 min. Wash the pellet three times in PBS and once with sterile water. Resuspend pellet in 50 μl of water, then dilute 1:10 with 1% Triton-X-100 and place in a boiling waterbath for 5 min. Place immediately on ice. Use 2 μl in the PCR mixture (Wang *et al.*, 1997).

6.1.6 DNA extraction from formalin-fixed, paraffin-embedded tissue

In many cases formalin-fixed paraffin-embedded tissue (FFPE) may be the only sample available for testing. However, extracting high-quality DNA of a suitable length is difficult due to the effects of fixation or tissue integrity. Degradation of nucleic acids can be affected by pre-fixation parameters such as tissue autolysis, tissue type, and fixation factors such as type of fixative, pH of fixative, temperature and duration of fixation. Post-fixation factors such as temperature and duration of storage also impact on DNA integrity. Formaldehyde as 10% neutral buffered formalin is the most widely used fixative in histopathology. Formaldehyde (HCHO) causes cross-linkages between DNA and protein, and RNA and protein, which reduces both the quantity and length of DNA available for amplification. Fragmentation of the DNA is affected by the age of the specimen and the pH of the fixative, which if unbuffered can be <1 (Gilbert *et al.*, 2007). DNA denaturation by formaldehyde is due to breakage of hydrogen bonds and unstacking of bases, and occurs at the AT-rich regions of double-stranded DNA. Formaldehyde reacts with DNA in four ways, as reviewed by Srinivasan *et al.* (2002) and some of the consequences lead to a high frequency of non-reproducible sequence, possibly caused by cross-linking of cytosine nucleotides. The cross-linked cytosine is not recognized by *Taq* polymerase, and adenine may be incorporated in place of the guanosine, thereby creating C-T or G-A mutations. Jumping between templates occurs with damaged DNA and is more likely when there are low cell numbers.

A study investigating the best methods for pre-treatment and DNA extraction from FFPE tissues produced the following findings and recommendations for improving the quality of DNA available for amplification (Gilbert *et al.*, 2007). Deparaffinization of tissues is not required. Critical point drying using alcohol washes did not improve amplifiable DNA. The use of lithium carbonate wash steps is not needed for Bouins-fixed tissues. Incubation at 98°C prior to enzymatic treatment did not improve amplifiable DNA. Proteinase K digestion time is optimal at 48 h, and although in some cases increased product yields were obtained at times above 48 h, this varied between samples and in some cases an adverse effect was noted at 96 h. Cross-linking is reversible with input of thermal energy, however thermal energy input also results in nucleic acid degradation. Digestion temperature is important, with 55°C recommended as the optimum, although some samples showed increased amplification product at 65°C. Higher temperatures of 75°C and 85°C did not improve the amount of amplifiable DNA and could have an adverse effect. The silica column commercial kit QIAamp® (Qiagen) improved the yield of DNA and its effect is thought to be due to certain propriety buffers that cause reversal of cross-linking. Some studies recommend hot-alkali incubation (120°C and buffer at pH 11.2 or use of 0.1 M NaOH) for high levels of amplifiable DNA, however DNA degradation also occurs with this treatment and there is a trade-off between cross-link reversal and degradation of DNA (Gilbert *et al.*, 2007). In another study that investigated four extraction methods, the length of product available for amplification was influenced by the DNA extraction method and pre-treatment. Proteinase K pre-treatment is a prerequisite for successful amplification from FFPE tissue. A product of 400 bp was successfully amplified with the QIAamp® DNA-blood-mini kit (Qiagen), EasyMAG® NucliSens® (bioMérieux) and heat treatment, whereas amplification of a product of a 600 bp was marginally successful, with the QIAamp® kit producing the best results. Amplification of 400 or 600 bp was not successful using Gentra Capture-Column™ kit (Huijsmans *et al.*, 2010).

Proteinase K pretreatment

A new microtome blade is used for each sample, and floater tissue removed from the waterbath to avoid contamination between samples. Paraffin blocks without tissue are processed between samples and used in the DNA extraction and PCR as negative controls. FFPE tissue is trimmed of excess paraffin and 3 μm thick sections are cut. A single section or 1–1.5 cm^2 of sectioned tissue is placed into 250 μl of digestion buffer consisting of 100 μl Proteinase K (20 mg/ml), and 10 μl of Tween 20 is added to 2 ml of TE buffer (1 mmol/l ethylenediaminetetraacetic acid, 10 mmol/l Tris-HCl buffer, pH 8.0). Digestion is carried out for 24–48 h at 55°C. Incubation time and temperature should be optimized between 24–72 h and 55–65°C, respectively. Proteinase K is inactivated at 100°C for 15–30 min. Samples are centrifuged at 10,000×*g* for 2 min. A 200-μl aliquot of supernatant is used for DNA extraction.

DNA extraction methods

Commercial kits QIAamp® DNA blood mini kit (Qiagen) and EasyMAG® NucliSens® (bioMérieux) produced the most amplifiable DNA following Proteinase K pre-treatment in comparison studies (Gilbert *et al.*, 2007; Huijsmans *et al.*, 2010). For each kit, 200 µl of proteinase K-treated supernatant is added to 200 µl AL buffer and 2 ml NucliSens lysis buffer, respectively, and the method followed according to the kit inserts.

DNA extraction methods for fungi

Some aquatic fungi are not true fungi but are more related to the brown algae and caution must be used to ensure the correct DNA extraction method is used to enable successful quantities of DNA to be achieved.

COMMERCIALLY AVAILABLE KITS. EZNA™-Fungi DNA miniprep kit (Omega Biotek, USA), QIAamp® DNA mini kit (Qiagen) and DNAzol® reagent (Invitrogen) have been successful for oomycete fungi.

Storage of DNA standards

DNA that is to be used as a standard for control in PCR methods should be stored at −20°C in a solution of 50% glycerol in double-distilled water (Röder *et al.*, 2010). The stability of the DNA depends on buffer composition, and the use of double-distilled water results in more stable DNA, as evidenced by reduced change in the threshold cycle value compared with storage of the DNA in PCR buffer. The use of glycerol appears to reduce the shearing of the DNA caused by the formation of ice crystals. DNA was stable for 100 days when stored in 50% glycerol and double-distilled water.

6.2 PCR Protocols

Any PCR that is introduced into the laboratory needs to be optimized for the *Taq* enzyme that is used by that laboratory and the primers that are prepared. Differences may occur between different batches of primers and this may be seen with the concentration of the dNTPs. For example, a pair of primers that may be optimal at a final concentration of 100 µM dNTP mix in the PCR reaction; when re-ordered a concentration of 200 µM may be optimal.

6.2.1 Method 1: Standard protocol

Table 6.2 suggests a standard PCR protocol that may be used to amplify from most target sequences. It is also a basic starting point from which to do optimizations. When optimizing the concentrations in the PCR mix, the suggested ranges for testing are: dNTP concentration (100–200 µM), primer concentration (0.1–1 µM), magnesium chloride concentration (1–4 mM) and *Taq* enzyme (1–2.5 Units). Too much or too little of any of these reagents will cause non-specific background, mispriming or insufficient product generated. In particular, high concentrations of dNTP, primer or $MgCl_2$ will lead to non-specific product (Saiki, 1989; Innis and Gelfand, 1990). A negative and positive control should be included with each PCR run.

Sometimes bovine serum albumin, gelatin, Tween 20 or DMSO can be added to the PCR

Table 6.2. Example of a standard PCR protocol.

Reagent (starting concentration)	x1 (µl)	x10 (µl)	Final concentration
Water	34.8	347.50	
PCR buffer (10×)	5.0	50.0	1×
dNTP mix (10 mM of each)	1.0	10.0	200 µM each
$MgCl_2$ (25 mM)	3.0	30.0	1.5 mM
Primer forward (50 µM)	0.5	5.0	0.5 µM
Primer reverse (50 µM)	0.5	5.0	0.5 µM
Taq enzyme (5 U/µl)	0.25	2.5	1.25 U/50 µl reaction
Total volume including volume of DNA to be added	50 µl		

reaction mix to help stabilize the *Taq* enzyme. DMSO may assist in preventing the inhibition of the function of the *Taq* enzyme caused by inhibitors present in tissue or from contamination by agar and agar-containing media such as Stuarts' transport medium. However, not more than 2% should be used because greater than this concentration is inhibitory to *Taq*.

It is suggested that the PCR be set up in an Excel spreadsheet (see Table 6.3) with the appropriate formulae in each cell. When a different number of samples are tested then the master mix is easily calculated.

Aliquot PCR master mix into 20 μl volumes. Add a drop (approximately 30 μl) of paraffin oil to each tube if thermocycler does not have a hot lid. Tubes can be stored at −20°C until required.

Add 5 μl of template (DNA) and place tubes on thermocycler.

Excel template for preparation of PCR master mix

An Excel spreadsheet can be used as a template to calculate the volumes of reagents to be added to ×1 single tube master mix and to calculate the volumes required for a bulk master mix according to the number of tubes required. Table 6.3 contains the formulae to be entered into each cell in the spreadsheet. The formulae is $C_1 \times V_1 = C_2 \times V_2$ where C_1 is the stock concentration, C_2 is the final concentration, V_1 is the unknown volume of stock or working solution (e.g. for primers) to be added and V_2 is the total volume of the mixture. Thus the unknown $V_1 = (C_2 \times V_2) \div C_1$. Table 6.3 suggests a master mix prepared using individual reagents. The reagents are adjusted in the spreadsheet depending on the type of PCR being performed.

The actual final concentrations of dNTPs, $MgCl_2$, primers, *Taq* and DNA are all subject to optimization.

6.2.2 Method 2: Commercial PCR reaction mixture

Most companies offer master mix for conventional, multiplex and real-time PCR that contains buffer, $MgCl_2$, *Taq* and dNTPs. The user adds the DNA template, primers (and fluorescent probe) and water. Some optimization may be needed for primer and probe and DNA template concentration. Always check the product insert from the manufacturer for *Taq* polymerase or a commercial master mix for the recommended concentrations and use this as a base for optimization. Some commercial master mixes are supplied with $MgCl_2$ as a separate component so that $MgCl_2$ can be optimized. Some companies supply master mixes that they claim have such buffering capacity that no optimization is required.

Table 6.3. Suggested Excel spreadsheet for preparing master mix.

	Reagent	Stock concentration (C_1 μM)	Volume for one tube (V_1 μl)	Volume for χ tubes (μl)	Final concentration (C_2 μM)
Row/column number in Excel spreadsheet		A	B	C	D
0				20	
1	Buffer for *Taq*	10	=D1×B9/A1	=B1×C0	1
2	dNTP Mix	25,000	=D2×B9/A2	=B2×C0	200
3	$MgCl_2$	25,000	=D3×B9/A3	=B3×C0	1500
4	Primer-F	50	=D4×B9/A4	=B4×C0	0.5
5	Primer-R	50	=D5×B9/A5	=B5×C0	0.5
6	*Taq* DNA polymerase	250 U	=D6×B9/A6	=B6×C0	1.25 U
7	DNA		5		
8	Water		=B9-SUM(B1:B7)	=B8×C0	
9	TOTAL volume (V_2 μl)		50	=B9×C0	

6.2.3 Thermocycling conditions

Cycling conditions will vary slightly depending on the type or model of thermocycler according to the ramp rate, that is, the time taken to reach the desired temperature, the annealing temperatures for the primers and the type of *Taq* enzyme, whether it is a hot-start *Taq* or not. A standard set of cycling conditions begins with one cycle at 95°C for 1–5 min to denature the DNA, followed by 25–35 cycles of denaturation at 95°C for 15–60 s, annealing at 50–68°C for 15–60 s and extension at 72°C for 30–60 s. The extension step is typically at 72°C but can range from 70 to 74°C and is held for 1–2 min with a final cycle programmed with an extension of 7–10 min. The thermocycler can be set to do a hold cycle at 4°C. The manufacturer's insert usually suggests a standard protocol to be used for their *Taq* enzyme or master mix, but may need to be adjusted according to optimization results.

The primer annealing temperature depends on the length of the primer, the GC content and concentration. A recommended annealing temperature is 5°C below the true T_m (melting temperature) of the primers. The T_m is usually provided on the data sheet sent with the primer. The T_m can be estimated by calculating 2°C for A or T, and 4°C for G or C in the primer sequence. The annealing temperature affects the specificity of the reaction. A lower than optimal annealing temperature may lead to mispriming of non-target sequence or the misextension of incorrect nucleotides at the 3′ end of the primers. A low extension temperature together with high dNTPs also favours mispriming and non-specificity (Saiki, 1989; Innis and Gelfand, 1990).

The optimal number of cycles is between 25 and 35. Increasing the number of cycles may lead to problems with the PCR because a plateau effect is reached where product is no longer amplified, resulting in preferential amplification of non-specific product. The plateau is reached according to the number of target copies initially present in the sample, and the amount of DNA synthesized. Also, reagent exhaustion (especially for *Taq*) occurs with an extended number of amplification cycles. The recommended number of cycles per starting material is 25–30 cycles (for 3×10^5 target molecules), 30–35 cycles (1.5×10^4 molecules), 35–40 cycles (1×10^3 molecules), 40–45 cycles (50 molecules) (Saiki, 1989; Innis and Gelfand, 1990).

6.2.4 Gel electrophoresis and detection of amplified product

A standard gel for the detection of amplified product is as follows. Prepare a 2% agarose gel using chromosomal grade agarose (Bio-Rad) by mixing 1.8 g agarose with 90 ml 0.5× TBE buffer for a wide mini gel bed (Bio-Rad) and dissolve using a microwave oven (3 min on medium). Pipette 5–10 µl of amplification product and 5–10 µl of 2× loading buffer into the wells. Electrophoresis is carried out at 5 volts/cm for 2 h. On a Bio-Rad PowerPac™ 300, 80 volts for 90 min produces well-separated bands. A molecular weight marker such as a 100-bp marker is always included with each run.

Staining of amplified product is done using fluorescent dyes such as SYBR®Safe (Invitrogen) or ethidium bromide that intercalate with double-stranded DNA. SYBR®Safe is a safer option than ethidium bromide, which is a carcinogen and mutagen. Tests suggest SYBR®Safe has no detectable amounts of mutagenic activity (SYBR® Safe datasheet). SYBR®Safe can be added to molten agarose (3 µl per 30 ml agarose solution). This has a time advantage because the amplified product can be detected at the completion of electrophoresis. Alternatively, the gel can be soaked in a solution of TAE or TBE buffer containing SYBR®Safe. The solution should be placed in a plastic container rather than glass because the dye can adhere to glass. The container should be covered or kept in the dark to reduce the loss of fluorescence.

For ethidium bromide, the gel is stained for 30 min to 1 h in 1 l of distilled water containing 50 µl of ethidium bromide. Ethidium bromide is a carcinogen, and therefore appropriate precautions need to be taken such as wearing a lab coats and gloves, and keeping the ethidium staining tank in a dedicated area. Ethidium-contaminated waste must not be washed down the sink. Ethidium must be neutralized using ethidium bromide destaining bags, also called

EtBr 'Green Bag' Disposal kit. The kit is available from companies such as Mo-Bio, Bio 100 Inc., Qbiogene, Amresco.

The stained gel is viewed over a UV transilluminator using a standard 300-nm transilluminator, a 254-nm epi- or transilluminator, or a blue-light transilluminator. There are a number of digital imaging systems available on the market. Appropriate fluorescent filters are required on the imaging system for the detection of amplified product stained by either ethidium bromide or SYBR®Safe. The filter used for ethidium bromide is not suitable for detection of amplified product stained with SYBR®Safe.

6.2.5 Reagent preparation of stock and working solutions

Definitions and terminology

Mole (abbreviation, mol) is the unit for measuring the amount of a substance, and is a measure of the mass (moles).

Molar (M) is defined as 1 mole of solute in 1 l of solution and is a measure of concentration (molarity).

Thus a 100-M solution is by definition 100 mol/l, and a 100-μM solution is 100 μmol/l. A 1-micromolar (1 μM) solution is 1 picomoles per microlitre (1 pmoles/μl).

Name	Abbreviation	Concentration (mol/l)
Millimolar	mM	10^{-3}
Micromolar	μM	10^{-6}
Nanomolar	nM	10^{-9}
Picomolar	pM	10^{-12}
Femtomolar	fM	10^{-15}

The SI unit prefixes are milli (m), micro (μ), nano (n), pico (p), femto (f) and atto (a).

Basic equation

To work out how much of each component to use in a PCR master mix, use the following formula.

$$C_1 \times V_1 = C_2 \times V_2$$

Where:
C_1 = stock or working solution concentration;
V_1 = is the volume required of the stock or working solution;
C_2 = is the final concentration required;
V_2 = is the total volume of the mixture.

Preparation of primers

A data sheet sent with each primer details the concentration and melting temperature of the primer. The tube is centrifuged briefly to collect the lyophilized contents into the bottom of the tube. A stock solution of 100 μM (100 pmol/μl) of a primer (oligonucleotide) is prepared from the lyophilized preparation by adding a volume of solvent (TE buffer or sterile ultrapure water) equal to 10 times the nmol amount. For example, if the primer data sheet states the primer is 31.2 nanomoles, add 312 μl of solvent to the lyophilized vial to get a 100 μM stock solution. Leave for 15–20 min to ensure the primer has dissolved completely. Primers prepared in TE buffer are more stable and may dissolve more easily than in water, however for sequencing reactions water is the preferred solvent because there are no interfering substances (Shane Herbert, Australian Genome Research Facility, personal communication). Primers prepared in either TE buffer or water are stable for at least 12 months. The concentration of prepared stocks can be checked spectrophotometerically.

A working solution of primer is prepared in sterile distilled ultrapure water in the range 10–50 μM (10–50 pmol/μl). Working solutions for probes are prepared in the range 2–10 μM. Aliquot and store frozen. Repeated freezing and thawing may lead to degradation. Store stocks and working solutions at −80°C or −20°C.

Converting from molarity (moles/l) to micrograms per microlitre (μg/μl)

Each of the units (moles and litres) is reduced according to the SI prefixes. SI refers to the International System of Units (abbreviated SI from French, Système international d'unités). 1 M $MgCl_2$=1 mol/l $MgCl_2$=1 mmol/ml $MgCl_2$=1 μmol/μl $MgCl_2$. For example, the molecular weight (MW) of $MgCl_2$ is 95.2 g/l. In 500 ml (0.5 l) the concentration of $MgCl_2$ is still 95.2 g/l but the total amount of $MgCl_2$ is reduced by half (0.5 mol) to 47.6 g.

If, for example, 5 pmol of primer is required in a PCR reaction, and the working concentration is 10 μM then 10 μM=10 μmol/l=nmol/ml=10 pmol/μl. Thus each 1 μl contains 10 pmol of primer, and 0.5 μl contains 5 pmol amount. Therefore, 0.5 μl of 10 μM working solution is added to the reaction to give a final amount of primer of 5 pmol per reaction (based on http://web.pdx.edu MolBioWorksheet).

Primer quantification by spectrophotometry

An accurate quantification can be done by spectrophotometer. Dilute the stock solution of primer 1:20 (i.e. 50 μl+950 μl of water).

The UV absorbance is read at 260 nm. The concentration of the primer or oligonucleotide in μg/ml=absorbance×dilution×weight per OD.

Single-stranded DNA (ss DNA) at a concentration of 33 μg/ml has an absorbance of 1. Therefore 33 μg of ssDNA=1 OD unit.

Therefore in the above example, the concentration of ssDNA μg/ml=absorbance×20×33.

Double-stranded DNA (ds DNA): $A_{260}=OD_{260}=1$ for a 50 μg/ml solution.

Oligomer – quantitation

For a 20-mer, a stock solution with $A_{260}=1$, contains 5 nM.

5 nM=33μg/(20×325).

For a 40-mer, a stock solution with $A_{260}=1$ contains 2.5 nM.

nM=33μg/(40×325).

Oligomer – conversion of pmoles of primer to μg of primer

Multiply pmoles by (length×325)/1,000,000. Example: 51,809.88 pmoles of a 20-mer (from data sheet).

(51,809.88×20×325)/1,000,000
=336.7 μg primer.

Conversion of μg of primer to pmoles of primer

Multiply by 1,000,000/(length×325).

Example: 365.73 μg of a 20-mer (from data sheet).

(365.73×1,000,000)/(20×325)
=56,266.15 pmoles primer.

Primer concentration for PCR amplification

Micromolar concentrations of primer=pmoles/μl. Thus 20 pmoles of primer in 100 μl PCR mixture =20 micromolar (20 μM).

Equation for estimating volume required from a stock concentration

A basic formula for estimating the required volume from a stock solution is as follows:

(initial concentration)×(volume needed)
=(final concentration)×(volume of sample).

Therefore, if you have a stock solution of 5 μM of primer with a final concentration required of 0.1 μM in a 25 μl reaction volume, the equation is:

$5\,\mu M \times \chi = 0.1\ \mu M \times 25\,\mu l$

$\chi = (0.1\,\mu M \times 25)/5\,\mu M$

$\chi = 0.5\,\mu l$. Therefore, add 0.5 μl of the 5 μM stock (or working solution) to a 25-μl reaction volume.

Storage of oligonucleotides

Oligonucleotides should be stored at −20°C. Store in aliquots to prevent multiple freeze and thaw of oligonucleotides.

Deoxynucleotide triphosphate (dNTP) concentration for PCR amplification

dNTPs can be added to the master mix as individual components, or as a stock mixture. A commercial dNTP solution mixture is an equimolar mixture of dATP, dCTP, dGTP and dTTP. If each nucleotide is present at a concentration of 10 mM, the mixture has a total concentration of 40 mM. If a final concentration of 200 μM (0.2 mM) is required for each dNTP, then 1 μl of 10 mM stock is added into a 50-μl reaction mix (or 0.5 μl for a final concentration of 100 μM).

Magnesium chloride

Usually supplied as a 25-mM stock solution. 1 µl in a 25-µl reaction will result in a 1-mM final concentration.

Taq DNA polymerase

A range of 1–1.25 units of *Taq* per 50 µl reaction is recommended. Manufacturers will recommend a concentration suitable for their product. For example, 1.25 units per 50 µl reaction might be recommended or 2.5 U/100 µl reaction.

DNA concentration

See under converting from molarity to micrograms for general explanation. Example. The concentration of extracted DNA based on either comparison in a gel against a molecular weight marker with bands of DNA at known concentrations, or calculated using a spectrophotometer (or nanodrop etc.) is 123 ng/µl. A concentration of 50 ng/µl is required in a volume of 50 µl. To find out how much DNA to add use $C_1V_1=C_2V_2$. $C_1=123$ ng/µl, V_1 is unknown, $C_2=50$ ng/µl, $V_2=100$ µl.

$V_1=C_2\times V_2/C_1.V_1=50\times 100/123$

$V_2=40.6$ µl and is the volume added to the reaction (based on http://web.pdx.edu Mol BioWorksheet).

Loop-mediated isothermal amplification method

Loop-mediated iosthermal amplification (LAMP) is a method that amplifies target DNA without the use of a thermocycler. Amplification of DNA occurs through a strand displacement method using a minimum of four inner and outer primers to produce 10^{-9} copies of DNA in less than an hour (Notomi *et al.*, 2000). LAMP is very specific due to the amplification occurring as result of hybridization of four highly specific primers to six regions on the target DNA. Designing the primers is complex compared with PCR. A pair of inner primers are usually greater than 40 mer in length with the outer primers being about 25 mer. The use of additional sets of lamp primers further reduces the reaction time. Primer design software is available at http://primer explorer.jp/lamp3.0.0. The reaction can be performed using non-denatured DNA, or DNA that has been denatured for 5 min at 95°C. The reaction proceeds at 60–65°C for 45–60 min. The reaction can be visualized using horizontal gel electrophoresis, or directly in the tube. On an electrophoresis gel a positive reaction is one that has smearing of DNA from the loading well to the base of the gel. Visualization in the tube can be done by adding intercalating dyes into the mixture and examining for fluorescence. Another in-tube detection method is to add magnesium pyrophosphate and centrifuge for the detection of a white precipitate. A negative reaction does not show smearing and does not produce a white precipitate.

LAMP is highly sensitive and may be susceptible to false positives due to contamination. Amplification and detection should be carried out in separate laboratory areas.

6.3 Assay Development, Verification and Validation of Tests for Nucleic Acid Detection by PCR Methods

6.3.1 Background

The introduction of a molecular test, usually PCR, into the suite of diagnostic tests performed at a diagnostic laboratory needs to undergo verification and validation before acceptance as a routine and/or accredited test. The following describes the verification and validation that must be undertaken for nucleic acid detection tests using PCR methods.

6.3.2 Introduction

Molecular methods are increasingly being used in the diagnostic laboratory and can enhance diagnostic capacity. However, many of these tests have not undergone sufficient standardization similar to the more traditional methods or commercial products.

The introduction of new PCR tests into the quality system of a laboratory may arise from a number of sources such as a method developed in-house, a published method or an OIE (World Organization for Animal Health) method. Each of these methods requires a certain amount of verification and validation depending on the source of the method. Methods developed in-house require the most verification and validation, published methods may require slightly less validation, whereas minimal validation may be required before adoption of a standardized method such as OIE methods.

As with all diagnostic tests, methods based on nucleic acid (RNA or DNA) detection have strengths and limitations and the results need to be interpreted in the appropriate context for which the test is used. Thus all molecular methods must undergo verification and validation within a laboratory to determine the strengths and limitations of the test.

Limitations can include the potential for false-positives resulting from contamination or cross-reactivity, false-negative results resulting from inhibition and inadequate specimen processing and the inability of the test to simultaneously detect multiple species or variants. Limitations for sensitivity of the test include the number of organisms in the blood or tissue samples. Infection with 10–100 organisms per 5–10 ml of blood may not be in the detectable range for PCR that has an analytical detection limit of one copy per reaction volume (e.g. 20 μl). Sensitivity is also affected by sample processing, which needs to ensure that inhibitors are removed from the specimen, complete lysis of the cell occurs during processing, efficient recovery of the nucleic acid is achieved after extraction, sufficient sample volume is used in the test reaction and that the nucleic acid is not degraded by the process. Specificity can be affected by false-positives due to exogenous contamination, carryover, particularly with high-throughput testing, non-specific signal detection and the presence of non-viable organisms.

6.3.3 Requirements for verification and validation of PCR methods

A laboratory must have a documented method for evaluating the accuracy and precision of new tests that are offered as a diagnostic test, or for verifying the accuracy and precision of a proposed new test before it can replace an older test. It also is a requirement that a laboratory using a commercially available test verify the same performance specifications established by the manufacturer can be achieved by the laboratory.

A laboratory must also have a process for continual review (validation) of established tests to assure consistency of results and to ensure personnel remain competent to perform tests and that new staff have undergone appropriate training.

Validation

The definition for validation may vary slightly depending on the standards. According to the OIE (2012a), validation is a process of determining a test's fitness for purpose and is applied to a test that has been developed, optimized and standardized. Validation is defined as 'confirmation through the provision of objective evidence that requirements for a specific intended use or application have been fulfilled' using the ISO 9000 standard, and according to the World Health Organization it is 'the action (or process) of proving that a procedure, process, system, equipment, or method used works as expected and achieves the intended result' (WHO-BS/95.1793).

Verification

Some laboratories use the term verification when they are 'validating' a commercial test or an in-house test. Whatever terminology is used, a laboratory must demonstrate that they can obtain performance specifications comparable to those established and reported by the manufacturer for commercially available PCR tests or for tests published in the literature.

6.3.4 Process for assay development and validation of nucleic acid detection by PCR

For a fully validated test the following criteria must be tested: fitness for purpose, optimization, standardization, repeatability including precision and accuracy, analytical sensitivity, analytical specificity, thresholds for negative, positive and intermediate test result categories, diagnostic sensitivity, diagnostic specificity and reproducibility. These criteria are discussed in part A: assay or test development and part B: test validation.

Part A: Assay (test) development

FITNESS FOR PURPOSE. PCR is used for direct detection of an infectious agent. It may be used to diagnose infection, to detect bacterial or viral DNA, to discriminate vaccine from field strains, to determine phylogenetic relationship, for fast diagnosis in infection outbreaks, to determine viral load, for monitoring of vaccinated animals, for detection of antimicrobial resistant mutant strains or for demonstrating freedom from infection in live animals or animal products.

During assay or test development, determining fitness for purpose of a test must take into consideration whether the test will be used as an individual animal test, a herd or region test; whether it is to be used to distinguish vaccinated animals from infected animals; and if a test that is analytically sound at the development phase will be problematic when applied to the target population.

VALIDATION OF EXTRACTION METHOD. Depending on the type of organism (bacterial cells, fungi, aquatic fungi, algae, viruses, protozoa, parasites, mycobacteria, etc.) being detected, the nucleic acid extraction method will vary according to what is optimal for the organism and the manufacturer's instructions. Nucleic acid extraction methods should be compared using spiked samples serially diluted, with each dilution extracted as an individual sample.

It is recommended that two or three DNA extraction methods be compared and the choice of extraction method will depend on its suitability for the type of sample. It has been found that the DNA extraction method can greatly affect the sensitivity and specificity of a PCR method, both conventional and real time PCR (Florisson, Ashley and Buller, unpublished).

The DNA extractions methods are tested using serial dilutions of microorganism seeded into the source material. Each dilution is treated as a sample and the DNA is extracted according to the respective extraction methods.

OPTIMIZATION. The reagents used in the master mix must be optimized. This will vary depending on the type of master mix. If it is an in-house master mix where all reagents (*Taq* enzyme, *Taq* buffer, $MgCl_2$, dNTPs and primers) are added individually, then all must be optimized.

1 Optimization of primer concentration

When designing primers, there are some rules to be taken into consideration, which will improve the specificity, sensitivity and robustness of the PCR. The types of PCR are increasing and include conventional PCR, Real Time PCR using Sybr Green or probes such as Fluorescent Resonance Energy Transfer (FRET), Five-Prime Exonuclease Probes, Molecular Beacons etc. Each of these technologies will have specifics for primer and probe design, and the information on the manufacturer's website should be consulted when designing or using these chemistries. The most common PCRs include conventional PCR (two primers, nested PCR, multiplex PCR), real-time PCR (using SybrGreen, TaqMan® probes) and reverse transcriptase PCR. Some basics for design of either primers or probes for these technologies are outlined below. Submissions for test validation must include details of test development, which includes optimization of reagents.

The number of microorganisms being sequenced is increasing at a logarithmic rate and the numbers of sequences entered into Genbank is increasing as a result. Therefore, it is advisable to check the specificity of primers (suggested annually) to ensure that specificity of PCR primers used in a diagnostic laboratory is maintained against unusual organisms. This is advisable for direct amplification from tissues, body fluids, soil, water etc., where the originating material may contain a vast number of different microorganisms.

1.1 Conventional PCR

- Design primers to produce an amplicon of between 100 and 300 bp for conventional PCR.
- Use G+C content equal to the template (ideally 50%).
- Avoid secondary structure at the 3′ end.
- Design primers that are non-complementary to each other.
- Design primers that are 20–30 bases long. Some recommendations are 15–30 and some recommendations are longer than 30 bases.

1.2 Real-time PCR

- Software, both proprietary and freeware, is available for designing TaqMan® primers and probes.

- Design probe first, then primers.
- Design primers to be close to the probe without overlapping it.
- Use a G+C content of 30–80%.
- Avoid runs of identical nucleotides, especially guanine (avoid runs of four or more Gs).
- Select primers that have a melting temperature (T_m) of 58–60°C.
- Design primers to have no more than two G or Cs for the last five nucleotides at the 3′ end.
- Long probes may be required for target DNA that has a low G+C content.
- Short probes may be optimal for DNA target that has high G+C content.

1.3 RNA PCRs
PCRs using RNA as a template are used mainly for detecting RNA viruses and are not covered in this manual. However, a point to remember if using an RNA PCR is that the PCR primers must be re-validated on a regular basis because of point mutations occurring in infectious agents from different geographical locations.

2 Optimization of *Taq* concentration
A range of concentrations is tested for in-house master mix preparations. *Taq* activity is related to $MgCl_2$ concentrations with 2 mM$MgCl_2$ maximally activating *Taq*, whereas high concentrations such as 10 mM $MgCl_2$ are inhibitory.

3 Optimization of $MgCl_2$ concentration
Test a range of concentrations between 1 and 4 mM $MgCl_2$.

4 Optimization of dNTP concentration
Test concentrations as 100 and 200 µM.

5 Optimization of annealing temperature
The annealing temperature depends on the primer length, GC content and concentration. In general the recommended annealing temperature is 5°C below the T_m of the primers. An annealing temperature lower than the optimum may result in mispriming of non-target sequence or mis-extension of incorrect nucleotides at the 3′ end of the primer.

6 Commercial master mix
Validating a PCR using a commercial master mix is done by optimizing primer concentration and annealing temperature. Some commercial mixes allow $MgCl_2$ concentration to be optimized.

7 Thermocycling parameters
Thermocycling parameters are dependent on the length and sequence of the template DNA, the sequence and per cent complementarity of the primers and the ramp times of the thermocycler used.

RAMP TIME. The ramp time is the time taken to achieve the desired temperature inside the reaction tube and is a crucial cycling parameter. The ramp time is usually longer than either the denaturation or primer annealing time. Ramp times are lower with thin-walled reaction tubes.

EXTENSION TIME. The extension time depends on the length of the target sequence. An extension time of 1 min/kb is used for target sequences >1 kb. For targets <100 bases in length, an extension time of 2 s is suitable.

CYCLE NUMBER. Cycle number depends on the efficiency of the reaction and the amount of DNA in the reaction. Generally cycle numbers of 30–35 are used. A starting amount of mammalian DNA of 100 ng, which is 10^4 cell equivalents, will produce a visible band on a ethidium bromide stained gel after 30 cycles. Increasing the cycle number to 40 can reduce *Taq* polymerase specific activity, increase the amplification of non-specific products and delete nucleotides or substrate. The last extension step is increased to 7 or 10 min to ensure all the PCR products are full length.

6.3.5 Quality control and standardization

The following factors are ones that may impact on successful test outcome and must be considered when the test is in the development phase.

Sample management

SAMPLE COLLECTION AND TRANSPORT. **Tissue**: Tissue samples should be collected using gloves to minimize exogenous DNA and prevent contamination of the sample. Transport on ice or at 4°C.

Blood: Blood samples for PCR testing should not be collected into heparin anticoagulant because

this is a potent inhibitor of PCR (as is heme). The recommended anticoagulants are ethylenediaminetetraacetic acid (EDTA) and acid citrate dextrose (ACD). Blood should not be lysed because lysed blood will contain PCR inhibitors.
Faeces: Transport on ice or at 4°C.
Semen: Chill immediately and transport to the laboratory maintained at 4°C.
Urine: The low pH and high urea content in urine rapidly denatures DNA, particularly at temperatures of 25°C and above.

SAMPLES FOR RNA STUDIES. Degradation of RNA and gene induction can occur with blood specimen collection, and within minutes of collection of tissue and body fluids. This leads to changes in the gene expression profile. Blood samples should be collected into tubes containing a RNA stabilizing agent or flash frozen in liquid nitrogen. Frozen specimens should be transported on dry ice and at the laboratory immediately placed at −70°C without thawing. Degradation of RNA will occur at −20°C, therefore storage for RNA should always be at −70°C.

SAMPLES AS FORMALIN-FIXED PARAFFIN-EMBEDDED TISSUE (FFPET). Formalin cross-links DNA and may prevent the amplification of high molecular weight amplicons. High molecular weight DNA from these samples may be obtained using the critical drying point method (1).

SAMPLE STORAGE. All samples containing microorganisms that will subsequently be detected using PCR should be refrigerated immediately on receipt in the laboratory. Samples not tested the same day should be stored at −20°C and at −70°C for long periods.

Sample preparation

CONCENTRATION OF THE PATHOGEN.
Concentration by centrifugation: This may be done by centrifugation where a 0.5 ml to 1-ml aliquot is centrifuged using high-speed centrifugation (24,000×*g* for 60 min for viruses). Centrifugation may also concentrate the proteins and other sample material that may contain PCR inhibitors.

Concentration of pathogen in faecal material: Faecal specimens are first diluted with a buffered (pH 7.4) solution and centrifuged and filtered to remove debris.

TISSUE FIXATIVES.
Formaldehyde (formalin): Ten per cent neutral buffered formalin decreases the solubility of DNA and can also lead to DNA degradation, even in short-term treatments of tissue in formalin, and therefore is generally not recommended for molecular tests.
Glutaraldehyde (1%): This is used as a standard fixative for electron microscopy. It does not degrade DNA to the same extent as 10% formalin.
Ethanol and methanol (100%): These are excellent fixatives for both DNA and RNA because they cause little chemical change in the nucleic acid. DNA is 'collapsed' in ethanol and methanol and reverts to the original form when the denatured DNA is rehydrated.

STORAGE OF PURIFIED DNA. Long-term storage of purified DNA should be at temperatures of less than 0°C and in polyallomer tubes that neither denature nor bind DNA.

Storage of DNA below 0°C minimizes the activity of enzymes that degrade DNA (DNAses). Containers should preferably contain a gasket so as to prevent evaporation. Polyallomer tubes are the recommended choice for storage because they do not bind or denature DNA, as happens with some types of tubes, but they are expensive. Specifically designed polypropylene tubes are suitable for long-term storage of DNA but need to be checked with the manufacturer because some types of polypropylene tubes are associated with DNA adsorption, particularly at high ionic strength. Polyethylene tubes are not to be used because DNA binds strongly to this surface.

Purified DNA can be stored in TE (Tris-EDTA) buffer at room temperature for 26 weeks, at 2–8°C for at least 1 year if free of contaminating DNAses, and for up to 7 years at −20°C and at least 7 years at −70°C or lower.

Freezers used to store DNA should not be 'frost-free' because the continual cycle of temperatures of these freezers causes the DNA to shear and therefore deteriorate.

Part B: Test validation

A validated test or assay means that it is capable of providing consistent results when detecting whether an animal is positive or negative for the presence of a microorganism and can predict with statistical certainty the exposure status of the animals. A validated test must have undergone previous development that included optimization and standardization to achieve the test performance parameters consistent for the fitness of purpose for which the test was developed.

A standardized method, such as OIE methods, must be performed exactly according to the method, otherwise validation will need to be undertaken according to the parameters that are changed from the standard method. Often, when introducing a standardized method into a laboratory the only component changed is the thermocycler. If this is the case, then validation must be undertaken on the new thermocycler by testing sensitivity, specificity and annealing temperature. The same sensitivity and specificity results should be obtained, and if they are different to the published method this must be recorded in the laboratory method.

A published method must be followed as published. Any changes to DNA/RNA extraction method, reagents, cycling conditions or thermocycler must have validation and optimization carried out to check the changes do not alter the sensitivity and specificity of the test.

6.3.6 Terminology and definitions

Most of these terms are ISO terms and used around the world.

Accuracy refers to the closeness of the agreement between the result of a measurement and a true value of the measureand.

Amplicon is the product from amplification of a nucleic acid target.

Analyte is the component represented in the name of a measurable quantity.

Analytical sensitivity in quantitative testing is the change in response of a measuring system or instrument divided by the corresponding change in the stimulus.

Analytical sensitivity in qualitative testing is the test method's ability to obtain positive results in concordance with positive results obtained by the reference method.

Analytical specificity in quantitative testing is the ability of a measurement procedure to determine only the component it purports to measure or the extent to which the assay responds only to all subsets of a specified analyte and not to other substances present in the sample.

Analytical specificity in qualitative or semi-qualitative testing is the method's ability to obtain negative results in concordance with negative results obtained by the reference method.

Diagnostic accuracy refers to the ability of a test system to obtain the correct result.

Diagnostic sensitivity (non-ISO term is clinical sensitivity). This is the fraction of clinically true positives divided by the sum of clinically true positives plus the clinically false-negative classifications.

Diagnostic specificity (non-ISO term is clinical specificity). This is the fraction of clinically true negatives divided by the sum of the clinically true negatives plus the clinically false-positives.

Equivocal is a test result within a specified range of the cut-off value, which cannot be interpreted as either positive or negative.

Inhibition arises when components within the specimen or exogenous substances introduced during specimen collection or processing reduce amplification or detection of the nucleic acid.

Interfering substances are substances in the specimen that can cause false-positive or false-negative results.

Internal amplification control is a non-target sequence present in the same sample tube, which is co-amplified so as to identify inhibition due to thermal cycler malfunction, suboptimal reagents or polymerase activity, or presence of inhibitory substances in the sample matrix. The internal amplification control can be a host gene or an artificial DNA molecule.

Limit of detection is also referred to as the minimum detectable concentration. It is the lowest amount of analyte in a sample that can be detected with probability.

Measurand is the particular quantity subject to measurement.

Measuring range (non-ISO term is reportable range) refers to a set of values of measureands for which the error of a measuring instrument is intended to lie within specified limits.

Repeatability (the non-ISO term is within run precision) describes the closeness of agreement between results of successive measurements of the same measurand carried out under the same conditions of measurement.

Sensitivity is the change in the response of a measuring instrument divided by the corresponding change in the stimulus.

Specificity is the ability of a measurement procedure to measure solely the measurand.

Trueness refers to the closeness of the agreement between the average value from a large series of measurements and to a true value of a measureand; the measurement of trueness is usually expressed in terms of bias.

Validation (ISO 9000 definition) is confirmation through the provision of objective evidence that requirements for a specific intended use or application have been fulfilled.

Validation (WHO definition) is the action or process of proving that a procedure, process, system equipment or method used works as expected and achieves the intended result. The components of validation are quality control, proficiency testing, validation of operator competency, instrument calibration and correlation with clinical findings.

Verification is confirmation through the provision of objective evidence that specified requirements have been fulfilled (ISO 9000). Verification is a process used to determine or confirm test performance characteristics before the test system is used for diagnostic testing.

6.4 Molecular Identification by 16S rDNA Sequencing

Sequencing the 16S rRNA has been used extensively to study bacterial evolution and phylogeny. With a vast number of 16S rRNA sequences available in the National Centre for Biotechnology Information (NCBI) and the Ribosomal Database Project (RDP) databases, sequencing the 16S rDNA is an essential tool in bacterial systematics and the identification of new species.

It is also a useful tool when used with biochemical tests for bacterial identification in the diagnostic laboratory. Ribosomal RNA contains variable and highly conserved regions that evolve very slowly and therefore can be specific to a genus.

The 16S rRNA is found in the ribosomes, which consist of proteins and RNA. In prokaryotes the ribosomes measure 70S, and consist of two subunits, 50S and 30S. S is the sedimentation rate, or Svedberg unit, during centrifugation. The 16S rRNA is found in the 30S subunit, whereas the 50S subunit contains the 5S and 23S rRNA molecules. 16S is approximately 1600 nucleotides, 23S, 3000 nucleotides and 5S is approximately 120 nucleotides.

Within the gene encoding for the 16S rRNA are areas of sequence that are conserved across the different genera, and it is these areas that have been used for the design of the universal eubacterial primers. Not all primers will bind to the DNA of the bacterium of interest; however, if a combination of primers are used then a section of DNA will be amplified that can then be sequenced for identification.

The theoretical threshold for delineation of a bacterial species is less than 97% similarity using 16S rRNA sequencing (Stackebrandt and Goebel, 1994).

Outline of steps for sequencing

Step	Method
1.	Extract DNA from bacterial colonies
2.	Amplify 16S rDNA using universal eubacterial primers
3.	Visualize DNA on agarose gel
4.	Clean up amplified product
5.	Quantitate DNA concentration
6.	Perform sequencing PCR reaction
7.	Sequencing (send to a sequencing facility)
8.	Analysis of sequence information using BioEdit
9.	BLAST search for sequence identification

6.5 Amplification of 16S rDNA Using Universal Eubacterial Primers

16S rRNA sequencing is used to delineate bacterial species. A sequence similarity of above 97% must be achieved for delineation of a bacterial species (Stackebrandt and Goebel, 1994).

The rDNA contains a number of sites that are conserved across genera, and as such there are a number of primers that can be used to amplify a part or all of the 16S rDNA for many genera. The universal primers A, B and C amplify regions that are universally conserved across prokaryotes and eukaryotes and were initially proposed by Lane *et al.* (1985). Since then a number of different variations have been proposed. Some primers are universal for bacteria only, some amplify the entire 16S rDNA, whereas others amplify different regions. The convention for identifying the primers is to name them according to the number of the *Escherichia coli* position at which the 3′ end of the primer anneals.

Some of the primers that are available to either amplify all or part of the 16S rDNA are detailed in Table 6.4. Not all primer combinations will be appropriate for all bacteria. Table 6.5 indicates primer combinations that have proved useful. The primers may be ordered from a company such as Qiagen or Invitrogen.

Primers A, B and C are universal for organisms from the three primary kingdoms, archaebacteria, eubacteria and eukaryotes (Lane *et al.*, 1985). Primer EUBB (7-26) of Suzuki *et al.* (1996) is the same as primer 27f of Lane (1991), except for a degeneracy (M) that possibly makes the primer more universal.

Primer 27f was modified because complementarity at the 3′ ends led to self-priming followed by primer depletion. The modified primer was referred to as POmod (Wilson *et al.*, 1990). Complementarity in primers and between primer pairs can lead to primer dimers, which is the product that is amplified in preference to the desired sequence on the DNA template (Watson, 1989).

The suggested pairs of primers of 63f and 1387r were found to improve on the primer pair of 27f and 1392r (Lane *et al.*, 1991), particularly with difficult DNA templates (Marchesi *et al.*, 1998).

The entire rDNA can be amplified with two sets of primer pairs. The primer pairs POmod and PC3mod amplify 789 bp of the rDNA and the pairs P3 and PC5 amplify the remainder of the rDNA of 721 bp (Wilson *et al.*, 1990). Alternatively, the entire 16S rRNA can be amplified with 27f or EUBB as the forward primer and 1525 or EUBA as the reverse primer (Weisburg *et al.*, 1991).

The universal primers 16/23S-F and 16/23S-R were used to amplify the intergenic spacer (IGS) region between the 16S rRNA and the 23S rRNA genes of the *Vibrio* spp. and this region was subsequently used to design specific primers for the detection of eight *Vibrio* spp. A standard PCR reaction mix, and thermocycling conditions with an annealing temperature of 55°C, was used (Lee *et al.*, 2002). See Table 6.1 for specific primers.

Once the entire 16S rDNA is amplified and purified, the sequencing primers are used in individual sequencing PCR reactions to sequence the 1.5-kbp product. For purification of the amplified product and for quantification of DNA, Table 6.7 details the sequencing master mix.

6.5.1 PCR master mix for universal primers

Use the standard PCR protocol, 6.1.4.1, as a guide, DMSO at no more than a 2% final concentration can be added to improve the specificity of hybridization of the primers (Wilson *et al.*, 1990). Prepare PCR master mix in a single tube. Aliquot 20 μl volumes to 0.6 ml or 0.2 ml microfuge tubes, depending on the requirements of the thermocycler. Add 20–30 μl of paraffin oil to each tube (if hot lid is not available on the thermocycler). Tubes can be labelled and stored at −20°C until required for use.

For the PCR add 5 μl of template DNA to a 20-μl reaction tube and place in thermocycler.

6.5.2 Thermocycling conditions for universal primers

The thermocycling conditions are much the same as for the universal PCR, except that a relaxed annealing temperature is used. Therefore, the cycling conditions begin with one cycle at 95°C for 1–5 min. Then, 25–35 cycles of denaturation at 95°C for 30 s, annealing at 45–55°C 30 s, extension at 72°C for 30 s. A final cycle is suggested as for the last cycle but for an extension time of 10 min. The thermocycler can be set to do a hold cycle at 4°C. The annealing temperature can be between 45 and 55°C.

6.5.3 Visualize DNA on agarose gel

To check that the DNA has been amplified, visualize the product on a 2% agarose gel.

Table 6.4. Universal eubacterial primers for 16S rDNA sequencing.

Primer (*E. coli* numbering)	Primer sequence 5′ to 3′	Target	Reference
27f (EUBB) (7–26)	AGAGTTTGATCMTGGCTCAG	Most Eubacteria	Lane (1991); Suzuki and Giovannoni (1996); Weisburg *et al.* (1991)
20F	AGAGTTTGATCATGGCTCAG	Eubacterial	Weisburg *et al.* (1991)
POmod (8–22) modified from 27f	AGAGTTTGATCMTGG	Eubacterial kingdom only	Wilson *et al.* (1990)
63f (43–63)	CAGGCCTAACACATGCAAGTC		Marchesi *et al.* (1998)
357r (339–357)	CTCCTACGGGAGGCAGCAG	Most Eubacteria	Lane (1991); Weisburg *et al.* (1991)
530f (515–530)	GTGCCAGCMGCCGCGG	Most Eubacteria, Eukaryotes, Archaebacteria	Lane (1991)
P3mod f (787–806)	ATTAGATACCCTDTAGTCC	Eubacterial kingdom only	Wilson *et al.* (1990)
519r (519–536)	GWATTACCGCGGCKGCTG	Primer A. Universal for all kingdoms	Lane *et al.* (1985); Lane (1991)
PC3mod r (787–806)	GGACTAHAGGGTATCTAAT	Eubacterial kingdom only	Wilson *et al.* (1990)
685r3	TCTRCGCATTYCACCGCTAC	Most Gram-positives, cyanobacteria	Lane (1991)
907r (907–926)	CCGTCAATTCMTTTRAGTTT	Primer B. Universal for all kingdoms	Lane *et al.* (1985); Lane (1991)
926f	AAACTYAAAKGAATTGACGG	Most Eubacteria, Eukaryotes, Archaebacteria	Lane (1991)
1100r (1100–1114)	GGGTTGCGCTCGTTG	Most Eubacteria	Lane (1991); Weisburg *et al.* (1991)
1114f	GCAACGAGCGCAACCC	Most Eubacteria	Lane (1991)
1387r (1387–1404)	GGGCGGWGTGTACAAGGC		Marchesi *et al.* (1998)
1392r (1392–1406)	ACGGGCGGTGTGTRC	Primer C. Universal for all kingdoms	Lane *et al.* (1985); Lane (1991)
1406f	TGYACACACCTCCCGT	Most Eubacteria, Eukaryotes, Archaebacteria	Lane (1991)
PC5 r (1492–1507) Modified from 1492r	ACCTTGTTACGACTT	Eubacterial kingdom only	Lane (1991); Wilson *et al.* (1990)
1492r (1492–1512)	TACGGYTACCTTGTTACGACTT	Most Eubacteria, Archaebacteria	Lane (1991)
1500R	GGTTACCTTGTTACGACTT	Eubacterial	Weisburg *et al.* (1991)
1525r (1525–1541)	AAGGAGGTGWTCCARCC	Universal for all kingdoms	Lane (1991); Weisburg *et al.* (1991)
1525r EUB A modified from 1525r	AAGGAGGTGATCCANCCRCA	Eubacterial kingdom only	Suzuki and Giovannoni (1996)
U8f	AGAGTTTGATCMTGGCTCAG	*Streptococcus* sp.	Gomila *et al.* (2005)
U1492r	TACGGYTACCTTGTTACGACTT	*Streptococcus* sp.	Gomila *et al.* (2005)
Str1	GTACAGTTGCTTCAGGACGTATC	*Streptococcus* sp.	Picard *et al.* (2004)
Str2	ACGTTCGATTTCATCACGTTG	*Streptococcus* sp.	Picard *et al.* (2004)

M=C:A, **R**=A:G, **K**=G:T, **W**=A:T (Lane *et al.* (1985). Other mixtures of nucleotides (known as degeneracies or wobbles) are **S**=C:G, **Y**=C:T, **V**=A:G:C, **H**=A:C:T, **D**=A:G:T, **B**=C:G:T, **N**=A:G:C:T. F=same sequence as rRNA. R=the complement of rRNA sequence. The primer positions correspond to the nucleotide numbering system of *E. coli* 16S rRNA (e.g. 9–27). The C in the name of the primers of Wilson *et al.* (1990) refers to the complementary sequence.

Table 6.5. Suggested primer pairs.

Forward primer 5′–3′	Reverse primer 5′–3′	Reference
27f GAGTTTGATCCTGGCTCAG	1392R ACGGGCGGTGTGTRC	Lane (1991)
63f CAGGCCTAACACATGCAAGTC	1387R GGGCGGWGTGTACAAGGC	Marchesi *et al.* (1998)
530F GTGCCAGCMGCCGCGG	1100R GGGTTGCGCTCGTTG	Lane (1991)
POmod AGAGTTTGATCMTGG	PC3mod. GGACTAHAGGGTATCTAAT	Wilson *et al.* (1990)
P3mod ATTAGATACCCTDTAGTCC	PC5. TACCTTGTTACGACTT	Wilson *et al.* (1990)
16/23S-F (1390-1407) TTGTACACACCGCCCGTC	16/23S-R (474-456) CCTTTCCCTCACGGTACTG	Lee *et al.* (2002)

Table 6.6. Universal primers for fungal species.

Primer name	E. coli position	Sequence 5′–3′	Reference
ITS1			White *et al.* (1990)
ITS2			
16F926	908–926	AACTCAAAGGAATTGACGG	Modified from primer 926f from Lane (1991)
16F1112	1093–1112	AGTCCCGCAACGAGCGCAAC	Thompson *et al.* (2001)
16F1241	1222–1241	GCTACACACGTGCTACAATG	Thompson *et al.* (2001)
16R339	358–339	ACTGCTGCCTCCCGTAGGAG	Modified from primer 342r from Lane (1991)
16R519	536–519	GTATTACCGCGGCTGCTG	Modified from primer 519r from Lane (1991)
16R1093	1112–1093	GTTGCGCTCGTTGCGGGACT	Thompson *et al.* (2001). Similar to 1100r Lane (1991)

Table 6.7. Sequencing PCR master mix.

Reagent	Half reaction (µl)	Full reaction (µl)
Water (ultra-pure)	0	0
Terminator ready mix	4	8
Primer F	0.5	1
DNA (10–30 ng/µl)	5.5	11
Volume µl (including DNA)	10	20

6.5.4 Clean-up amplified product

Before sequencing, the amplified product needs to be purified from the other components in the PCR reaction. Commercial kits such as QIAquick® PCR kit or QIAprep® Spin kit (Qiagen), PCR Kleen™ spin columns (Bio-Rad) or Wizard® PCR Preps (Promega) can be used for post-PCR clean up.

6.5.5 Quantitate DNA concentration

Once the DNA is free of other contaminants, it needs to be quantitated for the PCR sequencing reaction. This can be done spectrophotometrically by reading the absorbance at 260 nm. The concentration of double-stranded DNA in µg/ml= absorbance×dilution×weight per OD. Weight per OD for dsDNA is 50 µg/ml.

Alternatively, the concentration of DNA can be estimated by gel electrophoresis in a 2% agarose gel and estimating the amount of DNA against a quantitative DNA molecular weight marker, which has known molecular weights in nanograms per band. Examples are HyperLadder™ IV from Bioline and GeneRuler™ from MBI Fermentas.

For the sequencing PCR, the concentration of DNA needs to be 10–30 ng for a reaction volume of 10 µl. If the DNA needs to be concentrated, see section 6.5.9 on product purification.

6.5.6 Concentrating DNA, ethanol precipitation

If the DNA needs concentrating to obtain the desired concentration for sequencing, then an ethanol precipitation is used. If oil has been used as an overlay, remove as much as possible before the post-PCR clean up or ethanol precipitation. This method is from Sambrook *et al.* (1989).

1. Add a 1/10 volume of 3M sodium acetate pH 4.6.
2. Add an equal volume of 95% ethanol (non-denatured alcohol).
3. Centrifuge 13,000 rpm for 15 min at room temperature.
4. Carefully remove the supernatant.
5. Wash twice with 70% ethanol.
6. Wash once with 95% ethanol.
7. Air dry, or vacuum dry.
8. Add the desired volume of water or TE buffer to the dried pellet.

Note that 3M sodium acetate = 24.6 g sodium acetate (molecular weight = 82.03) in 100 ml of distilled water.

6.5.7 Perform PCR sequencing reaction

The sequencing PCR is done using the BigDye™ Terminator v3.0 kit from Applied Biosystems.

There is only one primer per sequencing reaction. Therefore, the universal forward primer used in the amplification PCR is used in one sequencing tube, and the reverse primer in a second tube. It is advisable to do each primer in duplicate or triplicate to check for errors due to mispriming.

This is the recommended protocol from Applied Biosystems for use with their BigDye™ V3.0 Cycle Sequencing kit.

The full reaction volume is that recommended by the supplier. The volume of water is used to adjust for the concentration of DNA. Sequencing results will not be obtained if the DNA is too dilute. For a thermocycler without a hot lid, overlay tubes with 30 μl mineral oil.

6.5.8 Thermocycling conditions for the sequencing PCR

Perform one cycle at 96°C for 2 min. Run 25 cycles of 96°C for 30 s, 50°C annealing for 30 s, 60°C for 4 min. Hold at 4°C. Times may be shorter, depending on the thermocycler used.

6.5.9 Product purification before sequencing

The product from the sequencing PCR reaction needs to be purified to remove salts and unincorporated dye terminators, otherwise the first 100 bases of the sequence will not be readable. Clean up can be done by using Centri-Sep™ columns from Applied Biosystems, Micro Bio-Spin™ 30 columns from Bio-Rad or QIAquick® PCR purification kit from Qiagen. Check with the sequencing facility to find out how they want the product to be presented to them for sequencing.

The following manual protocol gives good results and is based on an ethanol precipitation.

1. Into a 0.6-ml microfuge tube place 25 μl of 95% ethanol (non-denatured, absolute alcohol).
2. Add 1 μl of 3M sodium acetate, pH 4.6. Add entire PCR product (10 μl if a half reaction volume was used).
3. Place on ice for 10–20 min. Some methods recommend a strict 10 min only.
4. Centrifuge at 13,000 rpm at room temperature in a table-top centrifuge for 30 min.
5. Keep the hinges of the tube to the outside so that the DNA pellet can be more easily located.
6. Carefully remove supernatant without disturbing the pellet, which may not be visible.
7. Immediately add 125 μl of 70% ethanol and gently roll the tube.
8. Centrifuge at 13,000 rpm for 5 min.
9. Carefully remove the supernatant without disturbing the pellet.
10. Air dry or vacuum dry the pellet.
11. Send the dried product to a sequencing facility.

6.5.10 Sequencing

The sequencing reaction won't be covered in this manual. Usually, this reaction is carried out by a specialized laboratory. Check with the sequencing facility to find out how they want the product to be presented to them for sequencing.

6.5.11 Analysis of sequence information using BioEdit

The sequence information can be analysed by a number of different programs. Proprietary software from Applied Biosystems is available for purchase from the company; however, they also have freeware available on their website such as Sequence Scanner Software that allows sequence data to be viewed, edited and printed.

A number of other freeware programs are available and a comprehensive one is called BioEdit. It is available for download from http://www.mbio. ncsu.edu/BioEdit/bioedit.html, but the software is no longer being updated.

Citation: Hall, T.A. (1999) BioEdit: a user-friendly biological sequence alignment editor and analysis program for Windows 95/98/NT. *Nucleic Acids Symposium Series* 41, 95–98.

Using BioEdit

A brief explanation of the use of BioEdit for analysing sequence data is given here, but the reader is referred to the help notes under 'General use of BioEdit' contained with the program.

The sequence information is usually e-mailed to the client laboratory. Open the e-mail and save the attachments into a file on the hard drive.

Open the BioEdit program. Select OPEN SEQUENCE SET. This opens a 'open file' window. The sequence files will have the ending .ab1. Select the sequence to be examined and double click. This places the title of the sequence in the left-hand side of the BioEdit window, and the DNA sequence in the right-hand side window. A chromatogram window is also seen. Position the two windows so that both can be seen and check the sequence for inaccuracies. In many cases where a no base is called (N), checking the chromatograph can interpret the correct base. Do the same with the duplicate sequence. Once both are checked a consensus sequence can be obtained.

First select the duplicate sequence by clicking in the right-hand window to highlight the sequence name. Go to the menu bar and EDIT, COPY SEQUENCE. Then click on the window containing the first sequence. Go to the menu bar and EDIT, PASTE SEQUENCE. The primary and duplicate sequences will now be in the one window.

While holding down the shift key, click on both sequence names in the right-hand window to highlight both sequences. Go to menu bar SEQUENCE, PAIRWISE ALIGNMENT, ALIGN TWO SEQUENCES (optimal global alignment). A new window appears with the alignment result for the two sequences. From this window go to the menu bar and select ALIGNMENT, CREATE CONSENSUS SEQUENCE. A window now appears with the consensus sequence. To save the consensus sequence, highlight the consensus sequence, go to the menu bar EDIT, COPY SEQUENCE. Open a new word document and select PASTE in the Word menu bar. The consensus sequence will appear thus:

>Consensus

GGTACTGACC etc.

Re-name the sequence as appropriate. This format is known as 'fasta' format, and is the format required for a BLAST search.

Aligning two sequences using Clustal W

BioEdit also contains the Clustal program. To use Clustal within BioEdit to align two sequences, select the NEW ALIGNMENT icon. Go to the document with the sequence to be aligned. Make sure the sequence is in fasta format. Go back to the BioEdit NEW ALIGNMENT window. Under FILE, select IMPORT FROM CLIPBOARD. The sequence name and sequence will appear in the window. Do this for the second sequence. Under ACCESSORY APPLICATION, select CLUSTAL. Run Clustal. When it is finished, close the Clustal window to see the alignment in the BioEdit window.

If the reverse complement is needed, under SEQUENCE go to NUCLEIC ACID, then to REVERSE COMPLEMENT.

Clustal W citation: Thompson, J.D., Higgins, D.G. and Gibson, T.J. (1994) Clustal W: Improving the sensitivity of progressive multiple sequence alignment through sequence weighting, position specific gap penalties and weight matrix choice. *Nucleic Acids Research* 22, 4673–4680.

Web software for multiple alignments

http://dot.imgen.bcm.tmc.edu:9331/multi-align/multi-align.html

http://www.ebi.ac.uk/clustalw Clustal alignment program.

http://www.ncbi.nlm.nih.gov/blast/bl2seq/bl2.html Blast for alignment of two sequences.

http://www.mbio.ncsu.edu/Bioedit/bioedit BioEdit program for sequence analysis.

http://www.technelysium.com.au/chromas.html Chromas is a sequence editor.

or http://bioinfo.weizmann.ac.il/pub/software/chromas.

BLAST search for sequence identification

The National Centre for Biotechnology Information (NCBI) is located at http://www.ncbi.nlm.nih.gov. Select the Blast icon. There is a tutorial here. However, BLAST is very easy to use. Select Nucleotide Blast. In the search window paste the sequence in fasta format. Press 'Now Blast it'. Press 'Format' to get the results. The results are placed in a queue and it may take 1 to 2 min for the results to be downloaded to the desktop.
Citation for BLAST: Altschul, S.F., Madden, T.L., Schäffer, A.A., Zhang, J., Zhang, Z., Miller, W. and Lipman, D.J. (1997) Gapped BLAST and PSI-BLAST: a new generation of protein database search programs. *Nucleic Acids Research* 25, 3389–3402.

The Ribosomal Database Project (RDP) contains ribosomal sequence information at http://rdp.cme.msu.edu/html.

Phylogenetic analyses

Molecular phylogenetics is the analysis of hereditary differences in the DNA sequence to understand the evolution of the organism. The result of the analyses is the construction of a phylogenetic tree. Molecular systematics includes phylogenetics to assist taxonomy. To perform phylogenetic analysis from the results of sequencing house-keeping genes, pairwise alignment of sequences is done using Clustal W. A number of websites offer Clustal W. It is also contained within programs such as MEGA and BioEdit, although the latter program is no longer being maintained (http://www.mbio.ncsu.edu/BioEdit/bioedit.html). The MEGA (Molecular Evolutionary Genetics Analysis ([Kumar *et al.*, 2008]) software has been designed for biologists and a comprehensive manual and tutorial system plus the program for different computer platforms is available from the download website (Tamura *et al.*, 2011). Version 5 can be downloaded from http://www.megasoftware.net. This software infers evolutionary relationships of homologous sequences and estimates the neutral and selective evolutionary divergence. The Neighbour-joining (NJ) distance method (Saitou and Nei, 1987) and the maximum-parsimony character methods are used to construct trees and can be done within MEGA5.

The START2 (Sequence Type Analysis and Recombination Tests) software package (Jolley *et al.*, 2001) can be used to calculate recombination tests and the ratio of mean non-synonymous substitutions per synonymous site (d_s:d_{sn}) (http://pubmlst.org/software/analysis/start2). Split tree decomposition analysis can be performed using freeware (http://www.splitstree.org).

This application uses methods such as split decomposition, neighbor-net, consensus network, super networks methods to compute hybridization or simple recombination networks.

Concatenated sequences

There are different ways in which a phylogenetic tree can be constructed from sequence information, but two of the most commonly used methods are concatenation and consensus treatment of the sequence information when constructing the tree. Concatenation involves the alignment head to tail to form a super gene alignment of a number of housekeeping genes. Consensus phylogeny involves the alignment of individual genes. The concatenated approach for inferring species phylogenies from multiple genes was found to be more accurate than the consensus method when constructing phylogenetic trees (Gadagkar *et al.*, 2005). Within the *Vibrio* genus (and within other organisms) incongruence may occur between different genes analysed. Incongruence may occur using single or a small number of concatenated gene sequences. It has been suggested that 20 genes should be used to obtain an accurate fully resolved species tree (Rokas *et al.*, 2003). For example, incongruence was found for *Vibrio caribbeanicus* where data from *ftsZ*, *gyrB*, *mreB*, *pyrH* and *rpoA* showed similarity to the Orientalis clade, whereas *gapA*, *recA* and *topA* showed similarity to *V. alginolyticus* and *Aliivibrio fischeri* (Hoffman *et al.*, 2012a).

6.6 Fluorescent In Situ Hybridization (FISH)

Fluorescent in situ hybridization is used increasingly in clinical laboratories for the identification of bacteria from clinical samples (DeLong *et al.*, 1989; Jansen *et al.*, 2000; Hogardt *et al.*, 2000).

The same principal oligonucleotides or primers that are species-specific as shown by a BLAST search of experimentation can be used to develop sensitive FISH assays for aquatic bacteria. The primers listed in Table 6.1 would be a good starting point for developing specific FISH assays that could be applied to smears of colonies grown on agar plates. Although none of these primers have been tested as being suitable for specific probes, it was suggested that eight primers that were species-specific for *V. costicola*, *V. diazotrophicus*, *V. fluvialis*, *V. nigripulchritudo*, *V. proteolyticus*, *V. salmonicida*, *V. splendidus* and *V. tubiashii* would be suitable for use as specific probes (Lee *et al.*, 2002).

6.6.1 Outline of steps for FISH

Step	Method
1.	Prepare smear from bacterial colonies
2.	Hybridization
3.	Washing
4.	Microscopic examination

6.6.2 Oligonucleotide probes

Primers should be selected from Table 6.1 and a BLAST search conducted to ascertain whether the primers are species-specific. The single-stranded oligonucleotide are synthesized and covalently labelled with fluorescein isothiocyanate at the 5′ end.

A universal positive probe is prepared using EUB primer 5′-GCT GCC TCC CGT AGG AGT-3′. This sequence corresponds to positions 338 to 355 on the *Escherichia coli* numbering system. A universal negative probe is prepared using primer non-EUB 5′-ACT CCT ACG GGA GGC AGC-3′ (Amann *et al*, 1990; Jansen *et al.*, 2000). Fluorescein isothiocyanate is added to the 5′ end of each primer.

Prior to use, the probes are diluted to a concentration of 10 ng/ml in hybridization buffer.

6.6.3 Preparation of smears

Bacterial cells from a colony are emulsified in a drop (10–15 μl) of sterile saline or sterile distilled water within a marked area on a glass microscope slide. The suspension should not be too thick, but allow individual cells to be seen under the microscope. Prepare one drop per probe to be tested plus drops for the positive and negative controls. Allow to air dry.

Once dry, the cells are fixed to the slide by using a fixative solution of 4% formaldehyde in 96% ethanol. Gram-positive organisms must be permeabilized before application of the hybridization buffer. The fixed slide is placed into permeabilization buffer of 1 mg/ml of lysozyme for 5 min. Gram-negative organisms do not need the permeabilization step (Jansen *et al.*, 2000).

6.6.4 Hybridization

The fixed slides are hybridized in a hybridization buffer of 20 mM Tris-HCl, 0.9 M NaCl, 0.1% sodium dodecyl sulfate (SDS), pH 7.2 that contains the probe at a concentration of 10 ng/ml. The hybridization buffer (10–15 μl) can be placed on top of the fixed cells. Gram-negative organisms are hybridized for 45 min, and Gram- positives for 2 h at a temperature of 50°C (Jansen *et al.*, 2000).

One hybridization method pre-treated slides in ascending ethanol steps of 50, 80 and 96% for 3 min each step. The hybridization buffer then contained 30–40% formamide with 50 ng of probe. This concentration of formamide improved the specificity of the probe (Hogardt *et al.*, 2000). However, a concentration of 20% (vol/vol) of formamide did not show an increase in probe specificity (Jansen *et al.*, 2000).

6.6.5 Washing

The slides are washed in washing buffer of 20 mM Tris-HCl, 0.9 M NaCl, pH 7.2, at a temperature of 50°C for 10 min (Jansen *et al.*, 2000).

6.6.6 Examination of smears

The slides are mounted with VectaShield® (Vector Laboratories, Burlingame, California, USA), or Citifluor (Citifluor Ltd, London, UK).

The slides are examined for cells showing fluorescence (positive result), using a fluorescent microscope and filter set capable of detecting fluorescein. Fluorescence should be seen with the positive control, and no fluorescence with the negative control.

7

Preparation of Media for Culture and Identification

7.1 General Isolation and Selective Media

Acetate agar. See Rogosa medium under section 7.2 Biochemical Test Media

Alkaline peptone water (APW) (May be used as an enrichment medium for *Vibrio* spp., Furniss *et al.*, 1978)

Reagent	Amount	Preparation of media	Description of growth characteristics
Peptone	10.0 g	Dissolve reagents in distilled water and pH to 8.6. Dispense in 10 ml aliquots to McCartney bottles and autoclave at 121°C for 20 min.	May be used as an enrichment medium to isolate *Vibrio* spp. from contaminated samples such as faeces and polluted water. To be effective, the APW cultures should be subcultured at 6 h if incubated at 37°C and overnight if incubated at 18–20°C.
Sodium chloride	10.0 g		
Distilled water	1000 ml		

Amies Transport Medium (Oxoid)

This transport medium can be purchased ready-made as swabs in tubes of media, or prepared from a powdered medium available from Oxoid.

Formulae: 10.0 g charcoal (pharmaceutical), 3.0 g sodium chloride, 1.15 g sodium hydrogen phosphate, 0.2 g potassium dihydrogen phosphate, 0.2 g potassium chloride, 1.0 g sodium thioglycollate, 0.1 g calcium chloride, 0.1 g magnesium chloride, 4.0 g agar, 1000 ml distilled water, pH 7.2.

Anacker–Ordal agar (AO) (Used for growth of freshwater and marine *Cytophaga*, *Flavobacterium* and *Flexibacter* spp., Anacker and Ordal, 1955, 1959)

Reagent	Amount	Preparation of media	Description of growth characteristics
Bacto-tryptone (Difco)	0.5 g	Add all reagents to 1000 ml of distilled water and adjust pH to 7.2–7.4. Autoclave at 121°C for 15 min (15 lb/20 min). Cool to 50°C and pour media into Petri dishes. Store plates in sealed plastic bags at 4°C.	Using a cotton-tipped swab, collect samples from skin lesions and gills of fish and inoculate the plate. Colonies of *Flavobacterium columnare* appear at 2–5 days as yellow-pigmented, rhizoid colonies with a thin spreading growth. *See photographic section*. Early colonies may be viewed with the aid of a stereomicroscope. Subculture by cutting out a block of agar containing the colony, and invert onto new plate. A Pasteur pipette prepared in the shape of a 'hockey stick' with the end unsealed can be used to pick out a plug of agar.
Yeast extract (Difco)	0.5 g		
Sodium acetate	0.2 g		
Beef extract (Difco) or Lab-Lemco powder (Oxoid)	0.2 g		
Agar (Difco or Oxoid agar No. 1)	9 g /l		
Distilled water	1000 ml		

Sodium acetate inhibits the growth of cells and the medium may be improved by removing it from the formulation (Oplinger and Wagner, 2012).

Enriched Anacker–Ordal agar with skim milk and serum (EAO-Sh) (Improved medium for growth of *Flavobacterium psychrophilum*, Oplinger and Wagner, 2012)

Reagent	Amount	Preparation of media	Description of growth characteristics
Bacto-tryptone (Difco)	5.0 g	Add all reagents to 1000 ml of distilled water, and adjust pH to 7.2–7.4. Autoclave at 121°C for 15 min (15 lb/20 min). Cool to 50°C and pour media into Petri dishes. Store plates in sealed plastic bags at 4°C.	Using a cotton-tipped swab, collect samples from skin lesions and gills of fish and inoculate the plate. Colonies of *Flavobacterium psychrophilum* appear at 2–5 days as yellow-pigmented, rhizoid colonies with a thin spreading growth. Early colonies may be viewed with the aid of a stereomicroscope. Subculture by cutting out a block of agar containing the colony, and invert onto new plate. A Pasteur pipette prepared in the shape of a 'hockey stick' with the end unsealed can be used to pick out a plug of agar.
Yeast extract (Difco)	0.5 g		
Beef extract (Difco) or Lab-Lemco powder (Oxoid)	0.2 g		
Skim milk	2.0 g		
Horse serum	10 ml		
Agar (Difco or Oxoid agar No. 1)	9 g		
Distilled water	1000 ml		

This formulation was validated and optimized by Oplinger and Wagner (2012) and is based on AO medium (Anacker and Ordal, 1955, 1959), enriched AO medium that contains increased concentration of tryptone (Bernardet and Kerouault, 1989) and skim milk (Daskalov *et al.*, 1999).

Anacker–Ordal agar – Marine (AO-M) (Used for growth of marine *Cytophaga*, *Flavobacterium* and *Flexibacter* spp. and *Tenacibaculum* (*Flexibacter*) *maritimum*; Anacker and Ordal, 1955, 1959)

Add artificial seawater salts (Sigma) at 38 g/l. If using seawater, use sterile at 50–100% final concentration (Ostland *et al.*, 1999).

Enriched Anacker–Ordal (EAO) agar with activated charcoal and aromatic compounds (EAOcA) (Used for growth of *Flavobacterium psychrophilum*; Álvarez and Guijarro, 2007)

Prepare enriched AO (EAO) using 5 g/l tryptone (Merck), 5% (v/v) horse serum (Gibco) and the remaining reagents in AO.

To prepare EAOCa. To EAO medium add 0.5 g/l activated charcoal (Sigma), and aromatic compounds; 10 μmol/l of L-tyrosine, L-phenylalanine, L-tryptophan, *p*-aminobenzoic acid, *p*-hydroxybenzoic acid and 2,3-di-hydroxybenzoic acid (Sigma).

Anaerobe plates (ANA) (For growth of anaerobic Gram-positive and Gram-negative organisms; Oxoid manual)

Reagent	Amount	Preparation of media	Description of growth characteristics
Oxoid Wilkins–Chalgren Anaerobe agar	21.50 g	Suspend powder in water in a 1000-ml Schott bottle with magnetic flea. Autoclave at 121°C for 20 min then cool to 50°C in waterbath. Aseptically add blood. Pour into Petri dishes.	Inoculated plates are incubated at the appropriate temperature in an anaerobic atmosphere.
Distilled water	500 ml		
Sterile equine blood	30 ml		

Gram-negative anaerobe plates (ANA-GN) (For growth of anaerobic Gram-negative organisms; Oxoid manual)

Reagent	Amount	Preparation of media	Description of growth characteristics
Oxoid Wilkins–Chalgren Anaerobe agar	21.50 g	Dissolve agar in water. Autoclave at 121°C for 20 min then cool to 50°C in a waterbath. Reconstitute 1 vial of supplement with 10 ml sterile distilled water and add to base. Aseptically add blood. Pour into Petri dishes.	Inoculated plates are incubated at the appropriate temperature in an anaerobic atmosphere.
Distilled water	500 ml		
Sterile equine blood	15 ml		
Oxoid G-N selective supplement	10 ml		

Arcobacter selective broth (H BROTH) (Houf *et al.*, 2001)

Reagent	Amount	Preparation of media	Description of growth characteristics
Arcobacter-specific broth (Oxoid)	24 g	Dissolve agar in water. Autoclave at 121°C for 20 min then cool to 55°C in a waterbath. Aseptically add 50 ml of antibiotic solution (concentrations are final concentrations). Aliquot 9 ml volumes to sterile MacCartney bottles. Store at 4ºC.	1 g of material is inoculated to 9 ml of Arcobacter broth and incubated at 30°C (or appropriate temperature) in aerobic and microaerobic atmospheres for 24 h. 20 µl of enrichment broth is cultured to CC plates.
Distilled water	950 ml		
5-fluorouracil	100 mg/ml		
Amphotericin B	10 mg/ml		
Cefoperazone	16 mg/ml		
Novobiocin	32 mg/ml		
Trimethoprim	64 mg/ml		

***Arcobacter* selective plate (CC)** (Kemp *et al.*, 2005)

Reagent	Amount	Preparation of media	Description of growth characteristics
Modified charcoal cefoperazone deoxycholate agar (mCCDA) (Lab M, UK)	500 ml	Dissolve agar in water. Autoclave at 121°C for 20 min then cool to 50°C in a waterbath. Aseptically add 1 vial of CAT supplement to 500 ml base. Pour into Petri dishes. Dry and package and store at 4°C.	1 g of material is inoculated to 9 ml of Arcobacter broth and incubated at 30°C (or appropriate temperature) in aerobic and microaerobic atmospheres for 24 h. 20 µl of enrichment broth is cultured to CC plates.
CAT supplement (Oxoid) CAT supplement consists of vials containing Cefoperazone 4 mg/vial, Teicoplanin 2 mg/vial, Amphotericin B 5 mg/vial	1 vial		

Artificial seawater – ASW: sea salts

Reagent	Amount	Preparation of media	Description of growth characteristics
Sea salt (Sigma) #S 9883	38 g	Add salts to distilled water and pH to7.6. Autoclave at 121°C for 15 min.	May be added to media for growth of marine organisms. Add to AO medium for isolation of marine *Flavobacterium* and *Tenacibaculum* spp.
Distilled water	1000 ml		

Addition of 18.7 g/l makes a 50% seawater concentration of 17.5% salinity.

Artificial seawater medium (Lewin, 1974 – used for isolation of marine *Flexibacter*/*Flavobacterium* group. Baumann *et al.*, 1971 quoted MacLeod, 1968 and is the same medium but without the trace element mixture – used for marine *Vibrio* spp.)

Reagent	Amount	Preparation of media	Description of growth characteristics
NaCl	20.0 g	Dissolve reagents in distilled water. Add 1.0 ml of trace element solution. Adjust pH to 7.5 and autoclave at 121°C for 20 min.	Can be used as an inoculating medium for biochemical identification sets or as an initial isolation broth for marine organisms. Add agar at 15 g/l and pour into Petri dishes if a solid medium is required. Lewin used this medium for marine *Flavobacterium* and *Flexibacter* spp. The medium used by MacLeod and Baumann did not contain the trace element solution.
$MgSO_4.7H_2O$	5.0 g		
KCl	1.0 g		
$CaCl_2.2H_2O$	1.0 g		
Distilled water	1000 ml		
B (soluble salt of element)	0.5 mg/ml	Prepare a stock solution of trace elements so that the final concentration of each element is either 0.5 mg/l or 0.01 mg/l.	
Fe (soluble salt of element)	0.5 mg/ml		
Mn (soluble salt of element)	0.5 mg/ml		
Co (soluble salt of element)	0.01 mg/ml		
Cu (soluble salt of element)	0.01 mg/ml		
Mo (soluble salt of element)	0.01 mg/ml		
Zn (soluble salt of element)	0.01 mg/ml		

Blood agar – BA

Reagent	Amount	Preparation of media	Description of growth characteristics
Oxoid Columbia BA base	19.5 g	Suspend agar base in water. Autoclave at 121°C for 15 min. Cool in waterbath to 50°C. Add blood to cooled agar, mix well and pour into Petri dishes to depth of approximately 3 mm. Store at 4°C in sealed plastic bags (plates can be left overnight on the bench and sealed in plastic bags the following day. This prevents too much moisture build up once stored in the plastic bags).	Plates are inoculated with a swab of the specimen, and incubated at the appropriate temperature and atmosphere. Plates are examined daily for growth and haemolysis.
Distilled water	500 ml		
Sterile equine blood	15 ml		

Brain–heart infusion agar (BHIA)

Bordetella bronchiseptica selective agar – CFPA medium (Smith and Baskerville, 1979; Rutter, 1981; Hommez *et al.*, 1983)

Reagent	Amount	Preparation of media	Description of growth characteristics
Columbia agar base (Oxoid)	19.5 g	Add agars to distilled water and autoclave at 121°C for 15 min. Cool to 50°C. Add sterile blood to cooled agar mixture. Reconstitute 1 vial of Bordetella supplement with 2 ml of distilled water and add to mixture. Add 5 ml of antibiotic stock. Pour plates and store at 4°C.	Colonies are 1 mm at 48 h and may be haemolytic or non-haemolytic, opaque, smooth and pearl-like, or rough, translucent, raised in the centre with an undulating margin, depending on phase variation.
Agar technical No. 3 (Oxoid)	10.0 g		
Distilled water	500 ml		
Bordetella Pertussis supplement (Oxoid, Code SR082E)	2 ml		
Sterile equine blood	15 ml		
Penicillin	20 mg	Antibiotic stock: Add 20 mg of each antibiotic to 10 ml of normal saline. Store in fridge. Handle furaltadone with caution and use mask and gloves.	
Furaltadone	20 mg		
Normal saline	10 ml		

Burkholderia pseudomallei selective media – glycerol medium (Thomas *et al.*, 1979)

Reagent	Amount	Preparation of media	Description of growth characteristics
Agar No. 3 (Oxoid)	2.4 g	Add all reagents to distilled water (glycerol may be warmed to aid pipetting). Autoclave at 121°C	Plates are a mauve colour. Colonies appear at 24 h, smooth, mauve-coloured with slight metallic sheen. As incubation increases, colonies become wrinkled and umbonate. Incubate for 4 days.
Glycerol	6.0 ml		
Crystal violet stock solution	0.5 ml		
Distilled water	194 ml		
Crystal violet – Stock (1/5000 dilution)	0.5 g	Add 0.5 g of crystal violet to distilled water and stir to dissolved. Store at room temperature.	
Distilled water	100 ml		

Burkholderia pseudomallei selective broth (Modified from Thomas *et al.*, 1979)

Reagent	Amount	Preparation of media	Description of growth characteristics
MacConkey broth (purple) (Oxoid)	100 ml	Prepare MacConkey broth and add reagents except antibiotics. Autoclave at 121°C for 15 min. Cool to 50°C and add filter-sterilized antibiotics. Aseptically dispense into sterile McCartney bottles.	Detection of *Burkholderia* can be improved by incubating material in broth for 24 and 48 h, followed by subculture to plates.
Crystal violet	0.001 g		
Gentamycin	0.8 mg		
Streptomycin sulfate	5000 units		

Burkholderia pseudomallei **selective media – Ashdown's medium** (Ashdown, 1979a)

Reagent	Amount	Preparation of media	Description of growth characteristics
Tryptone soya agar	40 g	Add all ingredients except gentamycin to distilled water. Autoclave at 121°C for 15 min. Cool to 50°C. Add filter sterilized gentamycin. Mix well and pour into Petri dishes.	Colonies are flat, rough, wrinkled after 3 days. Colonies of *Burkholderia pseudomallei* absorb the neutral red after 3 days whereas *Pseudomonas* spp. have no dye uptake at 3 days.
Glycerol	40 ml		
Crystal violet	5 mg		
Neutral red	50 mg		
Gentamycin	4 mg		
Distilled water	1000 ml		

Brucella agar (Available from Difco or Oxoid media suppliers. Also see under Farrell's medium)

Carbon dioxide atmosphere (See under section 7.2 Biochemical Test Media)

Cellobiose–Colistin agar (For selective isolation of *Vibrio vulnificus*; Høi *et al.*, 1998; Massad and Oliver, 1987)

Reagent	Amount	Preparation of media	Description of growth characteristics
Solution 1		Adjust pH to 7.6. Autoclave at 121°C for 15 minutes and cool to 55°C.	The medium is olive green to light brown. Incubate plates at 40°C for 24–48 h. *Vibrio vulnificus* appears as yellow colonies surrounded by a yellow zone due to fermentation of cellobiose. *Vibrio cholerae* appears as purple colonies surrounded by a blue zone.
Bacto-peptone (Difco)	10 g		
Beef extract (Difco)	5 g		
NaCl	20 g		
Bromothymol blue	40 mg		
Cresol red	40 mg		
Agar	15 g		
Distilled water	900 ml		
Solution 2		Filter sterilize. Add to cooled reagents of solution 1 and mix. Dispense into Petri dishes.	
Cellobiose	15 g		
Colistin	0.03 mg/ml ($4\times10_5$ U/l)		
Distilled water	100 ml		

CFPA media (See under *Bordetella bronchiseptica* medium)

Dermatophilus **selective medium – polymyxin plates** (Abu-Samra and Walton, 1977)

Reagent	Amount	Preparation of media	Description of growth characteristics
Columbia agar base (Oxoid)	19.5 g	Add agar base to distilled water and autoclave at 121°C for 15 min. Cool to 50°C and aseptically add blood and polymixin. Polymixin B may be dissolved in sterile normal saline first.	Grind scab material using a sterile pestle and mortar. Transfer finely ground material to a bijou bottle and add two parts of distilled water. Shake thoroughly then allow to settle for 15 min. Take a loopful of surface material and plate to BA and polymixin plates. Incubate at 25 or 37°C depending on the habitat of the infected animal. Examine plates at 24 and 48 h for presence of adherent pitted dry or mucoid colonies. This medium is not completely selective and some contaminating bacteria will grow.
Distilled water	500 ml		
Sterile equine blood	50 ml		
Polymixin B (Use at a ratio of 1000 IU/ml of medium)	62.5 mg		

Dubos medium (For isolation of *Cytophaga hutchinsonii*; Bernardet *et al.*, 1989)

Reagent	Amount	Preparation of media	Description of growth characteristics
$NaNO_3$	0.5 g	Add all reagents except cellobiose, to distilled water. Adjust pH to 7.2 and autoclave at 121°C for 15 min. Cool to 50°C and add filter-sterilized cellobiose solution. Mix well and pour into Petri dishes.	
K_2HPO_4	1 g		
$MgSO_4.7H_2O$	0.5 g		
KCl	0.5 g		
$FeSO_4.7H_2O$	0.01 g		
Distilled water	1000 ml		
Agar	15 g		
D-cellobiose	30% w/v	Prepare cellobiose solution and filter-sterilize.	

For isolation of *Cellulomonas* spp. add 0.5 g yeast extract.

***Edwardsiella ictaluri* medium (EIM)** (For isolation of *Edwardsiella ictaluri*; Shotts and Waltman, 1990)

Reagent	Amount	Preparation of media	Description of growth characteristics
Bacto-tryptone (Difco)	10 g	Dissolve all reagents in distilled water and adjust the pH to 7.0–7.2. Autoclave at 121°C for 15 min. Cool to 50°C and add the 10 ml filter-sterilized solution 1 containing mannitol, colistin, bile salts and fungizone. Mix well and pour into Petri dishes.	This medium is used for the isolation of *Edwardsiella ictaluri*. Most Gram-negative bacteria are inhibited with the exception of *Proteus* spp., *Serratia marcescens*, *Aeromonas hydrophila* and *Yersinia ruckeri*. Gram-positive bacteria are inhibited with the exception of *Enterococci*. 90% of *E. tarda* isolates grow on this medium. *E. ictaluri* is seen as 0.5–1.0 mm green, translucent colonies at 48 h. Colonies of *E. tarda* have a similar size and appearance. *Proteus* spp. are 2–3 mm, brownish-green and may swarm. *S. marcescens* colonies are 2–3 mm and reddish brown. *Yersinia ruckeri* are 1–2 mm yellowish green, *A. hydrophila* are 2–3 mm yellowish-green opaque colonies and *Enterococci* appear as 0.5 mm yellowish colonies.
Yeast extract (Difco)	10 g		
Phenylalanine	1.25 g		
Ferric ammonium citrate	1.2 g		
Sodium chloride	5.0 g		
Bromothymol blue	0.03 g		
Agar (Difco)	17 g		
Distilled water	990 ml		
Solution 1			
Mannitol	3.5 g	Add reagents to 10 ml of distilled water and filter-sterilize.	
Colistin	10 mg		
Bile salts	1 g		
Fungizone	0.5 mg		
Distilled water	10 ml		

Edwardsiella tarda **(ET) selective medium** (Castro *et al.*, 2011)

Reagent	Amount	Preparation of media	Description of growth characteristics
MacConkey agar base	40.0 g	Add MacConkey agar base, yeast extract and additional agar to distilled water and autoclave autoclave at 121°C for 15 min. Cool to 50°C and add 100 ml of glucose solution and 10 ml of colistin solution. Total volume is 1.01 l. Pour plates and when dry store in sealed plastic bags at 4°C.	Used for the isolation of *Edwardsiella tarda*. Colonies are clear to whitish with a black centre when incubated at 25°C for 24–48 h.
Yeast extract	1.0 g		
Additional agar	4.5 g		
Distilled water	900 ml		
Glucose	2.0 g	Filter-sterilize.	
Sucrose	5.0 g		
Mannitol	5.0 g		
Xylose	5.0 g		
L-lysine	10.0 g		
Sodium thiosulfate	6.8 g		
Ferric ammonium sulfate	0.8 g		
Distilled water	100 ml		
Colistin	10.0 mg	Filter-sterilize.	
Distilled water	10 ml		

Electrolyte supplement (Added to biochemical test media to improve growth of organism when Na^+ alone is insufficient; Lee *et al.*, 1979)

Reagent	Amount	Preparation of media	Description of growth characteristics
NaCl	100 g	Add all reagents to distilled water and autoclave at 121°C for 15 min.	If addition of NaCl alone to biochemical identification media does not improve the growth of an organism, this electrolyte supplement may improve growth. It is added at a rate of 0.1 ml of supplement per 1.0 ml of medium.
$MgCl_2.6H_2O$	40 g		
KCl	10 g		
Distilled water	1000 ml		

EMJH medium for culture of *Leptospira*

Reagent	Amount	Preparation of media	Description of growth characteristics
Disodium hydrogen phosphate, (Na_2HPO_4)	1.0 g	Add all reagents to distilled water and autoclave at 121°C for 15 min.	If addition of NaCl alone to biochemical identification media does not improve the growth of an organism, this electrolyte supplement may improve growth. It is added at a rate of 0.1 ml of supplement per 1.0 ml of medium.
Potassium dihydrogen phosphate ($KH_2.PO_4$)	0.3g		
NaCl	1.0 g		
Ammonium chloride (NH_4Cl)	0.25 g		
Thiamine	0.005 g		
Glycerol	0.1 g		
Bovine albumin, fraction V	100.0 g		

Continued.

Calcium chloride ($CaCl_2.2H_2O$)	0.1 g		
Magnesium chloride ($MgCl_2.6H_2O$)	0.1 g		
Zinc sulfate ($ZnSO_4.7H_2O$)	0.04 g		
Copper sulfate ($CuSO_4.5H_2O$)	0.003 g		
Iron sulfate ($FeSO_4.7H_2O$)	0.5 g		
Vitamin B12	0.002 g		
Tween 80	12.5 g		
Distilled water	1000 ml		

This medium is available from some commercial suppliers such as Becton Dickinson (BD Diagnostics) or BioScientific Pty Ltd.

***Erysipelothrix* selective medium** (See 'Wood's broth' and 'Packer's plates')

Farrell's medium (For the growth of *Brucella* spp; Farrell, 1974)

Reagent	Amount	Preparation of media	Description of growth characteristics
Oxoid blood agar base	20.0 g	Add agar to distilled water and autoclave at 121°C for 15 min. Cool to 50°C and leave at this temperature while the other reagents are prepared.	Incubate plates at 37°C in 10% CO_2 for 14 days. Isolates from sea mammals appear after 4 days' incubation, whereas isolates from seals don't appear until day 10 of incubation, or do not grow. Incubation of plates should continue for 14 days. It has been suggested that the isolation of strains from seals may be improved by the decreasing the concentration or removing bacitracin and/or nalidixic acid from the medium (Foster *et al.*, 2002). Always inoculate a non-selective medium such as blood agar or serum dextrose medium. Colonies are 1–2 mm in diameter at 5 days' incubation, pale yellow, translucent, convex and round with entire edges.
Distilled water	500 ml		
Normal horse serum	25 ml	Inactivate the horse serum at 50°C for 35 min.	
25% dextrose (D-glucose)	20 ml	Add 125 g of D-glucose to 375 ml distilled water and dissolve by gentle heating. Dispense into 20 ml aliquots into McCartney bottles and with lids loose, autoclave at 121°C for 15 min. Store at 4°C.	
Oxoid Brucella selective supplement	10 ml	Dissolve Brucella supplement in 5 ml sterile distilled water and 5 ml methanol. Incubate for 10–15 min at 37°C and shake well to dissolve. For complete Farrell's medium aseptically combine all ingredients while stirring. Mix well and pour media into plates to an approximate depth of 3 mm.	
Final media preparation		Combine all prepared reagents (serum, dextrose, supplement) into agar. Mix well and pour into Petri dishes to a depth of 3 mm.	

The original medium uses the following antibiotic concentrations. Bacitracin (25 units/ml), vancomycin (20 µg/ml), polymixin B (5 units/ml), nalidixic acid (5 µg/ml), nystatin (100 units/ml) and cycloheximide (100 µg/ml) (Farrell, 1974). Farrell's medium is available from Oxoid as Brucella medium base, to which Brucella selective supplement (Oxoid code SR83) is added. The selective supplement is based on the Farrell formulation.

Farrell's modified medium (Recommended for the growth of *Brucella* spp. isolated from marine mammals; Foster *et al.*, 2002 Tryland *et al.*, 2005)

Reagent	Amount	Preparation of media	Description of growth characteristics
Oxoid blood agar base	20.0 g	Add agar to distilled water and autoclave at 121°C for 15 min. Cool to 50°C and leave at this temperature while the other reagents are prepared.	Incubate plates at 37°C in 10% CO_2 for 14 days. Isolates from most marine mammals appear after 4 days' incubation. Incubation of plates should continue for 14 days. Always inoculate a non-selective medium such as blood agar or serum dextrose medium. Colonies are 1–2 mm in diameter at 5 days' incubation, pale yellow, translucent, convex and round with entire edges.
Distilled water	500 ml		
Normal horse serum	25 ml	Inactivate the horse serum at 50°C for 35 min.	
25% dextrose (D-glucose)	20 ml	Add 125 g of D-glucose to 375 ml distilled water and dissolve by gentle heating. Dispense into 20 ml aliquots into McCartney bottles and with lids loose, autoclave at 121°C for 15 min. Store at 4°C.	
Antibiotics and antifungal agents Vancomycin (20 µg/ml) Polymixin B (5 units/ml) Nystatin (100 units/ml) Amphotericin (1.25 µg/ml)	10 ml	Dissolve antibiotics and antifungal agent in sterile distilled water. Filter-sterilize and add to rest of ingredients. For complete Farrell's medium aseptically combine all ingredients while stirring. Mix well and pour media into plates to an approximate depth of 3 mm	
Final media preparation		Combine all prepared reagents (serum, dextrose, supplement) into agar. Mix well and pour into Petri dishes to a depth of 3 mm.	

The isolation of strains from seals may be improved by decreasing the concentration or removing bacitracin and/or nalidixic acid from the medium (Foster *et al.*, 2002; Tryland *et al.*, 2005).

***Flavobacterium maritimus* media (FMM)** (May improve primary isolation for *T. maritimum*; Pazos *et al.*, 1996)

Reagent	Amount	Preparation of media	Description of growth characteristics
Peptone	5.0 g	Add reagents and powders to sterile seawater. Adjust pH to 7.2–7.4 and sterilize by autoclaving at 121°C for 20 min.	Colonies pale-yellow, flat, irregular colonies with uneven edges. Reduces growth of *Vibrio* spp. and *Aeromonas* spp. that may out-grow *Flexibacter* on media such as MSA-B or MA 2216.
Yeast extract	0.5 g		
Sodium acetate	0.01 g		
Agar	15 g		
Sterile seawater	1000 ml		

***Flavobacterium psychrophilum* medium (FPM)** (May improve isolation rate and colony size for *F. psychrophilum*; Daskalov *et al.*, 1999)

Reagent	Amount	Preparation of media	Description of growth characteristics
Tryptone T (Oxoid)	0.5 g	Add all reagents (except sugars and skimmed milk) to distilled water and autoclave at 121°C for 15 min. Cool in a 50°C waterbath. Prepare supplements (galactose, glucose, rhamnose and skim milk) as separate 10% (w/v) solutions, and filter by sterilization through a 0.22-µm pore size Millipore Millex porosity filter. Add filter-sterilized solutions to cooled agar medium. Pour into Petri dishes and store at 4°C.	Colonies grow after 3–6 days and are an intense yellow colour with a spreading and irregular shape.
Yeast extract (Oxoid)	0.5 g		
Beef extract (Oxoid)	0.2 g		
Sodium acetate trihydrate (Sigma)	0.2 g		
D(+) galactose (Sigma)	0.5 g		
D(+) glucose (BDH)	0.5 g		
L-rhamnose (Sigma)	0.5 g		
Skimmed milk (Oxoid)	0.5 g		
Water	1000 ml		
Agar – bacteriological agar no. 1 (Oxoid)	9.0 g		

This medium is based on Anacker Ordal (AO) agar (also known as Cytophaga agar) supplemented with galactose, glucose, rhamnose and skim milk. The size and number of colonies of *F. psychrophilum* isolated is increased compared with AO medium. Broth can be prepared without the addition of agar.

***Flexibacter polymorphus* media** (Lewin, 1974)

Reagent	Amount	Preparation of media	Description of growth characteristics
NaCl	20.0 g	Dissolve reagents in distilled water. Adjust pH to 7.5 and autoclave at 121°C for 20 min.	The organism is maintained in 5 ml of the broth medium and subcultured twice weekly. Filaments several hundred microns in length and 1.5 µm wide are produced. They have rounded ends, are flexuous, unbranched and cylindrical. The growing filaments may be peach-coloured. Cultures in late growth phase may show a refractile granule at each end, which is more easily seen after cell lysis. Cobalamin is essential for growth. Agar can be added to the medium for plate media. The organism also grows on agar medium prepared with seawater and yeast extract (10 mg/ml). Colonies may be peach coloured with a filamentous margin.
$MgSO_4.7H_2O$	5.0 g		
KCl	1.0 g		
$CaCl_2.2H_2O$	1.0 g		
Fe (soluble salt)	0.5 mg		
B (soluble salt)	0.5 mg		
Mn (soluble salt)	0.5 mg		
Co (soluble salt)	0.01 mg		
Cu (soluble salt)	0.01 mg		
Mo (soluble salt)	0.01 mg		
Zn (soluble salt)	0.01 mg		
Distilled water	1000 ml		

***Flexibacter* maintenance medium** (Lewin, 1974; Lewin and Lounsbery, 1969)

Reagent	Amount	Preparation of media	Description of growth characteristics
Tryptone (Difco)	1.0 g	Add all reagents to filtered seawater and adjust pH to 7.5. Dispense media into tubes. Autoclave at 121°C for 20 min. Optimum cobalamin is 0.3 µg/l.	The organisms are maintained in 5 ml of the broth medium, incubated between 22 and 32°C, and subcultured twice weekly. The Flexibacteria may show a range of pigmented filaments ranging from pink, orange or yellow colouration. On solid medium they may produce fimbriate margins. Agar can be added to the medium for plate media. A semi-solid medium can be produced using 0.3% agar, and viable filaments can be maintained for up to 1 month when grown in this medium at room temperature.
Casamino acids (Difco)	1.0 g		
Monosodium glutamate	5.0 g		
Sodium glycerophosphate	0.1 g		
Cobalamin	1.0 µg		
Fe (soluble salt)	0.5 mg		
B (soluble salt)	0.5 mg		
Mn (soluble salt)	0.5 mg		
Co (soluble salt)	0.01 mg		
Cu (soluble salt)	0.01 mg		
Mo (soluble salt)	0.01 mg		
Zn (soluble salt)	0.01 mg		
Filtered seawater	1000 ml		

Glycerol Lab Lemco broth (Used for storage of cultures at −80°C)

Reagent	Amount	Preparation of media	Description of growth characteristics
Lab Lemco Broth (Oxoid)	0.64 g	Add all reagents together. Pipetting of glycerol may be facilitated by pre-warming. Dispense into 2 ml volumes into sterile bottles. Autoclave at 121°C for 15 min. Store media at 4°C.	Used as −80°C storage medium. Using a sterile cotton-tipped swab, scrape bacterial growth from an agar plate and inoculate into 1 ml of Glycerol Lab Lemco medium in a nunc tube. Use an inoculum of approximately tube 5 McFarland opacity. Place tube in cryobox in −80°C freezer. This medium is also suitable for liquid nitrogen storage.
Glycerol	20 ml		
Distilled water	80 ml		

***Helicobacter* selective media** (Use Skirrow's medium, Also known as VPT media; Skirrow, 1977)

Reagent	Amount	Preparation of media	Description of growth characteristics
Blood agar base No. 2 (Oxoid)	20 g	The agar is dissolved in water and autoclaved at 121°C for 15 min. After cooling to 50°C add 15 ml of sterile equine blood. Reconstitute 1 vial of supplement using 2 ml of sterile distilled water and add to cooled media. Mix thoroughly. Pour media into plates to an approximate depth of 3 mm. Store plates at 4°C.	Other bases such as Columbia agar base (Oxoid), or Brucella medium base (Oxoid), or tryptone soya agar may be used to prepare this medium. However, it is suggested that blood agar base No. 2 is more nutritionally rich and also that trimethoprim is more active in this base. To reduce contamination from other organisms, the sample can be filtered through a 0.65-µm filter and the filtrate cultured to the plate (Butzler *et al.*, 1973). The plates are incubated at 37°C in a microaerophilic atmosphere of N_2, H_2, CO_2 (80:10:10) for 2–4 weeks. Gas-generating packs are available from commercial suppliers such as MGC Anaero Pak™ Campylo from Mitsubishi Gas Chemical Company. *Helicobacter* spp. will grow as pinpoint colonies and also appear as a thin spreading film across the plate.
Distilled water	500 ml		
Sterile equine blood	15 ml		
Oxoid Campylobacter supplement (Skirrow)	2 ml		

This medium was developed by M. B. Skirrow (1977) for the isolation of *Campylobacter* spp. It is also suitable for the isolation of *Helicobacter* spp. and is widely quoted in many references. It is also available as ready-made media from a number of media producers. The antibiotics are often supplied as a supplement and depending on the originators of the medium, the supplements will be named accordingly. For example, Oxoid produces 'Campylobacter selective supplement (Skirrow) for additions of vancomycin, polymyxin and trimethoprim, and Campylobacter selective supplement (Blaser–Wang) for additions of vancomycin, polymyxin, trimethoprim, amphotericin B, and cephalothin'. The additive containing the amphotericin B is recommended when fungal contamination is suspected. The original concentrations described by Skirrow (1977) are vancomycin (10 mg/l), polymyxin B (2.5 IU/ml) and trimethoprim (5 mg/l). The series of papers describing the isolation and identification of *Helicobacter cetorum* from dolphins and whales (Harper *et al.*, 2000, 2002a, 2002b) describes the use of TVP and CVA media from Remel. These are in essence Skirrow's media and Blaser–Wang media, respectively.

Hsu-Shotts agar – HS (For *Flavobacterium columnare, F. psychrophilum;* Bullock *et al.*, 1986)

Reagent	Amount	Preparation of media	Description of growth characteristics
Tryptone	2 g	Add all reagents to distilled water, except neomycin sulfate. Autoclave at 121°C for 15 min. Cool to 50°C. Add filter-sterilized neomycin sulfate, mix well and pour into Petri dishes. When dry, package and store at 4°C.	This is a semi-selective medium for the isolation of *Flavobacterium columnare*. At 48 h characteristic yellow, spreading colonies are seen.
Yeast extract	0.5 g		
Gelatin	3 g		
Agar	15 g		
Distilled water	1000 ml		
Neomycin sulfate	4.0 µg/ml	Prepare solution and filter-sterilize. Add to cooled mixture as above.	

Note: HSM. Addition of 18.7 g/l sea salts (Sigma) to make a 50% seawater concentration (17.5% salinity), and 200 IU/ml polymyxin B is suitable for isolation of *Tenacibaculum* (*Flexibacter*) *maritimum* (Chen *et al.*, 1995).

Inositol horse serum (Suspension medium for freeze-drying organisms)

Reagent	Amount	Preparation of media	Description of growth characteristics
Inositol	5 g	Dissolve inositol in horse serum. Filter through a 0.45-µm filter followed by further filtration through a 0.22-µm filter for sterilization. Check sterility. Dispense into McCartney bottles and store at 4°C.	Place 1 ml of sterile inositol horse serum into a freeze-dried vial (Wheaton serum bottle). Using a sterile cotton swab or loop, emulsify a heavy suspension of bacteria in the medium. Snap-freeze in liquid nitrogen and follow instructions for freeze-drying relevant to the equipment being used.
Horse serum	100 ml		

Meso-inositol nutrient broth suspension medium for freeze-drying *Vibrio* spp. (Carson *et al.*, 2009)

Reagent	Amount	Preparation of media	Description of growth characteristics
Nutrient broth No. 2 (Oxoid CM067B)	16 ml	Prepare nutrient broth and *meso*-inositol. Filter-sterilize by filtration through a 0.22-μm filter. Add aseptically to inactivated horse serum. Mix and dispense into McCartney bottles and store at 4°C.	Place 1 ml of sterile nutrient broth, *meso*-inositol horse serum solution into a freeze-dried vial (Wheaton serum bottle). Using a sterile cotton swab or loop, emulsify a heavy suspension of bacteria in the medium. Snap-freeze in liquid nitrogen and follow instructions for freeze-drying relevant to the equipment being used.
meso-inositol	2.4 g		
Horse serum (inactivated)	32 ml		

For reconstitution of *Vibrio* spp. use Vibrio recovery medium.

KDM2 (For growth and isolation of *Renibacterium salmoninarum;* Evelyn, 1977)

Reagent	Amount	Preparation of media	Description of growth characteristics
Tryptone	1 g	Add all reagents to distilled water and adjust pH to 6.5–6.8 (with NaOH). Dispense into tubes and autoclave at 121°C for 20 min. Store at 4°C for 1 month.	Incubate plates at 15°C for up to 2 months. Initial growth is visible between 2 and 8 weeks. Colonies are pinpoint to 2 mm in size. Old colonies may have a granular or crystalline appearance. To assist in prevention of overgrowth by fast-growing colonies, examine plates every few days and aseptically remove these colonies. Lot-to-lot variations in peptone may produce inconsistent growth rates. This can be overcome by the use of a nurse culture of *R. salmoninarum* or the addition of 10% v/v of sterilized spent culture medium termed KDM2 Ten-M medium (see below).
Yeast extract	0.05 g		
L-cysteine (chlorhydrate)	0.1 g		
Agar	1.5 g		
Distilled water	100 ml		
Fetal calf serum	5–10%	For use, heat tubes. Cool to 50°C and add calf serum. Pour into plates.	

KDM2 Ten-M (For growth and isolation of *Renibacterium salmoninarum*; Evelyn *et al.*, 1990; Teska, 1994; Chamber and Barker, 2006)

To improve the isolation rate, sterilized spent culture broth is added to the medium as a supplement at 10% v/v (KDM2 Ten-M). Spent culture broth is prepared by inoculating a culture of *R. salmoninarum* into 100 ml of KDM2 broth and incubating at 15°C for up to 18 days or until a turbidity of 0.83 optical density (OD) at 540 nm is reached. The culture is centrifuged and the supernatant is filter-sterilized using a 0.22-μm filter. The sterilized spent culture broth can be stored frozen. Incubation in a microaerobic atmosphere may increase colony size.

KDMC (For *Renibacterium salmoninarum*; Daly and Stevenson, 1985)

Reagent	Amount	Preparation of media	Description of growth characteristics
KDM2 medium		As above	Addition of charcoal acts as a detoxifying agent.
Activated charcoal (Difco)	0.1%	For use heat tubes. Cool to 50°C and charcoal. Pour into plates.	

Also see SKDM medium.

Marine salt agar with blood (MSA-B) (Used as a general-purpose medium for isolation of organisms from the marine environment)

Reagent	Amount	Preparation of media	Description of growth characteristics
Tryptone soya agar (Oxoid)	20.0 g	Dissolve TSA and NaCl in water and autoclave at 121°C for 15 min. Cool in waterbath to 50°C. Aseptically add 15 ml blood to cooled agar, mix well and pour into Petri dishes to depth of approximately 3 mm. Store at 4°C.	TSA with added blood has the advantage of providing differentiation between haemolytic and non-haemolytic *Vibrio* spp. Some *Vibrio* spp. will be haemolytic on blood agar and non-haemolytic on MSA-B despite the latter medium providing better growth for most marine *Vibrio* spp.
NaCl	7.5 g (= 2% w/v final conc)		
Distilled water	500 ml		
Sterile equine blood	15.0 ml		

Marine sea salt agar (MSSA)

Reagent	Amount	Preparation of media	Description of growth characteristics
Tryptone soya agar (Oxoid)	24.0 g	Dissolve TSA and sea salts in water and autoclave at 121°C for 15 min. Cool in waterbath to 50°C. Aseptically add 18 ml blood to cooled agar, mix well and pour into Petri dishes to depth of approximately 3 mm. Store at 4°C.	TSA with added blood and a mixture of sea salts rather than NaCl alone enhances the culture for bacteria that require salts other than NaCl. The addition of blood provides differentiation between haemolytic and non-haemolytic *Vibrio* spp., although the plates lyse after storage for a few days. Bacteria that grow on MA 2216 usually have an increased colony size on MSSA.
Sea salts (Sigma)	22.8 g		
Distilled water	600 ml		
Sterile equine blood	18.0 ml		

Marine 2216 agar (Difco) (ZoBell, 1941)

Reagent	Amount	Preparation of media	Description of growth characteristics
Bacto marine agar 2216 (Difco – Complete medium)	55.1 g	Add powder to distilled water and boil to dissolve. Autoclave at 121°C for 15 min. pH 7.6±0.2.	Plates are a light amber colour and slightly opalescent. Used for the culture of a wide range of heterotrophic marine bacteria including *Vibrio* spp. *Tenacibaculum soleae* grows on this medium.
Distilled water	1000 ml		

MA 2216 can be purchased from a number of companies including Difco and Cultimed (Panreac) as a broth or agar. Composition of MA 2216: 5.0 g Bacto peptone; 1.0 g Bacto yeast extract; 0.1 g Fe(III) citrate; 19.45 g NaCl; 5.9 g $MgCl_2$ (dried); 3.24 g $NaSO_4$; 1.80 g $CaCl_2$; 0.55 g KCl; 0.16 g Na_2CO_3; 0.08 g KBr; 34.0 mg $SrCl_2$; 22.0 mg H_3BO_3; 4.0 mg Na-silicate; 2.4 mg NaF; 1.6 mg $(NH_4)NO_3$; 8.0 mg Na_2HPO_4; 15.0 g agar; 1000 ml distilled water.

Marine agar (Jensen *et al.*, 1996)

Reagent	Amount	Preparation of media	Description of growth characteristics
Peptone	2.5 g	Add reagents to distilled water and mix to dissolve. Autoclave at 121°C for 15 min. pH 7.1	Used for the culture of a wide range of heterotrophic marine bacteria including *Pseudoalteromonas* and *Vibrio* spp.
Yeast extract	1.5 g		
Glycerol	1.5 ml		
Sea salts (Instant Aquarium Systems Inc.)	28.0 g		
Bacto agar	17.0 g		
Distilled water	1000 ml		

This is a high-nutrient medium reported for growth of *Pseudoalteromonas piscicida* (Nelson and Ghiorse, 1999).

Medium K (Used for isolation of *Chryseobacterium (Flavobacterium) scophthalmum*; Mudarris *et al.*, 1994)

Reagent	Amount	Preparation of media	Description of growth characteristics
Yeast extract (Oxoid)	1 g	Add all ingredients to seawater and adjust to pH 7.2. Autoclave 121°C for 15 min.	Orange-pigmented colonies are shiny, smooth, round, raised entire, 5–6 mm at 48 h, 25°C. On primary culture isolates may show slight gliding motility, but lose this after storage.
Beef extract (Oxoid)	5 g		
Casein (Oxoid)	6 g		
Tryptone (Oxoid)	2 g		
Anhydrous $CaCl_2$	1 g		
Agar (Oxoid No. 1)	15 g		
750 ml seawater (aged 30 days)			

Middlebrook 7H10-ADC medium (Used for isolation of *Mycobacterium abscessus*; Teska *et al.*, 1997)

Reagent	Amount	Preparation of media	Description of growth characteristics
NH_4SO_4	0.05 g	Add chemical reagents to distilled water. Autoclave 121°C for 15 min. Cool to 50°C. Aseptically add sterile bovine albumin, nalidixic acid and cycloheximide, pyridoxine and biotin. Media can be prepared as broths or 1.5 g of agar can be added for preparation of plates. All chemicals can be purchased from Sigma.	Incubate at 25°C for 14–28 days. Growth occurs in 7 days. Young colonies may exhibit a light blue-green tint. With age, colonies are off-white to tan in colour.
KH_2PO_4	0.15 g		
Na_2HPO_4	0.15 g		
Sodium citrate	0.04 g		
$MgSO_4$	0.0025		
$CaCl_2$	0.00005 g		
$ZnSO_4$	0.0001 g		
$CuSO_4$	0.0001 g		
L-glutamic acid	0.05 g		
Ferric ammonium citrate	0.004 g		
Pyridoxine	0.0001 g		
Biotin	0.00005 g		
Malachite green	0.025 g		
Glycerol	0.5 ml		
Nalidixic acid (35 µg/ml)	0.0035 g		
Cycloheximide (400 µg/ml)	0.04 g		
Bovine albumin V	0.5 ml		
D-glucose	0.2 g		
Catalase	0.0003%		
Distilled water	100 ml		

This medium is modified from Middlebrook's medium (Middlebrook *et al.*, 1960) by the addition of ADC (albumin, catalase and dextrose).

Mycoplasma medium – general purpose medium (Bacto PPLO agar and broth with supplements (Difco). Used for isolation of *Mycoplasmas* spp. This commercial medium, with the added supplements, is based on Hayflick's medium)

Reagent	Amount	Preparation of media	Description of growth characteristics
PPLO agar	35 g	Add dehydrated media to distilled water and autoclave 121°C for 15 min. Cool to 50–60°C. Aseptically add 300 ml mycoplasma supplement or 300 ml of mycoplasma supplement S. Mix well. Dispense into small (5 cm Petri dishes. Store in plastic bags at 4°C.	Mince or grind tissue and place a few loopfuls into a 3-ml broth. Incubate broth at 25 and 37°C. A change in pH results in a colour change from red to yellow, and may indicate growth of Mycoplasma. Bacterial contamination may also change the pH of the medium. The bacterial contamination can be filtered off by passing the broth through a 0.22-µm filter. Place drops of the filtered broth to a fresh broth and to a plate.
Distilled water	700 ml		
PPLO broth	21 g	Add dehydrated media to distilled water and autoclave 121°C for 15 min. Cool to 50–60°C. Add either supplement as for agar.	Subculture two drops from a broth culture to a PPLO agar plate every 3–4 days. Incubate plates in air and 5–10% CO_2 at 25 and 37°C. Subculturing on a Tuesday and a Friday may be a convenient time frame for the laboratory. Examine plates under a stereomicroscope for presence of typical 'fried-egg' type colonies. To differentiate mycoplasma colonies from contaminating bacterial colonies, stain colonies with Dienes stain. Mycoplasma colonies stain blue with a distinctly dense blue centre and a lighter blue periphery. Mycoplasma colonies retain the stain for 24 h, whereas bacterial colonies decolourize after 30 min. See Dienes stain in 'Tests' section.
Distilled water	700 ml		
Phenol red			
Mycoplasma supplement	1 vial	Rehydrate vial with distilled water and swirl to dissolve. Aseptically add 1 vial per 70 ml of PPLO agar or broth.	
Sterile distilled water	30 ml		
Mycoplasma supplement S	1 vial	Rehydrate vial with distilled water and swirl to dissolve. Aseptically add 1 vial per 70 ml of PPLO agar or broth.	
Sterile distilled water	30 ml		
DNA (optional)		Prepare 0.2% solution by dissolving 0.2 g DNA in 100 ml distilled water. Autoclave at 121°C for 15 min. Add 10 ml to 1000 ml of above medium to produce a final concentration of 1%.	
DNA (optional) (Calbiochem polymerized calf thymus DNA) 0.2% solution	1%		
Distilled water 100 ml			

Formulations

- PPLO agar=per litre contains 50 g Bacto beef heart infusion, 10 g Bacto peptone, 5 g sodium chloride, Bacto agar 14 g. The broth does not contain the agar component.
- Mycoplasma supplement=per 30 ml vial is Bacto yeast extract 0.01 g, 1.6 g desiccated horse serum.

- Mycoplasma supplement S=per 30 ml vial contains Bacto yeast extract 0.01 g, 1.6 g of desiccated horse serum, 55,000 units penicillin, 50 mg thallium acetate.
- Bacto Heart Infusion Broth (Difco) contains 500 g beef heart infusion, 10 g bacto-tryptone, 5 g sodium chloride.
- The addition of 0.2% DNA can be added to the above medium. DNA is recommended for primary isolation of the bovine mycoplasma *Mycoplasma bovigenitalium*, and is thought to stimulate the growth of other *Mycoplasma* spp. as well. Therefore it is recommended as an optional media component (Razin and Tully, 1983).

Mycoplasma medium – modified Hayflick medium (Chanock *et al.*, 1962; Modified and used for isolation of *Mycoplasma mobile*; Kirchhoff and Rosengarten, 1984. Also used for many animal *Mycoplasma* spp.)

Reagent	Amount	Preparation of media	Description of growth characteristics
Bacto PPLO broth (Difco)	16.8 g	Dissolve PPLO broth in distilled water and autoclave at 121°C for 15 min. Cool to 50°C for agar containing media, or room temperature for broths, and aseptically add other filter-sterilized ingredients. Dispense in 3 ml volumes to bijou tubes for broths, or 5 ml volumes to small Petri dishes for agar plates.	See above under Mycoplasma medium – General-purpose medium.
Horse or bovine serum	200 ml		
Yeast extract (50% w/v)	10 ml		
DNA	0.02 g		
Penicillin	2000 IU/ml		
Thallium acetate (1.25% w/v)	10 ml		
Distilled water	800 ml		

For agar plates add 1.0% w/v purified agar (Oxoid) or use PPLO agar. *Note*: Handle thallous acetate with caution. Use gloves and mask. Thallous acetate may be inhibitory to some *Mycoplasma* spp. (T-strain mycoplasma, which is *Ureaplasma urealyticum*) (Shepard and Lunceford, 1975) and an alternative to thallium acetate may be colistin sulfate at 37.5 mg/l.

Mycoplasma medium (Used for isolation of *Mycoplasma* from crocodiles; Kirchhoff *et al.*, 1997)

Reagent	Amount	Preparation of media	Description of growth characteristics
Brain–heart infusion broth (Oxoid)	37 g	Dissolve brain heart infusion, yeast extract and glycerol in distilled water. Glycerol can be warmed slightly to facilitate pipetting. Autoclave at 121°C for 15 min. Prepare other reagents and filter sterilize. Add each reagent to cooled autoclaved media. Dispense into 5 cm Petri dishes. Store in plastic bags at 4°C.	Incubate plates in a candle jar or similar CO_2-generating atmosphere, at 37°C.
Yeast extract (Oxoid)	2 g		
Glycerol	8 ml		
Thallium acetate (10% w/v)	2.5 ml		
Ampicillin (5% w/v)	2 ml		
NAD (1% w/v)	10 ml		
Distilled water	700 ml		

Kirchhoff *et al.* (1997) also used this medium as a base to study the metabolism of glucose, arginine hydrolysis, TTC and phosphatase activity.

Nutrient agar (Used as a general isolation medium)

Reagent	Amount	Preparation of media	Description of growth characteristics
Nutrient agar (Oxoid)	14.0 g	Add agar powder to distilled water. Autoclave at 121°C for 15 min. Cool to 50°C and dispense into Petri dishes. Store at 4°C.	
Distilled water	500 ml		

Nutrient agar supplemented with activated charcoal (NAC) (Used for isolation of *Flavobacterium psychrophilum*; Álvarez and Guijarro, 2007)

Reagent	Amount	Preparation of media	Description of growth characteristics
Nutrient agar (Oxoid)	15.0 g	Add agar powder and charcoal to distilled water. Autoclave at 121°C for 15 min. Cool to 50°C and dispense into Petri dishes. Store at 4°C.	May improve recovery rate of *Flavobacterium psychrophilum*
Activated charcoal (Sigma)	0.5 g		
Distilled water	1000 ml		

Organic growth requirements

Some strains of *Vibrio*, e.g. *Vibrio ordalii* (previously *V. anguillarum* type II), *Moritella* (*Vibrio*) *marina* and some strains of *Photobacterium* (*Vibrio*) *logei* need 0.05% yeast extract for growth (Baumann *et al.*, 1980). A 10× stock solution of yeast extract can be prepared and then 250 µl added to 5 ml of broth media.

Packer's plates (Selective medium for *Erysipelothrix*; Packer, 1943)

Reagent	Amount	Preparation of media	Description of growth characteristics
Columbia blood agar base (Oxoid)	20.0 g	Add agar base to water and autoclave at 121°C for 15 min. Cool to 50°C and aseptically add sodium azide, crystal violet and blood. Mix well and pour into Petri dishes.	Colonies are 0.5–1 mm at 24–48 h. It is suggested this medium is used in conjunction with Wood's broth. Approximately 1 g of chopped or minced tissue or sample material is placed into 10 ml of Wood's broth. Incubate at 25°C and subculture from broth to BA and Packer's plates at 24 and 48 h. Examine plates at 24 and 48 h for typical colonies. Colonies of *Erysipelothrix rhusiopathiae* are 0.5–1 mm, grey-green at 48 h, with slight α-haemolysis around the colony. They resemble α-haemolytic Streptococci. *See photographic section.*
Distilled water	500 ml		
Sterile equine blood	25 ml		
Crystal violet (stock = 0.25 g in 100 ml of distilled water)	2 ml	Prepare stock solutions of crystal violet and sodium azide. Autoclave at 121°C for 15 min.	
Sodium azide (NaH_3) (stock = 1 g in 100 ml of distilled water)	12.5 ml		

***Pasteurella multocida* selective media.** NB plates (Rutter *et al.*, 1984)

Reagent	Amount	Preparation of media	Description of growth characteristics
Columbia agar base (Oxoid)	7.8 g	Add agar to distilled water and autoclave at 121°C for 15 min. Cool to 50°C and aseptically add blood and 1 ml of antibiotic stock. Mix well and pour into Petri dishes.	This is a semi-selective medium. Colonies of *Pasteurella multocida* appear as grey non-haemolytic colonies 1–2 mm and may be slightly mucoid.
Distilled water	200 ml		
Sterile equine blood	12 ml		
Neomycin sulfate (2.0 µg/ml)	4.0 mg	Prepare antibiotic stock by adding neomycin and bacitracin to normal saline. Add 1 ml to cooled agar and blood mixture as described above.	
Bacitracin (3.5 µg/ml)	7.0 mg		
Normal saline	10 ml		

Peptone yeast medium – PY (For growth of *Cytophaga, Flavobacterium* and *Sphingobacterium* spp.; Takeuchi and Yokota, 1992)

Reagent	Amount	Preparation of media	Description of growth characteristics
Peptone	1 g	Add all reagents to distilled water and pH to 7.0. Autoclave at 121°C for 15 min. Cool to 50°C and dispense into Petri dishes. Can be prepared as a broth by omitting the agar.	For growth of *Pedobacter heparinus, Sphingobacterium spiritivorum, S. multivorum* and *Flavobacterium* spp. Incubate at 28°C for 2 days.
Yeast extract	0.2 g		
Glucose	0.2 g		
NaCl	0.2 g		
Agar	1.5 g		
Distilled water	100 ml		

Poly plates (See under *Dermatophilus congolensis*)

PYE medium (peptone yeast extract) (For growth of *Planococcus kocurii*; Van Hao and Komagata, 1985)

Reagent	Amount	Preparation of media	Description of growth characteristics
Polypeptone	10 g	Add all reagents to distilled water and pH to 7.5. Autoclave at 121°C for 15 min. Cool to 50°C and aseptically dispense into Petri dishes.	Incubate at 30°C for 24–48 h.
Yeast extract (Difco)	5 g		
NaCl	5 g		
Bacto agar (Difco)	20 g		
Distilled water	1000 ml		

For growth of *Planococcus okeanokoites*, this media was used with the addition of 1 g $MgSO_4.7H_2O$ and 30 g NaCl (Nakagawa *et al*., 1996).

PYS-2 medium (For growth of *Vibrio rumoiensis*; Yumoto *et al*., 1999)

Reagent	Amount	Preparation of media	Description of growth characteristics
Polypeptone	8 g	Add all reagents to distilled water and pH to 7.5. Autoclave at 121°C for 15 min. Cool to 50°C and aseptically dispense into Petri dishes.	Colonies of *Vibrio rumoiensis* are circular and colourless. Growth can occur between 2 and 34°C with an optimum at 27–30°C.
Yeast extract	3 g		
NaCl	5 g		
Agar	15 g		
Distilled water	1000 ml		

R2A agar (Oxoid CM 906, or Difco) (Used for isolation of Antarctic *Flavobacterium* spp.; McCammon and Bowman, 2000)

Reagent	Amount	Preparation of media	Description of growth characteristics
Agar CM906	18.1 g	Add powder to distilled water and adjust pH to 7.2. Boil to dissolve. Sterilize by autoclaving at 121°C for 15 min. Cool to 50°C and pour into Petri dishes.	This is a nutritionally reduced medium that enhances recovery of heterotrophic bacteria from treated waters, and assists in recovery of bacteria that are stressed or chlorine-tolerant. The sodium pyruvate increases the recovery of stressed cells.
Distilled water	1000 ml		

Composition g/l of Oxoid CM 906: yeast extract (0.5), tryptone (0.25), peptone (0.75), dextrose (0.5), starch (0.50, di-potassium phosphate (0.3), magnesium sulfate (0.024), sodium pyruvate (0.3), agar (15.0).

Rogosa acetate agar (Oxoid). See under 'Tests' section

Strontium chloride B enrichment broth (For isolation of *Salmonella* spp. and *Edwardsiella tarda*; Iveson, 1971)

Reagent	Amount	Preparation of media	Description of growth characteristics
Bacto tryptone (Difco)	0.5 g	Add all reagents to the distilled water. Dispense into 10 ml volumes in McCartney bottles and autoclave at 121°C for 20 min. Final concentration of strontium chloride is 3.4% and pH is 5.0–5.5.	Place macerated or chopped sample (0.5 ml) or swab of sample into broth. Incubate at 37°C for 24 and 48 h. Subculture from broth onto selective media of MCA or DCA. On MCA plates the colonies of *Edwardsiella tarda* and *Salmonella* spp. appear as non-lactose fermenting colonies. On DCA the colonies appear as pale pink to clear colonies with or without a black centre (H_2S). *E. tarda* colonies are slightly smaller in size that *Salmonella* and black centre takes longer to develop.
Sodium chloride	0.8 g		
Potassium dihydrogen phosphate	0.1 g		
Strontium chloride 60%	6.0 ml		
Distilled water	100 ml		

Seawater – artificial (See under artificial seawater)

Seawater peptone-yeast agar (Medium for *Moritella* spp.; Colwell and Morita, 1964)

Reagent	Amount	Preparation of media	Description of growth characteristics
Proteose peptone (Difco)	10 g	Add all reagents to distilled water. Autoclave at 121°C for 15 min.	
Yeast extract (Difco)	3 g		
Artificial seawater salts (Sigma)	38 g		
Agar (Difco)	2 g		
Distilled water	1000 ml		

Serum-dextrose agar (Non-selective medium for *Brucella* spp.; Alton and Jones, 1967)

Reagent	Amount	Preparation of media	Description of growth characteristics
Nutrient agar	95 ml	Prepare nutrient agar and cool to 50°C. Add 5 ml of a stock solution of serum-dextrose per 95 ml of nutrient agar. Mix and pour into Petri dishes.	After 4 days' growth, smooth colonies of *Brucella* are small, round, glistening and blue or blue-green in colour when examined by obliquely reflected light. Rough colonies are yellow-white in colour and have a dry granular appearance.
Serum-dextrose	5 ml to 95 ml of nutrient agar		
Stock solution of serum-dextrose	1 g of dextrose per 5 ml of serum	For a stock solution of serum-dextrose, dissolve pure dextrose (glucose) in inactivated serum, at a rate of 1 g of dextrose per 5 ml of serum. Filter-sterilize and store at 4°C or –20°C.	

This medium is available from Oxoid as Brucella medium base Code CM169. Inactivated horse serum (5%) is added to the autoclaved base medium.

Shieh medium + Tobramycin (SM-T) (Selective medium for *Flavobacterium columnare and F. psychrophilum*; Decostere *et al.*, 1997; Shieh, 1980)

Reagent	Amount	Preparation of media	Description of growth characteristics
Peptone (Difco)	5 g	Suspend all chemicals into 1000 ml of distilled water and pH to 7.2. Autoclave at 121°C for 15 min. Cool to 50°C and added filter-sterilized solution of tobramycin. Mix and pour media into Petri dishes. Ensure the moisture content remains in the media by storing plates at 4°C, in sealed plastic bags. Growth of *F. columnare* is optimal when the plates are fresh, but reduces as the surface moisture decreases.	Using a cotton-tipped swab, collect samples from skin lesions and gills of fish and inoculate the plate. Colonies of *F. columnare* appear at 2–5 days as yellow-pigmented, rhizoid colonies with a thin spreading growth. *See photographic section.* Early colonies may be viewed with the aid of a stereomicroscope and a plug of the colony cut out using an unsealed Pasteur pipette prepared in the shape of a 'hockey stick'. Subculture the colony plug to a fresh plate. Shieh medium plus tobramycin reduces the growth and colony size of *Aeromonas hydrophila* and prevents growth of *A. salmonicida*, *S. putrefaciens* and *P. fluorescens*.
Yeast extract (Difco)	0.5 g		
Sodium acetate	0.01 g		
$BaCl_2 (H_2O)_2$	0.01 g		
K_2HPO_4	0.1 g		
KH_2PO_4	0.05 g		
$MgSO_47H_2O$	0.3 g		
$CaCl_22H_2O$	0.0067 g		
$FeSO_47H_2O$	0.001 g		
$NaHCO_3$	0.05 g		
Tobramycin	0.5 μg		
Noble agar (Difco)	10 g		
Distilled water (pH 7.2)	1000 ml		

Note: The original method states 1 μg/ml of tobramycin, but 0.5 μg/ml is recommended for Australian strains of *F. columnare* because growth may be inhibited or reduced at 1 μg/ml of tobramycin (Dr Annette Thomas, Department of Primary Industries, Queensland, 2000, personal communication).

SKDM (Selective medium for *Renibacterium salmoninarum*; Austin *et al.*, 1983)

Reagent	Amount	Preparation of media	Description of growth characteristics
Tryptone	1.0 g	Add all reagents to distilled water and adjust pH to 6.8. Autoclave at 121°C for 15 min, then cool to 50°C.	Inoculate plates with infectious material and incubate in a humid atmosphere at 15°C for 12 weeks. Examine plates regularly for the presence of pinpoint colonies. Maximum size of colonies is 2 mm. Colonies are white or creamy, shiny, smooth, raised and round. Old colonies may have a granular or crystalline appearance.
Yeast extract	0.05 g		
Cycloheximide	0.005 g		
Agar	1.0 g		
Distilled water	100 ml		
Fetal calf serum	10.0 g	Filter-sterilize L-cysteine hydrochloride, D-cycloserine, polymyxin B sulfate and oxolinic acid. Add these plus sterile fetal calf serum to autoclaved media. Pour into plates.	
L-cysteine hydrochloride	0.1 g		
D-cycloserine	0.00125 g		
Polymyxin B sulfate	0.0025 g		
Oxolinic acid	0.00025 g		

Shewanella marine agar (SMA) (Used for isolation of *Shewanella woodyi*; Makemson *et al.*, 1997)

Reagent	Amount	Preparation of media	Description of growth characteristics
Bacto peptone (Difco)	5.0 g	Add all reagents to distilled water. Autoclave at 121°C for 15 min.	A medium suitable for growth of luminous marine bacteria, especially *Shewanella*. *Shewanella* spp. can also grow on MA 2216.
Bacto yeast extract (Difco)	1.0 g		
Bacto agar (Difco)	15.0 g		
1× sea salts (see below)	200 ml		
Distilled water	1000 ml		

5× sea salts stock: 2.58 M NaCl, 0.125 M $MgCl_2$, 0.125 M $MgSO_4$, 0.1 M KCl, distilled water 1000 ml, pH to 7.5.

Serum-dextrose agar (For *Brucella* spp.)

Reagent	Amount	Preparation of media	Description of growth characteristics
Nutrient agar	95 ml	Prepare nutrient agar and cool to 50°C. Add 5 ml of a stock solution of serum-dextrose per 95 ml of nutrient agar. Mix and pour into Petri dishes.	After 4 days' growth, smooth colonies of *Brucella* are small, round, glistening and blue or blue-green in colour when examined by obliquely reflected light. Rough colonies are yellow-white in colour and have a dry granular appearance.
Serum-dextrose	5 ml to 95 ml of nutrient agar		
Stock solution of serum-dextrose	1 g of dextrose per 5 ml of serum	For a stock solution of serum-dextrose, dissolve pure dextrose (glucose) in inactivated serum, at a rate of 5 g of dextrose per 5 ml of serum. Filter-sterilize and store at 4°C or –20°C.	

Siem selective medium (Selective medium for *Aerococcus viridans*; Stewart, 1972; Gjerde, 1984)

Reagent	Amount	Preparation of media	Description of growth characteristics
Glucose	6.5 g	Add powders to distilled water and dissolve. Adjust pH to 7.4. Autoclave.	Growth of Gram-positive cocci is promoted with a change of medium from purple to yellow because of acid production. Incubate at 25°C for 5 days. Subculture yellow-coloured broth to BA.
Yeast extract	4.5 g		
Tryptone	15.0 g		
NaCl	6.4 g		
Phenyl ethyl alcohol	2.5 g		
Bromocresol purple	0.008 g		
Distilled water	1000 ml		

Skirrow's medium (See under *Helicobacter* selective medium)

SWT (A seawater-based complex medium for growth of marine organisms; Nealson, 1978; Boettcher *et al.*, 1999)

Reagent	Amount	Preparation of media	Description of growth characteristics
Tryptone	0.5 g	Add reagents to distilled water and autoclave at 121°C for 15 min. Cool to 50°C and aseptically add filter-sterilized seawater. Pour into Petri dishes.	Used for cultivation of *A. logei*, *A. fischeri*, *Roseobacter* spp., *Stappia* spp. Seawater can be replaced with 1.5% sea salts.
Yeast extract	0.3 g		
Glycerol	0.3 g		
Seawater	70 ml		
Distilled water	30 ml		
Agar	1.2 g		

TCBS cholera medium (TCBS) (Semi-selective for growth of *Vibrio* spp.; Kobayashi *et al.*, 1963; Nicholls *et al.*, 1976)

Reagent	Amount	Preparation of media	Description of growth characteristics
TCBS cholera medium (Oxoid)	44.0 g	Add powder to distilled water. Autoclave at 121°C for 15 minutes. Cool to 50°C and dispense into Petri dishes. Store at 4°C.	This selective medium will grow the majority of *Vibrio* spp. Some strains, however, such as *Vibrio* (*Listonella*) *ordalii* do not grow on this medium. *Aeromonas* spp. and *Pseudomonas* will grow weakly as small colonies of approximately 1 mm at 24–48 h. *Vibrio* spp. that ferment sucrose appear as yellow colonies, whereas sucrose-negative isolates appear as green colonies.
Distilled water	500 ml		

Notes:

- Some *Vibrio* spp. grow poorly on TCBS and a few strains do not grow at all on this medium. Some brands of TCBS are more inhibitory than others. Eiken and Oxoid brand supported the growth of a greater number of *Vibrio* spp. than BBL or Difco brands, in particular for *V. cholerae* and non-cholera *Vibrios*. Also, it is recommended that each batch of medium is tested for growth of *Vibrio* spp. (Nicholls *et al.*, 1976).
- TCBS detects a lower number of *Vibrio* colonies than MSA-B when performing total bacterial counts on water. Colonies that grow on MSA-B and not on TCBS on primary culture often grow on TCBS when subcultured to that medium, and identify as *Vibrio* spp. biochemically.
- After prolonged incubation, sucrose-fermenting colonies that initially appeared yellow may turn green once the sucrose in the medium has been used up or the pH of the medium changes.
- Formulation: Yeast extract (5 g), peptone (10 g), sodium thiosulfate (10 g), sodium chloride (10 g), ox bile (8 g), sucrose (20 g), NaCl (10 g), ferric chloride (1 g), bromothymol blue (0.04 g), thymol blue (0.04 g), agar (1.4 g), water (1000 ml) to pH 8.6.

Tryptone soya agar (TSA) (Used as a general-purpose isolation medium)

Reagent	Amount	Preparation of media	Description of growth characteristics
Tryptone soya agar (Oxoid)	20.0 g	Dissolve TSA in water and autoclave at 121°C for 15 min. Cool in waterbath to 50°C and pour into Petri dishes to depth of approximately 3 mm. Store at 4°C.	Suitable as a general purpose medium. Add NaCl (7.5 g) to give a final concentration of 2% to isolate bacteria from marine sources.
Distilled water	500 ml		

Tryptone yeast extract salt medium (TYES) (Used for the growth of *Flavobacterium columnare*; Triyanto and Wakabayashi, 1999)

Reagent	Amount	Preparation of media	Description of growth characteristics
Tryptone	0.4 g	Dissolve reagents in water and autoclave at 121°C for 15 min. Cool in waterbath to 50°C and pour into Petri dishes to depth of approximately 3 mm. Store at 4°C.	Used for the growth of *F. columnare*. Incubate plates at 25°C for 24 h. Colonies appear as yellow-pigmented, rhizoid, mucoid or honey-comb shaped colonies. Has been used for culture of *F. psychrophilum*. Incubate at 15°C for 3–5 days.
Yeast extract	0.04 g		
$MgSO_4.7H_2O$	0.05 g		
$CaCl_2.10H_2O$	0.05 g		
Distilled water	100 ml		

Tryptone-yeast extract-glucose agar (TYG) (Used for growth of *F. psychrophilum*, *F. columnare*, *T. maritimum*; Cipriano *et al.*, 1996)

Reagent	Amount	Preparation of media	Description of growth characteristics
Tryptone	0.2 g	Add all reagents except neomycin to distilled water. Autoclave at 121°C for 15 min. Cool to 50°C, and aseptically add filter-sterilized neomycin. Pour into Petri dishes.	Colonies are yellow with a thin spreading margin, which indicates the ability of the organism to exhibit gliding motility. Optimal temperature for *F. psychrophilum* is 14–20°C, for *F. columnare* 22–30°C and 15–34°C for *T. maritimum*. Plates are examined daily for up to 7 days. Examination for colonies using a stereomicroscope may assist in early detection of characteristic colonies.
Yeast extract	0.05 g		
Glucose	0.3 g		
Agar	1.5 g		
Distilled water	100 ml		
Neomycin sulfate	0.4 mg		

Addition of 10 IU/ml of Polymyxin B sulfate may assist as a selective medium for *F. columnare* (Shamsudin and Plumb, 1996).

TYG-M (Addition of 4 μg/ml neomycin sulfate, 200 IU/ml polymyxin B and 18.7 g/l ASW can be used for isolation of *Tenacibaculum* [*Flexibacter*] *maritimum*; Chen *et al.*, 1995).

VAL (Selective and differential medium for *Vibrio alginolyticus*; Chang *et al.*, 2011)

Reagent	Amount	Preparation of media	Description of growth characteristics
Brain–heart infusion agar	37 g	Dissolve all reagents in distilled water, and boil. Cool to 55°C. Adjust pH to 7.4±0.2 with 5 M NaOH. Pour into Petri dishes. Store at 4°C.	The uninoculated medium is a violet-blue colour. After incubation at 37°C, *V. alginolyticus* colonies are round, flat, greenish-yellow with a diameter larger than 2 mm at 24–48 h. *Vibrio proteolyticus* and *Staphylococcus epidermidis* are able to grow on this medium. *Vibrio proteolyticus* produces blue colonies. *Staphylococcus epidermidis* produces 0.5–1 mm green-yellowish colonies with a similar appearance to *V. alginolyticus*. *Aeromonas hydrophila*, *anguillarum*, *V. fluvialis*, *V. harveyi*, *V. parahaemolyticus* and *V. vulnificus* are unable to grow on VAL medium.
Sucrose	15 g		
Oxgall (Difco)	1 g		
NaCl	85 g		
Bromocresol purple	0.02 g		
Agar	15 g		
Distilled water	1000 ml		

VAM (Presumptive differentiation medium for *Vibrio* [*Listonella*] *anguillarum*; Alsina *et al.*, 1994)

Reagent	Amount	Preparation of media	Description of growth characteristics
Sorbitol	15 g	Dissolve all reagents (except ampicillin) in distilled water, and boil. Cool to 50°C. Adjust pH to 8.6 with 5 M NaOH. Aseptically add ampicillin. Pour into Petri dishes. Store at 15°C. Plates are viable for 3 weeks. After this time the ampicillin begins to lose activity.	The uninoculated medium is a violet-blue colour. After 48 h incubation at 25°C, *V. anguillarum* colonies are flat, round and bright yellow with a diffuse yellow halo due to the fermentation of sorbitol. *Vibrio fluvialis*, *V. harveyi* and *V. metschnikovii* are able to grow as yellow colonies on VAM, and therefore must be differentiated biochemically from *V. anguillarum*. *Vibrio alginolyticus* grows as a non-swarming blue colony.
Yeast extract	4 g		
Bile salts	5 g		
NaCl	35 g		
Ampicillin	10 mg		
Cresol red	40 mg		
Bromothymol blue	40 mg		
Agar	15 g		
Distilled water	1000 ml		

Vibrio recovery medium (Carson *et al.*, 2009)

Reagent	Amount	Preparation of media	Description of growth characteristics
Peptone (Oxoid LP0037)	5.0 g	Dissolve all reagents in distilled water. Adjust pH to 7.5–7.6. Autoclave at 121ºC for 15 min. Dispense into MaCartney bottles and store at 4ºC.	For the recovery of *Vibrio* spp. from freeze-dried vials, aseptically add 1 ml of Vibrio recovery medium and allow to stand until dissolved. Culture to suitable agar plate media such as MSA, MSSA or MA 2216 for salt-requiring organisms and to BA, TSA or NA for freshwater organisms.
Yeast extract	1.0 g		
Ferrous sulfate ($FeSO_4.7H_2O$)	0.2 g		
Sodium thiosulfate ($Na_2S_2O_3.5H_2O$)	0.3 g		
Sodium pyruvate	1.0 g		
Bacteriological charcoal	2.0 g		
Sea salts (Sigma)	900 ml		
Distilled water	100 ml		

This medium has been modified by replacing aged seawater with Sigma sea salts (this manual).

VVM (Selective medium *for Vibrio vulnificus*; Cerdà-Cuéllar *et al.*, 2001)

Reagent	Amount	Preparation of media	Description of growth characteristics
D-cellobiose	15.0 g	Dissolve all reagents in distilled water and bring to boil. Cool to 50°C and adjust pH to 8.5 using 5 M NaOH. (This media does not require autoclaving).	VVM plates are violet-blue in colour. *Vibrio vulnificus* is seen as a bright yellow colony with a yellow halo of diffusion due to the fermentation of cellobiose. Other cellobiose fermenting *Vibrios* such as *V. campbellii*, *V. harveyi* and *V. navarrensis* will grow on VVM agar. Some strains of *V. aestuarianus*, *V. alginolyticus* and *V. anguillarum* may also grow on this medium.
NaCl	10.0 g		
Yeast extract	4.0 g		
$MgCl_2{\cdot}6H_2O$	4.0 g		
KCl	4.0 g		
Cresol red	40.0 mg		
Bromothymol blue	40.0 mg		
Polymyxin B	10^5 U/l		
Colistin methanesulfonate	10^5 U/l		
Agar	15.0 g		
Distilled water	1000 ml		

A modification of this medium, termed VVMc, has the same reagents except the polymyxin B, and recovery rates are virtually the same (Cerdà-Cuéllar *et al.*, 2001).

Wagatsuma agar (Medium for detection of beta-haemolysis from *Vibrio parahaemolyticus*, known as the Kanagawa Phenomenon; Miyamoto *et al.*, 1969)

Reagent	Amount	Preparation of media	Description of growth characteristics
Yeast extract	3 g	Wash red blood cells in equal or larger volume of saline by centrifuging at 4000 × *g* at 4°C for 15 min. Wash twice. After third wash resuspend cell pellet to original volume with saline. Dissolve all reagents (except blood) in distilled water, and boil. Do not autoclave. Adjust pH to 8.0±2. Steam for 30 min. Cool to 45–50°C. Add 50 ml of washed red blood cells to the cooled medium. Pour into Petri dishes. Dry plates. Use promptly.	Erroneous results for beta-haemolysis and detection of the Kanagawa Phenomenon will occur if the NaCl concentration is less than 5%, if horse blood is used or if the ingredients are autoclaved. Subculture *V. parahaemolyticus* to Wagatsuma agar and incubate at 37°C for 24–48 h. A KP positive result is a clear zone of beta-haemolysis. A KP negative result is either no haemolysis or alpha (green) haemolysis.
Bactopeptone (Difco)	10 g		
NaCl	70 g		
K_2HPO_4	5 g		
Mannitol	10 g		
Crystal violet	0.001 g		
Agar	15 g		
Distilled water	1000 ml		
Human or rabbit red blood cells (fresh, 24 h) with anticoagulant	50 ml		

Wood's broth (Selective broth for detection of *Erysipelothrix rhusiopathiae*; Wood, 1965)

Reagent	Amount	Preparation of media	Description of growth characteristics
Bacto tryptose (Difco)	7.5 g	Add tryptose, Lab Lemco and NaCl to distilled water, pH to 7.5 and autoclave at 121°C for 15 min. Cool to room temperature.	Place swab of material or chopped pieces of tissue into medium. Incubate at 37°C for 48 h. Subculture to BA at 24 and 48 h.
Lab Lemco powder (Oxoid)	1.5 g		
NaCl	2.5 g		
Distilled water	500 ml		
Sterile horse serum (not inactivated)	25 ml	Add antibiotics to 10 ml of sterile distilled water. Aseptically add horse serum and antibiotic solution to cooled base. Dispense 10 ml volumes to McCartney bottles and store at 4°C.	Colonies of *Erysipelothrix* are 0.5 mm to 1 mm, grey-green at 48 h, with slight α-haemolysis around the colony. They resemble α-haemolytic streptococci. *See photographic section.*
Kanamycin	200 mg		
Neomycin	25 mg		
Vancomycin	12.5 mg		

Wood and Packer (1972) modified this medium by replacing the distilled water with 0.1 M phosphate buffer (12.02 g Na_2HPO_4, 2.09 g KH_2PO_4 and 1000 ml distilled water).

Yersinia selective agar (*Yersinia ruckeri* does not grow on this medium)

Reagent	Amount	Preparation of media	Description of growth characteristics
Yersinia selective agar base (Oxoid)	29.0 g	Add Yersinia selective base to distilled water and autoclave at 121°C for 15 min. Cool to 50°C.	*Yersinia* spp. grow as pink colonies with a darker 'bull's eye' centre at 24–48 h.
Distilled water	500 ml		
Oxoid Yersinia selective supplement	2 ml	Reconstitute 1 vial of supplement by aseptically adding 1 ml distilled water and 1 ml ethanol. Mix to dissolve and aseptically add to base. Pour into Petri dishes.	

7.2 Biochemical Test Media

Aesculin

Reagent	Amount	Preparation of media	Description of test
Aesculin	0.2 g	Dissolve all chemicals, except aesculin, by boiling. Cool and add aesculin while stirring. Dispense 5 ml of media into tubes. Autoclave at 121°C for 15 min. Store tubes in the dark.	Bacterial cells are inoculated into aesculin broth and incubated at appropriate temperature for 24–48 h. Development of a black colour is positive. Because some bacteria, especially some *Vibrio* spp., cause blackening of the medium due to melanin production, true hydrolysis of aesculin must be tested for loss of fluorescence at 354 nm long-wave UV light (MacFaddin, 1980; Choopun *et al.*, 2002). Fluorescence indicates a negative reaction.
Ferric citrate	0.1 g		
Peptone water (Oxoid)	3.0 g		
Distilled water	200 ml		

Arcobacter growth and tolerance tests. Used for identification of *Arcobacter, Campylobacter* and *Helicobacter* (On and Holmes, 1991)

Reagent	Amount	Preparation of media	Description of test
Base medium	100 ml	Prepare 100 ml lots of base medium by mixing reagents in water and heat and stir to dissolve. For glycine growth test add 1.0 g glycine to 100 ml of base medium. For glucose growth test add 8.0 g glucose to 100 ml base medium. Follow amounts for other tests as indicated. Dispense 4 ml of media to 5 ml tubes and autoclave at 115°C for 10 min. Label tubes and store at 4ºC. Tests can also be prepared as BA plates containing the individual supplements.	Growth in 1% glycine, 8% glucose, nitrate, 1.5% NaCl etc. is used for the identification of *Arcobacter*, *Campylobacter* and *Helicobacter* spp. Inoculate the semisolid tubed media using a sterile straight wire. Incubate at the appropriate temperature. Examine daily for signs of growth in the tubes. This may be seen as an umbrella-shaped growth at the top of the tube, particularly for organisms that are microaerophilic. For the nitrate test add one drop each of nitrate reagent A and B to the nitrate tube. A red colour is positive. Growth at different temperatures or atmosphere is done on BA plates. Consistency of inoculum size is important.
Nutrient broth No. 2 (Oxoid)	2.9 g		
Agar	2 g		
Distilled water	100 ml		
Individual test reagents			
Glycine (1%)	1.0 g		
Glucose (8%)	8.0 g		
Potassium nitrate	0.1 g		
Sodium chloride (1.5% or 4%)	1.5 or 4 g		
Bile or oxgall (1%)	1.0 g		
Safranine O (0.05%)	0.05 g		
Cefoperazone	64 mg/l		
2,3,5-triphenyltetrazolium chloride (TTC) (0.04 or 0.01%)	0.04 g or 0.01 g		

On and Holmes (1991) found the use of nutrient agar supported the growth of a greater number of *Arcobacter*, *Campylobacter* and *Helicobacter* spp. and produced less aberrant results than the use of Brucella broth as a base. The use of BA plates with the added test reagents gave the most favourable results for growth and tends to be the medium of choice for tolerance tests such as tolerance to bile, where as the use of nutrient agar base gave the most reproducible results for tests such as safranine O.

Arginine dihydrolase (Moeller) (See under Decarboxylases)

Arginine dihydrolase (Thornley) (See under Decarboxylases)

Brucella. Assay of metabolic activity (Alton and Jones, 1967; Jahans *et al.*, 1997)

Reagent	Amount	Preparation of media	Description of test
L-alanine	1.25 g/l	Dissolve various metabolic substrates in sterile PBS, pH 7.2 at a concentration of 1.25 g/l. Sterilize each solution by filtration through 0.22-µm filter membrane. Store at 4°C. Prepare MTT solution in PBS at 1.0 g/l. Sterilize by filtration through a 0.22-µm filter.	Substrate mediated metabolic activity is assayed by the stimulation of tetrazolium reduction. Prepare a cell suspension in PBS and adjust the concentration to 10^{10} organisms per ml. In a flat-bottomed microtitre plate, place 100 µl of each substrate into each well. Place 50 µl of cell suspension in each well and incubate at 37°C in 10% CO_2 for 18 h. After incubation, place 50 µl of MTT in each well. Incubate 1 h at room temperature and add 50 µl of formaldehyde to each well. After 2–4 h, read the optical density at 630 nm. The metabolic index for each substrate is expressed as a ratio of the OD with substrate against the OD without substrate.
L-asparagine	1.25 g/l		
L-glutamic acid	1.25 g/l		
L-arginine	1.25 g/l		
DL-ornithine	1.25 g/l		
I-lysine	1.25 g/l		
D-galactose	1.25 g/l		
D-ribose	1.25 g/l		
D-xylose	1.25 g/l		
Meso-erythritol	1.25 g/l		
Urocanic acid	1.25 g/l		
3-[4,5-dimethylthiazol-2-yl]-2,5-diphenyltetrazolium bromide (MTT)	1.0 g/l		
Formaldehyde	40%		

Brucella: growth on substrate media (Alton and Jones, 1967; Jahans *et al.*, 1997)

Reagent	Amount	Preparation of media	Description of test
Basic fuchsin	20 µg/ml (1/50,000)	A 0.1% stock solution of each dye is made in distilled water, and sterilized by being held in boiling water for 1 h. Stock solution should be renewed after 3 months. Each dye is then added to a base media such as tryptone-soya agar or serum-dextrose agar. Alton and Jones (1967) suggest that a laboratory should determine the concentration of each dye that gives optimum results, the range being somewhere between 1:25,000 and 1:100,000 (10 to 40 µg of dye per ml of medium), using the FAO/WHO reference strains. The concentrations are suggested by Jahans *et al.* (1997). The required amount of the dye stock solution is added to the melted base medium, mixed, and poured into Petri dishes and allowed to set. Label each plate with appropriate dye identification.	Prepare a suspension of bacterial cells in 0.5 ml of sterile normal saline. The unknown culture should be inoculated onto the same plate along with the reference strains. Therefore mark off an appropriate number of areas on a plate. Using a loop of culture suspension, make five separate streaks without recharging the loop onto each specifically marked area. Also inoculate a control plate that contains basic medium only and no dye. Incubate plates at 37°C with and without CO_2 for 4 days. Record results of 'growth' or 'no growth' in the presence of each dye.
Safranin O	100 µg/ml (1/10,000)		
Thionin	20 µg/ml (1/50,000)		

Brucella selective media (*See Farrell's medium*)

Carbohydrate fermentation (General-purpose medium for detection of carbohydrate fermentation; Vera, 1948, 1950)

Reagent		Amount	Preparation of media	Description of test
Cystine tryptic medium (Gibco)		7.13 g	Basal medium preparation Boil distilled water and cysteine tryptic medium. Dispense 5 ml volumes into 10 ml tubes. Carbohydrate preparation To prepare the carbohydrates, make a 10% solution (3 g in 30 ml). For salicin use a 4% solution (1.2 g in 30 ml of distilled water). For carbohydrates that can be autoclaved Pipette 0.5 ml of carbohydrate solution per 5 ml of basal medium. Autoclave at 10 lb per 10 min. For filter-sterilization: Filter sterilize the carbohydrate solution through a 0.22-µm filter. Autoclave basal medium 5 ml in 10 ml tubes. Cool to 50°C and aseptically add the filter-sterilized carbohydrate solution, 0.5 ml to 5 ml of basal medium.	Fermentation is detected by a colour change from red to yellow. References may state carbohydrates are added at 0.5% to 1.0%. However, because reversion of the reaction may occur when 0.5% is used, using a concentration of 1% ensures against depletion of the carbohydrate by the organism.
Distilled water		250 ml		
Carbohydrate (see below)		Sterilization method		
Carbohydrate	Sterilization	1% final concentration		
Arabinose	Filter			
Aesculin	Autoclave			
Glucose	Autoclave			
Inositol	Autoclave			
Lactose	Autoclave			
Maltose	Filter			
Mannitol	Autoclave			
Mannose	Filter			
Salicin	Autoclave			
Sorbitol	Autoclave			
Sucrose	Filter			
Trehalose	Autoclave			
Xylose	Filter			

Cystine tryptic medium contains Bacto tryptose, L-cystine, sodium chloride, sodium sulfite, agar and phenol red. With the addition of yeast extract at 0.01%, this medium is also suitable for testing carbohydrate fermentation of the *Cytophaga*, *Flavobacterium*, *Flexibacter*, *Tenacibaculum* group, plus certain *Vibrio* spp. that require yeast extract, such as *V. ordalii*, *Moritella marinus* and some strains of *P. logei*. Baumann *et al.* (1980) add yeast extract at 0.05% for strains of *Vibrio* and *Photobacterium* spp. that require organic growth factors.

Notes:

- In the literature some original recipes used bromothymol blue as the pH indicator, but it may be toxic to a number of marine bacteria and so phenol red is the recommended pH indicator in carbohydrate fermentation tests (Leifson, 1963).
- Peptone-containing media may give misleading results for carbohydrate breakdown because weak acid production may be masked by peptone-breakdown products (Vera, 1950).
- Sucrose should be filter-sterilized, because it is thermolabile and will give false-positive results if sterilized by autoclaving (Stanier *et al.*, 1966).
- To prevent colour change of the medium caused by CO_2 build up in screw-capped tubes when inoculated with organism of interest, ensure lids are loosened for about an hour before results are read. Preparing tubed media to only contain half their volume in media also assists to prevent pH reversion caused by CO_2 build-up.
- Most carbohydrates can be stored at room temperature for 2–3 weeks, with the main stocks stored at 4°C for longer term storage. However, some carbohydrates should always be stored at 4°C to prevent unwanted colour change in the medium. These are arabinose, mannitol, xylose and also the oxidative-fermentative media and ONPG media.

Alternative methods for carbohydrate fermentation for specific bacteria

Carbohydrate fermentation media for *Flavobacterium* spp.

- Broth medium containing 0.05% tryptone, 0.05% yeast extract, 0.0018% phenol red and filter-sterilized carbohydrate at a final concentration of 0.2% (Wakabayashi *et al.*, 1986).
- AO medium was used as a basal medium for carbohydrate fermentation, but it was unsuitable for this test, because colour changes were seen in the negative control tube (Bernardet and Grimont, 1989).

Carbohydrate fermentation media for marine bacteria (Leifson, 1963, used by Gauthier, 1976)

Reagent	Amount	Preparation of media	Description of test
Casitone (Difco)	0.1 g	The ingredients are dissolved in half-strength seawater and the pH is adjusted with HCl to 7.5. Autoclave and re-check the pH, which should be about 8. Adjust with HCl if necessary. Filter-sterilize the carbohydrate and add aseptically to the base medium. Dispense 3 ml into 13×100 mm tubes.	Fermentation is indicated by a change in colour from red to yellow.
Yeast extract	0.01 g		
Ammonium sulfate	0.05 g		
Tris buffer	0.05 g		
Agar	0.3 g		
Phenol red	0.001 g		
Artificial seawater	Half-strength		
Carbohydrate	1%		

Einar Leifson (1963) modified this recipe from the original medium published in 1953 (Hugh and Leifson, 1953). The pH indicator in the 1953 medium (bromothymol blue) was found to be toxic for a number of marine bacteria and therefore phenol red is recommended. This media is also used to test for oxidation/fermentation. Add individual carbohydrates to the base medium for fermentation tests. This method is similar to the general-purpose medium.

Carbohydrate fermentation media for *Moritella viscosa* (Used by Lunder *et al.*, 2000)

Reagent	Amount	Preparation of media	Description of test
Nutrient broth (Difco)	2.7 g	The ingredients are dissolved in distilled water and the pH is adjusted with HCl to 7.5. Autoclave and re-check the pH, which should be about 8. Adjust with HCl if necessary. Filter-sterilize the carbohydrate and add aseptically to the base medium. Dispense 3 ml into 13×100 mm tubes.	Fermentation is indicated by a change in colour from red to yellow.
Proteose peptone (Difco)	5.0 g		
NaCl	20.0 g		
$K_2HPO_4 \cdot 2H_2O$	0.3 g		
Mineral salt solution	10 ml		
Phenol red (0.2%)	6 ml		
Distilled water	1 l		
Carbohydrate	0.7 g		

Mineral salt solution: 20 g $MgCl_2.6H_2O$, 5.7 g $CaCl_2.2H_2O$, 0.5 g $FeCl_2.4H_2O$, 0.5 g $ZnSO_4.7H_2O$, 0.25 g $CoCl_2.6H_2O$, 0.25 g $CuSO_4.5H_2O$, 0.25 g Na_2MoO_4, 0.5 g $MnSO_4.H_2O$. Dissolve in 1 H_2O.

Carbohydrate fermentation media for *Pseudomonas* and *Flavobacterium* spp. (Gilardi, 1983)

Reagent	Amount	Preparation of media	Description of test
OF basal medium (Difco)	9.4 g	Add reagent in distilled water and dissolve by boiling. Dispense into 100 ml aliquots. Autoclave at 121°C for 15 min. To each 100 ml aliquot aseptically add 10 ml of a 10% carbohydrate solution. Aseptically dispense each 100 ml of respective carbohydrate into 5 ml aliquots in 10 ml tubes.	Colour change to yellow indicates a positive fermentation reaction.
Distilled water	1000 ml		
Carbohydrate 10% solution	10 ml	Filter-sterilize and add as above.	

Composition of Difco OF medium: tryptone (2.0 g), sodium chloride (5.0 g), dipotassium phosphate (0.3 g), agar (2.0 g), bromothymol blue (0.08 g), in 1000 ml distilled water, pH 6.8.

Carbohydrate fermentation media using acid from ammonium salt sugar (ASS) (Used for *Sphingobacterium, Flavobacterium, Shewanella putrefaciens, Pseudomonas* spp. Used by Holmes *et al.* [1975] for testing *Pseudomonas* spp. Used by Bernardet and Grimont [1989] as an inoculation medium for API 50CH when testing *Flavobacterium*. Quoted in Cowan and Steel [1970] from the method of Smith *et al.* [1952])

Reagent	Amount	Preparation of media	Description of test
$(NH_4)_2HPO_4$	1.0 g	Add ingredients to distilled water and dissolve by boiling or steaming. Add indicator and autoclave 115°C for 20 min. Allow medium to cool to 60°C and add filter-sterilized carbohydrate solution. Dispense into tubes and incline tube so medium sets as a slope.	Inoculate and incubate at the appropriate temperature. Examine over 7 days. Colour change from purple to yellow is indicative of fermentation of carbohydrate.
KCl	0.2 g		
$MgSO_4.7H_2O$	0.2 g		
Yeast extract	0.2 g		
Agar	20 g		
Distilled water	1000 ml		
Bromocresol purple 0.2% solution	4 ml		
Carbohydrate 10% solution	100 ml		

Notes

- Bacteria that do not give reliable fermentation reaction with peptone-containing medium should be tested in ASS. In general, peptone-containing media give the least positive reactions and ammonium salt sugars the most positive reactions (Cowan and Steel, 1970).
- Reactions using this medium are not reported in this book. *Flavobacterium* spp. reactions reported in this book were the same by both the peptone-containing method (Carbohydrate general purpose medium) and ASS.

Carbohydrate fermentation media for *Vibrio* spp.

The following media were used by Baumann *et al.* (1971) for *Vibrio* spp. and *Beneckea* – adapted from the methods of Stanier *et al.* (1966)

- Basal medium (BM): 50 mM tris (hydroxymethyl)aminomethane (tris)-hydrochloride (pH 7.5), 190 mM NH_4Cl, 0.33 mM $K_2HPO_4.3H_2O$, 0.1 mM $FeSO_4.7H_2O$ and half-strength seawater.
- Basal medium agar (BMA): Mix equal volumes of double-strength MB with 20 g of Ionagar (Oxoid) per litre.
- Yeast extract broth (YEB): Add 5 g per litre of yeast extract (Difco) to BM.
- For Yeast extract agar (YEA) add 20 g of agar (Difco) per litre of YEB.
- Carbohydrate fermentation medium (F-2): To YEB add 100 mM Tris-hydrochloride (pH 7.5), 1 g/l of Ionagar (Oxoid), 1 g/l of sodium thioglycollate and 10 g/l of filter-sterilized glucose.
- Carbohydrate fermentation medium (F-3): To BM add 25 mM Tris-hydrochloride (pH 7.5), 0.5 g/l of yeast extract (Difco), 1 g/l of Ionagar (Oxoid), 2 ml/l of a 1.6% (w/v) alcoholic solution of bromocresol purple and 10 g/l of filter-sterilized carbohydrate.

Carbohydrate fermentation media for Nocardia spp. Basal inorganic nitrogen medium
(Gordon *et al*., 1974. Quoted from Ayers *et al*., 1919)

Reagent	Amount	Preparation of media	Description of test
$(NH_4)_2HPO_4$	1.0 g	Add all reagents except carbohydrate to distilled water and adjust pH to 7.0. Dispense 5 ml into tubes and autoclave 121°C for 15 min. Aseptically add 0.5 ml of a 10% solution of each carbohydrate (autoclaved separately) to tubes and allow to set on a slant.	Production of acid due to fermentation of the carbohydrate is seen as a colour change from purple to yellow. Incubate at 28°C and read from 7 to 28 days. With the addition of yeast extract, this medium is the same as ammonium salt sugars.
KCl	0.2 g		
$MgSO_4.7H_2O$	0.2 g		
Agar	15		
Distilled water	1000 ml		
Bromocresol purple 0.04% solution	15 ml		
Carbohydrates to be tested (10% solution)	0.5 ml		
Optional Yeast extract (Difco)			

When testing *N. seriolae*, add 2% w/v yeast extract (Difco) to medium (Kudo *et al*., 1988).

Carbon dioxide atmosphere (Used when incubation conditions require a carbon dioxide atmosphere)

For laboratories that do not have access to a carbon dioxide incubator, the following methods offer a cheap alternative.

Method 1: Candle jar method. Place agar plates in an airtight container. Place a lighted candle in the jar and seal the lid. This method produces 2% CO_2 (J. Lloyd, Department of Agriculture, Western Australia, 1985, personal communication) with other reports suggesting 6–8% CO_2 (Cottral,

1978). If the lid is plastic, then protect from the heat produced by the candle by using an aluminium foil-covered device.

Method 2: 'ENO salts' method (Lloyd, 1985). A total of 7.5 g of ENO salts per 10 l container generates 4% CO_2. Use an appropriate container that has a well-sealed lid, such as a household food container or Mitsubishi Gas Chemical Company anaerobe box. Place 3 g of ENO salts (Sigma) into 20 ml of water for a container that measures 22×22×8 cm or 10 g of ENO salts into 40 ml of water for a container that measures 30×30×14 cm. This technique needs to be performed quickly, therefore place salts into a small piece of tissue paper. Place the water into a 50-ml urine container and place into the incubation container. Quickly add the salts to the water and seal lid. Place box at appropriate temperature. ENO is a sparkling antacid that contains per 5 g, sodium bicarbonate (2.32 g), sodium carbonate (0.5 g), citric acid anhydrous (2.18 g) and can be purchased through SIGMA chemicals.

Catalase test (From Cowan and Steel, 1970)

Smear a colony onto a glass slide. Place a drop of 30% H_2O_2 on the smear. Presence of bubbles indicates a positive reaction. When picking growth from a blood agar plate make sure no blood-containing agar is placed on the slide because false positive results may occur.

Cellulose digestion (Wakabayashi *et al.*, 1989)

To a broth culture medium appropriate to the growth requirements of the organism being tested, add a piece of cellulose filter paper. Observe for disintegration of the filter paper. A characteristic of the *Cytophaga* spp. is that they digest cellulose, therefore when testing an organism suspected of being a *Cytophaga* spp. use a broth of Anacker–Ordal medium.

Coomassie brilliant blue agar (CBBA) (Used to detect the A+ protein layer of *Aeromonas salmonicida*; Udey, 1982; Evenberg *et al.*, 1985, Cipriano and Bertolini, 1988; Markwardt *et al.*, 1989)

Reagent	Amount	Preparation of media	Description of growth characteristics
TSA	44.0 g	Add Coomassie blue and TSA to distilled water. Autoclave 121°C for 15 min. Cool to 50°C and pour into Petri dishes.	Coomassie brilliant blue is a protein-specific dye that results in production of dark blue colonies of bacteria that have the A-layer protein. A-layer positive *A. salmonicida* stain deep blue. This media is not selective and other bacteria will produce blue-coloured colonies. However, it assists in the isolation and detection of *A. salmonicida*.
Coomassie brilliant blue R250 (Bio-Rad)	0.1 g		
Distilled water	1000 ml		

Citrate. Simmons' method (Simmons, 1926)

Reagent	Amount	Preparation of media	Description of test
Simmons' citrate (Difco)	3.63 g	Suspend media in distilled water, and bring to the boil. Dispense 3 ml of media into tubes. Autoclave at 121°C for 15 min. Lay on a slight incline to create slopes.	Development of a deep blue colour is positive. Simmons' citrate (Difco) contains 1% NaCl. Citrate is tested as a sole carbon source. (Christensen's citrate doesn't test for citrate as the sole carbon source because it contains other nutrients).
Distilled water	150 ml		

Congo red (Tests for presence of extracellular galactosamine glycan in the *Flavobacteriaceae*; Johnson and Chilton [1966] who quote personal communication with E.J. Ordal for this test; McCurdy, 1969)

Reagent	Amount	Preparation of media	Description of test
Congo red	10 mg	Weigh 10 mg of Congo red and add to 100 ml of distilled water. Final concentration is 0.01% aqueous. Mix well, label bottle and store at room temperature.	Place 1–2 drops of Congo red on a few isolated colonies growing on AO or Shiels medium. A positive reaction is red-colour development of colonies. This colour lasts for a number of hours. This test detects the presence of extracellular glucans.
Distilled water	100 ml		

Congo red agar (Used for detection of A protein layer of *A. salmonicida*; Ishiguro *et al.*, 1985)

Prepare tryptic soy agar (Difco) containing 30 g/ml of Congo red. Incubate at 20°C and examine for red-coloured colonies after 48 h.

Decarboxylases and arginine dihydrolase (Used in Cowan and Steel, 1970; West and Colwell, 1984)

Reagent	Amount	Preparation of media	Description of test
Decarboxylase base Moeller (Difco) (Moeller, 1955)	2.1 g	In separate bottles, dissolve each individual amino acid in the 200 ml of distilled water and decarboxylase base. Dispense 5 ml into 10 ml tubes. Autoclave at 121°C for 15 min.	A tube of basal medium without amino acid is always inoculated in parallel with the test media. Inoculate with heavy inoculum. Overlay all tubes with paraffin oil, and incubate for 7–14 days for some strains. Most strains can be read at 48 h. Check the bottom of tube and the cloudiness of the tube for signs of sufficient inoculum and growth. For salt-requiring strains, add 0.5 ml of 20% NaCl to tubes before inoculation with organism.
Distilled water	200 ml		
L-arginine HCl	2.0 g		
L-ornithine monohydrochloride	2.0 g		
L-lysine monohydrochloride	2.0 g		
Control – use decarboxylase base and distilled water only.			See under Paraffin oil for sterilization procedures for the oil.

Arginine dihydrolase (ADH) (Method of Thornley, 1960. Recommended for marine bacteria)

Reagent	Amount	Preparation of media	Description of test
Bacto-peptone (Difco)	0.1 g	Add reagents to distilled water and pH to 6.8. Dispense 5 ml into 10 ml tubes. Autoclave at 121°C for 15 min. Note: the original pH described by Thornley was 7.2, however pH 6.8 is recommended by West and Colwell (1984).	A tube of basal medium without amino acid is always inoculated in parallel with the test media. Inoculate with heavy inoculum. Overlay all tubes with paraffin oil, and incubate for 7–14 days for some strains. Most strains can be read at 48 h. Check the bottom of tube and the cloudiness of the tube for signs of sufficient inoculum and growth.
NaCl	0.5 g		
K_2HPO_4	0.03 g		
Agar	0.3 g		
Phenol red	0.001 g		
Arginine HCl	1.0 g		
Distilled water			

Some *Vibrio* spp. are negative in Moeller's method for ADH, but positive by the method of Thornley. These are; *Vibrio mediterranei*, *V. mytili*, *V. orientalis* and *V. splendidus* biogroup I and some strains of *V. tubiashii* (Macián *et al.*, 1996). The glucose in Moeller's medium appears to inhibit the reaction due to catabolite repression of the inducible ADH system (Macián *et al.*, 1996). Baumann *et al.* (1971) found that *Photobacterium* spp. can produce alkaline products in Thornley's medium, yet none possess a constitutive ADH system when tested with more sensitive analytical methods (West and Colwell, 1984). Different brands of peptone were found to give different results and the Difco Bacto-peptone was recommended (Thornley, 1960).

Thornley's medium was modified by Baumann and Baumann (1981) by using basal medium without the tris-HCl. See under alternative media for carbohydrate fermentation, media for *Vibrio* spp., for this recipe.

Dienes stain (Used to differentiate mycoplasma from bacteria; Dienes, 1939; Hayflick, 1965)

Reagent	Amount	Preparation of media	Description of test
Methylene blue	2.5 g	Dissolve all reagents in distilled water. Store in a screw-cap bottle.	Method I. Place some stain onto a coverslip and allow to dry. Agar blocks containing colonies are excised and placed onto a glass slide with colonies uppermost. Place coverslip onto agar block containing colonies and press coverslip lightly so that there is good contact between the colonies and the stain. Method II. Using a cotton wool swab directly apply the stain to the mycoplasma colonies. Mycoplasma colonies stain blue with a distinctly dense blue centre and a lighter blue periphery. Mycoplasma colonies retain the stain for 24 h, whereas bacterial colonies decolourize after 30 min.
Azure II	1.2 g		
Maltose	10 g		
Sodium carbonate (anhydrous)	0.25 g		
Distilled water	100 ml		

Some methods contain 0.2 g of benzoic acid.

Digitonin (Used to differentiate *Mycoplasma* spp. from *Acholeplasma* spp.)

Reagent	Amount	Preparation of media	Description of test
Digitonin (1.5% w/v)	1.5 g	Dissolve digitonin in ethanol. Warm to 37°C to dissolve. Place 20 µl of stock solution onto a 6-mm sterile filter paper disc (Oxoid). Dry at 37°C for 1 h.	Plate suspect *Mycoplasma* colony to a 5-cm Mycoplasma agar plate. This will either be from a broth or agar plate. When transferring colonies from an agar plate, use a sterile spatula or open-ended 'hockey stick' (made from a bent Pasteur pipette) to select a plug of agar containing a colony. Invert this agar plug, colony-side down, onto a fresh agar plate and rub over the surface. If inoculating the plate from a broth culture, pipette the broth culture onto the plate and then remove the excess. Allow the plate to dry for a few hours. Place the digitonin disc in the middle of the plate, and incubate at appropriate temperature under CO_2 for 3–4 days. Mycoplasmas are sensitive to digitonin and so a 4-mm zone of inhibition of growth is seen from the edge of the disc (zone of 14 mm diameter). *Acholeplasma* are resistant to digitonin and a zone of less than 1 mm is seen from the edge of the disc.
Ethanol	100 ml		

DNase (Used by West and Colwell (1984) and reported from all sources used in this manual)

Reagent	Amount	Preparation of media	Description of test
DNase test agar (Oxoid)	19.50 g	Suspend agar in water in 1000 ml. Autoclave at 121°C for 15 min then cool to 50°C in waterbath. Mix well and pour media into Petri dishes.	Spot-inoculate or streak as a single line, a loop of bacterial culture onto the plate. Incubate plates for 2–7 days at 24°C. Flood plate with 1% HCl to precipitate the DNA. A positive reaction is seen as a clear zone around the bacterial streak. *See photographic section*.
Distilled water	500 ml		

See under HCl for preparation of 1 M HCl.

Furunculosis agar (For detection of pigment production from *Aeromonas salmonicida*; Bernoth and Artz, 1989)

Reagent	Amount	Preparation of media	Description of test
Bacto-tryptone (Difco)	10 g	Add all ingredients to distilled water. Adjust pH to 7.3, and autoclave at 121°C for 15 min. Cool to 50°C and pour into Petri dishes.	Pigment production is best detected by incubating at 15 to 20°C for up to 7 days. At 25°C less pigment production is seen. Pigment is seen as a diffuse brown coloration around each colony.
Yeast extract (Difco)	5 g		
L-tyrosine (Merck)	1 g		
NaCl (Merck)	2.5 g		
Agar (Oxoid L11)	15 g		
Distilled water	1000 ml		

Furunculosis agar is recommended as a superior medium for detection of pigment production as a greater number of strains show pigment on FA than on TSA, NA or BHIA (Hirvelä-Koski *et al.*, 1994; Hänninen and Hirvelä-Koski, 1997). However, FA is not the best medium for primary isolation of *Aeromonas salmonicida* (Bernoth and Artz, 1989) and BA is recommended. (It may be possible that by adding blood to FA, the medium would be a suitable primary isolation medium).

Gelatinase (0% and 3% NaCl) (Smith and Goodner, 1958. Used by West and Colwell, 1984)

Reagent	Amount 0% NaCl	Amount 3% NaCl	Preparation of media	Description of test
Difco Bacto peptone	1.0 g	1.0 g	Add reagents to distilled water and autoclave at 121°C for 15 min. Cool to 50°C. Use split Petri dishes, and label plates 0% and 3% per side of each plate. Pour appropriate media into each half of each plate. Store at 4°C.	Spot inoculate the gelatin plate on both the 0% NaCl and 3% NaCl sides with approximately equal amounts of bacterial growth. Incubate at an appropriate temperature 24–48 h. Production of gelatinase is seen as either a cloudy or clear zone around the area of bacterial growth. The plate should be held up to the light and read against a darkish background. Chilling the plates may provide a sharper contrast between zones and the unaffected portions of the media. Flooding the plate with ammonium sulfate may assist with definition of zones of clearing.
Difco yeast extract	0.25 g	0.25 g		
Oxoid gelatin	3.75 g	3.75 g		
NaCl	–	7.50 g		
Oxoid agar No. 1	3.75 g	3.75 g		
Distilled water	250 ml	250 ml		

This plate method of Smith and Goodner (1958) is used because it detects a change in the composition of gelatin rather than its liquefaction and thus is a very sensitive method. Use a split plate with one side 0% NaCl and the other side containing 3% NaCl. This allows the salt requirement of the organism to be detected at the same time. Alternatively, use small (5 cm) Petri dishes for each NaCl concentration.

Glucose yeast extract agar (GYEA) (For colonial morphology and survival at 50°C of *Nocardia* spp.; Gordon *et al.*, 1974)

Reagent	Amount	Preparation of media	Description of test
Yeast extract	10 g	Add reagents to water and pH to 6.8. Autoclave at 121°C for 15 min. May be poured into plates, or prepared as tubed slants.	Plate media is used for observation of colonial morphology. Tubed slants are used for testing survival at 50°C.
Glucose	10 g		
Agar	15 g		
Tap water	1000 ml		

Haemolysis

Record from growth on BA or MSA-B within 7 days. For some *Vibrio* spp. haemolysis may be more pronounced on BA than MSA-B, even though their preferred growth medium is MSA-B.

HCl (1N) (For detection of hydrolysis of DNA in DNase medium)

Reagent	Amount	Preparation of media	Description of test
Concentrated HCl, 32%	9.85 ml	Add 9.85 ml of 32% acid to 80 ml water, then make up to 100 ml.	To a 24 or 48 h growth on a DNase plate flood with 1 N HCl. Wait 1 min and observe over a black tile for zone of clearing around bacterial growth. *See photographic section.*
Distilled water	80 ml		

Safety note: Always add acid to water, not vice versa. Other concentrations of HCl are available. To prepare from 35.4%, add 8.9 ml to 80 ml of water, and then make up to 100 ml. To prepare from 37%, add 8.5 ml of acid to 80 ml of water, and then make up to 100 ml.

Hippurate hydrolysis (Hwang and Ederer, 1975)

Reagent	Amount	Preparation of media	Description of test
Hippuric acid sodium salt	0.15 g	Dissolve hippuric acid in water. Dispense 0.4 ml per sterile, 5 ml yellow-capped plastic tube. Store at −20°C.	Inoculate medium with a large loopful of bacterial growth from culture plate. Incubate at appropriate temperature for 2.5 h. Add 200 µl of ninhydrin reagent and incubate a further 10 min. Development of deep purple colour is positive. (Original reference suggests a purple colour as deep as the colour of crystal violet used in the Gram stain). Used in identification of *Arcobacter*, *Campylobacter* and *Helicobacter.*
Distilled water	15 ml		

Reagent: Ninhydrin.

Ninhydrin reagent (Used for detection of hippurate hydrolysis. Hwang and Ederer, 1975)

Reagent	Amount	Preparation of media	Description of test
Ninhydrin	0.35 g	Dissolve ninhydrin in the acetone/butanol mixture. Dispense into 5 ml tubes and store in the dark.	To a mixture of organisms incubated in hippurate solution for 2.5 h, add 200 µl of ninhydrin reagent. Incubate 10 min and observe formation of purple layer.
Acetone	5 ml		
Butanol	5 ml		

Hydrogen sulfide production

Many of the references for the biochemical reactions use a method whereby a lead acetate strip (H_2S indicator) is suspended over a tube of growth medium containing cysteine (sulfur source). This is a very sensitive method. However, the preparation of the lead acetate strips is hazardous (for method see Cowan and Steel, 1970). Alternative methods are available but may not have the same sensitivity. The test can be performed in an API 20E or a triple sugar iron tube. A hydrogen sulfide Biostrip (catalogue number TM343) is available from MedVet Science. When using these media be aware of the growth requirements of the test organism and the sensitivity of the test.

Indole (Cowan and Steel method 2, 1970; Colwell and West, 1984; MacFaddin, 1985)

Reagent	Amount	Preparation of media	Description of test
Tryptone broth (Difco)	2.5 g	Combine reagents and pH to 7.5. Dispense 5 ml per tube and autoclave 121°C for 15 min.	Use a heavy inoculum and incubate 48 h at 25 or 37°C, depending on growth requirements of the bacterium. To read, add six or seven drops of Kovács' reagent and shake tube. A positive result is development of a cherry red colour in the upper reagent layer on top of the broth medium. No colour development indicates a negative result. For organisms that require salt, add 0.5 ml of 20% NaCl to 5 ml of medium. Even though a salt-requiring organism may show growth in the tube at 48 h, if NaCl is not present a false-negative reaction can still occur. *See photographic section.*
NaCl	1.25 g		
Distilled water	250 ml		

Results between API 20E and tryptone broth may differ. Add NaCl to a final concentration of 2% for marine isolates (i.e. add 500 µl of 2% NaCl to 5 ml of test medium).

Indoxyl acetate hydrolysis (Mills and Gherna, 1987)

Indoxyl acetate discs may be prepared by making a 10% (wt/vol) solution of indoxyl acetate in acetone. Add 25 µl to a sterile blank disc, 0.64 cm in diameter. Discs are available from Oxoid. Allow discs to dry and store at 4°C in an amber-coloured bottle. Shelf life of discs is approximately 6 months. A 25-µl volume compared with 50 µl was found to be more discriminatory for *Campylobacter* identification (On and Holmes, 1992). Discs are also available commercially from companies such as Remel.

Test: Bacterial growth from an agar plate is scraped onto an indoxyl acetate disc and a drop of distilled water added. A positive result is the development of a dark blue colour within 5–10 min. Alternatively, colonies from an agar plate can be emulsified in 0.3 ml of distilled water. Add an indoxyl acetate disc. A positive result is the development of a dark blue colour within 10–15 min (Mills and Gherna, 1987).

Kovács' indole reagent (Kovács' 1928; Cowan and Steel, 1970; MacFaddin, 1985)

Reagent	Amount	Preparation of media	Description of test
p-dimethylaminobenzaldehyde	5 g	Handle *p*-dimethylaminobenzaldehyde and HCl with caution. Dissolve *p*-dimethylaminobenzaldehyde by warming in a waterbath at 50°C. Only leave long enough to dissolve, otherwise it turns pink then dark brown. Cool and slowly add the HCl to the aldehyde-alcohol mixture. The reagent should be light yellow to light brown in colour. Store reagent in a brown glass stoppered bottle at 4°C.	See under 'Indole'.
Iso-amyl alcohol	75 ml		
Concentrated hydrochloric acid	25 ml		

KOH (Tests for presence of Flexirubin pigments; Reichenbach *et al.*, 1974, 1981)

Reagent	Amount	Preparation of media	Description of test
KOH	20%		Place one or two drops of KOH onto a few isolated young colonies on AO medium. Detection of a brown or red pigment is positive. Detects flexirubin pigment (Reichenbach *et al.*, 1981). Some reports indicate a purple colour as positive (Mudarris and Austin, 1989).

Pigment production may be influenced by factors in the growth medium such as yeast extract, which promotes pigment production through stimulated growth. Also, pH can affect pigment, where a lower pH may reduce pigment production (Reichenbach *et al.*, 1974) (Reichenbach states 5% KOH in methanol).

MacConkey agar

Reagent	Amount	Preparation of media	Description of test
MacConkey agar (Oxoid)	26.0 g	Suspend the agar in the distilled water and autoclave 121°C for 20 minutes. Cool to 50°C and pour into Petri dishes. Store plates at 4°C.	Organisms that ferment lactose are seen as dark pink colonies. Non-lactose fermenting colonies are seen as yellow-clear colonies. *Salmonella* spp. and *Edwardsiella tarda* will appear as non-lactose fermenting colonies.
Distilled water	500 ml		

Marine oxidative fermentative medium (MOF) (Leifson, 1963; See under Oxidative fermentative marine)

McFarland (nephelometer) standards preparation

McFarland Tube No.	Sulfuric acid 1% aqueous solution (ml)	Barium chloride, 1% aqueous solution (ml)	Corresponding density of bacteria – 10^6	International Units (IU) of opacity
1	9.9	0.1	300	3
5	9.5	0.5	1500	15

Method: Mix the 1% sulfuric acid and the 1% barium chloride according to the table. Use clean, clear glass tubes. Plug or cap the tubes and seal with parafilm. Store tubes upright.
Use: When comparing bacterial density against the tubes, ensure that the bacterial suspension is in a similar-sized glass tube to the standard. It may be best to prepare the standards in a tube or bottle that the bacterial suspensions will be normally prepared in.
Sensitivity to vibriostatic agent (0/129 discs), prepare lawn inoculum at McFarland tube 1.
API 20E, API 20NE, API 50CH. Inoculate with inoculum prepared to turbidity of McFarland tube 1.
API Zyme. Inoculate with inoculum prepared to turbidity of McFarland tube 5.

Taken from Difco manual, referenced in *Gradwohl's Clinical Laboratory Methods and Diagnosis*. A.C. Sonnenwirth and L. Jarett (eds). C.V. Mosby Company, 1980, p. 1363.

Motility – Hanging drop method (Recommended in West and Colwell, 1984)

Place a drop of suspension from an early stationary phase broth culture onto a coverslip. Place small dollops of vaseline or plasticine at the corners. Invert coverslip and place onto a glass slide so that the drop of culture hangs free of the glass slide. Examine under phase contrast, or bright-field with the substage condenser aperture closed down to give greater contrast. Weakly motile strains may not be detected in semi-solid motility medium. For some organisms, motility is temperature-dependent; therefore, ensure growth medium is incubated at the appropriate temperature.

MRVP test medium (Clark and Lubs, 1915; Cowan and Steel, 1970. Used by West and Colwell, 1984)

Reagent	Amount	Preparation of media	Description of test
MRVP medium (Oxoid)	3.75 g	Dissolve reagent in distilled water and autoclave 121°C for 15 min. Dispense 5 ml per tube.	Add NaCl for marine organisms. Must incubate for 2–3 days before testing. VP reaction: red colouration may develop up to 18 h after the addition of reagents.
Distilled water	250 ml		

For Voges–Proskaüer test, see under VP.

Methyl Red reagent

Reagent	Amount	Preparation of media	Description of test
Methyl Red	40 mg	Add Methyl Red reagent to ethanol and heat to 50°C in a waterbath to assist dissolution. Make up to 100 ml total with distilled water. More ethanol can be added if a precipitate forms.	After incubation at optimal temperature for at least 2 days, add three to four drops of methyl Red reagent. Persistence of a red colour indicates a positive result. *See photographic section.*
Ethanol 95%	40 ml		

Mycoplasma characterization tests (Aluotto *et al.*, 1970 with modifications)

Reagent	Amount	Preparation of media	Description of test
Heart infusion broth – stock		Add dehydrated media to distilled water. Using 5 N NaOH, adjust pH to 7.6.	Inoculate a test and substrate control tube with 1 ml of a 24-h culture grown in broth. Read tests daily for up to 2 weeks. A pH drop of 0.5 unit or more in the glucose tube compared with the appropriate substrate control tube is a positive reaction. A rise of 0.5 pH unit in the arginine or urea tubes compared with the appropriate substrate tubes indicates a positive reaction. The pH values are read by comparison with a set of standards ranging in pH from 5.6 to 8.4.
Heart infusion broth	25 g		
Distilled water	1000 ml		
Basal medium		Prepare yeast extract stock, and phenol red stocks. To 74 ml of HIB add horse serum, yeast extract, phenol red and test substrate. Using 5 N NaOH of 5 N HCl, adjust to appropriate pH depending on test substrate. Prepare substrate control tubes, that is, basal medium without test substrate. Filter-sterilize and dispense in 5 ml amounts to sterile bijou or other screw-capped bottle.	
Heart infusion broth (HIB) (Difco) stock	74 ml		
Horse serum (Heated 56°C for 30 min)	10 ml		
Yeast extract (Oxoid) stock solution of 10% w/v. Filter-sterilize	5 ml		
Phenol red (0.5% w/v) autoclaved	1 ml		
Test substrate			
Glucose (10% w/v)	10 ml	pH 7.6	
Arginine (0.2%)	10 ml	pH 7.0	
Urea (10% w/v)	10 ml	pH 7.0	
OF test		Dispense 1 ml of sterile medium into 15×45 mm tubes.	Inoculate duplicate test and substrate control tubes with 0.5 ml of an overnight broth culture. Overlay one test and one control tube with sterile paraffin oil. Incubate at appropriate temperature and examine daily for up to 2 weeks for pH change compared with pH standards. Fermentative organisms produce acid in both the aerobic and anaerobic tubes, whereas oxidative organisms produce acid in the aerobic tube only.
Basal medium	90 ml		
Glucose (10% w/v)	10 ml, pH 7.6		

Reagent	Amount	Preparation of media	Description of test
Tetrazolium (TTC) reduction		Aseptically add reagents to HIA stock and pour into 5 cm Petri dishes. TTC can also be tested in broth rather than on agar plates.	Inoculate duplicate plates with agar plugs taken from dense growth on culture plates. Agar plugs can be excised using a scalpel blade, alcohol flame-sterilized spatula or open-ended 'hockey stick' made from a bent Pasteur pipette. Slide the inverted agar plug over the surface of the TTC plates. Incubate one plate aerobically and the other anaerobically for up to 2 weeks. Development of a pink-red colour in 3–4 days is positive.
HIA stock	74 ml		
Horse serum	20 ml		
Yeast extract stock	5 ml		
TTC (2% w/v)	1 ml		
Phosphatase		Aseptically add reagents to HIA stock and pour into 5 cm Petri dishes. Phosphatase can also be tested in broth rather than on agar plates.	Inoculate three plates with a drop from a 24 h broth culture. Incubate plates at the appropriate temperature. On days 3, 7 and 14, flood plate with 5 N NaOH. A positive reaction is the appearance of a red colour.
HIA	74 ml		
Horse serum	20 ml		
Yeast extract stock	5 ml		
Phenolphthalein diphosphate (sodium salt) (1% w/v)	1 ml		
Film and Spots 1		Aseptically remove egg yolks and homogenize with an equal amount of sterile distilled water. Add the homogenate to HIA at a final concentration of 10%. Pour into 5 mm Petri dishes.	Inoculate the egg yolk medium with test organism and incubate in CO_2 at 37°C for up to 14 days. Examine macroscopically using reflected light. An iridescent or pearly film seen over areas of heavy growth indicates a positive reaction.
HIA	90 ml		
Egg yolk	10 ml		
Film and Spots 2		An alternative method is to use 20% horse serum in the medium.	Inoculate plates with test organism and incubate in CO_2 at 37°C for up to 14 days. The development of small spots in the medium indicates a positive reaction.
HIA	80 ml		
Horse serum	20 ml		
Catalase			Flood a plate containing 24–48 h growth with 30% hydrogen peroxide. Production of bubbles indicates a positive reaction.
HIA plate			
See also Digitonin and Dienes stain			

This is the original medium proposed by Aluotto *et al.* (1970). Bacto heart infusion broth (Difco) contains 500 g beef heart infusion, 10 g bacto-tryptone, 5 g sodium chloride. An alternative is to use Difco PPLO broth with mycoplasma supplement, to which the phenol red and test substrate is added. Some Mycoplasmas may be inhibited by 1% arginine originally proposed by Aluotto, therefore 0.2% is recommended. TTC = 2,3,5-triphenyltetrazolium chloride.

pH standards: Add phenol red to basal medium and dispense 3 ml lots into 5ml tubes. Adjust pH in tubes to give a pH range of 5.6–8.4.

Control organisms:

Glucose positive=*M. bovirhinis* ATCC 19884, and negative=*M. arthritidis* ATCC 19611.

Arginine hydrolysis positive=*M. arthritidis* ATCC 19611, and negative=*M. bovirhinis* ATCC 19884.

Urea positive=T-strain mycoplasma, and negative=*M. arthritidis* ATCC 19611.

OF test organisms: Oxidative=*M. pneumoniae* ATCC 15531, Fermentative=*M. bovirhinis* ATCC 19884.

TTC control organisms: Positive aerobic and anaerobic=*M. bovirhinis* ATTCC 19884. Negative for both is *M. arthritidis* ATCC 19611.

Phosphatase control organisms: Positive=*M. arthritidis* ATCC 19611. Negative=*M. bovirhinis* ATCC 19884.

Film and Spots control organisms: Positive= *M. gallinarum* ATCC 19708. Negative=*M. arthritidis* ATCC 19611.

Sterol requirement: Inoculate into serum-free basal medium, and basal medium with 5 mg bovine serum albumin per ml, 0.01 mg of palmitic acid per ml, and different concentrations of cholesterol (1.0, 5.0, 10.0, 20 mg/ml). Growth is determined by sedimenting the mycoplasma and testing the medium for protein content by the biuret method. Paper disc inhibition method, and use same formatting as *Sterol requirement*. Use a 1.5% w/v solution of digitonin (Sigma), and 5, 10, 20% w/v solutions of sodium polyanetholesulfonate (Koch-Light Labs, UK). The widths of the zones of growth inhibition are measured. Zones of inhibition are 5–10 mm wide.

Ninhydrin reagent (Used for detection of hippurate hydrolysis [from Cowan and Steel, 1970]. See under Hippurate hydrolysis)

Nitrate broth (Quoted in Crosby, 1967; Cowan and Steel, 1970; West and Colwell, 1984)

Reagent	Amount	Preparation of media	Description of test
KNO_3	0.25 g	Add reagents to distilled water and autoclave at 121°C for 15 min. Dispense 5 ml into tubes.	Inoculate media with organisms and incubate at the appropriate temperature. For organisms that require salt, add 0.5 ml of 20% NaCl to 5 ml of test media.
Nutrient broth (Oxoid)	3.25 g		
Distilled water	250 ml		

Reagent	Amount	Preparation of media	Description of test
Nitrate reagent A		Add sulfanilic acid to distilled water, then dissolve in the acetic acid.	After incubation for 24 or 48 h, add five drops of nitrate reagent A, and five drops of nitrate reagent B to nitrate broth. Appearance of a red colour is indicative of a positive reaction. Tubes that do not show a red colour should be tested with zinc dust (match head amount). Development of a red colour indicates a true negative result, whereas no colour indicates that nitrate is absent, indicating it was reduced by the organism to nitrite, which in turn was reduced – therefore, a positive result.
Sulfanilic acid	1.28 g		
Distilled water	110 ml		
Acetic acid	50 ml		
Nitrate reagent B		Handle with caution and use mask and gloves. Add dimethyl-α-naphthylamine to distilled water. Then add acetic acid and dissolve by heating to 50°C in a waterbath.	
Dimethyl-α-naphthylamine	0.96 ml		
Distilled water	110 ml		
Acetic acid	50 ml		

***Nocardia* spp. growth medium** (See Glucose yeast extract agar)

0/129. See under Vibrio discs for description of test.
0/129 = 2,4-diamino-6,7-diisopropyl pteridine (0/129) phosphate. Discs are available from Oxoid and other companies. Available at concentrations of 10 μg and 150 μg.

ONPG – o-nitrophenyl-β-D-galactosidase (Cowan and Steel (1970) uses the method of Lowe, 1962)

Reagent	Amount	Preparation of media	Description of test
Oxoid peptone water	0.9 g	Dissolve the peptone water in distilled water and autoclave at 121°C for 20 min.	Inoculate a tube of ONPG broth and incubate at the appropriate temperature for 24–48 h. A yellow colour indicates a positive result and indicates the presence of the enzyme β-galactosidase.
Distilled water	60 ml		
ONPG	0.15 g	Dissolve the ONPG in the phosphate solution at pH 7.5 and filter-sterilize. Aseptically add to the peptone water, and dispense 2.5 ml into sterile tubes. Store at 4°C and keep protected from the light.	
Na_2HPO_4	0.035 g		
Distilled water	25 ml		

ONPG = o-nitrophenyl-β-D-galactopyranoside. ONPG discs are available from Oxoid. Discs are recommended when testing *Flavobacteriaceae*.

Oxidase test (Kovacs, 1956. Used in Cowan and Steel, 1970; West and Colwell, 1984)

Reagent	Amount	Preparation of media	Description of test
Tetramethyl-*p*-phenylenediamine	1% aqueous solution	Prepare a 1% solution of oxidase reagent in water. Use immediately. The oxidase reagent must be stored in a stoppered dark glass bottle, protected from the light and stored in the fridge. It should not be used if it has become deep blue in colour.	Place a piece of filter paper in an empty Petri dish. Dampen with freshly prepared oxidase reagent. Using a platinum loop or wooden orange stick or toothpick, smear a streak of bacterial growth across the paper. Appearance of a deep purple colour in 10–30 s indicates a positive reaction. A purple colour after 2 min may be a false positive. Discard once the filter paper becomes a blue colour. Do not test for oxidase reaction from media that contains carbohydrate such as TCBS and MCA (Jones, 1981).

Commercial oxidase strips are also available and are recommended because they offer a standardized test method. The BD method has been shown to be the most reliable.

Oxidative-fermentative (O-F) media (Media of Hugh and Leifson, 1953. For marine organisms, use MOF medium of Leifson, 1963)

Reagent	Amount	Preparation of media	Description of test
Difco bacto tryptone	0.8 g	Mix all ingredients, except horse serum, in distilled water and boil for 1 min to dissolve. Cool slightly and pH to 7.1 using 10 N NaOH. Dispense 5 ml to tubes and autoclave at 121°C for 15 min. Cool to 50°C and aseptically add two drops of sterile horse serum per tube. Store at 4°C.	Fermentative organisms produce an acid reaction (yellow colour) in both the sealed and open tubes. Oxidative organisms produce acid at the surface of the open tube. Growth is seen at the surface of the medium with little or no growth at the bottom of the open tube, and no growth in the sealed tube. Organisms that fail to either ferment or oxidize glucose may produce an alkaline reaction (purple) at the surface of the open tube.
NaCl	2.0 g		
K_2HPO_4	0.12 g		
Glucose	4.0 g		
Oxoid agar No. 1	0.8 g		
Indicator (see below)	24 ml		
Sterile horse serum not inactivated	2 drops per 5 ml tube		
Distilled water	400 ml		

Notes from Hugh and Leifson (1953): Metabolism of carbohydrate involves two different mechanisms. One mechanism is called fermentation and occurs without oxygen and is therefore an anaerobic process. The other mechanism is called oxidation and occurs in the presence of oxygen and is therefore an aerobic process. The oxidative-fermentative (OF) medium from Oxoid is based on Hugh and Leifson (1953).

OF indicator

Reagent	Amount	Preparation of media	Description of test
Cresol red	0.15 g	Dissolve the bromothymol blue in water. 0.3 ml of a 1% solution is added to each 100 ml of medium.	
Bromothymol blue	0.10 g		
NaOH	0.20 g		
Distilled water	500 ml		

Marine oxidative fermentative medium for marine organisms (MOF) (Used for carbohydrate metabolism for marine organisms; Leifson, 1963)

Reagent	Amount	Preparation of media	Description of test
Casitone (Difco)	0.1 g	Dissolve ingredients in distilled water and adjust pH to 7.5. Autoclave at 121°C for 15 min. Autoclave artificial seawater separately and add to ingredients in distilled water. Aseptically add 10 ml filter-sterilized glucose per 100 ml. Aseptically dispense 5 ml to 10 ml tubes.	Inoculate two tubes of medium with organism and overlay the medium in one tube with a layer of sterile paraffin oil to approximately 1 cm depth, or 0.5 ml. Incubate tubes at appropriate temperature. Organisms that ferment carbohydrate acidify the medium in both tubes, whereas aerobic organisms acidify the medium in the 'open' tube only. For fermenting organisms record 'F' on results sheet. For aerobes or oxidizing organisms, record 'O' on results sheet.
Yeast extract	0.01		
Ammonium sulfate	0.05 g		
Tris buffer	0.05		
Agar	0.3 g		
Distilled water	50 ml		
Artificial seawater	50 ml		
Phenol red (0.1% stock)	1 ml	Phenol red: Used at a 0.001% final concentration. (Prepare a 0.1% solution and use 1 ml of this per 100 ml of medium).	
Glucose	1.0%	Prepare a 10% solution and filter-sterilize.	

The indicator bromothymol blue that is used in the conventional OF medium may be toxic to some marine organisms.

Paraffin oil (Used for overlay in decarboxylase tests and OF test)

Reagent	Amount	Preparation of media	Description of test
Paraffin oil	As required	Dispense into 100 ml bottles or any appropriate-sized bottle. Hot-air sterilize at 160°C for 60 min. Do not autoclave because the oil goes cloudy.	Used as overlay in ADH, LDC, ODC and OF tests. For ease of use, oil can be dispensed into a 1-l Schott bottle and capped with a Socorex 2-ml dispenser unit. Dispense 0.5 ml into 10 ml media tubes.

Physiological saline (See saline)

Rogosa agar (RA). Oxoid manual (Selective medium for *Lactobacilli*; Rogosa *et al*., 1951)

Reagent	Amount	Preparation of media	Description of test
Rogosa agar (Oxoid)	82 g	Suspend Rogosa agar in distilled water and bring to the boil to dissolve. Add glacial acetic acid and mix thoroughly. Heat to 90–100°C for 2–3 min with frequent agitation. Distribute into sterile tubes or Petri dishes. Do not autoclave.	May assist to differentiate *Lactobacilli* from other Gram-positive bacteria such as *Carnobacterium*, *Arcanobacterium* and *Vagococcus* spp.
Distilled water	1000 ml		
Glacial acetic acid	1.32 ml		

Ingredients: g/l: tryptone (10.0), yeast extract (5.0), glucose (20.0), sorbitan mono-oleate (1.0), potassium dihydrogen phosphate (6.0), ammonium citrate (2.0), sodium acetate (25.0), magnesium sulfate (0.575), manganese sulfate (0.12), ferric sulfate (0.034), agar (20.0).

Saline: physiological saline

Reagent	Amount	Preparation of media	Description of test
NaCl	0.85 g	Add salt to distilled water. Dispense 10 ml into McCartney bottles. Autoclave at 121°C for 15 min.	May be used as an inoculum for commercial kit tests and 'Biochem Set'.
Distilled water	100 ml		

Salt – 20% stock solution

Reagent	Amount	Preparation of media	Description of test
NaCl	100 g	Add NaCl to distilled water. Autoclave 121°C for 15 min.	For ease of use, 20% NaCl can be dispensed into a 1-l Schott bottle and capped with a Socorex 2-ml dispenser unit with 0.5 ml graduations. Dispense 0.5 ml per 5 ml of media to produce final NaCl concentration of 2% in test.
Distilled water	400 ml		

In most cases, the carbohydrate media tubes do not require NaCl in addition to the 2% NaCl inoculum fluid. However, liquid tube media such as aesculin, ADC, ODC, LDC, the decarboxylase control, MRVP, indole (TW) and nitrate do require that 500 µl of a 20% NaCl solution is added to the 5 ml of media.

Salt tolerance

0%, 3% NaCl: See Gelatin/NaCl split plate. Using a wire loop, spot inoculate equal amounts of the bacterium to the 0% NaCl side and the 3% NaCl side. After 1–2 days' incubation at the appropriate temperature, examine for areas of clearing or opacity. *See photographic section.*

10% NaCl: For tolerance to 10% salt, dispense equal volumes of TSB and 20% NaCl stock solution into a sterile bijou or 10 ml tube. Add bacteria to turbidity of 0.5 or tube 1 McFarland standard. Incubate at appropriate temperature for 24–48 h. Look for obvious increase in the growth of organism as seen by cloudiness in the medium.

Survival at 50°C (Used for *Nocardia* spp.; Gordon *et al.*, 1974)

Inoculate slants of glucose yeast extract agar, and incubate in a 50°C waterbath for 8 h. Remove from waterbath, cool quickly and incubate at 28°C for 3 weeks. Examine tubes for growth.

TDA reagent (For use in the API 20E kit. Can be purchased commercially)

Reagent	Amount	Preparation of media	Description of test
Ferric chloride	3.4 g	Dissolve the ferric chloride in 90 ml of water, and then make up to 100 ml.	Used in the API 20E kit. A positive is a brown colour. *Proteus* strains tend to be positive.
Distilled water	100 ml		

Triple sugar iron agar (TSI)

Reagent	Amount	Preparation of media	Description of test
Triple sugar iron agar (Oxoid)	9.75 g	Dissolve reagent in distilled water and pH to 7.4. Dispense in 5 ml aliquots to 10 ml tubes. Autoclave at 121°C for 15 min. Allow to cool and solidify in a slanted position to give deep butts.	Inoculate the tube using a straight wire. Stab into medium, and then zig-zag the slope. H_2S producers will turn the medium black. Fermentative organisms will acidify the TSI butt (yellow), whereas non-fermentative organisms will grow on the slant only and show either no pH change or an alkaline reaction, and rarely, an acidic reaction. For some organisms this test method may not be as sensitive as using a lead acetate paper strip. A hydrogen sulfide Biostrip (catalogue number TM343) is available from MedVet Science.
Distilled water	150 ml		

Urea (Christensen, 1946)

Part A

Reagent	Amount	Preparation of media	Description of test
Oxoid agar No 1	3.75 g	Add agar to distilled water and autoclave at 121°C for 15 min. Cool to 50°C.	Add 1% NaCl for marine organisms.
Distilled water	225 ml		

Part B

Reagent	Amount	Preparation of media	Description of test
BBL brand urea agar base	8.7 g	Add reagents to distilled water and filter-sterilize through a 0.22-µm Millipore filter. Add Part A to 25 ml of Part B. Aseptically dispense 5 ml volumes per tube. Allow to solidify in a sloped position.	Inoculate the slant heavily over the entire surface with a loopful of bacteria harvested from an agar medium. Bright pink indicates a positive reaction.
Distilled water	30 ml		

Urea slopes should be prepared with a deep butt and a short slant (Gilardi, 1983).

Vibrio discs

0/129 = 2,4-diamino-6,7-diisopropyl pteridine (0/129) phosphate, vibriostatic reagent (µg). The *Vibrio* spp. (including *Aliivibrio* spp., *Moritella* spp. and *Photobacterium* spp.) are sensitive to this compound, commonly called vibrio static agent (Shewan *et al.*, 1954). This test assists in the differentiation of *Vibrio* spp. from other Gram-negative rods, particularly *Aeromonas* spp., which are resistant to vibriostatic agent. Almost all *Vibrio* spp. are sensitive to 0/129 at 150 µg and some are sensitive to 10 µg. However, *Vibrio cholerae* 0139 has developed resistance to 0/129 at 150 µg concentration (Albert *et al.*, 1993; Islam *et al.*, 1994).

Treat this test in the same manner as 'sensitivity testing' used to test an organism as sensitive or resistant to an antimicrobial reagent. Prepare an inoculum in normal saline to the opacity of McFarland tube 1, a suspension that is just visible to the eye. Use a sterile cotton-tipped swab to lawn inoculate the plate. Use BA for freshwater organisms and MSA-B for marine organisms. Place the two discs onto the agar surface so that they are at least 4 cm apart. Invert the plate, and incubate at the appropriate temperature for 24 h. If there is insufficient growth, incubate for a further 24 h; however, normally the zones are recorded at 24 h because further growth, especially when testing *Vibrio* spp. that swarm, may show a false resistant result. Record zone sizes as 'sensitive' (S) or 'resistant' (R). For the 0/129 150 µg disc, a zone of 9 mm is susceptible for *Vibrio* spp. (Bernardet and Grimont, 1989). A zone size of 22 mm is considered sensitive for *Photobacterium damselae* ssp. *damselae* (Love *et al.*, 1981).

The discs are purchased from Oxoid or Rosco Diagnostics. The Oxoid codes for the two concentrations are DD14 for 0129 10 µg, and DD 15 for 150 µg.

Voges–Proskaüer reaction (See MRVP test; Clark and Lubs, 1915; Voges and Proskauer, 1898 as reported in Cowan and Steel, 1970. Used by West and Colwell, 1984)

Use for the detection of acetoin from glucose fermentation. The incubation time and temperature rather than the method influence the production of acetoin.

MRVP test medium (The commercial medium is based on Clark and Lubs medium, 1915)

Reagent	Amount	Preparation of media	Description of test
MRVP medium (Oxoid)	3.75 g	Dissolve reagent in distilled water and autoclave at 121°C for 15 min. Dispense 5 ml per tube.	Add NaCl for marine organisms. Must incubate for 2–3 days before testing. Add VP reagent A, and VP reagent B. Red colouration may develop up to 18 h after the addition of reagents.
Distilled water	250 ml		

MR=methyl red, VP=Voges–Proskaüer.

VP test reagents (Barritt, 1936)

Reagent	Amount	Preparation of media	Description of test
Reagent A		Dissolve α-naphthol in ethanol. Store at 4°C.	After incubation of MRVP medium for 48 h at appropriate temperature, place 1 ml into a test tube. Add 0.6 ml of reagent A, and 0.2 ml of reagent B. Examine for pink colour up to 4 h at room temperature. Tests for acetylmethylcarbinol (acetoin).
α-naphthol	5.0 g		
Absolute ethanol	100 ml		
Reagent B		Weigh KOH. Make up to 100 ml final volume with distilled water. Store at 4°C.	
Potassium hydroxide (KOH)	40.0 g		
Distilled water	Add to final 100 ml		

VP reagent I and II in the API 20E kit are also suitable. Transfer 250 µl of medium to a microfuge tube. Add 150 µl and 50 µl of reagent I and II, respectively, shaking after each addition. Read after 10–20 min.

7.3 Stains

Gram stain

Fite's method for ZN stain. The use of peanut oil instead of alcohol in the hydration and dehydration process reduced the risk of extraction of bacterial capsular lipids during the staining process (Luna, 1968).

Fite's acid-fast stain

Used to stain Mycobacteria that may stain weakly with the ZN stain. Fite's method uses peanut oil and xylene instead of alcohol in the hydration and dehydration process to minimize the bacterial cell-wall exposure to solvents which may remove lipids from the cell wall when treating and staining paraffin-embedded sections for the presence of acid-fast bacilli. Removal of lipids from the cell wall will result in loss of acid-fastness, leading to a false-negative result.

Reagent	Amount	Preparation of media	Procedure for stain
Xylene	50 ml	Prepare a solution of xylene and peanut oil 50:50 v/v. The solution is stable for 1 year. See ZN stain for preparation of carbol-fuchsin and methylene blue.	Deparaffinize sections in xylene/peanut oil (50:50) for 10 min, twice. Drain slides and blot to remove excess oil. Rinse in distilled water until oil clears. Stain in carbol-fuchsin for 30 min at room temperature. Wash in tap water. Dip or flood with acid alcohol until stain is removed. Wash in tap water. Counterstain in methylene blue for 30 s. Blot and air dry. Coverslip and examine. Acid-fast bacilli stain red against a blue background.
Peanut oil	50 ml		
Carbol-fuchsin			
Acid alcohol	1.00%		
Methylene blue			

Ziehl-Neelson acid-fast stain

Reagent	Amount	Preparation of media	Procedure for stain
Carbol-fuchsin		Add basic fuchsin to ethanol and mix to dissolve. Incubate at 37°C overnight. Use mask and gloves and add phenol to distilled water. Mix and dissolve. Add phenol mixture to basic fuchsin mixture. Filter. Store in a brown glass bottle. Stable for 1 year. Warning: carcinogenic. Phenol is toxic by absorption into the skin and inhalation.	Cut paraffin sections and place on glass slides. Remove paraffin through descending solutions of xylene. Hydrate to water. Heat carbol-fuchsin in coplin jar for 45 s in microwave at 80 power, or in oven at 60°C for 1 h. Place slides in hot carbol-fuchsin for 5 min (filter solution once a week). Wash slides in running tap water for 5 min. Flood with 1% acid alcohol until pink stain is removed. Wash slides in running tap water for 5 min. Rinse in distilled water. Place slides in working solution of methylene blue for 30 s. Rinse in water. Dehydrate, clear and coverslip. Acid-fast bacilli appear red against a blue-green background. For Mycobacteria that may be affected by xylene removing lipids from the cell wall, use Fite's modification of the ZN stain.
Basic fuchsin	2.5 g		
Distilled water	250 ml		
95% ethanol	25 ml		
Phenol crystals, melted	12.5 ml		
1% acid alcohol		Place alcohol into a measuring cylinder. In a well-ventilated area or chemical hood, add the HCl slowly to the alcohol while stirring. Store in glass bottle at room temperature. Stable for 1 year. Corrosive: HCl is corrosive	
Hydrochloric acid	10 ml		
70% alcohol	990 ml		
Methylene blue stock solution		Mix well. Filter. Store in glass or plastic bottle. Stable for 1 year. Warning: treat with caution. Animal reproductive effects.	
Methylene blue	0.7 g		
Distilled water	50 ml		
Methylene blue working solution		Pour into coplin jar or bottle. Renew after 2 months.	
Methylene blue stock	5 ml		
Distilled water	45 ml		

Further Reading and Other Information Sources

Books on Fish Diseases

Austin, B. and Austin D.A., (2012) *Bacterial Fish Pathogens: Disease of Farmed and Wild Fish*, 5th edn. Springer–Praxis Publishing, Chichester, UK.

Bergey's Manual of Systematic Bacteriology, Volumes 1 to 4 published from 2001 to 2011. Springer, London/Williams & Wilkins Co., Baltimore, Maryland.

OIE *Diagnostic Manual for Aquatic Animal Diseases*. These manuals are also available online: http://www.oie.int/en/international-standard-setting/terrestrial-manual/access-online

Woo, P.T.K. and Bruno, D.W. (eds) (2011) *Fish Diseases and Disorders*: *Viral, Bacterial and Fungal Infections*. Volume 3. CAB International, Wallingford, UK.

Books on Biochemical Identification Tests

Barrow, G.I. and Feltham, R.K.A. (1993) *Cowan and Steel's Manual for the Identification of Medical Bacteria*, 3rd edn. Cambridge University Press, Cambridge.

Cowan, S. and Steele, K. (1970) *Manual for the Identification of Medical Bacteria*. Cambridge University Press, Cambridge.

MacFaddin, J.F. (1985) *Biochemical Tests for Identification of Medical Bacteria*, 2nd edn. Williams and Wilkins, Baltimore, Maryland.

MacFaddin, J.F. (2000) *Biochemical Tests for Identification of Medical Bacteria*, 3rd edn. Williams and Wilkins, Baltimore, Maryland.

Other Biochemical Identification Schemes

Alsina, M. and Blanch, A. (1994) A set of keys for biochemical identification of environmental *Vibrio* species. *Journal of Applied Bacteriology* 76, 79–85.

Alsina, M. and Blanch, A. (1994) Improvement and update of a set of keys for biochemical identification of *Vibrio* species. *Journal of Applied Bacteriology* 77, 719–721.

Carson, J., Wagner, T., Wilson, T. and Donachie, L. (2001) Miniaturised tests for computer-assisted identification of motile *Aeromonas* species with an improved probability matrix. *Journal of Applied Microbiology* 90, 190–200.

Carson, J., Higgins, M.J., Wilson, T.K., Gudkovs, N. and Bryant, T.N. (2009) Identification of *Vibrionaceae* from Australian aquatic animals using phenotypic and PCR procedures. In: *Australia and New Zealand Standard Diagnostic Procedure: Vibrionaceae*. http://www.scahls.org.au/__data/assets/pdf_file/0007/1516516/Vibriosis.pdf

Schmidtke, L.M. and Carson, J. (1994) Characteristics of *Vagococcus salmoninarum* isolated from diseased salmonid fish. *Journal of Applied Bacteriology* 77, 229–236.

Journals

Applied and Environmental Microbiology	http://aem.asm.org
Aquaculture	http://www.elsevier.com/locate/aquaculture
Aquaculture Journal	http://www.elsevier.nl/inca/homepage/san/aqua/fp.sht
Bulletin of the European Association of Fish Pathologists	http://eafp.org/bulletin-archive
Bulletin of the Japanese Society of Scientific Fisheries (The Japanese Society of Fisheries Science, Nippon Suisan Gakkaishi)	https://www.jstage.jst.go.jp/browse/suisan/52/0/_contents
Current Microbiology	http://link.springer.de/link/service/journals/00284
Diseases of Aquatic Organisms	http://www.int-res.com/journals/dao
Fish Pathology	https://www.jstage.jst.go.jp/browse
International Journal of Systematic and Evolutionary Microbiology	http://ijs.sgmjournals.org
Journal of Applied Ichthyology	http://www.blackwell-synergy.com/Journals/issuelist.asp?journal=jai
Journal of Applied Microbiology	http://www.blackwell-synergy.com/Journals/issuelist.asp?journal=jam
Journal of Aquatic Animal Health	http://www.tandfonline.com/toc/uahh20/current#.UtwGzCvFJEg
Journal of Clinical Microbiology	http://jcm.asm.org
Journal of Fish Diseases	http://onlinelibrary.wiley.com/journal/10.1111/(ISSN)1365-2761
Journal of General and Applied Microbiology	http://www.journalarchive.jst.go.jp/english/jnllist_en.php
Veterinary Microbiology	http://www.elsevier.nl/locate/vetmic

Culture Collections

ACAM: Australian Collection of Antarctic Microorganisms, Antarctic CRC, University of Tasmania, Hobart, Australia.

AHLDA: Animal Health Laboratories, Department of Agriculture. 3 Baron-Hay Court, South Perth, Western Australia 6151.

ATCC: American Type Culture Collection. Corporate: ATCC, 10801 University Boulevard, Manassas, VA 20110-2209, USA. Products & Services Orders: ATCC, PO Box 1549, Manassas, VA 20108-1549, USA. http://www.atcc.org

CCUG: Culture Collection, University of Göteborg, Department of Clinical Bacteriology, Institute of Clinical Bacteriology, Immunology, and Virology, Guldhedsgatn 10A s-413, 46 Göteborg, Sweden.

CDC: Center for Disease Control, 1600 Clifton Rd, Atlanta, GA 30333, USA.

CECT: Coleccion Espanola de Cultivos Tipo, Universidad de Valencia, Burjassot, Spain.

CIP: Collection de l'Institut Pasteur, Institut Pasteur, 28 Rue du Docteur Roux, 75724 Paris Cedex 15, France.

CNCM: Collection Nationale de Culture de Microorganismes, Institut Pasteur, Paris, France.

DSMZ (DSM): Deutsche Sammlung von Mikroorganismen und Zellkulturen GmbH, Mascheroder Weg 1B, D-38124, Braunschweig, Germany. http://www.dsmz.de/dsmzhome.htm

Ecole Nationale Vétérinaire de Toulouse, 23 chemin des Capelles, F-31076 Toulouse cedex 03, France. http://www.bacterio.cict.fr/collections.html

GDOU: Guangdong Ocean University, China.
IAM: Institute of Molecular and Cellular Biosciences (formerly Institute of Applied Microbiology, Culture Collection – IAMCC), The University of Tokyo, Yayoi, Bunkyo-Ku, Tokyo, Japan.
IMCAS: Institute of Microbiology of Chinese Academy of Sciences, China.
KMM: Collection of Marine Microorganisms, Pacific Institute of Bioorganic Chemistry, Vladivostok, Russia.
NCFB: National Collection of Food Bacteria (previously named NCDO). Transferred from the IFR (Institute of Food Research) Reading to National Collections of Industrial, Food and Marine Bacteria, 23 Machar Drive, Aberdeen AB24 3RY, Scotland.
NCIMB: National Collection of Industrial and Marine Bacteria, National Collections of Industrial, Food and Marine Bacteria, 23 Machar Drive, Aberdeen AB24 3RY, Scotland. http://www.ncimb.co.ukncimb.htm
NCTC: National Collection of Type Cultures, Central Public Health Laboratory, Colindale Avenue, London NW9 5HT, UK. http://www.phls.co.uk
RVAU: Royal Veterinary and Agricultural University, Copenhagen, Denmark.
SCSIO: South China Sea Institute of Oceanology, Chinese Academy of Science, China.
UB: University of Barcelona, Barcelona, Spain.

Taxonomy and Nomenclature

Alga

http://www.algaebase.org/about

Bacteria

Bacterial nomenclature is continually updated at the following websites:
DSMZ (DSM): Deutsche Sammlung von Mikroorganismen und Zellkulturen GmbH, Mascheroder Weg 1B, D-38124, Braunschweig, Germany. http://www.dsmz.de/dsmzhome.htm
List of Bacterial Names with Standing in Nomenclature: http://www.bacterio.cict.fr
NCBI: http://www.ncbi.nlm.nih.gov/Taxonomy/taxonomyhome.html

Vibrio

The association of Vibrio Biologists, AviB, was formed in 2005 to promote collaboration on research into *Vibrio* species and the dissemination of information through the website and biannual meetings. http://www.vibriobiology.net.

Fungi, taxonomy and nomenclature

Mycobank. http://www.mycobank.org

Fish Disease Websites

FishDisease.net website is for aquatic animal health professionals and contains information on different diseases, an image library on parasites, notification of conferences, jobs and contacts. http://www.fishdisease.net
World Aquatic Veterinary Medical Association website offers information on a variety of topics of interest to aquatic veterinarians. http://www.wavma.org

FishBase

Accepted and valid scientific and common names of fish species; updated regularly. http://www.fishbase.org/home.htm and http://www.fishbase.org/search.html

Catalogue of Fishes

California Academy of Sciences website for names of fish species. http://research.calacademy.org/redirect?url=http://researcharchive.calacademy.org/research/Ichthyology/catalog/fishcatmain.asp

Fish Veterinary Society publishes the *Fish Veterinary Journal*. http://www.fishvetsociety.org.uk/publications.php

World Register of Marine Species

A database containing over 200,000 species; includes species names and synonyms. http://www.marinespecies.org/index.php

Common Name and Scientific Name of Aquatic Animals

Common name	Scientific name
Abalone	*Haliotis discus hannai*
Adriatic sturgeon	*Acipenser naccarii*
African or Black-footed penguin	*Spheniscus demersus*
Almorha loach	*Botia almorhae*
Amazon freshwater dolphin	*Inia geoffrensis*
Amberjack	*Seriola dumerili*
American alligator	*Alligator mississippiensis*
American crayfish	*Orconectes limosus, Pacifastacus leniusculus, Procambarus clarkii*
American eel	*Anguilla rostrata*
American plaice	*Hippoglossoides platessoides*
Angelfish, freshwater ornamental fish	*Pterophyllum scalare*
Antarctic fur seal	*Arctocephalus gazella*
Arctic char	*Salvelinus alpinus* L.
Atlantic bottlenose dolphin	*Tursiops truncatus*
Atlantic cod	*Gadus morhua* L.
Atlantic croaker	*Micropogonias* (*Micropogon*) *undulatus*
Atlantic menhaden	*Brevoortia tyrannus* Latrobe
Atlantic salmon	*Salmo salar* L.
Atlantic spotted dolphin	*Stenella frontalis*
Atlantic tomcod	*Microgadus* (*Gadus microgadus*) *tomcod*
Atlantic walrus	*Odobenus rosmarus rosmarus*
Atlantic white-sided dolphin	*Lagenorhynchus acutus*
Atlantic wolffish	*Anarhichas lupus*
Australian fur seal	*Arctocephalus pusillus doriferus*
Australian herring	*Arripis georgianus*
Australian oyster	*Saccostrea commercialis*
Australian sea lion	*Neophoca cinerea*
Ayu	*Plecoglossus altivelis altivelis* (Temminck and Schlegel)

Continued

Continued.

Common name	Scientific name
Ballan wrasse	*Labrus berggylta*
Balloon molly	*Poecilia*
Banana prawn	*Penaeus merguiensis*
Banded coral shrimp	*Stenopus hispidus*
Barbour's seahorse	*Hippocampus barbouri*
Barramundi	*Lates calcarifer* (Bloch)
Beaked whale	*Mesoplodon bidens*
Beluga whale	*Delphinapterus leucas*
Bighead carp	*Hypophthalmichthys* (*Aristichthys*) *nobilis*
Big-scale redfin	*Tribolodon hakonensis* (Gunther, 1877)
Black acara	*Cichlasoma bimaculatum*
Black-footed or African penguin	*Spheniscus demersus*
Black mullet or sea mullet	*Mugil cephalus*
Black scraper	*Thamnaconus* (*Navodon*) *modestus*
Black skirted tetra	*Hyphessobrycon* species
Blacksmith (Damselfish)	*Chromis punctipinnis*
Blacktip shark	*Carcharhinus limbatus*
Blenny, viviparous eelpout	*Zoarces viviparus*
Blue discus	*Symphysodon aequifasciatus*
Blue fish	*Pomatomus saltatrix*
Blue mackerel	*Scomber australasicus*
Blue manna crab	*Portunus pelagicus*
Blue mussel	*Mytilus edulis*
Blue shrimp	*Litopenaeus stylirostris*
Bony bream	*Nematolosa come* (Richardson)
Borneo or largescale mullet	*Chelon* (*Liza*) *macrolepis*
Bottlenosed dolphin	*Tursiops* (*gephyreus*) *truncatus*
Bottlenose whale	*Hyperoodon ampullatus*
Bowhead whale	*Balaena mysticetus*
Bream, common bream	*Abramis brama*
Brill	*Scophthalmus rhombus* (L.)
Brine shrimp	*Artemia salina*
Brook trout	*Salvelinus fontinalis* (Mitchill)
Brown bullhead	*Ameiurus* (*Ictalurus*) *nebulosus* (Lesueur)
Brown mussel	*Perna perna*
Brown shark	*Carcharhinus plumbeus*
Brown trout	*Salmo trutta trutta*
Burnett salmon	*Polydactylus* (*sheridani*) *macrochir* (Günther, 1987)
Californian sea lion	*Zalophus californianus*
Canadian shrimp or cleaner shrimp	*Lysmata amboiensis*
Carp	*Cyprinus carpio* Linnaeus
Carpet shell clam	*Venerupis* (*Ruditapes*) *decussata*
Catfish	*Clarias batrachus* L.
Chameleon cichlid	*Australoheros* (*Cichlasoma*) *facetus* (Jenyns)
Channel catfish	*Ictalurus punctatus* (Rafinesque)
Chinook salmon	*Oncorhynchus tshawytscha*
Chub	*Leuciscus cephalus*
Chub mackerel	*Scomber japonicus*
Chum salmon	*Oncorhynchus keta* (Walbaum)
Cichlid	*Oreochromis mossambicus*
Clam	*Venerupis* (*Ruditapes*) *philippinarum*, *Venerupis* (*pullastra*) *senegalensis*
Cleaner shrimp	*Lysmata amboiensis*

Continued

Continued.

Common name	Scientific name
Cod	*Gadus morhua*
Coho salmon	*Oncorhynchus kisutch*
Common carp	*Cyprinus carpio* L.
Common dolphin	*Delphinus delphis*
Common seal	*Phoca vitulina*
Common snook	*Centropomus undecimalis*
Common or Atlantic wolffish	*Anarhichas lupus*
Coral prawn	*Metapenaeopsis* species
Crucian carp	*Carassius carassius*
Cultured flounder	*Paralichthys olivaceus*
Cutthroat trout	*Oncorhynchus clarkii clarkii*
Cuttlefish	*Sepia officinalis*
Dab	*Limanda limanda*
Dace	*Leuciscus leuciscus* L.
Danube sturgeon	*Acipenser gueldenstaedtii*
Discus fish	*Symphysodon discus, S. aequifasciatus*
Dolphin (Amazon freshwater)	*Inia geoffrensis*
Dolphin	*Tursiops (gephyreus) truncatus*
Dolphin fish	*Coryphaena hippurus* L.
Dusky grouper	*Epinephelus marginatus*
Eastern freshwater cod	*Maccullochella ikei*
Eastern mosquitofish	*Gambusia holbrooki*
Eastern oyster	*Crassostrea virginica*
Eastern painted turtle	*Chrysemys picta picta*
Eel	*Anguilla japonica, A. reinhardtii*
Elephant seal	*Mirounga leonina*
European crayfish	*Astacus astacus, A. leptodactylus, A. pachypus, Austropotamobius pallipes, A. torrentium*
European eel	*Anguilla anguilla*
European flounder	*Platichthys flesus*
European otter	*Lutra lutra*
European seabass	*Dicentrarchus labrax* Serranidae
Fairy shrimps	*Anostraca* species, *Branchipus schaefferi* (Fisher), *Chirocephalus diaphanus* (Prévost), *Streptocephalus torvicornis* (Waga)
False killer whale	*Pseudorca crassidens*
Fathead minnows	*Pimephales promelas*
Fighting fish	*Betta splendens*
Finless porpoise	*Neophocaena phocaenoides*
Firemouth cichlid	*Cichlasoma meeki*
Flathead grey mullet	*Mugil cephalus* (L.)
Flat-tailed mullet	*Liza (dussumieri) subviridis* (Valenciennes)
Flounder	*Paralichthys olivaceus, P. flesus*
Four bearded rockling	*Enchelyopus cimbrius* L.
Freshwater catfish	*Pangasianodon hypophthalmus* Sauvage
Freshwater cod (Australian native fish)	*Maccullochella* sp.
Freshwater dolphin	*Inia geoffrensis*
Freshwater prawn	*Macrobrachium rosenbergii*
Fur seal	*Arctocephalus australis*
Gentoo penguin	*Pygoscelis papua*
Giant river prawn	*Macrobrachium rosenbergii*
Gilthead seabream	*Sparus aurata*
Golden shiner	*Notemigonus crysoleucas* (Mitchell)

Continued

Continued.

Common name	Scientific name
Goldfish	*Carassius auratus* L.
Goldsinny wrasse	*Ctenolabrus rupestris*
Gourami (three-spot)	*Trichopodus* (*Trichogaster*) *trichopterus*
Grass carp	*Ctenopharyngodon idella*
Grayling	*Thymallus thymallus* L.
Greasy grouper	*Epinephelus tauvina*
Great sand eel	*Hyperoplus lanceolatus* (Le Sauvage, 1824)
Greater weever	*Trachinus draco*
Greenback flounder	*Rhombosolea tapirina* Gunther
Green knife fish	*Eigemannia virescens*
Green moray eel	*Gymnothorax funebris*
Green sturgeon	*Acipenser medirostris*
Green turtle	*Chelonia mydas*
Grey seal	*Halichoerus grypus*
Grooved carpet shell clam	*Venerupis* (*Ruditapes*) *decussata*
Grouper	*Epinephelus* (*guaz*) *marginatus, E. coioides*
Gulf menhaden	*Brevoortia patronus*
Guppy	*Poecilia reticulata* (Peters)
Haddock	*Melanogrammus aeglefinus* L.
Halibut	*Hippoglossus hippoglossus* L.
Harbour porpoise	*Phocoena phocoena*
Harbour seal	*Phoca vitulina*
Hardhead sea catfish	*Ariopsis felis*
Harp seal	*Phoca* (*groenlandica*) *groenlandicus*
Hawaiian bobtail squid	*Euprymna scolopes*
Hooded seal	*Cystophora cristata*
Horse mackerel	*Trachurus trachurus*
Humboldt penguin	*Spheniscus humboldti*
Iberian or Spanish toothcarp	*Aphanius iberus*
Japanese abalone	*Haliotis discus hannai* (Ino, 1953)
Japanese eel	*Anguilla japonica*
Japanese flounder	*Paralichthys olivaceus*
Japanese or Pacific oyster	*Crassostrea gigas*
Japanese rice fish or medaka	*Oryzias latipes*
Jewel tetra	*Hyphessobrycon* (*callistus*) *eques* (Steindachner, 1882)
Johnston's freshwater crocodile	*Crocodylus johnstoni*
Killer whale	*Orcinus orca*
King prawn	*Penaeus latisulcatus*
Knife fish	*Gymnotus carapo*
Lake trout	*Salvelinus namaycush* (Walbaum, 1792)
Lamprey	*Petromyzon marinus, Lampetra tridentata*
Largemouth bass	*Micropterus salmoides*
Leatherback turtle	*Dermochelys coriacea*
Lemon shark	*Negaprion brevirostris*
Little blue penguin	*Eudyptes minor*
Little penguin	*Eudyptula minor*
Loach, pond loach	*Misgurnus anguillicaudatus* (Cantor, 1982)
Lobster	*Homarus gammarus* L.
Local or blue mussel	*Mytilus edulis*
Long-tom	*Strongylura krefftii* (*Tylosurus* [*macleayanus*] *gavialoides*) (Ogilby, 1886)
Mackerel	*Scomber scombrus*
Manila clam	*Venerupis* (*Ruditapes, Tapes*) *philippinarum*

Continued

Continued.

Common name	Scientific name
Marble trout	*Salmo marmoratus*
Marbled spinefoot, Rabbitfish	*Siganus rivulatus* (Forsskål)
Masou salmon	*Oncorhynchus masou*
Menhaden, gulf menhaden	*Brevoortia patronus*
Milkfish	*Chanos chanos*
Minke whale	*Balaenoptera acutorostrata*
Minnow	*Phoxinus phoxinus* L.
Molly	*Poecilia velifera* (Regan)
Mud crab	*Scylla serrata*
Mullet	*Mugil cephalus*
Murray cod	*Maccullochella peeli*
Mussel (Far-Eastern)	*Crenomytilus grayanus*
Mussel	*Leukoma* species (*Protothaca jedoensis*), *Mytilus edulis, Mytilus galloprovincialis*
Narrow-bellied seahorse	*Hippocampus angustus*
Narrow-clawed crayfish	*Astacus leptodactylus*
Narwahl whale	*Monodon monoceros*
Neon tetra	*Paracheirodon* (*Hyphessobrycon*) *innesi*
New Zealand fur seal	*Arctocephalus forsteri*
New Zealand mussel	*Perna canalicula*
New Zealand sea lion	*Phocarctos hookeri*
Nile tilapia	*Oreochromis niloticus*
North-east Atlantic mackerel	*Scomber scombrus*
Northern elephant seal	*Mirounga angustirostris*
Northern fur seals	*Callorhinus ursinus*
Northern pike	*Esox lucius* L.
Northern right whale dolphin	*Lissodelphis borealis*
Nurse shark	*Ginglymostoma cirratum*
Octopus	*Octopus vulgaris, O. joubini*
One-spot or white bream	*Diplodus sargus sargus*
Orange-spotted grouper	*Epinephelus coioides*
Ornamental fish, freshwater angelfish	*Pterophyllum scalare*
Ornate spiny lobster	*Panulirus ornatus*
Oscar	*Astronotus ocellatus*
Otter, European otter	*Lutra lutra*
Oyster	*Ostrea edulis*
Oyster, eastern oyster	*Crassostrea virginica*
Pacific herring	*Clupea pallasii pallasii*
Pacific lamprey	*Lampetra tridentata*
Pacific or Japanese oyster	*Crassostrea gigas*
Pacific salmon	*Oncorhynchus* species
Pacific sandlance	*Ammodytes personatus* Girard
Pacific staghorn sculpin	*Leptocottus armatus*
Pacific white shrimp	*Litopenaeus* (*Penaeus*) *vannamei*
Pacific white-sided dolphin	*Lagenorhynchus obliquidens*
Paradise fish	*Macropodus opercularis* (L.)
Pejerrey	*Odontesthes bonariensis*
Penaoid shrimp, prawn	*Litopenaeus vannamei*
Penguin	*Aptenodytes patagonicus, Eudyptes chrysocome, Pyoscelis papua, Spheniscus demersus, Spheniscus humboldti*
Perch	*Perca fluviatilis*
Pike, northern pike	*Esox lucius*
Pilchard	*Sardinops* (*neopilchardus*) *sagax*

Continued

Continued.

Common name	Scientific name
Pilot whale	*Globicephala (scammoni) macrorhynchus*
Pink salmon	*Oncorhynchus gorbuscha*
Pink snapper	*Chrysophrys (unicolor) auratus*
Pinkfish	*Lagodon rhomboides*
Pirarucu	*Arapaima gigas* Cuvier
Plaice, European plaice	*Pleuronectes platessa*
Pompano, Florida pompano	*Trachinotus carolinus* L.
Pond loach	*Misgurnus anguillicaudatus* (Cantor, 1982)
Prawn	*Penaeus monodon*
Rabbitfish, Marbled spinefoot	*Siganus rivulatus* (Forsskål)
Rainbow trout	*Oncorhynchus mykiss* (Walbaum); *Salmo gairdnerii*
Red abalone	*Haliotis rufescens*
Red algae	*Jania* sp.
Red claw crayfish	*Cherax quadricarinatus*
Red discus	*Symphysodon discus*
Red drum, Redfish	*Sciaenops ocellatus*
Red-eared slider turtle	*Trachemys (Chrysemys) scripta elegans*
Red seabream	*Chrysophrys major*
Red swamp crawfish or crayfish	*Procambarus clarkii*
Redtail catfish	*Phractocephalus hemiliopterus*
Ringed seal	*Pusa (Phoca) hispida*
Risso dolphin	*Grampus griseus*
Roach	*Rutilus rutilus* L.
Rohu	*Labeo rohita*
Rosy barb	*Puntius conchonius*
Rudd	*Scardinius erythrophthalmus*
Ruff, Australian herring, Australian salmon	*Arripis georgianus*
Sablefish	*Anoplopoma fimbria* (Pallas, 1814)
Salmon	*Oncorhynchus kisutch* (Walbaum)
Saltwater crocodile	*Crocodylus porosus*
Sand eel	*Ammodytes (lancea) tobianus* (Cuvier, 1829), *Hyperoplus lanceolatus* (Le Sauvage, 1824)
Sandlance, Pacific sandlance	*Ammodytes personatus* Girard
Sand whiting	*Sillago ciliata* Cuvier
Saratoga	*Scleropages leichardti*
Sardine or South American pilchard	*Sardinops (melanostictus) sagax*
Scallop	*Pecten maximus, Argopecten purpuratus*
Scaly mackerel fish	*Sardinella lemuru* (Bleeker, 1853)
Seabass	*Dicentrarchus labrax*
Seabream	*Sparus aurata*
Sea cucumber	*Apostichopus japonicus*
Seagull, silver gull	*Chroicocephalus (Larus) novaehollandiae*
Sea mullet	*Mugil cephalus* L.
Sea-urchin	*Paracentrotus lividus*
Shotted halibut	*Eopsetta grigorjewi*
Shubunkin	*Carassius* sp.
Siamese fighting fish	*Betta splendens* Regan
Signal crayfish	*Pacifastacus leniusculus*
Silver black porgy	*Sparidentex hasta (Acanthopagrus cuvieri)*
Silver carp	*Hypophthalmichthys molitrix* Valenciennes
Silver gull, seagull	*Chroicocephalus (Larus) novaehollandiae*
Silver molly	*Poecilia* sp.
Silver or white mullet	*Mugil curema* (Valenciennes)

Continued

Continued.

Common name	Scientific name
Silver perch	*Bidyanus bidyanus* (Mitchell)
Silver sea trout	*Cynoscion nothus*
Sind danio	*Danio devario* (Hamilton, 1822)
Small abalone	*Haliotis diversicolor supertexta*
Smallmouth bass	*Micropterus dolomieu*
Snakehead fish	*Channa striatus*
Snapper	*Lutjanus guttatus*
Snub-nose garfish	*Arrhamphus sclerolepis* (Gunther)
Sockeye salmon	*Oncorhynchus nerka* (Walbaum)
Softshell clam	*Mya arenaria*
Sole	*Solea solea*
South African abalone	*Haliotis midae*
South American fur seal	*Arctocephalus australis*
South American pilchard	*Sardinops (melanostictus) sagax*
South American sea lion	*Otaria flavescens*
Southern black bream	*Acanthopagrus butcheri*
Southern elephant seal	*Mirounga leonina*
Southern rockhopper penguin	*Eudyptes chrysocome*
Sowerby's beaked whale	*Mesoploden bidens*
Spanish or Iberian toothcarp	*Aphanius iberus*
Spanner crab	*Ranina ranina*
Speckled longfin eel	*Anguilla reinhardtii*
Spiny soft-shelled turtle	*Trionyx spinifera*
Spot croaker	*Leiostomus xanthurus*
Spotted dolphin	*Stenella (plagiodon) frontalis*
Spotted moray eel	*Gymnothorax moringa*
Spotted wolffish	*Anarhichas minor*
Squid – Longfin inshore squid	*Doryteuthis (Loligo) pealeii*
Squid – Bobtail squid	*Sepiola* sp.
Staghorn damselfish	*Amblyglyphidodon curacao* (Bloch)
Starfish	*Asterias rubens*
Steelhead trout	*Oncorhynchus mykiss* (rainbow trout); *Salmo gairdnerii*
Stingray	*Dasyatis pastinaca*
Striped bass	*Morone saxatilis (Walbaum)*
Striped catfish	*Pangasianodon hypophthalmus* Sauvage
Striped dolphin	*Stenella coeruleoalba*
Striped jack	*Pseudocaranx dentex*
Striped mullet	*Mugil cephalus*
Striped-neck musk turtle	*Sternotherus minor peltifer*
Sturgeon	*Acipenser naccarii*
Sunfish (Ocean sunfish or common mola)	*Mola mola*
Surf bream	*Acanthopagrus australis*
Tasmanian lobster, southern rock lobster, spiny rock lobster	*Jasus (novaehollandiae) edwardsii*
Tiger shark	*Galeocerdo cuvier*
Tilapia	*Tilapia mosambica*
Tilapia	*Oreochromis (Sarotherodon) aureus* (Steindachner), *Oreochromis niloticus*
Trout cod	*Maccullochella macquariensis*
Turbot	*Scophthalmus maximus* L.
Turkish crayfish	*Astacus leptodactylus*
Venustus hap or Giraffe hap (African cichlid)	*Nimbochromis venustus*

Continued

Continued.

Common name	Scientific name
Viviparous eelpout or blenny	*Zoarces viviparus*
Weakfish	*Cynoscion regalis* (Bloch & Schneider)
Weddell seals	*Leptonychotes weddellii*
Wedge sole	*Dicologoglossa cuneata* (Moreau)
Western rock lobster	*Panulirus cygnus*
White bream	*Blicca bjoerkna*
White catfish	*Ictalurus catus* L.
White clawed crayfish or Atlantic stream crayfish	*Austropotamobius pallipes*
White leg shrimp	*Litopenaeus vannamei*
White perch	*Morone americanus*
White's seahorse	*Hippocampus whitei*
White-spotted spinefoot	*Siganus canaliculatus*
White whale or beluga whale	*Delphinapterus leucas*
Whitefish	*Coregonus* sp.
Whiting	*Merlangius merlangus*
Yellow bass	*Morone mississippiensis*
Yellow catfish	*Tachysurus* [*Pelteobagrus*] *fulvidraco* Richardson
Yellow perch	*Perca flavescens*
Yellowfin seabream	*Acanthopagrus latus*
Yellow-spotted Amazon river turtle	*Podocnemis unifilis*
Yellow tail	*Seriola quinqueradiata*
Yellowtail amberjack	*Seriola lalandi*
Yellowtail clownfish	*Amphiprion clarkii* (Bennett)
Yesso or Giant Ezo scallop	*Patinopecten yessoensis*
Yoyo loach	*Botia almorhae*
Zebra danio	*Danio* (*Brachydanio*) *rerio*

Glossary of Terms

α or αH	(referring to greening of agar) Alpha haemolysis
A	Alkaline reaction
AAHRL	Australian Animal Health Reference Laboratory
AAI	Amino acid identity
ACAM	Australian Collection of Antarctic Micro-organisms
ADH	Arginine dihydrolase
Aes	Aesculin
AFB	Acid-fast bacteria
AHL	Animal Health Laboratories
AHLDA	Animal Health Laboratories Department of Agriculture Culture Collection
Ala	L-Alanine
ANA	Plate media for anaerobic bacteria
ANI	Average nucleotide identity
AO	Anacker–Ordal agar for *Flavobacteria*
AO-M	Anacker–Ordal agar with added sea salts for growth of marine *Flavobacteria*
API 50CH	API identification system from bioMérieux. Carbohydrate tests
API 20E	API identification system from bioMérieux. Fermentation and enzyme tests
API 20NE	API identification system. Utilization tests
API Rapid ID32 Strep	API system for *Streptococci* and other Gram-positives
API 20 Strep	API identification system for *Streptococci* and other Gram-positive organisms
API ZYM	API identification system. Enzyme tests
Arab	L-Arabinose
Arg	Arginine
Asp	L-Asparagine
ASW	Artificial sea water
AT	Annealing temperature
ATCC	American Type Culture Collection, Rockville, Maryland, USA
β	Beta (refers to clear zone or beta haemolysis)
BA	Blood agar
BCRC	Bioresource Collection and Research Centre

BGD	Bacterial gill disease
BHA	Bacterial haemorrhagic ascites
BHIA	Brain–heart infusion agar
BKD	Bacterial kidney disease
bp	Base pairs (in kilobases)
BRD	Brown ring disease
Brucella agar	Used for isolation of *Brucella* species
C	Number of cycles in the PCR reaction
Ca	Calcium
Cat	Catalase
CBBA	Coomassie brilliant blue agar
CCA	Cellobiose Colistin agar for isolation of *Vibrio vulnificus*
CCRC	Culture Centre for Research and Collection
CCUG	Culture Collection of the University of Göteborg, Department of Clinical Bacteriology, Göteborg, Sweden
CDA	Czapek–Dox agar
CDC	Centres for Disease Control and Prevention, Atlanta, USA
cDNA	Complementary DNA
CFPA	Selective media for isolation of *Bordetella bronchiseptica*
CFU	Colony forming unit
Cit	Citrate
CMA	Corn meal agar
CNCM	Collection Nationale de Culture de Microorganismes (Institut Pasteur, Paris, France)
CO_2	Carbon dioxide
CPC	Cellobiose Polymyxin B Colistin agar for isolation of *Vibrio vulnificus*
Cr	Coccoid rods
CR	Congo red
CSBA	Columbia sheep blood agar
cv	Curved rods
CZA	Czapek's agar
d	Days
DCA	Desoxycholate-citrate agar
DMSO	Dimethyl sulfoxide
DNA	Deoxyribonucleic acid
DNase	Test for the detection of hydrolysis of DNA
dNTPs	Deoxyribonucleoside triphosphates
dsDNA	Double-stranded DNA
DSM	Deutsche Sammlung von Mikroorganismen und Zellkulturen GmbH, Braunschweig, Germany
EAOCa	Enriched Anacker Ordal's medium with activated charcoal and aromatic compounds
EIM	*Edwardsiella ictaluri* Medium
EM	Electron microscopy
ERM	Enteric red mouth
Ery	*meso*-erythritol
ESC	Enteric septicaemia of catfish
F	Fermentative (facultative anaerobe)
FA	Furunculosis agar for *Aeromonas salmonicida*
FAO	Food and Agriculture Organization of the United Nations
FINE	Flounder necrotizing enteritis

FM	Farrell's medium for *Brucella* species
FPM	*Flavobacterium psychrophilum* medium
Fusiform	A spindle-like shape, meaning it is wider in the middle and tapers at the ends
G	Gelatin
Gal	D-Galactose
Glid	Gliding motility
Glu	Glucose
Glut	Glutamic acid
Gm	Gram stain reaction (blue colour = positive, red = negative)
GUD	Goldfish ulcer disease
h	Hours
H_2S	Hydrogen sulfide
HCl	Hydrochloric acid
HG	(DNA) hybridization group
Hip	Hippurate
HS	Hsu–Shotts agar for *Flavobacterium columnare*
HSM	For the isolation of *Tenacibaculum maritimum*
I	Inert reaction (in OF test)
ID	Identification
IGS	Intergenic spacer
Ind	Indole
Inos	Inositol
ISO	International Organization for Standardization
ISP No. 2	Yeast malt extract agar (Difco)
JOD	Juvenile oyster disease
KCl	Potassium chloride
KDM2	Medium for growth of *Renibacterium salmoninarum*
KDMC	Medium for growth of *Renibacterium salmoninarum*
KOH	Potassium hydroxide. 20%
KUMA	Kumamoto Prefectural Institute of Public Health – culture collection prefix
Lac	Lactose
LCD	Lethargic crab disease
LD_{50}	Mean lethal dose. The figure represents the number of bacteria needed to kill 50% of the inoculated fish (Reed and Muench, 1938)
LDC	Lysine decarboxylase
LJM	Lowenstein–Jensen Medium for *Mycobacteria* species
LMG	Laboratorium Microbiologie Gent Culture Collection, Universiteit Gent, Belgium
LPS	Lipopolysaccharide
Lys	Lysine
MA	Malt agar
MA 2216	Medium for growth of marine organisms. Commercially available
Malt	Maltose
Man	Mannitol
Mano	Mannose
MCA	MacConkey agar
MEA	Malt extract agar
Mg	Magnesium
Middlebrook's media	For the isolation of *Mycobacterium* species

min	Minute
MOF	Marine oxidative-fermentative medium
Monophyletic	A monophyletic taxon is a group of organisms descended from a single ancestor
Mot	Motility
MR	Methyl red
MRVP	Methyl red Voges–Proskaüer
MSA-B	Marine salt agar – blood: medium for growth of marine organisms
N	Negative
NA	Nutrient agar
NAC	Nutrient agar supplemented with activated charcoal
NaCl	Salt
NaCl 0/3	Plate media containing either 0% NaCl or 3% NaCl concentration
NB	Nutrient broth
NCFB	National Collection of Food Bacteria, Agricultural and Food Research Council (AFRC) Institute of Food Research, Reading Laboratory, Reading, Berkshire, UK (formerly National Collection of Dairy Organisms)
NCIMB	National Collection of Industrial and Marine Bacteria, Aberdeen, Scotland, UK
NCTC	National Collection of Type Cultures, Central Public Health Laboratory, London, UK
ND	Not done
Neg	Negative
NG	No growth
Nit	Nitrate
NK	Not known
NLF	Non-lactose fermenting
nm	Nanometre
nmoles	Nanomoles
nt	Not tested
NVI	National Veterinary Institute, Oslo, Norway
O	Oxidative (aerobic metabolism)
ODC	Ornithine decarboxylase
OF	Oxidative-fermentative test
ONPG	*o*-nitrophenyl-β-D-galactopyranoside
Orn	Ornithine
Ox	Oxidase
Packer's plates	Selective media for the isolation of *Erysipelothrix* species
PBS	Phosphate buffered saline
PCR	Polymerase chain reaction
PDA	Potato dextrose agar
Phylogenetics	The study of evolutionary relatedness among groups of organisms or species
pmoles	Picomoles
Polyphyletic	A taxon or a group of organisms descended from more than one ancestor
Poly plates	Media for the isolation of *Dermatophilus* species
Pos	Positive
PY	Peptone yeast medium for *Pedobacter* and *Sphingobacterium* species
PYS-2	Peptone yeast medium 2
R	Resistant
R2A	Medium for isolation of marine *Flavobacterium* and other organisms

RAA	Rogosa acetate agar
RBC	Red blood cells
rDNA	Recombinant deoxyribonucleic acid
Rib	D-Ribose
RNA	Ribonucleic acid
RT-PCR	Reverse transcriptase polymerase chain reaction
S	Sensitive
SAB	Sabouraud's medium for fungi
Sal	Salicin
Shieh medium	For the isolation of *Flavobacterium columnare*
Siem agar	Selective medium for isolation of *Aerococcus viridans*
SKDM	Selective medium for *Renibacterium salmoninarum*
Skirrow's medium	For the isolation of *Helicobacter* species
Sor	Sorbitol
SS	Media for the isolation of *Salmonella* and *Shigella*
Suc	Sucrose
SW	Swarming growth on agar plate
SWT	Seawater-based complex medium
Taxonomy	The classification, identification and naming of organisms
TB Lab	Tuberculosis Laboratory
TCBS	Thiosulfate–citrate–bile salts–sucrose agar
Temp	Temperature
THA	Todd–Hewitt agar
Tm	Melting temperature
Tre	Trehalose
TSA	Tryptone soya agar
TSA-B	Tryptone soya agar with added blood
TSA+NaCl	Tryptone soya agar with added salt (2% final concentration)
TSB	Tryptic soy broth
TYG	Tryptone yeast extract glucose agar
TYG-M	Tryptone yeast extract glucose agar with added NaCl for growth of *T. maritimum*
Uro	Urocanic acid
V	Variable reaction in literature
VAM	*Vibrio anguillarum* medium for isolation of *Vibrio anguillarum*
VP	Voges–Proskaüer test
VPT	Vancomycin, polymyxin, trimethoprim: Skirrow's medium for the isolation of *Helicobacter* species
Vs	Variable or slow reaction
UK	United Kingdom
USA	United States of America
UV	Ultraviolet light
VVM	*Vibrio vulnificus* medium
WHO	World Health Organization
Wood's Broth	Selective broth for the isolation of *Erysipelothrix* species
XLD	Xylose lysine deoxycholate agar for isolation of *Salmonella* species
Xyl	D-Xylose
Y	Yellow-coloured colony on TCBS plate
YSA	*Yersinia* selective agar
ZN	Ziehl–Neelsen
0%–3%	Split plate containing gelatin, and concentration of salt at 0% and 3%

0129	Vibriostatic agent – 2:4-diamino- 6:7-diisopropylpteridine phosphate
+	Positive reaction
−	Negative reaction
−α	Negative result, but may show α-haemolysis after a week
−cr	Gram-negative curved rod
+g+	Glucose fermentation positive, gas produced
+g−	Glucose fermentation positive, no gas produced
+rt	Positive at room temperature (25°C) but negative at 37°C
+s	Slow positive reaction. May take 2–4 days for reaction to occur
+w	Weak positive reaction

References

1 Abayneh, T., Colquhoun, D.J. and Sørum, H. (2013) *Edwardsiella piscicida* sp. nov., a novel species pathogenic to fish. *Journal of Applied Microbiology* 114(3), 644–654.

2 Abbott, S.L., Cheung, W.K., Kroske-Bystrom, S., Malekzadeh, T. and Janda, J.M. (1992) Identification of *Aeromonas* strains to the genospecies level in the clinical laboratory. *Journal of Clinical Microbiology* 30(5), 1262–1266.

3 Abbott, S.L., Moler, S., Green, N., Tran, R.K., Wainwright, K. and Janda, J.M. (2011) Clinical and laboratory diagnostic characteristics and cytotoxigenic potential of *Hafnia alvei* and *Hafnia paralvei* strains. *Journal of Clinical Microbiology* 49(9), 3122–3126.

4 Abbott, S.L., Cheung, W.K.W. and Janda, J.M. (2003) The genus *Aeromonas*: biochemical characteristics, atypical reactions, and phenotypic identification schemes. *Journal of Clinical Microbiology* 41(6), 2348–2357.

5 Abbott, S.L., Seli, L.S., Catino, M., Hartley, M.A. and Janda, J.M. (1998) Misidentification of unusual *Aeromonas* species as members of the genus *Vibrio*: a continuing problem. *Journal of Clinical Microbiology* 36(4), 1103–1104.

6 Abu-Samra, M.T. and Walton, G.S. (1977) Modified techniques for the isolation of *Dermatophilus* spp. from infected material. *Sabouraudia* 15(1), 23–27.

7 Acosta, F., Real, F., Caballero, M.J., Sieiro, C., Fernández, A. and Rodríguez, L.A. (2002) Evaluation of immunohistochemical and microbiological methods for the diagnosis of brown trout infected with *Hafnia alvei*. *Journal of Aquatic Animal Health* 14(1), 77–83.

8 Acuña, M.T., Díaz, G., Bolaños, H., Barquero, C., Sánchez, O., Sánchez, L.M., Mora, G., Chaves, A. and Campos, E. (1999) Sources of *Vibrio mimicus* contamination of turtle eggs. *Applied and Environmental Microbiology* 65(1), 336–338.

9 Adler, B. and de la Peña Moctezuma, A. (2010) *Leptospira* and leptospirosis. *Veterinary Microbiology* 140(3–4), 287–296.

10 Aguirre, A.A., Balazs, G.H., Zimmerman, B. and Spraker, T.R. (1994) Evaluation of Hawaiian green turtles (*Chelonia mydas*) for potential pathogens associated with fibropapillomas. *Journal of Wildlife Diseases* 30(1), 8–15.

11 Aguirre-Guzmán, G., Ascencio, F. and Saulnier, D. (2005) Pathogenicity of *Vibrio penaeicida* for white shrimp *Litopenaeus vannamei*: a cysteine protease-like exotoxin as a virulence factor. *Diseases of Aquatic Organisms* 67(3), 201–207.

12 Ahmet, Z., Stanier, P., Harvey, D. and Holt, D. (1999) New PCR primers for the sensitive detection and specific identification of group B beta-hemolytic streptococci in cerebrospinal fluid. *Molecular and Cellular Probes* 13(5), 349–357.

13 Aiso, K., Simidu, U. and Hasuo, K. (1968) Microflora in the digestive tract of inshore fish in Japan. *Journal of General Microbiology* 52(3), 361–364.

14 Akagawa, M. and Yamasato, K. (1989) Synonymy of *Alcaligenes aquamarinus*, *Alcaligenes faecalis* subsp. *homari*, and *Deleya aesta: Deleya aquamarina* comb. nov. as the type species of the genus *Deleya*. *International Journal of Systematic Bacteriology* 39(4), 462–466.

15 Akineden, O., Hassan, A.A., Alber, J., El-Sayed, A., Estoepangestie, A.T.S., Lämmler, C., Weiss, R. and Siebert, U. (2005) Phenotypic and genotypic properties of *Streptococcus equi* subsp. *zooepidemicus* isolated from harbor seals (*Phoca vitulina*) from the German North Sea during the phocine distemper outbreak in 2002. *Veterinary Microbiology* 110(1–2), 147–152.

16 Alber, J., El-Sayed, A., Lämmler, C., Hassan, A.A., Weiss, R. and Zschöck, M. (2004) Multiplex polymerase chain reaction for identification and differentiation of *Streptococcus equi* subsp. *zooepidemicus* and *Streptococcus equi* subsp. *equi*. *Journal of Veterinary Medicine. B, Infectious Diseases and Veterinary Public Health* 51(10), 455–458.

17 Albert, M.J., Ansaruzzaman, M., Bardhan, P.K., Faruque, A.S.G., Faruque, S.M., Islam, M.S., Mahalanabis, D., Sack, R.B. and Salam, M.A. (1993) Large epidemic of cholera-like disease in Bangladesh caused by *Vibrio*

cholerae 0139 synonym Bengal. *The Lancet* 342(8868), 387–390.
18 Alcaide, E. (2003) Numerical taxonomy of *Vibrionaceae* isolated from cultured amberjack (*Seriola dumerili*) and surrounding water. *Current Microbiology* 46(3), 184–189.
19 Alcaide, E., Amaro, C., Todolí, R. and Oltra, R. (1999) Isolation and characterization of *Vibrio parahaemolyticus* causing infection in Iberian toothcarp *Aphanius iberus*. *Diseases of Aquatic Organisms* 35, 77–80.
20 Alcaide, E., Gil-Sanz, C., Sanjuán, E., Esteve, D., Amaro, C. and Silveira, L. (2001) *Vibrio harveyi* causes disease in seahorse, *Hippocampus* sp. *Journal of Fish Diseases* 24, 311–313.
21 Alcaide, E., Herraiz, S. and Esteve, C. (2006) Occurrence of *Edwardsiella tarda* in wild European eels *Anguilla anguilla* from Mediterranean Spain. *Diseases of Aquatic Organisms* 73(1), 77–81.
22 Al Dahouk, S., Scholz, H., Tomaso, H., Bahn, P., Göllner, C., Karges, W., Appel, B., Hensel, A., Neubauer, H. and Nöckler, K. (2010) Differential phenotyping of *Brucella* species using a newly developed semi-automated metabolic system. *BMC Microbiology* 10, 269.
23 Alderman, D.J. (1981) *Fusarium solani* causing an exoskeletal pathology in cultured lobsters, *Homarus vulgaris*. *Transactions of the British Mycological Society* 76(1), 25–27.
24 Alderman, D.J. (1996) Geographical spread of bacterial and fungal diseases of crustaceans. *Revue Scientifique et Technique (International Office of Epizootics)* 15(2), 603–632.
25 Alderman, D.J. and Polglase, J.L. (1985) *Fusarium tabacinum* (Beyma) Gams, as a gill parasite in the crayfish, *Austropotamobius pallipes* Lereboullet. *Journal of Fish Diseases* 8(2), 249–252.
26 Alderman, D.J. and Polglase, J.L. (1986) *Aphanomyces astaci*: isolation and culture. *Journal of Fish Diseases* 9, 367–379.
27 Al-Duboon, A.H., Al-Mukhtar, M.A., Jassim, A.A.R., Bader, S.Q. and Al-Zawar, J.M. (2006) Saprolegniosis of *Barbus sharpeyi*, Gunther (Bunnei fish) in Basrah/Iraq. *Iraq Journal of Aquaculture* 1, 25–29.
28 Aleksić, S., Steigerwalt, A.G., Bockemühl, J., Huntley-Carter, G.P. and Brenner, D.J. (1987) *Yersinia rohdei* sp. nov. isolated from human and dog feces and surface water. *International Journal of Systematic Bacteriology* 37(4), 327–332.
29 Algöet, M., Bayley, A.E., Roberts, E.G., Feist, S.W., Wheeler, R.W. and Verner-Jeffreys, D.W. (2009) Susceptibility of selected freshwater fish species to a UK *Lactococcus garvieae* isolate. *Journal of Fish Diseases* 32(10), 825–834.
30 Ali, A., Carnahan, A.M., Altwegg, M., Lüthy-Hottenstein, J. and Joseph, S.W. (1996) *Aeromonas bestiarum* sp. nov. (formerly genomospecies DNA group 2 *A. hydrophila*), a new species isolated from non-human sources. *Medical Microbiology Letters* 5, 156–165.
31 Ali, E., Hashem, M. and Al-Salahy, M. (2011) Pathogenicity and oxidative stress in Nile tilapia caused by *Aphanomyces laevis* and *Phoma herbarum* isolated from farmed fish. *Diseases of Aquatic Organisms* 94(1), 17–28.
32 Alker, A., Smith, G. and Kim, K. (2001) Characterization of *Aspergillus sydowii* (Thom et Church), a fungal pathogen of Caribbean sea fan corals. *Hydrobiologia* 460(1–3), 105–111.
33 Allam, B., Paillard, C., Howard, A. and Le Pennec, M. (2000) Isolation of the pathogen *Vibrio tapetis* and defense parameters in brown ring diseased Manila clams *Ruditapes philippinarum* cultivated in England. *Diseases of Aquatic Organisms* 41(2), 105–113.
34 Allam, B., Paillard, C. and Ford, S.E. (2002) Pathogenicity of *Vibrio tapetis*, the etiological agent of brown ring disease in clams. *Diseases of Aquatic Organisms* 48(3), 221–231.
35 Allen, D.A., Austin, B. and Colwell, R.R. (1983) *Aeromonas media*, a new species isolated from river water. *International Journal of Systematic Bacteriology* 33(3), 599–604.
36 Alperi, A., Martínez-Murcia, A.J., Ko, W.-C., Monera, A., Saavedra, M.J. and Figueras, M.J. (2010a) *Aeromonas taiwanensis* sp. nov. and *Aeromonas sanarellii* sp. nov., clinical species from Taiwan. *International Journal of Systematic and Evolutionary Microbiology* 60(Pt 9), 2048–2055.
37 Alperi, A., Martínez-Murcia, A.J., Monera, A., Saavedra, M.J. and Figueras, M.J. (2010b) *Aeromonas fluvialis* sp. nov., isolated from a Spanish river. *International Journal of Systematic and Evolutionary Microbiology* 60(Pt 1), 72–77.
38 Alsina, M., Martínez-Picado, J., Jofre, J. and Blanch, A.R. (1994) A medium for presumptive identification of *Vibrio anguillarum*. *Applied and Environmental Microbiology* 60(5), 1681–1683.
39 Altinok, I., Kayis, S. and Capkin, E. (2006) *Pseudomonas putida* infection in rainbow trout. *Aquaculture* 261(3), 850–855.
40 Altmann, K., Marshall, M., Nicholson, S.E., Hanna, P.J. and Gudkovs, N. (1992) Glucose repression of pigment production in atypical isolates of *Aeromonas salmonicida* responsible for goldfish ulcer disease. *Microbios* 72(292–293), 215–220.
41 Alton, G.G. and Jones, L.M. (1967) *Laboratory techniques in brucellosis*. World Health Organization, Geneva, Switzerland.
42 Altwegg, M., Steigerwalt, A.G., Altwegg-Bissig, R., Lüthy-Hottenstein, J. and Brenner, D.J. (1990) Biochemical identification of *Aeromonas* genospecies isolated from humans. *Journal of Clinical Microbiology* 28(2), 258–264.
43 Aluotto, B.B., Wittler, R.G., Williams, C.O. and Faber, J.E. (1970) Standardized bacteriologic techniques for the characterization of *Mycoplasma* species. *International Journal of Systematic Bacteriology* 20(1), 35–58.
44 Álvarez, B. and Guijarro, J.A. (2007) Recovery of *Flavobacterium psychrophilum* viable cells using a charcoal-based solid medium. *Letters in Applied Microbiology* 44(5), 569–572.
45 Alvarez, J.D., Austin, B., Alvarez, A.M. and Reyes, H. (1998) *Vibrio harveyi*: a pathogen of penaeid shrimps and fish in Venezuela. *Journal of Fish Diseases* 21(4), 313–316.
46 Alvarez-Perez, S., Mateos, A., Dominguez, L., Martinez-Nevado, E., Blanco, J.L. and Garcia, M.E. (2010) Polyclonal *Aspergillus fumigatus* infection in captive penguins. *Veterinary Microbiology* 144(3–4), 444–449.
47 Amagliani, G., Omiccioli, E., Andreoni, F., Boiani, R., Bianconi, I., Zaccone, R., Mancuso, M. and Magnani, M.

(2009) Development of a multiplex PCR assay for *Photobacterium damselae* subsp. *piscicida* identification in fish samples. *Journal of Fish Diseases* 32(8), 645–653.

48 Amandi, A., Hiu, S.F., Rohovec, J.S. and Fryer, J.L. (1982) Isolation and characterization of *Edwardsiella tarda* from fall chinook salmon (*Oncorhynchus tshawytscha*). *Applied and Environmental Microbiology* 43(6), 1380–1384.

49 Amann, R.I., Binder, B.J., Olson, R.J., Chisholm, S.W., Devereux, R. and Stahl, D.A. (1990) Combination of 16S rRNA-targeted oligonucleotide probes with flow cytometry for analyzing mixed microbial populations. *Applied and Environmental Microbiology* 56(6), 1919–1925.

50 Amaro, C., Hor, L.I., Marco-Noales, E., Bosque, T., Fouz, B. and Alcaide, E. (1999) Isolation of *Vibrio vulnificus* serovar E from aquatic habitats in Taiwan. *Applied and Environmental Microbiology* 65(3), 1352–1355.

51 Anacker, R.L. and Ordal, E.J. (1955) Study of a bacteriophage infecting the myxobacterium *Chondrococcus columnaris*. *Journal of Bacteriology* 70(6), 738–741.

52 Anacker, R.L. and Ordal, E.J. (1959) Studies on the myxobacterium *Chondrococcus columnaris*. I. Serological typing. *Journal of Bacteriology* 78(1), 25–32.

53 Angka, S.L., Lam, T.J. and Sin, Y.M. (1995) Some virulence characteristics of *Aeromonas hydrophila* in walking catfish (*Clarias gariepinus*). *Aquaculture* 130(2–3), 103–112.

54 Anguiano-Beltrán, C., Searcy-Bernal, R. and Lizárraga-Partida, M. (1998) Pathogenic effects of *Vibrio alginolyticus* on larvae and postlarvae of the red abalone *Haliotis rufescens*. *Diseases of Aquatic Organisms* 33, 119–122.

55 Angulo, L., Lopez, J.E., Vicente, J.A. and Saborido, A.M. (1994) Haemorrhagic areas in the mouth of farmed turbot, *Scophthalmus maximus* (L.). *Journal of Fish Diseases* 17, 163–169.

56 Aoki, T., Park, C.-I., Yamashita, H. and Hirono, I. (2000) Species-specific polymerase chain reaction primers for *Lactococcus garvieae*. *Journal of Fish Diseases* 23(1), 1–6.

57 Arakawa, C.K., Fryer, J.L. and Sanders, J.E. (1986) Serology of *Mycobacterium chelonei* isolated from salmonid fish. *Journal of Fish Diseases* 9, 269–271.

58 Aravena-Román, M., Harnett, G.B., Riley, T.V., Inglis, T.J. and Chang, B.J. (2011) *Aeromonas aquariorum* is widely distributed in clinical and environmental specimens and can be misidentified as *Aeromonas hydrophila*. *Journal of Clinical Microbiology* 49(8), 3006–3008.

59 Arias, C.R., Verdonck, L., Swings, J., Aznar, R. and Garay, E. (1997) A polyphasic approach to study the intraspecific diversity amongst *Vibrio vulnificus* isolates. *Systematic and Applied Microbiology* 20(4), 622–633.

60 Arkwright, J.A. (1912) An epidemic disease affecting salmon and trout in England during the summer of 1911. *The Journal of Hygiene* 12(4), 391–413.

61 Aronson, J.D. (1926) Spontaneous tuberculosis in salt water fish. *The Journal of Infectious Diseases* 39(4), 315–320.

62 Ashburner, L.D. (1977) Mycobacteriosis in hatchery-confined chinook salmon (*Oncorhynchus tshawytscua* Walbaum) in Australia. *Journal of Fish Biology* 10(6), 523–528.

63 Ashdown, L.R. (1979a) An improved screening technique for isolation of *Pseudomonas pseudomallei* from clinical specimens. *Pathology* 11(2), 293–297.

64 Ashdown, L.R. (1979b) Identification of *Pseudomonas pseudomallei* in the clinical laboratory. *Journal of Clinical Pathology* 32(5), 500–504.

65 Ast, J.C., Urbanczyk, H. and Dunlap, P.V. (2009) Multigene analysis reveals previously unrecognized phylogenetic diversity in *Aliivibrio*. *Systematic and Applied Microbiology* 32(6), 379–386.

66 Atkins, D. (1954) Further notes on a marine member of the *Saprolegniaceae*, *Leptolegnia marine* n. sp., infecting certain invertebrates. *Journal of the Marine Biological Association of the United Kingdom* 33, 613–625.

67 Auling, G., Reh, M., Lee, C.M. and Schlegel, H.G. (1978) *Pseudomonas pseudoflava*, a new species of hydrogen-oxidizing bacteria: its differentiation from *Pseudomonas flava* and other yellow-pigmented, Gram-negative, hydrogen-oxidizing species. *International Journal of Systematic Bacteriology* 28(1), 82–95.

68 Austin, B. and Austin, D. (1999) *Bacterial Fish Pathogens: Disease of Farmed and Wild Fish*, 3rd rev. edn. Springer, London/Praxis Publishing, Chichester, UK.

69 Austin, B., Austin, D.A., Dalsgaard, I., Gudmundsdóttir, B.K., Høie, S., Thornton, J.M., Larsen, J.L., O'Hici, B. and Powell, R. (1998) Characterization of atypical *Aeromonas salmonicida* by different methods. *Systematic and Applied Microbiology* 21(1), 50–64.

70 Austin, B. (1993) Recovery of 'atypical' isolates of *Aeromonas salmonicida*, which grow at 37°C, from ulcerated non-salmonids in England. *Journal of Fish Diseases* 16(2), 165–168.

71 Austin, B., Austin, D.A., Blanch, A.R., Cerda, M., Grimont, F., Grimont, P.A.D., Jofre, J., Koblavi, S., Larsen, J.L., Pedersen, K., Tiainen, T., Verdonck, L. and Swings, J. (1997) A comparison of methods for the typing of fish-pathogenic *Vibrio* spp. *Systematic and Applied Microbiology* 20(1), 89–101.

72 Austin, B., Embley, T.M. and Goodfellow, M. (1983) Selective isolation of *Renibacterium salmoninarum*. *FEMS Microbiology Letters* 17(1–3), 111–114.

73 Austin, B., Gonzalez, C.J., Stobie, M., Curry, J.I. and McLoughlin, M.F. (1992) Recovery of *Janthinobacterium lividum* from diseased rainbow trout, *Oncorhynchus mykiss* (Walbaum), in Northern Ireland and Scotland. *Journal of Fish Diseases* 15(4), 357–359.

74 Austin, B. and Robertson, P.A.W. (1994) Recovery of *Streptococcus milleri* from ulcerated koi carp (*Cyprinus carpio* L.) in the UK. *Bulletin of the European Association of Fish Pathologists* 13(6), 207–209.

75 Austin, B., Rodgers, C.J., Forns, J.M. and Colwell, R.R. (1981) *Alcaligenes faecalis* subsp. *homari* subsp. nov., a new group of bacteria isolated from moribund lobsters. *International Journal of Systematic Bacteriology* 31(1), 72–76.

76 Austin, B. and Stobie, M. (1992a) Recovery of *Serratia plymuthica* and presumptive *Pseudomonas pseudoalcaligenes* from skin lesions in rainbow trout, *Oncorhynchus mykiss* (Walbaum), otherwise infected with enteric redmouth. *Journal of Fish Diseases* 15(6), 541–543.

77 Austin, B. and Stobie, M. (1992b) Recovery of *Micrococcus luteus* and presumptive *Planococcus* sp. from moribund fish during an outbreak of rainbow trout,

Oncorhynchus mykiss (Walbaum), fry syndrome in England. *Journal of Fish Diseases* 15(2), 203–206.
78 Austin, B., Zachary, A. and Colwell, R.R. (1978) Recognition of *Beneckea natriegens* (Payne *et al.*) Baumann *et al.* as a member of the genus *Vibrio*, as previously proposed by Webb and Payne. *International Journal of Systematic Bacteriology* 28(2), 315–317.
79 Austin, B., Austin, D., Sutherland, R., Thompson, F. and Swings, J. (2005) Pathogenicity of vibrios to rainbow trout (*Oncorhynchus mykiss*, Walbaum) and *Artemia nauplii*. *Environmental Microbiology* 7(9), 1488–1495.
80 Austin, B., McIntosh, D. and Murray, K.R. (1988) Infection of Atlantic salmon (*Salmo salar*) with *Planococcus* sp. *Bulletin of the European Association of Fish Pathologists* 8(2), 21–22.
81 Austin, D.A., McIntosh, D. and Austin, B. (1989) Taxonomy of fish associated *Aeromonas* spp., with the description of *Aeromonas salmonicida* subsp. *smithia* subsp. nov. *Systematic and Applied Microbiology* 11(3), 277–290.
82 Ausubel, F. (1987) *Current Protocols in Molecular Biology*. Greene Publishing Associates/Wiley Interscience, New York.
83 Avendaño-Herrera, R., Balboa, S., Doce, A., Ilardi, P., Lovera, P., Toranzo, A.E. and Romalde, J.L. (2011) Pseudo-membranes on internal organs associated with *Rhodococcus qingshengii* infection in Atlantic salmon (*Salmo salar*). *Veterinary Microbiology* 147(1–2), 200–204.
84 Avendaño-Herrera, R., Toranzo, A.E. and Magariños, B. and (2006) Tenacibaculosis infection in marine fish caused by *Tenacibaculum maritimum*: a review. *Diseases of Aquatic Organisms* 71(3), 255–266.
85 Aydin, S., Celebi, S. and Akyurt, I. (1997) Clinical, haematological and pathological investigations of *Escherichia vulneris* in rainbow trout (*Oncorhynchus mykiss*). *Fish Pathology* 32(1), 29–34.
86 Aydin, S., Gultepe, N. and Yildiz, H. (2000) Natural and experimental infections of *Campylobacter cryaerophila* in rainbow trout: gross pathology, bacteriology, clinical pathology and chemotherapy. *Fish Pathology* 35, 117–123.
87 Ayers, S., Rupp, P. and Johnson, W. Jr (1919) A study of the alkali-forming bacteria found in milk. Bulletin No. 782. US Department of Agriculture.
88 Ayling, R.D., Bashiruddin, S., Davison, N.J., Foster, G., Dagleish, M.P. and Nicholas, R.A.J. (2011) The occurrence of *Mycoplasma phocicerebrale*, *Mycoplasma phocidae*, and *Mycoplasma phocirhinis* in grey and common seals (*Halichoerus grypus* and *Phoca vitulina*) in the United Kingdom. *Journal of Wildlife Diseases* 47(2), 471–475.
89 Azandégbé, A., Garnier, M., Andrieux-Loyer, F., Kérouel, R., Philippon, X. and Nicolas, J. (2010) Occurrence and seasonality of *Vibrio aestuarianus* in sediment and *Crassostrea gigas* haemolymph at two oyster farms in France. *Diseases of Aquatic Organisms* 91(3), 213–221.
90 Backman, S., Ferguson, H.W., Prescott, J.F. and Wilcock, B.P. (1990) Progressive panophthalmitis in chinook salmon, *Oncorhynchus tshawytscha* (Walbaum), a case report. *Journal of Fish Diseases* 13(5), 345–353.
91 Bader, J.A., Shoemaker, C.A. and Klesius, P.H. (2003) Rapid detection of columnaris disease in channel catfish (*Ictalurus punctatus*) with a new species-specific 16–S rRNA gene-based PCR primer for *Flavobacterium columnare*. *Journal of Microbiological Methods* 52(2), 209–220.
92 Bader, J.A. and Shotts, E.B. (1998) Identification of *Flavobacterium* and *Flexibacter* species by species-specific polymerase chain reaction primers to the 16S ribosomal RNA gene. *Journal of Aquatic Animal Health* 10(4), 311–319.
93 Baeck, G.W., Kim, J.H., Gomez, D.K. and Park, S.C. (2006) Isolation and characterization of *Streptococcus* sp. from diseased flounder (*Paralichthys olivaceus*) in Jeju Island. *Journal of Veterinary Science* 7(1), 53–58.
94 Baharaeen, S. and Vishniac, H.S. (1982) *Cryptococcus lupi* sp. nov., an Antarctic Basidioblastomycete. *International Journal of Systematic Bacteriology* 32(2), 229–232.
95 Bai, C., Garner, T.W.J. and Li, Y. (2010) First evidence of *Batrachochytrium dendrobatidis* in China: discovery of chytridiomycosis in introduced American bullfrogs and native amphibians in the Yunnan Province, China. *EcoHealth* 7(1), 127–134.
96 Bai, F., Pang, L., Qi, Z., Chen, J., Austin, B. and Zhang, X.-H. (2008) Distribution of five vibrio virulence-related genes among *Vibrio harveyi* isolates. *The Journal of General and Applied Microbiology* 54(1), 71–78.
97 Baker, J.A. and Hagan, W.A. (1942) Tuberculosis of the Mexican Platyfish (*Platypoecilus maculatus*). *The Journal of Infectious Diseases* 70(3), 248–252.
98 Bakopoulos, V., Adams, A. and Richards, R.H. (1995) Some biochemical properties and antibiotic sensitivities of *Pasteurella piscicida* isolated in Greece and comparison with strains from Japan, France and Italy. *Journal of Fish Diseases* 18, 1–7.
99 Bakopoulos, V., Peric, Z., Rodger, H., Adams, A. and Richards, R. (1997) First report of Fish Pasteurellosis from Malta. *Journal of Aquatic Animal Health* 9(1), 26–33.
100 Balboa, S., Ferguson, H.W. and Romalde, J.L. (2007) Phenotypic, serological and genetic characterization of *Pseudomonas anguilliseptica* strains isolated from cod, *Gadus morhua* L., in northern Europe. *Journal of Fish Diseases* 30(11), 657–664.
101 Balboa, S., Dieguez, A.L., Doce, A., Barja, J.L. and Romalde, J.L. (2012) Evaluation of different culture media for the isolation and growth of the fastidious *Vibrio tapetis*, the causative agent of brown ring disease. *Journal of Invertebrate Pathology* 111(1), 74–81.
102 Balboa, S., Doce, A., Diéguez, A.L. and Romalde, J.L. (2011) Evaluation of different species-specific PCR protocols for the detection of *Vibrio tapetis*. *Journal of Invertebrate Pathology* 108(2), 85–91.
103 Balboa, S. and Romalde, J.L. (2013) Multilocus sequence analysis of *Vibrio tapetis*, the causative agent of Brown Ring Disease: description of *Vibrio tapetis* subsp. *britannicus* subsp. nov. *Systematic and Applied Microbiology* 36(3), 183–187.
104 Balcázar, J.L., Gallo-Bueno, A., Planas, M. and Pintado, J. (2010a) Isolation of *Vibrio alginolyticus* and *Vibrio splendidus* from captive-bred seahorses with disease symptoms. *Antonie Van Leeuwenhoek* 97(2), 207–210.

105 Balcázar, J.L., Pintado, J. and Planas, M. (2010b) *Bacillus galliciensis* sp. nov., isolated from faeces of wild seahorses (*Hippocampus guttulatus*). *International Journal of Systematic and Evolutionary Microbiology* 60(Pt 4), 892–895.

106 Balcázar, J.L., Pintado, J. and Planas, M. (2010c) *Vibrio hippocampi* sp. nov., a new species isolated from wild seahorses (*Hippocampus guttulatus*). *FEMS Microbiology Letters* 307(1), 30–34.

107 Balcázar, J.L., Vendrell, D., de Blas, I., Ruiz-Zarzuela, I., Gironés, O. and Múzquiz, J.L. (2007) Quantitative detection of *Aeromonas salmonicida* in fish tissue by real-time PCR using self-quenched, fluorogenic primers. *Journal of Medical Microbiology* 56(Pt 3), 323–328.

108 Balebona, M.C., Andreu, M.J., Bordas, M.A., Zorrilla, I., Moriñigo, M.A. and Borrego, J.J. (1998a) Pathogenicity of *Vibrio alginolyticus* for cultured gilt-head sea bream (*Sparus aurata* L.). *Applied and Environmental Microbiology* 64(11), 4269–4275.

109 Balebona, M.C., Zorrilla, I., Moriñigo, M.A. and Borrego, J.J. (1998b) Survey of bacterial pathologies affecting farmed gilt-head sea bream (*Sparus aurata* L.) in southwestern Spain from 1990 to 1996. *Aquaculture* 166(1–2), 19–35.

110 Ballesteros, I., Martín, M.P., Cerenius, L., Söderhäll, K., Tellería, M.T. and Diéguez-Uribeondo, J. (2007) Lack of specificity of the molecular diagnostic method for identification of *Aphanomyces astaci*. *Bulletin Français de la Pêche et de la Pisciculture* (385), 17–24.

111 Ballesteros, I., Martín, M. and Diéguez-Uribeondo, J. (2006) First isolation of *Aphanomyces frigidophilus* (*Saprolegniales*) in Europe. *Mycotaxon* 95, 335–340.

112 Bang, S.S., Baumann, P. and Nealson, K.H. (1978) Phenotypic characterization of *Photobacterium logei* (sp. nov.), a species related to *P. fischeri*. *Current Microbiology* 1(5), 285–288.

113 Bangyeekhun, E., Cerenius, L. and Söderhäll, K. (2001) Molecular cloning and characterization of two serine proteinase genes from the crayfish plague fungus, *Aphanomyces astaci*. *Journal of Invertebrate Pathology* 77(3), 206–216.

114 Banin, E., Israely, T., Kushmaro, A., Loya, Y., Orr, E. and Rosenberg, E. (2000) Penetration of the coral-bleaching bacterium *Vibrio shiloi* into *Oculina patagonica*. *Applied and Environmental Microbiology* 66(7), 3031–3036.

115 Banner, C.R., Long, J.J., Fryer, J.L. and Rohovec, J.S. (1986) Occurrence of salmonid fish infected with *Renibacterium salmoninarum* in the Pacific Ocean. *Journal of Fish Diseases* 9, 273–275.

116 Bano, N., DeRae Smith, A., Bennett, W., Vasquez, L. and Hollibaugh, J.S. (2007) Dominance of *Mycoplasma* in the guts of the Long-Jawed Mudsucker, *Gillichthys mirabilis*, from five California salt marshes. *Environmental Microbiology* 9(10), 2636–2641.

117 Baptista, B., Romalde, J.L. and Toranzo, A.E. (1996) First occurrence of pasteurellosis in Portugal affecting cultured gilthead seabream (*Sparus aurata*). *Bulletin of the European Association of Fish Pathologists* 16(3), 92–95.

118 Baran, I. and Soylu, E. (1989) Crayfish plague in Turkey. *Journal of Fish Diseases* 12, 193–197.

119 Barbeyron, T., L'Haridon, S., Corre, E., Kloareg, B. and Potin, P. (2001) *Zobellia galactanovorans* gen. nov., sp. nov., a marine species of *Flavobacteriaceae* isolated from a red alga, and classification of [*Cytophaga*] *uliginosa* (ZoBell and Upham 1944) Reichenbach 1989 as *Zobellia uliginosa* gen. nov., comb. nov. *International Journal of Systematic and Evolutionary Microbiology* 51(3), 985–997.

120 Barrionuevo, J.S., Aguayo, R. and Lavilla, E.O. (2008) First record of chytridiomycosis in Bolivia (*Rhinella quechua*; Anura: Bufonidae). *Diseases of Aquatic Organisms* 82(2), 161–163.

121 Barrionuevo, S. and Mangione, S. (2006) Chytridiomycosis in two species of *Telmatobius* (Anura: *Leptodactylidae*) from Argentina. *Diseases of Aquatic Organisms* 73(2), 171–174.

122 Barritt, M.M. (1936) The intensification of the Voges-Proskauer reaction by the addition of α-naphthol. *The Journal of Pathology and Bacteriology* 42(2), 441–454.

123 Baumann, L., Baumann, P., Mandel, M. and Allen, R.D. (1972) Taxonomy of aerobic marine *Eubacteria*. *Journal of Bacteriology* 110(1), 402–429.

124 Baumann, L., Bowditch, R.D. and Baumann, P. (1983) Description of *Deleya* gen. nov. created to accommodate the marine species *Alcaligenes aestus*, *A. pacificus*, *A. cupidus*, *A. venustus*, and *Pseudomonas marina*. *International Journal of Systematic Bacteriology* 33(4), 793–802.

125 Baumann, P. and Baumann, L. (1981) The marine Gram-negative eubacteria: genera *Photobacterium*, *Beneckea*, *Alteromonas*, *Pseudomonas*, and *Alcaligenes*. In: Starr, M., Stolp, H., Trüper, H., Balows, A. and Schegel, H. (eds) *The Prokaryotes*, Vol. II. Springer, Berlin, pp. 1302–1331.

126 Baumann, P., Baumann, L. and Mandel, M. (1971a) Taxonomy of marine bacteria: the genus *Beneckea*. *Journal of Bacteriology* 107(1), 268–294.

127 Baumann, P., Bang, S.S. and Baumann, L. (1978) Phenotypic characterization of *Beneckea anguillara* biotypes I and II. *Current Microbiology* 1, 85–88.

128 Baumann, P., Baumann, L., Bang, S.S. and Woolkalis, M.J. (1980) Reevaluation of the taxonomy of *Vibrio*, *Beneckea*, and *Photobacterium*: abolition of the genus *Beneckea*. *Current Microbiology* 4(3), 127–132.

129 Baumann, P., Baumann, L., Mandel, M. and Allen, R.D. (1971b) Taxonomy of marine bacteria: *Beneckea nigrapulchrituda* sp. n. *Journal of Bacteriology* 108(3), 1380–1383.

130 Baumann, P., Bowditch, R.D., Baumann, L. and Beaman, B. (1983) Taxonomy of marine *Pseudomonas* species: *P. stanieri* sp. nov.; *P. perfectomarina* sp. nov., nom. rev.; *P. nautica: and P. doudoroffii*. *International Journal of Systematic Bacteriology* 33(4), 857–865.

131 Baxa, D.V., Groff, J.M., Wishkovsky, A. and Hedrick, R.P. (1990) Susceptibility of nonictalurid fishes to experimental infection with *Edwardsiella ictaluri*. *Diseases of Aquatic Organisms* 8, 113–117.

132 Baya, A., Li, T., Lupiani, B. and Hetrick, F.M. (1992a) *Bacillus cereus*, a pathogen for striped bass. In: *Eastern Fish Health and American Fisheries Society Fish Health Section Workshop, 16–19 June 1992*. Auburn University, Auburn, Alabama, p. 67.

133 Baya, A.M., Lupiani, B., Bandín, I., Hetrick, F., Figueras, A., Carnahan, A., May, E.M. and Toranzo, A.E.

(1992b) Phenotypic and pathobiological properties of *Corynebacterium aquaticum* isolated from diseased striped bass. *Diseases of Aquatic Organisms* 14, 115–126.
134 Baya, A.M., Lupiani, B., Hetrick, F.M., Roberson, B.S., Lukacovic, R., May, E. and Poukish, C. (1990a) Association of *Streptococcus* sp. with fish mortalities in the Chesapeake Bay and its tributaries. *Journal of Fish Diseases* 13(3), 251–253.
135 Baya, A.M., Romalde, J.L., Green, D.E., Navarro, R.B., Evans, J., May, E.B. and Toranzo, A.E. (1997) Edwardsiellosis in wild striped bass from the Chesapeake Bay. *Journal of Wildlife Diseases* 33(3), 517–525.
136 Baya, A.M., Toranzo, A.E., Lupiani, B., Li, T., Roberson, B.S. and Hetrick, F.M. (1991) Biochemical and serological characterization of *Carnobacterium* spp. isolated from farmed and natural populations of striped bass and catfish. *Applied and Environmental Microbiology* 57(11), 3114–3120.
137 Baya, A.M., Toranzo, A.E., Lupiani, B., Santos, Y. and Hetrick, F.M. (1992c) *Serratia marcescens*: a potential pathogen for fish. *Journal of Fish Diseases* 15(1), 15–26.
138 Baya, A., Toranzo, A.E., Nunez, S., Barja, J. and Hetrick, F.M. (1990b) Association of a *Moraxella* sp. and a reo-like virus with mortalities of striped bass, *Morone saxatilis*. In: Perkins, F. and Cheng, T. (eds) *Pathology in Marine Science*. Academic Press, New York, pp. 91–99.
139 Beaz Hidalgo, R., Cleenwerck, I., Balboa, S., De Wachter, M., Thompson, F., Swings, J., De Vos, P. and Romalde, J.L. (2008) Diversity of *Vibrios* associated with reared clams in Galicia (NW Spain). *Systematic and Applied Microbiology* 31(3), 215–222.
140 Beaz Hidalgo, R., Cleenwerck, I., Balboa, S., Prado, S., De Vos, P. and Romalde, J.L. (2009a) *Vibrio breoganii* sp. nov., a non-motile, alginolytic, marine bacterium within the *Vibrio halioticoli* clade. *International Journal of Systematic and Evolutionary Microbiology* 59(Pt 7), 1589–1594.
141 Beaz-Hidalgo, R., Alperi, A., Figueras, M.J. and Romalde, J.L. (2009b) *Aeromonas piscicola* sp. nov., isolated from diseased fish. *Systematic and Applied Microbiology* 32(7), 471–479.
142 Beaz-Hidalgo, R., Diéguez, A.L., Cleenwerck, I., Balboa, S., Doce, A., de Vos, P. and Romalde, J.L. (2010a) *Vibrio celticus* sp. nov., a new *Vibrio* species belonging to the Splendidus clade with pathogenic potential for clams. *Systematic and Applied Microbiology* 33(6), 311–315.
143 Beaz-Hidalgo, R., Doce, A., Balboa, S., Barja, J.L. and Romalde, J.L. (2010b) *Aliivibrio finisterrensis* sp. nov., isolated from Manila clam, *Ruditapes philippinarum* and emended description of the genus *Aliivibrio*. *International Journal of Systematic and Evolutionary Microbiology* 60(Pt 1), 223–228.
144 Beaz-Hidalgo, R., Doce, A., Pascual, J., Toranzo, A.E. and Romalde, J.L. (2009c) *Vibrio gallaecicus* sp. nov. isolated from cultured clams in north-western Spain. *Systematic and Applied Microbiology* 32(2), 111–117.
145 Beijerinck, M.W. (1889) *Photobacterium luminosum*, a luminous bacterium in the North Sea. In: *Benchmark Papers in Microbiology, Vol. 11: Marine Microbiology*. Dowden, Hutchinson & Ross, Stroudsburg, Pennsylvania.
146 Beijerinck, M.W. (1901) On different forms of hereditary variation of microbes. *Huygens Institute – Royal Netherlands Academy of Arts and Science, Proceedings* 3, 352–365.
147 Bein, S.J. (1954) A study of certain chromogenic bacteria isolated from 'red tide' water with a description of a new species. *Bulletin of Marine Science* 4(2), 110–119.
148 Bej, A.K., Patterson, D.P., Brasher, C.W., Vickery, M.C.L., Jones, D.D. and Kaysner, C.A. (1999) Detection of total and hemolysin-producing *Vibrio parahaemolyticus* in shellfish using multiplex PCR amplification of *tl*, *tdh* and *trh*. *Journal of Microbiological Methods* 36(3), 215–225.
149 Bejerano, Y., Sarig, S., Horne, M.T. and Roberts, R.J. (1979) Mass mortalities in silver carp *Hypophthalmichthys molitrix* (Valenciennes) associated with bacterial infection following handling. *Journal of Fish Diseases* 2(1), 49–56.
150 Bell, G.R. and Traxler, G.S. (1986) Resistance of the Pacific lamprey, *Lampetra tridentata* (Gairdner), to challenge by *Renibacterium salmoninarum*, the causative agent of kidney disease in salmonids. *Journal of Fish Diseases* 9, 277–279.
151 Belland, R.J. and Trust, T.J. (1988) DNA:DNA reassociation analysis of *Aeromonas salmonicida*. *Journal of General Microbiology* 134(2), 307–315.
152 Ben-Dov, E.,Ben Yosef, D.Z., Pavlov, V. and Kushmaro, A. (2009) *Corynebacterium maris* sp. nov., a marine bacterium isolated from the mucus of the coral *Fungia granulosa*. *International Journal of Systematic and Evolutionary Microbiology* 59(Pt 10), 2458–2463.
153 Benediktsdóttir, E., Verdonck, L., Spröer, C., Helgason, S. and Swings, J. (2000) Characterization of *Vibrio viscosus* and *Vibrio wodanis* isolated at different geographical locations: a proposal for reclassification of *Vibrio viscosus* as *Moritella viscosa* comb. nov. *International Journal of Systematic and Evolutionary Microbiology* 50(Pt 2), 479–488.
154 Benediktsdóttir, E., Helgason, S. and Sigurjónsdóttir, H. (1998) *Vibrio* spp. isolated from salmonids with shallow skin lesions and reared at low temperature. *Journal of Fish Diseases* 24(1), 19–28.
155 Ben-Haim, Y. and Rosenberg, E. (2002) A novel *Vibrio* sp. pathogen of the coral *Pocillopora damicornis*. *Marine Biology* 141, 47–55.
156 Ben-Haim, Y., Thompson, F.L., Thompson, C.C., Cnockaert, M.C., Hoste, B., Swings, J. and Rosenberg, E. (2003a) *Vibrio coralliilyticus* sp. nov., a temperature-dependent pathogen of the coral *Pocillopora damicornis*. *International Journal of Systematic and Evolutionary Microbiology* 53(Pt 1), 309–315.
157 Ben-Haim, Y., Zicherman-Keren, M. and Rosenberg, E. (2003b) Temperature-regulated bleaching and lysis of the coral *Pocillopora damicornis* by the novel pathogen *Vibrio coralliilyticus*. *Applied and Environmental Microbiology* 69(7), 4236–4242.
158 Benisch, J. (1940) Artificially induced *Aphanomyces* infestation in chinese crabs. *Zeitschrift für Fischerei* 38, 71–80.
159 Benjamin, J., Leaper, S., Owen, R.J. and Skirrow, M.B. (1983) Description of *Campylobacter laridis*, a new species comprising the nalidixic acid resistant thermophilic *Campylobacter* (NARTC) group. *Current Microbiology* 8, 231–238.

160 Ben Kahla-Nakbi, A., Chaieb, K. and Bakhrouf, A. (2009) Investigation of several virulence properties among *Vibrio alginolyticus* strains isolated from diseased cultured fish in Tunisia. *Diseases of Aquatic Organisms* 86(1), 21–28.

161 Bercovier, H., Steigerwalt, A.G., Guiyoule, A., Huntley-Carter, G. and Brenner, D.J. (1984) *Yersinia aldovae* (formerly *Yersinia enterocolitica*-like Group X2), a new species of *Enterobacteriaceae* isolated from aquatic ecosystems. *International Journal of Systematic Bacteriology* 34(2), 166–172.

162 Bercovier, H., Ursing, J., Brenner, J.D., Steigerwalt, A.G., Fanning, G.R., Carter, G.P. and Mollaret, H.H. (1980) *Yersinia kristensenii*: a new species of *Enterobacteriaceae* composed of sucrose-negative strains (formerly called atypical *Yersinia enterocolitica* or *Yersinia enterocolitica*-like). *Current Microbiology* 4(4), 219–224.

163 Berger, L., Speare, R. and Humphrey, J. (1997) Mucormycosis in a free-ranging green tree frog from Australia. *Journal of Wildlife Diseases* 33(4), 903–907.

164 Berger, L., Speare, R., Daszak, P., Green, D.E., Cunningham, A.A., Goggin, C.L., Slocombe, R., Ragan, M.A., Hyatt, A.D., McDonald, K.R., Hines, H.B., Lips, K.R., Marantelli, G. and Parkes, H. (1998) Chytridiomycosis causes amphibian mortality associated with population declines in the rain forests of Australia and Central America. *Proceedings of the National Academy of Sciences* 95(15), 9031–9036.

165 Bergman, S., Selig, M., Collins, M.D., Farrow, J.A., Baron, E.J., Dickersin, G.R. and Ruoff, K.L. (1995) '*Streptococcus milleri*' strains displaying a gliding type of motility. *International Journal of Systematic Bacteriology* 45(2), 235–239.

166 Bernardet, J.F. (1989) '*Flexibacter columnaris*' – first description in France and comparison with bacterial strains from other origins. *Diseases of Aquatic Organisms* 6, 37–44.

167 Bernardet, J.F., Campbell, A.C. and Buswell, J.A. (1990) *Flexibacter maritimus* is the agent of 'black patch necrosis' in Dover sole in Scotland. *Diseases of Aquatic Organisms* 8, 233–237.

168 Bernardet, J.F. and Grimont, P.A.D. (1989) Deoxyribonucleic acid relatedness and phenotypic characterization of *Flexibacter columnaris* sp. nov., nom. rev., *Flexibacter psychrophilus* sp. nov., nom. rev. and *Flexibacter maritimus* Wakabayashi, Hikida, and Masumura 1986. *International Journal of Systematic Bacteriology* 39(3), 346–354.

169 Bernardet, J.F. and Kerouault, B. (1989) Phenotypic and genomic studies of '*Cytophaga psychrophila*' isolated from diseased rainbow trout (*Oncorhynchus mykiss*) in France. *Applied and Environmental Microbiology* 55(7), 1796–1800.

170 Bernardet, J.F. and Nakagawa, Y. (2006) An introduction to the family *Flavobacteriaceae*. In: Dworkin, M., Falkow, S., Rosenberg, E., Schleifer, K.H. and Stackebrandt, E. (eds) *The Prokaryotes: a Handbook on the Biology of Bacteria*, 3rd edn, Vol. 7. Springer, New York, pp. 455–480.

171 Bernardet, J.-F., Kerouault, B. and Michel, C. (1994) Comparative study on *Flexibacter maritimus* strains isolated from farmed sea bass (*Dicentrarchus labrax*) in France. *Fish Pathology* 29(2), 105–111.

172 Bernardet, J.-F., Segers, P., Vancanneyt, M., Berthe, F., Kersters, K. and Vandamme, P. (1996) Cutting a Gordian knot: emended classification and description of the genus *Flavobacterium*, emended description of the family *Flavobacteriaceae*, and proposal of *Flavobacterium hydatis* nom. nov. (basonym, *Cytophaga aquatilis* Strohl and Tait 1978). *International Journal of Systematic Bacteriology* 46(1), 128–148.

173 Bernardet, J.-F., Nakagawa, Y. and Holmes, B. (2002) Proposed minimal standards for describing new taxa of the family *Flavobacteriaceae* and emended description of the family. *International Journal of Systematic and Evolutionary Microbiology* 52(Pt 3), 1049–1070.

174 Bernoth, E.-M (1990) Autoagglutination, growth on tryptone-soy-Coomassie agar, outer membrane protein patterns and virulence of *Aeromonas salmonicida* strains. *Journal of Fish Diseases* 13(2), 145–155.

175 Bernoth, E.-M. and Artz, G. (1989) Presence of *Aeromonas salmonicida* in fish tissue may be overlooked by sole reliance on furunculosis agar. *Bulletin of the European Association of Fish Pathologists* 9(1), 5–6.

176 Berthe, F.C.J., Michel, C. and Bernardet, J.-F. (1995) Identification of *Pseudomonas anguilliseptica* isolated from several fish species in France. *Diseases of Aquatic Organisms* 21(2), 151–155.

177 Bian, B.Z. and Egusa, S. (1980) *Atkinsiella hamanaensis* sp. nov. isolated from cultivated ova of the mangrove crab, *Scylla serrata* (Forsskal). *Journal of Fish Diseases* 3(5), 373–385.

178 Biebl, H., Pukall, R., Lünsdorf, H., Schulz, S., Allgaier, M., Tindall, B.J. and Wagner-Döbler, I. (2007) Description of *Labrenzia alexandrii* gen. nov., sp. nov., a novel alphaproteobacterium containing bacteriochlorophyll *a*, and a proposal for reclassification of *Stappia aggregata* as *Labrenzia aggregata* comb. nov., of *Stappia marina* as *Labrenzia marina* comb. nov. and of *Stappia alba* as *Labrenzia alba* comb. nov. and emended descriptions of the genera *Pannonibacter*, *Stappia* and *Roseibium*, and of the species *Roseibium denhamense* and *Roseibium hamelinense*. *International Journal of Systematic and Evolutionary Microbiology* 57(5), 1095–1107.

179 Binesse, J., Delsert, C., Saulnier, D., Champomier-Vergès, M.-C., Zagorec, M., Munier-Lehmann, H., Mazel, D. and Le Roux, F. (2008) Metalloprotease vsm is the major determinant of toxicity for extracellular products of *Vibrio splendidus*. *Applied and Environmental Microbiology* 74(23), 7108–7117.

180 Biosca, E.G., Amaro, C., Larsen, J.L. and Pedersen, K. (1997) Phenotypic and genotypic characterization of *Vibrio vulnificus*: proposal for the substitution of the subspecific taxon biotype for serovar. *Applied and Environmental Microbiology* 63(4), 1460–1466.

181 Birkbeck, T.H., Laidler, L.A., Grant, A.N. and Cox, D.I. (2002) *Pasteurella skyensis* sp. nov., isolated from Atlantic salmon (*Salmo salar* L.). *International Journal of Systematic and Evolutionary Microbiology* 52(Pt 3), 699–704.

182 Bisharat, N., Agmon, V., Finkelstein, R., Raz, R., Ben-Dror, G., Lerner, L., Soboh, S., Colodner, R., Cameron, D.N., Wykstra, D.L., Swerdlow, D.L. and Farmer, J.J. III (1999) Clinical, epidemiological, and microbiological features of *Vibrio vulnificus* biogroup 3 causing

outbreaks of wound infection and bacteraemia in Israel. *Lancet* 354(9188), 1421–1424.

183 Bishop, E.J., Shilton, C., Benedict, S., Kong, F., Gilbert, G.L., Gal, D., Godoy, D., Spratt, B.G. and Currie, B.J. (2007) Necrotizing fasciitis in captive juvenile *Crocodylus porosus* caused by *Streptococcus agalactiae*: an outbreak and review of the animal and human literature. *Epidemiology and Infection* 135(8), 1248–1255.

184 Blackall, P.J. and Doheny, C.M. (1987) Isolation and characterisation of *Bordetella avium* and related species and an evaluation of their role in respiratory disease in poultry. *Australian Veterinary Journal* 64(8), 235–239.

185 Blaiotta, G., Pepe, O., Mauriello, G., Villani, F., Andolfi, R. and Moschetti, G. (2002) 16S-23S rDNA intergenic spacer region polymorphism of *Lactococcus garvieae*, *Lactococcus raffinolactis* and *Lactococcus lactis* as revealed by PCR and nucleotide sequence analysis. *Systematic and Applied Microbiology* 25(4), 520–527.

186 Blanco, M.M., Gibello, A., Vela, A.I., Moreno, M.A., Domínguez, L. and Fernandez-Garayzábal, J.F. (2002) PCR detection and PFGE DNA macrorestriction analyses of clinical isolates of *Pseudomonas anguilliseptica* from winter disease outbreaks in sea bream *Sparus aurata*. *Diseases of Aquatic Organisms* 50(1), 19–27.

187 Blaylock, R.B., Overstreet, R.M. and Klich, M.A. (2001) Mycoses in red snapper (*Lutjanus campechanus*) caused by two deuteromycete fungi (*Penicillium corylophilum* and *Cladosporium sphaerospermum*). *Hydrobiologia* 460, 221–228.

188 Blazer, V.S., Lilley, J.H., Schill, W.B., Kiryu, Y., Densmore, C.L., Panyawachira, V. and Chinabut, S. (2002) *Aphanomyces invadans* in Atlantic Menhaden along the East Coast of the United States. *Journal of Aquatic Animal Health* 14(1), 1–10.

189 Blazer, V.S., Vogelbein, W.K., Densmore, C.L., May, E.B., Lilley, J.H. and Zwerner, D.E. (1999) *Aphanomyces* as a cause of ulcerative skin lesions of menhaden from Chesapeake Bay Tributaries. *Journal of Aquatic Animal Health* 11(4), 340–349.

190 Bleicher, A., Neuhaus, K. and Scherer, S. (2010) *Vibrio casei* sp. nov., isolated from the surfaces of two French red smear soft cheeses. *International Journal of Systematic and Evolutionary Microbiology* 60(Pt 8), 1745–1749.

191 Bly, J.E., Lawson, L.A., Dale, D.J., Szalai, A.J., Durborow, R.M. and Clem L.W. (1992) Winter saprolegniosis in channel catfish. *Diseases of Aquatic Organisms* 13, 155–164.

192 Boeger, W.A., Pie, M.R., Ostrensky, A. and Patella, L. (2005) Lethargic crab disease: multidisciplinary evidence supports a mycotic etiology. *Memórias do Instituto Oswaldo Cruz* 100(2), 161–167.

193 Boerner, L., Nevis, K.R., Hinckley, L.S., Weber, E.S. and Frasca, S. Jr (2004) Erysipelothrix septicemia in a little blue penguin (*Eudyptula minor*). *Journal of Veterinary Diagnostic Investigation* 16(2), 145–149.

194 Boettcher, K.J., Barber, B.J. and Singer, J.T. (1999) Use of antibacterial agents to elucidate the etiology of juvenile oyster disease (JOD) in *Crassostrea virginica* and numerical dominance of an alpha-proteobacterium in JOD-affected animals. *Applied and Environmental Microbiology* 65(6), 2534–2539.

195 Boettcher, K.J., Barber, B.J. and Singer, J.T. (2000) Additional evidence that juvenile oyster disease is caused by a member of the *Roseobacter* group and colonization of nonaffected animals by *Stappia stellulata*-like strains. *Applied and Environmental Microbiology* 66(9), 3924–3930.

196 Boettcher, K.J. and Ruby, E.G. (1990) Depressed light emission by symbiotic *Vibrio fischeri* of the sepiolid squid *Euprymna scolopes*. *Journal of Bacteriology* 172(7), 3701–3706.

197 Boettcher, K.J, Geaghan, K.K., Maloy, A.P. and Barber, B.J. (2005) *Roseovarius crassostreae* sp. nov., a member of the *Roseobacter* clade and the apparent cause of juvenile oyster disease (JOD) in cultured Eastern oysters. *International Journal of Systematic and Evolutionary Microbiology* 55(Pt 4), 1531–1537.

198 Bondad-Reantaso, M.G., Lumanlan-Mayo, S.C., Natividad, J.M. and Phillips, M.J. (1992) Environmental monitoring of the epizootic ulcerative syndrome (EUS) in fish from Munoz, nueva Ecija in the Philippines. In: Shariff, M., Subasinghe, R.P. and Arthur, J.R. (eds) *Diseases in Asian Aquaculture*, Vol. I. Fish Health Section: Asian Fisheries Society, Manila, pp. 475–490.

199 Boomker, J., Henton, M.M., Naudé, T.W. and Hunchzermeyer, F.W. (1984) Furunculosis in rainbow trout (*Salmo gairdneri*) raised in sea water. *The Onderstepoort Journal of Veterinary Research* 51(2), 91–94.

200 Borg, A. (1948) *Studies on myxobacteria associated with diseases in salmonid fishes*. University of Washington, Washington, DC.

201 Borrego, J.J., Castro, D., Luque, A., Paillard, C., Maes, P., Garcia, M.T. and Ventosa, A. (1996) *Vibrio tapetis* sp. nov., the causative agent of the brown ring disease affecting cultured clams. *International Journal of Systematic Bacteriology* 46(2), 480–484.

202 Borteiro, C., Cruz, J.C., Kolenc, F. and Aramburu, A. (2009) Chytridiomycosis in frogs from Uruguay. *Diseases of Aquatic Organisms* 84(2), 159–162.

203 Borucinska, J.D. and Frasca, S. (2002) Naturally occurring lesions and micro-organisms in two species of free-living sharks: the spiny dogfish, *Squalus acanthias* L. and the smooth dogfish, *Mustelus canis* (Mitchill), from the north-western Atlantic. *Journal of Fish Diseases* 25(5), 287–298.

204 Bourke, J., Ohst, T., Gräser, Y., Böhme, W. and Plötner, J. (2011) New records of *Batrachochytrium dendrobatidis* in Chilean frogs. *Diseases of Aquatic Organisms* 95(3), 259–261.

205 Bouvet, O.M.M., Grimont, P.A.D., Richard, C., Aldova, E., Hausner, O. and Gabrhelova, M. (1985) *Budvicia aquatica* gen. nov., sp. nov.: a hydrogen sulfide-producing member of the *Enterobacteriaceae*. *International Journal of Systematic Bacteriology* 35(1), 60–64.

206 Bouvet, P.J.M. and Grimont, P.A.D. (1986) Taxonomy of the genus *Acinetobacter* with the recognition of *Acinetobacter baumannii* sp. nov., *Acinetobacter haemolyticus* sp. nov., *Acinetobacter johnsonii* sp. nov. and *Acinetobacter junii* sp. nov. and emended descriptions of *Acinetobacter calcoaceticus* and *Acinetobacter lwoffii*. *International Journal of Systematic Bacteriology* 36(2), 228–240.

207 Bowenkamp, K.E., Frasca, S. Jr, Draghi, A. II, Tsongalis, G.J., Koerting, C., Hinckley, L., De Guise, S., Montali, R.J., Goertz, C.E., St Aubin, D.J. and Dunn, J.L. (2001) *Mycobacterium marinum* dermatitis and panniculitis with chronic pleuritis in a captive white whale (*Delphinapterus leucas*) with aortic rupture. *Journal of Veterinary Diagnostic Investigation* 13(6), 524–530.

208 Bower, S.M., Goh, B. and Myer, G.R. (2005) Epizootiology and detection of nocardiosis in oysters. In: Walker, P., Lester, R. and Bondad-Reantaso, M.G. (eds) *Diseases in Asian Aquaculture*. Fish Health Section, Asian Fisheries Society, Manila, pp. 249–262.

209 Bowman, J.P., McCammon, S.A., Nichols, D.S., Skerratt, J.H., Rea, S.M., Nichols, P.D. and McMeekin, T.A. (1997) *Shewanella gelidimarina* sp. nov. and *Shewanella frigidimarina* sp. nov., novel Antarctic species with the ability to produce eicosapentaenoic acid (20:5 omega 3) and grow anaerobically by dissimilatory Fe(III) reduction. *International Journal of Systematic Bacteriology* 47(4), 1040–1047.

210 Bowman, J.P. and McMeekin, T.A. (2005) Order X. *Alteromonadales* ord. nov. In: Brenner, D.J., Krieg, N.R., Staley, J.T. and Garrity, G.M. (eds) *Bergey's Manual of Systematic Bacteriology*, 2nd edn, Vol. II *The Proteobacteria, part B, The Gammaproteobacteria*. Springer, New York, p. 443.

211 Bowman, J.P. and Nichols, D.S. (2002) *Aequorivita* gen. nov., a member of the family *Flavobacteriaceae* isolated from terrestrial and marine Antarctic habitats. *International Journal of Systematic and Evolutionary Microbiology* 52(Pt 5), 1533–1541.

212 Bowser, P.R., Rosemark, R. and Reiner, C.R. (1981) A preliminary report of vibriosis in cultured American lobsters, *Homarus americanus*. *Journal of Invertebrate Pathology* 37(1), 80–85.

213 Boyle, D.G., Boyle, D.B., Olsen, V., Morgan, J.A.T. and Hyatt, A.D. (2004) Rapid quantitative detection of chytridiomycosis (*Batrachochytrium dendrobatidis*) in amphibian samples using real-time Taqman PCR assay. *Diseases of Aquatic Organisms* 60(2), 141–148.

214 Boys, C.A., Rowland, S.J., Gabor, M., Gabor, L., Marsh, I.B., Hum, S. and Callinan, R.B. (2012) Emergence of epizootic ulcerative syndrome in native fish of the Murray-Darling river system, Australia: hosts, distribution and possible vectors. *PLoS ONE* 7(4).

215 Bozal, N., Tudela, E., Rosselló-Mora, R., Lalucat, J. and Guinea, J. (1997) *Pseudoalteromonas antarctica* sp. nov., isolated from an Antarctic coastal environment. *International Journal of Systematic Bacteriology* 47(2), 345–351.

216 Bragg, R.R., Huchzermeyer, H.F. and Hanisch, M.A. (1990) *Mycobacterium fortuitum* isolated from three species of fish in South Africa. *The Onderstepoort Journal of Veterinary Research* 57(1), 101–102.

217 Bransden, M.P., Carson, J., Munday, B.L., Handlinger, J.H., Carter, C.G. and Nowak, B.F. (2000) Nocardiosis in tank-reared Atlantic salmon, *Salmo salar L. Journal of Fish Diseases* 23(1), 83–85.

218 Brasher, C.W., DePaola, A., Jones, D.D. and Bej, A.K. (1998) Detection of microbial pathogens in shellfish with multiplex PCR. *Current Microbiology* 37(2), 101–107.

219 Brauns, L.A., Hudson, M.C. and Oliver, J.D. (1991) Use of the polymerase chain reaction in detection of culturable and nonculturable *Vibrio vulnificus* cells. *Applied and Environmental Microbiology* 57(9), 2651–2655.

220 Brayton, P.R., Bode, R.B., Colwell, R.R., MacDonell, M.T., Hall, H.L., Grimes, D.J., West, P.A. and Bryant, T.N. (1986) *Vibrio cincinnatiensis* sp. nov., a new human pathogen. *Journal of Clinical Microbiology* 23(1), 104–108.

221 Breed, R.S., Murray, E.G.D. and Smith, N. (1957) *Bergey's Manual of Determinative Bacteriology*. Williams & Wilkins Co., Baltimore, Maryland.

222 Brenner, D.J., Hickman-Brenner, F.W., Lee, J.V., Steigerwalt, A.G., Fanning, G.R., Hollis, D.G., Farmer, J.J., Weaver, R.E., Joseph, S.W. and Seidler, R.J. (1983) *Vibrio furnissii* (formerly aerogenic biogroup of *Vibrio fluvialis*), a new species isolated from human feces and the environment. *Journal of Clinical Microbiology* 18(4), 816–824.

223 Brenner, D.J., Müller, H.E., Steigerwalt, A.G., Whitney, A.M., O'Hara, C.M. and Kämpfer, P. (1998) Two new *Rahnella* genomospecies that cannot be phenotypically differentiated from *Rahnella aquatilis*. *International Journal of Systematic Bacteriology* 48(Pt 1), 141–149.

224 Brenner, D.J., Farmer, J.J., Fanning, G.R., Steigerwalt, A.G., Klykken, P., Wathen, H.G., Hickman, F.W. and Ewing, W.H. (1978) Deoxyribonucleic acid relatedness of *Proteus* and *Providencia* species. *International Journal of Systematic Bacteriology* 28(2), 269–282.

225 Brenner, D.J., Steigerwalt, A.G., Falcão, D.P., Weaver, R.E. and Fanning, G.R. (1976) Characterization of *Yersinia enterocolitica* and *Yersinia pseudotuberculosis* by deoxyribonucleic acid hybridization and by biochemical reactions. *International Journal of Systematic Bacteriology* 26(2), 180–194.

226 Brew, S.D., Perrett, L.L., Stack, J.A., MacMillan, A.P. and Staunton, N.J. (1999) Human exposure to *Brucella* recovered from a sea mammal. *The Veterinary Record* 144(17), 483.

227 Bricker, B.J., Ewalt, D.R., MacMillan, A.P., Foster, G. and Brew, S. (2000) Molecular characterization of *Brucella* strains isolated from marine mammals. *Journal of Clinical Microbiology* 38(3), 1258–1262.

228 Bromage, E.S., Thomas, A. and Owens, L. (1999) *Streptococcus iniae*, a bacterial infection in barramundi *Lates calcarifer*. *Diseases of Aquatic Organisms* 36(3), 177–181.

229 Brown, D.R., Farley, J.M., Zacher, L.A., Carlton, J.M., Clippinger, T.L., Tully, J.G. and Brown, M.B. (2001a) *Mycoplasma alligatoris* sp. nov., from American alligators. *International Journal of Systematic and Evolutionary Microbiology* 51(2), 419–424.

230 Brown, D.R., Nogueira, M.F., Schoeb, T.R., Vliet, K.A., Bennett, R.A., Pye, G.W. and Jacobson, E.R. (2001b) Pathology of experimental mycoplasmosis in American alligators. *Journal of Wildlife Diseases* 37(4), 671–679.

231 Brown, D.R., Whitcomb, R.F. and Bradbury, J.M. (2007) Revised minimal standards for description of new species of the class *Mollicutes* (division *Tenericutes*). *International Journal of Systematic and Evolutionary Microbiology* 57(Pt 11), 2703–2719.

232 Brown, G.R., Sutcliffe, I.C. and Cummings, S.P. (2001c) Reclassification of [*Pseudomonas*] *doudoroffii* (Baumann *et al.* 1983) into the genus *Oceanomonas*

gen. nov. as *Oceanomonas doudoroffii* comb. nov. and description of a phenol-degrading bacterium from estuarine water as *Oceanomonas baumannii* sp. nov. *International Journal of Systematic and Evolutionary Microbiology* 51(1), 67–72.
233 Brown-Elliott, B.A., Brown, J.M., Conville, P.S. and Wallace, R.J. Jr (2006) Clinical and laboratory features of the *Nocardia* spp. based on current molecular taxonomy. *Clinical Microbiology Reviews* 19(2), 259–282.
234 Bruno, D. (1989) Observations on a swim bladder fungal infection of farmed Atlantic salmon, *Salmo salar L. Bulletin of the European Association of Fish Pathologists* 9(1), 7–8.
235 Bruno, D.W., Griffiths, J., Mitchell, C.G., Wood, B.P., Fletcher, Z.J., Drobniewski, F.A. and Hastings, T.S. (1998a) Pathology attributed to *Mycobacterium chelonae* infection among farmed and laboratory-infected Atlantic salmon *Salmo salar*. *Diseases of Aquatic Organisms* 33(2), 101–109.
236 Bruno, D.W., Griffiths, J., Petrie, J. and Hastings, T.S. (1998b) *Vibrio viscosus* in farmed Atlantic salmon *Salmo salar* in Scotland: field and experimental observations. *Diseases of Aquatic Organisms* 34, 161–166.
237 Bruno, D.W., Hastings, T.S. and Ellis, A.E. (1986a) Histopathology, bacteriology and experimental transmission of cold-water vibriosis in Atlantic salmon *Salmo salar*. *Diseases of Aquatic Organisms* 1, 163–168.
238 Bruno, D.W. and Munro, A.L.S. (1986b) Uniformity in the biochemical properties of *Renibacterium salmoninarum* isolates obtained from several sources. *FEMS Microbiology Letters* 33(2–3), 247–250.
239 Bruno, D.W. and Stamps, D.J. (1987) Saprolegniasis of Atlantic salmon, *Salmo salar* L., fry. *Journal of Fish Diseases* 10, 513–517.
240 Bruno, D., Collet, B., Turnbull, A., Kilburn, R., Walker, A., Pendrey, D., McIntosh, A., Urquhart, K. and Taylor, G. (2007) Evaluation and development of diagnostic methods for *Renibacterium salmoninarum* causing bacterial kidney disease (BKD) in the UK. *Aquaculture* 269(1–4), 114–122.
241 Brunvold, L., Sandaa, R.-A., Mikkelsen, H., Welde, E., Bleie, H. and Bergh, Ø. (2007) Characterisation of bacterial communities associated with early stages of intensively reared cod (*Gadus morhua*) using Denaturing Gradient Gel Electrophoresis (DGGE). *Aquaculture* 272(1–4), 319–327.
242 Buck, J.D., Meyers, S.P. and Leifson, E. (1963) *Pseudomonas* (*Flavobacterium*) *piscicida* bein comb. nov. *Journal of Bacteriology* 86(5), 1125–1126.
243 Buller, N.B. (2008) *Crayfish Plague: Australian and New Zealand Standard Diagnostic Procedure*. Sub-Committee on Animal Health Laboratory Standards.
244 Buller, N.B. (n.d.) Unpublished data.
245 Buller, N. (2004) Molecular diagnostic tests to detect epizootic ulcerative syndrome (*Aphanomyces invadans*) and Crayfish plague (*Aphanomyces astaci*). *FRDC Project Number 2001/621*. Fisheries Research and Development Corporation, Deakin West, Australia.
246 Buller, N.B. (2007) Further research and laboratory trials for diagnostic tests for the detection of *A. invadans* (EUS) and *A. astaci* (Crayfish Plague). *FRDC Project Number 20004/091*. Fisheries Research and Development Corporation, Deakin West, Australia.
247 Bullock, G.L., Hsu, T.C. and Shotts, E.B. Jr (1986) Columnaris disease of fishes. *U.S. Fish and Wildlife Service Fish Disease Leaflet* 72, 1–9.
248 Bullock, G.L., Stuckey, H.M. and Chen, P.K. (1974) Corynebacterial kidney disease of salmonids: growth and serological studies on the causative bacterium. *Applied Microbiology* 28(5), 811–814.
249 Bullock, G.L., Stuckey, H.M. and Shotts, E.B. (1978) Enteric redmouth bacterium: comparison of isolates from different geographic areas. *Journal of Fish Diseases* 1(4), 351–356.
250 Burgents, J.E., Burnett, K.G. and Burnett, L.E. (2005) Effects of hypoxia and hypercapnic hypoxia on the localization and the elimination of *Vibrio campbellii* in *Litopenaeus vannamei*, the Pacific white shrimp. *The Biological Bulletin* 208(3), 159–168.
251 Burns, G.F. and Williams, H. (1975) *Clostridium botulinum* in Scottish fish farms and farmed trout. *The Journal of Hygiene* 74(1), 1–6.
252 Burr, S.E., Gobeli, S., Kuhnert, P., Goldschmidt-Clermont, E. and Frey, J. (2010) *Pseudomonas chlororaphis* subsp. *piscium* subsp. nov., isolated from freshwater fish. *International Journal of Systematic and Evolutionary Microbiology* 60(Pt 12), 2753–2757.
253 Bustos, P.A., Calbuyahue, J., Montana, J., Opazo, B., Entrala, P. and Solervicens, R. (1995) First isolation of *Flexibacter psychrophilus*, as causative agent of rainbow trout fry syndrome (RTFS), producing rainbow trout mortality in Chile. *Bulletin of the European Association of Fish Pathologists* 15(5), 162–164.
254 Butzler, J.P., Dekeyser, P., Detrain, M. and Dehaen, F. (1973) Related vibrio in stools. *Journal of Pediatrics* 82(3), 493–495.
255 Cabanillas-Beltrán, H., LLausás-Magaña, E., Romero, R., Espinoza, A., García-Gasca, A., Nishibuchi, M., Ishibashi, M. and Gomez-Gil, B. (2006) Outbreak of gastroenteritis caused by the pandemic *Vibrio parahaemolyticus* O3:K6 in Mexico. *FEMS Microbiology Letters* 265(1), 76–80.
256 Cai, J., Chen, H., Thompson, K.D. and Li, C. (2006a) Isolation and identification of *Shewanella alga* and its pathogenic effects on post-larvae of abalone *Haliotis diversicolor supertexta*. *Journal of Fish Diseases* 29(8), 505–508.
257 Cai, J., Han, H., Song, Z., Li, C. and Zhou, J. (2006b) Isolation and characterization of pathogenic *Vibrio alginolyticus* from diseased postlarval abalone, *Haliotis diversicolor supertexta* (Lischke). *Aquaculture Research* 37, 1222–1226.
258 Cai, S.H., Lu, Y.S., Wu, Z.H., Jian, J.C. and Huang, Y.C. (2009) A novel multiplex PCR method for detecting virulent strains of *Vibrio alginolyticus*. *Aquaculture Research* 41, 27–34.
259 Cai, S.H., Lu, Y.S., Wu, Z.-H., Jian, J.C., Wang, B. and Huang, Y.C. (2010) Loop-mediated isothermal amplification method for rapid detection of *Vibrio alginolyticus*, the causative agent of vibriosis in mariculture fish. *Letters in Applied Microbiology* 50(5), 480–485.
260 Cai, Y., Benno, Y., Nakase, T. and Oh, T.-K. (1998) Specific probiotic characterization of *Weissella hellenica* DS-12 isolated from flounder intestine. *The Journal of General and Applied Microbiology* 44(5), 311–316.

261 Caldwell, D.K., Caldwell, M.C., Woodard, J.C., Ajello, L., Kaplan, W. and McClure, H.M. (1975) Lobomycosis as a disease of the Atlantic bottle-nosed dolphin (*Tursiops truncatus* Montagu, 1821). *The American Journal of Tropical Medicine and Hygiene* 24(1), 105–114.

262 Callinan, R.B., Fraser, G.C. and Virgona, J.L. (1989) Pathology of red spot disease in sea mullet, *Mugil cephalus* L., from eastern Australia. *Journal of Fish Diseases* 12, 467–479.

263 Callinan, R.B., Paclibare, J.O., Bondad-Reantaso, M.G., Chin, J.C. and Gogolewski, R.P. (1995) *Aphanomyces* species associated with epizootic ulcerative syndrome (EUS) in the Philippines and red spot disease (RSD) in Australia: preliminary comparative studies. *Diseases of Aquatic Organisms* 21, 233–238.

264 Cammà, C., Ferri, N., Zezza, D., Marcacci, M., Paolini, A., Ricchiuti, L. and Lelli, R. (2010) Confirmation of crayfish plague in Italy: detection of *Aphanomyces astaci* in white clawed crayfish. *Diseases of Aquatic Organisms* 89(3), 265–268.

265 Campbell, S., Harada, R.M. and Li, Q.X. (2007) *Ferrimonas senticii* sp. nov., a novel gammaproteobacterium isolated from the mucus of a puffer fish caught in Kaneohe Bay, Hawai'i. *International Journal of Systematic and Evolutionary Microbiology* 57(Pt 11), 2670–2673.

266 Campbell, S., Harada, R.M. and Li, Q.X. (2008) *Chryseobacterium arothri* sp. nov., isolated from the kidneys of a pufferfish. *International Journal of Systematic and Evolutionary Microbiology* 58(Pt 1), 290–293.

267 Camus, A.C., Shewmaker, P.L., Mauel, M.J. and Wise, D.J. (2008) *Streptococcus ictaluri* arthritis, osteolysis, myositis, and spinal meningitis in channel catfish broodstock. *Journal of Aquatic Animal Health* 20(1), 54–62.

268 Candan, A., Kucker, M.A. and Karatas, S. (1996) Pasteurellosis in cultured sea bass (*Dicentrarchus labrax*) in Turkey. *Bulletin of the European Association of Fish Pathologists* 16(5), 150–153.

269 Cann, D.C. and Taylor, L.Y. (1982) An outbreak of botulism in rainbow trout, *Salmo gairdneri* Richardson, farmed in Britain. *Journal of Fish Diseases* 5, 393–399.

270 Cano-Gómez, A., Goulden, E.F., Owens, L. and Høj, L. (2010) *Vibrio owensii* sp. nov., isolated from cultured crustaceans in Australia. *FEMS Microbiology Letters* 302(2), 175–181.

271 Cano-Gómez, A., Høj, L., Owens, L. and Andreakis, N. (2011) Multilocus sequence analysis provides basis for fast and reliable identification of *Vibrio harveyi*-related species and reveals previous misidentification of important marine pathogens. *Systematic and Applied Microbiology* 34(8), 561–565.

272 Cao, H., Ou, R., He, S. and Yang, X. (2013) Identification of an *Achlya klebsiana* isolate as the causal agent of saprolegniosis of yellow catfish (*Pelteobagrus fuvidraco*) eggs and its control with herb extracts. *The Israeli Journal of Aquaculture – Bamidgeh* 65, 1–9.

273 Cao, H., Zheng, W., Xu, J., Ou, R., He, S. and Yang, X. (2012) Identification of an isolate of *Saprolegnia ferax* as the causal agent of saprolegniosis of yellow catfish (*Pelteobagrus fulvidraco*) eggs. *Veterinary Research Communications* 36(4), 239–244.

274 Cao, J., Xu, J., Zheng, Q. and Yan, P. (2010) Rapid detection of *Vibrio metschnikovii* in aquatic products by real-time PCR. *Folia Microbiologica* 55(6), 607–613.

275 Carleton, H.M., Drury, R.A.B. and Wallington, E.A. (1980) *Carleton's Histological Technique*. Oxford University Press, Oxford.

276 Carnahan, A., Fanning, G.R. and Joseph, S.W. (1991a) *Aeromonas jandaei* (formerly genospecies DNA group 9 *A. sobria*), a new sucrose-negative species isolated from clinical specimens. *Journal of Clinical Microbiology* 29(3), 560–564.

277 Carnahan, A.M., Chakraborty, T., Fanning, G.R., Verma, D., Ali, A., Janda, J.M. and Joseph, S.W. (1991b) *Aeromonas trota* sp. nov., an ampicillin-susceptible species isolated from clinical specimens. *Journal of Clinical Microbiology* 29(6), 1206–1210.

278 Carnahan, A.M., Marii, M.A., Fanning, G.R., Pass, M.A. and Joseph, S.W. (1989) Characterization of *Aeromonas schubertii* strains recently isolated from traumatic wound infections. *Journal of Clinical Microbiology* 27(8), 1826–1830.

279 Carson, J., Gudkovs, N. and Austin, B. (1993) Characteristics of an *Enterococcus*-like bacterium from Australia and South Africa, pathogenic for rainbow trout, *Oncorhynchus mykiss* (Walbaum). *Journal of Fish Diseases* 16(4), 381–388.

280 Carson, J., Higgins, M.J., Wilson, T.K., Gudkovs, N. and Bryant, T.N. (2009) *Identification of Vibrionaceae from Australian Aquatic Animals using Phenotypic and PCR Procedures*. Australia and New Zealand Standard Diagnostic Procedure. Sub-Committee on Animal Health Laboratory Standards.

281 Carson, J. and Wilson, T.K. (2009) *Yersiniosis in Fish*. Australia and New Zealand Standard Diagnostic Procedure. Sub-Committee on Animal Health Laboratory Standards.

282 Carson, J. and Handlinger, J. (1988) Virulence of the aetiological agent of goldfish ulcer disease in Atlantic salmon, *Salmo salar* L. *Journal of Fish Diseases* 11, 471–479.

283 Carson, J. and Gudkovs, N. (2001) *Aeromonas salmonicida*: a sleeping giant. *Microbiology Australia* (July), 28–30.

284 Carson, J., Schmidtke, L.M. and Munday, B.L. (1993) *Cytophaga johnsonae*: a puative skin pathogen of juvenile farmed barramundi, *Lates calcarifer* Bloch. *Journal of Fish Diseases* 16, 209–218.

285 Castro, D., Martinez-Manzanares, E., Luque, A., Fouz, B., Moriñigo, M., Borrego, J.J. and Toranzo, A.E. (1992) Characterization of strains related to brown ring disease outbreaks in southwestern Spain. *Diseases of Aquatic Organisms* 14, 229–236.

286 Castro, N., Toranzo, A.E., Barja, J.L., Núñez, S. and Magariños, B. (2006) Characterization of *Edwardsiella tarda* strains isolated from turbot, *Psetta maxima* (L.). *Journal of Fish Diseases* 29(9), 541–547.

287 Castro, N., Toranzo, A.E., Devesa, S., González, A., Nuñez, S. and Magariños, B. (2012) First description of *Edwardsiella tarda* in Senegalese sole, *Solea senegalensis* (Kaup). *Journal of Fish Diseases* 35(1), 79–82.

288 Castro, N., Toranzo, A.E., Nuñez, S. and Magariños, B. (2011) Evaluation of the selective and differential ET

medium for detection of *Edwardsiella tarda* in aquaculture systems. *Letters in Applied Microbiology* 53(1), 114–119.
289 Castro, N., Toranzo, A.E., Núñez, S., Osorio, C.R. and Magariños, B. (2010) Evaluation of four polymerase chain reaction primer pairs for the detection of *Edwardsiella tarda* in turbot. *Diseases of Aquatic Organisms* 90(1), 55–61.
290 Cavalier-Smith, T. (1998) A revised six-kingdom system of life. *Biological Reviews of the Cambridge Philosophical Society* 73(3), 203–266.
291 Cepeda, C., García-Márquez, S. and Santos, Y. (2003) Detection of *Flexibacter maritimus* in fish tissue using nested PCR amplification. *Journal of Fish Diseases* 26(2), 65–70.
292 Cerdà-Cuéllar, M., Permin, L., Larsen, J.L. and Blanch, A.R. (2001) Comparison of selective media for the detection of *Vibrio vulnificus* in environmental samples. *Journal of Applied Microbiology* 91(2), 322–327.
293 Cerdà-Cuéllar, M., Rosselló-Mora, R.A., Lalucat, J., Jofre, J. and Blanch, A. (1997) *Vibrio scophthalmi* sp. nov., a new species from turbot (*Scophthalmus maximus*). *International Journal of Systematic Bacteriology* 47(1), 58–61.
294 Cerenius, L. and Söderhäll, K. (1984) Repeated zoospore emergence from isolated spore cysts of *Aphanomyces astaci*. *Experimental Mycology* 8(4), 370–377.
295 Cerenius, L., Söderhäll, K., Persson, M. and Ajaxon, R. (1988) The crayfish plague fungus *Aphanomyces astaci* – diagnosis, isolation, and pathobiology. *Freshwater Crayfish* 7, 131–144.
296 Chambers, E. and Barker, G. (2006) Comparison of culture media for the isolation of *Renibacterium salmoninarum* from naturally infected rainbow trout (*Oncorhynchus mykiss*). *Bulletin of the European Association of Fish Pathologists* 26(3), 137–142.
297 Chambers, E.M., Nagel, D.A., Elloway, E.A., Addison, K.L., Barker, G.A., Verner-Jeffreys, D.W. and Stone, D.M. (2009) Polymerase chain reaction detection of *Renibacterium salmoninarum* in fish: validation of a modified protocol. *Aquaculture* 287(1–2), 35–39.
298 Chan, K.Y., Baumann, L., Garza, M.M. and Baumann, P. (1978) Two new species of *Alteromonas*: *Alteromonas espejiana* and *Alteromonas undina*. *International Journal of Systematic Bacteriology* 28(2), 217–222.
299 Chang, C.L., Jeong, J., Shin, J.H., Lee, E.Y. and Son, H.C. (1999) *Rahnella aquatilis* sepsis in an immunocompetent adult. *Journal of Clinical Microbiology* 37(12), 4161–4162.
300 Chang, C.-I., Lee, C.-F., Wu, C.-C., Cheng, T.C., Tsai, J.-M. and Lin, K.-J. (2011) A selective and differential medium for *Vibrio alginolyticus*. *Journal of Fish Diseases* 34(3), 227–234.
301 Chang, H.-W., Roh, S.W., Kim, K.-H., Nam, Y.-D., Jeon, C.O., Oh, H.-M. and Bae, J.-W. (2008) *Vibrio areninigrae* sp. nov., a marine bacterium isolated from black sand. *International Journal of Systematic and Evolutionary Microbiology* 58(Pt 8), 1903–1906.
302 Chang, T., Hsieh, C., Chang, C., Shen, Y., Huang, K., Tu, C., Chen, L., Wu, Z. and Tsai, S. (2006) Pathological and molecular studies on mycobacteriosis of milkfish *Chanos chanos* in Taiwan. *Diseases of Aquatic Organisms* 72(2), 147–151.
303 Chanock, R.M., Hayflick, L. and Barile, M.F. (1962) Growth on artificial medium of an agent associated with atypical pneumonia and its identification as a PPLO. *Proceedings of the National Academy of Sciences of the United States of America* 48(1), 41–49.
304 Chapman, P.F., Cipriano, R.C. and Teska, J.D. (1991) Isolation and phenotypic characterization of an oxidase-negative *Aeromonas salmonicida* causing furunculosis in coho salmon (*Oncorhynchus kisutch*). *Journal of Wildlife Diseases* 27(1), 61–67.
305 Chase, D.M. and Pascho, R.J. (1998) Development of a nested polymerase chain reaction for amplification of a sequence of the p57 gene of *Renibacterium salmoninarum* that provides a highly sensitive method for detection of the bacterium in salmonid kidney. *Diseases of Aquatic Organisms* 34(3), 223–229.
306 Chase, D.M., Elliott, D.G. and Pascho, R.J. (2006) Detection and quantification of *Renibacterium salmoninarum* DNA in salmonid tissues by real-time quantitative polymerase chain reaction analysis. *Journal of Veterinary Diagnostic Investigation* 18(4), 375–380.
307 Chen, D., Wang, K., Geng, Y., Wang, J., Huang, X. and He, M. (2011a) Pathological changes in cultured channel catfish *Ictalurus punctatus* spontaneously infected with *Streptococcus iniae*. *Diseases of Aquatic Organisms* 95(3), 203–208.
308 Chen, J.-D. and Lai, S.-Y. (1998) PCR for direct detection of *Edwardsiella tarda* from infected fish and environmental water by application of the hemolysin gene. *Zoological Studies* 37(3), 169–176.
309 Chen, M.E., Henry-Ford, D. and Groff, J.M. (1995) Isolation and characterization of *Flexibacter maritimus* from marine fishes of California. *Journal of Aquatic Animal Health* 7(4), 318–326.
310 Chen, Q., Yan, Q., Wang, K., Zhuang, Z. and Wang, X. (2008) Portal of entry for pathogenic *Vibrio alginolyticus* into large yellow croaker *Pseudosciaena crocea*, and characteristics of bacterial adhesion to mucus. *Diseases of Aquatic Organisms* 80(3), 181–188.
311 Chen, S.C., Lin, Y.D., Liaw, L.L. and Wang, P.C. (2001) *Lactococcus garvieae* infection in the giant freshwater prawn *Macrobranchium rosenbergii* confirmed by polymerase chain reaction and 16S rDNA sequencing. *Diseases of Aquatic Organisms* 45(1), 45–52.
312 Chen, S., Lee, J., Lai, C., Gu, Y., Wang, C., Chang, H. and Tsai, K. (2000) Nocardiosis in sea bass, *Lateolabrax japonicus*, in Taiwan. *Journal of Fish Diseases* 23(5), 299–307.
313 Chen, S.-C., Liaw, L.-L., Su, H.-Y., Ko, S.-C., Wu, C.-Y., Chaung, H.-C., Tsai, Y.-H., Yang, K.-L., Chen, Y.-C., Chen, T.-H., Lin, G.-R., Cheng, S.-Y., Lin, Y.-D., Lee, J.-L., Lai, C.-C., Weng, Y.-J. and Chu, S.-Y. (2002) *Lactococcus garvieae*, a cause of disease in grey mullet, *Mugil cephalus* L., in Taiwan. *Journal of Fish Diseases* 25(12), 727–732.
314 Chen, W.-M., Huang, W.-C., Young, C.-C. and Sheu, S.-Y. (2012a) *Flavobacterium tilapiae* sp. nov., isolated from a freshwater pond, and emended descriptions of *Flavobacterium defluvii* and *Flavobacterium johnsoniae*. *International Journal of Systematic and Evolutionary Microbiology*.
315 Chen, Y.F., Liang, R.S., Zhuo, X.L., Wu, X.T. and Zou, J.X. (2012b) Isolation and characterization of

Aeromonas schubertii from diseased snakehead, *Channa maculata* (Lacepède). *Journal of Fish Diseases* 35(6), 421–430.
316 Chen, Y.-G., Hu, S.-P., Tang, S.-K., He, J.-W., Xiao, J.-Q., Zhu, H.-Y. and Li, W.-J. (2011b) *Bacillus zhanjiangensis* sp. nov., isolated from an oyster in South China Sea. *Antonie Van Leeuwenhoek* 99(3), 473–480.
317 Chen, Y.-G., Zhang, L., Zhang, Y.-Q., He, J.-W., Klenk, H.-P., Tang, S.-K., Zhang, Y.-X. and Li, W.-J. (2011c) *Bacillus nanhaiensis* sp. nov., isolated from an oyster. *International Journal of Systematic and Evolutionary Microbiology* 61(Pt 4), 888–893.
318 Chien, C.H., Miyazaki, T. and Kubota, S. (1979) Branchiomycosis of the reared Japanese eel in Taiwan. *Fish Pathology* 13(4), 179–182.
319 Chien, M.-S., Gilbert, T.L., Huang, C., Landolt, M.L., O'Hara, P.J. and Winton, J.R. (1992) Molecular cloning and sequence analysis of the gene coding for the 57-kDa major soluble antigen of the salmonid fish pathogen *Renibacterium salmoninarum*. *FEMS Microbiology Letters* 96(2–3), 259–266.
320 Chimetto, L.A., Cleenwerck, I., Alves, N. Jr, Silva, B.S., Brocchi, M., Willems, A., De Vos, P. and Thompson, F.L. (2011a) *Vibrio communis* sp. nov., isolated from the marine animals *Mussismilia hispida, Phyllogorgia dilatata, Palythoa caribaeorum, Palythoa variabilis and Litopenaeus vannamei*. *International Journal of Systematic and Evolutionary Microbiology* 61(Pt 2), 362–368.
321 Chimetto, L.A., Cleenwerck, I., Moreira, A.P.B., Brocchi, M., Willems, A., De Vos, P. and Thompson, F.L. (2011b) *Vibrio variabilis* sp. nov. and *Vibrio maritimus* sp. nov., isolated from *Palythoa caribaeorum*. *International Journal of Systematic and Evolutionary Microbiology* 61(Pt 12), 3009–3015.
322 Chinabut, S., Roberts, R.J., Willoughby, G.R. and Pearson, M.D. (1995) Histopathology of snakehead, *Channa striatus* (Bloch), experimentally infected with the specific *Aphanomyces fungus* associated with epizootic ulcerative syndrome (EUS) at different temperatures. *Journal of Fish Diseases* 18(1), 41–47.
323 Choopun, N., Louis, V., Huq, A. and Colwell, R.R. (2002) Simple procedure for rapid identification of *Vibrio cholerae* from the aquatic environment. *Applied and Environmental Microbiology* 68(2), 995–998.
324 Chopra, A.K., Houston, C.W., Peterson, J.W. and Jin, G.F. (1993) Cloning, expression, and sequence analysis of a cytolytic enterotoxin gene from *Aeromonas hydrophila*. *Canadian Journal of Microbiology* 39(5), 513–523.
325 Chou, J.-H., Chou, Y.-J., Arun, A.B., Young, C.-C., Chen, C.A., Wang, J.-T. and Chen, W.-M. (2009) *Vogesella lacus* sp. nov., isolated from a soft-shell turtle culture pond. *International Journal of Systematic and Evolutionary Microbiology* 59(Pt 10), 2629–2632.
326 Chowdhury, M.A., Yamanaka, H., Miyoshi, S., Aziz, K.M. and Shinoda, S. (1989) Ecology of *Vibrio mimicus* in aquatic environments. *Applied and Environmental Microbiology* 55(8), 2073–2078.
327 Christensen, P. (1980a) *Flexibacter canadensis* sp. nov. *International Journal of Systematic Bacteriology* 30(2), 429–432.
328 Christensen, P. (1980b) Description and taxonomic status of *Cytophaga heparina* (Payza and Korn) comb. nov. (Basionym: *Flavobacterium heparinum* Payza and Korn 1956). *International Journal of Systematic Bacteriology* 30(2), 473–475.
329 Christensen, W.B. (1946) Urea decomposition as a means of differentiating *Proteus* and paracolon cultures from each other and from *Salmonella* and *Shigella* Types 1. *Journal of Bacteriology* 52(4), 461–466.
330 Chukanhom, K., Borisutpeth, P., Khoa, L.V. and Hatai, K. (2003) *Haliphthoros milfordensis* isolated from black tiger prawn larvae (*Penaeus monodon*) in Vietnam. *Mycoscience* 44(2), 123–127.
331 Chun, J., Seong, C.-N., Bae, K.S., Lee, K.-J., Kang, S.-O., Goodfellow, M. and Hah, Y.C. (1998) *Nocardia flavorosea* sp. nov. *International Journal of Systematic Bacteriology* 48(3), 901–905.
332 Cipriano, R.C., Schill, W.B., Pyle, S.W. and Horner, R. (1986) An epizootic in chinook salmon (*Oncorhynchus tshawytscha*) caused by a sorbitol-positive serovar 2 strain of *Yersinia ruckeri*. *Journal of Wildlife Diseases* 22(4), 488–492.
333 Cipriano, R. and Bertolini, J. (1988) Selection for virulence in the fish pathogen *Aeromonas salmonicida*, using Coomassie Brilliant Blue agar. *Journal of Wildlife Diseases* 24(4), 672–678.
334 Cipriano, R.C. and Bullock, G.L. (2001) Furunculosis and other diseases caused by *Aeromonas salmonicida*. *Fish Disease Leaflet* 66.
335 Cipriano, R.C., Schill, W.B., Teska, J.D. and Ford, L.A. (1996) Epizootiological study of bacterial cold-water disease in Pacific salmon and further characterization of the etiologic agent, *Flexibacter psychrophila*. *Journal of Aquatic Animal Health* 8(1), 28–36.
336 Clark, K.F. and Greenwood, S.J. (2011) *Aerococcus viridans* expression of Cpn60 is associated with virulence during infection of the American lobster, *Homarus americanus* Milne Edwards. *Journal of Fish Diseases* 34(11), 831–843.
337 Clark, W., Hollis, D., Weaver, R. and Riley, P. (1984) Identification of unusual pathogenic Gram-negative aerobic and facultatively anaerobic bacteria. US Department of Health and Human Services, Public Health Service, Centers for Disease Control, Washington, DC.
338 Clark, W.M. and Lubs, H.A. (1915) The differentiation of bacteria of the Colonaerogenes family by the use of indicators. *The Journal of Infectious Diseases* 17(1), 160–173.
339 Clarridge, J.E. and Zighelboim-Daum, S. (1985) Isolation and characterization of two hemolytic phenotypes of *Vibrio damsela* associated with a fatal wound infection. *Journal of Clinical Microbiology* 21(3), 302–306.
340 Clavareau, C., Wellemans, V., Walravens, K., Tryland, M., Verger, J.M., Grayon, M., Cloeckaert, A., Letesson, J.J. and Godfroid, J. (1998) Phenotypic and molecular characterization of a *Brucella* strain isolated from a minke whale (*Balaenoptera acutorostrata*) *Microbiology* 144(Pt 12), 3267–3273.
341 Cloeckaert, A., Grayon, M. and Grepinet, O. (2000) An IS*711* element downstream of the *bp26* gene is a specific marker of *Brucella* spp. isolated from marine mammals. *Clinical and Diagnostic Laboratory Immunology* 7(5), 835–839.

342 Cloeckaert, A., Verger, J.M., Grayon, M., Paquet, J.Y., Garin-Bastuji, B., Foster, G. and Godfroid, J. (2001) Classification of *Brucella* spp. isolated from marine mammals by DNA polymorphism at the *omp2* locus. *Microbes and Infection* 3(9), 729–738.

343 Coker, W.C. and Mathews, V.D. (1937) Blastocladiales, Monoblepharidales, Saprolegniales. *North American Flora* 2, 24.

344 Coker, W.C. (1923) *The Saprolegniaceae, with Notes on Other Water Molds*. University of North Carolina Press, Chapel Hill, North Carolina.

345 Colegrove, K.M., Lowenstine, L.J. and Gulland, F.M.D. (2005) Leptospirosis in northern elephant seals (*Mirounga angustirostris*) stranded along the California coast. *Journal of Wildlife Diseases* 41(2), 426–430.

346 Coleman, S.S., Melanson, D.M., Biosca, E.G. and Oliver, J.D. (1996) Detection of *Vibrio vulnificus* biotypes 1 and 2 in eels and oysters by PCR amplification. *Applied and Environmental Microbiology* 62(4), 1378–1382.

347 Coles, B.M., Stroud, R.K. and Sheggeby, S. (1978) Isolation of *Edwardsiella tarda* from three Oregon sea mammals. *Journal of Wildlife Diseases* 14(3), 339–341.

348 Collado, L., Cleenwerck, I., Van Trappen, S., De Vos, P. and Figueras, M.J. (2009) *Arcobacter mytili* sp. nov., an indoxyl acetate-hydrolysis-negative bacterium isolated from mussels. *International Journal of Systematic and Evolutionary Microbiology* 59(Pt 6), 1391–1396.

349 Collado, L. and Figueras, M.J. (2011) Taxonomy, epidemiology, and clinical relevance of the genus *Arcobacter*. *Clinical Microbiology Reviews* 24(1), 174–192.

350 Collins, M.D., Farrow, J.A., Phillips, B.A. and Kandler, O. (1983) *Streptococcus garvieae* sp. nov. and *Streptococcus plantarum* sp. nov. *Journal of General Microbiology* 129(11), 3427–3431.

351 Collins, M.D., Hoyles, L., Hutson, R.A., Foster, G. and Falsen, E. (2001) *Corynebacterium testudinoris* sp. nov., from a tortoise, and *Corynebacterium felinum* sp. nov., from a Scottish wild cat. *International Journal of Systematic and Evolutionary Microbiology* 51(Pt 4), 1349–1352.

352 Collins, M.D. and Lawson, P.A. (2000) The genus *Abiotrophia* (Kawamura *et al.*) is not monophyletic: proposal of *Granulicatella* gen. nov., *Granulicatella adiacens* comb. nov., *Granulicatella elegans* comb. nov. and *Granulicatella balaenopterae* comb. nov. *International Journal of Systematic and Evolutionary Microbiology* 50(Pt1), 365–369.

353 Collins, M.D., Martinez-Murcia, A.J. and Cai, J. (1993) *Aeromonas enteropelogenes* and *Aeromonas ichthiosmia* are identical to *Aeromonas trota* and *Aeromonas veronii*, respectively, as revealed by small-subunit rRNA sequence analysis. *International Journal of Systematic Bacteriology* 43(4), 855–856.

354 Collins, M.D., Ash, C., Farrow, J.A.E., Wallbanks, S. and Williams, A.M. (1989) 16S Ribosomal ribonucleic acid sequence analyses of lactococci and related taxa. Description of *Vagococcus fluvialis* gen. nov., sp. nov. *Journal of Applied Microbiology* 67(4), 453–460.

355 Collins, M.D., Farrow, J.A.E., Phillips, B.A., Ferusu, S. and Jones, D. (1987) Classification of *Lactobacillus divergens*, *Lactobacillus piscicola*, and some catalase-negative, asporogenous, rod-shaped bacteria from poultry in a new genus, *Carnobacterium*. *International Journal of Systematic Bacteriology* 37(4), 310–316.

356 Collins, M.D., Farrow, J.A.E., Katic, V. and Kandler, O. (1984) Taxonomic studies on streptococci of serological groups E, P, U and V: description of *Streptococcus porcinus* sp. nov. *Systematic and Applied Microbiology* 5(3), 402–413.

357 Collins, M.D., Hoyles, L., Foster, G. and Falsen, E. (2004) *Corynebacterium caspium* sp. nov., from a Caspian seal (*Phoca caspica*). *International Journal of Systematic and Evolutionary Microbiology* 54(Pt 3), 925–928.

358 Collins, M.D., Hoyles, L., Foster, G., Falsen, E. and Weiss, N. (2002a) *Arthrobacter nasiphocae* sp. nov., from the common seal (*Phoca vitulina*). *International Journal of Systematic and Evolutionary Microbiology* 52(Pt 2), 569–571.

359 Collins, M.D., Hutson, R.A., Foster, G., Falsen, E. and Weiss, N. (2002b) *Isobaculum melis* gen. nov., sp. nov., a *Carnobacterium*-like organism isolated from the intestine of a badger. *International Journal of Systematic and Evolutionary Microbiology* 52(Pt 1), 207–210.

360 Colorn, A., Diamant, A., Eldar, A., Kvitt, H. and Zlotkin, A. (2002) *Streptococcus iniae* infections in Red Sea cage-cultured and wild fishes. *Diseases of Aquatic Organisms* 49(3), 165–170.

361 Colorni, A., Ravelo, C., Romalde, J.L., Toranzo, A.E. and Diamant, A. (2003) *Lactococcus garvieae* in wild Red Sea wrasse *Coris aygula* (Labridae). *Diseases of Aquatic Organisms* 56(3), 275–278.

362 Colquhoun, D., Aase, I.L., Wallace, C., Baklien, A. and Gravningen, K. (2004) First description of *Vibrio ordalii* from Chile. *Bulletin of the European Association of Fish Pathologists* 24, 185–188.

363 Colwell, R.R. and Morita, R.Y. (1964) Reisolation and emendation of description of *Vibrio marinus* (Russell) Ford. *Journal of Bacteriology* 88(4), 831–837.

364 Company, R., Sitjà-Bobadilla, A., Pujalte, M.J., Garay, E., Álvarez-Pellitero, P. and Pérez-Sánchez, J.(1999) Bacterial and parasitic pathogens in cultured common dentex, *Dentex dentex* L. *Journal of Fish Diseases* 22(4), 299–309.

365 Conejero, M.J. and Hedreyda, C.T. (2003) Isolation of partial *toxR* gene of *Vibrio harveyi* and design of *toxR*-targeted PCR primers for species detection. *Journal of Applied Microbiology* 95(3), 602–611.

366 Copland, J.W. and Willoughby, L.G. (1982) The pathology of *Saprolegnia* infections of *Anguilla anguilla* L. elvers. *Journal of Fish Diseases* 5, 421–428.

367 Coquet, L., Cosette, P., Quillet, L., Petit, F., Junter, G.-A. and Jouenne, T. (2002) Occurrence and phenotypic characterization of *Yersinia ruckeri* strains with biofilm-forming capacity in a rainbow trout farm. *Applied and Environmental Microbiology* 68(2), 470–475.

368 Corbel, M.J. and Banai, M. (2005) Genus I. *Brucella* Myer and Shaw 1920, 173. In: Brenner, D.J., Krieg, N.R., Staley, J.T. and Garrity, G.M. (eds) *Bergey's Manual of Systematic Bacteriolog*, 2nd edn, Vol. II, *The Proteobacteria, Part C, The Alpha-, Beta-, Delta-, and Epsilonproteobacteria*. Springer, New York, pp. 307–386.

369 Corbel, M.J. and Morgan, W.J.B. (1975) Proposal for minimal standards for descriptions of new species and

biotypes of the genus *Brucella*. *International Journal of Systematic Bacteriology* 25(1), 83–89.
370 Cornick, J.W., Morrison, C.M., Zwicker, B. and Shum, G. (1984) Atypical *Aeromonas salmonicida* infection in Atlantic cod, *Gadus morhua* L. *Journal of Fish Diseases* 7(6), 495–499.
371 Costa, R., Mermoud, I., Koblavi, S., Morlet, B., Haffner, P., Berthe, F., Legroumellec, M. and Grimont, P. (1998) Isolation and characterization of bacteria associated with a *Penaeus stylirostris* disease (Syndrome 93) in New Caledonia. *Aquaculture* 164(1–4), 297–309.
372 Cottral, G.E. (1978) *Manual of Standardized Methods for Veterinary Microbiology*. Cornell University Press, Ithaca, New York.
373 Cousins, D.V. and Lloyd, J.M. (1988) Rapid identification of *Haemophilus somnus*, *Histophilus ovis* and *Actinobacillus seminis* using the API ZYM system. *Veterinary Microbiology* 17(1), 75–81.
374 Cousins, D.V., Bastida, R., Cataldi, A., Quse, V., Redrobe, S., Dow, S., Duignan, P., Murray, A., Dupont, C., Ahmed, N., Collins, D.M., Butler, W.R., Dawson, D., Rodríguez, D., Loureiro, J., Romano, M.I., Alito, A., Zumarraga, M. and Bernardelli, A. (2003) Tuberculosis in seals caused by a novel member of the *Mycobacterium tuberculosis* complex: *Mycobacterium pinnipedii* sp. nov. *International Journal of Systematic and Evolutionary Microbiology* 53(5), 1305–1314.
375 Cowan, D.F. (1993) Lobo's disease in a bottlenose dolphin (*Tursiops truncatus*) from Matagorda Bay, Texas. *Journal of Wildlife Diseases* 29(3), 488–489.
376 Cowan, S.T. and Steel, K.J. (1970) *Manual for the Identification of Medical Bacteria*. Cambridge University Press, Cambridge.
377 Coyne, V.E.and Al-Harthi, L. (1992) Induction of melanin biosynthesis in *Vibrio cholerae*. *Applied and Environmental Microbiology* 58(9), 2861–2865.
378 Creeper, J.H. and Buller, N.B. (2006) An outbreak of *Streptococcus iniae* in barramundi (*Lates calcarifera*) in freshwater cage culture. *Australian Veterinary Journal* 84(11), 408–411.
379 Creeper, J., Main, D., Berger, L., Huntress, S. and Boardman, W. (1998) An outbreak of mucormycosis in slender tree frogs (*Litoria adelensis*) and white-lipped tree frogs (*Litoria infrafrenata*). *Australian Veterinary Journal* 76(11), 761–762.
380 Crisafi, F., Denaro, R., Genovese, M., Cappello, S., Mancuso, M. and Genovese, L. (2011) Comparison of *16SrDNA* and *toxR* genes as targets for detection of *Vibrio anguillarum* in *Dicentrarchus labrax* kidney and liver. *Research in Microbiology* 162(3), 223–230.
381 Croci, L., Suffredini, E., Cozzi, L., Toti, L., Ottaviani, D., Pruzzo, C., Serratore, P., Fischetti, R., Goffredo, E., Loffredo, G. and Mioni, R. (2007a) Comparison of different biochemical and molecular methods for the identification of *Vibrio parahaemolyticus*. *Journal of Applied Microbiology* 102(1), 229–237.
382 Croci, L., Suffredini, E., Cozzi, L., Paniconi, M., Ciccaglioni, G. and Colombo, M.M. (2007b) Evaluation of different polymerase chain reaction methods for the identification of *Vibrio parahaemolyticus* strains isolated by cultural methods. *Journal of AOAC International* 90(6), 1588–1597.
383 Crosby, N.T. (1967) The determination of nitrite in water using Cleve's acid 1-naphthylamine-7-sulphonic acid. *Proceedings of the Society for Water Treatment and Examination* 16.
384 Crous, P., Gams, W., Stalpers, J., Robert, V. and Stegehuis, G. (2004) MycoBank: an online initiative to launch mycology into the 21st century. *Studies in Mycology* 50, 19–22.
385 Crow, G.L., Brock, J.A. and Kaiser, S. (1995) *Fusarium solani* fungal infection of the lateral line canal system in captive scalloped hammerhead sharks (*Sphyrna lewini*) in Hawaii. *Journal of Wildlife Diseases* 31(4), 562–565.
386 Crumlish, M., Dung, T., Ngoc, N. and Ferguson, H.W. (2002) Identification of *Edwardsiella ictaluri* from diseased catfish, *Pangasius hypophthalmus* (Sauvage), cultured in the Mekong Delta, Vietnam. *Journal of Fish Diseases* 25, 733–736.
387 Crumlish, M., Thanh, P.C., Koesling, J., Tung, V.T. and Gravningen, K. (2010) Experimental challenge studies in Vietnamese catfish, *Pangasianodon hypophthalmus* (Sauvage), exposed to *Edwardsiella ictaluri* and *Aeromonas hydrophila*. *Journal of Fish Diseases* 33(9), 717–722.
388 Cruz, J.C. (1938) *Mycobacterium fortuitum*. Um novo bacilo ácido-resistente patogênico para o homem. *Acta Medica do Rio de Janeiro* 1, 297–301.
389 Cruz, J., Saraiva, A., Eiras, J., Branco, R. and Sousa, J. (1986) An outbreak of *Plesiomonas shigelloides* in farmed rainbow trout, *Salmo gairdneri* Richardson, in Portugal. *Bulletin of the European Association of Fish Pathologists* 6, 20–22.
390 Cuellar, L. and Coll, M. (1983) Epizootiology of the crayfish plague (Aphanomycosis) in Spain. *Freshwater Crayfish* 5, 545–548.
391 Cunningham, A.A., Garner, T.W.J., Aguilar-Sanchez, V., Banks, B., Foster, J., Sainsbury, A.W., Perkins, M., Walker, S.F., Hyatt, A.D. and Fisher, M.C. (2005) Emergence of amphibian chytridiomycosis in Britain. *The Veterinary Record* 157(13), 386–387.
392 Cunningham, J.H. (1899) Experiments on *Saprolegnia ferax*, and their application to the trout hatchery. *Journal of the Boston Society of Medical Sciences* 4(3), 55–66.
393 Dakin, W.P., Howell, D.J., Sutton, R.G., O'Keefe, M.F. and Thomas, P. (1974) Gastroenteritis due to non-agglutinable (non-cholera) vibrios. *The Medical Journal of Australia* 2(13), 487–490.
394 Dalsgaard, A., Dalsgaard, I., Høi, L. and Larsen, J.L. (1996) Comparison of a commercial biochemical kit and an oligonucleotide probe for identification of environmental isolates of *Vibrio vulnificus*. *Letters in Applied Microbiology* 22(2), 184–188.
395 Dalsgaard, I., Gudmundsdóttir, B.K., Helgason, S., Høie, S., Thoresen, O.F., Wichardt, U.P. and Wiklund, T. (1998) Identification of atypical *Aeromonas salmonicida*: inter-laboratory evaluation and harmonization of methods. *Journal of Applied Microbiology* 84(6), 999–1006.
396 Dalsgaard, I., Høi, L., Siebeling, R.J. and Dalsgaard, A. (1999) Indole-positive *Vibrio vulnificus* isolated from disease outbreaks on a Danish eel farm. *Diseases of Aquatic Organisms* 35(3), 187–194.

397 Dalsgaard, I., Jürgens, O. and Mortensen, A. (1988) *Vibrio salmonicida* isolated from farmed Atlantic salmon in the Faroe Islands. *Bulletin of the European Association of Fish Pathologists* 8(3), 53–54.

398 Dalsgaard, I. and Paulsen, H. (1986) Atypical *Aeromonas salmonicida* isolated from diseased sand-eels, *Ammodytes lancea* (Cuvier) and *Hyperoplus lanceolatus* (Lesauvege). *Journal of Fish Diseases* 9, 361–364.

399 Daly, J.G. and Stevenson, R.M. (1985) Charcoal agar, a new growth medium for the fish disease bacterium *Renibacterium salmoninarum*. *Applied and Environmental Microbiology* 50(4), 868–871.

400 Daly, J.G. and Stevenson, R.M. (1993) Nutrient requirements of *Renibacterium salmoninarum* on agar and in broth media. *Applied and Environmental Microbiology* 59(7), 2178–2183.

401 Dang, H.T., Park, H.K., Myung, S.C. and Kim, W. (2012) Development of a novel PCR assay based on the 16S-23S rRNA internal transcribed spacer region for the detection of *Lactococcus garvieae*. *Journal of Fish Diseases* 35(7), 481–487.

402 Danley, M.L., Goodwin, A.E. and Killian, H.S. (1999) Epizootics in farm-raised channel catfish, *Ictalurus punctatus* (Rafinesque), caused by the enteric redmouth bacterium *Yersinia ruckeri*. *Journal of Fish Diseases* 22(6), 451–456.

403 Daoust, P.Y., Larson, B.E. and Johnson, G.R. (1989) Mycobacteriosis in yellow perch (*Perca flavescens*) from two lakes in Alberta. *Journal of Wildlife Diseases* 25(1), 31–37.

404 Darwish, A.M., Ismaiel, A.A., Newton, J.C. and Tang, J. (2004) Identification of *Flavobacterium columnare* by a species-specific polymerase chain reaction and renaming of ATCC43622 strain to *Flavobacterium johnsoniae*. *Molecular and Cellular Probes* 18(6), 421–427.

405 Daskalov, H., Austin, D.A. and Austin, B. (1999) An improved growth medium for *Flavobacterium psychrophilum*. *Letters in Applied Microbiology* 28(4), 297–299.

406 Daskalov, H., Stobie, M. and Austin, B. (1998) *Klebsiella pneumoniae*: a pathogen of rainbow trout (*Oncorhynchus mykiss, Walbaum*). *Bulletin of the European Association of Fish Pathologists* 18, 26–28.

407 Daura-Jorge, F.G. and Simões-Lopes, P.C. (2011) Lobomycosis-like disease in wild bottlenose dolphins *Tursiops truncatus* of Laguna, southern Brazil: monitoring of a progressive case. *Diseases of Aquatic Organisms* 93(2), 163–170.

408 David, J.C. and Kirk, P.M. (1997) *Index of Fungi*, Vol. 6. CAB International, Wallingford, UK.

409 Davies, R.L. (1990) O-serotyping of *Yersinia ruckeri* with special emphasis on European isolates. *Veterinary Microbiology* 22(4), 299–307.

410 Davies, R.L. (1991) Clonal analysis of *Yersinia ruckeri* based on biotypes, serotypes and outer membrane protein-types. *Journal of Fish Diseases* 14(2), 221–228.

411 Davies, R.L. and Frerichs, G.N. (1989) Morphological and biochemical differences among isolates of *Yersinia ruckeri* obtained from wide geographical areas. *Journal of Fish Diseases* 12(4), 357–365.

412 Davis, B.R., Fanning, G.R., Madden, J.M., Steigerwalt, A.G., Bradford, H.B., Smith, H.L. and Brenner, D.J. (1981) Characterization of biochemically atypical *Vibrio cholerae* strains and designation of a new pathogenic species, *Vibrio mimicus*. *Journal of Clinical Microbiology* 14(6), 631–639.

413 Davis, H.S. (1926) A new gill disease of trout. *Transactions of the American Fisheries Society* 56(1), 156–160.

414 Davis, H.S. (1922) A new bacterial disease of fresh-water fishes. *Bulletin of the Bureau of Fisheries, Washington* 38, 261–280.

415 Davis, J.W. and Sizemore, R.K. (1982) Incidence of *Vibrio* species associated with blue crabs (*Callinectes sapidus*) collected from Galveston Bay, Texas. *Applied and Environmental Microbiology* 43(5), 1092–1097.

416 Davison, N.J., Cranwell, M.P., Perrett, L.L., Dawson, C.E., Deaville, R., Stubberfield, E.J., Jarvis, D.S. and Jepson, P.D. (2009) Meningoencephalitis associated with *Brucella* species in a live-stranded striped dolphin (*Stenella coeruleoalba*) in south-west England. *The Veterinary Record* 165(3), 86–89.

417 Dawson, C.E., Perrett, L.L., Young, E.J., Davison, N.J. and Monies, R.J. (2006) Isolation of *Brucella* species from a bottlenosed dolphin (*Tursiops truncatus*). *Veterinary Record* 158, 831–832.

418 Dawyndt, P., Thompson, F.L., Austin, B., Swings, J., Koski, T. and Gyllenberg, M. (2005) Application of sliding-window discretization and minimization of stochastic complexity for the analysis of FAFLP genotyping fingerprint patterns of *Vibrionaceae*. *International Journal of Systematic and Evolutionary Microbiology* 55(Pt 1), 57–66.

419 Deane, E.E., Li, J. and Woo, N.Y.S. (2004) Modulated heat shock protein expression during pathogenic *Vibrio alginolyticus* stress of sea bream. *Diseases of Aquatic Organisms* 62(3), 205–215.

420 Dear, G. (1988) *Yersinia ruckeri* isolated from Atlantic salmon in Scotland. *Bulletin of the European Association of Fish Pathologists* 8, 18–20.

421 De Beer, H., Hugo, C.J., Jooste, P.J., Vancanneyt, M., Coenye, T. and Vandamme, P. (2006) *Chryseobacterium piscium* sp. nov., isolated from fish of the South Atlantic Ocean off South Africa. *International Journal of Systematic and Evolutionary Microbiology* 56(Pt 6), 1317–1322.

422 Debruyne, L., Broman, T., Bergström, S., Olsen, B., On, S.L.W. and Vandamme, P. (2010a) *Campylobacter subantarcticus* sp. nov., isolated from birds in the sub-Antarctic region. *International Journal of Systematic and Evolutionary Microbiology* 60(Pt 4), 815–819.

423 Debruyne, L., Houf, K., Douidah, L., De Smet, S. and Vandamme, P. (2010b) Reassessment of the taxonomy of *Arcobacter cryaerophilus*. *Systematic and Applied Microbiology* 33(1), 7–14.

424 Debruyne, L., On, S.L.W., De Brandt, E. and Vandamme, P. (2009) Novel *Campylobacter lari*-like bacteria from humans and molluscs: description of *Campylobacter peloridis* sp. nov., *Campylobacter lari* subsp. *concheus* subsp. nov. and *Campylobacter lari* subsp. *lari* subsp. nov. *International Journal of Systematic and Evolutionary Microbiology* 59(5), 1126–1132.

425 Decostere, A., Haesebrouck, F. and Devriese, L.A. (1997) Shieh medium supplemented with tobramycin for selective isolation of *Flavobacterium columnare* (*Flexibacter columnaris*) from diseased fish. *Journal of Clinical Microbiology* 35(1), 322–324.

426 Decostere, A., Haesebrouck, F. and Devriese, L.A. (1998) Characterization of four *Flavobacterium columnare* (*Flexibacter columnaris*) strains isolated from tropical fish. *Veterinary Microbiology* 62(1), 35–45.

427 De Grandis, S.A., Krell, P.J., Flett, D.E. and Stevenson, R.M.W. (1988) Deoxyribonucleic acid relatedness of serovars of *Yersinia ruckeri*, the enteric redmouth bacterium. *International Journal of Systematic Bacteriology* 38(1), 49–55.

428 De Hoog, G.S., Vicente, V.A., Najafzadeh, M.J., Harrak, M.J., Badali, H. and Seyedmousavi, S. (2011) Waterborne *Exophiala* species causing disease in cold-blooded animals. *Persoonia* 27, 46–72.

429 De Hoog, G.S., Zeng, J.S., Harrak, M.J. and Sutton, D.A. (2006) *Exophiala xenobiotica* sp. nov., an opportunistic black yeast inhabiting environments rich in hydrocarbons. *Antonie Van Leeuwenhoek* 90(3), 257–268.

430 Dekker, J. and Branda, J. (2011) MALDI-TOF spectrometry in the clinical laboratory. *Clinical Microbiology Newsletter* 33(12), 87–93.

431 De la Peña, L., Lavilla-Pitogo, C. and Paner, M. (2001) Luminescent vibrios associated with mortality in pond-cultured shrimp *Penaeus monodon* in the Philippines: species composition. *Fish Pathology* 36(3), 133–138.

432 Del Cerro, A., Mendoza, M.C. and Guijarro, J.A. (2002) Usefulness of a TaqMan-based polymerase chain reaction assay for the detection of the fish pathogen *Flavobacterium psychrophilum*. *Journal of Applied Microbiology* 93(1), 149–156.

433 De Ley, J., Cattoir, H. and Reynaerts, A. (1970) The quantitative measurement of DNA hybridization from renaturation rates. *European Journal of Biochemistry* 12(1), 133–142.

434 De Ley, J., Mannheim, W., Mutters, R., Piechulla, K., Tytgat, R., Segers, P., Bisgaard, M., Frederiksen, W., Hinz, K.H. and Vanhoucke, M. (1990) Inter- and intrafamilial similarities of rRNA cistrons of the *Pasteurellaceae*. *International Journal of Systematic Bacteriology* 40(2), 126–137.

435 Demarta, A., Küpfer, M., Riegel, P., Harf-Monteil, C., Tonolla, M., Peduzzi, R., Monera, A., José Saavedra, M. and Martínez-Murcia, A. (2008) *Aeromonas tecta* sp. nov., isolated from clinical and environmental sources. *Systematic and Applied Microbiology* 31(4), 278–286.

436 Denner, E.B.M., Vybiral, D., Fischer, U.R., Velimirov, B. and Busse, H.-J. (2002) *Vibrio calviensis* sp. nov., a halophilic, facultatively oligotrophic 0.2 μm filterable marine bacterium. *International Journal of Systematic and Evolutionary Microbiology* 52(Pt 2), 549–553.

437 Desolme, B. and Bernardet, J.F. (1996) Freeze-drying of *Flavobacterium columnare, Flavobacterium psychrophilum* and *Flexibacter maritimus*. *Diseases of Aquatic Organisms* 27(1), 77–80.

438 De Vos, P. and Truper, H.G. (2000) Judicial Commission of the International Committee on Systematic Bacteriology; IXth International (IUMS) Congress of Bacteriology and Applied Microbiology. *International Journal of Systematic and Evolutionary Microbiology* 50(6), 2239–2244.

439 Dewhirst, F.E., Chien, C.C., Paster, B.J., Ericson, R.L., Orcutt, R.P., Schauer, D.B. and Fox, J.G. (1999) Phylogeny of the defined murine microbiota: altered Schaedler flora. *Applied and Environmental Microbiology* 65(8), 3287–3292.

440 Diamant, A., Banet, A., Ucko, M., Colorni, A., Knibb, W. and Kvitt, H. (2000) Mycobacteriosis in wild rabbitfish *Siganus rivulatus* associated with cage farming in the Gulf of Eilat, Red Sea. *Diseases of Aquatic Organisms* 39(3), 211–219.

441 Dick, M.W., Vick, M.C., Gibbings, J.G., Hedderson, T.A. and Lopez Lastra, C.C. (1999) 18S rDNA for species of *Leptolegnia* and other *Peronosporomycetes*: justification for the subclass taxa *Saprolegniomycetidae* and *Peronosporomycetidae* and division of the *Saprolegniaceae sensu lato* into the *Leptolegniaceae* and *Saprolegniaceae*. *Mycological Research* 103(9), 1119–1125.

442 Diéguez, A.L., Beaz-Hidalgo, R., Cleenwerck, I., Balboa, S., de Vos, P. and Romalde, J.L. (2011) *Vibrio atlanticus* sp. nov. and *Vibrio artabrorum* sp. nov., isolated from the clams *Ruditapes philippinarum* and *Ruditapes decussatus*. *International Journal of Systematic and Evolutionary Microbiology* 61(Pt 10), 2406–2411.

443 Diéguez-Uribeondo, J. and Söderhäll, K. (1993) *Procambarus clarkii* Girard as a vector for the crayfish plague fungus, *Aphanomyces astaci* Schikora. *Aquaculture Research* 24(6), 761–765.

444 Diéguez-Uribeondo, J., Cerenius, L. and Söderhäll, K. (1994) Repeated zoospore emergence in *Saprolegnia parasitica*. *Mycological Research* 98(7), 810–815.

445 Diéguez-Uribeondo, J., Fregeneda-Grandes, J.M., Cerenius, L., Pérez-Iniesta, E., Aller-Gancedo, J.M., Tellería, M.T., Söderhäll, K. and Martín, M.P. (2007) Re-evaluation of the enigmatic species complex *Saprolegnia diclina–Saprolegnia parasitica* based on morphological, physiological and molecular data. *Fungal Genetics and Biology* 44(7), 585–601.

446 Diéguez-Uribeondo, J., Huang, T.-S., Cerenius, L. and Söderhäll, K. (1995) Physiological adaptation of an *Aphanomyces astaci* strain isolated from the freshwater crayfish *Procambarus clarkii*. *Mycological Research* 99(5), 574–578.

447 Diéguez-Uribeondo, J. and Söderhäll, K. (1999) RAPD evidence for the origin of an outbreak of crayfish plague in Spain. *Freshwater Crayfish* 12, 313–318.

448 Dienes, L. (1939) L organisms of Klieneberger and *Streptobacillus moniliformis*. *Journal of Infectious Diseases* 65(1), 24–42.

449 Dierckens, K.R., Vandenberghe, J., Beladjal, L., Huys, G., Mertens, J. and Swings, J. (1998) *Aeromonas hydrophila* causes 'black disease' in fairy shrimps (*Anostraca; Crustacea*). *Journal of Fish Diseases* 21(2), 113–119.

450 Diggles, B.K., Moss, G.A., Carson, J. and Anderson, C.D. (2000) Luminous vibriosis in rock lobster *Jasus verreauxi (Decapoda: Palinuridae)* phyllosoma larvae associated with infection by *Vibrio harveyi*. *Diseases of Aquatic Organisms* 43(2), 127–137.

451 Di Pinto, A., Ciccarese, G., Fontanarosa, M., Terio, V. and Tantillo, G. (2006) Detection of *Vibrio alginolyticus* and *Vibrio parahaemolyticus* in shellfish samples using collagenase-targeted multiplex PCR. *Journal of Food Safety* 26(2), 150–159.

452 Di Pinto, A., Ciccarese, G., Tantillo, G., Catalano, D. and Forte, V.T. (2005) A collagenase-targeted multiplex PCR assay for identification of *Vibrio alginolyticus, Vibrio cholerae*, and *Vibrio parahaemolyticus. Journal of Food Protection* 68(1), 150–153.

453 DiSalvo, L.H., Blecka, J. and Zebal, R. (1978) *Vibrio anguillarum* and larval mortality in a California coastal shellfish hatchery. *Applied and Environmental Microbiology* 35(1), 219–221.

454 Dodson, S.V., Maurer, J.J. and Shotts, E.B. (1999) Biochemical and molecular typing of *Streptococcus iniae* isolated from fish and human cases. *Journal of Fish Diseases* 22(5), 331–336.

455 Doménech, A., Fernández-Garayzábal, J.F., García, J.A., Cutuli, M.T., Blanco, M., Gibello, A., Moreno, M.A. and Domínguez, L. (1999) Association of *Pseudomonas anguilliseptica* infection with 'winter disease' in sea bream, *Sparus aurata* L. *Journal of Fish Diseases* 22(1), 69–71.

456 Doménech, A., Fernández-Garayzábal, J.F., Pascual, C., Garcia, J.A., Cutuli, M.T., Moreno, M.A., Collins, M.D. and Dominguez, L. (1996) Streptococcosis in cultured turbot, *Scopthalmus maximus* (L.), associated with *Streptococcus parauberis. Journal of Fish Diseases* 19(1), 33–38.

457 Donlon, J., McGettigan, S., O'Brien, P. and Ó'Carra, P. (1983) Re-appraisal of the nature of the pigment produced by *Aeromonas salmonicida. FEMS Microbiology Letters* 19(2–3), 285–290.

458 Dorsch, M., Lane, D. and Stackebrandt, E. (1992) Towards a phylogeny of the genus *Vibrio* based on 16S rRNA sequences. *International Journal of Systematic Bacteriology* 42(1), 58–63.

459 Dos Santos, N.M.S., do Vale, A., Sousa, M.J. and Silva, M.T. (2002) Mycobacterial infection in farmed turbot *Scophthalmus maximus. Diseases of Aquatic Organisms* 52(1), 87–91.

460 Douidah, L., De Zutter, L., Vandamme, P. and Houf, K. (2010) Identification of five human and mammal associated *Arcobacter* species by a novel multiplex-PCR assay. *Journal of Microbiological Methods* 80(3), 281–286.

461 Doukas, V., Athanassopoulou, F., Karagouni, E. and Dotsika, E. (1998) *Aeromonas hydrophila* infection in cultured sea bass, *Dicentrarchus labrax* L. and *Puntazzo puntazzo* Cuvier from the Aegean Sea. *Journal of Fish Diseases* 21(4), 317–320.

462 Drancourt, M., Bollet, C., Carta, A. and Rousselier, P. (2001) Phylogenetic analyses of *Klebsiella* species delineate *Klebsiella* and *Raoultella* gen. nov., with description of *Raoultella ornithinolytica* comb. nov., *Raoultella terrigena* comb. nov. and *Raoultella planticola* comb. nov. *International Journal of Systematic and Evolutionary Microbiology* 51(3), 925–932.

463 Drummond, L.C., Balboa, S., Beaz, R., Mulcahy, M.F., Barja, J.L., Culloty, S.C. and Romalde, J.L. (2007) The susceptibility of Irish-grown and Galician-grown Manila clams, *Ruditapes philippinarum*, to *Vibrio tapetis* and Brown Ring Disease. *Journal of Invertebrate Pathology* 95(1), 1–8.

464 Dunbar, S.A. and Clarridge, J.E. III (2000) Potential errors in recognition of *Erysipelothrix rhusiopathiae. Journal of Clinical Microbiology* 38(3), 1302–1304.

465 Duperthuy, M., Binesse, J., Le Roux, F., Romestand, B., Caro, A., Got, P., Givaudan, A., Mazel, D., Bachère, E. and Destoumieux-Garzón, D. (2010) The major outer membrane protein OmpU of *Vibrio splendidus* contributes to host antimicrobial peptide resistance and is required for virulence in the oyster *Crassostrea gigas. Environmental Microbiology* 12(4), 951–963.

466 Duremdez, R., Al-Marzouk, A., Qasem, J.A., Al-Harbi, A. and Gharabally, H. (2004) Isolation of *Streptococcus agalactiae* from cultured silver pomfret, *Pampus argenteus* (Euphrasen), in Kuwait. *Journal of Fish Diseases* 27, 307–310.

467 Dykstra, M.J., Noga, E.J., Levine, J.F., Moye, D.W. and Hawkins, J.H. (1986) Characterization of the *Aphanomyces* species involved with ulcerative mycosis (UM) in menhaden. *Mycologia* 78(4), 664.

468 Eaves, L.E., Ketterer, P.J., Anderson, I.G. and Beumer, J.P. (1990) The isolation and identification of *Edwardsiella tarda* from a diseased native Australian eel (*Anguilla reinhardtii*). *Australian Veterinary Journal* 67(9), 336–337.

469 Eaves, L.E. and Ketterer, P.J. (1994) Mortalities in red claw crayfish *Cherax quadricarinatus* associated with systemic *Vibrio mimicus* infection. *Diseases of Aquatic Organisms* 19, 233–237.

470 Eddyani, M., Ofori-Adjei, D., Teugels, G., De Weirdt, D., Boakye, D., Meyers, W.M. and Portaels, F. (2004) Potential role for fish in transmission of *Mycobacterium ulcerans* disease (Buruli Ulcer), an environmental study. *Applied and Environmental Microbiology* 70(9), 5679–5681.

471 Egan, S., Holmström, C. and Kjelleberg, S. (2001) *Pseudoalteromonas ulvae* sp. nov., a bacterium with antifouling activities isolated from the surface of a marine alga. *International Journal of Systematic and Evolutionary Microbiology* 51(Pt 4), 1499–1504.

472 Egidius, E., Soleim, Ø. and Andersen, K. (1984) Further observations on cold-water vibriosis or Hitra disease. *Bulletin of the European Association of Fish Pathologists* 4(3), 50–51.

473 Egidius, E., Wiik, R., Andersen, K., Hoff, K.A. and Hjeltnes, B. (1986) *Vibrio salmonicida* sp. nov., a new fish pathogen. *International Journal of Systematic Bacteriology* 36(4), 518–520.

474 Egusa, S. and Masuda, N. (1971) A new fungal disease of *Plecoglossus altivelis. Fish Pathology (in Japanese)* 6, 41–46.

475 Egusa, S. and Ohiwa, Y. (1972) Branchiomycosis of pond-cultured eels. *Fish Pathology* 7(1), 79–83.

476 Eissa, A.E., Elsayed, E.E., McDonald, R. and Faisal, M. (2006) First record of *Renibacterium salmoninarum* in the sea lamprey (*Petromyzon marinus*). *Journal of Wildlife Diseases* 42(3), 556–560.

477 Eissa, N.M.E., El-Ghiet, A., Shaheen, A.A. and Abbass, A. (2010) Characterization of *Pseudomonas* species isolated from tilapia '*Oreochromis niloticus*' in Qaroun and Wadi-El-Rayan Lakes, Egypt. *Global Veterinaria* 5(2), 116–121.

478 El Aamri, F., Padilla, D., Acosta, F., Caballero, M.J., Roo, J., Bravo, J., Vivas, J. and Real, F. (2010) First report of *Streptococcus iniae* in red porgy (*Pagrus pagrus*, L.). *Journal of Fish Diseases* 33(11), 901–905.

479 Eldar, A., Bejerano, Y., Livoff, A., Horovitcz, A. and Bercovier, H. (1995a) Experimental streptococcal meningoencephalitis in cultured fish. *Veterinary Microbiology* 43(1), 33–40.

480 Eldar, A., Ghittino, C., Asanta, L., Bozzetta, E., Goria, M., Prearo, M. and Bercovier, H. (1996) *Enterococcus seriolicida* is a junior synonym of *Lactococcus garvieae*, a causative agent of septicemia and meningoencephalitis in fish. *Current Microbiology* 32(2), 85–88.

481 Eldar, A., Goria, M., Ghittino, C., Zlotkin, A. and Bercovier, H. (1999) Biodiversity of *Lactococcus garvieae* strains isolated from fish in Europe, Asia, and Australia. *Applied and Environmental Microbiology* 65(3), 1005–1008.

482 Eldar, A., Bererano, Y. and Bercovier, H. (1994) *Streptococcus shiloi* and *Streptococcus difficile*: two new Streptococcal species causing a meningoencephalitis in fish. *Current Microbiology* 28, 139–143.

483 Eldar, A., Frelier, P., Assenta, L., Varner, P., Lawhon, S. and Bercovier, H. (1995b) *Streptococcus shiloi*, the name for an agent causing septicemic infection in fish, is a junior synonym of *Streptococcus iniae*. *International Journal of Systematic Bacteriology* 45(4), 840–842.

484 Elliott, D.G. and Shotts, E.B. Jr (1980) Aetiology of an ulcerative disease in goldfish, *Carassius auratus* (L.), experimental induction of the disease. *Journal of Fish Diseases* 3(2), 145–151.

485 Elliott, J.A. and Facklam, R.R. (1996) Antimicrobial susceptibilities of *Lactococcus lactis* and *Lactococcus garvieae* and a proposed method to discriminate between them. *Journal of Clinical Microbiology* 34(5), 1296–1298.

486 Elliott, R.F. (1968) Morphological variation in New Zealand *Saprolegniaceae*. *New Zealand Journal of Botany* 6(1), 94–105.

487 Ellis, A.E., Waddell, I.F. and Minter, D.W. (1983) A systemic fungal disease in Atlantic salmon parr, *Salmo salar* L., caused by a species of *Phialophora*. *Journal of Fish Diseases* 6(6), 511–523.

488 Elsayed, E.E., Eissa, A.E. and Faisal, M. (2006) Isolation of *Flavobacterium psychrophilum* from sea lamprey, *Petromyzon marinus* L., with skin lesions in Lake Ontario. *Journal of Fish Diseases* 29(10), 629–632.

489 Elston, R.A., Hasegawa, H., Humphrey, K.L., Polyak, I.K. and Häse, C.C. (2008) Re-emergence of *Vibrio tubiashii* in bivalve shellfish aquaculture: severity, environmental drivers, geographic extent and management. *Diseases of Aquatic Organisms* 82(2), 119–134.

490 Emmerich, R. and Weibel, C. (1894) Über eine durch Bakterien erzeugte Seuche unter den Forellen. *Archiv für Hygiene* Bd. XXI(1), 1–21.

491 Engelsma, M.Y., Roozenburg, I. and Joly, J.-P. (2008) First isolation of *Nocardia crassostreae* from Pacific oyster *Crassostrea gigas* in Europe. *Diseases of Aquatic Organisms* 80(3), 229–234.

492 Enger, Ø., Nygaard, H., Solberg, M., Schei, G., Nielsen, J. and Dundas, I. (1987) Characterization of *Alteromonas denitrificans* sp. nov. *International Journal of Systematic Bacteriology* 37(4), 416–421.

493 Esteve, C., Biosca, E.G. and Amaro, C. (1993) Virulence of *Aeromonas hydrophila* and some other bacteria isolated from European eels *Anguilla anguilla* reared in fresh water. *Diseases of Aquatic Organisms* 16, 15–20.

494 Esteve, C., Gutiérrez, M.C. and Ventosa, A. (1995a) *Aeromonas encheleia* sp. nov., isolated from European eels. *International Journal of Systematic Bacteriology* 45(3), 462–466.

495 Esteve, C., Gutiérrez, M.C. and Ventosa, A. (1995b) DNA relatedness among *Aeromonas allosaccharophila* strains and DNA hybridization groups of the genus *Aeromonas*. *International Journal of Systematic Bacteriology* 45(2), 390–391.

496 Esteve, C., Amaro, C., Biosca, E.G. and Garay, E. (1995c) Biochemical and toxigenic properties of *Vibrio furnissii* isolated from a European eel farm. *Aquaculture* 132(1–2), 81–90.

497 Esteve, C., Valera, L., Gutiérrez, C. and Ventosa, A. (2003) Taxonomic study of sucrose-positive *Aeromonas jandaei*-like isolates from faeces, water and eels: emendation of *A. jandaei* Carnahan *et al*. 1992. *International Journal of Systematic and Evolutionary Microbiology* 53(Pt 5), 1411–1419.

498 Euzéby, J.P. (n.d.) List of Bacterial Names with Standing in Nomenclature: a folder available on the Internet. *International Journal of Systematic and Evolutionary Microbiology*, 1997, 47, 590–592. (List of Prokaryotic Names with Standing in Nomenclature. http://www.bacterio.net).

499 Evans, J.J., Bohnsack, J.F., Klesius, P.H., Whiting, A.A., Garcia, J.C., Shoemaker, C.A. and Takahashi, S. (2008) Phylogenetic relationships among *Streptococcus agalactiae* isolated from piscine, dolphin, bovine and human sources: a dolphin and piscine lineage associated with a fish epidemic in Kuwait is also associated with human neonatal infections in Japan. *Journal of Medical Microbiology* 57(Pt 11), 1369–1376.

500 Evans, J.J., Pasnik, D.J., Klesius, P.H. and Al-Ablani, S. (2006) First report of *Streptococcus agalactiae* and *Lactococcus garvieae* from a wild bottlenose dolphin (*Tursiops truncatus*). *Journal of Wildlife Diseases* 42(3), 561–569.

501 Evans, J., Klesius, P.H., Gilbert, P.M., Shoemaker, C.A., Sarawi, M., Landsberg, J., Duremdez, R., Marzouk, A. and Zenki, S. (2002) Characterization of β-haemolytic Group B *Streptococcus agalactiae* in cultured seabream, *Sparus auratus* L. and wild mullet, *Liza klunzingeri* (Day), in Kuwait. *Journal of Fish Diseases* 25, 505–513.

502 Evelyn, T.P.T. (1971) An aberrant strain of the bacterial fish pathogen *Aeromonas salmonicida* isolated from a marine host, the sablefish (*Anoplopoma fimbria*), and from two species of cultured Pacific salmon. *Journal of the Fisheries Research Board of Canada* 28, 1629–1634.

503 Evelyn, T.P.T. (1977) An improved growth medium for the kidney disease bacterium and some notes on using the medium. *Bulletin de L' Office International des Epizooties* 87, 511–513.

504 Evelyn, T.P.T. (1993) Bacterial kidney disease – BKD. In: Inglis, V., Roberts, R.J. and Bromage, N.R. (eds) *Bacterial Disease of Fish*. Blackwell Scientific, London, pp. 177–195.

505 Evelyn, T.P.T., Prosperi-Porta, L. and Ketcheson, J.E. (1990) Two new techniques for obtaining consistent results when growing *Renibacterium salmoninarum* on KDM2 culture medium. *Diseases of Aquatic Organisms* 9, 209–212.

506 Evenberg, D., Versluis, R. and Lugtenberg, B. (1985) Biochemical and immunological characterization of the cell surface of the fish pathogenic bacterium *Aeromonas salmonicida*. *Biochimica et Biophysica Acta* 815(2), 233–244.

507 Ewalt, D.R., Payeur, J.B., Martin, B.M., Cummins, D.R. and Miller, W.G. (1994) Characteristics of a *Brucella* species from a bottlenose dolphin (*Tursiops truncatus*). *Journal of Veterinary Diagnostic Investigation* 6(4), 448–452.

508 Ewing, W.H. and Davis, B.R. (1972) Biochemical characterization of *Citrobacter diversus* (Burkey) Werkman and Gillen and designation of the neotype strain. *International Journal of Systematic Bacteriology* 22(1), 12–18.

509 Ewing, W.H. and Fife, M.A. (1972) *Enterobacter agglomerans* (Beijerinck) comb. nov. (the Herbicola-Lathyri bacteria). *International Journal of Systematic Bacteriology* 22(1), 4–11.

510 Ewing, W.H., McWhorter, A.C., Escobar, M.R. and Lubin, A.H. (1965) *Edwardsiella*, a new genus of *Enterobacteriaceae* based on a new species, *E. tarda*. *International Bulletin of Bacteriological Nomenclature and Taxonomy* 15(1), 33–38.

511 Ewing, W.H., Ross, A.J., Brenner, D.J. and Fanning, G.R. (1978) *Yersinia ruckeri* sp. nov., the redmouth (RM) bacterium. *International Journal of Systematic Bacteriology* 28(1), 37–44.

512 Facklam, R. (2002) What happened to the streptococci: overview of taxonomic and nomenclature changes. *Clinical Microbiology Reviews* 15(4), 613–630.

513 Faisal, M., Eissa, A.E. and Elsayed, E.E. (2007a) Isolation of *Aeromonas salmonicida* from sea lamprey (*Petromyzon marinus*) with furuncle-like lesions in Lake Ontario. *Journal of Wildlife Diseases* 43(4), 618–622.

514 Faisal, M., Elsayed, E., Fitzgerald, S.D., Silva, V. and Mendoza, L. (2007b) Outbreaks of phaeohyphomycosis in the chinook salmon (*Oncorhynchus tshawytscha*) caused by *Phoma herbarum*. *Mycopathologia* 163(1), 41–48.

515 Fall, J., Chakraborty, G., Kono, T., Maeda, M., Itami, T. and Sakai, M. (2008) Establishment of loop-mediated isothermal amplification method (LAMP) for the detection of *Vibrio nigripulchritudo* in shrimp. *FEMS Microbiology Letters* 288(2), 171–177.

516 Farkas, J. (1985) Filamentous *Flavobacterium* sp. isolated from fish with gill diseases in cold water. *Aquaculture* 44(1), 1–10.

517 Farmer, J.J. III and McWhorter, A.C. (1984) Genus *Edwardsiella*. In: Kreig, N.R. and Holt, J.G. (eds) *Bergey's Manual of Systematic Bacteriology*, Vol. I. Williams & Wilkins Co., Baltimore, Maryland, pp. 486–491.

518 Farmer, J.J., Hickman-Brenner, F.W., Fanning, G.R., Gordon, C.M. and Brenner, D.J. (1988) Characterization of *Vibrio metschnikovii* and *Vibrio gazogenes* by DNA–DNA hybridization and phenotype. *Journal of Clinical Microbiology* 26(10), 1993–2000.

519 Farmer, J.J. and Holmes, B. (2009) International Committee on Systematics of Prokaryotes; Subcommittee on the taxonomy of *Aeromonadaceae*, *Vibrionaceae* and related organisms. *International Journal of Systematic and Evolutionary Microbiology* 59(10), 2633–2637.

520 Farrell, I.D. (1974) The development of a new selective medium for the isolation of *Brucella abortus* from contaminated sources. *Research in Veterinary Science* 16(3), 280–286.

521 Farto, R., Armada, S.P., Montes, M., Guisande, J.A., Pérez, M.J. and Nieto, T.P. (2003) *Vibrio lentus* associated with diseased wild octopus (*Octopus vulgaris*). *Journal of Invertebrate Pathology* 83(2), 149–156.

522 Farto, R., Armada, S.P., Montes, M., Perez, M.J. and Nieto, T.P. (2006) Presence of a lethal protease in the extracellular products of *Vibrio splendidus–Vibrio lentus* related strains. *Journal of Fish Diseases* 29(12), 701–707.

523 Farto, R., Montes, M., Pérez, M.J., Nieto, T.P., Larsen, J.L. and Pedersen, K. (1999) Characterization by numerical taxonomy and ribotyping of *Vibrio splendidus* biovar I and *Vibrio scophthalmi* strains associated with turbot cultures. *Journal of Applied Microbiology* 86(5), 796–804.

524 Faury, N., Saulnier, D., Thompson, F.L., Gay, M., Swings, J. and Le Roux, F. (2004) *Vibrio crassostreae* sp. nov., isolated from the haemolymph of oysters (*Crassostrea gigas*). *International Journal of Systematic and Evolutionary Microbiology* 54(Pt 6), 2137–2140.

525 Ferguson, H.W., Collins, R.O., Moore, M., Coles, M. and MacPhee, D.D. (2004) *Pseudomonas anguilliseptica* infection in farmed cod, *Gadus morhua* L. *Journal of Fish Diseases* 27(4), 249–253.

526 Ferguson, H.W., Delannoy, C.M.J., Hay, S., Nicolson, J., Sutherland, D. and Crumlish, M. (2010) Jellyfish as vectors of bacterial disease for farmed salmon (*Salmo salar*). *Journal of Veterinary Diagnostic Investigation* 22(3), 376–382.

527 Fernández-Benéitez, M.J., Ortiz-Santaliestra, M.E., Lizana, M. and Diéguez-Uribeondo, J. (2008) *Saprolegnia diclina*: another species responsible for the emergent disease 'Saprolegnia infections' in amphibians. *FEMS Microbiology Letters* 279(1), 23–29.

528 Fernández-Garayzábal, J.F., Vela, A.I., Egido, R., Hutson, R.A., Lanzarot, M.P., Fernández-García, M. and Collins, M.D. (2004) *Corynebacterium ciconiae* sp. nov., isolated from the trachea of black storks (*Ciconia nigra*). *International Journal of Systematic and Evolutionary Microbiology* 54(Pt 6), 2191–2195.

529 Ferragut, C., Izard, D., Gavini, F., Kersters, K., Ley, J.D. and Leclerc, H. (1983) *Klebsiella trevisanii*: a new species from water and soil. *International Journal of Systematic Bacteriology* 33(2), 133–142.

530 Fidopiastis, P.M., Sørum, H. and Ruby, E.G. (1999) Cryptic luminescence in the cold-water fish pathogen *Vibrio salmonicida*. *Archives of Microbiology* 171(3), 205–209.

531 Fidopiastis, P.M., von Boletzky, S. and Ruby, E.G. (1998) A new niche for *Vibrio logei*, the predominant light organ symbiont of squids in the genus *Sepiola*. *Journal of Bacteriology* 180(1), 59–64.

532 Fields, P.I., Popovic, T., Wachsmuth, K. and Olsvik, O. (1992) Use of polymerase chain reaction for detection of toxigenic *Vibrio cholerae* O1 strains from the Latin American cholera epidemic. *Journal of Clinical Microbiology* 30(8), 2118–2121.

533 Figueiredo, H.C.P., Costa, F.A.A., Leal, C.A.G., Carvalho-Castro, G.A. and Leite, R.C. (2012) *Weissella* sp.

outbreaks in commercial rainbow trout (*Oncorhynchus mykiss*) farms in Brazil. *Veterinary Microbiology* 156(3–4), 359–366.
534 Figueiredo, H.C.P., Klesius, P.H., Arias, C.R., Evans, J., Shoemaker, C.A., Pereira, D.J. Jr and Peixoto, M.T.D. (2005) Isolation and characterization of strains of *Flavobacterium columnare* from Brazil. *Journal of Fish Diseases* 28(4), 199–204.
535 Figueras, M.J., Alperi, A., Saavedra, M.J., Ko, W.-C., Gonzalo, N., Navarro, M. and Martínez-Murcia, A.J. (2009) Clinical relevance of the recently described species *Aeromonas aquariorum*. *Journal of Clinical Microbiology* 47(11), 3742–3746.
536 Figueras, M.J., Alperi, A., Beaz-Hidalgo, R., Stackebrandt, E., Brambilla, E., Monera, A. and Martínez-Murcia, A.J. (2011a) *Aeromonas rivuli* sp. nov., isolated from the upstream region of a karst water rivulet. *International Journal of Systematic and Evolutionary Microbiology* 61(2), 242–248.
537 Figueras, M.J., Suarez-Franquet, A., Chacón, M.R., Soler, L., Navarro, M., Alejandre, C., Grasa, B., Martínez-Murcia, A.J. and Guarro, J. (2005) First record of the rare species *Aeromonas culicicola* from a drinking water supply. *Applied and Environmental Microbiology* 71(1), 538–541.
538 Figueras, M.J., Collado, L., Levican, A., Perez, J., Solsona, M.J. and Yustes, C. (2011b) *Arcobacter molluscorum* sp. nov., a new species isolated from shellfish. *Systematic and Applied Microbiology* 34(2), 105–109.
539 Figueras, M.J., Levican, A., Collado, L., Inza, M.I. and Yustes, C. (2011c) *Arcobacter ellisii* sp. nov., isolated from mussels. *Systematic and Applied Microbiology* 34(6), 414–418.
540 Filler, G., Ehrich, J.H., Strauch, E. and Beutin, L. (2000) Acute renal failure in an infant associated with cytotoxic *Aeromonas sobria* isolated from patient's stool and from aquarium water as suspected source of infection. *Journal of Clinical Microbiology* 38(1), 469–470.
541 Fisher, W.S., Nilson, E.H. and Shleser, R.A. (1975) Effect of the funqus *Haliphthoros milfordensis* on the juvenile stages of the American lobster *Homarus americanus*. *Journal of Invertebrate Pathology* 26(1), 41–45.
542 Flemming, L., Rawlings, D. and Chenia, H. (2007) Phenotypic and molecular characterisation of fishborne *Flavobacterium johnsoniae*-like isolates from aquaculture systems in South Africa. *Research in Microbiology* 158(1), 18–30.
543 Flores-Miranda, M. del C., Luna-González, A., Campa-Córdova, Á.I., González-Ocampo, H.A., Fierro-Coronado, J.A. and Partida-Arangure, B.O. (2011) Microbial immunostimulants reduce mortality in whiteleg shrimp (*Litopenaeus vannamei*) challenged with *Vibrio sinaloensis* strains. *Aquaculture* 320(1–2), 51–55.
544 Foo, J.T.W., Ho, B. and Lam, T.J. (1985) Mass mortality in *Siganus canaliculatus* due to streptococcal infection. *Aquaculture* 49(3–4), 185–195.
545 Forbes, L.B., Nielsen, O., Measures, L. and Ewalt, D.R. (2000) Brucellosis in ringed seals and harp seals from Canada. *Journal of Wildlife Diseases* 36(3), 595–598.
546 Foster, G., Jahans, K.L., Reid, R.J. and Ross, H.M. (1996a) Isolation of *Brucella* species from cetaceans, seals and an otter. *The Veterinary Record* 138(24), 583–586.
547 Foster, G., MacMillan, A.P., Godfroid, J., Howie, F., Ross, H.M., Cloeckaert, A., Reid, R.J., Brew, S. and Patterson, I.A.P. (2002) A review of *Brucella* sp. infection of sea mammals with particular emphasis on isolates from Scotland. *Veterinary Microbiology* 90(1–4), 563–580.
548 Foster, G., Ross, H.M., Malnick, H., Willems, A., Garcia, P., Reid, R.J. and Collins, M.D. (1996b) *Actinobacillus delphinicola* sp. nov., a new member of the family *Pasteurellaceae* Pohl (1979) 1981 isolated from sea mammals. *International Journal of Systematic Bacteriology* 46(3), 648–652.
549 Foster, G., Ross, H.M., Malnick, H., Willems, A., Hutson, R.A., Reid, R.J. and Collins, M.D. (2000) *Phocoenobacter uteri* gen. nov., sp. nov., a new member of the family *Pasteurellaceae* Pohl (1979) 1981 isolated from a harbour porpoise (*Phocoena phocoena*). *International Journal of Systematic and Evolutionary Microbiology* 50(Pt 1), 135–139.
550 Foster, G., Ross, H.M., Patterson, I.A., Hutson, R.A. and Collins, M.D. (1998) *Actinobacillus scotiae* sp. nov., a new member of the family *Pasteurellaceae* Pohl (1979) 1981 isolated from porpoises (*Phocoena phocoena*). *International Journal of Systematic Bacteriology* 48(Pt 3), 929–933.
551 Foster, G., Ross, H.M., Hutson, R.A. and Collins, M.D. (1997) *Staphylococcus lutrae* sp. nov., a new coagulase-positive species isolated from otters. *International Journal of Systematic Bacteriology* 47(3), 724–726.
552 Foster, G., Holmes, B., Steigerwalt, A.G., Lawson, P.A., Thorne, P., Byrer, D.E., Ross, H.M., Xerry, J., Thompson, P.M. and Collins, M.D. (2004) *Campylobacter insulaenigrae* sp. nov., isolated from marine mammals. *International Journal of Systematic and Evolutionary Microbiology* 54(6), 2369–2373.
553 Foster, G., McAuliffe, L., Dagleish, M.P., Barley, J., Howie, F., Nicholas, R.A.J. and Ayling, R.D. (2011) *Mycoplasma* species isolated from harbor porpoises (*Phocoena phocoena*) and a Sowerby's beaked whale (*Mesoplodon bidens*) stranded in Scottish waters. *Journal of Wildlife Diseases* 47(1), 206–211.
554 Foster, G., Osterman, B.S., Godfroid, J., Jacques, I. and Cloeckaert, A. (2007) *Brucella ceti* sp. nov. and *Brucella pinnipedialis* sp. nov. for *Brucella* strains with cetaceans and seals as their preferred hosts. *International Journal of Systematic and Evolutionary Microbiology* 57(Pt 11), 2688–2693.
555 Fouz, B., Larsen, J.L. and Amaro, C. (2006) *Vibrio vulnificus* serovar A: an emerging pathogen in European anguilliculture. *Journal of Fish Diseases* 29(5), 285–291.
556 Fouz, B., Larsen, J.L., Nielsen, B., Barja, J.L. and Toranzo, A.E. (1992) Characterization of *Vibrio damsela* strains isolated from turbot *Scophthalmus maxiumus* in Spain. *Diseases of Aquatic Organisms* 12, 155–166.
557 Fox, G., Wisotzkey, J. and Jurtshuk, P. (1992) How close is close: 16S rRNA sequence identity may not be sufficient to guarantee species identity. *International Journal of Systematic Bacteriology* 42(1), 166–170.
558 Fox, J.G., Yan, L.L., Dewhirst, F.E., Paster, B.J., Shames, B., Murphy, J.C., Hayward, A., Belcher, J.C. and Mendes, E.N. (1995) *Helicobacter bilis* sp. nov., a

novel *Helicobacter* species isolated from bile, livers, and intestines of aged, inbred mice *Journal of Clinical Microbiology* 33(2), 445–454.
559 Fox, J.G., Chien, C.C., Dewhirst, F.E., Paster, B.J., Shen, Z., Melito, P.L., Woodward, D.L. and Rodgers, F.G. (2000) *Helicobacter canadensis* sp. nov. isolated from humans with diarrhea as an example of an emerging pathogen. *Journal of Clinical Microbiology* 38(7), 2546–2549.
560 Fox, J.G., Handt, L., Xu, S., Shen, Z., Dewhirst, F.E., Paster, B.J., Dangler, C.A., Lodge, K., Motzel, S. and Klein, H. (2001) Novel *Helicobacter* species isolated from rhesus monkeys with chronic idiopathic colitis. *Journal of Medical Microbiology* 50(5), 421–429.
561 Fox, J.G., Taylor, N.S., Howe, S., Tidd, M., Xu, S., Paster, B.J. and Dewhirst, F.E. (2006) *Helicobacter anseris* sp. nov. and *Helicobacter brantae* sp. nov., isolated from feces of resident Canada geese in the Greater Boston Area. *Applied and Environmental Microbiology* 72(7), 4633–4637.
562 Francis-Floyd, R., Reed, P., Bolon, B., Estes, J. and McKinney, S. (1993) An epizootic of *Edwardsiella tarda* in largemouth bass (*Micropterus salmoides*). *Journal of Wildlife Diseases* 29(2), 334–336.
563 Franco, P.F. and Hedreyda, C.T. (2006) Amplification and sequence analysis of the full length *toxR* gene in *Vibrio harveyi*. *The Journal of General and Applied Microbiology* 52(5), 281–287.
564 Fraser, G.C., Callinan, R.B. and Calder, L.M. (1992) *Aphanomyces* species associated with red spot disease: an ulcerative disease of estuarine fish from eastern Australia. *Journal of Fish Diseases* 15(2), 173–181.
565 Fregeneda-Grandes, J.M., Rodríguez-Cadenas, F. and Aller-Gancedo, J.M. (2007) Fungi isolated from cultured eggs, alevins and broodfish of brown trout in a hatchery affected by saprolegniosis. *Journal of Fish Biology* 71(2), 510–518.
566 Friedman, C.S., Beaman, B.L., Chun, J., Goodfellow, M., Gee, A. and Hedrick, R.P. (1998) *Nocardia crassostreae* sp. nov., the causal agent of nocardiosis in Pacific oysters. *International Journal of Systematic Bacteriology* 48(Pt 1), 237–246.
567 Fringuelli, E., Savage, P.D., Gordon, A., Baxter, E.J., Rodger, H.D. and Graham, D.A. (2012) Development of a quantitative real-time PCR for the detection of *Tenacibaculum maritimum* and its application to field samples. *Journal of Fish Diseases* 35(8), 579–590.
568 Fryer, J.L. and Sanders, J.E. (1981) Bacterial kidney disease of salmonid fish. *Annual Review of Microbiology* 35(1), 273–298.
569 Fujino, T., Sakazaki, R. and Tamura, K. (1974) Designation of the type strain of *Vibrio parahaemolyticus* and description of 200 strains of the species. *International Journal of Systematic Bacteriology* 24(4), 447–449.
570 Fujioka, R.S., Greco, S.B., Cates, M.B. and Schroeder, J.P. (1988) *Vibrio damsela* from wounds in bottlenose dolphins *Tursiops truncates*. *Diseases of Aquatic Organisms* 4, 1–8.
571 Fujiwara-Nagata, E. and Eguchi, M. (2009) Development and evaluation of a loop-mediated isothermal amplification assay for rapid and simple detection of *Flavobacterium psychrophilum*. *Journal of Fish Diseases* 32(10), 873–881.
572 Fukuda, Y., Matsuoka, S., Mizuno, Y. and Narita, K. (1996) *Pasteurella piscicida* infection in cultured juvenile Japanese flounder [*Paralichthys olivaceus*]. *Fish Pathology* 31(1), 33–38.
573 Funke, G., Ramos, C.P., Fernández-Garayzábal, J.F., Weiss, N. and Collins, M.D. (1995) Description of human-derived Centers for Disease Control coryneform group 2 bacteria as *Actinomyces bernardiae* sp. nov. *International Journal of Systematic Bacteriology* 45(1), 57–60.
574 Funke, G., von Graevenitz, A. and Weiss, N. (1994) Primary identification of *Aureobacterium* spp. isolated from clinical specimens as '*Corynebacterium aquaticum*'. *Journal of Clinical Microbiology* 32(11), 2686–2691.
575 Furniss, A.L. and Donovan, T.J. (1974) The isolation and identification of *Vibrio cholerae*. *Journal of Clinical Pathology* 27(9), 764–766.
576 Furniss, A.L., Lee, J.V. and Donovan, T.J. (1977) Group F, a new *Vibrio*? *The Lancet* 310(8037), 565–566.
577 Furniss, A.L., Lee, J.V. and Donovan, T.J. (1978) *The Vibrios*. Public Health Laboratory Service Monograph Series 11.
578 Gadagkar, S.R., Rosenberg, M.S. and Kumar, S. (2005) Inferring species phylogenies from multiple genes: concatenated sequence tree versus consensus gene tree. *Journal of Experimental Zoology. Part B, Molecular and Developmental Evolution* 304(1), 64–74.
579 Gahlawat, S.K., Ellis, A.E. and Collet, B. (2009) A sensitive loop-mediated isothermal amplification (LAMP) method for detection of *Renibacterium salmoninarum*, causative agent of bacterial kidney disease in salmonids. *Journal of Fish Diseases* 32(6), 491–497.
580 Gales, N., Wallace, G. and Dickson, J. (1985) Pulmonary cryptococcosis in a striped dolphin (*Stenella coeruleoalba*). *Journal of Wildlife Diseases* 21(4), 443–446.
581 Gao, Z.-M., Xiao, J., Wang, X.-N., Ruan, L.-W., Chen, X.-L. and Zhang, Y.-Z. (2012) *Vibrio xiamenensis* sp. nov., a cellulase-producing bacterium isolated from mangrove soil. *International Journal of Systematic and Evolutionary Microbiology* 62(Pt 8), 1958–1962.
582 Garcia, M.T., Ventosa, A., Ruiz-Berraquero, F. and Kocur, M. (1987) Taxonomic study and amended description of *Vibrio costicola*. *International Journal of Systematic Bacteriology* 37(3), 251–256.
583 García-González, P., García-Lamas, N., Fuentes Edfuf, C. and Santos, Y. (2011) Development of a PCR method for the specific identification of the marine fish pathogen *Tenacibaculum soleae*. *Aquaculture* 319(1–2), 1–4.
584 Garland, C., Nash, G., Summer, C. and McMeekin, T. (1983) Bacterial pathogens of oyster larvae (*Crassostrea gigas*) in a Tasmanian hatchery. *Marine and Freshwater Research* 34(3), 483–487.
585 Garland, S., Baker, A., Phillott, A.D. and Skerratt, L.F. (2010) BSA reduces inhibition in a TaqMan assay for the detection of *Batrachochytrium dendrobatidis*. *Diseases of Aquatic Organisms* 92(2–3), 113–116.
586 Garland, S., Wood, J. and Skerratt, L.F. (2011) Comparison of sensitivity between real-time detection of a

TaqMan assay for *Batrachochytrium dendrobatidis* and conventional detection. *Diseases of Aquatic Organisms* 94(2), 101–105.

587 Garner, M.M., Lambourn, D.M., Jeffries, S.J., Hall, P.B., Rhyan, J.C., Ewalt, D.R., Polzin, L.M. and Cheville, N.F. (1997) Evidence of *Brucella* infection in *Parafilaroides* lungworms in a Pacific harbor seal (*Phoca vitulina* Richardsi). *Journal of Veterinary Diagnostic Investigation* 9(3), 298–303.

588 Garnier, M., Labreuche, Y. and Nicolas, J.-L. (2008) Molecular and phenotypic characterization of *Vibrio aestuarianus* subsp. *francensis* subsp. nov., a pathogen of the oyster *Crassostrea gigas*. *Systematic and Applied Microbiology* 31(5), 358–365.

589 Gaskins, J.E. and Cheung, P.J. (1986) *Exophiala pisciphila*. A study of its development. *Mycopathologia* 93(3), 173–184.

590 Gatesoupe, F.J., Lambert, C. and Nicolas, J.L. (1999) Pathogenicity of *Vibrio splendidus* strains associated with turbot larvae, *Scophthalmus maximus*. *Journal of Applied Microbiology* 87(5), 757–763.

591 Gauger, E.J. and Gómez-Chiarri, M. (2002) 16S ribosomal DNA sequencing confirms the synonymy of *Vibrio harveyi* and *V. carchariae*. *Diseases of Aquatic Organisms* 52(1), 39–46.

592 Gauger, E.J., Smolowitz, R., Uhlinger, K., Casey, J. and Gómez-Chiarri, M. (2006) *Vibrio harveyi* and other bacterial pathogens in cultured summer flounder, *Paralichthys dentatus*. *Aquaculture* 260, 10–20.

593 Gauthier, D.T. and Rhodes, M.W. (2009) Mycobacteriosis in fishes: a review. *Veterinary Journal* 180(1), 33–47.

594 Gauthier, G., Gauthier, M. and Christen, R. (1995a) Phylogenetic analysis of the genera *Alteromonas, Shewanella*, and *Moritella* using genes coding for small-subunit rRNA sequences and division of the genus *Alteromonas* into two genera, *Alteromonas* (emended) and *Pseudoalteromonas* gen. nov. and proposal of twelve new species combinations. *International Journal of Systematic Bacteriology* 45(4), 755–761.

595 Gauthier, G., Lafay, B., Ruimy, R., Breittmayer, V., Nicolas, J.L., Gauthier, M. and Christen, R. (1995b) Small-subunit rRNA sequences and whole DNA relatedness concur for the reassignment of *Pasteurella piscicida* (Snieszko *et al.*) Janssen and Surgalla to the genus *Photobacterium* as *Photobacterium damsela* subsp. *piscicida* comb. nov. *International Journal of Systematic Bacteriology* 45(1), 139–144.

596 Gauthier, M.J. (1976a) Morphological, physiological, and biochemical characteristics of some violet-pigmented bacteria isolated from seawater. *Canadian Journal of Microbiology* 22(2), 138–149.

597 Gauthier, M.J. (1976b) *Alteromonas rubra* sp. nov., a new marine antibiotic-producing bacterium. *International Journal of Systematic Bacteriology* 26(4), 459–466.

598 Gauthier, M.J. (1977) *Alteromonas citrea*, a new Gram-negative, yellow-pigmented species from seawater. *International Journal of Systematic Bacteriology* 27(4), 349–354.

599 Gauthier, M.J. (1982) Validation of the name *Alteromonas luteoviolacea*. *International Journal of Systematic Bacteriology* 32(1), 82–86.

600 Gauthier, M.J. and Breittmayer, V.A. (1979) A new antibiotic-producing bacterium from seawater *Alteromonas aurantia* sp. nov. *International Journal of Systematic Bacteriology* 29(4), 366–372.

601 Gavini, F., Ferragut, C., Izard, D., Trinel, P.A., Leclerc, H., Lefebvre, B. and Mossel, D.A.A. (1979) *Serratia fonticola*, a new species from water. *International Journal of Systematic Bacteriology* 29(2), 92–101.

602 Gavini, F., Mergaert, J., Beji, A., Mielcarek, C., Izard, D., Kersters, K. and Ley, J.D. (1989) Transfer of *Enterobacter agglomerans* (Beijerinck 1888) Ewing and Fife 1972 to *Pantoea* gen. nov. as *Pantoea agglomerans* comb. nov. and description of *Pantoea dispersa* sp. nov. *International Journal of Systematic Bacteriology* 39(3), 337–345.

603 Gay, M., Renault, T., Pons, A.-M. and Le Roux, F. (2004) Two *Vibrio splendidus* related strains collaborate to kill *Crassostrea gigas*: taxonomy and host alterations. *Diseases of Aquatic Organisms* 62(1–2), 65–74.

604 Gelev, I., Gelev, E., Steigerwalt, A.G., Carter, G.P. and Brenner, D.J. (1990) Identification of the bacterium associated with haemorrhagic septicaemia in rainbow trout as *Hafnia alvei*. *Research in Microbiology* 141(5), 573–576.

605 Geng, Y., Wang, K.Y., Huang, X.L., Chen, D.F., Li, C.W., Ren, S.Y., Liao, Y.T., Zhou, Z.Y., Liu, Q.F., Du, Z.J. and Lai, W.M. (2012) *Streptococcus agalactiae*, an emerging pathogen for cultured ya-fish, *Schizothorax prenanti*, in China. *Transboundary and Emerging Diseases* 59(4), 369–375.

606 Geng, Y., Wang, K., Chen, D., Huang, X., He, M. and Yin, Z. (2010) *Stenotrophomonas maltophilia*, an emerging opportunist pathogen for cultured channel catfish, *Ictalurus punctatus*, in China. *Aquaculture* 308(3–4), 132–135.

607 Geraci, J.R., Sauer, R.M. and Medway, W. (1966) Erysipelas in dolphins. *American Journal of Veterinary Research* 27(117), 597–606.

608 Germani, Y., Dauga, C., Duval, P., Huerre, M., Levy, M., Pialoux, G., Sansonetti, P. and Grimont, P.A. (1997) Strategy for the detection of *Helicobacter* species by amplification of 16S rRNA genes and identification of *H. felis* in a human gastric biopsy. *Research in Microbiology* 148(4), 315–326.

609 Getchell, R.G., Rohovec, J.S. and Fryer, J.L. (1985) Comparison of *Renibacterium salmoninarum* isolates by antigenic analysis. *Fish Pathology* 20(2/3), 149–159.

610 Gharaibeh, D.N., Hasegawa, H. and Häse, C.C. (2009) Development of a quantitative real-time PCR assay for detection of *Vibrio tubiashii* targeting the metalloprotease gene. *Journal of Microbiological Methods* 76(3), 262–268.

611 Ghiasi, M., Khosravi, A.R., Soltani, M., Binaii, M., Shokri, H., Tootian, Z., Rostamibashman, M. and Ebrahimzademousavi, H. (2010) Characterization of *Saprolegnia* isolates from Persian sturgeon (*Acipencer persicus*) eggs based on physiological and molecular data. *Journal de Mycologie Médicale/Journal of Medical Mycology* 20(1), 1–7.

612 Ghittino, C., Agnetti, F., Panzieri, C., Cabra, S., Colussi, S. and Prearo, M. (2004) *Vagococcus salmoninarum* infection in rainbow trout in Italy. *Ittiopatologia* 1, 25–33.

613 Gibello, A., Blanco, M.M., Moreno, M.A., Cutuli, M.T., Domenech, A., Domínguez, L. and Fernández-Garayzábal, J.F. (1999) Development of a PCR assay for detection of *Yersinia ruckeri* in tissues of inoculated and naturally infected trout. *Applied and Environmental Microbiology* 65(1), 346–350.
614 Gibello, A., Mata, A.I., Blanco, M.M., Casamayor, A., Domínguez, L. and Fernández-Garayzabal, J.F. (2005) First identification of *Streptococcus phocae* isolated from Atlantic salmon (*Salmo salar*). *Journal of Clinical Microbiology* 43(1), 526–527.
615 Gibson, L.F., Woodworth, J. and George, A.M. (1998) Probiotic activity of *Aeromonas media* on the Pacific oyster, *Crassostrea gigas*, when challenged with *Vibrio tubiashii*. *Aquaculture* 169(1–2), 111–120.
616 Giebel, J., Meier, J., Binder, A., Flossdorf, J., Poveda, J.B., Schmidt, R. and Kirchhoff, H. (1991) *Mycoplasma phocarhinis* sp. nov. and *Mycoplasma phocacerebrale* sp. nov., two new species from harbor seals (*Phoca vitulina* L.). *International Journal of Systematic Bacteriology* 41(1), 39–44.
617 Gil, P., Vivas, J., Gallardo, C.S. and Rodriguez, L.A. (2000) First isolation of *Staphylococcus warneri*, from diseased rainbow trout, *Oncorhynchus mykiss* (Walbaum), in Northwest Spain. *Journal of Fish Diseases* 23(4), 295–298.
618 Gilardi, G. (1983) *Identification of glucose non-fermenting Gram-negative rods*. American Society for Microbiology, Washington, DC.
619 Gilbert, M.T.P., Haselkorn, T., Bunce, M., Sanchez, J.J., Lucas, S.B., Jewell, L.D., Marck, E.V. and Worobey, M. (2007) The isolation of nucleic acids from fixed, paraffin-embedded tissues – which methods are useful when? *PLoS ONE* 2(6), e537.
620 Gilmartin, W.G., Vainik, P.M. and Neill, V.M. (1979) *Salmonellae* in feral pinnipeds off the Southern California coast. *Journal of Wildlife Diseases* 15(4), 511–514.
621 Gjerde, J. (1984) Occurrence and characterization of *Aerococcus viridans* from lobsters, *Homarus gammarus* L., dying in captivity. *Journal of Fish Diseases* 7(5), 355–362.
622 Gjessing, M., Davey, M., Kvellestad, A. and Vrålstad, T. (2011) *Exophiala angulospora* causes systemic inflammation in Atlantic cod *Gadus morhua*. *Diseases of Aquatic Organisms* 96(3), 209–219.
623 Glazebrook, J.S. and Campbell, R.S.F. (1990a) A survey of the diseases of marine turtles in northern Australia. I. Farmed turtles. *Diseases of Aquatic Organisms* 9, 83–95.
624 Glazebrook, J.S. and Campbell, R.S.F. (1990b) A survey of the diseases of marine turtles in northern Australia. II. Oceanarium-reared and wild turtles. *Diseases of Aquatic Organisms* 9, 97–104.
625 Goarant, C., Ansquer, D., Herlin, J., Domalain, D., Imbert, F. and De Decker, S. (2006a) 'Summer Syndrome' in *Litopenaeus stylirostris* in New Caledonia: pathology and epidemiology of the etiological agent, *Vibrio nigripulchritudo*. *Aquaculture* 253(1–4), 105–113.
626 Goarant, C. and Merien, F. (2006) Quantification of *Vibrio penaeicida*, the etiological agent of Syndrome 93 in New Caledonian shrimp, by real-time PCR using SYBR Green I chemistry. *Journal of Microbiological Methods* 67(1), 27–35.
627 Goarant, C., Reynaud, Y., Ansquer, D., de Decker, S. and Merien, F. (2007) Sequence polymorphism-based identification and quantification of *Vibrio nigripulchritudo* at the species and subspecies level targeting an emerging pathogen for cultured shrimp in New Caledonia. *Journal of Microbiological Methods* 70(1), 30–38.
628 Goarant, C., Reynaud, Y., Ansquer, D., de Decker, S., Saulnier, D. and le Roux, F. (2006b) Molecular epidemiology of *Vibrio nigripulchritudo*, a pathogen of cultured penaeid shrimp (*Litopenaeus stylirostris*) in New Caledonia. *Systematic and Applied Microbiology* 29(7), 570–580.
629 Goldenberger, D., Perschil, I., Ritzler, M. and Altwegg, M. (1995) A simple 'universal' DNA extraction procedure using SDS and proteinase K is compatible with direct PCR amplification. *PCR Methods and Applications* 4(6), 368–370.
630 Goldman, C.G., Loureiro, J.D., Matteo, M.J., Catalano, M., Gonzalez, A.B., Heredia, S.R., Zubillaga, M.B., Solnick, J.V. and Cremaschi, G.A. (2009a) *Helicobacter* spp. from gastric biopsies of stranded South American fur seals (*Arctocephalus australis*). *Research in Veterinary Science* 86(1), 18–21.
631 Goldman, C.G., Matteo, M.J., Loureiro, J.D., Degrossi, J., Teves, S., Heredia, S.R., Alvarez, K., González, A.B., Catalano, M., Boccio, J., Cremaschi, G., Solnick, J.V. and Zubillaga, M.B. (2009b) Detection of *Helicobacter* and *Campylobacter* spp. from the aquatic environment of marine mammals. *Veterinary Microbiology* 133(3), 287–291.
632 Goldman, C.G., Loureiro, J.D., Quse, V., Corach, D., Calderon, E., Caro, R.A., Boccio, J., Rodríguez Heredia, S., Di Carlo, M.B. and Zubillaga, M.B. (2002) Evidence of *Helicobacter* sp. in dental plaque of captive dolphins (*Tursiops gephyreus*). *Journal of Wildlife Diseases* 38(3), 644–648.
633 Goldman, C.G., Matteo, M.J., Loureiro, J.D., Almuzara, M., Barberis, C., Vay, C., Catalano, M., Heredia, S.R., Mantero, P., Boccio, J.R., Zubillaga, M.B., Cremaschi, G.A., Solnick, J.V., Perez-Perez, G.I. and Blaser, M.J. (2011) Novel gastric helicobacters and oral campylobacters are present in captive and wild cetaceans. *Veterinary Microbiology* 152(1–2), 138–145.
634 Gomez-Gil, B., Fajer-Avila, E. and García-Vargas, F. (2007) Vibrios of the spotted rose snapper *Lutjanus guttatus* Steindachner, 1869 from northwestern Mexico. *Journal of Applied Microbiology* 102(6), 1518–1526.
635 Gomez-Gil, B., Roque, A., Lacuesta, B. and Rotllant, G. (2010) Diversity of vibrios in the haemolymph of the spider crab *Maja brachydactyla*. *Journal of Applied Microbiology* 109(3), 918–926.
636 Gomez-Gil, B., Thompson, F.L., Thompson, C.C., Garcia-Gasca, A., Roque, A. and Swings, J. (2004a) *Vibrio hispanicus* sp. nov., isolated from *Artemia* sp. and sea water in Spain. *International Journal of Systematic and Evolutionary Microbiology* 54(Pt 1), 261–265.
637 Gomez-Gil, B., Thompson, F.L., Thompson, C.C. and Swings, J. (2003a) *Vibrio rotiferianus* sp. nov., isolated from cultures of the rotifer *Brachionus plicatilis*. *International Journal of Systematic and Evolutionary Microbiology* 53(Pt 1), 239–243.
638 Gomez-Gil, B., Thompson, F.L., Thompson, C.C. and Swings, J. (2003b) *Vibrio pacinii* sp. nov., from cultured

aquatic organisms. *International Journal of Systematic and Evolutionary Microbiology* 53(Pt 5), 1569–1573.
639 Gomez-Gil, B., Fajer-Avila, E., Pascual, J., Macián, M.C., Pujalte, M.J., Garay, E. and Roque, A. (2008) *Vibrio sinaloensis* sp. nov., isolated from the spotted rose snapper, *Lutjanus guttatus* Steindachner, 1869. *International Journal of Systematic and Evolutionary Microbiology* 58(Pt 7), 1621–1624.
640 Gomez-Gil, B., Roque, A., Chimetto, L., Moreira, A.P.B., Lang, E. and Thompson, F.L. (2012) *Vibrio alfacsensis* sp. nov., isolated from marine organisms. *International Journal of Systematic and Evolutionary Microbiology* 62(Pt 12), 2955–2961.
641 Gomez-Gil, B., Roque, A., Rotllant, G., Peinado, L., Romalde, J.L., Doce, A., Cabanillas-Beltrán, H., Chimetto, L.A. and Thompson, F.L. (2011) *Photobacterium swingsii* sp. nov., isolated from marine organisms. *International Journal of Systematic and Evolutionary Microbiology* 61(Pt 2), 315–319.
642 Gomez-Gil, B., Soto-Rodríguez, S., García-Gasca, A., Roque, A., Vazquez-Juarez, R., Thompson, F.L. and Swings, J. (2004b) Molecular identification of *Vibrio harveyi*-related isolates associated with diseased aquatic organisms. *Microbiology* 150(Pt 6), 1769–1777.
643 Gomez-Gil, B., Tron-Mayén, L., Roque, A., Turnbull, J.F., Inglis, V. and Guerra-Flores, A.L. (1998) Species of *Vibrio* isolated from hepatopancreas, haemolymph and digestive tract of a population of healthy juvenile *Penaeus vannamei*. *Aquaculture* 163(1–2), 1–9.
644 Gómez-León, J., Villamil, L., Lemos, M.L., Novoa, B. and Figueras, A. (2005) Isolation of *Vibrio alginolyticus* and *Vibrio splendidus* from aquacultured carpet shell clam (*Ruditapes decussatus*) larvae associated with mass mortalities. *Applied and Environmental Microbiology* 71(1), 98–104.
645 Gomila, M., Gascó, J., Busquets, A., Gil, J., Bernabeu, R., Buades, J.M. and Lalucat, J. (2005) Identification of culturable bacteria present in haemodialysis water and fluid. *FEMS Microbiology Ecology* 52(1), 101–114.
646 Gonzalez, S.F., Osorio, C.R. and Santos, Y. (2003) Development of a PCR-based method for the detection of *Listonella anguillarum* in fish tissues and blood samples. *Diseases of Aquatic Organisms* 55(2), 109–115.
647 Goodfellow, M., Embley, T.M. and Austin, B. (1985) Numerical taxonomy and emended description of *Renibacterium salmoninarum*. *Journal of General Microbiology* 131(10), 2739–2752.
648 Goodwin, A.E., Roy, J.S.J., Grizzle, J.M. and Goldsby, M.T.J. (1994) *Bacillus mycoides*: a bacterial pathogen of channel catfish. *Diseases of Aquatic Organisms* 18, 173–179.
649 Gordon, M.A. (1976) Characterization of *Dermatophilus congolensis*: its affinities with the *Actinomycetales* and differentiation from *Geodermatophilusi*. In: Lloyd, D.H. and Sellers, K.C. (eds) *Dermatophilus Infection in Animals and Man*. Academic Press, London, pp. 187–201.
650 Gordon, R.E., Barnett, D.A., Handerhan, J.E. and Pang, C.H.-N. (1974) *Nocardia coeliaca, Nocardia autotrophica*, and the Nocardin strain. *International Journal of Systematic Bacteriology* 24(1), 54–63.
651 Goris, J., Konstantinidis, K.T., Klappenbach, J.A., Coenye, T., Vandamme, P. and Tiedje, J.M. (2007) DNA–DNA hybridization values and their relationship to whole-genome sequence similarities. *International Journal of Systematic and Evolutionary Microbiology* 57(1), 81–91.
652 Goyache, J., Ballesteros, C., Vela, A.I., Collins, M.D., Briones, V., Hutson, R.A., Potti, J., García-Borboroglu, P., Domínguez, L. and Fernández-Garayzábal, J.F. (2003a) *Corynebacterium sphenisci* sp. nov., isolated from wild penguins. *International Journal of Systematic and Evolutionary Microbiology* 53(Pt 4), 1009–1012.
653 Goyache, J., Vela, A.I., Collins, M.D., Ballesteros, C., Briones, V., Moreno, J., Yorio, P., Dominguez, L., Hutson, R. and Fernández-Garayzábal, J.F. (2003b) *Corynebacterium spheniscorum* sp. nov., isolated from the cloacae of wild penguins. *International Journal of Systematic and Evolutionary Microbiology* 53(Pt 1), 43–46.
654 Grange, J.M. (1981) *Mycobacterium chelonei*. *Tubercle* 62(4), 273–276.
655 Green, S.L., Lifland, B.D., Bouley, D.M., Brown, B.A., Wallace, R.J. and Ferrell, J.E. (2000) Disease attributed to *Mycobacterium chelonae* in South African clawed frogs (*Xenopus laevis*). *Comparative Medicine* 50(6), 675–679.
656 Greenwood, A.G. and Taylor, D.C. (1978) Clostridial myositis in marine mammals. *Veterinary Record* 103(3), 54–55.
657 Griffin, P.J., Snieszko, S.F. and Friddle, S.B. (1953) A new adjuvant in the diagnosis of fish furunculosis caused by *Bacterium salmonicida*. *Veterinary Medicine* 48(7), 280–282.
658 Grimes, D., Gruber, S.H. and May, E.B. (1985) Experimental infection of lemon sharks, *Negaprion brevirostris* (Poey), with *Vibrio* species. *Journal of Fish Diseases* 8, 173–180.
659 Grimes, D.J., Stemmler, J., Hada, H., May, E.B., Maneval, D., Hetrick, F.M., Jones, R.T., Stoskopf, M. and Colwell, R.R. (1984) *Vibrio* species associated with mortality of sharks held in captivity. *Microbial Ecology* 10(3), 271–282.
660 Grimes, D.J., Jacobs, D., Swartz, D.G., Brayton, P.R. and Colwell, R.R. (1993) Numerical taxonomy of Gram-negative, oxidase-positive rods from carcharhinid sharks. *International Journal of Systematic Bacteriology* 43(1), 88–98.
661 Grimont, F. and Grimont, P.A.D. (2005) Genus XXXIV. *Serratia Bizio 1823, 288AL*. In: Brenner, D.J., Krieg, N.R., Staley, J.T. and Garrity, G.M. (eds) *Bergey's Manual of Systematic Bacteriology*, 2nd edn, Vol. II. Springer, New York.
662 Grimont, P.A.D., Grimont, F., Richard, C. and Sakazaki, R. (1980) *Edwardsiella hoshinae*, a new species of *Enterobacteriaceae*. *Current Microbiology* 4(6), 347–351.
663 Grocott, R.G. (1955) A stain for fungi in tissue sections and smears using Gomori's methenamine-silver nitrate technique. *American Journal of Clinical Pathology* 25(8), 975–979.
664 Groussaud, P., Shankster, S.J., Koylass, M.S. and Whatmore, A.M. (2007) Molecular typing divides marine mammal strains of *Brucella* into at least three groups with distinct host preferences. *Journal of Medical Microbiology* 56(Pt 11), 1512–1518.

665 Guerinot, M.L., West, P.A., Lee, J.V. and Colwell, R.R. (1982) *Vibrio diazotrophicus* sp. nov., a marine nitrogen-fixing bacterium. *International Journal of Systematic Bacteriology* 32(3), 350–357.

666 Gulland, F.M., Koski, M., Lowenstine, L.J., Colagross, A., Morgan, L. and Spraker, T. (1996) Leptospirosis in California sea lions (*Zalophus californianus*) stranded along the central California coast, 1981–1994. *Journal of Wildlife Diseases* 32(4), 572–580.

667 Gunn, B.A. and Colwell, R.R. (1983) Numerical taxonomy of Staphylococci isolated from the marine environment. *International Journal of Systematic Bacteriology* 33(4), 751–759.

668 Hada, H.S., West, P.A., Lee, J.V., Stemmler, J. and Colwell, R.R. (1984) *Vibrio tubiashii* sp. nov., a pathogen of bivalve mollusks. *International Journal of Systematic Bacteriology* 34(1), 1–4.

669 Hahnel, G.B. and Gould, R.W. (1982) Effects of temperature on biochemical reactions and drug resistance of virulent and a virulent *Aeromonas salmonicida*. *Journal of Fish Diseases* 5, 329–337.

670 Haldar, S., Neogi, S.B., Kogure, K., Chatterjee, S., Chowdhury, N., Hinenoya, A., Asakura, M. and Yamasaki, S. (2010) Development of a haemolysin gene-based multiplex PCR for simultaneous detection of *Vibrio campbellii, Vibrio harveyi* and *Vibrio parahaemolyticus*. *Letters in Applied Microbiology* 50(2), 146–152.

671 Haldar, S., Chatterjee, S., Sugimoto, N., Das, S., Chowdhury, N., Hinenoya, A., Asakura, M. and Yamasaki, S. (2011) Identification of *Vibrio campbellii* isolated from diseased farm-shrimps from south India and establishment of its pathogenic potential in an *Artemia* model. *Microbiology* 157(Pt 1), 179–188.

672 Hamada, M., Iino, T., Iwami, T., Harayama, S., Tamura, T. and Suzuki, K. (2010) *Mobilicoccus pelagius* gen. nov., sp. nov. and *Piscicoccus intestinalis* gen. nov., sp. nov., two new members of the family *Dermatophilaceae*, and reclassification of *Dermatophilus chelonae* (Masters *et al.* 1995) as *Austwickia chelonae* gen. nov., comb. nov. *The Journal of General and Applied Microbiology* 56(6), 427–436.

673 Handlinger, J., Soltani, M. and Percival, S. (1997) The pathology of *Flexibacter maritimus* in aquaculture species in Tasmania, Australia. *Journal of Fish Diseases* 20(3), 159–168.

674 Hänninen, M.L. and Hirvelä-Koski, V. (1997) Molecular and phenotypic methods for the characterization of atypical *Aeromonas salmonicida*. *Veterinary Microbiology* 56(1–2), 147–158.

675 Hansen, G.H., Bergh, O., Michaelsen, J. and Knappskog, D. (1992) *Flexibacter ovolyticus* sp. nov., a pathogen of eggs and larvae of Atlantic halibut, *Hippoglossus hippoglossus* L. *International Journal of Systematic Bacteriology* 42(3), 451–458.

676 Hansen, G.H., Raa, J. and Olafsen, J.A. (1990) Isolation of *Enterobacter agglomerans* from dolphin fish, *Coryphaena hippurus* L. *Journal of Fish Diseases* 13(1), 93–96.

677 Harf-Monteil, C., Flèche, A.L. , Riegel, P., Prévost, G., Bermond, D., Grimont, P.A.D. and Monteil, H. (2004) *Aeromonas simiae* sp. nov., isolated from monkey faeces. *International Journal of Systematic and Evolutionary Microbiology* 54(Pt 2), 481–485.

678 Harper, C.G., Feng, Y., Xu, S., Taylor, N.S., Kinsel, M., Dewhirst, F.E., Paster, B.J., Greenwell, M., Levine, G., Rogers, A. and Fox, J.G. (2002a) *Helicobacter cetorum* sp. nov., a urease-positive *Helicobacter* species isolated from dolphins and whales. *Journal of Clinical Microbiology* 40(12), 4536–4543.

679 Harper, C.M., Dangler, C.A., Xu, S., Feng, Y., Shen, Z., Sheppard, B., Stamper, A., Dewhirst, F.E., Paster, B.J. and Fox, J.G. (2000) Isolation and characterization of a *Helicobacter* sp. from the gastric mucosa of dolphins, *Lagenorhynchus acutus* and *Delphinus delphis*. *Applied and Environmental Microbiology* 66(11), 4751–4757.

680 Harper, C., Xu, S., Feng, Y., Dunn, J.L., Taylor, N.S., Dewhirst, F.E. and Fox, J.G. (2002b) Identification of novel *Helicobacter* spp. from a beluga whale. *Applied and Environmental Microbiology* 68(4), 2040–2043.

681 Harper, C., Xu, S., Rogers, A.B., Feng, Y., Shen, Z., Taylor, N.S., Dewhirst, F.E., Paster, B.J., Miller, M., Hurley, J. and Fox, J.G. (2003a) Isolation and characterization of novel *Helicobacter* spp. from the gastric mucosa of harp seals *Phoca groenlandica*. *Diseases of Aquatic Organisms* 57(1–2), 1–9.

682 Harper, C.G. Whary, M.T., Feng, Y., Rhinehart, H.L., Wells, R.S., Xu, S., Taylor, N.S. and Fox, J.G. (2003b) Comparison of diagnostic techniques for *Helicobacter cetorum* infection in wild Atlantic bottlenose dolphins (*Tursiops truncatus*). *Journal of Clinical Microbiology* 41(7), 2842–2848.

683 Harrell, L., Novotny, A., Schiewe, M.H. and Hodgins, H. (1976) Isolation and description of two vibrios pathogenic to Pacific salmon in Puget Sound, Washington. *Fishery Bulletin* 74(2), 447–449.

684 Hartley, J.W. and Pitcher, D. (2002) Seal finger – tetracycline is first line. *Journal of Infection* 45(2), 71–75.

685 Hartmans, S., De Bont, J.A.M. and Stackebrandt, E. (2006) The genus *Mycobacterium* – nonmedical. In: *The Prokaryotes*, 3rd edn, Vol. 3. Springer, New York, pp. 889–918.

686 Harwood, C.S. (1978) *Beneckea gazogenes* sp. nov., a red, facultatively anaerobic, marine bacterium. *Current Microbiology* 1(4), 233–238.

687 Hasegawa, H. and Häse, C.C. (2009) The extracellular metalloprotease of *Vibrio tubiashii* directly inhibits its extracellular haemolysin. *Microbiology* 155(Pt 7), 2296–2305.

688 Hasson, K.W., Wyld, E.M., Fan, Y., Lingsweiller, S.W., Weaver, S.J., Cheng, J. and Varner, P.W. (2009) Streptococcosis in farmed *Litopenaeus vannamei*: a new emerging bacterial disease of penaeid shrimp. *Diseases of Aquatic Organisms* 86(2), 93–106.

689 Håstein, T. and Holt, G. (1972) The occurrence of vibrio disease in wild Norwegian fish. *Journal of Fish Biology* 4(1), 33–37.

690 Håstein, T., Saltveit, S.J. and Roberts, R.J. (1978) Mass mortality among minnows *Phoxinus phoxinus* (L.) in Lake Tveitevatn, Norway, due to an aberrant strain of *Aeromonas salmonicida*. *Journal of Fish Diseases* 1(3), 241–249.

691 Hatai, K. and Hoshiai, G. (1992) Mass mortality in cultured coho salmon (*Oncorhynchus kisutch*) due to

Saprolegnia parasitica Coker. *Journal of Wildlife Diseases* 28(4), 532–536.
692 Hatai, K. and Kubota, S.S. (1989) A visceral mycosis in cultured masu salmon (*Oncorhynchus masou*) caused by a species of *Ochroconis*. *Journal of Wildlife Diseases* 25(1), 83–88.
693 Hatai, K., Fujimaki, Y., Egusa, S. and Jo, Y. (1986) A visceral mycosis in ayu fry, *Plecoglossus altivelis* Temminck & Schlegel, caused by a species of *Phoma*. *Journal of Fish Diseases* 9, 111–116.
694 Hatai, K. (1982) On the fungus *Haliphthoros milfordensis* isolated from temporarily held abalone (*Haliotis sieboldii*). *Fish Pathology* 17(3), 199–204.
695 Hatai, K. and Egusa, S. (1979) Studies on the pathogenic fungus of mycotic granulomatosis – III. Development of the medium for MG-fungus. *Fish Pathology* 13(3), 147–152.
696 Hatai, K., Egusa, S., Nakajima, M. and Chikahata, H. (1975) *Pseudomonas chlororaphis* as a fish pathogen. *Bulletin of the Japanese Society of Scientific Fisheries* 41, 1203.
697 Hatai, K., Egusa, S. and Nomura, T. (1977) *Saprolegnia australis* Elliott isolated from body surface lesions of rainbow trout fingerlings. *Fish Pathology* 11(4), 201–206.
698 Hatai, K. and Hoshiai, G.-I. (1992) Saprolegniasis in cultured coho salmon (*Oncorhynchus kisutch*). *Fish Pathology* 27(4), 233–234.
699 Hatai, K., Nakamura, K., Rha, S.A., Yuasa, K. and Wada, S. (1994) *Aphanomyces* infection in dwarf gourami (*Colisa lalia*). *Fish Pathology* 29(2), 95–99.
700 Hatai, K., Roza, D. and Nakayama, T. (2000) Identification of lower fungi isolated from larvae of mangrove crab, *Scylla serrata*, in Indonesia. *Mycoscience* 41(6), 565–572.
701 Haulena, M., Gulland, F.M.D., Lawrence, J.A., Fauquier, D.A., Jang, S., Aldridge, B., Spraker, T., Thomas, L.C., Brown, D.R., Wendland, L. and Davidson, M.K. (2006) Lesions associated with a novel *Mycoplasma* sp. in California sea lions (*Zalophus californianus*) undergoing rehabilitation. *Journal of Wildlife Diseases* 42(1), 40–45.
702 Hawke, J.P., McWhorter, A., Steigerwalt, A.G. and Brenner, D.J. (1981) *Edwardsiella ictaluri* sp. nov., the causative agent of enteric septicaemia of catfish. *International Journal of Systematic Bacteriology* 31(4), 396–400.
703 Hawke, J.P., Plakas, S.M., Minton, R.V., McPhearson, R.M., Snider, T.G. and Guarino, A.M. (1987) Fish pasteurellosis of cultured striped bass (*Morone saxatilis*) in coastal Alabama. *Aquaculture* 65(3–4), 193–204.
704 Hayashi, K., Moriwaki, J., Sawabe, T., Thompson, F.L., Swings, J., Gudkovs, N., Christen, R. and Ezura, Y. (2003) *Vibrio superstes* sp. nov., isolated from the gut of Australian abalones *Haliotis laevigata* and *Haliotis rubra*. *International Journal of Systematic and Evolutionary Microbiology* 53(Pt 6), 1813–1817.
705 Hayflick, L. (1965) Tissue cultures and mycoplasmas. *Texas Reports on Biology and Medicine, Supplement* 23, 285–303.
706 He, Y., Xu, T., Han, Y., Shi, X. and Zhang, X.-H. (2011) Phenotypic diversity of *Edwardsiella tarda* isolated from different origins. *Letters in Applied Microbiology* 53(3), 294–299.
707 Hebert, A.M. and Vreeland, R.H. (1987) Phenotypic comparison of halotolerant bacteria: *Halomonas halodurans* sp. nov., nom. rev., comb. nov. *International Journal of Systematic Bacteriology* 37(4), 347–350.
708 Heckert, R.A., Elankumaran, S., Milani, A. and Baya, A. (2001) Detection of a new *Mycobacterium* species in wild striped bass in the Chesapeake Bay. *Journal of Clinical Microbiology* 39(2), 710–715.
709 Hedlund, B. and Staley, J. (2001) *Vibrio cyclotrophicus* sp. nov., a polycyclic aromatic hydrocarbon (PAH)-degrading marine bacterium. *International Journal of Systematic and Evolutionary Microbiology* 51(1), 61–66.
710 Hedreyda, C.T. and Rañoa, D.R.E. (2007) Sequence analysis of the *Vibrio harveyi* ornithine decarboxylase (odc) gene and detection of a gene homologue in *Vibrio campbellii. The Journal of General and Applied Microbiology* 53(6), 353–356.
711 Hedrick, R., McDowell, T. and Groff, J. (1987) Mycobacteriosis in cultured striped bass from California. *Journal of Wildlife Diseases* 23(3), 391–395.
712 Heiberg, B. (1936) The biochemical reactions of Vibrios. *The Journal of Hygiene* 36(1), 114–117.
713 Hendrie, M.S., Hodgkiss, W. and Shewan, J.M. (1970) The identification, taxonomy and classification of luminous bacteria. *Journal of General Microbiology* 64(2), 151–169.
714 Hendrie, M.S., Hodgkiss, W. and Shewan, J.M. (1971a) Proposal that the species *Vibrio anguillarum* Bergman 1909, *Vibrio piscium* David 1927, and *Vibrio ichthodermis* (Wells and ZoBell) Shewan, Hobbs, and Hodgkiss 1960 be combined as a single species, *Vibrio anguillarum*. *International Journal of Systematic Bacteriology* 21(1), 64–68.
715 Hendrie, M.S., Hodgkiss, W. and Shewan, J.M. (1971b) Proposal that *Vibrio marinus* (Russell 1891) Ford 1927 be amalgamated with *Vibrio fischeri* (Beijerinck 1889) Lehmann and Neumann 1896. *International Journal of Systematic Bacteriology* 21(3), 217–221.
716 Henley, M.W. and Lewis, D.H. (1976) Anaerobic bacteria associated with epizootics in grey mullet (*Mugil cephalus*) and redfish (*Sciaenops ocellata*) along the Texas Gulf Coast. *Journal of Wildlife Diseases* 12(3), 448–453.
717 Henrichsen, J. (1972) Bacterial surface translocation: a survey and a classification. *Bacteriological Reviews* 36(4), 478–503.
718 Herbert, B. (2012) *Edwardsiella ictaluri* – a significant pathogen of fish. *Emergency Animal Disease Bulletin No. 108*. Department of Agriculture Fisheries and Forestry, Canberra.
719 Herbst, L.H., Costa, S.F., Weiss, L.M., Johnson, L.K., Bartell, J., Davis, R., Walsh, M. and Levi, M. (2001) Granulomatous skin lesions in moray eels caused by a novel *Mycobacterium* species related to *Mycobacterium triplex*. *Infection and Immunity* 69(7), 4639–4646.
720 Herr, R.A., Tarcha, E.J., Taborda, P.R., Taylor, J.W., Ajello, L. and Mendoza, L. (2001) Phylogenetic analysis of *Lacazia loboi* places this previously uncharacterized pathogen within the dimorphic *Onygenales*. *Journal of Clinical Microbiology* 39(1), 309–314.

721 Hickman, F.W., Farmer, J.J., Hollis, D.G., Fanning, G.R., Steigerwalt, A.G., Weaver, R.E. and Brenner, D.J. (1982) Identification of *Vibrio hollisae* sp. nov. from patients with diarrhea. *Journal of Clinical Microbiology* 15(3), 395–401.

722 Hickman-Brenner, F.W., Fanning, G.R., Arduino, M.J., Brenner, D.J. and Farmer, J.J.III (1988) *Aeromonas schubertii*, a new mannitol-negative species found in human clinical specimens. *Journal of Clinical Microbiology* 26(8), 1561–1564.

723 Hickman-Brenner, F.W., MacDonald, K.L., Steigerwalt, A.G., Fanning, G.R., Brenner, D.J. and Farmer, J.J. (1987) *Aeromonas veronii*, a new ornithine decarboxylase-positive species that may cause diarrhea. *Journal of Clinical Microbiology* 25(5), 900–906.

724 Hicks, C., Kinoshita, R. and Ladds, P. (2000) Pathology of melioidosis in captive marine mammals. *Australian Veterinary Journal* 78(3), 193–195.

725 Hidalgo-Vila, J., Díaz-Paniagua, C., Marchand, M.A. and Cunningham, A.A. (2012) *Batrachochytrium dendrobatidis* infection of amphibians in the Doñana National Park, Spain. *Diseases of Aquatic Organisms* 98(2), 113–119.

726 Hielm, S., Björkroth, J., Hyytiä, E. and Korkeala, H. (1998) Prevalence of *Clostridium botulinum* in Finnish trout farms: pulsed-field gel electrophoresis typing reveals extensive genetic diversity among type E isolates. *Applied and Environmental Microbiology* 64(11), 4161–4167.

727 Hijazin, M., Sammra, O., Ulbegi, H., Nagib, S., Alber, J., Lämmler, C., Kämpfer, P., Glaeser, S.P., Busse, H.-J., Kassmannhuber, J., Prenger-Berninghoff, E., Weiss, R., Siebert, U., Hassan, A.A., Abdelmawjood, A. and Zschöck, M. (2012) *Arcanobacterium phocisimile* sp. nov., isolated from harbour seals. *International Journal of Systematic and Evolutionary Microbiology*.

728 Hill, A.C. (1985) *Mycoplasma testudinis*, a new species isolated from a tortoise. *International Journal of Systematic Bacteriology* 35(4), 489–492.

729 Hill, J.E., Paccagnella, A., Law, K., Melito, P.L., Woodward, D.L., Price, L., Leung, A.H., Ng, L.-K., Hemmingsen, S.M. and Goh, S.H. (2006) Identification of *Campylobacter* spp. and discrimination from *Helicobacter* and *Arcobacter* spp. by direct sequencing of PCR-amplified *cpn60* sequences and comparison to *cpnDB*, a chaperonin reference sequence database. *Journal of Medical Microbiology* 55(4), 393–399.

730 Hill, W.E., Keasler, S.P., Trucksess, M.W., Feng, P., Kaysner, C.A. and Lampel, K.A. (1991) Polymerase chain reaction identification of *Vibrio vulnificus* in artificially contaminated oysters. *Applied and Environmental Microbiology* 57(3), 707–711.

731 Hirono, I., Masuda, T. and Aoki, T. (1996) Cloning and detection of the hemolysin gene of *Vibrio anguillarum*. *Microbial Pathogenesis* 21(3), 173–182.

732 Hirota, K., Nodasaka, Y., Orikasa, Y., Okuyama, H. and Yumoto, I. (2005) *Shewanella pneumatophori* sp. nov., an eicosapentaenoic acid-producing marine bacterium isolated from the intestines of Pacific mackerel (*Pneumatophorus japonicus*). *International Journal of Systematic and Evolutionary Microbiology* 55(Pt 6), 2355–2359.

733 Hirvelä-Koski, V., Koski, P. and Niiranen, H. (1994) Biochemical properties and drug resistance of *Aeromonas salmonicida* in Finland. *Diseases of Aquatic Organisms* 20, 191–196.

734 Hirvelä-Koski, V. (2008) The fish pathogen *Renibacterium salmoninarum*: growth in a microaerophilic atmosphere. *Veterinary Microbiology* 127(1–2), 191–195.

735 Hitchner, E.R. and Snieszko, S.F. (1947) A study of a microorganism causing a bacterial disease of lobsters. *Journal of Bacteriology* 54(1), 48.

736 Hiu, S.F., Holt, R.A., Sriranganathan, N., Seidler, R.J. and Fryer, J.L. (1984) *Lactobacillus piscicola*, a new species from salmonid fish. *International Journal of Systematic Bacteriology* 34(4), 393–400.

737 Hochwimmer, G., Tober, R., Bibars-Reiter, R., Licek, E. and Steinborn, R. (2009) Identification of two GH18 chitinase family genes and their use as targets for detection of the crayfish-plague oomycete *Aphanomyces astaci*. *BMC Microbiology* 9, 184.

738 Hoffmann, M., Monday, S.R., Fischer, M. and Brown, E.W. (2012a) Genetic and phylogenetic evidence for misidentification of *Vibrio* species within the Harveyi clade. *Letters in Applied Microbiology* 54(2),160–165.

739 Hoffmann, M., Monday, S.R., Allard, M.W., Strain, E.A., Whittaker, P., Naum, M., McCarthy, P.J., Lopez, J.V., Fischer, M. and Brown, E.W. (2012b) *Vibrio caribbeanicus* sp. nov., isolated from the marine sponge *Scleritoderma cyanea*. *International Journal of Systematic and Evolutionary Microbiology* 62(Pt 8), 1736–1743.

740 Hogardt, M., Trebesius, K., Geiger, A.M., Hornef, M., Rosenecker, J. and Heesemann, J. (2000) Specific and rapid detection by fluorescent in situ hybridization of bacteria in clinical samples obtained from cystic fibrosis patients. *Journal of Clinical Microbiology* 38(2), 818–825.

741 Høi, L., Dalsgaard, I., DePaola, A., Siebeling, R.J. and Dalsgaard, A. (1998) Heterogeneity among isolates of *Vibrio vulnificus* recovered from eels (*Anguilla anguilla*) in Denmark. *Applied and Environmental Microbiology* 64(12), 4676–4682.

742 Høie, S., Heum, M. and Thoresen, O.F. (1997) Evaluation of a polymerase chain reaction-based assay for the detection of *Aeromonas salmonicida* ss *salmonicida* in Atlantic salmon *Salmo salar*. *Diseases of Aquatic Organisms* 30, 27–35.

743 Holben, W.E., Williams, P., Gilbert, M.A., Saarinen, M., Särkilahti, L.K. and Apajalahti, J.H.A. (2002) Phylogenetic analysis of intestinal microflora indicates a novel *Mycoplasma* phylotype in farmed and wild salmon. *Microbial Ecology* 44(2), 175–185.

744 Hollis, D.G., Weaver, R.E., Baker, C.N. and Thornsberry, C. (1976) Halophilic *Vibrio* species isolated from blood cultures. *Journal of Clinical Microbiology* 3(4), 425–431.

745 Holman, J.D., Burnett, K.G. and Burnett, L.E. (2004) Effects of hypercapnic hypoxia on the clearance of *Vibrio campbellii* in the Atlantic blue crab, *Callinectes sapidus* Rathbun. *The Biological Bulletin* 206(3), 188–196.

746 Holmes, B. (1986) Identification and distribution of *Pseudomonas stutzeri* in clinical material. *The Journal of Applied Bacteriology* 60(5), 401–411.

747 Holmes, B., Lapage, S.P. and Malnick, H. (1975) Strains of *Pseudomonas putrefaciens* from clinical material. *Journal of Clinical Pathology* 28(2), 149–155.

748 Holmes, B. (1992) Synonymy of *Flexibacter maritimus* Wakabayashi, Hikida, and Masumura 1986 and *Cytophaga marina* Reichenbach 1989. *International Journal of Systematic Bacteriology* 42(1), 185–185.

749 Holmes, B., Owen, R.J., Evans, A., Malnick, H. and Willcox, W.R. (1977) *Pseudomonas paucimobilis*, a new species isolated from human clinical specimens, the hospital environment, and other sources. *International Journal of Systematic Bacteriology* 27(2), 133–146.

750 Holmes, B., Owen, R.J. and Hollis, D.G. (1982) *Flavobacterium spiritivorum*, a new species isolated from human clinical specimens. *International Journal of Systematic Bacteriology* 32(2), 157–165.

751 Holmes, B., Owen, R.J., Steigerwalt, A.G. and Brenner, D.J. (1984) *Flavobacterium gleum*, a new species found in human clinical specimens. *International Journal of Systematic Bacteriology* 34(1), 21–25.

752 Holmes, B., Owen, R.J. and Weaver, R.E. (1981) *Flavobacterium multivorum*, a new species isolated from human clinical specimens and previously known as group IIk, biotype 2. *International Journal of Systematic Bacteriology* 31(1), 21–34.

753 Holmes, B., Snell, J.J.S. and Lapage, S.P. (1977) Revised description, from clinical isolates, of *Flavobacterium odoratum* Stutzer and Kwaschnina 1929, and designation of the neotype strain. *International Journal of Systematic Bacteriology* 27(4), 330–336.

754 Holmes, B., Snell, J.J.S. and Lapage, S.P. (1978) Revised description, from clinical strains, of *Flavobacterium breve* (Lustig) Bergey *et al.* 1923 and proposal of the neotype strain. *International Journal of Systematic Bacteriology* 28(2), 201–208.

755 Holmes, B. (1998) Subcommittee on the taxonomy of *Vibrionaceae* Minutes of the meetings, 4 and 7 July 1994, Prague, Czech Republic. *International Journal of Systematic Bacteriology* 48(4), 1467–1469.

756 Hommez, J., Devriese, L.A. and Castryck, F. (1983) Improved media for the isolation of *Bordetella bronchiseptica*. In: Pedersen, K.B. and Nielsen, N.C. (eds) *Atrophic Rhinitis in Pigs*. Commission of the European Communities, Luxembourg, pp. 98–104.

757 Hoshina, T. (1962) On a new bacterium, *Paracolobactrum anguillimortiferum* n. sp. *Bulletin of the Japanese Society of Scientific Fisheries* 28(2), 162–164.

758 Houf, K., Tutenel, A., De ZutterL, L., Van Hoof, J. and Vandamme, P. (2000) Development of a multiplex PCR assay for the simultaneous detection and identification of *Arcobacter butzleri*, *Arcobacter cryaerophilus* and *Arcobacter skirrowii*. *FEMS Microbiology Letters* 193(1), 89–94.

759 Houf, K., Devriese, L.A., De Zutter, L., Van Hoof, J. and Vandamme, P. (2001) Development of a new protocol for the isolation and quantification of *Arcobacter* species from poultry products. *International Journal of Food Microbiology* 71(2–3), 189–196.

760 Howe, G.E. and Stehly, G.R. (1998) Experimental infection of rainbow trout with *Saprolegnia parasitica*. *Journal of Aquatic Animal Health* 10, 397–404.

761 Hoyles, L., Foster, G., Falsen, E., Thomson, L.F. and Collins, M.D. (2001a) *Facklamia miroungae* sp. nov., from a juvenile southern elephant seal (*Mirounga leonina*). *International Journal of Systematic and Evolutionary Microbiology* 51(Pt 4), 1401–1403.

762 Hoyles, L., Lawson, P.A., Foster, G., Falsen, E., Ohlén, M., Grainger, J.M. and Collins, M.D. (2000) *Vagococcus fessus* sp. nov., isolated from a seal and a harbour porpoise. *International Journal of Systematic and Evolutionary Microbiology* 50(Pt 3), 1151–1154.

763 Hoyles, L., Pascual, C., Falsen, E., Foster, G., Grainger, J.M. and Collins, M.D. (2001b) *Actinomyces marimammalium* sp. nov., from marine mammals. *International Journal of Systematic and Evolutionary Microbiology* 51(Pt 1), 151–156.

764 Hsieh, C.Y., Chang, T.C., Shen, Y.L., Chang, C.D., Tu, C., Tung, M.C., Chen, L.C. and Tsai, S.S. (2006) Pathological and PCR detection of mycobacteriosis in pond-cultured Chinese soft shell turtles, *Trionyx sinensis*. *Aquaculture* 261(1), 10–16.

765 Hsu, T.C., Waltman, W.D. and Shotts, E.B. Jr (1981) Correlation of extracellular enzymatic activity and biochemical characteristics with regard to virulence of *Aeromonas hydrophila*. *International Symposium on Fish Biologics: Serodiagnostics and Vaccines* 49, 101–111.

766 Huang, C.Y., Garcia, J.L., Patel, B.K., Cayol, J.L., Baresi, L. and Mah, R.A. (2000) *Salinivibrio costicola* subsp. *vallismortis* subsp. nov., a halotolerant facultative anaerobe from Death Valley, and emended description of *Salinivibrio costicola*. *International Journal of Systematic and Evolutionary Microbiology* 50(Pt 2), 615–622.

767 Huang, T., Cerenius, L. and Söderhäll, K. (1994) Analysis of genetic diversity in the crayfish plague fungus, *Aphanomyces astaci*, by random amplification of polymorphic DNA. *Aquaculture* 126(1–2), 1–9.

768 Huchzermeyer, K.D.A. and Van der Waal, B.C.W. (2012) Epizootic ulcerative syndrome: exotic fish disease threatens Africa's aquatic ecosystems. *Journal of the South African Veterinary Association* 83(1).

769 Hugh, R. (1964) The proposed conservation of the generic name *Vibrio* Pacini 1854 and designation of the neotype strain of *Vibrio cholerae* Pacini 1854. *International Bulletin of Bacteriological Nomenclature and Taxonomy* 14(2), 87–101.

770 Hugh, R. and Leifson, E. (1953) The taxonomic significance of fermentative versus oxidative metabolism of carbohydrates by various Gram negative bacteria. *Journal of Bacteriology* 66(1), 24–26.

771 Hughes, K.P., Duncan, R.B. and Smith, S.A. (2002) Renomegaly associated with a mycobacterial infection in summer flounder *Paralichthys dentatus*. *Fish Pathology* 37(2), 83–86.

772 Huijsmans, C.J., Damen, J., van der Linden, J.C., Savelkoul, P.H. and Hermans, M.H. (2010) Comparative analysis of four methods to extract DNA from paraffin-embedded tissues: effect on downstream molecular applications. *BMC Research Notes* 3, 239.

773 Hulvey, J.P., Padgett, D.E. and Bailey, J.C. (2007) Species boundaries within *Saprolegnia* (*Saprolegniales*, *Oomycota*) based on morphological and DNA sequence data. *Mycologia* 99(3), 421–429.

774 Humphrey, J. and Ashburner, L. (1993) Spread of the bacterial fish pathogen *Aeromonas salmonicida* after

importation of infected goldfish, *Carassius auratus*, into Australia. *Australian Veterinary Journal* 70(12), 453–454.
775 Humphrey, J.D., Lancaster, C.E., Gudkovs, N. and Copland, J.W. (1987) The disease status of Australian salmonids: bacteria and bacterial diseases. *Journal of Fish Diseases* 10(5), 403–410.
776 Humphrey, J.D., Lancaster, C., Gudkovs, N. and McDonald, W. (1986) Exotic bacterial pathogens *Edwardsiella tarda* and *Edwardsiella ictaluri* from imported ornamental fish *Betta splendens* and *Puntius conchonius*, respectively: isolation and quarantine significance. *Australian Veterinary Journal* 63(11), 369–371.
777 Humphry, D.R., George, A., Black, G.W. and Cummings, S.P. (2001) *Flavobacterium frigidarium* sp. nov., an aerobic, psychrophilic, xylanolytic and laminarinolytic bacterium from Antarctica. *International Journal of Systematic and Evolutionary Microbiology* 51(Pt 4), 1235–1243.
778 Huq, A., Alam, M., Parveen, S. and Colwell, R.R. (1992) Occurrence of resistance to vibriostatic compound 0/129 in *Vibrio cholerae* 01 isolated from clinical and environmental samples in Bangladesh. *Journal of Clinical Microbiology* 30(1), 219–221.
779 Huq, M.I., Alam, A.K., Brenner, D.J. and Morris, G.K. (1980) Isolation of *Vibrio*-like group, EF-6, from patients with diarrhea. *Journal of Clinical Microbiology* 11(6), 621–624.
780 Hussein, M.M.A. and Hatai, K. (1999) *Saprolegnia salmonis* sp. nov. isolated from sockeye salmon, *Oncorhynchus nerka*. *Mycoscience* 40(5), 387–391.
781 Hussein, M.M., Hatai, K. and Nomura, T. (2001) Saprolegniosis in salmonids and their eggs in Japan. *Journal of Wildlife Diseases* 37(1), 204–207.
782 Huys, G., Coopman, R., Janssen, P. and Kersters, K. (1996) High-resolution genotypic analysis of the genus *Aeromonas* by AFLP fingerprinting. *International Journal of Systematic Bacteriology* 46(2), 572–580.
783 Huys, G., Kämpfer, P., Altwegg, M., Coopman, R., Janssen, P., Gillis, M. and Kersters, K. (1997a) Inclusion of *Aeromonas* DNA hybridization group 11 in *Aeromonas encheleia* and extended descriptions of the species *Aeromonas eucrenophila* and *A. encheleia*. *International Journal of Systematic Bacteriology* 47(4), 1157–1164.
784 Huys, G., Kämpfer, P., Altwegg, M., Kersters, I., Lamb, A., Coopman, R., Lüthy-Hottenstein, J., Vancanneyt, M., Janssen, P. and Kersters, K. (1997b) *Aeromonas popoffii* sp. nov., a mesophilic bacterium isolated from drinking water production plants and reservoirs. *International Journal of Systematic Bacteriology* 47(4), 1165–1171.
785 Huys, G., Kämpfer, P. and Swings, J. (2001) New DNA–DNA hybridization and phenotypic data on the species *Aeromonas ichthiosmia* and *Aeromonas allosaccharophila: A. ichthiosmia* Schubert *et al.* 1990 is a later synonym of *A. veronii* Hickman-Brenner *et al.* 1987. *Systematic and Applied Microbiology* 24(2), 177–182.
786 Huys, G., Cnockaert, M., Abbott, S.L., Janda, J.M. and Vandamme, P. (2010) *Hafnia paralvei* sp. nov., formerly known as *Hafnia alvei* hybridization group 2. *International Journal of Systematic and Evolutionary Microbiology* 60(Pt 8), 1725–1728.
787 Huys, G., Cnockaert, M., Janda, J.M. and Swings, J. (2003a) *Escherichia albertii* sp. nov., a diarrhoeagenic species isolated from stool specimens of Bangladeshi children. *International Journal of Systematic and Evolutionary Microbiology* 53(Pt 3), 807–810.
788 Huys, G., Cnockaert, M. and Swings, J. (2005) *Aeromonas culicicola* Pidiyar *et al.* 2002 is a later subjective synonym of *Aeromonas veronii* Hickman-Brenner *et al.* 1987. *Systematic and Applied Microbiology* 28(7), 604–609.
789 Huys, G., Denys, R. and Swings, J. (2002a) DNA–DNA reassociation and phenotypic data indicate synonymy between *Aeromonas enteropelogenes* Schubert *et al.* 1990 and *Aeromonas trota* Carnahan *et al.* 1991. *International Journal of Systematic and Evolutionary Microbiology* 52(Pt 6), 1969–1972.
790 Huys, G., Kämpfer, P., Albert, M.J., Kühn, I., Denys, R. and Swings, J. (2002b) *Aeromonas hydrophila* subsp. *dhakensis* subsp. nov., isolated from children with diarrhoea in Bangladesh, and extended description of *Aeromonas hydrophila* subsp. *hydrophila* (Chester 1901) Stanier 1943 (approved lists 1980). *International Journal of Systematic and Evolutionary Microbiology* 52(3), 705–712.
791 Huys, G., Pearson, M., Kämpfer, P., Denys, R., Cnockaert, M., Inglis, V. and Swings, J. (2003b) *Aeromonas hydrophila* subsp. *ranae* subsp. nov., isolated from septicaemic farmed frogs in Thailand. *International Journal of Systematic and Evolutionary Microbiology* 53(Pt 3), 885–891.
792 Hwang, M.-N. and Ederer, G.M. (1975) Rapid hippurate hydrolysis method for presumptive identification of Group B Streptococci. *Journal of Clinical Microbiology* 1(1), 114–115.
793 Hyatt, A.D., Boyle, D.G., Olsen, V., Boyle, D.B., Berger, L., Obendorf, D., Dalton, A., Kriger, K., Heros, M., Hines, H., Phillott, R., Campbell, R., Marantelli, G., Gleason, F. and Colling, A. (2007) Diagnostic assays and sampling protocols for the detection of *Batrachochytrium dendrobatidis*. *Diseases of Aquatic Organisms* 73(3), 175–192.
794 Hyatt, A.D., Speare, R., Cunningham, A.A. and Carey, C. (2010) Amphibian chytridiomycosis. *Diseases of Aquatic Organisms* 92(2–3), 89–91.
795 Hyytiä, E., Hielm, S. and Korkeala, H. (1998) Prevalence of *Clostridium botulinum* type E in Finnish fish and fishery products. *Epidemiology and Infection* 120(3), 245–250.
796 Iida, T., Sakata, C., Kawatsu, H. and Fukuda, Y. (1997a) Atypical *Aeromonas salmonicida* infection in cultured marine fish. *Fish Pathology* 32(1), 65–66.
797 Iida, T., Suthienkul, O., Park, K.-S., Tang, G.-Q., Yamamoto, R., Ishibashi, M., Yamamoto, K. and Honda, T. (1997b) Evidence for genetic linkage between the *ure* and *trh* genes in *Vibrio parahaemolyticus*. *Journal of Medical Microbiology* 46(8), 639–645.
798 Ilardi, P., Abad, J., Rintamäki, P., Bernardet, J.-F. and Avendaño-Herrera, R. (2010) Phenotypic, serological and molecular evidence of *Chryseobacterium piscicola* in farmed Atlantic salmon, *Salmo salar* L., in Finland. *Journal of Fish Diseases* 33(2), 179–181.

799 Ilardi, P., Fernández, J. and Avendaño-Herrera, R. (2009) *Chryseobacterium piscicola* sp. nov., isolated from diseased salmonid fish. *International Journal of Systematic and Evolutionary Microbiology* 59 (Pt 12), 3001–3005.

800 Imai, D., Jang, S., Miller, M. and Conrad, P.A. (2009) Characterization of beta-hemolytic streptococci isolated from southern sea otters (*Enhydra lutris nereis*) stranded along the California coast. *Veterinary Microbiology* 136(3–4), 378–381.

801 Innis, M.A. and Gelfand, D.H. (1990) Optimisation of PCRs. In: Gelfand, D.H., Sninsky, J.J., Innis, M.A., and White, H. (eds) *PCR Protocols: A Guide to Methods and Applications*. Academic Press, San Diego, California, pp. 3–12.

802 Ishiguro, E.E., Ainsworth, T., Trust, T.J. and Kay, W.W. (1985) Congo red agar, a differential medium for *Aeromonas salmonicida*, detects the presence of the cell surface protein array involved in virulence. *Journal of Bacteriology* 164(3), 1233–1237.

803 Ishimaru, K., Akagawa-Matsushita, M. and Muroga, K. (1995) *Vibrio penaeicida* sp. nov., a pathogen of kuruma prawns (*Penaeus japonicus*). *International Journal of Systematic Bacteriology* 45(1), 134–138.

804 Ishimaru, K., Akagawa-Matsushita, M. and Muroga, K. (1996) *Vibrio ichthyoenteri* sp. nov., a pathogen of Japanese flounder (*Paralichthys olivaceus*) larvae. *International Journal of Systematic Bacteriology* 46(1), 155–159.

805 Ishimaru, K. and Muroga, K. (1997) Taxonomical re-examination of two pathogenic *Vibrio* species isolated from milkfish and swimming crab. *Fish Pathology* 32(1), 59–64.

806 Isik, K., Chun, J., Hah, Y.C. and Goodfellow, M. (1999) *Nocardia salmonicida* nom. rev., a fish pathogen. *International Journal of Systematic Bacteriology* 49(Pt 2), 833–837.

807 Islam, M.S., Hasan, M.K., Miah, M.A., Yunus, M., Zaman, K. and Albert, M.J. (1994) Isolation of *Vibrio cholerae* O139 synonym Bengal from the aquatic environment in Bangladesh: implications for disease transmission. *Applied and Environmental Microbiology* 60(5), 1684–1686.

808 Itano, T., Kawakami, H., Kono, T. and Sakai, M. (2006) Detection of fish nocardiosis by loop-mediated isothermal amplification. *Journal of Applied Microbiology* 100(6), 1381–1387.

809 Itoi, S., Abe, T., Washio, S., Ikuno, E., Kanomata, Y. and Sugita, H. (2008) Isolation of halotolerant *Lactococcus lactis* subsp. *lactis* from intestinal tract of coastal fish. *International Journal of Food Microbiology* 121(1), 116–121.

810 Itoi, S., Yuasa, K., Washio, S., Abe, T., Ikuno, E. and Sugita, H. (2009) Phenotypic variation in *Lactococcus lactis* subsp. *lactis* isolates derived from intestinal tracts of marine and freshwater fish. *Journal of Applied Microbiology* 107(3), 867–874.

811 Ivanova, E.P., Chun, J., Romanenko, L.A., Matte, M.E., Mikhailov, V.V., Frolova, G.M., Huq, A. and Colwell, R.R. (2000) Reclassification of *Alteromonas distincta* Romanenko *et al.* 1995 as *Pseudoalteromonas distincta* comb. nov. *International Journal of Systematic and Evolutionary Microbiology* 50(Pt 1), 141–144.

812 Ivanova, E.P., Kiprianova, E.A., Mikhailov, V.V., Levanova, G.F., Garagulya, A.D., Gorshkova, N.M., Vysotskii, M.V., Nicolau, D.V., Yumoto, N., Taguchi, T. and Yoshikawa, S. (1998) Phenotypic diversity of *Pseudoalteromonas citrea* from different marine habitats and emendation of the description. *International Journal of Systematic Bacteriology* 48(Pt 1), 247–256.

813 Ivanova, E.P., Sawabe, T., Gorshkova, N.M., Svetashev, V.I., Mikhailov, V.V., Nicolau, D.V. and Christen, R. (2001) *Shewanella japonica* sp. nov. *International Journal of Systematic and Evolutionary Microbiology* 51(Pt 3), 1027–1033.

814 Ivanova, E.P., Nedashkovskaya, O.I., Sawabe, T., Zhukova, N.V., Frolova, G.M., Nicolau, D.V., Mikhailov, V.V. and Bowman, J.P. (2004) *Shewanella affinis* sp. nov., isolated from marine invertebrates. *International Journal of Systematic and Evolutionary Microbiology* 54(Pt 4), 1089–1093.

815 Ivanova, E.P., Shevchenko, L.S., Sawabe, T., Lysenko, A.M., Svetashev, V.I., Gorshkova, N.M., Satomi, M., Christen, R. and Mikhailov, V.V. (2002) *Pseudoalteromonas maricaloris* sp. nov., isolated from an Australian sponge, and reclassification of [*Pseudoalteromonas aurantia*] NCIMB 2033 as *Pseudoalteromonas flavipulchra* sp. nov. *International Journal of Systematic and Evolutionary Microbiology* 52 (Pt 1), 263–271.

816 Ivanova, E.P., Kiprianova, E.A., Mikhailov, V.V., Levanova, G.F., Garagulya, A.D., Gorshkova, N.M., Yumoto, N. and Yoshikawa, S. (1996) Characterization and identification of marine *Alteromonas nigrifaciens* strains and emendation of the description. *International Journal of Systematic Bacteriology* 46(1), 223–228.

817 Iveson, J.B. (1971) Strontium chloride B and E.E. enrichment broth media for the isolation of *Edwardsiella*, *Salmonella* and *Arizona* species from tiger snakes. *The Journal of Hygiene* 69(3), 323–330.

818 Iwamoto, Y., Suzuki, Y., Kurita, A., Watanabe, Y., Shimizu, T., Ohgami, H. and Yanagihara, Y. (1995) *Vibrio trachuri* sp. nov., a new species isolated from diseased Japanese horse mackerel. *Microbiology and Immunology* 39(11), 831–837.

819 Iwanowicz, L.R., Griffin, A.R., Cartwright, D.D. and Blazer, V.S. (2006) Mortality and pathology in brown bullheads *Amieurus nebulosus* associated with a spontaneous *Edwardsiella ictaluri* outbreak under tank culture conditions. *Diseases of Aquatic Organisms* 70(3), 219–225.

820 Izard, D., Ferragut, C., Gavini, F., Kersters, K., Ley, J.D. and Leclerc, H. (1981) *Klebsiella terrigena*, a new species from soil and water. *International Journal of Systematic Bacteriology* 31(2), 116–127.

821 Izumi, S., Yamamoto, M., Suzuki, K., Shimizu, A. and Aranishi, F. (2007) Identification and detection of *Pseudomonas plecoglossicida* isolates with PCR primers targeting the *gyrB* region. *Journal of Fish Diseases* 30(7), 391–397.

822 Izumi, S. and Aranishi, F. (2004) Relationship between *gyrA* mutations and quinolone resistance in *Flavobacterium psychrophilum* isolates. *Applied and Environmental Microbiology* 70(7), 3968–3972.

823 Izumi, S. and Wakabayashi, H. (2000) Sequencing of *gyrB* and their application in the identification of

Flavobacterium psychrophilum by PCR. *Fish Pathology* 35, 93–94.

824 Izumikawa, K. and Ueki, N. (1997) Atypical *Aeromonas salmonicida* infection in cultured Schlegel's black rockfish. *Fish Pathology* 32(1), 67–68.

825 Izumiya, H., Matsumoto, K., Yahiro, S., Lee, J., Morita, M., Yamamoto, S., Arakawa, E. and Ohnishi, M. (2011) Multiplex PCR assay for identification of three major pathogenic *Vibrio* spp., *Vibrio cholerae*, *Vibrio parahaemolyticus*, and *Vibrio vulnificus*. *Molecular and Cellular Probes* 25(4), 174–176.

826 Jaffrès, E., Prévost, H., Rossero, A., Joffraud, J.-J. and Dousset, X. (2010) *Vagococcus penaei* sp. nov., isolated from spoilage microbiota of cooked shrimp (*Penaeus vannamei*). *International Journal of Systematic and Evolutionary Microbiology* 60(Pt 9), 2159–2164.

827 Jahans, K.L., Foster, G. and Broughton, E.S. (1997) The characterisation of *Brucella* strains isolated from marine mammals. *Veterinary Microbiology* 57(4), 373–382.

828 Janda, J.M. and Abbott, S.L. (2002) Bacterial identification for publication: when is enough enough? *Journal of Clinical Microbiology* 40(6), 1887–1891.

829 Janda, J.M. and Abbott, S.L. (2006) The genus *Hafnia*: from soup to nuts. *Clinical Microbiology Reviews* 19(1), 12–18.

830 Janda, J.M. and Abbott, S.L. (2010) The genus *Aeromonas*: taxonomy, pathogenicity, and infection. *Clinical Microbiology Reviews* 23(1), 35–73.

831 Jansen, G.J., Mooibroek, M., Idema, J., Harmsen, H.J.M., Welling, G.W. and Degener, J.E. (2000) Rapid identification of bacteria in blood cultures by using fluorescently labeled oligonucleotide probes. *Journal of Clinical Microbiology* 38(2), 814–817.

832 Janssen, W.A. and Surgalla, M.J. (1968) Morphology, physiology, and serology of a *Pasteurella* species pathogenic for white perch (*Roccus americanus*). *Journal of Bacteriology* 96(5), 1606–1610.

833 Jansson, E., Lindberg, L., Säker, E. and Aspán, A. (2008) Diagnosis of bacterial kidney disease by detection of *Renibacterium salmoninarum* by real-time PCR. *Journal of Fish Diseases* 31(10), 755–763.

834 Jasmin, A.M. and Baucom, J. (1967) *Erysipelothrix insidiosa* infections in the Caiman (*Caiman crocodilus*) and the American crocodile (*Crocodilus acutus*). *American Journal of Veterinary Clinical Pathology* 1, 173–177.

835 Jayaprakash, N.S., Pai, S.S., Philip, R. and Singh, I.S.B. (2006) Isolation of a pathogenic strain of *Vibrio alginolyticus* from necrotic larvae of *Macrobrachium rosenbergii* (de Man). *Journal of Fish Diseases* 29(3), 187–191.

836 Jensen, M., Tebo, B., Baumann, P., Mandel, M. and Nealson, K. (1980) Characterization of *Alteromonas hanedai* (sp. nov.), a nonfermentative luminous species of marine origin. *Current Microbiology* 3(5), 311–315.

837 Jensen, P.R., Kauffman, C.A. and Fenical, W. (1996) High recovery of culturable bacteria from the surfaces of marine algae. *Marine Biology* 126(1), 1–7.

838 Jensen, S., Samuelsen, O.B., Andersen, K., Torkildsen, L., Lambert, C., Choquet, G., Paillard, C. and Bergh, Ø. (2003) Characterization of strains of *Vibrio splendidus* and *V. tapetis* isolated from corkwing wrasse *Symphodus melops* suffering vibriosis. *Diseases of Aquatic Organisms* 53, 25–31.

839 Jing-jing, Z., Chang, C., Peng, L., Chun-hua, R., Xiao, J., Zhe, Z. and Chao-qun, H.(2011) SYBR Green I-based real-time PCR targeting the *rpoX* gene for sensitive and rapid detection of *Vibrio alginolyticus*. *Molecular and Cellular Probes* 25(2–3), 137–141.

840 Jiravanichpaisal, P., Miyazaki, T. and Limsuwan, C. (1994) Histopathology, biochemistry, and pathogenicity of *Vibrio harveyi* infecting black tiger prawn *Penaeus monodon*. *Journal of Aquatic Animal Health* 6, 27–35.

841 Jöborn, A., Dorsch, M., Olsson, J.C., Westerdahl, A. and Kjelleberg, S. (1999) *Carnobacterium inhibens* sp. nov., isolated from the intestine of Atlantic salmon (*Salmo salar*). *International Journal of Systematic Bacteriology* 49(Pt 4), 1891–1898.

842 Johnson, F.H. and Shunk, I.V. (1936) An interesting new species of luminous bacteria. *Journal of Bacteriology* 31(6), 585–593.

843 Johnson, J.L. and Chilton, W.S. (1966) Galactosamine glycan of *Chondrococcus columnaris*. *Science* 152(3726), 1247–1248.

844 Johnson, R., Colwell, R.R., Sakazaki, R. and Tamura, K. (1975) Numerical taxonomy study of the *Enterobacteriaceae*. *International Journal of Systematic Bacteriology* 25(1), 12–37.

845 Johnson, S.P., Jang, S., Gulland, F.M.D., Miller, M.A., Casper, D.R., Lawrence, J. and Herrera, J. (2003) Characterization and clinical manifestations of *Arcanobacterium phocae* infections in marine mammals stranded along the central California coast. *Journal of Wildlife Diseases* 39(1), 136–144.

846 Johnson, T.W. (1958) A fungus parasite in ova of the barnacle *Chthamalus fragilis denticulata*. *The Biological Bulletin* 114(2), 205–214.

847 Jolley, K.A., Feil, E.J., Chan, M.S. and Maiden, M.C. (2001) Sequence type analysis and recombinational tests (START). *Bioinformatics* 17(12), 1230–1231.

848 Jones, A. (1981) Effect of carbohydrate content of culture media on Kovac's oxidase test, with particular reference to *Vibrio* spp. *Medical Laboratory Sciences* 38, 133–137.

849 Jones, M.K. and Oliver, J.D. (2009) *Vibrio vulnificus*: disease and pathogenesis. *Infection and Immunity* 77(5), 1723–1733.

850 Jones, M.W. and Cox, D.I. (1999) Clinical disease in seafarmed Atlantic salmon (*Salmo salar*) associated with a member of the family *Pasteurellaceae* – a case history. *Bulletin of the European Association of Fish Pathologists* 19(2), 75–78.

851 Jores, J., Appel, B. and Lewin, A. (2007) *Vibrio navarrensis* biotype *pommerensis*: a new biotype of *V. navarrensis* isolated in the German Baltic Sea. *Systematic and Applied Microbiology* 30(1), 27–30.

852 Jørgensen, T., Midling, K., Espelid, S., Nilsen, R. and Stensvåg, K. (1989) *Vibrio salmonicida*, a pathogen in salmonids, also causes mortality in net-pen captured cod (*Gadus morhua*). *Bulletin of the European Association of Fish Pathologists* 9(2), 42–44.

853 Joseph, S.W., Colwell, R.R. and Kaper, J.B. (1982) *Vibrio parahaemolyticus* and related halophilic vibrios. *Critical Reviews in Microbiology* 10(1), 77–124.

854 Jung, M.Y., Chang, Y.-H. and Kim, W. (2010) A real-time PCR assay for detection and quantification of *Lactococcus garvieae*. *Journal of Applied Microbiology* 108(5), 1694–1701.

855 Kalina, G.P., Antonov, A.S., Turova, T.P. and Grafova, T.I. (1984) *Allomonas enterica* gen. nov., sp. nov.: deoxyribonucleic acid homology between *Allomonas* and some other members of the *Vibrionaceae*. *International Journal of Systematic Bacteriology* 34(2), 150–154.

856 Kaminski, G.W. and Suter, I.I. (1976) Human infection with *Dermatophilus congolensis*. *The Medical Journal of Australia* 1(13), 443–447.

857 Kämpfer, P., Lodders, N., Martin, K. and Avendaño-Herrera, R. (2012) *Flavobacterium chilense* sp. nov. and *Flavobacterium araucananum* sp. nov., isolated from farmed salmonid fish. *International Journal of Systematic and Evolutionary Microbiology* 62(Pt 6), 1402–1408.

858 Kämpfer, P. and Altwegg, M. (1992) Numerical classification and identification of *Aeromonas* genospecies. *Journal of Applied Microbiology* 72, 341–351.

859 Kämpfer, P., Fallschissel, K. and Avendaño-Herrera, R. (2011) *Chryseobacterium chaponense* sp. nov., isolated from farmed Atlantic salmon (*Salmo salar*). *International Journal of Systematic and Evolutionary Microbiology* 61(Pt 3), 497–501.

860 Kämpfer, P., Vaneechoutte, M. and Wauters, G. (2009) *Chryseobacterium arothri* Campbell *et al.* 2008 is a later heterotypic synonym of *Chryseobacterium hominis* Vaneechoutte *et al.* 2007. *International Journal of Systematic and Evolutionary Microbiology* 59(Pt 4), 695–697.

861 Kanamoto, T., Sato, S. and Inoue, M. (2000) Genetic heterogeneities and phenotypic characteristics of strains of the genus *Abiotrophia* and proposal of *Abiotrophia para-adiacens* sp. nov. *Journal of Clinical Microbiology* 38(2), 492–498.

862 Kaneko, K.-I. and Hashimoto, N. (1982) Five biovars of *Yersinia enterocolitica* delineated by numerical taxonomy. *International Journal of Systematic Bacteriology* 32(3), 275–287.

863 Kang, S.H., Shin, G.W., Shin, Y.S., Palaksha, K.J., Kim, Y., Yang, H., Lee, E., Lee, E., Huh, N., Ju, O.M. and Jung, T. (2004) Experimental evaluation of pathogenicity of *Lactococcus garvieae* in black rockfish (*Sebastes schlegeli*). *Journal of Veterinary Science* 5(4), 387–390.

864 Kang, S.-I., Her, M., Kim, J.W., Kim, J.-Y., Ko, K.Y., Ha, Y.-M. and Jung, S.C. (2011) Advanced multiplex PCR assay for differentiation of *Brucella* species. *Applied and Environmental Microbiology* 77(18), 6726–6728.

865 Kao, D.-Y., Cheng, Y.-C., Kuo, T.-Y., Lin, S.-B., Lin, C.-C., Chow, L.-P. and Chen, W.-J. (2009) Salt-responsive outer membrane proteins of *Vibrio anguillarum* serotype O1 as revealed by comparative proteome analysis. *Journal of Applied Microbiology* 106(6), 2079–2085.

866 Kapperud, G., Bergan, T. and Lassen, J. (1981) Numerical taxonomy of *Yersinia enterocolitica* and *Yersinia enterocolitica*-like bacteria. *International Journal of Systematic Bacteriology* 31(4), 401–419.

867 Karatas, S., Ercan, D., Steinum, T.M., Turgay, E., Memis, D. and Candan, A. (2010) First isolation of a *Flavobacterium johnsoniae* like bacteria from cultured Russian sturgeon in Turkey. *Journal of Animal and Veterinary Advances* 9(14), 1943–1946.

868 Kariya, T., Kubota, S., Nakamura, Y. and Kira, K. (1968) Nocardial infection in cultured yellowtails (*Seriola quinqueruiata* and *S. purpurascens*). I. Bacteriological study. *Fish Pathology* 3(1), 16–23.

869 Karthick, B. and Williams, D.M. (2012, February 25) The International Code for Nomenclature for algae, fungi and plants – a significant rewrite of the International Code of Botanical Nomenclature. *Current Science (Bangalore)* 102(4), 551–552.

870 Karunasagar, I., Karunasagar, I. and Pai, R. (1992) Systemic *Citrobacter freundii* infection in common carp, *Cyprinus carpio* L., fingerlings. *Journal of Fish Diseases* 15(1), 95–98.

871 Kasornchandra, J., Rogers, W.A. and Plumb, J.A. (1987) *Edwardsiella ictaluri* from walking catfish, *Clarias batrachus* L., in Thailand. *Journal of Fish Diseases* 10, 137–138.

872 Kawakami, H., Suzukawa, K., Yamashita, H., Kohno, Y. and Sakai, M. (1999) Mass mortality of wild largescale Blackfish *Girella punctata* caused by *Photobacterium damsela* subsp. *piscicida*. *Fish Pathology* 34(2), 87–88.

873 Kaznowski, A. (1998) Identification of *Aeromonas* strains of different origin to the genomic species level. *Journal of Applied Microbiology* 84(3), 423–430.

874 Ke, X.L., Wang, J.G., Gu, Z.M., Li, M. and Gong, X.N. (2009) Morphological and molecular phylogenetic analysis of two *Saprolegnia* sp. (*Oomycetes*) isolated from silver crucian carp and zebra fish. *Mycological Research* 113(5), 637–644.

875 Ke, X., Wang, J., Gu, Z., Li, M. and Gong, X. (2009) *Saprolegnia brachydanis*, a new oomycete isolated from zebra fish. *Mycopathologia* 167(2), 107–113.

876 Kemp, R., Leatherbarrow, A.J.H., Williams, N.J., Hart, C.A., Clough, H.E., Turner, J., Wright, E.J. and French, N.P. (2005) Prevalence and genetic diversity of *Campylobacter* spp. in environmental water samples from a 100-square-kilometer predominantly dairy farming area. *Applied and Environmental Microbiology* 71(4), 1876–1882.

877 Kent, M.L., Traxler, G.S., Kieser, D., Richard, J., Dawe, S.C., Shaw, R.W., Prosperi-Porta, G., Ketcheson, J. and Evelyn, T.P.T. (1998) Survey of salmonid pathogens in ocean-caught fishes in British Columbia, Canada. *Journal of Aquatic Animal Health* 10(2), 211–219.

878 Kent, M.L. (1982) Characteristics and identification of *Pasteurella* and *Vibrio* species pathogenic to fishes using API-20E (Analytab Products) multitube test strips. *Canadian Journal of Fisheries and Aquatic Sciences* 39(12), 1725–1729.

879 Kesarcodi-Watson, A., Kaspar, H., Lategan, M.J. and Gibson, L. (2009) Two pathogens of Greenshell mussel larvae, *Perna canaliculus: Vibrio splendidus* and a *V. coralliilyticus/neptunius*-like isolate. *Journal of Fish Diseases* 32(6), 499–507.

880 Keskin, O., Seçer, S., Izgür, M., Türkyilmaz, S. and Mkakosya, R. (2004) *Edwardsiella ictaluri* infection in rainbow trout (*Oncorhynchus mykiss*). *Turkish Journal of Veterinary and Animal Sciences* 28, 649–653.

881 Ketterer, P. and Eaves, L. (1992) Deaths in captive eels (*Anguila reinhardtii*) due to *Photobacterium*

(*Vibrio*) *damsela*. *Australian Veterinary Journal* 69(8), 203–204.

882 Keyes, M.C., Crews, F.W. and Ross, A.J. (1968) *Pasteurella multocida* isolated from a California sea lion (*Zalophus californianus*). *Journal of the American Veterinary Medical Association* 153(7), 803–804.

883 Khan, A.A. and Cerniglia, C.E. (1997) Rapid and sensitive method for the detection of *Aeromonas caviae* and *Aeromonas trota* by polymerase chain reaction. *Letters in Applied Microbiology* 24(4), 233–239.

884 Khan, A.A., Nawaz, M.S., Khan, S.A. and Cerniglia, C.E. (1999) Identification of *Aeromonas trota* (hybridization group 13) by amplification of the aerolysin gene using polymerase chain reaction. *Molecular and Cellular Probes* 13(2), 93–98.

885 Khan, M.H., Marshall, L., Thompson, K.D., Campbell, R.E. and Lilley, J.H. (1998) Susceptibility of five fish species (Nile tilapia, rosy barb, rainbow trout, stickleback and roach) to intramuscular injection with the oomycete fish pathogen, *Aphanomyces invadans*. *Bulletin of the European Association of Fish Pathologists* 18(6), 192–197.

886 Khashe, S. and Janda, J.M. (1998) Biochemical and pathogenic properties of *Shewanella alga* and *Shewanella putrefaciens*. *Journal of Clinical Microbiology* 36(3), 783–787.

887 Khoa, L.V. and Hatai, K. (2005) First case of *Fusarium oxysporum* infection in cultured kuruma prawn *Penaeus japonicus* in Japan. *Fish Pathology* 40(4), 195–196.

888 Khoa, L.V., Hatai, K., Yuasa, A. and Sawada, K. (2005) Morphology and molecular phylogeny of *Fusarium solani* isolated from kuruma prawn *Penaeus japonicus* with black gills. *Fish Pathology* 40(3), 103–109.

889 Khoo, L., Leard, A.T., Waterstrat, P.R., Jack, S.W. and Camp, K.L. (1998) Branchiomyces infection in farm-reared channel catfish, *Ictalurus punctatus* (Rafinesque). *Journal of Fish Diseases* 21(6), 423–432.

890 Khrulnova, S., Manukhov, I.V., Zarubina, A.P. and Zavilgelsky, G.B. (2010) *Aliivibrio logei* KCh1 (Kamchatka isolate), Biochemical and bioluminescence characteristics and cloning of the *lux* operon. *Microbiology* 79(3), 349–355.

891 Khulbe, R.D., Bisht, G.S. and Joshi, C. (1994) Epizootic infection due to *Achlya debaryana* in a catfish. *Mycoses* 37(1–2), 61–63.

892 Ki, J.-S., Zhang, R., Zhang, W., Huang, Y.-L. and Qian, P.-Y. (2009) Analysis of RNA polymerase beta subunit (*rpoB*) gene sequences for the discriminative power of marine *Vibrio* species. *Microbial Ecology* 58(4), 679–691.

893 Kiiyukia, C., Nakajima, A., Nakai, T., Muroga, K., Kawakami, H. and Hashimoto, H. (1992) *Vibrio cholerae* non-O1 isolated from ayu fish (*Plecoglossus altivelis*) in Japan. *Applied and Environmental Microbiology* 58(9), 3078–3082.

894 Kim, B.-J., Lee, S.-H., Lyu, M.-A., Kim, S.-J., Bai, G.-H., Kim, S.-J., Chae, G.-T., Kim, E.-C., Cha, C.-Y. and Kook, Y.-H. (1999a) Identification of mycobacterial species by comparative sequence analysis of the RNA polymerase gene (*rpoB*). *Journal of Clinical Microbiology* 37(6), 1714–1720.

895 Kim, D., Baik, K.S., Kim, M.S., Jung, B.-M., Shin, T.-S., Chung, G.-H., Rhee, M.S. and Seong, C.N. (2007) *Shewanella haliotis* sp. nov., isolated from the gut microflora of abalone, *Haliotis discus hannai*. *International Journal of Systematic and Evolutionary Microbiology* 57(Pt 12), 2926–2931.

896 Kim, D.-G., Bae, J.-Y., Hong, G.-E., Min, M.-K., Kim, J.-K. and Kong, I.-S. (2008) Application of the *rpoS* gene for the detection of *Vibrio anguillarum* in flounder and prawn by polymerase chain reaction. *Journal of Fish Diseases* 31(9), 639–647.

897 Kim, H.M., Hwang, C.Y. and Cho, B.C. (2010a) *Arcobacter marinus* sp. nov. *International Journal of Systematic and Evolutionary Microbiology* 60(Pt 3), 531–536.

898 Kim, J.H., Lee, J.K., Yoo, H.S., Shin, N.R., Shin, N.S., Lee, K.H. and Kim, D.Y. (2002) Endocarditis associated with *Escherichia coli* in a sea lion (*Zalophus californianus*). *Journal of Veterinary Diagnostic Investigation* 14(3), 260–262.

899 Kim, J.H., Gomez, D.K., Baeck, G.W., Shin, G.W., Heo, G.J., Jung, T.S. and Park, S.C. (2006) Pathogenicity of *Streptococcus parauberis* to Olive Flounder *Paralichthys olivaceus*. *Fish Pathology* 41(4), 171–173.

900 Kim, K.K., Kim, M.K., Lim, J.H., Park, H.Y. and Lee, S.-T. (2005) Transfer of *Chryseobacterium meningosepticum* and *Chryseobacterium miricola* to *Elizabethkingia* gen. nov. as *Elizabethkingia meningoseptica* comb. nov. and *Elizabethkingia miricola* comb. nov. *International Journal of Systematic and Evolutionary Microbiology* 55(Pt 3), 1287–1293.

901 Kim, K.-K., Kim, Y.-O., Park, S., Kang, S.-J., Nam, B.-H., Kim, D.N., Oh, T.-K. and Yoon, J.-H. (2011) *Shewanella upenei* sp. nov., a lipolytic bacterium isolated from bensasi goatfish *Upeneus bensasi*. *Journal of Microbiology* 49(3), 381–386.

902 Kim, Y.B., Okuda, J., Matsumoto, C., Takahashi, N., Hashimoto, S. and Nishibuchi, M. (1999b) Identification of *Vibrio parahaemolyticus* strains at the species level by PCR targeted to the *toxR* gene. *Journal of Clinical Microbiology* 37(4), 1173–1177.

903 Kim, Y.-O., Kong, H.J., Park, S., Kang, S.-J., Kim, K.-K., Moon, D.Y., Oh, T.-K. and Yoon, J.-H. (2010b) *Paracoccus fistulariae* sp. nov., a lipolytic bacterium isolated from bluespotted cornetfish, *Fistularia commersonii*. *International Journal of Systematic and Evolutionary Microbiology* 60(Pt 12), 2908–2912.

904 Kimura, B., Hokimoto, S., Takahashi, H. and Fujii, T. (2000) *Photobacterium histaminum* Okuzumi *et al.* 1994 is a later subjective synonym of *Photobacterium damselae* subsp. *damselae* (Love *et al.* 1981) Smith *et al.* 1991. *International Journal of Systematic and Evolutionary Microbiology* 50(Pt3), 1339–1342.

905 Kimura, H. and Kusuda, R. (1982) Studies on the pathogenesis of streptococcal infection in cultured yellowtails, *Seriola* spp.: effect of crude exotoxin fractions from cell-free culture on experimental streptococcal infection. *Journal of Fish Diseases* 5, 471–478.

906 Kimura, T. (1969) A new subspecies of *Aeromonas salmonicida* as an etiological agent of Furunculosis on ‘Sakuramasu’ (*Oncorhynchus masou*) and pink salmon (*O. gorbuscha*) rearing for maturity. *Fish Pathology* 3(2), 45–52.

907 Kingombe, C.I., Huys, G., Tonolla, M., Albert, M.J., Swings, J., Peduzzi, R. and Jemmi, T. (1999) PCR detection, characterization, and distribution of virulence

genes in *Aeromonas* spp. *Applied and Environmental Microbiology* 65(12), 5293–5302.
908 Kirchhoff, H., Beyene, P., Fischer, M., Flossdorf, J., Heitmann, J., Khattab, B., Lopatta, D., Rosengarten, R., Seidel, G. and Yousef, C. (1987) *Mycoplasma mobile* sp. nov., a new species from fish. *International Journal of Systematic Bacteriology* 37(3), 192–197.
909 Kirchhoff, H., Mohan, K., Schmidt, R., Runge, M., Brown, D.R., Brown, M.B., Foggin, C.M., Muvavarirwa, P., Lehmann, H. and Flossdorf, J. (1997) *Mycoplasma crocodyli* sp. nov., a new species from crocodiles. *International Journal of Systematic Bacteriology* 47(3), 742–746.
910 Kirchhoff, H. and Rosengarten, R. (1984) Isolation of a motile mycoplasma from fish. *Journal of General Microbiology* 130(9), 2439–2445.
911 Kiryu, Y., Blazer, V.S., Vogelbein, W.K., Kator, H. and Shields, J.D. (2005) Factors influencing the sporulation and cyst formation of *Aphanomyces invadans*, etiological agent of ulcerative mycosis in Atlantic menhaden, *Brevoortia tyrannus*. *Mycologia* 97(3), 569–575.
912 Kiszka, J., Van Bressem, M.-F. and Pusineri, C. (2009) Lobomycosis-like disease and other skin conditions in Indo-Pacific bottlenose dolphins *Tursiops aduncus* from the Indian Ocean. *Diseases of Aquatic Organisms* 84(2), 151–157.
913 Kita-Tsukamoto, K., Oyaizu, H., Nanba, K. and Simidu, U. (1993) Phylogenetic relationships of marine bacteria, mainly members of the family *Vibrionaceae*, determined on the basis of 16S rRNA sequences. *International Journal of Systematic Bacteriology* 43(1), 8–19.
914 Kitancharoen, N. and Hatai, K. (1995) A marine oomycete *Atkinsiella panulirata* sp. nov. from philozoma of spiny lobster, *Panulirus japonicus*. *Mycoscience* 36(1), 97–104.
915 Kitancharoen, N. and Hatai, K. (1997) *Aphanomyces frigidophilus* sp. nov. from eggs of Japanese char, *Salvelinus leucomaenis*. *Mycoscience* 38, 135–140.
916 Kitancharoen, N., Nakamura, K., Wada, S. and Hatai, K. (1994) *Atkinsiella awabi* sp. nov. isolated from stocked abalone, *Haliotis sieboldii*. *Mycoscience* 35(3), 265–270.
917 Kitao, T., Aoki, T. and Sakoh, R. (1981) Epizootic caused by β-haemolytic *Streptococcus* species in cultured freshwater fish. *Fish Pathology* 15, 301–307.
918 Kitao, T., Yoshida, T., Aoki, T. and Fukudome, M. (1985) Characterization of an atypical *Aeromonas salmonicida* strain causing epizootic ulcer disease in cultured eel. *Fish Pathology* 20(2/3), 107–114.
919 Klein, B., Kleingeld, D. and Bohm, K. (1993) First isolations of *Pleisiomonas shigelloides* from samples of cultured fish in Germany. *Bulletin of the European Association of Fish Pathologists* 13, 70–72.
920 Kloos, W.E. and Schleifer, K.H. (1975) Isolation and characterization of Staphylococci from human skin II. Descriptions of four new species: *Staphylococcus warneri*, *Staphylococcus capitis*, *Staphylococcus hominis*, and *Staphylococcus simulans*. *International Journal of Systematic Bacteriology* 25(1), 62–79.
921 Ko, Y.-M. and Heo, G.-J. (1997) Characteristics of *Flavobacterium branchiophilum* isolated from rainbow trout in Korea. *Fish Pathology* 32(2), 97–102.
922 Kobayashi, T., Enomoto, S., Sakazaki, R. and Kuwahara, S. (1963) A new selective medium for pathogenic vibrios TCBS Agar (modified Nakanishi's Agar). *Japanese Journal of Bacteriology* 18, 387–391.
923 Kodama, H., Nakanishi, Y., Yamamoto, F., Mikami, T., Izawa, H., Imagawa, T., Hashimoto, Y. and Kudo, N. (1987) *Salmonella arizonae* isolated from a pirarucu, *Arapaima gigas* Cuvier, with septicaemia. *Journal of Fish Diseases* 10, 509–512.
924 Kolby, J.E., Padgett-Flohr, G.E. and Field, R. (2010) Amphibian chytrid fungus *Batrachochytrium dendrobatidis* in Cusuco National Park, Honduras. *Diseases of Aquatic Organisms* 92(2–3), 245–251.
925 Kong, R.Y.C., Lee, S.K.Y., Law, T.W.F., Law, S.H.W. and Wu, R.S. S. (2002) Rapid detection of six types of bacterial pathogens in marine waters by multiplex PCR. *Water Research* 36(11), 2802–2812.
926 Königsson, M.H., Pettersson, B. and Johansson, K.E. (2001) Phylogeny of the seal mycoplasmas *Mycoplasma phocae* corrig., *Mycoplasma phocicerebrale* corrig. and *Mycoplasma phocirhinis* corrig. based on sequence analysis of 16S rDNA. *International Journal of Systematic and Evolutionary Microbiology* 51(Pt 4), 1389–1393.
927 Konstantinidis, K.T., Ramette, A. and Tiedje, J.M. (2006) The bacterial species definition in the genomic era. *Philosophical Transactions of the Royal Society B: Biological Sciences* 361(1475), 1929–1940.
928 Koppang, E.O., Fjølstad, M., Melgård, B., Vigerust, M. and Ørum, H.S. (2000) Non-pigment-producing isolates of *Aeromonas salmonicida* subspecies *salmonicida*: isolation, identification, transmission and pathogenicity in Atlantic salmon, *Salmo salar* L. *Journal of Fish Diseases* 23(1), 39–48.
929 Korczak, B.M., Stieber, R., Emler, S., Burnens, A.P., Frey, J. and Kuhnert, P. (2006) Genetic relatedness within the genus *Campylobacter* inferred from *rpoB* sequences. *International Journal of Systematic and Evolutionary Microbiology* 56(Pt 5), 937–945.
930 Korun, J. and Timur, G. (2005) The first pasteurellosis case in cultured sea bass (*Dicentrarchus labrax* l.) at low marine water temperatures in Turkey. *The Israeli Journal of Aquaculture – Bamidgeh* 57(3), 197–206.
931 Koski, P., Hirvelä-Koski, V. and Bernardet, J.F. (1993) *Flexibacter columnaris* infection in Arctic char (*Salvelinus alpinus* (L.)). First isolation in Finland. *Bulletin of the European Association of Fish Pathologists* 13(2), 66–69.
932 Kotlowski, R., Martin, A., Ablordey, A., Chemlal, K., Fonteyne, P.-A. and Portaels, F. (2004) One-tube cell lysis and DNA extraction procedure for PCR-based detection of *Mycobacterium ulcerans* in aquatic insects, molluscs and fish. *Journal of Medical Microbiology* 53(Pt 9), 927–933.
933 Kovacs, N. (1956) Identification of *Pseudomonas pyocyanea* by the oxidase reaction. *Nature* 178(4535), 703–703.
934 Kovács, N. (1928) Eine vereinfachte Methode zum Nachweis der Indolbildung durch Bakterien. *Zeitschrift fur immunitatsforschung – Immunobiology* 44, 311.
935 Kozinska, A., Figueras, M.J., Chacon, M.R. and Soler, L. (2002) Phenotypic characteristics and pathogenicity of *Aeromonas* genomospecies isolated from common

carp (*Cyprinus carpio* L.). *Journal of Applied Microbiology* 93(6), 1034–1041.
936 Kozubíková, E., Puky, M., Kiszely, P. and Petrusek, A. (2010) Crayfish plague pathogen in invasive North American crayfish species in Hungary. *Journal of Fish Diseases* 33(11), 925–929.
937 Kozubíková, E., Viljamaa-Dirks, S., Heinikainen, S. and Petrusek, A. (2011a) Spiny-cheek crayfish *Orconectes limosus* carry a novel genotype of the crayfish plague pathogen *Aphanomyces astaci*. *Journal of Invertebrate Pathology* 108(3), 214–216.
938 Kozubíková, E., Vrålstad, T., Filipová, L. and Petrusek, A. (2011b) Re-examination of the prevalence of *Aphanomyces astaci* in North American crayfish populations in Central Europe by TaqMan MGB real-time PCR. *Diseases of Aquatic Organisms* 97(2), 113–125.
939 Kraxberger-Beatty, T. and McGarey, D.J. (1990) *Vibrio harveyi*, an opportunistic pathogen of common snook, *Centropomus undecimalis* (Bloch), held in captivity. *Journal of Fish Diseases* 13, 557–560.
940 Kreger, A.S. (1984) Cytolytic activity and virulence of *Vibrio damsela*. *Infection and Immunity* 44(2), 326–331.
941 Krieg, N.R. (2010) Procaryotic domains. In: (eds) *Bergey's Manual of Systematic Bacteriology*, 2nd edn, Vol. II. Springer, New York, pp.
942 Krovacek, K., Huang, K., Sternberg, S. and Svenson, S.B. (1998) *Aeromonas hydrophila* septicaemia in a grey seal (*Halichoerus grypus*) from the Baltic Sea: a case study. *Comparative Immunology, Microbiology and Infectious Diseases* 21(1), 43–49.
943 Krugner-Higby, L., Haak, D., Johnson, P.T.J., Shields, J.D., Jones, W.M. III,, Reece, K.S., Meinke, T., Gendron, A. and Rusak, J.A. (2010) Ulcerative disease outbreak in crayfish *Orconectes propinquus* linked to *Saprolegnia australis* in big Muskellunge Lake, Wisconsin. *Diseases of Aquatic Organisms* 91(1), 57–66.
944 Kudo, T., Hatai, K. and Seino, A. (1988) *Nocardia seriolae* sp. nov. causing nocardiosis of cultured fish. *International Journal of Systematic Bacteriology* 38(2), 173–178.
945 Kuijper, E.J., Steigerwalt, A.G., Schoenmakers, B.S., Peeters, M.F., Zanen, H.C. and Brenner, D.J. (1989) Phenotypic characterization and DNA relatedness in human fecal isolates of *Aeromonas* spp. *Journal of Clinical Microbiology* 27(1), 132–138.
946 Kumar, D., Suresh, K., Dey, R.K. and Mishra, B.K. (1986) Stress-mediated columnaris disease in rohu, *Labeo rohita* (Hamilton). *Journal of Fish Diseases* 9, 87–89.
947 Kumar, N.R. and Nair, S. (2007) *Vibrio rhizosphaerae* sp. nov., a red-pigmented bacterium that antagonizes phytopathogenic bacteria. *International Journal of Systematic and Evolutionary Microbiology* 57(Pt 10), 2241–2246.
948 Kumar, S., Nei, M., Dudley, J. and Tamura, K. (2008) MEGA: A biologist-centric software for evolutionary analysis of DNA and protein sequences. *Briefings in Bioinformatics* 9(4), 299–306.
949 Kumar, S., Tamura, K. and Nei, M. (2004) MEGA3: Integrated software for molecular evolutionary genetics analysis and sequence alignment. *Briefings in Bioinformatics* 5(2), 150–163.
950 Küpfer, M., Kuhnert, P., Korczak, B.M., Peduzzi, R. and Demarta, A. (2006) Genetic relationships of *Aeromonas* strains inferred from 16S rRNA, *gyrB* and *rpoB* gene sequences. *International Journal of Systematic and Evolutionary Microbiology* 56(12), 2743–2751.
951 Kurup, P.V. and Schmitt, J.A. (1973) Numerical taxonomy of *Nocardia*. *Canadian Journal of Microbiology* 19(8), 1035–1048.
952 Kushmaro, A., Banin, E., Loya, Y., Stackebrandt, E. and Rosenberg, E. (2001) *Vibrio shiloi* sp. nov., the causative agent of bleaching of the coral *Oculina patagonica*. *International Journal of Systematic and Evolutionary Microbiology* 51(Pt 4), 1383–1388.
953 Kusrini, M.D., Skerratt, L.F., Garland, S., Berger, L. and Endarwin, W. (2008) Chytridiomycosis in frogs of Mount Gede Pangrango, Indonesia. *Diseases of Aquatic Organisms* 82(3), 187–194.
954 Kusuda, R., Kawai, K., Salati, F., Banner, C.R. and Fryer, J.L. (1991) *Enterococcus seriolicida* sp. nov., a fish pathogen. *International Journal of Systematic Bacteriology* 41(3), 406–409.
955 Kusuda, R., Kawakami, K. and Kawai, K. (1987) A fish-pathogenic *Mycobacterium* sp. isolated from an epizootic of cultured yellowtail. *Nippon Suisan Gakkaishi* 53(10), 1797–1804.
956 Kusuda, R., Dohata, N., Fukuda, Y. and Kawai, K. (1995) *Pseudomonas anguilliseptica* infection in striped jack. *Fish Pathology* 30(2), 121–122.
957 Kusuda, R. and Toyoshima, T. (1976) Characteristics of a pathogenic *Pseudomonas* isolated from cultured yellowtail. *Fish Pathology* 11(3), 133–139.
958 Kusuda, R., Toyoshima, T. and Nishioka, J. (1974) Characteristics of a pathogenic *Pseudomonas* isolated from cultured crimson sea breams. *Fish Pathology* 9, 71–78.
959 Kusuda, R. and Yamaoka, M. (1972) Etiological studies on bacterial pseudotuberculosis in cultured yellowtail with *Pasteurella piscicida* as the causative agent – I. On the morphological and biochemical properties. *Nippon Suisan Gakkaishi* 38(12), 1325–1332.
960 Kusuda, R., Yokoyama, J. and Kawai, K. (1986) Bacteriological study on cause of mass mortalities in cultured black sea bream fry. *Nippon Suisan Gakkaishi* 52(10), 1745–1751.
961 Kusunoki, S. and Ezaki, T. (1992) Proposal of *Mycobacterium peregrinum* sp. nov., nom. rev. and elevation of *Mycobacterium chelonae* subsp. *abscessus* (Kubica *et al.*) to species status: *Mycobacterium abscessus* comb. nov. *International Journal of Systematic Bacteriology* 42(2), 240–245.
962 Labeda, D.P. (2000) International Committee on Systematic Bacteriology; IXth International (IUMS) Congress of Bacteriology and Applied Microbiology. *International Journal of Systematic and Evolutionary Microbiology* 50(6), 2245–2247.
963 Labella, A., Vida, M., Alonso, M.C., Infante, C., Cardenas, S., Lopez-Romalde, S., Manchado, M. and Borrego, J.J. (2006) First isolation of *Photobacterium damselae* ssp. *damselae* from cultured redbanded seabream, *Pagrus auriga* Valenciennes, in Spain. *Journal of Fish Diseases* 29(3), 175–179.
964 Labreuche, Y., Le Roux, F., Henry, J., Zatylny, C., Huvet, A., Lambert, C., Soudant, P., Mazel, D. and

Nicolas, J.-L. (2010) *Vibrio aestuarianus* zinc metalloprotease causes lethality in the Pacific oyster *Crassostrea gigas* and impairs the host cellular immune defenses. *Fish and Shellfish Immunology* 29(5), 753–758.

965 Lacoste, A., Jalabert, F., Malham, S., Cueff, A., Gélébart, F., Cordevant, C., Lange, M. and Poulet, S.A. (2001a) A *Vibrio splendidus* strain is associated with summer mortality of juvenile oysters *Crassostrea gigas* in the Bay of Morlaix (North Brittany, France). *Diseases of Aquatic Organisms* 46(2), 139–145.

966 Lacoste, A., Jalabert, F., Malham, S.K., Cueff, A. and Poulet, S.A. (2001b) Stress and stress-induced neuroendocrine changes increase the susceptibility of juvenile oysters (*Crassostrea gigas*) to *Vibrio splendidus*. *Applied and Environmental Microbiology* 67(5), 2304–2309.

967 Ladds, P., Bradley, J. and Hirst, R. (1996) *Providencia rettgeri* meningitis in hatchling saltwater crocodiles (*Crocodylus porosus*). *Australian Veterinary Journal* 74(5), 397–398.

968 Laidler, L.A., Treasurer, J.W., Grant, A.N. and Cox, D.I. (1999) A typical *Aeromonas salmonicida* infection in wrasse (*Labridae*) used as cleaner fish of farmed Atlantic salmon, *Salmo salar* L., in Scotland. *Journal of Fish Diseases* 22(3), 209–213.

969 Lallier, R. and Higgins, R. (1988) Biochemical and toxigenic characteristics of *Aeromonas* spp. isolated from diseased mammals, moribund and healthy fish. *Veterinary Microbiology* 18(1), 63–71.

970 Lambert, C., Nicolas, J.L., Cilia, V. and Corre, S. (1998) *Vibrio pectenicida* sp. nov., a pathogen of scallop (*Pecten maximus*) larvae. *International Journal of Systematic Bacteriology* 48(2), 481–487.

971 Lan, J., Zhang, X.-H., Wang, Y., Chen, J. and Han, Y. (2008) Isolation of an unusual strain of *Edwardsiella tarda* from turbot and establish a PCR detection technique with the *gyrB* gene. *Journal of Applied Microbiology* 105(3), 644–651.

972 Lancefield, R. (1933) A serological differentiation of human and other groups of streptococci. *Journal of Experimental Medicine* 59, 441–158.

973 Lane, D. (1991) 16S/23S rRNA sequencing. In: Stackebrandt, E. and Goodfellow, M. (eds) *Nucleic Acid Techniques in Bacterial Systematics*. John Wiley & Sons, Chichester, pp. 115–147.

974 Lane, D.J., Pace, B., Olsen, G.J., Stahl, D.A., Sogin, M.L. and Pace, N.R. (1985) Rapid determination of 16S ribosomal RNA sequences for phylogenetic analyses. *Proceedings of the National Academy of Sciences* 82(20), 6955–6959.

975 Langdon, J.S. (1988) Fish diseases. Refresher course for Veterinarians. In: *Post-Graduate Committee in Veterinary Science,Proceedings No. 106*. University of Sydney, Sydney, Australia, pp. 225–259.

976 Langdon, J.S. and McDonald, W.L. (1987) Cranial *Exophiala pisciphila* infection in *Salmo salar* in Australia. *Bulletin of the European Association of Fish Pathologists* 7(2), 35–37.

977 Lansdell, W., Dixon, B., Smith, N. and Benjamin, L. (1993) Isolation of several *Mycobacterium* species from fish. *Journal of Aquatic Animal Health* 5(1), 73–76.

978 Lapage, S., Sneath, P., Lessel, E., Skerman, V., Seeliger, H. and Clark, W. (eds) (1992) *International Code of Nomenclature of Bacteria: Bacteriological Code, 1990 Revision*. ASM Press, Washington, DC.

979 Larsen, J.L. and Pedersen, K. (1996) A typical *Aeromonas salmonicida* isolated from diseased turbot (*Scophthalmus maximus* L.). *Acta Veterinaria Scandinavica* 37(2), 139–146.

980 Lasa, A., Diéguez, A.L. and Romalde, J.L. (2013) *Vibrio toranzoniae* sp. nov., a new member of the Splendidus clade in the genus *Vibrio*. *Systematic and Applied Microbiology* 36(2), 96–100.

981 Lau, K.W.K., Ren, J., Wai, N.L.M., Lau, S.C.L., Qian, P.-Y., Wong, P.-K. and Wu, M. (2006) *Marinomonas ostreistagni* sp. nov., isolated from a pearl-oyster culture pond in Sanya, Hainan Province, China. *International Journal of Systematic and Evolutionary Microbiology* 56(Pt 10), 2271–2275.

982 Laurent, F.J., Provost, F. and Boiron, P. (1999) Rapid identification of clinically relevant *Nocardia* species to genus level by 16S rRNA gene PCR. *Journal of Clinical Microbiology* 37(1), 99–102.

983 Laurent, F.J., Provost, F. and Boiron, P. (2000) Rapid identification of clinically relevant *Nocardia* species to genus level by 16S rRNA gene PCR. *Journal of Clinical Microbiology* 38(7), 2805.

984 Lawson, P.A., Collins, M.D., Falsen, E. and Foster, G. (2006) *Catellicoccus marimammalium* gen. nov., sp. nov., a novel Gram-positive, catalase-negative, coccus-shaped bacterium from porpoise and grey seal. *International Journal of Systematic and Evolutionary Microbiology* 56(Pt 2), 429–432.

985 Lawson, P.A., Falsen, E., Foster, G., Eriksson, E., Weiss, N. and Collins, M.D. (2001) *Arcanobacterium pluranimalium* sp. nov., isolated from porpoise and deer. *International Journal of Systematic and Evolutionary Microbiology* 51(Pt 1), 55–59.

986 Lawson, P.A., Foster, G., Falsen, E. and Collins, M.D. (2005) *Streptococcus marimammalium* sp. nov., isolated from seals. *International Journal of Systematic and Evolutionary Microbiology* 55(Pt 1), 271–274.

987 Lawson, P.A., Foster, G., Falsen, E., Davison, N. and Collins, M.D. (2004) *Streptococcus halichoeri* sp. nov., isolated from grey seals (*Halichoerus grypus*). *International Journal of Systematic and Evolutionary Microbiology* 54(Pt 5), 1753–1756.

988 Lawson, P.A., Foster, G., Falsen, E., Ohlén, M. and Collins, M.D. (1999a) *Vagococcus lutrae* sp. nov., isolated from the common otter (*Lutra lutra*). *International Journal of Systematic Bacteriology* 49(Pt 3), 1251–1254.

989 Lawson, P.A., Foster, G., Falsen, E., Ohlén, M. and Collins, M.D. (2000) *Atopobacter phocae* gen. nov., sp. nov., a novel bacterium isolated from common seals. *International Journal of Systematic and Evolutionary Microbiology* 50(5), 1755–1760.

990 Lawson, P.A., Foster, G., Falsen, E., Sjödén, B. and Collins, M.D. (1999b) *Abiotrophia balaenopterae* sp. nov., isolated from the minke whale (*Balaenoptera acutorostrata*). *International Journal of Systematic Bacteriology* 49(Pt 2), 503–506.

991 Lee, C.Y., Pan, S.F. and Chen, C.H. (1995) Sequence of a cloned pR72H fragment and its use for detection of *Vibrio parahaemolyticus* in shellfish with the PCR. *Applied and Environmental Microbiology* 61(4), 1311–1317.

992 Lee, J., Kim, J.S., Nahm, C.H., Choi, J.W., Kim, J., Pai, S.H., Moon, K.H., Lee, K. and Chong, Y. (1999) Two cases of *Chromobacterium violaceum* infection after injury in a subtropical region. *Journal of Clinical Microbiology* 37(6), 2068–2070.

993 Lee, J.V., Bashford, D.J., Donovan, T.J., Furniss, A.I. and West, P.A. (1982) The incidence of *Vibrio cholerae* in water, animals and birds in Kent, England. *Journal of Applied Microbiology* 52(2), 281–291.

994 Lee, J.V., Donovan, T.J. and Furniss, A.L. (1978) Characterization, taxonomy, and emended description of *Vibrio metschnikovii*. *International Journal of Systematic Bacteriology* 28(1), 99–111.

995 Lee, J.V., Hendrie, M.S. and Sherwan, J.M. (1979) Identification of *Aeromonas*, *Vibrio* and related organisms. In: Skinner, F.A. and Lovelock, D.W. (eds) *Identification Methods for Microbiologists. The Society of Applied Bacteriology. Technical Series No. 14*. Academic Press, London, pp. 152–166.

996 Lee, J.V., Shread, P., Furniss, A.L. and Bryant, T.N. (1981) Taxonomy and description of *Vibrio fluvialis* sp. nov. (synonym group F vibrios, group EF6). *The Journal of Applied Bacteriology* 50(1), 73–94.

997 Lee, K. and Ruby, E.G. (1995) Symbiotic role of the viable but nonculturable state of *Vibrio fischeri* in Hawaiian coastal seawater. *Applied and Environmental Microbiology* 61(1), 278–283.

998 Lee, K.-K., Yu, S.-R., Yang, T.-I., Liu, P.-C. and Chen, F.-R. (1996) Isolation and characterization of *Vibrio alginolyticus* isolated from diseased kuruma prawn, *Penaeus japonicus*. *Letters in Applied Microbiology* 22(2), 111–114.

999 Lee, S.E., Kim, S.Y., Kim, S.J., Kim, H.S., Shin, J.H., Choi, S.H., Chung, S.S. and Rhee, J.H. (1998) Direct identification of *Vibrio vulnificus* in clinical specimens by nested PCR. *Journal of Clinical Microbiology* 36(10), 2887–2892.

1000 Lee, S.K.Y., Wang, H.Z., Law, S.H.W., Wu, R.S.S. and Kong, R.Y.C. (2002) Analysis of the 16S-23S rDNA intergenic spacers (IGSs) of marine vibrios for species-specific signature DNA sequences. *Marine Pollution Bulletin* 44(5), 412–420.

1001 Lee, Y.S., Baik, K.S., Park, S.Y., Kim, E.M., Lee, D.-H., Kahng, H.-Y., Jeon, C.O. and Jung, J.S. (2009) *Tenacibaculum crassostreae* sp. nov., isolated from the Pacific oyster, *Crassostrea gigas*. *International Journal of Systematic and Evolutionary Microbiology* 59(Pt 7), 1609–1614.

1002 Leger, J.A.S., Begeman, L., Fleetwood, M., Frasca, S., Garner, M.M., Lair, S., Trembley, S., Linn, M.J. and Terio, K.A. (2009) Comparative pathology of Nocardiosis in marine mammals. *Veterinary Pathology Online* 46(2), 299–308.

1003 Lehmann, J., Mock, D. and Schäfer, W. (1999) Swim bladder infection of farmed Atlantic salmon (*Salmo salar* L.) by a fungus: a case report. *Bulletin of the European Association of Fish Pathologists* 19(2), 83–84.

1004 Lehmann, J., Mock, D., Stuerenberg, F.J. and Bernardet, J.F. (1991) First isolation of *Cytophaga psychrophila* from a systemic disease in eel and cyprinid. *Diseases of Aquatic Organisms* 10, 217–220.

1005 Leifson, E. (1962) The bacterial flora of distilled and stored water. III. New species of the genera *Corynebacterium, Flavobacterium, Spirillum* and *Pseudomonas*. *International Bulletin of Bacteriological Nomenclature and Taxonomy* 12(4), 161–170.

1006 Leifson, E. (1963) Determination of carbohydrate metabolism of marine bacteria. *The Journal of Bacteriology* 85(5), 1183–1184.

1007 LeJeune, J.T. and Rurangirwa, F.R. (2000) Polymerase chain reaction for definitive identification of *Yersinia ruckeri*. *Journal of Veterinary Diagnostic Investigation* 12(6), 558–561.

1008 Leonardo, M.R., Moser, D.P., Barbieri, E., Brantner, C.A., MacGregor, B.J., Paster, B.J., Stackebrandt, E. and Nealson, K.H. (1999) *Shewanella pealeana* sp. nov., a member of the microbial community associated with the accessory nidamental gland of the squid *Loligo pealei*. *International Journal of Systematic Bacteriology* 49(Pt 4), 1341–1351.

1009 Le Roux, F., Gay, M., Lambert, C., Nicolas, J.L., Gouy, M. and Berthe, F. (2004) Phylogenetic study and identification of *Vibrio splendidus*-related strains based on *gyrB* gene sequences. *Diseases of Aquatic Organisms* 58(2–3), 143–150.

1010 Le Roux, F., Goubet, A., Thompson, F.L., Faury, N., Gay, M., Swings, J. and Saulnier, D. (2005) *Vibrio gigantis* sp. nov., isolated from the haemolymph of cultured oysters (*Crassostrea gigas*). *International Journal of Systematic and Evolutionary Microbiology* 55(Pt 6), 2251–2255.

1011 Le Roux, F., Labreuche, Y., Davis, B.M., Iqbal, N., Mangenot, S., Goarant, C., Mazel, D. and Waldor, M.K. (2011) Virulence of an emerging pathogenic lineage of *Vibrio nigripulchritudo* is dependent on two plasmids. *Environmental Microbiology* 13(2), 296–306.

1012 Lescenko, P., Matlova, L., Dvorska, L., Bartos, M., Vavra, O., Navratil, S., Novotny, L. and Pavlik, I. (2003) Mycobacterial infection in aquarium fish. *Veterinární Medicína-Czech* 48(3), 71–78.

1013 Levi, M.H., Bartell, J., Gandolfo, L., Smole, S.C., Costa, S.F., Weiss, L.M., Johnson, L.K., Osterhout, G. and Herbst, L.H. (2003) Characterization of *Mycobacterium montefiorense* sp. nov., a novel pathogenic *Mycobacterium* from moray eels that is related to *Mycobacterium triplex*. *Journal of Clinical Microbiology* 41(5), 2147–2152.

1014 Levican, A., Collado, L., Aguilar, C., Yustes, C., Diéguez, A.L., Romalde, J.L. and Figueras, M.J. (2012) *Arcobacter bivalviorum* sp. nov. and *Arcobacter venerupis* sp. nov., new species isolated from shellfish. *Systematic and Applied Microbiology* 35(3), 133–138.

1015 Lévy-Frébault, V.V. and Portaels, F. (1992) Proposed minimal standards for the genus *Mycobacterium* and for description of new slowly growing *Mycobacterium* species. *International Journal of Systematic Bacteriology* 42(2), 315–323.

1016 Lewin, R.A. (1974) *Flexibacter polymorphus*, a new marine species. *Journal of General Microbiology* 82(2), 393–403.

1017 Lewin, R.A. and Lounsbery, D.M. (1969) Isolation, cultivation and characterization of Flexibacteria. *Journal of General Microbiology* 58(2),145–170.

1018 Li, G., Zhao, D., Huang, L., Sun, J., Gao, D., Wang, H., Tan, Y. and Liang, L. (2006) Identification and phylogenetic analysis of *Vibrio vulnificus* isolated from diseased *Trachinotus ovatus* in cage mariculture. *Aquaculture* 261(1), 17–25.

1019 Li, H., Qiao, G., Gu, J.-Q., Zhou, W., Li, Q., Woo, S.-H., Xu, D.-H. and Park, S.-I. (2010a) Phenotypic and genetic characterization of bacteria isolated from diseased cultured sea cucumber *Apostichopus japonicus* in northeastern China. *Diseases of Aquatic Organisms* 91(3), 223–235.

1020 Li, H., Qiao, G., Li, Q., Zhou, W., Won, K.M., Xu, D.-H. and Park, S.-I. (2010b) Biological characteristics and pathogenicity of a highly pathogenic *Shewanella marisflavi* infecting sea cucumber, *Apostichopus japonicus*. *Journal of Fish Diseases* 33(11), 865–877.

1021 Li, J., Zhou, L. and Woo, N. (2003) Invasion route and pathogenic mechanisms of *Vibrio alginolyticus* to Silver sea bream *Sparus sarba*. *Journal of Aquatic Animal Health* 15(4), 302–313.

1022 Li, L., Rock, J.L. and Nelson, D.R. (2008a) Identification and characterization of a repeat-in-toxin gene cluster in *Vibrio anguillarum*. *Infection and Immunity* 76(6), 2620–2632.

1023 Li, Z., Zhang, S., Wang, C. and Pang, Q. (2008b) Complement-mediated killing of *Vibrio* species by the humoral fluids of amphioxus *Branchiostoma belcheri*: implications for a dual role of O-antigens in the resistance to bactericidal activity. *Fish and Shellfish Immunology* 24(2), 215–222.

1024 Lightner, D., Redman, R.M., Mohney, L., Sinski, J. and Priest, D. (1988) A renal mycosis of an adult hybrid red tilapia, *Oreochromis mossambicus* X *O. hornorum*, caused by the imperfect fungus, *Paecilomyces marquandii*. *Journal of Fish Diseases* 11, 437–440.

1025 Lightner, D.V. and Redman, R.M. (1998) Shrimp diseases and current diagnostic methods. *Aquaculture* 164(1–4), 201–220.

1026 Lilley, J.H., Callinan, R.B., Chinabut, S., Kanchanakhan, S., MacRae, I.H. and Phillips, M.J. (1998) *Epizootic Ulcerative Syndrome (EUS) Technical Handbook*. The Aquatic Animal Health Research Institute, Bangkok.

1027 Lilley, J.H., Cerenius, L. and Söderhäll, K. (1997) RAPD evidence for the origin of crayfish plague outbreaks in Britain. *Aquaculture* 157(3–4), 181–185.

1028 Lilley, J.H. and Roberts, R.J. (1997) Pathogenicity and culture studies comparing the *Aphanomyces* involved in epizootic ulcerative syndrome (EUS) with other similar fungi. *Journal of Fish Diseases* 20, 135–144.

1029 Lin, B., Wang, Z., Malanoski, A.P., O'Grady, E.A., Wimpee, C.F., Vuddhakul, V., Alves, N. Jr, Thompson, F.L., Gomez-Gil, B. and Vora, G.J. (2010) Comparative genomic analyses identify the *Vibrio harveyi* genome sequenced strains BAA-1116 and HY01 as *Vibrio campbellii*. *Environmental Microbiology Reports* 2(1), 81–89.

1030 Lincoln, S.P., Fermor, T.R. and Tindall, B.J. (1999) *Janthinobacterium agaricidamnosum* sp. nov., a soft rot pathogen of *Agaricus bisporus*. *International Journal of Systematic Bacteriology* 49(Pt 4), 1577–1589.

1031 Linde, H.-J., Kobuch, R., Jayasinghe, S., Reischl, U., Lehn, N., Kaulfuss, S. and Beutin, L. (2004) *Vibrio metschnikovii*, a rare cause of wound infection. *Journal of Clinical Microbiology* 42(10), 4909–4911.

1032 Lio-Po, G.D., Albright, L.J., Michel, C. and Leaño, E.M. (1998) Experimental induction of lesions in snakeheads (*Ophicephalus striatus*) and catfish (*Clavias batvachus*) with *Aeromonas hydrophila*, *Aquaspirillum* sp., *Pseudomonas* sp. and *Streptococcus* sp. *Journal of Applied Ichthyology* 14(1–2), 75–79.

1033 Liston, J. (1957) The occurrence and distribution of bacterial types on flatfish. *Journal of General Microbiology* 16(1), 205–216.

1034 Liu, C.-H., Cheng, W., Hsu, J.-P. and Chen, J.-C. (2004a) *Vibrio alginolyticus* infection in the white shrimp *Litopenaeus vannamei* confirmed by polymerase chain reaction and 16S rDNA sequencing. *Diseases of Aquatic Organisms* 61(1–2), 169–174.

1035 Liu, G.F., Wang, J.Y., Xu, L.W., Ding, X. and Zhou, S.N. (2010a) Sensitive and rapid detection of *Vibrio corallilyticus* by loop-mediated isothermal amplification targeted to the alpha subunit gene of RNA polymerase. *Letters in Applied Microbiology* 51(3), 301–307.

1036 Liu, J.Y. and Li, A.H. (2012) First case of *Aeromonas schubertii* infection in the freshwater cultured snakehead fish, *Ophiocephalus argus* (Cantor), in China. *Journal of Fish Diseases* 35(5), 335–342.

1037 Liu, J.Y., Li, A.H., Ji, C. and Yang, W.M. (2009) First description of a novel *Weissella* species as an opportunistic pathogen for rainbow trout *Oncorhynchus mykiss* (Walbaum) in China. *Veterinary Microbiology* 136(3–4), 314–320.

1038 Liu, J.Y., Li, A.H., Zhou, D.R., Wen, Z.R. and Ye, X.P. (2010b) Isolation and characterization of *Edwardsiella ictaluri* strains as pathogens from diseased yellow catfish *Pelteobagrus fulvidraco* (Richardson) cultured in China. *Aquaculture Research* 41, 1835–1844.

1039 Liu, K.-C., Mai, L.-M. and Chien, C.-H. (1985) Electron microscope study of the reproductive stages of the gill rot pathogen, *Branchiomyces* sp., of fish. *Fish Pathology* 20(2/3), 267–272.

1040 Liu, P.C., Chen, Y.C., Huang, C.Y. and Lee, K.K. (2000) Virulence of *Vibrio parahaemolyticus* isolated from cultured small abalone, *Haliotis diversicolor supertexta*, with withering syndrome. *Letters in Applied Microbiology* 31(6), 433–437.

1041 Liu, P.C., Chuang, W. and Lee, K. (2003) Infectious gastroenteritis caused by *Vibrio harveyi* (*V. carchariae*) in cultured red drum, *Sciaenops ocellatus*. *Journal of Applied Ichthyology* 19, 59–61.

1042 Liu, P.C., Lee, K.-K., Yii, K.-C., Kou, G.-H. and Chen, S.-N. (1996) Isolation of *Vibrio harveyi* from diseased kuruma prawns *Penaeus japonicus*. *Current Microbiology* 33, 129–132.

1043 Liu, P.C., Lin, J.-Y., Chuang, W.-H. and Lee, K.-K. (2004b) Isolation and characterization of pathogenic *Vibrio harveyi* (*V. carchariae*) from the farmed marine cobia fish *Rachycentron canadum* L. with gastroenteritis syndrome. *World Journal of Microbiology and Biotechnology* 20, 495–499.

1044 Liu, P.-C., Lin, J.-Y., Hsiao, P.-T. and Lee, K.-K. (2004) Isolation and characterization of pathogenic *Vibrio*

alginolyticus from diseased cobia *Rachycentron canadum*. *Journal of Basic Microbiology* 44(1), 23–28.
1045 Llewellyn, L.C. (1980) A bacterium with similarities to the redmouth bacterium and *Serratia liquefaciens* (Grimes and Hennerty) causing mortalities in hatchery reared salmonids in Australia. *Journal of Fish Diseases* 3(1), 29–39.
1046 Lloyd, J. (1985) Estimation of amount of carbon-dioxide produced by two incubation methods. Department of Agriculture, Western Australia, unpublished.
1047 Loch, T.P., Xu, W., Fitzgerald, S.M. and Faisal, M. (2008) Isolation of a *Carnobacterium maltaromaticum*-like bacterium from systemically infected lake whitefish (*Coregonus clupeaformis*). *FEMS Microbiology Letters* 288(1), 76–84.
1048 Logan, N.A. (1989) Numerical taxonomy of violet-pigmented, Gram-negative bacteria and description of *Iodobacter fluviatile* gen. nov., comb. nov. *International Journal of Systematic Bacteriology* 39(4), 450–456.
1049 Longcore, J.E., Pessier, A.P. and Nichols, D.K. (1999) *Batrachochytrium dendrobatidis* gen. et sp. nov., a chytrid pathogenic to amphibians. *Mycologia* 91(2), 219.
1050 Lönnström, L., Wiklund, T. and Bylund, G. (1994) *Pseudomonas anguilliseptica* isolated from Baltic herring *Clupea harengus membras* with eye lesions. *Diseases of Aquatic Organisms* 18, 143–147.
1051 Lönnström, L.-G., Hoffrén, M.L. and Wiklund, T. (2008) *Flavobacterium psychrophilum* associated with mortality of farmed perch, *Perca fluviatilis* L. *Journal of Fish Diseases* 31(10), 793–797.
1052 López, J.R., Balboa, S., Núñez, S., de la Roca, E., de la Herran, R., Navas, J.I., Toranzo, A.E. and Romalde, J.L. (2011a) Characterization of *Vibrio tapetis* strains isolated from diseased cultured wedge sole (*Dicologoglossa cuneata* Moreau). *Research in Veterinary Science* 90(2), 189–195.
1053 López, J.R., de la Roca, E., Núñez, S., de la Herran, R., Navas, J.I., Manchado, M., Herrera, M. and Toranzo, A.E. (2009a) Identification of *Vibrio harveyi* isolated from diseased cultured wedge sole *Dicologoglossa cuneata*. *Diseases of Aquatic Organisms* 84 (3), 209–217.
1054 López, J.R., Hamman-Khalifa, A.M., Navas, J.I. and la Herran, R. (2011b) Characterization of ISR region and development of a PCR assay for rapid detection of the fish pathogen *Tenacibaculum soleae*. *FEMS Microbiology Letters* 324(2), 181–18.
1055 López, J.R., Núñez, S., Magariños, B., Castro, N., Navas, J.I., de la Herran, R. and Toranzo, A.E. (2009b) First isolation of *Tenacibaculum maritimum* from wedge sole, *Dicologoglossa cuneata* (Moreau). *Journal of Fish Diseases* 32(7), 603–610.
1056 López, J.R., Piñeiro-Vidal, M., García-Lamas, N., de la Herran, R., Navas, J.I., Hachero-Cruzado, I. and Santos, Y. (2010) First isolation of *Tenacibaculum soleae* from diseased cultured wedge sole, *Dicologoglossa cuneata* (Moreau), and brill, *Scophthalmus rhombus* (L.). *Journal of Fish Diseases* 33(3), 273–278.
1057 López-Goñi, I., García-Yoldi, D., Marín, C.M., de Miguel, M.J., Muñoz, P.M., Blasco, J.M., Jacques, I., Grayon, M., Cloeckaert, A., Ferreira, A.C., Cardoso, R., Corrêa de Sá, M.I., Walravens, K., Albert, D. and Garin-Bastuji, B. (2008) Evaluation of a multiplex PCR assay (Bruce-ladder) for molecular typing of all *Brucella* species, including the vaccine strains. *Journal of Clinical Microbiology* 46(10), 3484–3487.
1058 Lopez-Joven, C., de Blas, I., Ruiz-Zarzuela, I., Furones, M.D. and Roque, A. (2011) Experimental uptake and retention of pathogenic and nonpathogenic *Vibrio parahaemolyticus* in two species of clams: *Ruditapes decussatus* and *Ruditapes philippinarum*. *Journal of Applied Microbiology* 111(1), 197–208.
1059 López-Romalde, S., Magariños, B., Núñez, S., Toranzo, A.E. and Romalde, J.L. (2003) Phenotypic and genetic characterization of *Pseudomonas anguilliseptica* strains isolated from fish. *Journal of Aquatic Animal Health* 15(1), 39–47.
1060 Love, M., Teebken-Fisher, D., Hose, J.E., Farmer, J.J., Hickman, F.W. and Fanning, G.R. (1981) *Vibrio damsela*, a marine bacterium, causes skin ulcers on the damselfish *Chromis punctipinnis*. *Science* 214(4525), 1139–1140.
1061 Løvoll, M., Wiik-Nielsen, C.R., Tunsjø, H.S., Colquhoun, D., Lunder, T., Sørum, H. and Grove, S. (2009) Atlantic salmon bath challenged with *Moritella viscosa* – pathogen invasion and host response. *Fish and Shellfish Immunology* 26(6), 877–884.
1062 Lowe, G.H. (1962) The rapid detection of lactose fermentation in paracolon organisms by the demonstration of beta-*D*-galactosidase. *The Journal of Medical Laboratory Technology* 19, 21–25.
1063 Lü, A., Hu, X., Zheng, L., Zhu, A., Cao, C. and Jiang, J. (2011) Isolation and characterization of *Citrobacter* spp. from the intestine of grass carp *Ctenopharyngodon idellus*. *Aquaculture* 313(1–4), 156–160.
1064 Luan, X.-Y., Chen, J.-X., Zhang, X.-H., Jia, J.-T., Sun, F.-R. and Li, Y. (2007) Comparison of different primers for rapid detection of *Vibrio parahaemolyticus* using the polymerase chain reaction. *Letters in Applied Microbiology* 44(3), 242–247.
1065 Lucena, T., Ruvira, M.A., Arahal, D.R., Macián, M.C. and Pujalte, M.J. (2012) *Vibrio aestivus* sp. nov. and *Vibrio quintilis* sp. nov., related to Marisflavi and Gazogenes clades, respectively. *Systematic and Applied Microbiology* 35(7), 427–431.
1066 Ludwig, W., Schleifer, K.H. and Whitman, W.B. (2009) Revised road map to the phylum *Firmicutes*. In: De Vos, P., Garrity, G., Jones, D., Krieg, N.R., Ludwig, W., Rainey, F.A., Schleifer, K.-H. and Whitman, W.B. (eds) *Bergey's Manual of Systematic Bacteriology*, 2nd edn (Vol. III). Springer, New York, pp. 1–13.
1067 Lumanlan-Mayo, S.C., Callinan, R.B., Paclibare, J.O., Catap, E.S. and Fraser, G.C. (1997) Epizootic ulcerative syndrome (EUS) in rice-fish culture systems: an overview of field experiments 1993–1995. In: Flegel, T.W. and MacRae, I.H. (eds) *Diseases in Asian Aquaculture*, (Vol. III). Fish Health Section, Asian Fisheries Society, Manila, pp. 129–138.
1068 Lumsden, J.S., Ostland, V.E. and Ferguson, H.W. (1996) Necrotic myositis in cage cultured rainbow trout, *Oncorhynchus mykiss* (Walbaum), caused by *Flexibacter psychrophilus*. *Journal of Fish Diseases* 19(2), 113–119.

1069 Luna, G., Bongiorni, L., Gill, C., Biavasco, F. and Danovaro, R. (2010) *Vibrio harveyi* as a causative agent of the White Syndrome in tropical stony corals. *Environmental Microbiology Reports* 2(1), 120–127.

1070 Lunder, T., Evensen, Ø., Holstad, G. and Håstein, T. (1995) 'Winter ulcer' in the Atlantic salmon *Salmo salar*. Pathological and bacteriological investigations and transmission experiments. *Diseases of Aquatic Organisms* 23, 39–4.

1071 Lunder, T., Sørum, H., Holstad, G., Steigerwalt, A.G., Mowinckel, P. and Brenner, D.J. (2000) Phenotypic and genotypic characterization of *Vibrio viscosus* sp. nov. and *Vibrio wodanis* sp. nov. isolated from Atlantic salmon (*Salmo salar*) with 'winter ulcer'. *International Journal of Systematic and Evolutionary Microbiology* 50(Pt 2), 427–450.

1072 Luo, P. and Hu, C. (2008) *Vibrio alginolyticus gyrB* sequence analysis and *gyrB*-targeted PCR identification in environmental isolates. *Diseases of Aquatic Organisms* 82(3), 209–216.

1073 Lupiani, B., Baya, A.M., Magarinos, B., Romalde, J.L., Li, T., Roberson, B.S., Hetrick, F.M. and Toranzo, A.E. (1993) *Vibrio mimicus* and *Vibrio cholerae* non-01 isolated from wild and hatchery-reared fish. *Fish Pathology* 28(1), 15–26.

1074 Lynch, M., Taylor, T.K., Duignan, P.J., Swingler, J., Marenda, M., Arnould, J.P.Y. and Kirkwood, R. (2011) Mycoplasmas in Australian fur seals (*Arctocephalus pusillus doriferus*), identification and association with abortion. *Journal of Veterinary Diagnostic Investigation* 23(6), 1123–1130.

1075 MacDonell, M.T. and Colwell, R.R. (1985) Phylogeny of the *Vibrionaceae*, and recommendation for two new genera, *Listonella* and *Shewanella*. *Systematic and Applied Microbiology* 6(2), 171–182.

1076 MacDonell, M.T., Singleton, F.L. and Hood, M.A. (1982) Diluent composition for use of API 20E in characterizing marine and estuarine bacteria. *Applied and Environmental Microbiology* 44(2), 423–427.

1077 Macey, B.M., Achilihu, I.O., Burnett, K.G. and Burnett, L.E. (2008) Effects of hypercapnic hypoxia on inactivation and elimination of *Vibrio campbellii* in the Eastern oyster, *Crassostrea virginica*. *Applied and Environmental Microbiology* 74(19), 6077–6084.

1078 MacFaddin, J.F. (1980) *Biochemical Tests for Identification of Medical Bacteria*. Williams & Wilkins Co., Baltimore, Maryland.

1079 Macián, M.C., Arahal, D.R., Garay, E. and Pujalte, M.J. (2005) *Marinomonas aquamarina* sp. nov., isolated from oysters and seawater. *Systematic and Applied Microbiology* 28(2), 145–150.

1080 Macián, M.C., Garay, E., Grimont, P.A.D. and Pujalte, M.J. (2004) *Vibrio ponticus* sp. nov., a neighbour of *V. fluvialis–V. furnissii* clade, isolated from gilthead sea bream, mussels and seawater. *Systematic and Applied Microbiology* 27(5), 535–540.

1081 Macián, M.C., Garay, E. and Pujalte, M.J. (1996) The arginine dihydrolase (ADH) system in the identification of some marine *Vibrio* species. *Systematic and Applied Microbiology* 19, 451–456.

1082 Macián, M.C., Ludwig, W., Aznar, R., Grimont, P.A., Schleifer, K.H., Garay, E. and Pujalte, M.J. (2001a) *Vibrio lentus* sp. nov., isolated from Mediterranean oysters. *International Journal of Systematic and Evolutionary Microbiology* 51(Pt 4), 1449–1456.

1083 Macián, M.C., Ludwig, W., Schleifer, K.H., Garay, E. and Pujalte, M.J. (2000) *Vibrio pelagius*: differences of the type strain deposited at various culture collections. *Systematic and Applied Microbiology* 23(3), 373–375.

1084 Macián, M.C., Ludwig, W., Schleifer, K.H., Pujalte, M.J. and Garay, E. (2001b) *Vibrio agarivorans* sp. nov., a novel agarolytic marine bacterium. *International Journal of Systematic and Evolutionary Microbiology* 51(Pt 6), 2031–2036.

1085 MacKenzie, K. (1988) Presumptive mycobacteriosis in North-east Atlantic mackerel, *Scomber scombrus* L. *Journal of Fish Biology* 32(2), 263–275.

1086 MacKnight, K., Chow, D., See, B. and Vedros, N. (1990) Melioidosis in a macaroni penguin *Eudyptes chrysolophus*. *Diseases of Aquatic Organisms* 9, 105–107.

1087 MacLeod, R.A. (1968) On the role of inorganic ions in the physiology of marine bacteria. *Advances in Microbiology of the Sea* 1, 95–126.

1088 Madinabeitia, I., Ohtsuka, S., Okuda, J., Iwamoto, E., Yoshida, T., Furukawa, M., Nakaoka, N. and Nakai, T. (2009) Homogeneity among *Lactococcus garvieae* isolates from striped jack, *Pseudocaranx dentex* (Bloch & Schneider), and its ectoparasites. *Journal of Fish Diseases* 32(10), 901–905.

1089 Magariños, B., Romalde, J.L., Bandín, I., Fouz, B. and Toranzo, A.E. (1992) Phenotypic, antigenic, and molecular characterization of *Pasteurella piscicida* strains isolated from fish. *Applied and Environmental Microbiology* 58(10), 3316–3322.

1090 Maher, M., Palmer, R., Gannon, F. and Smith, T. (1995) Relationship of a novel bacterial fish pathogen to *Streptobacillus moniliformis* and the Fusobacteria group, based on 16S ribosomal RNA analysis. *Systematic and Applied Microbiology* 18(1), 79–84.

1091 Maiden, M.C., Bygraves, J.A., Feil, E., Morelli, G., Russell, J.E., Urwin, R., Zhang, Q., Zhou, J., Zurth, K., Caugant, D.A., Feavers, I.M., Achtman, M. and Spratt, B.G. (1998) Multilocus sequence typing: a portable approach to the identification of clones within populations of pathogenic microorganisms. *Proceedings of the National Academy of Sciences of the United States of America* 95(6), 3140–3145.

1092 Mainster, M.E., Lynd, F.T., Cragg, P.C. and Karger, J. (1973) Treatment of multiple cases of *Pasteurella multocida* and staphylococcal pneumonia in *Alligator mississippiensis* on a herd basis. *Annual Proceedings of the American Association of Zoo Veterinarians*, 34–36.

1093 Makemson, J.C., Fulayfil, N.R., Landry, W., Van Ert, L.M., Wimpee, C.F., Widder, E.A. and Case, J.F. (1997) *Shewanella woodyi* sp. nov., an exclusively respiratory luminous bacterium isolated from the Alboran Sea. *International Journal of Systematic Bacteriology* 47(4), 1034–1039.

1094 Maloy, A.P., Barber, B.J. and Boettcher, K.J. (2005) A PCR-based diagnostic assay for the detection of *Roseovarius crassostreae* in *Crassostrea virginica* affected by juvenile oyster disease (JOD). *Diseases of Aquatic Organisms* 67(1–2), 155–162.

1095 Manefield, M., Harris, L., Rice, S.A., de Nys, R. and Kjelleberg, S. (2000) Inhibition of luminescence and virulence in the black tiger prawn (*Penaeus monodon*) pathogen *Vibrio harveyi* by intercellular signal antagonists. *Applied and Environmental Microbiology* 66(5), 2079–2084.
1096 Maquart, M., Le Flèche, P., Foster, G., Tryland, M., Ramisse, F., Djønne, B., Al Dahouk, S., Jacques, I., Neubauer, H., Walravens, K., Godfroid, J., Cloeckaert, A. and Vergnaud, G. (2009) MLVA-16 typing of 295 marine mammal *Brucella* isolates from different animal and geographic origins identifies 7 major groups within *Brucella ceti* and *Brucella pinnipedialis*. *BMC Microbiology* 9, 145.
1097 Marchesi, J.R., Sato, T., Weightman, A.J., Martin, T.A., Fry, J.C., Hiom, S.J., Dymock, D. and Wade, W.G. (1998) Design and evaluation of useful bacterium-specific PCR primers that amplify genes coding for bacterial 16S rRNA. *Applied and Environmental Microbiology* 64(6), 2333–2333.
1098 Markwardt, N.M., Gocha, Y. and Klontz, G. (1989) A new application for Coomassie brilliant blue agar: detection of *Aeromonas salmonicida* in clinical samples. *Diseases of Aquatic Organisms* 6, 231–233.
1099 Martens, T., Heidorn, T., Pukall, R., Simon, M., Tindall, B.J. and Brinkhoff, T. (2006) Reclassification of *Roseobacter gallaeciensis* Ruiz-Ponte *et al.* 1998 as *Phaeobacter gallaeciensis* gen. nov., comb. nov., description of *Phaeobacter inhibens* sp. nov., reclassification of *Ruegeria algicola* (Lafay *et al.* 1995) Uchino *et al.* 1999 as *Marinovum algicola* gen. nov., comb. nov. and emended descriptions of the genera *Roseobacter*, *Ruegeria* and *Leisingera*. *International Journal of Systematic and Evolutionary Microbiology* 56(Pt 6), 1293–1304.
1100 Martin-Carnahan, A.M. and Joseph, S.W. (2005) Family *Aeromonadaceae* Colwell, MacDonell and De Ley 1986. In: Brenner, D.J., Krieg, N.R., Staley, J.T. and Garrity, G.M. (eds) *Bergey's Manual of Systematic Bacteriology*, 2nd edn, Vol. 2. Springer, New York, p. 556.
1101 Martínez-Murcia, A.J. (1999) Phylogenetic positions of *Aeromonas encheleia*, *Aeromonas popoffii*, *Aeromonas* DNA hybridization group 11 and *Aeromonas* group 501. *International Journal of Systematic Bacteriology* 49(Pt 4), 1403–1408.
1102 Martínez-Murcia, A.J., Esteve, C., Garay, E. and Collins, M.D. (1992) *Aeromonas allosaccharophila* sp. nov., a new mesophilic member of the genus *Aeromonas*. *FEMS Microbiology Letters* 70(3), 199–205.
1103 Martínez-Murcia, A.J., Figueras, M.J., Saavedra, M.J. and Stackebrandt, E. (2007) The recently proposed species *Aeromonas sharmana* sp. nov., isolate GPTSA-6T, is not a member of the genus *Aeromonas*. *International Microbiology: The Official Journal of the Spanish Society for Microbiology* 10(1) 61–64.
1104 Martínez-Murcia, A.J., Monera, A., Alperi, A., Figueras, M.-J. and Saavedra, M.-J. (2009) Phylogenetic evidence suggests that strains of *Aeromonas hydrophila* subsp. *dhakensis* belong to the species *Aeromonas aquariorum* sp. nov. *Current Microbiology* 58(1), 76–80.
1105 Martínez-Murcia, A.J., Monera, A., Saavedra, M.J., Oncina, R., Lopez-Alvarez, M., Lara, E. and Figueras, M.J. (2011) Multilocus phylogenetic analysis of the genus *Aeromonas*. *Systematic and Applied Microbiology* 34(3), 189–199.
1106 Martínez-Murcia, A.J., Saavedra, M.J., Mota, V.R., Maier, T., Stackebrandt, E. and Cousin, S. (2008) *Aeromonas aquariorum* sp. nov., isolated from aquaria of ornamental fish. *International Journal of Systematic and Evolutionary Microbiology* 58(Pt 5), 1169–1175.
1107 Martínez-Murcia, A.J., Soler, L., Saavedra, M.J., Chacón, M.R., Guarro, J., Stackebrandt, E. and Figueras, M.J. (2005) Phenotypic, genotypic, and phylogenetic discrepancies to differentiate *Aeromonas salmonicida* from *Aeromonas bestiarum*. *International Microbiology: The Official Journal of the Spanish Society for Microbiology* 8(4), 259–269.
1108 Massad, G. and Oliver, J.D. (1987) New selective and differential medium for *Vibrio cholerae* and *Vibrio vulnificus*. *Applied and Environmental Microbiology* 53(9), 2262–2264.
1109 Masters, A.M., Ellis, T.M., Carson, J.M., Sutherland, S.S. and Gregory, A.R. (1995) *Dermatophilus chelonae* sp. nov., isolated from chelonids in Australia. *International Journal of Systematic Bacteriology* 45(1), 50–56.
1110 Masumura, K. and Wakabayashi, H. (1977) An outbreak of gliding bacterial disease in hatchery-born red seabream (*Pagrus major*) and gilthead (*Acanthopagrus schlegeli*) fry in Hiroshima. *Fish Pathology* 12(3), 171–177.
1111 Mata, A.I., Blanco, M.M., Domínguez, L., Fernández-Garayzábal, J.F. and Gibello, A. (2004) Development of a PCR assay for *Streptococcus iniae* based on the lactate oxidase (*lctO*) gene with potential diagnostic value. *Veterinary Microbiology* 101(2), 109–116.
1112 Mauel, M.J., Miller, D.L., Frazier, K.S. and Hines, M.E. (2002) Bacterial pathogens isolated from cultured bullfrogs (*Rana castesbeiana*). *Journal of Veterinary Diagnostic Investigation* 14(5), 431–433.
1113 Mawdesley-Thomas, L.E. (1969) Furunculosis in the goldfish *Carassius auratus* (L.). *Journal of Fish Biology* 1(1), 19–23.
1114 Mayer-Scholl, A., Draeger, A., Göllner, C., Scholz, H.C. and Nöckler, K. (2010) Advancement of a multiplex PCR for the differentiation of all currently described *Brucella* species. *Journal of Microbiological Methods* 80(1), 112–114.
1115 McCammon, S.A. and Bowman, J.P. (2000) Taxonomy of Antarctic *Flavobacterium* species: description of *Flavobacterium gillisiae* sp. nov., *Flavobacterium tegetincola* sp. nov. and *Flavobacterium xanthum* sp. nov., nom. rev. and reclassification of [*Flavobacterium*] *salegens* as *Salegentibacter salegens* gen. nov., comb. nov. *International Journal of Systematic and Evolutionary Microbiology* 50(Pt 3), 1055–1063.
1116 McCammon, S.A., Innes, B.H., Bowman, J.P., Franzmann, P.D., Dobson, S.J., Holloway, P.E., Skerratt, J.H., Nichols, P.D. and Rankin, L.M. (1998) *Flavobacterium hibernum* sp. nov., a lactose-utilizing bacterium from a freshwater Antarctic lake. *International*

Journal of Systematic Bacteriology 48(Pt 4), 1405–1412.

1117 McCarthy, D.H. (1975) Fish furunculosis caused by *Aeromonas salmonicida* var. *achromogenes*. *Journal of Wildlife Diseases* 11(4), 489–493.

1118 McCarthy, D.H. (1977) The identification and significance of atypical strains of *Aeromonas salmonicida*. *Bulletin of the International Office of Epizootics* 87, 459–463.

1119 McCarthy, D.H. (1980) Some ecological aspects of the bacterial fish pathogen *Aeromonas salmonicida*. In: *Aquatic Microbiology. Symposium of the Society of Applied Bacteriology No. 6*, 299–324.

1120 McCarthy, D.H. and Johnson, K.A. (1982) A serotypic survey and cross-protection test of North American field isolates of *Yersinia ruckeri*. *Journal of Fish Diseases* 5(4), 323–328.

1121 McCurdy, H.D. (1969) Studies on the taxonomy of the *Myxobacterales*.I. Record of Canadian isolates and survey of methods. *Canadian Journal of Microbiology* 15(12), 1453–1461.

1122 McGinnis, M.R. and Ajello, L. (1974) A new species of *Exophiala* isolated from channel catfish. *Mycologia* 66(3), 518–520.

1123 McIntosh, D. and Austin, B. (1990) Recovery of an extremely proteolytic form of *Serratia liquefaciens* as a pathogen of Atlantic salmon, *Salmo salar*, in Scotland. *Journal of Fish Biology* 36(5), 765–772.

1124 McIntosh, D., Meaden, P.G. and Austin, B. (1996) A simplified PCR-based method for the detection of *Renibacterium salmoninarum* utilizing preparations of rainbow trout (*Oncorhynchus mykiss*, Walbaum) lymphocytes. *Applied and Environmental Microbiology* 62(11), 3929–3932.

1125 McKenzie, R.A. and Hall, W.T.K. (1976) Dermal ulceration of mullet (*Mugil cephalus*). *Australian Veterinary Journal* 52(5), 230–231.

1126 McLaughlin, R.W., Zheng, J.S., Chen, M.M., Zhao, Q.Z. and Wang, D. (2011) Detection of *Helicobacter* in the fecal material of the endangered Yangtze finless porpoise *Neophocaena phocaenoides asiaeorientalis*. *Diseases of Aquatic Organisms* 95(3), 241–245.

1127 McVicar, A.H. and White, P.G. (1979) Fin and skin necrosis of cultivated Dover sole *Solea solea* (L.). *Journal of Fish Diseases* 2(6), 557–562.

1128 Melero, M., Rubio-Guerri, C., Crespo, J.L., Arbelo, M., Vela, A.I., García-Párraga, D., Sierra, E., Domínguez, L. and Sánchez-Vizcaíno, J.M. (2011) First case of erysipelas in a free-ranging bottlenose dolphin (*Tursiops truncatus*) stranded in the Mediterranean Sea. *Diseases of Aquatic Organisms* 97(2), 167–170.

1129 Mellado, E., Moore, E.R., Nieto, J.J. and Ventosa, A. (1996) Analysis of 16S rRNA gene sequences of *Vibrio costicola* strains: description of *Salinivibrio costicola* gen. nov., comb. nov. *International Journal of Systematic Bacteriology* 46(3), 817–821.

1130 Mendes, E.N., Queiroz, D.M., Dewhirst, F.E., Paster, B.J., Moura, S.B. and Fox, J.G. (1996) *Helicobacter trogontum* sp. nov., isolated from the rat intestine. *International Journal of Systematic Bacteriology* 46(4), 916–921.

1131 Mendoza, L., Ajello, L. and Taylor, J.W. (2001) The taxonomic status of *Lacazia loboi* and *Rhinosporidium seeberi* has been finally resolved with the use of molecular tools. *Revista Iberoamericana de Micología* 18(3), 95–98.

1132 Merga, J.Y., Leatherbarrow, A.J.H., Winstanley, C., Bennett, M., Hart, C.A., Miller, W.G. and Williams, N.J. (2011) Comparison of *Arcobacter* isolation methods, and diversity of *Arcobacter* spp. in Cheshire, United Kingdom. *Applied and Environmental Microbiology* 77(5), 1646–1650.

1133 Merkel, J.R., Traganza, E.D., Mukherjee, B.B., Griffin, T.B. and Prescott, J.M. (1964) Proteolytic activity and general characteristics of a marine bacterium, *Aeromonas proteolytica* sp. n. *Journal of Bacteriology* 87(5), 1227–1233.

1134 Meyer, F.P. and Bullock, G.L. (1973) *Edwardsiella tarda*, a new pathogen of channel catfish (*Ictalurus punctatus*). *Applied Microbiology* 25(1), 155–156.

1135 Mian, G.F., Godoy, D.T., Leal, C.A.G., Yuhara, T.Y., Costa, G.M. and Figueiredo, H.C.P. (2009) Aspects of the natural history and virulence of *S. agalactiae* infection in Nile tilapia. *Veterinary Microbiology* 136(1–2), 180–183.

1136 Michel, C., Bernardet, J.F. and Dinand, D. (1992) Phenotypic and genotypic studies of *Pseudomonas anguilliseptica* strains isolated from farmed European eels (*Anguilla anguilla*) in France. *Fish Pathology* 27(4), 229–232.

1137 Michel, C., Faivre, B. and Kerouault, B. (1986) Biochemical identification of *Lactobacillus piscicola* strains from France and Belgium. *Diseases of Aquatic Organisms* 2, 27–30.

1138 Michel, C., Messiaen, S. and Bernardet, J.-F. (2002) Muscle infections in imported neon tetra, *Paracheirodon innesi* Myers: limited occurrence of microsporidia and predominance of severe forms of columnaris disease caused by an Asian genomovar of *Flavobacterium columnare*. *Journal of Fish Diseases* 25(5), 253–263.

1139 Michel, C., Nougayrde, P., Eldar, A., Sochon, E. and de Kinkelin, P. (1997) *Vagococcus salmoninarum*, a bacterium of pathological significance in rainbow trout *Oncorhynchus mykiss* farming. *Diseases of Aquatic Organisms* 30(3), 199–208.

1140 Middlebrook, G., Cohn, M.L., Dye, W.E., Russell, W.F. and Levy, D. (1960) Microbiologic procedures of value in tuberculosis. *Acta Tuberculosea Scandinavica* 38, 66–81.

1141 Migaki, G., Valerio, M.G., Irvine, B. and Garner, F.M. (1971) Lobo's disease in an Atlantic bottle-nosed dolphin. *Journal of the American Veterinary Medical Association* 159(5), 578–582.

1142 Mikalsen, J., Olsen, A.B., Tengs, T. and Colquhoun, D.J. (2007) *Francisella philomiragia* subsp. *noatunensis* subsp. nov., isolated from farmed Atlantic cod (*Gadus morhua* L.). *International Journal of Systematic and Evolutionary Microbiology* 57(Pt 9), 1960–1965.

1143 Miller, M.A., Byrne, B.A., Jang, S.S., Dodd, E.M., Dorfmeier, E., Harris, M.D., Ames, J., Paradies, D., Worcester, K., Jessup, D.A. and Miller, W.A. (2010) Enteric bacterial pathogen detection in southern sea

otters (*Enhydra lutris nereis*) is associated with coastal urbanization and freshwater runoff. *Veterinary Research* 41:01. DOI: 10.1051/vetres/2009049.
1144 Mills, C.K. and Gherna, R.L. (1987) Hydrolysis of indoxyl acetate by *Campylobacter* species. *Journal of Clinical Microbiology* 25(8), 1560–1561.
1145 Miñana-Galbis, D., Farfán, M., Fusté, M.C. and Lorén, J.G. (2004) *Aeromonas molluscorum* sp. nov., isolated from bivalve molluscs. *International Journal of Systematic and Evolutionary Microbiology* 54(Pt 6), 2073–2078.
1146 Miñana-Galbis, D., Farfán, M., Fusté, M.C. and Lorén, J.G. (2007) *Aeromonas bivalvium* sp. nov., isolated from bivalve molluscs. *International Journal of Systematic and Evolutionary Microbiology* 57(Pt 3), 582–587.
1147 Miñana-Galbis, D., Farfán, M., Gaspar Lorén, J. and Carmen Fusté, M. (2010a) Proposal to assign *Aeromonas diversa* sp. nov. as a novel species designation for *Aeromonas* group 501. *Systematic and Applied Microbiology* 33(1), 15–19.
1148 Miñana-Galbis, D., Farfán, M., Lorén, J.G. and Fusté, M.C. (2002) Biochemical identification and numerical taxonomy of *Aeromonas* spp. isolated from environmental and clinical samples in Spain. *Journal of Applied Microbiology* 93(3), 420–430.
1149 Miñana-Galbis, D., Farfàn, M., Lorén, J.G. and Fusté, M.C. (2010b) The reference strain *Aeromonas hydrophila* CIP 57.50 should be reclassified as *Aeromonas salmonicida* CIP 57.50. *International Journal of Systematic and Evolutionary Microbiology* 60(3), 715–717.
1150 Miñana-Galbis, D., Urbizu-Serrano, A., Farfán, M., Fusté, M.C. and Lorén, J.G. (2009) Phylogenetic analysis and identification of *Aeromonas* species based on sequencing of the *cpn60* universal target. *International Journal of Systematic and Evolutionary Microbiology* 59(Pt 8), 1976–1983.
1151 Misaka, N. and Suzuki, K. (2007) Detection of *Flavobacterium psychrophilum* in chum salmon *Oncorhynchus keta* and virulence of isolated strains to salmonid fishes. *Fish Pathology* 42(4), 201–209.
1152 Mitchell, A. and Goodwin, A. (2000) The isolation of *Edwardsiella ictaluri* with a limited tolerance for aerobic growth from channel catfish. *Journal of Aquatic Animal Health* 12, 297–300.
1153 Miura, M., Hatai, K., Tojo, M., Wada, S., Kobayashi, S. and Okazaki, T. (2010) Visceral mycosis in ayu *Plecoglossus altivelis* larvae caused by *Pythium flevoense*. *Fish Pathology* 45(1), 24–30.
1154 Miyamoto, Y., Kato, T., Obara, Y., Akiyama, S., Takizawa, K. and Yamai, S. (1969) *In vitro* hemolytic characteristic of *Vibrio parahaemolyticus*: its close correlation with human pathogenicity. *Journal of Bacteriology* 100(2), 1147–1149.
1155 Miyamoto, Y., Nakamura, K. and Takizawa, L. (1961) Pathogenic halophiles. Proposals of a new genus '*Oceanomonas*' and of the amended species names. *Japanese Journal of Microbiology* 5(4), 477–481.
1156 Miyashita, T. (1984) *Pseudomonas fluorescens* and *Edwardsiella tarda* isolated from diseased tilapia. *Fish Pathology* 19(1), 45–50.
1157 Miyata, M., Inglis, V. and Aoki, T. (1996) Rapid identification of *Aeromonas salmonicida* subspecies *salmonicida* by the polymerase chain reaction. *Aquaculture* 141(1–2), 13–24.
1158 Mizuki, H., Washio, S., Morita, T., Itoi, S. and Sugita, H. (2006) Distribution of a fish pathogen *Listonella anguillarum* in the Japanese flounder *Paralichthys olivaceus* hatchery. *Aquaculture* 261(1), 26–32.
1159 Mladineo, I., Miletic, I. and Bocina, I. (2006) *Photobacterium damselae* subsp. *piscicida* outbreak in cage-reared Atlantic bluefin tuna *Thunnus thynnus*. *Journal of Aquatic Animal Health* 18(1), 51–54.
1160 Mohanty, B.R. and Sahoo, P.K. (2010) Immune responses and expression profiles of some immune-related genes in Indian major carp, *Labeo rohita* to *Edwardsiella tarda* infection. *Fish and Shellfish Immunology* 28(4), 613–621.
1161 Mohi, M.M., Kuratani, M., Miyazaki, T. and Yoshida, T. (2010) Histopathological studies on *Vibrio harveyi*-infected tiger puffer, *Takifugu rubripes* (Temminck et Schlegel), cultured in Japan. *Journal of Fish Diseases* 33(10), 833–840.
1162 Mohney, L.L., Poulos, B.T., Brooker, J.H., Cage, G.D. and Lightner, D.V. (1998) Isolation and identification of *Mycobacterium peregrinum* from the Pacific white shrimp *Penaeus vannamei*. *Journal of Aquatic Animal Health* 10(1), 83–88.
1163 Mok, W.Y. and Carvalho, C.M. (1984) Occurrence and experimental infection of toads (*Bufo marinus* and *B. granulosus*) with *Mycobacterium chelonei* subsp. *abscessus*. *Journal of Medical Microbiology* 18(3), 327–333.
1164 Molitoris, E., Marii, M.A., Joseph, S.W., Krichevsky, M.I., Fanning, G.R., Last, G., El-Mishad, A.M., Batawi, Y.A.E. and Colwell, R.R. (1989) Numerical taxonomy and deoxyribonucleic acid relatedness of environmental and clinical *Vibrio* species isolated in Indonesia. *International Journal of Systematic Bacteriology* 39(4), 442–449.
1165 Mondal, S.K. and De, A.B. (2002) A fatal oomycotic disease of the fresh water fish *Aplocheilus panchax* in India caused by *Aphanomyces laevis*. *Mycopathologia* 154(1), 21–24.
1166 Montieri, S., Suffredini, E., Ciccozzi, M. and Croci, L. (2010) Phylogenetic and evolutionary analysis of *Vibrio parahaemolyticus* and *Vibrio alginolyticus* isolates based on *toxR* gene sequence. *The New Microbiologica* 33(4), 359–372.
1167 Mora, D., Scarpellini, M., Franzetti, L., Colombo, S. and Galli, A. (2003) Reclassification of *Lactobacillus maltaromicus* (Miller *et al.* 1974) DSM 20342(T) and DSM 20344 and *Carnobacterium piscicola* (Collins *et al.* 1987) DSM 20730(T) and DSM 20722 as *Carnobacterium maltaromaticum* comb. nov. *International Journal of Systematic and Evolutionary Microbiology* 53(Pt 3), 675–678.
1168 Moreira, A.P.B., Pereira, N. and Thompson, F.L. (2011) Usefulness of a real time PCR platform for GC content and DNA–DNA hybridization similarity estimations in vibrios. *International Journal of Systematic and Evolutionary Microbiology* 61(Pt 10), 2379–2383.
1169 Moreno, E., Cloeckaert, A. and Moriyón, I. (2002) *Brucella* evolution and taxonomy. *Veterinary Microbiology* 90(1–4), 209–227.

1170 Morris, J.G. Jr, Miller, H.G., Wilson, R., Tacket, C.O., Hollis, D.G., Hickman, F.W., Weaver, R.E. and Blake, P.A. (1982) Illness caused by *Vibrio damsela* and *Vibrio hollisae*. *Lancet* 1(8284), 1294–1297.

1171 Moyer, T.R. and Hunnicutt, D.W. (2007) Susceptibility of zebra fish *Danio rerio* to infection by *Flavobacterium columnare* and *F. johnsoniae*. *Diseases of Aquatic Organisms* 76(1), 39–44.

1172 Mudarris, M. and Austin, B. (1989) Systemic disease in turbot *Scophthalmus maximus* caused by a previously unrecognised *Cytophaga* like bacterium. *Diseases of Aquatic Organisms* 6, 161–166.

1173 Mudarris, M., Austin, B., Segers, P., Vancanneyt, M., Hoste, B. and Bernardet, J.F. (1994) *Flavobacterium scophthalmum* sp. nov., a pathogen of turbot (*Scophthalmus maximus* L.). *International Journal of Systematic Bacteriology* 44(3), 447–453.

1174 Muhvich, A.G., Reimschuessel, R., Lipsky, M.M. and Bennett, R.O. (1989) *Fusarium solani* isolated from newborn bonnethead sharks, *Sphyrna tiburo* (L.). *Journal of Fish Diseases* 12, 57–62.

1175 Müller, H.E. (1983) *Providencia friedericiana*, a new species isolated from penguins. *International Journal of Systematic Bacteriology* 33(4), 709–715.

1176 Müller, H.E., Fanning, G.R. and Brenner, D.J. (1995) Isolation of *Serratia fonticola* from mollusks. *Systematic and Applied Microbiology* 18(2), 279–284.

1177 Mullis, K. and Faloona, F.A. (1987) Specific synthesis of DNA in vitro via a polymerase-catalyzed chain reaction. In:Wu, R. (ed.) *Methods in Enzymology: Recombinant DNA*, Vol. 155. Academic Press, New York, pp. 335–350.

1178 Munchan, C., Kurata, O., Wada, S., Hatai, K., Sano, A., Kamei, K. and Nakaoka, N. (2009) *Exophiala xenobiotica* infection in cultured striped jack, *Pseudocaranx dentex* (Bloch & Schneider), in Japan. *Journal of Fish Diseases* 32(10), 893–900.

1179 Munday, B.L. and Peel, B.F. (1983) Severe ulcerative dermatitis in platypus (*Ornithorhynchus anatinus*). *Journal of Wildlife Diseases* 19(4), 383–385.

1180 Muñoz Perdiguero, G., Barrientos, A., Madrid, E. and Birkbeck, T.H. (2011) Isolation of *Rhodococcus erythropolis* from vaccinated Atlantic salmon, *Salmo salar* L., smolts in Chile. *Journal of Fish Diseases* 34(9), 715–717.

1181 Muraosa, Y., Lawhavinit, O. and Hatai, K. (2006) *Lagenidium thermophilum* isolated from eggs and larvae of black tiger shrimp *Penaeus monodon* in Thailand. *Fish Pathology* 41(1), 35–40.

1182 Muroga, K., Yamanoi, H., Hironaka, Y., Yamamoto, S., Tatani, M., Jo, Y., Takahashi, S. and Hanada, H. (1984) Detection of *Vibrio anguillarum* from wild fingerlings of ayu *Plecoglossus altivelis*. *Bulletin of the Japanese Society of Scientific Fisheries* 50(4), 591–596.

1183 Musa, N., Wei, L.S., Musa, N., Hamdan, R.H., Leong, L.K., Wee, W., Amal, M.N., Kutty, B.M. and Abdullah, S.Z. (2009) Streptococcosis in red hybrid tilapia (*Oreochromis niloticus*) commercial farms in Malaysia. *Aquaculture Research* 40(5), 630–632.

1184 Mutschmann, F., Berger, L., Zwart, P. and Gaedicke, C. (2000) Chytridiomycosis in amphibians-first report in Europe. *Berliner Und Münchener Tierärztliche Wochenschrift* 113(10), 380–383.

1185 Mutters, R., Ihm, P., Pohl, S., Frederiksen, W. and Mannheim, W. (1985) Reclassification of the genus *Pasteurella* Trevisan 1887 on the basis of deoxyribonucleic acid homology, with proposals for the new species *Pasteurella dagmatis*, *Pasteurella canis*, *Pasteurella stomatis*, *Pasteurella anatis*, and *Pasteurella langaa*. *International Journal of Systematic Bacteriology* 35(3), 309–322.

1186 Myhr, E., Larsen, J.L., Lillehaug, A., Gudding, R., Heum, M. and Håstein, T. (1991) Characterization of *Vibrio anguillarum* and closely related species isolated from farmed fish in Norway. *Applied and Environmental Microbiology* 57(9), 2750–2757.

1187 Nagai, T. and Iida, Y. (2002) Occurrence of bacterial kidney disease in cultured ayu. *Fish Pathology* 37(2), 77–81.

1188 Nagano, I., Oshima, S. and Kawai, K. (2011) *In vivo* analysis on the adherence and infection route of *Photobacterium damselae* subsp. *piscicida* in Yellowtail. *Fish Pathology* 46(2), 45–50.

1189 Nair, G.B. and Holmes, B. (1999) International committee on the systematic bacteriology subcommittee on the taxonomy of *Vibrionaceae*. *International Journal of Systematic Bacteriology* 49, 1945–1947.

1190 Najiah, M., Lee, K.L., Noorasikin, H., Nadirah, M. and Lee, S.W. (2011) Phenotypic and genotypic characteristics of *Mycobacterium* isolates from fighting fish *Betta* spp. in Malaysia. *Research in Veterinary Science* 91(3), 342–345.

1191 Naka, H. and Crosa, J.H. (2011) Genetic determinants of virulence in the marine fish pathogen *Vibrio anguillarum*. *Fish Pathology* 46, 1–10.

1192 Nakagawa, Y., Sakane, T. and Yokota, A. (1996) Emendation of the genus *Planococcus* and transfer of *Flavobacterium okeanokoites* Zobell and Upham 1944 to the genus *Planococcus* as *Planococcus okeanokoites* comb. nov. *International Journal of Systematic Bacteriology* 46(4), 866–870.

1193 Nakai, T., Fujii, N., Muroga, K., Arimoto, M., Mizuta, Y. and Matsuoka, S. (1992) *Pasteurella piscicida* infection in hatchery-reared juvenile striped jack. *Gyobo Kenkyu* 27, 103–108.

1194 Nakai, T., Hanada, H. and Muroga, K. (1985) First records of *Pseudomonas anguilliseptica* infection in cultured ayu, *Plecoglossus altivelis*. *Fish Pathology* 20(4), 481–484.

1195 Nakajima, K., Muroga, K. and Hancock, R.E.W. (1983) Comparison of fatty acid, protein, and serological properties distinguishing outer membranes of *Pseudomonas anguilliseptica* strains from those of fish pathogens and other pseudomonads. *International Journal of Systematic Bacteriology* 33(1), 1–8.

1196 Nakamura, K. and Hatai, K. (1994) *Atkinsiella parasitica* sp. nov. isolated from a rotifer, *Brachionus plicatilis*. *Mycoscience* 35(4), 383–389.

1197 Nakamura, K. and Hatai, K. (1995a) Three species of *Lagenidiales* isolated from the eggs and zoeae of the marine crab *Portunus pelagicus*. *Mycoscience* 36(1), 87–95.

1198 Nakamura, K. and Hatai, K. (1995b) *Atkinsiella dubia* and its related species. *Mycoscience* 36(4), 431–438.

1199 Nakamura, K., Nakamura, M., Hatai, K. and Zafran (1995) *Lagenidium* infection in eggs and larvae of

mangrove crab (*Scylla serrata*) produced in Indonesia. *Mycoscience* 36(4), 399–404.
1200 Nakamura, K., Wada, S., Hatai, K. and Sugimoto, T. (1994) *Lagenidium myophilum* infection in the coonstripe shrimp, *Pandalus hypsinotus*. *Mycoscience* 35(1), 99–104.
1201 Nakatsugawa, T. (1983) *Edwardsiella tarda* isolated from cultured young flounder. *Fish Pathology* 18(2) 99–101.
1202 Nam, Y.-D., Chang, H.-W., Park, J.R., Kwon, H.-Y., Quan, Z.-X., Park, Y.-H., Kim, B.-C. and Bae, J.-W. (2007) *Vibrio litoralis* sp. nov., isolated from a Yellow Sea tidal flat in Korea. *International Journal of Systematic and Evolutionary Microbiology* 57(Pt 3), 562–565.
1203 Nandi, B., Nandy, R.K., Mukhopadhyay, S., Nair, G.B., Shimada, T. and Ghose, A.C. (2000) Rapid method for species-specific identification of *Vibrio cholerae* using primers targeted to the gene of outer membrane protein OmpW. *Journal of Clinical Microbiology* 38(11), 4145–4151.
1204 Nealson, K.H. (1978) Isolation, identification, and manipulation of luminous bacteria. *Methods in Enzymology* 57, 153–166.
1205 Nedashkovskaya, O.I., Kim, S.B. and Bae, K.S. (2010) *Gramella marina* sp. nov., isolated from the sea urchin *Strongylocentrotus intermedius*. *International Journal of Systematic and Evolutionary Microbiology* 60(Pt 12), 2799–2802.
1206 Nedashkovskaya, O.I., Kim, S.B., Lysenko, A.M., Frolova, G.M., Mikhailov, V.V., Bae, K.S., Lee, D.H. and Kim, I.S. (2005) *Gramella echinicola* gen. nov., sp. nov., a novel halophilic bacterium of the family *Flavobacteriaceae* isolated from the sea urchin *Strongylocentrotus intermedius*. *International Journal of Systematic and Evolutionary Microbiology* 55(Pt 1), 391–394.
1207 Nedashkovskaya, O.I., Kim, S.B., Vancanneyt, M., Lysenko, A.M., Shin, D.S., Park, M.S., Lee, K.H., Jung, W.J., Kalinovskaya, N.I., Mikhailov, V.V., Bae, K.S. and Swings, J. (2006) *Echinicola pacifica* gen. nov., sp. nov., a novel flexibacterium isolated from the sea urchin *Strongylocentrotus intermedius*. *International Journal of Systematic and Evolutionary Microbiology* 56(Pt 5), 953–958.
1208 Neill, S.D., Campbell, J.N., O'Brien, J.J., Weatherup, S.T.C. and Ellis, W.A. (1985) Taxonomic position of *Campylobacter cryaerophila* sp. nov. *International Journal of Systematic Bacteriology* 35(3) 342–356.
1209 Neimark, H., Johansson, K.E., Rikihisa, Y. and Tully, J.G. (2001) Proposal to transfer some members of the genera *Haemobartonella* and *Eperythrozoon* to the genus *Mycoplasma* with descriptions of '*Candidatus Mycoplasma haemofelis*', '*Candidatus Mycoplasma haemomuris*', '*Candidatus Mycoplasma haemosuis*' and '*Candidatus Mycoplasma wenyonii*'. *International Journal of Systematic and Evolutionary Microbiology* 51(3), 891–899.
1210 Nelson, E.J. and Ghiorse, W.C. (1999) Isolation and identification of *Pseudoalteromonas piscicida* strain Cura-d associated with diseased damselfish (*Pomacentridae*) eggs. *Journal of Fish Diseases* 22(4), 253–260.
1211 Neogi, S.B., Chowdhury, N., Asakura, M., Hinenoya, A., Haldar, S., Saidi, S.M., Kogure, K., Lara, R.J. and Yamasaki, S. (2010) A highly sensitive and specific multiplex PCR assay for simultaneous detection of *Vibrio cholerae*, *Vibrio parahaemolyticus* and *Vibrio vulnificus*. *Letters in Applied Microbiology* 51(3), 293–300.
1212 Nesterenko, O.A., Nogina, T.M., Kasumova, S.A., Kvasnikov, E.I. and Batrakov, S.G. (1982) *Rhodococcus luteus* nom. nov. and *Rhodococcus maris* nom. nov. *International Journal of Systematic Bacteriology* 32(1), 1–14.
1213 Netto, L.N., Leal, C.A.G. and Figueiredo, H.C.P. (2011) *Streptococcus dysgalactiae* as an agent of septicaemia in Nile tilapia, *Oreochromis niloticus* (L.). *Journal of Fish Diseases* 34 (3), 251–254.
1214 Neville, S.A., Lecordier, A., Ziochos, H., Chater, M.J., Gosbell, I.B., Maley, M.W. and van Hal, S.J. (2011) Utility of matrix-assisted laser desorption ionization-time of flight mass spectrometry following introduction for routine laboratory bacterial identification. *Journal of Clinical Microbiology* 49(8), 2980–2984.
1215 Nguyen, H.T. and Kanai, K. (1999) Selective agars for the isolation of *Streptococcus iniae* from Japanese flounder, *Paralichthys olivaceus*, and its cultural environment. *Journal of Applied Microbiology* 86(5), 769–776.
1216 Nhung, P.H., Hata, H., Ohkusu, K., Noda, M., Shah, M.M., Goto, K. and Ezaki, T. (2007) Use of the novel phylogenetic marker *dnaJ* and DNA–DNA hybridization to clarify interrelationships within the genus *Aeromonas*. *International Journal of Systematic and Evolutionary Microbiology* 57, (6), 1232–1237.
1217 Nicholls, K.M., Lee, J.V. and Donovan, T.J. (1976) An evaluation of commercial thiosulphate citrate bile salt sucrose agar (TCBS). *The Journal of Applied Bacteriology* 41(2), 265–269.
1218 Nichols, D.S., Hart, P., Nichols, P.D. and McMeekin, T.A. (1996) Enrichment of the rotifer *Brachionus plicatilis* fed an Antarctic bacterium containing polyunsaturated fatty acids. *Aquaculture* 147(1–2), 115–125.
1219 Nicolas, J.L., Basuyaux, O. and Mazurié, J. (2002) *Vibrio carchariae*, a pathogen of the abalone *Haliotis tuberculata*. *Diseases of Aquatic Organisms* 50, 35–43.
1220 Nicolas, J.L., Corre, S., Gauthier, G., Robert, R. and Ansquer, D. (1996) Bacterial problems associated with scallop *Pecten maximus* larval culture. *Diseases of Aquatic Organisms*, 2767–2776.
1221 Nielsen, M.E., Høi, L., Schmidt, A.S., Qian, D., Shimada, T., Shen, J.Y. and Larsen, J.L. (2001) Is *Aeromonas hydrophila* the dominant motile *Aeromonas* species that causes disease outbreaks in aquaculture production in the Zhejiang Province of China? *Diseases of Aquatic Organisms* 46(1), 23–29.
1222 Nieto, T.P., López, L.R., Santos, Y., Núñez, S. and Toranzo, A.E. (1990) Isolation of *Serratia plymuthica* as an opportunistic pathogen in rainbow trout, *Salmo gairdneri* Richardson. *Journal of Fish Diseases* 13(2), 175–177.
1223 Nilsen, H., Johansen, R., Colquhoun, D.J., Kaada, I., Bottolfsen, K., Vågnes, Ø. and Olsen, A.B. (2011a) *Flavobacterium psychrophilum* associated with septicaemia and necrotic myositis in Atlantic salmon

Salmo salar: a case report. *Diseases of Aquatic Organisms* 97(1), 37–46.
1224 Nilsen, H., Olsen, A.B., Vaagnes, Ø., Hellberg, H., Bottolfsen, K., Skjelstad, H. and Colquhoun, D.J. (2011b) Systemic *Flavobacterium psychrophilum* infection in rainbow trout, *Oncorhynchus mykiss* (Walbaum), farmed in fresh and brackish water in Norway. *Journal of Fish Diseases* 34(5), 403–408.
1225 Nishibuchi, M., Doke, S., Toizumi, S., Umeda, T., Yoh, M. and Miwatani, T. (1988) Isolation from a coastal fish of *Vibrio hollisae* capable of producing a hemolysin similar to the thermostable direct hemolysin of *Vibrio parahaemolyticus*. *Applied and Environmental Microbiology* 54(8), 2144–2146.
1226 Nishibuchi, M. and Kaper, J.B. (1995) Thermostable direct hemolysin gene of *Vibrio parahaemolyticus*: a virulence gene acquired by a marine bacterium. *Infection and Immunity* 63(6), 2093–2099.
1227 Nishiguchi, M.K. and Nair, V.S. (2003) Evolution of symbiosis in the *Vibrionaceae*: a combined approach using molecules and physiology. *International Journal of Systematic and Evolutionary Microbiology* 53(Pt 6), 2019–2026.
1228 Nishimori, E., Hasegawa, O., Numata, T. and Wakabayashi, H. (1998) *Vibrio carchariae* causes mass mortalities in Japanese abalone *Suculus diversicolor supratexta*. *Fish Pathology* 33(5), 495–502.
1229 Nishimori, E., Kita-Tsukamoto, K. and Wakabayashi, H. (2000) *Pseudomonas plecoglossicida* sp. nov., the causative agent of bacterial haemorrhagic ascites of ayu, *Plecoglossus altivelis*. *International Journal of Systematic and Evolutionary Microbiology* 50(1), 83–89.
1230 Nishimura, Y., Kinpara, M. and Iizuka, H. (1989) *Mesophilobacter marinus* gen. nov., sp. nov.: an aerobic coccobacillus isolated from seawater. *International Journal of Systematic Bacteriology* 39(4), 378–381.
1231 Nitzan, S., Shwartsburd, B., Vaiman, R. and Heller, E.D. (2001) Some characteristics of *Photobacterium damselae* ssp. *piscicida* isolated in Israel during outbreaks of pasteurellosis in hybrid bass (*Morone saxatilis×M. chrysops*). *Bulletin of the European Association of Fish Pathologists* 21(2), 77–80.
1232 Noga, E.J. and Berkhoff, H.A. (1990) Pathological and microbiological features of *Aeromonas salmonicida* infection in the American eel (*Anguilla rostrata*). *Fish Pathology* 25, 127–132.
1233 Nogi, Y., Kato, C. and Horikoshi, K. (1998) *Moritella japonica* sp. nov., a novel barophilic bacterium isolated from a Japan Trench sediment. *The Journal of General and Applied Microbiology* 44(4), 289–295.
1234 Nogi, Y., Masui, N. and Kato, C. (1998) *Photobacterium profundum* sp. nov., a new, moderately barophilic bacterial species isolated from a deep-sea sediment. *Extremophiles: Life Under Extreme Conditions* 2(1), 1–7.
1235 Nomoto, R., Munasinghe, L.I., Jin, D.-H., Shimahara, Y., Yasuda, H., Nakamura, A., Misawa, N., Itami, T. and Yoshida, T. (2004) Lancefield group C *Streptococcus dysgalactiae* infection responsible for fish mortalities in Japan. *Journal of Fish Diseases* 27(12), 679–686.
1236 Notomi, T., Okayama, H., Masubuchi, H., Yonekawa, T., Watanabe, K., Amino, N. and Hase, T. (2000) Loop-mediated isothermal amplification of DNA. *Nucleic Acids Research* 28(12), E63.
1237 Novoa, B., Luque, A., Castro, D., Borrego, J.J. and Figueras, A. (1998) Characterization and infectivity of four bacterial strains isolated from Brown Ring Disease-affected clams. *Journal of Invertebrate Pathology* 71(1), 34–41.
1238 Novotny, L., Halouzka, R., Matlova, L., Vavra, O., Bartosova, L., Slany, M. and Pavlik, I. (2010) Morphology and distribution of granulomatous inflammation in freshwater ornamental fish infected with mycobacteria. *Journal of Fish Diseases* 33(12), 947–955.
1239 Nozue, H., Hayashi, T., Hashimoto, Y., Ezaki, T., Hamasaki, K., Ohwada, K. and Terawaki, Y. (1992) Isolation and characterization of *Shewanella alga* from human clinical specimens and emendation of the description of *S. alga* Simidu *et al.*, 1990, 335. *International Journal of Systematic Bacteriology* 42(4), 628–634.
1240 Nyaoke, A., Weber, E.S., Innis, C., Stremme, D., Dowd, C., Hinckley, L., Gorton, T., Wickes, B., Sutton, D., de Hoog, S. and Frasca, S. (2009) Disseminated phaeohyphomycosis in weedy seadragons (*Phyllopteryx taeniolatus*) and leafy seadragons (*Phycodurus eques*) caused by species of *Exophiala*, including a novel species. *Journal of Veterinary Diagnostic Investigation* 21(1), 69–79.
1241 Nybelin, O. (1936) Studies into the cause of the crayfish plague occurring in Sweden at present. *Reports from the Swedish State Institute of Freshwater Fishery Research, Drottningholm* 9.
1242 Nymo, I.H., Tryland, M. and Godfroid, J. (2011) A review of *Brucella* infection in marine mammals, with special emphasis on in the hooded seal (*Cystophora cristata*). *Veterinary Research* 42(1), 93.
1243 O'Brien, M. and Colwell, R.R. (1987) A rapid test for chitinase activity that uses 4-methylumbelliferyl-*N*-acetyl-β-*d*-glucosaminide. *Applied and Environmental Microbiology* 53(7), 1718–1720.
1244 Oakey, H.J., Gibson, L.F. and George, A.M. (1999) DNA probes specific for *Aeromonas hydrophila* (HG1). *Journal of Applied Microbiology* 86(2), 187–193.
1245 Obendorf, D.L., Carson, J. and McManus, T.J. (1987) *Vibrio damsela* infection in a stranded leatherback turtle (*Dermochelys coriacea*). *Journal of Wildlife Diseases* 23(4), 666–668.
1246 Obendorf, D.L., Peel, B.F. and Munday, B.L. (1993) *Mucor amphibiorum* infection in platypus (*Ornithorhynchus anatinus*) from Tasmania. *Journal of Wildlife Diseases* 29(3), 485–487.
1247 Odamaki, T., Yonezawa, S., Kitahara, M., Sugahara, Y., Xiao, J.-Z., Yaeshima, T., Iwatsuki, K. and Ohkuma, M. (2011) Novel multiplex polymerase chain reaction primer set for identification of *Lactococcus* species. *Letters in Applied Microbiology* 52(5), 491–496.
1248 Oh, M.-J., Kim, W.-S., Kitamura, S.-I., Lee, H., Son, B.W., Jung, T.-S. and Jung, S.-J. (2006) Change of pathogenicity in Olive flounder *Paralichthys olivaceus* by co-infection of *Vibrio harveyi*, *Edwardsiella tarda* and marine birnavirus. *Aquaculture* 257(1–4), 156–160.

1249 Oidtmann, B., Geiger, S., Steinbauer, P., Culas, A. and Hoffmann, R.W. (2006) Detection of *Aphanomyces astaci* in North American crayfish by polymerase chain reaction. *Diseases of Aquatic Organisms* 72(1), 53–64.

1250 Oidtmann, B., Schaefers, N., Cerenius, L., Söderhäll, K. and Hoffmann, R.W. (2004) Detection of genomic DNA of the crayfish plague fungus *Aphanomyces astaci* (Oomycete) in clinical samples by PCR. *Veterinary Microbiology* 100(3–4), 269–282.

1251 Oidtmann, B., Steinbauer, P., Geiger, S. and Hoffmann, R.W. (2008) Experimental infection and detection of *Aphanomyces invadans* in European catfish, rainbow trout and European eel. *Diseases of Aquatic Organisms* 82(3), 195–207.

1252 OIE (2000a) World Animal Health in 2000: Reports on the animal health status and disease control methods and tables on incidence of list A diseases. Office International des Epizooties, Paris.

1253 OIE (2000b) Bacterial kidney disease (*Renibacterium salmoninarum*). In:Diagnostic Manual for Aquatic Animal Diseases 2000. Office International des Epizooties, Paris, Chapter 2.2.6.

1254 OIE (2012a) Principles and methods of validation of diagnositc assays for infectious diseases. In: *Manual of Diagnostic Tests and Vaccines for Terrestrial Animals*, 7th edn. World Organisation for Animal Health (OIE), Paris, Chapter 1.1.5.

1255 OIE (2012b) Crayfish plague (*Aphanomyces astaci*). In: *Manual of Diagnostic Tests for Aquatic Animals*. World Organisation for Animal Health (OIE), Paris, Chapter 2.2.1.

1256 OIE (2012c) Infection with *Batrachochytrium dendrobatidis*. In: *Manual of Diagnostic Tests for Aquatic Animals*. World Organisation for Animal Health (OIE), Paris, Chapter 2.1.1.

1257 Okada, K., Iida, T., Kita-Tsukamoto, K. and Honda, T. (2005) Vibrios commonly possess two chromosomes. *Journal of Bacteriology* 187(2), 752–757.

1258 Okada, S. and Gordon, D.M. (2003) Genetic and ecological structure of *Hafnia alvei* in Australia. *Systematic and Applied Microbiology* 26(4), 585–594.

1259 Okuda, J., Nakai, T., Chang, P.S., Oh, T., Nishino, T., Koitabashi, T. and Nishibuchi, M. (2001) The *toxR* gene of *Vibrio* (*Listonella*) *anguillarum* controls expression of the major outer membrane proteins but not virulence in a natural host model. *Infection and Immunity* 69(10), 6091–6101.

1260 Okuzumi, M., Hiraishi, A., Kobayashi, T. and Fujii, T. (1994) *Photobacterium histaminum* sp. nov., a histamine-producing marine bacterium. *International Journal of Systematic Bacteriology* 44(4), 631–636.

1261 Olafsen, J.A., Mikkelsen, H.V., Giæver, H.M. and Hansen, G.H. (1993) Indigenous bacteria in hemolymph and tissues of marine bivalves at low temperatures. *Applied and Environmental Microbiology* 59(6), 1848–1854.

1262 Olivares-Fuster, O.,, Bullard, S., McElwain, A., Llosa, M. and Arias, C. (2011) Adhesion dynamics of *Flavobacterium columnare* to channel catfish *Ictalurus punctatus* and zebrafish *Danio rerio* after immersion challenge. *Diseases of Aquatic Organisms* 96(3), 221–227.

1263 Oliver, J.D., Warner, R.A. and Cleland, D.R. (1983) Distribution of *Vibrio vulnificus* and other lactose-fermenting vibrios in the marine environment. *Applied and Environmental Microbiology* 45(3), 985–998.

1264 Olivier, G. (1990) Virulence of *Aeromonas salmonicida*: lack of relationship with phenotypic characteristics. *Journal of Aquatic Animal Health* 2(2), 119–127.

1265 Olsen, A.B., Birkbeck, T.H., Nilsen, H.K., MacPherson, H.L., Wangel, C., Myklebust, C., Laidler, L.A., Aarflot, L., Thoen, E., Nygård, S., Thayumanavan, T. and Colquhoun, D.J. (2006a) Vaccine-associated systemic *Rhodococcus erythropolis* infection in farmed atlantic salmon *Salmo salar*. *Diseases of Aquatic Organisms* 72(1), 9–17.

1266 Olsen, A.B., Mikalsen, J., Rode, M., Alfjorden, A., Hoel, E., Straum-Lie, K., Haldorsen, R. and Colquhoun, D.J. (2006b) A novel systemic granulomatous inflammatory disease in farmed Atlantic cod, *Gadus morhua* L., associated with a bacterium belonging to the genus *Francisella*. *Journal of Fish Diseases* 29(5), 307–311.

1267 Olufemi, B.E., Agius, C. and Roberts, R.J. (1983) Aspergillomycosis in intensively cultured tilapia from Kenya. *The Veterinary Record* 112(9), 203–204.

1268 Olufemi, B.E. and Roberts, R.J. (1983) Method for the isolation of *Aspergillus* species pathogens of fish from clinical materials. *Veterinary Record* 112(1), 15.

1269 Olufemi, B.E. and Roberts, R.J. (1986) Induction of clinical aspergillomycosis by feeding contaminated diet to tilapia, *Oreochromis niloticus* (L.). *Journal of Fish Diseases* 9(2), 123–128.

1270 On, S.L. and Holmes, B. (1991) Reproducibility of tolerance tests that are useful in the identification of campylobacteria. *Journal of Clinical Microbiology* 29(9), 1785–1788.

1271 On, S.L. and Holmes, B. (1992) Assessment of enzyme detection tests useful in identification of campylobacteria. *Journal of Clinical Microbiology* 30(3), 746–749.

1272 Onarheim, A.M., Wiik, R., Burghardt, J. and Stackebrandt, E. (1994) Characterization and identification of two *Vibrio* species indigenous to the intestine of fish in cold sea water; description of *Vibrio iliopiscarius* sp. nov. *Systematic and Applied Microbiology* 17(3), 370–379.

1273 Oplinger, R.W. and Wagner, E.J. (2012) Effects of media ingredient substitution and comparison of growth of *Flavobacterium psychrophilum* among four media. *Journal of Aquatic Animal Health* 24(1), 49–57.

1274 Ordal, E.J. and Rucker, R.R. (1944) Pathogenic myxobacteria. *Proceedings of the Society for Experimental Biology and Medicine* 56(1), 15–18.

1275 Orieux, N., Bourdineaud, J.-P., Douet, D.-G., Daniel, P. and Le Hénaff, M. (2011) Quantification of *Flavobacterium psychrophilum* in rainbow trout, *Oncorhynchus mykiss* (Walbaum), tissues by qPCR. *Journal of Fish Diseases* 34, 811–821.

1276 Osorio, C.R., Barja, J.L., Hutson, R.A. and Collins, M.D. (1999) *Arthrobacter rhombi* sp. nov., isolated from Greenland halibut (*Reinhardtius hippoglossoides*). *International Journal of Systematic Bacteriology* 49(Pt 3), 1217–1220.

1277 Osorio, C.R. and Klose, K.E. (2000) A region of the transmembrane regulatory protein ToxR that tethers the transcriptional activation domain to the cytoplasmic membrane displays wide divergence among *Vibrio* species. *Journal of Bacteriology* 182(2), 526–528.

1278 Osorio, C.R., Toranzo, A.E., Romalde, J.L. and Barja, J.L. (2000) Multiplex PCR assay for *ureC* and 16S rRNA genes clearly discriminates between both subspecies of *Photobacterium damselae*. *Diseases of Aquatic Organisms* 40(3), 177–183.

1279 Osterman, B. and Moriyón, I. (2006) International Committee on Systematics of Prokaryotes; Subcommittee on the Taxonomy of *Brucella* Minutes of the Meeting, 17 September 2003, Pamplona, Spain. *International Journal of Systematic and Evolutionary Microbiology* 56(5), 1173–1175.

1280 Ostland, V.E., Byrne, P.J., Lumsden, J.S., MacPhee, D.D., Derksen, J.A., Haulena, M., Skar, K., Myhr, E. and Ferguson, H.W. (1999a) Atypical bacterial gill disease: a new form of bacterial gill disease affecting intensively reared salmonids. *Journal of Fish Diseases* 22(5), 351–358.

1281 Ostland, V.E., Ferguson, H.W. and Stevenson, R.M.W. (1989) Case report: bacterial gill disease in goldfish *Carassias auratus*. *Diseases of Aquatic Organisms* 6, 179–184.

1282 Ostland, V.E., Lumsden, J.S., Macphee, D.D. and Ferguson, H.W. (1994) Characteristics of *Flavobacterium branchiophilum*, the cause of salmonid bacterial gill disease in Ontario. *Journal of Aquatic Animal Health* 6(1), 13–26.

1283 Ostland, V.E., Morrison, D. and Ferguson, H.W. (1999b) *Flexibacter maritimus* associated with a bacterial stomatitis in Atlantic salmon smolts reared in net-pens in British Columbia. *Journal of Aquatic Animal Health* 11(1), 35–44.

1284 Otis, E.J., Wolke, R.E. and Blazer, V.S. (1985) Infection of *Exophiala salmonis* in Atlantic salmon (*Salmo salar* L.). *Journal of Wildlife Diseases* 21(1), 61–64.

1285 Otis, V.S. and Behler, J.L. (1973) The occurrence of Salmonellae and *Edwardsiella* in the turtles of the New York Zoological Park. *Journal of Wildlife Diseases* 9(1), 4–6.

1286 Ouellet, M., Mikaelian, I., Pauli, B.D., Rodrigue, J. and Green, D.M. (2005) Historical evidence of widespread chytrid infection in North American amphibian populations. *Conservation Biology* 19(5), 1431–1440.

1287 Oxley, A.P.A., Argo, J.A. and McKay, D.B. (2005) *Helicobacter* spp. from captive bottlenose dolphins (*Tursiops* spp.) and polar bears (*Ursus maritimus*). *The Veterinary Journal* 170, 377–380.

1288 Oxley, A.P.A. and McKay, D.B. (2004) Fecal shedding of *Helicobacter* spp. by co-housed Australian sea lions (*Neophoca cinerea*) and Australian fur seals (*Arctocephalus pusillus doriferus*). *Veterinary Microbiology* 101(4), 235–243.

1289 Oxley, A.P.A. and McKay, D.B. (2005) Comparison of *Helicobacter* spp. genetic sequences in wild and captive seals, and gulls. *Diseases of Aquatic Organisms* 65(2), 99–105.

1290 Oxley, A.P.A., Powell, M. and McKay, D.B. (2004) Species of the family *Helicobacteraceae* detected in an Australian sea lion (*Neophoca cinerea*) with chronic gastritis. *Journal of Clinical Microbiology* 42(8), 3505–3512.

1291 Pacha, R.E. (1968) Characteristics of *Cytophaga psychrophila* (Borg) isolated from outbreaks of bacterial cold-water disease. *Applied Microbiology* 16(1), 97–101.

1292 Packer, R.A. (1943) The use of sodium azide (NaN3) and crystal violet in a selective medium for Streptococci and *Erysipelothrix rhusiopathiae*. *Journal of Bacteriology* 46(4), 343–349.

1293 Paclibare, J.O., Evelyn, T.P.T., Albright, L.J. and Prosperi-Porta, L. (1994) Clearing of the kidney disease bacterium *Renibacterium salmoninarum* from seawater by the blue mussel *Mytilus edulis*, and the status of the mussel as a reservoir of the bacterium. *Diseases of Aquatic Organisms* 18, 129–133.

1294 Padgitt, P.J. and Moshier, S.E. (1987) *Mycobacterium poriferae* sp. nov., a scotochromogenic, rapidly growing species isolated from a marine sponge. *International Journal of Systematic Bacteriology* 37(3), 186–191.

1295 Padilla, D., Acosta, F., Vega, J., Sorroza, L., Román, L., Bravo, J., Real, F. and Vivas, J. (2010) Study of adherence, invasion and survival of *Hafnia alvei* in RTG-2. *Fish Pathology* 45(4), 179–182.

1296 Padilla, D., Real, F., Gómez, V., Sierra, E., Acosta, B., Déniz, S. and Acosta, F. (2005) Virulence factors and pathogenicity of *Hafnia alvei* for gilthead seabream, *Sparus aurata* L. *Journal of Fish Diseases* 28(7), 411–417.

1297 Paillard, C. (2004) A short-review of brown ring disease, a vibriosis affecting clams, *Ruditapes philippinarum* and *Ruditapes decussatus*. *Aquatic Living Resources* 17(04), 467–475.

1298 Paillard, C., Allam, B. and Oubella, R. (2004) Effect of temperature on defense parameters in Manila clam *Ruditapes philippinarum* challenged with *Vibrio tapetis*. *Diseases of Aquatic Organisms* 59(3), 249–262.

1299 Paillard, C., Gausson, S., Nicolas, J.L., le Pennec, J.P. and Haras, D. (2006) Molecular identification of *Vibrio tapetis*, the causative agent of the brown ring disease of *Ruditapes philippinarum*. *Aquaculture* 253(1–4), 25–38.

1300 Paillard, C., Maes, P. and Oubella, R. (1994) Brown ring disease in clams. *Annual Review of Fish Diseases* 4, 219–240.

1301 Palleroni, N.J. and Holmes, B. (1981) *Pseudomonas cepacia* sp. nov., nom. rev. *International Journal of Systematic Bacteriology* 31(4), 479–481.

1302 Palmer, R., Drinan, E.M. and Murphy, T. (1994) A previously unknown disease of farmed Atlantic salmon: pathology and establishment of bacterial aetiology. *Diseases of Aquatic Organisms* 19, 7–14.

1303 Palmgren, H., McCafferty, D., Aspán, A., Broman, T., Sellin, M., Wollin, R., Bergström, S. and Olsen, B. (2000) *Salmonella* in sub-Antarctica: low heterogeneity in *Salmonella* serotypes in South Georgian seals and birds. *Epidemiology and Infection* 125(2), 257–262.

1304 Palomino, J.C. and Portaels, F. (1998) Effects of decontamination methods and culture conditions on

viability of *Mycobacterium ulcerans* in the BACTEC system. *Journal of Clinical Microbiology* 36(2), 402–408.
1305 Panangala, V.S., Shoemaker, C.A., Van Santen, V.L., Dybvig, K. and Klesius, P.H. (2007) Multiplex-PCR for simultaneous detection of 3 bacterial fish pathogens, *Flavobacterium columnare, Edwardsiella ictaluri*, and *Aeromonas hydrophila*. *Diseases of Aquatic Organisms* 74(3), 199–208.
1306 Pang, L., Zhang, X.-H., Zhong, Y., Chen, J., Li, Y. and Austin, B. (2006) Identification of *Vibrio harveyi* using PCR amplification of the *toxR* gene. *Letters in Applied Microbiology* 43(3), 249–255.
1307 Paperna, I. and Smirnova, M. (1997) *Branchiomyces*-like infection in a cultured tilapia (*Oreochromis* hybrid, Cichlidae). *Diseases of Aquatic Organisms* 31, 233–238.
1308 Park, K.-I., Paillard, C., Le Chevalier, P. and Choi, K.-S. (2006) Report on the occurrence of brown ring disease (BRD) in Manila clam, *Ruditapes philippinarum*, on the west coast of Korea. *Aquaculture* 255(1–4), 610–613.
1309 Park, S.-C., Baik, K.S., Kim, M.S., Kim, S.S., Kim, S.R., Oh, M.-J., Kim, D., Bang, B.-H. and Seong, C.N. (2009) *Aequorivita capsosiphonis* sp. nov., isolated from the green alga *Capsosiphon fulvescens*, and emended description of the genus *Aequorivita*. *International Journal of Systematic and Evolutionary Microbiology* 59(4), 724–728.
1310 Park, S.-C., Nakai, T. and Yuasa, A. (2002) Relationship between *in vitro* motility of *Pseudomonas plecoglossicida* and clinical conditions in affected ayu. *Fish Pathology* 37(3), 141–144.
1311 Park, S.-I., Wakabayashi, H. and Watanabe, Y. (1983) Serotype and virulence of *Edwardsiella tarda* isolated from eel and their environment. *Fish Pathology* 18(2), 85–89.
1312 Pârvulescu, L., Schrimpf, A., Kozubíková, E., Cabanillas Resino, S., Vrålstad, T., Petrusek, A. and Schulz, R. (2012) Invasive crayfish and crayfish plague on the move: first detection of the plague agent *Aphanomyces astaci* in the Romanian Danube. *Diseases of Aquatic Organisms* 98(1), 85–94.
1313 Pascual, C., Foster, G., Alvarez, N. and Collins, M.D. (1998) *Corynebacterium phocae* sp. nov., isolated from the common seal (*Phoca vitulina*). *International Journal of Systematic Bacteriology* 48, 601–604.
1314 Pascual, J., Macián, M.C., Arahal, D.R., Garay, E. and Pujalte, M.J. (2009) Description of *Enterovibrio nigricans* sp. nov., reclassification of *Vibrio calviensis* as *Enterovibrio calviensis* comb. nov. and emended description of the genus *Enterovibrio* Thompson *et al.* 2002. *International Journal of Systematic and Evolutionary Microbiology* 59(Pt 4), 698–704.
1315 Pascual, J., Macián, M.C., Arahal, D.R., Garay, E. and Pujalte, M.J. (2010) Multilocus sequence analysis of the central clade of the genus *Vibrio* by using the 16S rRNA, *recA*, *pyrH*, *rpoD*, *gyrB*, *rctB* and *toxR* genes. *International Journal of Systematic and Evolutionary Microbiology* 60(Pt 1), 154–165.
1316 Pasquale, V., Baloda, S.B., Dumontet, S. and Krovacek, K. (1994) An outbreak of *Aeromonas hydrophila* infection in turtles (*Pseudemis scripta*). *Applied and Environmental Microbiology* 60(5), 1678–1680.
1317 Pavan, M.E., Abbott, S.L., Zorzópulos, J. and Janda, J.M. (2000) *Aeromonas salmonicida* subsp. *pectinolytica* subsp. nov., a new pectinase-positive subspecies isolated from a heavily polluted river. *International Journal of Systematic and Evolutionary Microbiology* 50(3), 1119–1124.
1318 Payne, W., Eagon, R. and Williams, A. (1961) Some observations on the physiology of *Pseudomonas natriegens* nov. spec. *Antonie van Leeuwenhoek* 27(1), 121–128.
1319 Pazos, F., Santos, Y., Macías, A.R., Núñez, S. and Toranzo, A.E. (1996) Evaluation of media for the successful culture of *Flexibacter maritimus*. *Journal of Fish Diseases* 19(2), 193–197.
1320 Pearson, M.D., Hirono, I., Aoki, T., Miranda, R. and Inglis, V. (2000) Virulence properties of motile aeromonads isolated from farmed frogs *Rana tigerina* and *R. rugulosa*. *Diseases of Aquatic Organisms* 40(3), 185–193.
1321 Pedersen, K., Austin, B., Austin, D.A. and Larsen, J.L. (1999a) Vibrios associated with mortality in cultured plaice *Pleuronectes platessa* fry. *Acta Veterinaria Scandinavica* 40(3), 263–270.
1322 Pedersen, K., Dalsgaard, I. and Larsen, J.L. (1997) *Vibrio damsela* associated with diseased fish in Denmark. *Applied and Environmental Microbiology* 63(9), 3711–3715.
1323 Pedersen, K., Grisez, L., van Houdt, R., Tiainen, T., Ollevier, F. and Larsen, J.L. (1999b) Extended serotyping scheme for *Vibrio anguillarum* with the definition and characterization of seven provisional O-serogroups. *Current Microbiology* 38(3), 183–189.
1324 Pedersen, K., Kofod, H., Dalsgaard, I. and Larsen, J.L. (1994) Isolation of oxidase-negative *Aeromonas salmonicida* from diseased turbot *Scophthalmus maximus*. *Diseases of Aquatic Organisms* 18, 149–154.
1325 Pedersen, K., Skall, H.F., Lassen-Nielsen, A.M., Bjerrum, L. and Olesen, N.J. (2009) *Photobacterium damselae* subsp. *damselae*, an emerging pathogen in Danish rainbow trout, *Oncorhynchus mykiss* (Walbaum), mariculture. *Journal of Fish Diseases* 32(5), 465–472.
1326 Pedersen, K., Verdonck, L., Austin, B., Austin, D.A., Blanch, A.R., Grimont, P.A.D., Jofre, J., Koblavi, S., Larsen, J.L., Tiainen, T., Vigneulle, M. and Swings, J. (1998) Taxonomic evidence that *Vibrio carchariae* Grimes *et al.* 1985 is a junior synonym of *Vibrio harveyi* (Johnson and Shunk 1936) Baumann *et al.* 1981. *International Journal of Systematic Bacteriology* 48(3), 749–758.
1327 Pedersen, O.A. and Langvad, F. (1989) *Exophiala psychrophila* sp. nov., a pathogenic species of the black yeasts isolated from farmed Atlantic salmon. *Mycological Research* 92(2), 153–156.
1328 Pereira, W.L.A., de Souza, A.J.S., Gabriel, á M., Cardoso, A.M.C., Monger, S.G.B., Seligmann, I.C.A., Pereira, A.C.A. and Queiroz, D.K.S. (2012) Branchiomycosis in tambaqui, *Colossoma macropomum* (Cuvier), from the eastern Brazilian Amazon. *Journal of Fish Diseases* 35(8), 615–617.
1329 Perera, R.P., Johnson, S.K., Collins, M.D. and Lewis, D.H. (1994) *Streptococcus iniae* associated with mortality of *Tilapia nilotica* × *T. aurea* hybrids. *Journal of Aquatic Animal Health* 6(4), 335–340.

1330 Pérez, T., Balcázar, J.L., Peix, A., Valverde, A., Velázquez, E., de Blas, I. and Ruiz-Zarzuela, I. (2011) *Lactococcus lactis* subsp. *tructae* subsp. nov. isolated from the intestinal mucus of brown trout (*Salmo trutta*) and rainbow trout (*Oncorhynchus mykiss*). *International Journal of Systematic and Evolutionary Microbiology* 61(8), 1894–1898.

1331 Pessier, A.P., Nichols, D.K., Longcore, J.E. and Fuller, M.S. (1999) Cutaneous chytridiomycosis in poison dart frogs (*Dendrobates* spp.) and White's tree frogs (*Litoria caerulea*). *Journal of Veterinary Diagnostic Investigation* 11(2), 194–199.

1332 Petrie, J., Bruno, D.W. and Hastings, T.S. (1996) Isolation of *Yersinia ruckeri* from wild, Atlantic salmon, *Salmo salar* L., in Scotland. *Bulletin of the European Association of Fish Pathologists* 16(3), 83–84.

1333 Phadee, P., Kurata, O., Hatai, K., Hirono, I. and Aoki, T. (2004) Detection and identification of fish-pathogenic *Aphanomyces piscicida* using polymerase chain reaction (PCR) with species-specific primers. *Journal of Aquatic Animal Health* 16(4), 220–230.

1334 Phuoc, L.H., Corteel, M., Nauwynck, H.J., Pensaert, M.B., Alday-Sanz, V., Van den Broeck, W., Sorgeloos, P. and Bossier, P. (2008) Increased susceptibility of white spot syndrome virus-infected *Litopenaeus vannamei* to *Vibrio campbellii*. *Environmental Microbiology* 10(10), 2718–2727.

1335 Picard, F.J., Ke, D., Boudreau, D.K., Boissinot, M., Huletsky, A., Richard, D., Ouellette, M., Roy, P.H. and Bergeron, M.G. (2004) Use of *tuf* sequences for genus-specific PCR detection and phylogenetic analysis of 28 streptococcal species. *Journal of Clinical Microbiology* 42(8), 3686–3695.

1336 Pickett, M.J. and Nelson, E.L. (1955) Speciation within the genus *Brucella*. IV Fermentation of carbohydrates. *Journal of Bacteriology* 69(3), 333–336.

1337 Pickett, M.J. and Pedersen, M. (1970) Characterization of saccharolytic nonfermentative bacteria associated with man. *Canadian Journal of Microbiology* 16, 351–362.

1338 Pidiyar, V.J., Jangid, K., Dayananda, K.M., Kaznowski, A., Gonzalez, J.M., Patole, M.S. and Shouche, Y.S. (2003) Phylogenetic affiliation of *Aeromonas culicicola* MTCC 3249(T) based on *gyrB* gene sequence and PCR-amplicon sequence analysis of cytolytic enterotoxin gene. *Systematic and Applied Microbiology* 26(2), 197–202.

1339 Pidiyar, V., Kaznowski, A., Narayan, N.B., Patole, M. and Shouche, Y.S. (2002) *Aeromonas culicicola* sp. nov., from the midgut of *Culex quinquefasciatus*. *International Journal of Systematic and Evolutionary Microbiology* 52(5), 1723–1728.

1340 Pie, M., Boeger, W., Patella, L., Vicente, V., Ribeiro, R. and Ostrensky, A. (2011) Specific primers for the detection of the black-yeast fungus associated with lethargic crab disease (LCD). *Diseases of Aquatic Organisms* 94(1), 73–75.

1341 Pier, G.B. and Madin, S.H. (1976) *Streptococcus iniae* sp. nov., a beta-hemolytic Streptococcus isolated from an Amazon freshwater dolphin, *Inia geoffrensis*. *International Journal of Systematic Bacteriology* 26(4), 545–553.

1342 Pier, G.B., Madin, S.H. and Al-Nakeeb, S. (1978) Isolation and characterization of a second isolate of *Streptococcus iniae*. *International Journal of Systematic Bacteriology* 28(2), 311–314.

1343 Pilarski, F., Rossini, A.J. and Ceccarelli, P.S. (2008) Isolation and characterization of *Flavobacterium columnare* (Bernardet *et al.* 2002) from four tropical fish species in Brazil. *Brazilian Journal of Biology* 68(2), 409–414.

1344 Piñeiro-Vidal, M., Carballas, C.G., Gómez-Barreiro, O., Riaza, A. and Santos, Y. (2008a) *Tenacibaculum soleae* sp. nov., isolated from diseased sole (*Solea senegalensis* Kaup). *International Journal of Systematic and Evolutionary Microbiology* 58(Pt 4), 881–885.

1345 Piñeiro-Vidal, M., Centeno-Sestelo, G., Riaza, A. and Santos, Y. (2007) Isolation of pathogenic *Tenacibaculum maritimum*-related organisms from diseased turbot and sole cultured in the Northwest of Spain. *Bulletin of the European Association of Fish Pathologists* 27(1), 29–35.

1346 Piñeiro-Vidal, M. and, Gijón, D. and, Zarza, C. and Santos, Y. (2012) *Tenacibaculum dicentrarchi* sp. nov., a marine bacterium of the family *Flavobacteriaceae* isolated from European sea bass. *International Journal of Systematic and Evolutionary Microbiology* 62(Pt 2), 425–429.

1347 Piñeiro-Vidal, M. and, Riaza, A. and Santos, Y. (2008b) *Tenacibaculum discolor* sp. nov. and *Tenacibaculum gallaicum* sp. nov., isolated from sole (*Solea senegalensis*) and turbot (*Psetta maxima*) culture systems. *International Journal of Systematic and Evolutionary Microbiology* 58(Pt 1), 21–25.

1348 Plumb, J.A., Schachte, J.H., Gaines, J.L., Peltier, W. and Carroll, B. (1974) *Streptococcus* sp. from marine fishes along the Alabama and northwest Florida coast of the Gulf of Mexico. *Transactions of the American Fisheries Society* 103(2), 358–361.

1349 Plumb, J. and Sanchez, D.J. (1983) Susceptibility of five species of fish to *Edwardsiella ictaluri*. *Journal of Fish Diseases* 6, 261–266.

1350 Pollard, D.R., Johnson, W.M., Lior, H., Tyler, S.D. and Rozee, K.R. (1990) Detection of the aerolysin gene in *Aeromonas hydrophila* by the polymerase chain reaction. *Journal of Clinical Microbiology* 28(11), 2477–2481.

1351 Pollock, F.J., Morris, P.J., Willis, B.L. and Bourne, D.G. (2010) Detection and quantification of the coral pathogen *Vibrio coralliilyticus* by real-time PCR with TaqMan fluorescent probes. *Applied and Environmental Microbiology* 76(15), 5282–5286.

1352 Popoff, M. and Véron, M. (1976) A taxonomic study of the *Aeromonas hydrophila–Aeromonas punctata* group. *Journal of General Microbiology* 94(1), 11–22.

1353 Portaels, F., Fonteyene, P.A., de Beenhouwer, H., de Rijk, P., Guédénon, A., Hayman, J. and Meyers, M.W. (1996) Variability in 3′ end of 16S rRNA sequence of *Mycobacterium ulcerans* is related to geographic origin of isolates. *Journal of Clinical Microbiology* 34(4), 962–965.

1354 Pot, B., Devriese, L.A., Hommez, J., Miry, C., Vandemeulebroecke, K., Kersters, K. and Haesebrouck, F. (1994) Characterization and identification of *Vagococcus fluvialis* strains isolated from domestic animals. *Journal of Applied Microbiology* 77(4), 362–369.

1355 Pourahmad, F., Cervellione, F., Thompson, K.D., Taggart, J.B., Adams, A. and Richards, R.H. (2008) *Mycobacterium stomatepiae* sp. nov., a slowly growing, non-chromogenic species isolated from fish. *International Journal of Systematic and Evolutionary Microbiology* 58(12), 2821–2827.

1356 Pourahmad, F., Thompson, K.D., Adams, A. and Richards, R.H. (2009) Detection and identification of aquatic mycobacteria in formalin-fixed, paraffin-embedded fish tissues. *Journal of Fish Diseases* 32(5), 409–419.

1357 Prado, S., Romalde, J.L., Montes, J. and Barja, J.L. (2005) Pathogenic bacteria isolated from disease outbreaks in shellfish hatcheries. First description of *Vibrio neptunius* as an oyster pathogen. *Diseases of Aquatic Organisms* 67(3), 209–215.

1358 Preheim, S.P., Timberlake, S. and Polz, M.F. (2011) Merging taxonomy with ecological population prediction: a case study of *Vibrionaceae*. *Applied and Environmental Microbiology* 77(20), 7195–7206.

1359 Prol, M.J., Bruhn, J.B., Pintado, J. and Gram, L. (2009) Real-time PCR detection and quantification of fish probiotic *Phaeobacter* strain 27–4 and fish pathogenic *Vibrio* in microalgae, rotifer, *Artemia* and first feeding turbot (*Psetta maxima*) larvae. *Journal of Applied Microbiology* 106(4), 1292–1303.

1360 Pu, Z.Y., Dobos, M., Limsowtin, G.K.Y. and Powell, I.B. (2002) Integrated polymerase chain reaction-based procedures for the detection and identification of species and subspecies of the Gram-positive bacterial genus *Lactococcus*. *Journal of Applied Microbiology* 93(2), 353–361.

1361 Pujalte, M.J. and Garay, E. (1986) Proposal of *Vibrio mediterranei* sp. nov.: a new marine member of the genus *Vibrio*. *International Journal of Systematic Bacteriology* 36(2), 278–281.

1362 Pujalte, M.J., Macián, M.C., Arahal, D.R. and Garay, E. (2005) *Stappia alba* sp. nov., isolated from Mediterranean oysters. *Systematic and Applied Microbiology* 28(8), 672–678.

1363 Pujalte, M.J., Ortigosa, M., Urdaci, M.-C., Garay, E. and Grimont, P.A.D. (1993) *Vibrio mytili* sp. nov., from mussels. *International Journal of Systematic Bacteriology* 43(2), 358–362.

1364 Pujalte, M.J., Sitjà-Bobadilla, A., Macián, M.C., Belloch, C., Alvarez-Pellitero, P., Pérez-Sánchez, J., Uruburu, F. and Garay, E. (2003) Virulence and molecular typing of *Vibrio harveyi* strains isolated from cultured dentex, gilthead sea bream and European sea bass. *Systematic and Applied Microbiology* 26(2), 284–292.

1365 Puttinaowarat, S., Thompson, K.D., Kolk, A. and Adams, A. (2002) Identification of *Mycobacterium* spp. isolated from snakehead, *Channa striata* (Fowler), and Siamese fighting fish, *Betta splendens* (Regan), using polymerase chain reaction–reverse cross blot hybridization (PCR–RCBH). *Journal of Fish Diseases* 25(4), 235–243.

1366 Pychynski, T., Malanowska, T. and Kozlowski, M. (1981) Bacterial flora in branchionecrosis of carp (particularly *Bacillus cereus* and *Bacillus subtilis*). *Medycyna Weterynaryjna* 37, 742–743.

1367 Pye, G.W., Brown, D.R., Nogueira, M.F., Vliet, K.A., Schoeb, T.R., Jacobson, E.R. and Bennett, R.A. (2001) Experimental inoculation of broad-nosed caimans (*Caiman latirostris*) and Siamese crocodiles (*Crocodylus siamensis*) with *Mycoplasma alligatoris*. *Journal of Zoo and Wildlife Medicine* 32(2), 196–201.

1368 Raguénès, G., Christen, R., Guezennec, J., Pignet, P. and Barbier, G. (1997) *Vibrio diabolicus* sp. nov., a new polysaccharide-secreting organism isolated from a deep-sea hydrothermal vent polychaete annelid, *Alvinella pompejana*. *International Journal of Systematic Bacteriology* 47(4), 989–995.

1369 Rahe, R. and Soylu, E. (1989) Identification of the pathogenic fungus causing destruction to Turkish crayfish stocks (*Astacus leptodactylus*). *Journal of Invertebrate Pathology* 54(1), 10–15.

1370 Ramasamy, P., Rajan, P.R., Jayakumar, R., Rani, S. and Brennan, G.P. (1996) *Lagenidium callinectes* (Couch, 1942) infection and its control in cultured larval Indian tiger prawn, *Penaeus monodon* Fabricius. *Journal of Fish Diseases* 19(1), 75–82.

1371 Rameshkumar, N., Fukui, Y., Sawabe, T. and Nair, S. (2008) *Vibrio porteresiae* sp. nov., a diazotrophic bacterium isolated from a mangrove-associated wild rice (*Porteresia coarctata* Tateoka). *International Journal of Systematic and Evolutionary Microbiology* 58(Pt 7), 1608–1615.

1372 Rameshkumar, N., Gomez-Gil, B., Spröer, C., Lang, E., Dinesh Kumar, N., Krishnamurthi, S., Nair, S. and Roque, A. (2011) *Vibrio plantisponsor* sp. nov., a diazotrophic bacterium isolated from a mangrove associated wild rice (*Porteresia coarctata* Tateoka). *Systematic and Applied Microbiology* 34(7), 487–493.

1373 Rameshkumar, N., Sproer, C., Lang, E. and Nair, S. (2010) *Vibrio mangrovi* sp. nov., a diazotrophic bacterium isolated from mangrove-associated wild rice (*Poteresia coarctata Tateoka*). *FEMS Microbiology Letters* 307(1), 35–40.

1374 Ramos, C.P., Foster, G. and Collins, M.D. (1997) Phylogenetic analysis of the genus *Actinomyces* based on 16S rRNA gene sequences: description of *Arcanobacterium phocae* sp. nov., *Arcanobacterium bernardiae* comb. nov. and *Arcanobacterium pyogenes* comb. nov. *International Journal of Systematic Bacteriology* 47(1), 46–53.

1375 Rand, T.G., Bunkley-Williams, L. and Williams, E.H. (2000) A hyphomycete fungus, *Paecilomyces lilacinus*, associated with wasting disease in two species of tilapia from Puerto Rico. *Journal of Aquatic Animal Health* 12(2), 149–156.

1376 Ransangan, J. and Mustafa, S. (2009) Identification of *Vibrio harveyi* isolated from diseased Asian seabass *Lates calcarifer* by use of 16S ribosomal DNA sequencing. *Journal of Aquatic Animal Health* 21(3), 150–155.

1377 Ransom, D.P., Lannan, C.N., Rohovec, J.S. and Fryer, J.L. (1984) Comparison of histopathology caused by *Vibrio anguillarum* and *Vibrio ordalii* in three species of Pacific salmon. *Journal of Fish Diseases* 7(2), 107–115.

1378 Rasheed, V., Limsuwan, C. and Plumb, J. (1985) Histopathology of bullminnows, *Fundulus grandis* Baird & Girard, infected with a non-haemolytic group B *Streptococcus* sp. *Journal of Fish Diseases* 8(1), 65–74.

1379 Rattanama, P., Srinitiwarawong, K., Thompson, J.R., Pomwised, R., Supamattaya, K. and Vuddhakul, V. (2009) Shrimp pathogenicity, hemolysis, and the presence of hemolysin and TTSS genes in *Vibrio harveyi* isolated from Thailand. *Diseases of Aquatic Organisms* 86(2), 113–122.

1380 Ravelo, C., Magariños, B., Romalde, J. and Toranzo, A.E. (2001) Conventional versus miniaturized systems for the phenotypic characterization of *Lactococcus garvieae* strains. *Bulletin of the European Association of Fish Pathologists* 21(4), 136–144.

1381 Razin, S. and Tully, J.G. (1983) Methods in Mycoplasmology. Vol. I: Mycoplasma characterisation. Academic Press, New York.

1382 Reddacliff, G.L., Hornitzky, M., Carson, J., Petersen, R. and Zelski, R. (1993) Mortalities of goldfish, *Carassius auratus* (L.), associated with *Vibrio cholerae* (non-01) infection. *Journal of Fish Diseases* 16(5), 517–520.

1383 Reddacliff, G.L., Hornitzky, M. and Whittington, R.J. (1996) *Edwardsiella tarda* septicaemia in rainbow trout (*Oncorhynchus mykiss*). *Australian Veterinary Journal* 73(1), 30.

1384 Reddy, C.A., Cornell, C.P. and Fraga, A.M. (1982) Transfer of *Corynebacterium pyogenes* (Glage) Eberson to the genus *Actinomyces* as *Actinomyces pyogenes* (Glage) comb. nov. *International Journal of Systematic Bacteriology* 32(4), 419–429.

1385 Reed, L.J. and Muench, H. (1938) A simple method of estimating fifty per cent endpoints. *American Journal of Epidemiology* 27(3), 493–497.

1386 Register, K.B., Sacco, R.E. and Foster, G. (2000) Ribotyping and restriction endonuclease analysis reveal a novel clone of *Bordetella bronchiseptica* in seals. *Journal of Veterinary Diagnostic Investigation* 12(6), 535–540.

1387 Rehulka, J., Marejková, M. and Petráš, P. (2012) Edwardsiellosis in farmed rainbow trout (*Oncorhynchus mykiss*). *Aquaculture Research* 43(11), 1628–1634.

1388 Reichelt, J.L. and Baumann, P. (1975) *Photobacterium mandapamensis* Hendrie *et al.*, a later subjective synonym of *Photobacterium leiognathi* Boisvert *et al. International Journal of Systematic Bacteriology* 25(2), 208–209.

1389 Reichelt, J.L., Baumann, P. and Baumann, L. (1976) Study of genetic relationships among marine species of the genera *Beneckea* and *Photobacterium* by means of *in vitro* DNA/DNA hybridization. *Archives of Microbiology* 110(1), 101–120.

1390 Reichenbach, H., Kleinig, H. and Achenbach, H. (1974) The pigments of *Flexibacter elegans*: novel and chemosystematically useful compounds. *Archives of Microbiology* 101(1), 131–144.

1391 Reichenbach, H., Kohl, W. and Achenbach, H. (1981) The flexirubin-type pigments, chemosystematically useful compounds. In: Reichenbach, H. and Weeks, O.B. (eds) The Flavobacterium-Cytophaga Group. Proceedings of the International Symposium on Yellow-Pigmented Gram-Negative Bacteria of the Flavobacterium-Cytophaga Group, 11 July, Braunschweig-Stöckheim, Germany). GBF Monograph Series no. 5.

1392 Reid, G.A. and Gordon, E.H. (1999) Phylogeny of marine and freshwater *Shewanella*: reclassification of *Shewanella putrefaciens* NCIMB 400 as *Shewanella frigidimarina*. *International Journal of Systematic Bacteriology* 49(Pt 1), 189–191.

1393 Reid, H., Duncan, H., Laidler, L.A., Hunter, D. and Birkbeck, T.H. (2003a) Isolation of *Vibrio tapetis* from cultivated halibut (*Hippoglossus hippoglossus* L.). *Aquaculture* 221, 65–74.

1394 Reid, H.I., Soudant, P., Lambert, C., Paillard, C. and Birkbeck, T.H. (2003b) Salinity effects on immune parameters of *Ruditapes philippinarum* challenged with *Vibrio tapetis*. *Diseases of Aquatic Organisms* 56(3), 249–258.

1395 Reid, H.I., Treasurer, J.W., Adam, B. and Birkbeck, T.H. (2009) Analysis of bacterial populations in the gut of developing cod larvae and identification of *Vibrio logei*, *Vibrio anguillarum* and *Vibrio splendidus* as pathogens of cod larvae. *Aquaculture* 288(1–2), 36–43.

1396 Renault, T., Haffner, P., Malfondet, C. and Weppe, M. (1994) *Vibrio damsela* as a pathogenic agent causing mortalities in cultured sea bass (*Lates calcarifer*). *Bulletin of the European Association of Fish Pathologists* 14(4), 117–119.

1397 Reuter, R.E., Hutchinson, W., Ham, J. and Davis, S. (2003) *Exophiala* sp. infection in captured King George whiting (*Sillaginodes punctata*). *Bulletin of the European Association of Fish Pathologists* 23(3), 128–134.

1398 Rhodes, L.D., Durkin, C., Nance, S.L. and Rice, C.A. (2006) Prevalence and analysis of *Renibacterium salmoninarum* infection among juvenile Chinook salmon *Oncorhynchus tshawytscha* in North Puget Sound. *Diseases of Aquatic Organisms* 71(3), 179–190.

1399 Rhodes, M.W., Kator, H., Kaattari, I., Gauthier, D., Vogelbein, W. and Ottinger, C.A. (2004) Isolation and characterization of mycobacteria from striped bass *Morone saxatilis* from the Chesapeake Bay. *Diseases of Aquatic Organisms* 61(1–2), 41–51.

1400 Rhodes, M.W., Kator, H., Kotob, S., van Berkum, P., Kaattari, I., Vogelbein, W., Quinn, F., Floyd, M.M., Butler, W.R. and Ottinger, C.A. (2003) *Mycobacterium shottsii* sp. nov., a slowly growing species isolated from Chesapeake Bay striped bass (*Morone saxatilis*). *International Journal of Systematic and Evolutionary Microbiology* 53(Pt 2), 421–424.

1401 Rhodes, M.W., Kator, H., McNabb, A., Deshayes, C., Reyrat, J.-M., Brown-Elliott, B.A., Wallace, R. Jr, Trott, K.A., Parker, J.M., Lifland, B., Osterhout, G., Kaattari, I., Reece, K., Vogelbein, W. and Ottinger, C.A. (2005) *Mycobacterium pseudoshottsii* sp. nov., a slowly growing chromogenic species isolated from Chesapeake Bay striped bass (*Morone saxatilis*). *International Journal of Systematic and Evolutionary Microbiology* 55(Pt 3), 1139–1147.

1402 Richards, G.P., Watson, M.A., Crane, E.J., Burt, I.G. and Bushek, D. (2008) *Shewanella* and *Photobacterium* spp. in oysters and seawater from the Delaware Bay. *Applied and Environmental Microbiology* 74(11), 3323–3327.

1403 Richards, R.H., Holliman, A. and Helgason, S. (1978) *Exophiala salmonis* infection in Atlantic salmon *Salo salar* L. *Journal of Fish Diseases* 1, 357–368.

1404 Ringø, E. and Gatesoupe, F.-J. (1998) Lactic acid bacteria in fish: a review. *Aquaculture* 160(3–4), 177–203.

1405 Ringø, E., Seppola, M., Berg, A., Olsen, R.E., Schillinger, U. and Holzapfel, W. (2002) Characterization of *Carnobacterium divergens* strain 6251 isolated from intestine of Arctic charr (*Salvelinus alpinus* L.). *Systematic and Applied Microbiology* 25(1), 120–129.
1406 Ringø, E.,, Sinclair, P.D., Birkbeck, H. and Barbour, A. (1992) Production of eicosapentaenoic acid (20:5 n-3) by *Vibrio pelagius* isolated from turbot (*Scophthalmus maximus* (L.) larvae. *Applied and Environmental Microbiology* 58(11), 3777–3778.
1407 Ringuet, H., Akoua-Koffi, C., Honore, S., Varnerot, A., Vincent, V., Berche, P., Gaillard, J.L. and Pierre-Audigier, C. (1999) hsp65 sequencing for identification of rapidly growing mycobacteria. *Journal of Clinical Microbiology* 37(3), 852–857.
1408 Ripabelli, G., Sammarco, M.L., Grasso, G.M., Fanelli, I., Caprioli, A. and Luzzi, I. (1999) Occurrence of *Vibrio* and other pathogenic bacteria in *Mytilus galloprovincialis* (mussels) harvested from Adriatic Sea, Italy. *International Journal of Food Microbiology* 49(1–2), 43–48.
1409 Riquelme, C., Toranzo, A.E., Barja, J.L., Vergara, N. and Araya, R. (1996) Association of *Aeromonas hydrophila* and *Vibrio alginolyticus* with larval mortalities of scallop (*Argopecten purpuratus*). *Journal of Invertebrate Pathology* 67(3), 213–218.
1410 Rivas, A.J., Balado, M., Lemos, M.L. and Osorio, C.R. (2011) The *Photobacterium damselae* subsp. *damselae* hemolysins damselysin and HlyA are encoded within a new virulence plasmid. *Infection and Immunity* 79(11), 4617–4627.
1411 Rivera-Posada, J.A., Pratchett, M., Cano-Gomez, A., Arango-Gomez, J.D. and Owens, L. (2011) Refined identification of *Vibrio* bacterial flora from *Acanthasther planci* based on biochemical profiling and analysis of housekeeping genes. *Diseases of Aquatic Organisms* 96(2), 113–123.
1412 Robert-Pillot, A., Guenole, A. and Fournier, J.M. (2002) Usefulness of R72H PCR assay for differentiation between *Vibrio parahaemolyticus* and *Vibrio alginolyticus* species: validation by DNA–DNA hybridization. *FEMS Microbiology Letters* 215(1), 1–6.
1413 Roberts, R.J., Willoughby, L.G. and Chinabut, S. (1993) Mycotic aspects of epizootic ulcerative syndrome (EUS) of Asian fishes. *Journal of Fish Diseases* 16(3), 169–183.
1414 Robertson, L.A., Figge, M.J. and Dunlap, P.V. (2011) Beijerinck and the bioluminescent bacteria: microbiological experiments in the late 19th and early 20th centuries. *FEMS Microbiology Ecology* 75(2), 185–194.
1415 Robinson, J.A. and Meyer, F.P. (1966) Streptococcal fish pathogen. *Journal of Bacteriology* 92(2), 512.
1416 Röder, B., Fruhwirth, K., Vogl, C., Wagner, M. and Rossmanith, P. (2010) Impact of long-term storage on stability of standard DNA for nucleic acid-based methods. *Journal of Clinical Microbiology* 48(11), 4260–4262.
1417 Rodríguez, J.M., López-Romalde, S., Beaz, R., Alonso, M.C., Castro, D. and Romalde, J.L. (2006) Molecular fingerprinting of *Vibrio tapetis* strains using three PCR-based methods: ERIC-PCR, REP-PCR and RAPD. *Diseases of Aquatic Organisms* 69(2–3), 175–183.
1418 Rodríguez, L.A., Gallardo, C.S., Acosta, F., Nieto, T.P., Acosta, B. and Real, F. (1998) *Hafnia alvei* as an opportunistic pathogen causing mortality in brown trout, *Salmo trutta* L. *Journal of Fish Diseases* 21(5), 365–370.
1419 Rogers-Talbert, R. (1948) The fungus *Lagenidium callinectes* Couch (1942) on eggs of the blue crab in Chesapeake Bay. *The Biological Bulletin* 95(2), 214–228.
1420 Roggenkamp, A., Abele-Horn, M., Trebesius, K.-H., Tretter, U., Autenrieth, I.B. and Heesemann, J. (1998) *Abiotrophia elegans* sp. nov., a possible pathogen in patients with culture-negative endocarditis. *Journal of Clinical Microbiology* 36(1), 100–104.
1421 Rogosa, M., Mitchell, J.A. and Wiseman, R.F. (1951) A selective medium for the isolation and enumeration of oral and fecal lactobacilli. *Journal of Bacteriology* 62(1), 132–133.
1422 Rokas, A., Williams, B.L., King, N. and Carroll, S.B. (2003) Genome-scale approaches to resolving incongruence in molecular phylogenies. *Nature* 425(6960), 798–804.
1423 Romalde, J.L., Magarinos, B., Fouz, B., Bandín, I., Núñez, S. and Toranzo, A.E. (1995) Evaluation of BIONOR Mono-kits for rapid detection of bacterial fish pathogens. *Diseases of Aquatic Organisms* 21(1), 25–34.
1424 Romalde, J.L., Ravelo, C., Valdés, I., Magariños, B., de la Fuente, E., Martín, C.S., Avendaño-Herrera, R. and Toranzo, A.E. (2008) *Streptococcus phocae*, an emerging pathogen for salmonid culture. *Veterinary Microbiology* 130(1–2), 198–207.
1425 Romalde, J.L. and Toranzo, A.E. (1991) Evaluation of the API-20E system for the routine diagnosis of the enteric redmouth disease. *Bulletin of the European Association of Fish Pathologists* 11(4), 147–149.
1426 Romansic, J.M., Diez, K.A., Higashi, E.M., Johnson, J.E. and Blaustein, A.R. (2009) Effects of the pathogenic water mold *Saprolegnia ferax* on survival of amphibian larvae. *Diseases of Aquatic Organisms* 83(3), 187–193.
1427 Ross, A.J. (1960) *Mycobacterium salmoniphilum* sp. nov. from salmonid fishes. *The American Review of Respiratory Disease* 81, 241–250.
1428 Ross, A.J. and Brancato, F.P. (1959) *Mycobacterium fortuitum* Cruz from the tropical fish *Hyphessobrycon Innesi*. *Journal of Bacteriology* 78(3), 392–395.
1429 Ross, A.J., Rucker, R.R. and Ewing, W.H. (1966) Description of a bacterium associated with redmouth disease of rainbow trout (*Salmo gairdneri*). *Canadian Journal of Microbiology* 12(4), 763–770.
1430 Ross, H.M., Foster, G., Reid, R.J., Jahans, K.L. and MacMillan, A.P. (1994) *Brucella* species infection in sea-mammals. *The Veterinary Record* 134(14), 359.
1431 Rowland, S.J., Landos, M., Callinan, R.B., Allan, G., Read, P., Misfud, C., Nixon, M., Boyd, P. and Tully, P. (2007) Development of a health management strategy for the silver perch aquaculture industry. FRDC Project Nos. 2000/267 and 2004/089. *NSW Department of Primary Industries*.
1432 Royo, F., Andersson, G., Bangyeekhun, E., Múzquiz, J.L., Söderhäll, K. and Cerenius, L. (2004) Physiological and genetic characterisation of some new *Aphanomyces* strains isolated from freshwater crayfish. *Veterinary Microbiology* 104(1–2), 103–112.

1433 Rozas, J. and Rozas, R. (1999) DnaSP version 3: an integrated program for molecular population genetics and molecular evolution analysis. *Bioinformatics (Oxford, England)* 15(2), 174–175.

1434 Ruangpan, L. and Kitao, T. (1991) *Vibrio* bacteria isolated from black tiger shrimp, *Penaeus monodon* Fabricius. *Journal of Fish Diseases* 14(3), 383–388.

1435 Ruby, E.G. and Nealson, K.H. (1976) Symbiotic association of *Photobacterium fischeri* with the marine luminous fish *Monocentris japonica*; a model of symbiosis based on bacterial studies. *The Biological Bulletin* 151(3), 574–586.

1436 Rucker, R.R. (1949) A streptomycete pathogenic to fish. *Journal of Bacteriology* 58(5), 659–664.

1437 Rüger, H.-J. and Tan, T.L. (1983) Separation of *Alcaligenes denitrificans* sp. nov., nom. rev. from *Alcaligenes faecalis* on the basis of DNA base composition, DNA homology, and nitrate reduction. *International Journal of Systematic Bacteriology* 33(1), 85–89.

1438 Ruhnke, H.L. and Madoff, S. (1992) *Mycoplasma phocidae* sp. nov., isolated from harbor seals (*Phoca vitulina* L.). *International Journal of Systematic Bacteriology* 42(2), 211–214.

1439 Ruimy, R., Breittmayer, V., Elbaze, P., Lafay, B., Boussemart, O., Gauthier, M. and Christen, R. (1994) Phylogenetic analysis and assessment of the genera *Vibrio, Photobacterium, Aeromonas*, and *Plesiomonas* deduced from small-subunit rRNA sequences. *International Journal of Systematic Bacteriology* 44(3), 416–426.

1440 Ruimy, R., Riegel, P., Carlotti, A., Boiron, P., Bernardin, G., Monteil, H., Wallace, R.J. Jr and Christen, R. (1996) *Nocardia pseudobrasiliensis* sp. nov., a new species of *Nocardia* which groups bacterial strains previously identified as *Nocardia brasiliensis* and associated with invasive diseases. *International Journal of Systematic Bacteriology* 46(1), 259–264.

1441 Ruiz-Ponte, C., Cilia, V., Lambert, C. and Nicolas, J.L. (1998) *Roseobacter gallaeciensis* sp. nov., a new marine bacterium isolated from rearings and collectors of the scallop *Pecten maximus*. *International Journal of Systematic Bacteriology* 48(Pt 2), 537–542.

1442 Ruiz-Zarzuela, I., de Bias, I., Gironés, O., Ghittino, C. and Múazquiz, J.L. (2005) Isolation of *Vagococcus salmoninarum* in rainbow trout, *Oncorhynchus mykiss* (Walbaum), broodstocks: characterization of the pathogen. *Veterinary Research Communications* 29(7), 553–562.

1443 Runyon, E.H. (1959) Anonymous mycobacteria in pulmonary disease. *The Medical Clinics of North America* 43(1), 273–290.

1444 Ruthig, G.R. (2009) Water molds of the genera *Saprolegnia* and *Leptolegnia* are pathogenic to the North American frogs *Rana catesbeiana* and *Pseudacris crucifer*, respectively. *Diseases of Aquatic Organisms* 84(3), 173–178.

1445 Rutter, J.M. (1981) Quantitative observations on *Bordetella bronchiseptica* infection in atrophic rhinitis of pigs. *The Veterinary Record* 108(21), 451–454.

1446 Rutter, J.M., Taylor, R.J., Crighton, W.G., Robertson, I.B. and Benson, J.A. (1984) Epidemiological study of *Pasteurella multocida* and *Bordetella bronchiseptica* in atrophic rhinitis. *The Veterinary Record* 115(24), 615–619.

1447 Ruwandeepika, H.A.D., Defoirdt, T., Bhowmick, P.P., Karunasagar, I., Karunasagar, I. and Bossier, P. (2011) *In vitro* and *in vivo* expression of virulence genes in *Vibrio* isolates belonging to the Harveyi clade in relation to their virulence towards gnotobiotic brine shrimp (*Artemia franciscana*). *Environmental Microbiology* 13(2), 506–517.

1448 Ruwandeepika, H.A.D., Defoirdt, T., Bhowmick, P.P., Shekar, M., Bossier, P. and Karunasagar, I. (2010) Presence of typical and atypical virulence genes in vibrio isolates belonging to the Harveyi clade. *Journal of Applied Microbiology* 109(3), 888–899.

1449 Saavedra, M.J., Figueras, M.J. and Martínez-Murcia, A.J. (2006) Updated phylogeny of the genus *Aeromonas*. *International Journal of Systematic and Evolutionary Microbiology* 56(Pt 10), 2481–2487.

1450 Saavedra, M.J., Perea, V., Fontes, M.C., Martins, C. and Martínez-Murcia, A. (2007) Phylogenetic identification of *Aeromonas* strains isolated from carcasses of pig as new members of the species *Aeromonas allosaccharophila*. *Antonie Van Leeuwenhoek* 91(2), 159–167.

1451 Sack, D.A., Sack, R.B., Nair, G.B. and Siddique, A. (2004) Cholera. *The Lancet* 363(9404), 223–233.

1452 Saeed, M.O. (1995) Association of *Vibrio harveyi* with mortalities in cultured marine fish in Kuwait. *Aquaculture* 136, 21–29.

1453 Saeed, M.O., Alamoudi, M.M. and Al-Harbi, A.H. (1987) A *Pseudomonas* associated with disease in cultured rabbitfish *Siganus rivulatus* in the Red Sea. *Diseases of Aquatic Organisms* 3, 177–180.

1454 Saha, P. and Chakrabarti, T. (2006) *Aeromonas sharmana* sp. nov., isolated from a warm spring. *International Journal of Systematic and Evolutionary Microbiology* 56(Pt 8), 1905–1909.

1455 Saha, R., Spröer, C., Beck, B. and Bagley, S. (2010) *Pseudomonas oleovorans* subsp. *lubricantis* subsp. nov. and reclassification of *Pseudomonas pseudoalcaligenes* ATCC 17440T as later synonym of *Pseudomonas oleovorans* ATCC 8062T. *Current Microbiology* 60(4), 294–300.

1456 Saiki, R. (1989) The design and optimisation of the PCR. In:PCR Technology: Principles and Applications for DNA Amplification. Stockton Press, New York.

1457 Saiki, R.K., Scharf, S., Faloona, F., Mullis, K.B., Horn, G.T., Erlich, H.A. and Arnheim, N. (1985) Enzymatic amplification of beta-globin genomic sequences and restriction site analysis for diagnosis of sickle cell anemia. *Science* 230(4732), 1350–1354.

1458 Saitou, N. and Nei, M. (1987) The neighbor-joining method: a new method for reconstructing phylogenetic trees. *Molecular Biology and Evolution* 4(4), 406–425.

1459 Sakai, M. and Kobayashi, M. (1992) Detection of *Renibacterium salmoninarum*, the causative agent of bacterial kidney disease in salmonid fish, from pen-cultured coho salmon. *Applied and Environmental Microbiology* 58(3), 1061–1063.

1460 Sakai, T., Iida, T., Osatomi, K. and Kanai, K. (2007) Detection of type 1 fimbrial genes in fish pathogenic

and non-pathogenic *Edwardsiella tarda* strains by PCR. *Fish Pathology* 42(2), 115–117.
1461 Sakata, T., Matsuura, M. and Shimokawa, Y. (1989) Characteristics of *Vibrio damsela* isolated from diseased yellowtail *Seriola quinqueradiata*. *Nippon Suisan Gakkaishi* 55(1), 135–141.
1462 Sakazaki, R. (1968) Proposal of *Vibrio alginolyticus* for the biotype 2 of *Vibrio parahaemolyticus*. *Japanese Journal of Medical Science and Biology* 21(5), 359–362.
1463 Sakazaki, R. (2005) *Hafnia alvei*. In: Brenner, D.J., Krieg, N.R., Staley, J.T. and Garrity, G.M. (eds) *Bergey's Manual of Systematic Bacteriology. The Proteobacteria*, 2nd edn, Vol. 2. Springer, New York, pp. 681–685.
1464 Sakazaki, R., Iwanami, S. and Fukumi, H. (1963) Studies on the enteropathogenic, facultatively halophilic bacteria, *Vibrio parahaemolyticus* I. Morphological, cultural and biochemical properties and its taxonomical position. *Japanese Journal of Medical Science and Biology* 16(4), 161–188.
1465 Sakurai, J., Matsuzaki, A., Takeda, Y. and Miwatani, T. (1974) Existence of two distinct hemolysins in *Vibrio parahaemolyticus*. *Infection and Immunity* 9(5), 777–780.
1466 Salati, F., Tassi, P. and Bronzi, P. (1996) Isolation of an *Enterococcus*-like bacterium from diseased Adriatic sturgeon, *Acipenser naccarii*, farmed in Italy. *Bulletin of the European Association of Fish Pathologists* 16(3), 96–100.
1467 Saleh, M., Soliman, H. and El-Matbouli, M. (2008) Loop-mediated isothermal amplification as an emerging technology for detection of *Yersinia ruckeri* the causative agent of enteric red mouth disease in fish. *BMC Veterinary Research* 4, 31.
1468 Salogni, C., Perantoni, P., Pitozzi, A. and Alborali, G. (2007) Outbreak of *Vagococcus salmoninarum* in rainbow trout (*Oncorhynchus mykiss*) broodstocks. *Ittiopatologia* 4, 59–66.
1469 Sambrook, J., Fritsch, E. and Maniatis, T. (1989) Molecular Cloning. A Laboratory Manual. Cold Spring Harbor Laboratory Press, Cold Spring Harbor, New York.
1470 Sandell, T.A. and Jacobson, K.C. (2011) Comparison and evaluation of *Renibacterium salmoninarum* quantitative PCR diagnostic assays using field samples of Chinook and coho salmon. *Diseases of Aquatic Organisms* 93(2), 129–139.
1471 Sanders, J.E. and Barros, M.J. (1986) Evidence by the fluorescent antibody test for the occurrence of *Renibacterium salmoninarum* among salmonid fish in Chile. *Journal of Wildlife Diseases* 22(2), 255–257.
1472 Sanders, J.E. and Fryer, J.L. (1980) *Renibacterium salmoninarum* gen. nov., sp. nov., the causative agent of bacterial kidney disease in salmonid fishes. *International Journal of Systematic Bacteriology* 30(2), 496–502.
1473 Sandlund, N., Torkildsen, L., Magnesen, T., Mortensen, S. and Bergh, O. (2006) Immunohistochemistry of great scallop *Pecten maximus* larvae experimentally challenged with pathogenic bacteria. *Diseases of Aquatic Organisms* 69(2–3), 163–173.
1474 Sandoz, M.D., Rogers, R. and Newcombe, C.L. (1944) Fungus infection of eggs of the blue crab *Callinectes sapidus* Rathbun. *Science* 99(2563), 124–125.
1475 Santacana, J.A., Conroy, D.A., Mujica, M.E., Marin, C. and Lopez, N. (1982) Acid-fast bacterial infection and its control in three-spot gouramies, *Trichogaster trichopterus* Pallas. *Journal of Fish Diseases* 5, 545–547.
1476 Santos, Y., Romalde, J.L., Bandín, I., Magariños, B., Núñez, S., Barja, J.L. and Toranzo, A.E. (1993) Usefulness of the API-20E system for the identification of bacterial fish pathogens. *Aquaculture* 116(2–3), 111–120.
1477 Sarmiento-Ramírez, J.M., Abella, E., Martín, M.P., Tellería, M.T., López-Jurado, L.F., Marco, A. and Diéguez-Uribeondo, J. (2010) *Fusarium solani* is responsible for mass mortalities in nests of loggerhead sea turtle, *Caretta caretta*, in Boavista, Cape Verde. *FEMS Microbiology Letters* 312(2), 192–200.
1478 Sato, N., Yamane, N. and Kawamura, T. (1982) Systemic *Citrobacter freundii* infection among sunfish *Mola mola* in Matsushima aquarium. *Bulletin of the Japanese Society of Scientific Fisheries* 48(11), 1551–1557.
1479 Satomi, M., Oikawa, H. and Yano, Y. (2003) *Shewanella marinintestina* sp. nov., *Shewanella schlegeliana* sp. nov. and *Shewanella sairae* sp. nov., novel eicosapentaenoic-acid-producing marine bacteria isolated from sea-animal intestines. *International Journal of Systematic and Evolutionary Microbiology* 53(Pt 2), 491–499.
1480 Satomi, M., Vogel, B.F., Gram, L. and Venkateswaran, K. (2006) *Shewanella hafniensis* sp. nov. and *Shewanella morhuae* sp. nov., isolated from marine fish of the Baltic Sea. *International Journal of Systematic and Evolutionary Microbiology* 56(Pt 1), 243–249.
1481 Satomi, M., Vogel, B.F., Venkateswaran, K. and Gram, L. (2007) Description of *Shewanella glacialipiscicola* sp. nov. and *Shewanella algidipiscicola* sp. nov., isolated from marine fish of the Danish Baltic Sea, and proposal that *Shewanella affinis* is a later heterotypic synonym of *Shewanella colwelliana*. *International Journal of Systematic and Evolutionary Microbiology* 57(Pt 2), 347–352.
1482 Saulnier, D., Avarre, J.C., Le Moullac, G., Ansquer, D., Levy, P. and Vonau, V. (2000) Rapid and sensitive PCR detection of *Vibrio penaeicida*, the putative etiological agent of syndrome 93 in New Caledonia. *Diseases of Aquatic Organisms* 40(2), 109–115.
1483 Saulnier, D., De Decker, S. and Haffner, P. (2009) Real-time PCR assay for rapid detection and quantification of *Vibrio aestuarianus* in oyster and seawater: a useful tool for epidemiologic studies. *Journal of Microbiological Methods* 77(2), 191–197.
1484 Savas, H., Altinok, I., Cakmak, E. and Firidin, S. (2006) Isolation of *Renibacterium salmoninarum* from cultured Black Sea salmon (*Salmo trutta labrax*), first report in Turkey. *Bulletin of the European Association of Fish Pathologists* 26(6), 238–246.
1485 Sawabe, T., Fujimura, Y., Niwa, K. and Aono, H. (2007a) *Vibrio comitans* sp. nov., *Vibrio rarus* sp. nov. and *Vibrio inusitatus* sp. nov., from the gut of the abalones *Haliotis discus discus*, *H. gigantea*, *H. madaka* and *H. rufescens*. *International Journal of Systematic and Evolutionary Microbiology* 57(Pt 5), 916–922.

1486 Sawabe, T., Hayashi, K., Moriwaki, J., Fukui, Y., Thompson, F.L., Swings, J. and Christen, R. (2004a) *Vibrio neonatus* sp. nov. and *Vibrio ezurae* sp. nov. isolated from the gut of Japanese abalones. *Systematic and Applied Microbiology* 27(5), 527–534.

1487 Sawabe, T., Hayashi, K., Moriwaki, J., Thompson, F.L., Swings, J., Potin, P., Christen, R. and Ezura, Y. (2004b) *Vibrio gallicus* sp. nov., isolated from the gut of the French abalone *Haliotis tuberculata*. *International Journal of Systematic and Evolutionary Microbiology* 54(Pt 3), 843–846.

1488 Sawabe, T., Kita-Tsukamoto, K. and Thompson, F.L. (2007b) Inferring the evolutionary history of vibrios by means of multilocus sequence analysis. *Journal of Bacteriology* 189(21), 7932–7936.

1489 Sawabe, T., Makino, H., Tatsumi, M., Nakano, K., Tajima, K., Iqbal, M.M., Yumoto, I., Ezura, Y. and Christen, R. (1998) *Pseudoalteromonas bacteriolytica* sp. nov., a marine bacterium that is the causative agent of red spot disease of *Laminaria japonica*. *International Journal of Systematic Bacteriology* 48(Pt 3), 769–774.

1490 Sawabe, T., Sugimura, I., Ohtsuka, M., Nakano, K., Tajima, K., Ezura, Y. and Christen, R. (1998) *Vibrio halioticoli* sp. nov., a non-motile alginolytic marine bacterium isolated from the gut of the abalone *Haliotis discus hannai*. *International Journal of Systematic Bacteriology* 48(Pt 2), 573–580.

1491 Sawabe, T., Tanaka, R., Iqbal, M.M., Tajima, K., Ezura, Y., Ivanova, E.P. and Christen, R. (2000) Assignment of *Alteromonas elyakovii* KMM 162T and five strains isolated from spot-wounded fronds of *Laminaria japonica* to *Pseudoalteromonas elyakovii* comb. nov. and the extended description of the species. *International Journal of Systematic and Evolutionary Microbiology* 50(Pt 1), 265–271.

1492 Saylor, R.K., Miller, D.L., Vandersea, M.W., Bevelhimer, M.S., Schofield, P.J. and Bennett, W.A. (2010) Epizootic ulcerative syndrome caused by *Aphanomyces invadans* in captive bullseye snakehead *Channa marulius* collected from south Florida, USA. *Diseases of Aquatic Organisms* 88(2), 169–175.

1493 Schiewe, M.H., Trust, T.J. and Crosa, J.H. (1981) *Vibrio ordalii* sp. nov.: a causative agent of vibriosis in fish. *Current Microbiology* 6(6), 343–348.

1494 Schikora, F. (1906) Die Krebspest. *Fischerei-Zeitung* 9, 529–532, 561–566, 581–583.

1495 Schleifer, K.H. and Kloos, W.E. (1975) Isolation and characterization of staphylococci from human skin I. Amended descriptions of *Staphylococcus epidermidis* and *Staphylococcus saprophyticus* and descriptions of three new species: *Staphylococcus cohnii, Staphylococcus haemolyticus*, and *Staphylococcus xylosus*. *International Journal of Systematic Bacteriology* 25(1), 50–61.

1496 Schloegel, L.M., Daszak, P., Cunningham, A.A., Speare, R. and Hill, B. (2010) Two amphibian diseases, chytridiomycosis and ranaviral disease, are now globally notifiable to the World Organization for Animal Health (OIE), an assessment. *Diseases of Aquatic Organisms* 92(2–3), 101–108.

1497 Schmidtke, L.M. and Carson, J. (1994) Characteristics of *Vagococcus salmoninarum* isolated from diseased salmonid fish. *The Journal of Applied Bacteriology* 77(2), 229–236.

1498 Schmidtke, L.M. and Carson, J. (1995) Characteristics of *Flexibacter pyschrophilus* isolated from Atlantic salmon in Australia. *Diseases of Aquatic Organisms* 21, 157–161.

1499 Schubert, R.H.W. (1967) The taxonomy and nomenclature of the genus *Aeromonas* Kluyver and van Niel 1936. *International Journal of Systematic Bacteriology* 17(3), 273–279.

1500 Schubert, R.H.W., Hegazi, M. and Wahlig, W. (1990a) *Aeromonas enteropelogenes* species nova. *Hygiene Medizin* 15, 471–472.

1501 Schubert, R., Hegazi, M. and Wahlig, W. (1990b) *Aeromonas ichthiosmia* species nova. *Hygiene Medicine* 15, 477–479.

1502 Segers, P., Vancanneyt, M., Pot, B., Torck, U., Hoste, B., Dewettinck, D., Falsen, E., Kersters, K. and De Vos, P. (1994) Classification of *Pseudomonas diminuta* Leifson and Hugh 1954 and *Pseudomonas vesicularis* Büsing, Döll, and Freytag 1953 in *Brevundimonas* gen. nov. as *Brevundimonas diminuta* comb. nov. and *Brevundimonas vesicularis* comb. nov., respectively. *International Journal of Systematic Bacteriology* 44(3), 499–510.

1503 Seibold, H.R. and Neal, J.E. (1956) Erysipelothrix septicemia in the porpoise. *Journal of the American Veterinary Medical Association* 128(11), 537–539.

1504 Seidler, R.J., Allen, D.A., Colwell, R.R., Joseph, S.W. and Daily, O.P. (1980) Biochemical characteristics and virulence of environmental group F bacteria isolated in the United States. *Applied and Environmental Microbiology* 40(4), 715–720.

1505 Selvin, J. and Lipton, A.P. (2003) *Vibrio alginolyticus* associated with white spot disease of *Penaeus monodon*. *Diseases of Aquatic Organisms* 57(1–2), 147–150.

1506 Serracca, L., Ercolini, C., Rossini, I., Battistini, R., Giorgi, I. and Prearo, M. (2011) Occurrence of both subspecies of *Photobacterium damselae* in mullets collected in the river Magra (Italy). *Canadian Journal of Microbiology* 57(5), 437–440.

1507 Seymour, R.L. (1970) The genus *Saprolegnia*. *Nova Hedwigia* 19, 1–124.

1508 Shah, K.L., Jha, B.C. and Jhingran, A.G. (1977) Observations on some aquatic phycomycetes pathogenic to eggs and fry of freshwater fish and prawn. *Aquaculture* 12(2), 141–147.

1509 Shah, K.L. and Tyagi, B.C. (1986) An eye disease in silver carp, *Hypophthalmichthys molitrix*, held in tropical ponds, associated with the bacterium *Staphylococcus aureus*. *Aquaculture* 55(1), 1–4.

1510 Shaheen, A.A., El-Sayed, E. and Faisal, M. (1999) Isolation of *Aphanomyces* sp(p). associated with skin lesions and mortalities in the striped (*Mugil cephalus*) and the thin lip (*Liza ramada*) grey mullets. *Bulletin of the European Association of Fish Pathologists* 19(2), 79–82.

1511 Shamsudin, M.N. and Plumb, J.A. (1996) Morphological, biochemical, and physiological characterization of *Flexibacter columnaris* isolates from four species of fish. *Journal of Aquatic Animal Health* 8(4), 335–339.

1512 Shepard, M.C. and Lunceford, C.D. (1975) Unusual colonies of *Ureaplasma urealyticum* (T mycoplasmas) in primary agar cultures of certain urine specimens. *Journal of Clinical Microbiology* 2(5), 456–458.
1513 Sheu, S.-Y., Chiu, T.F., Chou, J.-H., Sheu, D.-S., Arun, A.B., Young, C.-C., Chen, C.A., Wang, J.-T. and Chen, W.-M. (2009) *Andreprevotia lacus* sp. nov., isolated from a fish-culture pond. *International Journal of Systematic and Evolutionary Microbiology* 59(Pt 10), 2482–2485.
1514 Sheu, S.-Y., Chiu, T.-F., Young, C.-C., Arun, A.B. and Chen, W.-M. (2011a) *Flavobacterium macrobrachii* sp. nov., isolated from a freshwater shrimp culture pond. *International Journal of Systematic and Evolutionary Microbiology* 61(Pt 6), 1402–1407.
1515 Sheu, S.-Y., Jiang, S.-R., Chen, C.A., Wang, J.-T. and Chen, W.-M. (2010) *Vibrio stylophorae* sp. nov., isolated from the reef-building coral *Stylophora pistillata*. *International Journal of Systematic and Evolutionary Microbiology*.
1516 Sheu, S.-Y., Jiang, S.-R., Chen, C.A., Wang, J.-T. and Chen, W.-M. (2011b) *Vibrio stylophorae* sp. nov., isolated from the reef-building coral *Stylophora pistillata*. *International Journal of Systematic and Evolutionary Microbiology* 61(Pt 9), 2180–2185.
1517 Sheu, S.-Y., Lin, K.-Y., Chou, J.-H., Chang, P.-S., Arun, A.B., Young, C.-C. and Chen, W.-M. (2007) *Tenacibaculum litopenaei* sp. nov., isolated from a shrimp mariculture pond. *International Journal of Systematic and Evolutionary Microbiology* 57(Pt 5), 1148–1153.
1518 Shewan, J.M., Hodgkiss, W. and Liston, J. (1954) A method for the rapid differentiation of certain nonpathogenic, asporogenous bacilli. *Nature* 173(4396), 208–209.
1519 Shewmaker, P.L., Camus, A.C., Bailiff, T., Steigerwalt, A.G., Morey, R.E. and Carvalho, M. da G.S. (2007) *Streptococcus ictaluri* sp. nov., isolated from Channel Catfish *Ictalurus punctatus* broodstock. *International Journal of Systematic and Evolutionary Microbiology* 57(Pt 7), 1603–1606.
1520 Shieh, H.S. (1980) Studies on the nutrition of a fish pathogen, *Flexibacter columnaris*. *Microbios Letters* 13(51/52), 129–133.
1521 Shieh, W.Y., Chen, Y.-W., Chaw, S.-M. and Chiu, H.-H. (2003) *Vibrio ruber* sp. nov., a red, facultatively anaerobic, marine bacterium isolated from sea water. *International Journal of Systematic and Evolutionary Microbiology* 53(Pt 2), 479–484.
1522 Shieh, W.Y., Chen, A.L. and Chiu, H.H. (2000) *Vibrio aerogenes* sp. nov., a facultatively anaerobic marine bacterium that ferments glucose with gas production. *International Journal of Systematic and Evolutionary Microbiology* 50(Pt 1), 321–329.
1523 Shimahara, Y., Nakamura, A., Nomoto, R., Itami, T., Chen, S.-C. and Yoshida, T. (2008) Genetic and phenotypic comparison of *Nocardia seriolae* isolated from fish in Japan. *Journal of Fish Diseases* 31(7), 481–488.
1524 Shin, G.-W., Nho, S.-W., Park, S.-B., Jang, H.-B., Cha, I.-S., Ha, M.-A., Kim, Y.-R., Dalvi, R.S., Joh, S.-J. and Jung, T.-S. (2009) Comparison of antigenic proteins from *Lactococcus garvieae* KG- and KG+ strains that are recognized by olive flounder (*Paralichthys olivaceus*) antibodies. *Veterinary Microbiology* 139(1–2), 113–120.
1525 Shin, S.B., Yoo, M.H., Jeong, J.B., Kim, Y.M., Chung, J.-K., Huh, M.D., Komisar, J.L. and Jeong, H.D. (2005) Molecular cloning of the *gyrA* gene and characterization of its mutation in clinical isolates of quinolone-resistant *Edwardsiella tarda*. *Diseases of Aquatic Organisms* 67(3), 259–266.
1526 Shiose, J., Wakabayashi, H., Tominaga, M. and Egusa, S. (1974) A report on a disease of cultured carp due to a capsulated *Pseudomonas*. *Fish Pathology* 9, 79–83.
1527 Shnit-Orland, M., Sivan, A. and Kushmaro, A. (2010) *Shewanella corallii* sp. nov., a marine bacterium isolated from a Red Sea coral. *International Journal of Systematic and Evolutionary Microbiology* 60(Pt 10), 2293–2297.
1528 Shotts, E.B. and Waltman, W.D. (1990) A medium for the selective isolation of *Edwardsiella ictaluri*. *Journal of Wildlife Diseases* 26(2), 214–218.
1529 Shotts, E.B. Jr, Talkington, P.D., Elliott, D.G. and McCarthy, D.H. (1980) Aetiology of an ulcerative disease in goldfish, *Carassius auratus* (L.), characterization of the causative agent. *Journal of Fish Diseases* 3(3), 181–186.
1530 Siddall, M.E., Worthen, P.L., Johnson, M. and Graf, J. (2007) Novel role for *Aeromonas jandaei* as a digestive tract symbiont of the North American medicinal leech. *Applied and Environmental Microbiology* 73(2), 655–658.
1531 Silva-Rubio, A., Acevedo, C., Magariños, B., Jaureguiberry, B., Toranzo, A.E. and Avendaño-Herrera, R. (2008) Antigenic and molecular characterization of *Vibrio ordalii* strains isolated from Atlantic salmon *Salmo salar* in Chile. *Diseases of Aquatic Organisms* 79(1), 27–35.
1532 Simidu, U. and Hasuo, K. (1968) Salt dependency of the bacterial flora of marine fish. *Journal of General Microbiology* 52(3), 347–354.
1533 Simidu, U., Kita-Tsukamoto, K., Yasumoto, T. and Yotsu, M. (1990) Taxonomy of four marine bacterial strains that produce tetrodotoxin. *International Journal of Systematic Bacteriology* 40(4), 331–336.
1534 Simmons, G.C., Sullivan, N.D. and Green, P.E. (1972) Dermatophilosis in a lizard (*Amphibolurus barbatus*). *Australian Veterinary Journal* 48(8), 465–466.
1535 Simmons, J.S. (1926) A culture medium for differentiating organisms of typhoid-colon aerogenes groups and for isolation of certain fungi with colored plate. *Journal of Infectious Diseases* 39(3), 209–214.
1536 Simon, G. and Oppenheimer, C.H. (1968) Bacterial changes in sea water samples, due to storage and volume. *Zeitschrift für allgemeine Mikrobiologie* 8(3), 209–214.
1537 Simoncelli, F., Fagotti, A., Dall'Olio, R., Vagnetti, D., Pascolini, R. and Rosa, I.D. (2005) Evidence of *Batrachochytrium dendrobatidis* infection in water frogs of the *Rana esculenta* complex in central Italy. *EcoHealth* 2(4), 307–312.
1538 Skaar, I., Gaustad, P., Tønjum, T., Holm, B. and Stenwig, H. (1994) *Streptococcus phocae* sp. nov., a new species isolated from clinical specimens from seals.

International Journal of Systematic Bacteriology 44(4), 646–650.
1539 Skirrow, M.B. (1977) Campylobacter enteritis: a 'new' disease. *British Medical Journal* 2(6078), 9–11.
1540 Smith, A.W., Brown, R.J., Skilling, D.E., Bray, H.L. and Keyes, M.C. (1977) Naturally-occurring leptospirosis in northern fur seals (*Callorhinus ursinus*). *Journal of Wildlife Diseases* 13(2), 144–148.
1541 Smith, A.W., Brown, R.J., Skilling, D.E. and DeLong, R.L. (1974) *Leptospira pomona* and reproductive failure in California sea lions. *Journal of the American Veterinary Medical Association* 165(11), 996–998.
1542 Smith, H.L. (1970) A presumptive test for vibrios: the 'string' test. *Bulletin of the World Health Organization* 42(5), 817–818.
1543 Smith, H.L. and Goodner, K. (1958) Detection of bacterial gelatinases by gelatin-agar plate methods. *Journal of Bacteriology* 76(6), 662–665.
1544 Smith, I.M. and Baskerville, A.J. (1979) A selective medium facilitating the isolation and recognition of *Bordetella bronchiseptica* in pigs. *Research in Veterinary Science* 27(2), 187–192.
1545 Smith, I.M., Gordon, R.E. and Clark, F. (1952) Aerobic sporeforming bacteria. *Monograph No. 16*. US Department of Agriculture, Washington, DC.
1546 Smith, I.W. (1963) The classification of '*Bacterium salmonicida*'. *Journal of General Microbiology* 33(2), 263–274.
1547 Smith, S.K., Sutton, D.C., Fuerst, J.A. and Reichelt, J.L. (1991) Evaluation of the genus *Listonella* and reassignment of *Listonella damsela* (Love *et al.*) MacDonell and Colwell to the genus *Photobacterium* as *Photobacterium damsela* comb. nov. with an emended description. *International Journal of Systematic Bacteriology* 41(4), 529–534.
1548 Sneath, P.H.A. and Skerman, V.B.D. (1966) A list of type and reference strains of bacteria. *International Journal of Systematic Bacteriology* 16(1), 1–134.
1549 Snieszko, S.F. (1981) Bacterial gill disease of freshwater fishes. *US Fish and Wildlife Publications Paper 147*.
1550 Snieszko, S.F., Bullock, G.L., Dunbar, C.E. and Pettijohn, L.L. (1964a) Nocardial infection in hatchery-reared fingerling rainbow trout (*Salmo gairdneri*). *Journal of Bacteriology* 88(6), 1809–1810.
1551 Snieszko, S.F., Bullock, G.L., Hollis, E. and Boone, J.G. (1964b) *Pasteurella* sp. from an epizootic of white perch (*Roccus americanus*) in Chesapeake Bay tidewater areas. *Journal of Bacteriology* 88, 1814–1815.
1552 Snieszko, S.F. and Taylor, C.C. (1947) A bacterial disease of the lobster (*Homarus americanus*). *Science* 105(2732), 500.
1553 Snipes, K.P. and Biberstein, E.L. (1982) *Pasteurella testudinis* sp. nov.: a parasite of desert tortoises (*Gopherus agassizi*). *International Journal of Systematic Bacteriology* 32(2), 201–210.
1554 Söderhäll, K., Rantamäki, J. and Constantinescu, O. (1993) Isolation of *Trichosporon beigelii* from the freshwater crayfish *Astacus astacus*. *Aquaculture* 116(1), 25–31.
1555 Soffientino, B., Gwaltney, T., Nelson, D., Specker, J., Mauel, M. and Gómez-Chiarri, M. (1999) Infectious necrotizing enteritis and mortality caused by *Vibrio carchariae* in summer flounder *Paralichthys dentatus* during intensive culture. *Diseases of Aquatic Organisms* 38, 201–210.
1556 Soler, L., Yáñez, M.A., Chacon, M.R., Aguilera-Arreola, M.G., Catalán, V., Figueras, M.J. and Martínez-Murcia, A.J. (2004) Phylogenetic analysis of the genus *Aeromonas* based on two housekeeping genes. *International Journal of Systematic and Evolutionary Microbiology* 54(Pt 5), 1511–1519.
1557 Soltani, M., Ahmadi, M.R., Yavari, H. and Zargar, S.S.M. (2010) Red-pink colony-producing *Pseudomonas* sp. is the causative agent of mass mortality in larvae and post-larvae of *Litopenaeus vannamei* raised in hatcheries in south Iran. *Iranian Journal of Veterinary Medicine* 2, 89–94.
1558 Sonnenwirth, A.C. (1970) Bacteremia with and without meningitis due to *Yersinia enterocolitica*, *Edwardsiella tarda*, *Comamonas terrigena*, and *Pseudomonas maltophilia*. *Annals of the New York Academy of Sciences* 174(2), 488–502.
1559 Sørensen, U.B. and Larsen, J.L. (1986) Serotyping of *Vibrio anguillarum*. *Applied and Environmental Microbiology* 51(3), 593–597.
1560 Sorroza, L., Padilla, D., Acosta, F., Román, L., Grasso, V., Vega, J. and Real, F. (2012) Characterization of the probiotic strain *Vagococcus fluvialis* in the protection of European sea bass (*Dicentrarchus labrax*) against vibriosis by *Vibrio anguillarum*. *Veterinary Microbiology* 155(2–4), 369–373.
1561 Sosa, E.R., Landsberg, J.H., Kiryu, Y., Stephenson, C.M., Cody, T.T., Dukeman, A.K., Wolfe, H.P., Vandersea, M.W. and Litaker, R.W. (2007) Pathogenicity studies with the fungi *Aphanomyces invadans*, *Achlya bisexualis*, and *Phialemonium dimorphosporum*: induction of skin ulcers in striped mullet. *Journal of Aquatic Animal Health* 19(1), 41–48.
1562 Soto-Rodriguez, S.A., Gomez-Gil, B. and Lozano, R. (2010) 'Bright-red' syndrome in Pacific white shrimp *Litopenaeus vannamei* is caused by *Vibrio harveyi*. *Diseases of Aquatic Organisms* 92(1), 11–19.
1563 Soto-Rodriguez, S.A., Roque, A., Lizarraga-Partida, M.L., Guerra-Flores, A.L. and Gomez-Gill, B. (2003) Virulence of luminous vibrios to *Artemia franciscana* nauplii. *Diseases of Aquatic Organisms* 53(3), 231–240.
1564 Sousa, J.A.P., Romalde, J.L., Eiras, J.C. and Toranzo, A.E. (1997) Phenotypic, serological and molecular characterization of Portuguese isolates of *Aeromonas salmonicida* subsp. *salmonicida*. *Fish Pathology* 32(3), 177–178.
1565 Speare, D.J., Brocklebank, J., MacNair, N. and Bernard, K.A. (1995) Experimental transmission of a salmonid *Rhodococcus* sp. isolate to juvenile Atlantic salmon, *Salmo salar* L. *Journal of Fish Diseases* 18(6), 587–597.
1566 Speare, R., Thomas, A.D., O'Shea, P. and Shipton, W.A. (1994) *Mucor amphibiorum* in the toad, *Bufo marinus*, in Australia. *Journal of Wildlife Diseases* 30(3), 399–407.
1567 Spencer, J.F.T., Phaff, H.J. and Gardner, N.R. (1964) *Metschnikowia kamienskii*, sp. n., a yeast associated with brine shrimp. *Journal of Bacteriology* 88(3), 758–762.

1568 Springer, B., Stockman, L., Teschner, K., Roberts, G.D. and Böttger, E.C. (1996) Two-laboratory collaborative study on identification of mycobacteria: molecular versus phenotypic methods. *Journal of Clinical Microbiology* 34(2), 296–303.

1569 Srinivasan, M., Sedmak, D. and Jewell, S. (2002) Effect of fixatives and tissue processing on the content and integrity of nucleic acids. *The American Journal of Pathology* 161(6), 1961–1971.

1570 Stackebrandt, E., Frederiksen, W., Garrity, G.M., Grimont, P.A.D., Kämpfer, P., Maiden, M.C.J., Nesme, X., Rosselló-Mora, R., Swings, J., Trüper, H.G., Vauterin, L., Ward, A.C. and Whitman, W.B. (2002) Report of the ad hoc committee for the re-evaluation of the species definition in bacteriology. *International Journal of Systematic and Evolutionary Microbiology* 52(Pt 3), 1043–1047.

1571 Stackebrandt, E. and Goebel, B.M. (1994) Taxonomic note: a place for DNA–DNA reassociation and 16S rRNA sequence analysis in the present species definition in bacteriology. *International Journal of Systematic Bacteriology* 44(4), 846–849.

1572 Stamper, M.A., Gulland, F.M. and Spraker, T. (1998) Leptospirosis in rehabilitated Pacific harbor seals from California. *Journal of Wildlife Diseases* 34(2), 407–410.

1573 Stanier, R.Y. (1947) Studies on nonfruiting *Myxobacteria* I. *Cytophaga johnsonae*, n.sp., a chitin-decomposing myxobacterium. *Journal of Bacteriology* 53(3), 297–315.

1574 Stanier, R.Y., Palleroni, N.J. and Doudoroff, M. (1966) The aerobic pseudomonads: a taxonomic study. *Journal of General Microbiology* 43(2), 159–271.

1575 Starliper, C.E. (2001) Isolation of *Serratia liquefaciens* as a pathogen of Arctic char, *Salvelinus alpinus* (L.). *Journal of Fish Diseases* 24(1), 53–56.

1576 Starliper, C.E., Shotts, E.B. Jr and Brown, J. (1992) Isolation of *Carnobacterium piscicola* and an unidentified Gram-positive bacillus from sexually mature and post-spawning rainbow trout *Oncorhynchus mykiss*. *Diseases of Aquatic Organisms* 13, 181–187.

1577 Stebbing, P.D., Pond, M.J., Peeler, E., Small, H.J., Greenwood, S.J. and Verner-Jeffreys, D. (2012) Limited prevalence of gaffkaemia (*Aerococcus viridans* var. *homari*) isolated from wild-caught European lobsters *Homarus gammarus* in England and Wales. *Diseases of Aquatic Organisms* 100(2), 159–167.

1578 Steenbergen, J.F., Kimball, H.S., Low, D.A., Schapiro, H.C. and Phelps, L.N. (1977) Serological grouping of virulent and avirulent strains of the lobster pathogen *Aerococcus viridans*. *Journal of General Microbiology* 99(2), 425–430.

1579 Stephens, F.J., Raidal, S.R., Buller, N.B. and Jones, B. (2006) Infection with *Photobacterium damselae* subspecies *damselae* and *Vibrio harveyi* in snapper, *Pagrus auratus* with bloat. *Australian Veterinary Journal* 84(5), 173–177.

1580 Stevens, E., Lipscomb, T.P. and Gulland, F.M. (1999) An additional case of leptospirosis in a harbor seal. *Journal of Wildlife Diseases* 35(1), 150.

1581 Stevenson, R.M.W. and Airdrie, D.W. (1984) Serological variation among *Yersinia ruckeri* strains. *Journal of Fish Diseases* 7(4), 247–254.

1582 Stevenson, R.M.W. and Daly, J.G. (1982) Biochemical and serological characteristics of Ontario isolates of *Yersinia ruckeri*. *Canadian Journal of Fisheries and Aquatic Sciences* 39(6), 870–876.

1583 Stewart, J.E. (1972) The detection of *Gaffkya homari*, the bacterium pathogenic to lobsters (genus *Homarus*). *Fisheries Research Board of Canada. New Series* 43, 1–5.

1584 Stewart, J.E., Cornick, J.W., Zwicker, B.M. and Arie, B. (2004) Studies on the virulence of *Aerococcus viridans* (var.) *homari*, the causative agent of gaffkemia, a fatal disease of homarid lobsters. *Diseases of Aquatic Organisms* 60(2), 149–155.

1585 Stine, C.B., Jacobs, J.M., Rhodes, M.R., Overton, A., Fast, M. and Baya, A.M. (2009) Expanded range and new host species of *Mycobacterium shottsii* and *M. pseudoshottsii*. *Journal of Aquatic Animal Health* 21(3), 179–183.

1586 Strohl, W.R. and Tait, L.R. (1978) *Cytophaga aquatilis* sp. nov., a facultative anaerobe isolated from the gills of freshwater fish. *International Journal of Systematic Bacteriology* 28(2), 293–303.

1587 Strongman, D.B., Morrison, C.M. and McClelland, G. (1997) Lesions in the musculature of captive American plaice *Hippoglossoides platessoides* caused by the fungus *Hormoconis resinae* (Deuteromycetes). *Diseases of Aquatic Organisms* 28(2), 107–113.

1588 Stueland, S., Hatai, K. and Skaar, I. (2005) Morphological and physiological characteristics of *Saprolegnia* spp. strains pathogenic to Atlantic salmon, *Salmo salar* L. *Journal of Fish Diseases* 28(8), 445–453.

1589 Suanyuk, N., Kong, F., Ko, D., Gilbert, G.L. and Supamattaya, K. (2008) Occurrence of rare genotypes of *Streptococcus agalactiae* in cultured red tilapia *Oreochromis* sp. and Nile tilapia *O. niloticus* in Thailand: relationship to human isolates? *Aquaculture* 284(1–4), 35–40.

1590 Suárez, P., Contreras, M., Fernández-Delgado, M., Salazar, V., Peña, R., Michelangeli, F. and García-Amado, M.A. (2010) Detection of *Helicobacter* in the digestive tract of an Atlantic spotted dolphin (*Stenella frontalis*). *Journal of Wildlife Diseases* 46(2), 622–626.

1591 Suer, L. and Vedros, N. (1988) *Erysipelothrix rhusiopathiae*.I. Isolation and characterization from pinnipeds and bite/abrasion wounds in humans. *Diseases of Aquatic Organisms* 5, 1–5.

1592 Sugumar, G., Nakai, T., Hirata, Y., Matsubara, D. and Muroga, K. (1998) *Vibrio splendidus* biovar II as the causative agent of bacillary necrosis of Japanese oyster *Crassostrea gigas* larvae. *Diseases of Aquatic Organisms* 33(2), 111–118.

1593 Sukenda, and Wakabayashi, H. (2000) Tissue distribution of *Pseudomonas plecoglossicida* in experimentally infected ayu *Plecoglossus altivelis* studied by real-time quantitative PCR. *Fish Pathology* 35(4), 223–228.

1594 Sun, F., Chen, J., Zhong, L., Zhang, X., Wang, R., Guo, Q. and Dong, Y. (2008) Characterization and virulence retention of viable but nonculturable *Vibrio harveyi*. *FEMS Microbiology Ecology* 64(1), 37–44.

1595 Sung, H.H., Hwang, S.F. and Tasi, F.M. (2000) Responses of giant freshwater prawn (*Macrobrachium rosenbergii*) to challenge by two strains of *Aero-*

monas spp. *Journal of Invertebrate Pathology* 76(4), 278–284.
1596 Suzuki, K., Arai, H., Kuge, T., Katagiri, T. and Izumi, S. (2008) Reliability of PCR methods for the detection of *Flavobacterium psychrophilum*. *Fish Pathology* 43(3), 124–127.
1597 Suzuki, M., Nakagawa, Y., Harayama, S. and Yamamoto, S. (2001) Phylogenetic analysis and taxonomic study of marine *Cytophaga*-like bacteria: proposal for *Tenacibaculum* gen. nov. with *Tenacibaculum maritimum* comb. nov. and *Tenacibaculum ovolyticum* comb. nov. and description of *Tenacibaculum mesophilum* sp. nov. and *Tenacibaculum amylolyticum* sp. nov. *International Journal of Systematic and Evolutionary Microbiology* 51(Pt 5), 1639–1652.
1598 Suzuki, M.T. and Giovannoni, S.J. (1996) Bias caused by template annealing in the amplification of mixtures of 16S rRNA genes by PCR. *Applied and Environmental Microbiology* 62(2), 625–630.
1599 Svensson, E. and Unestam, T. (1975) Differential induction of zoospore encystment and germination in *Aphanomyces astaci*, oomycetes. *Physiologia Plantarum* 35(3), 210–216.
1600 Sweeney, J.C. and Ridgway, S.H. (1975) Common diseases of small cetaceans. *Journal of the American Veterinary Medical Association* 167(7), 533–540.
1601 Swenshon, M., Lämmler, C. and Siebert, U. (1998) Identification and molecular characterization of beta-hemolytic streptococci isolated from harbor porpoises (*Phocoena phocoena*) of the North and Baltic Seas. *Journal of Clinical Microbiology* 36(7), 1902–1906.
1602 Taborda, P.R., Taborda, V.A. and McGinnis, M.R. (1999) *Lacazia loboi* gen. nov., comb. nov., the etiologic agent of lobomycosis. *Journal of Clinical Microbiology* 37(6), 2031–2033.
1603 Takeuchi, M. and Yokota, A. (1992) Proposals of *Sphingobacterium faecium* sp. nov., *Sphingobacterium piscium* sp. nov., *Sphingobacterium heparinum* comb. nov., *Sphingobacterium thalpophilum* comb. nov. and two genospecies of the genus *Sphingobacterium*, and synonymy of *Flavobacterium yabuuchiae* and *Sphingobacterium spiritivorum*. *Journal of General Microbiology* 38, 465–482.
1604 Takuma, D., Sano, A., Wada, S., Kurata, O. and Hatai, K. (2010) A new species, *Aphanomyces salsuginosus* sp. nov., isolated from ice fish *Salangichthys microdon*. *Mycoscience* 51(6), 432–442.
1605 Takuma, D., Sano, A., Wada, S., Kurata, O. and Hatai, K. (2011) *Aphanomyces sinensis* sp. nov., isolated from juvenile soft-shelled turtle, *Pelodiscus sinensis*, in Japan. *Mycoscience* 52(2), 119–131.
1606 Talaat, A.M., Reimschuessel, R. and Trucksis, M. (1997) Identification of mycobacteria infecting fish to the species level using polymerase chain reaction and restriction enzyme analysis. *Veterinary Microbiology* 58(2–4), 229–237.
1607 Talaat, A.M., Trucksis, M., Kane, A.S. and Reimschuessel, R. (1999) Pathogenicity of *Mycobacterium fortuitum* and *Mycobacterium smegmatis* to goldfish, *Carassius auratus*. *Veterinary Microbiology* 66(2), 151–164.
1608 Tall, B.D., Fall, S., Pereira, M.R., Ramos-Valle, M., Curtis, S.K., Kothary, M.H., Chu, D.M.T., Monday, S.R., Kornegay, L., Donkar, T., Prince, D., Thunberg, R.L., Shangraw, K.A., Hanes, D.E., Khambaty, F.M., Lampel, K.A., Bier, J.W. and Bayer, R.C. (2003) Characterization of *Vibrio fluvialis*-like strains implicated in limp lobster disease. *Applied and Environmental Microbiology* 69(12), 7435–7446.
1609 Tamura, K., Peterson, D., Peterson, N., Stecher, G., Nei, M. and Kumar, S. (2011) MEGA5: molecular evolutionary genetics analysis using maximum likelihood, evolutionary distance, and maximum parsimony methods. *Molecular Biology and Evolution* 28(10), 2731–2739.
1610 Tamura, K., Sakazaki, R., McWhorter, A.C. and Kosako, Y. (1988) *Edwardsiella tarda* serotyping scheme for international use. *Journal of Clinical Microbiology* 26(11), 2343–2346.
1611 Tarr, C.L., Patel, J.S., Puhr, N.D., Sowers, E.G., Bopp, C.A. and Strockbine, N.A. (2007) Identification of *Vibrio* isolates by a multiplex PCR assay and *rpoB* sequence determination. *Journal of Clinical Microbiology* 45(1), 134–140.
1612 Taylor, P.W. and Winton, J.R. (2002) Optimization of nested polymerase chain reaction assays for identification of *Aeromonas salmonicida*, *Yersinia ruckeri*, and *Flavobacterium psychrophilum*. *Journal of Aquatic Animal Health* 14(3), 216–224.
1613 Teixeira, L.M., Carvalho, M.G., Merquior, V.L., Steigerwalt, A.G., Brenner, D.J. and Facklam, R.R. (1997) Phenotypic and genotypic characterization of *Vagococcus fluvialis*, including strains isolated from human sources. *Journal of Clinical Microbiology* 35(11), 2778–2781.
1614 Teixeira, L.M., Merquior, V.L., Vianni, M.C., Carvalho, M.G., Fracalanzza, S.E., Steigerwalt, A.G., Brenner, D.J. and Facklam, R.R. (1996) Phenotypic and genotypic characterization of atypical *Lactococcus garvieae* strains isolated from water buffalos with subclinical mastitis and confirmation of *L. garvieae* as a senior subjective synonym of *Enterococcus seriolicida*. *International Journal of Systematic Bacteriology* 46(3), 664–668.
1615 Telenti, A., Marchesi, F., Balz, M., Bally, F., Böttger, E.C. and Bodmer, T. (1993) Rapid identification of mycobacteria to the species level by polymerase chain reaction and restriction enzyme analysis. *Journal of Clinical Microbiology* 31(2), 175–178.
1616 Temprano, A., Yugueros, J., Hernanz, C., Sánchez, M., Berzal, B., Luengo, J.M. and Naharro, G. (2001) Rapid identification of *Yersinia ruckeri* by PCR amplification of *yruI–yruR* quorum sensing. *Journal of Fish Diseases* 24(5), 253–261.
1617 Tendencia, E. (2002) *Vibrio harveyi* isolated from cage-cultured seabass *Lates calcarifer* Bloch in the Philippines. *Aquaculture Research* 33, 455–458.
1618 Teshima, C., Kudo, S., Ohtani, Y. and Saito, A. (1992) Kidney pathology from the bacterium *Hafnia alvei*: experimental evidence. *Transactions of the American Fisheries Society* 121(5), 599–607.
1619 Teska, J.D. (1994) *In vitro* growth of the bacterial kidney disease organism *Renibacterium salmoninarum* on a nonserum, noncharcoal-based 'homospecies-metabolite' medium. *Journal of Wildlife Diseases* 30(3), 383–388.

1620 Teska, J.D., Twerdok, L.E., Beaman, J., Curry, M. and Finch, R.A. (1997) Isolation of *Mycobacterium abscessus* from Japanese Medaka. *Journal of Aquatic Animal Health* 9(3), 234–238.

1621 Thoen, C. and Schliesser, T. (1984) Mycobacterial infections in cold-blooded animals. In: Kubica, G.P. and Wayne, L.G. (eds) *The Mycobacteria: A Sourcebook*, Vol. II. Marcel Dekker, New York, pp.1297–1311.

1622 Thomas, A.D., Forbes-Faulkner, J. and Parker, M. (1979) Isolation of *Pseudomonas pseudomallei* from clay layers at defined depths. *American Journal of Epidemiology* 110(4), 515–521.

1623 Thompson, C.C., Thompson, F.L., Vandemeulebroecke, K., Hoste, B., Dawyndt, P. and Swings, J. (2004) Use of *recA* as an alternative phylogenetic marker in the family *Vibrionaceae*. *International Journal of Systematic and Evolutionary Microbiology* 54(Pt 3), 919–924.

1624 Thompson, C.C., Thompson, F.L., Vicente, A.C.P. and Swings, J. (2007a) Phylogenetic analysis of vibrios and related species by means of *atpA* gene sequences. *International Journal of Systematic and Evolutionary Microbiology* 57(Pt 11), 2480–2484.

1625 Thompson, F.L., Gevers, D., Thompson, C.C., Dawyndt, P., Naser, S., Hoste, B., Munn, C.B. and Swings, J. (2005a) Phylogeny and molecular identification of vibrios on the basis of multilocus sequence analysis. *Applied and Environmental Microbiology* 71(9), 5107–5115.

1626 Thompson, F.L., Gomez-Gil, B., Vasconcelos, A.T.R. and Sawabe, T. (2007b) Multilocus sequence analysis reveals that *Vibrio harveyi* and *V. campbellii* are distinct species. *Applied and Environmental Microbiology* 73(13), 4279–4285.

1627 Thompson, F.L., Hoste, B., Thompson, C.C., Goris, J., Gomez-Gil, B., Huys, L., De Vos, P. and Swings, J. (2002a) *Enterovibrio norvegicus* gen. nov., sp. nov., isolated from the gut of turbot (*Scophthalmus maximus*) larvae: a new member of the family *Vibrionaceae*. *International Journal of Systematic and Evolutionary Microbiology* 52(Pt 6), 2015–2022.

1628 Thompson, F.L., Hoste, B., Thompson, C.C., Huys, G. and Swings, J. (2001a) The coral bleaching *Vibrio shiloi* Kushmaro *et al*. 2001 is a later synonym of *Vibrio mediterranei* Pujalte and Garay 1986. *Systematic and Applied Microbiology* 24(4), 516–519.

1629 Thompson, F.L., Hoste, B., Vandemeulebroecke, K., Engelbeen, K., Denys, R. and Swings, J. (2002b) *Vibrio trachuri* Iwamoto *et al*. 1995 is a junior synonym of *Vibrio harveyi* (Johnson and Shunk 1936) Baumann *et al*. 1981. *International Journal of Systematic and Evolutionary Microbiology* 52(3), 973–976.

1630 Thompson, F.L., Hoste, B., Vandemeulebroecke, K. and Swings, J. (2001b) Genomic diversity amongst *Vibrio* isolates from different sources determined by fluorescent amplified fragment length polymorphism. *Systematic and Applied Microbiology* 24(4), 520–538.

1631 Thompson, F.L., Hoste, B., Vandemeulebroecke, K. and Swings, J. (2003a) Reclassification of *Vibrio hollisae* as *Grimontia hollisae* gen. nov., comb. nov. *International Journal of Systematic and Evolutionary Microbiology* 53(Pt 5), 1615–1617.

1632 Thompson, F.L., Li, Y., Gomez-Gil, B., Thompson, C.C., Hoste, B., Vandemeulebroecke, K., Rupp, G.S., Pereira, A., De Bem, M.M., Sorgeloos, P. and Swings, J. (2003b) *Vibrio neptunius* sp. nov., *Vibrio brasiliensis* sp. nov. and *Vibrio xuii* sp. nov., isolated from the marine aquaculture environment (bivalves, fish, rotifers and shrimps). *International Journal of Systematic and Evolutionary Microbiology* 53(Pt 1), 245–252.

1633 Thompson, F.L., Thompson, C.C., Dias, G.M., Naka, H., Dubay, C. and Crosa, J.H. (2011) The genus *Listonella* MacDonell and Colwell 1986 is a later heterotypic synonym of the genus *Vibrio* Pacini 1854 (Approved Lists 1980) – a taxonomic opinion. *International Journal of Systematic and Evolutionary Microbiology* 61(Pt 12), 3023–3027.

1634 Thompson, F.L., Thompson, C.C., Hoste, B., Vandemeulebroecke, K., Gullian, M. and Swings, J. (2003c) *Vibrio fortis* sp. nov. and *Vibrio hepatarius* sp. nov., isolated from aquatic animals and the marine environment. *International Journal of Systematic and Evolutionary Microbiology* 53(Pt 5),1495–1501.

1635 Thompson, F.L., Thompson, C.C., Li, Y., Gomez-Gil, B., Vandenberghe, J., Hoste, B. and Swings, J. (2003d) *Vibrio kanaloae* sp. nov., *Vibrio pomeroyi* sp. nov. and *Vibrio chagasii* sp. nov., from sea water and marine animals. *International Journal of Systematic and Evolutionary Microbiology* 53(Pt 3), 753–759.

1636 Thompson, F.L., Thompson, C.C., Naser, S., Hoste, B., Vandemeulebroecke, K., Munn, C., Bourne, D. and Swings, J. (2005b) *Photobacterium rosenbergii* sp. nov. and *Enterovibrio coralii* sp. nov., vibrios associated with coral bleaching. *International Journal of Systematic and Evolutionary Microbiology* 55(Pt 2), 913–917.

1637 Thompson, F.L., Thompson, C.C. and Swings, J. (2003e) *Vibrio tasmaniensis* sp. nov., isolated from Atlantic salmon (*Salmo salar* L.). *Systematic and Applied Microbiology* 26(1), 65–69.

1638 Thompson, J.D., Higgins, D.G. and Gibson, T.J. (1994) CLUSTAL W: improving the sensitivity of progressive multiple sequence alignment through sequence weighting, position-specific gap penalties and weight matrix choice. *Nucleic Acids Research* 22(22), 4673–4680.

1639 Thomson, R., Macpherson, H.L., Riaza, A. and Birkbeck, T.H. (2005) *Vibrio splendidus* biotype 1 as a cause of mortalities in hatchery-reared larval turbot, *Scophthalmus maximus* (L.). *Journal of Applied Microbiology* 99(2), 243–250.

1640 Thornley, M.J. (1960) The differentiation of *Pseudomonas* from other Gram-negative bacteria on the basis of arginine metabolism. *Journal of Applied Microbiology* 23(1), 37–52.

1641 Thyssen, A., Grisez, L., van Houdt, R. and Ollevier, F. (1998) Phenotypic characterization of the marine pathogen *Photobacterium damselae* subsp. *piscicida*. *International Journal of Systematic Bacteriology* 48(Pt 4), 1145–1151.

1642 Tindall, B.J., Rosselló-Móra, R., Busse, H.-J., Ludwig, W. and Kämpfer, P. (2010) Notes on the characterization of prokaryote strains for taxonomic purposes. *International Journal of Systematic and Evolutionary Microbiology* 60(1), 249–266.

1643 Tinsley, J.W., Austin, D.A., Lyndon, A.R. and Austin, B. (2011) Novel non-motile phenotypes of *Yersinia ruckeri* suggest expansion of the current clonal complex theory. *Journal of Fish Diseases* 34(4), 311–317.
1644 Tison, D.L., Nishibuchi, M., Greenwood, J.D. and Seidler, R.J. (1982) *Vibrio vulnificus* biogroup 2: new biogroup pathogenic for eels. *Applied and Environmental Microbiology* 44(3), 640–646.
1645 Tison, D.L. and Seidler, R.J. (1983) *Vibrio aestuarianus*: a new species isolated from estuarine waters and shellfish. *International Journal of Systematic Bacteriology* 33, 699–702.
1646 Tobback, E., Decostere, A., Hermans, K., Haesebrouck, F. and Chiers, K. (2007) *Yersinia ruckeri* infections in salmonid fish. *Journal of Fish Diseases* 30(5), 257–268.
1647 Tomomasa, M., Takashi, K., Norihisa, O., Kenrou, K. and Takaji, I. (2005) Pathogenicity of motile and non-motile *Edwardsiella tarda* to some marine fish. *Fish Pathology* 40(3), 133–135.
1648 Toranzo, A. and Barja, J. (1992) Fry mortality syndrome (FMS) in Spain. Isolation of the causative bacterium *Flexibacter psychrophilus*. *Bulletin of the European Association of Fish Pathologists* 13(1), 30–32.
1649 Toranzo, A., Barreiro, S., Casal, J.F., Figueras, A., Magariños, B. and Barja, J.L. (1991) Pasteurellosis in cultured gilthead seabream (*Sparus aurata*), first report in Spain. *Aquaculture* 99, 1–15.
1650 Toranzo, A.E. and Barja, J.L. (1990) A review of the taxonomy and seroepizootiology of *Vibrio anguillarum*, with special reference to aquaculture in the northwest of Spain. *Diseases of Aquatic Organisms* 9, 73–82.
1651 Toranzo, A.E., Baya, A.M., Roberson, B.S., Barja, J.L., Grimes, D.J. and Hetrick, F.M. (1987) Specificity of slide agglutination test for detecting bacterial fish pathogens. *Aquaculture* 61(2), 81–97.
1652 Toranzo, A.E., Baya, A.M., Romalde, J.L. and Hetrick, F.M. (1989) Association of *Aeromonas sobria* with mortalities of adult gizzard shad, *Dorosoma cepedianum* Lesueur. *Journal of Fish Diseases* 12, 439–448.
1653 Toranzo, A.E., Cutrín, J.M., Núñez, S., Romalde, J.L. and Barja, J.L. (1995) Antigenic characterization of *Enterococcus* strains pathogenic for turbot and their relationship with other Gram-positive bacteria. *Diseases of Aquatic Organisms* 21(3), 187–191.
1654 Toranzo, A.E., Cutrin, J.M., Roberson, B.S., Nunez, S., Abell, J.M., Hetrick, F.M. and Baya, A.M. (1994) Comparison of the taxonomy, serology, drug resistance transfer, and virulence of *Citrobacter freundii* strains from mammals and poikilothermic hosts. *Applied and Environmental Microbiology* 60(6), 1789–1797.
1655 Toranzo, A.E., Romalde, J.L., Nuñez, S., Figueras, A. and Barja, J.L. (1993) An epizootic in farmed, market-sized rainbow trout in Spain caused by a strain of *Carnobactenum piscicola* of unusual virulence. *Diseases of Aquatic Organisms* 17, 87–99.
1656 Torkko, P., Suomalainen, S., Iivanainen, E., Tortoli, E., Suutari, M., Seppänen, J., Paulin, L. and Katila, M.-L. (2002) *Mycobacterium palustre* sp. nov., a potentially pathogenic, slowly growing mycobacterium isolated from clinical and veterinary specimens and from Finnish stream waters. *International Journal of Systematic and Evolutionary Microbiology* 52(5), 1519–1525.
1657 Torrent, A., Déniz, S., Ruiz, A., Calabuig, P., Sicilia, J. and Orós, J. (2002) Esophageal diverticulum associated with *Aerococcus viridans* infection in a loggerhead sea turtle (*Caretta caretta*). *Journal of Wildlife Diseases* 38(1), 221–223.
1658 Tortoli, E., Bartoloni, A., Bozzetta, E., Burrini, C., Lacchini, C., Mantella, A., Penati, V., Tullia Simonetti, M. and Ghittino, C. (1996) Identification of the newly described *Mycobacterium poriferae* from tuberculous lesions of snakehead fish (*Channa striatus*). *Comparative Immunology Microbiology and Infectious Diseases* 19(1), 25–29.
1659 Toyama, T., Kita-Tsukamoto, K. and Wakabayashi, H. (1994) Identification of *Cytophaga psychrophila* by PCR targeted 16S ribosomal RNA. *Fish Pathology* 29(4), 271–275.
1660 Toyama, T., Kita-Tsukamoto, K. and Wakabayashi, H. (1996) Identification of *Flexibacter maritimus*, *Flavobacterium branchiophilum* and *Cytophaga columnaris* by PCR targeted 16S ribosomal DNA. *Fish Pathology* 31(1), 25–31.
1661 Travers, M.-A., Le Goïc, N., Huchette, S., Koken, M. and Paillard, C. (2008) Summer immune depression associated with increased susceptibility of the European abalone, *Haliotis tuberculata* to *Vibrio harveyi* infection. *Fish & Shellfish Immunology* 25(6), 800–808.
1662 Triyanto, Kumamaru, A. and Wakabayashi, H. (1999) The use of PCR targeted 16S rDNA for identification of genomovars of *Flavobacterium columnare*. *Fish Pathology* 34(4), 217–218.
1663 Triyanto and Wakabayashi, H. (1999) Genotypic diversity of strains of *Flavobacterium columnare* from diseased fishes. *Fish Pathology* 34(2), 65–71.
1664 Trust, T.J., Khouri, A.G., Austen, R.A. and Ashburner, L.D. (1980) First isolation in Australia of atypical *Aeromonas salmonicida*. *FEMS Microbiology Letters* 9(1), 39–42.
1665 Tryland, M., Derocher, A.E., Wiig, Y. and Godfroid, J. (2001) *Brucella* sp. antibodies in polar bears from Svalbard and the Barents Sea. *Journal of Wildlife Diseases* 37(3), 523–531.
1666 Tryland, M., Kleivane, L., Alfredsson, A., Kjeld, M., Arnason, A., Stuen, S. and Godfroid, J. (1999) Evidence of *Brucella* infection in marine mammals in the North Atlantic Ocean. *The Veterinary Record* 144(21), 588–592.
1667 Tryland, M., Sørensen, K.K. and Godfroid, J. (2005) Prevalence of *Brucella pinnipediae* in healthy hooded seals (*Cystophora cristata*) from the North Atlantic Ocean and ringed seals (*Phoca hispida*) from Svalbard. *Veterinary Microbiology* 105(2), 103–111.
1668 Tsukamura, M., Yano, I. and Imaeda, T. (1986) *Mycobacterium fortuitum* subspecies *acetamidolyticum*, a new subspecies of *Mycobacterium fortuitum*. *Microbiology and Immunology* 30(2), 97–110.
1669 Tubiash, H.S., Chanley, P.E. and Leifson, E. (1965) Bacillary necrosis, a disease of larval and juvenile bivalve mollusks I. Etiology and epizootiology. *Journal of Bacteriology* 90(4), 1036–1044.

1670 Tuffs, S. and Oidtmann, B. (2011) A comparative study of molecular diagnostic methods designed to detect the crayfish plague pathogen, *Aphanomyces astaci*. *Veterinary Microbiology* 153(3–4), 343–353.
1671 Tung, M.-C., Tsai, S.-S., Ho, L.-F., Huang, S.-T. and Chen, S.-C. (1985) An acute septicemic infection of *Pasteurella* organism in pond-cultured Formosa snake-head fish (*Channa maculata* Lacepede) in Taiwan. *Fish Pathology* 20(2/3),143–148.
1672 Udey, L.R. (1982) A differential medium for distinguishing Alr+ from Alr– phenotypes in *Aeromonas salmonicida*. In:Proceedings of the 13th Annual Conference and Workshop and 7th Eastern Fish Health Workshop. Presented at the International Association for Aquatic Animal Medicine, Baltimore, Maryland.
1673 Udey, L.R., Young, E. and Sallman, B. (1977) Isolation and characterization of an anaerobic bacterium, *Eubacterium tarantellus* sp. nov., associated with striped mullet (*Mugil cephalus*) mortality in Biscayne Bay, Florida. *Journal of the Fisheries Research Board of Canada* 34(3), 402–409.
1674 Uhland, F., Hélie, P. and Higgins, R. (2000) Infections of *Edwardsiella tarda* among brook trout in Quebec. *Journal of Aquatic Animal Health* 12, 74–77.
1675 Une, Y., Kadekaru, S., Tamukai, K., Goka, K. and Kuroki, T. (2008) First report of spontaneous chytridiomycosis in frogs in Asia. *Diseases of Aquatic Organisms* 82(2), 157–160.
1676 Unestam, T. and Ajaxon, R. (1976) Phenol oxidation in soft cuticle and blood of crayfish compared with that in other arthropods and activation of the phenol oxidases by fungal and other cell walls. *Journal of Invertebrate Pathology* 27(3), 287–295.
1677 Unestam, T. and Weiss, D.W. (1970) The host–parasite relationship between freshwater crayfish and the crayfish disease fungus *Aphanomyces astaci*: responses to infection by a susceptible and a resistant species. *Journal of General Microbiology* 60(1), 77–90.
1678 Unestam, T. (1965) Studies on the crayfish plague fungus *Aphanomyces astaci* I. Some factors affecting growth *in vitro*. *Physiologia Plantarum* 18(2), 483–505.
1679 Unestam, T. (1969) On the physiology of zoospore production in *Aphanomyces astaci*. *Physiologia Plantarum* 22(2), 236–246.
1680 Unestam, T. (1972) On the host range and origin of the crayfish plague fungus. *Report Institute of the Freshwater Research Drottningholm* 52, 192–198.
1681 Unestam, T. (1975) Defence reactions in and susceptibility of Australian and New Guinea freshwater crayfish to European crayfish plague fungus. *Australian Journal of Experimental Biological Medical Science* 53(5), 349–359.
1682 Urakawa, H., Kita-Tsukamoto, K. and Ohwada, K. (1999) Reassessment of the taxonomic position of *Vibrio iliopiscarius* (Onarheim *et al*. 1994) and proposal for *Photobacterium iliopiscarium* comb. nov. *International Journal of Systematic Bacteriology* 49(Pt 1), 257–260.
1683 Urakawa, H., Kita-Tsukamoto, K., Steven, S.E., Ohwada, K. and Colwell, R.R. (1998) A proposal to transfer *Vibrio marinus* (Russell 1891) to a new genus *Moritella* gen. nov. as *Moritella marina* comb. nov. *FEMS Microbiology Letters* 165(2), 373–378.
1684 Urbanczyk, H., Ast, J.C. and Dunlap, P.V. (2011) Phylogeny, genomics, and symbiosis of *Photobacterium*. *FEMS Microbiology Reviews* 35(2), 324–342.
1685 Urbanczyk, H., Ast, J.C., Higgins, M.J., Carson, J. and Dunlap, P.V. (2007) Reclassification of *Vibrio fischeri*, *Vibrio logei*, *Vibrio salmonicida* and *Vibrio wodanis* as *Aliivibrio fischeri* gen. nov., comb. nov., *Aliivibrio logei* comb. nov., *Aliivibrio salmonicida* comb. nov. and *Aliivibrio wodanis* comb. nov. *International Journal of Systematic and Evolutionary Microbiology* 57(Pt 12), 2823–2829.
1686 Urdaci, M.C., Chakroun, C., Faure, D. and Bernardet, J.F. (1998) Development of a polymerase chain reaction assay for identification and detection of the fish pathogen *Flavobacterium psychrophilum*. *Research in Microbiology* 149(7), 519–530.
1687 Urdaci, M.C., Marchand, M., Ageron, E., Arcos, J.M., Sesma, B. and Grimont, P.A. (1991) *Vibrio navarrensis* sp. nov., a species from sewage. *International Journal of Systematic Bacteriology* 41(2), 290–294.
1688 Ursing, J.B., Rossellö-Mora, R.A., García-Valdés, E. and Lalucat, J. (1995) Taxonomic note: A pragmatic approach to the nomenclature of phenotypically similar genomic groups. *International Journal of Systematic Bacteriology* 45(3), 604.
1689 Valdebenito, S. and Avendaño-Herrera, R. (2009) Phenotypic, serological and genetic characterization of *Flavobacterium psychrophilum* strains isolated from salmonids in Chile. *Journal of Fish Diseases* 32(4), 321–333.
1690 Valheim, M., Håstein, T., Myhr, E., Speilberg, L. and Ferguson, H.W. (2000) *Varracalbmi*: a new bacterial panophthalmitis in farmed Atlantic salmon, *Salmo salar* L. *Journal of Fish Diseases* 23(1), 61–70.
1691 Valsecchi, E., Glockner-Ferrari, D., Ferrari, M. and Amos, W. (1998) Molecular analysis of the efficiency of sloughed skin sampling in whale population genetics. *Molecular Ecology* 7(10), 1419–1422.
1692 Valtonen, E.T., Rintamäki, P. and Koskivaara, M. (1992) Occurrence and pathogenicity of *Yersinia ruckeri* at fish farms in northern and central Finland. *Journal of Fish Diseases* 15(2), 163–171.
1693 Vancanneyt, M., Segers, P., Hauben, L., Hommez, J., Devriese, L.A., Hoste, B., Vandamme, P. and Kersters, K. (1994) *Flavobacterium meningosepticum*, a pathogen in birds. *Journal of Clinical Microbiology* 32(10), 2398–2403.
1694 Vancanneyt, M., Segers, P., Torck, U., Hoste, B., Bernardet, J.-F., Vandamme, P. and Kersters, K. (1996) Reclassification of *Flavobacterium odoratum* (Stutzer 1929) strains to a new genus, *Myroides*, as *Myroides odoratus* comb. nov. and *Myroides odoratimimus* sp. nov. *International Journal of Systematic Bacteriology* 46(4), 926–932.
1695 Vandamme, P., Bernardet, J.F., Segers, P., Kersters, K. and Holmes, B. (1994) New perspectives in the classification of the flavobacteria: description of *Chryseobacterium* gen. nov., *Bergeyella* gen. nov. and *Empedobacter* nom. rev. *International Journal of Systematic Bacteriology* 44(4), 827–831.
1696 Vandamme, P., Devriese, L.A., Pot, B., Kersters, K. and Melin, P. (1997) *Streptococcus difficile* is a nonhemolytic group B, type Ib streptococcus.

International Journal of Systematic Bacteriology 47(1), 81–85.

1697 Vandamme, P., Falsen, E., Rossau, R., Hoste, B., Segers, P., Tytgat, R. and De Ley, J. (1991) Revision of *Campylobacter*, *Helicobacter*, and *Wolinella* taxonomy: emendation of generic descriptions and proposal of *Arcobacter* gen. nov. *International Journal of Systematic Bacteriology* 41(1), 88–103.

1698 Vandenberghe, J., Li, Y., Verdonck, L., Li, J., Sorgeloos, P., Xu, H. and Swings, J. (1998) Vibrios associated with *Penaeus chinensis* (Crustacea: Decapoda) larvae in Chinese shrimp hatcheries. *Aquaculture* 169(1–2), 121–132.

1699 Vandersea, M.W., Litaker, R.W., Yonnish, B., Sosa, E., Landsberg, J.H., Pullinger, C., Moon-Butzin, P., Green, J., Morris, J.A., Kator, H., Noga, E.J. and Tester, P.A. (2006) Molecular assays for detecting *Aphanomyces invadans* in ulcerative mycotic fish lesions. *Applied and Environmental Microbiology* 72(2), 1551–1557.

1700 Vaneechoutte, M., Kämpfer, P., De Baere, T., Avesani, V., Janssens, M. and Wauters, G. (2007) *Chryseobacterium hominis* sp. nov., to accommodate clinical isolates biochemically similar to CDC groups II-h and II-c. *International Journal of Systematic and Evolutionary Microbiology* 57(Pt 11), 2623–2628.

1701 Van Gelderen, R., Carson, J. and Nowak, B. (2010) Experimentally induced marine flexibacteriosis in Atlantic salmon smolts *Salmo salar*. I. Pathogenicity. *Diseases of Aquatic Organisms* 91(2), 121–128.

1702 Van Hao, M. and Komagata, K. (1985) A new species of *Planococcus*, *P. kocurii* isolated from fish, frozen foods, and fish curing brine. *Journal of General Microbiology* 31(5), 441–455.

1703 Van West, P. (2006) *Saprolegnia parasitica*, an oomycete pathogen with a fishy appetite: new challenges for an old problem. *Mycologist* 20(3), 99–104.

1704 Varaldo, P.E., Kilpper-Bälz, R., Biavasco, F., Satta, G. and Schleifer, K.H. (1988) *Staphylococcus delphini* sp. nov., a coagulase-positive species isolated from dolphins. *International Journal of Systematic Bacteriology* 38(4), 436–439.

1705 Vaseeharan, B., Sundararaj, S., Murugan, T. and Chen, J.C. (2007) *Photobacterium damselae* ssp. damselae associated with diseased black tiger shrimp *Penaeus monodon* Fabricius in India. *Letters in Applied Microbiology* 45(1), 82–86.

1706 Vattakaven, T., Bond, P., Bradley, G. and Munn, C.B. (2006) Differential effects of temperature and starvation on induction of the viable-but-nonculturable state in the coral pathogens *Vibrio shiloi* and *Vibrio tasmaniensis*. *Applied and Environmental Microbiology* 72(10), 6508–6513.

1707 Vedros, N.A., Quinlivan, J. and Cranford, R. (1982) Bacterial and fungal flora of wild northern fur seals (*Callorhinus ursinus*). *Journal of Wildlife Diseases* 18(4), 447–456.

1708 Vela, A.I., Fernández, A., Bernaldo de Quirós, Y., Herráez, P., Domínguez, L. and Fernández-Garayzábal, J.F. (2011) *Weissella ceti* sp. nov., isolated from beaked whales (*Mesoplodon bidens*). *International Journal of Systematic and Evolutionary Microbiology* 61(11), 2758–2762.

1709 Vela, A.I., Fernandez, A., Espinosa de los Monteros, A., Goyache, J., Herraez, P., Tames, B., Cruz, F., Domínguez, L. and Fernández-Garayzábal, J.F. (2008) *Lactobacillus ceti* sp. nov., isolated from beaked whales (*Ziphius cavirostris*). *International Journal of Systematic and Evolutionary Microbiology* 58(Pt 4), 891–894.

1710 Vela, A.I., Fernandez, A., Sánchez-Porro, C., Sierra, E., Mendez, M., Arbelo, M., Ventosa, A., Domínguez, L. and Fernández-Garayzábal, J.F. (2007) *Flavobacterium ceti* sp. nov., isolated from beaked whales (*Ziphius cavirostris*). *International Journal of Systematic and Evolutionary Microbiology* 57(Pt 11), 2604–2608.

1711 Vela, A.I., Vázquez, J., Gibello, A., Blanco, M.M., Moreno, M.A., Liébana, P., Albendea, C., Alcalá, B., Mendez, A., Domínguez, L. and Fernández-Garayzábal, J.F. (2000) Phenotypic and genetic characterization of *Lactococcus garvieae* isolated in Spain from lactococcosis outbreaks and comparison with isolates of other countries and sources. *Journal of Clinical Microbiology* 38(10), 3791–3795.

1712 Venkateswaran, K., Dohmoto, N. and Harayama, S. (1998) Cloning and nucleotide sequence of the *gyrB* gene of *Vibrio parahaemolyticus* and its application in detection of this pathogen in shrimp. *Applied and Environmental Microbiology* 64(2), 681–687.

1713 Venkateswaran, K., Moser, D.P., Dollhopf, M.E., Lies, D.P., Saffarini, D.A., MacGregor, B.J., Ringelberg, D.B., White, D.C., Nishijima, M., Sano, H., Burghardt, J., Stackebrandt, E. and Nealson, K.H. (1999) Polyphasic taxonomy of the genus *Shewanella* and description of *Shewanella oneidensis* sp. nov. *International Journal of Systematic Bacteriology* 49(Pt 2), 705–724.

1714 Ventura, M.T. and Grizzle, J.M. (1988) Lesions associated with natural and experimental infections of *Aeromonas hydrophila* in channel catfish, *Ictalurus punctatus* (Rafinesque). *Journal of Fish Diseases* 11, 397–407.

1715 Vera, H.D. (1948) A simple medium for identification and maintenance of the gonococcus and other bacteria. *Journal of Bacteriology* 55(4), 531–536.

1716 Vera, H.D. (1950) Relation of peptones and other culture media ingredients to the accuracy of fermentation tests. *American Journal of Public Health and the Nations Health* 40(10), 1267–1272.

1717 Vera, P., Navas, J.I. and Fouz, B. (1991) First isolation of *Vibrio damsela* from seabream (*Sparus aurata*). *Bulletin of the European Association of Fish Pathologists* 11(3), 112–113.

1718 Verdonck, L., Grisez, L., Sweetman, E., Minkoff, G., Sorgeloos, P., Ollevier, F. and Swings, J. (1997) Vibrios associated with routine productions of *Brachionus plicatilis*. *Aquaculture* 149(3–4), 203–214.

1719 Verger, J.-M., Grimont, F., Grimont, P.A.D. and Grayon, M. (1985) *Brucella*, a monospecific genus as shown by deoxyribonucleic acid hybridization. *International Journal of Systematic Bacteriology* 35(3), 292–295.

1720 Verschuere, L., Heang, H., Criel, G., Sorgeloos, P. and Verstraete, W. (2000) Selected bacterial strains protect *Artemia* spp. from the pathogenic effects of

Vibrio proteolyticus CW8T2. *Applied and Environmental Microbiology* 66(3), 1139–1146.

1721 Vey, A., Soderhall, K. and Ajaxon, R. (1983) Susceptibility of *Orconectes limosus* Raff to the crayfish plague, *Aphanomyces astaci* Schikora. *Freshwater Crayfish* V, 284–291.

1722 Vicente, V.A., Orélis-Ribeiro, R., Najafzadeh, M.J., Sun, J., Guerra, R.S., Miesch, S., Ostrensky, A., Meis, J.F., Klaassen, C.H., de Hoog, G.S. and Boeger, W.A. (2012) Black yeast-like fungi associated with Lethargic Crab Disease (LCD) in the mangrove-land crab, *Ucides cordatus* (Ocypodidae). *Veterinary Microbiology* 158(1–2), 109–122.

1723 Vieira, V., Teixeira, L., Zahner, V., Momen, H., Facklam, R.R., Steigerwalt, A.G., Brenner, D.J. and Castro, A.C.D. (1998) Genetic relationships among the different phenotypes of *Streptococcus dysgalactiae* strains. *International Journal of Systematic Bacteriology* 48(4), 1231–1243.

1724 Vigneulle, M. and Laurencin, F.B. (1995) *Serratia liquefaciens*: a case report in turbot (*Scophthalmus maximus*) cultured in floating cages in France. *Aquaculture* 132(1–2), 121–124.

1725 Vilar, P., Faílde, L.D., Bermúdez, R., Vigliano, F., Riaza, A., Silva, R., Santos, Y. and Quiroga, M.I. (2012) Morphopathological features of a severe ulcerative disease outbreak associated with *Tenacibaculum maritimum* in cultivated sole, *Solea senegalensis* (L.). *Journal of Fish Diseases* 35(6), 437–445.

1726 Vilela, R., Rosa, P.S., Belone, A.F.F., Taylor, J.W., Diório, S.M. and Mendoza, L. (2009) Molecular phylogeny of animal pathogen *Lacazia loboi* inferred from rDNA and DNA coding sequences. *Mycological Research* 113(Pt 8), 851–857.

1727 Villamil, L., Figueras, A., Toranzo, A.E., Planas, M. and Novoa, B. (2003) Isolation of a highly pathogenic *Vibrio pelagius* strain associated with mass mortalities of turbot, *Scophthalmus maximus* (L.), larvae. *Journal of Fish Diseases* 26(5), 293–303.

1728 Vishniac, H.S. (1958) A new marine phycomycete. *Mycologia* 50(1), 66.

1729 Voges, O. and Proskauer, B. (1898) Beitraege zur Ernaehrungsphysiologie und zur Differential Diagnose der Bakterien der hemmorrhagischen Septicamie. *Zentralblatt für Hygiene* 28, 20–22.

1730 Volokhov, D.V., Simonyan, V., Davidson, M.K. and Chizhikov, V.E. (2012) RNA polymerase beta subunit (*rpoB*) gene and the 16S–23S rRNA intergenic transcribed spacer region (ITS) as complementary molecular markers in addition to the 16S rRNA gene for phylogenetic analysis and identification of the species of the family *Mycoplasmataceae*. *Molecular Phylogenetics and Evolution* 62(1), 515–528.

1731 Von Graevenitz, A., Bowman, J., Del Notaro, C. and Ritzler, M. (2000) Human infection with *Halomonas venusta* following fish bite. *Journal of Clinical Microbiology* 38(8), 3123–3124.

1732 Vossen, A., Abdulmawjood, A., Lämmler, C., Weiss, R. and Siebert, U. (2004) Identification and molecular characterization of beta-hemolytic streptococci isolated from harbor seals (*Phoca vitulina*) and grey seals (*Halichoerus grypus*) of the German North and Baltic Seas. *Journal of Clinical Microbiology* 42(1), 469–473.

1733 Voyles, J., Berger, L., Young, S., Speare, R., Webb, R., Warner, J., Rudd, D., Campbell, R. and Skerratt, L.F. (2007) Electrolyte depletion and osmotic imbalance in amphibians with chytridiomycosis. *Diseases of Aquatic Organisms* 77(2), 113–118.

1734 Vrålstad, T., Knutsen, A.K., Tengs, T. and Holst-Jensen, A. (2009) A quantitative TaqMan MGB real-time polymerase chain reaction based assay for detection of the causative agent of crayfish plague *Aphanomyces astaci*. *Veterinary Microbiology* 137(1–2), 146–155.

1735 Vreeland, R.H., Litchfield, C.D., Martin, E.L. and Elliot, E. (1980) *Halomonas elongata*, a new genus and species of extremely salt-tolerant bacteria. *International Journal of Systematic Bacteriology* 30(2), 485–495.

1736 Vuddhakul, V., Nakai, T., Matsumoto, C., Oh, T., Nishino, T., Chen, C.H., Nishibuchi, M. and Okuda, J. (2000) Analysis of *gyrB* and *toxR* gene sequences of *Vibrio hollisae* and development of *gyrB*- and *toxR*-targeted PCR methods for isolation of *V. hollisae* from the environment and its identification. *Applied and Environmental Microbiology* 66(8), 3506–3514.

1737 Vuillaume, A., Brun, R., Chene, P., Sochon, E. and Lesel, R. (1987) First isolation of *Yersinia ruckeri* from sturgeon, *Acipenser baeri* Brandt, in south west of France. *Bulletin of the European Association of Fish Pathologists* 7, 18–19.

1738 Wada, S., Hatai, K. and Ishii, H. (1993) Mycotic gastritis of juvenile ayu (*Plecoglossus altivelis*) caused by *Saprolegnia diclina* type 1. *Journal of Wildlife Diseases* 29(4), 587–590.

1739 Wada, S., Nakamura, K. and Hatai, K. (1995) First case of *Ochroconis humicola* infection in marine cultured fish in Japan. *Fish Pathology* 30(2), 125–126.

1740 Waechter, M., Le Roux, F., Nicolas, J.-L., Marissal, E. and Berthe, F. (2002) Characterization of pathogenic bacteria of the cupped oyster *Crassostrea gigas*. *Comptes Rendus Biologies* 325(3), 231–238.

1741 Wahli, T., Burr, S.E., Pugovkin, D., Mueller, O. and Frey, J. (2005) *Aeromonas sobria*, a causative agent of disease in farmed perch, *Perca fluviatilis* L. *Journal of Fish Diseases* 28(3), 141–150.

1742 Wakabayashi, H. and Egusa, S. (1972) Characteristics of a *Pseudomonas* sp. from an epizootic of pond-cultured eels (*Anguilla japonica*). *Nippon Suisan Gakkaishi* 38(6), 577–587.

1743 Wakabayashi, H. and Egusa, S. (1973) *Edwardsiella tarda* (*Paracolobacterium anguillimortiferum*) associated with pond-cultured eel disease. *Bulletin of the Japanese Society of Scientific Fisheries* 39(9), 931–936.

1744 Wakabayashi, H., Hikida, M. and Masumura, K. (1986) *Flexibacter maritimus* sp. nov., a pathogen of marine fishes. *International Journal of Systematic Bacteriology* 36(3), 396–398.

1745 Wakabayashi, H., Horiuchi, M., Bunya, T. and Hoshiai, G. (1991) Outbreaks of cold-water disease in coho salmon in Japan. *Fish Pathology* 26(4), 211–212.

1746 Wakabayashi, H., Huh, G.J. and Kimura, N. (1989) *Flavobacterium branchiophila* sp. nov., a causative agent of bacterial gill disease of freshwater fishes.

International Journal of Systematic Bacteriology 39(3), 213–216.

1747 Wakabayashi, H., Kira, K. and Egusa, S. (1970) Studies on columnaris disease of pond-cultured eels – I. Characteristics and pathogenicity of *Chondrococcus columnaris* isolate from pond-cultured eels. *Bulletin of the Japanese Society of Scientific Fisheries* 36(2), 147–155.

1748 Wakabayashi, H., Sawada, K., Ninomiya, K. and Nishimori, E. (1996) Bacterial hemorrhagic ascites of ayu caused by *Pseudomonas* sp. *Fish Pathology* 31(4), 239–240.

1749 Wakabayashi, H., Toyama, T. and Iida, T. (1994) A study on serotyping of *Cytophaga psychrophila* isolated from fishes in Japan. *Fish Pathology* 29(2), 101–104.

1750 Waldenström, J., On, S.L.W., Ottvall, R., Hasselquist, D., Harrington, C.S. and Olsen, B. (2003) Avian reservoirs and zoonotic potential of the emerging human pathogen *Helicobacter canadensis*. *Applied and Environmental Microbiology* 69(12), 7523–7526.

1751 Waldman, B., van de Wolfshaar, K.E., Klena, J.D., Andjic, V., Bishop, P.J. and Norman, R.J. (2001) Chytridiomycosis in New Zealand frogs. *Surveillance* 28(3), 9–11.

1752 Wallace, L., White, F.H. and Gore, H.L. (1966) Isolation of *Edwardsiella tarda* from a sea lion and two alligators. *Journal of the American Veterinary Medical Association* 149, 881–883.

1753 Wallace, R.J. Jr, Brown, B.A., Tsukamura, M., Brown, J.M. and Onyi, G.O. (1991) Clinical and laboratory features of *Nocardia nova*. *Journal of Clinical Microbiology* 29(11), 2407–2411.

1754 Wallach, J.D. (1977) Husbandry and general. *International Zoo Yearbook* 17(1), 170–171.

1755 Wallbanks, S., Martinez-Murcia, A.J., Fryer, J.L., Phillips, B.A. and Collins, M.D. (1990) 16S rRNA sequence determination for members of the genus *Carnobacterium* and related lactic acid bacteria and description of *Vagococcus salmoninarum* sp. nov. *International Journal of Systematic Bacteriology* 40(3), 224–230.

1756 Walling, E., Vourey, E., Ansquer, D., Beliaeff, B. and Goarant, C. (2010) *Vibrio nigripulchritudo* monitoring and strain dynamics in shrimp pond sediments. *Journal of Applied Microbiology* 108(6), 2003–2011.

1757 Waltman, W.D. and Shotts, E.B.Jr (1984) A medium for the isolation and differentiation of *Yersinia ruckeri*. *Canadian Journal of Fisheries and Aquatic Sciences* 41(5), 804–806.

1758 Waltman, W.D., Shotts, E.B. and Hsu, T.C. (1986) Biochemical characteristics of *Edwardsiella ictaluri*. *Applied and Environmental Microbiology* 51(1), 101–104.

1759 Wang, F.-I. and Chen, J.-C. (2006) The immune response of tiger shrimp *Penaeus monodon* and its susceptibility to *Photobacterium damselae* subsp. *damselae* under temperature stress. *Aquaculture* 258(1–4), 34–41.

1760 Wang, H., Liu, J., Wang, Y. and Zhang, X.-H. (2011) *Vibrio marisflavi* sp. nov., isolated from seawater. *International Journal of Systematic and Evolutionary Microbiology* 61(Pt 3), 568–573.

1761 Wang, J.-T., Chou, Y.-J., Chou, J.-H., Chen, C.A. and Chen, W.-M. (2008a) *Tenacibaculum aiptasiae* sp. nov., isolated from a sea anemone *Aiptasia pulchella*. *International Journal of Systematic and Evolutionary Microbiology* 58(Pt 4), 761–766.

1762 Wang, P.-C., Chen, S., Tsai, M., Weng, Y., Chu, S., Chern, R. and Chen, S. (2009) *Nocardia seriolae* infection in the three striped tigerfish, *Terapon jarbua* (Forsskål). *Journal of Fish Diseases* 32(4), 301–310.

1763 Wang, P.-C., Lin, Y.-D., Liaw, L.-L., Chern, R.-S. and Chen, S.-C. (2008b) *Lactococcus lactis* subspecies *lactis* also causes white muscle disease in farmed giant freshwater prawns *Macrobrachium rosenbergii*. *Diseases of Aquatic Organisms* 79(1), 9–17.

1764 Wang, R.-F., Cao, W.-W. and Cerniglia, C.E. (1997) A universal protocol for PCR detection of 13 species of foodborne pathogens in foods. *Journal of Applied Microbiology* 83(6), 727–736.

1765 Wang, X., Chi, Z., Yue, L., Li, J., Li, M. and Wu, L. (2007) A marine killer yeast against the pathogenic yeast strain in crab (*Portunus trituberculatus*) and an optimization of the toxin production. *Microbiological Research* 162, 77–85.

1766 Wang, X.-Y., Zhou, Z.-C., Guan, X.-Y.,Jiang, B., Chen, Z., Dong, Y. and Yang, A.-F. (2010a) Rapid PCR detection for *Vibrio alginolyticus* and *Vibrio splendidus* in sea cucumber *Apostichopus japonicas* and its culturing environment. *Journal of Agricultural Science and Technology* 12(3), 125–130.

1767 Wang, Y., Zhang, X.-H., Yu, M., Wang, H. and Austin, B. (2010b) *Vibrio atypicus* sp. nov., isolated from the digestive tract of the Chinese prawn (*Penaeus chinensis* O'sbeck). *International Journal of Systematic and Evolutionary Microbiology* 60(Pt 11), 2517–2523.

1768 Ward, L.N. and Bej, A.K. (2006) Detection of *Vibrio parahaemolyticus* in shellfish by use of multiplexed real-time PCR with TaqMan fluorescent probes. *Applied and Environmental Microbiology* 72(3), 2031–2042.

1769 Wauters, G., Janssens, M., Steigerwalt, A.G. and Brenner, D.J. (1988) *Yersinia mollaretii* sp. nov. and *Yersinia bercovieri* sp. nov., formerly called *Yersinia enterocolitica* biogroups 3A and 3B. *International Journal of Systematic Bacteriology* 38(4), 424–429.

1770 Wayne, L.G., Brenner, D.J., Colwell, R.R., Grimont, P.A.D., Kandler, O., Krichevsky, M.I., Moore, L.H., Moore, W.E.C., Murray, R.G.E., Stackebrandt, E., Starr, M.P. and Truper, H.G. (1987) Report of the ad hoc committee on reconciliation of approaches to bacterial systematics. *International Journal of Systematic Bacteriology* 37(4), 463–464.

1771 Weerakhun, S., Aoki, N., Kurata, O., Hatai, K., Nibe, H. and Hirae, T. (2007) *Mycobacterium marinum* infection in cultured yellowtail *Seriola quinqueradiata* in Japan. *Fish Pathology* 42(2), 79–84.

1772 Wei, Y., Xu, K., Zhu, D.-Z., Chen, X.-F. and Wang, X.-L. (2010) Early-spring survey for *Batrachochytrium dendrobatidis* in wild *Rana dybowskii* in Heilongjiang Province, China. *Diseases of Aquatic Organisms* 92(2–3), 241–244.

1773 Weiner, R.M., Coyne, V.E., Brayton, P., West, P. and Raiken, S.F. (1988) *Alteromonas colwelliana* sp. nov., an isolate from oyster habitats. *International Journal of Systematic Bacteriology* 38(3), 240–244.

1774 Weiner, R.M., Segall, A.M. and Colwell, R.R. (1985) Characterization of a marine bacterium associated with *Crassostrea virginica* (the Eastern Oyster). *Applied and Environmental Microbiology* 49(1), 83–90.
1775 Weinstein, M.R., Litt, M., Kertesz, D.A., Wyper, P., Rose, D., Coulter, M., McGeer, A., Facklam, R., Ostach, C., Willey, B.M., Borczyk, A. and Low, D.E. (1997) Invasive infections due to a fish pathogen, *Streptococcus iniae*. *S.iniae* study group. *The New England Journal of Medicine* 337(9), 589–594.
1776 Weisburg, W.G., Barns, S.M., Pelletier, D.A. and Lane, D.J. (1991) 16S ribosomal DNA amplification for phylogenetic study. *Journal of Bacteriology* 173(2), 697–703.
1777 Weldon, C.,du Preez, L.H., Hyatt, A.D., Muller, R. and Spears, R. (2004) Origin of the amphibian chytrid fungus. *Emerging Infectious Diseases* 10(12), 2100–2105.
1778 Welker, T.L., Shoemaker, C.A., Arias, C.R. and Klesius, P.H. (2005) Transmission and detection of *Flavobacterium columnare* in channel catfish *Ictalurus punctatus*. *Diseases of Aquatic Organisms* 63(2–3), 129–138.
1779 West, P.A., Brayton, P.R., Bryant, T.N. and Colwell, R.R. (1986) Numerical taxonomy of Vibrios isolated from aquatic environments. *International Journal of Systematic Bacteriology* 36(4), 531–543.
1780 West, P.A., Brayton, P.R., Twilley, R.R., Bryant, T.N. and Colwell, R.R. (1985) Numerical taxonomy of nitrogen-fixing 'decarboxylase-negative' *Vibrio* species isolated from aquatic environments. *International Journal of Systematic Bacteriology* 35(2), 198–205.
1781 West, P.A. and Colwell, R.R. (1984) Identification and classification of *Vibrionaceae* – an overview. In: Colwell, R.R. (ed.) *Vibrios in the Environment*. John Wiley & Sons, New York, pp.285–363.
1782 West, P.A., Lee, J.V. and Bryant, T.N. (1983) A numerical taxonomic study of species of *Vibrio* isolated from the aquatic environment and birds in Kent, England. *The Journal of Applied Bacteriology* 55(2), 263–282.
1783 Whipps, C.M., Butler, W.R., Pourahmad, F., Watral, V.G. and Kent, M.L. (2007) Molecular systematics support the revival of *Mycobacterium salmoniphilum* (ex Ross 1960) sp. nov., nom. rev., a species closely related to *Mycobacterium chelonae*. *International Journal of Systematic and Evolutionary Microbiology* 57(11), 2525–2531.
1784 Whipps, C.M., Matthews, J.L. and Kent, M.L. (2008) Distribution and genetic characterization of *Mycobacterium chelonae* in laboratory zebrafish *Danio rerio*. *Diseases of Aquatic Organisms* 82(1), 45–54.
1785 White, F.H., Simpson, C.F. and Williams, L.E. (1973) Isolation of *Edwardsiella tarda* from aquatic animal species and surface waters in Florida. *Journal of Wildlife Diseases* 9(3), 204–208.
1786 White, T., Bruns, T., Lee, S. and Taylor, J. (1990) Amplification and direct sequencing of fungal ribosomal RNA genes for phylogenetics. In: Innis, M., Gelfand, D., Shinsky, J. and White, T. (eds) *PCR Protocols: A Guide to Methods and Applications*. Academic Press, New York, pp.315–322.
1787 Whittington, R.J. and Cullis, B. (1988) The susceptibility of salmonid fish to an atypical strain of *Aeromonas salmonicida* that infects goldfish, *Carassius auratus* (L.), in Australia. *Journal of Fish Diseases* 11, 461–470.
1788 Whittington, R.J., Djordjevic, S.P., Carson, J. and Callinan, R.B. (1995) Restriction endonuclease analysis of atypical *Aeromonas salmonicida* isolates from goldfish *Carassius auratus*, silver perch *Bidyanus bidyanus*, and greenback flounder *Rhombosolea tapirina* in Australia. *Diseases of Aquatic Organisms* 22(3), 185–191.
1789 Whittington, R.J., Gudkovs, N., Carrigan, M.J., Ashburner, L.D. and Thurstan, S.J. (1987) Clinical, microbiological and epidemiological findings in recent outbreaks of goldfish ulcer disease due to atypical *Aeromonas salmonicida* in south-eastern Australia. *Journal of Fish Diseases* 10, 353–362.
1790 Wiens, G.D., Rockey, D.D., Wu, Z., Chang, J., Levy, R., Crane, S., Chen, D.S., Capri, G.R., Burnett, J.R., Sudheesh, P.S., Schipma, M.J., Burd, H., Bhattacharyya, A., Rhodes, L.D., Kaul, R. and Strom, M.S. (2008) Genome sequence of the fish pathogen *Renibacterium salmoninarum* suggests reductive evolution away from an environmental *Arthrobacter* ancestor. *Journal of Bacteriology* 190(21), 6970–6982.
1791 Wieser, M., Denner, E.B.M.,Kämpfer, P., Schumann, P., Tindall, B., Steiner, U., Vybiral, D., Lubitz, W., Maszenan, A.M., Patel, B.K.C., Seviour, R.J., Radax, C. and Busse, H.-J. (2002) Emended descriptions of the genus *Micrococcus*, *Micrococcus luteus* (Cohn 1872) and *Micrococcus lylae* (Kloos *et al.* 1974). *International Journal of Systematic and Evolutionary Microbiology* 52(2), 629–637.
1792 Wiik, R., Egidius, E. and Goksøyr, J. (1987) Screening of the Norwegian lobsters *Homarus gammarus* for the lobster pathogen *Aerococcus viridans*. *Diseases of Aquatic Organisms* 3, 97–100.
1793 Wiik, R., Torsvik, V. and Egidius, E. (1986) Phenotypic and genotypic comparisons among strains of the lobster pathogen *Aerococcus viridans* and other marine *Aerococcus viridans*-like cocci. *International Journal of Systematic Bacteriology* 36(3), 431–434.
1794 Wiklund, T. and Bylund, G. (1990) *Pseudomonas anguilliseptica* as a pathogen of salmonid fish in Finland. *Diseases of Aquatic Organisms* 8, 13–19.
1795 Wiklund, T. and Bylund, G. (1993) Skin ulcer disease of flounder *Platichthys flesus* in the northern Baltic Sea. *Diseases of Aquatic Organisms* 17, 165–174.
1796 Wiklund, T. and Dalsgaard, I. (1998) Occurrence and significance of atypical *Aeromonas salmonicida* in non-salmonid and salmonid fish species: a review. *Diseases of Aquatic Organisms* 32(1), 49–69.
1797 Wiklund, T., Dalsgaard, I., Eerola, E. and Olivier, G. (1994a) Characteristics of 'atypical', cytochrome oxidase-negative *Aeromonas salmonicida* isolated from ulcerated flounders (*Platichthys flesus* (L.)). *The Journal of Applied Bacteriology* 76(5), 511–520.
1798 Wiklund, T., Kaas, L., Lönström, L. and Dalsgaard, I. (1994b) Isolation of *Cytophaga psychrophila* (*Flexibacter psychrophilus*) from wild and farmed rainbow trout (*Oncorrhynchus mykiss*) in Finland. *Bulletin of the European Association of Fish Pathologists* 14(2), 44–46.
1799 Wiklund, T., Madsen, L., Bruun, M.S. and Dalsgaard, I. (2000) Detection of *Flavobacterium psychrophilum*

from fish tissue and water samples by PCR amplification. *Journal of Applied Microbiology* 88(2), 299–307.

1800 Wiklund, T., Tabolina, I. and Bezgachina, T.V. (1999) Recovery of atypical *Aeromonas salmonicida* from ulcerated fish from the Baltic Sea. *ICES Journal of Marine Science: Journal du Conseil* 56(2), 175–179.

1801 Willems, A., Busse, J., Goor, M., Pot, B., Falsen, E., Jantzen, E., Hoste, B., Gillis, M., Kersters, K., Auling, G. and Ley, J.D. (1989) *Hydrogenophaga*, a new genus of hydrogen-oxidizing bacteria that includes *Hydrogenophaga flava* comb. nov. (formerly *Pseudomonas flava*), *Hydrogenophaga palleronii* (formerly *Pseudomonas palleronii*), *Hydrogenophaga pseudoflava* (formerly *Pseudomonas pseudoflava* and '*Pseudomonas carboxydoflava*'), and *Hydrogenophaga taeniospiralis* (formerly *Pseudomonas taeniospiralis*). *International Journal of Systematic Bacteriology* 39(3), 319–333.

1802 Williams, A.M., Fryer, J.L. and Collins, M.D. (1990) *Lactococcus piscium* sp. nov. a new *Lactococcus* species from salmonid fish. *FEMS Microbiology Letters* 68(1–2), 109–113.

1803 Williams, M.L. and Lawrence, M.L. (2010) Verification of an *Edwardsiella ictaluri*-specific diagnostic PCR. *Letters in Applied Microbiology* 50(2), 153–157.

1804 Williams, R.E.O., Hirch, A. and Cowan, S.T. (1953) *Aerococcus*, a new bacterial genus. *Journal of General Microbiology* 8(3), 475–480.

1805 Willoughby, L.G. and Roberts, R.J. (1994) Improved methodology for isolation of the *Aphanomyces* fungal pathogen of epizootic ulcerative syndrome (EUS) in Asian fish. *Journal of Fish Diseases* 17(5), 541–543.

1806 Willoughby, L.G., Roberts, R.J. and Chinabut, S. (1995) *Aphanomyces invaderis* sp. nov., the fungal pathogen of freshwater tropical fish affected by epizootic ulcerative syndrome. *Journal of Fish Diseases* 18(3), 273–276.

1807 Willumsen, B. (1989) Birds and wild fish as potential vectors of *Yersinia ruckeri*. *Journal of Fish Diseases* 12(3), 275–277.

1808 Wilson, B.W. and Holliman, A. (1994) Atypical *Aeromonas salmonicida* isolated from ulcerated shub *Leuciscus cephalis*. *The Veterinary Record* 135, 185–186.

1809 Wilson, K.H., Blitchington, R.B. and Greene, R.C. (1990) Amplification of bacterial 16S ribosomal DNA with polymerase chain reaction. *Journal of Clinical Microbiology* 28(9), 1942–1946.

1810 Wolters, W.R. and Johnson, M. (1994) Enteric septicemia resistance in blue catfish and three channel catfish strains. *Journal of Aquatic Animal Health* 6, 329–334.

1811 Won, K., Kim, K. and Park, S. (2009) *Vibrio harveyi* associated with deep skin ulcer in river puffer, *Takifugu obscures*, exhibited in a public aquarium. *Aquaculture Research* 40, 903–907.

1812 Wong, F.Y.K., Fowler, K. and Desmarchelier, P.M. (1995) Vibriosis due to *Vibrio mimicus* in Australian freshwater crayfish. *Journal of Aquatic Animal Health* 7(4), 284–291.

1813 Wood, R.L. (1965) A selective liquid medium utilizing antibiotics for isolation of *Erysipelothrix insidiosa*. *American Journal of Veterinary Research* 26(115), 1303–1308.

1814 Wood, R.L. and Packer, R.A. (1972) Isolation of *Erysipelothrix rhusiopathiae* from soil and manure of swine-raising premises. *American Journal of Veterinary Research* 33(8), 1611–1620.

1815 Wyatt, L.E., Nickelson, R. and Vanderzant, C. (1979) *Edwardsiella tarda* in freshwater catfish and their environment. *Applied and Environmental Microbiology* 38(4), 710–714.

1816 Xie, Z.Y., Hu, C.Q., Zhang, L.P., Chen, C., Ren, C.H. and Shen, Q. (2007) Identification and pathogenicity of *Vibrio ponticus* affecting cultured Japanese sea bass, *Lateolabrax japonicus* (Cuvier in Cuvier and Valenciennes). *Letters in Applied Microbiology* 45(1), 62–67.

1817 Xie, Z.-Y., Zhou, Y.-C., Wang, S.-F.,Mei, B., Xu, X.-D., Wen, W.-Y. and Feng, Y.-Q. (2009) First isolation and identification of *Elizabethkingia meningoseptica* from cultured tiger frog, *Rana tigerina rugulosa*. *Veterinary Microbiology* 138(1–2), 140–144.

1818 Xu, X.-W., Wu, Y.-H., Wang, C.-S., Oren, A. and Wu, M. (2009) *Vibrio hangzhouensis* sp. nov., isolated from sediment of the East China Sea. *International Journal of Systematic and Evolutionary Microbiology* 59(Pt 8), 2099–2103.

1819 Yabuuchi, E., Kaneko, T., Yano, I., Moss, C.W. and Miyoshi, N. (1983) *Sphingobacterium* gen. nov., *Sphingobacterium spiritivorum* comb. nov., *Sphingobacterium multivorum* comb. nov., *Sphingobacterium mizutae* sp. nov. and *Flavobacterium indologenes* sp. nov.: glucose-non-fermenting Gram-negative rods in CDC Groups IIK-2 and IIb. *International Journal of Systematic Bacteriology* 33(3), 580–598.

1820 Yamada, Y. and Wakabayashi, H. (1999) Identification of fish-pathogenic strains belonging to the genus *Edwardsiella* by sequence analysis of *sodB*. *Fish Pathology* 34(3), 145–150.

1821 Yáñez, M.A., Catalán, V., Apráiz, D., Figueras, M.J. and Martínez-Murcia, A.J. (2003) Phylogenetic analysis of members of the genus *Aeromonas* based on *gyrB* gene sequences. *International Journal of Systematic and Evolutionary Microbiology* 53(Pt 3),875–883.

1822 Yang, H., Baek, H., Speare, R., Webb, R., Park, S., Kim, T., Lasater, K.C., Shin, S., Son, S., Park, J., Min, M., Kim, Y., Na, K., Lee, H. and Park, S. (2009) First detection of the amphibian chytrid fungus *Batrachochytrium dendrobatidis* in free-ranging populations of amphibians on mainland Asia: survey in South Korea. *Diseases of Aquatic Organisms* 86(1), 9–13.

1823 Yang, J.-L., Zhan, W.-B., Zhou, L. and Xing, J. (2008) Purification, virulence and characterization of an extracellular peptidase produced by *Aeromonas salmonicida* ssp. *masoucida* isolated from cultured sea cucumber *Apostichopus japonicus*. *Diseases of Aquatic Organisms* 82(3), 223–229.

1824 Yang, W. and Li, A. (2009) Isolation and characterization of *Streptococcus dysgalactiae* from diseased *Acipenser schrenckii*. *Aquaculture* 294(1–2),14–17.

1825 Yang, Y., Yeh, L.P., Cao, Y., Baumann, L., Baumann, P., Tang, J.S. and Beaman, B. (1983) Characterization of marine luminous bacteria isolated off the coast of China and description of *Vibrio orientalis* sp. nov. *Current Microbiology* 8(2), 95–100.

1826 Yanong, R.P.E., Curtis, E.W., Terrell, S.P. and Case, G. (2003) Atypical presentation of mycobacteriosis in a collection of frogfish (*Antennarius striatus*). *Journal of Zoo and Wildlife Medicine* 34(4), 400–407.

1827 Yassin, A.F., Hupfer, H., Siering, C. and Schumann, P. (2011) Comparative chemotaxonomic and phylogenetic studies on the genus *Arcanobacterium* Collins *et al.* 1982 emend. Lehnen *et al.* 2006: proposal for *Trueperella* gen. nov. and emended description of the genus *Arcanobacterium*. *International Journal of Systematic and Evolutionary Microbiology* 61(Pt 6),1265–1274.

1828 Yasunaga, N., Hatai, K. and Tsukahara, J. (1983) *Pasteurella piscicida* from an epizootic of cultured Red Sea bream. *Fish Pathology* 18(2), 107–110.

1829 Ye, S., Li, H., Qiao, G. and Li, Z. (2009) First case of *Edwardsiella ictaluri* infection in China farmed yellow catfish *Pelteobagrus fulvidraco*. *Aquaculture* 292, 6–10.

1830 Yii, K.C., Yang, T.I. and Lee, K.K. (1997) Isolation and characterization of *Vibrio carchariae*, a causative agent of gastroenteritis in the groupers, *Epinephelus coioides*. *Current Microbiology* 35(2), 109–115.

1831 Yishan, L., Jiaming, F., Zaohe, W. and Jichang, J. (2011) Genotype analysis of collagenase gene by PCR-SSCP in *Vibrio alginolyticus* and its association with virulence to marine fish. *Current Microbiology* 62(6), 1697–1703.

1832 Yoon, J.-H., Yeo, S.-H., Kim, I.-G. and Oh, T.-K. (2004) *Shewanella marisflavi* sp. nov. and *Shewanella aquimarina* sp. nov., slightly halophilic organisms isolated from sea water of the Yellow Sea in Korea. *International Journal of Systematic and Evolutionary Microbiology* 54(6), 2347–2352.

1833 Yoshiura, Y., Kamaishi, T., Nakayasu, C. and Ototake, M. (2006) Detection and genotyping of *Flavobacterium psychrophilum* by PCR-targeted to peptidyl-prolyl cis-trans isomerase C gene. *Fish Pathology* 41(2), 67–71.

1834 Yoshizawa, S., Karatani, H., Wada, M., Yokota, A. and Kogure, K. (2010a) *Aliivibrio sifiae* sp. nov., luminous marine bacteria isolated from seawater. *The Journal of General and Applied Microbiology* 56(6), 509–518.

1835 Yoshizawa, S., Tsuruya, Y., Fukui, Y., Sawabe, T., Yokota, A., Kogure, K., Higgins, M., Carson, J. and Thompson, F.L. (2012) *Vibrio jasicida* sp. nov., a member of the Harveyi clade, isolated from marine animals (packhorse lobster, abalone and Atlantic salmon). *International Journal of Systematic and Evolutionary Microbiology* 62(Pt 8), 1864–1870.

1836 Yoshizawa, S., Wada, M.,Kita-Tsukamoto, K., Ikemoto, E., Yokota, A. and Kogure, K. (2009) *Vibrio azureus* sp. nov., a luminous marine bacterium isolated from seawater. *International Journal of Systematic and Evolutionary Microbiology* 59(Pt 7), 1645–1649.

1837 Yoshizawa, S., Wada, M., Yokota, A. and Kogure, K. (2010b) *Vibrio sagamiensis* sp. nov., luminous marine bacteria isolated from sea water. *The Journal of General and Applied Microbiology* 56(6), 499–507.

1838 Yu, J.-H., Park, K.S., Park, K.H. and Park, S.W. (2009) *Edwardsiella tarda* infection in Korean catfish, *Silurus asotus*, in a Korean fish farm. *Aquaculture Research* 41, 19–26.

1839 Yuasa, K., Kitancharoen, N. and Hatai, K. (1997) Simple method to distinguish between *Saprolegnia parasitica* and *S. diclina* isolated from fishes with Saprolegniasis. *Fish Pathology* 32(3), 175–176.

1840 Yuasa, K., Kitancharoen, N., Kataoka, Y. and Al-Murbaty, F.A., (1999) *Streptococcus iniae*, the causative agent of mass mortality in rabbitfish *Siganus canaliculatus* in Bahrain. *Journal of Aquatic Animal Health* 11(1), 87–93.

1841 Yue, X., Liu, B. and Sun, L. (2011) Isolation and characterization of a virulent *Vibrio* sp. bacterium from clams (*Meretrix meretrix*) with mass mortality. *Journal of Invertebrate Pathology* 106(2), 242–249.

1842 Yue, X., Liu, B., Xiang, J. and Jia, J. (2010) Identification and characterization of the pathogenic effect of a *Vibrio parahaemolyticus*-related bacterium isolated from clam *Meretrix meretrix* with mass mortality. *Journal of Invertebrate Pathology* 103(2), 109–115.

1843 Yumoto, I., Iwata, H., Sawabe, T., Ueno, K., Ichise, N., Matsuyama, H., Okuyama, H. and Kawasaki, K. (1999) Characterization of a facultatively psychrophilic bacterium, *Vibrio rumoiensis* sp. nov., that exhibits high catalase activity. *Applied and Environmental Microbiology* 65(1), 67–72.

1844 Yumoto, I., Kawasaki, K., Iwata, H., Matsuyama, H. and Okuyama, H. (1998) Assignment of *Vibrio* sp. strain ABE-1 to *Colwellia maris* sp. nov., a new psychrophilic bacterium. *International Journal of Systematic Bacteriology* 48(Pt 4), 1357–1362.

1845 Zamora, L., Fernández-Garayzábal, J.F., Svensson-Stadler, L.A., Palacios, M.A., Domínguez, L., Moore, E.R.B. and Vela, A.I. (2012a) *Flavobacterium oncorhynchi* sp. nov., a new species isolated from rainbow trout (*Oncorhynchus mykiss*). *Systematic and Applied Microbiology* 35(2), 86–91.

1846 Zamora, L., Vela, A.I., Palacios, M.A., Sánchez-Porro, C., Moore, E.R.B., Domínguez, L., Ventosa, A. and Fernández-Garayzábal, J.F. (2012b) *Chryseobacterium tructae* sp. nov., isolated from rainbow trout (*Oncorhynchus mykiss*). *Systematic and Applied Microbiology* 35(5), 315–319.

1847 Zappulli, V., Mazzariol, S., Cavicchioli, L., Petterino, C., Bargelloni, L. and Castagnaro, M. (2005) Fatal necrotizing fasciitis and myositis in a captive common bottlenose dolphin (*Tursiops truncatus*) associated with *Streptococcus agalactiae*. *Journal of Veterinary Diagnostic Investigation* 17, 617–622.

1848 Zeigler, D.R. (2003) Gene sequences useful for predicting relatedness of whole genomes in bacteria. *International Journal of Systematic and Evolutionary Microbiology* 53(6), 1893–1900.

1849 Zerihun, M.A., Nilsen, H., Hodneland, S. and Colquhoun, D.J. (2011) *Mycobacterium salmoniphilum* infection in farmed Atlantic salmon, *Salmo salar* L. *Journal of Fish Diseases* 34(10), 769–781.

1850 Zhang, X.-J., Qin, G.-M., Bing, X.-W., Yan, B.-L. and Liang, L.-G. (2011) Molecular and phenotypic characterization of *Vibrio aestuarianus*, a pathogen of the

cultured tongue sole, *Cynoglossus semilaevis* Günther. *Journal of Fish Diseases* 34(1), 57–64.
1851 Zhou, S., Hou, Z., Li, N. and Qin, Q. (2007) Development of a SYBR Green I real-time PCR for quantitative detection of *Vibrio alginolyticus* in seawater and seafood. *Journal of Applied Microbiology* 103(5), 1897–1906.
1852 Zhou, S.M., Fan, Y., Zhu, X.Q., Xie, M.Q. and Li, A.X. (2011) Rapid identification of *Streptococcus iniae* by specific PCR assay utilizing genetic markers in ITS rDNA. *Journal of Fish Diseases* 34(4), 265–271.
1853 Ziemke, F., Höfle, M.G., Lalucat, J. and Rosselló-Mora, R. (1998) Reclassification of *Shewanella putrefaciens* Owen's genomic group II as *Shewanella baltica* sp. nov. *International Journal of Systematic Bacteriology* 48(Pt 1), 179–186.
1854 Zlotkin, A., Eldar, A., Ghittino, C. and Bercovier, H. (1998a) Identification of *Lactococcus garvieae* by PCR. *Journal of Clinical Microbiology* 36(4), 983–985.
1855 Zlotkin, A., Hershko, H. and Eldar, A. (1998b) Possible transmission of *Streptococcus iniae* from wild fish to cultured marine fish. *Applied and Environmental Microbiology* 64(10), 4065–4067.
1856 ZoBell, C.E. (1941) Studies on marine bacteria.I. The cultural requirements of heterotrophic aerobes. *Journal of Marine Research* 4, 42–75.
1857 Zorrilla, I., Balebona, M.C.,Moriñigo, M., Sarasquete, C. and Borrego, J.J. (1999) Isolation and characterization of the causative agent of pasteurellosis, *Photobacterium damsela* ssp. *piscicida*, from sole, *Solea senegalensis* (Kaup). *Journal of Fish Diseases* 22, 167–172.

Index

Page numbers in *italic* type refer to figures and tables. Under named groups or genera of organisms, general topics are listed first, followed by the names of individual species. The page numbers for sources hosts and diseases for the individual species is listed under the genus heading